2009 IEEE 22nd International Conference on Micro Electro Mechanical Systems

(MEMS)

Sorrento, Italy
25-29 January 2009

Pages 1-562

IEEE Catalog Number: CFP09MEM-PRT
ISBN: 978-1-4244-2977-6

Copyright © 2009 by the Institute of Electrical and Electronic Engineers, Inc
All Rights Reserved

Copyright and Reprint Permissions: Abstracting is permitted with credit to the source. Libraries are permitted to photocopy beyond the limit of U.S. copyright law for private use of patrons those articles in this volume that carry a code at the bottom of the first page, provided the per-copy fee indicated in the code is paid through Copyright Clearance Center, 222 Rosewood Drive, Danvers, MA 01923.

For other copying, reprint or republication permission, write to IEEE Copyrights Manager, IEEE Service Center, 445 Hoes Lane, Piscataway, NJ 08854. All rights reserved.

This publication is a representation of what appears in the IEEE Digital Libraries. Some format issues inherent in the e-media version may also appear in this print version.

IEEE Catalog Number: CFP09MEM-PRT
ISBN 13:   978-1-4244-2977-6

ISSN: 1084-6999

Additional Copies of This Publication Are Available From:

Curran Associates, Inc
57 Morehouse Lane
Red Hook, NY 12571 USA
Phone: (845) 758-0400
Fax: (845) 758-2633
E-mail: curran@proceedings.com

TABLE OF CONTENTS

INVITED SPEAKER I

MEMS EPIPHANY 1
Benedetto Vigna

SESSION I – MEMS ACTUATORS

DEVELOPMENT OF CALIBRATION STANDARDS FOR THE OPTICAL MEASUREMENT OF IN-PLANE DISPLACEMENTS OF MICROMECHANICAL COMPONENTS 7
J. Gaspar, J. Held, G. Pedrini, W. Osten, and O. Paul

3D MAGNETIC MICROACTUATOR MADE OF NEWLY DEVELOPED MAGNETICALLY MODIFIED PHOTOCURABLE POLYMER AND APPLICATION TO SWIMMING MICROMACHINE AND MICROSCREWPUMP......................................11
K. Kobayashi and K. Ikuta

SESSION II – RF MEMS

NOVEL CONCEPT OF MICROWAVE MEMS RECONFIGURABLE 7X45º MULTI-STAGE DIELECTRIC-BLOCK PHASE SHIFTER..15
N. Somjit, G. Stemme, and J. Oberhammer

MICROMECHANICAL RESONANT DISPLACEMENT GAIN STAGES 19
B. Kim, Y. Lin, W.-L. Huang, M. Akgul, W.-C. Li, Z. Ren, and C.T.-C. Nguyen

EPITAXIAL SILICON MICROSHELL VACUUM-ENCAPSULATED CMOS-COMPATIBLE 200 MHz BULK-MODE RESONATOR..23
K.-L. Chen, H. Chandrahalim, A.B. Graham, S.A. Bhave, R.T. Howe, and T.W. Kenny

A NOVEL STRESS-GRADIENT-ROBUST METAL-CONTACT SWITCH 27
H. Sedaghat-Pisheh, J.-M. Kim, and G.M. Rebeiz

SESSION III – FUEL CELL POWER MEMS

MICROBIAL FUEL CELL BASED ON ELECTRODE-EXOELECTROGENIC BACTERIA INTERFACE 31
E. Parra and L. Lin

A MICRO HYDROGEN GENERATOR WITH A MICROFLUIDIC SELF-REGULATING VALVE FOR SENSORS AND FUEL CELLS..35
V.V. Swaminathan, L. Zhu, B. Gurau, R.I. Masel, and M.A. Shannon

SESSION IV- MEMS IN MICROFLUIDICS

MICROCHANNELS WITH SUBSTANTIAL FRICTION REDUCTION AT LARGE PRESSURE AND LARGE FLOW 39
C.F. Carlborg, G. Stemme, and W. van der Wijngaart

STARJET: PNEUMATIC DISPENSING OF NANO- TO PICOLITER DROPLETS OF LIQUID METAL 43
T. Metz, G. Birkle, R. Zengerle, and P. Koltay

CONTINUOUS MICRO-PARTICLE SEPARATION USING OPTICALLY-INDUCED DIELECTROPHORETIC FORCES 47
W.-Y. Lin, Y.-H. Lin, and G.-B. Lee

POWERFUL ACTUATION OF MAGNETIZED MICROTOOL BY FOCUSED MAGNETIC FIELD ON A DISPOSABLE MICROFLUIDIC CHIP 51
F. Arai, S. Sakuma, Y. Yamanishi, and K. Onda

INVITED SPEAKER II

NANOBIOPHOTONICS AND BIOASICS FOR BIOMEDICAL INNOVATIONS 55
Luke P. Lee

SESSION V- CELLS AND PARTICLES

RECONSTRUCTION OF 3D HIERARCHIC MICRO-TISSUES USING MONODISPERSE COLLAGEN MICROBEADS 56
Y. Morimoto, Y. Tsuda, and S. Takeuchi

LABEL-FREE CONTINUOUS MICRO CELL SORTER WITH ANTIBODY-IMMOBILIZED OBLIQUE GROOVES 60
S. Hashimoto, T. Nishimura, J. Miwa, Y. Suzuki, and N. Kasagi

FLOW-LYSOMETRY AND ITS BIOMEDICAL APPLICATION: CYTOSOLIC ANALYSIS OF SINGLE CELLS IN LARGE POPULATIONS 64
W.C. Lee, F.A. Kuypers, Y.-H. Cho, and A.P. Pisano

STRETCHABLE SUBSTRATES FOR THE MEASUREMENT OF INTRACELLULAR CALCIUM ION CONCENTRATION RESPONDING TO MECHANICAL STRESS 68
Y.J. Heo, E. Iwase, K. Matsumoto, and I. Shimoyama

SESSION VI – MICRO PROCESSING AND CHARACTERIZATION

MICRO CHEMICAL VAPOR DEPOSTION SYSTEM: DESIGN AND VERIFICATION 72
Q. Zhou and L. Lin

MEMS NANOREACTOR FOR ATOMIC-RESOLUTION MICROSCOPY OF NANOMATERIALS IN THEIR WORKING STATE 76
J.F. Creemer, S. Helveg, G.H. Hoveling, S. Ullmann, P.J. Kooyman, A.M. Molenbroek, H.W. Zandbergen, and P.M. Sarro

NOVEL MEMS APPARATUS FOR IN SITU THERMO-MECHANICAL TENSILE TESTING OF MATERIALS AT THE MICRO- AND NANO-SCALE 80
J. Han, M. Uchic, and T. Saif

A NEW MICRO-FOUR-POINT PROBE DESIGN FOR VARIOUS APPLICATIONS 84
J.-K. Kim, Y. Zhang, and D.-W. Lee

STICTION IN LOW HUMIDITY ENVIRONMENT 88
F. Sammoura, A. Sparks, W. Sawyer, M. Bhagavat, M. Judy, and K. Yang

SESSION VII – PHYSICAL MEMS

A NOVEL HIGHLY-TWISTABLE TACTILE SENSING ARRAY USING EXTENDABLE SPIRAL ELECTRODES...92
  M.-Y. Cheng, C.-M. Tsao, Y.-T. Lai, and Y.-J. Yang

A DIRECTIONAL CAPACITIVE MEMS MICROPHONE USING NANO-ELECTRODEPOSITS 96
  S.-S. Je, J. Kim, M.N. Kozicki, and J. Chae

NANOTEXTURE ELECTRODE ON NANOPOROUS AAO DIELECTRIC FOR MICRO TACTILE SENSING DEVICES..100
  C.-T. Hong, L.-A. Chu, A.-S. Chiang, and W. Fang

AN ELECTRICALLY DECOUPLED LATERAL-AXIS TUNING FORK GYROSCOPE OPERATING AT ATMOSPHERIC PRESSURE 104
  Z.Y. Guo, L.T. Lin, Q.C. Zhao, J. Cui, X.Z. Chi, Z.C. Yang, and G.Z. Yan

TERAHERTZ METAMATERIALS WITH SIMULTANEOUSLY NEGATIVE ELECTRIC AND MAGNETIC RESONANCE RESPONSES BASED ON BIMATERIAL POP UP STRUCTURES..108
  H. Tao, A. Strikwerda, K. Fan, W.J. Padilla, R.D. Averitt, and X. Zhang

INVITED SPEAKER III

DAMAGE-FREE PLASMA ETCHING PROCESSES FOR FUTURE NANOSCALE DEVICES...112
  Seiji Samukawa

SESSION VIII – NANOPROCESSING AND NEMS

ATOMIC LAYER DEPOSITION (ALD) TUNGSTEN NEMS DEVISES VIA A NOVEL TOP-DOWN APPROACH...120
  B.D. Davidson, Y.J. Chang, D. Seghete, S.M. George, and V.M. Bright

NANOFIBROUS SURFACE PATTERNING USING NANO-MESHED MICROCAPSULES INDUCED BY PHASE SEPARATION ASSISTED ELECTROSPRAY 124
  M. Ikeuchi, R. Tane, M. Fukuoka, and K. Ikuta

TUNABLE OPTICAL ENHANCEMENT FROM A MEMS-INTEGRATED TIO$_2$ NANOSWORD PLASMONIC ANTENNA 128
  B.D. Sosnowchik, P.J. Schuck, J. Chang, and L. Lin

MICROELECTROMECHANICAL METAL-AIR-INSULATOR-SEMICONDUCTOR (MEM-MAIS) DIODE: A NOVEL HYBRID DEVICE FOR ESD PROTECTION 132
  D. Acquaviva, D. Tsamados, Ph. Coronel, T. Skotnicki, and A.M. Ionescu

SESSION IX- FABRICATION AND PACKAGING

PACKAGING OF 11 MPIXEL CMOS-INTEGRATED SIGE MICRO-MIRROR ARRAYS 136
  A. Witvrouw, H.A.C. Tilmans, L. Bogaerts, P. De Moor, T. Bearda, S. Halder, L. Haspeslagh, B. Schlatmann, M. van Bommel, C. de Nooijer, J. Lauria, R. Vanneer, and B. van Drieenhuizen

ROBUST WAFER-LEVEL THIN-FILM ENCAPSULATION OF MICROSTRUCTURES USING LOW STRESS PECVD SILICON CARBIDE.................................140
  V. Rajaraman, L.S. Pakula, H.T.M. Pham, P.M. Sarro, and P.J. French

DIRECT ETCHING OF HIGH ASPECT RATIO STRUCTURES THROUGH A STENCIL 144
G. Villanueva, O. Vazquez-Mena, C. Hibert, and J. Brugger

INDIUM TIN OXIDE (ITO) TRANSPARENT MEMS SWITCHES 148
B.-K. Lee, Y.-H. Song, and J.-B. Yoon

ADVANCEMENTS IN TECHNOLOGY AND DESIGN OF BIOMIMETIC FLOW-SENSOR ARRAYS 152
C.M. Bruinink, R.K. Jaganatharaja, M.J. de Boer, E. Berenschot, M.L. Kolster, T.S.J. Lammerink,
R.J. Wiegerink, and G.J.M. Krijnen

SESSION X- OPTICAL MEMS

A WAFER-LEVEL MICRO MECHANICAL GLOBAL SHUTTER FOR A MICRO CAMERA 156
C.-H. Kim, K.-D. Jung, and W. Kim

MICROFABRICATED FLIPPING GLASS DISC FOR STEREO IMAGING IN ENDOSCOPIC VISUAL INSPECTION 160
W. Choi, M. Akbarian, V. Rubtsov, and C.-J. Kim

RESONANT FREQUENCY TUNING OF TORSIONAL MICROSCANNER BY MECHANICAL RESTRICTION USING MEMS ACTUATOR 164
J.-I. Lee, S. Park, Y. Eun, B. Jeong, and J. Kim

REFLECTIVE DISPLAY USING IONIC LIQUID 168
N. Binh-Khiem, K. Matsumoto, and I. Shimoyama

FORMATION AND INTEGRATION OF A BALL LENS UTILIZING TWO PHASES LIQUID TECHNOLOGY 172
C.-C. Lee, S.-Y. Hsiao, and W. Fang

TEMPERATURE-CONTROLLED TRANSFER AND SELF-WIRING FOR MULTI-COLOR LED DISPLAY ON A FLEXIBLE SUBSTRATE 176
E. Iwase, H. Onoe, A. Nakai, K. Matsumoto, and I. Shimoyama

SESSION XI – BIOMEMS

MICROFLUIDIC ODORANT SENSOR WITH FROG EGGS EXPRESSING OLFACTORY RECEPTORS 180
N. Misawa, H. Mitsuno, R. Kanzaki, and S. Takeuchi

SEQUENTIAL AND SELECTIVE SELF-ASSEMBLY OF MICRO COMPONENTS BY DNA GRAFTED POLYMER 184
T. Tanemura, G. Lopez, R. Sato, K. Sugano, T. Tsuchiya, O. Tabata, M. Fujita, and M. Maeda

PIEZORESISTIVE CANTILEVER-BASED FORCE-CLAMP SYSTEM FOR THE STUDY OF MECHANOTRANSDUCTION IN C. ELEGANS 188
S.-J. Park, B. Petzold, M.B. Goodman, and B.L. Pruitt

SUB-5μM DIAMETER SILICON MICROPROBE AND SILICON DIOXIDE MICROTUBE ARRAYS FOR LOW-INVASIVE ANIMAL EXPERIMENTS.......................192
K. Takei, T. Kawano, T. Kawashima, K. Sawada, H. Kaneko, and M. Ishida

MINIMALLY INVASIVE PARYLENE DUAL-VALVED FLOW DRAINAGE SHUNT FOR GLAUCOMA IMPLANT 196
J.C.-H. Lin, P.-J. Chen, B. Yu, M. Humayun, and Y.-C. Tai

GLASS MICROPROBE WITH EMBEDDED SILICON VIAS FOR 3D INTEGRATION 200
C.-W. Lin, C.-W. Chang, Y.-T. Lee, R. Chen, Y.-C. Chang, and W. Fang

POSTER/ORAL PRESENTATIONS

BIOMEDICAL APPLICATIONS

**FABRICATION PROCESSES OF INTEGRATED MULTI-ANALYTE BIOCHIP
SYSTEM FOR IMPLANTABLE APPLICATIONS** 204
*Y.-C. Tsai, N.-F. Chiu, P.-C. Liu, Y.-C. Ou, H.-H. Liao, Y.-J. Yang, L.-J. Yang, U. Lei, F.-S. Chao, S.-
S. Lu, C.-W. Lin, P.-Z. Chang, and W.-P. Shih*

**MICROSYSTEM-BASED STUDY OF POLLEN-TUBE ATTRACTANTS SECRETED
BY OVULES** 208
J.R. Cooper, Y. Qin, L. Jiang, R. Palanivelu, and Y. Zohar

**A BATCH-PATTERNED SELF-EXPANDING BILIARY STENT WITH CONFORMAL
MAGNETIC PDMS LAYER AND TOPOLOGICALLY-MATCHED WIRELESS
MAGNETOELASTIC SENSOR** 212
S.R. Green and Y.B. Gianchandani

**RADIO-CONTROLLED CYBORG BEETLES: A RADIO-FREQUENCY SYSTEM FOR
INSECT NEURAL FLIGHT CONTROL** 216
H. Sato, Y. Peeri, E. Baghoomian, C.W. Berry, and M.M. Maharbiz

**HOLLOW MICRONEEDLE ELECTRODE ARRAYS FOR INTRACELLULAR
RECORDING APPLICATIONS** 220
*J. Held, J. Gaspar, P.J. Koester, C. Tautorat, M. Hagner, A. Cismak, A. Heilmann, W. Baumann, P.
Ruther, and O. Paul*

**BATCH FABRICATION OF OUT-OF-PLANE, IC-COMPATIBLE, NANOSCALE-TIP
SILICON NEUROPROBE ARRAYS** 224
A. Goryu, A. Ikedo, K. Takei, K. Sawada, T. Kawano, and M. Ishida

**ULTRA-COMPACT INTEGRATION FOR FULLY-IMPLANTABLE NEURAL
MICROSYSTEMS** 228
G.E. Perlin and K.D. Wise

**CMOS-BASED HIGH-DENSITY SILICON MICROPROBE ARRAY FOR
ELECTRONIC DEPTH CONTROL IN NEURAL RECORDING** 232
*K. Seidl, S. Herwik, Y. Nurcahyo, T. Torfs, M. Keller, M. Schüttler, H. Neves, T. Stieglitz, O. Paul,
and P. Ruther*

**IN-PLANE BANDPASS REGULATION CHECK VALVE IN HEAT-SHRINK
PACKAGING FOR DRUG DELIVERY** 236
R. Lo and E. Meng

**ON-CHIP BLOOD VISCOMETER TOWARDS POINT-OF-CARE HEMATOLOGICAL
DIAGNOSIS** 240
H. Zeng and Y. Zhao

**IMPLANTABLE FLEXIBLE-COILED WIRELESS INTRAOCULAR PRESSURE
SENSOR** 244
P.-J. Chen, S. Saati, R. Varma, M.S. Humayun, and Y.-C. Tai

INTEGRATED WIRELESS NEUROSTIMULATOR 248
W. Li, D.C. Rodger, and Y.C. Tai

TRANSDERMAL POWER TRANSFER FOR IMPLANTED DRUG DELIVERY DEVICES USING A SMART NEEDLE AND REFILL PORT 252
A.T. Evans, S. Chiravuri, and Y.B. Gianchandani

A FLEXIBLE REGENERATION MICROELECTRODE WITH CELL-GROWTH GUIDANCES 256
R. Gojo, N. Kotake, T. Suzuki, K. Mabuchi, and S. Takeuchi

ADAPTIVE BLOOD CELL SEPARATION USING COOPERATION OF SERIALLY CONNECTED MEMBRANE FILTERS 260
T. Kobayashi, D. Kato, H. Koga, K. Morimoto, M. Fukuda, Y. Kinoshita, H. Yoshida, and S. Konishi

BIOCHEMICAL SENSORS

HIGH SENSITIVITY MICRO-THERMAL CONDUCTIVITY DETECTOR FOR GAS CHROMATOGRAPHY 264
B.C. Kaanta, H. Chen, G. Lambertus, W.H. Steinecker, O. Zhdaneev, and X. Zhang

CMOS-MEMS CAPACITIVE HUMIDITY SENSOR 268
N. Lazarus, S. Bedair, C. Lo, and G.K. Fedder

A THERMOELECTRIC GAS SENSOR BASED ON AN EMBEDDED TIN OXIDE CATALYST FOR DETECTING HYDROGEN AND NOX GASES 272
S.-I. Yoon, C.-i. Lee, and Y.-J. Kim

A NOVEL MICROFLUIDIC PLATFORM FOR CONTINUOUS DNA EXTRACTION AND PURIFICATION USING LAMINAR FLOW MAGNETOPHORESIS 276
M. Karle, J. Miwa, G. Roth, R. Zengerle, and F. von Stetten

AN EWOD DROPLET MICROFLUIDIC CHIP WITH INTEGRATED LOCALTEMPERATURE CONTROL FOR MULTIPLEX PROTEOMICS 280
W. Nelson, I. Peng, J.A. Loo, R.L. Garrell, and C.-J. Kim

FREQUENCY DRIFT COMPENSATION IN MASS-SENSITIVE CHEMICALSENSORS BASED ON PERIODIC STIFFNESS MODULATION 284
K.S. Demirci, J.H. Seo, S. Truax, L.A. Beardslee, Y. Luzinova, B. Mizaikoff, and O. Brand

μGC MULTICAPILLARY COLUMNS WITH MONO-LAYER PROTECTED GOLD AS A STATIONARY PHASE 288
M.A. Zareian-Jahromi, and M. Agah

A SELF-CONTAINED MINIATURIZED PCR SYSTEM USING ELECTROMAGNETIC ACTUATORS 292
B.T. Chia, S.-A. Yang, M.-Y. Cheng, C.-L. Lin, C.-W. Lin, and Y.-J. Yang

NANOSTRUCTURE-ENHANCED FIBER-OPTIC INTERFEROMETRY FOR LABEL-FREE IMMUNE SENSING 296
Y.-T. Tseng, C.-T. Chang, M.-H. Chen, and F.-G. Tseng

BIOSENSORS BASED ON ALL-POLYMER RESONANT MICROBEAMS 300
S. Schmid, P. Wägli, and C.H. Hierold

BEAD-IMMOBILIZED MOLECULAR BEACONS FOR HIGH THROUGHPUT SNP GENOTYPING VIA A MICROFLUIDIC SYSTEM 304
R.D. Sochol, A. Mahajerin, B.P. Casavant, P. Singh, M. Dueck, L.P. Lee, and L. Lin

A FULLY AUTOMATED MICRO-SOLID PHASE EXTRACTION CHIP FOR GENETIC SAMPLE PREPARATION SYSTEM 308
S.-I. Han, H. Lee, and K.-H. Han

SELF-POLARIZED PIEZOELECTRIC BIOSENSOR ARRAY FOR MULTIPLE IMMUNOASSYS APPLICATIONS 312
T. Xu, J. Miao, Z. Wang, L. Yu, and C.-M. Li

SELECTIVITY AND LONG-TERM RELIABILITY OF RESONANT EXPLOSIVE-VAPOR-TRACE DETECTION BASED ON ANTIGEN-ANTIBODY BINDING 316
A. Lin and E.S. Kim

DNA-DECORATED CARBON NANOTUBES AS SENSITIVE LAYER FOR AlN CONTOUR-MODE RESONANT-MEMS GRAVIMETRIC SENSOR 320
C. Zuniga, M. Rinaldi, S.M. Khamis, T.S. Jones, A.T. Johnson, and G. Piazza

MICROFLUIDICS WITHIN A SWAGELOK®: A MEMS-ON-TUBE ASSEMBLY 324
S. Unnikrishnan, H.V. Jansen, J.W. Berenschot, B. Mogulkoc, and M.C. Elwenspoek

NANO-THICKNESS INTEGRATED RESONANT CANTILEVERS WITH SURFACE-STIFFENING SCHEME FOR ULTRA-SENSITIVE DETECTION OF TRACE CHEMICALS 328
X. Xia, Y. Yang, and X. Li

SURFACE ACOUSTIC WAVE-DRIVEN ACTIVE PLASMONICS BASED ON DYNAMIC PATTERNING OF NANOPARTICLES IN MICROFLUIDIC CHANNELS 332
J. Shi, Y.B. Zheng, and T.J. Huang

PUPAL-STAGE INSERTED SILICON MICROPROBE NEURAL INTERFACE IN INSECTS FOR GAS SENSING 336
C.J. Shen, R. Gilmour, and A. Lal

BEAD-BASED MICROFLUIDIC PLATFORM INTEGRATED WITH OPTICAL DETECTION DEVICES FOR RAPID DETECTION OF GENETIC DELETION FROM SALIVA 340
K.-Y. Lien, C.-J. Liu, P.-L. Kuo, and G.-B. Lee

A MICROFLUIDIC AFFINITY COCAINE SENSOR 344
J.P. Hilton, T. Nguyen, R. Pei, M. Stojanovic, and Q. Lin

DEVELOPMENT OF NEW BIOCHEMICAL IC CHIP-SET FOR REAL-TIME PCR 348
K. Ikuta, T. Sasao, Y. Okuda, K. Watamura, and M. Ikeuchi

A MEMS SENSOR FOR CONTINUOUS MONITORING OF GLUCOSE IN SUBCUTANEOUS TISSUE 352
X. Huang, S. Li, J. Schultz, Q. Wang, and Q. Lin

MONOLITHIC CENTRIFUGAL MICROFLUIDIC PLATFORM FOR BACTERIA CAPTURE AND CONCENTRATION, LYSIS, NUCLEIC-ACID AMPLIFICATION, AND REAL-TIME DETECTION 356
J.L. Garcia-Cordero, I.K. Dimov, J. O'Grady, J. Ducree, T. Barry, and A.J. Ricco

CELLS AND PARTICLES

IN VITRO DETECTION OF NEURAL ACTIVITY WITH VERTICALLY GROWN SINGLE PLATINUM NANOWIRE 360
D. Choi, A. Fung, H. Moon, G. Villareal, Y. Chen, D. Ho, N. Presser, G. Stupian, and M. Leung

GENERATION AND SELECTIVE RETRIEVAL OF MICRO DROPLETS IN AN ARRAY FOR MICRO-PCR USING A GLASS-SILICON-GLASS CHIP 363
K.H. Park, H.G. Park, and S. Takeuchi

ARRAYING SINGLE ADHERENT CELLS BY MICROPLATE SELF-ASSEMBLY 367
H. Ishihara, K. Kuribayashi, and S. Takeuchi

A DYNAMIC MICROARRAY WITH PNEUMATIC VALVES FOR SELECTIVE TRAPPING AND RELEASING OF MICROBEADS 371
K. Iwai and S. Takeuchi

SIZE-DEPENDENT PARTICLE FILTERATION USING MAGNETICALLY DRIVEN MICROTOOL AND CENTRIFUGAL FORCE IN MICROCHIP 375
H. Maruyama, S. Sakuma, Y. Yamanishi, and F. Arai

CLUSTER DYNAMICS IN FLOW OF SUSPENDED PARTICLES IN MICROCHANNELS 379
T. Gudipaty, M.T. Stamm, L.S.L. Cheung, L. Jiang, and Y. Zohar

A MULTIFUNCTIONAL VERTICAL MICROSIEVE FOR MICRO AND NANO PARTICLES SEPARATION 383
C. Shen, H.T.M. Pham, and P.M. Sarro

CLASSIFICATION AND CONDENSATION OF NANO-SIZED AIRBORNE PARTICLES BY ELECTRICALLY TUNNING COLLECTION SIZE 387
Y.-H. Kim, S. Kwon, D. Park, J. Hwang, and Y.-J. Kim

DEVELOPMENT OF A CYTOMIC FORCE TRANSDUCER FOR EXPERIMENTAL MECHANOBIOLOGY 391
E. Dy and C.M. Ho

A PDMS PLANAR PATCH-CLAMP ARRAY CHIP WITH POLY (ETHYLENE GLYCOL)/SU-8 BASED CELL-PATCH INTERFACE 395
B.J. Xu, Z.B. Liu, J.J. Yu, Y.K. Lee, and M. Yang

WHOLE CELL IMPEDANCE ANALYSIS OF METASTATIC AND NON-METASTATIC CANCER CELLS 399
H.S. Kim, Y.H. Cho, A.B. Frazier, Z.G. Chen, D.-M. Shin, and A. Han

INTEGRATION OF SKELETAL MUSCLE CELL ONTO SI-MEMS AND ITS GENERATIVE FORCE MEASUREMENT 403
K. Shimizu, H. Sasaki, H. Hida, H. Fujita, K. Obinata, M. Shikida, and E. Nagamori

A STRETCHABLE CELL CULTURE PLATFORM WITH EMBEDDED ELECTRODE ARRAY 407
P. Wei, R. Taylor, Z. Ding, G. Higgs, J.J. Norman, B.L. Pruitt, and B. Ziaie

ASSESSMENT OF SINGLE CELL VIABILITY FOLLOWING LIGHT-INDUCED ELECTROPORATION THROUGH USE OF ON-CHIP MICROFLUIDICS 411
J.K. Valley, H.-Y. Hsu, S. Neale, A.T. Ohta, A. Jamshidi, and M.C. Wu

MANIPULATION OF BIOSAMPLES AND MICROPARTICLES USING OPTICAL IMAGES ON POLYMER DEVICES 415
W. Wang, Y.-H. Lin, T.-F. Guo, and G.-B. Lee

FABRICATION OF LIVING CELL STRUCTURE UTILIZING ELECTROSTATIC INKJET PHENOMENA 419
S. Umezu, T. Kitajima, H. Murase, H. Ohmori, K. Katahira, and Y. Ito

SIZE-CONTROLLED ISLET-CELL SPHEROIDS FOR GEOMETRIC ANALYSIS OF INSULIN SECRETION 423
Y. Tsuda, M. Kato-Negishi, T. Okitsu, and S. Takeuchi

A 3-D MICROFLUIDIC COMBINATORIAL CELL CULTURE ARRAY 427
M.C. Liu and Y.C. Tai

FLOW ACCELERATION EFFECT ON CANCER CELL DEFORMATION AND DETACHMENT 431

L.S.L. Cheung, X.J. Zheng, A. Stopa, J. Schroeder, R.L. Heimark, J.C. Baygents, R. Guzman, and Y. Zohar

MICROFLUIDICS AND FLOW SENSORS

LINEAR EXPANSION AND CONTRACTION OF PAIRED PNEUMATIC BALOON BENDING ACTUATORS TOWARD TELESCOPIC MOTION 435

N. Fujiwara, S. Sawano, and S. Konishi

LOW-COST MICROFLUIDIC SINGLE-USE VALVES AND ON-BOARD REAGENT STORAGE USING LASER-PRINTER TECHNOLOGY 439

J.L. Garcia-Cordero, F. Benito-Lopez, D. Diamond, J. Ducree, and A.J. Ricco

DROPLET MIXER BASED ON SIPHON-INDUCED FLOW DISCRETIZATION AND PHASE SHIFTING 443

R. Burger, N. Reis, J.G. Fonseca, and J. Ducreé

AN ALL-POLYMER AIR-FLOW SENSOR ARRAY USING A PIEZORESISTIVE COMPOSITE ELASTOMER 447

A.R. Aiyar, C. Song, S.H. Kim, and M.G. Allen

DESIGN AND CHARACTERISATION OF A HYDRAULIC MICROACTUATOR FABRICATED BY LITHOGRAPHY 451

M. De Volder, F. Ceyssens, D. Reynaerts, and R. Puers

BIOMIMIC LOW HYSTERESIS SURFACE PREPARED BY HIERARCHICAL MICRO/NANO TEXTURES 455

M.-H. Chen1, T.-H. Hsu1, and F.-G. Tseng1,2

WETTABILITY SWITCHING TECHNIQUE OF A BIOCOMPATIBLE POLYMER 459

Y. Takahashi, K.S. Teh, and Y.W. Lu

MICRO CORIOLIS MASS FLOW SENSOR WITH INTEGRATED CAPACITIVE READOUT 463

J. Haneveld, T.S.J. Lammerink, M.J. de Boer, and R.J. Wiegerink

MEMS TACTILE DISPLAY WITH HYDRAULIC DISPLACEMENT AMPLIFICATION MECHANISM 467

T. Ninomiya, K. Osawa, Y. Okayama, Y. Matsumoto, and N. Miki

HIGH PURITY SEPARATION OF RARE SPECIES IN DROPLET MICROFLUIDICS USING DROPLET-CONDUIT STRUCTURES 471

G.J. Shah and C.-J. Kim

A HIGHLY FLEXIBLE SUPERHYDROPHOBIC MICROLENS ARRAY WITH SMALL CONTACT ANGLE HYSTERESIS FOR DROPLET-BASED MICROFLUIDICS 475

M. Im, D.-H. Kim, X.-J. Huang, J.-H. Lee, J.-B. Yoon, and Y.-K. Choi

AMBIENT TEMPERATURE-GRADIENT COMPENSATED LOW-DRIFT THERMOPILE FLOW SENSOR 479

M. Dijkstra, T.S.J. Lammerink, M.J. de Boer, R.J. Wiegerink, and M.C. Elwenspoek

AN ELECTRICALLY-DRIVEN, LARGE-DEFLECTION, HIGH-FORCE, MICRO PISTON HYDRAULIC ACTUATOR ARRAY FOR LARGE-SCALE MICROFLUIDIC SYSTEMS 483

H. Kim and K. Najafi

**ELECTROHYDRODYNAMIC JET PRINTING CAPABLE OF REMOVING
SUBSTRATE EFFECTS AND MODULATING PRINTING CHARACTERISTICS**................487
J.-S. Lee, Y.-J. Kim, B.-G. Kang, S.-Y. Kim, J. Park, J. Hwang, and Y.-J. Kim

**HYBRID ON DEMAND JETTING SYSTEM FOR ULTRA FINE DROPLET BASED ON
ELECTROHYDRODYNAMIC AND PIEZOELECTRIC ACTUATION**................491
Y.-J. Kim1,2, J.-S. Lee1, S.-Y. Kim1, S.E. Park1, J. Hwang1, and Y.-J. Kim1

**MICRO-MACHINED STENT-TYPE FLOW SENSOR FOR EVALUATION OF NASAL
RESPIRATION**................495
T. Yokota, J. Naito, M. Shikida, and K. Sato

**EHD MICRO PUMP USING PYROLYZED POLYMER 3-D CARBON MESH
ELECTRODES**................499
D. Wakui, N. Imai, Y. Nagaura, H. Sato, T. Sekiguchi, S. Konishi, S. Shoji, and T. Homma

**NON-LINEAR FLUIDIC INTEGRATED CIRCUITS REALIZED BY PNEUMATIC-
FIELD EFFECT TRANSISTORS WITH CONTROLLABLE OUTPUT RESISTANCE**................503
H. Takao, N. Tanaka, M. Sugiura, K. Sawada, and M. Ishida

**HIGH-RESOLUTION PIEZO INKJET PRINTHEAD FABRICATED BY THREE
DIMENSIONAL ELECTRICAL CONNECTION METHOD USING THROUGH GLASS
VIA**................507
M. Murata, T. Kondoh, T. Yagi, N. Funatsu, K. Tanaka, H. Tsukuni, K. Ohno, H. Usami, R. Nayve, N.
Inoue, S. Seto, and N. Morita

**SELF-RESONANT FLOW SENSOR USING RESONANCE FREQUENCY SHIFT BY
FLOW-INDUCED VIBRATION**................511
K.H. Kim and Y.H. Seo

MICROFLUIDIC-BASED DISPENSERS FOR SUB-MICROLITER PIPETTING................515
S.-B. Huang and G.-B. Lee

**PRECISE TITRATION MICROFLUIDIC SYSTEM USING SUB-PICOLITER
DROPLET INJECTION**................519
K. Kawai, M. Fujii, and S. Shoji

**DROPLETS GENERATION METHOD FOR WATER–IN-OIL STATE IN THE
POLYDIMETHYLSILOXANE MICROCHANNEL WITH GROOVES**................523
J. Kim, D. Byun, J. Hong, and A.J. deMello

**A LOW-COST FLEXIBLE HOT-FILM SENSOR SYSTEM FOR FLOW SENSING AND
ITS APPLICATION TO AIRCRAFT**................527
R. Zhu, P. Liu, X.D. Liu, F.X. Zhang, and Z.Y. Zhou

**INTEGRATING EWOD WITH SURFACE RATCHETS FOR ACTIVE DROPLET
TRANSPORT AND SORTING**................531
T.A. Duncombe, M. Kumemura, H. Fujita, and K.F. Böhringer

**THERMAL CHARACTERIZATION OF MICROLITER AMOUNTS OF LIQUIDS BY
A MICROMACHINED CALORIMETRIC TRANSDUCER**................535
G. Pärr, E. Santagata-Iervolino, A.W. van Herwaarden, W. Wien, and M.J. Vellekoop

**TIR-BASED DYNAMIC LIQUID-LEVEL AND FLOW-RATE SENSING AND ITS
APPLICATION ON CENTRIFUGAL MICROFLUIDIC PLATFORMS**................539
J. Hoffmann, L. Riegger, D. Mark, F. von Stetten, R. Zengerle, and J. Ducreé

NANO DEVICES AND CARBON NANOTUBES

100 NANOMETER SCALE RESISTIVE HEATER-THERMOMETER ON A SILICON CANTILEVER 543
Z. Dai, K. Park, and W.P. King

MEMS DIFFRACTIVE OPTICAL NANORULER TECHNOLOGY FOR TIP-BASED NANOFABRICATION AND METROLOGY 547
N. Yoshimizu, A. Lal, and C.R. Pollock

INTEGRATION OF BRIDGING-STRUCTURAL SWNTS ON FLEXIBLE PDMS SHEET BY STAMPING TRANSFER 551
Y. Takei, T. Kan, E. Iwase, K. Matsumoto, and I. Shimoyama

GIANT PIEZORESISTANCE OF NANO-THICK SILICON INDUCED BY INTERFACE ELECTRON TRAPPING EFFECT 555
Y. Yang and X. Li

ELECTRICAL DISCHARGE BASED MICROFABRICATION ON ELECTROSPUN NANOFIBERS 559
H. Zeng and Y. Zhao

INTEGRATION OF CARBON NANOTUBES INTO MEMS DEVICES VIA DETERMINISTIC TRANSPLANTING ASSEMBLY 563
S. Kim, H.W. Lee, and S.G. Kim

SURFACE-TEMPERATURE CONTROL OF SILICON NANOWIRES IN DRY AND LIQUID CONDITIONS 567
S. Akiyama, Y.T. Cheng, J. Fattaccioli, N. Takama, P. Löw, C. Bergaud, and B.J. Kim

ELECTRON BEAM STIMULATED OXIDATION OF CARBON (EBSOC) 571
P.S. Spinney, S.D. Collins, and R.L. Smith

ULTRA SMALL SINGLE WALLED CARBON NANOTUBE PRESSURE SENSORS 575
T. Helbling, S. Drittenbass, L. Durrer, C. Roman, and C.H. Hierold

METALLIC NANOPARTICLE MANIPULATION USING OPTOELECTRONIC TWEEZERS 579
A. Jamshidi, H.Y. Hsu, J.K. Valley, A.T. Ohta, S. Neale, and M.C. Wu

THIN FILM TRANSISTOR (TFT) SENSING ELEMENTS FABRICATED IN SURFACE MICROMACHINED POLYMERMEMS FOR A DIFFERENTIAL CALORIMETRIC FLOW SENSOR 583
S.-H. Tsang, K. Simard, I.G. Foulds, H. Izadi, K.S. Karim, and M. Parameswaran

PLANAR MEMS SUPERCAPACITOR USING CARBON NANOTUBE FORESTS 587
Y.Q. Jiang, Q. Zhou, and L. Lin

PIEZOELECTRIC NANOSWITCH 591
D.C. Judy, J.S. Pulskamp, R.G. Polcawich, and L. Currano

A MOLECULAR MUSCLE BASED NANO-ELECTRO-MECHANICAL-SYSTEMS (NEMS) 595
B.K. Juluri, A.S. Kumar, Y. Liu, T. Ye, Y.W. Yang, A.H. Flood, L. Fang, J.F. Stoddart, P.S. Weiss, and T.J. Huang

MATERIALS AND DEVICE-CHARACTERIZATION

HIGHLY EFFICIENT EXTRACTION OF MECHANICAL AND LINEAR AND QUADRATIC PIEZORESISTIVE PROPERTIES OF POLY-SI FILMS USING WAFER-SCALE MICROTENSILE TESTING 599
M.E. Schmidt, J. Gaspar, J. Held, S. Kamiya, and O. Paul

DESIGN OF POWER-OPTIMIZED THERMAL CANTILEVERS FOR SCANNING PROBE TOPOGRAPHY SENSING 603
H. Rothuizen, M. Despont, U. Drechsler, C. Hagleitner, A. Sebastian, and D. Wiesmann

MECHANICAL CHARACTERIZATION OF BIOMOLECULES IN LIQUID USING SILICON TWEEZERS WITH SUBNANONEWTON RESOLUTION 607
C. Yamahata, E. Sarajlic, L. Jalabert, M. Kumemura, D. Collard, and H. Fujita

EFFECT OF ELECTRODE GEOMETRY AND SURFACE PASSIVATION ON CORROSION OF POLYCRYSTALLINE SILICON UNDER HIGH RELATIVE HUMIDITY AND BIAS 611
F. Liu, C. Roper, C. Carraro, and R. Maboudian

MECHANICAL RESPONSE EVALUATION OF HIGH-THERMALLY-STABLE-GRADE PARYLENE SPRING 615
C. Kamezawa, Y. Suzuki, and N. Kasagi

PIEZOELECTRIC PZT VERY THIN FILMS IN THE 100NM RANGE: A SOLUTION FOR ACTUATORS EMBEDDED IN LOW VOLTAGE DEVICES 619
E. Defay, G. Le Rhun, F. Perruchot, P. Rey, A. Suhm, M. Aïd, L.J. Liu, S. Pacheco, and M. Miller

CMOS-COMPATIBLE SURFACE-MICROMACHINED TEST STRUCTURE FOR DETERMINATION OF THERMAL CONDUCTIVITY OF THIN FILM MATERIALS BASED ON SEEBECK EFFECT 623
Z. Wang, P. Fiorini, and C. van Hoof

A BIOCOMPATIBLE AND FLEXIBLE RF CMOS TECHNOLOGY AND THE CHARACTERIZATION OF THE FLEXIBLE MOS TRANSISTORS UNDER BENDING STRESSES 627
C.-Y. Hsieh, J.-S. Chen, W.-A. Tsou, Y.-T. Yeh, K.-A. Wen, and L.-S. Fan

RADIATION SENSITIVITY OF OHMIC RF-MEMS SWITCHES FOR SPATIAL APPLICATIONS 630
A. Tazzoli, G. Cellere, E. Autizi, V. Peretti, A. Paccagnella, and G. Meneghesso

RELIABILITY OF RF MEMS CAPACITIVE SWITCHES AND DISTRIBUTED MEMS PHASE SHIFTERS USING ALN DIELECTRIC 634
M. Fernández-Bolaños, D. Tsamados, P. Dainesi, and A.M. Ionescu

TENSILE MEASUREMENT OF A SINGLE CRYSTAL GALLIUM NITRIDE NANOWIRE 638
J.J. Brown, A.I. Baca, and V.M. Bright

CELLULOSE-BASED COMPOSITE AS A RAW MATERIAL FOR FLEXIBLE AND ULTRA-LIGHTWEIGHT MECHANICAL SWITCH DEVICE 642
S. Couderc, B.J. Kim, and T. Someya

DETERMINATION OF DENSITY AND YOUNG'S MODULUS OF ATOMIC LAYER DEPOSITED THIN FILMS BY RESONANT FREQUENCY MEASUREMENTS OF OPTICALLY EXCITED NANOCANTILEVERS 646
B. Ilic, S. Krylov, and H. Craighead

THE EFFECTS OF THERMAL OXIDATION OF MEMS RESONATOR ON TEMPERATURE DRIFT AND ABSOLUTE FREQUENCY 650
C. van der Avoort, J. van Wingerden, and J.T.M. van Beek

NON-DESTRUCTIVE QUANTITATIVE MEASUREMENT METHOD FOR NORMAL AND SHEAR STRESSES ON SINGLE-CRYSTALLINE SILICON STRUCTURES FOR RELIABILITY OF SILICON-MEMS 653
M. Komatsubara, T. Namazu, N. Naka, S. Kashiwagi, K. Ohtsuki, and S. Inoue

PHASE LOCK LOOP BASED TEMPERATURE COMPENSATION FOR MEMS OSCILLATORS 657
J. Salvia, R. Melamud, S. Chandorkar, H.K. Lee, Y.Q. Qu, S.F. Lord, B. Murmann, and T.W. Kenny

DEGRADATION OF MECHANICAL STRENGTH AT SI/SIO$_2$ INTERFACE ON SOI WAFERS UNDER CYCLIC LOADING 661
T. Ando, T. Takumi, and K. Sato

FABRICATION PACKAGING ENCAPSULATION

PART-TO-PART AND PART-TO-SUBSTRATE MAGNETIC SELF-ASSEMBLY OF MILLIMETER SCALE COMPONENTS WITH ANGULAR ORIENTATION 665
S.B. Shetye, I. Eskinazi, and D.P. Arnold

LOW TEMPERATURE FABRICATION PROCESS FOR HIGH-ASPECT-RATIO AND MULTI-COMPLIANT MEMS 669
S. Kühne, R. Blattmann, and C.H. Hierold

FABRICATION OF NANOPILLARS BASED ON SILICON OXIDE NANOPATTERNS SYNTHESIZED IN OXYGEN PLASMA REMOVAL OF PHOTORESIST 673
H.Y. Mao, D. Wu, W.G. Wu, J. Xu, H.X. Zhang, and Y.L. Hao

VISUAL OBSERVATION OF PDMS TIP IN LIQUID MICROCONTACT PRINTING 677
J. Fattaccioli, A. Ikeda, J.G. Kim, N. Takama, and B.J. Kim

SUPERCRITICAL FLUID DEPOSITION (SCFD) TECHNIQUE AS A NOVEL TOOL FOR MEMS FABRICATION 681
T. Momose, T. Ohkubo, T. Uejima, T. Saito, M. Sugiyama, and Y. Shimogaki

ELECTRODEPOSITION OF PERMALLOY IN DEEP SILICON TRENCHES WITHOUT EDGE-OVERGROWTH UTILIZING DRY FILM PHOTORESIST 685
S.-W. Park, D.G. Senesky, and A.P. Pisano

FABRICATION OF BIODEGRADABLE MAGNETIC MICROCAPSULES UTILIZING 3D, SURFACE MODIFIED PDMS MICROFLUIDIC DEVICES 689
C.Y. Liao and Y.C. Su

NOVEL TECHNOLOGY FOR CAPACITIVE PRESSURE SENSORS WITH MONOCRYSTALLINE SILICON MEMBRANES 693
K. Knese, S. Armbruster, H. Weber, M. Fischer, H. Benzel, M. Metz, and H. Seidel

DEVELOPMENT OF MULTI-USER MULTI-CHIP SOI CMOS-MEMS PROCESSES 697
K. Takahashi, M. Mita, M. Nakada, D. Yamane, A. Higo, H. Fujita, and H. Toshiyoshi

MEMS-BASED BATCH-MODE MICRO-ELECTRO-DISCHARGE MACHINING USING MICROELECTRODE ARRAYS ACTUATED BY HYDRODYNAMIC FORCE 701
C.R. Alla Chaitanya, and K. Takahata

SILICON CARBIDE SURFACE MICROMACHINING TECHNOLOGY BY TETRAMETHYLSILANE-BASED ATMOSPHERIC VAPOR DEPOSITION 705
Y. Hatakeyama, M. Esashi, and S. Tanaka

DEVELOPMENT OF A MONOLITHIC PMMA COMB-DRIVE MICRO ACTUATOR UTILIZING HOT EMBOSSING AND ULTRA-PRECISION MACHINING 709
S. Amaya, D.V. Dao, and S. Sugiyama

OPTICALLY PROGRAMMABLE SELF-ASSEMBLY OF HETEROGENEOUS MICRO-COMPONENTS ON UNCONVENTIONAL SUBSTRATES 713
E. Saeedi, J.R. Etzkorn, L. Draghi, and B.A. Parviz

FABRICATION OF NEAR NET SHAPE ALUMINA NICKEL COMPOSITE MICRO PARTS USING AQUEOUS SUSPENSION 717
H.S. Hassanin and K. Jiang

VERTICALLY ALIGNED VARIOUS LENGTHS DOPED-SILICON MICROWIRE ARRAYS BY REPEATED SELECTIVE VAPOR-LIQUID-SOLID GROWTH 721
T. Kawano, A. Ikedo, T. Kawashima, K. Sawada, and M. Ishida

FABRICATION OF ELECTROSTATICALLY-ACTUATED, IN-PLANE FUSED QUARTZ RESONATORS USING SILICON-ON-QUARTZ (SOQ) BONDING AND QUARTZ DRIE 725
Y.-S. Hwang, H.-K. Jung, E.-S. Song, I.-J. Hyeon, Y.-K. Kim, and C.-W. Baek

ADJUSTABLE REFRACTIVE INDEX METHOD FOR COMPLEX MICROSTRUCTURE BY AUTOMATED DYNAMIC MODE MULTIDIRECTIONAL UV LITHOGRAPHY 729
J.K. Kim, T.-S. Yun, H. Jee, and Y.K. Yoon

A NEW HIGH-SENSITIVITY PACKAGE-LEAK TESTING METHOD FOR MEMS SENSORS 733
M. Fujiyoshi, Y. Nonomura, and H. Senda

SPATIALLY ARRANGED MICROELECTRODES USING WIRE BONDING TECHNOLOGY FOR SPATIALLY DISTRIBUTED CHEMICAL INFORMATION ACQUISION 737
W. Tonomura, K. Shimizu, and S. Konishi

WAFER SCALE ENCAPSULATION OF LARGE LATERAL DEFLECTION MEMS STRUCTURES 741
A.B. Graham, M. Messana, P. Hartwel, J. Provine, S. Yoneoka, B. Kim, R. Melamud, R.T. Howe, and T.W. Kenny

PARYLENE-POCKET CHIP INTEGRATION 745
R. Huang and Y.C. Tai

A LOW-POWER OVEN-CONTROLLED VACUUM PACKAGE TECHNOLOGY FOR HIGH-PERFORMANCE MEMS 749
S.-H. Lee, J. Cho, S.-W. Lee, M.F. Zaman, F. Ayazi, and K. Najafi

DIRECT WIRE-BONDING OF SILICON DEVICES WITHOUT METAL PADS 753
A. Hirshberg and D. Elata

PHYSICAL SENSORS

A LOW-NOISE OSCILLATOR BASED ON A MULTI-MEMBRANE CMUT FOR HIGH SENSITIVITY RESONANT CHEMICAL SENSOR 757
H.J. Lee, K.K. Park, P. Cristman, Ö. Oralkan, M. Kupnik, and B.T. Khuri-Yakub

PERFORMANCE OF PACKAGED PIEZOELECTRIC MICROSPEAKERS DEPENDING ON THE MATERIAL PROPERTIES 761
S.H. Yi, S.C. Ur, and E.S. Kim

CMOS INTEGRATED STRESS MAPPING CHIPS WITH 32 N-TYPE OR P-TYPE PIEZORESISTIVE FIELD EFFECT TRANSISTORS 765
P. Gieschke, Y. Nurcahyo, M. Herrmann, M. Kuhl, P. Ruther, and O. Paul

DEVELOPMENT OF FABRIC-TYPE OF TENSIONAL FORCE SENSOR 769
Y. Suzuki, D. Ogura, M. Shikida, Y. Hasegawa, and K. Sato

MEASUREMENT RESULTS OF THE FIRST TWO CHIP SILICON MICROPHONE WITH LOW STRESS NICKEL MEMBRANE COVERING FULL AUDIO RANGE 773
S. Junge, F. Jakobs, and W. Lang

TOWARDS PIEZORESISTIVE CMOS SENSORS FOR OUT-OF-PLANE STRESS 777
B. Lemke, K. Kratt, R. Baskaran, and O. Paul

FLEXIBLE TACTILE SENSOR SHEET WITH LIQUID FILTER FOR SHEAR FORCE DETECTION 781
K. Noda, K. Matsumoto, and I. Shimoyama

BIOMIMETIC DESIGN AND FABRICATION OF COMPONENTS FOR ARTIFICIAL HAIR CELL SENSOR 785
H.S. Lee, B.-K. Lee, and T.H. Kwon

A NOVEL CONFIGURABLE MEMS INERTIAL SWITCH USING MICROSCALE LIQUID-METAL DROPLET 789
K. Yoo and J. Kim

NOVEL MICRO CAPACITIVE INCLINOMETER WITH OBLIQUE COMB ELECTRODE AND SUSPENSION SPRING ALIGNED PARALLEL TO {111} VERTICAL PLANES OF (110) SILICON 793
D.-H. Jeong, S.-S. Yun, M.-L. Lee, G. Hwang, C.-A. Choi, and J.-H. Lee

A DIGITAL OUTPUT PIEZOELECTRIC ACCELEROMETER USING PATTERNED PB(ZR,TI)O3 THIN FILMS ELECTRICALLY CONNECTED IN SERIES 797
T. Kobayashi, H. Okada, T. Masuda, and T. Itoh

ACCELERATION COMPENSATION OF MEMS RESONATORS USING ELECTROSTATIC TUNING 801
S. Yoneoka, G. Bahl, J. Salvia, K.L. Chen, A.B. Graham, H.K. Lee, G. Yama, R.N. Candler, and T.W. Kenny

DESIGN AND IMPLEMENTATION OF A NOVEL CMOS-MEMS SINGLE PROOF-MASS TRI-AXIS ACCELEROMETER 805
C.-M. Sun, M.-H. Tsai, and W. Fang

A LATCHING ACCELERATION SWITCH WITH MULTI-CONTACTS INDEPENDENT TO THE PROOF-MASS 809
Z.Y. Guo, Z.C. Yang, L.T. Lin, Q.C. Zhao, H.T. Ding, X.S. Liu, X.Z. Chi, J. Cui, and G.Z. Yan

AN ANALYSIS TO IMPROVE STABILITY OF DRIVE-MODE OSCILLATIONS IN CAPACITIVE VIBRATORY MEMS GYROSCOPES 813
S.E. Alper, K. Sahin, and T. Akin

MODELING OF A CAPACITIVE Σ-Δ ACCELEROMETER SYSTEM INCLUDING THE NOISE COMPONENTS AND VERIFICATION WITH TEST RESULTS 817
B. Boga, I.E. Ocak, H. Külah, and T. Akin

AIR PRESSURE SENSOR ON AN INSECT WING 821
H. Takahashi, K. Matsumoto, and I. Shimoyama

A FULLY CMOS-COMPATIBLE MICRO-PIRANI GAUGE BASED ON A CONSTANT CURRENT CIRCUIT 825
J. Wang and Z. Tang

GLASS AND QUARTZ MICROSCINTILLATORS FOR CMOS COMPATIBLE MULTI-SPECIES RADIATION DETECTION 829
  R.P. Waguespack, H. Berry, S. Pellegrin, and C.G. Wilson

SEMICONDUCTING Y-BA-CU-O INFRARED MICROBOLOMETER ARRAY FABRICATED AND CHARACTERIZED WITH CCBDI ROIC 833
  S. Kumar1 and D.P. Butler2

A MICROMACHINED QUARTZ AND STEEL PRESSURE SENSOR OPERATING UP TO 1000°C AND 2000 TORR 837
  S.A. Wright and Y.B. Gianchandani

HIGH SPEED (GHZ), ULTRA-HIGH PRESSURE (GPA) SENSOR ARRAY FABRICATED IN INTEGRATED CMOS+MEMS PROCESS 841
  M. Okandan, R. Olsson, M. Baker, P. Resnick, T.A. Hill, C. Lackey, S. Pearson, J. Castaneda, W. Trott, and D. Jones

SINGLE WAFER SURFACE MICROMACHINED FIELD EMISSION ELECTRON SOURCE 844
  F. Santagata, C.K. Yang, J.F. Creemer, P.J. French, and P.M. Sarro

SUSPENDED MEMBRANE SINGLE CRYSTAL SILICON MICRO HOTPLATE FOR DIFFERENTIAL SCANNING CALORIMETRY 848
  J. Lee1, C.M. Spadaccini2, E.V. Mukerjee2, and W.P. King1

RF MEMS AND RESONATORS

SIMPLE AND ROBUST AIR GAP-BASED MEMS SWITCH TECHNOLOGY FOR RF-APPLICATIONS 852
  P. Ekkels, X. Rottenberg, P. Czarnecki, R. Puers, and H.A.C. Tilmans

RUTHENIUM/GOLD HARD-SURFACE/LOW-RESISTIVITY CONTACT METALLIZATION FOR POLYMER-ENCAPSULATED MICROSWITCH WITH STRESS-REDUCED CORRUGATED SIN/SIO$_2$ DIAPHRAGM 856
  F. Ke, J. Miao, and J. Oberhammer

A MEMS RECONFIGURABLE QUAD-BAND CLASS-E POWER AMPLIFIER FOR GSM STANDARD 860
  L. Larcher, R. Brama, M. Ganzerli, J. Iannacci, B. Margesin, M. Bedani, and A. Gnudi

A RF-MEMS BASED MONOLITHICALLY INTEGRATED WIDE RANGE TUNABLE LOW PASS FILTER FOR RF FRONT END APPLICATION 864
  T. Fujii, T. Irieda, K. Nakamura, K. Ota, K. Ihara, D.T. McCormick, and L. Lin

LOW ACTUATION VOLTAGE SPDT RF MEMS K BAND SWITCH USING A SINGLE GOLD MEMBRANE 868
  R. Robin, O. Millet, K. Segueni, and L. Buchaillot

PERFORMANCE ENHANNCEMENT BY SUBSTRATE PERFORATION FOR A WAFER-LEVEL ENCAPSULATED RF MEMS DC SHUNT SWITCH 872
  F. Ke, J. Miao, and C.W. Tan

A STRESS TOLERANT TEMPERATURE-STABLE RF MEMS SWITCHED CAPACITOR 876
  I. Reines1, B. Pillans2, and G.M. Rebeiz1

TEMPERATURE COMPENSATION IN SILICON-BASED MICRO-ELECTROMECHANICAL RESONATORS 880
  F. Schoen, M. Nawaz, T. Bever, R. Gruenberger, W. Raberg, W. Weber, B. Winkler, and R. Weigel

A 12-GHZ DPDT RF-MEMS SWITCH WITH LAYER-WISE WAVEGUIDE/ACTUATOR DESIGN TECHNIQUE..................884
D. Yamane, H. Seita, W. Sun, S. Kawasaki, H. Fujita, and H. Toshiyoshi

POST-FABRICATION ELECTRICAL TRIMMING OF SILICON BULK ACOUSTIC RESONATORS USING JOULE HEATING..................888
A.K. Samarao and F. Ayazi

RF MEMS HIGH-IMPEDANCE TUNEABLE METAMATERIALS FOR MILLIMETER-WAVE BEAM STEERING..................892
M. Sterner, D. Chicherin, A.V. Räisenen, G. Stemme, and J. Oberhammer

MONOLITHICALLY INTEGRATED PIEZOMEMS SP2T SWITCH AND CONTOUR-MODE FILTERS..................896
J.S. Pulskamp, D.C. Judy, R.G. Polcawich, R. Kaul, H. Chandrahalim, and S.A. Bhave

A LIQUID-METAL RF MEMS SWITCH WITH DC-TO-40 GHZ PERFORMANCE..................900
P. Sen and C.-J. Kim

DESIGN AND FABRICATION OF A GAUSSIAN-SHAPED AT-CUT QUARTZ CRYSTAL RESONATOR..................904
T. Abe and H. Kishi

GAS-LIQUID SEPARATED RESONATOR FOR BIO-CHEMICAL APLICATION..................908
H. Hida, M. Shikida, M. Okochi, H. Honda, and K. Sato

5 - 10 GHZ ALN CONTOUR-MODE NANOELECTROMECHANICAL RESONATORS..................912
M. Rinaldi, C. Zuniga, and G. Piazza

CHAOS IN ELECTROSTATICALLY ACTUATED RF-MEMS MEASURED AND MODELED..................916
J. Stulemeijer, R.W. Herfst, and J.A. Bielen

THERMOELASTIC FEM-BEM MODEL FOR MEMS RESONATOR LOSS SIMULATION..................919
D. Ekeom and L. Buchaillot

RESONANT PIEZOELECTRIC ALGAN/GAN MEMS SENSORS IN LONGITUDINAL MODE OPERATION..................923
K. Brueckner, F. Niebelschuetz, K. Tonisch, R. Stephan, V. Cimalla, O. Ambacher, and M.A. Hein

FULLY DIFFERENTIAL INTERNAL ELECTROSTATIC TRANSDUCTION OF A LAMÉ-MODE RESONATOR..................927
M. Ziaei-Moayyed, D. Elata, J. Hsieh, J.-W.P. Chen, E.P. Quévy, and R.T. Howe

ANALYTICAL MODELING AND NUMERICAL SIMULATION OF CAPACITIVE SILICON BULK ACOUSTIC RESONATORS..................931
G. Casinovi, X. Gao, and F. Ayazi

THERMAL ANALYSIS AND CHARACTERIZATION OF HIGH Q FILM BULK ACOUSTIC RESONATOR (FBAR) AS BIOSENSERS IN LIQUIDS..................935
X. Zhang, W. Xu, A. Abbaspour-Tamijani, and J. Chae

OPTICAL MEMS

EL DISPLAY PRINTED ON CURVED SURFACE..................939
T.-M. Lee, H.-C. Choi, J.-H. Noh, and D.-S. Kim

SOLUTION ELECTROCHEMILUMINESCENT MICROFLUIDIC CELL FOR FLEXIBLE AND STRETCHABLE DISPLAY..................943
R. Okumura, S. Takamatsu, E. Iwase, K. Matsumoto, and I. Shimoyama

MEMS BASED LASER IMAGER WITH DIAGONAL PROGRESSIVE SCANNING 947
T. Sandner, T. Grasshoff, T. Klose, H. Schenk, and J.L. Massieu

**POLYMERIC (SU-8) OPTICAL MICROSCANNER DRIVEN BY ELECTROSTATIC
ACTUATION** 951
S.K. Lee, M.G. Kim, J.Y. An, M.H. Jun, S. Yang, and J.H. Lee

**VACUUM WAFER LEVEL PACKAGED TWO-DIMENSIONAL OPTICAL SCANNER
BY ANODIC BONDING** 955
H. Tachibana, K. Kawano, H. Ueda, and H. Noge

**ACCURACY ENHANCEMENT IN THE IN-PLANE DYNAMIC MEASUREMENT OF
MEMS ACTUATORS USING A LASER DOPPLER VIBROMETER WITH A 45º-
ANGLED OPTICAL FIBER** 959
M.G. Kim1, K.W. Jo1, Y.S. Park2, W.G. Jang, and J.H. Lee1

MICRO LIQUID PRISM 963
Y. Yoshihata, A. Takei, N. Binh-Khiem, T. Kan, E. Iwase, K. Matsumoto, and I. Shimoyama

MEMS OPTICAL LOGIC NOR GATE USING INTEGRATED TUNABLE LASERS 967
B. Liu, H. Cai, X.M. Zhang, J. Tamil, Q.X. Zhang, and A.Q. Liu

**TILTED PARABOLOIDAL REFLECTIVE LENS FOR FAR INFRARED
FABRICATED BY MASK WITH RECTANGULAR OPENINGS** 971
T. Takahata, K. Matsumoto, and I. Shimoyama

**MEMS LASER WITH TUNABLE WAVELENGTH AND POLARIZATION USING
OPTICAL TUNNELING EFFECT** 975
W.M. Zhu, X.M. Zhang, H. Cai, J. Tamil, W. Zhang, B. Liu, T. Bourouina, and A.Q. Liu

**A NOVEL TELESCOPE WITH MICROMIRROR FOR OBSERVATION OF
TRANSIENT LUMINOUS EVENTS FROM SPACE** 979
J.H. Park, S. Nam, G. Garipov, J.A. Jeon, J.Y. Jin, B. Khrenov, J.E. Kim, M. Kim, Y.K. Kim, J. Lee,
G.W. Na, I.H. Park, Y.-S. Park, and B.W. Yoo

**NON-SPHERICAL SU-8 MICROLENS ARRAY FABRICATED UTILIZING A NOVEL
STAMPING PROCESS AND AN ELECTRO-STATIC PULLING METHOD** 983
S.-M. Kuo and C.-H. Lin

**MULTICHAMBER TUNABLE LIQUID MICROLENSES WITH ACTIVE
ABERRATION CORRECTION** 987
D. Mader, P. Waibel, A. Seifert, and H. Zappe

**OPTICAL MICROMIRROR ACTUATION USING THERMOCAPILLARY EFFECT IN
MICRODROPLETS** 991
R.K. Dhull, I. Puchades, L. Fuller, and Y.W. Lu

DUAL-AXES CONFOCAL MICROLENS FOR RAMAN SPECTROSCOPY 995
C.P.B. Siu, H. Wang, H. Zeng, and M. Chiao

**TUNABLE SCANNING FIBER OPTIC MEMS-PROBE FOR ENDOSCOPIC OPTICAL
COHERENCE TOMOGRAPHY** 999
K. Aljasem, A. Seifert, and H. Zappe

**CMOS-INTEGRABLE PISTON-TYPE MICRO-MIRROR ARRAY FOR ADAPTIVE
OPTICS MADE OF MONO-CRYSTALLINE SILICON USING 3-D INTEGRATION** 1003
M. Lapisa, F. Zimmer, F. Niklaus, A. Gehner, and G. Stemme

**MONOLITHIC MULTICOLOR TOTAL INTERNAL REFLECTION (TIR)-BASED
CHIP FOR HIGHLY-SENSITIVE MULTIFLUORESCENCE DETECTION AND
IMAGING** 1007
N.C.H. Le, D.V. Dao, R. Yokokawa, J. Wells, and S. Sugiyama

CMOS-COMPATIBLE 2-AXIS SELF-ALIGNED VERTICAL COMB-DRIVEN MICROMIRROR FOR LARGE FIELD-OF-VIEW MICROENDOSCOPES 1011
K. Kumar and X.J. Zhang

TOPOLOGY OPTIMIZATION FOR MICRO ROTATIONAL MIRROR DESIGN AND SAFE MANUFACTURING 1015
T. Chen, Z. Liu, J.G. Korvink, S. Krausse, and U. Wallrabe

STRETCHABLE YARN OF DISPLAY ELEMENTS 1019
S. Takamatsu, K. Matsumoto, and I. Shimoyama

A MICROMACHINED THERMO-OPTIC TUNABLE LASER 1023
H. Cai, B. Liu, X.M. Zhang, W.M. Zhu, J. Tamil, W. Zhang, Q.X. Zhang, and A.Q. Liu

MEMS ACTUATORS & POWER MEMS

MASS-ANALYSIS SCANNING FORCE MICROSCOPY WITH ELECTROSTATIC SWITCHING MECHANISM 1027
C.Y. Shao, Y. Kawai, T. Ono, and M. Esashi

A MULTIPLE DEGREES OF FREEDOM ELECTROTHERMAL ACTUATOR FOR A VERSATILE MEMS GRIPPER 1031
D.-S. Chen, C.-Y. Yin, R.-J. Lai, and J.-C. Tsai

BIDIRECTIONAL ELECTROTHERMAL ELECTROMAGNETICTORSIONAL MICROACTUATORS 1035
Y. Eun, H. Na, and J. Kim

A 2D ELECTRET-BASED RESONANT MICRO ENERGY HARVESTER 1039
U. Bartsch, J. Gaspar, and O. Paul

PZT MEMS ACTUATED FLAPPING WINGS FOR INSECT-INSPIRED ROBOTICS 1043
J.R. Bronson, J.S. Pulskamp, R.G. Polcawich, C.M. Kroninger, and E.D. Wetzel

A NOVEL UNDERWATER ACTUATOR DRIVEN BY MAGNETIZATION REPULSION/ATTRACTION 1047
H.-T. Su, T.-L. Tang, and W. Fang

DEVELOPMENT OF THE FORWARD-LOOKING ACTIVE MICRO-CATHETER ACTUATED BY TI-NI SHAPE MEMORY ALLOY SPRINGS 1051
M. Komatsubara, T. Namazu, H. Nagasawa, T. Miki, T. Tsurui, and S. Inoue

LOW-RESONANT-FREQUENCY MICRO ELECTRET GENERATOR FOR ENERGY HARVESTING APPLICATION 1055
M. Edamoto, Y. Suzuki, N. Kasagi, K. Kashiwagi, Y. Morizawa, T. Yokoyama, T. Seki, and M. Oba

DEVELOPMENT OF THE NOVEL ELONGATION-MEASUREMENT DEVICE WITH IN-PLANE BIMORPH ACTUATOR FOR THE TENSILE TEST 1059
H. Fujii, T. Namazu, and S. Inoue

MICROACTUATION UTILIZING WAFER-LEVEL INTEGRATED SMA WIRES 1063
D. Clausi, H. Gradin, S. Braun, J. Peirs, G. Stemme, D. Reynaerts, and W. van der Wijngaart

AN INTEGRATED SOLUTION FOR WAFER-LEVEL PACKAGING AND ELECTROSTATIC ACTUATION OF OUT-OF-PLANE DEVICES 1067
K.L. Chen, R. Melamud, S. Wang, and T.W. Kenny

AN ELECTROMAGNETIC MICRO POWER GENERATOR FOR LOW FREQUENCY ENVIRONMENTAL VIBRATIONS BASED ON THE FREQUENCY UP-CONVERSION TECHNIQUE 1071
I. Sari, T. Balkan, and H. Külah

ELECTROSTATIC ROTARY STEPPER MICROMOTOR FOR SKEW ANGLE COMPENSATION IN HARD DISK DRIVES 1075
  E. Sarajlic, C. Yamahata, M. Cordero, and H. Fujita

A SURFACE-TENSION-DRIVEN PROPULSION AND ROTATION PRINCIPLE FOR WATER-FLOATING MINI/MICRO ROBOTS 1079
  S.K. Chung, K. Ryu, and S.K. Cho

DEVELOPMENT OF SKEWED DRIE PROCESS AND ITS APPLICATION TO ELECTROSTATIC TILT MIRROR 1083
  M. Nakada, K. Takahashi, T. Takahashi, A. Higo, H. Fujita, and H. Toshiyoshi

DESIGN OF A PASSIVE AND PORTABLE DMFC OPERATING IN ALL ORIENTATIONS 1087
  N. Paust, S. Krumbholz, S. Munt, C. Müller, R. Zengerle, C. Ziegler, and P. Koltay

A ROTARY MICROACTUATOR SUPPORTED ON ENCAPSULATED MICROBALL BEARINGS USING AN ELECTRO-PNEUMATIC THRUST BALANCE 1091
  M. McCarthy, C.M. Waits, M.I. Beyaz, and R. Ghodssi

PASSIVATED ELECTRODE ACTUATOR WITH STABLE RESONANCE AMPLITUDE 1095
  G. Bahl, R.G. Walmsley, B.E. DeMartini, K.L. Turner, and P.G. Hartwell

FAST POSITIONING AND IMPACT MINIMIZING OF MEMS DEVICES BY SUPPRESSION MOTION-INDUCED VIBRATION BY COMMAND SHAPING METHOD 1099
  K.-S. Chen and K.-S. Ou

A MICRO DIRECT METHANOL FUEL CELL INTEGRATED WITH A TEMPERATURE CONTROL SYSTEM FOR EXTREME ENVIRONMENTS 1103
  Q. Zhang, X. Wang, Y. Zhu, Y. Zhou, X. Qiu, and L. Liu

LEVER-BASED CMOS-MEMS PROBES FOR RECONFIGURABLE RF IC'S 1107
  J. Liu, M. Noman, J.A. Bain, T.E. Schlesinger, and G.K. Fedder

A NEW EFFICIENT METHOD FOR SIMULATING THE DYNAMIC RESPONSE OF ELECTROSTATIC SWITCHES 1111
  V. Leus and D. Elata

Author Index

LETTER FROM THE CO-CHAIRS

We welcome you to the 22nd IEEE International Conference on Micro Electro Mechanical Systems (MEMS 2009) in Sorrento, Italy!

The IEEE MEMS Conference series has started in 1987 and is known as the IEEE International Conference on Micro Electro Mechanical Systems since 1999. Over the last two decades, the MEMS community has experienced significant growth in various areas of device science and technology. New activities have evolved into major research fields such as Bio-MEMS, Optical MEMS, RF MEMS, Power MEMS, NEMS and many more interesting fields. The enthusiasm of this community is evident from the creativity in relentlessly expanding the range of MEMS applications.

It is our great pleasure to announce that this year we received a record breaking number of 856 abstract! A total of 279 papers were carefully selected by the 24 experts of the Technical Program Committee (TPC). The presentations are arranged in a single-session format that includes three invited keynote presentations, 47 oral presentations and 232 posters. The poster sessions have always been a highlight of this Conference and follow the successful poster/oral presentation format pioneered in 2004. Poster presentations are divided into four sessions to facilitate interaction with the authors, who will have their posters on display throughout the Conference. All Conference papers, oral and poster, are included in this Technical Digest, which is distributed on an USB stick to all Conference attendees. For the first time in the MEMS Conference history, printed books are available on request, only.

We would like to express our sincere gratitude to all the authors of the submitted abstracts. Their high quality work serves as the foundation for the success of this Conference. The papers were selected by the TPC made up with equal representation from three regional divisions: America, Europe & Africa, and Asia & Oceania. Four sub-committees were formed in order to facilitate a careful review of the large number of submitted abstracts. Each abstract has been evaluated and rated by six expert members of the TPC. The committee recommendations on the acceptance or rejection of papers were taken as binding. We are grateful to all TPC members who volunteered their valuable time, including participation in a two-day on-site meeting in Zürich, Switzerland, for paper selection. We are also indebted to the International Steering Committee and the Advisory Co-Chairs, Oliver Brand and Yitshak Zohar, for generously sharing their experience. We gratefully acknowledge the industrial support groups, exhibitors and benefactors for their involvement in this Conference, and are very thankful to the IEEE Robotics and Automation Society for their continued support of this meeting. The dedicated effort of Ms. Katharine Cline and her team at PMMI in managing this Conference is greatly appreciated.

In closing, we hope you enjoy the Conference presentations, posters, exhibition booths and events this week in Sorrento!

Lina Sarro
General Co-Chair

Christofer Hierold
General Co-Chair

ORGANIZING COMMITTEE

General Co-Chairs

Christofer Hierold
ETH Zürich, SWITZERLAND

Pasqualina M. Sarro
Delft University of Technology,
THE NETHERLANDS

Technical Program Committee

Christian Bergaud
LAAS, FRANCE

Sunil Bhave
Cornell University, USA

Karl Böhringer
University of Washington, USA

Oliver Brand
Georgia Institute of Technology, USA

David Elata
Technion - Israel Institute of Technology, ISRAEL

Udo-Martin Gómez
Robert Bosch GmbH, GERMANY

Christofer Hierold
ETH Zürich, SWITZERLAND

Gwo-Bin Lee
National Cheng Kung University, TAIWAN

Qiao Lin
Columbia University, USA

Liwei Lin
University of California, Berkeley, USA

Jianmin Miao
Nanyang Technological University, SINGAPORE

Takahito Ono
Tohoku University, JAPAN

Beth Pruitt
Stanford University, USA

Pasqualina M. Sarro
Delft University of Technology,
THE NETHERLANDS

Kazuaki Sawada
Toyohashi University of Technology, JAPAN

Yuji Suzuki
University of Tokyo, JAPAN

Wouter van der Wijngaart
KTH Royal Institute of Technology, SWEDEN

Xiaohong Wang
Tsinghua University, CHINA

Man Wong
Hong Kong University of Science and Technology,
HONG KONG

Eric Yeatman
Imperial College, UK

Jun-Bo Yoon
Korea Advanced Institute of Science and Technology
(KAIST), SOUTH KOREA

Euisik Yoon
University of Minnesota, USA

Hans Zappe
University of Freiburg, GERMANY

Yitshak Zohar
University of Arizona, USA

ORGANIZING COMMITTEE, continued

International Steering Committee

Co-Chairs

Oliver Brand
Georgia Institute of Technology, USA

Yitshak Zohar
University of Arizona, USA

Members

Tayfun Akin
Middle East Technical University, TURKEY

Karl Böhringer
University of Washington, USA

Oliver Brand
Georgia Institute of Technology, USA

Jong-Uk Bu
SenPlus Inc., KOREA

Christofer Hierold
ETH Zürich, SWITZERLAND

Liwei Lin
University of California, Berkeley

Satoshi Konishi
Ritsumeikan University, JAPAN

Lina Sarro
Delft University of Technology,
THE NETHERLANDS

Yuji Suzuki
University of Tokyo, JAPAN

Man Wong
Hong Kong University of Science and Technology,
CHINA

Roland Zengerle
University of Freiburg - IMTEK, GERMANY

Yitshak Zohar
University of Arizona, USA

MEMS Benefactor's Group and Exhibitors

We gratefully acknowledge the support of this Conference from the following companies and institutions as of the printing of December 1, 2008:

ASML Netherlands B.V.

Conventor, Inc.

DIMES - Delft University of Technology

ETH Zürich

Heidelberg Instruments Mikrotechnik GmbH

IEEE

The IET

IOP Publishing

John Wiley & Sons, Ltd.

MICRONARC

NEC SCHOTT Components Corporation

POLYTEC GmbH

Primaxx, Inc.

Semefab (Scotland) Limited

ST Microelectronics

Surface Technology Systems

Tokyo Electron Ltd.

XACTIX, Inc.

Yole Développement

MEMS 2008 TECHNICAL PROGRAM COMMITTEE

1 - Qiao Lin
2 - Marcel Suter (recorder)
3 - Yuji Suzuki
4 - Moritz Mattmann (recorder)
5 - Euisik Yoon
6 - Man Wong
7 - Matthias Muoth (recorder)
8 - Gwo-Bin Lee
9 - Beth Pruitt
10 - Xiaohong Wang
11 - Kazuaki Sawada
12 - Karl Böhringer
13 - Jun-Bo Yoon
14 - Udo-Martin Gómez
15 - Oliver Brand
16 - Pasqualina M. Sarro (Co-Chair)
17 - Sunil Bhave
18 - Liwei Lin
19 - Hans Zappe
20 - Christofer Hierold (Co-Chair)
21 - Eric Yeatman
22 - Christian Bergaud
23 - Yitshak Zohar
24 - Takahito Ono
25 - Wouter van der Wijngaart
26 - David Elata
27 - Jianmin Miao

MEMS EPIPHANY

Benedetto Vigna
STMicroelectronics
MEMS and Healthcare, RF and Sensors Product Division
Via Tolomeo, 1 – 20010 Cornaredo – Milano, Italy
Benedetto.Vigna@st.com

ABSTRACT

This paper addresses the key drivers of the "MEMS Epiphany," i.e. the successful interplay between the Technology Push and Design-Driven innovation strategies for MEMS (Micro-Electromechanical Systems) success in the consumer market. Before 2005, MEMS were known only by technology experts and were not yet ready for the high-volume consumer market.

Only the marriage of a technological breakthrough in the MEMS world, pioneered by STMicroelectronics, and the creation and delivery of new applications for end users, driven by Nintendo and Apple, paved the way for the MEMS Epiphanies. This paper will focus on the technological challenges that STMicroelectronics faced to enable this historical landmark in the "MEMS Consumerization" era. And it will also discuss briefly the next possible MEMS Epiphanies in relation to new products and applications.

INTRODUCTION

The Micro-Electromechanical Systems (MEMS) world is very complex with many different facets. MEMS are silicon-based microsystems where electrons aren't the only things moving. Membranes, cantilevers, suspended masses, and fluids are the living actors of the micromechanical world. Today, MEMS is a sexy word attracting the attention of many customers and as a result, of many semiconductor suppliers.

MEMS can be divided in two wide classes:

1. Actuators
2. Sensors.

The best known actuators are the thermal inkjets for printers and the tiny micro-mirrors of video projectors using DLP™ technology. These products find widespread use in our offices and homes. The most widely used sensors are: accelerometers, pressure sensors, flow sensors, gyroscopes and microphones.

Actuators are mostly used in the consumer market, while the automotive, industrial and medical market segments represent today the biggest markets for all physical sensors.

All automotive active and passive safety systems, like Vehicle Dynamic Control and Air Bags, use micromachined inertial sensors to continuously protect human lives.

Tiny pressure sensors measure barometric pressure and the intake manifold pressure to optimize the fuel/air ratio and allow us to drive in a more environmentally efficient way by reducing gasoline consumption. Pressure sensors in the tires increase passenger safety by measuring pressure in the tires constantly and alerting the driver if it falls below a threshold.

Pressure sensors and flow sensors are also used in the industrial market for factory automation, preventive machine diagnostics and optimization of machine operating conditions.

In the medical market accelerometers are mounted in pacemakers to adapt the pacing signal frequency to the patient's physical activity and ultra miniaturized pressure sensors are used for temporary invasive pressure measurements in several parts of the human body.

But in November 2006, a MEMS Epiphany occurred in the video game industry and this was followed in June 2007, with another Epiphany that occurred in the mobile phone industry. Prof. Roberto Verganti describes Nintendo Epiphany in his upcoming book "Design Driven Innovation." Nintendo Wii™ and Apple iPhone™ and iTouch™, the most popular consumer gadgets of 2007, were introduced and they each contained a micro-machined accelerometer as their beating heart. After those Epiphanies, laptops, mobile phones, Digital Still Camera and Camcorders, Personal Multimedia Players, Personal Navigation Devices, and many other electronic gadgets began to make wide usage of micro-machined sensors. Free-fall detection, more intuitive user interfaces, health monitoring, noise cancellation and image stabilization are just a few of the many applications for these sensors in the consumer market.

The MEMS Consumerization wave is driving the development of inexpensive, tiny and ultra low-power sensors that, coupled with a wireless module, enable the deployment of wireless sensor networks. It's commonly believed that after the current "Nomadic Era," wireless sensor networks built around appropriate sensors will drive the next big commercial adoption wave for semiconductors.

Thanks to the field-proven reliability of several hundreds of millions of devices in the consumer market and to the well consolidated 8" MEMS manufacturing plants and related economy of scale, in the near future automotive, medical and industrial markets will all benefit.

MICROMACHINED SENSORS IN THE AUTOMOTIVE MARKET

Modern cars use many different micromachined sensors in safety, comfort and environmental applications. Almost all of these systems require the fusion and concurrent evaluation of signals coming from different sensors. Pressure sensors, one- and two-axis accelerometers, yaw and roll gyroscopes, and air-flow sensors are the MEMS most commonly used in the car.

The first commercial application of MEMS dates back to 1974, when silicon micromachined pressure sensors appeared in automotive electronic fuel-injection sys-

978-1-4244-2977-6/09 $25.00 © 2009 IEEE

tems to monitor manifold absolute pressure (MAP) and mass airflow. The micromachined mass airflow sensor measures the mass of air entering the engine. This data, combined with information coming from a non-MEMS oxygen sensor and a MEMS-based MAP, optimizes the air-to-fuel ratio, reducing fuel consumption.

The 1980's brought a proliferation of pressure sensors: from manifold absolute pressure to barometric pressure, fuel pressure, oil pressure and more. Barometric Pressure sensors measure the altitude, since the atmospheric pressure decreases as you go from sea level up into the mountains, automatically leaning the fuel mixture. The monitoring of barometric pressure is important for the car's engine to maintain efficient operating conditions.

In the 1980's, the U.S. government required that all cars produced after April 1, 1989 include a driver's side airbag. In 1993, the U.S. National Highway Transportation Safety Administration (NHTSA) mandated dual front airbags in all passenger vehicles, and a few years' later, side-impact airbags also became a requirement. The core of the airbag system is a micromachined accelerometer that detects sudden deceleration and, in conjunction with a proper algorithm, decides whether or not to fire an explosive charge for fast deployment of the airbag.

Accelerometers used for airbags located in the center of the car are able to detect accelerations from 20g to 70g, where g is the Earth's gravitational acceleration.

Unfortunately, the unmodulated explosive charge in first-generation airbags were lethal for child passengers and smaller drivers sitting close to the steering wheel. Later-generation airbags are safer, since they deploy at a speed matching the weight of the passenger. Some Passenger Occupant Detection (POD) systems use pressure sensors to detect and monitor the way the passenger is seated before the bag deploys.

In the 1990's, the number of accelerometers per car grew exponentially. Some automakers started to use more accelerometers in the periphery of the car (satellite accelerometers with a full-scale range from 100g to 500g) to better protect passengers in case of accidents. These sensors gave the microcontroller more time to decide whether to deploy the airbag or not.

Moreover, automobile manufacturers began to deploy active safety systems – Vehicle Dynamic Control (VDC™) or Electronic Stability Control (ESC™) – after 1996. In contrast to an airbag, which is a passive safety system because it is deployed after the accident event happens, ESC is an active safety system, because it operates to prevent accidents from happening.

ESC systems require high-grade low-g accelerometers (up to 5g), yaw-rate gyroscopes, and magnetic speed sensors at each wheel. They help drivers better handle a car on icy or wet roads by working to prevent dangerous under- or over-steering. While all accelerometers used for this application are manufactured in silicon, gyroscopes are both silicon and quartz based.

The ESC system was the precursor of anti-rollover systems (ARO). Both ARO and ESC use similar sensor technologies and, in fact, have now been combined into a single system.

Modern ESC systems currently are only used in some European vehicles. But in 2004 the NHTSA released a study showing a 35% reduction in single-vehicle crashes for cars with ESC, with a reduction of up to 67% for SUVs. Thus, in April 2006, NHTSA mandated the adoption of active safety systems ESC in all U.S. passenger vehicles, beginning with 2012 vehicles; a similar mandate for all European cars is in place for 2013 vehicles.

In 2000 Firestone had disastrous quality issues with its tires that resulted in more than 270 deaths and almost 1,000 injuries. As a result, tire pressure sensors started to gain momentum. These sensors measure pressure inside the tire and alert the driver if the pressure is too high or too low. All vehicles from 2008 will have a tire pressure monitoring system, which consists of a pressure sensor, a simple microcontroller or state machine, and a wireless link. The system also contains an accelerometer that wakes-up the system when the vehicle is moving; some systems also have a secondary accelerometer to identify the side of the car where the problem tire is mounted.

Another automotive application for MEMS is in navigation systems based on Global Positioning Systems, where accelerometers, gyros and magnetic sensors can help the driver by performing dead reckoning to compensate for urban canyons, tunnels and other blind spots, when satellite signals from the GPS can't be detected.

CARVING OUT A QUIESCENT MEANING FOR THE CONSUMER MARKET

"Near the Georgian border there is a spring, from which gushes a stream of oil in such abundance that a hundred ships may load there at once. This oil is not good to eat; but it is good for burning and as a salve for men and camels affected with itch or scab. Men come from a long distance to fetch this oil, and in all the neighborhood no other oil is burnt but this." -- Marco Polo. (From Prof. Roberto Verganti's "Design Driven Innovation")

"The statue is already embedded in the marble block, I'm just removing the marble in excess." -- Michelangelo Buonarroti.

As Professor Roberto Verganti describes in his forthcoming book "Design Driven Innovation," every technology embeds many quiescent meanings, some of which are potentially disruptive, although they may not be visible at first, like the fountains of petroleum that Marco Polo encountered in his travels to China.

Only leading companies, nimble and visionary enough, as Michelangelo was, are able to carve out the new meaning and thus have a Technology Epiphany: *the manifestation of the essential and more powerful meaning of that technology.*

The Nintendo Wii console effectively combines a radical innovation of meaning with a radical innovation of technology; it's a MEMS Epiphany. On the Meaning axis, Nintendo said, "Playing is believing," suggesting that passive immersion in a virtual world only targeted to young players was over. On the Technology side, Wii uses a breakthrough technology: an inexpensive high-performance three-axis micro-machined accelerometer that is able to sense any movement in space.

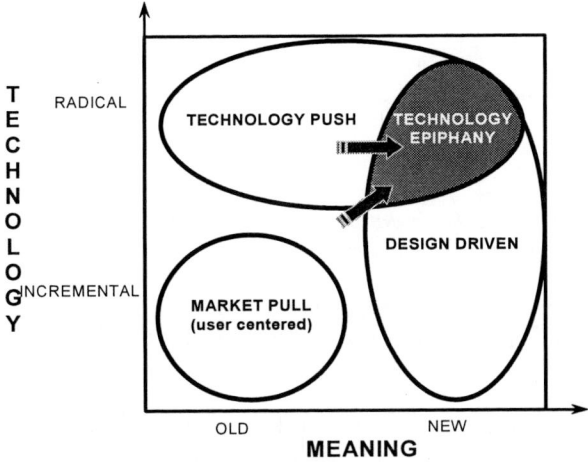

Figure 1: An innovation strategy comes from positioning between Meaning and Technology. (From Prof. Verganti's book "Design Driven Innovation")

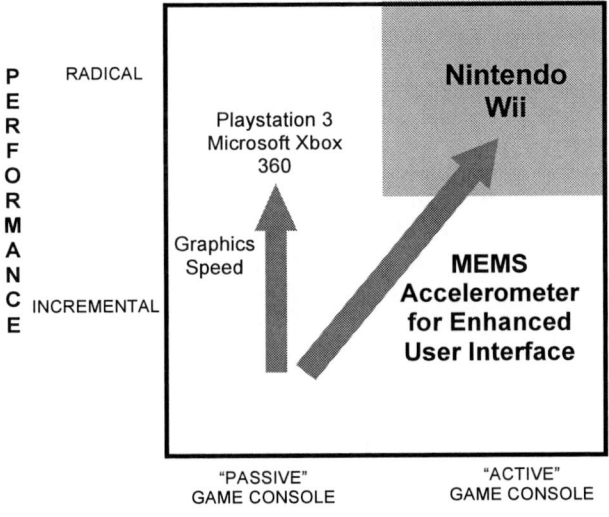

Figure 2: Comparison of innovation strategies of the main players of the game industry: Nintendo, Microsoft and Sony (From Prof. Verganti's book "Design Driven Innovation")

Nintendo's Wii console effectively combines a radical innovation of meaning with a radical innovation of technology; it's a Technology Epiphany.

Nintendo, Sony and Microsoft are the three major players in the $30 billion video-game market. Nintendo was the leader in late 1980's and early 1990's, when it brought new approaches to the industry. But with Sony's launch of the Playstation in 1995 and the Playstation 2 in 2000, and Microsoft's launch of the Xbox in 2001, Nintendo lost its leadership and fell into hard times.

Three years ago, Microsoft and Sony were pushing hard to deliver ever more processing power in their gaming systems, offering high-definition images and more complex games and graphics. Nintendo instead decided to play a different game by offering a radical change in meaning, compared with its competitors. Nintendo chose to deliver a physical experience that could be played, not with thumbs, but with the entire body, using natural movements common to sports and vigorous games.

And the same successful interplay between technology- push and design-driven innovation applies to the latest generation products from Apple, its iPhone™ and iTouch™ of 2007, to the iPhone, iPod™ and iTouch launched in the market just a few months ago.

All these new products demonstrate a radical new meaning to the movements the end-user makes while holding the portable equipment in their hands. And a MEMS accelerometer is the technology breakthrough enabling it.

Actually, in 2003, a few mobile-phone manufacturers had already approached the MEMS world to embed a wellness and health dimension into their phones by incorporating a pedometer function. But the perceived value to the end user was not so high. So the pedometer function did not become a MEMS Epiphany.

After June 2007, motion sensors began to become the new mouse for many handheld devices. The combination of micro-machined accelerometers and the appropriate application software eliminated the need for conventional switches or button and thumb wheels for scrolling, zooming and panning of web pages, e-books, and spreadsheets. This was an innovative way to solve the well-known "small button – big finger" problem that plagued many users. In fact, while small cell phones are convenient and easy to carry, their small display screens and limited graphic capabilities impair the total user experience. The sensor in these new-age mice in handheld devices detect basic human movements and use them as the input for display orientation, which in turn simplifies how the user views pictures or downloads Web pages. The user can simply navigate through web pages or pan through maps by simply tilting the device in the desired direction.

From mobile phones to Portable Multimedia Players and to remote controllers, the adoption of MEMS has been very easy and obvious. In fact, Apple's iTouch MP4 player uses a 3-axis accelerometer for a more intuitive user interface; the same is true for the latest generation of input devices from Logitech and Kodak.

In 2003, STMicroelectronics won an EDN Innovation Award as the first company to launch in production a three-axis accelerometer. After this historical landmark STMicroelectronics developed a complete family of ultra-low power, ultra-compact, highly manufacturable and inexpensive analog and digital accelerometers to support its customers (Fig. 3). Now, thanks to this Technology Epiphany, ST leads the consumer market for Motion Sensors.

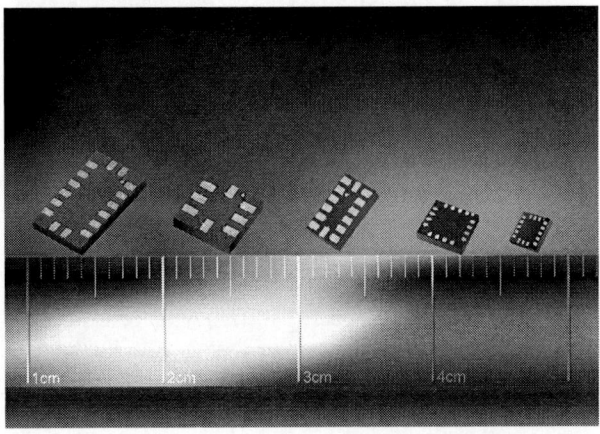

Figure 3: Land Grid Array three-axes accelerometers in different form factors from STMicroelectronics.

All STMicroelectronics' Motion Sensors for Consumer Market applications come in Land Grid Array (LGA) packages. Their footprint ranges from 35mm2 of the package on the extreme left in Fig. 3, to 9mm2 for the package on the extreme right. Thickness varies from 0.9mm to 1.5mm depending on the configuration of the two chips in the package itself, stacked or side-by-side and on the needs of the customer.

LIS3L02AL: THE TECHNOLOGY BREAKTHROUGH OF THE MEMS EPIPHANY

STMicroelectronics started high volume production for Nintendo Wii controllers in March 2006 with its LIS3L02AL. The sensor is a low-g micro-machined capacitive, three-axis high-performance accelerometer with analog output. It consists of two chips stacked in one package with eight leads (Fig. 3, second package from left).

ST faced many challenges in both technology and design of both chips (the MEMS mechanical transducer and an electronic interface). Among these were wafer manufacturing, product testing, and calibration. The remainder of this paper will focus on the challenges of the high-volume assembly and testing, since in 2004 no technique in use in the Automotive market was applicable to the Consumer Market.

High Volume Assembly

Accelerometers for automotive applications had been on the market for more than two decades and the assembly and packaging technology used for these devices made constant progress. But in 2004 those techniques were not yet ready to support the high volume production requirements of the consumer market. Some suppliers were packaging MEMS sensors one by one in their packaging line using pre-molded or ceramic packages. Other suppliers were sealing the tiny components directly in the cleanroom with a wafer-to-wafer bonding technology. These other suppliers were using bulky packages, like SO16 or SO24, with gel inside to guarantee good product performance.

ST selected low-melting-point glass, i.e. glass frit, to seal the micromechanical structure of its LIS3L02AL product at the wafer level, since this was the most commonly used technique. Unfortunately, no package was able to meet all the requirements of consumer-market high-volume production.

In 2003 STMicroelectronics pioneered the design and development of the LGA matrix platform to package the LIS3L02AL product. The metals layers and vias of the substrate were designed to reduce stresses on the sensing element. Due to the different physical properties of the material used in the package, thermal and mechanical stresses can be induced during manufacturing (i.e., molding and soldering of the devices) as well as during life in the field (i.e., thermal ageing, drops, shock damage). Extensive Finite Element Model (FEM) simulations during the design phase and customized experiments in the laboratory were used to tailor and optimize the design of the metal layers as well as the positions of the wire bonding and pads on the silicon dice.

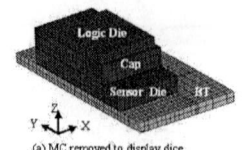 

(a) MC removed to display dice   (b) ½ Package model

Figure 4: FEM simulations of LIS3L02AL accelerometer assembled in LGA package 5x5x1.5mm³

Fig. 4 is an example of FEM simulation of LIS3L02AL. The sensor die is first bonded to a substrate with adhesive tape. The package footprint is minimized by stacking the logic die on top of the sensor die. After wire bonding, the devices on the strip are ready to be encapsulated via an injection molding process (see Fig. 5).

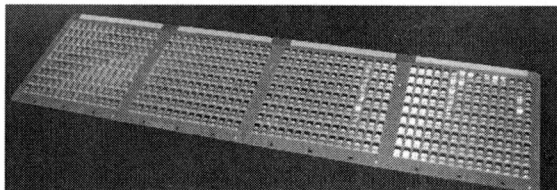

Figure 5: Strip populated of dice just before injection molding.

Still, the proper substrate design was not sufficient to guarantee robust and stable product. The use of low-stress glue to attach the silicon chip to the substrate or to the electrical chip and the selection of a proper mold material were additional key elements to obtain good and stable product performance. STMicroelectronics has worked, and continues to work, with materials suppliers to design dedicated low-stress materials.

Table 1: Properties of the materials important for the performance of the LIS3L02AL product.

Materials	E (GPa)	Tg (°C)	α₁ (ppm/°C)	α₂ (ppm/°C)
Silicon	131	-	2.8	-
Substrate Core	xy: 26 z:11	-	xy:15.2 z:40	-
Solder Mask	2.4	101	50	160
Copper	117	-	17.7	-
Die Attach	1.5	128	80	170
Mold Compound	23.5	125	8	34
Glass Frit	70	-	7	-

The thickness of the sensor, cap, and electronics dice also affect mechanical stress distribution and thus the final product's performances. In fact, FEM simulations show that compound molding warpage will result in sensor warpage, and thus ultimately influences the performance and stability of the final product. The warpage of the sensor substrate affects the signal performance of the accelerometer. Typically, the warpage effect after molding and post-mold curing is compensated by signal calibration prior to shipping products to customers. On the other hand, temperature variations over the sensor's lifetime also induce thermal mismatch and thus deformation of the sensor microstructures. These deformations would cause a drift of the sensor signal.

978-1-4244-2977-6/09 $25.00 © 2009 IEEE

Therefore, it is not sufficient to calibrate the device at a single temperature, but the dependence of the warpage on the temperature and its influence on the signal must be properly considered during the design of the entire product.

Production Testing and Calibration

Innovative technologies typically require dedicated novel testing equipment. In the early 2000's there was no high-volume testing equipment meant to calibrate a three-axis accelerometer, since the automotive industry was not using these products.

MEMS testing is performed on the final device after completion of all assembly-related process steps. A first screening of device function is done at the wafer level by means of electrical stimuli. This is called Electrical Wafer Sorting (EWS). Despite the moving structures present on a MEMS wafer, there is no difference between micro-system EWS and standard integrated-circuit EWS. MEMS testing at the packaging level aims at calibration of the devices and verification of the correct performance. Calibration and verification of the LIS3L02AL is performed to guarantee stability over the sensor's lifetime as well as to reduce drift versus temperature, i.e. about 0.5 mg/°C.

The specificity of LIS3L02AL testing arose from the need to stimulate the sensor in all three axes. In fact, key product parameters, such as sensitivity or zero-g level (output with no applied acceleration), had to be guaranteed for each component by means of spatial movements.

Figure 6: Testing equipment for MEMS accelerometers.

Fig. 6 shows that the test board for the LIS3L02AL is not fixed, but can move or shake in space in the X, Y and Z directions. The test board is typically decoupled mechanically from the other machine components. Furthermore, decoupling elements between the testing equipment and the testing floor reduces environmental vibrations and disturbances during product calibration. The handlers and sockets are also customized to minimize mechanical deformation and stress on the sensor element during calibration. LIS3L02AL insertion and its placement in the socket are carefully monitored because non-planar insertion or incorrect fixing in the socket could result in a calibration error due to an anomalous stress distribution.

High-volume testing methodologies, suitable to match the price target of customer, were also developed for the LIS3L02AL product. In fact, STMicroelectronics pioneered a novel testing strategy based on the strip format of the LGA package family: accelerometers were not tested one-by-one, but all together on a single strip, as in Fig. 5.

Strip testers present the following advantages:

- They reduce handling and stress issues related to the testing of single small packages, like the 3×3×0.9mm³ package used for the LIS331AL and LIS331DL products.
- They improve quality and speed up manufacturing via increased parallelism in going from 4/20 devices on classical robot handlers to 24/160 devices, representing a productivity increase of a factor of almost ten.
- They offer the possibility to switch easily between different classes of products. With a simple modification of the stimulus chamber, the same equipment can be used to test inertial MEMS (e.g., accelerometers, gyroscopes) and differential MEMS (e.g., pressure sensors, microphones).
- They enable the testing of sensor modules requiring the application of different physical stimuli.

NEW MEMS PRODUCTS

Today, accelerometers are broadly used in the consumer market and the accelerometer's product roadmap is meant to satisfy customer needs for many years.

The same is not yet true for multiple-axis gyroscopes, pressure sensors and high-performance digital microphones. MEMS suppliers, like STMicroelectronics, are actively developing low-power, ultra-compact, inexpensive and highly manufacturable devices. This is the first step towards the integration of different sensing functions in a single module.

STMicroelectronics is now producing a Coriolis-based yaw-axis gyroscope, with analog and digital output, in a package 5x5x1.5mm3 (Fig. 7)

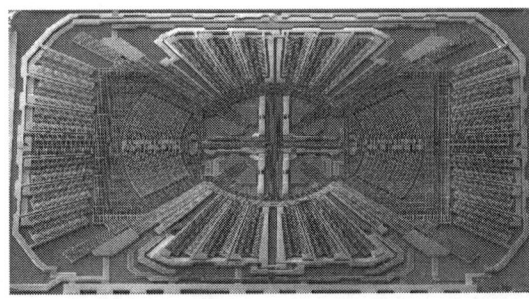

Figure 7: Mechanical Element of the Yaw axis gyroscope.

STMicroelectronics is also developing multiple-axis gyroscopes and compact pressure sensors using an innovative packaging solution (Fig. 8). In fact, STMicroelectronics patented a special package, Holed Land Grid Array (HLGA) that exploits all the manufacturing assets it has already installed for inertial sensors to offer customers in consumer markets very small and thin packages.

Figure 8: HLGA 3x3x1 mm³ pressure sensor without amplification and compensation IC, before and after injection molding (First Row); HLGA 5x5x1 mm³ pressure sensor with amplification and compensation IC, after die attach and wire bonding on substrate (second row, left picture) and after injection molding and singulation (right picture)

978-1-4244-2977-6/09 $25.00 © 2009 IEEE

THE NEXT EPIPHANY IN THE SENSOR WORLD

Microfabricated sensors are already used in numerous applications in industrial and medical markets. Smart factory automation, optimization of water and energy consumption in household appliances, preventive machine diagnostics, and continuous monitoring of patient activity are just a few examples of potential new epiphanies in markets beyond the consumer market. These applications could include fall-detection systems for elderly people that automatically activate an emergency call, context awareness for domestic robots to help with housework, asset tracking and monitoring for more efficient logistic services, security modules for rescuers such as firefighters, for people practicing different sports, and ambient assisted-living applications.

All of these new applications will likely require the deployment of wireless sensor networks, which today are still in their infancy. A wireless sensor network combines different motes, i.e., wireless sensor modules consisting of some combination of a sensor, an integrated circuit controller, a wireless receiver, an antenna and a battery, or even better, an energy scavenger.

Although it is difficult to predict when a new Technology Epiphany will occur, I'm sure that the combination of sensing function with wireless transmission (i.e. mote) will open new markets and new applications. It will drive us in the Wireless Sensing Network Era.

The potential market for motes is limited only by our imagination. Motes could, for example, find application in consumer markets, with solutions ranging from security and bio-detection to building and home automation, industrial control, pollution monitoring, and agriculture. Also, rising concerns for safety, convenience, entertainment, and efficiency, coupled with worldwide government regulation, could boost sensor usage to unprecedented levels, though not all of the sensors would necessarily be silicon-microfabricated.

Will there be another Epiphany like Nintendo's Wii™ or the Apple iPhone™? Today the answer is not clear, but suppliers must be ready to integrate different technologies in a modular format as well as to achieve high-volume manufacturability of motes.

REFERENCES

[1] Roberto Verganti, *Design Driven Innovation*. Boston, MA: Harvard Business Publishing, March 2009.

[2] Oboe R, Antonello R, Lasalandra E, et al. (2005) *Control of a z-axis MEMS vibrational gyroscope*. IEEE/ASME Transactions on Mechatronics 10:364-370.

[3] Vigna B. (2005), *More than Moore: micro-machined products enable new applications and open new markets*. Invited Talk, Proc. International Electron Devices Meeting, Washington, DC, USA, 8 pp.

[4] Bourne M. (2007) *A consumer's guide to MEMS and Nanotechnology*. Bourne Research LLC.

[5] www.st.com/mems.

[6] Johnson R. (2007) There's more to MEMS than meets the iPhone. EETimes, *http://www.eetimes.com/show Article.jhtml?articleID=200900669*.

[7] Toedosio M (2007) Making MEMS the new 'mouse'. EETimes Asia, *http://www.eetasia.com/ART_ 88004609 68_499495_NT_734b413b.HTM*.

[8] Marriott M (2006) At the heart of the Wii, micron-size machines. New York Times, *http://www.nytimes.com/2006/12/21/technology/21ho ww.html*. Accessed 04 Jan 08.

DEVELOPMENT OF CALIBRATION STANDARDS FOR THE OPTICAL MEASUREMENT OF IN-PLANE DISPLACEMENTS OF MICROMECHANICAL COMPONENTS

J. Gaspar[1], J. Held[1], G. Pedrini[2], W. Osten[2], and O. Paul[1]

[1]Department of Microsystems Engineering (IMTEK), University of Freiburg, GERMANY
[2]Institute for Technical Optics (ITO), University of Stuttgart, GERMANY

ABSTRACT

The goal of this work is to develop miniaturized reference standards for the optical measurement of in-plane displacements serving for the calibration of optical systems used in the production and characterization of MEMS. The proposed devices consist of SOI-based in-plane microactuators. The displacements resulting from both mechanical and electrostatic actuations are measured by means of optical techniques such as stroboscopic illumination, laser-deflection method and digital speckle interferometry.

INTRODUCTION

The trend towards miniaturization of microelectromechanical systems (MEMS) and micro-opto-electromechanical systems (MOEMS) continues to lead to more and more compact devices. Advantages lie in the integration with control electronics, lower power dissipation, higher sensitivity and better performance. Applications cover a wide range of fields, from optical telecommunications to medicine [1]. Part of the functionality and reliability of MEMS devices is based on the displacement and deformation of micromechanical parts under mechanical, thermal, magnetic or electrostatic loads. The measurement of the deformation of such systems is thus of great importance for confirming analytical and finite element models, accessing material and device properties, detecting potential defects and determining performance. These data can then be used in further device optimization. Since typical structures exhibit dimensions in the order of some micrometers, it is necessary to measure their deformation with accuracies down to the lower nanometer range. Standardized approaches and calibrated setups are therefore essential for the measurement of shape, displacement and deformation fields in the static and dynamic cases.

Several non-contact measuring procedures, suitable for the measurement of shape, deformation and strain of small objects, have been previously reported. Optical methods already applied in the inspection and measurement of microsystems include classic, grid, white light, digital holographic and digital speckle interferometries, electro-optical holography, digital speckle photography, laser-deflection technique, Doppler-vibrometry and digital image analysis and correlation [2]-[4]. Even though these techniques have been successfully tested in laboratory, their implementation in industrial micro-manufacturing environments has only taken place in few cases. A deficit is that calibration procedures are not available yet, or more precisely, have only been developed for macroscopic components through projects such as the standardization project for optical measurement of strain (SPOTS) [5], [6].

The work presented here aims at narrowing this gap by developing standards or norms for the use of different full-field optical techniques in the measurement of in-plane displacements of microsystems. The process itself involves the following steps:

(i) Development of micromechanical reference components designed to deform in a reproducible and precise way when submitted to known loads;

(ii) Calibration of the micromechanical reference devices by means of optical techniques, performed in laboratory-controlled conditions;

(iii) Determination of the measurement uncertainties according to the guide of expression of uncertainty in measurement (GUM [7]) and certification of the optical calibration setups by official bodies, allowing for traceability of the observed quantities to the international SI standards.

After development and calibration, the reference devices may then be used for the general calibration of optical measurement systems used in the MEMS industry.

This paper presents the fabrication and characterization of such micromechanical reference structures. The processing technology is based on the micromachining of silicon-on-insulator (SOI) substrates. The design tries to cover a broad variety of specifications found in micromechanical devices by including (a) several predefined displacement ranges tuned through mechanical demultiplication mechanisms, (b) surfaces with different optical characteristics (in terms of reflectivity, roughness and geometry patterns) and (c) more than one actuation mechanism on the same chip. The optical techniques used include stroboscopic illumination, laser-deflection method and speckle interferometry.

MICROMECHANICAL REFERENCE

The developed reference test structure is schematically shown in Fig. 1. It consists of a movable large body connected to anchors through parallel springs and mechanically coupled by a bending element to a smaller displacement structure, which is also stabilized by parallel elements. The bending element is responsible for demultiplying the displacement between the two movable bodies. The structural elements are made of crystalline silicon (c-Si) from the active layer of SOI substrates with a thickness t of 25 μm.

As a consequence of the arrangement and dimensions of the parallel springs, the test structure illustrated in Fig. 1 is relatively resilient in the x-direction and strongly constrained in the y- and z-directions, which allows for the in-plane displacement of the inner frame along the x-axis when a force is applied to it. Based on standard beam theory, any such spring acts with spring constants [8]

$$k_x = Ew^3t/L^3, \qquad (1)$$
$$k_y = Ewt/L^3, \text{ and} \qquad (2)$$

978-1-4244-2977-6/09 $25.00 © 2009 IEEE

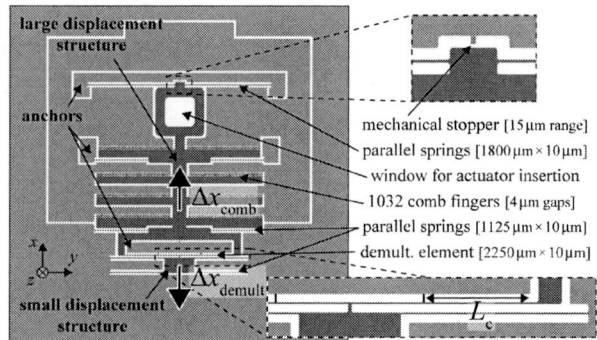

Figure 1: Schematics of micromechanical reference object with in-plane movable structures connected to anchors through parallel springs. The displacement demultiplication $\Delta x_{demult}/\Delta x_{comb}$ is set by the clamping length L_c.

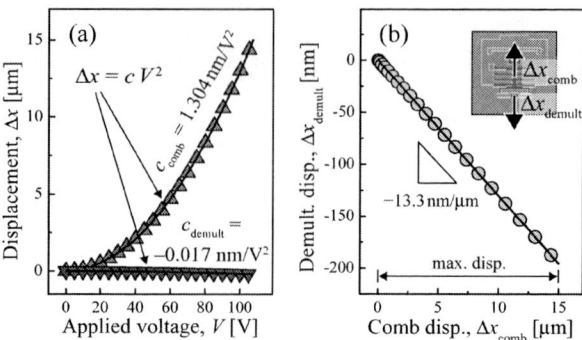

Figure 2: (a) FEM-simulated displacements of the comb structure and demultiplied motion body as a function of applied voltage and (b) resulting demultiplication characteristic curve for a microdevice with $L_c = 837.5\ \mu m$.

$$k_z = Ewt^3/L^3, \qquad (3)$$

along x, y and z, respectively, where L and w are the length and width of the spring, respectively, and $E = 168$ GPa is the elastic modulus along the $<100>$ direction. The contribution from the association of these elements is $k_x = 0.013$ mN/μm, $k_y = 196.0$ mN/μm, and $k_z = 0.083$ mN/μm for the structures guiding the large displacement mass and $k_x = 0.012$ mN/μm, $k_y = 150$ mN/μm, and $k_z = 0.074$ mN/μm for the body with demultiplied motion. This highlights the relative mechanical anisotropy of the proposed micromechanical reference. Moreover, the overall stiffness values are increased by the inclusion of the bending, demultiplication element.

The test structure can be actuated by mechanical and/or electrostatic means. In the case of mechanical excitation, the design includes a window, connected to the large displacement structure, allowing for the insertion of the pin of an external actuator. The on-chip electrostatic actuation is possible by applying a voltage V between the electrodes of a comb drive comprising a number n of 1008 gaps with a width g of 4 μm. The electrostatic force thus generated is [9]

$$F = n\varepsilon_r\varepsilon_0 tV^2/2g, \qquad (4)$$

where $\varepsilon_r \approx 1$ is the relative dielectric constant of air and ε_0 is the vacuum permittivity. Based on the layout shown in Fig. 1, and according to Eq. (4), F/V^2 amounts to 0.028 μN/V^2.

Mechanical stoppers have been implemented as well, limiting the displacement of the comb structure to target values of 15 μm. By construction, the displacements imposed to this body, Δx_{comb}, are demultiplied to amplitudes Δx_{demul} with opposite direction of the small displacement structure. Consequently, the demultiplied displacement body possesses a well-determined working range as well. Two different ranges are thus available on the same micromechanical device, enabling its use in optical calibration systems with different characteristics. The demultiplication factor $\Delta x_{demul}/\Delta x_{comb}$ can easily be tuned by varying the clamping length L_c: to designed values of L_c between 387.5 to 837.5 μm correspond modeled values of $\Delta x_{demul}/\Delta x_{comb}$ between −116.3 and −13.3 nm/μm. Figure 2 (a) shows the displacement of one device as a function of the voltage applied to the comb drive, obtain-

ed from finite element modeling (FEM) simulations, for both comb and demultiplied movements. The same displacements are plotted against each other in Fig. 2 (b).

FABRICATION

The processing of the micromechanical reference devices relies on the bulk micromaching of SOI substrates by deep reactive ion etching (DRIE) and the subsequent removal of the oxide layer in order to release the movable parts. The final stages of the fabrication process are summarized in Figs. 3 (a) to (c): after depositing and patterning thin films of low-temperature oxide (LTO), poly-silicon (poly-Si) and aluminum (Al), the 25-μm-thick SOI device layer is structured by DRIE, thus defining the geometry of the mechanical elements (movable components, springs, anchors, comb fingers), Fig. 3 (a). Next follows the rear DRIE of the 500-μm-thick silicon handle, with the 2-μm-thick SOI buried oxide acting as the etch stop, Fig. 3 (b). The final step, shown in Fig. 3 (c), consists of the oxide layer removal with concentrated 73% hydrofluoric (HF) acid, a process highly selective against the silicon structures, poly-Si films and Al metallization [10]. The poly-Si structures previously structured on top of the movable components, along with the bare c-Si areas and Al patterns serve as surfaces with different optical properties suitable for measurements using different optical techniques. The Al layer also serves the purpose of providing electrical contact to the comb drive. Variations of the fabrication process can be envisaged in order to integrate other films as optical ele-

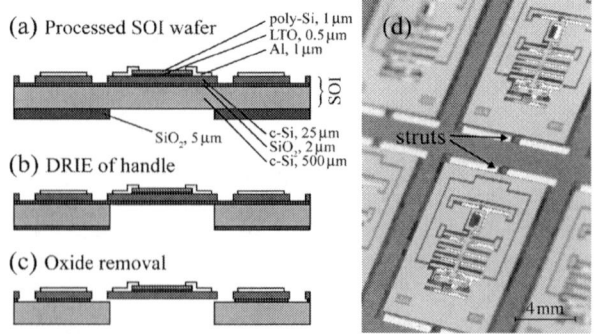

Figure 3: (a)–(c) Final steps of the fabrication process of the proposed micromechanical reference devices and (d) photograph of processed chips at the wafer level, after their release using a 73%-HF solution.

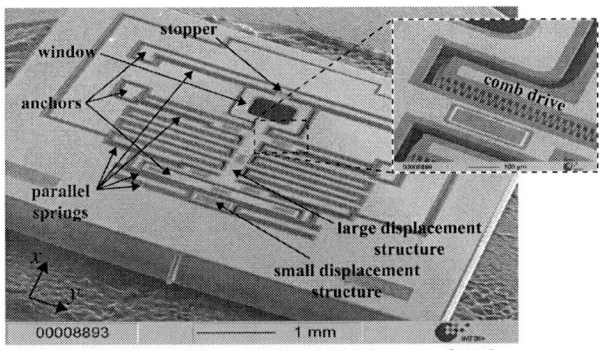

Figure 4: SEM micrograph of a fabricated reference test object. The displacement of the moving parts can be excited either mechanically or electrostatically. The close-up details the comb drive fingers comprising 1008 gaps.

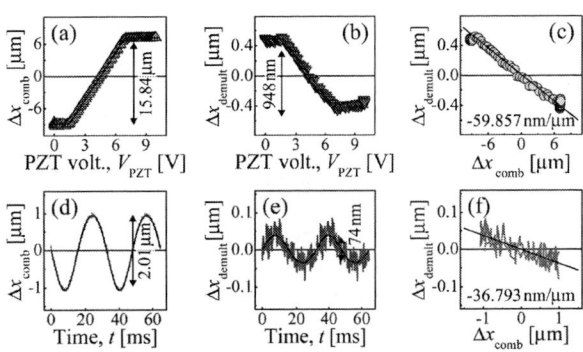

Figure 6: Displacements (a) Δx_{comb} and (b) Δx_{demult} as a function of the external PZT voltage and (c) demultiplication curve. (d)–(f) Data obtained from applied electrostatic comb loads [$V_{DC} = 4V_{AC} = 40$ V, $f = 31$ Hz].

ments. Figure 3 (d) shows a photograph of fabricated devices at the wafer level. Single, released devices can be extracted by gentle twisting that breaks the connecting struts, as developed elsewhere [11]. Scanning electron microscope (SEM) graphs of a fabricated microdevice are shown in Fig. 4.

EXPERIMENTAL

The different optical arrangements used to characterize the micromechanical reference are shown in Fig. 5. The mechanical actuation is provided by a comercial PZT actuator with a pin mounted on its tip, which is inserted into the window of the mechanical structure and enables its displacement along the x-direction. The electrostatic excitation is made possible by applying a voltage to the comb actuator via electrical connections to the socket into which the chip carrier with wirebonded device is mounted. The sample is placed inside a vacuum chamber with window, enabling optical measurements at pressures from 760 Torr down to 10^{-6} Torr.

Both laser-deflection method and stroboscopic video microscopy are implemented in the setup shown in Fig. 5 (a). In the case of the laser-deflection technique, a laser diode (658 nm) is partially focused on the microdevice, whose movement modulates the intensity of the reflected beam, which in turn is monitored with an avalanche photo diode (APD) detector [12]. The stroboscopic apparatus is a commercial system (Polytec MSA 400) that combines the acquisition of pictures using short light pulses synchronized with the object motion and

digital analysis, reaching resolutions down to 1 nm. The speckle interferometer is shown in Fig. 5 (b): the interference pattern of the speckle fields is produced by illuminating the sample with a Nd:YAG laser (532 nm) from two directions, and is recorded with a CCD camera [13], [14].

RESULTS AND DISCUSSION

Figure 6 shows typical measurement results and Fig. 7 presents a comparison between modeled and experimentally extracted results. In Fig. 6 (a), the displacement of the comb structure Δx_{comb} of a device with $L_c = 536.5$ µm clearly shows a well-defined range, 15.84 µm, as the displacement of the external PZT actuator is increased via the voltage applied to it, V_{PZT}. This value is close to the design value of 15 µm, limited by the contact point between the PZT and test structure and by the implemented mechanical stopper. The resulting demultiplied motion amplitude Δx_{demul} shown in Fig. 6 (b) is limited in this case to 948 nm, linear with V_{PZT} but, as expected, with negative slope. The resulting demultiplying characteristic, i.e. Δx_{demul} plotted as a function of Δx_{comb}, is shown in Fig. 6 (c), from which a factor of ~ -60 nm/µm is extracted. The measured electrostatic characteristics of a micromechanical reference with $L_c = 687.5$ µm are summarized in Figs. 6 (d) to (f). In this case, a voltage with DC and AC components, $V = V_{DC} + V_{AC} \sin(2\pi ft)$, with $V_{DC} = 40$ V, $V_{AC} = 10$ V, and $f = 31$ Hz, is applied to the comb drive. The frequency

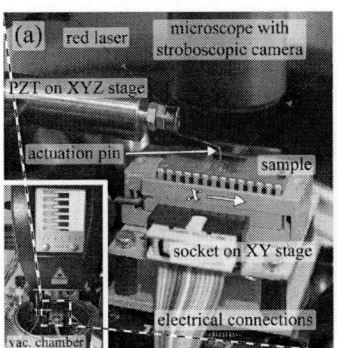

Figure 5: (a) Laser-deflection test setup and stroboscopic microscope for measurements in vacuum, and (b) digital speckle interferometer developed to characterize the displacements of the micromechanical references in air.

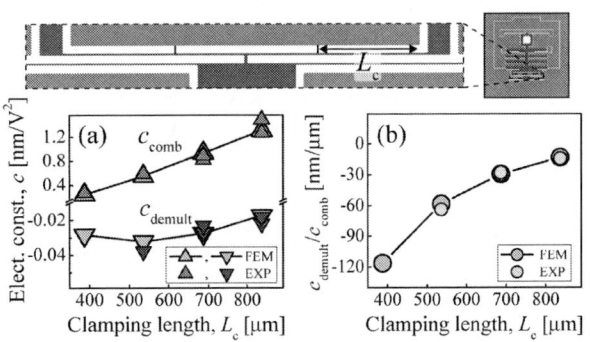

Figure 7: (a) Simulated and experimentally extracted electrostatic constants [$c = \Delta x/V^2$] for the comb and small displacement structures and (b) resulting demultiplication factor as function of the clamping length L_c.

f is far below the resonance frequency of the micro-mechanical devices and, therefore, such quasi-static conditions can be fully modeled by static mechanical models. Both Δx_{comb} and Δx_{demul} show a quadratic dependence on V, $\Delta x = cV^2$, Figs. 6 (b1) and (b2), respectively, and, expectedly, the demultiplied motion has a phase shift of 180° with respect to the comb drive. The constants c_{comb} and c_{demul} thus obtained are plotted in Fig. 7 (a) and the demultiplication factor is shown in Fig. 7 (b) as a function of L_c, showing good agreement between measured and FEM-modeled data. The micromechanical reference devices have resonance frequencies f_{res} in the kHz range, as shown in Fig. 8.

CONCLUSIONS

A micromechanical in-plane reference device for the calibration of MEMS optical measurement systems is proposed, fabricated and characterized. Both mechanical and electrostatic actuations are possible. Different coatings processed on the movable parts make it possible to characterize optical setups based on different principles using the same device. Moreover, a simple demultiplication mechanism extends the working range of the reference standard, thus covering a larger spectrum of displacements of MEMS devices.

ACKNOWLEDGMENTS

The authors gratefully acknowledge A. Baur (IMTEK RSC) for assistance in clean room processing, M. Pappalardo (IMTEK MML) for development of the laser-deflection setup in vacuum and T. Wu (ITO) for development of the speckle interferometer. This work was supported by the German Research Foundation (DGF) under grants PA792/4-1 and OS111/22-1.

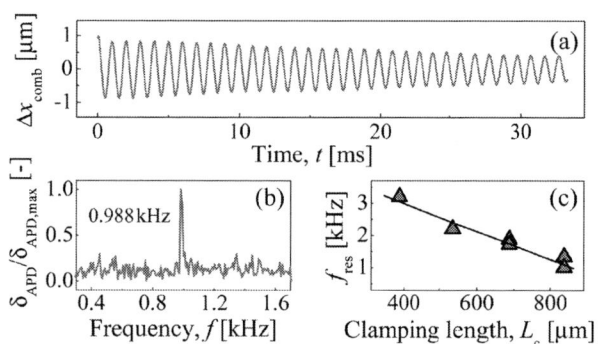

Figure 8: (a) Response of the comb structure in air to a voltage step, as a function of time, (b) resonance peak measured in vacuum (10^{-6} Torr) and (c) resonance frequency obtained as a function of the clamping length L_c.

REFERENCES

[1] J. Korvink, O. Paul, eds., *MEMS – A Practical Guide to Design, Analysis and Applications*, William Andrew Publishing, Norwich, NY, 2006.

[2] W. Osten, ed., *Optical Inspection of Microsystems*, CRC Press, Boca Raton, FL, 2006.

[3] W. Osten, ed., *Optical Microsystems Metrology I – Special Issue of Optics and Lasers in Engineering*, vol. 36, pp. 75-240, 2001.

[4] W. Osten, editor, *Optical Microsystems Metrology II – Special Issue of Optics and Lasers in Engineering II*, vol. 36, pp. 401-526, 2001.

[5] SPOTS: Standardization Project for Optical Techniques of Strain Measurement, *in* www.opticalstrain.org.

[6] M. Kujawinska, E. A. Patterson, R. L. Burgute, E. Hack, D. Mendels, T. Siebert, and M. Whelan, "Calibration and Assessement of Full-Field Optical Strain Measurement Procedures and Instrumentation," in *Proc. SPIE*, vol. 6341, 63410Q, 2006.

[7] International Standardization Organization (ISO), *Guide to the Expression of Uncertainty in Measurement (GUM)*, Geneva, 1995.

[8] S. Kamiya, J. H. Kuypers, A. Trautmann, P. Ruther, and O. Paul, "Process Temperature-Dependent Mechanical Properties of Polysilicon Measured Using a Novel Tensile Test Structure," *J. Microelectromech. Syst.*, vol. 16, pp. 202-212, 2007.

[9] H. J. De Los Santos, *Introduction to Microelectro-mechanical (MEM) Microwave Systems*, Artech House, Norwood, MA, 1999.

[10] P. T. J. Gennissen, and P. J. French, "Sacrificial Oxide Etching Compatible with Aluminum Metallization," in *Tech. Dig. Tranducers'97*, Chicago, June 16-19, 1997, pp. 225-228.

[11] J. Gaspar, M. Schmidt, and O. Paul, "Comparison of Improved Techniques for Mechanical Thin Film Characterization – Application to Polysilicon," in *Tech. Dig. Transducers'07*, Lyon, June 10-14, 2007, 575-578.

[12] J. Gaspar, V. Chu, and J. P. Conde, "Electrostatic microresonators from doped hydrogenated amorphous and nanocrystalline silicon thin films," *J. Microelectromech. Syst.*, vol. 14, pp.1082-1088, 2005.

[13] G. Pedrini *et al.*, "Calibration of optical systems for the measurement of microcomponents," *Optics and Lasers in Engineering*, 2008, in press.

[14] R. Jones, C. Wykes, *Holographic and Speckle Interferometry*, Cambridge University, NY, 1989.

3D MAGNETIC MICROACTUATOR MADE OF NEWLY DEVELOPED MAGNETICALLY MODIFIED PHOTOCURABLE POLYMER AND APPLICATION TO SWIMMING MICROMACHINE AND MICROSCREWPUMP

Kengo Kobayashi and Koji Ikuta

Department of Micro-Nano System Engineering, Nagoya University, Japan

E-mail: ikuta@mech.nagoya-u.ac.jp

ABSTRACT

This paper describes a newly developed photocurable polymer with ferromagnetic property for microstereolithography to fabricate magnetic microactuators with three dimensional (3D) structures and its application to a wirelessly controllable swimming micromachine and microscrewpump. This polymer was prepared by mixing a photocurable polymer with magnetic particles, together a viscosity-increasing agent to maintain the particle dispersion. It was demonstrated that this polymer were rapidly fabricated into various complicated 3D magnetic microstructures using microstereolithography. Finally, a swimming micromachine and microscrewpump driven by magnetic force wirelessly were developed using screw-type microactuators made of the polymer with a diameter of submillimeter order.

INTRODUCTION

Magnetic microactuators can generate large force and large displacement compared with electrostatic and piezoelectric ones. In addition, it can be driven by energy supplies without physical contact with components such as tethers. Because of this, they can be remotely controlled in closed spaces and in liquid. These advantages can be put to effective use in microactuators and micromachines for microelectromechanical systems (MEMS), microfluidic devices and minimally invasive surgery in the human body. Various microactuators and micromachines have already been developed [1]-[3].

Structure is a very important factor for machines because it decides functions and performance of them. For example, a fan has blades with 3D structure, and they convert rotational motion into force to pump fluid. Moreover, its pumping efficiency is dependant on the blade's structure. In microscale, the importance of structure is also same, and therefore appearance of microactuators and micromachines with arbitrary 3D structures is desired.

In the previous researches, the magnetic materials can be fabricated into microstructures using techniques such as electroplating, micromolding, and screen printing, but the magnetic microstructures fabricated by these techniques are generally only two dimensional [1],[2]. It is needed to assemble 3D magnetic microstructures from simply structured microparts such as a wire and cylinder [3]. Microassembly techniques, however, still lack the precision required for the dexterous handling and bonding of the microparts of miniaturized and complexly structured micromachines.

Microstereolithography, a technique based on rapid prototyping, can fabricate complex 3D microstructures within short process periods [4]-[6]. Yet as a process that can only be performed with photocurable polymers, microstereolitho-graphy has the drawback of poor material selectivity. Several groups have recently tried to solve this problem by developing new materials with specific functions [7],[8]. None of the materials developed, however, have had magnetic properties.

To overcome the structural limitation of magnetic microactuators and micromachines, we developed a magnetically modified photocurable (MPC) polymer for microstereolithography, and demonstrated the feasibility of 3D magnetic microstructures arbitrarily using this polymer [9]. For its application, we developed a wirelessly controllable swimming micromachine and microscrew-pump using screw-type microactuators made of the MPC polymer.

DEVELOPMENT OF MPC POLYMER

Photocurable polymer has no ferromagnetic property, because its primary components are organic. A material with both ferromagnetic and photocurable properties can be produced by mixing magnetic particles into photocurable polymer.

As it turned out, we were unable to obtain an MPC polymer with good dispersion of the magnetic particles merely by mixing the particles into the photocurable polymer. Figure 1 shows the dispersion of ferrite particles (mean particle size: 1.3 μm) of 10 wt% added into liquid photocurable polymer. After mixing a batch of the polymer with a mixer for 10 min, we dropped a portion onto a glass slide and observed it under an optical microscope.

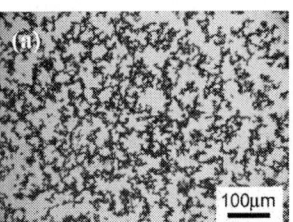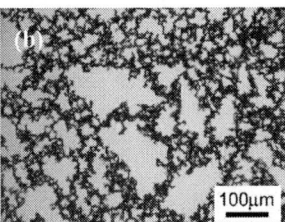

Figure 1: Aggregation of the magnetic particles in the liquid photocurable polymer matrix without additives (a) immediately after mixing and (b) 1 h later.

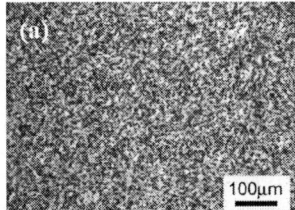

Figure 2: Dispersion of the magnetic particles in the liquid photocurable polymer matrix with the viscosity-increasing agent (a) immediately after mixing and (b) 1 h later.

As shown in Fig. 1, the magnetic particles gradually started to aggregate just after the mixing, and by 1 hour had thoroughly aggregated into a network-like structure. This was presumably caused by the magnetic attraction intrinsic to the individual particles. Aggregation such as this degrades the fabrication accuracy, while imparting an uneven magnetism in a structure. While a surfactant has proved to be effective for dispersing ceramic and metal particles [7],[8], the magnetic force of the particles used in our experiment countered the dispersive effect of the surfactant.

To stop the magnetic particles from aggregating, we tried adding an agent to increase the viscosity of the liquid photocurable polymer. Figure 2 shows the dispersion of the ferrite particles of 10 wt% in the liquid photocurable polymer with the viscosity-increasing agent of 5 wt% (identical to the mixed polymer shown in Fig. 1, but with the viscosity-increasing agent added). As shown in Fig. 2, the magnetic particles were dispersed uniformly after the mixing, and the dispersion was maintained for more than 1 hour. When a viscosity-increasing agent is added to a liquid polymer matrix, a highly viscous drag inhibits the aggregation of the magnetic particles caused by the magnetic attraction. Moreover, the plastic fluid property of the polymer matrix with the viscosity-increasing agent ensures that the good dispersion of the magnetic particles can be sustained for a long period. According to our experiments, the dispersion of the magnetic particles was sustained for more than 10 days.

3D MAGNETIC MICROSTRUCTURE

For the demonstration of 3D magnetic micro-structures, we fabricated many kinds of magnetic microstructures using MPC polymers and microstereo-lithography. Figure 3 shows microscrews with different diameters of 0.5 mm and 0.1 mm, and Fig. 4 shows a micro sirocco fan and a microsculpture of a beetle. The MPC polymer with 50 wt% magnetic particles was used to fabricate the microstructures in Figs. 3 and 4(a), and the MPC polymer with 30wt% magnetic particles was used for the microstructure in Fig. 4(b).

All of these microstructures were fabricated within 30 min. Figures 3 and 4 confirm that complex microstructures having high aspect ratios, overhangs, 3D curved surfaces, and ceilings can be easily fabricated using microstereo-lithography. Structures like these cannot be fabricated by conventional microfabrication techniques such as electroplating or micromolding.

SWIMMING MICROMACHINE

For an application of the magnetic microactuator fabricated by microstereolithography, we developed a wirelessly controllable swimming micromachine with the same size as a human hair.

Figure 5 shows the schematic diagram of the wirelessly controllable swimming micromachine. This micromachine consists of a magnetically driven screw-type microactuator as a locomotive part and two orthogonal Helmholtz coils to control the microactuator.

Driving mechanism of the screw-type microactuator as a swimming micromachine is as follows. The microactuator is magnetized in the radial direction, and then inserted into a capillary filled with water. The

capillary is located at the center of orthogonal Helmholtz coils which generate rotational magnetic field around the longitudinal axis of it. The microactuator is rotated by the rotainal magnetic field wirelessly, and its screw structure converts the rotational motion into the force to propel itself. Hence, the microactuator can swim in the capillary.

In our experiment, the screw-type microactuator made of the MPC polymer with a diameter of 0.1 mm shown in Fig. 3(b) was used as the locomotive part of the swimming micromachine, and glass tube with an inner diameter of 0.2 mm was used as the capillary. The rotational magnetic field of 1mT induced by the orthogonal Helmholtz coils was applied to the microactuator, and the relationship between the swimming velocity and the rotational frequency of magnetic field was measured.

Figure 6 shows time-series photographs where the screw-type microactuator was driven at a rotational frequency of 500 Hz. From these photographs, it was verified that the screw-type microactuator made of the MPC polymer can be wirelessly controlled by applying the external magnetic field to it and swim along the capillary. If the rotational direction of the magnetic field is reversed, the microactuator can swim in the opposite direction.

Figure 7 shows the relationship between the rotational frequency of the magnetic field and the swimming velocity of the screw-type microactuator. It was confirmed that the swimming velocity is linearly proportional to the rotational frequency.

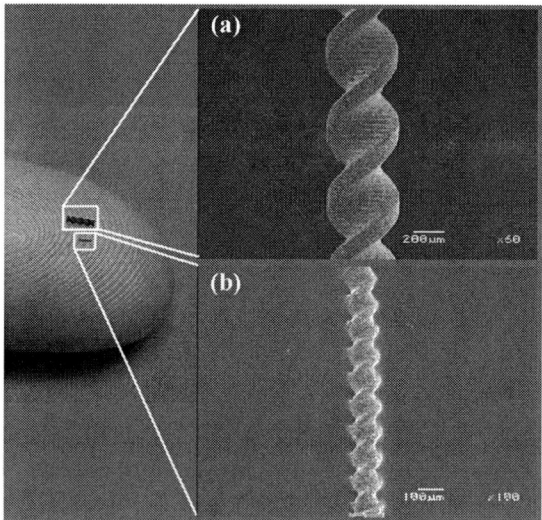

Figure 3: Magnetic microscrews fabricated by micro-stereolithography: (a) with a diameter of 0.5 mm, length of 2 mm, and spiral pitch of 1 mm and (b) with a diameter of 0.1 mm, length of 1 mm, and spiral pitch of 0.2 mm

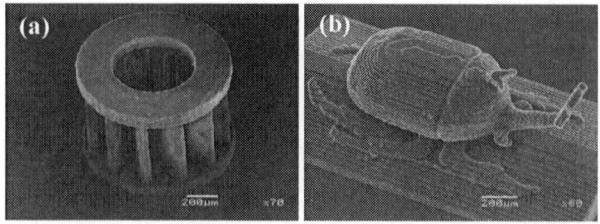

Figure 4: 3D magnetic microstructures fabricated by microstereolithography: (a) micro sirocco fan with a diameter of 1 mm and height of 0.7 mm; (b) beetle with a length of 2 mm, width of 1 mm, and height of 0.66 mm.

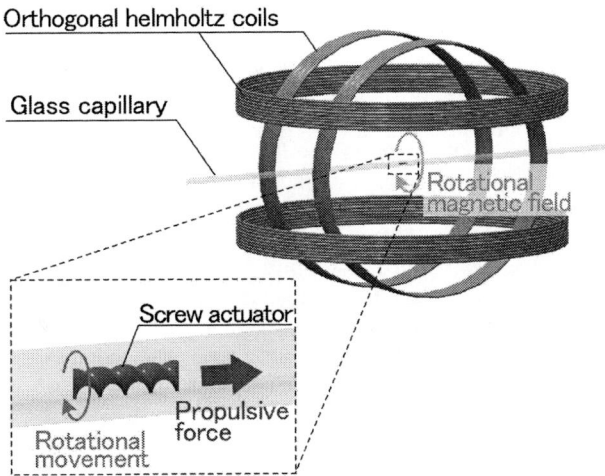

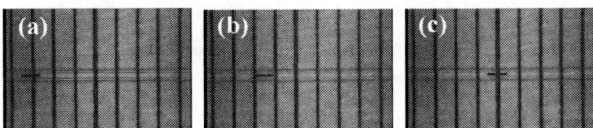

Figure 5: Schematic diagram of the wirelessly controllable swimming micromachine driven by magnetic force

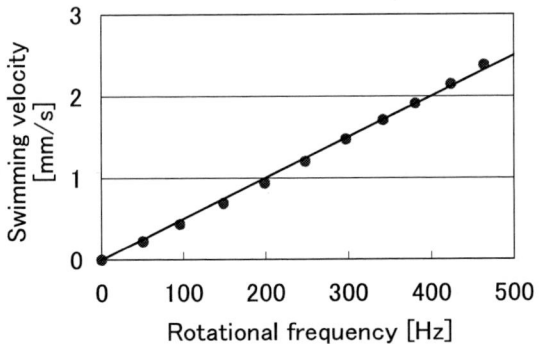

Figure 6: Photographs of the screw-type microactuator with a diameter of 0.1 mm swimming in the capillary with an inner diameter of 0.2 mm filled with water at a rotational frequency of 500Hz: (a) 0s, (b) 0.5 s and (C) 1 s later

Figure 7: Relationship between the rotational frequency of the magnetic field and the swimming velocity of the screw-type microactuator with a diameter of 0.1 mm

MICROSCREWPUMP

As mentioned above, the screw structure of the microactuator converts the rotational motion into the force to thrust fluid backward, and thereby the microactuator can swim forward. On the other hand, if the microactuator are prevented from swimming and only rotated at certain place in the capillary, it serves as a microscrewpump.

Magnetic trap was employed to prevent the screw-type microactuator from swimming away. In the magnetic trap, the gradient of magnetic field in the longitudinal direction of the capillary is generated by electromagnetic coils located around it, and magnetic attraction toward the electromagnetic coils is exerted on the microactuator. The attraction counters the propulsive force of the

microactuator, and prevents it from swimming away from the electromagnetic coils. When the rotational magnetic field is simultaneously generated by the electromagnetic coils, the microactuator is rotated near them, and consequently it can pump fluid (Fig. 8).

In our experiments, the screw-type mictoactuator with a diameter of 0.5 mm shown in Fig. 3(a) was magnetized in the radial direction and inserted into the capillary with an inner diameter of 0.53 mm. Silicon oils with different kinetic viscosities, 1 cSt, 10 cSt and 100 cSt, were filled in the capillary, respectively. The capillary is put horizontally and two pairs of electromagnetic coils were orthogonally located around it. These electromagnetic coils generated the gradient of the magnetic field in the longitudinal direction of the capillary and the rotational magnetic field around the axis of it. The strength of magnetic field at the center of the electromagnetic coils was 45 mT. The relationship between the flow rate and the rotational frequency was measured.

Figure 9 shows the relationships among the flow rate, the rotational frequency and the kinetic viscosity of fluid. From Fig. 9, it was found that the flow rate of the microscrewpump is linearly proportional to the rotational frequency regardless of the viscosity of fluid.

It was confirmed that the screw-type microactuator was released from the magnetic trap and swam away from the electromagnetic coils when the rotational frequency was higher than certain frequency and thereby it was not able to pump the fluids. The frequencies of swimming away were 240 Hz, 40 Hz and 4 Hz when the kinetic viscosities were 1 cSt, 10 cSt and 100 cSt, respectively.

This is because the propulsive force of the microactuator was higher than the magnetic attraction exerted on it by the electromagnetic coils. The propulsive force of the microactuator increases with increasing the rotational frequency as shown in Fig. 7. When the propulsive force is lower than the magnetic attraction, the microactuator is kept near the electromagnetic coils. On the other hand, when the propulsive force is higher than the magnetic attraction, the microactuator is released from the magnetic trap and hence it swims away from the electromagnetic coils. Moreover, because the propulsive force of the microactuator depends on the viscosity of fluid surrounding it, it swims away at lower rotational frequency when the viscosity is higher.

In microscrewpump, although the frequency of swimming away varies with change in the viscosity of fluid as mentioned above, the relationship between the flow rate and the rotational frequency, i.e. the gradient of the graph shown in Fig. 9, does not vary with change in the viscosity. Micro diaphragm pump which is representative of mechanical micropump, in principle, nonlinearly responses to the frequency, and its flow rate varies with change in the viscosity of fluid. Thus, when a fluid with uncertain viscosity is used in this pump, it is required to measure its viscosity or calibrate the flow rate of the pump in advance. Meanwhile, microscrewpump is more convenient owing to the characteristic that its flow rate is independent of the viscosity of fluid.

Although micro diaphragm pump with check valve or diffuser/nozzle can only pump fluid unidirectionally, the microscrewpump can pump fluid reversely because it does not have structural anisotropy. In addition, compared with

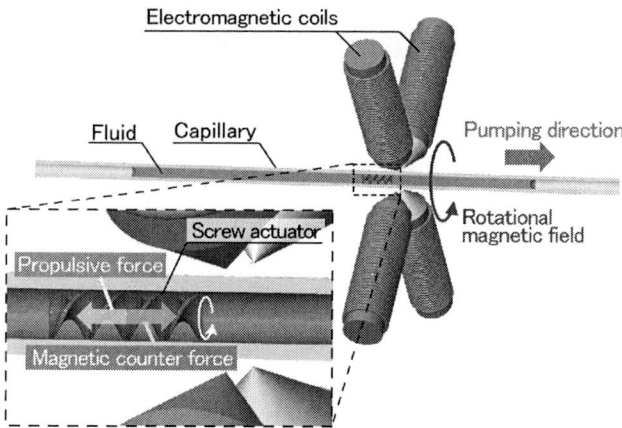

Figure 8: Schematic diagram of miroscrewpump

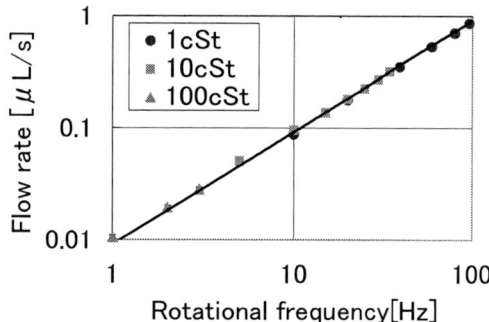

Figure 9: Relationship among the flow rate of microscrewpump, the rotational frequency of magnetic field and the kinetic viscosity of fluid

electroosmotic and electrohydrodynamic pump, many kinds of fluid are used in the microscrewpump because it pumps them mechanically.

It is not necessary for our microscrewpump to fabricate extra structures on a microfluidic chip. We have only to insert a screw-type microactuator into a microchannel on the chip and apply the magnetic field on it by electromagnetic coils located out of the chip. Thus, pump function can be arbitrarily added into microfluidic chips which do not intrinsically have it when used.

CONCLUSION

We developed special MPC polymer with well-dispersed magnetic particles and fabricated 3D magnetic microstructures using microstereolithography. As the result, it was verified that this achievement can realize a variety of 3D magnetic microactuators which are not feasible for conventional microfabrication techniques.

Using screw-type microactuators with a diameter of submillimeter order fabricated by microstereolithography, we developed a micromachine swimming in a capillary and microscrewpump. These micromachines do not need to have batteries in themselves, and external electromagnetic coils can supply energy with them and control them wirelessly. Thus, the swimming micromachine will be applicable to micro surgery microrobot driven in human body remotely.

Microscrewpump can be not only inserted into a capillary but also be implemented in a microfluidic chip. Thus, it will be applicable to implantable drug delivery systems controlled remotely in the future.

Our achievement is not limited to the feasibility of fabricating 3D magnetic microactuators. Because micro-stereolithography is based on computer-aided manufacturing, it easily cooperates with computer-aided analysis. Hence, it facilitates optimization of the design of magnetic microactuators and shortens development periods for them. It is expected that the swimming micromachine and microscrewpump will also be more efficient by optimizing their design for various applications.

REFERENCES

[1] S.Guan and B. J. Nelson, "Magnetic Composite Electroplating for Depositing Micromagnets", *J. Microelectromech. Syst.*, vol.15, pp.330-337, 2006

[2] W. C. Jackson, H. D. Tran, M. J. O'Brien, E. Rabinovich and G. P. Lopez, "Rapid prototyping of active microfluidic components based on magnetically modified elastomeric materials", *J. Vac. Sci. Technol. B*, vol.19, pp.596-599, 2001

[3] K. Kikuchi, A. Yamazaki, M. Sendoh, K. Ishiyama, and K. I. Arai, "Fabrication of a Spiral type Magnetic Micromachine for Trailing a Wire", *IEEE Trans. Magn.*, vol. 41, pp.4012-4014, 2005

[4] K. Ikuta and K. Hirowatari, "REAL THREE DIMESIONAL MICRO FABRICATION USING STEREO LITHOGRAPHY AND METAL MOLDING", *Proc. IEEE International workshop on Micro Electro Mechanical Systems (MEMS'93)*, Feb. 7-10, 1993, pp. 42-47

[5] K. Ikuta, T. Ogata, M. Tsubio and S. Kojima, "Development of mass productive micro stereo lithography (Mass-IH process)", in *Proc. IEEE the 9th Annual International workship on Micro Electro Mechanical Systems (MEMS'96)*, Feb 11-15,1996, pp.301-306

[6] S. Maruo and K. Ikuta, "Submicron stereolithography for the production of freely movable mechanisms by using single-photon polymerization", *Sens. Actuat. A*, vol.1, pp70-76, 2002

[7] X. Zhang, X. N. Jiang and C. Sun, "Micro-stereolithography of polymeric and ceramic micro- structures", *Sens. Actuat. A*, vol. 77, pp.149-156, 1999

[8] J. W. Lee, I. H. Lee and D.-W. Cho, "Development of micro-stereolithography technology using metal powder", *Microelectron. Eng.*, vo.83, pp.1253-1256, 2006

[9] K. Kobayashi and K. Ikuta, "Three-dimensional magnetic microstructures fabricated by microstereolithography", *Appl. Phys. Lett.*, vol.92, 262505, 2008

NOVEL CONCEPT OF MICROWAVE MEMS RECONFIGURABLE 7X45°
MULTI-STAGE DIELECTRIC-BLOCK PHASE SHIFTER

N. Somjit, G. Stemme, and J. Oberhammer
Microsystem Technology Lab., KTH-Royal Institute of Technology, Stockholm, SWEDEN

ABSTRACT

A novel concept of ultra-broadband multi-stage digital-type microwave MEMS phase shifters with the best performance optimized for W-band applications is introduced in this paper. The relative phase shift of 45° of a single stage is achieved by vertically moving a $\lambda/2$-long high-resistivity silicon dielectric block above a 3D micromachined coplanar waveguide (3D CPW) by electrostatic actuation, resulting in different propagation constants of the microwave signal for the up-state and the down-state. For full 360° phase-shift capability, seven stages are cascaded. The devices are fabricated and assembled by wafer-scale processes using bulk and surface micromachining. The measurement results of the first prototypes show that the W-band return and insertion loss of a single 45° stage is better than -15 dB and -1.7 dB, respectively, while the 7-stage phase shifter has a return loss better than -12 dB with an insertion loss less than -4 dB. The phase shifters also perform well from 1-110 GHz with the return loss better than -10 dB, an insertion loss of less than -1.5 dB and a fairly linear phase-shift over the whole frequency range and the actuation voltage is 30 V. To the knowledge of the authors this phase shifter is better than all previous works in term of insertion loss, return loss and phase shift per losses (°/dB) from 70-100 GHz.

INTRODUCTION

Phase shifters are one of the critical components in beam steering systems which can greatly improve the performances of modern radar and wireless communication systems. Microelectromechanical systems (MEMS) phase shifters in general are known for very low insertion loss, low parasitic effects and high linearity over a large bandwidth as compared to solid-state phase shifters. However, at millimeter wave frequencies, such as the 75-110 GHz W-band, switched true-time delay phase shifters are not suitable anymore because of high resistive and dielectric losses [1]. Distributed MEMS transmission line capacitive-loaded phase shifters (DMTL) realized by periodically loading a high-impedance transmission line using MEMS bridges in [2] perform well in the W-band but consist of thin metal bridges which cannot handle large induced current densities, and are therefore limited to low and medium power RF signals [3].

This paper introduces a novel concept of an ultra-broadband digital-type microwave MEMS phase shifters with best performance optimized for W-band applications. The proposed RF MEMS dielectric-block phase shifters are suitable to operate under elevated current densities since no thin moving metallic membranes are employed, and thus are targeting larger signal-power applications. Another advantage of the digital-type phase shifters is that they have less susceptibility to electrical noise because of only two different capacitance values, in the up and down state, respectively, compared to analog systems [4].

PHASE SHIFTER DESIGN

The dielectric-block digital-type microwave MEMS phase shifters consist of a 3D high-impedance (> 50Ω) micromachined CPW loaded by a high-resistivity silicon block. The relative phase shift between the up and the down state is achieved by vertically moving the dielectric block above the waveguide by electrostatic actuation, which results in different propagation constants of the microwave signal depending on the vertical displacement of the dielectric block. The phase shift of the digital-type operation of this phase shifter design is calculated to be:

$$\Delta\phi \,(rad/m) = \frac{2\pi f Z_0 \sqrt{\varepsilon_{r,eff}}}{c}\left(\frac{1}{Z_u} - \frac{1}{Z_d}\right) \qquad (1)$$

where f is the design frequency, Z_0 is the unloaded characteristic impedance of the high-impedance 3D CPW, $\varepsilon_{r,eff}$ is the effective dielectric constant, c is the speed of light in free space, Z_u and Z_d are the loaded impedance of the up and down state positions, respectively.

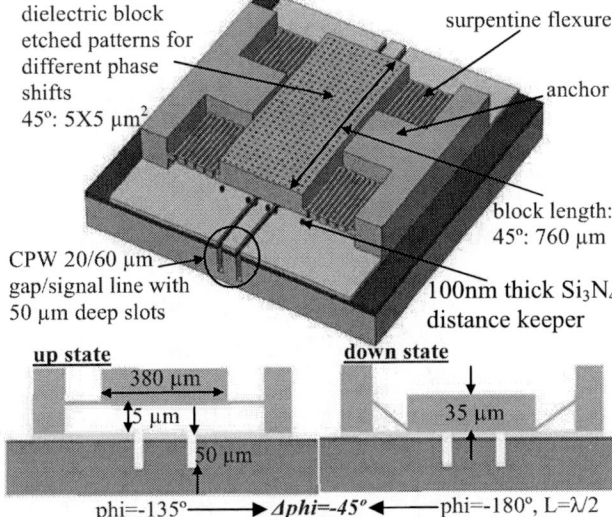

Figure 1: *Single stage phase shifter: cross-section and schematically functional drawing.*

Figure 1 shows schematic drawings of a single stage 45° dielectric-block phase shifter. A high-resistivity silicon block is suspended over the 20/60/20 μm 3D micromachined coplanar waveguide (CPW) with additional deep-etched slots of 50 μm, which decreases substrate loss and increases the sensitivity of the propagation speed to changes in the displacement of the dielectric block on a high-resistivity silicon substrate (HRSS) with dielectric constant of 11.9. The HFSS-modeled unloaded characteristic impedance Z_0 is 54.38 Ω. The length of the silicon block is chosen to $\lambda/2$ for the nominal frequency of 72 GHz to minimize return loss caused by reflected electromagnetic waves from both ends which are cancelling out for an electrical length of the

978-1-4244-2977-6/09 $25.00 © 2009 IEEE

block of 180°. The silicon block is suspended above the 3D micromachined CPW by four serpentine flexures. The relation between the electrostatic actuation voltage (V) and the effective spring constant (k) can be calculated from Eq. (2) which is used for numerical simulation and design of the serpentine flexures.

$$V = \sqrt{\frac{2k}{\varepsilon_0 A} g^2 (g_0 - g)} \qquad (2)$$

where $\varepsilon_0 = 8.8542 \times 10^{-12}$, A is the area of the actuation electrode, g is the distance between the dielectric block and the electrode and g_0 is the zero-bias block height.

Table 1: *HFSS simulated microwave circuit parameters of the single-stage 45° dielectric-block phase shifter and relative phase shift.*

design freq. (GHz)	Z_0 of 3D CPW (Ω)	$\varepsilon_{r,eff}$	Z_u (Ω)	Z_d (Ω)	$\Delta\emptyset$ HFSS (°)	$\Delta\emptyset$ Eq.(1) (°)
72	54.38	4.3	47.6	37.3	-44.2	-42.8

The phase shift of a single stage is tailor-made to 45° by artificially tuning the effective dielectric constant of the silicon block by varying the size of the etch-holes. The HFSS modeled microwave circuit parameters are found in Table 1, while Fig. 2 plots the HFSS simulation of the relation between phase shift and displacement of the block in analog tuning mode. For digital-type operation, an optimum operation point of high phase-shift sensitivity and medium deflection was chosen which requires an actuator displacement of 5 μm.

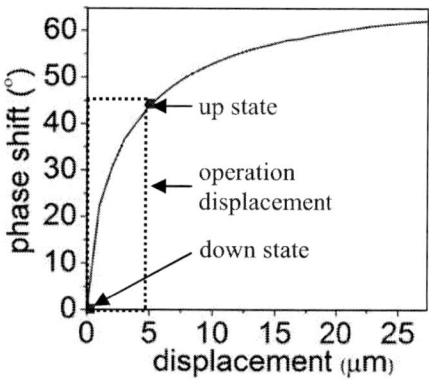

Figure 2: *HFSS simulation of phase shift corresponding to displacement of the dielectric block.*

Figure 3 shows a drawing of a 7-stage 45° phase shifter for full 360° phase-shift capability. The distance between each stage of the multi-stage phase shifter is 10 μm determined by HFSS simulation.

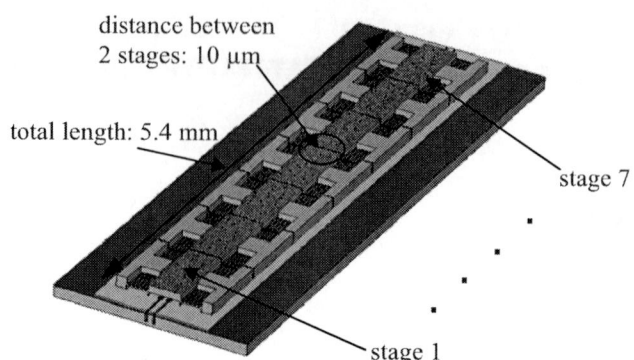

Figure 3: *Schematically drawing of 360° 7-stage phase shifter.*

FABRICATION

The process flow is schematically depicted in Fig. 4. The 1μm thick sputtered-gold CPW is patterned on a 500nm SiO_2 layer with thin adhesive Ti layer on the high resistivity silicon substrate. Above the gold, 100nm thick Si_3N_4 bumps are deposited as distance keepers to prevent stiction of the pulled-in block. The slots of the CPW are further etched by DRIE into the silicon substrate to create deep trenches. A high-resistivity silicon-on-insulator (SOI) wafer is bonded to this wafer by adhesive polymer bonding by mr-I 9250xp thermal nanoimprint lithography photoresist [5], and the SOI handle wafer is afterwards removed by plasma etching. The thin-film SiO_2 layer of the SOI wafer is used as oxide masks for further processing. The 35-μm thick device layer is patterned by two DRIE steps to shape the thick dielectric block with its release-etch holes and the 9μm thick serpentine flexures. Finally, the patterned structure is released by etching the photoresist sacrificial bonding layer in O_2 plasma. Fig.5 shows a top-view SEM picture of prototype of a 45° stage with 5-meander flexures, and Fig.6 shows a microscope picture of a full 7x45° phase shifter.

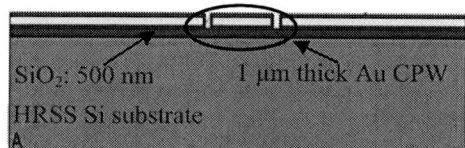

(a) depositing gold CPW (20/60/20 μm) on a 525-μm thick silicon substrate oxidized with 500-nm SiO_2.

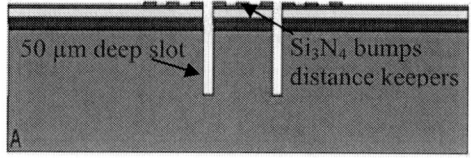

(b) deep-trench etch (DRIE : Borsch process) and depositing Si_3N_4 distance keeper array.

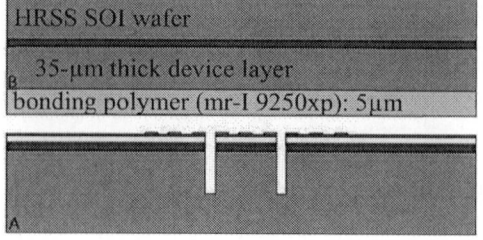

(c) adhesive polymer bonding with an SOI wafer.

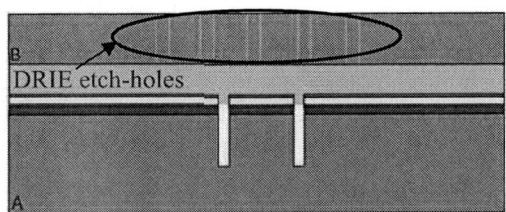

DRIE etch-holes

(d) patterning movable block and etch-holes.

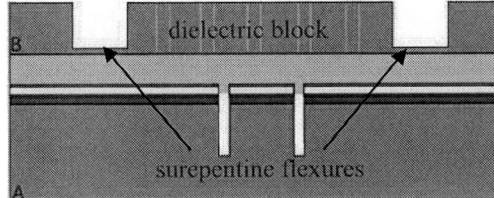

dielectric block

surepentine flexures

(e) patterning serpentine flexures.

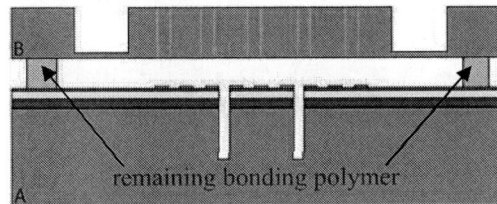

remaining bonding polymer

(f) release structure: O₂-plasma.

Figure 4: *Process flow.*

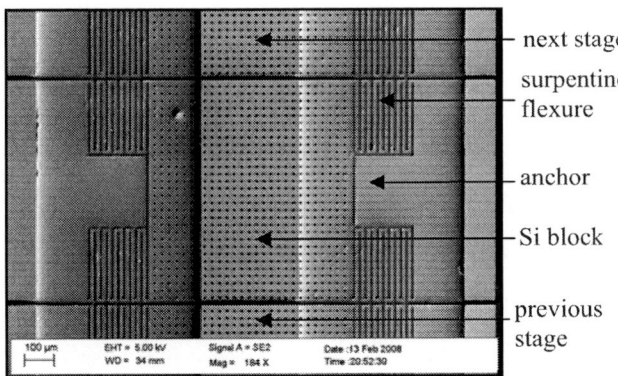

next stage

surpentine flexure

anchor

Si block

previous stage

EHT = 5.00 kV   Signal A = SE2   Date :13 Feb 2008
WD = 34 mm   Mag = 184 X   Time :20:52:30

Figure 5: *SEM picture of fabricated 45° phase shifter block within a 45° x7 multistage phase shifter.*

MEASUREMENTS

The RF measurement results for the single and multi-stage phase shifters are plotted in Fig. 7 and 8, respectively. At the nominal frequency, single-state phase shifters perform better than -0.7 dB of insertion loss and -20 dB of return loss and for the 7-stage phase shifter the return loss is better than -12 dB in the up and down state while the insertion loss is less than -4 dB. Over the W-band the return and insertion loss of a single 45° stage is better than -15 dB and -1.7 dB, respectively, while the multi-stage phase shifter has a return loss better than -12 dB with an insertion better than -5.1 dB. Thus, despite using a silicon block with non-ideal microwave properties for the first prototypes, the novel dielectric-block 7x45° phase shifter has a phase-shift of 74.1°/dB at 72 GHz and performs a linear phase shift of 592.1°/cm. To the knowledge of the authors this phase shifter is better than all previous works in term of insertion loss, return loss and phase shift per losses (°/dB) from 70-100 GHz. Even on an ultra-wide bandwidth, the return loss of a single stage is better than -12 dB from 10-110 GHz and for the 7x45° phase-shifter better than -11 dB from 1-110 GHz.

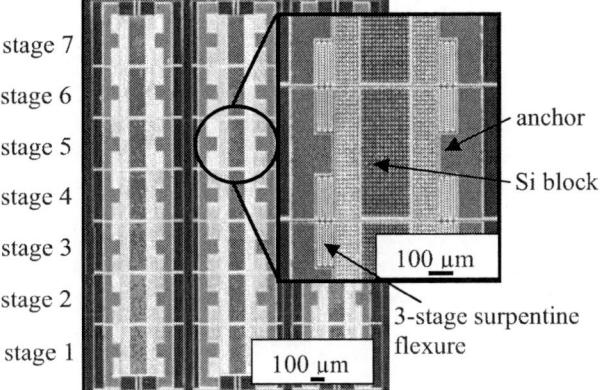

stage 7
stage 6
stage 5
stage 4
stage 3
stage 2
stage 1

anchor

Si block

3-stage surpentine flexure

100 μm

100 μm

Figure 6: *Microscope pictures of fabricated 7x45 ° phase shifters.*

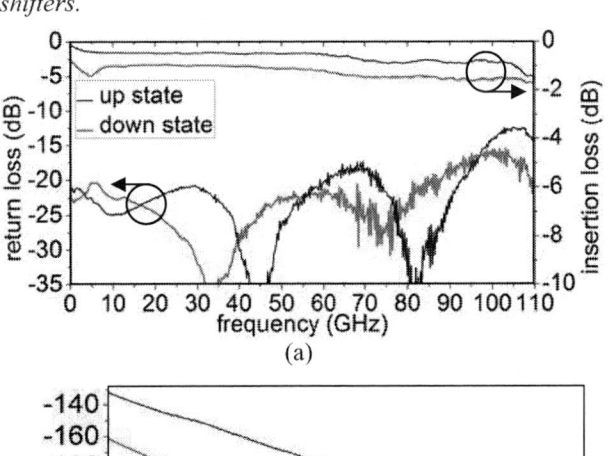

(a)

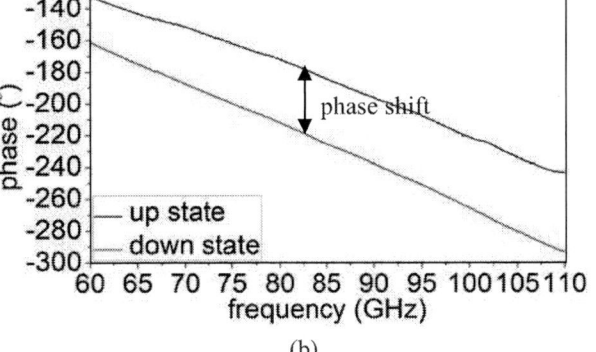

phase shift

(b)

Figure 7: *RF measurement of the single-stage phase shifter: (a) return and insertion loss, (b) phase of S_{21} for up and down state.*

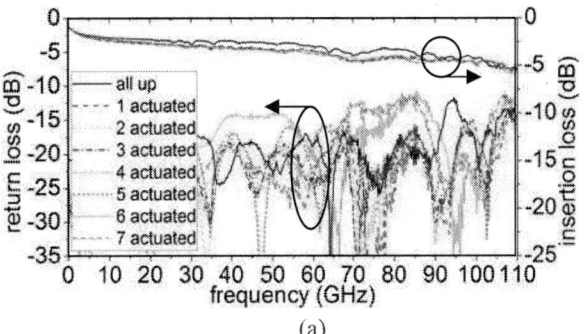

(a)

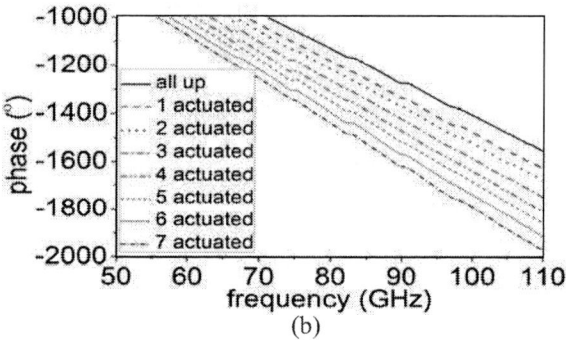

Figure 8: *RF measurement of the 7x45° multi-stage phase shifter: (a) return and insertion loss for actuating multiple stages, (b) phase of S_{21} for different configurations.*

The pull-in behavior of the electrostatic actuator was characterized by a white-light interferometric profilometer as shown in Fig. 9. The actuation curves, simulated by COMSOL Multiphysics (based on Eq. (2)) and measured by white-light interferometer, are plotted in Fig. 10 for a 3-stage and a 5-stage meander flexure; the pull-in voltage of the 3-stage serpentine flexure is 30 V. The effective spring constants determined by measurements are 36.67 N/m and 3.53 N/m for the 3-stage and 5-stage meander flexures, respectively.

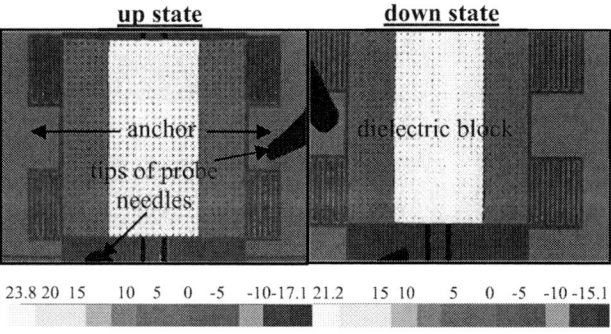

Figure 9: *Deflection of the block characterized by white-light interferometric profilometer: not actuated (left), actuated (right).*

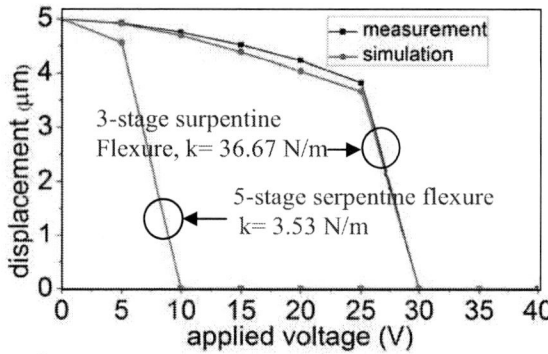

Figure 10: *Measured and simulated DC-actuation characteristics of 3-stage serpentine flexures. Only simulation results are given for the 5-stage meander flexure because the flexure is too soft and therefore too unstable for reproducible results.*

CONCLUSIONS

A novel ultra-broadband digital-type microwave MEMS phase shifter concept with best performance for W-band applications is introduced. The relative phase shift is controlled by vertically moving a dielectric block above the 3D micromachined coplanar waveguide. The single 45° stage and full 360° multi-stage phase shifters were fabricated and characterized. The RF measurements show that both single-stage and multi-stage phase shifters perform decent linear phase from 1-110 GHz with 74.1°/dB at 72 GHz and performs a linear phase shift of 592.1°/cm. The return and insertion losses of single-stage shifters are better than -15 dB and -1.7 dB while the cascaded 7-stage phase shifters perform better than -12 dB and -5.1 dB for the return and insertion loss, respectively. To the knowledge of the authors this phase shifter is better than all previous works in term of insertion loss, return loss and phase shift per losses (°/dB) from 70-100 GHz

REFERENCES

[1] N. S. Barker, G. M. Rebeiz, "Distributed MEMS True-Time Delay Phase Shifters and Wideband Switches," in *IEEE Trans. on Microwave Theory and Techniques*, VOL. 46, NO. 11, November 1998.

[2] N. S. Barker and G. M. Rebeiz, "Optimization of Distributed MEMS Transmission Line Phase Shifters U-Band and W-Band Designs," *IEEE Trans. Microwave Theory and Techn*, vol. 48, pp. 1957-1966 Nov. 2000.

[3] G. M. Rebeiz, "*RF MEMS: Theory, Design, and Technology,*" Wiley; 1st edition, June 15, 2002

[4] J. J. Hung, L. Dussopt and G. M. Rebeiz, "A Low-Loss Distributed 2-Bit W-Band MEMS Phase Shifter," in *33rd European Microwave Conference*, pp. 983-985, October 2003.

[5] M. Populin, A. Decharat, F. Nuklaus and G. Stemme, "Thermosetting Nano-Imprint Resists: Novel Materials for Adhesive Wafer Bonding," in *IEEE 20th International Conference on Micro Electro Mechanical Systems*, pp. 239-242, 21-25 January 2007.

MICROMECHANICAL RESONANT DISPLACEMENT GAIN STAGES

B. Kim, Y. Lin, W.-L. Huang*, M. Akgul, W.-C. Li, Z. Ren, C. T.-C. Nguyen

University of California, Berkeley, Department of EECS, Berkeley, California, USA

*Now at RF Micro Devices, Greensboro, North Carolina, USA

ABSTRACT

Micromechanical resonant displacement gain stages have been demonstrated that employ directionally engineered stiffnesses in resonant structures to effect displacement amplification from a driven input axis to an output axis. Specifically, the introduction of slots along the output axis of a 53-MHz wine-glass mode disk resonator structure realizes a single gain stage with a measured input-to-output displacement amplification of 3.08x. Multiple such mechanical displacement gain stages can then be cascaded in series via half-wavelength beam couplers to achieve multiplicative gain factors; e.g., two cascaded gain stages achieve a total measured gain of 7.94x. The devices have also been operated as resonant switches, where displacement gain allows impact switching via actuation voltages of only 400mV, which is 6x smaller than for previous resoswitches without displacement gain. The availability of such high frequency displacement gain strategies for resonant switches may soon allow purely mechanical periodic switching applications (such as power amplifiers and power converters) with much higher efficiencies than current transistor-based versions.

INTRODUCTION

Micromechanical switches operating at radio frequencies (RF) have exhibited superior performance over transistor-based counterparts in insertion loss, isolation, and power handling capability [1]. In particular, the small input capacitance of MEMS switches offers switch figure of merits (FOM's) several orders better than transistor counterparts, which could substantially enhance the efficiency of periodic switch applications, such as power amplifiers and converters. However, the majority of MEMS switches demonstrated so far still suffer from slow switching speeds, large actuation voltages, and poor long-term reliability. These unsolved issues greatly limit the use of these devices in RF applications, e.g., phased-array antennas, and outright prevents them from addressing higher volume switched-mode power amplifier or power converter applications.

To address these issues, the micromechanical resoswitch was recently introduced in [2, 3]. As shown in Figure 1a, this device comprises a wine-glass mode disk resonator similar to those previously used for oscillator and filter applications [4, 5], but now driven harder so that its conductive disk structure impacts surrounding electrodes. This then electrically shorts the disk and electrodes, thereby effecting periodic on/off switching at the disk's resonance frequency. By harnessing the resonance and nonlinear dynamical properties of their mechanical structures, resoswitches achieve significantly lower actuation voltage (~2.5V), much faster switching speed (rise time ~4ns), and substantially longer cycle lifetimes (>16.5 trillion cycles), than conventional MEMS switch counterparts, making them far more suitable for applications where periodic switching is needed [2, 3].

The resoswitch structure of [2, 3], however, was only a

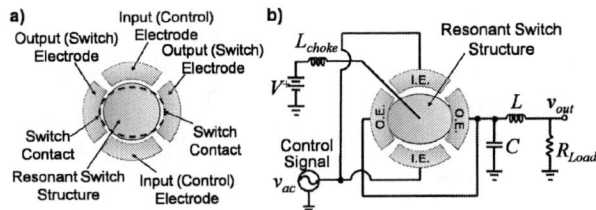

Figure 1: a) Schematic of a wine-glass disk "resoswitch" structure. When the structure resonates in the mode shape shown, it impacts select electrodes, effectively closing the switch. b) Example circuit diagram of a Class-E power amplifier utilizing this resoswitch.

demonstration vehicle, with numerous imperfections. Aside from its polysilicon (rather than metal) construction, it suffered from the fact that during vibration its structure impacts not only the output (switch) electrodes, but also the input (control) electrodes, and this input-impacting behavior greatly limits intended applications, such as the Class-E power amplifier illustrated in Figure 1b. The reason for impacting along both axes is simply that the tiny lateral electrode-to-disk gaps required for capacitive transduction of the device were formed not by lithography, but by a sacrificial spacer film deposition that sets all gap spacings, input and output, to one value (less than 100nm) [3, 4]. One obvious remedy to this problem is to use multiple sacrificial sidewall layers with appropriate masking steps to achieve different gap spacings along the input axis and the output axis. However, this approach entails a more complicated process flow, and thus, might sacrifice production yield and cost.

This work offers a much better solution that dispenses with the need for process modification and rather utilizes a design-centric approach to effect displacement amplification in the switch structure. In particular, via directional stiffness engineering, the resonator structure generates larger displacements along its output axis than its input, so that impact occurs only at the output electrodes. This solves a major issue with the resoswitch of Figure 1a and makes it more suitable for mechanical power converters and amplifier applications.

DISPLACEMENT GAIN STAGE DESIGNS
Disk Resonators with Slots

Figure 2 presents overhead schematics and ANSYS-simulated mode shapes for (a) a conventional wine-glass mode disk resonator; and (b) a new slotted disk resonator design that provides displacement amplification between orthogonal axes. The conventional disk resonator (Figure 2a) exhibits its maximum displacement along two orthogonal axes (*a1-axis* and *a2-axis*), where the magnitude of the displacement along these two axes is identical. In contrast, the new resonator design (Figure 2b) introduces slots along the *b2-axis*, which then reduces the stiffness along this axis relative to that along the *b1-axis*. As a result, at resonance, the displacement along the *b2-axis* becomes much larger than that along the *b1-axis*, effecting dis-

978-1-4244-2977-6/09 $25.00 © 2009 IEEE

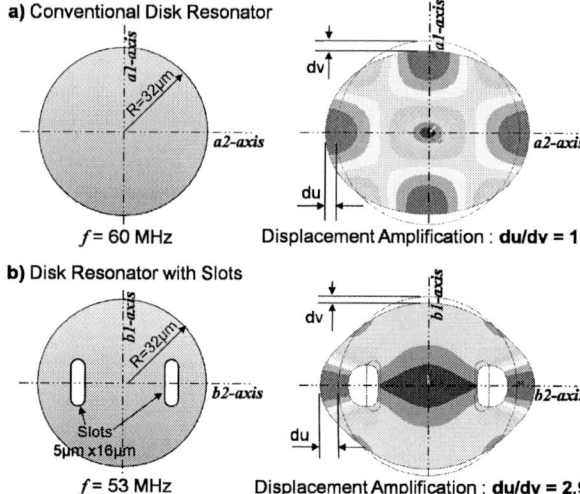

a) Conventional Disk Resonator

R=32μm

f = 60 MHz

Displacement Amplification : du/dv = 1

b) Disk Resonator with Slots

R=32μm

Slots
5μm x16μm

f = 53 MHz

Displacement Amplification : du/dv = 2.94

Figure 2: Design shapes and ANSYS FEA simulated mode shapes (wine-glass mode) of a) a conventional disk resonator and b) a disk resonator with slots. At resonance, the conventional disk has identical displacement along the a1-axis and the a2-axis, while the displacement along the slotted axis (b2-axis) is larger by 2.94x than the orthogonal axis (b1-axis) in the case of the disk resonator with slots.

placement amplification. ANSYS finite element simulation in fact predicts that the displacement along the slotted b2-axis is 2.94x larger than the displacement along the orthogonal b1-axis.

Figure 3 compares curves of normalized radial displacement ζ versus angular location along the perimeter θ for the two disk resonator designs operating in their wine-glass modes. Near the vertical axis ($0<\theta<45°$), both disk resonators exhibit similar mode shapes. The conventional disk resonator has a symmetric mode shape along the quasi-node point at $\theta=45°$. However, the relative radial displacement of the slotted disk resonator increases rapidly for angles past the nodal point and finally becomes 2.94x greater along the horizontal axis at $\theta=90°$ than along the vertical axis at $\theta=0°$.

Resonant Displacement Gain Stage Cascade

For further displacement amplification, displacement gain stages (disk resonators with slots) can be cascaded in series via beam couplers. These beam couplers become

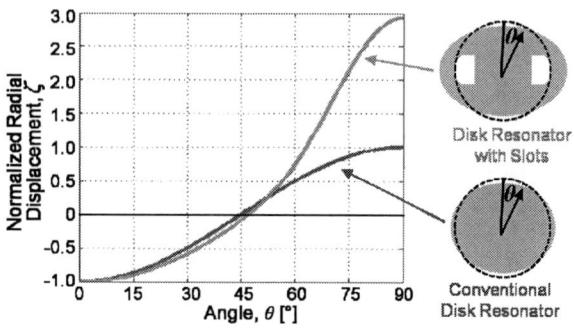

Figure 3: Mode shape comparison between a conventional disk resonator and the resonant gain stage. Here, radial displacements are normalized to the maximum displacement of the conventional disk resonator.

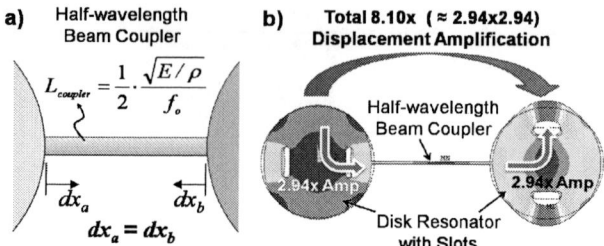

a) Half-wavelength Beam Coupler

$$L_{coupler} = \frac{1}{2} \cdot \frac{\sqrt{E/\rho}}{f_o}$$

$dx_a = dx_b$

b) Total 8.10x (≈ 2.94x2.94) Displacement Amplification

Half-wavelength Beam Coupler

2.94x Amp   2.94x Amp

Disk Resonator with Slots

Figure 4: a) The motions of two resonators connected by a beam coupler are synchronized when the beam length is half-wavelength. b) ANSYS FEA simulation predicts multiplication of displacement gains when two gain stages are series cascaded, yielding 8.10x total displacement gain.

most efficient when their dimensions correspond to extensional-mode half-wavelengths at the disk resonance frequency, at which point their dynamic stiffnesses become virtually infinite [6]. As a result, the motion of one resonator is effectively transferred to the next, as shown in Figure 4a [5]. In effect, coupling two slotted resonator displacement gain stages using a half-wavelength beam coupler, attaching to one at its small displacement point and the other at its large displacement point, multiplies the gains of each stage, yielding a much higher total gain. ANSYS finite element simulations for the two slotted disk cascade shown in Figure 4b predict 8.10x of total displacement amplification, further attesting to gain multiplication.

EXPERIMENTAL VERIFICATION
Fabrication

For experimental verification, micromechanical resonant gain stages were fabricated via a small lateral-gap

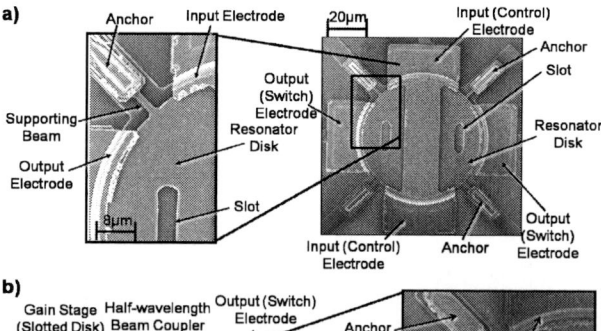

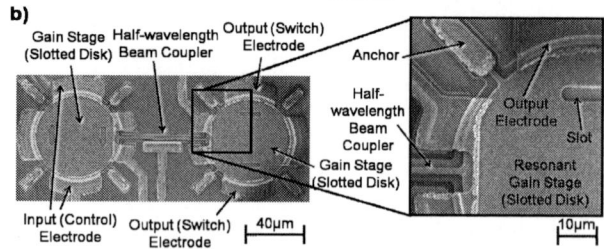

Figure 5: SEM's of fabricated micromechanical resonant gain stages. a) A single gain stage (disk resonator with sots). Slots (5μm x 16μm) were etched into a polysilicon disk resonator to realize the key amplifying features. b) A two slotted-disk resonator cascade, coupled via a half-wavelength beam coupler for further displacement amplification. The coupling beam connects a point along the slotted axis of the first resonator to a point along the un-slotted axis of the second to effect multiplication of gains.

978-1-4244-2977-6/09 $25.00 © 2009 IEEE

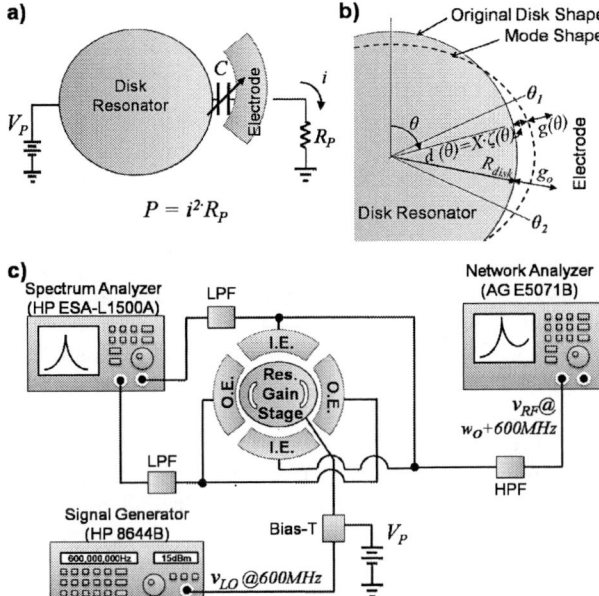

$$P = i^2 \cdot R_p$$

Figure 6: a) When a resonator vibrates, the gap between the resonator and the electrodes changes, resulting in current generation. b) Geometry schematic of a disk resonator. c) RF/LO mixing overtone test setup for simultaneous capacitive displacement measurement along each axis at resonance [7].

polysilicon surface micromachining process similar to those previously used [4]. Here, devices were patterned in a 3μm-thick polysilicon layer and electrodes were centered along the maximum displacement axes. A conformal sacrificial sidewall spacer high temperature oxide (HTO) deposition provided uniform lateral resonator-to-electrode gaps of 97nm. Slots were then patterned in the structure via a plasma etch. The devices were then released using hydrofluoric acid wet etching followed by critical point drying. SEM images of the fabricated devices are shown in Figure 5.

Capacitive Displacement Measurement

Determining the displacement gains in actual devices requires an ability to simultaneously measure and compare displacements along the different axes of a given device.

Since the devices used here are capacitively transduced, displacements are most conveniently measured by measuring the currents generated by the charged time-varying capacitors between the disks and their electrodes (Figure 6a). Using the normalized radial displacement ζ of a disk resonator operating in its wine-glass mode in Figure 3 and the amplitude of oscillation X, the current generated at an electrode i is

$$i = V_P \frac{dC}{dt} = X \frac{\varepsilon_o h R_{disk} V_P \omega_o}{g_o^2} \int_{\theta_1}^{\theta_2} \zeta d\theta \quad (1)$$

where, ε_o is the air permittivity, h is the disk thickness, ω_o is the angular resonant frequency, g_o is the static gap spacing, V_P is the bias voltage, R_{disk} is the disk radius, and θ_1 (overlap starts) and θ_2 (overlap finishes) are the overlap angles in Figure 6b.

By measuring the output power P at the electrode, the oscillation amplitude X can be obtained as

$$X = \sqrt{\frac{P}{R_P}} \frac{g_o^2}{\varepsilon_o h R_{disk} V_P \omega_o \cdot \int_{\theta_1}^{\theta_2} \zeta d\theta} \quad (2)$$

where R_p represents parasitic interconnect resistance. Given this, the displacement d at resonance at a given location θ is governed by

$$d(\theta) = X \cdot \zeta(\theta) = \sqrt{\frac{P}{R_P}} \frac{g_o^2 \cdot \zeta(\theta)}{\varepsilon_o h R_{disk} V_P \omega_o \cdot \int_{\theta_1}^{\theta_2} \zeta d\theta} \quad (3)$$

RF/LO Overtone Measurement

The output powers generated at the electrodes along each axis of each vibrating resonator under test were measured using the RF/LO overtone measurement setup illustrated in Figure 6c [7]. Using this setup, the motion of the resonator can be measured without interference from currents feeding through the static capacitors between the resonant structure and its electrodes. Here, a signal generator supplies an LO signal at 600MHz (which is away from the disk's resonance frequency) to the bias port via a bias-T, while a network analyzer provides an RF signal (with a frequency offset by ω_o from the LO signal). When this offset frequency, ω_o, matches the resonance frequency of the resonator, a force at ω_o is generated via the square-law voltage-to-force transfer function of the capacitive transducer that then drives the disk to resonance

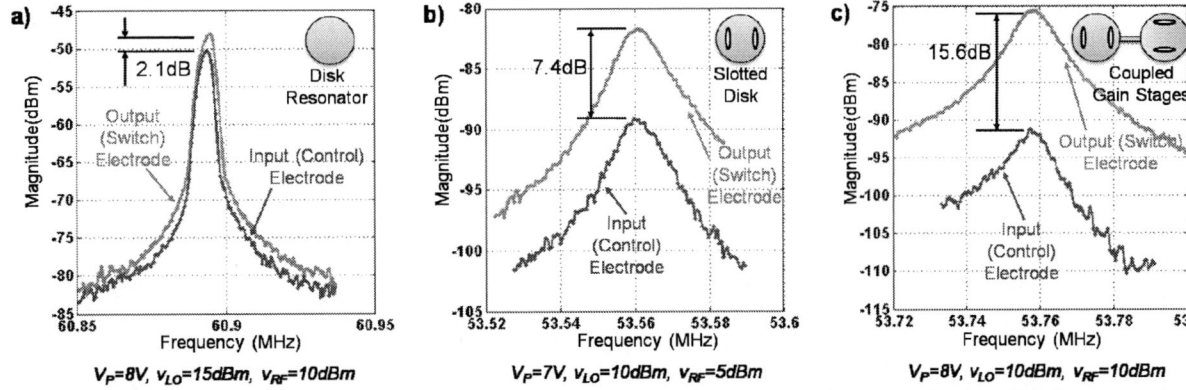

Figure 7: Comparison of measured power at each electrode of each resonator design using the RF/LO overtone measurement setup in Figure 6. The displacement along each axis can be extracted from the measured current out of each electrode. a) A conventional wine-glass mode disk resonator, b) Single resonant gain stage (slotted disk), c) Cascade of two slotted disk gain stages. The conventional disk resonator had virtually the same power generation within the measurement error range, while the single gain stage and two cascaded gain stages had significant differences.

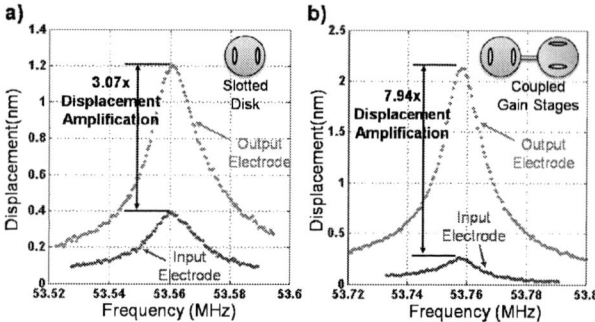

Figure 8: Comparison of displacement along each axis of displacement gain stages. These displacement data were extracted from Figure 7 using equation (3). a) The slotted disk resonator exhibited input and output axis displacements of 0.39nm and 1.20nm, respectively, providing an overall input-to-output displacement gain of 3.07x. b) In the case of two cascaded gain stages, the displacement was amplified by 7.94x, from 0.26nm (input) to 2.12nm (output).

vibration. Since the applied LO and RF signals do not coincide with the disk's resonance frequency, their feedthrough currents will not interfere with motional currents to be measured at resonance, allowing clean and accurate measurement by a spectrum analyzer.

Figure 7 shows the measured output power at each electrode for different designs. For the case of the conventional disk resonator, the measured power between the input and output electrodes were virtually identical, differing by less than the measurement error range. In contrast, significant differences are evident for the single resonant gain stage (slotted disk) and the cascaded two slotted disk gain stages.

Measured Displacement Gain

The displacement along each axis was extracted by plugging the resonator parameters (h=3μm, R_{disk}=32μm, g_o=97nm, and R_p=1.7kΩ) and the measured output power data (in Figure 7) into equation (3). Figure 8 plots the calculated displacements of each design. At resonance, the slotted disk (Figure 8a) exhibited input and output axis displacements of 0.39nm and 1.20nm, respectively, providing an input-to-output displacement amplification of 3.07x. The two cascaded gain stages (Figure 8b) exhibited an overall displacement gain of 7.94x, from 0.26nm (input) to 2.12nm (output). These displacement amplifications

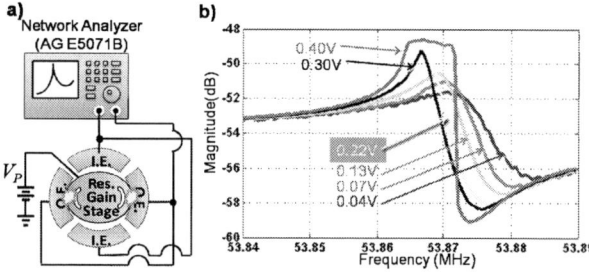

Figure 9: a) Test setup for switch-mode operation of resonant displacement gain stages. b) Frequency response plot of a preliminary resoswitch test of a single mechanical gain stage with V_{bias}=10V. The gain stage impacted the output electrodes and flattened the passband at V_{ac}=400mV.

match well with the predictions of the ANSYS finite element analysis shown in Figure 2b and Figure 4b.

Switch-Mode Operation of Displacement Gain Stages

To demonstrate the efficacy of displacement gain in resonant switch applications, a single resonant displacement gain stage (slotted disk) was tested in switch mode operation, as shown in Figure 9. Here, as the amplitude of the input voltage increases, the frequency response eventually flattens, indicating switching (impacting) of the resonant disk to the output electrode. This occurs when the input voltage amplitude is only 400mV$_{p-p}$, which is 6x smaller than for previous resoswitches without displacement gain [2, 3].

CONCLUSIONS

By strategic introduction of slots into their structures, the stiffnesses of disk resonators have been directionally engineered to yield more than 3x of displacement amplification from one axis to an orthogonal axis, and up to 7.94x when emplaced in two-stage cascades. Such gains have already led to 6x reductions in resoswitch actuation voltage. However, to truly reap the benefits of gain-enhanced resoswitches, further research is needed to lower their series resistance, e.g., by implementing them in metal (rather than polysilicon); and to improve their cycle lifetimes, where extensions from the current 16.5 trillion cycles to quadrillions of cycles are needed if such devices are to be used for practical power amplifier and power converter applications. Research on these topics is ongoing.

ACKNOWLEGEMENTS

The authors greatly appreciate the help and support of Li-Wen Hung, Inkyu Park, Hyuck Chou, and Matthew Wasilik from U.C.Berkeley for fabrication help and SEM image capture. This work was supported by DARPA.

REFERENCES

[1] P. D. Grant, et al., "A comparison between RF MEMS switches and semiconductor switches," in *ICMENS* 2004, pp. 515-521.

[2] Y. Lin, et al., "The micromechanical resonant switch ("Resoswitch")," in Hilton Head 2008, pp. 40-43.

[3] Y. Lin, et al., "A resonance dynamical approach to faster, more reliable micromechanical Switches," in *IEEE Frequency Control Symposium* Honolulu, Hawaii, USA, 2008, pp. 640-645.

[4] Y.-W. Lin, et al., "Series-resonant VHF micromechanical resonator reference oscillators," in *IEEE Journal of Solid-State Circuits*, vol. 39, pp. 2477-2491, 2004.

[5] S.-S. Li, et al., "Disk-array design for suppression of unwanted modes in micromechanical composite-array filters," in *IEEE MEMS'06*, Istanbul, Turkey, 2006, pp. 866-869.

[6] Y.-W. Lin, et al., "Low phase noise array-composite micromechanical wine-glass disk oscillator," in *IEDM 2005*, pp. 287-290.

[7] J. R. Clark, et al., "High-Q UHF micromechanical radial-contour mode disk resonators," in *IEEE Journal of Microelectromechanical Systems*, vol. 14, pp. 1298-1310.

EPITAXIAL SILICON MICROSHELL VACUUM-ENCAPSULATED CMOS-COMPATIBLE 200 MHz BULK-MODE RESONATOR

Kuan-Lin Chen[1], Hengky Chandrahalim[2], Andrew B. Graham[1],
Sunil A. Bhave[2], Roger T. Howe[1] and Thomas W. Kenny[1]
[1]Stanford University, Stanford, California, USA
[2]Cornell University, Ithaca, New York, USA

ABSTRACT

This paper shows the first successful combination of dielectrically-transduced 200 MHz resonators with the epi-silicon encapsulation process, and demonstrates a set of important capabilities needed for the construction of CMOS-compatible RF MEMS components. The result shows the resonant frequency of 207 MHz and a quality factor of 6,400. The high $f \cdot Q$ (1.2×10^{12} Hz) makes this encapsulated resonator an excellent candidate for applications in local oscillators and RF spectrum analyzers.

INTRODUCTION

Wireless communication has greatly impacted our daily lives since the first radio system was invented [1]. Applications, such as cellular phones, satellite television, GPS navigation, and wireless internet networks, are driving the development of RF components in the direction of being smaller, more inexpensive and requiring less power, and thus this topic has been one of the hottest research areas. MEMS resonators have great potential for replacing conventional resonators used in portable wireless applications because of their merits of small size, high quality factor (Q), and low power consumption [2]. There is also great interest in using coupled micro-resonators as band-pass filters and several research groups have already demonstrated exciting results [3-4].

Despite the advances in device performance, packaging for MEMS resonators remains a critical challenge. Because of their extreme sensitivity to the environment, MEMS resonators need vacuum encapsulation to achieve high quality factors (Q) and enable post-MEMS CMOS integration. The promising on-chip application also requires a CMOS compatible packaging process. Due to the stringent RF requirements, the electrical properties and hermeticity of the packaging are also very important.

This paper presents the design, fabrication, packaging design, and measurement of a 200 MHz, dielectrically-transduced, width-extensional mode resonator encapsulated in an epitaxial-grown silicon package. The paper also explains the challenges of merging two technologies into a promising solution of a high frequency MEMS-based device.

DESIGN OF RESONATOR

Dielectrically-transduced silicon bar resonators vibrating in the bulk-mode have superior linearity and high quality factors [5, 6]. The width-extensional mode resonance of a bar resonator depends on the width, W, with frequency

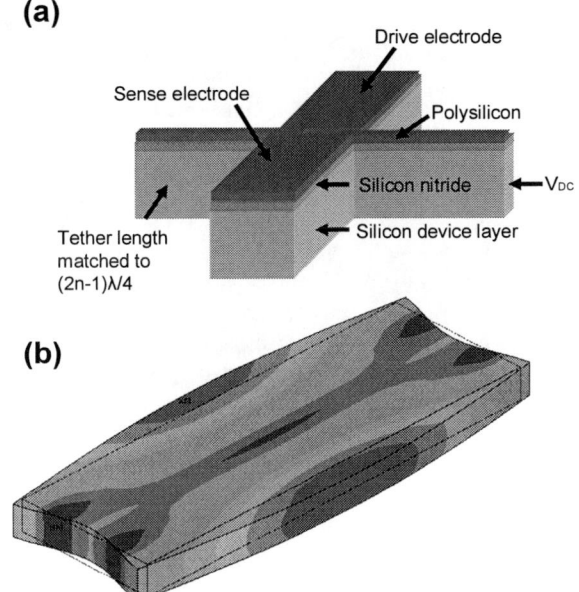

(a)

(b)

Figure 1: (a) Schematic diagram of the dielectrically-transduced, width-extensional mode resonator and (b) ANSYS contour plot of the width-extensional bar mode shape.

$$f = \frac{n}{2W}\sqrt{\frac{E}{\rho}}$$

where n is the mode number ($n = 1,3,\ldots$), and E and ρ are the effective elastic modulus for 2D expansion and the density of silicon, respectively. The width-extensional mode is excited by patterning electrodes in plate configuration on top of the resonator using dielectric transduction, as shown in Fig. 1(a). The main bar resonator consists of three layers - a 100nm polysilicon electrode layer, a 100nm silicon nitride transducer layer, and a 3μm single-crystal silicon resonator layer. As shown in Figure 1, the input and output electrodes are patterned on polysilicon and the silicon nitride layer is used as a high-κ dielectric to enhance the transduction (compared to an air-gap). The quarter-wavelength anchoring scheme is designed to minimize mass-loading of the resonator. A DC bias is superimposed on a small AC voltage using a Bias-T and applied to the drive electrode, while the silicon device layer is connected to RF ground. This time-varying voltage causes a squeezing force on the dielectric thin film. Due to the Poisson effect, the dielectric layer experiences a lateral strain. Since the silicon nitride layer is structurally integrated with the silicon bar, the lateral strain is transferred from the dielectric film to the silicon resonator layer. As the strain distributes through the resonator, the width-extensional mode is excited and the motional current is sensed through the sense electrode. The ANSYS fundamental mode shape of the width-extensional resonator is shown in Fig. 1(b).

EPITAXY-SILICON PACKAGING (MICROSHELL)

The fabrication process is the fusion of two technologies developed at Cornell and Stanford [6, 7]. The wafer-level encapsulation packaging developed at Stanford University has the advantage of a small foot print, superior long-term stability, and CMOS compatibility [8-10]. The epitaxial-grown silicon allows CMOS circuitry to be built on top of the package to be integrated with the MEMS device packaged inside.

Previously, there have been accelerometers and resonators working at 1.2–1.4 MHz or below, encapsulated in this packaging. However, there has not been an attempt to encapsulate devices with resonant frequencies higher than 20 MHz using this silicon package. A thorough study of the electrical performance of the packaging was investigated [11]. The paper indicates that the silicon vertical interconnect has less than 1 dB attenuation up to 6 GHz, which extends the utility of this packaging for IF to HF resonator applications.

FABRICATION

As shown in Fig. 2, the fabrication begins with a silicon-on-insulator (SOI) wafer with a device layer of 3μm. A stack of 100nm stoichiometric silicon nitride and 100nm heavily doped n+ polysilicon was deposited using low pressure chemical vapor deposition (LPCVD). The polysilicon layer was patterned to create electrodes. Part of the silicon nitride layer was etched to create electrical contact to the device layer, as shown in Fig. 2(b). The resonator was then patterned and etched through the nitride/silicon stack using deep reactive ion etching (DRIE). Low temperature oxide (LTO) was deposited as the sacrificial layer to create a cavity on top of the resonator. The LTO was then patterned and etched to create electrical contacts.

The main challenge is in creating low-loss vertical interconnects making contact to two different layers of the resonator. As shown in Fig. 2(d), the LTO was patterned and etched to create an electrical connection to the polysilicon layer as well as the device layer. Due to the thickness variation from the LTO deposition, stopping the etch on two different layers is extremely challenging using an academic-grade plasma etcher (insufficient etching selectivity, non-uniform etching across the wafer, etc.) However, this challenge can be greatly reduced with an industry-grade etcher.

Another challenging issue is the resistivity of the epi-polysilicon interconnects. A lot of characterization was done to ensure the resistivity of the epi-polysilicon was less than 10 mΩ-cm. Highly phosphorus doped epi-silicon was then grown on the top of the wafer to encapsulate the resonators. Vent holes were patterned to open access to the sacrificial LTO. A custom made HF vapor etcher in the Stanford Nanofabrication Facility was used to release the resonator. After releasing, vent holes were again sealed by LTO in a low pressure environment to vacuum-

encapsulate the resonators. Finally, an aluminum layer was used for electrical interconnects and bond pads. An SEM picture of a fabricated width-extensional mode resonator is shown in Fig. 3.

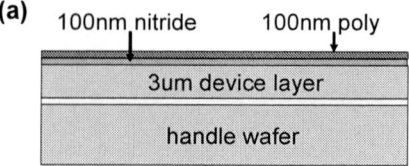

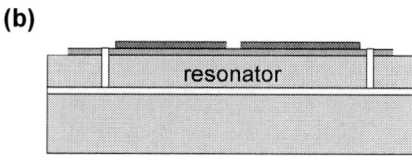

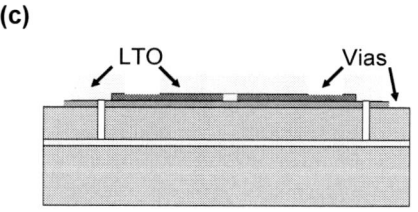

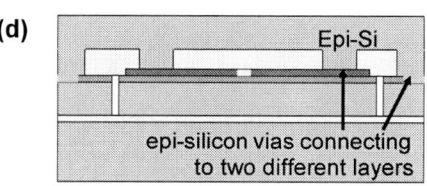

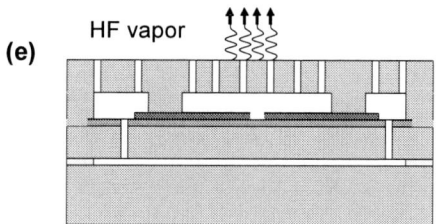

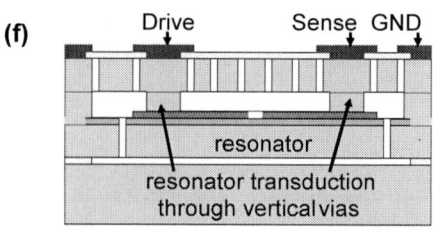

Figure 2: Epi-silicon microshell encapsulation process for dielectrically transduced resonators. Vertical epi-vias to the poly electrodes provide RF input/output while the vias to the silicon device layer provide a micro RF-cage providing a shielded microenvironment.

978-1-4244-2977-6/09 $25.00 © 2009 IEEE

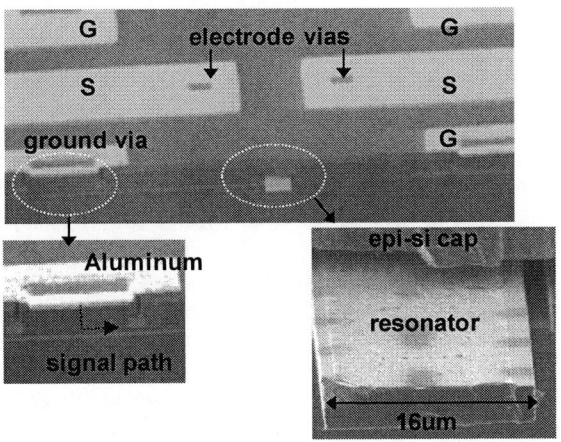

Figure 3: SEM of the fabricated resonator. The zoomed-in picture on the right shows the thin poly electrode on top of the resonator bar. The zoomed-in picture on the left shows the electrical contact between the epi-silicon and the device layer.

DIFFERENTIAL MEASUREMENT SETUP

As shown in Fig. 4, a simple two-port measurement gives a clear sharp dip in S21, indicating the electromechanical resonance is masked by the parasitic feedthrough. To overcome the capacitive feedthrough, the resonator was characterized using a differential measurement technique, inspired by [12, 13]. An RF signal from the network analyzer first goes into a splitter to create in-phase and out-of-phase signals, represented by $+RF_{in}$ and $-RF_{in}$, respectively. In order to induce resonance of the resonator, a DC bias, V_{DC}, was added into $+RF_{in}$ using a Bias-T. Two identical resonators were driven by $+RF_{in}$ and $-RF_{in}$. While $+RF_{in} + V_{DC}$ was driving the resonator, the pure AC $-RF_{in}$ was driving the dummy resonator, with only its parasitic capacitance being driven. The outputs from both structures were then mixed together, where the out-of-phase current from the parasitic capacitance cancels out the parasitic component in the signal from the resonator. The exact differential

measurement setup used in this work is express in Fig. 5. Although it would be ideal to have two identical resonators side-by-side when the differential measurement is conducted, the fabricated wafer only has a similar resonator by its side. Therefore, although the improvement of the signal to noise ratio is already greatly enhanced by using a similar structure as the dummy resonator, the feedthrough cancellation is not yet optimized.

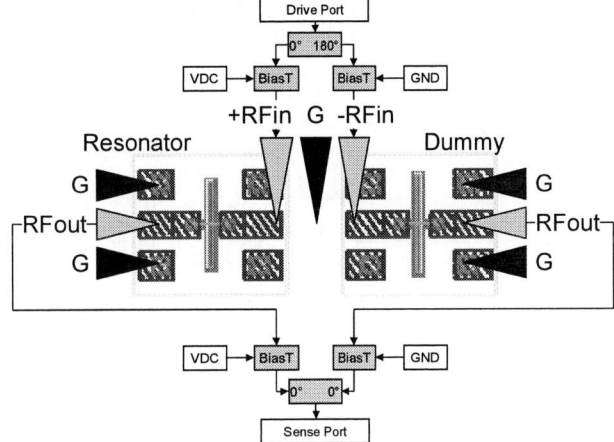

Figure 5: Schematic of the pseudo-differential measurement setup.

RESULT

From the SEM picture, one can observe that the vertical interconnect is firmly connected to the device layer, providing electrical ground to the resonator beam. One can also note that the thin (~13μm) epi-silicon is enough to provide mechanical stability for the low-pressure hermetic packaging. It is important to keep this layer thin because of the electrical requirements [11].

The fabricated resonator was measured using the differential measurement setup explained in the previous section. Fig. 6 shows the response using feedthrough cancellation on top of the response from the 2-port

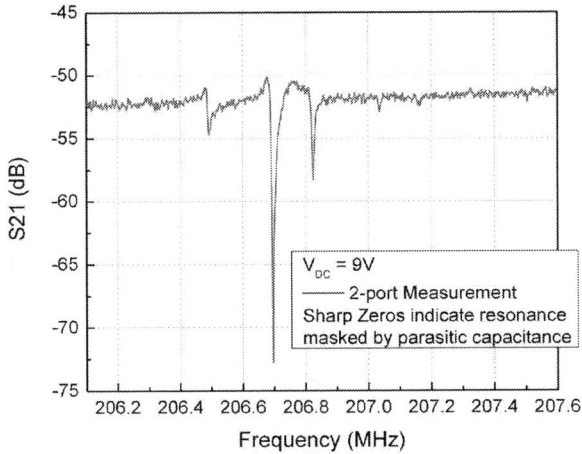

Figure 4: Transmission response using the 2-port measurement. Capacitive feedthrough masks the electromechanical resonance, resulting in a sharp dip at the target frequency.

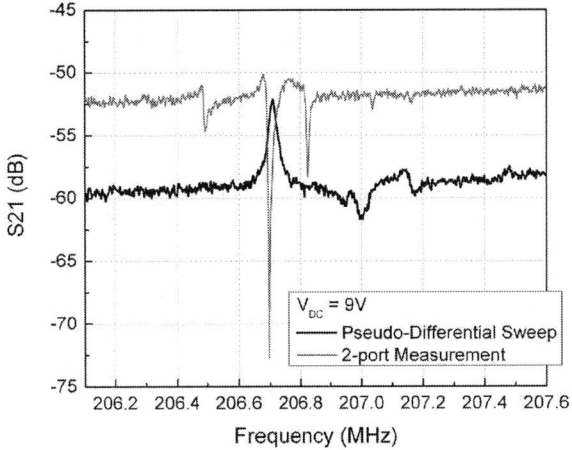

Figure 6: Transmission responses of the simple 2-port measurement and pseudo-differential measurement. It is clearly shown that the resonance peak matches the sharp dip shown in the 2-port measurement.

978-1-4244-2977-6/09 $25.00 © 2009 IEEE

measurement. The resonant frequency is 207 MHz and the quality factor is 6,400. This is the highest reported frequency for a MEMS resonator packaged in a microshell. One can also note that there is a -13 dB noise reduction due to the implementation of the differential measurement.

CONCLUSION

A fully encapsulated, width-extensional mode resonator was successfully fabricated with an $f \cdot Q$ product of 12×10^{11} Hz, making it an excellent candidate for drop-in insertion in local oscillator and RF spectrum analyzer applications. The performance of the epitaxial-silicon packaging is also verified to sufficiently package VHF devices.

In the future, we plan to introduce Hafnium dioxide (HfO$_2$) as the dielectric film to reduce the motional impedance. Furthermore, the length of the width-extensional mode resonator can be increased to enlarge the transduction area. Hence, the motional impedance of the encapsulated resonator could be reduced and alleviate the challenge of impedance matching.

ACKNOWLEDGEMENTS

The authors would like to thank the staff in the Stanford Nanofabrication Facility, Gary Yama from Robert Bosch LLC RTC, and other people who work in the SNF for technical support.

This work was supported by a subaward on the DARPA Analog Spectral Processors program, with Chris Conway of Rockwell Collins as PI. The authors gratefully acknowledge the financial support from the DARPA Analog Spectral Processors program for this project (contract #N00173-06-C-2055). Distribution Statement "A" (Approved for Public Release, Distribution Unlimited).

REFERENCES

[1] G. Marconi, "Wireless telegraphy," Smithsonian Institution, Washington, D.C.,Annual Report – 1901, 1902, pp. 287-298.

[2] Y.-W. Lin, S. Lee, S.-S. Li, Y. Xie, Z. Ren, and C. T.-C. Nguyen, "60-MHz wine glass micromechanical disk reference oscillator," Digest of Technical Papers, 2004 IEEE International Solid-State Circuits Conference, San Francisco, California, Feb. 15-19, 2004, pp. 322-323.

[3] S.-S. Li, Y.-W. Lin, Z. Ren, and C. T.-C. Nguyen, "A micromechanical parallel-class disk-array filter," Proceedings, 2006 IEEE Int. Frequency Control Symp., Geneva, Switzerland, May 29-June 1, 2007, pp. 1356-1361

[4] G.K. Ho, R. Abdolvand, and F. Ayazi, "Through-Support-Coupled Micromechanical Filter Array," Proc. IEEE Micro Electro Mechanical Systems Conference (MEMS '04), Maastricht, the Netherlands, Jan. 2004, pp.769-772.

[5] Hengky Chandrahalim and Sunil A. Bhave, "Digitally-tunable MEMS filter using mechanically-coupled resonator array," 21st IEEE International Conference on Micro Electro Mechanical Systems (MEMS 2008), Tucson, Arizona, January 13-17, 2008, pp. 1020-1023.

[6] Sunil A. Bhave and Roger T. Howe, "Silicon nitride-on-silicon bar resonator using internal electrostatic transduction," 13th International Conference on Solid-State Sensors, Actuators and Microsystems (Transducers'05), Seoul, Korea, June 5-9, 2005, pp. 2139-2142.

[7] R. Candler et al, "Single wafer encapsulation of MEMS devices," IEEE Trans. on Advanced Packaging 26(3) pp. 227-232, 2003

[8] B. Kim, T. Kenny et al, "Frequency stability of wafer-scale film encapsulated silicon based MEMS resonators," Sensors and Actuators A: Physical, 136(1) pp. 125-131, 2007

[9] R. Melamud, et al, "Temperature-compensated high-stability silicon resonators," Applied Physics Letters, 90 244107, 2007

[10] Park, W.T., et al., "Encapsulated submillimeter piezoresistive accelerometers." Journal of Microelectromechanical Systems, 15(3): p. 507-514., 2006

[11] K-L Chen, J. Silvia et al, "Performance Evaluation and Equivalent Model of Silicon Interconnects for Fully-Encapsulated RF MEMS Devices ," IEEE Trans. on Advanced Packaging, 2008, In press.

[12] K. Ekinci, M. Roukes et al, "Balanced electronic detection of displacement in nanoelectromechanical systems," Applied Physics Letters 81 2253, 2002

[13] P. Rantakari et al, "Reducing the effect of parasitic capacitance on MEMS measurements," Transducer, pp. 1556-1559, 2001

A NOVEL STRESS-GRADIENT-ROBUST METAL-CONTACT SWITCH

Hojr Sedaghat-Pisheh[1], Jung-Mu Kim[2] and Gabriel M. Rebeiz[1]
[1]University of California, San Diego, California, USA
[2]Chonbuk National University, Jeonju, Korea

ABSTRACT

This paper presents the design, fabrication and measurements on a new metal-contact RF MEMS switch with low sensitivity to stress gradients and temperature variations. The switch is based on a circular geometry with arc type springs, and results in high contact force, fast switching time, and has excellent microwave performance to > 40 GHz.

INTRODUCTION

RF MEMS offer several key advantages over solid-state switches, in terms of high isolation, high linearity, low loss, and low power consumption [1], and are used in applications such as switchable routing networks in RF system front-ends, connection to capacitor banks, true time-delay (TTD) phase shifters, tunable filters, reconfigurable antennas, high-resolution pulsed radar sensors, and space-based applications.

Metal-contact switches to-date are based on a very thick cantilever (6-10 um) with two contact points, and have demonstrated excellent performance and reliability [1-4]. The standard designs result in a high contact force but are sensitive to stress gradients which cause a large variation in the pull-down voltage across a wafer. This, in turn, results in a variation in the contact force and affects the switch reliability.

In this paper, we report a new design of single-pole/single-throw (SPST) RF MEMS metal contact switch with very low sensitivity to stress gradient. The insensitivity to stress gradients increases the yield from wafer to wafer, and process to process, and results in a higher performance switch.

DESIGN

The new switch design is based on a circular beam with arc-type springs such that the forces resulting from the stress-gradients are distributed in an arc-type fashion on the anchor and therefore cancel in the radial direction due to the symmetry of the design. This yields very low deflections even for large stress gradients. To demonstrate this, the novel switch and a standard rectangular cantilever were designed to result in the same spring constant of 120-140 N/m. Simulations done using CoventorWare [5] show that for the same stress gradient of 20 MPa/μm the standard rectangular cantilever deflects 41 times more than the new switch (0.41 μm for rectangular design vs. 0.01 μm for circular design).

The switch top view and cross-section are presented in Figure 1. The switch is cantilever-based and is fabricated on a 500 μm-thick quartz substrate with a 4 μm-thick circular beam (electroplated gold), a diameter of 100 μm, and a 120° circular anchor. PMMA A9 is used as the sacrificial layer to realize a 0.8-μm gap. Two 8-μm wide cutouts have been etched in the beam to define the springs, and the simulated spring constant is 167 N/m.

There are many parameters which can change the mechanical and RF performance of the switch. Cutouts (number, angle, center radius, width, and center angle), anchors (number and angle), diameter of the switch, thickness, gap, dimple (diameter and distance between) are among these parameters. To design the switch, we have considered the effect of those parameters on the pull-down voltage (Vp), collapse voltage, contact force, spring constant, isolation, stress gradient, resonant frequencies, switching time, etc.

Different anchors and spring designs were done and many resulted in excellent performance showing the robustness vs. different optimization parameters of this switch, but only one design is shown in this paper. Table 1 presents some key characteristics of the switch.

Table 1: Mechanical and electrical characteristics.

Parameter	Value
Switch Diameter (D)	100 μm
Beam Thickness (t)	4 μm
Gap (g_0)	0.8 μm
Spring constant (K)	167 N/m (Simulated)
Pull-Down Voltage (V_p)	64 V (Measured)
Release Voltage (V_r)	53 V (Measured)
Contact Force per Dimple @ 90V	105 μN (Simulated)
Resonant Frequency (f_0)	160 kHz (Measured)
Mechanical Q	1.5 (Measured)
Switching Time (up to down)	12 μs (Measured)
Switching Time (down to up)	3 μs (Measured)

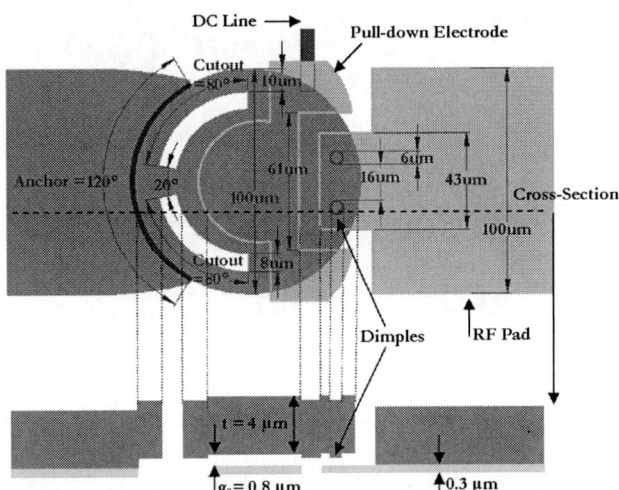

Figure 1: Schematics of the novel switch; (a) top view, (b) cross-section view.

The switch actuates by applying 65 V to the pull-down electrode using a high resistance bias-line (4 kΩ/sq). The contact force is 105 uN per dimple for an applied voltage of 90 V. For this work, the contacts are sputtered-gold to sputtered-gold and result in a very low resistance (but not necessarily the best long-term reliability).

The overlap area between RF pad and the beam

978-1-4244-2977-6/09 $25.00 © 2009 IEEE

should be minimized for a low up-state capacitance design. On the other hand, this may cause improper mechanical contact due to possible misalignments, under-etches, process variations, and minimum feature size limits.

FABRICATION

The fabrication steps are as follows: 1) A high resistive SiCr layer (1200 Å) is sputtered. This layer connects the DC pad to the pull-down electrode using a cut in the CPW ground plane, which is later connected using an air bridge in a subsequent process step. 2) A Ti/Au layer (200 Å/3000 Å) is sputtered and forms the bottom electrodes after patterning. 3) A 0.8 μm-thick PMMA A9 layer is used as the sacrificial layer. This layer is then etched selectively using RIE to form the anchor areas. To fabricate the dimples, an extra fabrication step is required. In this case, RIE is used to etch PMMA A9 for a depth of 0.4 μm. 4) A second metal layer consisting of Ti/Au/Ti (200 Å/3000 Å/200 Å) is sputtered, patterned, and electroplated (4 μm) as the structural material for the cantilever. 5) The second metal layer is etched and the cantilevers are released using a critical point dryer process.

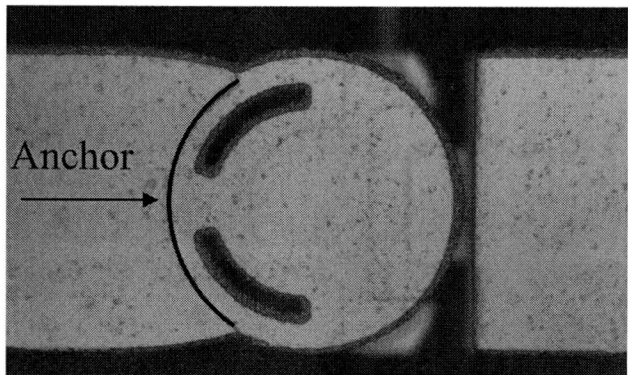

Figure 2: Optical image of the fabricated switch.

Figure 2 presents an optical image of a released switch while the microscope focused on top layer. The bottom metal and DC control line which are 5 μm under the top surface and are out of focus.

MEASUREMENT RESULTS
Flatness

An important characteristic of this design is the resulting switch flatness. An optical profilometer (VEECO) was used to capture two dimensional and three-dimensional data points of the fabricated switch. The 3D profile of a switch in a 50/100/50 μm CPW line shows that the switch is quite flat (Figure 3). 2D profile of some circular switches also verified this observation. A sample of 2D profiles as shown in Figure 4 illustrates that the average height difference between the tip (white arrows) and the middle (black arrows) is measured to be 0.02 μm. This result is achieved even with a measured stress gradient of 13-15 MPa/um (50-60 MPa for the 4 μm beam), and is much less than a standard cantilever.

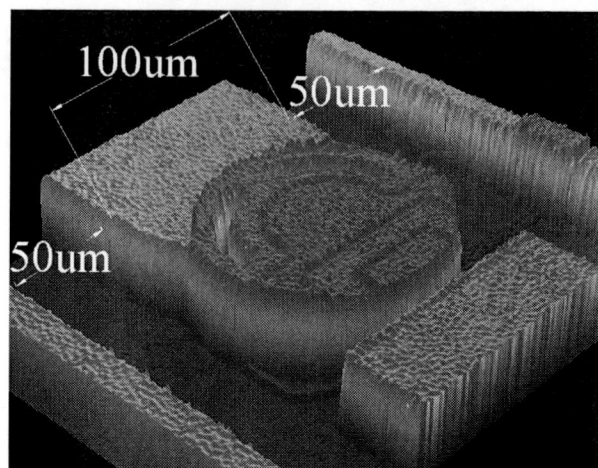

Figure 3: 3D profile of the fabricated switch.

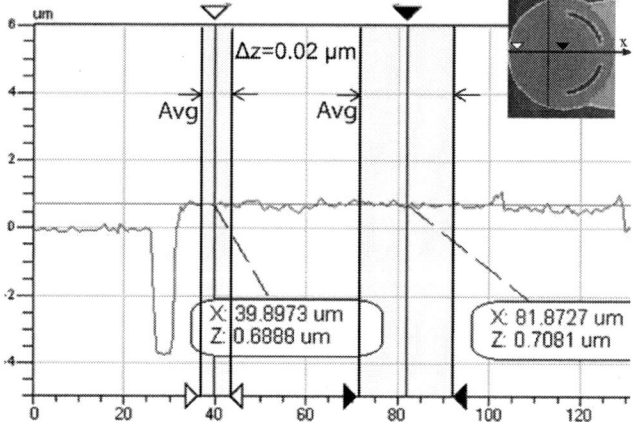

Figure 4: X-axis profile of the fabricated switch.

Pull-Down and Release Voltage vs. Temperature

Pull-down voltage (V_p) and release voltage (V_r) of the switch were measured for 13 different samples across the wafer for 5 different temperatures in heating and cooling cycles. Figure 9 shows V_p and V_r of two switches on the wafer. The pull down and release voltages are very stable with temperature and result in <2 V variation from 25-90 °C.

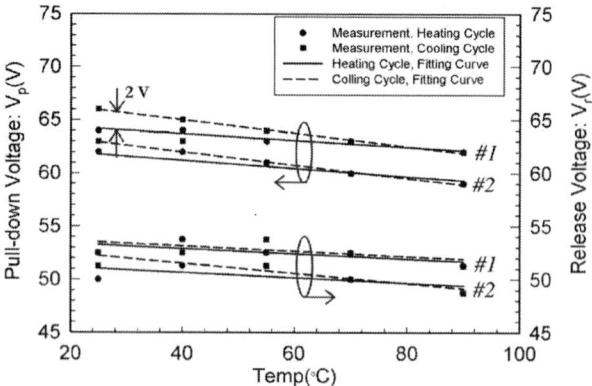

Figure 9: Measured pull-down voltage and release voltage versus temperature of two different switches.

Resonant Frequency

The resonant frequency was measured using AM

modulation on the DC electrode. A 10 GHz RF signal of was applied to the switch. Also, a sinusoidal voltage of 40 V peak to peak with an average of a 30 V is applied to the DC pad of the switch. This, results in AM sideband around the 10 GHz RF signal which can be easily measured using a spectrum analyzer. A resonant frequency of 160 kHz and a mechanical Q of 1.5 is obtained by fitting the measurement with a second order system (Figure 6).

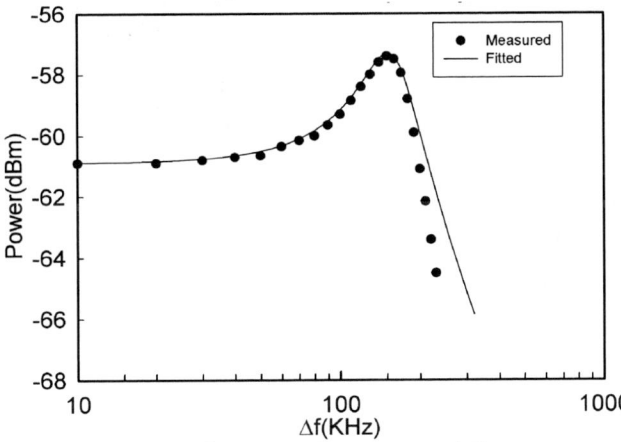

Figure 6: Measured resonant frequency and Q.

S-Parameter measurement and parameter extraction

The S-parameters of the switch were measured from 10 MHz to 40 GHz in the up- and down-state position. The calibration planes are at the tip of the CPW probes, and the measurements include 100 μm of CPW line on each side of the switch. The measurements fit almost perfectly with the expected values for C_u and R_s (Figure 5). The up-state capacitance is extracted by fitting the simulated S_{21} of a 13 fF capacitor to the measured S_{21} of the switch.

The actuation voltage is 64 V and the down-state measurement were done at 90 V. The down-state resistance and inductance are extracted by fitting the simulated S_{21} of a series R_s and L_s to the measured S_{21} of the switch. The fitted resistance is 0.4 Ω and the fitted inductance is 100 pH.

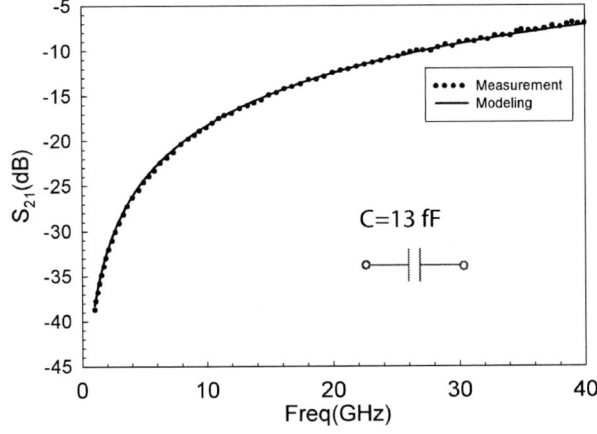

Figure 5 (a): Up-state S-parameter measurement and modeling.

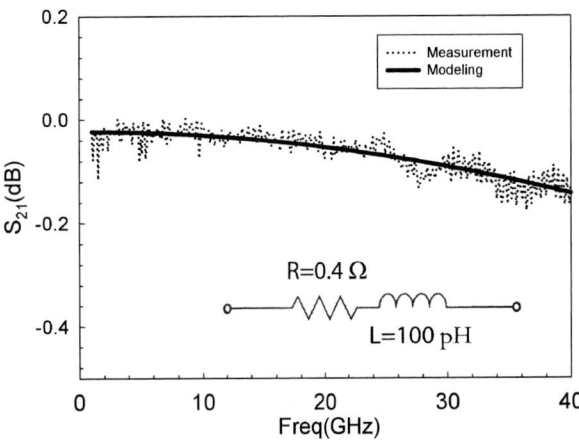

Figure 5 (b): Down-state S-parameter measurement and modeling.

Switching Time

The test setup for measuring the switching speed of a 2-port device is shown in Figure 7.

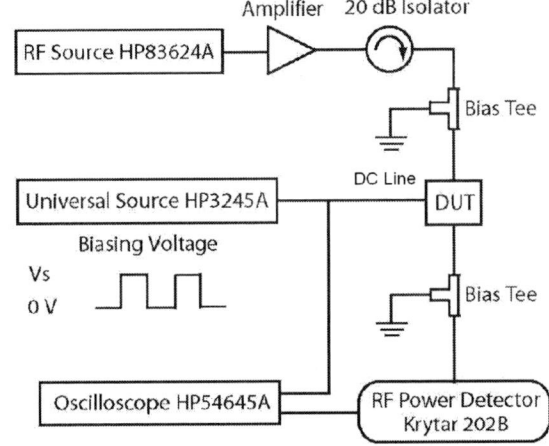

Figure 7: Switching speed measurement test setup.

The switch was actuated at a rate of 0.5 Hz using a 0-100 V unipolar voltage and injecting 0.3 mW of RF power at 10 GHz. Both actuation voltage and the output voltage of the diode detector are plotted in Figure 8 during the up-to-down and down-to-up switch transition. The measured switching speed is 12 μs from up to down and 3 μs from down-to-up transition.

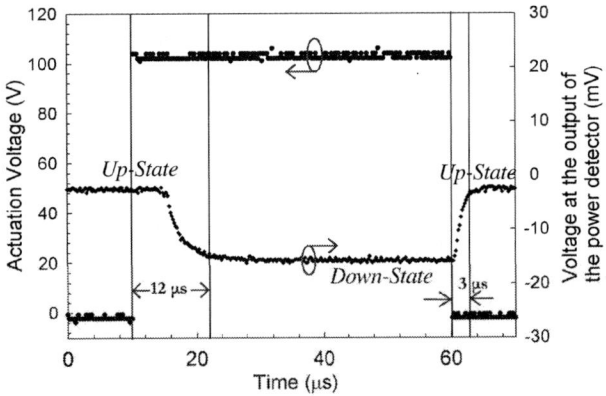

Figure 8: Measured switching time.

978-1-4244-2977-6/09 $25.00 © 2009 IEEE

CONCLUSIONS

A novel stress-gradient-robust metal-contact switch is designed, fabricated, and measured. The beam is very flat with < 0.02 μm deviation between the tip and the middle of the beam, and is therefore insensitive to stress gradients. V_p and V_r are very stable with temperature with < 2 V variation from 25-90 °C. A switching time of 3-12 μs was measured. S-parameter measurements indicate excellent isolation and insertion loss up to 40 GHz.

ACKNOWLEDGEMENTS

The authors would like to thank Isak Reines and Alex Grichener for valuable technical discussions. This work was supported by the UCSD/DARPA Center on RF MEMS Reliability and Design Fundamentals. The RF MEMS fabrication was done in CalIT^2 NANO3 facilities at UCSD.

REFERENCES

[1] Gabriel M. Rebeiz, *RF MEMS Theory, Design, and Technology*, John Wiley and Sons, Inc., 2003.

[2] S. Majumder, J. Lampen, R. Morrison, and J. Maciel, "A packaged, high-lifetime ohmic MEMS RF switch," in *IEEE MTT-S Int. Dig.*, Philadelphia, PA, June 2003, pp. 1935–1938.

[3] J. Costa, T. Ivanov, J. Hammond, J. Gering, E. Glass, J. Jorgenson, D. Dening, D. Kerr, J. Reed, S. Crist, T. Mercier, S. Kim, and P. Gorisse, "Integrated MEMS Switch Technology on SOI-CMOS," *Solid-State Sensors, Actuators, and Microsystems Workshop 2008*, pp. 18-21

[4] N. Nishijima, H. Juo-Jung Hung and G. M. Rebeiz, "Parallel-contact metal-contact RF-MEMS switches for high power applications," *17th IEEE International Conference on Micro Electro Mechanical Systems*, 2004, pp. 781-784.

[5] CoventorWare^{TM} version 2006, http://www.coventorware.com

MICROBIAL FUEL CELL BASED ON ELECTRODE-EXOELECTROGENIC BACTERIA INTERFACE

Erika Parra and Liwei Lin
Berkeley Sensor and Actuator Center
The University of California at Berkeley, USA

ABSTRACT

A microbial fuel cell with a microfabricated electrode and exoelectrogenic bacteria has been demonstrated. The superior compatibility of the organism's microbiology makes this approach more attractive and practical over previously micromachined microbial fuel cell designs. Both electrical performance and longevity of the device are greatly improved with continuous power output for over a week and enhanced performance over time. Using a $1mm^2$ anode, the fuel cell reaches and maintains $619 mV$ of open circuit voltage and delivers $0.12 \mu W$ maximum power using potassium ferricyanide at the cathode after 10 days of operation.

INTRODUCTION

The surge of oil prices along with concern for global warming has motivated the search for both short and long term alternatives to current energy technologies. Consequently, efficient and sustainable energy conversion is the focus of many research programs. Common areas of investigation in this field include improving the production of biofuels for combustion, solar cells, and hydrogen fuel cells. Conversely, a less studied biomimetic approach involves the direct conversion of simple sugars and alcohols into electricity using microorganisms. Termed microbial fuel cells (MFCs), these electrochemical devices utilize the enzymatic redox reactions of living microorganisms' metabolism to efficiently break down organic fuels and produce electric current. This is an attractive technology for it remedies both environmental and energy endeavors. The use of photosynthetic end products (biomass) as fuel, closes the carbon cycle, and the extraction of electricity directly through an electrochemical cell removes the Carnot efficiency limitation of thermal energy conversion. In addition, the use of inexpensive and self-renewable bacteria as the catalyst frees the demand of precious metals that are also a limited natural resource. In principle, potential applications of microbial fuel cells range from off-grid and portable electronics to industrial wastewater treatment, but the ultimate cost of materials and fabrication will dictate target markets. As we will show, MFCs are well suited for MEMS applications requiring steady, low level power sources with minimal maintenance.

Earlier MEMS-based MFCs have shown low energy conversion efficiency, low capacity, and short life time. Previously, we used *Saccharomyces cerevisiae* (baker's yeast) as the catalyst, and immobilized Thykaloids, to convert biomass into electrical energy [1-3]. Electron shuttles were required in these systems to carry electrons from living cells or enzymes to the inorganic electrodes. This results in low efficiency as current density is lost through the long diffusion process, and mediators compromise the microorganisms' viability. Here, we tackle these bottlenecks by adopting exoelectrogenic bacteria that break down various organic fuels into electrons and protons. These bacteria transport the separated charge extracellularly, eliminating the need for toxic exogenous compounds. In addition, the bacteria produce "organic nanowire" appendages used as electrical connections to transfer electrons directly to the electrode for enhanced efficiency [4-7].

PRINCIPLES OF OPERATION

We have adopted *Geobacter sulfurreducens* as our preferred organism because of its ability to thrive in highly anoxic (low oxygen) environments and transport electrons extracellularly through its intrinsic metabolic processes. In addition, these bacteria are capable of fully oxidizing compounds such as acetate, ethanol, and pyruvate to carbon dioxide with insoluble irons or electrodes as the electron acceptor for respiration, much like aerobic cells use oxygen. As illustrated in Fig. 1, electrical energy is harvested extracellularly from the bacteria through an electrochemical cell by inoculating an anode with the microorganisms that then colonize the electrode to form bacterial films that can reach over $40 \mu m$ in thickness [5].

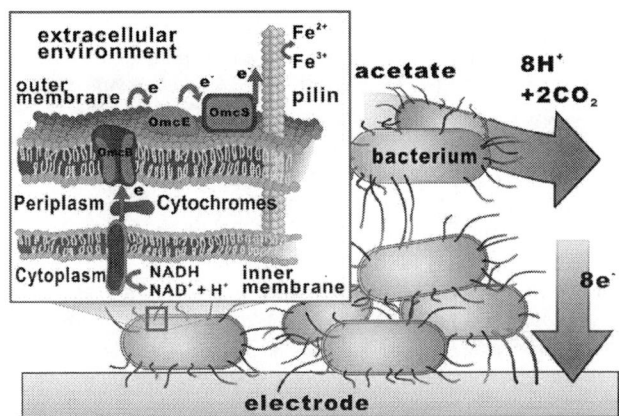

Figure 1. Exoelectrogenic bacteria membrane schematic (inset) illustrating electron externalization model and the pilins, "organic nanowires", of individual cells that form a network between cells and electrode. Here, bacteria break down acetate into electrons and protons to provide electricity.

In the case of acetate consumption, charged species (protons and electrons) and CO_2 are the metabolic byproducts, as shown in Equation 1. Electrons are transferred to the electrode via appendage nanowires that serve as electrical conduits or through direct electrode-cytochrome (membrane bound protein involved in charge transfer) contact between individual cells and the inorganic electrode [6-7].

$$CH_3COO^- + 2OH^- \rightarrow 2CO_2 + 8e^- + 5H^+ \quad (1)$$

To complete the circuit, protons released from the bacteria to regulate internal acidity travel through a proton exchange membrane (PEM) towards the cathode. At the cathode, protons combine with an oxidant as shown in Figure 2.

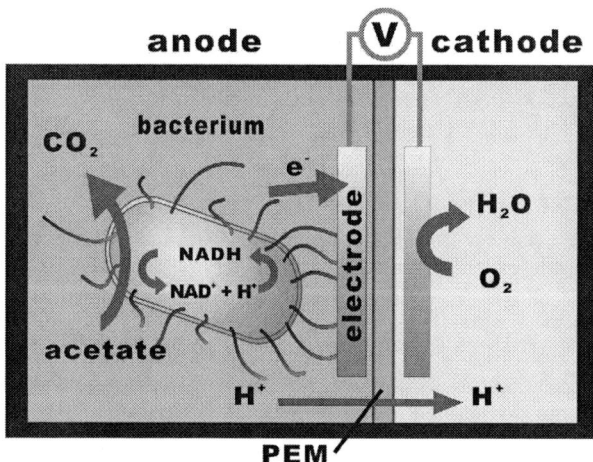

Figure 2. *Fuel cell using exoelectrogenic bacteria. Bacterial metabolism breaks down acetate into electrons, protons, and CO_2. Electrons are transferred to the anode while protons move to cathode through a PEM to complete the fuel cell operation.*

DESIGN AND FABRICATION

The prototype fuel cell anode geometry was designed to conform to the bacteria and biofilm dimensions. The anode consisted of a micro-patterned gold electrode that was 2 μm in width, the length of a single bacterium for single cell contact, arrayed at a 100 μm pitch to promote biofilm separation through the spacing of electrodes, as shown in Fig. 3a. The total anode surface area was 1 mm^2. The fabrication process started with the growth of a 500 nm oxide layer on a silicon wafer through a wet-furnace process, as shown in Fig. 3b. Next, photoresist was applied and patterned to define the electrodes using a lift-off process.

The electrochemical cell consisted of a two chamber flow-through configuration as shown in Fig. 4. The anodic chamber housed the bacteria in a 350 μL volume. The cathode consisted of a coiled gold wire with 100mm^2 surface area immersed in 200 μL of catholyte. The two chambers were separated by a 50 mm^2 Nafion 212 membrane. A miniature reference Ag/AgCl electrode was incorporated in the anolyte. Lastly, because the anodic

catalysis is of primary interest to this study, the cathodic reaction losses were mitigated by using potassium ferricyanide ($K_3[Fe(CN)_6]$) as an electron sink. The ferricyanide reduction to ferrocyanide proceeds as follows,

$$[Fe(CN)_6]^{3-} + e^- \rightarrow [Fe(CN)_6]^{4-} \quad (2)$$

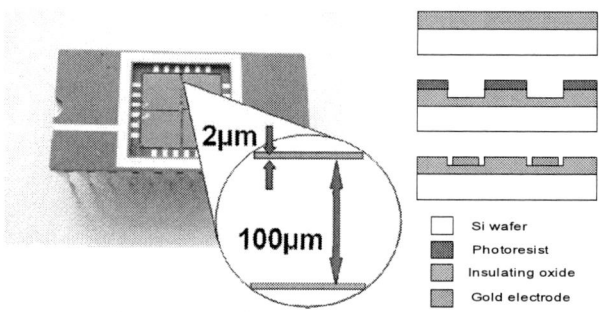

Figure 3. *(a) 2 μm wide patterned gold micro-electrodes with 100 μm spacing. (b) Electrode fabrication process. The process begins with a silicon wafer with insulating oxide on top. Photoresist is applied and patterned to define the electrodes with a lift-off process.*

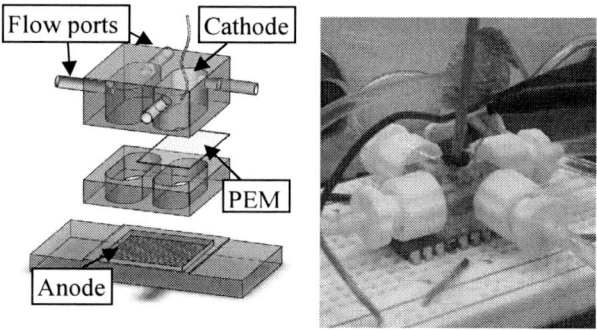

Figure 4. *(a) Exploded view of the MEMS MFC. Anolyte and catholyte circulate through corresponding chambers separated by a Nafion membrane. (b) Fuel cell experimental setup using potassium ferricyanide as electron sink at cathode.*

EXPERIMENTAL PROCEDURE

The bacteria, *Geobacter sulfurreducens,* was initially cultured using anaerobic media in a PIPES buffer using 20 mM acetate and 80 mM fumerate as electron acceptor at 30°C. The bacteria were then washed and added to the anolyte to colonize the electrode. Prior to taking all measurements, anaerobic media and acetate (5mM) were added to the anodic chamber. Quantitative measurements were acquired with a Gamry (Warminster, PA) Reference 600 potentiostat.

The catholyte consisted of a solution of 50mM phosphate buffer (pH 7) and 50 mM $K_3[Fe(CN)_6]$ with a redox potential of 436 mV (vs. SHE) at neutral pH. The reaction is quasi-reversible and the oxidized form can be renewed with aeration, but the set up used was anoxic and required continuous replenishment of the catholyte.

978-1-4244-2977-6/09 $25.00 © 2009 IEEE

RESULTS AND DISCUSSION

In contrast to our previous work, this system does not require an electron mediator. The *Geobacter* electrochemical cell generates electrical power, as shown in Fig. 2, by harnessing electrons from the bacterial metabolic breakdown of acetate. The bacteria retrieve energy from the fuel for biological maintenance functions or cell division, then release electrons as metabolic waste to be collected by an electrode as useful electrical energy. The mechanism is analogous to a standard fuel cell but *Geobacter* acts as the catalyst and acetate as the fuel, while the "organic nanowires" act as interfaces to the inorganic electrode, eliminating the need for an electron mediator.

Performance

Figure 5 shows the MFC polarization over time as the bacteria colonize the electrode surface. With increasing time and cell count on the electrode, the overpotential or connectivity losses are mitigated, and the open circuit voltage and current density increase. The maximum current obtained after 10 days of continuous operation, using acetate as fuel and potassium ferricyanide as the catholyte, was 1.4 μA/1 mm^2. SEM images of the anode indicated that only a monolayer of bacteria covered the electrode during the attainment of these measurements. The maximum current could be increased by using a catholyte with a more oxidizing redox potential (such as pure oxygen) and allowing the bacteria to cover the anode surface with a fully developed biofilm. Figure 6 depicts the power obtained as a function of current after ten days of bacterial respiration. In this case, a maximum power of 0.12 μW occurs at 0.61 μA.

The bacteria can produce a high potential upon inoculation of the anode due to a "pre-charging" of the membrane that occurs during test tube incubation, assuming that fuel has been metabolized without an electron acceptor. Prior studies estimate the redox potential of the electrons from the bacteria at -0.2 V (vs. SHE) [4], making the maximum open circuit voltage, V_{oc}, possible from the cell here presented roughly 630 mV. Figure 7 displays open circuit voltages under various operating conditions. Upon bacterial addition to the anode, a $V_{oc} \approx 630$ mV is obtained. Here, the cells were incubated in a test tube for two weeks prior to anolyte inoculation. Next, after an initial "discharging" of the bacteria, the V_{oc} reduces to 100 mV but increases over time as is illustrated by Fig. 5. After 10 days of respiration on the anode, V_{oc} values ranged from 550-600mV and maintained a low standard deviation (roughly 1 mV) over 10 minute intervals. The maximum steady state V_{oc} was 619mV or 98% of the maximum expected. Lastly, the control readings, taken with media and acetate but without bacteria, demonstrate negligible potential across the electrodes.

Figure 8 shows the MFC time response to different loads. These measurements were recorded 6 days after inoculation of the bacteria at a point where $V_{oc} \approx 0.2$ V. In each case, 90% of the V_{oc} was regained immediately upon load removal. In addition, the power output (voltage and current) was stable for all the loadings over time.

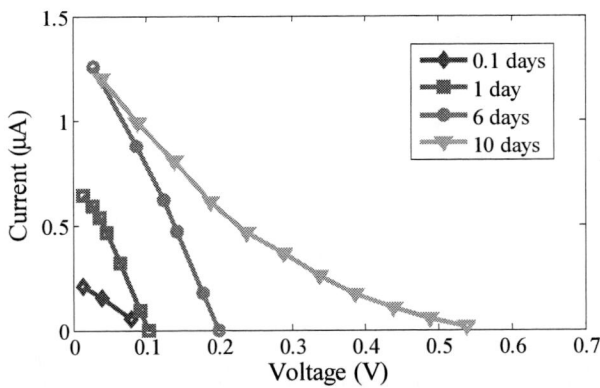

Figure 5. Polarization curves after 0.1, 1, 6, and 10 days, respectively, from the micro-patterned microbial fuel cell. Note that $V_{oc,max} \approx 630\,mV$ with potassium ferricyanide assuming that electrons from the bacteria are transferred extracellularly through c-type cytochromes with an estimated redox potential of -0.2 V (vs. SHE) [4].

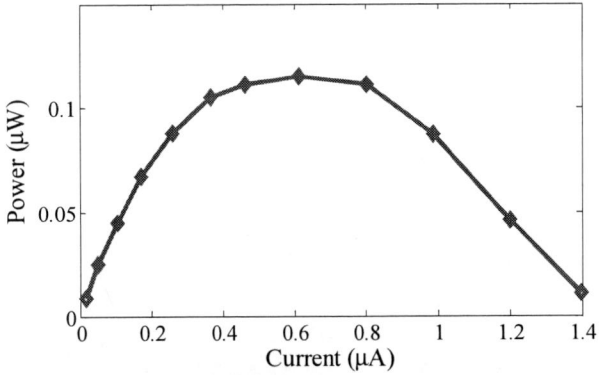

Figure 6. Power obtained at various loads after ten days of continuous bacterial respiration on the electrode.

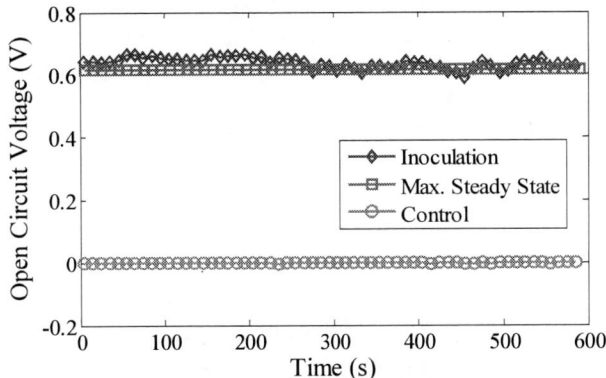

Figure 7. Open circuit voltage of MFC under various conditions. Upon inoculation of charged bacteria, the anode reaches the potential expected of metabolic enzymes. After 10 days of operation, steady state voltages ranged from 550 mV to 619 mV for several 10 minute intervals. Control sample with fuel but without catalyst showed negligible potential across the electrodes.

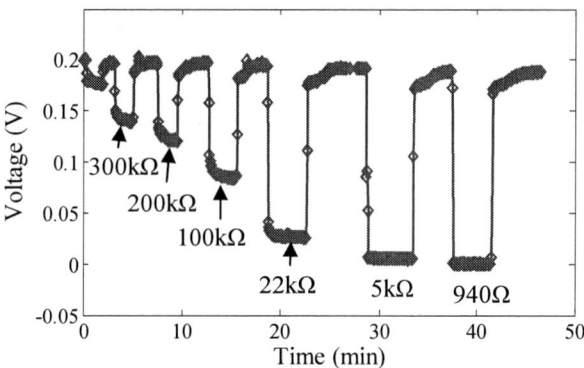

Figure 8. *MFC time response to different load resistors after 6 days of operation.*

Bacteria Viability

In addition to performance, the viability of the bacterial catalysts during respiration onto the electrode is also of interest. SEM images in Fig. 9 illustrate bacterial growth patterns on gold electrodes and on insulating oxide. After six days of respiration in the anodic chamber, bacteria more densely populate the electrode, as expected. Proximity to the conductive surface facilitates the release of electrons from charged enzymes used in metabolism and allows them to repeat the process efficiently. However, as is shown in Fig. 9b, cell division is common on the insulating surface making the current density contribution and connectivity means of these cells uncertain.

Figure 10 depicts the intricate rooting of bacteria on an electrode. Bacteria grow multiple appendages that form interconnects to the electrode and other bacteria [7].

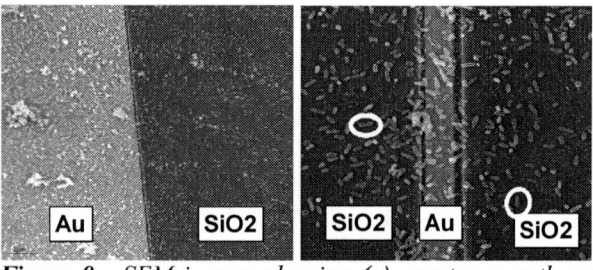

Figure 9. *SEM images showing (a) greater growth on gold electrode that insulating surface, and (b) white circles illustrating cell division occurring off the electrode.*

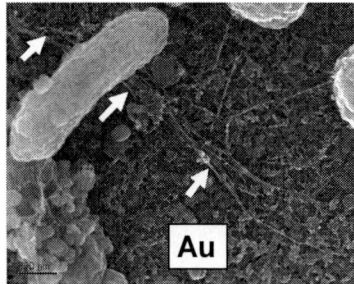

Figure 10. *SEM of bacteria illustrating significant appendages protruding into electrodes. Bacteria are roughly 300 nm in diameter and 1.5 μm long.*

CONCLUSIONS

Demonstrated here is a micro-fabricated MFC that uses *Geobacter sulfurreducens* as living catalyst and acetate as fuel. The system polarization (and power density) increases over time as the bacteria colonize the electrode surface. With a 1 mm² anode area, the fuel cell delivered $V_{oc} = 619$ mV and $P_{max} = 0.12$ μW using potassium ferricyanide as catholyte. We hypothesize that further performance improvements are possible by allowing the bacterial biofilm to fully cover the anode, increasing the electrode surface area, and enhancing the connectivity of the bacteria to the electrode.

In addition to promising performance, the presented approach of using exoelectrogenic catalysts offers a number of advantages over previous methods. Because the bacteria externalize electrons through their natural metabolic processes, no oxidizing electron shuttle is needed and viability is maintained. The system presented harvested energy for over a week. In addition, *Geobacter* appears to have the capability of transferring high energy electrons directly to the electrode via membrane contact or organic pilins.

ACKNOWLEDGEMENTS

We would like thank our collaborators, Dr. John Coates and Kelly Wrighton in the Plant and Microbial Biology, Dr. Haw Yang in Chemistry, Jui-Ming Yang and the Lin lab members, and the Microfabrication Laboratory at UC Berkeley. In addition, Dr. Cullen Buie, Tony Liu, and Kengo Suzuki for help reviewing the paper. This work is supported by the Sustainable Products and Solutions Program at UC Berkeley.

REFERENCES

[1] M. Chiao, K. Lam, and L. Lin, "Micromachined microbial and photosynthetic fuel cells," *JMicromech. and Microeng.*, Vol. 16, pp. 2547-2553, Dec. 2006.

[2] K. Lam, E. Johnson, and L. Lin, "A MEMS Photosynthetic Electrochemical Cell Powered By Sub-Cellular Plant Photosystems," *IEEE/ASME JMEMS,* Vol. 15. pp. 1243-1250, 2006.

[3] K. Lam, E. Irwin, K. Healy, and L. Lin, "Bioelectroanalytic Self-Assembled Thylakoids for Micro Power and Sensing Applications," *Sensors and Actuators - B*, Vol. 117, pp. 480-487, Oct. 2006.

[4] D. Bond and D. Lovley, "Electricity production by *Geobacter sulfurreducens* attached to electrodes," Appl Environ Microbiol 69(**3**):1548-1555, 2003.

[5] H. Richter, D. Lovley, et al., "Electricity generation by *Geobacter sulfurreducens* attached to gold electrodes," Langmuir 24(8):4376-4379, 2008.

[6] K. Weber, L. Achenbach, and J. Coates, "Microbes pumping iron: anaerobic microbial iron oxidation and reduction," Nature Reviews Microbiology 4:752-764.

[7] G. Reguera, D. Lovley, et al., "Extracellular electron transfer via microbial nanowires," Nature 435(7045):1098-1101, 2005.

A MICRO HYDROGEN GENERATOR WITH A MICROFLUIDIC SELF-REGULATING VALVE FOR SENSORS AND FUEL CELLS

Vikhram V. Swaminathan, L. Zhu, B. Gurau, R.I. Masel and M.A. Shannon[*]
University of Illinois at Urbana-Champaign, USA
[*]Corresponding Author: mshannon@illinois.edu

ABSTRACT

We have developed a fully integrated micro power generator, based on a hydrogen fuel cell. Key to the power generator is an on-board hydrogen production with a passive microfluidic control valve that provides self-regulating operation. Critical integration issues are identified and regimes established for efficient power generation. Preliminary results demonstrate the ability of our device to operate smoothly, with quick – order seconds – response for fluctuating power demands. Energy densities in excess of 150 Whr/L and peak powers of 369.2 W/L are promising for packaging the integrated hydrogen fuel cell into Power MEMS.

INTRODUCTION

Miniaturization of fuel-cell based power systems have been spurred by recent advances in microsystems. Among these, wireless sensors and bio-MEMS, for example, operate predominantly in the sub-milliwatt regime, with peak power requirement of about one milliwatt, and have duty cycles ranging from a few hours up to several days. Such systems require efficient power sources that can last over such periods of operation, and also meet the peak power demands in short bursts.

Fuel cells offer the promise of providing simultaneously higher power and energy densities than state-of-the-art batteries. However fuel cells have been successfully miniaturized down only to the centimeter scale, and focused on 1-10 W power generation for handheld electronic systems [1-3]. Scaling down further poses several challenges including fuel storage, handling, and control. In this paper, we demonstrate a fully integrated, millimeter-scale H_2-PEMFC, with an on-board hydrogen generator that readily supplies fuel through hydrolysis, which is controlled by a passive microfluidic valve that regulates water supply to the microreactor based on power demand encountered by the fuel cell.

Hydrogen Generator

Hydrogen generation is accomplished through controlled hydrolysis of calcium hydride (CaH_2), where

$$CaH_2 + 2H_2O \rightarrow Ca(OH)_2 + 2H_2\uparrow. \quad (1)$$

Numerous metallic hydrides have been identified as potentially high capacity sources of hydrogen [4]. The kinetics are influenced by the water-vapor partial pressure; the extent of reaction completeness; volume expansion; and solubility and porosity of the reaction products [5]. CaH_2 has a relatively high reaction rate and a hydrogen yield up to 90% [6], due to the hygroscopic, porous nature of by-product $Ca(OH)_2$. Allowance is made for volume expansion of 50% in designing the 4.4 mL capacity hydrogen generator, with a side-by-side hydride chamber and water reservoir configuration.

Fuel Cell

Hydrogen from the on-board CaH_2 source is used to run a miniaturized PEMFC. The membrane electrode assembly (MEA) for the H_2-PEMFC is comprised of a Nafion® 112 proton exchange membrane, and platinum black in a nafion-based ink as the catalyst. The MEA thickness is determined by factors such as resistivity, mass transport of protons and crossover of hydrogen, and strength against hydrogen pressure. In recent work done by our group [7], an air breathing cathode MEA showed peak power density as high as 280 mW/cm², which supplies the high-power density needed at the millimeter scale.

MEMS Control Valve

A microfabricated silicon control layer provides a fluidic connection between the water and hydride chambers to transport water for the hydrolysis reaction. It consists of holes aligned with the chambers', connected by a hydrophilic microchannel (150 μm wide × 5 μm deep × 2.2 mm long), and employs passive surface tension based pumping [8], due to capillary action (figure 1). The capillary pressure for a channel of width w and height h is described by the Young-Laplace equation

$$\Delta P = \rho \left(\frac{1}{r_1} + \frac{1}{r_2} \right), \quad (2)$$

with
$$2r_1 \cos\theta = h, \quad (3)$$

and
$$2r_2 \cos\theta = w, \quad (4)$$

where r_1 and r_2 are the principal radii of curvature for the meniscus, θ, the equilibrium contact angle, ρ, the surface tension of water, and ΔP, the pressure difference.

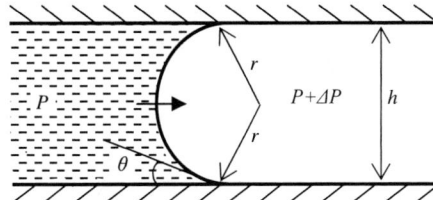

Figure 1: Surface tension pumping in a microchannel

A polyimide (PI) membrane bonded to the bottom of the control layer completes the channel geometry and forms a circular diaphragm valve below the water reservoir. A Cr/Au spot in the centre of the diaphragm provides water/hydrogen tight seal during valve closure. On the hydride side, a 300 μm hole creates a path for hydrogen to reach the MEA. A hydrophobic membrane, permitting water-vapor diffusion only, acts as a barrier to prevent liquid water from directly reacting with the CaH_2.

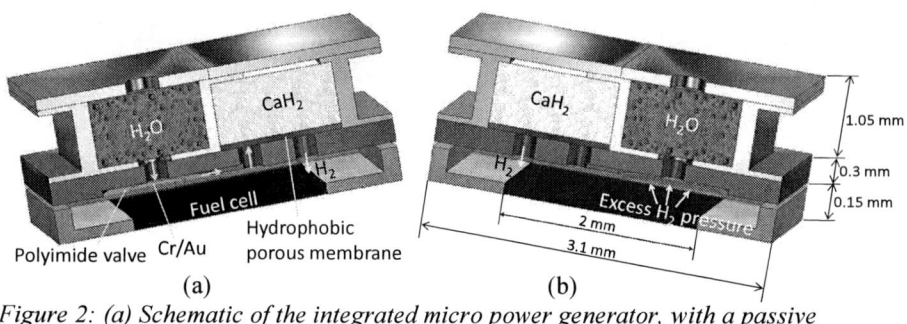
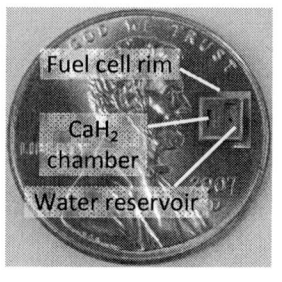

Figure 2: (a) Schematic of the integrated micro power generator, with a passive control valve, and (b) depicting the self-regulating action of the valve.

Figure 3: Photograph of the integrated device

The diameter and thickness of the diaphragm are estimated from plate theory with empirical modification [9], such that the pressure difference is

$$\Delta P = \frac{8}{3} \frac{Et}{a^4 (1 - \nu^2)} w_0^3 + 4 \frac{\sigma_0 t}{a^2} w_0, \qquad (5)$$

where $w(r)$ is the analytical shape of the membrane, w_0, the maximum centre deflection, t, the thickness, a, the radius, σ_0, the uniform residual stress, and E and ν, the Young's modulus and Poisson's ratio for PI, respectively. For a 1 mm diameter and 2 µm thick membrane, the theoretical closing pressure is 3.7 KPa, which is significantly lower than the capillary pressure of 24 KPa, cf. equations (2-4). Thus, the valve responds and closes first to prevent hydrogen from breaching into the microchannel and equalizing pressure [10]. A distinct advantage of the control layer placement exposing water-laden PI-Si microchannel to the MEA is maintaining humidity around the anode to prevent Nafion® dry out.

Figure 2 depicts a schematic of the integrated device, with the operation of the microfluidic control valve, and a photo is shown is figure 3 for size comparison. The overall volume of the millimeter-scale device is 12.7 µL.

Device operation

During power generation, water flows into the microchannel, vapor diffuses through the hydrophobic porous membrane, and reacts with CaH_2 to rapidly release hydrogen that is consumed at the anode. When the electrical load drops, hydrogen accumulates in the space between the MEA and the control layer, causing a pressure differential across the diaphragm. The valve responds to cut-off water needed for the hydrolysis, and hydrogen production stops, preventing any further rise in pressure, which is insufficient to displace water in the microchannel, thus sending the device into a suspended state. Conversely, when power demand increases, hydrogen is consumed, leading to a pressure drop. The valve opens fully, delivering more water to the hydride; the device now operates in a hydrogen limited regime, governed by water-vapor diffusion towards unreacted hydride. Thus, the microfluidic control layer acts as a simple regulator, switching operating regimes with varying demand, providing load control without using power itself.

FABRICATION

Figure 4 shows a fabrication sequence, broadly based on aspects specific to components described above.

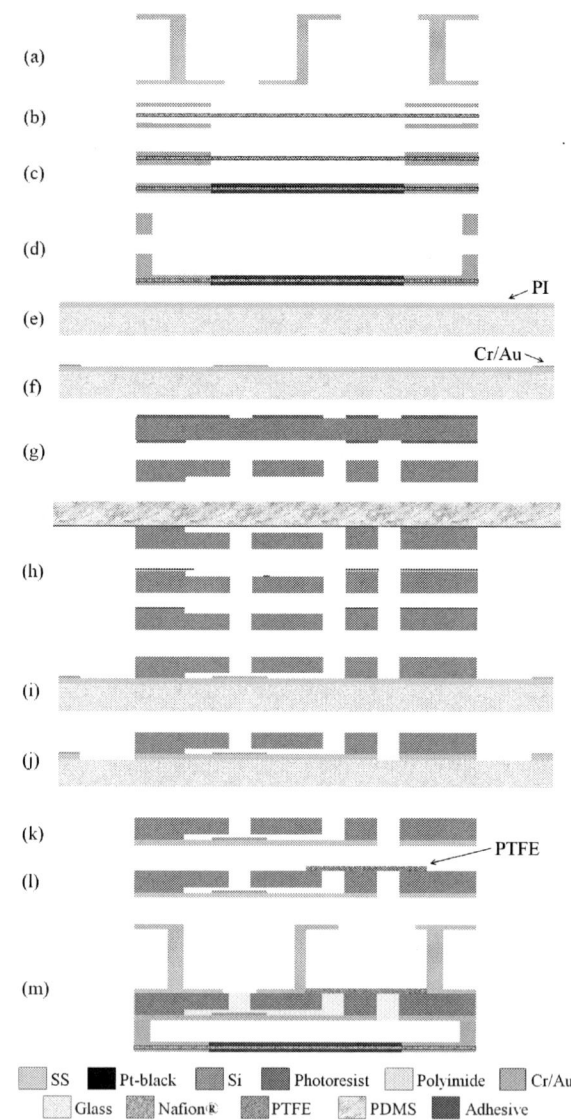

Figure 4: Fabrication flow sheet of the micro power generator.

The stainless steel (SS) hydrogen generator was electrical discharge machined, with rectangular water and hydride pockets. A 200-400 µm sized hole was provided at the bottom of the water chamber, whereas the hydride chamber bottom was left open with additional flange on the top to increase bonding surface. Material usage was economized and wall thickness reduced to 100 µm to increase volumetric energy density.

978-1-4244-2977-6/09 $25.00 © 2009 IEEE

The MEA (figure 4(b)-(d)) was made by sandwiching Nafion® 112 between two 25 µm SS sheets, each having a laser cut window of 4 mm². Bonding was done by adhesive transfer of glue, thermally activated at 110°C, and cured at 140°C under hot pressing to form a strong, water-resistant bond [11]. Pt-black based catalyst ink was painted directly onto the PEM square windows on either side, adequately contacting the SS sheet, to create electrodes. A 50 µm thick SS spacer was bonded around the MEA using 3-M 2216™ two-part epoxy resin.

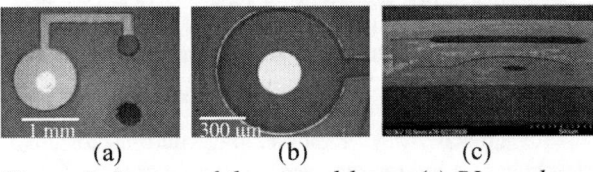

(a)   (b)   (c)

Figure 5: Images of the control layer. (a) PI membrane, microchannel and holes, (b) Cr/Au covering the water hole, and (c) SEM of Si control layer with a slot opening.

The control layer, shown in figure 5, was made using conventional microfabrication. The PI membrane was prepared through a spin-cure-release procedure (figure 4(e), (f)). A 25 mm square coverglass was treated by buffered HF and SC1 cleaning ($NH_4OH:H_2O_2:H_2O$ - 1:10:100 at 73°C for 30 min). HD Microsystems™ PI 2545 precursor was spun at 3000 rpm and baked on a hotplate at 90°C for 30 min. Imidization to 2.3 µm final thickness was achieved, by vacuum annealing in N_2, at 350°C for 3 hours. A 200 nm thick Au film on 5 nm Cr adhesion layer was sputtered onto the PI surface. Micromachined Si shadow masks were used with patterns of alignment marks and the 200/300 µm Cr/Au spot for covering the 100/200 µm water hole. The Si die was patterned using double sided photolithography and SF_6/C_4F_8 plasma etching in a PlasmaTherm™ ICP-DRIE (figure 4(g)). The bottom channel was etched 6.3 µm deep, and freshly masked by 100 nm sputtered Al, before etching through holes from the top. SC-1 cleaning was performed to make the channel walls hydrophilic ($\theta \leq 5°$) [12]. As shown in (figure 4(h)-(k)), the transfer bonding was done by a PDMS-Si contact printing process [11]. Using a shadow mask, adhesive in the channel was removed by isotropic O_2 plasma RIE. With precise alignment of the Cr/Au spot, Si was thermally bonded to PI at 140°C. PI covering the H_2 hole and around Si was etched away by O_2 plasma and shadow masking. The samples were released overnight in a DI water bath on a hotplate at 120°C. A 25 µm PTFE hydrophobic membrane was attached to cover the hydride holes, and finally, the three layers were stacked and bonded by epoxy resin (figure 4(l) & (m)). The device was loaded with powdered CaH_2 supplied by Aldrich Chemical Co. (St. Louis, MO) and capped with a SS cover using Loctite Super Glue™, in a Nitrogen Glovebox.

TESTS AND RESULTS

Three specimens of the fully integrated micro power generator were tested, using a SI 1287 Potentiostat (Solartron Analytical, Hampshire, UK), to obtain various performance metrics. At the start of each experiment, Millipore water was added to the water chamber using a syringe. The first device had a control layer with a slot,

and was used to obtain I-V characteristics. It was subsequently tested in switching mode, using ten, 10 min square waves of 0.8/0.3 V. The other two were tested for operating lifetime and energy density: the device with a 400 µm water delivery hole at 0.6 V; and the third with a 500 µm hole at 0.9 V. Results are depicted in figures 6-8 and summarized in Table 1.

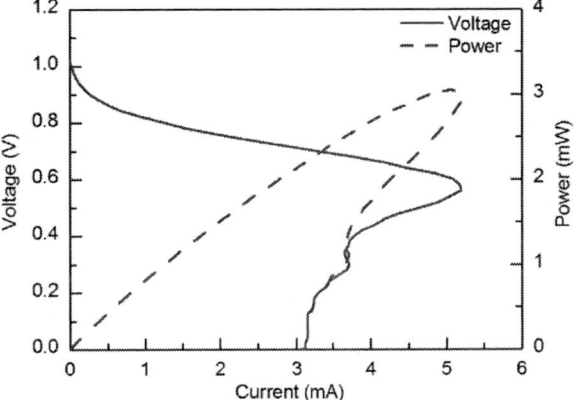

Figure 6: Polarization and theoretical power curve

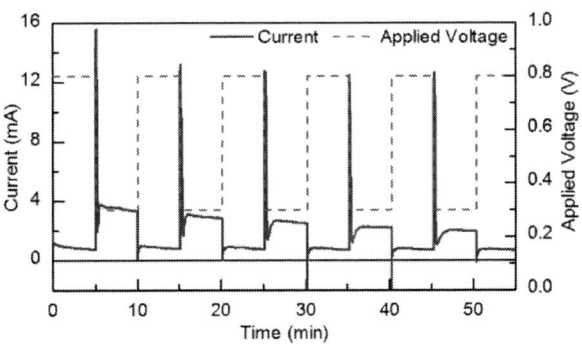

Figure 7: Square wave response between 0.8 and 0.3 V operating voltage; Peak Power density was 369.2 W/L

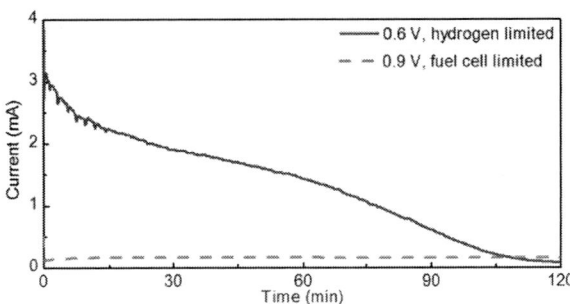

Figure 8: Chronoamperograms at two operating regimes

DISCUSSION

Numerous effects arise as a consequence of miniaturizing the H_2-PEMFC, evident from our results. The polarization and theoretical power curves suggest substantial drop in performance upon integration. The curves change direction when operated as high as 0.6 V, which indicates a H_2 mass transfer limit for the hydrogen generator. However, a high open circuit voltage of 1 V and the kinetic limit within 0.8 V imply the possibility of highly efficient power generation at sub-milliwatt capacities. The lifetime test on device 3 at 0.9 V generated

Table 1: Summary of test results for integrated devices

No.	Voltage (V)	Duration (hr)	Energy (Whr/L)	Peak Power (W/L)
1	0.8/0.3	1.67	83	369.2
2	0.6	2	152.1	159.4
3	0.9	10.5	86.75	31.86

a steady current during fuel cell limited operation. Running the device below 0.6 V for increased power is possible, but at reduced efficiency due to mass transport limitations.

Figure 7 depicts switching between high and low loads. The response shows the control of the water transport through the microfluidic channels to the hydride and the subsequent transport of the hydrogen to the MEA. The figure also shows the maximum power capacity of the device over a long period of dynamic operation. During the 5 min dwell times, we observed smooth current generation that gradually decreased over time due to a decrease in a mass transfer of H_2O into the generator and the H_2 gas out through the product hydrate. The nearly flat slope at 0.8 V shows fuel cell operating near the kinetic limit; but a small gradient arises from slowing of the hydrolysis reaction due to the increase in diffusion length for the water vapor to reach the hydride. The sharp peaks observed to be consistently in excess of 300 W/L while switching from 0.8 V to 0.3 V show that excess H_2 generation was sustained for long periods, and that the rapid change in load consumed the H_2 to meet demand. The drop to a lower value occurred as a consequence of a fully mass transport limited operation. However, current soon rose to levels predicted by the I-V curve, indicating the passive surface tension pumping of water in the microchannel to meet the H_2 demand. Similarly, the immediate drop upon switching back to a higher voltage, and settling into a steady operation with almost constant current, was made possible by successful closure of the valve, with its consequent increase in H_2 pressure.

The water-vapor diffusion limited H_2 generation theory was justified by a continuous, gradually decreasing current at 0.6 V operation. Figures 7 and 8 validate our approach and establish its versatility—the ability to generate power efficiently across different regimes, and respond to a dynamically varying demand. The energy density of the order of 150 Whr/L is on par with contemporary technologies, but high power capacity (369.2 W/L) and dynamic performance provide both high energy and power densities needed for many microsystems.

CONCLUSION

We have successfully demonstrated the concept of a self-regulating micro power generator with the capacity to generate H_2 on-demand for a fuel cell, with the ability to operate for long periods across multiple regimes, along with short bursts of high power density. The microfluidic control valve is sensitive to dynamically changing power requirements, without consuming power for control.

ACKNOWLEDGEMENTS

This research is funded by the Defense Advanced Research Projects Agency (DARPA) under U.S. Air Force grant FA8650-04-1-7121. Any opinions, findings, and conclusions or recommendations expressed in this manuscript are those of the authors and do not necessarily reflect the views of the US Government. Micro-Nano-Mechanical Systems Laboratory in University of Illinois at Urbana-Champaign provided the cleanroom facilities.

REFERENCES

[1] G.K. Pitcher, G.J. Kavarnos, "A Test Assembly for Hydrogen Production by the Hydrolysis of Solid Lithium Hydride", *Int. J. Hydrogen Energy*, vol. 22, no. 6, pp. 575-579, 1997.

[2] P.P. Prosini, P. Gislon, "A hydrogen refill for cellular phone", *J. Power Sources*, vol. 161, pp. 290-293, 2006.

[3] A.V. Pattekar, M.V. Kothare, "A Microreactor for Hydrogen Production in Micro Fuel Cell Application", *J. Microelectromech. S.*, vol. 13, no. 1, pp. 7-18, 2004.

[4] R. Aeillo, M.A. Matthews, D.L. Reger, J.E. Collins, "Production of Hydrogen Gas from Novel Chemical Hydrides", *Int. J. Hydrogen Energy*, vol. 23, no. 12, pp. 1103-1108, 1998.

[5] V.C.Y. Kong, D.W. Kirk, F.R. Foulkes, J.T. Hinatsu, "Development of Hydrogen Storage for Fuel Cell Generators II: Utilization of Calcium Hydride and Lithium Hydride", *Int. J. Hydrogen Energy*, vol. 28, pp. 205-214, 2003.

[6] L. Zhu, D. Kim, H. S. Kim, R.I. Masel, M.A. Shannon, "Hydrogen Generation from Hydrides in Millimeter Scale Reactors for Micro Proton Exchange Membrane Fuel Cell Applications", *J. Power Sources*, 2008, in press.

[7] B. Gurau, L. Zhu, H.S. Kim, R.D. Morgan, V. Vilasur Swaminathan, M.A. Shannon, R.I. Masel, "Integrated Miniature Fuel Cell – Hydrogen Generator for Portable Power Generation", *Proc. ECS PRiME*, 2008.

[8] G.M. Walker, D.J. Beebe, "A Passive Pumping Method for Microfluidic Devices", *Lab Chip*, vol. 2, pp. 131-134, 2002.

[9] J.C. Selby, M.A. Shannon, K. Xu, J. Economy, "Sub-micrometer Solid-state Adhesive Bonding with Aromatic Thermosetting Copolyesters for the Assembly of Polyimide Membranes in Silicon-based Devices", *J. Micromech. Microeng.*, vol. 11, pp. 672-685, 2001.

[10] S. Moghaddam, E. Pengwang, R.I. Masel, M.A. Shannon, "A Self-regulating Hydrogen Generator for Micro Fuel Cells", *J. Power Sources*, vol. 185, pp. 445-450, 2008.

[11] B.R Flachsbart,., K. Wong, J. M. Iannacone, E. N. Abante, R. L. Vlach, P. A. Rauchfuss, P. W. Bohn, J. V. Sweedler, and M. A. Shannon, "Design and Fabrication of a Multilayered Polymer Microfluidic Chip with Nanofluidic Interconnects via Adhesive Contact Printing," *Lab-On-A-Chip*, vol. 6 no. 5, pp. 667-674, 2006.

[12] D-H Eom, G-B Lim, J-G Park, A.A. Busnaina, "Reaction of Ozone and H_2O_2 in NH_4OH Solutions and Their Reaction with Silicon Wafers" *Jpn. J. Appl. Phys.*, vol. 43, no. 6A, pp. 3335-3339, 2004.

MICROCHANNELS WITH SUBSTANTIAL FRICTION REDUCTION AT LARGE PRESSURE AND LARGE FLOW

C.F. Carlborg, G. Stemme, and W. van der Wijngaart

Microsystem Technology Lab, KTH – Royal Institute of Technology, Stockholm, Sweden

ABSTRACT

This paper introduces and experimentally verifies a self-regulating method for reducing the friction losses in large microchannels at *high liquid pressures* and *large liquid flows*, overcoming limitations with regard to sustainable liquid pressure on a superhydrophobic surface.

Our design of the superhydrophobic channel creates an automatic adjustment of the gas pressure in the lubricating air layer to the local liquid pressure in the channel. This is achieved by pneumatically connecting the liquid in the microchannel to the air pockets trapped at channel wall trough a pressure feedback channel. When liquid enters the feedback channel it compresses the air and increases the pressure in the air pocket. This reduces the pressure drop over the air-liquid interface and increases the maximum sustainable liquid pressure. We define a dimensionless fluidic number, $W_F = P_L D_h / \gamma \cos\theta_c$, which expresses the fluidic energy carrying capacity of a superhydrophobic microchannel. We experimentally verified that our geometry can sustain several times higher liquid pressure before collapsing, and we measured better friction reducing properties at higher W_F values than in previous works. This method could be applicable for reducing near-wall laminar friction in both micro- and macroscale flows.

INTRODUCTION

The liquid mass flow through microchannels under a pressure drop ΔP_L is limited by the frictional losses induced by the no-slip boundary condition at the liquid-solid interface. Previous work [1-3] have demonstrated that the frictional losses can be significantly reduced by trapping pockets of air on a superhydrophobic surface at the channel wall, forming an air lubrication layer. Reduction of the frictional losses up to 40% in microchannels has been reported using superhydrophobic surfaces [1]. However, all previous efforts were limited by the integrity of the air pockets because the maximum liquid pressure, P_L, in the microchannel must remain below the Laplace collapse pressure,

$$P_{LC} = \frac{\gamma \cos\theta_c}{2a} \qquad (1)$$

of the superhydrophobic surface, with γ being the liquid-air surface energy, θ_c the liquid-solid contact angle and a being the characteristic cavity length of the superhydrophobic geometry.

The drag reduction in previous work is limited by the Laplace collapse pressure P_{LC} of the air filled cavities, because air pressure P_G is fixed at priming conditions and cannot be adjusted to increasing liquid pressure P_L:

$$P_L - P_G < P_{LC} = \frac{\gamma \cos\theta_c}{2a}. \qquad (2)$$

The Laplace collapse pressure is the change of the superhydrophobic surface from Cassie state (lubricating) to Wenzel state (wetted). Since the effective friction reduction of a superhydrophobic channel is proportional to the ratio a/D_h, where D_h is the hydrodynamic diameter of the channel, substantial drag reduction at high pressures can only be accomplished in very narrow channels [2].

In short, to achieve a *large effective friction reduction* in channels with *a large hydraulic diameter D_h* requires a *large air pocket size a* [4]. This will create an unstable air lubrication layer prone to collapse already at low liquid pressures P_L (Figure 1 A).

We previously reported the circumvention of these limitations using an active control of the lubricating air pressure [5], in which we connected the air pockets to external air pressure sources. In this paper we present a self-regulating method that does not need external pneumatic control for reducing the viscous drag in channels with *large hydraulic diameter D_h* and at *large pressures*, Figure 1B. This method could be applicable for near-wall laminar friction reduction in both micro- and macroscale flow, where laminar losses typically account for 100%, respectively 30% of overall flow losses.

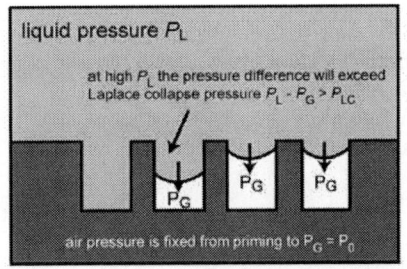

A) Standard superhydrophobic channel wall- P_G is fixed by the priming conditions

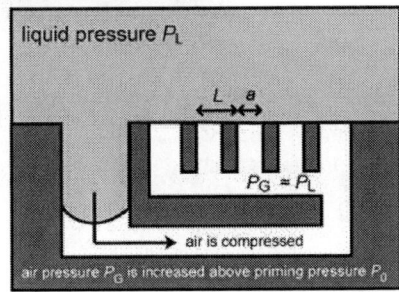

B) Self-regulating design - feedback channel makes air pocket pressure P_G increase with the local liquid pressure P_L

Figure 1 : In a normal superhydrophobic surface, the trapped air pocket pressure is fixed at the priming. When the liquid pressure is increased the pocket will eventually collapse. In the novel design, air pocket pressure is not fixed and is increased as the liquid pressure is increased, delaying significantly or altogether preventing collapse.

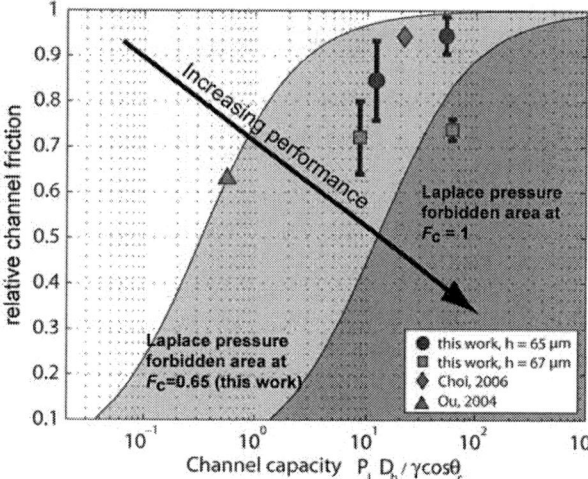

Figure 2: Plot of the relative friction, v, against the channel's energy capacity, W_F. The dark grey area corresponds to conditions that cannot be reached using previous methods, even assuming they could reach a 100% shear-free fraction $F_c=1$. The lighter gray area corresponds to conditions that cannot be reached with previous methods if they would have a 65% shear-free fraction $F_c=0.65$. Experimentally measured values of W_F are plotted for this work ($F_c=0.65$) and for work of Choi [2] ($F_c=0.78$) and Ou [1] ($F_c=0.97$).

ANALYSIS

We analyzed the (well studied) case of parallel plate flow between two ribbed superhydrophobic surfaces. The results of this analysis can be qualitatively generalized to other channel and surface roughness geometries. The relative friction in a ribbed superhydrophobic channel, i.e. the channel friction normalized against the channel friction of an equivalent channel at classical zero-slip condition, can be defined as

$$v = \frac{\text{slip flow losses}}{\text{zero - slip flow losses}} = \frac{f\,\mathrm{Re}}{f\,\mathrm{Re}|_{\mathrm{PP}}}, \qquad (3)$$

where $f\mathrm{Re}$ is the classical non-dimensional product of the Darcy friction factor f and the Reynolds number Re. We introduce a novel dimensionless number that expresses the fluidic energy carrying capacity of a superhydrophobic channel:

$$W_F = \frac{P_L D_h}{\gamma \cos \theta_c}, \qquad (4)$$

where P_L is the liquid pressure and D_h the channel's hydraulic diameter. The relation between the effective slip-length λ and the classical friction factor $f\mathrm{Re}$ can be expressed as [6]

$$\frac{\lambda}{D_h} = \frac{8}{f\,\mathrm{Re}} - \frac{1}{12}. \qquad (5)$$

The theoretically predicted effective slip length by Lauga et al. [4] is

$$\lambda = \frac{L}{2\pi} \ln(1/\cos(F_c \pi /2)), \qquad (6)$$

where $F_c = a/L$ is the "lubricated fraction", i.e. the liquid-air interface length, a, divided by the total liquid interface length along the channel wall L. Combining the equations (1), (5) and (6), shows that the relative friction reduction v in previous works is limited to values of W_F for which the relative channel friction is,

$$v > \left[1 + \frac{12}{\pi} \ln(\frac{1}{\cos(F_c \pi/2)}) \frac{1}{F_c W_F} \right]^{-1}. \qquad (7)$$

A plot of the relative friction factor v versus the dimensionless number W_F is shown in Figure 2. Lee et al. [3] showed for the first time values for $F_c > 0.99$, i.e. experimentally reaching the limits of standard superhydrophobic patterns

OPERATIONAL PRINCIPLE

We developed a self-regulating superhydrophobic grating geometry (Figure 1 and 3) that overcomes the Laplace burst limitation and thus allows for substantial drag reduction at high W_F values, i.e. high liquid pressures P_L and large channel sizes D_h. The lubricating properties of a normal superhydrophobic surface will disappear when the liquid pressure exceeds the Laplace collapse pressure of the superhydrophobic grating. Our geometry creates an automatic adjustment of the gas pressure P_G in the lubricating air layer to the local liquid pressure $P_L(x)$ in the channel, whereas in previous work, P_G is fixed after channel priming (Figure 1). This is accomplished by pneumatically connecting the main liquid channel to the air pockets via a pressure feedback channel that runs behind the superhydrophobic grating such that the liquid compresses the trapped air layer to nearly the same pressure as the local liquid pressure (Figure 1 B).

The experimental device is shown in Figure 3. After priming the main channel (Figure 3 B), the wider part of the feedback channel is filled with liquid, compressing the air behind the superhydrophobic grating. As the liquid pressure is increased far above the normal collapse pressure, (Figure 3 C), the liquid advances in the feedback channel and compresses the air behind the grating further, thus reducing the pressure drop over the air-liquid interface and preventing collapse of the lubricating air layer. The air layer pressure is approximately the same as the local liquid pressure minus the Laplace pressure at the gas-liquid interface in the feedback channel. The gas pressures in the air pockets, indicated in Figure 2, could be estimated by assuming isothermal compression of the gas and neglecting mass transport across the liquid-gas interface.

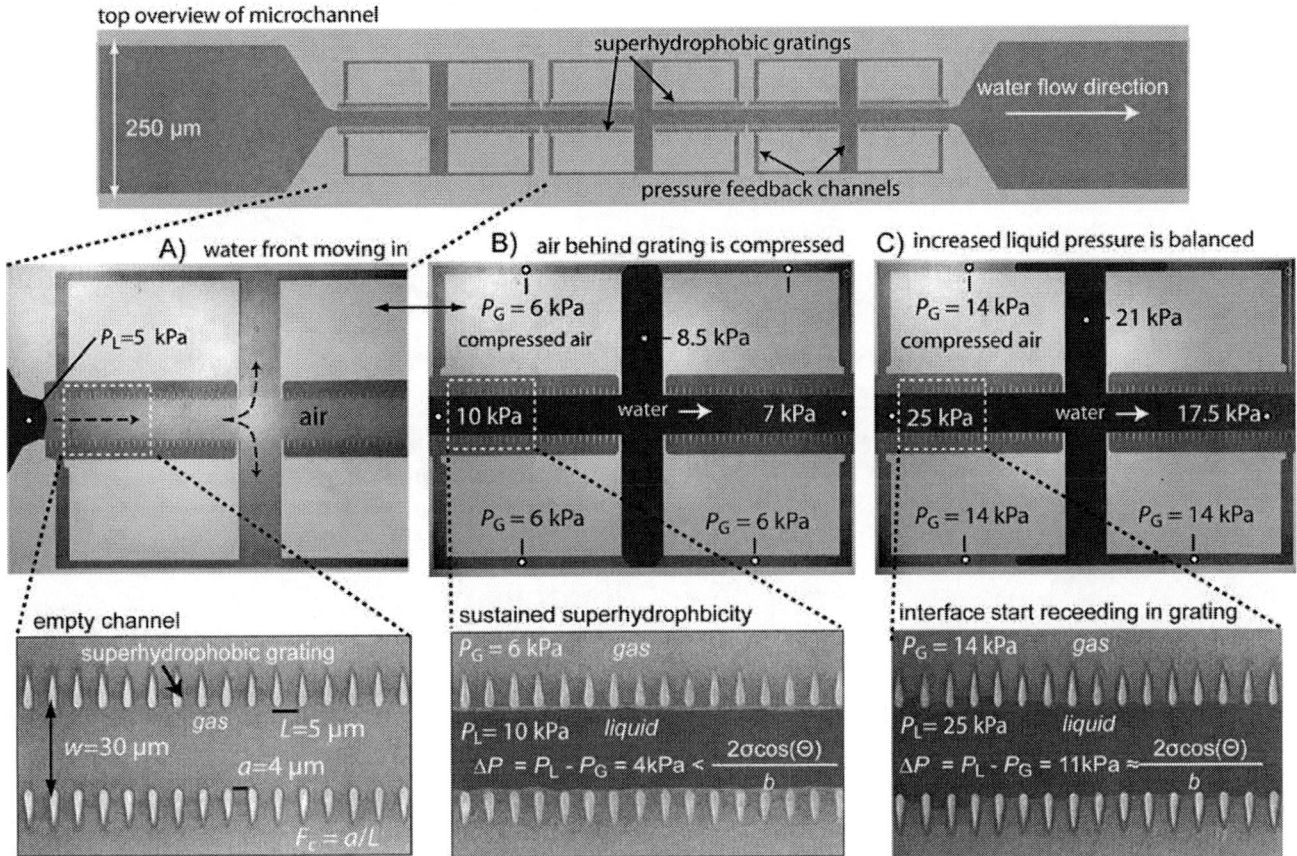

Figure 3 : Top-view photographs of the first cell (l =400 μm) in a L=1.2 mm long microchannel. Top: the entire superhydrophobic section of a test channel. Middle: Microscope pictures of (A) the channel prior to priming, (B) after priming the main channel, and (C) As the liquid pressure is increased far above the normal collapse pressure. For clarity, the contrast of the liquid has been enhanced.

EXPERIMENTS AND RESULTS

For experimental verification, we DRIE etched $h = 65$ μm and 67 μm deep and $w = 30$ μm wide channels ($D_h = 4A/S = 41$ μm) in silicon. Along most of the channel, the sidewalls consist of a pillar grating with pitch $L=5$ μm and an air-liquid interface length, a, between the pillars with 3 μm $< a <$ 5 μm (Figure 3). Each channel section of 400 μm length contains one feedback channel, with width < 25 μm, connected to an air pocket behind the pillar grating. The entire silicon surface was dip-coated with 0.5 % Teflon AF™ solution, spun at 3000 rpm and baked at 165 °C (measured θc=105±5°) and the structure was closed with an air tight hydrophobic tape.

In a first experiment we first liquid-filled the channels, after which we applied an equal ramping liquid pressure P_L to both the inlet and outlet (Figure 4 A). At high liquid pressures the liquid penetrates the perpendicular feedback channel and doing so compresses the air behind the superhydrophobic grating (Figure 3 B). The increased pressure in the air behind the grating makes it withstand higher liquid pressures P_L before collapsing. As the liquid pressure is increased far above the normal collapse pressure (Figure 3 C), the liquid continues to advance in the feedback channel and compresses the air in the lubricating layer further. The liquid-air interface at the grating was observed trough a microscope and remained intact for $0 < P_L < 45$ kPa. This value is between 2.6 and 9 times higher than the theoretical Laplace collapse

pressure of 16 kPa (b = 3 μm; θc=110°), respectively 5 kPa (b = 5 μm; θc=100°).

In a second experiment we coupled the ramping pressure source to the inlet while keeping the outlet at atmospheric pressure. Initially, the liquid pressure is not enough to penetrate the superhydrophobic channel (Figure 3 A), but at elevated pressures a flow is established through the main channel. At yet higher pressures the liquid penetrates the feedback channel and regulates the air pressure behind the grating (Figure 3 B and C). The liquid volume flow Q trough the channel and the pressure drop ΔP_L were recorded (Figure 4B and 5), and the relative friction in the microchannel, compared to a smooth channel with same hydraulic diameter D_h, could be calculated. The liquid-air interface was observed to remain intact up to almost twice (25 kPa) the maximum sustainable Laplace pressure before the first upstream air pocket collapsed. The loss of superhydrophobic properties could also be observed from the increased fluidic resistance (the slope of Figure 5).

With a shear free fraction of $F_c = 0.65$, estimated from Figure 3 (bottom), a normal superhydrophobic surface with the same roughness pitch is limited to the non-shaded part of Figure 2. With the current design we demonstrate that we can overcome this limitation and move into the Laplace pressure forbidden, shaded area, with increased performance in terms of "fluidic energy carrying capacity". The relative friction reduction is

978-1-4244-2977-6/09 $25.00 © 2009 IEEE   41

plotted in Figure 2, and compared with previous experimental works using superhydrophobic surfaces with different roughness pitch a. Note that in our experimental conditions, the influence of the zero-slip bottom and top walls of the channel probably cause a substantial deviation from the ideal parallel plate model investigated in the Analysis section. Our results show up to 5 times relative friction reduction for similar values of W_F [1] and up to 100 times larger W_F values while maintaining the same relative friction reduction [2].

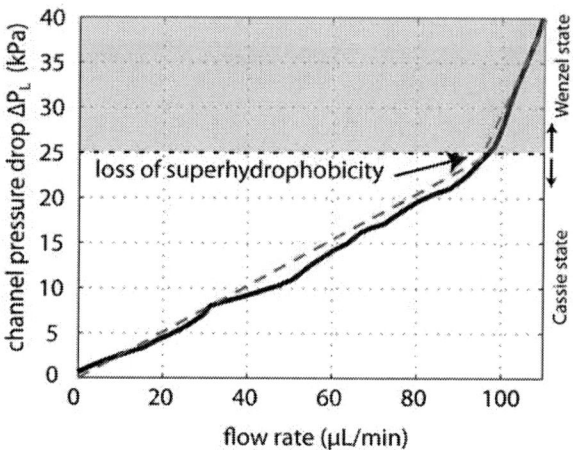

Figure 5: Pressure-flow curve when increasing the pressure drop over the channel. The bursting of air pockets is visible as a steeper slope (increased fluidic resistance) at flow rates exceeding 88 µL/min.

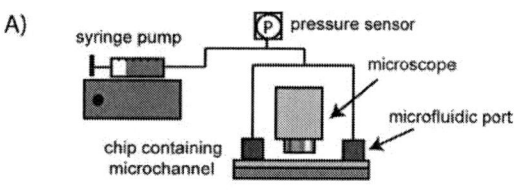

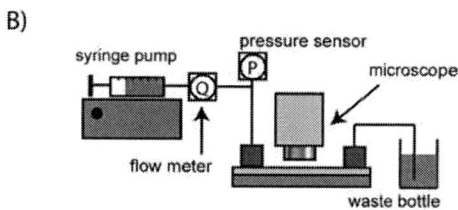

Figure 4 : Experimental setups A) Static conditions. Equal liquid pressure in the channel is ramped by a syringe pump. Pressure is monitored and air cavities are observed trough microscope. B) A liquid pressure drop is applied over the channel using a syringe pump. The flow is monitored by a flow meter and pressure is measured at the channel entrance.

CONCLUSION

We developed a novel design for microchannel friction reduction that is not limited in liquid pressure or channel size. Compared to previous work, our experiments show up to 5 times relative friction reduction for similar "fluidic energy carrying capacity" values W_F and up to 100 times larger W_F values while maintaining the same relative friction reduction.

REFERENCES

[1]. J. Ou, B. Perot and J.P. Rothstein, *Phys. Fluids*, vol 16, 2004, pp 4635-4643

[2] C-H Choi, U. Ulmanella, J Kim, C-M. Ho and C-J Kim, *Phys. Fluids*, vol 18, 2006, 087105

[3] C. Lee, C-H. Choi and C-J. Kim, *Phys Rev. Lett*, vol 101, 2008, 064501

[4] E. Lauga and H. A. Stone, *J. Fluid Mech.* 489, 2003, 55-77

[5] C.F Carlborg, M. Do-Quang, G.Stemme, G.Amberg, W. van der Wijngaart., *proc. IEEE MEMS 2008*, pp. 599-602.

[6] J. Davies, D. Maynes, B.W. Webb, B. Woolford, *Phys. Fluids*, Vol. 18, 2006, 0871110

STARJET: PNEUMATIC DISPENSING
OF NANO- TO PICOLITER DROPLETS OF LIQUID METAL

T. Metz[1], G. Birkle[2], R. Zengerle[1,2], and P. Koltay[2]
[1]HSG-IMIT, Villingen-Schwenningen, GERMANY
[2]IMTEK - University of Freiburg, Freiburg, GERMANY

ABSTRACT

In this work we present a novel, simple and robust, pneumatically actuated dispenser for nano- to picoliter sized droplets of liquid metals. The so called StarJet dispenser utilizes a star-shaped nozzle geometry that stabilizes plugs of liquid in the centre of the nozzle by capillary force. This minimizes the wall contact of the liquid plug and reduces contact line friction. Individual droplets of liquid metal can be pneumatically generated by interplay of the sheathing gas flow in the outer grooves of the nozzle and the liquid metal. The working principle was first discovered and studied by Computational Fluid Dynamic (CFD) simulations. For experimental validation silicon chips with the star-shaped geometry were fabricated by Deep Reactive Ion Etching (DRIE) and assembled into a printhead. With different nozzle chips volumes between 120 pl and 3.6 nl could be generated at natural frequencies of 90 Hz and 400 Hz. The StarJet can either be operated as drop on demand or as continuous droplet dispenser. We printed columns of metal with 0,5 to 1,0 mm width and 40 mm height (aspect ratio >40) to demonstrate the directional stability of the ejection.

INTRODUCTION

The contact free dispensing of tiny liquid volumes in the nano- to picoliter range is an important application field for MEMS technologies since more than twenty years, from inkjet printing [1], to the printing of microarrays [2] and for rapid prototyping methods [3]. A particularly challenging area is the dosage of liquid metals [4]. This is required for the generation of solder bumps [5] for flip chip bonding or for rapid prototyping of electric circuits [3] and metallic devices. Further the dosage of small droplets to arm MEMS fabricated mercury switches [6] is of interest.

The contact free dispensing of liquid metals is challenging for different reasons: The dosage device must operate above the typically high melting temperatures of the metal. The high temperatures and the temperature changes during heating and cooling can induce mechanical stress that complicates assembly of the devices and boosts oxidation. Preventing oxidation of the jetted droplet is an additional issue. Therefore a constant flow of inert gas around the liquid metal is required. If the droplet generators are driven by piezo ceramics [4], the piezos must be thermally insulated as their working range is limited by the Curie temperature that ranges normally between 150°C and 300°C. Finally, the high surface tension of liquid metals prohibits capillary priming of devices as it leads to very high contact angles on most materials.

The StarJet system presented in this work is a rather simple solution for the dosage of liquid metals and does not suffer from any of these problems. It applies the sheet gas flow required to prevent oxidation as actuation mechanism and takes advantage of the high contact angles between the liquid metal and the nozzle chip fabricated of silicon. The assembly of the nozzle chip into the printhead is kept simple so thermal stresses do not affect the system.

WORKING PRINCIPLE

The heart of the StarJet dispenser is a star shaped nozzle as shown in Fig 1a. The profile was first studied as geometry for minimizing gas bubble resistance in tubes [7] (StarTube). For the dispensing of liquid metals the nozzle is fabricated from silicon to provide thermal stability and a high contact angle ($\theta > 150°$) towards the liquid metal. In Fig. 2a/b SEM pictures are shown of the top and the interior of a StarJet nozzle chip fabricated by DRIE within this work. The star shaped nozzle profile leads to a centering of a droplet or plug of liquid metal in the nozzle as shown by a simulation using CFD [8] in Fig. 1b.

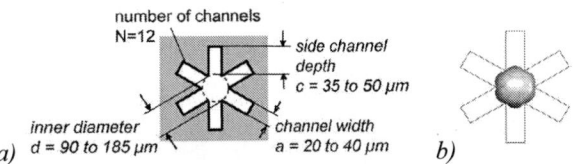

Figure 1: a) Schematic cross section of the star shaped nozzle. b) CFD simulation showing the centered droplet in the star shaped nozzle.

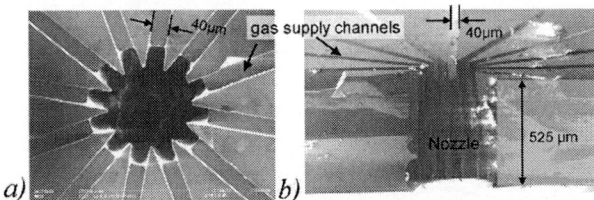

Figure 2: SEM pictures of fabricated StarJet nozzle chips

The droplet generation mechanism in the StarJet is explained in Fig. 3. It is somewhat similar to the droplet or bubble generation in a flow focusing device [9]. The liquid reservoir is placed directly above the inlet of the nozzle chip. The inner surface of the nozzle chip features gas supply channels (Fig. 2). These channels enter the nozzle in the side grooves of the star shaped profile. The liquid reservoir and the gas supply channels are in fluidic contact as shown in Fig. 3. When applying a pressure to the common inlet at top of the StarJet system, a gas flow through the gas supply channels is induced with a pressure drop Δp along the gas supply channels (Fig. 3a).

If this pressure drop Δp is high enough to overcome the capillary pressure of the nozzle centre, a liquid plug is forced into the star shaped nozzle since the driving pressure also acts on the liquid metal in the reservoir (Fig. 3b).

978-1-4244-2977-6/09 $25.00 © 2009 IEEE

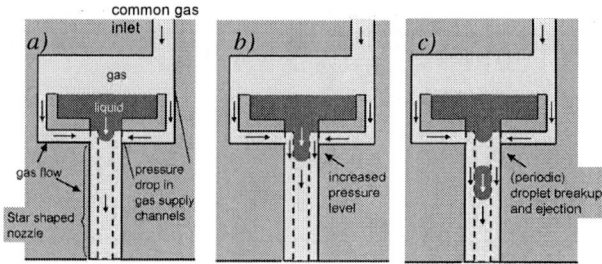

Figure 3: Working principle of the StarJet dispenser

As stated above, the liquid plug is confined in the center of the nozzle by capillary forces and the gas flow can still hold on in the outer grooves of the star shaped profile.

In response to the liquid plug entering the nozzle the gas flow resistance increases and the pressure level in the nozzle (Fig. 3b) raises. Like in an electrical voltage divider the pressure drop across the supply channels decreases. Therefore the liquid column collapses at the point of highest pressure near to the connection point of the gas supply channels. While the liquid column becomes necked a gas flow induced force onto the separating droplet towards the outlet supports the breakup. Finally the droplet is transported with the gas flow out of the nozzle (Fig. 3c). Until the droplet is ejected, the liquid column can not reenter the nozzle. Thus the system provides self controlled single droplet breakup. After ejection the sequence restarts.

A key feature of the StarJet dispenser is that the star-shaped profile reduces wall friction of droplets. This supports the transport decisively and reduces deflecting forces during ejection. The driving sheet gas protects the hot metal from oxidation as droplets during the generation and during free flight but also at rest in the reservoir by constantly rinsing the system at a low pressure level.

THEORY

Star shaped nozzle

A droplet is centered in the star shaped nozzle profile if the repellent capillary pressure of the outer fingers is too high for the droplet in the centre to move into the outer fingers. The effect depends on the number of fingers that form the nozzle and the wetting angle between the nozzle material and the liquid to be transported. The critical number of fingers necessary to enable a centered

droplet in dependence of the contact angle is given in the condition Eq. (1) that is illustrated in Fig. 4. Further details and a derivation of Eq. (1) are given in [7]

$$N > \pi \left(\arctan\left(\frac{2\cos^2\theta}{2\theta - \pi - 2\cos\theta\sin\theta} \right) \right)^{-1} \quad (1)$$

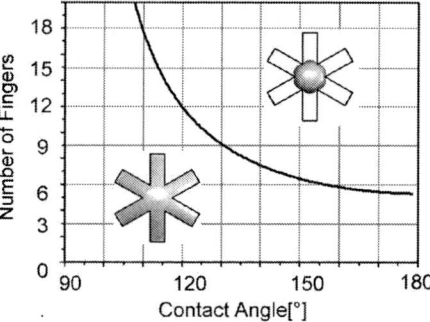

Figure 4: Number of side channels (fingers) required for centered droplets as function of the wetting angle.

Minimum actuation pressure

The minimum pressure difference Δp_{min} necessary to force a liquid column into the StarJet nozzle can be estimated by the inner diameter d of the nozzle (Fig. 1). The contact angle is virtually 180° as nearly no wall contact exists and the capillary pressure p_{cap} can be calculated from the surface tension σ of the liquid only by Eq. (2):

$$\Delta p_{min} > p_{cap} = \frac{4}{d}\sigma \quad (2)$$

As discussed above Δp_{min} has to drop along the gas inlet channels.

SIMULATIONS

The mechanism of droplet breakup was first studied by computational fluid dynamics (CFD) simulations using the software package CFD-ACE+ [8]. It allows the solution of the Navier-Stokes equations for two different fluids under consideration of surface tension and contact angles. In the simulation only the interior of the star shaped nozzle is considered with simplified boundary conditions: Where the gas supply channels enter the nozzle a constant gas flow is set. At the interface between liquid reservoir and nozzle a liquid inlet with a constant

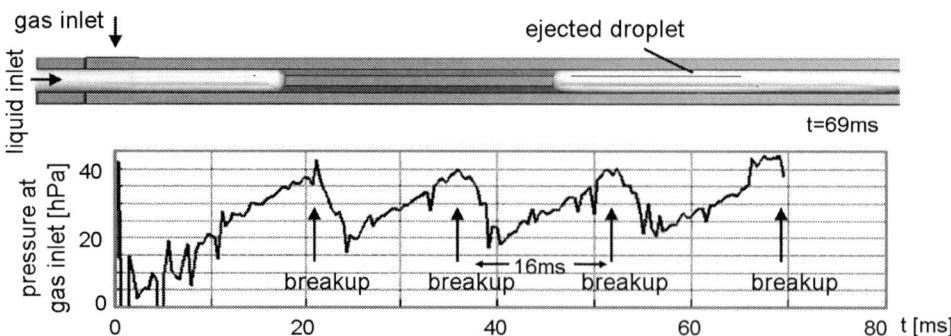

Figure 5: Result of CFD simulation with simplified model shows a periodic pressure signal at the gas inlet corresponding to a periodic droplet breakup. Boundary conditions: Constant pressure at liquid inlet (p=120hPa; mercury:$\rho = 1350 kgm^{-3}$; $\sigma = 484\ Nm^{-1}$; $\eta = 1,55\ Pa\ s$) and constant velocity at gas inlet (1 ms^{-1}, air) (Outlet: ambient pressure).

978-1-4244-2977-6/09 $25.00 © 2009 IEEE

pressure boundary condition is applied. Despite the constant boundary conditions the liquid column that enters the nozzle periodically breaks into plugs once the pressure gradient along the plug overcomes a critical threshold value as shown in Fig. 5.

FABRICATION
Chip fabrication

Nozzle chips were fabricated by etching the star-shaped nozzle profile and the gas channels into a 525 μm thick silicon wafer (Fig. 2). A thermal oxide layer was first structured with the geometry for the gas supply channels. The oxide layer was then covered by a resist film that was structured with the star shape. The nozzle profile was etched by DRIE into the wafers up to a residual thickness of 80 μm. The rest of the nozzle and the gas supply channels were etched using the oxide mask. The complete chip design additionally contains etched pin holes that serve for the alignment of the chips within the printhead. From Fig. 2a one can find that the etching of the star shaped profile is not perfectly straight but the inner diameter increases towards the outlet. This is caused by the DRIE process, but is not preventing the desired effect. In principle the tapering adds some additional pressure force onto the droplet breakup due to the narrowing of the gas channels.

Twenty different designs with parameters as defined in Fig. 1 were fabricated within one mask set. After examination of fabrication quality and some primary tests, two chip designs (SJet2, SJet14) were selected for experimental studies. The dimensions of these chips are given in Table 1 together with those used in the simulation shown in Fig. 5. The chip design named SJet14 has twice the central diameter d as the one named SJet2. The value p_{cap} is calculated according to Eq. 2.

Table 1: Parameters of StarJet nozzles studied.

	SJet2	SJet14	simulation
N	12	12	8
$a\ [\mu m]$	20	40	76
$c\ [\mu m]$	35	50	87
$d\ [\mu m]$	94	188	100
$p_{cap}\ [hPa]$	197	99	100

Experimental Setup

The printhead fluidically connecting the chips was designed as shown in Fig. 6 and fabricated in brass by milling. It was heated by a soldering iron and contains a solder reservoir and a gas inlet port for the pneumatic actuation. The nozzle chips are mechanically clamped into the printhead. A drilled hole in the center (Ø 400 μm) connects the reservoir to the liquid inlet of the chip. The alignment of the central hole is not critical as the same condition that prevents the liquid metal from moving into the outer channels of the star shaped profile protects the gas supply channels of the chip from being clogged by the liquid metal.

An external solenoid valve controls the gas flow of nitrogen supplied for pneumatic actuation. The valve switches between a low pressure level (50 hPa) for constant rinsing with sheet gas and an adjustable higher pressure level to actuate the system. The temperature is controlled during the experiments with a thermocouple

near the chip and was set to 255+/-5°C. For the observation of droplets the setup was placed in front of a stroboscopic camera as shown in Fig. 7. A microphone capsule in the pneumatic supply tubing was used to determine pressure disturbations during droplet breakup. In all experiments solder of grade Sn60Pb ($\rho = 860 kgm^{-3}$; $\sigma = 460\ Nm^{-1}$; $\eta = 2\ Pas$ at 255°C; T_{liq}~190°C), was used.

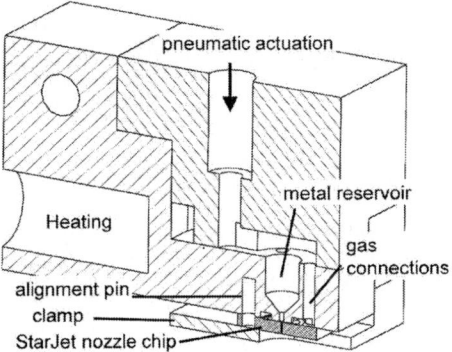

Figure 6: Design of the printhead used to heat and to fluidically connect the nozzle chips.

EXPERIMENTS

Experiments were performed with the two nozzle chips SJet2 and SJet14 to study the working principle. For a number of different pressure levels, the droplet size and breakup frequencies were studied. The minimum actuation pressure Δp_{min} was found to be 150 hPa for the larger nozzle SJet14 and 400 hPa for the smaller nozzle SJet2.

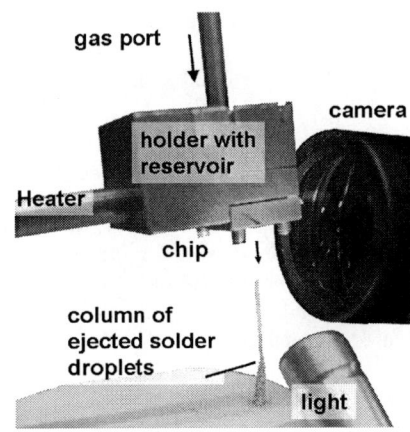

Figure 7: Photograph of the experimental setup.

When actuation with a static pressure a periodic droplet breakup could be observed like expected. In Fig. 8 stroboscopic pictures of the droplet breakup are shown. From these pictures it can be found that the solder leaves the nozzle already as preformed droplets. This proofs that the droplet breakup takes place already inside the nozzle.

The frequencies of droplet breakup were investigated from the signals recorded by the microphone. As shown in Fig. 9 the waveform is regular and periodic. For both chips the breakup frequencies close to Δp_{min} were determined to be around 95 Hz with a reproducibility of +/-15Hz. In single experiments the bandwidth was around +/- 3Hz.

978-1-4244-2977-6/09 $25.00 © 2009 IEEE

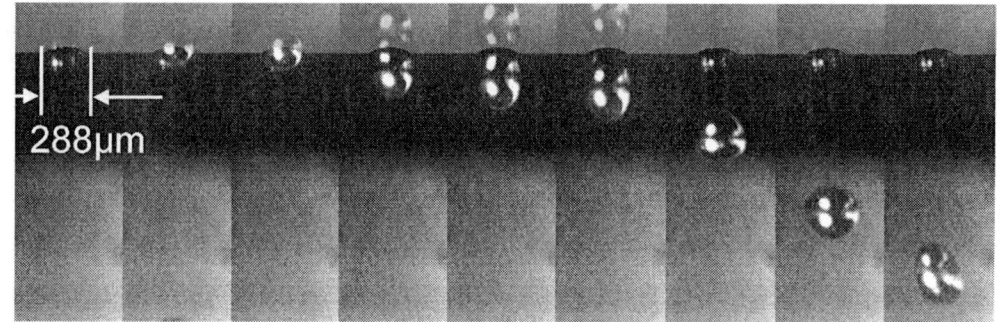

Figure 8: Stroboscopic pictures of droplet ejection are shown. Every picture relates to a new droplet.

For the SJet14 chips the frequency of the droplet breakup has been found to be stable up to 300 hPa. A further increase of the driving pressure raised the dispensing frequency strongly up to 400 Hz. When further increasing the pressure level towards 1500 hPa for SJet14 and towards 2000 hPa for SJet02 the system behavior changed towards spraying with a finite cone. For such high pressures the dynamic pressure in the gas flow becomes dominant and the described droplet generation principle fails.

By single actuation of the valve for only 6 ms with 200 hPa for Sjet14 dispensing of single droplets in a drop-on-demand mode could be achieved.

The size of droplets generated close to the minimum actuation pressure was determined by SEM pictures (Fig. 10). Due to the high heat capacity of the solder, the droplets do not solidify during flight but when meeting the substrate and get flattened. In contrast, nearly spherical droplets that do not merge could be achieved by dispensing from a distance of 30 cm into a water/soap mixture. The droplet diameter corresponds well to the inner diameter of the StarJet nozzles (Fig. 10). Droplet volumes were evaluated to be 120 pl and 3.6nl for SJet02 at 600 hPa and SJet14 at 200 hPa.

The columns built up during periodic dispensing on a constant position indicate the scattering of the droplets. Even for a height of up to 30 mm their width was only between 500 µm and 1 mm demonstrating the directional stability of the system. A reason for this might be the rather low contact area between droplet and nozzle chip. This prevents frictional forces at the nozzle outlet to deflect the ejected droplet.

CONCLUSIONS

The StarJet dispenser as presented in this work exhibits a new concept for droplet dispensing by pneumatic actuation that fits well to the requirements of the dosage of liquid metal. It is a promising principle for applications in this field. A prototype of the StarJet dispenser and a MEMS based fabrication process were successfully established. The working principle is basically understood. Three different working modes were found: continuous droplet generation, drop on demand and spray. The modes depend on the actuation pressure and duration which are externally controllable. The application of the StarJet dispenser must not be limited to liquid metal but can be applied to other liquids like aquaous solutions if nozzles are fabricated that meet the conditions given by Eq. 1 and Fig. 4 for that specific liquid.

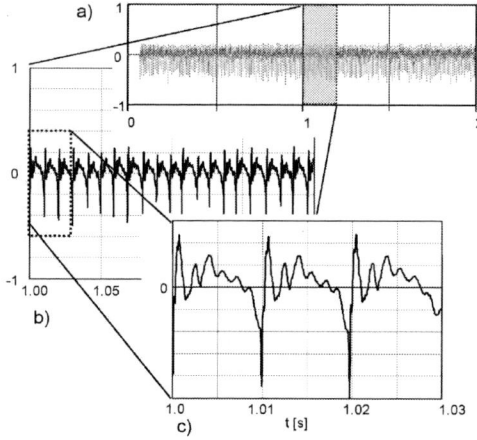

Figure 9: Acoustic signal of droplet breakup: a) full signal; b) zoom into 200ms of signal; c) zoom into 30ms of signal (SJet14, actuation pressure 150 hPa).

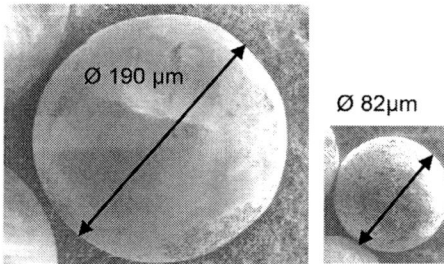

Figure 10: SEM pictures of droplets when dispensed from 30 cm distance into soap water.

REFERENCES

[1] J. Heinzl and C. H. Hertz, *Adv. Imag. Elec. Phy.*, vol. 65, pp. 91-171, 1985

[2] B. de Heij et al., *Anal. Bioanal. Chem.*, vol. 378, no. 1, pp. 119-122, Jan.2004

[3] M. Ession, et al., CA000002373149A1, 2000

[4] W. Wehl et al., Proc. IMAPS 2003

[5] D. Schuhmacher et al. *Proc. IEEE MEMS 2007*, pp. 357-360

[6] P. Sen and C. J. Kim, *Proc. IEEE MEMS 2007*, pp. 767-770

[7] T. Metz, et al. *Langmuir*, pp. 9204-9206, 2008

[8] ESI-Group, "CFD-ACE+ 2007," 2006

[9] S. L. Anna et al., *Appl. Phy. Let.* vol., 82, no. 3, pp. 364-366, Jan.2003.

CONTIUNOUS MICRO-PARTICLE SEPARATION USING OPTICALLY-INDUCED DIELECTROPHORETIC FORCES

Wang-Ying Lin [1], Yen-Heng Lin [1] and Gwo-Bin Lee [1,2]

[1]Department of Engineering Science, National Cheng Kung University, Tainan, Taiwan
[2]Medical Electronics and Device Technology Center, Industrial Technology Research Institute, Hsinchu, Taiwan

ABSTRACT

The current study presents new methods to quickly separate micro-particles with different sizes using optically-induced dielectrophoretic (ODEP) forces. It was found that the strength of the ODEP force induced on the hydrogenated amorphous silicon surface is determined by the wavelength of the illuminating patterns. Therefore two moving lines and one stationary illuminated line which produced the ODEP forces as virtual electrodes were first defined by projecting lights onto a photoconductive chip. The moving lines and one stationary illuminated line were used to generate a stronger and a weaker ODEP force, respectively. As the moving lines approached to the weaker line, micro-particles between them were concentrated by the higher ODEP forces and squeezed through the stationary electrode. Finally, lager micro-particles were subsequently separated from the smaller particles. With this approach, continuous particle separation can be automatically achieved in a short period of time. This developed method may be promising for a variety of applications such as cell-based assays and sample pre-treatment using micro-particles.

INTRODUCTION

Particle/bead and cell manipulation and separation are essential for a variety of applications, including sample pretreatment [1], cell manipulation [2], and diagnosis [3]. Recently, the separation/sorting of micro-sized biological and chemical particles has become popular. Biological samples usually consist of complicated compositions, which may have an extremely low trace of target cells, proteins, or nucleic acids. Targets of interest are usually expected to be separated from substances that may interfere with detection process such that the detection limit can be improved. A wide variety of techniques for the separation of cells and particles in bio-samples have been employed, utilizing their difference in sizes [4], electrical properties [5] and other physical properties. In the application of these techniques, several mechanisms such as electrical (electrophoresis or dielectrophoresis) [6], magnetic [7], and hydrodynamic forces [8] have been widely used. For instance, DEP force is commonly used for manipulation and separation [9] of dielectric particles and cells. Since DEP force provides a direct and convenient way for particle separation without complex sample preparation process, it has become one of the most attractive separation techniques. The conventional DEP forces manipulate/separate particles by using micro-fabricated electrodes to generate forces on dielectric particles. These fixed electrodes require delicate fabrication processes. Moreover, it is impossible to change their geometric configuration once they have been made.

Alternatively, optoelectronic tweezers (OET) [10] have been used to improve the efficiency in particle manipulations using virtual electrodes instead of fixed metal electrodes to spatially manipulate and separate particles. The virtual electrodes are formed by projecting electrode images onto photoconductive materials (e.g. hydrogenated amorphous silicon) to produce non-uniform electric fields, so that the virtual electrodes can generate DEP forces on dielectric particles.

In this study, it was found that the strength of the optically-induced dielectrophoresis (ODEP) force provided by a sandwiched photoconductive device can be determined by the wavelength of illuminating optical beams [11]. Under different wavelengths, the hydrogenated amorphous silicon has different abilities in creating corresponding electron-hole pairs to affect the strength of ODEP force. In addition to wavelength, changes in the line widths and brightness of patterned lines can also provide different ODEP forces on the amorphous silicon surface. In this study, virtual electrodes are then patterned with different wavelengths, widths and brightness to obtain two line-shaped lights with different ODEP forces. With this approach, new methods are reported for separation of micrometer-sized particles using these lines with different strengths to squeeze mixed particles with different sizes such that they can be sorted automatically with a high throughput.

DESIGN AND FABRICATION

The chip which generated ODEP forces is a sandwiched structure (Figure 1a), including a top layer made of an ITO glass, a liquid layer and a bottom layer with two thin films coated on another ITO glass. The bottom layer consists of an ITO glass substrate, a 10-nm molybdenum layer and a 1-μm thick amorphous silicon layer. An alternating current (ac) voltage is applied to the top and bottom layers to spatially produce an electric field on the chip. Since the bottom layer of the chip is a photoconductive surface, most of the voltage can drop across the illuminating spots, producing a non-uniform electric field at the light-patterned regions in the liquid layer. Therefore the micro-particles can be manipulated through the "virtual electrodes" generated by a commercial projector (PJ1172, Viewsonic, Japan). Micrometer-sized polystyrene (PS) beads (Duke Scientific, USA) are used to simulate the behavior of micro-particles in the liquid layer of the chip. Please note that cells or magnetic beads can be later conjugated on the polystyrene for further biological applications. The polystyrene beads of two sizes (10- and 20-μm diameters) are mixed in a liquid solution consisting of deionized (DI) water with 1% fetal bovine serum (FBS) to form a buffer solution with low conductivity (165 ~ 175 μs/cm). Then the mixed solution is injected into the liquid layer of the chip.

The experimental setup (Figure 1b) consists of an image generation system to project virtual electrodes onto the chip, and an image acquisition system to observe bead

978-1-4244-2977-6/09 $25.00 © 2009 IEEE

separation process. To construct the virtual electrodes, a commercial liquid crystal display (LCD) projector (PJ1172, Viewsonic, Japan) with a spatial resolution of 1024 pixels x 768 pixels is used as a light source to excite electron-hole pairs in the amorphous silicon layer. A 50× objective lens (Nikon, Japan) is mounted between the projector and the chip in order to focus and collimate the projector light onto the chip to form micro-scaled virtual patterns. The scanning movement of the line is accomplished using FLASH, a commercial computer software. The image acquisition system is placed on top of the chip to observe bead separation. It consists of a charge-coupled device (CCD, SSC-DC80, Sony, Japan) camera, an optical microscope (Zoom 125C, OPTEM, USA) and a computer equipped with an image acquisition interface card. The AC voltage is supplied by a function generator (Model 195, Wavetek, U.K.) and an amplifier.

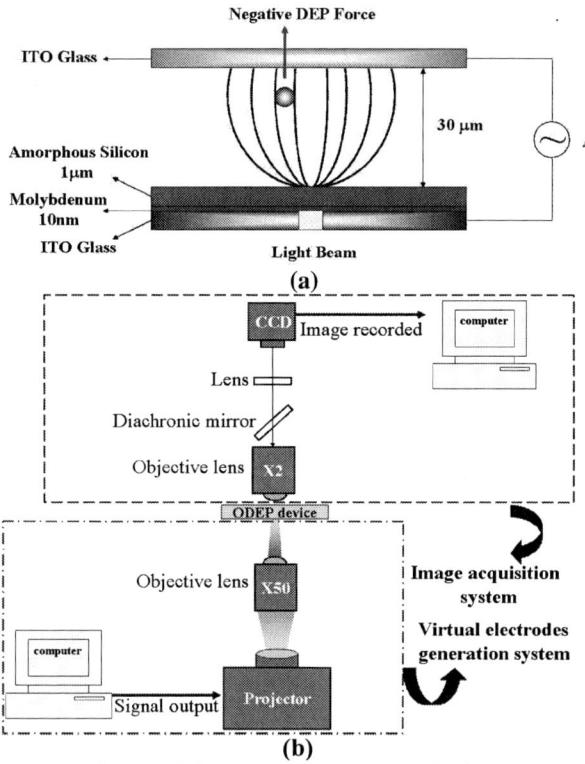

Figure 1: (a) Schematic illustration of the operating principle of the ODEP chip. (b) Experimental setup for micro-particles separating on optical induced DEP

Inducing ODEP force in different strength for separation

The strength of ODEP force is influenced by wavelengths, line widths, and brightness of projecting images. The new methods for particle separation generates two line-shaped virtual electrodes in different strength to sort beads with different sizes (10 and 20 µm). The separation process is performed in three major steps, including concentration, squeezing and separation, as shown in Figure 2. First, all the beads randomly located on the surface can be concentrated between two lines generating negative ODEP forces when the moving line with the stronger ODEP strength is scanned from left to right (Figure 2a-b). As the space between the two lines becomes smaller, it begins to squeeze the beads (Figure 2c). Finally, when the space is smaller than the size of the

larger beads, the larger beads jump to the right side of the stationary line, which has a smaller ODEP strength (Figure 2d). Note that this method can be easily extended to multiple lines with different strengths such that multiple-sized bead separation with high discrimination capability can be achieved.

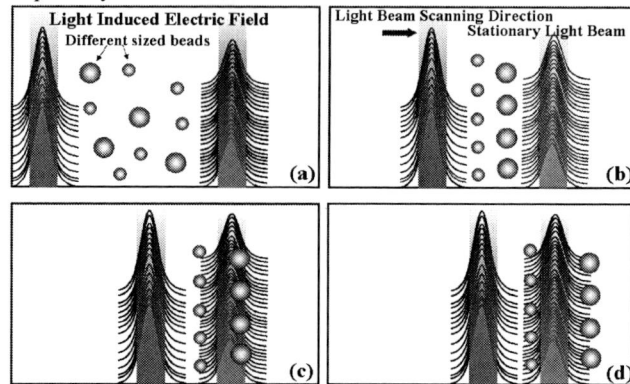

Figure 2: The top view of a new method for particle separation by using lines in different strengths.

Separation using different colors

It is found in our experimental results that green light induces a stronger ODEP force than the red and blue lights, because the amorphous silicon has better absorption coefficient at the wavelength of green lights [11]. Although the absorption of amorphous silicon for blue lights is as good as that of green light, most of the energy in the incident blue light was found to be dissipated as thermal energy and did not contribute to the photoconductivity effect. Therefore the exactly transformed photoconductivity effect within the amorphous silicon under different wavelengths, manifested as different colors should be expressed by the incident photon-to-current conversion efficiency (IPCE).

The DEP force is expressed as

$$F = 2\pi r^3 \varepsilon_m \, \text{Re}\,[K * (\omega)] \, \nabla(E^2) \qquad (1)$$

where r is the radius of the particle, ε_m is the permittivity of the surrounding buffer, Re $[K * (\omega)]$ is the real part of the Clausius-Mossotti (CM) factor which varies with the conductivity of the particle or the medium and the angular frequency of electric field, and E is the electric field strength. The induced ODEP forces are proportional to the gradient of the square of the electric field strength [$\nabla(E^2)$] which is directly related to the efficiency of the photoconductivity effect. According to the measured results in the IPCE of amorphous silicon, the peak is located around 550 to 570 nm [12-13] depending on the structure of amorphous silicon. A spectrum analysis of the light projector also confirms that the strength of the induced force for green light is the highest. Note that the wavelength of green light in our projector ranges from 500nm to 590 nm while the blue light is from 420nm to 500nm, and red light is from 590nm to 730nm.

In this case, red line is stationary to be the weaker virtual electrode and two green lines are scanned from left to right at a scanning velocity of 7.5 µm/s. The second line is designed for avoiding the missing of beads which are not separated by the first line. The final distance between the two lines is set to be about 13 µm such that 10- and 20-µm beads can be separated. The line-shaped lights here are all in the width of 21-μm and the light intensity of them are 145.5 W/cm². These virtual electrodes are powered under an ac voltage that is 36 V$_{pp}$, 100 KHz.

In addition to using the different colors to induce ODEP forces with different strengths, the other separating method by using different illumination line widths and brightness are also be applied for separation. Their separation procedures are identical to the one using different colors, as described above. In the operating mode of different line widths, the first and second scanning lines have the same width of 21 μm, and the stationary line has a width of 14 μm. All three lines are white lights with an intensity of 782.2 W/cm². Changing the brightness of lines to produce the difference in ODEP force is based on the fact that photoconductivity is proportional to the illumination intensity. The brightness of the two scanning lines and the stationary line are set at 255 (372.9 W/cm²) and 192 (145.5 W/cm²), respectively. All these three lines are in green lights. The line widths of these line-shaped lights are 21-μm-wide.

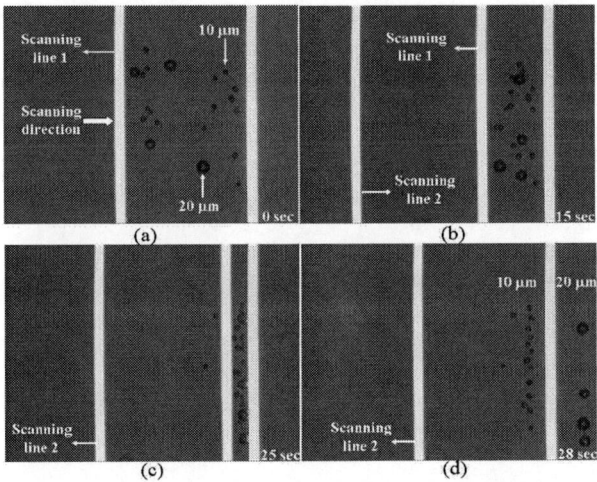

Figure 3: Separation of 10- and 20-μm PS beads by using two linear light segments with different wavelengths.

RESULTS AND DISCUSSION

The difference of wavelength is applied to generate line-shaped lights in different ODEP forces and two PS beads with different sizes can be separated successfully. As shown in Figure 3, PS beads with 10- and 20-μm are mixed and placed on the chip (Figure 3a). When the first green line is scanning on the mixed beads, these beads experience a negative DEP force and are pushed toward the red line (Figure 3b). Then the beads are squeezed to align along the line-shaped lights till the lager ones can not be contained inside the line space and then the lager size beads is pushed to the other side of the stationary line (Figure 3c). Then the first line disappears at the separating moment in order to avoid the space becoming smaller (Figure 3d). The separation process can be finished in 28 seconds.

In order to certify the magnitude of ODEP force generated by different wavelength, the terminal velocity of polystyrene beads hauled by line-shaped lights was measured. As shown in Table 1 and 2, the experimental results for the maximum velocity of 10- and 20-μm beads that are pushed under scanning lights of blue, green, and red. As expected, it can be observed that the green lights

push beads at a highest speed than the other two lights. According to the Stoke's law, the velocity of spherical beads and their DEP force are directly proportional when the ODEP force is balanced with the viscous drag force in fluid, i.e.,

$$F = 6\pi r \eta v \qquad (2)$$

where r is the radius of the spherical particle, η is the viscosity of the fluid (the viscosity of water at 20°C is 1.002×10-3 Ns/m²), and v is the maximum velocity of the particle. Thus the polystyrene beads experience lager force as they have a higher maximum velocity.

Table 1: Maximum velocity and calculated strength of ODEP forces under different illuminating wavelength with 20-μm PS beads

Wavelength (nm)	Maximum velocity (μm/s)	Induced force (pN)
Blue (420~500)	30.88	5.83
Green(500~590)	33.25	6.28
Red (590~730)	49.88	9.42

Table 2: Maximum velocity and calculated strength of ODEP forces under different illuminating wavelength with 10-μm PS beads

Wavelength (nm)	Maximum velocity (μm/s)	Induced force (pN)
Blue (420~500)	21.50	2.03
Green(500~590)	23.75	2.24
Red (590~730)	35.63	3.36

The successful separation processes using the other manipulating modes are shown in Figures 4 and 5, including the application of line widths and intensities. Under the same designed scanning speed, the entire separation process are all be finished in 28 seconds at a scanning speed of 7.5 μm /s. Note that the entire separation time is determined by the scanning speed.

Experimental results reveal that lights with different line widths and intensities can induce ODEP forces with different strengths. It was reported that changes in line widths and light intensities can generate different photoconductivity levels on the amorphous silicon surface. Similarly, the terminal velocities are used to calculate the induced ODEP forces. The induced ODEP forces on the 20-μm beads are 49.34 pN, 34.99 pN and 17.94 pN for 28-μm, 21-μm, and 14-μm lines, respectively. The trend for the 10-μm beads is similar to the one for 20-μm beads. Moreover, the illumination with a stronger intensity can generate more electron-hole pairs to make the voltage drop to the liquid layer more evident for particle manipulation. The calculated forces of the particles for the green lights with a brightness of 255 (372.9 W/cm²) is 23.98 pN for 20-μm beads and 7.70 pN for 10-μm beads. The induced forces when using green lines with a brightness of 192 (145.5 W/cm²) is 9.42 pN and 3.36 pN for 20 and 10-μm beads, respectively.

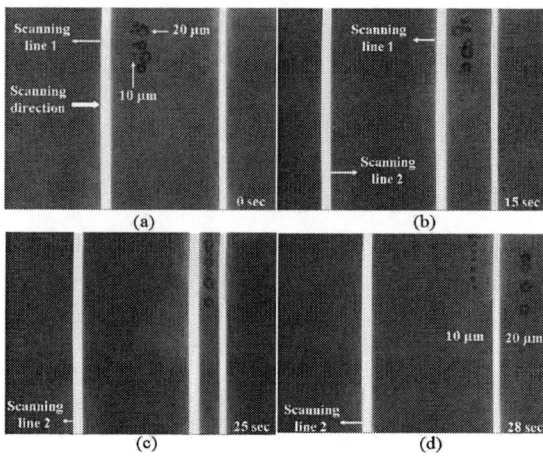

Figure 4: separation of 20- and 10-μm PS beads by using two linear light segments with different line widths.

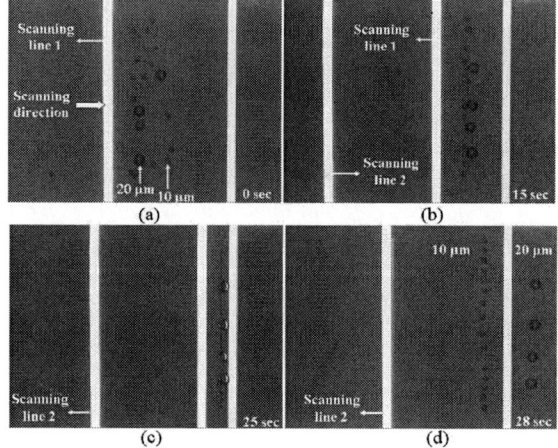

Figure 5: separation of 20- and 10-μm PS beads by using two linear light segments with different intensities.

CONCLUSIONS

The design presented in this study can manipulate micro-particles with different sizes by simply using thin-film deposition technology. The flexibility in this method is that particle separation can be conveniently achieved by simply changing light patterns with different wavelengths, intensities, and line widths. For particles with similar sizes, further modifications of this approach can be used for particle separation. For instance, multiple lines with different intensities can be used such that multiple bead separation with a high discrimination capability can be achieved. Particles with different sizes can eventually be separated if more scanning lines and stationary lines have been deployed.

ACKNOWLEDGEMENTS

The authors would like to thank Chi-Mei Optoelectronics Inc. for their financial support from grant number (96S036). Partial financial support provided to this study by the National Science Council of Taiwan is also greatly appreciated. Authors also thank Dr. T. F. Guo for valuable discussion.

REFERENCES

[1] A. J. de mello, N. Beard, "Dealing with 'real' samples:sample pre-treatment in microfluidic systems", *Lab Chip*, vol. 3, pp. 11N-19N, 2003.

[2] E. W. H. Jager, O. Inganas, I. Lundstrom, "Microrobots for micrometer-size objects in aqueous media: potential tools for single-cell manipulation", *Science*, vol. 288, pp. 2335-2338, 2000.

[3] P. Gascoyne, J. Satayavivad, M. Ruchirawat, "Microfluidic approaches to malaria detection", *Acta Trop.*, vol. 89, pp. 357-369, 2004.

[4] P. Vulto, G. Medoro, L. Altomare, G. A. Urban, M. Tartagni, R. Guerrieri N. Manaresi, "Selective sample recovery of DEP-separated cells and particles by phaseguide-controlled laminar flow", *J. Micromech. Microeng.*, vol. 16, pp. 1847-1853, 2006.

[5] C. Iliescu, G. L. Xu, P. L. Ong, K. J. Leck, "Dielectrophoretic separation of biological samples in a 3D filtering chip", *J. Micromech. Microeng.*,vol. 17, pp. S128–S136, 2007.

[6] S. Choi, J. K. Park, "Microfluidic system for dielectrophoretic separation based on a trapezoidal electrode array", *Lab Chip.*, vol. 5, pp. 1161-1167, 2005.

[7] J. M. Choi, C. H. Ahn, S. Bhansali, H. T. Henderson, "A new magnetic bead-based, filterless bio-separator with planar electromagnet surfaces for integrated bio-detection systems", *Sensors and Actuators B.*, vol. 68, pp. 34–39, 2000.

[8] M. Yamada, K. K. -Y. Tsud, J. Kobayashi, M. Yamato, M. Seki, T. Okano, "Microfluidic devices for size-dependent separation of liver cells", *Biomed Microdevices.*, vol. 9, pp. 637-645, 2006.

[9] M. P. Hughes, "Strategies for dielectrophoretic separation in laboratory-on-a-chip systems", *Electrophoresis Rev.*, vol. 23, pp. 2569-2582, 2002.

[10] P. Y. Chiou, A. T. Ohta, M. C. Wu, "Massively parallel manipulation of single cells and microparticles using optical images", *Nat.*, vol. 436, pp.370-372, 2005.

[11] D. E. Carlson, C. R. Wronski, "Amorphous silicon solar cell," *Appl. Phys. Lett.*, vol. 28, pp. 671-673, 1976.

[12] R. J. Loveland, W. E. Spear, A. Al-Sharboty, "Photoconductivity and absorption in amorphous Si", *J. Non-Cryst. Solids.*, vol. 13, pp. 55-68, 1973.

[13] S. D. Liu, S. C. Lee and M. Y. Chern, "Hydrogenated Amorphous Silicon-Germanium PIN X-Ray Detector", *IEEE Trans. Electron Devices.*, vol. 48, pp. 1564-1567, 2001.

POWERFUL ACTUATION OF MAGNETIZED MICROTOOL
BY FOCUSED MAGNETIC FIELD ON A DISPOSABLE MICROFLUIDIC CHIP

F. Arai, S. Sakuma, Y. Yamanishi and K. Onda
Tohoku University, Sendai, JAPAN

ABSTRACT

We succeeded in powerful noncontact actuation of magnetically driven microtool (MMT) by magnetizing it and focusing magnetic field in a microfluidic chip. Novelty of this paper is summarized as follows. (1) We employed neodium powder as the main component of MMT. The density of magnetic flux was improved about 100 times larger after magnetization. (2) We fabricated a pair of magnetic sharp needles in the chip by electroplating. MMT was placed between the needles and the density of magnetic flux was improved about 3 times larger. As a result, we succeeded in powerful actuation of MMT in the chip. Drive frequency was improved about 10 times faster (up to 180Hz).

INTRODUCTION

Sorting of microparticles in microchannels offers great potential in biology, chemistry, and environmental analysis. For example, cell sorters, which are based on flow cytometry, can sort cells in a continuous cell-laden flow, and can collect the sorted cells in cell suspensions [1]. Cell sorting is generally performed either electrically (by charging cells) or by mechanically moving the receiving dishes after a shot of laser light. The laser pulse is controlled such that a single cell is placed in a single droplet by thrusting the diluted cell suspension into an air phase. However, because this system tends to be large and expensive, a low-cost cell sorter has been developed on a microchip. There are many methods of microscale cell sorting, such as dielectrophoretic sorting [2]-[3], laser trap [4], magnetic isolation, and switching in microfluidic channels using microvalves [5]. It is necessary to design the system to fit the characteristics of the sorting object (e.g., size), the condition of the carrier liquid, and the sorting speed.

The most popular microactuators that can be applied in the confined space of microchannels are electrostatic microactuators, optical tweezers [6], and magnetic microactuators [7]. The Coulomb force has been often used in manipulating cells of the order of 10 μm, whereas it is necessary to apply a high voltage to manipulate particles of the order of 100 μm, which risks damaging cells such as oocyte by heat generation. The dielectrophoretic force can be adjusted by varying the squared value of the gradient of the electrical field; however, it is controllable only in the limited region adjacent to the electrodes, and requires higher voltages to sort larger objects. Optical tweezers can manipulate cells indirectly by non-contact actuation of microtools, thus reducing the risk of damaging cells during manipulation; however, the generated force is of the order of several pN, which is not suitable for manipulating cells of the order of 100 μm. On the other hand, the magnetic sorting offers a limited risk of cell contamination and it has been used in many studies because of its low cost [8]-[10].

We had developed magnetically driven microtools (MMT) in a previous research [11], [12], [13]. The basic concept of MMT was proposed in 2004 for noncontact manipulation of microscopic particles such as oocytes, cells and microbeads [11]. Then we developed polymer based MMT with magnetite powder by photolithography technique. This technique provides many microchannel functions, such as those of a valve, stirrer, and loader [12]-[13]. The developed MMT is flexible and biocompatible, and can be applied to cell sorting applications where there is a risk of contamination, by fabricating the microchannel out of a disposable material. It is particularly important to note that it can sort relatively large particles (of the order of 100 μm).

In developing the MMT, we especially focused on investigating a small actuation module to integrate many functions in the limited area on chip. First prototype developed is shown in Figure 1 (a). Permanent magnets have been reported to have significant advantages in miniaturizing systems effectively [7]. To actuate the soft magnetic MMT, our system has drive unit and permanent magnet unit to downsize the electromagnetic coil (drive unit). We used a permanent magnet that operated indirectly, and were able to downsize the drive unit significantly. The system consists of two modules - an upper module containing a disposable microchannel and a lower actuation module. The density of magnetic flux generated by the electromagnetic coil is amplified by the permanent magnet unit mounted between the microchannel and the magnetic circuit, and the MMT is moved by non-contact actuation. However the maximum actuation frequency of MMT was limited to about 18 Hz, because of the size of permanent magnet is significantly larger than MMT. Also, fluid resistance was remarkable in water and the power for actuation was not enough to exceed the measurement speed (30 fps).

For the current study, we solved these problems by focusing magnetic field of drive unit as shown in Figure 1 (b). To integrate of MMT and to obtain powerful actuation, we fabricated magnetized MMT using composite of neodymium powder and PDMS. Also, we fabricated magnetic needles in a chip to focus density of magnetic flux. The magnetized MMT was installed between them (Figure 2(a)).

Figure 2(b) shows concept image of sorting method and an actuation module. The direction of the current in the coil of the magnetic circuit can be switched to reverse the electromagnet's polarity, and the density of magnetic flux generated by the electromagnetic coil is focused by the magnetic needles fabricated in a chip, and the MMT is moved by non-contact actuation. Table 1 summarized the classifications of the MMTs. A permanent magnet unit was used to support the small magnetic power of the magnetite particles of MMT. The maximum actuation frequency of MMT was rate-limiting by the motion of permanent magnet. However the most effective interaction between MMT and

978-1-4244-2977-6/09 $25.00 © 2009 IEEE

the magnetic field can be obtained for the case of A-3 and B-2 in Table 1[7]. Therefore, the powder of permanent magnet was used to produce MMT, and MMT has also magnetized sufficiently to have the magnetic poles for the present work. Finally, with such a developed system, automating the sorting of different sizes of copolymer beads was completed using a real-time image processing system.

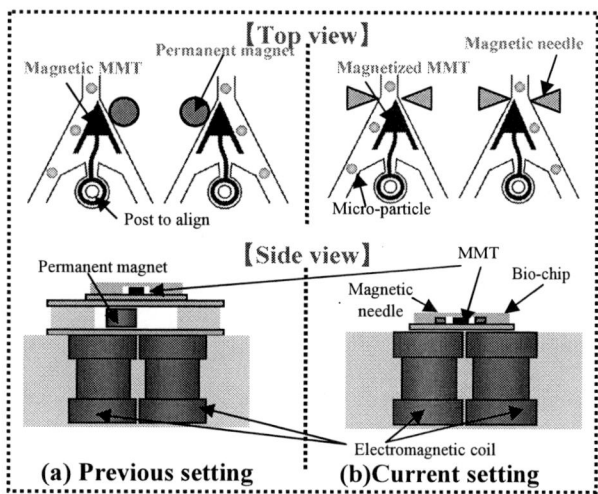

Figure 1: Schematic of experimental setup.

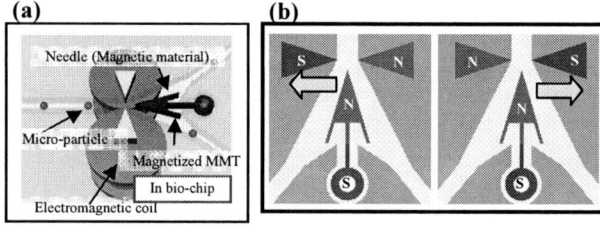

Figure 2: Concept and principle of actuation method of MMT sorting biochip.

PRODUCTION OF MMT

Figure 3(a) shows the fabrication process of the MMT, which may be summarized as follows: (1) pattern SU-8 over the silicon substrate, and an MMT mold was produced by photolithography, (2) a mixture of PDMS and neodium ($Nd_2Fe_{14}B$, 50 wt%) was spread over the patterned mold and baked on a hotplate (100℃, 15 min), (3) the layer of PDMS was spincoated over (2) and baked it in an oven (90℃, 10 min), (4),(5) separate MMT from the mold with PDMS, and peel MMT from PDMS, and (6) magnetize MMT. The surface of MMT was Teflon coated with CF_4 gas by plasma ashing method (Dischage Power: 130 W) for 30 minutes to avoid any stiction in microchannels. The use of magnetic powder in polymer has a great advantage for arbitral shape of fabrication. The baked PDMS (the catalyst was evaporated) is biocompatible and this is widely used as a material of biochip. In case of ferromagnetic materials, it is difficult to fabricate fine and complicated shape. To manipulate cells,

the shapes of actuator tend to be complicated for the accurate manipulations with protecting the sensitive material of cell. (The current shape of sorter MMT is not much complicated, however MMT itself has ability to fabricate more complicated shape and functions.)

Figure 3(b) shows the fabrication process of the magnetic needles, which may be summarized as follows: (1) sputter Cr/Au on a piece of cover glass, (2) pattern photoresist (OFPR800) and etched Cr/Au, (3) remove OFPR and pattern KMPR1050, (4) electro plated Ni, (5) remove KMPR1050 in a stripper liquid (Remover PG), and (6) assemble MMT into PDMS (micro-channel), and bond the cover glass together.

Table 1: Classification of driving method of MMT

Classifications		Characteristics
A. Control Method of Magnetic Field		
1. Fixed Electromagnet		Magnetic poles and magnetic field are controlled by the electric magnets
	1.1 Switching the magnetic pole	ON/OFF type
	1.2 Distribution of the magnetic field	Continuous type
2. Move of Electromagnet		Control the position of the electric magnet by an actuator and control the magnetic pole and magnetic field. (piezoelectric, AC/DC motor, etc..)
3. Move of Permanent magnet		Control the position of the permanent magnet by an actuator and control the magnetic pole and magnetic field. (piezoelectric, electromagnet, AC/DC motor, etc....)
B. Magnetic Characteristics of MMT		
1. VR-type		Use materials with a high magnetic permeability
2. PM-type		Use a permanent magnet
C. Degree of Freedom		
1DOF x, (y, z) - lateral		Continuous, ON/OFF
1DOF rotation		Continuous, step
2DOF xy-lateral, lateral + rotation		Continuous, ON/OFF, step
Multi-degree of Freedom		Continuous, ON/OFF, step

OPERATION OF MMT

Figure 4(a) shows the operation of the MMT using the actuation module. The needles are made by Ni electroplating. Images were captured by a CCD camera attached to an optical microscope. The depth of the channel was 200 μm and the height of MMT was 150 μm. The MMT was installed in the microchannel of the PDMS chip before the bonding between the cover glass and the PDMS chip. We applied 2.83 V and 0.15 A on the electromagnetic coils and the MMT was actuated at about 180 Hz (the velocity of edge of the MMT: 13.5 mm/s) in ethanol. It was confirmed that MMT was able to sort micro-particle of about 100 μm, and the edge of the MMT can be used to switch the microchannel. Drive frequency was improved about 10 times faster than previous work [14]. Figure 4(b) shows the outlook of sorting chip fabricated by the processes shown in Figure 3.

Production of MMT

1. Patterning SU-8

2. Filling in the composition

3. Spincoat PDMS

4. Separation MMT from the mold

5. Peeling MMT from PDMS

6. Magnetization

Production of Microchannel

1. Sputtering Cr/Au

2. Etching Cr/Au

3. Remove OFPR and Patterning KMPR

4. Ni electro plating

5. Remove KMPR

6. Assembly

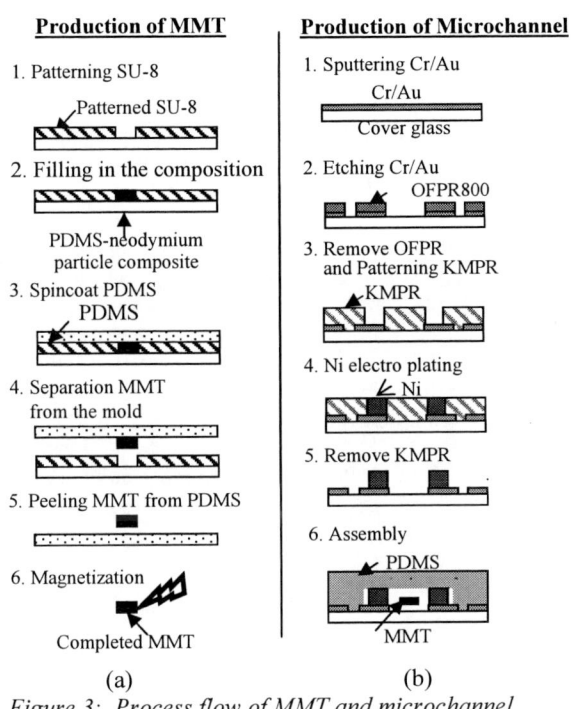

(a)   (b)

Figure 3: Process flow of MMT and microchannel

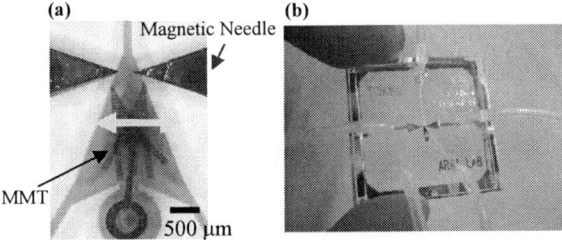

Figure 4: (a) Operation of sorting MMT and (b) Outlook of sorting chip

FEM ANALYSIS OF THE MAGNETIC FLUX

The distribution of density of magnetic flux near the MMT was calculated by FEM analysis software (COMSOL) and shown in Figure 5. The total number of turns of electromagnetic coil was 500 times, and the electric current of the coil was 0.12 A. Figure 5(a) shows a condition without magnetic needles, and (b) shows a condition with it to focus density of magnetic flux. It is evident that the magnetic flux was focused by magnetic needles fabricated in the chip, where the magnetic flux increased remarkably.

Figure 6 shows profiles of density of magnetic flux along x direction near the MMT (where y=2.25×10-3 [m]) of a case with magnetic needles. In case of the condition with needles, the density of magnetic flux at the position of MMT became about 2.7 times of the one without needles. It is clear to see that magnetic field is focused at the center of the needles.

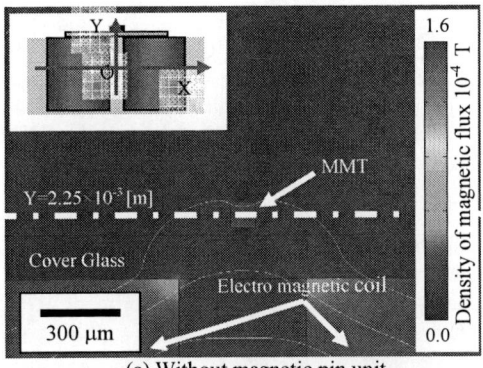

(a) Without magnetic pin unit

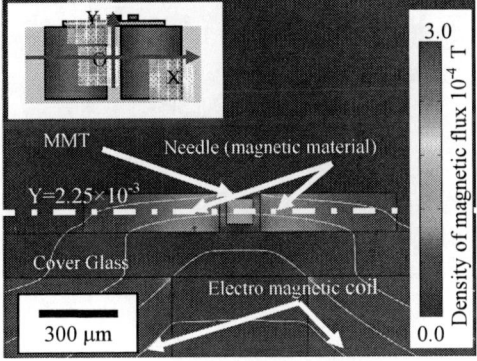

(b) With magnetic pin unit

Figure 5: Density of magnetic flux near the MMT.

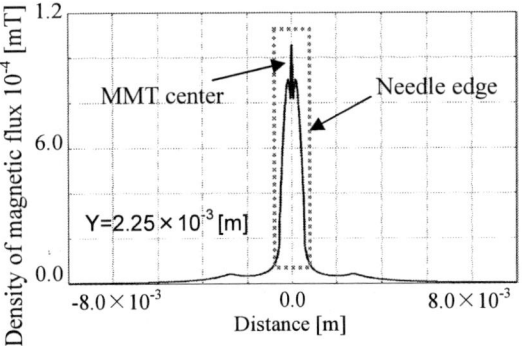

Figure 6: Profiles of density of magnetic flux.

DETECTION AND SEPARATION BY IMAGE PROCESSING

We demonstrated separation of copolymer beads in a chip, and beads of certain size are separated successfully (Figure 7). The sorting speed was 1 particles/ sec, which will be improved by sensor.

The procedures of the detection and separation are summarized as follows; the sensing system was calibrated using an image of representative sorting particles. The particles can then be sorted at a threshold, based on the results of the calculation of the coefficient of correlation between the CCD image data and the sample image data. A sensing area is then defined in the CCD image at the measurement. Then, the particle was detected and sorted by a result of correlation between the CCD image and the representative sorting particle one. A current, amplified by a power amplifier, is used to change the magnetic polarity of the electromagnet coil and the output of a D/A (Digital to Analogue) circuit is varied. This process was controlled by the results of image processing in the sensing area. We applied it to sort copolymer beads in a chip. The bead was successfully transported to the left microchannel by moving the MMT to the right wall of the microchannel. The maximum sorting rate for the current experiment was up to 15 Hz which was restricted by the image capturing system (Figure 8).

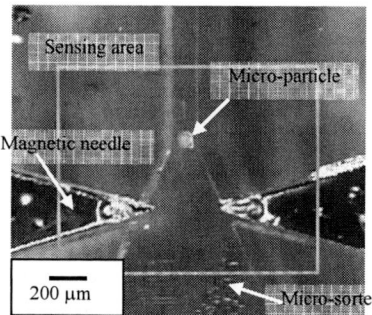

Figure 7: Operation of sorting MMT. .

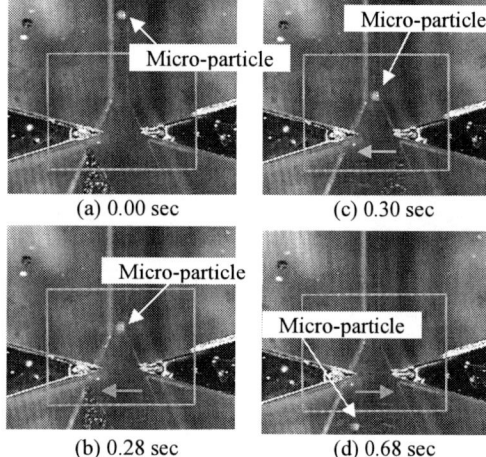

(a) 0.00 sec (c) 0.30 sec

(b) 0.28 sec (d) 0.68 sec

Figure 8: Demonstration of automatic sorting system.

CONCLUSIONS

We succeeded in powerful noncontact actuation of magnetically driven microtool (MMT) by magnetizing it and focusing magnetic field in a microfluidic chip. The density of magnetic flux was improved about 100 times larger after magnetization. We applied 2.83 V and 0.15 A on the electromagnetic coils and the maximum actuation speed of MMT was achieved up to 180 Hz (the velocity of edge of the MMT: 13.5 mm/s) in ethanol. We demonstrated separation of copolymer beads in a chip by image sensing, and beads of certain size are separated successfully

Proposed magnetic actuation method can be applied to install robotic components in a chip. This will lead us to develop a novel research field on "Robotics on a Chip".

ACKNOWLEDGMENT

This work was financially supported by the Research and Development Program for New Bio-industry Initiatives and the Ministry of Education, Culture, Sports, Science and Technology Grants-in-Aid for Scientific Research (17040017)& 19016004). The authors would like to thank Prof. S. Sugimoto, Tohoku University, for use of the magnetization facilities.

REFERENCES

[1] M. R. Melamed, T. Lindmo and M. L. Mendelsohn, "Flow Cytometry and Sorting", second edition, Wiley-Liss, New York, USA, 1991.

[2] G. Fuhr, R. Hagedorn, et al., "Linear motion of dielectric particles and living cells in microfabricated structures induced by traveling electric fields", Proc. of IEEE Micro Electro Mechanical systems, p.259–264, 1991.

[3] T. Schnelle, et al., "Paired Microelectrode System: Dielectrophoretic Particle Sorting and Force Calibration", Journal of Electrostatics, Vol.47, p.121-132, 1999.

[4] F. Arai, A. Ichikawa, M. Ogawa, T. Fukuda, K. Horio and K. Itoigawa, "High-speed separation systems of randomly suspended single living cells by laser trap and dielectrophoresis", Electrophoresis, Vol.22, p.283–288, 2001.

[5] Y. Shirasaki, et al., "A Novel Biomolecule Sorter Using Thermosensitive Hydrogel in Micro Flow System", Proc. of the Micro Total Analysis Systems 2002 (μ-TAS2002), p. 925–927, 2002.

[6] A. Ashkin and J. M. Dziedzic, "Optical trapping and manipulation of viruses and bacteria", Science, Vol. 235, p.1517, 1987.

[7] O. Cugat, J. Delamare and G. Reyne, "Magnetic Micro-Actuators and Systems" IEEE Transactions on magnetics, Vol.39, No.5, p.3607–3612, 2003.

[8] J. J. Abbott, Z. Nagy, F. Beyeler and B. J. Nelson, "Robotics in the Small", IEEE Robotics & Automation Magazine, p.92–103, June, 2007.

[9] G. A. Mensing, T. M. Pearce, M. D. Graham and D. J. Beebe, "An Externally driven magnetic microstirrer", Phil. Trans. R. Soc. Lond. A 362, p.1059–1068, 2004.

[10] M. Barbic, J. J. Mock, A. P. Gray and S. Schults, "Electromagnetic micromotor for microfluidics applications", Applied Physics Letters, Vol.79(9), p.1399–1401, 2001.

[11] H. Maruyama., F. Arai and T. Fukuda, "On-Chip microparticle handling using magnetically driven microdevice", μ-TAS2005, p.1422–1424, 2005.

[12] Y. Yamanishi, Y. C. Lin and F. Arai: "Magnetically Modified PDMS Devices for Active Microfluidic Control", μ-TAS2007, p.883–885, 2007.

[13] Y. Yamanishi, S. Sakuma and F. Arai: "Magnetically Modified Soft Micro Actuator for Oocyte Manipulation", IEEE International Symposium on Micromechatronics and Human Science (MHS), p.442–447, 2007.

[14] Y. Yamanishi, S. Sakuma, K. Onda, F. Arai, "Biocompatible Polymeric Magnetically Driven Microtool for Particle Sorting", Journal of Micro - Nano Mechatronics, (2008) (online).

978-1-4244-2977-6/09 $25.00 © 2009 IEEE

NANOBIOPHOTONICS AND BIOASICS FOR BIOMEDICAL INNOVATIONS

Luke P. Lee

Department of Bioengineering, University of California, Berkeley
Biomolecular Nanotechnology Center, Berkeley Sensor and Actuator Center
Email: lplee@berkeley.edu, URL: http://biopoets.berkeley.edu

ABSTRACT

In this talk, I will discuss quantitative biology and medicine by nanobiophotonics and BioASICs. Using new paradigms of biological inspiration and understanding of electron transfer mechanism in biological systems, we have created quantized Plasmon Resonance Energy Transfer (PRET) nanospectroscopy for molecular imaging of living cells (Fig. 1). For the remote optical control of gene regulation and protein expression, we have developed Oligonucleotides on a Nanoplasmonic Carrier Optical Switch (ONCOS). ONCOS allows on-demand gene silencing with nanometer-scale spatial resolution and localized temperature disturbance in living cells. The ONCOS and PRET will be used for experimental system biology, molecular/cellular diagnostics, and therapeutic applications since it will provide us precise spatial and temporal information of living cellular mechanism. Bionanophotonic molecular ruler is accomplished to measure the dynamics of DNA and protein interactions. *In-vivo* Surface Enhanced Raman Spectroscopy (SERS) probes, *in-vitro* integrated nanofluidic SERS, and optofluidic microprocessors are developed for label-free molecular diagnostics and drug discovery.

In order to accomplish physiologically relevant cell culture platforms, we have developed Biological Application Specific Integrated Circuits (BioASICs) for patch clamp array (Fig. 2) single cell biophysics, quantitative cell biology, and cell-based diagnostics by connecting novel microfluidics and nanofluidic circuits, which can impact on high-speed and high-content repeatable biology, quantitative medicine, and biophysics in new ways. We are creating a library of these "building blocks" to develop innovative single cell array, physiologically relevant dynamic cell culture array (Fig. 3), and biological microprocessors with integrated optical controls and detections capability. For biologically inspired electronics, we are elucidating the electron transfer mechanism of natural photosynthesis to develop artificial photosynthesis on chip. Finally, I will discus the importance of cell culture revolution by BioASICs (i.e. physiologically relevant dynamic cell culture platform) and quantitative real time nucleic acid detection by *i*NASBA (Integrated Nucleic Acid Sequence-Based Amplification), which can impact on the future quantitative biology, molecular medicine, and medical diagnostics.

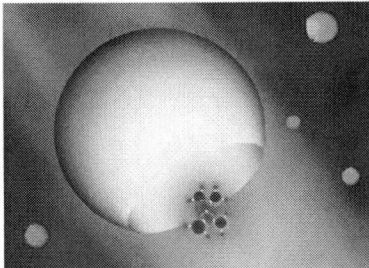  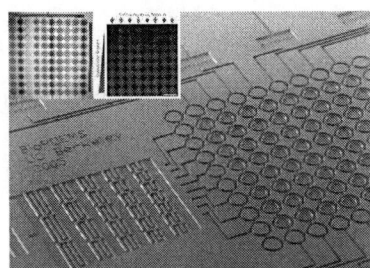

Fig.1. Nanobiophotonic Satellites   *Fig. 2. Patch-clamp Array*   *Fig. 3. Dynamic Cell Culture Array*

RECONSTRUCTION OF 3D HIERARCHIC MICRO-TISSUES USING MONODISPERSE COLLAGEN MICROBEADS

Y. Morimoto[1], Y. Tsuda[1, 2], and S. Takeuchi[1, 2, 3]

[1]Center for International Research on Micromechatronics (CIRMM), Institute of Industrial Science (IIS), The University of Tokyo, Japan

[2]Bio Electromechanical Autonomous Nano Systems (BEANS) Project, Ministry of Economy, Trade and Industry, Japan

[3]PRESTO, Japan Science and Technology, Japan

ABSTRACT

We successfully established a three-dimensional (3D) hierarchic cell co-culture system using micro-collagen beads. Our reconstructed tissues have double cell-layers that provide an *in vivo*-like micro-environment. To prepare the hierarchic tissue structures, we used an axisymmetric flow-focusing device (AFFD) that allows us to encapsulate HepG2 cells within monodisperse collagen beads. We then seeded 3T3 cells on the surface of the collagen beads. We observed that HepG2 cells and 3T3 cells successfully self-organized into hierarchical cell structures of uniform size. We believe the *in vitro* 3D co-culture beads are useful for the studies of *in vivo*-like cell-cell interactions with various cells. Moreover, the monodispersity of the 3D cell cultured beads facilitates on-chip assays of chemicals/drugs.

INTRODUCTION

Heterotypic 3D co-culture is essential to mimic tissues and organs since various types of cell-cell interactions are required *in vivo*. Reconstruction of 3D tissue co-culture of two different types of cells provides an *in vivo*-like environment. Although conventional 2D co-culture systems has revealed that cell-cell interactions affect the activation of cellular functions, they did not observe the same phenomena occurring within organisms [1, 2] In addition, it is difficult to form 3D co-cultures because they tend to aggregate [3], and consequently it is nearly impossible to dictate the orientation of each kind of cell.

Here, we arrange different types of cells within and on monodisperse collagen beads; these beads are biocompatible spherical scaffolds to reform their native structures (Fig. 1(a)). This co-culture method is able to control the location of each kind of cell and thus generate hierarchically layered cells; in these beads, the inner cells are sufficiently confined by outer cells. 3D co-culture of monodisperse collagen beads facilitates handling of these mobile tissues and provides a convenient experimental platform for chemical/drug assays.

To produce monodisperse collagen beads containing cells, we used the AFFD fabricated by stereolithography (Fig. 1(b)) [4, 5]. The use of AFFD has two advantages: (i) droplets size can be controlled with respect to the flow rate ratio of the outer fluid to the inner fluid; (ii) we can form droplets regardless of the components of the solutions as droplets do not contact the surface of the AFFD [6-8].

In this paper, using monodisperse collagen beads, we fabricated 3D hierarchic co-culture micro-tissues with HepG2 cells and 3T3 cells. HepG2 cells are able to secrete albumin and secretion of albumin increase when HepG2 cells are co-cultured with 3T3 cells [1, 3]. We

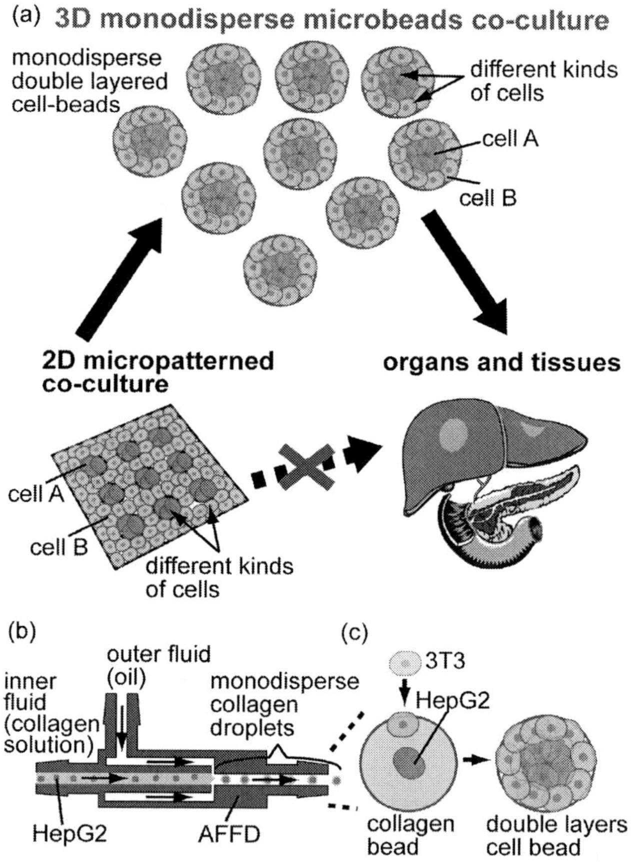

Figure 1. (a) Concept of 3D co-culture microtissues. These tissues provide microenvironment similar to organs and tissues rather than conventional 2D co-culture. (b) Schematic diagram of an AFFD fabricated by stereolithography to produce monodisperse collagen droplets. (c) Process flow of producing 3D co-culture systems based collagen beads using 3T3 cells and HepG2 cells.

demonstrated the incrementation of albumin secretion of HepG2 in our 3D tissue co-culture system using collagen, and this demonstration shows that our system has the ability to apply mimicking tissue structures and functions *in vitro*.

MATERIALS AND METHODS
Materials

Corn oil, lecithin (corn oil) and mineral oil were purchased from Wako Pure Chemical Industries, Ltd. Span80, Span20, Tween20 and hexadecane were purchased from Kanto Chemical Co., Inc. Dulbecco's

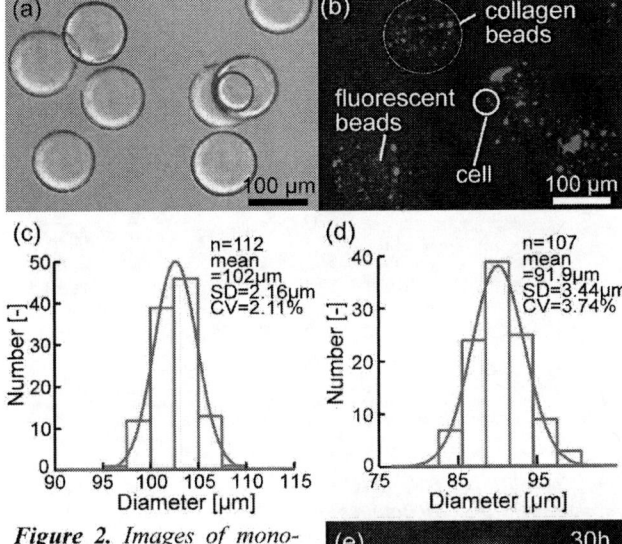

Figure 2. *Images of monodisperse (a) collagen droplets in corn oil and (b) collagen beads in culture medium. Collagen dropletss were produced using AFFD. In (b), we visualized collagen beads by subnano-sized fluorescent beads (red) and cell nuclei with Hoechst 33342 (blue). (c) and (d) Diameter distribution of the collagen droplets in (a) and the collagen beads in (b). Our method for the collagen gelation remains the monodispersity of collagen droplets supplied by AFFD. (e) Confocal microscopy of a collagen bead encapsulating cells after 30-hour incubaiton. In this image, we visualized live cells (Green) by Live/Dead Assay kit. This image show that most of the cells are alive in collagen beads fabricated by our method and grow along the shape of collagen beads.*

modified eagle's medium (DMEM) and phosphate buffered saline (PBS) were purchased from SIGMA-Aldrich. Neutral collagen solution in DMEM (2 mg/ml) was purchased from KOKEN Co., Ltd. Other chemicals were purchased from Kanto Chemical Co., Inc, Nacalai Tesque and Wako Pure Chemical Industries. Unless otherwise specified, all water used in the experiments refers to ultra pure water obtained from a Millipore system having a specific resistance of 18 MΩ·cm.

Cell culture

Adherent cells used here were 3T3 cells, mouse fibroblast like cell, and HepG2 cells, human hepatoma cell line, respectively. Both cell types were cultured at a condition of 37 °C and 5 % CO_2 atmosphere in DMEM with 10%(v/v) fatal bovine serum (Japan Bioserum Co Ltd.), and 5% Trypsin-EDTA solution (SIGMA-Aldrich) as a culture medium. Cells were fluorescently labeled with Hoechst33342 (Molecular Probes Inc.), Live/Dead assay (LIVE/DEAD® Viability/ Cytotoxicity kit, Invitrogen Co.) and Cell tracker (Cell Tracker Green CMFDA and Cell Tracker Red CMTPX, Invitrogen Co.) by a protocol provided from the manufacturer.

Formation of collagen sol droplets with AFFDs

Using the AFFD fabricated by stereolithography, we produced monodisperse collagen sol droplets. The mechanism of monodisperse droplets formation and the fabrication process of the AFFD have been previously reported elsewhere [4]. Briefly, the device comprises of two concentric hollow cylinders. Each cylinder has a connection port that separately guides fluid such as oil and water into the device. These fluids are immiscible and break up into droplets when the inner fluid streams through the exit of the orifice. The inner fluid is always surrounded by the outer fluid and the droplets formed are confined to the central axis of the microchannel. As they do not contact the channel surface, we can avoid the problem of wetting on the inner-wall even when various kinds of droplets incorporated with cells are produced. By varying the flow rate ratio of the inner fluid and the outer fluid, the size of droplets can be controlled. We designed the device with a 3D modeling software (Rhinoceros, AppliCraft) and, to fabricate the AFFD, used a commercial stereolithography modeling machine (Perfactory, Envision Tec, Germany) with photoreactive acrylates resin ("R11, 25-50 μm layers", Envision Tec) that consists of acrylic oligomer, dipentaerythritol pentaacrylate, propoxylated trimethylol-propane triacrylate, photoinitiator and stabilizers.

The size of droplets is affected by the orientation of the AFFD [6], thus here we oriented the device vertically for all the experiments to produce collagen sol droplets. Tefzel® tubes attached to the inlet and outlet ports using silicone rubber tubes and we infused inner and outer fluids into the device through these tubes using syringe pumps (KDS-210, KD Scientific Inc., USA). We allowed the system to stabilize for one minute before droplets were collected.

Gelation of collagen sol droplets encapsulating cells

Monodisperse droplets produced by the AFFD are platforms for the formation of monodisperse collagen gel beads encapsulating cells in viable conditions. In the AFFD, corn oil with lecithin (2 wt%) is used for the outer fluid and neutral collagen solution in DMEM with cells is used for the inner fluid, respectively. After forming monodisperse collagen sol droplets encapsulating cells by the AFFD, we collected them into microtubes including the mixture oil of corn oil with lecithin (2 wt%) and mineral oil with Span20 (2 wt%). Since neutral collagen solution gelates at a warmed condition, they are incubated in a water bath at 37 °C for 45 minutes to induce gelation of collagen solution.

After collecting and heating collagen gel beads in the mixture oil, we extracted these collagen gel beads from the oil and transferred them to DMEM using the method reported in another paper [9] as follows. We aspirated off the mixture oil around collagen gel beads after they were deposited at the bottom. We subsequently introduce hexadecane with Span80 (2 wt%) into the microtube to dissolve the remaining the mixture oil for protecting the surface of collagen gel beads from the oil sticking. We again aspirated off hexadecane with the mixture oil and added the DMEM with Tween20 (0.1 wt%) to separate collagen gel beads from oil. After suspending collagen gel beads in DMEM with Tween20, we collected these beads by centrifuging. We aspirated off the supernatant and resuspended these beads in DMEM. We repeated centrifuging and resupending in DMEM for rinsing. Then we obtained monodisperse collagen gel beads in DMEM.

RESULTS AND DISCUSSION

Monodisperse collagen gel beads with cells

We produced monodisperse collagen sol droplets and gel beads using the AFFD fabricated by stereolithography. Figure 2(a) and (b) show images of monodisperse collagen sol droplets encapsulating cells and collagen beads containing cells and fluorescent beads after gelation, respectively. We produced both droplets and gel beads with 9 μl/min and 60 μl/min with the flow rate of collagen solution and corn oil, respectively. Figure 2(c) and (d) show the diameter distribution of the collagen sol droplets and gel beads in Figure 2(a) and (b). Both collagen sol droplets and gel beads are monodisperse since the coefficient of variation (CV), which is defined as the ratio between the standard deviation and the mean, is less than 5% [6]. In addition, we can maintain the viability of the encapsulated cells and they can grow for 30 hours (Fig. 2(e)). The advantage of our gelation method allows cells encapsulated in collagen gel beads to be alive. When corn oil surrounds collagen sol droplets for 45 minutes, although monodispersity of droplets keeps, cells in these droplets die as corn oil does not have much oxygen permeability. Whereas, when mineral oil surrounds collagen sol droplets for 45 minutes, although cells keep alive, the droplets of collagen become polydisperse. Using the mixture oil of corn oil and mineral oil allow us to take the advantage of these oil, therefore we can obtain monodisperse collagen gel beads encapsulating cells in keeping the viability of them.

Mono-culture using collagen beads

3T3 cells on the surface of collagen gel beads

We seeded 3T3 cells on the surface of monodisperse collagen gel beads. When the 3T3 cells adhered on the collagen gel beads, they grew and gradually migrated over the surface of the collagen (Fig. 3(a)). Figure 3(b) shows the image of monodisperse 3T3 coated collagen beads. This image indicates that although the traction force of cell adhesion may transform the shape of collagen gel beads, monodispersity of collagen gel beads maintain after the 3T3 cell adhesion. 3T3 cells on the surface of collagen gel beads gradually migrated and became the layer of 3T3 cells (Fig. 3(c) and (d)). The layer of 3T3 cells was formed after a 30 hours incubation period and the inside of this bead is hollow (Fig. 3(e)).

HepG2 cells in the collagen gel beads

We cultured HepG2 cells encapsulated in collagen gel beads produced by our method. After 30-hour incubation period, most of HepG2 in the collagen gel beads were viable and monodispersity of collagen gel beads was maintained. This result shows that, for inner cells as HepG2 cells in this report, our collagen gelation method is no harm and keeps their activity.

3D tissue co-culture of 3T3 cells and HepG2 cells

On the basis of these results, when we seeded 3T3 cells on the surface of collagen gel beads encapsulating HepG2

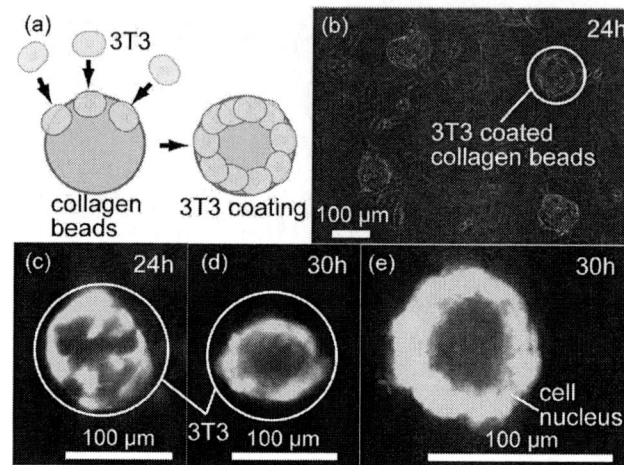

Figure 3. (a) Concept of 3T3 coated collagen beads. (b) Images of 3T3 cells coated collagen beads after 24-hour incubation. (c) and (d) Images of the beads coated with 3T3 cells after 24 hours (c) and 30 hours (d) incubation. We visualized 3T3 cells by Live/Dead Assay kit (Green). 3T3 cells adhered on the collagen beads grow and cover the surface of beads gradually. (e) Confocal microscopy of 3T3 cells after 30 hours incubation period. We visualized 3T3 cells by Live/Dead Assay kit (Green) and cell nuclei by Hoechst 33342 (blue). 3T3 cells make a layer on the surface of collagen beads and almost 3T3 cells were alive after 30 hours.

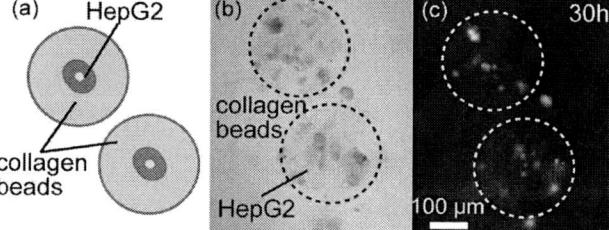

Figure 4. (a) Concept of HepG2 encapsulated in collagen beads. Images of HepG2 cells within collagen beads; (b) bright field image, (c) visualized HepG2 cells using Live/Dead Assay kit (Green) and HepG2 cell nuclei by Hoechst33342 (blue). Most of the HepG2 cells encapsulated in collagen beads were alive through our gelation method.

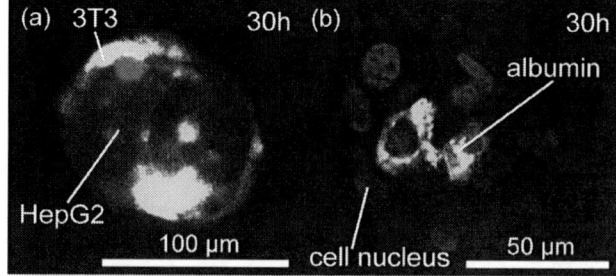

Figure 5. (a) Confocal microscopy of 3D co-culture based collagen beads after 30-hour incubation period; HepG2 cells stained red and 3T3 cells stained green. 3T3 cells sufficiently confined HepG2 cells. (b) Image of the hepatic albumin secretion by HepG2 cells. Albumin is visualized using an immunostaining method (Green). It was confirmed that HepG2 co-cultured with 3T3 cells secreted albumin.

cells and cultured them during 30 hours, we succeeded in forming a 3D co-culture system consisting of HepG2 cells confined by a layer of 3T3 cells (Fig. 5(a)). In this 3D co-culture system, we observed that HepG2 cells secreted albumin (Fig. 5(b)); albumin secreted from HepG2 cells was visualized by an immunostaining [10]. We confirmed the albumin secretion rate of HepG2 cells was enhanced due to the presence of 3T3 cells over that of just the mono-culture system of HepG2 cells; this *in vitro* phenomenon mimics *in vivo* hepatic function.

CONCLUSION

Using the AFFD, we can produce monodisperse collagen sol droplets encapsulating cells. Our gelation method using the mixture oil consisting of corn oil and mineral oil can produce monodisperse collagen gel beads and maintain the viability of encapsulated cell. We believe these collagen gel beads will be a convenient system for the analysis and experiment of 3D tissue co-cultures. For the demonstration of this system, we succeeded in producing the hierarchic 3D co-culture of 3T3 cells and HepG2 cells and confirmed that the albumin secretion rate of HepG2 cells increase due to the presence of 3T3 cells. By comparing this result with the albumin secretion rate of 2D co-culture in the same condition, we would understand the mechanism of 3D hierarchic cell-cell interactions of organs. In addition, monodispersity of collagen beads is important when making an array of them using beads-based microfluidic array systems that allow us to achieve quantitative analysis [11-13]. We believe that our 3D tissue co-culture method will be an economical and convenient tool for *in vitro* study of *in vivo*-like microenvironments and cell-cell interactions.

ACKNOWLEDGEMENTS

This work was supported by Bio Electromechanical Autonomous Nano Systems (BEANS) Project from Ministry of Economy, Trade and Industry, by Core Research for Evolutional Science and Technology (CREST) from Japan Science and Technology (JST) Agency, and by Grants-in-Aid for Scientific Research from Ministry of Education, Culture, Sports, Science and Technology Japan.

REFERENCES

[1] S. N. Bhatia, U. J. Baris, M. L. Yarmush and M. Toner, "Effect of cell-cell interactions in preservation of cellular phenotype: cocultivation of hepatocytes and nonparenchymal cells", *FASEB J.*, vol. 13, pp. 1883-1900, 1999.

[2] Y. Tsuda, A. Kikuchi, M. Yamato, G. Chen and T. Okano, "Heterotypic cell interactions on a dually patterned surface", *Biochem. Biophys. Res. Comm.*, vol. 348, pp. 937-944, 2006.

[3] A. Ito, H. Jitsunobu, Y. Kawabe and M. Kamihira, "Construction of heterotypic cell sheets by magnetic force-based 3-D coculture of HepG2 and NIH3T3 cells", *J. Biosci. Bioeng.*, vol. 104, pp. 371-378, 2007.

[4] Y. Morimoto, W.-H. Tan and S. Takeuchi, "Three-dimensional axisymmetric flow-focusing device using stereolithography", *Biomed. Microdev.*, in press.

[5] Y. Morimoto, W.-H. Tan and S. Takeuchi, ""Housing" for cells in monodisperse microcages", *Proc. of MEMS 2008*, pp.304-307.

[6] S. Takeuchi, P. Garstecki, D. B. Weibel and G. M. Whitesides, "An axisymmetric flow-focusing micro-fluidic device", *Adv. Mater.*, vol. 17, pp. 1067-1072,

2005.

[7] A. Luque, F. A. Perdigones, J. Esteve, J. Montserrat, A. M. Ganan-Calvo and J. M. Quero, "Silicon microdevice for emulsion production using three-dimensional flow focusing", *J. Microelectromech. Syst.*, vol. 16, pp. 1201-1208, 2007.

[8] A. S. Utada, E. Lorenceau, D. R. Link, P. D. Kaplan, H. A. Stone and D. A. Weitz, "Monodisperse double emulsions generated from a microcapillary device", *Science*, vol. 308, pp. 537-541, 2005.

[9] W.-H. Tan and S. Takeuchi, "Monodisperse alginate hydrogel microbeads for cell encapsulation", *Adv. Mater.*, vol. 82, pp. 364-366, 2003.

[10] A. Yamasaki, H. Jameson, C. Schonhoff, C. Leveille-Webster and M. S. Anwer, "Cyclic reverses tautolithocholate-induced inhibition of taurocholate uptake in human hepatoma cell line", *Hepatology*, vol. 44, pp. 381A, 2006.

[11] D. Di Carlo, N. Aghdam and L. P. Lee, "Single-cell enzyme concentrations, kinetics, and inhibition analysis using high-density hydrodynamic cell isolation arrays", *Anal. Chem.*, vol. 78, pp. 4925-4930, 2006.

[12] W.-H. Tan and S. Takeuchi, "A trap-and-release integrated microfluidic system for dynamic microarray applications", *Proc. Natl. Aca. Sci. USA*, vol. 104, pp. 1146-1151, 2007.

[13] W.-H. Tan and S. Takeuchi, "Dynamic microarray system with gentle retrieval mechanism for cell-encapsulating hydrogel beads", *Lab Chip*, vol. 8, pp. 259-266, 2008.

LABEL-FREE CONTINUOUS MICRO CELL SORTER
WITH ANTIBODY-IMMOBILIZED OBLIQUE GROOVES

Shin-ichi Hashimoto, Takahiro Nishimura, Junichi Miwa, Yuji Suzuki and Nobuhide Kasagi
Department of Mechanical Engineering, The University of Tokyo
Hongo 7-3-1, Bunkyo-ku, Tokyo 113-8656, Japan

ABSTRACT

We report development of a novel label-free continuous cell separation method for the extraction of rare cells from a limited amount of sample. The separation method is based on specific adhesion between target cells and antibody immobilized on oblique micro grooves etched into the channel wall. Due to asymmetric adhesive force, cross-flow displacement of the target cells is induced. Label-free lateral separation of micro beads as the cell model has been successfully achieved. Cross-flow displacement of human umbilical-vein endothelial cells (HUVEC) with the CD31 antibody is also demonstrated.

INTRODUCTION

Stem cell therapy is a rapidly evolving biomedical technology, in which multipotent stem cells are cultured in vitro and transplanted to regenerate damaged or deficit tissue. Since use of embryonic stem cells often encounters ethical issues, extraction of adult stem cells such as mesenchymal stem cell (MSC) from terminal blood attracts much attention. However, stem cells are very rare and difficult to identify using their physical properties. Thus, development of an efficient and accurate cell separation method of rare cells from cell mixture is necessary.

Cell separation methods often employ specific binding between cell surface antigen and its corresponding antibody. For example, antibody-conjugated fluorescence molecule and magnetic beads are respectively used in fluorescence activated cell sorting [1-3] and magnetic cell sorting [4-6] for labeling target cells. However, labeling cell is problematic for regenerative medicine because of unexpected effect of the labels in cultivation and transplantation.

In the present study, we propose a continuous label-free cell separation method, where the target cells are separated in the cross-flow direction using the asymmetric antibody pattern on the channel surface.

DESIGN

Miwa et al. [7] has proposed a label-free cell separation method using specific adhesion between target cells and antibody immobilized on the wall. Separation of cell mixture can be accomplished by simply flowing the sample plug through an antibody-immobilized micro channel without any pre-/post-processing. In their device, the target cells rolling on the channel wall are decelerated and separated from the other cells in the direction of main flow. They successfully separate a mixture of human umbilical-vein endothelial cells (HUVEC) and human leukocytes (HL60). However, its throughput remains low, since it is batch process.

Figure 1 shows the schematic of the present cell separation method. At the entrance of the separation channel, cell suspension is hydrodynamically focused to one side of the channel using buffer fluid. The channel wall has oblique grooves with immobilized antibody, which work as adhesive/non-adhesive stripes. When the cells roll on the surface, the target cells are displaced in the cross-flow direction by the asymmetric adhesive force, and collected from the outlet on the other side.

In the present study, it is required that the cells roll along the antibody-immobilized channel wall. Thus, the channel depth is chosen as 40 μm, which is approximately twice the diameter of a typical monocyte. Channel width is 200μm. The oblique grooves 0.2 μm in depth 1 μm in width are etched into the channel surface as the non-adhesive oblique patterns; when the cells located above a groove, the cell surface is displaced from the antibody on the bottom of the groove. The pitch between the grooves is 2 μm.

MICROFABRICATION

Figure 2 shows the fabrication process of the present micro cell separator. Firstly, micro channel structures are etched into a silicon substrate using DRIE with an oxide mask and fluidic ports are ultrasonically drilled (Fig. 2a). The silicon substrate is coated with 2 μm-thick parylene-C (Fig. 2c). Oblique grooves are patterned with EB lithography and etched with CHF_3 plasma into a Pyrex substrate (Fig. 2b). Then, anino-functionalized parylene diX-AM (Kisco) 0.1 μm in thickness is deposited on the Pyrex substrate to form dense amino group on its surface [7]. This is followed by parylene-parylene thermal bonding [8] between silicon and Pyrex substrate in vacuum for an hour (Fig. 2d). The bonding pressure and temperature are respectively 5 MPa and 200°C. Finally, inlet and outlet ports made of PDMS blocks are firmly

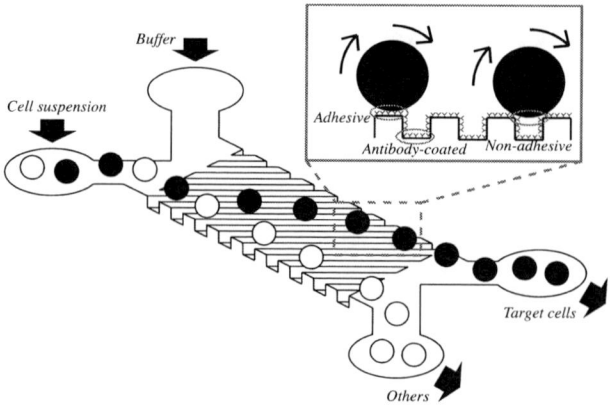

Figure 1: Cell separation principle using oblique antibody patterns.

bound to the silicon substrate after oxygen plasma treatment (Fig. 2e).

Figure 3 shows a photo of the present device and a SEM image of the oblique grooves. The oblique pattern locates in the straight part of the channel. Total number of the grooves is 10,000. Sample cell suspension is focused in the center of the channel by buffer fluid to avoid unwanted effect of velocity gradient near the sidewall.

Biomolecule immobilization on the amino group of diX-AM (Fig. 4) is described in Miwa et al [7]. It is performed by successive introduction and incubation of biomolecule solutions with a syringe pump; the diX-AM surface is firstly biotinylated by conjugating NHS-LC-LC-biotin to the surface amines. NHS-LC-LC-biotin is successively dissolved into dimethylsulfoxide and bicine buffered saline (pH 8). The biotin solution is then introduced into the device, and incubated for one hour at 30 °C. Streptavidin and biotin-conjugated CD31 antibody solutions, are separately dissolved into PBS (pH 7.4), and successively incubated. Although CD31 is used in the present study, various kinds of proteins and other biomolecules can be immobilized using the present procedure.

EXPERIMENTAL RESULTS

As proof-of-concept, we employ polystyrene micro beads in order to examine performance of the present device. Streptavidin-functionalized surface, which specifically adhere the biotin-coated beads, is formed by conjugation of NHS-LC-LC-biotin and streptavidin on the diX-AM surface. Biotin-coated Yellow and streptavidin-coated Nile Red fluorescence beads (Spherotec) are used respectively as the target and non-target model cells. The diameter of both beads is 7.66 ± 0.33 μm. Under a fluorescent microscope, the beads type can be distinguished by using different filter set. The beads are suspended in PBS (pH 7.4) at the number density of 4×10^6 beads/mL. Because the only difference between the two kinds of beads is the surface chemistry, any difference in the motion of the beads should be attributed to the difference in the adhesive force with the surface antibody.

We characterize the beads motion on the oblique grooves through microscopic measurements. The focal plane of the microscope is set at the bottom wall of the micro channel, and beads images are taken at 25 fps. The beads velocity is calculated from the displacement of the beads center between successive images. The number of beads analyzed is 30 for each beads type.

The flow ratio of the beads mixture and PBS buffer fluid is set 1:1 to focus the beads within 100 μm of the center of the channel, where the flow velocity is almost uniform. The angle between oblique grooves and the main flow is 45°. Figure 5 shows superimposed sequential images of a biotin-coated bead flowing in the micro channel. The cross flow position gradually shifts to the lower side of the image, which shows the oblique antibody pattern does induce cross-flow adhesive force.

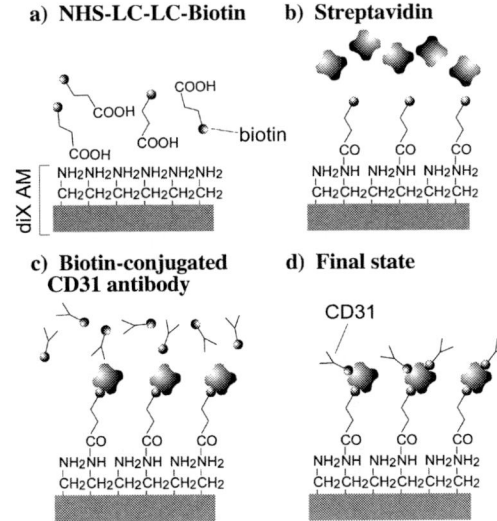

Figure 4: Schematic of the CD31 antibody immobilization process on diX-AM surface [7].

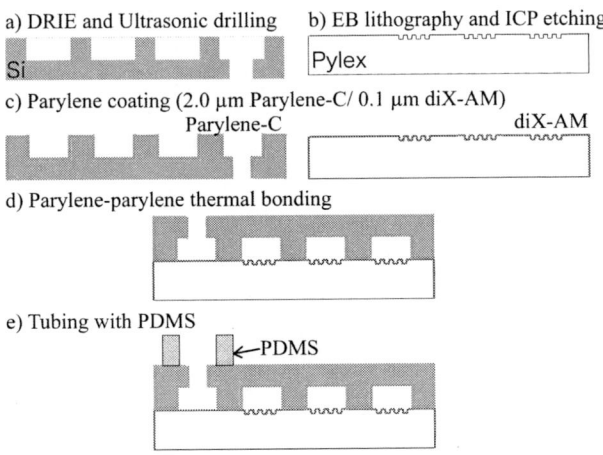

Figure 2: Microfabrication process.

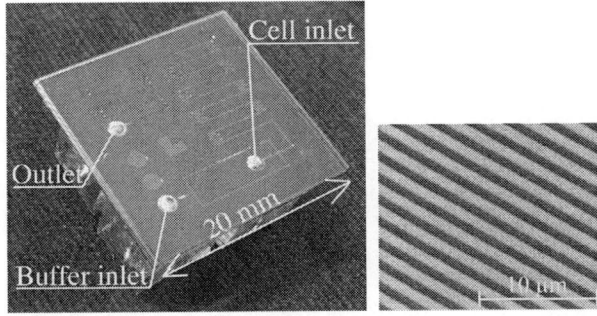

Figure 3: (a) Photo of the present device, (b) SEM image of the oblique grooves.

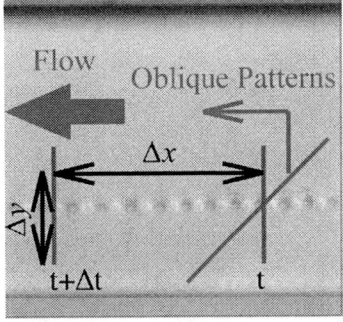

Figure 5: Microscopic measurement of micro beads velocity and displacement.

Figure 6 shows the streamwise velocity of beads u at by the bulk mean velocity u_b of 0.5 mm/s. All the beads are almost uniformly distributed in the cross-flow direction within the middle half of the channel. While the velocity of the streptavidin-coated beads is somewhat larger than u_b, the biotin-coated beads travels at about 0.4 u_b, showing effective deceleration due to the biotin-streptavidin interaction.

Figure 7 shows the cross-flow displacement Δy versus the beads traveling speed u. The biotin-coated beads are displaced in the cross-flow direction, and the mean value is 0.8% of the traveling length Δx. On the other hand, the cross-flow displacement of streptavidin-coated beads is less than 0.2%, and scattered around $\Delta y = 0$. Therefore, the adhesive force on the oblique patterns should be responsible for the cross-flow displacement, and the effect of the secondary flow or flow inside the grooves is negligible.

Figure 8 shows the mean cross-flow displacement for different bulk mean velocities. The mean cross-flow displacement is about 0.8% for all the cases examined, and independent of the flow velocity. This result indicates that the cell separation using the oblique antibody pattern can be accomplished in a wide range of flow velocity.

The beads separation is demonstrated using a beads mixture. The separation performance is evaluated by counting the number of beads passing the measurement

volume. The length of patterned area is 20 mm (10,000 grooves). The biotin- and streptavidin-coated beads are mixed at 1:1 with 2×10^4 beads/mL total concentration. The flow ratio of beads mixture and the PBS buffer fluid is set 1:2.5 to focus the beads within 50 μm of center of the channel. The bulk mean velocity is 1.0 mm/s.

Figure 9 shows histograms of cross-flow position of the beads. The sample beads are successively separated into each type of the beads after 10,000 oblique grooves, showing the effectiveness of present cell separation method.

Finally, we employ human umbilical vein endothelial cells (HUVEC) to evaluate performance of the present device for actual cells. HUVEC has mean diameter of 20 μm and nearly 100% expression of CD31 (Fig. 10).

The cells are stained with SYTO24 and suspended in

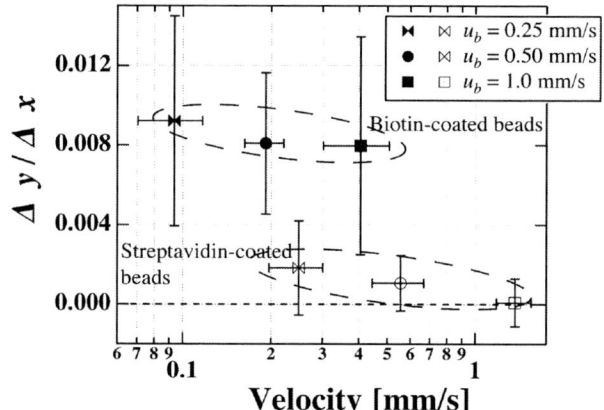

Figure 8: Dependence of the mean cross-flow displacement on the flow velocity. The error bars correspond to twice the standard deviation for $n = 30$.

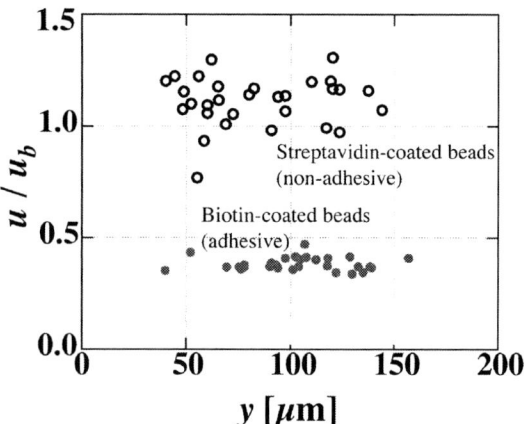

Figure 6: Streamwise velocities of individual beads at $u_b = 0.5$ mm/s.

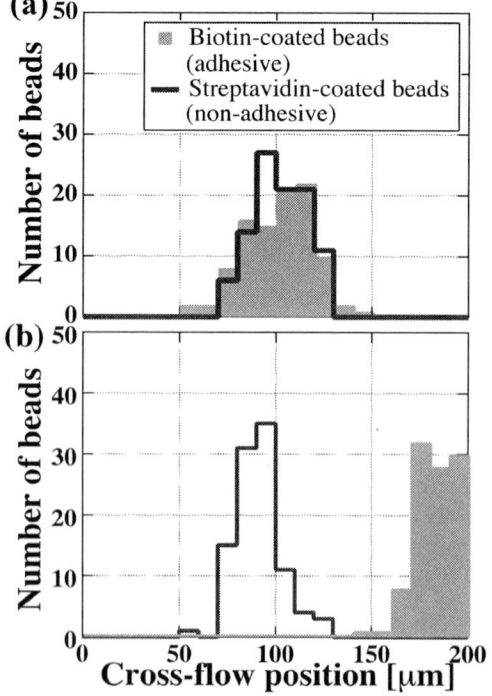

Figure 9: Histograms of the cross-flow position of the beads at $u_b = 1.0$ mm/s. (a) At the inlet, (b) After passing through 10,000 oblique streptavidin patterns.

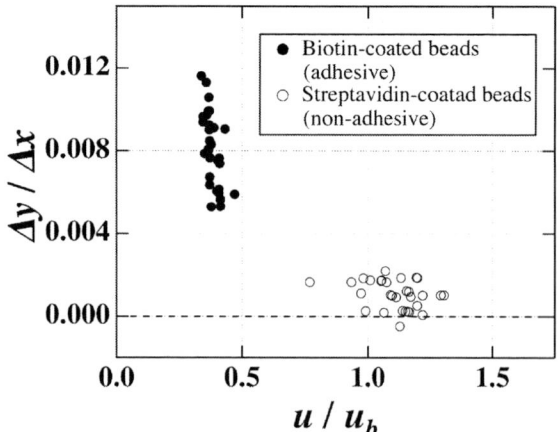

Figure 7: The cross-flow displacement versus the beads velocity at $u_b = 0.5$ mm/s.

PBS (pH 7.4) at the number density of 2×10^6 cells/mL, which corresponds to the volume concentration of approximately 0.25 %. The sample cell suspension is continuously introduced with a syringe pump. The bulk mean velocity is 1.0 mm/s. The flow ratio of beads mixture and the PBS buffer fluid is set 1:2.5.

Figure 9 shows histograms of the cross-flow positions of HUVEC before and after passing the CD31 oblique patterns. The histogram downstream the patterns is shifted in the cross-flow direction, but their displacement is smaller than the beads. Longer channel should be necessary for complete separation of HUVEC, partially due to small deceleration of the cells. Figure 12 shows HUVEC velocity near the outlet. Based on Miwa et al. [7], HUVEC are expected to decelerate to $0.6\ u_b$ in antibody immobilized micro channel. However, many cells travel faster than $0.6\ u_b$ in the present device.

CONCLUSIONS

A novel label-free continuous cell separation method has been developed for accurate extraction of rare cells from a limited amount of sample. The separation method is based on specific adhesion between target cells and oblique antibody patterns, which are realized through immobilization of antibodies on oblique micro grooves etched into the channel wall. It is found that, on the streptavidin-coated surface, biotin-coated beads are decelerated by 60 % due to the specific adhesion. The adhesion force induces cross-flow motion of the beads, and the cross-flow displacement of about 0.8 % of the beads' traveling distance is achieved. Complete separation of beads mixture in a 20 mm-long micro channel is also demonstrated. When HUVEC is employed as the target cell with the CD31 antibody, cross-flow distribution of HUVEC is also shifted in the cross-flow direction.

The authors thank Professors T. Ushida and K. Furukawa of the University of Tokyo. Photomasks are made using the University of Tokyo VLSI Design and Education Center (VDEC)'s 8-inch EB writer F5112+ VD01 donated by ADVANTEST Corporation.

REFERENCE

[1] W. A. Bonner, H. R. Hulett, R. G. Sweet, and L. A. Heizenberg, "Fluorescence activated cell sorting," Rev. Sci. Instrum, Vol. 43, pp. 404-409, 1972.

[2] A. Y. Fu, C. Spence, A. Scherer, F. H. Arnold and S. R. Quake, "A Microfabricated fluorescence-activated cell sorter," Nature Biotech., Vol. 17, pp. 1109-1111, 1999.

[3] A. Wolff, I. R. Perch-Nielsen, U. D. Larsen, P. Friis, G. Goranovic, C. R. Poulsen, J. P. Kutter and P. Telleman, "Integrating advanced functionality in a micro-fabricated high-throughput fluorescence-activated cell sorter," Lab Chip, Vol. 3, pp. 22-27, 2003.

[4] M. Berger, J. Castelino, R. Huang, M. Shah, and R. H. Austin, "Design of a microfabricated magnetic cell separator," Electrophresis, Vol. 22, pp. 3883-3892, 2001.

[5] W. H. Tan, Y. Suzuki, N. Kasagi, N. Shikazono, K. Furukawa and T. Ushida, "Lamination micro mixer for micro-immunomagnetic cell sorting," JSME Int. J. Ser. C, Vol. 48, pp. 425-435, 2005.

[6] K.-H. Han, and A. B. Frazier, "Paramagnetic capture mode magnetophoretic microseparator for high efficiency blood cell separations," Lab Chip, Vol. 6, pp. 265-273, 2006.

[7] J. Miwa, Y. Suzuki and N. Kasagi, "Adhesion-based cell sorter with antibody-coated amino-functionalized-parylene surface," J. Microelectro-mech. Syst., Vol. 17, pp. 611-621, 2008.

[8] H. Kim, and K. Najafi, "Characterization of low-temperature wafer bonding using thin-film parylene," J. Microelectromech. Syst., Vol. 14, pp. 1347-1355, 2005.

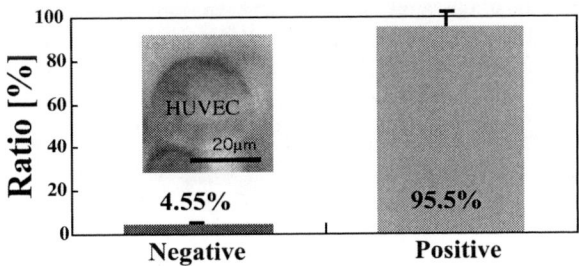

Figure 10: CD31 expression of HUVEC [5].

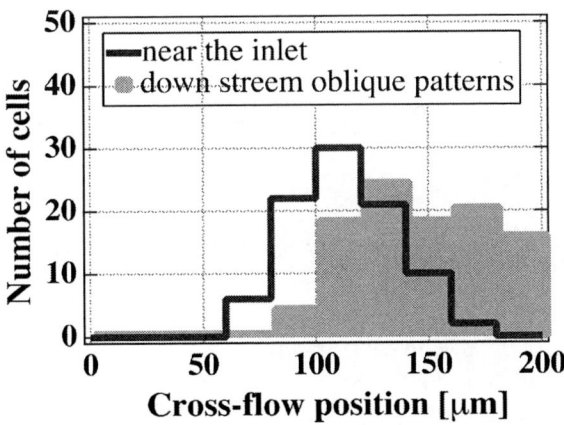

Figure 11: Histograms of cross-flow positions of HUVEC at u_b = 1.0 mm/s. (a) Near the inlet, (b) After passing through 10,000 oblique CD31 patterns.

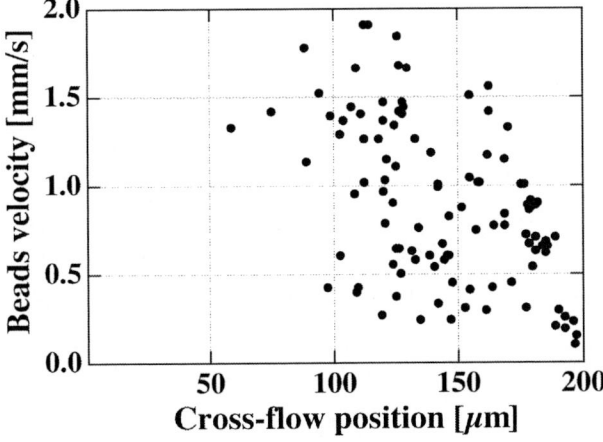

Figure 12: Traveling velocity of HUVEC versus the cross-flow position near the outlet.

FLOW-LYSOMETRY AND ITS BIOMEDICAL APPLICATION: CYTOSOLIC ANALYSIS OF SINGLE CELLS IN LARGE POPULATIONS

W.C. Lee[1,3], F.A. Kuypers[1], Y.-H. Cho[2], and A.P. Pisano[3]

[1]Red Cell Membrane Laboratory, Children's Hospital Oakland Research Institute, California, USA
[2]Digital Nanolocomotion Center, KAIST, Deajeon, Republic of Korea
[3]Berkeley Sensor & Actuator Center, UC Berkeley, California, USA

ABSTRACT

We present a novel single-cell analyzer, flow-lysometry, to measure cytosolic components in large cell populations. The present flow-lysometry performs and synchronizes three functions (cell-detection, cell-lysis, and component-sensing) in a continuous microfluidic channel, thus achieving high-throughput measurements (20 cells/min) of wide target components. In the experimental study, we measure the cytosolic component (Ca^{++} ion) from each single RBC and verify population analysis of single cells in mixed cell populations. Thus, this work shows that the present flow-lysometry is useful to characterize complex cell populations, which is required for various biomedical studies.

INTRODUCTION

Single-cell analysis has generated a wealth of information in cell population studies [1]. Using single-cell analysis, the characteristics of cells in a heterogeneous population can be analyzed without the loss of information that results from averaging the population as a whole. Until now, most single-cell analysis, such as flowcytometry and automated microscopy, has been focused on cell surface properties due to the lack of proper cytosolic analysis tools. Studies that report on single-cell analysis of cytosolic components can be categorized by the access methods to cytosolic components: biochemical [2] or physical [3,4]. In the biochemical access methods (Figure 1a), fluorescence markers penetrate through the plasma membrane into the cytosol or are expressed in the cytosol. Subsequently, using continuous fluidics, the automated microscope or flowcytometer successively measures the fluorescent light from individual cells as a measure of the target cytosolic component. However, only limited markers, such as fluorescent probes for calcium, can penetrate the membrane and report on the cytosol. For other components, the cell membrane needs to be made permeable in order to give access to compounds such as antibodies, while maintaining the cytosolic component to be measured inside the cell. In other approaches, the cell membranes are pierced (Figure 1b) or removed to release cytosol (Figure 1c) using MEMS devices that can manipulate single cells individually. In these methods, however, the cells are maintained in a stationary or quasi-stationary liquid to prevent diffusion of the cytosolic component. This in turn does not allow fast successive measurements of a significant number of cells, important to analyze cell populations. Thus, Many important cytosolic components are difficult to measure in single cells in a heterogeneous population. While these cellular compounds can be sensitively measured in a (lysed) cell population as a whole, the measurement of individual cells in a heterogeneous population is difficult.

The proposed flow-lysometer technique (Figure 1d) achieves high-throughput measurements using the continuous flow microchannel, while maintaining the target flexibility of the physical access methods. We propose to successively lyse single cells one by one, and sense the released cytosolic component in a continuous flow channel. Since the cell lysis in continuous flow causes unstable measurements, we also add a synchronizing cell detector. The cell-detector enables the synchronization of the cytosolic sensing signals with detected cell positions. Thus, the proposed flow-lysometry technique can analyze cytosolic components in single cells from heterogeneous cell populations.

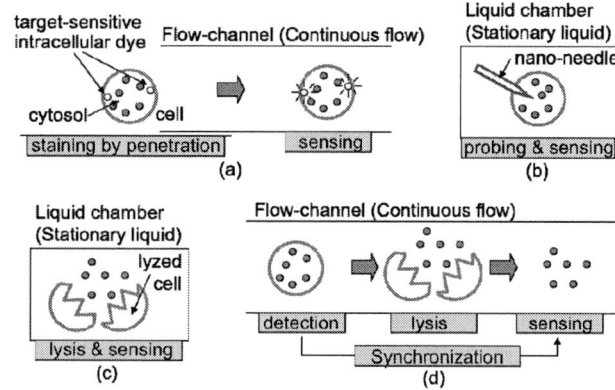

Figure 1: Single-cell cytosol analyzers: (a) the conventional method based on intracellular staining [2]; (b) the conventional method based on single-cell probing [3]; (c) the conventional method based on single-cell lysis [4]; (d) the proposed flow-lysometry.

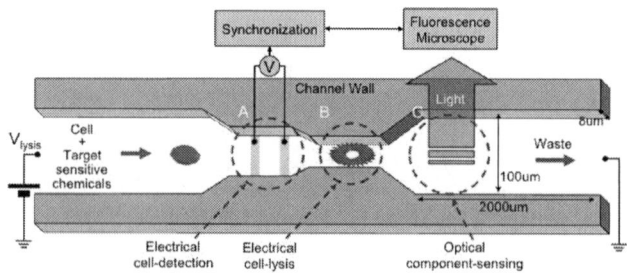

Figure 2: Top view of the proposed flow-lysometry device.

WORKING PRINCIPLES

The flow-lysometer (Fig.2) consists of three parts (cell-detection, cell-lysis, and component-sensing) along a single microfluidic channel. Initially, RBCs (red blood cells) and target-sensitive dyes are introduced with a syringe pump. The injected RBCs are detected between two electrodes (A in Fig.2) using the coulter counting principles [5]. Then, the detected RBCs are lysed by the focused electric field (B in Fig.2) [6] and cytosolic components are released. The released cytosolic components react with target-sensitive dyes and are measured by optical (fluorescence or chemiluminescence) methods (C in Fig.2). Finally, this optical signal is synchronized to the electrical pulses from the cell-detector, thus the flow-lysometer measuring the amount of cytosolic components (measured optical intensity) in each single-cell (each electrical pulse).

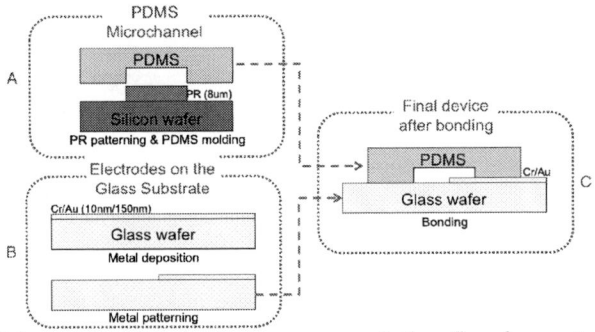

Figure 3: Fabrication process of the flow-lysometry device.

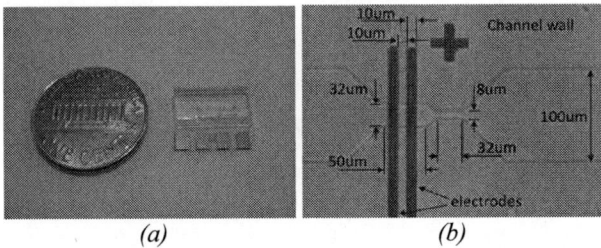

Figure 4: Fabricated device: (a) overall view; (b) an enlarged view of A, B, and C in Fig.2.

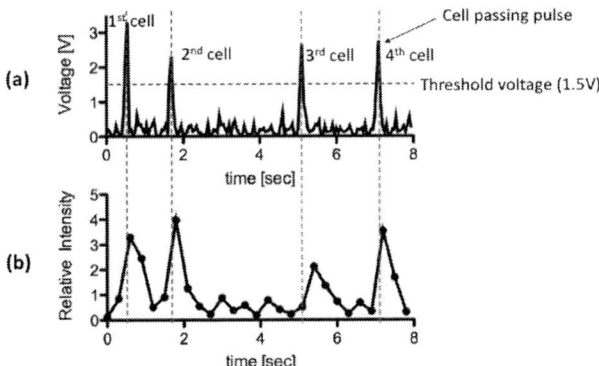

Figure 5: Synchronization between cell-detection and component-sensing signals: (a) measured electrical signals with cell passing; (b) relative optical intensity measured from Fig.7.

EXPERIMENTAL RESULTS

In order to test the fabricated device (Figs. 3 and 4), we measured Ca^{++} ion in single RBCs after Ca^{++} ion loading. First, three parts (cell-detection, cell-lysis, and component-sensing) were experimentally verified. 1) The cell-detector generated electrical pulses (Fig.5a) from the resistance change when a cell passes through the channel between two electrodes. 2) The electrical cell-lyzer successively broke red and white blood cells (Fig.6). We verified that cells passing through the channel did not lyse when no electrical field was applied (Fig.6a-I). Increasing the electric field from 0kV/cm (Fig.6a-I) to 1.2kV/cm (Fig.6a-II) led to a complete disruption of all cells in the orifice. Figure 6b also shows that lysis in an electrical field is not unique to red blood cells. 3) To verify the component sensing, we measured Ca^{++} ion in a single RBC. Ca^{++} ion and Fluo-4 (Ca^{++} ion sensitive fluorescence dye) were loaded to RBC samples and RBC samples were introduced to a microchannel where RBCs were lyzed. Figure 7 shows that the fluorescence light from the released cytosol is detected at 0.28sec and is reduced after 0.28sec due to the diffusion of the released cytosol.

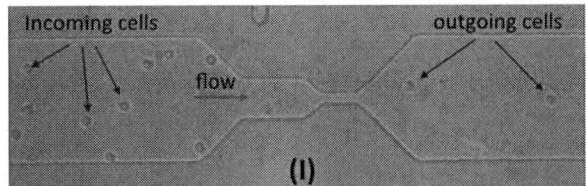

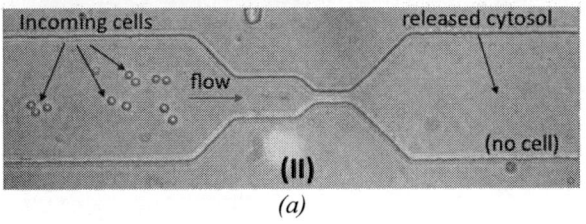

(a)

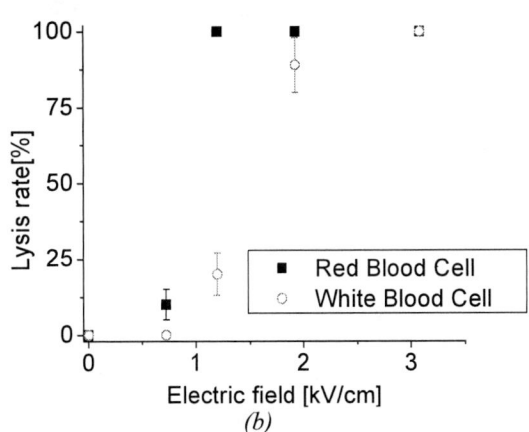

(b)

Figure 6: Movie clips and a graph for the cell-lysis characterization: (a-I) no RBC lysis in the absence of an electric field (0kV/cm); (a-II) RBC lysis in the electric field of 1.2kV/cm (42.6V applied). Note that cells lyse only when they enter the orifice; (b) Comparison of lysis of RBC and white blood cells (WBC) exposed to electrical fields of different strengths.

978-1-4244-2977-6/09 $25.00 © 2009 IEEE

Second, the amount of Ca^{++} ion in each single RBC was measured by the synchronization between electrical and optical signals. The fluorescence intensity signal in Fig.5b was obtained from fluorescence light in the sensing window of Fig.7. The optical intensity peaks (Fig5b) shows the synchronization to cell-passing peaks (Fig.5a) as shown in Fig.5 and Table 1. These synchronization means that the flow-lysometry measures the amount of cytosolic components (measured optical intensity) in each single-cell (each electrical pulse). Thus, the relative light intensity summarized in Table 1 also indicates the relative amount of Ca^{++} ion in each single RBC of Fig.5. Using the same procedure, we analyzed each of 105 cells from 5 min measurement (Fig.8), thus showing enough throughput (20 cells/min) for cell-population analysis.

Third, we verified the measurement capability of the single-cell cytosol in mixed cell populations. This capability is important for the flow-lysometry to expand its application to single-cell studies of complex cell populations. Figure 9 shows histograms of measured Ca^{++} ion from a mixture of normal and Ca^{++} loaded RBCs (1:1 and 1:3 ratios). The measured number of cells, which had low and high optical signals, were 46cells / 57cells and 27cells / 80cells, respectively, and these numbers coincided well with the original normal and Ca^{++} loaded cell ratios. This data indicates that normal and Ca^{++} loaded cells can be easily distinguished by the present flow-lysometry.

Table 1: Temporal properties and signal intensities of electrical and optical peaks. The relative optical intensities indicate the relative Ca^{++} amounts of each cell in Fig.5.

Cell	1st	2nd	3rd	4th
Time at the electrical peak	0.55s	1.70s	5.10s	7.10s
Time at the optical peak	0.60s	1.80s	5.45s	7.20s
Time leg between peaks	0.05s	0.10s	0.35s	0.10s
Relative optical intensity	3.29	3.98	2.12	3.54

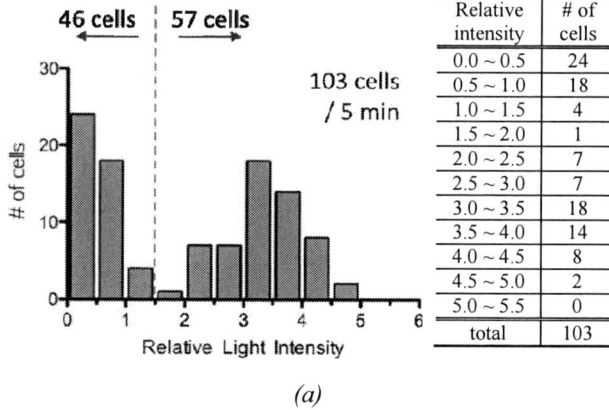

Relative intensity	# of cells
0.0 ~ 0.5	24
0.5 ~ 1.0	18
1.0 ~ 1.5	4
1.5 ~ 2.0	1
2.0 ~ 2.5	7
2.5 ~ 3.0	7
3.0 ~ 3.5	18
3.5 ~ 4.0	14
4.0 ~ 4.5	8
4.5 ~ 5.0	2
5.0 ~ 5.5	0
total	103

(a)

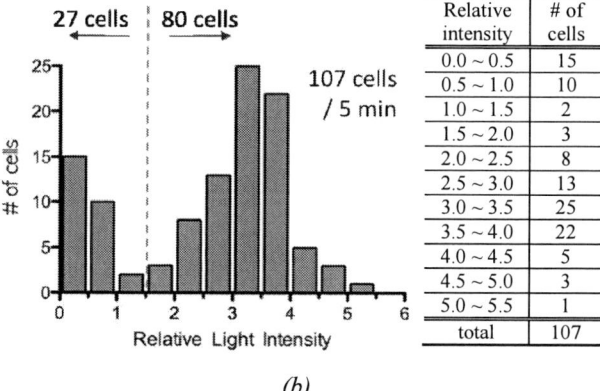

Relative intensity	# of cells
0.0 ~ 0.5	15
0.5 ~ 1.0	10
1.0 ~ 1.5	2
1.5 ~ 2.0	3
2.0 ~ 2.5	8
2.5 ~ 3.0	13
3.0 ~ 3.5	25
3.5 ~ 4.0	22
4.0 ~ 4.5	5
4.5 ~ 5.0	3
5.0 ~ 5.5	1
total	107

(b)

Figure 9: Histograms of the relative light intensity that indicates single-cell Ca^{++} in mixtures of normal and Ca^{++} loaded RBCs in the ratio of: (a) 1 to 1; (b) 1 to 3.

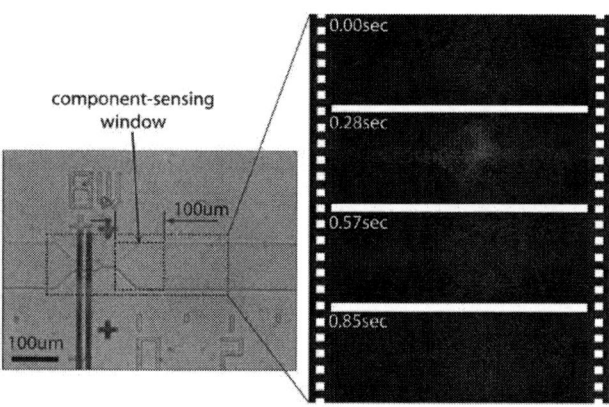

Figure 7: Fluorescence images of the component-sensing part: Ca^{++} ion released from cytosol is detected by Fluo-4 (fluorescence dye).

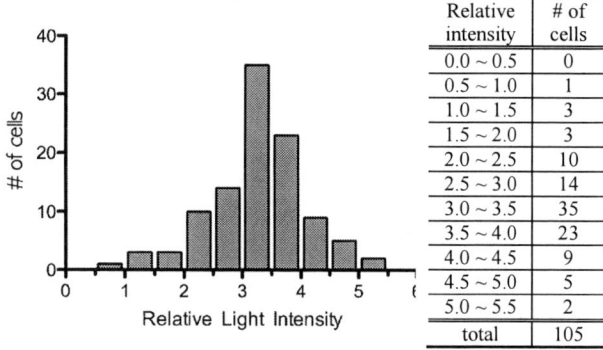

Relative intensity	# of cells
0.0 ~ 0.5	0
0.5 ~ 1.0	1
1.0 ~ 1.5	3
1.5 ~ 2.0	3
2.0 ~ 2.5	10
2.5 ~ 3.0	14
3.0 ~ 3.5	35
3.5 ~ 4.0	23
4.0 ~ 4.5	9
4.5 ~ 5.0	5
5.0 ~ 5.5	2
total	105

Figure 8: Histogram of the relative light intensity that indicates Ca^{++} amount in single RBCs.

CONCLUSIONS

In this paper, we presented and tested the flow-lysometry, the cytosolic analyzer of single-cells in large populations. First, we verified three fundamental functions (cell-detection, cell-lysis, and component-sensing) separately in the fabricated devices. Second, we synchronized these three functions in one device, thus experimentally showing the capability of single-cell analysis. Finally, we verified population analysis of single cells with mixed population of red blood cells. These experimental verifications showed the ability to the high-throughput measurements (20 cells/min) of wide target components. We envision that the flow-lysometry can be applied to various areas such as hematology, oncology, and immunology which require to characterize complex cell populations.

ACKNOWLEDGEMENTS

This work has been supported by the Exploratory /Developmental Grant (R21) of the National Institutes of Health (NIH), the National Creative Research Initiative Program of the Ministry of Science and Technology (MOST) and the Korea Science and Engineering Foundation (KOSEF), and the Korea Research Foundation Grant funded by the Korean Government (MOEHRD). (KRF-2007-357-D00025)

REFERENCES

[1] K. Cottingham, K., "The single-cell scene", *Analytical Chemistry*, vol. 76(13), pp. 235a-238a, 2004.

[2] M.G. Macey, *Flow Cytometry: Principles and Applications*, Humana Press, 2007.

[3] H. Andersson and A. Berg, "Microtechnologies and nanotechnologies for single-cell analysis", *Current opinion in biotechnology*, vol. 15, pp.44-49, 2004

[4] R.A. Yotter, *et. al.*, "Sensor technologies for monitoring metabolic activity in single cells - Part I: Optical, methods", *IEEE Sensors Journal*, vol. 4(4), pp.395-411, 2004.

[5] D.W. Lee, *et. al.*, "A flow-rate independent cell counter using a fixed control volume between double electrical sensing zones" *Digest Tech. Papers IEEE MEMS'05 Conference*, Jan. 30 - Feb. 3, 2005, pp.678- 681.

[6] D.W. Lee and Y.-H. Cho, "A continuous electrical cell lysis device using a low dc voltage for a cell transport and rupture", *Sensors and Actuators B-Chemical*, vol. 124(1), p. 84-89, 2007.

[7] F.A. Kuypers, *et. al.*, "Interaction of an Annexin V Homodimer (Diannexin) with Phosphatidylserine on Cell Surfaces and Consequent Antithrombotic Activity", *Thrombosis and Haemostasis*, vol. 97(101), pp. 478-486, 2007.

STRETCHABLE SUBSTRATES FOR THE MEASUREMENT OF INTRACELLULAR CALCIUM ION CONCENTRATION RESPONDING TO MECHANICAL STRESS

Y. J. Heo, E. Iwase, K. Matsumoto, and I. Shimoyama

The University of Tokyo, Japan

ABSTRACT

This paper presents the design, fabrication, and the characterization of a stretchable substrate, achieving the change of intracellular calcium ion concentration by mechanical stress. We propose the stretchable substrate integrated with the air chambers of the pneumatic actuator. We have measured the areal strain depending on input pressure and the intracellular calcium ion concentration increase in response to the mechanical stress. The present stretchable substrate is able to provide the areal strain of 5.21%~12.3% for the input pressure of 34.5kPa~103kPa. We also verified that the present stretchable substrate is able to stimulate cells by the mechanical stress applied through integrins, showing potential feasible application for monitoring the calcium ion influx caused by the mechanical stress.

INTRODUCTION

The researches for cell responses caused by mechanical stresses have been studied to understand age related disease and tissue degeneration [1]. The mechanical stress applied to the cells leads to intracellular change, and thereby influence to the control of many cell behaviors, such as migration, growth, differentiation, apoptosis and stem cell lineage switching. Many of the mechanical stresses are applied to cells through their extracellular matrix (ECM) adhesions. The mechanical stress entering via integrins, cell surface receptors with ECM, affects to not only the immediate cellular responses but also the later adaptive responses [2].

Intracellular calcium ion concentration change is the one of the immediate cellular responses due to the mechanical stress application to integrins. The human endothelial cells have the mechanosensitive ion channels [3]. The mechanosensitive calcium ion channels provide a mechanism for rapid transduction to an electrophysiological response induced by the applied force. Therefore, calcium ions function as an intracellular mediator responding to extracellular mechanical stress. Furthermore, the frequency of the calcium ion oscillation is translated into a frequency dependent cell response [4].

The previous devices for stimulating cells on collagen matrix using a transducer [1] and a needle [5] are not able to stimulate live cells due to the fixation process. However, stretchable substrates are able to supply the mechanical stress to cells in the live state. Therefore, the stretchable substrates have been applied to researches about both the rapid and slow cellular responses. The previous stretchable substrate using rotors [6] and frames [7] are not possible to provide periodic force to cells due to their force input principle. Their force input principle also has a limit to integrate to microchips. However, the present stretchable substrates using the pneumatic actuators provide the automatically driven periodic force and the feasibility of integrating to microchips. In this work, we coat the present

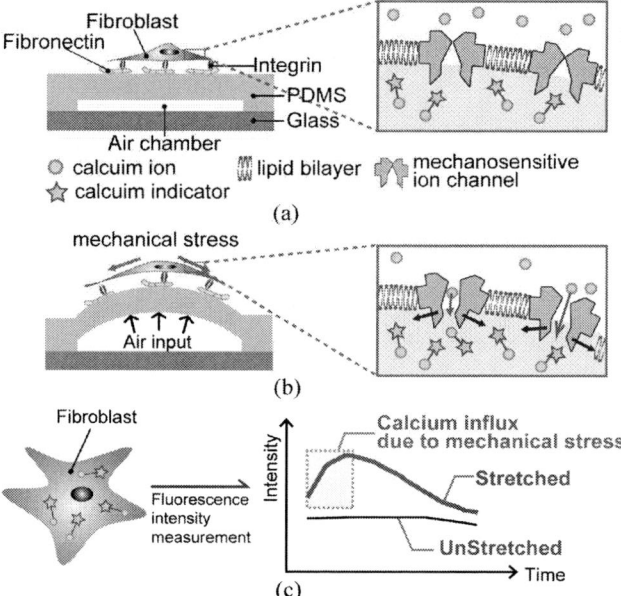

Figure 1: Principle of intracellular calcium ion concentration change responding to mechanical stress on the stretchable substrate: (a) illustrates a cell attached on the present device; (b) illustrates the stretched cell in (a) for input pressure; (c) illustrates the fluorescence intensity increase due to the mechanical stress.

stretchable substrate with fibronectin, an extracellular protein binding to integrins, and thereby measure the intracellular response modulated by the mechanical stress application to integrins. Thus, we are able to measure the intracellular calcium ion concentration, $[Ca^{2+}]_i$, applying the mechanical stress to the cells attached on the present stretchable substrates.

WORKING PRINCIPLE AND DESIGN

Figure 1a illustrates the stretchable substrate integrated with the air chamber of the pneumatic actuator. The present stretchable substrate consists of the patterned Polydimethylsiloxane (PDMS) and a glass, thereby providing transparent view. By using pneumatic actuators, the present devices are capable of providing mechanical stress only, without electrical, chemical, and thermal stresses. Figure 1 illustrates the working principle of the present stretchable substrate. When the pressure is inputted to the air chamber of the pneumatic actuator under the substrate, the substrate is stretched, which results in generating the mechanical stress to the cells attached on the substrates. The applied mechanical stress to the human endothelial cells causes the mechanosensitive ion channels of the cells to activate. Calcium ions enter into the cells through the activated ion channels, and then the intracellular calcium ion concentration increases.

978-1-4244-2977-6/09 $25.00 © 2009 IEEE

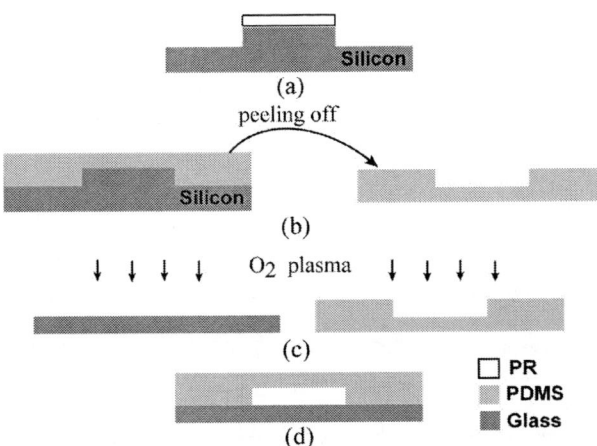

Figure 2: Microfabrication process: (a) PR patterning and silicon etching; (b) PDMS spin coating and peeling off; (c) treatment of O₂ plasma (d) increasing bonding strength at 120°C.

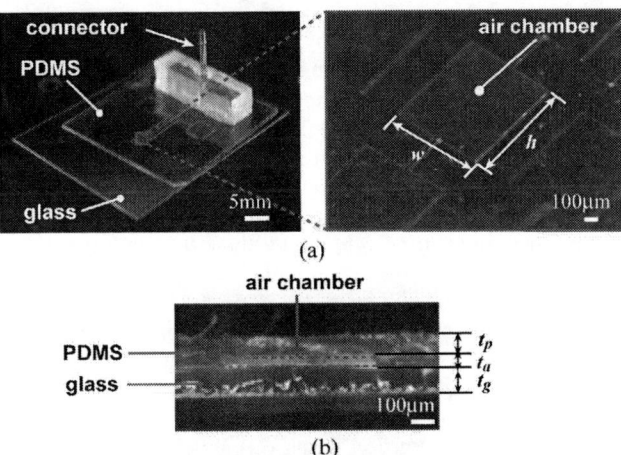

Figure 3: Fabricated device: (a) overall view and enlarged view of the air chambers; (b) cross sectional view along line AA' in (a).

If the cells are loaded with fluorescent calcium indicator, we can measure the intracellular calcium ion concentration. The fluorescence intensity increases in stretched cells due to the calcium ion influx, which does not in unstretched cells (Fig.1c).

FABRICATION PROCESS

Figure 2 illustrates the microfabrication process of the present stretchable substrate. We patterned photo resist (OFPR) on silicon wafer, then etched the silicon wafer using DRIE. We poured PDMS on the developed silicon mold. After curing process at 110°C for 10min, we peeled off the cured PDMS from the silicon mold. And then, PDMS mold and the glass were treated with oxygen plasma. The patterned PDMS substrate was carefully stacked on the glass. The stacked present device is cured again at 120°C for over 30 min for the irreversible bonding process [8]. In order to insert pressure to the air chambers, we attached a connector on the present device. We link the connector to a pressure source through a silicone tube. Figure 3 shows the photographs of the overall structure and the air chambers of the fabricated stretchable substrate. Table 1 lists the dimension of the present device.

Table 1: Dimension of the fabricated device in Fig.3

Dimension	Air chambers	Substrates
Width (w)	670.8±1.6μm	N/A
Height (h)	675.0±2.7μm	N/A
Thickness (t)	t_a = 41.9±3.0μm	t_p = 114±25μm t_g = 145±25μm

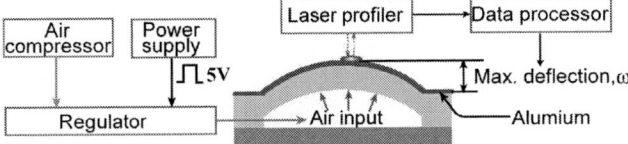

Figure 4: Experimental setup to measure the maximum deflections of the fabricated device for varying input pressure.

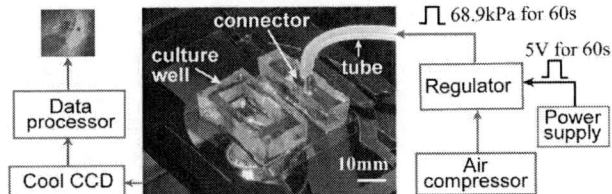

Figure 5: Experimental setup for the measurement of the intracellular calcium ion concentration responding to the mechanical stress.

EXPERIMENTAL RESULTS

From the fabricated devices, we have characterized the areal strain depending on the input pressure and the intracellular calcium ion concentration in response to the mechanical stress. The areal strain of the present device is estimated from the maximum deflection of the substrate for the input pressure. We have coated aluminum on the substrate for measuring the deflection of the substrate due to the input pressure using a laser profiler (Fig.4). We measured the maximum deflection of the top substrate for the input pressure of 34.5~138kPa.

The intracellular calcium ion concentration has been measured using human foreskin fibroblasts (ATCC CRL-2097). We attached PDMS culture well on the present device to fill culture medium. For coating fibronectin, the present device was wetted with Earle's balanced salt solution (EBSS; GIBCO). And then, we added human fibronectin derived from human donor plasma (CHEMION FC010) of 2~10μl/cm² into EBSS. After over 4 h at room temperature, fibronetin having the sites for cell binding [8] absorbed onto the fabricated substrate. The fibroblasts were seeded onto the present substrate coated with fibronectin. The fibroblasts were cultured in minimum essential medium (Eagle) with 2mM L-glutamine and bicarbonate, 0.1mM non-essential amino acids, and 1.0mM sodium pyruvate (ATCC) with 10% fetal bovine serum (GIBCO) and 1% antibiotics (GIBCO). After 2h from seeding the fibroblasts onto the substrate, the cells were incubated with 10μM Oregon Green BAPTA1-AM (OGB1-AM; Molecular probe). We stimulated the human foreskin fibroblasts by applying pressure to the air chambers under the present stretchable substrate. After the mechanical stress input, we obtained the fluorescence images using a fluorescent microscope (ZEISS, AxioVert200), as shown in Fig.5.

978-1-4244-2977-6/09 $25.00 © 2009 IEEE

Table 2: Maximum deflections and the areal strains in Portion A (Fig.7) for input pressures.

Input pressure, P_{in}	Max. deflection, ω	Areal strain, $\Delta A/A_i$
34.5kPa	56.6±1.2µm	5.21%
68.9kPa	82.5±3.9µm	8.30%
103kPa	119±4µm	12.3%
138kPa	152±1µm	16.1%

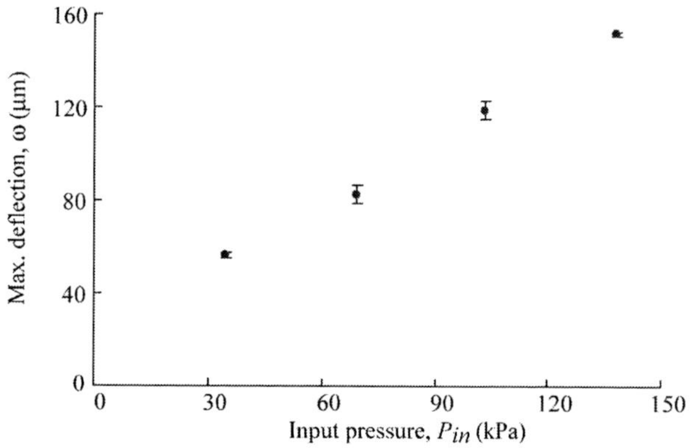

Figure 6: Maximum deflections of the present device for increasing input pressure.

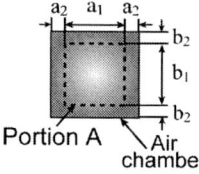

a_1	413.7µm
a_2	128.6µm
b_1	396.0µm
b_2	139.5µm

Figure 7: Stretched area on the air chamber, Portion A (left) and the dimension of Portion A (right).

Areal strain depending on the input pressure

We have performed 3 repeated measurement using 2 fabricated present devices. Figure 6 illustrates the maximum deflections of the present device for the input pressure of 34.5kPa~138kPa. From the results of the Fig.6, we calculated the changed area for the input pressure, A_f, of the fabricated device using FEM software (ANSYS). Then, we calculated the areal strain, $\Delta A/A_i$, using the following equation,

$$\Delta A/A_i = (A_f - A_i)/A_i \qquad (1)$$

where A_i indicates initial area.

The portion A of Fig.7 indicates the part of the substrate on the air chamber, whose areal strain is larger than 0. The fabricated substrate shows the areal strain of 5.21%~16.1% for the input pressure of 34.5kPa~138kPa in the Portion A (Table 2). Therefore, we experimentally verified that the present stretchable substrates have reasonable areal strain of the 5.12%~12.3% for the input pressure of 34.5~105kPa to demonstrate the cellular change in response to the mechanical stress [5].

Intracellular calcium ion concentration change in response to the mechanical stress

The input pressure of 68.9kPa was inserted to the air chambers under the stretchable substrate for 60s, which results in the areal strain of 8.30%. We obtained the fluorescence images of the intracellular calcium ion after stretching the substrates for about 5min (Fig.8).

The pseudoratio, $\Delta F/F$, is indicated by the following formula

$$\Delta F/F = (F - F_{base})/(F_{base} - B) \qquad (2)$$

where F is the measured fluorescence intensity of the calcium indicator, F_{base} is the fluorescence intensity of the calcium indicator in the cell before stimulation, and B is the background signal determined from the average of areas adjacent to the cells. In this work, we estimated the fluorescence intensity of calcium ion indicators from $\Delta F/F$.

Figure 8a illustrates the fluorescent images of the intracellular calcium ion in a whole cell after 0s from the force application. The bright portion in the cell is supposed to be endoplasmic reticulum, the calcium store in cells. We calculated $\Delta F/F$ of the cell using MATLAB. Then, we inspected $\Delta F/F$ change in the areas having the high fluorescence intensity as a function of time. Figure 8b illustrates the enlarged view of the portion enclosed with dotted line in Fig.8a. Figure 8c illustrates $\Delta F/F$ after 65.5s from the force application. The areas having fluorescence intensity of larger than 0.8 and 0.6 show 17.5% and 4.38% increases after 65.6s from the force application, compared to them after 0s from the force application (Fig.8b and c). Portion B and Portion C are defined as the areas, whose fluorescence intensities are larger than 0.8 and 0.6 in Fig.8b, respectively.

The fluorescence intensity, $\Delta F/F$, in the Portion B and Portion C of the stretched cell increases compared to an unstretched cell (Fig.9). The fluorescence intensity in the unstretched cell decreases due to the photobleaching of the fluorescent indicator [9]. The peak value of $\Delta F/F$ after the force application in Portion B is 1.6 times larger than that in Portion C (Fig.9). The fluorescence intensity pseudoratio, $\Delta F/F$, increase in the stretched cells, showing peak value after 70.1s and 61.1s from the force application in Portion B and Portion C, respectively. And then, the fluorescence intensity of the Portion B and Portion C decreases to the base line. Based on these results, we verified that the mechanical stress causes calcium ions to enter into the cells. The cells also responded to the second stress input, however, the influx of the calcium ion is less than the first response (Fig.10). Therefore, we verified the applied mechanical stress via integrins can activate mechanosensitive ion channels, which results in entering calcium ion into the fibroblasts.

CONCLUSIONS

In this paper, we presented the stretchable substrate integrated with the pneumatic actuators, designed to be able to demonstrate the intracellular calcium ion concentration applying the mechanical stress to integrins. In experimental study, we designed, fabricated, and tested the present stretchable substrate. From the measured maximum deflection, we calculated areal strain of the present stretchable substrate. The areal strain of the present device shows the reasonable values of 5.21%~12.3% to measure the cellular responses for the input pressure of 34.5kPa~103kPa.

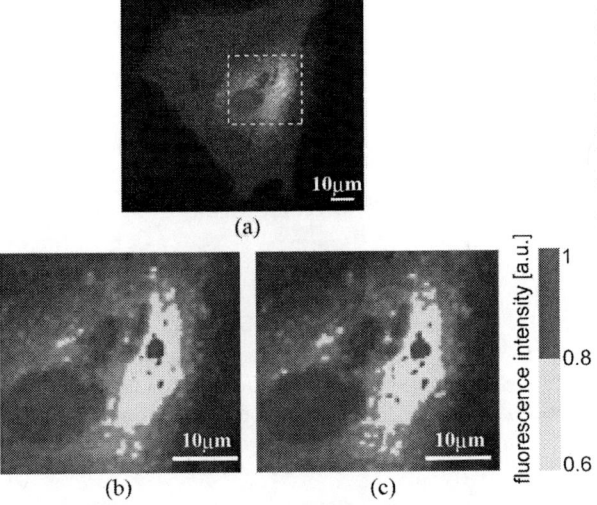

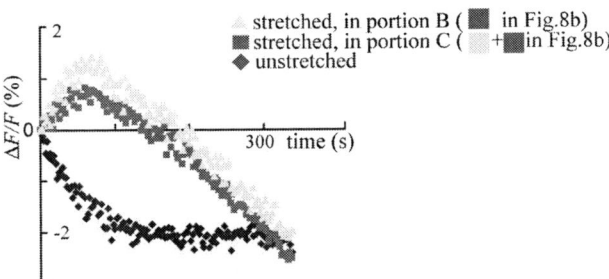

Figure 8: Fluorescence images of the cell: (a) whole cell image after 0s from the force application; (b) illustrates enlarged view of the cell portion enclosed with dotted line in (a); (c) illustrates the fluorescence image after 65.6s from the force application.

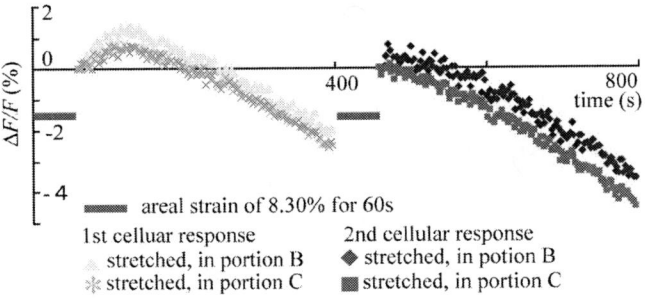

Figure 9: Fluorescence intensity of the stretched cell compared to the unstretched cell.

Figure 10: Fluorescence intensity in the stretched cell after first and second mechanical stress input.

We also verified that the intracellular calcium ion concentrations in the portions having the fluorescence intensity of larger than 0.8 and 0.6 increase responding to the mechanical stress, showing the peak value of after 70.1s and 61.1s from applying the areal strain of 8.30%. Therefore, we verified the applied mechanical stress via integrins can activate mechanosensitive ion channels, which results in entering calcium ion into the endothelial cells. On these experimental bases, we can conclude that the present stretchable substrate is able to stimulate cells by the mechanical stress applied through integrins, showing potential feasible application for monitoring the calcium ion influx caused by the mechanical stress.

ACKNOWLEDGEMENTS

The photolithography masks were made using the University of Tokyo VLSI Design and Education Center (VDEC)'s 8 inch EB writer F5112 + VD01 donated by ADVANTEST Corporation. This research is supported by Special Coordination Funds for Promoting Science and Technology, "IRT Foundation to Support Man and Aging Society."

REFERENCES

[1] R. A. Brown, et al., "Tensional homeostasis in dermal fibroblasts: Mechanical responses to mechanical loading in three-dimensional substrates," *J. Cellular Physiology*, Vol. 175, pp. 323-332, 1998.

[2] B. D. Matthews, et al., "Cellular adaptation to mechanical stress: role of integrins, Rho, cytoskeletal tension and mechanosensitive ion channels," *J. of Cell Science*, Vol. 119, pp. 508-518, 2006.

[3] S. Yano, et al., "Mechanical Stretching In Vitro Regulates Signal Transduction Pathways and Cellular Proliferation in Human Epidermal Keratinocytes," *J. Investigative Dermatology*, Vol. 122, pp. 783-790, 2004.

[4] B. Alberts, et al., *Molecular biology of the cell*, Garland Science, New York, 2008.

[5] W. M. Petroll, M. Vishwanath, and L. Ma, "Corneal Fibroblast Respond Rapidly to Changes in Local Mechanical Stress," *Investigative Ophthalmology & Visual Science*, vol. 45, pp. 3466-3474, 2004.

[6] M. chiquet, et al., "How do fibroblasts translate mechanical signals into changes in extracellular matrix production," *Matrix biology*, Vol. 22, pp. 73-80, 2003.

[7] P. F. Davies and S. C. Tripathi, "Mechanical Stress Mechanisms and the Cell : An Endothelial Paradigm," *Circulation Research*, Vol. 72, pp. 239-245, 1993.

[8] O. C. Jeong and S. Konishi, "All PDMS Pneumatic Microfinger with Bidirectional Motion and Its Application," *J. MEMS*, Vol. 15, pp.896-903, 2006.

[9] A. Takahashi, et al., "Measurement of Intracellular Calcium," *Physiological Reviews*, Vol. 79, No. 4, pp. 1089-1125, 1999.

MICRO CHEMICAL VAPOR DEPOSITION SYSTEM: DESIGN AND VERIFICATION

Q. Zhou and L. Lin

Berkeley Sensor and Actuator Center, Department of Mechanical Engineering, University of California, Berkeley

ABSTRACT

The conventional chemical vapor deposition system has been miniaturized to the micro scale, leading to several potential advantages for the synthesis of nanostructures. First, minute heat capacity leads to fast temperature stabilization. Second, tiny chamber volume helps for rapid gas species exchanges. Third, small Reynolds number ensures laminar flow for better control of deposition sources. Forth, small diffusion length near the chemical reaction surface enhances efficient gas mass transfer. As a demonstration of principle, high-quality single-walled carbon nanotubes (SWNTs) are synthesized while similar experimental parameters in a large scale system fail to construct good quality SWNTs.

INTRODUCTION

Chemical vapor deposition (CVD) has been widely used in microfabrication processes to produce thin films on substrates and has been an important method in the synthesis of various nanostructures [1]. The thermal decomposition or reaction of gaseous compounds is the key deposition sources in CVD. Atmosphere pressure CVD (APCVD) is performed at atmosphere pressure. As the mean free path for gas molecules is quite small (tens of nanometers), high flow rates are usually required to ensure efficient mass transfer of reactant gas. LPCVD (Low pressure CVD) system, on the other hand, typically operates at 30 to 250Pa with much lower flow rate due to the increased mean free path for gas molecules. CVD reactions can also take place with the assistance of plasma. Formation of the plasma could usually lower the reaction temperature, which is a primary advantage of plasma-enhanced CVD (PECVD) processes.

APCVD is easy to set up and does not require vacuum equipment, so it is widely used in the synthesis of nanostructures. The process is usually performed by forcing the reactant gas to flow through a heated tube in a furnace, as shown in Figure 1. After entering into the tube, the gas temperature inside the tube gradually approaches external furnace temperature. CVD reaction takes place either with or without the assistance of catalyst. For example, carbon nanotubes (CNTs) have been previously synthesized with the assistance of iron nanoparticles with methane as the carbon source. [2]

One may indentify several challenges, especially during the synthesis of nanostructures, for a conventional CVD system. First, large heat capacity means that both heating up and cooling down process will take a long time. Second, large volume also implies that it takes a long time to switch from one gas to another gas inside the reaction tube. Third, in APCVD, high flow rate is required to ensure efficient mass transport of reactant gases but the flow rate is also limited to laminar flow requirement for high quality thin film or nanostructures. High flow rate can cause turbulent flow, which has a non-stable flow field which would greatly affect the quality of synthesized materials [1, 5].

We propose and demonstrate a Micro-CVD system to address the aforementioned challenges in a large CVD system. By shrinking the conventional CVD system from the scale of meters to the scale of tens of micrometers, one can take the advantage of scaling effects to address issues described above. Theoretical analyses are conducted with experimental verifications to show the scaling effects in the CVD process.

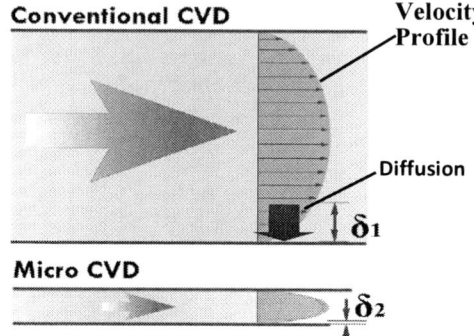

Figure 1: Comparing the conventional CVD system with the Micro-CVD system.

THEORETICAL MODELING

The proposed Micro-CVD system replaces the quartz tube in the conventional CVD system with microchannels as illustrated in Fig. 1. The gas is heated up to the CVD reaction temperature by heating up the microchannels via joule heating. In short, micro-CVD is like a miniaturized conventional CVD system with only tens of micrometers in length and several micrometers in diameter. Several possible advantages are analyzed below.

Quick temperature stabilization

Due to the large heat capacity of the conventional furnace system, it takes half an hour or more for the system to warm up and cool down and lots of unwanted reaction can happen. For example, the synthesis of ZnO (Zinc Oxide) nanowires prefers fast temperature rise and fall to avoid the formation of ZnO thin film or micro-crystals in low temperature region [3]. In order to analyze the time response of the system, we estimate the behavior of the system heat capacity C_h and the heat resistance R_h so that the time constant can be estimated by $\tau_h = R_h C_h$. The heat capacity of the system scales with volume or L^3, where L is the dimension of the system. The heat resistance, on the other hand, scales with L^{-1}. This is because the total heat flow

$$Q = k \oint_{system} \mathbf{dA} \cdot \nabla T \sim kL^2(\Delta T / L) \propto \Delta T \cdot L$$

where A is the area of heat flow, k is thermal conductivity, and T is temperature. ΔT represents the temperature difference between the heating zone and the environment.

The heat resistance is then $R_h = \Delta T/Q$ and scales with L^{-1}. Based on the previous analysis, the time constant of the system $\tau_h = R_h C_h$ should scale with L^2. Shrinking the conventional CVD system from several meters in dimension down to tens of micrometers will result in an approximately 10^{-8} reduction of response time. This means that the system temperature stabilization time will go down from tens of minutes to several microseconds and unwanted side reactions can be avoided by this super-fast temperature control.

Rapid gas species exchange

To control a CVD process, switching from one type of gas to another is often advantageous for the process flexibility. In the example of CNT synthesis process, carrier gases (such as argon) continue to flow during the furnace heating up process. After the furnace temperature is stabilized, the carrier gas is switched to reactant gas (such as methane). During the switching process, the gas concentrations continue to change as the concentration of the carrier gas drops and that of the reactant gas rises gradually. It is desirable to minimize this gas exchange period as it may affect the quality of the synthesis process with lower concentration of the reactant gases to starting reacting in the early stage. The time needed for the gas exchange period can be estimated by dividing the total volume of the reactant chamber by the flow rate of the gas. For a conventional furnace, this time can be on the order of one minute (The volume is 0.003 m^3, assuming a 5-feet long, 2-inch in diameter tube, and assume a flow rate of 3000 sccm). This gas exchange period is undesired because one minute could readily grow CNTs up to 10μm long in a typical process [4].

The micro-CVD system provides fast gas exchange because of the reduced reaction chamber volume. If one continues with the dimensional analysis, the volume of the reaction chamber scales with L^3. Assuming the same gas flow speed, the flow rate scales with L^2 due to the reduction of the cross sectional area. The gas exchange time should scale with L or about 10^{-4} reduction due to miniaturization for the proposed micro-CVD system. Therefore, the gas exchange process will be 10^4 quicker and only takes several milliseconds.

Small Reynolds number for Laminar flow

In a CVD process, laminar flow is often preferred. Reynolds number Re is often used to characterize different flow regimes. Laminar flow in a tube requires $Re<2000$ [5]. For a conventional CVD system, the Reynolds number is estimated to be

$$Re = \frac{\rho v_m D}{\mu} = \frac{\rho Q D}{\mu A} \approx \frac{1\text{Kg/m}^3 \times 3000\text{sccm} \times 0.05\text{m}}{1 \times 10^{-5}\text{Pa}\cdot\text{s} \times 0.002\text{m}^2} = 125$$

where ρ is the gas density, v_m is the mean gas speed, μ is the gas viscosity, A is the tube cross sectional area, and D is the tube diameter. Numbers are for methane (as it is used in the experimental section). We see that Re is already near the upper limit of laminar flow [6].

Reynolds number for the micro-CVD system, on the contrast, is as small as 0.01 assuming the same gas speed. With a small Re about 5 orders of magnitude away from the upper limit of laminar flow, we might increase the flow rate easily without worrying about turbulent flow.

Enhancement on reactant gas mass transfer

Mass transfer is one of the most important issues in a CVD system. Fresh reactant gas is brought onto the substrate surface to produce the desired deposit, and the by-products are carried away by the flow of gas. As such, efficient gas transfer is necessary to produce high quality products from the CVD process. The mass transfer mechanism in an APCVD system is usually diffusion [1]. To understand this, we notice that the gas flows in parallel layers (laminar flow condition), and that the velocity of the layer at the substrate surface where the deposition happens is actually zero (due to the fluid viscosity, see Fig. 1). As a result, reactant gas does not directly flow onto the substrate surface, but the gas diffuses from the flowing layers through the stagnant layer to the deposition surface. The diffusion length δ is therefore used to describe the distance over which the diffusion takes place. Short δ is desired because the gas can easily diffuse in and out. Obviously, δ is inversely proportional to the velocity gradient near the reaction surface such that a higher gas speed is preferred while no turbulence flow should be generated – a typical dilemma in the conventional CVD system.

The micro-CVD system greatly enhances mass transfer by increasing the gas velocity gradient near the reaction surface. For gas flow in a tube with circular cross section, the velocity profile is given by [7]:

$$v(r) = 2v_m \cdot \left[1 - \left(\frac{r}{r_0}\right)^2\right]$$

where r_0 is the radius of the tube. This gives a parabolic velocity profile. The velocity gradient near the surface of the tube is

$$\left.\frac{\partial v(r)}{\partial r}\right|_{r=r_0} = \frac{4v_m}{r_0}$$

It is observed that the velocity gradient scale with r_0^{-1}, so there would be a 10^4 increase in velocity gradient following out prototype example if the gas has the same mean velocity v_m. This large velocity gradient can ensure an efficient mass transfer.

Short gas warming up distance

One concern about the micro-CVD system is that whether the gas is fully heated up by flowing through such a short-distance in a microchannel. A simple heat transfer model is established to address this concern.

$$-v_m A \rho c_p \frac{dT(x)}{dx} + k_W A \frac{d^2 T(x)}{dx^2} + h(T_w - T(x))\pi D = 0$$

$$\Rightarrow \quad T(x) = T_w - (T_w - T_0)e^{-\frac{x}{s}}$$

where $s = \frac{2k_W}{v_m \rho c_p}\left(\sqrt{1 + \frac{16hk_W}{D(v_m \rho c_p)^2}} - 1\right)^{-1}$, c_p is the specific gas heat capacity, T is the gas temperature, T_w is the temperature of the tube, T_0 is the temperature of gas at the inlet, and h is the convection heat transfer coefficient. It is observed that s is a "temperature approaching length", that is, after the gas flows a distance of s, the temperature difference between the gas and the tube goes to $1/e$ of the previous value. Let's estimate the value s for the Micro-CVD system. For laminar flow in a tube, h can be estimated by dividing the thermal conductivity of a gas by the diameter of the tube [7]. This gives

978-1-4244-2977-6/09 $25.00 © 2009 IEEE

$$h \approx Nu \frac{k_W}{D} \approx 3.66 \times \frac{0.1W/m \cdot K}{5\mu m} = 7 \times 10^4 W/m^2 K$$

, so

$$s \approx 1\mu m$$

The "temperature approaching length" is so small such that within several micrometers, the gas temperature will reach that of the microchannels. This benefit comes from the great velocity gradient in the microchannels, which gives big convection heat transfer coefficient h (For ordinary forced convection, h only has a value between $50 \sim 250 W/m^2 K$). Figure 3 is the temperature profile from the inlet into the microchannel with the assumptions based on the prototype synthesis conditions. It is found that within several micrometers, the gas temperature reaches that of the sidewall at $900^\circ C$. This is an important advancement in temperature control as compared with the previous "Local synthesis" demonstration from our group where temperature control is rather difficult using microheaters in a cold reactant gas chamber [8].

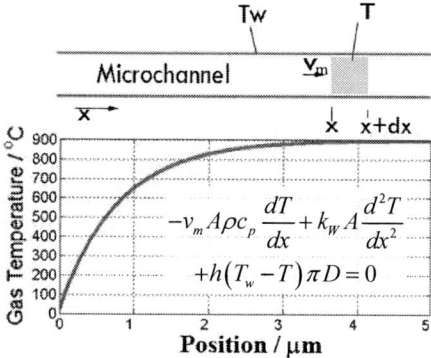

Figure 2: Analysis on the temperature profile of the gas within first several micrometers into the microchannel.

DEVICE DESIGN AND FABRICATION

Figure 3 illustrates the cross sectional view of the prototype Micro-CVD system consisting of a suspended heating stage with built-in vertical microchannels as the chemical reaction chambers. These microstructures are constructed on a SOI wafer. The prototype dimension of the microchannel is 5µm in diameter and 50µm in length. Reactant gas is introduced through the backside hole (100µm in diameter and 400µm in length) etched by a though-hole deep reactive-ion etching (DRIE) step. A scanning electron microscope (SEM) image of the device can be found in Fig. 4. The stage is supported by a pair of flexible beams anchored to the substrate. These beams also function as the electrical heaters to provide the heating power to the stage.

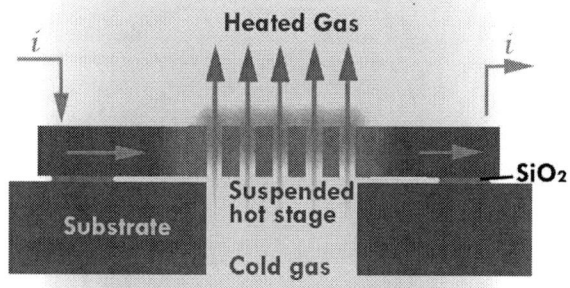

Figure 3: The concept and design of a Micro-CVD system.

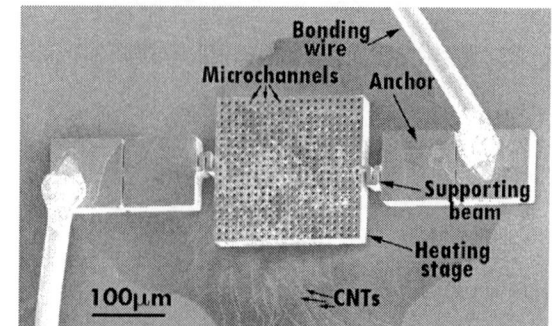

Figure 4: A SEM picture shows an overview of the device

Two pressure gauges are installed in the gas lines before and after the Micro-CVD chip to monitor the gas pressure drop along the microchannels. The flow speed of the reactant gas flowing through the microchannels can then be modulated by changing this pressure. Although the two pressure gauges are not mounted directly at the beginning and the end of the microchannels, the pressure drop mainly happens at the microchannels since it has the smallest diameter and thus the largest flow resistance. The flow speed v_m can be calculated by:

$$v_m = -\frac{r_0^2}{8\mu}\frac{dp}{dx} = \frac{r_0^2}{8\mu}\frac{\Delta p}{L_{mc}}$$

where p is the gas pressure, L_{mc} is the length of the microchannels. In order have a v_m on the order of centimeters per second (which is the typical velocity for conventional CVD system), Δp should be on the order of tens of pascals. The micro-CVD chip is then connected with electrical and gas interfaces as shown in Fig 5. By applying electric current through the beams, the stage can be heated up via joule heating.

Figure 5: Electrical and gas interfaces.

EXPERIMENTAL VERIFICATION

Figure 6 illustrates the process flow chosen for the verification of the concept of Micro-CVD by using the synthesis of single-walled carbon nanotubes as the demonstration example. A simple SOI process is conducted as stated previously to make the Micro-CVD platform with finished chips as illustrated in Fig. 3. A 0.5 µm-thick thermal oxide layer is grown as the barrier layer to prevent the iron nanoparticles (serving as the catalyst in CVD process) from diffusing into the silicon (Fig. 6a). A

978-1-4244-2977-6/09 $25.00 © 2009 IEEE   74

photolithography is performed and the oxide on the contact pads is etched away to allow electric contact (Fig. 6b). Iron nanoparticles are then deposited onto it as the catalysts following reference [2] (Fig. 6c). The CVD process is then performed at approximately 750°C (by observing the glowing color of the heating stage under microscope). The voltage applied is 16.65V and the current is 128mA in the prototype experiment. Under the gas inputs of hydrogen and ethylene (volume ratio 20:3), and the pressure drop measuring 10KPa across the microchannels, SWNTs are expected to grow from the iron nanoparticles (Fig. 6d).

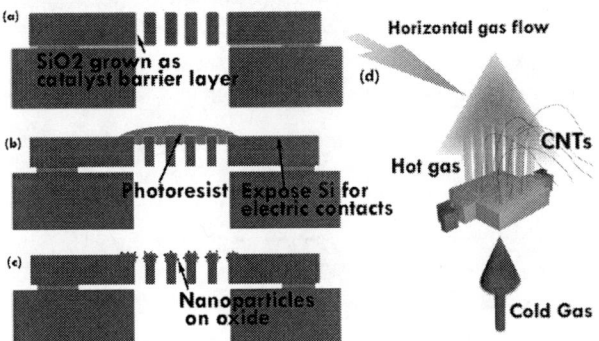

Figure 6: Process flow for single-walled CNT growth.

Figure 7a shows typical SWNTs synthesized by the Micro-CVD system. These SWNTs measure approximately 100 μm long and are grown in 5 minutes. It is suspected that these CNTs were floating in the gas environment during the growth process and later dropped onto the substrate as observed. It is also found that the CNT growth direction can be affected by a horizontal gas flow above the stage, which forces the CNTs to bend following the flow direction (Fig. 6d). The CNTs fall onto the substrate after the growth process, partially retaining this direction (Fig. 7b, indicated by arrows). These CNTs are verified by atomic force microscopy (AFM) measurements to be single-walled as their diameter is within 1~2 nm (Fig. 7c). They appear to be much thicker under SEM due to charge effects [9]. We believe that this is the first successful report to grow good quality single-walled CNTs using solution cast iron nanoparticles as catalysts, and hydrogen and ethylene as gas sources. A control experiment is performed using conventional thermal CVD with the same catalyst and gas under various temperatures between 600-850°C. Only short and curly CNTs are acquired as shown in Fig. 7d (growth temperature at 780°C). This is probably because amorphous carbon is formed on the iron nanoparticles due to the inefficient mass transfer of the conventional CVD system and the amorphous carbon blocks the CNT growth [10]. The good quality SWNTs acquired by the micro-CVD system is a direct verification of the superior operation of the Micro-CVD system as compared with the conventional one.

CONCLUSION

The concept of micro-CVD system is proposed and verified experimentally with distinctive advantages as compared with the conventional atmospheric pressure CVD system, including agile temperature control, fast gas species exchange, steady laminar gas flow and efficient

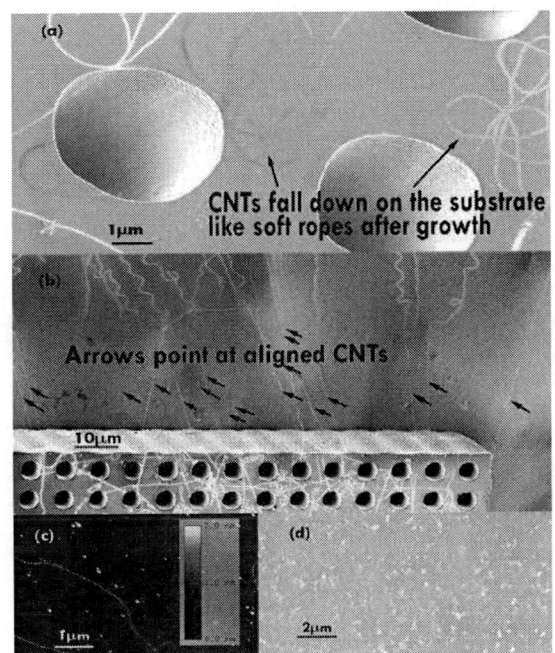

Figure 7: (a) CNTs grown by Micro-CVD. (b) Aligned growth. (c) AFM image. (d) CNTs by furnace growth.

gas mass transfer. Experimental results using a MEMS platform show the successful growth of good quality single-walled carbon nanotubes with approximately 100 μm in length and 1~2 nm in diameter. The control test using conventional furnace CVD system with the same setup was not able to produce high quality CNTs.

REFERENCES

[1] Chemical Vapor Deposition, edited by J. H. Park and T. S. Sudarshan, Materals Park, OH : ASM International, Ch. 2, 2001.

[2] H. C. Choi, S. Kundaria, D. Wang, A. Javey, Q. Wang, M. Rolandi, and H. Dai, "Efficient Formation of Iron Nanoparticle Catalysts on Silicon Oxide by Hydroxylamine for Carbon Nanotube Synthesis and Electronics", Nano Lett., Vol. 3, pp. 157-161 2003.

[3] S.H. Dalal, D. L. Baptista, et al. "Controllable Growth of Vertically Aligned Zinc Oxide Nanowires Using Vapour Deposition", Nanotechnology, Vol. 17, 4811-4818, 2006.

[4] W. Kim, H. C. Choi, et al. "Synthesis of Ultralong and High Percentage of Semiconducting Single-walled Carbon Nanotubes", Nano Lett., Vol. 2, No. 7, 703-708, 2002.

[5] J. P. Holman, Heat Transfer, McGraw-Hill, 2002, pp. 207.

[6] B. H. Hong, J. Y. Lee, T. Beetz, Y. Zhu, P. Kim, and K. S. Kim, "Quasi-Continuous Growth of Ultralong Carbon Nanotube Arrays", J. Am. Chem. Soc., Vol. 127, 15336-15337, 2005.

[7] F. P. Incropera and D. P. DeWitt, Fundamentals of Heat and Mass Transfer, 5th Edition, Wiley, pp. 472-487.

[8] O. Englander, D. Christensen, and L. Lin, "Local Synthesis of Silicon Nanowires and Carbon Nanotubes on Microbridges", Appl. Phys. Lett., Vol. 82, No. 26, pp. 4797-4799, 2003.

[9] Y. Homma, S. Suzuki, Y. Kobayashi, and M. Nagase, "Mechanism of Bright Selective Imaging of Single-walled Carbon Nanotubes on Insulators by Scanning Electron Microscopy", Appl. Phys. Lett., Vol. 84, 1750-1752, 2004.

[10] N. R. Franklin, Y. Li, R. J. Chen, A. Javey, and H. Dai, "Patterned Growth of Single-walled Carbon Nanotubes on Full 4-inch Wafers", Appl. Phys. Lett., Vol. 79, No. 21, 4571-4573, 2001.

978-1-4244-2977-6/09 $25.00 © 2009 IEEE

MEMS NANOREACTOR FOR ATOMIC-RESOLUTION MICROSCOPY OF NANOMATERIALS IN THEIR WORKING STATE

J.F. Creemer[1], S. Helveg[2], G.H. Hoveling[1], S. Ullmann[2], P.J. Kooyman[1], A.M. Molenbroek[2],
H.W. Zandbergen[1], P.M. Sarro[1]

[1]Delft University of Technology, Delft, NETHERLANDS
[2] Haldor Topsøe A/S, Kgs. Lyngby, DENMARK

ABSTRACT

We present a MEMS nanoreactor that for the first time enables transmission electron microscopy (TEM) of nanomaterials at the atomic scale during exposure to reactive gases at ambient pressure. This pressure exceeds that of existing in-situ TEM systems by a factor of one hundred. The reactor integrates a shallow flow channel (35 µm high) with a microheater and an array of robust electron-transparent windows of only 10 nm thickness. The Pt heater is embedded in a SiN_x membrane. The reactor is integrated into a dedicated TEM specimen holder. Its performance is demonstrated by the *live* formation of a nanostructured catalyst that is normally used for the production of methanol. The formation of Cu nanoparticles on the ZnO support crystals is imaged at 1.2 bar H_2 and up to 500 °C with very low thermal drift and with a spatial resolution of 0.18 nm.

INTRODUCTION

The interaction of nanostructured materials with gases is a subject that has been studied extensively in e.g. catalysis. The properties of nanomaterials are quite different from their bulk counterparts. They depend on the size, the shape, and the atomic arrangement of the material, but also on the working state: the pressure, the temperature, and the species of the gaseous environment [1].

A powerful and versatile tool for the imaging of nanostructured materials is transmission electron microscope (TEM) [2-7]. This instrument can combine high resolution with real-time imaging of both the atomic-scale structure and the dynamics.

However, the application of the TEM to study gas-solid interactions is extremely demanding because of the high vacuum conditions (10^{-6} mbar) it normally requires. Even slight amounts of gas molecules lead to scattering of the electron beam, resulting in a loss of resolution and contrast. In so-called environmental TEM (ETEM) gas is introduced deliberately around the specimen. Nevertheless, it is a key requirement to limit the gas pressure as well as the thickness of the layer through which the electron beam travels. This confinement of the gas can be obtained by two principles [2]: by differentially pumped vacuum systems [3-5] and by window cells [6,7].

Today, ETEM provides images with atomic lattice fringes below 0.2 nm at temperatures up to 900 °C [3-7]. However, the pressure is limited to about 10 mbar. This is a factor one hundred below atmospheric pressure, which is about the minimum pressure to which nanomaterials are exposed to in many applications. Information from images obtained at low-pressure conditions should therefore be extrapolated with great care.

Here we present a MEMS nanoreactor that for the first time enables electron microscopy with atomic-scale resolution at ambient pressure (1 bar) and up to 500 °C. It is based on the window cell principle but realized with silicon-based microsystems technology. Recently, we have shown the microscopic performance of this reactor [8]. In this paper we focus on the design and fabrication.

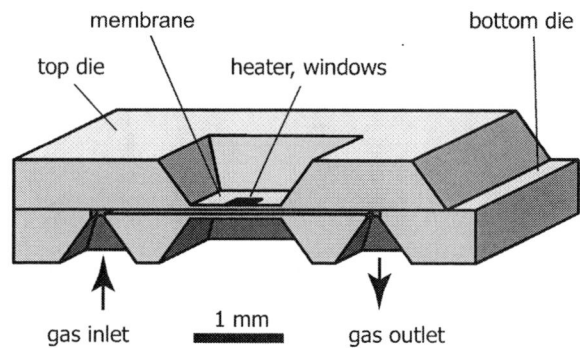

Figure 1: Schematic cross-section of the nanoreactor.

DESIGN

The major challenge in our reactor design is to combine high electron transparency with the robustness to withstand 1 bar of pressure difference, with heating up to 500 °C, and in the presence of reactive gases.

The reactor therefore consists of a shallow channel enclosed by two SiN_x membranes of 1x1 mm wide and 1.2 µm thick (Fig. 1). The channel height is set at a minimum of 4 µm by spacers. However, in the centre it can increase to about 40 µm by pressure-induced bulging [9]. In this case the gas density along the electron beam is limited to $2x10^3$ atoms/nm^2 at 1 bar and room temperature. In comparison, a state-of-the-art differentially pumped system has a gas layer thickness of 5 mm, corresponding to a density of $3x10^3$ atoms/nm^2 at only 10 mbar.

The good gas confinement comes at the expense, however, of a set of electron-transparent windows that in principle introduce additional electron scattering (Fig. 2). Our windows are 10 nm thick, adding $2x10^3$ atoms/nm^2 in the beam direction. In order to withstand a pressure difference inside the microscope, the width of the windows is limited to 9 µm. They are also made of SiN_x, as this material introduces little scattering by the low atomic number of its constituents [10]. In addition, it is amorphous. Therefore the remaining scattering doesn't create a periodic diffraction pattern, but adds only a small amount of noise to the microscopy images.

In the areas between the windows a Pt heater is embedded in the lower membrane incorporates in the form of a spiraled resistive wire (see Fig. 2) [11,12]. This resistor can heat up to 500 °C over a prolonged period of

978-1-4244-2977-6/09 $25.00 © 2009 IEEE

time. The heated area is 0.34x0.34 mm, limiting the dissipation to 30 mW in an evacuated reactor (10^{-6} mbar) and up to 0.1 W in the presence of a gas flow and 1 bar of pressure. During operation, the resistance of this wire is monitored by a four-point measurement to determine the actual temperature. The heater is driven and read out by custom-made instrumentation amplifiers. These amplifiers are controlled by a LabView program, which also stores the readings as a function of time.

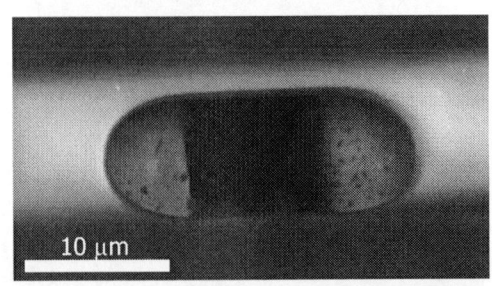

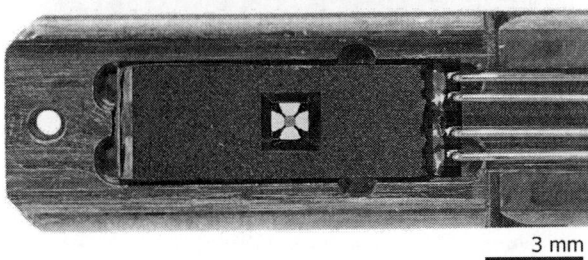

Figure 2: (a) Optical microscope image of the reactor membranes. The white spiral is the Pt heater. The ovaloids are the SiN_x electron-transparent windows. The circles are the spacers. (b) SEM image of a pair of superimposed electron-transparent windows of 10 nm thick, through which small catalytic particles are visible. (c) Nanoreactor in the tip of the dedicated TEM specimen holder, on the right hand side connected by four electrical spring contacts.

The reactor interface consists of a specially designed TEM specimen holder. This holder is a rod of approximately 25 cm that hosts the reactor in its tip (Fig. 2c). It provides four electrical spring contacts and gas in- and outlet tubing. The holder fits into regular TEMs of Philips/FEI Company, thereby avoiding the need for reconstruction of the base instrument. The holder has

been designed such that the reactor can be interchanged within a few minutes.

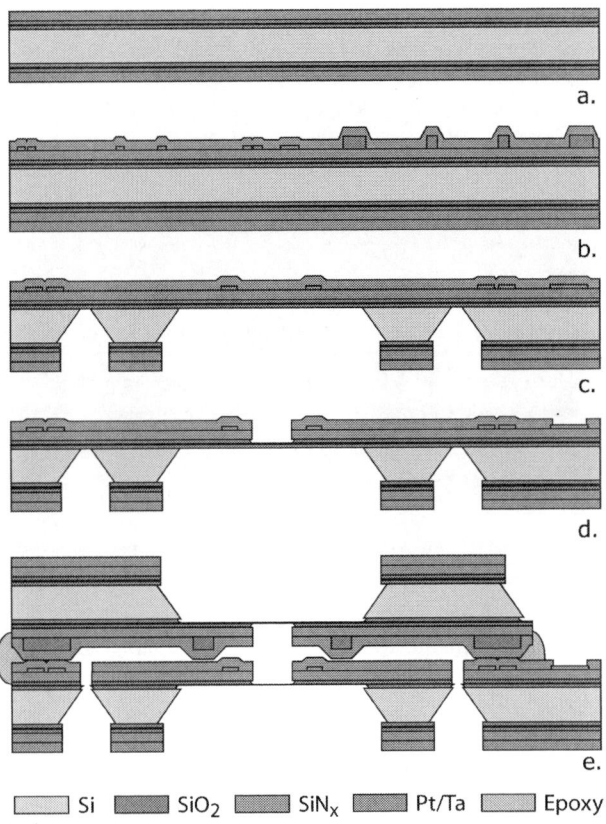

Figure 3: Schematic cross sections of the nanoreactor representing the fabrication sequence. (a) Layer stack forming the membrane and the windows. (b) Pt/Ta and LPCVD TEOS, patterned for the heater and the spacers. (c) KOH etch for membrane release and gas in- and outlet. (d) Opening of the contact pads and the windows. (e) Assembly, sealing and manual opening of the gas channel and the electrical connections.

FABRICATION

The reactors are fabricated by the sequence shown schematically in Fig. 3. The starting material consists of a Si (100) wafer of 100 mm in diameter and 525 µm thickness. The wafer is covered by 200 nm SiO_2 by wet thermal oxidation. Then, 15 nm of low-stress LPCVD SiN_x is deposited to form the electron-transparent windows. This nitride is covered by 200 nm of LPCVD TEOS oxide which later acts as an etch stop layer. This is followed by 500 nm of low-stress LPCVD SiN_x to form the membrane. Next, 185 nm of Pt is e-beam evaporated on top of a 15-nm thick Ta adhesion layer [11]. This metal stack is patterned into the shape of the heater by lift-off. The spacers are formed in a layer of 1 µm thick LPCVD TEOS. Spacers and heater are covered by another 500 nm of low-stress LPCVD SiN_x to prevent exposure to the gases and liquids during reactor operation. The back side of the wafers is then patterned by lithography and plasma etching, and etched in KOH to define the membrane and the gas in- and outlet. The electrical contact pads are opened by fluorine plasma. The electron transparent windows are released by a combination of dry and wet etching. Finally, the wafers are diced into devices of 10x3.33 mm. The top and bottom halves of the reactors are assembled and sealed with epoxy. Punching the covering membranes with a needle opens the gas in- and outlet and the electrical connections. The temperature dependence of the resistance is characterized on a hotplate in the range from 0 to 200 °C. For these experiments the dependence is assumed to be constant at higher values.

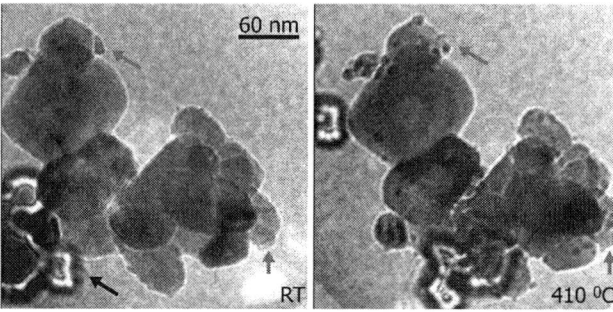

Figure 4: Cu nanocrystal growth: at higher temperatures the ZnO support particles become speckled with black dots of Cu that form the active sites of the catalyst. The blue arrows indicate some areas that have changed. The black arrow indicates out-of-focus particles on the opposite window.

EXPERIMENTS

The strength of the membrane and the windows was tested in a special chamber that can create pressure differences up to 2.7 bar across a reactor half. The large majority of the devices withstood those pressures, at both room temperature as well as during heating. The chamber was also used to verify the leak-tightness of the windows, of which the dimensions are too close to the resolution limit of light microscopy to inspect them optically.

The performance of the entire nanoreactor was demonstrated by the *live*, time-resolved monitoring of the formation of a nanostructured catalyst that is normally used in fuel cells and in the large-scale production of methanol. The catalyst precursor consists of CuO dispersed on ZnO nanocrystallites that act as support particles (Fig. 4a).

The experiment was performed in a Philips/FEI CM-300ST FEG TEM equipped with a differentially pumped environmental cell. This microscope was used for its proven gas supply system, which was now connected to the nanoreactor specimen holder [5]. The environmental cell was left unused. TEM images were recorded with a Tietz FastScan-F114 CCD camera. Electron energy loss spectra (EELS) were acquired with a Gatan Image Filter (GIF-2000). The pressure on the reactor in- and outlet was monitored by an MKS 910 DualTrans piezo transducer.

The catalyst precursor was loaded into the reactor, inserted into the microscope, and exposed to H_2 with pressures up to 1.2 bar. Images and EEL spectra were recorded from this starting situation. Then, the precursor was heated up to 500 °C within 68.5 s in eight quasi-instantaneous steps. Meanwhile, a time-lapsed image series was recorded with a frame rate of 6.9 frames/s.

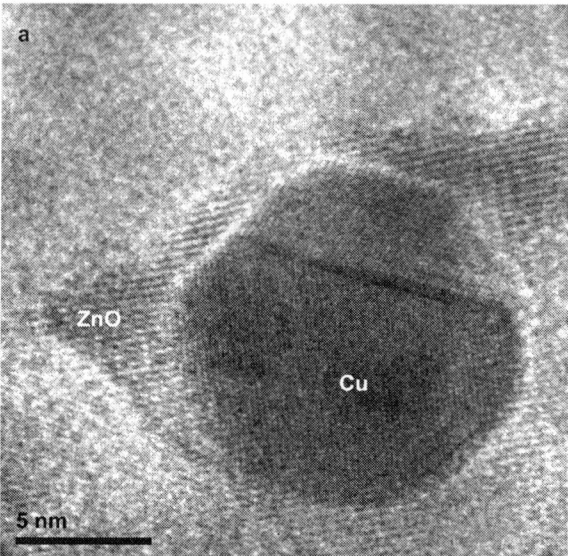

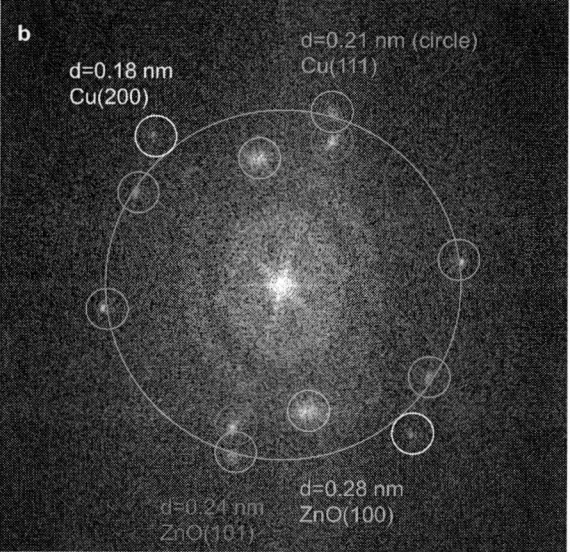

Figure 5: (a) High-resolution TEM image of a catalytic Cu nanoparticle on a ZnO support during exposure to 1.2 bar H_2 at 500 °C. (b) Fourier transform of the previous image. The bright spots represent the atomic lattice fringes. The smallest observable spacing is 0.18 nm, corresponding to the Cu (2 0 0) lattice planes. Adapted from [8].

RESULTS

For pressures above 0.4 bar, the EEL spectra show a distinct hydrogen K-edge peak at 13 eV. It confirms the presence of gas in the reactor and indicates the amount of gas molecules along the beam path. The peak height increases roughly quadratically with pressure. This is probably due to the combined effects of increasing gas density and the bulging of the membranes. By defocusing, the maximum gas channel height at 1.2 bar could be determined at 35 μm. This is in good agreement with the plate theory calculations. The channel height does not change noticeably during heating.

When the sample temperature exceeded 200 °C, it could be clearly observed in real time how the CuO was broken up and small Cu nanoparticles were formed all over the ZnO support (Fig. 4). At higher magnifications the nanoparticles show lattice fringes of the Cu (200) crystal planes with a spacing of only 0.18 nm (Fig. 5). This is close to the resolution of 0.14 nm of the microscope in vacuum. It is also small enough to be able to assess the important class of transition metal catalysts.

By autocorrelation of the image series it was found that the thermal drift of the system is very low compared to conventional, non-MEMS specimen holders with heating capability. Two drift effects are observed. The first one has a magnitude of 0.40 nm/°C with a stabilization time of less than an image frame (145 ms). This effect could be attributed to the expansion of the MEMS heater, which has a very small thermal mass. The other effect has an initial drift rate of 3.3 nm/s at 500 °C and decays exponentially with a time constant of 137 s. This could be attributed to the thermal expansion of the specimen holder. This holder has a much larger thermal mass than the MEMS heater. In the same time it evacuates the majority of the heat produced by the heater, since natural convection is absent in the vacuum of the microscope and because below 500 °C radiation losses are much lower that thermal conduction. The evacuation necessarily leads to a temperature gradient throughout the holder, thereby causing slow thermal expansion (drift).

Overall, the drift is so small that during the fast heating sequence the specimen in Fig. 3 could be easily kept in view although the viewing area was only 314x314 nm. The difference with conventional heating holders can be attributed to the miniaturization of the heater, which consumes about fifty times less power at 500 °C.

CONCLUSIONS

Our MEMS nanoreactor for *in situ* TEM enables atomic-scale imaging of nanomaterials during exposure to reactive gases at 1.2 bar, up to 500 °C, and with very modest thermal drift. We therefore believe that this reactor will particularly enable important new insights into the field of heterogeneous catalysis. Moreover, it is applicable to many other functional nanomaterials in their working state. It will be a valuable tool in a variety of areas such as crystal growth, nanofabrication, materials science and biology.

ACKNOWLEDGEMENTS

This work is supported by STW, applied science foundation of NWO and the Ministry of Economic Affairs with financial contributions from FEI Company, Haldor Topsøe A/S, and the European ESTEEM project. We thank the DIMES Technology Centre, the Nanofacility, IMT Neuchâtel and D. Briand for support in the cleanroom fabrication, F. Tichelaar, and U. Ziese for TEM characterization. FEI Company, and especially M. Stekelenburg, we thank for valuable technical input. Haldor Topsøe A/S acknowledges CTCI Foundation, Taiwan for participation in establishment of its ETEM facility.

REFERENCES

[1] N.I. Jaeger, "Bridging gaps and opening windows", *Science* 293 (2001) 1601.

[2] E.P. Butler, K.F. Hale, "Dynamic experiments in the electron microscope", in *Practical Methods in Electron Microscopy,* vol. 9, North Holland, Amsterdam, 1981.

[3] E.D. Boyes, P.L. Gai, "Environmental high resolution electron microscopy and applications to chemical science", *Ultramicroscopy* 67 (1997) 219.

[4] R. Sharma, P.A. Crozier, "Environmental Transmission Electron Microscopy in Nanotechnology", in *Microscopy in Nanotechnology*, Kluwer Academic, New York, 2005.

[5] P.L. Hansen, S. Helveg, A.K. Datye, "Atomic-scale imaging of supported metal nanocluster catalysts in the working state", *Adv. Catal.* 50 (2006) 77.

[6] G.M. Parkinson, "High resolution, in-situ controlled atmosphere transmission electron microscopy (CATEM) of heterogeneous catalysts", *Catal. Lett.* 2 (1989) 303.

[7] S. Giorgio, S. Sao Joao, S. Nitsche, D. Chaudanson, G. Sitja, C.R. Henry, "Environmental electron microscopy (ETEM) for catalysts with a closed E-cell with carbon windows", *Ultramicroscopy* 106 (2006) 503.

[8] J.F. Creemer, S. Helveg, G.H. Hoveling, S. Ullmann, A.M. Molenbroek, P.M. Sarro, and H.W. Zandbergen, "Atomic-scale electron microscopy at ambient pressure", *Ultramicroscopy* 108 (2008) 993.

[9] J. Yang, O. Paul, "Fracture properties of LPCVD silicon nitride thin films from the load-deflection of long membranes", *Sens. Actuators A* 97-98 (2002) 520.

[10] T. Doll, M. Hochberg, D. Barsic, A. Scherer, "Micro-machined electron transparent alumina vacuum windows", *Sens. Actuators A* 87 (2000) 52.

[11] D. Briand, A. Krauss, B. van der Schoot, U. Weimar, N. Barsan, W. Göpel, N.F. de Rooij, "Design and fabrication of high-temperature micro-hotplates for drop-coated gas sensors", *Sens. Actuators B* 68 (2000) 223.

[12] R.M. Tiggelaar, P. van Male, J.W. Berenschot, J.G.E. Gardeniers, R.E. Oosterbroek, M.H.J.M. de Croon, J.C. Schouten, A. van den Berg, M.C. Elwenspoek, "Fabrication of a high-temperature microreactor with integrated heater and sensor patterns on an ultrathin silicon membrane", *Sens. Actuators A* 119 (2005) 196.

NOVEL MEMS APPARATUS FOR IN SITU THERMO-MECHANICAL TENSILE TESTING OF MATERIALS AT THE MICRO- AND NANO-SCALE

J. Han[1], M.D.Uchic[2] and T. Saif[1]

[1]University of Illinois at Urbana-Champaign, Illinois, USA

[2]Air Force Research Laboratory, Materials & Manufacturing Directorate, Wright-Patterson AFB, Ohio, USA

ABSTRACT

We present, for the first time, a MEMS-based test methodology that potentially enables elevated-temperature mechanical tensile testing of nano- and micro-scale samples within a SEM or TEM (T > 500°C). Importantly, the test methodology allows for the samples to be fabricated separately from the MEMS-apparatus, a significant advancement from other test devices developed by some of the present authors [1]. Therefore the test methodology should be applicable to the study of a wide range of materials. Other advancements found in the methodology include a co-fabricated force calibration device, and a built-in thermocouple sensor to measure the stage temperature close to the sample.

INTRODUCTION

At the micro- and nano-scale, materials behavior can strongly depend on the scale of both internal microstructural features as well as the external dimensions of the structure in question. Microstructural length scales may include features such the grain size, the distance between impurities [2], and the layer thickness in a multilayered material, to name a few. With regards to the sample geometry, decreasing size results in an increase in the surface-to-volume ratio of the material. These surfaces and interfaces can affect the active deformation mechanisms and their dynamics, which may introduce new mechanisms of deformation. These surface- dominated properties may also be time and temperature dependent. As a result, bulk material properties cannot be extrapolated to the diminutive size-scales that are of interest to MEMS and NEMS, and thus the mechanical properties of materials with micro- or nano-scale dimensions need to be directly measured. A major challenge lies in measuring such properties, and also in understanding the mechanisms that control size-dependent mechanical properties.

In an earlier study, some of the present authors developed a MEMS stage that allowed for uniaxial tensile testing of ultra-thin free standing films (30 nm thick and larger) that was capable of operation within either a SEM or TEM [3]. This previously-developed MEMS-stage enabled the measurement of the stress-strain response of the films while also allowing for direct visualization of the films during testing. The force on the sample was determined by image-based measurement of the deflection of a force-sensing beam, while the sample deformation was determined by image-based measurement of the change of position of two gauges attached to the ends of the sample. A unique feature of the stage is that any misalignment error in loading is reduced by five orders of magnitude through the use of alignment flexures. This ensures precise uniaxial loading of the sample. The stage was employed to test aluminum and gold film samples with grain sizes ranging from 10-300 nm, and film thickness ranging from 30-300 nm. The experiments revealed unusual properties in nano-grained metal films, and helped uncover new fundamental mechanisms of deformation that occur in nanoscale microstructures; for example, the recovery of plastic strain with time [1], and non-linear elasticity in nanocrystalline metals [3].

However, the stage described above has some limitations because the sample and the stage are co-fabricated. First, this limits the choice of materials to be tested to those that can be deposited or grown on silicon and patterned. Also, the sample thickness is constrained to the typical range of thin films. Furthermore, each stage can only be used once. The new MEMS apparatus presented in this paper overcomes these limitations.

THE NEW TESTING APPARATUS

The test methodology described herein allows for testing of samples that are fabricated separately from the MEMS stage. The test apparatus consists of a MEMS tensile stage and a heating stage. Figures 1a and 1b show the Si MEMS tensile stage without a sample. The stage contains two grips with etched or focused ion beam (FIB)-fabricated wedges to hold a dog-bone shaped sample that is manufactured separately. One of the wedge grips is attached to the force sensing beam, while the other is attached to a set of beams and a U-beam that ensure alignment and uniaxial loading of the sample. The tensile sample is placed into the wedge grips using a micro-manipulator. The large etched holes on the outer ends of the MEMS-stage connect the stage to a macroscopic piezo-electric actuated test frame using rigid pins. In order to apply uniaxial tension to the sample, one end of the stage is held fixed while the other end is moved by the piezoelectric actuator. Force on the sample is measured from the deflection of the force sensing beam (Figure 2). There are two gauges in the stage. One provides the displacement of the force sensing beam, the other provides the relative displacement between the two wedge grips.

Procedure for calibration of spring constant of the force sensing beam

978-1-4244-2977-6/09 $25.00 © 2009 IEEE

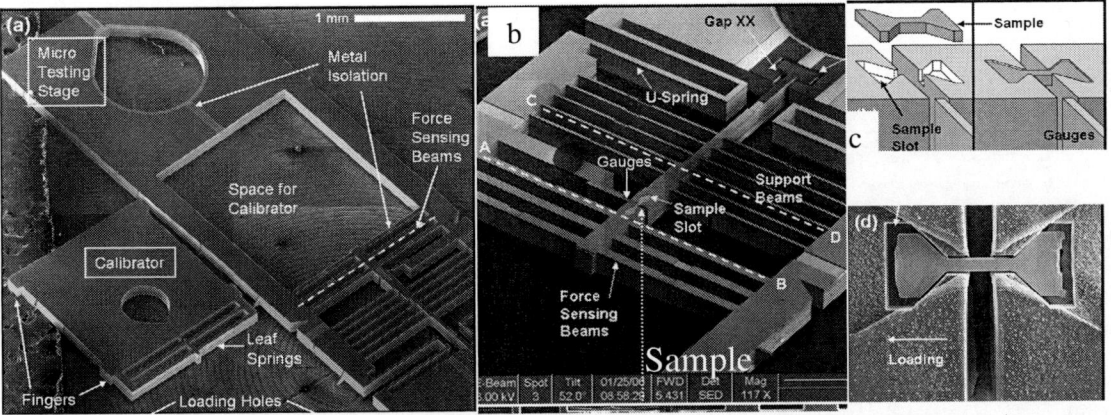

Figure 1. (a) SEM image of the MEMS tensile stage and the calibration device. The spring constant of the leaf spring of the calibrator is measured by a nano indenter. (b) Close-up image of the stage. (c) Schematic of sample loading. (d) Top view of the loading stage and the nickel-base superalloy sample fabricated by FIB.

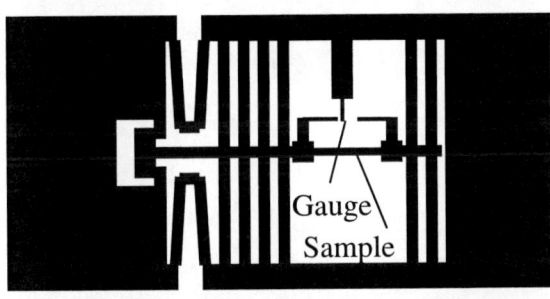 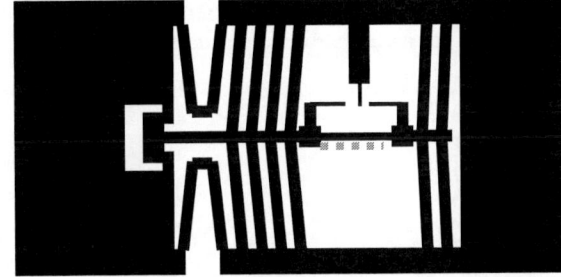

Figure 2. An unloaded (left) and a loaded (right) sample. The dotted line on the right diagram shows the original length of the sample. The force on the sample and the strain are measured from the change in the gauge positions.

The spring constant of the force sensing beam is calibrated using a co-fabricated calibration device (Figure 1a). The calibrator consists of a leaf spring. Its spring constant can be readily determined using a commercial nanoindentation system. Once the spring constant of the calibrator is determined, the calibrator is placed in the designated space of the MEMS stage where it can be placed in direct contact with the force sensing beam. Next, the calibrator is pushed by a micromanipulator to displace the force sensing beam. The spring constant of the force sensing beam is simply determined from the force imparted by the calibrator divided by the relative deflection of the force sensing beam.

Heating stage

In order to provide a capability to measure material properties at elevated temperatures, a micro-heater has been designed and tested that works with the MEMS testing stage (Figure 3). The micro-heater consists of a Si substrate with a patterned Cr micro-coil for Joule heating. For an elevated temperature test, the MEMS-stage is placed on top of the heating stage, and both the stage and sample are heated primarily by conduction. The temperature of the MEMS stage is measured by a built-in thermocouple. It consists of a silicon cantilever coated on one side by Cr. The mismatch between the thermal expansion coefficients of Si and Cr results in bending of the cantilever with increasing temperature, and the temperature is calculated from the relative deflection of the cantilever tip. The thermocouple is calibrated by measuring the tip deflection when the stage is heated to known temperatures. For calibration experiments, the stage temperature can be easily varied by using a hot plate or a convection oven. In a separate experiment, we have qualitatively verified the accuracy of the cantilever-based temperature measurement by melting materials with known melting temperatures. These materials include solder wire and aluminum, and the results from these experiments are shown in Figure 3. In order to further push the temperature capabilities of this type of micro-heater device, a silicon carbide substrate was substituted for the Si substrate. Here silicon carbide also served as the resistor for Joule heating. This heater was used to melt glass (Soda lime glass, softening temperature: 720C, Gold Seal Microslide) (placed on the heater), indicating that temperature can exceed 700°C.

Fabrication

The stage is fabricated from single crystal silicon by deep reactive ion etching. The process flow is given in Figure 4 with the details listed in the figure caption.

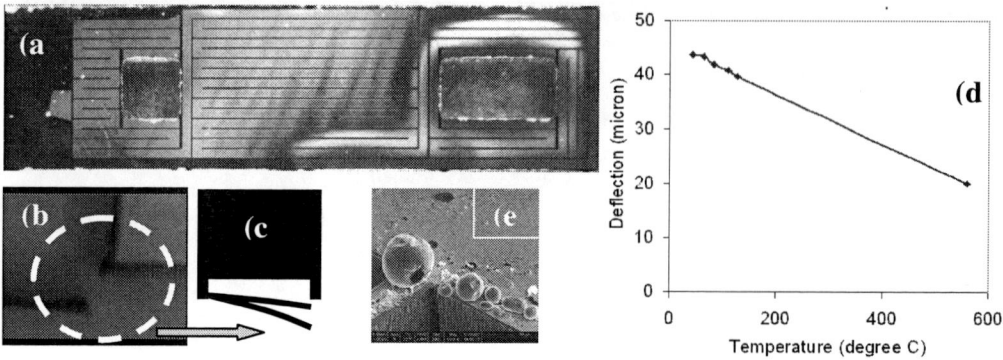

Figure 3. (a) The micro heater with Cr coils on a silicon substrate. The micro tensile stage is placed on top of the heater for uniaxial experiment. (b, c) Bimaterial (Si/Cr) temperature sensor built- in to the loading stage. The temperature is measured from the deflection of the beam (d). Temperature calibration of the sensor is shown (e) Al droplets produced by melting.

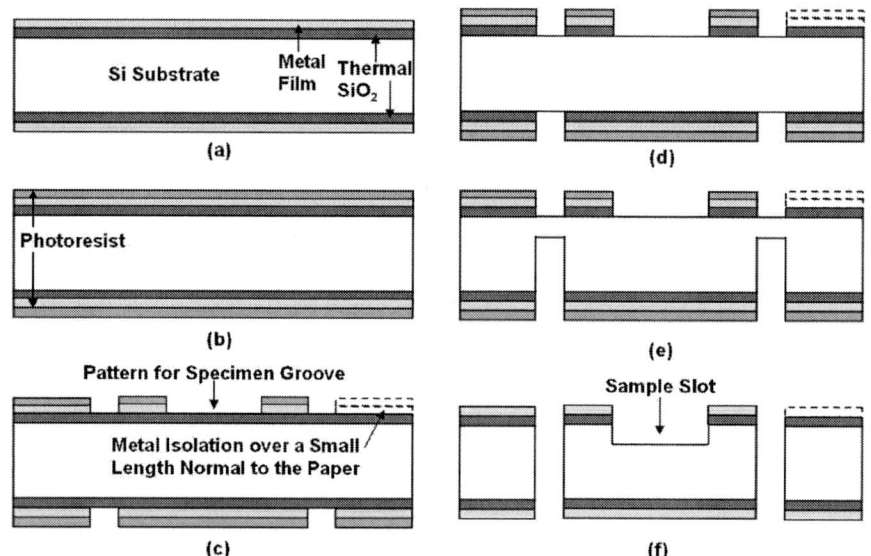

Figure 4. Schematic of process flow of the tensile stage: (a) Thermal growth of SiO_2 and deposition of aluminum layers on both sides of Si wafer; (b) Photoresist spin-coating on both sides of Si wafer; (c) Patterning of both sides by photolithography and liquid etching of metal layer (Here, the pattern of the bottom surface is a mirror image of the top pattern except for the patterns of the sample slot and the center gauge on top.); (d) Dry etching of SiO_2 layer on top and bottom by RIE; (e) Deep Si etching from bottom by ICP DRIE; (f) Shallow Si etching from top to form the sample slot and the center gauge by ICP DRIE and removal of Photoresist on top and bottom by O_2 plasma. Note that the metal layer is intended to serve as conductors for possible resistivity measurement of the sample.

EXPERIMENTS WITH THE STAGE

As a demonstration of the applicability of the stage, a nickel superalloy sample was tested under uniaxial loading. The specimen was made by FIB milling from a bulk crystal using techniques similar to those shown in [4]. The dimensions of the specimen gage section were approximately 25 μm in length, 10 μm in thickness, and 5 μm in width. The tensile axis of the sample was oriented parallel to the <123> direction, which corresponds to a single-slip orientation at the beginning of the experiment. Once the specimen was separated from the bulk crystal, it was transferred onto the wedge grips of the MEMS stage via micromanipulation within the FIB. The tension experiment was carried out in an SEM. The specimen and the displacement gauges were

Figure 5. In situ, room temperature uniaxial test of a nickel-base superalloy sample using the MEMS-stage.

978-1-4244-2977-6/09 $25.00 © 2009 IEEE

captured in the same SEM image so that both the stress-strain data and images of the deforming sample were captured together.

Figure 5(a-e) show images of the sample under increasing strains. After yielding, there is significant plastic flow with little strain hardening up to about 25% strain. In-situ SEM observation reveals that many slip lines are formed along the gage section that are consistent with activity from the primary slip system (white arrows). The sample then begins to strain harden until about 40% strain, at which time the stress begins to drop. The peak load coincides with the initiation of a second slip system (Figure 5e).

ACKNOWLEDGEMENTS

The work was supported by the US Air Force grants USAF–5212-STI-SC-0004, USAF-5215-FT1-SC-0032, and the National Science Foundation grants DMR 0237400, ECS 0304243. The tensile testing stage was fabricated at the Micro-Miniature Systems Lab and Micro-Nano Technology Laboratory of University of Illinois at Urbana-Champaign.

REFERENCES

[1] J. Rajagopalan, J. Han, and M. T. A. Saif. "Plastic Deformation Recovery in Freestanding Nanocrystalline Aluminum and Gold Thin Films", *Science,* 315, pp1831, 2007.

[2] E. Arzt, "Size effects in materials due to microstructural and dimensional constraints: a comparative review," Acta Materialia, 46, 5611, 1998.

[3] A. Haque, A., M. T. A. Saif. "Deformation Mechanisms in Free-standing Nano-scale Thin Films: A Quantitative In-situ TEM Study", *Proceedings of the National Academy of Science*, vol 101 (17), 6335-6340, (2004).

[3] J. Rajagopalan, J. H. Han, M. T. A. Saif. "Bauschinger Effect in Unpassivated Freestanding Nanoscale Metal Films", *Scripta Materialia*, Volume 59 , Issue 7 , Pages 734 – 737, 2008.

[4] M. D. Uchic and D. M. Dimiduk, "A methodology to investigate size scale effects in crystalline plasticity using uniaxial compression testing," *Materials Science and Engineering A*, 400-401, 268, 2005.

A NEW MICRO-FOUR-POINT PROBE DESIGN
FOR VARIOUS APPLICATIONS

Ji-Kwan Kim, Yan Zhang, and Dong-Weon Lee
MEMS & Nanotechnology Laboratory, School of Mechanical Systems Engineering
Chonnam National University, Gwangju, KOREA

ABSTRACT

In this paper, we proposed a new micro-four-point-probe (μ4PP) with sharp tips arranged in squire and spacing 20μm. The μ4PP consists of a main cantilever and four sub-cantilevers. A thermal actuator based on the bimorph effect is integrated on each sub-cantilever. The four-terminal configuration affords to versatile applications to measure electrical properties. Alternatively, the modified μ4PP structure can be also used as a micro-gripper to move micro/nano-objects. A spring constant of the fabricated sub-cantilevers is less than 1.5N/m which is suitable for fragile materials. Resistivity measurement on a metal particle is successfully performed using μ4PP.

INTRODUCTION

Usually, the four-point probe (4PP) is a much simple approach for sheet resistance measurement, which is available in several geometries. The most common is collinear and the Van Der Pauw method [1]. In the former case, current is passed between the two outer probes and the voltage is measured across the inner pair. The measured ratio of the voltage drop to the forced current is the sheet resistance of the sample. The result is multiplied by a geometric correction factor that depends on the probe geometry and the ratio of the probe spacing to the thickness of the conductive region of the sample to be evaluated [2, 3]. While in the Van Der Pauw method, to improve the accuracy, the probe connections are rotated 90°, positioned on the circumference of the sample and the average resistance is calculated through repeatable measured data.

As the development of advanced techniques in semiconductor, ultra small electronic, mechanical devices have been built for large scale integration and high performance. The analytical techniques for process control and characterization of these fabricated system is quite important and difficult, even a challenge in the further scaled feature size, especially for property investigation of thin films to understand the properties in nanoworld. Pengpeng Zhang, E P Nordberg et. al., evaluated the electrical conductivity in silicon nanomembranes using STM and the Van Der Pauw 4PP method [4]. Recently, Dirch H. Petersen et. al., performed Hall Effect measurement to determine carrier mobility, sheet carrier density and sheet resistance on ultra-shallow junctions in silicon and germanium by using the fabricated μ4PP under various experimental conditions with reproducibility of less than 1.5% [5]. In addition to electrical property evaluation, in-situ thermal behavior investigation for micro-heaters is also performed by using microcantilever with the principle of resistance measurement for temperature rise evaluation [6, 7, 8]. The test structure varies in numbers, geometry, dimensions of the probe as well as materials utilized etc. [9, 10]. Actually, characteristics of the fabricated 4PPs are also influenced by the measurement performance and the apparatus used, such as the sufficiently small contacts between the probe and the sample and other contact conditions. Especially the damage problem is an increasing issue with the scaled dimensions, also mentioned in reports [11,12]. It can be circumvented by selection of electrode materials to minimize adhesion, reduce the contact pressure and improve the techniques of either fabrication process or practical operation during the functionalization etc. And several attempts based on silicon micromachining have been made to improve 4PP characteristics, including a desire to decrease its probe-to-probe spacing for further applications. When the distance decreases, there are some advantages. The device miniaturizes, measurements get detailed and the damage of the substrate minimizes. However, there is the fundamental limitation to reduce the spacing of probe tips because the tips are lined up. Also, it is difficult to measure the electrical properties of the substrate or impurities existing in a curve.

In this paper the main focus for innovation is the design of smart μ4PP to minimize the damage to samples and provide accurate measured results for small contact area, especially for nano-scaled samples. A novel process is proposed for the fabrication of very sharp tips in plane. In addition, the modified new μ4PP can be employed as either grippers or micro-heaters.

DESIGN AND FABRICATION

Design consideration

Our smart μ4PP shown in Fig. 1 employs very sharp tips at each sub-cantilever end symmetrically arranged in two diagonals of a square frame extended from a suspended main cantilever as a support beam to position the four probes and the path of the electrical wire leads connecting the ones on the substrate platform to control the sub-cantilevers either independently or collectively depending on various applications of interest.

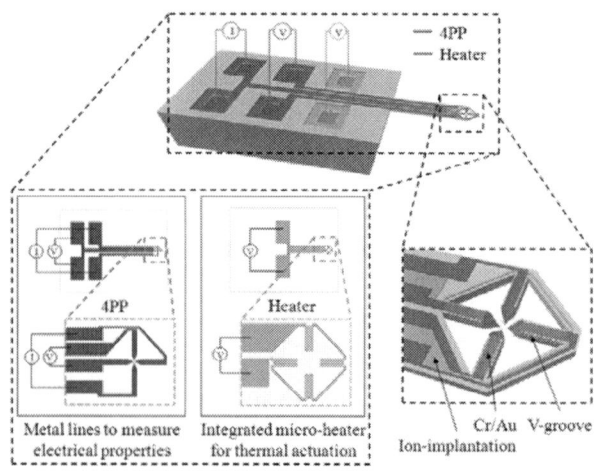

Figure 1: A schematic diagram of a new type μ4PP

978-1-4244-2977-6/09 $25.00 © 2009 IEEE

Four sub-cantilevers symmetrically suspended from the square frame at the end of the main cantilever are the sensitive components and should be protected from destruction or other contaminations. Take one of the electrodes for example, it consists of 3μm thick silicon with 0.5μm SiO_2 grown on, and coated with Cr/Au layer of 50/150 nm for electrical conducting. Each of the sub-cantilever has dimensions l=140μm, w=40μm, and t=3μm, with the V-groove pattern further reducing the stiffness for less damage to the test sample [13]. The probe spacing is 20μm even less varying from applications. Additionally, the thermal heater formed by either lightly doped silicon or deposited metal on the top layer as the thermal resistor is only placed on the end of the sub-cantilever. One of the probes can be used as a heater and the other probes can be for the thermal transfer behavior investigation through electronic method. Thermal induced expansion will lead to a little deflection downward, which can form a small angle between the heater and the sample. The sharp tip is desired for less damage and heating only small fragile elements. The test samples would also be designed to match the probes configuration for a reliable and efficient test process.

The thermal bimorph actuators based gripper is a common device in micro/nano-electro-mechanical system (M/NEMS), which has two flexible cantilevers forming the lateral contact part to make the target object move. The designed μ4PP can also be used as a gripper consists of four contact points for a reliable holding of the element. The V grooves pattern on each sub-cantilever is desired for large thermal expansion of metal, and also flexible in fitting the element geometry. The small contact area enables holding light objectives with minimized damage to the material.

Figure 2 shows the process flow of key steps in the fabrication of the μ4PP. First a 200nm thick layer of SiO_2 is grown by wet oxidation on a silicon-on-insulator (SOI) wafer with 3μm top silicon on 1μm buried oxide on 300μm bulk silicon. Then photoresist (AZ5214) is spun on top of the SiO_2 layer, and is exposed to UV light through a chromium mask defining the V-groove pattern on the wafer front surface. The V-groove pattern is then etched in tetramethylammonium hydroxide (TMAH) solution at 80℃. SiO_2 layer is grown again to define the cantilever and tip structure. The shape of the in-plane tip is defined in SiO_2 by photolithography. Then, the SiO_2 is wet-etched until the neck width of the base-shaped mask is 0. After removing the photoresist in acetone, silicon layer is dry-etched using an SF-based reactive ion etching (RIE). Finally, the SiO_2 is removed in a solution of buffered hydrofluoric acid (BHF), which defines the in-plane tip at the free end of the sub-cantilever. This novel process for tip fabrication can fabricate the ultra sharp tip end. Also this process can be fabricated simple and inexpensive without special equipment like an electron beam (EB) lithography system. Figure 3 shows optical and scanning electron microscope (SEM) images of the fabricated in-plane tips. For other parts, we use conventional micromachining techniques. The μ4PPs are finally released from the substrate by removing of an intermediate oxide layer of the SOI wafer in BHF. Figure 4 shows SEM images of the fabricated smart μ4PP and the sub-cantilever is initially curled by thermal stress of a thin metal layer deposited on the sub-cantilevers. Initial probe-to-probe spacing of the devices is about 20μm and can be decreased to 500nm by using EB or ion beam lithography. Further decrease of the initial distance is desired for the samples with small contact area.

Fabrication

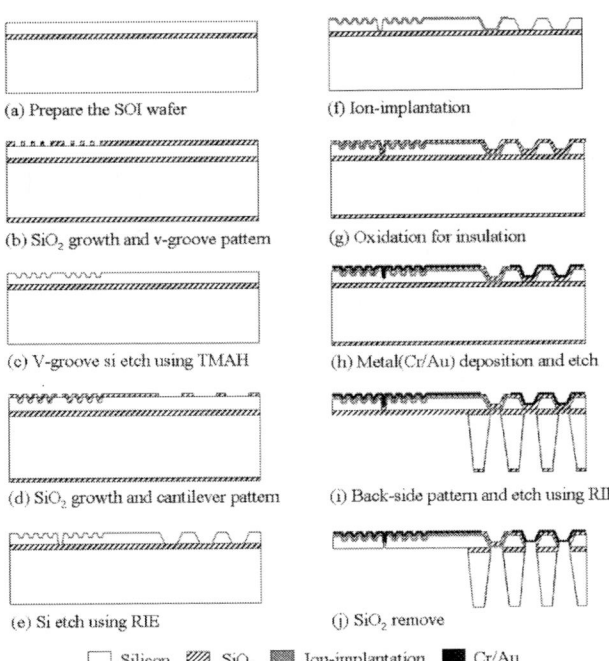

Figure 2: Process flow of key steps in the fabrication of μ4PP

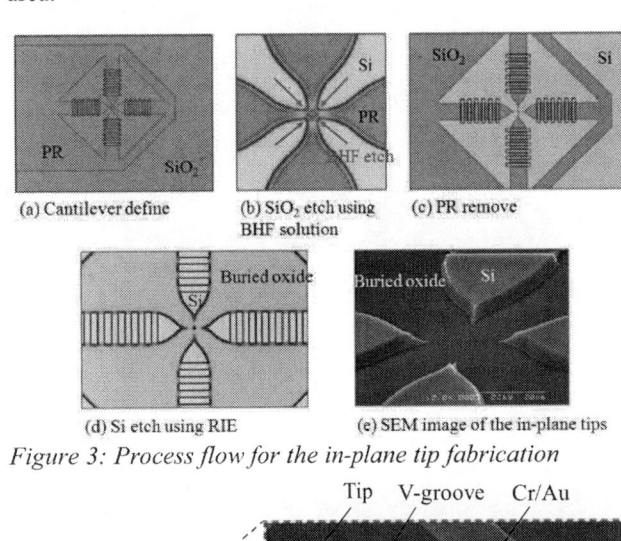

Figure 3: Process flow for the in-plane tip fabrication

Figure 4: SEM images of the fabricated smart μ4pp

To evaluate the spring constant of the sub-cantilevers, the

μ4PP was excited by a sine-wave voltage signal from a function generator. The deflection of the sub-cantilever is measured with a laser vibrometer, a laser spot is focused on the free end of the sub-cantilever. A spectrum analyzer records the input voltage across the cantilever and the deflection signal from the laser vibrometer. A spring constant of the fabricated sub-cantilever is less than 1.5N/m, which is suitable for fragile materials.

ELECTRICAL AND THERMAL PROPERTIES CHARACTERIZATION

To perform the electrical property measurement by using the fabricated μ4pp, an experiment setup is established combined with other micro-manipulates, shown in Fig. 5 (a). A current of 0.1~10mA is applied to two of the sub-cantilevers and then the voltage difference is measured through the other ones. The μ4PP can be moved away from the sample by using linear motion probe station during sample preparation. Then the μ4PP is made to approach the sample with the coarse motion feed through, and final fine positioning is performed. And then, the two outer electrodes act as the current source and drain, while the inner ones measure the voltage drop using a 2182A nanovoltmeter from Keithley. A constant current is applied in our experiments in order to minimize the influence of contact resistance between the electrode tips and the sample surface. Fig. 5 (b) shows the experimental I-V relations of silver paste particle samples.

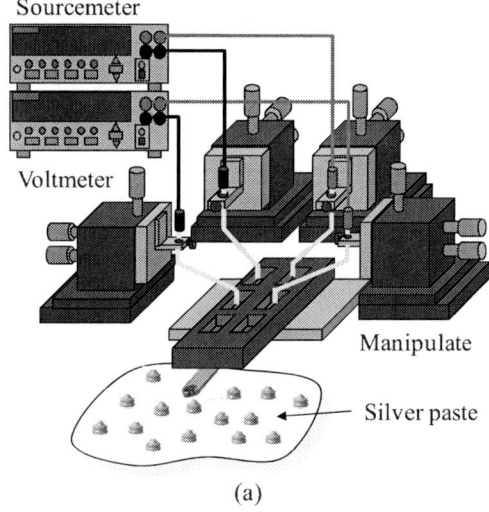

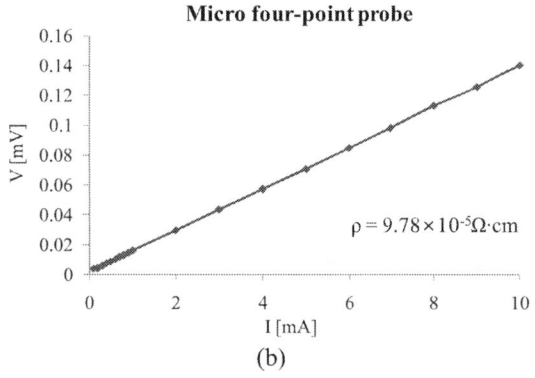

$\rho = 9.78 \times 10^{-5} \Omega \cdot cm$

(b)

Figure 5: (a) An experimental setup to measure electrical property and (b) resistivity measurement using the fabricated μ4pp

Another application is the contact resistance measurement between doped silicon denoted by O in Fig. 6 (a) and deposited metal layer in order to form ohm contact. Usually the cross pattern is defined with the purpose of accurate contact resistance measurement using the 4PP method rather than the digital volt meter with only two contact points at A and B. In Fig. 6 (a), a current source contacts electrode A and B, and then we drop the other probes at point C and D respectively to measure the potential difference there. In this case, we do not care the contact resistance between the current source probes and the contact metal whatever it is thanks to the feedback circuit in the current source to filter the undesired signals. And there is no current flow through the voltage meter, the influence such as the parasitic and Schottky junction etc. can be zero to the first order, and the measured contact resistance here should be V/I. To perform the Van Der Pauw method for the resistivity measure, the pitch of the contact pads A, B, C, and D of the test structure fixed along the circumference can be designed to match the ones of the μ4PP as shown in Fig. 6 (b). We assume that all the contacts are infinitely small and must be aligned along the circumference of the sample as the boundary conditions for precise measured results based on the numerical analysis. The specific resistance ρ of the material can be obtained numerically from equation (1) [1],

$$\exp\left(-\frac{\pi R_{AB,CD}d}{\rho}\right) + \exp\left(-\frac{\pi R_{BC,DA}d}{\rho}\right) = 1 \quad (1)$$

Where d is the thickness of the sample, $R_{AB,CD}$ denotes the measured resistance when the current enters the sample through the contact of A and leaves it through B, then we measure the potential difference of contact C and D, so is the definition of $R_{BC,DA}$. The resistivity is derived in the form

$$\rho = \frac{\pi d(R_{AB,CD} + R_{BC,DA})}{2\ln 2} \cdot f\left(\frac{R_{AB,CD}}{R_{BC,DA}}\right) \quad (2)$$

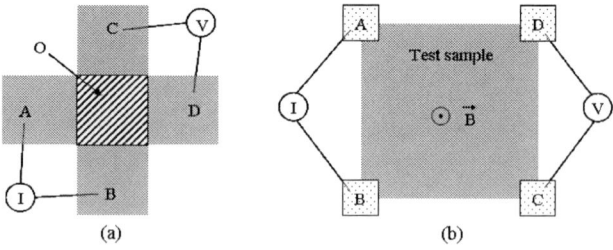

Figure 6: (a) The schematic diaphragm of contact resistance measurement and (b) the schematic representation for the van der pauw method and hall effect measurement.

Where f is a function of $R_{AB,CD}/R_{BC,DA}$ and must satisfy the equation

$$\frac{R_{AB,CD} - R_{BC,DA}}{R_{AB,CD} + R_{BC,DA}} = f \cdot \text{arccos} \, h\left(\frac{\exp(\ln 2/f)}{2}\right) \quad (3)$$

978-1-4244-2977-6/09 $25.00 © 2009 IEEE

When there is a magnet field applied perpendicular to the sample shown in Figure 6 (b), the Hall mobility can be derived from the resistance change $\Delta R_{BD,AC}$, given by

$$\mu_H = \frac{d\Delta R_{BD,AC}}{B\rho} \qquad (4)$$

Both of the measured electric resistivity and hall mobility would be influenced by either the contact area or the position of the sub-cantilever and suitable for the sharp tips utilization.

Furthermore, the thermal property such as the thermal coefficient of resistance ξ (TCR) can be evaluated from the resistance change measurement and derived from the general expressions for the relations between the resistance and the temperature gradient as follows:

$$R = R_0(1 + \xi \cdot \Delta T) \qquad (5)$$

Two of the sub-cantilevers can provide current flow through the test structure and the generated Joule heating power will lead a temperature rise through the test sample. The other two sub-cantilevers measure the potential difference and the temperature coefficient of resistance can be obtained from the governing equations of thermal transfer mechanisms varying from conditions of the test structures.

CONCLUSIONS

A novel μ4PP is optimized and fabricated for electrical and thermal properties evaluation of metals and semiconductors. The configuration and modified sharp tips at each sub-cantilever end enable good contact conditions to reduce damages for reliable electrical and thermal characterization, such as ohm contact resistance between metals and semiconductors, also suitable for Van Der Pauw method, in turn Hall Effect measurement. In addition, the modified μ4PP can be also used as grippers based on thermal actuation and micro-heater. The correspondingly theory analysis for the multi-function μ4PP is established for the data process during the experiment, to provide high efficiency and accuracy of the measured results to study both of the material property with variable conditions and the exact performance in practical functionalizations. Further study is being extended in the Micro/Nano application scopes of physics, chemistry and biology.

ACKNOWLEDGEMENTS

This work was supported by Korea Science & Engineering Foundation through the NRL Program (Grant R0A-2007-000-10157-0).

REFERENCES

[1] L. J. Van der Pauw, "A Method for Measuring the Specific Resistivity and Hall Effect of Discs of Arbitrary Shape", *Phillips Res. Rep.*, vol. 13, 1, 1958.

[2] L.B. Valdes, "Eesistivity Measurements on Germanium for Transistors", *in Proc. IRE*, vol. 42, 420, 1954.

[3] M. Yamashita and M.Agu, "Geometrical Correction Factor of Semiconductor Resistivity Measurement by Four Point Probe Method," *Japan J.Appl.Phys.*, vol. 23, 1499, 1984.

[4] Pengpeng Zhang, E P Nordberg, B-N Park, et. al., "Electrical Conductivity in Silicon Nanomembrane" *New Journal of Physics*, vol. 8, 200, 2006.

[5] Dirch H. Petersen, Ole Hansen, et. al,. "Micro-Four-Point Probe Hall Effect Measurement Method" *J. Appl. Phys.* vol. 104, 013710, 2008.

[6] Friedemann Volklein and Henry Baltes, "A Microstructure for Measurement of Thermal Conductivity of Polysilicon Thin Films," *J. Microelectromech. Syst.*, Vol. 1, pp. 193-196, 1992.

[7] Nicholas W. Botterill and David M. Grant, "Novel Micro-thermal Characterisation of Thin Film NiTi Shape Memory Alloys", *Materials Science & Engineering A*, vol. 378, pp. 424-428, 2004.

[8] Xu Gaobin and Huang Qing'an, "On-line Extraction for Thermal Conductivity of Surface-micromachined Polysilicon Thin Films", in *Tech. Papers IEEE Sensors*, Vienna, October 24-27, 2004, pp. 1593-1596.

[9] Christian Leth Petersen et. al., "Multi-Point Probe", *Patent US 7,323,890 B2*, Jan. 29, 2008.

[10] M. Baroncini, et. al., "Characterization of an Embedded Micro-heater for Gas Sensors Applications", in *Tech. Papers VLSI Technology, Systems, and Applications*, Hsinchu, April 18-20, 2001, pp. 164-167.

[11] R Lin, M Bammerlin, et. al., "Micro-Four-Point-Probe Characterization of Nanowires Fabricated Using the Nanostencil Technique", *Nanotechnology*, vol. 15, pp. 1363-1367, 2004.

[12] S. Keller, et. al., "Microscopic Four-Point Probe Based on SU-8 Cantilevers", *Review of scientific instruments*, vol. 76, 125102, 2005.

[13] Dong-Weon Lee, Christoph Gerber, et. al., "Switchable Cantilever Fabrication for a Novel Time-of-Flight Scanning Force Microscope", *Microelectronic Engineering*, vol. 67-68, pp. 635-643, 2003.

STICTION IN LOW HUMIDITY ENVIRONMENT

Firas Sammoura, Andrew Sparks, William Sawyer, Milind Bhagavat, Michael Judy, and Ken Yang
Micromachined Product Division, Analog Devices, Cambridge, MA, USA

ABSTRACT

The susceptibility of MEMS devices, treated with anti-stiction coatings, to stiction in a low humidity environment has been investigated for the first time. Wafer-level testing with a pull-in/pull-out voltage technique and current compliant source was used to detect stiction on capped and uncapped wafers. Devices capped in dry Nitrogen environment were found to be sticky at both the wafer-level pull-in/pull-out test as well as the packaged part tap test. Although uncapped devices did not show stiction in a room environment using the pull-in/pull-out detection technique, successive drops in pull-out voltage were detected as the conditions of the test control chamber became drier. The ability of the anti-stiction coating to retain charge was analyzed using KLA-Tencor Quantox XP system where the deposited charge decayed in 6 minutes when measured in room air conditions. Successive pull-in/pull-out cycles separated by a delay of 15 seconds showed decay in pull-out voltage of 22mV/cycle. This unprecedented demonstration opens up new opportunities for understanding stiction in MEMS devices and implementing the necessary solutions.

INTRODUCTION

Analog Devices has recently developed a three-axis accelerometer where the sensor and ASIC chips are fabricated separately and integrated in a plastic package. The MEMS sensor is hermetically sealed using lead borosilicate glass. In order to address the susceptibility of these MEMS devices to stiction problems arising during operation caused by tapping and shaking, several studies were done to evaluate the role of capping environment on stiction.

Stiction constitutes an important failure mechanism for MEMS devices and can emanate from two sources: (a) process-related stiction occurring during the release process and (b) operation-related stiction during device use [1]. Much research has been done to understand and resolve stiction. Self-assembled monolayers (SAMs) as anti-stiction coatings for MEMS have been investigated [2]. Although SAM treatments desensitized the MEMS actuator to the environment conditions, in particular to humidity, permanent stiction was still observed in OTS-coated devices. This was attributed to surface charging due to a higher resistivity layer. In addition, aligned carbon nanotubes (ACNTs) were grown perpendicular to the surface using chemical vapor deposition (CVD) in order to increase the surface roughness [3]. Models that predict the effect of humidity on MEMS stiction were developed [4, 5]. However, they were limited to high relative humidity regimes where stiction is likely to occur due to surface tension. In contrast to the previous works on stiction, this is the first paper that investigates stiction of MEMS devices coated with phenyl-silicone anti-stiction coatings [6] in a low humidity environment.

Two distinct achievements have been accomplished in this work: (1) pull-out voltage analysis versus control chamber humidity condition as well pull-in/pull-out cycling, and (2) surface charge decay study at room conditions.

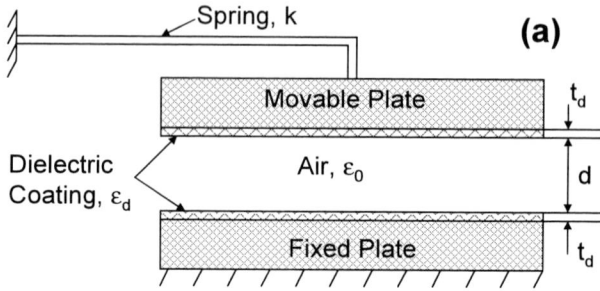

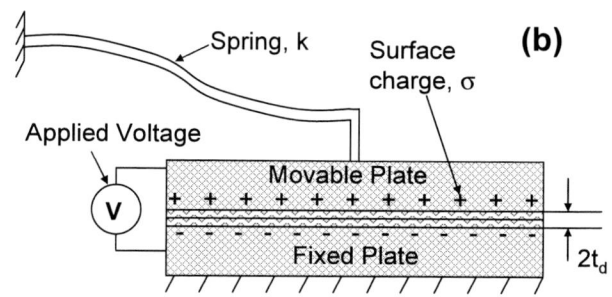

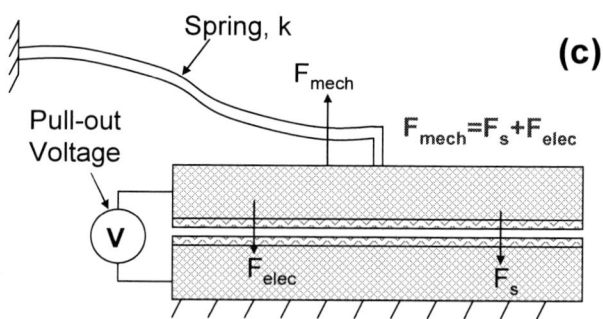

Figure 1: Schematic diagram showing test structures (a) at initial position, (b) after pull-in, and (c) at pull-out

THEORY

Figure 1(a) is a schematic diagram showing the test structure at initial position. The test structure resembles a parallel plate capacitor with a movable plate and fixed plate. The former is connected to a mechanical spring, while the latter is anchored to the substrate. The plates are coated by an anti-stiction coating, which is modeled as a dielectric material. In order to mimic mechanical forces operating on the beam during use, a DC voltage is applied between the plates in order to generate an electrostatic force as shown in Figure 1(b). As the voltage is ramped up, the movable plate gets closer to the fixed plate. Equilibrium between the electrostatic force and the mechanical restoring force is maintained as long as the voltage is less than the pull-in voltage. As the voltage exceeds the pull-in voltage level, the electrostatic force always exceeds the mechanical restoring force and the movable plate is snapped into the fixed plate. The pull-in voltage, V_{pin}, can be calculated using [7]:

$$V_{pin} = \sqrt{\frac{8kd^3}{27\varepsilon_0 A}} \qquad (1)$$

where d is the gap distance, k is the spring constant, ε_0 is the permittivity of free space, and A is the area of the drive electrode.

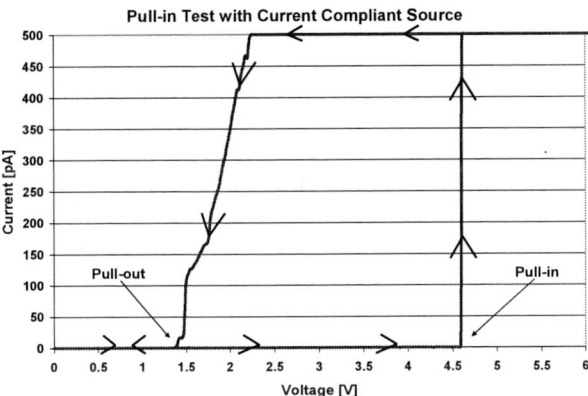

Figure 2: Hysteresis curve showing pull-in and pull-out of a parallel plate MEMS capacitor.

After pull-in, the electrostatic force becomes very large as the dielectric thickness is very small. The applied voltage is then reduced until the mechanical restoring force, F_{mech}, balances the electrostatic force, F_{elec}, and the stiction force, F_s, as shown in Figure 1(c). The pull-out voltage, V_{pout}, can be calculated using a similar approach undertaken by Basu al. [8]:

$$V_{pout} = \sqrt{\frac{2(kd - F_s)(2t_d)^2}{\varepsilon_0 \varepsilon_d A}} \qquad (2)$$

where t_d and ε_d are the thickness and the permittivity of the dielectric material covering the electrodes, respectively. The stiction force arises from many factors including surface charging, van der Waals forces, and capillary forces. The stiction force arising from the surface charge accumulated on the surface of a MEMS parallel plate capacitor can be approximated by assuming that equal and opposite charge density have accumulated on the two sides of the capacitor formed by the movable and fixed plate with the coating material as the dielectric [9] :

$$F_s \cong \frac{1}{2} \frac{A}{\varepsilon_0} \sigma^2 \qquad (3)$$

where σ is the surface charge density trapped on the dielectric material. Figure 2 shows a hysteresis curve illustrating the pull-in and pull-out voltage of the movable plate to the fixed plate [1]. The current source limit was set at 500pA to avoid excessive heating and thus welding. It is worth mentioning that for sticky structures, no real pull-out voltage is recorded and the capacitance or current values do not drop to pre-pull-in levels. Figure 3(a) and 3(b) show a set of lateral movable plate/fixed plate capacitors just before and after pull-in, respectively.

RESULTS AND DISCUSSION

The capped and uncapped test structures had different stiction performance (i.e. pull-in/pull-out voltages) as measured by electrical wafer-level probing. Uncapped wafers showed better stiction performance in a typical lab environment. In order to investigate the different behavior, residual gas analysis (RGA) techniques were adopted to study the gas composition inside the capped device. The primary difference in the RGA results between the sticky and non-sticky die was that the water content was lower for the sticky die. As shown in Figure 4, the stiction-free dies had higher water content, around 4%, compared to 1% water content in the sticky dies. Higher humidity is a precursor for stiction due to capillary surface tension effects, which has been shown in many of the stiction studies. Operation of the MEMS device in low humidity regimes gives rise to charging. The inability of surfaces, in a dry-air capped environment, which were coated with an anti-stiction coating to bleed off surface charges, is due to lower surface conduction leading to device failure due to stiction.

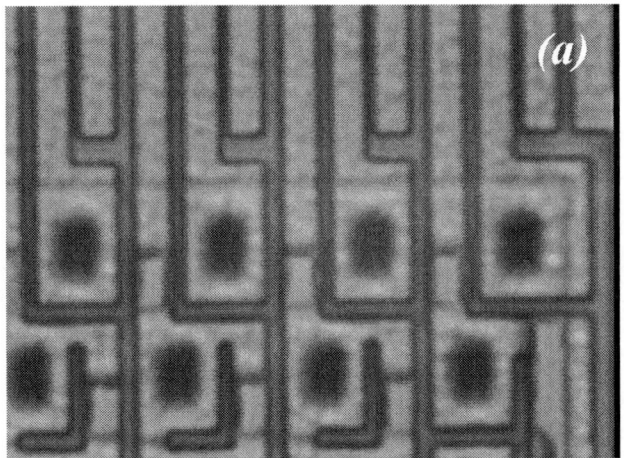

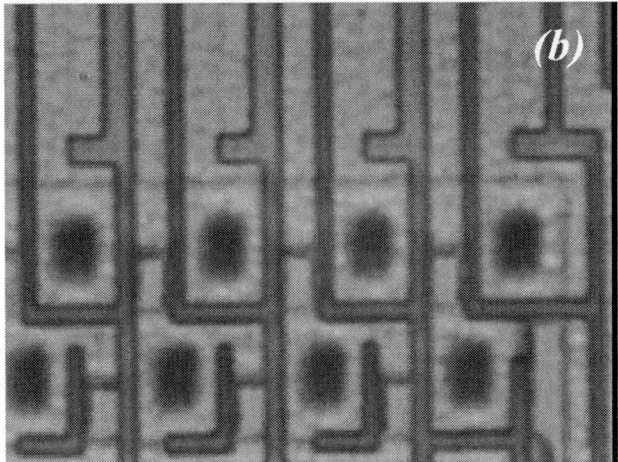

Figure 3: Digital picture showing series of lateral movable plates (a) Just before pull-in and (b) Just after pull-in to the fixed plates.

In order to verify the RGA findings, uncapped wafers were tested in a controlled chamber at various temperatures and nitrogen flow rates in an attempt to modulate the humidity. The humidity decreases as the wafer temperatures and the nitrogen flow rate increases. Figure 5 demonstrates that the average pull-out voltage across the wafer dropped from 1.5V to 0.25V as the chuck temperature was raised from -40°C to 150°C. Introducing Nitrogen further reduced the pull-out voltage. For instance, at 100°C, the average pull-out voltage across the wafer dropped from 1.25V to 0.5V as the nitrogen flow rate increased from 0sccm to 8sccm. Figure 6 reports the percentage of stuck dies across an uncapped wafer with respect to both temperature and nitrogen flow rate modulation. The percentage was as high as 65% across the wafer for the driest condition of 150°C and 8sccm nitrogen flow rate. A precise humidity control chamber such as Rigaku (Humidity Control TG-DTA) was missing in this analysis to better correlate the RGA data to the experimental findings. However, the results collected from modulating the humidity using Nitrogen flow rate and temperature modulation gave a clear signal pertaining to the effect of relative humidity on stiction behavior.

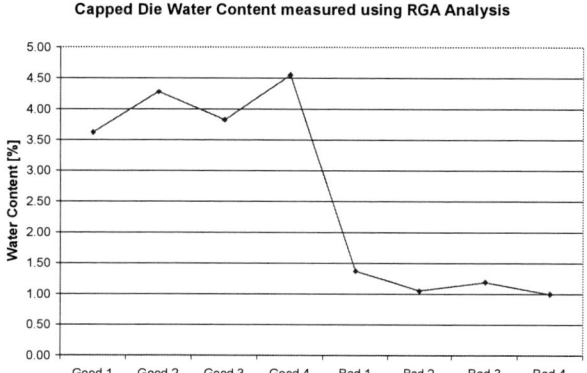

Figure 4: RGA data showing water content difference between sticky and non-sticky dies.

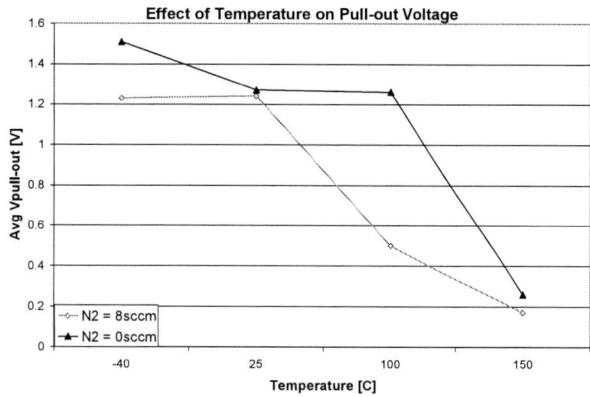

Figure 5: Effect of temperature and nitrogen flow rate on pull-out voltage.

The ability of surfaces coated with anti-stiction coating to retain charge in room air conditions was further investigated. Analysis was done using the KLA Tencor Quantox XP system with ACTIV technology control. The Quantox is a quasi-static measurement tool, where surface charge bias, Q, is applied using a high voltage to create corona ions. The surface voltage is then measured using a vibrating Kelvin probe. A baseline surface voltage 0.424V was measured. A surface charge of $8 \times 10^{-6} C/cm^2$ was then deposited resulting in an instantaneous increase of surface voltage to 0.462V as shown in Figure 7. The surface voltage was monitored every 40secs where it decayed with time until it returned to baseline levels after 360secs. The findings indicate that the anti-stiction coating is not a perfect dielectric because it is not capable of retaining charge for an extended period of time. The anti-stiction coating is capable of holding parasitic charges formed during movable and fixed plate contact due to static electricity as well as the applied electric potential for an extended period of time. The surface charge decay time constant of the anti-stiction coating was measured in room air conditions which could vary. It is envisioned that a controlled Quantox atmosphere will better characterize the dielectric properties of the anti-stiction coating and its ability to retain charge.

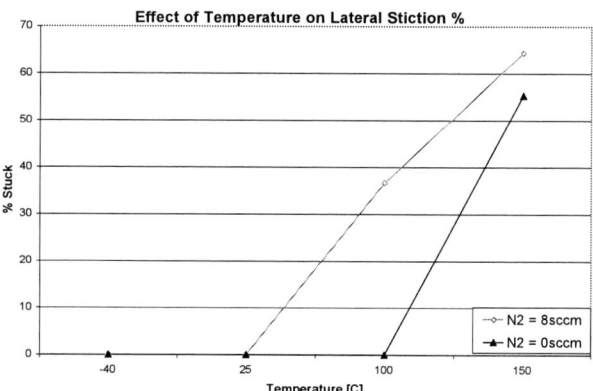

Figure 6: Percentage of stuck dies vs. temperature and nitrogen flow rate.

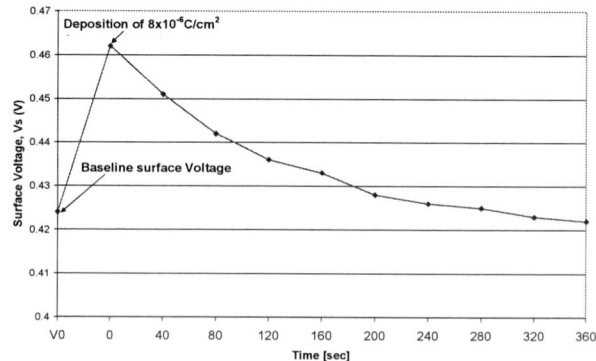

Figure 7: Surface voltage (Vs) decay vs. time. $8 \times 10^{-6} C/cm^2$ of charge was deposited. V0 indicates the baseline surface voltage. The decay was monitored for 360sec.

Finally, the capped test structure pull-out voltage drift with pull-in/pull-out cycling was examined. A total of 48 cycles were successively repeated separated by 15 second pauses. As observed in Figure 8, the pull-out voltage dropped as the number of pull-in/pull-out cycles increased. For instance, the pull-out voltage dropped from an initial value of 1.2V to around 0.1V in 48cycles or

22mV/cycle. Some test structures with an initial pull-out voltage of less than 1.2V showed stiction behavior after 25cycles. The data indicates that stressing the test structures causes gradual surface charge accumulation which ultimately leads to device failure due to stiction. It is foreseen, that modulating the pause time between cycles and repeating the measurement on uncapped wafers will further assess the surface charge accumulation/decay phenomenon of the anti-stiction coating.

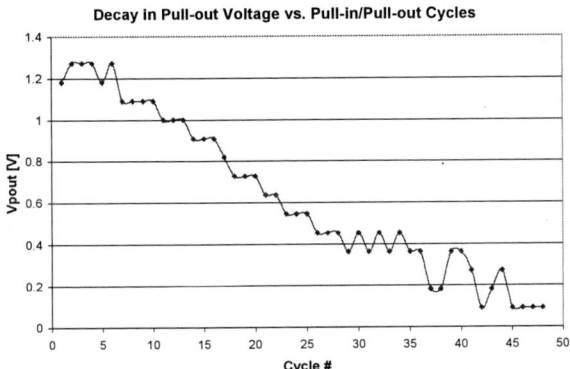

Figure 8: The decay of pull-out voltage versus pull-in/pull-out cycles.

CONCLUSION

Stiction in low humidity conditions has been investigated for the first time. Parallel plate stiction test structures made up of a movable plate connected to a restoring spring and a fixed plate anchored to the substrate were fabricated and coated with phenyl-silicone anti-stiction coatings. The test structures were capped using borosilicate glass. An electrical wafer-level stiction test using a pull-in/pull-out technique with 500pA current limiting was used to study stiction on capped as well as uncapped wafers. Stiction was detected on capped wafers using both electrical wafer-level testing as well as mechanical testing after packaging. On the other hand, no stiction was detected on uncapped wafers using the electrical testing technique. RGA analysis of sticky and non-sticky capped dies showed a distinctive higher humidity on the non-sticky dies. A water content of 4% was collected on all good dies, versus 1% on the bad dies. Modulating the relative humidity of a control chamber using Nitrogen gas flow rate and chuck temperature was done to verify the RGA findings. Uncapped wafers tested under various conditions inside the control chamber showed that stiction behavior is reproduced as the humidity levels decrease in the control chamber. A pull-out voltage drop from 1.3V to 0.5V was observed as the chuck temperature increased from 100°C to 150°C under 8sccm of Nitrogen flow rate. The ability of the anti-stiction coating to retain surface charge was studied using KLA Tencor Quantox XP system using surface voltage measurement. An 8×10^{-6} C/cm^2 of deposited surface charge vanished in 360 seconds indicating a relatively high charge decay time constant at room conditions. In addition, charge accumulation due to successive voltage stress cycling during pull-in/pull-out was analyzed. A 22mV/cycle decay in pull-out voltage was observed with a 15 second pause between cycles. It is envisioned that this novel study will resolve many issues related to the capping environment as well as anti-stiction coatings in order to better help in designing and fabricating MEMS devices that are robust to stiction failures.

ACKNOWLEDGEMENTS

The authors would like to thank Ms Kelley Halchuk and Mr. Bill Henderson from KLA Tencor for their help in surface charge characterization. The authors would also like to acknowledge Dr. Jack Martin, Dr. Mark DaSilva, and Dr. Steve Bart from the Micromachined Product Division at Analog Devices for their valuable discussion and their support with RGA data analysis.

REFERENCES

[1] E. Bhattacharya, S. Basu, and A. Prabhakar, "Stiction Force Estimation from Attachment Length and Electrostatic Measurement on Cantilever Beams," *Proc. of SPIE*, vol. 6111, 611101, (2006).

[2] R. Maboudian, W. R. Ashurst, and C. Carraro, "Self-Assembled Monolayers as Anti-Stiction Coatings for MEMS: characteristics and recent developments," *Sensors and Actuators* 82 (2000) 219-223.

[3] L. Zhu, J. Xu, Z. Zhang, D. W. Hess, and C. P. Wong, "Lotus Effect Surface for Prevention of Microelectromechanical System (MEMS Stiction," *Proceedings of the Electronic Components and Technology Conference*, vol. 2 (2005), 1798-1801.

[4] W. M. Van Spengen, R. Puers, and I. De Wolf, "A Physical Model to Predict Stiction in MEMS," *J. Micromech. Microeng.* 12 (2002) 702-713.

[5] M. P. de Boer, J. A. Knapp, T. M. Mayer, and T. A. Michalske, "The Role of Interfacial Propoerties on MEMS Performance and Reliability," *Invited paper for SPIE EOS Conference on Microsystems and Inspection*, Munich, June 15, 1999.

[6] J. Martin, "Process for Wafer Level Treatment to Reduce and Passivate Micromachined Surfaces and Compounds used therefore," *United States Patent Application Publication: US 2007/0196945 A1*.

[7] L. A. Rocha, E. Cretu, and R. F. Wolffenbuttel, "Compensation of Temperature Effects on the Pull-in Voltage of Microstructures," *Sensors and Actuators A: Physical*, vol. 115, issue 2-3 (September 2005), 351-356.

[8] S. Basu, A. Prabhakar, and E. Bhattacharya, "Estimation of Stiction Force from Electrical and Optical Measurements on Cantilever Beams," Journal of Microelectromechanical Systems, vol. 16, no. 5 (October 2007), 1254-1262.

[9] J. Wibbeler, G. Pfeifer, and M. Hietschold, "Parasitic Charging of Dielectric Surfaces in Capacitive Microelectromechanical systems (MEMS)," *Sensors and Actuators A: Physical*, vol. 71 (1998), 74-80.

A NOVEL HIGHLY-TWISTABLE TACTILE SENSING ARRAY USING EXTENDABLE SPIRAL ELECTRODES

M.-Y. Cheng, C.-M. Tsao, Y.-T. Lai, Y.-J. Yang
Department of Mechanical Engineering, National Taiwan University, Taipei, TAIWAN

ABSTRACT

This paper presents a novel design of highly twistable tactile sensing array which will be as artificial skin for robot applications. The proposed artificial skin, which employ extendable spiral electrodes, are highly flexible and durable so that it can conform to more complex surfaces without damaging the skin structure and the metal interconnects on the sensing array. Polydimethylsiloxane (PDMS) is employed as the main structure material of the skin, and conductive polymer is used as the tactile sensing elements. The electrical characteristic of conductive polymer is measured. By using an 8x8 sensing array, the tactile images induced by PMMA stamps of different shapes are successfully captured.

INTRODUCTION

Recently, the research on humanoid robots has progressed rapidly around the world. In order for ensuring effective and safe interactions between robots and humans, many sensing capabilities, such as tactile, temperature, vision and auditory senses, are indispensable. Flexible artificial skins with tactile array sensing capability are essential for robots to detect physical contact with humans/environment. Various approaches of realizing tactile sensing arrays have been proposed. For example, in [1], a flexible tactile sensing array is created by stitching thin metal wires on a large-area flexible conductive polymer sheet. The stitched wires are carefully arranged to function as the sensing electrodes and the signal interconnects of the sensing array. Yang *et al.* [2] dispensed individual conductive polymer bumps on pre-defined interdigital copper electrodes and showed that the electrical isolation between adjacent sensing elements can be effectively achieved. Lee *et al.* [3] proposed a capacitive tactile sensing array using PDMS elastomer with electroplated copper electrodes. The expandability of the sensing array for large area deployment is also demonstrated. Furthermore, Chen *et al.* [4] developed a flexible capacitive sensing array with interwoven structure using weaving-by-lithography micromachining process.

In those aforementioned approaches, the metal interconnects, which are essential for retrieving signals from each sensing element, are usually quite vulnerable when they are stretched to cover complex surfaces. As a result, many recent research works have been proposed to enhance the reliability and durability of the conductors or interconnects on flexible substrate under large deformation. Khang *et al.* [5] fabricated wavelike conductors which can be safely stretched and compressed to large levels of strain. The work in [6] presented a rubber-like stretchable active matrix with elastic conductors using single-walled carbon nanotubes composite film coated with PDMS rubber.

In this work, we present a novel approach to realize a highly-twistable and reliable artificial skin by using extendable spiral electrodes as the sensing electrodes and the interconnects. The proposed spiral electrode is made by winding copper wires around an elastic nylon line, and

can be easily mass-produced. Each tactile sensing element is formed by dispensing conductive polymer on the spiral electrodes. Also, the characteristic of the sensing material and the performance of the sensing array will be measured and discussed.

DESIGN

Figure 1(a) shows the schematic of the proposed skin which covers a sphere. The simultaneous incorporation of flexible polymer-based substrate and extendable conductive wires is the key to realize the highly-twistable tactile sensing array. In order to make the conductive wires highly extendable and flexible, copper wires of 0.25 mm in diameter are wound to a soft elastic nylon line of 0.3mm in diameter, as shown in Figure 1(b). The schematic diagram of the proposed artificial skin is showed in Figure 1(c). In the proposed design, PDMS elastomer and conductive polymer are employed as main structure material and sensing material of the skin, respectively.

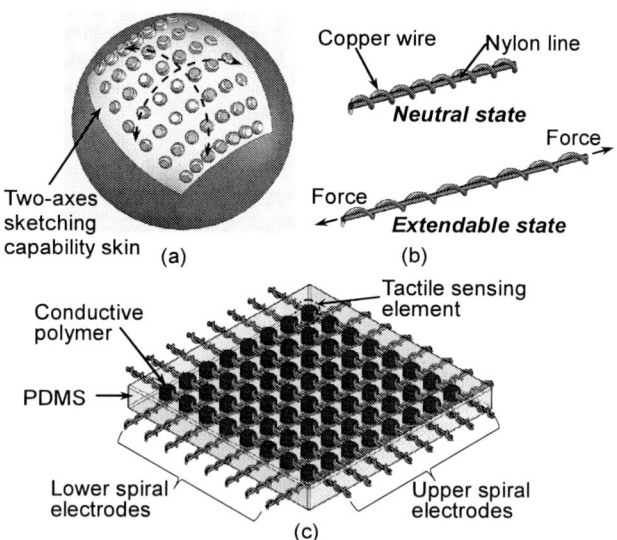

Figure 1. (a) The two-axes sketching capability skin. (b) Extendable spiral electrode. (c) The schematic diagram of the proposed artificial skin.

Figure 2 describes the tactile sensing principle of the design. Two spiral metal wires (i.e., spiral electrodes) are aligned perpendicularly without contacting each other. The sensing element is formed by dispensing conductive polymer [2] to create a resistor between the two spiral electrodes. When pressure is applied, the conductive polymer deforms and the resistance between the metal wires decreases so that the pressure can be detected.

Figure 3 shows the detailed structure and dimensions of a proposed tactile sensing element. The gap between two spiral electrodes is 500μm. The size and thickness of sensing element are 2500×2500 μm^2 and 2600μm, respectively. The helix pitch of spiral copper wires winding on the nylon line is 1200μm.

978-1-4244-2977-6/09 $25.00 © 2009 IEEE

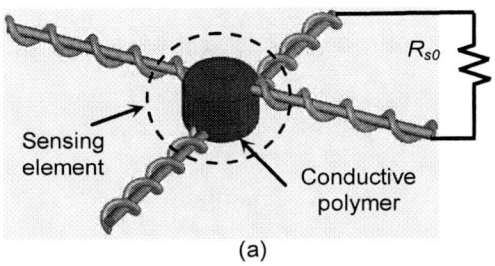

No pressure applied

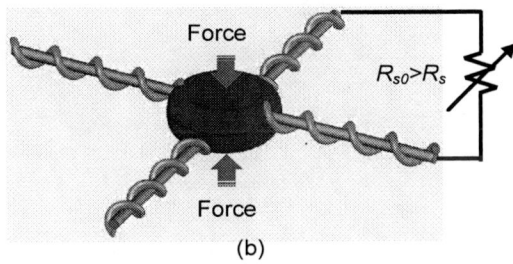

Pressure applied

Figure 2. The schematic of a tactile sensing element without and with applied pressure. Note that R_s and R_{s0} are the resistances of the sensing element with and without applied pressure, respectively.

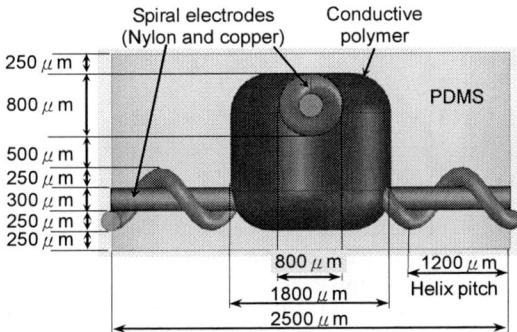

Figure 3. The detailed illustration and dimensions of the proposed tactile sensing element. This figure is not to scale.

FABRICATION

The fabrication process is shown in Figure 4. PDMS prepolymer and curing agent are mixed at 10:1 ratio. The prepared PDMS mixture is then poured in a PMMA handling mold (Figure 4(a) and 4(b)). Spiral electrodes, which can be easily mass-produced by a winding machine (Figure 5), are pre-fixed on a PMMA frame (Frame I). Frame I is then placed on the PMMA handling mold to form the row interconnect of sensing array (Figure 4(c)). After PDMS curing, conductive polymer is dispensed on appropriate locations (Figure 4(d)). Then, another PMMA frame (Frame II) with pre-fixed spiral electrodes is placed on Frame I to form the column interconnect (Figure 4(e)). Conductive polymer is dispensed again to ensure the electrical connection between the conductive polymer and the spiral electrodes. Then, the prepared PDMS mixture is filled into the mold (Figure 4(f)) and degassed. Figure 4(g) shows the flexible sensing array after curing PDMS and removing the PMMA mold/frames.

The conductive polymer material is formed by introducing nano carbon black (80nm in average size,

10.5%wt), silver nano powder (30nm in average size, 8.1%wt) and copper powder (45μm in average size, 32.5%wt) into PDMS prepolymer. The dispersant such as cyclohexane is also used for increasing the fluidity of the PDMS prepolymer and improving the mixture uniformity during the blending procedure (30 min). Then the PDMS curing agent is well mixed into the composites before all the mixture is degassed in a vacuum chamber for 30 minutes to evaporate the volatile cyclohexane. During the dispensing step, the unpolymerized conductive polymer material is put into a dispensing machine and is selectively dispensed on the copper spiral electrodes.

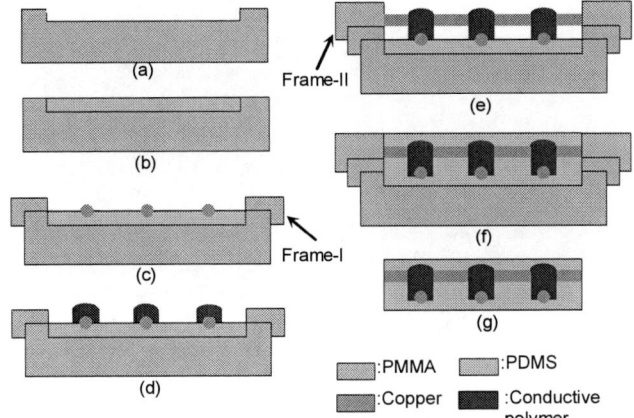

Figure 4. Fabrication processes of the proposed tactile sensor array.

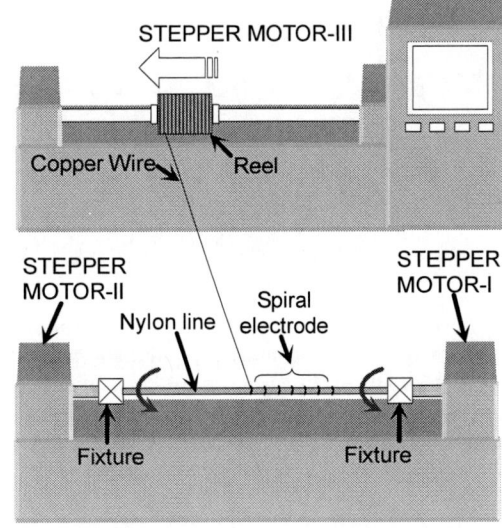

Figure 5. Schematic of the winding machine for fabricating the spiral electrode.

The schematic of the winding machine for fabricating the spiral electrode is shown in Figure 5. Firstly, both ends of the nylon line are fixed on the fixtures of Stepper-Motor-I (SM-I) and Step-Motor-II (SM-II). Also, one end of the copper wire is fixed on the fixtures of SM-I. During the winding process, SM-I and SM-II operate synchronously, and the copper-wire reel is laterally shifted by Stepper-Motor-III (SM-III), which is used to control the helix pitch distance of spiral copper wire on the nylon line.

Figure 6(a) shows the fabricated extendable spiral electrode. Figure 6(b) shows the process of dispensing conductive polymer materials using an automatic dispenser. By controlling X-Y table of the dispensing system, the conductive polymer can be efficiently dispensed on the appropriate positions. Figure 6(c) shows the dispensed conductive polymer array on the row interconnect (i.e., the spiral electrode). The fabricated highly-twistable 8×8 tactile sensing array with an effective area of 20 by 20 square millimeters is shown in Figure 6(d). As shown in Figure 6(e) and 6(f), the excellent flexibility and twistability of the sensing array are also demonstrated.

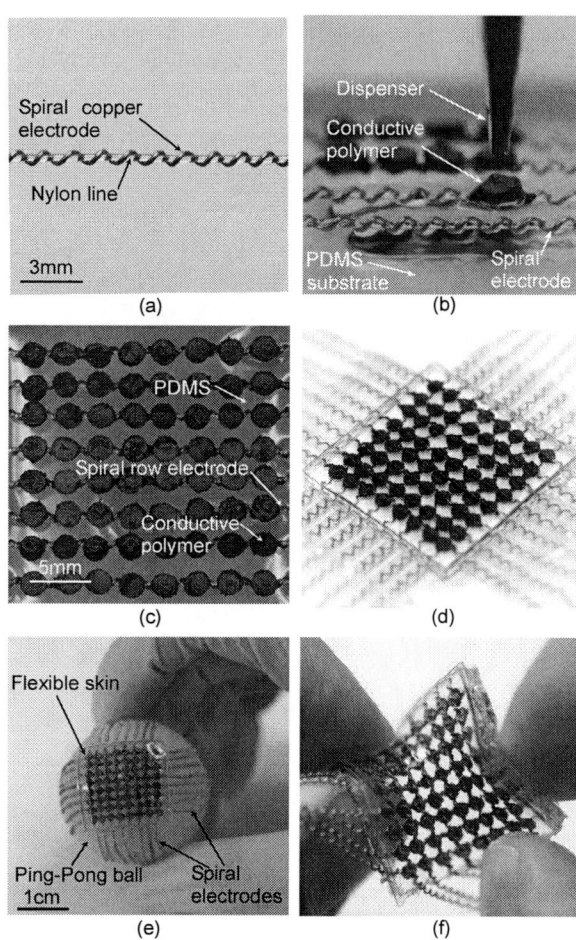

Figure 6. (a) The fabricated extendable spiral electrode. (b) The process of dispensing conductive polymer materials using an automatic dispenser. (c) The dispensed conductive polymer array on the row interconnect. (d)~(f) The pictures of the fabricated tactile sensing array.

MEASUREMENT AND DISCUSSION

The sensing array system, which consists of a sensing array and a scanning circuitry, is shown in Figure 7. For row scanning, a set of multiplexers provides the driving power to sensing elements. For column scanning, another set of multiplexers is used to receive the data outputs from each sensor. The scanning circuit measures the resistance change of each sensing element. A constant voltage is supplied, and the change of the current flowing through the resistor can be detected by the current-to-voltage converter, as shown in the figure. The scanned output voltages of tactile sensing arrays are transferred to their corresponding MCU by an Analog-to-digital converter. Scanning requests for the sensing array can be made by a PC through serial communication port (RS-232). Also, the

scanned data can be transferred to the PC through RS-232 for visualization and data analysis. The maximum scanning rate is greater than 3,000 elements per second.

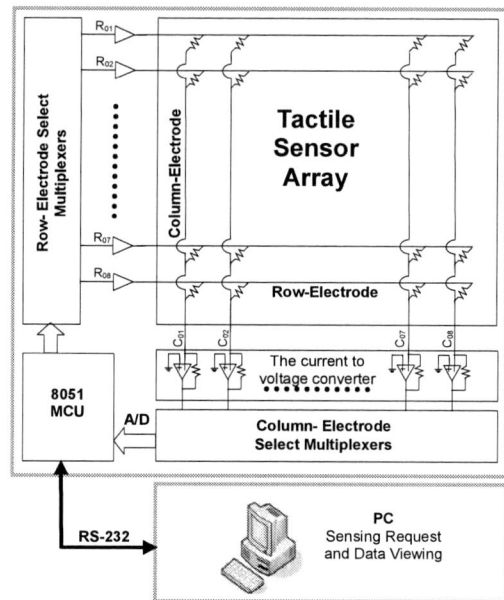

Figure 7. Schematic of the sensing-array system.

In order to characterize the sensing properties, the conductive polymer is trimmed as a circular shape with 3mm in diameter and 0.5mm in thickness. The conductive polymer film is then sandwiched by a copper plate and a metal probe. The diameter of the metal probe is also 3mm. The copper plate and the metal probe are connected to a multimeter for resistance measurement. The measured relationship of electrical resistance vs. applied pressure for various tactile sensing elements is shown in Figure 8. The curve in the figure is the average result by measuring one sample sheet 10 times with different applied pressure. The error bars indicate the measured maximum and minimum values. The resistance of the conductive polymer converges to a steady value as the applied pressure is greater than 450kPa.

Figure 9 shows the measured pressure distributions by pressing solid PMMA stamps of different alphabetic patterns on the proposed 8x8 sensing array. The figure also shows the pictures and dimensions of these stamps. Obviously, these shapes are clearly resolved by the sensing array. Also, the external applied pressure is produced by placing a 1kg weight on the top of each stamp.

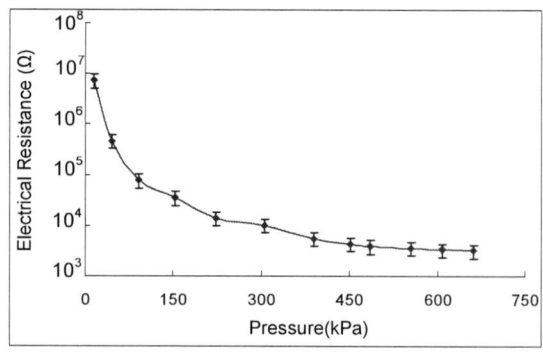

Figure 8. Measured response of the fabricated tactile sensing element with different applied pressure.

Figure 9. The measured pressure distributions using solid stamps of different patterns. The acrylic stamps are also shown.

CONCLUSION

The development of a novel highly-twistable 8×8 tactile sensing array was presented in the paper. Extendable spiral electrodes were employed as elastic interconnects of sensing array so that it can withstand significant deformation without damaging the sensing array. The tactile sensing elements were formed by dispensing conductive polymer on the extendable spiral electrodes. The corresponding scanning circuit for the sensing array was designed and implemented. The characteristic of the sensing element was also measured. By using the fabricated 8x8 sensing array, tactile images induced by PMMA stamps of different shapes were also demonstrated.

ACKNOWLEDGEMENT

This work was supported in part by the National Science Council, Taiwan, R.O.C. (Contract No: NSC 97-2628-E-002 -049 -MY3). The authors would like to thank Prof. Cheng-Wen Ma for help on the device packaging.

REFERENCES

[1] M. Shimojo, A. Namiki, M. Ishikawa, R. Makino, and K. Mabuchi, "A tactile sensor sheet using pressure conductive rubber with electrical-wires stitched method," *Ieee Sensors Journal*, vol. 4, pp. 589-596, Oct 2004.

[2] Y. J. Yang, M. Y. Cheng, W. Y. Chang, L. C. Tsao, S. A. Yang, W. P. Shih, E. Y. Chang, S. H. Chang, and K. C. Fan, "An integrated flexible temperature and tactile sensing array using PI-copper films," *Sensors and Actuators A-Physical*, vol. 143, pp. 143-153, May 2008.

[3] H. K. Lee, S. I. Chang, and E. Yoon, "A flexible polymer tactile sensor: Fabrication and modular expandability for large area deployment," *Journal of Microelectromechanical Systems*, vol. 15, pp. 1681-1686, Dec 2006.

[4] N. Chen, J. Engel, S. Pandya, and C. Liu, "Flexible skin with two-axis bending capability made using weaving-by-lithography fabrication method," *Proceedings of the IEEE International Conference on Micro Electro Mechanical Systems (MEMS'06)*, 2006, pp. 330-333.

[5] D. Y. Khang, H. Q. Jiang, Y. Huang, and J. A. Rogers, "A stretchable form of single-crystal silicon for high-performance electronics on rubber substrates," *Science*, vol. 311, pp. 208-212, Jan 2006.

[6] T. Sekitani, Y. Noguchi, K. Hata, T. Fukushima, T. Aida, and T. Someya, "A rubberlike stretchable active matrix using elastic conductors," *Science*, vol. 321, pp. 1468-1472, Sep 2008.

A DIRECTIONAL CAPACITIVE MEMS MICROPHONE USING NANO-ELECTRODEPOSITS

Sang-Soo Je, Jeonghwan Kim, Michael N. Kozicki, and Junseok Chae
Department of Electrical Engineering,
Arizona State University, 650 E. Tyler Mall, Tempe, AZ 85287, USA

ABSTRACT

We present an *in-situ* non-volatile tuning method for adjusting MEMS microphone sensitivity using integrated nano-electrodeposits to achieve high directionality in hearing aids. Using a DC bias at room temperature, nano-electrodeposits are electrochemically formed on a Ag-Ge-Se solid electrolyte film integrated with a microphone diaphragm. The *in-situ* growth mechanism generates mass/stress redistribution on the diaphragm, tuning microphone sensitivity to incoming acoustic sources. Acoustic measurements demonstrate the technique can achieve a tuning range of 1.67 dB (24%), corresponding to a 1.17-dB Directivity Index (DI) improvement, which is sufficient to enable acoustic directionality in high accuracy microphone arrays.

INTRODUCTION

Minimizing the sensitivity mismatch among multiple microphones in modern hearing aids is a challenge since a very small amount of mismatch destroys directionality [1]. High directionality is desirable to hearing aid users particularly in environments in which speech sound is heavily masked by noise [2]. Many methods currently exist to increase the directionality and signal-to-noise ratio of microphones in the hearing aids. Most research and development has focused on software calibration methods (electronic method), including noise-reduction algorithms, and second-order directional microphones [3-4]. However, these require significant complexity and power, which reduces the battery lifetime and substantially increases the overall cost of the hearing aids. Some research has been performed in the area of hardware manipulation for resonant frequency shift applications, including PECVD / LPCVD localized mass depositions. However, these approaches require high temperatures and specific gas environments during tuning, and are therefore not reversible in the field [5-6]. In our work, we present a hardware manipulation technique using nano-electrodeposits integrated with a microphone membrane for directional hearing aid applications. The technique operates at room temperature, consumes very little power, and is both non-volatile and reversible.

DEVICE DESIGN AND OPERATING PRINCIPLE

Device Design

The directional microphone is composed of two omni-directional microphones fabricated side by side. The microphones have two unique components: a multiple-layer diaphragm (parylene/Gold/parylene/SiO₂/Ag-doped GeSe from bottom to top layer) and cathode/anode electrodes to grow/retract nano-electrodeposits on the solid electrolyte. *Figure 1 (a)* shows a fabricated dual microphone. The air gap between top and bottom electrodes is 1 μm, defined by a SiO_2 sacrificial layer. An oxidizable Ag electrode (anode) and electrochemically-inert Ni electrode (cathode) are placed on a silver ion-containing solid GeSe electrolyte film. *Figure 1 (b) and (c)* show SEM and Veeco NT9800 optical images of grown Ag nano-electrodeposits from the cathode tip, respectively, on the parylene diaphragm. The height of nano-electrodeposits is a maximum of 90 nm.

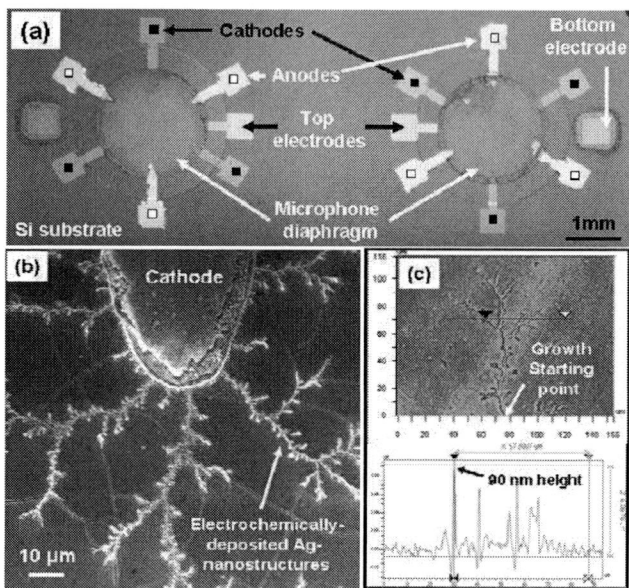

Figure 1: (a) Top view of a fabricated dual capacitive omni-directional microphone with a Ag-doped GeSe solid electrolyte on a suspended parylene diaphragm. (b) Grown nano-electrodeposits from a cathode on the parylene diaphragm. (c) An image produced by an optical profilometer; the height of the nano-electrodeposits is a maximum of 90 nm.

Operating Principle

Figure 2 illustrates the operating principle of electrodeposition on a Ag-doped GeSe solid electrolyte. The process is analogous to aqueous electrochemical processes such as electroplating but uses silver metal ions dissolved in a high-ion-mobility thin film instead of solution chemistry. The solid electrolyte film is heavily doped with silver ions; a Ag film is deposited by thermal evaporation on a $Ge_{30}S_{70}$ base glass and then dissolved into the solid electrolyte film by photodissolution. The photodissolution process causes the silver to diffuse into the film and breaks the GeSe chemical backbone, freeing Se to form superionic Ag_2Se, which is the source of Ag^+ ions in the electrolyte. When a forward DC bias is applied (the oxidizable Ag electrode is made positive and the electrochemically-inert nickel electrode is negative), an electrochemical reaction allows oxidation of the Ag electrode (anode) and rapidly draws Ag⁺ ions through the electrolyte across the diaphragm to the Ni electrode (cathode). Electrodeposition only occurs if the applied bias is sufficient to overcome the chemical potential barrier (a few hundred mV) at the electrode. When the polarity is reversed (oxidizable Ag electrode is negative and the nickel electrode positive), the grown electrodeposits dissolve via oxidation. Structural asymmetry is essential for repeated nano-electrodeposit growth (electrodeposition) and retraction (electrodissolution) [7]. The growth path of the nano-electrodeposits is dendritic, branching out laterally as they grow toward the anode.

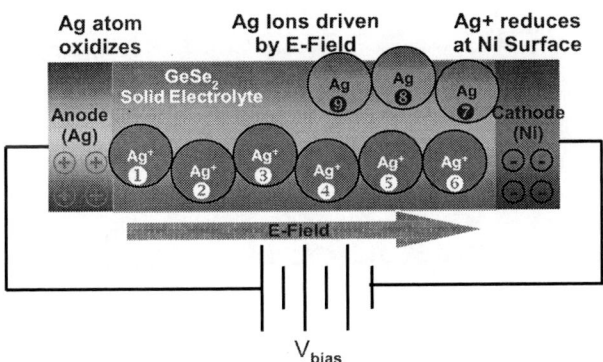

Figure 2: Operating principle of the nano-electrodeposition mechanism in a Ag-doped GeSe solid electrolyte.

FABRICATION AND EXPERIMENTAL SETUP

Fabrication

The fabrication process of a microphone combined with a Ag-doped GeSe solid electrolyte is shown in *Figure 3*. The structure requires a seven mask process, including deposition of the Ag-doped GeSe solid electrolyte on the parylene diaphragm suspended over the silicon substrate. All metal patterning is performed by lift-off processing.

(a) There are two parylene diaphragm coatings for the suspended diaphragm. The first parylene layer (3000Å) is coated on top of a 1-μm thick sacrificial oxide layer on a silicon wafer to provide electrical isolation between the top and bottom electrodes. The parylene layer is etched in an oxygen plasma for 1.5 min at an etch rate of 2000 Å/min under 15-sccm oxygen, 100-W RF power, and 500-V bias. A 3000-Å thick top electrode (Cr/Au) is then deposited and patterned on this first parylene layer. The second parylene layer (3 μm) is coated on the top electrode and patterned in oxygen plasma for 15 min. This second parylene layer is the main structural layer of the diaphragm. (b) GeSe base glass (2400 Å) and Ag layers (800 Å) are thermally evaporated and patterned on the diaphragm. The optimized ratio of GeSe to Ag is approximately 3:1 according to previous studies [8]. (c) Immediately after the deposition, photo-dissolution is performed using a 15-min UV exposure to diffuse Ag into the GeSe layer to form the solid electrolyte. (d) The anode (Ag) and cathode (Ni) are separately evaporated and patterned on the diaphragm. (e) The bottom electrode is formed by removing the 1-μm thick oxide by buffered oxide etchant for 50 min, and Cr/Au metal films are evaporated and patterned to access the silicon substrate. (f) Finally, the diaphragm is defined

by Deep Reactive Ion Etch (DRIE) using an etch rate of 3 μm/min on the back side of the silicon wafer. The diaphragm is then released using concentrated hydrofluoric acid for 15 min. to remove the sacrificial oxide layer.

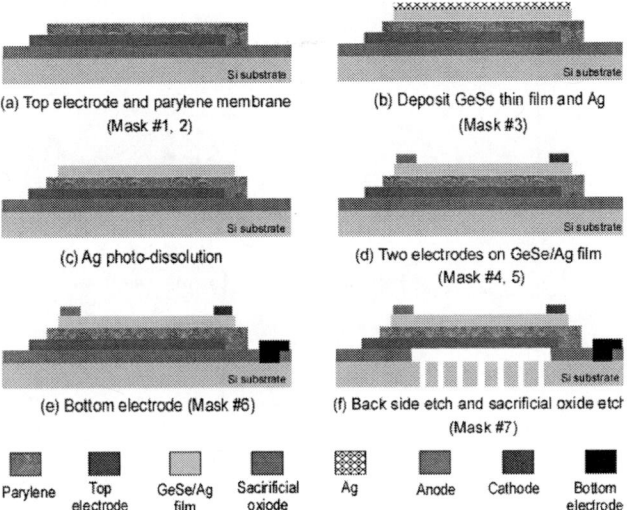

Figure 3: Microphone fabrication process.

Experimental Setup

Figure 4 (a) shows the experimental setup uses to test the directionality of the microphone. The microphone is mounted on a PCB along with capacitive readout circuits. The PCB is placed inside a Faraday cage to avoid electromagnetic interference. In order to minimize 60 Hz noise, Li-ion batteries are used to grow/retract the nano-electrodeposits and to power the circuits. The capacitive readout has a passband from 100 Hz to 100 kHz, with a peak capacitance to voltage conversion of 300 mV/pF. The entire experiment is performed in an anechoic chamber to ensure minimization of unwanted acoustic signals. The output of the readout circuit is connected to a spectrum analyzer (Agilent 35670A) and measured in an Electromagnetic Interference (EMI)-shielded test fixture [9]. An external loudspeaker is mounted on a rotation stage, 1-m away from the setup and driven by a burst sine wave at 3 kHz. *Figure 4 (b)* shows the calibration of the microphone; acoustic sound amplitude from the speaker in Sound Pressure Level (SPL) versus the output of the readout circuit. The response is linear before and after nano-electrodeposit growth, showing a regression of 1 and 0.9999, respectively.

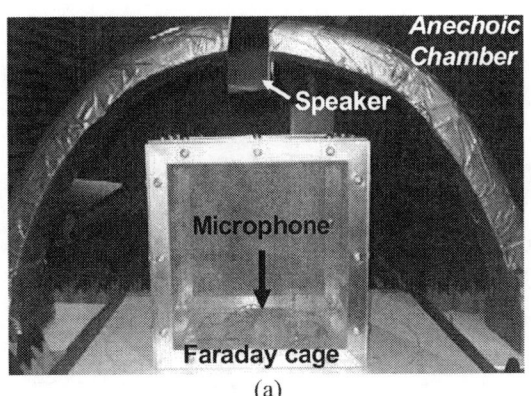

(a)

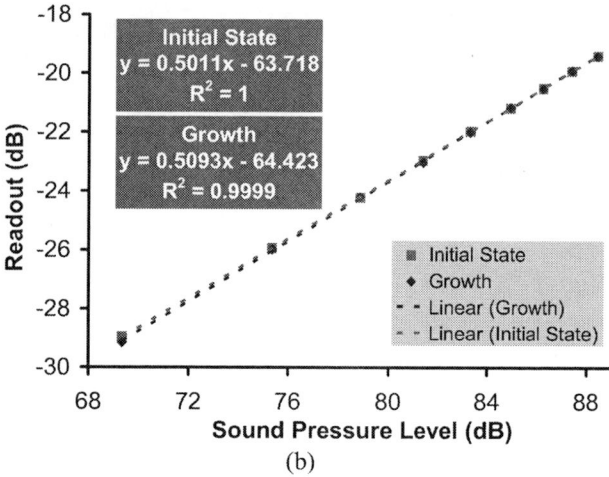

(b)

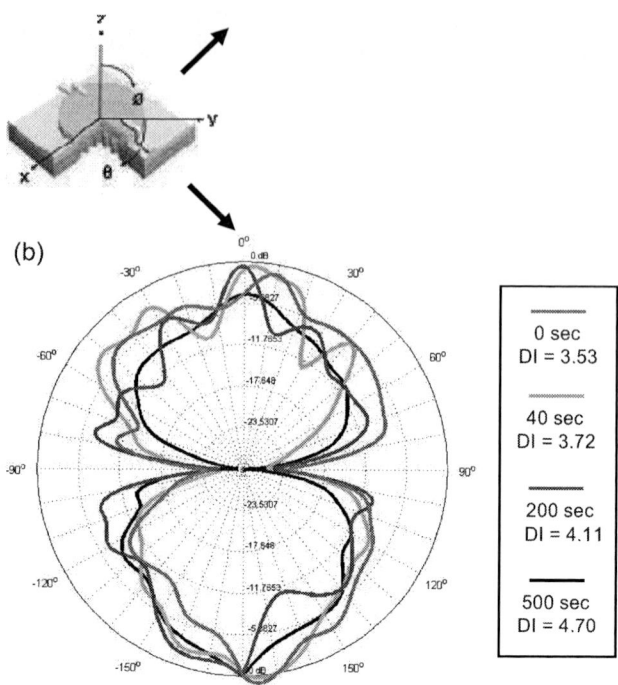

(b)

Figure 4: (a) Experimental setup of the microphone in a shielded Faraday cage. (b) The calibration curve of the microphone: Sound Pressure Level (SPL) versus the output of the readout circuit.

CHARACTERIZATION

Sensitivity tuning is performed by growing and retracting nano-electrodeposits on the solid electrolyte layer of the microphone. Using an external DC bias at room temperature, the nano-electrodeposits can be grown to precisely tune the microphone's sensitivity to incoming acoustic sources. *Figure 5* (a) presents omni-directional responses of individual microphones on a polar plot, showing the sensitivity mismatch of the omni-directional microphones. An acoustic source (3 kHz) rotates out of plane on a y-z axis (Ø) by 180°. The initial microphone sensitivity mismatch between two microphones is approximately 3-dB maximum, which occurs at 50° from the left ground position. *Figure 5* (b) shows the directionality of the microphone on a polar plot, showing a directionality change upon the growth of nano-electrodeposits. The acoustic source rotates in plane on a x-y axis (Θ) by 360° upon nano-electrodeposit growth up to 500 sec at 3 V. As the nano-electrodeposits grow, the polar plot patterns become closer to a "figure 8", indicating high directionality. During the growth period, power consumption is less than 100 µW. The sensitivity mismatch decreases by 1.67 dB (24%), and DI improves by 1.17 dB.

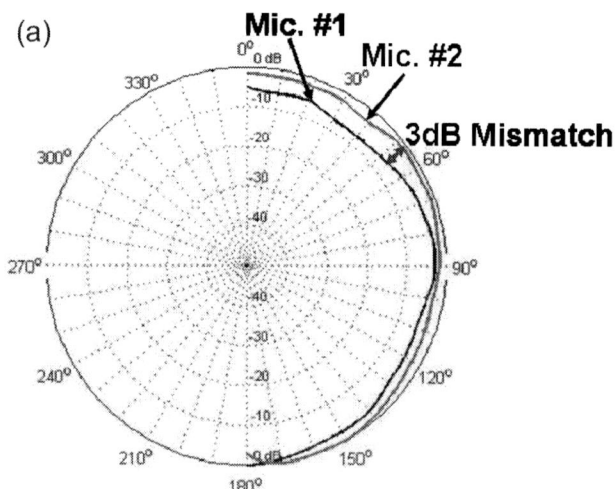

(a)

Figure 5: (a) Omni-directional responses of individual microphones on a polar plot at 3 kHz: rotation of Ø. (b) Directionality of the dual microphone on a polar plot at 3 kHz: rotation of Θ. DI is enhanced following the growth of Ag nano-electrodeposits.

Figure 6 plots the directionality improvement upon nano-electrodeposit growth at 3 kHz by comparing sensitivity mismatch and DI. The measurement results and theoretical calculations are in excellent agreement within 10%. Generally, DI is categorized by three performance ranges: 0–2.0 dB for poor directionality, 2.0–4.5 dB for fair directionality, and 4.5–6.0 dB for excellent directionality [10]. As the nano-electrodeposits grow, the mismatch reduces and consequently DI increases; directionality performance improves from fair to excellent during the nano-electrodeposit growth. Also, the error bars of the characterization curve demonstrate very small measurement errors. The theoretical calculation of microphone DI is given by

$$DI = 10\log\left|\frac{2[R(0)]^2}{\int_0^\pi [R(\theta)]^2 \sin\theta d\theta}\right|$$

, where $R(\theta)$ is the directivity pattern and a function of incident sound angle. The pattern, $R(\theta)$, is given by

$$R(\theta) = s_{-1}(1+\delta)\cdot e^{-j\frac{kd\cos\theta}{2}} + s_1\cdot e^{j\frac{kd\cos\theta}{2}}$$
$$\approx (s_{-1}+s_1) + j\frac{kd}{2}\cos\theta(s_{-1}+s_1)$$
$$\approx (A + B\cos\theta) + \frac{\delta}{2}(A + B\cos\theta) + j\frac{\delta B}{kd}$$

, where s_{-1} is the incorrect microphone sensitivity by δ, s_1 the correct microphone sensitivity, k is the acoustic wave number given by $k = \frac{2\pi}{\lambda}$, λ is the wavelength of the sound, d is the distance between microphones, θ is the angle between the line joining the microphones and the propagation direction of the incoming wavefront.

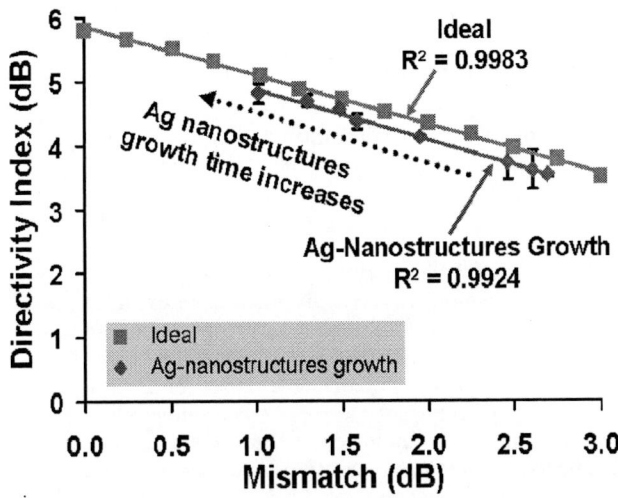

Figure 6: DI vs. mismatch analysis at 3 kHz: measured data shows DI improvements upon the growth of nano-electrodeposits (red solid line is an ideal response)

CONCLUSION

We have demonstrated an *in-situ* non-volatile tuning method for adjusting MEMS microphone sensitivity to achieve high directionality in hearing aids. The technique uses electrochemically formed nano-electrodeposits on a Ag-doped GeSe solid electrolyte integrated with the microphone diaphragm. The nano-electrodeposits are controlled by a DC bias at room temperature and consume less than 100 μW during tuning, with no power consumption following adjustment. The tuning can be used to reduce a dual microphone's mismatch to achieve better directionality by the way of mass and stress redistribution on the microphone diaphragm. In our studies, the sensitivity mismatch decreased by 1.67 dB, and DI improved by 1.17 dB. The success of the *in-situ* method described in this paper suggests that precise mechanical manipulation of a microdevice diaphragm is possible, and we believe that the results are a significant step toward realizing a practical application in MEMS devices, which could also include sensitivity matching, post-package trimming, and self calibration.

ACKNOWLEDGMENTS

The authors thank Mr. Yongmo Yang, Ms. Rhonda Steele, Knowles Electronics, and the staff at the Center for Solid State Electronics Research (CSSER) at Arizona State University for their help. We also appreciate the US National Science Foundation supporting this work under grant #0627777 and #0652136.

REFERENCES

[1] J. Tchorz, "Effects of Microphone Mismatch in Directional Hearing Instruments," *The Hearing Review*, 2001.

[2] S. Kochkin, "Customer Satisfaction with Single and Multiple Microphone Digital Hearing Aids," *The Hearing Review*, 2000.

[3] Gennum Corp., *Obtaining Directionality Using FRONTWAVE™ Multi-Microphone Hybrid.* 2000.

[4] V. Hamacher, *et al.*, "Signal Processing in High-End Hearing Aids: State of the Art, Challenges, and Future Trends," *J. of Applied Signal Processing*, vol. 18, pp. 2915–2929, 2005.

[5] S. C. Jun, *et al.*, "Electrother- mal frequency tuning of a nano-resonator," *Electronics Letters*, vol. 45, no. 25, pp. 1484-1485, 2006.

[6] D. Joachim *et al.*, "Characterization of selective polysilicon deposition for MEMS resonator tuning," *J. of MEMS*, vol. 12, no. 2, pp. 193-200, 2003.

[7] M. N. Kozicki *et al.*, "Mass transport in chalcogenide electrolyte films – materials and applications," *J. of Non-Crystalline Solids*, vol. 352, no. 6-7, pp. 567-577, 2006.

[8] M. Mitkova *et al.*, "Silver incorporation in Ge–Se glasses used in programmable metallization cell devices," *J of Non-Crystalline Solids*, vol. 299, no. 302, pp. 1023-1027, 2002.

[9] S-.S. Je, *et al.*, "In-situ Tuning of Omnidirectional microelectromechanical-systems microphones to improve performance fit in hearing aids," *Applied Physics Letters*, vol. 93, 123501, 2008.

[10] B. E. Walden *et al.*, "Comparison of benefits provided by different hearing aid technologies," *J. of the American Academy of Audiology*, vol. 11, no. 10, pp. 540-560, 2000.

NANOTEXTURE ELECTRODE ON NANOPOROUS AAO DIELECTRIC FOR MICRO TACTILE SENSING DEVICES

Chi-Tsung Hong[1], Li-An Chu[3], Ann-Shyn Chiang[3], and Weileun Fang[1,2]

[1] Institute of NanoEngineering and MicroSystems, [2]Department of Power Mechanical Engineering, and [3]Institute of Biotechnology, National Tsing Hua University, Hsinchu, Taiwan

ABSTRACT

This study reports the integration of nano-texture electrode and self-assembly anodic aluminum oxide (AAO) dielectric layer on Si-substrate for capacitive-type micro tactile sensing devices. The advantages of these nano-structure micro devices are as follows, (1) employing AAO as dielectric film for its large dielectric constant (2.5-fold higher than oxide), and high electrical resistance; and (2) employing the AAO with nanoporous as template to form the nano-texture on metal electrode film to increase sensing area. Thus, the sensitivity of tactile sensor is increased. In application, the MOS (*Au-AAO-Si*) and the parallel-plate (*Au-AAO-Ti*) capacitors are fabricated on Si-substrate and then characterized. These two capacitors also show their capability as a tactile sensing device for tiny objects by detecting the contact of *Drosophila* and ant.

1. INTRODUCTION

The anodic aluminum oxide (AAO) film with uniform-distributed and high-density nanoscale pore (several microns in thickness and several nanometers in diameter) is a promising material for MEMS/NEMS. Such nanoporous AAO film finds many applications in nano structures and devices [1-3]. The AAO can also integrate with metal film using planar fabrication process to realize various micro sensors [4-7]. In these applications, the AAO film acting as a template to enable the formation of nano-texture metal film. The film with nano-texture shows outstanding material properties for micro sensors. Thus, the integration of nano-texture thin film and AAO nanostructure on Si offers an alternative for MEMS applications. Moreover, the integration of AAO film with other nano-structures, such as nanotubes, nanowire, nanoparticle, and nanoaggregates has also been extensively reported. For instance, the existing works include atomic layer deposition (ALD) [6], thermal CVD synthesis [7-9], electrochemical deposition [10]. However, complicated as well as high temperature processes are required for these approaches.

This study presents a simple process to realize tactile sensor by integrating nano-texture metal electrode and nanoporous AAO dielectric layer on Si. The nanoporous AAO has higher dielectric constant than SiO2. In addition, the sensing area is significantly increased by the metal electrode with nano-texture. Thus, the sensitivity of the tactile sensor is improved. In application, the MOS and parallel-plate capacitors formed by the nano-structure are realized and characterized. The detecting of *Drosophila* and ant also demonstrate the capability of these two capacitors as a tactile sensing device for tiny objects.

2. DESIGN CONCEPT

This study exploits the AAO nano-structure to realize two different capacitor designs: (a) the MOS capacitor, and (b) the parallel-plate capacitor, and further employs these capacitor for the application of tactile sensor. Fig.1 illustrates the device architecture of the present capacitors with AAO nano-structure. In this architecture, the AAO is fabricated on top of Si substrate. The metal film is further deposited and patterned on top of nanoporous AAO. Thus, metal film with nano-texture surface topography is also achieved. Since AAO acts as an insulator, the architecture in Fig.1a consisted of *Au-AAO-Si* forms a metal–insulator-semiconductor (MIS) electronic device. On the other hand, the architecture in Fig.1b consisted of *Au-AAO-Ti* forms a metal–insulator-metal (MIM) electronic device. This study employs the MIS and MIM architectures to respectively fabricate MOS and parallel-plate capacitors on Si substrate using the AAO film. As a result, the metal-film with nano-texture surface topography has high surface-to-volume ratio and can significantly improve the sensitivity of the presented tactile sensor.

3. FABRICATION PROCESS AND RESULTS

This study has established the fabrication processes shown in Fig.2 to implement the presented capacitance sensing devices using AAO film. Firstly, the silicon substrate was respectively evaporated with Al film and Al/Ti films, as shown in Fig.2a. As in Fig.2b, the AAO film with uniform and high density nanoporous was fabricated through anodization process in 0.3M oxalic acid (at 40V and 10°C).

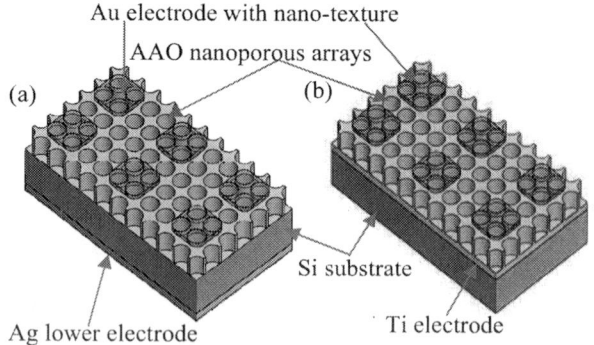

Fig. 1 Scheme of the proposed capacitive-type micro tactile sensing devices with nanoporous AAO and nano-texture Au electrode, (a) the MOS capacitor, and (b) the parallel-plate capacitor.

978-1-4244-2977-6/09 $25.00 © 2009 IEEE

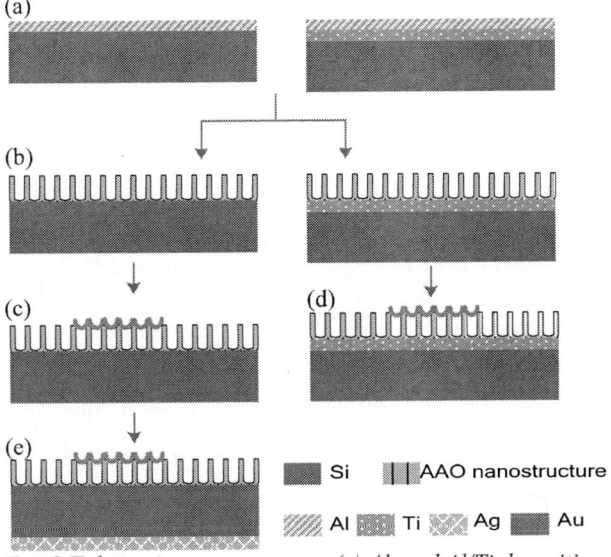

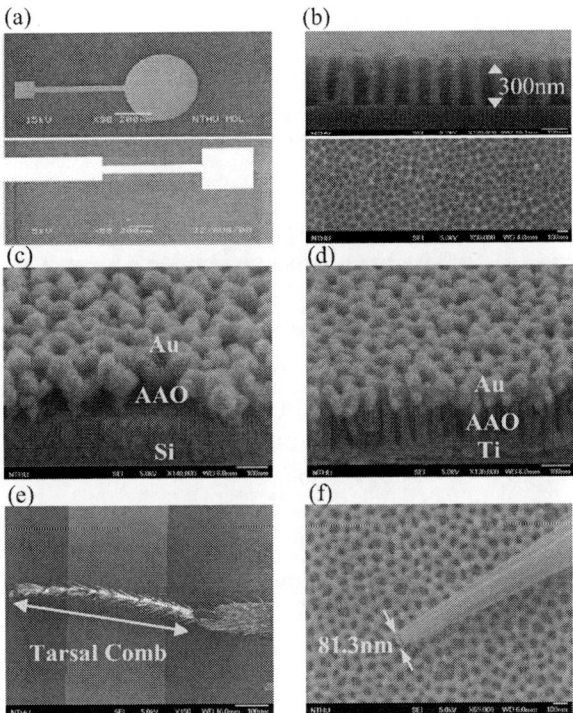

Fig. 2 Fabrication process steps, (a) Al, and Al/Ti deposition on Si substrate, (b) alumina nanoporous arrays anodization step, (c-d) Au film deposition on AAO to form nano-texture electrode for capacitors, other area of AAO film was etched by phosphoric acid, and (e) Ag deposition for the lower electrode of MOS capacitor.

As illustrated in Fig.2c-d, the Au film was evaporated and patterned on AAO film. Thus, the AAO acted as a template to enable the formation of nano-texture on Au film. The devices in Fig.2c and Fig.2d respectively form the MOS capacitor (*Au-AAO-Si*), and the parallel-plate capacitor (*Au-AAO-Ti*). The Ag film was deposited on the backside of MOS capacitor to act as the lower electrode, as shown in Fig.2e. Fig.3a shows Au electrode patterns can be easily defined on top of AAO. The top one is a circular Au electrode with 200μm in radius, and the bottom shows a 500μm×500μm square electrode. Fig.3b shows the side-view of AAO nanoporous (above); and the top-view of nano-texture on Au film (below). The nanoporous AAO contains vertical pore of 300nm in thickness. The top-view photo indicates the AAO template successfully enables the formation of nano-texture on Au film. Fig.3c shows the films stacking of the MOS capacitors. The Au nano-texture formed on top of the nanoporous AAO can be clearly observed. Figs.3d shows the films stacking of the parallel-plate capacitor. Fig.3e shows the Tarsal Comb of *Drosophila* leg on top of Au sensing electrode. The zoom-in micrograph in Fig.3f further shows one of the ~80nm wide bristle on Tarsal Comb contacting with the Au sensing electrode. The nano-texture on the Au film is clearly observed in this micrograph.

4. TESTING AND RESULTS

• Characterization of the AAO capacitors

The capacitors were characterized using the apparatus in Fig.4a. The C-V curve measured using the LCR meter was used to evaluate the electronic characteristics of nanoporous

Fig. 3 FE-SEM micrographs of typical fabrication results, (a) top view of Au electrode, (b) side view of AAO nanoporous and top view of Au nano-texture, (c) zoom-in of the Au-AAO-Si MOS capacitor, (d) zoom-in of the Au-AAO-Ti parallel-plate capacitor, (e) Tarsal Comb of Drosophila on Au electrode, and (f) a ~80nm wide bristle on Tarsal Comb contacting with the nano-texture Au electrode.

AAO and nano-texture metallic films. During the C-V measurement, the input signals contain a DC voltage with sweeping range of ±15V and a 100mV AC signal of different frequencies. Fig.4b shows the typical C-V curves measued from MOS capacitors. The triangle dots represent the measurement results from the *Au-AAO-Si* capacitor. In comparison, the measurement on the *Au-SiO$_2$-Si* capacitor was also performed, as depicted by the circle dots. The results show the capacitance of *Au-AAO-Si* MOS capacitor is much higher than that of the oxide one, as the bias DC voltage larger than +5V. The higher capacitance is mainly contributed by the higher dielectric constant of AAO film as well as the larger surface area of nano-texture Au electrode.

The measured C-V curve in Fig.4c indicates the *Au-AAO-Ti* parallel-plate capacitor has a stable capacitance C=75pF at different driving voltage. The dielectric constant ε of AAO film can be expressed as,

$$\varepsilon = \frac{C \cdot t}{\varepsilon_0 \cdot A} \qquad (1)$$

where A is the area of Au electrode, t is the thickness of AAO film, ε_0 is permittivity in the air. Thus, from the measured capacitance C and Eq.(1), the dielectric constant of AAO film was determined to be ε=10.03 (at 100kHz), which is 2.5-fold higher than that of SiO$_2$ (ε=3.9).

978-1-4244-2977-6/09 $25.00 © 2009 IEEE   101

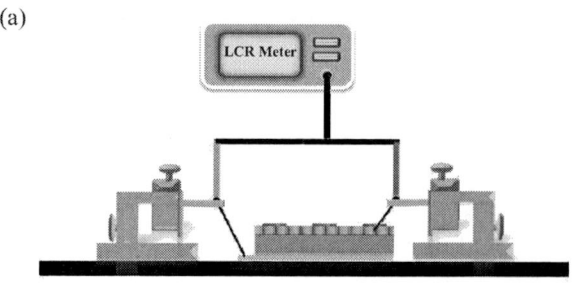

(a)

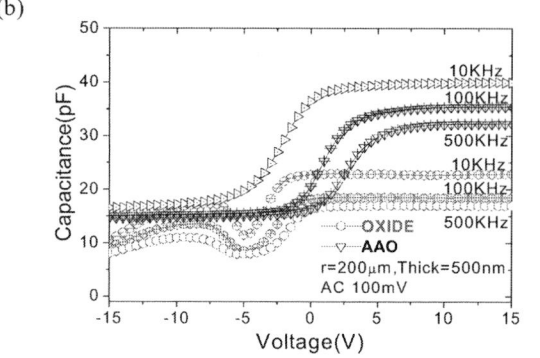

(b)

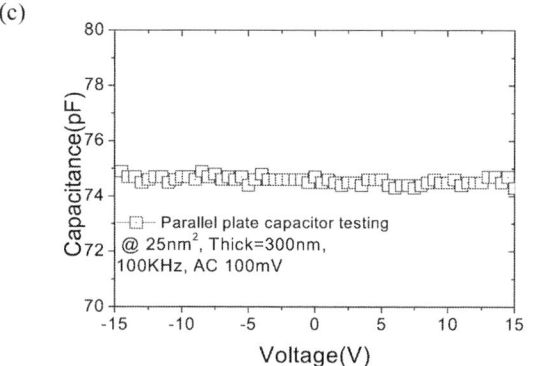

(c)

Fig.4 C-V characterization of the presented capacitors, (a) the test setup, (b) the C-V curve for MOS capacitor, and (c) the C-V curve for parallel-plate capacitor.

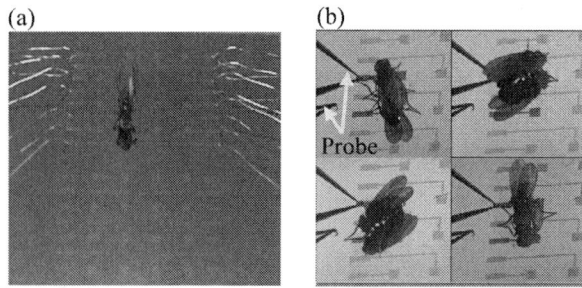

Fig.5 Micro tactile sensing tests using Drosophila, (a) a Drosophila on a sensing chip with wire bonding as interconnection, and (b) a Drosophila walking on a sensing chip with probe as interconnection.

● **Tactile sensing test of AAO capacitors by *Drosophila***

The performances of the AAO-based capacitors as the tactile sensor for tiny objects were also characterized. The *Drosophila* on employed as the sample for the micro tactile sensing tests. Fig.5a shows the *Drosophila* on a MOS capacitor sensing chip. The C-V curve of capacitor has a small perturbation as the *Drosophila* contact the sensing pixel on chip. Since the *Drosophila* was special treated, it could only walk on the chip surface but not fly, as shown in Fig.5b. Fig.6 shows two typical measurement results from MOS capacitor. The solid dots in Fig.6a show a 6% (in average) capacitance change on *Au-AAO-Si* MOS capacitors due to the contact of ant. In comparison, the hollow dots in Fig.6a show a 1.9% (in average) capacitance change on *Au-SiO₂-Si* MOS capacitors. It demonstrates the sensitivity of the present capacitance tactile sensor is successfully improved by the nano-structure. Moreover, Fig.6b shows a 2% (in average) capacitance change on *Au-AAO-Si* MOS capacitor due to the contact of *Drosophila*. The Pre-Tarsus of ant (more than 70μm in length) is larger than that the Tarsal Comb of *Drosophila* (near 30μm in length). Thus, the ant has a larger contact area with the sensing electrode, and the capacitance change due to the contact of ant is larger than that of *Drosophila*.

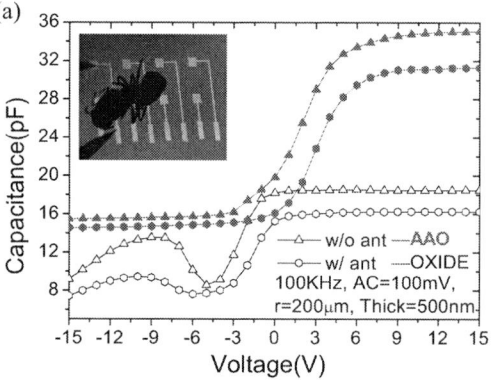

(a)

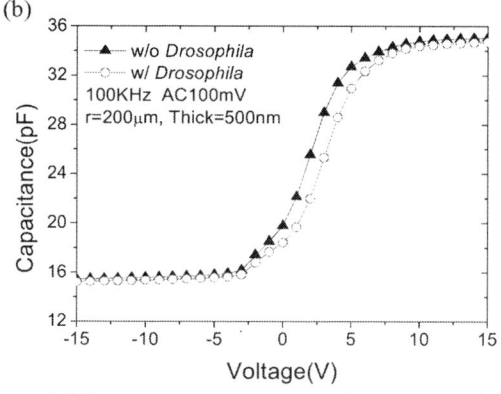

(b)

Fig.6 MOS capacitor micro tactile testing, (a) MOS capacitor C-V detecting for ant contact, (b) MOS capacitor C-V detecting for Drosophila contact.

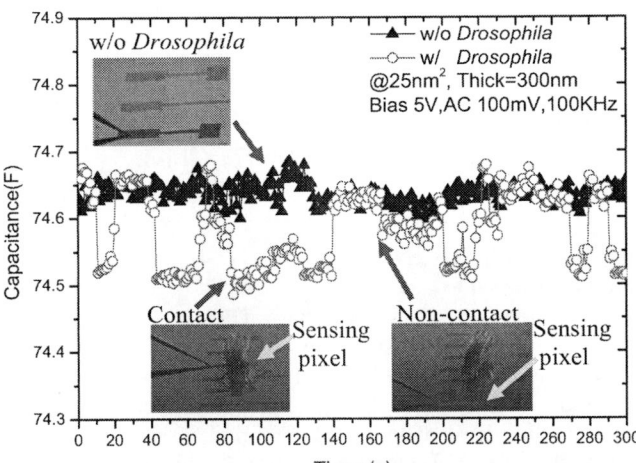

Fig.7 The measured capacitance change of Parallel-plate capacitor resulted from contact/non-contact of Drosophila during the micro tactile testing.

Fig.7 shows the capacitance recorded from the *Au-AAO-Ti* parallel-plate capacitor during the 5-minutes *Drosophila* contact and non-contact test. As show in Fig. 5, the *Drosophila* was walking on the sensing chip during the test. Thus, the *Tarsal Comb* of *Drosophila* intermittently contacted with the sensing electrode and leaded to the capacitance change. The solid dots in Fig.7 show the signal without *Drosophila* as a reference. The hollow dots indicate the capacitance change caused by the intermittent contact of *Drosophila*. Since the capacitance change varies with the total contact area between *Tarsal Comb* and sensing electrode, the capacitor has a 0.1~0.15pF capacitance change when *Drosophila* touching the sensing-plate.

5. CONCLUSIONS

This study reports the integration of nano-texture electrode and nanoporous AAO dielectric layer on Si-substrate for capacitive-type micro tactile sensing devices. These sensing devices can be batch fabricated on Si substrate. In applications, the MOS capacitor and parallel-plate capacitor have been successfully implemented and characterized. According to the higher dielectric constant (ε=10.03 (at 100kHz), and is 2.5-fold higher than that of SiO_2) of AAO film as well as the larger surface area of nano-texture Au electrode, the AAO based capacitors have superior electronic performances. The detecting of *Drosophila* and ant also demonstrate the capability of these two capacitors as a tactile sensing device for tiny objects. Moreover, the MOS capacitor is a fundamental CMOS device. Based on the presented approach, the nanoporous AAO film and the nano-texture metal film can be easily integrate with the existing micro fabrication processes on Si substrate. Thus, the superior AAO based MOS capacitor can further integrate with other electronic devices for more applications.

6. ACKNOWLEDGEMENTS

This paper was partially supported by Brain Research Center at the National Tsing Hua University, and by Nation Science Council, Taiwan, under contract NSC 95-2221-E-007-068-MY3. The authors would also like to appreciate National Tsing Hua U. (Taiwan), and National Chiao Tung U. (Taiwan) in providing fabrication facilities. The authors also deeply appreciate Prof. T.-K. Sang and Prof. K.-C. Hwang of National Tsing Hua U. for the preparing of *Drosophila* and FE-SEM micrographs, respectively.

REFERENCES

[1] P.S. Lee, J. Lee, N. Shin, K.H. Lee, D. Lee, S. Jeon, D. Choi, W. Hwang, and H. Park, "Microcantilevers with Nanochannels," *Adv. Mater.*, pp. 1732-1737, 2008.

[2] X.D. Dang, W. Plieth, S. Richter, M. Plotner, and W.J. Fischer, "Aluminum Oxide Film as Gate Dielectric for Organic FETs: Anodization and Characterization," *Phys. Stat. Sol. (a)*, pp. 626-632, 2008.

[3] J. Miao, Z. Xu, X. Zhang, N. Wang, Z. Yang, and P. Sheng, "Micropumps Base on the Enhanced Electroosmotic Effect of Aluminum Oxide Membranes," *Adv. Mater.*, pp. 4234-4237, 2007.

[4] D. Ding, and Z. Chen, "A Pyrolytic Carbon-Stabilized, Nanoporous Pd Film for Wide-Range H_2 Sensing," *Adv. Mater.*, pp. 1996-1999, 2007.

[5] L.M. Hagelsieb, B. Foultier, G. Laurent, R. Pampin, J. Remacle, J.P. Raskin, and D. Flandre, "Electrical Detection of DNA Hybridization: Three Extraction Technology Based on Interdigiated Al/Al_2O_3 Capacitor," *Biosens. Bioelectron.*, pp. 2199-2207, 2007.

[6] W.H. Kim, S.J. Park, J.Y. Son, and H. Kim, "Ru Nanostructure Fabrication Using an Anodic Aluminum Oxide Nanotemplate and Highly Conformal Ru Atomic Layer Deposition," *Nanotechnology*, 2008.

[7] J.I. Sohn, Y.S. Kim, C. Nam, B.K. Cho, and T.Y. Seong, "Fabrication of High-Density Arrays of Individually Isolated Nanocapacitors Using Anodic Aluminum Oxide Templates and Carbon Nanotubes," *Appl. Phys. Lett.*, 123115, 2005.

[8] N.V. Quy, N.D. Hoa, M. An, Y. Cho, and D. Kim, "A High-Performance Triode-Type Carbon Nanotube Field Emitter for Mass Production," *Nanotechnology*, 345201, 2008.

[9] N.D. Hoa, N.V. Quy, Y. Cho, and D. Kim, "An Ammonia Gas Sensor Based on Non-catalytically Synthesized Carbon Nanotubes on an Anodic Aluninum Oxide Template," *Sens. Actuators B*, pp. 447-454, 2007.

[10] M.S. Sander, and L.S. Tan, "Nanoparticle Arrays on Surface Fabrication Using Anodic Alumina Film as Template," *Adv. Funct. Mater.*, pp. 393-397, 2003.

AN ELECTRICALLY DECOUPLED LATERAL-AXIS TUNING FORK GYROSCOPE OPERATING AT ATMOSPHERIC PRESSURE

Z.Y. Guo, L.T. Lin, Q.C. Zhao, J. Cui, X.Z. Chi, Z.C. Yang, G.Z. Yan

Institute of Microelectronics, Peking University, Beijing, 100871 China

National Key Laboratory of Micro/Nano Fabrication Technology, Beijing, 100871 China

ABSTRACT

In this paper, a bulk micromachined lateral axis TFG (tuning fork gyroscope) with torsional sensing comb capacitors is presented. Both driving and sensing modes of the gyroscope are dominated by slide film air damping, then it can work even at atmosphere. Novel driving comb capacitors are used to electrically decouple the mechanical coupling from sensing mode to driving mode. The process for this gyroscope is also compatible with z-axis gyroscope, which makes it potentially to realize low coast monolithic MIMU (miniature inertial measurement unit) without vacuum packaging. The TFG was fabricated and tested at atmosphere. The sensitivity is 17.8mV/°/s while the nonlinearity is 0.6%. The bias stability is 0.05°/s (1σ) and the noise floor is $0.02°/s/Hz^{1/2}$.

INTRODUCTION

Though there are still great gaps between the performances of a micromachined gyroscope and the traditional gyroscopes such as ring laser gyroscopes, fiber optic gyroscopes and tuned rate gyroscopes, the micromachined gyroscopes draw more and more attentions and greatly extend the fields of gyroscopes' applications because of their extremely low costs, small size and low power consumption [1]. Micromachined gyroscopes have experienced rapid progress in the past two decades especially since the comb driver was used [1][2].The performance improving factor by 10× every two years has being kept on[3][4]. As for structures, a tuning fork gyroscope is one of the most successful structures of all the efforts to improve the performances. Tuning fork gyroscopes can realize differential sensing to improve the sensitivity and suppress the translational accelerations [2][5]. However, most of the previous reported TFGs used parallel plate capacitors to sense the out of plane motion, which needs high cost vacuum packaging to ensure good performance. Furthermore, although the CFDTS (coupling from driving mode to sensing mode) was suppressed by structure designs for most of the TFGs, the CFSTD (coupling from sensing mode to driving mode) was not well compensated.

In this paper, by adopting the torsional sensing comb capacitors [6] and special driving comb capacitors, a lateral axis TFG which can not only electrically decouple the mechanical coupling from sensing mode but also operate at atmospheric pressure was proposed. Moreover, folded bending sensing beams are used to provide the freedom of out-of-plan torsional movements for the gyroscope. Because of its symmetric structure and the arrangement of the torsional sensing comb capacitors [6], the TFG is immune to acceleration disturbances along x, y and z axes. The TFG was fabricated using modified silicon on glass (SOG) process [6] which is compatible with z-axis gyroscope process.

STRUCTURE DESIGN

Structure of the TFG

The schematic of the presented TFG is shown in Figure 1. It is a symmetric structure with two proof-masses. The movable combs of the novel comb drivers are rigidly connected with the proof-masses which are connected with the gimbal through driving beams. The gimbal with which the movable sensing combs are rigidly connected is connected with the anchors by the sensing beams. Two proof-masses of the TFG are electrostatically actuated to vibrate in opposite directions along x axis. When there is a rotation along y axis, the induced Coriolis forces will force the proof-masses to vibrate reversely along z axis and the torsional comb capacitors can differentially sense the movement.

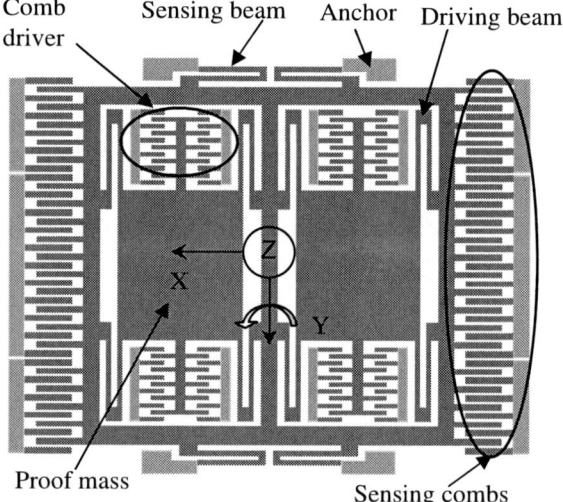

Figure 1: Schematic Diagram of the TFG.

Special Comb Driver Design

The schematic of proposed comb driver is shown in Figure 2 and Figure 3. The height of the movable driving comb h_m is symmetrically higher than the height of the fixed combs h_f by Δh on both top and bottom sides. By deliberately designing Δh and when the out of plane displacement is within a certain range, there is no disturbing electrostatic force between the fixed and the movable driving combs along z axis induced by the drive voltage V. Thus, the CFSTD is electrically decoupled.

According to reference [7], the electrostatic force along z axis per unit length of one unit of the comb driver, F_z, can be calculated using the following equation:

978-1-4244-2977-6/09 $25.00 © 2009 IEEE

$$F_z = \frac{\varepsilon_0 \pi^2}{4g(a-b)}(a\frac{b-d}{a-d} - b\frac{a-c}{b-c})$$

$$\times \frac{V^2}{[F(\delta(u=0),q) + F(k(u=0),q)]^2} \quad (1)$$

Where g is the gap between the adjacent movable and fixed comb fingers and

$$a = e^{\pi(-\frac{h_f}{2} + \frac{h_m}{2} + D_z)/g}$$

$$b = e^{\pi(-\frac{h_f}{2} - \frac{h_m}{2} + D_z)/g}$$

$$c = -e^{\pi(-h_f)/g}$$

$$d = -1 , \quad (2)$$

D_z is the out-of-plane displacement of the movable finger relative to the fixed finger along Z axis.

F denotes the incomplete elliptic integration of the first kind,

$$q = \sqrt{\frac{(b-d)(a-d)}{(a-c)(b-a)}}$$

$$\delta = \sin^{-1}\sqrt{\frac{(b-d)(a-d)}{(a-c)(a-d)}}$$

$$k = \sin^{-1}\sqrt{\frac{(a-c)(b-u)}{(b-c)(a-u)}} \quad (3)$$

It can be seen that F_z is a function of the parameters of the comb and the out-of-plane displacement, h_m, h_f, g and D_z. Considering $\Delta h = \frac{1}{2}(h_m - h_f)$ and taken that the height of the movable combs h_m and the gap g keep constant, for example they are 80 μm and 4μm respectively, F_z has two independent variables Δh and D_z.

At the same time, the desired electrostatic driving force per unit height which is along x axis, defined as F_{dr} here, can be evaluated by [7]

$$F_{dr} = \frac{\varepsilon_0 V^2}{g} \quad (4)$$

Thus the normalized driving force $F_{dr}/(\varepsilon_0 V^2) = 0.25\,\mu m^{-1}$ for g=4 μm.

Figure 4 shows the normalized force versus normalized out of plane displacement under different values of Δh.

As shown in Figure 4, the normalized force, $F_z/(\varepsilon_0 V^2)$, for the combs with equal height ($\Delta h = 0$) is inconstant and substantially large when compared with the desired driving force, which will deteriorates the performances of a gyroscope if it is not suppressed

efficiently.

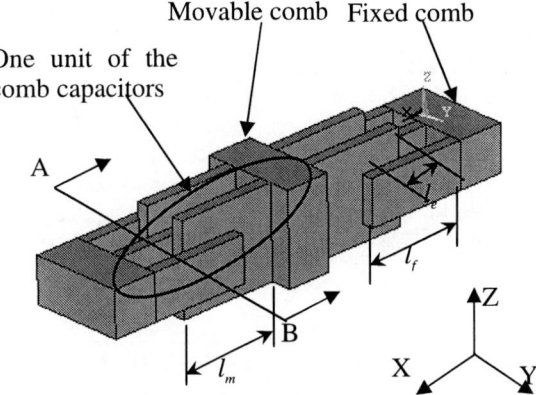

Figure 2: Schematic Diagram of the comb driver.

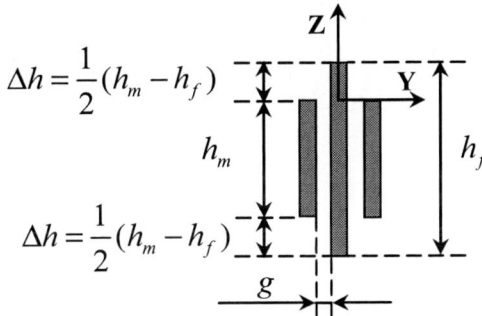

Figure 3: Sectional view of one unit of the comb driver.

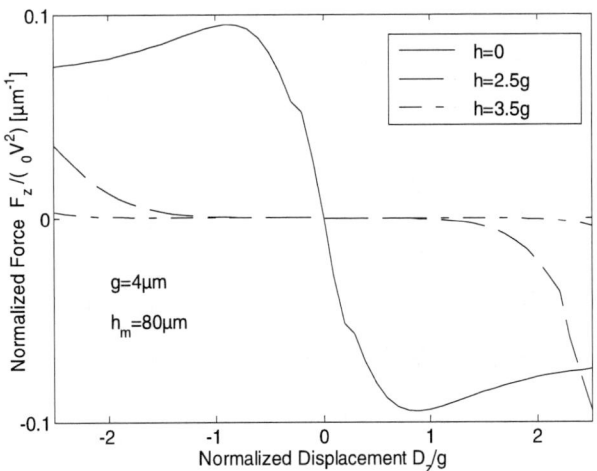

Figure 4: Normalized Z Axis force versus normalized out-of-plane displacement

However, with the increase of Δh, $F_z/(\varepsilon_0 V^2)$ decreases and tends to zero when the normalized displacement, D_z/g, is within a certain range. Furthermore, the range increases with Δh. So the disturbing electrostatic force, F_z, can be efficiently suppressed by reasonable design of comb driver. For example, when $\Delta h \geq 2.5g$ and $D_z/g \in (-1,1)$, F_z can be taken as zero.

Sensing Beams

Folded beams have been widely used in z axis in-plane motion micromachined gyroscopes [4]. However, for lateral out-of-plane rotational motion gyroscopes, double clamped torsional beams are commonly used, which will deteriorate the linearity of a gyroscope [2][5][6].

In our design, folded bending sensing beams are used to provide the freedom of out-of-plane rotational movements for the gyroscope. Folded bending sensing beams are very important for the TFG not only to get high linearity even for large full scale but also to accquire good bias stability bescase they can release residual stress [1][2][8].

FABRICATION

The TFG with novel driving and sensing combs can be realized by the process has been proposed in our previous work [6]. The last step of the fabrication process is shown in Figure 5. Figure 6, 7 are the SEM photo of close-up view of the proposed comb driver and the torsional comb capacitors of the TFG respectively. Figure 8 shows a photo of the fabricated gyroscope.

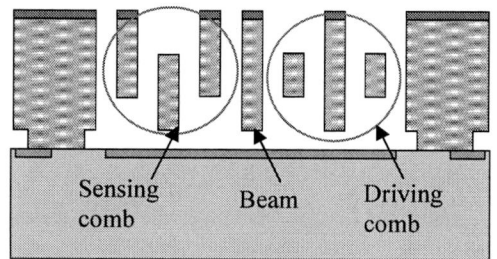

Figure5: The last step of Fabrication process of the TFG.

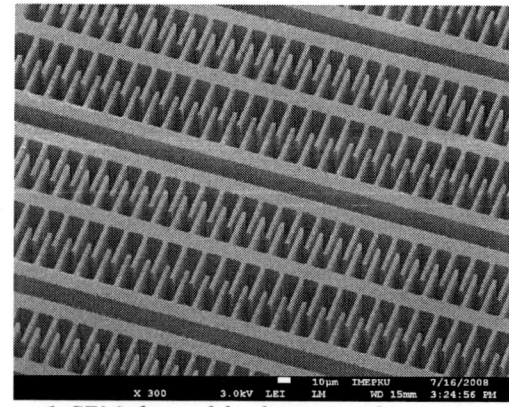

Figure 6: SEM photo of the driving combs.

Figure 7: SEM photo of the sensing combs.

Figure 8: Picture of the whole device.

CHARACTERIZATION

The TFG was tested at atmospheric pressure. Figure 9 shows the test result of the coupling between the two modes. The coupling from driving mode to sensing mode is -41dB while the coupling from sensing mode to driving mode is -50dB. Figure 10 shows the outputs under different input angular rate with full scale of 1000 °/s. The sensitivity is 17.8mV/°/s and the non-linearity is 0.6%. As shown in Figure 11, the bias stability of the TFG for 30 minutes is 0.05°/s (1 σ). Figure 12 shows the noise spectrum analysis. The noise floor is 0.02°/s/Hz$^{1/2}$.

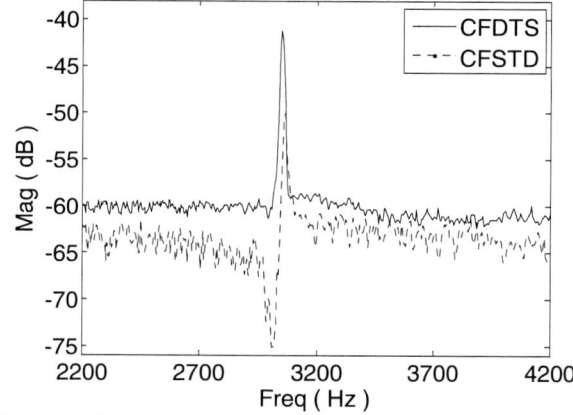

Figure 9: Coupling test of the TFG.

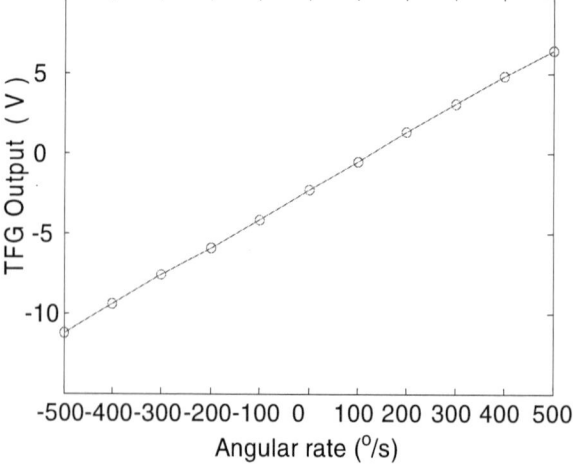

Figure 10: Output of the TFG versus angular rate input.

978-1-4244-2977-6/09 $25.00 © 2009 IEEE

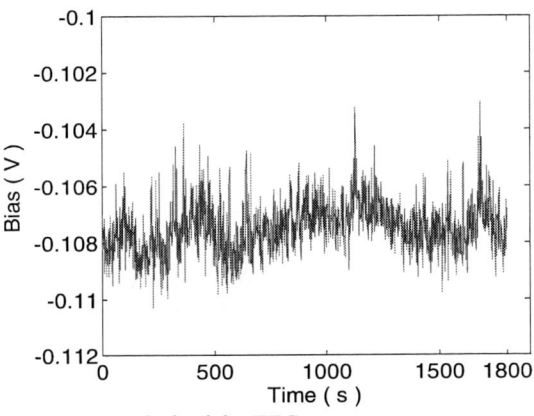

Figure 11: Bias drift of the TFG.

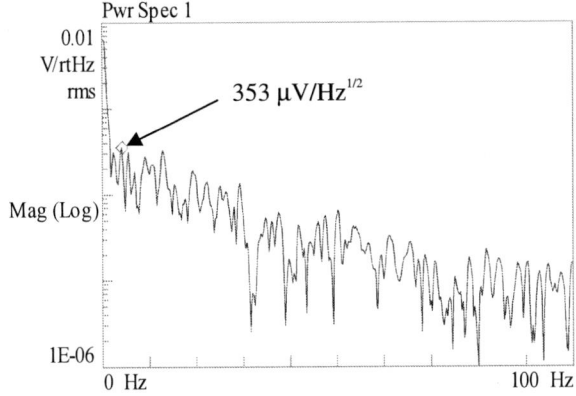

Figure 12: Noise spectrum analysis of the TFG.

CONCLUSION

An electrically decoupled lateral axis micromachined TFG with torsional sensing comb capacitors is presented. Both driving and sensing modes of the gyroscope are dominated by slide film air damping, then it can work even at atmosphere. The novel comb driver efficiently suppressed the coupling from sensing mode to driving mode. The measured CFSTD is -50dB. The process for this gyroscope is compatible with z-axis gyroscope. The sensitivity is 17.8mV/°/s while the nonlinearity is 0.6%. The bias stability is 0.05°/s (1σ) and the noise floor is 0.02°/s/Hz$^{1/2}$.

ACKNOWLEDGEMENT

The authors would like to thank the technical staffs of National Key Laboratory of Micro/Nano Fabrication Technology for the support on device fabrication. The work was partially supported by NSFC (Grant No. 50575001) and National High Technology Research and Development Program of China (Grant No. 2006AA04Z371).

REFERENCES

[1] W.C. Tang, T.H. Nguyen, and R.T. Howe, "Laterally driven polysilicon resonant microstructures," *Sensors and Actuators*, vol. 20, pp. 25-32, 1989.

[2] J. Bernstein, S. Cho, A. T. King, A. Kourepenis, P. Maciel, and M. Weinberg, "A Micromachined Comb-Drive Tuning Fork Rate Gyroscope," in *MEMS'93 Conference*, Fort Lauderdale, FL, February 13-17, 1993, pp. 143-148.

[3] N. Yazdi, F. Ayazi, and K. Najafi, "Micromachined Inertial Sensors," *Proc. of the IEEE*, pp.1640-1659, Aug. 1998.

[4] Ajit Sharma, Mohammad Faisal Zaman, Mark Zucher and Farrokh Ayazi, "A 0.1°/hr Bias Drift Electronically mathed Tunning Fork Microgyroscope," in *MEMS'08 Conference*, Tucson, January 13-17,2008, pp.6-9.

[5] A. Kourepenis, J. Borenstein, J. Connelly, R. Elliott, P. Ward, M. Weinberg, "Performance of MEMS Inertial Sensors," in *IEEE Position Location and Navigation Symposium*, Palm Springs, CA, 1998, pp.1-8.

[6] X.S. Liu, Z.C. Yang, X.Z. Chi, J. Cui, H.T. Ding, Z.Y. Guo, B. Lv, G.Z. Yan, "A X-axis Micromachined Gyroscope with Doubly Decoupled Oscilation Modes," in *MEMS'08 Conference*, Tucson, January 13-17, 2008, pp.860-863.

[7] J.-L. Andrew Yeh, E, Chung-Yuen Hui, and Norman C. Tien, "Electrostatic Model for an Asymmetric Combdrive," *J. Microelectromech. Syst.*, vol. 9, pp. 126-135, 2000.

[8] Francesco Braghin , Ferruccio Resta, Elisabetta Leo, Guido Spinola, "Nonlinear dynamics of vibrating MEMS", *Sensors and Actuators A*, vol.134 (2007), pp. 98–108.

TERAHERTZ METAMATERIALS WITH SIMULTANEOUSLY NEGATIVE ELECTRIC AND MAGNETIC RESONANCE RESPONSES BASED ON BIMATERIAL POP UP STRUCTURES

H. Tao[1], A. Strikwerda[1], K. Fan[1], W. J. Padilla[2], R. D. Averitt[1] and X. Zhang[1]

[1]Boston University, Boston, Massachusetts, USA
[2]Boston College, Chestnut hill, Massachusetts, USA

ABSTRACT

In this paper, we present a metamaterial with simultaneously negative electric and magnetic resonance responses in the terahertz (THz) frequency range. The structure is composed of in-plane split resonator rings (SRRs) for desired electric response and out-of-plane SRRs realized by bimaterial pop up structures for magnetic response. Experimental results conducted using THz Time Domain Spectroscopy (THz-TDS) show that the presented metamaterial has both electric and magnetic resonance peaks at 0.5 THz. The performance of the metamaterial can be further improved by optimization of the structure's geometry and the operating frequency of the metamaterial can be tuned by scaling dimensions.

INTRODUCTION

Terahertz Gap

Nearly all optoelectronic devices function on their responses to electromagnetic waves, and the electromagnetic spectrum covers a range of wavelengths and photon energies. However, the electromagnetic response is not evenly distributed across the electromagnetic spectrum due to the difference in working particles interacting with the electromagnetic waves. At lower frequencies such as a few hundred GHz and lower, electrons are the principle particles and the waves can be generated by electronic devices such as those found in radio and cell phones. On the other hand, at higher frequencies such as infrared through optical wavelengths, the lights are generated by optical devices where photon is the fundamental particle of choices [1]. There is one region of the electromagnetic spectrum comparatively devoid of material response, and was found difficult to be reached by

either end. This is the terahertz (THz) region, alternatively called the far-infrared, which lies below visible and infrared wavelengths and above microwave wavelengths, ranges from frequencies of about 100 GHz to 10 THz (1THz = 10^{12} Hz). Since it is least developed and therefore the least understood, this region is commonly referred to as the "Terahertz gap" [2]. Metamaterials are expected to play an important role in exploring this gap.

Metamaterials

Recently, artificially structured electromagnetic materials have become an extremely active research area because of the possibility of creating materials which exhibit novel electromagnetic responses not available in natural materials, such as negative refractive index [3]. Such electromagnetic composites, often called metamaterials, are sub-wavelength composites where the electromagnetic response originates from oscillating electrons in highly conducting metals such as gold or copper allowing for a designed specific resonant response of the electrical permittivity (μ) or magnetic permeability (ε). The advent of metamaterials has given rise to numerous electromagnetic functionalities previously unimaged, including superlensing [4-7], cloaking [8-10], and more generally, coordinate transformation materials design [11].

This is especially important for the technologically relevant THz frequency regime which is difficult to be reached due to lack of functional sources and detectors. Many metamaterials were initially implemented at microwave frequencies due to ease of fabrication and characterization [12-13]. However, the fabrication of sub-wavelength unit cells becomes increasingly

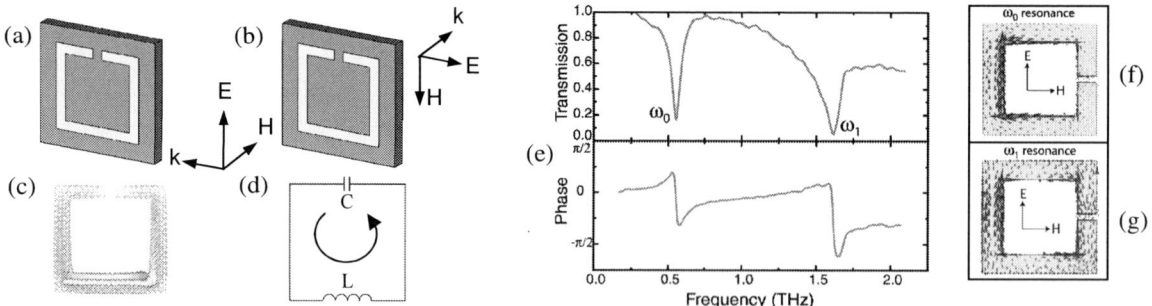

Figure 1: Split-ring resonator (SRR) element for construction of metamaterials: (a) a single SRR with an external alternating magnetic field incident upon it, which induces a circulating current in the ring; (b) the same SRR structure with an external electric field perpendicular to the gap, which also excites a circulating current, as shown in (c), and it has been proven that these two resonances have same dependence on the geometrical parameters; (d) equivalent circuit of SRR with the gap as the capacitance and the current path as the inductance. (e) shows the measured transmission(top) and phase(bottom) spectrum of planar SRR array. Two resonances were observed: the first one (ω_0) comes from the circulating current due to the split gap as shown in (f), which is our aim resonance; the second resonance (ω_1) comes from the dipole coupling from the side bars along the direction of electric field, as shown in (g).

978-1-4244-2977-6/09 $25.00 © 2009 IEEE

challenging in moving from the microwave to higher frequencies such as THz frequencies and even higher infrared and visible frequencies [14]. Since 1 THz corresponds to 300 μm in wavelength, the critical geometric dimensions of the sub-wavelength unit cells are tens of microns or smaller, where MEMS technologies happen to show extreme power and flexibility in terms of fabrication.

Split ring resonators (SRRs)

Split ring resonator, first theoretically introduced by Pendry et al. in 1999 [15] and experimentally verified by Smith et al. in 2000 [16], has been the element typically used for the narrow band resonance response to the electromagnetic field. The SRR structure can be tuned to show either negative electric or magnetic response by being exposed to different orientation of incident radiation and directions of electric and magnetic fields.

The electromagnetic metamaterial can be designed independently to respond to the electric or magnetic component of an electromagnetic wave. When a varying magnetic field polarized normal to the plane of the SRR, as shown in Fig. 1a, circulating currents will be induced within the ring. Due to the split gap, the induced circulating currents will result in a build up of charge in the gap, which concentrates the electric field and stores the energy as a capacitance. When the frequency of externally varying magnetic field is below ω_0, which is the resonance frequency of the SRR due to the gap, the induced currents in the SRR are able to keep up with the driving force, and thus a positive response is observed. However, as the frequency increases, the currents can no longer keep up and begin to lag eventually, which results in an out of phase or negative *magnetic response*.

Recently, there have been further advances in the development of *electric metamaterials* with similar SRR structures that were initially designed to act as magnetic metamaterials, by applying different electromagnetic polarization to the SRR plane [17], as shown in Fig. 1b. It has been shown that the response induced by electric field has the same dependence on the geometrical parameters as the magnetically-driven response [18].

Furthermore, the resonance frequency can be tuned by geometry because of the universality of metamaterial response over many decades of frequency; the operating region of negative values can be made to occur at nearly any frequency, from microwave to optical. There has been considerable effort to develop various metamaterial structures at THz frequencies, however, most THz metamaterials built on planar structures show only electric response due to fabrication and characterization limitations, which will be explained in the following section.

In this paper, we present the first metamaterial with simultaneously negative electric and magnetic response in the terahertz frequency range using MEMS-based bimaterial pop up structures.

DESIGN

As mentioned above, SRR can show either electric or magnetic response, which depends on the orientations of incident light and electromagnetic fields. However, it is difficult to get magnetic response for the planar THz SRR structures under normal incident light, as in this case the wave has to propagate along the SRR plane with magnetic field normal to the SRR plane to excite the magnetic response, which is difficult to be experimentally fabricated and characterized in the terahertz range since the incident light is usually limited normal to the SRR plane to get reasonable signal strength level for measuring the S-parameters. So to get both electric and magnetic response simultaneously for the planar THz SRR structures, the wave has to propagate so that electric field is perpendicular to the SRR's gap to excite the electric LC response and magnetic field is normal to, or at least has some portion projected into, the SRR plane to excite the magnetic response at the same time, which is barely possible for available planar SRRs.

Here we present a new metamaterial structure showing both negative electric and magnetic resonance responses. One single unit cell (εμ-SRR) consists of two in-plane SRR elements (ε-SRRs) for electric response and another two out-of-plane SRR elements (μ-SRRs) supported by bimaterial cantilever legs for magnetic response, as shown in Fig. 2. The bimaterial cantilever legs consist of seemingly arbitrary 200 nm thick gold (Au) and 400 nm thick silicon nitride (SiN$_x$) layers. Both ε-SRRs and μ-SRRs are supported on the 400 nm thick SiNx thin film, which is highly transparent to THz radiation. By polarizing electric filed *E* perpendicular to the ε-SRRs's gap, we are able to couple to the electric LC response. A magnetically-driven response with the same frequency depended features as the electrically-driven response could be obtained by popping up the μ-SRRs such that there is portion of magnetic field *H* projected into the μ-SRRs plane.

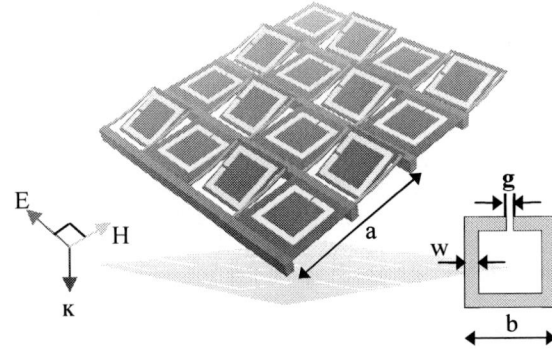

Figure 2: Schematic of the metamaterial structure with simultaneously electric and magnetic resonance responses. The corresponding dimensions are listed in Table I.

*Table I: Dimensions of the presented metamaterial structure (all units in μm): **a**, unit cell; **b**, outer dimension; **w**, line width, **g**, gap distance, t_{Au}, thickness of gold; t_{SiNx}, thickness of silicon nitride.*

Parameters	a	b	w	g	t_{Au}	t_{SiNx}
Dimensions	200	72	8	4	0.2	0.4

FABRICATION

The metamaterial structure was fabricated using a surface micromachining process which typically consists of four steps: 1) deposition of the SiN$_x$ films, 2) patterning of the Au layer for SRRs and bimaterial bending legs, 3) patterning the SiN$_x$ film for supporting plate and bending legs, 4) release of the structure by KOH etching, as shown in Fig. 3a.

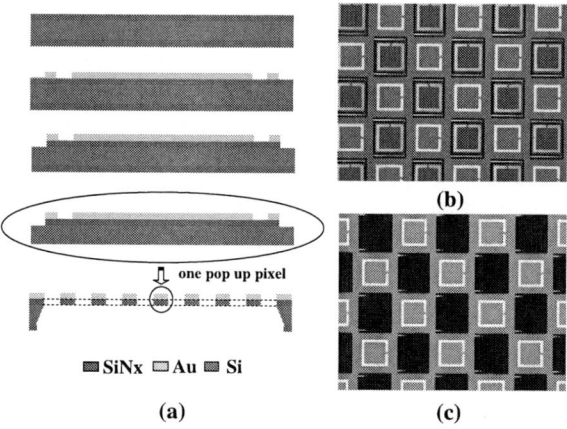

Figure 3: (a) Fabrication process; (b) microscopic image of the as-fabricated sample; (c) microscopic image of the sample after rapid thermal annealing (RTA). (Dark area: out-of-plane pixels/μ-SRRs)

The process started with a 4" silicon wafer with 400 nm thick SiN$_x$ films deposited on both sides. The initial bending angle of the pop up structure originates from residual stress in or between the films. It is well known that the residual stress in LPCVD SiN$_x$ film depends strongly on the deposition conditions such as temperature, pressure and the ratio of the flow rate of source gases during deposition. We purchased 4" (100) 500 μm thick silicon wafers with super low stress LPCVD SiN$_x$ films deposited on both sides (Addison Engineering, Inc.). The SiN$_x$ film on the front side serves as the supporting plate for SRRs and the bottom layer of bimaterial legs whereas the SiNx film on the back side serves as the etching windows for structure release in the last step. Then the SRRs and the top layer of the bimaterial legs were fabricated using standard photolithographic methods and electron beam evaporated deposition of 200 nm of gold following a 10 nm chromium adhesion layer, then undertaking a lift-off process. The next step is to lithograph photoresist mask and it is followed by the removal of unwanted SiNx layer employing reactive ion etching (RIE) technique. Next, the back side SiN$_x$ etching window is patterned by photolithography followed by RIE etching. Finally, the Si substrate beneath the pop up metamaterial elements was etched in KOH solution. The as-fabricated structures were near flat with the initial bending angle under 5° (Fig. 3b), and the bimaterial legs supported SRR elements were then popped up at around 40° by placing the samples in the rapid thermal annealing (RTA) oven at 400°C for 10 minutes (Fig. 3c).

CHARACTERIZATION

THz Time Domain Spectroscopy (THz-TDS) was used to characterize the metamaterial sample performances; the experimental set up is shown in Fig. 4. In THz-TDS, the time-varying electric field of the impulsive THz radiation is recorded, and the electric field spectral amplitude and phase are directly obtained by performing Fourier analysis.

The transmission of the THz electric field was measured for the sample and a reference, which in the present case is simply air. Prior to measurements, the metamaterial samples were diced into 1 cm × 1 cm squares. All experiments were performed at room temperature in a dry air atmosphere (< 0.1% humidity). The THz beam diameter was ~ 3 mm, which was safely covered by our samples.

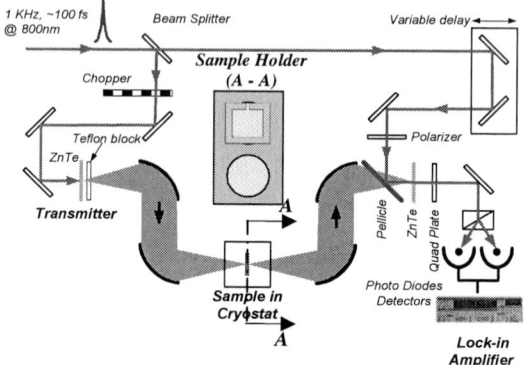

Figure 4: THz-TDS set up for the metamaterial sample performance characterization. (Top hole: sample; bottom hole: reference/air)

The samples, including εμ-SRR, pure ε-SRR and pure μ-SRR, were mounted on the sample holder with air as the reference, and were oriented with THz radiation normal to the sample plane. The electric field is perpendicular to the ε-SRR's gap and parallel to the μ-SRR.

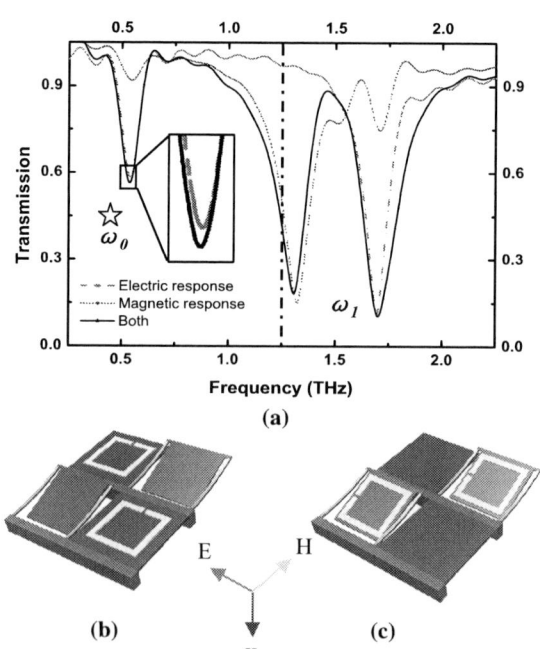

Figure 5: (a) Experimentally measured transmission for the μ-SRRs (Black Solid) along with pure in-plane -SRRs (Red Dash) and pure out-of-plane μ-SRRs (Blue Dot) for comparison; (b) schematic of pure in-plane -SRRs; (c) schematic of pure out-of-plane μ-SRRs.

The transmission spectra were measured and all samples showed resonance responses at ~ 0.5 THz, which originated from the induced circulating currents in the SRRs. εμ-SRR showed the strongest resonance response while μ-SRR showed the weakest as shown in Fig. 5, which verified our expectation since the strength of magnetic response relies on the magnetic field portion incident to the SRR plane decided by the out-of-plane angle, namely ~ 40° in this case. The angle could be further increased by optimizing RTA parameters and geometry of bimaterial legs. A stronger resonance should be expected by increasing the angle and 90° should be the ideal angle for maximizing the magnetic response.

CONCLUSION

In summary, we have experimentally demonstrated a new THz metamaterial structure that shows simultaneously negative electric and magnetic response, based on the MEMS-based bimaterial pop up structure. We anticipate such metamaterials will increasingly find wide variety of applications in the THz frequency range, such as superlens, wave plates, band-pass filters and etc.

ACKNOWLEDGEMENTS

This project has been supported in part by the DOD/Army Research Laboratory through grant W911NF-06-2-0040, National Science Foundation through grant ECCS-0802036, and the Defense Advanced Research Projects Agency through grant HR0011-08-1-0044. The authors would like to thank the Photonics Center at Boston University for all the technical support throughout the course of this research.

REFERENCES

[1] H. Tao, N. I. Landy, C. M. Bingham, X. Zhang, R. D. Averitt, and W. J. Padilla, "A metamaterial absorber for the terahertz regime: Design, fabrication and characterization," *Optics Express,* vol. 16, pp. 7181-7188, May 12 2008.

[2] H. T. Chen, W. J. Padilla, J. M. O. Zide, A. C. Gossard, A. J. Taylor, and R. D. Averitt, "Active terahertz metamaterial devices", *Nature,* vol. 444, pp. 597-600, 2006.

[3] V.G. Veselago, "The Electrodynamics of substances with simultaneously negative values of epsilon and mue", *Sov. Phys. Uspekhi.,* vol. 10, pp. 509-514, 1968.

[4] K. Aydin and E. Ozbay, "Left-handed metamaterial based superlens for subwavelength imaging of electromagnetic waves," *Applied Physics a-Materials Science & Processing,* vol. 87, pp. 137-141, 2007.

[5] N. Fang, H. Lee, C. Sun, and X. Zhang, "Sub-diffraction-limited optical imaging with a silver superlens," *Science,* vol. 308, pp. 534-537, 2005.

[6] N. Fang and X. Zhang, "Imaging properties of a metamaterial superlens," *Applied Physics Letters,* vol. 82, pp. 161-163, 2003.

[7] I. I. Smolyaninov, Y. J. Hung, and C. C. Davis, "Magnifying superlens in the visible frequency range," *Science,* vol. 315, pp. 1699-1701, 2007.

[8] W. S. Cai, U. K. Chettiar, A. V. Kildishev, and V. M. Shalaev, "Designs for optical cloaking with high-order transformations," *Optics Express,* vol. 16, pp. 5444-5452, 2008.

[9] J. Wood, "Mew metamaterial may lead to a magnetic cloak," *Materials Today,* vol. 11, pp. 8-8, 2008.

[10] D. Schurig, J. J. Mock, B. J. Justice, S. A. Cummer, J. B. Pendry, A. F. Starr, and D. R. Smith, "Metamaterial electromagnetic cloak at microwave frequencies," Science, vol. 314, pp. 977-980, 2006.

[11] J. B. Pendry, D. Schurig, and D. R. Smith, "Controlling electromagnetic fields," *Science,* vol. 312, pp. 1780-1782, 2006.

[12] K. Guven and E. Ozbay, "Near field imaging in microwave regime using double layer split-ring resonator based metamaterial," *Opto-Electronics Review,* vol. 14, pp. 213-216, 2006.

[13] R. P. Liu, T. J. Cui, B. Zhao, X. Q. Lin, H. F. Ma, D. Huang, and D. R. Smith, "Resonant crystal band gap metamaterials in the microwave regime and their exotic amplification of evanescent waves," *Applied Physics Letters,* vol. 90, 2007.

[14] D. Maystre, S. Enoch, B. Gralak, and G. Tayeb, "Metamaterials: from microwaves to the visible region," *Comptes Rendus Physique,* vol. 6, pp. 693-701, 2005.

[15] J. B. Pendry, A. J. Holden, D. J. Robbins, and W. J. Stewart, "Magnetism from conductors and enhanced nonlinear phenomena," *IEEE Transactions on Microwave Theory and Techniques,* vol. 47, pp. 2075-2084, 1999.

[16] D. R. Smith, W. J. Padilla, D. C. Vier, S. C. Nemat-Nasser, and S. Schultz, "Composite medium with simultaneously negative permeability and permittivity," *Physical Review Letters,* vol. 84, pp. 4184-4187, 2000.

[17] D. Schurig, J. J. Mock, and D. R. Smith, "Electric-field-coupled resonators for negative permittivity metamaterials," *Applied Physics Letters,* vol. 88, p. 041109, 2006.

[18] W. J. Padilla, "Group theoretical description of artificial electromagnetic metamaterials," *Optics Express,* vol. 15, pp. 1639-1646, 2007.

DAMAGE-FREE PLASMA ETCHING PROCESSES FOR FUTURE NANOSCALE DEVICES

S. Samukawa[1]

[1]Institute of Fluid Science, Tohoku University, Sendai, JAPAN

ABSTRACT

For the past 30 years, plasma etching technology has led in the efforts to shrink the pattern size of ultra-large-scale integrated (ULSI) devices. However, inherent problems in the plasma processes, such as charge buildup and UV photon radiation, limit the etching performance for nanoscale devices. To overcome these problems and fabricate sub-10-nm devices in practice, neutral-beam etching has been proposed. In this paper, we introduce the ultimate etching processes using neutral-beam sources and discuss the fusion of top-down and bottom-up processing for future nanoscale devices. Neutral beams can perform atomically damage-free etching and surface modification of inorganic and organic materials. This technique is a promising candidate for the practical fabrication technology for future nano-devices.

INTRODUCTION

Recent ultralarge-scale integration (ULSI) production processes have involved fabricating sub-50-nm patterns on Si wafers. High-density plasma sources, such as inductively coupled plasma (ICP) and electron-cyclotron- resonance (ECR) plasma, are key technologies for developing precise etching processes. The disadvantages of these technologies include several types of radiation damage caused by the charge buildup of positive ions and electrons [1-4] or by the radiation of ultraviolet (UV), vacuum ultraviolet (VUV), and X-ray photons [5-12] during etching. The voltages generated by charge build-up distort ion trajectories and lead to breakage of thin gate oxide films, stoppage of the etching, and pattern dependence of the etching rate. Additionally, high-density crystal defects are generated by UV or VUV photons radiating from the plasma to the etching surface. These serious problems must be overcome in the fabrication of future nanoscale devices. These problems strongly degrade electrical characteristics of the devices and increase the critical dimension losses in the etching process. For sub-10-nm devices, defect-free and charge-free atomic layer etching processes are indispensable. To achieve a breakthrough and overcome these problems and achieve accurate nanoscale patterning, a high-performance neutral-beam etching system has been investigated. However, previous neutral-beam sources have not been practical for etching processes [13,14] because of low neutralization efficiency and high-energy neutral beams; low etching rates and low etching selectivity resulted. Consequently, few studies have been carried out on the highly efficient generation of neutral beams with controllable beam energy ranging from a few tens to a few hundred electron volts (eV) for high-performance damage-free etching.

Our group has developed a highly efficient neutral-beam source to accomplish the ultimate top-down etching for future nanoscale devices. In this paper, I describe the problems in the conventional plasma etching processes and introduce the ultimate etching processes for future devices from 50 nm to sub-10 nm in using our new neutral beam sources.

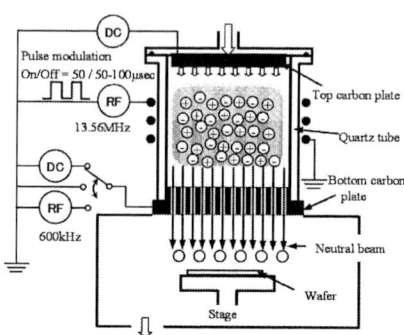

Figure 1. The newly developed neutral-beam etching system.

NEUTRAL BEAM SOURCE

Our newly developed neutral-beam source [17-19] is shown in Fig. 1. It consists of an ICP and parallel carbon plates. The quartz plasma chamber is 10 cm in diameter and 25 cm long. The process chamber is separated from the plasma chamber by a carbon plate fitted at the bottom, because carbon has the lowest sputtering yield under high-energy bombardment and does not contaminate semiconductor devices. Numerous apertures in the bottom carbon plate extract neutral beams from the plasma in the process chamber. The apertures account for 50% of an area 100 mm in diameter on the bottom carbon plate. The apertures are 1 mm in diameter and 10 mm long. A pulse-time-modulated RF power (13.56 MHz, on-time/ off-time=50 μs/50 μs) was applied to generate a large number of negative ions in the after-glow during the period when the plasma was off. [20] Negative ions are more easily neutralized as they pass through the aperture than positive ions, because the detachment energy of electrons from negative ions is much smaller than the charge transfer energy of positive ions. The differences in neutralization efficiency between negative (Cl^-) and positive (Cl_2^+) ions passing through the carbon aperture in the pulsed Cl_2 plasma are shown in Fig. 2. Neutral Cl beams with a neutralization efficiency of almost 100% could be generated by using Cl^- at any negative acceleration voltage, whereas a low neutralization efficiency (max. 60%) was observed for Cl_2^+. Furthermore, neutral F beams could be efficiently generated using F^- in either F_2- or SF_6-pulsed plasma by the same mechanism. Consequently, high-flux (more than 1 mA/cm^2) and highly efficient (neutralization efficiency: 98%) neutral beams could be extracted using negative ions (Cl^-, F^-) from a pulse-time-modulated plasma (Cl_2-, F_2- or SF_6-pulsed plasma) through the apertures. In this system, a DC or RF (600 kHz) bias is applied to the bottom carbon plate to control the neutral beam energy precisely. The gas pressures are 1.0 Pa in the plasma chamber and 0.1 Pa in the process chamber. Under these conditions, a low-energy neutral beam (Cl, F beam) ranging from 10 to 100 eV can be generated with a high-density flux of about 1 mA/cm^2.

978-1-4244-2977-6/09 $25.00 © 2009 IEEE

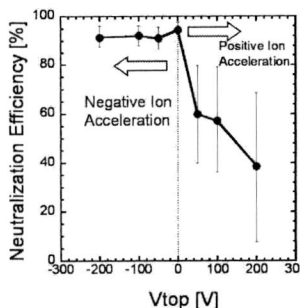

Figure 2. Neutralization efficiency of positive (Cl_2^+) and negative (Cl^-) ions in the pulsed Cl_2 plasma after passing through carbon aperture.

ULTIMATE NANOSCALE ETCHING
Influence of etching damage on less than 20 nm generation CMOS devices

Recently, nonplanar double-gate metal-oxide-semiconductor field emission semiconductor (MOSFETs) have provided a potential solution for nanoscale complementary MOS (CMOS) technology thanks to their ability to control leakage while maintaining a high drive current. [21-23] However, the fabrication of vertical Si fins is challenging. With conventional plasma etching, the generation of defects by the irradiation of charged particles and UV/VUV photons during processing seriously affects device performance and reliability. Surface damage and mobility degradation of the plasma-etched sidewall have already been reported. [24-25] To overcome these problems, a damage-free neutral beam etching process is very effective. In this section, to find a potential solution for damage-free nanoscale CMOS fabrication, we demonstrate the impact of neutral beam etching through the formation and investigation of performance of Fin type FETs.

A three-dimensional schematic diagram of Fin FETs with vertical Si fins is shown in Fig. 3. The starting materials were (100)- and (110)-oriented silicon-on-insulator (SOI) wafers, and the channel orientations of the fabricated Fin type FETs were (110) and (111), respectively. The process flow of the n-channel FinFET for this investigation of performance is summarized in Fig. 4. An electron beam resist was used to pattern Si fins or transfer the patterns to a SiO_2 hard mask. Our neutral- beam system was used to fabricate the Si fin structure. Cl_2 gas was used and pulse- time-modulated RF (13.56 MHz) power was applied to generate the plasma. Then, Cl^- ions were accelerated through the carbon aperture by a negative DC bias and a neutral Cl beam was generated efficiently. Under these conditions, the peak energy of the neutral Cl beam was estimated to be about 10 eV. We have already shown that neutral-beam processes can markedly eliminate charge buildup and photon radiation damage during etching. We also used conventional plasma etching with ICP for comparison. After Si-fin etching, a 2.5 nm gate oxide was formed at 850°C and then covered with n^+ poly-Si. Some Fin FETs were also fabricated on the (110) SOIs with an additional sacrificial oxidation step before gate oxidation. After the gate electrode formation using electron beam lithography and ICP etching, a shallow implantation into the extension of the source/drain was performed, followed by gate-sidewall formation, source/drain implantation, activation annealing, and electrode metalization. Finally, the devices were sintered in 3% H_2 ambient at 450°C for 30 min.

Using low-energy neutral Cl beams, we successfully fabricated charge-free and photon-radiation-free anisotropic Si fins with an ideally rectangular shape using a SiO_2 hard mask, as shown in Fig. 5. [26] We also fabricated similar rectangular Si fins by conventional plasma etching for comparison.

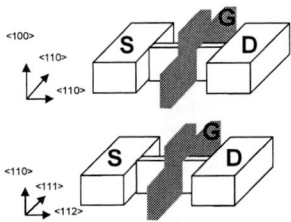

Figure 3. Three-dimensional schematic diagram of Fin FETs with vertical Si fins.

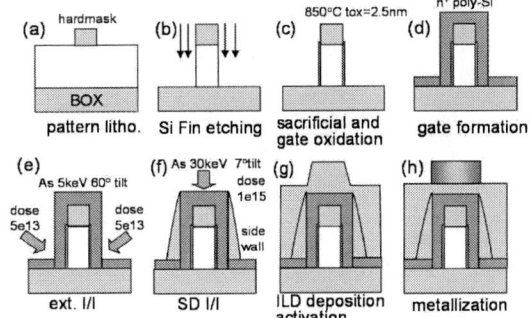

Figure 4. Process flow of n-channel Fin FET for this investigation of performance.

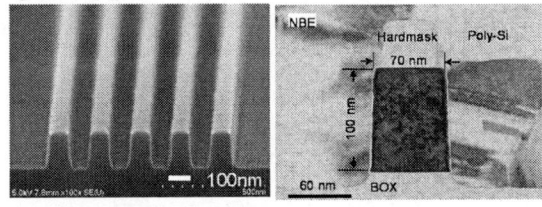

Figure 5. SEM and TEM images of Si fin etching profile. Using low-energy neutral Cl beams, we successfully fabricated charge-free and photon radiation-free anisotropic Si fins with an ideally rectangular shape using a SiO_2 hard mask.

The n-channel Fin FETs with fin width (tfin) = 50 nm and gate length (Lg) = 110 nm fabricated by neutral beam etching have superior subthreshold characteristics compared with the conventional plasma etching ones at the same Vth (Fig. 6).[27,28] However, it is noteworthy that there is a large difference in the drain on-currents: the drain current of the FinFET processed by neutral-beam etching was significantly improved. To investigate this difference in detail, we fabricated multiple Fin FETs with 80 channels and Lg = 50 μm [27,28] and examined the carrier mobility. The electrical characteristics of the multiple Fin FET also showed excellent subthreshold characteristics (Fig. 7). It is apparent from Fig. 8[27,28] that the electron mobility of the Si channel processed by neutral-beam etching is superior to that of the one processed by conventional plasma etching and is almost equivalent to that of the bulk planar (110) MOSFET. [29] The superiority of neutral-beam etching remained unchanged for different values of t_{fin} (Fig. 9). [27,28] Cross-sectional high-resolution TEM images of the Si fins produced using the neutral-beam and conventional plasma etching processes clearly show that the neutral-beam process creates defect-free, atomically smooth sidewall surfaces, whereas the conventional plasma etching processes cause deep crystal defects and significant roughness (Fig. 10). It is speculated that the roughness produced using plasma etching is enhanced by the UV and VUV radiation. Thus, the remarkable contrast in the electric characteristics between devices produced by neutral-beam and conventional plasma etching processes can be explained by the atomic-level roughness and damage to the etched channels. In future nanoscale devices, the atomic-level roughness and damage must be controlled, and the ultimate etching processes enabled by our developed neutral beam will be indispensable.

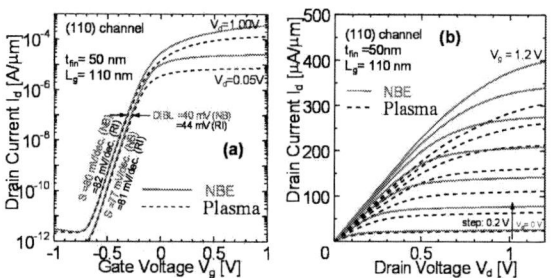

Figure 6. Subthreshold characteristics of Fin FET for neutral-beam etching and conventional plasma etching. The n-channel Fin FETs with fin width (tfin) = 50 nm and gate length (Lg) = 110 nm fabricated by neutral beam etching have superior subthreshold characteristics compared with the conventional plasma etching ones and the same Vth.

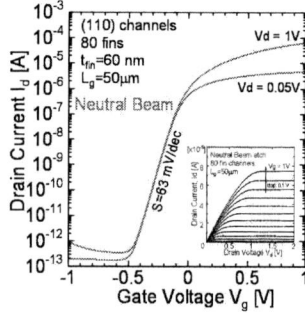

Figure 7. Electrical characteristics of multiple Fin FET in the case of neutral-beam etching.

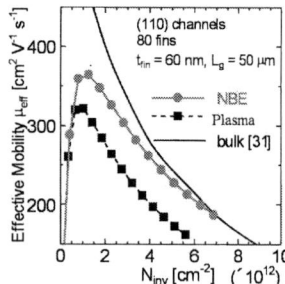

Figure 8. Electron mobility of Si channel processed by neutral-beam etching. It is superior to that of the Si channel processed by conventional plasma etching and is almost equivalent to that of the bulk planar (110) MOSFET.

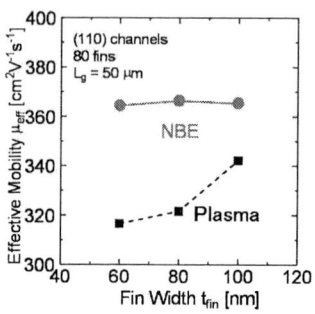

Figure 9. Dependence of electron mobility of Si channel processed by neutral beam etching and plasma etching as function of fin width. The superiority of the neutral beam etching remained unchanged for different values of fin width.

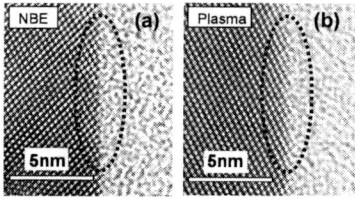

Figure 10. TEM image of sidewall surface roughness after (a) neutral beam etching and (b) plasma etching.

Defect-free sub-10-nm Si nanocolumn etching

Semiconductor devices are being further miniaturized in keeping with Moore's Law. Within the next decade, we expect that the smallest line width of devices will be less than 25 nm. However, it is difficult for conventional optical lithographic techniques to draw patterns smaller than the wavelength of light. Electron beam lithography is widely used for research purposes to fabricate nanometer-sized structures, but it is not suitable for mass production. [30] Scanning-probe microscope lithography has also been widely researched, but it is too time consuming. [31]

Colloidal lithography and natural lithography are candidates for fabricating nanometer-sized structures called nanopillars. [30, 32-41] These lithographies use colloidal particles, such as gold, as etching masks. Structures as small as approximately 10 nm are reported to have been fabricated. Such structures are expected to be used in single-electron devices, quantum dot lasers, and photonic crystals. However, colloids have a distributed size range, so the final nanocolumn structures will also have a range. In addition, normal plasma etching cannot be applied to the fabrication of high-aspect-ratio nanometer-scale structures. Some researchers have reported the fabrication of sub-10-nm nanopillars using plasma etching. [34, 41] However, some of this research has resulted in poor aspect ratios [41] and in other reports structural details were not given. [33] We propose a nanoprocess technology that will break through these barriers by combining protein-derived etching masks with the neutral-beam etching process.

Because proteins are synthesized on the basis of the information in DNA, a large number of atomically identical protein molecules can easily be produced. Hence, proteins are an excellent candidate for making

978-1-4244-2977-6/09 $25.00 © 2009 IEEE

nanometer-sized templates, such as uniform nanoblocks for making nanostructures. Furthermore, some proteins exhibit biomineralization, which enables them to make inorganic materials on their surfaces. This ability can be used to fabricate inorganic nanostructures. [37] One such protein is ferritin, a spherical protein with a 7-nm-diameter cavity. Ferritin can biomineralize iron as hydrated iron oxide ($5Fe_2O_3 \cdot 9H_2O$) in vivo. The iron core can also be produced in vitro using native or recombinant ferritin. This enables us to use the ferritin molecule as a highly accurate template for inorganic structures.

Yamazaki et al. studied plasma etching with a ferritin iron core mask. [43] They succeeded in fabricating cone-shaped structures, but the iron core mask broke during the etching process, making it impossible to etch for a long period of time. We think that this occurred because the high-energy photons in the plasma damaged the iron core. To achieve accurate and fine etching, we used a neutral beam ICP source [17-19, 44] and neutralized ions in the plasma using a carbon electrode with numerous apertures. Because the neutral beam does not contain charged particles or high-energy photons, [17-19, 37, 45, 46] it enables etching using masks like the ferritin iron core that can be easily damaged by conventional etching processes.

The experimental procedure is roughly illustrated in Fig. 11. The experimental details have been given elsewhere. [46] Recombinant ferritin with an artificially hydrated iron oxide core was prepared. A cleaned silicon wafer was treated using a UV/ozone stripper (SAMCO International Inc. UV-1) to remove any organic contamination and make the surface hydrophilic. A ferritin solution was dropped onto the wafer. The wafer was stored in a centrifuge tube and then centrifuged at 10,000 rpm. Less than 1 monolayer of ferritin molecules was adsorbed on the wafer by this process. An UV/ozone treatment was performed again to remove the ferritin protein shell and leave the iron core.

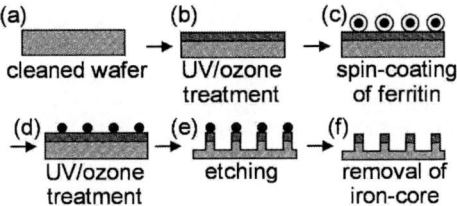

Figure 11. Flow of nanoprocess technology combining protein-derived etching masks with our neutral-beam etching process.

The ferritin iron core was used as a mask for etching Si nanocolumns. Cl_2 gas flowed into the quartz cylindrical chamber at a rate of 40–60 sccm. When the Cl_2 gas flow rate was 40 sccm, the pressure in the chamber was about 10 mTorr. We applied a 13.56 MHz RF power of 800 W to a three-turn inductive coil antenna to generate a high-density plasma. The RF power was modulated at a pulse timing of 50 μs to generate a large number of negative ions in the plasma. [47, 48] A negative DC bias voltage of –100 V was applied to the top electrode to accelerate the negative ions toward the etching chamber. Under this condition, we optimized the neutral-beam energy. The bottom electrode was connected to the ground level or to a 600 kHz RF bias, ranging from 20 to 40 W, to control the neutral-beam

energy incident to the substrate. The etching time was calculated so that the etching depth would be 50 nm after performing experiments to determine the etching rate.

To clarify the advantages of neutral beam etching, we also investigated ECR plasma etching for nanocolumn structures for comparison. About 30 sccm of Cl_2 gas was introduced into the chamber at a pressure of about 2 mTorr. ECR plasma at a power of 500 W was discharged. We supplied 20 W RF power to the sample stage during etching to produce the same ion bombardment energy as in the neutral-beam etching experiments. The sample was cooled to –20°C during etching. The etching time was determined so that the etching depth of silicon would be about 50 nm.

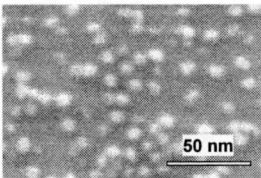

Figure 12. SEM image of wafer after centrifuging and UV/ozone treatment. The 7-nm uniform iron cores dispersed on the wafer were clearly visible.

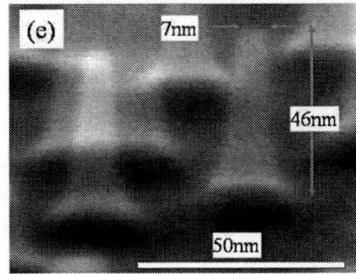

Figure 13. Dependence of etching profile obtained using neutral beams on beam energy. By increasing RF power, the taper angle of the Si nanocolumn etching profile was improved. (e) shows the optimum condition for fabricating anisotropic 7 nm Si nano-columns.

The sample was immersed in 5 mol/L HCl for 10 min to remove the iron mask after neutral beam and ECR plasma etchings. SEM (Scanning Electron Microscopy) images were taken using a LEO-982 microscope (LEO Electron Microscopy) with a 30 kV acceleration voltage. X-ray photoelectron spectroscopy (XPS) measurements were used to determine the etching selectivity of the iron core mask to the Si. An SEM image of a wafer after centrifuging and UV/ozone treatment is shown in Fig. 12. The 7 nm uniform iron cores dispersed on the wafer are clearly visible.

SEM images of Si nanocolumn etching profiles under

various RF bias powers at the bottom electrode of the neutral beam source are shown in Fig. 13. When the RF bias at the bottom electrode was increased, the beam energy increased. Nanocolumns with more vertical structures were observed in all images.

An SEM image of a Si nanocolumn etching profile with no RF bias at the bottom electrode is shown in Fig. 13(a). The top diameter and height of the nanocolumn were about 7 and 41 nm, respectively. The profile had a slight taper of about 77°. We think this is due to the relatively low collimation of the neutral beam under these conditions. On the other hand, the XPS iron peak on the substrate was hardly decreased at all even after the etching. Thus, the etching depth of the iron core was calculated to be 0.083 nm from the XPS results. Since the silicon etching depth was 41 nm, the calculated etching selectivity was about 500.

When RF bias power (600 kHz) was applied to the bottom electrode [Figs. 13(b)-13(d)], the etching anisotropy increased with RF power. This suggests that the neutral beam was collimated by increasing the RF bias power. When the RF bias power was 20 W, the taper angle of the etching profile was about 85°. The top diameters of these nanocolumns were about 7 nm, which coincides with the diameter of the iron core mask. The heights of the nanocolumns were about 46, 45, and 47 nm, under conditions with 20, 30, and 40 W of RF bias power at the bottom electrode, respectively. Many short columns were observed when the bottom electrode RF bias was 40 W (Fig. 13(d)). This suggests that some of the iron cores were completely etched or removed from the sample surface during Si etching. The etching selectivity of silicon to the iron core was calculated to be about 80, 40, and 30 under conditions with 20, 30, and 40 W of RF bias power at the bottom electrode, respectively. It is clear that the etching selectivity of silicon to the iron core strongly depended on the bias condition at the bottom electrode. A lower-energy neutral beam had higher selectivity. The iron core etching rate corresponded to the beam energy. This result is consistent with the low etching selectivity in the case of higher RF bias power, as shown in Fig. 13(d). As a result, we concluded that the RF power of 20 or 30 W applied to the bottom electrode was the optimum etching condition for obtaining high etching selectivity and high anisotropic etching at the same time [Fig.13 (e)].

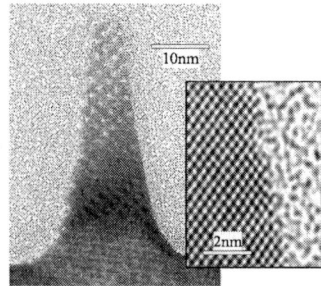

Figure 14. Cross-sectional TEM image of Si nanocolumn structure formed using neutral beam etching.

TEM observation was performed using the high-density area of the sample. A cross-sectional TEM image of a Si nanocolumn is shown in Fig. 14. The top diameter and the height were about 7 and 10 nm, respectively. In the image, the silicon lattice is clearly seen

both in the substrate and in the nanocolumn. The lattice constant was the same as that of silicon, and no distortion was observed even close to the surface. These results show that our newly developed neutral beam carried out performed ultimate defect-free atomic-layer etching because it is free of charged particles and UV photons with a beam energy as low as around 10 eV. We have been eliminating the influences of native silicon oxide on the silicon surface to improve the etching profile by optimizing the beam energy and mixing ratio between Cl and F beams. Based on these results, it is speculated that our developed neutral beam can accomplish a few nm size patterning.

Based on the high aspect Si nanocolumn structure, we have been developing new quantum effects devices.

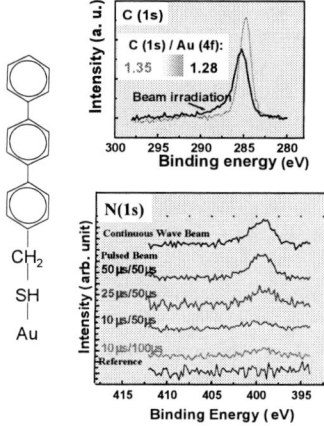

Figure 15. XPS peak of C 1s and N 1s on surface of terphenlys on the gold substrate after N_2 neutral beam irradiation. Using N_2 neutral beam, the surface nitridation of terphenlys could be accomplished with keeping the bulk molecular structure.

FUTURE PROSPECTS OF NEUTRAL BEAM PROCESSES

Future sub-10-nm devices will require extremely precise anisotropic etching processes. Any charge buildup and UV photon radiation must be completely using collimated and accelerated low-energy neutral particles. Charge buildup distorts ion trajectories, and crystal defects cause large critical-dimension loss and low etching selectivity. Furthermore, we think that bio-supermolecules (e.g., DNA and proteins), organic molecules [e.g., self-assembling monolayers (SAMs)] and carbon nanotube will be also used for active areas on silicon in future nano-devices. In these devices, the fusion of top-down and bottom-up processes will be necessary to integrate these new materials on the silicon substrate. A preliminary investigation of nitridation for organic materials, such as SAMs fabricated from terphenyls [49] (TP1) on gold substrates, has been attempted using our neutral beam to control electrical characteristics. A N_2 gas pulsed plasma was used to generate the neutral beam. The N_2 neutral beam energy was periodically modulated at 10 eV during the pulse-on time and at 1 eV during the pulse-off time. [50] Basically, the SAM could be nitrided only during the pulse-on time. Under these conditions, for the first time, we found that the surface nitridation of the SAM could be precisely controlled by changing the pulse-on time while maintaining the bulk structure, [51] as shown in Fig. 15.

This result suggests that the neutral-beam process can provide a damage-free top-down process even for weak organic materials. The neutral-beam process should enhance the practical development of future nanoscale devices.

CONCLUSIONS

Inherent problems in plasma processes, such as charge buildup and UV photon radiation, limit the etching performance for nanoscale devices. To overcome these problems and fabricate nanoscale devices in practice, neutral-beam etching has been proposed. Neutral beams carry out atomically damage-free etching and surface modification of inorganic and organic materials. This technique is a promising candidate for an ultimate top-down etching for future nanodevices.

ACKNOWLEDGMENT

I would like to thank Mr. Katsunori Ichiki and Dr. Hirokuni Hiyama of Ebara Research Co., Ltd., Dr. Shuichi Noda of Oki Electric Industry Co.,Ltd., Dr. Misturu Okigawa and Mr. Yoshinari Ichihashi of Sanyo Electric Co.,Ltd., Dr. Tadashi Shimmura of Toshiba Corporation, Mr.Seiichi Fukuda of Sony Corporation, Dr. Kenji Ishikawa of Fujitsu Laboratories, Dr. Hiroto Ohtake and Dr. Yukinori Ochiai of NEC Corporation, Dr. Kazuhiko Endo, Dr. Eiichi Suzuki, Dr. Satoshi Yamazaki and Dr. Takao Ishida of National Institute of Advanced Industrial Science and Technology (AIST), Professor Yukiharu Uraoka, Professor Takashi Fuyuki and Professor Ichiro Yamashita of Nara Institute of Science and Technology, Prof. T. Ono, Prof. K. Toji of Tohoku University, Prof. Yasuo Ohno of Toshima University, Professor Demetre Economou, and Professor Vincent Donnelly of University of Huston for useful advice, discussions and collaboration.

REFERENCES

[1] T. Nozawa, T. Kinoshita, T. Nishizuka, A. Narai, T. Inoue, A. Nakaue, "The Electron Charging Effects of Plasma on Notch Profile Defects", *Jpn. J. Appl. Phys.* vol. 34, pp. 2107-2133 ,1995.

[2] T. Kinoshita, M. Hane and J. P. McVittie, "Notching as an example of charging in uniform high density plasmas", *J. Vac. Sci. Technol. B*, vol. 14, pp. 560-565, 1996.

[3] H. Ootera, T. Oomori, M. Tsuda and K. Namba, "Simulation of Ion Trajectories near Submicron-Patterned Surface Including Effects of Local Charging and Ion Drift Velocity toward Wafer", *Jpn. J. Appl. Phys.* vol. 33, pp. 4276-4280, 1993.

[4] H. Ohtake and S. Samukawa, "Microloading-free Si Trench Etching in Pulse-time Modulated ECR Plasma with 600 kHz RF bias", in *Proc. 17th Dry Process Symp.*, Tokyo, November, 1995, pp. 45-50.

[5] T. Okamoto, T. Ide, A. Sasaki, K. Azuma, and Y. Nakata, "Irradiation Damage in SiO2/Si System Induced by Photons and/or Ions in Photo-Oxidation and Plasma- Oxidation" *Jpn. J. Appl. Phys.*, vol. 43, pp. 8002-8006, 2004.

[6] K. Yonekura, K.Goto, M. Matsuura, N.Fujiwara and K. Tsujimoto, "Low-Damage Damascene Patterning Using Porous Inorganic Low-Dielectric-Constant Materials", *Jpn. J. Appl. Phys.*, vol. 44, pp. 2976-2981,

2005.

[7] K. P. Cheung and C. S. Pai, "Charging damage from plasma enhanced TEOS deposition", *Electr. Dev. Lett.* vol. 16, pp. 220-222, 1995.

[8] J. Carrere, J-C. Oberlin and M. Haond, "Topographical dependence of charging and new phenomenon duringinductively coupled plasma (ICP) CVD process", in *Proc. 5th Int. Symp. on Plasma Process-Induced Damage*, Santa Clara, May 22-24, 2000, pp. 164-167.

[9] T. Dao and W. Wu, "Charging of Underlayer at Via Etch causing Slow Down in Oxide Etch Rate", in *Proc. 1st Int. Symp. on Plasma Process-Induced Damage*, Monterey, May 13-14, 1996, pp. 54-57.

[10] M. Joshi, J. P. McVittee and K. Sarawat, "Direct experimental determination and modeling of VUV induced bulkconduction in dielectrics during plasma processing", in *Proc. 5th Int. Symp. on Plasma Process-Induced Damage*, Santa Clara, May 22-24, 2000, pp. 157-160.

[11] C. Cismura, J. L. Shohet and J. P. McVittee, "Plasma vacuum ultraviolet emission in a high density etcher" in *Proc. 5th Int. Symp. on Plasma Process-Induced Damage*, Santa Clara, May 22-24, 2000, pp. 192-195.

[12] J. R. Woodworth, M. G. Blain, R. L. Jarecki, T. W. Hamilton and B. P. Aragon, "Absolute intensities of the vacuum ultraviolet spectra in a metal-etch plasma processing discharge", *J. Vac. Sci. Technol. A*, vol. 17, pp. 3209-3217, 1999.

[13] T. Mizutani and S. Nishimatsu, "Sputtering yield and radiation damage by neutral beam bombardment", *J. Vac. Sci. Technol. A*, vol. 6, pp. 1417-1420, 1988.

[14] F. Shimokawa, "High-power fast-atom beam source and its application to dry etching", *J. Vac. Sci. Technol. A*, vol. 10, pp. 1352-1357, 1992.

[15] D. Lee, B. Park and G. Yeom, "Effects of Axial Magnetic Field on Neutral Beam Etching by Low-Angle Forward-Reflected Neutral Beam Method" *Jpn. J. Appl. Phys.*, vol. 44, pp. L63-L66, 2005.

[16] S. Panda, D. J. Economou, L. Chen, "Anisotropic etching of polymer films by high energy (~100s of eV) oxygen atom neutral beams", *J. Vac. Sci. Technol. A*, vol. 19, pp. 398-404, 2001.

[17] S. Samukawa, K. Sakamoto and K. Ichiki, "Generating high-efficiency neutral beams by using negative ions in an inductively coupled plasma source", *J. Vac. Sci. Technol. A*, vol. 20, pp. 1566-1573, 2002.

[18] S. Samukawa, K. Sakamoto and K. Ichiki, "High-Efficiency Neutral-Beam Generation by Combination of Inductively Coupled Plasma and Parallel Plate DC Bias", *Jpn. J. Appl. Phys.*, vol. 40, pp. L779-L782, 2001.

[19] S. Samukawa, Y. Minemura and S. Fukuda, "Ultrathin Oxynitride Films Formed by Using Pulse- Time-Modulated Nitrogen Beams", *Jpn. J. Appl. Phys.*, vol. 42, pp. L795-L797, 2003.

[20] S. Samukawa and T. Mieno, "Pulse-time modulated plasma discharge for highly selective, highly anisotropic and charge-free etching", *Plasma Sources Sci. Technol.*, vol. 5, pp. 132-138, 1996.

[21] T. Sekigawa, "Calculated threshold- voltage

characteristics of an XMOS transistor having an additional bottom gate", *Solid State Electron.*, vol. 27, pp. 827-828, 1984.

[22] M. Masahara, Y. Liu, K. Ishii, H. Takashima, E. Sugimata, T. Matsukawa, H. Yamauchi and E. Suzuki, "15-nm-thick Si channel wall vertical double-gate MOSFET", in *Proc. IEDM Tech. Deg.*, 2002, pp. 949-951.

[23] Y. Liu, M. Masahara, K. Ishii, H. Takashima, E. Sugimata, T. Matsukawa, H. Yamauchi and E. Suzuki, "Flexible threshold voltage FinFETs with independent double gates and an ideal rectangular cross-section Si-Fin channel", in *Proc. IEDM Tech. Deg.*, 2003, pp. 986-989.

[24] G. S. Oehrlein, R. M. Tromp, Y. H. Lee and J. Petrillo, "Study of silicon contamination and near-surface damage caused by CF_4/H_2 reactive ion etching", *Appl. Phys. Lett.*, vol. 45, pp. 420-422, 1984.

[25] C. J. Petti, J. P. McVittie, J. D. Plummer, "Characterization of surface mobility on the sidewalls of dry-etched trenches" , in *Proc. IEDM Tech. Deg.*, 1988, pp. 104-107.

[26] K. Endo, S. Noda, T. Ozaki, S. Samukawa, M. Masahara,Y. Liu, K. Ishii, H. Takashima, E. Sugimata, T. Matsukawa, H. Yamauchi, Y. Ishikawa, and E. Suzuki, "Damage-free neutral beam etching technology for high mobility FinFETs", in *Ext. Abstr. Int. Microprocesses and Nanotechnology Conference*, Tokyo, 2005.

[27] K. Endo, S. Noda, M. Masahara, T. Ozaki, S. Samukawa, Y. Liu, K. Ishii, H. Takashima, E. Sugimata, T. Matsukawa, H. Yamauchi, Y. Ishikawa and E. Suzuki, "Damage-free neutral beam etching technology for high mobility FinFETs", in *Proc. IEDM Tech. Deg.*, 2005, pp. 840-843.

[28] K. Endo, S. Noda, M. Masahara, T. Ozaki, S. Samukawa, Y. Liu, K. Ishii, H. Takashima, E. Sugimata, T. Matsukawa ,H. Yamauchi, Y. Ishikawa and E. Suzuki, "Fabrication of FinFET by Damage-free Neutral Beam Etching Technology", *IEEE Trans. Elec. Dev.*, vol.53, pp. 1826-1833, 2006.

[29] H. Irie, K. Kita, K. Kyuno and A. Toriumi, "In-plane mobility anisotropy and universality under uni-axial strains in nand p-MOS inversion layers on (100), [110], and (111) Si", in *Proc. IEDM Tech. Deg.*, 2004, pp. 225-228.

[30] V. Poborchii, T. Tada and T. Kanayama, "Subwavelength-Resolution Raman Microscopy of Si Structures Using Metal-Particle-Topped AFM Probe", *Jpn. J. Appl. Phys.*, vol. 38, pp. L202-L204, 2005.

[31] L. M. Demers, D. S. Ginger, S.-J. Park, Z. Li, S.-W. Chung and C. A. Mirkin, "Direct Patterning of Modified Oligonucleotides on Metals and Insulators by Dip-Pen Nanolithography", *Science*, vol. 296, pp. 1836-1838, 2002.

[32] H. W. Deckman and J. H. Dunsmuir, "Natural lithography", *Appl. Phys. Lett.* vol. 41, pp. 377-379, 1982.

[33] T. Sato, D. G. Hasko and H. Ahmed, "Nanoscale colloidal particles: Monolayer organization and patterning", *J. Vac. Sci. Technol. B*, vol. 15, pp. 45-48, 1997.

[34] P. A. Lewis, H. Ahmed and T. Sato, "Silicon nanopillars formed with gold colloidal particle masking", *J. Vac. Sci. Technol. B*, vol. 16, pp. 2938-2941, 1998.

[35] P. A. Lewis, H. Ahmed and B. W. Alphenaar, "Colloidal gold natural lithography technique for fabricating GaAs nanopillars", *Microelectron. Eng.* vol. 57-58, pp. 925-930, 2001.

[36] M .A. Wood, M. Riehle, and C. D. W. Wilkinson, "Patterning colloidal nanotopographies", *Nanotechnology*, vol.13, pp. 605-609, 2002.

[37] M. Bale, A. J. Turner and R. E. Palmer, "Fabrication of ordered arrays of silicon nanopillars at selected sites", *J. Phys. D: Appl. Phys.*, vol. 35, pp. L11-L14, 2002.

[38] M. Haupt, S. Miller, A. Ladenburger, R. Sauer, K. Thonke, J. P. Spatz, S. Riethmüller, M. Möller and F. Banhart, "Semiconductor nanostructures defined with self-organizing polymers", *J. Appl. Phys.*, vol. 91, pp. 6057-6059, 2002.

[39] K. Seeger and R. E. Palmer, "Fabrication of ordered arrays of silicon nanopillars", *J. Phys. D: Appl. Phys.*, vol. 32, pp. L129-L132, 1999.

[40] C.-W. Kuo, J.-Y. Shiu, P. Chen, and G. A. Somorjai, "Fabrication of Size-Tunable Large-Area Periodic Silicon Nanopillar Arrays with Sub-10-nm Resolution", *J. Phys. Chem. B*, vol. 107, pp.9950-9953, 2003.

[41] Y.-K. Hong, J. H. Bahng, G. Lee, H. Kim, W. Kim, S. Lee, J.-Y. Koo, J.-I. Park, W.-R. Lee, and J. Cheon, "Facile fabrication of 2-dimensional arrays of sub-10 nm single crystalline Si nanopillars using nanoparticle masks", *Chem. Commun.*, pp. 3034-3035, 2003.

[42] I. Yamashita, "Fabrication of a two-dimensional array of nano-particles using ferritin molecule", *Thin Sold Films* vol. 393, pp. 12-18, 2001.

[43] G. Yamazaki, Y. Uraoka, T. Fuyuki and I. Yamashita, "Nano-etching Using Nanodots Mask Fabricated by Bio-nano-process", *J. Photopolym. Sci. Technol.* vol. 16, pp. 439-444, 2003.

[44] S. Samukawa, K. Sakamoto and K. Ichiki, "High-Efficiency Low Energy Neutral Beam Generation Using Negative Ions in Pulsed Plasma", *Jpn. J. Appl. Phys.*, vol. 40, L997-999, 2001.

[45] S. Noda, H. Nishimori, T. Ida, T. Arikado, K. Ichiki, and S. Samukawa, "Neutral Beam Etching for Damage-free 50 nm Gate Electrode Patterning", in *Ext. Abst. Int. Conf. Solid State Devices and Materials*, Tokyo, 2003, pp. 472-473.

[46] T. Kubota, T. Baba, S. Samukawa, H. Kawashima, Y. Uraoka, T. Fuyuki, and I. Yamashita, "A 7nm-Nanocolum Structure Fabricated by Using a Ferritin Iron-Core Mask and Low Energy Cl Neutral Beams.", *Appl. Phys. Lett.*, vol. 84, pp. 1555-1557, 2004.

[47] S. Samukawa and T. Mieno, "Pulse-time modulated plasma discharge for highly selective, highly anisotropic and charge-free etching", *Plasma Sources Sci. Technol.*, vol. 5, pp. 132-138, 1996.

[48] S. Samukawa, H. Ohtake, and T. Mieno, "Pulse–time-modulated electron cyclotron resonance plasma discharge for highly selective, highly anisotropic, and charge-free etching", *J. Vac. Sci. Technol. A*, vol. 14, pp. 3049-3058, 1996.

[49] T. Ishida, M. Sano, H. Fukushima, M. Ishida and S. Sasaki, "Stability of Terphenyl Self-Assembled Monolayers Exposed under UV Irradiation", *Langmuir* vol. 18, pp. 10496-10499, 2002.

[50] S.Samukawa, Y.Minemura and S. Fukuda, "Control of nitrogen depth profile in ultrathin oxynitride films formed by pulse-time-modulated nitrogen beams", *J. Vac. Sci. Technol. A*, vol. 22, pp. 245-249.

[51] Y. Ishikawa, T. Ishida and S. Samukawa, "The Low Damage Surface Modification of the Self-assembled monolayer by the N_2 Neutral Beam Irradiation", in *Int. Symp. AVS*, 2004, PS+BI-FrM9.

ATOMIC LAYER DEPOSITION (ALD) TUNGSTEN NEMS DEVICES VIA A NOVEL TOP-DOWN APPROACH

B.D. Davidson[1], Y.J. Chang[1], D. Seghete[2], S.M. George[2], and V.M. Bright[1]
[1]Department of Mechanical Engineering
[2]Departments of Chemical and Biological Engineering and Chemistry
University of Colorado, Boulder, CO USA

ABSTRACT

In this paper we present a novel low temperature, CMOS compatible, direct top-down nano-fabrication process employing ALD tungsten (WALD) as a structural material for nano-electro-mechanical systems (NEMS). Using this process doubly clamped suspended NEMS devices have been successfully fabricated and demonstrated. The devices have been observed to operate comparably to 2-terminal electrostatic carbon nanotube (CNT) switches, and MEMS tunneling devices. A lifetime in excess of 660,500 cycles has been observed under low-current-limited operating conditions. Under these conditions the device behavior is stable, reproducible and hysterisis free, resembling that of MEMS tunneling devices.

1. INTRODUCTION

The past 20 years has seen the rapid growth in MEMS from a field of novel research to a consumer market. Today MEMS devices are commonplace in a range of products, from cars, to watches, phones, and video game controllers. In the past decade or so the field of MEMS has branched into the very active research field of NEMS as the effort to miniaturize micro-mechanical devices to take advantage of some of the superior properties of nano-mechanical systems, such high natural frequencies, low mass, and low operational power.

The effort to miniaturize and improve well-established MEMS devices – such as mass/chemical sensors and electromechanical switches, has subsequently led to the improvement of existing processing techniques. Nanofabrication processes are generally categorized as top-down (direct, indirect) or bottom-up [1]. Processes that have been borrowed from MEMS and applied to NEMS are generally top-down. This implies the fabrication of structures by deposition of material layers, and then selectively etching one or more of these layers to define geometries, vias, interconnects or to release suspended devices such as mechanical resonators or switches whose functionality may rely on out-of-plane motion. Bottom-up, or self-assembly, processes entail employing the self-assembly of molecules in fabrication of a device. Examples of this are NEMS devices that incorporate either single-walled or multi-walled CNTs. Jang *et al* have demonstrated NEMS switches with vertically aligned multi-walled CNTs fabricated using a combination of top-down and bottom-up nanofabrication processing techniques by growing multi-walled CNTs on top of Nb electrodes created using electron beam lithography (EBL) [2].

We introduce here a novel low temperature, CMOS compatible, top-down nano-fabrication process employing WALD [3-4]. Our process is based on surface micromachining techniques with selective etching to create suspended devices having out of plane functionality. The process uses ALD thin film tungsten as a structural material. Formerly, ALD alumina (Al_2O_3) technology was incorporated in the fabrication of MEMS devices for use as protective and wear resistant coatings [5-6], and as a structural material [7-9].

The majority of NEMS to date have been based on CNTs [10]. Because our devices are made using ALD tungsten they should not be as stiff as their CNT counterparts. For electrostatically actuated device applications this is advantageous because a WALD device will have a lower actuation bias compared to a similarly scaled CNT device. A number of different electromechanical switch designs have been fabricated and demonstrated, ranging from horizontally oriented cantilevers [11] and doubly-clamped devices [12-14] suspended above actuation electrodes, to vertically grown CNTs whose mechanical structures double as actuation electrodes [2]. The devices in [2] are fabricated combining top-down and bottom-up techniques that require high temperature CVD deposition processes (600-850°C) for CNT synthesis [2,12,13]. To our knowledge tunable tunneling NEMS devices as demonstrated by our group are unique.

2. DESIGN AND FABRICATION

Process

The fabrication process for nano-scale WALD structures is a top-down surface micro-machining approach illustrated in Figure 1. In steps 1-3 alignment marks used for all EBL steps are formed. The substrate is coated by a 1.8 μm thick layer of AZP 4210 photoresist and then patterned by photolithography to define the alignment marks. After patterning the photoresist a 5nm thick Ti adhesion layer followed by a 50nm thick layer of Au are deposited by thermal evaporation, and the alignment marks defined by lift-off.

In steps 3-5 the actuation electrodes for the device are fabricated. The substrate is coated with a 150 nm thick layer of PMMA, and the device actuation electrodes patterned via EBL. Next, a 5 nm thick adhesion layer of Ti is deposited by thermal evaporation followed by a 10 nm thick layer of Au, and the actuation electrode then formed via lift-off. Here Au has been chosen as the conducting material for the electrode because Au is inert and thus resistant to oxidation, which can have deleterious effects on device performance [15]. Very thin electrodes are necessary in this process due to their being formed by lift-off.

In step 5, following the electrode lift-off, a 50nm thick sacrificial layer of Ni is deposited via thermal evaporation. Next the structural layer of WALD is deposited; first 3 nm of ALD alumina (Al_2O_3), used as a

978-1-4244-2977-6/09 $25.00 © 2009 IEEE

seed layer, followed by a 30 nm thick layer WALD are grown at a temperature of 120 °C.

In steps 6-7 the substrate is coated with a 150 nm thick layer of PMMA, and the device geometry patterned via EBL. A hard mask used to define the geometry of the device is formed after lift-off of a 30 nm thick thermally evaporated Ni layer. The NEMS structures are defined after etching the ALD W and alumina layers in an O_2:CF_4 RIE environment.

In steps 8-9 the process is completed; the substrate is once again coated by a 1.8 μm thick layer of AZP 4210, which is then patterned by photolithography to define alignment marks for a 100 nm thick layer of Au thermally evaporated and defined by lift-off. This layer will create large electrodes that connect to the actuation electrodes, as well as WALD devices allowing for probing, and if made thicker, wire bonding. These large probing/bonding pads also act to fix the device in place in the event that the sacrificial layer is over etched. The substrate is then immersed into TFB nickel etchant for ~ 60 seconds. The wet-etch serves to release the WALD NEMS structures, remove the hard mask, and define the device anchors via a partial under-etching of the WALD layer. Finally, the device is dried in a CO_2 critical point dryer in order to prevent device stiction.

Design/Results

The design of our WALD NEMS devices is based on that of a two terminal electrostatically actuated switch with out-of-plane functionality. The design is similar to CNT NEMS switches and CNT non-volatile memory devices reported elsewhere [12-14], in that the device is suspended over a single actuation electrode. The fabricated device is shown in Figure 2. The lateral dimensions of the device are 2,000 x 700 x 30 nm, with a gap height of 50 nm from actuation electrode to the WALD beam, and an electrode/beam overlap area is 1,500 x 700 nm^2. The thickness of the WALD and ALD alumina layers was measured to be 32 nm by x-ray refraction.

3. OPERATING PRINCIPLE

The device is designed to actuate by application of electrostatic forces. By modeling the device as a parallel plate capacitor, using a lumped model, and approximating the spring constant by applying simply-supported boundary conditions to the beam, we can derive the well known relation between voltage and displacement, given by Equation 1, where k_{eff} is the effective spring constant of the system, z the displacement, and d_o the initial gap height. Although the device is designed with doubly clamped boundary conditions, Chang *et al* demonstrated that because the anchor is defined by an under-etch of the sacrificial material it is more appropriate to model the boundary conditions as simply supported [7].

$$V_{app} = \sqrt{\frac{2k_{eff}z(d_0 - z)^2}{\varepsilon_o A_{overlap}}} \qquad (1)$$

Assuming a perfectly rectangular cross section, the modulus of elasticity of bulk tungsten, WALD thickness of 32 nm (as measured), and pull-in at $z \sim d_o/3$, pull-in is expected to occur at $V_{pull-in} \sim 4.33$ Volts.

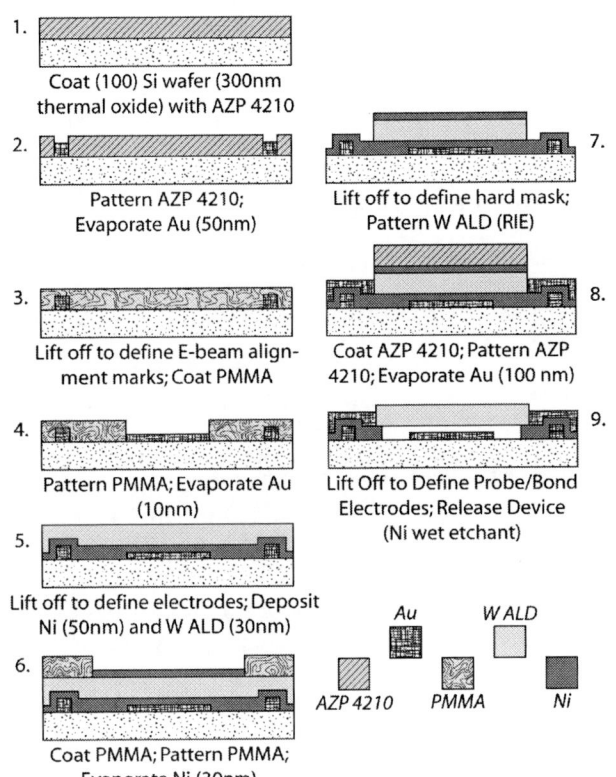

Figure 1: *WALD Device Process, 10nm Au electrodes (EBL), 50nm air gap, 30nm W ALD grown at 120°C at 3.3 Å/cycle, patterned via EBL, etched in RIE. Wet etch release of sacrificial Ni.*

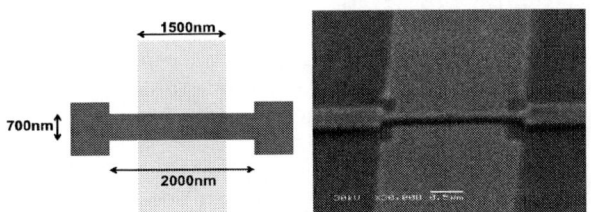

Figure 2: *(a) 2-d schematic of the device (not drawn to scale): length = 2um, width = 700nm, actuation overlap = 1500x700nm, beam thickness = 30nm, gap =50nm. (b) Successfully fabricated two terminal WALD NEMS device.*

4. TESTING AND RESULTS

The pull-in voltages of the WALD NEMS switches were characterized by the measurement of I-V curves. The measurements were performed using an HP 4145B semiconductor parameter analyzer, MC systems 8806 probe station, and custom Lab View VI used to control the parameter analyzer. The curves were measured by sweeping the bias voltage in 20 mV intervals from 0 to V_{max}. Here V_{max} represents a value typically greater than observed actuation, which is defined by an exponential increase in measured current. For these sweeps current was limited to avoid burn-out. Similar testing techniques have been reported to characterize CNT switching behavior [2,11-14].

The average pull-in voltage for our WALD devices was found to be 4.93 ± 0.31 Volts, which is within ~13% of the theoretically predicted value of 4.33 Volts, and comparable to a similarly dimensioned CNT device [11].

978-1-4244-2977-6/09 $25.00 © 2009 IEEE

A plot of pull-in voltage with current limited to 500 nA is shown in Figure 3a. Hysteresis of the pull-in behavior was examined by sweeping the bias voltage forward from 0 – 6 volts and then backward from 6 – 0 volts in 20 mV intervals, and is displayed in Figure 3b.

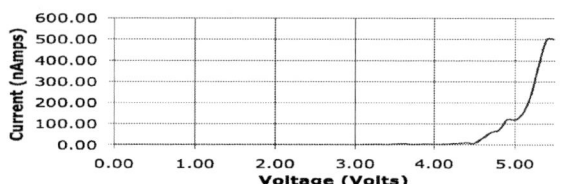

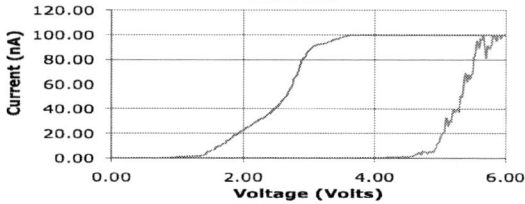

Figure 3: (a) I-V curve for a WALD device, current limited at 500 nA, pull-in ~ 5.4 Volts. (b) I-V hysteresis for WALD device, current limited at 100nA, pull-in ~ 5.5 Volts, pull-out ~ 3 Volts.

Pull-in testing revealed device life-times < 300 cycles when operated with a current limit between 250-500 nA. During pull-in characterization testing it was frequently observed that low levels of current < 20 nA were measured for applied biases < 3V, and that in this region the I-V curves shared a striking resemblance to I-V curves used to characterize MEMS tunneling accelerometers [15]. Lee *et al* attributed the nonlinear increase in current

to tunneling behavior induced as the device is drawn towards the actuation electrode [11].

Tunneling in our WALD NEMS was investigated using the same methods described above; however, the bias maximum was 2 V and current was limited at 10nA, results are shown in Figure 4a. By following scanning tunneling microscopy (STM) theory that has been previously applied to MEMS tunneling accelerometers [15-16], tunneling current is given by Equation 2. Here V_b = bias voltage, h = gap height, ϕ = barrier height, and $\alpha = 1.025\text{eV}^{-1/2}\text{Å}^{-1}$.

$$I_t \propto V_b \exp(-\alpha\sqrt{\phi}h) \qquad (2)$$

Next, following the example set by Cui *et al*, a semi-log plot was created of the measured current vs. displacement, Figure 4b [16]. In equation 3 displacement is approximated by assuming higher order terms of length to be insignificant when compared to first order terms in equation 1. This approximation introduces some error into the analysis, and in the future a more refined approximation for displacement should be used.

$$z = \sqrt{\frac{\varepsilon_o A_{overlap}}{2k_{eff}d_o}}V_{app} \qquad (3)$$

Using equation 3 and the semi-log plot of measured current, the barrier height is found to be $\phi \sim 0.0036$ eV. Hysteresis for tunneling behavior was also investigated and is shown in Figure 4c. For this limited range of operation no hysteresis was observed implying predictable and controllable device behavior. The data presented suggest that within a limited region (before snap-through) our WALD NEMS device may behave as a tunable tunneling device – tunable in respect that the gap height and subsequently measured current can be adjusted via applied voltage.

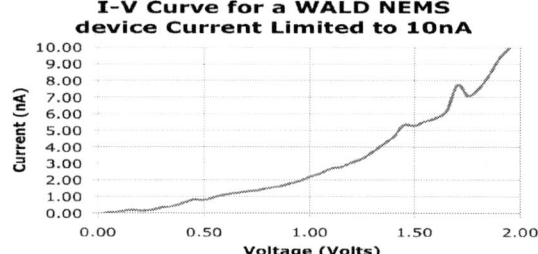

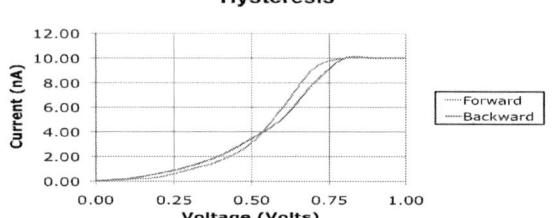

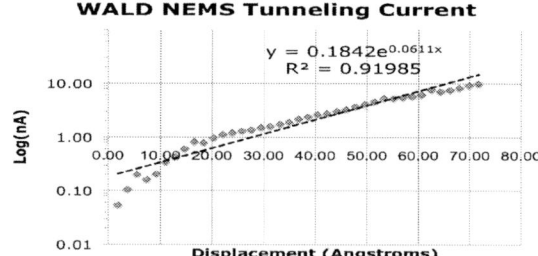

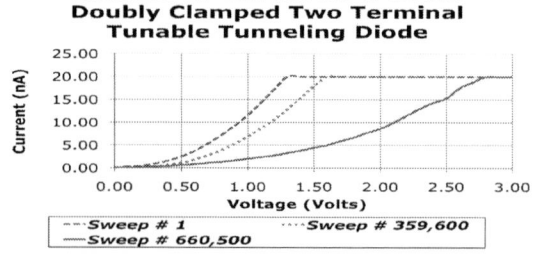

Figure 4: (a) Low current I-V. (b) Log(Current) vs. displacement, $\phi \sim 0.0036$ eV. (c) Hysteresis analysis of tunneling current. (d) I-V curve of lifetime testing for a doubly clamped device shown to have a lifetime > 660,500 cycles. Plot shows 3 sweeps at different cycle numbers during the lifetime testing of the NEMS WALD tunneling device.

The experimental procedure used to characterize pull-in behavior was modified to accommodate lifetime testing for low current operation. Current was limited to 20nA and actuation voltage was switched directly from 0 to a value large enough to induce electron tunneling. An I-V code was written in LabView to record one sweep for every one hundred cycles. After several thousand cycles tunneling behavior was examined via an I-V curve to ensure the device had not failed. Using this method we have demonstrated that our two-terminal tunable NEMS WALD tunneling device has a lifetime greater than 660,500 cycles, Figure 4d.

5. CONCLUSION

A novel low-temperature CMOS compatible top-down nano-fabrication process using ALD has been introduced, and devices successfully fabricated. The WALD NEMS devices we have developed display switching characteristics that are comparable to existing CNT NEMS switches. From Figure 3a,b the 'contact resistance' of our devices in contact mode is measured to be ~ 10-50 MΩ. The large contact resistance can be attributed to native W oxide present on the tungsten beam. In the future the devices should be either coated with additional thin layer of gold or operated in oxygen free environment to avoid W oxide formation.

When operating as a tunneling device at low voltages our device suggests a barrier height of ~ 0.0036 eV. Here we have measured what we believe to be tunneling current over a much larger distance than one would expect, as well as a relatively large current. The effect of surface asperities on device behavior should be considered. Very sharp features can drastically increase the magnitude of the localized electric field inducing field emission, which effectively lowers the energy required to free an electron from its metal. We can be certain this occurs to some degree at the edge of the actuation electrode because of the consequences of the lift-off process used for fabrication in our process. Devices operated in this fashion have demonstrated a lifetime > 660,500 cycles.

The fabrication process and WALD devices introduced present WALD as a capable structural material for use in the field of NEMS. The process extends design techniques established in MEMS into the nanoscale. Devices fabricated using this process show promise for a new class of devices. The tunneling behavior observed is of great interest but must be studied in more depth before any final conclusions about the physics involved can be made. Finally, the switching behavior observed offers great promise for the development of WALD NEMS switches capable for use in digital logic and memory applications.

ACKNOWLEDGEMENTS

This research was supported by DARPA NEMS (SPAWAR Contract No.: N66001-07-1-2033), DARPA Award # HR0011-06-1-0048, the DARPA Focus Center for Integrated Micro/NanoMechanical Transducers (iMINT Center) at the University of Colorado, Boulder and the Air Force Office of Scientific Research.

REFERENCES

[1] Z. Cui, C. Gu, "Nanofabrication challenges for NEMS", *1st IEEE Int. Conf. on Nano/Micro Eng. and Mol. Sys.*, Zhuhai, China, Jan 18-21, 2006, pp. 340-344.

[2] J. Jang, D. Kang, D. Hasko, *et al*, "Nanoelectromechanical switches with vertically aligned carbon nanotubes", *Appl. Phys. Lett.*, vol. 87, 3114, 2005.

[3] S. George, *et al*, "Surface chemistry for atomic layer growth," *J. Phys. Chem.*, vol. 100, no. 31, pp. 13121-13131, 1996.

[4] R. Grubbs, *et al*, "Gas phase reaction products during tungsten atomic layer deposition using WF_6 and Si_2H_6", *J. Vac. Sci. Technol. B*, vol. 22, no. 4, pp. 1811-1821, 2004.

[5] N. Hoivik, J. Elam, *et al.*, "Atomic layer deposited protective coatings for micro-electromechanical systems," *Sensors and Actuators A - Physical*, vol. A103, no. 1-2, pp. 100-8, 2003.

[6] T. M. Mayer, *et al.*, "Atomic-layer deposition of wear-resistant coatings for microelectromechanical devices," *App. Phys. Lett.*, vol. 82, no. 17, pp. 2883-2885, 2003.

[7] Y. Chang, *et al*, "Atomic layer deposited alumina for micromachined resonators", *21st IEEE Int. Conf. MEMS 2008, Tucson, USA, Jan.13-17, 2008.* pp. 387-390.

[8] M. Tripp, C. Herrmann, *et al*, "Ultra-thin multilayer nanomembranes for short wavelength deformable optics", *17th IEEE Int. Conf. MEMS 2004 Technical Digest.* pp. 77-80.

[9] M. K. Tripp, C. Stampfer, *et al*, "The mechanical properties of atomic layer deposited alumina for use in micro- and nano-electromechanical systems," *Sensors and Actuators A - Physical*, vol. 130, pp. 419-429, 2006.

[10] S. Demoustier *et al*, Comptes Rendus Physique, vol. 9, pp. 53-66, Jan-Feb 2008.

[11] S. Lee, D. Lee, *et al*, "A three-terminal carbon nannorelay", *Nano Lett.*, vol. 4, no. 10, pp. 2027-2030, 2004.

[12] A. Kaul, E. Wong, *et al*, "Electromechanical carbon nanotube switches for high-frequency applications", *Nano Lett.*, vol. 6, no. 5, pp. 942-947, 2006.

[13] J. Ward, M. Meinhold, *et al*, "A non-volatile nanoelectromechanical memory element utilizing a fabric of carbon nanotubes", *Proc. 2004 Non-Volatile Mem. Tech. Symp.*, Orlando, Nov 15-17, 2004, pp. 34-38.

[14] E. Dujardin, V. Derycke, *et al*, "Self-assembled switches based on electroactuated multiwalled nanotubes", *App. Phys. Lett.*, vol. 87, 3107, 2005.

[15] P. Hartwell, F. Bertsch, *et al*, "Single mask lateral tunneling accelerometer", *Proc. IEEE MEMS '98*, Heidelberg, Germany, Jan 25-29, 1998, pp. 340-344.

[16] T. Cui, J. Wang, "Polymer-based wide-bandwidth and high-sensitivity micromachined electron tunneling accelerometers using hot embossing", *J. Microelectromech. Syst.*, vol. 14, no. 5, pp. 895-902, 2005.

NANOFIBROUS SURFACE PATTERNING USING NANO-MESHED MICROCAPSULES INDUCED BY PHASE SEPERATION ASSISTED ELECTROSPRAY

Masashi Ikeuchi, Ryosuke Tane, Muneaki Fukuoka and Koji Ikuta,
Department of Micro-Nano Systems Engineering, Nagoya University, Aichi, Japan

ABSTRACT

This is the first report on a nanofibrous surface patterning process using nano-meshed polymer microcapsules for biological applications. The nano-meshed microcapsules were formed using a method named phase separation assisted electrospray. Features of the nano-meshed microcapsules were found to be tunable by adjusting process conditions. Contrary to non-woven mats formed of continuous nanofibers via conventional electrospinning, nano-meshed microcapsule reported herein enabled a single step nanofibrous surface patterning by electrostatic focusing following the microcapsule formation. The biodegradable polylactic acid made microcapsules were patterned onto a glass substrate with up to 500μm resolution and human hepatocyte cells were cultured on the patterned areas to confirm biocompatibility of the microcapsule.

INTRODUCTION

Currently nanofibrous surface modification has been attracting a lot of attention, especially in biological and medical applications. Nanofibrous surfaces composed of biocompatible tiny polymer fibers ~φ100nm can effectively interact with cell surfaces and promote cell proliferation [1]. Among several methods to fabricate nanofibrous surfaces, electrospinning has been widely explored because of its simplicity and versatility [2]. In electrospinning, a polymer solution is ejected from a charged nozzle toward a target electrode substrate of opposite polarity (Fig.1). After being ejected from the nozzle tip the polymer solution is elongated and simultaneously splits into many thinner fibers due to external electrostatic attractive forces and internal electrostatic repulsion. Thus, a nanofiber mesh is obtained on the target electrode.

For tissue engineering applications, micro-patterned nanofibrous surfaces are highly required to guide cell proliferation in defined configuration. However, direct micro-patterning of nanofiber meshes formed by electrospinning is of great difficulty in principle because of the fibers' continuous form. Although there are some reports on patterning electrospun nanofibers on a substrate by patterning electrodes on the target substrate, the patterned and un-patterned areas are not fully separated due to the fibers' continuity.

In electrospinning, the viscosity of the polymer solution affects the morphology of the product created. By decreasing the viscosity of the polymer solution, nanofibers gradually thins until a point where the fiber starts to break up and particulate. At such sufficiently low viscosities, micro-particles below φ1μm are formed – a process called electrospraying differentiated from electrospinning. The morphological change of the electrospinning products depending mainly on the solution conditions has been thoroughly investigated to obtain uniform nanofibers [3]. Compared to the explosive increase in researches in recent years on electrospun nanofibers, a lot less attention has been paid to the microparticle formation process since mere microparticle can be produced using other methods.

One foggy morning in spring 2006, one of the authors M.F. found porous particulated structures in the products of electrospun nanofibers. Since then, we have extensively studied the formation process of the porous structures, and found the ambient humidity is the key factor among numerous parameters for the porous particle formation. By fine tuning the solution conditions and the ambient humidity, we can now control the morphology of the products – nanofiber, particle-nanofiber complex, normal particle and porous particle (Fig.2).

In this report, the formation conditions of the porous particle named "nano-meshed microcapsule", was investigated, and a novel direct surface patterning method of the nano-meshed microcapsule using electrostatic focusing was described. Finally, the biocompatibility of the patterned surface was examined by culturing human hepatocyte cells for future tissue engineering applications.

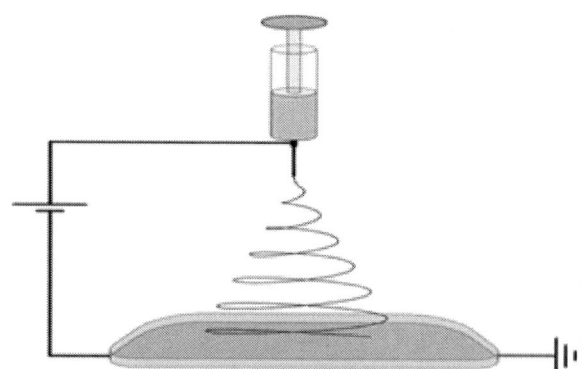

Figure 1: Schematic of the electrospinning process. A polymer solution is fed to a metal capillary by a syringe. A high voltage is applied to the capillary, forcing the polymer to eject towards the target electrode

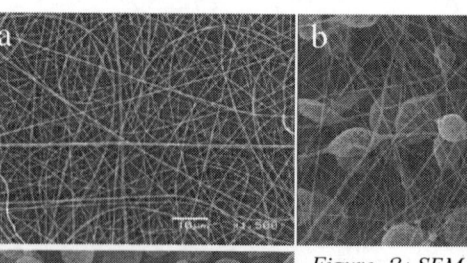
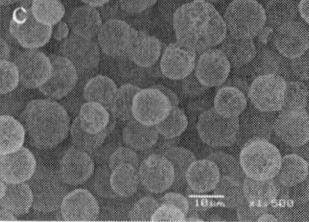

Figure 2: SEM image of (a) nanofibers, PLA 2.0%, Chloroform:Ethanol = 80:20 (w/w), (b) particle-nanofiber complex, PLA 2.0%, Chloroform:Ethanol = 90: 10, (c) "nano-meshed microcapsule", PLA 0.5%, Chloroform:Ethanol = 95:5

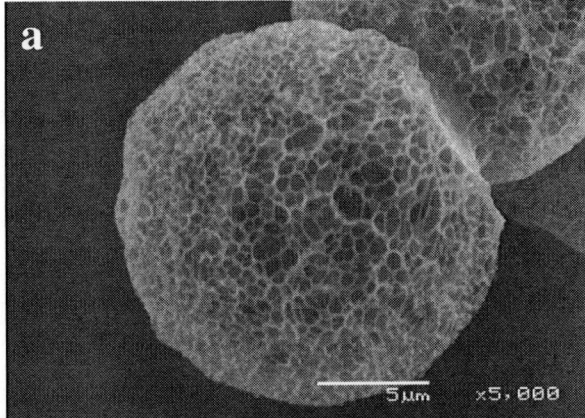

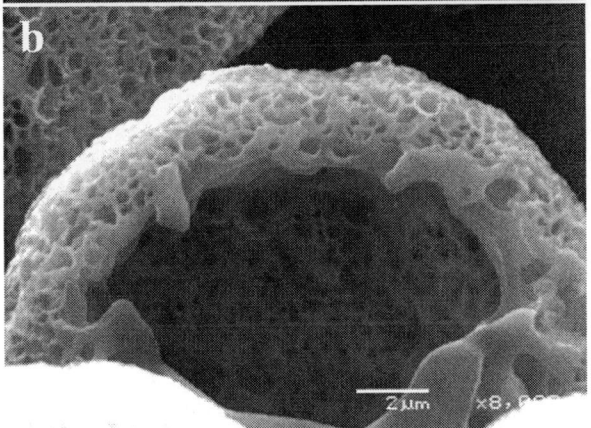

Figure 3: (a) SEM image of a "nano-meshed microcapsules" induced by phase separation assisted electrospray (b) Cross-section of a microcapsule cut with an excimer laser: The edge was slightly deformed due to heat generated by excimer laser ablation.

FORMATION OF THE NANO-MESHED MICROCAPSULE

In electrospinning or electrospraying, there are numerous parameters needed to be adjusted to control the final product. Among these parameters one of the most critical must be the composition of the polymer solution. Usually polymer solutions prepared for electrospinning or electrospraying are a mixture of polymer and solvents.

In this report, the polymer solution is constituted of 1) polylactic acid (PLA): a synthetic biodegradable polymer commonly used for biomedical applications, 2) chloroform: a volatile liquid used as a solvent, 3) ethanol: used as a secondary solvent to adjust the viscosity of the solution. By changing the ratio of the above three constituents, a wide range of morphologies can be created ranging from nanofibers to micro particles. It has been reported that the key factors of polymer solution that influence the formation of various morphologies are viscosity, elasticity, conductivity, and surface tension. When ratio of constituents is altered, all of the above properties of the solution are changed resulting in different morphologies for final products.

Nanofibers $\phi100\sim200$nm was formed with polymer solution constituted with an excessive amount of ethanol (Fig.2a). When the ratio of ethanol was decreased, nanofibers started to contain irregular particles (Fig.2b).

Further decreasing of ethanol along with the decreasing of polylactic acid resulted in the disappearance of fibers

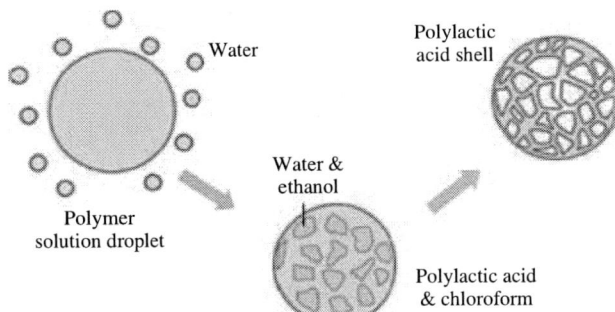

Figure 4: Schematic representation of the formation process of the nano-meshed microcapsule induced by phase separation assisted electrospray

and the formation of uniform sized microparticles. We found the surface of the microparticle became porous (Fig.2c), when the process was taken place in high ambient humidity. The porous microparticle, named "nano-meshed microcapsule", was found to have a hollow spherical shell composed of nanofibers $\phi100\sim200$nm. The nanofibers formed penetrating microporous structure in the shell (Fig.3).

The possible mechanism of the nano-meshed microcapsule formation is as follows (Fig.4): 1) the polymer solution jet electrosprayed from the nozzle breaks up into tiny droplets due to instability of the surface and electrostatic repulsion, 2) The droplet subjected to high ambient conditions, and water vapor condenses on the surface of the microparticle causing phase separation between the water and solvent phases, 3) the droplet of the polymer solution expands to a spherical shell due to electrostatic repulsion, 4) the water and solvent rapidly evaporates, leaving the polymer to remain to form a nano-meshed structure.

CONTROL OF THE DIMENSIONS OF THE NANO-MESHED MICROCAPSULE

The porosity of the nano-meshed surface and the diameter of the microcapsules can be tuned by adjusting the ambient humidity during the electrospray process and the flow rate of the polymer solution respectively.

The porosity of the surface was sensitive to ambient humidity, and increased by electrospraying at higher humidity (Fig.5). This is presumably because the amount of the condensed water on the surface of the electrosprayed

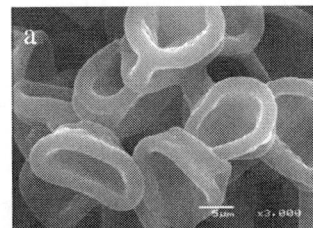

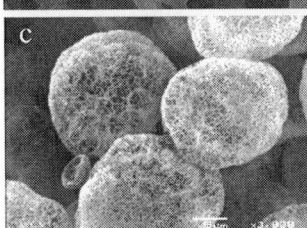

Figure 5: SEM image of nano-meshed microcapsules (a) humidity = 15%, (b) humidity = 50%, (c) humidity = 80%. Other conditions were same for each experiment: temperature = 25℃, flow rate = 3.6ml/h, PLA 1.0%, Chloroform: Ethanol = 95:5

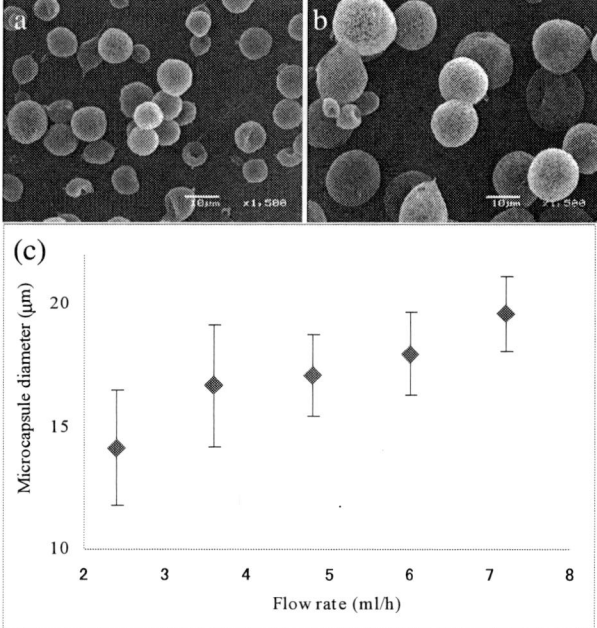

Figure 6: (a) SEM image of nano-meshed microcapsules: flow rate = 2.4ml/h, (b) SEM image of nano-meshed microcapsules: flow rate = 6.0ml/h, (c) Plot of microcapsule diameter as a function of the flow rate of the solution in the ejection nozzle. Experimental conditions except flow rate were same for each experiment: temperature = 25℃, humidity = 80%, PLA 1.0%, Chloroform: Ethanol = 95:5. Voltage = DC20kV

droplet of the polymer solution increases under high humidity, leaving larger pores on the final products through the formation of larger hydrophilic domain during phase separation process.

The diameter of the nano-meshed microcapsule was tunable from ~φ10μm to φ20μm by adjusting the flow rate of the solution to the nozzle. The diameter was increased with faster flow rates (Fig.6). A faster flow rate seems to cause the increase of the initial diameter of the electrosprayed jet of the solution from the nozzle, resulting in larger initial droplets formation by instability of the surface.

DIRECT SURFACE PATTERNING OF THE NANO-MESHED MICROCAPSULE

The nano-meshed microcapsules have the characteristics of both a nanofiber and a microparticle. That is, the surface of the microcapsule is composed of nanofibers ~φ100nm, and at the same time, can be treated as an individual particle. In this section we introduce a method using an electrostatic lens placed concentric to the ejection nozzle in order to focus the nano-meshed microcapsules. This set up can realize microcapsule formation and simultaneous direct patterning of the microcapsules on a substrate, whereas conventional electrospun nanofibers are difficult to be directly patterned due to its continuous form.

Schematic setup of the improvised electrospray and direct patterning system is shown in Fig.7a,b. All the process is done in a closed chamber capable of temperature and humidity control. The polymer solution is supplied to the nozzle from a motorized syringe, and the grounded target substrate is mounted onto a PC-controlled X-Y-Z stage. The electrosprayed nano-meshed microcapsules are

focused at the bottom of the lens, and arbitrary shapes can be patterned in a single step by moving the stage (Fig.7c). In the following experiment, DC20kV was applied to ejection nozzle and DC15kV was applied to the electrostatic lens.

To verify the formation and simultaneous electrostatic focusing of the nano-meshed microcapsule, microcapsules were converged to a single spot on a glass substrate (Fig.8a). The height was increased depending on the electrospray duration, whereas the bottom width was kept constant due to electrostatic focusing (Fig.8b). The diameter of the pillar was φ~500μm at the bottom, and was decreased to φ~200μm in the upper part. By moving the X stage, a straight line composed of nano-meshed microcapsule was drawn on the glass substrate (Fig.8c). The depicted line was observed to be ~500μm in width. Using programmable motorized X-Y stage, arbitrary shape can be pattered on the substrate (Fig.8d). Contrary to the previously reported patterning methods of electrospun nanofibers by fabricating electrodes on the target substrate, the direct patterning method developed herein has high flexibility in modifying the pattern, which is suitable for fabricating customized products in medical applications.

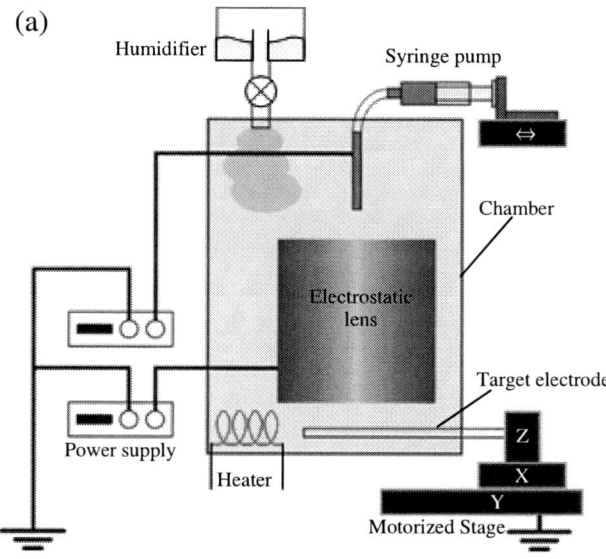

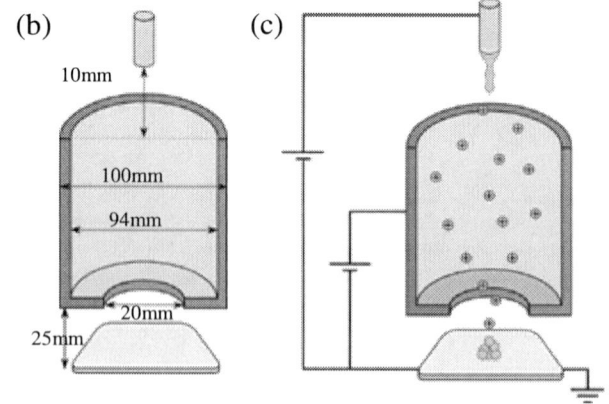

Figure 7: (a) Schematic of the improvised phase separation assisted electrospray patterning system (b) Dimensions of the setup with an electrostatic lens (c) Schematic of the formation and the focusing process of nano-meshed microcapsules using electrostatic lens.

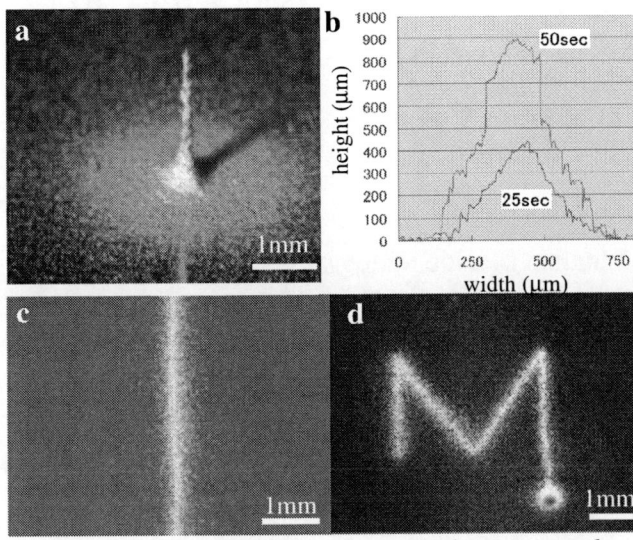

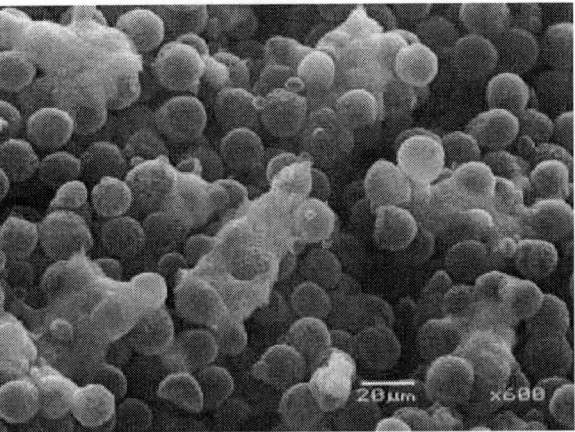

Figure 9: Pseudo-colored SEM image of human hepatocyte cells (green) cultured on the patterned nano-meshed microcapsules (purple) for 48h

Figure 8: (a) Converged nano-meshed microcapsules using the electrostatic lens, (b) Cross-sectional profile of the deposited structure composed of microcapsules: The height was increased depending on the electrospray duration, whereas the bottom width was kept constant due to electrostatic focusing. (c) SEM image of a line drawn using the patterning system, (d) "M" shaped 2D pattern drawn using the patterning system as a demonstration

BIOCOMPATIBILITY OF THE NANO-MESHED MICROCAPSULE

In recent years, there has been an enormous rise in interest for biomedical research, having been one of the most beneficial fields owing to the advancement in nanotechnology. Among various biomedical researches, tissue engineering is advancing rapidly towards the realization of a totally new medicine. One of the challenges in tissue engineering is to design an ideal scaffold capable of both providing structural support and supplying nutrients for cells.

Electrospun nanofibers have been explored extensively as components of scaffolds because their large surface area and high porosity are suitable for cell adhesion and liquid circulation. However, there are still difficulties lying in the micro-patterning of electrospun nanofibers to create an arbitrary scaffold because of its continuous form.

One of the most attractive applications of the nano-meshed microcapsule is for usage as a component to create a scaffold, since it has fine nanofibrous surface like electrospun nanofibers, and at the same time, it can be patterned in an arbitrary shape as a particle.

To verify the application of the nanofibrous surface patterning method using nano-meshed microcapsules for tissue engineering, human hepatocyte cell was cultured on the nano-meshed microcapsules. Good cell adhesion and proliferation was observed after 48 hours of cultivation (Fig.9). It can be speculated that the topological characteristics of the nano-meshed microcapsules are favorable for cell adhesion and that the hollow structure of the capsules will allow nutrients and gases to circulate within the structures suitable for three-dimensional cell proliferation. Thus, the biocompatibility of the nano-meshed microcapsules was confirmed.

CONCLUSIONS

This paper has presented a new morphology "nano-meshed microcapsule" created by phase separation assisted electrospray. The microcapsule has excellent uniformity in size and shape and possesses a nanofibrous surface which is ideal for cell culture. Using the microcapsules, nanofibrous surface patterning was realized by applying a secondary electric field created by an electrostatic lens. The patterning resolution was as high as 500μm. It was also confirmed that this surface patterning exhibited good cell adhesion indicating the possibility for further applications in biomedical and tissue engineering fileds.

ACKNOWLEDGEMENTS

The authors thank Dr. Akira Yamada, and Kuniho Kasugai for their contribution in starting up the electrospinning project.

REFERENCES

[1] F. Causaa, P.A. Nettib, and L. Ambrosiob, Biomaterials 28 (2007), pp.5093-5099

[2] W.E. Teo and S. Ramakrishna, Nanotechnology 17 (2006), pp.R89-R106

[3] K.H.Lee, H.Y.Kim, H.J.Bang, Y.H.Jung and S.G.Lee, Polymer 44 (2003), pp.4029-4034

[4] Z. Maa, M. Kotakia and S. Ramakrishna, J.Membrane Science 272 (2006), pp.179-187

[5] D. Yang, B. Lu, Y. Zhao and X. Jiang, Advanced Materials 19 (2007), pp.3702-3706

[6] A. Jaworek, Powder Technology 176 (2007), pp.18-35

[7] J.M. Deitzel, J.D. Kleinmeyer, J.K. Hirvonen, N.C. Beck Tan, Polymer 42 (2001), pp.8163-8170

[8] Y.K. Hwang, U. Jeong, E.C. Cho, Langmuir 24 (2008), pp.2446-2451

[9] Z. Huang, Y. Zhang, M. Kotaki, S. Ramakrishna, Composites Science and Technology 63 (2003), pp.2223-2253

[10] H. Fong, I. Chun, D.H. Reneker, Polymer 40 (1999), 4585-4592

TUNABLE OPTICAL ENHANCEMENT FROM A MEMS-INTEGRATED TIO₂ NANOSWORD PLASMONIC ANTENNA

Brian D. Sosnowchik[1], P. James Schuck[2], Jiyoung Chang[1], and Liwei Lin[1]

[1]Berkeley Sensor & Actuator Center, University of California at Berkeley, Berkeley, CA, USA
[2]Imaging & Manipulation Facility, Lawrence Berkeley National Laboratory, Berkeley, CA, USA

ABSTRACT

In this work, we present the fabrication and testing of a MEMS-integrated, variable-gap titanium dioxide nanosword plasmonic antenna. Two-photon photoluminescence (TPPL) testing was performed, and enhancements as large as 251-fold were observed at the sharp, tapered nanosword tip. The tunable nanosword antenna successfully and repeatedly demonstrated increases in the TPPL intensity as the gap was reduced below 50nm. Dark-field scattering experiments were performed at the tip to quantify the shape-dependent plasmonic resonance of the gold-coated nanosword tip. It is believed that this device, with further improvements to the enhancement, may enable previously unattainable functionality for surface-enhance Raman scattering (SERS) analysis.

INTRODUCTION

Surface-enhanced Raman spectroscopy (SERS) is a widely used, versatile technique for the detection of small concentrations of complex molecules [1]. Improvement in SERS detection depends, in part, on the ability to *reproducibly* generate strongly enhanced, concentrated electromagnetic fields in a localized region of analysis. Generation of these fields arises from the substrate's physical features (like a sharp tip), plasmonic coupling, or both. Two common approaches to creating substrates that contain localized regions of enhanced SERS signal, or "hot-spots," are through the roughening of silver surfaces [2] or the dispersion of gold nanoparticles [3]. However, such random methods are limited in the ability to pre-determine the location and optical properties of the "hot-spot." Hence, lithographically-patterned nanoantennas are significantly advantageous [4, 5]. The previously reported "bowtie" geometries take advantage of structural tapering, sharp tips, and small gaps which are known to improve the electric field enhancement and plasmonic coupling [6, 7]. Further improvements to this design may arise from the ability to change the gap of the optical antenna [8], enabling the tuning of the plasmonic resonance [9].

At *MEMS 2007*, we reported a method to rapidly synthesize a variety of nanostructured materials [10, 11], and at *MEMS 2008*, we presented the discovery of a previously unreported titanium dioxide nanostructure called a "nanosword" [12]. The nanoswords have a repeatable geometry which is broad and flat, tapering to a sharp tip with a radius of curvature as low as 2nm. In this work, these TiO₂ nanoswords were used to fabricate a MEMS-integrated, variable-gap plasmonic antenna.

Figure 1 illustrates the operating principle of the device. Nanoswords are precisely positioned and aligned on opposing platforms of an electrostatic MEMS actuator and coated with a thin layer of gold. When the nanoswords are far apart, the optical field enhancement

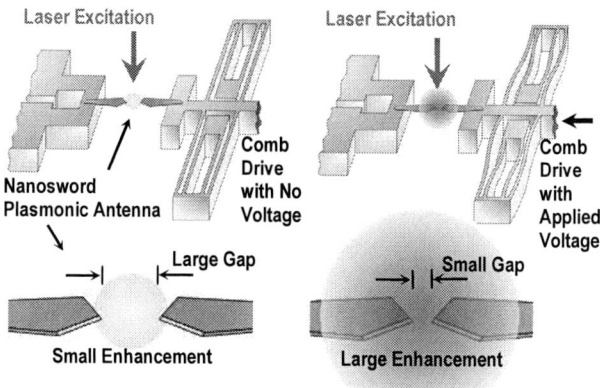

Titanium Dioxide Nanosword Plasmonic Antenna

Figure 1: Operating principle of a nanosword plasmonic antenna. Nanoswords are positioned on opposing sides of a MEMS actuator and coated with gold. When far apart, the optical enhancement is small. When the gap between the tips is reduced, large enhancements can occur. Such enhancements are vital for surface-enhanced Raman scattering analyses.

results solely from the sharpness of the tips (i.e. – the "lightning rod effect"). However, as the gap closes, the nanostructures become optically coupled, resulting in increased enhancement. In theory, this shape-dependent enhancement can rise several orders of magnitude as the separation distance drops below 50nm [5].

SAMPLE PREPARATION & ANALYSIS

The nanoswords were synthesized using the method described previously [12]. Briefly, a 200nm layer of titanium was thermally evaporated onto a copper TEM grid, which was placed atop a 3x3mm² titanium heater chip and positioned in a quartz tube within an induction coil. Power was initiated rapidly heating both the TEM grid and the titanium chip for a period of 3-10 minutes in an acetylene environment resulting in the synthesis of the vapor-solid-grown nanoswords. The dimensions of the nanoswords used in this work had widths between 500-800nm, thicknesses of 55-60nm, lengths of 5-8μm, and had a taper angle of approximately 81°.

Plasmonic tests were performed on two different nanosword specimen as shown in Figure 2. For the case of isolated nanoswords, a tungsten probe tip on a micropositioner was used to acquire nanoswords from the growth substrate and position them cantilevered (Figure 2a) onto the underside of a copper TEM grid as shown in Figure 2b. Samples were cleaned for 15sec in an oxygen barrel plasma, and a thin film of 2nm chromium and 10nm or 20nm gold was thermally evaporated onto the surface of the nanosword.

978-1-4244-2977-6/09 $25.00 © 2009 IEEE

Fabrication of the tunable plasmonic antenna devices occurred as follows. A 50µm-thick device layer silicon-on-insulator (SOI) wafer was used to fabricate electrostatic actuators as shown in the scanning electron microscope (SEM) image in Figure 2c. Attached to the rotor is a moving platform which, upon actuation, is driven closer to a stationary platform. Prior to the placement of the nanoswords, the moving platform displacement vs. voltage data were obtained using optical and scanning electron microscopes. Two opposing nanoswords were subsequently positioned and aligned on the platforms. After oxygen-plasma cleaning, a thin-film of 2nm chromium and 20nm gold was then thermally evaporated onto the sample. It should be noted that all SEM images were taken after TPPL testing to prevent carbon contamination on the gold film.

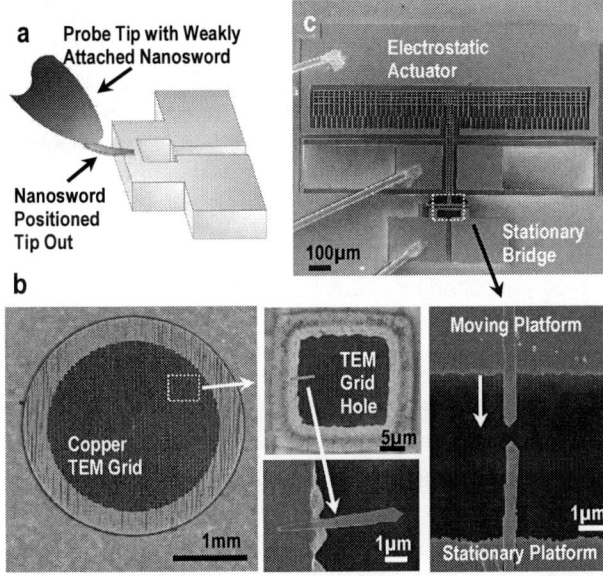

Figure 2: Sample preparation for the two-photon photoluminescence tests. a) Using a tungsten probe tip, nanoswords were positioned onto target structures to observe the TPPL resulting from two different nanosword specimen. b) Isolated nanoswords were positioned cantilevered on a copper TEM grid testing platform. c) SEM of the plasmonic antenna device. Opposing nanoswords are positioned and aligned on the two platforms of the SOI-fabricated device. Electrostatic actuation closes the gap between the nanosword tips, enabling the tuning of the resulting optical intensity.

EXPERIMENTAL SETUP

A 120fs, 832nm, 75MHz Ti:sapphire laser was used for optical excitation with typical average powers being on the order of 100µW at the sample. Laser light polarized axially with the nanoswords was focused through an inverted optical microscope setup to a diffraction-limited spot on the specimen. Emitted photons were measured using an avalanche photodiode while backscattered laser light was effectively blocked. For the antenna device, the packaged MEMS structures were actuated with a DC power supply.

A phenomenon called two-photon photoluminescence (TPPL) was measured to observe the field enhancements of the isolated nanoswords and the antenna. For noble metals like gold, sharp tips and roughened surfaces give rise to localized field enhancement which, when excited with the appropriate wavelength, can lead to two-photon photoluminescence, an emission continuum that results from interband electron transitions. More importantly, the TPPL resulting from a gold nanostructure is very sensitive to the incident excitation [13] making it an appropriate metric for quantifying local field enhancement. In our case, TPPL emission from a test sample demonstrated a quadratic dependence on pump power, showing clearly that emission originates from a two-photon excitation process.

RESULTS

Figure 3 illustrates the TPPL tests for the case of isolated nanoswords on TEM grids. The nanosword shown in Figure 3a resulted in the TPPL profile mapped in Figure 3b. From the mapping, it is clear that the enhancement was strongest at the tip region of the nanosword, with a TPPL enhancement of 10.8-fold compared to the midsection of the nanosword. Such an increased enhancement at the tip region was consistent for all nanoswords tested. In Figure 3c, the radius of curvature was estimated to be ~10nm. It should be mentioned that the bright "hot spots" occurring at the base of the nanoswords may be attributed to localized plasmonic coupling of nanogaps in the roughened copper grid and the gold on the nanosword. Alternatively, the TPPL mapping of the nanosword in Figure 3d showed a 251-fold increase in the photoluminescent intensity with respect to the midsection of the nanosword, shown in Figure 3e. This significant enhancement may be attributed to the gold nanoparticle which formed at the tip shown in Figure 3f.

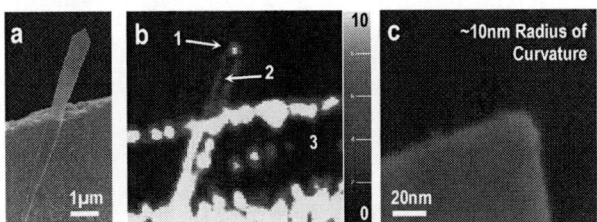

2nm/10nm Cr/Au: 1) Tip Intensity 10.8kHz, 2) Mid-Section Intensity 1kHz, 3) Smooth Grid Base Intensity 0.5kHz

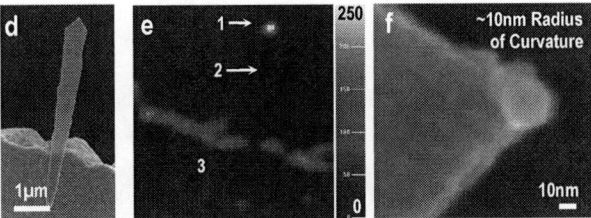

2nm/20nm Cr/Au: 1) Tip Intensity 251kHz, 2) Mid-Section Intensity 1kHz, 3) Smooth Grid Base Intensity 0.5kHz

Figure 3: TPPL tests of two isolated nanoswords. a-c) Enhancement of 10.8-fold was observed at the tip, while higher intensities occurred when gold particle formed at the tip (d-f).

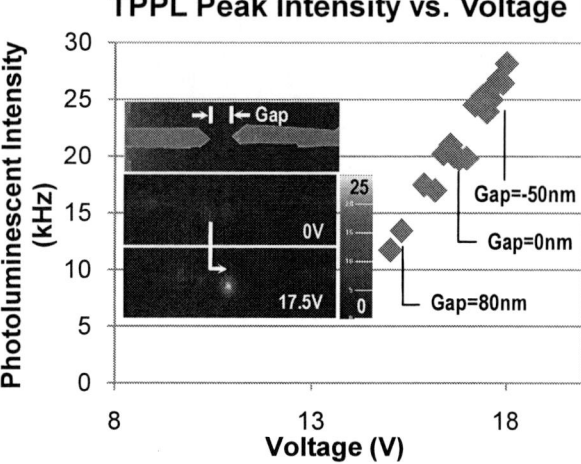

TPPL Peak Intensity vs. Voltage

Figure 4: TPPL peak intensity vs. actuator voltage. As the gap between the nanostructures was reduced, the intensity increased significantly. The inset illustrates the TPPL mapping of the nanoswords with an open and closed gap.

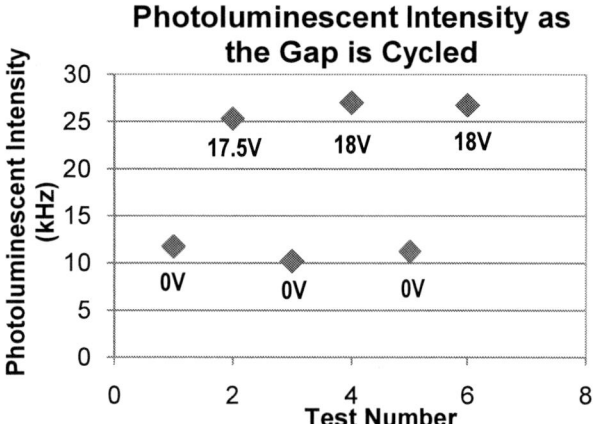

Photoluminescent Intensity as the Gap is Cycled

Figure 5: The photoluminescent intensity for the plasmonic antenna was repeatable when cycled several times between 0 and 17.5-18V.

For the nanosword plasmonic antenna device, TPPL profiles were obtained as the gap between the opposing nanoswords was incrementally closed. The initial gap was observed to be 568nm. As the tips converged upon one another, Figure 4 illustrates the observed increase in the TPPL intensity (in units of photon yield rate) with decreasing gap between the nanoswords. From a gap of over 80nm through an overlap of 50nm, the intensity increased by approximately 2.5x (more on this below). The inset of Figure 4 illustrates a scanning electron micrograph of the opposing nanoswords and the corresponding TPPL profiles for two cases – first, for the case of no actuation (with a maximum tip-to-tip gap at 0V), and second, for the case of a ~50nm tip overlap (at 17.5V). For visual convenience, all three images have been scaled to the same size. Evidence of plasmonic coupling may be observed from the TPPL at the tips. When apart, each tip had a photoluminescent intensity of ~5kHz. However, when the tips were brought into close proximity, the intensity rose significantly beyond the sum

of the two (10kHz), clearly demonstrating the effect of plasmonic coupling.

To demonstrate the repeatability of the nanosword antenna, the device underwent several cycles between the maximum gap and a ~50nm tip-to-tip overlap. The results are shown in Figure 5, where the intensity at 0V represents the sum of the two tips. The data illustrate a repeatable trend in the enhancement at the tips and further demonstrate the plasmonic coupling. It should be noted that cycling was discontinued after six tests to prevent unnecessary damage to the antenna.

DISCUSSION

Several points of note should be mentioned about the device fabricated in this work. As each opposing arm of the antenna, the nanosword presents an ideal geometry for this device. With its structural tapering, the device resembles the previously reported "bowtie" designs. Additionally, the tip sharpness, at a 2nm radius of curvature, is comparable to state of the art e-beam lithographically patterned designs. More importantly, the crystal dictates the geometry, so a repeatable taper angle and tip is frequently achieved. By using MEMS as a platform, not only can the optical intensity be tuned as demonstrated herein, but in theory, the plasmonic resonance and enhancement should be able to be tuned in real time. Such tuning is important, since SERS detection of different molecules depends on the plasmon resonance. Finally, such a device could provide a measureable and unambiguous approach to understanding to how the gap of an antenna influences the plasmonic resonance.

Scattering Spectrum of Nanosword Tip

Figure 6: Dark-field scattering spectrum near the tip region of a gold-coated nanosword.

While we have shown an unprecedented degree of plasmonic device control, we feel there still exist some questions that must be addressed. For one, why was a greater enhancement not observed for the initial device? Perhaps some insight can be gained when one considers that significant enhancements are known to arise through the excitation at a plasmon resonant wavelength. To obtain a greater understanding of the plasmonic resonances of the nanoswords, isolated nanoswords with a 2nm/20nm Cr/Au thin film were placed onto a quartz substrate, and dark-field scattering spectra were obtained using a 100W halogen lamp. The normalized spectrum is shown in Figure 6 from the tip in the inset. Two resonant peaks are visible. The first occurs at around 550nm, which is believed to arise from grains of gold along the edge of the nanosword [14]. The second peak at 870nm is believed to be the surface plasmon resonance attributed to the thin gold film deposited on, and in the shape of, the nanosword. If the peak resonance red shifts as the tips are brought into close proximity [4], the resonant wavelength will extend further from the excitation wavelength possibly resulting in lower enhancements like those observed. Improvements to the enhancement may arise by tuning the laser wavelength or structurally modifying the nanosword, which will be addressed shortly. Misalignment should also play a role – the tips of the nanoswords in this antenna were laterally misaligned by 71nm, and although theoretically straight, vertically misalignment cannot be conclusively ruled out. Greater insight may be provided through finite difference time domain simulation and further experimentation.

There are several strategies, however, which we plan to employ to improve the enhancement of the device. One approach is to use a focused ion beam (FIB) to etch a geometry on the gold similar to the bowtie structures used in [5], which are known to resonate with laser excitation at 832nm. Alternatively, FIB may be used to etch plasmonic gratings along the length of the nanosword [15]. Using such an approach, resonance can be structurally induced for a particular laser wavelength, allowing the propagation of surface plasmons to the tip of the nanosword. Finally, lateral misalignment may be solved, in the short-term, by using a micropositioner integrated in an SEM to precisely align the tips of the nanoswords, while in the long-term, a MEMS device with both lateral and transverse motion should solve this problem.

CONCLUSION

In this work, we present the fabrication and testing of a variable-gap TiO_2 nanosword plasmonic antenna. Two-photon photoluminescence was measured on both isolated, gold-coated nanoswords as well as the plasmonic antenna. For isolated nanoswords, enhancements as large as 251-fold were observed at the tips of the nanoswords. For the antenna device, an SOI MEMS actuator enabled the controlled closing of the gap between the antenna's nanosword arms. The tunable, localized optical field was observed to depend on the gap between the opposing nanoswords. Dark-field scattering spectra were taken at the nanosword tip illustrating a peak resonance of 870nm. Future testing will involve the aforementioned efforts to further improve the enhancement and subsequent SERS testing.

ACKNOWLEDGEMENTS

The authors would like to acknowledge the help of Frank Ogletree, Ed Wong, Lei Luo, Jong-Yoon Ha, and Shiwei Wu. Samples used in this work were fabricated as part of MEMS Exchange Run 4119 by Sia Parsa, Antal Kovats, and Attila Szabo in the UC Berkeley Microfabrication Laboratory. This project is supported in part by the DARPA N/MEMS Fundamental Sciences program. Work at the Molecular Foundry was supported by the Office of Science, Office of Basic Energy Sciences, of the U.S. Department of Energy under Contract No. DE-AC02-05CH11231.

REFERENCES

[1] A. Campion and P. Kambhampati, "Surface-Enhanced Raman Scattering," *Chem. Soc. Rev.*, vol. 27, pp. 241-250, 1998.

[2] M. Fleischmann, P. J. Hendra, and A. J. McQuillan, "Raman Spectra of Pyridine Adsorbed at a Silver Electrode," *Chem. Phys. Let.*, vol. 26, pp. 163-166, 1974.

[3] S. Nie and S. R. Emory, "Probing Single Molecules and Single Nanoparticles by Surface-Enhanced Raman Scattering," *Science*, vol. 275, pp. 1102-1106, 1997.

[4] D. P. Fromm, et al., "Gap-Dependent Optical Coupling of Single "Bowtie" Nanoantennas Resonant in the Visible," *Nano Let.*, vol. 4, pp. 957-961, 2004.

[5] P. J. Schuck, et al., "Improving the Mismatch Between Light and Nanoscale Objects with Gold Bowtie Nanoantennas," *Phys. Rev. Let.*, vol. 94, pp. 017402-1-4, 2005.

[6] K. B. Crozier, et al., "Optical Antennas: Resonators for Local Field Enhancement," *J. App. Phys.*, vol. 94, pp. 4632-4642, 2003.

[7] M. I. Stockman, "Nanofocusing of Optical Energy in Tapered Plasmonic Waveguides," *Phys. Rev. Let.*, vol. 93, pp. 137404-1-4, 2004.

[8] K. Iwami, T. Ono, and M. Esashi, "Optical Near-Field Probe Integrated With Self-Aligned Bow-Tie Antenna and Electrostatic Actuator for Local Field Enhancement," *J. Microelectromech. Sys.*, vol. 15, pp. 1201-1208, 2006.

[9] J. Merlein, et al., "Nanomechanical Control of an Optical Antenna," *Nature Photonics*, vol. 2, pp. 230-233, 2008.

[10] B. D. Sosnowchik and L. Lin, "Rapid Synthesis of Carbon Nanotubes by Bulk and Localized Inductive Heating," presented at the 20th IEEE MEMS Conference, Kobe, Japan, 2007.

[11] B. D. Sosnowchik and L. Lin, "Rapid Synthesis of Carbon Nanotubes via Inductive Heating," *App. Phys. Let.*, vol. 89, pp. 193112, 2006.

[12] B. D. Sosnowchik, et al., "Rapid, Localized Synthesis of Titanium-Based Nanosword on MEMS," presented at the 21st IEEE MEMS Conference Tucson, AZ, 2008.

[13] G. T. Boyd, Z. H. Yu, and Y. R. Shen, "Photoinduced Luminescence from the Noble Metals and Its Enhancement on Roughened Surfaces," *Phys. Rev. B*, vol. 33, pp. 7923-7936, 1986.

[14] S. A. Maier and H. A. Atwater, "Plasmonics: Localization and Guiding of Electromagnetic Energy in Metal/Dielectric Structures," *J. App. Phys.*, vol. 98, pp. 011101, 2005.

[15] C. Ropers, et al., "Grating-Coupling of Surface Plasmons onto Metallic Tips: A Nanoconfined Light Source," *Nano Let.*, vol. 7, pp. 2784-2788, 2007.

MICROELECTROMECHANICAL METAL-AIR-INSULATOR-SEMICONDUCTOR (MEM-MAIS) DIODE: A NOVEL HYBRID DEVICE FOR ESD PROTECTION

D. Acquaviva[13], D. Tsamados[1], Ph. Coronel[2], T. Skotnicki[3] and A. M. Ionescu[1]

[1]Ecole Polytechnique Fédérale de Lausanne, Lausanne, SWITZERLAND
[2]LITEN-CEA, Grenoble, FRANCE
[3]ST Microelectronics, Crolles, FRANCE

ABSTRACT

In this paper we propose and experimentally validate the concept of Microelectromechanical Metal-Air-Insulator-Semiconductor Diode (MEM-MAIS) as a novel hybrid device for ESD protection fabricated on SOI. The proposed ESD switch has unique figures of merit inherited from the MEM part: the *in-series* air gap guaranties record low leakage current ($\sim$10fA/μm) practically independent on temperature up to 100°C in off-state and a parasitic capacitance of few fF. The scalable electro-mechanical design of the actuation voltage makes the new proposed switch suitable for voltage domains ranging from few volts to tens of volts. A Fowler-Nordheim conduction is demonstrated for the on-state of the MEM-MAIS diode. Basic ESD HBM and MM experiments with MEM-MAIS diodes are reported and the versatility of the new device demonstrated.

INTRODUCTION

Electrostatic discharge (ESD) protection circuits are becoming a challenging design problem both for RF and submicron CMOS integrated circuits (IC) [1], because of the trends towards higher speed and smaller device size. In order to obtain high ESD robustness with a low impact on the circuit performances, on-chip ESD protection circuits have to be added into the IC products [2]. Diodes, latching thyristors and bipolar snap back structures (or networks) [3] still remain the most used solutions able to absorb the total energy of an ESD pulse, redesigned and rearranged in different configurations to satisfy the new technology requirements. Especially diodes have been proposed in many solutions, both for bulk and SOI technologies, in single and stacked configurations. Low leakage diode string for RF-ESD application has already been demonstrated in [4] with a leakage current of 10^{-13}A at room temperature and a parasitic capacitance of 20fF. Even if SOI technology precludes the simple integration of common bulk ESD protection device, due to the presence of the buried oxide which leads to increased self-heating, several SOI ESD protection devices have been also proposed: Salman has reported in [5] a new field effect diode for ESD protection with a leakage current density of 10^{-12}A/μm and a parasitic capacitance density of few fF/μm.

In this work we present a novel device for ESD protection, which combines the bistability of a MEMS switch with the exponential characteristics of a MIS diode. The MEM-MAIS (Microelectromechanical Metal-Air-Insulator-Semiconductor) diode could represent a unique solution to all the ones already reported in literature since the load dump can be set by MEM switch

design and the presence of the air-gap guaranties quasi-zero leakage current ($\sim$fA) up to the *pull-in* event and an extremely very low parasitic capacitance ($\sim$fF), nevertheless an area layout saving. Moreover, the voltage operation of the device can be imposed only by the design of the MEM pull-in voltage, ranging from few volts to tens (or hundreds) of volts and it is based on a single device concept in a single technology with potential applications in standard CMOS or automotive.

DESIGN AND FABRICATION

Fig. 1(a) and 1(c) show the SEM images of the fabricated structure and the device cross section, respectively. The suspended beam or membrane forming the movable electrode of the diode (Fig. 1(b)) has a double clamped crab configuration; it has been designed keeping constant the contact area ($W \times w = 20\mu m \times 10\mu m$) and varying the arms length in order to demonstrate flexibility in defining actuation voltages (pull-in) between 20 and 40V [6], with an air gap of 300nm, as might be required by automotive applications. It is worth noting that by scaling the gap, the metal thickness and the device area one can practically scale the pull-in voltage from tens or hundreds of volts down to $\sim$1V.

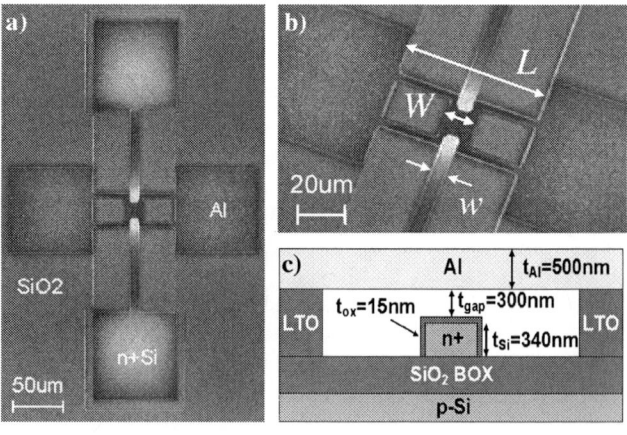

Figure 1: a) SEM top view of the fabricated MEM-MAIS diode. b) SEM image of the suspended membrane tilted of 20°. L, W and w represent the distance between two aluminum anchors, the width of the membrane central region, the width of the bottom contact, respectively. c) Device cross-section.

The device is fabricated (Fig. 2) on SOI wafer with 340nm thick, n-doped ($\sim 10^{20}$cm^{-3}) Si bottom contact. On the top of the Si contact a 15nm thick oxide is thermal grown, in order to control the surface quality and serve as a tunneling insulator. The suspended electrode is made of

978-1-4244-2977-6/09 $25.00 © 2009 IEEE

500nm thick Al anchored on LTO regions. Polyimide is used as sacrificial layer to release the suspended diode membrane. A detailed device design and fabrication process is reported in [7].

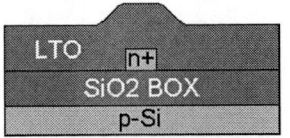
1. n-doping of Si layer; LTO deposition

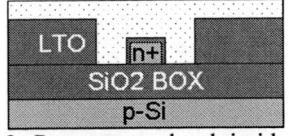
3. Promoter and polyimide (PI) spinning

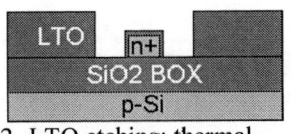
2. LTO etching; thermal oxide growth on Si

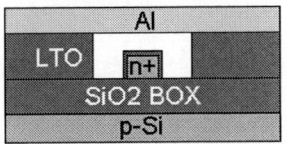
4. CMP; Al deposition; PI removal by O₂ plasma

Figure 2: Main steps of the MEM-MAIS diode fabrication process. Step 1: n-doping of Si layer by PoCl3 followed by the patterning of the n+Si regions based on chlorine chemistry. Deposition of thick low thermal oxide (LTO) on the too of the silicon contacts. Step 2: Opening of trenches in the LTO by reactive ion etching (RIE). Growth of a thin thermal oxide on the silicon beams. Step3: Spinning of the sacrificial layer: adhesion promoter plus polyimide (PI). Step 4: Chemical mechanical polishing (CMP) to planarize the surface and to define the air gap. Aluminum deposition by sputtering and chlorine-based pattering. Structures releasing (PI removal) by plasma oxygen.

ELECTRICAL CHARACTERIZATION

The typical log-linear I-V characteristics of an experimental MEM-MAIS device with a length of 30μm and a stiffness constant k_l=150.8N/m, are reported in Fig.3. The measurements are performed using a HP4156C semiconductor parameter analyzer in medium sampling mode; the silicon bottom contact is grounded and the voltage is varied on the suspended membrane.

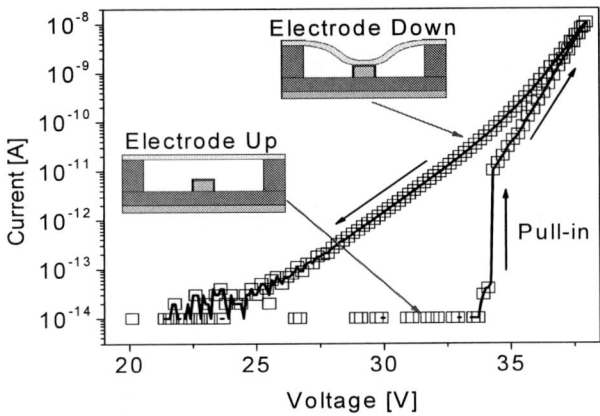

Figure 3: Log scale I-V characteristics of a MEM-MAIS switch with L=30um, W= 20um and a stiffness constant k_l=150.8N/m.

The pull-in occurs above 30V with slope that is quasi-infinite, in agreement to the simulation (the actuation voltage is evaluated according to the analytical model reported in [6] and predicts the electromechanical pull-in

at 32.9V, really close to the measured 34.3V). As expected, the off-state current is in the order of 10fA (the resolution of the equipment), because of the in-series air-gap. When the pull-in occurs, an extremely abrupt current increase over 3 decades with a slope better than 20mV/dec (limited by the step-voltage of the measurement equipment) is observed. Following contact between metal and oxide, a local field of the order of 10-15MV/cm develops, inducing a Fowler-Nordheim (FN) tunneling through the thin 15nm SiO₂ layer. The calculated pull-out occurs at 1V and is not visible because it is much lower than the 25V reported in Fig. 3 at which the current reaches the leakage floor.

In the reverse bias we confirm the quasi-perfect isolation of the device, due to the in-series air-gap, with an experimental current always in the order of fA.

In order to demonstrate the FN tunneling conduction, temperature measurements have been performed. According to what it is reported in [8], the current density (J) generated by the induced electric field (E), has the following expression (WKB approximation which ignores the temperature dependence and the lowering of the barrier by the Schottky effect):

$$J \propto K_1 E^2 exp(-K_2 / E) \qquad (1)$$

where K_1 and K_2 are two constants proportional to the potential barrier at the metal/oxide interface and the effective electron mass. In its not approximated form, FN conduction law is linked to the temperature via the Fermi-Dirac distribution; the temperature increase produces a decreasing of the effective barrier at metal/oxide interface, inducing the increasing of the current density.

Fig. 4 shows the FN log plot of the current density (normalized to the square of the electric field) versus the inverse of the electric field for different temperatures (0°C-100°C), after the pull-in event. As expected, the temperature increase provokes the shift of the current density, without changing the slope of all the curves.

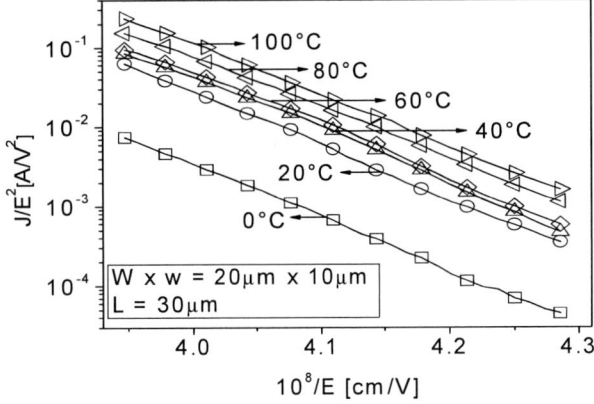

Figure 4: Fowler-Nordheim (FN) log scale plots of the current density versus the inverse of the electric field, for different temperatures, confirming basic FN conduction mechanisms in MEM-MAIS switch. The current density is normalized to the square of the electric field.

Fig. 5 better explains and confirms the FN tunneling conduction with respect to the temperature. In fact, the temperature dependence is more pronounced at high temperatures and at low field because the separation between subbands is smaller at lower applied voltages so that the occupation of high energy subbands is larger [9].

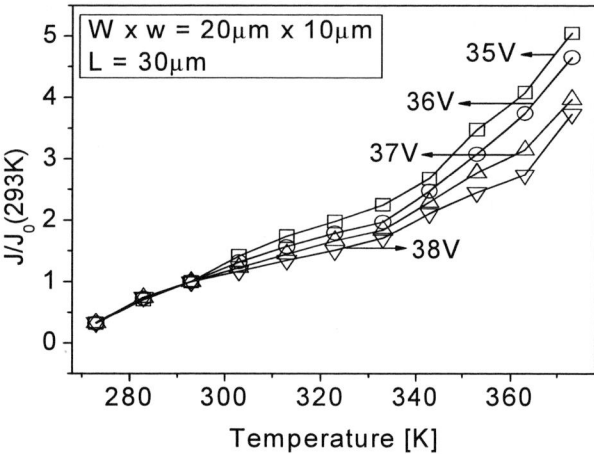

Figure 5: FN current density versus temperature for different applied voltages. The current density is normalized to the current density at room temperature.

Furthermore the temperature increasing provokes a decreasing of the pull-in voltage as well as an increasing of the current for the same voltage value (Fig. 6). Thanks to the air-gap, *the leakage current is not at all influenced by the temperature* and remains constant at few fA. Concerning the capacitance in the up-position (before the pull-in event) calculations provides a value of 5fF that is masked in C-V measurements by the pad capacitance that is in the order of 300fF.

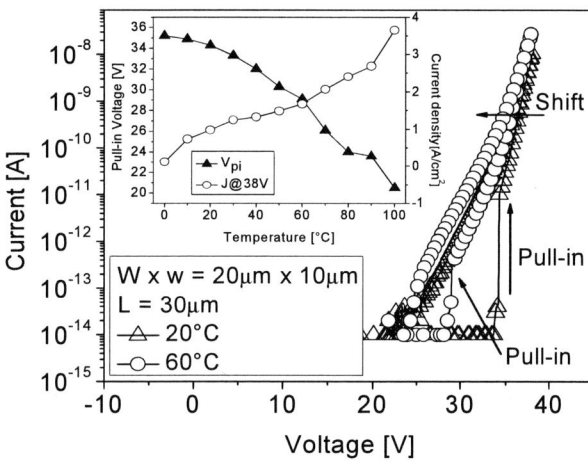

Figure 6: Log scale I-V characteristics for the same device tested at room temperature and at 60°C, respectively. The small graph shows the pull-in voltage and the current density versus the temperature. The current density has been collected after the pull-in event, at 38V.

ESD APPLICATION OF MEM-MAIS DIODE

In this section we propose and investigate a novel application of the MEM-MAIS diode switch. The low capacitance and the temperature-independent low leakage current make the MEM-MAIS diode switch suitable for ESD protection in both low voltage CMOS and automotive applications. The ESD appropriateness of the new switch has been estimated with respect to the Human Body Model (HBM - 100pF, 1.5kΩ) for a 1kV and 2kV pulse and the Machine Model (MM - 200pF) for a 200V

short pulse. Fig. 7 shows the experimental transient current response of a MEM-MAIS structure for standard HBM and MM set-up's, respectively. Fast current rise followed by longer current decays are observed.

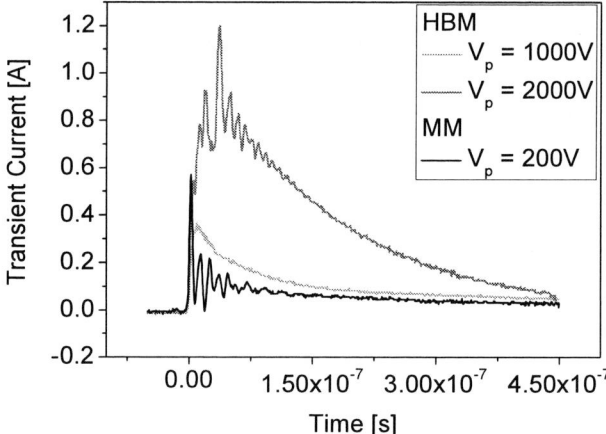

Figure 7: Transient current response of a 200V, 1kV and 2kV ESD event of the device 30um long (L) and 20um wide (W), with respect to the Machine Model (MM) and the Human Body Model (HBM).

Fig. 8 reports the comparison between the current transient of HBM tests performed at 1kV pulse on the diode switch and on short-circuit wire reference. The diode current response has a rise time t_r (defined by the 10%-90% of the current peak) of 7.5ns and a decay time t_d (36.8% of the current peak) of 100ns.

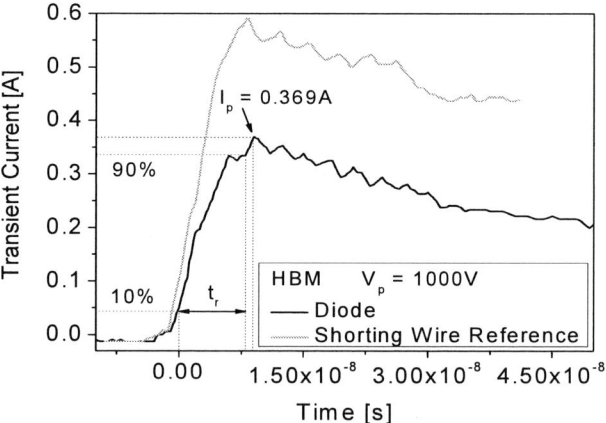

Figure 8: Comparison between the transient current responses of the MEM-MAIS diode (k_l=150.8N/m) and the shorting wire reference of a 1kV ESD event for the HBM.

Surprisingly even devices with large metal membranes are able to react as ultra-fast ESD switches in sub-100ns short time, essentially due to the extremely high voltage pulse. This is explained by the extremely high ESD charge pulse applied in few tens of ns, which induces an equivalent DC signal and a resulting very high actuation force that pulls-down the diode membrane (similarly with the DC component resulting at high level of RF power on MEM capacitors [10]).

In Fig. 9, the comparison between the diode characteristics before and after three successive 200V ESD pulses is reported; degradations of I_{off} and of pull-in voltage are observed due to charge injection/trapping in the oxide.

978-1-4244-2977-6/09 $25.00 © 2009 IEEE

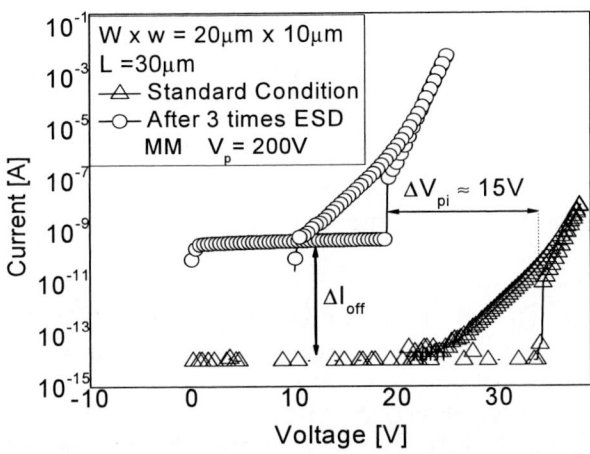

Figure 9: Comparison between the log scale I-V characteristics of the diode before and after the ESD event. Shift is due to the charging of the oxide, maintaining the same characteristics.

CONCLUSION

A novel MEM-MAIS diode switch using a movable conductive electrode over a fixed oxidized and highly-doped semiconductor has been presented and its operation was experimentally validated. The air-gap diode offers a very low off-current, practically independent of temperature and parasitic capacitance of the order of few fFs. The conduction is based on FN tunneling.

The new switch device is versatile in terms of the scaling of actuation voltage that can be varied by appropriate electro-mechanical design. The use of the MEM-MAIS switch for ESD protection applications has been investigated based on standard MM and HBM experiments and it is concluded that the MEM-MAIS diode shows promise for ESD applications.

AKNOWLEDGMENT

The author wants to thank Dr Didier Bouvet and Mr Giovanni Salvatore to have helped him in the fabrication of the MEM-MAIS diode.

REFERENCES

[1] G. Chen, IEEE Electron Device Letters, Vol. 25, May 2004, pp. 323-325.

[2] M. Ker and W. Lo, IEEE Trans. on Solid State Circuits, Vol.35, N.4, April 2000.

[3] R.K. Williams, Int. Symp. Power Semic. Dev. & IC's, 1997, pp. 357-360.

[4] S. Chen, IEEE Electron Device Letters, Vol. 24, Sept. 2003, pp. 595-597.

[5] A.A. Salman, Tech. Digest IEDM 2006.

[6] G.M Rebeiz, "RF MEMS theory, design and technology", John Wiley & Sons, New Jersey, 2003.

[7] D. Acquaviva, et al., MNE08

[8] Y. Khlifi, M. J. Condensed Matter, Vol.3, N.1, July 2000, pp. 53-57.

[9] J. Sune, IEEE Trans. on El. Devices, Vol. 40, May, 1993, pp. 1017-1019.

[10] H. Nieminen, J. Micromech. Microeng. 12 (2002) pp. 1-10.

PACKAGING OF 11 MPIXEL CMOS-INTEGRATED SIGE MICRO-MIRROR ARRAYS

A. Witvrouw[1], H. A.C. Tilmans[1], L. Bogaerts[1], P. De Moor[1], T. Bearda[1], S. Halder[1], L. Haspeslagh[1], B. Schlatmann[2], M. van Bommel[2], C. de Nooijer[2], J. Lauria[3], R. Vanneer[4] and B. van Drieenhuizen[4]

[1]IMEC, Kapeldreef 75, 3001 Leuven, Belgium,
[2]Philips Applied Technologies, Eindhoven, The Netherlands
[3]ASML, Wilton Connecticut, USA, [4]ASML, Veldhoven, The Netherlands

ABSTRACT

This article reports for the first time on the packaging of 10 cm² 11 MPixel SiGe micro-mirror arrays, intended to be used as spatial light modulator (SLM). Due to very stringent requirements on mounted die flatness (<0.01 mrad), first-level packaging of the SLM die is done using specially designed SiC holders. To avoid trapped particles between die and holder, which would jeopardize the flatness spec, special backside cleaning of the dies (≤ 1 0.8 µm particle/cm²) is needed before mounting the SLM die on the holder. To enable this backside cleaning and to avoid front-side particles during dicing, handling and wire bonding, a temporary wafer-level or 0-level packaging cap which can be placed and removed at room temperature was developed. Dynamic white light interferometer measurements of packaged dies showed that 99.5% of 123,648 mirrors tested are within spec.

INTRODUCTION

Monolithically integrated micro-mirror arrays are well established devices in applications such as video projection [1, 2] and mask writers [3]. These integrated micro-mirrors are mostly Al-based [1, 2], often giving rise to reliability problems such as creep (hinge memory effect [4]), thermal elastic or plastic deformation and oxidation. Replacing Al by silicon (Si) solves many of these problems, but, because of the high thermal budget of poly-Si processing, integrating the Si mirrors with the CMOS driving circuitry can only be accomplished in a hybrid fashion through wafer- or die-bonding [4]. Poly-SiGe offers much better integration possibilities: it can be deposited at temperatures compatible with above-CMOS processing, while it has comparable properties to poly-Si and can be processed using similar state-of-the-art tools [5].

Recently very reliable 10 cm² 11 MPixel SiGe-based micro-mirror arrays, fabricated on top of standard 0.18µm analog-CMOS wafers featuring 6 interconnect levels, have been developed (Fig. 1) [6]. These arrays, consisting of 8µm x 8µm pixels with a density which is almost double compared to the state-of-the-art [2], are intended to be used as spatial light modulators (SLM) [7, 8]. The required grayscales can be created through the programmable deflection angle of the mirrors.

This article reports for the first time on the packaging of these 10 cm² arrays of 11 million SiGe micro-mirrors. Due to very stringent requirements on mounted die flatness (<0.01 mrad or <100 nm peak-to-valley surface deviation over 1 cm), first level packaging of the SLM die is done using specially designed SiC holders. Next to the mechanical interface, this holder also provides electrical, purging and cooling interfaces (Fig. 2). Trapped particles between die and holder will jeopardize the flatness specifications and therefore special backside cleaning of the dies (≤ 1 0.8 µm particle/cm²) is needed before mounting the SLM die onto the holder. To enable this backside cleaning and to avoid front-side particles during dicing, handling and wire bonding, a temporary wafer-level or 0-level packaging cap which can be placed and removed at room temperature was developed.

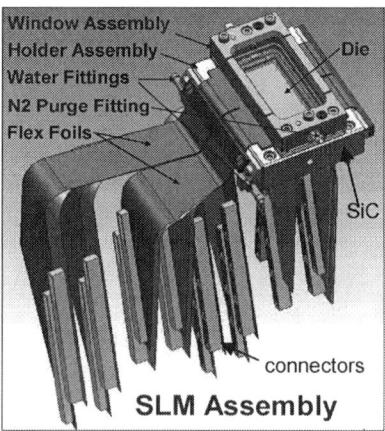

Figure 2: Schematic of SiC SLM holder with SLM die and flexible electrical connection.

MEMS PROCESSING

A schematic view of the front-end process flow is shown in Table 1. On top of standard 0.18µm analog-CMOS wafers, fabricated by NXP and featuring 6 metal interconnect levels, 8µm pitch micro-mirrors, spaced 0.3 µm apart, were fabricated using SiGe as structural material and SiO₂ as sacrificial layer (Fig. 1). To prevent attack of the CMOS during the vapor HF (VHF) or wet HF release, it is protected with a SiC layer. This layer is perforated with more than 50 million W-vias per SLM die

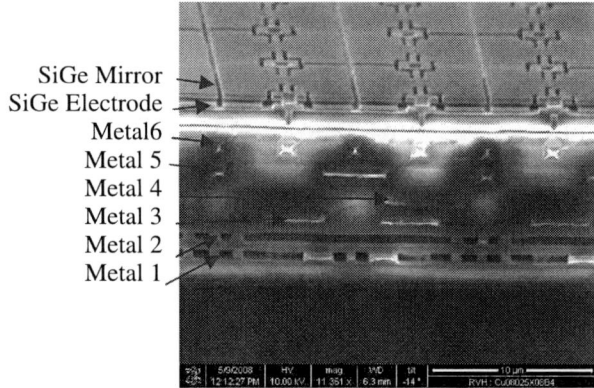

Figure 1: Cross-sectional view of the integrated micro-mirror array, showing the mirrors on top of the 6 layers of Al interconnect [6].

to connect the mirrors to the underlying driving circuitry (Fig. 3).

Table 1: Schematic process flow of micromirror module.

Start: CMOS base wafer (from NXP)
Top level metal planarisation + passivation
SiC protection layer
W-vias
SiGe electrode + planarisation
Sacrificial oxide deposition
Mirror hinge formation
SiGe mirror layer deposition + CMP
Phasestep etch (optional)
Al coating (optional)
Mirror etch
Oxide protection + Al bondpads
Release etch mirrors

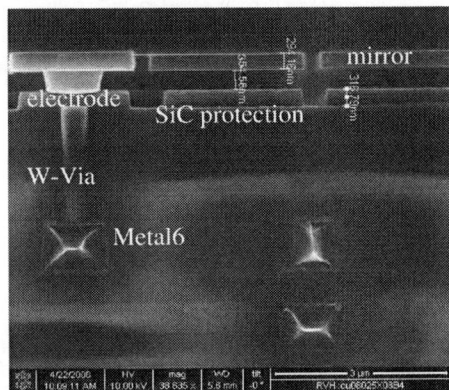

Figure 3: Cross-section of mirror showing W-via, SiC protection layer, SiGe electrodes, SiGe mirror layer.

A 300nm hydrogenated microcrystalline SiGe film deposited by PECVD [5] is used for the structural layer and to fill the 1 μm supporting posts in the sacrificial layer. This layer has an optimised strain gradient smaller than 4.10^{-4} μm^{-1}. The RMS roughness of the structural layer was reduced from 5 nm as-deposited to around 0.3 nm by a SiGe CMP process. Optionally a phase step is created in each mirror by removing an additional stress-free SiGe layer selectively from half the mirror area by a timed etch [6]. An Al coating can also optionally be added [6]. The front-end process is finalized by adding Al bondpads and releasing the mirrors either in buffered HF + CPD (Critical Point Drying) or VHF.

PACKAGING AND CLEANING

Temporary 0-level packaging

To protect the SLM die (in particular the fragile released mirrors) during wafer dicing, handling and 1- and 2-level packaging (mounting on the SiC holder, wire-bonding, etc...), a temporary wafer-level or 0-level package has been developed. The 0-level package relies on a chip-capping approach, implemented either in a die-to-wafer (D2W) or a die-to-die (D2D) fashion. In both implementations, a glass cap, laminated with a 110 μm thick double-sided UV-releasing adhesive tape (SELFA-BG tape from Sekisui), is bonded to the mirror wafer (or

die). In the D2D solution the UV-tape is patterned forming a ring around the released mirrors (see Fig. 4(b)).

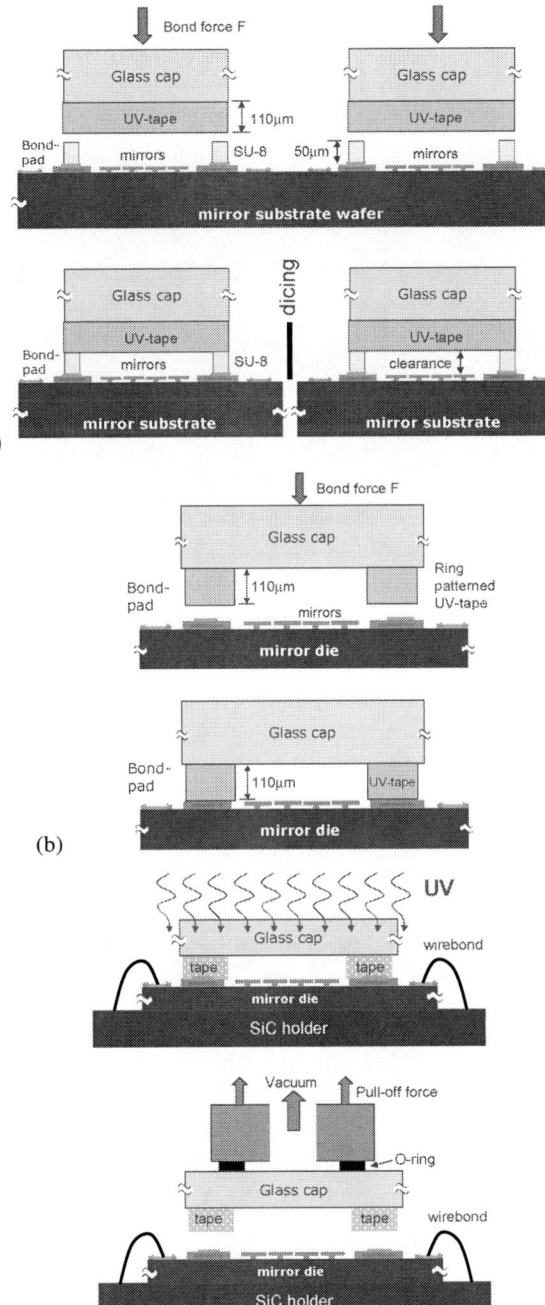

Figure 4: Illustration of temporary 0-level packaging concept. (a) For D2W; (b) for D2D, and (c) Decapping.

In the D2W solution the uniform UV tape is placed on top of a 50 μm thick, 250 μm wide SU-8 spacer ring, which is deposited and patterned before release of the mirrors (see Fig. 4(a)). The SU-8 ring (or the patterned UV-tape for that matter) ensures a sufficient clearance between the glass cap and the top surface of the mirror chip. After placing the temporary protecting cap, the wafer (in case of D2W) can be safely diced, the backside of the 0-level packaged SLM die is cleaned and the SLM die is mounted and wire-bonded to the SiC holder, (see Fig. 4(c)). Next, the temporary cap is removed at room temperature with a vacuum pick-up tool after UV-irradiation of the adhesive tape, which weakens the bond. Immediately following the removal of the temporary cap, a permanent window is placed (see below).

While leak-tight 0-level packages were obtained after dicing of test wafers (Fig. 5) for the D2W approach, the D2D approach was finally chosen due to SU-8 adhesion problems on the device wafers during release.

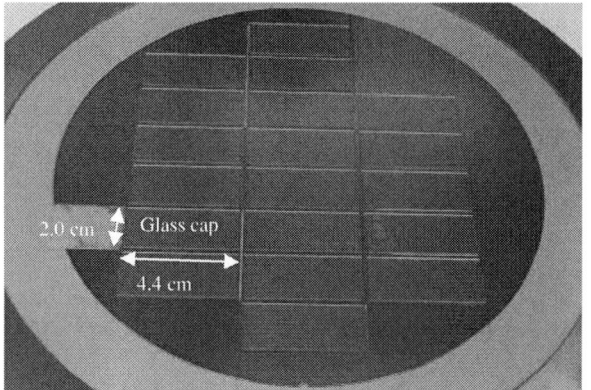

Figure 5: Diced SiGe test wafers capped die-to-wafer. No water was observed inside the cavity after dicing.

Backside cleaning

After dicing, the backside of the die is contaminated with particles and glue residues. A spinning tool with small vacuum chuck was built to enable backside cleaning of individual dies. An ammonia/peroxide/water (APM)

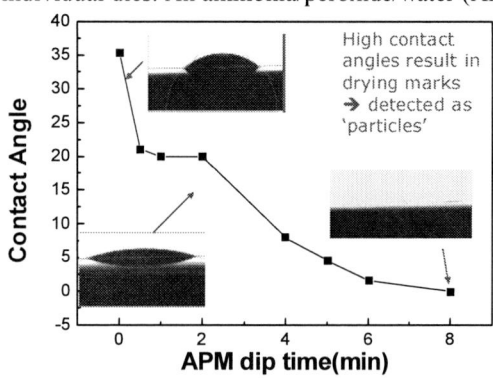

Figure 6: Contact angle of water droplets a.f.o APM dip time. Organic contamination results in a high contact angle.

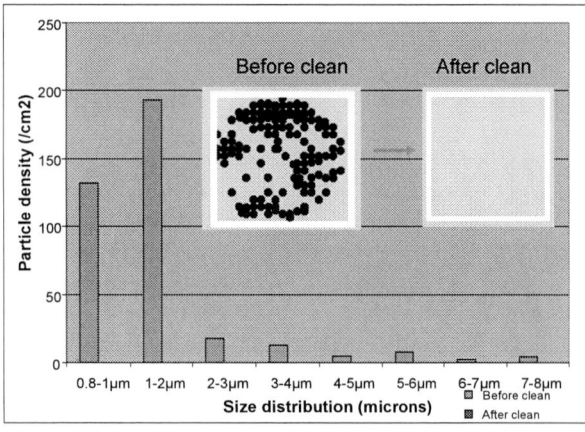

Figure 7: Particle size distribution and maps before and after APM clean.

mixture was found to be effective in removing both types of contaminants as demonstrated by contact angle measurements of water droplets (Fig. 6) and particle measurements (Fig. 7). After 10'APM followed by a 1' rinse and spin-dry, no particles > 0.8 μm were found by light scattering on the five tested dies, thus fulfilling the specification.

First/second level packaging

The cleaned SLM die is then attached to the SiC SLM holder by dispensing small dots of a special two-component glue (with minimal shrinkage upon curing) in between a 10-20 μm high pillar structure in the SiC holder (Fig. 8). This structure was designed such that the SLM flatness (Fig. 9) and thermal specification can be met without having to impose even more stringent specifications on the backside cleanliness.

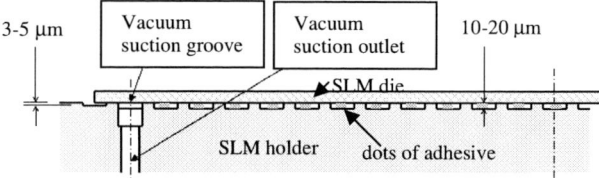

Figure 8: Schematic layout of pillar structure of SiC holder.

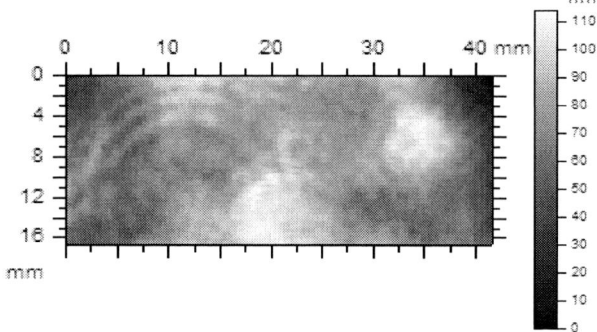

Figure 9: Flatness measurement of the 16.8x41.6 mm² mirror area of a test die glued on the SiC holder showing a max. surface deviation of 115 nm peak-to-valley.

After glueing the die to the holder, 8 multi-layer flex foils, 4 stacked on each side of the die, and a decoupling board are placed on the SLM holder. The flex foils are designed such that over 2500 interconnections can be made by wire bonding to 36 μm pitch bondpads with high yield (Fig. 10).

Figure 10: Schematic and actual impression of the wire bonding principle.

Right before mounting a permanent window, the temporary cap could easily be removed by lifting it with a vacuum tool after bond weakening by UV exposure (Fig. 4c). The packaged SLM die is maintained in an inert atmosphere, by purging with clean N_2, which is an additional feature integrated in the SLM assembly.

Figure 11: Fully packaged and wire-bonded SLM assembly.

CALIBRATION AND TESTING

The packaged dies (Fig.11) are then ready for calibration and testing. The physical attributes of the individual SLM mirrors are evaluated on a white light interferometer system (Zygo™ NewView model 5032). This system supports stroboscopic measurements of the MEMS mirrors, allowing dynamic measurements at a 6 kHz frame rate. The mirror geometry measured through tilt angle provides data for the dynamic mirror tilt response, the mirror shape, the mirror-to-mirror differences, and the undesirable piston motion and tilt [7].

The mirror bias voltage that provides the correct electrical charge needed to activate the mirrors over a

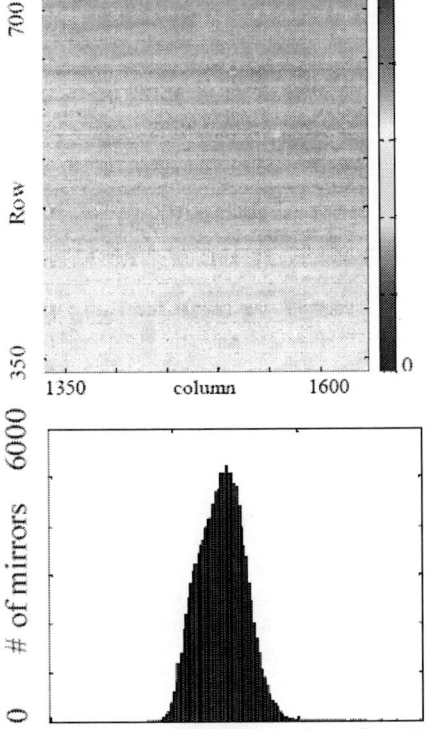

*Figure 12: 2D overview (top) and distribution (bottom) of fitted slope of driving voltage (in DAC counts) versus tilt response (in mrad) for 123,648 mirrors of un-calibrated **fully packaged** SLM (see Fig. 11).*

desired tilt range of ±10 mrad is determined experimentally. Patterns of specific voltage inputs are applied and the data is collected and analyzed. The output of this analysis is tilt response (Fig. 12) and geometrical data as a function of applied control voltage. A summary of the mechanical testing results is provided in Table 2.

Table 2: Mechanical testing results for123,648 mirrors

Item	Specification	Results
Good Mirrors %	/	99.5 %
Mean Cupping	< 10 nm	< 5.4 nm
Residual Piston	< 1nm	< 1 nm

CONCLUSIONS

$10 \, cm^2$ 11 MPixel CMOS-integrated micro-mirror arrays were for the first time successfully packaged. Packaging of the micro-mirror dies is done using specially designed SiC holders. To avoid trapped particles between die and holder, special backside cleaning of the dies is needed before mounting them on the holder. To enable this backside cleaning and to avoid front-side particles during dicing, handling and wire bonding, a temporary 0-level packaging cap which can be placed and removed at room temperature by using UV-sensitive tape was developed. Dynamic testing demonstrated that the packaged SLM device is working with 99.5% of 123,648 tested mirrors within spec.

REFERENCES

[1] L.J. Hornbeck, Proc. SPIE **3013**, 27, 1997.

[2] L.J. Hornbeck, "Combining Digital Optical MEMS, CMOS and Algorithms for Unique Display Solutions" IEDM 2007, pp. 17-24.

[3] T. Sandner et al., "Highly reflective coatings for micro-mechanical mirror arrays operating in the DUV and VUV spectral range", Proc. SPIE 5721, 72, 2005.

[4] B. Völker, M. Friedrichs, D. Rudloff, T. Bakke, "Drift Free, Highly Planar Silicon Micromirror Arrays", Proc. MME 2005, 223.

[5] M.Gromova, L. Haspeslagh, A.Verbist, B. Du Bois, R. Van Hoof, B. Eyckens, B. Sijmus, I. De Wolf, V. Simons, Ph. Muller, T. Lauwagie, M. Willegems, S. Locorotondo, W. Boullart, K. Baert and A. Witvrouw, "Highly reliable and extremely stable SiGe micro-mirrors", Proc. IEEE MEMS 2007, 759.

[6] L Haspeslagh, J. De Coster, O. Varela Pedreira, I. De Wolf, B. Du Bois, A. Verbist, R. Van Hoof, M. Willegems, S. Locorotondo, G. Bryce, J. Vaes, B. van Drieenhuizen and A. Witvrouw, "Highly reliable CMOS-integrated 11MPixel SiGe-based micro-mirror arrays for high-end industrial applications", accepted for publication in Proc. IEDM 2008.

[7]] J. Lauria, R. Albright, O. Vladimirsky, L. Chen, M. Hoeks, R. Vanneer, B. van Drieenhuizen, L. Haspeslagh, A. Witvrouw, B. Schlatmann, "SLM Device for 193 nm Lithographic Applications", accepted for publication in Proc. MNE 2008.

[8] J.Suzuki, A. Komai, Y. Ohuchi, Y. Tezuka, H. Konishi, M. Nishiyama, Y. Suzuki and S. Owa, "Micro-mirror On Ribbon-actuator (MOR) for high speed spatial light modulator", Proc. IEEE MEMS 2008, 763.

ROBUST WAFER-LEVEL THIN-FILM ENCAPSULATION OF MICROSTRUCTURES USING LOW STRESS PECVD SILICON CARBIDE

V. Rajaraman[1], L.S. Pakula[1], H.T.M. Pham[2], P.M. Sarro[2] and P.J. French[1]

[1]Electronic Instrumentation Laboratory, [2]ECTM Laboratory, DIMES, Dept. of Microelectronics, Faculty of EEMCS, Delft University of Technology, The Netherlands

ABSTRACT

This paper presents a new low-cost, CMOS-compatible and robust wafer-level encapsulation technique developed using a stress-optimised PECVD SiC as the capping and sealing material, imparting harsh environment capability. This technique has been applied for the fabrication and encapsulation of a wide variety of surface- and thin-SOI microstructures that included microcavities, RF switches and various accelerometers. Advantages of our technique are its versatility, smaller footprint, reduced chip thickness and process complexity, post-CMOS batch processing capability and added functionality due to the possibility of integrating additional electrodes for MEMS. Besides fabrication details, this work also discusses related design aspects for large-area MEMS and demonstrates the encapsulation results. Successfully encapsulation of device geometries as large as $955 \times 827 \mu m^2$ has been achieved.

INTRODUCTION

Using a meticulously developed IC-compatible microfabrication technology, MEMS can in principle be monolithically integrated with CMOS control/detection circuitry. However, a key distinguisher between the MEMS and IC is that the former contains free-standing and/or moving components with gaps and feature sizes in the range of a few microns that requires protection against dust, contamination and handling during backend processing such as dicing and assembly. Besides, some MEMS devices like pressure sensors, resonators, gyroscopes, infrared sensors, etc. also require a stable and controlled environment, like vacuum for instance, for reliable operation. These challenging demands are usually fulfilled by means of packaging MEMS devices on the wafer-level using either wafer bonding [1,2] or thin-film chip-scale encapsulation [1-6], also known as zero-level packaging.

Typical requirements for wafer-level MEMS packaging are low cost processing, CMOS compatibility (in terms of materials employed, process development and device compatibility), hermeticity, robustness, stability, good sealing characteristics and an ability to withstand harsh environments. Although widespread, zero-level packaging using wafer bonding technologies suffers from limitations in terms of size, cost and complexity. Hence, thin-film encapsulation technique offers an interesting alternative for zero-level packaging of MEMS.

A generic representation of thin-film packaged MEMS is illustrated in Fig. 1. This technique involves the deposition and patterning of a sacrificial material, over the microstructure, above which the encapsulation material is applied. Later, the sacrificial material is etched through the encapsulation via etch-windows to form a microcavity that is then sealed with a second material, the sealing layer, which can either be a different material or the same as the one used for encapsulation. Some key advantages of wafer level chip-scale packaging of MEMS by thin-film encapsulation are a smaller footprint, reduced chip thickness, reduced process complexity and lower costs. Thin-film packaging technique specifically is more appropriate for surface micromachined and thin-SOI micromachined devices due to the possibility of producing ultra compact MEMS devices, both integrated and standalone.

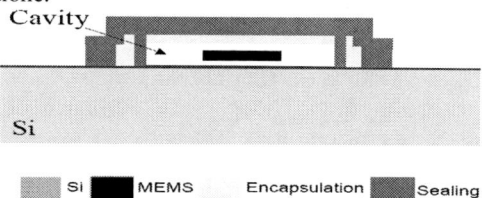

Figure 1: Schematic representation of a thin-film encapsulated MEMS device

Several thin-film encapsulation techniques, using different materials and processes, are reported in the literature [1-6], each having their distinctive pros and cons. However, a majority of the reported techniques have shortcomings like being costly and exclusive, requiring high temperature processing, employing less reliable materials, involving process complexity, etc. In this work, we have developed two thin-film encapsulation processes, for small- and large-area MEMS, respectively, using a stress-optimised PECVD SiC for encapsulation and sealing that is compatible with post-CMOS processing. Moreover, the use of PECVD SiC for packaging brings in benefits such as mechanical robustness, chemical inertness and an ability to operate at high temperatures, making it suitable for harsh environment applications [7-9].

GENERAL CONSIDERATIONS

There are several issues to consider in the design and development of a wafer-level thin-film encapsulation process, some of which are discussed below.

Choice of the Encapsulation Material

The material being used for encapsulation should be mechanically robust with a minimal residual stress, crack-resistant and pin-hole-free for achieving hermitic sealing. Also the material should preferably exhibit low tensile stress such that the encapsulation does not deform and is held intact when suspended. This aspect has been taken care of in our encapsulation process by choosing a pin-hole-free PECVD SiC, which is a mechanically robust material with a high Young's modulus (460GPa), and by optimising its

978-1-4244-2977-6/09 $25.00 © 2009 IEEE   140

residual tensile stress level to 65MPa [9].

Deflection of the Encapsulation Layer

The maximum deflection caused due to the pressure difference between the sealed cavity and the ambience acting across the encapsulation layer has to be optimised as it also introduces stress along the edges. This can be done by choosing an appropriate thickness for the encapsulation layer as well as the sacrificial layer that will later define the height of the microcavity on removal. This value was determined both numerically and verified by FEM simulation using COMSOL for rectangular and circular PECVD SiC encapsulation geometries, to ensure that the encapsulation does not collapse onto the microstructures. An example simulation result shown in Fig. 2 suggests that a 4μm thick, 200x200μm^2, self-supported PECVD SiC encapsulation undergoes a maximum deflection that is around 100nm. Based on the FEM results, the thickness of the PECVD SiC was fixed between 4μm-8μm in the encapsulation processes, discussed in the next section.

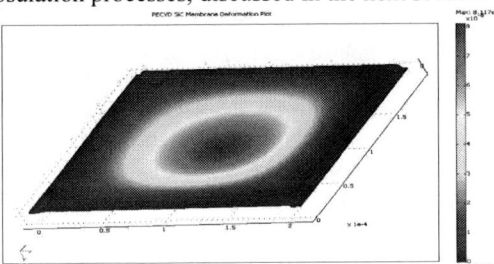

Figure 2: Simulation of the maximum deformation of a 4μm thick, 200x200μm^2 PECVD SiC encapsulation under atmospheric pressure loading

Inclusion of Etch-Windows and Anchors

The third design issue concerns the geometry of the etch-windows that need to be etched in the encapsulation layer for removing the sacrificial material so that the encapsulation and the MEMS can be released simultaneously and made free-standing. These etch-windows could be located either around the periphery of smaller microstructures or on top of the larger microstructures, enabling quicker release, depending on the device design. According to [6] the use of slit shaped etch-windows instead of circular ones results in a smaller principal stress on the encapsulation layer. So in this work, horizontal and vertical 1x5μm^2 slit shaped etch-windows that were placed apart by 10μm, representing a unit cell, were designed. Multiple cells were placed all over on top of the encapsulation layer with 10μm spacing, in order to facilitate quicker release of the encapsulation layer.

Yet another and important design issue is the inclusion of support-pillars or anchors in the MEMS device design that can assist the suspended encapsulation layer to spread over large-areas without deformation. Anchors are especially required for accommodating large-area MEMS structures, like accelerometers for instance, into the thin-film package. Thus, 40x40 μm^2 anchor points were included in the design of large-area thin SOI-MEMS structures like inertial sensors that were placed apart by 150μm-225μm.

Encapsulated Geometries

A variety of circular and rectangular encapsulations with varying dimensions were designed to accommodate a wide range of microstructures that included cantilevers, resonators, micro-bridges, switches, accelerometers and gyroscopes. The diameter of the circular encapsulations intended for small-area microstructures ranged between 50μm-150μm. The minimum and maximum dimensions of the rectangular encapsulations ranged from a meagre 13x63μm^2 for small area-microstructures through to 955x827μm^2 for large-area thin-SOI microstructures. In the design of encapsulation for large area microstructures, the effective area of the rectangular encapsulations was minimised at fixed intervals to facilitate the mechanical stability of the long encapsulation layer.

ENCAPSULATION PROCESSES

Two wafer-level thin-film encapsulation processes, with integrated MEMS fabrication, were developed for surface micromachined and thin-SOI microstructures on 100mm wafers. A 65MPa low tensile stress PECVD SiC layer, used for encapsulation and sealing, was obtained in a Novellus Concept One plasma deposition system that allows post-CMOS processing. The RIE of SiC was performed in an Alcatel GIR 300 plasma etch system. The key process parameters are summarised in Table 1. Also, proper etch selectivity was achieved against inorganic and organic sacrificial materials (like PECVD oxides and polyimide), Al and Si using different etch chemistries.

Table 1: Key process parameters for low-stress PECVD SiC

Parameter		Deposition	Etching
Gas Flow	SiH$_4$	250 sccm	-
	CH$_4$	3000 sccm	-
	CF$_4$	-	70 sccm
	SF$_6$	-	10 sccm
	O$_2$	-	10 sccm
Temperature		400 °C	-
Pressure		2 Torr	37.5 Torr
HF Power		450 W	60 W
LF Power		150 W	-

Surface Micromachining with Integrated Packaging

The post-CMOS surface micromachining fabrication process with integrated PECVD SiC encapsulation is illustrated in Fig. 3. In this process, Al was micromachined and used as the mechanical material as well as the electrodes, PECVD SiC was used as the encapsulation and sealing material and PECVD oxide was used as the sacrificial layer. The fabrication starts with the sputtering of Al(1% Si) on a Si substrate pre-isolated with nitride or oxide followed by the deposition of a 100nm PECVD SiC thin-film layer and then both layers are patterned (Fig.3a-a). The PECVD SiC thin-film layer is used to prevent stiction and to provide isolation between the electrodes and the microstructure. Next, 1.5μm of PECVD oxide sacrificial layer is deposited and patterned (Fig. 3a-b). This step is followed by sputtering of a 2.5μm thick Al(1% Si) mechanical layer that is micromachined to form the MEMS structure (Fig. 3a-c). Later, a second PECVD oxide sacrificial layer is deposited and patterned (Fig. 3a-d). On the top of the sacrificial layer, again a 100nm PECVD SiC isolation layer is deposited that is followed by sputtering of a 0.6μm second Al(1% Si) layer. Both layers are then patterned forming a second set of electrodes. A 4 μm PECVD SiC is then used for encapsulation. At first, a 2μm PECVD SiC is deposited and patterned to create the first layer of encapsulation. Simultaneously, etch-windows are opened for sacrificial layer etching (Fig. 3a-e). The structure is then released by removing the PECVD oxide sacrificial layer in 73% HF, followed by a freeze-drying procedure to prevent stiction. Finally, another 2μm of PECVD SiC layer is deposited to seal the microcavity and to strengthen the encapsulation (Fig. 3a-f).

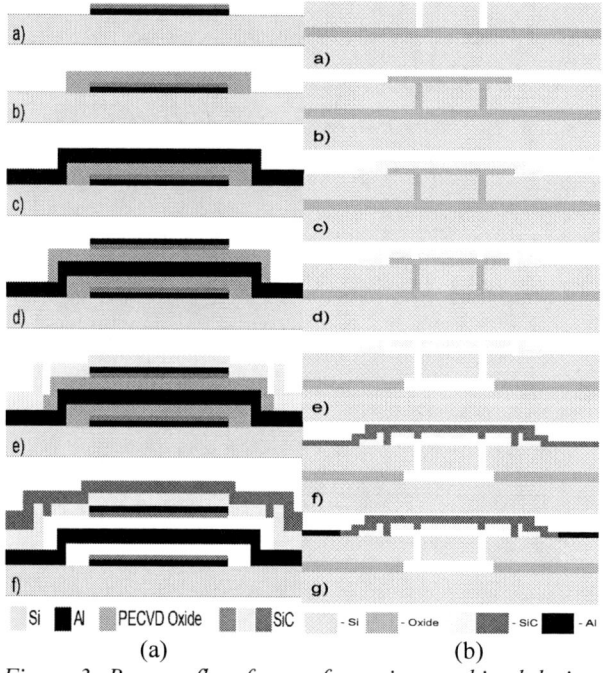

(a) (b)

Figure 3: Process flow for: surface micromachined devices (a) and thin-SOI micromachined devices (b)

Thin-SOI Micromachining with Integrated Packaging
The fabrication process for thin-SOI microstructures and their PECVD SiC encapsulation is presented in Fig. 3b. Here, an SOI substrate with a thin active Si (thin-SOI) layer and a reasonably thick buried oxide (BOX) layer was used. PECVD oxide was used as the sacrificial layer and PECVD SiC is used for encapsulation and sealing as with the other process. Processing starts with RIE of the active Si layer to define the microstructures (Fig. 3b-a) over which a 5μm-7μm thick sacrificial PECVD oxide layer is deposited and patterned (Fig. 3b-b). Then a 4μm PECVD SiC encapsulation layer is deposited and patterned (Fig. 3b-c). Now etch-windows are patterned on the PECVD SiC encapsulation layer for accessing and etching the sacrificial layer together with the buried oxide (Fig. 3b-d). The sacrificial layer is then etched in 40% HF that is succeeded by a freeze-drying procedure to prevent stiction, thereby simultaneously releasing the PECVD SiC encapsulation layer together with the thin-SOI microstructure (Fig. 3b-e). Another 4μm of PECVD SiC sealing layer is then deposited

over the suspended PECVD SiC encapsulation layer and patterned, thus sealing the microcavity (Fig. 3b-f). The layer also adds to the first encapsulation layer making it mechanically more robust. Now for devices requiring electrical contacts, the 8μm thick PECVD SiC layer is patterned and etched followed by the sputtering and patterning of 0.6μm Al (1% Si) for the bond pads (Fig. 3b-g) completing the encapsulation process.

RESULTS

Several circular PECVD SiC encapsulations, suitable for small-area microstructures, ranging in diameters from 50μm-150μm were realised using the presented process in Fig. 3a. Fig. 4a presents an array of free-standing circular microcavities after removal of the sacrificial layer. The etch-windows that were situated around the periphery of the encapsulations can also be seen. A close-up cross-sectional view of one such encapsulation is visualised in Fig. 4b. The cross-sectional schematic and the top view of the fabricated surface micromachined RF MEMS switch are shown in Fig. 5 and the electrical characterisation results of the RF switch are tabularised in Table 2. In this device, the function of PECVD SiC is two-fold. Firstly, it has to hold the top electrode that is embedded beneath the PECVD SiC encapsulation layer and secondly, it has to encapsulate and seal the microcavity defining the zero-level packaged environment for device operation.

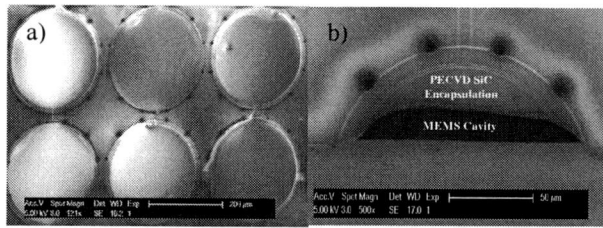

Figure 4: An array of free-standing circular SiC encapsulations (a); cross-sectional view of a microcavity (b)

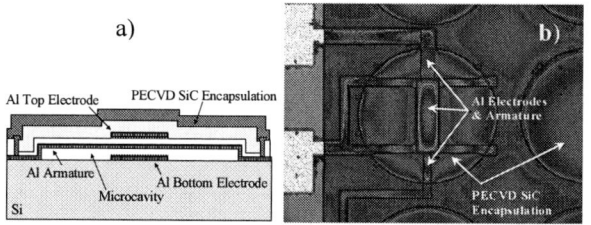

Figure 5: Schematic cross-section (a) and top view (b) of the surface micromachined RF switch

Table 2: Characterisation results of the RF Switch

Device Parameter	Values
Resonant frequency	204 kHz
Quality factor	687.5
Pull-in voltage	38 V
Pull-off voltage	1.18 V / 0.8 V
Capacitance (down/up)	2.66 pF / 14.7 fF
Operating voltage	5V

In the thin-SOI micromachining process, illustrated in Fig. 3b, PECVD SiC thin-film encapsulation of a variety of small- and large-area microstructures were achieved. The encapsulated small-area microstructures included cantilevers, resonators, micro-bridges, etc. Amongst large area microstructures, see Fig. 6, a range of accelerometers micromachined on thin-SOI before the deposition of the sacrificial oxide layer. The largest area encompassing the proof-mass and the comb electrodes requiring a successful encapsulation was, $955\times827\mu m^2$, that of the left-most accelerometer seen in Fig. 6a. The cross-sections in Fig. 7a reveals the trenches that were first etched into Si and later filled with PECVD sacrificial oxide, over which the PECVD SiC encapsulation was applied. In order to reveal the interface of the contrasting layers, these samples were diced and then etched using HF. Fig. 7b shows an accelerometer structure encapsulated with PECVD SiC and patterned with etch-windows before removal of the sacrificial PECVD oxide and the buried oxide.

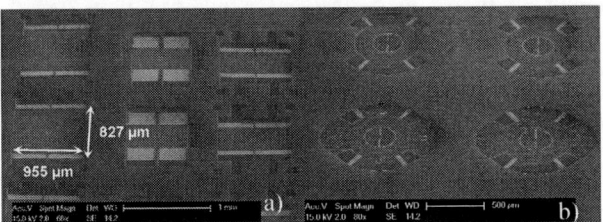

Figure 6: Thin-SOI accelerometers structures before encapsulation

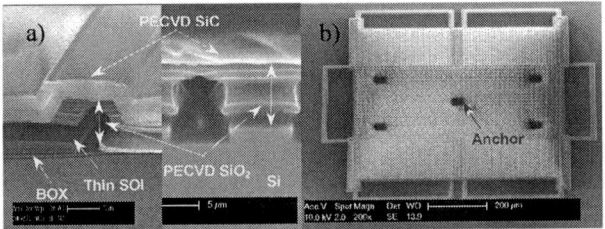

Figure 7: Cross-sectional view revealing contrasting layers after deposition of PECVD SiO₂ and PECVD SiC layers (a); top view of an encapsulated accelerometer with patterned etch-windows and before the sacrificial oxide etching (b)

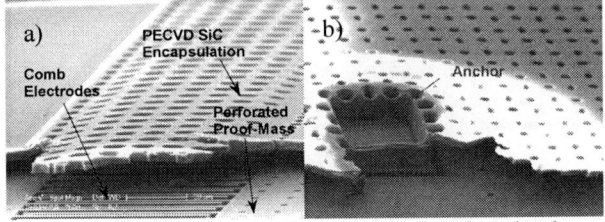

Figure 8: Detailed view of an encapsulated and released accelerometer showing various parts after removal of the sacrificial and buried oxide.

Fig. 8a presents a detailed view of a released device that was later cleaved showing the free-standing PECVD SiC encapsulation layer with etch-windows, the comb electrodes and the perforated proof-mass. Fig. 8b shows the PECVD SiC encapsulation layer anchored to one of the support-pillars that is meant to improve the reliability of the PECVD SiC encapsulation layer. A series of encapsulated, sealed and fully released accelerometers are shown in Fig. 9a. The

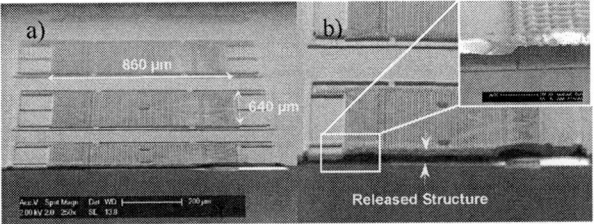

Figure 9: Encapsulated and released accelerometers (a) and a detailed cross-sectional view (b)

close-up SEM image in Fig. 9b shows the final free-standing 8μm thick PECVD SiC encapsulation obtained after sealing the etch-windows. Thus, the above results demonstrate the versatility of this encapsulation technique to accommodate a wide range of MEMS devices.

CONCLUSIONS

We presented two encapsulation processes, related design issues and the results of a low-cost, robust and reliable wafer-level thin-film encapsulation for surface micromachined and thin-SOI MEMS devices using a 4μm-8μm thick stress-optimised PECVD SiC layer. A wide range of microstructures can be encapsulated in this process and results were demonstrated for microcavities, RF-switches, and accelerometers. The largest encapsulated geometry was about $955\times827\mu m^2$. The presented approach is CMOS-compatible and moreover, it enables a smaller footprint and reduced chip thickness compared to wafer bonding. PECVD SiC based thin-film encapsulation technique for wafer-level MEMS packaging is very interesting for automotive, industrial and medical applications where devices are often subjected to harsh environments requiring a stable, durable and reliable encapsulation layer.

ACKNOWLEDGEMENT

The authors, V. Rajaraman and P.J. French, would like to gratefully acknowledge the financial support provided by NXP Semiconductors, The Netherlands. The authors wish to thank: Dr. H. Boezen and Dr. J.-J. Koning of NXP Semiconductors, Nijmegen, for their support, the staff of DIMES ICP Group, TU Delft, for their assistance during microfabrication, Dr. K.A.A. Makinwa of EI, TU Delft, for his support, Q. Li, Dr. J.F.L. Goosen of 3ME, TU Delft, for the interesting discussions, and Dr. H.W. van Zeijl of DIMES, TU Delft, for his assistance in investigating the CMOS compatibility of PECVD SiC layers.

REFERENCES

[1] M. Esashi, *J. Micromech. Microeng.*, 18, 2008, pp.1-13.
[2] K. Najafi, Proc. *SPIE Micromachining and Microfabrication Process Tech.*, Vol. 4979, 2003, pp. 1-19.
[3] A. Höchst et al., *Sensors and Actuators A*, 114, 2004, pp. 355-361.
[4] D. Reuter et al., *Sensors and Actuators A*, 145-146, 2008, pp.316-322.
[5] P. Monajemi et al., *J. Micromech. Microeng.* 16, 2006, pp.742-750.
[6] Y. Shimooka et al., *Proc. ECTC 2008*, Florida, 2008, pp. 824-828.
[7] P.M. Sarro, *Sensors and Actuators A*, 82, 2000, pp. 210-218.
[8] H.T.M. Pham, *PhD Thesis*, Delft Univ. of Tech., NL, 2004.
[9] L.S. Pakula et al., *J. Micromech. Microeng.*, 14, 2004, pp. 1478-1483.

978-1-4244-2977-6/09 $25.00 © 2009 IEEE

DIRECT ETCHING OF HIGH ASPECT RATIO STRUCTURES THROUGH A STENCIL

G. Villanueva, O. Vazquez-Mena, C. Hibert and J. Brugger
EPFL, Lausanne, Switzerland

ABSTRACT

This paper reports the feasibility of the fabrication of high aspect ratio structures on substrates via dry etching through a stencil mask placed onto the sample. It demonstrates the possibility to use standard equipment and processes with this novel masking technique, which allows the patterning of fragile and pre-structured surfaces, and avoids the use of resist or additional coating of the sample, reducing costs and processing time. Aspect ratios as high as 13:1 and pattern transfer with a gap of 100 μm are demonstrated.

INTRODUCTION

High Aspect Ratio structures

One of the major challenges of microtechnology fabrication has been, and still is in some cases, the definition of high aspect ratio (HAR) structures. LIGA technology [1, 2] is a very powerful technique allowing incredibly high aspect ratios but with the inconvenient of huge cost, due to the fact that X-Rays are necessary. A similar but much cheaper solution is the use of SU-8 [3, 4], which has been used in many different MEMS applications.

However, silicon has always been the most used material for MEMS [5] and it was not until almost 15 years ago that the invention of the Deep Reactive Ion Etching process (Bosch® process) [6] provided a major advancement in the field of silicon technology. This etching technique combines two different gases in alternating steps of etching (with SF_6) and polymer deposition (with C_4F_8). The polymer is deposited both in the sidewalls and in the horizontal surfaces but removed at a much slower rate from the vertical walls, which eventually yields an anisotropic process. In addition, the etch rates are quite high, allowing to perform bulk micromachining of a standard Si wafer in less than 1 hour (a major breakthrough when compared to previous techniques like KOH or TMAH etching [7]).

Therefore, DRIE allows defining structures in silicon with Aspect Ratios (AR) higher than 15:1 (AR depends on the exact etching conditions and can be 100:1 or 150:1) and with a high etch rate. However, a lithographic step to define the mask is always necessary prior to the etching, which implies the use of photoresist and consequently coating, exposure, development and removal of the resist. This also imposes certain restrictions regarding the materials and substrates that can be patterned, i.e. substrates sensitive to chemicals and non-planarized are incompatible with standard techniques.

As an alternative, we present here the use of DRIE processes using a micro-stencil as a reusable mask for the definition of HAR structures. This work is inspired by the already described dry etching of thin films with sub-micrometer resolution [8, 9] but with a focus for MEMS applications.

Stencil Lithography

Stencil Lithography (SL) has been widely used in the last years to locally deposit metals on a substrate, providing sub-micrometer resolution [10, 11]. The metals deposited can be used either to create electronic structures [12] or to act as a mask for posterior etching of the substrate [13]. Although this technique has mainly been used for the selective deposition of metal, it has also been used to perform ion implantation [14] and etching of different substrates in the several hundreds of microns regime [15, 16] and in the sub-micrometer range [8, 9].

EXPERIMENTS

Stencil fabrication

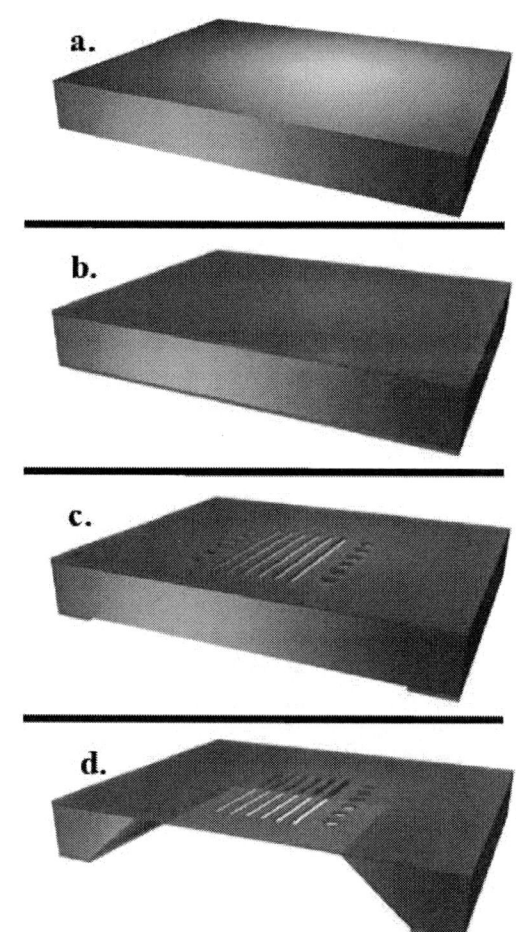

Figure 1: Schematics of the fabrication of a stencil. a) Double side polished wafer, b) deposition of Low Stress Silicon Nitride, c) pattern of apertures in the front-side and windows in the back-side, d) KOH etching to release the membranes.

For this work, low stress silicon nitride (LS-SiN) membranes, 500 nm thick and ~1x1 mm² in lateral size, were fabricated in a p-type Si wafer (100 mm). The fabrication involves standard micro-technology processes and a detailed description can be found elsewhere [4]. It starts with a double sided polished Silicon wafer (100 mm in diameter) (Figure 1.a). A layer of low stress silicon nitride (LS-SiN) (100 to 500 nm thick) is deposited (Figure 1.b) on both sides of the wafer. Then, the nitride layer is patterned opening functional apertures on the front-side and windows for bulk micromachining on the backside (Figure 1.c). Finally, the release of the membranes is made using a KOH etching (Figure 1.d). The dimensions of the different apertures in the membranes used for the experiments presented here were ranging between 2 µm and 200 µm. Some examples of membranes with apertures are presented in Figure 2.

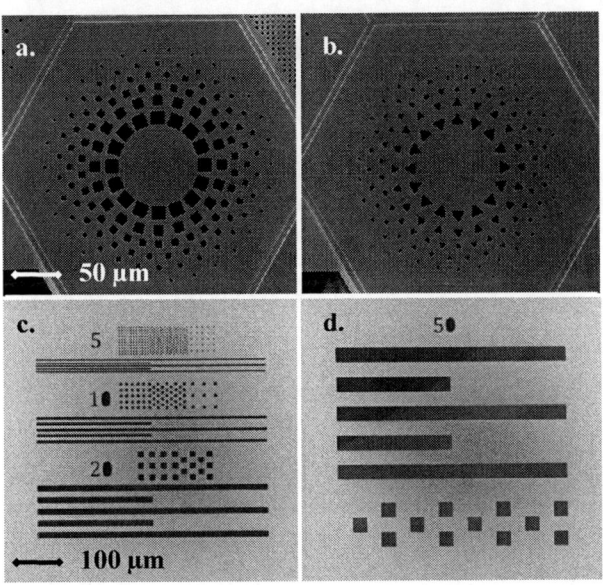

Figure 2: SEM micrographs of different membranes with holes after release. a) Squares and b) triangles ranging from 2 to 10 µm. c, d) squares and lines from 5 to 50 µm.

Once the membranes are released and prior to their use as shadow masks for etching of silicon, both sides of the membranes are covered by a thin Al layer (around 50 nm) in order to avoid damaging the membrane and allow their re-usability. Aluminum is chosen because of its high selectivity with respect to Si in Bosch process.

Table 1: Process conditions in Alcatel ICP-601E used in the presented work

Process	Si-etch		Poly-coat
Gas	SF_6	$c\text{-}C_4F_8$	$c\text{-}C_4F_8$
Flux (sccm)	300	200	50
Cycle Time (s)	7	2	-
Pressure	41% (6.5 Pa)	41% (4.5 Pa)	3 Pa
Source Power (W)	1800	1800	1800
Platten Power (W)	90	90	0

Substrates preparation

In order to prove that this technique was suited for the patterning of substrates with pre-defined topography, several wafers were prepared with square-shaped wells (Figure 3.a). This topography was defined either by DRIE (vertical sidewalls, conditions in Table I) or by KOH etching (sloped sidewalls). Gaps between 15 and 100 µm were defined.

Etching experiment

Once the membranes were protected, the stencils were placed on top of a p-type Si wafer (substrate, Figure 3.a) and a dry etching using a standard Bosch process (Table 1, "Si etch") was performed in an Alcatel ICP-601E (Figure 3.b). After etching, the stencil is released from the substrate and the resulting patterns can be observed (Figure 3.c). It is worth stressing at this point that the etching recipe used was the same as the one used for silicon etching using conventional etching masks like photoresist, silicon oxide or aluminum.

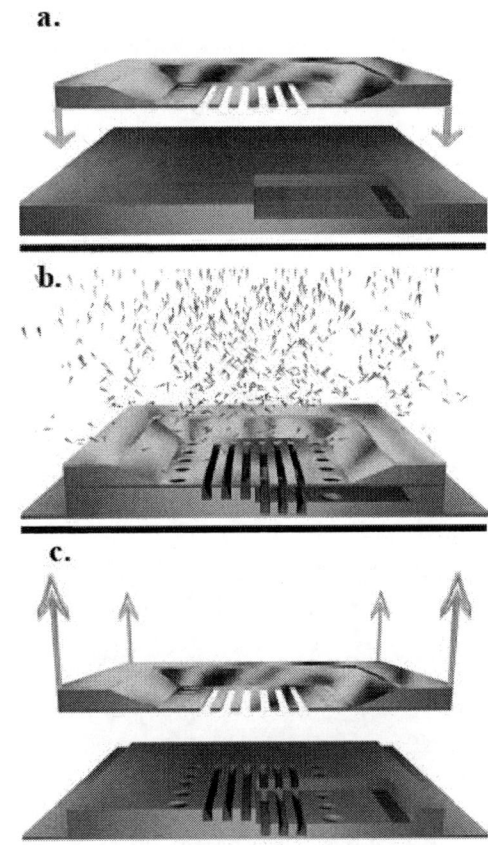

Figure 3: Steps to fabricate deep trenches through stencil: a) Stencil covered with an Al layer is placed on the substrate, c) DRIE is performed, d) stencil is released and the structures have been transferred.

RESULTS

Flat substrate

When etching on a flat substrate, all the patterns included in the membranes were transferred (Figure 4.a and b) yielding a maximum measured AR of 13:1 (for a typical trench of 2.5 µm in width and 33 µm in depth), proving therefore the feasibility of the process. The value for the maximum AR must be compared with the maximum value attained with the same process but using an aluminum layer as a mask, which is 15:1. This means a reduction of approximately 15% of the maximum AR. A reduction of the etch rate was also observed.

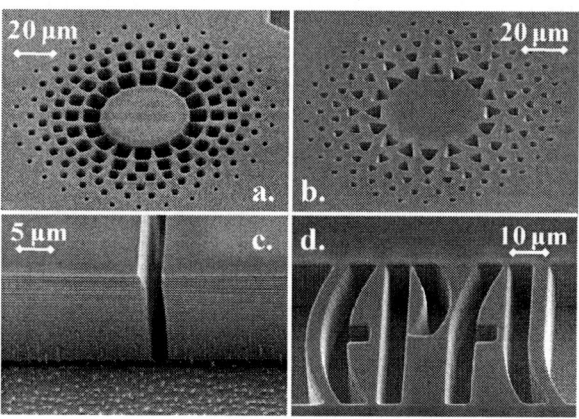

Figure 4: Results after 10 minutes etching of silicon on a flat substrate. Squares (a) and triangles (b) from 2 to 10 µm in size transferred into Si. These patterns correspond to the stencils in Figure 2.a and b. c) Si trench with an AR of 10:1. d) Transfer of arbitrary shaped structures.

Endurance tests were performed on the masks and, due to the Al protective layer and its high selectivity towards Si, it was possible to etch-through a full wafer (525 µm) twice without noticeably degrading the stencil.

However, it is important to note that when Al was not covering both parts of the membrane the corrosion was fast. This is due to the neutral etching species which perform a purely chemical etching in an isotropic way. After going through the apertures in the membrane, those compounds disperse freely in the gap between the membrane and the substrate, damaging the membrane if it is unprotected.

Stepped substrates

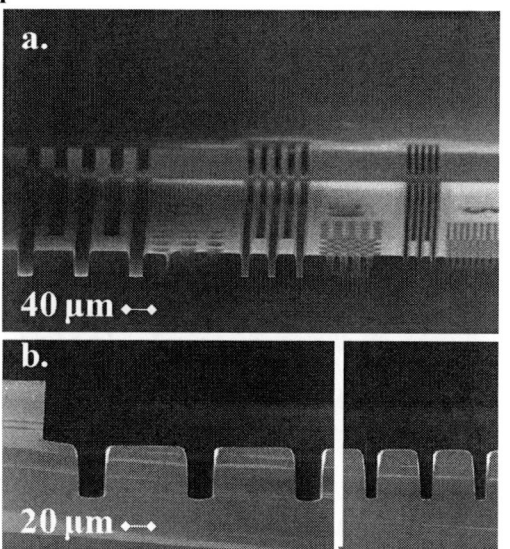

Figure 5: a) and b) 40 µm deep trenches etched in the Si over a gap of 50 µm. Widths from 10 to 50 µm were correctly transferred.

To verify the structuring of non-planar, pre-patterned substrates, we performed etching through openings widths

ranging from 10-50 µm over different gaps. Some typical results for 50 µm gaps are shown in Figure 5.

In this case, we observed differences between standard and shadow mask process that increase with the gap, such as: surface damage that can be caused by the neutral species (slight etching, huge extension. Figure 6.a) or by charged particles which are not completely perpendicular to the mask and therefore increase the exposed area (small size, serious damage to corners, Figure 6.b and c). Also loss of Aspect Ratio (Figure 6.d) and reduction on the etch rate (Figure 7) are observed.

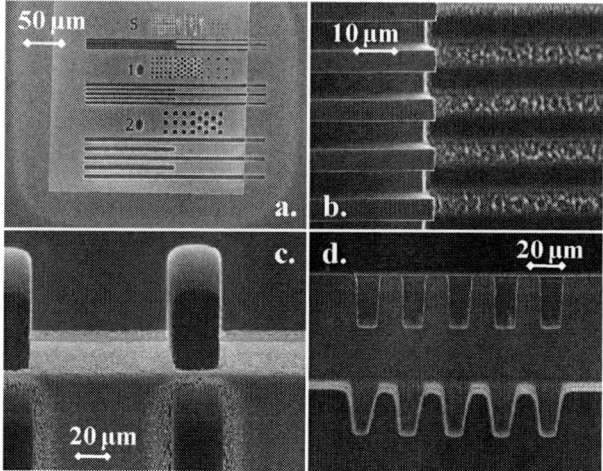

Figure 6: a) Pattern transfer corresponding to the stencil in Figure 2.c. The zone with the bright halo presents damage only in the surface. b) and c) Huge damage of the region close to the corners. d) 20 µm trench opened with and without gap. AR of the structures is smaller after the gap.

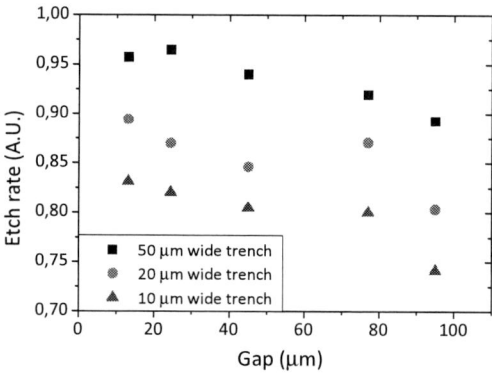

Figure 7: Relative etch rate for different widths as a function of the gap between the stencil and the substrate. The difference in the etch rate for different widths is due to the difference AR. The difference due to the gap is due to the dispersion of the etching species in the chamber after they go through the membrane apertures.

Finally, a novel effect was observed when etching on sloped walls. Figure 8 is showing a SEM micrograph and a schematic of the cross section. It is possible to see that in the etched and sloped region, silicon grass is formed, whereas in the etched and flat regions, clean results are obtained. This behavior is due to the fact that the gas has locally more surface to etch than in the flat case. Then the balance between the two gases is broken and therefore spikes appear. A possible solution to this issue is to increase the Platen Power, which makes the etching species more energetic and capable of removing those spikes while processing.

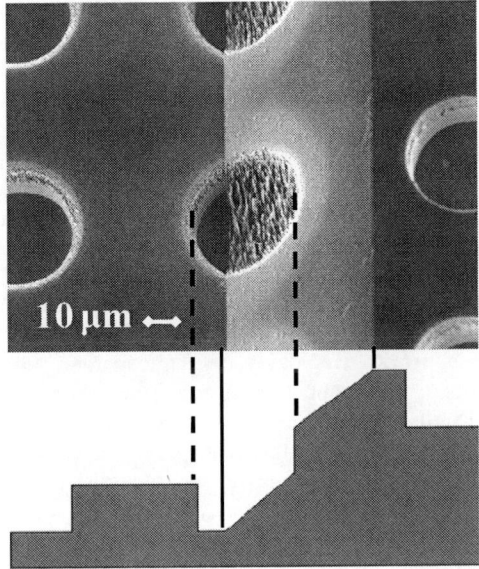

Figure 8: Circles 40 μm deep defined at two different gaps and in a KOH slope simultaneously. In the slope, the effective area is larger, which implies that the etching gas is not enough using a standard recipe. Higher platen power can be used to minimize that effect.

CONCLUSIONS

We have shown in this paper the feasibility of the use of membranes with apertures as a hard mask to perform DRIE on silicon substrates. The gap between the mask and the wafer has a direct effect on the surface damage (increasing when the gap increases), the achievable aspect ratio and the etch rate (both decreasing when the gap increases).

We have achieved AR as high as 13:1, full wafer etching without noticeable degradation of the stencil, definition of structures after a gap of 100 μm and correct pattern into sloped surfaces.

This is therefore a very promising technique which we believe can broaden the use of DRIE techniques for MEMS applications.

REFERENCES

[1] E.W. Becker, W. Ehrfield, P. Hagmann, A. Maner, and D. Münchmeyer, "Fabrication of microstructures with high aspect ratios and great structural heights by synchrotron radiation lithography, galvanoforming, and plastic moulding (LIGA process)", *Microelectron. Eng.*, 1986, 4(1), pp. 35-56.

[2] C.K. Malek and V. Saile, "Applications of LIGA technology to precision manufacturing of high-aspect-ratio micro-components and -systems: a review", *Microelectronics Journal*, 2004, 35(2), pp. 131-143.

[3] M. Despont, H. Lorenz, N. Fahrni, J. Brugger, P. Renaud, and P. Vettiger. "High-aspect-ratio, ultrathick, negative-tone near-UV photoresist for MEMS applications", in *Proceedings - IEEE MEMS '97*, 1997, pp. 518-522.

[4] H. Lorenz, M. Despont, N. Fahrni, N. LaBianca, P. Renaud, and P. Vettiger, "SU-8: a low-cost negative resist for MEMS", 1997, 7(3), pp. 121-124.

[5] K.E. Petersen, "Silicon as a Mechanical Material", *Proceedings of the IEEE*, 1982, 70(5), pp. 420-457.

[6] F. Laermer and A. Schilp, Robert Bosch Gmbh (Bosch), 1994, Patent number: WO9414187-A, US5501893-A.

[7] G.T.A. Kovacs, N.I. Maluf, and K.E. Petersen, "Bulk micromachining of silicon", *Proceedings of the IEEE*, 1998, 86(8), pp. 1536-1551.

[8] S.W. Pang, M.W. Geis, W.D. Goodhue, N.N. Efremow, D.J. Ehrlich, R.B. Goodman, and J.N. Randall, "Pattern Transfer by Dry Etching through Stencil Masks", *J Vac Sci Technol B*, 1988, 6(1), pp. 249-252.

[9] G. Villanueva, O. Vazquez-Mena, M.A.F. van den Boogaart, K. Sidler, K. Pataky, V. Savu, and J. Brugger, "Etching of sub-micrometer structures through Stencil", 2008, 85(5-6), pp. 1010-1014.

[10] M.M. Deshmukh, D.C. Ralph, M. Thomas, and J. Silcox, "Nanofabrication using a stencil mask", *Appl Phys Lett*, 1999, 75(11), pp. 1631-1633.

[11] J. Brugger, J.W. Berenschot, S. Kuiper, W. Nijdam, B. Otter, and M. Elwenspoek, "Resistless patterning of sub-micron structures by evaporation through nanostencils", *Microelectron Eng*, 2000, 53(1-4), pp. 403-405.

[12] O. Vazquez-Mena, G. Villanueva, V. Savu, K. Sidler, M.A.F. van den Boogaart, and J. Brugger, "Metallic Nanowires by Full Wafer Stencil Lithography", 2008.

[13] J. Arcamone, G. Rius, G. Abadal, J. Teva, N. Barniol, and F. Perez-Murano, "Micro/nanomechanical resonators for distributed mass sensing with capacitive detection", *Microelectron Eng*, 2006, 83(4-9), pp. 1216-1220.

[14] J.N. Randall, D.C. Flanders, N.P. Economou, J.P. Donnelly, and E.I. Bromley, "High-Resolution Ion-Beam Lithography at Large Gaps Using Stencil Masks", *Appl Phys Lett*, 1983, 42(5), pp. 457-459.

[15] K. Kolari, "Deep plasma etching of glass with a silicon shadow mask", *Sens and Act A-Phys*, 2008, 141(2), pp. 677-684.

[16] K. Kolari and A. Hokkanen, "Tunable hydrophilicity on a hydrophobic fluorocarbon polymer coating on silicon", *J Vac Sci Technol A*, 2006, 24(4), pp. 1005-1011.

INDIUM TIN OXIDE (ITO) TRANSPARENT MEMS SWITCHES

Byung-Kee Lee, Yong-Ha Song, and Jun-Bo Yoon

School of Electrical Engineering and Computer Science, KAIST, Republic of Korea

ABSTRACT

This paper presents transparent conductive oxide (TCO) switches on glass substrates using indium tin oxide (ITO) for the first time. Mechanical properties of the sputtered ITO film on the glass (density : 7.3gcm^{-3}, Young's modulus : ~190GPa) were measured by X-ray reflectivity (XRR) and nano-indentation analysis. Improvement of transmittance, sheet resistance and wet-etching of 1-μm-thick sputtered ITO was obtained through thermal annealing at 230°C. The fabricated MEMS switches made of ITO with a length of 50μm and width of 10μm showed the pull-in voltage of 23V and pull-out voltage of 20V, and an excellent ON/OFF current ratio of ~10^6. In addition, up to 10μA current could flow through the transparent MEMS switch.

INTRODUCTION

Recently, transparent electronic devices have attracted tremendous attention in the fields of flat panel displays and solar cells. Accordingly, a considerable number of researches have been conducted on the transparent oxide thin film transistor (TFT) using TCO and transparent oxide semiconductor (TOS). Compared with amorphous silicon (a-Si) TFT, oxide TFTs have powerful advantages of high mobility (>10cm^2 V^{-1} s^{-1}) and high ON/OFF ratio (>10^6), which are suitable for active-matrix organic light-emitting diodes (AMOLEDs) application. Furthermore, their transparency could improve light efficiency of liquid crystal displays (LCDs), AMOLEDs and other types of displays [1, 2]. Thus, fully transparent oxide TFT array was successfully integrated onto the color filter array and positioned at the viewing side of the electronic paper (E ink) display [3].

For a few decades, the MEMS switches have been actively researched for radio frequency (RF) applications, because of various advantages such as low insertion loss, high isolation, and negligible power consumption [4]. Especially, the electrostatic MEMS switches have been well developed and used mainly for the wireless communication thanks to their low on-resistance of several Ω, long lifetime (cold-switched) exceeding 10 billion cycles, fast response time of several micro-seconds and high current driving capability [5]. These advantages are also necessitated in the TFT field. We previously reported the possibility and superiority of MEMS switches over conventional a-Si TFT in fulfillment of the higher current

requirements for the large display applications and possibility of being integrated on the low cost plastic substrate [6]. However, MEMS switches made of metal materials cannot obtain the transparency unless TCOs are used.

As a good candidate for transparent MEMS switch material, ITO is the one of most widely used TCOs. It consists of indium oxide (In$_2$O$_3$) doped with tin (Sn). It is extensively exploited for optoelectonic devices [7], displays [8] and solar cells [9] because it has a low resistivity (~10^{-4}Ωcm) and high transparency (>80%) to visible light and it reflects IR radiation [10-12].

In this work, we developed transparent MEMS switches using ITO and characterized the mechanical, optical and electrical properties as a substitute for oxide TFTs. Figure 1 illustrates the conceptual schematic view of the transparent MEMS switches integrated into the transparent display. Light can pass through the transparent MEMS switch because the beam and bottom electrodes are made of transparent material on the glass substrate.

MATERIALS AND FABRICATION

Mechanical property

Mechanical properties of the ITO film such as density and Young's modulus should be known to predict the pull-in voltage and resonant frequency of the ITO MEMS switches. However, many researches about the density and Young's modulus of ITO have not reported because ITO has been used only for transparent electrodes. The density of 1000-Å-thick sputtered ITO on the glass substrate (7.3gcm^{-3}) was measured using XRR analysis. Young's modulus of the 5000-Å-thick sputtered ITO on the glass substrate was measured using nano-indentation analysis [13]. Figure 2 shows the load-displacement curve of ITO nano-indentation test. The first step in the analysis procedure is to determine the contact stiffness (S = dP/dh) which is the slope of initial portion of the unloading curve. Young's modulus is related to the contact stiffness by

$$S = \frac{2}{\sqrt{\pi}} \beta \left(\frac{1-v_s^2}{E_s} + \frac{1-v_i^2}{E_i} \right)^{-1} \sqrt{A} \qquad (1)$$

where β is a constant that depends on the geometry of indenter (β = 1.034 for a Berkovich indenter) and A is the projected contact area. E$_s$ and v$_s$ are Young's modulus and

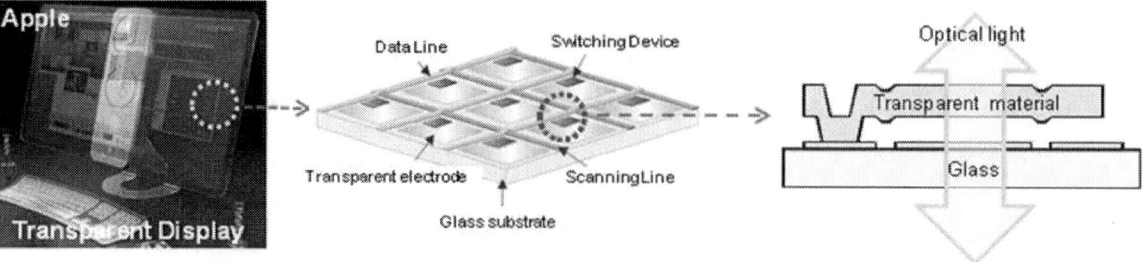

Figure 1: Conceptual schematic view of the transparent MEMS switches integrated into the transparent display.

978-1-4244-2977-6/09 $25.00 © 2009 IEEE

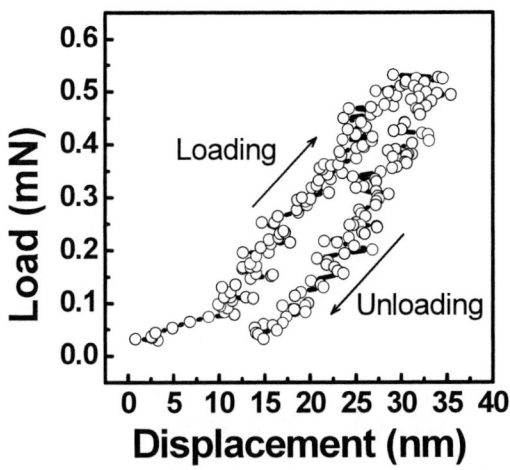

Figure 2: Load-displacement curve of ITO nano-indentation test.

Poisson's ratio respectively, and E_i and v_i are the same quantities for the indenter. For a diamond nano-indenter tip, E_i and v_i are 1140 GPa and 0.07 respectively. Poisson's ratio ranges between 0 and 0.5 for most material, and v_s is assumed as around 0.3. From the Eq. (1), the Young's modulus of ITO (E_s) is calculated into ~190GPa, which is similar to that of Ni.

Thermal annealing and wet-etching

In order to use ITO as a MEMS structural material, wet-etching property of the thick ITO film should be considered. Figure 3 shows the optical images of the wet-etched 1-μm-thick ITO with photoresist mask after (a) no thermal annealing, and (b) annealing at 150°C, (c) 200°C and (d) 230°C for 1 hour. Wet etching was conducted using 18% HCl-4.5% HNO$_3$ at 60°C. No annealed ITO film was torn from the substrate in the etching process, but 230°C-annealed ITO was etched very well during the wet etching.

XRD and crystallinity

In order to obtain good optical and electrical properties of the ITO film, high crystallinity is necessary. Most widely used technique is sputtering ITO films with the substrate temperature of 200 - 400°C. In this work, 1-μm-thick ITO was sputtered at room temperature, and improving the crystallinity of ITO was done by post

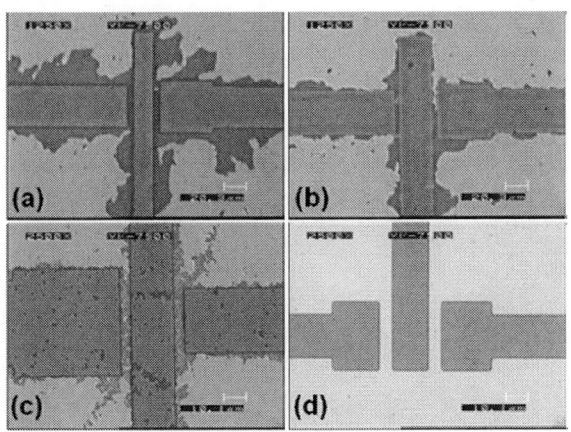

Figure 3: optical images of wet-etched 1-μm-thick ITO with photoresist mask after (a) no thermal annealing, (b) 150°C, (c) 200°C, and (d) 230°C for 1 hour.

thermal annealing at 230°C for 1 hour. Figure 4 shows x-ray diffraction (XRD) spectra of ITO films before and after thermal annealing. Peak identification confirms these sputtered films to be Sn-doped In$_2$O$_3$. Both amorphous and polycrystalline ITO were obtained by room temperature sputtering. After the thermal annealing, Sn-doped In$_2$O$_3$ peak intensity increased and broad peak from 25 to 35° disappeared, which indicate that amorphous ITO was transformed into polycrystalline ITO.

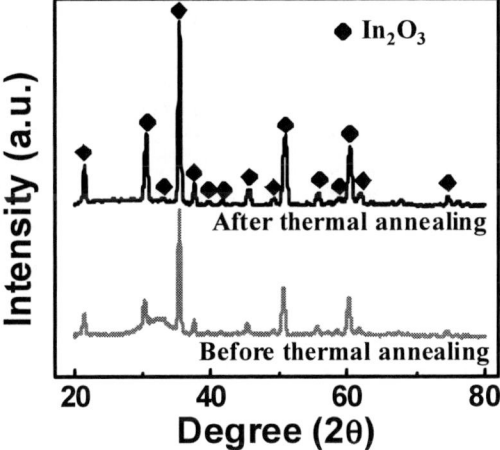

Figure 4: XRD spectra of 1-μm-thick sputtered ITO films on the glass substrate before and after thermal annealing.

Optical and electrical property

The optical transmittance of the 1-μm-thick ITO before and after 230°C thermal annealing is shown in Fig. 5. The average optical transmittance (@550nm wavelength) of the ITO film was improved from 34 to 57% by annealing process. The inset shows 1-μm-thick ITO on the glass substrate before (left) and after thermal annealing (right). The sheet resistance of the sputtered 1-μm-thick ITO decreased from 6.8 to 2.3 Ω/□ after annealing, which also indicates the crystallinity of ITO was improved by thermal annealing. This improvement is because of activation of Sn. Before the thermal annealing, Sn is not activated in the amorphous state and is not able to dope

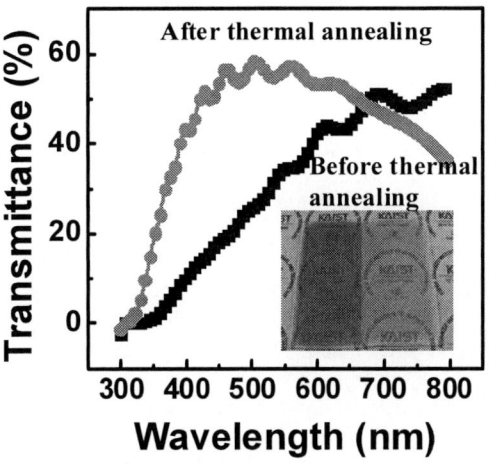

Figure 5: Optical transmittance spectrum of 1-μm-thick ITO on glass substrates before (left inset) and after thermal annealing (right inset).

978-1-4244-2977-6/09 $25.00 © 2009 IEEE   149

In_2O_3 effectively. In the amorphous state, most of the electrical free carriers arise from oxygen vacancy-like state. However, after annealing, additional carriers are provided by substitutional Sn, which is activated. Thus low resistance of polycrystalline ITO was obtained.

Fabrication process

Figure 6 illustrates the fabrication process flow of the transparent ITO MEMS switch. All switch structures including bottom electrodes were simply made of ITO on the glass substrate. The process started with the 1450-Å-thick ITO coated glass, and (a) ITO was patterned to form bottom electrodes. (b) 500-nm-thick Cr was sputtered and patterned as a sacrificial layer. Then, (c) 1-μm-thick ITO was sputtered and annealed at 230°C for 1 hour to improve crystallinity. After the formation of ITO beam, (d) the Cr sacrificial layer was removed (d) and finally released using critical point dryer (CPD) to prevent stiction.

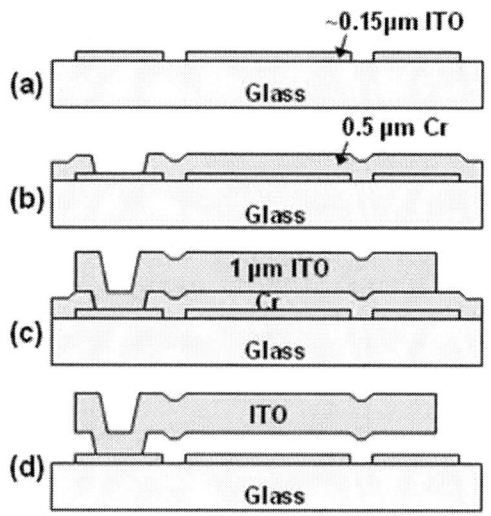

Figure 6: Schematic fabrication process of the ITO MEMS switch.

RESULTS

The SEM images of the fabricated ITO switch with a length of 30 μm and width of 10 μm are shown in Fig. 7. ITO residues were observed after wet-etching of 1-μm-thick ITO because there still existed amorphous ITO phases after annealing. Moreover, beam thinning phenomenon was observed in Fig. 7(b). It was led by the penetration of ITO wet etchant into ITO/Cr interface due to poor adhesion.

Figure 8 shows the fabricated ITO switches arrays including bottom electrodes on the 3cm-by-3cm glass substrate. Backside images can be clearly seen through the sample in Fig. 8(a) and ITO switches made visible by adjusting the angle of illumination in Fig. 8(b).

I-V characteristic of the fabricated ITO switch with a length of 50 μm and width of 10 μm is shown in Fig. 9. Agilent 4156C precision semiconductor parameter analyzer was used. It shows the pull-in and pull-out voltages of 23V and 20V respectively, and an excellent drain current ON/OFF ratio of ~10^6. In addition, ON

current of up to 10μA could flow through the ITO MEMS switch when the drain and source voltages of 1V and 0V are applied, respectively.

CONCLUSION

In this work, highly transparent switches were newly fabricated using ITO, and the mechanical, optical and electrical properties were observed. Improvement of transmittance, sheet resistance and wet-etching of the 1-μm-thick sputtered ITO is obtained through thermal annealing. The ITO switches with a length of 50μm and width of 10μm showed the pull-in voltage of 23V and pull-out voltage of 20V, and an excellent ON/OFF current ratio of ~10^6. Moreover, up to 10μA current could flow through the transparent MEMS switch. The proposed switches are expected to be the promising substitute for oxide TFTs, and with them fully transparent displays are on the verge of realization.

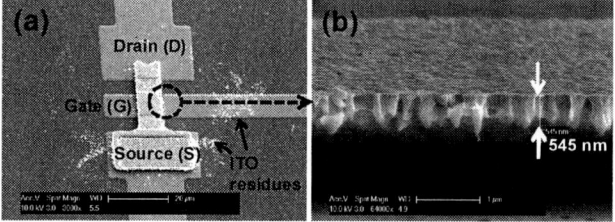

Figure 7: SEM images of (a) fabricated ITO switch with a length of 30 μm and width of 10 μm and (b) side view of ITO beam.

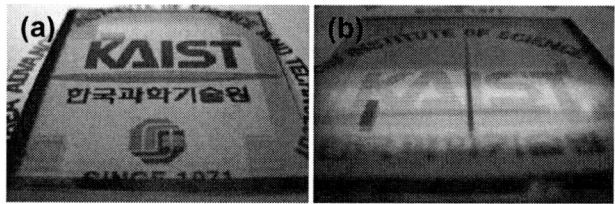

Figure 8: Photos of (a) the transparent ITO switches on the glass substrate and (b) ITO switches made visible by adjusting the angle of illumination.

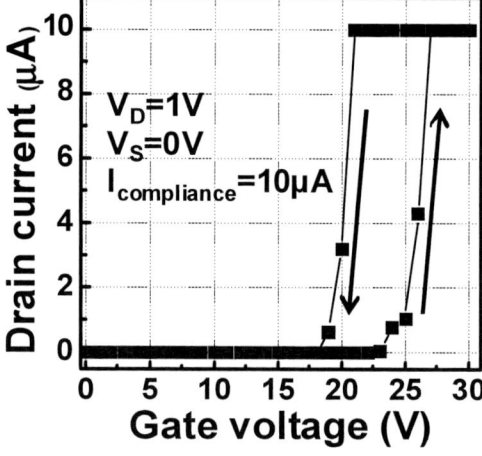

Figure 9: I-V characteristic of the fabricated ITO switch with a length of 50 μm and width of 10 μm.

ACKNOWLEDGEMENTS

The authors would like to thank Devices and Materials Laboratory, LGE Advanced Research Institute for sputtering ITO. This work was supported by the Korea Science and Engineering Foundation (KOSEF) grant funded by the Korea government (MEST) (No. R11-2007-045-03003-0).

REFERENCES

[1] H. Hosono, "Recent progress in transparent oxide semiconductors: Materials and device application", *Thin Solid Films*, vol. 515, pp. 6000-6014, 2007.

[2] G. S. Chae, "Modified transparent conducting oxide for flat panel displays only", *Jpn. J. Appl. Phys.*, vol. 40, pp. 1282-1286, 2001.

[3] M. Ito, M. Kon, C. Miyazaki, N. Ikeda, M. Ishizaki, Y. Ugaji and N. Sekine, "Front drive display structure for color electronic paper using fully transparent amorphous oxide TFT array", *IEICE TRANS. ELECTRON.*, E90-C, pp. 2105-2111, 2007.

[4] G. M. Rebeiz and J. B. Muldavin, "RF MEMS switches and switch circuits", *IEEE Microw. Mag.*, vol. 2, pp. 59-71, 2001.

[5] http://www.radantmems.com/radantmems/switchperformance.html

[6] J. O. Lee, H. H. Yang and J. B. Yoon, "A new method of driving an AMOLED with MEMS switches", *proc. MEMS 2008*, pp. 132-135.

[7] T. Margalith, O. Buchinsky, D. A. Cohen, A. C. Abare, M. Hansen, S. P. DenBaars and L. A. Coldren, "Indium tin oxide contacts to gallium nitride optoelectronic devices", *Appl. Phys. Lett.*, vol. 74, pp. 3930- 3932, 1999.

[8] K. H. Choi, J. Y. Kim, Y. S. Lee and H. J. Kim, "ITO/Ag/ITO multilayer films for the application of a very low resistance transparent electrode", *Thin Solid Films*, vol. 341, pp. 152-155, 1999.

[9] J. Yang, A. Banerjee, and S. Guha, "Triple-junction amorphous silicon alloy solar cell with 14.6% initial and 13.0% stable conversion efficiencies", *Appl. Phys. Lett.*, vol. 70, pp. 2975-2977, 1997

[10] C. G. Granqvist and A. Hultåker, "Transparent and conducting ITO films: new developments and applications", *Thin solid Films*, vol. 411, pp. 1-5, 2002.

[11] J. L. Vossen and W. Kern, *Thin Film Processes*, Academic Press, New York, 1978.

[12] I. A. Rauf, "Structure and properties of tin-doped indium oxide thin films prepared by reactive electron-beam evaporation with a zone-confining arrangement", *J. Appl. Phys.* Vol. 79, pp. 4057-4065, 1996.

[13] W. C. Oliver and G. M. Pharr, "An improved technique for determining hardness and elastic modulus using load and displacement sensing indentation experiments", *J. mater. Res.*, vol. 7, pp. 1564-1583, 1992.

ADVANCEMENTS IN TECHNOLOGY AND DESIGN OF BIOMIMETIC FLOW-SENSOR ARRAYS

C.M. Bruinink, R.K. Jaganatharaja, M.J. de Boer, E. Berenschot, M.L. Kolster, T.S.J. Lammerink, R.J. Wiegerink, and G.J.M. Krijnen

Transducers Science and Technology, MESA$^+$ and IMPACT Research Institutes, University of Twente, P.O. Box 217, NL-7500 AE Enschede (The Netherlands)

ABSTRACT

This paper reports on recent developments to increase the performance of biomimetic flow-sensor arrays by means of several technological advancements in the fabrication procedures and corresponding sensor design optimizations. Advancements include fabrication procedures with higher process latitude and geometrical modifications of several parts of the flow sensor. The conclusive measurements in this paper support our sensor-model predictions for a 100-fold increase in acoustic sensitivity (down to oscillating flow amplitudes in the order of 1 mm·s^{-1}) translating to substantially higher capacitive outputs in comparison to our first-generation biomimetic flow-sensor arrays.

1. INTRODUCTION

Crickets can extract aerodynamic information from many filiform flow-sensitive hairs (of various length, diameter and directivity) on two conical appendages (cerci) at the rear of their abdomen (Figure 1).

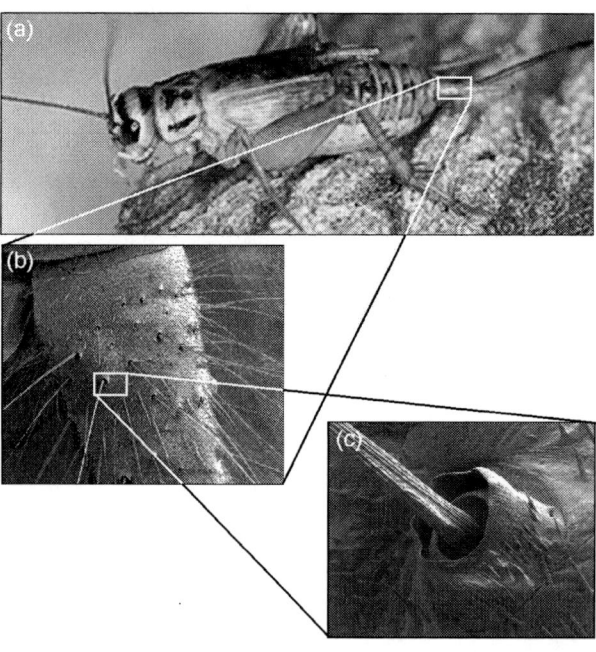

Figure 1: (a) Photograph of the Acheta Domesticus and (b,c) SEM images of the base part of a cercus and a close-up of a filiform flow-sensitive hair (courtesy of J. Casas, IRBI, University of Tours).

The longer filiform hairs (500-1000 μm) are highly perceptive to low-frequency sound waves [1] and can sense air velocities as small as 30 μm·s^{-1} [2], enabling crickets to identify and escape from predators.

Several research groups have been working on modeling, fabrication and optimization of biomimetic flow-sensing devices consisting of arrays of many flow-sensors. MEMS fabrication technology is very practical in the realization of these devices for its capabilities of parallel fabrication and integration of flow sensors into large sensor arrays. The additional value of this type of arrays is that the flow-sensitive hairs allow measurement of flows with high spatial resolution and therefore could facilitate complex flow pattern measurements. Several types of biomimetic flow-sensors can be found in literature incorporating various transduction mechanisms. These include piezoresistive sensing [3], capacitive sensing [4] and switching [5].

This paper describes the successful continuation of the fabrication and design of flow-sensor arrays with capacitive sensing [4] for reasons of high intrinsic sensitivity in combination with low power dissipation and promising performance.

2. SENSOR-MODEL DESIGN

In our sensors, capacitive transduction relies on the tilting of a membrane by drag forces acting on the receptive hair of the sensor (Figure 2). The metallic electrodes on top of the membrane form capacitors with the underlying common electrode. A tilt at the rotational axis of the membrane by the deflection of the drag-force receptive hair causes a differential change in capacitance and therefore provides a means of measuring air flows.

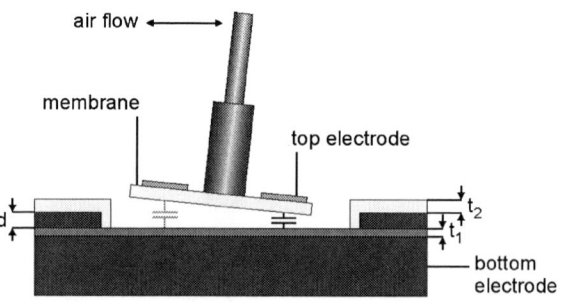

Figure 2: Schematic representation of the operational principle of capacitive sensing.

Application of the aerodynamic models for the filiform hairs of crickets [1] to our biomimetic flow sensors results in several essential design considerations for optimal sensor performance [6]. For comparison of the performance of the biomimetic flow sensors with the filiform hairs of crickets, we define a figure of merit (FoM) [6] as the product of usable bandwidth and the low-frequency sensitivity:

$$FoM = bandwidth \cdot sensitivity \equiv \omega_0 \cdot \frac{T_d}{S} \qquad (1)$$

(S: torsional spring constant)

978-1-4244-2977-6/09 $25.00 © 2009 IEEE

The bandwidth of the flow sensors is proportional to the mechanical resonance frequency given by:

$$\sqrt{\frac{S}{I}} \propto \sqrt{\frac{S}{\rho \cdot L^3 \cdot D^2}} \qquad (2)$$

(L: hair-length; D: hair-diameter; ρ: density of the hair; I: moment of inertia)

For hair lengths above the boundary layer thickness (e.g. at a frequency of 100 Hz, this translates to a minimum hair length of 440 μm) and for the frequency (10 – 1000 Hz), hair-length (100 – 1000 μm) and hair-diameter (5 – 50 μm) ranges of interest, the sensitivity of the flow sensors is roughly proportional to [6]:

$$\frac{L^2 \cdot D^{1/3}}{S} \qquad (3)$$

By combining equations 2 and 3, the FoM becomes proportional to:

$$FoM \propto \sqrt{\frac{L}{\rho \cdot S \cdot D^{4/3}}} \qquad (4)$$

Equation 4 provides clear directions for (design) modifications to enhance the overall performance of the first-generation flow-sensor arrays (see Figure 3a) [4].

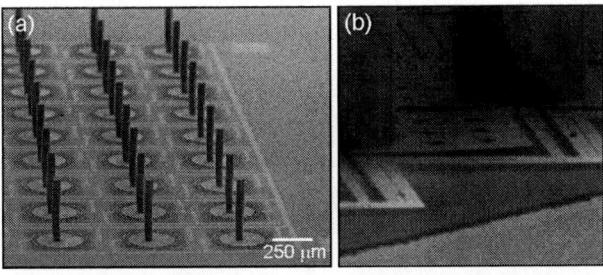

Figure 3: (a) SEM image of the first-generation flow-sensors arrays with 450 μm-long hairs and spiral suspensions. (b) SEM image of the membrane curvature (δ ≈ 2.5 μm) as a result of the high tensile stress in the chromium electrodes on top of the membrane.

These modifications include the incorporation of a two-step photolithographic procedure for the fabrication of longer SU-8 hairs (to increase drag-torque pick-up), smaller inter-electrode gaps (to increase the capacitive sensitivity) and adjustments in the lateral dimensions of the torsion beams (to reduce torsional stiffness). Additionally, experimental findings illustrate the need of low-stress electrode materials in combination with a more optimal electrode design to counteract for the membrane curvature (Figure 3b). The effect of the curvature (δ) on the sensitivity η is given by [6]:

$$\eta = \frac{w \cdot l^2}{(d_0 + \delta) \cdot d_0} \qquad (5)$$

with $d_0 = d \cdot \dfrac{t_1 + t_2}{e_r}$ (see Figure 2)

(δ: curvature at the rim of the membrane; d_0: dielectric thickness (Figure 2); w: membrane-width; l: membrane-length; ε_r: dielectric constant of silicon nitride)

The effects of these modifications on the sensitivity are shown as (theoretical) sensitivity improvement factors in Table 1.

Table 1: Technological sensitivity improvement factors for various design parameters.

Design parameter		Factor
inter-electrode gap	1.0 μm → 0.6 μm	2.7
hair length	450 μm → 900 μm	4.0
hair diameter	50 μm → 25 μm	0.8
membrane curvature	2.5 μm → no bending	4.3
membrane shape	circular → rectangular	1.5
torsion beams		1.8
- length	75 μm → 100 μm	
- width	10 μm → 5 μm .	

The adjustment of the hair-diameter (Table 1) comes at the penalty of a slight reduction in drag-force, however, the smaller diameter considerably decreases the moment of inertia (by about 65%), allowing smaller torsional stiffness while still maintaining sufficient bandwidth.

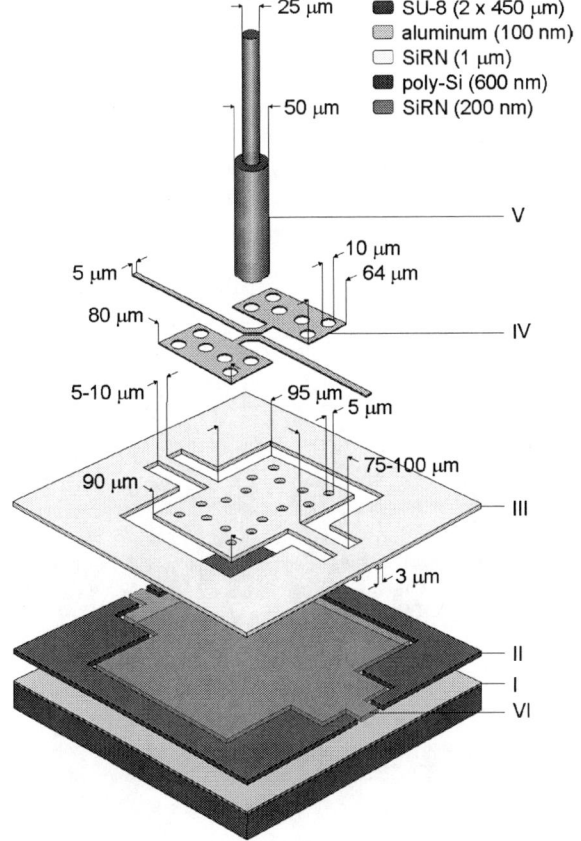

Figure 4: Schematic representation of the fabrication of the artificial hair sensor, including the dimensions.
(I) LPCVD of the protective SiRN layer; (II) LPCVD of the sacrificial poly-Si layer and patterning of the protection trenches; (III) LPCVD of the SiRN layer and patterning of the membrane/torsion beam structures; (IV) sputtering of aluminum and patterning of the top electrodes; (V) two-step SU-8 processing of the hair; (VI) sacrificial layer etching.

978-1-4244-2977-6/09 $25.00 © 2009 IEEE

3. FABRICATION

A thin silicon-rich nitride (SiRN, 200 nm) layer was deposited on a highly conductive silicon wafer (as the common bottom electrode) by low-pressure chemical vapor deposition (LPCVD, Figure 4-I) for protection of the wafer during later sacrificial layer etching. The sacrificial poly-silicon (poly-Si) layer was deposited to a thickness of 600 nm by LPCVD and patterned by reactive-ion etching (RIE) to form protection trenches (Figure 4-II). A second 1 μm-thick SiRN layer was deposited by LPCVD to form the actual membrane and torsion beams after etching by RIE (Figure 4-III) and to complete the protection of specific areas in the poly-Si layer. From past experiments it is known that the deposition and design of the electrode systems is of prime importance for the sensor performance (due to membrane curvature by the internal stress in the metal layer). After performing several test runs for stress and resistance measurements of the electrode systems, the thickness of the aluminum layer was set to 100 nm resulting in low-stress electrode systems. After sputtering the low-stress aluminum by low-power sputtering at room temperature, the electrode systems were patterned by wet-etching in standard resist developer (Figure 4-IV). Use of the developer has several advantages over the conventional aluminum etchants, (low etch rate, room temperature etching) and results in the definition of the electrode systems with high fidelity (< 1% deviation with lateral dimensions of the resist mask). For the fabrication of 900 μm-long SU-8 hairs, processing was done by a sequential exposure procedure of two 450 μm-thick SU-8 layers imposed by the maximum exposure thickness of only 700 μm due to UV light adsorption.

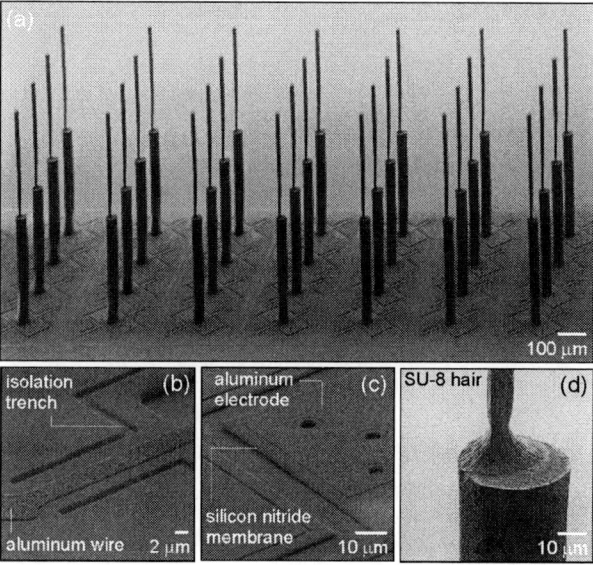

Figure 5: Series of SEM images illustrating in detail the final second-generation flow-sensor array: (a) part of the array with 900 μm-long hairs, (b) part of the aluminum wiring, (c) close-to-flat membrane with aluminum electrode, and (d) connection between two segments of the SU-8 hair. Note: charging during SEM imaging is causing parts of the membrane to stick to the substrate (image c).

Additionally, this two-step photolithographic procedure

enables the fabrication of hairs consisting of two segments of different diameter (Figure 4-V). Finally, the sacrificial layer etching of the poly-silicon for releasing the sensor from the substrate (Figure 4-VI) was done on the front-side using a glass cover over the entire wafer to protect the sensor parts (membrane, electrodes and SU-8 hair) from ion-bombardment during etching.

The series of SEM images in Figure 5 illustrate in detail the resulting second-generation flow-sensor array, consisting of 124 flow-sensors in a parallel configuration with 45° rotation with respect to the longitudinal axis of the sensor array. The significant reduction of the membrane curvature, from 2.5 μm (in case of chromium in the first-generation sensors) to about 100 nm (light interference microscopy data not shown), by using low-stress aluminum as the electrode material is seen in Figure 5c.

Overall, the present fabrication modifications result in second-generation flow sensors with close-to-flat membranes, longer hairs and a smaller inter-electrode gap.

4. ACOUSTIC MEASUREMENTS

Acoustic characterization of this flow-sensor array was carried out with a directional flow-sensitivity measurement set-up (Figure 6) using oscillating flows with amplitudes in the range of 1-100 mm·s⁻¹ and measuring the output at different source angles with respect to the rotational axis of the flow sensor.

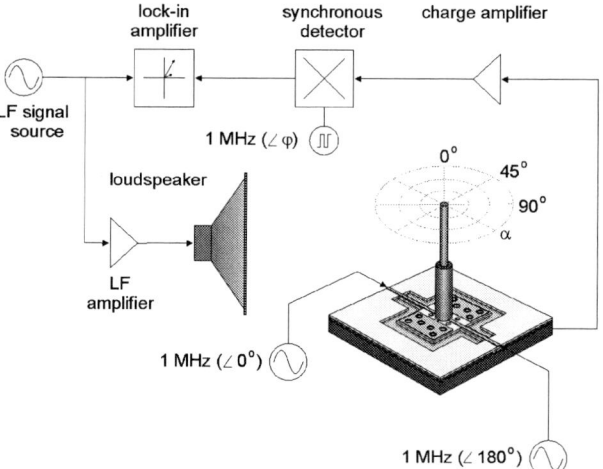

Figure 6: Schematic representation of our acoustic flow measurement set-up, illustrating the essential components for performing acoustic measurements.

This measurement set-up allows the characterization of the flow-sensor arrays with near field acoustic flows from a loudspeaker by employing capacitive amplitude modulation of two 1 MHz electrical signals [7]. The electrical signals are taken 180° out of phase for acquiring a differential-mode rotational signal of the flow sensor. After the charge amplifier and synchronous detector (by a multiplier circuit), the resulting base-band signal is fed to a lock-in amplifier to obtain the LF signal. Figure 7 shows the results of such a differential-mode rotational measurement on our present flow-sensor arrays. The signal follows to a large extend a figure-of-eight, indicating the preferential sensitivity of the flow sensor in the direction perpendicular to the rotation axis.

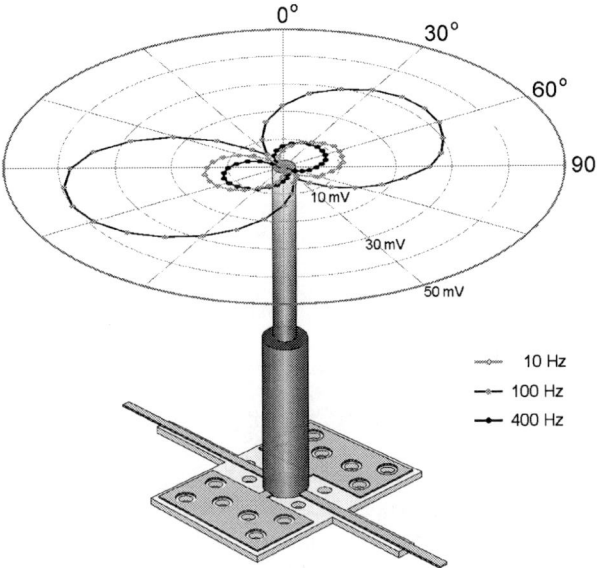

Figure 7: Superposition of the acoustic measurements at three different frequencies on the flow-sensor.

Sensitivity measurements were taken by using the same measurement set-up, including appropriate filtering and noise amplitude measurements (Figure 8). The lower flow detection limit is given by the intersection of the (1 Hz bandwidth) measured signals with the larger bandwidth noise values. Depending on signal frequency and operating bandwidth, minimum flow amplitudes of about 2 mm·s^{-1} can be obtained. This measurement supports the (theoretical) 100-fold increase in sensitivity (see Table 1) due to the present modifications in the second-generation flow-sensor arrays.

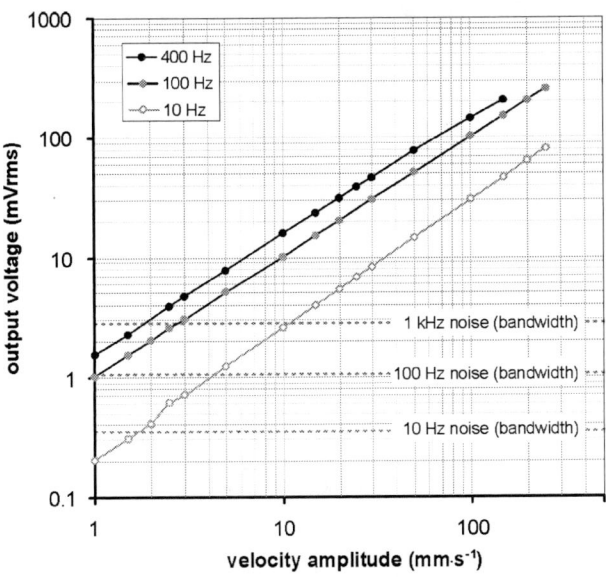

Figure 8: Sensitivity measurements at three different frequencies (dots), with the dashed red lines indicating the RMS noise level for three bandwidths.

Significant improvements in the noise levels are foreseen by further optimization of the electronics.

5. CONCLUSIONS

We have shown significant advancements in sensor technology and design that enable us to fabricate bio-mimetic flow-sensor arrays with sensitivities on the order of 1 mm·s^{-1}. Future research will concentrate on further enhancements by incorporation of different torsion beam constructions (for lowering the torsional spring stiffness), searching for optional materials to decrease the density of the hairs, and by the definition of separate bottom electrodes (for reduction of parasitic capacitances) in order to match the performance of our biomimetic flow-sensor arrays to the sensory hairs of crickets with sensitivities down to 30 μm·s^{-1}.

REFERENCES

[1] (a) J.A.C. Humphrey, R. Devarakonda, I. Iglesias, F.G. Barth, "Dynamics of arthropod filiform hairs. I. Mathematical modeling of the hair and air motions" *Phil. Trans. R. Soc. Lond. B*, 340, pp. 423-444, 1993. (b) T. Shimozawa, T. Kumugai, Y. Baba, "Structural scaling and functional design of the cercal wind-receptor hair of crickets", *J. Comp. Physiol. A*, 183, pp. 171-186, 1998.

[2] T. Shimozawa, J. Murakami, T. Kumagai, *Sensors and Sensing in Biology and Engineering (Chapter 10)*, ed. F.G. Barth, J.A.C. Humphrey and T.W. Secomb, Springer, Vienna, 2003.

[3] Y. Ozaki, T. Ohyama, T. Yasuda, I. Shimoyama, "Air flow sensor modeled on wind receptor hairs of insects", in *Proc. IEEE Int. Conf. MEMS*, Miyazaki, 2000, pp. 531-536.

[4] J.J. van Baar, M. Dijkstra, R. Wiegerink, T.S.J. Lammerink, R. de Boer, G.J.M. Krijnen, "Arrays of cricket-inspired sensory hairs with capacitive motion detection", in *Proc. IEEE Int. Conf. MEMS*, Miami, 2005, pp. 646-649.

[5] W.J. Kuipers, J.J. van Baar, M. Dijkstra, R.J. Wiegerink, T.S.J. Lammerink, J.H. de Boer, G. J.M. Krijnen, "Drag force actuated bistable microswitches for flow sensing", in *Proc. IEEE Int. Conf. MEMS*, 2006, pp. 658-661.

[6] G.J.M. Krijnen, A. Floris, M. Dijkstra, T.S.J. Lammerink, R. Wiegerink, "Biomimetic micro-mechanical adaptive flow sensor arrays", in *SPIE Microtechnologies for the New Millennium*, Gran Canaria, 2007, pp. 6592-6616.

[7] M. Dijkstra, J.J. van Baar, R.J. Wiegerink, T.S.J. Lammerink, J.H. de Boer, G.J.M. Krijnen, "Artificial sensory hairs based on the flow sensitive receptor hairs of crickets", *J. Micromech. Microeng.*, 15, pp. S132-S138, 2005.

ACKNOWLEDGEMENTS

The authors want to thank Mark Smithers for high-resolution SEM imaging, our colleagues in the EU project CILIA (www.cilia-bionics.org) for stimulating discussions and input to this work, the EU for financial support by the Future and Emergent Technologies arm of the IST Programme in the 6th Framework Programme and NWO/STW for financial support in the framework of VICI project BioEARS.

A WAFER-LEVEL MICRO MECHANICAL GLOBAL SHUTTER FOR A MICRO CAMERA

Che-Heung Kim, Kyu-Dong Jung, and Woonbae Kim
CTO-SAIT, Samsung Electronics, Co., Ltd., Yongin-City, KOREA

ABSTRACT

A novel wafer-level manufactured micro mechanical global shutter utilizing thin-film roll actuators is presented for a micro mobile camera. The aperture of the shutter is 2.2 mm in diameter and covered with 36 triangular roll actuators whose radius of curvature is designed to 235 μm. The stress induced rolling of thin composite layers and their pull-in behaviors are analyzed and experimented. A 0.8 mm-thick wafer-level shutter array is successfully implemented using batch processes. The fabricated shutter can follows 500 Hz-square wave signal of 30 V

INTRODUCTION

There is always a typical contradiction in the field of a mobile camera module: adding more functions at slimmer dimensions and lower cost. Traditional part-assembly manufacturing technology is getting saturated and required a breakthrough to satisfy new needs.

Recently, the wafer-level processed micro camera is being considered as the one of the most promising approaches to reduce the manufacturing cost and dimensions simultaneously. Both of a precise passive alignment and batch processes are the inherent power of the wafer-level manufacturing technology. Hence many industrial players, for example, Haptagon, Anteryon, and Tessra, have already presented the wafer-level array lens and their stacking. These researches sufficiently show the very possibility of a low-price fixed focus module. However mechanical components such as an auto-focus lens and a mechanical shutter are indispensable for the high quality camera module which is expected to dominate the market in near future. Unlike the wafer-level lens, no one has provided the wafer-level solution of these mechanical components so far.

Therefore authors propose a novel wafer-level MEMS shutter based on the surface micromachined roll actuator in this paper. The surface micromachined curved beam is a prevalent structure in number of applications due to its simple fabrication process and large stroke at a low driving voltage [1-3]. However a near 360° over-rolled and millimeter scale one utilized in this paper is seldom detected. Chow et al. at PARC studied a comparable scale stationary metallic roll for a chip scale interconnection [4]. Distinctively this paper discusses the integration of millimeter scale thirty-six roll actuators on the transparent electrode and their electrostatic driving characteristics.

DESIGN

A 1.4mm-long triangular roll actuator is a typical

element of the proposed shutter structure (Figure 1a, b). It is fabricated on the transparent electrode and initially rolled up with a specific radius of curvature (ROC); hence it is a 'normally open' switch. The ROC of the actuator is determined by the stress difference between two composing layers (e.g. Aluminum and Silicon Nitride). When a voltage is applied to the actuator, it is pulled in the grounded transparent electrode and then the switch is 'closed' (Figure 1b). Thirty-six roll actuators are combined to make a single 36-side equilateral polygon aperture. Each of roll corresponds to each side of polygon (Figure 1c). As seen in figure 1d, if all actuators are pulled in the substrate simultaneously, then the shutter is closed.

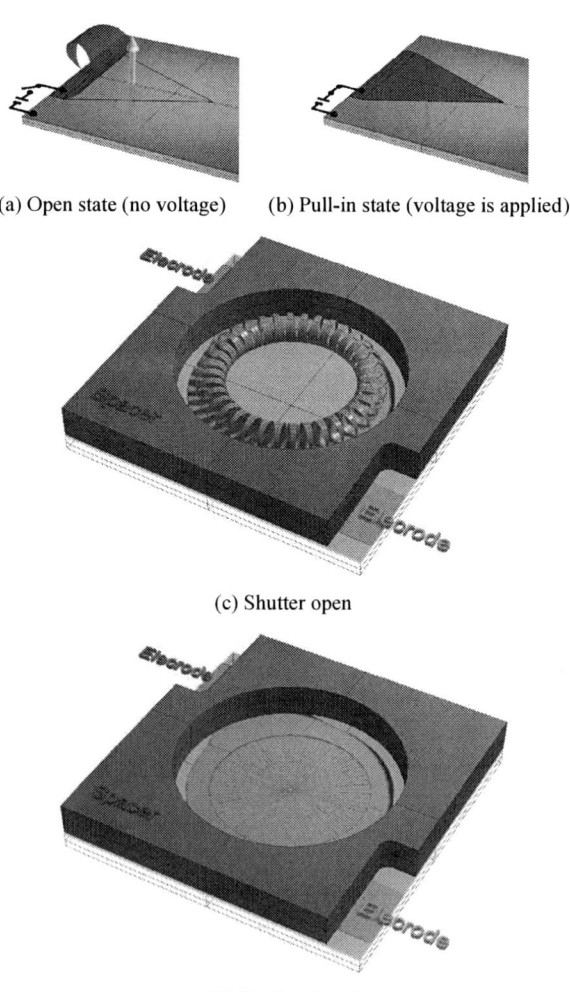

(a) Open state (no voltage) (b) Pull-in state (voltage is applied)

(c) Shutter open

(d) Shutter closed

Figure 1: Structures and principles of the proposed shutter

The diameter of the aperture is determined to 2.2 mm by the optical format of a ¼″ CMOS image sensor, the field of view (FOV) of the first lens (30°), and the distance from the image sensor. The ROC of the roll is herein designed to 235 μm considering an arc angle of 1.95π radians (Figure. 2). The larger arc angle results in the smaller ROC and the thinner shutter but requires the higher voltage to pull in.

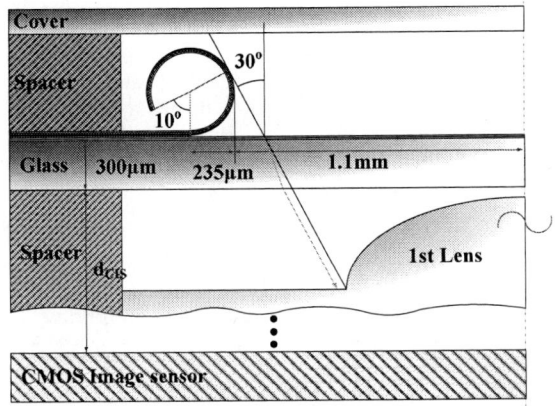

Figure 2: Left half cross-section view of the shutter module

Generally the ROC of n-composite layered beam member is expressed as an equation (1) with given residual stresses and thicknesses.

$$\rho = \frac{\overline{EI}}{M} = \frac{\sum_{i=1}^{n} w E_i' t_i \left[\frac{1}{12} t_i^2 + \left(\frac{1}{2} t_i + \sum_{j=1}^{i} t_{j-1} - \overline{z} \right)^2 \right]}{\frac{w}{2E_1'} \sum_{i=1}^{n} \sigma_i E_i' \left[\left(\sum_{j=1}^{i} t_j - \overline{z} \right)^2 - \left(\sum_{j=1}^{i-1} t_{j-1} - \overline{z} \right)^2 \right]} \quad (1)$$

Where, ρ is the ROC of the roll, $\overline{E}$ and $\overline{I}$ are an equivalent elastic modulus and moment of inertia, respectively. M is the bending moment generated from the stress difference. w is the width of the beam. E' is an effective elastic modulus, $\overline{z}$ is the neutral position of the fixture, σ and t are the residual stress and thickness of each layers, respectively.

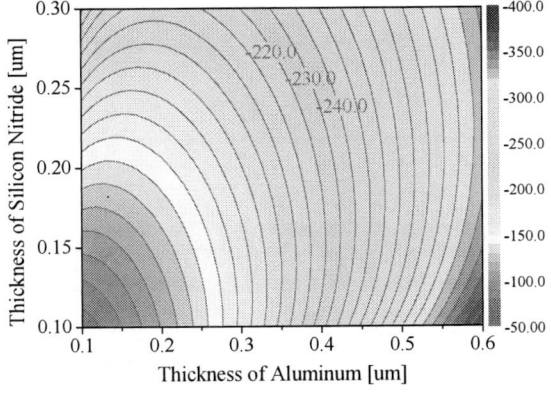

Figure 3: Thickness design window of the roll actuator

The Figure 3 is a design window to obtain the desired ROC (235 μm) for a composite layer used in this experiment. Residual stresses are assumed as 170 MPa and 40 MPa, respectively. As shown in figure 3, the thickness margin of the Aluminum layer is smaller than that of the Silicon Nitride layer, which means that the Aluminum is more effective to control the ROC, however it requires more reliable process conditions.

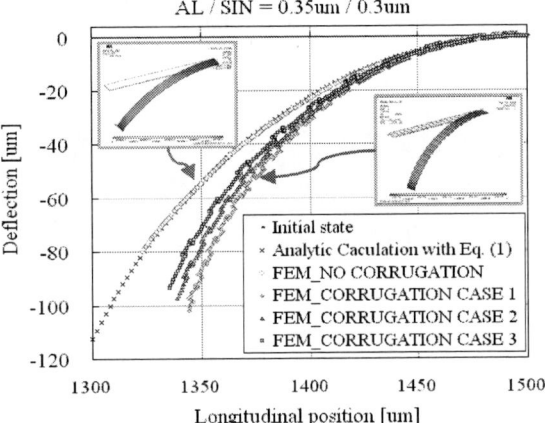

Figure 4: The Effect of the corrugation on the actuator

The result of the FEM simulation (ANSYS9.0) is compared with that of analytical calculation in figure 4. The thicknesses of Aluminum and Silicon Nitride in this case are 3500Å and 3000Å, respectively. Two methods are almost exactly matched to each other.

The corrugation is indispensable with this thin but large area actuator to initiate the roll direction. Forming dense corrugations makes the ROC small in figure 4.

The electrostatic pull-in voltage is simulated utilizing the FEM with respect to the ROC of the actuator in figure 5. The gap in the graph notes the initial gap formed by the thickness of the sacrificial layer. The expected pull-in voltages are about 27 V for a 1500Å gap and 37 V for a 3000Å gap.

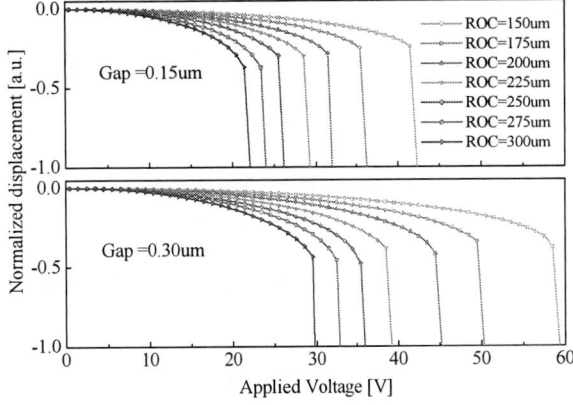

Figure 5: The simulated pull-in behavior with respect to the ROC and initial gap

FABRICATION

A 1500 Å Indium Tin Oxide (ITO) on a 300 μm glass wafer is used as a starting substrate. Thin Perylene sacrificial layer, Silicon Nitride, and Aluminum are deposited and patterned in succession. Finally, the patterned Aluminum and Silicon Nitride are released by the oxygen plasma. The change of the ROC with respect to the Aluminum thickness is presented in Figure 6. The ROC increases with increasing the thickness of Aluminum because the increased thickness stiffens the roll against the stress.

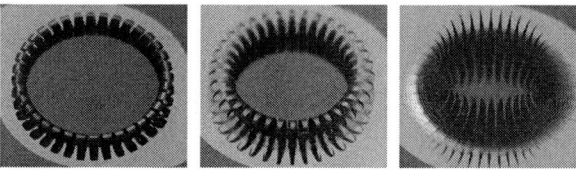

Figure 6: The ROC of the actuators with respect to Aluminum thickness: from left (Al/SiN_x) 2000Å/3000Å, 3500Å/3000Å, 5500Å/3000Å

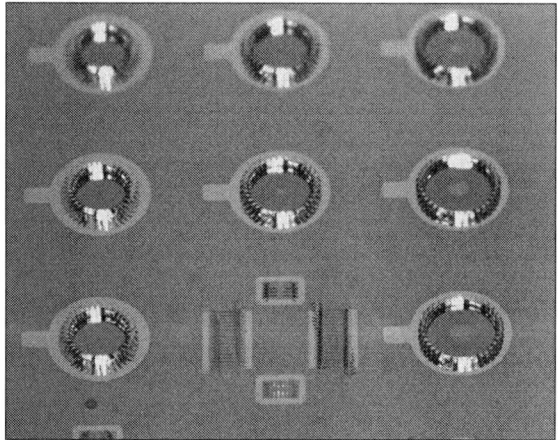

(a) Actuator only array on ITO glass

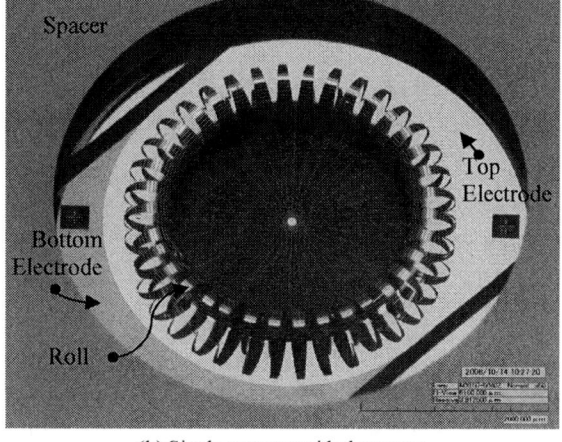

(b) Single actuator with the spacer

Figure 7: The view of the fabricated wafer-level shutter

The fabricated wafer-level micro shutter array is presented in the figure 7a. Among three columns of the array, the rightmost one contains most corrugations. Figure 7b is the enlarged view of a single shutter with the integrated spacer and its ROC is about 224 μm. Total thickness of the shutter module is measured about 0.8 mm.

MEASUREMENT

The harmonic and step response of the unit roll actuator is measured in Figure 8. The spot of the laser doppler vibrometer (LDV) is positioned on the 50 μm inner point from the anchor (not the roll tip) and a low driving voltage (under the pull-in voltage) is applied. The actuator doesn't touch down in this case. The dominant resonant frequencies of the desired actuation are 1.34 kHz and 4.84 kHz in Figure 8a. The figure 8b shows a highly under damped response at applying 20 V-square wave. The first crossing point is measured about 0.2 msec in zoomed graph.

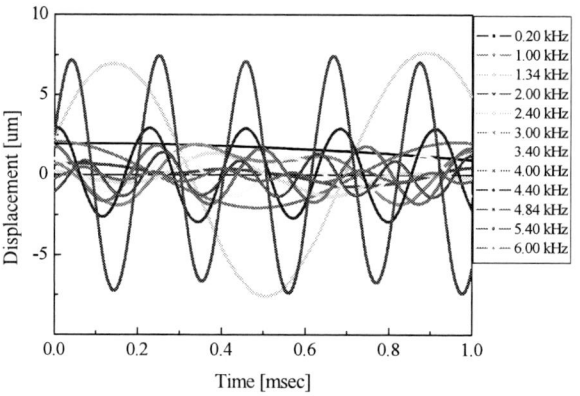

(a) Harmonic response

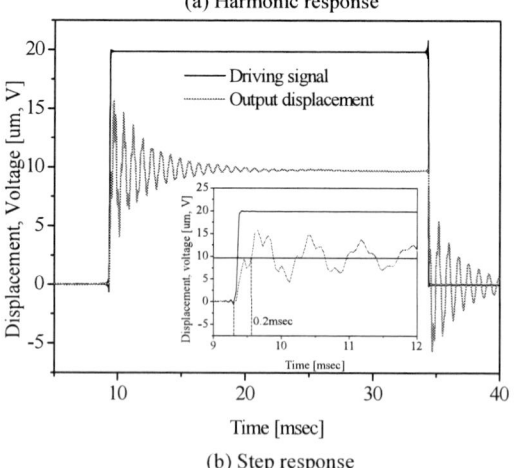

(b) Step response

Figure 8: Harmonic and Step response of the unit roll actuator

The apparatus for the pull-in actuation and response of shutter are provided in the figure 9. The spot of the LDV is focused on the roll tip position in this case. The measurable displacement was limited by the FOV of the LDV. The zero-displacement region reasons open or closed state, and bidirectional peaks are a transient period between open and closed states. The actuator well follows the driving signal (30 V-square wave) until 500 Hz. Non-zero regions in the 1000 Hz graph means that the actuator cannot follow the driving signal any more. Therefore the maximum shutter speed is estimated 1/500s in this case. The squeeze damping in the pull-in actuation is the most possible reason to reduce the speed over the step response in figure 8.

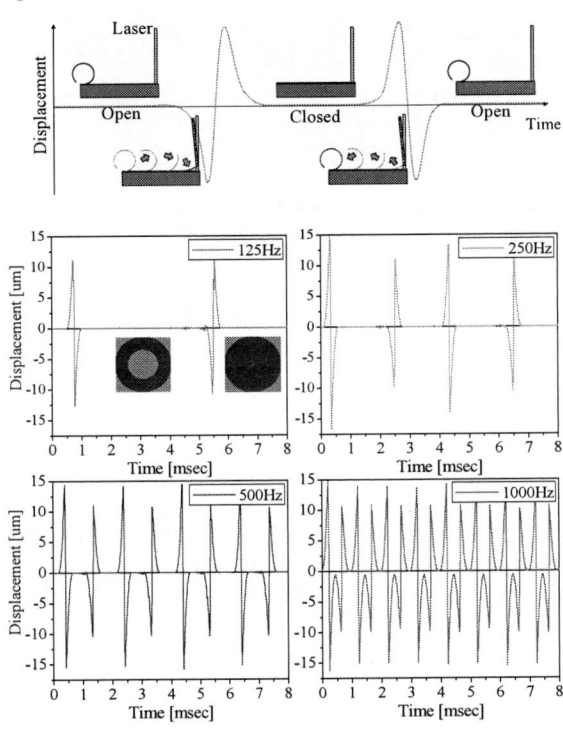

Figure 9: Pull-in actuation of the shutter with respect to the driving frequency

CONCLUSION

A novel wafer-level micro shutter utilizing the stress induced roll actuator was proposed and discussed. Ultra thin, 0.8mm-thick shutter was implemented using the surface micromachining. The driving voltage and the shutter speed were measured to 30 V and 1/500 sec, respectively. The improvement of the shutter speed and uniformity should be performed as a future work. This study proposed a distinctive approach to make a very thin, wafer-level mechanical shutter and is applicable to reflective/transmissive displays, optical sensors, and reconfigurable passives and so on.

REFERENCES

[1] S. Duffy et al., Microwave and Wireless comp. Lett., Vol.11, No. 3, pp. 106-108.

[2] C. L. Chua et al., Proceedings of SPIE, Vol. 4981(2003), pp. 150-155

[3] M. Pizzi. et al., Optical MEMS,2003 IEEE/LEOS Int. Conf., pp. 173-174.

[4] E. M. Chow et al., Electronic Components and Technology Conference; 2003 May 27-30, pp. 1714-1717.

MICROFABRICATED FLIPPING GLASS DISC FOR STEREO IMAGING IN ENDOSCOPIC VISUAL INSPECTION

Wook Choi[1], Minoo Akbarian[2], Vladimir Rubtsov[2] and Chang-Jin "CJ" Kim[1]

[1]Mechanical and Aerospace Engineering Department, University of California, Los Angeles (UCLA)
[2]Intelligent Optical Systems, Inc. (IOS), Torrance, CA, U.S.A.

ABSTRACT

In many fields, including medical and manufacturing, endoscopy is a powerful tool for remote visual inspection. However, images obtained by an endoscope are mostly planar and provide no accurate size or distance information about the object of interest. Use of stereo images to solve these problems greatly increases the performance of such inspection tools due to its three dimensional measurability. In this study, a method to obtain stereo images using a single "flipping" glass disc device for endoscopic applications, and its realization into a miniaturized device are presented.

1. INTRODUCTION

Endoscopy is a powerful tool for remote visual inspection in difficult-to-access areas and widely used in industrial and medical fields. It is mainly used for non-destructive inspection of machinery or minimally invasive diagnosis on body organs. However, most of such endoscopes, e.g., endoscopic crack inspection tools for quality assurance in industry, are able to obtain only planar images, with which size and distance information of objects of interest cannot be read easily. To overcome such limitations and improve the overall optical inspection performances of the endoscopic tools, stereo imaging methods using single lens [1]-[4] and multiple lens systems [5]-[7] have been introduced with more accurate evaluation capabilities. Such stereo images are used for precise measurement and three dimensional reconstruction of the object of interest using various triangulation techniques, which is never possible with the planar images.

In this study, a stereo imaging method using a single electrostatically "flipping" transparent plate is introduced for endoscopic applications. Design, fabrication, and operation of the microdevice that implements the imaging method are reported.

2. PRINCIPLE

The proposed stereo image generation utilizes refraction of light when the light passes through two different media, explained by Snell's law. Figure 1 illustrates the stereo image generation by a single transparent flipping plate without using multiple lenses, as in [3]. As the transparent plate tilts counterclockwise or

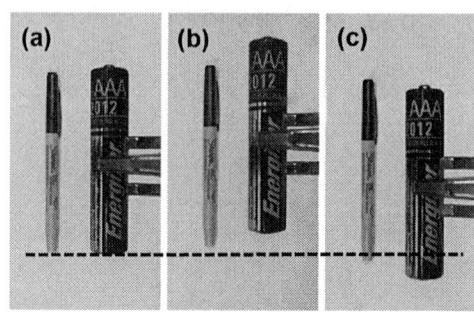

Figure 2: Demonstration of how to measure the distance from image shift. A 14 cm-long pen and a 4.5 cm-long AAA battery are seen through a 2.5 cm-thick Plexiglas® plate. (a) When the plate is still. (b) When the plate tilts forward, and (c) when the plate tilts backward. As the transparent plate tilts, the image of the closer object shifts more.

clockwise, the image of an object shifts up or down, respectively, due to the refraction of light when passing through the tilted plate. The amount of the image shift is determined by the index of refraction, tilting angle and thickness of the transparent plate, and can be calculated by

$$\Delta L = t \sin \theta \left(1 - \sqrt{\frac{1 - \sin^2 \theta}{\left(\frac{n_2}{n_1}\right)^2 - \sin^2 \theta}} \right) \quad (1)$$

where t is the thickness of the tilting plate, θ is the tilting angle, and n_1 and n_2 are refractive indexes of surrounding media and the tilting plate material, respectively.

Figure 2 demonstrates how the distance between the object and the observer can be evaluated from image shift by plate tilting. A 14 cm-long pen and a 4.5 cm-long AAA battery at different distances are seen through a 2.5 cm-thick Plexiglas® plate. Although smaller, the battery appears the same in size as the pen because it is closer to the observer, illustrating that size and distance cannot be read from the planar image. Conventional endoscopic devices using a single camera setup can only generate planar images in which operators sometimes cannot evaluate the right size or distance of the object of interest. When the plate tilts forward (Figure 2b) and backward (Figure 2c), the images of the objects shift up and down, respectively, as a result. What can be used at this point to evaluate the distance information is the fact that each object image shows a different shifting amount with respect to the distances. When the plate tilts, an object closer to the observer generates a larger image shift than an object further away due to the difference in distance; the

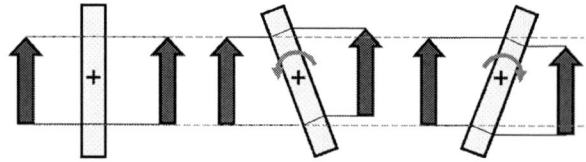

Figure 1: Image shift achieved by plate tilting. Due to the refraction of light, the image of an object shifts up or down by tilting the transparent plate.

978-1-4244-2977-6/09 $25.00 © 2009 IEEE

battery closer to the observer shifts more than the pen in the back. By using this difference in image shift, the distance and size of each object can be evaluated, once the image shift at a certain working distance is calibrated.

3. DESIGN AND FABRICATION

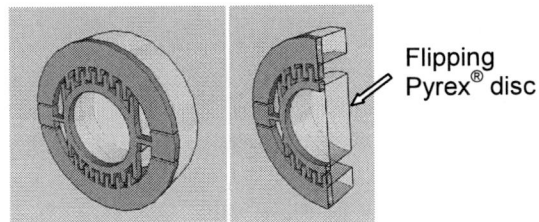

Figure 3: Schematic of the flipping disc device. (Left) Perspective view. (Right) Cut-out view showing the glass disc in the middle and the ring frame.

Device design

Figure 3 is a schematic drawing of the microfabricated flipping disc device (not to scale). A 420 µm-thick transparent disc 1300 µm in diameter is attached to a 40 µm-thick silicon layer with a 1000 µm-diameter opening in the center to expose the glass, and suspended by two 10 × 40 × 800 µm torsion bridges. The disc is electrostatically actuated (i.e., flipped) by two interdigitated sets of silicon comb drives. While photosensitive polymers (e.g., SU-8) are widely used to make lenses in optical MEMS field [8]-[10], Pyrex® is chosen as the transparent disc material due to its much higher resistance to degradation over time and discoloration during high temperature fabrication processes than the photosensitive polymers. To achieve the required shapes of the transparent disc and the surrounding structures without using expensive and time consuming glass machining methods, glass molding [11] is used at an elevated temperature (> 800ºC), reflowing Pyrex® into a patterned silicon mold.

Fabrication process

The process flow for the flipping disc device is shown in Figure 4. A silicon-on-insulator (SOI) wafer with heavily doped device layer is used as the starting material (Figure 4a). The wafer has 100 µm-thick device layer, 1 µm-thick buried oxide, and 500 µm-thick handling layer. The handling layer of the wafer with polished surface is patterned and etched down to the oxide layer using deep reactive ion etching (DRIE) to make a mold for glass filling, and the exposed oxide is removed (Figure 4b). The patterned SOI wafer is anodically bonded with a Pyrex® wafer in vacuum (Figure 4c).

The bonded wafers are thermally annealed at 850°C (above the glass transition temperature of the Pyrex®) so that the Pyrex® flows to fill in the Si mold (Figure 4d) [11], followed by additional anodic bonding to ensure the strong bond between the filled Pyrex® and the Si device layer. Because of the vacuum state inside the cavities, the Si mold is filled by melted Pyrex® using the atmospheric process pressure in the heating chamber without any external force applied during the molding process. Unfilled Pyrex® is then removed by chemical mechanical polishing (CMP) revealing the Si / Pyrex® surface (Figure 4e). The surface is additionally polished for smoothness to ensure

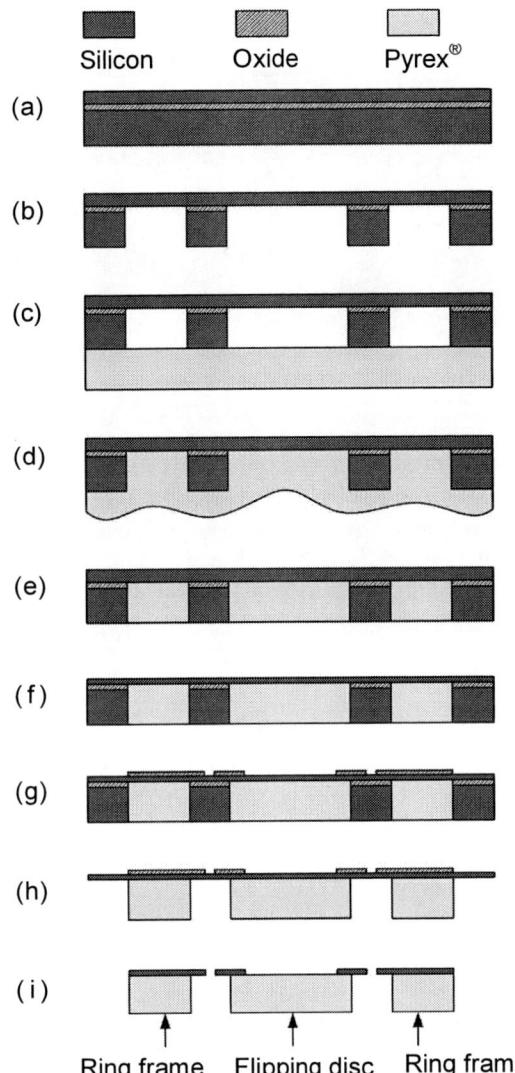

Figure 4: Process flow for tilting glass plate device; (a) SOI wafer as starting material, (b) device layer etch and exposed oxide removal, (c) anodic bonding of Pyrex® wafer to patterned SOI wafer, (d) glass molding into the Si mold followed by 2nd anodic bonding, (e) glass removal and polishing by CMP, (f) topside Si layer thinning by DRIE, (g) oxide deposition on top and patterning, (h) backside Si removal, and (i) comb patterning by DRIE, releasing the device at the same time, followed by oxide mask removal

transparency of the glass structures. The device layer is then thinned down to 40 µm using DRIE (Figure 4f) followed by oxide masking layer deposition and patterning on the thinned Si surface (Figure 4g). After the Si mold in the handling layer is removed (Figure 4h), the device layer is patterned using DRIE to form comb drive structures and torsion springs, releasing the device at the same time. The fabrication is completed by removing the oxide mask on the device layer (Figure 4i).

Figure 5 shows the fabricated flipping disc device, showing the 1.3 mm-diameter Pyrex® disc suspended by two torsion springs. Instead of using additional metal wires, the patterned silicon device layer of the SOI wafer is used as electrodes for device operations.

978-1-4244-2977-6/09 $25.00 © 2009 IEEE

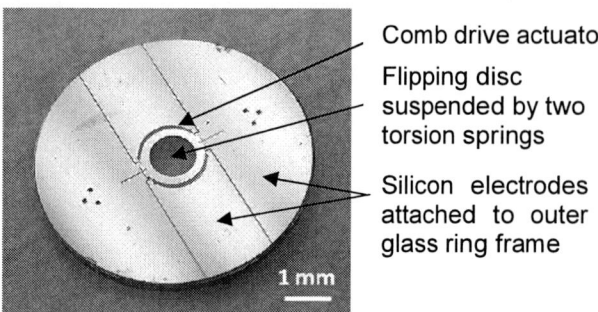

Figure 5: Fabricated flipping disc device.

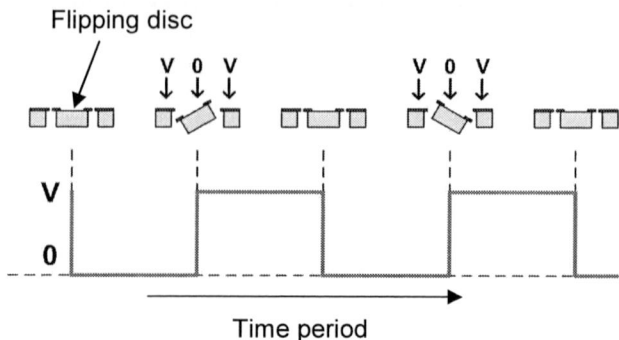

Figure 6: Cross section of the flipping disc (Top), and the corresponding driving signal (Bottom). A rectangular wave signal with a duty cycle of 50% at twice the flipping disc's mechanical resonant frequency is applied to both comb drives.

4. DEVICE TEST AND RESULT

The suspended Pyrex® disc is driven by two sets of interdigitated comb drives at resonant frequency. The actuation is similar to [12]-[14] and occurs through fringe-field electrostatic attraction vertical to the silicon layer. A rectangular pulse with a duty cycle of 50% at twice the mechanical resonant frequency of the flipping disc is used for the tilting. Figure 6 shows the cross section of the flipping disc and the corresponding driving signal. Initial excitation for device operation is achieved either by having asymmetric disc shape for small initial deflection [13] or by utilizing fabrication induced asymmetry [12]. Figure 7 is the experimental result using the fabricated disc device with 20 V applied to the comb drives in an open lab environment (1 atm). Driving signal is swept up and down by increments of 1 Hz to find the optimum driving frequencies. The maximum tilting is observed when the signal frequency is swept down to 994 Hz with 7° of total tilting, which is equivalent to the disc's mechanical resonant frequency of 497 Hz. With the disc thickness of 420 μm, the image shift with 7° tilting is measured to be 16.2 μm by image analysis.

A schematic of the experiment setup for stereo image generation is shown in Figure 8. The fabricated Pyrex® disc device flips back and forth by electrostatic actuation, and the resulting shift of the images are observed through the disc by a camera connected to a microscope and illuminating lights. The camera records image shifts at 300 frames per second to obtain overlapped images showing the image shifting distance. Figure 9 is the computer screen image of a cross-shaped pattern captured by the camera. The pattern is 180 μm (wide) × 180 μm (high) in size, and 3 mm in front of the flipping disc. From the grabbed image, the shifting distance can be read on the screen (as in Figure 9) and defined as "D". The relative distance D on the screen represents the already measured real image shifting distance of 16.2 μm. That is, the distance D on the screen can be used as a reference for objects at 3 mm distance. If the same disc and test setup is used, various distances of different objects can be estimated by comparing the relative distance on the screen with the distance D. If an observed image shift on the screen is larger or smaller than this value D, it means that the object lies closer or farther to the flipping disc than the reference distance of 3 mm. In this way, size information can also be obtained without knowing the real size of viewed objects. Because the distance D on the screen is already known to be 16.2 μm in real dimensions,

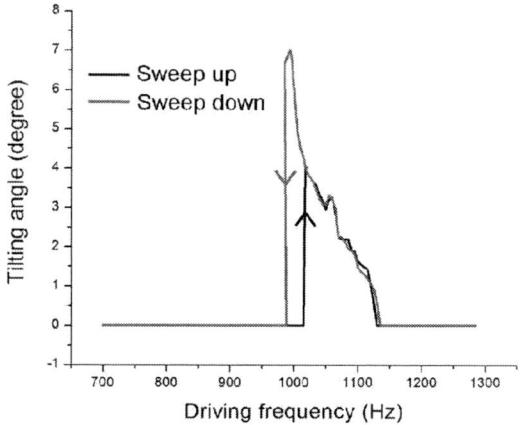

Figure 7: Dynamic characterization of the flipping disc by frequency sweeping using 20 V_{AC} of rectangular waveform with 50% duty cycle. A total tilting angle of 7° has been obtained at 994 Hz in 1 atm air.

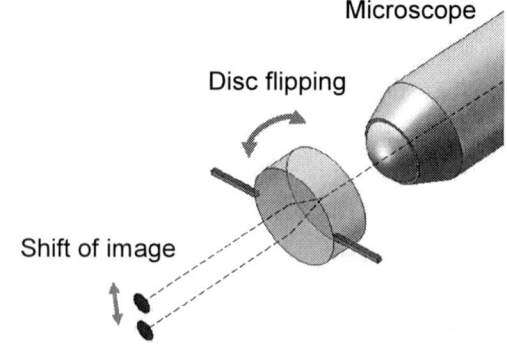

Figure 8: Schematic of image shift experiment using the flipping disc.

the object size can be calculated by comparing the size of the objects on the screen with D. For example, the height of the cross pattern on the screen (as in Figure 9) is approximately 11 times D, which means that the height of the cross pattern is in fact 180 μm.

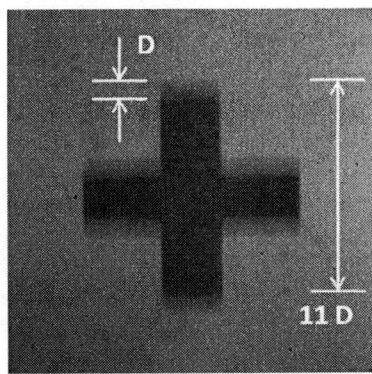

Figure 9: Image shift captured through the flipping disc. A 180 μm (wide) × 180 μm (high) cross pattern at 3 mm distance makes a vertical image shift of "D" on the screen which is equivalent to 16.2 μm of measured image shift. The size of the cross pattern on the screen is about 11×"D", equivalent to 180 μm of real pattern height.

5. CONCLUSION

Use of stereo images greatly increases the performance of endoscopic devices by its three-dimensional measurement capabilities. This paper presented a stereo imaging method using a single transparent flipping disc miniaturized and operated by MEMS technologies. The developed transparent flipping disc device has confirmed the feasibility of being used in front of single-lens endoscopic devices to generate such stereo images for non-contact dimensional measurements in difficult-to-access areas, minimizing the complexity of the overall device setup.

6. ACKNOWLEDGEMENT

The authors would like to thank Dr. Igor Ternovskiy at Intelligent Optical Systems (IOS) for his help measuring and confirming the image shift using his image analysis software. The work was partially supported by the National Science Foundation (NSF) Small Business Innovative Research (SBIR) program.

7. REFERENCES

[1] W. Teoh and X. D. Zhang, "An inexpensive stereoscopic vision system for robots," *Proc. Int. Conf. Robotics and Automation*, vol. 1, pp. 186-189, 1984.

[2] C. Gao and N. Ahuja, "Single camera stereo using planar parallel plate," *Proc. Int. Conf. Pattern Recognition*, vol. 4, pp. 108-111, 2004.

[3] Y. Nishimoto and Y. Shirai, "A feature-based stereo model using small disparities," *Proc. Computer Vision and Pattern Recognition*, pp. 192-196, 1987.

[4] A. Goshtasby and W. A. Gruver, "Design of a single-lens stereo camera system," *Pattern Recognition*, vol. 26, pp. 923–937, 1993.

[5] Example of stereo inspection systems found and accessed on 10 Oct 2008 at http://rvi.olympus-global .com/en/special/iplexfx.

[6] A. F. Durrani and G. M. Preminger, "Three-dimensional video imaging for endoscopic surgery," *Comput. Biol. Med.*, vol. 25, pp. 237-247, 1995.

[7] H. Becker, A. Melzer, M. O. Schurr, and G. Buess, "3-D video techniques in endoscopic surgery," *Endosc. Surg. Allied Technol.*, vol. 1, pp. 40-46, 1993.

[8] S. Moon, N. Lee, and S. Kang, "Fabrication of microlens array using micro-compression molding with electroformed mold insert," *J. Micromech. Microeng.*, vol. 13, pp. 98–103, 2003

[9] A. Tuantranont, V. M. Bright, J. Zhang, W. Zhang, J. A. Neff, and Y. C. Lee, "Optical beam steering using MEMS-controllable microlens array," *Sensors and Actuators A*, vol. 91, pp. 363-372, 2001.

[10] R. Yang and W. Wang, "Out-of-plane polymer refractive microlens fabricated based direct lithography of SU-8," *Sensors and Actuators A*, vol. 113, pp. 71–77, 2004.

[11] P. Merz, H. J. Quenzer, H. Bernt, B. Wagner, and M. Zoberbier, "A novel micromachining technology for structuring borosilicate glass substrates," *Transducers*, vol. 1, pp. 258–261, 2003.

[12] H. Schenk, P. Durr, T. Haase, D. Kunze, U. Sobe, H. Lakner, and H. Kuck, "Large deflection micromechanical scanning mirrors for linear scans and pattern generation," *J. Select. Topics Quantum Electron.*, vol. 6, pp. 715–722, 2000.

[13] K. N. Lee, Y. H. Jang, H. Kim, Y. S. Lee, and Y. K. Kim, "Monolithic fabrication of optical benches and scanning mirror using silicon micromachining," *J. Micromech. Microeng.*, vol. 15, pp. 747–755, 2005.

[14] C. Lee, "Design and fabrication of epitaxial silicon micromirror devices," *Sensors and Actuators A*, vol. 115, pp. 581–90, 2004.

RESONANT FREQUENCY TUNING OF TORSIONAL MICROSCANNER BY MECHANICAL RESTRICTION USING MEMS ACTUATOR

Jae-Ik Lee, Sunwoo Park, Youngkee Eun, Bongwon Jeong and Jongbaeg Kim
School of Mechanical Engineering, Yonsei University, Seoul, Republic of Korea

ABSTRACT

We have demonstrated resonant frequency-tuning of electrostatic torsional microscanners actuated by the staggered vertical comb (SVC) sets. Mechanical restriction unit composed of thermal actuator, scissor mechanism and shaft-holder is used for continuous and reversible resonant frequency tuning. The microscanner and tuning mechanism are fabricated on single crystal silicon of double Silicon-On-Insulator (DSOI) wafer in order to make vertically self-aligned comb sets and electrically insulate the electrostatic microscanner from the thermally actuated tuning unit to prevent charge leakage. The microscanner is actuated at a resonant frequency of 1.698 kHz under driving voltages of $5V_{ac}$ and $10V_{dc}$ without frequency tuning. As the stiffness of a torsional spring is modified gradually through the shaft-holding flexure operated by thermal actuator, the resonant frequency of the torsional microscanner is shifted up to 1.880 kHz showing the maximum tuning ratio of 10.7%.

INTRODUCTION

Torsional micromirror scanners have been utilized in various applications such as telecommunication [1], medical and biological metrology [2], display and image processing [3]. However, due to the inevitable dimensional errors existing in the microfabrication processes and the changes of operational conditions such as pressure or temperature, it has been an important issue to achieve the desired resonant frequency of the microscanners especially for those operated at resonant mode. To tune the deviated resonant frequency of the comb drive MEMS resonator to the intended value, previous works include increasing the vibrating mass by the laser ablation [4] or platinum deposition using focused ion beam [5]. Stiffness adjustment is another method to tune the resonant frequency, and increasing the stiffness of the mechanical spring by localized polysilicon deposition [6] has been demonstrated. While these methods are based on post processing that are not reversible and result in permanent change of resonance, numbers of reversible frequency tuning methods have been studied utilizing the electrostatic spring effect [7, 8] and localized thermal stressing effect [9]. All the aforementioned works have been applied only to laterally driven comb actuators. On the other hand, while the vertical comb-based electrostatic torsional actuators are widely chosen in many light-scanning applications for their large scanning angles without a mechanical contact or an obstruction unlike the parallel plate type resonators [10], no frequency tuning method has been developed yet.

In this paper we present a novel approach for tuning the resonant frequency of torsionally driven vertical comb actuators by mechanically restricting the torsional spring using integrated tuning actuator. Staggered vertical comb-drive actuator is fabricated for the torsional microscanner, and the chevron thermal actuator [11] is designed right beside it as an actuator to mechanically restrict the torsional spring of microscanner. The torsional microscanner is driven by electrostatic force between two sets of comb structures at different vertical positions and the tuning unit is operated by chevron type thermal actuator that generates linear displacement with large force. The stiffness of the torsional spring changes by the stiffening effect of the holding flexure along with the friction generated between the torsional spring and the holding flexure. It enables continuous and repetitive tuning based on reversible holding and releasing mechanism. This scheme can not only compensate the deviation from the designed resonant frequency but also be applied to the resonator popularly used in sensing, signal filtering and microscanning in order to achieve high sensitivity, selective band pass filter characteristic and adjustable scanning speed, respectively.

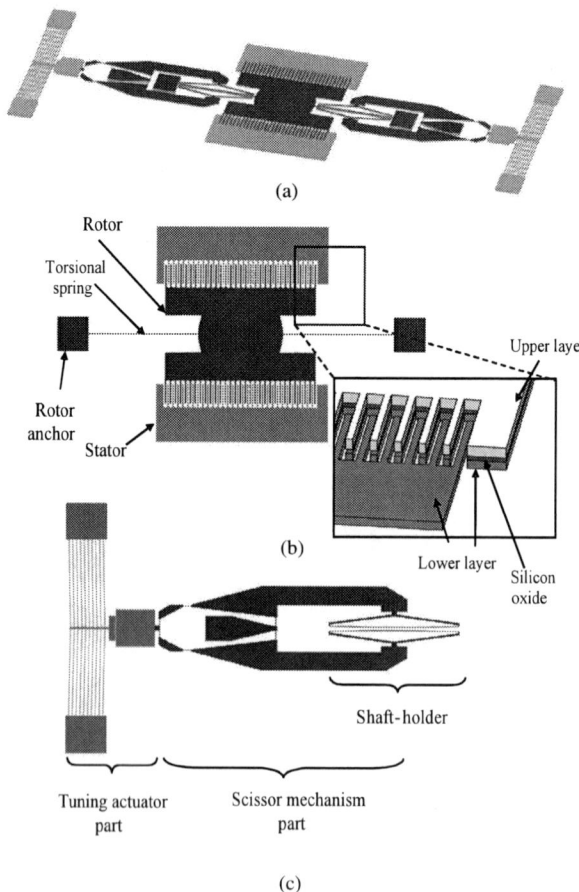

Figure 1: Schematic view of microscanner and resonant frequency tuning unit. (a) Overview of torsional microscanner integrated with frequency tuning unit. (b) Vertical comb-based torsional microscanner. (c) The frequency tuning unit.

978-1-4244-2977-6/09 $25.00 © 2009 IEEE

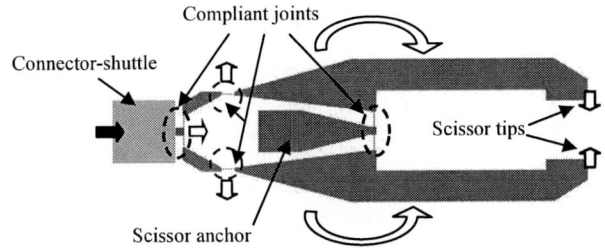

Figure 2: Motion of scissor mechanism with respect to the linear motion from the chevron thermal actuator.

STRUCTURE AND PRINCIPLE

The schematic view of vertical comb-based torsional microscanner and frequency tuning unit are shown in Figure 1. The torsional microscanner actuated by SVC sets is fabricated on two different levels of single crystalline silicon layers of DSOI with fixed combs on upper layer and moving combs and mirror on lower layer. The usage of silicon double layers allows self-alignment between fixed and moving combs with simple fabrication [12]. Bias voltage between fixed and moving combs induces electrostatic force, rotating the mirror with respect to torsional springs.

The frequency tuning unit is composed of three parts: actuator, scissor mechanism and shaft-holder. To generate a linear displacement and large force to operate the scissor mechanism, chevron thermal actuator is used. The scissor mechanism with three pairs of compliant hinges is designed to amplify the displacement generated by thermal actuator and to transform one-directional linear motion of chevron thermal actuator to symmetrical angular motion of shaft holders. The motion amplification is necessary for wide rage frequency tuning, since the thermal actuators typically generate large force and small displacement. As depicted in Figure 2, when the left end is pushed rightward by the thermal actuator, the scissor mechanism is closed at the right end approaching the shaft holders attached to scissor tips to torsional spring of micromirror scanner. Another critical aspect to consider for the design of the scissor mechanism is the insulation of electrical current which may flow from chevron thermal actuator to electrostatically driven torsional actuator when the shaft holder contacts with torsional spring. Since the scissor mechanism and the shaft holders are driven by thermal actuator, the electrical charges in electrostatic microscanner may leak when the shaft holders make contact with torsional spring of the electrostatic actuator resulting in malfunction of the scanner. In order to prevent this charge leakage, two isolated structural silicon layers of DSOI wafer are employed as shown in Figure 3. The chevron thermal actuator is defined on the upper layer of DSOI and the connector-shuttle is composed of both upper and lower layer plates that are mechanically connected by buried oxide, but electrically isolated. When the upper layer chevron thermal actuator pushes the connector- shuttle, it makes contact only with upper layer of the connector-shuttle. The lower layer of connector-shuttle is also pushed, actuating the scissor mechanism and closing the scissor tip, but the scissor mechanism defined on the lower layer is electrically isolated from the thermal actuator. Then the shaft holders attached to the scissor tips approach the torsional spring of microscanner and come into contact

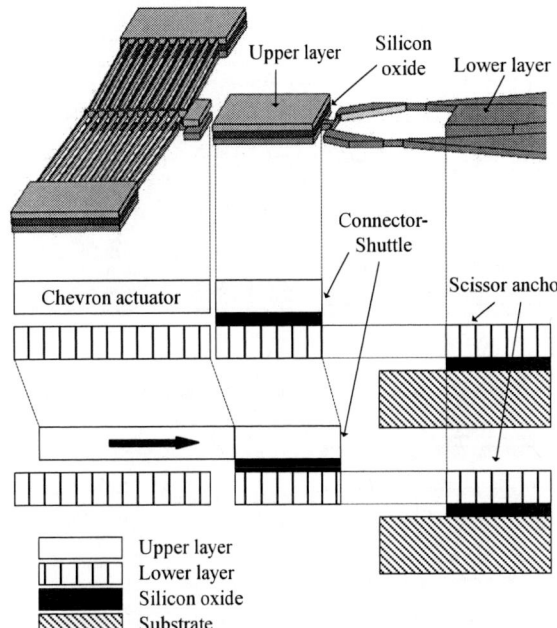

Figure 3: Electrical isolation design to avoid short circuit formation and charge leakage caused by mechanical contact.

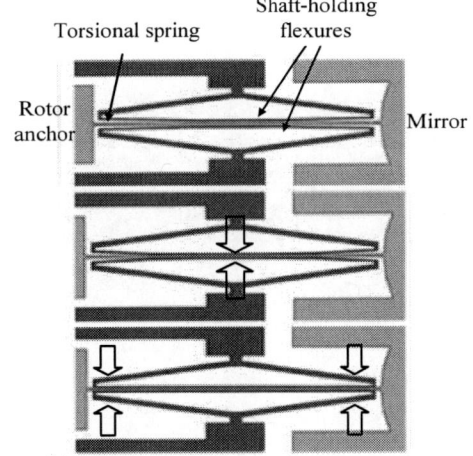

Figure 4: Schematic view of shaft-holding mechanism.

with it as shown in Figure 4. Thus, the contact of shaft holders to torsional springs does not cause any charge leakage from the electrostatic actuator. The torsional spring, mirror, moveable combs and rotor anchor are all defined on the lower layer and connected. Rotor anchor is electrically grounded for electrostatic actuation, while the upper layer fixed combs are biased.

The shaft-holding flexure that directly touches the torsional spring of the microscanner has a bow-like curved shape, and it elastically deforms when the constraining process begins, expanding the contact area as the stroke of thermal actuator increases. The thickness of shaft-holding flexure should be taken into consideration since a thin flexure would result in less change of the stiffness of torsional spring, whereas a thick flexure would require more force to carry out elastic deformation. In our design, finite element analysis (FEA) is conducted to obtain an optimized value for the thickness of the shaft-holding

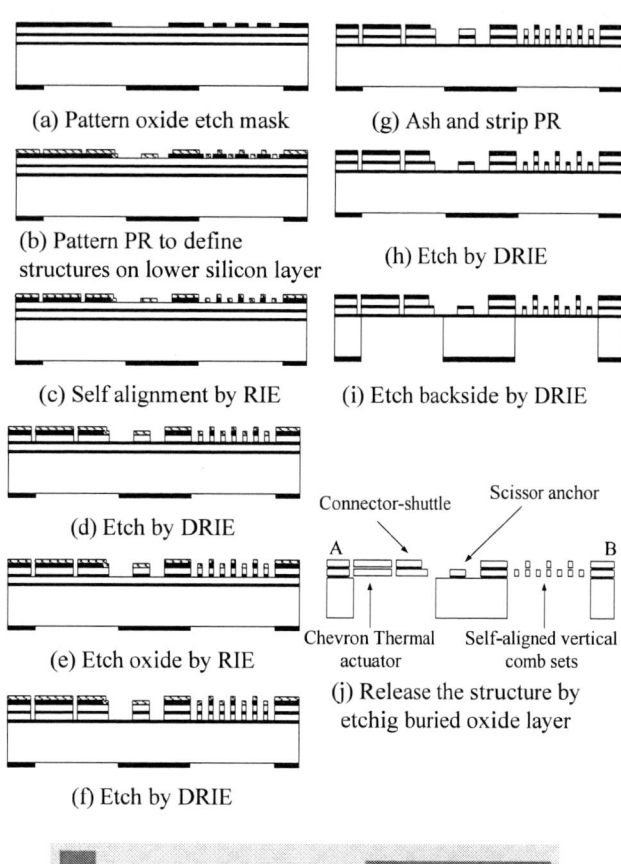

(a) Pattern oxide etch mask

(g) Ash and strip PR

(b) Pattern PR to define structures on lower silicon layer

(h) Etch by DRIE

(c) Self alignment by RIE

(i) Etch backside by DRIE

(d) Etch by DRIE

(e) Etch oxide by RIE

(j) Release the structure by etchig buried oxide layer

(f) Etch by DRIE

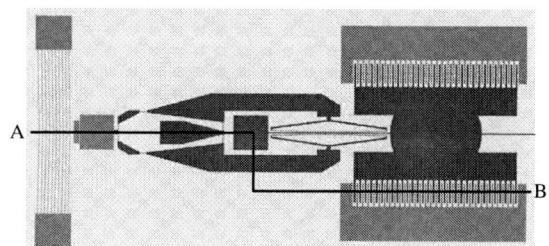

Figure 5: Fabrication process using DSOI.

flexure. As the contact area between shaft-holding flexure and torsional spring of scanner increases, the stiffness of torsional spring also increases. The continuous resonant frequency tuning is achieved by gradually increasing or decreasing this contact area using tuning actuator.

FABRICATION

The device is fabricated on double Silicon-On-Insulator (DSOI) wafer with two silicon device layers on the 425 μm thick base substrate, each of which is separated by two layers of buried oxide. Each silicon device layer is 20 μm thick, and each buried oxide layer is 1 μm thick. Figure 5 shows the fabrication process for the vertical comb-based torsional microscanner and frequency tuning unit. Firstly, silicon dioxide on top of upper silicon device layer is patterned to be used as an etch mask for silicon etch as in (a). Second photolithography is performed to pattern PR as in (b), and this pattern will be transferred to lower layer of DSOI later after the following silicon etch steps. Between step (a) and (b), rough alignment is acceptable since the following oxide etch will form self-alignment between oxide and PR, as shown in (c). During the following steps, (d) ~ (g), both upper and lower silicon

layers and buried oxide are etched by deep reactive ion etching (DRIE) and reactive ion etching (RIE). After unnecessary upper layer silicon are removed by DRIE in step (h), backside holes are formed as shown in (i) by etching the substrate to give space for the torsional motion of microscanner. Finally, buried oxide and etch mask oxide layers are etched in HF as in (j) such that the buried oxides under the large area silicon structures are remained for anchoring and mechanical connection between upper and lower layers, while the buried oxides under narrow silicon structures and etch mask oxides are completely removed.

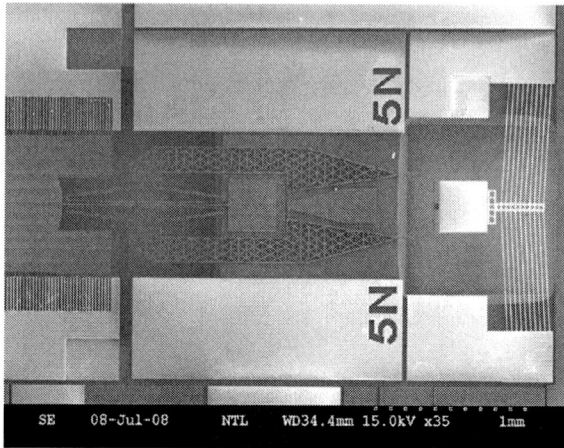

Figure 6: SEM image of fabricated microscanner and frequency tuning unit.

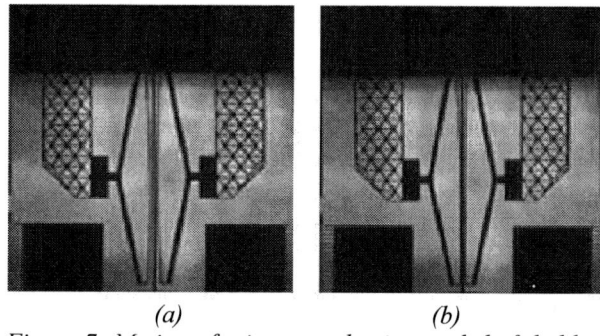

(a)   (b)

Figure 7: Motion of scissor mechanism and shaft-holding flexure before (a) and after (b) the actuation of tuning actuator.

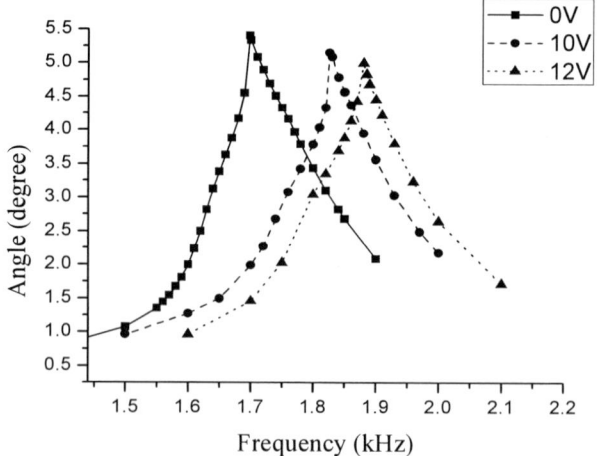

Figure 8: Frequency response change for different tuning bias on tuning actuator.

978-1-4244-2977-6/09 $25.00 © 2009 IEEE   166

Table 1: Resonant frequency tuning ratio for different tuning bias.

Tuning voltage (V)	Resonant frequency (kHz)	Tuning ratio (%)
0V	1.698	-
8V	1.749	3.03
10V	1.826	7.05
12V	1.880	10.7

EXPERIMENTAL RESULT

Figure 6 shows a scanning electron microscope (SEM) image of the fabricated device. The darker structures in the image are on the lower silicon layer, and the brighter structures are on upper layer. Before the chevron thermal actuator is operated, the shaft-holding flexure and the torsional spring are separated as shown in Figure 7 (a). As the driving bias on tuning actuator increases, the scissor tips are closed approaching the shaft-holding flexures to the torsional spring as compared in Figure 7 (b). Once the shaft-holding flexures make contact with torsional spring, they mechanically restrict the rotational motion of torsional spring increasing the torsional stiffness. It is observed as the operating voltage in tuning actuator increases, the contact area is also enlarged augmenting the torsional stiffness.

The shaft-holding flexures started to make contact with the torsional spring at 7V and are at full contact when 12V is applied. Between 7V and 12V, the mechanical restriction process is completely reversible, and continuous frequency tuning up and down is achieved. The experimental measurement results of resonant frequency tuning are presented in Table 1 while the driving voltages of $5V_{ac}$ and $10V_{dc}$ are biased on the electrostatic microscanner. When the tuning DC voltage increases to 8V, 10V and 12V, the resonant frequency is shifted to 1.749 kHz, 1.826 kHz and 1.880 kHz, respectively from the untuned resonant frequency of 1.698 kHz. Figure 8 shows how the frequency response of microscanner is changed from the untuned state to differently tuned states. As the resonant frequency is tuned up, due to the stiffened torsional spring, the measured optical scan angles have been decreased to 5.14 ° at 1.826 kHz and 4.99 ° at 1.880 kHz, respectively from the initial rotational angle of 5.40 ° at 1.698 kHz, without altering the driving voltage of electrostatic microscanner. The tuning ratios are 3.03%, 7.05% and 10.7%, at the tuning bias of 8V, 10V and 12V, respectively.

CONCLUSION

In this research, resonant frequency tuning of the electrostatic torsional microscanner has been demonstrated by mechanically restricting the motion of torsional spring using MEMS actuator. The continuous and reversible tuning is performed based on the holding and releasing motion of shaft-holding flexures that causes the stiffness modification of the torsional spring. The frequency tuning units are integrated to the microscanner, and there is no necessity of additional post fabrication process for frequency-tuning. When the tuning voltage applied to the chevron type thermal actuator in tuning unit increases to 12V, the resonant frequency is shifted up to 1.880kHz, from the untuned resonant frequency of 1.698kHz showing the maximum tuning ratio of 10.7%. Future work will focus on optimizing the structural design in order to improve the maximum tuning ratio.

ACKNOWLEDGEMENT

This work was supported by Seoul R&BD Program (11032).

REFERENCES

[1] H. Toshiyoshi and H. Fujita, "Electrostatic micro torsion mirrors for an optical switch matrix," *J. Microelectrom- echanical Systems,* vol. 5, pp. 231-237, 1996.

[2] W. Jung, D. McCormick, J. Zhang, L. Wang, N. Tien and Z. Chen, "Three-dimensional endoscopic optical coherence tomography by use of a two-axis microelectromechanical scanning mirror," *Applied Physics Letters,* vol. 88, pp. 163901, 2006.

[3] Y. Ko, J. Cho, Y. Mun, H. Jeong, W. Choi, J. Kim, Y. Park, J. Yoo and J. Lee, "Eye-type scanning mirror with dual vertical combs for laser display," *Sensors and Actuators A: Physical,* vol. 126, pp. 218-226, 2006.

[4] M. Chiao and L. Lin, "Post-packaging frequency tuning of microresonators by pulsed laser deposition," *J. Micromechanics and Microengineering,* vol. 14, pp. 1742-1747, 2004.

[5] S. Enderling, J. Hedley, L. Jiang, R. Cheung, C. Zorman, M. Mehregany and A. Walton, "Characterization of frequency tuning using focused ion beam platinum deposition," *J. Micromechanics and Microengineering,* vol. 17, pp. 213-219, 2007.

[6] D. Joachim and L. Lin, "Characterization of selective polysilicon deposition for MEMS resonator tuning," *J. Microelectromechanical Systems,* vol. 12, pp. 193-200, 2003.

[7] K. Lee and Y. Cho, "A triangular electrostatic comb array for micromechanical resonant frequency tuning," *Sensors and Actuators: A. Physical,* vol. 70, pp. 112-117, 1998.

[8] K. Lee, L. Lin and Y. Cho, "A closed-form approach for frequency tunable comb resonators with curved finger contour," *Sensors & Actuators: A. Physical,* vol. 141, pp. 523-529, 2008.

[9] T. Remtema and L. Lin, "Active frequency tuning for micro resonators by localized thermal stressing effects," *Sensors and Actuators: A. Physical,* vol. 91, pp. 326-332, 2001.

[10] K. Petersen, "Silicon torsional scanning mirror," *IBM J. Res. Develop,* vol. 24, pp. 631-637, 1980.

[11] C. D. Lott, T. W. McLain, J. N. Harb and L. L. Howell, "Modeling the thermal behavior of a surface-micromachined linear-displacement thermo-mechanical microactuator," *Sensors and Actuators A: Physical,* vol. 101, pp. 239-250, 2002.

[12] Y. Mizoguchi and M. Esashi, "Design and fabrication of a pure-rotation microscanner with self-aligned electrostatic vertical combdrives in double SOI wafer," in *Digest Tech. Papers Transducers'05 Conference,* Seoul, Korea, June 5-9, 2005, pp. 65-68.

REFLECTIVE DISPLAY USING IONIC LIQUID

Nguyen Binh-Khiem[1], Kiyoshi Matsumoto[1], and Isao Shimoyama[1]
[1] The University of Tokyo, Tokyo, JAPAN

ABSTRACT

We propose a reflective display device using an array of encapsulated micro droplets of a room temperature ionic liquid mixed with pH indicator dyes. The array is embedded in and addressed by a passive matrix. Parylene-on-liquid deposition is employed to encapsulate the liquid droplet arrays. When applying current to each droplet, the liquid mixture changes its color, providing the ability to display a pixel of graphics information. Once changed, the droplets can be tuned back to their initial color by reversing the direction of the applied electric current. The on-off switching time is less than 0.5s. The response speed of the devices is at least 2Hz.

INTRODUCTION

Recently, research on reflective display devices has been given intensive attention. Instead of emitting light as in emissive ones, reflective display devices usually have an array of pixels, which can change their appearance color according to electrical input signals. The change in appearance color can be achieved by changing the shape of a colored liquid droplet as in electrowetting-based devices [1], or by changing the distribution of small colored particles inside the pixels as in electrophoretic-based devices [2]. However, in reported devices, improving color contrast is still an unsolved issue.

Chemists have long been living in a colorful world of countless colored chemicals. Many chemicals have their colors sensitive to ambient changes. pH indicators are chemical compounds that change their colors in solutions of different pH values. This happens because pH indicator molecules can react easily with H^+ or OH^- ions existing in the solution. These reactions slightly modify the molecules so that their absorption spectra change; therefore, the compounds show different colors. pH indicators are often used in analytic chemistry to roughly determine the extent of chemical reactions.

In this paper, we propose a "chemical approach" to achieve changeable colors in reflective display devices. Our device composes of encapsulated droplets that are filled with a pH indicator solution. Each droplet has two contact electrodes. Upon applying voltage to these electrodes, the pH value on the electrode surface changes thus changes the color of the droplet solution.

The major issue arising in this approach is the use of liquid solutions, which have been traditionally aqueous ones. The volatility of aqueous solutions, their narrow electrochemical window, and the lack of suitable packaging technique are their main drawbacks.

We propose the use of room temperature ionic liquids (RTIL) [3] and Parylene-on-liquid deposition (PoLD) [4] to solve the "liquid solution issue". RTILs are known to have negligible volatilities because they possess very low vapor pressures. Thus, devices using RTILs can stably work for long periods of time without evaporation. RTILs also have considerably wide electrochemical windows that allow the devices to be operated by electrochemical methods. PoLD provides an encapsulation method for

A) Conceptual structure

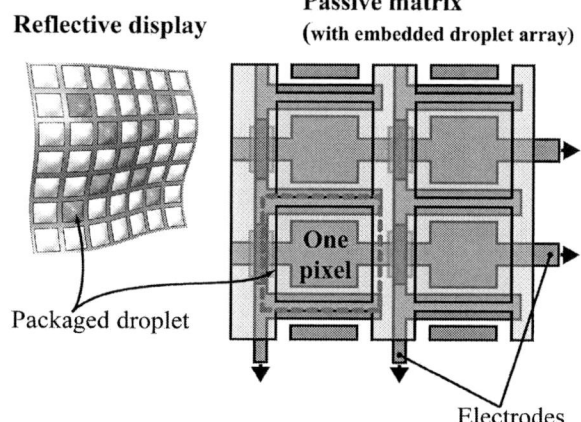

B) Droplet color change

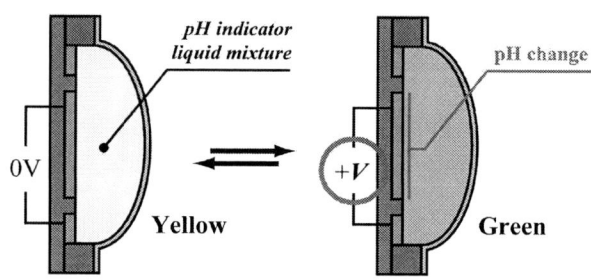

Figure 1. Conceptual structure of the reflective display. A) Conceptual structure. B) Droplet color change.

micro devices using liquids [4-8]. Droplet arrays with droplet sizes less than 100µm fabricated by PoLD have been reported [9].

DEVICE DESIGN

Our device conceptual structure is shown in Figure 1. Display units, the droplet, are filled with solutions of pH indicator bromothymol blue in 1-butyl-1-methyl-pyrrolidinium bis(trifluoromethylsulfonyl)imide, an RTIL (Figure 2A). Each droplet has two contact gold electrodes; one has a larger area than the other. The surface of this larger electrode is where the color change occurs. Upon applying voltage to these electrodes of each droplet, pH change induced on the electrode surface changes the liquid mixture color.

Our device has an array of these droplets embedded in a passive matrix structure. Each droplet can be considered as one member resistor of the passive matrix. One droplet can be addressed at a time with this passive matrix.

Bromothymol blue is a pH indicator that is yellow in acidic solution and green in slightly alkaline one. The pH threshold for color changing of bromothymol blue is between 6 and 7.6. Figure 2B shows the optical spectrum of the two colors of this liquid mixture with different pH condition.

978-1-4244-2977-6/09 $25.00 © 2009 IEEE

A) Bromothymol blue (1) and 1-butyl-1-methyl-pyrrolidinium bis(trifluoromethylsulfonyl)imide (2)

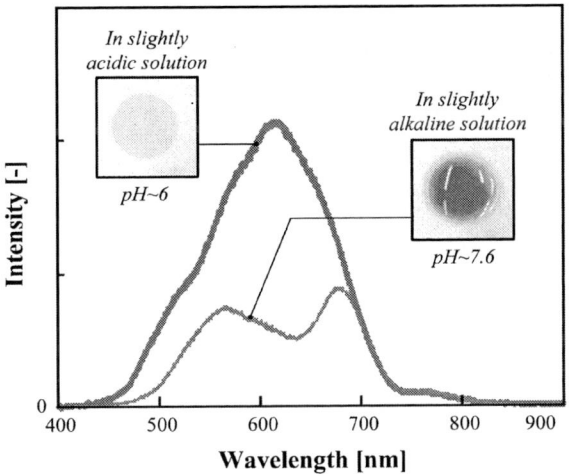

B) Optical spectrum of the liquid mixture

In slightly acidic solution

pH~6

In slightly alkaline solution

pH~7.6

Figure 2. Colors of the mixture of room temperature ionic liquid (RTIL) and pH indicator. A) pH indicator bromothymol blue and RTIL 1-butyl-1-methyl-pyrrolidinium bis (trifluoromethylsulfonyl)imide. B) Optical spectrum and colors of the liquid mixture.

FABRICATION PROCESS

Figure 3 shows the fabrication process of the device. This process creates the passive matrix and embeds the liquid droplet array into the matrix structure. In the process, firstly, we form gold electrode patterns on a base substrate (glass wafers or polyimide sheets) by metal deposition and photolithography. These electrode patterns serve as contact gold electrodes of each droplet. A layer of polyimide is spincoated, patterned by photolithography, and wet-etched to form the insulation islands between the electrodes at cross-points in the matrix. Another metal layer (copper) is deposited and patterned to form the conducting bridges on insulation islands to complete the electrode pattern of the passive matrix. Cytop, an amorphous fluorocarbon polymer (Asahi Glass Co., Ltd.), is then spincoated onto the wafer and patterned by oxygen plasma etching to form liquophobic/liquophilic array patterns for the droplet array. The RTIL is deposited onto the substrate. Due to the CYTOP liquophobic/liquophilic patterns formed on the substrate surface in the previous step, liquid deposited on this surface is automatically shaped into the array pattern by liquid surface tension. Finally, a thin Parylene film is formed on the substrate by chemical vapor deposition. The Parylene deposition completely encapsulates both the solid and the liquid parts of the device.

Figure 4 shows the fabrication results. The passive matrix structure with a droplet size of 500μm is formed on

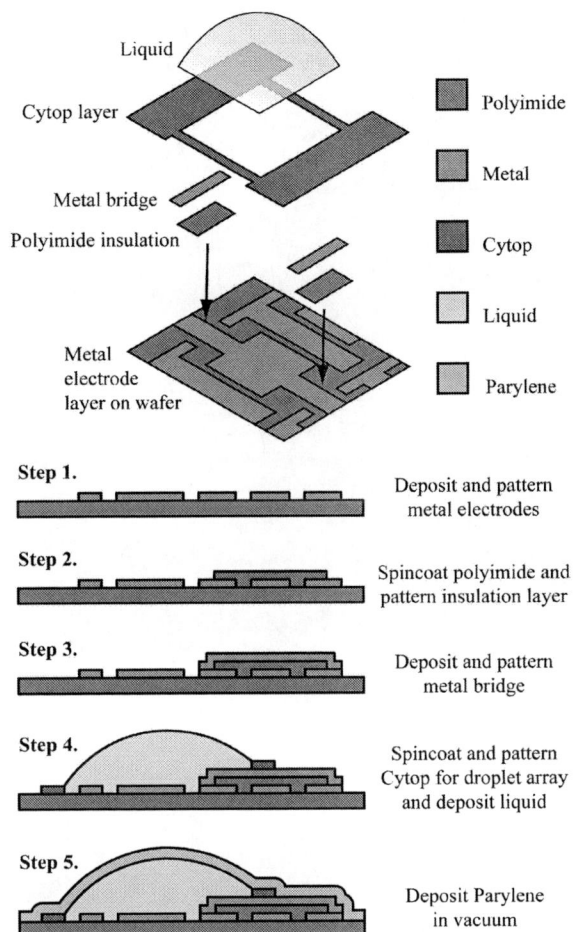

Figure 3. Fabrication process flow.

a glass substrate with good uniformity. Figure 4A(1) shows the matrix structure before liquid deposition. At every electrode cross-point in the matrix, an island of polyimide covers the underneath electrode to insulate it from the top side electrodes which is connected by a copper bridge. Figure 4A(2) shows the matrix structure after liquid deposition. Liquid droplets are precisely shaped and positioned into the designed array arrangement. Figure 4B shows devices fabricated on a solid glass wafer and on a flexible polyimide sheet. The device fabricated on a polyimide sheet is bendable. In the case of devices fabricated on polymer sheets, the device structure is made of mainly polymers and liquids, thus is suitable for large-area and low-cost devices.

EXPERIMENTS AND RESULTS

Figure 5 shows the optical spectrum of the liquid mixture when voltage is applied. Optical spectrum was taken by a USB2000 Ocean Optics Spectrometer with an Ocean Optics Tungsten Halogen Light Source. Voltages from 0V to 4V with a step of 0.2V were applied to the fabricated device. When applied voltage changed, the optical spectrum changed, which caused the liquid mixture color to change from yellow to green. The device should be operated within the electrochemical window 3V of 1-butyl-1-methyl-pyrrolidinium bis(trifluoromethylsulfonyl)imide.

Figure 6 shows the reflectivity and the sum-reflectivity

A) Passive matrix structure

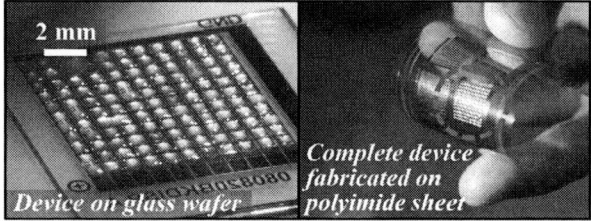

Figure 5. *Optical spectrum of the liquid mixture upon applied voltage. Voltage range: 0 – 4V, step: 0.2V.*

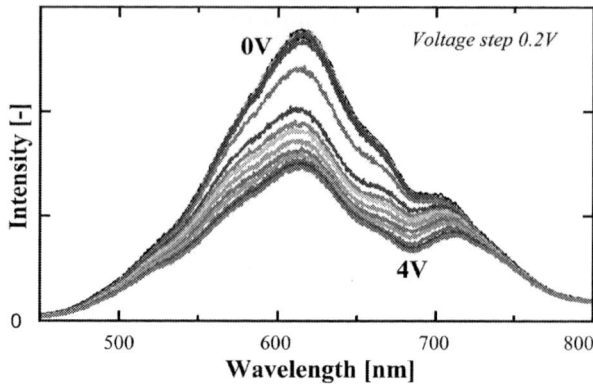

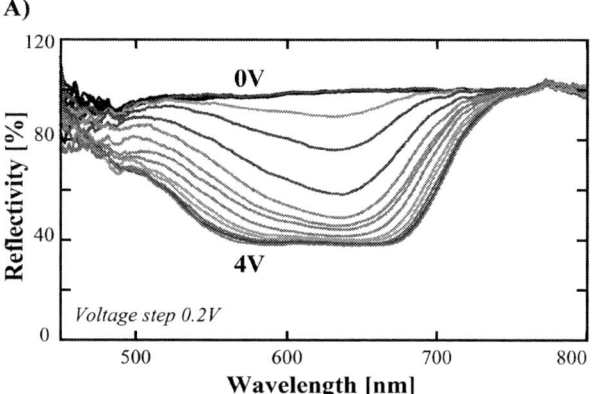

Figure 4. *Fabricated devices. A) Passive matrix structure. The matrix electrodes are completed with bridge parts, which are composed of a polyimide insulation island and a copper bridge. B) Fabricated devices. The device can be fabricated on solid substrates such as glass wafers, or flexible substrates such as polyimide sheets.*

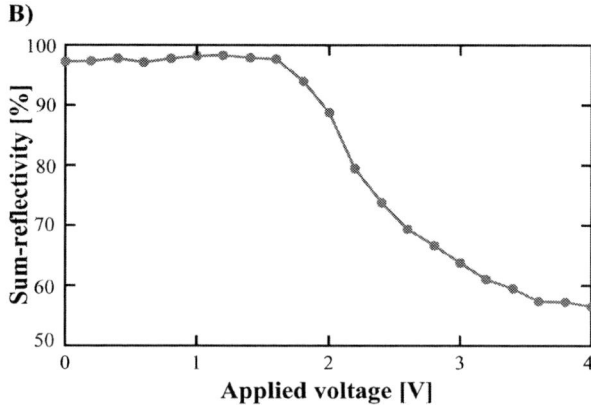

Figure 6. *A) Reflectivity and B) Sum-reflectivity (in visible range) of the device's pixels upon applied voltage. Voltage range: 0 – 4V, step: 0.2V.*

(in visible range) of the reflected light from the device, which also changed when voltages were applied. When applied voltage changed in between 0V to 4V, visible light with wavelengths from about 550nm to 700nm had the greatest reflectivity change, which decreased from 100% to less than 40%. In the whole visible range, the sum-reflectivity also decreased from 100% to about 50%. Both the change in optical spectrum and the change in reflectivity contributed to the color contrast of the display device, thus provided the device with a good color contrast (Figure 7).

Figure 7 shows two display demonstrations. When applying voltages to simple devices with patterned electrodes, color change was sufficient to display the patterns. A rectangular voltage (0 – 2.5V) was used. The on-off switching time was less than 0.5s. The response speed of the devices was at least 2Hz. The patterns were displayed in green color on a yellow background in devices using pH indicator bromothymol blue. Other colors should be similarly available due to the many colors and ways of changing color that pH indicators have [10].

Our present device can response to an input signal of at least 2Hz, but can be made faster by miniaturizing the droplet size. Our design uses mainly low-cost polymers and liquids, thus is suitable for large-area, low-cost, and flexible display devices — an application field that still has very few presently available technologies.

CONCLUSION

We proposed the use of liquid mixtures of pH indicator bromothymol blue in room temperature ionic liquid 1-butyl-1-methyl-pyrrolidinium bis(trifluoromethyl-sulfonyl)imide in a reflective display device. The device has a droplet array embedded in and addressed by a passive

978-1-4244-2977-6/09 $25.00 © 2009 IEEE

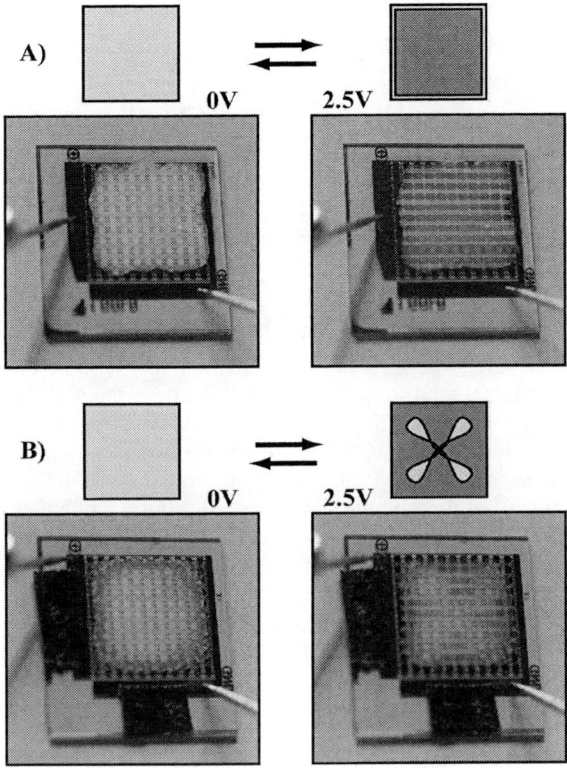

Figure 7. Display demonstrations. When applying voltages to simple devices with patterned electrodes, color change was sufficient to display the patterns. 0 – 2.5V rectangular voltage was used. The on-off switching time was less than 0.5s. The response speed of the devices was at least 2Hz.

matrix. The device is encapsulated in a thin Parylene film fabricated by Parylene-on-liquid deposition. One droplet at a time can be addressed with the passive matrix. When being applied current to, the liquid mixture changes its color to display a pixel of graphics information. The droplet color can be tuned back to its initial state by reversing the direction of the applied electric current. The on-off switching time is less than 0.5s. The response speed of the devices is at least 2Hz.

ACKNOWLEDGMENTS

The photolithography masks were made using the University of Tokyo VLSI Design and Education Center (VDEC)'s 8 inch EB writer F5112 + VD01 donated by ADVANTEST Corporation.

REFERENCES

[1] R. A. Hayes, B. J. Feenstra, "Video-speed electronic paper based on electrowetting", *Nature*, vol. 425, pp. 383-385, 2003.

[2] B. Comiskey, J. D. Albert, H. Yoshizawa, J.Jacobson, "An electrophoretic ink for all-printed reflective electronic displays", *Nature*, vol. 394, pp.253-255, 1998.

[3] F. Endres, S. Z. E. Abedin, "Air and water stable ionic liquids in physical chemistry", Phys. Chem. Chem. Phys., vol. 8, pp. 2101-2116, 2006.

[4] N. Binh-Khiem, K. Matsumoto, I. Shimoyama, "Polymer thin film deposited on liquid for varifocal encapsulated liquid lenses", *Appl. Phys. Lett.*, vol. 93, pp. 124101-3, 2008.

[5] T. Kan, N. Binh-Khiem, K. Matsumoto, I. Shimoyama, "Tunable SPR coupler by flexible polymer grating", in *21th IEEE International Conference on Micro Electro Mechanical Systems*, Tucson, USA, January 13-17, 2008, pp. 774-777.

[6] Y. Yoshihata, N. Binh-Khiem, A. Takei, E. Iwase, K. Matsumoto, I. Shimoyama, "Scanning micromirror using deformation of a parylene-encapsulated liquid structure", in *21th IEEE International Conference on Micro Electro Mechanical Systems*, Tucson, USA, January 13-17, 2008, pp. 770-773.

[7] H. Terae, B.-K. Nguyen, T. Takahata, E. Iwase, K. Matsumoto, I. Shimoyama, "Tapered waveguide by liquid for a coupler of optical fibers to MEMS devices", in *21th IEEE International Conference on Micro Electro Mechanical Systems*, Tucson, USA, January 13-17, 2008, pp. 794-797.

[8] I. Shimoyama, N. Binh-Khiem, Y. Yoshihata, "MEMS devices as display", in *14th International Display Workshops*, Sapporo, Japan, December 6, 2007, pp. 1325-1327.

[9] N. Binh-Khiem, K. Matsumoto, I. Shimoyama, "Active Parylene-encapsulated droplets for displays", in *American Chemical Society 235th National Meeting*, New Orleans, USA, April 6-10, 2008.

[10] N. Binh-Khiem, K. Matsumoto, I. Shimoyama, "Multicolor flexible display with a Parylene-encapsulated droplet array", in *4th Asia-Pacific Conference on Transducers and Micro-Nano Technology 2008*, Tainan, Taiwan, June 22-25, 2008.

FORMATION AND INTEGRATION OF A BALL LENS UTILIZING TWO PHASE LIQUID TECHNOLOGY

Chih-Chun Lee[1], Sheng-Yi Hsiao[2], and Weileun Fang[1,2]

[1]NEMS Inst., [2]Power Mechanical Engineering, National Tsing Hua University, HsinChu, Taiwan

ABSTRACT

This study presents a novel microlens formation technology to implement a polymer ball lens in liquid medium. The highly symmetric ball lens is achieved due to the lower gravity influence in liquid. The microlens' diameter is controlled by the volume of the dispensed polymer. Moreover, it is easy to dispense, form, and align the liquid-phase polymer microlens with MEMS structures while processing in liquid. Thus, the microlens can either be a discrete optical component, or directly integrated with MEMS structures during the process to form a SiOB (silicon-optical-bench). To date, ball lenses with diameter ranging 200μm~600μm and surface roughness <10nm are fabricated. A typical sphericity is 8.6μm for a 525μm diameter ball lens. The integration of such polymer micro ball lens on SiOB is also demonstrated.

1. INTRODUCTION

Micro ball lenses are commercially available and commonly used for light coupling, such as LED or laser-to-fiber and fiber-to-fiber coupling. The smaller ball lens has an advantage in miniaturizing the optical system. The ball lens is needed to be precisely assembled in a designed optical system. However, assembly of micro scale components with MEMS devices is not straightforward. The approaches to fabricate planar micro lens with out-of-plan optical axis have been extensively reported. For example, the etching selectivity techniques, such as boron diffusion, have been exploited to form a silicon microlens [1, 2]. The reflow process is commonly applied to micro lens manufacturing, especially in industry. For example, the photoresist reflow lens has been reported in [3, 4]. In [5, 6], the LIGA technology is used to define the resist shape (usually PMMA) with deep x-rays and then applying thermal reflow process to form lens. Comparing to reflow processes, [7] employs a nozzle to dispense the liquid polymer droplet onto heated substrate. However, above lens fabricating technologies can not be directly applied in in-plane optical applications. A further approach to implement the in-plane-axis optical system is to lift up and assemble the micro components such as thin film grating structure, HOE, Fresnel lens, etc [8, 9]. Yet, in order to realize component assembly, complicate structures and fabrication processes are required. In addition, accuracy, yield, and cost problems become challenges.

This study reports a new approach to fabricate discrete micro ball lens in a gravity-free and non-wetting liquid ambient, and further integrates the ball lens with MEMS device. Thus, the proposed micro lens is free from handling as well as assembly. In this approach, the lens diameter can be easily controlled by simply changing the dispensing polymer volume. Furthermore, such ball lens device is easily applied for micro-optical-system with optical-axis in in-plane and out-of-plane directions. In addition, the ball lens fabrication process is a room temperature post-process, and is compatible to most semiconductor devices.

2. CONCEPT AND PROCESS DESIGN

The existing liquid-phase formed polymer lenses are fabricated on a solid surface in air medium. In this regard, the final shape of a lens is defined by the balance of three forces, including surface tension force, gravity, and supporting force by the supporting surface. The lens curvature is mainly determined by the volume and surface tension of polymer. However, the gravity will lead to the asymmetry of lens shape. To prevent the influence of gravity, this study presents the approach to form polymer lens in liquid, as shown in Fig.1. In short, the liquid polymer is dispensed into immiscible buffer liquid to form a ball lens. As shown in Fig.1a, the dispenser is inserted into the immiscible liquid and the liquid phase polymer is dispensed. The surface tension of the immiscible liquid is employed to separate the polymer droplet and the dispenser, as shown in Fig.1b. As a result, droplet is finally immersed in buffer liquid. The lens floats in the buffer liquid and the supporting force is thus contributed by buoyancy.

In order to achieve the formation of a ball lens, a buffer liquid has the same density as that of dispensed polymer is selected. Thus, the gravity and buoyancy applied to the

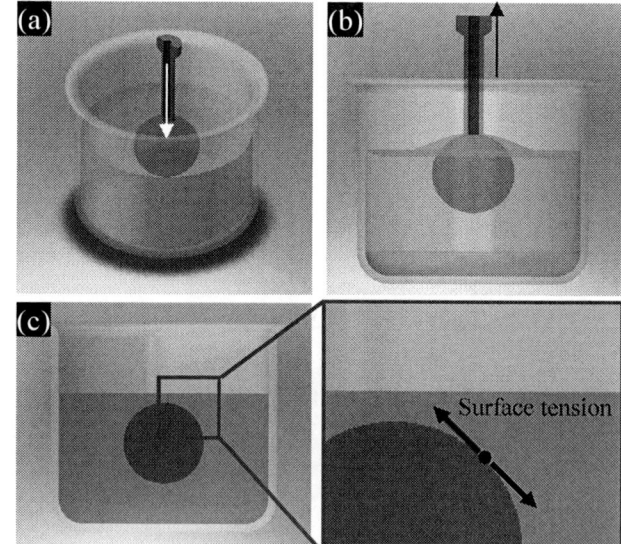

Figure 1: (a-b) Discrete micro ball lens fabricated by dispensing polymer in immiscible liquid, and (c) Polymer droplet in liquid medium is gravity-free. Spherical ball lens is formed to satisfy the minimum energy.

978-1-4244-2977-6/09 $25.00 © 2009 IEEE

polymer droplet is balanced, and the surface tension becomes the only force remained on droplet, as shown in Fig.1c. As a result, the surface area of the droplet is minimized by the surface tension. The droplet is then shaped into a perfect sphere which is with minimum surface area versus volume ratio. Moreover, the diameter of micro ball droplet can be properly controlled by the volume of dispensed polymer. The process parameters such as dispensing pneumatic pressure, and dispense time can be controlled by a commercial pneumatic dispensing system.

The aforementioned discrete ball lens formation technique is a room temperature process, and can integrate with various semiconductor devices. Thus, the ball lens can be integrated with other MEMS structure, such as frames and actuators to build a SiOB. Figs.2a-d further shows a typical example for the integration of polymer lens with MEMS structure in buffer-liquid. Fig.2a shows a Si-substrate containing a micromachined frame, and bonded with a Pyrex-glass. The substrate is immersed in a buffer liquid. Fig.2b shows the UV-curable polymer droplet was dispensed between the frame and Pyrex-glass. As shown in Fig.2c, the polymer is then solidified by means of UV-curing in buffer-liquid. Finally, the polymer ball lens is formed and confined by the frame and Pyrex-glass on Si substrate, as illustrated in Fig.2d. The lens can also be confined by micromachined frame and Si-substrate (Fig.2e), or by two frames (Fig.2f), to offer the key-components for SiOB. In conclusion, it is easy to integrate the ball lens with micromachined structures of different sizes and shapes since the lens is formed from the liquid-phase polymer.

In application, due to the feature of fully symmetric shape, these ball lenses can be applied without concerning of tilting problem. The misalignment is only caused by the off-center of ball lens. A ball lens can be applied both in in-plane and out-of-plane optical directions. For instance, typical

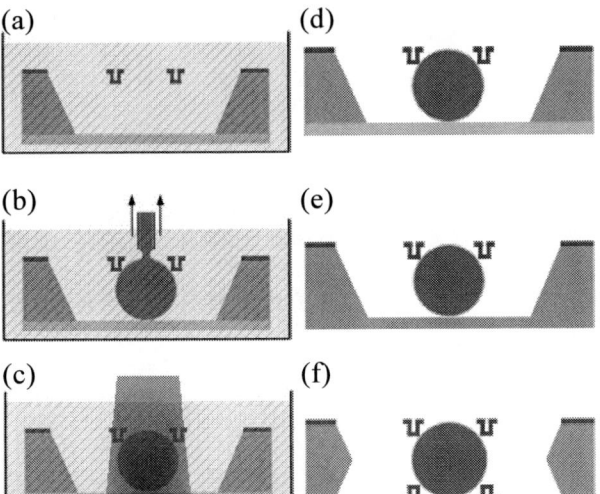

Figure 2: (a-c) Formation and integration of ball lens for SiOB; the SiOBs formed by (d) MEMS structure, ball lens, and glass substrate, (e) MEMS structure and ball lens, and (f) double-side MEMS structures and ball lens.

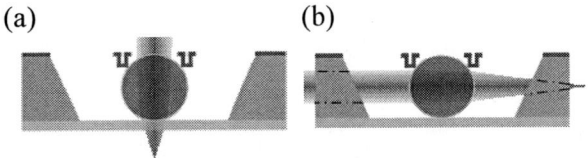

Figure 3: Ball lens for (a) out-of-plane, and (b) in-plane light condensing/collimating applications.

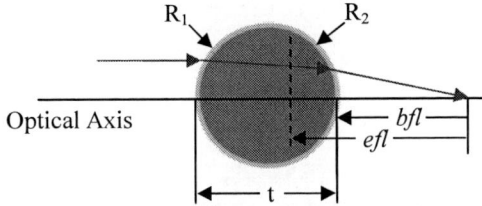

Figure 4: Optical parameters of a micro lens.

in-plane and out-of-plane condensing/collimating optical applications are shown in Fig.3. The vertical direction of light path as shown in Fig.3a is suitable for integration of stacking chips. The in-plane optical axis shown in Fig.3b is benefic to achieve a SiOB within one chip. Moreover, this study employs the optical power, ϕ, to represent the optical properties of the ball lens. As shown in Fig.4, a lens with thickness t, surface radii of curvature R_1, R_2, and material refractive index n, the relative optical power can be expressed as [10],

$$\phi = \frac{1}{f} = (n-1)\left[\frac{1}{R_1} - \frac{1}{R_2} + \frac{t(n-1)}{R_1 R_2 n}\right] \tag{1}$$

where f is the effective focal length as depicted in Fig.4. For a ball lens of radius R, Eq.(1) can be rewritten as,

$$\phi = \frac{2(n-1)}{nR} \tag{2}$$

Moreover, the optical power ϕ can also be represented by the back focal length bfl and the radius R of a ball lens,

$$\phi = \frac{1}{(bfl + R)} \tag{3}$$

As a result, the optical property of the dispensed ball lens is determined by measuring the bfl and R.

3. FABRICATION RESULTS

In experiment, the polymer employed in this study is Norland Optical Adhesive 63 (NOA63) with n=1.56, and specific gravity of 1.2. The glycerol with specific gravity of 1.26 is selected as the immiscible buffer liquid. Since the immiscible buffer liquid has almost the same density as the dispensed polymer, the influence of gravity force is ignored. First, the liquid phase NOA63 was dispensed in the glycerol

978-1-4244-2977-6/09 $25.00 © 2009 IEEE

to form discrete ball lens. The liquid polymer lens was then solidified by UV light curing process. Further cleaning and drying steps were applied to complete the ball lens process. Fig. 5 shows the results of various discrete micro ball lenses. Fig. 5a shows dispensing process under the buffer liquid. As shown in Fig.5b, the typical lens diameter ranging from 200μm to 600μm is properly controlled by polymer volume using a commercial pneumatic dispensing system. Fig. 5c shows the clear image "NTHU", in back of the micro ball lens, observed through the ball lens from microscope. Fig. 5d shows the integration of the discrete ball lens with MEMS device by assembly. Because of the fully symmetric character, a ball lens can be assembled without tilting concern. Thus, the assembly process is easy in application.

Figure 5: (a) Photo of applying two phase liquid technology to dispensing a ball lens under buffer liquid, (b) typical fabricated ball lens of different diameter, (c) image "NTHU" observed from ball lens, and (d) assembly of the discrete ball lens on MEMS actuator.

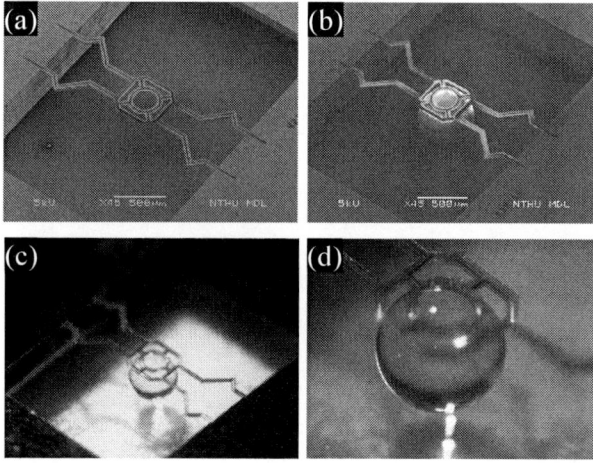

Figure 6: The SEM micrographs of a device (a) before and (b) after the integration of ball lens, (c) the OM photo of this integrated device, and (d) zoom-in of the ball lens. The lens is confined by micromachined silicon nitride frame and Pyrex glass substrate.

The results in Figs.6-7 demonstrate the integration of ball lens with MEMS devices, as illustrated in Fig.2d and Fig.2f. The SEM micrographs in Fig.6a-b respectively show a micromachined Si_3N_4 frame before and after the integration of polymer ball lens. The ball lens with a diameter larger than the edge length of lens frame is directly dispensed, formed, self-aligned, and cured in place. The device is further demonstrated by the photo images in Fig.6c-d. The glassy ball lens confined by transparent glass substrate and thin Si_3N_4 frame is clearly observed. Fig. 7 shows a ball lens being supported by two-side micromachined Si_3N_4 lens frames indicated in Fig.2f. In these two cases, the light can pass through the ball lens and the glass substrate, so as to realize stacking of SiOB chips vertically. Moreover, a fiber

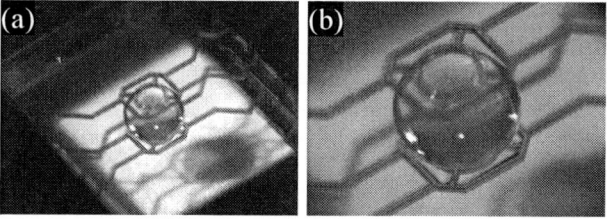

Figure 7: The OM photos of (a) ball lens confined by double side Si_3N_4 frames, and (b) zoom-in of the ball lens.

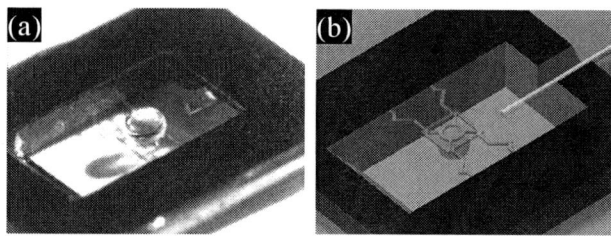

Figure 8: (a) Photo, and (b) illustration of a SiOB containing the integrated ball lens device in Fig.6 coupling with an optical fiber.

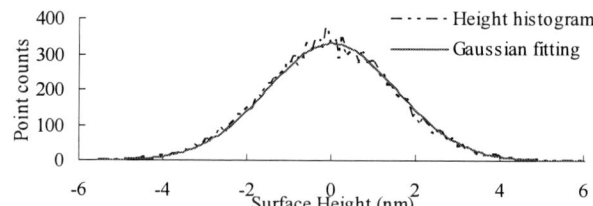

Figure 9: Surface height histogram of a ball lens (curvature removed). The 95% surface heights distribution is within ±3.05nm from the mean value.

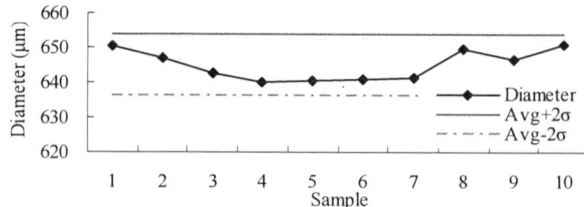

Figure 10: The statistics and deviation of the measured spherical lens diameter. The deviation is within ±2σ.

can be assembled in a wet etched V-groove and integrated with the proposed ball lens to construct an in-plane SiOB as shown in Fig. 8. shows a typical measured surface height histogram, after removing the curvature of ball lens. It indicates that 95% of the surface heights distribution is within ±3.05nm from the mean value. In general, average surface roughness of dispensed ball lens is lower than 10nm. Ten micro ball lenses were dispensed and measured to verify the repeatability of lens formation. Fig.10 shows the deviation of the micro ball lenses diameters are within ±2σ for these 10 samples. The two phase liquid technique is proved with reasonable repeatability.

Moreover, the sphericity and the misalignment of the integrated ball lens were evaluated using the image processing approach by commercial software. As shown in Fig.11, the typical off-center misalignment with the Si$_3$N$_4$ frame is about 3.4μm for a ball lens of 525μm in diameter. In addition, the typical sphericity of the ball lens is 8.6μm. Fig. 12 shows the experiment setup for lens optical test. A 632.8nm He-Ne Laser beam incidents into the specimen, moveable OM is used to measure back focal length. The typical measured back focal length of the micro ball lens with diameter 560μm (R=280μm) is about *bfl*=120μm. Based on the measured *bfl*, the optical power of the lens extracted from Eq.(3) is 2.50mm^{-1}. As a comparison, since the lens material

NOA63 has the refractive index of n=1.56, the optical power predicted from Eq.(2) is 2.56mm^{-1}.

5. CONCLUSIONS

This study presents a novel ball lens formation and integration technology utilizing the two phase liquid approach. The lens is formed in highly symmetric shape due to the gravity-free condition in buffer liquid. The diameter of the ball lens is controlled by the volume of the dispensed polymer. In application, the formation of NOA63 polymer ball lens in glycerol is demonstrated. The typical ball lens diameter ranges from 200μm to 600μm, and its surface roughness is near 10nm. Polymer ball lens of different materials can also be realized using the same approach. The ball lens formation process is a room temperature post-process, and is compatible with most semiconductor devices. Furthermore, the lens fabrication process has a merit of self-alignment when integrating with MEMS structures. The integration of such polymer micro ball lens with MEMS components to form SiOB is also demonstrated. In summary, the present two phase liquid ball lens formation and integration approach is a promising technology for micro optical systems.

ACKNOWLEDGEMENTS

This work was supported partially by the Ministry of Economic Affairs, Taiwan, under contract no. 96-EC-17-A-07-S1-011 and by the Nation Science Council, Taiwan, under contract no. NSC-96-2628-E-007-008-MY3.

REFERENCES

[1] Y. S. Kim, J. Kim, J. S. Choe, Y. G. Roh, H. Jeon, and J. C. Woo, *IEEE Photon. Tech. Lett.*, vol. 12, p. 507-509, 2000.

[2] C. S. Lee and C. H. Han, *Sensors & Actuators A*, vol. 88, pp. 87-90, 2001.

[3] P. Heremans, J. Genoe, M. Kujik, R. Vounckx, G. Borghs, and L. Imec, *IEEE Photon. Tech. Lett.*, vol. 9, pp. 1367-1369, 1997.

[4] H. S. Alhokail, in Proc. *10th Int. Conf. Microelectronics 1998*, Monastir, Tunisia, Dec. 14-16, 1998, pp. 49-52.

[5] S. K. Lee, K. C. Lee, and S. S. Lee, J. *Micromech. Microeng.*, vol. 12, pp. 334-340, 2002.

[6] S. Kim, S. S. Y. Lee, and S. Kwon, in Proc. *Optical MEMS 2002*, Lugano, Switzerland, Aug. 20-23, 2002 , pp. 77-78.

[7] D. L. MacFarlane, V. Narayan, J. A. Tatum, W. R. Cox, T. Chen, and D. J. Hayes, *IEEE Photon. Technol. Lett.*, vol. 6, pp. 1112-1114, 1994.

[8] L. Y. Lin, S. S. Lee, K. S. J. Pister, and M. C. Wu, *Electron. Lett.*, vol. 30, pp. 448-449, 1994.

[9] J. Y. Chang, C. M. Wang, C. C. Lee, H. F. Shih, and M. L. Wu, *IEEE Photon. Technol. Lett.*, vol. 17, pp. 214-216, 2005.

[10] W. Smith, *Modern Optical Engineering 3rd edition*, McGraw Hill, New York, 2000.

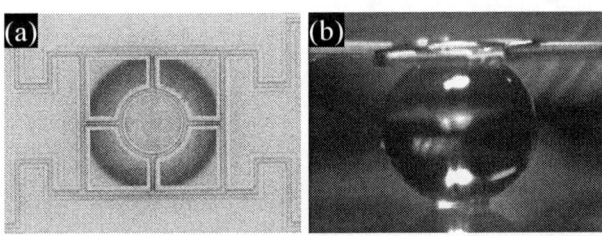

Figure 11: Top view and side view of ball lens integration with MEMS structure. (a) Misalignment between frame and polymer is 3.4μm, and (b) circular fitting is 8.6μm.

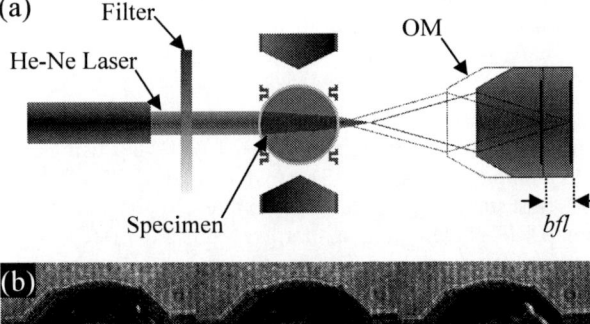

Figure 12: (a) Experiment setup for lens focal test, and (b) Laser beam spots recorded by movable OM. OM captures spots at different position from lens surface to focal point.

TEMPERATURE-CONTROLLED TRANSFER AND SELF-WIRING FOR MULTI-COLOR LED DISPLAY ON A FLEXIBLE SUBSTRATE

Eiji Iwase[1], Hiroaki Onoe[2], Akihito Nakai[1], Kiyoshi Matsumoto[1], and Isao Shimoyama[1]

[1] Depertment of Mechano-Informatics, University of Tokyo
[2] Center for International Research on MicroMechatronics, Institute of Industrial Science, University of Tokyo

ABSTRACT

We propose an integration method for arranging LED bare chips on a flexible substrate to fabricate a multi-color LED display. LED chips (240 μm × 240 μm × 75 μm) which were arrayed on an adhesive sheet were transferred to a flexible circuit substrate using our temperature-controlled transfer (TCT) and self-wiring (SW) method. Using these methods, we demonstrated a 5-by-5 LED flexible device and a two-color (blue and green) LED device, and observed light emission from the LED chips.

INTRODUCTION

Although inorganic LEDs have excellent brightness, high energy efficiency and long life time, there have been no consumer-use full-color LED displays for televisions or monitors of personal computers so far. The main reason is a difficulty of fabricating different color LEDs on a single substrate. Micro-scale self-assembly has been examined for heterogeneous integration of LED chips to create LED displays [1-3], but it needs assembly-control systems such as sequential heating using microheaters [4] to arrange different-color LED chips by micro self-assembly. Moreover, it has been difficult to obtain high chip density by the micro self-assembly because LED chips obstruct each other. On the contrary, our proposed method here enables us to fabricate a multi-color LED device by simply repeating the TCT process for each color of LED, and to achieve high chip density due to our stamping transfer apparatus with precise alignment (< 2 μm) [5,6].

TCT and SW have been advanced from our previous stamping transfer techniques using a PDMS (poly-(dimethylsiloxane)) stamping sheet [5,6]. The TCT is a control method of adhesion force using thermal transition of liquid/solid phases of adhesive layers. We used polyethylene glycol (PEG) and low melting point (LMP) bismuth-based solder as adhesive layers. Both the PEG and the LMP-solder layers are solid at room temperature, but they melt at 46 °C and 47 °C, respectively. Therefore, we can obtain sufficiently strong adhesion force to hold LEDs at room temperature (23 °C), and release the LEDs to transfer at high temperature (80 °C). By the TCT, the LEDs that were strongly bonded to the adhesive sheet can be released to a receiver substrate. We evaluate the TCT and SW and demonstrate a multi-color LED flexible display in this paper.

PRINCIPLE AND FABRICATION

The principle of our TCT and SW method is summarized in Figure 1. We chose commercially-available LED bare chip array on an adhesive sheet as an initial material for our transfer. Diced 240 μm × 240 μm × 75 μm LEDs were originally arranged and fixed on an adhesive sheet. We transferred the LEDs to a flexible circuit substrate and made electrical connection as follows; (step 1) releasing

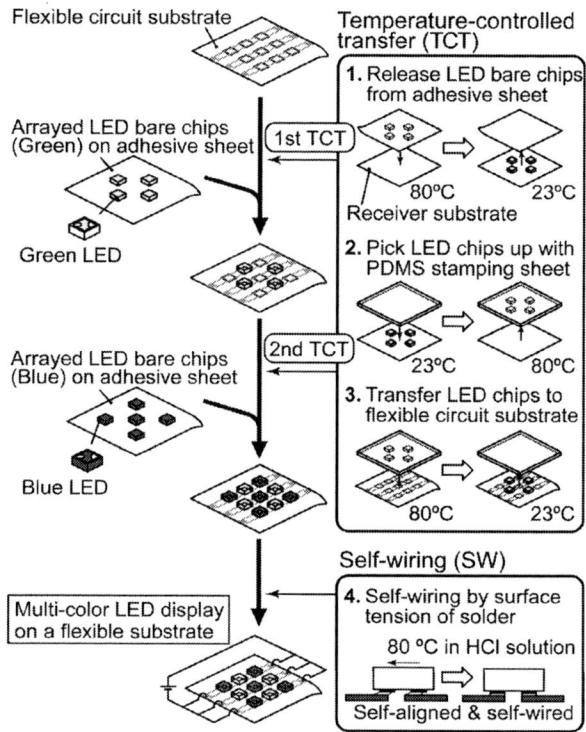

Figure 1: Schematic illustration of our temperature-controlled transfer (TCT) and self-wiring (SW) method for multi-color LED display.

the arrayed LEDs from the adhesive sheet to a receiver substrate, (step 2) picking the released LEDs up using a PDMS stamping sheet, (step 3) transferring the LEDs from the PDMS stamping sheet to a flexible circuit substrate where LMP-solder was patterned on, and (step 4) self-wiring between the LEDs and the circuit substrate by using the surface tension of the solder. During the successive transfer processes (step 1-3), adhesion force between the LEDs and substrates were controlled by temperature.

Basically, in order to transfer the chips from one substrate (donor substrate) to another substrate (receiver substrate), the adhesion force between the chips and the receiver substrate must be stronger than that between the chips and the donor substrate,

$$F_{\text{donor}} < F_{\text{receiver}}, \qquad (1)$$

where F is adhesion force between the LEDs and a substrate, and the subscript means the name of the substrate. Thus, we need to satisfy the relationship (1) at each step of our release, pick-up, and transfer processes.

In the release process, we adopted a glass slide with

978-1-4244-2977-6/09 $25.00 © 2009 IEEE   176

(A) Release process

(1) Heat receiver substrate at 80°C to melt PEG and LMP-solder. Press adhesive sheet to receiver substrate.

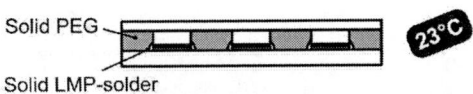

(2) Cool receiver substrate at R.T. to solidizing PEG and LMP-solder.

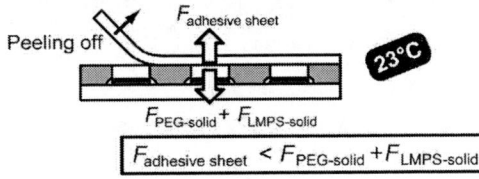

(3) Peel adhesive sheet off. Release LED bare chips.

(4) Wash PEG away by DI water.

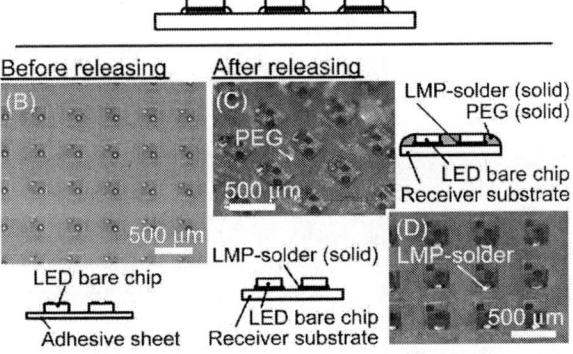

Figure 2: The release process. (A) Outline of the release process. (B) Arrayed LED bare chips on an adhesive sheet provided by manufacturer. (C) Released LEDs on a receiver substrate. (D) Released LEDs after washing the PEG layer out.

PEG/LMP-solder double layers as a receiver substrate to release the LEDs from the adhesive sheet as illustrated in Figure 2(A)-(1). After melting the PEG/LMP-solder double layer by heating at 80 °C, the LEDs are buried in the melted PEG and LMP-solder layers (Figure 2(A)-(2)), and then the substrates is cooled to solidify both PGE and LMP-solder layers. Since the solid PEG and LMP-solder layers are able to hold the LEDs structurally, the adhesion force between the LEDs and the solid PEG/LMP-solder double layer becomes larger than that between the chips and the adhesive sheet as follows,

$$(F_{LMPS-solid} <) \ F_{adhesive\,sheet} < F_{PEG-solid} + F_{LMPS-solid}. \quad (2)$$

Thus, the LED chips remains in the PEG and LMP-solder layers after peeling the adhesive sheet (Figure 2(A)-(3)). After that, the PEG layer is dissolved away by water for the next pick-up step (Figure 2(A)-(4)). The adhesion force of only LMP-solder layer is not sufficient to hold LEDs as

(A) Pick-up process

(1) Contact PDMS stamping sheet to LED chips.

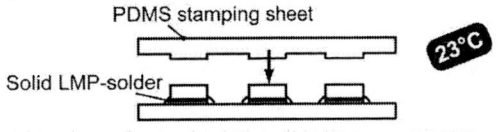

(2) Heat receiver substrate at 80°C to melt LMP-solder.

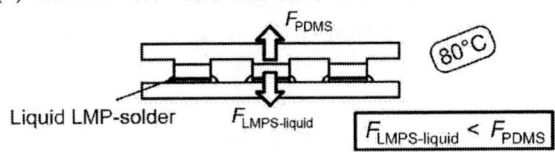

(3) Pick LED chips up with PDMS stamping sheet.

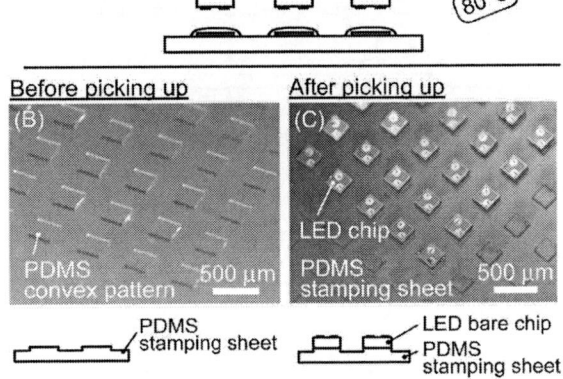

Figure 3: The pick-up process. (A) Outline of the pick-up process. (B) A PDMS stamping sheet before picking up. (C) LEDs that were picked up with the PDMS stamping sheet.

shown in the relationship (2). The PEG can not be patterned for a precise positioning at a transfer. Therefore, we used the PEG/LMP-solder double layer.

In the pick-up and transfer processes shown in Figure 3(A) and Figure 4(A), the released LED array is picked up and then transferred to the final circuit substrate using a PDMS stamping sheet. As shown in Figure 3(B), the PDMS stamping sheet has convex micro-patterns on its surface that enables us to selectively pick the LEDs up from the receiver substrate. To control the adhesion force during the pick-up and transfer processes, LMP-solder was used as a temperature-controllable adhesive layer by melting and solidifying it.

When the LMP-solder is liquefied by heating, the adhesion force between the LEDs and the PDMS stamping sheet is greater than that of the chips and the liquid LMP-solder. On the contrary, when the LMP-solder is solid, the adhesion force between the LEDs and the PDMS is smaller than that between the chips and the solid LMP-solder. Therefore, the relationship of the adhesion force can be described as,

$$F_{LMPS-liquid} < F_{PDMS} < F_{LMPS-solid}. \quad (3)$$

Using the relationship (3), the LEDs can be picked up from the receiver substrate with the PDMS stamping sheet by melting the LMP-solder as the pick-up process (Figure 3(A)). Meanwhile, we are able to transfer the LEDs from

(A) Transfer process

(1) Heat circuit substrate at 80°C to melt LMP-solder. Press LED bare chips to circuit substrate.

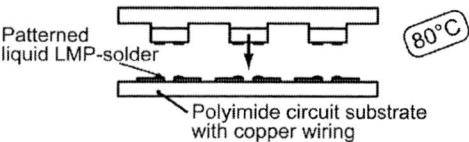

(2) Cool circuit substrate at R.T. to solidize LMP-solder.

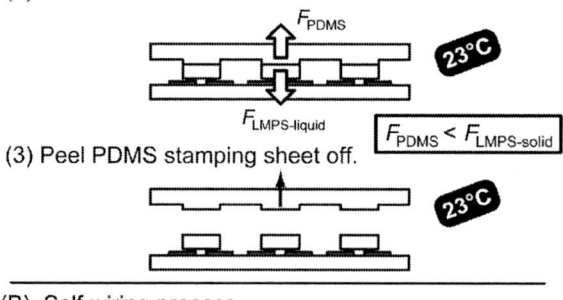

(3) Peel PDMS stamping sheet off.

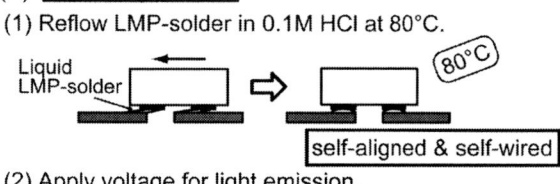

(B) Self-wiring process

(1) Reflow LMP-solder in 0.1M HCl at 80°C.

(2) Apply voltage for light emission.

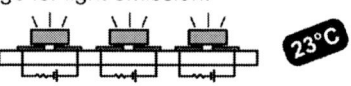

Figure 4: The transfer process and the self-wiring process.

the PDMS stamping sheet to the LMP-solder-patterned circuit substrate by solidifying the LMP-solder as the transfer process (Figure 4(A)). These temperature-controlled release, pick-up and transfer processes are repeatable for arranging multiple colors of LEDs on the circuit substrate.

Finally, the transferred LEDs are electrically connected to the circuit substrate by reflow of the LMP-solder in 0.1 M hydrochloric acid (HCl) at 80 °C as the self-wiring process (Figure 4(B)). Due to surface tension of liquid LMP-solder, the liquid LMP-solder moves spontaneously to only metal contact pads (e.g. copper or gold) rather than to other areas such as glass or polymers. Thus, electrical wiring between the LEDs and the circuit substrate is achieved automatically by reflow of the LMP-solder. In addition, the LEDs are dragged and precisely self-aligned to the contact pads on the circuit substrate by the surface tension of the liquid LMPS at the same time. Consequently, the transferred LEDs are self-aligned to the circuit substrate as well as the self-wiring, and the LEDs emit light by applying electrical current via the circuit substrate.

EXPERIMENTS AND RESULTS

We used GaN-based blue (wavelength of emission: λ = 454 nm) and green (λ = 550 nm) LEDs for our experiment. Both LEDs were 240 μm × 240 μm × 75 μm in size and arrayed on an adhesive sheet in gap distance of 500 μm. Each LED has both anode and cathode contact

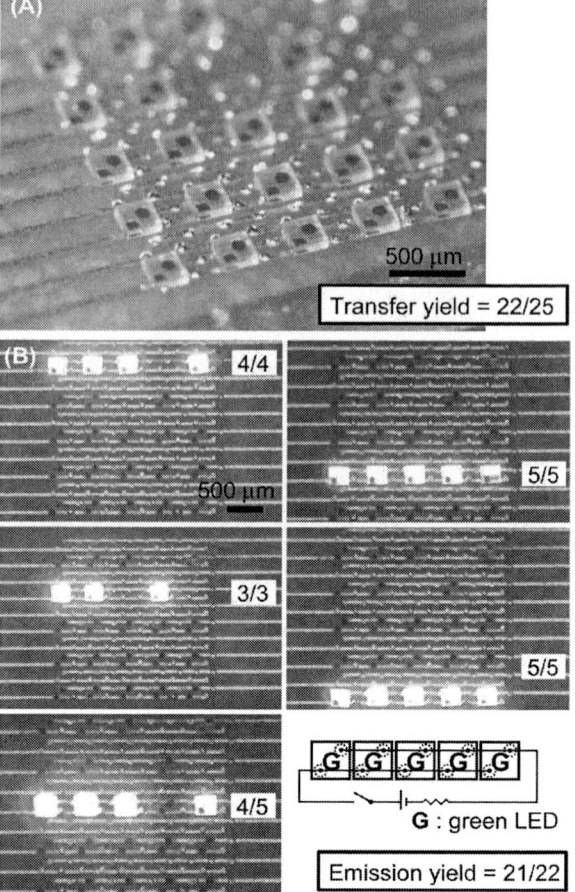

Figure 5: Photographs of the 5-by-5 LED flexible device. (A) Transferred LEDs on a flexible circuit substrate. The transfer yield of the LED chips was 88 % (22/25). (B) Light emission from the each line of the LED device. The emission yield of the LED chips was 95 % (21/22).

Table 1. Positioning accuracy and yield.

LED bare chips	x pitch [μm] ave.	x pitch [μm] s.d.	y pitch [μm] ave.	y pitch [μm] s.d.	Yield
on adhesive sheet (before "release")	500*	—	500*	—	—
on receiver substrate (after "release")	505	6.7	498	7.7	149/149 (100%)
on PDMS stamp sheet (after "pick-up")	501	6.4	495	7.7	25/25 (100%)
on circuit substrate (after "transfer")	505	6.6	500	6.1	22/25 (88%)

ave.: average value, s.d.: standard deviation, * original design

pads on the top surface of the chip. Since the base substrate of the LED was made of transparent sapphire, emitted light can be observed at both top and bottom sides of the chip.

We fabricated a 5-by-5 LED flexible device (Figure 5) and evaluated the yield and position accuracy (Table 1). Yields of the release, pick-up, and transfer step were 100 % (149/149), 100 % (25/25), and 88 % (22/25), respectively. The pitches between the LEDs were 505 ± 6.6 μm for

978-1-4244-2977-6/09 $25.00 © 2009 IEEE   178

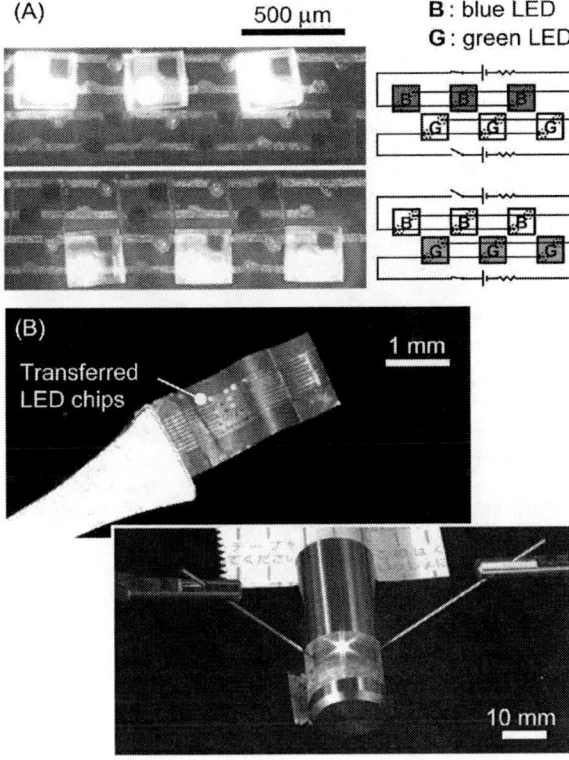

B : blue LED
G : green LED

Figure 6: Light emission from LED devices. (A) Emission of blue and green light from the transferred LED bare chips. (B) Light emission from bent device. The radius of curvature is 6 mm.

x-axis and 500 ± 6.1 μm for y-axis after the transfer step. Yield of the electrical connection of the transferred LED chips after the SW was 95 % (21/22) as shown in Figure 5.

We also demonstrated fabrication of multi-color LED deice by repeating two sets of the TCT processes, and performing the SW process at the end. As shown in Figure 6(A), three blue and three green LEDs were arranged on a flexible circuit substrate. The device was sufficiently thin and strong for bending deformation. As shown in Figure 6(B), the light emission from the bent device attached on a cylinder of 6 mm in the radius was also observed.

CONCLUSIONS

We proposed the TCT and SW method for arranging LEDs on a flexible substrate. In our method, we utilized temperature-controlled adhesive layers, PEG and LMP-solder, to adjust the adherent force of the LEDs during the TCT processes. And we made electrical connections between the transferred LEDs and the circuit substrate by reflow of the LMP-solder during the SW processes. A flexible LED device and multi-color LED device were demonstrated using our method.

ACKNOWLEDEMENT

The photolithography masks were fabricated using the EB lithography apparatus of the VLSI Design and Education Center (VDEC) of the University of Tokyo.

REFERENCES

[1] H. O. Jacobs, A. R. Tao, A. Schwartz, D. H. Gracias, and G. M. Whitesides, "Fabrication of a cylindrical display by patterned assembly," *Science*, vol. 296, pp. 323-325, 2002.

[2] K. F. Bohringer, "Surface modification and modulation in microstructures: controlling protein adsorption, monolayer desorption and micro-self-assembly," *J. Micromech. Microeng.*, vol. 13, pp. S1-S10, 2003.

[3] W. Zheng, P. Buhlmann, and H. O. Jacobs, "Sequential shape-and-solder-directed self-assembly of functional microsystems," *Proc. Natl. Acad. Sci. U.S.A.*, vol. 101, pp. 12814-12817, 2004.

[4] J. H. Chung, W. Zheng, T. J. Hatch, and H. O. Jacobs, "Programmable reconfigurable self-assembly: Parallel heterogeneous integration of chip-scale components on planar and nonplanar surfaces," *J. Microelectromech. Syst.*, vol. 15, pp. 457-464, 2006.

[5] H. Onoe, E. Iwase, K. Matsumoto, I. Shimoyama, "Three-dimensional integration of heterogeneous silicon micro-structures by liftoff and stamping transfer," *J. Micromech. Microeng.*, vol. 17, pp. 1818-1827, 2007.

[6] E. Iwase, H. Onoe, K. Matsumoto, I. Shimoyama, "Hidden vertical comb-drive actuator on PDMS fabricated by parts-transfer," *Proc. MEMS 2008*, pp. 116-119, 2008.

MICROFLUIDIC ODORANT SENSOR WITH FROG EGGS EXPRESSING OLFACTORY RECEPTORS

Nobuo Misawa[1, 2], Hidefumi Mitsuno[3], Ryohei Kanzaki[1, 3], and Shoji Takeuchi[1, 2, 4, 5]
[1]IRT Research Initiative, The University of Tokyo, JAPAN
[2]Institutehe of Industrial Science, The University of Tokyo, JAPAN
[3]Research Center for Advanced Science and Technology, The University of Tokyo, JAPAN
[4]PRESTO, JST, Center for Advanced Science and Technology
[5]BEANS Project, Life BEANS Center, The University of Tokyo

ABSTRACT

This study describes a membrane-protein based odorant sensor device consisting of microfluidic channels and frog eggs (*Xenopus laevis* oocyte) expressing olfactory receptors. These receptors are from silkmoth's receptors, BmOR1 and BmOR3 excited by specific pheromones, bombykol and bombykal, respectively. We employ a conventional two electrode voltage clamp method for the signal measurement of the oocytes in our microfluidic system. We have succeeded in selectively detecting these two types of odorant-like chemicals and the two current traces can be recorded by parallel measurement in this system. It implies that our suggested devices can be applied to multichannel detection as a chemical sensor using biological reactions.

INTRODUCTION

Various approaches to a chemical sensor as an odorant detector so-called "an artificial nose" are reported. These technologies are based on a metal-oxide-semiconductor transistor [1], a micromechanical cantilever array [2], a modified quartz crystal microbalance [3, 4], a nanostructured material [5], and a living cell's or biological response [6-9]. Chemical sensors using cells and biomolecules are especially studied actively due to their high specificity and sensitivity for target chemicals, accumulable property and based on scientific curiosity about mechanisms of the biological responses. In particular, using recombinant cells as sensing element of a chemical sensor gains much attention. *Xenopus laevis* oocyte (*X. laevis* oocyte), among others, has high versatility as a host cell for such a system since expression systems of many membrane proteins including G-protein coupled receptor which is a large family of plasma proteins have been constructed with *X. laevis* oocyte.

X. laevis oocyte is often used as platforms as test cells in chemical sensor applications and in drug screening [10-12]. Here, we describe a microfluidic device for measuring the electrophysiological activity of an oocyte that expresses artificially pheromone receptors of silkmoth. The receptors are called BmOR1 and BmOR3 which belong to seven-transmembrane receptor and involved in regulation of ion flows by recognizing selectively pheromones bombykol and bombykal, respectively [13]. In our system, the fluidic channels have a cell-trap region and two electrodes for measurement of a cell's response to a pheromone by two electrode voltage clamp method as shown in Figure 1.

In this paper, we demonstrated the device that can distinguish two similar chemicals, bombykol and bombykal, due to the receptors' specificity and sensitivity.

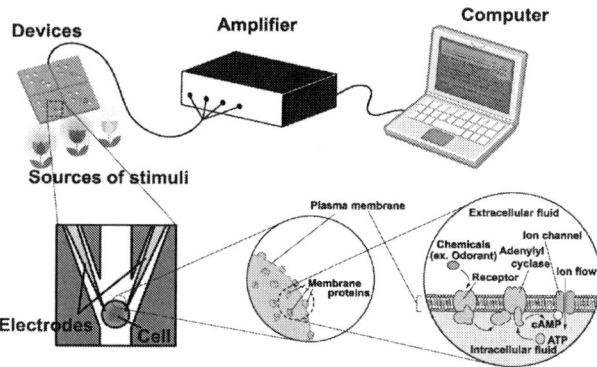

Figure 1. Conceptual image of chemical sensor using cells. The system is composed three parts: computer, amplifier and devices with cells expressed membrane proteins of receptors artificially. The signal is measured with two electrode voltage clamp method. General olfactory sense is shown as a model system here.

MATERIALS AND METHODS

Device fabrication

The device is composed of (i) Pyrex® glass substrate 2.5 cm × 2.5 cm × 2 mm (#7740, borosilicate glass, Corning Inc.), (ii) fluidic channel of polydimethylsiloxane (PDMS) (SILPOT 184, Dow Corning TORAY), (iii) two glass capillaries 1 mm outer diameter, 0.8 mm inner diameter, and 90 mm whole length (G-100, borosilicate glass tube, NARISHIGE Co., Ltd.) manually bended L-shaped by a gas burner, and (iv) two Ag/AgCl electrodes 0.4 mm diameter (AG-401355, The Nilaco Co.) prepared preliminarily by soaking in commercial bleach overnight. The glass capillary is pulled by a puller (PB-7, NARISHIGE Co., Ltd.) for making the tip outer diameter is less than about 50 μm.

Fabricated acrylic substrates are used for the mold of PDMS. The PDMS fluidic channel and the Pyrex glass substrate are bonded through oxygen plasma treatment performed with capacitively-coupled plasma (CCP) (Compact Etcher FA-1, SAMCO) at about 10 Pa with output power 75 W for 5 sec, frequency of 13.56 MHz and 20 ml/min oxygen flow rate.

As shown in Figure 2, two Ag/AgCl electrodes are inserted into the two capillaries for measurements of oocyte's responses. The distance of two capillaries tips is less than about 1 mm since diameter of *X. laevis* oocyte is generally 1-1.5mm. The capillaries are filled with 3 M KCl solution preliminarily. Furthermore, the center channel is filled with 5 mM HEPES/NaOH buffer solution pH 7.5 containing 96 mM NaCl, 2mM KCl, 1.8 mM CaCl$_2$, and 1.6 mM MgCl$_2$. The buffer solution is perfused constantly using a peristaltic pump.

978-1-4244-2977-6/09 $25.00 © 2009 IEEE

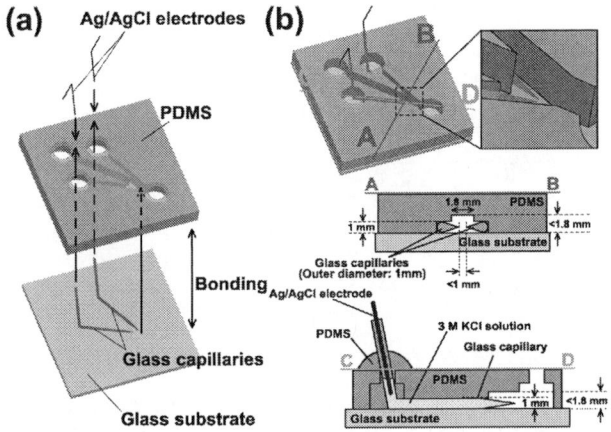

Figure 2. Schematic views of fluidic channel for cell trapping and electrical measurement of cell signals. (a) Components of a device. (b) Cross-section diagrams of the fluidic channels with glass capillaries and Ag/AgCl electrodes.

Cell trapping and silkmoth's pheromone receptors

Figure 3 shows image of the device and *X. laevis* oocyte. The buffer solution flows in center channel from near side (inlet) to far side (outlet) with 10-20 ml/min flow rate as seen in Figure 3(a) and 3(b). Oocyte can be trapped by two capillaries on the stream as shown in Figure 3(e) and 3(f).

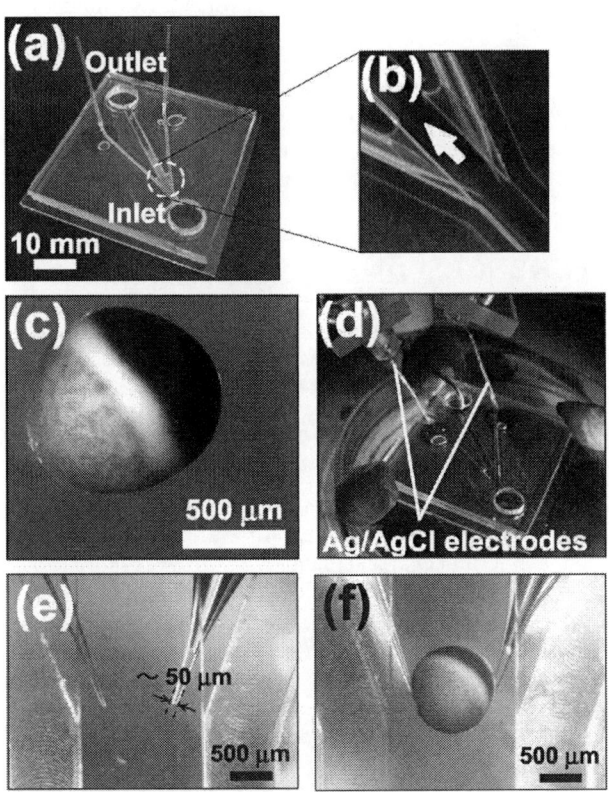

Figure 3. Pictures of the device and frog egg (Xenopus laevis oocyte). (a) Whole image of the device. (b) Close-up picture of cell trap region. (c) Picture of X. laevis oocyte. (d) Picture of the actual measurement system. (e) Picture of cell trap region without oocyte and (f) with oocyte.

Figure 4 schematically illustrates pheromone receptor of silkmoth for (E,Z)-10,12-hexadecadien-1-ol (bombykol) and (E,Z)-10,12-hexadecadien-1-al (bombykal). The receptors are called BmOR1 and BmOR3, respectively. They work as ion channels with ligand recognitions.

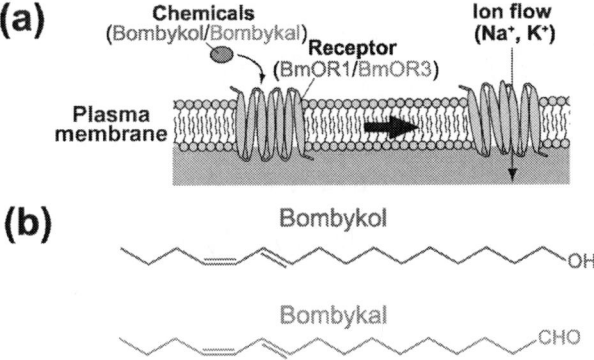

Figure 4. (a) Schematic image of silkmoth's pheromone receptor, BmOR1 and BmOR3 for bombykol and bombykal respectively. (b) Chemical structures of bombykol and bombykal.

Receptor expression in *Xenopus laevis* oocyte and electrophysiological recording

Oocytes are microinjected with RNA of BmOR1 or BmOR3 after collagenase treatment. Injected oocytes were incubated a few days in Barth's solution (88 mM NaCl, 1 mM KCl, 0.3 mM Ca(NO$_3$)$_2$, 0.4 mM CaCl$_2$, 0.8 mM MgSO$_4$, 2.4 mM NaHCO$_3$, 15 mM HEPES/NaOH buffer solution pH 7.6).

Oocytes currents are recorded with a two electrode voltage clamp method using a custom-built multichannel amplifier (Triton, Tecella LLC.). The current are monitored every 100 μsec and hold voltage is -80 mV.

RESULTS AND DISCUSSION
Selectivity of receptors for pheromone types

Figure 5(a) as a negative control shows an evidence of our system is activated by the oocyte. No peak appears in any case of applying buffer solution, bombykol and bombykal in Figure 5(a). Figure 5(b) and 5(c) are results of selective responses of the oocyte for bombykol and bombykal, respectively. Although intensities of responses depend on oocyte's condition, we find that our suggested device can detect odorant-like chemicals obviously. These results are in good agreement with results taken by conventional measuring system.

This system does not require microscopes and micromanipulators to insert electrodes into target cells since the device can trap the cell with inserting electrodes spontaneously. Furthermore, these two different responses are recorded at the same time with independent device as shown in an inset picture of Figure 5. It is possible to record four different oocyte responses in this system due to a multichannel amplifier we use.

In our system, there is no need any shield boxes and any vibration isolated tables in contrast with conventional electrophysiological recording systems. In addition, the amplifier is very compact. So, this system needs only three elements: computer, amplifier and microfluidic device with oocytes.

978-1-4244-2977-6/09 $25.00 © 2009 IEEE

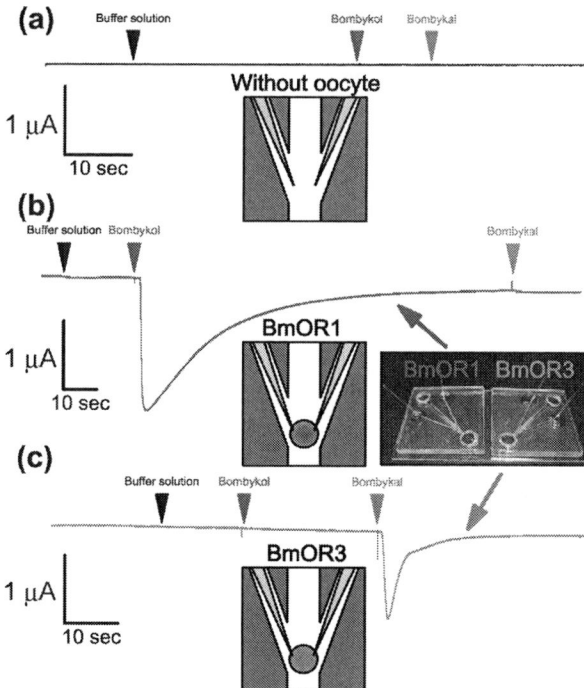

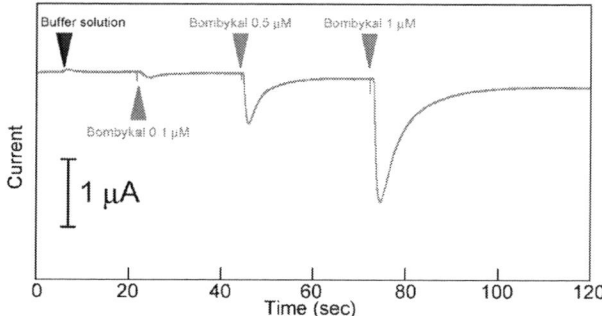

Figure 7. Bombykal responses of Xenopus oocyte expressing BmOR3. Buffer solution 20 μl, 1 μM, 5 μM and 10 μM bombykal 20 μl were applied at the time indicated by arrowheads.

CONCLUSIONS

We demonstrated a device that can distinguish two similar odorant-like chemicals, bombykol and bombykal, due to receptors' specificity and sensitivity.

This microfluidic device thus is useful as a chemical odorant sensor, easy to use, portable, and is easily multiplexed showing the potential of this odorant detection system as a next generation chemical detection sensor.

ACKNOWLEDGMENTS

The authors would like to thank Prof. M. Asashima of The University of Tokyo for kindly giving the X. laevis oocytes. We also really appreciate Dr. Y. Tanaka of Tecella LLC for customizing of a multichannel amplifier, Triton, for our experiment. This research is partly supported by Special Coordination Funds for Promoting Science and Technology, "IRT Foundation to Support Man and Aging Society".

Figure 5. (a) Current trace without Xenopus oocyte as a negative control. Buffer solution 20 μl, 10 μM bombykol 20 μl and 10 μM bombykal 20 μl were applied at the time indicated by arrowheads. (b) Responses of oocyte expressing BmOR1. Buffer solution 20 μl, 10 μM bombykol 20 μl and 10 μM bombykal 20 μl were applied at the time indicated by arrowheads. (c) Responses of oocyte expressing BmOR3. Buffer solution 20 μl, 0.5 μM bombykol 20 μl and 0.5 μM bombykal 20 μl were applied at the time indicated by arrowheads. (b) and (c) were measured in parallel.

Relativity of receptors for pheromone concentrations

Figure 6 and Figure 7 are pheromones' concentration dependent responses of the oocyte for bombykal and bombykol. According to these results we confirmed that our suggested device can detect a difference in chemicals concentration with enough sensitivity.

REFERENCES

[1] D. Kohl, L. Heinert, J. Bock, Th. Hofmann, P. Schieberle, "Systematic studies on responses of metal-oxide sensor surfaces to straight chain alkanes, alcohols, ketones, acids and esters using the SOMMSA approach", Sensors and Actuators B, vol. 70, pp. 43-50, 2000.

[2] H.P. Lang, M.K. Baller, R. Berger, Ch. Gerber, J.K. Gimzewski, F.M. Battiston, P. Fornaro, J.P. Ramseyer, E. Meyer, H.J. Güntherodt, "An artificial nose based on a micromechanical cantilever array", Analytical Chimica Acta, vol. 393, pp. 59-65, 1999.

[3] K. Yano, U.T. Bornscheuer, R. D. Schmid, H. Yoshitake, H-S. Ji, K. Ikebukuro, Y. Masuda, I. Karube, "Development of an odorant sensor using polymer-coated quartz crystals modified with unusual lipids", Biosensors & Bioelectronics, vol. 13, pp. 397-405, 1998

[4] H-S. Ji, S. McNiven, K. Ikebukuro, I. Karube, "Selective piezoelectric odor sensors using molecularly imprinted polymers", Analytical Chimica Acta, vol. 390, pp. 93-100, 1999.

[5] X-J. Huang, Y-K. Choi, "Chemical sensors based on nanostructured materials", Sensors and Actuators B, vol. 122, pp. 659-671, 2007.

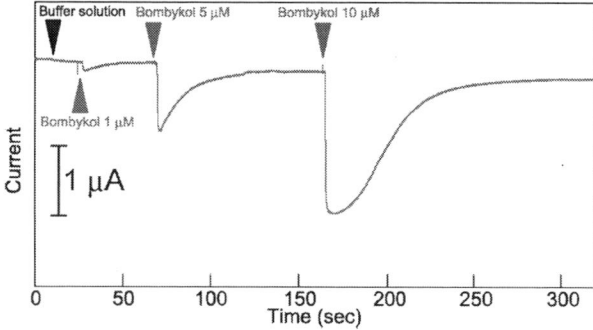

Figure 6. Bombykol responses of Xenopus oocyte expressing BmOR1. Buffer solution 20 μl, 1 μM, 5 μM and 10 μM bombykol 20 μl were applied at the time indicated by arrowheads.

[6] X. Feng, J. Castracane, N. Tokranova, A. Gracias, G. Lnenicka, B.G. Szaro, "A living cell-based biosensor utilizing G-protein coupled receptors: Principles and detection methods", *Biosensors and Bioelectronics*, vol. 22, pp. 3230-3237, 2007.

[7] V. Radhika, T. Proikas-Cezanne, M. Jayaraman, D. Onesime, J.H. Ha, D.N. Dhanasekaran, "Chemical sensing of DNT by engineered olfactory yeast strain", *Nature chemical biology*, vol. 3, pp. 325-330, 2007.

[8] Ch. Ziegler, W. Göpel, H. Hämmerle, H. Hatt, G. Jung, L. Laxhuber, H-L. Schmidt, S. Schütz, F. Vögtle, A. Zell, "Bioelectronic noses: a status report. Part II", *Biosensors & Bioelectronics*, vol. 13, pp. 539-571, 1998

[9] X. Feng, J. Castracane, N. Tokranova, A. Gracias, G. Lnenicka, B.G. Szaro, "A living cell-based biosensor utilizing G-protein coupled receptors: Principles and detection methods", *Biosensors and Bioelectronics*, vol. 22, pp. 3230-3237, 2007.

[10] P.R. Joshi, A. Suryanayanan, M.K. Schulte, "A vertical flow chamber for Xenopus oocyte electrophysiology and automated drug screening", *Journal of Neuroscience Methods*, vol. 132, pp. 69-79, 2004.

[11] E. Dahan, V. Bize, T. Lehnert, J-D. Horisbeger, M.A.M Gijs, "Integrated Microsystems for no-invasive electrophsiological measurements on Xenopus oocytes", *Biosensors and Bioelectronics*, vol. 22, pp. 3193-3202, 2007.

[12] R. Pantoja, J.M. Nagarah, D.M. Starace, N.A. Melosh, R. Blunck, F. Bezanilla, J.R. Heath, "Silicon chip-based patch-clamp electrodes integrated with PDMS microfluidics", *Biosensors and Bioelectronics*, vol. 20, pp. 509-517, 2004.

[13] T. Sakurai, T. Nakagawa, H. Mitsuno, H. Mori, Y. Endo, S. Tanoue, Y. Yasukochi, K. Touhara, T. Nishioka. "Identification and functional characterization of a sex pheromone receptor in the silkmoth *Bombyx mori*", *Proceeding of the National Academy of Science of the USA*, vol. 101, pp. 16653-16658, 2004.

SEQUENTIAL AND SELECTIVE SELF-ASSEMBLY OF MICRO COMPONENTS BY DNA GRAFTED POLYMER

T. Tanemura[1], G. Lopez[1], R. Sato[1], K. Sugano[1], T. Tsuchiya[1], O. Tabata[1], M. Fujita[2] and M. Maeda[2]
[1]Kyoto University, Kyoto, JAPAN
[2]RIKEN, Saitama, JAPAN

ABSTRACT

We demonstrated that Deoxyribonucleic Acid (DNA) can be utilized as an intelligent adhesive activated by temperature control to realize sequential and selective self-assembly of 5 mm squared silicon micro components on a substrate. Two kinds of complementary DNA pairs whose melting transition temperature (T_m) and base sequence were designated appropriately were utilized to verify the feasibility of the proposed approach. It was successfully confirmed that sequential and selective assembly of micro components on the substrate can be realized below each T_m. Furthermore, DNA grafted polymer was proposed for the first time as a flexible spacer between micro components and a substrate to improve self-assembly yield and speed.

INTRODUCTION

An integration of multiple heterogeneous functional components made from different materials such as silicon (Si), metal, ceramic, polymer and compound semiconductor components with size ranging from a few μm to a few tenths nm with Micro Electro Mechanical Systems (MEMS) as a single platform becomes an essential technology to extend MEMS functionality [1]. A robotic assembly using pick-and-place methods to manipulate and place macro-scale components such as IC chips and passive components individually is one of the widely used approaches to integrate various parts. However, the robotic assembly is not suitable to assemble micro- to nano-scale components in large quantities since it is not a parallel process and components handling becomes harder when their sizes decrease resulting in insufficient alignment accuracy and low assembly yield.

One of the key-technologies to address the integration of various micro- or nano-scale components and MEMS in parallel with higher alignment accuracy and higher yield is self-assembly. It applies the principle of potential energy minimization to realize ordered structures with mass-production and low cost. Several approaches have been proposed for self-assembly of micro- or nano-scale components driven by capillary force, hydrophobic force, electrostatic force and so on. In self-assembly, authors are paying much attention on sequential assembly of micro- to nano-scale components since the sequence control is the essential assembly step to build a complex system composed of multiple components with different functions as is the case with a macro scale assembly process [2]. Although a paper about sequential assembly of micro components using interactive forces control by changing the pH of the solution [3] was reported, its possible number of sequence control was limited. The method using DNA hybridization was proposed to realize a more complex sequence control and it was applied to assemble Au nanoparticles with diameter of 15 nm [4]. However, sequential and selective self-assembly process with larger components up to a few micro-scales and a substrate has

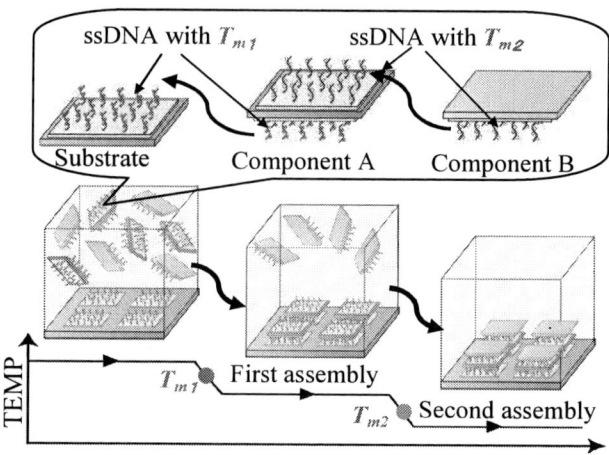

Figure 1: The concept of DNA mediated sequential assembly

not been realized yet. In MEMS 2008, we reported the validity of sequential assembly of Au nano particles and a substrate using DNA [2]. However, the applicability of this technique to micro-scale components was not proved yet.

In this paper, by revising process conditions and devising DNA grafted polymer techniques, we succeed to demonstrate the validity of hybridization of DNA and DNA grafted polymer for sequence and selectivity control of the self-assembly process with micro–scale components and a substrate.

DNA FOR SEQUENTIAL SELF-ASSEMBLY
Characteristics of DNA

In the proposed process, the characteristics of DNA such as hybridization specificity of complementary pairs of ssDNA into dsDNA as well as dependency of hybridization reaction on ambient temperature were taken as an advantage. A ssDNA consists of nucleotides and phosphate acid as a backbone. The nucleotides have four types of molecules called bases, Adenine (A), Thymine (T), Cytosine(C) and Guanine (G). A ssDNA is the sequence of these four bases along the backbone and each type of base on a strand can form a hydrogen bond with a specific type of base, such as A-T (two hydrogen bonds) or G-C (Three hydrogen bonds), to make a double helix structure (dsDNA). Therefore, when two entirely complementary ssDNA immobilized to different surfaces establish contact with each other, all bases can bond and form hybridized dsDNA. It works as an adhesive between two surfaces. Hybridization occurs below a specific temperature T_m called melting temperature, which can be designated by a length and a configuration of the complementary base sequence.

Sequential and selective self-assembly

Sequence and selectivity control of self-assembly can be achieved by taking advantage of the forementioned characteristics of DNA hybridization. The concept is shown in Fig. 1. Components of two types and a substrate

978-1-4244-2977-6/09 $25.00 © 2009 IEEE

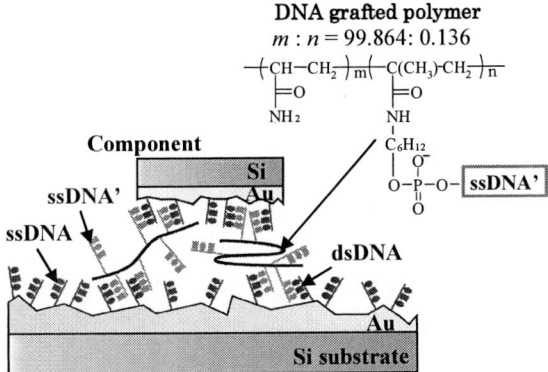

DNA grafted polymer
$m : n = 99.864 : 0.136$

Figure 2: DNA grafted polymer and its application for spacer between components and substrate

Table 1: DNA sequence and melting temperature T_m.

DNA	sequence 5' to 3'	T_m [°C]
D21	(SH)-AAA AAA **TTA AGA CGA GGC AAT CAT GCA**	65.4
D21'	(SH)-AAA AAA **TGC ATG ATT GCC TCG TCT TAA**	65.4
D9	(SH)-AAA AAA AAA AAA **ATT GCC TGC**	48.5
D9'	(SH)-AAA AAA AAA AAA **GCA GGC AAT**	48.5

containing sites are prepared and their surfaces are modified by one of two pairs of complementary ssDNA with melting temperatures of T_{m1} and T_{m2} ($T_{m1} > T_{m2} >$ Room Temperature (RT)). The self-assembly process starts from high temperature above T_{m1} and then decreasing to RT. As temperature decreases, the more stable ssDNA sequences with T_{m1} hybridize first, followed by the hybridization of less stable sequences with T_{m2} at lower temperature.

DNA grafted polymer

Figure 2 schematically shows an assembled component on a substrate using DNA grafted polymer. The DNA grafted polymer was made by copolymerizing acrylamid and methyl methacrylate including ssDNA' [5]. The surfaces of the components and the substrate were functionalized with the same ssDNA which have a complementary sequence of ssDNA' on the polymer. Since the DNA grafted polymer works as a flexible linker to infill the gap between two surfaces of components and the substrate, it was expected to provide a higher self-assembly yield compared to direct self-assembly using DNA hybridization without any intermediary.

MATERIALS AND PROCEDURES FOR THE EXPERIMENTS

Synthesized DNA and grafted polymer

In order to demonstrate the feasibility of above concept, two pairs of complementary DNA, D21-D21' and D9-D9', were synthesized with a desired length and sequence in Table 1 by Sigma-Aldrich, Japan. The continuous adenines work as a spacer to elongate their length and equalize the hybridized DNA length between two complementary DNA pairs. The T_m was calculated by the Nearest Neighbor method [6]. The thiol on the 5' edge of ssDNA was attached for covalent bonding to gold surface. DNA grafted polymer with D9' was synthesized to verify its effect on assembly yield and speed.

Fabrication and modification with DNA for micro components and substrate

A micro components and a substrate for assembly experiments were fabricated by using conventional micro fabrication process. As a first step, Cr and Au layers of 20 and 80 nm, respectively were deposited on silicon on insulator (SOI) wafer (active layer 2 μm, silicon dioxide 0.5 μm, handle wafer 450 μm) for the micro components and Si wafer for assembly sites. Au and Cr were patterned by lithography and etched in wet etchant. For assembly sites, the Si wafer was diced into 5 mm square chips. Each

chip has 100 assembly sites (200 x 200 μm²). The chips were cleaned with Piranha ($H_2SO_4:H_2O_2=4:1$) and immersed into a solution with 500 ul of 0.1 M NaCl, 10mM phosphoric acid, 0.01 % (w/w) Tween 20 and 0.5 μM thioled ssDNA for 24 hours. With this procedure, the patterned Au film was modified with ssDNA by covalent bonding. The substrate was then carefully rinsed with buffer (0.1 M NaCl, 10mM phosphoric acid and 0.01 % (w/w) Tween 20) for one minute to remove non-attached ssDNA. Figure 4 shows the fabricated assembly sites where Au film is patterned to define the same areas of Si and Au to compare the number of components as will be described in the section of results and discussion.

After patterning of Au and Cr, the device layer of SOI wafer was etched by ICP-RIE. Subsequently, diced chips (5 mm x 5 mm) were immersed in acetone and subsequently Piranha to remove photoresist and clean the Au film surface. The sacrifice silicon dioxide layer was

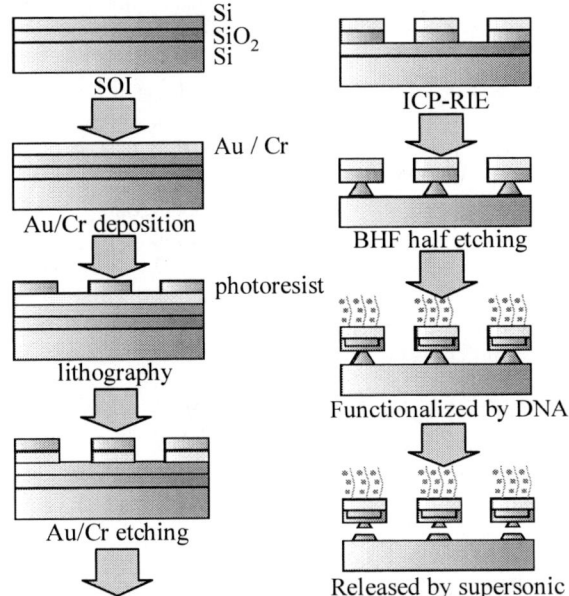

Figure 3: Fabrication process flow for micro components

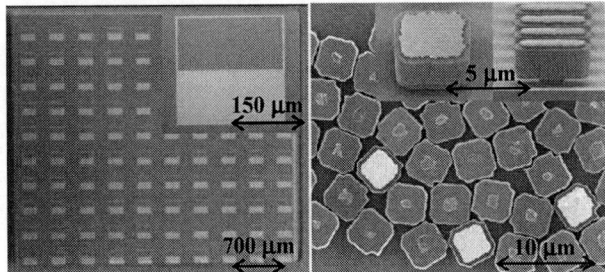

Figure 4: Assembly sites (left), fabricated micro components and components before release (right)

half-etched with buffered hydrofluoric acid solution (BHF). After modification with DNA using the same solutions mentioned above, micro components (5 μm x 5 μm x 2 μm) were released into the NaCl and phosphoric acid solution by supersonic. Figures 3 and 4 show the fabrication process flow and the fabricated components, respectively.

Self-assembly experiments

The substrate with assembly sites and the micro components of about 10^4 pieces are immersed into the test tube containing a buffer solution. Six kinds of samples containing components and substrate modified with different DNAs were utilized. The combination of DNAs were (D21-D21'), (D9-D9'), (D21-D9'), (D9-D21'), (no DNA) and (D9-DNA grafted polymer-D9). During the assembly process, the test tube containing samples was rotated with 4 rpm as agitation. First, the test tubes were denatured to unbind hybridized dsDNA and stretch ssDNA, then by decreasing the temperature from 90 °C to 60 °C, components with D21 self-assembled on Au film with D21' on substrate. With further decrease temperature to RT, the samples of (D9-D9') and (D9-DNA grafted polymer-D9) with low T_m showed self-assembly. During this process, the other 3 samples did not show self-assembly.

In the observation after the self-assembly process, the micro components attached on the substrate were categorized into four conditions; on Au film with their Au face up or down and on Si substrate with their Au face up or down. Much more components were observed with Au face down on Au film than any other three conditions. From this observation, it was concluded that the adhesion force due to hybridization is large enough compared to the hydrodynamic force induced by agitation which acts to release the micro components from the substrate. After a decided time, we counted the number of components of each condition to evaluate an assembly yield. Two kinds of experiments were carried out. (1) Keep six kinds of samples mentioned above rotating at 60 °C and RT for 24 hours to demonstrate sequence and selective control. (2) Assembly experiments using complementary samples (D21-D21', D9-D9', D9-DNA grafted polymer-D9) by changing the assembly time to confirm the assembly speed dependency on base sequence and DNA grafted polymer at RT.

RESULTS AND DISUCUSSIONS
Selectivity and sequence control and effect of DNA grafted polymer

Figures 5 and 6 show the experimental results. Only the combination of (D21-D21') at 60 °C and the three samples of (D21-D21'), (D9-D9') and (D9-DNA grafted polymer-D9) at RT showed a great amount of

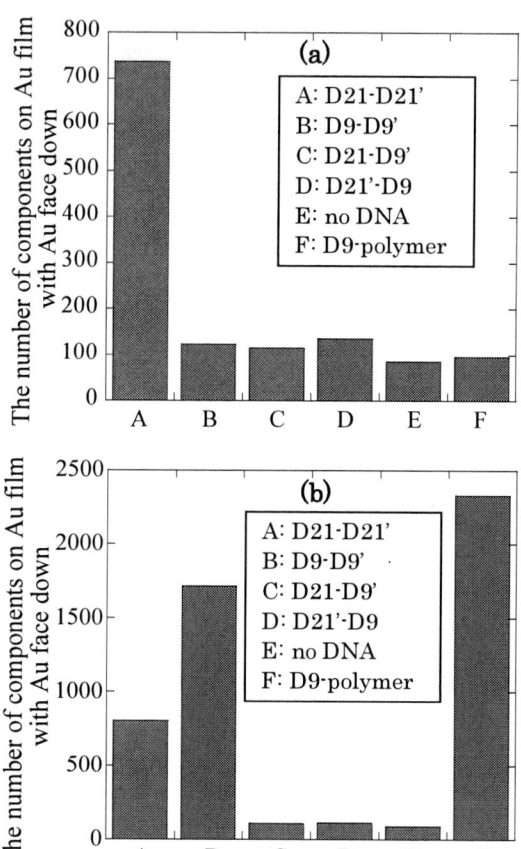

Figure 5: The number of micro components self-assembled on Au film with Au face down (a) at 60 °C, (b) at RT.

self-assembled components. This means, self-assembly occurred only for the samples with complementary DNA combination between the components and the substrate, and thus the number of self-assembled components on Au film with Au face down was much larger than any other three conditions when the temperature was below their T_m. The other combination has few components in every condition such as in Figure 6(b). Comparing D21-D21' to D9-D9' in Figure 5(b), there was much difference about the number of self-assembled components on Au film with Au face down although they are both below their T_m. The superiority of (D9-D9') can be explained by the difference of the length of hybridized parts in D9 and D9'. Only nine bases in D9 need to hybridize with their complementary bases in D9'. Therefore, it is easier to make dsDNA in the top nine bases than twenty one bases for the pair of (D21-D21'). On the other hand, the bonding strength of self-assembled components of (D9-D9') is weaker than that of (D21-D21'), but it seems to be enough to overcome the agitation force induced by rotation in order to keep Au

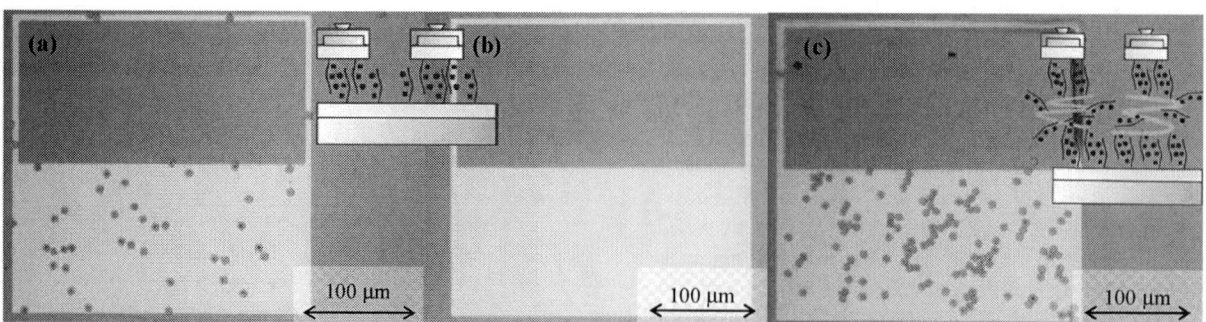

Figure 6: Optical photomicrograph of sites after assembly experiments at RT; (a) D9-D9' (b) D21-D9' (c) D9-DNA grafted polymer with D9'

978-1-4244-2977-6/09 $25.00 © 2009 IEEE

face down components on their assembled condition. When DNA grafted polymer containing D9' was applied as a spacer between components and the substrate, both modified with D9, the assembly yield was improved as shown in samples B and F in Figure 5. The reason was that the DNA grafted polymer worked well as a spacer to infill the gap between two surfaces. From these experiments, it is successfully confirmed that the DNA hybridization can work as a useful adhesive to control the sequence and selectivity of self-assembly between micro components and the substrate. As a future work, the influence of a percentage of the grafted DNA (0.136 % in this experiment) of whole polymer and number of components inside the buffer on yield will be investigated.

Time dependency on yield

Another assembly experiments were done only for complementary samples of (D21-D21'), (D9-D9') and (D9-DNA grafted polymer-D9) at RT with parameterizing the time during the rotation of test tubes with 4 rpm. Figure 7 shows the number of assembled components on Au film modified with complementary ssDNA with Au face down. The assembly process could be finished within six hours in the case of (D21-D21') and three hours for (D9-D9') and (D9-DNA grafted-polymer-D9). From the results, it was revealed that the length of hybridized parts in ssDNA not only affects the final yield but also the assembly speed. A shorter ssDNA is preferable for high yield and speed self-assembly. When DNA grafted polymer was used, it showed higher and faster yield than (D9-D9').

Figure 8 shows the transition of the number of components attached on the substrate with time. These four images were taken from the same substrate. After one hour, some components on both areas of Au film and Si substrate were observed. One hour later, all components on Au film in Figure 8(a), surrounded with a square, were remained on Au film as shown in Figure 8(b) keeping their orientation. Also, a greater amount of components were self-assembled on Au film. However, the two components on Si, surrounded with a circle in Figure 8(a) and 8(b), were released because of agitation. In Figure 8(c), two components were stuck with their Au side facing up. One of them was on Au film, surrounded with a triangle in Figure 8(c) and the other was on Si substrate, surrounded with a diamond shape. Both were removed as can be seen in Figure 8(d). During the assembly process, the number of self-assembled components on Au film with Au face down increased steadily until six hours have passed. From these results, it was expected that almost all unexpected components can be released by adjusting the strength of agitation.

CONCLUSIONS

In this paper, we proposed and demonstrated sequential and selective self-assembly using the characteristics of DNA hybridization, such as specific bonding laws and self-assembly sequence control by ambient temperature. The sample of (D9-D9') showed a higher and faster self-assembly at RT than (D21-D21') because of the difference in length of the DNA involved in the hybridization process. Furthermore, DNA grafted polymer used as a spacer in between components and the substrate improved the final assembly yield and speed than the samples of (D9-D9').

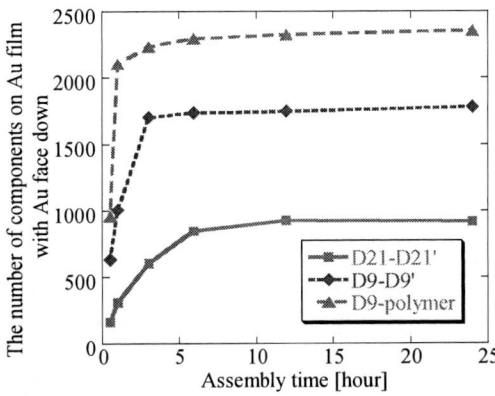

Figure 7: The dependency of assembly yield on time

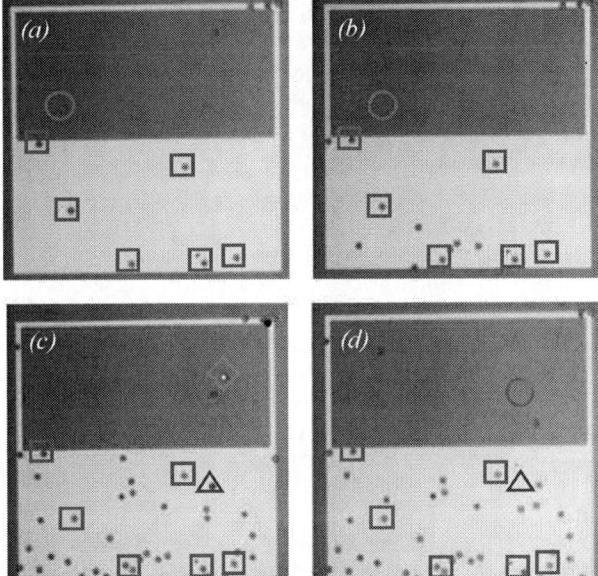

Figure 8: The transition of components on a assembly site. The assembly process after (a) 1hour (b) 2hours (c) 6 hours (d) 12 hours

ACKNOWLEDGEMENTS

The authors would like to acknowledge Prof. Noji and Prof. Tabata of Osaka University for the knowledge of DNA and Mr. Kinoshita of Kyoto University for taking SEM images.

REFENENCES

[1] C. J. Morris, S. A. Stauth and B. A. Parviz, "Self-Assembly for Microscale and Nanoscale Packaging: Steps Toward Self-Packaging," *IEEE ADVANCED PACKAGING*, Vol.28, No.4, pp.600-611, 2005.

[2] T. Kusakabe, T. Tanemura, Y. Higuchi, K. Sugano, T. Tsuchiya and O. Tabata, "DNA Mediated Sequential Self-assembly of Nano/Micro Components", *MEMS'08*, pp. 1052-1055, 2008.

[3] H. Onoe, K. Matsumoto, and I. Shimoyama, "Sequential Self-Assembly by Controlling Interactive Forces between Microparticles", *MEMS' 04*, pp.5-8, 2004.

[4] L. M. Dillenback, G. P. Goodrich, and C. D. Keating, "Temperature-Programmed Assembly of DNA: Au Nanoparitcle Bioconjugates", *Nano Lett.*, 6, pp.16-23, 2006.

[5] M. Maeda, "Sequence-Specific Aggregation Behavior of DNA-Carrying Colloidal Nanoparticles Prepared from Poly (N-isopropylacrylamide)-graft-Oligodeoxyribonucleotide", *Polymer Journal*, Vol. 38 No. 11, pp. 1099-1104, 2006

[6] Sigma-Aldrich technical information, *http://www.Sigma aldrich.com/Brands/Sigma_Genosys/Custom_DNA/Key_R esources/Oligos_Melting_Temp.html*

PIEZORESISTIVE CANTILEVER-BASED FORCE-CLAMP SYSTEM FOR THE STUDY OF MECHANOTRANSDUCTION IN C. ELEGANS

S.-J. Park[1], B. Petzold[1], M.B. Goodman[2], and B.L. Pruitt[1]

[1]Department of Mechanical Engineering, Stanford University, Stanford, CA, USA
[2]Department of Molecular and Cellular Physiology, Stanford University, Stanford, CA, USA

ABSTRACT

Understanding how the mechanoreceptor neurons of *Caenorhabditis elegans* mediate mechanotransduction can unravel how touch works, but new tools are required to quantitatively analyze the relationship between mechanical loading and the physiological response. Here we present a piezoresistive cantilever-based force clamp system that can apply user-defined force profiles to *C. elegans*. We present a novel MEMS force-clamp system and demonstrate a piezoresistive cantilever with low 1/f noise, low noise floor and high force resolution suitable for these measurements. Initial studies enabled by the system are also discussed.

INTRODUCTION

Cellular mechanotransduction, the conversion of force into to an electrical or biochemical signal, is a fundamental process essential to normal life, including hearing, touch and balance. Among these, touch sensation is the least understood. The nematode *Caenorhabditis elegans* is one of the most powerful model organisms in which to analyze the mechanism of touch sensation [1]. *C. elegans* is extremely simple compared to humans: it has 302 neurons and about 1000 cells. Six touch receptor neurons that innervate the length of its body are responsible for touch sensitivity. The classical method to test for touch sensitivity in *C. elegans* is to score its response to applied mechanical stimuli. One of us (Goodman) previously used micro-von Frey hairs to touch to freely moving worms until buckling was observed. Each different size probe has different buckling force, so each probe was used to measure touch sensitivity at different force levels. However, this method has the intrinsic issue of precision and accuracy: buckling of the probe is an unstable deformation mode. In addition, it is incapable of delivering forces <10μN which is at least an order of magnitude larger than the likely threshold for touch-sensitivity. Therefore, we developed new tools for quantitative evaluation of touch of *C. elegans*.

Piezoresistive cantilevers are well-suited to this application: they cover the relevant range of forces (nN to 100 μN), displacements (nm to 10 μm), and offer sufficient bandwidth (up to kHz).. Designing appropriate piezoresistive cantilevers is challenging, however, since numerous coupled design parameters should be chosen carefully to improve sensitivity while minimizing noise. Thus, we developed analytical models to predict cantilever sensitivity by analyzing the results of Tsuprem4 doping profile simulations and the observed sensitivity of experimental cantilevers [2]. Here, we demonstrate a piezoresistive cantilever with low 1/f noise (1 Hz corner frequency), low noise floor (4 nV/Hz$^{1/2}$) and high force resolution (< 500 pN over 0.1 to 10 kHz). By integrating the cantilever with a fast real-time controller, we present a novel MEMS force-clamp system that can apply user-

defined force profiles (e.g. constant, sinusoidal) to *C. elegans*. We also present initial mechanotransduction studies enabled by the system which reveal new insights into the inner working of the sense of touch in this tiny nematode.

METHOD

Piezoresistive cantilever

We designed and fabricated custom microcantilevers of several sizes (1.7 to 6 mm long, 30 to 400 μm wide, 7 to 50 μm thick cantilever with a 200 to 350 μm long and 10 to 20 μm wide U-shaped piezoresistor) using previously discussed microfabrication techniques [3, 4]. We glued completed cantilevers to custom printed circuit boards with epoxy (Devcon, Glenview, IL) and created aluminum wire interconnects between the device bond pads and the circuit board contacts using a wirebonder (Figure 1A). To control the contact area, a glass bead (10μm diameter, Duke Scientific, Palo Alto, CA) was attached to the end of each cantilever with an ultraviolet adhesive (Loctite 352, Henkel Technologies, Germany) (Figure 1B).

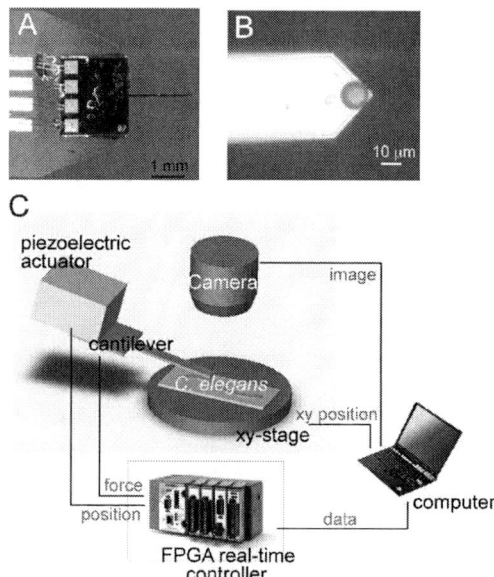

Figure 1: Piezoresistive microcantilever (2mm long, 30μm wide, 7μm thick) glued on printed circuit board (A) with 10-μm-diameter glass bead (B) and schematic of force and displacement clamp system using proportional-integral-derivative (PID) field programmable gate array (FPGA) controller.

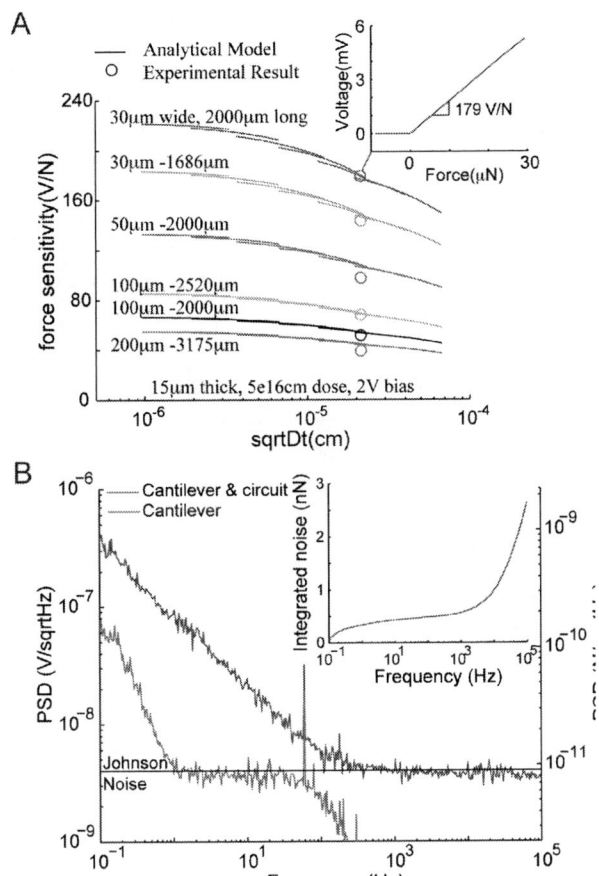

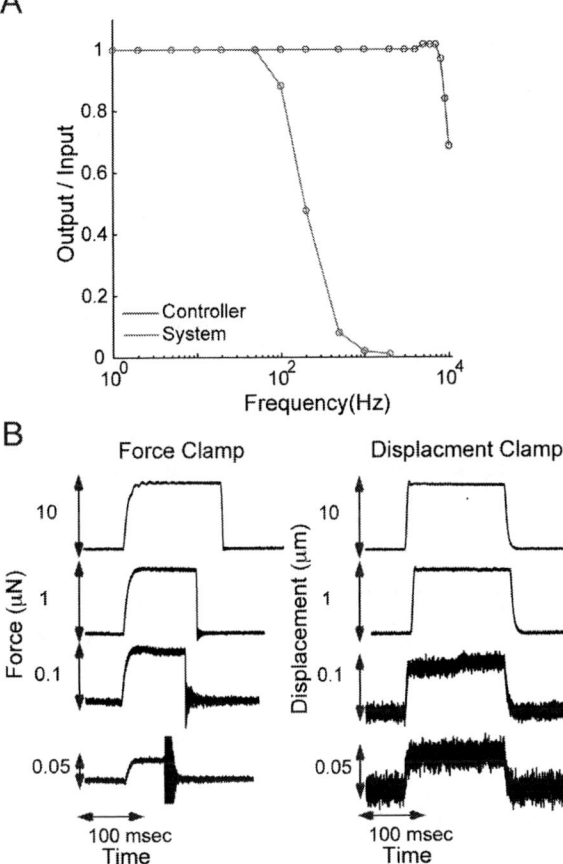

Figure 2: (A) Force sensitivity of piezoresistive cantilevers with various cantilever dimensions and process conditions. The force sensitivity is linear up to a few tens of μN (inset). (B) Noise power spectral density (PSD) for a cantilever and associated conditioning circuitry and integrated noise from 0.1Hz (inset).

We determined the sensitivity of the mounted device by utilizing a previously described resonant frequency technique which incorporates a piezoelectric shaker and a laser doppler vibrometer [3, 4]. To characterize the system noise, we implemented conditioning circuitry and an HP3562A dynamic analyzer. A TI INA103 instrumentation amplifier (Johnson noise floor 1 nV/Hz$^{1/2}$, 1/f noise corner frequency of 100 Hz) was used to remove the 1/f noise of instrumentation amplifier, allowing the noise of the cantilever itself to be quantified [5].

Force-clamp system

The force-clamp system consists of a piezoresistive cantilever, a piezoelectric actuator with a built-in capacitive sensor (PIHera P-622.Z, Physik Instrumente) to move the cantilever, and a real-time controller (CompactRIO Field Programmable Gate Array (FPGA) System, National Instruments) (Figure 1C). Resistance changes associated with applied force and associated cantilever deflection are measured using a Wheatstone bridge and an instrumentation amplifier to magnify the resultant voltage signal (AD8221, Analog Devices, for piezoresistor with high resistance, and INA103, Texas Instruments with low resistance). The position of the

Figure 3: Force-clamp system: (A) Frequency response of the controller vs. that of the system. (B) Performance in force and displacement clamp modes exhibit large dynamic range (10 to 0.05 μN and 10 to 0.05 μm) and fast response (<20msec rise time). The attraction force between the cantilever and sample resulted in ringing upon removal of force clamp stimulus (region not critical for experimental purposes).

cantilever is controlled and monitored by a piezoelectric actuator with a built-in capacitive sensor. To correct the error between the voltage signal from the piezoresistive cantilever and a desired setpoint, the real-time controller calculates and adjusts the driving voltage of the piezoelectric actuator with a 100 kHz feedback loop frequency, utilizing proportional, integral, derivative (PID) logic. During a measurement, the force signal from piezoresistive cantilever, displacement signal from actuator built-in sensor, and driving voltage of the actuator are recorded with a data-acquisition card (DAQCard 6036E, National Instruments) and LabView, enabling post hoc analysis of force-displacement relationships.

C. elegans behavioral study

An age-synchronized population of wild-type (N2) *C. elegans* nematodes was obtained by cultivating animals on standard 2% NGM agar plates for 48-52 hours at 20°C until animals reached the L4, young adult stage.

To precisely control the probe position and touch a selected spot (head or tail) on a moving *C. elegans*, we implemented a vision-based x-y tracking system in MATLAB® to operate in parallel with the force-clamp system. The tracking system allows the force probe to be rapidly positioned over an animal of interest so that the desired force to be applied precisely to a selected spot on the animal. In addition to the tracking system, we also developed MATLAB based image processing program, which can calculate the position and moving speed of head or tail before and after force is applied.

RESULTS AND DISCUSSION
Piezoresistive cantilever

We fabricated piezoresistive cantilevers with high force sensitivity (few 100 V/N at 2V bias) while maintaining low noise (few 100 nV, few nN integrated noise at 0.1 Hz to 10 kHz). The analytical model agrees well with the experimentally determined force-sensitivity (<20% error, Figure 2A). Our experimental results also show that the cantilevers maintain linear force sensitivity up to tens of μN (Figure 2A, inset).

Our typical piezoresistive cantilever (2000 x 50 x 7 μm, 1 kOhm piezoresistor fabricated with a 5E16 cm^{-2} boron dose followed by 15 min. oxidation and 10 min. inert annealing at 1000 °C) has a 4 nV/Hz$^{1/2}$ (10 pN/Hz$^{1/2}$) Johnson noise floor and 1 Hz 1/f corner frequency. The 1/f noise of cantilever (0.1 Hz to 10 kHz) is lower than that of the TI INA 103 for a 2 V bias (Figure 2B). This suggests that the bias voltage could potentially be increased in the future to induce higher sensitivity (sensitivity is proportional to bias voltage) until the 1/f noise of cantilever becomes dominant over that of instrumentation amplifier (1/f noise is also proportional to bias voltage). Utilizing this cantilever, the resulting system force resolution is ~500 pN over 0.1 Hz to 10 kHz band (Figure 2B, inset).

Force-clamp system

We integrated the piezoresistive cantilever and circuit with an FPGA-PID, real-time controller and a piezoelectric actuator. We characterized the output frequency response of the system to input signals of varying frequency driven at the maximum amplitude of the actuator (250 μm). Since the resonant frequency of the cantilever and corner frequency of the controller are ~ 5 kHz, the system is limited by the actuator (50 Hz roll-off) (Figure 3A). We tested the force and displacement clamping modes by holding scaled force or displacement at a desired magnitude (Figure. 3B). The demonstrated large dynamic range (50 nN to 10 μN) and short rise time (< 20 msec) of the system are suitable for analysis of touch sensitivity in freely-moving and immobilized *C. elegans* worms.

C. elegans behavioral study

To study the behavioral response of *C. elegans* to touch stimuli, we utilized the system in force-clamping mode. We applied a 1 μN force to freely-moving worms (Figure. 4). Image processing successfully determined the head position (Figure 3A and B) and calculated the moving speed (Figure 3C) of the animal. Of 25 wild-type worms tested, 100% responded to 1 μN force, i.e. they reversed direction and accelerated in response to such stimuli. The behavioral result agrees with our prior work which suggests that electrical responses to touch saturate near a force threshold of 1 μN [7]. Thus, *C. elegans* nematodes are more sensitive to touch than are humans, whose threshold for touch-sensitivity on the palm of the hand is several hundred milli-Newtons [8].

CONCLUSION

We have demonstrated a piezoresistive cantilever with low noise and high force sensitivity and a force-clamp system characterized by a fast response and a large dynamic range which is suitable for analysis of *C. elegans* touch-sensitivity. We also conducted several initial experiments with unprecedented force resolution. To find the behavioral threshold of touch, we will improve the force resolution of the system in the future by increasing bias voltage, making softer cantilevers, and by integrating a low-pass filter to remove high frequency noise. To

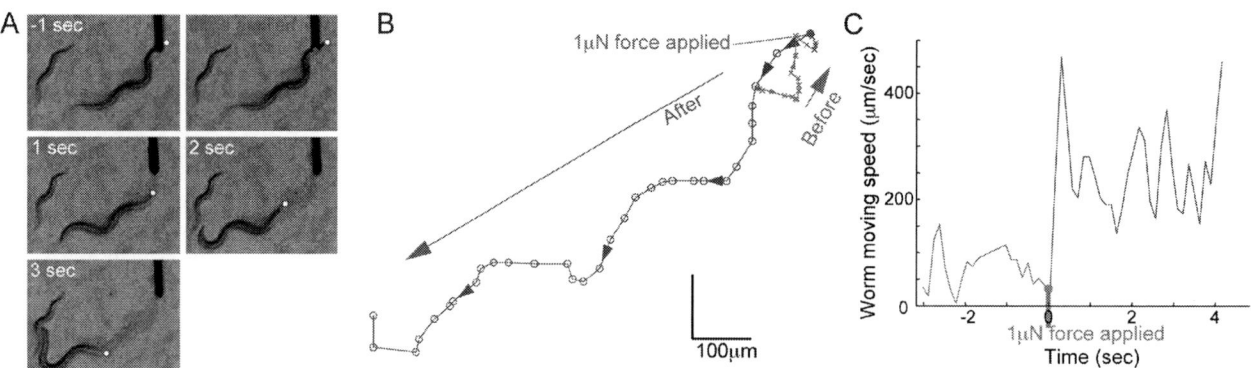

Figure 4: Behavioral study of C. elegans touch sensation. (A) Image sequence of C. elegans response to a 1μN force. The white and red dots correspond to the head position and contact location, respectively. (B) Sample head trajectory and (C) worm moving speed data, sampled every 130 msec, indicating that the animal reversed direction and accelerated in response to a 1 μN stimulus. This hard-wired behavior, characteristic of normal touch response in C. elegans, has previously only been studied for forces down to 10 μN. Our MEMS-based system allows unprecedented resolution and precision in touch sensitivity screening.

resolve the effects of mutations on touch sensitivity, we will evaluate sensitivity of wild type and mutant worms which lack key proteins known to be required for proper touch sensation. Future applications include integrating the force-clamp system with electrophysiological measurements of force-gated ion channels in *C. elegans* touch receptor neurons, to observe the transient electrochemical signals which propagate in response to applied mechanical stimuli. These behavioral and physiological data will be correlated with each other, to unravel mystery of mechanotransduction mechanism.

ACKNOWLEDGEMEMTS

We thank A. Rastegar for help with noise measurements. This work was performed in part at the Stanford Nanofabrication Facility (a member of the National Nanotechnology Infrastructure Network) which is supported by the National Science Founding under Grant 9731293, its lab members, and the industrial members of the Stanford Center for Integrated Systems. Funding was also provided under NSF ECCS-0449400, ECCS-0425914, and NIH R01EB006745. S-JP was supported by a Samsung fellowship.

REFERENCES

[1] M. B. Goodman, E. Schwarz, "Transducing Touch in Caenorhabditis elegans", *Annual Review of Physiology* vol. 65, pp. 429-452, 2003.

[2] S. –J. Park, A. J. Rastegar, T. H. Fung, A. A. Barlian, J. R. Mallon Jr., and B. L. Pruitt, "Optimization of Piezoresistive Cantilever Performance", in *Digest Tech. Sensors and Actuators Workshop, Hilton Head Island, SC*, June 1-5, 2008, pp. 98-101.

[3] B. L. Pruitt, T. W. Kenny, "Piezoresistive Cantilevers and Measurement System for Characterizing Low Force Electrical Contacts", *Sens Actuators A: Phys.*, vol. 104, pp. 68-77, 2003.

[4] S. –J. Park, M. B. Goodman, B. L. Pruitt, "Analysis of Nematode Mechanics by Piezoresistive Displacement Clamp", *Proc. Natl. Acad. Sci. USA.*, vol. 104, pp. 17376-17381, 2007.

[5] J. R. Mallon Jr., A. J. Rastegar, A. A. Barlian, M. T. Meyer, T. H. Fung, B. L. Pruitt, "Low 1/f Noise, Full Bridge, Microcantilever with Longitudinal and Transverse Piezoresistors", *Appl Phys Lett*, vol. 92 033508, 2008.

[6] I Hope, *C. elegans: A Practical Approach*, Oxford Univ Press, Oxford, 1999.

[7] R. O'Hagan, M. Chalfie, M. B. Goodman, "The MEC-4 DEG/ENaC Channel of Caenorhabditis elegans Touch Receptor Neurons Transduces Mechanical Signals", *Nat Neurosci*, vol. 8, pp. 43-50, 2005.

[8] R. S. Johansson, A. B. Vallbo, G. Westling, "Thresholds of Mechanosensitive Afferents in the Human Hand as Measured with von Frey Hairs", *Brain Res*, vol. 184, pp. 343-351, 1980.

SUB-5µM DIAMETER SILICON MICROPROBE AND SILICON DIOXIDE MICROTUBE ARRAYS FOR LOW-INVASIVE *IN-VIVO* ANIMAL EXPERIMENTS

Kuniharu Takei[1,3], Takeshi Kawano[1,4], Takahiro Kawashima[1], Kazuaki Sawada[1,4], Hidekazu Kaneko[2], Makoto Ishida[1,4]

[1]Toyohashi University of Technology, Toyohashi, JAPAN
[2]National Institute of Advanced Industrial Science and Technology, Ibaraki, JAPAN
[3]Research Fellow of the Japan Society for the Promotion of Science, Tokyo, JAPAN
[4]Japan Science and Technology Agency – CREST, Saitama, JAPAN

ABSTRACT

We report applications of sub-5 µm diameter silicon microprobes and silicon dioxide microtubes fabricated by vapor-liquid-solid method and microfabrication techniques, to use in *in-vivo* animal experiments. Before animal experiments, electrical impedance and test signal detection via the microprobe electrode were investigated. According to the results, electrical impedance at 1 kHz is approximately 2 MΩ, and microprobe detected the test signals ($80 \mu V_{p-p}$ at 1kHz) in saline solution. For microtube array, liquid flow through the microtube is confirmed at injection pressures of 25-45 kPa to get sub-µL/min flow rates. Finally, the microprobe and the microtube were applied to tibial muscle and sciatic nerves of rats, respectively. From these animal experiments, we could observe the electrical and chemical reactions of muscle and nerves using both sub-5 µm diameter microprobe and microtube successfully.

INTRODUCTION

A large number of microfabricated devices such as microneedles have been addressed *in-vivo/in-vitro* recording of neural activities, and played an important role in understanding neural behaviors [1-3]. However these microneedle structures are still proposed and improved to realize more useful device structures and further functionalities, for various recordings of neural responses [4-5]. One of the candidates for the devices is heterogeneous integration of microelectrodes for recording and microtubes for drug delivery. Several integrated devices have already been proposed, that consists of needle structures with the diameter of over 100 µm [6-7]. Even though, the proposed devices give good opportunity to study neural behaviors by observing electrical and chemical reactions of neurons, the spatial resolution of the analyses of neural behaviors is still low and the invasiveness is large because of their diameters. Edell *et al.* have reported the needle size for minimum invasiveness against neurons [8]. According to the report, needle diameter should be smaller than 10 µm.

We have proposed a concept that a microchip realizes electrical recording of neurons via the microprobe while drug is released into the neuronal tissue from another microtube. To achieve the probe-tube device, we have reported a heterogeneous integration of silicon microprobe and silicon dioxide microtube arrays with on-chip NMOSFETs circuitry, fabricated by IC process followed by a vapor-liquid-solid (VLS) method and microfabrication processes using silicon microprobe templates as shown in Fig. 1 [9]. The probe and tube array

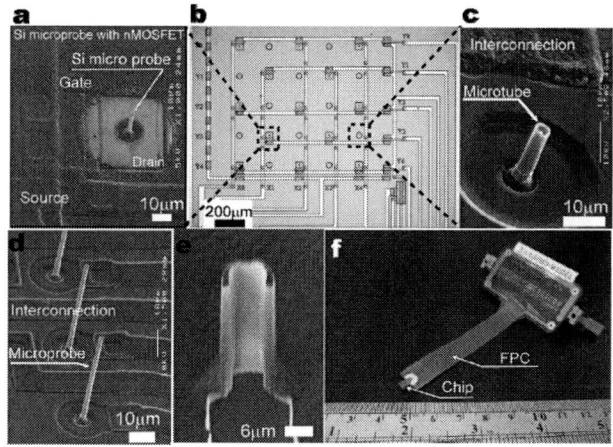

Figure 1: Microprobe and microtube arrays integrated with NMOSFET circuitry. a: microprobe on a drain region of NMOSFET, b: microscope image of 5×5 array of microprobe and microtube with NMOSFET circuitry, c: microtube integrated with microprobe and circuitry, d: microprobe array without circuit for test recording, e: cross sectional image of microtube fabricated by a focus ion beam, f: in-vivo packaging image.

is very fine diameter of less than 5 µm, which realizes low invasiveness [8] and high density with less than 100µm pitch. In this paper, we present both the electrical properties of the microprobe and chemical-flow properties of the microtubes. Here we also report *in-vivo* animal (rat) experiments, focusing on the recording of the electrical responses of myoelectric potential via the microprobes and the drug delivery into a sciatic nerve through the microtubes. The proposed device will be able to provide simultaneous electrical recording and chemical delivery at localized area in tissue with minimum invasiveness against the tissues, due to the integrated sub-5µm diameter probe/tube arrays.

ELECTRICAL PROPERTIES

Firstly we investigate the electrical properties of microprobes to apply to neural recording. To discuss the neural electrode, electrical impedance is very important factor, because the neural signal is very small such as the amplitude of less than 1 mV (extracellular signals). Electrode impedance and test signal recording of the microprobe are measured in saline environment after the device was packaged (Fig.1f). Figure 2a indicates the impedance property of a microprobe, which has 2 µm in a diameter and 30 µm in a length, as a function of signal

---

978-1-4244-2977-6/09 $25.00 © 2009 IEEE   192

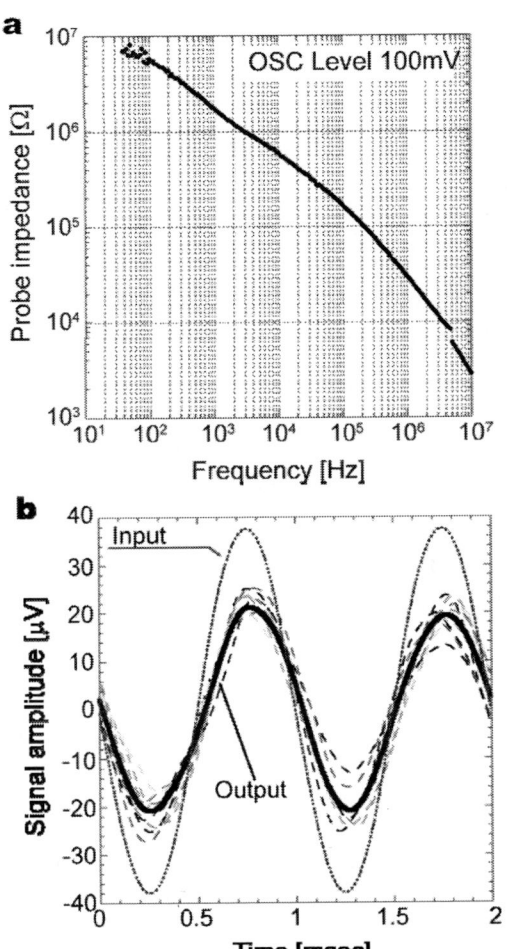

Figure 2: Electrical properties of microprobe measured in saline solution. a: electrical impedance as a function of frequency, b: test signal recording through silicon microprobe electrode in saline solution. Input signal is 1 kHz frequency and approximately 80 μV_{p-p} signal amplitude

frequency. The impedance at 1 kHz, which frequency is well known as same as neural signals, is approximately 2 MΩ. The value is small compared to commercial microneedle (Example: 10 MΩ, a few micrometer diameter of tungsten needle). Figure 2B shows the signal output results with input test sinusoidal waves (80μV_{p-p} at 1kHz), and this indicates that the microprobe can detect the test signals. The output/input signal ratio was found to be about 60% (Fig. 2b), and the reason is due to the parasitic capacitances existing in the substrate [10]. This is good agreement why we also observe the low impedance value compared to the above commercial needle. However, very fine microprobe can detect enough signal amplitude over the noise level (approximately 20 μV_{p-p}) to record the neural response.

LIQUID FLOW THROUGH MICROTUBE

Figure 3 shows the flow behavior of the microtube, that liquid (water) injections were observed through a 4.1μm inner diameter microtube. The flow rate of 1.1 μL min^{-1} was achieved using the flow pressure of 35 kPa.

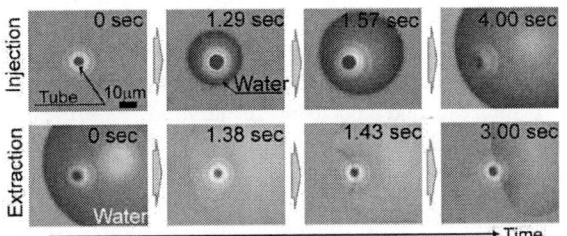

Figure 3: Liquid flow experiments through a microtube, which has 4.1 μm inner diameter. Upper figures are injection from the microtube, and lower figures are extraction from the microtube

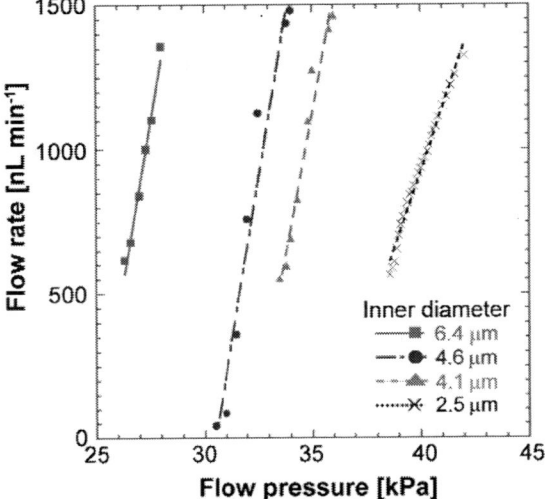

Figure 4: Liquid (DI water) flow rates through 2.5 μm to 6.4 μm inner diameter microtubes, as a function of flow pressure

Figure 3 indicates that the microtube has capability to liquid flow through the microtube without breaking after injection and extraction. The detail of flow rate as a function of flow pressure is shown in Fig. 4. The results show the linear relationship between flow pressure and flow rate, similar to the Hagen-Poiseuille law, even though the sizes of the microtubes are only a few micro meters. However, the observed values of liquid flow rates through our microtubes are not consistent with the theoretical values estimated by the Hagen-Poiseuille law. This discrepancy is probably due to eddy flow occurred at the connection between the bottom of microtube and the top of the silicon channel. In any case, microtubes with 2.5 ~ 6.4 μm inner diameters allow us to control the flow rate level of sub-μL min^{-1} at a flow pressure on the order of 10^5 Pa. The understanding of these flow properties will be usefull in efforts to precisely control the dose of drug and to design some micropumps to integrate with the microtube devices.

ANIMAL EXPERIMENTS

We applied each microprobe and microtube to nerves of rats (Sprague-Dawley (SD), male) in-vivo for the application of neuroscience. Figure 5 shows the schematic image of animal experiments. For in-vivo recording, the microprobes are penetrated into the anterior tibial muscle to record the evoked myoelectric potential (Fig. 5a). To record them, a coaxial electrode for stimulation was inserted in a sciatic nerve, and we used the stimulation parameters of 1.6 µA stimulatin current, 3 pulses trains, 20 ms pulse interval, and 200 µs/pulse duration. Figure 5b indicates an image of lidocain injection through the microtube. The microtubes are penetrated into a sciatic nerve for the drug administration. As a model agent for drug delivery, we used lidocaine, which temporarily disrupts the conduction of the nerve action potentials. To confirm the drug delivery, we observe evoked myoelectric potentials using Ag-AgCl electrodes, which penetrated into the anterior tibial muscle, while recording before/after the drug delivery. As same as recording of myoelectric potential, electrodes for stimulation of sciatic nerve are applied on the sciatic nerve. For both experiments, the branches of the sciatic nerve were cut except for the tibial nerve branch under anesthesia. All experimental procedures were based on the guidelines from the National Institutes of Health of the United States (1996) and the Japan Neuroscience Society.

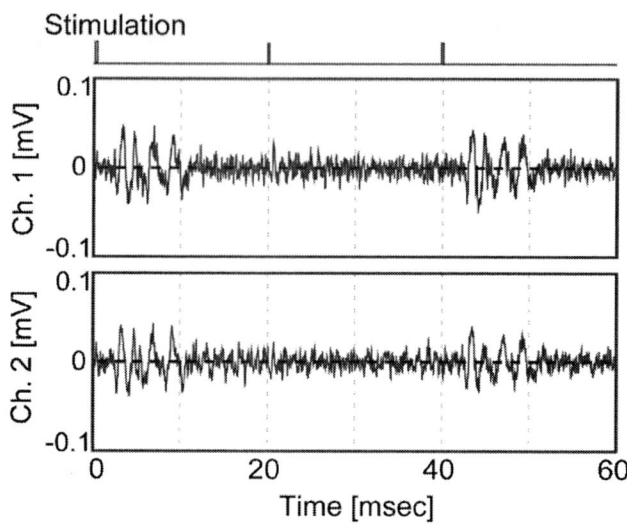

Figure 6: Evoked myoelectric potential from the anterior tibial muscle recorded by the microprobe array. The impedances of microprobes of Ch.1 and Ch.2 are 1.91 MΩ and 6.16 MΩ, respectively.

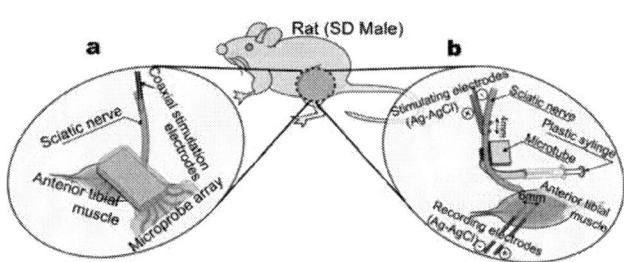

Figure 5: Schematic image of animal experiments. a:evoked signal recording using the microprobe array from an anterior tibial muscle stimulated by sciatic nerve, b: drug delivery to sciatic nerve using the microtube.

Myoelectric potential recording via microprobes

Microprobe arrays, which have 2 µm in a diameter and 60µm in a length, are used to record the evoked myoelectric potential. Electrical impedances of two-channel microprobes are 1.91 MΩ and 6.16 MΩ at 500 Hz, respectively. Figure 6 indicates the results of evoked potentials via two-channel microprobes. Myoelectric potentials were observed via both microprobes in 1 ~ 2 msec after stimulation of the sciatic nerve. We could also observe the no signal between 20 msec and 40 msec, even though the sciatic nerve was stimulated with the second pulse, as shown in Fig. 6. The reason is that the recorded potentials obeyed "all-or-none" law of nervous activity. This is good agreement to record myoelectric potential via microprobe array successfully. In this paper, we found the signal attenuation via microprobe because of high impedance of the electrode and high parasitic capacitance existing in the substrate, as shown in Fig. 2b. However, amplitude of myoelectric potential is much higher than the noise level. This result proves that a few micron diameter microprobe arrays can apply to myoelectric potential recording devices.

Lidocaine injection through microtube

The reactions of sciatic nerve against lidocaine administration injected through a microtube (4.6 µm in a diameter and 20 µm in a length) are observed. We confirmed that before releasing the drug, the tibial muscle contracted synchronously with the stimulation, which threshold of the stimulation current was 3.0 µA. In experiments, we set the stimulation current to 3.5 µA for the nerve-block test. The volume of injected lidocaine is less than 5 µL in 1.5 min. Figure 7a-c result that evoked potentials before and after lidocaine injection through the microtube. The myoelectric potential was disappeared in 10.5 min after the release of lidocaine (Fig. 7b), and the evoked muscle potentials recovered ~ 140 min after lidocaine injection (Fig. 6c).

We also measured threshold currents for inducing muscle evoked potentials. After lidocain injection, threshold current was drastically increased as shown in Fig. 7d. Next the threshold current decrease from 50 µA to 9 µA between 55 min and 73 min after the injection. The amplitude then recovered to level similar to that before the lidocaine administration in 140 min after injection. The results prove that drug administration through the sub-5µm diameter microtube was successfully implemented to the sciatic nerve [11].

CONCLUSION

In this paper, we have indicated and proved the possibilities of microprobe and microtube devices, which have sub-5 µm in a diameter, for applications to *in-vivo* analysis of neural networks with electrical recordings and chemical reactions. In future, the integrated device might be applicable to the electrical recordings of localized neuronal tissues as well as single neurons to specific bio-reactions during agent administration.

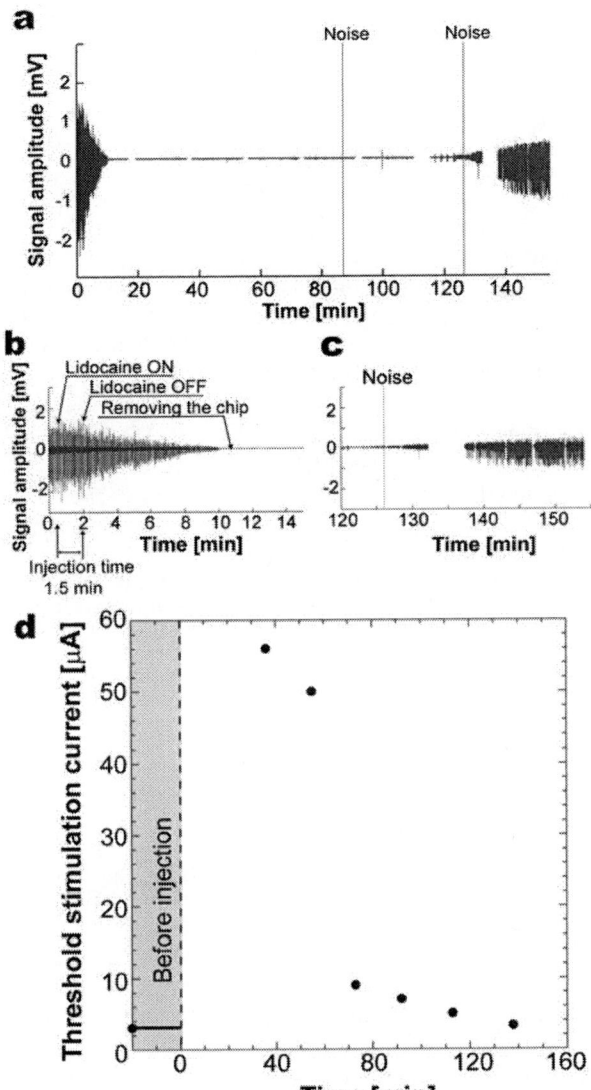

Figure 7: Evoked myoelectric potential before and after lidocain injection into the sciatic nerve through the microtube. a: signal amplitude before/after lidocaine injection through microtube as a function of time, b: initial 14 min, c: 115 min after lidocaine injection, d: changes in the threshold of stimulation current

ACKNOWLEDGEMENTS

The authors would like to thank Mr. Mitsuaki Ashiki for his assistance with fabrication processes. This work was supported by a Grand-in-Aid for Scientific Research (S) (MI) and for JSPS Fellows (KT), by the global COE Program "Frontiers of Intelligent Sensing" from the Ministry of Education, Culture, Sports, Science and Technology of Japan (MI), and by grants from the National Institute of Advanced Industrial Science Technology (HK).

REFERENCES

[1] K. D. Wise, J. B. Angell, A. Starr, "An Integrated-Circuit Approach to Extracellular Microelectrodes", *IEEE Trans. Biomedical Engineering*, vol. BME-17, pp. 238-247, 1970.

[2] P. K. Campbell, K. E. Jones, R. J. Huber, K. W. Horch, R. A. Normann, "A Silicon-Based, Three-Dimensional Neural Interface: Manufacturing Processes for an Intracortical Electrode Array", *IEEE Trans. Biomedical Engineering*, vol. 38, pp. 758-768, 1991.

[3] L. Lin, A. P. Pisano, "Silicon-processed microneedles", *J. Microelectromech. Syst.*, vol. 8, pp. 78-84, 1999.

[4] T. Kawano, Y. Kato, R. Tani, H. Takao, K.Sawada, M. Ishida, "Selective vapor-liquid-solid epitaxial growth of micro-Si probe electrode arrays with on-chip MOSFET on Si (111) substrates", *IEEE Trans. Electron Devices*, vol. 51, pp. 415-420, 2004.

[5] K. Takei, T. Kawashima, K. Sawada, M. Ishida, "Out-of-Plane Microtube Arrays for Biomedical Sensors Using Vapor-Liquid-Solid Growth Method", *IEEE Sensors Journal*, vol. 8, pp. 470-475, 2008.

[6] S. Takeuchi, D. Ziegler, Y. Yoshida, K. Mabuchi, T. Suzuki, "Parylene flexible neural probes integrated with microfluidic channels", *Lab Chip*, vol. 5, pp. 519-523, 2005.

[7] J. Chen, K. D. Wise, J. F. Hetke, S. C. Bledsoe, "A multichannel neural probe for selective chemical delivery at the cellular level", *IEEE Trans. Biomedical Engineering*, vol. 44, pp. 760-769, 1997.

[8] D. J. Edell, V. V. Toi, V. M. McNeil, L. D. Clark, "Factors Influencing the Biocompatibility of Insertable Silicon Microshafts in Cerebral Cortex", *IEEE Trans. Biomedical Engineering*, vol. 39, pp. 635-643, 1992.

[9] K. Takei, T. Kawashima, T. Kawano, H. Takao, K. Sawada, M. Ishida, "Integration of out-of-plane silicon dioxide microtubes, silicon microprobes and on-chip NMOSFETs by selective vapor-liquid-solid growth", *J. Micromech. Microeng.*, vol. 18, p. 035033, 2008.

[10] D. A. Robinson, "The electrical properties of metal microelectrodes", Proc. IEEE, vol. 56, pp. 1065-1071, (1968).

[11] K. Takei, T. Kawashima, T. Kawano, H. Kaneko, K. Sawada, M. Ishida, "Out-of-plane microtube arrays for drug delivery – Liquid flow properties and an application to the nerve block test – ", to be submitted.

MINIMALLY INVASIVE PARYLENE DUAL-VALVED FLOW DRAINAGE SHUNT FOR GLAUCOMA IMPLANT

Jeffrey Chun-Hui Lin[1], Po-Jui Chen[1], Brian Yu[1], Mark Humayun[2], Yu-Chong Tai[1]
[1]California Institute of Technology, USA
[2]University of Southern California, USA

ABSTRACT

A parylene-enabled microvalved shunt implant for glaucoma drainage is presented in this paper. Enabled by the dual-checkvalve operation, this device can physically drain the extra intraocular fluid and regulate the intraocular pressure (IOP) within the normal range (15-20 mmHg). Improved surgical features, in addition to the functional/microfluidic components, such as parylene-tube carrier and anchors, are also incorporated in such device to realize minimally invasive suture-less implantation, suitable for practical *in vivo* use. With the optimized micromachining and post-fabrication process procedures, the developed implant is the first checkvalved glaucoma drainage device (GDD), which is passive, consumes no additional power, and functions without any circuit involved to pursue its medical application.

INTRODUCTION

It is estimated 66.8 million people worldwide have glaucoma, most of whom are associated with abnormally high intraocular pressure (IOP) (> 20 mmHg) [1]. The most common way to treat glaucoma is using eye drops, which either reduce the generation of aqueous humor, or help the eye drain out fluid. Usually, this kind of medication needs patients to take the eye drop regularly every morning and evening, making it inconvenient and easily forgotten. On the other hand, some patients may get refractory responses or become allergic to the eye drops. In 1907, Rollet first proposed the idea of implanting a horse-hair thread subconjuctivally trying to drain out the excess eye fluid. Subsequently, other people tried many different kinds of materials [2]. These previous hollow tubes didn't get too much success because of bio-incompatibility and migration of the implants. In addition, the hollow tube, with neither threshold pressure constraint nor high-pressure protecting mechanism, causes excess fluid to leak out of the eye, which leads to hypotony. Thus, it is desirable to have a passive device to regulate the IOP to be below 20 mmHg, along with a protecting mechanism saving the eye from unexpected high pressure (e.g. >50 mmHg). This paper presents the first integrated parylene-tube-type microvalved glaucoma drainage device (GDD) incorporating microflow control system, parylene-tube carrier and anchors, which is designed to fit in a needle-implantable form factor for suture-less minimally invasive implantation through subconjunctival needle injection.

It is well known that a successful GDD must have continuous and reliable IOP control capability. In recent works, we have demonstrated that such requirements can be achieved by using dual checkvalves with flow regulation in required ranges compatible with IOP regulation specification [3]. However, their integration with appropriate surgical components for fixation is also necessary for its practical applications. After proving the concept of using the surgical features in the proposed device in MicroTAS07 [4] and demonstrating the preliminary assembly in HiltonHead08 [5], we have optimized the fabrication and packaging of a new device to create a parylene-tube GDD that targets subconjunctival implantation with needle-inserted and suture-less surgical procedures (Figure 1).

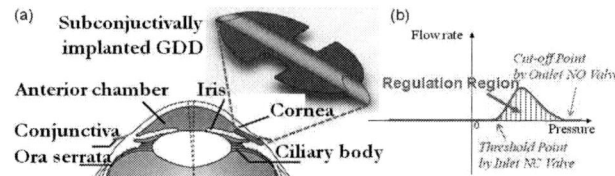

Figure 1: (a) Subconjunctival implantation concept of GDD; (b) Concept of IOP band passes regulation behavior.

DESIGN

The complete implant comprises a dual-valve microflow regulation system in a tube with integrated flexible tissue anchors (i.e., fixation). Dual back-to-back microvalves with one normally closed (NC) but open at 20 mmHg and the other normally open (NO) but closed beyond 50 mmHg are designed (Figure 2). A stiction pre-stressed NC valve is developed featuring spontaneous bonding from drying process. Few holes are designed on stiction-bonding parts where epoxy can be used to further ensure bonding strength and to prevent parylene from de-lamination after repeated operations. Trenches are added in the free-standing membrane of the NO valves to avoid stiction and to improve its high pressure sealing behavior.

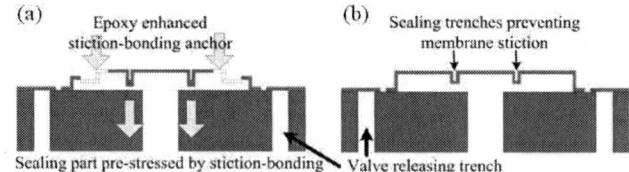

Figure 2: A dual valve micro-flow regulation system: (a) NC vavle; (b) NO valve.

To accommodate two microvalves, hollow tubes made of thick parylene are utilized and the length is chosen as 6 mm long (Figure 3). Both ends of these parylene tube carriers are slanted at 30 degrees to facilitate the surgical implantation and guard the GDD against iris retraction. To prevent the implanted GDD from dislodging, rollable/foldable anchors with hemispherical recesses are also designed. The anchor has a wingspan larger than the O.D. of 19-gauge hypodermic needles and can stretch out after needle retraction for robust fixation. The radius of hemispherical recesses is 300 μm, fitting and covering the parylene tube O.D. well during later assembly. Every parts of the GDD, (microvalves, parylene tube carriers, parylene anchors), is fabricated separately and finally integrated into one system. The fluid can flow from NC valve to NO valve before the pressure reaches the limit of NO valve, while the opposite direction is forbidden due to the closed NC valve.

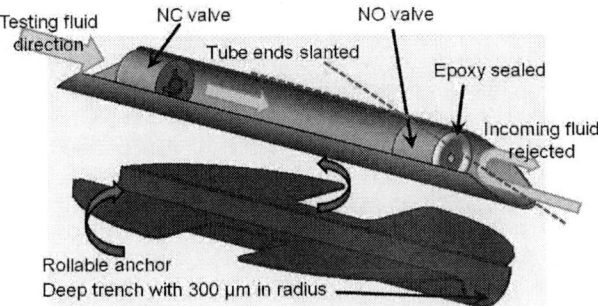

Figure 3: Full GDD system consisting of a dual-valve micro-flow regulation system, a parylene-tube carrier, and a rollable/foldable anchor.

FABRICATION AND PACKAGING

The microvalve fabrication started with backside DRIE to leave a 50-μm-thick membrane (Figure 4-(a)). The circular boundary of the valve seats were also defined here with diameter to be 500 μm, which can be fit into parylene carrier I.D. smoothly. XeF_2 was used to roughen the front side surface to improve the adhesion between parylene and the silicon valve seat. Lithography, which defines the valves' height difference was then performed and done by a 3-step exposure. After parylene coating, RIE was used to pattern coated parylene and then through holes were opened by backside DRIE. The wanted stiction parts of the circumferential of NC valve bonded to silicon surface simultaneously after the sacrificial photoresist were stripped and dried. Tiny epoxy drops were then applied onto these NC valves' stiction-bonding parts to make sure the parylene won't de-laminate after several repeated operations. The fabrication results of microvalves are shown in Figure 5-(a)-(c).

As for the fixation parts of the system, oxide was first patterned for later XeF_2 etching (Figure 4-(b)). The wafer was isotropically etched to create semi sphere recesses with radius to be 300 μm in depth. 20 μm thick parylene was then coated onto the wafer. Aluminum, used as the oxygen plasma etching mask, was deposited and then patterned by dry film photolithography. Dry film is one kind of negative photoresist that is widely used in printed circuit board (PCB) industry. The process provides the capability to pattern circuit boards with small holes without any photoresist collapsing. Hence it becomes a very good material that helps people pattern aluminum and parylene with a very deep trench. The final anchors were released by soaking in DI water. Different anchor shapes can be designed and patterned, depending on the surgical requirement. Results of the fabricated anchors are shown in Figure 5-(f). Parylene tube carrier was made by coating 35 μm thick parylene onto glass tubing with 530 μm in O.D., which were then cut into desired lengths of 6 mm with slanted ends on both sides and later free-released in BHF (Figure 4-(c)). Result of fabricated parylene tube carrier is shown in Figure 5-(d).

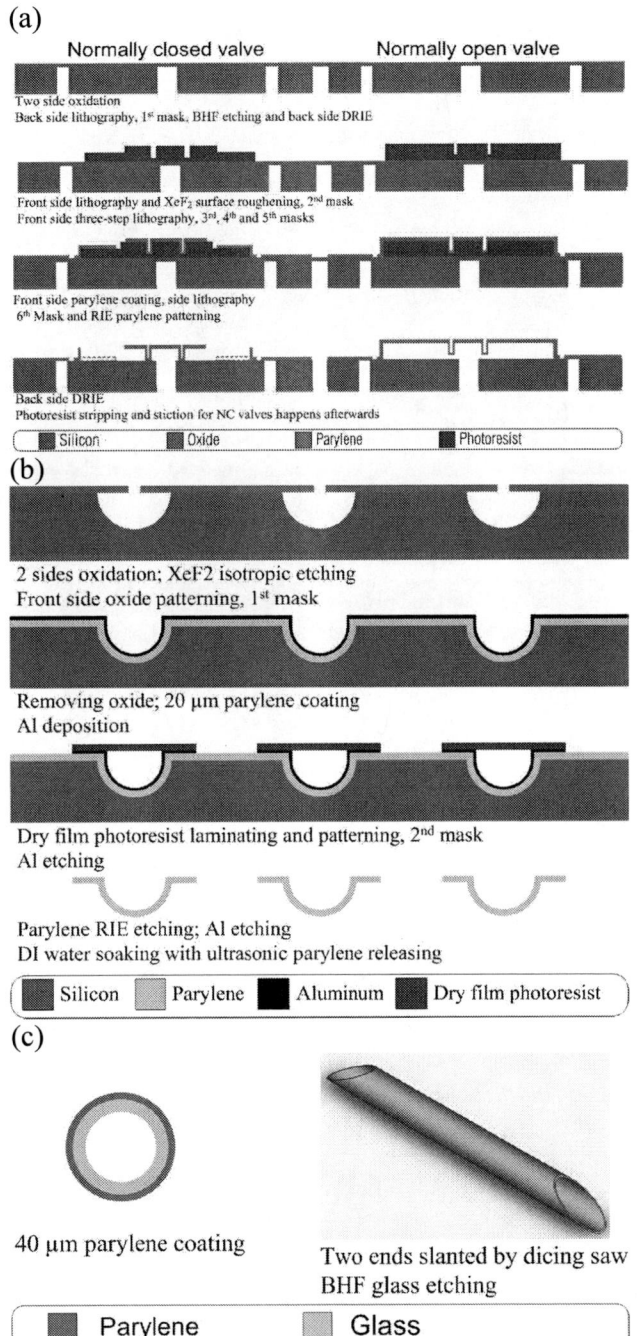

Figure 4: Fabrication procedures of: (a) Microvalves, (b) Anchors, and (c) Parylene tube carriers.

To package the entire GDD system, one NC valve and one NO valve were released and then inserted into either ends of the parylene tube carrier. The gap between the microvalves' seats and the tube carrier inner wall was sealed by epoxy (Figure 5-(e)). To mount the finished valved tube onto parylene fixation anchors, tiny epoxy drops was first wiped on the semi sphere recess. Then, the valved tube was assembled onto the recess. Figure 5-(g) shows the final assembled GDD with different anchor shapes. Since the O.D. of the parylene tube carrier is 600 μm (530 μm + 70 μm parylene coating), it fixes the fixation anchor nicely after assembled. The anchor is so flexible that it can be rolled/folded and insert into the testing Teflon tube thereafter (figure 5-(h)).

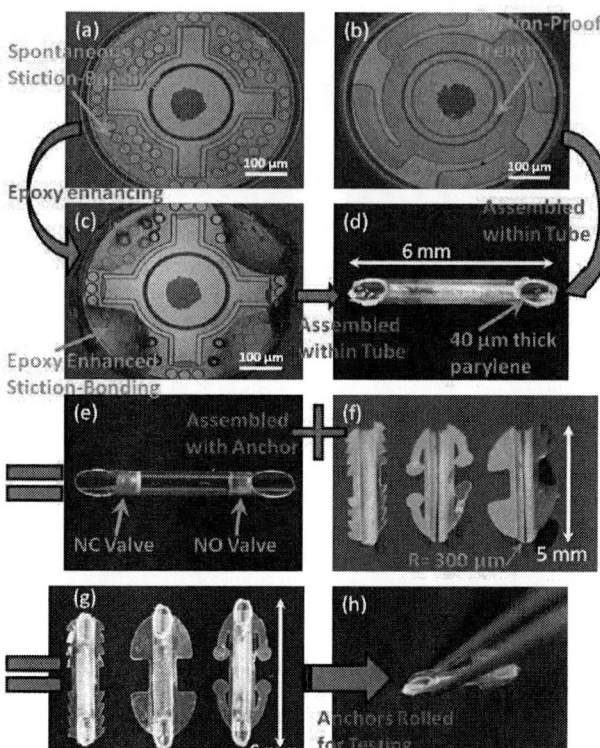

Figure 5: Fabrication results and packaging procedures: Micro graph of (a) NC valve, (b) NO vavle, (c) Epoxy enhanced NC vavle, (d) Hollow parylene tube carrier, (e) One NC and one NO valves sealed in tube carrier (transparent glass tube used here for clarity), (f) Anchors with trenches of 300 μm in radius, different anchor shapes designed to facilitate the surgical convenience and future GDD fixation: (left: ragged anchor; middle: foldable anchor; right: rollable squeeze-tail anchor), (g) Complete assembled GDD in top view, (h) Anchors can be rolled/folded for testing convenience. Valves are first sealed in the carrier, which is then assembled onto anchors.

BENCH TOP TESTING AND DISCUSSION

To test the completed GDD, one wingspan was folded by tweezers and inserted into Teflon FET tubing, which has I.D. of 750 μm. The gap between the GDD and the Teflon tubing was sealed with photoresist and dryied in the air (Figure 6). Water was chosen as the working fluid to mimic the anterior chamber environment. The Teflon tubing was connected to a pressure gauge with resolution up to 0.01 psi (~5 mmHg). The working fluid was pushed by air, which is regulated by a pressure regulator connected between the air source and pressure gauge. Marching speed of the working fluid advancement inside the Teflon tube was measured so that the flow rate can be calculated by knowing the fluid flow speed and the diameter of the Teflon tubing.

Both NC and NO valves were tested separately and the full GDD was then characterized with experimental verification that its behavior meets the standard IOP regulation requirements (Figure 7). The single NC valve starts to pop open at 0.2-0.3 psi (10-15 mmHg), and no obvious flow rate is observed before the cracking pressure. This result meets our simulation expectation and provides the evidence that stiction does provide the required pre-stress force to create the cracking pressure. Fluid in single NO valve flows smoothly before pressure reaches the high limit. The flow rate starts to decrease when the pressure goes up to 0.5 psi (~25 mmHg) and almost closes after 1.0 psi (~50 mmHg). The leak rate is less than 5 μL/min. This result proves that the sealing trench designed on NO valve does seal the NO valve opening. The small amount of leak rate may come from the non-flat bottom under the sealing trench which comes from the top surface of the sacrificial photoresist layer. For the dual valve GDD testing, the NC valve successfully opens at 0.33 psi (~17 mmHg) and NO valve starts to function at 1.1 psi (~57 mmHg). The reason that off pressure of NO valve in dual valve system is larger than single valve systems is that part of the energy of the flowing fluid is consumed before NO valve by friction and tether deformation of the NC valve so that only some energy is left to push the NO valve membrane to close the valve. This off pressure delay makes the off pressure of the dual valve system just meet our design requirement to close after 50 mmHg. The reported aqueous formation rate is 2-3 μL/min [6], which is satisfied perfectly by our GDD.

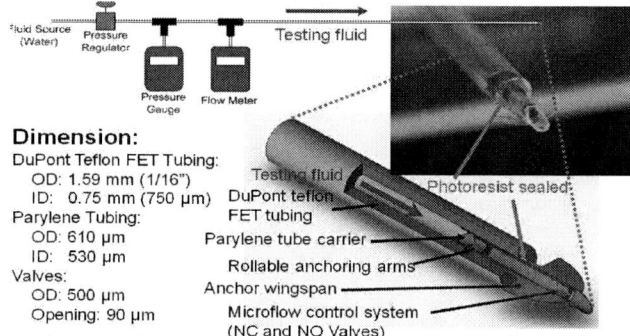

Figure 6: Testing setup of the GDD: Photoresist is painted in the gap between GDD and Teflon tubing for sealing.

Before every GDD is sent to animal surgical testing, the devices are prepared and tested first. After testing is done, the functional GDD is released by soaking into acetone again to ensure functionality of the device. The device wingspan is also stretched back to its original shape after pulling the GDD out of the testing Teflon tubing. The working device is then sent to be sterilized in order to prepare it for later *in vivo* functionality verification. In vivo animal trial is currently underway to verify the device's bioefficacy in the intraocular environment.

CONCLUSION

A fully passive, non-power integrated GDD combining micro flow regulation systems, parylene tube carrier and anchors is designed and developed. The parylene anchor with a very deep semi sphere recess is fabricated utilizing the dry film photolithography technique. With this improved surgical features and the newly developed packaging techniques, the overall finished GDD form factor not only still keeps minimally invasive implantation possible, but also prevents dislodging after subconjuctival implantation.

978-1-4244-2977-6/09 $25.00 © 2009 IEEE

(a)

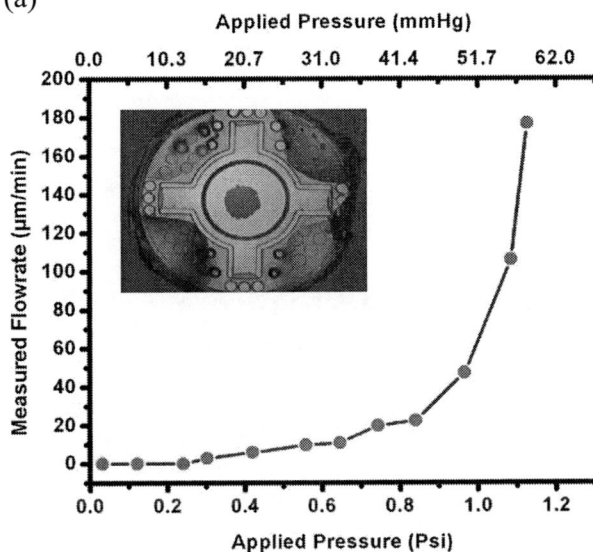

(b)

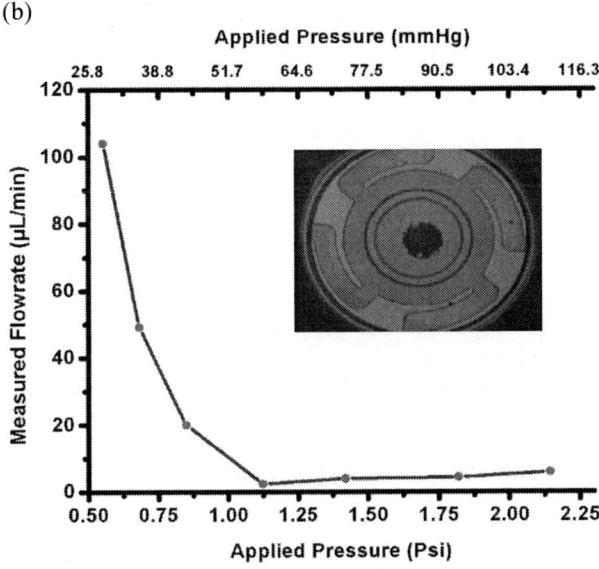

(c)

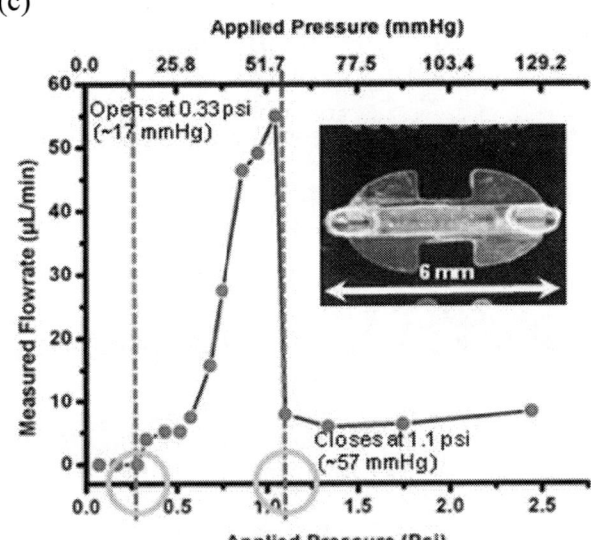

Figure 7: Testing results of (a) Single NC valve, (b) Single NO valve, (c) Dual valved GDD. In (c), the fluid started to flow after 0.33 psi (~17 mmHg), and closed at 1.1 psi (~57 mmHg). Water was chosen as working fluid.

A pressure-bandpass configuration is characterized by on bench verification which illustrates that the results meet medical requirements to treat glaucoma patients. It also shows that the epoxy and photoresist sealing during the packaging and testing procedures secure the system from water leakage very well. In terms of the cracking pressure of the NC valve, it can be trimmed, depending on the surgical requirements, by changing the number of tethers, length of tethers, parylene thickness, and sacrificial photoresist height.

ACKNOWLEDGEMENTS

This work was supported by Bausch and Lomb. The authors would also like to thank Mr. Trevor Roper for his valuable fabrication assistance.

REFERENCES

[1] http://www.ahaf.org/glaucoma/about/glabout.htm
[2] C.-H. Hong, A. Arosemena, D. Zurakowski, R. S. Ayyala, "Glaucoma Drainage Devices: A Systematic Literature Review and Current Controversies", *Surv Ophthalmol* vol. 50, pp. 48-60, 2005
[3] P.-J. Chen, D. C. Rodger, E. M. Meng, M. S. Humayun, and Y.-C. Tai, "Surface-Micromachined Parylene Dual Valves for On-Chip Unpowered Microflow Regulation," *J. Micoelectromech. Syst.*, vol. 16, pp. 223–231, 2007.
[4] P.-J. Chen, D. C. Rodger, S. Saati, J. C. Altamirano, C.-H. Lin, R. Agrawal, R. Varma, M. S. Humayun, and Y.-C. Tai, "Implementation of Microfabricated Sutureless Flexible Parylene Tissue Anchors on Minimally Invasive Biomedical Implants", in *Proc. MicroTAS 2007*, Paris, France, pp. 518-520.
[5] J. C.-H. Lin, P.-J. Chen, S. Saati, R. Varma, M. Humayun, Y.-C. Tai, "Implantable Microvalve-Packaged Glaucoma Drainage Tube", in *Tech. Digest 13th Solid State Sens., Actuators, and Microsyst. Workshop*, Hilton Head, SC, Jun. 1-5, 2008, pp. 146–149.
[6] http://www.nyee.edu/pdf/solomonaqhumor.pdf
[7] N. Tas, T. Sonnenberg, H. Jansen, R. Legtenberg, and M. Elwenspoek, "Stiction in Surface Micromachining," *J. Micromech. Microeng.*, vol. 6, pp. 385-397, 1996.
[8] H. Zhang, S. Wang, and Y. Xu, "Study and Applications of A Parylene Self-Sealing Structure," Technical Digest, in *Tech. Digest 19th IEEE International Conference on MicroElectroMechanical Systems (MEMS 2006)*, Istanbul, Turkey, Jan. 22-26, pp. 282-285.
[9] C. T. Leondes, *MEMS/NEMS Handbook Technigues and Applications Volume 5*, Springer Science+Business Media, Inc., New York, 2006.
[10] K. Dietz, *Dry Film Photoresist Processing Technology*, Electrochemical Publications Ltd, Port Erin, British Isles, 2001.

GLASS MICROPROBE WITH EMBEDDED SILICON VIAS FOR 3D INTEGRATION

Chiung-Wen Lin[1], Chih-Wei Chang[2], Yu-Tao Lee[1],
Rongshun Chen[1,2], Yen-Chung Chang[1,3], and Weileun Fang[1,2]
[1]Institute of NanoEng. and Microsystems, [2]Power Mechanical Eng. Dept. ,
[3]Institue of Molecular Medicine,
National Tsing Hua University, Taiwan

ABSTRACT

This study presents a novel process to realize glass 2D-microprobe array. The through silicon vias (TSVs) can also be integrated with the glass 2D-microprobe using the micromachining process. The vertical integration of chips containing glass 2D-microprobe array is realized using these silicon TSVs. The 3D-microprobe array and the signal processing IC can be easily .implemented after vertical assembly of 2D-microprobe chips using bonding. In application, the 2D glass microprobe is fabricated with a low impedance of 439kΩ at 1kHz; and the neural recording experiment has also been successfully performed on a rat.

INTRODUCTION

The probe array is a useful tool for neurophysiology to record neural or cell signals. Presently, the microfabrication is regarded as a promising approach to realize probe array. There are several advantages for the microfabricated probe array. Since the microprobe has the inherent characteristic of smaller size, the damage of neural tissue during penetration can be reduced. A microprobe with well located electrodes also helps recording multiple signals from cells. The existing microprobes are mainly made of single crystal silicon (SCS) as in [1-4]. According to the characteristic of semiconductor material, the single crystal silicon microprobe suffers from signal coupling during neural recording. Compared to silicon, glass is suitable for signal isolation and can provide better biocompatibility. Moreover, the fracture toughness of SCS varies with different crystal planes and ranges 0.62-1.34MPa×m$^{1/2}$ (glass is 0.77MPa×m$^{1/2}$ [5]). The SCS microprobe could be easily broken along a particular crystal plane. However, the high-aspect-ratio glass micromachining has rarely been investigated [6,7].

The vertical integration of planar microfabricated probe is a key technology to realize 3D microprobe array. As reported in [4,5], the integration is typically composed by another silicon carrier and complex process. Presently, the through-silicon via (TSV) technology is a promising approach to realize the vertical integration of chips [8-10]. The technique in [10] is capable of micromachining glass and embedded vertical silicon vias to achieve 3D stacking. This study presents a novel process to realize 2D glass microprobe array. In addition, the silicon vias can also be integrated with the 2D glass microprobe using the micromachining process. The vertical integration of chips containing 2D glass microprobe array is realized using these silicon TSVs. Thus, the 3D microprobe array can be implemented.

CONCEPT AND PROCESS STEPS

The designs of the present 2D and 3D glass microprobe arrays are illustrated in Fig. 1. In Fig. 1a, the 2D glass microprobe is fabricated using the glass reflow process. Fig. 1b further shows the 2D glass microprobe with embedded silicon vias. The low resistivity silicon vias are surrounded by the glass which acts as the insulator for vias. Electrical signal from electrode sites or other probe can be conducted through these vias. As shown in the left figure of Fig. 1c, upper set of vias are used to transmit signal from upper microprobe, while signal from electrode sites is connected to lower set of vias (cover with rectangular metal pad) and transmitted to backside of this microprobe. Thus the 3D microprobe arrays, as shown in the right figure of Fig. 1c, can be easily implemented after the vertical assembly of 2D microprobe chips using solder. Moreover, the signal processing IC can also be integrated with the microprobe array using the same manner. As a result, the present glass microprobe arrays with embedded silicon vias has the following merits: (1) Arbitrary shape of high-aspect-ratio glass microprobe can be fabricated using the present technology; (2) Drawbacks like the cross-talk of sensing signals and the opto-electronic effect in silicon microprobe can be prevented; (3) Vertical silicon vias can be fabricated for 3D stacking of glass microprobe and control IC.

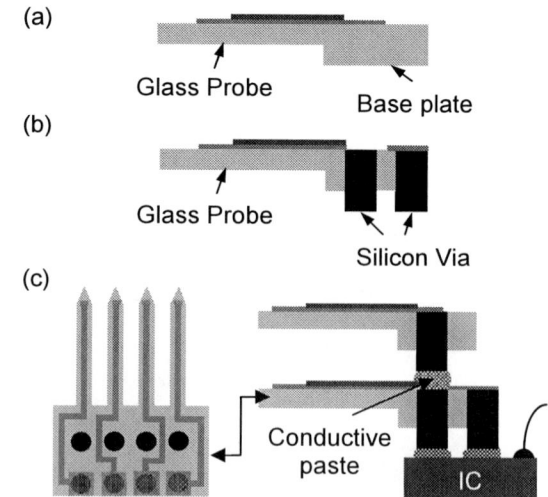

Figure 1: The concept of present glass microprobe, (a) Glass microprobe, (b) Glass microprobe with embedded silicon vias, and (c) 3D integration of microprobes.

978-1-4244-2977-6/09 $25.00 © 2009 IEEE

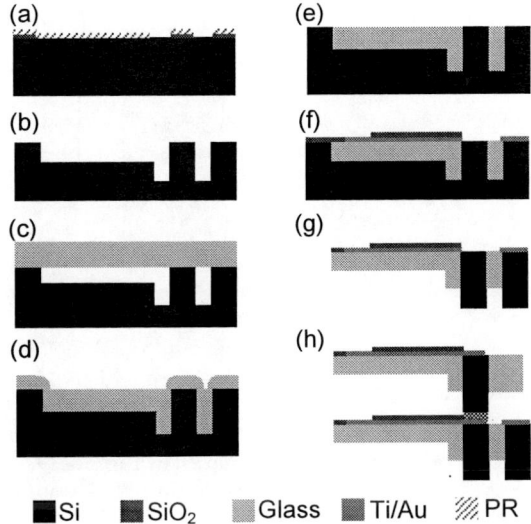

Figure 2: Process flow for silicon vias embedded glass microprobe.

The fabrication process illustrated in Fig. 2 has been established to implement concept in Fig. 1. As shown in Fig. 2a, two DRIE etching masks (PR and SiO$_2$) were deposited and patterned on low-resistivity silicon substrate. At the first DRIE etching of Si substrate, a deep trench for via insulation is defined. Afterward, PR was stripped and then followed by the second DRIE etching which defined the thickness of glass probe, as in Fig. 2b. In Fig. 2c, a Pyrex7740 glass wafer was then anodic bonded to the Si substrate in a vacuum chamber. After that, the bonded Si and Pyrex7740 glass were annealed at 750°C for 7 hours. By the assisting of pressure load, melting glass was filled into the cavity, as shown in Fig. 2d. A lapping process was performed to remove excessive glass until the expose of silicon via, as shown in Fig. 2e. The exposed silicon via was employed to act as the electrical interconnection. After lapping, metal thin film(Ti/Au) and SiO$_2$ were deposited and patterned on glass probe for electrical signal conducting from pad to silicon via, as indicated in Fig. 2f. As shown in Fig. 2g, backside DRIE etching of silicon was performed to fully release the glass probe. And 3D microprobes array can be accomplished by stacking 2D planar microprobes in Fig. 2h by conductive material such as solder or silver paste for vertical signal transmission. Direct integration of signal processing circuit on the backside of the microprobe can also be achieved. A typical 2D glass microprobe without embedded silicon vias can also be fabricated in the same process.

EXPERIMENT AND RESULTS

To demonstrate the feasibility of this novel glass micromachining in Fig. 2, two different types of glass microprobes (with or without embedded silicon vias) are fabricated. Various tests regarding to the performances of the glass microprobe, such as the impedance measurement and the neural signal recording, are performed.

- **Fabrication result**

The SEM micrographs in Fig. 3 show the silicon master mold prepared by the 2-steps DRIE etching of Si substrate illustrated in Fig. 2b. The silicon master mold is used for the following glass molding. The silicon mold consisted of probe array and the base plate is clearly observed in the left micrograph. In addition, the mold also contains conductive silicon vias. The zoom-in micrograph in Fig. 3 shows two different depths defined by the 2-steps DRIE. The thickness of base plate defined by both of the first and the second DRIE etching is about 330μm. The thickness of the microprobes defined by only the second DRIE etching is about 70μm. After completing the glass molding and lapping processes, metal films are deposited and patterned as shown in Fig. 4.

The close-up view of the glass microprobe, base plate, and metal conduction lines are clearly observed in the micrographs of Fig. 5. Fig. 5a shows a typical glass microprobe with base plate. The micrograph in Fig. 5b shows a typical glass microprobe with embedded silicon vias for vertical signal transmission. Electrical signal from electrode sites or different layer of stacking 3D microprobes can be transmitted by these vias as shown in Fig. 1c.

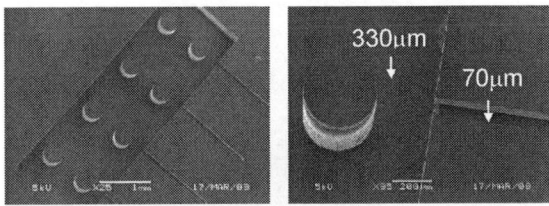

Figure 3: The fabricated silicon master mold for glass molding.

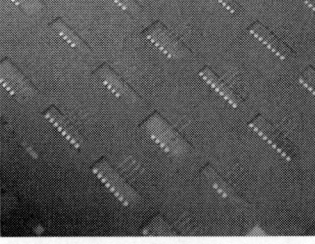

Figure 4: The fabricated glass microprobe with metal interconnections on silicon wafer.

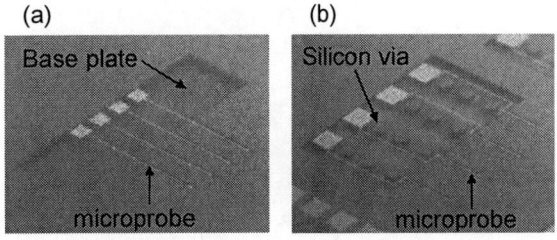

Figure 5: The fabricated glass microprobe with metal interconnections on silicon wafer, (a) all-glass microprobe, and (b) glass microprobes with embedded silicon via.

The SEM micrograph in Fig. 6 shows the cross-section of the glass/silicon compound substrate before the microprobe was fully released, as indicated in Fig. 2f. The through wafer silicon vias embedded inside the base plate of microprobe are clearly observed. It also shows the molded glass is properly filled inside the deep cavity on Si mold after reflow.

The micrographs in Fig. 7 show two different 2D glass microprobes after fully released from silicon substrate. Fig. 7a shows the typical all-glass 2D microprobe array with no embedded silicon vias. The inset SEM micrograph shows the close-up view of probe tips with metal lines and sensing electrodes. In this case, the probe thickness determined by the shallow trench of Si mold is 70μm. Fig. 7b shows the fabricated glass 2D microprobe with embedded silicon vias. The inset micrograph in Fig. 7b shows the zoom-in view of silicon vias inside the glass. After directly stacking of the 2D microprobe chip, as illustrated in Fig. 2h, the 3D microprobe array was achieved. Fig. 8 shows a typical result of assembled 3D-probe (2×3) by vertical stacking of 2D-probe (1×3) chips.

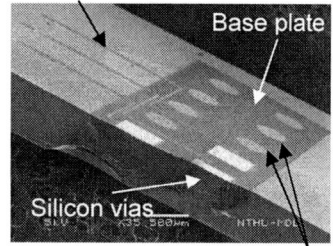

Figure 6: The fabricated glass microprobe, vertical silicon vias can be observed clearly on the sidewall of this substrate.

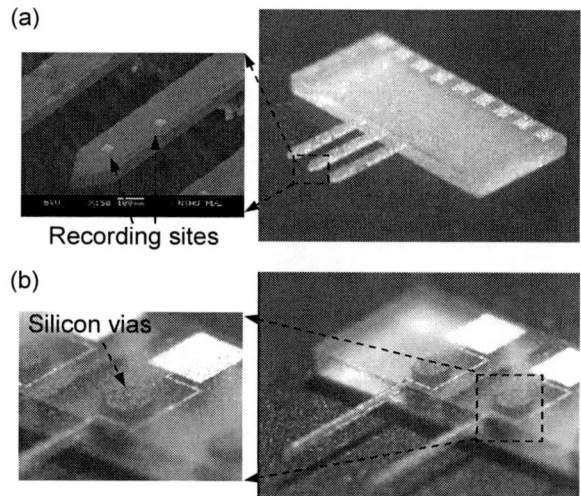

Figure 7: The optical images of fully released glass microprobe, (a) all-glass microprobe, and (b) glass microprobe with embedded silicon vias.

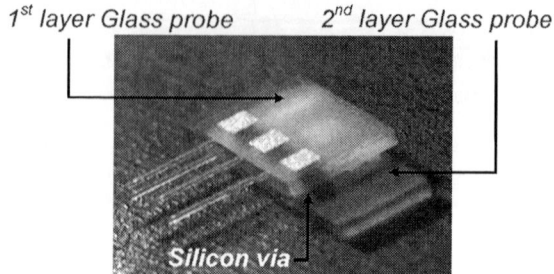

Figure 8: The 3-D glass probe array (2×3) assembled by stacking of 2D-probe (1×3) chip.

● **Tests**

The impedance of the glass microprobe was characterized by the experiment setup shown in Fig. 9. A fully released glass probe is wire bonding on a PCB board for impedance measurement, as shown in the inset of Fig. 9. The base plate of glass microprobe is encapsulated with UV-curable polymer for insulation. The tested microprobe is then immersed into the PBS solution. The measured impedance and phase variation at different frequency is shown in Fig. 10. The typical impedance of the present glass probe is 51.03kΩ at 1kHz.

The neural recording experiment was carried out on a rat. The rat was secured on a stereotaxic frame. A craniotomy was made to access the M1 cortex, and the present glass probe is inserted into the cortex for spontaneous spikes recording. The signal was measured by commercial

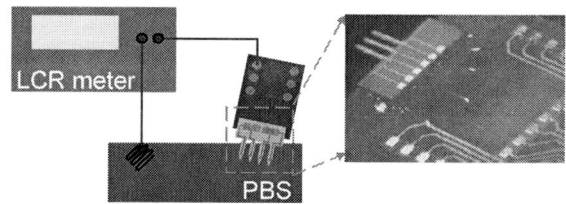

Figure 9: Impedance measurement set-up of glass microprobe, and the inset figure shows the glass microprobe wire bond to PCB.

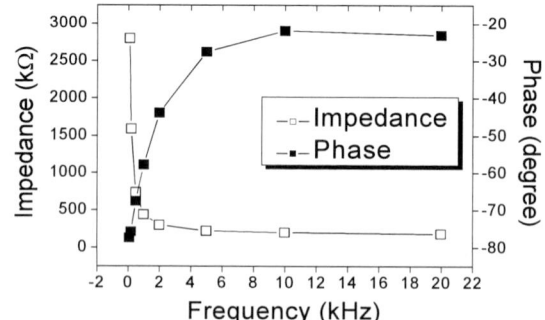

Figure 10: The measured impedance and phase of the glass microprobe at different frequency.

978-1-4244-2977-6/09 $25.00 © 2009 IEEE

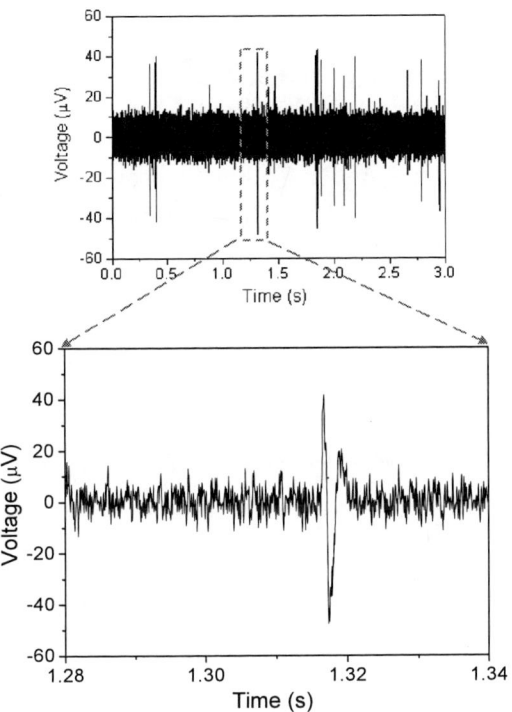

Figure 11: The measured spontaneous spikes of rat's cortex.

16-channels neuron recording systems (Plexon Inc., Dallas, Texas). The measured results are shown in Fig. 11. Upper figure shows the recorded neural signal for 3 seconds duration. The lower figure further shows the recorded signal near a spike at 1.317 second. The spike amplitude of 90μV (peak-to-peak) with SNR of 4.5 is successfully recorded. This result successfully demonstrated the in vivo neural recording of the presented glass microprobe.

CONCLUSION

A novel glass probe fabricated by glass reflow process is successfully demonstrated. Drawbacks like the cross-talk of sensing signals and the opto-electronic effect in silicon microprobe can be prevented. The typical impedance of this glass probe is 51.03kΩ at 1kHz. Successfully spontaneous spikes recording of rat's cortex is also demonstrated. The vertical silicon vias can be embedded into the glass microprobe chip by microfabrication process. Electrical signals between each 2D probes can be transmitted by these silicon vias. 3D glass probe arrays can be implemented by vertical stacking of 2D probes. Further integration with signal processing IC with this 3D probe array can be accomplished.

ACKNOWLEDGEMENTS

This research is sponsored in part by the NSC of Taiwan under grant of NSC-96-2627-E-007-002 and NSC-96-2628-E-007-007-MY3. The authors would like to appreciate the Nano Facility Center of National Tsing Hua University, the NEMS Research Center of National Taiwan University, and the NSC National Nano Device Laboratory (NDL) in providing the fabrication facilities. The author also would like to thank Yung-Chen Chen and Prof. Hsin Chen of National Tsing Hua University in providing the impedance measurement and Prof. You-Yin Chen of National Chiao Tung University in providing the rat's cortex measurement facility.

REFERENCES

[1] C. Moldovan, V. Llian, Ghe. Constantin, R. Iosub, M. Modreanu, I. Dinoiu, B. Firtat, and C. Voitincu, "Micromachining of a silicon multichannel microprobe for neural electrical activity recording," *Sensors and Actuators A*, Vol. 99, pp. 119-124, 2002.

[2] P. K. Campbell, K. E. Jones, R. J. Hubert, K. W. Horch, and R. A. Normann, "A silicon-based, three-dimensional neural interface: Manufacturing processes for an intracortical electrode array," *IEEE Transaction on Biomedical Engineering*, Vol. 38, pp. 758-768, 1991.

[3] Q. Bai, K. D. Wise, and D. J. Anderson, "A high-yield microassembly structure for three-dimensional microelectrode arrays," *IEEE Transaction on Biomedical Engineering*, Vol. 47, pp. 281-289, 2000.

[4] Y. Yao, M. N. Gulari, J. A. Wiler, and K. D. Wise, "A microassembled low-profile three-dimensional microelectrode array for neural prosthesis applications," *J. Microelectromech. Syst.*, vol. 16, pp. 977-988, 2007.

[5] Corning Pyrex Properties, http://www.ferroceramic.com/corning_table.htm.

[6] X. Li, T. Abe, and M. Esashi, "Deep reactive ion etching of Pyrex glass using SF6 plasma," *Sensors and Actuators A*, Vol. 87, pp. 139-145, 2001.

[7] A. Baram, and M. Naftali, "Dry etching of deep cavities in Pyrex for MEMS applications using standard lithography," *J. Micromech. and Microeng.*, Vol. 16, pp. 2287-2291, 2006.

[8] C.-W. Lin, H.-A. Yang, W. C. Wang, and W. Fang, "Implementation of three-dimensional SOI-MEMS wafer-level packaging using through-wafer interconnections," *J. Micromech. and Microeng.*, Vol. 17, pp. 1200-1205, 2007.

[9] J. Chae, J. M. Giachino, and K. Najafi, "Fabrication and characterization of a wafer-level MEMS vacuum packages with vertical feedthroughs," *J. Microelectromech. Syst.*, vol.17, 2008, pp. 193-200.

[10] C.-W. Lin, C.-P. Hsu, H.-A. Yang, W. C. Wang, and W. Fang, "Implementation of silicon-on-glass MEMS devices with embedded through-wafer silicon vias using the glass reflow process for wafer-level packaging and 3D chip integration," *J. Micromech. and Microeng.*, Vol. 18, 025018.

978-1-4244-2977-6/09 $25.00 © 2009 IEEE

FABRICATION PROCESS OF INTEGRATED MULTI-ANALYTE BIOCHIP SYSTEM FOR IMPLANTABLE APPLICATION

Y. -C. Tsai[1], N.-F. Chiu[2], P.-C. Liu[2], Y.-C Ou[3], H.-H. Liao[1], Y.-J. Yang[1], L.-J. Yang[3],
U. Lei[4], F.-S. Chao[2], S.-S. Lu[5], C.-W. Lin[2], P.-Z. Chang[4], and W. -P. Shih[1]

[1]Department of Mechanical Engineering, National Taiwan University, Taipei, Taiwan
[2]Institute of Biomedical Engineering, National Taiwan University, Taipei, Taiwan
[3]Department of Mechanical and Electro-Mechanical Engineering, Tamkang University, Tamsui, Taiwan
[4]Institute of Applied Mechanics, National Taiwan University, Taipei, Taiwan
[5]Department of Electrical Engineering, National Taiwan University, Taipei, Taiwan

ABSTRACT

This paper presents the integration of the fabrication processes for an implantable multi-analyte biochip system. The implantable biochip system has a microfluidic channel in which an electrochemical biosensor and a dielectrophoresis (DEP) micropump are fabricated by surface micromachining. The enzyme-based biosensor consists of three electrodes and detects the glucose level instantaneously by utilizing glucose oxidase (GOD). The DEP force resulting from the four-phase voltage exchange of the interdigitated electrodes drives the blood flow in the microchannel. By utilizing the MEMS fabrication and the electric circuit design, a compact implantable biochip system is achieved.

INTRODUCTION

In order to continuously and simultaneously monitor the physiological parameters such as glucose concentration, it is necessary to develop stable and accurate implantable biosensor systems. Micromachining techniques have been widely used for developing diverse biosensors. The flexible biocompatible polymer glucose sensors [1,2] and thermal metabolic sensor integrated with microfluidics [3] have been reported. Silicon nanochannel biological field effect transistors have also been developed for measuring the glucose concentration, and a linear response has been obtained [4]. In order to avoid the damage of the enzyme in the biosensor during the fabrication processes, the enzyme is usually applied at the last process step. For example, the enzyme membrane can be applied through the microholes on top of a fabricated microchannel [5]. This method can effectively resolve the incompatibility between the biomaterials and the MEMS processes. However, the sensitivity of the biosensor will decrease due to the leakage of the microholes. The development of an amperometric biochip that is designed for continuously in-*vivo* monitoring physiological analytes has been reported [6,7]. But the proposed device has not been miniaturized for implantable applications. The flexible polymer tube lab-chip integrated with microsensors has been proposed and applied for developing smart microcatheter [8]. The main problems in the biosensor chip fabrication include: (1) the incompatibility between the biomaterials and the MEMS processes and (2) system miniaurization. The miniaturized and packaged implantable biochip systems of multi-functions are still rare.

In this paper a multi-analyte implantable biochip system has been developed. It consists of a biocompatible package,

a control and wireless module, a battery, an inductive power coupling module, and a PDMS microchannel (Figure 1). The PMMA package is coated by 5μm-thick parylene-C for biocompatibility. It has circular shape to avoid the tissue injury during implantation. The control module is to drive the DEP micropump and to supply the working voltage to the electrochemical electrodes. Its wireless sub-module includes a microcontroller unit (MCU), an amplifier, and an RF transmission section for signal transmission and process. The battery supplies a 3.7V voltage for the biochip system and can be recharged via inductive coupling. The microchannel is connected to blood vessels for in-situ biological detections in order to increase the performance of the measurement. The programmatic target in this paper is to fabricate the microchannel which contains the micropump and the biosensors to apply implantable multi-analyte biochip system.

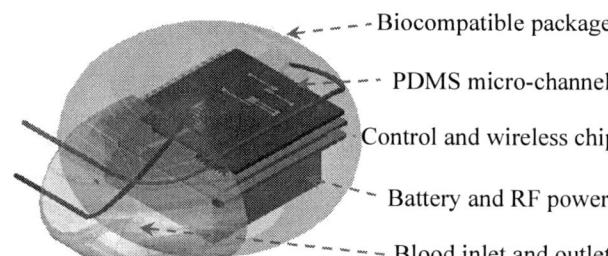

Figure 1: Schematic of the wireless implantable biochip system.

The DEP micropump and electrochemical electrodes are embedded in the microchannel (Figure 2). The voltage difference between each electrode in the DEP micropump can drive the electrolytes in the blood. Simultaneously, the motion of the electrolytes can urge the blood flow in the microchannel [9]. The low supply voltage of the DEP micropump is suitable for implantable biochip system. The GOD enzyme is coated on the gold working electrode by electrochemical polymerization for electrochemical detection [10]. In addition to the working electrode, the platinum counter electrode and silver reference electrode are fabricated in the microchannel for improving the sensor lifetime and accuracy.

978-1-4244-2977-6/09 $25.00 © 2009 IEEE

FABRICATION

Electrode/Glass Chip

The electrodes of the DEP micropump and the biosensors are fabricated on the glass substrate by using surface micromachining. The electrodes are deposited by e-beam evaporation and patterned by using lift-off process. The DEP micropump has 36 gold electrodes. The electrode thickness is 180nm. The area of each DEP electrode is 500μm*15μm, and the gap between adjacent DEP electrodes is 15μm. The area of each electrochemical electrode in the biosensor is 500μm*500μm, and the gap is 500μm. The thickness of the gold working electrodes, the platinum counter electrode, and silver reference electrode is 180nm, 200nm, and 250nm, respectively. Underneath all of the electrodes is a 10nm chromium adhesion layer.

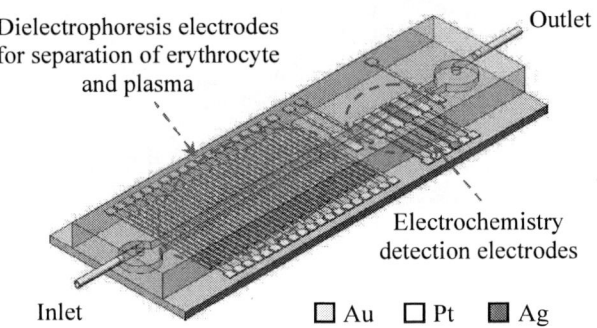

Figure 2: Schematic of the PDMS microchannel which contains dielectrophoresis electrodes and chemical electrodes.

PDMS Microchannel

To fabricate the PDMS microchannel, the SU-8 mold on a silicon substrate is first patterned by photolithography. Then PDMS is applied and cured on the SU-8 mold. The cured PDMS is peeled off the SU-8 mold and then bonded on electrode/glass substrate. The length, width, and height of the PDMS microchannel are 12mm, 500μm, and 40μm, respectively. The inlet/outlet section in the microchannel has a large circular area in which the interconnection can be easily implemented. To make a compact package, the interconnection passage in the PDMS is drilled in the form of an "L" shape.

Biochip Integration

The process for fabricating the implantable biosensor system should be carefully designed to resolve the incompatibility of the biomaterials and to achieve the system miniaturization. The electrode/glass chip is fixed on the printed circuit board (PCB) by AB glue (Figure 3(a)). The electrodes on the glass substrate are connected to the corresponding pads on the PCB by applying wire-bonding. The wires are then covered by the AB glue which serves as the protection layer in the following process steps (Figure 3(b)). The wire-bonding process should be carried out prior to bonding the PDMS microchannel on the glass substrate. Otherwise, the PDMS microchannel would hinder the view for aligning the electrodes with the bonding pads in the wire-bonding process. The AB glue must not cover the electrodes on the electrode/glass chip for the success of bonding the PDMS microchannel. Before bonding the PDMS microchannel, the Ag/AgCl reference electrode is chloridized. The enzyme polymerization on the Au working electrode should also be carried out prior to bonding the PDMS microchannel because it is difficult to proceed in the microchannel. The electroplating method is used to chloridize the Ag/AgCl reference electrode. A 0.1M NaCl solution is applied on the electrode/glass chip. Meanwhile, the platinum and silver electrodes are protected. In the electroplating process, the current density is 40μA/cm^2. The processing time is 10 minutes. The polypyrrole (from Merck) and GOD mixture is electrochemically polymerized on the working electrode by controlling the voltage from 0V to 1.2V in a cyclic voltammetry. After the GOD polymerization, the working electrode becomes dark brown.

To bond the PDMS microchannel onto the glass substrate, the bonding surfaces should be modified by using oxygen plasma. It should be noted that the working electrode and the Ag/AgCl reference electrode should be covered for avoiding direct contact with the oxygen plasma (Figure 3(c)). Otherwise, the enzyme on the working electrode will be damaged, and the Ag/AgCl reference electrode will be oxidized. The microchannel which is bonded on the glass substrate is illustrated in Figure 3(d). Finally, the tubes are connected with PDMS microchannel. The interconnection is sealed by AB glue (Figure 3(e)). The polyethylene tube has 0.86mm inner diameter and 1.27mm outer diameter. In order to make the compact interconnection, the "L" shape passage of 1.25mm diameter is used. The diameter of the passage is slightly smaller than the outer diameter of the tube. The elastic PDMS passage can tightly clamp the tube and avert the AB glue from permeating into the microchannel by capillary force.

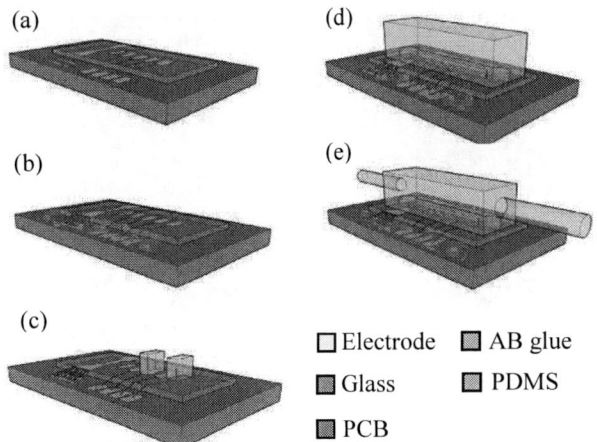

Figure 3: Chip assembly process. (a) Fixing the electrode/glass chip on the PCB by AB glue. (b) Wire-bonding and applying AB glue on the wires for protection. (c) Placing the PDMS lump on the working and reference electrodes, respectively. (d) Bonding the PDMS microchannel by oxygen plasma. (e) Connecting the tubes to the microchannel and then welding ICs on the PCB.

978-1-4244-2977-6/09 $25.00 © 2009 IEEE

After the microchannel and the electrode/glass substrate are assembled, the DEP control ICs (TPS2010 from Texas Instruments and 74ACT04 from Fairchild Semiconductor) and electrochemistry control ICs (OPA4336 from Texas Instruments) are welded on the PCB (Figure 4). The DEP control ICs are to supply the constant voltage and exchange the voltage phase to drive the DEP micropump. The electrochemistry control ICs supply the required voltages to the working electrodes and the reference electrode. They also measure the current change between the working electrode and the counter electrode. The integration of the microchannel electrodes and the control ICs enables the multi-analyte detection on a single chip and has made the implantable system practicable.

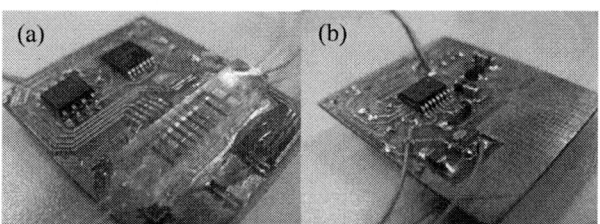

Figure 4: Pictures of the electrochemistry detection system and the DEP control module. (a) Front side: micro-channel and DEP control ICs. (b) Back side: circuit for electrochemical detection.

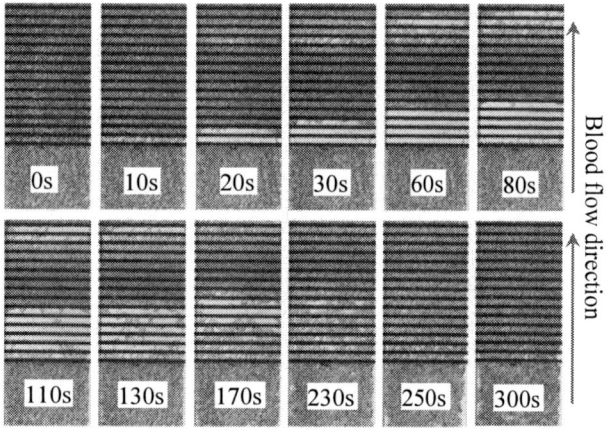

Figure 5: Optical images of the blood flow driven by the four-phase DEP micropump.

TEST RESULT AND DISCUSSION

In this paper, whole human blood is used to test the transport capability of the DEP micropump. In the DEP micropump test, the blood is injected into the microchannel firstly. Then the driving signal is applied once the static state of the blood flow has been obtained. The driving signal of the DEP micropump is a 25MHz, 5V four-phase sine wave. The blood flow driven by the DEP micropump is observed though microscopy. Initially, all of the electrodes are uniformly covered by the red blood cells. Hence the red color is observed. Once the DEP micropump is activated, the red blood cells drift in response to the asymmetric electric field, resulting in the blood flow (Figure 5). The measured blood

flow velocity is 14μm/s. This driving condition requires low electric voltage so that the low power consumption of the DEP micropump can be achieved and be suitable for the implantable biochip system.

Before testing the amperometric responses of the implantable biochip system, a cyclic voltammetry is used to measure the peak potential of the glucose solution. The phosphate buffer solution (PBS) and different glucose concentration solution are injected into microchannel individually for measurement. The measured peak potential for different glucose concentration is about 0.25V (Figure 6). The current response increases with the increasing glucose concentration at 0.25V.

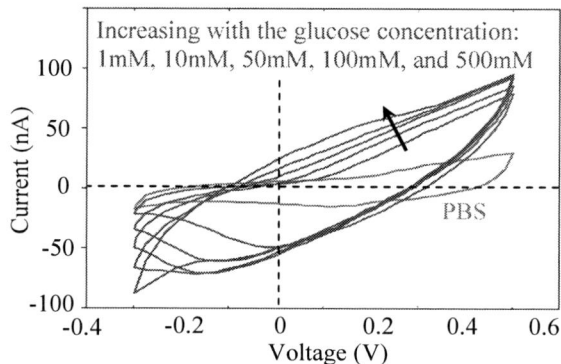

Figure 6: Cyclic voltammetric measurement result of different glucose concentrations.

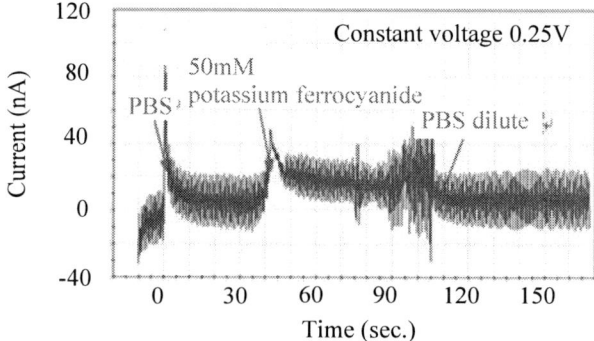

Figure 7: Measurement result of the current response after injecting 50mM potassium ferrocyanide..

The working voltage is maintained at 0.25V for detecting the glucose. The fabricated biosensor exhibits reversible sensing capability. The time domain of the detection result shows that an apparent current response about 50nA is obtained after the mixture solution of the glucose and 50mM potassium ferrocyanide ($C_6FeK_4N_6$ from Sigma) mediator is injected into the microchannel (Figure 7). The potassium ferrocyanide mediator serve as a means of shuttling electrons between theimmobilized glucose oxidase enzyme and the electrode surface during amperometric signal recordings. After the mixture solution passes by, the current response decreases with time due to the decreasing concentration. Then the phosphate buffer solution is injected to the microchannel, the current level returns to its initial value. The measurement can be supplied by a 3.7V battery, controlled by the ICs, and obtained by a wireless transmission system.

CONCLUSION

The integration of an implantable multi-analyte system for continuously in-*vivo* monitoring important physiological signals has been proposed and demonstrated. By protecting the biomaterials during the oxygen plasma bonding, multiple biosensors can be easily integrated in the microchannel. The proposed process sequences of the implantable biosensor system can overcome the difficulty in the system miniaturization. The glucose detection is achieved by using electrochemical method, and the function of the DEP micropump is verified. The electrochemistry and DEP control circuits are integrated with the inductive power coupling and wireless communication modules. A fully packaged miniature system with the total volume of 13c.c. is achieved (Figure 8). In addition, the GOD enzyme can be replaced by other enzymes or other bio-marker for different implantable applications.

Figure 8: Picture of the wireless implanted biochip system.

ACKNOWLEDGEMENT

This project has been supported by the Minister of Economy and National Science Council, Taiwan.

REFERENCES

[1] H. Kudo, T. Sawada, E. Kazawa, H. Yoshida, Y. Iwasaki, K. Mitsubayashi, "A flexible and wearable glucose sensor based on functional polymers with Soft-MEMS techniques", *Biosensors and Bioelectronics*, vol. 22, pp. 558-562, 2006.

[2] K. Mitsubayashi, S. Iguchi, T. Endo, S. Tanimoto, D. Murotomi, "Flexible glucose sensors with a film-type oxygen electrode by microfabrication techniques", in *Digest Tech. Papers Transducers '03 Conference*, Boston, June 8-12, 2003, pp. 1201-1204.

[3] L. Wang, D. M. Sipe, Y. Xu, Q. Lin, "A MEMS thermal biosensor for metabolic monitoring applications", *Journal of Microelectromechanical Systems*, vol. 17, no.

2, pp. 318-327, 2008.

[4] X. Wang, Y. Chen, K. A. Gibney, S. Erramilli, P. Mohanty, "Silicon-based nanochannel glucose sensor", *Applied Physics Letters*, vol. 92, 013903, 2008.

[5] J. Wu, J. Suls, W. Sansen, "The glucose sensor integratable in the microchannel", *Sensors and Actuators B*, vol. 78, pp. 221-227, 2001.

[6] A. Guiseppi-Elie, S. Brahim, G. Slaughter, K. R. Ward, "Design of a subcutaneous implantable biochip for monitoring of glucose and lactate", *IEEE Sensors Journal*, vol. 5,no. 3, pp. 345-355, 2005.

[7] M. Pepper, N. S. Palsandram, P. Zhang, M. Lee, H. J. Cho, "Interconnecting fluidic packages and interfaces for micromachined sensors", *Sensors and Actuators A*, vol. 134, pp. 278-285, 2007.

[8] C. Li, P.-M. Wu, J. Han, C. H. Ahn, "A flexible polymer tube lab-chip integrated with microsensors for smart microcatheter", *Biomed Microdevices*, vol. 10, pp. 671-679, 2008.

[9] C. Y. Yang, U. Lei, "Dielectrophoretic force and torque on an ellipsoid in an arbitrary time varying electric field", *Applied Physics Letters*, vol. 90, 153901, 2007.

[10] P. A. Fiorito, S. I. Cordoba de Torresi, "Glucose amperometric biosensor based on the Co-immobilization of glucose oxidase (GOx) and Ferrocene in poly(pyrrole) generated from ethanol/water mixtures", *J. Braz. Chem. Soc.*, vol. 12, no. 6, pp. 729-733, 2001.

MICROSYSTEM-BASED STUDY OF POLLEN-TUBE ATTRACTANTS SECRETED BY OVULES

J.R. Cooper[1], Y. Qin[2], L. Jiang[3], R. Palanivelu[2], and Y. Zohar[1]
[1]Dept. Aerospace and Mechanical Engineering, [2]Dept. Plant Sciences, [3]College of Optical Sciences
University of Arizona, Tucson, AZ, USA

ABSTRACT

Microdevices are developed to resemble the *in-vivo* micro-environment of ovule fertilization by pollen tubes in model research plant *Arabidopsis thaliana*. The PDMS-based microdevices are filled with pollen germination medium (PGM) providing pollen tubes with a proper growth environment. Pollen tubes are found to grow within the microgrooves with an average rate of 130μm/hr, reaching a final tube length of about 450μm. Targeting ovules by pollen tubes has also been tested with an observed efficiency of about 67%. Both the pollen tube growth rate and the ovule targeting efficiency in microdevices are similar to those obtained using *in-vitro* plate experiments. Finally, initial results indicate that pollen tube bundles preferentially turn toward ovule containing chambers, suggesting that the pollen tubes respond to the attractants secreted by unfertilized ovules.

INTRODUCTION

Research on pollen tube guidance has received significant attention in the past decade because of its importance in both biological and biotechnological research. In nature, a pollen grain lands on the surface of the pistil, it absorbs water from the stigma and forms a pollen tube; a long polar process that transports all of the cellular contents, including sperm. Pollen tubes grow through stigma and style, traveling through the transmitting tract, where they find guidance cues that modify their direction and lead them to the ovule micropyle [1]. Typically, only one tube enters the ovule, terminating its journey within a synergid cell and bursting to release sperm cells.

A schematic diagram in Figure 1 illustrates the *in-vivo* micro-environment of pollen tube targeting of ovules within an *A. thaliana* flower. The remarkable precision in the process of a pollen tube approaching an ovule micropyle is controlled by a series of molecular signals involving pollen tubes, ovule tissues and the female gametophytes [2-6]. Our earlier study using an *in-vitro* assay defines three signaling events that regulate pollen tube guidance in A. thaliana: i) contact-mediated competence conferred by the stigma and style, ii) diffusible ovule-derived attractants, and iii) repellents exuded from recently-targeted ovules. The species specificity and diffusion properties of the ovule attractant are consistent with a protein signal, while the abrupt transmission and response to the repellent suggests the activity of a small molecule, a peptide or post-translational modifications to signals synthesized prior to fertilization [5].

The *in-vitro* studies were conducted using a plate assay involving a petri dish with pollen germination medium (PGM) filled to a thickness of about 3mm. A cut portion of the pistil and a few ovules are placed on the PGM to recapitulate the *in-vivo* events. A schematic diagram illustrating the *in-vitro* plate assay is shown in Figure 2. Pollen tubes emerge from the cut end of the pistil, and travel across the PGM before entering the excised ovules. The plate assay is performed on a an open PGM surface without structures of the transmitting tract and the ovule chambers, which physically confine the path of the attractant diffusion emitted from the ovules in the *in-vivo* environment. The attractant diffusion is believed to be directional; thus, the model assay should be anisotropic rather than isotropic. Therefore, the isotropic model of the plate assay is not suitable for quantitative investigation of ovule-derived attractants in the *in-vivo* micro-environment.

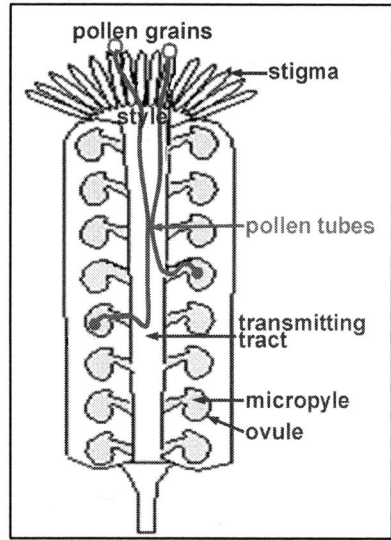

Figure 1: A diagram illustrating *in-vivo* pollen tube growth and guidance within a flower.

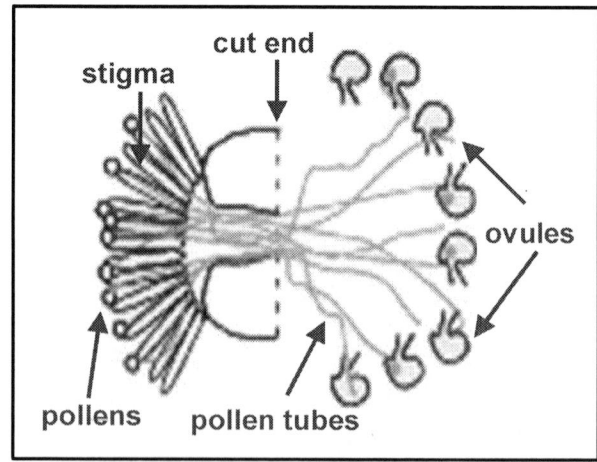

Figure 2: A diagram illustrating pollen tube growth and ovule targeting in an *in-vitro* plate assay.

978-1-4244-2977-6/09 $25.00 © 2009 IEEE

Microsystem technology is potentially an ideal tool for such investigations because it has the capacity of constructing complex configurations similar to those found in the *in-vivo* micro-environment. Three dimensional structures can be realized with a length scale ranging from 1μm to 1000μm. Indeed, microdevices using various materials, such as glass and polymers, have been developed for numerous biological applications [7]. In this work, microdevices are fabricated to resemble the micro-environment of ovule fertilization by pollen tubes in model research plant *A thaliana*. An *in-vitro* assay using the microdevices is performed. The results demonstrate the advantages and the potential of pursuing further research on pollen tube guidance, in response to ovule secretions, using microsystem technology.

DEVICE DESIGN AND FABRICATION

In order to model the micro-environment of ovule fertilization by pollen tubes, a microfluidic device is designed consisting of a main groove and two ovule chambers on each side of the groove. To facilitate *in-vitro* pollen growth in a microdevice, the pollen tubes and the ovules have to be in contact with PGM to obtain necessary nutrients. Therefore, the groove and the chambers must be filled with PGM thick enough to meet the pollen tube growth requirements. In addition, transparent devices are desired to allow direct inspection using conventional microscopy techniques.

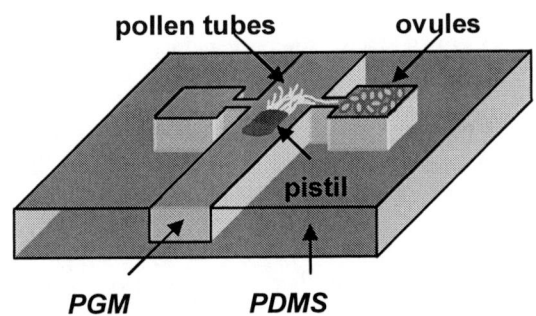

Figure 3: A schematic diagram of the designed micro-system consisting of a main groove and two side chambers.

Silicon-based organic polymer Polydimethylsiloxane (PDMS) is chosen to serve as the substrate material for the microdevices because it is bio-compatible and transparent. Furthermore, device fabrication using PDMS is simple, fast and cheap. A schematic device layout is shown in Figure 3, consisting of a long pistil groove and two square ovule chambers. The fabrication process starts with a mold formation using photosensitive polymer SU-8. To achieve a thickness of 500μm, SU-8 2100 is spin-coated on a silicon wafer using a two-layer coating technique. After soft-baking at temperatures 65°C and 95°C, the SU-8 film is exposed to UV light using an ABM aligner and a photo-mask. Following post-baking at the same temperatures, the SU-8 film is developed; thus, transferring the device pattern from the mask. After rinsing the mold with de-ionized water and drying it in nitrogen, omni agent is coated on the Si/SU-8 mold, which is then baked at 100°C for 20min allowing easy release of PDMS replicas. A PDMS mixture of base and curing agent with a ratio of 10:1 is poured onto the mold and cured for 2 hours at 60°C. Microdevices are obtained by peeling off the PDMS replicas from the mold. The groove width is about 1000μm to accommodate a cut pistil, whereas the area of the side chambers varies from 250μm×250μm to 1000μm×1000μm to allow placement of ovules increasing in number. Both the width and length of these capillaries vary in the range of 50μm-250μm to control the distance and, consequently, the concentration of the ovule attractants, mimicking the micropyles in the *in-vivo* environment.

PGM is prepared exactly as described in [5]. The pH of the solution is adjusted to be 7.0. Then 0.5% noble agar (Difco) is added to the solution, which is heated until the agar is completely dissolved. Simultaneously, a PDMS device is exposed to oxygen plasma to render the surfaces of the groove and chambers hydrophilic. A needle syringe is used to inject the liquid PGM at one end of the groove. The PGM fills up the groove and chambers due to capillary forces, and solidifies as it cools down to room temperature. A picture of a fabricated PDMS microdevice filled with PGM is shown in Figure 4. The device is placed on a 3mm thick PGM, in a covered Petri dish, to prevent PGM dry-out in the microdevice; at this point, it is ready for placement of plant tissues.

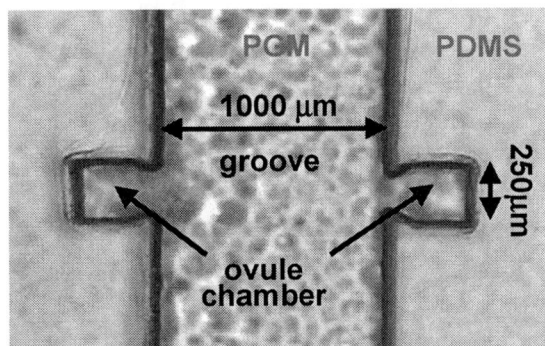

Figure 4: A microscopic photograph of a fabricated PDMS device filled with PGM.

POLLEN TUBE GROWTH RATE

Pollen tube growth in fabricated microdevices is first investigated and as a control, a plate assay is carried out using 35mm diameter Petri dishes. Freshly prepared cut pistils from young *A. thaliana* flowers are utilized. The pistils are cut at the junction of style and ovary, and the cut portion with stigma is used for the experiments. In both cases, the cut pistil is laid on the PGM surface and gently pushed down to achieve good contact with PGM as illustrated in Figure 3. The image in Figure 5 shows a typical bundle of pollen tubes fanning out of the cut end of a pistil in a PGM plate assay. Similarly, a cut pistil is placed on the PGM in a microgroove. The pistil is carefully aligned along the groove, at the middle between the two side walls, such that the pistil cut edge is perpendicular to groove axis. As shown in Figure 6, pollen tubes emerge from the cut end of a pistil and grow on the PGM in the microdevice. The pollen tube activity in both assays is very similar, as shown in Figs. 5 & 6, indicating that microdevices can provide reliable data about the growth of pollen tubes.

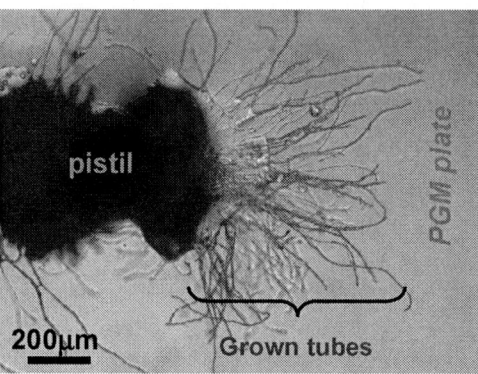

Figure 5: An image of pollen tube growth through a cut pistil placed on a PGM layer in a plate.

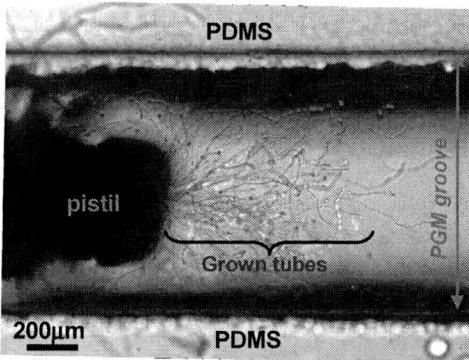

Figure 6: An image of pollen tube growth through a cut pistil placed on a PGM layer in a fabricated microdevice.

In order to quantify the pollen tube growth rate in both environments, tube length is measured at time intervals of 1 hour during the growth period. The growth of a typical pollen tube in a microdevice is compared in Figure 7 to the growth of a typical tube in a plate. In both cases, it takes about 4 hours for pollen tubes to emerge from the cut edge of a pistil; this is the time needed for pollen grains to germinate, form tubes and grow through the stigma and style. Furthermore, the pollen tube growth rate decreases with time, such that the tubes asymptotically attain their maximum length in about 8-9 hours. The average growth rate is estimated to be about 130μm/hr.

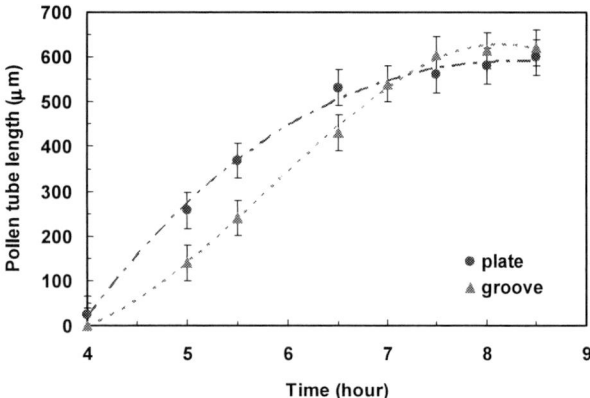

Figure 7: A comparison between pollen tube growth with time in a microdevice and in a plate.

As expected, significant variations in their final length are observed not only between bundles but among pollen tubes in the same bundle as well. Therefore, the final length of more than 100 pollen tubes has been measured in each assay, microdevice and plate. The resulting probability distributions of the pollen tube final length are plotted in Figure 8; the probability distribution of each curve is normalized such that the area underneath the curve is unity. In both assays, the measured average of pollen tube final length is about 450μm with a standard deviation of 200μm. This is a quantitative confirmation that the *in-vitro* microdevice assay provides a micro-environment for pollen tube growth at least as adequate as the one found in the plate assay.

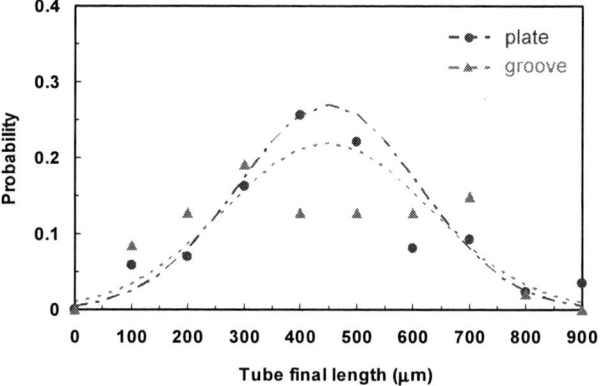

Figure 8: Probability distribution of the final length of pollen tubes grown in a microdevice and in a plate.

POLLEN TUBE TARGETING OF OVULES

A critical aspect of an *in-vitro* assay is the interaction between pollen tubes and ovules. Therefore, to investigate targeting of ovules by pollen tubes, *A. thaliana* flower ovules are freshly picked and gently distributed on the PGM facing the pistil cut edge in both assays, plate and micro-device, as shown in Figures 9(a) and (b). Since the measured average final length of pollen tubes is about 450μm, the ovules are placed within a distance of 400μm from the cut pistil edge ensuring that at least some pollen tubes can reach the ovules. When pollen tubes enter micropyles, they are usually obscured by the opaque ovule integument cells. Thus, to facilitate continuous pollen tube observation, the flower plants are transformed with a GFP reporter under the control of a pollen-specific LAT52 promoter [8]. Upon reaching a synergid cell, these tubes burst and GFP rapidly diffuses throughout a relatively large area within the ovule, conveniently marking targeted ovules. The images taken under a fluorescent microscope, shown in Figures 9(a) and (b), depict the GFP-marked pollen tubes as they approach the ovules. The red arrows in the images point to pollen tube burst events within the ovule indicating successful targeting of ovules. Of the tubes that sufficiently grow in length to reach the placed ovules, 60% (38/64) successfully entered the micropyle in plates compared to 67% (39/58) in micro-devices. Although this may statistically be negligible, the slightly higher efficiency in microdevices could be attributed to higher attractant concentration secreted from the ovules due to directional diffusion in confined PGM.

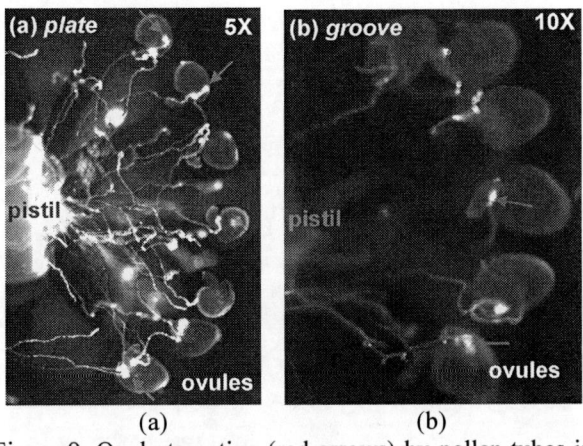

Figure 9: Ovule targeting (red arrows) by pollen tubes in a: (a) plate, and (b) microdevice.

POLLEN TUBE ATRACTANTS SECRETED BY OVULES

Using the PGM-filled PDMS microdevices, pollen tube response to attractants secreted by ovules is investigated. A device, with a microgroove and two identical chambers at each side of the groove, is used for this purpose. A cut pistil is placed in the microgroove near the chambers, while several ovules are placed in one chamber leaving the other one empty. The pollen tubes emerge from the cut edge of the pistil, as shown in Figure 10, grow initially along the groove axis and then preferentially turn toward the chamber containing the ovules. This phenomenon has been observed in more than 50% of the tested samples.

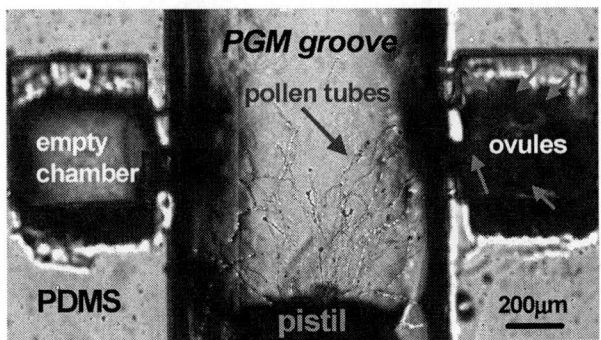

Figure 10: Pollen tubes grown in a microdevice turning preferentially towards the ovule-containing chamber.

It is believed that the unfertilized ovules release attractants, which diffuse via the connecting capillary filled with PGM. The pollen tubes, while growing, detect the attractants and adjust their growth orientation towards the concentration gradient of the secreted attractants.

These results suggest that utilizing PGM-filled PDMS-based microdevices, experimental investigation of pollen tube growth guidance can be conducted in a controlled manner. Indeed, these microsystems offer greater promise in attempting to identify individual attractants and quantifying their effects on pollen tube guidance than current assays. Investigating the existence of growth biases due to ovule secretions in *A. thaliana* has not been successfully performed using plate assays thus far.

CONCLUSIONS

Recapitulation of *in-vivo* pollen tube guidance to excised *A. thaliana* ovules in fabricated microdevices has been demonstrated. The measured growth rate and final length of pollen tubes in microdevices and plates are about the same. The targeting efficiency of ovules by pollen tubes is also similar in both assays. These results show that micro-environments for the study of *A. thaliana* pollen tube growth (and other species that contain extremely thin tubes) are promising. Initial results indicate preferential turning of pollen tubes toward unfertilized ovules secreting attractants. This encourages further studies such as measurements of attractant micro-gradients secreted by ovules as well as repulsion signals, which are impossible to conduct *in-vivo* and difficult to control using *in-vitro* plate assays.

ACKNOWLEDGEMENTS

This work is partially supported by NSF IOS-0723421

REFRENCES

[1] J.P. Mascarenhas, "The biochemistry of angiosperm pollen development", *Bot Rev* vol. 41, pp.259–314, 1975.

[2] R. Malho, "Pollen tube guidance-the long and winding road", *Sex Plant Report*, vol. 11, pp.242–244, 1998.

[3] V.E. Franklin-Tong, "Signaling and pollination", *The Plant Cell*, vol.11, pp.490-495, 1999.

[4] T. Higashiyama, H. Kuroiwa and T. Kuroiwa, "Pollen-tube guidance: beacons from the female gametophyte", *Current Opinion in Plant Biology*, vol. 6, pp.36-41, 2003.

[5] R. Palanivelu and D. Preuss., "Distinct short-range ovule signals attract or repel Abrabidopsis thaliana pollen tubes in vitro", *BMC Plant Biology*, vol. 6, pp. 1-9, 2006.

[6] A. Geitmann and R. Palanivelu, "Fertilization requires communication: signal generation and perception during pollen tube guidance", *Floriculture and Ornamental Biotechnology*, vol. 1 (2), pp77-89, 2007.

[7] L.M. Lee, R.L. Heimark, R. Guzman, J.C. Baygents, and Y. Zohar, "Low melting point agarose as a protection layer in photolithographic patterning of aligned binary proteins", *Lab on a Chip*, vol.6, pp.1080-1085, 2006.

[8] D. Twell, R. Wing, J. Yamaguchi and S. MaCormick, "Isolation and expression of an anther-specific gene from tomato", *Mol Gen Genet*, vol. 217 (2-3), pp. 240-245, 1989.

A BATCH-PATTERNED SELF-EXPANDING BILIARY STENT WITH CONFORMAL MAGNETIC PDMS LAYER AND TOPOLOGICALLY-MATCHED WIRELESS MAGNETOELASTIC SENSOR

Scott R. Green and Yogesh B. Gianchandani
University of Michigan, Ann Arbor, Michigan, USA

ABSTRACT

This paper presents a system for wirelessly monitoring the accumulation of sludge within biliary stents. The system comprises a sensor and biasing permanent magnet layer that conform to the meshed topology and tubular curvature of a biliary stent. The sensors have an active area of 7.5 mm x 29 mm and a mass of 9.1 mg. Annealing the sensor at 375°C results in reducing the required biasing magnetic field from 6 Oe to 2 Oe. The integrated system shows a 38% decrease in resonant frequency (from 61.6 kHz to 38.2 kHz) after an applied mass load of 20.9 mg, or 2.3x the mass of the sensor. The system architecture allows the mechanical properties of the stent to be maintained while adding important monitoring capabilities to the implanted device.

INTRODUCTION

Stents are mesh tubular structures used to impart and maintain patency in a variety of vessels and ducts that have become constricted as a result of stenotic pathology. Though the act of implanting a stent relieves symptoms caused by the constriction, in-stent restenosis – a reappearance of the narrowing, typically due to the reaction of the body to the presence of the stent – is a risk associated with all stenting procedures.

An example of a stent application area – and the focus of this work – is the bile duct. Restenosis can occur in a biliary stent over an unpredictable timeframe of 2-12 months via formation of a bacterial matrix known as biliary "sludge" [1]. Current diagnostic techniques are indirect or invasive. As such, a direct, non-invasive method of diagnosis – such as that shown in Fig. 1 – would enable timely intervention, minimize patient discomfort, and eliminate unnecessary procedures.

We have previously reported on magnetoelastic wireless sensing of sludge accumulation utilizing externally applied AC interrogative and stent-integrated DC biasing magnetic fields [2-3]. The magnetic fields cause a magnetoelastic sensor integrated with the stent to resonate at a frequency that changes as local viscosity increases and as sludge accumulates. The mechanical resonance generates an oscillating magnetic flux that can be measured with an external pick-up coil. Our previous work utilized discrete neodymium magnets to optimally bias the anisotropy of a ribbon sensor – i.e. a rectangular strip of magnetoelastic material – that was similar in design to sensors used in industrial/environmental applications [4-6]. Although this combination was shown to be effective in benchtop testing, the discrete nature of the magnet and sensor components leaves room for improvement – especially with regards to maintaining important distributed flexibility of the biliary stent. Components that conform to or mimic the open, flexible structure of the stent would lead to a system that is better able to withstand and accommodate the deformations required during catheter-based delivery, as well as lead to a system that preserves the structural functionality of the

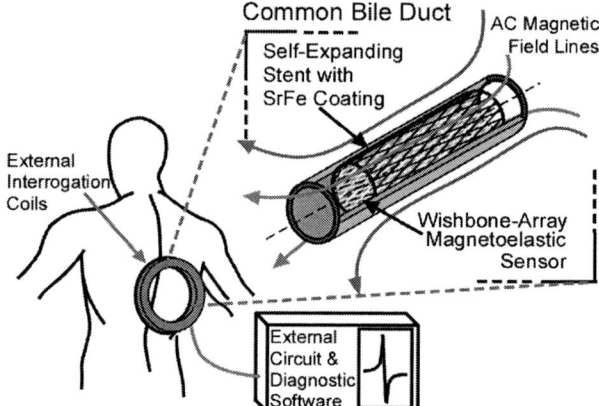

Figure 1: Conceptual diagram of in vivo magnetoelastic sensing of sludge accumulation for biliary stents.

stent. With this viewpoint, this work focuses on an integrated magnetoelastic system with a sensor and biasing permanent magnet layer that match the meshed topology and tubular curvature of a biliary stent. Further, we investigate the impact of sensor shaping and material optimization via thermal annealing.

DESIGN AND MODELING

Stent – Biliary stents generally reach their final *in situ* diameter via an elastic self-expansion. This is in opposition to the plastic expansion of typical balloon-assisted cardiac stents. The need for large elastic diameter recovery in biliary stents leads to not only the utilization of materials with superior elastic properties (e.g. chrome-nickel Elgiloy or nickel-titanium Nitinol) but also to the use of open diamond-shaped patterns. Often these patterns are formed by braiding filaments into a tubular shape, or by serially cutting the pattern from a metal tube.

In this work, we have investigated the batch-patterning of self-expanding stents from a planar Elgiloy foil utilizing a photochemical machining (PCM) process [2]. As shown in Figure 2, an elongated wishbone-array pattern is used; this pattern allows high elastic expandability for the stent. By mechanically joining the long sides of the planar pattern together, a tubular stent is formed. Two methods for joining the sides of the stent are described in the fabrication section.

Wishbone-Array Sensor – In keeping with the philosophy of mimicking the design of the stent with the design of the magnetoelastic sensor, we would like to use a material with superior elastic properties and to shape the material in diamond-shaped patterns. Fortunately, Metglas[TM] alloys are materials with excellent magnetostrictive properties as well as excellent elastic properties. For instance, the 2826MB alloy as used in this work is reported to have a yield strain of 1.6% [7], which is even higher than most cold-reduced Elgiloy yield strains of ~1% [8]. Metglas[TM] is readily available in foil form and suitable for batch-patterning using PCM. Again,

978-1-4244-2977-6/09 $25.00 © 2009 IEEE   212

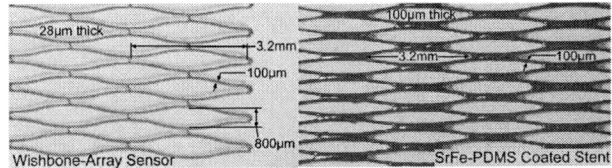

Figure 2: LEFT: A portion of a wishbone-array sensor. RIGHT: Stent with SrFe-PDMS coating.

the elongated wishbone-array pattern is used to improve the expandability of the structure and to avoid plastic strain in the sensor during deformation that is required during catheter-based delivery (Fig. 2).

Because the wishbone-array pattern represents a significant departure from typical ribbon sensors – which are analyzed for this application in detail in [2,9] – we developed an FEA tool that is appropriate for estimating mode shapes and expected signal amplitudes from sensors with complicated structures. Linearized piezomagnetic equations [10] are implemented in this work utilizing COMSOL Multiphysics and coupled time-harmonic (frequency response) induction current and stress-strain modes. A detailed look at an FEA implementation for magnetostrictive materials is in [11]; the approach used in this work is modified for application to resonant sensors. In Figure 3, the calculated mode shape at 61.6 kHz results in the largest response signal amplitude. Note that the mode shapes combine significant longitudinal and transverse motion, whereas mode shapes of traditional ribbon sensors are limited to longitudinal motion.

Conformal Magnetic Layer – To achieve optimal magnetomechanical coupling, the magnetoelastic material must be biased with a DC magnetic field. This field offsets the as-fabricated anisotropy of the magnetic domains in the material, and sets the operating point of the sensor in response to the small AC interrogative signal.

While sensor performance is improved when the bias field is as uniform as possible, this uniformity is difficult to achieve with integrated discrete magnets because the field strength will necessarily decay as the distance from the magnets increases. In addition, localized peaks and high magnetic field gradients lead to undesirable magnetic forces. The distributed magnet is chosen in this work to be a layer of strontium ferrite (SrFe) particles (~1 μm average diameter, Hoosier Magnetics) suspended in polydimethylsiloxane (PDMS, Sylgard 184, Dow Corning). This choice is made again in keeping with minimally altering the functionality and structure of the biliary stent with the additional components. In this case, the polymer-suspended particles can be applied in a thin, flexible layer conforming exactly to the stent structure (Fig. 2).

Other polymers have been used as a base for SrFe particles in microfabricated magnets described elsewhere [12]. SrFe particles have the advantages of being chemically inert (owing to their ceramic nature), and of being widely and inexpensively available in very small particle sizes. The chemical inertness is especially

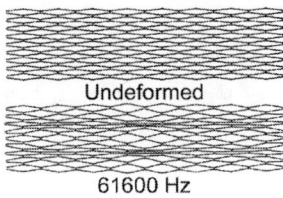

Undeformed

61600 Hz

Figure 3: FEA calculated mode shape and frequency prediction.

valuable in our implantable application. PDMS is chosen in this work due to its generally accepted biocompatibility and due to processing ease.

FABRICATION

Stent – The stent is batch fabricated from a 100 μm thick foil of Elgiloy using the PCM process. The feature sizes and patterns are identical to those of the sensor (Fig. 2). The overall stent size is 4 mm (dia.) x 40 mm.

Wishbone-Array Sensor – The wishbone-array sensors for this work are batch fabricated from a 28 μm thick foil of 2826MB Metglas™ (a NiFeMoB alloy) utilizing the PCM process. Feature sizes of the individual struts are 100 μm, which is near the feature size limit for the technology. The overall size of the active portion of the sensor (not including the anchor areas discussed later) is 7.5 mm x 29 mm, with a mass of 9.1 mg.

PCM is a planar process, so the as-fabricated sensors are also planar. Because the stent application calls for a tubular shape, and the lateral dimension of the sensor is larger than the diameter of the stent, the sensor must be curved into a tubular or semi-tubular shape to best match the stent geometry. Rolling the sensor into a tube inside the stent resulted in degradation of the signal amplitude. As such, the tubular shape is achieved in this work by placing the sensor against the inner wall of a fixture tube and annealing for 30 minutes at 375 °C. Various final radii can be achieved by either changing the fixture tube radius or by changing the anneal temperature.

Conformal Magnetic Layer – To form the conformal magnetic layer, the PDMS is first mixed in a 10:1 base-to-curing-agent ratio. Subsequently, the SrFe particles are introduced in 1:1, 3:1, or 1:3 SrFe-to-PDMS by weight ratios and mixed in by hand until the mixture is consistent (usually about 1 minute of mixing time). The mixture is then poured or spread into a mold containing the stent. The stent is then peeled out of the mold, with a conformal layer of the magnetic suspension adhered. The layer is then cured for 30 minutes at 60°C. Thicker layers can be built up by repeating the process. Finally, the layer is magnetized uniformly along the long axis of the stent using a benchtop pulse magnetizer. In general, the 1:1 SrFe:PDMS ratio offered the best combination of workability and remnant strength of the ratios tested.

System Assembly – Lateral portions of the wishbone-array sensor are connected to the active area with single struts. These areas act as anchors, and the single struts isolate the vibrating active area from the anchors. The anchors are bonded to the stent with a thin layer of PDMS.

Subsequently, the stent is rolled into a tubular shape and the resulting seam where the edges of the stent adjoin is mechanically joined. This joining process is achieved with two methods in this work (Fig. 4). In one method, the edges are brought into alignment with a fixture that also acts as a mold. PDMS is poured into the mold and cured, encasing the seam edges. In the other method, a wire is formed in the pattern on one edge of the stent. The wire is then woven through both edges to join the seam. The entire assembly process is shown in Figure 5, and an assembly is shown in Figure 6.

EXPERIMENTAL METHODS AND RESULTS

Stent – Important characteristics for biliary stents

978-1-4244-2977-6/09 $25.00 © 2009 IEEE

include deliverable diameter, expanded diameter, and radial stiffness. The ratio of the expanded diameter and deliverable diameter is a measure of the expandability of the stent. This ratio should be high to minimize invasiveness during stent placement. Radial stiffness should be high so that the stent can act as a mechanical scaffold to prop the duct open.

To establish deliverable and expanded diameter, stents were passed through tubes of known diameter and measured after passing through the tube with calipers (Fig. 7). To evaluate radial stiffness, the stent was placed in a semi-cylindrical fixture and probed with a force gauge (Imada, Inc.). The force gauge was equipped with a blade probe that allowed a localized (5 mm x 1 mm) application of force to best simulate a local bile duct lesion. The stent was probed at the mid-length and end. The stent was also tested in two different orientations – with the force applied to the seam or with the force applied 90° from the seam – to evaluate stiffness changes due to the seam. The testing was applied to uncoated stents, coated stents, stents with a bonded seam, stents with a woven seam, and to a commercially available biliary stent. Results are summarized in Table I.

Sensor Annealing – The optimal bias field – with which the amplitude of the response is largest (10 mVp-p) – is around 6 Oe for as-cast sensors (Fig. 8). The frequency and amplitude show repeatable performance across the tested sensors, indicating a repeatable PCM fabrication process.

The sensors were thermally treated either above (375°C) or below (325°C) the material Curie temperature (353°C) and either remained planar or were given curvature. Post-treatment evaluation showed lower optimal biasing field (~2 Oe, Fig. 8) and improved signal level (up to 13.5 mVp-p). This important result shows that thermal treatment facilitates thinner SrFe-PDMS layers, which simplifies fabrication and minimizes

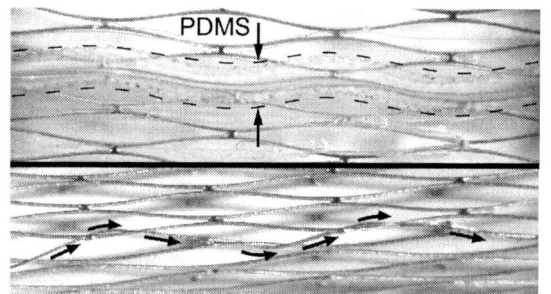

Figure 4: Seam joining techniques. TOP: PDMS seam joint (PDMS extent denoted by dashed line). BOTTOM: Woven wire seam (wire path denoted by arrows).

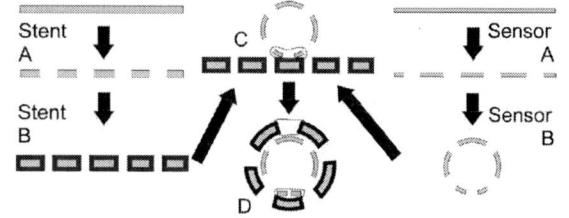

Figure 5: Fabrication process. A) PCM patterning of Elgiloy (stent) and Metglas^TM (sensor). B) Stent coated in SrFe-PDMS layer and magnetized. Sensor annealed in a tube. C) Sensor anchors bonded to stent with PDMS. D) Stent seam bonded with PDMS or joined by threading.

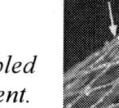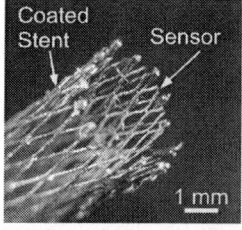

Figure 6: Assembled sensor and coated stent.

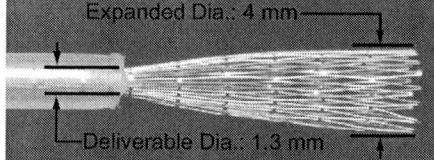

Figure 7: The batch-patterned stents are capable of 3:1 self-expansion from the deliverable diameter, similar to commercially available braided stents. The conformal coating requires higher forces to reach the deliverable diameter but does not change the elasticity.

concerns about chronically implanted magnetic fields.

Integrated System – The integrated system consists of a curved wishbone-array sensor and a SrFe-PDMS coated stent. For all tests, a swept-frequency network analyzer signal was amplified and sent through a transmit coil, while the same analyzer measured the EMF generated on a receive coil. The sensors were located concentrically with these coils. All biasing of the sensor was provided by the conformal SrFe-PDMS magnetic layer. For the integrated system, sensitivity to viscosity over a physiologically appropriate range was measured even as mass was added. This experimental process showed that the normalized frequency response of the sensor to viscosity changes was not significantly affected by mass buildup (Fig. 9). Application of the acrylate terpolymer sludge simulant as a mass load showed that the frequency and signal amplitude of the integrated sensor reacted to mass loads similarly to those of the isolated sensors (Fig. 10).

DISCUSSION

Three important advantages of the wishbone-array sensor over typical ribbon sensors in this application are made clear by this work. First, the fine features sizes and large open area of the pattern present little obstruction to bile flow, which is the primary objective of a biliary stent.

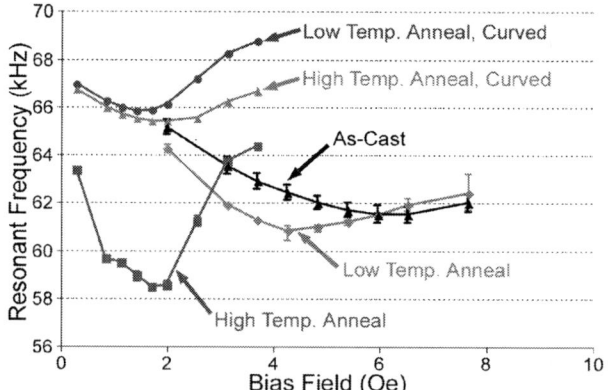

Figure 8: Thermal treatment of the sensor reduces the required biasing field from 6 Oe to 2-4 Oe depending on the treatment. The required biasing field is located at the minimum of each curve.

Second, the sensors are much more accommodating of the large deformations required for catheter-based delivery. Third, the sensors have a higher sensitivity to viscosity changes, which is a clinically relevant parameter in many pathological conditions [13]. The principal disadvantage of the wishbone-array sensor, at least with the present design, is the smaller signal amplitude. However, preliminary results show that the signal amplitude scales with the overall sensor length and width, so this disadvantage may be mitigated in future designs.

Prior to *in vivo* testing of the system, further evaluation of the mechanical properties of the stent must be done. Analytical models imply that the sludge simulants used in this work represent a worst-case scenario, as biofilms like sludge are likely to be less stiff than the test materials [14].

CONCLUSION

This work integrates a flexible wishbone-array magnetoelastic sensor and conformal magnetic layer with a biliary stent as a wireless system that monitors the stent environment. The system is sensitive to physiologically appropriate viscosity changes, showing a 7% decrease in resonant frequency in 10 cP fluid. The system also is capable of measuring mass buildup that is associated with sludge accumulation, showing a 38% decrease in the resonant frequency after an applied mass load of 20.9 mg, or 2.3x the mass of the sensor. The mechanical properties of the batch-patterned stents compare favorably with those of commercially available stents. Annealing the sensor results in control over the shape of the sensor and a reduced required biasing field. The integrated system is

Table I: Stent Mechanical Property Comparison

	Commercial	Batch-Patterned
Deliverable Dia.	1.2-1.3 mm	1.3-1.4 mm
Expanded Dia.	4 mm	4 mm
Radial Stiffness	0.07-0.15 N/mm	0.06-0.18 N/mm

robust to deformations required for delivery and provides a uniform biasing layer that minimally affects stent mechanics. With appropriate scaling, the sensing methodology may be applicable in any stent, including cardiovascular and esophageal stents. Additionally, the improved viscosity sensitivity of the wishbone-array sensor may find use in industrial applications like monitoring oil refinement.

ACKNOWLEDGEMENTS

The authors acknowledge Dr. Grace Elta and Dr. Richard Kwon for discussions regarding stent usage. Mark Richardson assisted with test setup design and implementation. Metglas Inc., Hoosier Magnetics, and Dow Corning provided samples for this project. This work was supported in part by a NSF Graduate Research Fellowship, the NSF ERC for Wireless Integrated Microsystems (WIMS), and the University of Michigan. Y. Gianchandani acknowledges support through the IR/D program while working at the National Science Foundation. The findings do not necessarily reflect the views of the NSF.

REFERENCES

[1] G. Donnelli, et al., "Plastic Biliary Stent Occlusion: Factors Involved and Possible Preventive Approaches," Clinical Medicine & Research, Vol. 5, No. 1, 2007, pp. 53-60.

[2] S. R. Green, et al., "Photochemically Patterned Biliary Stents with Integrated Permanent Magnets and Deformable Assembly Features for Wireless Magnetoelastic Tissue Growth Sensing," Transducers 2007, pp. 213-7.

[3] S. R. Green, et al., "Wireless Biliary Stent System with Wishbone-Array Resonant Magnetoelastic (WARM) Sensor and Conformal Magnetic Layer," *Hilton Head 2008: A Solid-State Sensors, Actuators, and Microsystems Workshop*, pp. 158-161.

[4] C. Grimes, et al., "Magnetoelastic Microsensors for Environmental Monitoring," Transducers 2001, pp. 278-81.

[5] M. Jain, et al., "A Wireless Micro-sensor for Simultaneous Measurement of pH, Temperature, and Pressure," Smart Mat. and Struc., Vol. 10, 2001, pp. 347-53.

[6] C. Grimes, et al., "Simultaneous measurement of liquid density and viscosity using remote query magnetoelastic sensors," Rev. Sci. Inst., Vol. 7, Issue 10, 2000, pp. 3822-4.

[7] J.-J. Lin, et al., "Embrittlement of Amorphous Fe40Ni38Mo4B18 Alloy by Electrolytic Hydrogen," Met. and Mat. Trans. A, Vol. 26, No. 1, 1995, pp. 197-201.

[8] Elgiloy Specialty Metals. Available online at www.elgiloy.com

[9] M.T. Richardson, et al., "Magnetoelastic Wireless Sensing of Tissue Growth for Self-Expanding Biliary Stents," IEEE MEMS 2007, pp. 469-472.

[10] G. Engdahl (ed.), *Handbook of Giant Magnetostrictive Materials*, Academic Press, 2000.

[11] J. Benatar, "FEM Implementations of Magnetostrictive-Based Applications", MS thesis, Univ. of Maryland, 2005.

[12] L. Lagorce, et al, "Magnetic and Mechanical Properties of Micromachined Strontium Ferrite/Polyimide Composites," JMEMS, Vol. 6, No. 4, 1997, pp. 307-312.

[13] H. Zhang et al., "Role of Bile Mucin in Bacterial Adherence to Biliary Stents," J. Lab Clin Med, 139(1), pp. 28-34, Jan 2002.

[14] A.W. Cense, et al., "Mechanical Properties and Failure of *Streptococcus mutans* Biofilms, Studied Using a Microindentation Device," J. Microbiological Methods, Vol. 67, 2006, pp. 463-472.

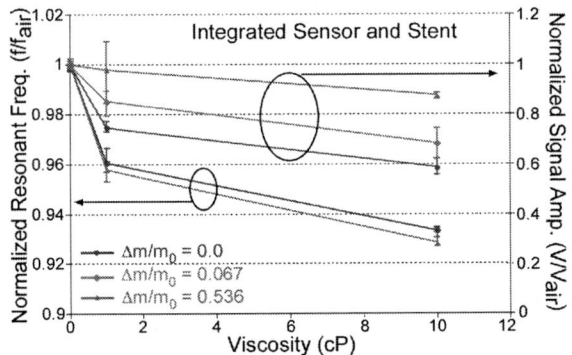

Figure 9: Integrated sensor response to physiologically relevant viscosity changes. The sensor was tested even as mass was added.

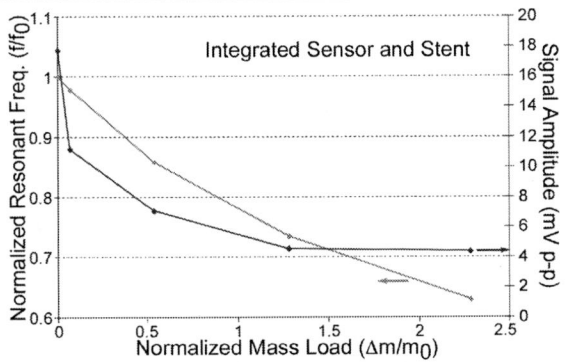

Figure 10: Mass was added to the fully integrated sensor with acrylate terpolymer. The signal amplitude and frequency of the integrated sensor responds to mass loading in a manner similar to isolated sensors.

RADIO-CONTROLLED CYBORG BEETLES:
A RADIO-FREQUENCY SYSTEM FOR
INSECT NEURAL FLIGHT CONTROL

H. Sato[1], Y. Peeri[1], E. Baghoomian[1], C.W. Berry[2], M.M. Maharbiz[1,2]

[1]Electrical Engineering and Computer Science, University of California, Berkeley, CA, USA
[2]Electrical Engineering and Computer Science, University of Michigan, Ann Arbor, MI, USA

ABSTRACT

We present the first report of radio control of a cyborg beetle in free-flight. The microsystem (Figs. 1, 2) consisted of a radio-frequency receiver assembly, a micro battery and a live giant flower beetle platform (Mecynorhina polyphemus or Mecynorhina torquata). The assembly had six electrode stimulators implanted into the left and right optic lobes, brain, posterior pronotum (counter electrode), right and left basalar flight muscles. Initiation and cessation of flight were accomplished by optic lobe stimulation while muscular stimulation of either right or left basalar flight muscles (referenced to the posterior pronotum electrode) elicited left or right turns, respectively. Flight commands were wirelessly transferred to the beetle-mounted system (running BeetleBrain v1.0 code) via an RF transmitter operated by a laptop running custom software (BeetleCo mmander v1.0) through a USB/Serial interface.

1. INTRODUCTION

Micro air vehicles (MAV's) which can navigate into locations not easily accessible to humans have been the subject of much recent research [1]. However, man-made MAV's are still limited in size, payload, distance and performance. In contrast, many insects have as-yet unmatched flight performance and increasingly understood muscular and nervous systems [2]. Additionally, some insects undergo complete metamorphosis (i.e. form pupae) and are amenable to implantation and internal manipulation during pupation. In light of this, there have been recent efforts by several groups to implant microsystems into insects to control their flight [3-8].

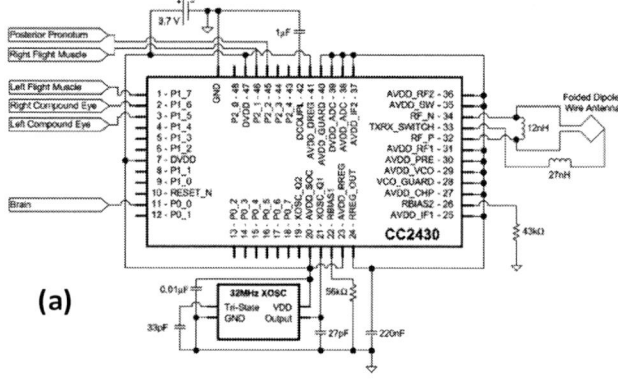

(a)

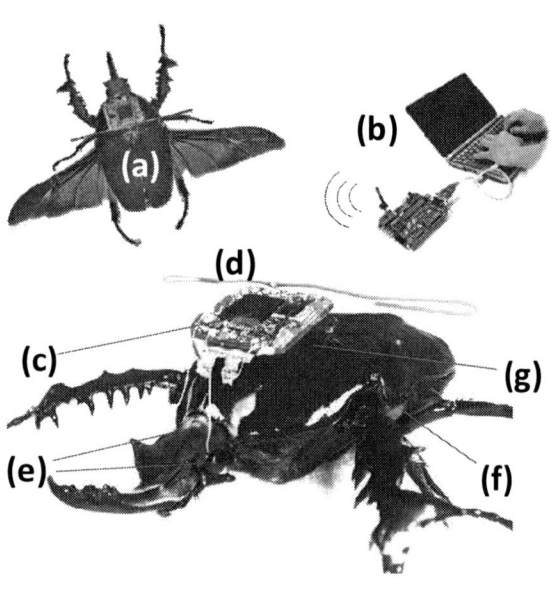

Figure 1: Photographs of a radio-controlled cyborg Mecynorhina beetle; (a) live beetle platform (Mecynorhina, 4 - 10 g, 4 - 8 cm), (b) RF transmitter (CC2431 microcontroller mounted on a Chipcon Texas Instruments SmartRF 04EB) operated by user's personal computer via USB/Serial-interface. (c) RF receiver assembly (Chipcon Texas Instruments CC2431, 2.4 GHz, on custom PCB, see also Fig. 2), (d) half-wave dipole antenna, (e) stimulating electrode terminals at both optic lobes, (f) basalar flight muscle (left), (g) posterior pronotum (counter electrode).

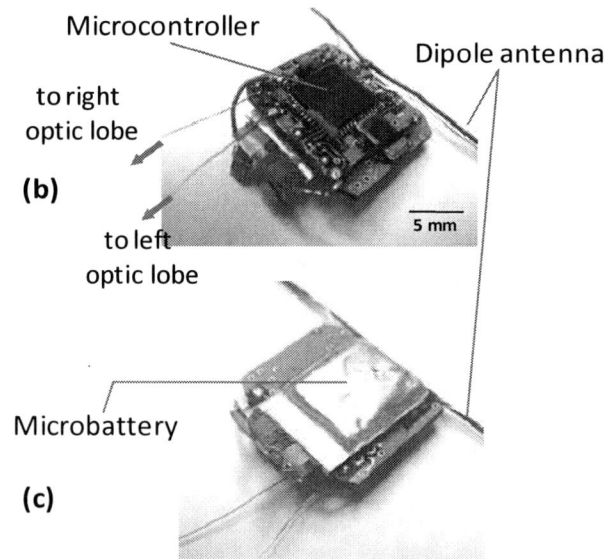

Figure 2: RF receiver assembly (a) circuit diagram, (b) front and (c) backside photographs. Total weight was 1331 mg; rigid PCB + misc components = 687 mg, microbattery (8.5 mAh, Micro Avionics) = 350 mg, antenna = 74 mg, microcontroller = 130 mg, insect adhesive = 90 mg.

978-1-4244-2977-6/09 $25.00 © 2009 IEEE

If it were possible to remotely control the flight of insects, and receive information from on-board sensors, there would be many applications. In biology, the ability to control insect flight would be useful for studies of insect communication, mating behavior and flight energetics, and for studying the foraging behavior of insect predators such as birds, as has been done with terrestrial robots [9]. In engineering, electronically-controllable insects could be useful models for insect-mimicing M/NAV's [10-12]. Furthermore, tetherless, electrically-controllable insects themselves could be used as M/NAV's and serve as couriers to locations not easily accessible to humans or terrestrial robots.

We previously reported flight control of live beetles including initiation, cessation, elevation and turns using a conventional neural wire stimulators driven by a pre-programmed microcontroller, powered by a microbattery, mounted on the anterior pronotum [6, 7].

In this study, we extended this concept to include a microcontroller with an RF transceiver, an antenna and a slightly bigger microbattery. This allowed us to wirelessly remote control insect flight initiation, cessation and turning in free flight.

2. EXPERIMENTAL RESULTS

RF System

The remote control system used two Chipcon Texas Instruments CC2431 microcontrollers (6 x 6 mm, 130 mg, 2.4 GHz); one acting as the beetle-mounted RF receiver and one as the computer-driven RF transmitter base station. Based on the circuit diagram shown in Fig. 2(a), we designed and manufactured custom PCB's (printed circuit boards, 16 x 13 mm, FR4 (rigid) version: 500 mg, polyimide (flexible) version: 70 mg) for the receiver. After programming, the microcontroller and the other components were assembled on the PCB as shown in Fig. 2(b). The microcontroller was powered by a rechargeable micro lithium-polymer-battery (Micro Avionics, 4 V, 8.5 mAh, 350 mg) which was attached on the backside of PCB with a piece of double adhesive tape as shown in Fig. 2(c) and electrically connected to the PCB when used.

We employed *Mecynorhina polyphemus or Mecynorhina torquata* beetle (4 – 10 g, 4 – 8 cm) as the insect platform. The assembly was mounted on the beetle's posterior pronotum (Fig. 1) and glued with beeswax. The terminals of 6 output wires from the assembly were inserted into the left and right optic lobes, brain, posterior pronotum, left and right basalar flight muscles (Fig. 3).

Flight commands were generated by custom control software (BeetleCommander v1.0) running on a personal computer interfaced via a USB port with the transmitter (CC2431 microcontroller mounted on a Chipcon Texas Instruments SmartRF 04EB). BeetleCommander v1.0 allowed for in-flight control of stimulation parameters including frequency, number and duty cycle of control voltage pulses to stimulated sites. Signals were transmitted using the CC2431's built-in 2.4 GHz IEEE 802.15.4 compliant transmitter broadcasting on a single channel (1A, 2.480 GHz) using direct sequence spread spectrum RF modulation. The transmitter sent a command to the receiver every 1 ms for 300 ms when instructed to do so. The flight commands were mapped to appropriate voltage pulse trains at the beetle's neural stimulators by custom

signal generating software (BeetleBrain v1.0) running on the receiver. To adjust the applied potential to a value other than the 4 V supplied from the lithium-polymer battery, surface mount resistors were soldered in parallel as voltage dividers to each output pin.

The working range of the beetle-mounted wireless system was ~10 m indoors in a modern office environment; outside, the range is 2 – 5 x greater depending on line of sight and objects present. At full power, the receiver consumes ~77 mW. Cycling sleep and receive modes, consumption is 10.95 mW for operation. The optic lobe and basalar flight muscle stimulations consumed ~500 μW and ~20 μW, respectively as shown in Fig. 4.

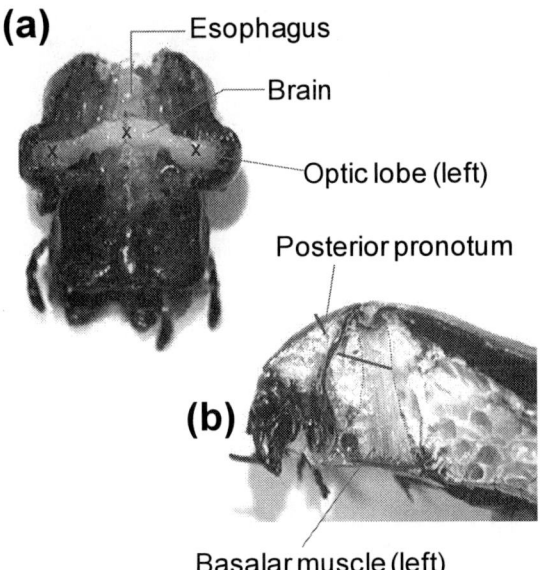

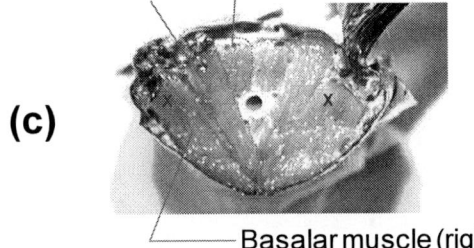

Figure 3: (a) front view of dissected beetle head showing the brain and optic lobe implant sites and relevant internal regions. (b) sagittal section of thorax showing the posterior pronotum implant (counter electrode), the basalar muscle implant (muscular stimulator) and relevant internal regions. (c) cross-section of mesothorax also showing the basalar muscle stimulator site and relevant internal regions. The basalar muscle stimulator was inserted rostral-caudal on either side of the insect, about midway between sternum and notum of mesothorax to a depth of approximately 1 cm. The letters X and bars indicate implant sites and approximate implant length, respectively. Green clay was used to support the objects in (c). Cotinis texana (which has nearly identical, scaled anatomy to the Mecynorhina beetle) was used for these anatomical images.

Flight Control

Flight initiation was triggered by applying a 4 V, 100 Hz, 20 % duty cycle, alternating positive and negative potential pulses (Figs. 4(a), 4(b)) to the two neural stimulators implanted into the optic lobes (Figs. 1 and 3(a)) via the mounted receiver as shown in Fig. 5 and Movie 1 in [13]. The response time was < 1 sec (N = 9); a typical, untethered beetle responded to the flight initiation command in 270 ms, as determined by frame by frame analysis in 30 fps video as shown in Figs. 5 (c) – (e). Flight cessation was triggered by a single 4 V, 1 sec pulse applied between the optic lobes.

Turn could be elicited in free flight by applying 2 V, 100 Hz positive potential pulses (Figs. 4(c), (d)) to either left or right basalar flight muscle (working electrode) with respect to posterior pronotum (counter electrode). The beetle turned in a direction opposite to stimulated side: left turn was, for example, elicited by stimulating the right basalar flight muscle. Representative turn control is shown in Fig. 6 and Move 2 in [13].

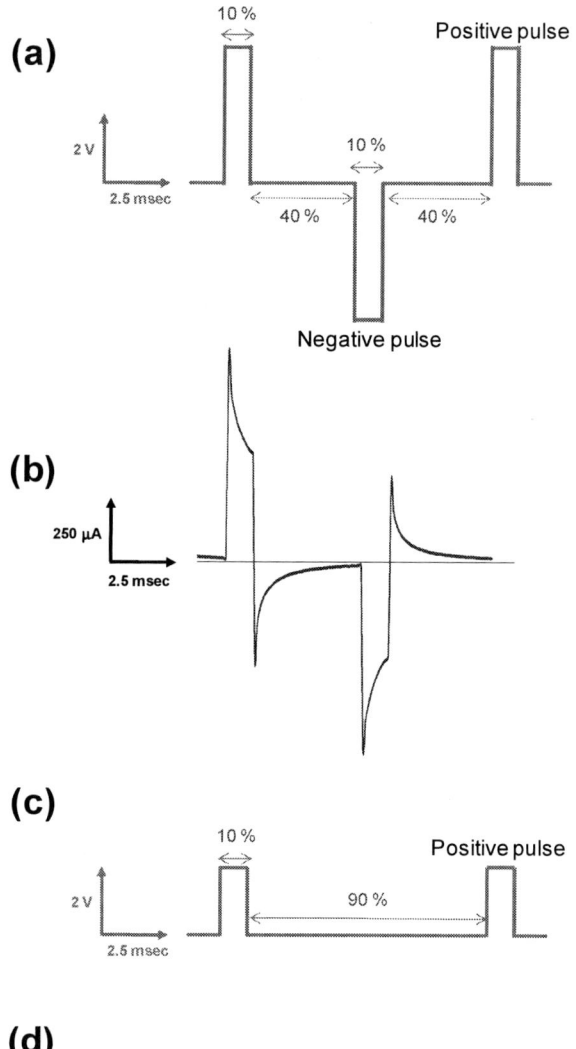

Figure 4: (a) alternating positive and negative potential pulse trains (100 Hz) applied between the left and right optic lobes for initiation of flight. (b) typical current wave monitored when applying (a), (c) positive potential pulse trains (100 Hz) applied between either left or right basalar flight muscle and the posterior pronotum (counter electrode) for eliciting turns, (d) typical current wave monitored when applying (c).

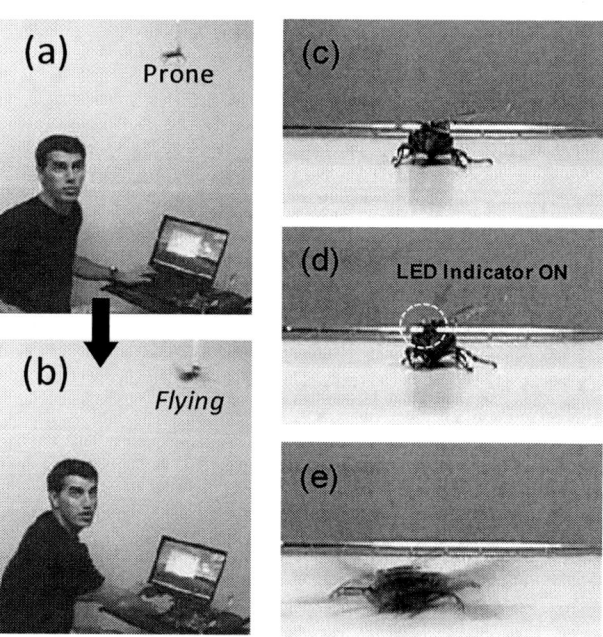

Figure 5: Photographs of flight initiation induced by the optic lobe simulator (100 Hz, 4 V amplitude, alternating positive and negative potential pulse trains between left and right optic lobes). The stimulator was wirelessly operated by user via the CC2431's built-in 2.4 GHz IEEE 802.15.4 compliant transmitter broadcasting on a single channel (1A, 2.480 GHz) using direct sequence spread spectrum RF modulation. (a) user signaled for initiation of flight, (b)beetle initiated flight. (c)-(e) close-up views of un-tethered initiation of flight. (c) standing position, (d) the LED indicator showed the stimulator turned on, (e) wings unfolded and flight started 0.27 sec after control signal was sent (b). See Movie 1 in [13].

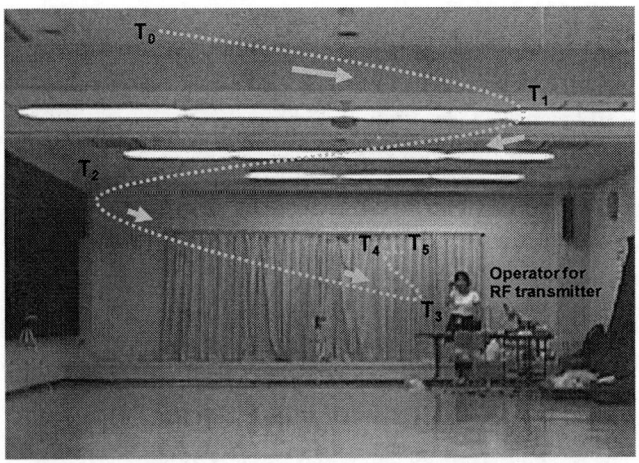

Figure 6: Flight path of a flying beetle wirelessly stimulated for turn control. The beetle was initially flying towards the operator. T_0 (0.00 sec) is the start time of the filming. At T_1 (0.6 sec), the operator signaled a left turn from the base station (right basalar muscle stimulation). At T_2 (1.6 sec), the operator switched the stimulated side from the right to the left basalar flight muscle, and the beetle turned right. At T_3 (3.1 sec), the right basalar flight muscle was stimulated (left turn). At T_4 (4.2 sec), the left basalar flight muscle and turning right again. At T_5 (4.8 sec), the beetle touched on the curtain and stopped the flight. See Movie 2 in [13] for detail.

3. CONCLUSION

We present the first-ever wireless flight control microsystem using a small RF receiver mounted on a live beetle and an RF transmitter operated from a base station. Flight initiation and cessation were accomplished by neural stimulation of both optic lobes while turns in free flight were elicited by muscular stimulation of basalar flight muscle on either side.

4. ACKNOWLEDGMENT

We thank Professor Jon Harrison (Arizona State University) for his helpful advice on insect biology and entomology. This work was financially supported by Defense Advanced Research Projects Agency (DARPA) Hybrid Insect MEMS program.

5. REFERENCES

[1] D. J. Pines, F. Bohorquez, "Challenges Facing Future Micro-Air-Vehicle Development", *J. Aircraft*, vol. 43, pp. 290-305, 2006.

[2] M. Dickinson, G. Farman, M. Frye, T. Bekyarova, D. Gore, D. Maughan, T. Irving, "Molecular Dynamics of Cyclically Contracting Insect Flight Muscle *in vivo*", *Nature*, vol. 433, pp. 330-333, 2005.

[3] J. Marshall, "The Fly Who Bugged Me: Labs around the World Are Hatching a New Breed of Cyborg Animal Spies", *New Scientist*, March, pp. 41-43, 2008.

[4] A. Bozkurt, A. Paul, S. Pulla, A. Ramkumar, B. Blossey, J. Ewer, R. Gilmour, A. Lal, "Microprobe Microsystem Platform Inserted During Early Metamorphosis to Actuate Insect Flight Muscle", *Proc. IEEE MEMS 2007*, pp. 405-408.

[5] A. Bozkurt, R. Gilmour, D. Stern, A. Lal, "MEMS Based Bioelectronic Neuromuscular Interfaces for Insect Cyborg Flight Control" *Proc. IEEE MEMS 2008*, pp. 160-163.

[6] H. Sato, C. W. Berry, B. E. Casey, G. Lavella, Y. Yao, J. M. VandenBrooks, M. M. Maharbiz, "A Cyborg Beetle: Insect Flight Control Through an Implantable, Tetherless Microsystem", *Proc. IEEE MEMS 2008*, pp. 164-167.

[7] H. Sato, C. W. Berry, M. M. Maharbiz, "Flight Control of 10[+] Gram Insects by Implanted Neural Stimulators", *Proc. Solid-State Sens., Actuators, and Microsyst. Workshop Hilton Head 2008*, pp. 90-91.

[8] W.M. Tsang, Z. Aldworth, A. Stone, A. Permar, R. Levine, J.G. Hildebrand, T. Daniel, A.I. Akinwande, J. Voldman, "Insect Flight Control by Neural Stimulation of Pupae-Implanted Flexible Multisite Electrodes", *Proc. μTAS 2008*, pp. 1922-1924.

[9] A. Michelsen, B. B. Andersen, W. H. Kirchner, M. Lindauer, "Honey Bees Can be Recruited by a Mechanical Model of a Dancing Bee" *Naturwiss*, vol. 76, pp. 277-280, 1989.

[10] W. C. Wu, L. Schenato, R. J. Wood, R. S. Fearing, "Biomimetic Sensor Suite for Flight Control of a Micromechanical Flying Insect: Design and Experimental Results", *Proc. IEEE Int. Conf. Rob. Autom.*, vol. 1, pp. 1146-1151, 2003.

[11] R. J. Wood, "The First Takeoff of a Biologically Inspired At-Scale Robotic Insect", *IEEE Trans. Rob.* vol. 24, pp. 341-347, 2008.

[12] L. Schenato, W. C. Wu and S. Sastry, "Attitude Control for a Micromechanical Flying Insect via Sensor Output Feedback", *IEEE Trans. Robs. Autom.*, vol. 20, pp. 93-106, 2004.

[13] Movie files are downloadable from: ftp://ftp.eecs.umich.edu/people/maharbiz/CotinisFlightMovies/MEMS2009.

HOLLOW MICRONEEDLE ELECTRODE ARRAYS FOR INTRACELLULAR RECORDING APPLICATIONS

J. Held[1], J. Gaspar[1], P. J. Koester[2], C. Tautorat[2], M. Hagner[3], A. Cismak[4],
A. Heilmann[4], W. Baumann[2], P. Ruther[1], and O. Paul[1]

[1]Department of Microsystems Engineering (IMTEK), University of Freiburg, Germany
[2]Department of Biophysics, University of Rostock, Germany
[3]Department of Physics, University of Konstanz, Germany
[4]Department of Biological Materials and Interfaces, Fraunhofer Institute, Halle, Germany

ABSTRACT

This paper reports on the fabrication of hollow microneedle electrodes with fluidic channels arranged in 8×8 arrays. Features of these electrodes include (i) an increased surface area for improved intracellular potential measurements with simultaneous membrane cell poration capabilities, (ii) their potential use in highly parallel patch-clamp applications and (iii) the ability to efficiently inject reagents and extract cytoplasm into and from the cell interior, respectively. Three different fabrication processes to realize hollow microneedle electrode arrays with incorporated microfluidic components, as well as initial experiments with cell cultures, are presented.

INTRODUCTION

A fundamental method in drug screening and disease studies is the electrical recording of the specific cell response to an external stimulus in living cell cultures. The classical patch-clamp method enables the measurement of intracellular membrane potentials [1]. However, due to the sophisticated equipment used, this is a time-consuming technique that can only be applied to single cells or to a small number of cells in a cluster. On the other hand, patch-on-chip systems allow the parallel examination of a larger number of cells, but only for cells in suspension [1]. In case of adherently growing cell cultures, classical CMOS-based electrode chips allow the measurement of extracellular potentials only [2].

To provide a method that enables the measurement of intracellular cell potentials of adherently growing cells, a novel cell chip design comprising microneedle-based electrodes was recently developed by the authors [3,4]. These microneedle chip systems are further developed here to obtain hollow microneedle chips, which possess an increased surface area for improved intracellular

recording applications. The chip contains an array of 64 hollow microneedle-based electrodes with diameters in the sub-micron range and heights below 10 µm. Different cell types are cultivated on these three-dimensional electrodes, which are subsequently introduced into the cytoplasm using local micro invasive needle electroporation (LOMINE), schematically shown in Fig. 1 [3,5-9]. Due to the low radius of curvature at the needle tip, a voltage pulse on the order of a few hundred millivolts up to a few volts is sufficient for this purpose [5,6].

FABRICATION

The proposed hollow microneedle processes are schematically summarized in Fig. 2, and the respective layout of one chip is given in Fig. 3. Each chip includes 64 microneedles over an area of approximately 1×1 mm², one reference electrode and a substrate contact. The total chip area of 10×10 mm². The three different process sequences are described in the following.

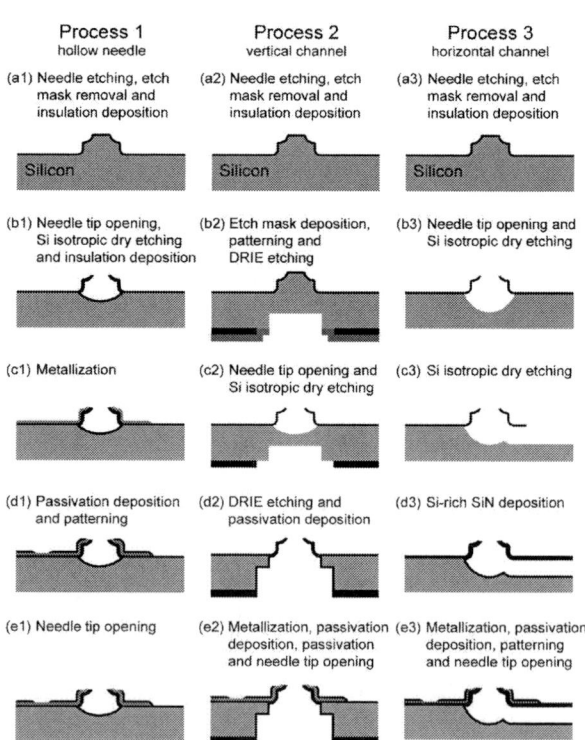

Figure 2: Schematics of the three different fabrication processes of hollow microneedle arrays for intracellular recording applications.

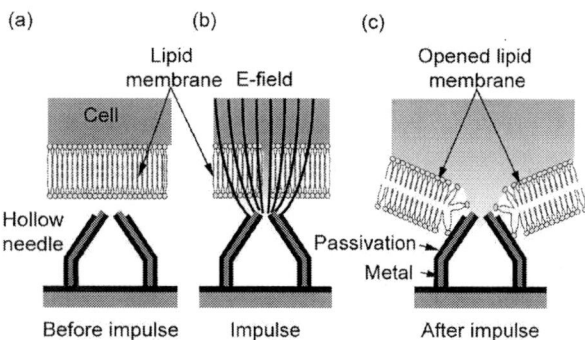

Figure 1: (a-c) Schematic sequence of local electroporation resulting in an opening of the lipid membrane.

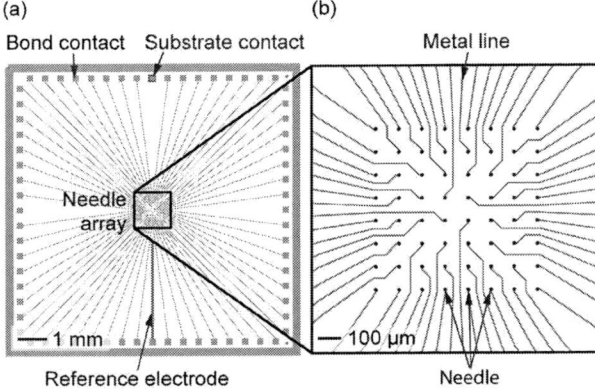

Figure 3: (a) Chip layout for the four-masks-process, and (b) close-up of the needle array with metal leads.

Process 1 – Hollow Microneedles

As described in Fig. 2, the fabrication process 1 performed on single-side polished (100) silicon wafers starts with the deposition of a 300-nm-thick low-temperature oxide (LTO) layer serving as an etch mask. This oxide layer is patterned by reactive ion etching (RIE) using a commercial RIE etcher (STS RIE Multiplex). The definition of the three-dimensional shape of the hollow microneedles is obtained by sequences of isotropic, anisotropic, and isotropic silicon dry etching, as explained in detail elsewhere [3,4]. The etching steps are performed in an inductively coupled plasma (ICP) etcher from STS. By varying the etch mask diameter and process parameters, different needle profiles can be realized [4]. After etching of the microneedles, the LTO mask is removed, followed by the deposition of a 400-nm-thick SiO layer serving as a etch mask as illustrated in Fig. 2 (a1) [3]. The hollow needle is realized by opening the oxide layer at the needle tip by maskless lithography and RIE. The exposed silicon is then etched by isotropic dry etching using the ICP system, Fig. 2 (b1), followed by the deposition of a 400-nm-thick SiO insulation layer on the wafer surface. This oxide layer is patterned by RIE for the substrate contacts. For electrical connection of the hollow microneedles, a thin metallization layer is sputter deposited and patterned using lift-off. This metallization consists of a stack of titanium tungsten (TiW, 15 nm) acting as an adhesion layer and platinum (Pt, 150 nm) for the electrical contact [Fig. 2 (c1)]. In the lift-off process, a bi-layer photoresist scheme using LOR3A and AZ1518 resists is employed. This bi-layer process is required since the standard lift-off process using AZ5214E resist only in combination with sputter deposition creates small sidewalls at the borders of the metal leads. As passivation of the metal, a 400-nm-thick silicon nitride (SiN) layer is deposited by plasma enhanced chemical vapor deposition (PECVD) using a mixed frequency (MF) process. The SiN is patterned by RIE to open the bond pads and the reference electrode, as shown in Fig. 2 (d1). For measurement purposes, the metal has to be exposed at the needle tip. This is again done using a maskless process similar to the one described in Fig. 2 (b1). At the end of the process, the SiN layer is removed from the tip of the hollow microneedle by RIE [Fig. 2 (e1)].

A hollow microneedle realized using process 1 is shown in Fig. 3. It may be connected to out-of-plane and in-plane fluidic channels as proposed in processes 2 and 3, respectively.

Process 2 – Hollow Microneedles with Vertical Channels

The fabrication of vertical out-of-plane channels using process 2 starts with a 380-µm-thick double-side polished (100) silicon wafer. The three-dimensional definition of the needles is realized similar to process 1 [see Figs. 2 (a1) and (a2)]. The vertical out-of-plane channels of process 2 are obtained by using a dry etching process performed in three steps: (i) deep reactive-ion etching (DRIE) of the wafer rear, (ii) isotropic dry etching of the hollow microneedle on the front side, and (iii) a second DRIE step on the wafer rear. This requires two additional lithography masks. At first, the etch mask required in the second DRIE step is defined depositing a PECVD SiO layer at high frequency (HF) patterned using RIE. This is followed by the definition of the etch mask for the first DRIE using photoresist. This second photoresist mask with typical feature sizes of 7.5 µm overlaps the preceding SiO masking layer. After etch mask definition, holes with a diameter of 8 µm are etched to a depth of 80 µm using DRIE as illustrated in Fig. 2 (b2). The following process steps include resist removal at the wafer rear and etching of the hollow needles as described in process 1 [Fig. 2 (c2)]. The final DRIE step concludes the wafer through-etch with a diameter of 50 µm from the wafer rear. This enlarged diameter is intended to increase the etch rate during fabrication [Fig. 2 (d2)]. After etching of the vertical channels, the silicon substrate is insulated using a SiO

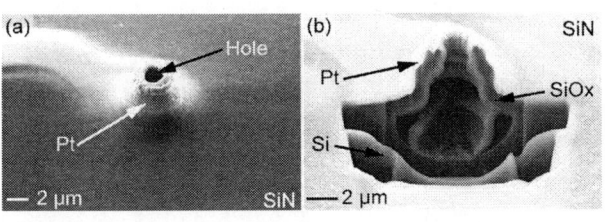

Figure 3: SEM micrograph of (a) hollow needle fabricated using process 1 and (b) respective FIB cross section.

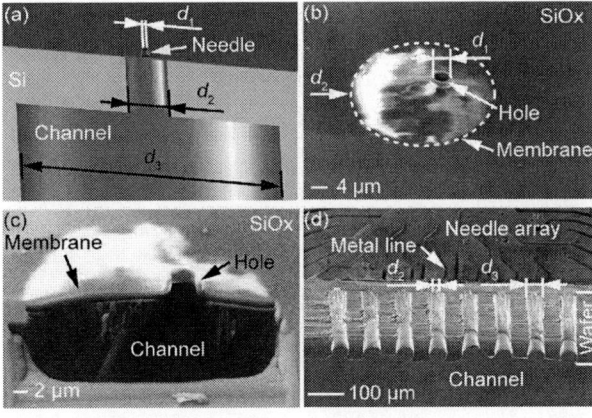

Figure 4: (a) Schematics of hollow needle with vertical channel fabricated using process 2; SEM micrograph of fabricated structures: (b) overview, (c) FIB cross section showing remaining membrane, hollow needle and the channel, and (d) overview on vertical channels.

978-1-4244-2977-6/09 $25.00 © 2009 IEEE

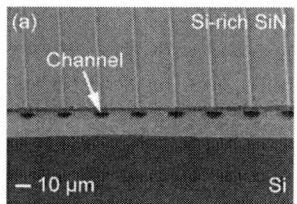 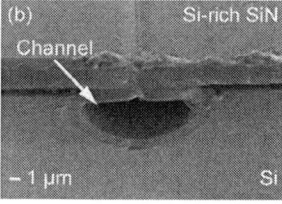

Figure 5: SEM micrograph of horizontal channels, fabricated using process 3; (a) overview and (b) close-up.

layer, the metallization is deposited and patterned and the final passivation is deposited and selectively removed as illustrated in Fig. 2 (e2).

The layout of the etch mask leads to narrower channel diameters at the tip of the hollow needle, as shown in Fig. 4.

Process 3 – Hollow Microneedles with Horizontal Channels

In process 3, horizontal fluidic channels are realized in order to connect the hollow needles. Again, as described in process 1, the proposed fabrication starts with the definition of the microneedle as show in Fig. 2 (a3), followed by the etching of the hollow needle structure [Fig. 2 (b3)]. The horizontal channels are realized by first patterning a SiO layer by RIE. The profile of the channel is then defined by isotropic dry etching using a commercial ICP etcher as illustrated in Fig. 2 (c3). The etch openings are then closed by conformal deposition of a 3-μm-thick silicon-rich SiN layer by low pressure chemical vapor deposition (LPCVD) as described in Fig. 2 (d3) and [10]. After the channels have been sealed, the metallization is deposited and patterned, followed by the deposition, patterning and selective removal of the passivation layer as shown in Fig. 2 (e3) and described in process 1. Scanning electron micrographs (SEM) of fabricated horizontal channels are shown in Fig. 5.

EXPERIMENTAL RESULTS

The measurement setup used for local micro invasive needle electroporation (LOMINE) and intracellular recording is shown in Fig. 6. It is possible to apply voltages and to measure intracellular potentials from 16 hollow microneedles in parallel by using a data acquisition card (NI 6221 National Instruments). Due to the specific interconnection scheme on the hollow needle chip and its carrier, a consecutive measurement of all hollow needles is feasible by rotating the chip carrier in steps of 90°. The cell position and attachment on the chip are examined optically using water immersion

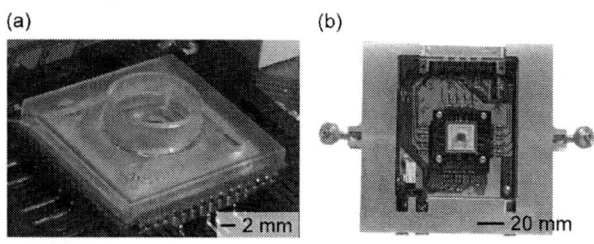

Figure 6: (a) Hollow microneedle chip wire bonded to a commercial leadless chip carrier and (b) measurement setup.

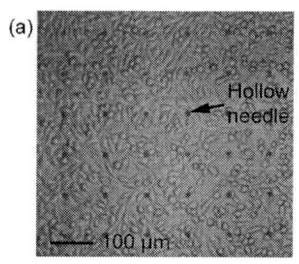 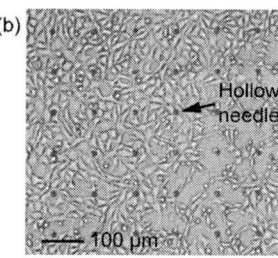

Figure 7: Micrograph of L929 fibroblasts on an array of hollow microneedles; (a) before and (b) after poration tests and chemical fixation.

microscopy. The successful seeding of a cell culture of murine L929 fibroblasts on a hollow needle chip, before and after poration tests is shown in Fig. 7.

To investigate the interface between the hollow needle and a cell either grown on or electroporated with the microneedle electrode, focused ion beam (FIB) technology together with SEM is applied. The samples are prepared using the usual procedures for SEM investigations of biological samples [11]: (i) primary fixation in glutardialdehyde, (ii) post-fixation with osmium tetroxide, (iii) dehydration and drying with critical point technology. The samples are finally coated with a thin platinum layer to prevent surface charging. Figure 8 (a) shows a hollow needle chip with cultivated L929 fibroblasts with an enlarged view in Figure 8 (b). The FIB preparation shown in Fig. 8 (c) demonstrates that the cell is in contact with the hollow needle tip. The interface between the cell and the hollow needle is illustrated in the FIB cross-section in Fig. 8 (d). As it can be seen, the cell is in direct contact with the tip of the hollow needle. A good electrical contact is thus expected as well.

Promising measurement results of the intracellular potential of L929 cells using hollow structures are shown in Fig. 9. The diagram shows the measured potentials at two hollow needle electrodes as a function of time. In case

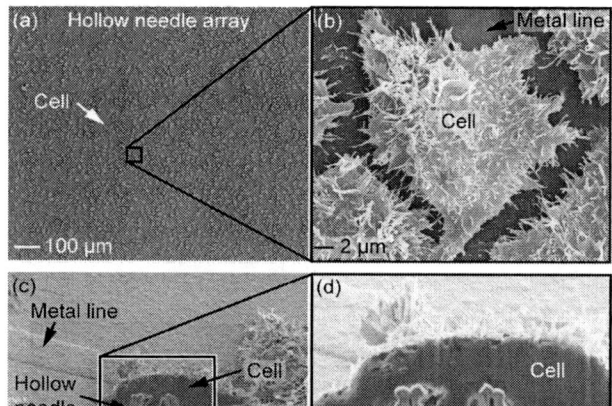

Figure 8: SEM micrographs of L929 fibroblasts on a hollow needle chip; (a) overview, (b) close-up, (c) FIB cross section showing a fibroblast on top of a hollow needle tip, and (d) interface between hollow needle and cell.

978-1-4244-2977-6/09 $25.00 © 2009 IEEE

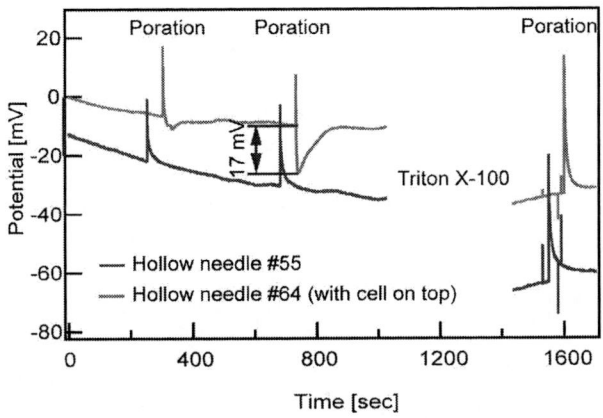

Figure 9: Measurement of potential with hollow microneedle chip. The dashed line indicated the needle without cell; the solid line with a cell on top.

of electrode #64 (solid line) a L929 cell is located on the electrode while no cells are present close to needle #55 (dashed line). An electroporation signal is applied to both needles. After the second electroporation pulse a potential drop of around −17 mV is measured at needle #64 indicating the successful poration. Prior to the third electroporation signal triton X-100 is added to the cell culture to kill the cells. In this case, potential measurements at needle #64 indicate the same behavior as needle #55 without cells.

CONCLUSION

Three fabrication processes to realize hollow silicon-based microneedle electrodes with fluidic channels for intracellular recording application are demonstrated. Hollow microneedles with opening diameters smaller than 2 μm, and heights around 5 μm are able to penetrate membranes through LOMINE. Current work involves performing electroporation experiments, as previously demonstrated with solid silicon microneedles [3]. Further development of the hollow needle chips involves the integration of dielectrophoretic electrodes for cell positioning, as previously demonstrated with solid microneedle chips [12,13].

ACKNOWLEDGMENTS

The authors gratefully acknowledge A. Baur and M. Reichel (IMTEK cleanroom service center) for useful discussions and help in cleanroom fabrication. The authors also thank BMBF/VDI/VDE for funding the project MIBA – *Mikrostrukturen und Methoden für die intrazelluläre Bioanalytik* (project number 16SV2337).

REFERENCES

[1] O. P. Hamil, A. Marty, E. Neher, B. Sakmann, F. J. Sigworth, "Improved patch-clamp techniques for high-resolution current recording from cells and cell-free membranes patches," *Eur. J. Physiol.*, 1981, Vol. 406, pp. 73-82.

[2] W. Baumann, E. Schreiber, G. Krause, S. Stüwe, A. Podssun, S. Homma, H. Anlauf, I. Freund, R. Rosner, M. Lehmann, "Multiparametric neurosensor microchip," *Proc. Eurosensors XVI*, 2002, pp. 1169-1172.

[3] J. Held, J. Gaspar, P. J. Koester, C. Tautorat, A. Cismak, A. Heilmann, W. Baumann, A. Trautmann, P. Ruther, O. Paul, "Microneedle arrays for intracellular recording applications," *Dig. Tech. Papers IEEE MEMS Conf.*, 2008, pp. 268-271.

[4] J. Held, J. Gaspar, P. Ruther, M. Hagner, A. Cismak, A. Heilmann, O. Paul, "Systematic characterization of DRIE-based fabrication process of silicon microneedles," in *Mat. Res. Soc. Symp. Proc*, 2007, 1052-DD07-07.

[5] P. J. Koester, C. Tautorat, J. Held, J. Gaspar, P. Ruther, O. Paul, A. Cismak, A. Heilmann, J. Gimsa, W. Baumann, "Local mirco-invasive needle electroporation (LOMINE) of single cells attached on silicon chips", *Proceedings IBS'08 Conference*, 2008, in press.

[6] C. Tautorat, P. J. Koester, J. Held, J. Gaspar, P. Ruther, O. Paul, A. Cismak, A. Heilmann, J. Gimsa, H. Beikirch, W. Baumann, "Intracellular potential measurements of adherently growing cells using microneedle arrays", *Proceedings μTAS'08 Conference*, 2008, pp. 1777-1780.

[7] P. van Stiphout, T. Knott, T. Danker, A. Stett, "3D Microfluidic chip for automated patch-clamping," *MST-Kongress*, 2005, pp. 435-438.

[8] T. Y. Tsong, "Electroporation of cell membranes," *Biophys. J.*, 1991, Vol. 60, pp. 297-306.

[9] M. P. Rols, J. Teissié, "Electropermeabilization of cells to macromolecules," *Biophys. J.*, 1998, Vol. 75, pp. 1415-1423.

[10] M. Dijkstra, M. J. de Boer, J. W. Berenschot, T. S. J. Lammerink, R. J. Wiegerink, M. Elwenspoek, "Miniaturized flow sensor with planar integrated sensor structures on semicircular surface channels," *J. Micromech. Microeng.*, Vol. 17, 2007, pp. 1971-1977.

[11] A. Heilmann, F. Altmann, A. Cismak, W. Baumann, M. Lehmann, "Investigation of cell-sensor hybrid structures by focused ion beam (FIB) technology," in *Mat. Res. Soc. Proc.*, 0983-LL03-03, 2007.

[12] P. J. Koester, C. Tautorat, A. Podssun, J. Gimsa, W. Baumann, "Dielectrophoretic positioning of cells for the measurement of intracellular potentials using kidney-shaped electrodes", *Proceedings MEA Meeting*, 2008, pp. 317-318

[13] J. Held, J. Gaspar, P. Ruther, O. Paul, "Microneedle electrode arrays with dielectrophoresis electrodes for intracellular recording applications", *Proceedings MEA Meeting*, 2008, pp. 295-296.

BATCH FABRICATION OF OUT-OF-PLANE, IC-COMPATIBLE, NANOSCALE-TIP SILICON NEUROPROBE ARRAYS

A. Goryu, A. Ikedo, K. Takei, K. Sawada, T. Kawano and M. Ishida

Department of Electrical and Electronic Engineering, Toyohashi University of Technology,
1-1 Hibarigaoka Tempaku-cho, Toyohashi, Aichi, JAPAN, 441-8580
(Tel:+81(532)-44-6746 / Fax:+81(532)-44-6757, E-mail: goryu-a@dev.eee.tut.ac.jp)

ABSTRACT

We developed a batch-fabrication of nanoscale-tip silicon microprobe arrays for use in multipoint nanoscale investigations of cell/neuron in-vivo/in-vitro. Sharpened tips, less than 100nm diameter, can be formed at the tips of out-of-plane three-dimensional silicon microprobe (length >10µm) arrays, by silicon wet etching-based batch-process within only 1-3min, providing precisely controlled tip angles ranging from 15° to 50°. The penetration capability of the nanoscale-tip microprobes was demonstrated, using finite element modeling (FEM) simulations and penetration tests with a gelatin as tissue/cell.

INTRODUCTION

Multichannel penetrating microelectrode will accelerate the study of neuron and brain science. The microelectrode is needed for electrical recordings as well as for stimulations of a large number of neurons in tissue. For this field, we have proposed penetrating microscale silicon probe arrays integrated with on-chip signal processing circuitry, by integrated circuits (ICs) process followed by selective Vapor-Liquid-Solid (VLS) growth [1] [2].

For example, the previous microprobe arrays, with the diameter of 2-4µm, have been used for electrophysiological recordings of outer cell (extracellular) potentials [3]. However, the recording of inside cell (intracellular) requires the probe diameter less than 1 µm, in order to penetrate the probe inside the cell (Fig. 1a). Recent advances in the MEMS and Nanotechnology have realized the intracellular probe devices, such as using sharpened AFM probe, nanoscale-tip metals as well as nanowire/nanotube [4-6]. Although VLS growth can make nanoscale probe [7], the nanoprobe with the length of several ten to hundred microns must decrease in its mechanical strength, resulting in the probe breaking during the penetration (Fig. 1b). To realize the multipoint nanoscale probe in tissue, one possible approach is that the probe body is formed to be in several ten/hundred microns as the mechanical support and only tip section is formed to be in nanoscale.

Here, we propose 2-4µmdiameter, 50-100µm length microprobe arrays, consisting of sharpened sub-100nm diameter tips, which can be integrated with on-chip microelectronics (IC, MEMS). The nanoscale-tip can be formed at the tip of the preliminary fabricated micro diameter probe, by spray coating of Photoresist around the probe and subsequent isotopic wet etching of the probe-tip. The tip-angle of the probe is controllable with the etching rate and the etching time. For the probe capability to the neuronal penetration, the probe penetration into a gelatin is discussed.

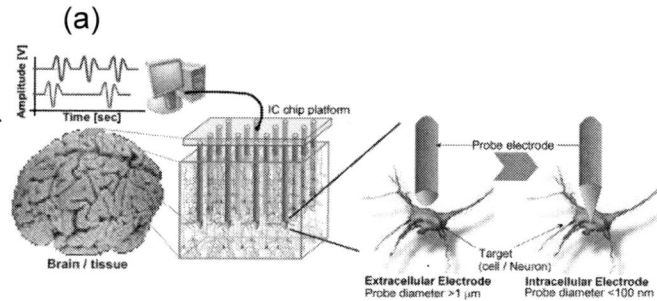

(a)

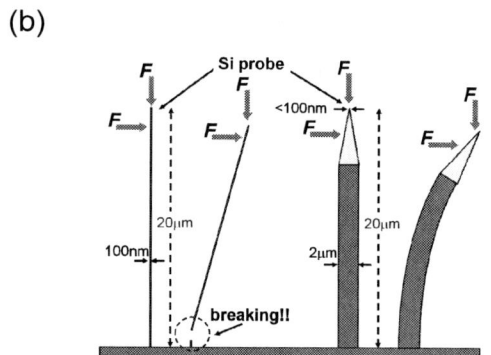

(b)

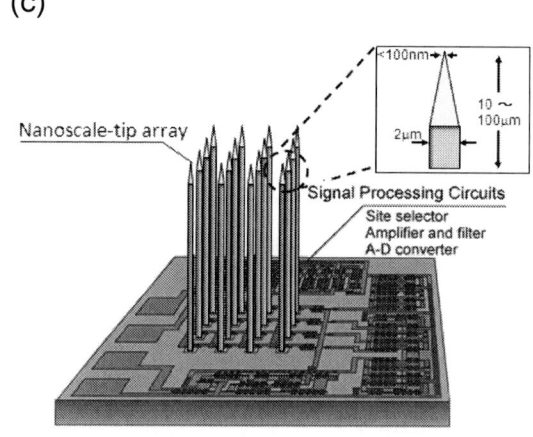

(c)

Figure 1: a) An application of probe array to neuronal tissue penetration, and comparison between mcroscale-tip and nanoscale-tip probe electrodes for neuron. b) Difference in the mechanical property between nanoprobe (e.g. φ=100nm) and microprobe (e.g. φ=2µm) under external forces applied. c) Schematic image of proposed nanoscale-tip silicon microprobe array device.

978-1-4244-2977-6/09 $25.00 © 2009 IEEE

FABRICATION OF NANOSCALETIP SILICON MICROPROBE ARRAYS

We first start with a silicon (111) substrate consisting of the silicon microprobes integrated by the IC-compatible selective VLS growth method. Figure 2 shows subsequent batch-process to form nanoscale-tip probes: a-b) after the selective VLS growth of silicon microprobe arrays, c) silicon probe covered with conformal 3-μm-thick Photoresist by the spray coating, d-e) exposing the probe-tip from the Photoresist by oxygen plasma etching, and subsequent isotopic wet etching of the probe-tip with HF+HNO$_3$ solution at 40 °C, f) Photoresist removed, and g) after the process completion. Sharpened microprobe exhibited the tip diameter of less than 100nm, which was confirmed by SEM observation (Fig. 2g). In this work, the length and diameter of used silicon probes were 30μm and 2μm, respectively (cylindrical-shape). In addition, the spray coating of Photoresist can be applicable to probe structure with further high aspect ratio, such as more than 100μm length

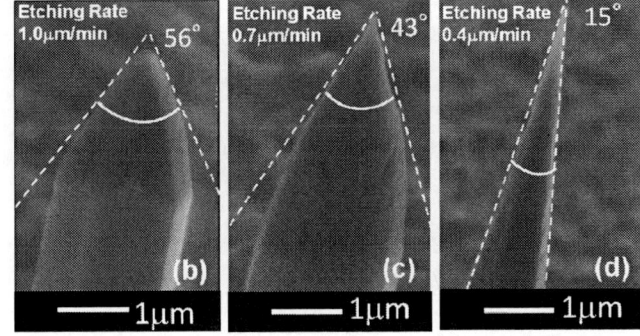

Before

Table 1: Etching Parameters.

	Solution HF:HNO$_3$:H$_2$O	Etching Time (sec)	Etching Rate (μm/min)	Angle (deg)
b	1:50:0	60	1.0	56
c	1:100:0	85	0.7	43
d	1:50:25	180	0.4	15

All etchings carried out at 40°C

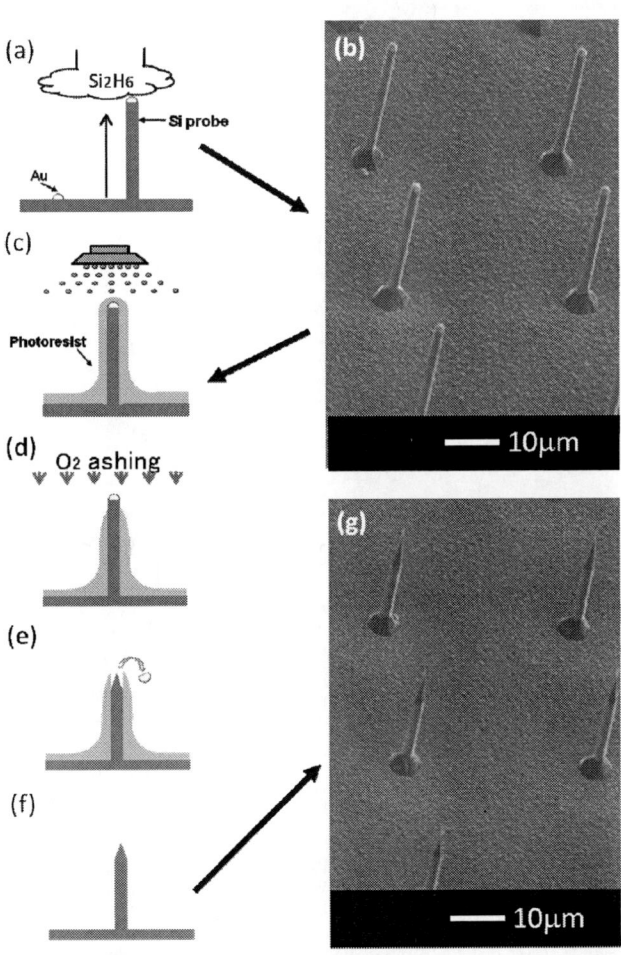

Figure 2: Process sequence: a-b) after VLS growth process, c) Photoresist coated, d) O$_2$ plasma etching of the Photoresist at the probe-tip, e) isotopic wet etching of the probe-tip, f) Photoresist removal, g) SEM image of the sharpened probe array after process completion.

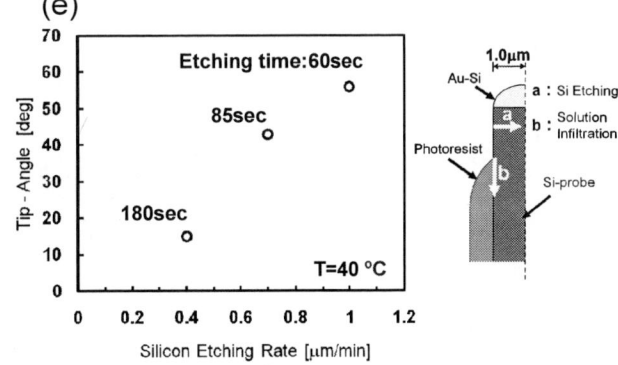

Figure 3: Tip angles controlled by etching parameters (solution and time). a) Before the etching, sharpened angle of b) 56°, c) 43°, d)15°, and e) etching rate-tip angle plots for the etching of 1μm radius silicon probe, shown in (a).

This wet etching-based process is repeatable and fast with etching time of only 1-3 min (etching rates=0.4-1μm/min). In addition, the process can precisely control the tip angle, which depends on the ratio of HF in the solution and the etching time, providing various angles ranging from 15° to 50° (Fig. 3 a-d). We found that the tip angle becomes to be sharpened when both slower etching rate and longer etching time were used. For example, the angle of 15° can be formed with the etching solution of HF:HNO$_3$:DIW=1:50:25 (etching rate=0.4μm/min) and the etching time for 3min. Figure 3e shows silicon etching rate – tip angle plots, taken from the experimental results, shown in Table 1. To complete the nanoscale-tip formation for 1μm radius silicon probe (Fig. 3a), each etching rates of 1.0μm/min, 0.7μm/min and 0.4μm/min required etching times for 1.0min, 1.4min and 2.5min, respectively, and the differences in etching parameters result in the change in the tip-angle of the probe. Actually,

978-1-4244-2977-6/09 $25.00 © 2009 IEEE

we found another etching behavior is that the etching solution infiltrates into the interface between the Photoresist and the surface of the silicon probe, as shown Fig. 3e. This behavior also influences the shape of the probe tip.

Figure 4 shows the TEM image of a typical result of nanoscale-tip silicon probe with the angle of 15°, shown in Figure 3d. We observed that sharpened microprobe exhibited the radius of the curvature of less than 50nm. The inset image shows the crystal structure of the tip, confirming that the probe tip is crystalline structure of silicon because of epitaxial VLS growth process. This result promises that the tip is mechanically robust during above application to probe penetration.

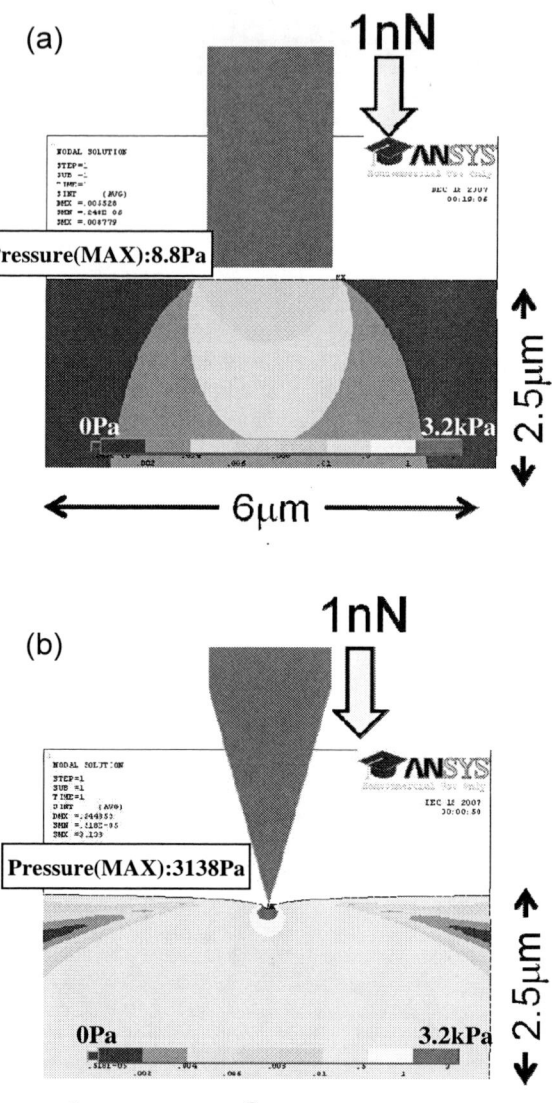

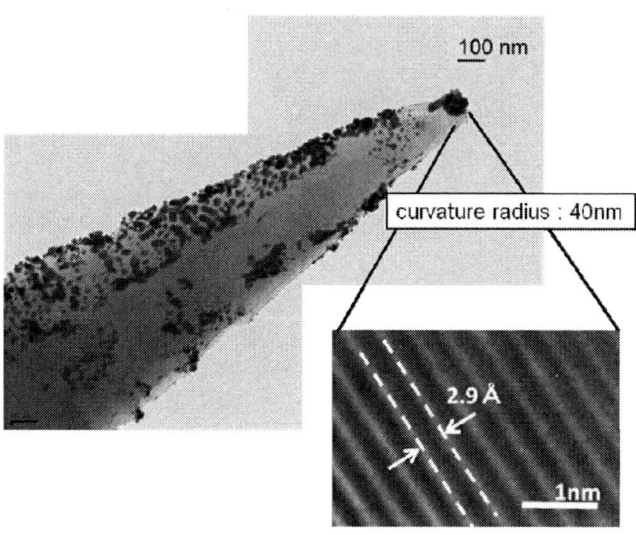

Figure 4: TEM image of a typical sharpened probe (15°). Inset image shows crystalline structure of silicon with the lattice plane distance of 2.9Å. Note that the small black dots around the silicon probe are gold particles prepared for previous SEM observation.

Figure 5: FEM simulation of pressure distribution while probe reaching cell model with 1nN force. a) tip-diameter of 2μm and b) 100nm.

NANO-TIP PROBE PENETRATION
FEM simulation

We also simulated pressure distributions of nanoscale-tip probe (100nm-diameter) while probe is reaching a cell model. It is also compared to that with microprobe (diameter is 2μm). Figure 5 show simulation results by finite element method (ANSYS), where a total force of 1nN [8] is vertically applied to the cell through the probe (cell model: Poisson ratio=0.3, Young's modulus=5×10^3Pa) [9]. As we expected, this confirms that the pressure at the cell surface can locally be increased with the nanoscale-tip probe. The increased pressure is 3kPa, and this value is about 350 times larger than that with the microprobe with the tip diameter of 2μm (8.8Pa). Based on results of localized stress and probably minimal deformation of a target (tissue/cell) during the penetration, the penetrating characteristics of the microprobe can significantly be improved by the nanoscale-tip formation.

Penetration tests

In investigations of neurons in tissue, the silicon probe will be penetrated into the tissue, which causes the compressive and/or bending forces, as shown in Fig. 1b. To confirm the penetration capability and reliability of the probe, the penetration experiment was demonstrated using a gelatin with the hardness similar to neuronal tissue. Figure 6 shows the penetration experiment on two nanoscale-tip probes. The used probe has the same tip angle of 15°, shown in Fig. 3d. Consequently, we demonstrated easy penetration and extraction without breaking of probe in continuous several tests (more than 10times). Therefore, we confirmed that this probe has the capability to use in the neuronal tissue penetration for cell/neuron investigations.

CONCLUSIONS

In summary, we have proposed a fabrication process to realize nanoscale-tip silicon microprobe arrays, using fast and repeatable batch-process, based on the spray coating of Photoresist and isotopic wet etching of silicon probe. We observed that sharpened microprobe tip exhibited the radius of the curvature of less than 50nm. Finally, the probe penetration into a gelatin was conducted, and this result indicated that the nanoscale-tip probe has the capability to penetrate the probes into neuronal tissue and investigate cell/neuron with the nanoscale-tip probes.

ACKNOWLEDGEMENTS

The authors gratefully acknowledge K. Mayumi, T. Harimoto, A. Fujishiro, and A. Okugawa for useful discussions regarding VLS growth, K. Muramoto for his TEM work, Professor A. Ishihara at the Chukyo University and Professor S. Usui at the RIKEN for discussions regarding electrophysiological measurement of neuron. All devices in this work were fabricated at the Electron Device Research Center (EDRC) at the Toyohashi University of Technology. This work was supported by the grant for young research project of Research Center for Future Technology (Toyohashi University of Technology), by a Grant-in-Aid for Scientific Research (S), by the global COE Program "Frontiers of Intelligent Sensing" from the Ministry of Education, Culture, Sports, Science and Technology of Japan.

REFERENCES

[1] R. S. Wagner and W. C. Ellis, "Vapor-liquid-solid mechanism of single crystal growth", Appl. Phys. Lett. Vol. 4, No. 5, pp. 89-90, March 1964.

[2] T. Kawano, Y. Kato, M. Futagawa, H. Takao, K. Sawada, and M. Ishida, "Fabrication and properties of ultrasmall Si wire arrays with circuits by vapor liquid solid growth," Sensors and Actuators, Vol. 97-98, pp. 709-715, 2002.

[3] T. Kawano, A. Ishihara, T. Harimoto, H. Takao, K. Sawada, S. Usui and M. Ishida, "Three-dimensional Multichannel Si Microprobe Electrode Array Chip for Analysis of the Nervous System", International Electron Device Meeting (IEDM) 2004, pp. 1013-1016.

[4] Y. Hanein, C. G. J. Schabmueller, G. Holman, P. L¨ucke, D. D. Denton and K. F. B¨ohringer, "High-aspect ratio submicrometer needles for intracellular applications", Journal of Micromechanics and Microengineering 13, pp91–95, 2003.

[5] T. Kawano, Chung Yeung Cho, Liwei Lin, "Carbon Nanotube-based Nanoprobe Electrode", Nano/Micro Engineered and Molecular Systems (NEMS) 2007, pp. 895-898.

[6] J. R. Freedman, D. Mattia, G. Korneva, Y. Gogotsi, G. Friedman and A. K. Fontecchioa, "Magnetically assembled carbon nanotube tipped pipettes", Appl. Phys. Lett. 90, pp.103-108, 2007.

[7] M. Ishida, T. Kawano, M. Futagawa, Y. Arai, H. Takao & K. Sawada, "A Si nano-micro-wire array on a Si (111) substrate and field emission device applications", Superlattices and Microstructures 34, pp.567-575, 2004.

[8] Ikuo Obataya, Chikashi Nakamura, SungWoong Han, Noriyuki Nakamura and Jun Miyake, "Nanoscale Operation of a Living Cell Using an Atomic Force Microscope with a Nanoneedle", Nano letters Vol. 5, No. 1, pp.27-30, 2005.

[9] H. Kojima, A. Ishijima, and T. Yanagida, "Direct measurement of stiffness of single action filaments with and without tropomyosin by in vitro nanomanipulation", Proc. Natl. Acad. Sci.Vol. 91, pp. 12962-12966, 1994.

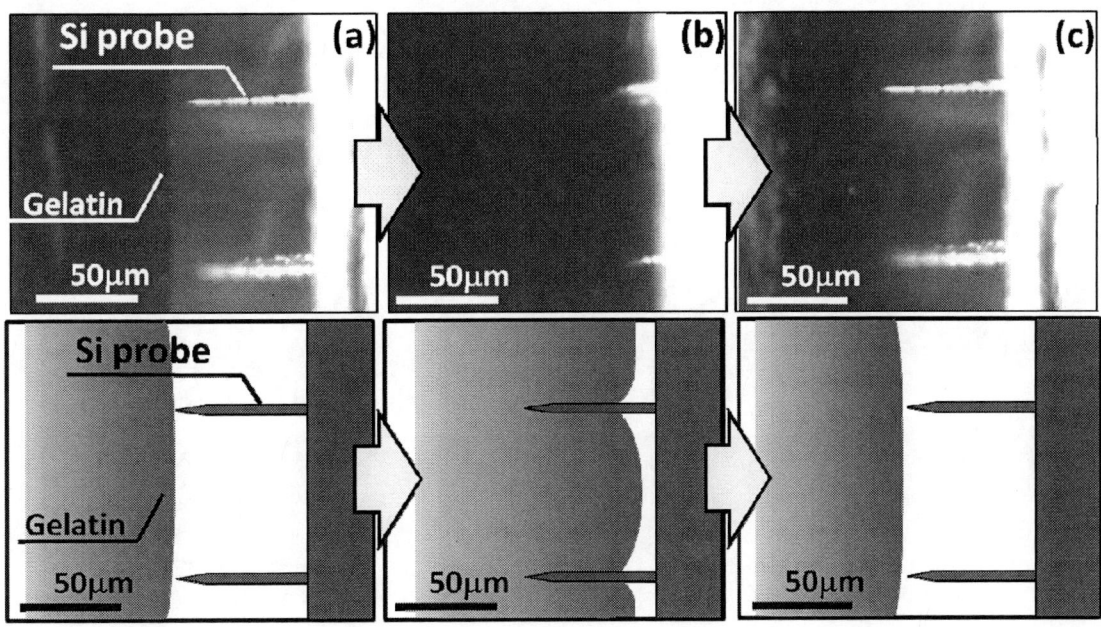

Figure 6: Penetration experiment using a gelatin: a) before penetration, b) state of penetration, and c) after extraction

ULTRA-COMPACT INTEGRATION FOR FULLY-IMPLANTABLE NEURAL MICROSYSTEMS

G.E. Perlin and K.D. Wise
Center for Wireless Integrated Microsystems,
The University of Michigan, Ann Arbor, MI, USA

ABSTRACT

A new approach to microsystem integration to replace conventional area-consuming platform architectures with an overlay integration cable is presented. A parylene cable carrying interconnect lines is used to integrate a 3-D array of silicon microelectrodes with a custom-designed signal conditioning chip to realize a neural recording microsystem in its most compact form. This low-profile integrated front-end was implanted in a guinea pig and used to obtain discriminable neural activity.

INTRODUCTION

One of the key challenges in realizing fully-implantable microsystems for neural prosthesis and neuroscience applications is the integration and packaging of their various components, especially the electrodes and their interface circuitry. Approaches based on flip-chip bonding are being pursued to vertically integrate such components into commercial 2-D neural electrode arrays, and suitable packages are in development [1]; however, such micro-systems must deal with the size and metallurgical constraints of flip-chip bonding and the increased thickness which results from the stacking of components. A low profile is especially important to allow the dura to be replaced over the implant, decoupling it from the skull and allowing it to remain free of the skull. Previously-reported integration approaches that embed the components in a silicon platform achieve a lower vertical rise, but traditionally at the expense of a larger lateral area. For example, wire bonding between the components and platform bonding pads as shown in Fig. 1(a) has been used to electrically integrate a neural recording microsystem [2]; however, the wire bonds have

appreciable height, and for high-channel-count systems the interconnect routing and bonding pads occupy significant area (48% of the total platform area for the neural recording microsystem in [2]).

This work presents a new approach to microsystem integration which results in nearly zero vertical-rise above the platform while simultaneously minimizing the size overhead in the lateral dimensions. In this approach, shown in Fig. 1(b), a silicon platform is micromachined with dry etched cavities to recess the components, creating a planar surface. A lithographically-defined polymer cable is used to carry interconnect lines that terminate in lead transfer tabs. The cable is aligned on top of the recessed components and ultrasonically bonded to achieve electrical integration of the microsystem. There are no bonding pads or interconnect lines on the platform so the components can be efficiently placed very close to each other. Interconnect lines run directly on top of the components and do not consume extra area. The cable thickness, on the order of few micrometers, is negligible in terms of the vertical dimension of the micro-system. For very complex multi-channel interconnect routing, multiple cables can be stacked, again with negligible penalty in vertical rise.

DESIGN

The new integration approach has been used to develop a low-profile compact front-end for a neural recording microsystem. The front-end includes a 64 channel 3-D array of penetrating silicon microelectrodes that sense microvolt-level neural activity and signal-conditioning circuitry that selects active channels and amplifies and filters the neural activity. In neural recording applications, it is critical that the signal-

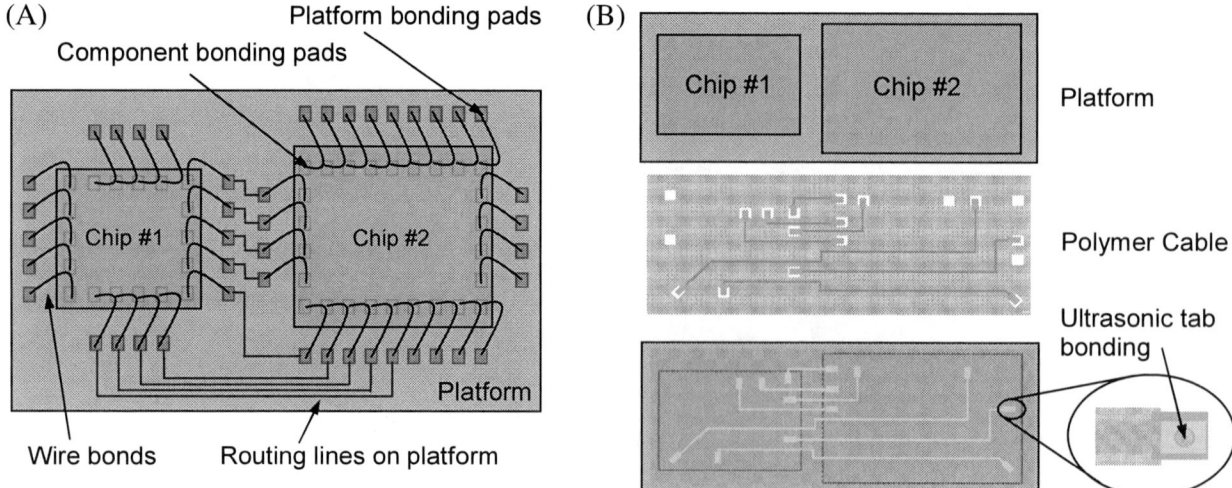

Fig. 1: (a) Conventional system integration approach on a platform, and (b) microsystem integration approach of this work.

conditioning circuitry be as close to the source of the signal as possible to avoid noise corruption of the microvolt signals.

The 3-D array consists of four parallel 2-D planar silicon probes assembled into dry etched slots in a silicon platform [3]. The back-end of each 16-channel 2-D probe (4 sites per shank) has bendable lead transfer tabs that fold orthogonal to the shanks onto the surface of the platform, resulting in a planar surface. A custom-designed signal-conditioning chip containing 16 amplifiers, each with programmable gain (up to 60dB) and tunable bandwidth [4], is recessed into a 300μm deep cavity (corresponding to the thickness of the IC) formed using DRIE in the silicon plat-form. The integration cable is a parylene-metal-parylene structure carrying the electrical interconnect lines between the 3-D array and the signal conditioning chip. Parylene-C was selected as the structural material for the cable due to its compatibility with low-temperature deposition, lithographic patterning, mechanical flexibility and biocompatibility [5]. Following the assembly of all components in the platform recesses, the parylene cable is placed over the components, aligned to their bonding pads, and electrically connected using ultrasonic bonding. The design of the neural re-cording platform and integration cable are shown in Fig. 2.

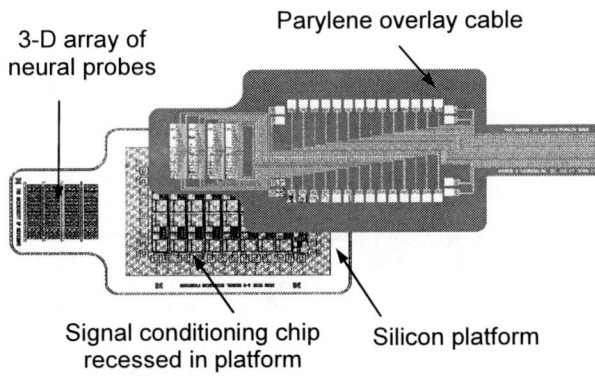

3-D array of neural probes

Parylene overlay cable

Signal conditioning chip recessed in platform

Silicon platform

Fig. 2: Integration of a neural recording microsystem front-end, including a 3-D array of neural probes and a custom designed signal-conditioning chip.

FABRICATION

The fabrication and assembly of the 3-D array are detailed in [3]; the custom-designed signal-conditioning chip was fabricated at a foundry in 0.5μm technology [4]. The fabrication details of the silicon platform and integration cable are discussed here.

The fabrication of the package starts with a standard silicon wafer approximately 500μm thick. The first step is to lithographically pattern the front side of the wafer and then to deposit and liftoff chromium (300Å) and gold (5000Å) for the bonding pads between the slots of the 3-D array. Then, the chip cavity is patterned and DRIE etched from the front side to a depth of about 300μm, which is the thickness of the chip. This cavity opening, measuring 2.2mm x 1.5mm, includes a 150μm tolerance all around to account for chip-to-chip size differences and for positioning during assembly. Next, the wafer is patterned

in the 3-D array region to define the front-side slot and perimeter openings. The challenge here is to conformally coat the photoresist around a 300μm deep cavity. Since the cavity opening is relatively large, conformal coating was achieved using a very viscous photoresist (AZ 9260) with a slow and long spread time, followed by a spin/dry step. This technique was characterized to allow the edges of the cavity to remain protected with resist. The slots and perimeter of the package are then DRIE etched to a depth of 300μm, which is the height of the probe backend. Since the front-side slot/perimeter depth and the chip cavity depth are the same in this design, a single-step lithography for the front-side etch (cavity and slots) was explored. However, due to the significant differences in the mask openings and aspect ratios, an optimal DRIE etch recipe was not achieved. The difficulty is that the chip cavity was found to etch at nearly twice the rate as the narrow slots, requiring a two-step front-side etch. The final step is to pattern the back side of the wafer and DRIE etch the slots and perimeter (~200μm) until the platform is released from the wafer. The platforms are soaked in acetone to remove the photoresist and then cleaned with IPA.

The fabrication of the integration cable begins with a silicon wafer having a sacrificial layer on the front side. In this work, three sacrificial layers were explored: PECVD oxide (5000Å), evaporated titanium (300Å) and native oxide on bare silicon. Next, the first layer of parylene is deposited at room temperature using the Specialty Coating System PDS 2010. Approximately 5μm of parylene is deposited on both sides of the wafer. This first layer of parylene is patterned on the front side of the wafer using ~15μm thick photoresist (AZ 9260) and dry etched in an oxygen plasma (100sccm, 100mTorr, 105W) to define the outline of the cable and tab cutout regions. Following patterning of the parylene, the interconnect lithography and definition take place. The interconnect metal is a stack of chromium (300Å), gold (3500Å), chromium (300Å) which is defined using liftoff. A top layer of chromium is used since a second layer of parylene will be later deposited, which has better adhesion to chromium than to gold [5]. The next step is to open the tab regions with lithography and sputter an electroplating seed layer: Cr (300Å), Au (2000Å). Due to the 5μm step height between the wafer and top surface of the parylene, it is critical that sputter metallization, due to its conformal coverage, rather than evaporation be used to deposit an electrically continuous seed layer. For the same reason, the tab regions and the inter-connect metallization are defined in two steps rather than one. A single-step metallization will result in poor lithography near the edge of the step. After deposition of the seed layer, lithography is again used to open the tab regions for electroplating. Before electroplating, an oxygen plasma ash (250mT, 250W, 1min) is used to modify the resist surface so that it is hydrophilic to avoid wetting voids in the electroplating solution. The tab regions are then electroplated with gold at a current density of 3mA/cm^2 to a thickness of 4.5μm to 5μm. The electroplating resist is stripped in acetone along with the liftoff of the seed layer. The top parylene layer is deposited (~5μm), similar to the first layer, patterned and dry etched in the field and tab regions. The final step is to release the individual cables from the wafer. As

978-1-4244-2977-6/09 $25.00 © 2009 IEEE 229

mentioned, three different sacrificial layers were explored (PECVD oxide: 5000Å, Ti: 300Å, native oxide on silicon: 10Å-20Å) as well as parylene deposition with and without adhesion promoter (2.5ml A-174 silane, 250ml IPA, 250ml DI H_2O). These wafers were successfully released in 1:1 HF:DI H_2O. The PECVD oxide sacrificial layer was not only the longest release method (overnight) but also caused the cables to stick to the wafer even after all oxide had been undercut in the HF solution, causing low yield. Using a thin titanium sacrificial layer or bare silicon provides the quickest release (~30min) and highest yield. Obviously, using adhesion promoter in the deposition of the first parylene layer makes it more difficult to release the cables, requiring significant agitation. In this fabrication sequence it was found that adhesion promoter is not critical since contact lithography was successful without significant peeling or bubbling of the parylene layer.

The details of a released parylene cable are shown in the SEM images of Fig. 4. The free-hanging tabs overlaying the 3-D probe array (4 x 16), shown in Fig. 4(a), are 30μm wide, spaced at 40μm pitch. The tabs overlaying the IC, shown in Fig. 4(b), are 75μm wide, and the parylene cutout region is 100μm on each side, leaving about a 12μm gap on three sides of the tab. The interconnect traces on the cable are 10μm wide and spaced at 20μm pitch.

RESULTS

The assembled and integrated front-end of the neural recording microsystem is shown on an index finger in Fig. 5. The ultrasonic bonding of the electroplated tabs on the cable to the bonding pads on the chip is shown in Fig. 6. This low-profile package with its planar top surface is robust for handling during implantation.

The integrated front-end was implanted into guinea pig auditory cortex and used to successfully record discriminable neural activity. The pictures of the implanted front-end are shown in Fig. 7. The package was handled with tweezers and implanted manually. The picture in Fig. 7(a) shows partial insertion of the probe shanks into tissue. The tail-end of the parylene cable was connected to a custom printed circuit board to transfer power to the chip and the recorded signals from the chip to an oscilloscope as shown in Fig. 7(b). Using this setup, both stimulated and spontaneous neural activity were recorded with the gain of the amplifiers on the chip set at 60dB and the bandwidth set from 100Hz to 9kHz. In Fig. 7(c), two different channels are shown, recording neural responses to an acoustic noise burst (an 80dB log up-sweep from 500Hz to 16kHz with a duration of 164ms) played into the animal's ear. The spontaneous activity obtained is shown in Fig. 7(d) on two separate channels.

CONCLUSIONS

In summary, this paper presents a new integration method for fully-implantable microsystems. This method eliminates the interconnect routing conventionally formed on the supporting platform, allowing components to be arranged in the most compact configurations. The surface area (size) of the microsystem can be significantly reduced and is only limited by the number/size of the components themselves. The electrical lines are carried

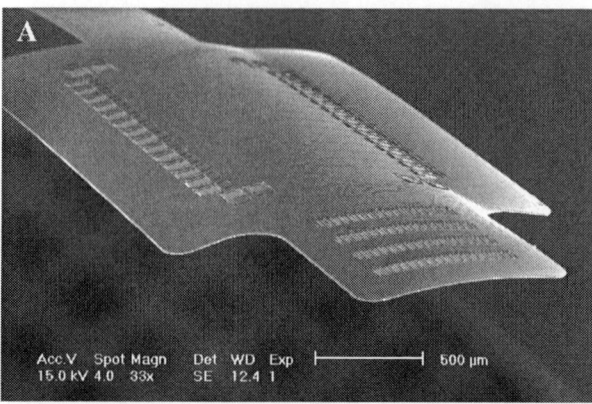

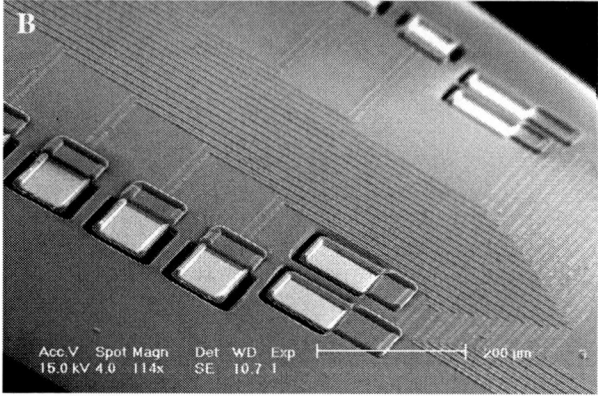

Fig. 4: (a) Fabricated parylene integration cable, and (b) details of the electroplated bonding tabs and interconnect lines in the chip region.

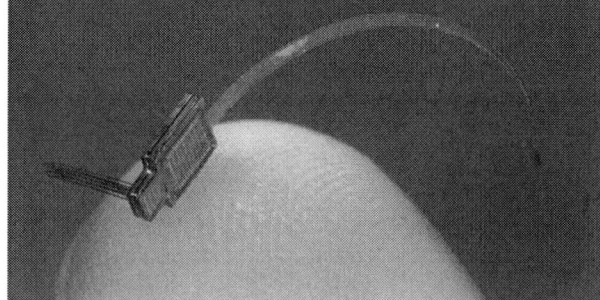

Fig. 5: Integrated neural recording microsystem front-end on an index finger.

Fig. 6: Ultrasonic tab bonding of the integration cable tabs aligned over the chip bonding pads.

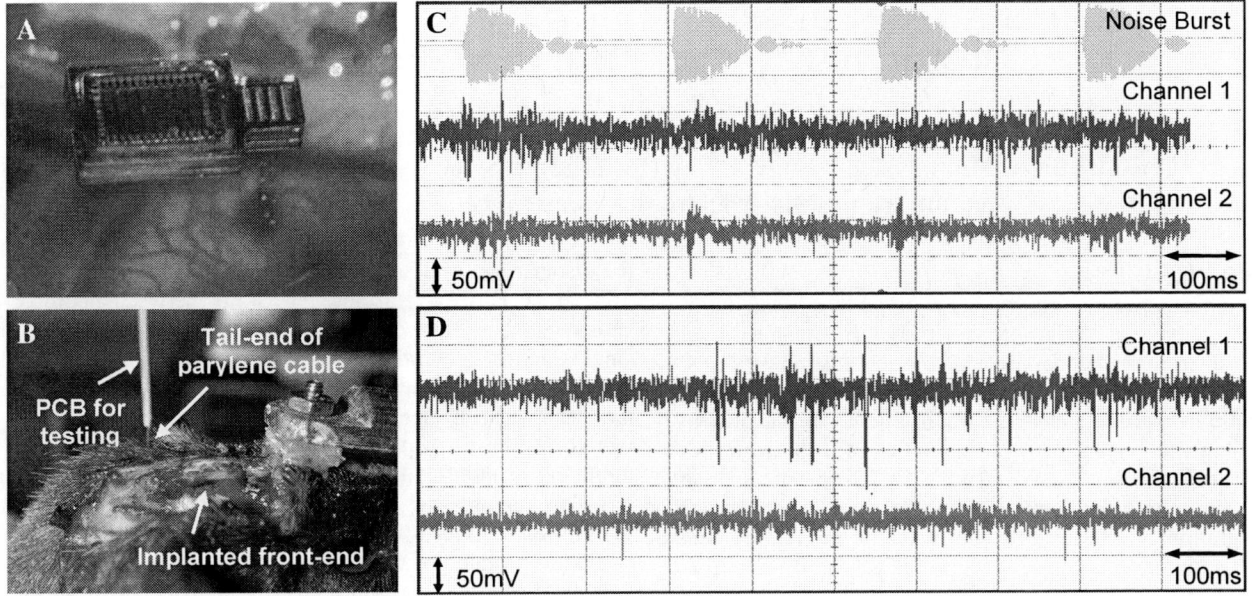

Fig. 7: (a) Partial insertion of the integrated front-end in guinea pig cortical tissue, (b) in-vivo test setup using a PCB connector, (c) stimulated neural activity, and (d) spontaneous neural activity recorded from guinea pig inferior colliculus using the integrated neural recording front-end.

by a flexible polymer cable (Parylene-C in this work) that is placed directly on top of the components. The interconnect lines on the cable terminate in lead tabs that are ultrasonically bonded to the component bonding pads. This method was used to integrate a neural recording front-end, including the 3-D array architecture presented in [3] and the signal conditioning circuitry presented in [4]. The ultrasonically bonded overlay cable approach was validated in-vivo by recording neural signals using silicon microelectrodes connected the chip while the power and data transfer to and from the chip were carried by the parylene cable. This integrated front-end achieves an ultra-compact low-profile fully-implantable microsystem with near zero-rise above the surface of the platform.

FUTURE WORK

In future work, a complete wireless neural microsystem that includes signal processing and telemetry circuitry in addition to the front-end will be integrated on a separate (satellite) platform using a similar approach. The parylene cable will be extended to accommodate the integration of the satellite platform with the front-end platform. Multiple overlay cables can be stacked to aid in simplifying the design and bonding of high-channel-count microsystems with single/multiple components, resulting in a negligible increase in vertical rise. This planned two-platform architecture will allow not only compact lateral integration but also a physical separation, reducing the electromagnetic interference generated by the wireless components in the microsystem from the sensitive analog front-end. Both platforms, except for the electrodes, will finally be encapsulated with a biocompatible material such as parylene.

ACKNOWLEDGMENTS

The authors would like to thank Brendan Casey for his assistance with assembly and Jim Wiler for his assistance with the in-vivo experiments.

REFERENCES

[1] M. Töpper et al., "Biocompatible hybrid flip chip microsystem integration for next generation wireless neural interfaces," *Proc. of the Electronic Components and Technology Conf. (ECTC)*, 2006, pp.705-708.

[2] A. M. Sodagar, G. E. Perlin, Y. Yao, K. D. Wise, and K. Najafi, "An implantable microsystem for wireless multi-channel cortical recording," *Proc. Solid State Sensors, Actuators, and Microsystems Conf. (Transducers)*, 2007, pp. 69-72.

[3] G. E. Perlin, and K. D. Wise, "A compact architecture for three-dimensional neural microelectrode arrays," *Proc. IEEE International Conf. of the Engineering in Medicine and Biology Society (EMBS)*, 2008, pp. 5806-5809.

[4] G. E. Perlin, A. M. Sodagar and K. D. Wise, "A neural amplifier with high programmable gain and tunable bandwidth," *Proc. IEEE International Conf. of the Eng. in Medicine and Biology Society (EMBS)*, 2008, pp. 3154-3157.

[5] Tze-Jung Yao, Parylene for MEMS applications, *Ph.D. Dissertation*, California Institute of Technology, 2002.

CMOS-BASED HIGH-DENSITY SILICON MICROPROBE ARRAY FOR ELECTRONIC DEPTH CONTROL IN NEURAL RECORDING

K. Seidl[1], S. Herwik[1], Y. Nurcahyo[1], T. Torfs[2], M. Keller[1], M. Schüttler[1],
H. Neves[2], T. Stieglitz[1], O. Paul[1], and P. Ruther[1]

[1]Department of Microsystems Engineering (IMTEK), University of Freiburg, Germany
[2]Interuniversity Microelectronics Centre (IMEC), Belgium

ABSTRACT

This paper reports on a novel CMOS-based high-density silicon microprobe array for intracortical recording applications. In contrast to existing systems, CMOS multiplexing units are integrated on the slender, needle-like probe shaft 160 µm in width. In the present implementation an unequaled number of 188 electrodes (diameter 20 µm, pitch 40 µm) are arranged in two columns along the 4-mm-long probe shaft. The on-shaft integration of electronics is motivated by the requirement to discriminate single action potentials in extracellular recordings necessitating a close proximity between the neuron of interest and the recording electrode. Instead of a mechanical movement of the probe shaft carrying the electrode, the position adjustment during long-term recordings is performed by switching between different electrodes arranged in a dense array. The paper presents the probe concept, the post-CMOS fabrication process and initial measurements evaluating electrode impedance and probe performance during measurements in saline solution.

INTRODUCTION

Recording of single neuron activity with high spatial resolution in the cerebral cortex is required for a basic understanding of neural processes and for an improved control of neural prostheses [1-4]. Currently this goal can only be achieved using silicon based MEMS arrays with slender probe shafts comprising multiple electrodes. Advantages are precise one-dimensional (1D) [5, 6], two-dimensional (2D) [5, 7-9] as well as three-dimensional (3D) electrode configurations [4, 10, 11] and the possibility to integrate electronic circuitry, e.g. for buffering, multiplexing, amplification, signal processing and telemetry. So far, the electronics have been integrated either in the connecting part [4, 12], i.e. the interface between probe and a cable or a platform, or on a separate chip attached to the backbone of the arrays [13].

Despite recent advances in the development of these systems, current microelectrode arrays for intracortical applications comprise a comparably small number of electrodes per shaft only. This is determined by geometrical constraints, i.e. minimal line width and spacing of internal leads defined by lithography versus the tolerable shaft dimensions; further constraints are the acceptable number of bonding pads on the connecting part, and the choice of technology, i.e. in-plane [4, 7-9] versus out-of-plane concepts [10, 13]. Further, the integration of electronics on the probe shaft enabling an increased number of electrodes with minimal contact leads is often incompatible with the specific process technologies [4, 10].

An issue during long-term recordings is the occurrence of micromotions of the probe shaft increasing the distance between a recording electrode and the neuron of interest. However, a close proximity between neuron and recording electrode is mandatory to discriminate single action potentials. In contrast to single wire electrodes a manual adjustment of the electrode position is not feasible with electrode arrays, as all recording sites would be moved in parallel. Further, a probe movement might cause additional tissue trauma.

To overcome the restrictions of existing systems, i.e. the limited number of electrodes on MEMS-based probes and the desirable position control of the recording sites, a novel system concept is presented in this paper. It applies the idea of electronic depth control recently proposed in [14] by integrating CMOS circuitry on the probe shafts. The on-shaft circuitry reduces the number of connecting lines and allows to select a subset of recording sites from an unequaled number of electrodes. Over time, a possible drift of the probe shaft in the tissue is compensated by switching to nearby electrodes. The paper presents the probe design and its functionality, the CMOS-compatible processing matching the technological approach used throughout the NeuroProbes project [15] and initial experimental results.

DESIGN AND FABRICATION

Probe Design

Figure 1 illustrates the active, CMOS-based neural probe in comparison with the passive probe of similar geometrical dimensions developed within the *NeuroProbes* project [8, 9]. The active probe comprises a slender, 100-µm-thick and 4-mm-long probe shaft realized using deep reactive ion etching of silicon. The sharp tip has an opening angle of 19°. The 300-µm-thick connecting part is 880×670 µm^2 in size and carries 13

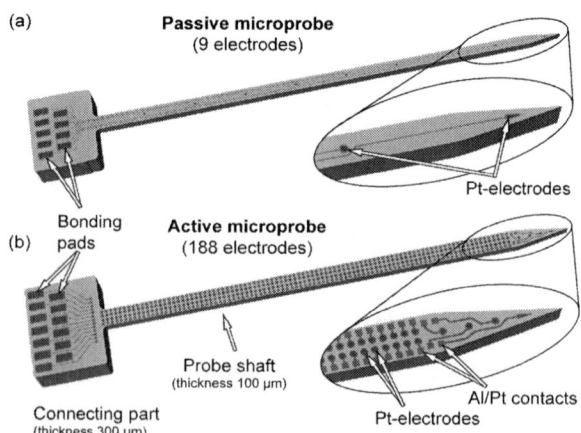

Figure 1: Schematic of passive probe shaft (a) with 9 electrodes with a pitch of 500 µm in comparison with active probe shaft (b) comprising rows of electrodes with a pitch of 40 µm.

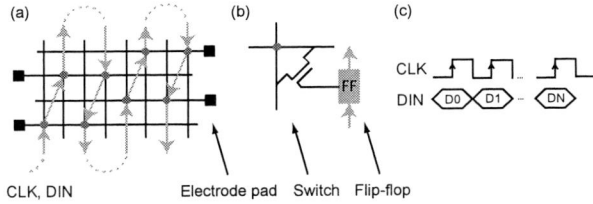

Figure 2: Probe functionality: Electrodes are selected via a shift register comprising D-type flip-flops (FF). (a) Elementary cell of switch matrix, (b) single node with switch and D-FF and (c) control signals CLK and DIN.

bonding pads: Eight analog output lines and five additional electronic control lines.

Two rows of 92 electrodes with a pitch of 40 μm and four electrodes in the tip region are distributed along the probe shaft resulting in a total number of 188 recording sites. Similar to the passive probes with nine electrodes per shaft [8, 9], the open area of the electrodes is 20 μm in diameter. Due to biocompatibility issues and metal corrosion in the brain tissue the aluminum metallization of the CMOS circuitry cannot be used as electrode material. Therefore, platinum is deposited as electrode material in a post-CMOS process which contacts metal 3 of the CMOS circuitry. To improve the electrode stability in a physiological environment, the electrode sites are shifted by 40 μm with respect to the aluminum contact pads with the interconnecting leads completely sealed by a passivation layer similar to that reported in Ref. 16.

Electronics Design

The CMOS electronics of the neural probes are fabricated using the commercial 0.6 μm CMOS process XC06 (triple-metal, double-poly, 5 V power supply) from X-FAB Semiconductor Foundries AG (Erfurt, Germany). They comprise a switch matrix on the probe shaft to select in parallel eight recording sites from the total 188 electrodes. Each electrode except for those in the tip region of the shaft can be switched to two of the eight analog output lines A1 to A8. This switching matrix contains a shift register formed by a chain of 372 D-type flip-flops which allows a serial programming of the switches using two control lines (data input, DIN, and clock, CLK) in combination with two lines for power supply (VSS and VDD). The probe functionality is illustrated in Fig. 2 with the probes configurable in any combination of two tetrodes (2×2 electrodes) or certain combinations of eight single electrodes. The switches comprise NMOS and PMOS transistors of equal size (width 2.6 μm and length 0.6 μm). The on-resistance was measured between 1.5 and 3.5 kΩ for an input voltage range of 0 to 5 V.

System Design

As illustrated in Fig. 3, the control signals (DIN, CLK) for the microprobe are generated by a *controller* connected to a host computer on which the electrode selection is made. Currently an FPGA (XC3S200, XILINX Spartan) is used as controller.

The *recording hardware* consists of a preamplifier stage, bandpass filter and amplifier. The A/D converter card PCIe 6259 from National Instruments is used for

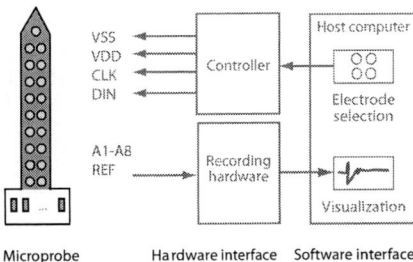

Figure 3: System design: Electrode selection is transferred from a host computer via a controller to the microprobe. Neural signals are recorded and visualized.

digitizing the neural signals. It has a sampling rate of 31.25 kHz per channel, up to 32 analog inputs, a resolution of 16 bit, an internal amplification of 100 and 48 digital I/O lines that can be used for communication with the controller.

The *software interface* makes it possible to acquire, process, display and save the neural signals and to communicate with the hardware controller of the microprobe. It includes a graphical user interface (GUI) with visual electrode selection programmed under the free multiplatform wxWidgets framework using DialogBlocks and is based on C++.

Post-CMOS Fabrication

The post-CMOS compatible fabrication technology is based on the process of passive probes developed in the framework of the NeuroProbes project [8,9]. The 4-mask process for the deposition and patterning of the electrode and bonding pad metallization and the structuring of the probe shaft is summarized in Fig. 4. First, the CMOS substrates are ground to a thickness of 300 μm (Fig. 4 (a)). The passivation layer (SiN) of the CMOS substrate with a thickness of 800 nm is opened to vias of 15 × 15 μm² to connect to the electrodes and bonding pads realized in the post-CMOS metallization. This metallization (300-nm-thick Pt sandwiched between 30-nm-thick Ti adhesion layers) is sputter-deposited and patterned using a lift-off process (Fig. 4 (b)) which applies a bi-layer photoresist of LOR10A (thickness of 350 nm, MicroChem Corp.) and AZ1518 (thickness of 2.1 μm, MicroChemicals GmbH). While the AZ-resist is processed as usual, an improved undercut between substrate and AZ4533 is achieved by the isotropic development of the LOR resist. In this way, a

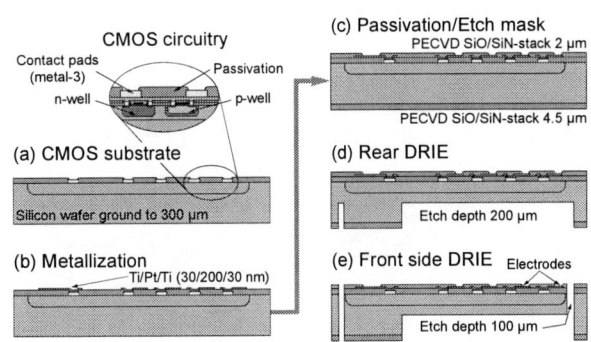

Figure 4: Post-CMOS fabrication process based on DRIE of silicon. The CMOS substrates are ground to 300 μm.

978-1-4244-2977-6/09 $25.00 © 2009 IEEE

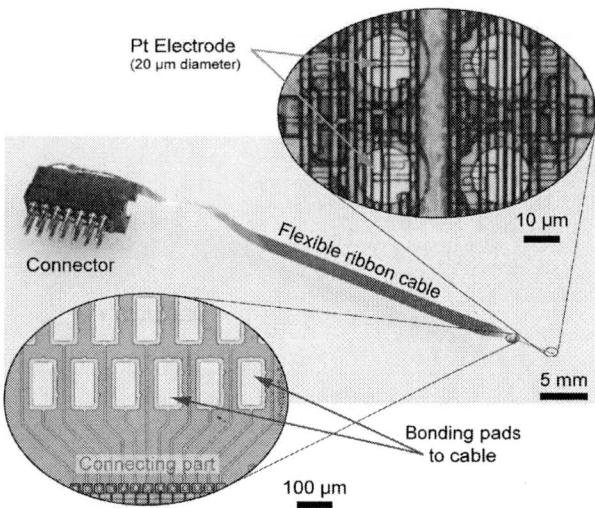

Figure 5: Assembled probe with flexible polyimide ribbon cable and connector with details showing post-processed Pt electrodes and the connecting part.

precise definition of electrodes and leads with minimum space of 4 µm is achieved in the sputter process.

Following the post-CMOS metallization, stress-compensated silicon oxide (SiO)/silicon nitride (SiN) layer stacks are deposited on both sides of the wafer by plasma enhanced chemical vapor deposition (PECVD) (Fig. 4 (c)). These layer stacks serve as a passivation and etch mask for subsequent deep reactive ion etching (DRIE). The electrodes and bonding pads on the wafer front are opened by patterning the 2-µm-thick post-CMOS passivation using reactive ion etching (RIE). The titanium layer of the post-CMOS metallization serving as adhesion promoter is selectively removed in 1% hydrofluoric acid (HF) to expose the Pt-electrodes.

The 4.5-µm-thick SiO/SiN-stack on the wafer rear is patterned using RIE followed by DRIE using an inductively coupled plasma (ICP) etcher (ICP Multiplex from STS) to define the probe geometry and adjust the shank thickness to 100 µm (Fig. 4 (d)). Finally, to release the probes from the CMOS substrate, a second DRIE step is performed from the wafer front. Due to the mechanical fragility of the 300-µm-thin substrates, the processing steps (b) to (e) are performed on sections of the 6-inch wafers. These sections are mounted on a handle wafer during RIE and DRIE steps using Cool-Grease (AI Technology Inc.).

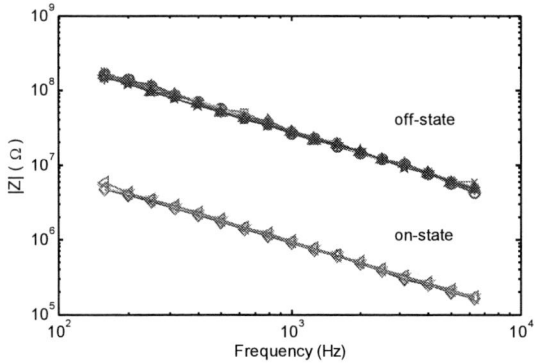

Figure 6: Electrode impedance vs. frequency, with the electrode switched on and off.

Figure 5 shows an assembled probe with flexible polyimide ribbon cable attached as described in [8] and details of the Pt recording electrodes on top of the CMOS circuitry and close to the CMOS contact pads.

EXPERIMENTAL RESULTS

The electrode impedances are characterized at frequencies between 10^2 and 10^4 Hz using a three-electrode setup with the selected microelectrode, a Pt counter electrode and a Ag/AgCl reference electrode stored in 0.9% saline solution. Typical results are shown in Fig. 6 for the on- and off-switched state. The Pt electrodes (diameter 20 µm) have impedances between 0.9 to 1.4 MΩ at 1 kHz comparable to our passive electrodes [8]. In the off-state, impedances of 25 to 27 MΩ are measured, clearly indicating that the electrodes are switched off.

The electrical performance of the probes is determined using the experimental setup illustrated in Fig. 7. It comprises a stimulating sharp tungsten electrode (parylene coated wire with a diameter 100 µm and exposed tip) translated in steps of 50 µm along the probe shaft using a z-stage. The stimulating electrode is aligned in a distance closer than 100 µm from the microprobe using the xy-stage shown in Fig. 7. A periodic current (100 µA peak-to-peak @ 1 kHz) is injected between reference and counter electrode using the current source HP3245A. One electrode of the microprobe is selected for recording and the measured potential is amplified by a factor of 500 (input impedance 1 GΩ) with the amplitude measured using a digital multimeter (34401A, Agilent

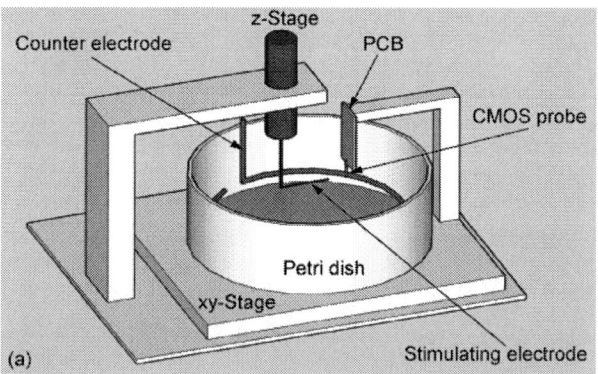

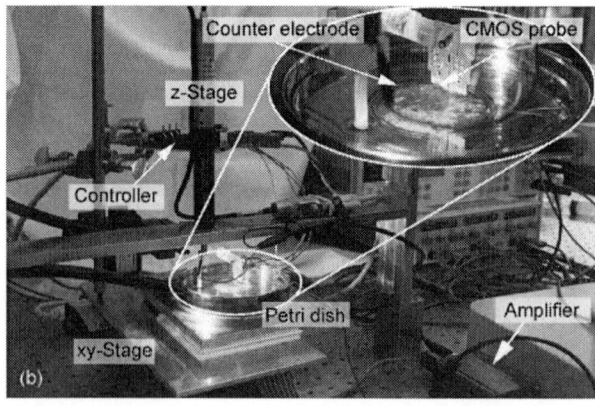

Figure 7: Experimental setup with automated xy- and z-stages moving the probe and Petri dish relative to the stimulating wire electrode relative. (a) Schematic and (b) photograph.

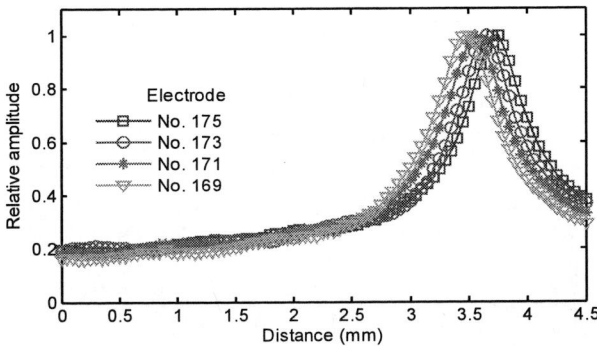

Figure 8: Potential measurement scan along the shaft with 4 different activated electrodes (step size 50 μm, stimulating current 100 μA at 1 kHz, distance between activated electrodes 80 μm).

Technologies). Figure 8 shows normalized electrode signals recorded during the scan of the stimulating electrode. As expected, the distance between the measured signal maxima reflect the electrode pitch.

CONCLUSION AND OUTLOOK

Silicon microprobe arrays with integrated multiplexing units directly on the probe shaft were fabricated using CMOS technology and post-processing. An unequaled number of electrodes was achieved allowing an electronic depth control with close proximity between neurons and electrodes by selecting the nearest electrode. The electrodes were characterized *in vitro* and showed the desired electrode properties and performance. Current work involves performing *in vivo* experiments and extension to 2D and 3D arrays, i.e. integration into platform, as successfully demonstrated with our passive probes [8, 9, 11].

ACKNOWLEDGEMENTS

The authors gratefully acknowledge A. Baur and M. Reichel (IMTEK cleanroom service center) for useful discussions and technical support in cleanroom fabrication. The work was performed in the frame of the Information Society Technologies (IST) Integrated Project NeuroProbes of the 6th Framework Program (FP6) of the European Commission (Project number IST-027017).

REFERENCES

[1] R.C. Kelly, M.A. Smith, J.M. Samonds, A. Kohn, A.B. Bonds, J.A. Movshon, T.S. Lee, "Comparison of Recordings from Microelectrode Arrays and Single Electrodes in the Visual Cortex", *J. Neurosci.*, vol. 27, pp. 261-264, 2007.

[2] T.J. Blanche, M.A. Spacek, J.F. Hetke, N.V. Swindale, "Polytrodes: High-Density Silicon Electrode Arrays for Large-Scale Multiunit Recording", *J. Neurophysiol,* vol. 93, pp. 2987-3000, 2005.

[3] J. Csicsvari, D. Henze, B. Jamieson, K. Harris, A. Sirota, P. Barthó, K. Wise, G. Buzsáki, "Massively Parallel Recording of Unit and Local Field Potentials With Silicon-Based Electrodes", *J. Neurophysiol.*, vol. 90, pp. 1314-1323, 2003.

[4] R.H. Olsson, K. Wise, "A Three-Dimensional Neural Recording Microsystem With Implantable Data Compression Circuitry", *IEEE J. Solid-State Circ.*, vol. 40, pp. 2796-2804, 2005.

[5] NeuroNexus Technologies, 3985 Research Park Dr. Suite 100, Ann Arbor, MI 480108, USA.

[6] K. Najafi, K. Wise, T. Mochizuki, "A High-Yield IC-Compatible Multichannel Recording Array", *IEEE Trans. Electr. Dev.*, vol. 32, pp. 1206-1211, 1985.

[7] R.J. Vetter, R.M. Miriani, B.E. Casey, K. Kong, J.F. Hetke, D.R. Kipke, "Development of a Microscale Implantable Neural Interface (MINI) Probe System", 27th *Int. IEEE EMBS Conf.*, Shanghai, Sept. 1-4, 2005, pp. 7341-7444.

[8] S. Kisban, S. Herwik, K. Seidl, B. Rubehn, M.A. Umilta, L. Fogassi, T. Stieglitz, O. Paul, P. Ruther, "Microprobe Array With Low Impedance Electrodes and Highly Flexible Polyimide Cables for Acute Neural Recording", 29th *Int. IEEE EMBS Conf.*, Lyon, Aug. 23-27, 2007, pp. 175-178.

[9] S. Herwik, S. Kisban, A.A.A. Aarts, K. Seidl, G. Girardeau, K. Benchenane, M. Zugaro, S. Wiener, H. Neves, O. Paul, P. Ruther, "Fabrication Technology for Silicon Based Microprobe Arrays Used in Acute and Subchronic Neural Recording", in *Techn. Dig. 19th MicroMechanics Europe Workshop*, Aachen, Sept. 28-30, 2008, pp. 57-60.

[10] R. Bhandari, S. Negi, L. Rieth, R. Normann, F. Solzbacher, "A Novel Method of Fabricating Convoluted Shaped Electrode Arrays for Neural and Retinal Prosthesis", *Dig. Tech. Papers Transducers '07 & Eurosensors XXI*, Lyon, June 10-14, 2007, pp. 1231-1234.

[11] A. Aarts, H.P. Neves, I. Ulbert, L. Wittner, L. Grand, M.B.A. Fontes, S. Herwik, S. Kisban, O. Paul, P. Ruther, R.P. Puers, C. Van Hoof, "A 3D Slim-Base Probe Array for In Vivo Recorded Neuron Activity", *Proc. 30th Annual Int. Conf. of the IEEE Eng. in Medicine and Biology Society (EMBC)*, Aug. 20-24, 2008, pp. 5798-5801.

[12] Q. Bai, K. Wise, "Single-Unit Neural Recording With Active Microelectrode Arrays", *IEEE Trans. Biomed. Eng.*, vol. 48, pp. 911-920, 2001.

[13] R.R. Harrison, P.T. Watkins, R.J. Kier, R.O. Lovejoy, D.J. Black, B. Greger, F. Solzbacher, "A Low-Power Integrated Circuit for a Wireless 100-Electrode Neural Recording System", *IEEE J. Solid-State Circ.*, vol. 42, pp. 123-133, 2007.

[14] H.P. Neves, T. Torfs, R.F. Yazicioglu, J. Aslam, A.A.A. Aarts, P. Merken, P. Ruther, and C. Van Hoof, "The NeuroProbes Project: A Concept for Electronic Depth Control", 30th *Int. IEEE EMBS Conf.*, Vancouver, Aug. 20-24, 2008, p. 1857.

[15] P. Ruther, A. Aarts, O. Frey, S. Herwik, S. Kisban, K. Seidl, S. Spieth, A. Schumacher, M. Koudelka-Hep, O. Paul, T. Stieglitz, R. Zengerle, H. Neves, "The NeuroProbes Project - Multifunctional Probe Arrays for Neural Recording and Stimulation", *Biomed. Tech.*, vol. 53, no. 1, 2008, pp. 238-240.

[16] S. Hafizovic, F. Heer, W. Franks, F. Greve, A. Blau, C. Ziegler, A. Hierlemann, "CMOS Bidirectional Electrode Array for Electrogenic Cells", in *Dig. Tech. Papers IEEE MEMS Conf.*, 2006, pp. 4-7.

IN-PLANE BANDPASS REGULATION CHECK VALVE IN HEAT-SHRINK PACKAGING FOR DRUG DELIVERY

R. Lo and E. Meng

University of Southern California, Los Angeles, California, USA

ABSTRACT

The first check valve featuring dual regulation of in-plane flow and heat-shrink tubing packaging is presented. This modular design is optimized for integration into low-profile fluidic devices requiring flow control, such as drug delivery devices. Theoretical and finite-element-modeling (FEM) analyses were performed to guide valve design and these results were confirmed experimentally. The valve regulates flow between 150-900 mmHg (20-120 kPa) and withstands >500 mmHg (66.7 kPa) of reverse pressure. The heat-shrink packaging scheme does not require adhesives and is extremely robust (>2000 mmHg without leakage).

INTRODUCTION

The out-of-plane orientation of typical MEMS check valves can complicate the process of fluidic packaging. We previously presented an ocular drug delivery device with a simple check valve [1]. The device comprised a drug reservoir for storage of pharmaceutical solutions, a flexible cannula for directed delivery to diseased tissues, and a flow regulating check valve integrated at the tip of the cannula. This check valve prevented bodily fluids from backflowing into the drug reservoir but lacked over-pressure protection to prevent accidental dosing. Also, its out-of-plane orientation could result in contact with tissues limiting its practical *in vivo* implementation. Furthermore, the valve was integrated into a cannula having rectangular geometry which prevented tight seals with the tissue at the incision site after suturing.

Thus, we propose a new modular valve paradigm that incorporates both a pressure limiting safety feature and surgically-friendly medical grade heat-shrink tubing packaging scheme. The round heat-shrink tubing is also the durg delivery cannula (Figure :1). This valve paradigm is easily adapted for use in other microfluidic systems.

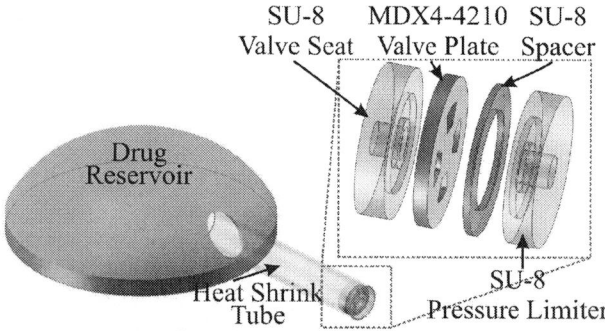

Figure :1 Surgical model of a MEMS ocular drug delivery device featuring a valve packaged in a biocompatible heat-shrink tube. The valve is comprised of four modular components (inset): valve seat, pressure responsive valve plate, a spacer, and pressure limiter.

DESIGN

The modular valve consists of four stacked disks: valve seat, pressure responsive valve plate, spacer plate, and pressure limiter (Figure 2).

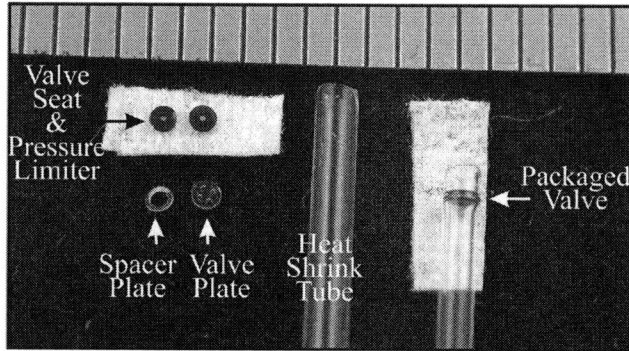

Figure 2: Photo of the valve components (valve seat, valve plate, spacer plate, and pressure limiter), heat–shrink tube, and fully assembled valve (ruler divisions = 1 mm)

The valve operates in a manner analogous to a bandpass filter and allows fluid flow when the forward pressure exceeds the valve cracking pressure; flow ceases when the closing pressure is reached. A SU-8 spacer plate defines the distance between the movable valve plate and pressure limiter plate and thus the operating pressure range (opening and closing pressures). Valve components are stacked together and packaged into a biocompatible 22G fluorinated ethylene propylene (FEP) heat-shrink tube (Figure 3). The circular tube facilitates incision/tube sealing with sutures. This packaging method is extremely robust and does not require any adhesives.

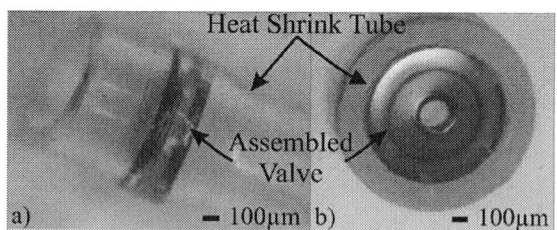

Figure 3: a) Side view and b) top view of the packaged valve in FEP heat-shrink tube. The valve was placed inside the tube utilizing a custom-made jig and then heated to 215 °C at 1.5 °C/min, and cooled at the same rate to room temperature.

The valve dimensions were selected to meet the surgical requirements; a 1 mm incision is permitted to insert the cannula/valve. Therefore, a 900 µm diameter valve was chosen leaving 100 µm for the packaging. The thickness of the spacer and valve plates was determined by FEM analysis and using the relationship governing large-deflections in a flexible plate of uniform thickness (eq. 1,2). Maximum deflection (w_{max}) can be calculated from plate thickness (t), applied pressure (p), plate radius

978-1-4244-2977-6/09 $25.00 © 2009 IEEE

(*a*), and flexural rigidity (*D*). Flexural rigidity is a function of Young's modulus (*E*), plate thickness (*t*), and Poisson's ratio (υ) [2].

$$w_{max}\left(1 + 0.486\frac{w^2_{max}}{t^2}\right) = \frac{pa^4}{64D} \qquad (1)$$

$$D = \frac{Et^3}{12(1-\upsilon^2)} \qquad (2)$$

Three different valve plate designs (hole, straight arm, and s-shape arm) were investigated (Figure 4). Each design yielded different deflection behaviors and thus differing bandpass flow regulating characteristics (*e.g.* opening and closing pressure and flow resistance). FEM estimations and theoretical analyses were used to assign valve geometries such that the operational pressure range would be limited at the lower bound by normal intraocular pressure (IOP), <35 mmHg (4.67kPa), and an upper bound of 1000 mmHg (133.3kPa) (Table 1).

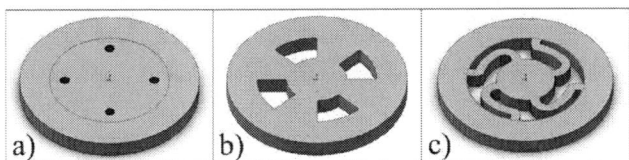

Figure 4: Three different valve plate designs a) hole, b) straight arm, and c) s-shape arm.

Table 1: Dimensions of valve components, including the three valve designs (hole, straight arm, s-shape arm). All components are 900 µm in diameter.

	Valve Seat/ Pressure Limiter	Hole Valve Plate	Straight Arm Valve Plate	S-Shape Arm Valve Plate	Spacer Plate
Material	SU-8	MDX4-4210	MDX4-4210	MDX4-4210	SU-8
Thickness [µm]	200	75	75	75	40

FABRICATION

SU-8 Valve Seat and Pressure Limiter

The SU-8 valve seat and SU-8 pressure limiter were fabricated using a two-layer SU-8 process (Figure 5). First, a soda-lime wafer (Mark Optics, Santa Ana, CA) was pretreated with Omnicoat (MicroChem, Newton, MA), a SU-8 release layer. Three layers of Omnicoat were applied (3000 rpm, 30 sec) with a bake step (1 min at 200 °C) performed after each coat (Figure 5a). Three coats facilitated component release from the substrate; these extra layers of Omnicoat resulted in a reduction in the time and temperature necessary to release the SU-8. A 160 µm layer of SU-8 2100 (MicroChem, Newton, MA) formed the base section (Figure 5b,c). A 40 µm layer was added to form the features in the valve seat and pressure limiter (Figure 5d-f). The wafer was immersed in Remover PG (MicroChem, Newton, MA) for 5 minutes to separate the SU-8 components from the wafer (Figure 5g). The components were rinsed (IPA and DI H₂O) and then hardbaked at 215 °C for 1 hour. This final step annealed

the SU-8 components to improve thermal resistance for the subsequent heat-shrink packaging process.

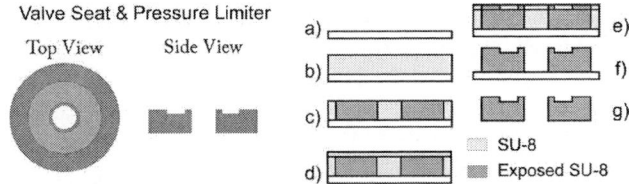

Figure 5: Fabrication process for the valve seat and pressure limiter plates.

SU-8 Spacer

The 40 µm thick spacer plate was also fabricated on an Omnicoat-coated soda lime wafer using SU-9 2050 (Figure 6a-d). Spacer plates were released from the substrate using Remover PG and rinsed using IPA and DI H₂O (Figure 6e).

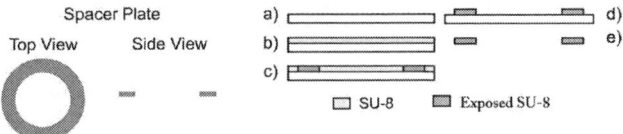

Figure 6: Fabrication process for the SU-8 spacer plate.

Valve Plate

The valve plate was fabricated by casting medical grade silicone (MDX4-4210, Dow Corning, Midland, MI) against a SU-8 master. The SU-8 master was created on a soda lime wafer using SU-8 2050. 4 µm of Parylene C (Specialty Coating Systems, Inc., Indianapolis, IN) was vapor deposited onto the wafer to prevent the SU-8 from delaminating from the wafer due to thermal mismatch with the substrate [3]. A 75 µm layer of SU-8 2050 defined the valve plate thickness (Figure 7a-c). MDX4-4210 (10:1 base to curing agent ratio), was poured onto the mold and degassed under vacuum. Excess silicone was removed with a metal squeegee (Figure 7d) [4]. The silicone was cured at room temperature for 48 hours to minimize shrinkage. Then valve plates where separated from the mold (Figure 7e).

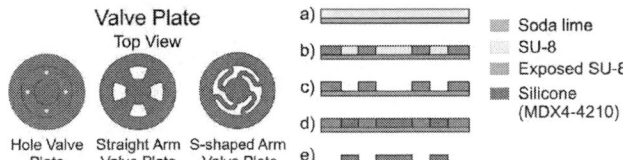

Figure 7: Fabrication process for the valve plate using an SU-8 master mold.

Valve Packaging in Heat-Shrink Tubing

The valve was packaged in the heat-shrink tubing using a custom-made Teflon jig with a stainless steel post (813 µm diameter). A 22G (inner diameter prior to shrinkage: 914.4 µm, maximum wall thickness: 254 µm) 1.3:1 shrink ratio FEP heat-shrink tube (Zeus Industrial Products Inc., Orangeburg, SC) was placed around the post followed by the four valve components (valve seat, valve plate, spacer plate, pressure limiter) (Figure 8a). A second Teflon block with a matching and adjustable stainless steel post was aligned and secured such that the two posts held the valve assembly in place while the FEP tubing was slipped it (Figure 8b). The entire jig was placed in an oven and heated to 215 °C at a rate of 1.5 °C/min; baked for 30 minutes, and then cooled to room temperature at the same rate to limit the thermally induced stress on SU-8 (Figure 8c). The jig was disassembled and the packaged valve was removed from the posts (Figure 8d).

The final outer diameter of the valve and packaging was 1.23 ± 0.004 µm (n=7, mean ± SE); while the outer diameter of the tube surrounding the valve was 1.04 ± 0.006 µm (n=21, mean ± SE). The post size was chosen to provide the largest possible surface area on which to balance the valve, however the final dimension for the tube outer diameter was limited by the stainless steel post.

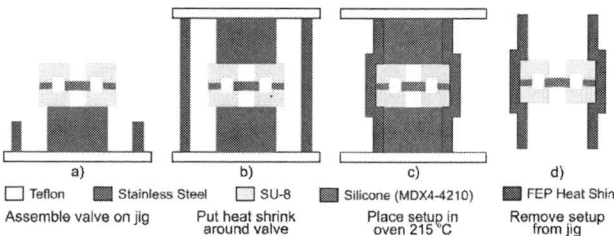

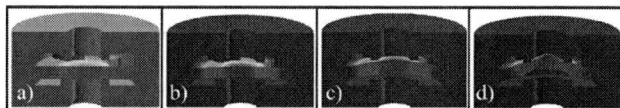

Figure 8: Process for packaging valve in heat-shrink tubing.

EXPERIMENTS AND RESULTS
Valve Plate Deflection and Stress Analysis

Valve plate deflection and stress distribution for various stages of valve operation were modeled using FEM analysis (Figure 9). These stages include the valve at rest, opening under application of forward pressure, and closure at higher forward pressures.

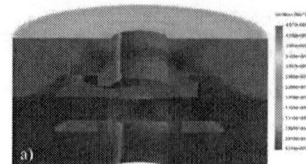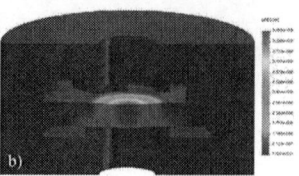

Figure 9: FEM images of valve plate deflection under a) negligible forward pressure, b) 100 mmHg, c) 500 mmHg, and d) 10000 mmHg (used to visually exaggerate the valve closing mechanism). Forces between the valve seat and plate were not modeled; therefore the valve opened for any non-zero applied forward pressure.

Under forward applied pressure (1000 mmHg, 133.3 kPa), the maximum stress on the valve plate (0.99 MPa) was concentrated at the outer edge of contact with the valve seat. The stress was <20% of MDX4-4210 tensile strength (5 MPa) and significantly less than the tensile strength of SU-8 (60 MPa). Reverse pressure (500 mmHg, 66.7 kPa) analysis verified the stress on the valve (0.46 MPa) was at least an order of magnitude less than the tensile stress of MDX4-4210 or SU-8 (Figure 10b). Additionally, the valve plate deflected <7 µm under reverse pressure, maintaining an effective seal between the valve plate and valve seat (Figure 10b). Normal IOP ranges are 5-35 mmHg (4.67 kPa), thus the valve can withstand reverse pressure conditions greater than 10 times the IOP without failing.

Figure 10: FEM analysis of 500 mmHg reverse pressure on the assembled valve. a) Maximum stress (0.46 MPa) was significantly lower than the tensile stress of the valve materials. b) Deflection of the valve plate was <7µm.

The deflection for each valve plate design under increasing forward pressure was measured using a microscope. The valve plate was clamped along the outer edge and pressurized air was applied to the plate. The results were compared to the theory (large deflection equations of a clamped membrane with uniform thickness (eq. 1, 2)) (Figure 11).

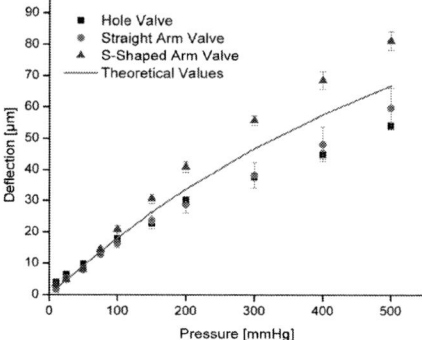

Figure 11: Comparison of calculated valve deflection values using theoretical equations versus experimentally obtained values for all three designs (n=4, mean ± SE).

As expected, the hole and straight arm valves had similar performance while the s-shaped arm valve had the greatest deflection [5]. The s-shaped tethers allow the valve plate to twist as it deflects away from the valve seat. Furthermore, the s-shape arm valve presents less fluidic resistance than the other two designs.

Heat-Shrink Tubing Packaging Method

The biocompatible heat-shrink tubing provides an extremely robust package and eliminates the use of adhesives [6, 7]. To ensure even shrinkage and prevent cracking, the tube is uniformly heated and cooled to/from 215 °C (1.5 °C/min) and room temperature in a digitally-controlled oven (Model VO914A, Lindberg/ Blue, Asheville, NC).

To quantify the fluidic integrity of this packaging method, a solid 200 µm thick SU-8 disk with the same diameter as the valve (900 µm) was packaged in heat-shrink tubing. Pressurized water was applied through the heat-shrink tubing to one side of the packaged SU-8 disk. A 100 µL calibrated pipette (Clay Adams, Parsippany, NJ, USA) was placed at the outlet to measure leakage of water between the disk and heat-shrink tubing. The disk remained in its packaged position and the entire system was leak-tight up to 2000 mmHg (266.6 kPa) which is the pressure limit of our testing apparatus. Pressurized N₂ gas was applied to the system, with the tubing outlet immersed in water to visualize any bubbles due to

978-1-4244-2977-6/09 $25.00 © 2009 IEEE   238

leakage. The packaged system was also able to withstand up to 2000 mmHg (266.6 kPa).

Valve Operation

The behavior of a packaged valve (hole valve plate) was determined using a custom-made jig and pressure system. Pressurized water (0-1000 mmHg, 0-133.3 kPa) was applied in incremental steps to the valve inlet. The flow rate from the packaged valve was measured using a 100 µL calibrated pipette placed at the valve outlet. The system was held at each test pressure set point for 5 minutes to allow the system to equilibrate.

The valve cracking pressure was 150 mmHg (20 kPa) and closed at a pressure of 900 mmHg (120 kPa) for the hole valve plate design (Figure 12). Minimal leakage, less than 18 times peak flow, was observed after valve closure. The valve was able to withstand reverse pressures in excess of 500 mmHg (66.7 kPa) without leaking. The valve operating range is much greater than normal and abnormal IOP values, preventing the valve from opening due to normal eye pressures or transient fluctuations (*e.g.* as a result of flying or sneezing).

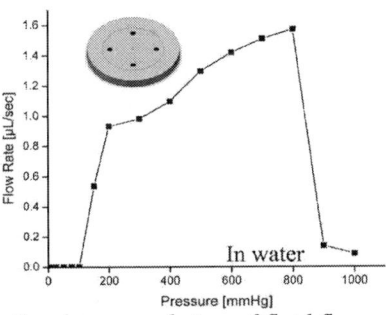

Figure 12: Bandpass regulation of fluid flow was verified on a packaged valve (hole valve plate). Pressurized DI water was applied to the inlet of a package valve and flow rate was measured using a 100 µL calibrated pipette.

CONCLUSION

A bandpass regulation, in-plane check valve packaged within biocompatible heat-shrink tubing without the use of adhesives is presented. The valve achieved bandpass regulation of pressurized water with a cracking pressure of 150 mmHg (20 kPa) and closing pressure of 900 mmHg (120 kPa). The valve was able to withstand a reverse pressure of 500 mmHg (66.7 kPa) without leaking. The package is very robust and can withstand water and N_2 gas pressures in excess of 2000 mmHg (266.6 kPa). Packaged valves were incorporated into a surgical model of the drug delivery device (made of MDX4-4210 with a stainless steel ring and PEEK baseplate) that will be used for *ex vivo* and *in vivo* validation of valve operation (Figure 13).

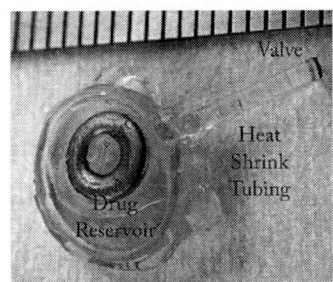

Figure 13: The valve was incorporated into a surgical model containing a drug reservoir for in vivo testing (ruler divisions = 1 mm). The valve is modular and easily replaced with another valve.

ACKNOWLEDGEMENTS

This work was funded by the NIH/NEI under award number R21EY018490. The authors would like to thank Dr. Donghai Zhu, Benjamin Lee, and the members of the USC Biomedical Microsystems Laboratory for their assistance with this project.

REFERENCES

[1] R. Lo, P. Y. Li, S. Saati, R. Agrawal, M. S. Humayun, and E. Meng, "A refillable microfabricated drug delivery device for treatment of ocular diseases," Lab on a Chip, vol. 8, pp. 1027-1030. 2008.

[2] A. C. Ugural, Stresses in Plates and Shells, 2nd ed. New York: McGraw-Hill, 1999.

[3] R. Lo and E. Meng, "Integrated and reusable in-plane microfluidic interconnects," Sensors and Actuators B: Chemical, vol. 132, pp. 531-539. 2008.

[4] R. Kee Suk, W. Xuefeng, K. Shaikh, and L. Chang, "A method for precision patterning of silicone elastomer and its applications," Journal of Microelectromechanical Systems, vol. 13, pp. 568-575. 2004.

[5] X.-Q. Wang, Q. Lin, and Y.-C. Tai, "Parylene micro check valve," Proceedings of the IEEE Micro Electro Mechanical Systems (MEMS) Orlando, Jan. 17-21, 1999, pp. 177-182.

[6] T. Pan, A. Baldi, and B. Ziaie, "A reworkable adhesive-free interconnection technology for microfluidic systems," Journal of Microelectromechanical Systems, vol. 15, pp. 267-272. 2006.

[7] T. Pan, J. D. Brown, and B. Ziaie, "An Artificial Nano-Drainage Implant (ANDI) for Glaucoma Treatment," presented at the 28th Annual International Conference of the IEEE Engineering in Medicine and Biology Society, 2006. EMBS '06., New York, Aug. 30- Sept. 3, 2006, pp. 3174-3177.

ON-CHIP BLOOD VISCOMETER
TOWARDS POINT-OF-CARE HEMATOLOGICAL DIAGNOSIS
H. Zeng and Y. Zhao
Department of Biomedical Engineering
The Ohio State University, Columbus, Ohio, USA

ABSTRACT

Blood viscosity is an important hematological parameter which is widely used for the diagnosis of atherosclerosis, thrombosis and stroke. Currently used Couette rheometer drives the blood flow at a certain shear rate and measures the viscosity from the resistance toque towards the rotational shaft. Although effective, conventional rheometer is limited due to the complex configuration and the relatively large sample volume. More important, the rotating components hinder the miniaturization for point-of-care and unattended diagnosis. To address this, a microchip is reported for measuring the blood viscosity from the electrical impedances of the blood flowing inside a microchannel. This microdevice is advantageous over rotational viscometers in less sample consumption, rapid response and simple configuration. Considering the vital role of blood viscosity in cardiovascular disorders, this microchip provides a promising start point for on-chip hematological diagnosis.

INTRODUCTION

Cardiovascular research has continued to provide evidences that the blood viscosity highly correlates with a wide array of cardiovascular diseases, such as thrombosis, hypertension, and strokes [1], and can be used as a measure of various cardiovascular conditions. For example, an increased blood viscosity usually requires an elevated pressure generated by the heart (or the assistive heart device) to pump the blood to organs. The changes of blood viscosity may also suggest the adhesion of different protein species on the sidewall of blood vessels, which may increase the chance of thrombosis formation [2]. More clinically important, in some critical surgeries, the change of blood viscosity may indicate heart attack, which needs to be *in situ* monitored throughout the entire surgery [3]. Therefore, it is clear that precise and real time monitoring of blood viscosity is significant in both basic cardiovascular research and clinical practice.

As a non-Newtonian fluid, blood exhibits varying viscosity at different flowing shear rates. Investigation has shown that at high shear rates, blood exhibits a constant viscosity. At lower shear rates, however, the blood viscosity significantly increases as the shear rate decreases [4]. This is mainly attributed to the large amount of floating RBCs and their aggregation under different flowing conditions. Rheometer (rotational viscometer) is the most commonly used equipment for determining the blood viscosity under different shear rates. In the viscosity measurement, the blood sample is driven to a defined shear rate within a Couette flow. The viscosity is derived from the rotational torque that is required to maintain the shear rate. The blood viscosity can also be determined by maintaining a constant rotational toque and measuring the spin rate of the blood sample under such a toque when the motion becomes stable. The blood viscosity measurement using the rheometer has shown that blood with a higher hematocrit value (volume ratio of RBCs in the blood) exhibits larger viscosity than blood with lower hematocrit values, while other physiological conditions are kept constant [5]. Under a constant hematocrit value, the RBCs aggregation becomes the predominant factor. At low shear rates, the RBCs tend to aggregate to form rouleaux, a structure likes a column of coins stacking loosely. Such structure leads to the increased blood viscosity [6]. RBC deformability is another important factor determining the blood viscosity, which becomes increasing important as the shear rate increases to above 1100 s^{-1} [7]. Besides these three primary parameters, namely: hematocrit, RBCs aggregation, and RBCs deformability, other secondary parameters such as blood temperature and the concentration of macromolecules in the blood also affect blood viscosity [8].

It should be noted that although the rheometer is effective for measuring the blood viscosity, it is incompatible with point-of-care diagnosis in many clinical conditions. Despite different configurations, the rheometer usually requires extraction of a relatively large amount of blood (often on the order of a few ml) to be sheared between the parallel positioned driving plate and the sensing plate. The blood sample under test is in Couette flow, different from what it experiences in native blood vessels (circular Poiseuille flow). Moreover, the complicated configuration and the limited access of the costly equipment hinder the wide adaptation of the rheometer into a portable device for quick and self-operative diagnosis. It is also difficult to perform intra-operative monitor of blood viscosity using a rheometer. Therefore, it is clear that alternative solutions of measurement are increasingly needed to address the above issues.

DESIGN OF ON-CHIP BLOOD VISCOMETER

It is interesting to note that most of the predominant parameters determining the blood viscosity are also the determinants of the electrical impedance of the whole blood. Research has shown that the electrical resistivity of the whole blood is highly dependent on the hematocrit value [9]. The electrical impedance of blood increases at low shear rates because of the RBCs aggregation [10]. Such behavior is due to different electrical properties of the RBCs and the blood plasma: the blood plasma can generally be regarded as a conductive material, while the cell membranes of RBCs are more dielectric [11, 12].

In this work, a microfluidic channel with a rectangular cross section was developed (Figure 1). In order to minimize the wall effect induced by the cell free zone [13], the lateral width of the microchannel was designed as 500 μm. Two electrodes were patterned on the bottom surface of the microchannel as shown in the figure. The width of

978-1-4244-2977-6/09 $25.00 © 2009 IEEE

each electrode is 500 µm. The two electrodes are separated by 500 µm. The electrodes extend over the entire width of the microchannel. As the blood flow was perfused into the microchannel, the electrical impedance between the two electrodes was recorded as a measure of the blood rheological behavior.

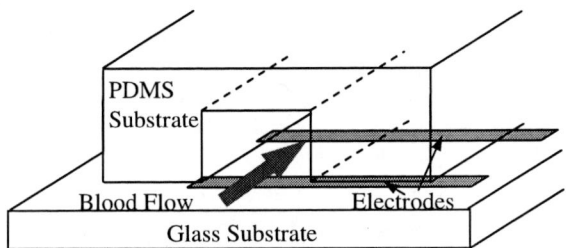

Figure 1: Schematic view of the electrical impedance based microfluidic device.

The distribution of the electric field generated by the two measuring electrodes was estimated using finite element method. The results showed that the electric field density decreases with the vertical distance from the planar electrodes. At the point 500 µm distant from the electrodes, the electrical field strength drops about 70%. In other words, this region includes the majority of the electrical energy induced by the two measuring electrodes. Therefore, in this work, the height of the microchannel was designed as 500 µm. The small change of rheological behavior of the blood can thus be monitored from electrical impedance measurement.

MICROCHIP FABRICATION

Figure 2 shows the fabrication process of the microchip. PDMS is selected as the channel material because its biocompatibility and the mechanical properties similar to human tissues [14]. The measuring electrodes are designed to be 30 mm away from the inlet to ensure the sufficiently developed blood flow at the point of measurement. The channel surface was treated using oxygen plasma to eliminate the influence of surface charges on PDMS substrate. The treatment also helps to convert the polymer surface from hydrophobic to hydrophilic.

Since the microfluidic device is used to perfuse blood sample with a wide range of shear rates, it should be able to withstand high hydraulic pressure occurring at high shear rates. This requires strong bonding between PDMS and glass substrates. To achieve this, the surfaces of PDMS and glass substrates were processed with oxygen plasma. After this treatment, a modest pressure was applied for bonding. Finally, the device was baked on the hotplate at 95°C overnight. After oxygen plasma processing and heating-assisted bonding, the PDMS and glass substrates were bonded firmly. Fluid leakage was not observed throughout the entire experiment.

BLOOD PREPARATION

The blood samples used in this study were obtained by drawing fresh whole blood from adult cattle. Heparin sodium at the rate of 40 U/ml of blood was used as anticoagulant and mixed thoroughly with the flesh blood immediately after the extraction. The blood was

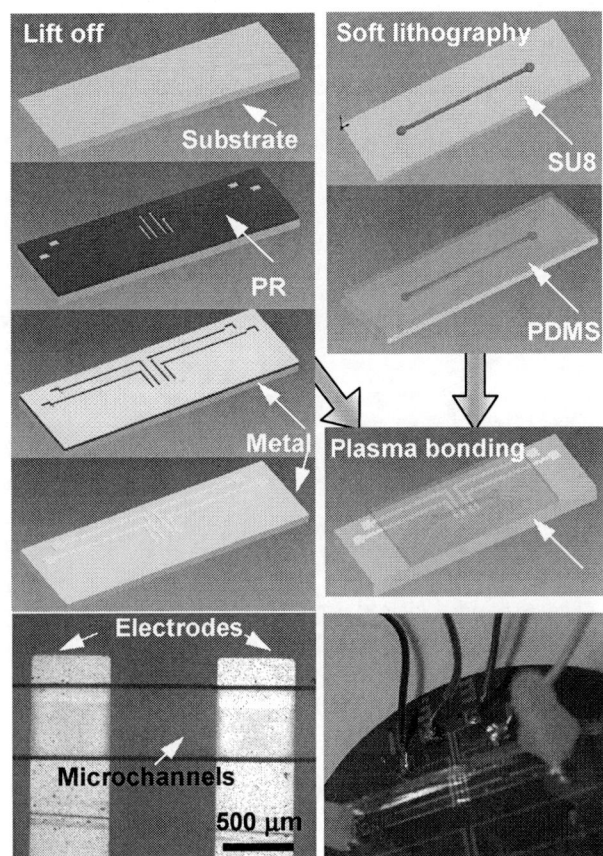

Figure 2: The photolithographic and softlithographic processes were used to fabricate the electrical impedance based microflduic device. The bottom left subfigure shows the top view of the microchannels with the patterned microelectrode. The bottom right subfigure shows the overview of the microchip.

centrifuged at 3400 rpm for 15 minutes to separate plasma, WBCs and RBCs. The plasma was carefully aspirated without disturbing the WBCs in the intermediate layer. Then the WBCs were slowly aspirated using a transfer pipette. Finally, the RBCs in the bottom layer were collected. The plasma and RBCs were mixed at different volume ratios to formulate blood samples with different hematocrit values (20%, 50% and 80%). Given the fact that the WBCs play a relatively minor role in determining the blood viscosity and the electrical impedance, the intermediate layer was removed from the recomposed blood samples for analysis simplicity.

VISCOSITY MEASUREMENT

The blood viscosity was measured using a commercialized Couette rheometer ARES LS II (TA Instruments, Inc.). The blood sample was sheared into two parallel plates placed close to each other. The driving plate propelled the blood to a constant shear rate and the viscosity was obtained by the rotational toque measured from the opposing plate. The measurement was not performed until the toque became stable. Figure 3 shows the viscosity measurement of the blood samples with the hematocrit of 20%, 50%, and 80%, under the shear rate from $10s^{-1}$ to $1000s^{-1}$. The measurements were performed at room temperature (25 °C). The results showed that the

978-1-4244-2977-6/09 $25.00 © 2009 IEEE   241

viscosity increases with the hematocrit, and decreases with the shear rate.

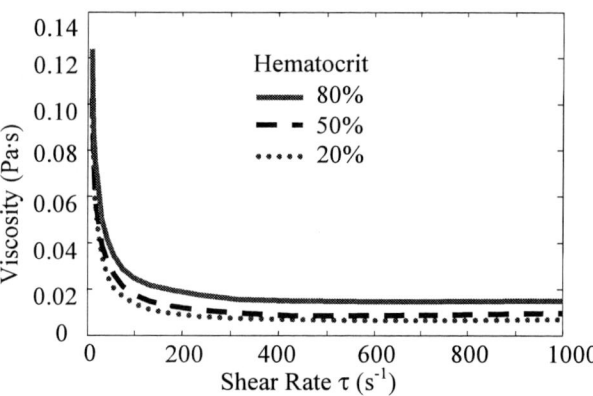

Figure 3: Blood viscosity varies as a function of hematocrit value and the flowing shear rate.

EXPERIMENTAL SETUP

The experimental setup of the electrical impedance measurement is shown in Figure 4. A precision syringe pump was used to precisely control the flow rate and therefore regulate the shear rates. An electrical impedance analyzer was used to measure the impedance of the blood sample flowing through the microchannel. The measurements were performed under the frequency of 20 kHz, where a small stimulating current was used in order not to induce RBCs deformation. To avoid the artifact due to channels clotting, optical monitoring was performed by placing the microchip on the stage of a measuring microscope, with the microchannels right beneath the objective.

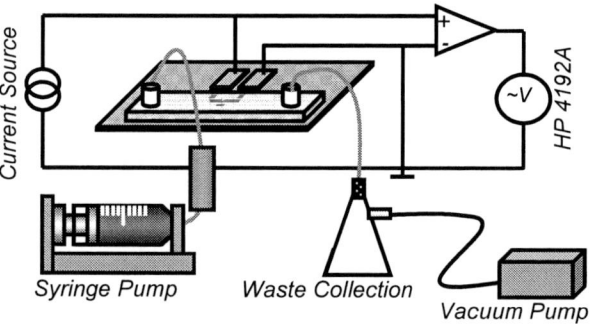

Figure 4: The experimental setup for electrical impedance measurement of the flowing blood sample using the microchip.

IMPEDANCE MEASUREMENT

The viscosity of blood samples at defined hematocrit values and shear rates is obtained from above measurements using rotational viscometer, whilst the electrical impedance of the blood samples under the same conditions is acquired using the microchip. The electrical impedance measurements were performed after 2 minutes from the start of the perfusion. This is to ensure a steady flow rate and to avoid the influence of RBCs sedimentation.

In this investigation, the electrical impedance of blood was measured at a stimulating frequency of 20 kHz in the physiological regime of shear rate from 10 s^{-1} to 500 s^{-1}. The real and imaginary parts of the electrical impedance at this frequency were derived from the measurement and were plotted against two important physiological parameters: hematocrit and shear rate. Figure 5(a) illustrates the real part of the measured electrical impedance (resistance) as a function of hematocrit and shear rate. It is seen from the figure that the electrical resistance increases with hematocrit and decreases with the shear rate. Figure 5(b) shows that the imaginary part of the electrical impedance keeps constant under a certain hematocrit. This indicates that the imaginary part of electrical impedance is relatively insensitive to the shear rate and thus can be used as a good predictor of blood hematocrit.

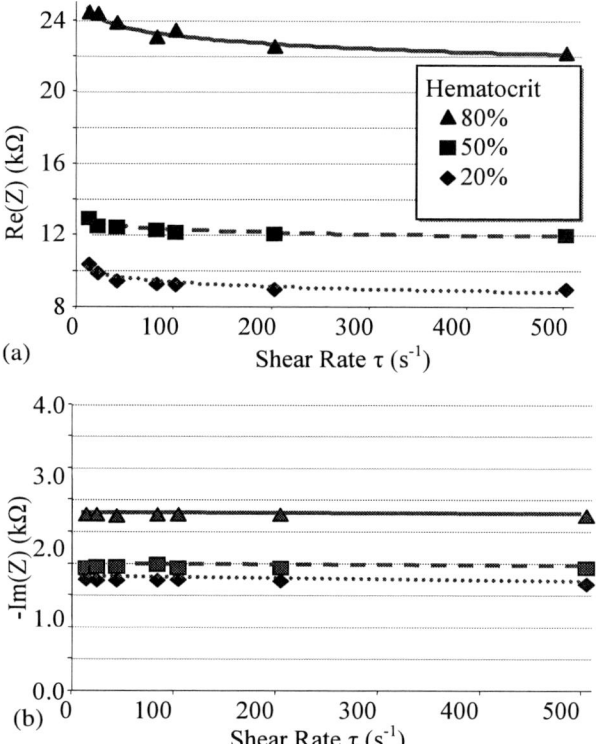

Figure 5: The measurement by the microchip shows that the electrical impedance of blood is a function of hematocrit and flowing shear rate. (a) real part of the electrical impedance changes with hematocrit and shear rate; (b) imaginary part of the electrical impedance changes with hematocrit, but barely varies with the shear rate.

In order to establish a qualitative correlation between the electrical impedance and the viscosity of blood at a defined hematocrit value, second-order regression models were employed (Figure 6). The statistical analysis showed the correlation coefficients for blood samples at all the three hematocrit values are higher than 90%. These results showed that electrical impedance measurement using this microchip is a viable method to predict the viscosity of blood samples. Furthermore, the predicted values of the blood viscosity were compared with additional viscosity measurements using a Couette rheometer, where a good match was observed (Figure 7).

978-1-4244-2977-6/09 $25.00 © 2009 IEEE   242

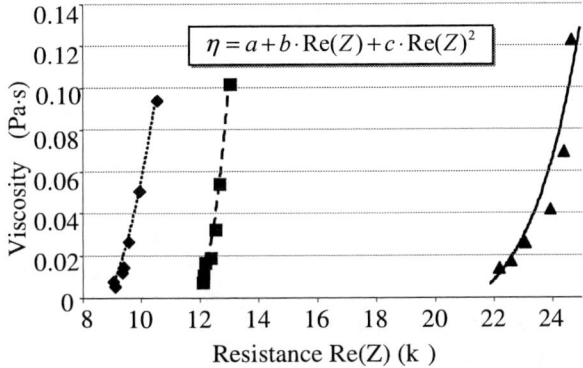

$$\eta = a + b \cdot \mathrm{Re}(Z) + c \cdot \mathrm{Re}(Z)^2$$

Figure 6: Correlation between blood viscosity and electrical impedance at 20 kHz. The measurements at three hematocrit values were fitted using second-order polynomial functions.

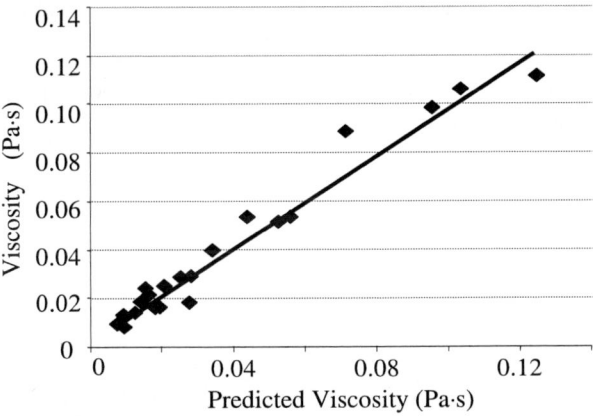

Figure 7: The blood viscosity measured using the microchip (predicted viscosity) is in well accordance with the measurement using a Couette rheometer.

CONCLUSION

In this work, we report the design, fabrication and test of a microfluidic device for assessing blood viscosity. This device avoids the rotational components in conventional rheological measurements, and thus allows for miniaturization. The viscosity is predicted from the electrical impedance measured by in-channel microelectrodes, and compared with the measurements by a conventional viscometer. The results show that the on-chip blood viscometer is a viable and simpler solution for determination of rheological behavior of blood samples. The digital readout and small sample consumption demonstrate the potential for point-of-care and unattended diagnosis.

ACKNOWLEDGEMENTS

The authors would like to thank Dr. Kurt Koelling and his laboratory at Ohio State University for the facility support and technical help on the blood viscosity measurements.

REFERENCES

[1] W. Lowe, "Understanding Disuse Atrophy," Massage Today, vol. 5, 2005.

[2] M. C. Rozenberg, "Blood viscosity and thrombosis in vitro in patients with previous myocardial infarction," Angiology, vol. 19, pp. 527-533, 1968.

[3] H. B. Aronson, S. Cotev, F. Magora, J. B. Borman, and G. Merin, "Blood viscosity and open heart surgery. A comparison of values in systemic and pulmonary blood vessels," Br J Anaesth, vol. 46, pp. 722-725, 1974.

[4] S. Chien, "Shear dependence of effective cell volume as a determinant of blood viscosity," Science, vol. 168, pp. 977-979, 1970.

[5] R. E. Wells, Jr. and E. W. Merrill, "Influence of flow properties of blood upon viscosity-hematocrit relationships," J Clin Invest, vol. 41, pp. 1591-1598, 1962.

[6] S. Shin, Y. Ku, M. S. Park, and J. S. Suh, "Measurement of red cell deformability and whole blood viscosity using laser-diffraction slit rheometer," Korea-Australia Rheology Journal, vol. 16, pp. 85-90, 2004.

[7] P. W. Rand, E. Lacombe, H. E. Hunt, and W. H. Austin, "Viscosity of normal human blood under normothermic and hypothermic conditions," J Appl Physiol, vol. 19, pp. 117-122, 1964.

[8] T. Zhao, B. Jacobson, and T. Ribbe, "Triple-frequency method for measuring blood impedance". Physiol Meas, vol. 14(2), pp. 145-156, 1993.

[9] K.R. Visser, "Electric properties of flowing blood and impedance cardiography". Annals of Biomedical Engineering, vol. 17(5), pp. 463-473, 1989.

[10] D.W. Hill, and F.D. Thompson, "The effect of haematocrit on the resistivity of human blood at 37°C and 100kHz". Med Biol Eng, vol. 13(2), pp. 182-185, 1975.

[11] H.D. Fuller, "The electrical impedance of plasma: A laboratory simulation of the effect of changes in chemistry". Ann Biom Eng, vol. 19(2), pp. 123-129, 1991.

[12] A.R. Pries, D. Neuhaus, and P. Gaehtgens, "Blood viscosity in tube flow: dependence on diameter and hematocrit". Am J Physiol Heart Circ Physiol, vol. 263(6), pp. H1770-H1778, 1992.

[13] H.P. Schwan, "Electrode polarization impedance and measurements in biological materials". Ann. New York Acad. Science, vol. 148, pp. 191-209, 1968.

[14] M.C. Belanger, and Y. Marois, "Hemocompatibility, biocompatibility, inflammatory and in vivo studies of primary reference material low-density polyethylene and polydimenthylsiloxane: A review". Journal of Biomedical Materials Research Part B: Applied Biomaterials, vol. 58(5), pp. 467-477, 2001.

IMPLANTABLE FLEXIBLE-COILED WIRELESS INTRAOCULAR PRESSURE SENSOR

Po-Jui Chen[1], Saloomeh Saati[2], Rohit Varma[2,3], Mark S. Humayun[2,3], and Yu-Chong Tai[1]
[1]California Institute of Technology, Pasadena, CA, USA
[2]Doheny Eye Institute, Los Angeles, CA, USA
[3]University of Southern California, Los Angeles, CA, USA

ABSTRACT

This work presents an implantable wireless passive pressure sensor for long-range continuous intraocular pressure (IOP) monitoring of glaucoma patients. The sensor is microfabricated with use of parylene C (poly-chloro-p-xylylene) to create a flexible coil substrate that can be folded during implantation for suture-less minimally invasive surgery, while stretched back without damage for enhanced inductive sensor-reader coil coupling and the corresponding sensing signal. Extensive device characterizations including on-bench testing and *in vivo* and *ex vivo* animal studies verify the device feasibility in both engineering (1 mmHg pressure sensing accuracy and 2 cm sensing distance) and surgical (robust fixation to the iris and long-term biocompatibility in the intraocular environment) aspects, all meeting specifications for future practical implementation of such IOP sensing technology.

INTRODUCTION

Passive telemetric sensing has been widely used as one of the viable methods to accomplish continuous and accurate non-contact IOP measurements of glaucoma patients [1, 2]. It enables straightforward IOP sensing by utilizing a transensor implant that registers environmental pressure variations inside the eye, so that the IOP can be directly measured by using an external reader wirelessly interrogating the implant. As illustrated in Fig. 1, this sensing methodology requires an external reader to interrogate IOP variations electrically registered by an implanted sensor through a wireless inductive coupling link. Accordingly, long sensing distance between sensor and reader coils in reasonable arrangements is crucial for practical IOP measurements. Previously, we developed a miniaturized implantable IOP sensor for preliminary feasibility study [3]. In spite of successful wireless sensing demonstration with achieving high pressure sensitivity, the sensing distance of the entire telemetric system was unsatisfactory (~2 mm) due to low quality factor and dimensional constraints of the sensor. This fact seriously hinders its application especially when considering the glasses-to-iris distance (1.5-2 cm) in practice, and thus needs further improvement.

Based on applying the impedance phase-dip technique as the wireless measurement method [1-3], the sensor implant is designed to have a pressure-sensitive electrical LC-tank resonant circuit with a corresponding resonant frequency represented as

$$f_s = \frac{1}{2\pi}\sqrt{\frac{1}{L_s C_s} - \frac{R_s^2}{L_s^2}} \cong \frac{1}{2\pi\sqrt{L_s C_s}} \quad \text{if } R_s^2 \ll \frac{L_s}{C_s},$$

where L_s, C_s, and R_s are respectively the inductance, capacitance, and resistance of the sensor. When the sensor is excited at resonance, the phase of the equivalent impedance Z_{eq} to be measured drops to the minimum with a difference approximated as

$$\Delta\phi \cong \tan^{-1}\left(k^2 Q_s\right),$$

where k is the coupling coefficient of the inductive link totally dependent on physical geometries such as the planar size of the sensor and reader coils and the separation distance between the coils, and $Q_s = R_s^{-1}(L_s C_s^{-1})^{1/2}$ the quality factor of the sensor at resonance. As long as the impedance phase dip is detectable in the frequency scan, the resonant frequency of the sensor varied by pressure conditions can be accurately characterized, and the resultant IOP can be obtained in real time. Therefore, the objective of this work is to develop a microfabricated implantable wireless pressure sensor with improved electromagnetic coil coupling and enhanced quality factor in order to increase the signal strength with more distinguishable phase dip, so that the overall sensing distance can be increased for necessarily longer-range IOP detection, which eventually accommodates the iris-to-glasses distance for appropriate implementation of the glasses-type reader in the future.

DESIGN

It is deduced from the aforementioned description that both the integrated coil size and quality factor of the LC-tank should be increased as much as possible in order to increase the k^2Q factor. On the other hand, the entire device needs to be in a small form factor suitable for minimally invasive implantation in order to minimize associated surgical difficulty and complications. As a result, the proposed wireless IOP microsensor incorporates the flexible coil paradigm reported in the literature [4] to best meet the both requirements. Fig. 2 shows design of the microsensor comprising a pressure-sensitive parallel-plate variable capacitor embedded in a deformable diaphragm chamber, a spiral metal wire to serve as a planar inductor, and a flexible/foldable disk substrate to support the coil. The disk diameter is designed to be 4 mm to accommodate a substantially large coil while still fitting the iris rim width in normal conditions. The disk is also designed to have with sufficient mechanical flexibility so as to maintain a

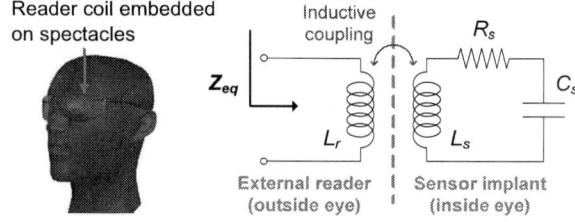

Figure 1. *Proposed IOP monitoring in practice: (left) Continuous sensing using the glasses reader paradigm; (right) equivalent circuit schematic of the sensing system.*

978-1-4244-2977-6/09 $25.00 © 2009 IEEE

small required incision (< 2 mm) during device implantation. These structures are integrated along with a rigid piece having air/gas cavity to provide pressure reference. The width of the backing piece is predetermined 1.5 cm so that the coiled disk can be folded to smaller than 4 mm × 2 mm for suture-less intraocular implantation as well as for ease of realization and handling of the parallel-plate variable capacitor. Because this prototype also needs another attachment piece to create the pressure reference in gauge pressure sensing, surgical anchoring features are directly created on the non-electronic sealing silicon piece for suture-less implant fixation on the iris. Electrical/mechanical design, material selection, and micromachining processes of the pressure-sensitive resonant circuit structures are analogous to those in the previous prototype [3]. The mechanical behavior of the parallel-plate variable capacitor is designed specifically in the small deflection regime to better estimate the pressure-sensitive performance of the transensor. All involving materials in the fabrication processes are in implantable grade. Parylene C is selected as the diaphragm and disk substrate material because of its flexibility (Young's modulus ~4 GPa), CMOS/MEMS processes compatibility, and biocompatibility (USP Class VI grade). Additionally, its low water permeability (0.08 g-mm/m^2-day) and low water absorption (< 0.06% after 24 hr) favor stable behavior of such device when immersed in the aqueous humor of the eye after implantation. Moreover, besides being used to create the top parallel capacitor plate, the second metal layer is arranged to be with all boundaries of the pressure-sensitive parylene diaphragm chamber to act as a strong barrier to transmissions/permeations of water vapor and gas in between the encapsulated air/gas and the ambience of the device. This inclusion can effectively prevent the sensor implant from substantial performance drift, which supports their long-term use in the intraocular environment.

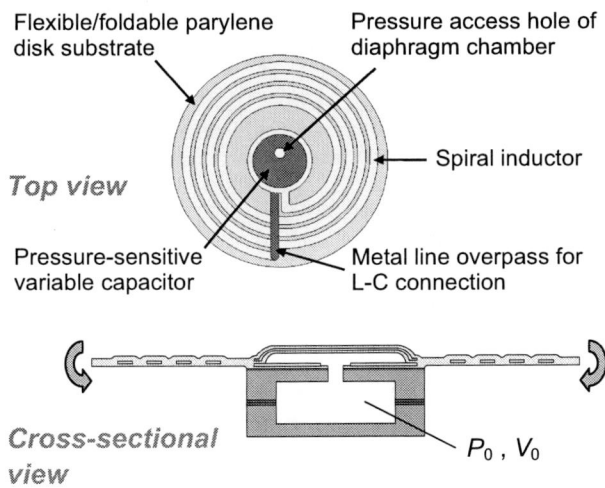

Figure 2. *Pressure sensor design schematics.*

FABRICATION

The sensor is monolithically microfabricated as depicted in Fig. 3 facilitated by low-temperature multi-layer parylene micromachining and deep silicon etching technologies. The fabrication processes started by thermally growing and patterning 2 μm oxide on a double-side-polished silicon wafer. Deep reactive-ion etching (DRIE) was performed on the backside of the wafer to define device release boundaries and access hole of the pressure diaphragm chamber until leaving approximately 50 μm silicon for ease of through-wafer etching using the remaining oxide as the mask. In frontside processes, a photoresist layer was created not only as a sacrificial layer for coil disk release, but also as a mask to gas-phase XeF$_2$ silicon roughening for physically strengthened parylene-silicon adhesion of the device. Afterwards, multiple parylene and metal layers were deposited and patterned to create the flexible-coiled pressure sensor structures along with using another sacrificial photoresist layer to define the free-standing pressure diaphragm chamber. The bottom parylene layer mechanically supports the metal coil windings and electrically isolates the other device layers from the substrate. The metals were e-beam-evaporated for better material quality purposes while thick deposition was required to obtain low overall resistance from the metal lines. Two thick titanium/gold layers (200 Å/3 μm and 200 Å/0.5 μm) were patterned using standard metal etching techniques. As previously stated, the sandwiched parylene-metal-parylene layers were arranged at the chamber sidewalls to minimize liquid/gas permeation from/to the encapsulated air cavity. In fact, this metal layer extended to the surrounding of the diaphragm chamber to contribute additional capacitance by interaction with the underneath parylene-metal layers, which facilitates control the device resonant frequency falling into reader scan range of interest. Devices were finally released with a backside recess for larger closed volume of pressure reference after performing another DRIE followed by photoresist stripping with acetone. Fig. 4 shows the air-dried microfabricated sensor before device packaging.

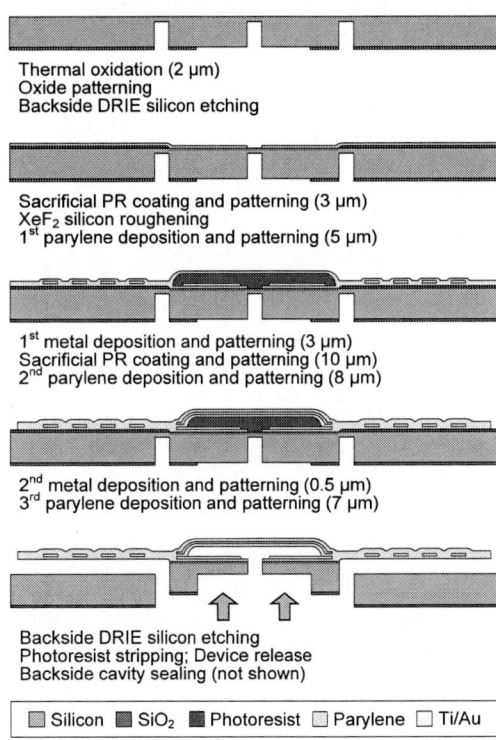

Figure 3. *Fabrication process flow. The arrows indicate the outcome of device release after fabrication processes.*

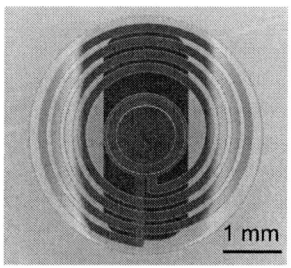 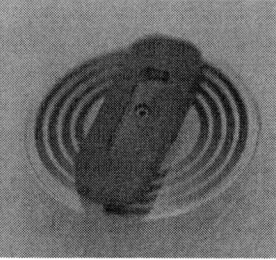

Figure 4. *Full-scale photographs of the microfabricated with its top (left) and bottom (right) views before packaged. The diameter of the flexible coil disk is 4 mm.*

The post-microfabrication sensor packaging was conducted in air under atmospheric pressure at room temperature as the conditions of the encapsulated air. A non-electronic piece was attached to the bottom of the microfabricated sensor using epoxy to form an air cavity inside the device for pressure reference in gauge pressure sensing. The overall size was measured at $\phi 4$ mm $\times$ 1 mm and reduced to a form factor of 4 mm $\times$ 1.5 mm $\times$ 1 mm by folding the coil disk as shown in Fig. 5, suitable for minimally invasive intraocular implantation. Given the high yield strain (~3%) characteristic of parylene C, the flexible disk can be stretched back to its original circular shape without severe permanent deformation or other damages after this extent of folding, and the inductor characteristics were hence unvaried.

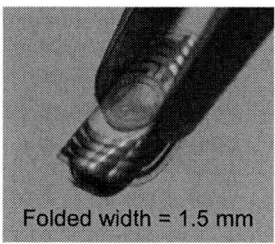 

Figure 5. *Device flexibility demonstration: (left) Coil disk folding; (right) Coil disk stretched back after folding.*

RESULTS

Device testing was conducted using a hand-wound coil connected to a HP 4195A network/spectrum analyzer to serve as the external reader for electrical measurements. Electrical parameters of the fabricated microsensor were first obtained by analyzing the measurement data from both the actual device with the external wireless readout method and several test structures with on-chip probing. Table I lists the experimental results which were in good agreement with device design estimates.

Table I *Electrical parameters of the fabricated sensor*

Parameters	Values
Inductance	57 nH
Capacitance	3.6 pF
Resistance	4.2 Ω
Resonant frequency	~350 MHz
Quality factor	~30

In on-bench wireless pressure sensing test, the device was placed inside a customized chamber connected to a pressurization setup to simulate the environmental pressure variation surrounding the device. The wireless pressure sensing behavior of the device was characterized with the measured phase-dip curves as shown in Fig. 6 with an approximately 455 ppm/mmHg sensitivity analyzed using the derived pressure response of the normalized resonant frequency [3] as

$$\frac{f_{\min}(\Delta P)}{f_{\min}(\Delta P = 0)} = \frac{\frac{1}{2\pi\sqrt{L_s(C_s + \Delta C_s)}}}{\frac{1}{2\pi\sqrt{L_s C_s}}} \cong (1 - \alpha \Delta P)^{\frac{1}{2}},$$

where α is the fitting parameter. In order to meet the IOP sensing requirement, the allowed frequency fluctuation in measurements must be lower than the shift contributed by 1 mmHg pressure difference. Consequently, a supportive data-processed readout method [5] was utilized to achieve low-noise, high signal-to-noise ratio (SNR) phase-dip detections associated with long-range pressure sensing. Testing results indicate that SNR > 10 is the criterion of controlling the frequency shift noise lower than 400 ppm for applicable measurements. Given the minimally obtained $\Delta\phi_{\text{noise}} \sim 0.001°$ using the developed readout method, the maximum sensing distance was confirmed to be 2 cm using a 2.2-cm-diameter external reading coil, appropriate for the proposed glasses-type reader paradigm in practical wireless IOP monitoring with 1 mmHg pressure sensing accuracy. These investigations verify the device feasibility in engineering aspect.

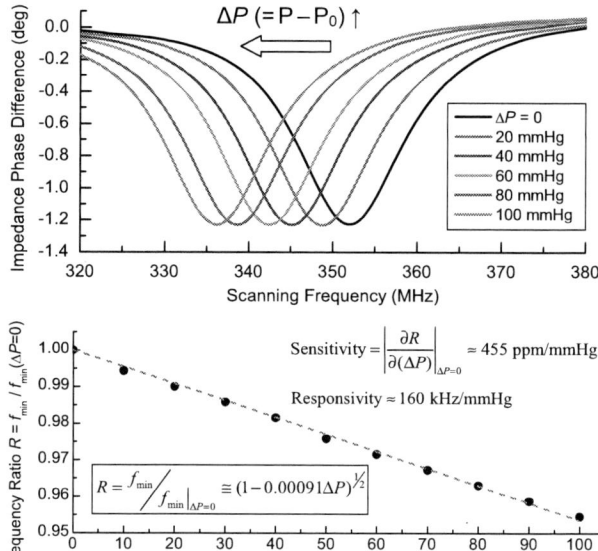

Figure 6. *On-bench wireless pressure sensing test: (top) Overlay plot of impedance phase-dip curves; (bottom) Pressure response characterization.*

In surgical aspect, a commercially available flexible ophthalmic iris retractor (Alcon/Grieshaber AG, Schaffhausen, Switzerland) was modified using a razor blade and then attached to the non-electronic sealing piece using epoxy as shown in Fig. 7 to serve as the tissue anchor for suture-less implant fixation. This surgical attachment

978-1-4244-2977-6/09 $25.00 © 2009 IEEE

and the microfabricated sensor were assembled as the complete packaged device. After being introduced to the anterior chamber, the tapered feature at the end of the retractor hook penetrates the iris stratum to provide sufficient anchoring force to the device. This anchoring process is reversible if the implant needs to be removed from the eye. The packaged device could be conformally coated by a thin parylene layer to ensure its long-term biocompatibility in the intraocular environment.

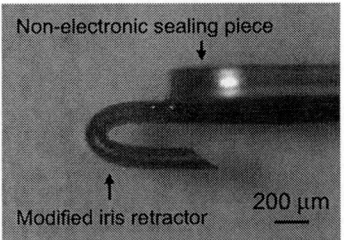

Figure 7. *Representative iris anchor assembly result (side view). The tapered retractor can hook the iris stratum to enable minimally invasive device fixation inside the eye.*

Chronic animal study was conducted with live rabbit model as shown in Fig. 8 to investigate surgical/biological compatibility of the implant. The entire surgery was remarkably completed within 15 min given its minimally invasive nature. In addition, the specific implant location has a tremendous advantage in aligning sensor and reader coils so that signal loss through coil misalignment can be minimized. Observation confirms that no device dislocation or post-operative complications (e.g., inflammatory response, tissue encapsulation/fibrosis, etc.) were found over 4 months, verifying surgical feasibility and bioefficacy of the device. Follow-up study is still in progress to collect long-term (e.g., 6-month) data.

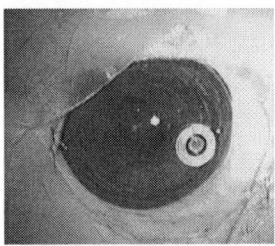 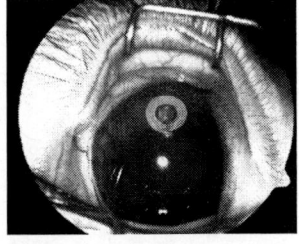

Figure 8. *Surgical feasibility test: (left) Implanted sensor inside the rabbit eye model as observed through the cornea; (right) Fundus photograph showing no device dislocation or post-operative complications in the intraocular environment after 3-month follow-up study.*

Finally, acute animal study was conducted to demonstrate *in vivo* wireless pressure sensing using the sensor implant inside a rabbit eye with the setup as shown in Fig. 9. *Ex vivo* testing using an enucleated porcine eye with the same setup was also accomplished to obtain quantitative results of sensor characterization in the simulated intraocular environment. A medical manometer was used in the infusion-based pressurization setup of this experiment to achieve precise IOP measurements as compared with the electrical readouts from the sensor. The wireless sensing results were analyzed and confirmed in good agreement with those in on-bench testing as shown in

Fig. 6, indicating that the pressure response of the sensor was consistent when situated in different environments, regardless of different electrical performance (e.g., quality factor drop when the sensor was inside the eye) due to the medium effect which needs further study. Advanced surgical protocols and tools more suitable for such implant paradigm will also be investigated to better meet the future requirements in practice.

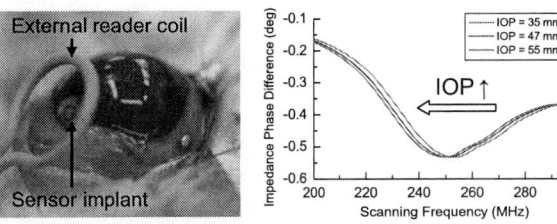

Figure 9. *In vivo wireless pressure sensing test: (left) Experimental setup. (right) Overlay plot of measured impedance phase-dip curves.*

CONCLUSION

An implantable wireless pressure sensor implementing passive telemetric sensing technology has been successfully developed as a solution to IOP measurements. Featuring a parylene-based flexible coil disk, the sensor achieves a satisfactory sensing distance suitable for practical glasses-type reader realization, as well as maintains a small form factor suitable for minimally invasive implantation. Engineering and surgical/biological performance of the sensor were characterized in a complete suite of on-bench, *ex vivo*, and *in vivo* experiments, and the results provide substantial evidence that the sensor has great potential for fulfilling continuous, real-time, reliable, and convenient IOP monitoring of glaucoma patients.

ACKNOWLEDGEMENTS

This work was supported in part by Bausch and Lomb. The authors especially thank Dr. Wen Li for her assistance on data acquisition of electrical impedance measurements, Dr. Damien Roger for his comments on surgical procedures and pressure testing with animal models, and Mr. Trevor Roper for his fabrication assistance.

REFERENCES

[1] C. C. Collins, "Miniature Passive Pressure Transensor for Implanting in the Eye," *IEEE Transactions on Biomedical Engineering*, BME-14(2), pp. 74-83, 1967.

[2] K. C. Katuri, S. Asrani, and M. K. Ramasubramanian, "Intraocular Pressure Monitoring Sensors," *IEEE Sens. J.*, vol. 8, pp. 12-19, 2008.

[3] P.-J. Chen, D. C. Rodger, S. Saati, M. S. Humayun, and Y.-C. Tai, "Implantable Parylene-Based Wireless Intraocular Pressure Sensor," *Proc. MEMS 2008*, pp. 58-61.

[4] M. A. Fonseca, M. G. Allen, J. Kroh, and J. White, "Flexible Wireless Passive Pressure Sensors for Biomedical Applications," *Proc. Hilton Head 2006*, pp. 37-42.

[5] P.-J. Chen, "Implantable Wireless Intraocular Pressure Sensors," *Ph.D. dissertation*, California Institute of Technology, 2008.

INTEGRATED WIRELESS NEUROSTIMULATOR

W. Li[1], D.C. Rodger[2], and Y.C. Tai[1]

[1]California Institute of Technology, Pasadena, California, USA

[2]Keck School of Medicine of the University of Southern California, Los Angeles, California, USA

ABSTRACT

This paper presents the design, fabrication and functional testing of a fully implantable, flexible, Parylene-enabled neurostimulator that features single channel wireless stimulation capability. The system comprises a CMOS stimulator chip, a fold-and-bond RF coil, two platinum electrodes, and discrete capacitors. The MEMS components are fabricated with a Parylene-metal skin technology, and the system assembly is achieved by interconnecting individual components together on a Parylene substrate with silver epoxy. The functionality of the integrated system is verified using a telemetry link setup, and single-phase pulses with amplitudes ranging from 7 to 8.5 V are detected.

INTRODUCTION

In recent decades, integrated wireless microsystems have provided tremendous opportunities in neural prostheses by establishing an artificial interface to the central or peripheral neural system. For example, retinal implants have been studied for the treatment of outer retinal degenerative diseases, such as age related macular degeneration (AMD) and retinitis pigmentosa (RP) [1]. However, many technologies are still in development and few have actually been transferred to the clinical practice due to constraints in material biocompatibility, device miniaturization, and flexibility [2, 3].

It has always been our goal to develop flexible wireless neural implants, employing Parylene-based MEMS devices integrated with other discrete components, such as application specific integrated circuits (ASICs) and chip capacitors. A Parylene-metal skin technology has been developed previously, which allows us to microfabricate RF coils and multielectrode arrays in a process compatible way [4, 5]. In this paper, we will integrate such a coil and an electrode array with a single channel stimulator chip to build a fully functional system for neural stimulation. For the first prototype, MEMS devices are fabricated separately, and assembled with other system components using biocompatible silver epoxy. Parylene-C serves as a substrate and packaging material, and therefore the final system is highly flexible and biocompatible for medical implantation.

SYSTEM DESIGN

Figure 1 depicts the system schematic of a single channel stimulator, which consists of a BION1-3 CMOS chip, two capacitors, a specially designed RF MEMS coil for wireless power and data transmission from an external personal trainer, as well as a carrier substrate. The flexible carrier has interconnection leads and contact pads for system assembly, and two electrodes of which one serves as the stimulating electrode while the other is the floating ground. The overall physical dimensions of the MEMS coil and the carrier substrate are to meet specifications used in retinal implantation, which are determined based on

surgical implantation results in canine eyes.

The BION chip is a single-channel stimulator which is initially developed to mimic muscle spindle function, and to treat patients suffering from muscle paralysis. Once implanted into paralyzed muscles, the chip can receive both power and command signals through inductive coupling over a 480 kHz ~ 500 kHz power carrier generated in an external unit, allowing the chip to emit precisely timed stimulation pulses with highly regulated amplitude and pulse-width [6]. The BION chip has physical dimensions of ~ 1 mm in width, ~ 2.33 mm in length, and ~ 257 µm in thickness. There are five pads on the chip with sizes of approximately 120 µm × 74 µm. The distance between two adjacent pads is about 110 µm.

Two ceramic chip capacitors (AVX Corporation, Myrtle Beach, SC, USA) are incorporated in the integration system to tune the circuit. As seen from the circuit layout (Figure 1), C_1 is a frequency tuning capacitor in parallel with the receiving coil to achieve a resonant frequency of ~ 500 kHz. C_2 is a charge storage capacitor which has a capacitance of ~ 22 nF, provided by Dr. Loeb's group at the University of Southern California. These capacitors are about 1 mm × 0.5 mm × 0.56 mm in size.

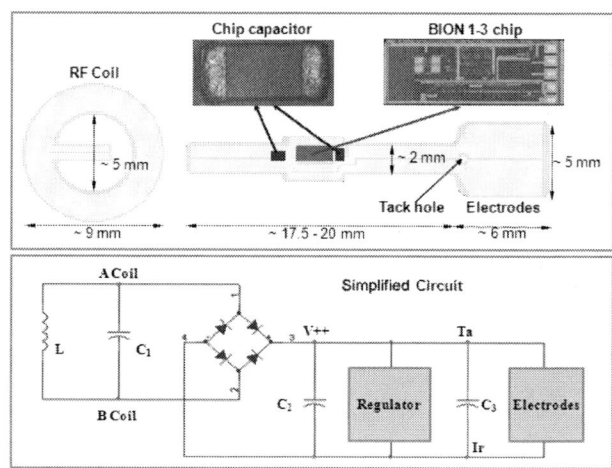

Figure 1: System schematic of the single channel stimulator and the corresponding circuit layout, showing the connections between individual components.

FABRICATION

The fabrication is divided into three steps: carrier substrate fabrication, RF MEMS coil fabrication, and final system assembly and packaging. To build the flexible substrate, a 200 nm layer of platinum is e-beam evaporated on a Parylene-C coated silicon substrate. Then the metal is patterned using lift-off to form connection pads, interconnection leads, and electrode sites [5]. Platinum is selected as the electrode material for optimal simulation capability. No adhesion metal layer is need in this case because platinum and Parylene are known to have good adhesion. After metal patterning, another layer of

978-1-4244-2977-6/09 $25.00 © 2009 IEEE   248

Parylene-C is deposited to seal the entire structure, followed by oxygen plasma etching in a reactive ion etching (RIE) system with a photoresist mask to define the outer geometry of the substrate, as well as to open the electrode sites and the contact vias. Finally, the device is peeled off from the silicon substrate in a water bath. Figure 2 illustrates the detailed process for making the carrier substrate, where steps (b)-(d) describe the lift-off technology for platinum patterning.

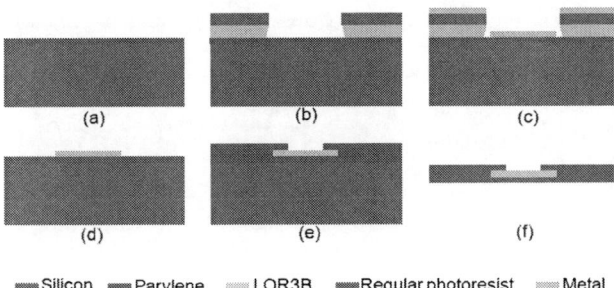

▬▬ Silicon ▬▬ Parylene ▭ LOR3B ▬▬ Regular photoresist ▬▬ Metal

Figure 2: Fabrication process of the Parylene-based carrier substrate.

Figure 3 presents a fabricated carrier substrate. The microscope image shows a special chip site design where Parylene ribbons are etched in a way such that the chip can be held in place and self-aligned to the contact vias on the substrate during system assembly. The array site contains a 450 μm diameter tack hole so that the array can be attached to the retina using a retinal tack.

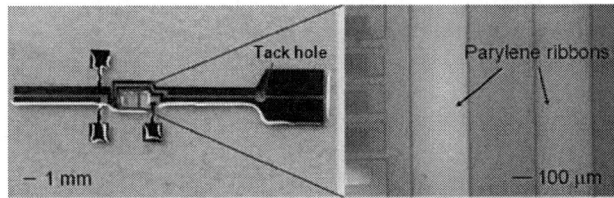

Figure 3: A fabricated carrier substrate with electrodes, interconnection leads and contacts. The right image shows Parylene ribbons for holding the BION chip.

To fabricate the MEMS coil, a fold-and-bond technology is involved [7], in which two coils are placed in series, and made of a single 3 μm layer of gold to achieve low resistance and high Q factor. Then the device is folded into two layers and stacked together with the assistance of two glass slides. Aluminum sheets are inserted between the Parylene surface and the class slides to prevent Parylene sticking on the glass. After that, the stacked coil is placed in a vacuum oven with a chamber pressure of ~ 10 Torr for bonding. The oven temperature ramps from room temperature to 250°C. The device is then soaked at the bonding temperature for 2 days, followed by a slow cool-down to room temperature. Nitrogen backfill is introduced during the thermal bonding process to equalize the chamber temperature.

A fabricated coil is shown in Figure 4, comprising two layers of metal with 10 turns in each layer. In-and-out leads are connected to the carrier substrate from the center, facilitating the surgical procedure. Through vias are designed to overlap with the contact pads, so that the interconnections to the carrier substrate can be formed from either side. The electrical properties of this coil are measured, showing an inductance of approximately 2.24 μH and a DC resistance of approximately 15.82 Ω, which results in a Q factor of 0.45 at 500 kHz.

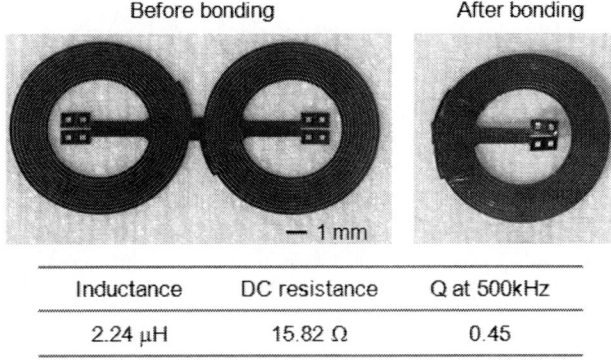

Inductance	DC resistance	Q at 500kHz
2.24 μH	15.82 Ω	0.45

Figure 4: A fabricated fold-and-bond coil with two layers of metal. The electrical characteristics are measured and given in the table.

For hybrid system assembly, individual components (the BION chip, the coil and the capacitors) are aligned to the corresponding interconnection vias on the carrier substrate. A small amount of biocompatible silver epoxy EPO-TEK H20E (Epoxy Technology, Billerica, MA, USA) is then applied on the contacts and cured at 80°C for 3 hours in a convection oven. The conductive epoxy serves two purposes: to form the interconnections between the components as well as to bond the components onto the substrate. Finally, Parylene-C is coated on the entire system just leaving the electrode sites open in order to protect the circuitry from the corrosive eye fluids and to improve the durability of the epoxy contacts. Figure 5 gives an assembled single channel stimulator system, and close-up views on the interconnections for each component.

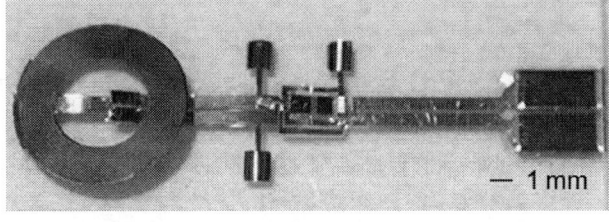

Figure 5: (a) A assembled BION system. (b) Interconnects formed with biocompatible conductive silver epoxy: (left) coil contacts, (right) chip and capacitor contacts.

Because the BION chip has only 5 pads, hand assembly using epoxy interconnection is quick and easy. However, a main problem encountered during the fabrication is the lack of control of the epoxy amount when applied by hand. Short circuits can be created if too much epoxy is applied. Epoxy reflow can also happen during high temperature curing, resulting in short circuits. For chips with high density pad layouts, this hand assembly is no longer applicable, and thus a wafer level integration technology is necessary [8].

TEST RESULTS AND DISCUSSION

After the assemble system is made, its functionality is verified using a telemetry link setup (Figure 6(a)). The primary stage (Figure 6(b)), which comprises a personal trainer unit, a class-E coil driver and a hand-wound transmitting coil, can generate a power carrier of approximately 500 kHz. The personal trainer stores command programs personalized for individual subjects, records the time and duration of treatment, and transfers this information to an external computer for real time monitoring [9]. Up to three programs can be preloaded into the memory of the personal trainer. The coil driver is connected to the personal trainer through a custom made adapter. The transmitting coil has an inductance of ~ 46.4 μH and a Q factor of ~ 118 at 500 kHz. Litz wires 1025-44 SPN are used for winding the primary coil to reduce the skin effect and proximity effect losses. The transmitting coil is built in a solenoid shape to establish a more uniform electromagnetic field inside the coil. Additionally, ferrite cores are inserted in the transmitting coil to magnify the electromagnetic field and therefore to improve voltage transfer efficiency.

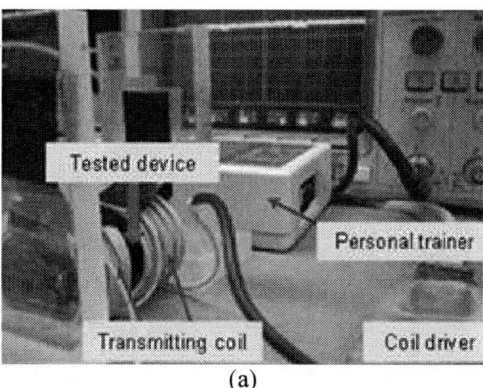

(a)

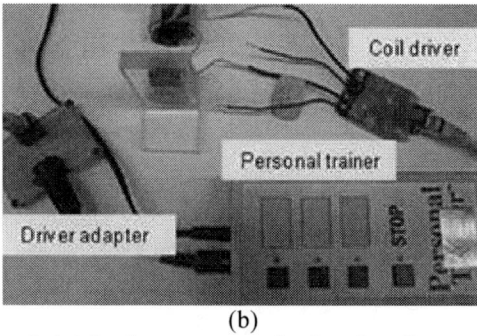

(b)

Figure 6: (a) A telemetry setup for functionality test of the assembled BION system. (b) A personal trainer and other peripheral accessories of the primary stage.

In Vitro measurements have been conducted using this setup, and output signals of the integrated stimulator are monitored by connecting two electrodes directly to an HP 54645A oscilloscope (HP/Agilent Technologies Inc., Santa Clara, CA, USA). The distance between the two coils is varied during testing, and a maximum detectable range of ~ 4 mm is found. The recorded stimulating pulses at the different separation distances are given in Figure 7, showing a pulse width of approximately 500 μs, and amplitudes varying from 7 to 8.5 V.

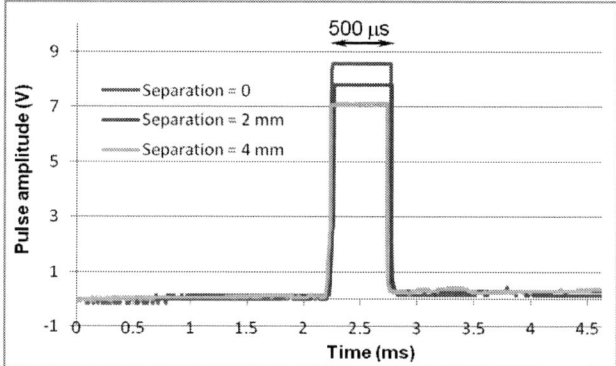

Figure 7: Recorded stimulation pulses at different coil separation distances.

In order to estimate the power transfer capability of the MEMS coil, the voltage across the receiving coil terminals and the current delivered to the chip are measured. Figure 8 shows typical waveforms of the transferred voltage and current, indicating a resonant frequency of ~ 505 kHz and a ~25 degree phase drift between voltage and current. As mentioned earlier, the tuning capacitor is a commercially available chip capacitor, which has limited options of capacitance values. Therefore, it is difficult to fine tune the resonant circuit to achieve precise synchronization. The delivered power at the different separation distances are also investigated, as shown in Figure 9. It can be seen that our MEMS coil can transfer a maximum power of ~ 43 mW through this inductive link at the separation distance of 1 mm. As the separation distance increases to 2 mm, the power drops by 62%, which is mainly limited by the low Q factor of the receiving coil.

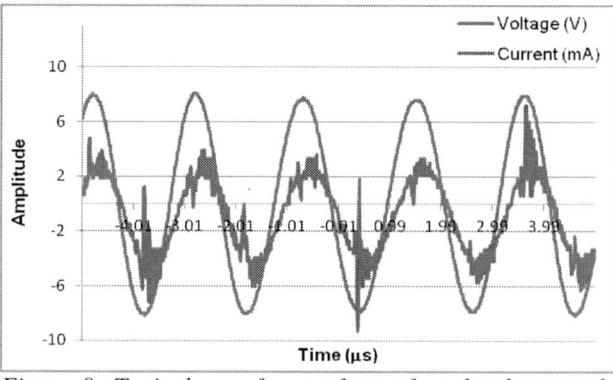

Figure 8: Typical waveforms of transferred voltage and current to the chip.

978-1-4244-2977-6/09 $25.00 © 2009 IEEE

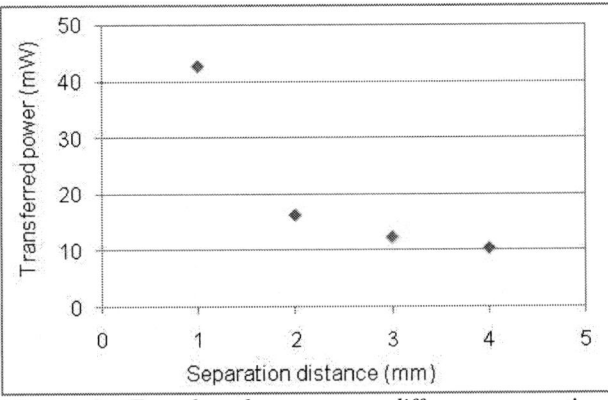

Figure 9: Transferred power at different separation distances between two coils.

The functionality of the integrated system has been successfully demonstrated in air. However, the short detectable distance would restrict the practicality of this system in real applications. Further improvement can be achieved by optimizing coil design, such as increasing metal thickness and/or the number of metal layers, in order to enhance the coil Q factor and the power transfer efficiency. Current efforts lie in the verification of system functionality through implantation in rabbit eyes.

CONCLUSION

A single channel neural stimulator has been designed, and the first prototype has been successfully fabricated and tested. Preliminary test results demonstrate that the BION chip can be driven by the MEMS coil within a 4 mm separation distance. Output pulses with a pulse width of ~ 500 μs and amplitudes of more than 7 V are measured from the simulating electrode, indicating that system can be operated *in vitro*. We are expecting to implant the device in animal subjects to further characterize the system performance *in vivo*.

ACKNOWLEDGMENTS

This work is supported in part by the Engineering Research Center Program of the National Science Foundation under Award Number EEC-0310723 and by a fellowship from the Whitaker Foundation (D.R.). The authors would also like to thank Dr. Gerald Loeb and his group members for providing BION chips and testing units. My gratitude also goes to Mr. Trevor Roper and all other members of the Caltech Micromachining Laboratory for assistance with fabrication.

REFERENCES

[1] M. Javaheri, D.S. Hahn, R.R. Lakhanpal, J.D. Weiland, and M.S. Humayun, "Retinal Prostheses for the Blind", *Anneal Academy of Medicine*, vol. 35, pp. 137-144, 2006.

[2] W. Mokwa, "Medical Implants Based on Microsystems", *Measurement Science and Technology*, vol. 18, pp. 47-57, 2007.

[3] P. Federspil and P.K. Plinkert, "Restoring Hearing with Active Hearing Implants", *Biomed. Tech.*, vol. 49, pp. 76-82, 2004.

[4] W. Li, D.C. Rodger, E. Meng, J.D. Weiland, M.S. Humayun, and Y.C. Tai, "Flexible Parylene Packaged Intraocular Coil for Retinal Prostheses," in *Proc. 4th International IEEE-EMBS Special Topic Conference on Microtechnologies in Medicine and Biology*, Okinawa, Japan, May 9-12, 2006, pp. 105-108.

[5] D.C. Rodger, A.J. Fong, W. Li, H. Ameri, I. Lavrov, H. Zhong, S. Saati, P. Menon, E. Meng, J.W. Burdick, R.R. Roy, V.R. Edgerton, J.D. Weiland, M.S. Humayun, and Y.C. Tai, "High-density Flexible Parylene-based Multielectrode Arrays for Retinal and Spinal Cord Stimulation", in *Digest Tech. Papers Transducers'07 Conference*, Lyon, France, June 10-14, 2007, pp. 1385-1388.

[6] N.A. Sachs and G.E. Loeb, "Development of a BIONic Muscle Spindle for Prosthetic Proprioception", *IEEE Transaction on Biomedical Engineering*, vol. 54, pp. 1031-1041, 2007.

[7] P.J. Chen, W.C. Kuo, W. Li, Y.J. Yang, and Y.C. Tai, "Q-enhanced Fold-and-bond MEMS Inductors", in *Proc. IEEE Int. Conf. on Nano/Micro Engineering and Molecular Systems*, Sanya, China, January 6-9, 2008, pp. 869-872.

[8] W. Li, D.C. Rodger, and Y.C. Tai, "Implantable RF-coiled Chip Packaging", in *Proc. IEEE Int. Conf. on Micro Electro Mechanical System*, Tucson, USA, January 13-17, 2008, pp. 108-111.

[9] A.C. Dupont, S.D. Bagg, J.L. Creasy, C. Romano, D. Romano, F.J.R. Richmond, and G.E. Loeb, "First Clinical Experience with BION Implants for Therapeutic Electrical Stimulation", *Neuromodulation*, vol. 7, pp. 38-47, 2004.

TRANSDERMAL POWER TRANSFER FOR IMPLANTED DRUG DELIVERY DEVICES USING A SMART NEEDLE AND REFILL PORT

Allan T. Evans[1], Srinivas Chiravuri[2], and Yogesh B. Gianchandani[1]

[1]Department of Electrical Engineering and Computer Science, University of Michigan, Ann Arbor,
[2]Department of Anesthesiology, University of Michigan, Ann Arbor

ABSTRACT

This paper describes a pair of micromachined components that work together for transdermal power (and data) transfer into implantable drug delivery devices. In particular, it describes a smart refill port that is electrically accessed with a mating, multi-pole, plug-in needle. These components are fabricated entirely from molded PDMS, Parylene, Kapton, and micro-electro-discharge machined stainless steel. The component pair demonstrates power transfer with low current leakage into both dry and saline ambient, and it has been used to recharge batteries with currents ranging from 10 mA up to 500 mA while temperature change is monitored.

INTRODUCTION

Motivation

Implantable drug delivery devices (IDDDs) are used to treat medical conditions, such as spasticity or chronic pain, through targeted delivery of medications like baclofen or local anesthetics. Conventional programmable IDDDs use 25-50% of the implanted device volume for a battery that is intended to last the entire duration (5-10 years) of the implant. However, in typical IDDDs [1] medication is typically refilled every 10 weeks by transdermal injection into a subcutaneous refill port (Figure 1). This research is motivated by the consideration that the overall volume efficiency of an IDDD, which is critical to its placement and usability (particularly in pediatric cases), can be improved substantially if the conventional battery is replaced with a smaller battery that is recharged. It is preferable that the recharging occurs in the same session that the drug reservoir is refilled, although not necessarily at precisely the same time [2]. While wireless power transfer is possible for very low-power applications [3], DC recharge capability [4] offers high current levels and may be more suitable for IDDDs (Fig. 2).

Power Transfer

While it is relatively easy to transfer data across most electrical connections, it is more difficult to transfer high levels of power. One problem is resistive heating in the conductors. The heat energy generated (Q) is due to the power (P) transferred through the conductor over a set amount of time (t). The power is related to the current (I) being transferred through a conductor with a particular resistance (R). The resistance is due to the resistivity of the conductor (ρ), the Length (L), and the cross sectional area (A) of the conductor. Assuming a constant material and length, the heat generated for any particular recharging current is inversely proportional to the cross sectional area of the conductor (Eqn. 1).

$$Q = Pt = (I^2 R)t = (I^2 \rho L t)/A \qquad (1)$$

In order to maximize the amount of power that can be transferred across a refill port, the conductor through the

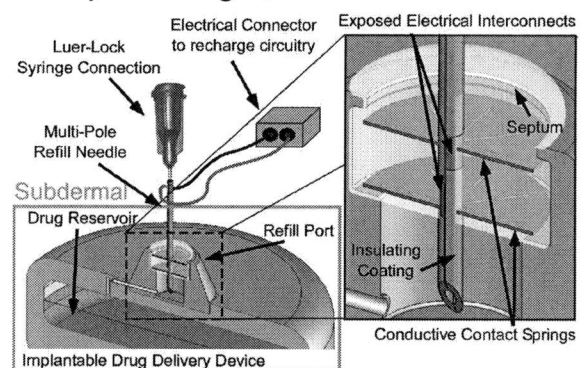

Figure 1: The system view: A two-pole needle is inserted into the refill port of a drug delivery device. Inset: A close view of the two needle halves making electrical contact with springs inside the septum.

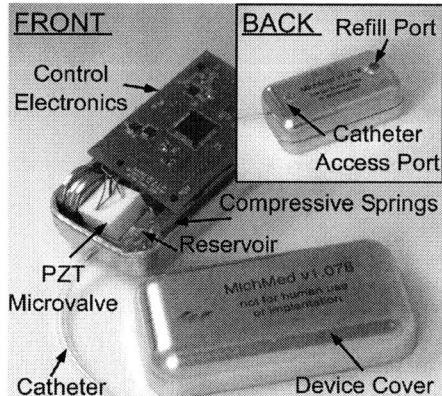

Figure 2: A photo of the front of an assembled microvalve-regulated drug delivery device with the back side refill port shown inset [8].

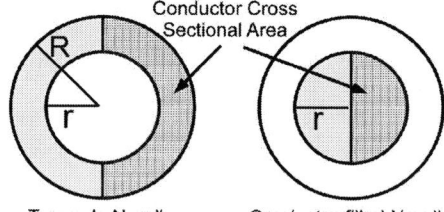

Figure 3: Two possible needle configurations. Either the needle is split (left) or it is filled with a conductor (right).

needle should have the largest possible cross sectional area. The two methods of creating conductors with the largest cross sectional area are either using the needle itself, or filling the needle with a conductor (Fig. 3).

Typical refill needles used in IDDDs range in size from 22 gauge to as narrow as 28 gauge. The ratio of the inner diameter (r) to outer diameter (R) ranges from 0.55 to 0.6 across this needle range. Even in the worst case, the cross sectional area of the conductor of the two-pole needle is 1.78 times that of a filled needle. This means that a two-pole needle would generate about half of the

978-1-4244-2977-6/09 $25.00 © 2009 IEEE   252

heat a needle with a filled conductor would for the same recharge current. For both this reason and the retained ability to transport liquids, the transdermal power transfer system is designed to provide a mating scheme that utilizes a multi-pole needle.

Device Concept

The refill ports of most implantable drug delivery devices are composed of a polymer septum through which the needle enters the device. The polymer is relatively thick (2-5 mm), and is designed to reseal itself after the refill needle is removed from the device. Below the septum is a small open volume that is connected to the reservoir of the implant. The thickness and insulating properties of the septum make it an appealing candidate for modifications that would allow power transfer.

A two-pole needle and a power transfer refill port (Fig. 1) are designed to mate within the septum. The two-pole needle is designed so that each half is an independent conductor that is isolated from the other half, the exterior, and the interior lumen of the needle through which the drug would flow. The needle halves are exposed on the exterior at two separate locations along its length. These locations correspond to the position of two contact springs within the septum.

The ports in conventional IDDDs are typically accessed by non-coring needles that are curved at the tip (Huber needles). The system mates when a multi-pole, non-coring needle punctures the silicone septum of the refill port and is advanced until the tip of the needle reaches the metal base plate at the bottom of the port. (This metal base plate is electrically floating.) This design architecture has several advantages. The location of the springs and the exposure "windows" of the needle are designed so the windows align with the springs when the needle is fully inserted because the tip presses against the bottom of the refill port. Also, by using two separate springs located at different heights, no rotational alignment of the needle is necessary to make electrical contact. This prevents the need for medical professionals to twist the needle upon insertion, and it also prevents mating incorrect conductors to each other. This topology is also particularly safe because the springs located in the non-conductive septum and the conductors of the needle are electrically isolated from the casing of the port, the patient's body, and the drug being refilled.

DESIGN AND FABRICATION

Two-pole Needle

The two major system components are the non-coring needle and the refill port. Presently, implantable drug delivery refill ports are designed to be accessed with Huber needles ranging from 22 gauge to 28 gauge. Two-pole Huber needles (700 µm ø, 26 gauge) for use with a smart refill port are fabricated by lapping two stainless steel needles in half using an oil-based diamond slurry (Fig. 4). The needle halves are completely coated in 2.8 µm thick Parylene to electrically isolate the halves from each other and from the environment. Parylene is selectively removed from areas near the needle tip to create the contact windows that mate in the refill port and from the back of the needle to allow contact with the power source. The needle halves are then aligned and bonded [5] (Fig. 5), with further insulation and sealing

strength provided by inserting the needle into a Kapton tube.

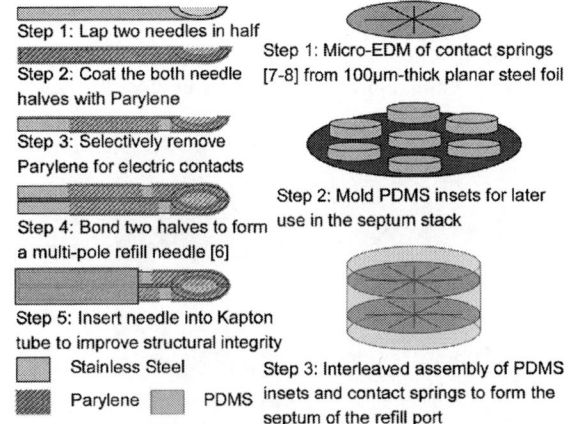

Figure 4: The needle and septum are fabricated from biologically compatible materials. The fabrication process aligns the conductive needle windows to the springs so they make contact upon insertion.

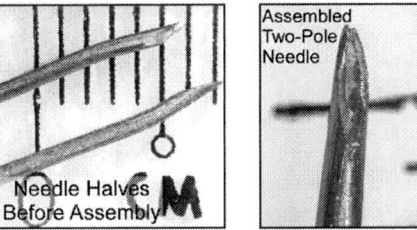

Figure 5: Photographs of 26 gauge coated needle halves before and after assembly taken on a white ruler.

Septum Springs

The refill port requires a septum designed with contact springs at specific heights. The electrical contact springs are fabricated from 100 µm thick stainless steel by micro-electro-discharge machining (µEDM) [6-7]. The contact springs are 5.2 mm in diameter with four quadrants separated by 300 µm wide slots (Fig. 6). These springs press against the needle as it is inserted. The pressure forces the needle toward the middle of the refill port, and it also improves the lead transfer conductance by maintaining pressure at the spring/needle junction. The symmetrical nature of the contact springs prevents the need for needle alignment. Additionally, the springs are supported by the septum polymer and return to their initial positions after the needle is removed. This allows for multiple recharging sessions to occur using a single port. The springs are located at specific heights within the polymer septum for needle alignment purposes.

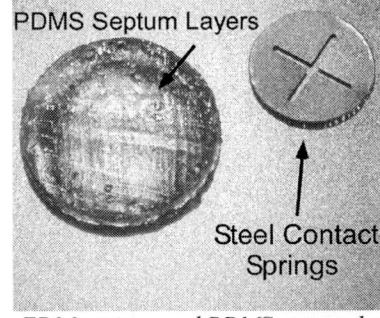

Figure 6: µEDM springs and PDMS septum layers.
Polydimethylsiloxane (PDMS; or alternatively silicone) is

used as the septum material. It is shaped using circular molds of specific heights and cured for 30 min. at 70°C to form septum layers with the correct spring spacing. The PDMS and springs are stacked and inserted into the refill port (Fig. 7) to allow for electrical coupling between a needle and the implantable drug delivery device (Fig. 2).

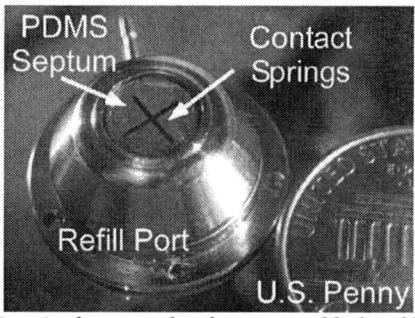

Figure 7: A photograph of an assembled refill port in which the top contact spring of the septum is clearly visible. The port is pictured here with a US Penny.

EXPERIMENTAL RESULTS
Insertion Testing

The quality of the electrical connection between the two-pole needle and the port can be evaluated by measuring the resistances between the two elements in both dry and wet environments. In mating both normal Huber needles and two-pole needles with assembled refill ports, testing can identify if electrical connections are made as expected (Fig. 8). Resistances between the two poles of the needle (A and B), the two septum springs (C and D), and the electrically floating metal base plate were measured as both types of needles were advanced into the septum. Typical resistances for a fully inserted normal needle were about 0.5 Ω from A to C and D; and 1 Ω between the A and E. The two-pole needle has slightly higher resistances from A to C and B to D (~0.7 Ω) but maintains isolation from A to B (1.6 MΩ) and from A to E (>10 MΩ). The resistance tests also confirm that the C and D are electrically insulated from each other and from the E. The slightly higher resistances attributed to the

two-pole needle are likely due to the decreased cross sectional area of the needle, and the decreased contact area between the conductors of the split needle and the metallic contact springs. A number of combinations of contact points were tested during multiple insertions, and the results exhibited expected contact properties.

In order to further evaluate the performance of the mating pair, saline was introduced into the needle lumen and the port cavity. Insertion tests, similar to those above, were conducted in the wet ambient. As shown in Figure 8, resistances from A to C and B to D were low (< 2 Ω), and electrical isolation was maintained from B to C and A to E (> 2 MΩ). These results indicate that the parasitic current leakage paths are minimal. Additionally, no electrolysis was observed in either the needle or in the mated port during characterization.

Battery Recharging

Recharging batteries using the needle-port system generates resistive heating in the needle and at interface between the needle and the contact springs. Two 1.2 V NiMH AA batteries were recharged using various currents at room temperature in a dry environment. The air environment restricts the thermal conductance of heat away from the refill port as compared to a refill port located subcutaneously *in vivo*. Additionally, a lack of liquid located within or flowing through the lumen, reduces the thermal capacitance of the system and increases the temperature change of the device for any particular power transfer. Typical room temperature recharging with power transfer rates ranging from 10-500 mA demonstrate the expected temperature rises at needle-septum interface (Fig. 9) and out on the port housing (Fig. 10) of the refill port. The temperature changes at the needle septum entry point exhibit much higher swings because the PDMS is not thermally conductive, so the principal mechanism of heat dissipation occurs via conduction through the air. This is a location in the system that will experience one of the largest temperature

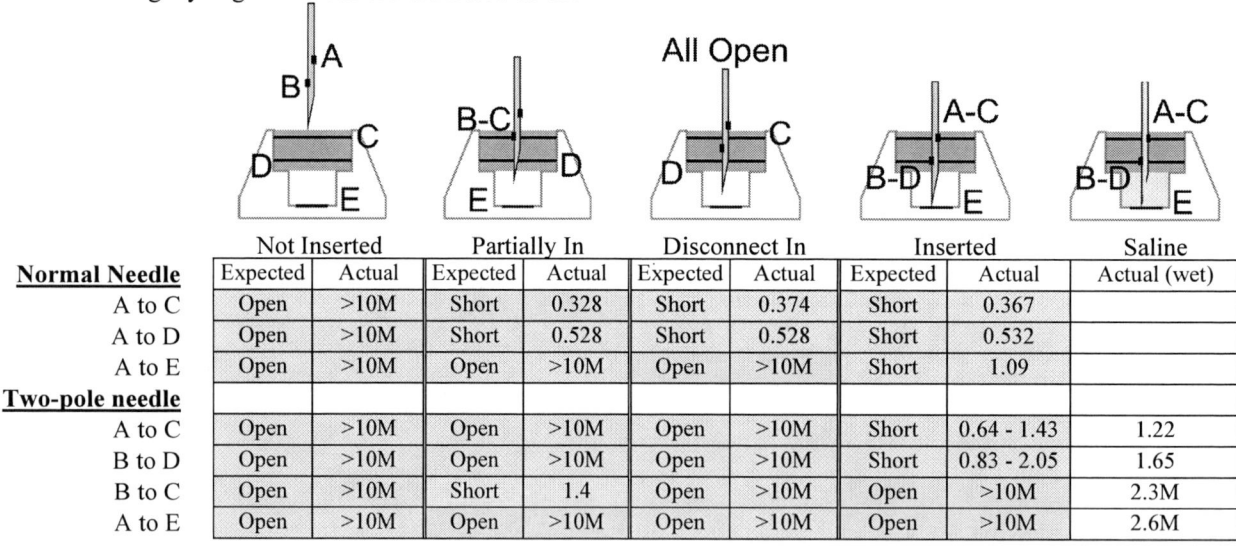

	Not Inserted		Partially In		Disconnect In		Inserted		Saline
Normal Needle	Expected	Actual	Expected	Actual	Expected	Actual	Expected	Actual	Actual (wet)
A to C	Open	>10M	Short	0.328	Short	0.374	Short	0.367	
A to D	Open	>10M	Short	0.528	Short	0.528	Short	0.532	
A to E	Open	>10M	Open	>10M	Open	>10M	Short	1.09	
Two-pole needle									
A to C	Open	>10M	Open	>10M	Open	>10M	Short	0.64 - 1.43	1.22
B to D	Open	>10M	Open	>10M	Open	>10M	Short	0.83 - 2.05	1.65
B to C	Open	>10M	Short	1.4	Open	>10M	Open	>10M	2.3M
A to E	Open	>10M	Open	>10M	Open	>10M	Open	>10M	2.6M

Figure 8: Stages of insertion with resistances color coded to expected states for normal and split needles both dry and completely filled with saline after insertion. (*All resistances are in Ω)

increases for a specific recharging current. The battery voltage was monitored (Fig. 11) along with temperature for varying current to confirm the batteries were recharging at relative rates. No measureable heating was observed for currents of 100 mA or less. The battery voltage increases coupled with the low heating levels at higher currents, like 500 mA, (< 15°C septum and < 2°C casing) represent power transfer rates necessary to fully recharge a battery used in an implantable drug delivery device during the reservoir refill session.

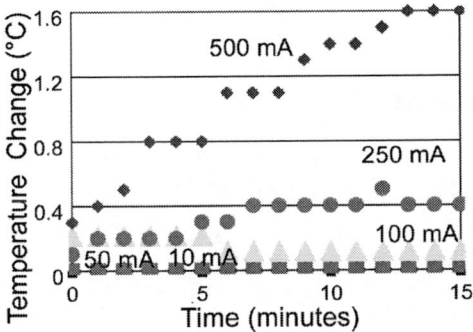

Figure 9: Temperature change at the septum entry point over time for recharging currents from 10-500 mA. The test was conducted in an air ambient environment with a baseline temperature of 22.4 °C.

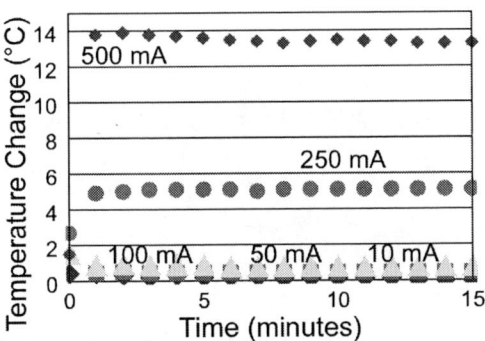

Figure 10: The temperature change of the exterior of the port housing for battery charging currents ranging from 10-500 mA. The temperature increase has resolution of 0.1°C and was conducted in ambient air with a temperature of 22.4 °C.

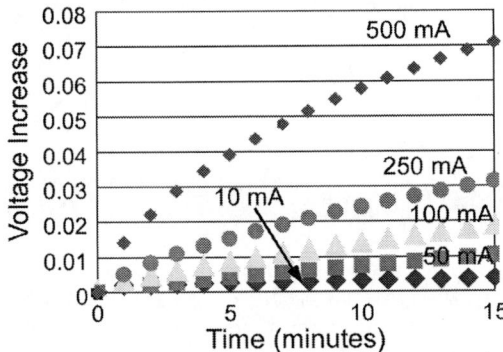

Figure 11: The voltage increase of a NiMH 1.2V AA battery as it is being recharged across a refill port with various charging currents. Charging profiles match expected values, and demonstrate power transmission with acceptable heating rates for a smart needle and port to be used in IDDDs.

CONCLUSIONS

A mechanism for transferring power across the needle through the refill port of an IDD has been explored to maximize the power transfer rate. A two-pole needle was fabricated from bio-compatible components, and a refill port was designed with stainless steel contact springs to mate with the needle upon insertion.

During the various stages of needle insertion, the two-pole needle exhibited expected resistive characteristics with little current leakage when filled with saline. The refill port springs self-aligned with the needle and made ohmic contact when the needle was fully inserted without additional alignment. NiMH batteries were recharged using power transfer across a refill port mated with a two-pole needle. Recharging currents ranging from 10 – 500 mA were transferred for typical refill times of 15 minutes while changes in temperature were measured at various points of the system. The relatively low increases in temperature for even the highest current levels indicate that this topology can be used in implants. This allows for IDDDs to be made smaller because they will not need batteries designed to power the device for 5 - 10 years, but they will be able to use batteries that power the device for 26 weeks. With more sophisticated needle and port designs, high speed parallel data transfer may also be possible. The reduced communication circuitry and battery sizes could result in a total IDDD size reduction of up to 40%.

REFERENCES

[1] S.J. Hassenbusch, et al., "Polyanalgesic Consensus Conference 2003: an update on the management of pain by intraspinal drug delivery-- report of an expert panel," *J. Pain Symptom Manage*, 2004. **27**(6): p. 540-63.

[2] M. Carmichael, "The Changing Science of Pain," *Newsweek*, June 4, 2007 pp. 40-47

[3] Boveja, et al, "Method and system for providing pulsed electrical stimulation to provide therapy for erectile/sexual dysfunction, prostatitis, prostatitis pain, and chronic pelvic pain," U.S. Patent 7,330,762, Feb. 12, 2008

[4] R. Vipul, "Vipul's lifetime lifeline permanent pacemaker and implantable cardioverter-defibrillator," U.S. Patent 7,239,917, Jul. 3, 2007

[5] H. Kim, and K. Najafi, "Characterization of Parylene-assisted wafer bonding: Long-term stability and influence of process chemicals," *Transducers '05*, pp. 2015-2018, Seoul, Korea, June, 2005

[6] K. Takahata, and Y.B Gianchandani, "Batch Mode Micro-Electro-Discharge Machining", *IEEE/ASME J. Microelectromechanical Systems*, 11(2), pp. 102-110, 2002

[7] M.T. Richardson, and Y.B Gianchandani, "Achieving precision in high density batch mode micro-electro-discharge-machining", *J. Micromech. Microeng*, 18 015002 (12pp), January 2008

[8] A.T. Evans, J.M. Park, S. Chiravuri, and Y.B. Gianchandani "Dual drug delivery device for chronic pain management using micromachined elastic metal structures and silicon microvalves," *Micro Electro Mechanical Systems, 2008. pp.* 252-55

A FLEXIBLE REGENERATION MICROELECTRODE WITH CELL-GROWTH GUIDANCE

Riho Gojo[1], Naoki Kotake[2], Takafumi Suzuki[3], Kunihiko Mabuchi[2,3] and Shoji Takeuchi[1,4,5]

[1]Institute of Industrial Science, The University of Tokyo,
[2] Dept. of Advanced Interdisciplinary Studies, Grad. School of Engineering, The University of Tokyo,
[3]Dept. of Information Physics & Computing, Grad. School of Info. Sci. & Tech., The University of Tokyo
[4]BEANS project, METI and [5]PRESTO, JST, JAPAN,

ABSTRACT

In this study we fabricate a cell-guide on a thin disc-shaped probe with numerous microholes to be implanted between the severed stumps of axons; the cell-guide directs the regenerating axons though the holes in our probe (Fig. 1(a)). Since cell growth is greatly influenced by the topography of the substrate [1-3] we design a SU8 microfluidic cell-guide with a particular shape (Fig. 1(b)) to prevent the random spreading of cells by directing cell growth through the holes. This guide may reduce the chance of short-circuits of the probe by cells that span two different electrodes. Furthermore, as the cells are directed to grow through the holes, we can measure their electrical potential with gold electrodes that we micro-fabricate.

1. INTRODUCTION

When a peripheral nerve fiber is severed or badly damaged, the body begins to repair the damaged region between the two nerve stumps with different processes from specific directions. Initially the end of the distal region begins a process called Wallerian degeneration where the end of the distal nerve begins to break-down, while simultaneously forming a fibrinous bridge; this bridge is made from fibrin and eventually links the two nerve stumps. It is thought that axon growth is inhibited until this bridge reaches the proximal nerve stump. Once connected, by the bridge, axons begin to grow down along it towards the distal nerve stump [4].

Measuring the current from single nerve cells is extremely difficult due to the inability to isolate individual cells and because of the difference in length-scale between the cells and the electrodes. Therefore, since each individual nerve cell must grow linearly to bridge the gap between the severed stumps during nerve fiber regeneration, we have a unique opportunity to isolate individual cells. By placing a microfabricated electrode in the path of the regenerating nerve cells, we can then isolate individual cells and measure the current through them once the nerve fiber has healed.

The advantage of this technique is that the electrode is in position to make continuous measurements throughout the regeneration process. Furthermore, since the electrode is fully integrated into the nerve bunch, we can both stimulate the nerve cells by applying voltage pulse and take long-term measurements.

Another method of interfacing nerves to electrodes entails using a 2D planar silicon sieve to enable the regeneration between the two sides of a severed nerve [5-7].

The sieve enables nerve fibers to regenerate through metalized holes; these holes serve as the electrodes. However, since these sieve electrodes are 2D, they provide

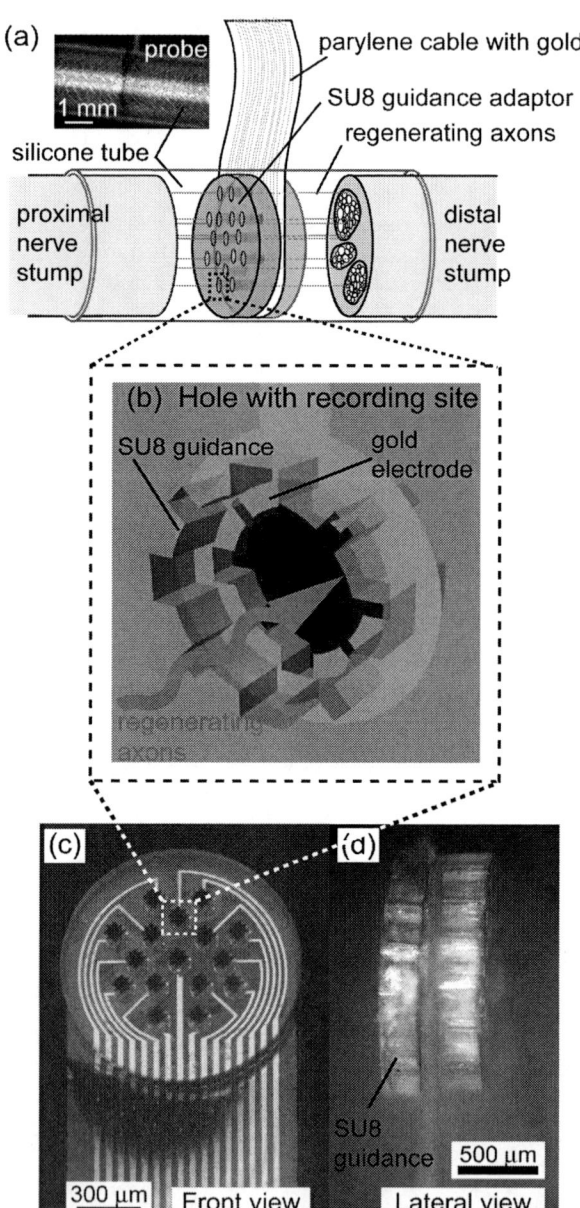

Figure 1: Schematics and photographs of the cell-growth guidance probe. (a) We designed a probe with 16 micrometer-sized holes in order to direct regenerating axons at a severed gap of stumps through the holes. (b) The gear-like shape of the hole for the regenerating axons, to enhance cell growth by the topographical advantage. (c-d) Photographs of the fabricated probe.

no guidance for nerve cells that reach a region of the electrode without holes. The cells simply grow randomly and find the holes by chance. In addition, since these electrodes and cables are made from hard materials, such as silicon and glass [8-9] as well as the devices is bigger than the target nerve, the chance for causing further damage to the surrounding tissue is high.

Here, we present a technique to fabricate an electrode capable of simultaneously enabling current measurements and application of stimulation to nerve cells during and after nerve cell regeneration. In addition, we incorporate guidance channels into the surface of the probe to facilitate the cell's rapid discovery of the through-holes.

2. DESIGN AND FABRICATION

Target nerve

We focus on the tibial nerve. Because the tibial nerve in rats is a useful nerve to target as it controls the manner in which the rat walks. Normal rats walk on their toes, however, if the tibial nerve is severed, it begins to walk flat-footed. Upon recovery of the tibial nerve, the rat once again begins to walk normally. We can use a walking track analysis of the rats to evaluate the function of the peripheral tibial nerve in rats [10] and as a result we can confirm whether the rat's nerve has healed without operating.

Cell-Growth Guidance Design

It is well known that cells recognize and are affected by the topography of the substrate on which they are growing [1-3]. In addition, open microfluidic grooves patterned in SU8 have shown utility in cultivating neural cells to guide axon outgrowth [13]. Utilizing these results, we designed the recording site holes with topographical microfluidic guidance (Fig. 1(b)). The SU8 adaptor guides the direction of cell growth and simultaneously prevents short-circuits. The guidance was designed according to the size of tibial nerve (1.2 mm in diameter).

Fabrication Process

Our flexible probe is composed of a thin biocompatible polymer film (Parylene) with 16 holes, within which we encase an electrode [11-12]. We sandwich this probe between two cell-guides made from SU8 (Fig. 2).

To fabricate our electrode, we first pattern a gold layer on a parylene film (Fig. 2(a-b)). The thickness of the gold layer is 400 nm. Gold and parylene are patterned with photo resist (S1818) with standard photolithography. Parylene is a thin, biocompatible polymer film. Next, we deposit and pattern a second parylene film (Fig. 2(c-e)). Each parylene film has a 20 µm thickness.

Then we fabricate the SU8 guidance holes on one side using a photolithography process (Fig. 2(f-g)) and peel the electrode off from the wafer (Fig. 2(h)). Finally we complete the process by fabricating the guidance holes on the other side (Fig. 2(i-l)).

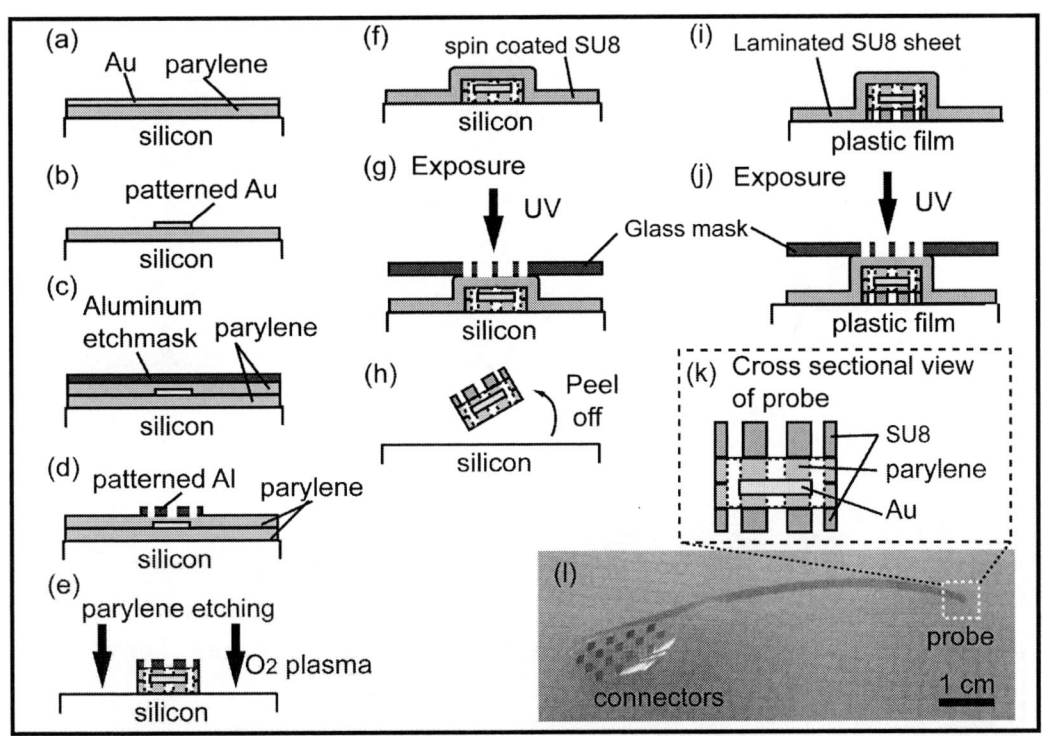

Figure 2: Schematic diagrams of fabrication process. (a-b) First, a gold layer was patterned on a parylene film and (c-e) next the second parylene film was deposited and patterned. (f-g) Then the SU8 guidance was fabricated on one side with a photolithography process, (h) followed by peeling off from the wafer. (i-l) Finally the process is completed by fabricating the other side of the guidance.

978-1-4244-2977-6/09 $25.00 © 2009 IEEE   257

3. EXPERIMENTAL RESULTS

We observed our probe using a scanning electron microscope (SEM), which allowed us to evaluate the quality of the probe's features; an image of the probe is shown in Fig.3(a). The impedance of the probe's 16 channels are approximately 1.6 MΩ and this value is unaffected by the presence of SU8 (Fig. 3(b)); however, there were channel-to-channel differences in impedance. As for the electrode with SU8, the interelectrode impedance is high. This implies that the interelectrode electric leakage is low (Fig. 3(b)) since there is sufficient distance between neighboring recording sites. Furthermore, each recording site is insulated enough by the SU8 guide.

We examined the adhesive properties of 3T3 cells on the SU8 cell-guide (Fig 4(a-b)). 3T3 cells are mouse fibroblast-like cells. We cultured the cells at 37 °C with 5 % CO_2 in Dulbecco's modified eagle medium (DMEM, 11885, Gibco), 10 % fetal bovine serum (FBS, F7524-500 ml, SIGMA), and 5 % L-Glutamine-Penicillin-Streptomycin. We find that without a guide, the cells simply move about randomly on the probe surface (Fig. 4(a)). However, with our guide, we find that the cell motion is directed through the holes (Fig. 4(b)). Utilizing the cell-guide, we are currently conducting *in vivo* experiments with our implanted electrode in a severed rat tibial nerve (Fig. 5). We first place the electrode in a silicone tube and the nerve and the silicone tube are sewn

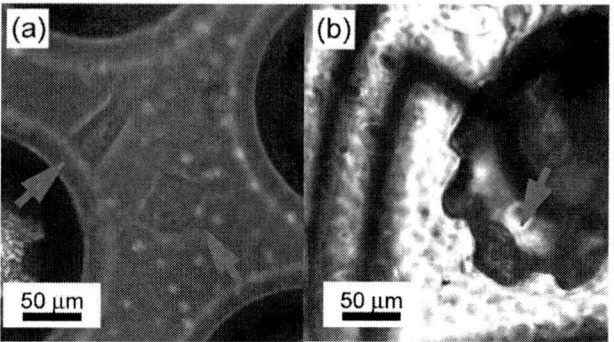

(b)	vs the external reference electrode	between the two electrodes on one probe
A	1.58 ± 1.00	2.19 ± 1.05
B	1.60 ± 0.61	3.21 ± 1.14

Figure 3: *Influence of the SU8 guidance. (a) SEM image of the SU8 surface. (b) Impedance data of the electrodes: Row A shows the impedance without SU8. Row B shows the impedance with SU8.*

Figure 4: *Cell-adhesive images by a phase contrast microscopy. (a) Without the SU8 guidance, the flat topography of the gold/parylene substrate let a random spread of cells. (b) With the SU8, cells adhere to the corner of the walls, resulting in directing the cells through the holes.*

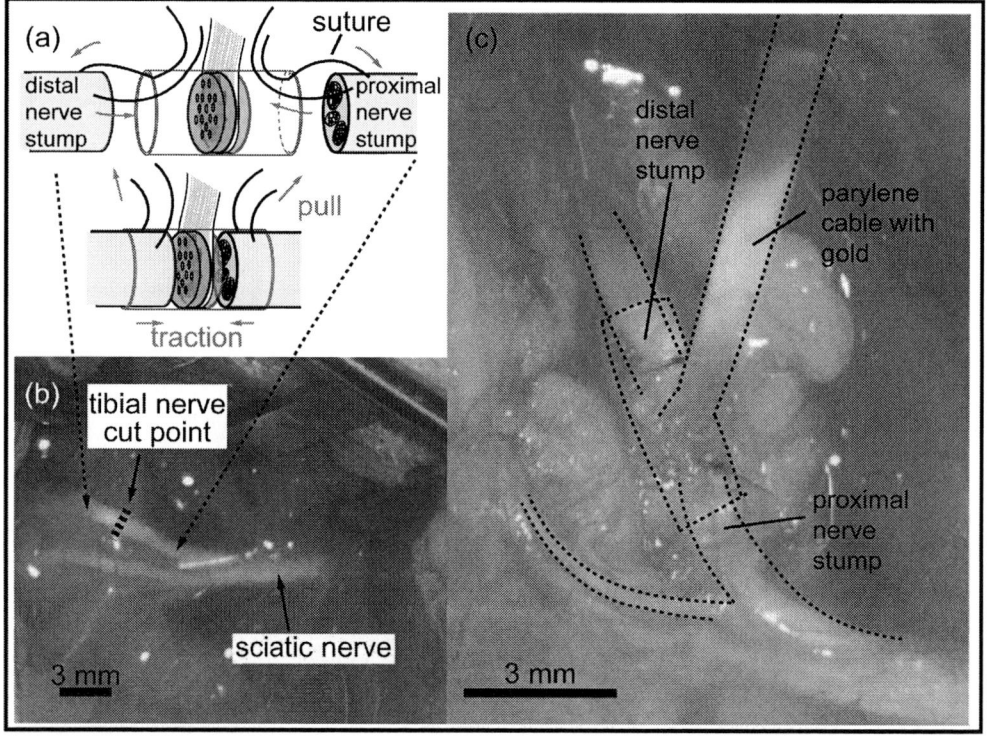

Figure 5: *Photographs of the implant experiment. (a) Schematic illustration of the implant process. (b) Image of the tripartite division of a rat sciatic nerve. The tibial nerve was cut and led into the silicone tube. (c) Image after the implant of the probe.*

together with Nylon suture string. Since the electrode is very thin it can be stored inside of the body without undue stress. Furthermore, although the probe is only 40 μm thick, it is strong enough to withstand the operation without breaking.

Finally, we conceptually describe a future manifestation of our probe. We plan to increase the number of simultaneous measurements that we can take by stacking four electrodes end-to-end (Fig. 6(a-b)). Furthermore, since we fabricate the probe with a constant thickness, we can measure the potential at equal intervals within each passage. One further step that we take is the incorporation of additional microfluidic channels perpendicular to while still connected to the original cell-guide holes; these channels are for the injection of chemicals and for microfluidic neural-cell guidance (Fig. 6(c)).

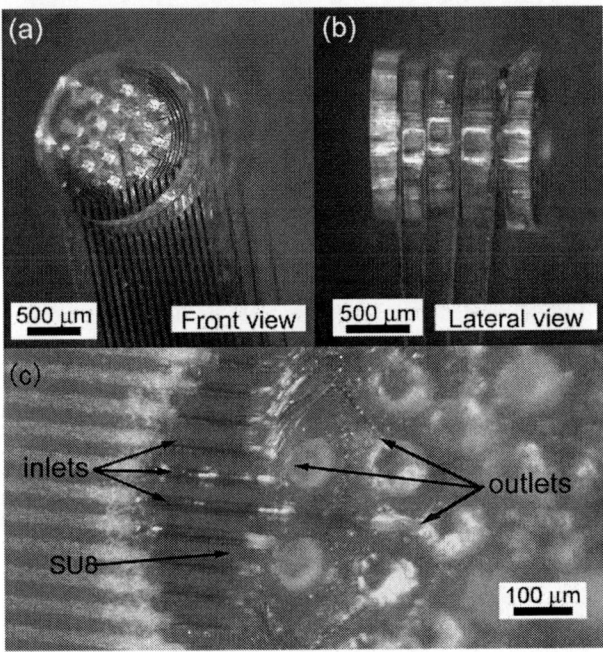

Figure 6: Applications and perspectives. (a-b) The fabricated probes are able to pile-up to increase the number of the sensing electrodes easily. (c) Newly proposed internal microfluidic-channels, along with the gold leads, allow injection of drugs/chemicals through the cell-growth guidances.

4. CONCLUSION

In summary, we propose a regeneration-type electrode that shows important advances over current neural electrodes. By incorporating microfluidic cell-guides, we may reduce the potentiality of short-circuits of the probe by cells that span two different electrodes and can ensure that the cells are directed to grow through the holes. Furthermore, these probes are made using micro-fabrication with additional channels that are easily incorporated, and they can be mass-produced.

ACKNOWLEDGEMENT

This study was partly supported by Health and Labour Sciences Research Grant (H20-Nano-general-003) from the Ministry of Health, Labour and Welfare of Japan.

REFERENCES

[1] J. Y. Wong, et al., Balance of chemistry, topography, and mechanics at the cell－biomaterial interface: Issues and challenges for assessing the role of substrate mechanics on cell response, Surface Science, vol. 570, pp. 119-133, 2004.

[2] Y. Tsuda, et al., Thermoresponsive microtextured culture surfaces facilitate fabrication of capillary networks, Adv. Mater., vol. 19, pp. 3633-3636, 2007.

[3] Y.-C. Lo, et al., Neural guidance by open-top SU-8 microfluidic channel, IEEE MEMS, Maastricht, January, pp. 671-674, 2004.

[4] W.L.C. Rutten, Selective electrical interfaces with the nervous system, Annu. Rev. Biomed. Eng., vol. 4, pp. 407-452, 2002.

[5] Q. Zhao, et al., Rat sciatic nerve regeneration through a micromachined silicon chip. J. Biomater., vol. 18, pp. 75-80, 1997.

[6] A. Blau, et al., Characterization and optimization of microelectrode arrays for in-vivo nerve signal recording and stimulation. Biosens. Bioelectron., vol. 12, pp. 883-92, 1997.

[7] X. Navarro, et al., Stimulation and recording from regenerated peripheral nerve through polyimide sieve electrodes. J. Peripher. Nerv. Syst., vol. 2, pp. 91-101, 1998.

[8] G. T. A. Kovacs, et al., Regeneration microelectrodes array for peripheral nerve recording and stimulation, IEEE Trans. on BME., vol. 9, pp. 893-902, 1992.

[9] T. Akin, et al., A micromachined silicon sieve electrode for nerve regeneration applications. IEEE Trans. on BME., vol. 41, pp. 305-313, 1994.

[10] G. M. T. Hare, et al., Walking track analysis: a long-term assessment of peripheral nerve recovery. Plast Reconstr Surg, vol. 89, pp. 251-258, 1992.

[11] T. Suzuki, et al., Flexible microelectrode for interfacing regenerating peripheral nerves, Proc. of 19th International IEEE EMBS., 1997 (CD-ROM).

[12] S. Takeuchi, et al., 3D Flexible multichannel neural probe array, Journal of Micromechanics and Microengineering, vol. 14, pp. 104-108, 2004.

ADAPTIVE BLOOD CELL SEPARATION USING COOPERATION OF SERIALLY CONNECTED MEMBRANE FILTERS

T. Kobayashi[1], D. Kato[1], H. Koga[1], K. Morimoto[1]
M. Fukuda[2], Y. Kinoshita[2], H. Yoshida[2], and S. Konishi[1]
[1]Ritsumeikan University, Kusatsu, Shiga, JAPAN
[2] Research & Development Laboratory, Nipro corporation, Kusatsu, Shiga, JAPAN

ABSTRACT

This paper proposes a cooperative operation of serially connected membrane filters toward adaptive blood cell separation system in order to overcome a restriction of a single membrane filter. A cooperation of serially connected membrane filters allows that downstream filters extract blood plasma from residual blood at upstream filters. Consequently, it becomes possible to adapt filtering characteristics to changing properties of blood. We focus on trans-membrane pressure (TMP) in order to prevent hemolysis. Our strategy can be realized as a miniaturized PDMS fluidic chip. Our laboratory experiment using a prototype shows that plasma extraction efficiency is improved from 34% to 75%. Toward an integrated system, this paper successfully demonstrates multiple filters integrated into a PDMS fluidic chip.

INTRODUCTION

Blood tests are vitally important for medical diagnostics. A device for blood analysis should have properties such as low invasiveness, high-throughput, and high accuracy for patient quality of life (QOL). It is important to facilitate quantitative analysis from the smallest possible volume of the whole blood in minimally invasive blood testing. Wide distribution for commercial production will require to save the manufacturing cost of such devices.

Micro total analysis systems (µ-TAS)-based technology shows promise for establishing a highly efficient way of separating blood cells and plasma. Many types of blood separation devices have been developed by µTAS. A filter-based blood cell separation device has been proposed in our group [1]. Horiike et al. [2] achieved both separation and transportation of blood on one chip using centrifugation. Khumpuang et al. [3] reported plasma separation on a silicon (Si) chip using a capillary effect that did not require electricity or mechanical pressure. Miki et al. [4] developed a separation device consisting of multi-layered porous filters with 2-nm-diameter pores, which functions as a small artificial kidney. In the present study, a blood cell separation device using serially connected membrane filters is proposed. The experimental results are shown that the particular emphasis is put on the importance of adjusting filtration characteristics to blood flow properties.

CONCEPTUAL DESIGN
Blood Cells Separation Device using Serially Connected Membrane Filters

A schematic view of the proposed separation device is shown in Figure 1. The device consists of several single filters and fluidic channels as shown in Fig. 1(b). It is one problem of existing devices that cell clogging inhibits both blood flow and further filtration. This device makes repeated use of the residual blood un-separated at the previous filtration step to improve collection efficiency of separated plasma. The ratio of separated volume to un-separated volume can be adjusted by controlling the ratio of P_l(negative pressure in lower channels) to P_u (negative pressure in upper channels) as shown in Fig. 1(b), since the proposed device has upper and lower channels. This method allows the simultaneous separation of the plasma and transportation of the residual blood.

Adjusting Filtration Characteristics to Blood Flow Properties

In this method, the concentration of the whole blood increased with the number of filtration sections, because the plasma is continuously collected from the lower outlet and un-separated blood is ejected from the upper outlet into the next filter device. High hematocrit blood tends to undergo hemolysis and cause filter clogging because the shear stress, trans-membrane pressure (TMP), increased due to high viscosity. Colton et al. [5] derived an equation of the relationship between the specifications of hollow-fiber filter and plasma flow rate or shear stress at the wall of membrane pores, based on the assumption that the plasma-separating flow rate is defined by the polarization of red blood cell concentration. With high TMP, red blood cells are more likely to flow into the pores of the filters, and the hemolysis is induced. To prevent hemolysis during filtration, shear stress of blood flow should be maintained lower than ~1000 Hg (6.6 - 9.2 kPa) [6-8] , and TMP should be lower than 50-70 mm.

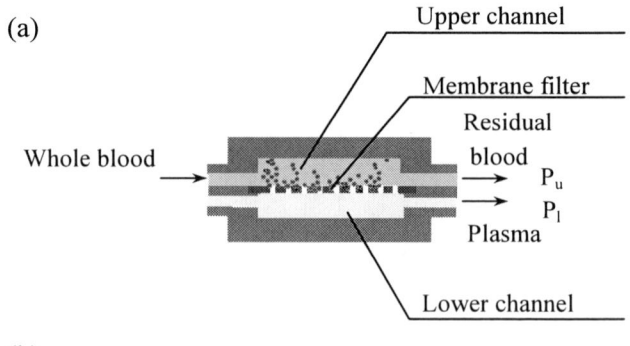

Figure 1: Principle of our proposed blood cells separation device.

Therefore, a filter device must be designed to adapt to the greater viscosity at each filtering section. Clogged cells result in not only the decrease of filtration fluxes but also the increase of TMP. Thus, to avoid hemolysis, the suppression of cell clogging is important.

EXPERIMENTS & RESULTS

Rapid prototype

As shown in Fig. 2, prototypes of the separation device were prepared for preliminary testing. The prototype was fabricated using rapid prototype equipment (Eden, Fasotec Co. Ltd.). A structural design of the prototype is shown in Fig. 2(a). The prototype contained an inlet, outlet, and valves corresponding to the upper and lower fluidic channels. Valves were integrated to allow the evaluation of blood conditions or volume at each section. Membrane filters (filter paper No.2, Advantec) were set into the prototype devices. The pseudo-whole blood consisted of 55% pseudo plasma and 45% blood cells. The 40% glycerin diluted in water was used as pseudo plasma. Sephadex G-25(GE Healthcare Bio-Sciences KK, Sephadex G-25 Superfine, $\Phi15\sim88$ μm) was used as pseudo-blood cells.

Change in Blood Flow Properties in Proposed Device and Influence on Separation Process

First, hematocrit was evaluated before and after separation. Results are shown in Fig. 3. Fig. 3(a) shows the change of hematocrit level in filtration. As shown in Fig. 3(b), whole blood was sampled at regions A and B using capillary tubes for hematocrit measurement and the height of sediment regions. The hematocrit level increased from 20% to 35%. The pseudo-whole blood was introduced into the upper and lower fluidic channels with the use of vacuum. The negative pressures supplied (P_u and P_l) were approximately 20 kPa and 40 kPa, respectively. Future devices need to be designed so that filtration volume and operation time can be adapted to TMP of each filter under the hemolysis threshold value using effective area, total number of hollow fibers, or the radius of the hollow fibers of each filter device.

Evaluation of Sampled Volume of Plasma

The dependence of the number of connected filters on sampled plasma volume is shown in Figure 4. The effective area and chamber volume of each filter were 30 mm^2 and 180 μl for a single connected type, 15 mm^2 and 90 μl for a double connected type, and 10 mm^2 and 60 μl for a triple connected type. The total effective area of each type was set to the same value. The upper chamber became clogged with pseudo-blood cells followed by inhibition of flow through the channels, when the supplied pressure P_u = 0 kPa and P_l = 40 kPa. The volume of the separated plasma was quite low (101 μl). The percentage of clogging cells was reduced and the dead volume was pumped to the next filter by negative pressure, when P_u = 20 kPa and P_l = 40 kPa. The volume of the sampled plasma was increased according to the total number of filters without increasing the effective area of the filters. The amount of plasma collected was 144 μl and 208 μl in the double connected type and triple connected type, respectively. The collection efficiency improved from 34% to 75%.

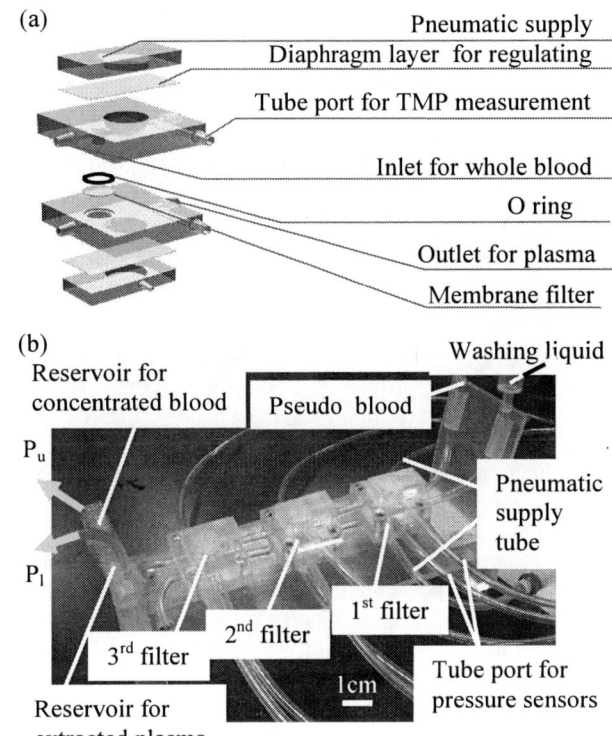

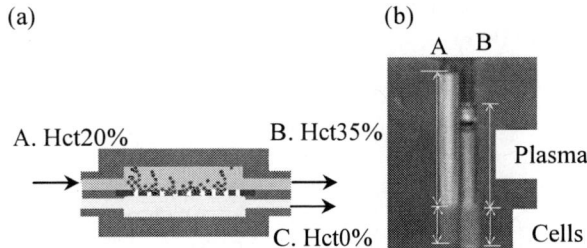

Figure 2: Prototype of separation device using serially-connected filters: (a) Composition of prototyped device, (b)Three membrane filters developed by rapid prototyping.

Figure 3: The change of hematocrit level in filtration: (a) The
schematic view of the changing hematocrit, (b) Sedimentation of pseudo blood sampled at A and B.

Monitoring of TMP for Adaptive Separation to Changing Blood Properties

Monitoring and controlling TMP at each filter under the hemolysis threshold value can be realized using our proposed device. Profiles of TMP at each filtering sections are changed by controlling the ratio of P_l to P_u as shown in Fig. 5. TMP was monitored using several ports for pressure sensors connected to upper and lower channel of a double-connected prototype. ΔP1 and ΔP2 are TMP of 1st filter section and 2nd filter section, respectively. 1st filter would be clogged faster than 2nd filter according to Fig. 5(a) that shows TMP of 1st filter was increased faster than 2nd filter when the supplied only P_l=40kPa. Fig. 5(b) shows that TMP of 2nd filter was increased faster than 1st filter when the supplied P_u=20kPa and P_l=40kPa, 2nd filter was clogged faster than 1st filter and smaller change than Fig. 5(a). Fig. 5(c) shows that TMP of 1st and 2nd filter changed slowly, those filters were clogged slower than Fig5(a) and Fig5(b) when the supplied pressure $P_u=40kPa$ and $P_l=40kPa$, most of blood was not separated but was transported to upper channel of next filter. It is possible to change value of TMP by actuating diaphragm integrated in fluidic chip according to Fig. 5(d). Adjustment of filtration characteristics can be realized by regulating the channels in our device.

Integration of Separation Function into a PDMS Fluidic Chip

As described previously, the serially connected filters improved collection efficiency. This study demonstrated that high-efficiency separation could be achieved from less than 50 μL of whole blood. It is important to integrate the separation function into the miniaturized fluidic chip. A schematic view of the separation device integrated into a PDMS chip is shown in Figure 6.

The structure of the separation device was designed to be as simple as possible by arranging the consecutive filters horizontally. The device developed here contains both separation and pumping components, including valves. Photographs of developed chip and its operation are shown in Fig. 7. As shown in Fig. 7(a), filters are integrated into fluidic channel structure. Fig. 7(b) is a micro scale photograph during regulation by actuating a diaphragm regulator. Filters are embedded by PDMS films as diaphragm layer as shown in Fig. 7(c). Fig. 7(d) and (e) is micro scale photographs that show separation is successful on a fluidic chip. Based on these results, we are trying to adapt each filters to fluidic properties as viscosity on a micro fluidic chip.

CONCLUSIONS

A conceptual design of adaptive blood cells separation device was proposed, with filtration characteristics that can be adjusted based on spatially changing blood flow properties. Through laboratory experiments using a prototype of serially connected filters, the effect on improvement of the collection efficiency of plasma was examined that plasma collection efficiency was improved from 34% to 75%. Sampling plasma without hemolysis is necessary to ensure accuracy in quantitative analyses. This device allows the adjustment of filtering characteristics to regulate the channels for avoiding hemolysis. The filters and diaphragm regulators were integrated into a PDMS fluidic chip. The separation of pseudo-plasma was successfully demonstrated and the usefulness of the proposed device demonstrated.

Future work will involve the development of a micro fluidic system with sensing and feedback control of TMP for high-efficiency blood cell/plasma separation without hemolysis and the optimization of the present separation device for blood sampled from a living body.

ACKNOWLEDGEMENTS

This research was partially supported by the Ministry of Education, Science, Sports and Culture, Grant-in-Aid for Cooperation of Innovative Technology and Advanced Research in City Area Program in the Southern Area of Lake Biwa, 2007 - 2009, coordinated by Shiga Prefectural Industrial Support Center.

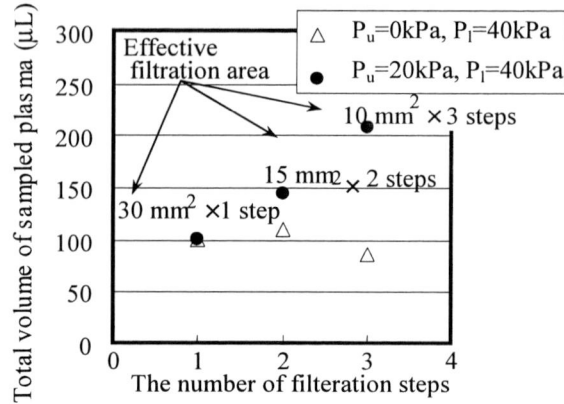

Figure 4: Effects of the number of filtration on the total extracted plasma volume.

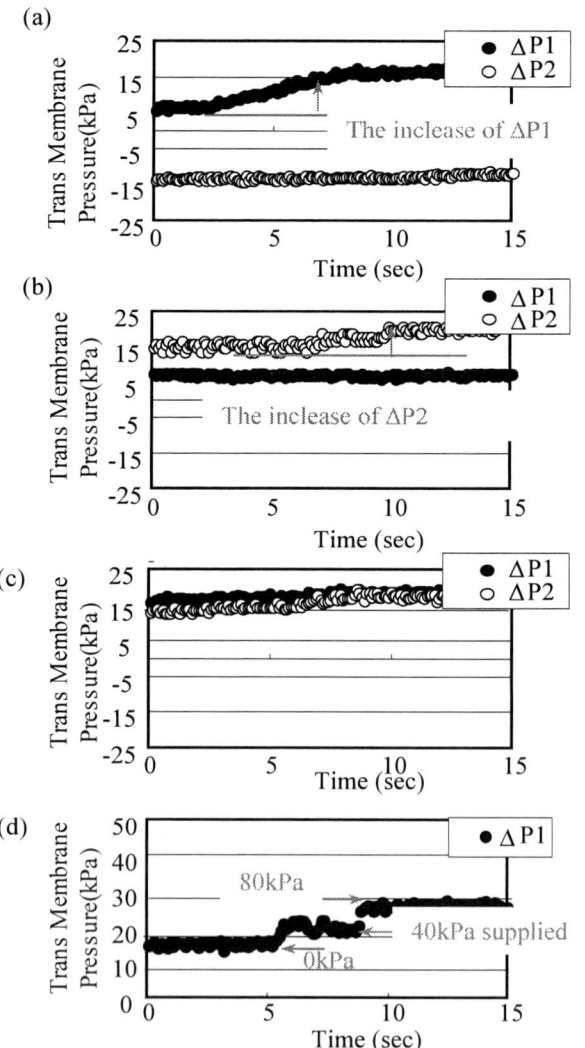

Figure 5: Measurement of TMP at various conditions: (a) $P_u=0kPa$, $P_l=40kPa$, (b) $P_u=20kPa$, $P_l=40kPa$, (c) $P_u=40kPa$, $P_l=40kPa$, (d) Regulation of TMP by changing the supplied pressure to a diaphragm regulator.

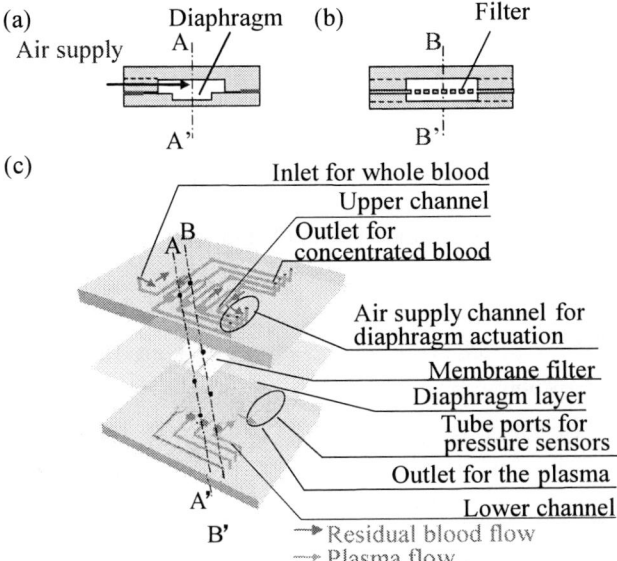

Figure 6: Schematic of integrated blood cells separation chip: (a) Diaphragm regulator,(b) Filter section, (c) Composition of separation chip.

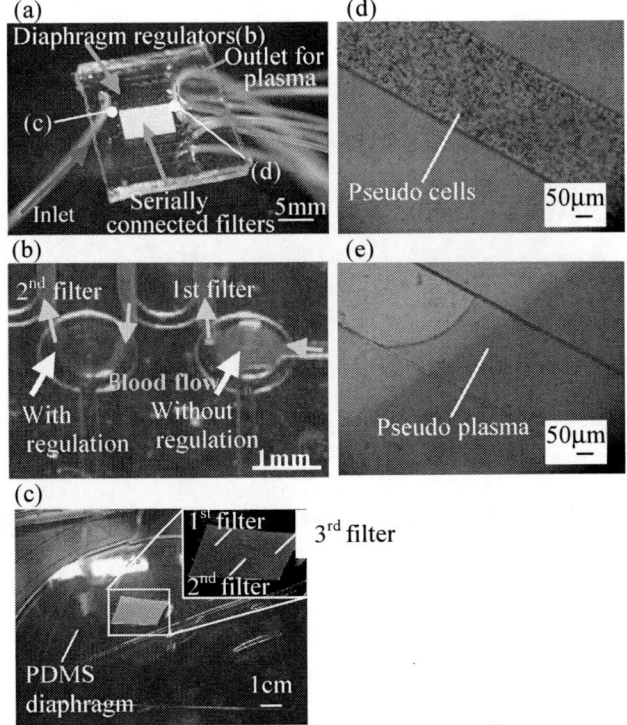

Figure 7: Photographs of developed chip and its operation: (a) Photograph of a separation chip, (b) Regulation of pressure using diaphragm regulators, (c) Filter embedded by PDMS film, (d) Before separation, (e) After separation.

REFERENCES

[1] S. Konishi, Y. Kinoshita, JPP2008-125708, May 13,2008.

[2] Y. Horiike, H. Koda, S.-H. Chang, R. Ogawa, S. Hashioka, M. Nagai, H. Ogawa, "Calorimetric Measurement Clinical Chip For Home Medical Diagnosis", *Micro TAS 2006*, pp1558-1560, Japan, 2006

[3] S. Khumpuang, T. Tanaka, F. Aita, Z. Meng, K. Ooe, M. Ikeda, Y. Omori, K. Miyamura, H. Yonezawa, K. Matsumoto, S. Sugiyama, "Blood Plasma Separation Device Using Capillary Phenomenon ", *14th International Conference on Solid-State Sensors , Actuators and Microsystems*, pp1957-1970, France, 2007

[4] Y. Gu, and N. Miki, "A Multilayered Microfilter with High performances applicable for Wearable Artificial Kidneys", *Asia-Pacific Conference of Transducers and Micro-Nano Technology*, 3C1-6, Taiwan, 2008

[5] A.I. Zydney, C.K. Colton, *Trans.Am.Soc.Artif. Intern. Intern. Organs.*, Volume 28, pp408, 1982

[6] L.B. Leverett, J.D. Hellums, C.P. Alfrey, E.C. Lynch, Red Blood Cell Damage by Shear Stress, *Biophys. J.12*, pp258-273

[7] J-H. Yeh, W-H. Chen, H-C. Chiu, C-H. Bai, "Hemolysis in Double-Filtration Plasmapheresis", *American Journal of Clinical Pathology*, Volume 127, number 1, pp76-80, 2007

[8] P.M. Reimann, P.D. Mason, "Plasmapheresis: technique and complications", *Intensive Care Med*, Volume16, pp3-10, 1990

HIGH SENSITIVITY MICRO-THERMAL CONDUCTIVITY DETECTOR FOR GAS CHROMATOGRAPHY

B.C. Kaanta[1], H.Chen[2], G. Lambertus[2], W.H. Steinecker[2], O. Zhdaneev[2], X. Zhang[1]
[1]Boston University, Boston, MA, USA
[2]Schlumberger-Doll Research, Cambridge, MA, USA

ABSTRACT

Design, fabrication and testing of a micro thermal conductivity detector (μTCD) with a <u>detection limit of 260 parts per billion (hexane in helium)</u>, no excess dead-volume, low power consumption, fast response time (150 μsec) and low sensitivity to flow rate is described herein. Significantly, this μTCD is shown to have a linear response and high sensitivity across a 10^4-fold concentration range; to the best of our knowledge this linearity and sensitivity has not been previously shown for a μTCD. The thermal conduction properties of several configurations are examined to identify their influence on performance.

1. INTRODUCTION

The more we can know about our environment, the safer, healthier and more efficiently we can live. In analytical chemistry, gas chromatography is the premier tool for separation and analysis of gases and volatile liquids; at present, it is used primarily for laboratory applications. Since the mid-1970's scientists and engineers have been working to expand the uses of gas chromatography by developing micro gas chromatography systems (μGC) [1]. μGC systems have the potential of being more portable, far more robust and, due to batch fabrication and economies of scale, much cheaper to produce relative to traditional GC system. This will enable applied gas chromatography to make an impact on fields ranging from heath services and homeland security, to industry process control and geological exploration. The goal of this work is the development of a keystone technology for μGC: a sensitive, reliable and robust thermal conductivity detector.

A thermal conductivity detector (TCD) measures the physical properties of a fluid as it flows over an electrically heated resistive element. Traditionally, the resistive heating element is located in a detection chamber. When substances with different thermal conductivities fill the chamber, the heat flux from the element is altered, producing varying electrical outputs.

In the push for novel gas sensing and detection mechanisms, little effort has been put into enhancing TCDs, one of the oldest gas sensors. However, TCDs are uniquely suited for miniaturization, since they are sensitive to the concentrations of substances within the mixture, not the total mass of a sample, as with flame ionization detectors (FIDs). Therefore, miniaturization of GC systems employing TCDs can maintain functional sensitivity while processing smaller sample masses, and can simultaneously reduce power consumption and increase mechanical robustness.

In the limited μTCD literature, some reports have indicated attempts to maximize thermal isolation, and therefore, sensitivity, to gas concentration. Wu et al.

fabricated a solid membrane over a cavity and a membrane between two flow channels in an attempt to increase thermal isolation [2]. However, even a thin contiguous membrane is significantly more thermally conductive than a gas, causing a huge loss in potential sensitivity. A suspended square silicon nitride pad was reported by Cruz et al. but this structure was released by etching a pyramidal cavity, creating a large dead volume, which would affect performance in gas chromatography [3].

Some work has also been done in modeling the thermal performance of a μTCD. Preliminary tests confirm some of the theoretical work of Huang & Bau [4], and Chen & Wu [5], which concluded that heat loss from forced convection is proportional to thermal entry length. However, their calculations were based on the assumption that the heater pad is isothermal, which is invalid for most situations.

Figure 1: SEM of a thermally isolated, low mass heater pad suspended in a 150μm wide channel. On top of the support structure is a 5μm wide meandering Ni self-heating resistor.

This paper reports a new generation of μTCDs, which incorporates a high level of thermal isolation, with the heater pad supported by thin legs at the four corners in the center of the flow channel, with an understanding of heat flux pathways. Design, fabrication, and testing of a detector with a fast time constant, a low limit of detection and a linear dynamic range that improves upon previous literature is presented (Figure 1).

2. MODELING AND DESIGN

The principle of operation for a thermal conductivity detector is simple: energy is put into heating a mass and the amount of thermal energy carried away is measured. Combined with an injection source and a separation column, a TCD can identify a chemical species and its concentration within a sample. However, the fluidic and thermodynamic properties involved in the TCD response are much more complex. Modeling allows for the optimization of detector sensitivity and dynamic range

978-1-4244-2977-6/09 $25.00 © 2009 IEEE

through variation of materials as well as the dimensions of the filament and the surrounding channel.

Thermal energy travels by three methods: conduction, radiation and convection. In macro-scale devices, heat is carried away by the fluid as it passes the heating element, a process controlled by convective heat transfer. On the micro-scale, a thermal boundary layer begins to fill the area between the heating element and the channel wall. Once this layer has developed, the thermal conductivity of the gas dominates. However, the flow rate, size and shape of the channel, as well as the surface roughness of the heated element and channel wall all play a role in how this layer forms.

Lumped Element Modeling

To understand and simulate the operation of a μTCD, the system can be modeled using electrical circuit elements. The thermal energy domain can be mapped to the electrical domain, [6] where temperature (T) is equivalent to voltage (V), thermal heat flux (Q) replaces current (I) and thermal resistance (R_T) is comparable to electrical resistance (R) (Figure 2).

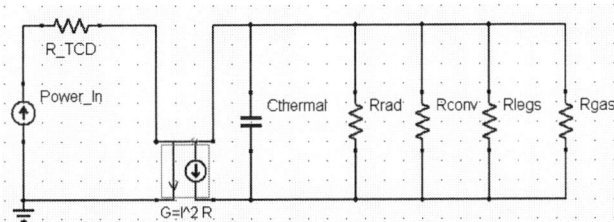

Figure 2: Electric to Thermal circuit model. Model includes heat storage ($C_{thermal}$) and all heat loss pathways (R_{rad}, R_{conv}, R_{legs}, R_{gas}).

Thermoresistance can be measured by applying a known voltage and current to the μTCD filament and calculating its electrical resistance. Power dissipation in the filament causes joule heating (measured in watts), increasing the electrical resistance of the filament. When the filament temperature is known, the heat flow and resistance to heat flow can be calculated. To perform these calculations, input power was adjusted until the desired temperature was reached.

FEM Simulations

Existing modeling of μTCD operation assumes that the heater pad has a uniform temperature.[3][4] However, it is known that this is not the case in a system where heat is generated through joule heating. In this case, heat flux modeling proves more applicable. The temperature near the support legs was found to be much lower, creating a temperature gradient along the length of the heater pad. Varying temperature along the heater length prevents the development of isotherms parallel to the pad. Figure 3 shows how heat flux from the heater element creates isotherms. This region lacking parallel isotherms causes the detector output to be sensitive to flow rate since variation in gas velocity will affect the amplitude of convective heat flux.

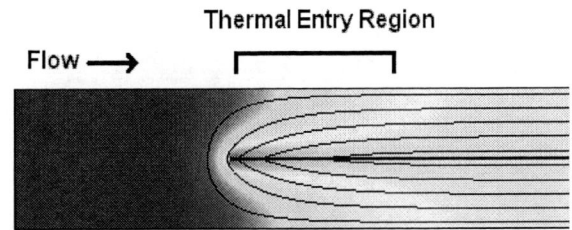

Figure 3: Simulation of heat flux showing temperature isotherm development after the flow contacts the heater pad. 'Thermal Entry Region' is where the isotherms are not parallel to flow and convection cannot be discounted as a heat flow pathway. Modeling preformed using Comsol Multiphysics.

3. FABRICATION

The presented μTCD was fabricated by depositing nickel over a chromium adhesion layer. This metal layer was supported on a 0.5μm silicon nitride film. Before the metal layer was deposited, the SiN layer was pre-etched 150nm in the pattern of the mask for the metal layer. The metal was then deposited, filling these pre-etched grooves (Figure 4a). This process creates a flush surface, which is critical for bonding a sealing Pyrex cover over the flow channel. The nickel layer connects contact pads with 100um-wide traces to 5um-wide resistive heaters.

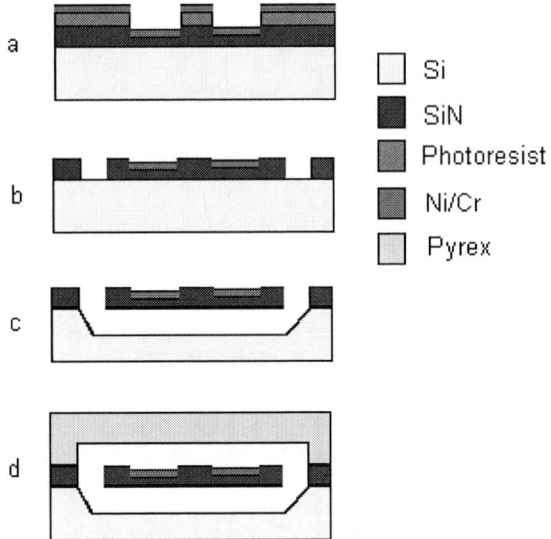

Figure 4: Device fabrication sequence for suspended μTCD. a) Pre-Etch of SiN and lift-off of metal layer. b) Etch of SiN film to define heater pad and channels. c) Release of heater pad with KOH etch. d) Device is sealed with Pyrex covers with anodic bonding.

After lift-off, the flow channels and support pad forming the resistive heating element were defined by patterning and etching trough the SiN layer (Figure 4b). Several versions of the μTCD were fabricated, with pad length as the only variable parameter (Figure 5).

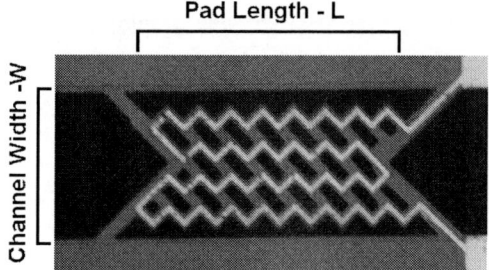

Figure 5: Top view of µTCD structure. The critical dimension changed in fabricated devices was the support pad length (L). All other dimensions were kept constant.

The Si bulk material was etched with a 30% KOH solution at 80°C, to a channel depth of 80µm (Figure 4c). The final etch depth was verified with a profilometer. The design of the suspended heating element is such that the maximum lateral undercut required during the wet etch release of the structure is 7um.

After release, the flush surface on the Si wafer allows the system to be sealed with a Corning Pyrex 7740 cover (Figure 4d). The Pyrex cover is fabricated with the mirror image of the channels on the Si wafer without the heating element.

Prior to bonding between the top and bottom wafers alignment of the flow channels is required. Alignment was achieved with the use of keys placed in the etched channels. Examination of the bonded devices showed a maximum alignment error of 5µm using this method.

Anodic bonding was performed by heating the aligned wafers to 450°C and applying 1000V to the hotplate and a contact probe. Using these parameters the Pyrex was successfully bonded through the 0.5µm SiN film.

4. TESTING AND RESULTS

Electrical Resistance

Characterizing the electrical resistance change of the µTCD as a function of temperature was achieved by placing the device in a temperature-controlled oven. For any material this change in resistance is proportional to its Thermal Coefficient of Resistance (TCR).

Nickel was selected for the resistive heater because of its high TCR (bulk TCR of ~0.006/°C at room temperature). However, some chromium nickel alloys can have TCRs 50 times lower. Since the combination of metals in the filament possessed an unknown TCR, this was quantified. TCRs for these devices were calculated from the resistance measurements and found to be between 0.0041/°C and 0.0044/°C.

Thermal Resistance

The thermal resistances of several different devices were measured and compared at different flow rates. Results indicated that the thermal resistance under conditions of no gas flow was proportional to the heater pad area.

When exposed to fluid flow, it can be seen that the smaller pads were more affected by the changing flow rate (Figure 6). This is due to the fact that the thermal entry length on each design is similar, making the thermal entry region a larger portion of the device's total pad area for smaller designs.

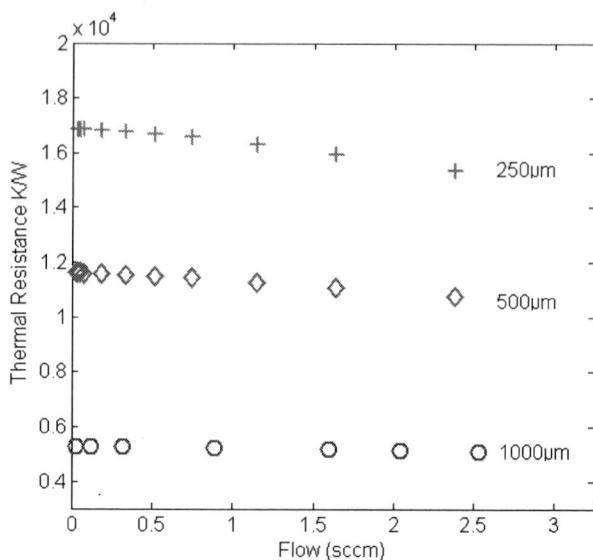

Figure 6: Total thermal resistance for µTCDs with heater pads 250µm, 500µm, and 1000µm in length tested in nitrogen. Smaller devices are more affected by changing flow rate then larger structures.

Electrical Measurements

All electrical measurements where performed by placing the fabricated µTCD in a wheatstone bridge circuit. The differential voltages on the two sides of the bridge were amplified by 100X and passed to a data acquisition card.

Step Response

The time constant for the system is very important to insure that that output of the detector accurately represents the presence of a gas in the detector.

If a narrow solute peak elutes from the GC column but passes the detector before the device has time to adjust, at best, inaccurate data will be collected and at worst, no detection will occur.

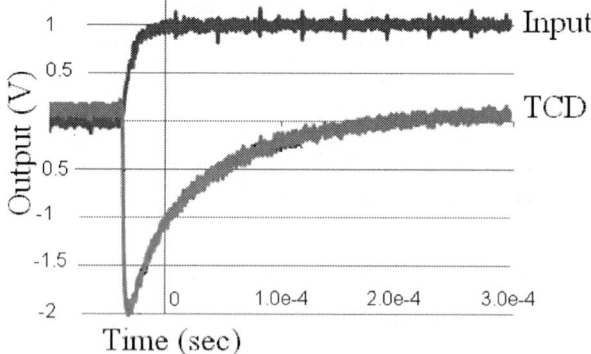

Figure 7: Step response to one volt input. µTCD output is 90% settled in under 150µsec

The time constant of the µTCD was measured by measuring the voltage step response (Figure 7). Before the step input, the TCD was at ambient temperature. Immediately after the step, the difference between the TCD resistance and the reference resistor creates an output spike. As the TDC heats up, its resistance changes

978-1-4244-2977-6/09 $25.00 © 2009 IEEE   266

and approaches the reference value, causing the output to settle. For our devices the 90% settling time was ~150μsec.

Calibration Curves - Sensitivity

Devices with 1000μm and 500μm pad lengths were tested downstream of a chromatographic column using hexane in helium over a varying concentration range to determine sensitivity and detection limits. Samples were prepared by serial dilutions of hexane into octane. Fixed volumes of the prepared samples were injected into an Agilent 7890 GC system with an integrated flame ionization detector (FID). Hexane concentration was calculated based on calibration from the known input mass and the FID output. The maximum linear detection range was limited by overloading of the GC column stationary phase, not the μTCD.

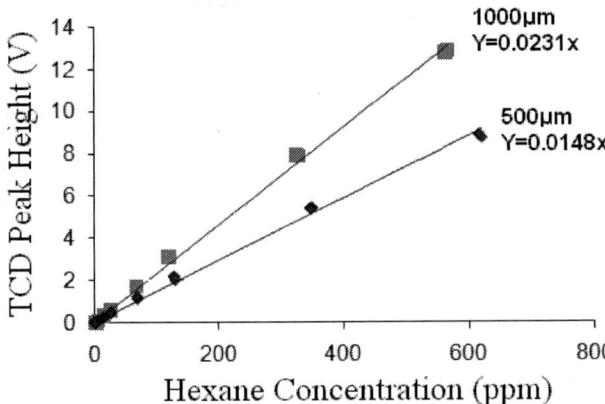

Figure 8: Plot of μTCD output vs. concentration. Heater pads with lengths 1000μm and 500μm. Heater pads were set at 50°C above the 100°C gas temperature. The limit of detection is defined as 3 times the rms noise divided by sensitivity, which was 260ppb and 410ppb for the 1000μm and 500μm designs respectively.

The device with the 1000μm pad length had a sensitivity of 23.1mv/ppm while the device with a 500μm pad length had a sensitivity of 14.8mv/ppm (Figure 8). The larger devices had higher sensitivity due to a larger portion of the thermal energy in the system being conducted from the heater pad through the gas (R_{gas} in Figure 2), than for the smaller devices.

Limit of Detection

The limit of detection (LOD), for detectors with a calibration curve with a zero y-intercept, is defined as the concentration of analyte yielding a signal to noise ratio of three, shown by the equation:

$$LOD = 3 N_{rms} \frac{1}{S} \tag{1}$$

where S is the device sensitivity and N_{RMS} is the RMS noise. For the detection circuitry used, the output has 2mv RMS noise. Therefore, the LOD for the 1000μm heater element is 260ppb, while the 500μm device has a LOD of 410pbb.

5. CONCLUSION

The design, fabrication and testing of several highly sensitive μTCDs has been discussed. The heat flux in the system has been modeled and tested experimentally to aid in understanding system thermodynamics and improving the device sensitivity. This work has focused only on the detector for a μGC system. Combing this detector with a micro-fabricated separation column and injector would create a complete μGC system. If a pre-concentrator was added to the front end of this system, its limit of detection would be decreased by the pre-concentration factor (typically 500-1000X), resulting in parts per trillion LODs.

6. ACKNOWLEDGEMENT

We would like thank Schlumberger-Doll Research and the Boston University Photonics Center for their support of this research.

REFERENCES

[1] S. C. Terry, J. H. Jerman and J. B. Angell, "A gas chromatographic air analyzer fabricated on a silicon wafer," IEEE Trans. Electron Devices, 1979, ED–26, 1880.

[2] Y.E. Wu, K. Chen, C.W. Chen, K.H. Hsu, "Fabrication and characterization of thermal conductivity detectors (TCDs) of different flow channel and heater designs," Sensors and Actuators A 100, pp 37-45, 2002.

[3] D. Cruz et al. "Microfabricated thermal conductivity detector for the micro-ChemLab," Sensors and Actuators B 121, pp. 414–422, 2007.

[4] Y. Huang, H. H. Bau, "The Effects of Forced Convection on the Power Dissipation of Constant-Temperature Thermal Conductivity Sensors," Transactions of the ASME, Vol. 119, 1997.

[5] Chen, K., and Wu, Y. E., "Thermal Analysis and Simulation of the Microchannel Flow in Miniature Thermal Conductivity Detectors," Sensors and Actuators A, vol. 79, pp. 211–218, 2000.

[6] S. D. Senturia, "Microsystem Design," Kluwer Academic Publishers, 4th ed, 2002.

CMOS-MEMS CAPACITIVE HUMIDITY SENSOR

N. Lazarus[1], S. Bedair[2], C. Lo[3] and G. Fedder[1]
[1]Carnegie Mellon University, Pittsburgh, Pennsylvania, USA
[2]U.S. Army Research Laboratory, Adelphi, Maryland, USA
[3]Maxim Integrated Products, San Jose, California, USA

ABSTRACT

This paper reports a method for improving the sensitivity of integrated capacitive chemical sensors by removing the underlying substrate. The sensor is integrated with CMOS testing electronics using mask-less post-processing followed by inkjet deposition of sensitive polymer. This technique is demonstrated with a humidity sensor but could be used for other types of analytes. The measured sensitivity is 0.18% change in capacitance per percent relative humidity, a factor of four improvement over previous integrated capacitive sensors.

INTRODUCTION

Background

Capacitive sensing is commonly used for humidity sensing due to the high dielectric constant of water. Large capacitance changes have been demonstrated using vertical parallel plate sensors consisting of a polyimide film sandwiched between two metal electrodes (Figure 1a). Sensitivities reported are approximately 0.2% change in capacitance for every 1% change in relative humidity [1]. This structure is difficult to integrate with CMOS circuitry, so past integrated capacitive sensors have consisted of interdigitated electrodes coated with adsorbent material (Figure 1b). Since the electrodes sit on the substrate, a parallel capacitance through the substrate exists that can be as large as or larger than through the adsorbent material, resulting in significantly lower sensitivity. CMOS chips also contain a layer of oxide between the metal electrodes, decreasing the sensitivity even further. In [2], the second layer of metal in CMOS is used to force more electric field lines to pass through the sensitive polymer to improve the sensitivity of the interdigitated electrode approach (Figure 1c). Despite this technique, a large parallel capacitance remains. In [2], the sensing capacitance is 1.4 pF in parallel with a substrate capacitance of 6.4 pF. Since 18% of the total capacitance is affected by analyte, the sensitivity is at most 18% of that of the parallel plate sensor, or about 0.04% change in capacitance per percent relative humidity.

This paper demonstrates a technique to eliminate the parallel substrate capacitance by removing the underlying substrate and releasing the structure (Figure 1d). By using the metal layers available in CMOS to create a horizontal parallel plate capacitor, a comparable sensitivity to the vertical parallel plate can be integrated with CMOS.

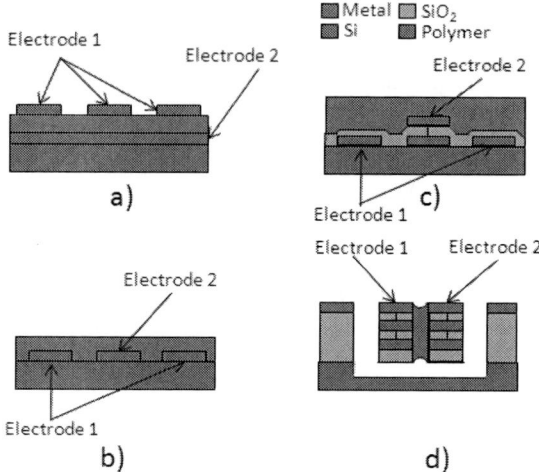

Figure 1: Designs of capacitive humidity sensors

Fabrication

The sensor is integrated with detection electronics and released using a previously developed technique that uses the top layer of metal in CMOS as a release mask (Figure 2) [3]. The process begins with a standard CMOS chip (Figure 2a) fabricated in the Jazz Semiconductor 0.35 μm BiCMOS technology. An anisotropic vertical oxide etch (Figure 2b) is followed by an isotropic silicon etch to release the MEMS structure (Figure 2c). Sensitive polymer is added by using a custom drop-on-demand inkjet system[4] to deposit polymer in solution to an attached well, with polymer drawn into the released structure using capillary forces; this technique was previously demonstrated for gravimetric sensing (Figure 2d)[5]. Wicking allows well controlled repeatable material deposition without damaging the fragile released structure.

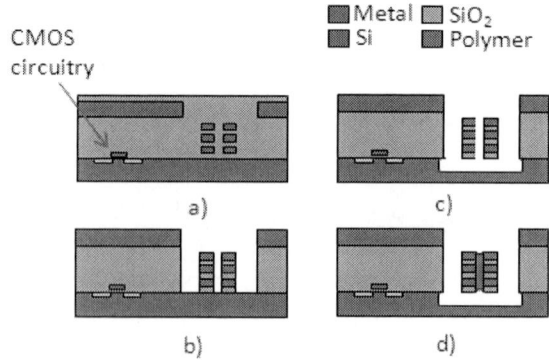

Figure 2: Fabrication process flow

Polyimide was chosen as an adsorbent material because its common usage in humidity sensing allows comparison with other humidity sensors. A formulation of polyimide in solution (HD Microsystems PI 2556) was diluted by a factor of 24:1 with the solvent HD Microsystems T-9039, a 50:50 mixture of n-methoyl-2-pyrrolidone and methoxy propanol to obtain a low enough viscosity to deposit with the inkjet. Figure 3 shows the structure before inkjet deposition; Figures 4 a) and b) show the wicking channel before and after deposition.

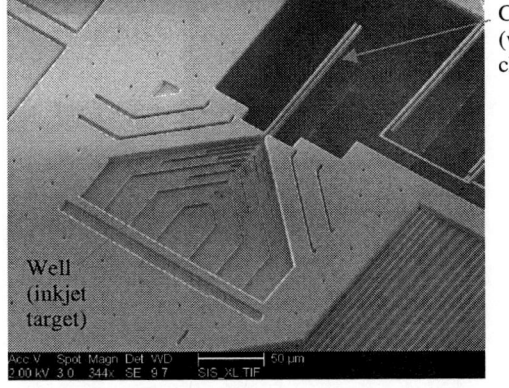

Figure 3: SEM of capacitive humidity sensor

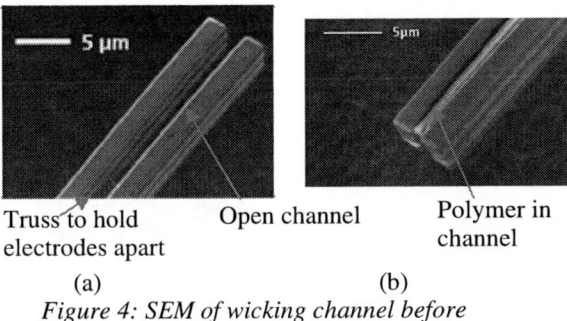

(a)   (b)

Figure 4: SEM of wicking channel before (a) and after (b) polymer deposition

The capacitor was tested using a charge-based capacitance measurement (CBCM) circuit [6] (Figure 5a).

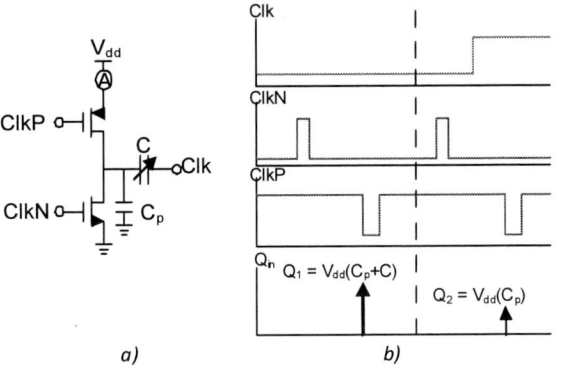

a)   b)

Figure 5: a) Charge-based capacitance measurement circuit b) Timing diagram

The circuit consists of two switch transistors, a PMOS and an NMOS; C is the chemical sensitive capacitor, C_p is a parasitic capacitance to ground. Figure 5b) shows a timing diagram of the testing circuit. A measurement is first taken with Clk set to 0 V; charge is drained from the central node to ground, using the NMOS transistor. The charge necessary to bring the central node to V_{dd} is measured using an ammeter, giving the charge Q_1 equal to $V_{dd}(C_p+C)$. The central node is then reset to zero charge; Clk is then set to V_{dd}, followed by turning on the PMOS

transistor, bringing the central node to V_{dd}. The charge measured is now equal to $V_{dd}C_p$, allowing the parasitic capacitance to be subtracted away to obtain the value of C. By repeating this measurement over time, changes in capacitance, and thus chemical absorbed, can be measured.

Theoretical Model

The capacitor can be modeled as a parallel plate capacitor with an additional fixed fringe capacitance. The dielectric constant ε of the polyimide film is given by [7]:

$$\varepsilon = \left\{ \gamma(\varepsilon_{H2O}^{1/3} - \varepsilon_{PI}^{1/3}) + \varepsilon_{PI}^{1/3} \right\}^3 \quad (1)$$

where γ is the volume fraction of water in the film, and ε_{PI} and ε_{H2O} are the dielectric constants of the polyimide and water respectively. Expansion was ignored because lateral motion was constrained by trusses between the capacitor beams and the volume coefficient of polyimide is only 60-75 PPM/%RH [8].

COMSOL Simulation

A series of COMSOL simulations were performed to compare the sensor performance to past integrated sensors similar to that in [2]. The simulations were also intended to investigate which of the changes from that type of design have the most significant effect on sensitivity. The first significant change is the removal of the oxide between the metal electrodes, allowing the most direct electric field path to be filled with polyimide. In [2], the oxide between the electrodes was left in place, with a coating of sensitive material added on the top surface of the oxide. The CMOS-MEMS structure is also able to use multiple layers of the metal in the stack, allowing a taller structure with more capacitance through the polyimide relative to the amount of substrate capacitance. The release then reduces the parallel substrate capacitance by removing the substrate. Figure 6 shows the capacitor behavior changes due to each of these changes.

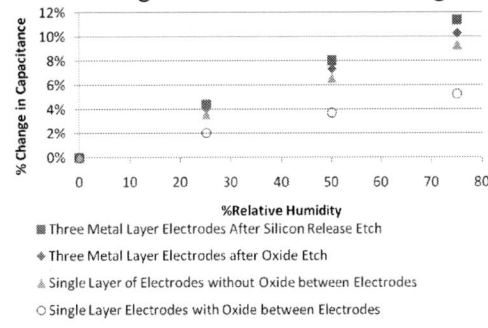

■ Three Metal Layer Electrodes After Silicon Release Etch
♦ Three Metal Layer Electrodes after Oxide Etch
▲ Single Layer of Electrodes without Oxide between Electrodes
○ Single Layer Electrodes with Oxide between Electrodes

Figure 6: Simulated sensor performance for different electrode configurations

The removal of the oxide between the electrodes resulted in a factor of two increase in sensitivity; the sensitivity improvements due to the taller structure and the release were 10.3% and 10.6% respectively.

978-1-4244-2977-6/09 $25.00 © 2009 IEEE   269

RESULTS

The sensor was tested using the flow system shown in Figure 7.

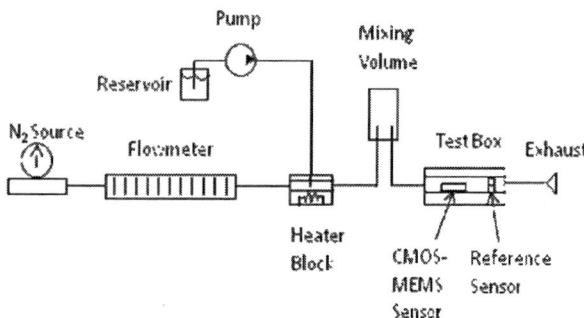

Figure 7: Humidity sensor test setup

A Milligat M6 solvent pump was used to pump liquid water into a dry nitrogen flowstream, controlled by a flow meter to a flow rate of 1 L/min. This occurs in a brass heater block heated to a temperature of 120° F, to increase the volatility of the water. The flow then continues into a 0.5 L mixing volume to improve the stability of concentration and allow the flow time to return to room temperature, after which it flows over the CMOS-MEMS humidity sensor. In order to obtain the actual relative humidity, a Honeywell HIH-4000 commercial humidity sensor was used as a reference.

The capacitance of the device was measured before and after deposition, and compared to COMSOL simulations (Table 1). The deviation for the post-inkjet value arises from polyimide not completely filling the gap between the electrodes due to the surface tension profile.

Table 1: Simulated and Measured Capacitance of Sensor

	Simulation	Measured
Pre-deposition Capacitance	7.93 fF	6.15 fF
Post-deposition Capacitance	17.79 fF	16.4 fF

A larger device consisting of five parallel wicking channels, shown in Figure 8, was used to test the humidity response. The larger sensing capacitance was more easily resolved from the fixed parasitic capacitance to ground, allowing for more accurate measurements.

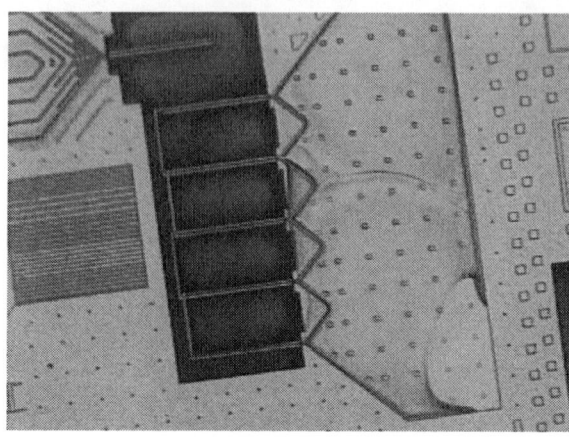

Figure 8: Five channel capacitive humidity sensor

The capacitance of the sensor was measured for a range of humidity values and compared to the theoretical model (Figure 9).

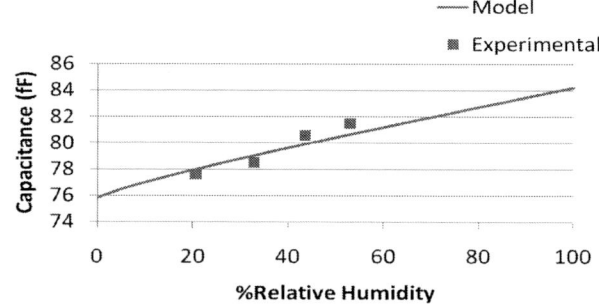

Figure 9: Capacitance vs. Relative Humidity

The sensitivity is 0.16% change in capacitance for every 1% change in humidity, slightly lower than the 0.2% reported for vertical capacitors; this is due to the capacitance through the air above and below the polymer.

One possible method of lowering the parallel air capacitance is to jet directly on the top surface of a wicking channel. Top surface of the electrodes will be coated with several microns of polyimide, moving the air on the top of the channel far enough from the electrode that the parallel capacitance from the air will be significantly smaller. Capillary forces will also pull material down into the wicking channel, filling the gap between the two electrodes. Only the parallel air capacitance below the capacitor will remain.. A device with containing 87 parallel channels was designed and tested (Figure 10). The wicking channels are fixed to the substrate at both ends, giving the mechanical stability necessary to survive an inkjet drop.

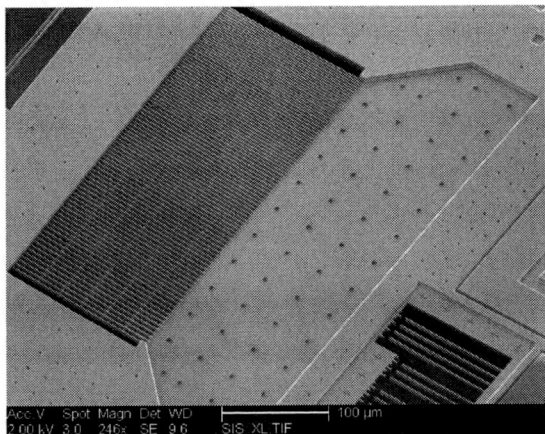

Figure 10: 87 channel capacitor before inkjet deposition

After inkjetting, the measured capacitance of the structure was 1.1 pF. Figure 11 shows a plot of the sensor output as the humidity concentration is pulsed to different values.

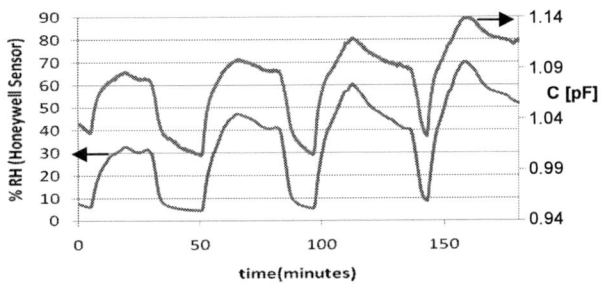

Figure 11: Humidity test with 87 channel capacitor

Figure 12 shows a plot of the response to changes in relative humidity for both the five channel and 87 channel capacitive sensors.

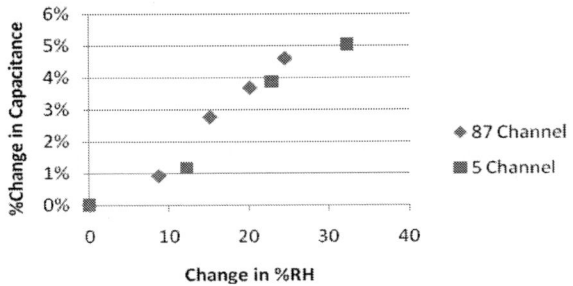

Figure 12: Change in capacitance vs. change in %RH for five and 87 channel devices

The measured sensitivity of the 87 channel device was 0.18 % change in capacitance per percent change in relative humidity. The sensitivity of the 87 channel device appears slightly more sensitive than the 5 channel wicked structure, as expected, although this may also be due to measurement uncertainty.

CONCLUSION

The measured results demonstrate a capacitive humidity sensor integrated with CMOS circuitry that is able to achieve sensitivities as high as 0.18% change in capacitance per percent change in humidity. Two different methods for depositing adsorbent polymer are considered, using a drop-on-demand inkjet to jet directly on the structure and using capillary forces to pull polymer into the capacitive gaps from a larger inkjet target. The experimental results show that both approaches give comparable sensitivity. This sensitivity is four times that of the unreleased interdigitated capacitor [2], which is the best design previously integrated with CMOS.

ACKNOWLEDGEMENT

The authors thank Lawrence Shultz and Lee Weiss for advice and assistance with inkjetting and Suresh Santhanam and the Carnegie Mellon Nanofabrication facility for the post CMOS processing. This work is funded in part by NIOSH/CDC (contract 200-2002-00528) and AFOSR grant FA9550-07-1-0245.

REFERENCES

[1] P. Harrey, B. Ramsey, P. Evans, D. Harrison, "Capacitive-type humidity sensors fabricated using the offset lithographic printing process", *Sensors and Actuators B*, vol. 87 pp. 226-232, 2002.

[2] A. Hierlemann, *Integrated Chemical Microsensor Systems in CMOS Technology*, Springer, Germany, 2005.

[3] G. K. Fedder, S. Santhanum, M. L. Reed, S. C. Eagle, D. F. Guillou, M. S. C. Lu, and L. R. Carley, "Laminated High-Aspect-Ratio Microstructures in a Conventional CMOS Process", in *Proc.of IEEE MEMS 1996*, San Diego, California, pp. 13-18, February 1996 .

[4] L. Weiss, L. Schultz, and E. Miller, "Inkjet deposition system with computer vision-based calibration for targeting accuracy," Technical report CMU-RI-TR-06-15, Carnegie Mellon University, March, 2006.

[5] S. Bedair and G. K. Fedder, "Polymer Mass Loading of CMOS/MEMS Microslot Cantilever for Gravimetric Sensing", in *Proc. Of IEEE Sensors 2007*, Atlanta, Georgia, pp. 1164-1167, October 2007.

[6] Y. Chang, H. Chang, T. Lu, Y. King, W. Ting, Y. Ku, and C. Lu, "Charge-Based Capacitance Measurement for Bias-Dependent Capacitance", *IEEE Electron Device Letters*, Vol. 27, No. 5, pp. 390-392, 2006.

[7] H. Shibata, M. Ito, M. Asakura, and K. Watanabe, "A Digital Hygrometer Using a Capacitance-to-Frequency Converter", *Proc. of IEEE IMTC 1995* Waltham, Massachusetts, pp. 100-106, April 1995.

[8] K. Sager, A. Schroth, and G. Gerlach, "Humidity-Dependent Mechanical Properties of Polyimide Films and Their Use for IC-Compatible Humidity Sensors", in *Tech. Dig. 8th Int. Conf. on Solid-State Sensors and Actuators (Transducers '95)*, Stockholm, Sweden, pp. 736-739, June 1995.

A THERMOELECTRIC GAS SENSOR BASED ON AN EMBEDDED TIN OXIDE CATALYST FOR DETECTING HYDROGEN AND NOX GASES

Seung-Il Yoon, Chung-il Lee, and Yong-Jun Kim*

School of Mechanical Engineering, Yonsei University, Seoul, REPUBLIC OF KOREA

ABSTRACT

This paper reports a novel gas sensing method by using a thermoelectric device, thermopile in this case, with an embedded tin oxide catalyst. By using a thin catalyst film and gas adsorption process instead of absorption, the response time and recovery time were remarkably improved. And the fabricated sensor was characterized through detecting a hydrogen gas. Finally, NOx gas was successfully detected by using the proposed sensor. The respond and recovery time were shorter than 1 minute. The output signal change according to the concentrations of the NOx gas was $1.5 \times 10^{-1} \mu V/ppm$.

INTRODUCTION

As an interest to the air quality has been increased, demanding for monitoring or detecting harmful gases have been also increased. Especially, a demanding for detection of hydrogen and nitrogen oxide (NOx) gases, which are air pollutants, released form combustion exhaust of automobile engines, home heaters, furnaces and plants have become more serious all over the world in recent years [1]. Because those gases at low concentrations affect human health seriously, many efforts based on gas sensors with metal oxide focused on high sensitivity, short response time, and continuous monitoring.

In these view points, various gas sensors such as resistor type [2] and nano-structure type [3] have been researched. Major issues for gas sensor researches would be classified into two parts; one is sensing materials, and the other is sensing methods. But although there are many efforts for improving the sensing materials, a resistance measurement type has dominated over all other sensing methods. Of course, the resistance measurement type has advantages such as simple structure and moderated sensitivity, but it is limited to be applied to real-time and continuous detection. The previous researches related to the resistance type are based on electrical property change by gas **_"absorption"_** to a thick catalyst layer, which results delayed response time and recovery time (~mins) [4]. And also, since the gas is accumulated in the catalyst layer during the measurement, its output signal is unstable and nonlinear to the concentrations of the gas in the case of long duration monitoring.

For real-time and continuous gas detection, a thermoelectric gas sensor based on a thermopile can be a strong candidate. It can produce electrical potential when thermal energy is released or absorbed due to chemical reactions between target gases and catalyst. And the gas sensor based on the thermopile does not require any power consumption. Furthermore, when using gas **_"adsorption"_** instead of gas absorption, response and recovery times can be improved because the adsorption reaction occurs on the surface of the catalyst and a penetration process of the gas into the catalyst is not required.

In this paper, a new gas sensing method based on **_"adsorption"_** instead of absorption is proposed. By using a thermopile, the proposed sensor can detect the target gas by measuring the reaction heat between the gas and a thin catalyst film. The proposed gas sensor was fabricated by using MEMS technology, and characterized through detecting a hydrogen and NOx gases.

MATERIALS AND METHODS

Sensing Principle

The proposed gas sensor consists of sensing and reference thermopiles with bismuth and chrome pairs, Detailed dimensions are as follows: (1) overall dimensions; 28 mm (l) × 23 mm (w); (2) width of the thermopile: 50 μm; (3) length of the thermopile: 5 mm; (4) the number of thermocouples: 120; (5) size of the catalyst layer: 150 μm (l) × 200 μm (w) × 1 μm (h).

The thermopile can generate an electrical potential proportional to temperature difference between hot and cold junctions without any power consumption as shown in Figure 1(a). For detecting the target gas, the catalyst film is embedded under only the hot junctions of the thermopile. And for removing thermal noises, a reference thermopile, which does not have a catalyst layer, is integrated with the sensing thermopile as shown in Figure 1(b). By using a differential amplifier, output signal changes only due to the gas reaction can be extracted.

When the gas is adsorbed to the catalyst surface, the reaction heat is generated immediately. Since the adsorption process is much faster than the absorption process, response time and recovery time can be improved. And because the adsorption process of the catalyst occurs at the lower temperature, the working temperature for detecting the target gases can be decreased. Furthermore, since the generated reaction heat is proportional to the gas concentrations, linear output signals can be obtained.

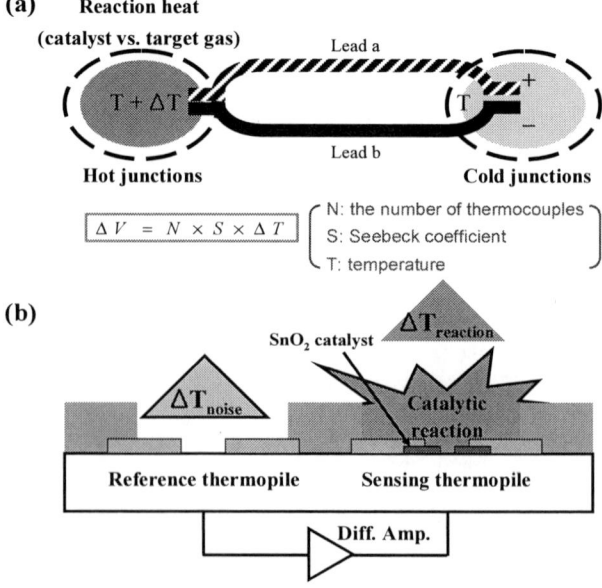

Figure 1: Gas sensing principle of the proposed sensor

Fabrication Process

The proposed sensor was realized on a Pyrex substrate. The electrical potential measured by the thermopile is proportional to the temperature differences between the hot and cold junctions of the thermopile. For higher sensitivity, the temperature differences between the junctions should be maximized. So it is important to reduce heat transfer through the substrate. Desirably, heat loss itself from the thermopile to the substrate as well as the heat transfer should be minimized. In this view point, the Pyrex substrate, which has lower thermal conductivity than that of a silicon substrate, was used for the substrate instead of the silicon wafer.

First of all, 2.3 μm-thick photoresist is spin-coated on the Pyrex substrate and patterned for defining the catalyst pattern. And then, 0.3 μm-thick SnO_2 layer is formed on the Pyrex substrate by using a lift-off process. Secondly, 1 μm-thick Titanium-bismuth layer is deposited on the catalyst layer by using a sputtering system and patterned. After this patterning, 1 μm-thick Chrome is deposited and patterned as crossed with Titanium-bismuth patterns. The fabricated thermopile includes 120 hot and cold junctions, respectively. Figure 3 shows the fabricated thermoelectric gas sensor.

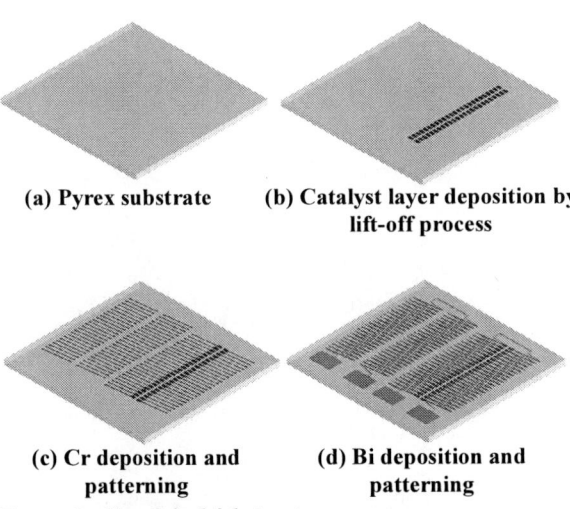

(a) Pyrex substrate   **(b) Catalyst layer deposition by lift-off process**

(c) Cr deposition and patterning   **(d) Bi deposition and patterning**

Figure 2: Simplified fabrication processes

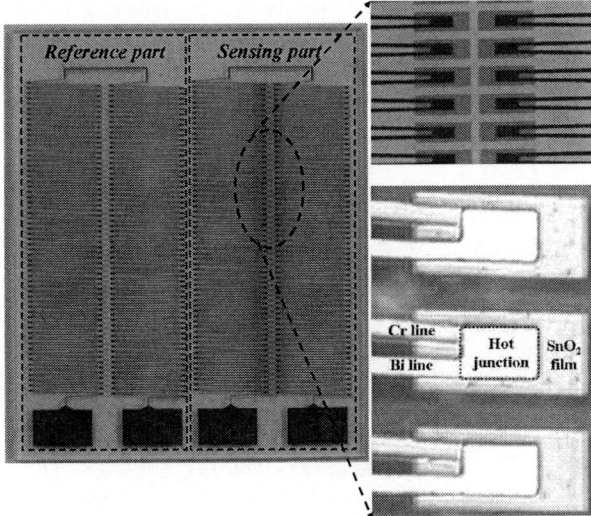

Figure 3: Optical views of the fabricated gas sensor

Thermal Characteristics of the Proposed Gas Sensor

For checking thermal characteristics of the proposed sensor, output signals depending on temperature difference between junctions were measured by using N_2 gas. The N_2 gas with various temperatures, which are 5 ℃ to 35 ℃ higher than the room temperature, was injected only to hot junctions of the sensing thermopile, and the output voltage changes due to the applied heat were measured.

The applied heat and the sensitivity of the sensor can be calculated by using Eq. (1) and Eq. (2).

$$q = C \times \dot{m} \times \Delta T \qquad (1)$$

q : Applied heat (W)

$\dot{m}$: Mass flow rate of N_2 gas (g/sec)

ΔT : Temperature difference (℃)

C : Specific heat (cal·g^{-1}· ℃$^{-1}$)

$$Sensitivity = \Delta V / q \qquad (2)$$

ΔV : Measured output voltage according to temperature difference between junctions (V)

The output signal changes of the sensor was 3.01 mV/℃ (=4.61 V/W) according to the temperature difference between hot and cold junctions, as shown in Table 1 and Figure 4.

Table 1: Output signals of the proposed gas sensor according to temperature differences between hot and cold junctions

Temperature difference (℃)	Applied thermal energy (mW)	Measured output voltages (mV)		Sensitivity (V/W)
		Average	Standard variation	
5	178.04	12.38	1.35	4.51
15	534.12	35.38	1.52	4.29
25	890.20	68.94	3.41	5.02
35	1246.28	81.96	2.75	4.26

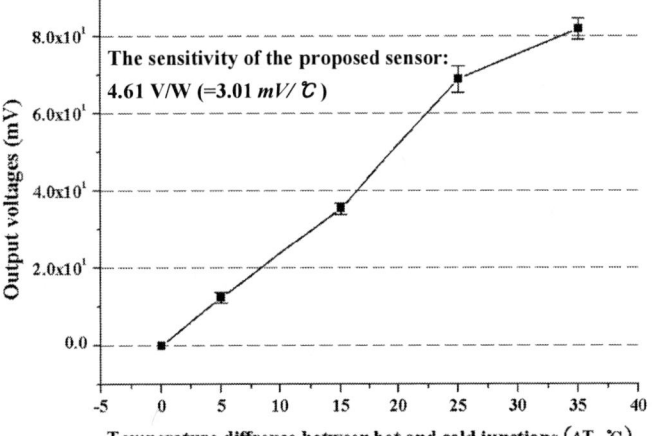

Figure 4: Output voltages according to temperature differences between hot and cold junctions

EXPERIMENT AND RESULT

The fabricated gas sensor was characterized through detecting hydrogen and NOx gases. Figure 5 shows the measurement setup.

Three kinds of gases, N_2, H_2, and NO_2 in this case, were used for the measurement. The N_2 gas, which was dehydrated by using a HEPA filter, was used as the reference gas and the mixture gas for maintaining specific concentrations of the target gases. Each gases was precisely controlled by using a mass flow controller. And by using an environmental chamber, the surroundings of the sensor were maintained at the fixed temperature and humidity. Output signal changes due to chemical reaction between the target gases and SnO_2 catalyst were recorded by an I-V meter (Semiconductor Characterization System 4200, Keithley).

Initially, the reference gas, the dehydrated N_2 gas, was supplied into the chamber at a constant flow rate. During this step, the output voltages were changed due to not reaction, but the gas flow itself. This noise was compensated by comparing output voltages between the sensing and reference thermopiles. And then, by switching the reference gas to the target gases, the target gases with various concentrations were introduced to the chamber. After a specific elapsed time for the gas reaction, the reference gas was re-supplied into the chamber instead of the target gases. All experiments have been performed several times, and the chemical and physical alteration of the catalyst was not observed.

Hydrogen Gas Detection

The concentrations of the hydrogen gas were varied from 1000 to 10000 ppm. Figure 6 shows the results of the hydrogen gas detection at room temperature (27 ℃). The response of the sensor for the hydrogen gas sensing took less than 10 seconds. And the recovery time were ranged from 18 to 38 seconds. The output signal change according to the concentrations of the hydrogen gas was measured to $1.06 \times 10^{-1} \mu V/ppm$.

NOx Gas Detection

For detecting NOx gas, NO_2 gas with various concentrations from 100 to 1000 ppm was used. Figure 7 shows the results of the NOx gas detection at 100 ℃. The respond and recovery time were shorter than 1mins. The output signal change according to the concentrations of the NOx gas was $1.5 \times 10^{-1} \mu V/ppm$.

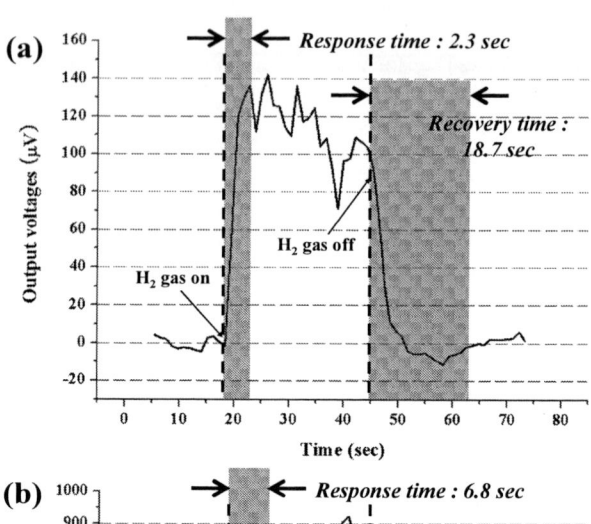

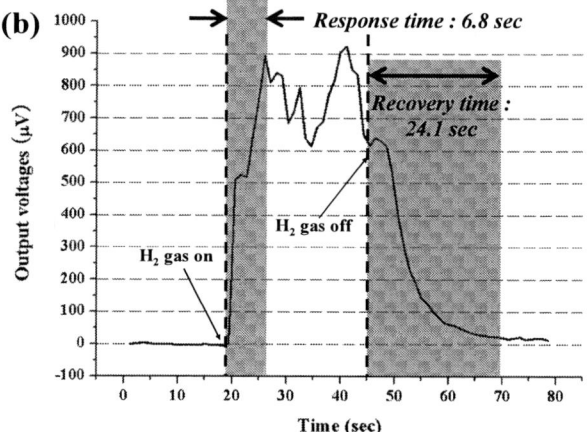

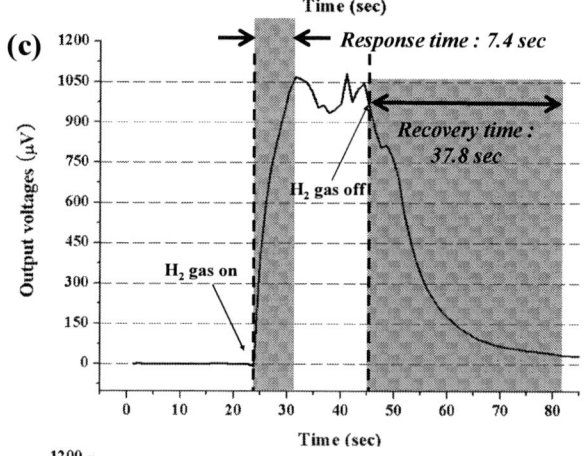

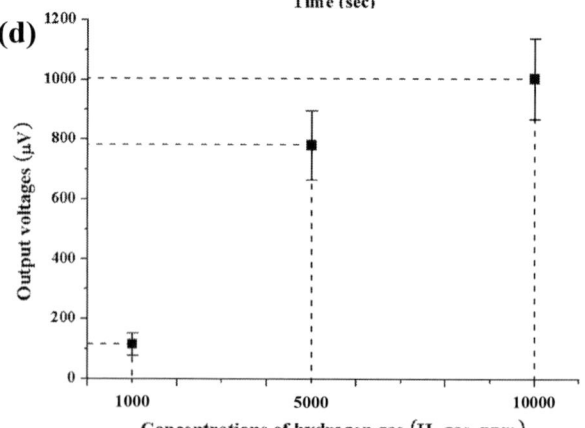

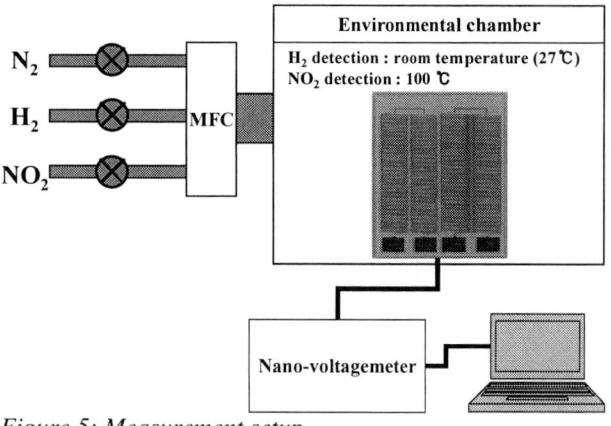

Figure 5: Measurement setup

Figure 6: Results of H_2 gas detection, (a) 1000 ppm, (b) 5000 ppm, (c) 10000 ppm, (d) Output voltages according to the concentrations of H_2 gas

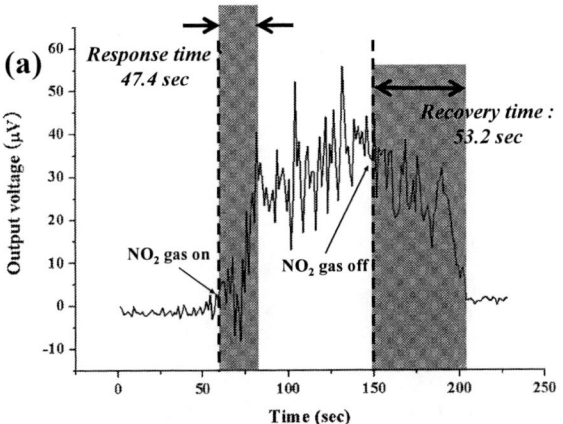

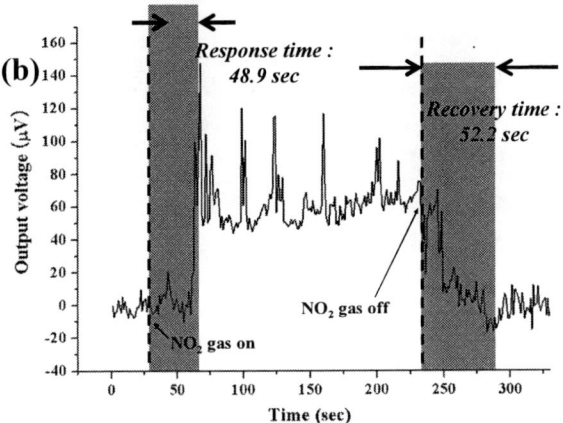

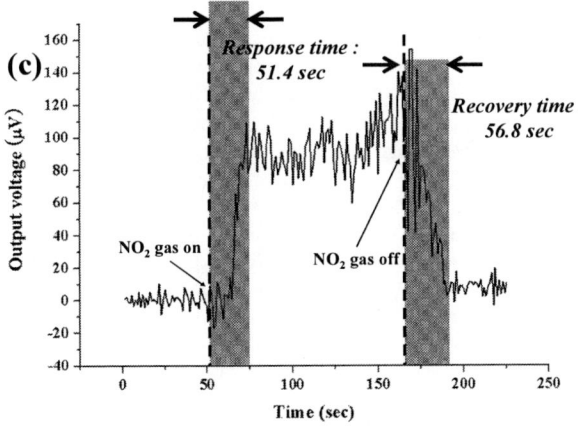

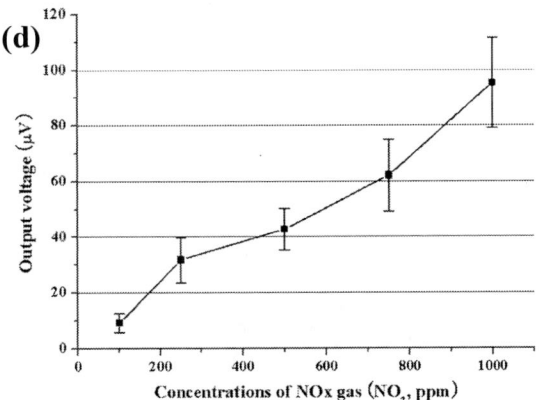

Figure 7: Results of NOx(NO₂) gas detection, (a) 250 ppm, (b) 750 ppm, (c) 1000 ppm, (d) Output voltages according to the concentrations of NOx gas

CONCLUSION

Gas detection researches have focused on high sensitivity, good selectivity, and fast detection. Sensitivity and selectivity can be achieved depending on the type of the used catalysts. Therefore, with respect to the detection method itself, improvement of the detection time has emerged as an important issue. In the case of the resistor type based on an thick catalyst layer, the gas absorption process require several minutes at the least, which makes fast response and recovery impossible. For real-time detection, a novel gas sensing method based on gas adsorption with a thermoelectric device was proposed. By using a thin catalyst film, the response time and recovery time were remarkably improved. The fabricated gas sensor was characterized through detecting hydrogen and NOx gases. In H_2 and NOx gas detection, the measured response and recovery times (~secs) were much less than those of the previous gas sensors (~mins). And the output signals show good linearity to the concentrations of the target gases.

ACKNOWLEDGEMENT

This work was supported by the Seoul R&BD program and the National Core Research Center (NCRC) of Yonsei University.

REFERENCES

[1] Ch.Y. Wang, M. Ali, Th. Kups, C.-C. Rohlig, V. Cimalla, Th. Stauden, and O. Ambacher, "NO_x sensing properties of In_2O_3 nanoparticles prepared by metal organic chemical vapor deposition", *Sens. Actuator B-Chem.*, vol. 130, pp. 589-593, 2008.

[2] J. Tamaki, J. Niimi, S. Ogura, and S. Konishi, "Effect of micro-gap electrode on sensing properties to dilutechlorine gas of indium oxide thin film microsensors", *Sens. Actuator B-Chem.*, vol. 117, pp. 353-358, 2006.

[3] E. Comini, "Metal oxide nano-crystals for gas sensing", *Anal. Chim. Acta.*, vol. 568, pp. 28-40, 2006.

[4] S. Shukla, P. Zhang, H. Choa, S. Seal a, and L. Ludwig, "Room temperature hydrogen response kinetics of nano–micro-integrated doped tin oxide sensor", *Sens. Actuator B-Chem.*, vol. 120, pp. 573-583, 2007.

A NOVEL MICROFLUIDIC PLATFORM FOR CONTINUOUS DNA EXTRACTION AND PURIFICATION USING LAMINAR FLOW MAGNETOPHORESIS

Marc Karle[1], Junichi Miwa[2], Günter Roth[2], Roland Zengerle[1,2], Felix von Stetten[1,2]
[1]HSG-IMIT, Villingen-Schwenningen, Germany
[2]IMTEK - University of Freiburg, Freiburg, Germany

ABSTRACT

We present a novel microfluidic platform using laminar-flow magnetophoresis for combined continuous extraction and purification of DNA. All essential unit operations (DNA binding, sample washing and DNA elution) are integrated on one single chip. The key function is the motion of magnetic beads given by the interplay of laminar flow and time-varying magnetic field. The time for extraction was 1 minute. The device is a central part of a complete biochemical system for continuous monitoring of cell-growth in bioreactors. The novel platform allows continuous purification of DNA, but is also applicable to purification of RNA, proteins or cells, including their subsequent real-time analysis in general.

INTRODUCTION

Monitoring of biological agents, including the pathogenic microorganisms, protein and free nucleic acids, is highly relevant in the field of security (B-detection), blood monitoring and process control in pharmaceutical fermentations. Key requirement to perform these monitoring tasks is the continuous processing of biochemical assays. Such continuously working and automated monitoring systems are currently not available.

Continuous on-chip PCR systems for amplification and detection of DNA have already been developed [1] including a transcription into cDNA for RNA detection [2]. A concept for continuous molecular enrichment using segmented flow has also been proposed [3]. However, a method for continuous DNA or RNA extraction from any sample such as whole blood or cell cultures was not achieved. Therefore we develop a novel microfluidic platform in order to realize a fully integrated microfluidic continuous extraction and analysis system for on-line monitoring of cell growth in bioreactors.

Among various techniques for DNA purification, we use DNA adsorption onto superparamagnetic beads. Different microfluidic approaches using this method in a batch-wise manner have already been reported [4-6]. In contrast our device allows the continuous operation in a flow through manner. Until now the only existing continuous magnetophoretic devices are for single-step cell separation [7;8], while we now report well-controlled manipulation of magnetic beads through at least three subsequent assay steps necessary for nucleic acid purification.

WORKING PRINCIPLE

The working principle is depicted in Fig. 1. The first step of DNA extraction and purification after sampling and lysis of the cells is binding of the DNA onto the superparamagnetic beads in order to separate the DNA from impurities such as enzymes and/or cell debris. After

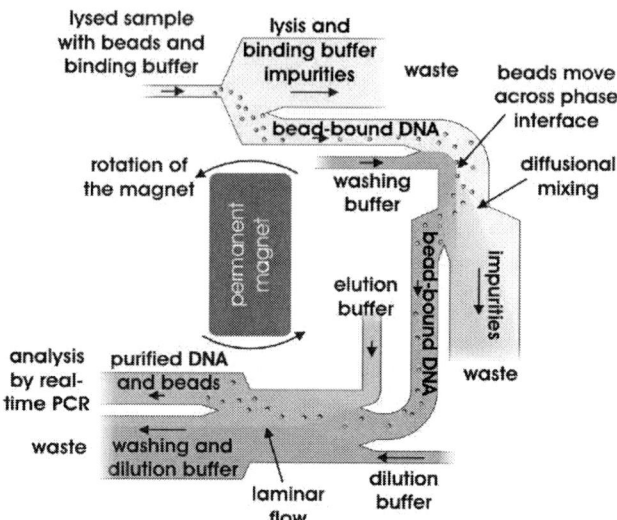

Figure 1: Schematic view of the microfluidic chip for continuous DNA extraction. The rotation of the permanent magnet is opposite to the direction of the buffer flow.

the separation a washing step is included to remove the high salt buffers required for DNA binding as well as any residual impurities. As a last step the beads are transferred into the elution buffer where the DNA dissociates from the beads to allow real-time analysis.

Chip design

Our microfluidic chip applies magnetophoresis to control superparamagnetic beads in different buffer solutions under continuous laminar flow (Fig. 1).

Bead-bound DNA samples flow through three chambers for impurity separation, sample washing, and DNA elution from the beads. The circular arrangement of the microfluidic structure (Fig. 2), together with a single rotating permanent magnet positioned below the chip center, allows the transfer of magnetic beads from one reagent to the other in each separation chamber. Subsequently, the magnetic beads transport the DNA through three separation chambers for the separation of impurities, sample washing and the final elution of DNA elution.

The separation chambers consist each of one or more inlets and two outlets which are different in diameter. This is a key design feature for the division of the buffer flow into two fractions of different flow rates; a fraction of small volume containing a high concentration of magnetic beads and thus high concentration of sample, and a large fraction without any magnetic beads but most of the buffer solution.

978-1-4244-2977-6/09 $25.00 © 2009 IEEE

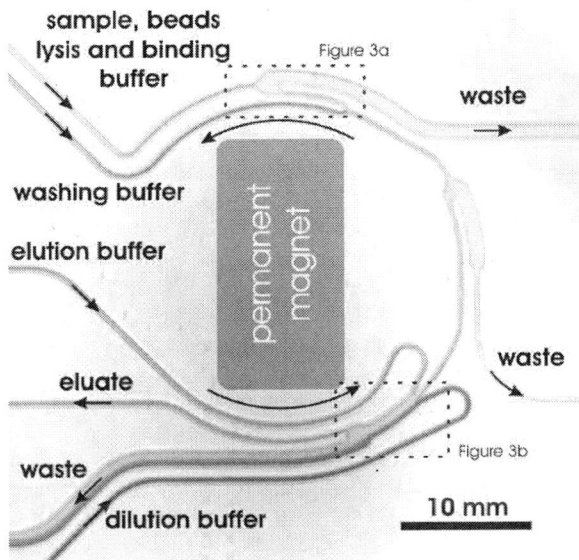

Figure 2: Polycarbonate chip with micro-milled microchannels. This chip design was used for the DNA extraction experiment. To illustrate the extraction process dyed buffer solutions were injected into the chip. Yellow: Lysis and binding buffer, blue: washing solution, green: elution buffer, red: dilution buffer.

The magnetic beads exit the separation chambers through the small outlet and are transferred to the next section whereas the large fraction of the buffer solution leaves through the wider outlet that directly leads to the waste.

Time-varying magnetic field

The circular arranged microchannels guide different buffer flows from different inlets of the chip around a single external permanent magnet located beneath the center of the chip. The permanent magnet is rotated slowly using a stepping motor to provide the required periodically changing magnetic field to attract the magnetic beads in the radially inward direction. According to the oscillating strength of the magnetic field, the beads are attracted and stick to the side-wall during the strong field phase but are released during the weak field phase. With the circular microfluidic design shown in Fig. 2, the DNA-bound beads are attracted by sufficiently strong magnetic force during the strong field phase at each of the separation chambers. The average magnetophoretic velocity of beads is determined by the balance between the flow rate and the strength of magnetic field, the latter of which can be controlled by altering the magnet rotation speed. We have also observed that by rotating the magnet rotation in the direction opposite to the flow, the streamwise bead velocity is larger. Thus, we realize efficient separation of the magnetic beads without immobilizing them on the walls of the microchannels.

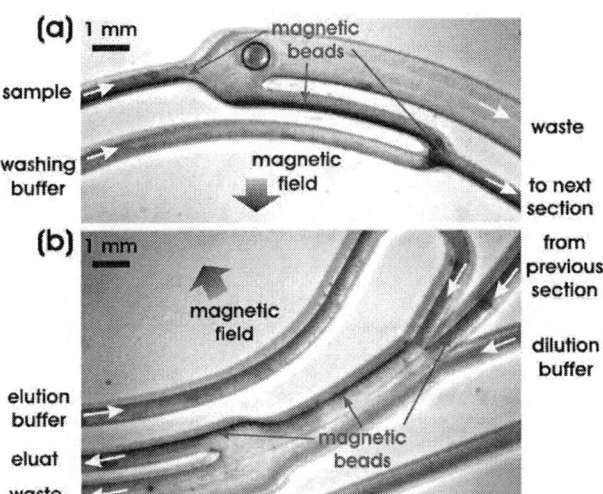

Figure 3: Instantaneous image of (a) the first separation chamber and (b) the third separation chamber. The beads are strongly attracted by the magnetic field and flow in the vicinity of the inner side wall throughout the device. The inlet flow velocity of each stream is 12.5 mm·s^{-1} and the average velocity of beads estimated from consecutive images is approximately 1.6 mm·s^{-1}.

Buffer flow protocol

A mixture of cell lysate, binding buffer and superparamagnetic beads, which are already bound to the DNA, enters the chip through the 1st inlet (Fig. 2, top left). In the first separation chamber the beads are concentrated by the central permanent magnet and separated from most of the buffer solution containing impurities (Fig. 3a).

During the transfer to the next section, washing buffer is added forming a laminar flow with the remaining binding buffer. The permanent magnet attracts the beads to the inner part of the channel system and ensures the transfer of the magnetic beads across the phase interface.

In the second separation chamber the magnetic beads are again concentrated and most of the buffer is transferred to the waste, while the magnetic beads leave through the small outlet towards the third separation chamber (Fig. 3b).

Here the beads are transferred to the elution buffer, in which the DNA is released from the bead surface. The eluate is collected for further analysis. A key feature in the third separation chamber is the introduction of a dilution buffer.

The washing buffer contains a high concentration of ethanol, which is a PCR inhibitor. However, the hygroscopic nature of ethanol leads to rapid dispersion throughout the sample flow, and it is thus not possible to establish a clear laminar interface between the washing and elution buffers. Therefore we introduce a dilution buffer that is injected from the opposite side. The purpose of the dilution buffer is to assimilate the ethanol from the washing buffer and therefore reduce the ethanol concentration in the eluate or even exclude ethanol from the eluate.

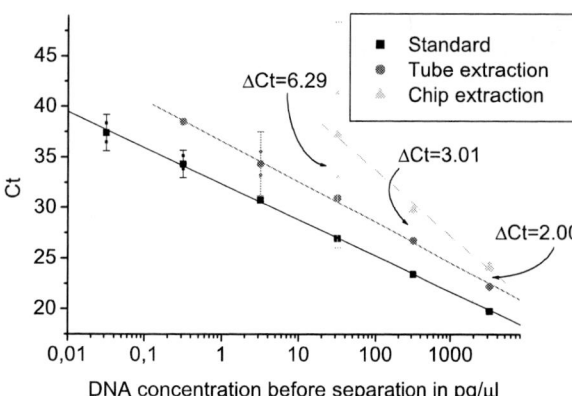

Figure 4: Real-time PCR of extracted genomic E. coli DNA. Shown in the graph is the result of a DNA extraction experiment performed both in a test tube and on the microfluidic chip. The genomic E. coli DNA was diluted and spiked with herring sperm DNA (therefore the gap between the standard curve and the tube extraction). At E. coli DNA concentrations of 3.15 ng·μl^{-1} the on-chip purification achieves 25% of the reference in the test tube.

On the other hand, DNA molecules and magnetic beads are large enough that they do not diffuse in the sample solutions so rapidly. Thus, after the mixing of elution and washing buffer on one side of the third separation chamber and the mixing of washing and dilution buffer on the opposite side, the purified DNA remains in the elution buffer stream to be collected at the outlet without any loss.

CHIP PRODUCTION

The 2D-layout of the microfluidic channels was designed with CAD software and converted to G-code compatible to a precision milling machine. For chip production polycarbonate was used. The mircochannels were milled directly into the chip using a 400 μm mill, with a standard width and depth of 400 μm. After the milling process the channels were sealed with adhesive foil.

RESULTS

The flow of sample and buffers of a commercially available DNA purification kit was controlled by syringe pumps (neMESYS, cetoni GmbH). The permanent magnet was mounted below the chip system close to the surface and rotated by a stepping motor with a rotational frequency of 1 Hz.

We extracted and purified a sample of genomic DNA from *E. coli* DH5αZ1 [9] bacteria on the chip in a flow-through manner. An inlet flow of 12.5 mm·s^{-1} lead to an average bead velocity of 1.6 mm·s^{-1} and a sample transition time of approximately 1 minute.

The extracted DNA was successfully amplified off-chip via real-time PCR to demonstrate that PCR inhibitors included in the buffers are sufficiently diluted and/or excluded during the on-chip purification. At a starting concentration of 3.15 ng·μl^{-1} the fluorescence signal of the chip-extracted total DNA crossed the threshold value only

2 cycles later than the reference extraction performed in a test tube (Fig. 4).

CONCLUSION

We established a microfluidic platform for continuous DNA purification by magnetophoretic manipulation of magnetic beads in a time-varying magnetic field. The platform closes the gap between continuous cell lysis and continuous DNA amplification to set up powerful microsystems for the monitoring of biological agents, pathogenic microorganisms, protein, free nucleic acids and cell growth in bioreactors.

The magnetic beads enable the transport of the DNA across the interfaces between co-flowing laminar streams in a circular channel arrangement around a central rotating permanent magnet inducing time-varying magnetic field, which prevents the beads from sticking to the channel walls and enables controlled transfer of beads between different extraction reagents. An inlet flow velocity of 12.5 mm·s^{-1} lead to an average bead velocity of 1.6 mm·s^{-1}. The sample transition time is approximately 1 minute. Compared to a reference extraction in a test tube, 25% of DNA could be detected by Real-Time PCR. However, we expect significant improvements in the recovery rate and purity by optimizing the magnet motion and fluid resistance at each section of the microfluidic structure.

Applications of this concept for the extraction and purification of biomolecules are the monitoring of cell growth in fermenters or continuous safety monitoring of drinking water against biological contaminations. Other potential applications are in the field of continuous protein purification, immuno- and cell based assays or other biological assays that are compatible with target-binding to magnetic beads.

ACKNOWLEDGEMENT

The present study is funded by the German Federal Ministry of Education and Research (16SV3528).

REFERENCES

[1] M. U. Kopp, A. J. de Mello, and A. Manz, "Chemical Amplification: Continuous-Flow PCR on a Chip," in *Science 280* 1998, pp. 1046-1048.

[2] J. Felbel, A. Reichert, M. Kielpinski, M. Urban, N. Hafner, M. Durst, J. M. Kohler, J. Weber, and T. Henkel, "Technical concept of a flow-through microreactor for in-situ RT-PCR," *Engineering in Life Sciences*, vol. 8, no. 1, pp. 68-72, Feb.2008.

[3] O. K. Castell, C. J. Allender, and D. A. Barrow, "Continuous molecular enrichment in microfluidic systems," *Lab Chip*, vol. 8, pp. 1031-1033, May2008.

[4] U. Lehmann, C. Vandevyver, V. K. Parashar, and M. A. M. Gijs, "Droplet-based DNA purification in a magnetic lab-on-a-chip," *Angewandte Chemie-International Edition*, vol. 45, no. 19, pp. 3062-3067, 2006.

[5] Y. K. Cho, J. G. Lee, J. M. Park, B. S. Lee, Y. Lee, and C. Ko, "One-step pathogen specific DNA extraction from whole blood on a centrifugal microfluidic device," *Lab Chip*, vol. 7, no. 5, pp. 565-573, Feb.2007.

978-1-4244-2977-6/09 $25.00 © 2009 IEEE

[6] M. Shikida, N. Nagao, R. Imai, H. Honda, M. Okochi, H. Ito, and K. Sato, "A palmtop-sized rotary-drive-type biochemical analysis system by magnetic bead handling," *J. Micromech. Microeng.*, vol. 18, no. 3 Mar.2008.

[7] N. Pamme and A. Manz, "On-chip free-flow magnetophoresis: Continuous flow separation of magnetic particles and agglomerates," *Anal. Chem.*, vol. 76, no. 24, pp. 7250-7256, Dec.2004.

[8] P. H. Shih, J. Y. Shiu, P. C. Lin, C. C. Lin, T. Veres, and P. Chen, "On chip sorting of bacterial cells using sugar-encapsulated magnetic nanoparticles," *Journal of Applied Physics*, vol. 103, no. 7 Apr.2008.

[9] R. Lutz and H. Bujard, "Independent and tight regulation of transcriptional units in Escherichia coli via the LacR/O, the TetR/O and AraC/I-1-I-2 regulatory elements," *Nucleic Acids Research*, vol. 25, no. 6, pp. 1203-1210, Mar.1997.

AN EWOD DROPLET MICROFLUIDIC CHIP WITH INTEGRATED LOCAL TEMPERATURE CONTROL FOR MULTIPLEX PROTEOMICS

Wyatt Nelson[1], Ivory Peng[2], Joseph A. Loo[2], Robin L. Garrell[2], and Chang-Jin "CJ" Kim[1]
[1]Mechanical and Aerospace Engineering Department
[2]Department of Chemistry and Biochemistry
University of California, Los Angeles (UCLA), California, U.S.A.

ABSTRACT

We present a microfluidic chip for multiplex thermal processing of discrete micro- and nano-liter sample droplets driven by electrowetting-on-dielectric (EWOD). The system features local temperature control integrated on chip in addition to the basic fluidic functionalities previously reported for EWOD devices [1]. Multi-functioning indium-tin-oxide (ITO) electrodes serve as optically transparent resistive heaters, temperature sensors, and EWOD actuation pads. With a 1 μL sample, heaters reach programmed set points in a few seconds and stabilize to ±1 °C. We demonstrate automated on-chip proteomics sample processing and direct characterization by matrix-assisted laser desorption/ionization mass spectrometry (MALDI-MS). The results, showing increased insulin disulfide reduction efficiency with heater temperatures from 22 to 130 °C, highlight the efficacy of the chip.

INTRODUCTION

One challenging goal of biochemical microfluidics is to build thumbnail chips capable of automatically performing assays that would otherwise require laboratory equipment, technicians, and hours of processing time. This means shrinking the dimensions of everything from fluid samples to ovens and mixers by several orders of magnitude. Micro-electro-mechanical systems (MEMS) manufacturing techniques enable fabrication of a vast array of on-chip transducers, and thus we achieve seamless integration of electrical and fluid handling components. To realize complete laboratory-on-a-chip miniaturization, a device should not only control the location and composition of sub-microliter samples, but also the temperature, pressure, and electrical signals in and around working fluids.

Microfluidic devices have been used successfully for miniaturizing biochemical assay protocols that require thermal cycling such as polymerase chain reaction (PCR) [2]. Oft-cited advantages of using microscale fluid volumes include lower waste and reagent usage, faster processing time (*e.g.,* rapid heating and cooling, shorter diffusion length), potentially higher throughput, efficiency, and levels of automation. Resistive heating and temperature-sensing elements are easily integrated into microfluidic chips, often as thin-film platinum wires [3]. While many reported lab-on-a-chip systems use integrated heating elements and temperatures sensors to eliminate the need for macroscale thermal components (which add bulk and thermal crosstalk), they commonly require external pumps and valves for pressure-driven fluid handling. Interfacing macroscale tubes with microfluidic chips inhibits scalability and parallelization.

Driving mechanisms such as externally applied pressure and electroosmosis can provide excellent control of flow rates in continuous flow microfluidic channels, but problems arise due to excessive power consumption, analyte dispersion, and, for electrokinetic mechanisms, electrolysis and Joule heating in the working fluid. Droplet-based, or digital, systems *excluding two-phase channel flows* alternatively use mechanisms including electrowetting, dielectrophoresis, or thermocapillarity to drive discrete droplets without physical pumps or valves. For example, EWOD refers to the electromechanical force that pulls a conductive liquid toward an electric field applied across an underlying dielectric layer. In this way, droplet manipulation is enacted via electrical signals.

Droplet (or digital) microfluidic (DMF) chips using EWOD actuation can create, transport, merge, and split nano- and picoliter droplets individually in reconfigurable pathways. They are usually manufactured using simple thin-film processes derived from integrated circuits fabrication steps, which can be redesigned in order integrate on-chip transducers for localized temperature control with EWOD fluidic handling. Such advancement is rather obvious; the key virtue we pursue here is simplicity, *i.e.* integrated functionalities are added without complicating the fabrication process.

Our proposed EWOD DMF chip performs chemical reactions in individual microscale samples under precisely controlled heating cycles. The design features arrays of multi-functional electrodes, each one serving as an individually addressable EWOD actuation pad, Joule heater, and temperature sensor. While reported droplet-based electrowetting devices have been mostly limited to room-temperature assays, ours is the first such device with integrated localized temperature control, which enables optimal on-chip conditions for chemical processing.

EWOD AND TEMPERATURE CONTROL

All functionalities of EWOD with local Joule heating and thermistor sensing are controlled by electrical signals, *i.e.* upon sample loading the chip operates solely according to wire-borne commands. The complete system diagram is shown in Fig. 1.

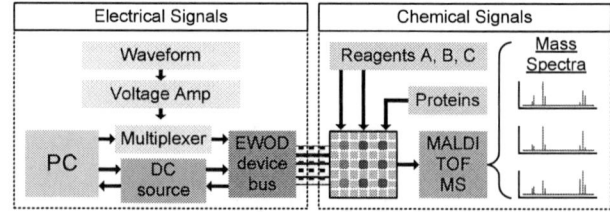

Figure 1: System diagram: A personal computer (PC) runs a custom LabView program that simultaneously controls multiplex EWOD actuation and temperature controlled heating. Chemical samples are processed on the EWOD chip, which is loaded directly into the MALDI-MS chamber.

A custom LabView program generates electrical signals for multiplex EWOD actuation and heating. Actuation signals are administered by a DAQ-controlled home-built multiplexer (National Instruments DAQPad 6507). Heater temperatures are maintained by proportional-integral-derivative (PID) control based on resistance measurements, sampled at 10 Hz by the DC source (Keithley 2425 SourceMeter). Chemical samples are generated, transported, mixed, and heated on the EWOD chip, which is loaded directly into the MALDI time-of-flight mass spectrometer chamber for characterization (Voyager-DE STR MALDI-TOF Mass Spectrometer).

EWOD actuation and heating/temperature sensing can be controlled separately by adjusting electrical bias levels. Each multi-functioning pad has two leads, which are at the same potential during actuation, and biased during heating. The operating principles for EWOD actuation and localized heating are depicted in Fig. 2. To move, voltage is applied to a substrate electrode, grounded through the droplet by the top plate. To heat, voltage is applied across a single patterned electrode, which serves as a resistive heater.

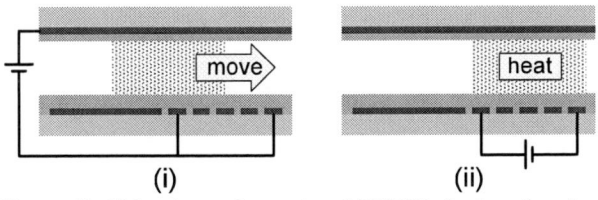

Figure 2: Side view schematic of EWOD device showing the operating principals for multi-function electrodes: (i) droplet movement by EWOD occurs when the pad is equipotential and grounded through the drop by the top plate, and (ii) resistive heating occurs when the pad is biased. Droplets are separated from the bottom and top electrodes by thick and thin hydrophobic dielectric layers, respectively.

DEVICE FABRICATION

Our fabrication is typical for planar EWOD devices, requiring no extra steps due to integrated heaters. Fig. 3 illustrates the process, with a top view of the ITO and gold layers. The substrate is a 700 μm thick glass wafer that has been coated with 140 nm ITO by the manufacturer (TechGophers Co.).

After evaporating 200 nm gold upon a 10 nm chromium adhesion layer, photoresist (AZ 5214) is coated and patterned by UV exposure through a mask defining the ITO electrode layout. Gold and chromium are wet etched, and the latter serves as the masking layer for the following wet etch of ITO in 5 %wt oxalic acid (see Fig. 3.1). With another photoresist mask, gold and chromium are etched again to form heater leads (see Fig. 3.2). For accurate temperature sensing, i.e. resistance measurement, it is critical to provide low-resistance electrical paths to the ITO heaters in order to ensure negligible power dissipation along lead wires; gold leads are about 100 times more conductive than ITO heaters, with approximate resistances of 10 and 1000 ohms, respectively. Plasma-enhanced chemical vapor deposition (PECVD) is used to deposit 1 μm silicon nitride (see Fig. 3.3). A hydrophobic polymer, 6 %wt Cytop®, is spin-

coated at 2000 rpm to yield a 200 nm layer, which is in some cases etched by oxygen plasma to form hydrophilic patches on heaters (see Fig. 3.4). Device top plates (see Figure 2) are fabricated by spin-coating 200 nm Cytop® onto an ITO-coated glass wafer.

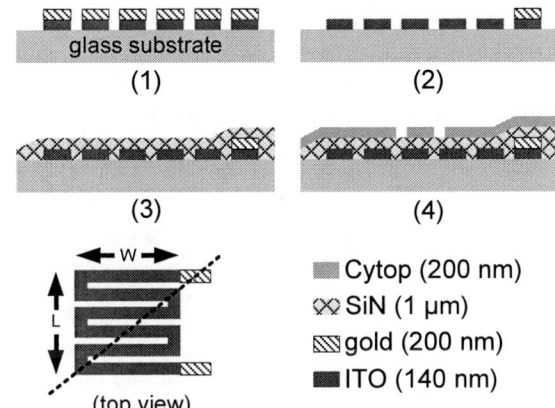

Figure 3: Fabrication process flow shown across the dotted line in the top view: (1) ITO etch in oxalic acid with a Cr/Au mask; (2) Au wet etch with a photoresist mask; (3) PECVD deposition of SiN; (4) Spin coat and RIE Cytop® with a photoresist mask

DEVICE CHARACTERIZATION

ITO heating elements are designed such that they are about 100 times more resistive than any other electrical connections in the system. Therefore, they double as accurate thermistors, or resistance temperature detectors, which rely on the material property temperature coefficient of resistance (TCR) to relate electrical resistance to heater temperature. The inset in Fig. 4 is a representative TCR calibration curve for our ITO heaters. For the testing range of 22 °C to 152 °C resistance varies linearly with temperature.

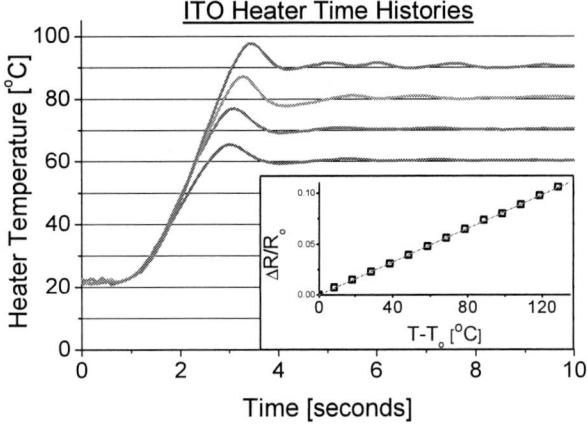

Figure 4: ITO heater time histories at set points 60 °C, 70 °C, 80 °C, 90 °C with 1 μL droplet; inset: TCR calibration curve for 22 °C to 152 °C.

Fig. 4 has temperature time histories from experiments using 3 x 3 mm (dimensions are L x W of Fig. 3) heaters under 1 μL water droplets sandwiched by a top plate placed 100 μm above the substrate. Set points were reached within a few seconds and maintained to

978-1-4244-2977-6/09 $25.00 © 2009 IEEE   281

within about ± 1 °C once stable, about 8 seconds after the initial rise.

EXPERIMENTAL PROCEDURE

Macroscale proteomics sample preparation steps, *e.g.* reduction and alkylation, are often performed at elevated temperatures, usually between 30 and 70 °C. Microscale EWOD devices for room temperature sample preparation and direct analysis by MALDI-MS have demonstrated protein purification [4] and enzyme reactions with rapid quenching [5]. We utilized EWOD with localized temperature control to perform an automated protocol for insulin disulfide reductions at various temperatures. The chip has three heating sites (see Fig. 5), which also serve as MALDI targets after matrix crystallization. A top view schematic of the device and video frames depicting an automated thermal processing sequence are shown in Fig. 5. Prior to installation of the device top plate, sample reservoirs A, B, and C were loaded with working fluids by pipette. After top plate placement, sample preparation steps were carried out automatically by a LabView program controlling EWOD actuation (40 to 60 rms volts at 1 kHz) and thermal cycling (direct current). During step 1, ~500 nL droplets of insulin (from reservoir A) and reducing agent dithiothreitol (DTT) in 50% aqueous acetonitrile (from reservoir B) were created and merged, forming a 1 μL droplet having initial concentrations of 0.5 μM insulin and 25 mM DTT. The combined sample was mixed by moving back and forth three times along the path shown by black arrows in Fig. 5.2. In step 3, the mixed sample was loaded onto a heating site. During heating for 180 seconds at a heater temperature of 70 °C, the sample evaporated rapidly, reducing in volume from ~1 μL to ~60 nL. This resulted in a ~95% increase in concentration.

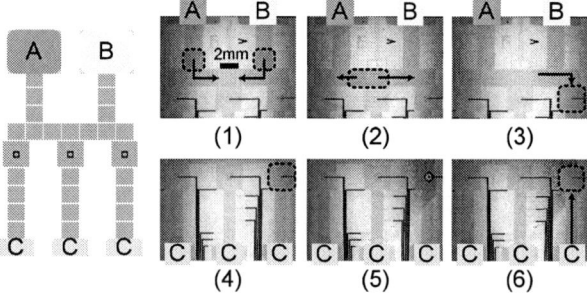

Figure 5: Left: Top view schematic of EWOD chip for proteomics sample preparation. Heating sites are three large gray boxes in the middle, containing small dark square rings, which depict hydrophilic regions. Right: Video frames of MALDI-MS sample preparation by EWOD with localized heating: (1) create and merge ~500 nL solutions of insulin (A) and DTT (B); (2) mix; (3) move ~1 μL sample to heating site; (4-5) heat; (6) add matrix. Black dotted lines and arrows indicate droplet locations and moving paths, respectively.

Note that Fig. 5.5 shows that the heated sample remains centered on the heating site. This position control was obtained by the hydrophilic patches (see small dark square rings in the schematic), which keep the droplets pinned during evaporation. In Fig. 5.6, a 700 nL droplet of MALDI matrix, 2,5-dihydroxybenzoic acid (DHB, 3

mg/mL) in 50% aqueous acetonitrile and 0.1% trifluoroacetic acid (TFA), was moved from the far-right reservoir C to the heating site. After preparing two more samples by the same process, the top plate was removed and crystallization occurred during solvent evaporation.

Merging the acidic DHB solution with the heated sample effectively quenches the disulfide reduction, providing a well-controlled reaction time. This quenching effect was confirmed by a control experiment showing no significant reduction for a sample in which all solutions were combined with DHB before heating at 70 °C, a temperature which yields high reduction efficiency using the protocol shown in Fig. 5.

Hydrophilic ring patterns centered on heater electrodes play a central role in device design by providing control of heated sample location and enhanced matrix crystal growth. Droplet centering was accomplished by both permanent and switchable hydrophilic regions; the former were formed by etching Cytop®, and the latter were formed by having individually addressable EWOD pinning electrodes centrally located in the heating site. Desirable shard-like morphologies were observed on permanent hydrophilic patches. Figure 6 shows a clean heater with ring pattern and the same heater with DHB crystals grown at room temperature.

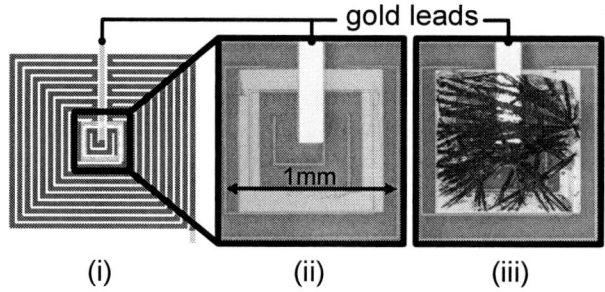

Figure 6: Matrix crystal growth on hydrophilic ring: (i) heater schematic showing field of view; (ii) clean heater; (iii) dry DHB crystal grown over ring.

We observed that in most cases, crystals grown on square rings originate from nucleation sites on hydrophilic regions and grow over surrounded hydrophobic patches. The final structure (see Fig. 6.iii) provides a good MALDI target, *i.e.* shard-like crystals yielded good spectra.

A series of insulin disulfide reduction experiments were carried out using the protocol of Fig. 5 in order to observe how reaction efficiency varies with heater temperature. Samples were processed three per chip using different temperatures, from 22 to 70 °C. After room temperature crystallization, which took approximately 15 minutes, the chip was loaded directly into the mass spectrometry chamber on a normal sample holder that had been milled to compensate for the wafer thickness.

Insulin disulfide reductions were performed at 130 °C in 50% DMSO, which in its pure form has a boiling point of 189 °C. Due to its extremely low evaporation rate compared to water, room temperature crystallization was too slow. Therefore, upon top plate removal, heaters were maintained at 90 °C until crystals were observed. This hot crystallization yielded shard-like morphologies much like those grown at room temperature from water. We did not confirm that DMSO was completely evaporated before loading the chip into the mass spectrometry chamber, but the crystals yielded very good spectra. It is possible that any remaining liquid evaporated in the MALDI chamber, which is under vacuum.

RESULTS AND DISCUSSION

Disulfide reduction of insulin breaks the molecule into its constitutive a- and b-chain polypeptides. For a sample containing insulin and a reducing agent, the reduction efficiency can be approximated using MALDI-MS spectra by comparing signal intensities of intact insulin and b-chain peaks. We define an intensity ratio, which is equal to the intact molecule peak intensity divided by that of the b-chain peak. Fig. 7 has a representative spectrum for the 50 °C case showing b-chain and intact insulin peaks; they were observed at approximately 5740 Da and 3400 Da, respectively. MALDI-MS data for 1 µL samples on heaters at temperatures from 22 to 70 °C for 180 seconds is summarized by a plot of intensity ratios in Fig. 7; it shows improved disulfide reduction efficiency with increasing temperature.

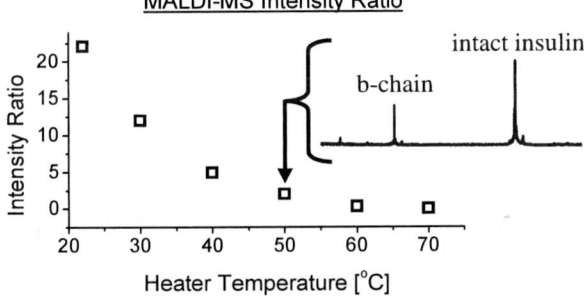

Figure 7: MALDI-MS spectra intensity ratio ($I_{intact\ insulin}/I_{b-chain}$) vs. heater temperature for 1 µL samples heated for 180 seconds. A representative spectrum from the 50 °C case shows b-chain and intact insulin peaks.

The above results represent a simple temperature study in which samples were prepared automatically in multiplex fashion by EWOD with local integrated heating. It demonstrates the ability to thermally cycle discrete droplets surrounded by air, with control over heating and chemical reaction times.

Elevated temperatures are necessary for not only thermally cycling, but also evaporating high boiling point solvents within practical timeframes, which is desirable for MALDI-MS characterization. Disulfide reductions were performed using the protocol shown in Fig. 5 with 0.25 µM insulin and 12 mM DTT in 50 % DMSO instead of aqueous acetonitrile. Fig. 8 shows a representative heater time history. A set point of 130 °C was maintained for about 10 seconds during disulfide reduction and 90 °C for about 250 seconds during hot crystallization.

Fast insulin disulfide reductions in DMSO yielded encouraging results, showing that local high-temperature cycling of droplets makes it possible to prepare samples for MALDI-MS characterization in about 5 minutes.

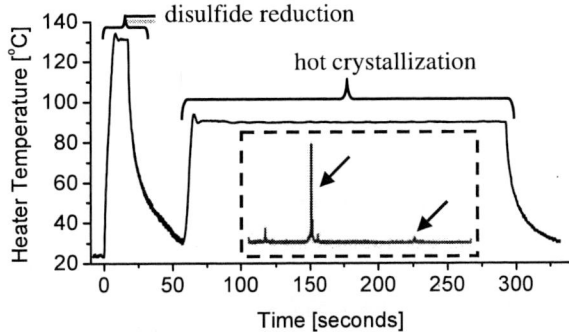

Figure 8: Time history for 1 µL 50% DMSO sample with reaction at 130 °C and crystallization at 90 °C. The corresponding MALDI-MS spectrum is shown inside the dashed box and indicates a nearly complete reduction, evidenced by a large b-chain peak (left arrow) and a tiny intact insulin peak (right arrow).

CONCLUSIONS

The work introduces and confirms the potency of an EWOD DMF chip with integrated local heating. By simple MEMS fabrication, we built a scalable device that performs multiplex thermal cycling of discrete microscale droplets. An automated temperature study of insulin disulfide reductions in DTT was carried out on the chip, which was loaded directly into a MALDI-MS chamber for characterization. Results verified that reaction efficiency increases with temperature.

AWKNOWLEDGEMENTS

Special thanks to Dr. Prosenjit Sen. This work was supported by NIH (1 R01 RR020070:01A2) and NSF Integrative Graduate Education and Research Traineeship (IGERT) through the UCLA Materials Creation Training Program (MCTP).

REFERENCES

[1] S. K. Cho, H. Moon, and C.-J. Kim, "Creating Transporting, Cutting, and Merging Liquid Droplets by Electrowetting-Based Actuation for Digital Microfluidic Circuits," *J. Microelectromech. Syst.*, 12 (2003), pp. 70-80.

[2] C. Zhang and D. Xing, "Miniaturized PCR chips for Nucleic Acid Amplification and Analysis: Latest Advances and Future Trends." *Nucleic Acids Research*, vol. 35, no. 13 (2007), pp. 4223-4237.

[3] Z. Q. Niu, W. Y. Chen, S. Y. Shao, X. Y. Jia, and W. P. Zhang, "DNA Amplification on a PDMS-glass Hybrid Microchip." *J. Micromech and Microeng.*, 16 (2006) pp. 425-433.

[4] H. Moon, A. R. Wheeler, R. L. Garrell, J. A. Loo, and C.-J. Kim, "An Integrated Digital Microfluidic Chip for Multiplexed Proteomic Sample Preparation and Analysis by MALDI-MS," *Lab Chip*, 6 (2006), pp. 1213-1219.

[5] K. P. Nichols and J. G. E. Gardeniers, "A Digital Microfluidic System for the Investigation of Pre-Steady-State Enzyme Kinetics Using Rapid Quenching with MALDI-TOF Mass Spectrometry," *Anal. Chem.*, 79 (2007), pp. 8699-8704.

FREQUENCY DRIFT COMPENSATION IN MASS-SENSITIVE CHEMICAL SENSORS BASED ON PERIODIC STIFFNESS MODULATION

K. S. Demirci[1], J. H. Seo[1, 3], S. Truax[1], L.A. Beardslee[1], Y. Luzinova[2], B. Mizaikoff[2], and O. Brand[1]

[1] School of Electrical & Computer Engineering, Georgia Institute of Technology, Atlanta, GA, USA

[2] School of Chemistry & Biochemistry, Georgia Institute of Technology, Atlanta, GA, USA

[3] current affiliation: Qualcomm MEMS Technologies, Qualcomm Inc., San Jose, CA 95134, USA

ABSTRACT

The successful compensation of frequency drift in a mass-sensitive chemical microsensor is demonstrated. The proposed compensation method uses a periodic stiffness modulation, generated by a second feedback loop, to monitor the microresonator's quality factor (Q-factor). The Q-factor is solely obtained from frequency measurements and monitored along with the measurand-induced frequency shift during normal closed-loop sensor operation. This simultaneous measurement of Q-factor and frequency shift enables the compensation of frequency drift induced by environmental disturbances using the extracted Q-factor. The feasibility of drift compensation has been demonstrated by implementing the compensation scheme into a closed-loop chemical sensing system and performing gas-phase chemical measurements.

INTRODUCTION

In mass-sensitive chemical microsensors [1-3], a microresonator is covered with a sensitive layer, and the change in its resonance frequency in response to mass changes during analyte sorption is monitored. Although microresonators show high mass-sensitivities because of their small dimensions, their frequency output is also affected by aging and environmental disturbances, i.e. variations in temperature, humidity and atmospheric pressure. Frequency drift caused by these environmental changes limits the resolution of mass-sensitive microsensors, especially in long-term monitoring applications with slowly changing analyte signatures. Therefore, without compensating for the frequency drift caused by environmental variations, a high mass-resolution cannot be achieved.

In resonant sensors, the Q-factor is one of the key parameters, which affects the (short-term) frequency stability and, thus, the sensor resolution. The Q-factor is related to the mechanical properties of the resonator, namely its stiffness, mass, and damping. Therefore, any variation in these properties in response to a measurand causes a Q-factor change. In particular, this means that the response of a resonant sensor to environmental disturbances can be monitored via the Q-factor. This fact has been exploited by using the Q-factor to sense e.g. temperature [4], fluidic properties such as viscosity and density [4, 5], and pressure [6].

In addition to monitoring environmental variations, the Q-factor can be used to compensate for frequency drift caused by them. In [7], the Q-factor has been utilized as a temperature sensor and used to compensate for temperature-induced frequency drift. The Q-factor has been obtained from the resonator's oscillation amplitude, requiring precise A/D conversion, in which case the advantages of a resonant sensor are not fully utilized.

In the present work, we demonstrate the compensation of frequency drift induced by environmental disturbances of a mass-sensitive chemical microsensor using frequency measurements only. The compensation method uses a stiffness modulation, generated by a feedback loop, to track the resonator's Q-factor and has been previously applied to temperature compensation [8]. Here, this compensation scheme has been successfully implemented into a closed-loop chemical sensing system. Moreover, an improved scheme using both positive and negative stiffness modulation is used to accurately track the Q-factor irrespective of phase-tuning errors in the feedback loop. Therefore, the Q-factor is solely obtained from frequency measurements without need for additional sensing devices and monitored along with the measurand-induced frequency shift during normal closed-loop sensor operation.

PRINCIPLE OF DRIFT COMPENSATION

For the application as a mass-sensitive sensor, the microresonator is generally incorporated in an amplifying feedback loop as the frequency-determining element. In this feedback loop (see main loop in Fig. 1), the resonator output signal is amplified and phase matched with the vibration velocity of the resonator; the resulting signal is amplitude-limited using a comparator and fed back to the excitation terminal via a buffer amplifier. This primary excitation signal compensates for the resonator damping and a constant-amplitude oscillation is sustained at the mechanical resonance frequency. This way, the measurand induced frequency change can be simply detected by using a frequency counter.

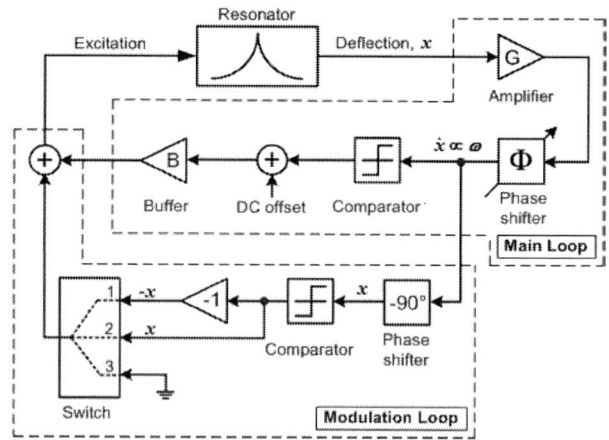

Figure 1: Schematic diagram of feedback circuitry for resonator with main loop and modulation loop for the implemented drift compensation method.

978-1-4244-2977-6/09 $25.00 © 2009 IEEE

Besides the main loop, a second electronic feedback loop generates a signal in-phase with the deflection of the resonator (see modulation loop in Fig.1). This signal is applied to the resonator excitation elements in addition to the signal generated by the main loop and modulates the effective stiffness of the resonator [9]. By controlling the polarity of the modulation signal, the effective stiffness of the resonator can be either reduced or increased. As a result, the resonance frequency of the harmonic oscillator is changed in response to the modulated effective stiffness.

The force-deflection relation of a mechanical resonator when modeled by a simple second-order system consisting of mass m, damper b and spring k is given by Eq. (1). The motion of the resonator considered in this work with main loop and enabled modulation loop can be described by Eq. (2) using the describing function of a comparator, defined as the complex ratio of the fundamental harmonic component of the comparator output signal to its input sinusoidal signal [10].

$$m\ddot{x} + b\dot{x} + kx = F \tag{1}$$

$$m\ddot{x} + b\dot{x} + kx = \psi_{main}\frac{\dot{x}}{|\dot{x}|} \mp \psi_{mod}\frac{x}{|x|} \tag{2}$$

The force amplitudes ψ_{main} and ψ_{mod} in Eq. (2) are defined as

$$\psi_{main} = \chi\gamma\frac{4M_{main}}{\pi}, \quad \psi_{mod} = \chi\gamma\frac{4M_{mod}}{\pi} \tag{3}$$

Here, χ is the transfer function of the generated mechanical force to an applied electrical excitation signal, γ is the transfer function of the electrical read-out signal to the resonator deflection, and M_{main} and M_{mod} are the amplitudes of the square wave signals generated by the comparators in the main and modulation loop, respectively. Combining Eqs. (2) and (3), the system equation for the resonator shown in Fig. 1 is expressed by

$$m\ddot{x} + \left[b - \frac{\psi_{main}}{|\dot{x}|}\right]\dot{x} + \left[k \pm \frac{\psi_{mod}}{|x|}\right]x = 0 \tag{4}$$

Under steady-state oscillation, the force generated by the main loop is the same as the damping force of the resonator. In this case, the magnitudes of the steady-state vibration velocity $\dot{x}$ and the vibration amplitude x of the resonator embedded in the feedback circuit are obtained from Eq. (4) by forcing the effective damping force (second term in Eq. (4)) to zero, i.e.

$$\dot{x} = \frac{\psi_{main}}{b}, \quad x = \frac{\psi_{main}}{b\omega_o} \tag{5}$$

As a result, the equation of motion becomes:

$$m\ddot{x} + \left[k \pm \frac{\psi_{mod}}{\psi_{main}}b\omega_o\right]x = 0 \tag{6}$$

Using the relation $Q = k\,b^{-1}\omega_0^{-1}$, Eq. (6) can be expressed in terms of Q-factor by

$$m\ddot{x} + \left[k\left(1 \pm \frac{\psi_{mod}}{\psi_{main}}\frac{1}{Q}\right)\right]x = 0 \tag{7}$$

Eq. (7) can be simplified to Eq. (8), if the amplitudes of the excitation signals generated by the main and modulation loops are equal, i.e. $\psi_{main} = \psi_{mod}$

$$m\ddot{x} + \left[k\left(1 \pm \frac{1}{Q}\right)\right]x = 0 \tag{8}$$

Eq. (8) reveals that the relative stiffness change induced by the modulation loop is directly related to the Q-factor of the resonator. In fact, the stiffness is modified by a factor of $(1 \pm Q^{-1})$ by the modulation loop in Fig. 1.

It should be noted that Eqs. (2) - (8) only hold when the excitation signals generated by main and modulation loop are exactly phase matched with the vibration velocity $\dot{x}$ and the deflection of the resonator x, respectively. In a real system, the phase-tuning is done by an electronic circuit, such as an all-pass filter, and there is a possibility of a tuning error, which will introduce an error in the estimated Q-factor. To minimize this phase tuning error, the polarity of the signal generated by the modulation loop is altered to obtain both positive and negative stiffness modulations, resulting in an increase or decrease in the resonance frequency. In this case, the resonance frequencies with positive and negative stiffness modulation, respectively, become

$$\omega_{pos} = \sqrt{\frac{k(1 + 1/Q)}{m}}, \quad \omega_{neg} = \sqrt{\frac{k(1 - 1/Q)}{m}} \tag{9}$$

From the ratio of the frequencies $\omega_{pos}/\omega_{neg}$, the Q-factor of the resonator can be calculated as

$$Q = \frac{\alpha + 1}{\alpha - 1}, \quad \alpha = \left(\frac{\omega_{pos}}{\omega_{neg}}\right)^2 \tag{10}$$

Therefore, by measuring the resonance frequency changes caused by the periodically enabled modulation loop of Fig. 1, the Q-factor (or the relative Q-factor change) of a microresonator can be extracted and tracked during closed-loop sensor operation.

If the influence of the Q-factor on the resonance frequency is known, the measured Q-factor data can be used to compensate for frequency drift. In the present work, an initial calibration step establishes the relation between the relative Q-factor dQ/Q and frequency $d\omega_0/\omega_0$ changes.

MEASUREMENT SETUP

To demonstrate the feasibility of the drift compensation scheme for mass-sensitive chemical microsensors, gas-phase chemical measurements have been performed using the setup shown in Fig. 2.

978-1-4244-2977-6/09 $25.00 © 2009 IEEE

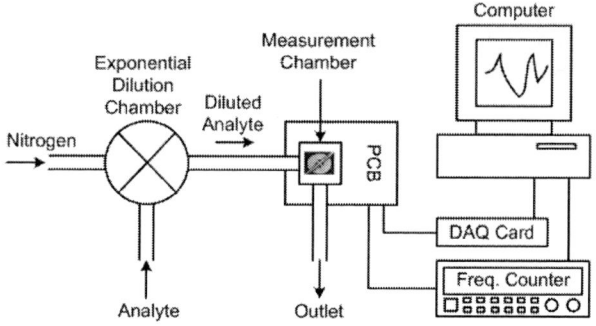

Figure 2: Schematic diagram of measurement setup. The PCB includes the feedback circuitry, and the DAQ card controls the switch position (see Fig. 1).

In the chemical measurements, a known volume of analyte is injected into an exponential dilution chamber [11] and continuously diluted by nitrogen flowing through the chamber at a constant rate of 30ml/min. The time-dependent analyte concentration in the measurement chamber is given by Eq. (11), where C_o is the initial analyte concentration inside the dilution chamber and α is the ratio of the nitrogen flow rate to the dilution chamber volume. The diluted analyte is drafted onto the resonator.

$$C_{analyte} = C_o e^{-\alpha t} \qquad (11)$$

A custom-made printed circuit board includes the feedback circuitry shown in Fig. 1. A LabView program controls the switch position shown in Fig. 1 via a National Instruments data acquisition (DAQ) card and reads the resonance frequency from a frequency counter.

The measurements have been performed with silicon-based disk-type microresonators [12] featuring integrated electrothermal excitation and piezoresistive detection elements (see Fig. 3). The microresonator is spray-coated with polyisobutylene (PIB) as chemically sensitive polymer film. The film thickness was measured as 0.3μm using a profilometer. Thereafter, the resonator is placed inside the air-tight measurement chamber.

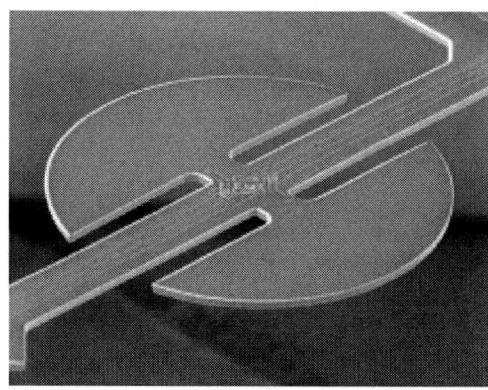

Figure 3: SEM photograph of silicon-based disk-type resonator with electrothermal excitation and piezoresistive detection elements.

RESULTS

During the chemical measurements, the negative and positive stiffness modulation (switch position 1 and 2 in Fig. 1) are periodically enabled for 2 seconds each, every 54 seconds (see Fig. 4). The disk-type resonator with a 533.4 kHz resonance frequency exhibits a frequency change (f_{pos}-f_{neg}) of approximately 241 Hz by the enabled modulation loop. Generally, a larger Q-factor yields a smaller frequency change (f_{pos}-f_{neg}). To prevent variations in resonance frequency by different static power dissipations in the excitation resistors with enabled and disabled modulation loop, the static power dissipation produced by the electrothermal excitation elements is kept constant during the experiment [8].

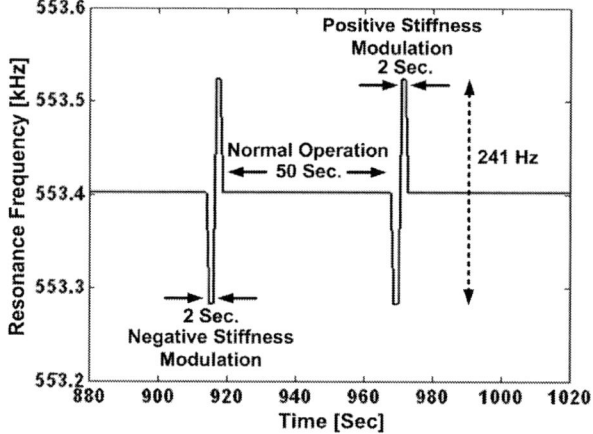

Figure 4: Resonance frequency of disk-type resonator with periodically enabled positive and negative stiffness modulation.

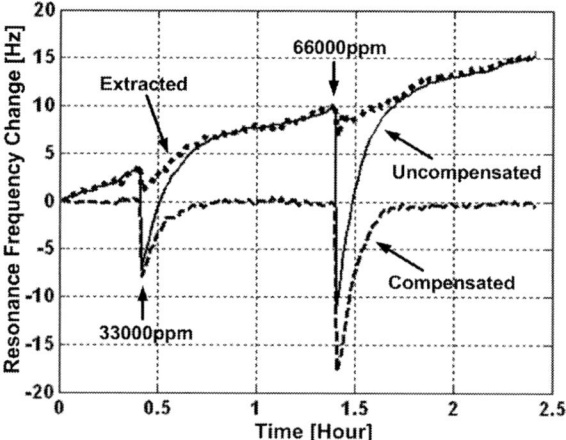

Figure 5: Resonance frequency change with (dashed line) and without (solid line) compensation during a gas-phase chemical measurement; the resonator (f=553.4kHz) is subsequently exposed to different concentrations of ethyl chloride. The dotted line shows the extracted frequency change based on the Q-factor measurement.

Fig. 5 shows the resonance frequency change with and without compensation during a gas-phase chemical measurement without temperature control, with two subsequent analyte injections into the exponential dilution chamber with different ethyl chloride concentrations. The

Q-factor is extracted using Eq. (10) and shown in Fig. 6. Using this extracted Q-factor, the frequency change stemming from the Q-factor change is estimated using the relation $dQ/Q = 360 \, d\omega_0/\omega_0$ obtained in the initial calibration step. This estimated frequency change is shown as the dotted line in Fig. 5 and is used to compensate for frequency drift. The final compensated frequency signal (dashed line in Fig. 5) is simply the difference of the measured (solid line) and estimated frequency (dotted line). In the measurement shown in Fig. 5, the observed frequency drift of 15Hz is caused by a roughly 1°C temperature decrease during the entire measurement. After compensation, the frequency is stable to within ±0.3Hz over the 2.5 hour measurement time, and the frequency decreases during analyte exposure are clearly visible.

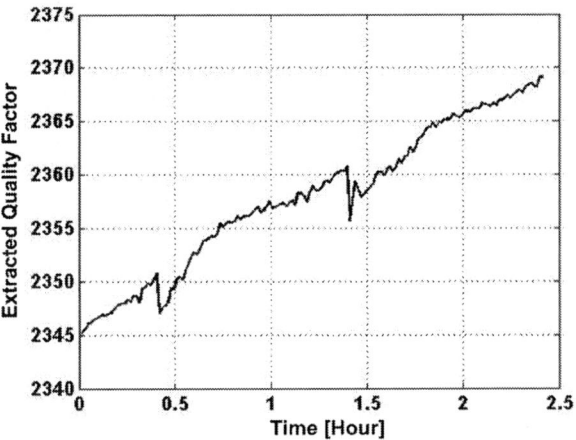

Figure 6: Extracted Q-factor from Eq. (10) during the gas-phase chemical measurement shown in Fig. 5.

CONCLUSIONS

A method for compensation of frequency drift induced by environmental disturbances in a mass-sensitive chemical microsensor is proposed and verified with gas-phase chemical measurements. The proposed method only requires frequency measurements and uses a periodic stiffness modulation that can be implemented easily and cost effectively using a simple electronic feedback loop. With this scheme, the frequency drift induced by environmental disturbances can be compensated during the normal mass-sensing operation in which the measurand-induced frequency shift is detected.

ACKNOWLEDGEMENTS

The authors would like to thank the staff of the Georgia Tech Microelectronics Research Center for their assistance during device fabrication. This work has been funded in part by the National Science Foundation under award 0601467.

REFERENCES

[1] D. Lange, C. Hagleitner, A. Hierlemann, O. Brand, H. Baltes, "Complementary metal oxide semiconductor cantilever arrays on a single chip: mass-sensitive detection of volatile organic compounds," *Analytical Chemistry*, vol. 74, pp. 3084-3095, 2002.

[2] C. Vancura, I. Dufour, S. M. Heinrich, F. Josse, A. Hierlemann, "Analysis of resonating microcantilevers operating in a viscous liquid environment," *Sensors and Actuators A*, vol. 141, pp. 43-51, 2008.

[3] S. Truax, K. S. Demirci, J. H. Seo, P. Kurzawski, Y. Luzinova, A. Hierlemann, B. Mizaikoff, O. Brand, "Gas and liquid phase sensing of volatile organics with disk microresonator," in *Proc. IEEE Conf. Micro Electro Mechanical Systems (MEMS 2008)*, pp. 220-223, 2008.

[4] M. K. Jain, S. Schmidt, C. A. Grimes, "Magneto-acoustic sensors for measurement of liquid temperature, viscosity and density," *Applied Acoustics*, vol. 62, pp. 1001-1011, 2001.

[5] S. J. Martin, R. W. Cernosek, J. J. Spates, "Sensing liquid properties with shear-mode resonator sensors," in *Proc. Transducers '95*, vol. 2, pp. 712-715, 1995.

[6] S. Bianco, M. Cocuzza, S. Ferrero, E. Giuri, G. Piacenza, C. F. Pirri, A. Ricci, L. Scaltrito, D. Bich, A. Merialdo, P. Schina, R. Correale, "Silicon resonant microcantilevers for absolute pressure measurement," *J. Vacuum Science Technology B*, vol. 24, pp. 1803-9, 2006.

[7] M. A. Hopcroft, M. Agarwal, K. K. Park, B. Kim, C. M. Jha, R. N. Candler, G. Yama, B. Murmann, T. W. Kenny, "Temperature compensation of a MEMS resonator using quality factor as a thermometer," in *Proc. IEEE Conf. Micro Electro Mechanical Systems (MEMS 2006)*, pp. 222-225, 2006.

[8] J.H. Seo, K. S. Demirci, A. Byun, S. Truax, O. Brand, "Temperature compensation method for resonant microsensors based on a controlled stiffness modulation," *Journal of Applied Physics*, v 104, pp. 014911-1-9, 2008.

[9] R. Sunier, T. Vancura, Y. Li, K.-U. Kirstein, H. Baltes, O. Brand, "Resonant magnetic field sensor with frequency output," *J. Microelectromechanical Systems*, vol. 15, pp. 1098-1107, 2006.

[10] K. Ogata, *Modern Control Engineering*, 4th ed., Prentice Hall, 2002.

[11] C. Charlton, B. Temelkuran; G. Dellemann, B. Mizaikoff, "Mid-infrared sensors meet nanotechnology: trace gas sensing with quantum cascade lasers inside photonic band-gap hollow waveguides," *Applied Physics Letters*, vol. 86, 2005, pp. 194102-1-3.

[12] J.H. Seo, O. Brand, "High Q-factor in-plane mode resonant microsensor for gaseous/liquid environment," *J. Microelectromechanical Systems*, vol. 17, pp. 483-493, 2008.

µGC MULTICAPILLARY COLUMNS WITH MONO-LAYER PROTECTED GOLD AS A STATIONARY PHASE

M. A. Zareian-Jahromi, M. Agah

MEMS Lab, Virginia Tech, Blacksburg, Virginia, USA

ABSTRACT

This paper reports the first MEMS-based stationary phase coating technique for ultra-narrow single capillary (SCC) and multicapillary (MCC) microfabricated gas chromatography (µGC) columns yielding the highest separation performance reported to date. The conventional coating methods are mainly based on the solvent evaporation of a polymeric solution. These methods, which are developed for fused silica capillary tubing, cannot afford conformal coating in the rectangular-shaped µGC columns with high-aspect-ratio structures. The proposed wafer-level coating technique overcomes this traditional challenge by merging MEMS and nanotechnology. The new µGC stationary phase has been achieved by depositing a uniform functionalized gold layer with an adjustable thickness (150nm-2µm) in 25µm-wide single columns as well as in four-capillary MCCs.

INTRODUCTION

Gas Chromatography is a unique and versatile technique for separation and identification of volatile and semi-volatile organic mixtures. As an analytical tool, which was introduced in 1950s, GC has a wide range of applications in the pharmaceutical industry, environmental monitoring, petroleum distillation, clinical chemistry, and food processing.

In a GC system, the mixture to be separated and analyzed is vaporized and injected into a column whose walls are coated by a polymer called stationary phase. The mixture components traverse the length of the column in a mobile phase (i.e. carrier gas) at rates determined by their retention in the stationary phase. If the column length and difference in the retention times are sufficient, a complete separation of components is possible. The separated components pass over a detector such as a flame ionization detector (FID), which generates a signal corresponding to the mass of the component present in the sample [1, 2].

Conventional GCs provide accurate analysis of complex mixtures but at the expense of using large, power-hungry, and relatively expensive table-top instruments. MEMS technology has already demonstrated the possibility of realizing micro gas chromatography systems (µGC) which exhibit faster analysis times, lower power consumption, and higher portability compared to bulky conventional GCs [2-4]. A µGC system is a hybrid integration of several micromachined modules such as a preconcentrator (PC), separation column, gas detector, micro valves and pumps. The design and optimization of each individual module is the current trend in µGC development [5-7].

The development of high-speed and high-performance separation columns as the heart of the system is of paramount importance in the world of µGC for on-field applications such as environmental monitoring, breath analysis, and public safety. µGC columns are generally fabricated in silicon by anisotropic deep reactive ion etching (DRIE) and then are sealed by a Pyrex wafer. The columns are then coated by a thin layer of a polymeric stationary phase using conventional static or dynamic techniques. In HiltonHead 2008, we introduced the concept of MCCs and showed theoretically that these columns as well as ultra-narrow SCCs can achieve fast separation of gas mixtures with high resolution suitable for in-field applications [8]. However, since the early µGC work [9], stationary phase coating of MEMS columns has been a challenge mainly because the conventional techniques relying on solvent evaporation can be utilized after separation channels are sealed. These methods, which are developed for fused silica capillary tubing, cannot afford conformal coating in the rectangular-shaped µGC columns. They suffer from pooling effects [1] and can plug the column during solvent evaporation especially as the column width is reduced to enhance the separation efficiency [8]. Moreover, even for wide µGC columns, the thickness and surface properties are not easily adjustable. The presented technique shifts the coating step before column sealing and takes advantage of microfabrication techniques along with nanotechnology in order to obtain stationary phases with high uniformity, reproducibility, and robustness. The following sections describe the proposed fabrication steps and experimental evaluation of the separation performance.

FABRICATION

Column Fabrication

In order to study the proposed coating method, SCC columns with capillary widths ranging from 25µm to 250µm as well as MCCs comprising four 63µm-wide capillaries were designed and fabricated. All the columns had a length of 25cm. The fabrication of both silicon-glass SCCs and MCCs started with cleaning a silicon wafer in a Piranha solution. Then, a 10µm-thick layer of AZ9260 photoresist was spun on the wafer at 3krpm for 60 seconds. After patterning the photoresist, the wafer was etched to a depth of 250µm using DRIE to form channels with aspect ratios varying between 1 (for 250µm-wide columns) and 10 (for 25µm-wide columns) (Fig. 1a).

Coating Method

The surface treatment includes four steps: deposition, patterning, stabilizing and functionalizing.

1) Deposition

In order to deposit a conformal layer in high-aspect-ratio channels, we used pulse electroplating technique. The proposed seedless gold electroplating on heavily phosphorous (n$^+$) doped silicon affords a variable thickness of gold layer which would be easily patterned after deposition due to low adhesion strength between gold and silicon. First, Silicon wafer with etched channels is doped by solid source diffusion of phosphorous at 950°C for 6 hours. The 2μm-deep doped region created on the silicon surface is used as the cathode. Before dipping the channels in electroplating bath, the native oxide is etched by HF. The electroplating bath is maintained at a temperature of 55°C with a stirring rate of 200rpm. The pulse electroplating has been adjusted based on theoretical models for maximum pulse duration (<2sec) and experimental modification for minimum time period (>45sec) to deposit a conformal layer on all the column surfaces (Fig. 2). Unlike solvent evaporation method, the stationary phase thickness in this new method is independent of the column size and is adjustable to maximize the separation performance [8].

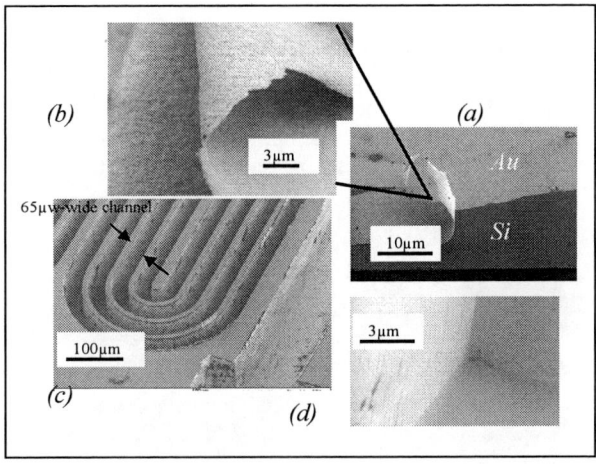

Fig. 2: SEM images show the deposition quality, (a) gold layer folded by mechanical force demonstrating its low adhesion (b) cross-section of 150nm-thick gold layer showing the uniformity of plated layer (c) 4-channel MCC after removing gold from the top flat surface, (d) showing the complete gold coverage on the corner. .

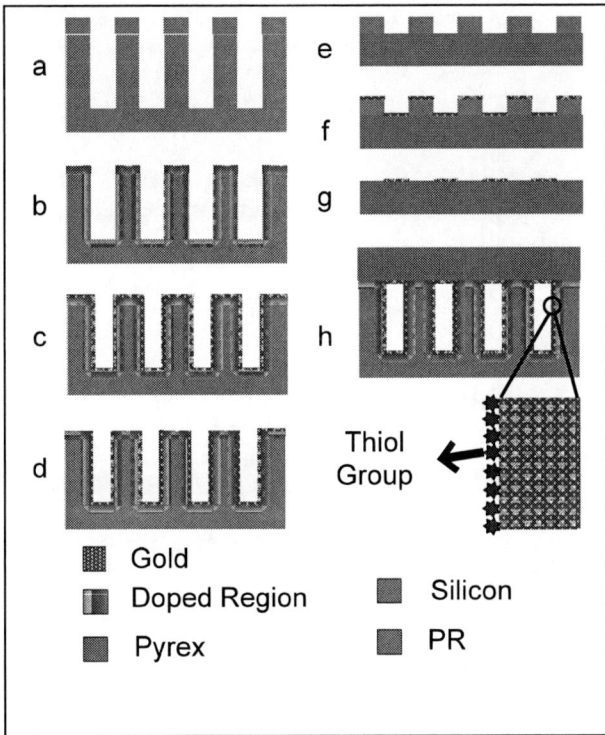

Fig. 1: Fabrication process flow of the MEMS μGC column, (a) DRIE, (b) phosphorous diffusion (c) gold electroplating, (d) gold patterning, (e-g) Ti/Au deposition and patterning (h) anodic bonding

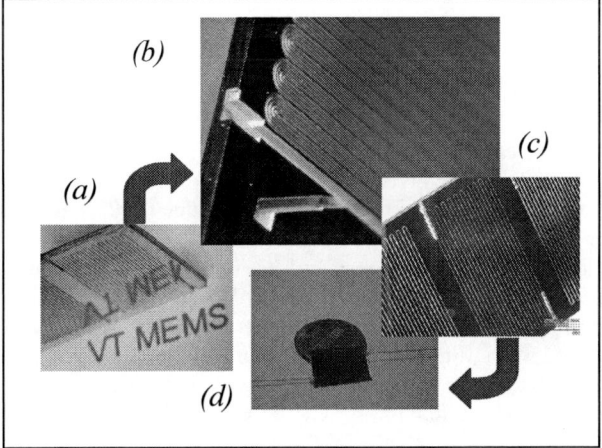

Fig. 3: (a) 25μm-wide, 25cm-long, 250μm-deep SCC coated by 150nm gold layer; the reflective surface shows the quality of the deposition (b) 4-channel MCC after gold patterning (c) 25μm- and 100μm wide SCCs after gold plating (d) a μGC column die.

2) Patterning

As Fig. 2 and Fig. 3 show, the seedless electroplated gold has very low but adequate adhesion to silicon. This is a critical characteristic making it easily removable from the top flat surface by an adhesive tape without affecting the coverage on the column sidewalls. Furthermore, the high value of silicon Young's modulus causes its perseverance to high vertical stress. This clearly demonstrates that it is possible to pattern the gold layer even for very tiny structures without applying any shear stress. After this patterning step, the silicon wafer was anodically bonded to a Pyrex 7740 substrate patterned with evaporated Ti/Au (Fig. 1e-g). The bonding temperature and pressure were 350°C and 2KPa, respectively, and applying a DC voltage of 1250V for 20 minutes established a permanent bonding.

3) Stabilizing

Thermal annealing during the bonding process at temperatures less than the Si-Au eutectic temperature (365°C) stabilizes the electroplated gold by significantly improving its adhesion to the doped silicon surface [10]. The gold coating was proved to be stable under GC operating temperature (350°C) and flow conditions (1ml/min). After bonding, the wafer was diced and fused silica capillary tubing with 167-μm-O.D. and 100-μm-I.D. was attached to fluidic ports.

4) Functionalizing

The last step is performed by filling the column with a solution of octadecylthiol ($C_{18}H_{35}SH$) in hexane for six hours. This functionalizing process attached the alkane group to the electroplated gold surface by self assembly

978-1-4244-2977-6/09 $25.00 © 2009 IEEE

and resulted in a monolayer-protected gold (MPG) layer that could be used as a stationary phase.

Fig. 4a shows an AFM image of the gold surface topology and its nanometer-scale grain size. The XPS analysis, Fig. 4b, proves the existence and the high uniformity of sulfur (thiol monolayer) on the gold surface.

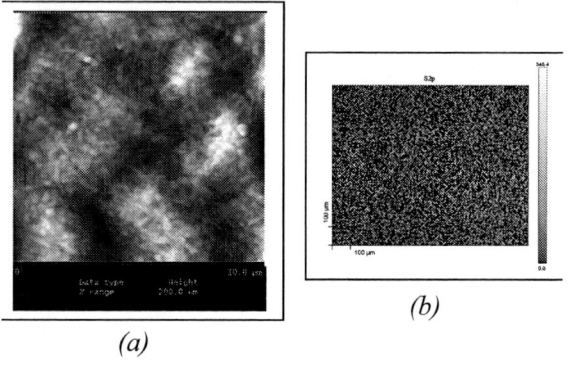

(a)   (b)

Fig. 4: Surface Analysis, (a) AFM image shows the topology of the 200nm-thick pulse electroplated gold (b) XPS image of MPG coated silicon surface shows the existence and uniformity of the sulfur (white dots)

EXPERIMENTS AND RESULTES

In order to evaluate the capabilities of the microfabricated columns, several experiments were conducted using conventional GC system (5890 Series II GC, Hewlett Packard) having a split inlet (200:1) and flame ionization detector (FID). The capillary column inside the GC was replaced by the MEMS column. The autosampler (7673B, Hewlett Packard) was used to inject the samples. Air was used as a carrier gas. All gases were purified with filters for water vapor and hydrocarbons. The injector and FID temperatures were both maintained at 280°C. The efficiency of GC columns is expressed by the height-equivalent-to-a-theoretical-plate (HETP). Dodecane was used to determine HETP because of having a retention factor more than 10, thus making HETP less dependent on the compound selected (k) and experimental conditions (mainly temperature) [8].

HETP was measured by injecting a 1μl mixture of dodecane ($C_{12}H_{20}$) and methylene chloride (CH_2Cl_2). The dodecane retention time t_R was used to determine the HETP experimentally based on:

$$HETP = \frac{L}{5.54(\frac{t_R}{w_h})^2} = \frac{1}{Plate\ Number}$$

where L and w_h are the length and width of the peak at the half height. Only symmetric peaks (symmetry of 0.9-1.1) were used to measure HETP.

As Table 1 indicates, 25μm-wide columns with MPG stationary phase have yielded 20,000 plates/m, which is the world-record separation efficiency. In addition, this is the first time that we have been able to successfully coat MEMS-based MCCs to achieve high-performance separations. Also, the separation performance of the SCCs and MCCs have been explored by separating a gas mixture containing n-alkanes (C_9-C_{10}-C_{12}-C_{14}-C_{16})

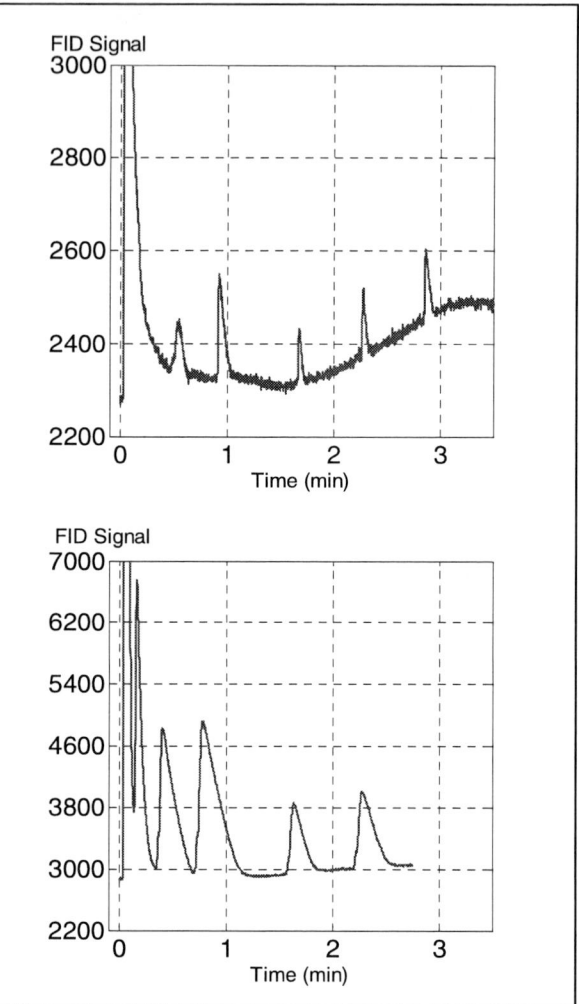

Fig. 5: Chromatogram of five alkanes (in order from left to right C_9, C_{10}, C_{12}, C_{14}, C_{16}), obtained by 25μm-wide SCC (top) and 4-capillary MCC (bottom) at temperature rate of 30°C/min.

(Sigma-Aldrich Inc., MO, US), in the conventional GC oven. Fig 5 shows the gas separations obtained by the high performance ultra-narrow columns (25μm-wide SCCs) and high sample capacity columns (4-channel MCCs with ~65μm-wide capillaries) at a 30°C/min temperature programming rate. It is worth mentioning that the sample concentration in MCC case is almost ten times more than the sample used for SCC to show the higher sample capacity of MCC design. The SCC column was saturated at this sample concentration and could not produce any separation.

Conclusion

This paper reports for the first time a high-yield MEMS-based surface treatment method for high-aspect ratio μGC separation channels. This new stationary phase coating technique is based on seedless gold electroplating and then functionalizing the gold layer with thiol groups using self assembly. The conformal deposition of gold with varying thickness and ability to be patterned on 3D structures is the main advantage of this method. Furthermore, the inherent properties of gold such as biocompatibility, conductivity and chemical stability are

978-1-4244-2977-6/09 $25.00 © 2009 IEEE

another important feature of the technique presented in this paper. It is notable that a wide variety of thiol-based organic molecules can be self-assembled to tune the polarity of this new class of μGC stationary phases.

Table 1: Experimental results, shows the column performance evaluated by measuring plate number (N), using dodecane (retention factor, k>10) at 65°C isothermal chromatogram.

Column Design	Plate number
	MPG
100μm-SCC	3000
25μm-SCC	20000
4-MCC	7700

ACKNOWLEDGMENT

The authors would like to thank Mr. Vaibhav Jain and Professor James Heflin for their technical assistance. This work has been supported primarily by the National Science Foundation under Award Number ECCS-0601456.

REFERENCES

[1] R. L. Grob and E. F. Barry, *Modern Practice of Gas Chrmatography* fourth ed. Hoboken, NJ: Wiey-Interscience, 2004.

[2] M. Agah and K. D. Wise, "Low-mass PECVD oxynitride gas chromatographic columns," *Journal of Microelectromechanical Systems,* vol. 16, pp. 853-860, 2007.

[3] P. R. Lewis, R. P. Manginell, D. R. Adkins, R. J. Kottenstette, D. R. Wheeler, S. S. Sokolowski, D. E. Trudell, J. E. Byrnes, M. Okandan, J. M. Bauer, R. G. Manley, and G. C. Frye-Mason, "Recent advancements in the gas-phase MicroChemLab," *IEEE Sensors Journal,* vol. 6, pp. 784-795, Jun 2006.

[4] M. Agah, J. A. Potkay, G. Lambertus, R. Sacks, and K. D. Wise, "High-performance temperature-programmed microfabricated gas chromatography columns," *Journal of Microelectromechanical Systems,* vol. 14, pp. 1039-1050, 2005.

[5] B. Alfeeli, M. Ashraf-Khorassani, L. T. Taylor, and M. Agah, "Multi-Inlet/Outlet Preconcentrator With 3-Dμ-Structures Coated By Inkjet Printing of Tenax TA," in *North American Solid-State Sensors, Actuators, and Microsystems Workshop, Hilton Head Island, SC,* 2008, pp. 118-121.

[6] E. T. Zellers, S. Reidy, R. A. Veeneman, R. Gordenker, W. H. Steinecker, G. R. Lambertus, K. Hanseup, J. A. Potkay, M. P. Rowe, Z. Qiongyan, C. Avery, H. K. L. Chan, R. D. Sacks, K. Najafi, and K. D. Wise, "An Integrated Micro-Analytical System for Complex Vapor Mixtures," in *Solid-State Sensors, Actuators and Microsystems Conference, 2007. TRANSDUCERS 2007. International,* 2007, pp. 1491-1496.

[7] S. Ali, M. A. Zareian-Jahromi, M. Ashraf-Khorassani, L. T. Taylor, and M. Agah, "Semi-Packed Micro Gas Chromatography Columns," in *11th International Conference on Miniaturized Systems for Chemistry and Life Sciences (MicroTAS 2007),* Paris, France, 2007, pp. 239-241.

[8] M. A. Zareian-Jahromi, M. Ashraf-Khorassani, L. T. Taylor, and M. Agah, "Design, Modeling and Fabrication of MEMS-Based Multicapillary Gas Chromatography Columns," in *North American Solid-State Sensors, Actuators, and Microsystems Workshop,* Hilton Head Island, SC, 2008, pp. 256-259.

[9] S. C. Terry, J. H. Jerman, and J. B. Angell, "A gas chromatographic air analyzer fabricated on a silicon wafer," *Electron Devices, IEEE Transactions on,* vol. 26, pp. 1880-1886, 1979.

[10] T. Fujita, S. Nakamichi, S. Ioku, K. Maenaka, and Y. Takayama, "Seedlayer-less gold electroplating on silicon surface for MEMS applications," *Sensors and Actuators A: Physical,* vol. 135, pp. 50-57, 2007.

A SELF-CONTAINED MINIATURIZED PCR SYSTEM USING ELECTROMAGNETIC ACTUATORS

Bonnie T. Chia[1], Sheng-An Yang[1], Ming-Yuan Cheng[1],
Chun-Liang Lin[1], Chii-Wann Lin[2], and Yao-Joe Yang[1]
[1]Department of Mechanical Engineering, [2]Department of Biomedical Engineering,
National Taiwan University, Taipei, TAIWAN

ABSTRACT

In this paper, the development of a portable polymerase chain reaction (PCR) device is presented. The fully integrated self-contained system consists of four major parts: a disposable chamber chip with micro-channels and pumping membranes, a heater chip with micro-heaters and temperature sensors, a linear array of electromagnetic actuators, and a control/sensing circuit. The system can be fully operated with a 5V DC voltage, and does not require any external air compressor or bulky power supply. The size of the whole system is 67 mm × 66 mm × 25 mm, and is smaller than a PDA cell-phone. The miniaturized PCR system not only has the advantage of smaller size, less consumption of DNA solution, but also can effectively reduce the PCR process time into one-third of the time required by typical commercial PCR system.

INTRODUCTION

The concept of lab-on-a-chip (L.O.C) has made great impacts on biological and medical research by minimizing the system size, reducing the sample volume, and shortening the reaction time. The researches on miniaturized PCR chips for nucleic acid amplification have grown rapidly because of the widespread applications to DNA sequencing, medical diagnosis, disease assay, and so on. In terms of heating process, there are two typical types of PCR chips: the stationary chamber-based type and the dynamic continuous-flow-based type [1]. For the chamber-based device, the PCR process is performed by keeping DNA solution in a reaction chamber which is cycled between different temperatures [2]-[4]. In [5], virtual chambers, which are in fact DNA sample encapsulated with mineral oil, are proposed for PCR process to accelerated the cooling and heating due to smaller thermal capacity. For continuous-flow-based devices, miniaturized channels used to convey DNA solution through zones of different temperatures for specific reaction duration [6]-[8]. This approach does not require heating or cooling each zone to different temperature at different time, and therefore might speed up the whole PCR process.

In this work, we develop a self-contained miniaturized chamber-based PCR system which employs reliable electromagnetic actuators for pushing DNA solution between the reaction chambers made by PDMS polymer. The proposed device has three chambers of different temperatures. During PCR process, DNA solutions are pushed among these chambers by tiny electromagnetic actuators. A micro heater and a micro temperature sensor are monolithically fabricated in each chamber. A controller (MCU) is used for controlling the actuation sequences and the chamber temperatures.

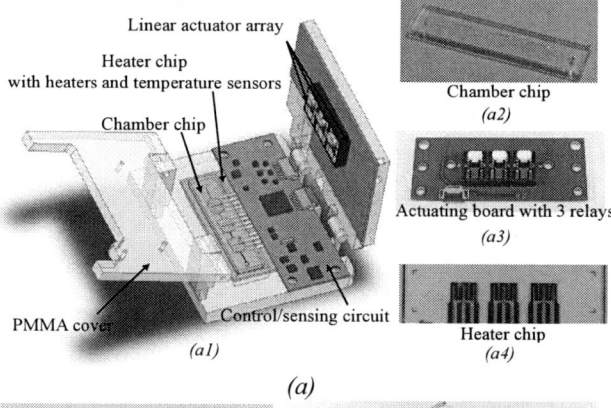

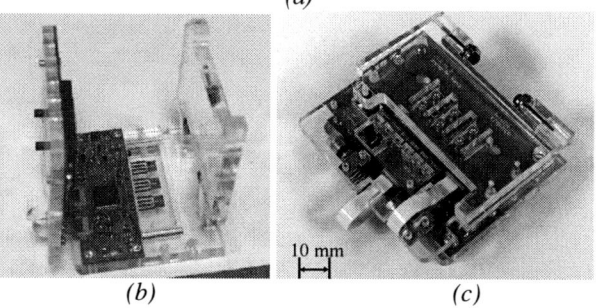

Figure 1: Miniaturized PCR system. The size of the integrated system is 67 mm × 66 mm × 25 mm.

DESIGN

Figure 1(a) shows schematic of the proposed PCR system. The detailed views of the components are shown in Figure 1(a2-a4). Figure 1(b) and 1(c) is the photographs of the assembled PCR system. The miniaturized PCR system consists of four major parts: a disposable *chamber chip* with micro-channels and pumping membranes, a *heater chip* with micro-heaters and temperature sensors, a *linear actuator array*, and a *control/sensing circuit*.

Figures 2(a) and 2(b) describe the operational principle of the device. These figures are the cross-sectional views of the $\overline{AA}$ line shown in Figure 2(c). Note that only the chamber chip, which consists of a *PDMS layer* and a thin *glass layer*, is shown in the figure. Figure 1(a2) is the picture of the fabricated chamber chip. The flexible membrane on chamber chip can be easily deformed by the compression force generated from a mini-actuator on top of each chamber. The chambers are connected by micro-channels.

As shown in Figure 2(a), these three chambers from left to right are the denaturation chamber, the extension chamber, and the annealing chamber. The temperature of each chamber is controlled by the heater chip (not shown) below the *glass layer* of the chamber chip. During the PCR process, two of the three actuators push the membranes of the corresponding chambers, and squeeze the DNA solution to the only un-squeezed chamber for

978-1-4244-2977-6/09 $25.00 © 2009 IEEE

reaction at certain temperature for a specific time period. The DNA solution is pushed back and forth among these chambers for denaturation, annealing and extension reactions for 20 or 30 amplification cycles.

Furthermore, these three actuators can also generate specific peristaltic sequences, as shown in Figure 2(b), for pumping fluid. In other words, the device can also serve as a three-phase peristaltic micro pump. Moreover, the low-cost PDMS chip, which is designed to be disposable, is the only contaminated part where DNA solution is injected and manipulated. Figure 1(a3) shows the linear actuator array on which three mini-actuators are soldered. Commercially -available mini-relays (Panasonic TQ2-L2-5V) are employed as the mini-actuators.

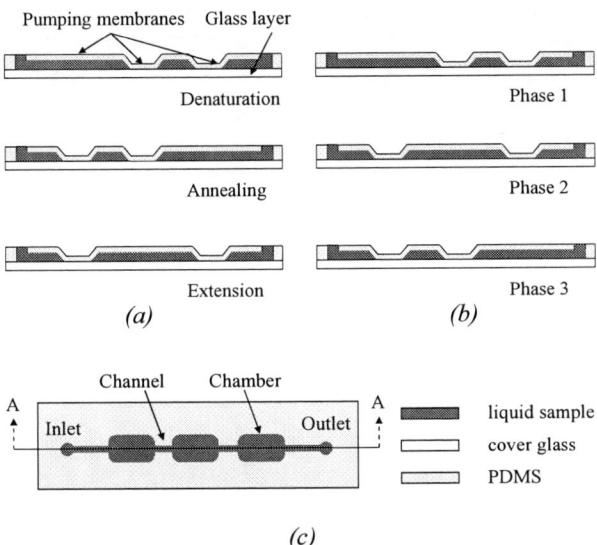

(a)   *(b)*

(c)

Figure 2: Working sequence of the actuators (a) for PCR amplification procedures; (b) as a three-phase peristaltic micro-pump. (c) The top view of the chamber chip.

For the *heater chip*, micro heaters and micro temperature sensors are realized by depositing and patterning platinum film on a glass substrate. Figure 1(a4) shows the fabricated heater chip. Real-time temperature of each chamber can be detected by the temperature sensor. By means of PID feedback control technique, the temperature in the chamber is maintained at the target level. Figure 3 shows the schematic illustration of a heater and a temperature sensor for each chamber. Since the temperature on the edge of the heater is usually lower than that in the center, the width of the serpentine wires near the edge of the chamber is narrower so that the heat generation is larger (under constant electric current). By this mean, temperature compensation around the edges can be achieved, and the temperature inside the whole chamber could be more uniform.

The control/sensing circuit controls the actuators and the heaters, and read the signals from the temperature sensors. The self-contained system can be fully operated with a 5V DC voltage, and does not require any external air compressor or bulky power supply. Six LEDs are also placed on the circuit board to indicate the operating status.

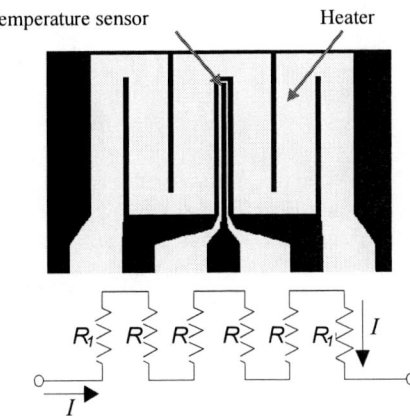

Figure 3: Design of a heater incorporated with a temperature sensor. The equivalent circuit of the heater illustrates the effect of temperature compensation.

FABRICATION

Chamber Chip

Figure 4(a) illustrates the fabrication process of the chamber chip. The chamber chip with micro-channel and micro-chamber structures is fabricated using a SU-8 mold. First, SU-8 (GM 1070) is spin-coated and patterned on a silicon wafer as the mold for the channel and chamber structures with a thickness of 100 μm. Then, PDMS replicates are created using the mold. On each PDMS replicate, two holes are punched as the inlet and outlet for loading DNA solution. Finally, the PDMS layer and a cover glass are bonded together by oxygen plasma treatment of 90 seconds.

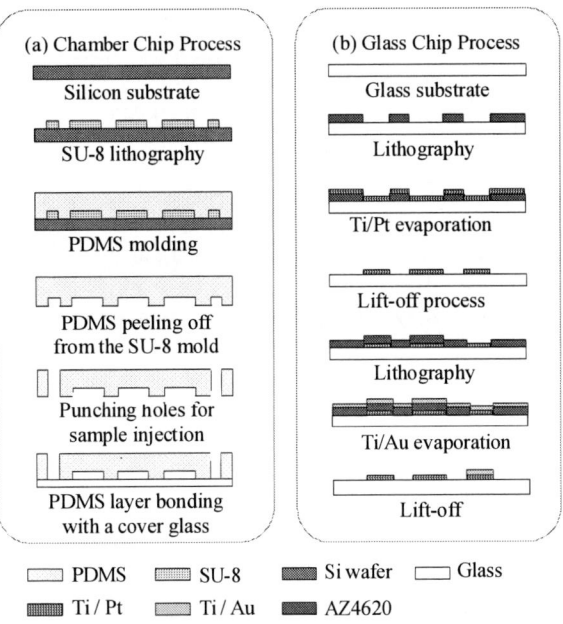

Figure 4: Fabrication process of the PDMS chip and the glass chip.

Heater Chip

Figure 4(b) describes the fabrication process of the heater chip. Micro-heaters and temperature sensors are fabricated on a glass substrate by thin-film deposition and lift-off process. First, lithography of AZ4620 as photo-resist is carried out. A 80 nm thin platinum (Pt) layer is deposited with a 20 nm titanium (Ti) adhesion layer by e-beam evaporation. The platinum pattern is defined by lift-off process in acetone solution for the temperature sensors and the micro-heaters. Again, the lithography process and lift-off process are carried out for defining the electrical leads with a 400 nm evaporated aurum (Au) layer and a 20 nm Ti adhesion layer.

RESULTS

Figure 5 illustrates the measured transient temperatures at the centers of the three chambers detected by an infrared thermometer. These three chambers are heated to 94 ℃, 54 ℃, and 72 ℃ for denaturation, annealing, and extension, respectively. The temperatures in the annealing and extension chambers reach the target levels within 15 seconds, while the denaturation chamber reaches its target level around 40 seconds. PWM method is used to feed power to the heater.

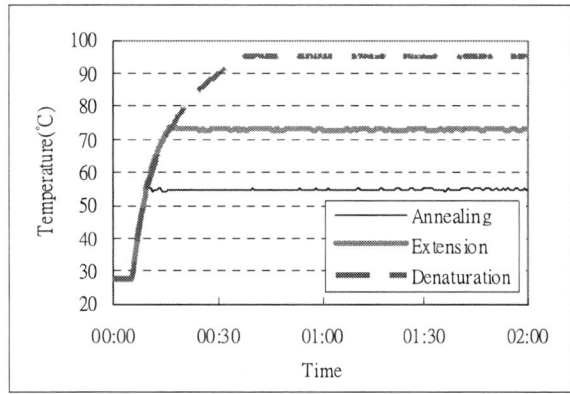

Figure 5: Temperature measured during heating by an infrared thermometer of the 3 chambers respectively.

Figure 6 shows the temperature distribution captured by an infrared imager. Figure 6(a) shows the infrared image in which all three chambers are heated simultaneously at the designated temperatures. Figure 6(b), (c), and (d) are the infrared images when one of chambers is heated to its corresponding temperature. Even though thermal interaction occurs between the three chambers, the temperatures of the three chambers are successfully controlled at the target level.

The measured flow rates at different frequencies are presented in Figure 7 by using the working sequence illustrated in Figure 2(b). Obviously, the flow rate reaches a maximum value of 25 μL/ min as the operating frequency is around 40 Hz, and then decay rapidly as the frequency increases.

During the PCR process, the chamber chip, which is loaded with DNA solution, is fastened above the heater chip by folding the PMMA covers which is integrated with the linear actuator array, as shown in Figure 1(c). By using the miniaturized system, successful PCR is performed for the MCF-7/adr cell line (122 bp segment). Primer sets of GAPDH (forward primer: 5'-AGT CAA CGG ATT TGG TCG TA-3'; reverse primer: 5'-GAA ACA TGT AAA CCA TGT AG-3') are chosen. The 8 μL DNA solutions,

which contains DNA samples, double distilled water (ddH$_2$O), and Gotaq® Green Master Mix, is injected into the PDMS chip for the PCR amplification. Firstly, the DNA solution is warmed up at 94 ℃ for 90 seconds. Then, 30 thermal cycles of different temperatures are performed. In each cycles, the set point temperatures are 94 ℃ for 25 seconds, 54 ℃ for 25 seconds and 72 ℃ for 30 seconds. Finally, temperature of 72 ℃ is kept for 5 minutes.

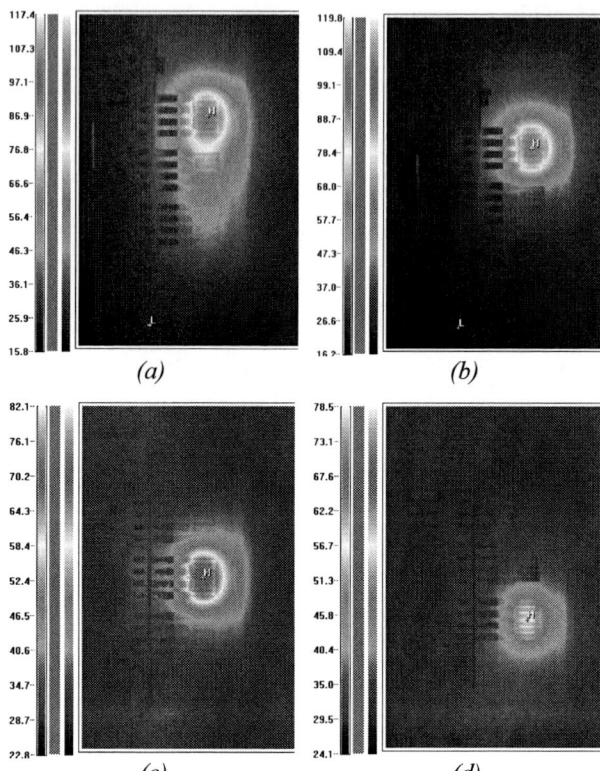

Figure 6: Infrared thermal image of the three chambers.

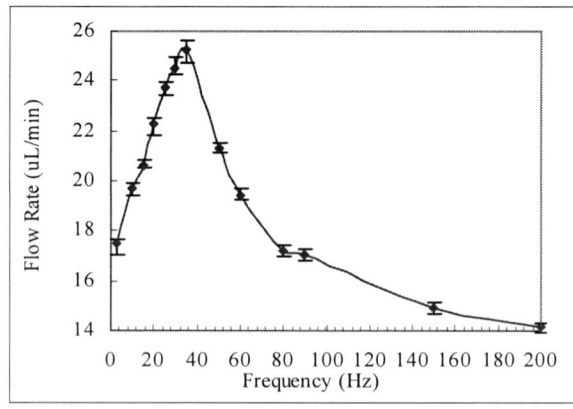

Figure 7: The measured flow rate of the peristaltic pump actuated by three electromagnetic actuators.

Figure 8 shows the photograph of slab gel electrophoresis of the PCR products. Lane M is the DNA ladder. Lane 1 is the result by a commercial PCR machine (Mastercycler gradient, Eppendorf®). The result of our miniaturized PCR system is shown in Lane 2. For both cases, the 122 bp PCR products are clearly observed, and no other specific products are found. Also, the results by our system are consistent with the result by the commercial

PCR machine. Our PCR system takes 46 min and 40 sec for the whole PCR process, while the commercial PCR machine takes about 2 hours and 30 minutes. Note that the required DNA sample for commercial PCR machine is 25 µL. Therefore, our PCR system only required less than 1/3 of the DNA sample.

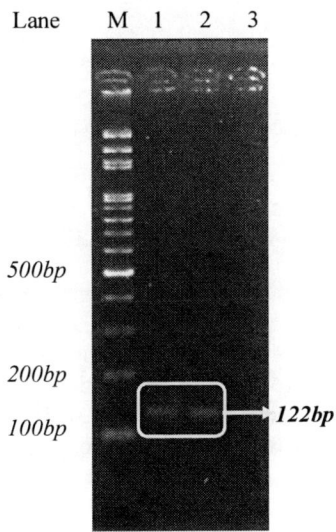

Figure 8: The observed fluorescence photograph of slab gel-electrophoresis of the PCR products.

CONCLUSION

A self-contained PCR system is developed in this work. Three chambers are designed for different reactions at specific temperatures. DNA solution is driven by electromagnetic mini-actuators between these chambers. Micromachined heaters and temperature sensors are also implemented in the system. The PWM method is used to feed the power to the heaters, and the PID technique is employed to control the temperatures at the target levels. Successful PCR amplification is performed. The DNA sample required by the miniaturized PCR device is less than one third of that required by a commercial PCR machine. Also, the proposed PCR system takes about 47 minutes for the whole PCR reaction, which is about one third of that by a commercial PCR machine. The self-contained system, whose size is 67 mm × 66 mm × 25 mm, can be fully operated with a 5V DC voltage, and does not require any external air compressor or bulky power supply.

ACKNOLEDGEMENTS

This project is sponsored by the National Science Council, Taiwan, ROC (contract no: NSC 96-2323-B-002-009). The authors would like to thank Prof. A.-B Wang for providing the infrared imaging system.

REFERENCES

[1] C. S. Zhang and D. Xing, "Miniaturized PCR Chips for Nucleic Acid Amplification and Analysis: Latest Advances and Future Trends" *Nucleic Acids Research*, vol. 35, pp. 4223-4237, 2007.

[2] C. Ke, A. M. Kelleher, H. Berney, M. Sheehan, and A. Mathewson, "Single Step Cell Lysis/PCR Detection of Escherichia Coli in an Independently Controllable Silicon Microreactor," *Sensors and Actuators B-Chemical*, vol. 120, pp. 538-544, 2007.

[3] K. Y. Lien, W. C. Lee, H. Y. Lei, and G. B. Lee, "Integrated Reverse Transcription Polymerase Chain Reaction Systems for Virus Detection", *Biosensors & Bioelectronics*, vol. 22, pp. 1739-1748, 2007.

[4] C. H. Chien and H. M. Yu, "The Design and Fabrication of Polymerase Chain Reaction Platform", *Microsystem Technologies: Micro and Nanosystems-Information Storage and Processing Systems*, vol. 13, pp. 1523-1527, 2007.

[5] P. Neuzil, C. Y. Zhang, J. Pipper, S. Oh, and L. Zhuo, "Ultra Fast Miniaturized Real-time PCR: 40 cycles in less than six minutes", *Nucleic Acids Research*, vol. 34, e77, 2006.

[6] S. Li, D. Y. Fozdar, M. F. Ali, H. Li, D. B. Shao, D. M. Vykoukal, J. Vykoukal, P. N. Floriano, M. Olsen, J. T. McDeitt, P. R. C. Gascoyne, and S. C. Chen, "A Continuous-flow Polymerase Chain Reaction Microchip with Regional Velocity Control", *Journal of Microelectromechanical Systems*, vol. 15, pp. 223-236, 2006.

[7] J. Grover, R. D. Juncosa, N. Stoffel, M. Boysel, A. I. Brooks, M. P. McLoughlin, and D. W. Robbins, "Fast PCR Thermal Cycling Device", *IEEE Sensors Journal*, vol. 8, pp. 476-487, 2008.

[8] N. Crews, C. Wittwer, and B. Gale, "Continuous-flow Thermal Gradient PCR", *Biomedical Microdevices*, vol. 10, pp. 187-195, 2008.

NANOSTRUCTURE-ENHANCED FIBER-OPTIC INTERFEROMETRY FOR LABEL-FREE IMMUNE SENSING

Yuan-Tai Tseng[1], Ching-Tsen Chang[2], Ming-Hung Chen[1] and Fan-Gang Tseng[1, 2, 3*]

[1] Institute of NanoEngineering and MicroSystems (NEMS), National Tsing Hua University,
Hsinchu 30013, Taiwan R.O.C.
[2]Department of Engineering and System Science, National Tsing Hua University,
Hsinchu 30013, Taiwan R.O.C.
[3]Division of Mechanics, Research Center for Applied Sciences, Academia Sinica,
Taipei 11574, Taiwan R.O.C.
*Corresponding author: Fan-Gang Tseng, NTHU ESS, #101, Sec. 2, Kuang-Fu Rd., Hsinchu,
Tawian ROC, Tel: +886-3-5715131 ext. 34270, e-mail: fangang@ess.nthu.edut.tw

ABSTRACT

This paper proposes a nanostructure-enhanced fiber-optic interferometry (NSFOI) biosensor, combining the abilities of high sensitivity, real-time monitoring and *in-situ* measurement for label-free immunological detection, by employing high-aspect-ratio pillar-like nanostructures on the tip of sensor surface. The nanostructure layer, fabricated by a simple etching process on resonant cavity without mask patterning, was employed not only for the improvement of the detection limit to reach 1ng/ml for rabbit-IgG (~6.7pM) due to the increase of the biomolecular binding surface, but also for the enhancement of the optical path length to enlarge the maximum magnitude of fringe shift which was approximated to 1.4nm by both experimental and theoretical results.

INTRODUCTION

Since the significant meaning of the *in-situ* and *in-vivo* detection of proteins in biomedical research, an immune sensing tool combining the powers of needle-type configuration, high sensitivity, and reasonable frequency response to monitor real-time changes of protein concentration is highly required. Traditional label-based techniques has been well developed for *in-vivo* protein analysis, such as using enzyme-linked immunosorbent assay (ELISA) coupled with the microdialysis sampling method [1], yielding considerably low detection limit and high lateral resolution. Yet, the labeling procedures were usually costly and time consuming due to the labeling process and efficiency, respectively, need to be properly controlled for quantitative analysis. To overcome those obstacles, label-free immune sensors based on optical principles were widely established on optic-fiber schemes [2], which promised the advantages of small size, rapid response, immunity from electrical or electromagnetic interference, and resistance to corrosive environments .

Fiber-optic interferometry (FOI) sensor is one of the fiber-based optical sensing methods with considerable robustness and sensitivity for the application of bio-sensing. The sensing mechanism stems from the measurement of the altering on either the refractive index [3] or the length of the resonant cavity by the affiliated biomolecules [4]. For example, Villar et al. and Arregui

et al. demonstrated the detection for ionic self-assembly monolayers stacked on fiber tip [5]. A similar sensor was also applied to detect the immunological activities of immunoglobulin G [6], giving a sensitivity of 25µg/ml. However, the result was less sensitive than that of the chip-based counterparts which can approach to 0.07µg/ml [3] due to the combination of nanoporous structures on the surface. In addition, the detected signals were only demonstrated for static rather than kinetic due to the measurement of fringe shift in atmosphere instead of aqueous environment. Therefore, a novel designed nanostructure-enhanced fiber-optic interferometry (NSFOI) biosensor was proposed in this paper to demonstrate the *in-situ* and real-time immune detection. Besides, by fabricating a layer of pillar-like nanostructures on the fiber tip, the detection limit and the magnitude of signals can be enhanced due to a large increase of surface for immune protein conjugation as well as optical path length were also verified by theoretical and experimental results.

DESIGN AND METHODS

Experimental Setup

The setup of the NSFOI sensing system is shown in figure 1(a). A beam light is generated by a broadband edge-emitting light emitting diode (ELED) driven by a current controller with a peak wavelength of 1550nm, spectrum half-width of 70nm, and emission power of 15µW. The waveband that is employed can avoid interference from visible lights and has a low loss for a long transmission distance. The light emitted from the ELED was transmitted by a single mode optical fiber with 125µm fiber and 8µm core diameter, respectively, and directed into the immune-sensing fiber to carry the interferometry signal through the coupling of the splice. A three-port optical circulator was utilized to guide the interfered light to an optical spectrum analyzer to record and analyze the interference spectrum.

Fabrication Process of NSFOI

The fabrication process of the sensing fiber, as shown in figure 1(b), was started from depositing a 3nm thin gold film on a fiber tip by e-gun evaporation. A resonant cavity was then produced by a dip-coated polydimethylsiloxane (PDMS) and curing it at 90°C for

10 minutes. The nanostructures layer on the PDMS surface was fabricated by self-masked high-aspect-ratio nano-pillar (SMHAR) fabrication method [7], using reactive ion etching process with a dummy material (cover glass) from which the nanomask was sputtered onto PDMS surface and then etched by gas ratios of 5:3 (CF_4:O_2) with RF power of 150W for 30min. The SEM image of the sensor tip after etching process was shown in figure 1(c). The height and width of nanopillar is averagely 1μm and 0.2μm, respectively (figure 1(d)).

Next, the sensing surface was hydrophilic treated by O_2 plasma for the immobilization of aminopropyltrimethoxysilane (APTS, 5% diluted in 99.5% alcohol). By immersing in APTS solution for 2hr and then annealing at 80°C oven for 2hr, the APTS molecules will form a self-assembled monolayer binding on the PDMS surface and to bind with proteins by the outside amino group. Therefore, after coating with 1mg/ml anti-rabbit-IgG (Anti-Rabbit IgG (whole molecule), antibody produced in goat, Sigma) solution as antibody for 2hr and blocking the uncoated sites by bovine serum albumin (BSA) solution, the sensor probe can be used to detect the concentration of rabbit-IgG (IgG from rabbit serum, ≥95%, Sigma) as antigen in the solution.

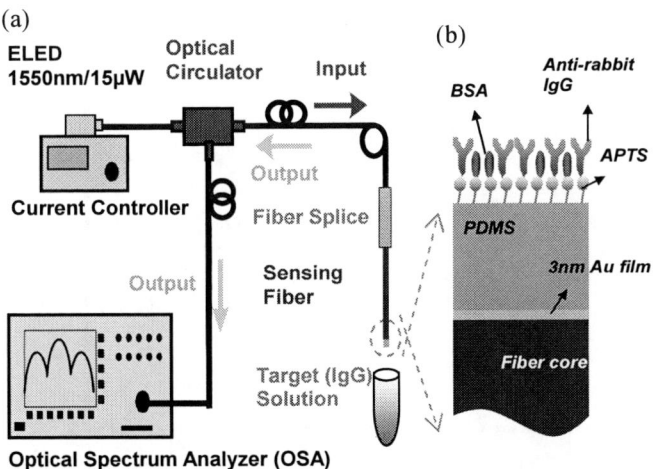

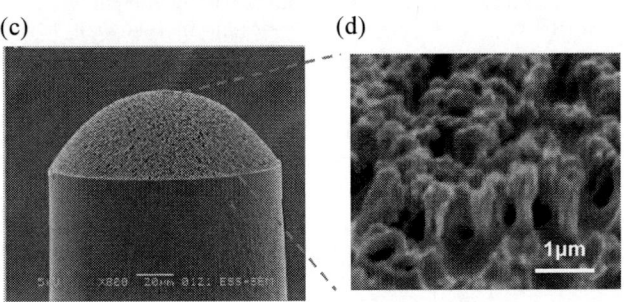

Figure 1: (a) Schematic diagram of NSFOI detection system and the SEM image of (c) the tip of the sensing fiber and (d) the nanopillars on the surface.

Sensing Theory

To quantitatively analyze the fringe shift from NSFOI sensor, a simplified theoretical approximation is developed. The sensing mechanism is illustrated in figure 2. As an incident light (I_0) passing through the resonant cavity (figure 2(a)), two reflections of R_1 and R_2 are reflected from the interfaces yield the interference spectrum as shown in figure 2(c) due to the phase difference (φ) between the two reflections R_1 and R_2. The phase difference φ, which is represented by constructive/destructive peak/valley of wavelength (λ) in the spectrum, can be expressed as:

$$\varphi = \frac{4\pi(n_P L_P + n_{NS} L_{NS})}{\lambda} \tag{1}$$

where the n_P, n_{NS}, L_P and L_{NS} indicate the refractive indices and lengths of the PDMS and nanostructure layer, respectively.

When proteins are added to the sensor surface, as shown in figure 2(b), the interfering fringe will have a shift (Δλ) (figure 2(d)) due to the change of optical path length from the second reflection (R_2'). The increase of optic path length can be represented by the increase in the effective length (ΔL_{NS}) and refractive index (Δn_{NS}) of the nanostructure layer. Therefore, the equation (1) can be expressed as the following equation:

$$\varphi = \frac{4\pi[n_P L_P + (n_{NS} + \Delta n_{NS})(L_{NS} + \Delta L_{NS})]}{\lambda + \Delta\lambda} \tag{2}$$

By tracing the same phase in equations (1) and (2), the interference fringe shift can be obtained as follows:

$$\Delta\lambda = \frac{n_{NS}\Delta L_{NS} + \Delta n_{NS} L_{NS}}{n_P L_P + n_{NS} L_{NS}}\lambda \tag{3}$$

where the term $\Delta n_{NS}\Delta L_{NS}$ is too small and so can be neglected in the equation.

Furthermore, the nanostructure layer is actually consisted by the nanopillers, void space and affiliated biomolecules. Therefore, the increase of effective length can be regarded as the length of biomolecules, such as IgG protein has a height of 10nm as binding on the surface. Beside, the refractive index (n_{NS}) and the effective refractive index change (Δn_{NS}) of the nanostructure layer can be expressed by the volume fractions (f_{pillar}, f_{void}, and f_{bio}) and refractive indices (n_{pillar}, n_{void}, and n_{bio}) of each part:

$$n_{NS} = f_{pillar} n_{pillar} + f_{void} n_{void} \tag{4}$$

$$\Delta n_{NS} = f_{bio}(n_{bio} - n_{void}) \tag{5}$$

where $f_{pillar} + f_{void} = 1$, and f_{bio} is related to the density of proteins which conjugated on the sensor surface instead of filling with the void space.

In our case, the $f_{pillar}=0.9$ and $f_{void}=0.1$ can be estimated by calculating the diameter of each pore (D_P) is 200nm averagely. Besides, the void space is filled with water solution before binding, thus $n_{void}=1.33$. When the surface of NSFOI sensor is fully coated with IgG protein, $f_{bio}=0.193$ can be obtained by calculating the volume ratio of protein in the nanostructure. Therefore, the maximum fringe shift of 1.4nm can be obtained by considering $L_P=30$μm, $L_{NS}=1$μm, $\Delta L_{NS}=10$nm, and $n_{pillar}=n_P=1.45$. However, for the FOI sensor without nanostructure fabrication on the surface, the maximum fringe shift is 0.5nm as saturate conjugation with proteins due to the signal is only contributed from the increase of the effective length of resonant cavity.

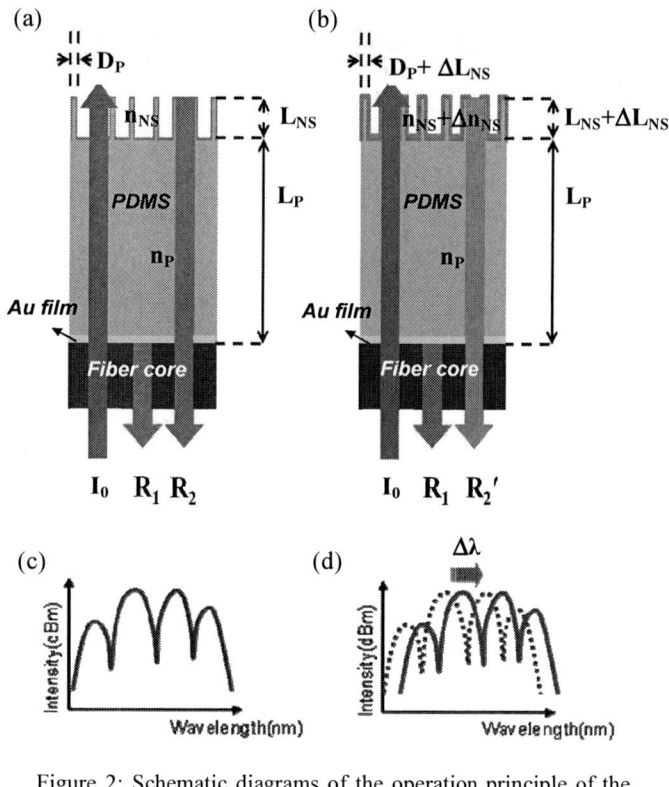

Figure 2: Schematic diagrams of the operation principle of the interferometry and related interference spectrum (a),(c) before and (b),(d) after immobilizing biomolecules on the sensor surface.

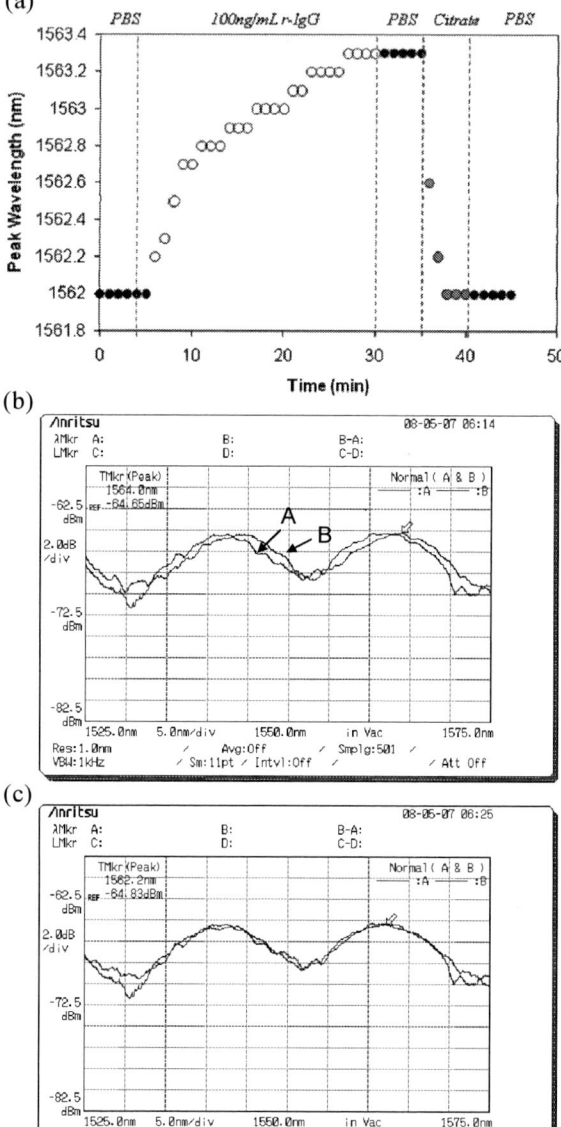

Figure 4: (a) Real-time employment of the immuno-detection; (b) the magnitudes of interference shift (from A to B) caused by immersing in 100ng/ml and (c) after the elution process.

RESULTS AND DISCUSSIONS

The *in-situ* and real-time monitoring of the immuno-conjugation was carried out by using NSFOI sensor to detect rabbit IgG solutions with constant volume of 100μl. During the measurement, the environmental temperature was controlled strictly at 22°C to avoid the thermal expansion of resonant cavity. The result was shown in figure 4(a), a stable signal by immersing the sensor tip in phosphate-buffered saline (PBS, pH 7.2) solution was first recorded as a baseline. Then, the probe was introduced into the concentration of 100ng/ml rabbit-IgG solution. The gradually increase of fringe shift can be observed to reveal the protein conjugation. Figure 4(b) shows the comparison of the interfering spectrums before and after immersing in rabbit-IgG solution for 20min, where the fringe shifted to the right side of 1.3nm was apparent due to the saturated conjugation of rabbit-IgG with anti-rabbit-IgG on the sensor surface. In addition, the immune reaction was verified by employing pH 3.0 sodium citrate buffer solution for the elution process. Since the acid solution can disrupt the ionic bonds of immune complexes by high concentration of protons, most of the rabbit-IgG molecules will be dissociated from the surface despite of the rarely non-specific binding proteins. As shown in figure 4(c), the elution efficiency can be observed from the fringe shifted toward left side and finally almost overlapped with initial fringe, after putting the sensor into the acid buffer for 5min.

The minimum detectable concentration was also verified in this study by demonstrating the enhancement of nanostructures. By following the detection process mentioned above, 1ng/ml and 10ng/ml of rabbit-IgG solution was also detectable as using NSFOI sensor to obtain a 0.3nm and 0.5nm shift, respectively. Yet, the flat surface of FOI sensor without nanostructures layer was only approaching the detection limit of 100ng/ml of rabbit-IgG solution as shown in figure 5. It indicates the NSFOI sensor can reach 2 orders of improvement for the detection limit as comparing with FOI sensor without nanostructures. Besides, the maximum shifts of the NSFOI and FOI sensor were measured as 1.3nm and 0.6nm, respectively, which were close to the magnitude estimated by theoretical prediction, also demonstrated the enhancement of the signals by the increase of optical path length from nanostructures as saturated conjugation of proteins.

Although the nanostructures layer was successfully used to enhance the fringe shift, the pillar-like surface also brought out the scattering effect on the second reflection, which will decrease the visibility of signal. As shown in figure 3, comparing interference fringes without (red curve) and with (blue curve) nanostructures on the sensor tip, the sharper peak and valley of fringe represented the visibility of smooth surface (figure 3(a)) was better than nanopillar-covered surface (figure 3(b)). The lower visibility of signal will cause the constructive peak, which was employed to monitor the fringe shift, become broader and easier to be interfered by the small perturbations. To diminish the noise, the peak values recorded within 10sec was averaged to obtain a mean value during the measurement, such as each dot appeared in figure 4(a).

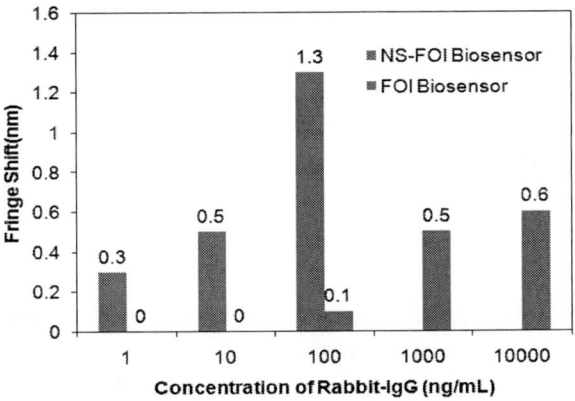

Figure 5: Comparison of the sensing signal of the sensing surface with and without nanostructures modification.

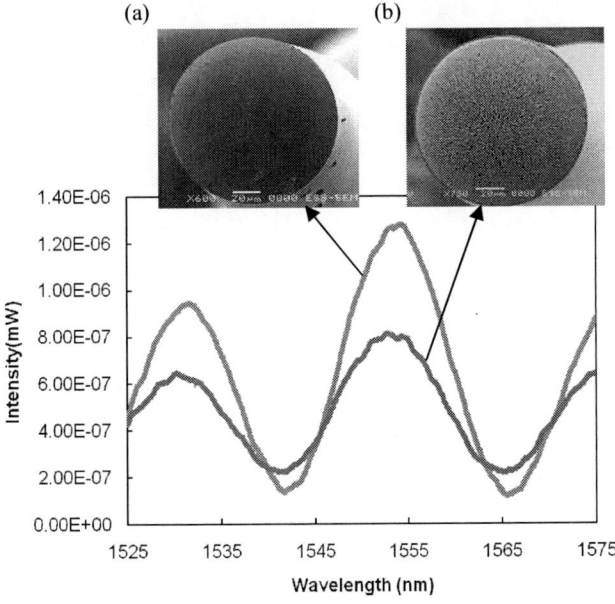

Figure 3: Comparison of the interfering spectrums of fiber sensor tip (a) without and (b) with the fabrication of nanopillars layer.

CONCLUSION

A novel design of nanostructure-enhanced fiber-optic interferometry (NSFOI) biosensor was proposed in this paper to demonstrate the abilities of high sensitivity, real-time monitoring, label-free and *in-situ* immune detection. By employing a pillar-like nanostructure layer on the sensor surface, the detection limit of 1ng/ml for rabbit-IgG (~6.7pM) can be approached due to the increase of the protein binding surface, which is two orders of improvement rather than the FOI sensor with smooth surface. Beside, the maximum fringe shift 1.4nm can be estimated from theoretical simulation and shows a highly agreement with experimental result. The other tests such as competition detection of different kinds of proteins and repetition detection on one sensing fiber are still verifying in progress.

Acknowledgements

The authors would gratefully acknowledge the financial support from VUST project (Veterans General Hospital and University System of Taiwan), NHRI (National Health Research Institutes, Taiwan), and National Nanoscience and Nanotechnology program NSC 96-2120-M-007-010 by National Science Council, Taiwan.

REFERENCES

[1] J. Hillman, O. Aneman, M. Persson, C. Andersson, C. Dabrosin, P. Mellergard, "Variations in the response of interleukins in neurosurgical intensive care patients monitored using intracerebral microdialysis", *J. Neurosurg.*, vol. 106, pp. 820–825, 2007.

[2] A. Leung, P. M. Shankar, R. Mutharasan, "A review of fiber-optic biosensors", *Sensor. Actuat. B-Chem.*, vol. 125, pp. 688–703, 2007.

[3] J. C. Lee, J. Y. An, B. W. Kim, "Application of anodized aluminium oxide as a biochip substrate for a Fabry–Perot interferometer" *Journal of Chemical Technology & Biotechnology*, vol. 82 , pp. 1045-1052, 2007.

[4] P. I. Nikitin, M. V. Valeiko, B. G. Gorshkov, "New direct optical biosensors for multi-analyte detection" *Sensor. Actuat. B-Chem.*, vol. 90, pp. 46-51, 2003.

[5] F. J. Arregui, I. R. Matias, Y. Liu, K. M. Lenahan, R. O. Claus, "Optical fiber nanometer-scale Fabry–Perot interferometer formed by the ionic self-assembly monolayer process", *Opt. Lett.*, vol. 24, pp. 596-598, 1999.

[6] Y. Zhang, H. Shibru, K. L. Cooper, A. Wang, "Miniature fiber-optic multicavity Fabry–Perot interferometric biosensor", *Opt. Lett.*, vol. 30, pp. 1021-1023, 2005.

[7] M. H. Chen, T. H. Hsu, Y. J. Chuang, P. H. Chen, F. G. Tesng, "Self-formed high-aspect-ratio polymer nanopillars by RIE", in *Digest Tech. Papers Transducers'07 Conference*, 2007, pp.563-566.

BIOSENSOR BASED ON ALL-POLYMER RESONANT MICROBEAMS

S. Schmid, P.Wägli, and C. Hierold

Micro and Nanosystems, Department of Mechanical and Process Engineering, ETH Zurich,
Switzerland

ABSTRACT

We present all-polymer resonant microbeams which can detect the adsorption of streptavidin and DNA on the surface in aqueous solution with a high selectivity. The polymer surface allows the direct functionalization with the receptor biomolecules without intermediate Au-layer. The highly sensitive resonant microbeams offer a low-cost system for disposable and label-free biosensing applications. Minimal analyte (streptavidin) concentrations of 0.025µg/ml were detected causing an oscillation frequency shift of 1.9%. 0.13 nmol/ml of 17-mer single-stranded DNA could be selectively detected causing a frequency change of 0.9%.

INTRODUCTION

Microcantilever based biosensors are typically made of silicon microbeams which change their resonant frequency or deflection as a cause of analyte molecules adsorbing on the surface [1]. For the application of the deflection method, only one side of the cantilever has to be functionalized with the receptor molecules. This can be done by coating the upper surface with a gold layer which can be functionalized by thiol-conjugated receptor molecules [2]. Besides gold, functional layers of chemical sensors but also such for biosensing applications often consist of polymer coatings [3]. Polymer microcantilevers with and without a gold layer have been used as deflecting biosensors because of their softness which results in high deflection for a induced surface stress, detecting streptavidin [4] and DNA [5]. Tamayo et al. [6] predicted a superior sensitivity of soft polymeric resonant cantilevers based on an increasing stiffness due to the adsorbed molecular layer resulting in a positive frequency shift. Furthermore, oscillating polymer microbeam sensors can be directly functionalized with receptor biomolecules [7], thereby covering the complete surface which simplifies the pre-treatment procedure.

In water, the quality factor of resonant microbeams is dominated by the damping of the aqueous environment. The quality factors of silicon and polymer micro cantilevers in water have similar low values of $Q = 1.8$ at 13 kHz for a silicon nitride cantilever [8] and $Q = 1.4$ at 7.3 kHz for SU-8 cantilevers [5]. In order to overcome the high damping in water, the microbeams are typically actuated in a self-excited oscillation by a positive feedback-loop [8] which has shown to be also feasible for polymer microbeams in aqueous environments [9]. Consequently, resonant polymer microbeams have a high potential as low-cost system for disposable and label-free biosensing applications. In this work, single- and double-clamped resonant all-polymer microbeams made from SU-8 are used for the detection of streptavidin and DNA. The polymer microstructures are functionalized by immersing them into the solution containing the receptor biomolecules which directly bind to the polymer surface.

EXPERIMENTAL

The microbeams are fabricated from SU-8 2002 (Microchem Corp.), an epoxy type photoresist, by a sacrificial layer technique using lift-off resist (LOR, Microchem Corp.) as the sacrificial layer material. The SU-8 and the LOR film have a thickness of 1.45 µm and 3.0 µm, respectively. The substrate with the electrodes is coated with a 120 nm thick aluminum oxide layer in order to prevent the Cr/Au electrodes from fast degradation. Fig. 1 shows pictures of the single- and double-clamped SU-8 microbeams. The structures are electrostatically actuated by means of the Kelvin polarization force [10]. The polymer microbeams are placed over an electrode gap producing a inhomogeneous electric field. The highly polar water is attracted towards the electric field intensity maximum at the electrodes' gap, pushing away the polymer microbeam. The oscillation is monitored with a laser-Doppler vibrometer. The resulting velocity signal is used to drive the polymer microresonator by means of a positive feedback [9]. The amplitude response of a streptavidin measurement is shown in Fig. 2. The frequency of the self-sustained oscillation during the measurement is monitored with a frequency counter.

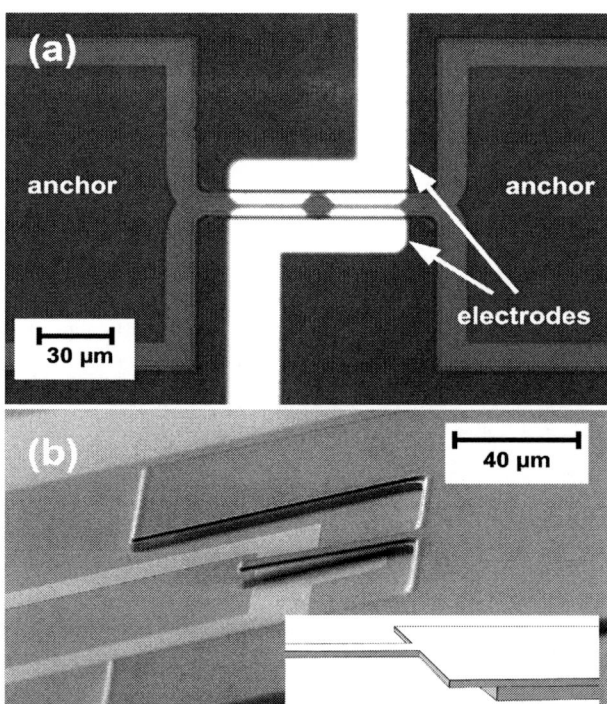

Figure 1: (a) Microscope image of double-clamped (d-c) and (b) SEM image of a single-clamped (s-c) SU-8 microbeam. The insert in (b) shows a schematic of the process related suspended anchor. (Taken from [11]).

978-1-4244-2977-6/09 $25.00 © 2009 IEEE

Protocol of streptavidin detection

The receptor molecule for the detection of streptavidin is biotin which was bound to the polymer surface via a carrier molecule which is known to bind to the SU-8 surface. In order to functionalize the SU-8 beams prior to the measurement, the polymer microstructures are fully immersed into a phosphate buffered saline (PBS) containing 0.46 mg/ml biotin conjugated anti-mouse immunoglobulin G (IgG). The sample was then stored in the solution at 37°C for 60 min. Each antibody is conjugated with approximately 5 biotin molecules. The orientation of the functional layer is therefore not an issue. To remove unspecific bonds, the samples were rinsed for 20 min with tris buffered saline with tween 20 (TBST). After replacing the TBST with water, active binding sites on the SU-8 were blocked with bovine serum albumin (BSA) (5 µg/ml). After rinsing with water, for the negative control, the sample was exposed to an aqueous solution containing 5 µg/ml of BSA. For the positive control, the sample was exposed to an aqueous solution containing various concentrations of streptavidin.

Protocol of DNA detection

To detect DNA strands of a specific sequence, the polymer has to be functionalized with the complementary single DNA strand (ssDNA). Streptavidin is used as interface layer to which later biotinylated ssDNA (receptor molecules) can be attached by means of a strong biotin streptavidin binding. Therefore, the SU-8 samples were stored in an aqueous solution containing 20 µg/ml streptavidin for 60 min at 37°C. After rinsing with water, free binding sites are blocked with a solution of 1 mg/ml PLL-g-PEG (60 min at room temperature) which enables protein resistance of the remaining beam surface. After rinsing, the samples are functionalized with the receptor DNA. Therefore, the sample is immersed in an aqueous solution containing biotinylated ssDNA (0.13 nmol/ml). For the negative test, the samples were exposed to an aqueous solution containing 0.13 nmol/ml of a wild type ssDNA. For the positive test, the samples were exposed to an aqueous solution containing 0.13 nmol/ml of the complementary ssDNA.

RESULTS

Streptavidin detection

Fig. 3 shows the monitored oscillation frequency of a SU-8 double-clamped microbeam during the sequential exposure to aqueous solutions containing 5 µg/ml of BSA as a negative test and 3.0 µg/ml of streptavidin as a positive test causing a frequency shift of 6.1%. The sample was functionalized prior to the measurement with immunoglobulin G conjugated with biotin which has a high affinity to streptavidin. Minimum streptavidin concentrations of 0.025µg/ml were detected with a corresponding frequency shift 1.9% (Fig. 4).

Fig. 5 shows the relative frequency change of single-clamped (s-c) and double-clamped (d-c) microbeams for the detection of different concentrations of streptavidin. Maximum frequency shifts of 15.6% for the positive test were measured whereas the maximum measured frequency shift of the negative test was 0.4%.

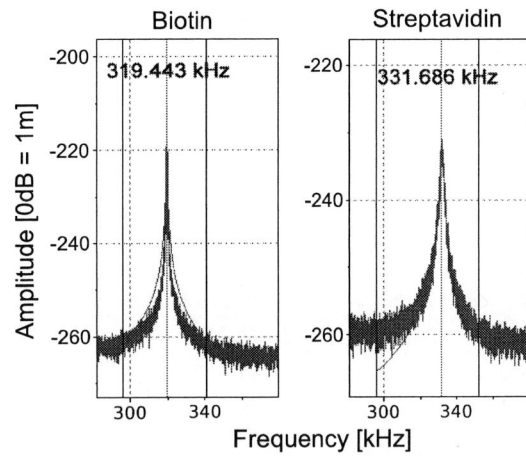

Figure 2: Frequency response before and after streptavidin detection (5.0 µg/ml) with a biotinylated SU-8 resonant double-clamped microbeam (sample: d-c 1).

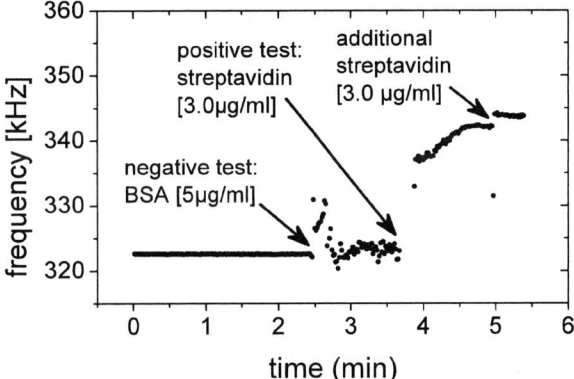

Figure 3: Detection of 3 µg/ml streptavidin with a double-clamped SU-8 microbeam (length=80 µm, thickness=1.45 µm, width=14 µm) at 24°C (sample: d-c 2).

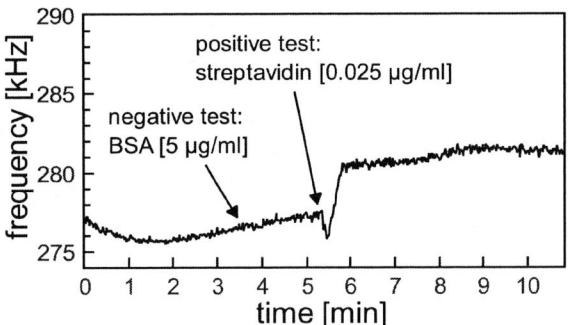

Figure 4: Detection of 0.025 µg/ml streptavidin with a double-clamped SU-8 microbeam (length=200 µm, thickness=1.45 µm, width=14 µm) at 25°C (sample: d-c 4).

DNA detection

Fig. 6 shows the monitored oscillation frequency of a SU-8 single-clamped microbeam during the sequential exposure to aqueous solutions containing 0.13 nmol/ml of biotinylated 17-mer single-stranded DNA (ssDNA), wild type 17-mer ssDNA as a negative test and the complementary ssDNA as a positive test. A frequency shift due to the adsorption of the complementary ssDNA of 0.9% was monitored.

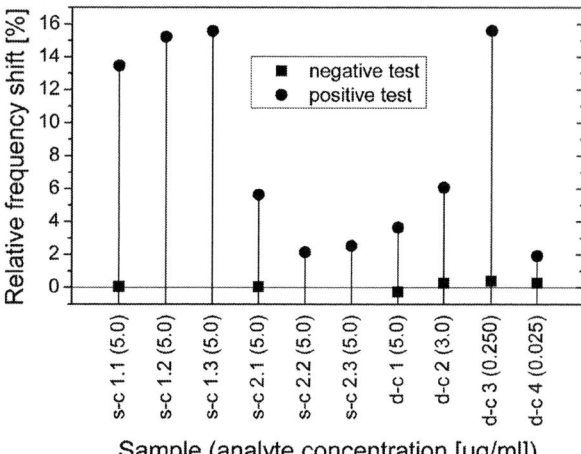

Figure 5: Relative frequency shift of single-clamped (s-c) and double-clamped (d-c) SU-8 microbeams for the detection of streptavidin. The oscillation frequencies vary between 272-322 kHz.

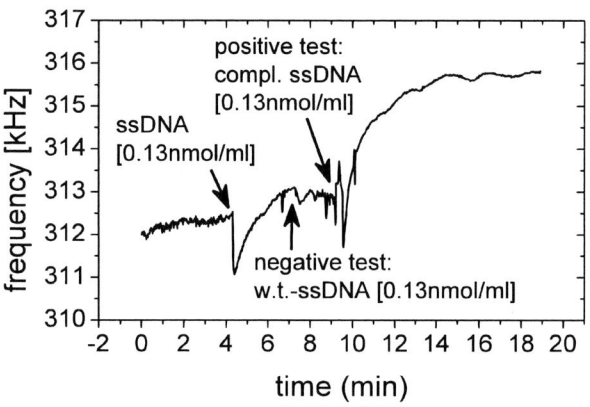

Figure 6: Detection of complementary 17-mer ssDNA with a single-clamped SU-8 microbream (length = 55 μm, thickness = 1.45 μm, width = 14 μm) at 23°C.

DISCUSSION

Streptavidin detection

To verify, that the analyte molecules actually bind to the biotinylated SU-8 surface, a histogram of the particle size on the beam surface before the measurements and after the streptavidin detection was calculated from the topographical information of an AFM scan (Fig. 7). The majority of the particles after the streptavidin detection have a diameter of 6 nm which corresponds to the size of the analyte molecules. Streptavidin builds dimers [12], entities of two equal streptavidin monomers, which have a diameter of around 10-13 nm. This explains the increased levels of particle diameters of this dimension in the histogram. It is concluded that the shift of the oscillation frequency is caused by the adsorption of the analyte molecules.

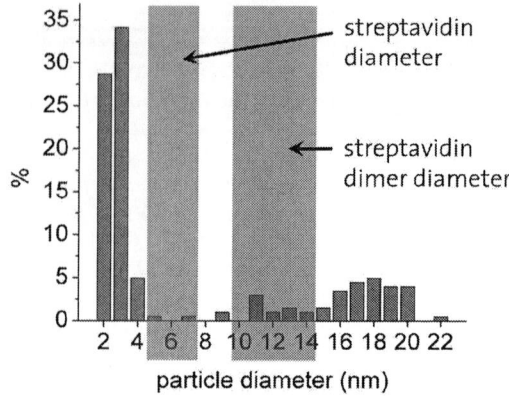

SU-8 surface before streptavidin detection

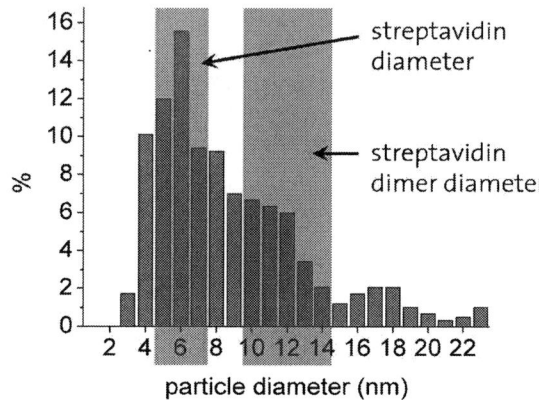

SU-8 surface covered with streptavidin

Figure 7: AFM histogram of particle diameter on SU-8 surface before and after streptavidin detection.

As shown in Fig. 5, the frequency shift of the streptavidin detection is always positive which might be due to an increase of the flexural rigidity of the soft polymer microbeams caused by the adsorbate stiffness as explained in [6]. In air, the d-c microbeams are pre-stressed and they would be less sensitive to a change in the flexural rigidity. But in water the polymer swells, the pre-stress is released and the d-c structures are sensitive to changes of the flexural rigidity. For a similar experiment where 0.2 mg/ml neutravidin was detected with a biotinylated silicon nitride cantilever with a thickness between 0.7 to 1 μm the resonant frequency showed a negative shift of 19% after the adsorption of the analyte molecules [8] due to the mass uptake. The silicon nitride microbeam is too stiff to be sensitive to a change of the adsorbate stiffness.

Another effect which can induce a shift of the oscillation frequency comes from the self-sustained actuation by the positive feedback. If the phase of the driving signal does not perfectly match the phase of the sensor response, the oscillation frequency will be different from the resonant frequency of the sensor, as explained in [13]. For a constant phase offset this causes the oscillation frequency to be sensitive to the system damping. The damping ratio of the system is defined as [14]

$$\zeta = \frac{c}{2\sqrt{k \cdot m}}$$

where c is the coefficient of damping force, m is the mass and k is the spring constant. An additional mass would thus decrease the system damping which induces a shift of the oscillation frequency. But from the fact, that all measured frequency shifts depicted in Fig. 5 are positive, it can be assumed, that the influence of the feedback, being sensitive to the damping, is not destructive. Furthermore, it can be seen from Fig. 5, that the frequency shifts of the sensors from the same chip (chip 1 and 2) are similar. This suggests that the sensor sensitivity mainly depends on the sample quality and not on the feedback control. Furthermore, no systematic difference in the sensitivity could be observed between the single- and double-clamped beams.

In Fig. 2 it can be seen, that additional analyte only slightly increased the resonant frequency. This behavior was observed at several measurements. After a first flush with 3.0 or 5.0 µg/ml of streptavidin, the streptavidin molecules seem to almost cover the whole surface of the beam, such that only little free biotin molecules remain available. Hence, for the second injection of streptavidin, there are not enough biotin binding sites available anymore.

DNA detection

For the streptavidin detection and for the DNA detection measurements, the observed frequency shifts were always positive. In Fig. 6 it can be seen, that the actual functionalization with the biotinylated receptor ssDNA causes a frequency drop, which slightly gets neutralized by a subsequent increase to a constant frequency. The addition of the wild-type ssDNA triggered no significant frequency reaction, whereas the addition of the complementary ssDNA results in a positive frequency shift. The exact mechanisms causing this specific behavior are not fully understood yet. The different behavior of the ssDNA and the complementary ssDNA adsorption possibly comes from a different effect of the ssDNA- and the double-stranded DNA-layer on the microbeam resonant frequency.

CONCLUSION

In conclusion, we have demonstrated the feasibility of polymer microbeam resonators as biosensors. The experimental results show that streptavidin and DNA can be detected with a high sensitivity. The main advantage of a resonant polymer microbeam biosensor is the possibility to directly adsorb the receptor biomolecules on the polymer without the need of an additional functional coating. The presented biosensors are not suitable to measure a specific concentration of a certain analyte. But it can determine if there are analyte molecules present in a test solution with a measured minimal concentration of 0.025 µg/ml of streptavidin. All-polymer resonant microbeams have thus a high potential as low-cost system for disposable and label-free biosensing applications.

ACKNOWLEDGEMENTS

We would like to thank Ronald Grundbacher and Andreas Leimbacher for the work in the cleanroom. We also acknowledge the biochemistry expert discussions with Ivo Zemp from Ulrike Kutay group. This project was partly founded by the Swiss National Science Foundation (Project No. 200020-113350).

REFERENCES

[1] C. Ziegler, "Cantilever-based biosensors", Anal. Bioanal. Chem., vol. 379, pp. 946-959, 2004.

[2] N. V. Lavrik, M. J. Sepaniak, and P. G. Datskos, "Cantilever transducers as a platform for chemical and biological sensors", Review of Scientific Instruments, vol. 75, pp. 2229-2253, 2004.

[3] P. Bergese, et al., "Investigation of a biofunctional polymeric coating deposited onto silicon microcantilevers", Applied Surface Science, vol. 253, pp. 4226-4231, 2007.

[4] M. Calleja, et al., "Low-noise polymeric nano-mechanical biosensors", Applied Physics Letters, vol. 88, pp. 113901, 2006.

[5] M. Calleja, et al., "Highly sensitive polymer-based cantilever-sensors for DNA detection", Ultramicroscopy, vol. 105, pp. 215-222, 2005.

[6] J. Tamayo, D. Ramos, J. Mertens and M. Calleja, "Effect of the adsorbate stiffness on the resonance response of microcantilever sensors", Applied Physics Letters, vol. 89, pp. 224104, 2006.

[7] R. Marie, et al., "Immobilisation of DNA to polymerized SU-8 photoresist", Biosensors & Bioelectronics, vol. 21, pp. 1327–1332, 2006.

[8] A. Mehta, et al., "Manipulation and controlled amplification of Brownian motion of microcantilever sensors", Applied Physics Letters, vol. 78, pp. 1637-1639, 2001.

[9] S. Schmid, P. Senn and C. Hierold, "Electrostatically actuated nonconductive polymer microresonators in gaseous and aqueous environment", Sensors and Actuators A: Physical, vol. 145-146, pp. 442-448, 2008.

[10] S. Schmid, M. Wendlandt, D. Junker, and C. Hierold, "Nonconductive polymer microresonators actuated by the Kelvin polarization force", Applied Physics Letters, vol. 89, pp. 163506, 2006.

[11] S. Schmid and C. Hierold, "Daming mechanisms of single-clamped and prestressed double-clamped resonant polymer microbeams", Journal of Applied Physics, accepted.

[12] Y. Pazy, et al., "Dimer-tetramer transition between solution and crystalline states of streptavidin and avidin mutants", J. Bacteriology, vol. 185, pp. 4050-4046, 2003.

[13] U. Dürig, H.R. Steinauer, and N. Blanc, „Dynamic force microscopy by means of the phase-controlled oscillator method", J. Appl. Phys., vol. 80 (8), pp. 3641-3651, 1997.

[14] M. Bao, "Analysis and Design Principles of MEMS Devices", Elsevier, 2005.

BEAD-IMMOBILIZED MOLECULAR BEACONS FOR HIGH THROUGHPUT SNP GENOTYPING VIA A MICROFLUIDIC SYSTEM

R.D. Sochol, A. Mahajerin, B.P. Casavant, P. Singh, M. Dueck, L.P. Lee, and L. Lin
Berkeley Sensor and Actuator Center
The University of California at Berkeley, USA

ABSTRACT

This paper reports the first demonstration of single nucleotide polymorphism (SNP) genotyping via molecular beacon probes immobilized on polystyrene microbead substrates within a dynamic microfluidic system. Additionally, we present an optimized bead immobilization technique, micropost array trapping (µPAT), for high-density and high-throughput arraying of beads. Quantitative detection was achieved at room temperature for three label-free DNA oligonucleotide sequences based on the genome of the Hepatitis C Virus (HCV) in humans. SNP detection technology is integral for the realization of personalized medicine and genetic disease identification, and this system offers a promising technique for fast, sensitive, and cost-effective SNP genotyping.

INTRODUCTION

Single Nucleotide Polymorphism (SNP) Genotyping via Molecular Beacon Probes

With the human genome now mapped, it is important to quickly and accurately detect variations in the gene sequences of individuals [1]. The most common form of DNA variation is a single nucleotide polymorphism (SNP) (Fig. 1A), occurring once per every 100-1000 base pairs, and molecular beacon probes (MBs) offer a powerful tool for detecting SNPs [1, 2]. MBs are self-complimentary single-stranded DNA sequences with a fluorescent molecule on one end and a quencher on the other (Fig. 1B) [2]. The MB is designed to maintain a 'hairpin' conformation, which restricts fluorescence due to the proximity between the fluorophore and quencher molecules through a Förster Resonance Energy Transfer (FRET) interaction. However, in the presence of a target DNA oligonucleotide sequence, the MB hybridizes with the target DNA. This conformational change causes a loss of FRET due to the increased distance between the fluorophore and quencher molecules. Hence, the resulting unquenched fluorescence of the fluorophore can be detected. The advantages of MB-based genotyping methods have been demonstrated previously, including high signal-to-noise ratios, minimal required DNA for detection, reaction kinetics at room temperature, and the capability of genotyping unlabeled DNA targets [2, 3].

Microfluidic Dynamic Microarrays

MB genotyping applications have been primarily limited to static arrays, where MBs are immobilized on a static substrate; however, microfluidic bead-based arraying techniques are superior to static microarrays due to faster reaction kinetics, higher bead surface areas, reduced background noise, lower costs, and the ability to 'mix-and-match' beads corresponding to different screenings [4, 5]. One caveat of bead-based microfluidic techniques is that microbeads must be immobilized for

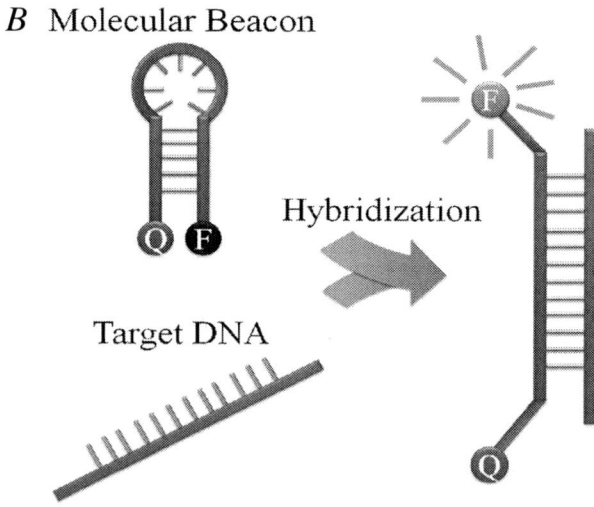

Figure 1: Conceptual illustration of (A) a SNP, and (B) molecular beacon genotyping

visualization and signal detection. This has been a significant limitation of previous MB-based dynamic arraying techniques, necessitating the physical placement of beads into array positions during fabrication [4].

DESIGN

Bead-based Molecular Beacon Genotyping in a Microfluidic System

A schematic of our microfluidic system is shown in Figure 2. Simultaneously, homogenous solutions of biotinylated MBs and streptavidin-coated polystyrene microbeads with an extended biological linker are

978-1-4244-2977-6/09 $25.00 © 2009 IEEE

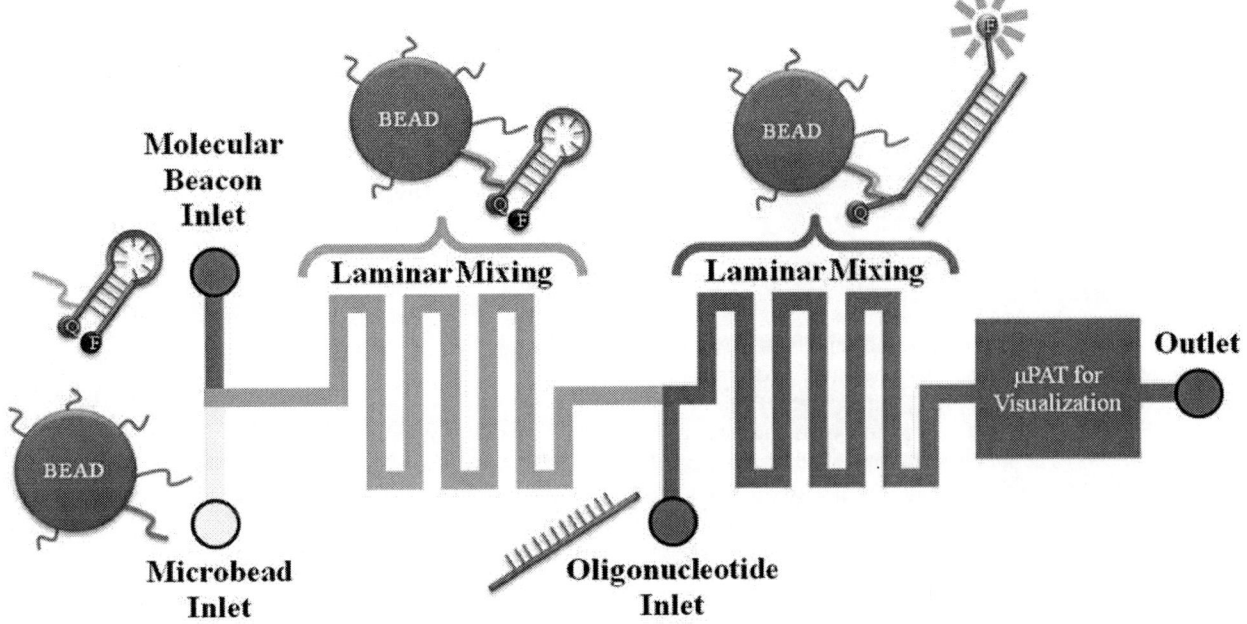

Figure 2: Conceptual illustration of the microfluidic system for bead-based molecular beacon genotyping

introduced via the 'Molecular Beacon Inlet' and the 'Microbead Inlet,' respectively. Laminar mixing in the microfluidic channel enables MBs to attach to the microbead linkers while maintaining the non-fluorescent hairpin conformation. A homogenous solution containing one of the DNA oligonucleotide sequences is introduced via the 'Oligonucleotide Inlet' and mixed with the MB-microbead solution. If the *Target DNA* oligonucleotides are present, they will hybridize with the MBs. Finally, the microbeads are immobilized for fluorescence visualization and quantification using micropost array trapping (μPAT).

Micropost Array Trapping (μPAT)

μPAT consists of arrays of rectangular microposts, which are spaced to form trapping sites as shown in Figure 3. Initially, higher flow rates between the microposts cause beads to be directed to and immobilized at these trapping sites. As each trapping site of the array is filled, the flow rate becomes higher at the subsequent trapping site. This chain reaction continues until all of the trapping sites are occupied by beads and flow is diverted around the traps.

Molecular Beacons and DNA Oligonucleotide Sequences

Previously published MB and DNA oligonucleotide sequences were used for testing the microfluidic system (Table 1), including three DNA oligonucleotide sequences based on the genome of the Hepatitis C Virus (HCV) in humans: (i) *Target Sequence*, (ii) *One Mismatch (SNP)*, and (iii) *Negative Control* [6]. The MB was designed to hybridize with the *Target Sequence*, as well as to be self-complimentary prior to hybridization, with a fluorophore (FAM) on the 5' end of the probe, and a quencher (BHQ-1) on the 3' end.

Fabrication
Microfluidic Devices

The devices were fabricated using a one-mask soft

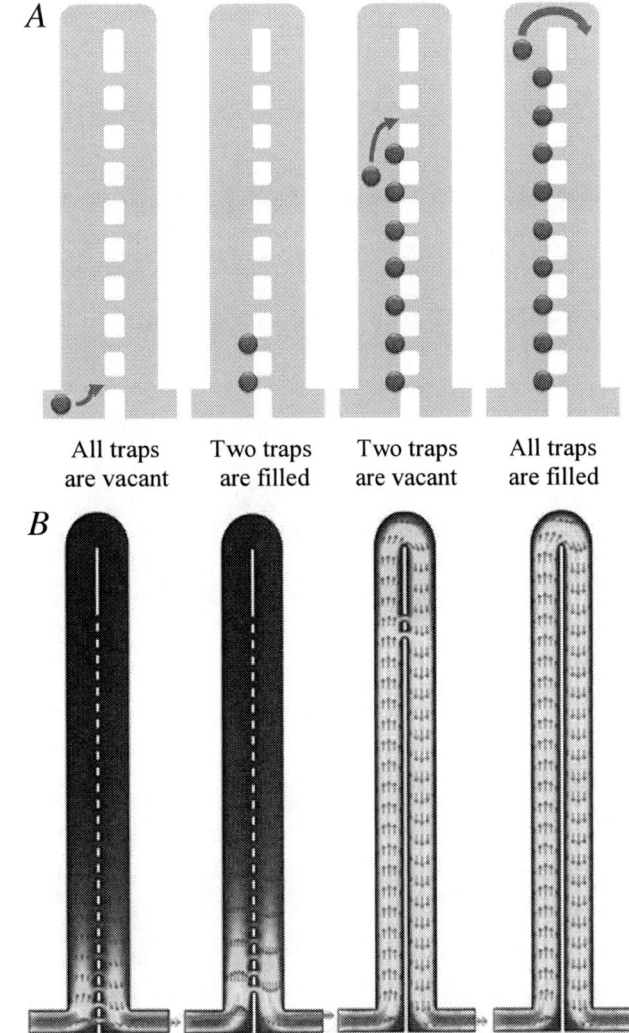

Figure 3: Micropost array trapping (μPAT): (A) Conceptual illustration and (B) COMSOL fluid velocity field simulations

Table 1: Names and sequences (5'-3' orientation) of the molecular beacon and DNA oligonucleotides.

Name	Sequence
Molecular Beacon (MB)	6-FAM-GCGAGCCACCGGAATTG CCAGGACGACCGCTCGC-BHQ-1
Target Sequence	GGTCGTCCTGGCAATTCCGGTG
One Mismatch (SNP)	GGTCGTCC**G**GGCAATTCCGGTG
Negative Control	GAGGGGCGGCGACTGGTGAG

lithography process for micromolding channels, followed by glass slide bonding. Briefly, SU-8 negative photoresist was spin-coated onto a clean silicon (Si) wafer. Microchannel patterns were UV-defined onto the wafer using a photomask. The wafer was developed, removing unpolymerized photoresist, to become a positive master. The silicone elastomer, Poly(dimethylsiloxane) (PDMS), was mixed and then poured onto the master. After curing, the PDMS was demolded, producing devices. The PDMS devices were cut and then punched with holes at inlet and outlet locations. Glass slide substrates and PDMS devices were UV ozone-treated in preparation for permanent bonding. Finally, the PDMS devices were thermally bonded to the glass substrates. The fabricated microfluidic system is shown in Figure 4.

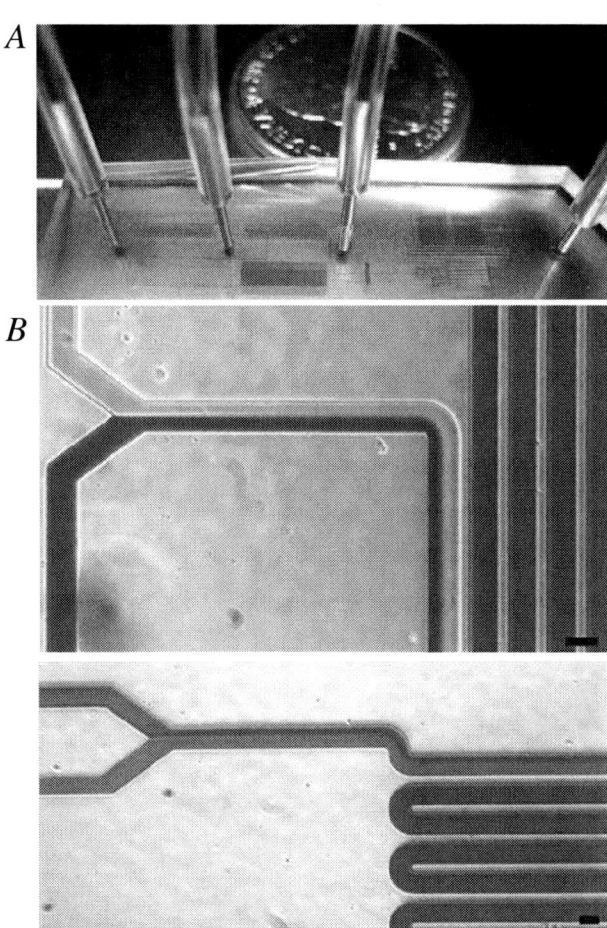

Figure 4: (A) Photograph of the complete microfluidic system for bead-based molecular beacon genotyping, and (B) Micrographs of the laminar mixing channels. Scale Bars = 50 μm

Microbead Biological Linker Treatment

Prior work has investigated the amplified fluorescent signal of MBs as a function of their separation distance from immobilization surfaces, which is regulated by biological linkers [3]. The biological linker is designed to be an elongated attachment between the streptavidin-coated polystyrene microbead and the biotinylated MB. To achieve this, streptavidin-coated microbeads were mixed with biotinylated bovine serum albumin (BSA) and subsequently washed to remove excess. The resulting biotinylated BSA-streptavidin-coated microbeads were then mixed with avidin to complete the biological linker. This was followed by another wash to remove unbound proteins. The MBs bound to the avidin-biotinylated BSA-streptavidin-coated microbeads fluoresced at higher intensities relative to MBs bound to streptavidin-coated microbeads without biological linkers.

RESULTS

Single Nucleotide Polymorphism Genotyping

Fluorescence results corresponding to the three DNA oligonucleotide sequences are shown in Figure 5. The fluorescence intensities of immobilized 21 μm diameter microbeads were quantified using the software, Image J, and verified off-chip via fluorescence images taken with a real-time polymerase chain reaction machine. Statistical significance was calculated using unpaired Student's *t*-tests. Each of the DNA oligonucleotide sequences fluoresced at distinct intensities during experimental testing. For the *Negative Control* case, immobilized microbeads fluoresced at approximately 44% of the

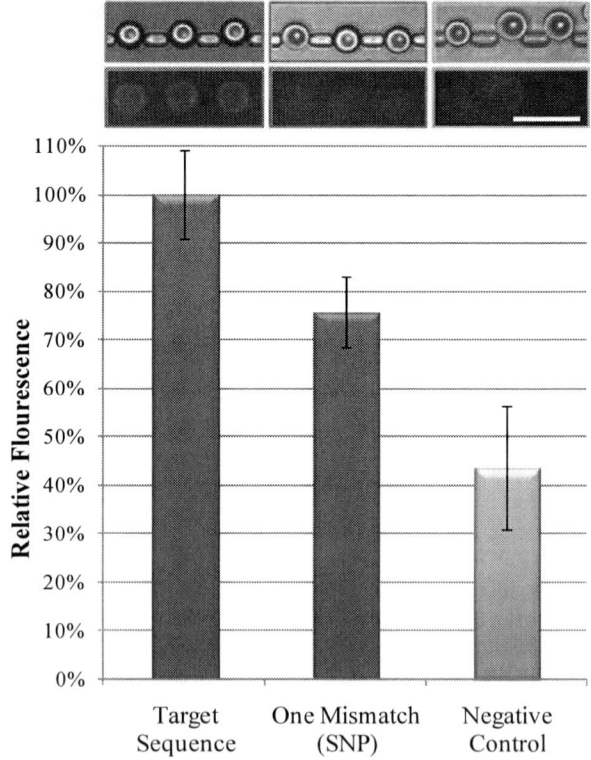

Figure 5: Relative fluorescence intensity of immobilized 21 μm diameter microbeads corresponding to different DNA oligonucleotide sequences. Error bars represent 95% confidence intervals. Scale Bar = 50 μm

fluorescence intensity of the *Target Sequence* case (p < 0.0001). Microbeads from the *One Mismatch (SNP)* experiments fluoresced at approximately 76% of the *Target Sequence* intensity and were successfully distinguished from both the *Target Sequence* (p < 0.001) and the *Negative Control* (p < 0.0001) cases.

Micropost Array Trapping

Three different sizes of microbeads were successfully arrayed using the µPAT technique: 15 µm, 21 µm, and 30 µm diameter microbeads. Up to 30 microbeads were successfully arrayed in adjacent traps. While microbeads generally arrayed in trapping sites sequentially (Fig. 6A), this was not always the case as different flow streamlines altered this sequential nature (Fig. 6B). Nonetheless, the specific order in which the traps were filled did not have any impact on the fluorescence data.

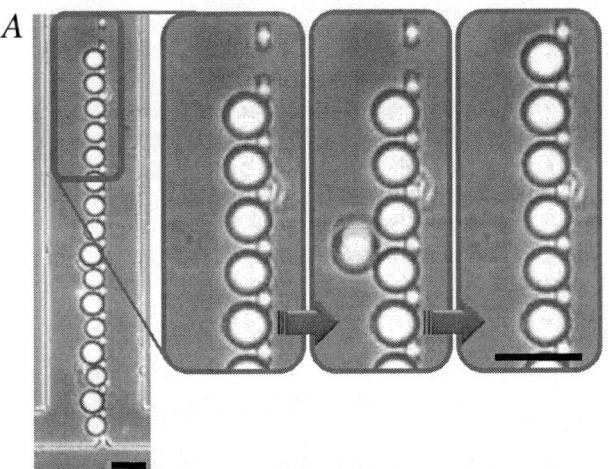

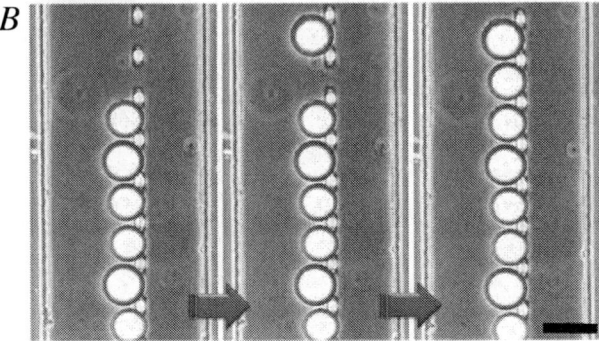

Figure 6: Micropost array trapping results for 30 µm diameter microbeads: (A) Sequential bead trapping, and (B) non-sequential bead trapping. Scale Bars = 50 µm

CONCLUSIONS

Our microfluidic system offers a fast, simple and sensitive technique for genotyping SNPs. By employing MBs for detection, time-consuming DNA labeling processes were avoided, thereby reducing the overall experiment time. Additionally, by integrating MBs into a bead-based microfluidic system, the experiment time was further reduced by taking advantage of the rapid diffusion times and enhanced reaction kinetics inherent to dynamic microfluidic arrays. We successfully distinguished between three DNA oligonucleotide sequences, including sequences that only differed by a single base. This demonstrates our achievement of the necessary sensitivity required for SNP detection, which was enhanced due to both a reduced background noise from using a microfluidic device as well as the high signal-to-noise ratios of MBs. For micro-particle trapping, µPAT was successfully employed to array a large number of beads within a limited area. This establishes the potential for µPAT as an effective tool for high-density and high-throughput micro-particle arraying, especially in automated systems.

Future work should concentrate on developing systems that apply microfluidic bead-based MB genotyping to distinguish multiple DNA oligonucleotide sequences simultaneously on a single microfluidic chip. This would result in a decrease in the overall time it takes to genotype multiple DNA oligonucleotide sequences since the current device was designed to genotype only one oligonucleotide sequence at a time per device. Furthermore, the increased sensitivity inherent to this method could be applied to distinguish not only *Target Sequence*, *One Mismatch (SNP)*, and *Negative Control* DNA oligonucleotide sequences, but also DNA sequences of two, three or other degrees of mismatches. While there are a number of potential directions for applying this technology, we have demonstrated that the integration of MB genotyping techniques with bead-based microfluidic arrays provides a promising basis for realizing fast, high-throughput, and cost-effective SNP genotyping.

ACKNOWLEDGEMENTS

The authors would like to acknowledge Adrienne Higa, Brian Sosnowchik, Paul Lum, Hansang Cho, John Waldeisen, J. Tanner Nevill, Edward Ha, Frankie Myers, Sisi Chen, the Liwei Lin Lab, and the Luke Lee Lab for their help, guidance, and support.

REFERENCES

[1] B.M. Paegel, C.A. Emrich, G.J. Weyemayer, J.R. Scherer, R.A. Mathies, "High throughput DNA sequencing with a microfabricated 96-lane capillary array electrophoresis bioprocessor", *PNAS*, vol. 99, pp. 574-579, 2002.

[2] Y. Baaj, C. Magdelaine, V. Ubertelli, C. Valat, L. Talini, F. Soussaline, E. Khomyakova, B. Funalot, J.-M. Vallat, F.G. Sturtz, "A highly specific microarray method for point mutation detection", *BioTechniques*, vol. 44, pp. 119-126, 2008.

[3] D. Horejsh, F. Martini, F. Poccia, G. Ippolito, A. Di Caro, M.R. Capobianchi. "A molecular beacon, bead-based assay for the detection of nucleic acids by flow cytometry", *Nucleic Acids Research*, vol. 33, 2005.

[4] W.-H. Tan, S. Takeuchi, "A trap-and-release integrated microfluidic system for dynamic microarray applications", *PNAS*, vol. 104, pp. 1146–1151, 2007.

[5] X.B. Zuo, X.H. Yang, K.M. Wang, W.H. Tan, J.H. Wen, "A novel sandwich assay with molecular beacon as report probe for nucleic acids detection on one-dimensional microfluidic beads array", *J. Analytica Chimica Acta*, vol. 587, pp. 9-13, 2007.

[6] J.H. Yang, J.-P. Lai, S.D. Douglas, D. Metzger, X.-H. Zhu, W.-Z. Ho, "Real-time RT-PCR for quantitation of hepatitis C virus RNA", *J. Virol. Methods*, vol. 102, pp. 119–128, 2002.

A FULLY AUTOMATED MICRO-SOLID PHASE EXTRACTION CHIP FOR GENETIC SAMPLE PREPARATION SYSTEM

Song-I Han, Hwanyoung Lee, and Ki-Ho Han*

School of Nano Engineering, Inje University, South Korea

ABSTRACT

This paper presents a fully automated micro-solid phase extraction (micro-SPE) chip for extracting genomic DNA from white blood cells. For realizing automated microfluidic control, high pressure dome-shaped microvalves are monolithically integrated. To increase the DNA extraction efficiency, silica beads were packed in the extraction microchannel involving two weir structures. Experimental results show that DNA extraction efficiency of the automated micro-SPE chip was 63%, averaged from five times successive measurement, in the first 8 μl elution solutions through automated SPE procedures. Furthermore, through the PCR-amplification of the β-globulin gene, the extracted genomic DNA is verified that it is a good quality for downstream genetic assays.

INTRODUCTION

Much research [1, 2] has focused on the development of miniaturized genetic sample preparation systems integrating various biological/chemical functionalities, including DNA purification, amplification by PCR, and analysis using CE. The miniaturized genetic sample preparation systems have inherent benefits, such as smaller geometric size, faster analysis times, less sample/reagent consumption, and disposability. To obtain a rapid and accurate performance of the miniaturized genetic sample preparation, extraction of high-quality DNA from crude biological samples is a significant functionality. Conventional DNA purification procedures are generally performed by silica-based SPE methods, whose efficiency depends on the interaction between DNA and the solid support, involving either electrostatic or hydrogen bonding. The SPE method has the advantages of ease of implementation for the complicated and miniaturized genetic analysis systems and the capacity to integrate possibly with genetic-based assays in micro format downstream. In the SPE procedure, a biological sample is loaded to a solid support and the DNA binds to the support. Next, other biological substances on the surface of the support are washed out, and the bound DNA is released from the support using a low-ionic-strength solution. To obtain high-quality DNA from the SPE procedure, it is essential to remove inhibitors in the biological sample that interfere with DNA binding to the solid support.

Although, previously developed micro-SPE chips have been reported for their capacities, the complicated manipulation and the labor-intensive nature of the approach due to their manual procedures are troublesome. In addition, they may limit the potential of the manual micro-SPE chips for incorporation with other biological functionalities into an on-chip complex microfluidic system. One practical method to solve the limitation is to integrate microvalves into the complicated microfluidic systems for automatic fluid control.

For over two decades, a number of microvalves based on silicon and glass materials have been developed for manipulating and controlling fluid flow in microsystems. Typically, these microvalves have been usually realized as hybrid type devices and their performance demonstrated. Due to their hybrid nature, their relatively large flow / dead volumes have led to limited success for their application to a complicated microfluidic system. To overcome this geometric scaling limitation, elastomeric microvalves have been implemented. However, microvalves using elastomeric material have issues with the hydrophobicity and porosity of native elastomer surfaces and auto-fluorescence. Furthermore, the low operating pressure (< 50 kPa) of elastomeric microvalves is a critical problem for their application to SPE procedures, which require a high fluidic pressure (> 200 kPa) for passage through a densely packed microchannel with solid support, such as silica beads, sol-gel, or solid column. Hence, microvalves, which can control the high fluidic pressure and can be monolithically implemented, are mandatory to realize an on-chip integrated microfluidic system.

In this study, we present a fully automated micro-SPE chip involving monolithically integrated high-pressure microvalves [3] for genetic sample preparation. Through a fluid flow test using three colored inks, the micro-SPE procedure, including loading, washing, and elution steps, was confirmed with automatic control. The extraction efficiency and reproducibility of the automated micro-SPE chip was evaluated using λ-phage DNA, and its capability was verified by extracting genomic DNA from human WBCs. Furthermore, PCR-amplification of the 260-bp β-globulin gene confirmed that the extracted genomic DNA is of sufficient quality to be used directly in downstream genetic analyses.

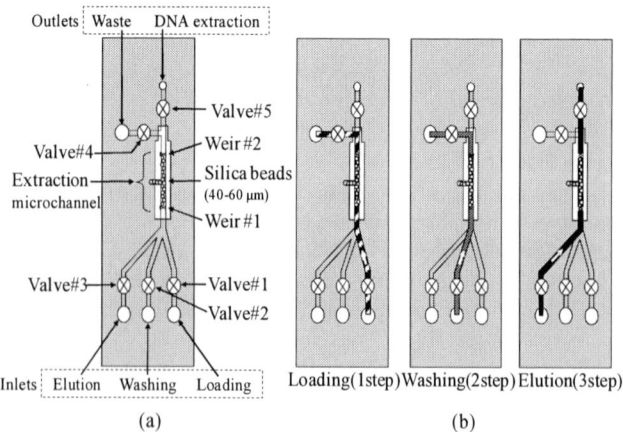

Figure 1: A schematic view of the automated micro-SPE chip and fluidic flow of loading, washing, elution steps for DNA extraction.

DESIGN

The automated micro-SPE chip consisted of three inlets (loading, washing, and elution), an extraction microchannel that contains the silica beads for binding DNA, and two outlets (for DNA extraction and waste), as shown in Fig. 1(a). Figure 1(b) illustrates the presenting fluid path of the loading (hatched), washing (gray), and elution (black) buffers as the automatic micro-SPE procedure was controlled using five monolithically integrated microvalves. To hold the silica beads within the extraction microchannel, two weirs, fabricated by chemical etching, were formed at the ends of the extraction microchannel. As a gap of about 10 µm is created between the weir and the top glass, the silica beads are held between two weirs and sub-cellular biological substances can pass through them.

Using silica beads in the extraction microchannel, the effective binding surface for DNA adsorption was increased over 60-fold compared to the microchannel without beads. Using the silica beads increased the DNA extraction efficiency dramatically. We used bare silica beads to avoid some of the problems are encountered with a solid-phase matrix. For instance, the solid-phase matrix requires a long time for curing and has the potential to release, which may cause contamination to other compartments integrated with on-chip microfluidic systems.

FABRICATION

The glass chip of the micro-SPE chip was fabricated by chemical etching and glass-to-glass thermal bonding processes, as depicted in Figure 2. The bottom glass (Borofloat™ glass, 0.7 mm thick, Howard Glass Co., MA, USA) was etched to a depth of 150 µm using 25% HF solution. To form the weir structure at both ends of the extraction microchannel, lines 280 µm wide were patterned on the microchannel. By chemical etching the bottom glass, the weir was formed separated from the top glass by about 10 µm. The dome-shaped valve holes placed on the top glass were etched using 25% HF solution for approximately 10 h, until holes of the appropriate diameter were etched through the bottom of the top glass (Fig. 2(a)).

To fabricate the through-holes in the top glass used for fluid interconnections, a mechanical drill was used along with a stereolithography (SLA) alignment jig, as shown in Fig. 2(b). The top and bottom glasses were bonded together by glass-to glass thermal bonding at 685°C for 3.5 h (Fig. 2(c)). A nitrile rubber O-ring (size 001-1/2, McMaster-Carr, GA, USA) was used to avoid clogging of the microchannels with the low- viscosity UV adhesive (1187-M, DYMAX Co., USA) during bonding of the microfluidic system interface (MSI) [4] to the underlying glass chip. Another nitrile rubber O-ring (Size-001, McMaster-Carr) was used to provide a tight fluid seal between the external capillary tubing and the microfluidic interconnects. A 120-µm-thick latex membrane was used to form the active element of the microvalve (Fig. 2(d)). The MSI and the glass chip were aligned and placed in a clamped jig. Next, the UV adhesive was placed into 'bonding' vias formed into the MSI, as shown in Fig. 2(d). Capillary force caused the UV adhesive to fill the gap between the MSI and the glass chip. The UV adhesive was then cured by placing it under a UV light for 30 min, completing fabrication of the glass-based automatic micro-SPE chip, as shown in Fig. 2(e). Figure 3 shows the fabricated glass-based automatic micro-SPE chip (a) before and (b) after the bonding of the MSI.

RESULTS

Monolithic integrated microvalves for automatic fluid control

Since microfluidic devices fabricated by glass-to-glass bonding technology result in a permanent mechanical bonding between adjoining layers, the technology lends itself to microfluidic applications requiring mechanical rigidity, high fluidic pressures, or relatively high operating temperatures. Considering these advantages of the glass-based microfluidic devices, the automated micro-SPE chip herein was fabricated based on glass substrates, and completed by assembling the glass device and the MSI, including the monolithic high pressure microvalves. As a result, the monolithic microvalves integrated in the glass-based micro-SPE chip can be operated under a high pressure. It is critical to manipulate the fluid flow for the micro-SPE procedure.

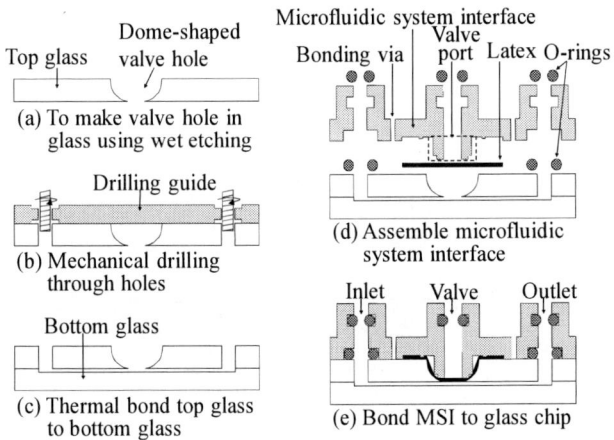

Figure 2: Fabrication and assembly process used to realize the automated micro-SPE chip involving microvalves.

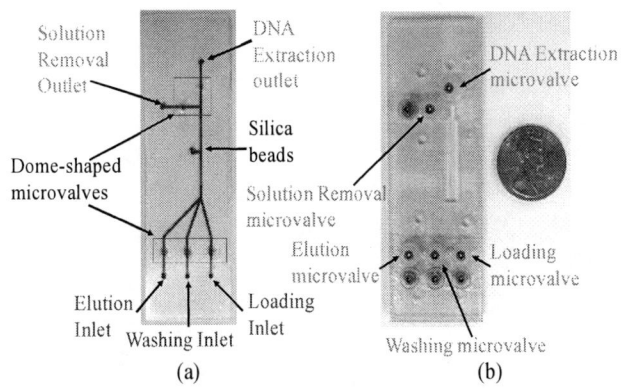

Figure 3: Fabricated the automated micro-SPE chip before and after bonding with the microfluidic system interface. (a) Glass chip. (b) Micro-SPE chip.

978-1-4244-2977-6/09 $25.00 © 2009 IEEE

The SPE procedure is automatically controlled for the loading, washing, and elution steps by the on/off sequential pneumatic control of the monolithic high-pressure microvalves. At the loading step, microvalves #2, #3, and #5 are closed and microvalves #1 and #4 are open. Then, the loading buffer is passed through the loading channel and flows out the waste outlet, as shown in Fig. 4(a). For the washing step, microvalves #1, #3, and #5 are closed and microvalves #2 and #4 are open. The wash buffer is passed through the washing channel and flows out the waste outlet to remove unwanted biological substances from the extraction microchannel, as shown in Fig. 4(b). For the elution step, microvalves #1, #2, and #4 are closed and microvalves #3 and #5 are open. The elution buffer flows from the elution inlet to the DNA extraction outlet, and the purified DNA is eluted and collected, as shown in Fig. 4(c). For the each step, fluid is driven through the densely packed silica beads in the extraction microchannel, thereby generating a high fluidic pressure (> 200 kPa). Under the high fluidic pressure, the proposed high-pressure microvalves successfully controlled the fluid flow, which was delineated by three different colored inks representing each step, as shown in Fig. 4. In the experiment, the dead volume of the microvalves was measured as only 22.5 nl, which is small enough for microfluidic manipulation. In addition, during the SPE procedure, no fluid leaks occurred through the closed microvalves, and therefore the possibility of biological cross-contamination between the inlet and outlet reservoirs was not a concern. Due to the characteristics of the monolithic microvalves, the developed automated micro-SPE chip has the ability to be integrated easily into on-chip microfluidic systems with other biological functionalities.

DNA extraction efficiency

The DNA extraction capability of the automated micro-SPE chip was evaluated by the extraction efficiency and reproducibility of the DNA purification. A 10 μl loading solution containing 1 ng/μl λ-DNA was passed through the extraction microchannel packed with silica beads at a flow rate of 180 μl/h. During the elution step, effluent from the DNA extraction outlet were collected every 2 μl for 10 fractions, and assayed using

the PicoGreen® fluorescence method to obtain the extraction traces, as illustrated in Fig. 5(a). The result indicated that 75.87% (i.e., 4.2 ng) of the extracted DNA came out in the first 8 μl of the eluted solution. This means that most of the extracted DNA can be enriched into several tens of microliters. Conventional PCR-amplification processes are usually implemented with > 0.1 ng DNA. In the case of our DNA extraction process using the automated micro-SPE chip, the quantity of purified DNA was more than 4 ng, an amount sufficient for PCR techniques. In addition, the fact that most of the DNA was eluted in the first 8 μl of solution means that the automated micro-SPE can be integrated easily into miniaturized genetic analysis systems, which typically manipulate trace amounts of DNA.

In a new SPE procedure, used silica beads were refreshed by washing with elution buffer for 20 min, followed by distilled water for 30 min, and then dried in a 40°C dry oven for 5 h. The DNA recovery rate from the micro-SPE chip, in which 10 ng λ-DNA was loaded, was measured as 63% averaged from five successive extractions (Fig. 5(b)). From these results, we can predict that the micro-SPE chip involving bare silica beads can be reused five times without any loss in DNA extraction performance.

Application to genomic sample preparation

Sub-cellular biological samples contain not only nucleic acids, but also, for instance, proteins, lipids, metabolites, and inorganic ions. These other compounds can interfere with DNA binding to the surfaces of silica beads. The unwanted biological substances may also act as inhibitors in genetic-based analysis procedures (e.g., PCR and base sequencing). Therefore, the extraction of high-purity genomic DNA from the crude biological sample is a critical technology for the reliability of the genetic-based analysis procedures.

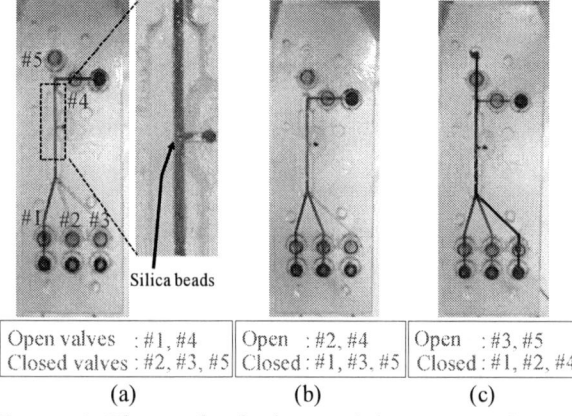

Figure 4: The result of ink test of the automated micro-SPE chip. (Back side) (a) Loading step. (b) Washing step. (c) Elution step.

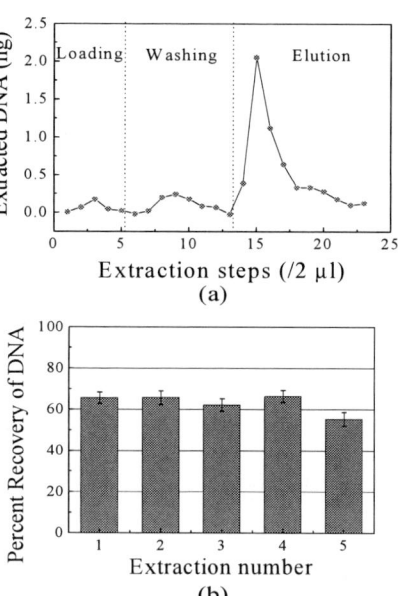

Figure 5: (a) λ-DNA extraction profile of the automated micro-SPE chip and (b) extraction efficiency according to extraction number.

As a biological sample, human WBCs were prepared from 20 µl fresh human whole blood by centrifugation at 3000 g for 30 min using a centrifugation media. The prepared WBCs were lysed with 100 µl loading buffer (6 M GuHCl in TE buffer) containing 10 µl proteinase K (20 mg/ml), warmed in a water bath at 70°C for 10 min, and then mixed with 90 µl ethanol to increase the genomic DNA extraction efficiency. Loading (12 µl), washing (16 µl), and elution (20 µl) buffers were consecutively passed through the extraction microchannel at a flow rate of 180 µl/h, with the total SPE procedure steps taking ~16 min. During the SPE procedure, effluent from the DNA extraction outlet was collected every 2 µl and fluorescently assayed using the PicoGreen® method (Fig. 6(a)). Similar to the extraction profile of λ-DNA (Fig. 5(a)), the genomic DNA was released at the start of the elution step; the most DNA came out in the second fraction. After the peak, the amount of eluted DNA tapered off slowly compared to the extraction profile of λ-DNA. The difference in elution profiles may be due to the larger size of the WBC genomic DNA compared to the λ-DNA, or the fact that the total amount loaded was greater than that of the λ-DNA. Using a spectrophotometer (NanoDrop ND-1000, MA, USA), the purity of the eluted sample; absorbance ratio of DNA to protein (A_{260}/A_{280}) was 1.56. The measured absorbance ratio indicates that nucleic acids percentage is 30%, the other side; protein is 70% in the eluted solution.

To verify the feasibility of human genomic DNA extraction from the micro-SPE chip, it was used for PCR with primers for amplification of a 260-bp β-globulin gene. Sequences of the forward and reverse primers were 5′-CAACTTCATCCACGTTCACC-3′ and 5′-GAAGAGCCAAGGACAGGTAC-3′, respectively. The 20 µl PCR mixture contained 4 µl of 5 × taq-PCR mix, 2 µl forward/reverse primers (10 pmol/µl), and 1 µl eluted DNA sample as a template, and 11 µl distilled water. Using a commercial thermocycler (PTC-200, MJ Research, USA), a standard PCR protocol was performed: initial denaturation at 94°C for 5 min, followed by 40 cycles of denaturation at 94°C for 30 s, annealing at 55°C for 30 s, primer extension at 72°C for 30 s, followed by a final extension at 72°C for 7 min. The PCR products were analyzed by 1.2% agarose gel electrophoresis with ethidium bromide staining followed by observation under UV light to compare the size using fluorescently-labeled DNA standards, and processed by ImageJ software (NIH), as shown in Fig. 6(b). Although, the percentage of nucleic acids of the eluted sample was less than the protein, the successful PCR-amplification of the β-globulin gene demonstrates that the genomic DNA extracted from the micro-SPE chip is pure enough to be used directly in subsequent genetic analysis procedures.

CONCLUSIONS

A fully automated micro-SPE chip was realized using monolithic integrated high-pressure microvalves. The micro-SPE chip was fabricated by glass-to-glass thermal bonding and microfluidic system interface technologies. Although a high fluidic pressure was generated for passage through the extraction microchannel in the micro-SPE chip, the monolithic microvalves operated successfully to control the fluid flow for the SPE procedure. An SPE experiment with λ-DNA showed that the extraction efficiency of the micro-SPE chip using bare silica beads was 75.87% (i.e., 4.2 ng) averaged using five successive extractions, and the efficiency was sustained up to 63% for the five extractions. Human genomic DNA extracted from the micro-SPE chip was used for PCR-amplification of a 260-bp β-globulin gene and its product was identified successfully by gel electrophoresis. These results have demonstrated that the purified genomic DNA is of sufficient quality to be used directly in subsequent genetic analyses in micro format. Further optimization of the automated micro-SPE chip will be investigated for reducing the extraction time, increasing the extraction efficiency, and improving methods to simplify on-line coupling of SPE, PCR, and CE in a µTAS device.

REFERENCES

[1] R. Zhong, D. Liu, L. Yu, N. Ye, Z. Dai, J. Qin, and B. Lin, "Fabrication of two-weir structure-based packed columns for on-chip solid-phase extraction of DNA", *Electrophoresis*, vol. 28, pp. 2920-2926, 2007.

[2] M.C. Breadmore, K.A. Wolfe, I.G. Arcibal, W.K. Leung, D. Dickson. B.C. Giordano, M.E. Power, J.P. Ferrance, S.H. Feldman, P.M. Norris, and J.P. Landers, "Microchip-Based Purification of DNA from Biological Samples", *Analytical Chemistry*, vol. 75, pp. 1880-1886, 2003.

[3] S.-I. Han, Y. Jung, K.-H. Han, and A.B. Frazier, "On-Chip Integratable Elastomeric Dome Valves for Glass Microfluidic Systems", *Sensors and Materials*, vol. 20, pp. 35-43, 2008.

[4] K.-H. Han, R.D. McConnell, C.J. Easley, J.M. Bienvenue, J.P. Ferrance, J.P. Landers, and A.B. Frazier, "An active microfluidic system packaging technology", *Sensors and Actuators B*, vol. 122, pp. 337-346, 2006.

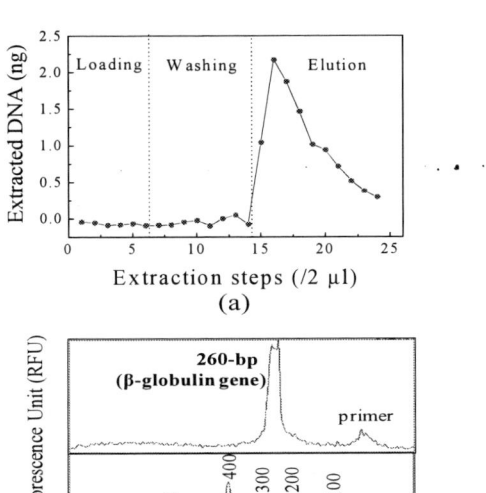

Figure 6: (a) Genomic DNA extraction profile from human WBCs lysate and (b) electropherograms of PCR product after amplification of the β-globulin gene (260-bp).

SELF-POLARIZED PIEZOELECTRIC BIOSENSOR ARRAY FOR MULTIPLE IMMUNOASSYS APPLICATIONS

Ting Xu, Jianmin Miao, Zhihong Wang, Ling Yu and Chang-Ming Li
Nanyang Technological University, SINGAPORE

ABSTRACT

This paper reports a novel microfabricated self-polarized piezoelectric biosensor array for simultaneous detection of 3 different types of antigens. This array consists of 8 individual circular sensors with a diameter of 1 mm and self-polarization because of the stress accumulated in the piezoelectric membrane during the fabrication process. HBsAg, HBcAg and α-Fetoprotein (AFP) were immobilized onto sensing surface as the probe molecules. Due to the mass sensitive nature of these sensors, their resonant frequencies were depressed after the target anti-bodies were captured by the probe molecules. A clear frequency depression was observed and the mass sensitivity of the device is estimated to be about 4.80 Hz/ng. The preliminary results demonstrate that the developed piezoelectric biosensor array has potential applications for multiple immunoassys.

INTRODUCTION

Micro-machined piezoelectric biosensors attract much research attention due to their compact size, high sensitivity, and rapid response [1-3]. However, development of biosensors array for simultaneously detecting multi-biomaterials to significantly shorten the diagnosis time and cost is still under exploration. To achieve the multi-biomaterial detecting function, the sensing device needs to be designed in an array form. Although quartz-crystal microbalance (QCM) type biosensor has many advantages, such as high reliability and sensitivity, and suitable for both gas and liquid media sensing [4], it is not practical for miniaturization and multiple sensing applications due to its fragile material property. Many researchers attempted to fabricate the piezoelectric silicon based cantilever biosensor array using micro-machining technology because it has high sensitivity and can be fabricated in the array form [2,5]. However, cantilever biosensor is not suitable to be used in liquid because of its low quality factor (Q value) and also low fabrication yield.

To overcome the problems arising from the QCM and cantilever biosensors, micro-machined piezoelectric membrane based sensor array is one of the alternative candidates [6]. We previously reported a piezoelectric biosensor array for detecting 2 types of antigens simultaneously [7]. In this work, a novel microfabricated self-polarized piezoelectric biosensor array for simultaneous detection of 3 different types of antigens with a sensitivity of 4.80 Hz/ng was reported. To our best knowledge, it is the first time to report on the self-polarized piezoelectric biosensor for the biomaterial detection.

FABRICATION

Fabrication Processes for the Sensor Array

The piezoelectric sensor array was fabricated by micro-fabrication technology. As illustrated in Figure 1, five main steps were involved during the fabrication processes. A silicon-on-insulator (SOI) wafer <100> was used as the main substrate. Firstly, a thin TiO$_2$/Pt (15 nm/200 nm) film was deposited as the bottom electrode by sputtering at high temperature of 250 °C. Secondly, a thin Pb(Zr$_{0.52}$Ti$_{0.48}$)O$_3$ (PZT) film (~0.7 μm) was deposited by the sol-gel deposition technique [8]. To open the access to the bottom electrode pad, in the third step, the PZT film was wet etched in diluted HCl : HF solution. Afterwards, a Si$_3$N$_4$ layer (350 nm) was deposited by plasma enhanced chemical vapor deposition (PECVD) to serve as an insulation layer to minimize parasitic capacitance induced by the patterned electrode wiring. The top electrode (10 nm Ti/200 nm Pt) was sputtered and patterned by using lift-off technique. Finally, the backside silicon holes used as the reaction chamber during the bio-immobilization process was etched by deep reactive ion etching (DRIE).

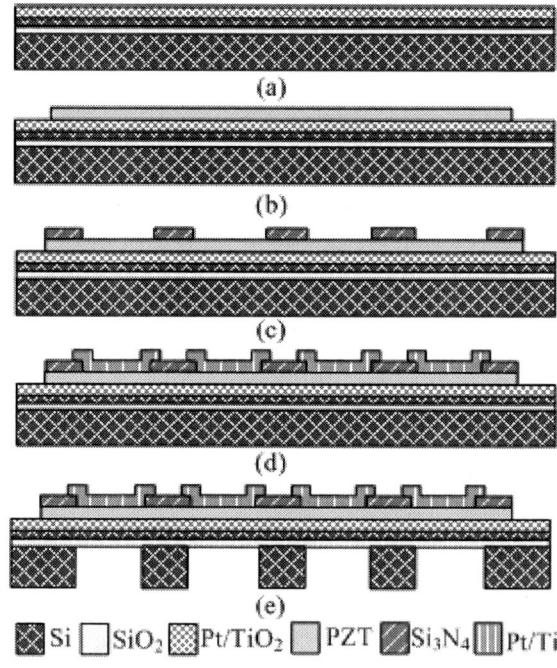

Figure 1: Sketched fabrication process flow of the biosensor array.

Fabrication Results

The optical and SEM images of the fabricated sensor array are presented in Fig. 2. As shown in Fig. 2 (a), the two rectangular blocks are the bottom electrode pads. A total of 8 individual sensors are located in the two parallel rows in the array. The individual sensor is formed by one circular PZT membrane and one square top electrode pad. The backside of the sensor array is shown in Fig. 2 (b), where the 8 holes with a diameter of 1 mm serve as the reaction chambers during the bio-immobilization process. The inset in Fig. 2 (b) is the SEM image of one chamber.

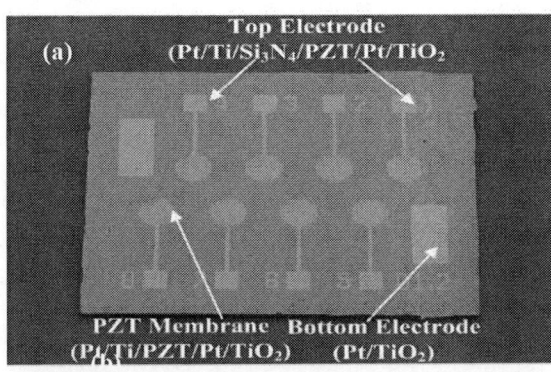

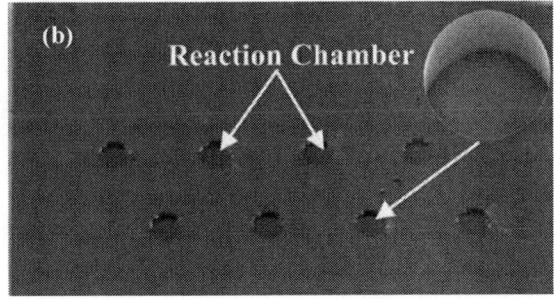

Figure 2: Optical images of the fabricated sensor array. (a) Front view. (b) Backside view.

CHARACTERIZATION

Piezoelectric Characterization of the Sensor Array

The measured P-E hysteresis loop of the sensor is shown in Fig. 3 (a), it can be seen that the thin PZT film (~ 0.7 μm thick) as the sensor membrane exhibits good ferroelectric properties after fabrication. The piezoelectric membrane based sensor was characterized using the impedance spectrum. An Agilent 4294A impedance analyzer was used to test the fabricated sensors in this work. A probe station was used to connect the analyzer to the wiring pads of the sensors. Fig. 3 (b) shows the impedance spectrum of a sensor at the first resonant mode without any prior polarization. With the Q factor definition reported in our previous work [9], the obtained Q value is about 59 at a quite low operating frequency of around 71 kHz. The results present in Fig. 3 (b) reveal that the thin PZT membrane deposited in our work is self-polarized, induced by the stresses accumulated during the fabrication processes.

Fig. 3 (c) shows the optical image of the fabricated PZT membrane. It clearly shows the color difference on the different membrane surfaces, indicating that the membrane was deformed after the fabrication process due to the accumulated stresses. The possible stresses come from three sources. Firstly, the residual thermal stress exists inside the SOI wafer due to the deposition of SiO_2 layer at high temperature. The second thermal stress source comes from the PZT deposition process, where the sample went through several runs of high temperature heating (~ 450 °C) and annealing (600 °C) processes. The thermal stress was certainly generated in this process because of the different coefficients thermal expansions (CTE) of different materials. The last one may be the mechanical stress generated during the thin membrane release process by DRIE.

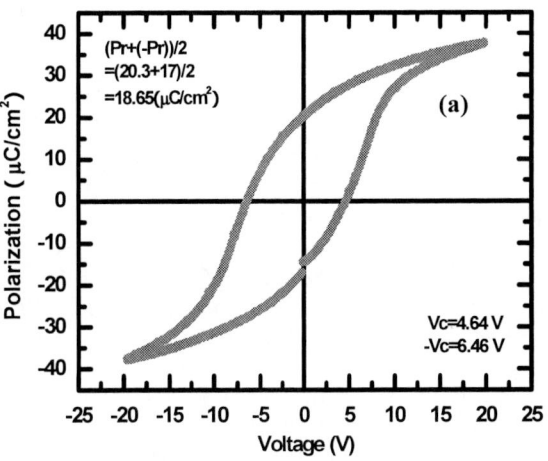

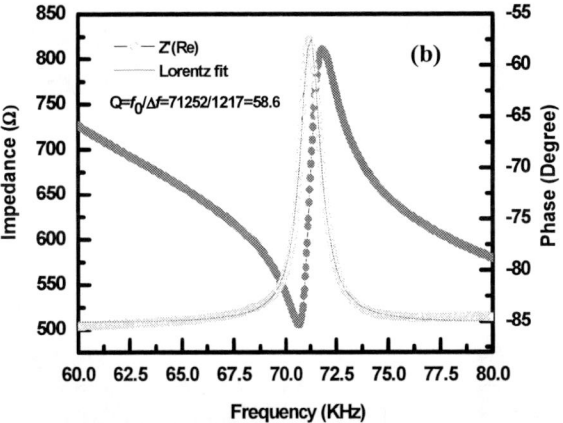

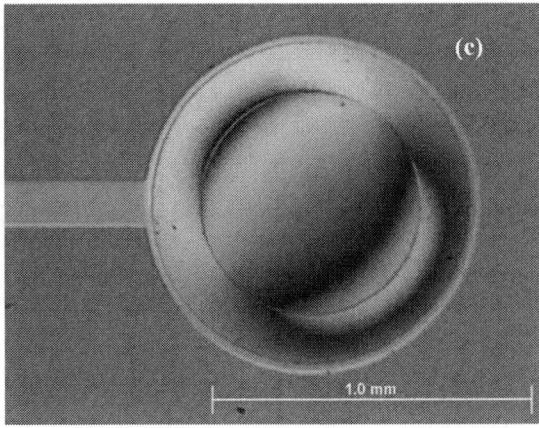

Figure 3: Characterization of the sensor array. (a) P-E hysteresis loop of the PZT layer. (b) Impedance spectrum of a sensor at the first resonant mode without any prior polarization. (c) Optical image of the PZT membrane on one sensor.

The continuous ion bombardment on the surface could also cause the mechanical deformation of the very thin membrane. Comparing with normal piezoelectric sensor, this self-polarized PZT membrane based sensor offers a much easier and faster fabrication process, it also lower the fabrication cost because the normal polarization process was omitted.

Immobilization of Biomaterials

Before using the piezoelectric sensor as a biosensor, the sensing surface should be first immobilized with bio-receptors. In this experiment, antigens were fixed onto the sensor surface to detect the specific antibodies. To realize the multi-detecting function of the developed piezoelectric sensor array, three different biological entities, HBsAg, HBcAg and α-Fetoprotein (AFP), were used to verify the feasibility of applying this sensor as an immunosensor. The sketch of biological entities immobilization processes is depicted in Fig. 4. The sensing surface, the bottom surface of the reaction chamber in this work, is the SiO_2 layer which facilitates the protein immobilization. Hence, in the first step, HBsAg, HBcAg and AFP (Sigma, USA) in phosphate buffered saline (PBS, pH7.4) were applied into the individual reaction chambers with the same concentration (100 µg/ml) in sensor 1 and 5, 2 and 6, 3 and 7, respectively.

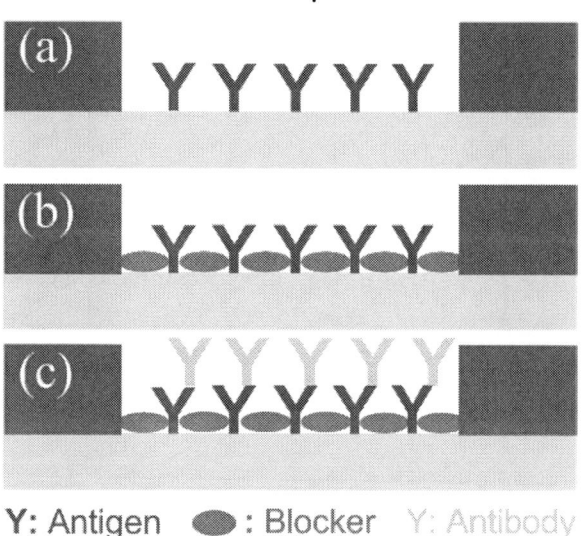

Figure 4: Schematic processes of immobilizing. (a) Attaching the antigens onto the PZT membrane. (b) Blocking the open surface by Blocker TM Casein in TBS. (c) Hybridization of antigens and antibodies.

One pair of the sensors was immobilized with the same biomaterials to ensure the accuracy and repeatability of the measurements. To investigate the effect of washing processes as well as the non-specific biomaterials absorption on the sensors' resonant frequency, one pair of sensors (4 and 5) was used as the reference which underwent all the same washing and drying process but no biomaterial was immobilized onto them. After 30 minute deposition at room temperature, the reaction chambers were washed by tris-buffered-saline (TBS washing buffer, pH8.0) washing buffer and deioned (DI) water, they were dried under nitrogen (N_2) air flow. In the second step, Blocker ^{TM} Casein in TBS (1 mg/µl) was added into the reaction chambers and incubated in the culture chamber (~ 37 °C) for 30 min to block the open space around immobilized proteins. The excess blockers were washed away by TBS and DI water. Lastly, a mixed solution (10 µg/ml) containing anti-AFP and anti-HBsAg was added into all the reaction chambers in the sensor array for the hybridization between the antigens and antibodies. This process took about 45 min at a temperature of 37 °C. Again, the un-bonded antibodies were cleaned by TBS and DI water. The sample was then dried by N_2 air flow. The sample was undergone the resonant frequency measurement after each immobilization process. Each measurement took less than 5 min so that all the proteins were in active conditions.

Results and Discussions

Due to the mass sensitivity nature of the piezoelectric biosensor, the resonant frequency of the sensor decreases after the external mass is loaded. In order to study this relationship in detail, the resonant frequencies of the sensors were measured after each immobilization or reaction process. All the measurements were conducted in gas phase with a temperature of 30 °C. After the first immobilization, frequency depression among the sensor array varied from 520 to 1800 Hz. Fig. 5 (a) shows the frequency shift of the sensor 3 after the anti-AFP was captured by the AFP. It can be clearly seen that the frequency was decreased. Detailed frequency shift after each immobilization process for the sensor array is presented in Fig. 5 (b).

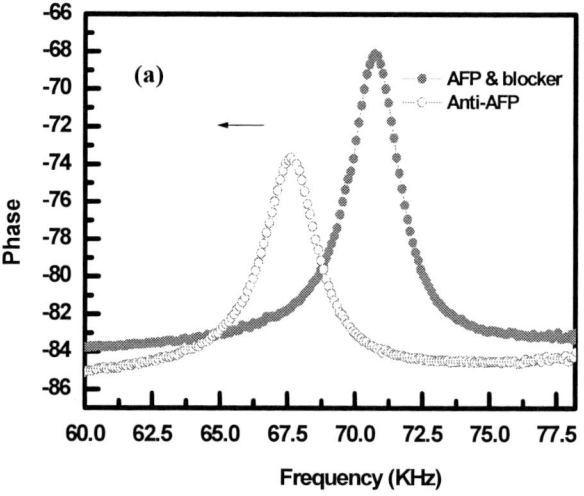

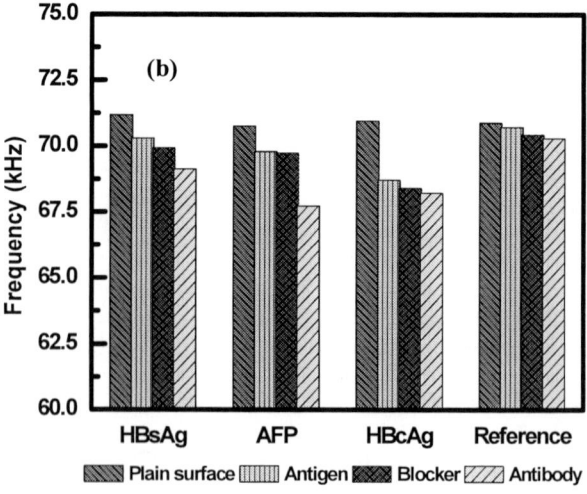

Figure 5: Frequency shift of the biosensor array. (a) After anti-AFP was captured. (b) After each immobilization process.

The frequency shifts were calculated based on the average frequency values of the one pair of sensors after 5 measurements. The first measurement shows that the basic resonant frequencies of the 4 pairs of sensors before the bio-immobilization were almost the same, indicating the uniform performance of the individual sensors within the array. Although the concentration and volume of the immobilized probe proteins, such as HBsAg, AFP and HBcAg, were the same in the first step, the depressions of the frequency in the second measurement was not the same. This implied that absorbed masses by each sensor were different. The possible reason may be attributed to the different physical properties of proteins, such as density and mobility. Thus the amount of proteins successfully immobilized onto the individual sensor surface was different. As shown in Fig. 5 (b), the frequency shift was very small after immobilizing the blockers; the small frequency variation revealed that most of the sensing surface was occupied by the added antigens. In the third immobilization process, the applied mixed solutions only contain anti-HBsAg and anti-AFP. The data hence presented in Fig. 5 (b) shows a significant change in the sensors with HBsAg and AFP. The frequency change in the sensors with HBcAg was negligible. These results indicate that the frequency depressions of the biosensors with HBsAg and AFP in the last step were mainly due to the mass of the captured antibodies by the antigens through the coupling effect. The frequency of the reference sensor has very small variations (< 80Hz) during the whole immobilization processes, showing that the washing processes and the non-specific biomaterial absorption had negligible effect on the sensors' resonant frequency. By using our previously reported calculation method, the sensitivity of mass per unit area for the sensor array was calculated to be - 45.33 m^2/kg, which is more than 3 times higher than that obtained in our previous work [7]. This is mainly because the thickness of the deposited PZT film (0.7 μm) is thinner than that in the last work (3.5 μm). Another sensitivity in the unit of Hz/ng can also be calculated by considering the reference resonant frequency (~ 71 kHz) and the surface area (7.85×10^{-7} m^2). It was calculated to be 4.80 Hz/ng and 4.79 Hz/ng for detecting anti-HBsAg and anti-AFP, respectively. This result shows the biosensor has almost the same resolution for detecting the antibodies of HBsAg and AFP. It also demonstrates that the sensor array developed in this work has the capability of detecting the bio-materials in quantity.

CONCLUSIONS

In this study, we presented a novel microfabricated self-polarized piezoelectric biosensor array which has the capability of detecting of 3 different types of antigens at the same time. These sensors were self-polarized because of the stress in the piezoelectric membrane accumulated during the fabrication process. Anti-HBsAg, anti-HBcAg and anti-AFP were successfully detected by the immobilized probe molecules.

A clear frequency depression was observed and the frequency depression among the sensor array varied from 520 to 1800 Hz. The mass sensitivity of the device is estimated to be about 4.80 Hz/ng. The preliminary results demonstrate that the developed piezoelectric biosensor array can be potentially used for multiple immunoassys in low cost medical diagnosis applications.

REFERENCE:

[1] K.S. Hwang, J.H. Lee, J. Park, D.S. Yoon, J.H. Park, T.S. Kim, "In-situ quantitative analysis of prostate-specific antigen (PSA) using nanomechnical PZT cantilever in liquid cell", *Lab Chip*, vol. 4, pp.547-552, 2004.

[2] R. Raiteri, M. Grattarola, H.J. Butt, P. Skladal, "Integrated flow sensor for in situ measurement and control of acoustic streaming in flexural plate wave micropumps", *Sens. Actuators B,* vol. 79, pp. 115-126, 2001.

[3] Y. Lee, G. Lim, W. Moon, *Sens. Actuators A,* 130 (2006), pp. 105-110.

[4] X. T. Mo, Y. P. Zhou, H. Lei, L. Deng," Microbalance-DNA probe method for the detection of specific bacteria in water", *Enzyme Microb. Technol. Enzyme.* vol.30, pp.583-589, 2002.

[5] L. G. Carrascosa, M. Moreno, M. Alvarez, L. M. Lechuga, "Nanomechanical biosensor s: a new sensing tool", *TrAC, Trends Anal. Chem* .vol. 25, pp. 196-206, 2006.

[6] L. Nicu, M. Guirardel, F. Chambosse, P. Rougerie, S. Hinh, E. Trevisiol, *et al.* "Resonating piezoelectric membranes for microelectromechanically based bioassay: detection of streptavidin-gold nanoparticles interaction with biotinylated DNA", *Sens. Actuators B*, vol. 110, pp. 125-136, 2005.

[7] T. Xu, Z.H. Wang, J.M. Miao, L. Yu, C.M. Li, "Micro-machined piezoelectric membrane based immunosensor array", *Biosens. Bioelectron.* vol. 24, pp.638-643, 2008.

[8] C. Zhao, Z. Wang, W. Zhu, X. Yao and W. Liu, "PZT Thick Films Fabrication Using a Sol-Gel Based 0-3 Composite Processing", *Int. J. Mod. Phys. B.* vol.16, pp.242-248, 2002.

[9] Zhihong Wang, Jianmin Miao, Ting Xu, Ling Yu; Chang Ming Li, Xiaofeng Chen, "Biosensors Based on Flexural Mode Piezo-Diaphragm", In *Digest Tech. Papers on Nano/Micro Engineered and Molecular Systems Conference*, Jan 6-9, 2008, pp. 374-378.

SELECTIVITY AND LONG-TERM RELIABILITY OF RESONANT EXPLOSIVE-VAPOR-TRACE DETECTION BASED ON ANTIGEN-ANTIBODY BINDING

Anderson Lin and Eun Sok Kim

Department of Electrical Engineering-Electrophysics

University of Southern California

Los Angeles, CA 90089-0271, USA

ABSTRACT

This paper reports experimental results on the selectivity and long-term reliability of the explosive-vapor-trace detection based on antigen-antibody binding on a film bulk acoustic resonator (FBAR). An FBAR coated with specific antibody was shown to be able to detect vapor trace of TNT (Trinitrotoluene) or RDX (Cyclotrimethylenetrinitramine) without any pre-concentrator [1], but its false alarm rate and long-term reliability have never been experimentally investigated. This paper confirms that the anti-TNT coated FBAR sensor indeed responds to TNT vapor very selectively with permanent resonant frequency shift, as expected since an antigen binds only to its specific antibody. Also experimentally obtained is a surprisingly long lifetime (up to a month) of the anti-TNT at room temperature.

INTRODUCTION

In the fight against terrorism, the ability to selectively detect traces of explosives in their vapor phase is critical for the safety of all citizens. Among the possible approaches to achieve such detection, an approach that is highly selective, cheap and portable is highly desirable. In our previous work, we have demonstrated selective mass sensing by exploiting the known high binding specificity between an antibody and its targeted antigen. We have shown that an FBAR coated with specific antibody is able to detect vapor trace of TNT or RDX without any pre-concentrator [1].

In order to investigate the practicality of such sensing method, two issues are of great importance; namely, the false alarm rate and long-term reliability of the mass sensor. With our choice of sensing method based on antigen-antibody binding, we expect the false alarm problems to be minimized. However, there may be adsorption/absorption issue that has to be considered, which will be discussed in our selectivity results reported in this paper.

For biological experiments, antibodies are usually stored in either a freezer or refrigerator to ensure its integrity over a length of time. However, such a constraint may be quite inconvenient for our mass sensors when put into practical usage. Thus, one would want to find out if the antibodies coated on the gold surface of the FBAR sensor would be able to maintain its functionality when placed in a dry environment at room temperature. As will be presented in this paper, our experimental results on the antibody lifetime are quite interesting as the anti-TNT we tested was able to withstand for up to a month at room temperature.

In this paper, we will report our experimental results and offer comments on the selectivity and long-term reliability of the explosive-vapor-trace detection based on antigen-antibody binding on an FBAR-based mass sensor.

DESIGN AND FABRICATION

FBAR Fabrication

FBAR fabrication was done by first depositing low stress LPCVD silicon nitride on a (100) silicon wafer. Silicon nitride on the backside of the wafer was patterned, and the silicon substrate was etched in KOH solution to produce the supporting silicon nitride diaphragm on the front side. To produce the sandwich structure of FBAR, bottom electrodes were first deposited in an e-beam metal evaporator, with Cr used as an adhesion layer for Au. After patterning of the bottom electrodes, piezoelectric ZnO film was sputter-deposited and patterned. Top electrodes were then formed using lift-off technique after depositions of Au/Cr in an e-beam metal evaporator. Then, a gold layer was deposited on the backside of the silicon nitride diaphragm for antibody immobilization for selective mass sensing. The fabricated FBARs consist of Au/ZnO/Au/SiN/Au layers as shown in Fig. 1.

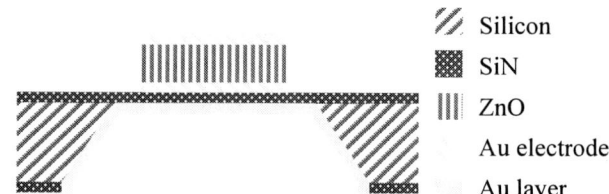

Fig. 1: Cross sectional view of the FBAR sensor with gold deposited on the backside.

Antibody Immobilization

To immobilize antibody on the gold layer, protein A was adsorbed to the gold, to form a polarized bond between the gold layer and antibody [2]. Protein A (isolated from microbial pathogen Staphylococcus aureus) was designed to recognize and bind with high affinity to the Fc portion of an IgG subclass antibody, leaving its antigen binding sites free. Protein A ensures that the antibody is properly oriented, facing sources of antigens.

Here in this work, we have optimized the antibody immobilization process [3] which now consists of the following: (1) immerse the FBAR backside in 1.2 M NaOH for 10 minutes followed by DI water wash, (2) immerse the FBAR backside in 1.2 M HCl for 10 minutes followed by DI water wash, (3) wash the FBAR backside with ethanol and DI water, (4) soak the FBAR backside in protein A solution (diluted in Phosphate Buffered Saline (PBS)) for 30 minutes at room temperature, followed by washing with PBS and DI water to remove unbound protein A, and (5) soak the FBAR backside in antibody solution (diluted with PBS) for one hour at room temperature, followed by washing with PBS and DI water to remove unbound

978-1-4244-2977-6/09 $25.00 © 2009 IEEE

antibodies. The sample is then allowed to dry, and this procedure results in an antibody layer immobilized on the FBAR gold surface as shown in Fig. 2.

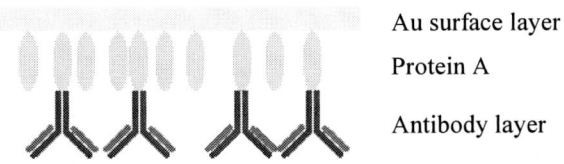

Au surface layer

Protein A

Antibody layer

Fig. 2: Antibody immobilization on the gold surface of the backside of an FBAR sensor.

SELECTIVITY
Testing Setup

The fabricated FBAR mass sensor coated with anti-TNT described above is expected to have outstanding selectivity, since the detection is based on antigen-antibody binding. To investigate the selectivity issue, we set up an experimental apparatus (Fig. 3) that consists of the following: (1) a testing platform where the FBAR can be exposed to vapors flowing into the platform, (2) an air pump that pumps air through a tubing, which brings air to the testing platform at a rate of 2 liters per minute (LPM), (3) a T-connector, connecting two tubes, that allows TNT or other chemicals to be put through the flowing air. The end of the T-connector where we placed chemicals was sealed with a tape throughout the experiment for a constant air-flow through the tubing.

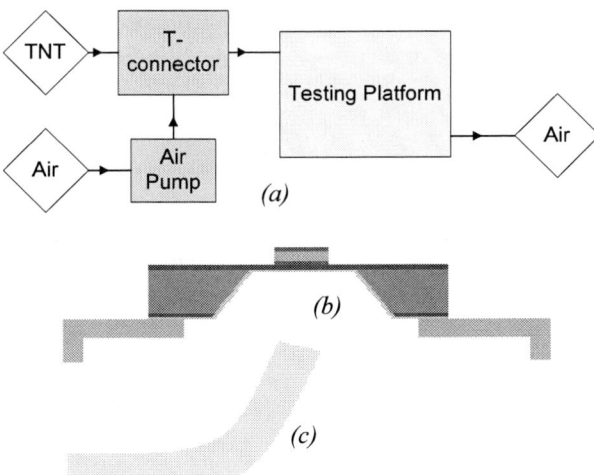

Fig. 3: (a) Testing schematic, with FBAR sensor situated on the testing platform. (b) Cross-sectional view of the testing platform with the FBAR sensor. (c) The tube from the T-connector.

The FBAR's impedance characteristics were measured with HP8753D network analyzer, and processed with LabVIEW on a computer. We have developed a LabVIEW program to track the resonant frequency where the Q is highest. A typical FBAR sensor coated with the antibody was measured to have a Q of 500 - 600 at 1.3 – 1.7 GHz with a noise floor of 1 ppm.

To narrow down the frequency range to monitor, we took the following steps in sequence: (1) measure FBAR's impedance with the network analyzer and choose the

frequency range to monitor, (2) calibrate the network analyzer, reconnect the network analyzer to the FBAR, and start the LabVIEW tracking program. Then after waiting until the frequency stabilizes, we placed chemicals into the T-connector (by opening the sealing tape briefly), and read out the resonant frequency shift.

Results

We investigated the anti-TNT coated sensor's selectivity to TNT over cyclotrimethylenetrinitramine (RDX), water, perfume, acetone, methanol and isopropyl. When we injected 20 µl RDX (the same volume as for TNT testing) once at a specific time, we did not observe any significant response as shown in Fig. 4, as expected since the sensor was coated with anti-TNT. Since RDX has a very low vapor pressure of 4.6 x 10^{-9} torr at room temperature [4], its mass loading on the FBAR surface through adsorption would be negligible.

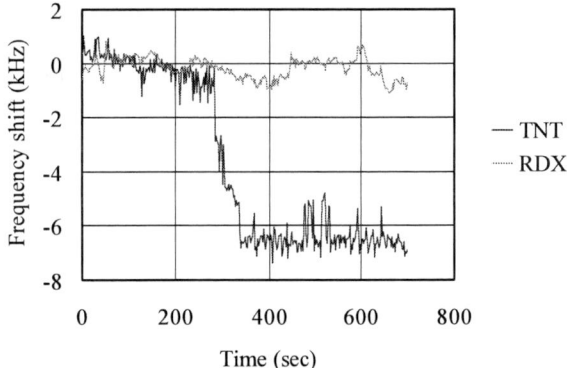

Fig. 4: The FBAR sensor's selective response to TNT over RDX vapor. The FBAR was coated with anti-TNT. If the sensor had anti-RDX immobilized on its surface, it would have responded selectively to RDX, not TNT.

Also, the anti-TNT coated sensor showed very little response for water and perfume, when a volume of 20 µl was injected into the T-connector once at a specific time. As shown in Fig. 5, the response from water or perfume was more noticeable, though temporary, than that from RDX, due to higher vapor pressure that makes more adsorption on the sensor surface or absorption inside the coated layers.

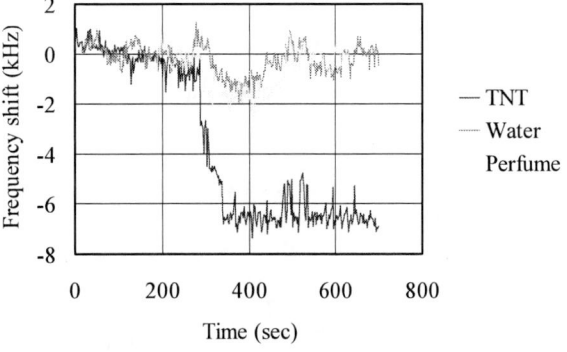

Fig. 5: FBAR sensor's responses to TNT, water and perfume vapors.

978-1-4244-2977-6/09 $25.00 © 2009 IEEE   317

When we injected 20 μl of acetone, methanol, or isopropyl into the T-connector once at a specific time, we observed some temporary frequency shifts due to those vapors as shown in Fig. 6. The vapors produced some frequency shifts at the point of injection, likely due to adsorption (and some absorption inside the coated layers of the sensor) of the vapors. As the adsorbed/absorbed vapors desorbed from the sensor, the loaded mass on the sensor was removed, and the resonant frequency returned to the original level prior to the chemical exposure. Thus, the anti-TNT coated FBAR sensor responded to organic chemicals of high vapor pressure, but only temporarily, unlike to TNT that produced a permanent shift in the resonant frequency due to antigen-antibody binding.

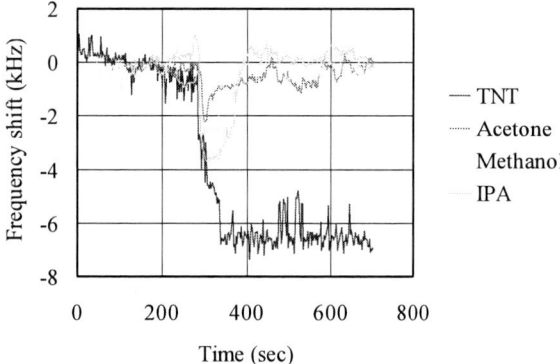

Fig. 6: FBAR sensor's responses to TNT, acetone, methanol, and isopropyl.

The anti-TNT coated FBAR sensor's permanent frequency shifts have been consistently obtained in the presence of TNT, as shown in Fig. 7.

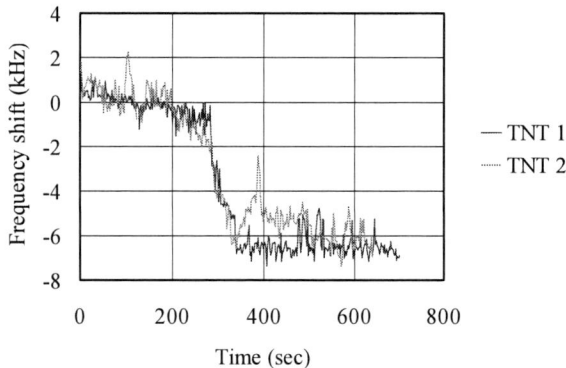

Fig. 7: Repeatability of the FBAR sensor to TNT at two different times.

LONG-TERM RELIABILITY

To investigate the long-term reliability of the FBAR coated with antibody, we tested the lifetime of anti-TNT at room temperature, using a technique similar to dot blotting [1]. For this, we immobilized anti-TNT over gold surface on silicon substrate, following the exact same procedure as stated in the antibody immobilization section. We prepared many samples at one time and stored the samples in a Petri dish at room temperature. Then as days go by, the

samples were pulled out at different points in time and put through a procedure to confirm the integrity of the antibody.

This procedure, similar to dot blotting, consists of the following steps: (1) soak the sample in a secondary antibody solution, which contains HRP (horseradish peroxidase) conjugated goat IgG anti-mouse, for 30 minutes, (2) wash the sample multiple times with buffer solutions for 30 minutes (3) apply chemiluminescent substrate to the sample, (4) expose the sample to a X-ray film. The theory is that if the anti-TNT (primary antibody) is in its working condition, then the secondary antibody will bind to the primary antibody situated on the gold surface. If the secondary antibody binds to the primary antibody and remains after a washing cycle, the HRP on the secondary antibody when treated with chemiluminescent substrate will produce light. The intensity of this light is then a direct indication of the integrity of the primary antibody. An illustration of this concept is shown in Fig. 8.

Fig. 8: A 1° antibody (lower upright Y) is situated on gold surface via protein A after antibody immobilization. Then a 2° antibody (upper slanted inverted Y) specific to the Fc (stems of the Y structures) region of the 1° antibody and conjugated with HRP (star) is bound and rinsed for specificity. Chemiluminescent substrate (small rectangles) is then exposed to the sample. Wherever there is HRP, the substrate will be cleaved and emit light (spreading lines). 1° antibody = mouse IgG anti-TNT, 2° antibody = HRP conjugated goat IgG anti-mouse, chemiluminescent substrate = SuperSignal West Femto (From Pierce Biotechnology).

The emitted light that is captured with X-ray film will present observable difference in luminance level from sample to sample depending on the antibody condition. Some examples that were used to determine the lifetime of the anti-TNT at room temperature are shown in Fig. 9.

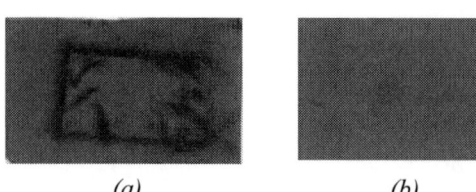

(a)   (b)

Fig. 9: Light from the HRP on the gold samples captured on X-ray film: (a) strong signal, (b) weak signal.

We summarize the results obtained with the samples over a period of time as shown in Fig. 10. Our results show that the integrity of the anti-TNT gradually decays as time passes, but can be considered to be good for up to a month at room temperature.

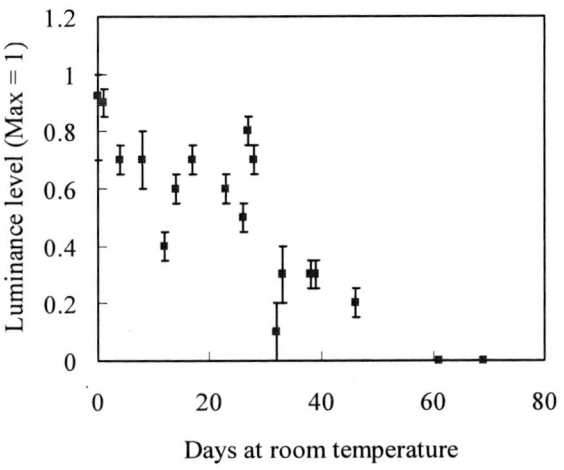

Fig. 10: Room temperature lifetime test of the anti-TNT immobilized on gold surface. The integrity of the anti-TNT was measured by the luminance level when the anti-TNT was treated with HRP-conjugated secondary antibody followed by chemiluminescent substrate.

Additionally, to shorten the length of time we need to carry out the lifetime experiments, we performed accelerated lifetime tests at elevated temperatures, and obtained similar conclusion, as shown in Fig. 11. For this experiment, we raised the temperature 30°C above room temperature, and estimated the days at room temperature, according to the "10 degree rule" that is commonly used for predicting shelf-life of medical products. The "10 degree rule" states that the lifetime of a product is expected to decrease by half per 10°C increases in temperature [5].

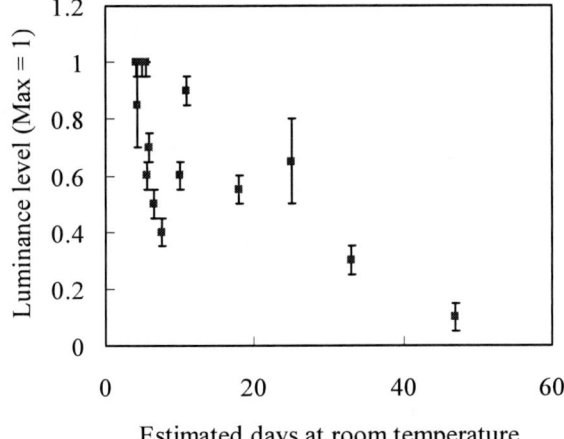

Fig. 11: Accelerated lifetime test of the anti-TNT on gold surface at an elevated temperature (i.e., 30 °C above room temperature). The measured data at elevated temperatures were converted to equivalent room temperature data by using the commonly observed fact that the lifetime decreases by a half per 10°C temperature increase.

CONCLUSION

In our previous work, we demonstrated that an FBAR coated with specific antibody is able to detect vapor trace of TNT or RDX without any pre-concentrator. However, we did not have sufficient experimental data to support the selectivity and long-term reliability of our FBAR mass sensors. In this paper, we present, for the first time, the experimental results on the selectivity and long-term reliability of an antibody-coated resonant mass sensor. The anti-TNT coated FBAR sensor was shown to produce a permanent frequency shift in response to TNT only among many other vapors. Other chemicals with high vapor pressure produced some temporary frequency shifts due to adsorption and/or absorption (which eventually resulted in desorption), but did not produce a permanent frequency shift. Thus, permanent frequency shifts were observed only with TNT due to the antigen-antibody binding. Also, we have measured the lifetime of anti-TNT coated on a gold surface with a technique similar to dot blotting, and report that the anti-TNT was functionally good up to 30 days at room temperature.

ACKNOWLEDGEMENTS

This material is based upon work supported by a grant from KalScott Engineering, Inc. Special thanks to Prof. Steven D. Goodman and Michael S. Waters from the Department of Molecular and Computational Biology at University of Southern California on providing the laboratory for antibody works as well as consultation and suggestions.

REFERENCES

[1] A. Lin, H. Yu, M.S. Waters, E.S. Kim, and S.D. Goodman, "Explosive Trace Detection with FBAR-Based Sensor," *IEEE International Micro Electro Mechanical Systems Conference*, Tucson, AZ, January 13 – 17, 2008, pp. 208-211.

[2] L. Ghitescu and M. Bendayan, "Immunolabeling Efficiency of Protein A-Gold Complexes," *J. Histochemistry and Cytochemistry*, vol. 38, no. 11, pp. 1523-1530, 1990.

[3] R.-J. Pei, J.-M. Hu, Y. Hu, and Y. Zeng, "A Piezoelectric Immunosensor for Complement C4 Using Protein A Oriented Immobilization of Antibody," *J. Chem. Technol. Biotechnol.*, vol. 73, pp. 59-63, 1998.

[4] J. Morgan, W. Bryden, J. Miragliotta, and L. Aamodt, "Improved Detection of Explosive Residues by Laser Thermal Desorption," *Johns Hopkins APL Technical Digest*, vol. 20, no. 3, pp. 389-395, 1999.

[5] D.W.L. Hukins, A. Mahomed, and S.N. Kukureka, "Accelerated aging for testing polymeric biomaterials and medical devices," *Medical Engineering & Physics,* in press, 2008

DNA-DECORATED CARBON NANOTUBES AS SENSITIVE LAYER FOR AlN CONTOUR-MODE RESONANT-MEMS GRAVIMETRIC SENSOR

C. Zuniga[1], M. Rinaldi[1], S.M. Khamis[2], T.S. Jones[1], A.T. Johnson[2] and G. Piazza[1]
[1]Department of Electrical and System Engineering
[2]Department of Physics and Astronomy
University of Pennsylvania, Philadelphia, PA.

ABSTRACT

In this work a nano-enabled gravimetric chemical sensor prototype based on single-stranded DNA (ss-DNA) decorated single-walled carbon nanotubes (SWNT) as nano-functionalization layer for Aluminun Nitride (AlN) contour-mode resonant-MEMS gravimetric sensors has been demonstrated. Two resonators fabricated on the same silicon chip and operating at different resonance frequencies, 287 and 450 MHz, were functionalized with this novel bio-coating layer to experimentally prove the capability of two distinct single strands of DNA bound to SWNT to enhance differently the adsorption of volatile organic compounds such as dinitroluene (DNT, simulant for explosive vapor) and dymethyl-methylphosphonate (DMMP, a simulant for nerve agent sarin).

The introduction of this bio-coating layer addresses the major drawbacks of recovery time (50% recovery in less than 29 seconds has been achieved) and lack of selectivity associated with gas sensor based on polymers and pristine carbon nanotube functionalization layers.

I. INTRODUCTION

The development of miniaturized smart sensors for real time detection of chemical and biological species has recently been a research topic of primary interest due to the growing number of applications ranging from national safety to medical diagnostics and industrial emission monitoring.

The implementation of high performance gravimetric gas sensors for detection of multiple Volatile Organic Chemicals (VOCs) requires both a high quality transducer and an excellent chemically active adsorbent layer.

Among the wide variety of transducers suggested for chemical sensing, gravimetric sensors based on piezoelectric resonators represent an established category suitable for a broad range of analytes including less polar chemical vapors that generally cannot be effectively sensed by conductance based sensors [1].

MEMS/NEMS devices show great potential as gravimetric sensors since their reduced size and frequency of operation higher than conventional quartz crystal microbalance (QCM) permit them to achieve unprecedented value of mass sensitivity and limit of detection. Although promising, there are still some unsolved challenges related to their effective employment as VOC sensors. For instance, nanocantilever beam based sensors suffer from the greatly reduced size of the area dedicated to the adsorbent layer, which limits the overall mass that can be adsorbed in the presence of a given analyte concentration in the environment. Thin film bulk acoustic resonators (FBAR) can attain high values of sensitivity to adsorbate concentration thanks to their high working frequency and wider area dedicated to sensing, but they cannot provide for different frequencies of operation on the same die, therefore posing a limitation in in the realization of a micromechanical nose with extended sensitivity range.

In this perspective, the AlN MEMS resonant transducer of this work (Fig.1) represents a combination of the key advantages previously demonstrated in MEMS/NEMS technologies. In fact, the AlN resonators have shown very high experimentally-verified sensitivity [2], enhanced effective area dedicated to sensing and the ability to achieve a very wide dynamic range (up to 8 orders of magnitude in sensitivity) by having multiple frequencies on the same chip.

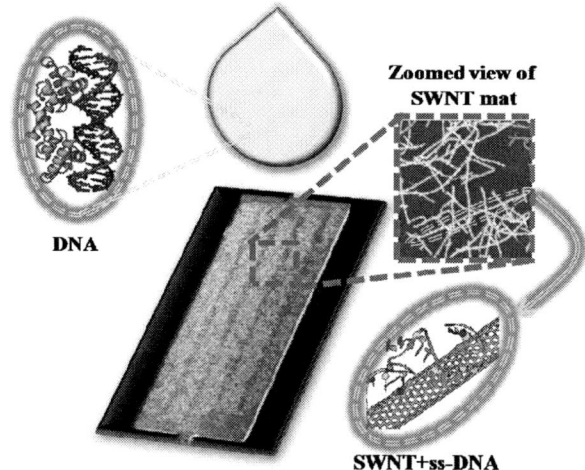

Figure 1: SEM of SWNTs grown on AlN contour mode resonator and schematic representation of the nano-functionalization process of the adsorbent layer.

In regards to the adsorbent layer, a wide range of nanoengineered materials have been extensively studied and employed [1]. Among those, Single Wall Carbon Nanotubes (SWNTs) stand out as prime candidates for sensing applications because their peculiar hollow structure and large surface to volume ratio greatly enhances their adsorptive capacity. In particular, SWNT bundles offer a great advantage over polymeric films, since they provide the transducer with a larger number of adsorbing sites distributed along two dimensions and therefore enable a quick and reversible adsorption process that solely involves the device surface. In general, this cannot be achieved with polymers, for which the adsorption process occurs within the bulk of the film and is consequently associated with a slower response time.

Generally, the use of bare carbon nanotubes as nanosensitive layer is not sufficient to address the problem of selectivity, since SWNTs are very sensitive to a broad target of molecules. In order to overcome the lack of specificity posed by pristine SWNTs, covalent and

978-1-4244-2977-6/09 $25.00 © 2009 IEEE

non-covalent functionalization layers have been employed. SWNTs have been coated by different layers such as chemo-selective polymers [3], enzyme [4], and metal nanoclusters [5]. Although promising, most of these techniques cannot be applied on a large scale and do not allow functionalizing adjacent transducers on the same chip with different chemical coatings. These factors pose an essential limitation to the realization of an on-chip electronic nose.

A novel nano-functionalizing bio-coating constituted by single stranded DNA (ss-DNA) has been successfully demonstrated for SWNT-based FET devices [6]. This technology sets a very promising pathway for the deployment of electronic noses, but it has never been investigated for gravimetric sensor.

In this paper, in order to overcome the common limitations of gas sensor based on pristine CNTs, DNA-decorated SWNT are employed as adsorbent layer for AlN countor-mode resonant sensors. Two resonators fabricated on the same die and operating at different resonance frequencies, 287 and 450 MHz, were functionalized with this novel bio-coating layer consisting of SWNTs grown by CVD and two distinct sequences of ss-DNA. Such sensors were tested for gas species such as dinitroluene, DNT, a simulant for explosive vapor and dymethyl-methylphosphonate, DMMP, a simulant for nerve agent sarin, demonstrating a net enhancement in the adsorption of both chemical agents. The analytes employed in this work responded differently to the same strand leading up to an 8 fold enhancement in DNT detection using one sequence of ss-DNA and a 2-3 times increase in both species detection when a second sequence was used.

II. DESIGN DETAILS AND FABRICATION

Sensor Transduction Mechanism

In an AlN contour-mode resonator, where W, L and T are respectively width, length and thickness of each sub-device (Fig. 2), the application of an electric field across the film thickness produces an in plane dilatation through the equivalent d_{31} coefficient and excites the resonator into lateral contour-extensional vibrations. The frequency of operation is approximately set by the width of a single sub-resonator, W:

$$f_0 = \frac{1}{2W}\sqrt{\frac{E_0}{\rho_0}} \qquad (1)$$

where E_0 and ρ_0 represents the Young's modulus and the density of the resonator.

As previously demonstrated [2], the resonator is best operated as a sensor and attains highest sensitivity when the adsorption layer is placed on its top surface. The large area available renders the resonator extremely sensitive to analyte concentration according to the following equation:

$$S_{eff} = \frac{-f_0^2}{\sqrt{E_0 \cdot \rho_0}} \cdot \frac{W}{T} \qquad [\text{kHz}\cdot\mu\text{m}^2/\text{fg}] \qquad (2)$$

where S_{eff} is the effective sensitivity to analyte concentration of a higher order contour mode resonator. For this kind of piezoelectric transducers the W/T ratio

plays a key role, since it is generally greater than 1 and clearly advantageously improves the device sensitivity.

In this work, two devices with frequencies of operation of 287 MHz ($T = 2\mu\text{m}$, $W = 15\mu\text{m}$, $L = 200\mu\text{m}$,

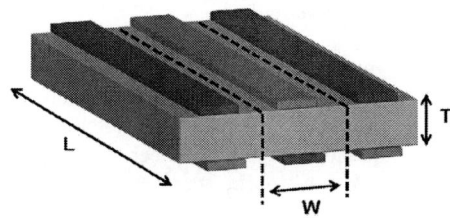

Figure 2: Basic geometry of a higher order laterally vibrating AlN resonator. In this case the number of sub-resonators, n, is 3.

$n = 5$) and 450 MHz ($T = 2\mu\text{m}$, $W = 10\mu\text{m}$, $L = 200\mu\text{m}$, $n = 5$) were fabricated on the same chip and tested by exposing them to different concentration of analytes. The devices have experimentally confirmed sensitivities to analyte concentration of 17.1 and 27.4 kHz·µm²/fg, which are in close agreement with analytically derived values (17.3 kHz·µm²/fg and 28 kHz·µm²/fg, respectively). With a device frequency accuracy of 1 ppm (ultimately affected by device Q) a minimum mass density of 16 ag/µm² can be resolved by these devices.

Furthermore, recent experimental work [8] on AlN contour-mode resonators has demonstrated devices operating at 7.23 GHz with Q of 630, proving that even higher sensitivities and Figure of Merit (FOM= fQ = $4\cdot10^{12}$) can be achieved.

Fabrication

The fabrication process employed to realize the nano-enabled sensor essentially consists in the combination of top-down microfabrication techniques for the resonator [9] and bottom-up growth of SWNTs by catalytic Chemical Vapor Deposition (CVD) as presented in Figure 2.

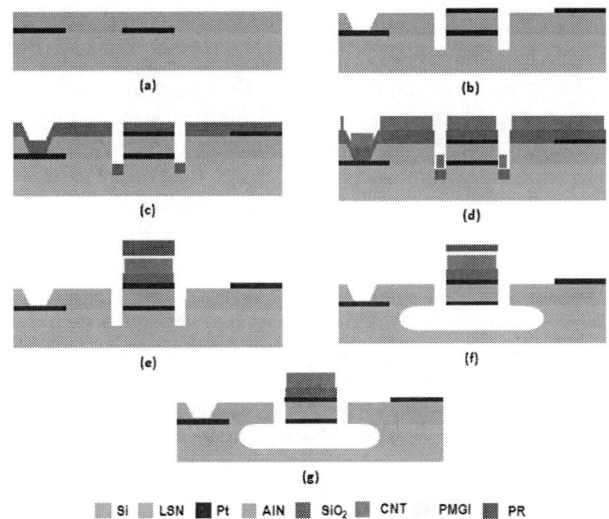

Figure 3: Fabrication process: a) sputter deposition of Pt bottom electrode and AlN on Si and LSN; b) open via in AlN, lithographic patterning of sputtered Pt as top electrode, dry etching of AlN in Cl₂ based chemistry; c)

978-1-4244-2977-6/09 $25.00 © 2009 IEEE   321

sputter deposition of oxide and evaporation of Iron catalyst; d) SWNTs growth by catalytic CVD at 900 °C in methane containing atmosphere; e) spinning of PMGI + photoresist, development and dry etching of SiO₂ in CF₄ based chemestry; f) XeF₂ dry release of AlN resonator; g) PMGI + photoresist removal.

The direct integration of SWNTs was performed at the die level immediately before the release of the device from the Si substrate by XeF_2 dry etching. PMGI patterned by standard lithographic techniques protected the body of the resonator during the oxide etching in CF_4 chemistry. This step removed the oxide seed layer and SWNTs from every other region of the die but the device body.

The process of integration described above has been performed at the die level because the dimension of the available CVD reactor cannot accommodate a 4-inch wafer substrate, but it can be easily extended to large scale manufacturing [7].

The functionalization of the adsorbent layer represents the last step of the process and it took place after the device fabrication. The single stranded DNA sequences chosen as functionalizing agents were:

Seq.1: 5' GAC TCT GTG GAG GAG GTA GTC 3'
Seq.2: 5' CTT CTG TCT TGA TGT TTG TCA AAC 3'

and were applied respectively to the 450 MHz and 287 MHz resonators. The functionalizing agents were obtained from Invitrogen (Carlsbad, CA) and diluted in distilled water to make a solution of 100 µM. The DNA solution was applied on the device with a micropipette for 45 minutes in a humid environment and hence dried in a nitrogen stream (Fig.1, 3).

III. EXPERIMENTAL RESULTS

The sensors were exposed to varying concentrations of DNT and DMMP before and after the ss-DNA functionalization of the SWNTs adsorbent layer.

Mass flow controllers regulated the flow of high purity argon carrier gas through two lines, one dedicated solely to the carrier, and a second one merged into a bubbler through which the analyte flow was generated. The two lines recombined the flow immediately before the testing chamber inlet so that different concentrations of the analytes were produced by controlling the relative flow of the two lines. The total flow delivered in the testing chamber was kept at a constant rate of 1000 sccm. The sensors were refreshed by solely flowing argon inside the testing chamber at the end of each analyte delivery phase. A complete testing cycle was hence formed by sequencing both analyte and refresh phases of identical duration (285 seconds each). A Labview controlled system of two and three-way solenoid valves switched the flows so that the two phases were automatically alternated.

The die under test was attached to a custom designed PCB that is provided with on-board calibration standards and served as a lid for the testing chamber. The resonators were wirebonded to 50 Ω lines and the analytes-induced frequency shift were monitored by an Agilent® N5230A Network Analyzer connected to the PCB.

As mentioned above, the effect of two distinct single strands of DNA as functionalizing agents for SWNTs have been investigated. Analyte concentrations equal to 10, 25 and 50 % of the saturated vapor pressure were delivered on both sensors.

In order to observe the effect of the nano-functionalization layer, the sensor response was characterized for the same analytes with just bare SWNTs and after the application of ss-DNA .

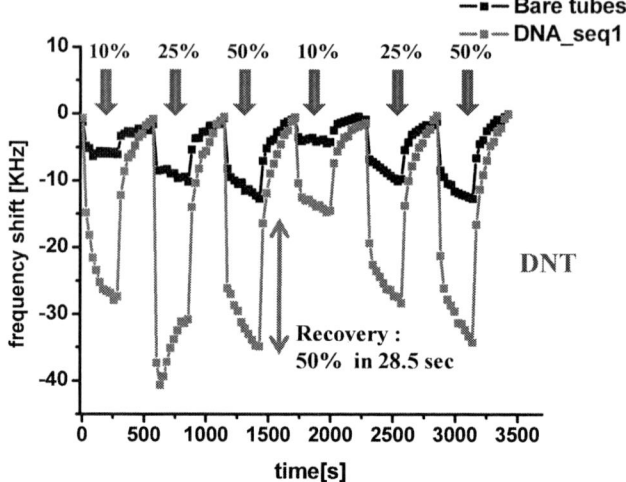

Figure 4: *Response to DNT of a 450 MHz AlN contour-mode resonator with and without DNA seq.1 functionalization. The sensor is refreshed to its original state by flowing Ar. No external heating is required and 50% recovery is achieved in 28.5 seconds.*

Figure 4 shows the frequency shift that was recorded for the 450 MHz resonator after the application of ss-DNA sequence 1. The data demonstrate a net enhancement in the adsorption of DNT with respect to bare SWNTs. As for DMMP, an adsorption enhancement was also recorded, but it was just a factor of 2-3 X compared to the 8 fold increase obtained for DNT.

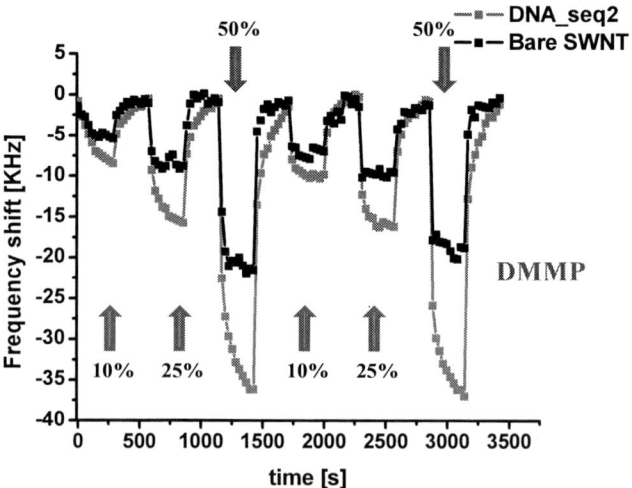

Figure 5: *Response to DMMP of a 287 MHz AlN Contour-Mode resonator with and without DNA seq.2 functionalization*

978-1-4244-2977-6/09 $25.00 © 2009 IEEE

The ss-DNA nano-enabled resonant sensor at 287 MHz and functionalized with sequence 2 exhibited a clear improvement in the response to both chemical agents. Differently, in this case, no substantial diversification in the response to DNT and DMMP was recorded; in fact, both species adsorption was amplified by a factor of nearly 2-3 folds over the bare SWNT-coated resonator.

During the refresh phase both sensor responses recovered at least 50% of their initial value in approximately 28.5 seconds, demonstrating that the sensor is reversible.

Although the adsorption mechanism is not fully understood and further investigations are needed, from the previous results it can be inferred that the DNA nano-coating layer increases the binding affinity between the analyte and the SWNT adsorbent layer, therefore enhancing the amount of adsorbed mass for a given concentration.

The amount of mass adsorbed on the nano-sensitive layer for each analyte concentration has been estimated from the change in the motional inductance of the extracted MBVD circuit (Fig.6). The adsorbed mass, Δm, is modeled by a series motional inductance, Lm^*, proportional to the adsorbed mass and a series resistance, Rm^*, that takes into consideration the small additional damping introduced by the adsorption process:

$$\Delta m = 2 \cdot n \cdot Lm^* \cdot \eta^2 \qquad (3)$$

where η is the electromechanical coupling coefficient of the resonator ($\eta = 2 \cdot d_{31} E_o L$).

The estimated VOC mass adsorbed for 10 % concentration of both analytes is reported in Figure 7, where the diversification in the affinity to different chemical species clearly points out that the adsorption process is DNA sequence dependent.

After the application of the DNA solution, a decrease in Q factor of approximately 10% (considering the value of Q after flowing Ar) was recorded for both devices. The partially aqueous environment created by the addition of DNA is considered the primary cause of this additional damping. No change in the off-resonance capacitance was observed, while Rm slightly increased (approximately 10% of its original value) reflecting the corresponding drop in Q.

the resonator surface. A total DNA mass of 230 pg was attached to the resonator. The same technique was employed to extract the added mass due to adsorption of VOC.

IV. CONCLUSION

A novel nano-enabled gravimetric chemical sensor prototype based on ss-DNA decorated SWNTs on AlN contour-mode MEMS resonator has been designed, fabricated and tested.

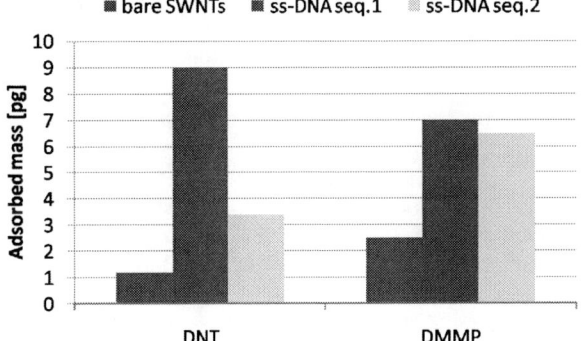

Figure 7: *Comparison between DNT and DMMP response with different ss-DNA functionalization sequences for 10% analyte concentration. The chart shows the enhanced mass adsorption occurred because of the functionalization process and the ability to selectively distinguish between DNT and DMMP.*

The novel functionalization layer permits to achieve up to an 8 fold increase in sensitivity to DNT over bare nanotubes with 1 sequence of DNA, and a 2-3 fold enhancement to both DNT and DMMP when a second sequence is employed. This work constitutes the first experimental verification towards the demonstration of a highly selective nanomechanical nose, whose realization will be made possible by gaining access to a vast ss-DNA library.

REFERENCES

[1] T. Zhang, S. Mubeen, N. V. Myung, M. Deshusses, "Recent progress in carbon nanotube-based gas sensors", *Nanotechnology* 19, 2008.

[2] M. Rinaldi, C. Zuniga et al. "Gravimetric chemical sensor based on the direct integration of SWNTs on AlN contour-mode MEMS resonators", *IEEE IFCS* 2008 proceedings, pp. 443-448.

[3] A. Star, T. R. Han, V. Joshi, J. C. Gabriel, and G. Gruner, *Adv. Mater. 16*, 2004.

[4] K. Besteman et al. *Nano Lett.*, vol. 3, No.6, pp. 727-30, 2003.

[5] M. Penza et al., *Appl. Phys. Lett.* 90, 173123, 2007.

[6] C. Staii, A. T. Johnson, M. Chen, A. Gelperin, *Nano Lett.*, vol. 5 No. 9, pp. 1774-1778.

[7] N.R. Franklin, Y. Li, R.J. Chen, "Patterned growth of single walled carbon nanotubes on a full 4-inch wafer", *Appl. Phys. Lett.* 79, No. 27, 2001.

[8] M. Rinaldi, C. Zuniga and G. Piazza, "5-10 GHz AlN contour mode Nanoelectromechanical MEMS resonators", *IEEE MEMS* 2009 proceedings.

[9] G. Piazza, P. J. Stephanou, A. P. Pisano, J. of Microel. Syst., Vol.15, Dec.2006, pp1406-1418.

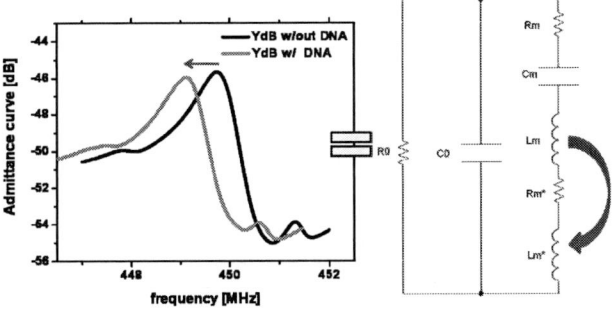

Figure 6: *The admittance curve of the resonant device was measured before and after the DNA deposition and both curves were fitted to the Modified Butteworth van Dyke (MBVD) model to extract the added mass of DNA on*

MICROFLUIDICS WITHIN A SWAGELOK®:
A MEMS-ON-TUBE ASSEMBLY
S. Unnikrishnan, H.V. Jansen, J.W. Berenschot, B. Mogulkoc, and M.C. Elwenspoek
MESA+ Institute for Nanotechnology – University of Twente, Enschede, THE NETHERLANDS

ABSTRACT

A novel packaging cum interfacing technique for microfluidic devices is reported. Unlike the conventional approach towards packaging in which the MEMS is first developed and finally packaged, a reverse approach is shown here that integrates the package with the MEMS either at the beginning or within the fabrication process. This new method employs standard glass tubes as substrates on which microfluidic components are fabricated. The tubular-substrate directly translates into a package and an interface, leading to 'plug-n-play' devices. Maintaining the total size of the MEMS device within the circumference of the glass tube enables this *MEMS-on-tube* assembly to be encapsulated within standard Swagelok® connectors.

INTRODUCTION

Microfluidics is a tremendously growing field with applications in gas-separation, filtration, microreactors, lab-on-a-chip systems etc. [1-4]. There has been a continuous quest to miniaturize various fluidic components including pumps, flow-channels, valves and sensors within a microchip so as to minimize the size, cost and dead volume of the system. Such researches are mostly focused on having a complete system on a chip (SOC). SOC demands a complicated fabrication process scheme and also faces a tough challenge of hermetic packaging and interfacing to the external world. It is not only important to properly interface the final microfluidic device to the macro-world, but the intermediate characterization of its individual components is also sometimes essential. A reliable package must consist of a robust support and a suitable interface to the equipment where it would be implemented. Providing leak-proof connections to such microfluidic chips is non-trivial. Usually, interconnections to such microchips are made via mechanical clamping or by gluing [5,6]. To attain hermetic sealing, mechanical clamps exert forces on the delicate microchips, which could lead to breakage. Glued connections, on the other hand, could block channels and/or capillaries and moreover cannot withstand harsh thermal and chemical environments. Presented in this paper is a convenient solution to hermetically package microfluidic components using tubular-substrates, which at the same time solves the interfacing issue.

MEMS has been mostly based on a two-dimensional microfabrication methodology involving processes being carried out on planar silicon or glass substrates. One of the reasons for this has been the chip-oriented approach owing to the fact that MEMS has been derived from planar CMOS fabrication technologies used for making Integrated-Circuit chips. In this paper, we show a new micromachining method using three-dimensional tubular substrates. The tubular-substrates used here are commercially available standard Duran® or Pyrex® glass tubes. Micromachining devices on such substrates lead directly to a package - MEMS-on-tube assembly - that is interfacable to standard Swagelok® connectors. The glass tube acts as a support for the relatively smaller MEMS device, while also being a functional connection to the macro-world. Moreover, it absorbs vibrations or shocks during connecting or operating, leaving the fragile MEMS device undisturbed.

FABRICATION

The basic fabrication scheme involves, 1) Preparation of the micro(fluidics) component 2) Preparation of the tubular substrate 3) Assembly of the micro(fluidics) component onto the tubular substrate 4) Continuing micromachining of the tube assembled MEMS - if required 5) Direct mounting of the MEMS-on-tube assembly within the Swagelok® 6) Test and usage of the MEMS device. Based on this fabrication scheme, there is a class of various tube assembled devices possible like thin-film membranes, particle-filter, and gas-separators, which are discussed later on in this section.

The integration of the tubular glass substrate with the MEMS can be done at various stages of the fabrication process depending on the intended application. By assembling directly on a tube, the microfluidic component transforms into a usable device that can be connected using standard Swagelok® connectors. By maintaining the total size of the MEMS device within the circumference of the glass tube, it is possible to mount this MEMS-on-tube assembly onto various equipments by a double-Swagelok® technique (see figure 1), which secures the device within the connector. The glass tube is tightened inside the Swagelok® using teflon ferrules instead of the usual stainless steel ferrules, which could break the glass.

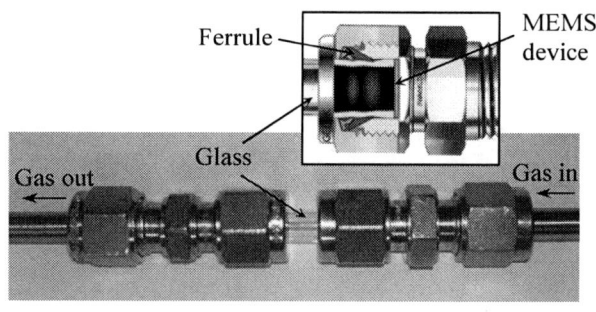

Figure 1: The double-Swagelok® connection technique for tube-assembled MEMS devices. Cut-sectional representation showing the MEMS device on one end of the glass tube.

The glass tubes are integrated to the wafer using a fusion bonding technique [7]. After placing the glass tubes on the wafer in an oven, they are heated up to 800°C, where the glass begins to soften. Since the viscosity of the glass is lowered at this high temperature, it starts to reflow thereby covering the glass-silicon interface by capillarity. Given enough time for glass to flow at elevated temperature, homogenous coverage of the bonding region

978-1-4244-2977-6/09 $25.00 © 2009 IEEE   324

is obtained. When the oven is cooled down below its transition temperature, glass solidifies to form a stable bond.

Next, few examples of micromachined devices assembled on a tube are described.

Tube assembled *thin-film membrane*

For many applications in the field of acoustics or pressure sensing [8], a free hanging thin membrane is useful. Such a micromachined membrane can be easily supported and interfaced using a glass tube as explained below.

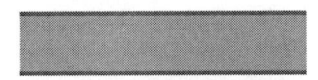

i) LPCVD growth of 1µm Si_3N_4 on silicon wafer

ii) Plasma etching of the Si_3N_4 layer on one side

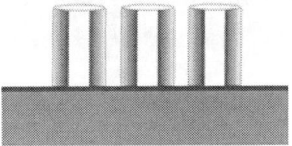

iii) Fusion bonding of Duran® glass tubes at 800°C

iv) Releasing the tube-assembled membranes by KOH etching of the silicon wafer

Figure 2a: Fabrication process for making free-hanging silicon nitride membrane assembled on a glass tube.

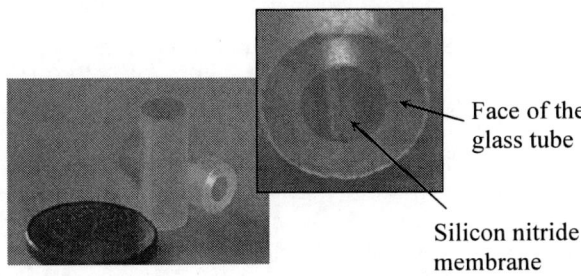

Face of the glass tube

Silicon nitride membrane

Figure 2b: 1µm thick silicon nitride membrane packaged on a glass tube of 3mm internal diameter.

The fabrication process (see figure 2a) involves batch-bonding (figure 3) of the glass tubes to a wafer having a uniformly deposited thin-film material; in this case a 1µm thick LPCVD silicon nitride layer. Subsequently, the silicon wafer is dissolved in KOH solution, which results in the release of the membrane free-hanging on the glass tubes. The tubes are easily separated from each other by breaking it from the weak silicon nitride membrane between them. Shown in figure 2b is the 1µm thick free-hanging silicon nitride membrane packaged onto a glass tube of 1.5mm wall-thickness and 3mm inner-diameter. The glass tubes can be of any desired diameters and wall thicknesses which are defined based upon the desired strength of the membrane.

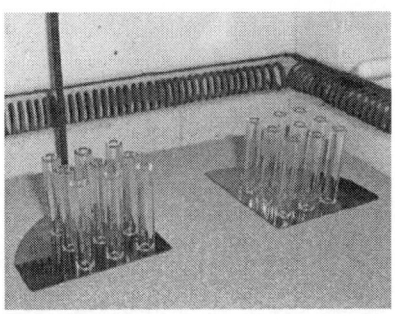

Figure 3: Fusion bonding of a batch of glass tubes of 30mm length onto silicon wafer

Tube assembled *particle filter*

Like dense membranes, it is also possible to assemble perforated membranes on a glass tube. Described in figure 4a is the fabrication process for making a tube assembled particle filter. After photolithography of a hexagonally packed pattern of Ø5µm microholes on a silicon wafer, they are plasma etched 90µm deep (see figure 4a(ii)). After stripping of the mask and proper cleaning, the wafer is oxidized to grow a etch stop layer (see figure 4a(iii)). Next, the entire wafer is plasma back-etched till the stop-layer is exposed, which is then stripped in hydrofluoric acid, thus resulting in a perforated silicon membrane sieve (figure 4a(iv)). Subsequently, as shown in figure 4a(v), glass tubes of desired size are fusion bonded as a batch onto the wafer and each of the tubes are just cleaved out of the wafer aided by the orderly microholes pattern.

i) Photolithography of the Ø5µm microsieve mask pattern

ii) DRIE Etching Ø5µm holes 90µm into silicon wafer

iii) Oxidation of the wafer to grow SiO_2 stop-layer

iv) Wafer back-etching and stripping of the stop-layer in HF

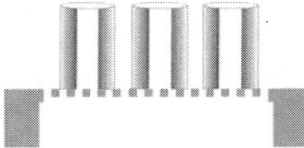

v) Fusion bonding of Duran® glass tubes at 800°C

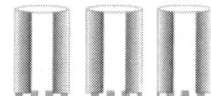

vi) Breaking away of the tube-bonded particle filter

Figure 4a: Fabrication process flow for making a glass tube assembled particle filter membrane.

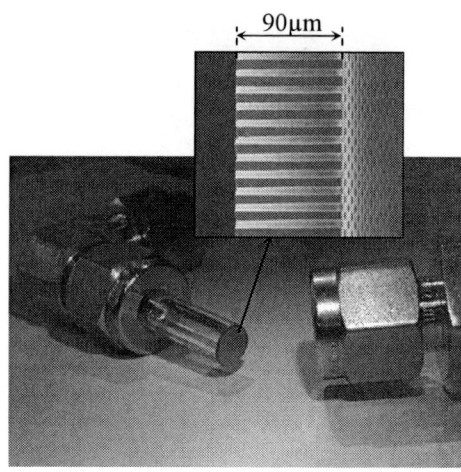

Figure 4b: Picture of a silicon microsieve particle filter (with Ø5μm pores) on a glass tube within a double-Swagelok®

Seen in figure 4b is the 90μm thick silicon particle filter with Ø5μm pores and 18% porosity, assembled on a glass tube. In a fluidic system it is possible to use a series of these sieves with different pore sizes for stage-wise retention of particles of various sizes.

Tube assembled *gas separators*

For gas separation and reaction applications, ultra-thin membranes are desired [1] because thinner membranes have lower resistance to permeation. Inorganic membranes like silicon dioxide and palladium are usually used for selective gas permeation applications. Using micromachining techniques it is possible to create defect free, low flow resistance ultra-thin membranes. By incorporating the MEMS-on-tube assembly technique, such delicate membranes can be easily packaged. But thin delicate membranes have to be supported by a porous membrane like the silicon microsieve (described in figure 4a) for mechanical strength. These supported ultra-thin membranes can either be flat or corrugated as shown in figure 5 and 6 respectively.

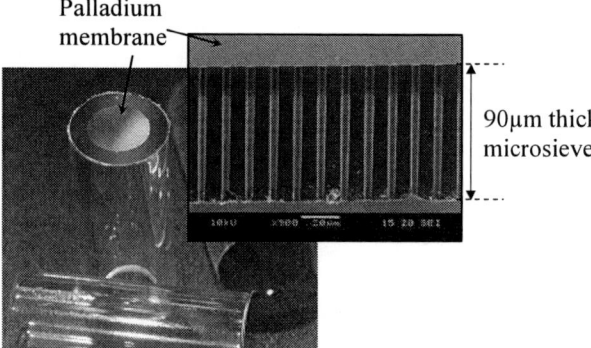

Figure 5: A supported flat palladium membrane assembled on Ø8mm inner diameter glass tube. SEM picture shows the cross-section of the membrane stack.

Figure 5 shows a 150nm thick tube-assembled palladium membrane (for hydrogen separation) supported

on a silicon-microsieve. The process of nano-membrane fabrication involves thin-film transfer technique which has been previously described by the authors [9].

Corrugated gas permeation membranes have the advantage of having large surface area which thereby results in a higher permeate flux. An example of a corrugated silica membrane of 50nm thickness supported on a silicon microsieve (assembled on a glass tube) can be seen in figure 6. These are fabricated directly on a silicon microsieve support using a similar process flow as in figure 4a, but with the difference that instead of removing the oxide etch stop-layer, it is retained while assembly on the glass tube.

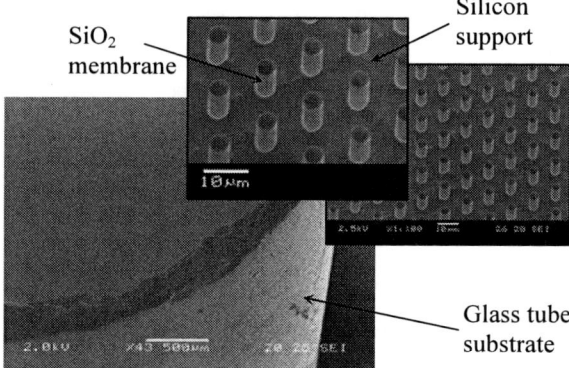

Figure 6: SEM picture of a microsieve supported corrugated SiO_2 membrane packaged on glass tube of 3mm internal diameter.

EXPERIMENTS AND RESULTS

The glass tubes assemblies were tested for leak using a helium gas flow set-up consisting of a gas chromatograph. All the leak tests were performed at different temperatures ranging from 30°C until 200°C ambient temperature. The silicon nitride membrane shown in figure 2b was tested for helium leak till 0.1bar transmembrane pressure. The silicon microsieve supported corrugated silica membrane shown in figure 6 being stronger, was tested for leak under 5bar helium transmembrane pressure. During high pressure tests (>3bars), for safety, the surface of the tube was slightly roughened to avoid its slipping off the ferrule grip. No helium leak was detected by the gas chromatograph for any of the membrane samples, confirming that the bond-interface is hermetically sealed.

It is good to mention here that the homogeneity of bond across the mating surfaces of the glass tube and silicon is crucial for the hermeticity of the bond. For this reason the preparation of the glass tube prior to bonding plays an important role. Apart from hermeticity, bond-strength is another parameter which defines the quality of the fusion-bond. Bond-strength tests done by Fazal [7] with glass tubes of 3mm internal diameter and 1.5mm wall thickness bonded to a plain silicon wafer of 525μm thickness revealed water burst strength of 65bar. But perforated membranes like the microsieve can break at a lower pressure. Tests showed a burst strength of 7bars for a microsieve of 90μm thickness with 18% porosity. In both cases, the silicon membrane broke and not the bond, thus proving the robustness of the bond. The resistance of the MEMS-on-tube assembly to harsh chemical environments was also found to be good after testing for ca. 30 minutes in aggressive solutions like hot concentrated HNO_3 (69% at 95°C) or Piranha (96% H_2SO_4 + 31% H_2O_2 at 100°C).

DISCUSSION

Although the MEMS-on-tube assemblies can withstand higher temperatures of operation, due to the usage of teflon-ferrules inside the Swagelok® connectors, the temperature of device operation is limited to ~220°C. This limitation can be overcome by using graphite-ferrules that can withstand higher temperatures.

The glass tube preparation is extremely important for the quality of the fusion bond. After dicing the glass tubes to appropriate lengths, they are polished to optical grade. Another method of preparation is to smoothen the surface by a pre-heat treatment. During batch bonding, the precise placing of an array of glass tubes over the wafer is a delicate job and wetting agents like isopropanol or ethanol can help to retain them in a particular position against vibrations.

It was suggested by Fazal [7] that glass tubes can be bonded to a microfluidic chip to act as fluidic interconnects. The problem of a chip with multiple non-concentric tube connections is that the tubes cause material failure at their interface due to external handling loads (torsional and bending forces) while tightening connectors like Swagelok®. In our approach, this issue is solved by confining the MEMS device within the glass tube's circumference. By this means, the external handling forces on the microfluidic system are redirected from the MEMS device and the bond-interface towards the glass tube, and therefore device robustness is substantially improved.

For using the double-Swagelok® technique, the glass tube must have a minimum length of 30mm, which ensures the proper connectability of two Swagelok® couplings on either end of the tube. Longer tubes are better in this respect, but they have the limitation that during fusion bonding process, they could bend or curve due to the pull of gravity. A mechanical bonding support could help for straight bonding of high aspect ratio tubes. This bending of the tube if not properly managed could be a disadvantage for Swagelok® coupling.

For certain MEMS materials, the fusion bonding temperature of 800°C could be too high, like metal layers. Due to this reason the palladium membrane shown in figure 5 was deposited after the glass tube bonding step. For such cases it is also possible to lower the bonding temperature and increase the bonding-time instead. Extremely smooth glass tube surfaces could be helpful in this regard. The other option is to use tubes made of low-melting point glass, which will soften at a lower temperature.

CONCLUSIONS

A new methodology for a one-step packaging cum interfacing technique for microfluidic devices has been demonstrated. Instead of the usual 2D micromachining approach, a 3D perspective is adopted, which uses tubular glass substrates. While in conventional packaging, the MEMS is first developed and finally packaged, a 'turn-around' approach is adopted here by starting from the package or integrating it with the MEMS device during the fabrication process itself. Using this new technique, it is also possible to characterize independently the sub-components of microfluidic systems. A double-Swagelok® technique has been shown here, which enables the MEMS-on-tube assembly to be mounted onto various equipments. Swagelok® connectors which are normally used for fluidic interconnections can now contain a smart MEMS device within them. This novel technique enables the easy implementation of microfluidic devices into various applications like air-sterilization, emulsification, fluid-filtration, gas permeation, microreactors, cell samplers etc. Overall, the MEMS-on-tube assembly acts as a versatile platform for microfluidic packaging and interfacing.

ACKNOWLEDGEMENTS

The authors would like to thank the Dutch Technology Foundation STW for funding this project.

REFERENCES

[1] H.D. Tong, F.C. Gielens, J.G.E. Gardeniers, H.V. Jansen, J.W. Berenschot, M.J. de Boer, J.H. de Boer, C.J.M. van Rijn, and M.C. Elwenspoek., "Microsieve Supporting Palladium-Silver Alloy Membrane and Application to Hydrogen Separation", *Journal of MEMS*, Vol. 14, No. 1, February 2005

[2] C.J.M. van Rijn, W. Nijdam, S. Kuiper, G.J. Veldhuis, H. van Wolferen and M. Elwenspoek, "Microsieves made with laser interference lithography for micro-filtration applications", *J. Micromech. Microeng.* 9 (1999), pp. 170–172

[3] S.V. Karnik, M.K. Hatalis and M.V. Kothare, "Towards a Palladium Micro-Membrane for the Water Gas Shift Reaction", *Journal of MEMS*, Vol. 12, No. 1, February 2003

[4] B.H. Weigl, R.L. Bardell, C.R. Cabrera, "Lab-on-a-chip for drug development", *Advanced Drug Delivery Reviews* 55 (2003), pp. 349–377

[5] B.L. Gray, D. Jaeggi, N.J. Mourlas, B.P. van Drieenhuizen, K.R. Williams, N.I. Maluf, G.T.A. Kovacs, "Novel interconnections technologies for integrated microfluidics", *Sensors Actuators A* 77 (1999), pp. 57–65

[6] T.J. You, S. Lee, W. Fang and Y.C. Tai, "Micromachined rubber O-ring microfluidic couplers", *Proc. MEMS 2000*, pp. 624–7

[7] I. Fazal, E. Berenschot, R. de Boer, H. Jansen and M. Elwenspoek, "Bond strength tests between silicon wafers and Duran tubes", *Proc. TRANSDUCERS 2005*, pp. 936-939

[8] W.P. Eaton and J.H. Smith, "Micromachined pressure sensors: review and recent developments", *Smart Mater. Struct.* 6 (1997), pp. 530–539.

[9] S. Unnikrishnan, H. Jansen, E. Berenschot and M. Elwenspoek, "Wafer scale nano-membrane supported on a silicon microsieve using thin-film transfer technology", *J. Micromech. Microeng.* 18 (2008), 064005, pp. 7

NANO-THICKNESS INTEGRATED RESONANT CANTILEVERS WITH SURFACE-STIFFENING SCHEME FOR ULTRA-SENSITIVE DETECTION OF TRACE CHEMICALS

Xiaoyuan Xia, Yongliang Yang, Xinxin Li

State Key Lab of Transducer Technology, Shanghai Institute of Microsystem and Information Technology, Chinese Academy of Science, Shanghai 200050, China (Email: xxli@mail.sim.ac.cn)

ABSTRACT

The paper reports an ultra-thin silicon resonant cantilever sensor with a novel through-cantilever doped piezoresistive sensing element and a Lorentz-force resonance exciting element integrated. Torsion-mode resonance is used where shear-stress piezoresistance can be used for the 95nm-thick silicon nano-cantilever. Thanks to the ultra-thin cantilever where surface sensing effect becomes dominant, molecule specific-adsorption induces a compressive surface-stress that further causes axial-force-induced spring-stiffening of the nano-cantilever. With the surface modified by specific SAM (self-assembled monolayer), the surface-stiffening-induced frequency-shift exhibits an much higher sensitivity to trace trimethylamine vapor, compared to the conventional mass-loading frequency-shift.

1. INTRODUCTION

Mass-loading frequency-shift has been used in micro-thickness resonant-cantilever sensors for bio-molecule detection [1]-[5]. However, for gaseous chemical molecules, which possess much smaller molecule-weight, the direct mass-loading scheme frequently feels difficult due to inadequate mass adsorption.

Instead of the mass-loading induced frequency-shift, surface adsorption is expected to build double-sided surface-stress that generates an axial-force-induced spring stiffening and an increase in resonant-frequency. To make the surface sensing effect a dominant one, here the cantilever is thinned into nano-scale.

2. SENSOR DESIGN AND FORMATION

In traditional bending resonance-mode, heavy and shallow impurity doping for piezoresistance integration is difficult for nano-thickness cantilever [6] as the doping depth cannot exceed the middle plane of the cantilever cross-section. Otherwise the bending-stress-induced piezoresistance signal at the top half cantilever will counteract with that at the bottom half, which will decrease the piezoresistive signal. If the piezoresistor is doped throughout the cantilever depth, the counteracting effect will cause a null sensing output that is illustrated in Figure 1(c).

As well known, the widely used piezoresistive effect works by detecting the stress at the piezoresistor location [7], [8]. Tensile-stress induced piezoresistive signal is opposite in sign to compressive-stress induced one. Therefore, when integrated in silicon diagram pressure sensors or beam-mass accelerometers, the piezoresistors are doped at one side of the neutral plane (i.e. the middle plane in most cases) of the sensing structures for detecting either tensile or compressive bending stress. If the doped

piezoresistor layer reaches or gets across the neutral plane, the stress will change its value from tensile to compressive (or vice versa) and, thus, the piezoresistive signal will be weakened by the different stresses [9]. If the piezoresistor layer is doped throughout the silicon structure thickness, most likely the piezoresistive signal will be counteracted by the different stresses into almost null. In conventional micromechanical sensors, the worst case mentioned above was difficult to happen, as the piezoresistive doping layer was normally much thinner than the half thickness of the silicon sensing structures. In recent years, however, the piezoresistive sensing elements have been becoming very tiny.

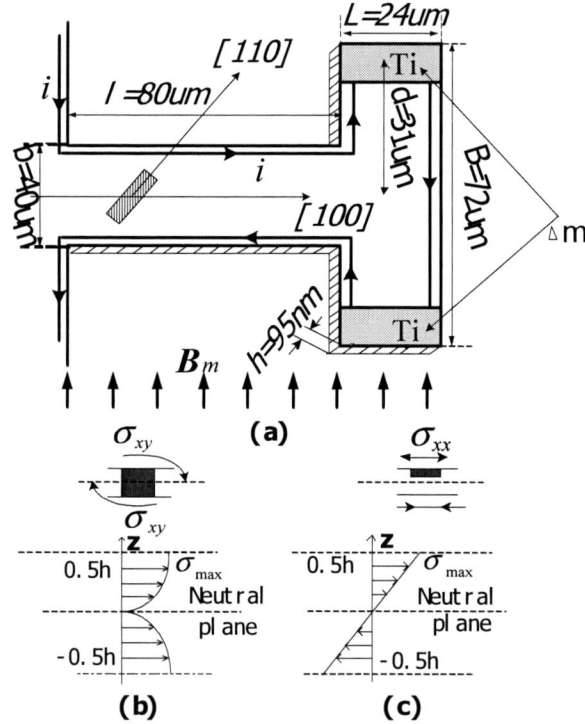

Figure 1: (a) Top-view schematic of the torsion-mode nano-cantilever. The schematic of the through-thickness piezoresistance doping scheme and the shear stress profile along the cantilever thickness in torsion-mode is shown in (b), which is compared with the piezoresistance situation of the conventional flexure-mode cantilever shown in (c).

For a flexural cantilever where normal stress is measured, it is difficult to breakthrough the neutral-plane limited piezoresistance design rule. However, in either static or dynamic cantilever applications like the microscale torsion-mode resonant cantilever sensor in Ref. [10], shear stress piezoresistive detection can be used to replace the bending-stress detection. Previous studies have revealed that torsion-mode resonant cantilevers feature superior resonance performance and better mass-sensing resolution compared to the conventional flexure-mode resonant cantilevers. In Ref. [10], the developed torsion-mode resonant cantilever was with a thickness of about 3μm, where the neutral-plane limitation is not a severe problem for piezoresistor integration. In this research, however, the silicon thickness of the nano-cantilever will be below 100nm. To our best knowledge, the previously developed resonant cantilevers were with the thinnest silicon cantilever as about 30nm [2], [5]. In those cantilevers, however, neither a resonant exciting element nor a sensing element was integrated on the cantilever. Actuation of the resonance and sensing to the specific-adsorbed mass was performed by using the piezoelectric scanner and the laser light bending readout of an SPM equipment. In this study, for conveniently realizing on-the-spot portable applications, Lorentz-force resonance exciting and piezoresistive sensing elements are all integrated on the nano-thickness cantilevers.

Herein we design a T-shaped nano-cantilever resonating in torsion-mode, where shear stress is used for piezoresistive signal-readout. Illustrated in Figure 1(b), the shear-stress can be efficiently sensed by a through-cantilever-doped p-type piezoresistor that is optimally designed along <110> orientation, while the cantilever is along <100>. Sketched in Figure 1(a), the through-thickness piezoresistance design-rule for the torsional resonant nano-cantilevers is largely different from that for the traditional flexure-mode piezoresistive cantilevers [10].

To translate the through-thickness shear stress into an effective piezoresistive signal, a specific design rule for both the cantilever orientation and the piezoresistor direction, which never previously reported, is elucidated in the present study. For p-type doped piezoresistance, we use n-type (100) silicon wafers. According to the theory in Ref. [9], the resistance change can be expressed as:

$$\Delta R / R = \sigma_s [(\pi_{11} - \pi_{12}) \sin 2\beta - (\pi_{11} - \pi_{12} - \pi_{44}) \sin 2\alpha \cos 2(\alpha - \beta)]$$

where α and β are the angle between the piezoresistor orientation and <100> silicon crystal coordinate, as well as, the angle between the piezoresistor orientation and the cantilever longitudinal direction, respectively. π_{11}, π_{12} and π_{44} are the three components of silicon piezoresistive coefficient tensor. By solving the equation, the maximum piezoresistive sensitivity value of $(\Delta R/R)_{max} = \pi_{44} \sigma_s$ is obtained when $\alpha = \beta = \pi/4$, i.e. the cantilever is along <100> orientation and the piezoresistor is along <110> direction. This through-thickness piezoresistance design rule for torsional resonant nano-cantilevers is quite different from that for the traditional flexural piezoresistive cantilevers [10].

The resonance of the proposed cantilever is excited by using Lorentz force. As is sketched in Figure 1(a), a Cr/Au line is integrated on the cantilever to flow a looped AC electric current, i. Under an external magnetic field supplied by a tiny NdFeB magnet that is mounted in the sensor package, an actuating force-moment is induced to excite the resonance. Based on the parameters denoted in Figure 1(a), the generated maximum shear stress can be calculated as

$$\sigma_{s\ max} = \frac{T}{k_2 b h^2} = \frac{2 B_m i (lb + LB)}{k_2 b h^2} = \frac{2 B_m i l (1 + \lambda \gamma)}{k_2 h^2}$$

where T is the trosional force-moment, B_m represents the magnetic field, k_2 is a constant relative to the cantilever geometric dimensions of b and h. When $b >> h$, $k_2 \approx 1/3$. $\lambda = L/l$ and $\gamma = B/b$ are used as the figures of merit that can be varied for optimal mass-loading sensitivity. For the through-thickness piezoresistance design, the piezoresistive output corresponds to the average shear stress and can be expressed as $\Delta R / R = Q \pi_{44} \sigma_{s\max} / \sqrt{2}$, where Q is the resonance quality-factor.

Shown in Figure 1(a), when a mass of Δm is adsorbed on the two sensing locations at the cantilever end-paddles, the sensitivity of the sensor can be expressed as

$$S = \frac{\Delta f}{\Delta m} \approx \frac{1}{2} \frac{f}{I_\phi} d^2$$

where I_Φ is the torsion-mode inertial moment, f is the resonant frequency and d is the distance from the adsorbed mass center to the cantilever's central line. With the structure and material parameters substituted, the mass sensitivity can be further expressed as

$$S = \frac{1}{2\pi} \sqrt{\frac{27 G k_1 \gamma^4}{\rho^3 l^4 b^4 \left(1 + \lambda \gamma^3\right)^3}}$$

where G is shear modulus, ρ is density, and k_1 is a constant relative to the ratio of b/h. When $b >> h$, $k_1 \approx 1/3$, it can be analyzed from the equation that the sensitivity will increase when λ is decreased. However, a compromise must be made for the end-paddle design. Since an adequate mass adsorption area is needed, the end-paddle length should not be very small. After λ is fixed, the highest sensitivity can be obtained when $\gamma^3 = 4/(5\lambda)$.

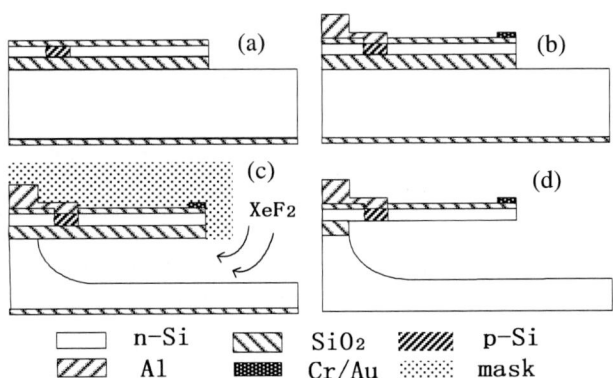

Figure 2: Cantilever processes. (a) p-type doping throughout the nano-thick Si layer of SOI wafers followed by RIE shaping the cantilever. (b) Interconnection for the doped piezoresistor and formation of metal loop for Lorenta-force generation. (c)-(d) Cantilever release by XeF2 dry etching followed by backside BOX layer removing.

The torsion-mode resonant nano-cantilever is formed by using micromachining technique. With the main process steps sketched in Figure 2, the fabrication starts from n-type (100) silicon-on-insulator (SOI) substrate. 35nm-thick SiO2 is grown on the surface and the remained top-layer silicon is about 95nm-thick. Reactive ion etch is used to form the silicon cantilever shape. Boron ion-implantation is used to dope the piezoresistor throughout the cantilever thickness. Then aluminum

978-1-4244-2977-6/09 $25.00 © 2009 IEEE

interconnection is formed. For the interconnection segment on the cantilever, the Al line is very thin. When the metal line is led out to the frame, a much thicker Al layer is used for a better electric interconnection. Then about 40nm-thick Cr/Au loop for Lorentz-force excitation is evaporated and patterned. Finally, XeF_2 isotropic dry etching is used to laterally excavate the silicon substrate beneath the cantilever [11], [12] and the buried oxide (BOX) layer under the silicon cantilever is removed by buffered HF. Figure 3 shows the SEM image of the fabricated cantilever. Shown in the SEM inset the thickness of the cantilever is measured, where the cantilever is with an oblique angle of 45°from the image plane. Covered by a 35-nm-thick SiO_2 layer, the silicon of the cantilever is about 95nm in thickness.

Figure 3: EM image of the fabricated integrated torsion-mode nano-cantilever, with the cantilever thickness illustrate in the SEM inset.

3. SENSING EXPERIMENTS

Resonance performance of the fabricated cantilever sensor is characterized with a network analyzer. Most of the sensors feature the resonance frequency ranged in 30-40kHz and the open-looped Q-factor value around 12-15 under air atmosphere. The non-uniformity of frequency is probably due to the non-uniformity of the cantilever silicon thickness that is originated from the SOI wafers.

Firstly the traditional mass-loading sensitivity of the sensor is tested. An extra 15nm-thick Ti-film is added on top of the Lorentz-force-generating Cr/Au lines. After the 114pg Ti pattern is stripped off by H_2O_2, a resonant frequency increase due to the mass un-loading is tested as 4.125kHz. This mass removing process can be considered as the reversed process of a mass loading on the cantilever. The testing results are illustrated in Figure 4, with the mass-loading sensitivity is experimentally obtained as 36Hz/pg.

Then the surface-stiffening sensing method is validated by experiment. Both the front-side and the backside SiO_2 surfaces of the cantilever is functionalized with the SAM of carbon-oxy-ethyl-silane-triol, sodium salt, and the SAM is simultaneously acidized to form -COOH sensing terminals. 6ppm tri-methylamine vapor (well known as the mark of fish freshness) is introduced and the target molecules are specific captured by the SAM within

30 seconds. With the testing results demonstrated in Fig. 5, the surface-stiffening effect induced frequency increase is measured as high as 4.101kHz.

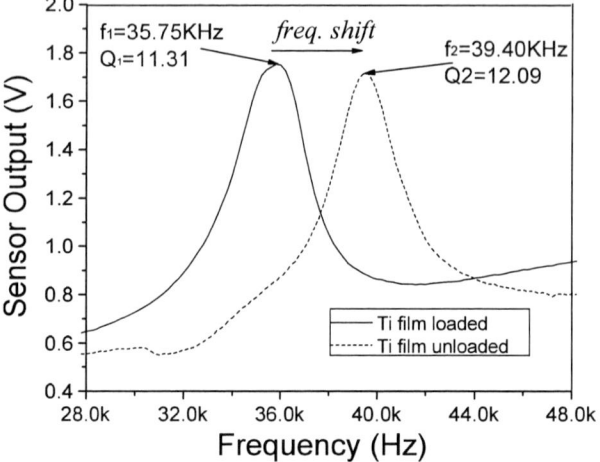

Figure 4: Measured resonant frequency shift according to a mass unloading of a thin-film Ti on top of the metal loop, resulting in sensitivity of 36Hz/pg.

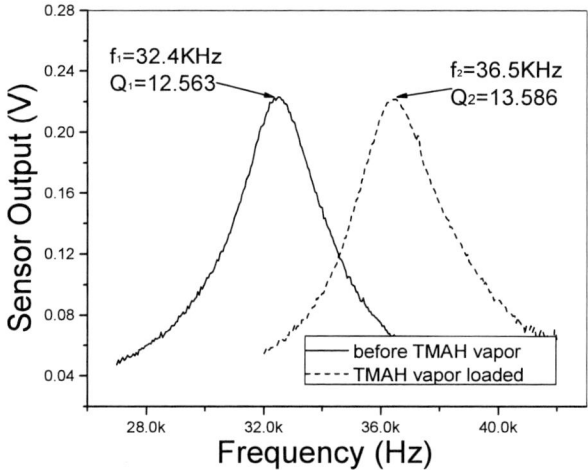

Figure 5: Tested frequency shift induced by tri-methylamine molecule-adsorption generated surface-stiffening effect, achieving an ultra-high sensitivity of 774Hz/pg.

Assuming that the -OH density at the SiO_2 surface is high enough, we estimate the theoretically maximum mass adsorption of tri-methylamine would be 4.77pg, i.e. the equivalent sensitivity is as high as about 860Hz/pg. Considering that the practically adsorbed molecule number would be much less than the theoretically maximum number, the surface-stress spring-stiffening sensing method indeed achieves a higher than one order of magnitude sensitivity compared to the conventional gravimetric sensing method where the frequency-shift is induced by mass-load.

4. CONCLUSION & ACKNOWLEDGMENT

A specific molecule reaction induced surface-stiffening sensing scheme is proposed to nano-thickness resonant cantilever sensors for ultra-high bio/chemical detecting sensitivity. A torsion-mode integrated silicon resonant cantilever is designed and fabricated with the silicon thickness as 95nm. Trace vapor specific detecting experiment has approved that the surface-stiffening sensing scheme exhibits an ultra-high sensitivity, compared to the conventional adsorbed-mass directly weighting scheme.

This research is supported by the NSFC Project under Contract No. 60725414, the Chinese 973 Program (2006CB300405) and Chinese 863 Project (2006AA04Z365). The third author thanks to the support of the NSFC Project of No. 60721004.

REFERENCES

[1] P. S. Waggoner, H. G. Craighead, "Micro- and nanomechanical sensors for environmental, chemical, and biological detection", *Lab on a Chip*, Vol. 7, pp. 1238-1255, 2007.

[2] A. Gupta, P. Nair, D. Akin, M. Ladisch, S. Broyles, M. Alam and R. Bashir, "Anomalous resonance in a nanomechanical biosensor", *PNAS* Vol. 103, pp. 13362-13367, 2006.

[3] B. Ilic, D. Czaplewski, H. G. Craighead, P. Neuzil, C. Campagnolo and C. Batt, "Mechanical resonant immunospecific biological detector", *Applied Physics Letter*, Vol. 77, pp. 450-452, 2000.

[4] B. Ilic, D. Czaplewski, M. Zalalutdinov, H. G. Craighead, P. Neuzil, C. Campagnolo and C. Batt, "Single cell detection with micromechanical oscillators", *Journal Vacuum Scientist and Technology B*, Vol. 19, pp. 2825-2828, 2001.

[5] A. Gupta, D. Akin and R. Bashir, "Single virus particle mass detection using microresonators with nanoscale thickness", *Applied Physics Letter*, Vol. 84, pp. 1976-1978, 2004.

[6] J. A. Harley and T. W. Kenny, "High-sensitivity piezoresistive cantilevers under 1000 Å thick", *Appl. Phys. Lett.*, Vol. 75, pp. 289-291, 1999.

[7] C. S. Smith, "Piezoresistance effect in germanium and silicon", *Phys. Rev.*, Vol. 94, pp. 42-49, 1954.

[8] Y. Kanda, "A graphical representation of the piezoresistance coefficients in silicon", *IEEE Trans. Electron Devices*, ED-29, pp. 64-70, 1982.

[9] M. H. Bao, *Analysis and Design Principles of MEMS Devices*, Elsevier, Amsterdam, 2005.

[10] D. Jin , Xinxin Li, H. Bao, Z. Zhang, Y. Wang, H. Yu, G. Zuo, "Integrated cantilever sensors with a torsional resonance mode for ultraresoluble on-the-spot bio/chemical detection", *Appl. Phys. Lett.*, Vol. 90, 041901, 2007.

[11] Xinxin Li, T. Ono, Y. Wang, M. Esashi, "Ultrathin single-crystalline-silicon cantilever resonators: Fabrication technology and significant specimen size effect on Young's modulus", *Appl. Phys. Lett.*, Vol. 83 pp. 3081-3083, 2003.

[12] Peng Li and Xinxin Li, "A single-sided micromachined piezoresistive SiO2 cantilever sensor for ultra-sensitive detection of gaseous chemicals", *J. Micromech. Microeng.*, Vol. 16, pp. 2539-2546, 2006.

SURFACE ACOUSTIC WAVE-DRIVEN ACTIVE PLASMONICSBASED ON DYNAMIC PATTERNING OF NANOPARTICLES IN MICROFLUIDIC CHANNELS

Jinjie Shi, Yue Bing Zheng and Tony Jun Huang[*]

Department of Engineering Science and Mechanics, The Pennsylvania State University
University Park, PA 16802, USA
*E-mail: junhuang@psu.edu, Tel: 1-814-863-4209

ABSTRACT

The nanoparticles redistribution in a microfluidic channel through a dynamic patterning process will locally change the refractive index of the medium around the nanodisk arrays, as well as the LSPR. The dynamic patterning of nanoparticles was achieved by a standing surface acoustic wave (SSAW), from which the acoustic force generated and force the particles to the pressure nodes. The SSAW were formed through two parallel interdigital transducers (IDTs) on a LiNbO$_3$ substrate, on which the gold nanodisk arrays were fabricated through a nanosphere lithography. A PDMS microchannel was aligned with the IDTs and bonded with the substrate to cover the nanodisk arrays. A solution of fluorescent polystyrene beads (diameter: 320 nm) was injected into the channel through a pressure driven flow. When SSAW was on, the original peak (λ=694 nm) is split into two peaks at λ_1=662 nm and λ_2=746 nm, respectively.

INTRODUCTION

The field of surface plasmon-based photonics, or "plasmonics", has both the capacity of photonics and the miniaturization of electronics, offering the potential to merge photonics and electronics at nanoscale dimensions [1]. Most plasmonic devices demonstrated thus far, including light sources, filters, waveguides, and polarizers, are *passive* structures [2, 3]. Before the potential of plasmonics can be realized, *active* plasmonic devices such as switches and modulators have to be developed.

Recently, several active plasmonic mechanisms and materials have been developed for usage in applications. Leroux showed that the LSPR of gold nanoparticles embedded in polyaniline thin films can be actively tuned by an electrochemical method [4]. Wang demonstrated in experiment the electrochemical tuning of LSPR of WO$_3$ sol-gel modified silver nanoparticle arrays [5]. Due to their large birefringence, Liquid crystals (LC's) have been widely used in optical switches and modulators, particularly in LC displays. By applying the electric field, the molecular orientation of liquid crystal can be tuned, thus changing the refractive index, resulting in the refractive index modulation and further tuning the LSPR of gold nanoparticles embedded in the LCs [6]. These mechanisms and techniques, however, all relay on the changes of the characteristics of the medium, either by charging the nanoparticles, or rotating the molecular orientation of the medium. These mechanisms require complex experimental setup or high electric field, which are inconvenient for most of applications and hinders the fast response and good reversibility. All limitations of present mechanisms call for a novel design which can achieve active LSPR tuning in a simple manner.

WORKING MECHANISM

LSPR is strictly dependent on the refractive index of the surrounding medium, such that changing the surrounding medium is the first and mostly adopted method to achieve active LSPR. In the recent work by Shi *et al.*, a standing surface acoustic wave (SSAW) has been demonstrated as a powerful tool to achieve dynamic patterning of micro/nano particles in a microfluidic channel [7, 8], which is believed to provide a promising method to locally change the refractive index of the fluid medium through the beads redistribution occurred in patterning. Here we report an integrated device combining a SAW-based dynamic patterning microfludic system with a gold nanodisk array on a piezoelectric substrate.

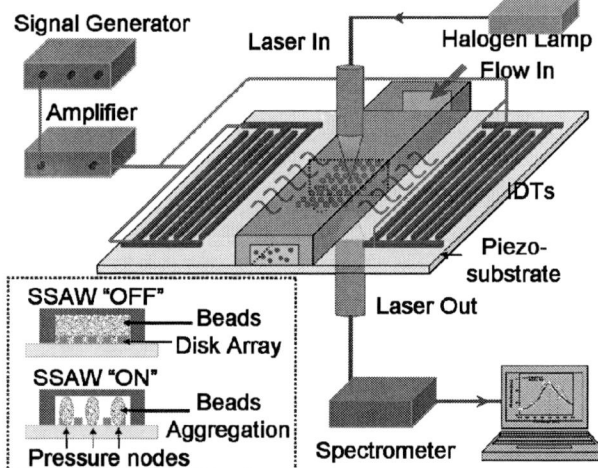

Fig. 1: Schematic drawing of the device's operation mechanism: light generated from a halogen lamp passes through the microchannel and the gold disk array, and is collected by the other fiber head and recorded by a spectrometer. Two parallel IDTs are used to generate SAW. Inset showing the cross section of the microchannel.

Figure 1 depicts the working mechanism of the SSAW-based active LSPR device. It is well known that SAW is a composite wave of longitudinal and transverse modes, which propagates only along the surface of a substrate, with most of its energy localized within one to two wavelengths beneath the surface [7]. Therefore, it has low propagation loss and high sensitivity to the surface modification, a property that has been utilized to form a uniform high frequency standing wave on the surface of the substrate. The SSAW is generated by activating two parallel IDTs with identical period deposited on a piezoelectric substrate. A microfluidic channel is aligned with these two parallel IDTs and tightly bonded with the substrate. A bead solution is infused into the channel using a pressure driven flow. When SAW is off, the beads are uniformly distributed inside the channel (Figure 1 inset upper). When the two IDTs are applied with identical radio frequency (RF) signals (generated from a signal generator and amplified by a power amplifier), two series of SAWs generate and propagate in opposite directions toward the bead solution inside the channel. The interference of the two SAWs generates a SSAW field in the region where the channel is bonded, resulting in the periodic distribution of pressure nodes and anti-nodes along the surface of the substrate. When SSAW encounters the liquid medium, a longitudinal leakage waves will be generated inside the liquid, resulting in the pressure fluctuation of the medium. The primary acoustic radiation forces originated from the pressure fluctuations drive the suspended particles toward the pressure nodes or anti-nodes (depending on the comparison of densities and compressibility between the particles and the liquid medium) [9], thereby achieving the patterning of beads inside the microchannel (Figure 1 inset lower, three pressure nodes are covered by the channel, resulting in three bead aggregations). Because of the laminar nature of the flow inside the channel, the patterned beads maintain their positions even after passing the SSAW region.

The primary acoustic force exerted on the suspended particles in a standing wave field can be expressed as [10]:

$$F_a = -(\pi p_0^2 V_c \beta_l / 2\lambda) \cdot \phi(\beta, \rho) \cdot \sin(2kx) \qquad (1)$$

$$\phi = (5\rho_p - 2\rho_l)/(2\rho_p + \rho_l) - \beta_p/\beta_l \qquad (2)$$

where V_c, p_0, λ, β, ρ are the particle volume, acoustic pressure, SAW wavelength, the compressibility and the density, respectively. Subscripts l and p denote liquid medium and particles, respectively. The notation of Φ decides whether the particle will be trapped at the pressure nodes (positive Φ) or the pressure anti-nodes (negative Φ), which is dependent on the density and compressibility of the medium and the particles. In this experiment, polystyrene beads with positive Φ is used and they are expected to be patterned at pressure nodes.

EXPERIMENT SETUP
Device fabrication

The fabrication procedure (as shown in Figure 2) for the SSAW focusing device includes four steps [8]: (1) fabrication of gold nanodisks on the piezo-substrate using nanosphere lithography; (2) the fabrication of IDTs for SAW generation; (3) the fabrication of a PDMS-based microchannel; (4) the bonding of the PDMS microchannel to the SAW substrate. Firstly, a thin layer of photoresist (PR) SPR3012 (MicroChem, Newton, MA) was spin-coated on a Y+128° X-propagation lithium niobate (LiNbO$_3$) wafer, patterned with a UV light source, and developed in a photoresist developer (MF CD-26, Microposit). A double metal layer (Ti/Au, 30Å/300Å) was subsequently deposited on the wafer using an e-beam evaporator (Semicore Corp), followed by a lift-off process to form the metal layer for nanodisk array fabrication. A self-assembly monolayer (SAM) of polystyrene beads

(diameter of 320 nm) is formed on the metal thin film, followed by an oxygen plasma etching to tune the bead size to 170 nm in diameter. A argon plasma is then used to etch the gold thin film using the polystyrene beads as the mask. After removing the PR in a followed lift-off process, the device is immersed in the toluene solution to remove the polystyrene beads with the help of ultrasound [11, 12]. Secondly, a PR layer is deposited and the second lithography is used to give the opening of IDTs, followed by a metal deposition (50Å/800Å, Cr/Au) and lift-off, finally the SAW substrate integrated IDTs and gold nanodisk arrays is obtained.

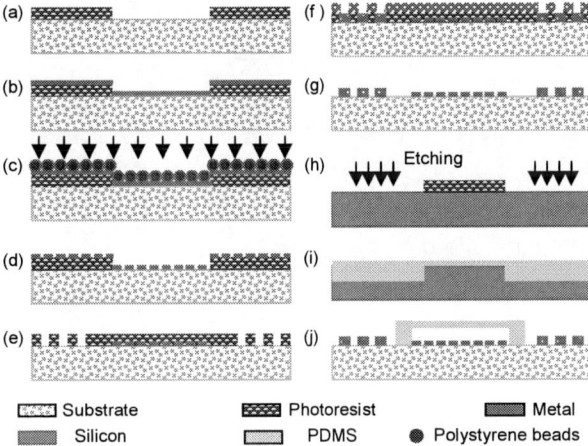

Fig. 2: Fabrication process of the SAW substrate with gold nanodisk and the PDMS channel: (a)~(g) nanosphere lithography was used to form gold nanodisk array on a piezo-substrate, followed by the formation of two parallel gold IDTs for SAW generation. (h)~(j) A silicon substrate was patterned and etched, serving as the channel mold, followed by the PDMS coating and curing. After removing from the mold, the PDMS channel was aligned and bonded with the piezo-substrate, covering the gold nanodisks.

The polydimethylsiloxane (PDMS) microchannels were fabricated with standard soft-lithography and mold-replica techniques [8]. The silicon mold for the microchannel was patterned by photoresist Shipley 1827 (MicroChem, Newton, MA) and etched by a Deep Reactive Ion Etching (DRIE, Adixen, Hingham, MA) process. After DRIE, the silicon mould was coated with 1H,1H,2H,2H-perfluorooctyl-trichlorosilane (Sigma Aldrich, St. Louis, MO) in a vacuum chamber to reduce surface energy and hence the damage to the PDMS channel during the demolding process. Sylgard™ 184 Silicone Elastomer Base and Sylgard™ 184 Silicone Elastomer Curing Agent (Dow Corning, Midland, MI) were mixed at an 11:1 (weight) ratio, cast onto the silicon mould, and cured at 70 °C in an oven for 30 minutes (varied for different hardness). After peeled from the silicon mold, inlets and outlets were created using a silicon carbide drill bit.

At last, surfaces on both the PDMS microchannel and the SAW substrate were activated with oxygen plasma (oxygen flow rate: 50 sccm; chamber pressure: 750 mTorr; and power: 150 W) to facilitate the bonding process. The alignment of the SAW substrate and the PDMS channel were conducted manually under the microscope. A drop of ethanol was placed in between the

SAW substrate and the microchannel serving as lubricant, so that the PDMS channel could slide on top of the SAW substrate during the aligning process. Following that, ethanol was removed by leaving the aligned device in a vacuum chamber at 50°C for 15 min. Finally, polyethylene tubings (Becton Dickson, Franklin Lakes, NJ) were inserted into the inlet and outlet of the channel and connected the device to a syringe pump (KDS 210, KD scientific, Holliston, MA). Figure 3(a) shows the bonded device (channel is filled with ink for visualization), and zoom-in figures in (b), (c) and (d) indicate the fabricated IDTs, AFM and SEM of gold nanodisk arrays (diameter ~170nm, period of 320 nm).

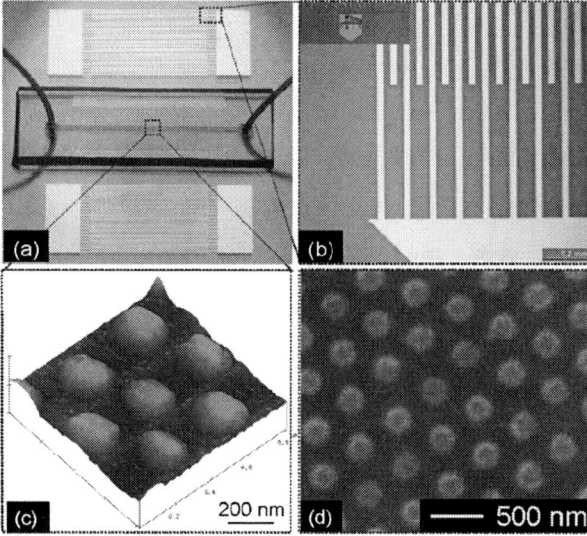

Fig. 3: (a) PDMS channel and the bonded device. The channel is drilled and connected with two silicone tubings such that the bead solution can be infused inside the channel. (b) Fabricated IDTs, (c) and (d) are the AFM and SEM figures of the gold nanodisks fabricated using the nanosphere lithography, respectively.

Measurement setup

The bonded device was firstly mounted on an inverted microscope (Nikon TE2000U) to visualize the beads pattern when SSAW is applied. Nano polystyrene beads (Bangs Laboratories) were injected to the channel through a pressure driven flow using a syringe pump with a flow rate of 7 μl/min. Following that, an RF signal was generated from a signal generator (Agilent E4422B) and amplified by a power amplifier (Amplifier Research 100A250A). The output signal from the amplifier was then split into two identical signals, which were applied to the IDTs and generated SAWs. The applied AC signal was set to be 38.2 MHz (the resonance frequency for SAW on $LiNbO_3$) and the input power is 25 dBm (~310 mW). In order to observe the active tuning of LSPR of the gold nanodisk arrays in the beads patterning process, a halogen lamp was equipped with the microscope and the transmitted white light through the device is collected by a spectrometer (HR 4000, Ocean Optics), where the absorption spectrum is recorded.

RESULTS AND DISCUSSION

Figure 4(a) shows the beads distribution in the microfluidic channel when SAW was absent. When SAW was applied and SSAW formed, the beads redistribution

happened in the channel and bead aggregations were recorded (Figure 4(b)). Figure 4(c) is the zoom-in figure of the bead aggregation region, showing the bead concentration increasing. As the bead material has a higher refractive index than that of water (solvent of beads solution), we observed that the original LSPR peak (within uniform beads solution medium) is split into two discrete peaks when SAW is applied. This is due to the re-distribution of the beads inside the channel: at the bead aggregation regions (pressure nodes), the refractive index increases, resulting in the red shift (~52 nm) of the LSPR of the local gold nanodisks; on the other hand, at other regions, the reduced bead concentration results in the refractive index decreasing and blue shift (~31nm) of the LSPR. Consequently, two new peaks are generated and the original peak intensity decreases (Figure 4(d)). As stated in our former works, the SSAW-based dynamic patterning process has a fast response (~ 5ms), thus this active tuning method is believed to provide a fast response comparing with other methods. Furthermore, by splitting one peak to two, this method provides more choices in plasmonics communications.

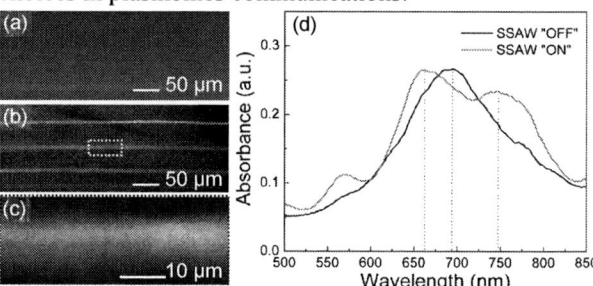

Fig. 4: (a) Uniform bead distribution in channel when the SAW was off. (b) Bead redistribution at multiple lines when SSAW was applied, resulting in periodic RI change. (c) The inset shows bead aggregations at pressure nodes in SAW. (d) When SAW is on, the original peak (=694 nm) is split into two peaks at $_1$=662 nm and $_2$=746 nm, respectively.

CONCLUSION

In summary, we have demonstrated an active plasmonics device including a SSAW-based microfluidic system and a gold nanodisk array. Patterning of nanoparticles helps build a period distribution of refractive index in a microfluidic channel, which inversely affected the LSPR of the gold nanodisk arrays integrated with the SAW substrate and covered by the channel. Compared with other techniques in active plasmonics [1, 6], our technique is simple, efficient, fast response (~5 ms), and have good reversibility. We believe that this technique provides a promising solution for plasmonic switches and modulators, important components for plasmonics-based optical communications.

ACKNOWLEDGEMENTS

The authors thank Dr. Bernhard R. Tittmann for help with experiment. This research was supported in part by the Grace Woodward Grants for Collaborative Research in Engineering and Medicine, and the NSF NIRT grant (ECCS-0609128). Components of this work were conducted at the Penn State node of the NSF-funded

National Nanotechnology Infrastructure Network.

REFERENCES

[1] E. Ozbay, "Plasmonics: Merging Photonics and Electronics at Nanoscale Dimensions", Science, vol. 311, pp. 189–193, 2006.

[2] S. A. Maier, M. D. Friedman, P. E. Barclay, and O. Painter, "Experimental demonstration of fiber-accessible metal nanoparticle plasmon waveguides for planar energy guiding and sensing", Appl. Phys. Lett., vol. 86, pp. 071103, 2005.

[3] W. Nomura, M. Ohtsu, and T. Yatsui, "Nanodot coupler with a surface plasmon polariton condenser for optical far/near-field conversion", Appl. Phys. Lett., vol. 86, pp. 181108, 2005.

[4] Y. R. Leroux, J. C. Lacroix, K. I. Chane-Ching, C. Fave, N. Felidj, G. Levi, J. Aubard, J. R. Krenn, A. Hohenau, "Conducting Polymer Electrochemical Switching as an Easy Means for Designing Active Plasmonic Devices", J. Am. Chem. Soc., vol. 127, pp. 16022–16023, 2005.

[5] Z. C. Wang, G. Chumanov, "WO3 Sol-Gel Modified Ag Nanoparticle Arrays for Electrochemical Modulation of Surface Plasmon Resonance", Adv. Mater., vol. 15, pp. 1285–1289, 2003.

[6] K. C. Chu, C. Y. Chao, Y. F. Chen, Y. C. Wu, C. C. Chen, "Electrically Controlled Surface Plasmon Resonance Frequency of Gold Nanorods", Appl. Phys. Lett., vol. 89, pp.103107, 2006.

[7] J. Shi, D. Ahmed, X. Mao, and T. J. Huang, "Surface acoustic wave (SAW) induced patterning of micro beads in microfluidic channels", MEMS 2008, Tucson, AZ, USA, January 13-17, pp. 26–29, 2008.

[8] J. Shi, X. Mao, D. Ahmed, A. Colletti, and T. J. Huang, "Focusing microparticles in a microfluidic channel with standing surface acoustic waves (SSAW)", Lab Chip, 8, 221–223, 2008.

[9] F. Petersson, A. Nilsson, H. Jo1nsson, T. Laurell, "Carrier Medium Exchange through Ultrasonic Particle Switching in Microfluidic Channels", Anal. Chem., vol. 77, pp. 1216–1221, 2005.

[10] K. Yosioka and Y. Kawasima, "Acoustic radiation pressure on a compressible sphere." Acustica, vol. 5, pp. 167–173, 1955.

[11] Y. B. Zheng, B. K. Juluri, X. Mao, T. R. Walker and T. J. Huang, "Systematic Investigation of Localized Surface Plasmon Resonance of Long-Range Ordered Au Nanodisk Arrays", J. Appl. Phys., vol. 103, pp. 014308, 2008.

[12] Y. B. Zheng, and T. J. Huang, "Surface Plasmons of Metal Nanostructure Arrays: from Nanoengineering to Active Plasmonics", J. Assoc. Lab. Autom., Vol. 13, pp. 215-226, 2008.

PUPAL-STAGE INSERTED SILICON MICROPROBE NEURAL INTERFACE IN INSECTS FOR GAS SENSING

C.J. Shen[2], R. Gilmour[2,3], A. Lal[1,2]

*Sonic*MEMS Laboratory

[1] School of Electrical and Computer Engineering
[2] Department of Biomedical Engineering
[3] Department of Biomedical Sciences

Cornell University, Ithaca, NY, USA

ABSTRACT

This paper reports on the first-ever hybrid bio-electromechanical sensor realized using Early Metamorphosis Implantation Technology (EMIT). Using EMIT, microprobes are inserted in *Manduca sexta* moth pupae early in their developmental cycle. The probes electrically report the neural reactions of pupae olfactory response. By forming direct neural interfaces to insect olfactory biological circuits, this approach taps into the high specificity and sensitivity (as low as several parts per quadrillion) developed in insects over millions of years of evolution. Here we demonstrate the ability to detect tobacco leaves and pheromone components with high sensitivity of at least 6 parts per billion (ppb) in an integrated system, and the potential for tapping into the much higher sensitivities intrinsic in insect olfactory processing systems.

1. INTRODUCTION

Insect biosensors that detect increased neural activity from single olfactory receptor neurons (ORNs) on the antenna or from electroantennograms have been previously investigated [1, 2]. Studies have demonstrated some species, such as the moth *Bombyx mori*, have olfactory neurons capable of detecting pheromones with concentrations as low as several parts per quadrillion [3]. With traditional neural recording equipment using *in vivo* preparations of insects to test responses to chemical odorants, concentrations of parts per trillion can be detected, comparable to or surpassing sensitivities of large proton-transfer reaction-mass spectrometers [4]. However, these biosensors have limitations in their practical effectiveness. Some respond similarly to the presence of any odor, thus not demonstrating specificity, while others only sample molecules which diffuse close to a single ORN (with a cross-section of several μm^2) limiting reliability and portability by requiring accurate

and stable probe placement. These sensors require that microelectrodes be positioned accurately on the antenna surface, or near a specific ORN.

Studies have demonstrated that the range of responsiveness of moths can be expanded beyond those that moths naturally respond to using Pavlovian conditioning. For example, one study trained adult moths to respond to explosives odors with a feeding response [5]. This particular approach to achieve a gas sensor, which is based on behavioral responses, has limitations in its reliability and specificity since behavioral responses can vary with other stimuli (e.g. light and sound) [5]. However, together with another recent paper showing that moths can also be trained to respond to certain odors prior to the pupal stage, during the larvae stage (Fig. 1), and retain memory of their training into the adult stage [6], these studies demonstrate the possible range of applicability of an integrated insect sensor.

For an effective sensor system, it is desirable to have reliability, specificity, and sensitivity. It has been demonstrated that surgically implanted substrates for microsystems can be implanted in pupal stage insects without greatly affecting their adult lifespan [7], suggesting promise for utilizing a pupal-stage implanted microelectrode in the pupal brain for an integrated gas sensor. As shown in Fig. 1, probes inserted during the pupal stage of the *Manduca sexta* life span provide the best electrical and mechanical integration. The insect microsystem could provide enhanced reliability, specificity, and sensitivity by tapping into the high sensitivity and specificity of an insect's olfactory system, coupled with pupal stage insertions allowing for good tissue integration and reliability. Additionally, further studies have demonstrated that moths can be maintained in the pupae stage and prevented from becoming an adult by injecting juvenile hormone, allowing a hybrid sensor to remain active indefinitely [8].

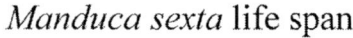

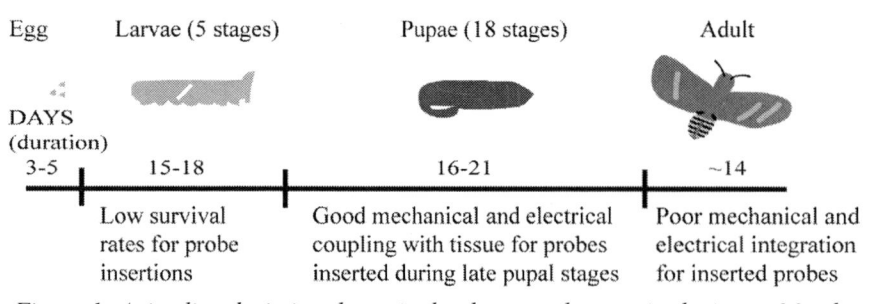

PUPAL STAGE
Olfactory Development

STAGE

9   Antennal lobe (AL) neurons respond to antennal nerve shock

11   AL neurons respond with action potentials to antennal nerve shock

15   Antenna sensilla begin responding to odors

16   First measured antennal lobe response to tobacco odor with our neural sensor

Figure 1: A timeline depicting the main developmental stages in the insect Manduca sexta, as well as results of substrate insertions at various point during development.

2. METHODS, DEVICE FABRICATION AND CHARACTERIZATION

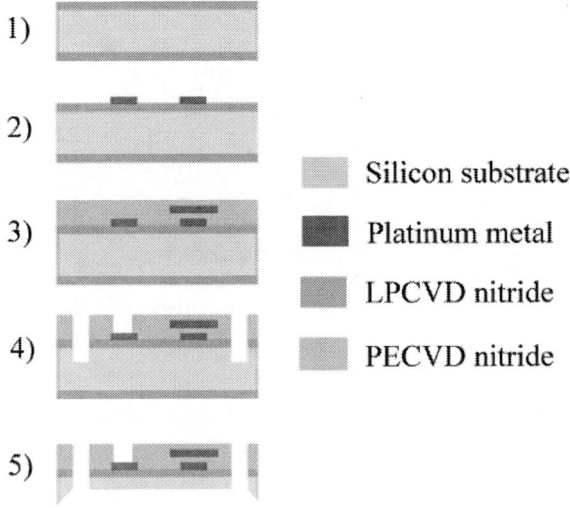

Figure 2: Process flow for microfabricating neural probes illustrated using cross-section of probe tip. 1) Deposit LPCVD nitride on silicon wafer. 2) Deposit photoresist and evaporate and pattern metal using lift-off. 3) Deposit an insulating layer of nitride, evaporate and pattern another ground layer of metal using lift-off, then passivate surface with another layer of silicon nitride. 4) Etch openings to bond and recording sites. DRIE silicon to define probe boundaries. 5) KOH etch to release probes.

Microfabricated Probe Fabrication and Characterization

The fabrication process is as presented in Fig. 2. The devices consist of two evaporated platinum layers insulated by silicon nitride. The fabricated probe and equivalent circuit of the electrode-electrolyte interface are shown in Fig. 3. The interface can be modeled as consisting of spreading resistance, R_s, charge transfer resistance, R_{ct}, and interfacial capacitance Z_{CPA} [9]. Using a Gamry FS2 potentiostat, cyclic voltammetry followed by electrochemical impedance spectroscopy (EIS) were performed on the microfabricated electrodes. During the tests, the electrodes were immersed in insect saline to mimic the hemolymph inside the insect. Averaging three consecutive EIS measurements together yields the experimentally measured interface impedance shown in Fig. 3(c). The simplified equivalent circuit model shown in Fig. 3(b) was fit to the measured results yielding the fitted parameters shown in Fig. 3.

Experimental Design and Device Integration

Male *Manduca sexta* late stage pupae reared at the Boyce Thompson Institute, Cornell University were used for the experiments. Short pulses of air mixed with target molecules were directed at the antenna of pupal stage male moths. One antenna was freed from the pupal cuticle, and the resulting opening in the pupal cuticle was sealed with wax.

As shown in Fig. 4(a), probes were designed such that the 10 μm by 10 μm recording sites matched the cell body diameter of 60% of the neurons in the lateral cluster

in the antennal lobe [10]. Microprobes were inserted with a micromanipulator during the 18 stages of pupal development. As shown in Fig. 4(b), probes were directed towards the antennal lobe ipsilateral to the freed antenna. Fig. 4(c) shows a dissection of an adult sphinx moth revealing the brain structure [11], and Fig. (d) demonstrates the location that was targeted with the probe insertions. These insertions were performed during the later stages of pupal development, as a cuticle does not form around the antennae for robustness while freeing it from inside the pupal cuticle until pupal stage twelve [12]. After initial insertions, the electrodes are coarse and fine positioned during development to find the strongest signals in the target brain volume.

To apply the olfactory stimuli, a setup similar to that presented in [10] was used. Charcoal filtered air (approximately 0.05 LPM) was normally directed through a glass cartridge aimed at the insect antenna. To apply each stimulus, small pieces of filter paper with stimulus dissolved were placed in a separate glass cartridge also directed towards the antenna. Using a relay controlled pinch valve, 300-400 ms pulses of air were diverted from the charcoal filtered normal air flow to the stimulus

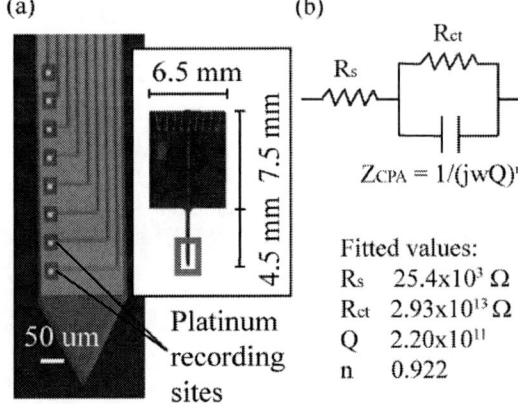

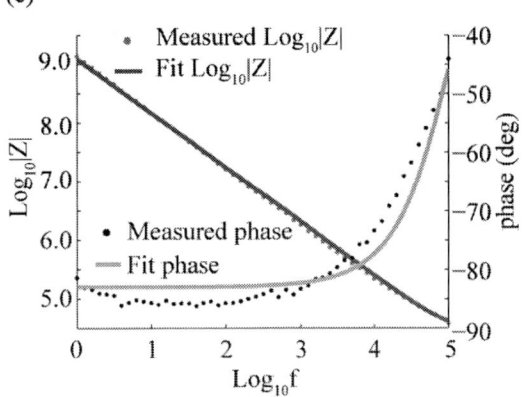

Figure 3: (a) Microfabricated silicon neural probe with eight platinum recording lines insulated with silicon nitride, and square 10 μm recording sites along the length of the probe for sensor (b) Simplified equivalent circuit and calculated equivalent parameters from fit of equivalent circuit model to measured impedance shown in (c). (c) Measured impedance data averaged over 3 consecutive electrochemical impedance spectroscopy (EIS) measurements and fit to the simplified model.

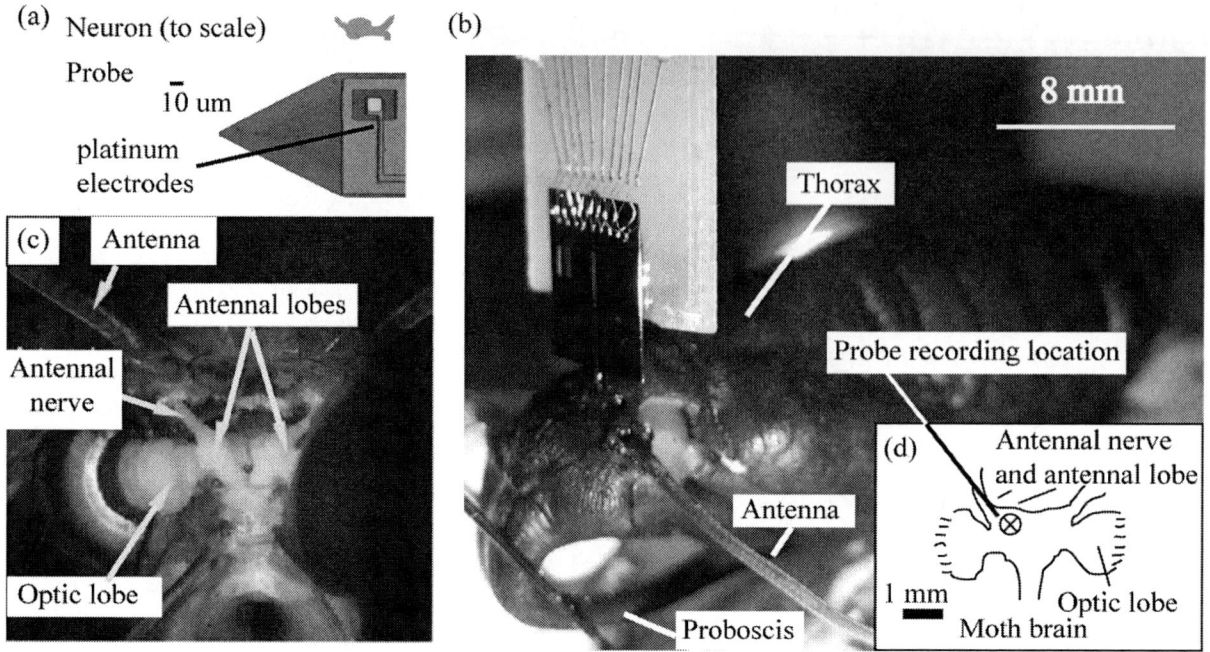

Figure 4: (a) Size comparison of probe tip and representation of neuron drawn to scale (b) Photograph of hybrid bio-electromechanical sensor system composed of microprobe inserted in Manduca sexta pupae using Early Metamorphosis Insertion Technology (EMIT) (c) Frontal view of brain of adult sphinx moth (adapted from [11]) (d) Diagram of insect brain.

cartridge. Stimuli included *Manduca sexta* host plant tobacco, major pheromone component E10Z12-16Al (Bombykal) [13], a mimic of another major pheromone component E11Z13-15Al ('C-15') [13], and clean air. Pheromone components were generously supplied by Prof. Jocelyn Millar (UC Riverside). Small pieces of tobacco leaves were placed directly into the cartridges for tobacco tests.

Electrophysiological signals were amplified and filtered with a notch filter at 60 Hz using a commercial A-M systems Model 1800 amplifier, and digitized at 10 kHz using a data acquisition system from National Instruments. Data were recorded using a PC with custom LabVIEW code.

3. RESULTS AND DISCUSSION

Pupal stage insertions allowed fine and coarse positioning of microprobes to determine locations with the strongest response signals. Probes inserted in the 14th pupal stage (P14) and maintained in a given location did not demonstrate antennal lobe tobacco responsiveness until the moth reached P16, indicating olfactory response development between the 14th and 16th stages (Fig. 1, right). This is supported by studies investigating the onset of antennal lobe responsiveness to various natural stimuli [12].

Recordings from P16 inserted probes in response to 300-400 ms air pulses containing various stimuli are shown in Fig. 5. The sensor demonstrates high specificity for tobacco, but is also weakly responsive to 6 ppb concentrations of bombykal, demonstrating a tobacco signal-to-noise ratio of greater than 10. This signal-to-noise ratio is defined as the ratio of amplitudes of the tobacco response to any other stimulus response. As

shown in Fig. 6, the insect response is sensitive to the concentration and/or history of applied stimuli, as the magnitude of the response decreases over consecutive

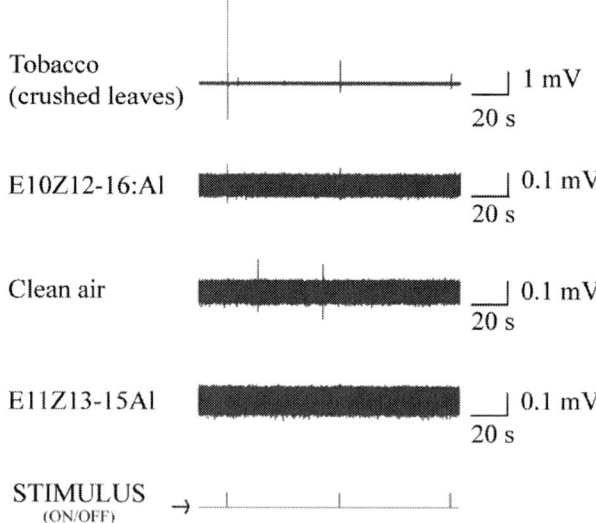

Figure 5: Recorded neural activity (blue, top traces) in response to indicated stimulus pattern (red, bottom trace) consisting of three 300-400 ms pulses of air mixed with target molecules. Measurements were recorded from male 16th pupal stage Manduca sexta. Strong responses to tobacco were observed at least 21 times, and weak-to-no responses to the pheromone and pheromone-mimic components E10Z12-16:Al and E11Z13-15:Al were observed, respectively. Measurements showed a tobacco response amplitude to non-tobacco response amplitude greater than 10.

978-1-4244-2977-6/09 $25.00 © 2009 IEEE

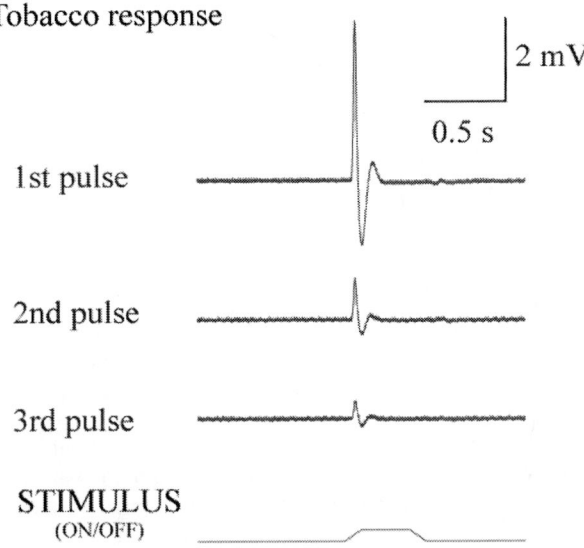

Figure 6: Responses of 3 odor pulses separated by 60 s each. Decreasing amplitude of responses were seen with each consecutive pulse.

tobacco pulses. However, applying further tobacco pulses after removal of the stimulus and waiting a period of several minutes yields similar initial response magnitudes and response patterns, lending support to the idea that some olfactory neuron adaptation may be occurring.

Our method provides greatly enhanced performance in reliability, specificity, and sensitivity by: 1) Measuring neural signals directly from the insect's first odor processing areas, the antennal lobes. These signals convey integrated information from neurons spanning the entire antenna (~2-3 cm long, 1 mm wide) that are receptive to a specific odor. 2) EMIT pupal stage probe insertions which improve insect and neural integration. That is, EMIT enables development of the insect brain *around* the microprobes, rather than insertion into fully developed brains. Due to tissue-anchoring, EMIT avoids physical damage to the adult neural circuitry, and uses the highly active brain development during metamorphosis to heal any damage. As further brain growth occurs around probes, even the small pupae motion does not shift the position of the electrodes away from the desired neurons, making signal capture robust and reliable.

4. CONCLUSION

We have demonstrated the feasibility of a new hybrid electromechanical sensor paradigm by recording long term from a pupal inserted microfabricated neural interface in response to the presentation of various odors. The range of chemical responsiveness can be modified in the future by targeting neurons responsive to other odors of interest.

ACKNOWLEDGEMENTS

The authors would like to thank the members of the *Sonic*MEMS laboratory for useful discussions, technical assistance, and advice. The authors would like to thank in particular Abhishek Ramkumar, Alper Bozkurt, and Ayesa Paul for their help. We also thank Prof. Jocelyn Millar (UC Riverside) for graciously providing insect pheromone components. This research was supported by DARPA-MTO, and part of the work was conducted at the Cornell Nanofabrication Facility, Cornell University.

REFERENCES

[1] M.J. Huotari, "Biosensing by insect olfactory receptor neurons", *Sensors and Actuators B*, vol. 71, pp. 212-222, 2000.

[2] M.J Schoning, S. Schutz, P. Schroth, B. Weissbecker, A. Steffen, P. Kordos, H.E. Hummel, H. Luth, "A BioFET on the basis of intact insect antennae", *Sensors and Actuators B*, vol. 47, pp. 235-238, 1998.

[3] R.L. Metcalf, "Ultramicrochemistry of Insect Semiochemicals", *Mikrochim. Acta*, vol. 129, pp. 167-180, 1998.

[4] C. Lindinger, P. Pollien, A. Santo, C. Yeretzian, I. Blank, T. Mark, "Unambiguous identification of volatile compounds by proton-transfer reaction mass spectrometry coupled with GC/MS", *Anal. Chem.* vol. 77, pp. 4117-4124, 2005.

[5] T.L. King, F.M Horine, K.C. Daly, B.H. Smith, "Explosives detection with hard-wired moths", *IEEE Trans. on Instrum. and Meas.*, vol. 53(4), pp. 1113-1118, 2004.

[6] D.J. Blackiston, C.E. Silva, M.R. Weiss, "Retention of memory through metamorphosis: can a moth remember what it learned as a caterpillar?", *PLoS ONE*, vol. 3(3): e17, 2008.

[7] A. Paul, A. Bozkurt, J. Ewer, B. Blossey, A. Lal, "Surgically implanted micro-platforms in *Manduca sexta* moths", in *2006 Solid State Sensor and Actuator Workshop*, Hilton Head Island, June 2006, pp. 209-211.

[8] H.F. Nijhout, C.M. Williams, "Control of Moulting and Metamorphosis in the Tobacco Hornworm, *Manduca sexta (L.)*: Cessation of Juvenile Hormone Secretion as a Trigger for Pupation", *J. Exp. Bio.* vol. 61, pp. 493-501, 1974.

[9] G.T.A. Kovacs, "Introduction to the theory, design, and modeling of thin-film microelectrodes for neural interfaces", *Enabling technologies for Cultured Neural Networks*, Academic Press, 1994.

[10] S.G. Matsumoto, J.G. Hildebrand, "Olfactory mechanisms in the moth *Manduca sexta*: response characteristics and morphology of central neurons in the antennal lobes", *Proc. of the Royal Society of London. Series B, Biological Sciences*, vol. 213 (1192), pp. 249-277, 1981.

[11] T. Michaelis, T. Watanabe, O. Natt, S. Boretius, J. Frahm, S. Utz, J. Schachtner, "In vivo 3D MRI of insect brain: cerebral development during metamorphosis of Manduca sexta", NeuroImage, vol. 24, pp. 596-602, 2005.

[12] E.S. Schweitzer, J.R. Sanes, J.G. Hildebrand, "Ontogeny of electroantennogram responses in the moth, Manduca sexta", J. Insect Physiol., vol. 22, pp. 955-960, 1976.

[13] T.A. Christensen, J.G. Hildebrand, "Male-specific, sex pheromone-selective projection neurons in the antennal lobes of the moth Manduca sexta", J. Comp. Physiol. A, vol. 160, pp. 553-569, 1987.

BEAD-BASED MICROFLUIDIC PLATFORM INTEGRATED WITH OPTICAL DETECTION DEVICES FOR RAPID DETECTION OF GENETIC DELETION FROM SALIVA

Kang-Yi Lien [1], Chien-Ju Liu [2], Pao-Lin Kuo [3] and Gwo-Bin Lee [1, 2, 4]

[1]Institute of Nanotechnology and Microsystems Engineering, [2]Department of Engineering Science, [3]Department of Obstetrics and Gynecology, National Cheng Kung University, Tainan, Taiwan
[4]Medical Electronics and Device Technology Center, Industrial Technology Research Institute, Hsinchu, Taiwan

ABSTRACT

This study presents a new magnetic bead-based microfluidic platform for genomic DNA (gDNA) extraction and rapid detection of genetic deletion from saliva samples by using micro-electro-mechanical-systems (MEMS)-based technologies. With the incorporation of several microfluidic devices, a charge-switchable, magnetic bead was used to extract gDNA released from the buccal cells in the saliva samples and collected by a built-in microcoils array. A new normally-closed micropump for liquid delivery and a new self-compensated micro temperature control module for on-chip PCR were all integrated into a single chip. Successful demonstration for detection of the patients with genetic deletion has been achieved by using the proposed microfluidic platform. Consequently, the integrated miniature saliva-based microfluidic system can provide a promising platform for rapid DNA extraction and fast detection of genetic diseases in a shorter period of time automatically.

INTRODUCTION

Point-of-care systems capable of sample preparation, transportation, separation, reaction and detection have been explored extensively in the recent years. Fewer manual operation steps are required and all the experimental processes from pre-treatment through analysis can be carried out with little human intervention in a lab-on-a-chip (LOC) systems [1-2]. Not only do the integrated micro-total-analysis-systems (μ-TAS) make a great impact on the development of miniaturized instruments for biomedical analysis, but they also provide an innovative technique to access the genetic information at a molecular level.

Recently, the development of preventive medicine and the genetic analysis of target genes have been widely explored. The deletion and mutation of genetic genes would cause serious illness such as cardiovascular or genetic diseases that may interfere with physical and mental health of human bodies and have attracted considerable interests. Therefore, the genotyping of nucleotide polymorphism and the analysis of genetic gene have become one of the most concerned issues in the clinical diagnosis. Among them, extraction of gDNA from clinical samples plays an important role in the genetic analyses [3-4]. Nonetheless, the invasive and painful venepuncture and several tedious purification steps of gDNA extraction from clinical whole blood samples are inevitable. Besides, well-trained personnel are normally required to perform the entire time-consuming process in using the organic reagents.

Saliva samples have therefore attracted considerable interest as a potential alternative source of human gDNA in genetic epidemiology [5-6]. A variety of biological substances including enzymes, hormones, immunoglobulins, and bio-molecules have already been successfully quantified in saliva [7]. The usage of saliva samples would be more convenient compared to the whole blood sample. High-quality gDNA can be extracted and purified from the saliva for subsequent biological applications. Recently, DNA extraction using magnetic bead-based techniques from clinical bio-samples has been extensively explored [8-9]. The magnetic beads which may be surface-modified with functional groups or specific probes would bind the DNA onto the beads with a high-efficient affinity. Nevertheless, additional block heaters and bulk PCR machines are always needed to perform the subsequent biomedical genetic analyses. In addition, optical detection using a bulky and costly apparatus is always required afterwards. As a result, the development of the LOC integrated with multi-functional modules to perform rapid gDNA extraction, nucleic acid amplification and on-line optical detection of genetic genes is still in great needs.

The study therefore presents an integrated bead-based microfluidic platform incorporated with an optical detection device for rapid detection of genetic deletion from saliva. The high-quality gDNA can be automatically isolated and extracted from the saliva samples utilizing the charge-switchable magnetic particles, followed by amplifying the target region of genetic deletion with specific primers and detecting a fluorescence-based DNA probe using an on-chip PCR module. With this approach, the genetic deletion of gDNA can be rapidly analyzed within a shorter period of time in an automatic format.

MATERIALS AND METHODS

Experimental principle

Figure 1 shows a schematic illustration for gDNA extraction and optical detection of genetic deletion. Fresh human saliva is first incubated with charge-switchable magnetic beads suspended in cell lysis buffer solution (Figs. 1a-1b). Taking advantage of bead-based technology that provides a charge-switchable surface of the magnetic beads, the gDNA with negative charges would be bound onto the surface of the magnetic beads with positive charges in the surrounding buffer (Fig. 1c). The DNA-bound magnetic complex can be further purified by a magnetic field generated by an on-chip circular microcoils array (Fig. 1d). After washing out the other interference substances in sample solution, the isolated gDNA can be eluted from the magnetic beads to the buffer (Fig. 1e). Subsequently, the released gDNA was amplified utilizing a self-compensated temperature control module that can

978-1-4244-2977-6/09 $25.00 © 2009 IEEE

provide a uniform temperature distribution during PCR process (Fig. 1f-1g), followed by on-line optical detection utilizing a photomultiplier tube (PMT) module (Fig. 1h).

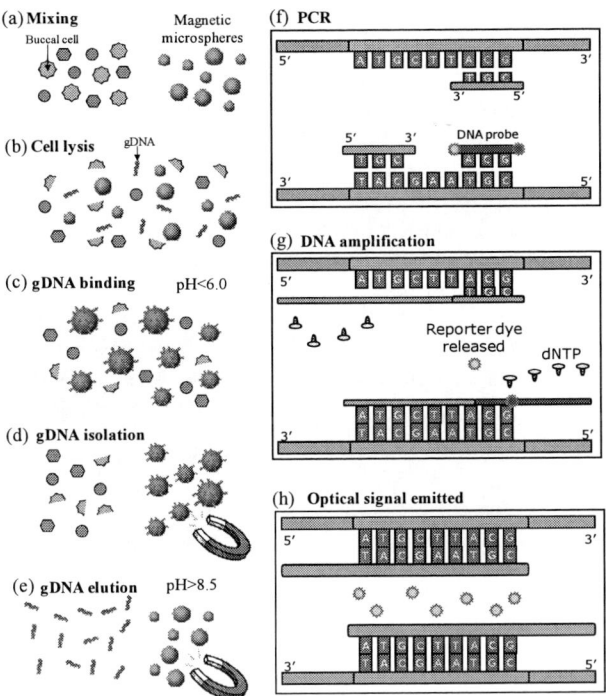

Figure 1: Schematic illustration of the operation process of gDNA purification, amplification and detection of genetic mutation.

Chip design

The integrated saliva-based microfluidic system is composed of three components including a DNA extraction module, a self-compensated PCR module and an optical detection module. Figure 2 shows the layout of the assembled bead-based microsystem with dimensions of 42 mm × 27 mm and scanning electron microscope (SEM) images of each component. From the illustration, the system is made of one glass substrate and two polydimethylsiloxane (PDMS) structure layers. The DNA extraction module consists of normally-closed pneumatic micropumps, a sample loading chamber, reaction reservoirs, microfluidic channels and a microcoils array. The normally-closed micropump is designed for liquid delivery with a high pumping rate (Fig. 2b-2). A required magnetic field for purification of the DNA-bead complex can also be generated by the built-in microcoils array (Fig. 2b-1). A new self-compensated PCR module composed of two symmetric array-type microheaters and three temperature sensors is developed to improve the temperature uniformity in the reaction chambers for rapid nucleic acid amplification of genetic genes (Fig. 2b-3). The platinum resistors are deposited and used for microheaters/temperature sensors, followed by depositing gold metallization as electrical leads of the micro temperature sensors and the array-type microheaters. Please note that all modules can be precisely controlled by using a digital application specific integrated circuit (ASIC) controller.

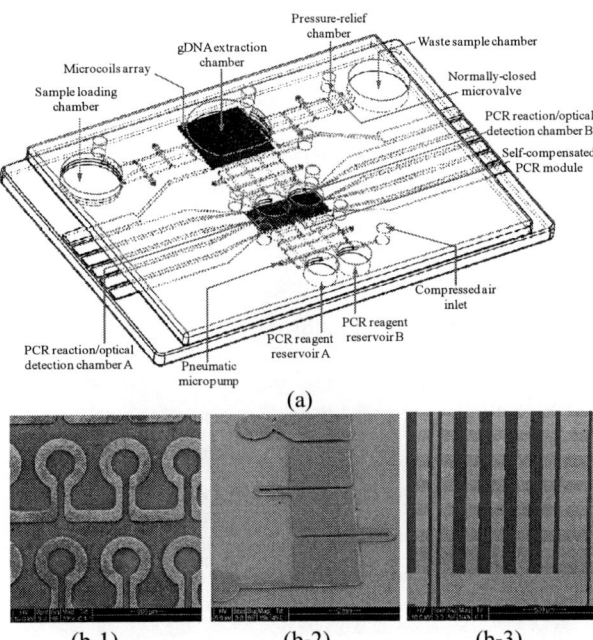

Figure 2: (a) Schematic diagram of the miniature microfluidic system. (b) SEM photos of each modules in the microfluidic system.

A fluorescence-based PCR technique is employed in the microfluidic system for rapid detection of genetic deletion. The fluorescent reporter dye labelled on the 5'-end of specific TaqMan® DNA probe can be emitted and detected by an end-point optical detection module while the 5'-nucleotide of the probe is cleaved and released by the Taq DNA polymerase during the extension phase of PCR amplification process. Experimental setup of the microfluidic chip and the end-point optical detection module is shown in Fig. 3. The module is composed of a PMT device (Hamamatsu, Japan), a mercury lamp (Nikon Corp., Japan), a set of optical components with three fluorescence filters (Nikon B-2A, Nikon Corp., Japan), and one objective (Nikon Corp., Japan). With this approach, highly sensitive detection of genetic deletion can be achieved and the genetic disease can be rapid diagnosis in a shorter period of time by utilizing this end-point optical detection module in the miniature system.

RESULTS AND DISCUSSION

Characterization of the microfluidic modules

The normally-closed pneumatic micropump proposed in the current study is designed for sample delivery and transportation with a high pumping rate. Figure 4a shows the relationship between the flow pumping rate and the driving frequency (f_d) of EMVs at an air pressure of 20 psi. The time-phased expansion of successive PDMS membranes along the microfluidic channel can generate a peristaltic effect to drive the fluids moving forwards. The normally-close valve of the micropump is a PDMS-based floating block structure located inside the sample flow microchannel, which can be activated by hydraulic pressure caused by the peristaltic activation of the PDMS

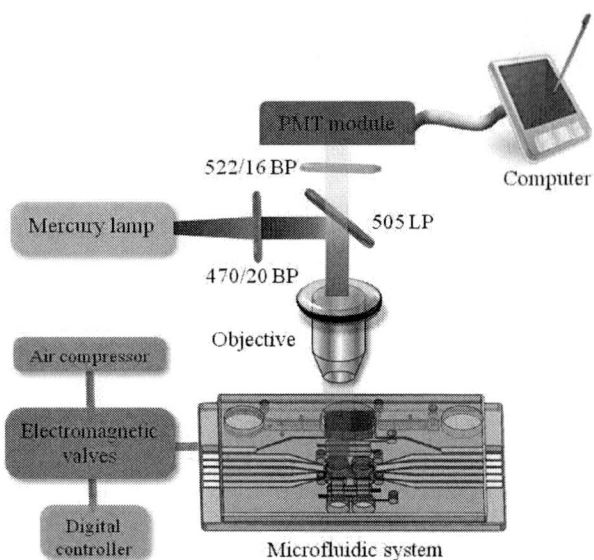

Figure 3: Experimental setup of the optical detection module and the microfluidic chip.

membranes. The valve can be used to increase pumping rates since it can be utilized to prevent the back flow. Three sets of pneumatic micropumps including a pneumatic micropump without a valve in the single microchannel, a micropump with a normally-closed valve in the single microchannel and a normally-closed micropump in the multi-channels are tested. It can been seen that a high pumping rate of 866 µl/min can be achieved for the micropump integrated with a normally-closed valve (f_d=81.7 Hz) while the maximum pumping rate is only 288 µl/min (f_d =34.8 Hz) for the micropump without a valve, both for the single microfluidic channel. The micropump with a higher pumping rate can be realized by integrating multiple microfluidic channels. A pumping rate of 1381 µl/min (f_d =72 Hz) can be achieved for multi-channel layouts. In addition, the built-in microcoils array can generate a magnetic field to attract the magnetic beads bound with target gDNA during the purification process. A separation test is then performed by flowing 4.5-µm DNA-bound magnetic beads through the extraction chamber while a DC current is applied through the microcoils array. Magnetic beads are observed to move freely when the DC current is not applied on the microcoils (Fig. 4b-1, t= 0 s). When a magnetic field with a magnitude up to 65 gauss is generated by the microcoils array at an applied current of 150 mA, the collection of beads is achieved. The magnetic beads are then successfully collected on the bottom of the extraction chamber (Fig. 4b-2, t= 35 s).

gDNA extraction

The capability of gDNA extraction from saliva is first demonstrated in the developed microfluidic system. The extraction of gDNA from saliva is performed. Two methods, including a manual operation of the magnetic particles for gDNA extraction and an automatic extraction process using the magnetic particles in the developed microfluidic system, are performed and compared. The yield of extracted gDNA is then verified by measuring optical intensity (OD) of UV absorbance at 260 nm and 280 nm utilizing a spectrophotometer. A high-quality

gDNA with an average concentration of 52.36 ng/µl can be isolated and purified automatically from 100 µl of saliva samples in the proposed microfluidic system. It reveals that the microfluidic system provides an excellent potential for automatic extraction of gDNA from the clinical saliva while compared with the manual extraction of gDNA (48.96 ng/µl). The entire extraction procedure in the microfluidic system can be completed within 15 minutes in an automatic fashion while the manual methods may need more than 45 minutes. Hence, the proposed microfluidic system can perform high-efficient purification of gDNA from the saliva sample without any invasive venepuncture. The high-purity gDNA can be further used for the subsequent biomedical applications.

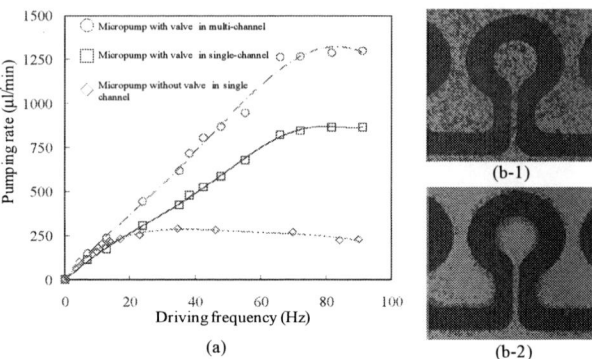

Figure 4: (a) The pumping rate of the microfluidic system; (b) the DNA-bound magnetic beads collected by the circular microcoils array (b-1) t=0 s; (b-2) t=35 s.

Table 1: Comparison of two methods for gDNA extraction

Method	Saliva volume (µl)	Extracted gDNA conc. (ng/µl)	OD$_{260/280}$	Total gDNA (µg)
Manual extraction using magnetic beads	100	48.96	1.80	4.896
On-chip extraction using the integrated system	100	52.36	1.83	5.236

Detection of α-thalassemia deletion

Genetic deletion associated with genetic diseases is explored by using the proposed microfluidic system. Alpha-thalassemia is a genetic disorder resulting from impaired or absent α-globin production. Genetic deletion of this gene has been investigated to verify the capability of the proposed microfluidic system in the current study. Two sets of primers specific for α-thalassemia-1 with SEA-type deletion including S1/S2 (for the detection of wild-type α-globin gene alleles) and S1/S3 (for the detection of α-thalassemia-1 with SEA-type deletion). A specific fluorescence-labeled TaqMan® DNA probe is used to verify the genetic deletion of α-thalassemia. For the cases without the SEA-type deletion, only normal α-globin gene clusters would be presented and amplified with the primer set S1/S2 whereas no signals are produced from the primer set S1/S3. Two types of patients including a normal case (lane 3-4, 7-8) and an abnormal case with carriers of α-thalassemia (lane 1-2, 5-6) are tested using the on-chip PCR module, followed by analyzing the amplified PCR products utilizing traditional slab-gel electrophoresis (Fig. 5a) and the end-point optical detection module (Fig. 5b), respectively. The experimental results show that the specific florescent signals with a wavelength of 508 nm can be emitted and detected by the PMT module while the filter of the objective is on (t=30 sec). From the comparable results, patients with genetic deletion can be rapid diagnosed automatically with a high sensitivity and specificity using the developed microfluidic system. The detection limit is measured to be 12 pg/μl. In addition, the entire diagnostic procedure including PCR and the optical detection of genetic deletion can be completed within 45 minutes in an automatic fashion.

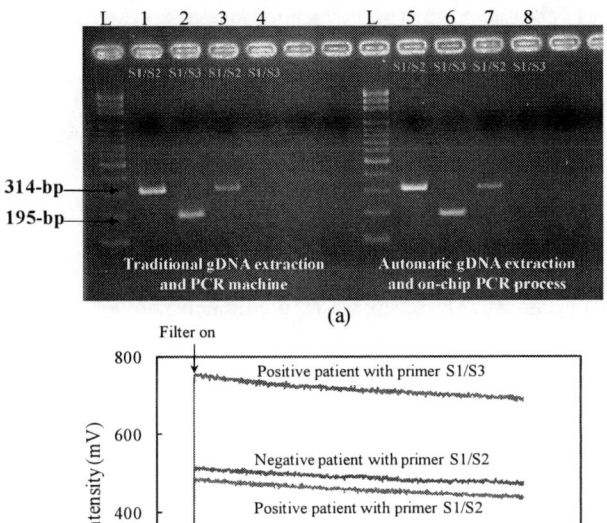

Figure 5: (a-1) Slab-gel electropherograms of positive patients detected by using the traditional methods and the microfluidic system. (a-2) The optical signals of the positive/negative patients with α-thalassemia deletion.

CONCLUSIONS

The study demonstrated a new magnetic-bead-based microfluidic system for rapid gDNA extraction, fast nucleic acid amplification and on-line optical detection of genetic diseases. The gDNA can be successfully purified and enriched from saliva in an automatic fashion by using the magnetic beads and the microfluidic module. The fast nucleic acid amplification utilizing the on-chip PCR module can be also performed. The developed micro-system may provide a powerful tool for automatic gDNA extraction and fast diagnosis of genetic diseases.

ACKNOWLEGEMENTS

The authors would like to thank financial support from the National Science Council in Taiwan (NSC 96-2120-M-006-008). The authors would also like to acknowledge Dr. Tsung-Min Hsieh for his assistance on the digital micro controllers and for discussion on the design of micro PCR module.

REFERENCES

[1] D. R. Reyes, D. Iossifidis, P. A. Auroux, A. Manz, "Micro total analysis systems 1. Introduction, theory, and technology," *Anal. Chem.*, vol. 74, pp. 2623-2636, 2002.

[2] C. D. Chin, V. Linder, S. K. Sia, "Lab-on-a-chip devices for global health: past studies and future opportunities," *Lab Chip*, vol. 7, pp. 41-57, 2007.

[3] S. M. Aljanabi, I. Martinez, "Universal and rapid salt-extraction of high genomic DNA for PCR-based techniques," *Nucleic Acids Res.*, vol. 25, pp. 4692-4693, 1997.

[4] S. Godreuil, M. -N. Didelot, C. Perez, A. Leflèche, P. Boiron, J. Reynes, F. Laurent, H. Jean-Pierre, H. Marchandin, "*Nocardia veterana* Isolated from Ascitic Fluid of a Patient with Human Immunodeficiency Virus Infection," *J. Clin. Microbiol.*, vol. 41(6), pp. 2768-2773, 2003.

[5] D. P. K. Ng, D. Koh, S. G. L. Choo, . V. Ng, Q. Fu, "The use of phosphate buffered saline for the recovery of cells and spermatozoa from swabs," *Clinica Chimica Acta*, vol. 343, pp. 191-194, 2004.

[6] D. Quinque, R. Kittler, M. Kayser, M. Stoneking, I. Nasidze, "Evaluation of saliva as a source of human DNA for population and association studies," *Anal. Biochem.*, vol. 353, pp. 272-277, 2006.

[7] C. F. Streckfus, L. R. Bigler, "Saliva as a diagnostic fluid," *Oral Dis.*, vol. 8, pp. 69-76, 2002.

[8] M. Fuentes, C. Mateo, A. Rodriguez, M. Casqueiro, J. C. Tercero, H. H. Riese, R. Fern´andez-Lafuente, J. M. Guis´an, "Detecting minimal traces of DNA using DNA covalently attached to superparamagnetic nanoparticles and direct PCR-ELISA," *Biosens. Bioelectron.*, vol. 21, pp. 1574-1580, 2006.

[9] K. -Y. Lien, C. -J. Liu, G. -B. Lee, "Magnetic bead-based microfluidic systems for detection of genetic diseases," in *Digest Tech. Papers MEMS '08 Conference*, Tucson, January 13-17, 2008, pp. 66-69.

A MICROFLUIDIC AFFINITY COCAINE SENSOR

John P. Hilton[1], ThaiHuu Nguyen[1], Renjun Pei[2], Milan Stojanovic[2], and Qiao Lin[1]
[1]Department of Mechanical Engineering,
[2]Department of Medicine, Columbia University, New York, USA

ABSTRACT

We present a novel microfluidic sensor that is capable of detecting trace cocaine concentrations and is fully regenerable at modest temperatures. The sensor exploits affinity aptamers (synthetic DNA/RNA oligonucleotides), labeled with a fluorophore and immobilized onto polymer microbeads as a highly sensitive cocaine receptor medium. The device demonstrates the capability of detecting native cocaine concentrations as low as 100 pM, with an eight-orders-of-magnitude dynamic range. By way of enriching rarefied cocaine samples, this detection limit is further improved to 10 pM. Additionally, the device utilizes temperature-dependent reversibility of aptamer-analyte binding to regenerate the sensor surface at a relatively modest temperature (40°C). This feature allows repeated reuse of the sensor without loss in functionality.

1. INTRODUCTION

Detection of trace amounts of illicit compounds such as cocaine is important in the fields of law enforcement and clinical medicine. Conventional approaches to cocaine detection include gas chromatography-mass spectrometry (GC-MS), and presumptive testing using reagents such as cobalt thiocyanate. Unfortunately, these methods are often expensive to operate, time consuming and can be non-specific. GC-MS, for example, offers high sensitivity but is costly due to complicated equipment and often time-prohibitive as a result of the high level of expertise required for operation. In addition, GC-MS systems generally utilize physisorption media for cocaine recognition such as hydrophobic or ion-exchange gels, which possess the negative side-effect of extracting impure compounds with similar physical or chemical properties as the target [1]. Presumptive testing is a very non-specific colorimetric process in which a reagent is exposed to a possible sample of cocaine [2]. This type of test produces only a discrete yes/no signal in response to a high concentration of cocaine. In addition, cobalt thiocyanate is known to produce false positives in response to materials such as lidocaine, a local anesthetic, which is similar to cocaine in structure and functionality. An alternative approach to cocaine detection is the use of affinity aptamer receptors as specific sensing molecules.

Aptamers are synthetic oligonucleotides (DNA or RNA) that bind to chemical and biological analyte targets via affinity interactions. Through an *in vitro* generation process, aptamers can be developed for a variety of analytes, such as small molecules, proteins, and cells. Aptamers exhibit high target selectivity by using secondary conformational structures that specifically fit the target. Interestingly, aptamer-target binding is also reversible and depends strongly on external stimuli such as pH and temperature [3]. Furthermore, aptamers exhibit advantages over standard affinity receptors such as antibodies, which are developed in vivo and therefore suffer from batch-to-batch variability depending on host immunogenicity and analyte/host compatibility. In this regard, they offer enhanced long-term stability, relatively straight forward synthesis, and the capability of readily modifiable end-chains to facilitate labeling or immobilization.

Conventional aptamer-based sensors (aptasensors) have been developed for cocaine using a variety of transduction methods, including electrochemical and optical sensing [4, 5]. However, these sensors generally produce excessive reagent waste while requiring bulky and expensive equipment to operate. Efficient MEMS-based cocaine sensors have been developed, which reduce sample and reagent consumption and allow smaller platform systems. Unfortunately, such devices rely on presumptive testing indicators that suffer from the inherent limitation of non-specificity, or antibodies, which can be undesirable receptors as mentioned above [2, 6]. No cocaine sensors have been developed to date that incorporate both the benefits of miniaturization and high specificity aptamer-based analyte recognition.

Here, a microfluidic cocaine aptasensor is demonstrated which utilizes a signal-on fluorescence scheme (requires no cocaine labeling) to detect low concentrations of cocaine. The device surfaces are functionalized with a cocaine-specific DNA aptamer that selectively binds native cocaine. The device employs thermally induced reversibility of aptamer-cocaine binding to controllably release cocaine, which occurs at 38 °C. Since this process is reversible, extended reuse of the device is allowed. In addition, device operation is simplified as detection and release occur in a homogeneous medium without altering solution composition or polarity. Although demonstrated with cocaine, this platform is applicable to other illicit substances as well.

2. PRINCIPLE AND DESIGN

Microfluidic Design.

The device consists of a microchamber packed with aptamer-immobilized microbeads for cocaine recognition, a microheater and temperature sensor for thermally induced cocaine release, and microchannels for introduction and removal of solutions [7]. The hexagonal 1 μL sample chamber is accessible via 3 inlet/outlet channels, two of which are equipped with passive weirs for retaining the microbeads (Fig. 1). The third inlet/outlet is used for bead introduction and permanently sealed before testing begins to ensure the retention of beads in the sample chamber. On-chip heaters and sensors lie directly beneath and around the microchamber. These are placed so as to create a uniform temperature distribution inside the chamber while providing accurate temperature sensing at the chamber center.

Signal-On Fluorescence Mechanism.

The signal-on fluorescence scheme in our device employs carboxyfluorescein (FAM), which is

978-1-4244-2977-6/09 $25.00 © 2009 IEEE

characterized by a peak absorption wavelength of 492 nm (blue light) and a peak emission wavelength of 520 nm (green light). Signal-on techniques are advantageous since a fluorescently labeled aptamer-quencher pair, free from the analyte, is used for signal transduction rather than labeling the analyte directly (Fig. 2). Quenchers such as dabcyl are chromophores which reduce the fluorescence emission of fluorophores like FAM when brought into close proximity. In sensors that use surface-immobilized aptamers, the quencher can be placed on a short complementary oligonucleotide. Initially this short oligonucleotide is hybridized to the aptamer, bringing the quencher into close proximity with the fluorophore and reducing visible fluorescence (Fig. 2b). Upon exposure of the aptamer to a target molecule, the stronger affinity of the aptamer to its target results in competitive displacement of the complementary strand, resulting in an increase in visible fluorescence (Fig. 2c) [8].

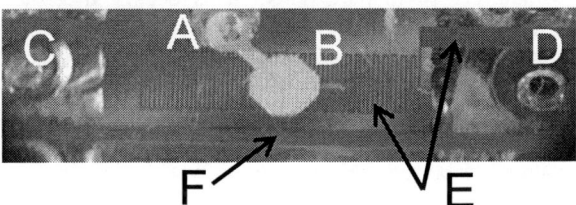

Figure 1: Photo of a packaged device. One inlet is used for bead insertion (A), while weirs (not visible) prevent the beads from flowing out of the chamber (B) through the sample inlets and outlets (C, D). An on-chip heater (E) and sensor (F) ensure uniform heating and accurate temperature sensing.

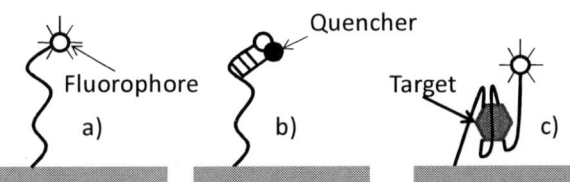

Figure 2: The fluorescent sensing principle. A fluorophore-modified aptamer is immobilized to the sensing surface (a). Quencher-labeled complementary strands are introduced, resulting in a loss of fluorescence (b). Introduction of target molecules causes competitive displacement and an increase in fluorescence (c).

3. EXPERIMENTAL

Materials and Equipment.

Biotinylated, FAM-labeled cocaine aptamers and dabcyl-labeled complementary strands are purchased from Integrated DNA Technologies. Streptavidin-coated beads, made of a co-polymer of bis-acrylamide and azlactone, are purchased from Pierce Biotechnology. Samples are diluted in a buffer containing pH 7.4 Tris-HCL, 150mM NaCl, 5mM KCl, and 2mM $MgCl_2$. Fluorescence measurements are conducted using an inverted fluorescence microscope (Nikon Diaphot 300).

Device Fabrication.

The device is fabricated according to a previously described approach [7]. Briefly, the microfluidic chamber and channels are defined by utilizing soft lithography.

SU-8 photoresist is spun onto a silicon wafer, exposed, and developed into a mold for PDMS casting. PDMS pre-polymer is cast onto this mold and cured before it is used for bonding. Meanwhile, a chrome/gold bilayer is thermally evaporated onto glass slides and patterned to create the on-chip heaters and temperature sensors. The slides are then bonded to a thin PDMS layer, which serves as an electrical passivation layer between the on-chip electronics and the microfluidic channels. Bonding to the PDMS channels follows. Having fabricated the device, the sensors are characterized in a convection oven and the chambers are packed with beads prior to experimentation.

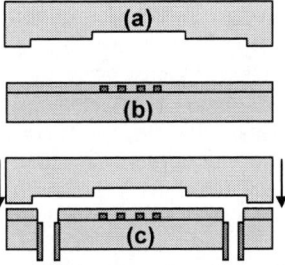

Figure 3: Device fabrication flowchart. PDMS channels (a) are defined using an SU-8 mold, gold heaters and sensors are patterned onto a glass slide and passivated (b), and the two pieces are then bonded and packaged (c).

Experimental Procedure.

Testing on the sensor followed a defined procedure. In each case, after bead packing is completed and the bead inlet/outlet is sealed, the device is initially washed with buffer solution (10 µL, all following experimental washes similar). Five microliters of 2 µM aptamer solution are then flowed into the chamber and allowed to incubate with the beads. Streptavidin-biotin attachment occurs within seconds, resulting in a very bright image under fluorescent excitation. Having immobilized the aptamers, 5 µL of 5 µM quencher solution is flowed into the chamber and allowed to incubate, resulting in a decrease in fluorescence. This procedure is repeated as necessary to achieve a fully quenched system. Having quenched the chamber, the sensor is now ready for cocaine sample introduction. In tests of the sensor's ability to respond to a particular cocaine concentration, protocol consisted of introduction of the cocaine, followed by an incubation period, followed by a wash. This procedure is additionally followed for tests of a control substance. When testing the ability of the sensor to concentrate cocaine, the wash step was omitted.

A photo is taken during each experimental step for fluorescence analysis. Fluorescence excitation during the data collection steps is completed rapidly to minimize the effects of photobleaching. This was achieved by opening the excitation light source only when taking a photo, and closing it immediately afterward. The sensor was otherwise kept in a dark environment. Fig. 4 indicates the effect of photobleaching on the fluorescence response of the sensor. This data was collected by introducing aptamers to the microbeads and taking digital photos over a 6 minute period. We observed negligible reduction of fluorescence signal during this time period, suggesting little effect from photobleaching.

The value of brightness intensity for every data point

978-1-4244-2977-6/09 $25.00 © 2009 IEEE

is normalized against the maximum and minimum readings using a dimensionless parameter, phi. Phi is equal to the difference of the measured fluorescence f and the minimum fluorescence f_{min} (measured during quenching), divided by the different between the minimum and maximum fluorescence, f_{max} (measured after aptamer incubation), for the sensor. Hence, $\Phi = (f - f_{max}) / (f_{max} - f_{min})$.

4. RESULTS AND DISCUSSION

We initially characterized the time-dependent response of the sensor to cocaine introduction to understand the kinetic behavior of the cocaine-aptamer system. Specifically, a sample of buffer solution containing 100 μM cocaine was introduced to a quenched sensor and the sensor response was measured over a period of eight minutes (Fig. 5). The observed fluorescence intensity with respect to time appears to reach an asymptotic value after six minutes, indicating that the time constant for signal response is roughly four minutes. For experimental purposes, all subsequent tests required a cocaine sample incubation period of four minutes to ensure that thorough binding of cocaine to the sensor had occurred before acquiring fluorescence data.

We then tested the sensor's response to 100 pM, 100 nM, 100 μM, and 10 mM concentrations of cocaine (Fig. 6). In addition, the sensor's response to a negative control substance, deoxycholic acid (DCA) was also tested in varying concentrations (1 pM, 1 nM, 1μM, and 1 mM).

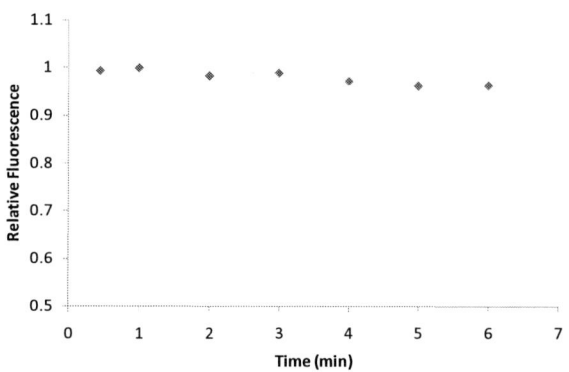

Figure 4: Example of the effect of photobleaching on fluorescent intensity. This graph indicates negligible change in fluorescent intensity after coating microbeads with fluorophore-labeled aptamers.

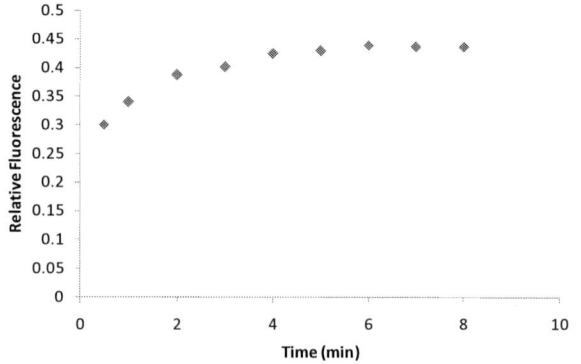

Figure 5: Transient response of the sensor to introduction of cocaine solution.

The sensor responds linearly as the concentration of cocaine is increased from 100 pM to 10 mM and indicates an 8 order of magnitude dynamic range. The sensor responds minimally to incubation with DCA. Although the lowest detected value of pure cocaine was 100 pM, we concentration (as demonstrated below). This result indicated that the sensor successfully detects cocaine at practically and physiologically relevant levels and compares well with reported cocaine aptasensors [4, 5], [9].

To further increase the sensitivity of the device and lower the detection limit, we attempted to concentrate rarefied cocaine solutions through continuous injection of a dilute sample into the sensing chamber. Specifically, 10 pM cocaine solution was introduced to the chamber continuously at 10 μL/min using a syringe pump (Fig. 7). At each measured time interval, fluorescence intensity was measured, which increased until a saturation value of approximately 0.19 relative units. This occurs after 95 minutes and approximately corresponds to a single injection signal generated by a 1 nM cocaine sample. Hence, a concentration factor of approximately 100 is determined for this experiment.

In addition to specific sensing, it may be critically important for captured cocaine to be properly released for subsequent analysis. This is accomplished in our device by thermally activated release and isocratic elution of analytes from aptamer-functionalized surfaces. To demonstrate this, a 100 μM cocaine sample was injected into the sensor which results in a relatively high initial fluorescence signal. Following this, the on-chip heaters and sensors were used to heat the sample intermittently from room temperature to approximately 38°C while pure buffer was being flowing through the sensor at 10 μL/min (Fig. 8). As the temperature is raised, a sharp increase in relative fluorescence occurs at approximately 33°C as the aptamer begins to release the cocaine and leftover quencher strands. Signal increase continues until fluorescence saturates at approximately 38°C. This

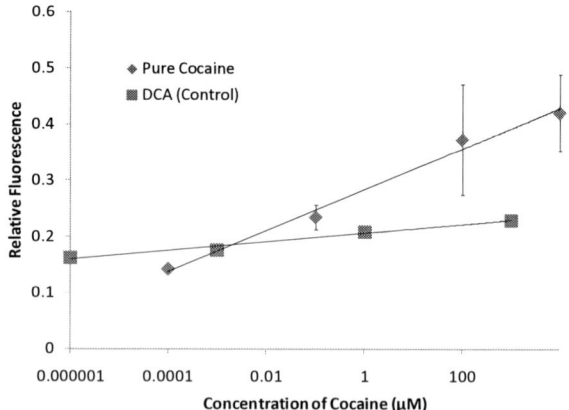

Figure 6: Fluorescence Response to Concentration of Cocaine and a control molecule, Deoxycholic Acid. Results indicate eight orders of magnitude of linear response.

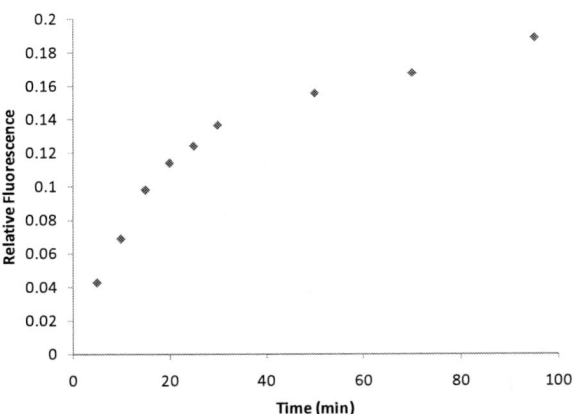

Figure 7: Relative Fluorescence vs. time for continuous flow of 10 pM cocaine. Fluorescence intensity indicates concentration of 10 pM cocaine to a fluorescence level corresponding to greater than 1 nM cocaine.

indicates that the sensor can be heated to a relatively modest temperature (38°C) to effectively release cocaine using a single homogeneous phase.

Thermal release of cocaine from the sensor surface additionally regenerates the surface functionality and renders the device reusable, a practically beneficial attribute. Regeneration of our device is demonstrated by subjecting the sensor to consecutive detection and thermal release cycles (Fig. 9). During each detection cycle, a distinct cocaine concentration is used to elucidate the effects of sample variation on device performance. In each case, the sensor achieves maximum fluorescence before use, minimum fluorescence after quenching, and an appropriate level of fluorescence (corresponding to Fig. 5) after exposure to a particular concentration of cocaine. This indicates that the sensor performance is consistent following regeneration and is independent of the concentration of cocaine being tested. Unlike previous microfluidic cocaine sensors [2, 6], mild heating and washing grants our device repeated use without loss of functionality.

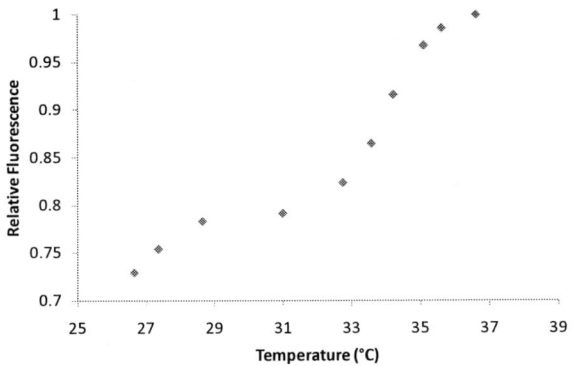

Figure 8: Relative Fluorescence during sensor regeneration as a function of temperature. On-chip heaters were used to heat the sample while flowing buffer at 10uL/min. On-chip sensor measured temperature as resistance, which was converted to temperature measurements.

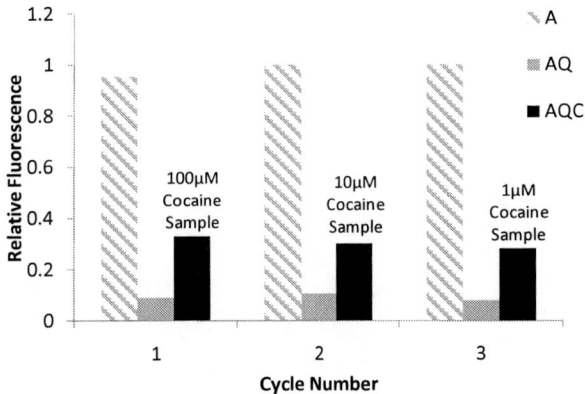

Figure 9: Relative Fluorescence for 3 regeneration cycles, each consisting of aptamers only (A), quencher aptamers (AQ), and the addition of cocaine (AQC). The cycles are separated by heating at approximately 40°C, causing the cocaine and quencher to break off of the aptamer.

5. CONCLUSION

A novel microfluidic aptasensor has been demonstrated that can detect physiologically relevant concentrations of cocaine. The sensor has a detection limit of 10 pM and a linear range of 8 orders of magnitude. In addition, the device is fully regenerable for extended reuse.

ACKNOWLEDGMENTS

We gratefully acknowledge financial support from the National Science Foundation (Grant Nos. CBET-0693274, ECCS-0707748, and NSF-0324845).

REFERENCES

[1] Washton, A. and Gold, M., eds. *Cocaine: A Clinician's*. The Guilford Press, New York. 1987.

[2] Bell, S. and Hanes, R. A Microfluidic Device for Presumptive. *J. Forensic Sci.* vol 52, 2007.

[3] S. Jayasena, "Aptamers: an emerging class," *Clin. Chem.*, vol. 45, pp. 1628-1650, 1999.

[4] Baker, B. R., R. Y. Lai, et al. (2006). An electronic, aptamer-based small-molecule sensor for the rapid, *J. Am. Chem. Soc* 128(10): 3138-3139.

[5] Li, Y., H. L. Qi, et al. (2007). "Electrogenerated chemiluminescence aptamer-based biosensor *Electrochemistry Communications* 9(10): 2571-2575.

[6] Frisk, T. et. al. A micromachined interface for airborne sample-to-liquid *Lab Chip*, 6, 2006.

[7] Nguyen, T., R. Pei, et al. (2007). Programmed Affinity Extraction of Molecules on a Microfluidic Platform. *Int. Conf. Nano/Micro Engineered and Molecular Systems (NEMS '07)*, Bangkok, Thailand.

[8] Nutiu, R. and Y. Li (2003). "Structure-Switching Signaling" *J. Am. Chem. Soc.* 125(16): 4771-4778.

[9] Javaid, J, et. al. Cocaine Plasma Concentration: Relation to Physiological and Subjective Effects in Humans. *Science*, Vol. 202, 1978.

DEVELOPMENT OF NEW BIOCHEMICAL IC CHIP-SET FOR REAL-TIME PCR

Koji Ikuta, Toshio Sasao, Yuya Okuda, Kenji Watamura and Masashi Ikeuchi
Department of Micro/Nano systems Engineering, Nagoya University, Aichi, Japan

ABSTRACT

We have developed new biochemical IC chip-set which realizes real-time PCR in the finger top size. Unlike conventional μ-TAS approach, our chip-set has great versatility and portability since all chips are designed in standardized micro-architecture and the fabrication process can be combined with conventional MEMS processes. Combining this chip-set with existing Biochemical IC chip-set family, researchers can easily develop order-made palm-top system for proteomics research, on-site diagnosis, and tailor-made medicine.

INTRODUCTION

In recent years, the infection of SARS, Bird Flu and other infectious diseases have been spreading over the world. To prevent the pandemic of serious diseases, rapid detection and on-site analysis of the infection is highly required. The key technology for rapid diagnosis of infection is real-time PCR. Real-time PCR is a technique which is used to amplify and simultaneously quantify the target DNA. Since it enables DNA amplification and simultaneous quantification by comparing the fluorescence signal from the product to a standard curve calibrated by PCR of a known DNA, we can diagnose more rapidly compared to conventional PCR. There is a problem, however, that real-time PCR apparatuses are so large to carry out on-site diagnosis. Even the recently advancing lab-on-chip and μ-TAS technologies cannot be applied directly to the on-site diagnosis, because they only focus on the miniaturization of specific function –e.g. reaction and detection [1], and not the whole device. Therefore, miniaturization of the whole system for real-time PCR is strongly needed.

BIOCHEMICAL IC CHIPS

To overcome the above issue, we have been proposing and developing Biochemical IC chips. The Biochemical IC, which is proposed by Ikuta [2-4] in 1993, is a group of standardized micro chemical device with various functions for biological experiments. Each chip has a specific function, and arbitrary micro-systems for biological analysis can be realized by assembling these chips (Figure 1). To date, we have developed chips for reaction [5], concentration [6], pumping [7], homogenization [8], cell-free protein synthesis [9], protein analysis (protein separation and collection) etc.

Recently we have developed "μ-Total Protein Analysis System" by combining these chips, and we can do the experiment from cell homogenization to collection of cell protein in the finger top [10]. These applications are the necessary techniques for the post-genome research such as proteomics research.

In this paper, development of a new Biochemical IC chip-set for real-time PCR is reported. Using the chip-set, accurate temperature control over an hour with high reproducibility was verified, and we have successfully amplified specific DNA and simultaneously quantified the products by fluorescence intensity with a CCD camera.

CONCEPT OF REAL-TIME PCR CHIP-SET

The following is the required specifications to design new Biochemical IC chip-set for real-time PCR.

(1) A precise and high speed temperature control which can realize real-time PCR thermal cycling
(2) Highly compatible and extensible structure suitable for module concept
(3) Detection of the fluorescence at multiple points in the reactor chip to realize a number of real-time PCR reactions in parallel.
(4) Fluorescence excitation and detection

Here is the schematic of Real-time PCR chip-set concept (Figure 2).

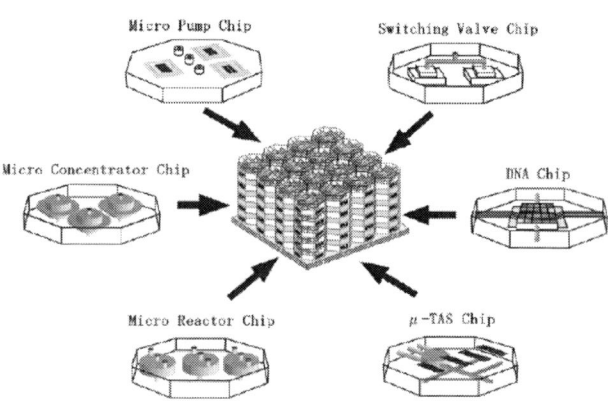

Figure 1: Module Concept of Biochemical IC

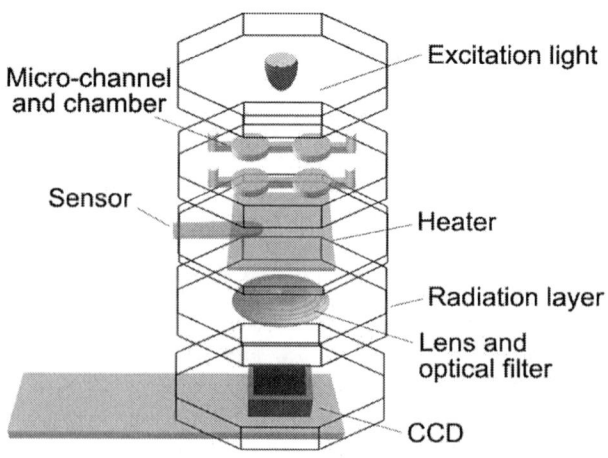

Figure 2: Concept of Real-time PCR chip-set

978-1-4244-2977-6/09 $25.00 © 2009 IEEE   348

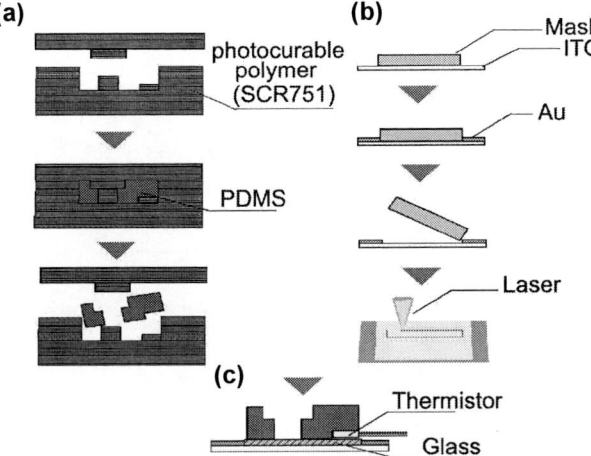

Figure 3: Fabrication process of the real-time PCR chip-set
(a) Fabrication of the micro-reactor
(b) Fabrication of the temperature control module
(c) Assembly of the two parts by PDMS bonding

We adopt a transparent conductive film since it can transmit the excitation light and fluorescence. Besides, it can work as a heater at the same time. A micro-reactor that has micro-channels and chambers is used to fulfill the 2nd and the 3rd specification stated above. To realize the 4th specification, Blue LED is set above the micro-reactor as an excitation light source. The fluorescence from the marker specific to double strand DNA in the sample is focused by an objective lens, and detected by a CCD sensor through a band-pass optical filter.

FABRICATION OF THE REAL-TIME PCR CHIP-SET

The micro-reactor must have PCR compatibility and transparency, and the temperature control module should be capable of controlling PCR thermal cycling for a number of times. We adopted PDMS as a material of the micro-reactor since it is widely used in biochemical and medical devices, and the more, it has fine transparency. The fabrication process of the micro-reactor and the temperature control module are described below (Figure 3).

Fabrication process of the micro-reactor
(Figure 3 (a))
(1) Fabricate molds made of photocurable polymer (SCR751, D-MEC) by microstereolithography and heat it for 12 hours at 200℃, then perform fluorine coating
  (dip CTX-109AE (ASAHI GLASS) diluted by CT-Solv. 100E (ASAHI GLASS), and heat for 10 min at 80℃, then dip again and heat for 1 hour at 180℃)
(2) Mix PDMS (SYLGARD 184 SILICON ELASTOMER, TORAY/DOW CORNING) by BASE: CURING AGENT =10:1 and inject the premix into the mold
(3) Heat for 8 hours at 80℃ after removing bubbles *in vacuo*

Fabrication of the temperature control module
(Figure 3 (b), (c))
(1) Cut the ITO (Indium tin oxide coated PET sheet, 35Ω/sq surface resistivity, Sigma-Aldrich.Inc) to the size of 6(mm)×13(mm)
(2) Place a stencil mask on the ITO to pattern the electrodes
(3) Fabricate a pair of electrodes by sputtering Au on the ITO/PET substrate
(4) Remove the mask
(5) Pattern the ITO layer by excimer laser (the line width is about 25 μm)

Assembly of the micro-reactor and the temperature control module
(1) Coat premixed PDMS on the surface of a cover glass (the thickness is about 0.15 (mm)) and combined with the micro-reactor by heating at 80℃ for 8 hours
(2) Attach the lower surface of the cover glass and the ITO using PDMS
(3) Remove the protective layer of the thermistor (104JT, ISHIZUKA ELECTORONICS) to improve the thermal response. A fine enamel wire is used to connect the thermistor and the voltmeter to decrease the heat capacity of the thermistor module. The module is then inserted into a preformed space for the thermistor in the micro-reactor, and enclosed by PDMS.
(4) Insert (3) into a holder to complete the real-time PCR chip-set.

Prototype of real-time PCR chip-set
The Real-time PCR chip-set developed consists of a PCR chip and a spacer chip (Figure 4). PCR chip is an assembly of a holder chip, a temperature control module and a micro-reactor made of PDMS.

The temperature control module has an ITO/PET substrate as a heater and a thermistor for temperature measurement. The ITO allows DNA amplification, irradiation of the excitation light, and real-time detection of

Figure 4: Fabricated Real-time PCR chip-set
(a) Assembly in a holder unit (side view)
(b) Assembly in a holder unit (top view)
(c) PCR chip (d) Spacer chip

978-1-4244-2977-6/09 $25.00 © 2009 IEEE

fluorescence simultaneously due to its transparency. In addition, since we can observe from its top or bottom, there is a lot of flexibility to design optical system and it is easier to miniaturize the whole system.

TEMPERATURE CONTROL OF THE MICRO-REACTOR

The PCR chip and the spacer chip fabricated by the above process were set in the holder unit. We controlled PCR thermal cycle using this chip-set and measured the temperature on the ITO surface by thermograph. As a result, we confirmed that the temperature gradient is so steep for heat concentration at the center of the heater when we use the ITO formed on the PET uniformly. The temperature gradient becomes steeper as temperature rises. PCR needs to repeat the specified cycle for the DNA amplification, so this problem made it difficult to perform PCR in the micro-reactor and calibration was also difficult since a slight displacement causes big differences of temperature.

Improvement of the temperature distribution

To solve the problem, the ITO layer was patterned using excimer laser to make a uniform temperature distribution (Figure 5). As a result, we verified the uniform temperature distribution and successfully extended the PCR effective area.

In addition, we improved the structure of the spacer chip and modified the temperature distribution. The spacer chip fixes the temperature control module and the micro-reactor at a prescribed distance. Micro-holes are patterned on the spacer chip to correct the difference of heat conductivity in the temperature control module caused by the thermistor and homogenize the temperature distribution in the micro-reactor.

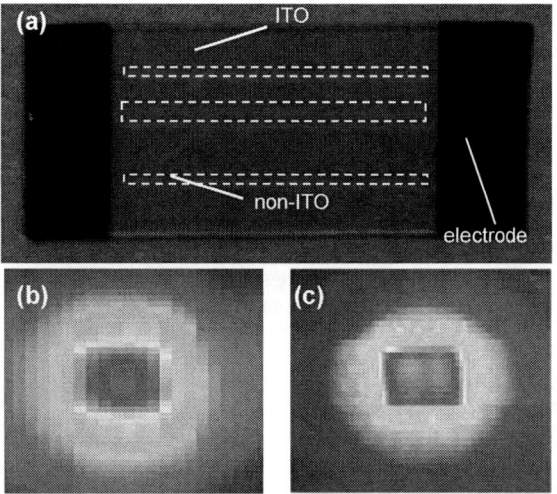

Figure 5: (a) Patterned ITO/PET substrate (b) Thermograph of the ITO/PET substrate before excimer laser patterning. (c) Thermograph of the ITO/PET substrate after excimer laser patterning

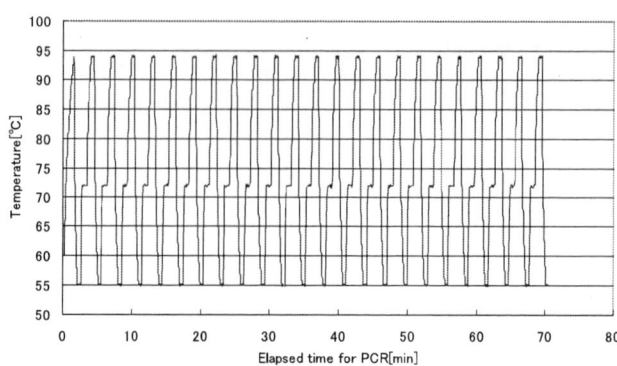

Figure 6: Temperature transition of the PCR cycle using the developed temperature control module

Verification of the uniform temperature distribution using the patterned ITO

To verify the temperature control using the chip-set, temperature cycles of real-time PCR (94℃・55℃・72℃) is demonstrated. Labview was used for PCR thermal cycle control. The results showed accurate temperature control over an hour with high reproducibility (Figure 6)

EXPERIMENTAL VERIFICATIONS OF REAL-TIME PCR FUNCTION

Experimental methods

We performed real-time PCR using the developed chip-set for the verification. We adopted the intercalater method for real-time PCR. The intercalater is a molecule that produces fluorescence when it combined with double-strand DNA. By adding the intercalater to the sample solution, the amount of the amplified unknown DNA can be estimated by comparing the measured fluorescence intensity and a standard curve calibrated by PCR of a known DNA.

Here, we performed 40 cycles of real-time PCR by assembling these chips in the holder unit, adding the intercalater to PCR sample, and injecting mineral oil over sample for preventing evaporation of sample.

And we obtained fluorescence images of the micro-reactor at each cycle (72℃・30sec) with a CCD camera to record the transition of fluorescence intensity. The excitation light was irradiated using a conventional fluorescence microscope instead of Blue LED. After the PCR reaction finished, the sample was taken out and electrophoresed using agarose gel. The fluorescence of the bands was observed using a transilluminator.

Experimental conditions

・ Sample: 3(μl)

  intercalater (SYBR-Green II , TAKARA BIO): F primer 2μM: R primer 2μM: template DNA 2g/l: sterile water =1.8:1:1:1:1.2

・ Mineral oil: 3(μl)

  mineral oil for molecular biology, light oil (SIGMA)

・ Fluorescence reagent for electrophoresis

  EnVISION DNA as Loading Buffer (Ameresco Inc)

・ Excitation light / Florescence: Blue / Green

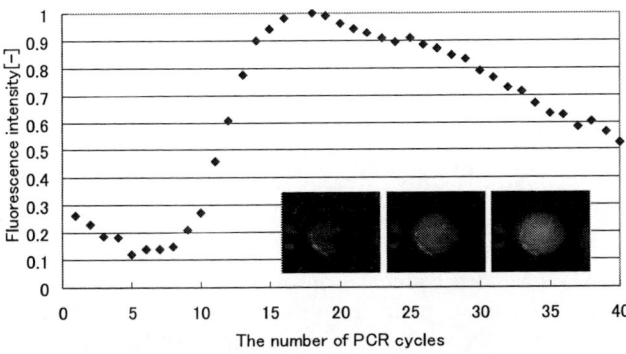

Figure 7: Transition of the fluorescence intensity during real-time PCR using the developed chip-set (the inset shows the CCD image of 8th, 12th and 16th cycle)

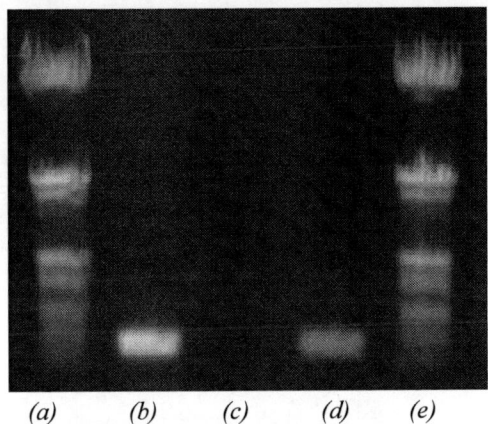

Figure 8: Electrophoresis of the products (a,e) marker (b) positive control (c) negative control (d) real-time PCR chip-set

PCR thermal cycle was as shown below.
94 ℃ (5sec)→[55 ℃ (30sec)→72 ℃ (45sec)→94 ℃ (30sec)]×40→72℃ (5min)

The result of real-time PCR

Figure 7 shows the transition of fluorescence signal from the micro-reactor. The fluorescence intensity increased in proportion to DNA amplification until it reached to the depletion of PCR resources at 16th cycle. Figure 8 shows the result of electrophoresis of the final products. Figure 8(b) shows positive control obtained by the conventional PCR apparatus and Figure 8(d) shows the result of the developed chip-set. The DNA band was detected at the same position as positive control. These results proved successful amplification of the target DNA and simultaneous quantification of the products in the new real-time PCR chip-set.

CONCLUSIONS

We have developed new Biochemical IC chip-set for real-time PCR. And we demonstrated successful amplification of the target DNA and simultaneous quantification of the products in the new real-time PCR chip-set. This chip-set has great versatility and portability since all chips are designed in standardized micro-architecture and the fabrication process can be combined with conventional MEMS processes. This chip-set together with other Biochemical IC family should be a powerful tool for on-site rapid detection of infections, tailor-made medicine and proteomics research.

REFERENCES

[1] S.-K.Hsiung and G.-B.Lee, "A controllable micro-lens structure for bio analytical applications," Proc.of MEMS07, pp.763-766 (2007)

[2] K.Ikuta, K.Hirowatari and T.Ogata, "Tree dimensional micro integrated fluid system (MIFS) fabricated by stereolithography," Proc.of MEMS 94, 1-6 (1994)

[3] K.Ikuta, T.Ogata, M.Tsuboi, S.Kojima, "Development of mass productive micro stereolithography (Mass-IH process)," Proc.of MEMS 96, pp.301-306 (1996)

[4] K.Ikuta, "3D micro integrated fluid system toward living LSI-Challenge for artificial cellular device-,"vArtificial Life V (MIT press), 17-24 (1997)

[5] K.Ikuta, S.Maruo, Y.Fukaya and T.Fujisawa, "Biochemical IC chip toward cell free DNA protein synthesis," Proc.of MEMS 98,131-136 (1998)

[6] K.Ikuta, S.Maruo, T.Fujisawa and A.Yamada, "Micro Concentrator with Opto-sense Micro Reactor for Biochemical IC Chip Family -3D Composite Structure and Experimental Verification -," Proc.of MEMS 99,376-381 (1999)

[7] K.Ikuta, T.Hasegawa, T.Adachi and S.Maruo, "Fluid drive chips containing multiple pumps and switching valve for biochemical IC chip family - Development of SMA drive 3D micro pumps and valves in leak-free polymer package -," Proc.of MEMS 2000,739-744 (2000)

[8] K.Ikuta, A.Talahashi, K,Ikeda and S.Maruo, "Fully integrated micro biochemical laboratory using Biochemical IC Chips - Cell-free protein synthesis by using a built-in micropump chip -," Proc.of MEMS 03,451-454 (2003)

[9] K.Ikuta, Y.Sasaki, H.Maegawa, and S.Maruo, "Biochemical IC Chip for pretreatment in biochemical experiments – Micro ultrasonic homogenizer chip made by hybrid microstereolithography -," Proc. of MEMS 03,343-346 (2003)

[10] K.Ikuta, N.Satake, T.Ohashi and M.Shibata, "Finger-top Total Protein Analysis System based on New Biochemical IC Chips-," Proc of MEMS 08, (2008)

A MEMS SENSOR FOR CONTINUOUS MONITORING OF GLUCOSE IN SUBCUTANEOUS TISSUE

X. Huang[1], S. Li[2], J. Schultz[3], Q. Wang[2], and Q. Lin[1]

[1]Mechanical Engineering Department, Columbia University, New York, NY, USA
[2]Chemistry and Biochemistry Department, University of South Carolina, Columbia, SC, USA
[3]Bioengineering Department, University of California, Riverside, CA, USA

ABSTRACT

We present a MEMS sensor for continuous glucose monitoring for diabetes management. The device consists of a microcantilever, which is driven by remote magnetic field and situated in a microchamber separated from the sensing environment by a semi-permeable membrane. As glucose concentration varies, viscosity changes induced by glucose/copolymer binding in a poly(acrylamide-*ran*-3-acrylamidophenylboronic acid) (PAA-*ran*-PAAPBA) copolymer solution produce a measurable change in cantilever vibration. The device has been used to measure physiologically relevant glucose concentrations from 0 to 324 mg/dL. The response time of the sensor to glucose concentration changes was 3 minutes and can be further improved with optimized device designs.

INTRODUCTION

Continuous glucose monitoring system (CGM) is highly desirable for diabetes treatment. This is commonly achieved by subcutaneously implanted enzymatic electrochemical sensors such as, the MiniMed Paradigm, Freestyle Navigator CGMS, and DexCom STS CGMS. These FDA approved commercial products detect glucose by enzyme-catalyzed reactions. While electrochemical methods allow sensitive glucose detection, they also incur significant drawbacks, such as irreversible glucose consumption, drift from hydrogen peroxide production, and interference from electrode-eroding chemicals. As a result, electrochemical CGM sensors generally exhibit large drifts over time, and require frequent calibration, making long term operation difficult [1].

Miniaturized sensors, based on MEMS technology, have provided low-cost non-invasive or minimally-invasive glucose monitoring based on methods including electrochemistry, impedance, hydrogel swelling, optics, glucose binding protein, and synthetic glucose-responsive polymers [2]. We previously reported a MEMS glucose detecting device [3] adopting concanavalin A, a glucose-binding protein which suffered from immunogenicity and cytotoxicity [4]. In addition, due to material limitations and osmotic pressure imbalances, it showed limited mechanical reliability, poor reversibility, and significant drift. Here, we present a new MEMS affinity glucose sensor which exploits a biocompatible synthetic polymer for glucose detection and parylene-based construction to address problems such as immunogenicity, cytotoxicity, mechanical reliability, and reversibility. Additionally, due to its MEMS-based fabrication, the sensor is amenable to subcutaneous implantation for long-term, stable CGM applications.

PRINCIPLE AND DESIGN

Device design and operational principle. The MEMS viscometric glucose sensor is based on a cantilever situated inside a microchamber (Fig.1). The parylene cantilever is anchored onto the substrate at one end, and suspended over a cavity. A permalloy thin film is electroplated onto the cantilever at its free end, and subsequently passivated by an additional parylene layer. The microchamber is formed between the substrate and a cellulose acetate (CA) semi-permeable membrane, and is filled with a polymer solution for glucose sensing. The membrane is semi-permeable so that environmentally present glucose can permeate through the membrane and bind with the polymer inside the microchamber. This interaction between the glucose and the polymer increases the viscosity of the fluid in the microchamber, thereby damping cantilever vibration. As a result, decreased cantilever vibration amplitude along with a vibration phase shift occurs.

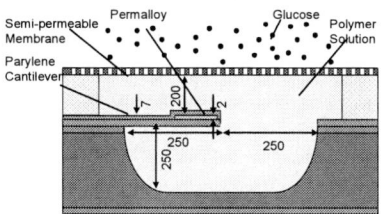

Fig. 1. *Schematic of the MEMS viscometric glucose sensor.*

This microcantilever is actuated by an electromagnetic field-producing solenoid. The electromagnetic field generates a torque on the magnetized permalloy film with a magnitude proportional to the product of permalloy volume, magnetic field, and magnetization along the length of the permalloy. This torque, distributed along the width of the cantilever, causes cantilever bending. Thus, a time-dependent electromagnetic field generated by an AC voltage actuated solenoid produces a time-dependent torque, leading to the vibration of the cantilever. This vibration is directly related to the solution viscosity. Therefore, by measuring the damped vibration, the viscosity of the sensing solution can be determined.

Glucose sensitive copolymer. The polymer sensing solution consists of PAA-*ran*-PAAPBA copolymer which is a stable and biocompatible glucose sensitive system compared to those conventionally-based on reactive proteins and enzymes. The boronic acid moiety in the polymer forms reversible, strong bonds with glucose, resulting in crosslinking of the polymer and an increase in viscosity (**Fig.2**). The biocompatible poly(acrylamide) (PAA) segment in this synthesized copolymer was chosen to improve the solubility of the sensing segment PAAPBA. Furthermore, PAA provides additional neighbor stabilizing effects besides the hydrogen bonding effect (e.g. the boronic acid moiety and amide coordination via B-N

978-1-4244-2977-6/09 $25.00 © 2009 IEEE

interaction), which could enhance the binding of boronic acid to carbohydrates.

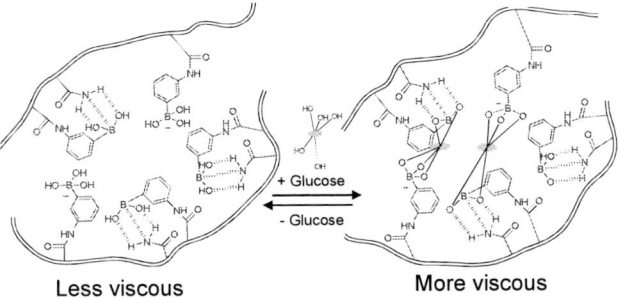

Fig. 2. *The sensing principle of the biocompatible, glucose-sensitive copolymer PAA-ran-PAAPBA.*

FABRICATION

The device fabrication process started with etching an anchor layer onto a silicon dioxide (SiO₂)-coated wafer to increase the adhesion between the wafer and the parylene layer (**Fig. 3**). After the deposition of a 5 μm parylene layer, a 100 nm copper layer was deposited as a seed layer for electroplating. S1818 photoresist was then spin-coated and patterned to form the electroplating area. Permalloy was then electroplated to a thickness of 1.5 μm at a deposition rate of approximately 150 nm/min for ten minutes with a 50 mA pulse current. This was followed by the removal of the photoresist and the copper seed layer. The second parylene layer, with a thickness of 2 μm, was then deposited to seal the permalloy. Next, the parylene layer was patterned to form a cantilever (250 μm × 250 μm). The exposed SiO₂ layer was removed by HF wet etching. Gas-phase XeF₂ etching was then used to release the cantilever by etching the silicon directly underneath it forming a cavity approximately 250 μm in height. The SiO₂ beneath the cantilever was then removed via another HF wet etch. The microcantilever was then fixed in a poly(dimethylsiloxane) (PDMS) microchamber and sealed by a CA semi-permeable membrane (MWCO 3500) forming the sensing chamber (30 μL). Another PDMS microchamber, for introduction of glucose samples, was attached on top of the CA membrane and functioned as a test cell. The inlet and outlet for the test cell and microchamber were provided via adhesive-affixed Teflon tubes (**Fig. 4**)

EXPERIMENTAL METHOD AND SETUP

Materials. 1 M glucose solution was obtained by dissolving glucose (1.8 g) in 10 mL of Phosphate Buffered Saline (PBS). A series of glucose concentrations (27 mg/dL, 54 mg/dL, 108 mg/dL, 216 mg/dL, and 324 mg/dL) were prepared by diluting the 1 M glucose with PBS. 1.9% PAAPBA was prepared using a method in [5].

Experimental method. Experiments were performed according to two methods. The first method was referred to as "premixed measurement", in which the microchamber and test cell were each filled with premixed samples of glucose and copolymer at the same concentration. The second method, referred to as "permeation measurement," was intended to simulate the environment of an implanted glucose sensor. The glucose concentration in the microchamber was changed according to physiologically relevant glucose levels, while the copolymer concentration

inside the test cell remained unchanged. The glucose concentration in the test chamber and microchamber reached equilibrium due to diffusive transport of glucose across the permeable membrane.

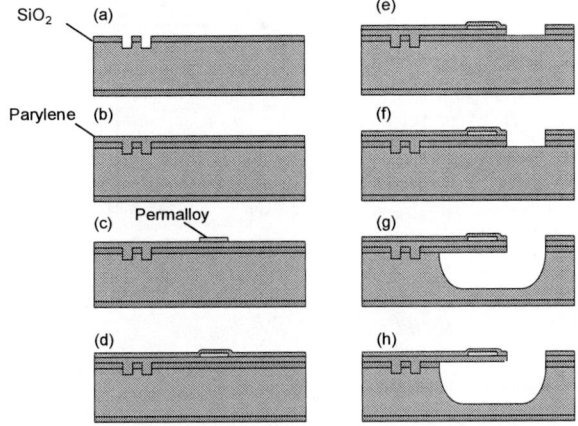

Fig. 3. *Fabrication process for the sensor chip: (a) Anchor layer etching; (b) First parylene layer deposition; (c) Permalloy electroplating; (d) Second parylene layer deposition; (e) Parylene and permalloy layer patterning; (f) SO2 layer around the cantilever removal to open a window for Si etching; (g) XeF₂ dry etching; (h) SO₂ layer beneath the cantilever removal.*

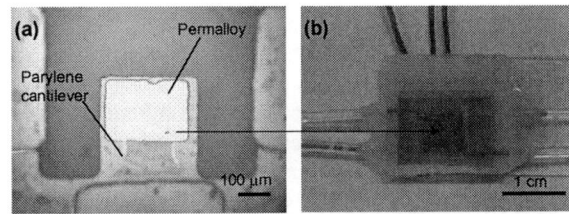

Fig. 4. *Images of the MEMS sensor: (a) before, and (b) after packaging.*

The experimental setup is shown in **Fig. 5** [3]. The cantilever vibration was driven by a home-made solenoid (2000 turns of a 200 um diameter copper wire on a 2.5 um diameter plastic core), which under a driving voltage of 5 V$_{rms}$, produced a magnetic field strength of about 500 kA/m perpendicular to the cantilever surface. A permanent magnet with field strength of 500 A/m was placed parallel to the cantilever surface to magnetize the permalloy film. The vibration of the cantilever was detected by an optical-lever.

Ultra-thin kapton heat-film was used to affix a K-type thermal couple to the bottom of the device. The thermocouple was connected to a multimeter (Agilent 34420A Nano Volt/Micro Ohm meter) to obtain temperature measurements. These values were then transmitted to a computer to control the voltage output of the DC power supply (Agilent E3631A DC power supply) connected to the heater. All experiments were conducted at 37°C with closed-loop temperature control. This temperature simulated a suitable in-vivo glucose monitoring environment.

978-1-4244-2977-6/09 $25.00 © 2009 IEEE

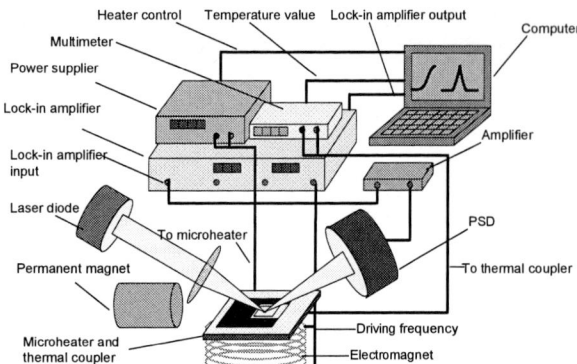

Fig. 5. *Experimental setup for characterization of the MEMS glucose sensor. The cantilever vibration was measured with an optical lever system.*

EXPERIMENTAL RESULTS

Time response. To characterize the basic measuring function of our device and obtain the time constant of its response, the chamber of the device was initially filled with PBS buffer, and was then exposed to 108 mg/dL glucose solution. The cantilever vibration amplitude, at a fixed frequency (28 Hz), was obtained over time (**Fig. 6**). We observed a gradual decrease in solution viscosity inside the test cell, corresponding to glucose permeation through the membrane followed by copolymer binding. Amplitude response leveled as glucose concentration between the microchamber and test cell equilibrated. The time constant, which represented the time consumption for glucose permeation into the chamber and equilibrium binding with the copolymer, was then determined to be 3 minutes. It was adequate for CGM applications [6], considering a 5-15 minutes detection period for current commercial CGM products.

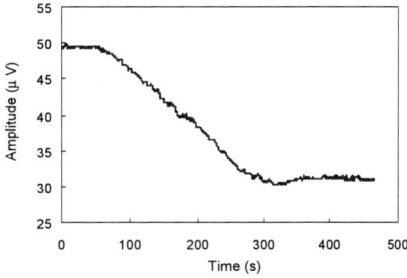

Fig. 6. *Time history of the cantilever vibration amplitude at 28 Hz.*

Evaluation of equilibrium binding through the membrane. To study the efficiency of molecular permeation through the membrane, an experiment comparing the cantilever response in premixed measurements and permeation measurements was carried out. The premixed measurement involved placing samples consisting of the same glucose and copolymer concentration (108 mg/dL glucose solution and 1.9% copolymer) in both the microchamber and the test cell. In the permeation measurement, the glucose concentration was initially zero in the microchamber and 108mg/dL in the test cell. The glucose in the test cell would then penetrate though the membrane and bind with the copolymer in the microchamber until the concentrations reached equilibrium. The resulting cantilever vibration

responses during premixed and permeation conditions were compared (**Fig. 7**). The resonance frequency was 27 Hz in premixed measurements and 27.2 Hz in permeation measurements, while the vibration amplitude (from 10 Hz to 40 Hz) was almost identical. The likely cause of the negligible resonance frequency shift (0.2Hz) was inconsistencies during sample preparation. The good agreement between the premixed measurement and the permeation measurement indicated efficient glucose molecular transmembrane diffusion and interaction with the copolymer.

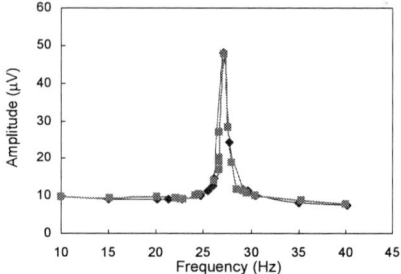

Fig. 7. *Comparison of the device's amplitude frequency response after glucose concentration was equilibrated between the two sides of the membrane (diamonds) with that from an experiment where glucose concentration was set to 108 mg/dL from the start (squares).*

Sensor response at varying glucose concentrations. To model the measurement of physiological glucose concentration in interstitial fluid (ISF), harmonic cantilever vibrations were measured at varying glucose concentrations (27 mg/dL, 54 mg/dL, 108 mg/dL, 216 mg/dL, and 324 mg/dL) using the permeation experimental protocol (**Fig. 8**). These experiments compared the signal changes in vibration and phase spectrum of each glucose solution. As the glucose concentration increased from 27 to 324 mg/dL, the vibration amplitude decreased accordingly with an observed total reduction of nearly 70%. This was accompanied by a shift of vibration resonance frequency from 27.54 to 26.77 Hz and a drop of Q-factor from approximately 29 to 7, which indicated a significant increase in vibrational damping and hence, viscosity of the polymer solution. Moreover, there was a significant change in the phase shift for the cantilever vibration (**Fig. 9**). For example, at 10 Hz, the phase shift increased from 2.2° at 27 mg/dL to 28.3° at 324 mg/dL. The results of harmonic vibration measurements demonstrated the capability of measuring different glucose solutions by recording either the phase shift or vibration amplitude at a fixed frequency where different glucose solutions are most distinguished.

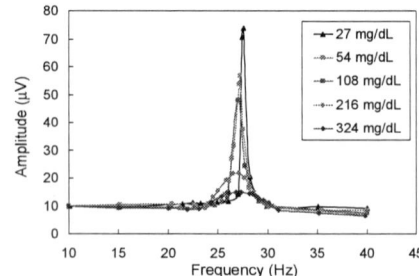

Fig. 8. *Frequency dependent amplitude of the cantilever vibration at physiologically relevant glucose concentrations.*

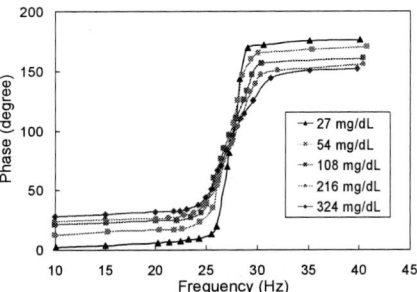

Fig. 9. Frequency dependent phase shift of the cantilever vibration at physiologically relevant glucose concentrations.

Reversibility. The reversibility of the sensor determines the potential of our device for long-term glucose monitoring without repeated calibration. We tested and observed reversibility of the device with respect to glucose concentration changes by alternatively measuring 0 and 108 mg/dL glucose solutions (**Fig. 10**). The measured vibration amplitude at 28 Hz repeatedly alternated between 37 and 43 μV. This result reflected an excellent reversibility of the device, indicating its ability for long-term continuous monitoring of glucose in subcutaneous tissue free of laborious recalibration.

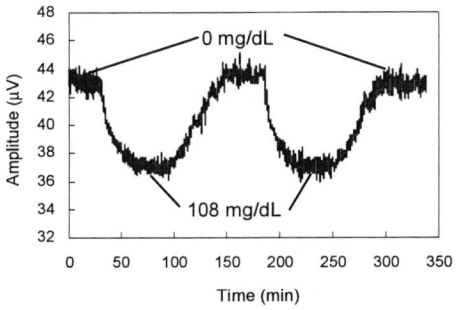

Fig. 10. Reversibility of the MEMS sensor to glucose concentration changes. (The noise shown reflects environmental disturbances to the optical setup.)

Drift. Drift is also a common obstacle with continuous glucose monitoring devices. Here, we assessed the drift in the device as it was exposed to glucose (108 mg/dL) over a long duration of time (**Fig. 11**). We observed a consistent vibration amplitude measurement of 37 V and found that there was virtually no drift of this measured signal over a preliminary measurement period of 5 hours. Compared with our previous system [3], this result indicated a significant improvement and highlighted the excellent stability of the current device.

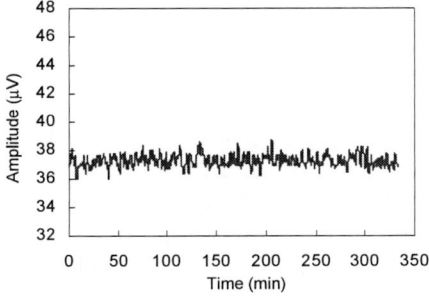

Fig. 11. evaluation of drift for the MEMS glucose sensor.

CONCLUSIONS

In this paper, a MEMS viscometer utilizing a novel biocompatible glucose-responsive copolymer is presented. The device, consisting of a magnetically driven parylene microcantilever coated with a permalloy thin film, is located in a PDMS microfluidic chamber. The sensing fluid, consisting of PAAPBA copolymer, exchanges glucose with the fluid outside the device through a CA semi-permeable membrane. Glucose concentration can be determined by detecting viscosity changes induced by the binding of glucose to PAAPBA, which affects the vibration characteristics of the cantilever. The dynamic characteristics of the device were characterized with the cantilever vibration in PBS buffer and 108 mg/dL glucose solution at 28Hz. The device time response was also found to be 3 minutes, comparing well to previously reported glucose sensors. In addition, the device response to physiologically relevant glucose concentrations (27 mg/dL to 324 mg/dL) was obtained. Moreover, the device exhibited excellent reversibility and negligible drift during long experimental time (~5 hours). These measurements have shown that the MEMS viscometer using PAAPBA to specifically detect glucose has the potential for long-term, stable, continuous monitoring of ISF glucose.

ACKNOWLEAGEMENTS

We gratefully acknowledge financial support from NSF (grant # ECCS-0702101) and the Columbia Diabetes and Endocrinology Research Center (NIH grant # DK63068-05). X.H. has been supported in part by a National Scholarship from the China Scholarship Council.

REFERENCES

[1] J. C. Pickup et al., "In vivo glucose monitoring: the clinical reality and the promise", *Biosens. Bioelectron.*, 20:1897-1902, 2005.

[2] K. Kataoka et al., "Totally synthetic polymer gels responding to external glucose concentration: Their preparation and application to on-off regulation of insulin-release", *J.Am.Chem.Soc.*, 120: 12694-12695, 1998.

[3] Y. Zhao et al., "A MEMS viscometric sensor for continuous glucose monitoring", *J. Micromech. & Microeng.*, 17: 2528-2537, 2007.

[4] T. Kataoka et al., "Immunogenicity and amplifier cell production by tumor vaccines enhanced by concanavalin A", *Gann*, 73:193-205, 1982.

[5] S. Li et al., "Development of novel glucose sensing fluids with application to MEMS-based continuous glucose monitoring", *J. Diabetes Science & Tech.*, (in press)

[6] J. Reifman et al., "Predictive monitoring for Improved management of glucose levels", *J. Diabetes Science & Tech.*, 1: 478-486, 2007.

MONOLITHIC CENTRIFUGAL MICROFLUIDIC PLATFORM FOR BACTERIA CAPTURE AND CONCENTRATION, LYSIS, NUCLEIC-ACID AMPLIFICATION, AND REAL-TIME DETECTION

J.L. García-Cordero[1], I.K. Dimov[1,2], J. O'Grady[1,3], J. Ducrée[1], T. Barry[1,3], and A.J. Ricco[1]

[1]Biomedical Diagnostics Institute, Dublin City University, Dublin, IRELAND
[2]Department of Biomedical Engineering, Universidad de Valparaiso, CHILE
[3]Biomedical Diagnostics Institute Programme, National Centre for Biomedical Engineering Science, National University of Ireland, Galway, IRELAND

ABSTRACT

We report the design, fabrication, and characterization of a polymer centrifugal microfluidic system for the specific detection of bacterial pathogens. This single-cartridge platform integrates bacteria capture and concentration, supernatant solution removal, lysis, and nucleic-acid sequence-based amplification (NASBA) in a single unit. The unit is fabricated using multilayer lamination and consists of five different polymer layers. Bacteria capture and concentration are accomplished by sedimentation in five minutes. Centrifugation forces also drive the subsequent steps. A wax valve is integrated in the cartridge to enable high-speed centrifugation. Oil is used to prevent evaporation during reactions requiring thermal cycling. Device functionality was demonstrated by real-time detection of *E. coli* from a 200-μL sample.

INTRODUCTION

Rapid detection and precise identification of bacterial pathogens is critical in food and water monitoring and in clinical diagnostics [1, 2]. Detection becomes more challenging because of the low concentration of pathogens (10 – 100) in large-volume samples [2]. Traditional techniques such as visual microscopy, plating, or culture enrichment can require times in the order of days, require access to unique facilities (bio-hazard containment), and need skilled technicians who may have to perform many repetitive steps.

Detection assays based on nucleic acid amplification methods have made possible the precise identification of low numbers of pathogens in relative short periods of time (tens of minutes to hours) [3]. However, compared to traditional methods, there are more process steps involved to achieve sensitive detection of pathogens, such as lysis of the organism and in some cases nucleic acid purification. Also, reagents (lysis buffer, primers, and enzymes) are needed to conduct these reactions. Sample concentration may be required for the assay to be most effective and to reduce reagent consumption. In addition, processing involves manual handling, expensive instrumentation, and liquid transfer to different containers, which can reduce the limits of detection or lead to false positives. Thus, there is a need for an automated platform that integrates the different steps involved in nucleic-acid amplification-based assays.

Microfluidics and lab-on-a-chip (LOC) technologies are enabling the integration and automation of sophisticated diagnostic tests, traditionally carried out in clinical laboratories, into nearly-autonomous monolithic systems the size of a microscope slide or a credit card [2, 4-13]. These devices offer reductions in reagent and sample volumes, as well as shorter assay times.

Several microfluidic devices have been reported for the detection of bacteria [4-10] and viruses [11-14]. Magnetic microbeads coated with specific antibodies against the target organism are one popular technique to capture, isolate, and concentrate pathogens [4-13]. However, this technique adds costs to the assay because specific antibodies are needed for each target organism. Dielectrophoresis is another technique used to capture and concentrate bacteria; however, it has low throughput because low flow rates are needed to efficiently capture bacteria [6-7].

Operation of these devices requires the combination of electrophoretic and thermopneumatic pumps [4], whereas other microdevices require expensive syringe pumps and pneumatic lines to move liquids through the device [5-7, 10-12]. Centrifugal microfluidics offers inherent advantages over these and other microfluidic pumping and actuation mechanisms: it can pump liquids in a wide range of flow rates, mixing can be easily accomplished by alternating the direction of rotation, and valving can be implemented by carefully controlling the rotation speed [14].

Centrifugation is widely utilized in many laboratories to capture and isolate bacteria. Because most bacteria have a higher density than the media in which they are suspended, they can be sedimented and thus concentrated and captured. Our device uses similar principles, combining the advantages of centrifugal microfluidics to control liquid transportation and to implement liquid metering. Our monolithic system executes two basic operations: bacteria are sedimented in a container and concentrated; successive reaction steps (lysis, sample conditioning, and NASBA) that otherwise would require manual pipetting are integrated on the disc using a microfluidic channel that allows removal of unneeded materials.

DESIGN AND FABRICATION

Design

The design of the centrifugal microfluidic cartridge is shown in Figure 1. The device consists of a main chamber in the form of a funnel that holds a sample of 200 μL (scalable to hold larger volumes). The bottom of the funnel features a rectangular-shaped vessel where the various reactions are carried out. The volume of this vessel is 5μL, and as with the funnel, it can also be scaled with corresponding scaling of reagent consumption.

A 5-cm-long microfluidic channel connects the waste chamber to the main funnel-shaped chamber; its purpose is to remove reactants not needed and to define the vessel

978-1-4244-2977-6/09 $25.00 © 2009 IEEE

volume. The entrance to the microfluidic channel is initially sealed with wax. The waste chamber holds 100 µL more volume than the main chamber, which is needed for the subsequent steps where other reagents are loaded into the cartridge. The waste chamber features a hole that functions as a vent, equalizing it with atmospheric pressure. Alignment holes are included on each layer to facilitate alignment.

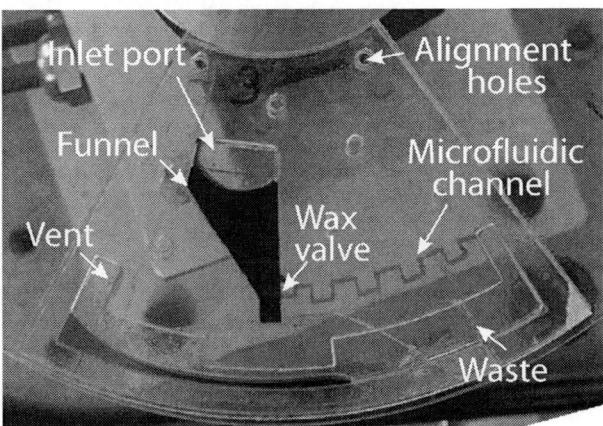

Figure 1: Centrifugal cartridge for the detection and analysis of bacteria.

Fabrication

Devices were fabricated using multi-layer lamination as shown in Figure 2. A CO_2 laser system (Laser Micromachining LightDeck, Optec, Belgium) was used to cut the polymer layers. To laminate the layers, a thermal roller (Titan-110, GBC Films, USA) was used. The funnel structure and waste camber were cut from a 1.2-mm thick poly(methylmethacrylate) sheet (GoodFellow, UK) previously laminated with an 80-µm thick layer of pressure-sensitive adhesive, PSA (AR9808, Adhesives Research, Ireland). The microfluidic channel, cut from the same PSA, had a measured width of approximately 400 µm.

All layers except the cover were carefully aligned and passed through the laminator. The valve was formed by manually loading a small amount of solid wax at the intersection of the microfluidic channel with the main chamber. The device was then warmed on a hot plate at 70 °C to melt the wax, which then filled the entrance of the channel by capillary action. Finally, the cover layer was laminated onto the device. The final assembled device was mounted on a transparent compact disc (Åmic, Sweden) using the same PSA (AR9808).

MATERIALS AND METHODS

Figure 3 presents the sequence of steps performed on the microfluidic cartridge. A brushless DC motor with an integrated optical encoder (Series 4490, Faulhaber, Switzerland) is used to rotate the disc through the various steps.

Sample is loaded in the main chamber and spun for 10 min at 6000 rpm to sediment the bacteria to the bottom of the funnel, Figure 3A. Although higher speeds could decrease the sedimentation time, the adhesive layer does not withstand pressures above this speed, and the funnel would eventually rupture from the bottom. Sealing of the

different layers with thermal lamination could prevent this problem.

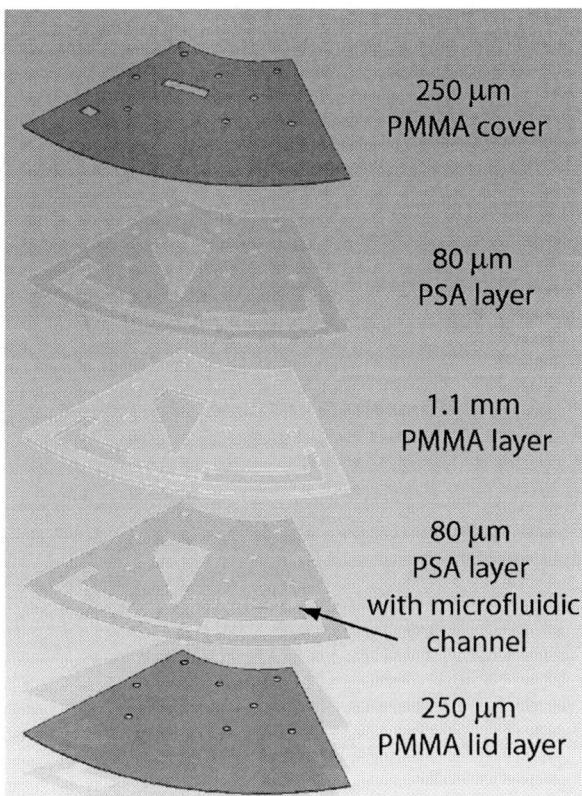

Figure 2: Five different plastic layers are used to build the centrifugal cartridge.

The disc is stopped and the wax valve opened by heating it to 70° C using a Thermofoil flexible heater (Minco, USA), Figure 3B. The disc is spun again at 700 rpm for 1 min to discharge the supernatant solution through the microfluidic channel into the waste chamber. The remaining 5 µL of liquid in the vessel includes a "pellet" that contains most of the bacteria, Figure 3C. Platform capture efficiency is characterized by applying different concentrations of *E. coli*, genetically modified to express green fluorescent protein (GFP), and assaying the captured bacteria by fluorescence.

Next, 10 µL of oil is introduced through the inlet port as shown in Figure 3D. The disc is spun and the oil covers the contents of the reaction vessel and also seals the microfluidic channel, preventing any evaporation during the various reactions steps.

Next, 20 µL of chemical lysis solution (microLysis Plus, Microzone, UK) are introduced through the inlet and, by spinning, the lysis reagent sinks and mixes with the bacteria pellet, Figure 3E. This mixture is thermally cycled as per the manufacturer's instructions (65°C: 15 min, 96°C: 2 min, 65°C: 4 min, 96°C: 1 min, 65°C: 1 min, 96°C: 30 sec) to lyse the bacteria, releasing nucleic acids including the RNA target. Thermal cycling also continuously mixes both solutions, Figure 3F.

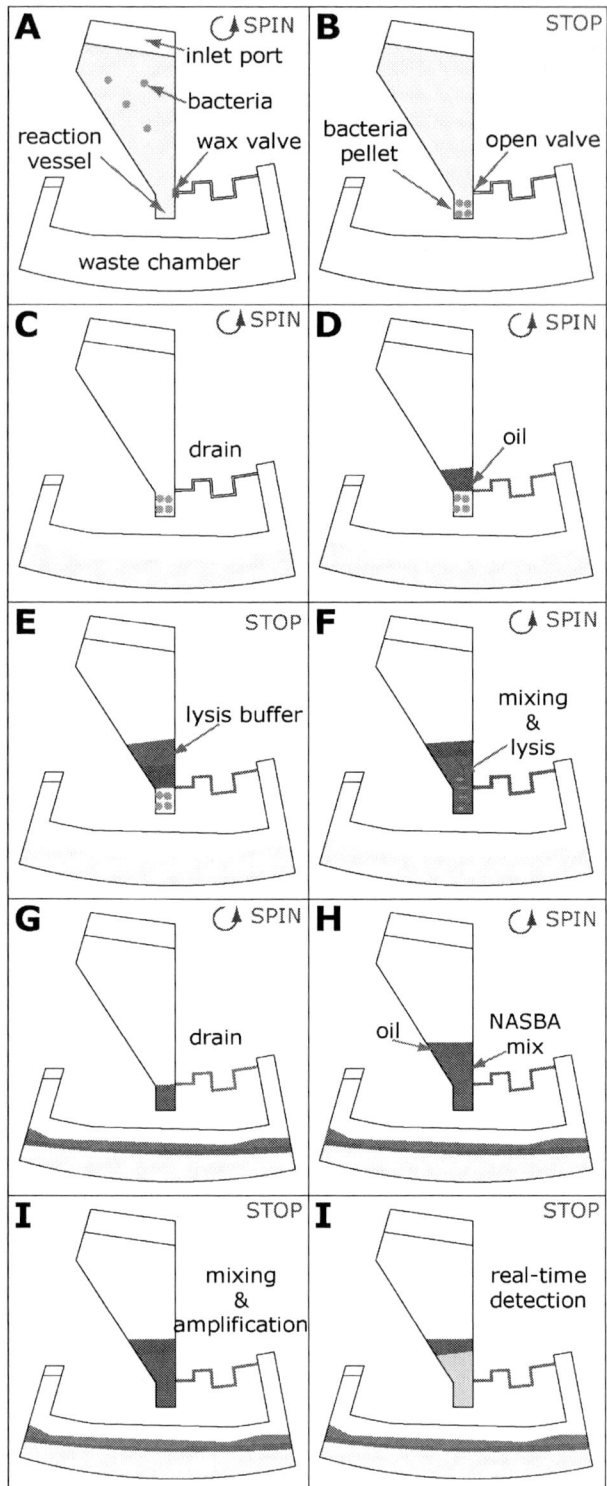

tmRNA target is in real-time via molecular beacons whose fluorescence is unquenched when hybridized to complementary RNA amplicons [15-16]. Figure 5 shows the NASBA-based detection of *E. coli* on this platform.

RESULTS AND DISCUSSION

Concentrations of GFP *E. coli* ranging from $10^1 – 10^8$ cfu/mL were prepared. Then, 100 μL of each concentration were loaded into a microwell plate. The plate was read with a fluorescent plate reader (Infinite 200, Tecan, USA), revealing a linear correlation of fluorescent intensity to bacteria concentration from $10^4 – 10^8$ bacteria/mL. These data enabled characterization of the bacteria capture efficiency of the funnel.

The density of *E. coli* is 1.088 g/cm^3, while that of the sample buffer was ~1.022 g/cm^3. This difference allows the sedimentation and concentration of bacteria. Microfluidic units were loaded with different concentrations, valves opened, and supernatant discarded into the waste chamber. 100 μL of supernatant were extracted from the waste chamber with a pipette, loaded into the microwell plate, and read with the fluorescent scanner.

To calculate the capture percentage efficiency of the devices, the supernatant concentration was subtracted from the initial concentration and divided by 100. Typically, more than 80% of the cells were captured in the compartment, as shown in Table 1. Experiments were repeated in triplicate. Figure 4 shows bacteria captured on one of the funnel from a solution containing 10^8 GFP *E.coli*/mL.

Table 1: Bacteria capture efficiency of the centrifugal microfluidic platform.

Initial Concentration (ml)	Supernatant Concentration (ml)	Capture percentage (%)
10^8	1.4×10^7	85
10^7	7.5×10^5	92
10^6	1.9×10^5	80

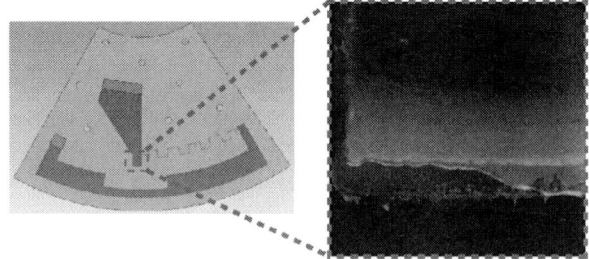

Figure 4: GFP E. coli *captured at the bottom of the funnel.*

The unit was characterized using water spiked with a concentration of 10^5 *E. coli*/mL, performing all the steps described in the Methods section. Figure 5 shows the NASBA-based detection of *E. coli* on this platform.

Figure 3: Steps involved in the detection of bacteria in the centrifugal microfluidic platform. See texts for details.

The disc is spun again, and some of the lysate goes into the waste, but 5 μL of the lysate remain in the vessel. The lysate is covered by a fresh layer of oil, and 5 μL of NASBA enzymes are introduced into the compartment by spinning and allowed to react with the lysate for 5 min at 61°C. Finally, 15 μL of NASBA mixture, including primers and molecular beacons for the target bacterial tmRNA fragment, are loaded into the platform and the solution incubated at 41°C. Detection of amplified

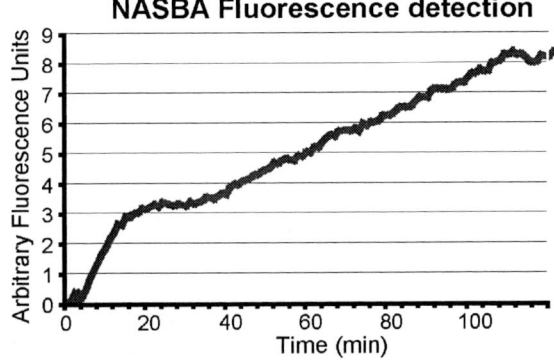

Figure 5: NASBA real-time detection of a positive sample containing 10^5 bacteria per mL.

CONCLUSIONS

We have demonstrated that the detection of bacteria can be accomplished in a centrifugal microfluidic cartridge. We integrated sample preparation and concentration, lysis of pathogens, and nucleic acid sequence-based amplification into a monolithic polymer platform that is inexpensive to manufacture. We are working to integrate reagent storage and laser-activated valves on this platform to enable a fully automated, self-contained unit to detect pathogenic bacteria.

ACKNOWLEDGMENTS

This work was supported by the Science Foundation Ireland under Grant No. 05/CE3/B754. We thank Majella Maher, Terry Smith, Paul Daly, and Claus Poulsen for helpful discussions and Niamh Gilmartin for providing GFP *E. coli*.

REFERENCES

[1] S. Bhattacharya, J. Jang, L. Yang, D. Akin, R. Bashir, "Biomems and nanotechnology-based approaches fro rapid detection of biological entities", *J Rapid Met Autom Microbiol*, vol 15, pp.1-32, 2007.

[2] A.K. Balasubramanian, K.A Soni, A. Beskok, S.D. Pillai, "A microfluidic device for continuous capture and concentration of microorganisms from potable water", *Lab Chip*, vol 7, pp.1315-1321, 2007.

[3] L.A. Jaykus, "Challenges to developing real-time methods to detect pathogens in food", *ASM News*, vol 69, pp.341-347, 2003.

[4] R.H. Liu, J. Yang, R. Lenigk, J. Bonanno, P. Grodzinski, "Self-contained, fully integrated biochip for sample preparation, polymerase chain reaction amplification, and DNA microarray detection", *Anal Chem*, vol 7, pp.1824-1831, 2004.

[5] E.T. Lagally, J.R. Scherer, R.G. Blazej, N.M. Toriello, B.A. Diep, M. Ramchandani, G.F. Sensabaugh, L.W. Riley, R.A. Mathies, "Integrated portable genetic analysis microsystem for pathogen/infectious disease detection", *Anal Chem*, vol. 76, pp. 3162-3170, 2004

[6] E.T. Lagally, S. Lee, H.T. Soh, "Integrated microsystem for dielectrophoretic cell concentration and genetic detection", *Lab Chip*, vol 5, pp.1053-1058, 2005

[7] B.H. Lapizco-Encinas, B. A. Simmons, E.B. Cummings, Y. Fintschenko, "Insulator-based dielectrophoresis for the selective concentration and separation of live bacteria in water", *Electrophoresis*, vol. 25, pp. 1695-1704, 2004

[8] C.G. Koh, W. Tan, M.Q. Zhao, A.J. Ricco, Z.H. Fan, "Integrating polymerase chain reaction, valving, and electrophoresis in a plastic device for bacterial detection", *Anal Chem*, vol. 75, pp. 4591-4598, 2003

[9] L.C. Waters, S.C. Jacobson, N. Kroutchinina, J. Khandurina, R.S. Foote, J.M. Ramsey, "Microchip device for cell lysis, multiplex PCR amplification, and electrophoretic sizing", *Anal Chem*, vol. 70, pp. 158-162, 1998

[10] Z. Chen, M.G. Mauk, J. Wang, W.R. Abrams, P.L. Corstjens, R.S. Niedbala, D. Malamud, H.H. Bau, "A microfluidic system for saliva-based detection of infectious disease", *Ann N Y Acad Sci*, vol. 1098, pp. 429-436, 2007

[11] R. Pal, M. Yang, R. Lin, B.N. Johnson, N. Srivastava, S.Z. Razzacki, K.J. Chomistek, D.C. Heldsinger, R.M. Haque, V.M. Ugaz, P.K. Thwar, Z. Chen, K. Alfano, M.B. Yim, M. Krishnan, A.O. Fuller, R.G. Larson, D.T. Burke, M.A. Burns, "An integrated microfluidic device for influenza and other genetic analyses", *Lab Chip*, vol 5, pp.1024-1032, 2005.

[12] C.J. Easey, J.M. Karlinsey, J.M. Bienvenyue, L.A. Legendre, M.G Roper, S.H. Feldman, M.A. Hughes, E.L. Hewlett, T.J. Merkel, J.P. Ferrance, J.P. Landers, "A fully integrated microfluidic genetic analysis system with sample-in-answer-out capability", *Proc Natl Acad Sci USA*, vol 103, pp19272-19277, 2006

[13] J. Pipper, M. Inoue, L. Ng, P. Neuzil, Y. Zhang, L. Novak, "Catching bird flu in a droplet", *Nat Met*, vol. 13, pp. 1259-1263, 2007

[14] J. Ducrée, S. Haeberle, S. Lutz, S. Pausch, F. von Stetten, R. Zengerle, "The centrifugal microfluidic platform", *J Micromech Microeng*, vol. 17, pp. S103-S115, 2007

[15] I. Dimov, J.L.Garcia-Cordero, J.O'Grady, C.R. Poulsen, C. Viguier, L. Kent, P. Daly, B. Lincoln, M. Maher, R. O'Kennedy, T.JSmith, A.J. Ricco, L.P. Lee, "Integrated microfluidic tmRNA purification and real-time NASBA device for molecular Diagnostics", *Lab Chip*, in press.

[16] J. O'Grady, S. Sedano-Balbas, M. Maher, T.J Smith, T. Barry, "Real-time PCR detection of *Listeria monocytogenes* in enriched food samples based on the ssRNA gene, a novel diagnostic target", *Food Microbiology*, vol. 25, pp. 75-84, 2008

IN VITRO DETECTION OF NEURAL ACTIVITY WITH VERTICALLY GROWN SINGLE PLATINUM NANOWIRE

D. Choi[1], A. Fung[2], H. Moon[3], G. Villareal[4], Y. Chen[2], D. Ho[5], N. Presser[6], G. Stupian[6], and M. Leung[6]

[1]University of Idaho at Moscow, Idaho, USA
[2]University of California at Los Angeles, California, USA
[3]University of Texas at Arlington, Texas, USA
[4]Galenea Corporation, Cambridge, Massachusetts, USA
[5]Northwestern University, Evanston, Illinois, USA
[6]The Aerospace Corporation, El Segundo, California, USA

ABSTRACT

We present a process for fabricating a nanoscale probing device for *in vitro* sensing of neural activity. Single vertical platinum (Pt) nanowires were fabricated on a microelectrode array by focused ion beam (FIB) – chemical vapor deposition (CVD) in order to improve the spatial resolution of recording and to minimize damage to the cells. Electrodes in contact with cells detected prominent spontaneous electrical activity. Such small geometry of vertical nanowires may enable probing multiple sites within a cell.

INTRODUCTION

The detection of electrical and chemical signaling within neural networks is critical in neuroscience [1-3]. Conventional approaches towards neurobiological studies, such as patch clamping and large probe insertion, have led to the basic understanding of neural activity as well as implanted neural detection systems. These approaches have been limited by their reliance on high operational power and laborious/invasive techniques that can induce cell damage [4, 5]. In many cases, the cell-probe interaction induces a cell stress response [6]. Patch clamping involves the formation of a seal between the cell membrane and the mouth of a pulled glass pipette electrode, and the subsequent rupture of the cell membrane. The mechanics of such an interface severely hinder simultaneous probing at multiple locations on a single cell, and limits the capability to monitor a population of cells. Furthermore, the cellular damage caused by whole-cell probing is unsuitable for chronic measurements. The modulation of intracellular physiology in turn precludes neural population studies based upon the inability to chronically monitor neural activity. Multi-electrode arrays (MEAs) are an alternative approach to electrophysiology that obtains parallel extracellular recordings from a distribution of micro-electrodes. However, the planar metal electrode does not approximate a physiological surface with which a cell may form an intimate interface. As the gap between the electrode and cell increases, the signal represents less a specific intracellular potential and more a delocalized field potential. The integration of high-aspect-ratio nanostructures in MEAs holds great potential to MAE with minimal invasive cell-electrode interface. We present here a vertical Pt nanowire probe fabricated by focused ion beam (FIB) and chemical vapor deposition (CVD). An array of these nanoprobes was used to record potential waveforms from cultured rat neuroblasts.

DESIGN AND FABRICATION

The arrays of interconnection microelectrodes were fabricated using conventional photolithography (Figure 1). A 4x4 array of electrodes with an area of 10 μm x 10 μm and a thickness of 10 nm Ti/100 nm Cr/50 nm Au and their metal interconnections were fabricated on a glass substrate by e-beam evaporation and lift-off process. A 500 nm thick silicon oxide was deposited on the substrate by CVD (STS MESC) to passivate the electrodes. The passivative oxide layer was removed from the area of electrical contact pads by dipping the edges of the chip in buffered oxide etchant (BOE).

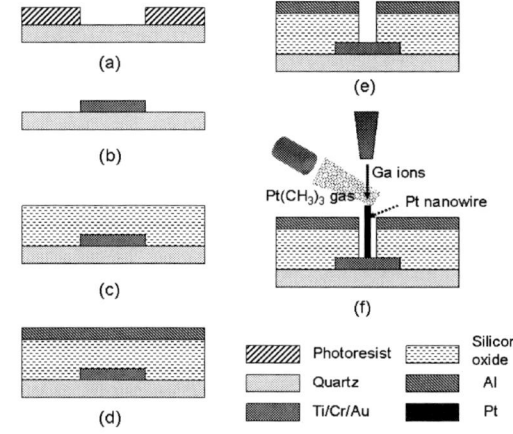

Figure 1: Schematic diagrams of fabrication of nanowire-based neuron probing device; (a) patterning photoresist on a glass substrate by optical lithography; (b) formation of Ti/Cr/Au electrodes by e-beam evaporation and a subsequent lift-off process; (c) deposition of silicon oxide by PECVD; (d) deposition of an Al layer by e-beam evaporation; (e) formation of a hole by FIB; (f) growth of single Pt nanowire in the hole by a FIB-CVD process.

Single vertical Pt nanowires were fabricated at the center of each micro-electrodes through the passivative oxide layer by FIB and CVD. . FIB has the flexibility to fabricate the nanostructures at a desired position on a substrate without the time-consuming lithographic processes [7-9]. Both

978-1-4244-2977-6/09 $25.00 © 2009 IEEE

nanoscale holes and vertical Pt nanowires were fabricated using a dual-beam (electron beam and focused gallium ion beam) Strata 235 FIB system (FEI, Inc.). First, a hole with 200 nm in diameter was created in a microelectrode by FIB milling in the energy range of 20 keV to 30 KeV and beam currents in the range of 0.2 pA to 3 nA. The exposure time was in the range between a few seconds and 200 seconds. A hole was milled down to the Au layer at each electrode site while monitoring a change in surface currents. An approximately 10 nm diameter ion beam at 25 KeV was incident on the surface and $Pt(CH_3)_3$ (Dimethyl-platinum-hexafluotaacetyacetonate) that was fed towards the surface with a small capillary tube. As a result of controlling beam currents in the range of 0.2 pA to 1nA, a Pt nanowire was grown in the hole to form a "*nanobud*". The smallest feature size from FIB deposition is a few tens of nanometers due to the area of dispersion of the secondary electrons generated by ion beam incidence on the substrates during FIB - CVD [10,11]. Figure 2 (A) shows the scanning electron microscope (SEM) image of microelectrode arrays fabricated on the glass substrate. A SEM image of a hole created on a microelectrode by the FIB milling process is shown in Figure 2 (B) and (C) as progressively zoomed-in images. Using the point beam irradiation, the applied voltage and emitter current were set at 10 keV and 2 nA, respectively. The beam diameter was approximately 8 nm and the actual beam current was 2 pA during the growth of nanowires. Under the same conditions, a vertical single Pt nanowire was fabricated in the hole located in the center of a microelectrode. Figure 2 (D) shows the scanning electron microscope image of vertical single Pt nanowire with a 50 nm in diameter and 1 μm in length grown in the hole located in the center of a microelectrode.

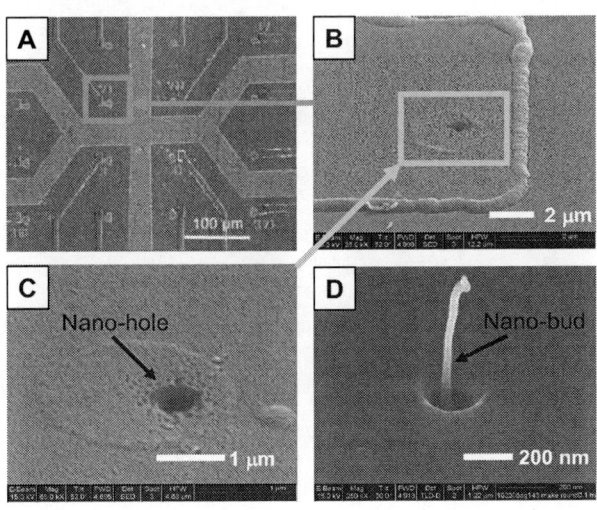

Figure 2: FESEM images of; (A) top view of microelectrode arrays; (B) a hole with about 200 nm in diameter generated by FIB milling; (C) the hole in the zoomed area pointed in (B); (D) single vertical Pt nanowire grown in the hole by FIB-CVD process.

A culture chamber was formed around the active electrode area with a polypropylene ring sealed to the chip with silicone grease. The chip was sterilized with ethanol, diluted with water and then Dulbecco's Modified Eagle's Medium (DMEM) to four times the volume of the culture chamber. The apparatus was allowed to equilibrate in the cell incubator for 15 minutes before the introduction of cells. Rat neuroblasts (ATCC,CRL-2754) were cultured in DMEM with 10% fetal bovine serum at 37 °C and 5% CO_2. The culture was grown to approximately 80% confluence for three passages with a change of medium every 2 to 3 days. The culture was then trypsinized and transferred to the electrode array, where they attached and remained in culture for 1 day. The array was then inspected by optical microscopy to verify which electrodes were overlaid with cells (Figure 3). Tungsten probes and micromanipulators were used to contact two selected electrodes on the chip. Each probe was connected to the input of a head stage, then to a bridge circuit of an electrophysiology microelectrode amplifier. The bath was grounded with an Ag/AgCl electrode immersed in the culture medium. The voltage signal was monitored on two channels simultaneously.

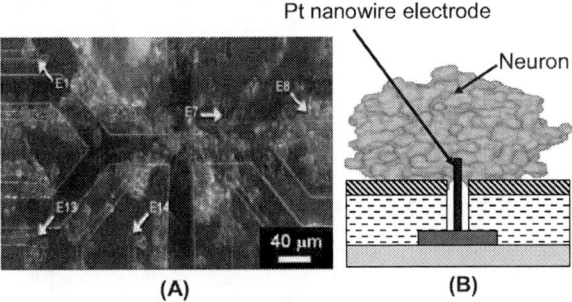

Figure 3: Optical micrograph of the nanowire electrode arrays with cultured B35 neurons; (A) Arrow and numeric designations indicate some of the electrodes that were used for recording; (B) drawing of a neuron located on the top of the nanowire electrodes.

RESULTS

Several recordings of at least 25 seconds were acquired on two channels that had no overlaying cells. The data showed a baseline signal with no events (Figure 4a). Simultaneous acquisitions from an electrode without cells and one with cells showed a series of spontaneous negative deflections (Figure 4b). These waveforms were typically 800 to 900 ms in duration, with an abrupt falling edge and a relatively slow rise time. While these signals are too slow to be action potentials, we believe that they represent a cellular signal modulated by slow charge transfer characteristics that are expected for a high surface-area-to-volume ratio polarizable electrode. The 3-dimensional aspect of the nanowire facilitates close contact with a cell. A valuable question should be explored concerning how a neuron interacts with an adjacent vertical nanowire. Previous studies showed that released nanowires were readily internalized by B35 in culture [12]. If the same phenomenon occurred with the

978-1-4244-2977-6/09 $25.00 © 2009 IEEE

anchored nanowires of these electrodes, it could provide minimally invasive access to the intracellular potential. We envision that biomolecular functionalization of the nanowire could be used to direct the cell-electrode interaction to improve signal quality and reduce movement artifacts of motile neurons in culture.

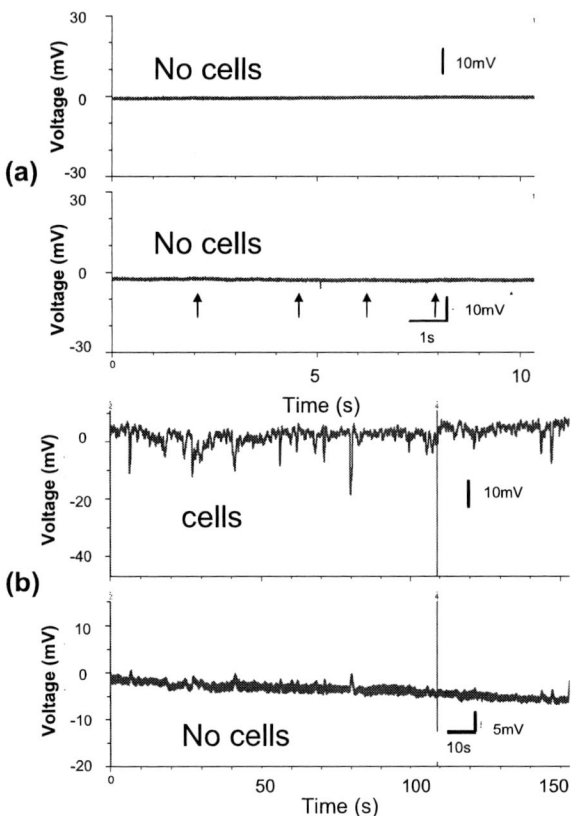

Figure 4: Voltage recordings from nanowire electrodes; (a) baseline signals of electrodes in the absence of cells; (b) signals from electrodes with cells (upper trace) and without cells (lower trace).

The small geometry of the nanowires makes it possible to probe multiple sites within a cell. In future work, the performance of the single Pt nanowire electrode should be clarified with further characterization of its electrochemical properties. Future studies will quantify the device impedance spectrum and crosstalk.

CONCLUSIONS

We demonstrated *in vitro* detection of neural activity with single vertical Pt nanowires (50 nm in diameter, 1 μm in length) fabricated by a novel method of FIB-CVD. These Pt nanowires were connected with an array of individually addressable microelectrodes and used to probe rat neuroblasts in culture. Voltages recorded by the nanowires indicated spontaneous activity of the cells

ACKNOWLEDGEMENTS

The authors gratefully acknowledge support for this work by the Director's Research & Development Fund, UI New Faculty Start-up Fund and the Korea Science and Engineering Foundation grants (Pioneer Program). AF gratefully acknowledges a NSERC postgraduate scholarship for study abroad.

REFERENCES

[1] S. Marom, G. Shahaf, *Q. Rev. Biophys*. 35, 63, 2002.

[2] W. L. C. Rutten, *Annu. Rev. Biomed. Eng*. 4, 407, 2002.

[3] P. Fromhertz, *Chem. Phys. Chem*. 3, 276, 2002.

[4] G. W. Gross, E. Rieske, G. W. Kreutzberg, and A. Meyer, *Neurosci. Lett*. 6, 101, 1977.

[5] J. Pine, *J. Neurosci. Meth*. 2, 19, 1980.

[6] T. Stieglitz, M. Schuettler, and K. Koch, IEEE Eng. Med. Biol. Magn 24, 58, 2005.

[7] S. Matsui, T. Kaito, J. Fujita, M. Komuro, K. Kanda, and Y. Haruyama, *J. Vac. Sci. Technol. B* 18, 3181, 2000.

[8] A. Stanishevsky, *Thin Solid Films* 398, 560, 2001.

[9] J. Fujita, M. Ishida, T. Ichihashi, Y. Ochiai, and S. Matsui, *Nuclear Instrum, Methods Phys. Res. B* 206, 472, 2003.

[10] B. Park, K. Jung, W. Song, B. Oh, and S. Ahn, *Adv. Mater*. 18, 95, 2006.

[11] G. Dearnaley, J. Freeman, R. Nelson, J. Stephen, *Ion Implantation* (North Holland Publishing Company, Amsterdam) Chap. 4, 1973.

[12] D. Choi, A. Fung, H. Moon, D. Ho, Y. Chen, E. Kan, Y. Rheem, B Yoo, N Myung, *Biomed. Microdevices* 9:143–148, 2007.

GENERATION AND SELECTIVE RETRIEVAL OF MICRO DROPLETS IN AN ARRAY FOR MICRO-PCR USING A GLASS-SILICON-GLASS CHIP

K.H. Park[1,2], H.G. Park[2], and S. Takeuchi[1,3,4]

[1]Institute of Industrial Science (IIS), The University of Tokyo, Tokyo, JAPAN
[2]Korea Advanced Institute of Science and Technology, Daejeon, KOREA
[3]BEANS Project, METI, Tokyo, JAPAN
[4]PRESTO, Japan Science and Technology, Tokyo, JAPAN

ABSTRACT

In this study, we fabricate a glass-silicon-glass chip for the generation of a water-droplet array. Using this array, we succeeded in performing polymerase chain reaction (PCR) within the hundreds of trapped water droplets. Then, we selectively retrieve a specific PCR droplet with a bubble formed by heating an aluminum micro pad with an infrared laser.

INTRODUCTION

The most popular nucleic acid amplification method is polymerase chain reaction (PCR), and this simple method is in advance adapted to a microfluidic field. Developments of on-chip PCR devices are focused in several important factors such as substrate materials and fluidic designs. Biocompatibility and thermal conductivity of the substrates are crucial in the efficiency of PCR. Designs of the PCR chip is also very important factors for the reaction speed and cross-contamination.

To fabricate PCR devices, many kinds of materials have been used as the substrates such as polymer and silicon. Due to the biocompatibility, excellent optical transparency and flexibility, polydimethylsiloxane (PDMS) have increasingly been used in a biochip field. Despite these advantages, PDMS has low chemical resistance to organic solvents and sometimes shows undesired absorption of biomolecules and fluorescent dyes. Also, the permeability of PDMS is not attractive when performing PCR; evaporation of the solution through the PDMS causes the change of the concentration. On the other hand, glass and silicon show high thermal conductivity, good chemical resistance and a non-absorption property of non-specific materials. Although the fabrication process needs expensive equipment compared to that with PDMS, the fabricated devices of silicon and glass can be easily recycled with a simple washing step [1,2].

Designs of PCR chips can be divided in two types: stationary chamber-based chips and continuous flow chips. Compared to stationary PCR chips, continuous flow chips are useful for fast PCR [3]. However, they usually have problems on (i) a cross-contamination, (ii) the poor controllability of the flow and temperature gradient, and (iii) the difficulty in the continuous observation of a specific target [4].

On the other hand, in the stationary chips, simultaneous observation of all the targets is possible. Also, separated chambers of a stationary chip are free from cross-contamination. However, most of recent stationary chips have been fabricated with open-chip architectures composed of etched chambers that needs large volume of PCR solution and covers for the protection of solution evaporation. Therefore, various type of research has been performed to construct stationary chips having a small volume microfluidic architecture without any covering process [5].

Here, we propose a glass-silicon-glass microfluidic chip for PCR. The 'Fishbone-like' etched silicon slide is designed to construct separated stationary chambers of PCR solution. Small microchambers are used as small batch reactors for PCR. Closed microchip and oil can protect evaporation of PCR mixture. Oil types, injection speed are optimized and selective retrieval of a specific product is successfully performed.

EXPERIMENTAL METHODS

Chip fabrication

We prepare three slides for the fabrication of the chip. The upper glass slide contains inlet and outlet holes. The middle silicon wafer is etched to make the channel and chamber walls. The bottom glass is prepared having an aluminum spots for droplet retrieval. To fabricate aluminum patterns, firstly, the S1818 photoresist is spread on the bottom glass and arrays of aluminum patterns are manufactured using a photolithography. After the lithography, extra aluminum is deposited onto the glass and etched. Then, three substrates are bonded using the anodic bonding process. Figure 1 shows scheme images of all components, a real image of the completed chip and a scanning electron micrograph (SEM) of the architecture inside of the device. After bonding of slides, PDMS molds for easy insertion and removal of solutions using silicon tubes are bonded by oxygen plasma treatment at the inlet and outlet regions of the upper glass.

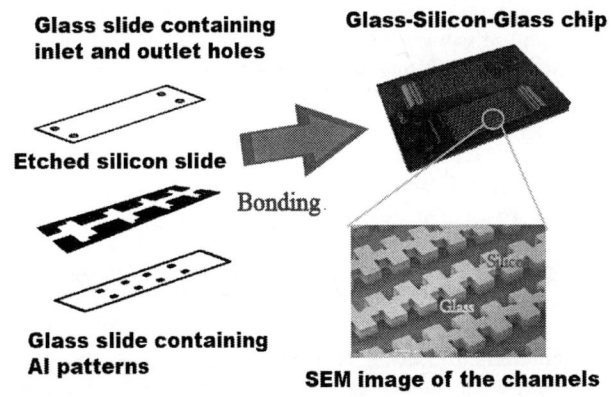

Figure 1. Fabrication of a glass-silicon-glass chip for PCR and selective retrieval. A SEM image shows the shapes of the silicon structure and glass surface. The size of each chamber is 100 x 100 x 115μm.

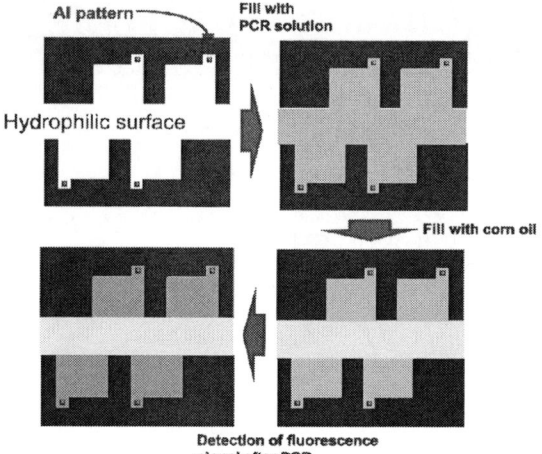

Figure 2. Scheme of all process in hydrophilic chip.

PCR amplification

The pUC19 plasmid DNA having 1.2bp is used as a template. The solution is composed of 50000 unit of template, 0.2 mM of dNTPs, 2.5 μM of each primer (forward: 5`-GTTTTCCCAGTCACGACGTTG-3` and reverse: 5`-ATGACCATGATTACGCCAAGC-3`), 2 mM MgSO₄, 2 unit/μL of Taq DNA polymerase (Invitrogen) in a volume of 50 μL. 1 μL SybrGreen I (Invitrogen) which intercalates into the double strand DNA and shows green fluorescence for the detection of PCR products is added to the mixture. After denaturing at 94°C for 7 min, the PCR is carried out for 30 cycles of 94 °C for 15 s, 55°C for 30 s, and 70°C for 1 min followed by final extension for 5 min at 70°C. PCR amplification is carried out on the 'Peltier-device' composed of Pt-100 heater, cooler, control panel and DC adapter purchased from 'VICS Japan'. Temperature is controlled using 'Tera-term' computer program.

Hydrophilic chip

Figure 2 shows the whole process in hydrophilic chip. First, prepared PCR mixture is injected in the channels and inside of the chip is fully filled. Then, oil is pumped into the channels for fabrication of a PCR array. We use two kinds of oil; corn oil, and mineral oil. Also we diversify of oil condition using 0.1% a surfactant, Span-80. And then the microfluidic chip is set on the 'Peltier-device' and PCR is carried out. During the PCR, we observe the midterm results of fluorescence intensity.

Hydrophobic chip

Figure 3 shows the whole process in hydrophobic chip. Before introducing of PCR solution, inside channels and chambers are rendered hydrophobic with octadecytimethoxysilane treatment. Modification solution is composed of 2% octadecytrimethoxy silane, and 0.5% butylamine and toluene. The modifier is injected into the chip and incubated for 20 min at room temperature. Modification solution is blown out following washing step with pure toluene. After the modification, PCR mixture solution is injected into the chip. Because of hydrophobic property of the chip, air in the chambers is maintained without flushing situation by PCR solution. Remained air in the chambers should be excluded for the formation of PCR mixture array.

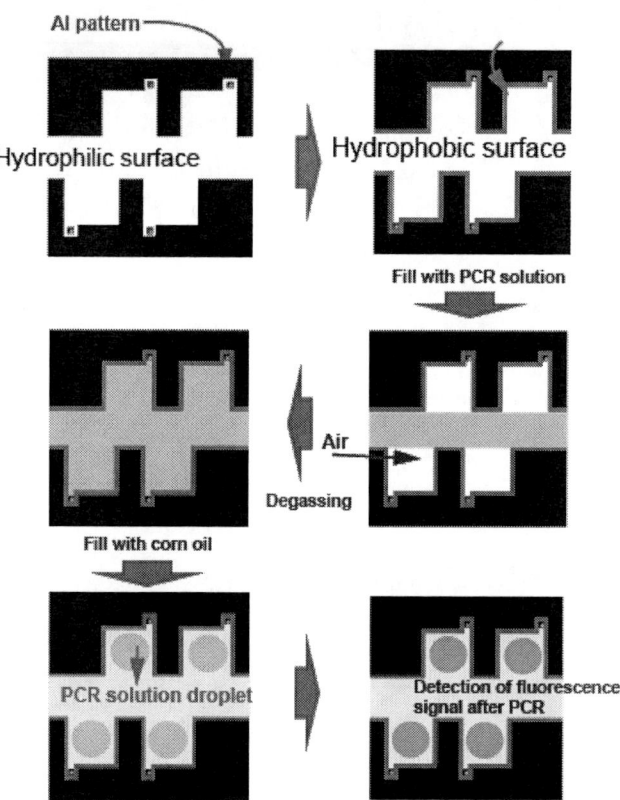

Figure 3. Scheme of all process in hydrophobic chip.

Figure 4 shows the degassing step in this chip. PCR mixture is injected into the channels and extra solution is located at the output. Inlet tube is plugged using a clip and aspiration is performed using a vacuum chamber. During the reducing the air pressure, air bubbles and PCR solution are removed from the chip.

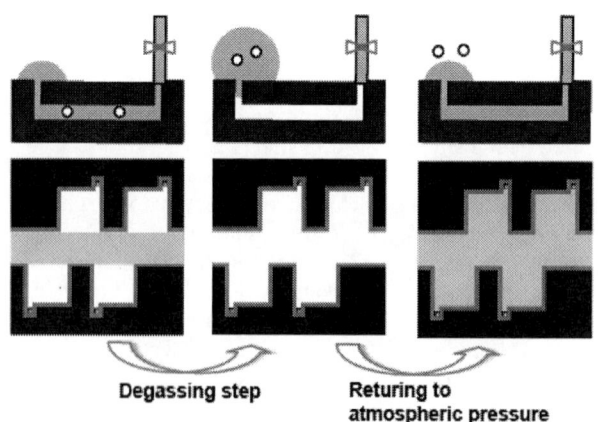

Figure 4. Degassing step. Sufficient water is injected into the channels and extra solution is located at the output. After blocking of inlet, the chip is degassed. Air and solution inside of the channels are removed during the decreasing of pressure. During the returning of pressure, water solution is inserted into the whole volume of the chip.

978-1-4244-2977-6/09 $25.00 © 2009 IEEE   364

After empting the inside of the chip, pressure is recovered to atmospheric pressure. During the pressure recovering, only the PCR solution at the outlet region is flowed into the chambers. Therefore, all the channels and the chambers are filled with PCR solution. Following this step, oil is injected into the chip, and whole the walls of the channels and chambers are soaked with oil inducing fabrication of water droplets at the chambers. PCR is performed with the same method of hydrophilic chip.

Retrieval of water droplets

After PCR, we selectively retrieve a specific water droplet. The chip is located onto a microscope and laser controlled using a manipulator is focused on a specific aluminum pattern[6]. After focusing, we irradiate an infrared laser (near 1064 nm, 2.2 W) to the pattern and heating result generates an air bubble that pushes water-bubble containing PCR products. The specific bubble is then carried to the end of channel with the main stream and collected.

RESULTS

Fabrication of PCR array, PCR results and the retrieval process in hydrophilic chip

Because of hydrophilic property of the channels, the water solution is confused inside of the chamber not forming water droplets. Figure 5 (a) shows the optical image of water-oil phase. As shown in a fluorescent image of Figure 5 (b), there is little fluorescence at all area before PCR. After more than 30 thermal cycles of temperature change, observed fluorescence intensity at only water-phase regions is outstandingly increased as shown in Figure 5 (c).

Fluorescence intensity is increased after 25 cycles at this experiment. Increase of fluorescence intensity is affected with the quantitative amount and length of double helix DNA, irradiation time of exciting light, and product size. In this experiment, we obtained a shot product (111bp); therefore it takes a large number of cycles to observe

increased signal compared to that before reaction. By controlling of various factors such as product size and observing conditions, the time to decide the accomplishment of the PCR can be optimized

After confirmation of PCR results, we carry out a retrieval process of one specific product in a specific chamber using laser heating method. However, during the air bubble formation inside of the chamber, the PCR solution is not escaped from the chamber. Because the silicon and glass substrate is hydrophilic, water runs up the hydrophilic walls. This phenomenon brings out the cross-contamination of PCR droplets. In conclusion, in hydrophilic chip, it is impossible to archive a specific product.

Fabrication of water droplets, PCR results and the retrieval process in hydrophobic chip

Figure 6 shows the results containing from preparation to observation steps. In Figure 6(b), air bubbles which remained after PCR mixture insertion of Figure 6(a) are removed after degassing step. Figure (c) and (d) show the fabricated water droplets after corn-oil injection. Compared to water shape of hydrophilic chip, oil is flowed through the whole walls and glass substrates. Therefore, oil wrap up the PCR solution in each chamber, and water droplets are fabricated.

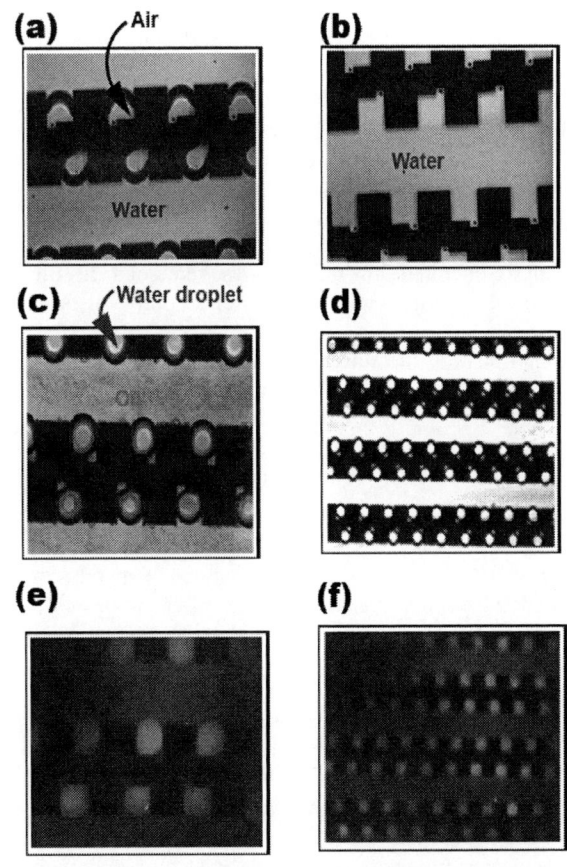

Figure 6. Optical/fluorescence images in hydrophobic channels: (a) An optical image after injection of water. (b) An optical image after degassing. (c) and (d) Optical images after oil injection. (e) and (f) Fluorescence images after PCR.

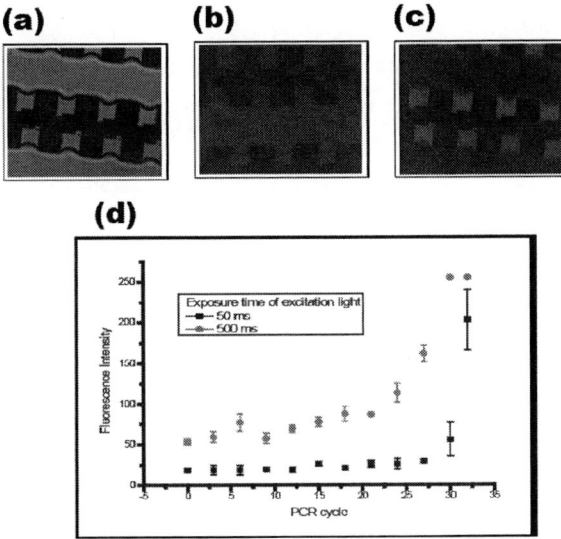

Figure 5. Results in hydrophilic channels: (a) An Optical image of water-oil phase. (b) A fluorescence image before PCR. (c) A fluorescence image after PCR. (d) A graph of the intensity vs. PCR cycle.

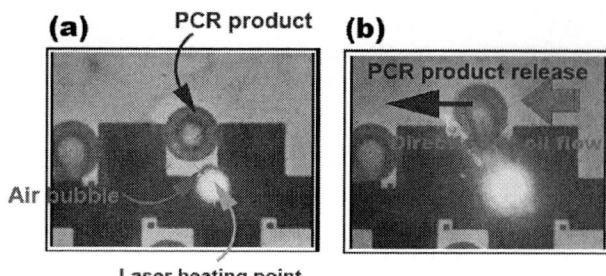

Figure 7. Selective retrieval of a specific PCR product from a microchamber: (a) Fabrication of air bubble by heating an aluminum spot using laser. (b) Retrieval of PCR product by air bubble.

Fabricating condition and size of water droplets are involved precisely with the pumping speed of oil. Water in all chambers is pushed away from the chambers below than 7 μm/min. More than above speed, mono-disperse water droplets are formed, and the size of water droplet is decreased as the speed of oil injection is increased.

PCR is performed with same method with hydrophilic chip after water droplet fabrication. Figure 6 (e) and (f) show the PCR result of water droplets. Detected fluorescent intensities of water droplets are not much in it of the hydrophilic chip.

After observation of PCR results, we try to retrieve one specific water droplet using a laser. In mineral oil, air bubble is not generated because of its high boiling point. On the other hand, in corn oil, air bubble is generated well as shown in Figure 7. The air bubble is fabricated on the Al pattern and it takes about 30 sec for fully removal of a water droplet from the chamber. Removed selected PCR product in the corn-oil containing 0.1% surfactant is acquired at the output, while water droplet is fused with a droplet of the next chamber in the corn oil without surfactant. Generated air bubble was remained in the chamber replaced of water bubble after stopping of laser irradiation.

After finishing the selection step, the chip is washed using 90% chloroform and 10% methanol solution. By simple heating for complete removal of remained washing solution, we reuse the chip to perform same process, and all the steps are successfully achieved again.

CONCLUSION

Our work shows various advantages over current micro-PCR stationary systems; (i) The microfluidic chip needs small amount of mixture, (ii) selective retrieval of a specific product is possible, and (iii) there is no evaporation without additional covering step since the PCR solutions are contained within the device and surrounded by oil. More efficient real-time, rapid and simultaneous detection is possible by controlling the experimental conditions such as chamber size and fluorescence related factors. Based on this result, we can analyze arbitrary biomolecules using our retrieval method combined with additional analytical tools such as primer-labels or aptamer- conjugated particles.

ACKNOWLEDGEMENT

We acknowledge to Dai Nippon Printing, Co. Ltd. (DNP) for the technical assistance in chip fabrication. This work was partly supported by Core Research for Evolutional Science and Technology (CREST) from Japan Science and Technology (JST), and Grants-in-Aid for Scientific Research from Ministry of Education, Culture, Sports, Science and Technology Japan. K.H. Park was supported by Brain Korea 21 (BK 21) Program from Ministry of Education, Science and Technology Korea,

REFERENCES

[1] C.S. Zhang, J.L. Xu, W.L. Ma, W.L. Zheng, "PCR Microfluidic Devices for DNA Amplification", *Biotechnol. Adv.*, vol. 24, pp. 243-284, 2005

[2] S.H. Kim, J. Noh, M.K. Jeon, K.W. Kim, L.P. Kim, S.I. Woo, "Micro-Raman Thermometry for Measuring the temperature distribution inside the microchannel of a polymerase chain reaction chips", *J. Micromech. Microeng.*, vol. 16, pp. 526-530, 2006.

[3] M.U. Kopp, A.J. de Mello, A. Manz, "Chemical Amplification: Continuous-Flow PCR on a Chip", *Science*, vol. 280, pp. 1046-1048, 1998.

[4] J.S. Marcus, W.F. Anderson, S.R. Quake, "Parallel Picoliter RT-PCR Assays using Micrlofluidics", *Anal. Chem.,*. vol. 78, pp. 956-958, 2006.

[5] C. Zhang, D. Xing "Miniaturized PCR Chips for Nucleic Acid Amplification and Analysis: Latest advances and future trends", *Nucleic Acid Research*, vol. 35, pp. 4223-4237, 2007.

[6] W.-H. Tan, S. Takeuchi, "A Trap-and-release Integrated Microfluidic System for Dynamic Microarray Application", *PNAS*, vol. 104, pp. 1146-1151, 2007

ARRAYING SINGLE ADHERENT CELLS BY MICROPLATE SELF-ASSEMBLY

Hirotaka Ishihara[1], Kaori Kuribayashi[1], and Shoji Takeuchi[1, 2, 3]
[1]Institute of Industrial Science, The University of Tokyo, Japan
[2]PRESTO, JST, Japan, [3]BEANS Project, METI, Japan

ABSTRACT

This paper describes a method for handling a single adherent cell using a self-assembly technique. We produce microplates using Parylene with both hydrophilic and hydrophobic sides. We culture the cells on the hydrophilic side, and utilize the hydrophobic side to self-assemble the plates onto the hydrophobic regions. Culturing a single cell on each Parylene microplate facilitates their handling since we can manipulate them as floating cells, and easily relocate them without the loss of cellular activity. In our experiment, we assemble 50 μm circle microplates with adherent cells in an array.

INTRODUCTION

Handling and arranging cells are essential techniques in tissue engineering and bioscience [1-3]. However, there is a problem caused by a difference of cell types: floating cells and adherent cells. Floating cells are easy to handle, while adherent cells are much more difficult to handle and rearrange once they have spread on a substrate since they cannot live without adhering to the substrate. When we use the adherent cells, typically, a chemical treatment with trypsin is used, which enables the re-dispersal of adherent cells but this process reduces cell activity/cell cycles due to its toxicity. To overcome this problem, in the previous report, we presented a method for easily handling adherent cells without the use of toxic chemicals by using mobile microplates [4]. The microplates are guides, and we can freely handle adherent cells like floating cells or micro-components.

Here, we further advance this method by making a platform for arraying the cell-adhered mobile microplates with a self-assembly technique utilizing the hydrophobic interaction [5]. Figure 1(a) shows a concept image of microplates with the self-assembly of cells. We produce the microplates with Parylene and culture adherent cells on the plates. We also produce a glass substrate patterned the lubricant on self-assembled monolayers (SAMs) of alkanethilates on gold, presenting hydrophobic surface [6]. We collect the microplates by pipette, and flush into the glass substrate with an array of lubricant drops. When the microplates with the cells contact the lubricants, capillary force occurs resulting from hydrophobic interaction between the lubricant and the bottom surface of the microplate with hydrophobicity (Figure 1(b)). Finally, the microplates immobilize at the lubricant area. In this paper, we carry out experiments of self-assembly of microplates without and with the adherent cells.

ADHERENT CELLS

Conditions of Cell Culturing

Adherent cells we use were NIH/3T3, mouse fibroblast like cell. Cell culturing condition was 37 °C, 5% CO_2 using Dulbecco's modified eagle medium (DMEM,

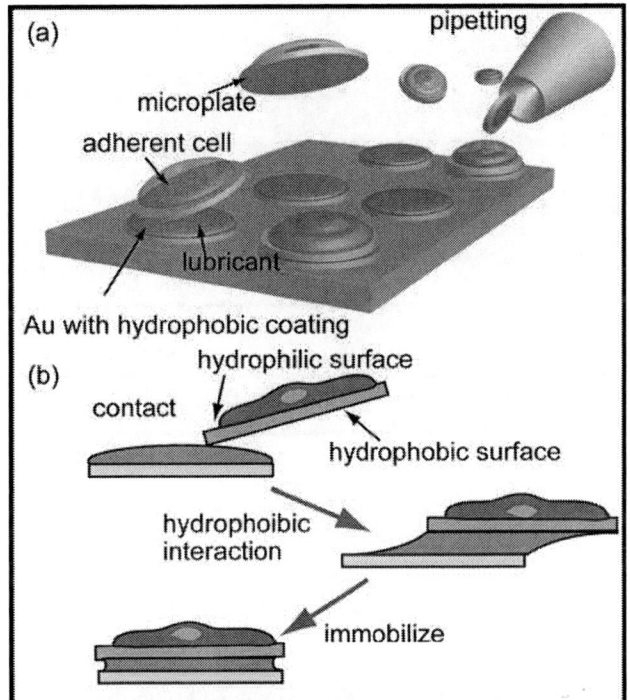

Figure 1 : *(a) Conceptual image of self-assembly of microplates with adherent cells. The adherent cells are cultured on the microplates. Each microplate acts as a single microcomponent of the self-assembling device. Using hydrophobic interaction, we can assemble adherent cells on a patterned area. (b) Sequential images of self-assembly.*

11885, Gibco) with 10% fatal bovine serum (FBS, F7524-500ML, SIGMA), and 5% L-Glutamine-Penicillin-Streptomycin (G1146- 100ML, SIGMA) as a culture medium.

Adherent Properties of the Cells

Figure 2(a) shows microphotographs of NIH/3T3 cells on both glass and Parylene-C (DPX-C, Speedline Technology) substrates with or without treating O_2 plasma treatment (Compact Etcher FA-1, Samco International Inc., 25 W, 10 ml s^{-1} oxygen, 30 sec). The O_2 plasma treatment is for making the surface of the Parylene substrate hydrophilic [7]. Before culturing the cells, we rinsed all substrates with 99 % ethanol and exposed UV light for 30 min. We dispersed 3T3 cell solution into the substrates. After 3 hours when 3T3 cells adhered on the substrates, we rinsed them by potassium buffered saline (PBS).

From the image of Parylene-C without the plasma treatment, which usually hydrophobic property, the adherent cells attached less on its surface with compared to the other substrates. In the image of Parylene-C substrate treated by the plasma, the cells adhere on the substrate since the surface of the Parylene-C is temporarily modified into hydrophilic. Figure 2(b) summarizes the adherent

978-1-4244-2977-6/09 $25.00 © 2009 IEEE   367

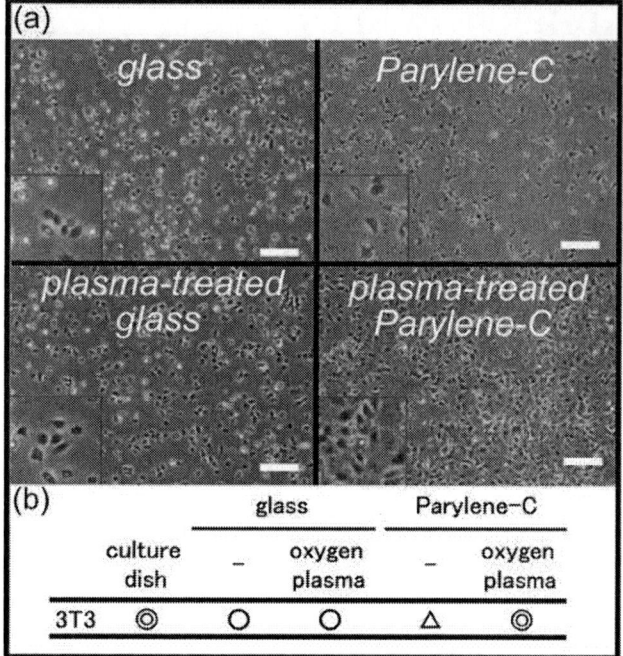

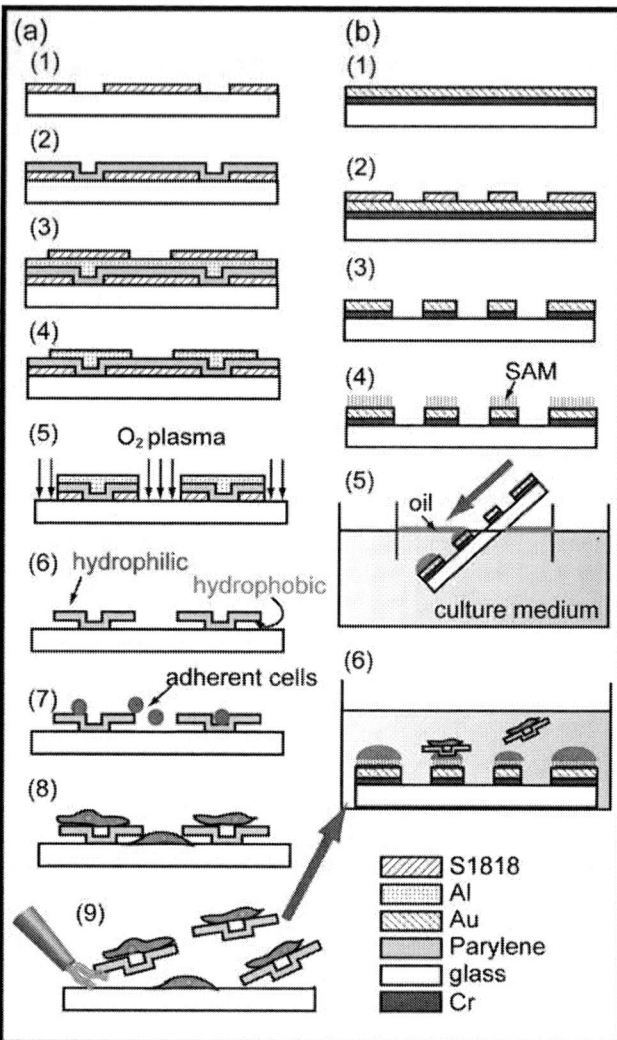

Figure 2: *Adherent property of 3T3 cells. (a) Optical Micrograph of 3T3 on glass and parylene-C surface with plasma treatment. Scale bar = 100 µm. (b) Table of adherent property of cells.*

property of the cells on both glass and Parylene-C substrates. The Parylene-C substrate treated with O_2 plasma has a better adherent property than other substrates. Therefore, we believe the O_2 plasma treatment is necessary for handling adherent cells.

METHODS

Fabrication of the Microplates for Culturing the Cells

To achieve handling a single adherent cell using a self-assembly technique, we prepared microplates that have both hydrophilic (upper) and hydrophobic (bottom) sides (Figure 3(a6)). The hydrophilic side is used for the cells to adhere onto the microplate, while the hydrophobic surface is essential for self-assembly. In this experiment, we produce the microplates with Parylene-C since it can be fabricated and changed the surface properties easily by O_2 plasma. In addition, the Parylene is biocompatible and suitable for bio applications using cells.

Figure 3(a1-6) shows the fabrication process of the microplates. On a glass substrate, we patterned S1818 photoresist (PR, Shipley) as a sacrificial layer to produce small contact areas between the glass and microplates (Figure 3(a1)), which allows us to release the microplates from the glass substrate easily after culturing the cells (Figure 3(a9)). The thickness of the S1818 layer is about 3 µm. Next, we deposited 5 µm thick Parylene-C on the substrate (Figure 3(a2)) using a Labcoater PDS2010 chemical deposition system (Specialty Coating Systems, Indianapolis). We then deposited aluminum (Al) and patterned S1818 atop the Parylene-C by a standard lithography technique (Figure 3(a3)). After that, we etched unmasked Al layer by Al etchant (Figure 3(a4)). The Al serves as a mask for patterning the Parylene-C, and we etched the Parylene and PR with O_2 plasma (Compact Etcher FA-1, Samco, International Inc. , 25 W, 10 ml s^{-1}

Figure 3: *(a1-6) Fabrication process of microplates. (a7-9) Culturing adherent cells and release of the microplates. (b1-4) Patterning of Au and SAM-coating of 1-octadecanethiol (ODT) on Au surface. (b5) Lowering the substrate into oil phase before immersing into culture solution. (b6) Microplates are injected though jetting flow and the self-assembly occurs.*

oxygen, 18 min) (Figure 3(a5)). We completely etched the unmasked the Parylene-C layer. Finally, we removed Al mask on the microplates by Al etchant, and treated the surface of the plate with O_2 plasma (Figure 3(a6)). We then dispersed 3T3 cells over the microplates and cultured them (Figure 3(a7, 8)).

Fabrication of the Substrates Patterned Lubricant Drops

Figure 3(b1-4) shows a process of producing patterned hydrophobic areas. We evaporated chrome and gold (Cr/Au) on a glass substrate, (Figure 3(b1)). Then we patterned photolithographically and etched the Cr/Au into circle, which is a same as a shape of the microplate (Figure 3(b2, 3). Before coating SAM, we treated the substrate with O_2 plasma (25 W, 10 ml s^{-1} oxygen, 30sec), rinsed it with ethanol and dried it by nitrogen gas for uniform deposition of SAM on Au. The SAM solution we used was 1 mM 1-octadecanetiol ($CH_3(CH_2)_{17}SH$, ODT, 98 %,

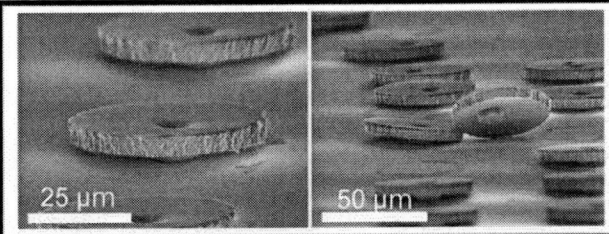

Figure 4: SEM images of parylene microplates (diameter: 50 μm, thickness: 5 μm). The small contact area between the microplate and glass substrate allows the microplates to release easily.

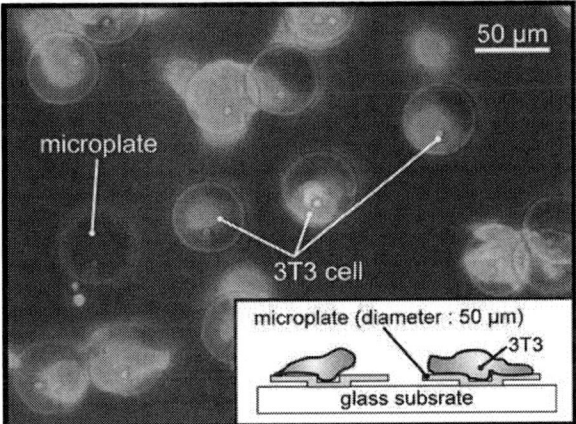

Figure 5 : Superimposed fluorescent and phase contrast images of single cell-cultured microplates. 3T3 cells are labelled green fluorescent.

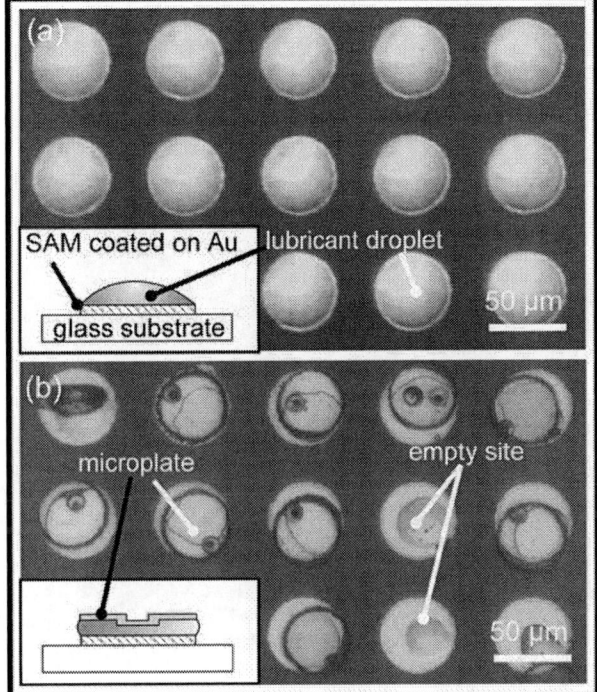

Figure 6 : Microscopic images of self-assembled microplates on patterned ODT-SAM in water before culturing cells. (a) Lubricant (squalene) wetted at SAM coated sites. (b) The microplates are self-assembled on the patterned lubricant droplets.

Aldrich) in ethanol. We immersed the substrate into the SAM solution for about 30 min and rinsed it by ethanol several times. The substrate formed hydrophobic monolayers (Figure 3 (b4)).

Figure 3(b5) shows a process of generating lubricant drops on the patterned area of SAM. At first, we filled a culture solution into a container. Secondly, a pour lubricant immiscible with the solution was added to a wall made of 1 mm thick silicone lubber. We used squalene as a lubricant of hydrophobic oil. Then, we inserted the patterned SAM substrate into the lubricant phase slowly at the speed of ~1 cm/s. As the substrate passed into the lubricant phase, lubricant droplets assembled only the areas coated with the hydrophobic SAM.

Arraying of Self-Assembly of the Microplates

After the cell culturing on the microplates, they were physically released from the glass substrate using a manipulator (TransferMan NK2, eppendorf) with glass capillary tube (Figure 3(a9)). We collected them, and subsequently introduced them to the patterned lubricant drops (Figure 3(b6)); consequently, they self-aligned and immobilized.

EXPERIMENTS AND RESULTS

Figure 4 shows SEM images of the microplates (50 μm in diameters). Since the microplates were attached on a glass substrate with small area (10 μm in diameter) at the glass substrate, we are able to release them easily after culturing the cells.

Figure 5 shows an image of 3T3 cells after 1 day culturing on Parylene microplates treated with O_2 plasma. The image was made by superimposing a fluorescent image onto a phase contrast image of the cells. We labeled the cells using Live Assay (L3224, Invitrogen) according to a protocol provided by the manufacturer. We successfully mounted a single 3T3 cell onto each Parylene microplate. After 30 min of the culturing, we observed that most of adherent cells adhered on only the Parylene microplates. We rinsed the dish with PBS twice and added culture medium so that rest of the 3T3 cells, which did not adhere to Parylene microplate, cannot disturb a self-assembly experiment.

The lubricant of squalene wetted only at hydrophobic SAM area. We have successfully generated an array of the lubricant drops on patterned SAM on Au in water (Figure 6(a)). For the self-assembly, we released and collected the microplates without the cells by using a pulled glass pipette and then dispersed the microplates on the lubricant drops. After agitating the water several times, we achieved that the microplates self-aligned on lubricant sites due to hydrophobic interaction between the lubricant and the hydrophobic side of the microplate (Figure 6(b)). We found that the amount of the lubricant was one of important parameters for the self-assembly.

We have also tried to align the microplates with the single 3T3 cell onto the patterned the lubricant droplet. Figure 7 shows sequential images of the experiment. We released the microplate by jetting of glass capillary in the culture medium (Figure 7(a)). The microplate sank down to the substrate and contacted to the lubricant (Figure 7(b)). When the microplate contacted the

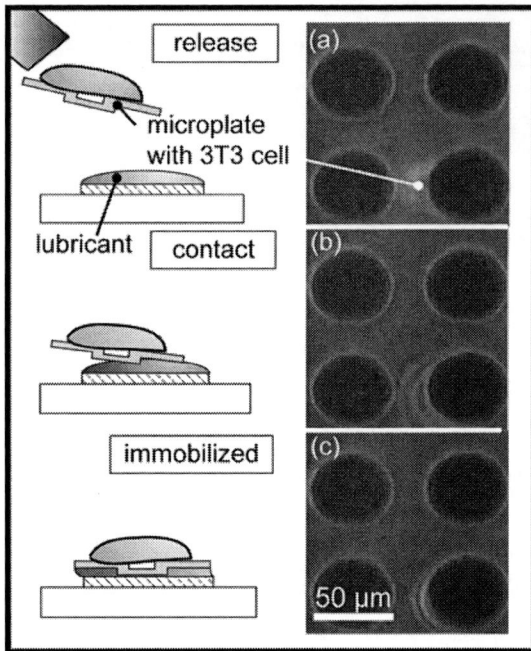

Figure 7 : Sequential images of self-assembly of 3T3 cell mounted on the micrplate. The microplate is released by a glass cappilary tube. Lubricant patterned on SAM catch the microplate due to the hydrophobic interaction. Flame rate is 0.5 fps.

lubricant, hydrophobic interaction occurs, and we then could immobilize the microplate (Figure 7(c)).

CONCLUSION

We applied self-assembly technique to arraying adherent cells using microplate handling. We chose Parylene-C as a material of microplates by modifying surface property using plasma treatment to promote cell adhesion. This assembly technique would be attractive as a tool of cell studies. By adding other functions to Parylene-C microplate, we would perform further analysis of cell behavior and cell-cell interactions.

ACKNOWLEDGEMENTS

This work was partly supported by Core Research for Evolutional Science and Technology (CREST) from Japan Science and Technology Agency and by Grants-in-Aid for Scientific Research from Ministry of Education, Culture, Sports, Science, and Technology Japan.

REFFERENCES

[1] N. Kojima, T. Matsuo, and Y. Sakai, "Rapid Hepatic Cell Attachment onto Biodegradable Polymer Surfaces without Toxicity Using an Avidin-Biotin Binding System," *Biomaterials*, Vol. 27, pp. 4904-4910, 2006

[2] P. Panda, S. Ali, E. Lo, B. G. Chung, T. A. Hatton, A. Khademhosseini, and P. S. Doyle, "Stop-flow lithography to generate cell-laden microgel particles," *Lab Chip*, Vol. 8, pp. 1056-1061, 2008

[3] W. H. Tan, and S. Takeuchi, "Dynamic microarray system with gentle retrieval mechanism for cell-encapsulating hydrogel beads," *Lab Chip*, Vol. 8, pp. 259-266, 2008

[4] H. Onoe, and S. Takeuchi, "Microfabricated Mobile Microplates for Handling Single Adherent Cells," *Journal of Micromechanics and Microengineering*, Vol. 18, pp. 095003, 2008

[5] U. Srinivasan, D. Liepmann, and R. T. Howe, Fellow, IEEE, "Microstructure to Substrate Self-Assembly Using Capillary Forces," *J. Microelectromech. Syst.*, Vol. 10, pp. 17-24, 2001

[6] H. A. Biebuyck, and G. M. Whitesides, "Self-Organization of Organic Liquids on Patterned Self-Assembled Monolayers of Alkanethiolates on Gold," *Langmuir*, Vol. 10, pp. 2790-2793, 1994

[7] T. Y. Chang, et al., "Cell and Protein Compatibility of Parylene-C Surfaces," *Langmuir*, Vol. 23, pp. 11718-11725, 2007

A DYNAMIC MICROARRAY WITH PNEUMATIC VALVES FOR SELECTIVE TRAPPING AND RELEASING OF MICROBEADS

K. Iwai[1] and S. Takeuchi[1, 2, 3]

[1]Center for International Research on Micromechatoronics, Institute of Industrial Science,
The University of Tokyo, Tokyo, Japan
[2]PRESTO, JST
[3]BEANS, Ministry of Economy, Trade and Industry, Japan

ABSTRACT

This paper describes a dynamic microarray device with pneumatic valves for trapping and releasing microbeads selectively. We fabricated thin membranes of polydimethylsiloxane (PDMS) inside our conventional dynamic microfluidic device. The membrane works as a pneumatically driven valve that changes the fluidic resistance, which ultimately determines the modes of the device: trapping, passing, or releasing. Using this device, we successfully controlled these three modes by using 100 μm polystyrene microbeads. Moreover, we succeeded in arraying two different types of microbeads alternately in a single device by selectively activating and deactivating the pneumatic valves.

INTRODUCTION

In cellomics studies, drug screening and detection/diagnostic applications, multifunctional microfluidic devices are highly needed to transport and immobilize various kinds of small particles such as microbeads coated with proteins or living cells. Such devices need to trap and array particles to infuse reagents, to continuously observe the chemical reactions, to retrieve selected particles, and to release the trapped particles after experiments for the repeated usage.

We have reported a dynamic microfluidic device for trapping and making an array of microbeads, retrieving selected microbeads and quickly releasing all of the trapped microbeads [1-3].

These devices, however, were not able to make an alternating array of different types of microbeads. When we introduce different types of particles into these devices, the particles are trapped in a random order preventing the formation of an alternating array. Although it is possible to make such an alternating array by retrieving selected particles with our previously reported optical retrieval system, it is time-consuming and impractical for high density microarrays to retrieve numbers of trapped particles one-by-one. Microarrays with selectively trapped particles enable us to conduct the experiments with various kinds of objects at the same time and to acquire large amount of data at once for meaningful conclusions.

Here, we demonstrate a simple dynamic microarray device with pneumatically driven valves [4, 5] controlling the flow rate and changing the modes for trapping, passing and releasing, which allows us to make an array of microbeads in a selected order.

DESIGN AND FABRICATION

Design

The schematic images of our microfluidic device with pneumatically driven valves are shown in Figure 1. Our device has three different modes depending on both the direction of the flow and the on/off state of the valves. When we introduce a forward flow and deactivate the valves the fluidic resistance through the trapping route is smaller than through the bypass route between point A and B (Fig. 1a) [6] and consequently infused microbeads are trapped in an array. When we pressurize the valves, they swell and reduce the flow rate of trapping stream causing the microbeads to pass through the bypass route (Fig. 1b). Using these functions, it is possible to trap microbeads selectively since they are only trapped in the deactivated valves. When we reverse the flow the activated valves also reduce the trapping stream and we can release all the trapped microbeads allowing us to reuse the device (Fig. 1c).

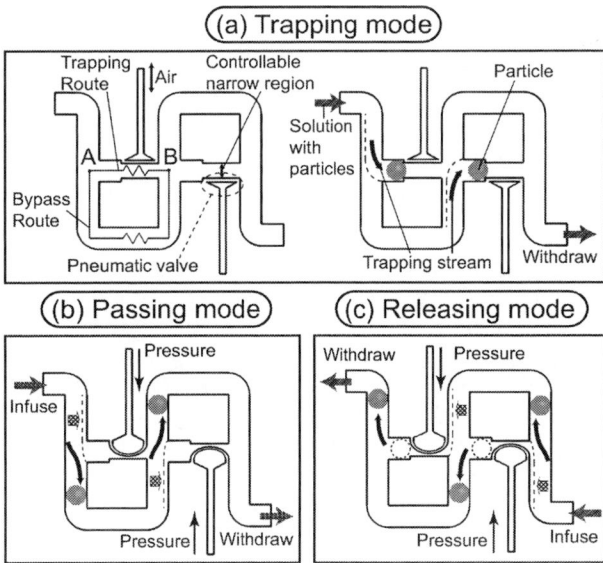

Figure 1: Schematic images of our dynamic microarray with pneumatic valves. (a) Trapping mode: When the pneumatic valve is not activated the fluidic resistance of the trapping route is smaller than that of the bypass route and particles are trapped. (b) Passing mode: When the pneumatic valve is activated, fluidic resistance of trapping route increases and particles are not trapped. (c) Releasing mode: When backward flow is infused, activated pneumatic valves prevent the particles from becoming stuck at the outlet of the traps and the particles are released.

978-1-4244-2977-6/09 $25.00 © 2009 IEEE   371

Fabrication

We fabricated our microfluidic device with pneumatic valves using standard soft lithography [7]. The device was fabricated from PDMS that was permanently bonded to glass by O_2 plasma treatment. The design configuration and the fabricated device are shown in Fig. 2(a) and 2(b), respectively. The pneumatic valves are 180 μm in length and 20 μm in width, respectively, and the channel height is approximately 115 μm.

We introduced pure water with 1 μm polystyrene beads into the channel to check for leaks in the pneumatic valves. The flow of the particles causes the channels to take on a darker hue compared to the pneumatic valve channels. This indicates that there are no leaks in the valves.

EXPERIMENTAL SETUP

The colloidal suspension that we infused into our channels contained 10% w/v of two different diameter polystyrene beads (ϕ = 98.7 μm and ϕ = 0.967 μm, coefficient of variation < 3%, 10%w/v aqua dispersion, Microparticles GmbH, Berlin, Germany) and were suspended to 0.1% w/v in ultrapure water (specific resistance of 18 MΩcm) with 0.5% of Tween 20 (Kanto Chemical, Tokyo, Japan).

We used syringe pumps (Micro 4; World Precision Instruments, Sarasota, FL) and Hamilton Gastight syringes (1700 series, TLL) to introduce the microbeads solution and for activating the valves. A pressure sensor (AP-V80 series; Keyence, Japan) and microscopic system (VH-900; Keyence, Japan) were used to measure the valve displacement. We used a Hamamatsu CCD camera mounted on an inverted microscope (Eclipse, TE300; Nikon, Tokyo) to image the process.

RESULTS

Controlling Flow

The valves were activated with 100 kPa pressure as shown in Figure 3a. We measured the distortion of valves by changing the applied pressure. The results of this measurement are shown in Figure 3b. We can observe that the displacement of the valves has dramatically increased when we applied high pressure. At 80 kPa, the membrane is deformed to the extent that it reduces the cross-sectional area of the channel by half.

Utilizing these valves, we successfully conducted the trapping, passing and releasing 100 μm polystyrene microbeads. Figure 4 shows the superimposed images of these three modes. When we introduced the beads in the forward direction, the beads were sequentially trapped one-by-one (Fig. 4a). When we applied pressure into the pneumatic channels and activated the valves the beads were no longer trapped in the activated valves (Fig. 4b). After all the traps are filled with beads we reversed the flow and the beads are released with activated valves (Fig. 4c).

These results indicate that the pneumatic valves are able to control the three modes of our microfluidic device by changing the fluidic resistance of the narrow region.

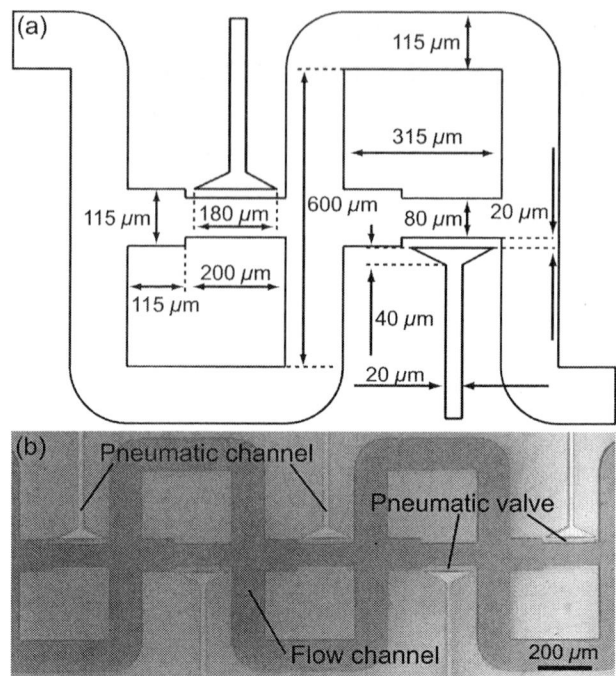

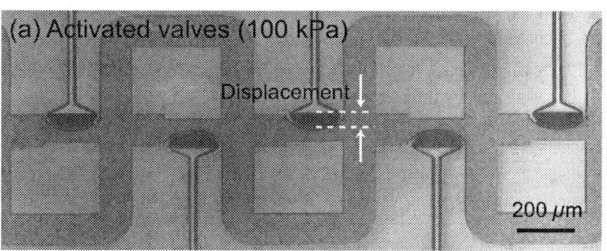

Figure 2: Design configuration and fabricated PDMS microfluidic channel. (a) The dimensions of the fluidic channel and pneumatic valves are indicated. (b) Image of the fabricated channels. The height of these channels is 115 μm. The microchannels have a darker color than the pneumatic valves due to the 1 μm polystyrene beads that we introduced into the channel to check for leakages.

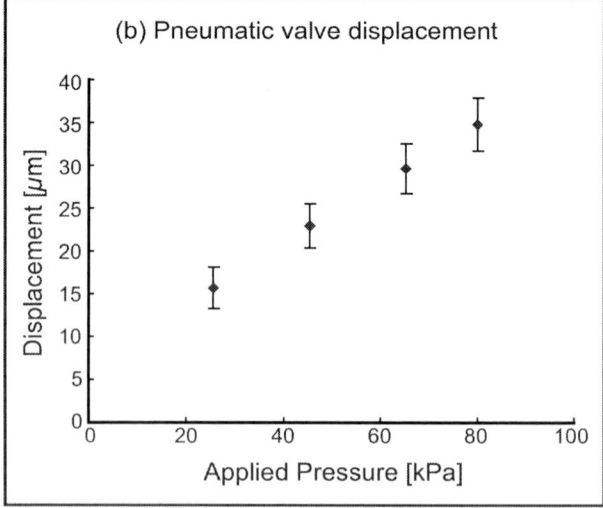

Figure 3: Pneumatically driven Valves. (a) Image of activated valves at 100 kPa. (b) Measured valve displacement when pressure varies (N = 100).

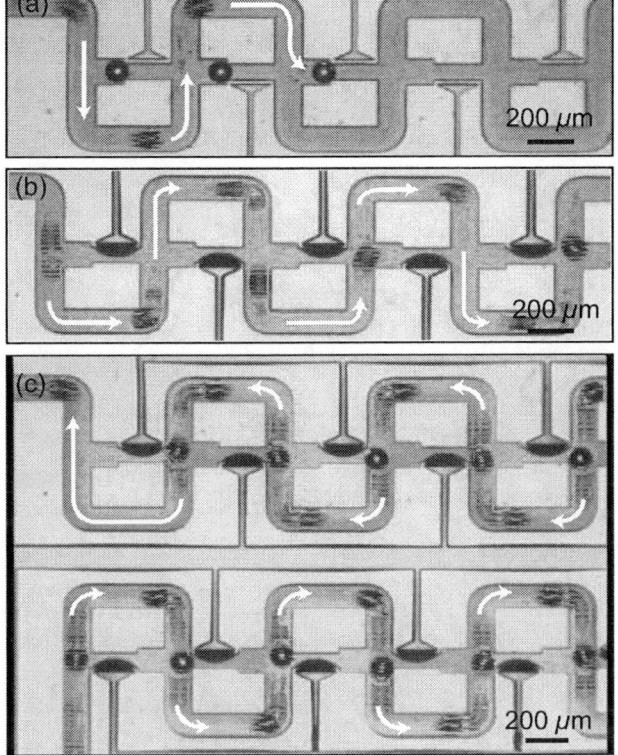

Figure 4: Superimposed images of (a) trapping, (b) passing, and (c) releasing of 100 μm polystyrene microbeads.

Selective Trapping

We fabricated a device with alternately working pneumatic valves for trapping and making an array of two different microbeads. Pneumatic valves on one side of the fluidic channels are connected to the single air inlet and pneumatic valves on the other side are connected to the other inlet.

We first activated valves on one side and trapped microbeads shown in blue with a white dashed circle in Figure 5a. Then, we deactivated all of the valves and trapped different microbeads shown in red with a white circle in Figure 5b. We added this false color to easily demonstrate that we maintain the order that the beads are trapped. We confirm this by imaging and tracking the beads as they enter the channels.

Selective Releasing

For the selective releasing of alternately trapped beads, we added small valves at the traps to hold the trapped particles. Figure 6 shows the fabricated valves and activated valves. The dimensions of small valves are 100 μm in length and 20 μm in width, respectively.

When valves are activated, the small valves work only to hold the trapped microbeads and do not affect the control of fluidic resistance at the large valves since the displacement of the larger valve controls the resistance in the channel. This configuration allows us to release only the selected particles by activating the valves alternately.

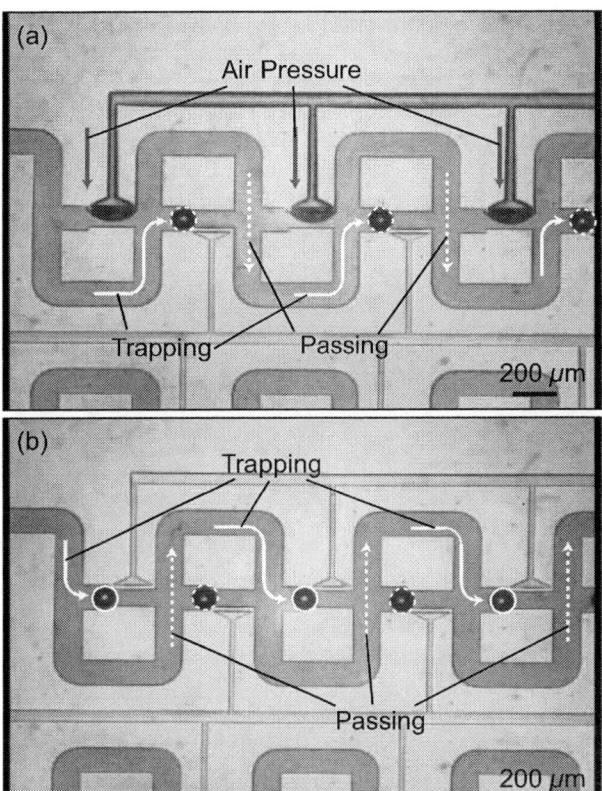

Figure 5: Selective trapping (false color was added to the beads in these images). (a) The beads that were introduced first were trapped only in the traps with deactivated valves. (b) After filling the traps, the next set of infused beads was trapped in the remaining traps.

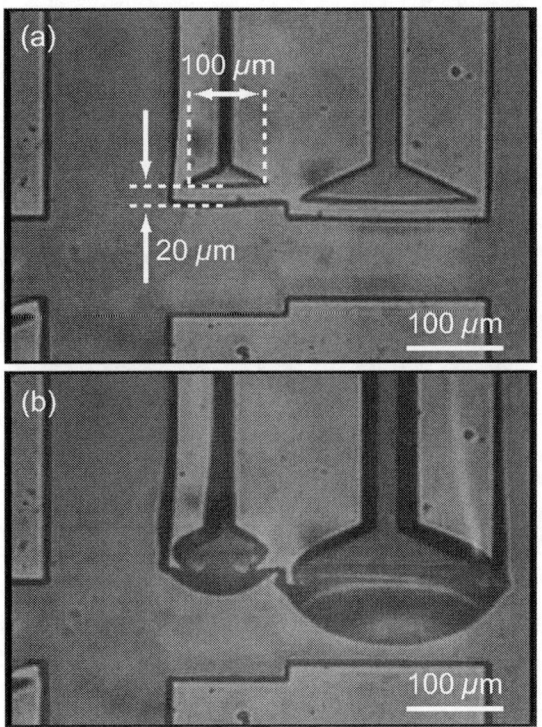

Figure 6: Small valve for selective release. (a) Image of fabricated valves for selective releasing. (b) Image of activated valves. Small valve on left side holds the trapped particles while the large valve on right side controls the fluidic resistance.

978-1-4244-2977-6/09 $25.00 © 2009 IEEE   373

Utilizing these pneumatic valves, we successfully released two types of microbeads separately after making an alternating array of the beads. Figure 7a-f shows the process of selective release. Alternately activated valves held the trapped particles (Fig. 7a, b) and reversed flow retrieves the particles at deactivated valves (Fig. 7c, d). After this selective release, the remaining beads were released by repeating the same process (Fig. 7 e, f).

These results indicate that our device is applicable to trap and release multiple microbeads in a single device. This function enables it to make an array of different types of particles selectively, to enable continuous observation under chemical stimulus, to selectively release a certain subset of the trapped particles separately, and to be able reset the device for reuse.

CONCLUSION

We successfully developed a dynamic microarray system for selective trapping and releasing. We achieved selective trapping and releasing functions by utilizing pneumatic valves that control the fluidic resistance. Using this dynamic microarray device with pneumatic valves, arraying and observation of different types of particles such as microbeads or living cells will be easier and simpler and we believe that our device will aid in the advancement of research in high throughput screening.

ACKNOWLEDGEMENTS

This work was partly supported by Grants from Research and Development Program for New Bio-industry Initiatives, by Core Research for Evolutional Science and Technology (CREST) from Japan Science and Technology Agency, and by Grants-in-Aid for Scientific Research from Ministry of Education, Culture, Sports, Science, and Technology Japan.

REFERENCES

[1] W. H. Tan, S. Takeuchi, "A Trap-and-release Integrated Microfluidic System for Dynamic Microarray Applications", *Proc. Natl. Acad. Sci. U.S.A.*, vol. 104, no. 4, pp. 1146-1151, 2007.

[2] W. H. Tan, S. Takeuchi, "Dynamic Microarray System with Gentle Retrieval Mechanism for Cell-encapsulating Hydrogel Beads", *Lab Chip*, vol. 8, Issue. 2, pp. 259-266, 2008.

[3] K. Iwai, W. H. Tan, S. Takeuchi, "A Resettable Dynamic Microfluidic Device", in *Digest Tech. Papers Microelectromech. Syst. '08 Conference*, Tucson, January 13-16, 2008, pp. 649-652.

[4] M. A. Unger, H. Chou, T. Thorsen, A. Scherer, S. R. Quake, "Monolithic Microfabricated Valves and Pumps by Multilayer Soft Lithography", *Science*, vol. 288, pp. 113-116, 2000.

[5] V. Studer, G. Hang, A. Pandolfi, M. Ortiz, W. F. Anderson, S. R. Quake, "Scaling Properties of a Low-actuation Pressure Microfluidic Valve", *J. Appl. Phys.*, vol. 95, no. 1, pp. 393-398, 2004.

[6] C. Amador, A. Gavriilidis, P. Angeli, "Flow Distribution in Different Microreactor Scale-out Geometries and the Effect of Manufacturing Tolerances and Channel Blockage", *Chem. Eng. J.*, vol. 101, pp. 379-390, 2004.

[7] D. C. Duffy, J. C. McDonald, O. J. A. Schueller, G. M. Whitesides, "Rapid Prototyping of Microfluidic Systems in Poly (dimethylsiloxane)", *Anal. Chem.*, vol. 70, no. 23, pp. 4974-4984, 1998.

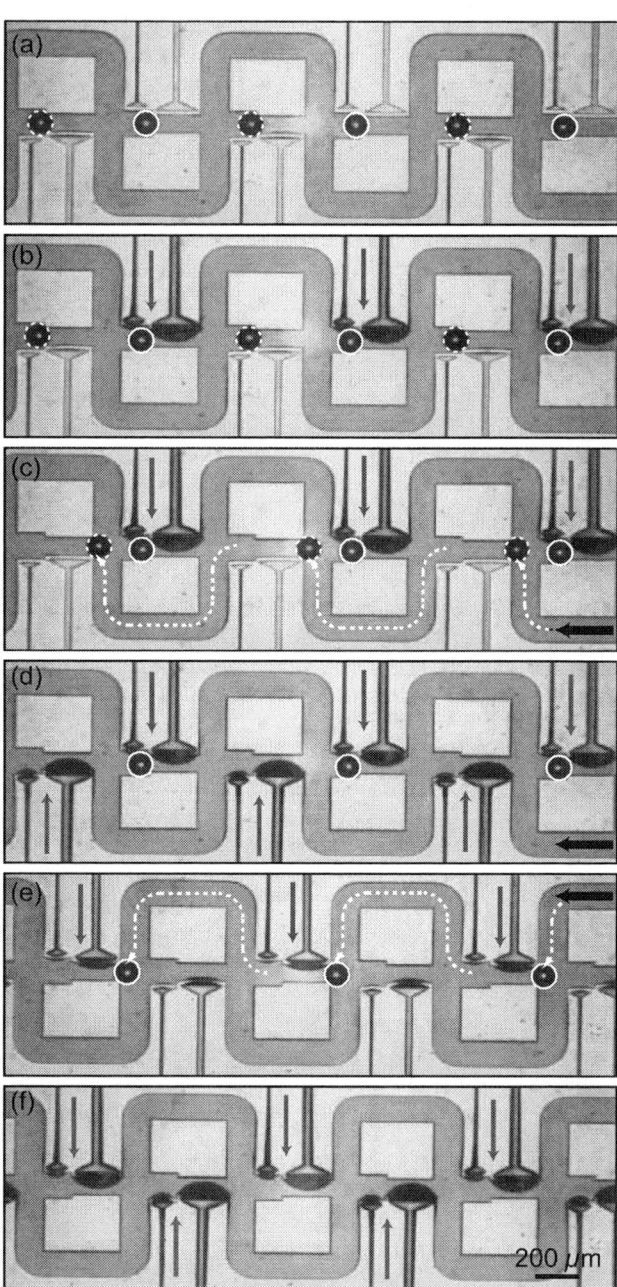

Figure 7: Selective releasing (false color was added to the beads in these images). (a) After trapping beads selectively, (b) we held the first set of beads by pushing the small valves and then (c, d) released the first set of beads. (e, f) By repeating the same procedure, the second set of beads was released and collected separately from the first microbeads.

SIZE-DEPENDENT PARTICLE FILTERATION USING MAGNETICALLY DRIVEN MICROTOOL AND CENTRIFUGAL FORCE IN MICROCHIP

H. Maruyama, S. Sakuma, Y. Yamanishi and F. Arai
Department of Bioengineering and Robotics, Tohoku University, Sendai, Miyagi, Japan

ABSTRACT

We succeeded in size-dependent filtration of microparticles by rotation of magnetically driven microtool (MMT) and centrifugal force in a microchip. Novelties of this paper are summarized as follows. (1) Filtering efficiency was improved than filtration by solely centrifugal force by MMT rotation. (2) Clogging of microparticles was avoided by swirling flow generated by rotation of 3D-MMT with fins. (3) This filtration is robust against pressure fluctuation in a microchip by mechanical particle separation using internal walls. Microparticles with different sizes flow in spiral microchannels and are separated according to their sizes by pass through under each sidewall of microchannels by centrifugal force. MMT is set inside the microchamber and rotated by external magnetic force. Rotation of MMT avoids the clogging of the microparticles and enhances the sorting efficiency. We demonstrated filtration of the microparticles in a microchip using 3D-MMT rotation and centrifugal force.

1. INTRODUCTION

Size-dependent filtration of microparticles such as copolymer beads and cells (from several μm to a hundred μm) is a very important microfluidic process for various research fields[1, 2]. In these days, many types of particle separation methods are developed for the separation of particles and cells, employing acoustic force, optical tweezers, electrostatic force and magnetic force [3-7]. These methods require external forces and modification of the particles prior to filtration, which may cause particle damage and complicate sample collection procedures. It is very important for biological and chemical analysis in a microchip to keep the target particles or cells intact during separation.

Recently, separation techniques based solely on the particle size and the hydrodynamic forces have been developed [8, 9]. Hydrodymanic particle separation does not cause damage to particles and can handle various particles with different properties such as electric permittivity and magnetism. Therefore, this approach eliminates the need for external fields or bead attachment. Research on size-based separation using hydrodymanic force has led to several techniques, including pinched flow fractionation (PFF) and deterministic lateral displacement. Both techniques allow for continuous particle separation, which is crucial to achieve large scale processing. Although these sorting techniques are capable of high efficiency separation, hydrodynamic force is not robust against pressure fluctuation inside microchannel.

In this paper, we proposed an on-chip particle filtration based on the centrifugal force and MMT rotation. We employed a microchip with three spiral microchannels. These microchannels are divided by internal walls with different clearances between the sidewalls and glass. Particles are separated according to their sizes by centrifugal force and internal walls. Our proposed method is robust against the pressure fluctuation inside microchannel because particles are separated mechanically by the sidewalls. 3D-MMT is employed for avoiding the clogging of particles and enhancing the particle separation. Size-dependent partilce filtration was performed successfully.

2. FILTERATION OF MICROPARTICLES USING MAGNETICALLY DRIVEN MICROTOOL AND CENTRIFUGAL FORCE

Principle of particle filtration

Figure 1 shows a schematic of a microchip. Figure 2 shows a principle of particle filtration. Microchip has one microchamber and three spiral microchannels. These microchannels function as particle filters. Each microchannel is connected to next microchnnel by the clearance between glass and bottom of sidewalls. Microparticles smaller than the clearance pass though the sidewall by centrifugal force, while particles bigger than it couldn't pass the sidewall. Particles with different sizes are separated according to their sizes. This filtration is robust against pressure fluctuation since particles are separated mechanically by sidewalls. When pressure is fluctuated, bigger particles do not move to the outside channel. Even if small particles move to the inner channel, these particles move to the outside channel again by centrifugal force.

Moreover, we employed 3D-MMT with fins to prevent microparticles from clogging and enhance the filtering. In microchamber, the rotation of the 3D-MMT generates swirling flow. These flow stirrer particles and avoid clogging of particles. Purity of the filtered particle can be improved by repetition of the filtration because the construction of circulating system is easy in the proposed method.

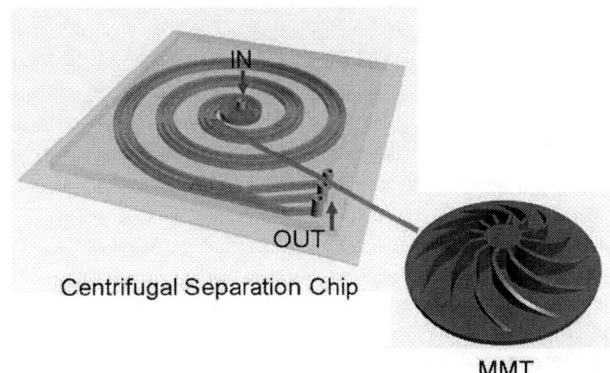

Figure 1: Schematic of size-dependent microparticle filtration using magnetically driven microtool and centrifugal force.

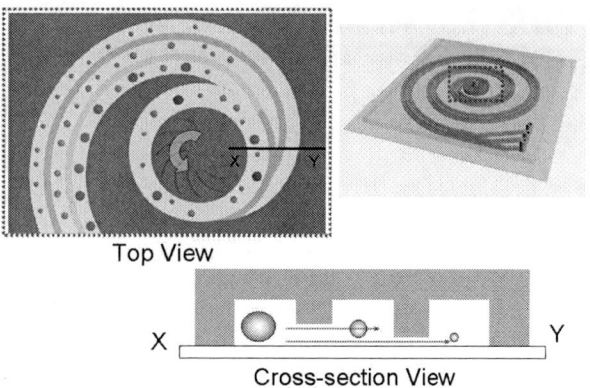

Top View

Cross-section View

Figure 2: Principle of size-dependent particle sorting.

Fabrication of PDMS microchip

PDMS microchip with three spiral microchannels was fabricated using photolithography and replica molding techniques. The fabrication process of the PDMS microchip was shown in Figure 3. We employed SU-8 sheets and multi-exposure to make the microchip with the different height area. Each microchannel is 200 μm in wide and 128 mm in deep. The width of sidewall is 100 μm. The clearance of inner sidewall is 24 μm and that of outer sidewall is 58 μm.

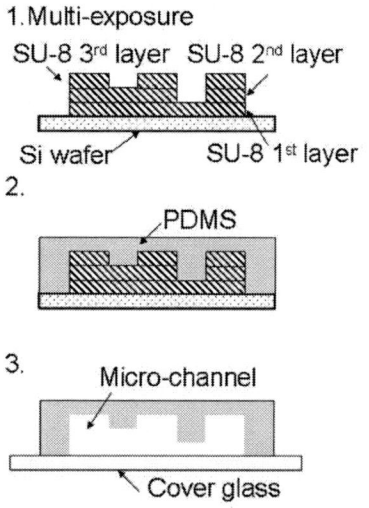

Figure 3: Fabrication process of PDMS microchip using multi-exposure.

Fabrication of 3D-MMT

Figure 4 shows the fabrication process of a 3D-MMT [10]. First, we spin coat KMPR (Kayaku MicroChem Co., Lt) resist on a glass substrate (thickness is about 120 μm), then use ultraviolet exposure to pattern the resist. We employed grayscale photolithography to make MMT three-dimensional structure [11]. In this process, backside exposure is required and a gray-scale mask is positioned at the opposite side of the glass substrate coated with KMPR. Exposure and development are the same processes as conventional photolithography.

Then, we coat a mixture of PDMS and magnetite (50

wt% of Fe_3O_4). After the bake (80 degrees of C), we use a striper liquid to remove the completed MMT with three-dimensional structure. Figure 5 shows the 3D-mold made of KMPR. By using replica molding method, the 3D-MMT made of PDMS mixed with magnetite is completed as shown in Figure 6. Moreover, we can produce a 3D-MMT having smooth curves within the only one exposure step.

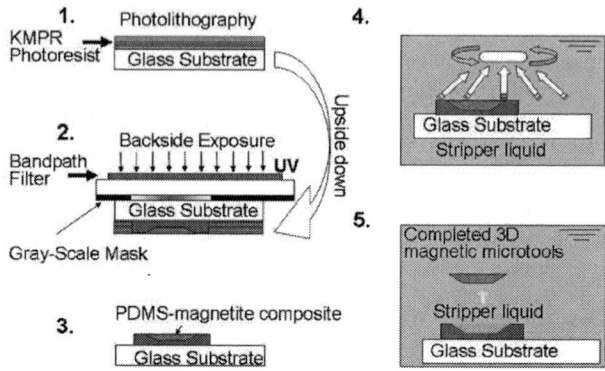

Figure 4: Fabrication process of 3D-MMT using gray scale lithography.

Figure 5: Photographs of 3D-molds made of KMPR.

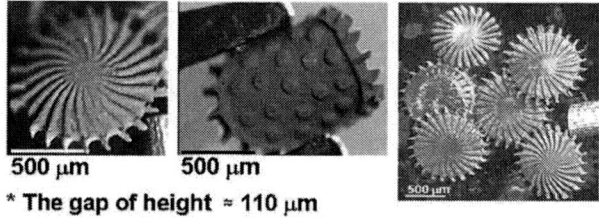

* **The gap of height ≈ 110 μm**

Figure 6: Photographs of 3D-MMTs.

Analysis of swirling flow generated by rotation of 3D-MMT

The rotation of the 3D-MMT with fins generates the swirling flow in the microchamber. This swirling flow is useful not only to avoid clogging of the particles in a microchamber but also to enhance the centrifugal separation of particles. Figure 7 shows the FEM analysis result of distribution of velocity around the rotating 2D (Flat)-MMT and 3D (Tapered)-MMT. Particles move obliquely downward on the 3D-MMT. On the other hand, particles on the 2D-MMT receive lateral force only. We confirmed that 3D-MMT can avoid deposition of the particles on the MMT and pump the particles to the spiral microchannels continuously.

978-1-4244-2977-6/09 $25.00 © 2009 IEEE

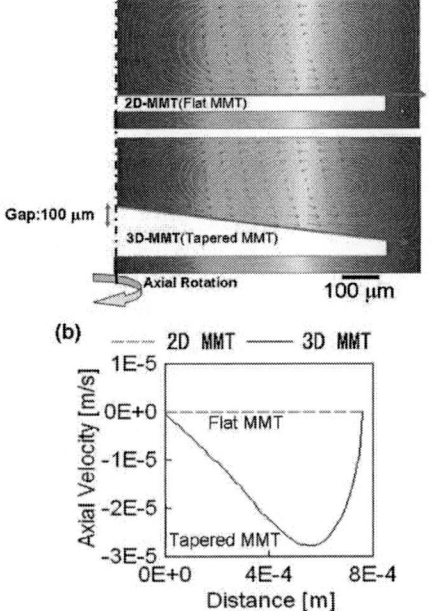

Figure 7: Distribution of velocity around the rotating 2D (Flat) and 3D (Tapered)-MMT. (a) Velocity distribution, (b) Profiles of axial velocity along the lateral component.

EXPERIMENTATAL

Experimental setup

Figures 8 and 9 show the experimental setup. The PDMS microchip was fabricated by photolithography and replica molding techniques. Small particles move to the outside microchannel through the sidewalls as shown in Fig. 8 (b). 3D-MMT with fins was made by grayscale photolithography as shown in Figure 8 (c) and positioned in the microchamber. MMT is 3 mm in diameter and 80 µm in thickness. MMT is rotated by the external rotational magnetic field as shown in Figure 9.

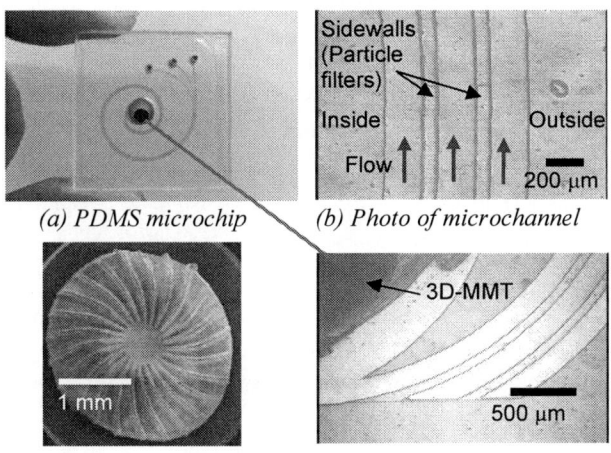

(a) PDMS microchip   *(b) Photo of microchannel*

(c) SEM image of 3D-MMT   *(d) MMT in microchip*
Figure 8: Experimental setup.

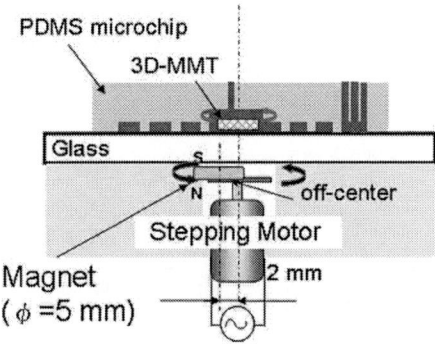

Figure 9: Actuation system of 3D-MMT.

Particle filtering using single-size particle

We used polystyrene particles for experiments. The sizes of microparticles are 20 µm, 50 µm, and 70 µm. Each particle is expected to be separated to the outer, middle, and inner microchannel. Particles were introduced from microchamber by using a syringe pump.

First, we introduced single-size particles to confirm the function of our proposed particle filter. We succeeded in filtration of the particles according to their sizes by centrifugal force and MMT rotation as shown in Figure 10. Particles which are bigger than the clearance between the sidewall and glass did not pass through the sidewall even if pressure was distributed. This system could separate more than 360 particles per second. From these results, we confirmed that the proposed method can be employed as continuous particle filter.

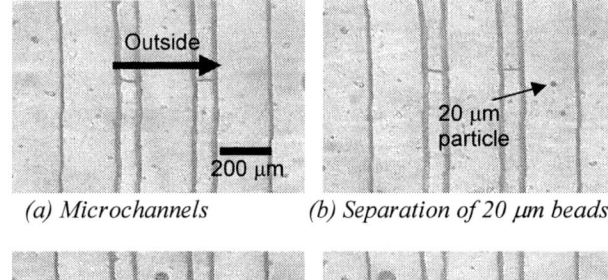

(a) Microchannels   *(b) Separation of 20 µm beads*

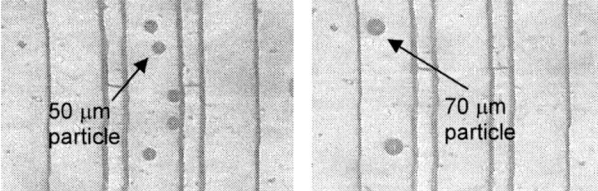

(c) Separation of 50 µm beads   *(d) Separation of 70 µm beads*
Figure 10: Experimental result of particle filtering.

Particle filtering using multi-size particles

Then, we performed the particle filtration using a solution including particles of different size. We used polystyrene particles for this experiment. The sizes are 20 µm, 50 µm, and 70 µm. Experimental results are shown in Figure 11. We succeeded in the filtration of the particles of two sizes by centrifugal force and MMT rotation as shown in Figures 11 (a), 11(b). We also succeeded in the filtration of the particles of three sizes as shown in Figures 11 (c), 11(d). From Table 1, MMT rotation improves sorting efficiency. Microchip and MMT are low cost and disposal. Therefore, proposed system is economical and suited for biomedical applications.

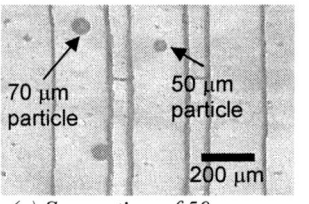 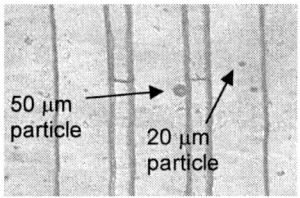

| *(a) Separation of 50 μm and 70 μm beads* | *(b) Separation of 20 μm and 50 μm beads* |

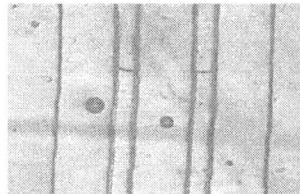 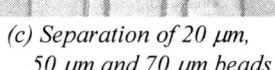 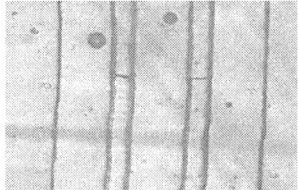

| *(c) Separation of 20 μm, 50 μm and 70 μm beads* | *(d) After separation* |

Figure 10: Experimental result of particle filtering.

Table 1: Comparison of filtering efficiency between with

Particle size	20 μm		50 μm		70 μm	
MMT rotation	With out	With	With out	With	With out	With
Outer channel	24%	38%	0%	0	0%	0%
Middle channel	43%	32%	64%	81%	0%	0%
Inner channel	33%	30%	36%	19%	100%	100%

MMT rotation and without MMT rotation.

CONCLUSIONS

Size-dependent filtration of microparticles by rotation of magnetically driven microtool and centrifugal force in a microchip was developed. We succeeded in size-dependent filtration of particles of multi sizes. Rotation of 3D-MMT avoided the clogging of the particles in the microchamber and enhanced the efficiency of filtration. 3D-MMT is fabricated by grayscale lithography. Rotation of 3D-MMT with fin generates the swirling flow. Microparticles in the microchamber move to the microchannel without clogging by this flow. This flow also enhances the centrifugal separation of microparticles. PDMS microchip with three spiral microchannels and one microchamber was fabricated by using multi-exposure method. Each sidewall between each microchannel has different clearance to glass surface. Particles are separated to each microchannel by passing through the clearance. Our method is robust against pressure disturbance because particles are separated mechanically by sidewalls. We can increase the number of the flittering size by increasing the number of the spiral microchannel and adjusting the clearance between sidewall and glass surface.

Microchip and MMT are low cost and disposal. Proposed system is economical and suited for biomedical applications. This technique for continuous particle filtration will make great contributions for cell biology and chemistry.

ACKNOWLEDGEMENT

This research is supported by The Ministry of Education, Culture, Sports, Science and Technology Grant-in-Aid for Scientific Research (19016004) and by the Sasakawa Scientific Research Grant from The Japan Science Society.

REFERENCES

[1] C. R. Cabrera, P. Yager, "Continuous concentration of bacteria in a microfluidic flow cell using electrokinetic techniques", Electrophoresis, 22, 2, pp.335-362, 2001.

[2] P. Kin Wong, C.-Yang Chen, T.-Huei Wang, and C.-Ming Ho, "Electrokinetic Bioprocessor for Concentrating Cells and Molecules", Analytical chemistry, 76, 23, pp.6908-6914, 2006.

[3] P. Gascoyne, C. Mahidol, M. Ruchirawat, J. Satayavivad, P. Watcharasit and F. F. Becker, "Microsample preparation by dielectrophoresis: isolation of malaria", Lab on a Chip, 2, pp.70-75, 2002.

[4] P. Yu Chiou, A. T. Ohtaand Ming C. Wu, "Massively parallel manipulation of single cells and microparticles using optical images", Nature, 436, pp.370-372, 2005.

[5] M. Yoshida, K. Thoda, M. Gratzl, "Hydrodynamic Micromanipulation of Individual Cells onto Patterned Attachment Sites on Biomicroelectromechanical System Chips", Anal. Chem, 75, pp. 4686-4690, 2003.

[6] F. Arai, A. Ichikawa, M. Ogawa, T. Fukuda, K. Horio, K. Itoigawa, "High Speed Separation System of Randomly Suspended Single Living Cells by Laser Trap and Dielectrophoresis", Electrophoresis, 22, 2, pp.283 – 288, 2001.

[7] S. Vankrunkelsven, D. Clicq, et. al., "A novel microstep device for the size separation of cells", Electrophoresis, 25, 10-11, pp.1714-1722, 2004.

[8] M. Yamada, M. Seki, "Microfluidic Particle Sorter Employing Flow Splitting and Recombining", Anal. Chem., 78, pp. 1357-1362, 2006.

[9] S. Sunahiro, M. Yamada, M. Yasuda1 and M. Seki, "CENTRIFUGAL MICRODEVICE FOR CONTINUOUS AND SIZE-DEPENDENT SEPARATION OF PARTICLES", Proc. of μTAS2007, pp. 898-900, 2007.

[10] Y. Yamanishi, S. Sakuma, K. Onda and F. Arai, "Biocompatible polymeric magnetically driven microtool for particle sorting", Journal of Micro - Nano Mechatronics (online).

[11] R. Mori, K. Hanai, Y. Matsumoto, "Three dimensional micro fabrication of photoresist and resin materials by using gray-scale lithography and molding", T.IEEE Japan, 124-E, 10, pp.359-363, 2004.

CLUSTER DYNAMICS IN FLOW OF SUSPENDED PARTICLES IN MICROCHANNELS

T. Gudipaty[1], M.T. Stamm[1], L.S.L. Cheung[1], L. Jiang[2], and Y. Zohar[1]

[1]Department of Aerospace and Mechanical Engineering, [2]College of Optical Sciences
University of Arizona, Tucson, AZ, USA

ABSTRACT

Cluster dynamics in microchannels due to flow of dilute suspensions of polystyrene spherical particles has been studied experimentally. Cluster-cluster interaction as well as the functional dependence of cluster growth rate on several control parameters has been studied. Destructive, e.g. cluster collision, and constructive modes, e.g. cluster merging, of cluster-cluster interaction have been observed. Cluster growth rate with time is found to increase with either particle concentration or shear strain rate, and decrease with channel-height to particle-diameter ratio.

INTRODUCTION

The utility and the life-time of microfluidic systems depend critically on their ability to maintain flow without interruption. Hence, the elucidation of potential conditions under which blockages can occur is important. However, despite their importance, the physical mechanisms that lead to clogging of porous media are exceptionally complicated and still not well understood. The simplest cause of clogging is that of pure size exclusion, where blockage occurs only if particles enter a pore that is smaller than their diameter. However, clogging can be observed even when the pores are much larger than the suspended particles. Indeed, adhesive particulates have been identified as a leading cause of failure in microfluidic devices, including those currently being developed for biological and chemical applications.

Possible mechanisms suggested for clogging include the random occurrence of a local particle concentration large enough to result in a physical blocking of pores [1] or arch formation within a channel [2]. Based on a geometrical analysis, it was shown that once the particles are in the arching configuration, lateral forces induced by the shear on the arch can hold the particles in place and stabilize the arch.

Model descriptions of clogging often focus on different length scales than do typical experimental investigations [3]. A soft-sphere discrete element model was used to examine particle aggregate formation and particle capture by walls in a laminar channel flow [4]. Particle capture by a substrate surface was examined with a DEM model using a 'frozen deposit' assumption [5]. Shear flow of particulate suspension was analyzed using a kinetic clustering model and direct numerical simulations [6].

Aggregation processes inevitably generate clusters, which often result in clogging. Hence, it is important to understand the aggregation process. Shear flows bring particles close to each other faster than Brownian motion, thus enhancing the aggregation rate. The cluster formation rate, induced by van der Waals attractive interaction, should be proportional to the shear rate for large Peclet number. The clusters break up under large shear stresses, at high shear, and the aggregation rate levels off or even decreases. Several experiments investigating these phenomena qualitatively confirmed such expectations [7,8].

The difficulty in conducting aggregation experiments, based on van der Waals interactions, lies in the sensitivity of the attractive forces to variations on the particle surfaces and in the solution. However, systems employing electric or magnetic attractive forces are less sensitive to such variations. Aggregation of paramagnetic particles in microchannels has recently been reported [9]. The growth mechanism for linear chain was observed to consist of the accumulation of isolated particles or small clusters onto existing chains, which were all moving at different speeds. At short times, the chain length increased linearly and had a growth rate that increased as a power law with the shear; while, at longer times, the chain length saturated.

A description of clogging may require an understanding of the blockage mechanism at a single pore level. In such a study, clogging was found to be independent of particle flow rate and volume fraction [10]. The occurrence of sticking events was determined by the balance of the repulsive forces and the drag force towards the wall. A critical distance of minimum approach was identified corresponding to a length comparable to the attractive well of the interaction potential that stabilizes particles against aggregation. This new scale was found to be comparable to the typical thickness of an electrostatic double layer at low ionic concentration.

In our previous work, the formation and evolution of single clusters due to a flow of particle suspension in a microchannel has been reported [11]. However, as clusters grow in number and size, cluster-cluster interaction becomes the dominant mechanism for channel blockage. Here we discuss these interactions, and quantify the dependence of the cluster area growth rate on the dominant control parameters.

EXPERIMENTAL ARRANGEMENTS

Microdevice Fabrication

The fabrication process starts with thermal growth and patterning of a 3000Å-thick oxide on a silicon wafer to serve as a mask for silicon etching using TMAH. The oxide etch mask is stripped away after the formation of a complimentary microchannel mold on the substrate about 3cm in length, 1mm in width and 15μm in height. A fresh 1μm-thick oxide layer is then thermally grown on the Si wafer to complete the fabrication of the microchannel mold. Next, a polydimethylsiloxane (PDMS) mixture of base and curing agent with a ratio of 10:1 is poured onto the mold. After curing at 60°C for 5 hours, the PDMS substrate with a microchannel replica is peeled off. Inlet/outlet holes are then drilled mechanically through the PDMS substrate, followed by attachment of two adaptors at both ends of the microchannel. The bonding surface of the PDMS substrate is treated in a plasma reactor rendering it hydrophilic.

978-1-4244-2977-6/09 $25.00 © 2009 IEEE   379

Finally, the treated PDMS substrate is bonded to a flat oxidized silicon wafer by pressing together the two substrates to complete the construction of the microchannel device.

Experimental Set-up

A schematic of the experimental set-up, with a photograph of a packaged microchannel device, is shown in Figure 1. Commercially-available, electrosterically stable polystyrene microparticles (Phosphorex, Inc.) were diluted in distilled water to obtain the suspension solutions. In this work polystyrene particles 1.5, 5 and 7.5μm in diameter were used. Three levels of void fraction were tested: 0.001, 0.005 and 0.01, providing one order of magnitude in range. The suspensions, sonicated immediately before an experiment to obtain uniform particle distribution, were forced through the microchannels using a syringe pump at a flow rate of 0.5, 1 or 2ml/hr. A gauge pressure transducer was installed between the syringe pump and the microchannel inlet to monitor the inlet pressure as an indicator of a complete channel blockage. All experiments were conducted under an optical microscope equipped with a CCD camera connected to a DVD recorder to facilitate further analysis of instantaneous images captured during the experiments.

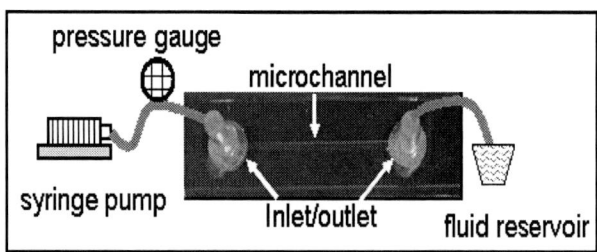

Figure 1, A schematic of the experimental set-up with a photograph of a packaged microchannel device.

CLUSTER DYNAMICS

Incipience of nucleation sites, serving as seeds for particle aggregation to form clusters on a microchannel bottom surface, is random both in space and time. Clusters initially evolve independently, with no cluster-cluster interaction, through aggregation of flowing isolated particles. Under most conditions tested in this work, the cluster shape resembles that of a wake behind a 2-D cylinder placed in a uniform flow as shown in Figure 2a. However, shown in Figure 2b, a chain-like cluster shape appears when the particle diameter is about half the height of the channel.

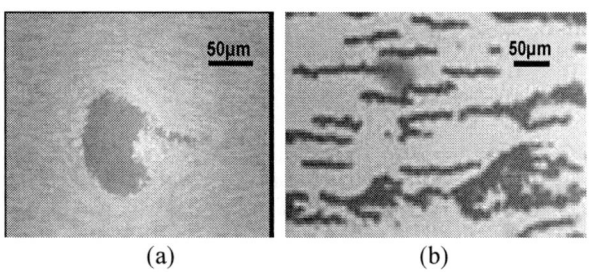

Figure 2, Images of cluster shape under similar flow conditions for: (a) $h/d=10$, and (b) $h/d=2$.

As the clusters continue to grow, cluster-cluster interactions start to play a prominent role. Several modes of cluster-cluster interactions are observed and, in general, they can be classified into either constructive or destructive modes. A most obvious example of constructive mode is cluster merging. This occurs when stationary neighboring clusters continuously grow in place until the gap between them is filled with particles recruited from the main stream. It can also occur when an upstream cluster is pushed under flow-induced stress, as depicted in Figures 3a-b, combining together with a downstream cluster to form a larger cluster.

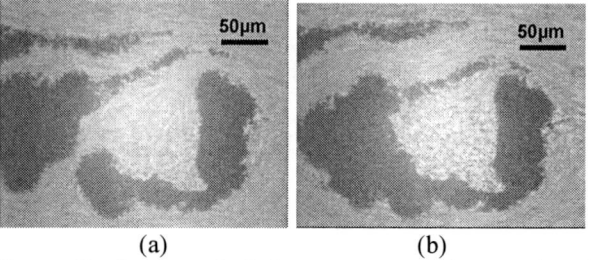

Figure 3, Images depicting a constructive mode of cluster-cluster interaction: (a) two clusters before merger, and (b) a single cluster after merger.

In contrast, a simple example of a destructive mode is the detachment of a whole cluster from the channel surface due to hydrodynamic forces sweeping it downstream. This mode also occurs when a flowing particle aggregate collides with a stationary cluster, shown in Figures 4a-b, resulting in the removal of the stationary cluster.

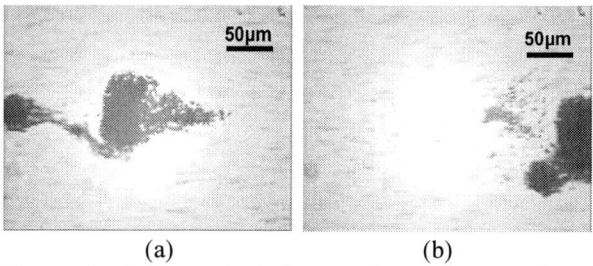

Figure 4, Images depicting a destructive mode of cluster-cluster interaction: (a) a collision between a flowing particle aggregate and a stationary cluster, and (b) cluster detachment following the collision.

CLUSTER EVOLUTION

Prior to onset of cluster-cluster interaction, single particle-cluster interaction is the dominant mechanism for cluster growth. To characterize this growth, the area and number of clusters in three microchannel zones: inlet, center and outlet, are determined using recorded instantaneous images. The average cluster area A_Z can then be calculated for each zone, and the results are summarized in Figure 5. The cluster area grows almost linearly with time, with a growth rate depending on its location along the microchannel. Similarly, we have previously reported that both the number of clusters and the cluster area are the lowest at the microchannel center in comparison with those values at the channel inlet and outlet regions [11].

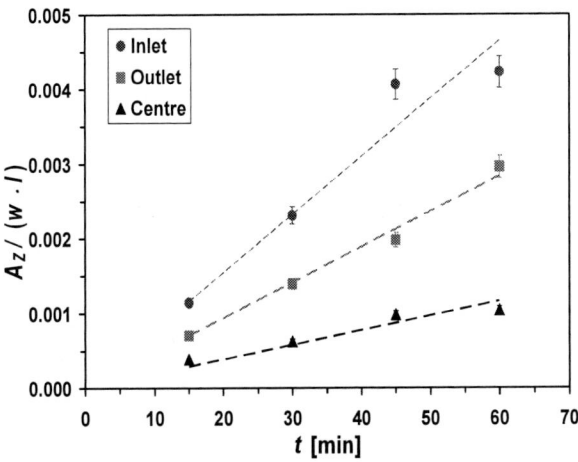

Figure 5, Cluster-area growth in time for the inlet, center and outlet regions of the microchannel.

To characterize cluster dynamics in the microchannel, the total area of all clusters is required. The time-dependent total cluster area measured under various flow rate, particle diameter and void fraction conditions is plotted in Figure 6. In each case, the growth rate is constant for at least 1 hour.

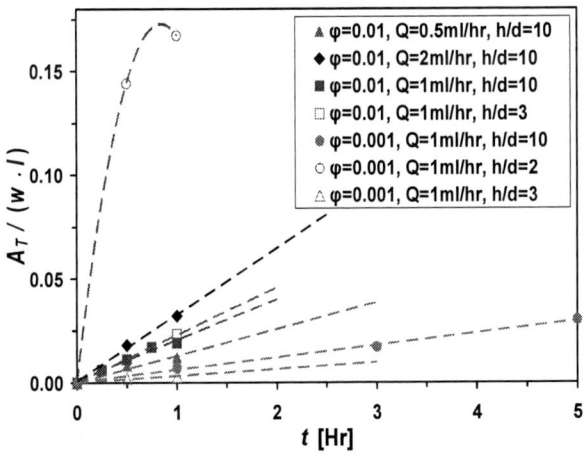

Figure 6, Growth of the total cluster area with time in the microchannel for various experimental conditions.

CLUSTER GROWTH RATE ANALYSIS

The first treatment of flocculation, aggregation of particles dispersed in a solution to form larger-size clusters, was introduced by Smoluchovski. His theoretical model can be simplified significantly assuming that the population is composed of two elements: $N(t)$ isolated particles, and N_C clusters; furthermore, the isolated particle population in the suspension decays as particles join existing clusters. Under the assumptions, the Smoluchovski equations reduce to [9]:

$$\frac{dN}{dt} = -N(t)N_C K \qquad (1)$$

where the kernel K represents the rate of cluster formation. Assuming that the isolated particles randomly aggregate into close-packed, two-dimensional clusters leads to:

$$\varphi_R \frac{dV_C}{dt} = -\frac{V_P}{N_C}\frac{dN}{dt} \qquad (2)$$

where $V_P = \pi d^3/6$ is the volume of a single particle, $V_C = h \cdot A_C$ is the cluster volume, and $\varphi_R = 0.64$ is the volume fraction filled by mono-disperse spherical objects in random close packing; here, d is the particle diameter, h the channel height, and A_C is the cluster area. Equations (1) and (2) can now be combined to obtain:

$$\frac{dA_C}{dt} = -\frac{\pi d^3}{6\varphi_R h N_C}\frac{dN}{dt} \approx \frac{\pi d^3}{6h\varphi_R}N(t)K \qquad (3)$$

Particle Concentration Effect

Equation (3) suggests that, for weakly varying particle concentration $N(t)$, the cluster area growth is linear in time; this is consistent with the results shown in Figure 6. Hence, the initial area growth can be expressed as follows [9]:

$$A_C(t) = \beta N(t)t \qquad (4)$$

If the particle concentration $N(t)$ is quasi steady, then β represents the growth rate with time of the cluster volume/area. Equation (4) indicates that the cluster area is directly proportional to particle concentration. Using the data in Figure 6, the cluster growth rate is plotted in Figure 7 as a function of the particle void fraction, φ, which is directly proportional to the particle concentration ($\varphi = N \cdot V_P$). Indeed, the cluster growth rate increases linearly with particle concentration (void fraction) as suggested by Equation (4).

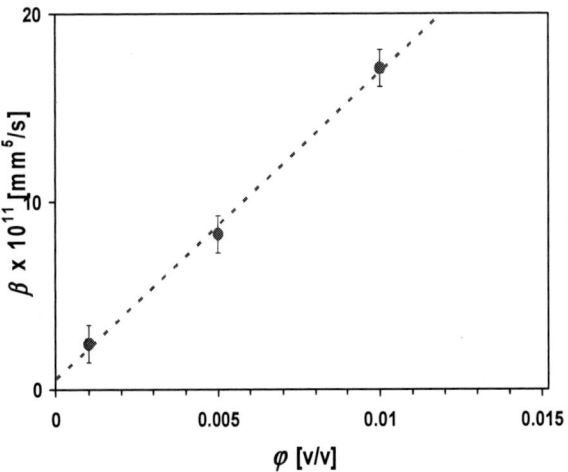

Figure 7, Cluster-area growth rate dependence on the particle void fraction, φ.

Shear Strain Rate Effect

In orthokinetic flocculation, i.e. particle aggregation due to fluid motion, K is equal to the shear strain rate S multiplied by a capture volume on the order of λ^3; where λ is a characteristic scale depending on the particle dynamics. Thus, combining Equations (3) and (4) leads to:

$$\beta \propto \frac{d^3}{h}S\lambda^3 \qquad (5)$$

The wall shear strain rate for a plane channel is given by: $S = 6Q/wh^2$; where Q is the volume flow rate, and w is the channel width. Since the dependence of the scale λ on the shear S is complicated and unknown at this stage, it is not possible to directly discern the relationship between the cluster growth rate and the flow shear strain rate. However, dimensional analysis suggests a power law relationship:

$\beta \sim S^a$ with an exponent a. The measured growth rate dependence on the wall shear strain rate, plotted in Figure 8, yields an empirical estimate for $a \cong 0.65$. In comparison, an exponent of 0.25 has been reported for aggregation of paramagnetic particles in a magnetic field forming linear chains [3].

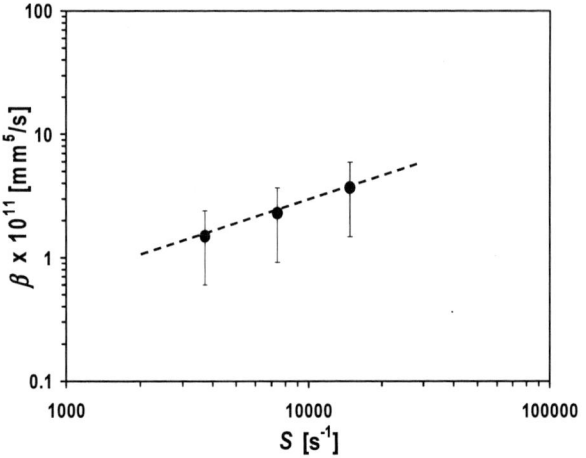

Figure 8, Cluster-area growth rate dependence on the wall shear strain rate, S.

Particle diameter to Channel Height Ratio Effect

Equation (5) indicates that the growth rate depends not only on the shear strain rate but also on the channel-height to particle-diameter ratio: $\beta \sim (d/h)^3 \lambda^3$. Indeed, based on experimental results, the critical particle concentration needed for blockage of microtubes has been reported to depend on $(d/h)^{-3}$ [12]. However, just as in the strain rate case, λ also depends on d and h. In this work, while the channel height is kept constant at h=15μm, experiments were conducted using particles with diameter of d=1.5, 5, and 7.5μm. The cluster growth rate measurements are plotted in Figure 9 as a function of h/d. The current results yield a best fit power law of $\beta \sim (h/d)^{-4}$. This suggests that the characteristic scale λ is perhaps directly proportional to the particle-diameter to channel-height ratio, $\lambda \sim (d/h)^{1/3}$; however, more experiments are required to confirm this relationship.

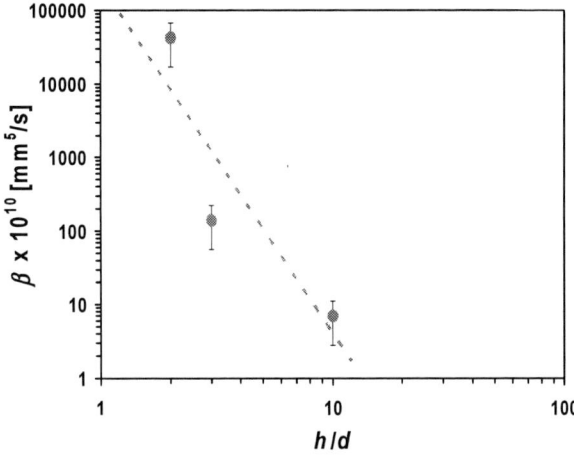

Figure 9, Cluster-area growth rate dependence on the channel-height to particle-diameter ratio h/d.

CONCLUSIONS

Microdevices based on PDMS-to-silicon bonding technology have been utilized to study cluster growth due to flow of polystyrene-particle suspensions through micro-channels. Initially, clusters grow almost linearly with time by the capture of isolated flowing particles. As the clusters keep growing, constructive and destructive modes of cluster-cluster interaction have been observed, such as cluster collision and merging, leading to a non-linear cluster growth rate. The initial cluster growth rate is found to increase linearly with the particle concentration, as suggested by a simplified theoretical model. Furthermore, fitting a power law to these preliminary data, the cluster growth rate is empirically found to increase with the wall shear strain rate (to the power of 0.65), and decrease with channel-height to particle-diameter ratio (to the power of -4).

ACKNOWLEDGEMENTS

This work is supported by Arizona Biomedical Research Commission grant 06-080.

REFERENCES

[1] G. H. Goldsztein and J. C. Santamarina, "Suspension extraction through an opening before clogging", *Appl. Phys. Lett.*, vol. 85, pp. 4535-4537, 2004.

[2] K. V. Sharp and R. J. Adrian, "On flow-blocking particle structures in microtubes", *Microfluidics and Nanofluidics*, vol. 1, pp. 376-380, 2005.

[3] C.J. Koh *et al.*, "An experimental investigation of concentrated suspension flows in a rectangular channel", *J. Fluid Mech.*, vol. 266, pp. 1-32, 1994.

[4] J.S. Marshall, "Particle aggregation and capture by walls in a particulate aerosol channel flow", *J. Aerosol Sci.*, vol. 38, pp. 333-351, 2007.

[5] A. Konstandopoulos, "Deposit growth dynamics: Particle sticking and scattering phenomena", *Powder Technology*, vol. 109, pp. 266-277, 2000.

[6] P. Raiskinmaki *et al.*, "Clustering and viscosity in a shear flow of a particulate suspension", *Physical Review E*, vol. 68, 061403 1-5, Dec 2003.

[7] M. Vanni and G. Baldi, "Coagulation efficiency of colloidal particles in shear flow", *Adv. Colloid Interface Sci.* vol. 97, pp. 151-177, 2002.

[8] H. Mousa *et al.*, "Experimental Investigation of the Orthokinetic Coalescence Efficiency of Droplets in Simple Shear Flow", *J. Colloid Interface Sci.*, vol. 240, pp. 340-348, 2001.

[9] E. Brunet *et al.*, "Aggregation of paramagnetic particles in the presence of a hydrodynamic shear", *J. Colloid Interface Sci.*, vol. 282, pp. 58–68, 2005.

[10] H.M. Wyss *et al.*, "Mechanism for clogging of microchannels", *Physical Review E*, vol. 74, pp. 061402 1-4, 2006.

[11] T. Gudipaty *et al.*, "Cluster formation and evolution in particle-laden microchannel flow", *Proc. MicroTAS'08*, pp. 110-112, San Diego, CA, 12-16 October, 2008.

[12] E. Yamaguchi and R.J. Adrian, *XXI ICTAM*, Warsaw, Poland, 15-21 August, 2004.

A MULTIFUNCTIONAL VERTICAL MICROSIEVE FOR MICRO AND NANO PARTICLES SEPARATION

C.Shen, H.T.M.Pham, P.M.Sarro

Delft University of Technology, DIMES-ECTM, Delft, the Netherlands

ABSTRACT

This paper presents the design, fabrication and characterization of a new concept of microsieves with vertical nano-perforated walls. A new approach is applied to realize these walls accurately without sophisticated or non-conventional lithography while preserving IC-compatibility. By simply changing the deposition and etch time in the fabrication process, different pore size microsieves are fabricated. Further, by combining several walls with specific design modifications, a range of functionalities (filtration, separation, anti-choking, etc) can be integrated on one chip. Microsieves with 100 nm and 1μm pore size are fabricated. Separation and anti-choking functionalities are successfully demonstrated.

INTRODUCTION

Filtration or separation of particles in fluids, based on their size, is quite common and very useful in laboratory analyses. A lot of applications can be found in medical, pharmaceutical and food industry. For example, by sieving and identifying specific bacteria from blood sample, some diseases can be determined faster and more accurately [1].

These applications usually deal with different particle sizes, need an anti-choking function to extend the lifetime of the separation device and are less sensitive to flow rate variations. Several kinds of devices can be found in literature or on the market.

Cross-linked porous polymer or glass membranes are the most widely used filtration devices. These devices can be of different shape, with different pore size and are already commercially available at a relatively low cost thanks to volume production [2-3]. However, their disadvantage is also clear: the big particles will be trapped inside and cannot be easily retrieved even after filtration. This results in a limited lifetime for the sieving devices. Besides, the pore size distribution of these porous materials is poor compared to other micromachined filtration devices.

Micromachined Brownian ratchets [4-7] overcome the lifetime problem. Brownian ratchets make use of the difference in Brownian motion between small and big particles to separate them. However, Brownian motion is sensitive to many factors, like particles density, fluids temperature, and even surface properties of the device [7]. Changes in these factors may also affect the separation accuracy and efficiency.

When particles are carrying electronic charges, electrical or magnetic fields can be used to help the separation. An example is the Anisotropic NanoFilter Array (ANA) [8]. But to use these devices, particles must be electrically charged.

In the search towards separation devices for uncharged particles with a sharp separation profile and good reliability, mechanical separation using micromachined microsieves is reported. Like a miniaturized flour sieve in kitchen, microsieves usually have a thin horizontal membrane with micro pores patterned with photolithography and etched through [1]. Through these pores, small particles can penetrate the membrane while big particles cannot.

Microsieves have highly uniform pore size, simple structure and possibility to trap particles for further analysis. However, previously reported lateral mechanical sieving devices [9] face three challenges: nanosieve fabrication, anti-choking and sorting particles with different sizes in one device. It's difficult to fabricate <100 nm pores on a membrane with conventional lithography. Using advanced lithography would be too expensive for most MEMS applications and with non-conventional photolithography technologies -- like e-beam [10] and laser interference [9] -- either fabrication takes a long time or process conditions are quite critical. Another challenge is to have anti-choking for continuous separation and to extend device lifetime. Particle may accumulate on the surface or inside the microsieve and block the pores which might inhibit the fluid flow through the microsieve [9]. Efficient removal of these particles is also quite challenging. The last challenge is because one lateral microsieve can have only one pore size, the integration of several lateral microsieves on one chip is very difficult.

Our vertical microsieve reported here provides a simple but versatile solution to achieve a range of functionalities (separation, filtration, anti-choking, etc) in a wide range (sub 100 nm to several microns) particle size, which addresses the above mentioned challenges.

DESIGN AND FABRICATION

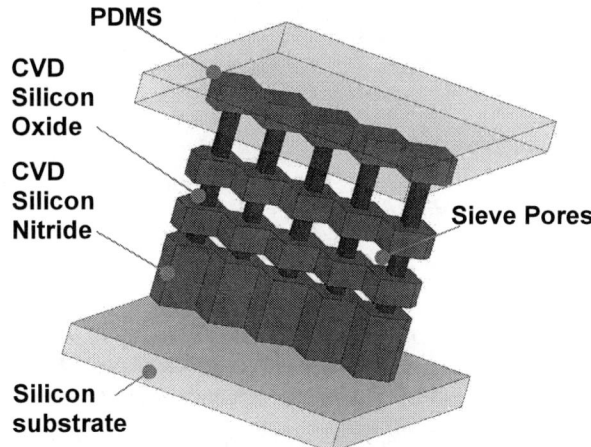

Figure 1: A schematic view of our versatile vertical wall microsieve concept: different pore sizes on a single wall are fabricated by changing the sacrificial layers thickness.

The key to the versatile character of our microsieves is the perforated walls which form separated fluid channels (Fig.1). Only particles smaller than the pore size can penetrate the walls and reach the adjacent channels. The walls are composed of easy-to-etch sacrificial layers and hard-to-be-removed structural layers. By using one single wet chemical etch step after all the vertical walls are defined, flat rectangular pores are formed on all the sacrificial layers at once. The pore's height, which is smaller than the width and controls the particle size, is determined by the sacrificial layers thickness. By changing the thickness, we can customize the sieve, realizing different pore sizes on each layer and on each wall (Fig.1).

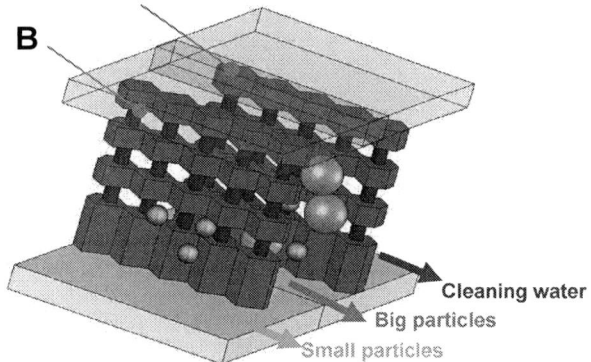

Figure 2: Anti-choking function with two identical sieve walls: Wall A is flushed by cleaning water while wall B is functioning as sieve. By alternating these modes, a non-stop anti-choking microsieve is obtained.

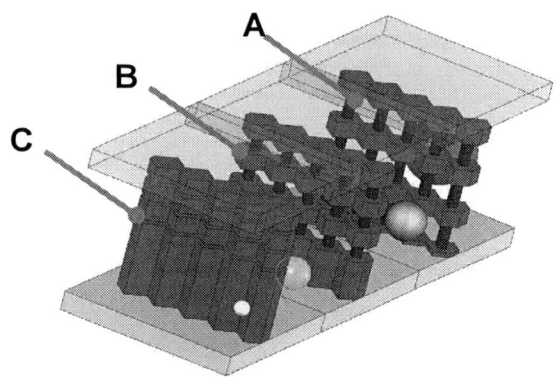

Figure 3: Multiple sievewalls with different pore sizes (A, B, C in the drawing) can be used to separate particles into different size ranges.

The inherent versatility of this concept allows having many vertical walls, different in shape and pore sizes integrated on the same chip. Consequently, many types of microsieves with different functions are realized. With one single wall (Fig.1) we have a binary separator or a filter. Fig.4 shows (close-up) a fabricated demonstrator for sieves walls with different pore sized from 0.5 μm to 1.5 μm. With two identical sieve walls placing in parallel, the anti-choking function (Fig.2) can be built as shown in the results and discussion section. With more walls, a particle separation system that covers a certain range of particle sizes is obtained (Fig.3). Fig.5 shows such a demonstrator with multiple sieve walls.

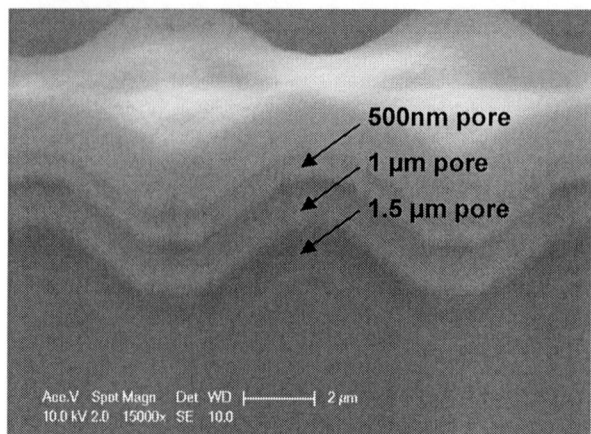

Figure 4: SEM picture of a microsieve with different sized pores.

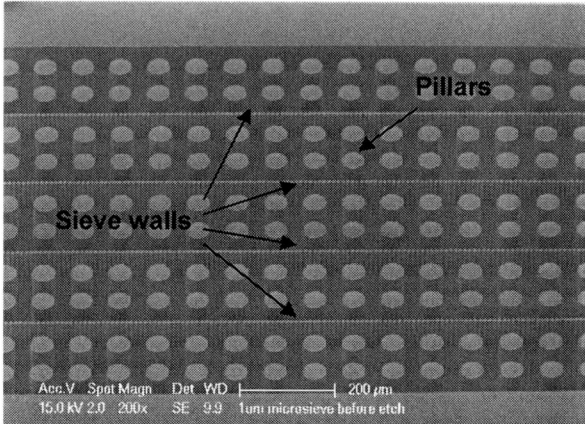

Figure 5: SEM picture of a microsieve with 4 sieve walls (top view). The pillars are designed to provide additional support to the PDMS film used to seal the microsieve.

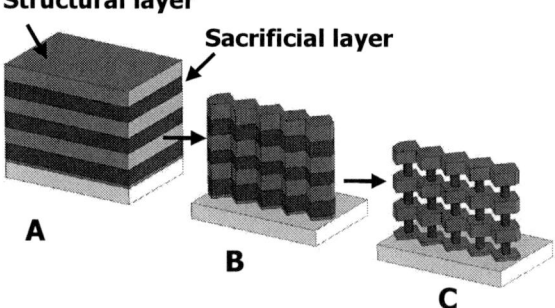

Figure 6: Schematic drawing of the sieve wall main fabrication steps: a) deposition of the layer stack by CVD, b) dry etching of the stack to build the zigzag shaped wall, c) wet etching of sacrificial layers to get the perforated sieve structure.

The fabrication process is simple as the structure is optimized in the design phase. After the deposition of the sacrificial and structural layers by chemical vapor deposition (CVD), the sieve walls are patterned and the channels etched. The walls are designed in a zigzag shape with the narrowest part being only 1μm wide. During etching of the sacrificial layer, the narrowest part will be rapidly etched through to form pores while the rest will be only partly etched leaving sufficient amount to support the structure (Fig.6). A porous structure is thus formed. Poly-

978-1-4244-2977-6/09 $25.00 © 2009 IEEE   384

dimethyl-siloxane (PDMS) sheets are then bonded on the top of the wafer to seal the microsieves. Finally, a surface treatment like oxygen plasma exposure and ethanol immersion is used to improve the wetting for small pores. Since there is no separate patterning for each sacrificial layer, the whole fabrication process can be done with only one lithography step. A detailed description of the fabrication process is reported in [11].

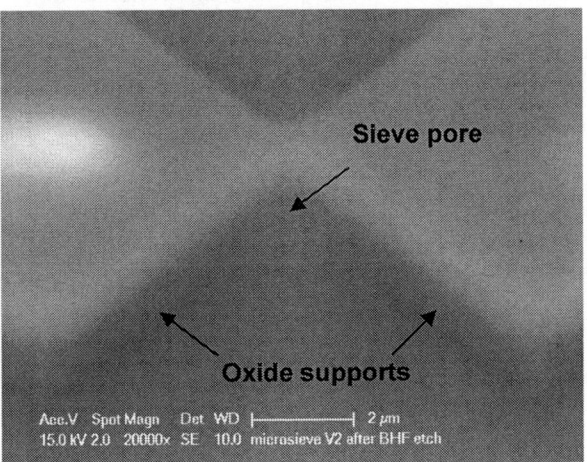

Figure 7: Close-up view of a pore structure showing oxide supports next to a pore.

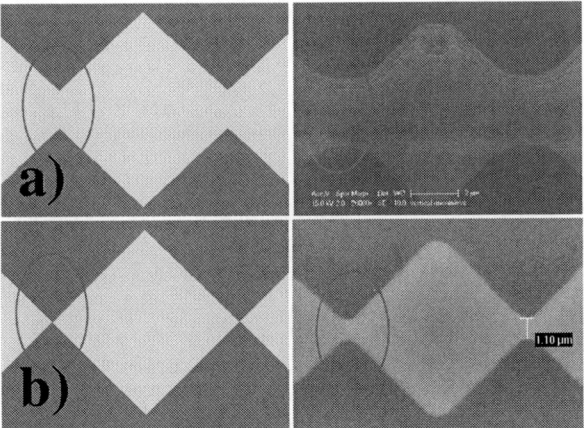

Figure 8: Comparison between the two layouts used and the corresponding result after exposure: a) initial design used in [11] with 2-3 μm in the narrowest part of sieve wall; b) new layout with only 1 μm in the narrowest part.

In order to improve the process as well as the performance of the sieves we have modified the layout of the walls. Compared to our previous design [11], the main change is in the reduction of the narrowest area of the shape of the zigzag walls (see Fig.7). In the initial design the narrowest part of the sieve wall is about 2-3 μm. After redesign, the narrowest part has been reduced to 1 μm. This change can shorten the distance particles have to travel through the sieve wall, which is important in reducing the choking phenomena as small particles may stick to the inner surface of the sieve. Moreover, the sacrificial layer etch time is also greatly reduced. Especially for the small sized (100 nm) microsieves where the wet etchant needs a long time to penetrate into the pores, this can be a significant improvement.

RESULTS AND DISCUSSION

To test and characterize the proposed concept of vertical wall sieves, two types of microsieves are fabricated. The geometric parameters are as listed in Table-1. The type I is used to test the filtration result for 100 nm microsieves. Type II is used to test the filtration of 1 μm microsieves and the anti-choking functionality.

Table 1: Geometric parameters of two fabricated microsieves. Type I: 100nm sieve Type II: 1μm sieve.

Geometric parameters	Type I	Type II
Sieve wall width (max)	10 μm	10 μm
Sieve wall width (min)	2-3 μm	1 μm
Lateral interval between two pores	9 μm	10 μm
Sacrificial layer thickness	100 nm	1 μm
Structure layer thickness	500 nm	1 μm
Sacrificial layer number	3	3
Pore width	2-3 μm	2-3 μm
Pore height	100 nm	1 μm

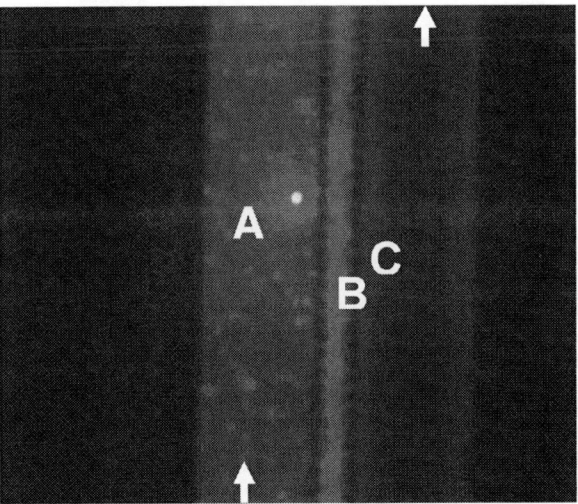

Figure 9: Microscope pictures of a 100 nm microsieve during operation: 200nm and 2μm particles (A) are separated from coloured water (C) by two 100 nm sieve walls (B).

In the experiment with 100 nm microsieves (Type I) fluorescent beads larger than 200 nm and water dyed with Rhodamine, are used. We inject both particles and water from the bottom of the left channel and observed successful blocking of 200nm particles in the left channel and water in the right channel (Fig.9).

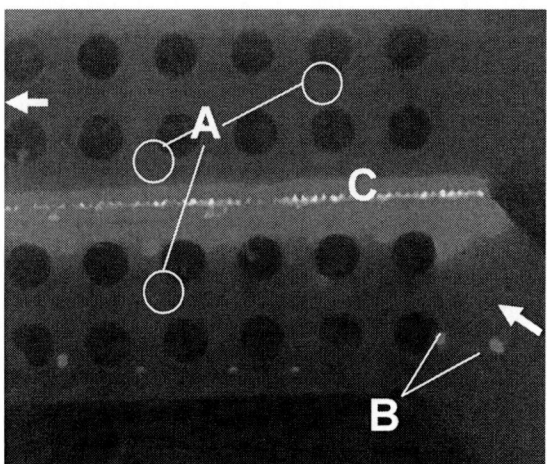

Figure 10: Microscope picture of 1μm microsieve during operation: 200 nm particles (A) are separated from 2μm particles (B) by one 1μm sieve wall (C). It's clear to see particles accumulate at the sieve wall(C).

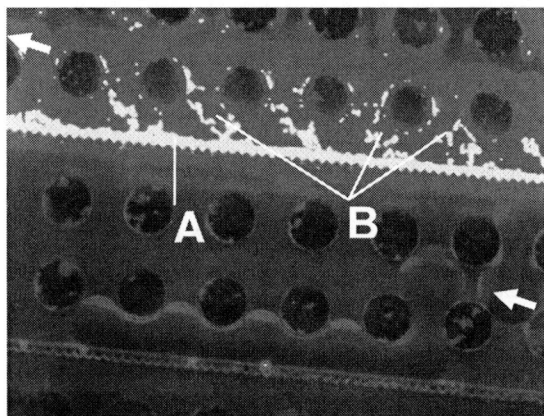

Figure 11: Microscope pictures for anti-choking test: particle choked on upper surface of sieve wall (A) are pushed away (B) by water flushed from backside of the wall (following white arrow direction).

In the experiment with the 1μm microsieve (type II), 2μm and 200nm particles are separated by a 1μm microsieve. The 2 μm particles start to accumulate on the sieve wall right after the particles are injected (Fig.10). After cleaning water flows from the backside of the wall, particles that choke the sieve wall are visibly pushed away (Fig.11). This demonstrates that the proposed anti-choking mechanism is working.

One limitation of the design shown here is the flow rate. According to the particle velocity measured in our experiments, the flow rate for a 10mm long sieve wall with 1μmx3μm pores is 0.972μl/hr. Larger flow rates are possible using higher pressure or longer walls.

CONCLUSIONS

In this paper we have presented a new concept of microsieves that uses vertical nano-perforated walls. The flexibility of this concept allows different sized microsieves to be fabricated with the same mask set, by simply changing the thickness of the structural and sacrificial layers in the fabrication process. Moreover, modifications at design level allow combining several types of sieve walls thus integrating several functionalities on one chip.

Both 100 nm and 1 μm microsieves have been designed, fabricated and tested. Successful separation of water, 200 nm and 2 μm particle beads has been observed with these microsieves.

The proposed anti-choking functionality has been tested as well. By flushing water reversely on the sieve wall, particles choked on the sieve can be flushed away. By placing two sieve walls in parallel, it is therefore possible to keep one sieve working while the other is being flushed. In this way, a non-stop anti-choking microsieve can be fabricated.

ACKNOWLEDGEMENTS

The authors would like to thank the DIMES IC Process Group and Bio-devices Group for technical support. This project is funded by MicroNed (SMACT 2-D).

REFERENCES

[1] http://www.fluxxion.com
[2] http://www.spg-techno.co.jp
[3] http://www.Millipore.com
[4] Chia-Fu Chou, Olgica Bakajin, Stephen W. P. Turner, Thomas A. J. Duke, Shirley S.Chan, Edward C. Cox, Harold G. Craighead, Robert H. Austin "Sorting by diffusion: An asymmetric obstacle course for continuous molecular separation" *PNAS* Vol. 96, pp.13762-13765, 1999
[5] R.H. Austin, N. Darnton, R. Huang, J. Sturm, O. Bakajin, T. Duke "Ratchets: the problems with boundary conditions in insulating fluids "*Appl. Phys. A: Mater. Sci. Process*, Vol.75, pp.279, 2002
[6] James W Dufty,J Javier Brey " Brownian motion in a granular fluid". *New J. Phys.* 7, pp.20, 2005
[7 Lotien Richard Huang, Pascal Silberzan, Jonas O. Tegenfeldt, Edward C. Cox, James C. Sturm, Robert H. Austin, Harold Craighead "Role of Molecular Size in Ratchet Fractionation" *Phys.Rev.Lett.* 89, pp.178301, 2002
[8] Jianping Fu, Reto B. Schoch, Anna L. Stevens, Steven R. Tannenbaum, Jongyoon Han "A patterned anisotropic nanofluidic sieving structure for continuous-flow separation of DNA and proteins" *Nature nanotechnology* vol 2, pp. 121-128
[9] Stein Kuiper, Henk van Wolferen, Cees van Rijn, Wietze Nijdam, Gijs Krijnen, Miko Elwenspoek Stein Kuiper, "Fabrication of microsieves with sub-micron pore size by laser interference lithography" *J. Micromech. Microeng.* 11, pp. 33-37, 2001
[10] M. J. Kim, M. Wanunu, D. C. Bell, A. Meller, "Rapid Fabrication of Uniformly Sized Nanopores and Nanopore Arrays for Parallel DNA Analysis" *Adv. Mater.*vol. 18, pp. 3149, 2006
[11] C.Shen, H.T.M.Pham, P.M.Sarro, "Vertical-Wall Microsieves for Nano Particle Separation" *Proc. Eurosensors XXII*, pp.1357, 2008

CLASSIFICATION AND CONDENSATION OF NANO-SIZED AIRBORNE PARTICLES BY ELECTRICALLY TUNNING COLLECTION SIZE

Yong-Ho Kim, Soon-myoung Kwon, Dongho Park, Jungho Hwang, and Yong-Jun Kim
School of Mechanical Engineering, Yonsei University, Seoul, Republic of Korea

ABSTRACT

This paper demonstrates a hybrid particle classification and condensation device based on both aerodynamics and electrostatics using a micro virtual impactor (µVI). The µVI is capable of classifying the nanoparticles according to their size and condense their number concentration that we are interested. The µVI was fabricated using polymer micromachining. Its classification efficiency was examined using solid particles, polystyrene latex (PSL) ranging from 80 to 250 nm in diameter. Thereafter, specific-sized nanoparticles, NaCl of 50 nm in diameter, were condensed by applying an electric potential of 1.1 kV. As a result, the electric signal detected was amplified by 4 times higher than that when no electric potential was applied. The output signal was amplified by 4 times (before condensation: 4 fA, after condensation: 16 fA).

INTRODUCTION

Air contains a variety of airborne particles, including particulate matter, micro-organisms and gaseous compounds, which range in size from several nanometers to tens of micrometers [1]. Airborne particles are classified according to size into fine (particles with an aerodynamic diameter larger than 2.5 µm), ultrafine (particles larger than 100 nm), and nano (particles larger than 50 nm) particles. Airborne particles are a threat to human health and have been reported to be a cause of climate change [2–3].

In particular, ultrafine and nano particles are considered to be a significant threat to human health [4–6]. For monitoring those ultrafine and nano particles, we need to consider two aspects, size-dependant classification and detection. Several methods have been used for particle classification, which include inertial classification, gravitational sedimentation, centrifugation, and thermal precipitation [7].

Among these techniques, a virtual impactor is widely used for particle sampling on account of its high performance and real-time classification capability [8, 9]. In a virtual impactor, the airborne particles are classified by separating them into a straight low-velocity flow (minor flow) and a perpendicular high-velocity flow (major flow) as shown in Fig. 1. The influx flow containing polydisperse particles is accelerated through a converging nozzle known as the injection nozzle. The flow is controlled at more than 90% to the major flow and less than 10% to the minor flow. Large-inertia particles in the flow move to the minor channel and small-inertia particles follow the major flow with a turning angle of 90°. However, existing VIs have following limitations: (1) for size-classification of particles < 100 nm, the particles need to be aerodynamically accelerated to a sonic range. This can causes particle fragmentation. Also, it is not suitable for miniaturization of a VI because a very tight packaging method is required due to the large pressure drop [10], and (2) the cut-off diameter

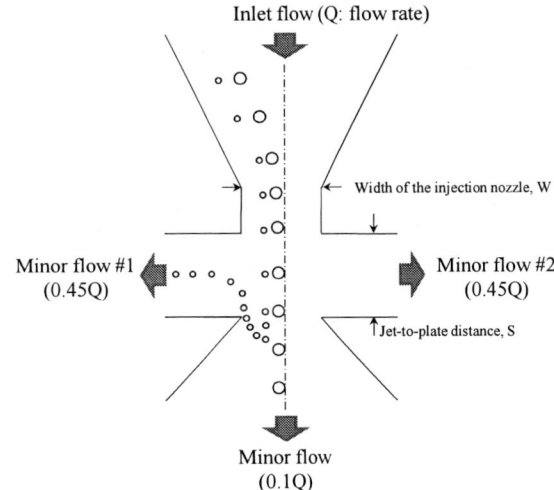

Figure 1: Schematic view of trajectories of particles in aVI.

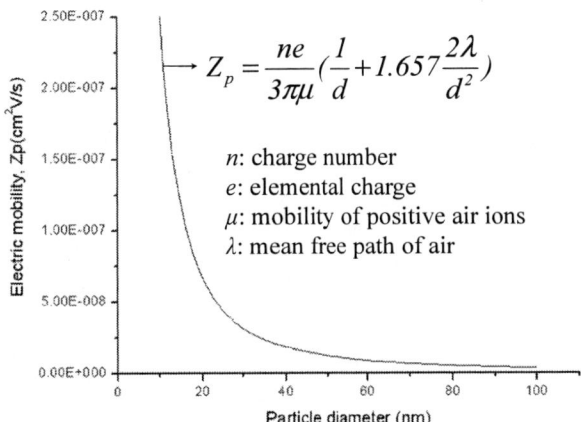

$$Z_p = \frac{ne}{3\pi\mu}(\frac{1}{d} + 1.657\frac{2\lambda}{d^2})$$

n: charge number
e: elemental charge
µ: mobility of positive air ions
λ: mean free path of air

Figure 2: Theoretical electrical mobility of particles

is defined by the geometry of a VIs.

For detection of nanoparticles, charging those particles and measuring their electric current are the most popular method. The charging value of those particles is proportional to square root of their diameter [1]. So, the value becomes dramatically deceases with decreasing of the particles' diameter. This results in difficulty on measurement.

To overcome the two limitations of the existing VIs and also enhance the detection capability, we propose a µVI that is capable of (1) classifying nanoparticles by electrostatically accelerating them to a sufficient high velocity, (2) electrically tuning the cut-off diameter (not determined by the geometry), and (3) detecting the particles of lower concentration by condensating them. Following sections will presents details of design, fabrication, experimental methods and results.

MATERIALS AND METHODS

Design

Since the electric mobility (Z_p) of nanoparticles is inversely proportional to square of their diameter (Fig. 2), they could be accelerated to be different velocities by applying an electric potential to the acceleration electrode without adverse effects for miniaturization. By adjusting the electric potential, nanoparticles interested was collected at the minor port. The particles were also concentrated by controlling the flow rates, 90% to the major flow and 10% to the minor flow. Accordingly, a µVI was designed as shown in Fig. 3.

The geometric cut-off diameter was calculated by considering the Stokes number (Stk), Reynolds number (Re), flow rate (Q), velocity (U), and injection nozzle width (W). The design of the virtual impactor was based on a trial-and-error method using the following procedure: 1) determination of a target cut-off diameter (d_p), flow rate, channel thickness (t) and Stk; 2) calculation of the slip-correction factor (SCF, Cc) using Eq. (1); 3) calculation of W using Eq. (2); and 4) verification of Re to determine if the resulting flow is within the laminar flow regime ($< 1,000$).

First, a target cut-off diameter of 1.0 µm was chosen according to the geometry of the virtual impactor. The upstream flow rate was 0.30 l/min, which was split into 0.27 and 0.03 l /min for the major and minor flow, respectively. The thickness of the microchannel was 200 µm.

Second, the Stk and the SCF were considered. The Stk is the ratio of the particle-stopping-distance at an average air velocity (U) to half the nozzle width ($W/2$), as expressed in Eq. (1). In Eq. (1), τ is the relaxation time, ρ_p is the particle density (1 g/cm3), d_p is the diameter of a particle, and µ is the dynamic viscosity of a particle (1.18×10-5 Pa·sec/m2). The Stk depends on the geometry of the flow channel, and Stk_{50} was considered to be the design rule for the Stk. The Stk_{50} is the Stk for a 50% collection efficiency of a particle. According to previous reports, Stk_{50} was chosen to be 0.229 because the proposed virtual impactor has a square channel [1, 11]. The SCF (C_c) was defined in Eq. (2), where λ is the mean free path of air (0.066 µm). The SCF is considered only when the particle diameter is smaller than 1.0 µm. Otherwise, the value is assumed to be 1. The SCF was assumed to be 1 because the cut-off diameter of the virtual impactor was 1.0 µm. The injection nozzle width was calculated using Eq. (3) by applying the Stk_{50} and SFC.

Finally, the Re was examined to ensure that it is within the laminar flow regime by applying all the variables related to Eq. (4). The flow in the channels needs to be laminar in order to achieve a sharp collection efficiency curve. The jet-to-plate distance (S) was determined to be 1.5 times larger than the width of the injection nozzle (W) [12]. Table 1 shows parameters for design of the δµVI and results.

$$Stk = \frac{\tau U}{W/2} = \frac{\rho_p d_p^2 U C_c}{9\mu W} \tag{1}$$

$$C_c = 1 + 1.257\frac{2\lambda}{d_p} + 0.40\frac{2\lambda}{d_p}\exp(-1.10\frac{d_p}{2\lambda}) \tag{2}$$

$$W = \sqrt{\frac{\rho_p d_p^2 q C_c}{9\mu Stk_{50} t}} \tag{3}$$

$$Re = \frac{4\rho WhU}{\mu(W+t)} \tag{4}$$

Fabrication

Fabrication of a virtual impactor begins with the definition of a 2 µm-thick Al electrode pair on a glass wafer. 3000 Å-thick silicon dioxide was patterned for electrical insulation of the Al electrode. Finally, a 200 µm-thick microchannel was defined using SU-8 (SU-8 2100, MicroChem Corp.). Fig. 4 shows a cross-section and fabrication result of the µVI. Polymethylmethacrylate (PMMA) plates and polydimethylsiloxane (Sylgard 184, Dow Corning Corp.) gaskets were used for tight sealing.

Table 1: Geometrical parameters and results of the µVI.

	Description	Value
Input parameter	Designed cut-off diameter	160 nm
	Flow rate (Inlet, major and minor flow)	0.22, 0.2, 0.02 ℓ/min
Output parameter	Width of the injection nozzle	120 µm
	Jet-to-plate distance	180 µm
	Microchannel thickness	150 µm

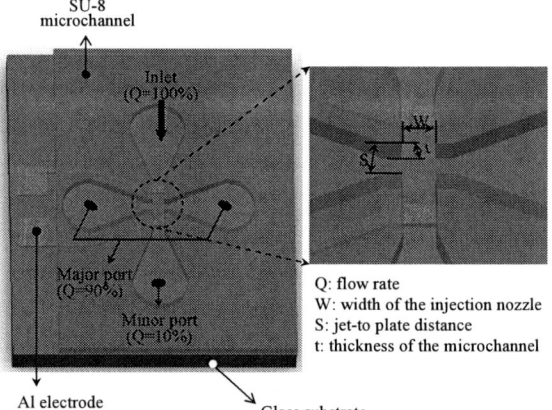

Figure 3: Schematic diagram of the proposed virtual impactor.

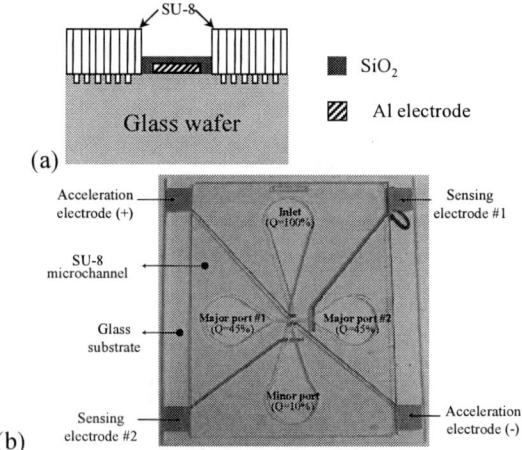

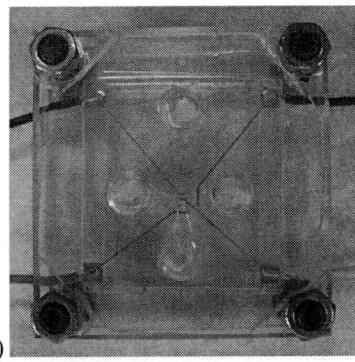

(c)

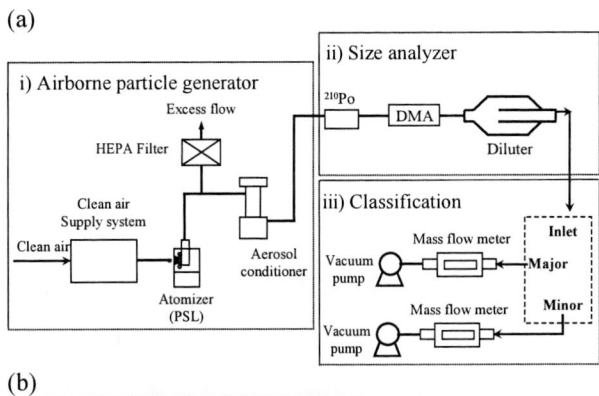

(a)

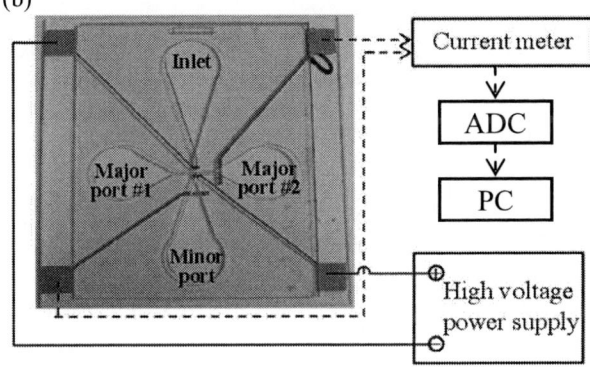

(b)

Figure 4: (a) Cross-section (b) bird's eye view of the μVI, and (c) packaged device.

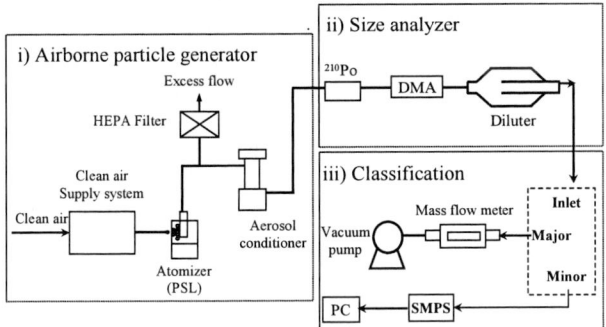

Figure 5: Experimental setup for generation of PSL ranging 80 to 250 nm: In section i) polydisperse PSL were generated, In section ii) PSL used in classification, ranging from 80 to 250 nm, were generated by filtering away other-size particles, In section iii), PSL ranging from 80 to 250 nm were classified using the proposed μVI.

EXPERIMENTAL

Geometric cut-off diameter

Fig. 5 shows the experimental setup for examining the cut-off characteristics of the proposed μVI and results. The setup consisted of an aerosol generator, a size analyzer, and a classifier with an efficiency measurement apparatus.

In section i), PSL were generated. Initially, ambient air was introduced into the clean air supply system, where it was dried, filtered and compressed. The cleaned air was then supplied to an atomizer. In the atomizer, a mixture of isopropyl alcohol (IPA) and PSL was sprayed into the air. The PSL aerosol ranging in size from 100 to 600 nm was encapsulated by IPA. Subsequently, the IPA was removed from the PSL aerosol in the aerosol conditioner (TSI 3072, USA).

PSL particles were then moved to section ii), the size analyzer. PSL except 80 to 250 nm were filtered away using a differential mobility analyzer (DMA, TSI 3081).

In section iii), the PSL were classified using the proposed μVI. The size distribution of the classified particles was measured again by a scanning mobility particle sizer (SMPS, TSI 3936). Flow rates of the μVI were controlled by the SMPS and were 0.30, 0.27 and 0.03 ℓ/min for the inlet, major and minor flow, respectively.

Figure 6: (a) monodisperse NaCl particles of 50 nm in diameter was generated, condensed and detected, and (b) Condensation and detection flow of NaCl particles of 50 nm in diameter.

Condensation of number concentration and detection

Monodisperse NaCl solid particles of 50 nm in diameter were used for the classification. Fig. 6(a) shows the experimental setup. Polydisperse NaCl particles were generated by supplying clean air to a toner chamber, which were then singly charged by a neutralizer, $^{210}P_0$. The particles generated were then filtered to have a monodisperse distribution of 50 nm using a DMA (TSI 3081). The number concentration of the NaCl particles was approximately 10^3 particle/cm^3. Finally, the monodisperse NaCl particles were classified using the proposed μVI by applying voltages of 1.1 kV to the acceleration electrodes (Fig. 6(b)). Current of the NaCl particles were detected by the sensing electrodes. The flow rate was 0.30, 0.27 and 0.03 ℓ/min for the inlet, major flow and minor flow, respectively.

RESULTS AND DISCUSSION

Geometric cut-off diameter

The geometric cut-off diameter (d_p)of μVI was measured to 190 nm as shown in Fig. 7, while the designed was d_p 160 nm. Percent deviation of the measured d_p from the designed d_p was approximately 19%. This deviation may be originated from resolution of fabrication and operational condition including flow rates. From this measurement results, we can expect that particles < d_p will be collected in the major port (side channel) of the μVI, but particles < d_p will be collected in the minor port (straight channel).

978-1-4244-2977-6/09 $25.00 © 2009 IEEE   389

Condensation of number concentration and detection

According to Fig. 7, particles < 190 nm were supposed to follow the major flow with 50% probability. For example, the possibility that the particles <50 nm were collected at the minor port were less than 10%. By virtue of the electric potential of 1.1 kV applied to the acceleration electrodes, the cut-off diameter was tuned. As a result, NaCl particles of 50 nm were collected at the minor port and also detected though the sensing electrode #2 (Fig. 8). Before applying the electric potential, the output signal was 4 fA. But the signal was amplified to be 16 fA by applying the electric potential. This means that the concentration of the 50-nm particles was amplified by 4 times.

CONCLUSION

We have demonstrated a size-dependant particle detection chip which is capable of (1) classifying nanoparticles by electrically accelerating them to a sufficient high velocity, (2) electrically tuning the cut-off diameter (not determined by the geometry), and (3) detecting the nanoparticles of lower concentration by condensating them.

A maximum charging value of nanoparticles is very small. This is a main obstacle on detection of nanoparticles. In this study, the condensation of those particles has been successfully achieved as well as the size-classification. Due to enhanced detection capability of nanoparticles, related researches including analysis of harmful effects of nanoparticles to human health could be facilitated.

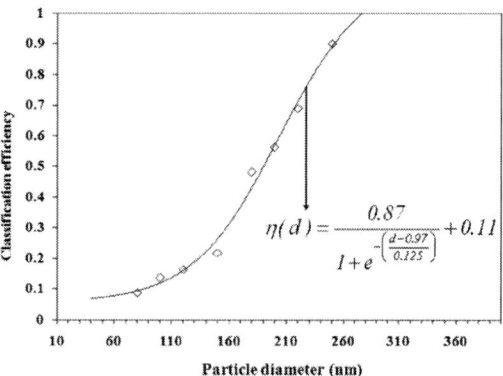

Figure 7: Classification result of PSL using the proposed μVI: geometric cut-off diameter was calculated to 190nm.

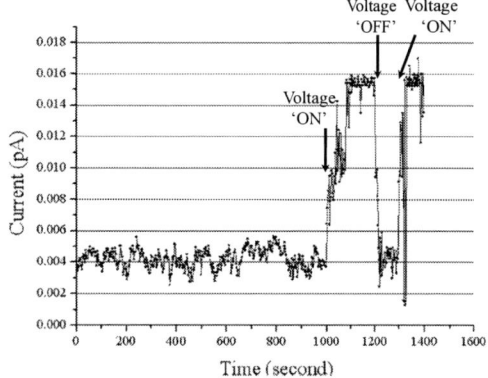

Figure 8: Output signal detected at the sensing electrodes. Signal level was amplified by approximately 4 times as a result of the electric potential of 1.1 kV.

ACKNOWLDGEMENTS

This work was supported in part of the Seoul R&BD program and in part of the National Core Research Center (NCRC) of Yonsei University.

REFERENCES

[1] W.C. Hinds, *Aerosol Technology: Properties, Behavior, and Measurement of Airborne Particle*, 2nd ed., Wiley, New York, 1999.

[2] M.Z. Jacobson, "Strong radiative heating due to the mixing state of black carbon in atmospheric aerosols," *Nature*, vol. 409, pp. 695–697, 2001.

[3] R. Wilson and J.D. Spengler, *Particles in our air: concentration and health effects*, Harvard University Press, Cambridge, 1996.

[4] J. Pekkanen, K.L. Timonen, J. Ruuskanen, A. Reponen, and A. Mirmes, "Effects of ultrafine and fine particles in urban air on peak expiratory flow among children with asthmatic symptoms," *Environ. Res.*, vol. 74, pp. 24–33, 1997.

[5] F. Alessandrini, H. Schulz, S. Takenaka, B. Lentner, E. Karg, H. Bhrendt and T. Jakob, "Effects of ultrafine carbon particle inhalation on allergic inflammation of the lung," *J. Allergy Clin. Immunol.*, vol. 117, pp. 824–830, 2006.

[6] K. Konaldson, X.Y. Li and W. MacNee, "Ultrafine (nanometer) particle mediated lung injury," *J. Aerosol Sci.*, vol. 29, pp.553–560, 1998.

[7] P.A. Baron and K. Willeke, *Aerosol Measurement*, 2nd ed., Wiley, New York, 2001.

[8] J.Y. Maeng, D. Park, Y.H. Kim, J. Hwang and Y.J. Kim, Micromachined cascade virtual impactor for aerodynamic size classification of airborne particles, *Proc. The 20th IEEE MEMS*, 2007, pp. 619–622.

[9] Y.H. Kim, J.Y. Maeng, D. Park, J. Hwang and Y.J. Kim, A micromachined cascade virtual impactor with a flow rate distributor for wide range airborne particle classification, *Appl. Phys. Lett.*, vol. 91, , 043512, 2007.

[10] C. Sioutas, P. Koutrakis and B.A. Olson, "Development and evaluation of a low cut-point virtual impactor," *Aerosol Sci. Technol.*, vol. 3, pp. 223-235, 1994.

[11] J.S. Hauglund, A.R. McFarland, "A circumferential slot virtual impactor," *Aerosol Sci. Technol.*, vol. 38, pp. 664-674, 2004.

[12] B.W. Loo and C.P. Cork, "Development of high efficiency virtual impactors," *Aerosol Sci.Technol.*, vol. 9, pp. 976-985, 1988.

DEVELOPMENT OF A CYTOMIC FORCE TRANSDUCER FOR EXPERIMENTAL MECHANOBIOLOGY

E. Dy and C.M. Ho

University of California Los Angeles, California, USA

ABSTRACT

In this work electrostatic actuation in ionic liquid environments was achieved through a unique atmospheric pressure packaging scheme in conjunction with Trichloro(1H,1H,2H,2H-Perfluorooctyl)Silane surface modification. This technique avoids common problems of electrolysis, charge blocking, and current leakage without the need for any drive signal considerations. When combined with cellular self-assembly onto sacrificial polymers, this breakthrough opens the possibility of creating a cytomic force transduction system through which mechanobiological experiments can be conducted on a multitude of cell phenotypes *in vitro*. Testing of the device in liquid demonstrated actuation with as little as 15V and continuous operation in liquid was successful for over two weeks.

INTRODUCTION

All cells in our body sense and respond to their surroundings through cell signaling which governs basic cellular activities and coordinates cell actions. The microenvironment that cells respond to consists of chemical, electrical and mechanical stimulations that are coordinated by neighboring cells. Mechanobiology is the interdisciplinary study that is specifically concerned with the cell's biological response to mechanical loads and the mechanisms by which these loads are transduced into a cascade of cellular and molecular events [1]. The application of MEMS to mechanobiology is intriguing for a variety of reasons.

The similar scale of cells and microdevices eliminate the need for large transducers or transducer interfaces whose large mass will deleteriously affect measurement frequency range and introduce compliance and drift issues. The ability of MEMS to sense and actuate with forces on the same magnitude of cells will result in higher resolution than most systems can achieve. Additionally, advances in fabrication techniques that provide means of fabricating MEMS and integrated circuits simultaneously opens the possibility of creating force and electrical feedback in a closed loop system. The application of MEMS force sensors and actuators to mechanobiology can help provide answers to many fundamental questions regarding the pathogenesis of disease.

Creating a biocompatible cytomic force transduction system will provide scientists with the necessary tools to carry out systematic studies on adherent cell phenotypes with versatile range of operating parameters. However, the conditions intrinsic to biology are often inimical to MEMS operation. Cells require ionic liquid environments with a constant temperature of 37°C to survive, while most MEMS devices operate exclusively in dry conditions under a wide range of temperatures. Looking specifically at electrostatic actuators, which possess the ideal force range and response times for mechanobiological studies, an array of problems arise which prevent easy application of these well characterized devices to cell studies.

The application of electrostatic actuators in aqueous mediums has received little attention up to this point largely due to the complications which arise from operating in liquid environments. Aside from surface tension and stiction problems which are common to all MEMS in liquid, high electric fields in water cause a completely different set of problems. Electrolysis, which occurs as low as 2V, charge blocking due to the formation of an electric double layer, and current leakage further add to the complications [2]. Previous attempts to apply electrostatic in liquid have all required special high frequency drive signals, which would complicate and possibly limit the range of force and frequency applied to cells [3]. Despite these challenges, development of electrostatic actuators capable of functioning in ionic aqueous environments opens the possibility of manipulating cellular behavior with unprecedented fidelity.

In order to generate this biotic/abiotic system a unique packaging system was combined with advanced cell patterning techniques on comb drive actuators. The work I have conducted in developing this system can be broken down into three areas: microfabrication of a comb drive actuator and the complimentary device components, device assembly for actuation in liquid, and development of compatible cell-silicon interface bridges.

DESIGN

The atmospheric pressure packaging system employed in this design effectively blocks liquid from contacting the folded flexure support beams or interdigitated comb fingers, while still keeping the actuator shuttle exposed to the external liquid for cellular contact and adhesion. The packaging scheme consists of two steps: microfabrication and bonding of a silicon lid and surface treatment with Trichloro(1H,1H,2H,2H-Perfluorooctyl) Silane. To begin with though let's first consider the comb drive actuator.

The actuator consists of a folded flexure spring (k=2N/m) that supports a 220 x 568μm shuttle with 139 fingers. Some features unique to this system worth mentioning are the cell adhesion anchor and the capillary pressure stops (figure 1).

978-1-4244-2977-6/09 $25.00 © 2009 IEEE

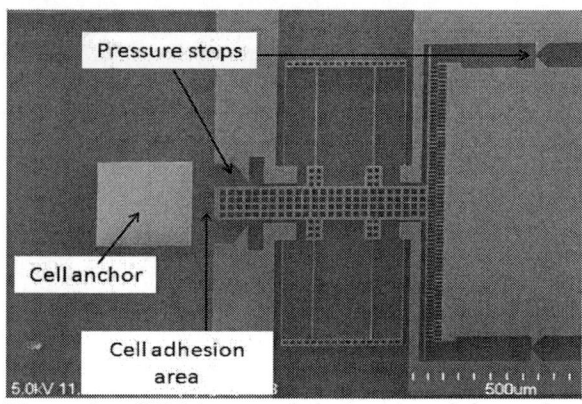

Figure 1: SEM image of device viewed from above, select parts are labeled.

The capillary pressure stops function as passive valves to stop the flow of liquid using a capillary pressure barrier that develops when the channel cross section changes abruptly [4]. This helps prevent liquid from leaking into the protected comb drive cavity.

The packaging cap is fabricated from a 200µm thick silicon substrate that is backside etched to eliminate contact with the actuator and thus prevent friction and though etched on the topside to provide access to the actuator shuttle for cellular attachment. To further reduce the ability of the surrounding liquid to penetrate into the protected comb drive cavity, vapor phase surface treatment with Trichloro(1H,1H,2H,2H-Perfluorooctyl) Silane is used post lid bonding. This creates an inert monolayer of silane molecules over all exposed surfaces rendering them hydrophobic [5].

One last design consideration is development of a cell-silicon interface to promote cellular adhesion between the cell anchor and actuator shuttle. Adherent cells require a solid substrate onto which they can attach and network with surrounding cells to form tissue bundles, but upon maturation the cells should be free to move. To meet these two restrictions a biocompatible thermoresponsive sacrificial hydrogel, poly-N-isopropylacrylimide (PNIPAAm) is deposited with a microdeposition system. Subsequent shadow mask patterning of the solidified hydrogel layer with a positive adhesion promoter such as Au further promotes selective cellular attachment between the actuator shuttle and an anchor pad 60µm away (figure 2) [6].

FABRICATION

The fabrication of comb drive actuators has been well characterized and is straightforward. The use of a low resistivity SOI wafer (500µm handle/25µm device/2µm sac ox) further simplified processing since the entire device could be defined through one mask using deep reactive ion etching (DRIE). Similarly the fabrication of both the packaging cap and shadow mask are fabricated with simple photolithography and thru etching with DRIE. Release of the patterned device was carried about by soaking the SOI substrate in methanol overnight,

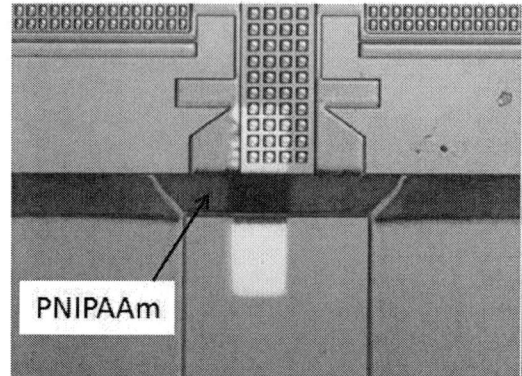

Figure 2: PNIPAAm cell-silicon interface patterned with Au positive cellular adhesion layer. Cells will selectively adhere to gold area

followed by a sacrificial oxide etch in BOE and critical point drying.

Following lid fabrication the device substrate and lid are aligned with a Karl Suss MA6/BA6 using bond align mode, and subsequently bonded together using an intermediate layer of AZ5214 photoresist. After hard baking the packaged device, the assembly is treated with O_2 plasma to clean and functionalize the surface and then placed under vacuum in a bell jar with 50uL of Trichloro(1H,1H,2H,2H-Perfluorooctyl) Silane for 15 minutes. This vapor phase surface treatment creates a layer of silane over all exposed surfaces (figure 3).

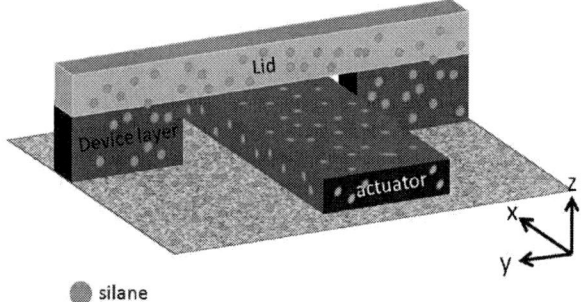

Figure 3: Schematic view of unpackaged area which will be exposed to vapor phase Trichloro-silane treatment

To create the solid sacrificial layer for cellular adhesion previously discussed, PNIPAAm is deposited using a microdeposition system consisting of a syringe pump, Tygon tubing, and a micromanipulator fitted with a syringe tip. Post deposition the excess PNIPAAm is removed from the top of the cell anchor and actuator shuttle with O_2 plasma in DRIE. Finally the solid sacrificial layer is patterned with a Au positive cellular adhesion layer deposited via CHA e-beam evaporation.

RESULTS

Once fully assembled the device is ready for testing and immersion in liquid (figure 4). Prior to submerging in liquid the device was characterized in air by applying

978-1-4244-2977-6/09 $25.00 © 2009 IEEE

varying DC voltages and tracking the displacement of the actuator shuttle. This was accomplished thru the use of a probe station, high voltage DC power supply and CCD camera with video recording software. Tracking of individual dies performance shows great uniformity in actuator performance, additionally the experimental actuator performance also closely matched the theoretical values (figure 5). In total a full 4" wafer yielded slightly over 90% functional dies.

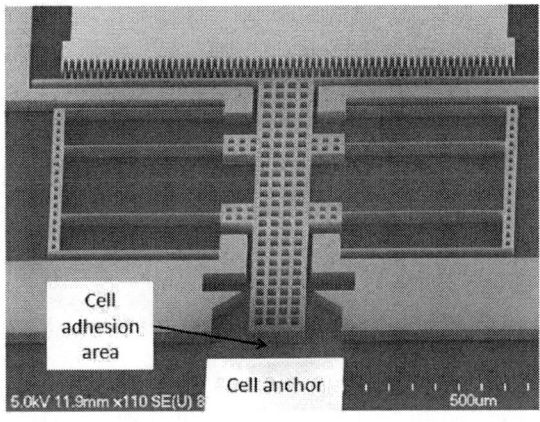

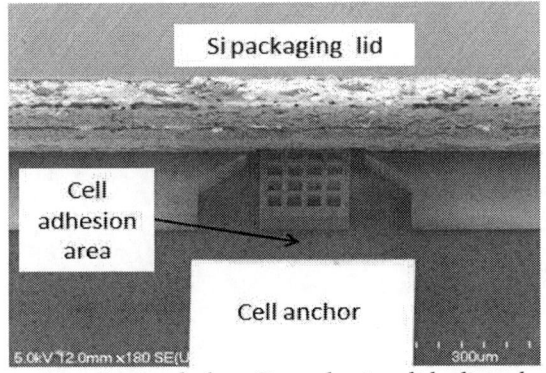

Figure 4: Device before Si packaging lid aligned and bonded to SOI wafer (above), and device post packaging with actuator shuttle still exposed to external environment (below).

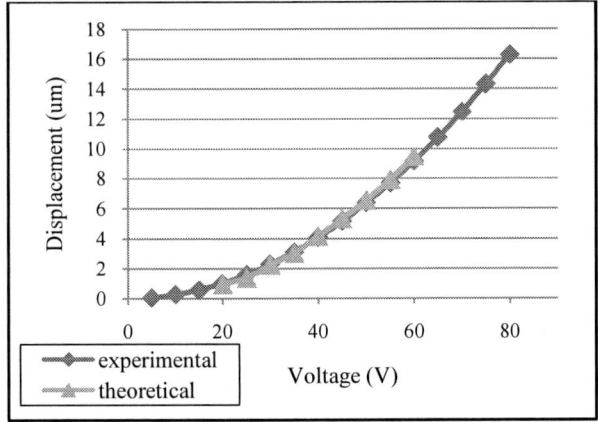

Figure 5: Average value of displacement vs. voltage for all device tested in air

Following characterization in air, the devices were placed in Petri dishes and covered in culture media for 24 hours and then retested to determine how performance in liquid deviated from that in air. Displacement of the actuator in liquid was successful and actuation was achieved with as little as 15V. The actuator performance in liquid was similar to that which was seen in air; however the measured displacement per unit voltage was less (figure 6). This is to be expected as there are additional forces at play which counteract the electrostatic force generated between the comb fingers.

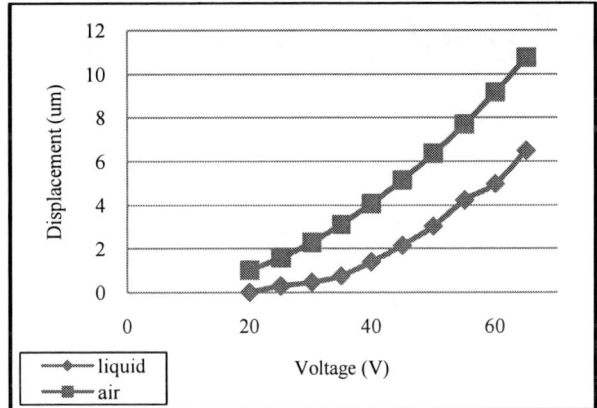

Figure 6: Difference in displacement vs. voltage for device operating in air vs. liquid

A force balance carried out on the portion of the rotor shuttle that is exposed to the liquid environment reveals the source of decreased actuator movement. In addition to the driving electrostatic force, the spring force, drag force, and surface tension forces act to inhibit device movement. Drag force can be neglected since the actuator moves to a static position and thus the velocity drops to zero and the force will not affect the final displacement. The main contributing factor to the decreased displacement then is surface tension.

After two weeks of continuous actuation while submerged in liquid the device still maintained displacement to voltage values similar to that found after 24 hours indicating stable hydrophobic treatment and maintenance of a liquid free comb drive cavity. This was supported by contact angle testing; silane treatment rendered the exposed surfaces hydrophobic effectively changing the contact angle for culture media from 57°-102°. Testing of the contact angle several weeks after treatment produced the same results further supporting the assumption that the layer of silane is inert.

Cell adhesion testing onto the Au patterned sacrificial PNIPAAm cell-silicon interface produced mixed results. Cell culture onto test structures supported cell adhesion and proliferation as can be seen by the neonatal rat ventricular myocyte cells fixed and stained after one week of culture (figure 7). However, adhesion of cells to the device are still lacking as cell plating protocols need to be refined to better target cell deposition onto the adhesion area.

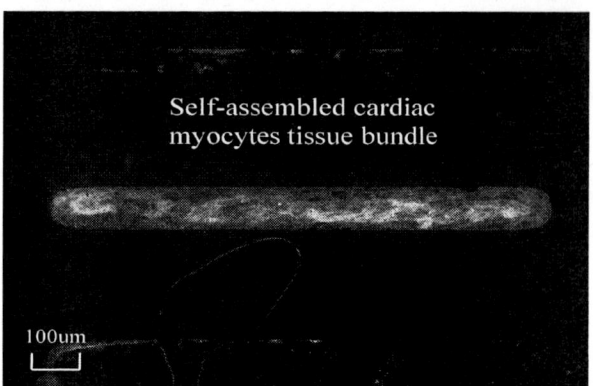

Figure 7: Fluorescently stained cardiac myocytes selectively adhered to a rectangular gold pattern on PNIPAAm

CONCLUSION

The cytomic force transduction system developed here utilizes electrostatic actuation in ionic liquid environments necessary for cellular proliferation. A cell anchor point and actuator shuttle exposed to the liquid environment provide an area for cellular adhesion; while the comb drive actuator and folded flexure support structure operate in a liquid free environment through a packaging scheme that employs capillary pressure stops in combination with hydrophobic vapor phase surface treatment with Trichloro(1H,1H,2H,2H-Perfluorooctyl) Silane. With this packaging method actuation in liquid can be achieved for extended period of times without the need for any special drive signals. This device provides scientists with the necessary tools to carry out systematic force vs. frequency studies on any array of adherent cell phenotypes with an unlimited number of dimensions.

ACKNOWLEDGMENTS

Special thanks to the UCLA Nanolab staff for their support in completion of this project, Katya Butylkova of the MacLellan Lab for supplying me with a steady supply of cardiac myocytes, and the funding sources: National Science Foundation "NSEC: Center for Scalable and Integrated Nano Manufacturing (SINAM)" & NIH/NDC "Center for Systemic Control of Cyto-Networks (CSCC)"

REFERENCES

[1] Ingber, D.E., *Cellular Basis of Mechanotransduction*. Biol Bull, 1998. 194(3): p. 323-327.

[2] Sameomoto, A.a.G.E., *Operation of electrothermal and electrostatis MUMPs microactuators underwater*. J Micromech Microeng, 2004. 14: p. 1359-1366.

[3] Sounart, T.L., T.A. Michalske, and K.R. Zavadil, *Frequency-dependent electrostatic actuation in microfluidic MEMS*. Microelectromechanical Systems, Journal of, 2005. 14(1): p. 125-133.

[4] Man, P.F., et al. *Microfabricated capillarity-driven stop valve and sample injector*. in *Micro Electro Mechanical Systems, 1998. MEMS 98. Proceedings., The Eleventh Annual International Workshop on*. 1998.

[5] Inc, G., *Hydrophobicity, Hydrophilicity and Silane Surface Modification*. 2006.

[6] Xi, J., J.J. Schmidt, and C.D. Montemagno, *Self-assembled microdevices driven by muscle*. 2005. 4(2): p. 180-184.

A PDMS PLANAR PATCH-CLAMP ARRAY CHIP WITH POLY (ETHYLENE GLYCOL)/SU-8 BASED CELL-PATCH INTERFACE

B.J. Xu[1], Z.B. Liu[1], J.J. Yu[1], Y.K. Lee[2] and M. Yang[1]

[1]Biomedical Engineering Division, the Hong Kong Polytechnic University, Hong Kong
[2]Mechanical Engineering, the Hong Kong University of Science & Technology, Hong Kong

ABSTRCT

This paper presents a new PDMS planar patch-clamp array chip integrated with microfluidic units and a poly (ethylene glycol)/SU-8 mixture based cell-patch interface for whole cell recordings. In our experiments, the PEG/SU-8 mixture with the weight ratio of 1:1 has been proven to have improved properties to fabricate perfect apertures compared with PDMS for cell-patch interface, such as large Young's modulus and good hydrophilic surface properties. The integrated chip was tested to trap 20μm polystyrene microsphere onto 2μm aperture by a negative pressure, and the variation of impedance spectra in trapping process demonstrated that the excellent sealing resistance has been achieved. Our present results indicate the potential of a planar patch-clamp array chip system with PEG/SU-8 base cell-patch interface for high-throughput whole cell recording.

INTRODUCTION

The patch-clamp method has proven to be an accepted standard for many fundamental studies of ion channel proteins and the discovery of drugs that affect them. There are several critical drawbacks in the conventional pipette-based technology, such as, costly, delicate and cumbersome micropositioning hardware, non-portable fluidic handling setups, fine manual dexterity, burdensome preparation, and hard to exchange the solution inside the micropipette [1]. Furthermore, this traditional method is not practical for high-throughput screening because it requires a skilled operator to manually manipulate the glass pipette onto the cell. In addition, the irreversible nature of the sealing process requires a new pipette be used for each recording [2].

Recently, many efforts have been taken to improve the chip-based patch-clamp system using glass, silicon, or elastomeric substrates [2-4]. Among them, PDMS has the advantage of low cost, and the air-molding technique proves to be a simple way for planar patch-clamp aperture fabrication [5]. However, its flexible property may block the small aperture under high negative pressure, as well as its hydrophobic surface may prevent the establishment of a high-resistance seal with cell membrane [6]. As UV sensitive materials, both PEG and SU-8 can be involved in chip fabrication like PDMS by soft lithography method. PEG/SU-8 compound has some advantages for cell-patch interface fabrication, such as, large Young's modulus, well hydrophilic surface, appropriate transparency and fine liquid compatibility.

MATERIALS AND METHODS
Design of planar patch-clamp array chip

Figure 1 shows the mechanism of this planar patch-clamp array chip integrated with PEG/SU-8 based aperture, gold micro-electrodes, and microfluidic units including inlet, channel, outlet and reservoir. Positive pressure on the flow channel was used to force the electrode (intracellular) solution to fill the inlet, the outlet, the channel, the cavity below the cell-patch interface and the aperture. And then, a bath (extracellular) solution which contained microspheres or cells was dropped onto the planar patch partition. Subsequently, a gentle suction was used to draw a nearby microsphere or cell to the aperture. Next, a platinum wire connected the bath solution to ground. The planar Au electrode was connected to the signal input of the amplifier (patch clamp amplifier, HEKA electronics Inc.). The signal was recorded with multi-channel data acquisition software using the patch master (both from HEKA electronics Inc.). Once a gigaohm seal (gigaseal) was observed, the suction was released and recording protocol could be started for whole cell configuration.

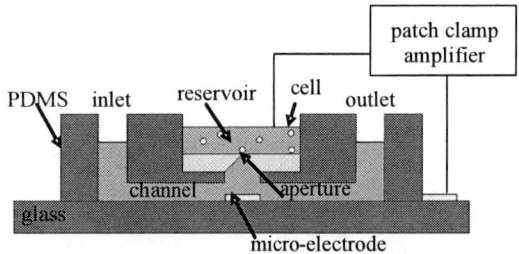

Figure 1: Schematic view of the planar patch-clamp array chip with PEG/SU-8 based aperture

Fabrication of PEG/SU-8 based cell-patch interface

For the experimental validation of PEG/SU-8 based cell-patch interface, polymer prototype was fabricated by the standard MEMS (microelectromechanical system) technologies. The micro-aperture which was the functional element of the interface was realized by UV photolithograph, KOH wet-etching, twice soft lithography for PDMS and PEG/SU-8. The applied fabrication process is illustrated in detail in Figure 2.

A 4 inch silicon wafer (Wafernet Inc., San José, CA, USA) with the orientation of <1,0,0> was oxidized double sides by a wet oxidation process to make a mask to protect silicon against Potassium hydroxide (KOH) etching as shown in Figure 2a. Then, the oxide layer on one side was lithographically patterned to expose a 2×2 array of square dots each measuring 400μm in width with 2000μm center-to-center spacing. The exposed silicon oxide was removed by a wet etching process using buffered hydrofluoric acid (HF) and the photoresist was then removed by acetone and ethanol. KOH (40 wt. %, Aldrich) heated to 50oC was then applied to the wafer to etch inverted pyramid-shaped holes. Etching occurred along the crystal plane to form 54.7o-tapered walls terminating in a sharp point. Thus, the silicon master for the PDMS slice with sharp tips was prepared as shown in Figure 2b.

PDMS (Sylgard 184, silicone elastomer, Dow Corning) was mixed at a 10:1 ratio of silicone base to curing agent by weight. After mixing, the PDMS was poured onto the above silicon master, and placed into a vacuum chamber at about 15 Torr for 30min to allow air trapped in the pyramid holes to escape the mixture as shown in Figure 2c. After baking at 80oC in conventional oven for 2 hours, PDMS replica with sharp tips could be easily peeled from the master at 200°C on the hotplate.

Next, the backside of PDMS slice was tightly adhered onto the transparent and rigid chamber, and arranged the array of sharp tips in the center of open window as shown in Figure 2d, and the tips side was covered with the planar glass slide coating a layer PDMS in order to avoiding stress difference between two sides. After pouring the compound mixed by PEG (Poly(ethylene glycol) diacrylate, Aldrich, 455008-100ml) and SU-8 2050 (MicroChem Corp., MA) onto the PDMS tips, the contact between the flexible pyramid tips and PDMS on the rigid glass can be precisely controlled by the inner pressure of chamber under an invert microscope observation to control the aperture size, as shown in Figure 2e.

Subsequently, the PEG/SU-8 compound was exposed by UV light in adequate energy from two sides as shown in Figure 2f, since all of the glass, PDMS, and the rigid chamber were transparent.

Finally, the PEG/SU-8 compound chip with apertures could be easily released from the PDMS mold, and it should be baked at 95°C for 30min to enhance the mechanical property as shown in Figure 2g.

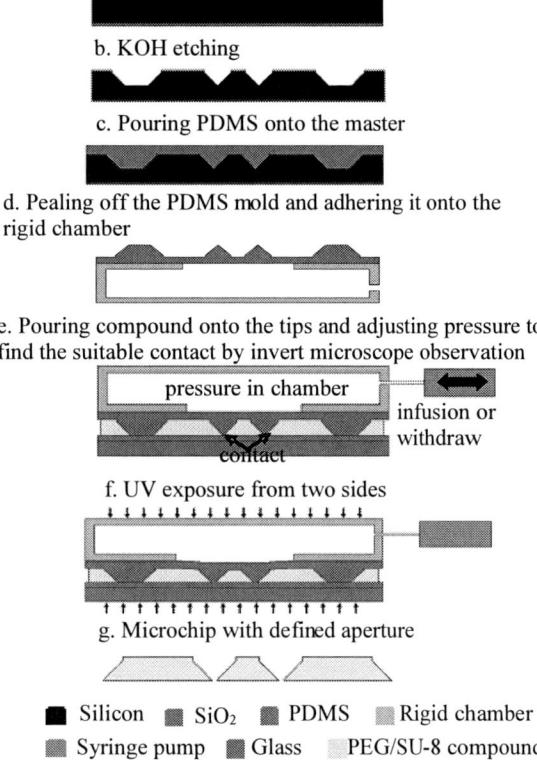

Figure 2: Schematic view of fabrication process of PEG/SU-8 compound based aperture

Assembly of the whole patch clamp array chip

The process flow of making the planar patch-clamp array chip system is shown in Figure 3. Two layers of PDMS were molded individually from SU-8 masters and bonded together after oxygen plasma treatment of the surfaces. A 75μm layer of SU-8 2050 was patterned using a 20000 dpi resolution transparency film (Kernel electronics Co., Hong Kong) as a mask. Casting and peeling off the PDMS gave a microchannel structure with a high-fidelity negative replication of the photoresist pattern. The reservoir, the cavity below aperture, the inlet and the outlet were punched with the sharpened needles with different gauges.

Planar Au electrodes were fabricated with metal evaporation and "lift-off" technology. A shadow mask (Kernel electronics Co., Hong Kong) was used to selectively coat nickel (50Å) and gold (0.5μm) on the cleaned glass slide. After cleaned with acetone and ethanol, both PDMS microfluidic part and the glass with electrodes were treated by O2 plasma for bonding together. The methanol could help the alignment between the cavities and the electrodes. Planar PEG/SU-8 cell-patch interfaces, also treated by oxygen plasma, were placed onto the PDMS microfluidic layer and carefully aligned to the below cavities. The edges of the PEG/SU-8 partition could use 1:1 epoxy glue to strengthen the bonding intensity.

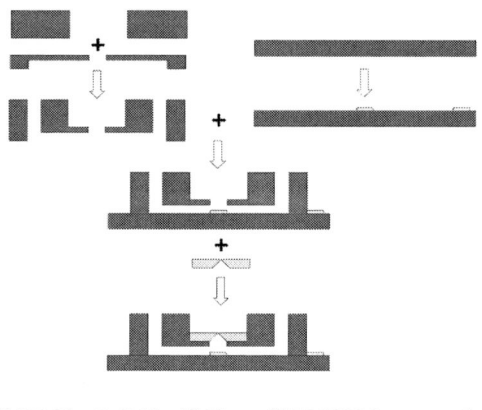

■ PDMS ▨ Gold ■ Glass ☐ PEG/SU-8 compound

Figure 3: Schematic view of the fabrication procedure of planar patch-clamp array chip

RESULTS

Properties of material for aperture fabrication

As we know, PDMS is a flexible silicone elastomer, and it is easy to deform under high negative pressure in microfluidic channel. The PDMS aperture may be blocked in suction process. At the same time, the PDMS hydrophobic surface prevents the establishment of a high-resistance seal with cell membranes, which should reduce the suction success greatly. In order to determine the feasibility of PEG/SU-8 mixture as an interface material for patch clamp measurements, the different materials including PDMS, PEG, PEG/SU-8(3:1, wt.), PEG/SU-8(1:1, wt.), PEG/SU-8(1:3, wt.), and SU-8 were compared. Figure 3a showed their surface chemistry and mechanical properties. The higher is the PEG concentration, the less is the contact angle. However, the chip which contained high PEG is easily deformed when its surface contacts the liquid buffer. At the same time, the higher is SU-8 concentration, the higher is Young's modulus. However, the SU-8 chip is not transparent and it's hard to align in the next packing process. It is proved that the mixture of PEG and SU-8 with the weight ratio of 1:1 is suitable for cell-patch interface fabrication because it has the increased Young's modulus and surface hydrophillicity, and also the appropriate transparency and liquid compatibility.

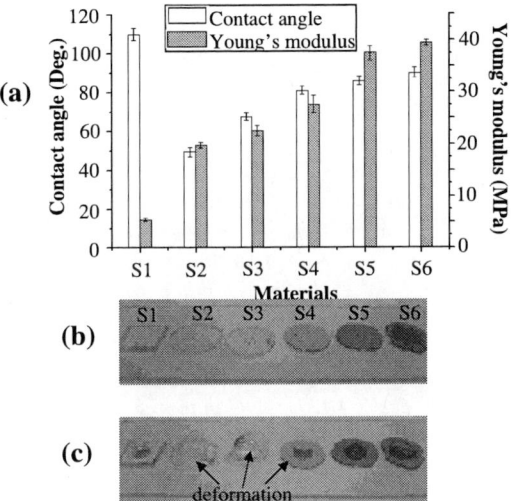

Figure 4: Materials properties for aperture fabrication, (a) Contact angle and Young's modulus, (b) transparency and color, (c) compatibility with liquid buffer. The samples in(a), (b) and (c): S1 is PDMS, S2 is PEG, S3 is PEG:SU-8(3:1wt.), S4 is PEG:SU-8(1:1 wt.), S5 is PEG:SU-8(1:3wt.), S6 is SU-8

Relationship between the pressure and aperture's size

The tips in PDMS mold were closely attached onto the glass side due to gravity if the experimental setup was arranged in Figure 2e. Small apertures could be fabricated by adjusting the negative pressure in chamber. Syringe pump was used to adjust the pressure by withdrawing movement. Different aperture sizes from 0.5µm-22.1µm have been achieved by modulating the pressure from 270 Pa to 0 Pa as shown in Figure 5. It suggested that the higher negative pressure could result in smaller apertures because the contact between PDMS tips and PDMS on the glass became smaller and the excluded PEG/SU-8 mixture would be less. The pressure of 225 Pa was chosen to make holes ~2µm in size.

Prototypes

PDMS mold with sharp tips was successfully fabricated. The tip's minimal size could reach about 0.5µm as shown in Figure 6a, so the minimal resolution of aperture in the cell-patch interface could be about 0.5µm. That could meet the requirement of patch clamp system for whole cell recording. Figure 6b shows the SEM image of the fabricated aperture with the diameter of 2µm. Due to inner stress effect, the exposed PEG/Su-8 mixture may have a little adjustment to form circle aperture instead of rectangle one in the subsequent baking process. Figure 5c shows 2×2 array of microfluidic units with separate channel and cavity. Figure 5d shows the total planar patch-clamp array chip integrated with microfluidic units, PEG/SU-8 compound based cell-patch interface, and

microelectrodes.

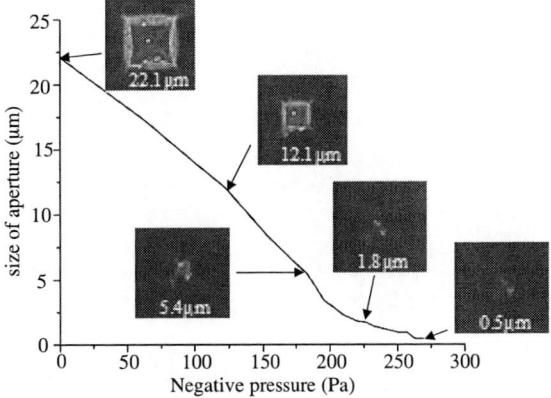

Figure 5: Controlling of aperture size by modulating pressure on the flexible PDMS mold

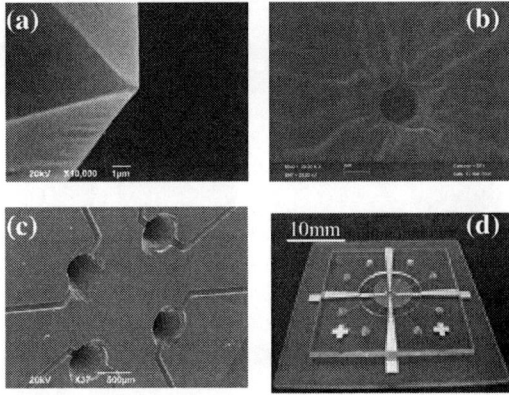

Figure 6: (a) SEM image of the tip of the pyramid PDMS mold, (b) SEM image of fabricated aperture with the diameter of 2µm, (c) 4 individual channels and chambers for buffer inletting and suction, (d) 2×2 array planar patch-clamp chip

Microsphere trapping testing

Both bath solution and electrode solution are phosphate buffered saline (PBS, Sigma-Aldrich). Figure 7 shows the dynamic process of a 20µm microsphere (Fluoresbrite™ plain YG 20.0 micron microsphere, Polysciences, Inc., Warrington, PA) capturing by the integrated chip when a gentle negative pressure of 500Pa. Compared to bottom microspheres in the bath solution, suspended microspheres can be easily trapped to the 2µm aperture.

Simulation validation

To validate the above trapping experiment, a 2D model has been established to see the effect of height on trapping efficiency for the CFD-simulations as shown in Figure 8a. Figure 8b shows the velocity vector mapping of five cells (d=20 µm) with the same distance of 100 µm but different heights to the aperture. Figure 8c illustrates the curve of average velocity magnitudes of these five cells, which demonstrates the above trapping experiments. It has been shown that the velocity of bottom cells is much smaller than that of the suspended cells, and the latter has much greater tendency to move to the aperture than the former under the same suction conditions. Hence, the microspheres or cells had better be suspended in the bath solution before the suction process was performed.

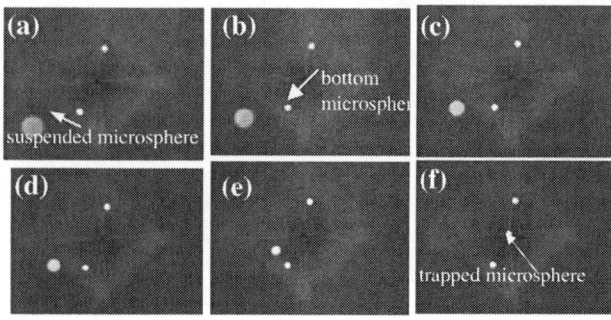

Figure 7: (a)-(f) Six sequential images showing fluorescence labeled microsphere with the diameter of 20μm being trapped by suction

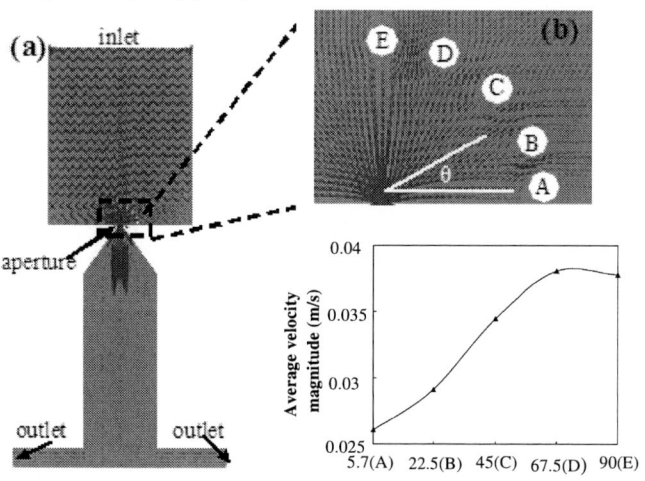

Figure 8: CFD-simulation shows the tendency of five trapped cells (A-E) with different heights in the buffer. Boundary conditions: pressures of inlet and outlet are constants; the left outlet's velocity is -0.1m/s, and the right one is 0.1m/s; cell diameter is 20μm, and the aperture size is ^

Giga-seal testing

Impedance spectra was performed using the impedance analyzer VersaSTAT3 (Princeton Applied Research, USA) controlled by a personal computer. Impedance of the electrode in the bath solution and the planar gold electrode in the electrode solution was measured at a fixed frequency (10 kHz) every minute during the impedance sweep interval to obtain real time recording. Figure 9 shows the resistance change for the trapping process of 20μm microsphere using the planar patch-clamp array chip with PEG/SU-8 based aperture. At first, there is an air bubble between two electrodes, so the testing circuit is open and the highest impedance of about 1.5Mohm can be achieved. When the electrode buffer overflows to the reservoir via the aperture, the impedance will be sharply decreased because it shows the buffer's impedance as low as about 0.1Mohm. When the microsphere is captured by suction process, the impedance will be increased according to the applied pressure. The syringe pump increases the negative pressure slowly, and the patched gap between the microsphere and the aperture reduces slowly as well, so the impedance increases tardily. Once the impedance becomes steady to about 1.1Mohm,

the syringe pump must be held and the pressure will be kept. This status is the giga-sealing. If the negative pressure is further increased, bubble will appear in the cavity below the cell-patch interface, and the impedance will be enlarged to that of open circuit.

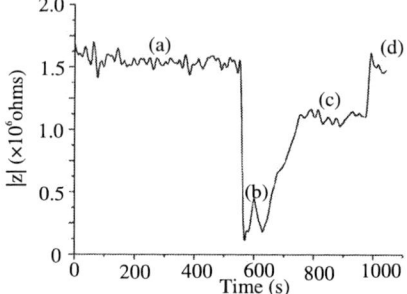

Figure 9: Impedance spectrum shows the trapping process of 20μm microsphere onto 2μm aperture by suction, (a) buffer inletting through micro-channel, (b)buffer via aperture, (c)microsphere captured by aperture, (d)bubble formation between two electrodes. Testing conditions: buffer is PBS, AC frequency on two micro-electrodes is 10 kHz, DC potential is 0.5V

REFERENCES

[1] C. Chen, A. Folch, "A High-Performance Elastomeric Patch Clamp Chip", *Lab on a Chip*, vol. 6, pp. 1338-1345, 2006.

[2] K.G. Klemic, J.F. Klemic, M.A. Reed, F.J. Sigworth, "Micromolded PDMS Planar Electrode", *Biosensors & Bioelectronics*, vol. 17, pp. 597-604, 2002.

[3] X.H. Li, K.G. Klemic, M.A. Reed, F.J. Sigworth, "Microfluidic System for Planar Patch Clamp Electrode Arrays", *Nano Letters*, vol. 6, pp. 815-819, 2006.

[4] J. Seo, C. Zanetti, J. Diamond, R. Lal, L.P. Lee, "Integrated Multiple Patch-Clamp Array Chip via Lateral Cell Trapping Junctions", *Applied Physics Letters*, vol. 8, pp. 1973-1975, 2004.

[5] K.G. Klemic, J.F. Klemic, F.J. Sigworth, "An Air-Molding Technique for Fabricating PDMS Planar Patch-Clamp Electrodes", *Pflügers Archiv European Journal of Physiology*, vol. 449, pp. 564-572, 2005.

[6] A.Y. Lau, P.J.Hung, A.R. Wu, L.P. Lee, "Open-Access Microfluidic Patch-Clamp Array with Raised Lateral Cell Trapping Sites", *Lab on a Chip*, vol. 6, pp. 1510-1515, 2006.

WHOLE CELL IMPEDANCE ANALYSIS OF METASTATIC AND NON-METASTATIC CANCER CELLS

H.S. Kim,[1] Y.H. Cho,[1] A.B. Frazier,[2] Z.(G). Chen,[3] D.M. Shin,[3] and A. Han[1]
[1]Texas A&M University, College Station, Texas, USA
[2]Georgia Institute of Technology, Atlanta, GA, USA
[3]Emory University, Atlanta, GA, USA

ABSTRACT

We have developed a micro electrical impedance spectroscopy (µEIS) system and utilized it for analyzing highly-metastatic and poorly-metastatic head and neck cancer (HNC) cell lines. The device is composed of 16 arrays of impedance analysis sites, with each site capable of capturing a single cell and analyzing its electrical impedance spectrum. Electrical impedance measurements on the highly-metastatic HNC cell line 686LN-M4e and the poorly-metastatic HNC cell line 686LN were conducted over a frequency range of 40 Hz to 10 MHz. The phase part of the impedance spectrums from the 686LN-M4e and 686LN cells showed clear difference. This result indicates that the µEIS system can be used to distinguish the metastatic status of HNC cell lines.

1. INTRODUCTION

Quantifying heterogeneous tumor cell populations has been one of the major challenges in understanding tumorigenesis and improving cancer therapy. Detection of cancer cells with metastatic potential at early stages of cancer or evaluation of patient response to drug therapies has significant clinical implications. Personalized therapies will be possible if an instrument can be developed which can quantify the heterogeneity of tumor cell populations and how such heterogeneous tumor cell populations change in response to therapies.

Electrical impedance spectroscopy (EIS) is a technique used for distinguishing different types of tissues/cells or pathological tissues/cells from normal tissues/cells, by measuring the electrical properties of tissues or cells in the frequency spectrum [1]. The non-invasive nature of the measurement makes this technology suitable when further analysis of the target sample is needed, and makes the overall analysis process simple. Among various applications of the EIS, distinguishing different types of tissues and studying pathological tissues, especially cancer tissues, have been of great interest [2]. However, impedance measurements at tissue level can vary significantly because of the complex composition and structure of tissues. As a result, impedance measurements on cells instead of tissues have drawn wide interest, and various impedance measurement methods targeting cells have been developed.

Recent advances in microfabrication and lab-on-a-chip technology allow electrical impedance spectroscopy devices to detect and analyze a single cell rapidly and accurately, using integrated electrode arrays [3-5]. Measurements on individual cells instead of population of cells by EIS have many advantages. The single-cell approach does not require time-consuming preparation of a uniform population of cells and can be used to quantify a heterogeneous mixture of cells. Moreover, theoretical analysis becomes relatively simple

compared to analyzing population of cells.

We have previously developed a micro EIS (µEIS) system for differentiating and quantifying human breast cancer cells from various cancer stages with single cell resolution [6]. One of the drawbacks of this device was that the cell trap part of the device composed of a thin silicon dioxide membrane was not mechanically robust, and resulted in frequent device breakdown as well as making it challenging to visualize the cell trapping process. In addition, measurements on large numbers of cells for obtaining statistically relevant results were challenging using this device since the fabrication steps to make the device were complicated. Here, we present a µEIS system having arrays of horizontal cell traps to overcome such limitations. The microsystem can be fabricated with vastly simpler processing steps. We used this µEIS device to analyze and compare the highly-metastatic head and neck cancer (HNC) cell line 686LN-M4e and the poorly-metastatic HNC cell line 686LN.

2. EXPERIMENT

Design of the µEIS system

The µEIS system is composed of two cell flow channels and 16 parallel impedance analysis sites. Each analysis site consists of a cell trap into which a single cell can be captured and an electrode pair for impedance measurements (Figure 1). Each cell trap site has a constricting channel structure so that a single cell flowing through the cell flow channel can be captured inside this microfluidic trap by applying negative pressure via the cell trapping side channel. The integrated electrodes were designed so that once a single cell is trapped, the electrodes naturally form a tight seal between the electrodes and the cell, with the electrodes being positioned at the middle of

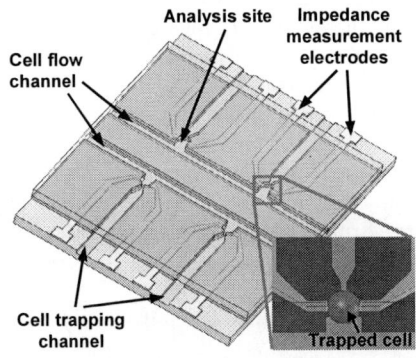

Figure 1: Schematic of the µEIS system showing four impedance analysis sites along the two cell flow channels. Close-up view illustrates a single cell trapped inside an impedance analysis site by applying negative pressure through a cell trapping side channel. The trapped cell is positioned between the two opposing impedance electrodes.

978-1-4244-2977-6/09 $25.00 © 2009 IEEE

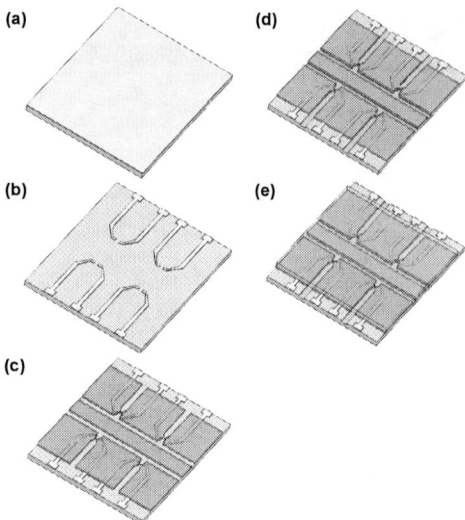

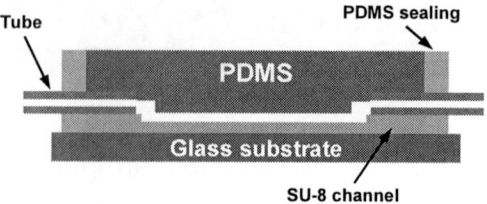

Figure 3: Schematic illustration of the lateral tubing interconnection scheme.

Figure 2: Schematic of the overall fabrication steps. (a) Cr/Au deposition, (b) Electrode patterning, (c) SU-8 microfluidic structure patterning, (d) Glass etching, (e) PDMS packaging and lateral tubing interconnects.

the cell both vertically and horizontally.

Fabrication of the μEIS system

The overall fabrication steps are illustrated in Figure 2. A Cr/Au layer was evaporated to a thickness of 20 and 250 nm, respectively, on a piranha-cleaned glass slide (Microslides 2947-75 x 50 mm, Corning Inc., NY). The Cr/Au layer was patterned using S1818 photoresist (Microposit™, MicroChem Corp., MA) to make 16 pairs of opposing electrode arrays for impedance measurement. Diluted Au etchant (Au etchant:DI water = 1:5, Gold etch – Type TFA, Transene Company Inc., MA) was used for Au etching since the fast etching rate of the undiluted etchant caused significant undercut to the electrodes. The resulting electrodes had a width of 6 μm, and gaps of 2 μm or 4 μm between the opposing electrodes.

A 7 μm thick SU-8 was spun over the electrode layer and patterned to form the two parallel cell flow channels, 16 cell trapping channels perpendicular to the cell flow channel, and 16 cell traps positioned at the interface between the cell flow channel and the cell trapping channel. The SU-8 pattern served both as a microfluidic channel as well as an etch mask for the subsequent glass etching process. After SU-8 patterning, the backside of the substrate was spin-coated with S1818 photoresist to protect the backside, followed by immersing the glass substrate into Buffered Oxide Etch (BOE, 7:1, J.T. Baker). Six minutes in BOE resulted in 2~3 μm deep glass etching. From this etching step, 9~10 μm deep microfluidic channel structure were formed on the glass substrate with the impedance electrodes positioned at the middle of the captured cell instead of at the bottom of the cell. The S1818 photoresist was removed using isopropyl alcohol.

Microfluidic Interconnect

Utilizing the μEIS device for single cell impedance measurement requires the cell trapping and releasing process to be observed under an upright microscope for validation purpose. The conventionally used vertical tubing interconnection scheme, where tubes are vertically inserted into inlet and outlet holes punched out in microchannel structure, is not compatible due to the space limitation between the device and the objective lens. To overcome this limitation, we developed a novel lateral fluidic interconnection scheme with arrays of tubing connecting laterally to the device rather than vertically (Figure 3). The lateral fluidic interfacing layer was composed of a poly(dimethylsiloxane) (PDMS) microchannel starting out large enough so that a tubing can be inserted from the side, and then shrinking down to a smaller channel so that the inserted tubing can be stopped. A tubing having an outer diameter of 1.6 mm and inner diameter of 0.76 mm was used. To smoothly slide in this tubing and to provide mechanical support for the inserted tubing, the entrance part of the microchannel dimension was defined to be 1.6 mm wide, 6.5 mm long, and 1.6 mm high. This channel size was then reduced to a channel having width, length, and height of 0.7 mm, 3 mm, and 0.7 mm, respectively, so that the inserted tubing can be stopped and create a fluidic interconnect between the tubing and the microchannel on the device.

A desktop Subtractive Rapid Prototyping (SRP) tool (MDX-40 milling machine, Roland ASD.) was first used to fabricate an acrylic master mold having two different steps for the fluidic interface layer having two different microchannel depths. PDMS was used to replicate the fluidic interface layer from this acrylic master.

Bonding between the PDMS channel layer and the SU-8 structure was achieved through air plasma treatment for 90 seconds. To align the two layers, water was used as a lubricant since commonly used methanol damaged the SU-8 structure. After alignment, the two layers were clamped at a pressure of 3.2 kPa and bonded at 80 °C for one hour.

Following bonding, fluorinated ethylene propylene (FEP) tubes (1.6 mm outer diameter, Upchurch Scientific Inc., USA) were inserted into the PDMS channel. To achieve leak free fluidic interconnects, the whole interface was sealed by applying a PDMS mixture (mixing ratio of base solution to curing agent is 10:2) followed by curing at 85 °C for 15 minutes.

Head and Neck Cancer (HNC) cell line preparation

Poorly-metastatic HNC cell line 686LN and highly-metastatic HNC cell line 686LN-M4e were used for the impedance measurement. The highly-metastatic 686LN-M4e cell line was established through *in vivo* selections from an SCCHN cell line 686LN by Dr. Chen at the Winship Cancer Institute at Emory University (Atlanta, GA) [7]. Cells were cultured in Dulbecco's Modified Eagle Medium (DMEM/F12, Invitrogen Corp., CA) supplemented with 10 % heat-inactivated fetal bovine serum (FBS) and antibiotics (Streptomycin, Penicillin G and Amphotericin B) as a monolayer at 37 °C, 5 % CO_2

environment, inside a humidified incubator. These cells were dissociated from the culture plate using Cellstripper, a non-enzymatic solution from Mediatech (Hermdon, VA), and re-suspended in phosphate buffered saline (PBS, Invitrogen Corp., CA). All cells were used within 4 hours of dissociation and at a concentration of 5×10^5 cells/μL.

Impedance Measurement Protocol

The μEIS system was placed under a microscope (Eclipse ME600, Nikon Corp., NY) and a syringe pump was used to load the suspended cells into the device. As cells flowed through the cell flow channel and passed by the impedance analysis sites, negative pressure was applied through the cell trapping channel so that a single cell can be trapped inside the microfluidic cell trap. Since the cell is larger than the size of the cell trap, the captured cell slightly deformed inside and formed a tight electrical contact with the electrodes. The electrical impedance measurement using a precision impedance analyzer (4294A, Agilent Technology, Inc., CA) was carried out at an operating voltage of 500 mV and over a frequency range of 40 Hz to 10 MHz. Two hundred data points were measured over the frequency range, with each data point representing an eight-point average measurement. The analyzer was set to obtain sixteen sweep average measurements at which point the measurement result was saved. The microdevice having two different electrode designs, electrode to electrode distance of 2 and 4 μm, were used to obtain the electrical impedance spectrum.

3. RESULTS AND DISCUSSION

Micrographs of the fabricated μEIS system are shown in Figure 4. The μEIS device has two cell flow channels, each having eight impedance analysis sites connected to eight cell trapping side channels. All structures including the cell flow channels, the side channels, and the analysis sites were defined with SU-8 patterning and glass etching. The overall height of the channels were 9~10 μm, resulting from the 7 μm thick SU-8 channel and the 2~3 μm deep glass etching. These two fabrication steps resulted in the pre-defined impedance electrodes on the glass substrate to be naturally positioned 2~3 μm above from the bottom of the channel due to the glass etching. Unlike flat electrodes located at the bottom of the devices, this suspended electrode scheme allows tight sealing between the trapped cell and the electrodes to prevent any influence from the surrounding media on the impedance measurements. One of the advantages of this μEIS system is its simple fabrication process so that many devices could be easily fabricated to obtain sufficient number of cell impedance data. Figure 4(b) and 4(c) show micrographs of a single analysis site having two integrated electrodes and the close-up view of electrodes sticking out from the SU-8 structure, respectively. The design to have the impedance electrodes sticking out 1 μm from the microfluidic trap was to assure a direct electrical contact between the trapped cell and the electrodes for impedance measurements.

The device has two inlets and two outlets connected to a syringe pump through the lateral tubing. The 16 cell measurement sites were laterally connected to the side channel in order to control each analysis site independently using a gas-tight glass syringe (1705, Hamilton, USA). A single cell was trapped inside the analysis site by applying

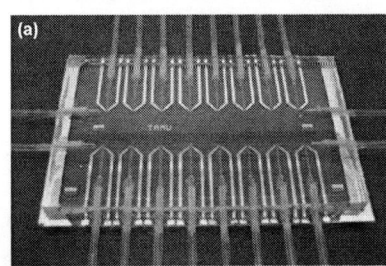
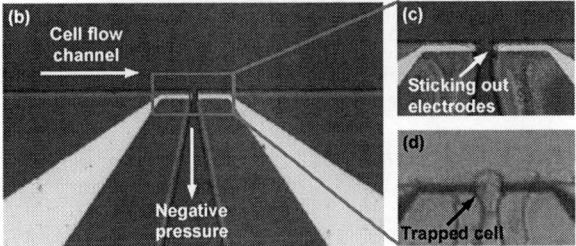

Figure 4: Micrographs of the fabricated μEIS system. (a) The completely assembled device with laterally connected tubings, (b) the enlarged view of a single analysis site, (c) impedance electrodes sticking out into the cell trap, (d) a single cell trapped and forming a tight contact with the pair of impedance electrodes.

negative pressure through the side channel, which is perpendicular to the cell flow channel and connected to the cell trap. Once a cell was trapped inside the analysis site and naturally positioned in the middle of the impedance electrodes, the syringe pump was stopped. The 10~20 μm and 5~10 μm diameter 686LN cells and 686LN-M4e cells fit tightly into the 4-6 μm wide cell trap (Figure 4(d)). This tight fit enabled cells not only to form tight contacts with the electrodes but also to cover the exposed portion of the electrodes completely, enabling impedance measurements not to be affected by any surrounding buffer solution.

The average electrical impedance phase spectrums for the 686LN and 686LN-M4e cells measured using the two different electrode designs (gap: 2 and 4 μm) are shown in Figure 5. For both electrode designs, the 686LN cell line showed higher phase value than the 686LN-M4e cell line. It could be also observed that the phase decreased as the frequency increased up to around 3 MHz, but then began to fluctuate, and finally converge to similar values with almost no difference between the 686LN cells and the 686LN-M4e cells. This is possibly due to the membrane capacitive component which plays negligible effects on the impedance value. The standard deviation of the impedance was relatively large since the measured impedance was affected by various factors, such as variability in channel and electrode dimensions among each of the fabricated devices, the electrode and electrolyte interface impedance, and the shunt path impedance formed by the saline solution surrounding the cell [3]. Each impedance phase spectrum shown in Figure 5 was an average of more than 40 cells tested (686LN-M4e, n=48; 686LN, n=39). It took approximately 4 seconds to carry out a frequency sweep impedance measurement to obtain 200 data points with eight-point average measurements over a frequency range of 40 Hz to 10 MHz. This measurement time is dependant on the number of data points taken over the frequency range and the number of point-averages. Overall, 33 electrodes from 15 devices having a 2 μm electrode gap

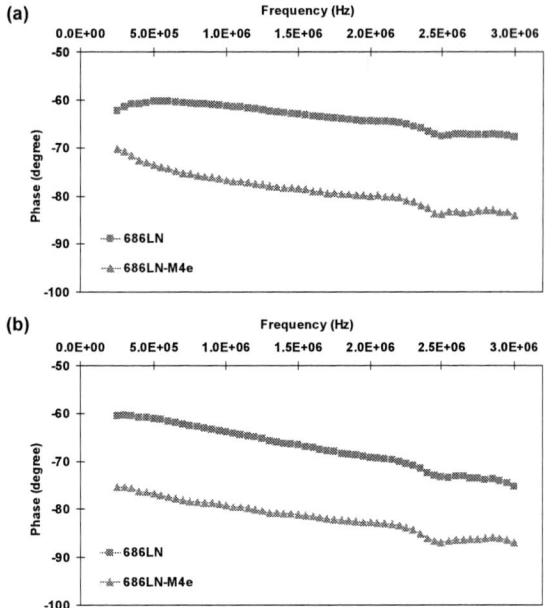

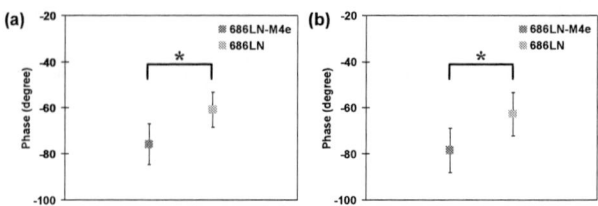

Figure 6: T-test result between the 686LN-M4e and 686LN cells at 0.8 MHz for impedance phase using the (a) 2 μm gap design and (b) 4 μm gap design. (*P < 0.001).

Figure 5: Impedance phase spectrum of the 686LN-M4e and 686LN cells using the (a) 2 μm gap design and the (b) 4 μm gap design over a frequency range of 200 kHz to 3.0 MHz.

and 30 electrodes from 19 devices having a 4 μm electrode gap were used for the impedance measurements shown in Figure 5.

Pooled two sample T-test was used to statistically analyze the differences between the 686LN-M4e and 686LN cells. Figure 6 shows the result of the T-test conducted between the phase values of the 686LN-M4e and 686LN cells measured at 0.8 MHz. Average and standard deviation of the impedance phase for the 686LN-M4e and 686LN cells measured using the 2 μm gap electrode design were -75.8 ± 8.7 ° and -60.9 ± 7.6 °, respectively. Similarly, the average and standard deviation of the 686LN-M4e and 686LN cells measured by the 4 μm gap electrode design was -78. 5 ± 9.7 ° and -62.8 ± 9.5 °, respectively. For both electrode designs, there were statistically significant differences between the 686LN-M4e cells and 686LN cells (P < 0.001). This difference in phase could be derived from the differences in morphology, distinct protein expression, and genetic alteration between the 686LN cell and the 686LN-M4e cell. At frequencies below 3.5 MHz, T-test analysis of the impedance phase data showed P < 0.05 between the two cell lines. However, the difference in impedance magnitude between the two cell lines was less distinct. When using the 2 μm gap electrode design, statistical differences (P < 0.05) could be only observed at frequencies between 2.6 and 3.4 MHz, and using the 4 μm gap electrode design, only at frequencies below 800 kHz.

Through these analyses, we can select a particular frequency showing the largest impedance differences between the two cancer cell lines. Based on this information, we can conduct impedance measurements on two different HNC cell lines at that fixed frequency and quantify the percentage of each cell type in a heterogeneous mixture. As a result, we can use a discrete, multi-point frequency analysis which takes less than 20 ms rather than performing a full frequency sweep over a wide range of frequencies.

4. CONCLUSION

This paper presents a μEIS system having arrays of 16 parallel impedance analysis sites with single cell resolution. Each of the sites can be independently controlled to trap a single cell for impedance interrogation. The impedance spectrums of the poorly-metastatic HNC cell line 686LN and the highly-metastatic HNC cell line 686LN-M4e were analyzed using the μEIS system to examine the capability of differentiating HNC cells from different pathological stages. The results show significant differences in the impedance phase between the two HNC cell lines. It is expected that this system will serve as a powerful tool for future detection and quantification of cancer cells from various tumor stages, and offer a platform to examine the effect of drug therapies on cancer cells or to customize drug therapies for cancer cells.

REFERENCES

[1] B. Rigaud, J.P. Morucci, N. Chauveau, "Bioelectrical Impedance Techniques in Medicine. Part I Bioimpedance Measurement. Second Section: Impedance Spectrometry", *Crit. Rev. Biomed. Eng.*, vol. 24, pp. 257-351, 1996.

[2] M.E. Valentinuzzi, J.P. Morucci, C.J. Felice, "Bioelectrical Impedance Techniques in Medicine. Part II: Monitoring of Physiological Events by Impedance", *Crit. Rev. Biomed. Eng.*, vol. 24, pp. 353-466, 1996.

[3] H.E. Ayliffe, A.B. Frazier, R.D. Rabbitt, "Electric Impedance Spectroscopy Using Microchannels with Integrated Metal Electrodes", *J. Microelectromech. Syst.*, vol. 8, pp. 50–57, 1999.

[4] S. Gawad , L. Schild and Ph. Renaud, "Micromachined Impedance Spectroscopy Flow Cytometer for Cell Analysis and Particle Sizing", *Lab Chip*, vol. 1, pp. 76–82, 2001.

[5] Y.H. Cho, T. Yamamoto, Y. Sakai, T. Fujii, B.J. Kim, "Development of Microfluidic Device for Electrical/physical Characterization of Single Cell", *J. Microelectromech. Syst.*, vol. 15, pp. 287–295, 2006.

[6] A. Han, L. Yang, A.B. Frazier, "Quantification of the Heterogeneity in Breast Cancer Cell Lines Using Whole-cell Impedance Spectroscopy", *Clin. Cancer Res.*, vol. 13, pp. 139-143, 2007.

[7] X. Zhang, Y. Liu, M.Z. Gilcrease, X.H. Yuan, G.L. Clayman, K. Adler-Storthz, Z. Chen, "A Lymph Node Metastatic Mouse Model Reveals Alterations of Metastasis-related Gene Expression in Metastatic Human Oral Carcinoma Sublines Selected from a Poorly Metastatic Parental Cell Line", *Cancer*, vol. 95, pp. 1663–1672, 2002.

978-1-4244-2977-6/09 $25.00 © 2009 IEEE

INTEGRATION OF SKELETAL MUSCLE CELL ONTO SI-MEMS AND ITS GENERATIVE FORCE MEASUREMENT

K.Shimizu[1], H. Sasaki[2], H. Hida[2], H. Fujita[1], K. Obinata[1], M. Shikida[2], and E. Nagamori[1]

[1]Toyota Central R&D Labs., Inc., Aichi, JAPAN
[2]Nagoya University, Aichi, JAPAN

ABSTRACT

We propose skeletal muscle cells as a new material for Bio-MEMS actuator, and developed the fabrication process to integrate it onto Si-MEMS devices. As the first trial of the integration of the skeletal muscle cells onto the Si-MEMS device, we used a cantilever-based force measurement Si-MEMS device for evaluating the generative force of the muscle cells. Murine skeletal muscle cell line C2C12 myoblasts were patterned on the device by using poly-N-isopropylacrylamide as a sacrificed layer. The cell-patterned device was immersed into the cell culture medium and cultivated for 7 days. The results of the immunological staining of the muscle specific protein, α-actinin, confirmed that the patterned myoblasts successfully differentitated into myotubes on the device. Then, we applied the electrical stimulation to the myotubes, evaluated the active tension generated from the myotubes, and revealed that its tetanic value was approximately 1.2 μN.

INTRODUCTION

Recent advancement of microfabrication technology offers new possibilities to develop microscale device actuated by bio-actuator. Motor protein molecules such as myosin, kinesin, dynein, and F1-ATPase which generate force by use of adenosine triphosphate (ATP) as an energy source have been investigated to employ for powering the microdevices [1-4]. Although these molecules are an attractive source to alternate conventional microactuators, they can generate only a few piconewtons per molecule. Hence, another material for bio-actuators which can generate sufficiently robust force to actuate microstructures has been expected. Cardiac muscle cells which are typical microsized bio-actuator powered by renewable glucose energy resource can generate a few micronewtons per cell [5, 6]. In these several years, it was succeeded to use the cardiac muscle cells for actuating some microdevices [7-10]. However, the cardiac muscle cells have some drawbacks for the practical use. The biggest problem is that they are driven spontaneously. As a result, it was difficult to apply the cardiac muscle cells for creating controllable actuators. In addition, the inability of the cardiac muscle cells to enter mitosis is an unavoidable difficulty. In fact, since the cardiac muscle cells used in the previous studies were primary cultured cells, the sacrifice of laboratory animals is continuously needed to fabricate their microdevices. To avoid these issues, we propose skeletal muscle myoblast cell lines as a new material for bio-actuator. Unlike the cardiac muscle cells used previously, the myoblasts can proliferate infinitely and differentiate into myotubes which generate active tension only when the electrical or chemical stimulation is applied externally. In this manner, we focused on the usage of the skeletal muscle cells, which have rarely been used so far as bio-actuators.

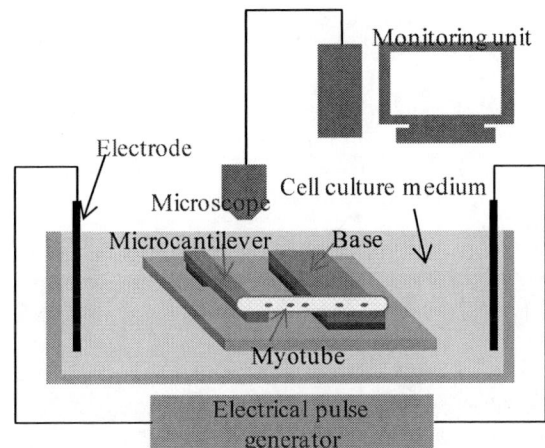

Figure 1: Schematic drawing of the device and the stimulating and monitoring units.

In the present study, we developed fabrication process to integrate the skeletal muscle cells onto Si-MEMS device. We used murine skeletal muscle cell line, C2C12, which were established in 1970s and has been used all over the world as a common model to investigate muscle biology and physiology [11]. However, the force generated by the cultured C2C12 myotube has not been clearly understood. Therefore, as a first trial of the integration, we used a cantilever-based force measurement Si-MEMS device to measure the generative force of the myotubes (Fig. 1). We fabricated the device consisting of a microcantilever and a base, and the Si-MEMS device and the myotubes were assembled. The myotubes bridged the gap between the cantilever and base. The integrated device was immersed into a cell culture medium, and the myotubes were actuated by the electrical signals. The value of the active tension generated by the myotubes was calculated by measuring the displacement of the cantilever bending.

FABRICATION

We used a cantilever-based force measurement Si-MEMS device in this study (Fig. 1). The fabrication process of the device is shown in Fig. 2. The device was fabricated using a (100) silicon on insulator wafer consisting of a 10 μm thickdevice layer with a 1 μm thick buried oxide (Fig. 2a). Positive photoresist (OFPR-800) was spincoated onto a wafer and patterned with a photomask (Fig. 2b and c). The wafer was etched using deep reactive ion etching, followed with buffered HF etch to remove the buried oxide layer (Fig. 2d and e). Then, the device with microcantilever was yielded (Fig. 2f). SEM image of the fabricated device was shown in Fig. 3a (10 μm in width, 1000 μm in length). The magnified image of the front edge of the microcantilever revealed that the microcantilever did not stick to the surface of Si-MEMS (Fig. 3b). We designed several dimensional cantilevers

978-1-4244-2977-6/09 $25.00 © 2009 IEEE 403

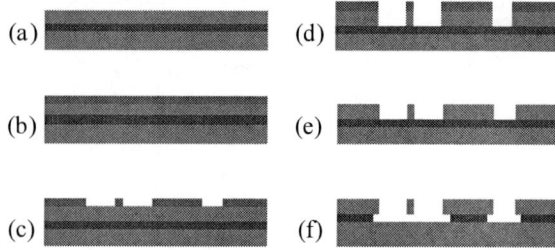

Figure 2: Fabrication process of the cantilever-based Si-MEMS device. a) SIO wafer, b) Spin-coating of photoresist layer, c) Microcantilever structure patterning, d)Deep-RIE etching, e) Removing photoresist, f) Buffered HF etching.

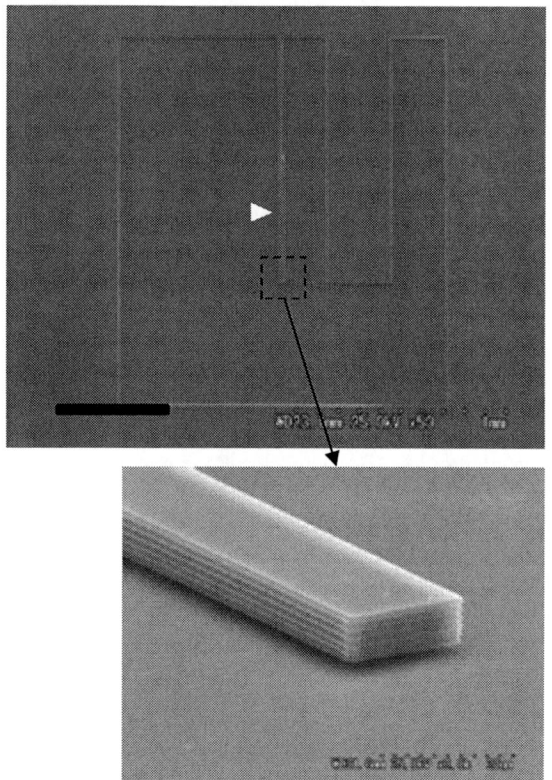

Figure 3: SEM of the device with microcantilever (white arrow head). Magnified image (bottom) represents tip of the microcantilever. The dimension of microcantilever is 10 μm in width and 1000 μm in length. Scale bar: 500 μm.

having different spring constant (k= 0.005 to 20) to increase of the force measurement range.

EXPERIMENTAL METHODS

We attempted to integrate skeletal muscle C2C12 cells onto the cantilever-based Si-MEMS device and measure the active tension generated by the myotubes cultured on the device as shown in Fig. 1. To position the myotubes over the gap between the cantilever and the base at low-temperature process, we used poly-N-isopropylacrylamide (PNIPAAm) as a sacrificed layer as reported previously with some modifications (Fig. 4 and 5) [7]. The PNIPAAm solves to water when the

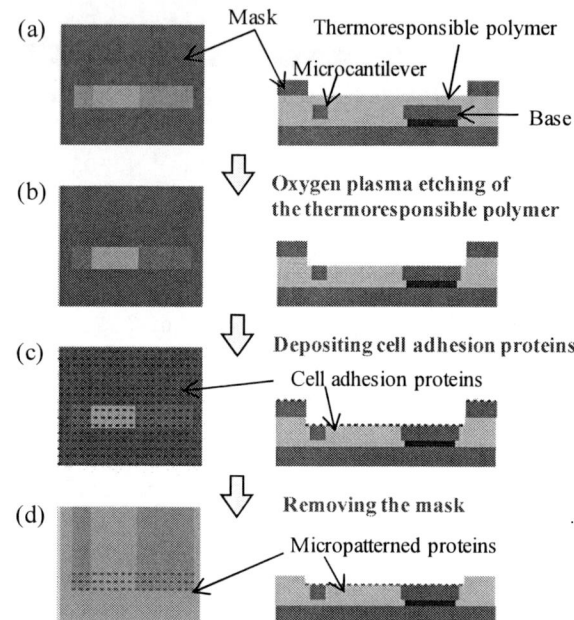

Figure 4: Schematic drawing of plasma etching of polymer and patterning of proteins on the device.

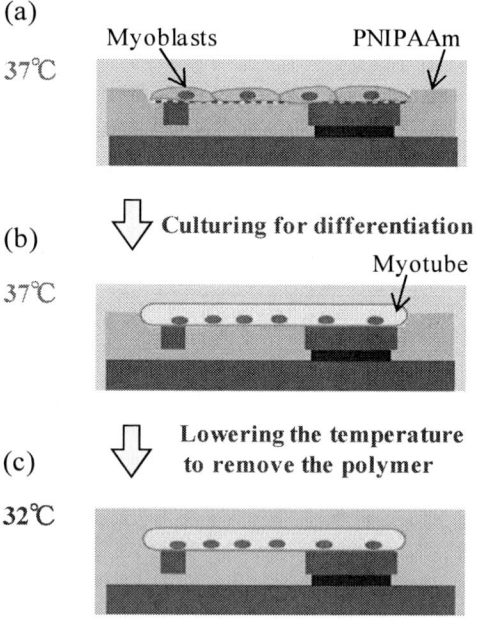

Figure 5: Schematic illustration of the process flow. PNIPAAm was used as a sacrificed layer.

temperature is below 32 °C. Thus, PNIPAAm can be removed without adding any thermal damages to the cells. This process can be divided into 2 steps as described below.

Plasma etching of polymer and patterning of proteins

First, the Si-MEMS device was covered with 5% of PNIPAAm in ethanol and dried at room temperature. Consequently, the microcantilever and the base were buried by PNIPAAm (Fig. 4a). Second, to remove the PNIPAAm selectively, oxygen plasma etching of

(a)
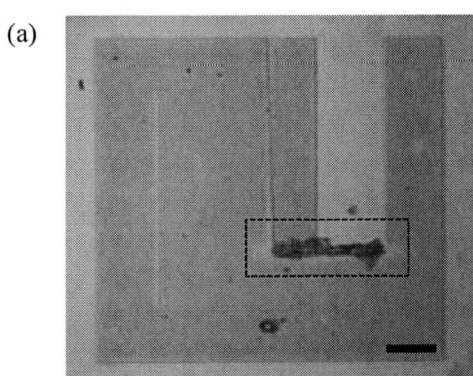

(b)
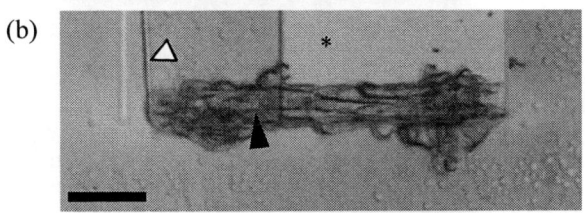

(c)
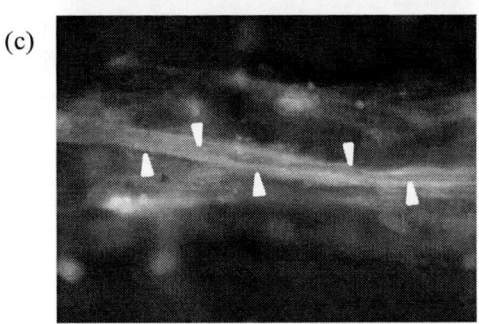

(a)
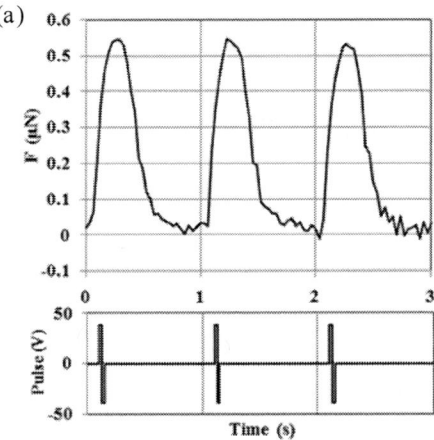

(b)
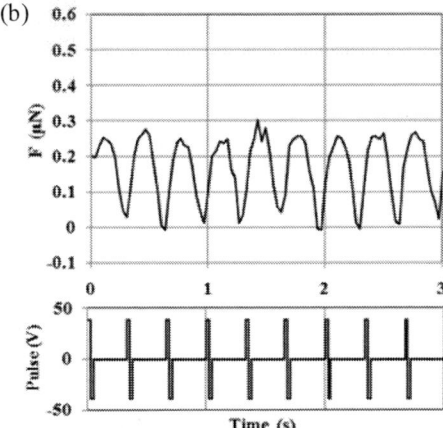

Figure 6: a) A bright field image of the device with myotubes. Scale bar: 200 μm. b) Magnification of the dotted box in Fig. 6a. Black arrow head: myotubes, white arrow head: microcantilever, asterisk: base, Scale bar: 100 μm. c) Immunological staining of the cells cultured on the device. Green; α-actinin (white arrow heads), red; F-actin, blue; nucleus.

PNIPAAm was performed with a stencil mask made by a polyimide tape (Fig. 4b). Third, the aqueous solution of cell adhesive proteins (laminin and Cell-Tak) was casted, dried, and patterned on the device (Fig. 4c). The proteins aid the attachment of the myoblasts to the polymer, enabling full differentiation into myotubes in the next step. This process was carried out at 37 °C to prevent solving the polymer into the aqueous solution. Finally, the stencil mask was removed, and then the patterned proteins were obtained (Fig. 4d).

Culturing of C2C12 cells on the device

At the second step, C2C12 myoblasts were cultured on the device (Fig. 5). The myoblasts were seeded onto the device which the cell adhesive proteins were patterned on (Fig. 5a). The C2C12 myoblasts selectively attached over the gap between the microcantilever and the base where the proteins were patterned because mammalian cells including myoblasts do not adhere to the intact surface of PNIPAAm [9]. Then, the device was immersed into

Figure 7: The myotubes cultured on the device for 7 days were stimulated by the electrical pulse at 1 Hz (a) and 3 Hz (b). The bipolar square pulse, voltage: 0.64 V/mm, pulse width: 24 ms.

Doulbecco's modified eagle medium supplemented by 2% horse serum (DM) to differentiate into myotubes (Fig. 5b). After 7 days cultivation in DM, the cantilever and the myotubes were released by removing the sacrificial layer of the PNIPAAm (Fig. 5c).

EXPERIMENTS

Integration of myotubes onto the device

Fig. 6 shows the C2C12 cells cultured on the cantilever-based device for 7 days. The myotubes were positioned only at the gap between the microcantilever and the base (Fig. 6a). The magnified image revealed that the myotubes elongated successfully (Fig. 6b). The differentiation of myoblasts into myotubes on the device was confirmed by the immunological staining of α-actinin which is one of the muscle specific proteins (Fig.6c, green; α-actinin, red; F-actin, blue; nucleus). Thus, we accomplished the integration of skeletal muscle cells onto the Si-MEMS device.

Measurement of active tension

Nextly, we evaluated the actuation performance of the cultured myotubes on our device by applying the electrical stimulation. At first, the myotubes were stimulated at 1 Hz

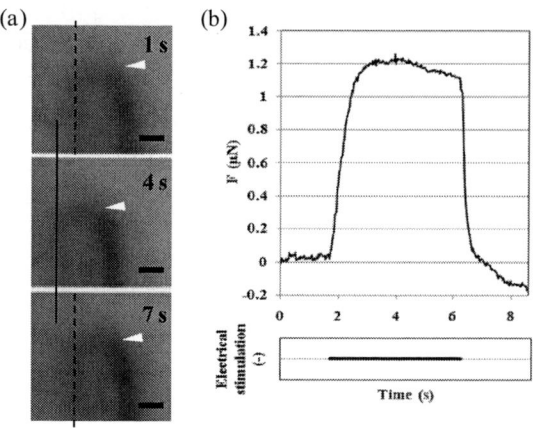

Figure 8: Tetanic active tension of the myotubes cultured on the device for 7 days. a) Movement of the edge of microcantilever was monitored by microscope. White arrw head: the edge of microcantilever. Black line and dotted line indicate the value of displacement. Scale bar: 5 μm. b) The tetanic active tension generated by the myotubes cultured on the device.

(bipolar square pulse, voltage: 0.64 V/mm, pulse width: 24 ms). The movement of the cantilever tip driven by the myotubes were monitored by microscope and recorded by digital video camera (Fig. 1). As shown in Fig. 7a, the myotubes on the device contracted and generated the active tension in response to the electrical pulse. Since the myotubes also responded to other frequency of the electrical pulse (Fig. 7b), it was suggested that the myotubes could be used as a controllable actuator for the Si-MEMS device.

Finally, we measured the tetanic active tension generated by the myotubes cultured on our device. Tetanic tension is a maximum tension which is generated by skeletal muscle cells when they are stimulated at 10-100 Hz of electrical pulse. The myotubes cultured on the device for 7 days were stimulated by the electric pulse at 20 Hz (bipolar square pulse, voltage: 0.64 V/mm, pulse width: 24 ms). Fig. 8a shows the movement of the cantilever tip driven by the myotubes. From the value of the cantilever displacement, we calculated the generated active tension, and obtained the value of approximately 1.2 μN per myotube (Fig. 8b).

CONCLUSION

In the present paper, we propose the skeletal muscle cell lines as a new material for bio-actuator of the MEMS devices and reported the fabrication process to integrate the cells onto Si-MEMS device. We succeeded to differentiate the murine skeletal muscle cell line C2C12 myoblasts into myotubes on the microcantilever-based Si-MEMS device, regulate the movement of the myotubes by applying the electrical stimulation, and measure the active tension of the myotubes. Therefore, the skeletal muscle cell lines have a great potential as a bio-actuator for the microdevices, and the developed integration process and the acquired value of tetanic active tension (approx. 1.2 μN) are useful to design new Si-MEMS devices actuated by the skeletal muscle cells.

REFERENCES

[1] H. Suzuki, A. Yamada, K. Oiwa, H. Nakayama, and S. Mashiko, "Control of actin moving trajectory by patterned poly(methyl methacrylate) tracks", *Biophys. J.*, vol.72, pp.1997-2001, 1997.

[2] L. Limberis and R.J. Stewart, "Toward kinesin-powered microdevices", *Nanotechnology*, vol.11, pp.47-51, 2000.

[3] R.K. Soong, G.D. Bachand, H.P. Neves, A.G. Olkhovets, H.G. Craighhead, C.D. Montemagno, "Powering an inorganic nanodevice with a biomolecular motor", *Science*, vol.290, pp.1555-1558, 2000.

[4] Y. Hiratsuka, T. Tada, K. Oiwa, T. Kanayama, and T.Q. Uyeda, "Controlling the direction of kinesin-driven microtubule movements along microlithographic tracks", *Biophys J.* vol.81, pp.1555-1561, 2001.

[5] S. Yasuda, S. Sugiura, N. Kobayakawa, H. Fujita, H. Yamashita, K. Katoh, Y. Saeki, H. Kaneko, Y. Suda, R. Nagai, H. Sugi, "A novel method to study contraction characteristics of a single cardiac myocyte using carbon fibers", *Am. J. Physiol, Heart Circ. Physiol.*, vol.281, pp.1442-1446, 2001.

[6] M.G. Garcia-Webb, A.J. Taberner, N.C. Hogan, I.W. Hunter, "A modular instrument for exploring the mechanics of cardiac myocytes", *Am. J. Physiol, Heart Circ. Physiol.*, vol.293, pp.866-874, 2007.

[7] J. Xi, J.J. Schmidt, C.D. Montemagno, "Self-assembled microdevices driven by muscle", *Nat. Mater.*, vol.4, pp.180-184, 2005.

[8] Y. Tanaka, K. Morishima, T. Shimizu, A. Kikuchi, M. Yamato, T. Okano, T. Kitamori, "An actuated pump on-chip powered by cultured cardiomyocytes", *Lab. Chip.*, vol.6, pp.362-368, 2006.

[9] J. Kim, J. Park, S. Yang, J. Baek, B. Kim, S.H. Lee, E. Yoon, K. Chun, S. Park, "Establishment of a fabrication method for a long-term actuated hybrid cell robot", *Lab. Chip.*, vol.7, pp.1504-1508, 2007.

[10] A.W. Feinberg, A. Feigel, S.S. Shevkoplyas, S. Sheehy, G.M. Whitesides, K.K. Parker, "Muscular thin films for building actuators and powering devices", *Science*, vol.317, pp.1366-1370, 2007.

[11] D. Yaffe, and O. Saxel, "Serial passaging and differentiation of myogenic cells isolated from dystrophic mouse muscle", *Nature Lond.*, vol.270, pp.725-727, 1977.

A STRETCHABLE CELL CULTURE PLATFORM WITH EMBEDDED ELECTRODE ARRAY

P. Wei[1,2], R. Taylor[4], Z. Ding[2,3], G. Higgs[4], J.J. Norman[4,5] B.L. Pruitt[4], B. Ziaie[1,2]

[1]School of Electrical and Computer Engineering
[2]Birck Nanotechnology Center
[3] Department of Physics
Purdue University, West Lafayette, IN, USA
[4]Department of Mechanical Engineering
Stanford University, Stanford, CA, USA
[5]Department of Pediatric Cardiology
Stanford Medical School, Stanford, CA, USA

ABSTRACT

In this paper, we present a stretchable electrode array for studying cell behavior subjected to mechanical strain. The electrode array consists of four gold nail-head pins (250μm tip diameter and 1.75mm spacing) inserted into a polydimethylsiloxane (PDMS) platform (25.4x25.4mm^2). Fusible indium alloy (liquid at room temperature) filled microchannels are used to connect the electrodes to the outside, thus providing the required stretchability. The electrode platform is biocompatible and can withstand strains of up to 40%. We tested these electrodes by repeatedly (100 times) subjecting them to 35% strain and did not notice any failure. We also successfully cultured mice cardiomyocytes onto the platform and performed electrical pacing.

INTRODUCTION

Stretchable electrodes as cell culture platforms have recently garnered particular attention for their utility in studying cellular behavior central to several important pathologies such as traumatic brain injury, cardiomyopathy, and vascular disorders. A central theme common to these diseases is the subjection of tissue to strain. The ability to record and stimulate nerve and muscle cell populations while subjecting them to mechanical strain can provide insights into mechanism of the aforementioned diseases. In addition, stretchable electrodes can be used to study stem cell differentiation since it is widely believed that mechanical cues are important in this process.

Most reported stretchable electrodes are based on the evaporation of a thin gold layer on a PDMS substrate. While many of these electrodes can be stretched to tens of percent, they are not very robust (break after a few rounds of stretch) [1-5]. In this work, we present a stretchable cell culture platform with embedded electrode array that alleviates the abovementioned problems by using room temperature liquid-alloy filled microchannels as interconnects and miniature gold nail-head pins as the electrodes.

STRETCHABLE PLATFORM

Figure 1 shows a schematic of the cell culture platform which consists of two PDMS layers. The top layer incorporates sub-surface liquid-alloy-filled microchannels and provides a biocompatible exterior surface for cell culture. The bottom layer is bonded to the top layer,

sealing the microchannels. Gold coated nail-head pins acting as electrodes are placed in the channels with their sharp head punched through the top PDMS layer and their flange-shaped nail-head bottom flush against the PDMS sealing the junction against the leakage of alloy into the culture medium.

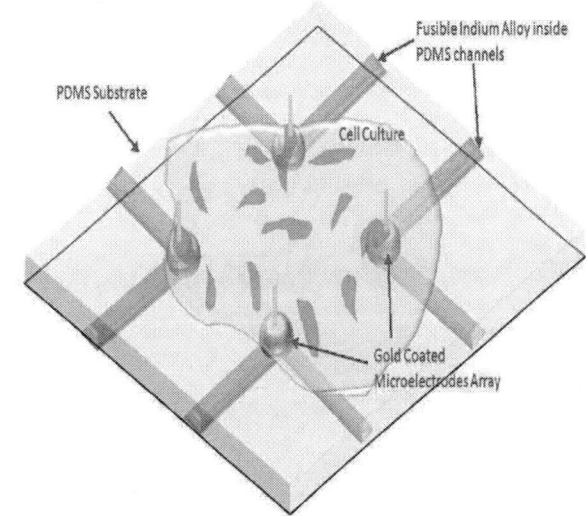

Figure 1. Schematic of the stretchable cell culture platform with embedded electrodes.

FABRICATION PROCESS

Figure 2 shows the fabrication process. It starts with a SU8 (SU8-2150, MicroChem Inc.) mold on a silicon wafer defining the microchannels (600μm height, 500μm width) and electrode placement areas, Figure 2a. Subsequently, uncured PDMS (800μm thickness) is cast against the mold with polyimide tapes (200μm) as spacers, Figure 2b. After release, the gold coated nail-head pins (250μm tip diameter and 750μm head diameter, Mill-Max MGF. Corp.) are inserted at electrode locations, Figure 2c. Another thin PDMS layer (300μm in thickness) treated with a high frequency charges generator (BD-10A, Electro-Technic Products, Inc.) is bonded to the top layer, Figure 2d. The bonded layers are kept at room temperature under atmospheric pressure for 24 hours to ensure a strong interfacial bond. Finally, room-temperature-liquid indium alloy (Gallium/Indium = 75.5/24.5, liquidus temperature ≥ 15.7°C) is injected into the channels and thin connecting

978-1-4244-2977-6/09 $25.00 © 2009 IEEE

wires are inserted into the inlet and outlet ports for electrical connections, Figure 2e.

Figure 3 shows a photograph of the fabricated device. The device has total area of 25.4x25.4mm² with four electrodes of 1.75mm spacing. The inlet and outlet ports are encapsulated with small PDMS droplets to prevent the leakage and oxidation of fusible indium alloy.

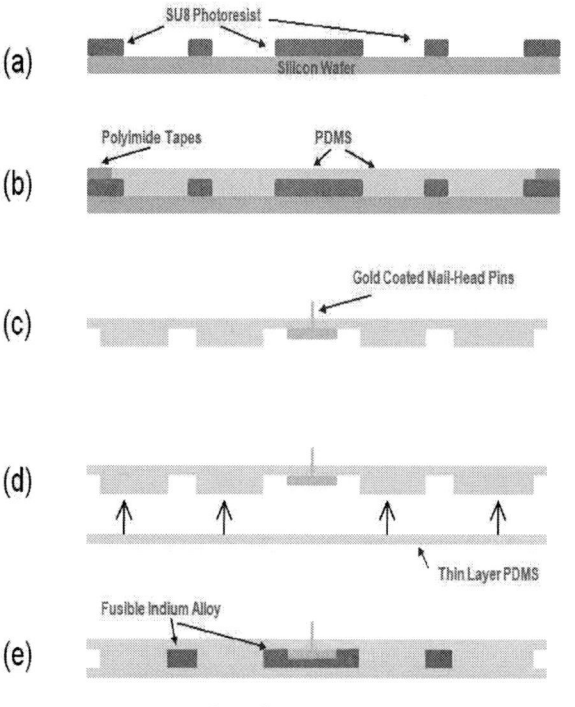

Figure 2. Fabrication Process.

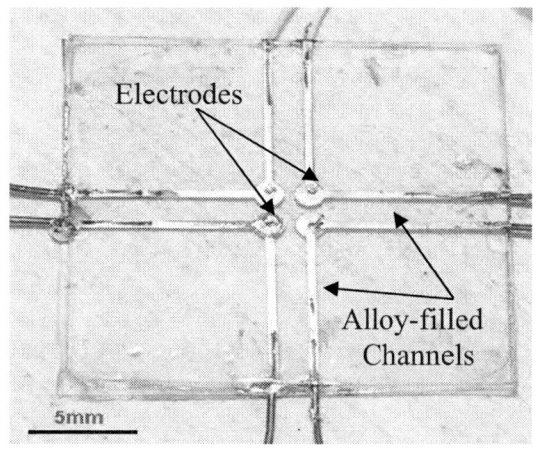

Figure 3. Photograph of a fabricated device.

DEVICE SIMULATION

We performed finite element simulations (COMSOL) to identify stress concentration points and platform behavior under applied strain. Figure 4 shows the displacement and stress distribution assuming a Young's modulus and tensile strength of 750kPa and 2.24MPa for PDMS, respectively. The device was fixed at the left end

and the red arrows indicate the direction of applied displacement (4.4mm, 22% strain) in the x-direction. As expected, the displacement distribution increases linearly in the x-direction. This results in a uniform stress distribution across most of the platform except the channel boundary regions which are subjected to an increased stress (30 times increase as compared to other areas), Figure 4. In this case (22% strain) the maximum strain at the channel boundary exceeds the tensile strength indicating a breakage point.

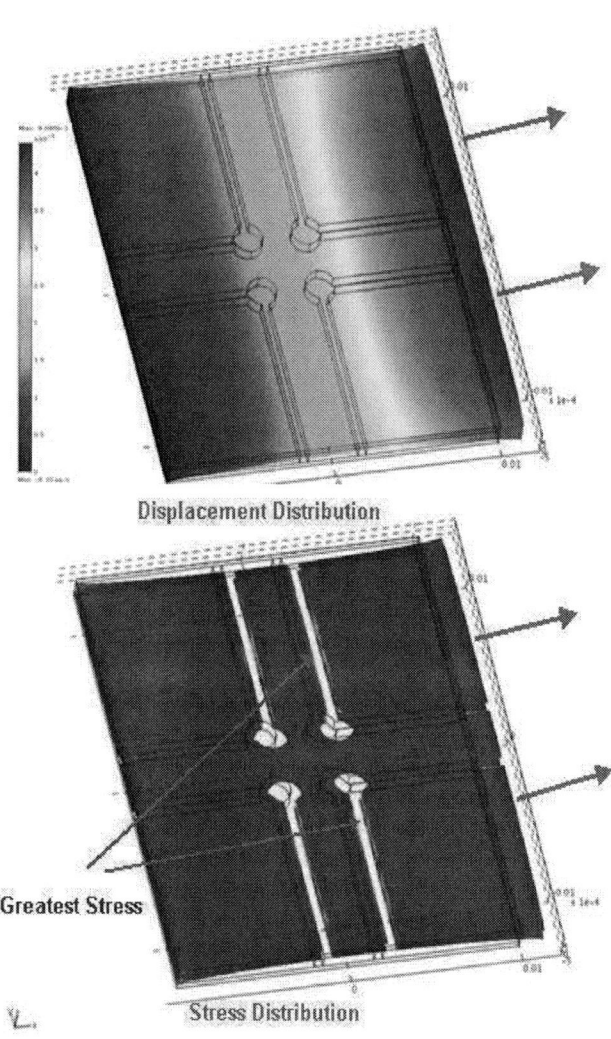

Figure 4. Simulation results for displacement and stress distribution of the stretchable electrode array platform.

ELECTRICAL CHARACTERIZATION

The stretchable electrode array, without cell culture, was stretched with micro-manipulator to evaluate its electrical properties, i.e., interconnect resistance as a function of applied strain. Figure 5 illustrates the stretch test experimental set-up. The device was clamped to a pair of micro-manipulators with a droplet of phosphate buffer saline (pH = 7.4) placed on the device. The impedance was measured between two electrodes at 1kHz with the platform stretched by 5% strain increments and decrements. Figure 6 shows the strain vs. impedance

978-1-4244-2977-6/09 $25.00 © 2009 IEEE   408

measurements results showing a constant impedance of ~ 1.1kΩ with applied strains of up to 35%. We performed reliability tests by subjecting the platform to 100 cycles of stretch-release, after which, we did not observe any alloy leakage or electrode breakage.

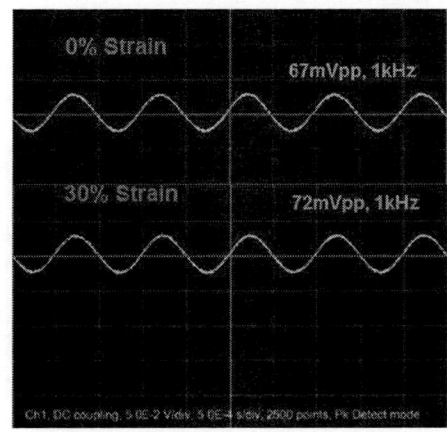

Figure 7. Recorded signals through an agarose gel (top: signal recorded at 0% strain; bottom: signal recorded at 35% strain)

CELL CULTURE EXPERIMENTS

Cell culture experiments with live/dead assay (Invitrogen L3224) were performed to confirm hermeticity of indium alloy inside the PDMS channels. PDMS retaining rings on the devices created a sealed region for cell culture media (Figure 8). Devices were prepared by coating them with 2% gelatin after O_2 plasma surface activation. Human aortic smooth muscle cells were then plated on the devices. Inset in Figure 6 shows the fluorescent microscopic image of the cells after two days with the green being the cell cytoplasm and the red dots the nuclei of dead cells (blue is the bright field image). The presence of live cells is indicated by green cell bodies without red nuclei. These images indicate that cells can survive near and even on the electrodes.

Figure 9 shows cells stretched to 10% strain. As can be seen, cells remained attached to the PDMS surface demonstrating their viability for mechanical studies.

Figure 5. Stretch test setup using a micro-manipulator.

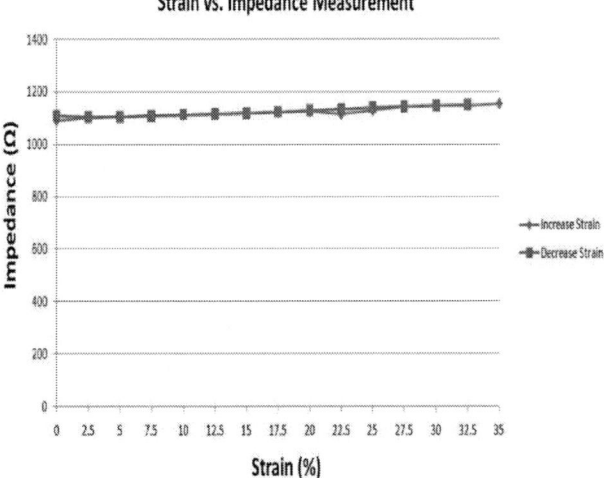

Figure 6. Strain vs. impedance measurement results.

We also performed in vitro tests using agarose gel (0.2% w/v) to mimic a brain tissue slice [6]. The agarose gel was placed on the stretchable electrode array platform within a PDMS ring container. The entire platform was clamped to a pair of micro-manipulators with a probe inserted into the agarose gel. A sinusoidal signal (100mVpp, 1kHz) was applied to the probe and transmitted voltages were recorded through gold coated nail-head electrodes. Figure 7 shows the recorded waveforms at 0 and 30% strain. The results clearly demonstrate the recording capability of the stretchable electrode array platform within a tissue medium during mechanical deformation.

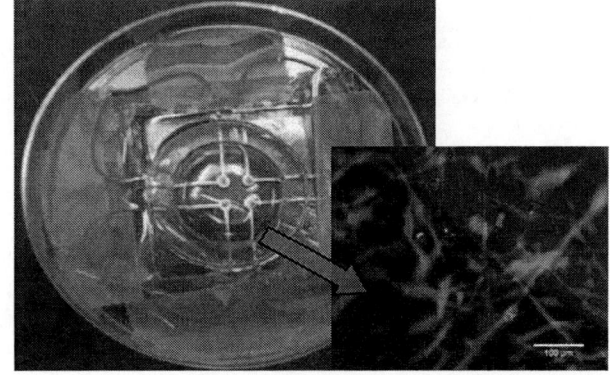

Figure 8. Stretchable cell culture platform populated with human aortic cells enclosed by PDMS retaining ring after 2 days of incubation.

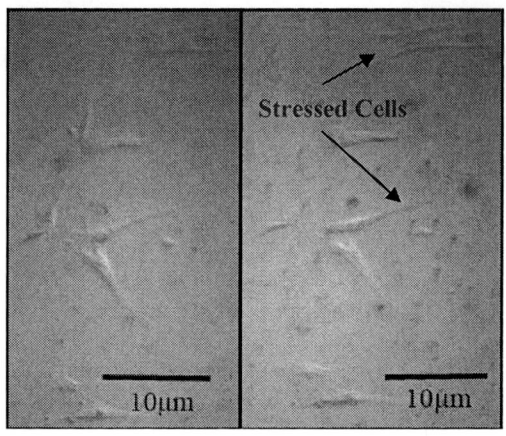

Figure 9. Cells remain attached to the platform and stretch with it upon the application of 10% strain.

In order to demonstrate the platform functionality with plated cells, we performed electrical pacing of mouse cardiomyocytes. The cell population was prepared and deposited on the device allowing 30 to 90 minutes for cell attachment. Then buffer media containing non-attached cells was aspirated and Tyrode's salt solution was added to the PDMS ring wells shown in Figure 6. Following the preparation of cardiomyocytes, devices wires were connected to a commercial electrical pacer (MyoPacer, IonOptix LLC). A square wave signal (30Vpp, 1Hz) was then applied resulting in contraction of cardiomyocytes on the surface, demonstrating that the device could be utilized for *in-vivo* testing, Figure 10.

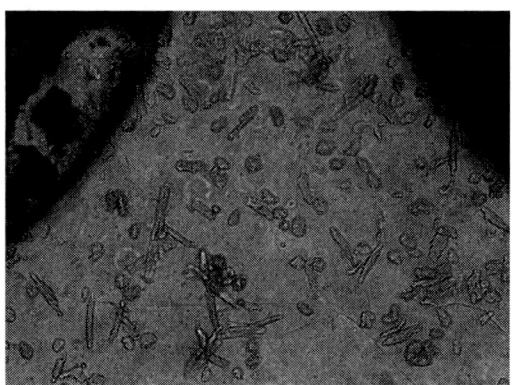

Figure 10. Electrically paced mice cardiomyocytes

CONCLUSIONS

We designed a simple and cost effective process for fabricating a stretchable cell culture platform with embedded electrode array using PDMS, room temperature-liquid-alloy, and gold coated nail-head pins. The device was fabricated and fully characterized. It was demonstrated that the platform is biocompatible to human aortic muscle cells and mice cardiomyocyte. The cells adhered well to the substrate during mechanical strain and survived near and on the electrodes after two days of incubation. The platform also maintained its electrical capabilities after being subjected to repeated cycles of mechanical deformation.

REFERENCES

[1] S. P. Lacour, C. Tsay, S. Wagner, Y. Zhe, and B. Morrison, B., *2005 IEEE Sensors*, pp. 617-620.

[2] J. Blundo *et al.*, *Proceedings of the ASME 2007 BioEng. Conf.*, June 2007, Colorado USA.

[3] D.-H. Kim et al., *Science*, 320 (2008), pp.507-511

[4] J. Xiao et al, *Appl. Phys. Lett.*, 93 (2008), pp013109-1-3.

[5] M. Maghribi et al., *2002 IEEE-EMBS*, pp. 80-83.

[6] Z. Chen, G.T. Gillies, P.P. Fatouros, *J. Neurosurg.*, Vol. 101 (2004), pp. 312 – 322.

ASSESSMENT OF SINGLE CELL VIABILITY FOLLOWING LIGHT-INDUCED ELECTROPORATION THROUGH USE OF ON-CHIP MICROFLUIDICS

Justin K. Valley[1], Hsan-Yin Hsu, Steven Neale, Aaron T. Ohta, Arash Jamshidi, and Ming C. Wu
[1]Berkeley Sensor & Actuator Center, Department of Electrical Engineering and Computer Sciences
University of California, Berkeley, USA

ABSTRACT

The high throughput electroporation of single cells is important in applications ranging from genetic transfection to pharmaceutical development. Light-induced electroporation using optoelectronic tweezers (OET) shows promise towards achieving this goal. However, cell viability following light-induced electroporation has yet to be shown. Here we present a novel OET device which incorporates microfluidic channels in order to assess the viability of single cells following light-induced electroporation. Monitoring of single cell electroporation and viability is achieved through the use of fluorescent dyes which are exchanged using the integrated fluidic channels. The successful *reversible* electroporation of HeLa cells is shown.

INTRODUCTION

Electroporation is a widely used technique for the introduction of exogenous molecules across the cell membrane through the use of external electric fields. If the field the cell is subjected to is large enough, the cell's membrane will porate, allowing molecular exchange with the external environment. The field can be properly tuned so that these pores can then reseal. This technique is largely used for genetic transfection and fluorescent tagging.

Conventional electroporation techniques are, however, limited by either low throughput or limited selectivity. Due to this, there has been increasing interest in creating a system capable of performing high throughput electroporation with single cell selectivity for on-chip transfection and cell-monitoring studies [1-3]. We have recently reported on the use of OET to achieve light-induced electroporation [4]. As shown in Figure 1, in this scheme, we use light-induced dielectrophoresis to manipulate individual cells in parallel and, then, by increasing the bias, we selectively electroporate the illuminated cells. By combing OET with electroporation, a high throughput, high selectivity assay can be performed.

Optoelectronic tweezers uses patterned light to alter the conductivity of a photosensitive film to create localized electric field gradients. These gradients result in a dielectrophoretic (DEP) force on particles in the vicinity. Figure 2 shows the electric field distribution created by illuminating a region of the photosensitive film. Note the concentration of electric field and creation of localized electric field gradients at the center of the x-axis where the 20 μm light spot is incident. Because of the low light power necessary for actuation, compared to the more traditional optical tweezers, thousands of simultaneous traps can be created and manipulated in parallel [5].

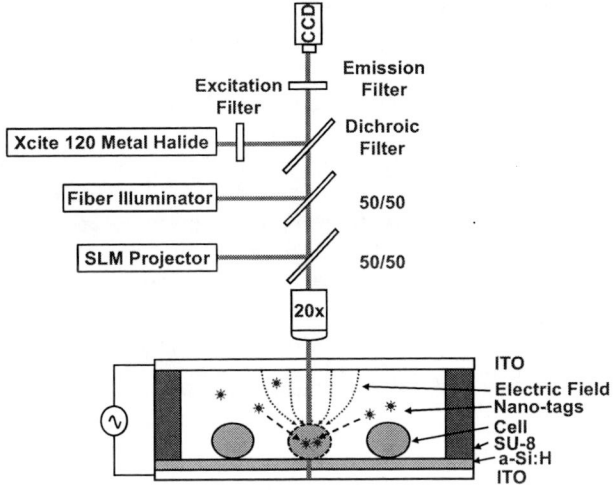

Figure 1: Experimental setup and schematic showing electroporation mechanism. Electric field is concentrated across the illuminated cell resulting in electroporation. A series of filters are used to select the appropriate fluorescent dye to monitor. Fluid exchange occurs through the use of a syringe pump (not pictured).

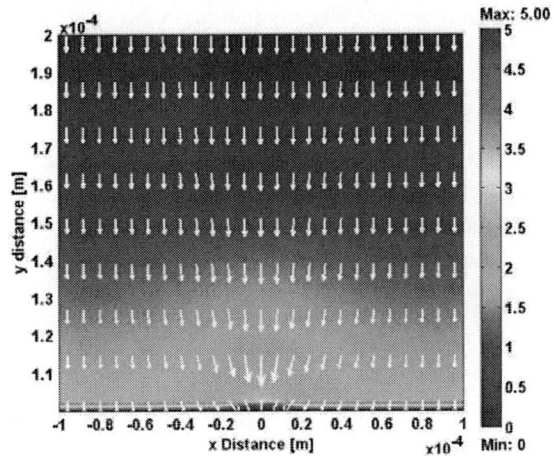

Figure 2: Electric field profile created by illumination in the OET device at center of x-axis. Arrows correspond to electric field, while the surface plot corresponds to electric potential for a 5 V amplitude signal. Note how localized electric field gradients and electric field concentration occurs in the illuminated region.

When a cell is illuminated by the projected light in the OET device, the electric field is concentrated across it. If the electric field is large enough, the cell's membrane will form nanoscopic pores allowing exogenous molecules to enter the cell (Figure 1). In this manner, one

978-1-4244-2977-6/09 $25.00 © 2009 IEEE

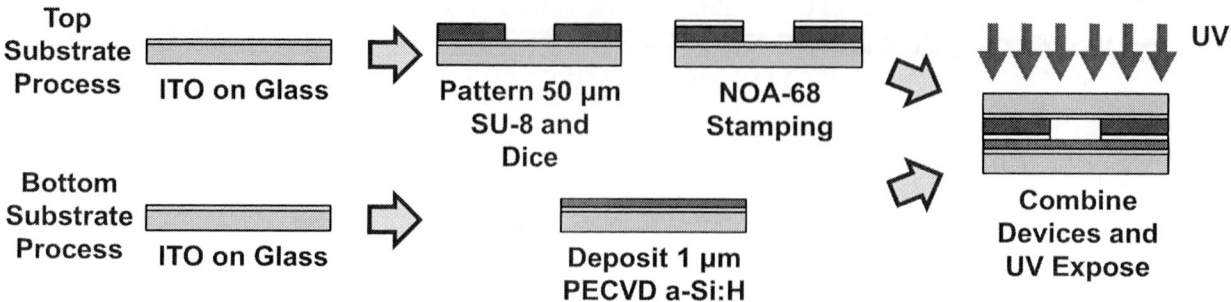

Figure 4: Fabrication of OET with integrated microfluidic channels. Channels are defined in SU-8 on the topside OET substrate and bonded to the bottom OET substrate using a UV-curable epoxy.

can individually select and electroporate cells in parallel.

However, until now, the reversible electroporation of cells has not been demonstrated in this device. Reversible electroporation refers to the ability of the pores in an electroporated cell to reseal. This process typically takes on the order of minutes to tens of minutes [6] and is highly dependent on the electric field dose applied during the electroporation process. A common method for the investigation of reversible electroporation is the use of fluorescent dyes [7]. First a cell is electroporated in the presence of a dye which causes successfully electroported cells to fluoresce. Next, the solution surrounding the cells is replaced with a solution containing another dye which indicates whether the electroporated cell is viable. For cells not adhered to the surface, this procedure requires the use of on-chip microfluidic channels. Here we present a process by which the OET device is integrated with lithographically defined channels and demonstrate its capability by showing that reversible electroporation can be obtained.

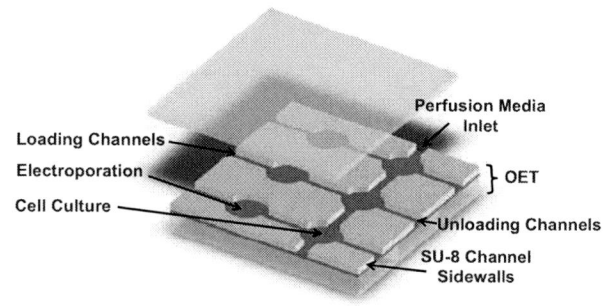

Figure 3: Concept overview of OET-based electroporation assay. Cells are electroporated in parallel with single cell selectivity in lithographically defined compartments and then moved into perfusion chambers allowing for subsequent cell culture.

FABRICATION

The integration of microfluidic circuits with the OET device will allow for applications ranging from on-chip sorting to on-chip cell culture (Figure 3). We have developed a process that allows lithographically defined channels to be integrated with the OET electroporation device. This process needs to allow for arbitrary top and bottom substrates. This is because the OET device requires a transparent conductive top surface and a photosensitive bottom substrate. Additionally, in order to accommodate cells, the channel height must be relatively thick (~50 μm) and the channel material must be non-conductive. SU-8 lends itself well to this task due to its ability to easily form high-aspect ratio, permanent, structures. The process is depicted in Figure 4.

Indium tin oxide (ITO) (300 nm) coated glass serves as the top and bottom OET surface. The bottom substrate is coated with a 1 μm layer of PECVD a-Si:H (100 sccm 10%SiH$_4$:Ar, 400 sccm Ar, 900 mTorr, 350°C, 200 W). The top substrate is patterned with 50 μm of SU-8 (Microchem, SU-8 2050) which defines the channel geometry. Both substrates are then diced into 2x2 cm chips using a dicing saw (Esec 8003). Fluidic input and output ports are then drilled into the topside device with a drill press and 0.75 mm drillbit.

A 20 μm layer of UV-curable epoxy (Norland, OA-68) is then spin coated (4000 RPM, 2 min.) on a dummy wafer. A block of polydimethlysiloxane is used to stamp the NOA-68 from the dummy wafer to the SU-8 channels. This results in a ~15 μm layer of NOA-68 on the top of the SU-8 channel sidewalls. Finally, the top and bottom OET substrates are combined and the epoxy is cured using a handheld UV gun (Norland, Opticure-4, 10sec). No alignment is needed in this case as the bottom substrate is featureless. The combination of the two substrates can be accomplished by hand or through the use of a flip-chip bonder. Fluidic access connections are then attached to the ports on the topside of the device with additional NOA-68.

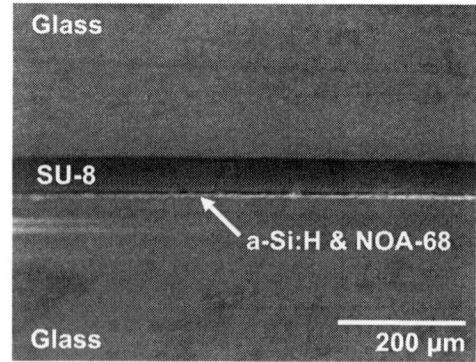

Figure 5: SEM of cross section of OET device with integrated SU-8 microfluidic channels.

Figure 5 shows an SEM cross-section of the described device showing the top and bottom substrates, SU-8, a-Si:H, and NOA-68. Figure 6 is a picture of the completed device.

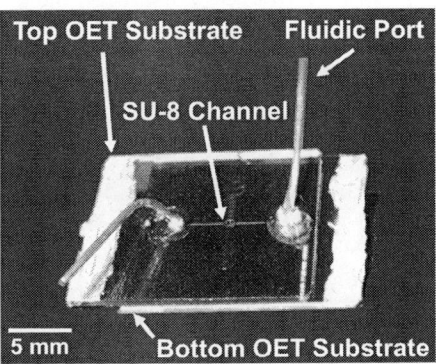

Figure 6: Picture of completed device. The top and bottom OET substrates are bonded together using a UV curable epoxy resulting in sealed microfluidic channels. Fluids are introduced through fluidic ports.

EXPERIMENTAL

For many applications (e.g. transfection), the cell must remain viable following electroporation (i.e. reversible electroporation). For a demonstration of the integrated device, we assess the viability of a single cell following light-induced electroporation. In order to show this, we use the technique mentioned before using two cellular dyes. The first dye, Propidium Iodide (PI) (Invitrogen), is a membrane impermeant dye. However, if the dye enters the cell, it binds to DNA and fluoresces red. It is an indicator of successful electroporation. The second dye, Calcein AM (CaAM) (Invitrogen), passively diffuses across the cell membrane. Once in the cytosol, it interacts with enzymes in the cell and produces a green fluorescent molecule which is membrane impermeant. Strong green fluorescence in the presence of CaAM means that the cell's membrane is intact and that the intracellular contents contain the necessary enzymes to catabolize the CaAM. If these two conditions are met, it is a strong indicator of cellular viability.

Since CaAM is membrane permient it is necessary to introduce it following electroporation, necessitating the use of microfluidic channels. To demonstrate successful reversible electroporation, we first select a cell, electroporate it in the presence of PI, and then exchange the media with a solution containing CaAM. If the cell fluoresces both red and green at the completion of the experiment, the cell has been reversibly electroporated.

HeLa cells were washed three times in and suspended in electroporation buffer (Cytopulse® Cytoporation Medium, 10 mS/m) at a cellular concentration of 2×10^6/mL. Due to the presence of large electrical fields during electroporation (kV/cm), joule heating in the liquid is an issue. Small temperature changes can reduce cellular viability significantly. Since joule heating is directly proportional to liquid conductivity, it is advantageous to use lowly conducting media.

PI dye was added to the cellular solution to achieve a concentration of 2 μM. CaAM dye at 6 μM was also prepared to be introduced later. The cell and PI solution was then introduced into the device via a syringe pump. The CaAM solution was also connected to the device to be introduced later.

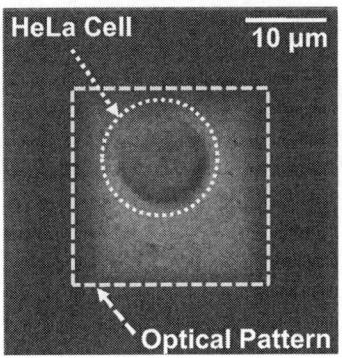

Figure 7: Bright-field image of HeLa cell under test in OET device with applied optical pattern. Cells are suspended in Cytopulse® Electroporation media (8 mS/m).

An individual HeLa cell is selected and a light pattern is generated to illuminate it (Figure 7). After initial positioning at low voltage (4 Vppk), the cell does not exhibit any fluorescence (Figure 8). This is because the fields the cell experiences during movement are less than that required to porate the cell membrane. However, once the electroporation bias is applied (20 Vppk), the critical field required to porate the membrane is surpassed and the cell uptakes PI and fluoresces accordingly. No CaAM fluorescence is seen at this stage as this dye is not currently present in the solution (Figure 8). Next the media surrounding the cells is exchanged, via the fluidic channels using a syringe pump, with the CaAM solution. After fluid exchange, the cells are incubated for 15 minutes to allow the CaAM to diffuse into the cells. After incubation, the cell now exhibits both PI and CaAM uptake (Figure 8). This means that the cell's membrane has resealed following electroporation and that it contains the appropriate enzymes necessary to catabolize the CaAM into its fluorescent derivative. Therefore, successful reversible electroporation has occurred.

CONCLUSION

We have demonstrated the integration of microfluidic chambers onto the OET device to demonstrate *reversible* light-induced electroporation. This process can be extended to allow for on-chip gene

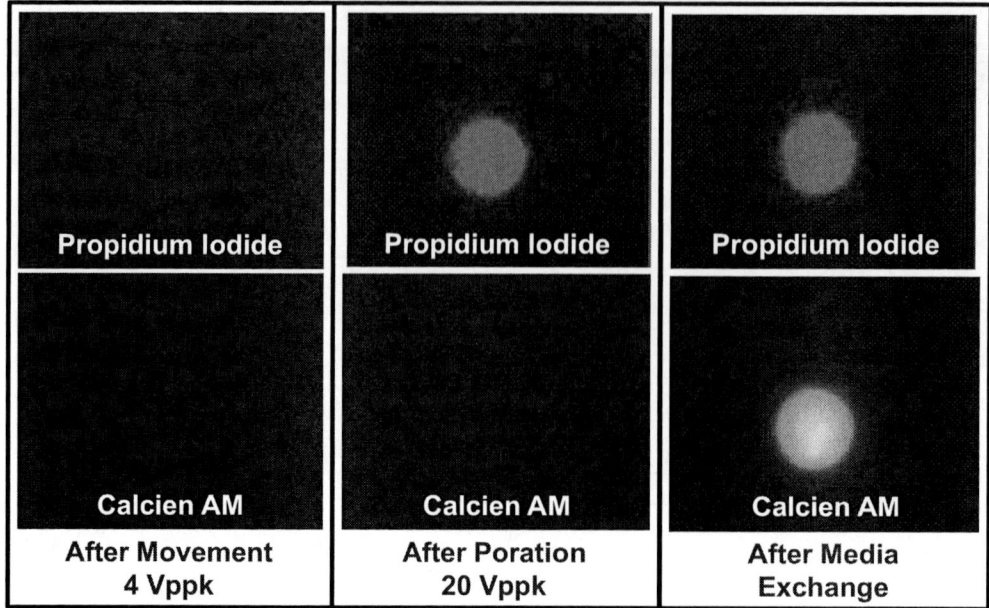

Figure 8: Fluorescent response of cell in Figure 7 for both PI and CaAM. Initially, the cell is suspended in a solution only containing PI (6µM). In the first panel, no dye uptake is observed following positioning of the cell with OET at 4 Vppk. In the second panel, the cell is subjected to the electroporation bias resulting in PI dye uptake. In the third panel, the media is exchanged using microfluidic channels with a solution containing CaAM (2µM). The cell now exhibits CaAM and PI response verifying successful reversible electroporation.

transfection and cell culture, eventually leading to a high throughput electroporation assay with single cell selectivity.

ACKNOWLEDGEMENTS

The authors would like to thank Ann Fischer of the UC Berkeley Cell Culture Facility for providing the cells. This work was funded by the Center for Cell Control, a National Institute of Health Nanomedicine Development Center, Grant # PN2 EY018228.

REFERENCES

[1] H. Q. He, D. C. Chang, and Y. K. Lee, "Using a micro electroporation chip to determine the optimal physical parameters in the uptake of biomolecules in HeLa cells," *Bioelectrochemistry*, vol. 70, pp. 363-368, May 2007.

[2] H. Y. Wang and C. Lu, "Microfluidic electroporation for delivery of small molecules and genes into cells using a common DC power supply," *Biotechnology and Bioengineering*, vol. 100, pp. 579-586, Jun 2008.

[3] M. Khine, C. Ionescu-Zanetti, A. Blatz, L. P. Wang, and L. P. Lee, "Single-cell electroporation arrays with real-time monitoring and feedback control," *Lab on a Chip*, vol. 7, pp. 457-462, 2007.

[4] J. K. Valley, H. Y. Hsu, A. T. Ohta, S. Neale, A. Jamshidi, and M. C. Wu, "In-situ Single Cell Electroporation Using Optoelectronic Tweezers," in *IEEE/LEOS Optical MEMS*, Freiburg, Germany, 2008.

[5] P. Y. Chiou, A. T. Ohta, and M. C. Wu, "Massively parallel manipulation of single cells and microparticles using optical images," *Nature*, vol. 436, pp. 370-372, Jul 21 2005.

[6] S. W. Hui, "Effects of Pulse Length and Strength on Electroporation Efficiency," in *Animal Cell Electroporation and Electrofusion Protocols*. vol. 48, J. A. Nickoloff, Ed.: Humana Press, 1995, pp. 29-40.

[7] J. S. Soughayer, T. Krasieva, S. C. Jacobson, J. M. Ramsey, B. J. Tromberg, and N. L. Allbritton, "Characterization of Cellular Optoporation with Distance," *Anal. Chem.*, vol. 72, pp. 1342-1347, 2000.

MANIPULATION OF BIOSAMPLES AND MICROPARTICLES USING OPTICAL IMAGES ON POLYMER DEVICES

Wei Wang [1], Yen-Heng Lin [1], Tzung-Fang Guo [2] and Gwo-Bin Lee [1,3]

[1]Department of Engineering Science, [2]Department of Electro-optical Science and Engineering,
National Cheng Kung University, Tainan, Taiwan
[3]Medical Electronics and Device Technology Center, Industrial Technology Research Institute,
Hsinchu, Taiwan

ABSTRACT

This paper presents a new method using a polymer material to generate optically-induced dielectrophoretic (ODEP) forces for particle manipulation. Instead of using the common material such as amorphous silicon, a polymer (P3HT/PCBM) which has excellent light absorption, was used in this study. Without using delicate and high-temperature thin-film process, the entire fabrication process was performed at room temperature. As a light beam was illuminated onto the polymer surface, electron-hole pairs was generated such that a non-uniform electric field was induced when an AC voltage was applied on the ODEP chip, thus generating an ODEP force. It can be used to manipulate particles for several functions such as concentration, transportation and separation of particles. The concentration of the polymer was found to play an important role on the magnitude of the ODEP force. Experimental data showed that polystyrene beads can be moved around using the developed chip. The maximum ODEP force was experimentally found to be around 35.2 pN. The new material may provide a cost-effective approach for mass-production of the ODEP chips.

INTRODUCTION

Recently, the manipulation of particles by using DEP forces [1] has been extensively explored for biological applications. Conventional DEP manipulation requires complicated photolithography and delicate thin-film processes to form driving electrodes. Different geometric configuration of the electrodes requires tedious process to redesign and fabricate a new chip. Alternatively, ODEP [2-3] technology using optical images to define various kinds of electrodes for particle manipulation has attracted considerable interests. When a light beam is illuminated onto the surface of a photoconductive material (usually amorphous silicon), a non-uniformed electric field is formed to induce dielectrophoretic forcs such particles can be manipulated while an AC voltage is applied on the chip. The flexibility to dynamically change the layouts of the driving electrodes using ODEP provides a new approach for particle manipulation. However, the deposition of amorphous silicon requires a delicate vapor deposition machine and also a high deposition temperature. It may hinder their practical applications when ODEP devices are integrated with other polymer-based microfluidic components.

ODEP platforms fabricated on silicon wafers with an amorphous silicon layer to manipulate particles have been recently explored extensively. It can simplify the complicated micro-electrode fabrication process by using "virtual" electrodes formed by light illumination to generate dielectrophoretic (DEP) forces on particles. However, the manufacturing process of amorphous silicon devices uses a plasma-enhanced chemical vapor deposition (PECVD) method, which requires relatively high temperature. Besides, it is still costly and complex manufacture procedure. As a result, it remains a critical issue to simplify ODEP manufacturing process and also replace the amorphous silicon layer. In this study, we report a new polymer material using a room-temperature process for ODEP applications. By using computer flash programs and a digital projector, the polymer chip can generate ODEP forces to manipulate particles.

In the past decades, polymers have been known as excellent materials for fabricating microfluidic devices and systems for biological applications. Among them, conjugated polymers [4] have drawn a lot of attention recently for their unique optoelectronic properties, low-cost for production, light weight, band gap tailorability and solubility in organic solvents. Taking all these characters into account, the development of conjugated polymers therefore leads to optoelectronic applications. Two major mechanisms for these materials, including a light generation mechanism and light harvesting mechanism, have been applied for a variety of industrial applications. For example, organic light emitting diodes (OLEDs) use the light generation mechanism while organic solar cells use the light harvesting mechanism. Similarly, organic field effect transistors (OFETs) are commonly used for semiconductor applications. Among the polymers for a thin-film photovoltaic (PV) market, regioregular poly (3-hexylthiophene) (rr-P3HT) [5] is one of the most promising polymer materials due to its low band gap, good solubility, highly crystalline morphology, good charge mobility and thermal stability. When incorporated with phenyl-C61-butyric acid methyl ester (PCBM), it is reported that a blend film with P3HT (as the standard fabrication of plastic solar cells) can achieve an excellent light absorption, high drift mobility and also semiconducting properties.

The polymer thin film layer, a blend film of P3HT and PCBM, is a well-known bulk heterojunction structure [6-7]. It has been reported to have excellent optoelectronic transmission efficiency. Meanwhile, the polymer thin film chip can be mass fabricated at room temperature. It is also feasible to fabricate ODEP devices on flexible plastics if using this new material. Therefore, the current study presents an innovative platform by using the P3HT/PCBM blend film patterned onto the indium-tin-oxide (ITO) substrate for particle manipulation by utilizing the generated ODEP forces from the illuminated pattern of the light.

978-1-4244-2977-6/09 $25.00 © 2009 IEEE

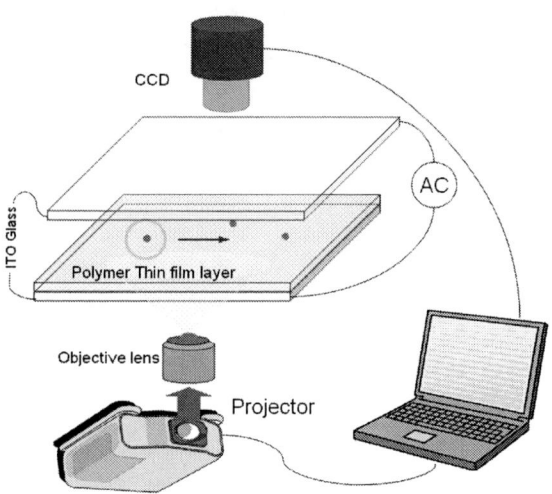

Figure 1: Schematic illustration of polymer thin-film ODEP manipulation platform.

hours to perform a "slow solvent vapor treatment" process [9]. Finally, another ITO glass is placed on top of the ITO substrate and bonded by a double-sided tape (30 μm) as a space layer to form a complete polymer ODEP chip.

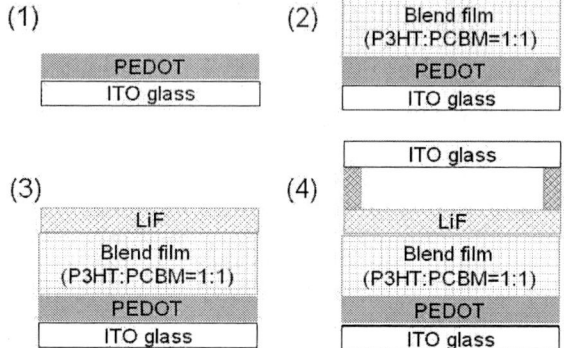

Figure 2: Fabrication process of the polymer thin-film ODEP chip

EXPERIMENTAL

Design

Figure 1 shows a schematic illustration about the operation principle of polymer-based ODEP chip. The illuminated pattern induces electron-hole pairs and makes the applied voltage drop across the bottom layer, producing a non-uniform electric field at the light-patterned regions in the liquid layer and thus inducing the ODEP force. The projected image then provides a similar effect as metal electrodes for generating DEP forces to manipulate particles.

Fabrication

The polymer thin-film ODEP chip has a sandwiched structure composed of a top ITO (RITEK Corp., 15Ω□) glass, a space layer and a bottom ITO glass with photoconductive polymer layers. The bottom ITO glass has several layers including a poly(3,4-ethylenedioxythiophene):poly(styrene- sulfonate) (PEDOT:PSS) film, a P3HT/PCBM blend film, and a lithium fluoride (LiF) layer (50 Å). Figure 2 shows a simplified fabrication process. Prior to fabrication, the ITO glass substrate is cleaned by ultrasonic treatment using detergent, deionized water, acetone and isopropylalcohol. Then the PEDOT:PSS (Baytron P, Bayer AG, Germany) layer is spin-coated at 4000 rpm to modify the ITO surface. Please note that the following procedure has to be performed in a nitrogen-filled glove box to form the active layer. After baking at 150°C for 30 min, a blend solution of P3HT (Rieke Metals, Inc., USA) and PCBM at a ratio of 1:1 is dissolved in 1,2-dichlorobenzene [6, 8] with a concentration around 3 wt% to 5wt% was prepared by stirring for at least 24 hours at room temperature. Then the solvent is filtered to prevent the aggregation of particles. The active layer is formed by spin-coating the blend solution with different concentrations. Please note that different spin rates have been tested. Finally, LiF is thermally deposited onto the P3HT: PCBM film to prevent water and oxygen in the air from destroying the polymer film in practical applications.

In order to ensure the purity of the crystalline, every wet film is immediately sealed in a dichlorobenzene vapor-filling Petri glass dish to slowly evaporate for four

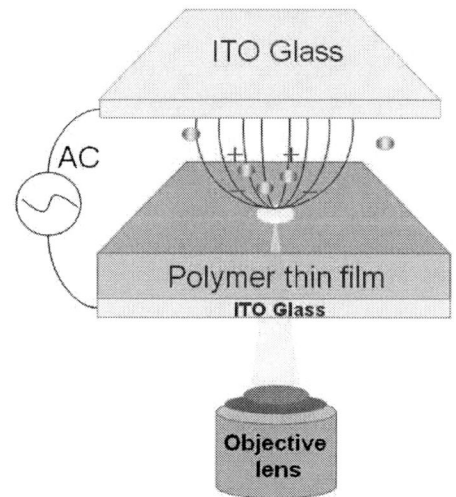

Figure 3: Schematic illustration of the operation principle for particle manipulation using the ODEP force. When a light beam is focused and projected onto the polymer layer under an AC voltage, ODEP forces is formed to manipulate polystyrene beads.

Particle manipulation

As shown schematically in Figure 3, liquid that contains polystyrene beads is sandwiched between the top ITO glass and the bottom ITO glass with a photoconductive polymer layer. An AC voltage (24 V_{pp}, 100 kHz) is applied across the liquid layer. The illumination source is a poly-Si TFT micro-lens projector (ViewSonic PJ1172, Japan) with a spatial resolution of 1024 pixels x 768 pixels. The programmable flash image is first focused and then projected onto the polymer layer through a 50 X objective lens (Nikon, Japan). The excited electron-hole pairs may induce a non-uniform electric field across top and bottom layers at the illuminated area. A negative DEP force is then generated by the "virtual" electrode formed by the illuminated area to push or confine polystyrene beads. By using different patterns of illumination, polystyrene beads can be moved around. The AC voltage is supplied by a function generator (Model 195, Wavetek, U.K.) and a power amplifier. A computer is used to generate the images shown on the illuminated area and also acquire images of particle manipulation using a charge-coupled-device (CCD, SSC-DC80, Sony, Japan) camera.

RESULTS AND DISCUSSION

A manipulation of polystyrene beads (20 µm) has been successfully demonstrated using the developed chip, as illustrated in Fig. 4. The polystyrene beads can be manipulated to move around a square. This indicates that the ODEP technology can be realized on our polymer thin film chips. Furthermore, the trapping area can be tailored to match different bead sizes and patterns, which can be used to manipulate, separate and sort biosamples and microparticles for further biological applications.

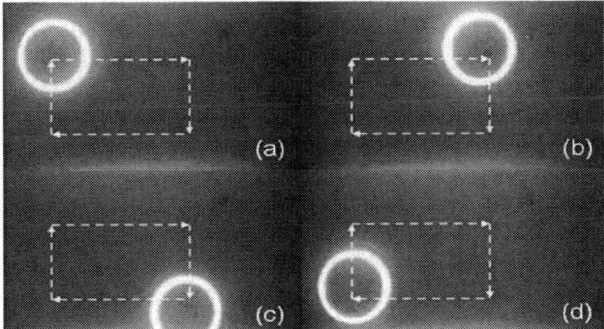

Figure 4: Manipulation of polystyrene beads to move around a square by using a projected circle.

The optical properties of the polymer films have been characterized. Absorption spectra of the P3HT: PCBM blend film is investigated by using a spectrophotometer (Model U-4100, Hitachi, Japan). Each film is measured by using a single-beam UV-Vis spectrophotometer spectrum. The P3HT/PCBM blend films with a ratio of 1:1 are tested. The concentrations of the blend films range from 1 wt% to 6% with an increment of 1%. Please note that each solvent has to be filtered before spin-coating to prevent the aggregation of particles. All drying procedure of the wet films has to be performed under slow solvent vapor treatment. Figure 4 shows the relationship between the absorption and the concentration of the blend film. From the UV-Vis spectrum, two peaks have been observed. The first peak appears at 335 nm corresponding to the PCBM, while the other peak (500–550 nm) represents the contribution of P3HT. As shown in this figure, the two peaks do not shift by the filtering process. It is further verified that the morphology does not change with the increase of concentrations in the slow-thermo evaporation process

In Figure 4, it is clearly seen that increasing concentration corresponds to higher UV-Vis absorption spectra. The absorption level reaches a peak for 5wt%. Nevertheless, the absorption level starts to diminish for 6wt%. It is due to the fact that P3HT/PCBM is aggregated in this high concentration even if it has been filtered.

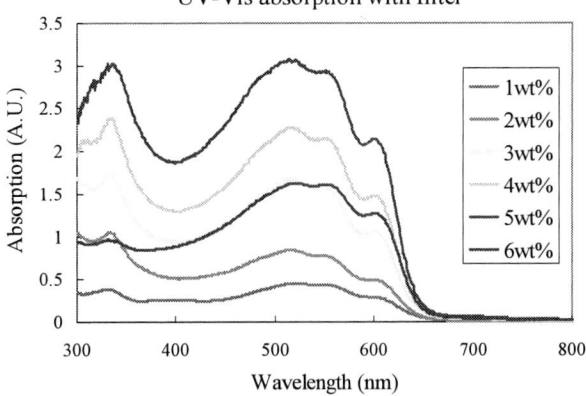

Figure 5: The relationship between the absorption and the wavelength of the projected light at different concentrations of the polymer films. The maximum peak occurs at a concentration of 5wt%.

The magnitude of the ODEP forces has been characterized by dragging the beads at their maximum velocity. According to the Stoke's law, the velocity of spherical beads and their DEP force are directly proportional when the ODEP force is balanced with the viscous drag force in fluid. A flash program is designed to drag a polystyrene bead. Figure 6 shows the relationship between the dragging velocity of polystyrene beads and the thickness of the polymer films. Please note that the thickness of the polymer film is measured by an alpha step profilometer (DETAK 6M, Veeco, UK). The experiments are performed under 24 V_{pp} and 100 kHz. It is observed that the thicker the film is, the higher the ODEP force is. The maximum velocity for 5 wt% films to manipulate a polystyrene bead (20 µm) is measured to be 186 µm/second. The ODEP force is then calculated to be 35.2 pN by using the Stokes' law.

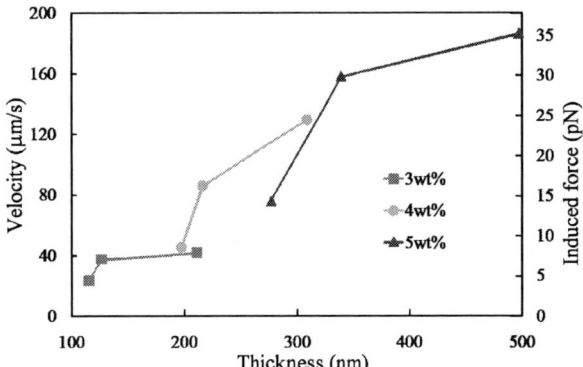

Figure 6: The relationship between the dragging velocity of polystyrene beads and the thickness of the polymer films. The maximum velocity to manipulate a polystyrene bead (20 µm) is 186 µm/s.

CONCLUSIONS

In this study, we report a new polymer material on ODEP applications. Instead of using amorphous silicon, the P3HT/PCBM blend film was used to manipulate particles. Experimental data shows that the thickness of the polymer thin film plays an important role for generating ODEP forces. The film with a concentration of 5wt% was found to have the maximum thickness in

978-1-4244-2977-6/09 $25.00 © 2009 IEEE   417

slow-thermo evaporation procedure, thus generating the maximum ODEP force.

ACKNOWLEGEMENTS

The authors would like to thank financial support from the National Science Council in Taiwan (NSC 96-2120-M-006-008).

REFERENCES

[1] M. P. Hughes, "Strategies for dielectrophoretic separation in laboratory-on-a-chip systems", *Electrophoresis Review*, vol. 23, pp. 2569-2582, 2002.

[2] P. Y. Chiou, A. T. Ohta, and M. C. Wu, "Massively parallel manipulation of single cells and microparticles using optical images", *Nature*, vol. 436, pp. 370-372, 2005.

[3] A. T. Ohta, P. Y. Chiou, T. H. Han, J. C. Liao, U. Bhardwaj, E. R.B. McCabe, Y. Fuqu, S. Ren, M. C. Wu, "Dynamic Cell and Microparticle Control via Optoelectronic Tweezers", *J. Microelectromech. Syst.*, vol. 16, pp. 491-499, 2007.

[4] N. S. Sariciftci, A. J. Heeger, *Handbook of Organic Conductive Molecules and Polymers*, JOHN WILEY & SONS , New York ,1997

[5] H. Sirringhaus, N. Tessler, R. H. Friend, "Integrated Optoelectronic Devices Based on Conjugated Polymers", *Science*, vol. 280, pp. 1741-1744, 1998.

[6] F. Padinger, R. S. Rittberger, N. S. Sariciftci, "Effects of Postproduction Treatment on Plastic Solar Cells", *Adv. Funct. Mater.* vol. 13, pp. 85-88, 2003.

[7] Y. Kim, S. A. Choulis, J. Nelson, D. D. C. Bradley, S. Cook, J. R. Durrant, "Device annealing effect in organic solar cells with blends of regioregular poly 3-hexylthiophene and soluble fullerene", *Appl. Phys. Lett.* vol. 86, pp. 063502, 2005.

[8] V. Dyakonov, "Mechanisms controlling the efficiency of polymer solar cells", *Appl. Phys. A: Mater. Sci. Process.*, vol. 79, pp. 21-25, 2004.

[9] G. Li, V. Shrotriya, J. S. Huang, Y. Yao, T. Moriarty, K. Emery and Y. Yang, "High-efficiency solution processable polymer photovoltaic cells by self-organization of polymer blends", *Nat. Mater.*, vol. 4, pp.864-868, 2005.

FABRICATION OF LIVING CELL STRUCTURE UTILIZING ELECTRO-STATIC INKJET PHENOMENA

S. Umezu[1], T. Kitajima, H. Murase, H. Ohmori, K. Katahira and Y. Ito
[1] RIKEN, Institute of Chemistry and Physics

ABSTRACT

We fabricated living cell lines, cell wall and cell cylinder utilizing electrostatic inkjet phenomena. The inkjet phenomena have two merits, higher resolution than commercial printer and ability to eject highly viscous liquid. These merits were preferable to print liquid with cells precisely and to eject liquid with scaffolds that were relatively high viscosity. In spite that high voltage was applied in case of the electrostatic inkjet phenomena, cells were living because current did not flow through cells but around cells.

INTRODUCTION

The object of this study is to fabricate 3-Dimensional cell structures utilizing electrostatic inkjet phenomena. It is preferable to perform laboratory experiments with 3D cell structures in tissue engineering and artificial organ. However it is difficult to fabricate 3D cell structures because own weight of cell is above the bonding force between cells.

A fabrication method was suggested [1]. When 3D scaffolds were put into liquid with cells, cells were attached to the surface of the scaffolds. However it was difficult to control density of the attached cells and there was much waste of cells. Another fabrication method [2] was carried out to clear these problems. Commercial piezo inkjet technology was applied for 3D positioning of alginate capsule which contained cells because liquid with scaffolds was difficult to eject due to high viscosity. Because alginate capsules were easy to stick each other, 3D positioning of cells was succeeded. However cells could not contact each other by the wall of the alginate capsule. To clear this problem, inkjet technology should be more powerful to eject highly viscous liquid.

We have investigated mechanism and fundamental characteristics of the electrostatic inkjet phenomena and now been applying for new printing technology of high image quality and 3D printing technology of glass paste [3]. The phenomena have two merits, higher resolution than commercial printer and ability to eject highly viscous liquid [4]. We were able to eject glass paste that viscosity was 30000 mPas. In this paper, we applied the inkjet phenomena for printing living cells. We investigated fundamental characteristics of printing cells utilizing the inkjet phenomena. We fabricated living cell structures.

ELECTROSTATIC INKJET PHENOMENA
Experimental Set-up

An experimental set-up illustrated in Fig. 1 was constructed to investigate characteristics of the formation of droplets in the electric field. A tube filled with ink was mounted perpendicular to a plate electrode made of stainless steel. DC voltage was applied by a function gen-

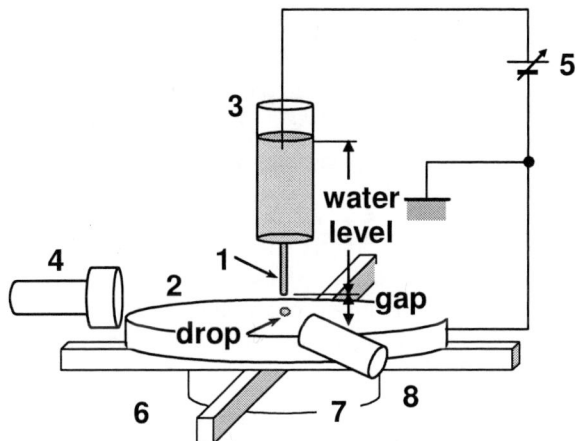

Figure 1: Experimental set-up of single nozzle inkjet system. (1: water pin electrode, insulative capillary tube filled with ink, 2: metal plate electrode, 3: ink tank, 4: high speed camera, 5: high voltage amplifier and function generator, 6: linear stages, x and y directions, 7: mechanical z-stage, 8: light)

erator (Iwatsu, Tokyo, SG-4105) and a high voltage amplifier (Matsusada Precision Inc, HEOP-10B2). The formation of the droplet was observed with a high-speed microscope camera (Photron Inc., Japan, FAST-CAM-MAX 120K model 1) with a light (Sanei Electric Inc., Japan, XEF-501S).

Mode of Drop Formation and Print Demonstration

Figure 2 shows the current-voltage characteristics of the water pin electrode [5]. This figure indicated that the formation of the droplet was classified into the following three modes corresponding to the discharge modes. In MODE 1, the diameter of the drop was several times larger than that of the tube diameter and the drop period was long, more than a second. In MODE 2, a Taylor cone [6] was formed at the end of the tube and the tip of the cone periodically separated from the cone to form a very small droplet of the order of several tens of microns in diameter. This MODE 2 is suitable for inkjet printing. In MODE 3, the Taylor cone changed to hemispherical and the droplet became relatively large, nearly the same as the tube diameter.

Figure 3 shows the print sample utilizing MODE 2 region. Maximum resolution of these samples was approximately 1550 dpi. Resolution of print samples utilizing the phenomena was higher than that of print samples utilizing commercial printer.

The electrostatic inkjet phenomena had a potential to realize three-dimensional printing with emulsified liquid that consisted largely of nano-particles, because highly viscous liquid, more than 30,000 mPa.s, could be ejected

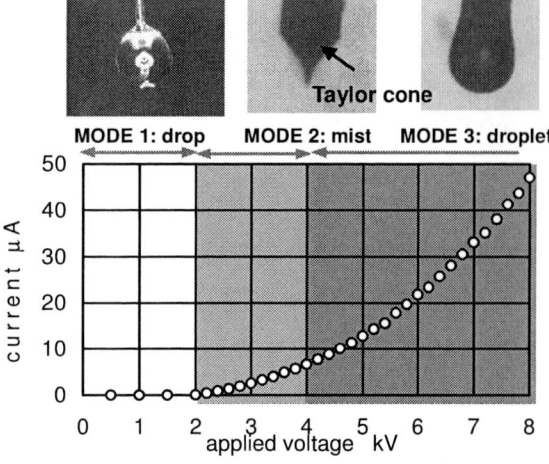

Figure 2: V-I curves in pin-to-plate electrode system. (ϕ 100 μm inner tube diameter, ϕ 100 μm metal pin diameter, 3 mm air gap)

機
Original image : 64×64 pixel

resolution	635 dpi	847 dpi	1270 dpi	1588 dpi
Dyes ink	機	機	機	
Pigment ink	機	機	機	機
	1.0 mm	1.0 mm	1.0 mm	1.0 mm

Figure 3: Print samples of Chinese character "mecha". (material: dyes ink and pigment ink)

by the phenomena. Compositions of liquid we have synthesizes for the 3D printing are 20 % alumina nano-particles (440 nm), 3 % binder (PVA), 0.2 % dispersing agent, and 77 % water. The viscosity was 12 mPa.s and the contact angle was 70 deg. Figure 4 shows demonstrated 3D patterns.

CELL PATTERNING

Experimental Set-up

An experimental set-up shown in figure 5 was constructed to print cell structures utilizing the electrostatic inkjet phenomena. A capillary tube made of silica coated by polyimide (inner diameter of the tube: 100 micron meters, outer diameter of the tube: 170 micron meters, Polymicro Technologies, Phoenix, AZ) was equipped with a bottom of a syringe. The tube filled with the liquid which contained cells was hanged down perpendicular to a dish filled with medium. Voltage was applied between the syringes and the dish by a power supply (voltage range: -5kV ~ +5kV, Matsusada Precision Inc, Tokyo, HVR-10P). The air gap was adjusted by a z-stage and the plate electrode was moved in x and y directions with two linear motors. Voltage application and motion of the linear stages were controlled by a PC.

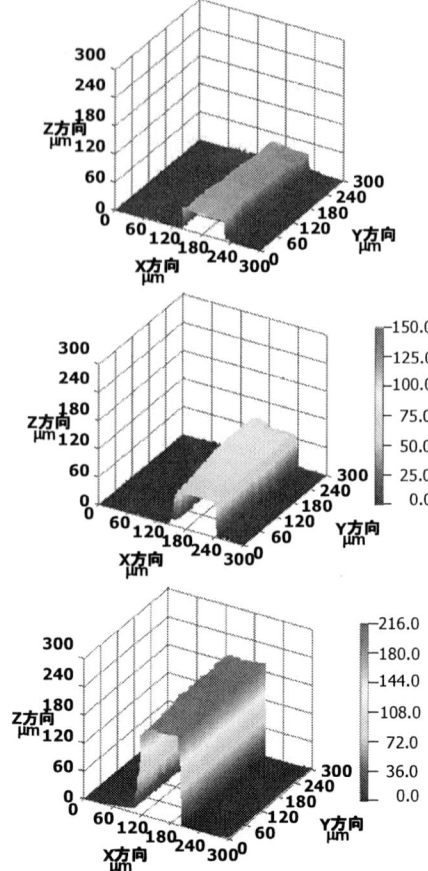

Figure 4: Print samples of 3D line. (material: glass paste)

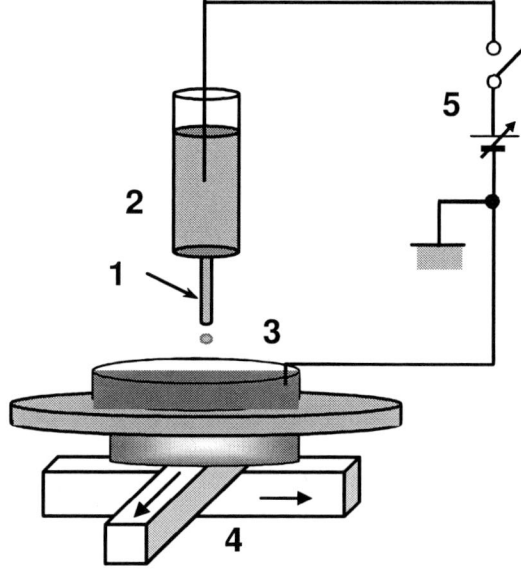

Figure 5: Experimental set-up of cell patterning utilizing electrostatic inkjet. (1: water pin electrode, insulative capillary tube was mounted at tip of syringe, 2: tank, filled with liquid of cells, 3: dish, filled with medium, 4: xyz linear stages, 5: DC high voltage power supply)

Results

Figure 6 and 7 show pictures of printed cells when one day has passed after cells were ejected utilizing the electrostatic inkjet phenomena. Figure 8 shows the enlarged view of the printed cells. Cells were living in spite of high voltage application because current was not flowed through cells but around cells. These pictures indicated that printing cell utilizing the electrostatic inkjet phenomena was suitable for practical use.

Figure 9 shows the printed cell line utilizing the electrostatic inkjet phenomena. The line width was less than 100 micron meters. When liquid with cells and liquid with scaffolds were ejected alternately, 3D cell structures, cell wall and cell cylinder were fabricated shown in figure 10 and 11.

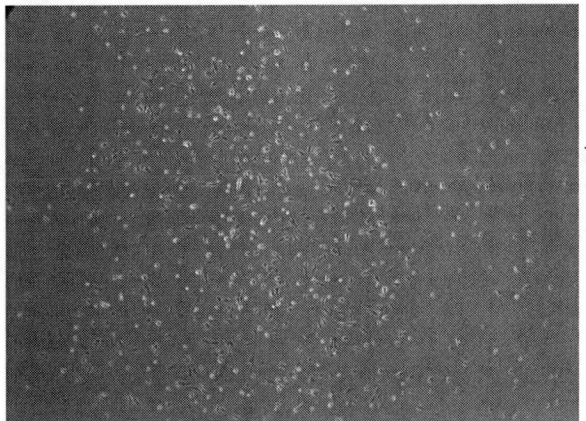

Figure 6: Picture of printed cells.
(applied voltage: 2.2 kV)

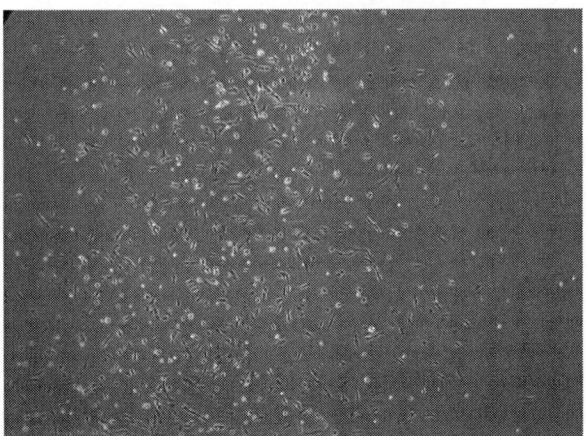

Figure 7: Picture of printed cells.
(applied voltage: 3.2 kV)

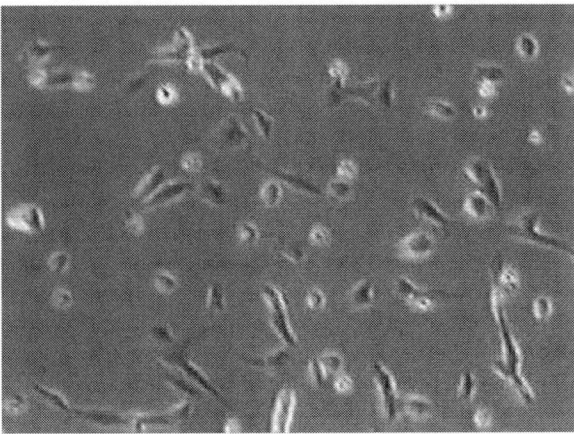

Figure 8: Enlarged view of printed cells.

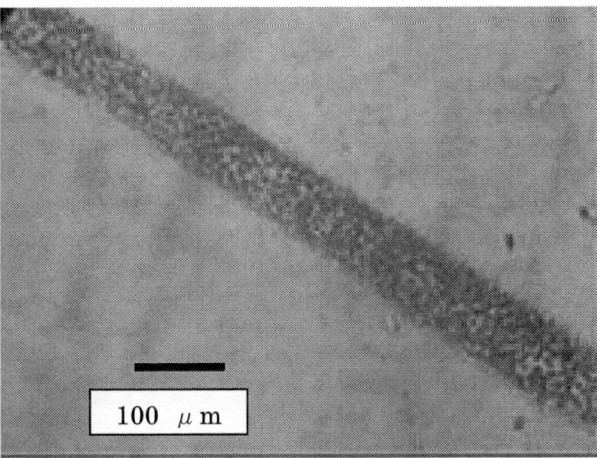

Figure 9: Picture of printed cell-line.

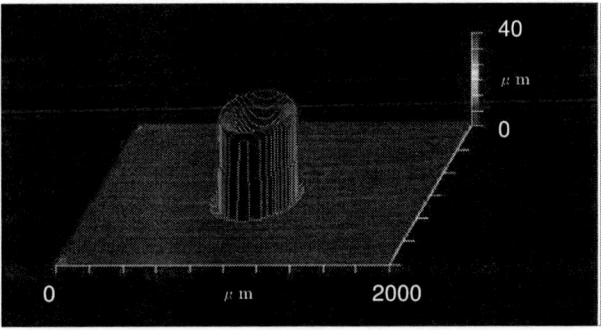

Figure 10: Picture of cell cylinder.

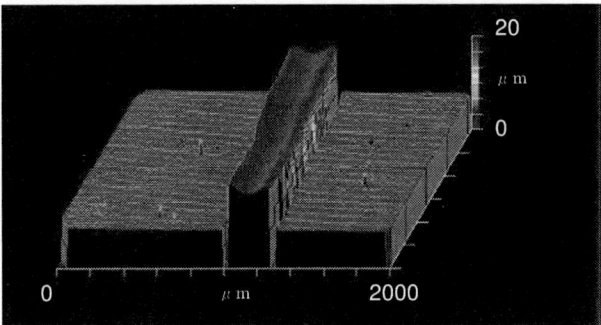

Figure 11: Picture of cell cylinder.

CONCLUSION

We applied the electrostatic inkjet phenomena for printing cells. It is possible to eject liquid with living cells utilizing the electrostatic inkjet phenomena in spite of high voltage application because current did not flowed through cells but flowed around cells. It is possible to eject liquid with scaffolds because the inkjet phenomena have potential to eject highly viscous liquid. We demonstrated living cell three dimensional cell structures, cell cylinder and cell wall utilizing liquid with cells and liquid with scaffolds.

ACKNOWLEDGEMENTS

The authors would like to express their thanks to Prof. Kawamoto for his help of useful advice. This work is supported by Grant-in-Aid for Young Scientists (B) of Japan Society for Promotion of Science and Special Grants of RIKEN.

REFERENCES

[1] X. Yang, R. S. Tare, K. A. Partridge, H. I. Roach, N. M. Clarke, S. M. Howdle, K. M. Shakesheff and R. O. Oreffo, Induction of Human Osteoprogenitor Chemotaxis, Proliferation, Differentiation and Bone Formation by Osteoblrast Stimulating Factor-1/Pleiotrophin: Osteoconductive Biomimetic Scaffolds for Tissue Engineering, *J. Bone and Mineral Research*, 18, 1, pp. 47-57 (2003).

[2] C. Henmi, M. Nakamura, Y. Nishiyama, K. Yamaguchi, S. Mochizuki, K. Takiura and H. Nakagawa, Development of an effective three dimensional fabrication technique using inkjet technology for tissue model samples, *AATEX*, 14, Special Issue, pp. 689-692 (2007).

[3] S. Umezu, K. Katahira and H. Ohmori, New Micro Fabrication Techniques Utilizing Electrostatic Inkjet Phenomena, *Proc. 10th anniversary int'l conference EUSPEN 2008*, pp. 443-447 (2008).

[4] A. Bailey, *Electrostatic Spraying of Liquids*, Research Studies Press, Tounton, Somerset, England (1988).

[5] H. Kawamoto, S. Umezu and R. Koizumi, Fundamental Investigation on Electrostatic Inkjet Phenomena in Pin-to-Plate Discharge System, *J. Imaging Sci. Technol*, 49, 1, pp.19-27 (2005).

[6] G. Taylor, Disintegration of water drops in an electric field, *Proc. Roy. Soc. London*, A 280, pp. 383–397 (1964).

SIZE-CONTROLLED ISLET-CELL SPHEROIDS
FOR GEOMETRIC ANALYSIS OF INSULIN SECRETION

Y. Tsuda[1,2], M. Kato-Negishi[1], T. Okitsu[2,3] and S. Takeuchi[1,2,4]

[1]Institute of Industrial Science (IIS), The University of Tokyo, Tokyo, JAPAN

[2]BEANS Project, METI, JAPAN

[3]Transplant Unit, Kyoto University Hospital, Kyoto, JAPAN

[4]PRESTO, Japan Science and Technology, JAPAN

ABSTRACT

We developed a method to prepare size- and shape-controlled islet-cell spheroids. We cultured β-cell line (MIN6-m9) in PDMS-microwells and enabled to construct their spheroids with the various sizes corresponding to those of the microwells. We demonstrated the geometric analysis of insulin secretion of the cells in the spheroids. We found that these spheroids respond to glucose and subsequently secrete insulin. Also, immunocytochemistry and the real-time live cell imaging of the Ca2+ oscillation revealed for the first time that the only cells approximately 36 μm apart from the periphery showed higher insulin secretion in the spheroid. This result indicates that our MEMS technique controlling the size and the shape of the spheroids is useful to construct islet-like tissue that functions efficiently and is applicable to transplantation therapy targeted to insulin dependent diabetes mellitus.

INTRODUCTION

Recently, clinical islet transplantation has shown promise as a means of curing insulin dependent diabetes mellitus. However, the performance of this approach is dependent on the availability of donor tissue. Therefore, regeneration of islets from cultured cells is desired. Regenerating islets requires the cells to form aggregation structures, called "spheroids", because cells in a spheroid secrete more insulin compared to dispersed cells [1]. The size of the spheroid is considered as an important factor for regenerated islets to function [2]. Many of 3D culture system to make a spheroid [3, 4] have been reported previously, but there has been no study dealing with precisely controlling the size of the spheroid, nor evaluating the function in situ of the cells forming the spheroid (Fig.1).

In this study, we developed the methodology to construct size- and shape-controlled beta-cell spheroids using PDMS-microwells, and investigated geometric insulin production of the cells in the 3D spheroid structures at various sizes with immunocytochemistry and cytoplasmic free calcium ion imaging.

MATERIALS AND METHODS

Materials

Commercial materials used in cell culture experiments were obtained as follows: trypsin-EDTA solution, and antibiotics (streptomycin and penicillin), Dulbecco's-modified Eagle's medium (DMEM), bovine serum albumin fraction V (BSA) and phosphate buffered saline (PBS) were from SIGMA-Aldrich (St. Louis, MO, USA), fetal bovine serum (FBS) from Japan Bioserum Co Ltd., polyclonal Guinea pig anti-swine insulin was from Dako, Alexa Fluor 488 goat anti-guinea pig IgG, Alexa Fluor 568-conjugated phalloidin, Hoechst 33342, LIVE/DEAD Viability/Cytotoxicity Kit and Fluo-4/AM were from Invitrogen.

PDMS-microwell fabrication

Polydimethylsiloxane (PDMS)-cell culture microwells are generated using conventional photolithography and soft lithography techniques. Briefly, SU-8 negative photoresist (MicroChem Corp., Newton, MA, USA) were span onto a silicon wafer and soft baked. The resist was exposed to UV light through a patterned photomask using a mask aligner and developed using SU8-developer solution (MicroChem Corp.). This process creates an SU8 master that could be used to mold PDMS. PDMS-microwells were formed by pouring PDMS (1:1 prepolymer: curing agent, (Silpot 184, Dow Corning, Tokyo, Japan)) onto the master and curing for 2h at 75°C. Removal of the PDMS from the wafers yielded PDMS-microwells for cell culture. Topological feature of both SU8 masters and PDMS-microwells were observed using a 3D laser scanning microscope (VK9700 Generation II; Keyence Corp., Osaka, Japan).

(a) Importance of 3D culture and geometrical analysis

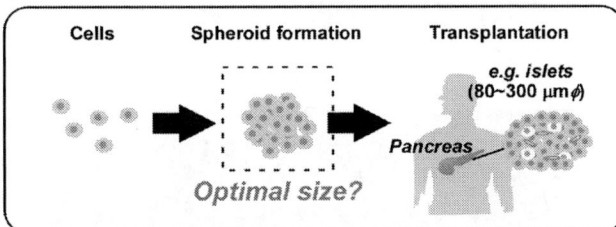

(b) Spheroid formation

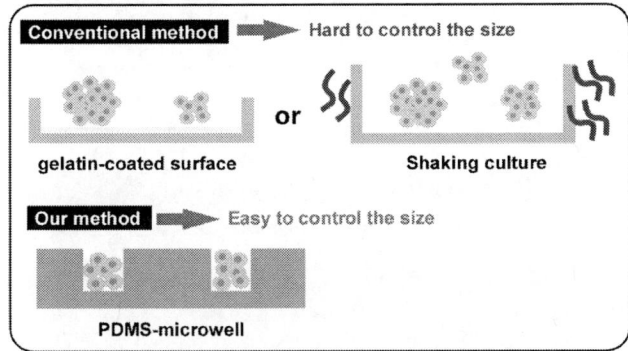

Figure 1. Concept of this study. (a) Importance of 3D culture system. (b) Various methods for the spheroid formation.

Cells and cell culture

Glucose-responsive pancreatic β-cell line, MIN6-m9 [3], were kindly provided from Dr. Seino (Kobe University, Kobe, Japan). Cells were maintained in DMEM supplemented with 10% heat-inactivated FBS, 100 U/ml penicillin and 100 mg/ml streptomycin at 37°C under a humidified atmosphere of 5% CO_2 and 95% air. Cells were harvested with 0.25% trypsin and 0.02% EDTA in PBS and subcultured at a 1:5 split ratio after reaching approximately 70~80% confluence. Cultured cells were observed with a phase-contrast microscope.

Spheroid formation using microwells

MIN6-m9 cells were seeded onto the PDMS-microwells at a density of 2~3 x 10^5 cells/cm^2. Cells which non-dropped in the well were removed after 30 min by changing the media (Fig.2). The cells were maintained in DMEM supplemented with 10% FBS at 37°C under a humidified atmosphere of 5% CO_2 and 95% air.

Cell viability assay

Cell viability was assessed using the LIVE/DEAD Viability/Cytotoxicity Kit based on a simultaneous determination of living and dead cells with two probes, calcein-AM for intracellular esterase activity and ethidium

homodimer-1 (EthD-1) for plasma membrane simultaneous determination of living and dead cells with two probes, calcein-AM for intracellular esterase activity and EthD-1 for plasma membrane integrity. In brief, MIN6-m9 spheroids were incubated in a working solution (2 mM calcein-AM and 5 mM EthD-1) for 10 min at 37°C. The stained cells were observed under fluorescent microscopy and confocal laser scanning microscopy.

Immnocytochemistry

Formed MIN6-m9 spheroids on PDMS-microwells were washed three times with PBS and fixed with 4% paraformaldehyde solution. After permeabilization with 0.5% Triton X-100 in PBS, cells were blocked with 5% BSA in PBS and reacted with a 1:200 dilution of polyclonal Guinea pig anti-swine insulin [5] at 4°C overnight, followed by incubation with a 1:500 dilution of Alexa Fluor 488 goat anti-guinea pig IgG at 25°C and finally rinsed with in PBS. For F-actin staining [6], cells were double-stained with a 1:200 dilution of Alexa568-conjugated phalloidin (200 U/ml) at 25°C. For cell nuclear staining, cells were triple-stained with a 1:1000 dilution of DNA-binding dye, Hoechst 33342 (1 mg/ml) at 25°C for 5 min. These stained cells were observed under a fluorescence microscope and confocal fluorescence laser-scanning microscope.

Cytoplasmic free calcium ion imaging

Intracellular calcium ion of 2D-cultured MIN6-m9 cells and 3D cultured MIN6-m9 spheroids were measured using calcium imaging system [7, 8]. Cells were incubated with glucose-free HEPES-balanced Krebs' ringer bicarbonate (HKRB) buffer (129 mM NaCl, 5 mM $NaHCO_3$, 4.7 mM KCl, 1.2 mM KH_2PO_4, 1.2 mM $MgSO_4$, 2 mM $CaCl_2$, 10mM HEPES, 0.1% bovine serum albumin, pH 7.4) and 3 μM Fluo-4/AM for 30 min at 37°C. Cells were washed twice with HKRB buffer, and mediated with

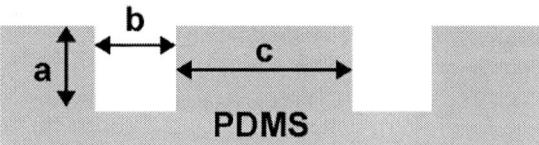

1. Cell seeding

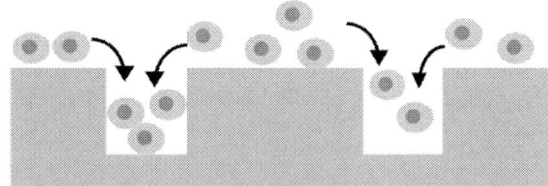

2. Cells drop into the microwells

3. Wash non-dropped cells

Figure 2. Procedure of MIN6-m9 spheroids formation in PDMS-microwells.

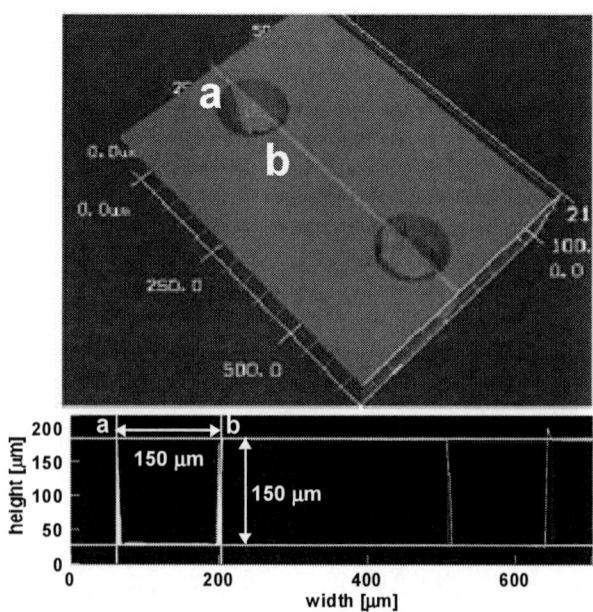

Figure 3. Topological feature of PDMS-microwell observed by 3D laserscanning microscope.

HKRB buffer containing 20 mM glucose. The concentration of intracellular Ca^{2+} concentration ($[Ca^{2+}]_i$) was observed with a laser scanning confocal microscope system using an excitation wavelength of 488 nm and an emission wavelength of 526 nm. We observed $[Ca^{2+}]_i$ oscillations on spheroids of cultured MIN6-m9 cells.

RESULTS AND DISCUSSION

MIN6-m9 spheroid formation and release

Glucose-responsive pancreatic β-cell line, MIN6-m9, were seeded on the PDMS microwells. The seeded cells settled down into the PDMS-microwells within 30 min, and the cells located on the top of PDMS surface were removed by changing media, then exposed to culture at 37°C (Fig.4a). PDMS surface prevented adhesion of the seeded cells, and the cells in the PDMS-microwells formed spheroid-like structures. As MIN6-m9 cells show a rapid cell-to-cell contact within 60 min on the non-adherent surfaces. Various shapes of spheroids can be prepared when suspended cells are collected in a distinct space (Fig.4b and c). Using this system, we demonstrated spherical and ring-shaped spheroids (Fig.4b). These spheroids can be easily released from the substrates by just pipetting or aspiration, and collected spheroids would be possible to use for the further applications of tissue assembly.

3D insulin distribution

Five-day cultured spheroids in microwells in both 150 μmφ and 50 μmφ are determined to be alive by Live/Dead Assay (Fig.5). MIN6-m9's specific function, insulin secretion which is a specific function of MIN6-m9 was examined by immunostaining of intracellular insulin and measurement of cytoplasmic free calcium ion ($[Ca^{2+}]i$) imaging using fluo-4/AM. Insulin-secreting cells at an inner part of the larger spheroids (approximately 120-μmφ

size) were not observed (Fig.6). On the other hand, cells located 36 ± 14.7 μm (n=8) apart from the periphery showed the insulin secretion, and most of the cells in smaller spheroids (approximately 30-μmφ size) are positive for the insulin secretion. Cytoplasmic free calcium ion imaging of spheroids revealed that MIN6-m9 cells in spheroid shows calcium-oscillation, indicating cells secrete insulin responding to glucose (Fig.7).

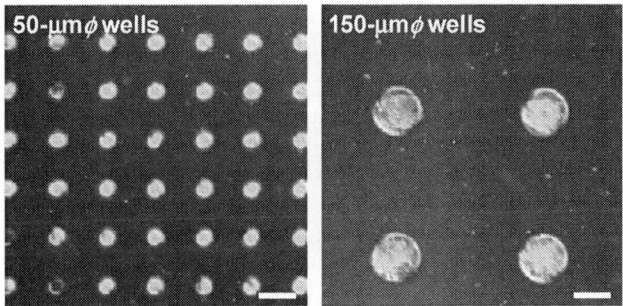

Figure 5. *Live/Dead Assay for MIN6-m9 spheroids formed in microwells. Dead cells stained red and living cells stained green. Most of the cultured spheroids were alive. Scale bars: 100 μm.*

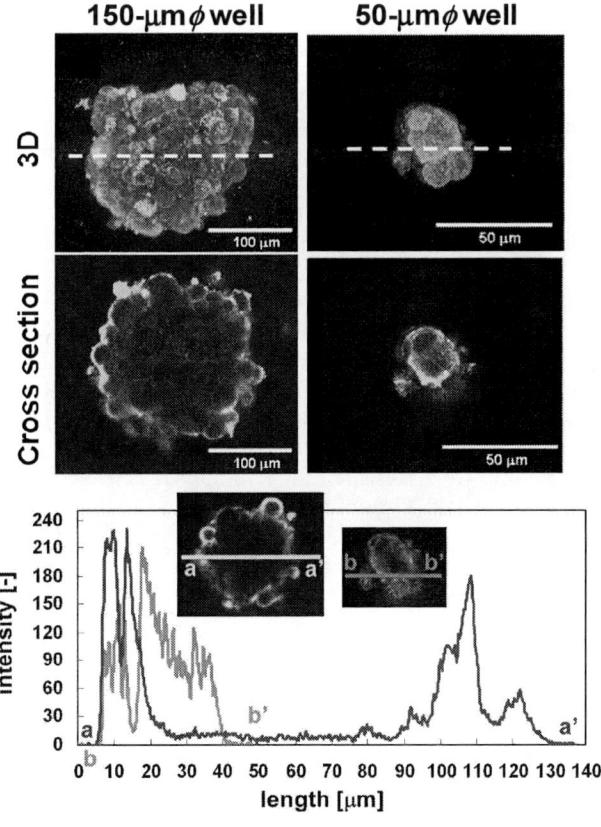

Figure 6. *Immunocytochemistry of MIN6-m9 spheroids at different size. Intracellular insulin stained green, F-actin stained red and nuclei stained blue. Cells located 36 ± 14.7 μm (n=8) apart from the periphery showed the insulin secretion, and most of the cells in smaller spheroids (approximately 30-μmφ size) are positive for the insulin secretion.*

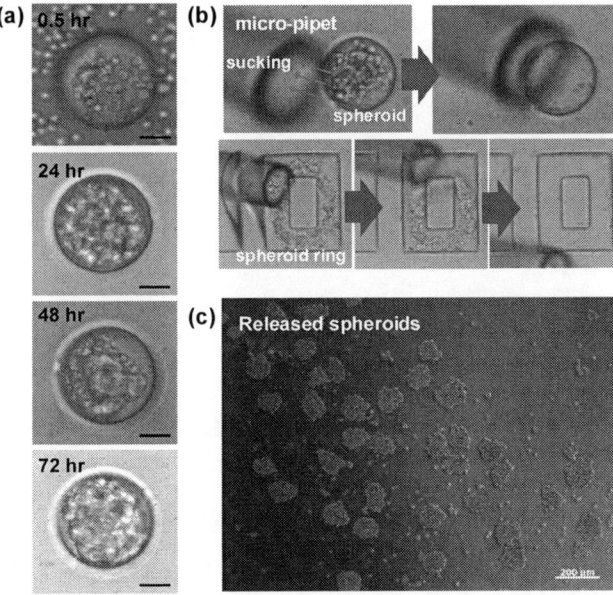

Figure 4. *. Formation of MIN6-m9 spheroid in PDMS-microwells (a). Releasing spheroids with spherical/ringt shape by sucking using micro-pipets (b). Spheroids released from 150-μmφ microwells (c).*

978-1-4244-2977-6/09 $25.00 © 2009 IEEE   425

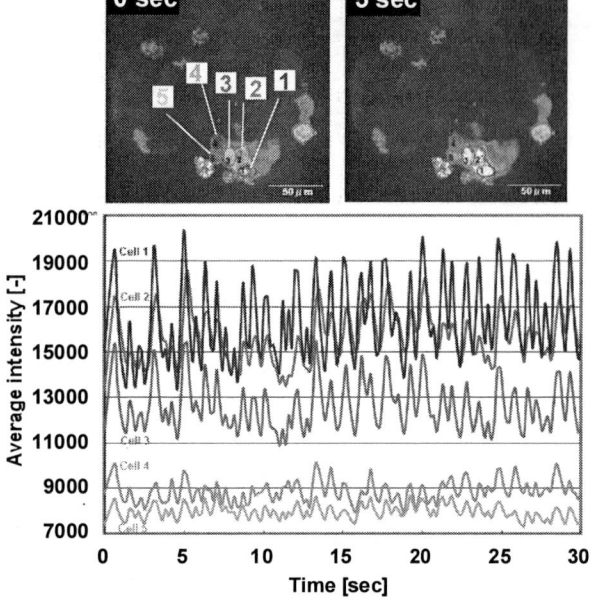

Figure 7. *Real-time Ca^{2+} imaging of cultured MIN6-m9 spheroid.*

CONCLUSIONS

We present the easy and convenient method to prepare the size- and shape-controlled MIN6-m9 spheroids which secrete insulin. Also, this represents the first effort of the geometric analysis for insulin secretion on MIN6-m9 spheroids. This technique holds promise for the further study of fundamental cell-to-cell communication and mechanism in 3D cultured cells, and reconstruction of viable tissues for the regenerative medicine in diabetes.

ACKNOWLEDGEMENTS

We acknowledge to Ms. H. Kato and Mr. M. Yamamoto (Leica Microsystems Corp., Tokyo, Japan) for their technical assistance in confocal microscopy. We also acknowledge to Prof. Seino (Kobe University, Kobe, Japan) for providing MIN6-m9 cells. This work was partly supported by Core Research for Evolutional Science and Technology (CREST) from Japan Science and Technology (JST) Agency, and by Grants-in-Aid for Scientific Research from Ministry of Education, Culture, Sports, Science and Technology Japan.

REFERENCES

[1] M.J. Luther, E. Davies, D. Muller, M. Harrison, A.J. Bone, S.J. Persaud, P.M. Jones, "Cell-to-cell contact influences proliferative marker expression and apoptosis in MIN6 cells grown in islet-like structures", Am. J. Physiol. Endocrinol. Metab., Vol. 288, pp. E502- E509, 2005.

[2] R. Lehmann, R.A. Zuellig, P. Kugelmeier, P.B. Baenninger, W. Moritz, A. Perren, P.A. Clavien, M. Weber, G.A. Spinas, "Superiority of small islets in human islet transplantation.", Diabetes. Vol. 56, pp. 594-603, 2007.

[3] J. Miyazaki, K. Araki, E. Yamato, H. Ikegami, T. Asano, Y. Shibasaki, Y. Oka, K. Yamamura, "Establishment of a pancreatic beta cell line that retains glucose-inducible insulin secretion: special reference to expression of glucose transporter isoforms", Endocrinoloygy, Vol. 127, pp.126-132, 1990.

[4] K. Minami, H. Yano, T. Miki, K. Nagashima, C.Z. Wang, H. Tanaka, J.I. Miyazaki, S. Seino, "Insulin secretion and differential gene expression in glucose-responsive and -unresponsive MIN6 sublines.", Am. J. Physiol. Endocrinol. Metab., Vol. 279, pp. E773-E781, 2000.

[5] T. Ibii, H. Shimada, S. Miura, E. Fukuma, H. Sato, H. Iwata, "Possibility of insulin-producing cells derived from mouse embryonic stem cells for diabetes treatment.", J. Biosci Bioeng., Vol. 103, pp. 140-146, 2007.

[6] Y. Tsuda, A. Kikuchi, M. Yamato, G. Chen and T. Okano, "Heterotypic cell interactions on a dually patterned surface", Biochem. Biophys. Res. Comm., vol. 348, pp. 937-944, 2006.

[7] K. Iizuka, H. Nakajima, A. Ono, K. Okita, J. Miyazaki, J. Miyagawa, M. Namba, T. Hanafusa, Y. Matsuzawa, "Stable overexpression of the glucose-6-phosphatase catalytic subunit attenuates glucose sensitivity of insulin secretion from a mouse pancreatic beta-cell line.", J. Endocrinol., vol. 164, pp. 307-314, 2000.

[8] A. Calabrese, M. Zhang, V. Serre-Beinier, D. Caton, C. Mas, "Connexin 36 controls synchronization of Ca2+ oscillations and insulin secretion in MIN6 cells", Diabetes, Vol. 52, pp. 417-424, 2003.

A 3-D MICROFLUIDIC COMBINATORIAL CELL CULTURE ARRAY

M. C. Liu and Y. C. Tai
California Institute of Technology, Pasadena, CA, USA

ABSTRACT

We present the development of a three-dimensional (3-D), on-glass combinatorial cell culture array chip featured with integrated three-input, eight-output combinatorial mixers and cell culture chambers. The device is designed to simultaneously screen for the effects of multiple compounds and concentrations on cultured cells. Experimentally, we first developed a precise way to characterize the combined compound concentration profile at each chamber with a fluorescence method. We then successfully demonstrated the functionality of the cell assay by culturing neuron cells on this device and screening the ability of three chemicals to attenuate cell death caused by cytotoxic hydrogen peroxide. Based on the same technology, the number of inputs and outputs of the combinatorial mixer can be scaled-up to construct lab-on-chip devices for performing high-throughput cell-based assay and highly parallel and combinatorial chemical or biochemical reactions with reduced labors, reagents, and time.

INTRODUCTION

Microfluidic devices have the potential to become inexpensive platforms for high-throughput cell-based screenings, and recently, many devices with cell culturing and assaying abilities have been developed [1,2]. However, most of the cell culturing devices can only screen for a single compound at once [3,4]. A microfluidic device that can screen for the combinatorial effect of multiple compounds on cells is valuable because cells are sustained in complex environments and cell fates are dictated by the integration of numerous extracellular signals. For example, combinatorial effects are evident in cellular processes such as gene regulation [5]. Also, certain combinations of anti-tumor compounds can interact in a synergistic manner on cancer cells [6]. In our previous works, we have demonstrated the method to monolithically fabricate three-dimensional (3-D) microfluidic networks on silicon substrate, and based on this fabrication technology, we have constructed a cell culture device with an integrated combinatorial mixer [7]. In this work, we extended such fabrication method to construct 3-D microfluidic networks on glass substrate and demonstrated a cytotoxic assay using this chip. Although the technology can be extended for more compounds, this work demonstrated a device for three compounds as a feasibility study.

DESIGN

Figure 1 shows the layout of the 1.7 cm by 1.7 cm chip. Compounds A, B, and C are flown into eight chambers with different combinations and concentrations depending on the flow resistance of the channels. The overpass structure allows one microfluidic stream to crossover another, and a control channel that receives none of the three inputs is included. The concentration of each compound inside the chamber is determined by the fluidic resistance of the channels. Assuming that the channel width is much larger than the channel height, the resistance can be modeled using the simplified formula for rectangular channel [8]:

$$R = \frac{12 \mu L}{w h^3}$$

R is the resistance, μ is the fluid viscosity, and L, w, and h are the length, width and height of the channel, respectively. Chambers receiving multiple compounds are designed to have equal dilution of the input compounds.

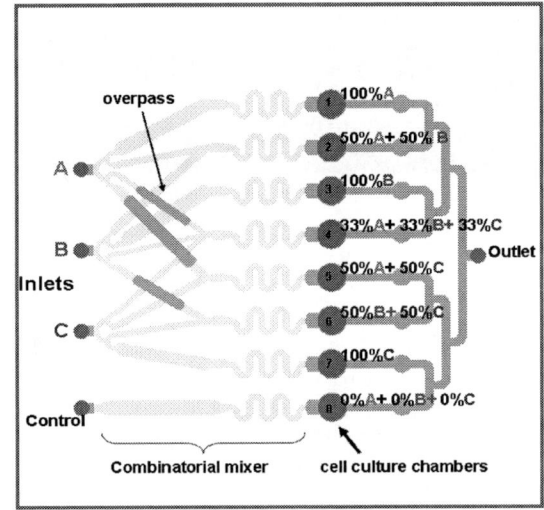

Figure 1. Design Layout of Device

FABRICATION

The fabrication process for constructing 3-D microfluidic networks is shown in Figure 2. Briefly, a first thin adhesion layer of parylene C was deposited (3 μm) on soda lime glass wafer. The first sacrificial photoresist layer was spin-coated (14 μm) and patterned to define the first-level channels. A second layer of parylene C (10 μm) was then deposited to cover the sacrificial photoresist, forming the first-level channels. Parylene C was patterned using oxygen plasma so the areas where the overpass structures would be joined were exposed. This parylene patterning also opened the area where the mixer and the culture chamber would be connected. A second sacrificial photoresist was spin-coated (25 μm) and patterned to define the overpass structure and the culture chamber. This second sacrificial photoresist covered the areas that were etched open, and the overpasses spanned several of the first-level microfluidic channels. A third layer of parylene C (2 μm) was deposited. The whole chip was planarized with thick SU8 (90μm) and parylene C was patterned to define the access holes. Finally, the chips were soaked in acetone to dissolve the sacrificial photoresist.

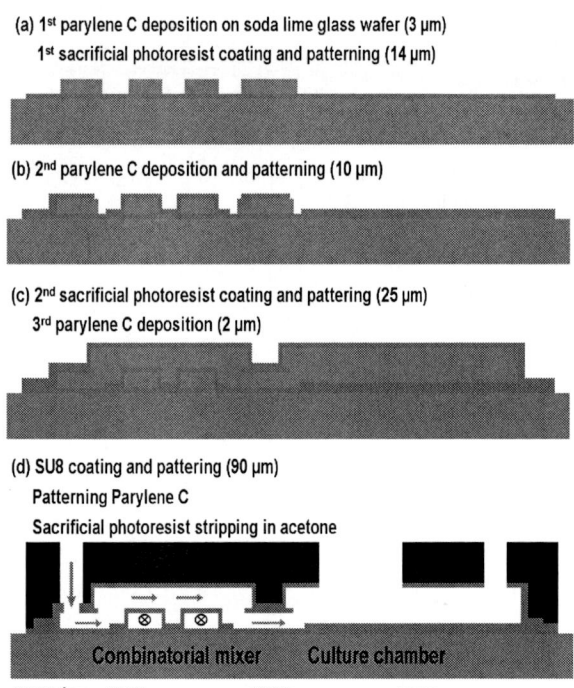

(a) 1st parylene C deposition on soda lime glass wafer (3 μm)
1st sacrificial photoresist coating and patterning (14 μm)

(b) 2nd parylene C deposition and patterning (10 μm)

(c) 2nd sacrificial photoresist coating and pattering (25 μm)
3rd parylene C deposition (2 μm)

(d) SU8 coating and pattering (90 μm)
Patterning Parylene C
Sacrificial photoresist stripping in acetone

Combinatorial mixer   Culture chamber

glass   Parylene C   Photoresist   SU8

Figure 2. Fabrication process flow. The overpass allows one fluid stream (→) to crossover another (⊗), as shown in (d).

EXPERIMENTAL

Figure 3 shows the packaging method. A transparent polyester membrane with 0.4 μm pores was cut out from Corning® Transwell® cell culture insert, and a rectangular piece of the membrane was placed on top of the chambers. A piece of PDMS (Polydimethylsiloxane) with 100 μm height microfluidic networks was fabricated by replicate molding from a SU8 mold. The PDMS was aligned onto the chip and the assembly was clamped together by acrylics. Figure 4 shows the layout of the PDMS piece, which has punched holes that match the locations of the inlet and outlet holes on the parylene microfluidic chip. The PDMS piece also has a cell-seeding inlet and a one-to-eight manifold that distributes cells into eight chambers, which align to the location of the culture chambers of the parylene microfluidic chip. With the membrane, the culture chamber turns into a two-level configuration, and the cell culture level is separated from the fluid delivery level. Syringes were connected with Teflon tubes, which were plugged into the holes of the piece of PDMS.

To characterize the concentration of the three input compounds inside each chamber, we first injected a red fluorescence solution (Sulforhodamine 101) into input A, while injecting DI water into input B, input C and control, and the fluorescence image of each chamber was captured. Then, the fluorescence intensity was measured using ImageJ. Fluorescence solution was then switched to input B, while injecting DI water into other inlets. The same procedure was repeated for input C. The injection was controlled by syringe pump and was set at 5 μL/min for duration of 20 minutes.

This microfluidic device was used to assay for the ability of 1,5-dihydroxyisoquinoline (ISO), deferoxamine (DFO), and 3-aminobenzoic acid (ABA) to alleviate cytotoxicity of hydrogen peroxide, which is generated during brain ischemia and reperfusion injury and various neuropathological conditions. The device operation for the cell assay experiment is shown in Figure 5. Suspension of rat neuronal cell (B35 rat neuroblastoma cells) was injected, and cells were efficiently collected by and grown on the membrane. Maintaining cells on the membrane and separating them from the fluid delivery channel also mimics the *in-vivo* tissue environment. Following three hours of incubation after cell seeding, each compound was mixed in media containing 1.5 mM hydrogen peroxide and injected at 5 μL/min for 30 minutes, followed by incubation with normal media for three hours. Live/dead stains (Calcein AM/propidium iodide) were injected, and cytotoxicity was determined by the ratio of dead cells to total cells.

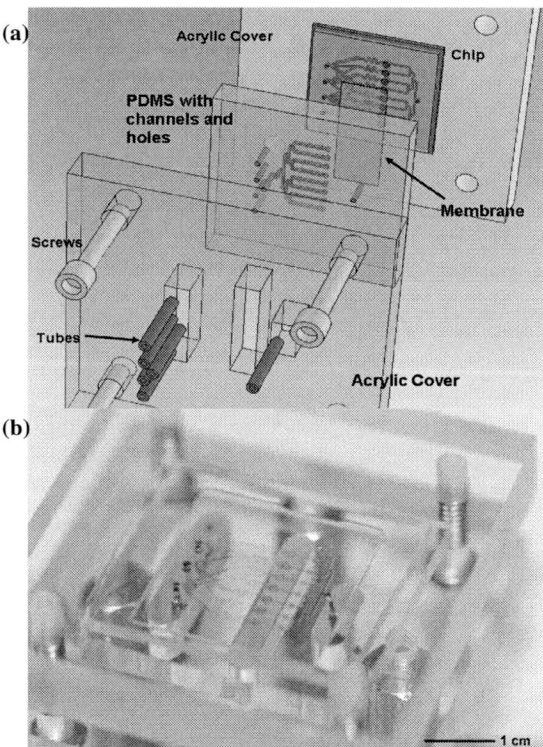

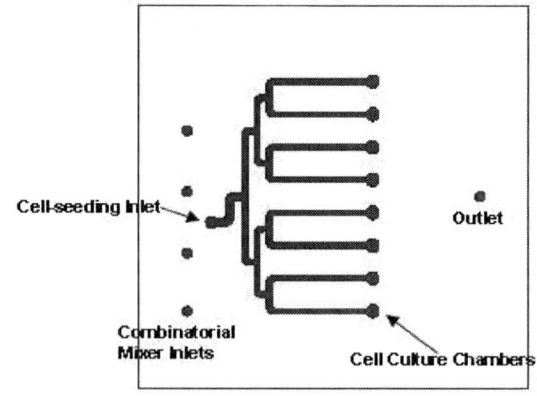

Figure 3. Device packaging. (a) Device packaging schematic. (b) Packaged device with various food coloring solutions injected.

Figure 4. Layout of the PDMS piece.

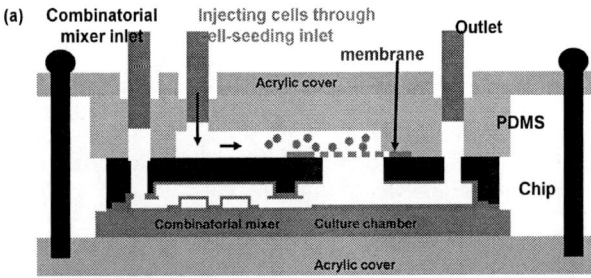

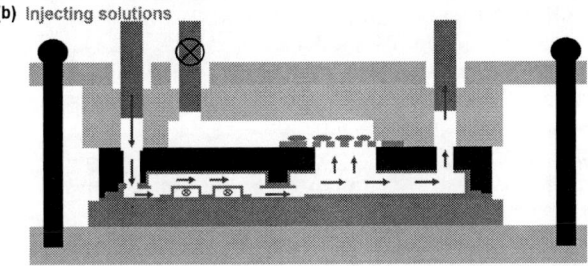

Figure 5. Device operation for cell-based assay. (a) Cell seeding by injecting cell suspension into the cell-seeding inlet. The one-to-eight manifold on the PDMS splits the cell suspension and cells flow into the chamber area where they are trapped by the membrane. (b) Injecting various solutions through the combinatorial mixer inlets for cell assay.

RESULTS AND DISCUSSION

Figure 6 shows the concentration of the three input compounds at each chamber. The combinatorial mixer was able to generate the correct combinations of the compound solutions. The concentration of the compounds can deviate from the design value because of dimensional inaccuracy in processing.

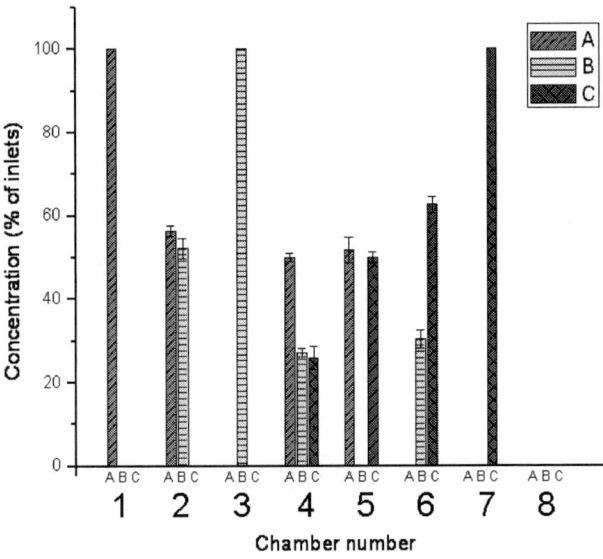

Figure 6. Concentration of the three input compounds at each chamber.

Integrating a porous membrane into the device packaging greatly facilitated the on-chip cell assay experiment. Cell seeding was done without high cell loss as in other microfluidic culture devices that load cells by flowing them through channels. This setup also

eliminated the need for coating the channels with cell attachment factors such as collagen or fibronectin. The device was used to assay for effect of three compounds on reducing the cytotoxicity of hydrogen peroxide. Figure 7 shows the result of the cytotoxicity experiment, and Figure 8 shows the arrays of cells after the cells were stained by live/dead stains. 1,5-dihydroxyisoquinoline and deferoxamine showed cytoprotective effect against hydrogen peroxide. 1,5-dihydroxyisoquinoline reduces cell death by inhibiting Poly(ADP-Ribose) polymerase, whose over-production in response to hydrogen peroxide can cause cell death [9]. Deferoxamine reduces cell death by decreasing the conversion of hydrogen peroxide to hydroxyl radicals that damage DNA. 3-aminobenzoic acid, structurally similar to 1,5-dihydroxyisoquinoline, showed minimal effect on hydrogen peroxide cytotoxicity. In general, this device successfully demonstrated that, even with just three inputs, eight distinct cytotoxic experiments were done with different combinatorial compounds using a single chip.

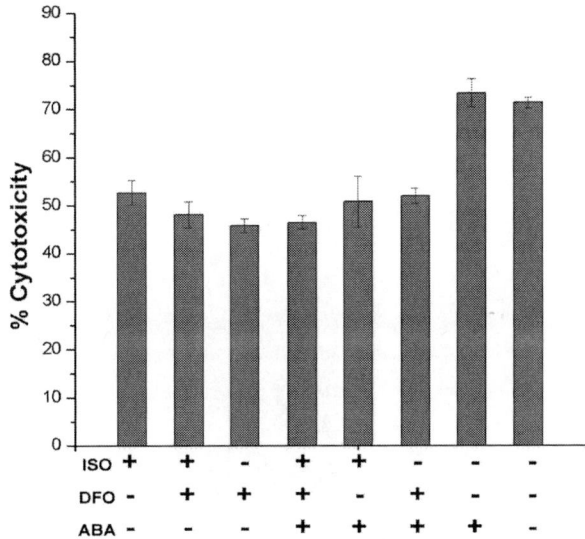

Figure 7. Cytotoxicity as a result of each treatment. The effects of three compounds, 1,5-dihydroxyisoquinoline (ISO), deferoxamine (DFO), and 3-aminobenzoic acid (ABA) on alleviating cell death caused by hydrogen peroxide were tested. Each compound was mixed in media containing 1.5 mM hydrogen peroxide.

CONCLUSION

In this work, we demonstrated the fabrication of a cell culture chip on glass with integrated combinatorial mixer, and showed a cell-based screening of different combinations of compounds. The combinatorial mixer correctly delivered the distinct combinations of input compounds into the chambers and the concentration of each input compound at the chamber was measured. A porous polyester membrane was incorporated into the culture chamber, and a cytotoxicity experiment was done using this chip. By scaling-up this technology, chip with high-density cell array and integrated high-input combinatorial mixer can be constructed. Such chip will enable inexpensive high-throughput cell-based assay, and significantly benefits research in a spectrum of fields including drug screening and systems biology.

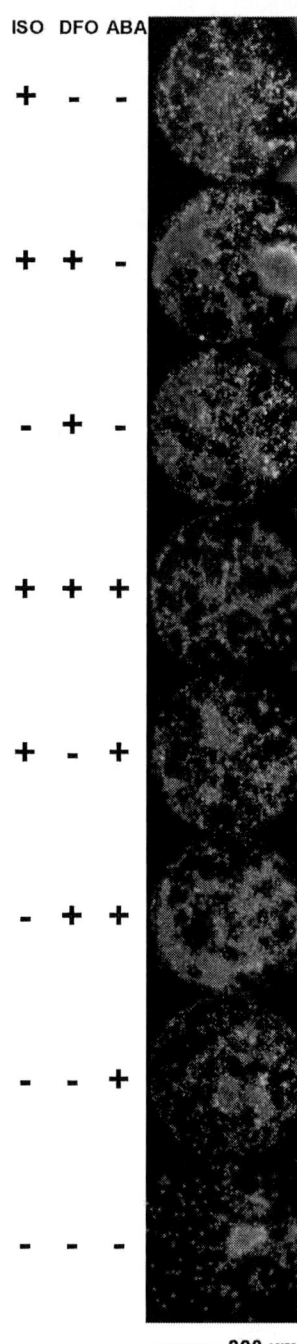

Figure 8. Cells in each chamber after live (green)/dead (red) staining. Some early apoptotic cells are stained by both dyes.

REFERENCES

[1] L. Kim, Y.C. Toh, J. Voldman and H. Yu, "A practical guide to microfluidic perfusion culture of adherent mammalian cells", *Lab Chip,* vol. 7, pp. 681-694, 2007.

[2] P. J. Lee, P. J. Hung, V. M. Rao and L. P. Lee, "Nanoliter scale microbioreactor array for quantitative cell biology", *Biotechnol. Bioeng.,* vol. 94, pp. 5-14, 2006.

[3] K. R. King, S. Wang, D. Irimia, A. Jayaraman, M. Toner and M. L. Yarmush, "A high-throughput microfluidic real-time gene expression living cell array", *Lab Chip*, vol. 7, pp. 77-85, 2007.

[4] I. Barbulovic-Nad, H. Yang, P. S. Park and A. R. Wheeler, "Digital microfluidics for cell-based assays", *Lab Chip,* vol. 8, pp. 519-526, 2008.

[5] S.J. Morrison, N.M. Shah, D.J. Anderson, "Regulatory mechanisms in stem cell biology", *Cell,* vol. 88, pp. 287-298, 1997.

[6] J. A. Menéendez, S. Ropero, M. M. Barbacid, S. Montero, M. Solanas, E. Escrich, H. Cortés-Funes and R. Colomer, "Synergistic interaction between vinorelbine and gamma-linolenic acid in breast cancer cells", *Breast Cancer Res. Treat.*, vol. 72, pp. 203-219, 2002.

[7] M. C. Liu, D. Ho and Y.C. Tai, "Monolithic fabrication of three-dimensional microfluidic networks for constructing cell culture array with an integrated combinatorial mixer", *Sens. Actuators B.,* vol. 129, pp. 826-833, 2008.

[8] G. T. A. Kovacs, *Micromachined Transducers Sourcebook,* McGraw-Hill Companies, Inc., New York, 1998.

[9] J. Bowes, J. Piper and C. Thiemermann, "Inhibitors of the activity of poly (ADP-ribose) synthetase reduce the cell death caused by hydrogen peroxide in human cardiac myoblasts", *Br. J. Pharmacol.,* vol. 124, pp. 1760-1766, 1998.

FLOW ACCELERATION EFFECT ON CANCER CELL DEFORMATION AND DETACHMENT

L.S.L. Cheung[1], X.J. Zheng[1], A. Stopa[2], J. Schroeder[3], R.L. Heimark[4]
J.C. Baygents[2], R. Guzman[2], and Y. Zohar[1]

[1]Dept. Aerospace and Mechanical Engineering, [2]Dept. Chemical and Environmental Engineering,
[3]Dept. Molecular and Cellular Biology and [4]Dept. Surgery, University of Arizona, Tucson, USA

ABSTRACT

The effect of flow acceleration, rather than just the flow rate, on the response of an attached cancer cell is for the first time reported. Selective binding of prostate cancer cells to a surface functionalized with anti-N-cadherin antibodies utilizing a microfluidic system under flow conditions has been studied [1]. Here, the behavior of a captured cell under a time-dependent flow field is investigated experimentally and numerically. Under slowly increasing flow rate, the cell deformation is more pronounced resulting in lower drag force on attached cells. Furthermore, the contact area between the cell and the functionalized surface is larger, potentially enhancing the cell adhesion force. Consequently, a higher flow rate is required to detach cells exposed to such a flow field. Numerical simulations have been utilized in effort to quantify the required detachment force. The results confirm that to obtain a similar shear stress, a higher flow rate is needed for attached cells under lower flow acceleration.

INTRODUCTION

Circulating tumor cells (CTCs) have been identified in peripheral blood from cancer patients, and are probably the origin of intractable metastatic disease [2]. These CTCs are very rare in blood and, thus, their isolation presents a formidable technical challenge [3]. Nonetheless, these cells represent a potential alternative to invasive biopsies for monitoring of non-haematologic cancers [4]; hence, the ability to characterize circulating tumor cells could further the discovery of cancer bio-markers and enhance the understanding of metastasis. Microfluidic systems provide unique opportunities for cell sorting and rare-cell detection; they have been applied for flow cytometry, continuous size-based separation, and adhesion-based separation [5]. In a landmark study, using EpCAM antibody coating, a microchannel with microposts has been developed for selective separation of viable circulating tumor cells from peripheral whole blood samples with approximately 50% purity [6].

Cadherins make up a family of trans-membrane cell adhesion molecules, and one cadherin sub-type only interacts with its counter receptor. Typically, prostate cancer cells down-regulate E-cadherin and up-regulate N-cadherin. A common technique to measure adhesion forces between such molecules involves immobilizing a single cell on a micro cantilever beam and mechanically pulling it away by contact to another cell. The binding between E-cadherins is reported to be stronger than N-cadherins. Also, the rupture force of E-cadherin shows higher dependence on loading velocity [7].

Requiring less expensive equipment and providing better visualization, microfluidic devices have been extensively used to study cell response to different flow conditions. In experiments of cell detachment due to shear flow, the fraction of adherent cells depends not only on the shear stress but also on time when the shear stress is applied in step-wise increments. Under higher shear stress, cells detached more rapidly in the initial few minutes and then approach a steady state level [8]. In a study about cell-surface interaction, the dissociation rate is reported to depend on hydrodynamic loading [9]. By estimating the fluidic force and moment on a biological cell, the detachment force is calculated and the dissociation rate is found to increase with the force. Recently, the attachment and detachment of cells on modified microchannel surfaces has been reported, including a theoretical model for fitting the experimental detachment data. The model predicts that the number of attached cell decreases exponentially with increasing shear stress. Two fitting parameters: decay rate and maximum detachment level are determined empirically for systems varying in cell type and surface coating [10].

It is difficult to obtain three dimensional geometry of an attached cell in a conventional parallel flow chamber. Therefore, calculations of flow-induced drag force and actual binding area can be underestimated. A side-view imaging technique has been proposed for studying cell motion and deformation on an antibody coated surface. Observations show that both the cell height and length change due to fluid flow, and a three dimensional model is suggested to describe the geometry of a deformed cell [11]. Numerical simulations have been used to study the deformation of normal and cancer cell under flow conditions. In comparison to a rigid spherical model, cell deformation reduces the hydrodynamic force by about 30%, while the contact area increases by factor of 2.5 to 4 [12]. The deformation mechanism is therefore critical for cell detachment. In our pervious work, we proposed that flow acceleration affects cell detachment [1]. Here, we explore the details of flow acceleration effect on cell deformation and detachment using both experimental measurements and numerical computations.

DEVICE AND FABRICATION

Device fabrication starts with the construction of a microchannel mold. A 0.3μm thick oxide layer is thermally grown on a 4" <100> P-type silicon wafer about 500μm in thickness. The microchannel pattern is transferred to the oxide etch mask utilizing standard photolithography and etching techniques; the microchannel length is about L=32mm and its width about W=1mm. The microchannel mold, about H=100μm in height, is etched in the silicon substrates using 25% TMAH. After stripping the oxide etch mask, de-molding agent (OmniCoat[TM] from MicroChem Corp.) is spin-coated to suppress adhesion force. PDMS solution, with a mixture of 10:1 base to curing agent, is next poured on the fabricated mold and

978-1-4244-2977-6/09 $25.00 © 2009 IEEE

cured at 100°C for 1hr. After peeling it off the mold, the cured PDMS substrate is treated in air plasma to render its bonding surface hydrophilic. The treated PDMS substrate is then bonded to another silicon wafer with a 3000Å-thick thermal oxide layer. Finally, after drilling holes through the PDMS substrate, inlet/outlet adaptors are glued to complete the process. A photograph of a fabricated microdevice is shown in Figure 1.

A standard immunoassay is used for derivatizing the channel surface with antibodies as previously described [1]. Briefly, the microchanel is filled with 1% (vol/vol) 3-aminopropyltrioxysilane (APTES)-acetone solution for 15min at room temperature. The APTES-coated surface is activated with 2% (vol/vol) glutaraldehyde in water (Fluka) for 2 hours at room temperature to promote a Schiff-base reaction. After flushing the microchannel with excess of DI water and 1× phosphate buffer saline (PBS) solution, Protein G from E.coli (Zymed Lab Inc.), 50 µg/ml in 1×PBS, is incubated on the activated surface overnight at 4°C. In order to block excess of silane sites, the channel is filled with bovine serum albumin (BSA) solution (2mg/ml in 1×PBS) for 1 hour at room temperature. Anti-N-cadherin (anti-N-Cad from Sigma) antibodies from mouse, 100 µg/ml in 1×PBS are then incubated on the protein G layer at room temperature for 1 hour. The immunoassay is finally completed by washing the microchannel in PBS solution. The surface bio-activity is tested by incubating the microchannel in anti-mouse IgG antibodies conjugated with Cy3 labels in 1×PBS. The uniform intensity of the emitted red light, observed under a fluorescent microscope as shown in Figure 1, indicates uniform antibody activity on the microchannel surface.

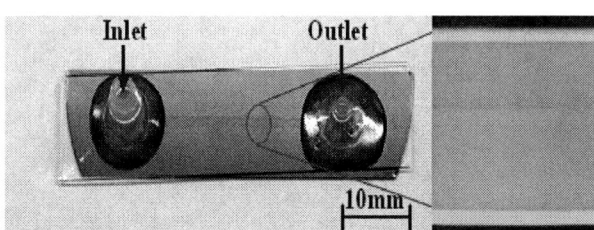

Figure 1, A photograph of a packaged microdevice and a fluorescent microscope image of a microchannel segment following labeled-antigen/antibody interaction indicating the antibody layer uniform bioactivity.

CANCER CELL ATTACHMENT

The N-cadherin expressing PC3N prostate cancer cell line (about 20µm in diameter) is maintained and grown in 1×DMEM/F12 (Invitrogen) with 10% fetal bovine serum (Cellgro) and 1% penicillin-streptomycin (Invitrogen) at 37°C in 5% CO2 in humid environment. The growth medium is aspirated and the cells are incubated with 4mM EDTA in 1×calcium free PBS (CMF-PBS) for 15mins for detachment. After centrifugation and solution removal, the cells are re-suspended in 1×CMF-PBS, while the concentration is determined by cell counting with a haemocytometer. The PC3N cell suspension is immediately loaded into a functionalized microchannel ready for experiment.

To test cell binding, PC3N prostate cancer cells are incubated in an anti-N-cadherin coated microchannel followed by washing. As shown in Figure 2, the number of

attached cells increases rapidly within 5min of incubation time before reaching 100% in less 10min. Hence, in this work, 15min incubation time has been used to capture cells.

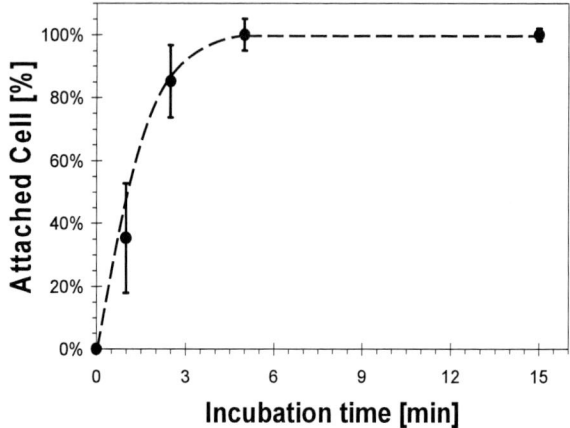

Figure 2, The effect of incubation time on the percentage of cancer cells captured on an anti-N-cadherin coated microchannel surface.

EXPERIMENTAL CONDITIONS

Once the binding between the cells and the functionalized surface is established, a certain level of fluid energy is required to detach cells. However, the same flow rate can be obtained after different time periods depending on the pump setting. Hence, two flow acceleration levels have been selected: dQ/dt=0.32ml/min^2 (High, denoted by **A**) and 0.032ml/min^2 (Low, denoted by **B**) to study the acceleration effect. Three flow rates, Q=0.165ml/min, 0.33ml/min and 0.66ml/min, have been tested in this work. Furthermore, once a design flow rate has been obtained using a syringe pump, the flow rate was kept constant for about 5min extra to obtain a steady state. All twelve experimental conditions are summarized in Figure 3 as A_{jk} and B_{jk}; where the subscript j=1,2,3 denotes the flow rate level, and k=1,2 denotes the start and finish of the constant flow rate interval.

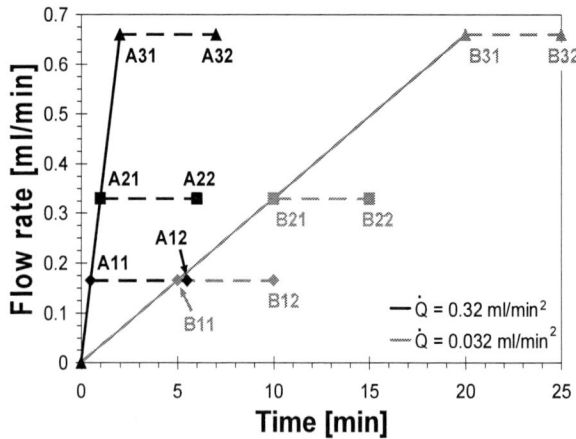

Figure 3, A summary of the experimental conditions under which cell detachment has been studied. Flow applied under high (Black color) and low (gray color) acceleration.

CELL DEFORMATION

As the flow accelerates, an attached cell experiences flow-induced shear stress increasing with time; in response, the cell is deformed. To show the effect of flow acceleration, a series of cell images have been recorded. Using a Matlab image processing program, the cell planar area is calculated. Under <u>low</u> acceleration, up to B_{31}, the cells continuously deform and elongate as shown in Figure 4. Initially, the cell contact area increases almost linearly up to 130% of the area under no-flow condition at about $Q=0.2$ml/min. Then, the deformation rate is slowly leveled off approaching a steady state contact area of about 140% at $Q=0.5$ml/min. The cell deformation is mainly along the flow direction resulting in a 'tear' shape. In sharp contrast, under <u>high</u> acceleration up to A_{31}, the cell practically retains the same spherical shape and contact area. The contact area increases slightly to 115% at about $Q=0.5$ml/min. Most of the cells are detached by the time the flow rate of 0.66ml/min is reached; thus, there isn't enough data to calculate the contract area.

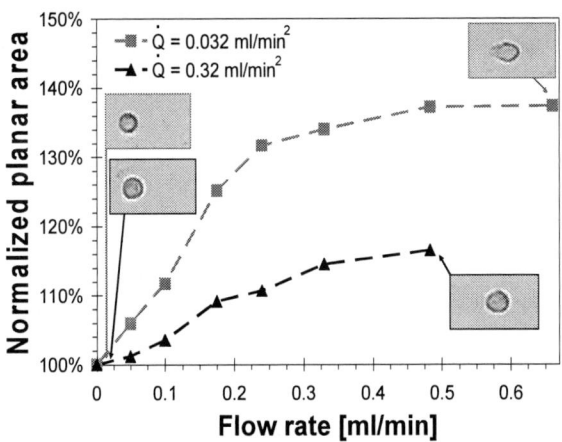

Figure 4, Cell deformation and contact area dependence on flow rate for both low and high acceleration.

CELL DETACHMENT

Once a tested flow rate is reached, points A_{j1} and B_{j1} with j=1-3, the number of attached cells is counted. The pump is programmed to maintain the flow rate at a constant level for about 5min extra, points A_{j2} and B_{j2} with j=1-3, until the number of attached cells reached a steady state. Normalized by the number at points A_{k1} and B_{k1}, the time-dependent number of attached cells is plotted in Figure 5. For <u>low</u> flow acceleration, the number of attached cells is about the same, close to 100%, independent of time and flow rate. Since the flow rate increases slowly, the attached cells have enough time to deform in essentially a quasi-steady process resulting in a significantly lower shear stress. Hence, the number of attached cells is independent of time during the constant flow rate period, from B_{j1} to B_{j2} with j=1-3. However, for <u>high</u> flow acceleration, the number of attached cell decreases with time except for the lowest flow rate level (A_{11} to A_{12}). Under the lowest flow rate, $Q=0.165$ml/min, the shear stress is too small for cell detachment; thus, while keeping it constant, the number of attached cells also stays constant. For the higher flow rates, the shear stress increases rapidly such that the cells do not have enough time to deform

keeping their spherical shape. As a result, cells continue to detach under constant high shear stress in a time-dependent manner, from A_{j1} to A_{j2} with j=2-3.

The final number of attached cells at points A_{j2} and B_{j2} is plotted in Figure 6 as a function of the flow rate. The percentage of attached cells is similar when the final flow rate is small. However, the two curves significantly diverge with increasing flow rate. Since there is more time for cell adjustment under low acceleration, the percentage of attached cells for low acceleration is higher than that for high acceleration. The results summarized in Figure 6 indicate that, under the same flow rate, the fraction of detached cancer cells strongly depends on the preceding flow evolution. This can be viewed as a balance between two competing time scales; a detaching-force time scale determined by the externally-applied flow acceleration and a cell-deformation time scale depending on the inherent cell physiology. If the flow acceleration is sufficiently high, the cell deformation is negligible. Thus, at a low flow rate, the hydrodynamic forces are large enough to overcome the adhesion forces leading to cell detachment. Numerical simulations will be used to compute these hydrodynamic forces since it is difficult to measure them experimentally.

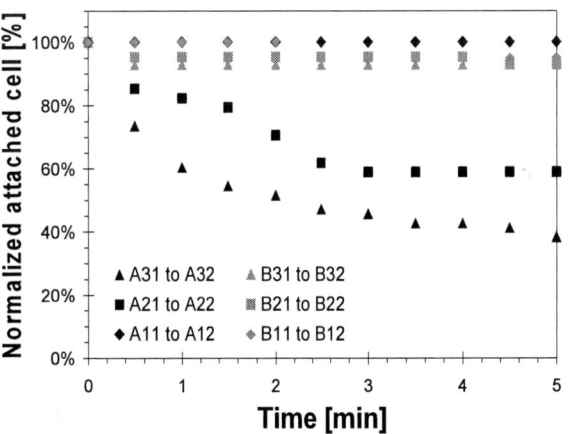

Figure 5, Time-dependent number of attached cells under different flow rates for both flow accelerations.

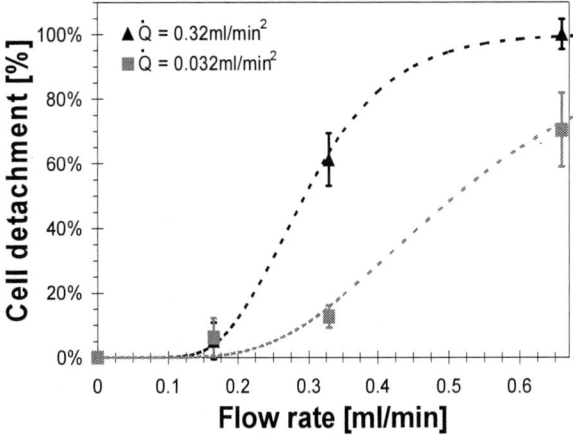

Figure 6, The percentage of attached cells on anti-N-cadherin microchannel surface as a function of the applied flow rate for both low and high flow accelerations.

978-1-4244-2977-6/09 $25.00 © 2009 IEEE

NUMERICAL SIMULATION

In order to estimate the hydrodynamic loads acting on an attached cell deformed due to fluid flow, a commercial CFD package (CFD-ACE) is used to numerically simulate the 3-D flow field. The shape of the deformed cell is modeled by the following equation [11]:

$$\frac{y^2}{b^2} + \frac{z^2}{c^2} = \left(1 - \frac{x^2}{a^2}\right)\left(1 + \frac{\beta x}{a}\right)^2 \qquad (1)$$

where a is the half length of a deformed cell; the other constants b, c and β are parameters determined by fitting the measured contact area in Figure 4 to the simulation model. The computed velocity vectors corresponding to conditions A_{31} and B_{31} are shown in Figure 7. Although the cell volume is the same in both cases, the cell under <u>high</u> acceleration is about double in height compared to the cell under <u>low</u> acceleration. Consequently, the pressure and shear stress developed on the deformed cell under low acceleration are lower. Accounting for the force and moment distribution, the resultant hydrodynamic force working against the cell adhesive forces can be computed as plotted in Figure 8. The estimated detachment force for both cases increase linearly with increasing flow rate but at different slopes. Indeed, under the same flow rate, the resultant force is higher for high flow acceleration. Consistent with the present experiments, the simulations suggest that cells under higher flow acceleration experience higher hydrodynamic loads resulting in a higher fraction of detached cells.

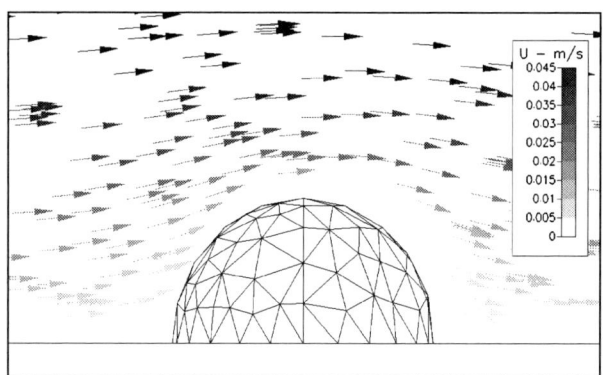

(a)

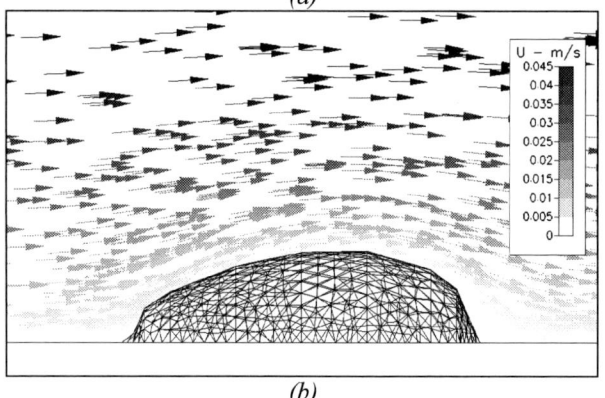

(b)

Figure 7, *3-D CFD simulations of the velocity field around an attached cell under the same flow rate but with (a)* <u>*high*</u>, *point A_{31}, and (b)* <u>*low*</u>, *point B_{31} flow acceleration.*

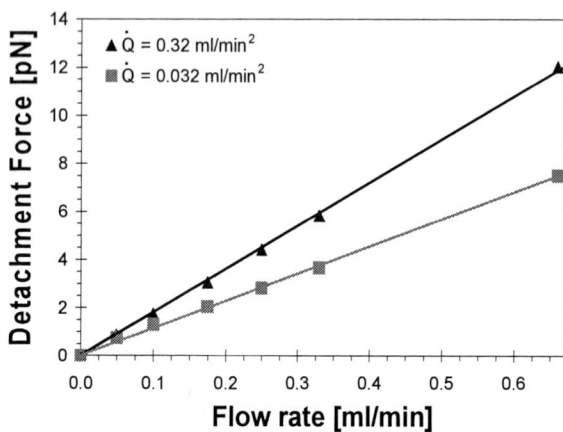

Figure 8, *Numerical computations of the hydrodynamic resultant force on an attached cell as a function of the flow rate for both low and high flow accelerations.*

CONCLUSIONS

Flow acceleration effect on the detachment of cancer cell immobilized on an antibody-coated surface has been quantitatively characterized using a microfludic device. Under low flow acceleration, in a quasi-steady process, attached cancer cells are significantly deformed in shape resulting in larger contact area and lower profile. In contrast, under high acceleration, the deformation of attached cells is negligible. Consequently, under the same flow rate, the fraction of immobilized cells detaching from a functionalized surface increases with increasing flow acceleration. CFD simulations compliment the experimental observations; the hydrodynamic loads are larger for higher flow acceleration, under the same flow rate, leading to more detached cells.

ACKNOWLEDGEMENTS

This work is supported by Arizona Biomedical Research Commission grant 06-080.

REFERENCES

[1] L.S.L. Cheung *et al.*, *Proc. MicroTAS'08 Conf.*, San Diego, Oct 12-16, 2008, pp.1159-1161.

[2] M. Cristofanilli *et al.*, *N. Engl. J. Med.*, vol 351, pp. 781-791, 2004.

[3] H.J. Kahn *et al.*, *Breast Cancer Res. Treast.*, vol 86, pp. 237-247, 2004.

[4] J.B. Smerage *et al*, *Br. J. Cancer*, vol 94, pp. 8-12, 2006.

[5] W.C. Chang *et al.*, *Lab Chip*, vol 5, pp. 64-73, 2005.

[6] S. Nagrath *et al.*, *Nature*, vol 450, pp. 1235-1239, 2007.

[7] P. Panorchan *et al.*, *J. Cell Sci.*, vol 119, pp. 66-74, 2005.

[8] H. Lu *et al.*, *Anal. Chem.*, vol 76, pp. 5257-5264, 2004.

[9] R. Alon *et al.*, *Nature*, vol 374, pp. 539-542, 1995.

[10] X. Zhang *et al.*, *Chem. Eng. J.*, vol 135S, pp. S82-S88, 2008.

[11] J. Cao *et al.*, *Microvasc. Rec.*, vol 55, pp. 124-137, 1998.

[12] S.P. Wankhede *et al.*, *Biotechnol. Prog.*, vol 22, pp. 1426-1433, 2006.

LINEAR EXPANSION AND CONTRACTION OF PAIRED PNEUMATIC BALOON BENDING ACTUATORS TOWARD TELESCOPIC MOTION

[1]N.Fujiwara, [1]S. Sawano, and [2]S. Konishi

[1]Graduate School of Science and Engineering, Ritsumeikan University, Kusatsu, Shiga, JAPAN
[2]Department of Micro System Technology, Ritsumeikan University, Kusatsu, Shiga, JAPAN

ABSTRACT

This paper proposes a novel actuation mechanism for linear expansion and contraction, or telescopic motion by pneumatic balloon actuator (hereafter PBA). PBA is characterized by small, soft, and safe (S^3) features because of its soft and flexible structure and safe driving principle. A single PBA bends due to a tensile force generated by a swelling balloon. Bending motion of PBA is transformed into the telescopic motion by using a paired PBA. Two opposing PBAs are bonded at both ends so as to compose a paired PBA. The paired PBA transforms the bending motion into the telescopic motion. An elemental telescopic motion PBA composed of a paired PBA (2 mm × 11 mm × 400 μm) can generate 15mN contraction force at 150kPa. It is also possible to design serial or parallel connection of the telescopic motion PBA. This paper presents design, fabrication, and characterization of proposed telescopic motion PBA. Medical forceps is demonstrated as promising medical application of the telescopic motion PBA.

INTRODUCTION

Recently, soft structure and flexible motion become important to deal with living organisms in bio and medical. It is expected to provide precise tools suitable for tiny living organisms in biotechnology and advanced medical tools for low inversion medical operation.

Minimally invasive surgery is expected to reduce the suffering of patient in medical fields [1,2]. We can find various actuators for small and flexible surgical tools. We proposed pneumatic balloon actuator as a soft or flexible micro actuator (hereafter PBA). PBA is featured by small, soft, and safe (S^3) characteristics because of its soft and flexible structure and safe driving principle [3-5]. Molding process based on MEMS technologies allows small structures. The current design of PBA employing all PDMS (polydimethylsiloxane) structure is the third generation. PDMS allows soft and flexible structure. Pneumatic driving method can provide safe operation especially for medical application. All PDMS PBA is composed of two different PDMS layers so as to form cavities for balloons and pneumatic supply channels. Bending motion of PBA uses a tensile force generated by pressure-dependent swelling motion of balloon on PBA. We have reported various devices using the PBA such as [6-9]. Most of developed devices are based on bending motion of PBA.

The motion of expansion and contraction is widely used actuation mechanism as much as the bending motion. Furthermore, telescopic motion actuators such as wire-driven actuators and artificial muscle actuators have been extensively developed and applied in robotics and medical applications. Pneumatic driving principle is attractive for actuators in medical application compared with conventional electrical or electromagnetic mechanism.

We have developed PBA to provide pneumatically driven actuators for robotics and especially medical applications. In this paper, we propose a telescopic motion PBA which can transform the bending motion into the telescopic motion by taking advantage of S^3 features of PBA. Proposed telescopic motion PBA can generate linear expansion and contraction motion by bonding a pair of opposing PBA at both ends. It is also possible to connect elemental telescopic motion PBAs in series or parallel in order to improve output force or displacement.

This paper describes a mechanism of telescopic motion of a paired PBA transformed from bending motion of a single PBA. The telescopic motion PBA is designed and developed based on results of characterization. Device performances such as output force and displacement are improved by connecting in series.

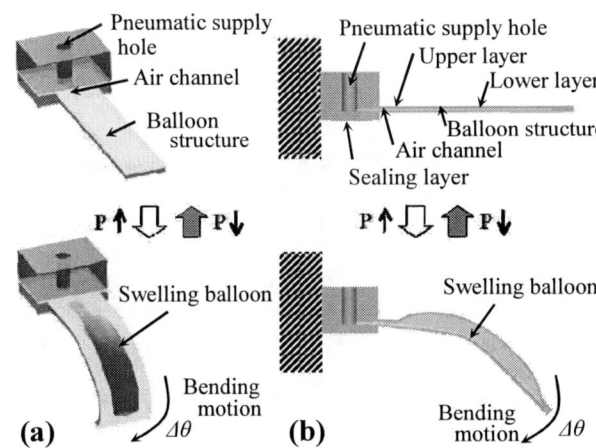

Figure 1: Schematic drawing of bending motion PBA. (a) whole structure and (b) cross sectional view.

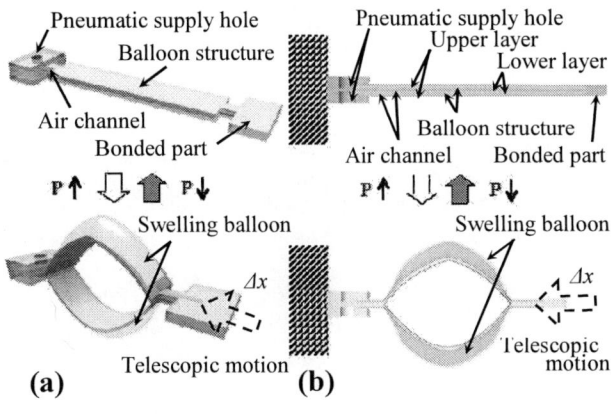

Figure 2: Schematic drawing of a paired PBA. (a) whole structure and (b) cross sectional view.

978-1-4244-2977-6/09 $25.00 © 2009 IEEE

DESIGN AND FABRICATION OF TELESCOPIC MOTION PBA

Bending Motion PBA

PBA is made of PDMS and has a balloon structure and an air channel as show in Fig. 1. PBA consists of two flexible PDMS diaphragms having different thickness and hardness. The upper one is a soft and thin diaphragm while the lower one is a hard and thick diaphragm. The lower one has the bottom structure for a balloon actuator and air flow channels. PBA uses a pressure-dependent swelling and shrinking motion of the upper diaphragm as a bending principle. If the pressure is applied to the balloon actuator, the thin top diaphragm is firstly inflated and the bending motion is generated. As the applied pressure increases, the upward bending motion restricted by the pulling force of the thick bottom diaphragm. The bending direction of PBA is changed from upward to downward since the generated pulling force of the thick bottom diaphragm is dominate. And then, the bending motion increases as the applied pressure increases.

Telescopic Motion of A Paired PBA

Telescopic motion PBA is composed of two opposing PBAs bonded at both ends as shown in Fig. 2. The paired PBA transforms the bending motion into the telescopic motion. Bonded PBAs bend by applying pressure in the opposite direction (see Fig. 2). A paired PBA transforms the opposite bending motion of two PBAs into the telescopic motion, linear expansion and contraction motion as shown in Fig. 2. Telescopic motion PBA is fabricated by bonding of two PBAs at both ends. Individual PBA is fabricated as follows. PDMS-A (8:1 mixture of the base polymer and the curing agent, Sylgard 184, Dow Corning Inc.) for the balloon structure as the lower layer is spin-coated over SU-8 mold on a silicon substrate. PDMS-A is cured at 75 °C for 20 minutes. PDMS-B (12:1 mixture) as the upper layer is coated on a glass substrate and cured at 75 °C for 4 minutes. Prepared two PDMS films A and B are bonded together. Bonded structure is cut into the final shape and implemented by pneumatic connector. Finally, two PBAs were bonded at both ends as shown in Fig. 3. Developed paired PBA for telescopic motion is shown in Fig. 4. Figure 4 shows photographs of developed device and corresponding drawings.

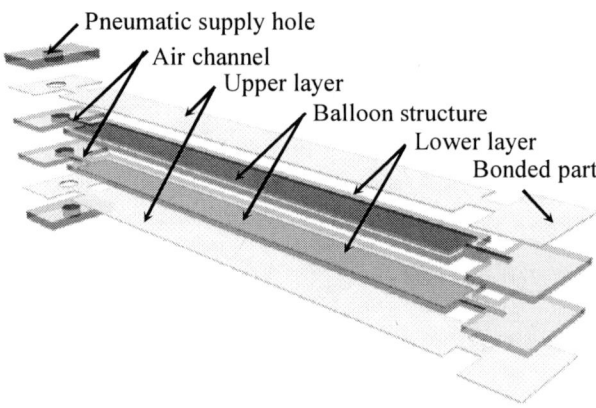

Figure 3: Schematic drawing of a paired PBA.

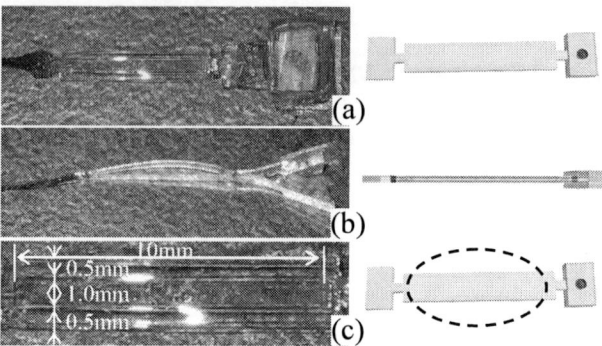

Figure 4: Photographs of implemented telescopic motion PBA. (a) top view, (b) side view, (c) magnified view of balloon structure.

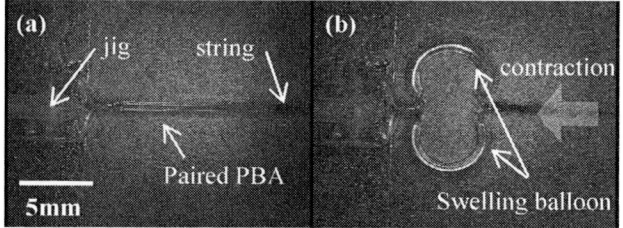

Figure 5: Operation of developed telescopic motion PBA. (a) before action, (b) after actuation.

OPERATION AND CHARACTERIZATION

A Single Telescopic Motion PBA

Figure 5 shows an operation of developed device. A paired PBA contracts as a result of opposite bending motion of bonded PBAs. The telescopic motion of paired PBA was successfully demonstrated as shown in Fig. 5. Telescopic motion of developed device was evaluated from the aspect of deformation and generated force. Longitudinal and transverse displacement (ΔX and ΔZ, respectively) were estimated depending on the length of PBA. The width and thickness of balloon were fixed in the evaluation. Applied pressure was fixed at 150 kPa. Figure 6 shows evaluation results of longitudinal (ΔX) and transverse (ΔZ) displacement. Evaluation results show that both ΔX and ΔZ increase according to applied pressure.

Generated tensile force was estimated depending on the length of PBA. The width and thickness of PBA were fixed as 1mm and 400μm, respectively. Applied pressure was fixed at 150 kPa while generated force increased according to applied pressure. Measured result on generated force is shown in Fig. 7. Figure 7 shows the existence of the optimum design in terms of ratio between length and width of PBA. Telescopic motion PBA (1 mm × 10 mm) could generate 15mN as the largest force in Fig. 7. In this study, 2 mm × 11 mm × 400 μm actuator was employed as a standard design in order to understand performance of telescopic motion PBA.

Serial Connection of Telescopic Motion PBA

It is possible to connect elemental telescopic motion PBA in series or parallel. Serial connection of actuators would contribute to improve output force and displacement. Parallel connection of actuators could improve output force without increasing the length of device. Serial or parallel connection of actuators were

978-1-4244-2977-6/09 $25.00 © 2009 IEEE

shown in Fig. 8(a) and (b). This paper reports serial connection of telescopic motion PBA by taking account of application of wire-driven actuators. Serially connected telescopic motion PBAs were developed as shown in Fig. 8 (c). Developed actuator was characterized in Fig. 9. Figure 9 tells that a serial combination of actuators could increase output force.

DEMONSTRATIONS

Medical forceps for laparoscopically assisted surgery was demonstrated as one of typical applications of wire-driven actuators. The wire-driven mechanism for forceps was replaced with developed serially connected telescopic motion PBAs. PBA based actuator is attractive for medical application because of its S^3 features. Telescopic motion PBA generates transversal deformation as well as longitudinal deformation while the abdominal port should be minimized to reduce inversion. Slits were designed for outer cover of forceps in order to allow the transverse deformation of telescopic motion PBA without increasing the diameter of the cover. Telescopic motion PBA is operated after introducing in an abdominal cavity. Abdominal cavity can accommodate structures protruding from slits of the outer cover. Required force to operate a commercialized forceps was measured.

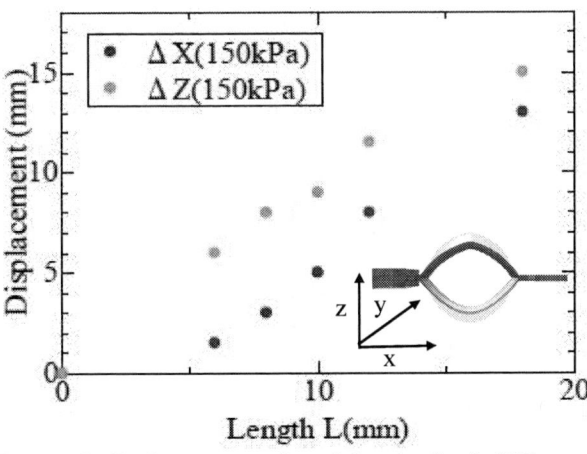

Figure 6: Evaluation results of longitudinal (ΔX) and transverse (ΔZ) displacement.

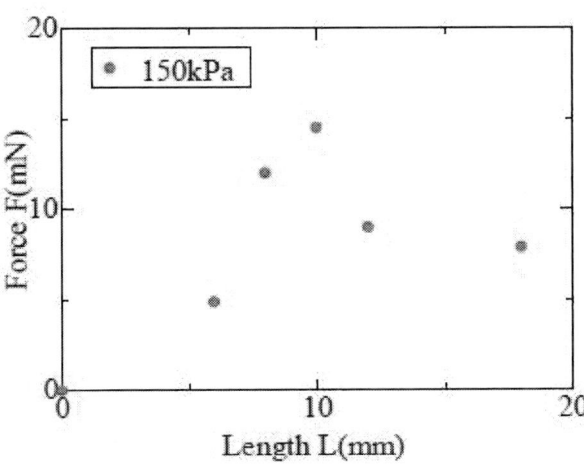

Figure 7: Measured characteristics of generated force.

It was estimated about 90 mN to drive a forceps. Twenty one elemental telescopic motion PBAs were connected in series to satisfy a requirement for the driving force. Figure 10 shows a forceps equipped with developed serial telescopic motion PBAs. The forceps initially opens (upper photo) and pinches when driving pressure is applied (lower photo). Telescopic motion PBAs contracts by applied pressure and operates the forceps based on wire-driven mechanism. In vitro experiment was executed by the forceps to estimate possibility of medical application. Figure 11 shows operation result to grasp a chicken liver (1g). The result says that we can use the telescopic motion PBA to drive forceps instead of conventional wire-driven mechanism.

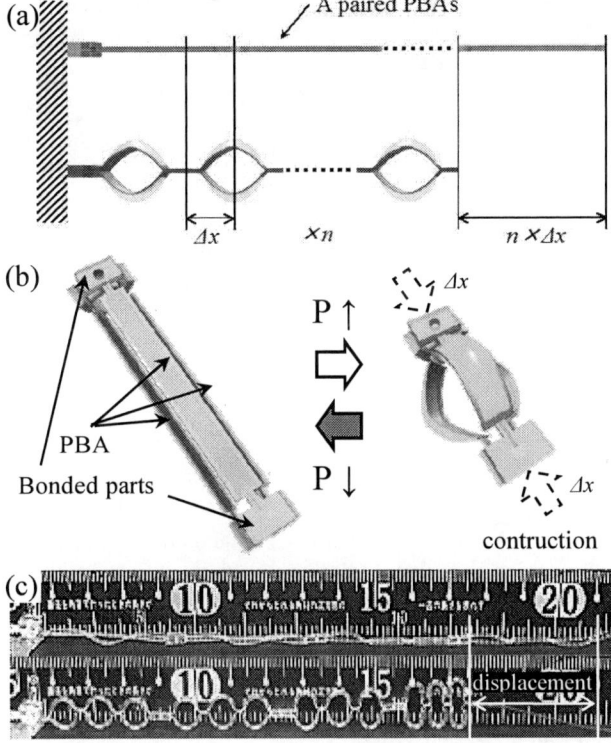

Figure 8: Serial or parallel connection of telescopic motion PBAs. (a) serial connection, (b) parallel connection, (c) demonstration of serial connection.

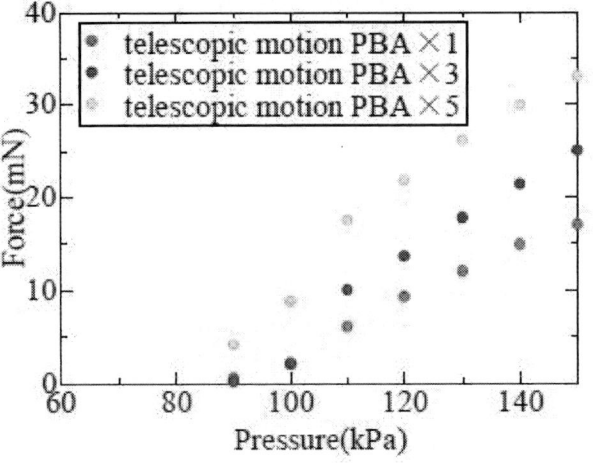

Figure 9: Characterization of serially connected telescopic motion PBA.

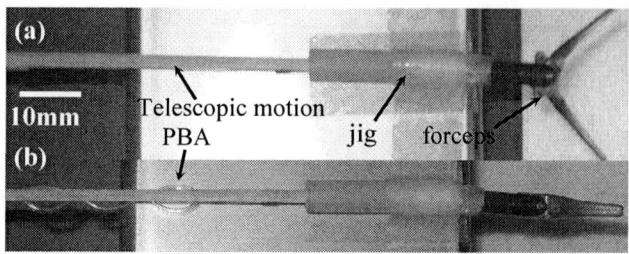

Figure 10: Operation of forceps equipped with telescopic motion PBAs. (a) before operation, (b) after operation.

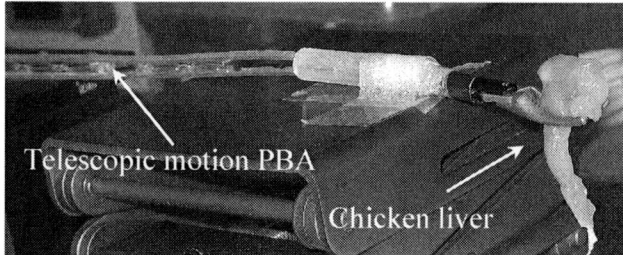

Figure 11: Operation result to grasp a chicken liver.

CONCLUSION

This paper proposed a novel actuation mechanism for linear expansion and contraction transformed from the bending motion of pneumatic balloon actuator (PBA). The telescopic motion PBA was composed of a paired PBA bonded at both ends. The paired PBA could successfully generate telescopic motion as designed. Developed telescopic motion PBA (1 mm × 10 mm) could generate 15 mN at 150 kPa. Operation of forceps by telescopic motion PBA was also demonstrated. Commercialized forceps requires 90 mN to drive. Serial connection of elemental telescopic motion PBA could improve output force and displacement. We could operate medical forceps by connecting telescopic motion PBAs in series. Linear expansion and contraction by PBA will enhance potential application of PBA in various fields, especially medical benefiting by its S^3 features.

ACKNOWLEDGEMENT

This research was partially supported by the Ministry of Education, Science, Sports and Culture, Grant-in-Aid for Cooperation of Innovative Technology and Advanced Research in City Area Program in the Southern Area of Lake Biwa, 2007-2009, coordinated by Shiga Prefectural Industrial Support Center.

REFERENCES

[1] S. Konishi, M. Nokata, O. C. Jeong, S. Kusuda, T.Sakakibara, M. Kuwayama and H. Tsutsumi, "Pneumatic Micro hand and Miniaturized Parallel Link Robot for Micro Manipulation Robot System", *Proc. of the 2006 IEEE International Conference on Robotics and Automation*, Orlando, Florida, May, 2006, pp. 1036-1041.

[2] K. Suzumori, T. Hama and T. Kanda, "New Pneumatic Rubber Actuators to Assist Colonoscope Insertion", *Proc. of the 2006 IEEE International Conference on Robotics and Automation*, Orlando, Florida, May, 2006, pp. 1824-1829.

[3] S. Konishi, F. Kawai and P. Cusin "Thin flexible end-effecter using pneumatic balloon actuator", *Proc. of IEEE International Conference of MEMS 2000*, Miyazaki, Japan, Jan. 2000, pp. 391-396.

[4] O. C. Jeong, S. Kusuda and S. Konishi, "All PDMS PNEUMATIC BALLOON ACTUATORS FOR BIDIRECTIONAL MOTION OF MICRO FINGER", *Proc. of IEEE International Conference of MEMS 2005*, Miami beach, Florida, USA, Jan. 30 –Feb. 3, 2005, pp. 407-410.

[5] O. C. Jeong, S. Konishi and S. Kusuda, "All PDMS Pneumatic Microfinger With Bidirectional Motion and Its Application", *Journal of MICROELECTROMECAHNICAL SYSTEMS, VOL.15, NO.4*, August, 2006, pp. 896-903.

[6] S.Kusuda, S.Sawano and S.Satoshi, "FLUID-RESISTIVE BENDING SENSOR HAVING PERFECT COMPATIBILITY WITH FLEXIBLE PNEUMATIC BALLOON ACTUATOR" *Proc. of IEEE International Conference of MEMS 2007*, pp. 615-618.

[7] S. Konishi, "SMALL, SOFT AND SAFE ACTUATOR BY USING MEMS-BASED PNEUMATIC BALLOON ACTUATOR", *Asia-Pacific Conference of Transducers and Micro-Nano Technology –APCOT 2006*, pp. 144-147, 2006.

[8] Yusaku Watanabe, Masanori Maeda, Naoko Yaji, Ryoichi Nakamura, Hiroshi Iseki, Masayuki Yamato, Teruo Okano, Sadao Hori, and Satoshi Konishi, "SMALL, SOFT AND SAFE MICROACTUATOR FOR RETINAL PIGMENT EPITHELIUM TRANSPLANTATION", *20th IEEE International Conference on Micro Electro Mechanical Systems MEMS2007*, pp. 659-662.

[9] K.Takemura and S.Yokota, "A Micro Artificial Muscle Actuator using Electro-conjugate Fluid", *Proceeding of the 2005 IEEE International Conference on Robotics and Automation*, Barcelona, Spain, April 2005.

[10] K.Takemura and S.Yokota, "A Micro Artificial Muscle Actuator using Electro-conjugate Fluid", *Proceeding of the 2005 IEEE International Conference on Robotics and Automation*, Barcelona, Spain, April 2005. pp. 532-537.

[11] T.Hasuo, T.Suzuki, E.Aoki, E.Kobayashi, K.Konishi, M.Hashizume, I.Sakuma, " Application of super elastic alloy to wire-driven multi-DOFs bending forceps", *Proceedings of the 2005 JSME Conference on Robotics and Mechatronics*, 2005, No.05-4

[12] Elizabeth V. Mangan, Dan A. Kingsley, Roger D. Quinn and Hillel J. Chiel, "Development of a Peristaltic Endoscope", *Proceedings of the 2002 IEEE International Conference on Robotics and Automation*, Washington, DC, May, 2002. pp. 347-352.

LOW-COST MICROFLUIDIC SINGLE-USE VALVES AND ON-BOARD REAGENT STORAGE USING LASER-PRINTER TECHNOLOGY

J.L. García-Cordero[1], F. Benito-Lopez[2], D. Diamond[1,2], J. Ducrée[1], and A.J. Ricco[1]
[1]Biomedical Diagnostics Institute and [2]Adaptive Sensors Group,
National Centre for Sensor Research, Dublin City University, Dublin, IRELAND

ABSTRACT

We report for the first time the laser-printer-based fabrication of vapor- and pressure-resistant microfluidic single-use valves and their implementation on a centrifugal microfluidic "lab-on-a-disc platform". As an extension of this technology, we implemented long-term storage of liquids for up to one month with no signs of evaporation. This simple technology is compatible with a range of polymer microfabrication technologies and should facilitate the design and fabrication of fully integrated and automated lab-on-a-chip cartridges that require pressure-resistant valves or long-term reagent storage.

INTRODUCTION

Point-of-care (POC) diagnostic devices will revolutionize and improve global public health by diagnosing diseases in a timely manner, preventing epidemics, controlling chronic health conditions, tailoring treatments, and decreasing national health system costs. POC systems for deprived-resource settings require portability, disposability, low-cost, simplicity of use, ruggedness, and temperature independence [1-3]. They also need to deliver assay results with similar sensitivity, reproducibility, and selectivity to centralized laboratory tests [1-3]. Finally, POC devices should operate with minimal, non-expert operator attention.

Of the different technologies that currently exist to address this issue, microfluidic and lab-on-a-chip technologies have the potential and the toolset to make POC diagnostic systems a reality [3-4]. The lab-on-a-chip vision is to miniaturize clinical laboratory processes, integrating them onto disposable units the size of a credit card using minute amounts of complex samples and precious reagents. These autonomous and integrated chips would consist of different modules or components that would handle a complex sample such as blood, preparing it and mixing it with the necessary reagents to produce a signal that can be read by a miniaturized, even an on-chip, detection system.

Although the field of microfluidics has produced several components that in theory can help to realize this vision, the complexities of integrating and fabricating them at low cost are many and the challenges are daunting. We address one key component of this challenge by introducing a technology for the low-cost production of valves that can, among other tasks, enable on-chip long-term wet reagent storage.

Microfluidic valves and pumps are ubiquitous in integrated microfluidic systems, but fluid actuation and control can greatly add to the fabrication costs of integrated microsystems: external actuators may be needed to drive them [5] or their implementation into manufacturing processes may be a costly engineering challenge. Therefore, pumping and valving control for

POCT should be easy to integrate into the manufacturing processes at minimum cost while still offering maximum flexibility in design and miniaturization of diagnostic integrated systems.

On-chip long-term reagent storage will be necessary for market success of many microfluidic point-of-care devices. Although both wet and dry reagent storage in microfluidic compartments has been reported [6-8] a key issue remains: delivering the reagents after an extended storage time, in a well-controlled fashion. Linder and colleagues demonstrated storage of reagents inside plastic tubing in liquid plugs separated by air gaps [8], but this and other methods that do not provide a sealed physical barrier are ill-suited to storage beyond a few hours due to migration of water in the vapor phase. Furthermore, this approach will not necessarily work in more complex, integrated microfluidic systems; alternatives are needed.

We report and demonstrate the fabrication of a single-use valve based on laser printing technology. The valves are 'opened' with a single laser shot or pulse. As an application of the same technology, we also demonstrate a system for the storage of liquid reagents in sealed reservoirs for up to 30 days with no significant evaporation. This simple technology is compatible with polymers and fabrication techniques such as hot embossing and multilayer plastic lamination.

Compared to other technologies [9], our approach requires lower laser powers and uses transparent foils (enabling addressing of valves on multiple fluidic levels). The electronics and software-control algorithms to operate the valves are simpler and the precision of positioning the laser spot less demanding since a general raster of the laser beam in the vicinity of the valve opens it.

DESIGN AND FABRICATION
Design

The schematic of the microfluidic single-use valves is shown in Figure 1. The device consists of three layers. The top layer features the main microfluidic channel, which connects to other microfluidic modules. Its continuity is interrupted at places where valves are necessary. The bottom layer contains a connecting microfluidic channel that links the segments of the top microfluidic channel. A plastic foil with laser-printed dots is sandwiched between these two layers. Two dots are aligned at the intersections of the top and bottom channels (Figure 1).

The purpose of each laser-printed dot is to absorb optical energy from the laser diode, rapidly heating and thus perforating the plastic foil by melting it, while clear areas of the foil remain unaffected. This reduces the required accuracy of aiming the laser, provided it is scanned over an area that encompasses the valve spot.

978-1-4244-2977-6/09 $25.00 © 2009 IEEE   439

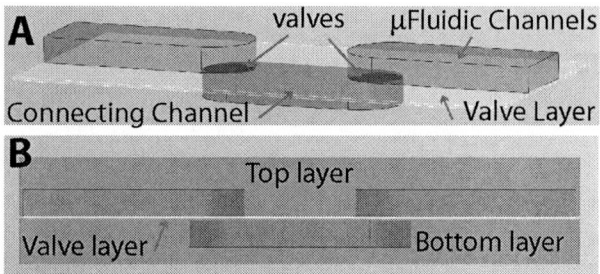

Figure 1: Schematic representation of the single-use valves. Upper microfluidic channels are connected through the bottom connecting channel (A). Between the two microfluidic channel layers is a thin film with laser-printed spots that define the valves as shown in the cross sectional view (B).

Operation of the valves is illustrated in Figure 2. Liquid flows into the upper left microfluidic channel. A laser diode is positioned to point at the first valve; a short pulse of light melts the plastic foil. The laser diode is then moved to the next dot and the operation repeated. Liquid then can be moved through the bottom channel into the upper right channel.

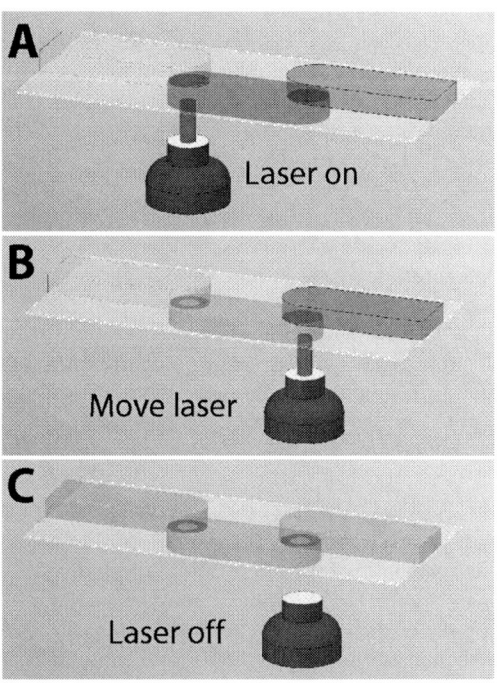

Figure 2: Operation of the single-use valves. Liquid (green) is loaded into the upper left channel (A). The laser diode is activated, sending a short pulse of light and melting the plastic (A), connecting the two channels. The laser diode is then moved to the next valve and the operation repeated (B, C).

The need for two dots per valve may not be obvious: one of the two could be perforated before assembly, reducing the complexities of positioning and control of the laser, but adding an additional step to the fabrication process. Pre-perforation of one dot would also eliminate the redundancy against leakage or slow permeation of water vapor afforded by two dots/valve. This trade-off is being explored and assessed.

A design for long-term reagent storage reservoirs for microfluidic devices, based on the functionality of the laser valves, is demonstrated in Figure 3A-C. A reservoir is defined in the upper layer and can be designed to hold any fluid volume. The valve for this storage reservoir is laser-printed at its peripheral end and at the intersection with a microfluidic channel that connects it to the rest of the microfluidic system. The barrier properties of the foil to store liquid reagents are exploited.

The operation of the device is shown in Figure 3D-E. After assembly of the device, reagents are loaded into the reservoir and encapsulated using pressure-sensitive adhesive (PSA) film. This film seals tightly the storage reservoir and prevents evaporation. Liquid from the container can be cleanly released into the channel by centrifugal or capillary actuation.

Fabrication

Devices shown in Figures 5 and 6 were fabricated using multi-layer lamination. A CO_2 laser (Laser Micromachining LightDeck, Optec, Belgium) system was used to cut the various polymer layers. To laminate the plastic layers, a thermal roller laminator (Titan-110, GBC Films, USA) was used. A laser-printer (resolution: 600dpi, LaserJet 4050 Series, HP, USA) was used to print dots onto a transparency film.

Connecting channels were cut from an 80-μm thick layer of PSA (AR9808, Adhesives Research, Ireland) and laminated onto a 250-μm poly(methylmethacrylate), PMMA, support layer (GoodFellow, UK). The width of the connecting microfluidic channel was measured to be approximately 400 μm. This assembly of channels constituted the connecting layer in both devices.

The upper chambers shown in Figure 5 were laser-cut from a 250-μm PMMA sheet; those in Figure 6 were cut from a 1.2-mm-thick polycarbonate layer (GoodFellow, UK). These layers were then laminated onto the connecting layer. Finally, a layer of PSA with laser-cut holes that function as vents was laminated on top of the chambers.

EXPERIMENTAL SETUP

Individual devices were mounted in a disc. A brushless DC motor with an integrated optical encoder (Series 4490, Faulhaber, Switzerland) was used to rotate the disc. We used a laser diode with similar characteristics to those used in commercial DVD-RW players (wavelength: 650 nm, power: 150 mW, Wicked Lasers, USA).

RESULTS AND DISCUSSION

Figure 4 shows a laser-printed dot on a transparency film. A laser pulse melts the plastic in less than one second. The minimum size of the spot is determined by the resolution of the laser printer. We printed dots with diameters down to 150 μm.

It is important to note that the transparency film was passed through the printer three times to impregnate the film with a high density of carbon ink microparticles and increase absorption of light from the laser diode.

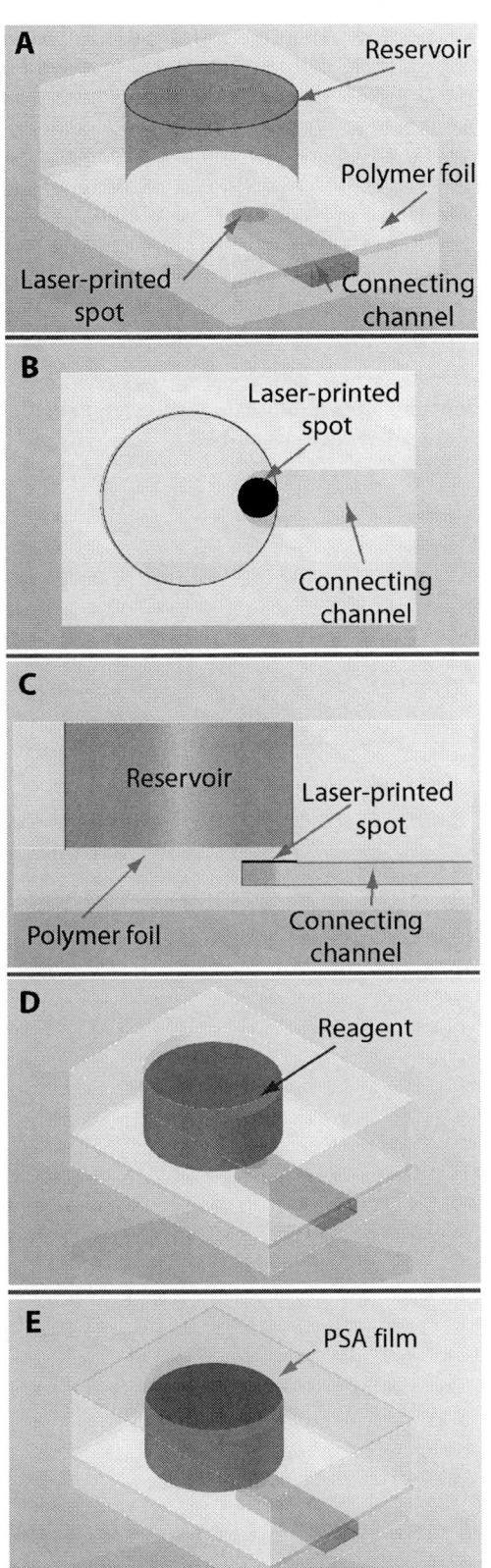

Experiments performed with foils printed in a single pass did not absorb enough energy to melt the plastic. An increase in the laser power would melt the plastic for a single-pass laser-printed film.

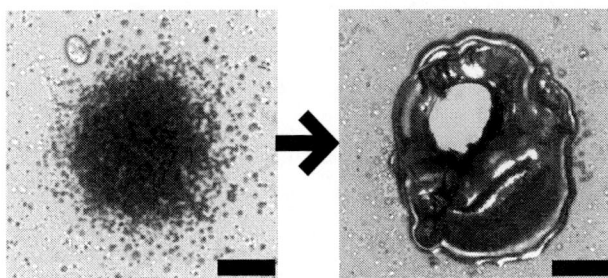

Figure 4: Laser-printed spot on a poly(ethylene terephthalate) (PET) substrate (left). A laser pulse melts the plastic foil (right) and creates a hole that allows communication between two channels. Scale bar: 400 μm.

To demonstrate the laser valve concept, we fabricated a centrifugal microfluidic "lab-on-a-disc" cartridge with two chambers connected by a microfluidic channel (Figure 5). The solution is initially loaded into the upper chamber. The disc was rotated at different speeds and no leakage was observed through the valve even while spinning at 5000 rpm. The disc was then stopped and light from the laser diode aimed at the laser-printed area, creating a communication port in less than 1 sec. The disc was spun again and the solution was fully transferred to the bottom chamber.

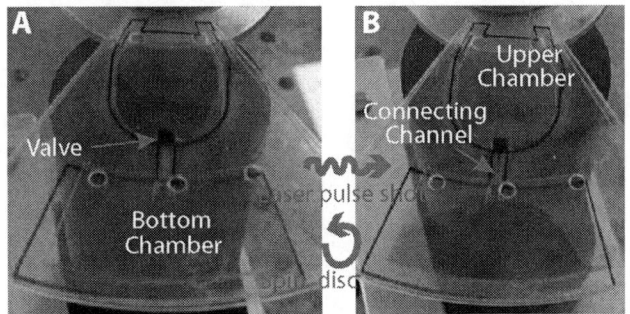

Figure 5: A centrifugal microfluidic system consisting of two chambers connected by a channel is shown. The valve is the laser-printed black dot (A). After the valve is melted, the disc is spun for a few seconds and all the liquid is transferred to the bottom chamber.

Figure 6 illustrates the system design for the on-board reagent-storage device. In this system, two reservoirs are defined near the center of the disc. Two solutions are loaded into the reservoirs and sealed with PSA-coated film. The valves are then opened and the disc spun to displace the liquid into a mixing chamber. Stored solutions did not evaporate for a period of 30 days, and suitable polymers could extend this significantly. The valves prevent fluid leakage at rotation rates of at least 5000 rpm (corresponding to 840x**g**).

Figure 3: Illustration of on-board reagent storage using laser-printed valves. The polymer foil is placed between the microfluidic connecting channel and the reservoir. Cross sectional (B) and frontal (C) views are shown. Reagent is loaded in the reservoir (D) and covered by a pressure-sensitive adhesive film (E).

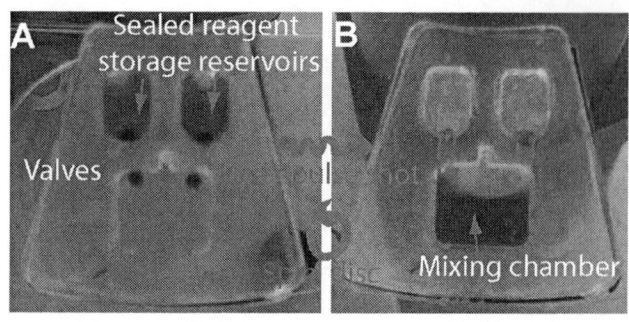

Figure 6: On-board storage of two solutions. Two solutions are initially loaded in two different compartments and sealed to prevent evaporation (A). After the laser valves are opened, the two solutions flow into the mixing chamber by spinning the disc (B).

CONCLUSIONS

A new laser-printed valve technology facilitates the design and fabrication of fully integrated and automated lab-on-a-chip cartridges that require pressure-resistant valves or long-term reagent storage. One key advantage is the absence of mechanical components in the valve and its actuation, facilitating its manufacture and use.

This technology can be adapted to multilevel microfluidics where layers of microfluidic channels are separated by valving layers. As long as the laser-printed spots do not overlap, the appropriate valve can be selected on demand and channels on different layers connected at will.

Future work will involve the full characterization of these laser valves. Melting temperatures, laser powers, and the effects of any (bio)chemically active residue of the melting process will be investigated.

ACKNOWLEDGMENTS

This work was supported by Science Foundation Ireland under Grant No. 05/CE3/B754. FBL acknowledges support from the Irish Research Council for Science, Engineering and Technology (IRCSET) under Fellowship No 2089.

REFERENCES

[1] P. Yager, T. Edwards, E. Fu, K. Helton, K. Nelson, M.R. Tam, B.H. Weigl, "Microfluidic diagnostic technologies for global public health", *Nature*, vol. 442, pp. 412-418, 2006

[2] P. Yager, G.J Domingo, J. Gerdes, "Point-of-Care diagnostics for global health", *Annu. Rev. Biomed.Eng.*, vol. 10, pp. 107-144, 2008

[3] C.D Chin, V. Linder, S.K. Sia, "Lab-on-a-chip devices for global health: Past studies and future opportunities", *Lab Chip*, vol. 7, pp. 41-57, 2007

[4] G. Whitesides, "The origin and future of microfluidics", *Nature*, vol. 442, pp. 368-373, 2006

[5] Z. Hua, R. Pal, O. Srivannavit, M. A. Burns and E. Gulari, "A light writable microfluidic "flash memory": Optically addressed actuator array with latched operation for microfluidic applications", *Lab Chip*, vol. 8, pp. 488-491, 2008

[6] E. Garcia, J.R. Kirkham, A.V. Hatch, K.R. Hawkins, P. Yager, "Controlled microfluidic reconstitution of functional protein from an anhydrous storage depot", *Lab Chip*, vol. 4, pp. 78-82, 2004

[7] R. Seetharam, Y. Wada, S. Ramachandran, H. Hess, P. Satir, "Long-term storage of bionanodevices by freezing and lypholilization", *Lab Chip*, vol. 6, pp. 1239-1242, 2006

[8] V. Linder, S.K. Sia, G.M. Whitesides, "Reagent-loaded cartridges for valveless and automated fluid delivery in microfluidic devices", *Anal. Chem.*, vol. 77, pp. 64-71, 2005

[9] P. Zucchelli, B. Van de Vyver, "Devices and methods for programmable microscale manipulation of fluid", U.S. Patent 7,152,616 (2006)

DROPLET MIXER BASED ON SIPHON-INDUCED FLOW DISCRETIZATION AND PHASE SHIFTING

R. Burger[1,2,3], N. Reis[1], J. G. Fonseca[1], and J. Ducrée[3]

[1]Biosurfit SA, Lisbon, Portugal, [2]IMTEK, Freiburg, Germany and [3]Biomedical Diagnostics Institute, Dublin City University, Ireland

ABSTRACT

We present a novel mixing principle for centrifugal microfluidic platforms. Siphon structures are designed to disrupt continuous flows in a controlled manner into a sequence of discrete droplets, displaying individual volumes as low as 60 nL. When discrete volumes of different liquids are alternately issued into a common reservoir, a striation pattern of alternating liquid layers is obtained. In this manner diffusion distances are drastically decreased and a fast and homogeneous mixing is achieved. Efficient mixing is demonstrated for a range of liquid combinations of varying fluid properties such as aqueous inks or saline solutions and human plasma. Volumes of 5 μL have been mixed in less than 20 s to a high mixing quality. One-step dilutions of plasma in a standard phosphate buffer solution up to 1:5 are also demonstrated.

INTRODUCTION

Mixing and diluting are essential steps in many assay procedures and constitute important unit operations for lab-on-a-chip platforms. In particular for point-of-care applications, mixing and dilution methods need to be fast. In contrast to macroscopic systems where liquid mixing can be easily achieved by stirring, shaking or other methods promoting turbulence in the liquid system, mixing in microfluidic systems is more challenging. Due to the small characteristic dimensions of microfluidic devices the flow is typically laminar and microfluidic mixers have to rely on diffusion and chaotic advection. Several microfluidic mixing principles have been introduced in the past [1] [2]. Among these are lamination mixers where liquids are laminated in a common channel to decrease diffusion distances. This can be further enhanced by placing obstacles in the channel or introducing curvatures and abrupt changes in cross-sectional area of the channels to promote chaotic advection or vortex mixing. Other mixers, especially suited for centrifugal microfluidics explore the Coriolis force present in rotating systems to induce secondary flows and promote mixing [3] or use periodically changing angular accelerations to perform batch mixing [4].

The novel mixing strategy presented here, performs mixing of two liquids by alternating the injection of discrete liquid volumes into a common reservoir under the impact of a centrifugal field. This approach enables mixing within short timescales (typically seconds) by generating an alternating pattern of thin lamellae which reduce the diffusion length , and also the kinetic impact of the discrete liquid volumes on the predeposited liquid volumes.

Given a liquid medium with a heterogeneous spatial distribution of species i.e. a concentration gradient ∇C , there will be a flux Φ of the species following a gradient in concentration. This flux is described by Fick's first law:

$$\Phi = -D \cdot \nabla C \qquad (1)$$

where D is the diffusion coefficient of the species in the particular medium. The time t_D required for a species to travel a distance l_D by diffusion is given by [5]:

$$t_D = \frac{l_D^2}{D} \qquad (2)$$

Equation 2 shows that diffusion times are proportional to the square of the diffusion distances and hence minimizing the diffusion length is a requirement for fast mixing. For discrete alternating layers of different liquids such as the approach being proposed, minimizing the thickness of the layers and maximizing the contact area has a significant impact in reducing the mixing time.

For continuous flow mixing this multilayer lamination approach has already been demonstrated. [1] [6]. The main advantages of the present approach is the simplicity for forming multiple thin layers from only two inlet channels plus the enhanced mixing effect of the kinetic impact of the alternate liquid volumes on a pre-deposited liquid. Moreover, the strategy is also highly suitable for mixing unequal liquid volumes such as required for dilutions.

MIXING MECHANISM

In order to create a fine striation pattern of liquids from two different inlets in a common outlet reservoir, two steps are necessary:
1. Liquid discretization in small volumes
2. Issuing the discrete volumes in alternating fashion into a common chamber to create a multi layer stack.

The first requirement is achieved using siphon-induced flow discretization which is illustrated in Figure 1.

The basic mechanism can be described as follows: a given liquid enters a reservoir featuring a siphon outlet at flow rate Q_i (Figure 1 a). Under the influence of a density dependent volume force such as centrifugal force and given that the rotational frequency is sufficiently high to suppress capillary priming of the siphon channel, liquid entering the chamber will accumulate in the reservoir (Figure 1 b). With time, the liquid levels in the reservoir and siphon outlet rise. Provided the liquid level is below the crest point, the siphon acts as a cloesed valve, retaining the liquid while the chamber is filling. Once the liquid level is above the crest point, the liquid flows outwards through the siphon channel at an outgoing flow rate Q_o (Figure 1 c). If the outgoing flow rate is higher than the incoming flow rate, the liquid level in the chamber will decrease until the liquid column breaks and gas enters the siphon (Figure 1 d). When the liquid column is disrupted, the same process resumes, thus inducing the periodic emission of droplets.

978-1-4244-2977-6/09 $25.00 © 2009 IEEE

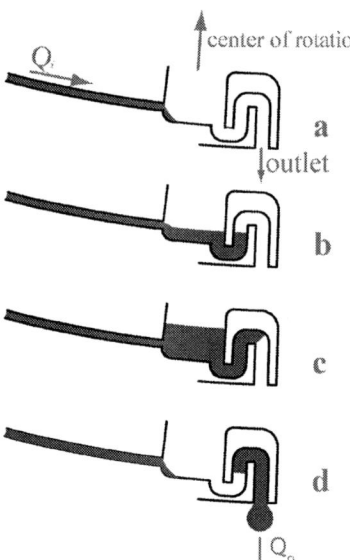

Fig 1: Working principle of the siphon-induced flow discretization. Liquid fills the discretization chamber and is held back by the outlet siphon until the crest point of the siphon is reached (a and b). When the crest point is passed (c) liquid starts flowing out at a flow rate Q_o, higher than the inlet flow rate Q_i, thus emptying the chamber (d). When the chamber is empty, the process is resumed.

Since a discrete volume / droplet is issued once the liquid in the discretization chamber reaches the crest point of the siphon, the volume of each droplet can easily be tuned varying the geometry and dimensions of the discretization chamber. The drop volume is given by $V_s + V_c$, where V_s is the volume of the siphon channel until the crest point and V_c is the volume of the chamber (see Figure 2 a). Discrete liquid volumes can be further reduced to the siphon volume alone (Figure 2 b) by eliminating the chamber provided that the relation between incoming and outgoing flow rates is maintained.

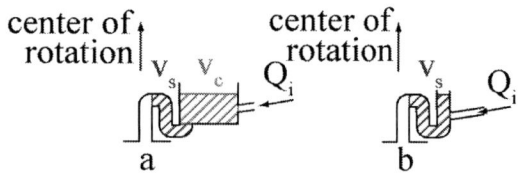

Fig 2: Definition of droplet volumes by geometrical properties of the discretization chambers.

To fulfil the second requirement (alternately issuing droplets into a common reservoir) the outlets of two of the above described discretization structures are brought together in a common reservoir, while the inlet channels are connected to different supply reservoirs (Figure 3).

The combined action of the centrifugally induced artificial gravity field as well as the kinetic impact make the droplets spread upon impact, thus forming thin lamellae (Figure 3). An alternating striation pattern evolves in the mixing chamber as a result of an arbitrary phase shift between the action of the two siphons. Non-

uniform mixing ratios can be enforced by an asymmetry between the fluidic structures defining the two discretizing structures.

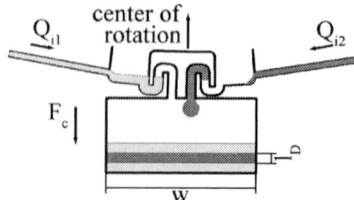

Fig 3: Schematic drawing of the siphon-induced flow discretization mixing structure.

EXPERIMENTAL

MATERIALS AND FABRICATION

Disk based microfluidic devices were fabricated by standard photolithography procedures using Dry Film Resists of different thickness (Ordyl P50000 and SY300 series, Elga Europe, Nerviano, Italy). Reservoirs and mixing structures were fabricated with depths of 100 or 120 μm (P50000 series) on top of 2 mm thick PMMA substrates (Repsol Glass, Repsol, Spain). Disk shapes and fluidic connections such as IO ports were machined by CO_2 laser ablation of the PMMA substrates. Channels were produced with depths of 30 or 55 μm (SY300series) on top of flat, 0.6 mm thick PC substrates consisting of blank DVD halves. After developing and etching the structures, disks containing reservoirs and channels were aligned and bonded by thermo-lamination. Micro fluidic devices for experimental characterisation of mixing performance contained either reservoir depths of 100 μm and channel depths of 30 μm (for aqueous ink mixtures) or reservoir depths of 120 μm and 55 μm deep channels (for human blood plasma and PBS mixtures or dilutions). PBS buffer was obtained from Invitrogen, Carlsbad, CA, USA. Blood plasma was obtained from Hytest, Turuk, Finland.

METHODS

The experimental setup for testing the fluidic structures consisted of a PC controlled servo-motor for spinning the disks (cool muscle 2, muscle corp., Osaka, Japan), a flood illumination system and a fire wire camera (Unibrain 501b, Unibrain Inc, San Ramon, Ca, USA). The camera was externally triggered by a signal from the motor and acquiring one image per rotation. This allowed for recording images of a specific position of the disk while spinning.

Mixing quality was assessed by two different methods. When mixing aqueous inks, a colorimetric method was used. After the liquids have been mixed, an image of the resulting mixture was acquired, and converted to a 8 bit grey scale image. Using an image processing software (ImageJ, NIH, USA), the area which contains the mixture was selected and the standard deviation of the histogram calculated. To compensate systematic errors due to the experimental setup, the standard deviation of a reference mixture was measured, and subsequently used to normalize the other standard deviations. The reference

978-1-4244-2977-6/09 $25.00 © 2009 IEEE

mixture was prepared by mixing the aqueous inks in the same ratios using a vortex mixer. After the mixing was performed, the same volume of liquid as obtained during the experiments was pipetted to an identical disk device and a reference image was acquired. The experimental setup consisted of an illumination table, a surrounding box to keep the illumination conditions constant and a digital camera (D 80, Nikon, Japan).

The assessment of mixing quality using plasma and PBS was done using a spectrophotometer (Nanodrop 1000, Thermo Scientific, Waltham, MA, USA). After mixing was performed, the mixture was split in 1,5 µL aliquots. The concentration of plasma proteins in these aliquots was subsequently determined using absorbance measurements at 280 nm wavelength. A perfect mixture would have the same concentration of proteins in all aliquots.

In order to collect the aliquots an additional microfluidic structure was employed. (Figure 4).

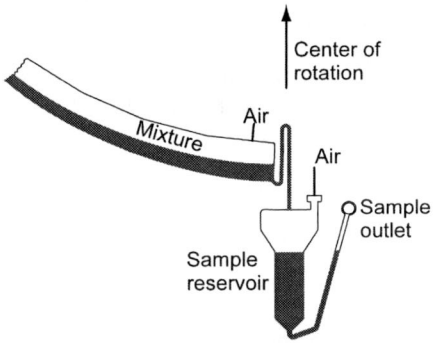

Fig 4: sample collection structure used in experiments where plasma and PBS was mixed. Under rotation liquid from the mixing reservoir is transferred to to the sample reservoir. Whenever a liquid level corresponding to 1,5 µl is reached, the disk is stopped and the liquid is collected. This process is repeated until all liquid in the mixing reservoir is collected.

Briefly, the aliquot collection procedure is done as follows: after the mixing step is performed, the disk is stopped and the siphon at the outlet of the mixing reservoir primes. When the disk resumes spinning, liquid from the reservoir is transferred into the sample reservoir. The rise of the liquid level in the sample reservoir is observed using a magnification lens coupled to the camera. The disk is stopped when the liquid level reaches a point which is known to correspond to a volume of 1,5 µL. This volume is then collected via the sample outlet using a standard micro-pipette. The process is repeated until all liquid from the sample reservoir is collected. Typically 7 to 9 aliquots were collected per mixing trial.

RESULTS

All results reported here concern mixing experiments which were carried out at a rotation frequency of 40 Hz. When mixing unequal liquid volumes (dilutions), the flow rates were adjusted via changes to the hydraulic resistance of the inlet channels in order to achieve a simultaneous emptying of both supply reservoirs. The mixing quality was evaluated immediately after performing the experiments.

Experiments using equal volumes of aqueous inks employed three different structure designs to enable discretization of volumes with 60, 180 and 360 nL. Additionally the discretization structures emptied directly either at the centre or at the side of the receiving reservoir. In all experiments 5 µl of each ink were used and mixing has been performed in less than 20 s. As expected reducing the discretization volumes leads to an increased homogeneity of mixing (Figure 5).

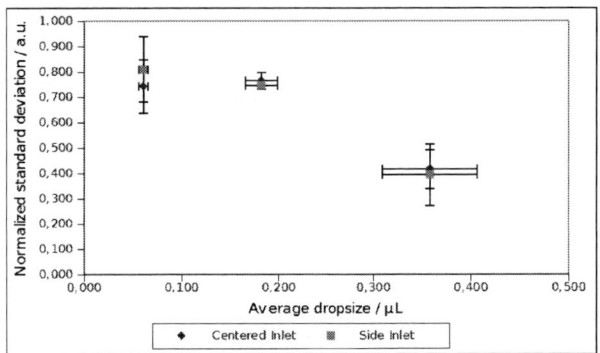

Fig 5: Normalized standard deviation (1 = perfect mixing) of mixed solutions as a function of the discrete volume and type of mixing structure. The measured standard deviation of the mixtures were normalized using the standard deviation of a sample mixed using a vortexer.

Moreover, the mixture performed with 360 nL discrete volumes took more than 40 minutes in order to reach the mixing quality obtained with 60 nL discrete volumes directly after mixing. It is also noteworthy that mixing quality of the lower discretization volumes is already close to that of the reference mixture. The position at which the discretization chamber enters the mixing reservoir was observed to be of minor importance for the mixing quality.

Figure 6 shows images collected during the mixing process with aqueous inks. There is clear evidence of liquid lamellae which are concentric around the point where the discrete volumes impinge the liquid pre-deposited in the mixing reservoir. With increasing distance from the center the boundaries between lamellae become fuzzier, indicating an ongoing mixing process.

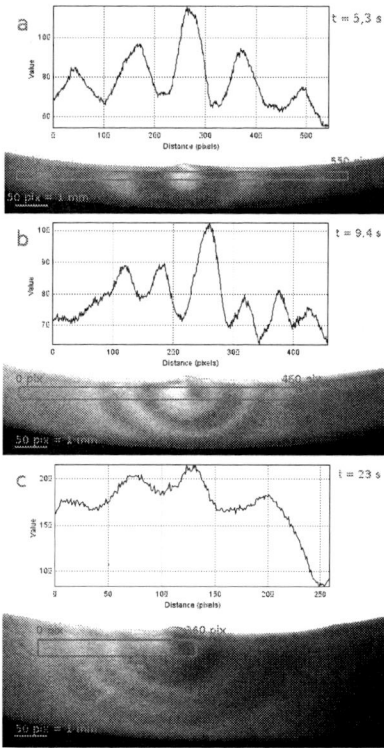

Fig 6: Image sequence of the mixing of two differently colored aqueous inks. Displayed is the center of the mixing chamber, where the droplets are issued and the histograms of the areas enclosed by the rectangles. The black part on the right side of image c is due to deficient illumination.

Since the mixing structures designed to obtain 60 nL discrete volumes showed superior mixing quality, the same discretization level was chosen for diluting plasma in PBS in ratios of 1:2.6 and 1:5. Figure 7 shows the results obtained for several samples of the two dilutions, taking into account all aliquot determinations per sample as shown in the standard deviation error bars. These results show that the approach proposed is suitable for diluting human plasma in PBS.

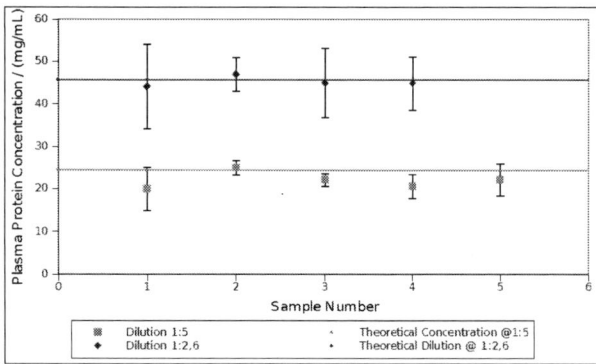

Fig 7: Results of the plasma dilution experiments. Shown is the concentration of plasma proteins in different samples. A homogeneous mixture would have the concentration which is indicated by the horizontal lines in all samples.

CONCLUSIONS

A novel mixing structure based on siphon-induced flow discretization has been developed and implemented on a centrifugal microfluidic platform. The structure is suitable for mixing liquids at volume ratios between unity and 1:5. Additionally, our mixing technology is also suitable for mixing liquids of significantly different hydrodynamic characteristics such as human blood plasma and standard buffer solutions. Furthermore, the mixer is simple to integrate and run in centrifugal microfluidic platforms, being also amenable to large scale manufacturing materials and processes such as plastic injection molding. Moreover, it does not require any surface modification processes and mixing is achieved in a fast, one-step centrifugation procedure.

REFERENCES

[1] N. T. Nguyen, Z. Wu, Micromixers – a review, J. Micromech. Microeng., vol. 15 R1-R16, 2005.

[2] A. P. Sudarsan, V. M. Ugaz, Multivortex micromixing, PNAS, vol. 103, pp. 7228-7233, 2006.

[3] S. Haeberle, T. Brenner, H. P. Schlosser, R. Zengerle, J. Ducrée, Centrifugal Micromixer, Chem. Eng. Technol., vol. 28, pp. 613-616, 2005.

[4] M. Grumann, A. Geipel, L. Riegger, R. Zengerle, J. Ducrée, Batch-mode mixing on centrifugal microfluidic platforms, Lab Chip, vol. 5, pp. 560-565, 2005.

[5] J. P. Brody, P. Yager, Low Reynolds Number Micro-fluidic Devices, Technical Digest of the IEEE Solid State Sensor and Actuator Workshop, Hilton Head Island, SC, June 3-6, pp. 105-108, 1996.

[6] J. Branebjerg, P. Gravesen, J. P. Krog, C. R. Nielsen, Fast mixing by lamination, Proc. IEEE MEMS Workshop, San Diego, CA, pp. 441-446, 1996.

AN ALL POLYMER AIR-FLOW SENSOR ARRAY USING A PIEZORESISTIVE COMPOSITE ELASTOMER

A.R. Aiyar, C. Song, S.H. Kim and M.G. Allen
Georgia Institute of Technology, Atlanta, GA, USA

ABSTRACT

This paper presents an out-of-plane micromachined piezoresistive flow sensor array based on laser micromachining of polymer films, microstencil printing, and stress-engineered curvature. The developed process is suitable for low cost, large-area sensor array fabrication, and can leverage traditional flex-circuit fabrication. Each device is composed of an out-of-plane curved microtuft formed from laser-machined Kapton® polyimide and PECVD-deposited SiO_2, and a conductive elastomer piezoresistor with a measured gage factor of 7.3 located at the base of the microtuft. The fabrication and performance of a prototype array and a fabrication sequence for large-area arrays on flexible substrates is demonstrated, for flow field mapping across an airfoil. The fabrication sequence also enables backside interconnects without adding further process complexity, which facilitates integration and enables the sensing of airflow with minimum interference due to the sensing circuitry. Individual microtufts as small as 1.5mm in length and 0.4mm in width, with 70 μm wide piezoresistor lines have been fabricated. Wind tunnel testing demonstrated sensitivities as high as 66 Ω/(m/s).

1. INTRODUCTION

Flow sensors are of paramount importance in flight control applications involving unmanned aerial vehicles (UAVs). They essentially consist of out-of-plane structures, such as artificial cilia, or in-plane shear stress sensors [1-3]. Most of these devices are fabricated using silicon, which limits the achievable height of the cilia to well below the boundary layer thickness. This gives little or no information about the flow fluctuations in the main stream [4].

The design of the sensor is based on the extensively studied and reported wind receptor hair in insects [5-6]. Air flow around the hair causes a drag force induced deflection, which in turn activates a sensory response, providing the insect with information about variations in the local flow field. Tall out-of-plane microtufts based on multiple fabrication approaches have been previously demonstrated for this application [5-7].

In this work, we report on the fabrication of a microcantilever based all-polymer flow sensor array with high flow sensitivity, while simultaneously minimizing the need for complex processes. The established fabrication process flow mimics many of the materials and processes in flex-circuit manufacturing, integrating polymer micromachining and laser MEMS techniques. This approach combines the advantages of planar processing with the 3-D capability of a tall out-of-plane flow sensing element in large area arrays on flexible substrates. The elastomer composite piezoresistive material used in the process is also discussed, with an emphasis on its mechanical and electrical characteristics. The fabrication

technique also enables a backside interconnection, which results in minimum intrusion of circuitry into the wind flow.

2. DESIGN AND OPERATION

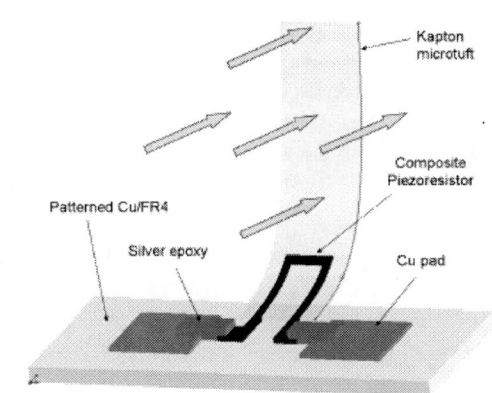

Figure 1: Sensing principle of the microtuft based flow sensor

Figure 1 shows the sensing principle of a prototype sensor. The microtuft itself is fabricated using a thin film of Kapton® (7.6 μm thick, Kapton® 30HN obtained from Dupont). The sensing element is a commercially-available conductive carbon black elastomer with a high piezoresistive gage factor (Elastosil® LR3162, obtained from Wacker Chemie AG), as is demonstrated in this paper. When air flows across the microtuft, it causes deformation of the beam structure and this in turn induces a strain in the piezoresistor. The strain eventually changes the resistance as a function of the applied air flow.

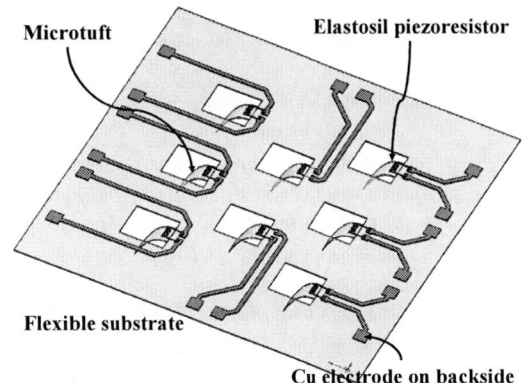

Figure 2: Array of sensors with backside interconnections

Figure 2 illustrates a 3×3 array of the all-polymer flow sensor. The microtuft will have a curvature into the substrate due to the stress-gradient in the material while all the electrical interconnections are placed on the substrate, leading to a backside interconnection scheme.

3. MATERIAL CHARACTERIZATION OF ELASTOSIL® ELASTOMER COMPOSITE

The Elastosil® composite used in the process is a silicone based elastomer loaded with carbon black particles, with a reported volume resistivity of 9 Ω-cm. It is a good candidate for the piezoresistor since it combines a low value of Young's modulus that ensures high values of strain for a given stress and electrical conductivity. In this paper, the Young's modulus and gage factor of the Elastosil® composite material are characterized before implementing the flow sensor array.

Young's Modulus characterization

The Young's modulus of the material is characterized using an Instron (Instron Engineering Corp.) tensile testing instrument.

In order to prepare samples for the tensile testing, a stencil mask is patterned from 12.7 μm thick Kapton®, using a CO_2 laser (Gravograph Newhermes), in the form of a rectangular strip of 5 mm width. The LS500XL series CO_2 laser operates at a wavelength of 10.6 μm and has a beam spot size of 160 μm. A low level of energy as well as cutting speed is used to prevent charring. The stencil is then adhered onto a 125 μm thick sheet of FEP Kapton® (FEP is Fluorinated ethylene propylene), using a spray adhesive (3M Corp.). The Elastosil® composite material itself, supplied as a two part system, is mixed in a 1:1 ratio of Part A and Part B and cured at a temperature of about 120-130℃ for 2 hours. The two ends of the 5mm wide strip are then placed on two small pieces of FR4 boards and secured in place using a quick cure 2 part epoxy (Loctite Inc.). The FR4 boards are used to prevent any damage to the sample itself by metal clamps in the Instron tool.

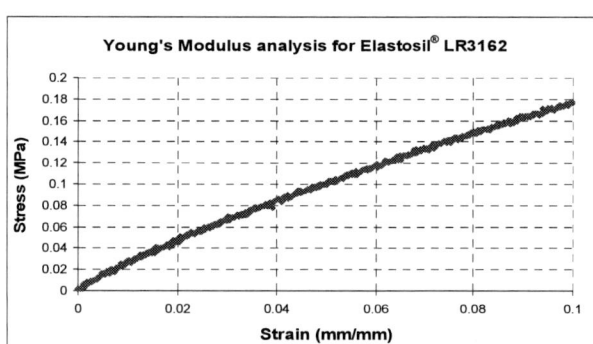

Figure 3: Stress-strain curve for a 5mm wide composite elastomer sample. The thickness of the material is 97μm and the unstrained length is 21mm. The applied strain is 10%.

Once the samples are mounted between the two clamps of the Instron, the tool is programmed to apply a constant strain rate of 0.21 mm/min. A typical stress-strain curve extracted from the load-deflection data obtained from the Instron is shown in Figure 3. The graph indicates that the stress-strain behavior of the composite elastomer is non-linear in nature. The line of best fit has an estimated slope of about 1.9 MPa and it is the approximate Young's modulus of the material.

Characterization of piezoresistive gage factor

The sample making procedure is the same as described in the previous section, except that connection between the material strip and the Cu-FR4 PCB is made using 2 part conductive silver epoxy (Epotek, Ted Pella Inc.). Wires are then soldered onto the PCB to connect the sample to an external resistance measurement unit, which consists of a multimeter (Keithely 196 system DMM), interfaced with a computer that records the resistance real-time, using a data acquisition software (HP-VEE).

The gage factor measurement was performed using a microscope (Nikon MM-40), in conjunction with a multi-axis dimensional measurement instrument (Quadra-Chek 200). The sample is taped onto the side of the microscope such that strains in the elastomer can be applied by rotating the z-axis knob. The corresponding elongation of the sample can be measured using the Quadra-Chek 200. The measurement setup is shown in Figure 4.

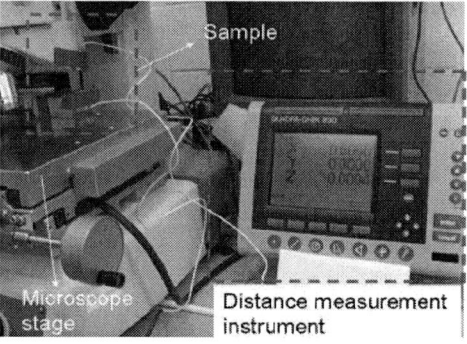

Figure 4: Experimental setup showing sample undergoing tensile testing

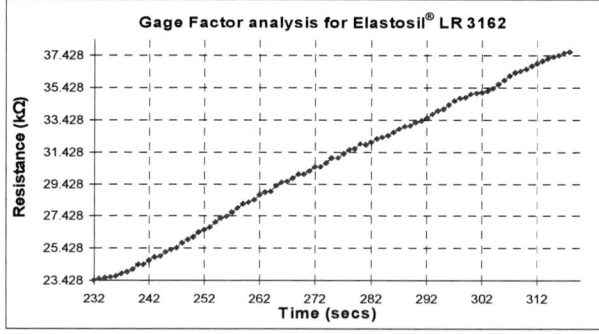

Figure 5: Loading profile for a maximum strain of 10%.

Testing is conducted on a virgin sample strip that has never seen strain. The experiment is started by first ensuring that the sample strip is just taut. Substantial time is allowed for the resistance to stabilize before any strain is applied to the sample. The sample is then continuously strained from 0% to the maximum strain, using the z-focus knob of the microscope. Tests are conducted at 1%, 5% and 10% strains and the corresponding gage factors calculated. A typical tensile test conducted on a 23mm × 5mm × 0.08mm (LxWxt) is shown in Figure 5.

The gage factor for the elastomer is calculated from the loading profile, as show in Figure 5, using the equation relating the change of resistance to the applied strain [8].

$$\frac{\Delta R}{R} = G\varepsilon$$

The average gage factor (G) calculated from a series of measurements at the above mentioned strains is about 7.3, which is considerably higher than that for metal thin films. It must be mentioned though, that errors could arise in the calculation of the gage factor, since strain values corresponding to each value of resistance are not recorded and the strain rate used is an average value.

4. DEVICE FABRICATION

The process starts with the fabrication of the base or substrate structure. The material for the base is either Kapton® 500HN (for the 1×3 test array) or 300 FN021 (for the larger 3×3 or 5×5 array). The former consists of a 125 µm thick Kapton® layer, while the latter consists of a 50.8 µm thick Kapton® layer and a 25.4 µm thick FEP (Fluorinated ethylene propylene) adhesive layer for lamination purposes.

The first step involves machining of square cavities in the base Kapton®, which act as release holes for releasing the microtufts in the final step (Figure 6(a)). This is achieved by ablating the Kapton® using the CO_2 laser.

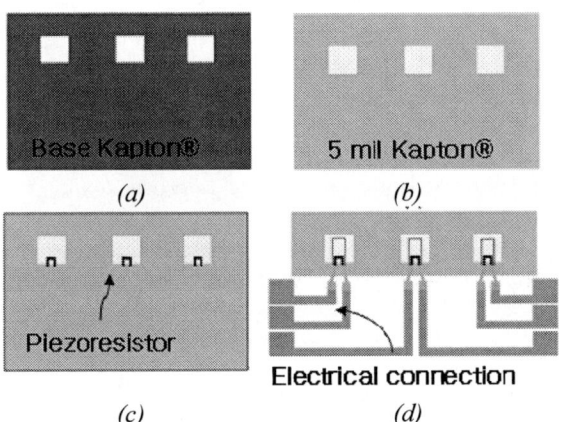

Figure 6: Fabrication process sequence for 1×3 array

The substrate structure or the base is then laminated to the 7.6 µm thick device Kapton® layer (Kapton® 30HN) at a temperature of 280℃ and a pressure of 1.4 MPa, using a bench top lamination press (Carver Inc.). For the 1×3 array, the substrate layer is adhered to the device layer using a spray adhesive (3M Corp.) (Figure 6(b)).

The next step in the process is patterning the piezoresistive composite material on the laminated substrate. This is achieved by a microstenciling process, whereby the elastomer material is doctor-bladed through a stencil mask onto the substrate. The stencil mask itself is fabricated by the Excimer laser ablation of 12.7 µm thick Kapton®. The LPX2000 series excimer laser (Lambda Physik) operates at a wavelength of 248 nm and a spot size of 10 µm is obtained through the demagnification optics. After stencil printing of the conductive elastomer, it is cured for about 2 hours at a temperature of 130℃ (Figure 6(c)). Interconnections between the conductive elastomers and the external circuit are obtained either using silver epoxy, in the case of the 1×3 array, or photolithography-based metal deposition for the larger area arrays.

In order to improve the flow-structure interaction, a layer of SiO_2 is deposited on the film using plasma enhanced chemical vapor deposition (PECVD). The residual stress induced due to the oxide deposition produces a curvature in the Kapton®, upon release in the final step of the process (Figure 6(d)).

The final release of the microtufts is achieved using the Excimer laser described before. The laser spot is aligned to features on the substrate and the structure is released by ablating the film. The residual stress causes the microtuft to bend out of or into the plane depending on which side of the substrate the oxide is initially deposited. Figure 7 shows photographs of two devices fabricated with microtuft lengths of 3.5 and 1.5 mm respectively.

The fact that the direction of the curvature (either into or out-of-plane) can be controlled simply by changing the face on which the SiO_2 is deposited, can be used to our advantage for the implementation of the proposed backside interconnects. In this scheme, shown in Figure 8, which is used for the large area arrays, the active circuitry, including all piezoresistors are on the backside of the chip, while microtufts are the only structures that are exposed to the wind flow, causing minimal disturbance of the flow profile.

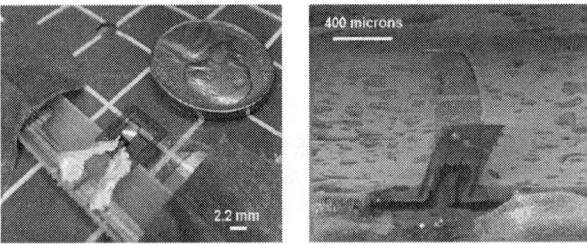

Figure 7: Devices fabricated using the "1×3" array fabrication process

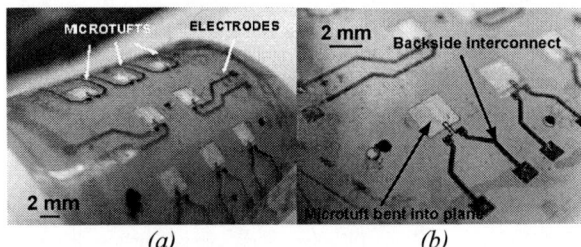

Figure 8: (a) 3×3 flexible sensor array (b) close up demonstrating the backside interconnects. The microtufts are bending into the plane of the paper

5. WIND TUNNEL TESTING

The fabricated devices are tested in a bench top wind tunnel (ST 180 Scantek 2000), shown in Figure 9. The mean free stream velocity in the wind tunnel is measured using a thermal anemometer (Omega FMA-605-I) with a range of 0-5000 SFPM (0-25 m/s), operating between 4-20 mA and this is used as a reference. All the data are recorded in the form of a resistance measurement using a digital multimeter (Keithley).

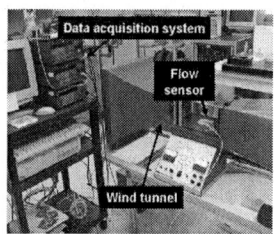 

Figure 9: Wind tunnel testing of the fabricated flow sensors

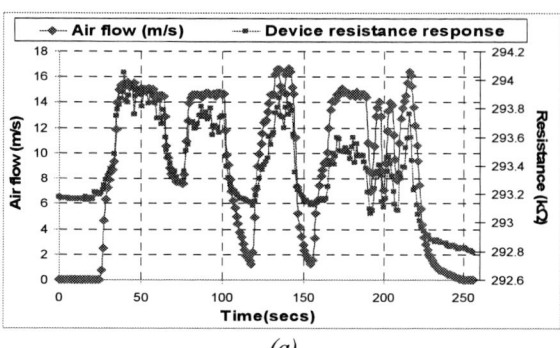

(a)

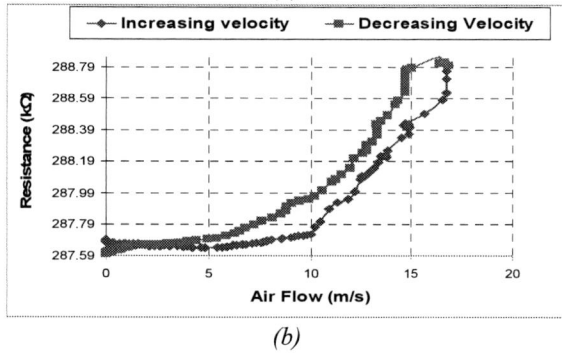

(b)

Figure 10: (a) Device response showing correlation between the wind velocity and the resistance profile (b) Device response with increasing and decreasing wind velocity. The device sensitivity calculated from the second graph is approximately 66 Ω/(m/s).

The sensor has a maximum resistance change of about 1.1 kΩ leading to a device sensitivity of approximately 66 Ω/(m/s) as shown in Figure 10. The device shows good time response characteristics, though there is resistance hysteresis and drift associated with the piezoresistive response, which may well be the signature of carbon-black elastomer composite materials, as has been extensively reported by other researchers [9,10]. These issues are currently being investigated.

6. CONCLUSION

The proof-of-concept array demonstrates the simple fabrication, low cost, and significant sensor response offered by this approach. Viable deployment, in concert with interface circuitry on the backside, is further supported by its backside interconnect scheme, keeping the circuit elements from interfering with the measured flow, while simultaneously affording the extra protection of the sensor and circuitry. The fabrication has also been extended to large area (3×3) arrays on flexible circuits.

Table 1: Characteristics and performance of the microtuft based flow sensor

Device dimensions	1.5 mm (l) × 0.4 mm (w) × 7.6 µm (t)
Maximum change in device response	~ 1.1 kΩ
Maximum change in windflow	16.9 m/s
Sensitivity	66 Ω/(m/s)

ACKNOWLEDGEMENTS

This work was supported in part by the U.S. Air Force under the Multidisciplinary University Research Initiative.

REFERENCES

[1] A. L. Huang, J. Tai, and C. M. Ho, "Microsensors and actuators for macrofluidic control," *Sensors Journal, IEEE*, vol. 4, pp. 494-502, 2004.

[2] F. Jiang, G. B. Lee, Y. C. Tai, and C. M. Ho, "A flexible micromachine-based shear-stress sensor array and its application to separation-point detection," *Sensors & Actuators: A. Physical*, vol. 79, pp. 194-203, 2000.

[3] J. Chen, Z. Fan, J. Zou, J. Engel, and C. Liu, "Two-Dimensional Micromachined Flow Sensor Array for Fluid Mechanics Studies," *Journal of Aerospace Engineering*, vol. 16, pp. 85, 2003.

[4] Z. Fan, J. Chen, J. Zou, D. Bullen, C. Liu, and F. Delcomyn, "Design and fabrication of artificial lateral line flow sensors," *Journal of Micromechanics and Microengineering*, vol. 12, pp. 655-661, 2002.

[5] Y. Ozaki, T. Ohyama, T. Yasuda, and I. Shimoyama, "An air flow sensor modeled on wind receptor hairs of insects," *Micro Electro Mechanical Systems, 2000. MEMS 2000. The Thirteenth Annual International Conference on*, pp. 531-536, 2000.

[6] R. J. F. Wiegerink, A. Jaganatharaja, R. K. Izadi, N. Lammerink, T. S. J. Krijnen, and J. M. Gijs, "Biomimetic Flow-Sensor Arrays Based on the Filiform Hairs on the Cerci of Crickets," *Sensors, 2007 IEEE*, pp. 1073-1076, 2007.

[7] Y. H. Wang, C. Y. Lee, and C. M. Chiang, "A MEMS-based Air Flow Sensor with a Free-standing Micro-cantilever Structure," *Sensors*, vol. 7, pp. 2389-2401, 2007.

[8] S. Senturia, *Microsystem Design*, Kluwer Academic Press, 2002.

[9] X. W. Zhang, Y. Pan, Q. Zheng, and X. S. Yi, "Time dependence of piezoresistance for the conductor-filled polymer composites," *Journal of Polymer Science Part B Polymer Physics*, vol. 38, pp. 2739-2749, 2000.

[10] K. Yamaguchi, J. J. C. Busfield, and A. G. Thomas, "Electrical and mechanical behavior of filled elastomers. I. The effect of strain," *Journal of Polymer Science, Part B, Polymer Physics*, vol. 41, pp. 2079-2089, 2003.

978-1-4244-2977-6/09 $25.00 © 2009 IEEE

DESIGN AND CHARACTERISATION OF A HYDRAULIC MICROACTUATOR FABRICATED BY LITHOGRAPHY

M. De Volder[1], F. Ceyssens[2], D. Reynaerts[1] , and R. Puers[2]

[1]Katholieke Universiteit Leuven, Dept. Mechanical Engineering, PMA, Leuven, BELGIUM
[2]Katholieke Universiteit Leuven, Dept. Electrical Engineering, MICAS, Leuven, BELGIUM

ABSTRACT

To improve the force output of microactuators, this work focuses on actuators driven by pressurized gasses or liquids. Despite their well known ability to generate high actuation forces, hydraulic actuators remain uncommon in microsystems. This is both due to the difficulty of fabricating these microactuators with the existing micromachining processes and to the lack of adequate microseals. This paper describes how to overcome these limitations with a combination of anisotropic micromachining, UV definable polymers and low temperature bonding. The functionality of these actuators is proven by extensive measurements which showed that actuation forces of 0.1 N can be achieved for actuators with an active cross-section of 0.15 mm^2. This is an order of magnitude higher than what is reported for classic MEMS actuators of similar size.

INTRODUCTION

Future microrobotic applications require actuators that can generate a high actuation force and stroke in a limited volume. Recent research revealed that hydraulic and pneumatic microactuators can develop high force and power densities at microscale [1-10]. Despite these promising characteristics, hydraulic actuators are rare in microsystem technology. Inflatable balloon or membrane actuators are the only hydraulic actuators that are readily used in microsystems [6-10]. These actuators are relatively easy to fabricate, but they are limited to short actuation strokes, and show non-linear actuation characteristics. In large scale applications, piston-cylinder pneumatic actuators are used because they can achieve relatively large strokes and their actuation force is a linear function of the applied pressure. Nevertheless, appropriate fabrication and seal technologies need to be developed in order to miniaturize these actuators. In the past we reported on how these actuators can be sealed relying on surface tension [1], ferrofluids [2], lipseals [3] or clearance seals [4, 5]. These actuators were fabricated by piece-wise production methods such as micromilling and micro-EDM [1-4] (see figure 1a and 1b). In this paper, we discuss how these actuators can be fabricated in batch using lithography (see figure 1c.).

The functionality of these actuators is proven by extensive measurements which confirm their exceptionally high force densities. Therefore, hydraulics is possibly the key technology to develop new microdevices that require high actuation forces. Particularly microrobots for assembly and inspection applications, microfactories, tactile displays, and tools for minimally invasive surgery can benefit from these new microactuators. For these applications we target an actuator with a cross section of 0.5 mm^2 (including the packaging) that can generate forces in the order of 0.1 N. Nevertheless, smaller

actuators with a cross section of 0.15 mm^2 have been fabricated as well. In what follows the design, fabrication and characterization of these devices will be discussed into more details.

Figure 1: (a.) Hydraulic actuator with an outside diameter of 1.5 mm (b.) Tungsten carbide microactuator with an outside diameter of 125 µm (c.) Microactuator with a piston of 180 by 400 µm fabricated in batch using lithography.

ACTUATOR DESIGN AND FABRIACTION

The fabrication of piston-cylinder actuators by lithographic processes requires a redesign of the actuator geometry. For instance, pistons with a square cross section instead of a circular one are used as illustrated in figure 2. The "cylinder" is in this case the cavity in the substrate that contains the piston. The pressure supply channels are fabricated in the same substrate as the actuator. In figure 2, the middle channel is used to push the piston out of the "cylinder", while the two outer channels can be pressurized to return the piston in the "cylinder". The latter channels are arranged symmetrically around the "cylinder" in order to prevent side-loading of the piston.

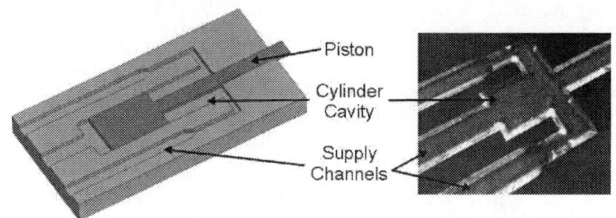

Figure 2: Left: Overview of the developed hydraulic actuators (cover not shown). Right: Substrate with a "cylinder" cavity and supply channels.

An important component of these actuators is the seal that prevents the driving fluid from leaking. In this case, clearance seals are used which rely on a microsized gap between the piston and the "cylinder" to control the leakage [4]. An important advantage of clearance seals is that a trade-off can be made between the fabrication accuracy and leakage of the actuator. For instance, in several medical applications a certain amount of leakage is not detrimental if the actuator is driven by physiological

978-1-4244-2977-6/09 $25.00 © 2009 IEEE

water [10]. Consequently, for those applications, clearances of a few micrometers between the piston and the "cylinder" are acceptable. However, the leakage can virtually be reduced to zero by reducing the piston-cylinder clearance.

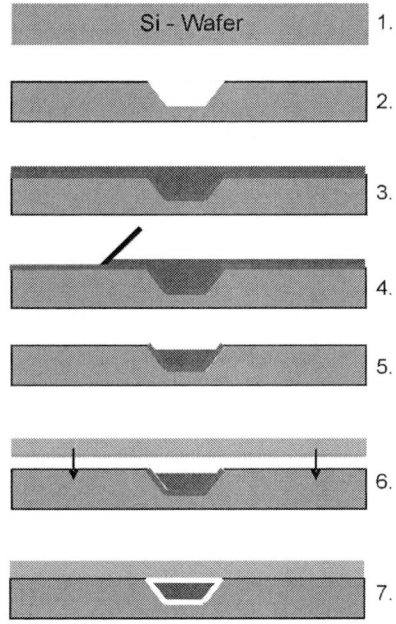

Figure 3: Overview of the production process. (1. Si wafer with thermal SiO$_2$. 2. KOH etch with convex corner compensation. 3. Molybdenum sputtering and Ormocomp moulding. 4. Ormocomp planarisation. 5. O$_2$ plasma etch. 6. Low temperature lamination bonding. 7. Dicing and sacrificial layer removal).

In what follows, the fabrication process of the actuators will be discussed. As illustrated in figure 3, the "cylinder" and supply channels are defined by a KOH etch. Corner compensation structures are designed similar to Fan et al. to prevent under etching of convex corners [11, 12]. Figure 4 shows a picture of mask and the substrate after a KOH etch of 150 μm deep for different compensation structure sizes.

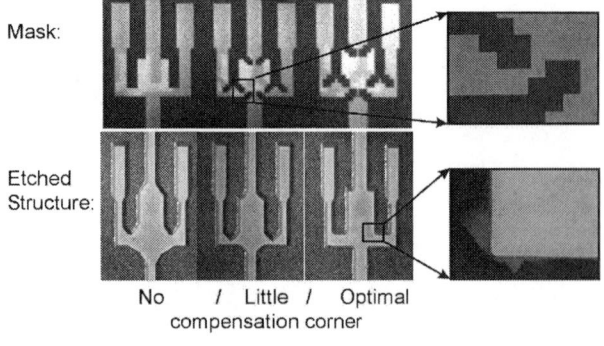

Figure 4: Illustration of the design of corner compensation structures (top), and their influence on the "cylinder" geometry after etching (bottom).

Next, a sacrificial molybdenum (Mo) layer of about 3 μm thickness is sputtered on the wafer. This layer will define the gap between the piston and the "cylinder" and therefore also the leakage rate. Next the pistons are fabricated by molding and patterning Ormocomp [13] in the KOH trenches. In step 5, a plasma etch is used in order to reduce the height of the pistons so that they will not make contact with the cover glass plate during the consecutive bonding process (see step 6). A short H$_2$O$_2$ dip is used to remove the Mo sacrificial layer form the wafer. This dip is short so that the Mo is not removed from under the pistons (see step 5 in figure 3). As a result, the pistons stay attached to the substrate during further processing. The plasma etch and the subsequent H$_2$O$_2$ etch ensure that the wafer surface is clean for bonding. In step 6, the wafer is bond to a glass plate using a low temperature thermo-compression process with a partially cross-linked Ormocomp layer at 150 °C and 100 kPa during 3 hours. This bonding process is strong enough to resist the final dicing saw treatment, and is a key for the successful fabrication of these hydraulic actuators since the pistons cannot resist the high temperatures needed for instance in anodic bonding. Finally, the pistons are released from the "cylinder" by etching the sacrificial Mo layer with H$_2$O$_2$. Different actuator sizes have been fabricated, but a typical piston has a height of 150 μm and a width of 1 mm. Figure 3 shows a prototype actuator, and the inset is showing that the piston can be moved in and out the "cylinder". This is the first report of a double acting piston-cylinder actuator fabricated by lithography.

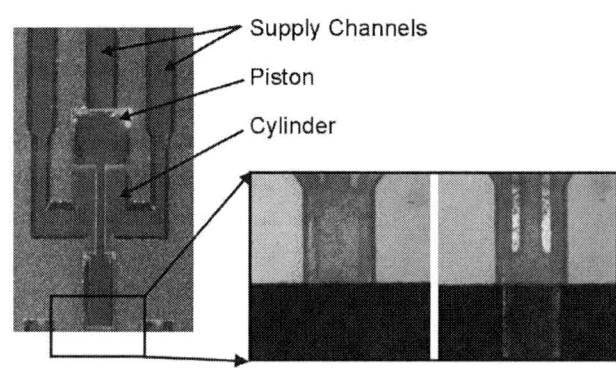

Figure 5: Left: Actuator with an Ormocomp "cylinder" before bonding and dicing. Right: Illustration of the piston moving in an operational actuator.

Two alternative routes for fabricating hydraulic microactuators have been investigated, and will be discussed briefly. In the first alternative approach, the actuator has the same geometry as depicted in figure 2, but the piston is made out of Ni by electroplating. This fabrication process is illustrated in figure 6 left: First, the "cylinder" and supply channels are realized by KOH etching. In the second step, a Ti-Cu-Ti trilayer is sputtered onto the substrate to form a seed layer for electroplating. Next the "cylinder" is filled with SU8. The SU8 layer is exposed and developed in such a way that the entire wafer is covered with SU8 except for the area where the piston needs to be electroplated. This SU8 masking method allows to electroplate Ni pistons in the "cylinder" as illustrated in step 3 of figure 6. The above approach allows fabricating a piston that perfectly matches the dimensions of the "cylinder". In step 4, the piston is

978-1-4244-2977-6/09 $25.00 © 2009 IEEE

planarized by polishing and the SU8 layer is removed. Finally, the actuator is packaged similarly to the actuator described above and the piston is released by etching the *Ti-Cu-Ti* layer. Several attempts have been made to obtain operational devices by this fabrication process. However, the polishing step revealed to be particularly delicate because it pulls the piston out of the *KOH* trenches resulting in a very low yield.

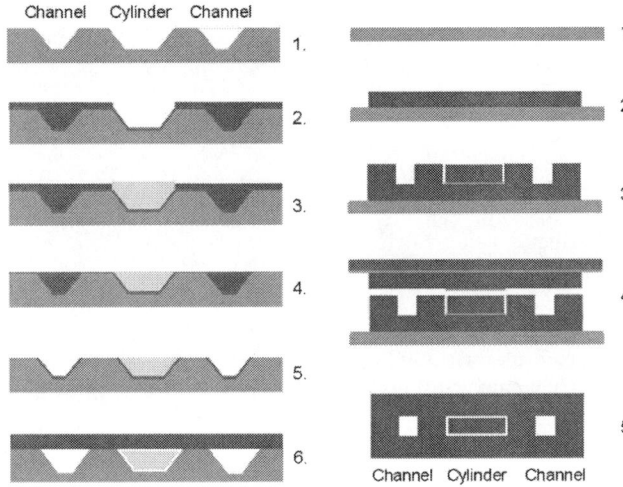

Figure 6: Left: Fabrication process based on electroplating in a KOH trench masked by SU8. Right: Fabrication process for actuators entirely made out of SU8 epoxy.

A second, more successful alternative fabrication process is shown on the right hand side of figure 6. This approach is based entirely on SU8 processing eliminating the need for dicing. In step 1, a substrate (here a *Si* wafer) is covered by a sacrificial *Mo* layer. Next three SU8 layers are sandwiched to form the piston and "cylinder". The first layer (see step 2) is a SU8 layer to support the device. On that layer a second sacrificial layer is deposited, enabling the release of moving parts later on. On top of this layer, a second SU8 layer that defines the "cylinder", the piston and the supply channels is deposited (step 3). The geometry of the piston-cylinder is shown in the left hand side of figure 7. Next, a sacrificial layer (LOR10B) and a third SU8 layer are spun on a second substrate and patterned (step 4). Then, a 3 µm thick SU-8 spacer layer is added. This second wafer is bonded to the first by thermocompression bonding at 3 bar and 150 °C, sealing the actuator. Finally, the piston and then the entire structure is released by etching the sacrificial layers. At this point the actuators are detached from the substrate and no further dicing steps are required. Figure 7, right shows a prototype hydraulic microactuator fabricated entirely out of SU8 by the production method described above. Moving actuators with square 100 micron wide "cylinders" and side gaps down to 5 microns have been produced.

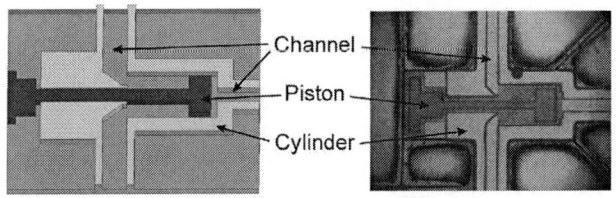

Figure 7: Left: Overview of an hydraulic actuator made entirely out of SU8. Right: Prototype actuator after step 3 of the production process.

MEASUREMENTS

In what follows, we will focus on the characterization of the KOH-Ormocomp actuators fabricated following the procedure described in figure 3. In a first experiment, the stroke of the actuator is determined to be 750 µm using an optical microscope. Next, the actuation force is measured using both pressurized air and water up to pressures of 1.6 MPa. The same test setup as described earlier on in [3] has been used to characterize the actuators. This setup consists of a data acquisition system that records simultaneously the actuation force measured by a custom made high sensitivity strain gauge bridge, and the pressure monitored by a Si-based commercial pressure sensor. The actuator is positioned on a micrometer stage in order to be able to measure the actuation force for different strokes of the actuator. These measurements result in force-pressure plots illustrated in figure 8. This figure shows that actuation forces of 0.1 N have repeatedly been reached at pressures of 1.6 MPa and that there is a relatively large amount of hysteresis on the force output. This hysteresis is mainly due to friction between the piston and the "cylinder". These measurements were obtained with an active piston area of 0.15 mm² and a cross section of 0.5 mm² inclusive the packaging. Similar results have been achieved using pressurized water as driving fluid.

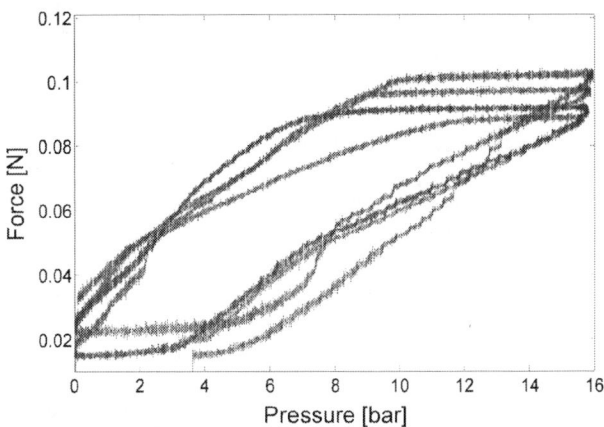

Fig. 8. Output force as a function of the applied function measured for 4 different strokes of the actuator using pressurized air as driving fluid.

Figure 9 compares the output force per cross section area as a function of the actuator volume for different actuation technologies reported in literature. This figure shows that the output force of the actuator developed in this paper is an order of magnitude higher than what is achieved by classic MEMS actuators [5-10, 14-17]. Since this actuator operates at high pressures, higher actuation forces are achieved than typically reported for pneumatic microactuators in literature. High pressure hydraulics is therefore possibly the key technology to develop new microdevices that require high actuation forces. On-going research is investigating the integration of more advanced sealing technologies [1-3] in the above actuators and the incorporation of microvalves such as [18] in the same substrate as the actuator.

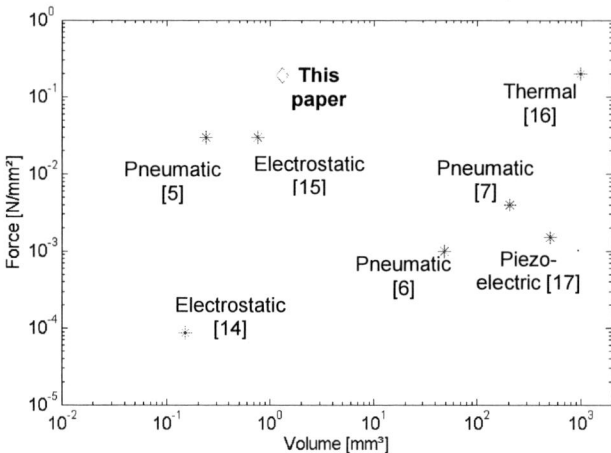

Fig. 9. *Comparison of the output force per cross section area as a function of the actuator volume for different actuation technologies.*

CONCLUSIONS

This paper introduces a new batch fabrication process for piston-cylinder hydraulic and pneumatic microactuators. The actuators consist of KOH trenches filled with pistons made out of Ormocomp or alternatively actuators made entirely out of SU8. The production process is validated by fabricating and testing several actuator prototypes. Actuation strokes of 750 μm and the actuation forces of 0.1 N have been achieved at supply pressures of 1.6 MPa. This actuation force is higher than that of typical electrostatic and thermal microactuators with similar dimensions. Furthermore, the actuation force can still be increased by using higher driving pressures and reducing piston-cylinder friction. Particularly microrobots for assembly and inspection applications, microfactories, tactile displays, and tools for minimally invasive surgery can benefit from these hydraulic microactuators.

ACKNOWLEDGEMENTS

This research was supported by the Fund for Scientific Research – Flanders (FWO). The authors would also like to thank J. Avau for his help (master thesis student at ESAT).

REFERENCES

[1] M. De Volder, J. Peirs, D. Reynaerts, J. Coosemans, R. Puers, O. Smal, B. Raucent, "A Novel Hydraulic Microactuator Sealed by Surface Tension", *Sensors and Actuators A,* Vol. 123-124, pp. 547-554, 2005.

[2] M. De Volder, D. Reynaerts, "A Ferrofluid seal technology for fluidic microactuators", *Proc. of SPIE Vol. 6415*, pp. 64150X-1, 2006.

[3] M. De Volder, F. Ceyssens, D. Reynaerts, R. Puers, "A PDMS lipseal for hydraulic and pneumatic microactuators", *J. Micromech. Microeng.* Vol 17, pp. 1232-1237, 2007.

[4] M. De Volder, J. Coosemans, R. Puers, D. Reynaerts, "Characterisation and control of a pneumatic microactuator with an integrated inductive position sensor", *Sensors and Actuators A*, Vol. 141, pp. 192 – 200, 2008.

[5] A. A. Norton, M. A. Minor, "Pneumatic microactuator powered by the deflagration of sodium azide", *Journal of Microelectromechanical systems,* Vol. 15, pp. 344 - 354, 2006.

[6] S. Buetefish, V. Seidemann, S. Buttgenbach, "Novel micro-pneumatic actuator for MEMS", *Sensors and Actuators A* 97-98, pp. 638-645, 2002.

[7] S. Konishi, F. Kawai, P. Cusin, "Thin flexible end-effector using pneumatic balloon actuator", *Sensors and Actuators A,* Vol 89, pp. 28-35, 2001.

[8] Y.-C. Su, L. Lin, "A water-powered osmotic microactuator", *J. Microelectromech. Syst.*, Vol. 11, pp. 736 - 742, 2002.

[9] K. Takemura, S. Yokota, K. Edamura, "Development and Control of a micro artificial muscle cell using electro-conjugate fluid", *Sensors and actuators A*, Vol. 133, pp. 493-499, 2007.

[10] H. Okayasu, J. Okamoto, M.G. Fujie, M. Umezu, H. Iseki, "Development of a hydraulic-driven flexible manipulator for neurosurgery", *International Congress Series 1256*, pp. 607 - 612, 2003.

[11] W. Fan, D. Zhang, "A simple approach to convex corner compensation in anisotropic KOH etching on a (100) silicon wafer", *J. Micromech. Microeng.* Vol. 16, pp. 1951 – 1957, 2006.

[12] B. Puers, W. Sansen, "Compensation structures for convex corner micromachining in silicon", *Sensors & Actuators*, Vol. A, 23, pp.1026-1041, 1990.

[13] M. Popall et al., "ORMOCER - New Photo-Patternable Dielectric and Optical Materials for MCM-Packaging", *IEEE Electronic Components and Technology Conference*, pp. 1018 – 1025, 1998.

[14] R. Yeh, S. Hollar, K. S. J. Pister, "Single mask, large force, and displacement electrostatic linear inchworm motors", *Proc. MEMS 2001 conf.*, pp. 260 – 264, 2001.

[15] T. A. Saif, N.C. MacDonald, "A millinewton microloading device", *Sensors and Actuators A*, Vol. 52, pp. 65 – 75, 1996.

[16] J. Peirs, D. Reynaerts, H. Van Brussel, "Design of a shape memory actuated endoscopic tip", *Sensors and Actuators A* 70, pp. 135-140, 1998.

[17] Website of the company Klocke Nanotechnik: www.nanomotor.de

[18] M. De Volder, K. Yoshida, S. Yokota, D. Reynaerts, "The use of liquid crystals in microsystems: model and measurements", *J. Micromech. Microeng.*, pp. 612 – 619, 2006.

BIOMIMIC LOW HYSTERESIS SURFACE PREPARED BY HIERARCHICAL MICRO/NANO TEXTURES

M. -H. Chen[1] , T. -H. Hsu[2], F. -G. Tseng[1,2,3]

[1]Institue of NanoEngineering and MicroSystems, National Tsing Hua University, Hsinchu 30013,

[2]Department of Engineering and System Science, National Tsing Hua University, Hsinchu 30013,

[3] Research Center for Applied Sciences, Academia Sinica, Taipei 11529,

TAIWAN R.O.C

ABSTRACT

In this work, a low hysteresis surface prepared by hierarchical textures for the reduction of liquid droplet adhesion is proposed. This surface is important to provide a platform for droplet-based biological reactions with minimum surface interferences. The hierarchical textures, mimicking lotus leaf surface, were composed of two length scales (nano-/micro- textures). Hereby, a high contact angle (160°) and low hysteresis (~2.7°) was obtained and comparable to the surface properties of lotus leaf. In dynamic contact angle analysis by droplet impinging onto the surface, multi-metastable states were categorized and showed a high correlation to the multi-length scale that would be noticeable for contact angle control.

INTRODUCTION

In recent years, superhydrophobic (SHP) surface was noticed on nature "lotus effect" [1] and introduced to many applications due to the self-cleaning phenomenon provided by its extremely large water contact angle and low hysteresis [2,3]. Recently, research interests were very much focused on the fabrication of SHP surface [4-7] and the theoretical analysis of static droplet on nano- or micro- textures [8-10]. However, the understanding of dynamic process for droplet action on SHP surface is more useful for real applications [11-13], such as droplet based bioassay for lab on a chip systems, but lacking much research. *Z. Wang* and coauthors proposed their observation of dynamic droplet impinging on carbon nanotube arrays composed by mono-length scale nanotextures [14]. Nature lotus leaf, however, showed a hierarchical surface constructed with multi-length scale textures. *L. Gao* and coauthors showed a preliminary result that hysteresis on surface with two length scales topography could be obviously decreased due to the increasing of trapping air [15]. However, those surfaces can not guarantee the bouncing back when droplets are wetting into either the micro or nano structures during actuation, greatly losing droplet control capability. Therefore, this work studies the dynamic process of droplet impacting on a biomimic SHP surface with micro/nano hierarchical textures and wish to find out controllable conditions for droplet manipulation on such a surface without wetting and much surface interaction.

THEORY

The theoretical analysis to model the apparent contact angle of sessile droplet on a rough surface was proposed by Wenzel [16] and Cassie [17]. In Wenzel model, it is assumed that the rough surface is fully wetted by liquid and the apparent contact angle θ_W is given by

$$\cos \theta_W = r \cos \theta_0 \tag{1}$$

where r is a ratio of the actual contact area of solid-liquid interface to the projected area on the horizontal plate and θ_0 is the contact angle of the liquid droplet on the flat surface.

In Cassie model, it is assumed that the droplet is placed on the rough surface but partially wetted the surface, means the droplet stay on the top of texture and air pocket is trapped under droplet so as to form a composite interface. The apparent contact angle θ_C is then given by

$$\cos \theta_C = f \cos \theta_0 + (1-f) \cos \theta_{air} \tag{2a}$$

where f is the area fraction of solid-liquid contact and θ_{air} is the contact angle in free space of 180°. Consequently, the Eq (2a) can be rewritten as

$$\cos \theta_C = f(1 + \cos \theta_0) - 1 \tag{2b}$$

A SHP surface, defined with high contact angle larger than 150° and low hysteresis less than 10°, is followed by the Cassie model due to the air trapping underlying droplet. A droplet in Wenzel state, however, dramatically increases the solid-liquid contact area and induces a large hysteresis that is quite different from "lotus effect". As we gently place a droplet on such a rough surface, the composite layer is obtained so that the fraction of solid-liquid contact area is decreased. In order to perform a high contact angle surface, the lower f and larger θ_0 are required. The θ_0 of flat surface, however, can reach only 120° and limit the construction of SHP surface. Nature system such as plant leaf of lotus and bamboo achieve SHP property with different strategy which the surface roughness is composed with two length-scale textures. The utility of multi-length scale can stabilize a SHP surface to prevent the dynamic impinging from environment which is more practical problem as applying a SHP property [15,18].

As considering the wetting process on a two length scales induced SHP surface, we suggest that the droplet enforced into textures may show a hierarchical step wetting owing to the existing of different roughness dimensions. The impalement Laplace pressure P_{imp} is given by

$$P_{imp} = \gamma C \tag{3}$$

where γ is liquid-air surface tension and C is the curvature of the liquid-air meniscus which is correlated to the pitch of textures. As a result, impalement Laplace pressure of microtextures may be obtained lower than that of nanotextures and it is suggested that there should be divided into three wetting schemes: non-wetting, micro-wetting, and nano-wetting (fully wetting).

EXPERIMENTAL
The Preparation of Biomimic SHP (BSHP) Surface

The biomimic SHP (BSHP) surface was composed of nanopillars superposed on microbumps as a hierarchical roughness. The fabrication process is depicted in Figure 1.

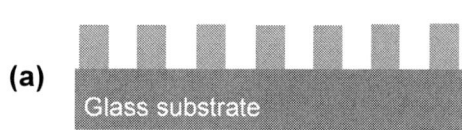

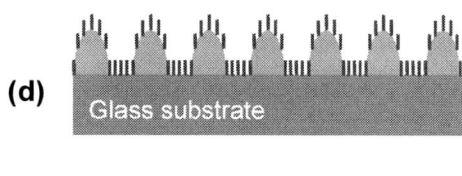

Figure 1: Fabrication process of SHP chip. (a) definition of AZ-9260 microtexture; (b) thermal reflow to rounded the microtexture; (c) Parylene C conformal coating; (d) after special RIE etching process [19], nanopillars was superposed on microbumps.

First, micro-texture of positive-tone AZ-9260 photoresist was defined on 4-inch glass substrate with photolithography and followed by thermal reflow in oven at 150°C for 3 min to form top-rounded microbumps. The pillar-like nano-texture was fabricated by reactive ion etching (RIE) with dummy material (cover glass) from which the nanomasks were sputtered onto Parylene C film during RIE etching [19], called "self-masked high-aspect-ratio polymer nanopillars fabrication process" (SMHAR process). The plasma was composed with CF_4 and O_2 in a ratio of 5:3 as well as the RF-power and pressure were 150 W and 150 mTorr, respectively. The formation of nanopillars on microbumps by SMHAR process was rapid and conformal that meets the needs as preparing the BSHP surface in this study.

The SEM image of the BSHP chip with 20-minute RIE etching is shown in Figure 2. The topography of

microbumps was 13.8 μm both in height and width with rounded top of 6.9 μm in radius as well as the nanopillar was 1.2 μm in length and 120 nm in diameter. The BSHP surface revealed a high water contact angle (CA) of 160° and a low hysteresis (θ_{hys}) of ~2.7°. From Eq(2b), f of the BSHP surface is 0.2 and θ_0 is replaced by the contact angle on nanopillars θ_{nano}, where $\theta_{nano} = 146°$, and the we can get a theoretical contact angle of 165°, compared to the measured result.

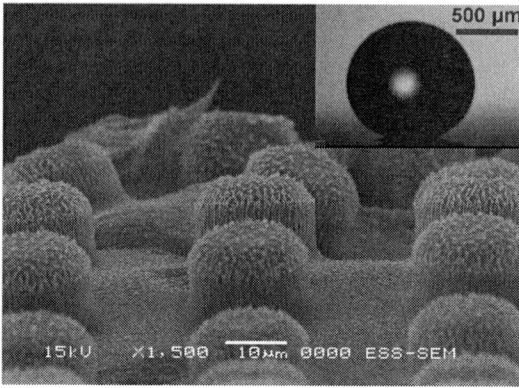

Figure 2: SEM photo of biomimetic SHP surface. The inset photo shows the static contact angle of 160° on this artificial surface.

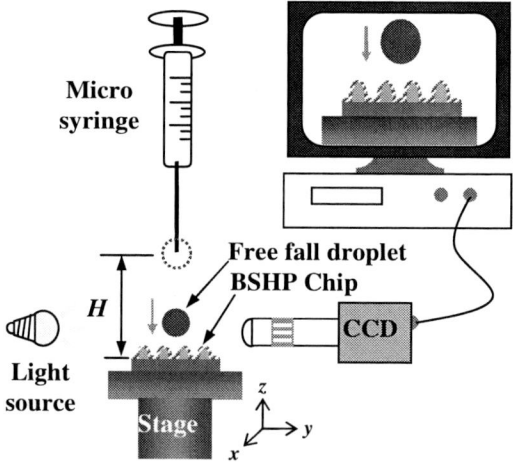

Figure 3: The setup of observation system. The dynamics of free fall droplet impinging on the SHP chip was recorded by high speed CCD at 6000 frames/s.

The Setup of Impinging Test

In the impinging test, as schematically shown in Figure 3, a free fall droplet of 5 μL DI water was precisely released from a glass microsyringe. The impinging energy (kinetic energy) of the droplet was controlled by the potential energy at various releasing heights (H) ranging from 1 cm to 5 cm. The correlation of H and impinging velocity (V) can be easily represented as $V^2 = 2gH$, where g is the gravity constant (g = 980 cm/s²). The impact dynamics was observed by a high speed CCD (Ultima APX Fastcam, Photron LTD., Japan) at an image capture rate of 6000 frames/s.

RESULTS AND DISCUSSION

The time-lapse snapshots of water droplet impinging on the BSHP chip are presented in Figure 4. As $H \leq 3$ cm, the impinging droplet spread on the surface and then rebounded back to take off the surface as a whole (donated as "*rebounding case*"). As $H \geq 3.5$ cm, the droplet also spread as the former case but fragmentation was occurred that droplet was partially stuck on the surface by residual droplets through necking and separating processes (donated as "*sticky case*").

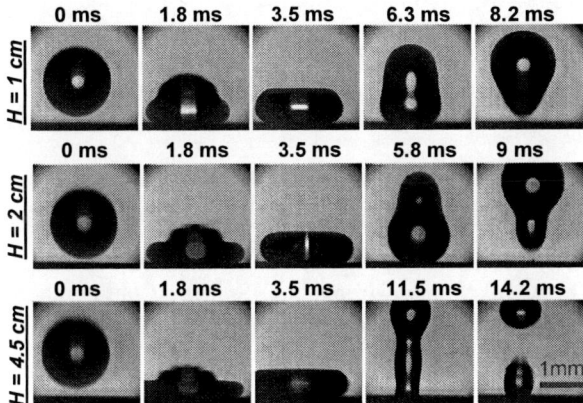

Figure 4: The snapshots of free fall droplets released from various heights. As releasing from lower height (H ≤ 3 cm), the impinging droplet spread and then rebound off the surface and donated as rebounding case (in upper and middle series); as releasing from higher height (H ≥ 3.5 cm), the droplet necked and separated into two parts after rebound which was donated as sticky case (in lower series).

For further understanding, goniometric analysis of these snapshots was performed and the time-varied contact angles were measured as shown in Figure 5. In all cases, the advancing duration times (τ_{adv}) of droplets were almost the same ($\tau_{adv} \sim 3.5$ ms) but the receding duration times (τ_{rec}) were different, with longer ones for the sticky cases.

In rebounding case, interestingly, it can be further categorized into two different states. One is the fully non-wetting state (0 cm < H ≤ 1.4 cm) that dynamic contact angle remained in ~145° at all time within a little contact angle variation of ~5°, which implied a homogeneous and low interaction on surface during droplet impingement and droplet remained on the vertex of the micro/nano textures.

This BSHP surface shows a better stability in dynamic contact angle compared to a difference of ~20° in contact angle proposed in Ref. 15. The overall duration time was 7.5 ms ($\tau_{adv} = 3.5$ ms and $\tau_{rec} = 4$ ms). As a result, a similar dynamics for contact angle evolution was observed no matter the difference in H. On the other hand, a micro-wetting state was observed at 1.45 cm ≤ H ≤ 3 cm that larger kinetic energy squeezed the liquid partially into the micro-texture but not nano-texture. In this state, the contact angle in receding process due to the limited increasing of liquid-solid interface (the droplet still contacted with the tips of nanostructures). This squeeze also induced an oscillation of the droplet and the duration time was slightly increased to 8.5 ms ($\tau_{adv} = 3.5$ ms, $\tau_{rec} =$

5 ms). The 25° contact angle variation can be further manipulated for droplet movement laterally without much hysteresis and surface interaction. Similarly, in this state, the impinging droplet can be reassembled back thoroughly without retention on the surface. The dynamics of droplet was also independent of H.

However in sticky case (nano-wetting state), much more impinging energy enforced the liquid penetrate into not only the micro structures but also the nano-textures, so as to stick the droplet on the surface due to high hysteresis inside the nanostructures. As the height increased to $H = 3.5$ cm, droplets rebound after hitting surface in most experiments but left tiny amount of liquid on the surface. The contact angle was decreased to 100° in receding process and the duration time was obviously increased to 10.2 ms ($\tau_{adv} = 3.5$ ms, $\tau_{rec} = 6.7$ ms). As H further increased to 5 cm, the droplet was deformed violently and fully resided on the surface after impinging without bouncing back. The duration time was also increased to 14.5 ms ($\tau_{adv} = 3.5$ ms, $\tau_{rec} = 11$ ms) and the contact angle further decreased to 90°.

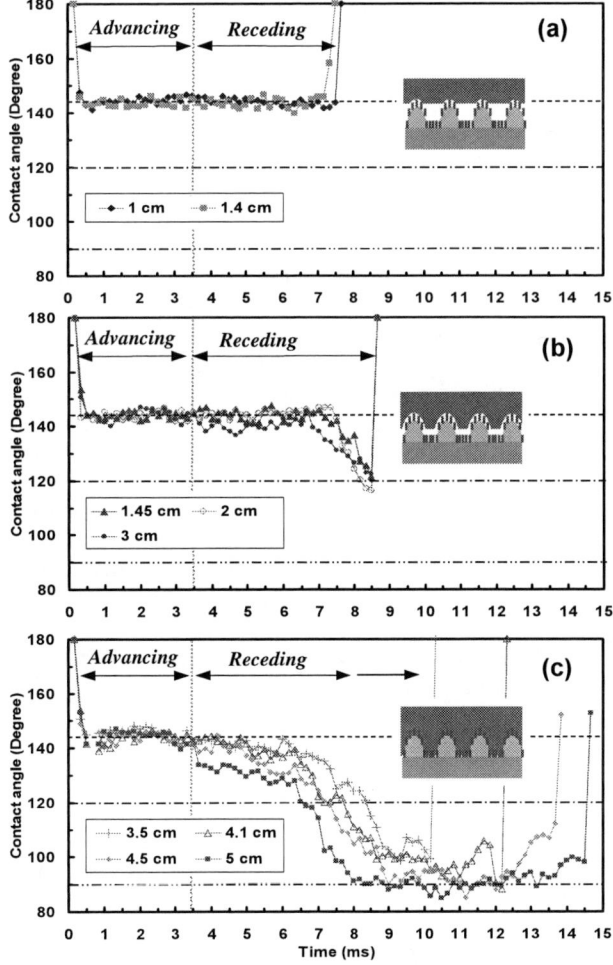

Figure 5. Contact angle variation of dynamic droplet with different H. (a) As 0 cm < H ≤ 1.4 cm, liquid front of droplet keeps on the micro- and nano-texture vertex; (b) as 1.45 cm ≤ H ≤ 3 cm, liquid front partially wets the micro-texture but not nano-texture; (c) as 3.5 cm ≤ H, liquid front start to penetrate into nano-texture and finally the droplet sticks on surface.

After the oscillation of the residual droplet stopped, the contact angle of this static droplet was ~105°, indicating a fully wetting (Wenzel) condition.during impinging history was accordingly decreased to 120° from 145°.

CONCLUSIONS

In summary, these results showed that multi-length scale surface can stabilize the liquid droplet in a middle state to allow contact angle change but minimize droplet wetting and hysteresis conditions, which is not possible to be reached by mono-scale (i.e. either micro or nano) structures. Furthermore, the wetting behaviors of impinging droplet with various energy levels tuned by releasing height (H) can clearly depict the characteristics of dynamic contact angle and duration time on BSHP surface. The surface conditions can be clearly categorized into three different states, including fully non-wetting, micro-wetting, and nano-wetting (sticky). As a result, discrete energy states could be designed by superposing textures in different length scales for droplet manipulation with low hysteresis and surface interactions.

ACKNOWLEDGEMENT

The authors greatly appreciate the financial support from National Science Council of Taiwan ROC through National Nanotechnology and Nanoscience Program under Contract NSC- 96-2120-M-007-010.

REFERENCES

[1] W. Barthlott and C. Neinhuis, "Purity of the sacred lotus, or escape from contamination in biological surfaces", *Planta*, vol. 202, pp. 1-8, 1997.

[2] R. Blossey, "Self-Cleaning Surface: virtual realities", *Nat. Mater.*, vol. 2, pp. 301-306 ,2003.

[3] R. Fürstner, W. Barthlott, C. Neinhuis, P. Walzel,"Wetting and Self-Cleaning Properties of Artificial Superhydrophobic Surfaces", *Langmuir*, vol. 21, pp. 956-961, 2005.

[4] C. -H. Choi and C. -J. Kim, "Fabrication of Dense Array of Tall Nanostructures over a Large Sample Area with Sidewall Profile and Tip Sharpness Control", *Nanotechnology*, vol.17, pp. 5326-5333, 2006.

[5] O. -H. Park, J. Y. Cheng, M. W. Hart, T. Topuria, P. M. Rice, L. E. Krupp, R. D. Miller, H. Ito ,and H. -C. Kim, "High Aspect-Ratio Cylindrical Nanopore Arrays and Their Use for Templating Titania Nanoposts", *Adv. Mater.*, vol. 20, pp. 738-742, 2008.

[6] C. L. Cheung, R. J. Nikolic, C. E. Reinhardt, and T. F. Wang, "Fabrication of Nanopillars by Nanosphere Lithography", *Nanotechnology*, vol. 17, pp. 1339-1343, 2006.

[7] Y. Zhang, C. -W. Lo, J. A. Taylor, and S. Yang, "Replica Molding of High-Aspect-Ratio Polymeric Nanopillar Arrays with High Fidelity", *Langmuir*, vol. 22, pp. 8595-8601, 2006.

[8] M. Nosonovsky and B. Bhushan, "Patterned Nanadhesive Surfaces: Superhydrophobicity and Wetting Regime Transitions", *Langmuir*, vol. 24, pp. 1525-1533, 2008.

[9] C. W. Extrand, " Criteria for Ultralyophobic Surface", *Langmuir*, vol. 20, pp. 5013-5018, 2004.

[10] A. Marmur, "The Lotus Effect: Superhydrophobicity and Metastability", *Langmuir*, vol. 20, pp. 3517-3519, 2004.

[11] D. Richard, C. Clanet, and D. Quéré, "Surface Phenomena: Contact Time of A Bouncing Drop", *Nature*, vol. 417, pp. 811, 2002.

[12] B. Liu and F. Lange, "Pressure Induced Transition between Superhydrophobic States: Configuration Diagrams and Effect of Surface Feature Size", *J. Colloid Interface Sci.*, vol. 298, pp. 899-909, 2006.

[13] J. G. Fan, D. Dyer, G. Zhang, and Y. P. Zhao, "Nanocarpet Effect: Pattern Formation During the Wetting of Vertically Aligned Nanorod Arrays", *Nano Lett.*, vol. 4 , pp. 2133-2138, 2004.

[14] Z. Wang, C. Lopez, A. Hirsa, and N. Koratkar, "Impact Dynamics and Rebound of Water Droplets on Superhydrophobic Carbon Nanotube Arrays", *Appl. Phys. Lett.*, vol. 91, 023105, 2007.

[15] L. Gao and T. J. McCarthy, "The "Lotus Effect"Explained: Two Reasons Why Two Length Scales of Topography Are Important", *Langmuir*, vol. 22, pp. 2966-2967, 2006

[16] R. N. Wenzel, "Resistance of Solid Surface to Wetting by Water", *Ind. Eng. Chem.*, vol. 28, pp. 988-994, 1936.

[17]A. B. D. Cassie and S Baxter, "Wettability of Porous Surface", *Trans. Faraday Soc.*, vol. 40, pp. 546-561, 1944.

[18] M. Nosonovsky and B. Bhushan, "Biologically Inspired Surfaces: Boradening the Scope of Roughness", *Adv. Funct. Mater.*, vol. 18, pp.843-855, 2007.

[19] M.-H. Chen, T.-H. Hsu, Y.-J. Chuang, P.-H. Chen, and F.-G. Tesng," Self-formed High-Aspect-Ratio Polymer Nanopillars by RIE", in *Digest Tech. Papers Transducers' 07 conference*, Lyon, France, June 10-14, 2007, pp.563-566.

WETTABILITY SWITCHING TECHNIQUE OF A BIOCOMPATIBLE POLYMER

Y. Takahashi[1], K. S. Teh[2], and Y.W. Lu[1]

[1]Rochester Institute of Technology, Rochester, New York, USA
[2]San Francisco State University, San Francisco, California, USA

ABSTRACT

In this paper we present data of protein adsorption on topographically altered active polymeric films, investigated using atomic force microscopy (AFM), fluorescence microscopy, and contact angle measurement. The surface wettability of biocompatible polypyrrole (PPy) films was altered via low applied currents, and we have demonstrated switchable hydrophobic-hydrophilic surfaces, used as protein adhesion substrates. The preliminary results showed that both BSA and fibronectin have higher affinities for and adhere preferentially to hydrophobic surfaces.

INTRODUCTION

In the development of bio-MEMS, bio-microelectromechanical systems, for a new generation of biomedical device, one of the major issues is the possibility of protein fouling. Creating a smart surface that controls protein adhesion can be of great benefit for microscale devices, which have higher surface to volume ratios. The ability to reconfigure surfaces wettability or even to switch between hydrophobic and hydrophilic provides an effective strategy for promoting and preventing protein adsorption [1]. For example, bovine serum albumin (BSA) strongly adsorbs by cell adhesive proteins on hydrophobic self assembled monolayer (SAM) [2]. In microfluidics, BSA detachment with a shear flow can be improved by utilizing superhydrophobic surfaces [3]. Considerable research effort has been invested on reversible modification of surface wettability by means of high voltage, light, chemical, and mechanical techniques [4 - 6]. For example, a fully reversible photoswitchable wetting surface was achieved via alternation of ultraviolet irradiation on ZnO nanorod films and dark storage [7]. While an applied electrical potential was used in order to induce conformational transitions of a low-density SAM which has a hydrophobic chain capped by a hydrophilic functional group, effectively switching surface wettability [8].Our technique which differs from earlier work is reversible and does not require a high electrical potential, and can be applied to larger area, and also easy to apply in device developments. In this paper, we describe recent progress on mechanism realization, material characterization, and biological application of active nanostructuring of doped-polypyrrole (PPy) films.

Our topography alteration mechanism is based on creating nanoscale structures on the polymeric surface, as described in elsewhere [9]. The creation of surface nanostructures is induced by a low current electrochemical process. Our polymer substrate used is a biocompatible conjugated conducting polymer – polypyrrole (PPy). The doing level of conducting polymer

is reversible and can be controlled by reduction and oxidation processes. Our previous study found that there is a surface topographical modification associated with the change in doping level of PPy, and this property can be exploited to make reconfigurable hydrophobic/hydrophilic smart surface.

The surface roughening mechanism was discussed in elsewhere [9]. Briefly, as the electric potential is applied on the PPy film, cations (Na^+) are being expelled from the surface, while the bulky anions (DBS^-) remain immobilized in the polymer, as shown in **Figure 1**. The net exodus of Na^+ causes reorientation of the polymer network and creates nanoscale structures on the surface. Such surface roughening effect contributes in decreasing the net surface energy, as described by Cassie model [10]. We capitalize on this mechanism by developing a galvanostatic redox method to keep the current constant through the electrolytic cells, and to control the amount of ions in/out the film.

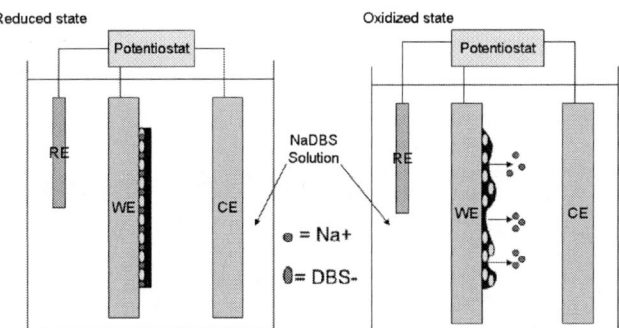

Figure 1. Schematic diagram of surface nanostructuring mechanism. (a) During reduction, cations (Na^+) enter the polymer matrix on the as-deposited film. (b) During oxidation cations (Na^+) leaves the polymer matrix of the film, creating the nanostructures and roughening the surface. [9]

EXPERIMENTAL
Doped-PPy substrate deposition

Doped PPy layer was synthesized by an electropolymerization process. Instead of using a potentiostatic method described elsewhere [9], we used a current-controlled method (galvanostatic) to deposit doped PPy used in this experiment. 0.1M of aqueous pyrrole solution was made by dissolving pyrrole monomer (MW: 67.09) in 0.1M aqueous sodium dodecylbenzene sulfonate, NaDBS (MW: 348.5) at room temperature. Two gold-coated glass slides (25 x 75 mm^2, Fisher Scientific) were cleaned in acetone, methanol and DI water, and were used as the working electrode (WE) and

978-1-4244-2977-6/09 $25.00 © 2009 IEEE   459

the counter electrode (CE). To ensure a uniform electric field upon electrolysis, a 10 mm x 10 mm window, which was to be the reaction site, was made using masking tapes on WE. A low current density at $0.2mA/cm^2$ for 5 minutes was employed to yield the doped PPy film with good uniformity and adhesion to the gold substrate. The galvanostatic method exhibited an improved consistency than the previously used potentiostatic method.

Doped-PPy redox and Surface Reconfiguration

As-deposited PPy film was placed in a 0.1M aqueous NaDBS solution and was subjected to a +/- $0.1mA/cm^2$ applied current to induce the redox process. The galvanostatic redox method is utilized to control the amount of ions in/out the film and the nanostructurings. The PPy samples were then rinsed in DI water and blow-dried by nitrogen. Contact angle measurement and atomic force microscopy (AFM) were utilized to characterize surface wettability and roughness (RMS). The relationship between the redox duration time, the PPy surface contact angle, and the average surface roughness at this positive current density ($0.1mA/cm^2$) was studied, and the results are illustrated in **Figure 2**. Both the contact angle and surface roughness were proportional to the redox duration. As redox duration increases, the number of cations expelled from the surface increases, causing both contact angles and surface roughness (RMS) to increase. After the redox duration reached ~800 to 1000 seconds, PPy started to peel off, and the process was discontinued. **Table 1** summarizes the experimental condition for this study.

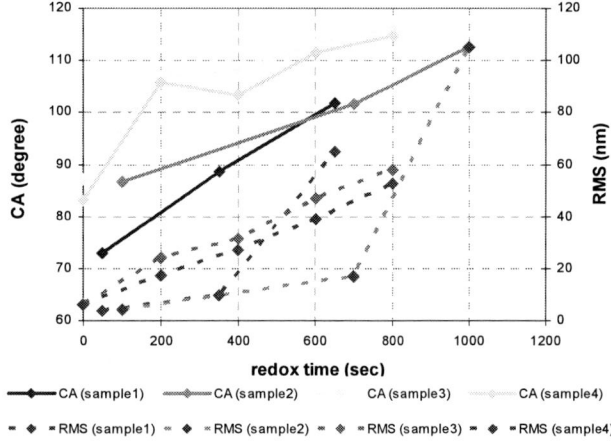

Figure 2. Contact angle and surface roughness (RMS) vs. redox duration time.

Meanwhile, the wetting reversibility was investigated. The negative applied current was used to induce switching of surface wettability. The as-deposited PPy film was subjected a redox process at a positive current for 600 seconds in NaDBS solution. Subsequently a negative current density was applied to the PPy film. The contact angle was increased to ~120° via the positive redox current, and the negative redox current reduced the surface contact angle by 40° to 50° as shown in **Figure 3**. When the order of positive and negative currents were reversed (a negative current, followed by a positive

current), the contact angle initially reduced by between 10° to 20°, and it then increased to approximately 120° once again.

Table 1. *Experimental conditions for nanostructuring test. These samples are the same set of samples used in* **Figure 2**.

Sample Set	Film Deposition Condition	Nanostructuring	
		Current	Time
1	$+0.2$ mA/cm^2, 300 sec	$+0.1$ mA/ cm^2	0 to 650 sec
2		$+0.1$ mA/ cm^2	0 to 1000sec
3		$+0.1$ mA/ cm^2	0 to 800 sec
4		$+0.1$ mA/ cm^2	0 to 800 sec

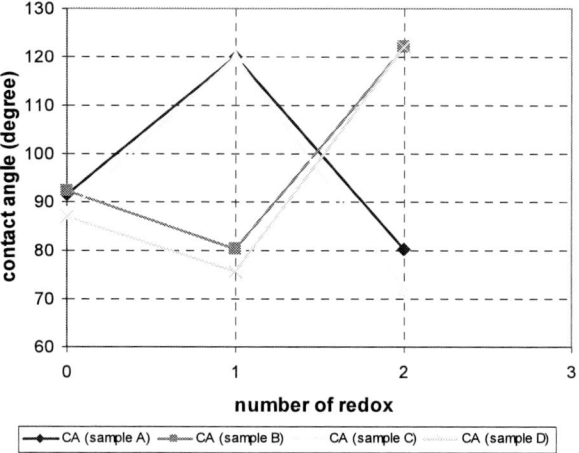

Figure 3. *Contact angle switching via positive and negative currents redox.*

BSA and Fibronectin adhesion

Bovine serum albumin (BSA) with fluorescein, and fibronectin from bovine plasma were dissolved in phosphate buffer solution (PBS) to make 5mg/ml and 0.1mg/ml solutions respectively. Each solution was directly placed on the wettability controlled PPy films. After 2 hours of incubation, the protein was washed off with PBS, and the surface was characterized using green florescent light (395nm) and AFM.

RESULTS AND DISCUSSION

Surface Reconfiguration and Wetting Mechanism

By altering the direction of applied current on the redox process, the wettability of PPy film can be switched: changing between hydrophilic and hydrophobic. In the case of **Figure 4 (a)**, the cations first diffused in, and then were repelled from the film, while in the case of **Figure 4 (b)**, the cations were first repelled from, and then diffused into the film. In the former case, the RMS changed from 5.756 nm, to 5.519 nm, and to 102.89 nm – the surface roughened and became hydrophobic only when a positive current was applied. In the latter case, the

RMS changed from 6.089 nm, to 80.046 nm, and to 73.021 nm – the surface became roughened and hydrophobic when a positive current was first applied. It then became hydrophilic and less roughened when a negative current was applied. Once the surface is subjected to a positive current and nanostructuring is formed in high degree, however, it is unable to reverse the topography completely with a negative current redox process (**Figure 4. b**). Although the surface nanostructures still existed, the experimental result exhibited a significant decrease in the contact angle measurement. **Table 2** summarizes the experimental conditions.

Table 2. Experimental conditions for wettability switching.

Sample Set	Film Deposition Condition	Nanostructuring	
		1st redox (600 sec)	2nd redox (600 sec)
Figure 4. (a)	+0.2 mA/cm^2, 300 sec	-0.1 mA/ cm^2	+0.1 mA/ cm^2
Figure 4. (b)	+0.2 mA/ cm^2, 300 sec	+0.1 mA/ cm^2	-0.1 mA/ cm^2

The influence of roughness on the wettability of solid surfaces has been widely investigated, showing that surface roughness may enhance non-wetting. **Figure 5** explains the possible mechanism with a positive current redox followed by one with a negative current, based on previous literature [11]. The hierarchical structures of the substrate play an important role in determining the surface wettability. If the contact angle of the substrate material is less than $\pi/2$, the liquid should at least overhang across the smallest structures, which can result from the Laplace pressure of the liquid surface. As the polymer is subjected to the positive current, the surface structures are formed with smaller indentations, where less steepness is required for structures to support a liquid surface. When the polymer is negatively reduced, the cations enter the film, which flatten the smaller indentations. On a larger scale, these structures are no longer able to support the liquid droplet without smaller indentations.

(a)   (b)

Figure5. Schematics of wetting mechanism of nano-scale structures for (a) hydrophobic and (b) hydrophilic states.

BSA and Fibronectin adhesion

The effect of wettability on protein adsorption was studied. The PPy films in **Table 2** were incubated in 5mg/ml of BSA solution. **Figure 6** shows the fluorescent images (GFP) of BSA on the PPy samples, where the fluorescein-dyed BSA appears as white dots. The wettability modified hydrophobic surface in **Figure 6 (b)** exhibited a significantly more protein adsorption than the hydrophilic PPy surface in **Figure 6 (a)**.

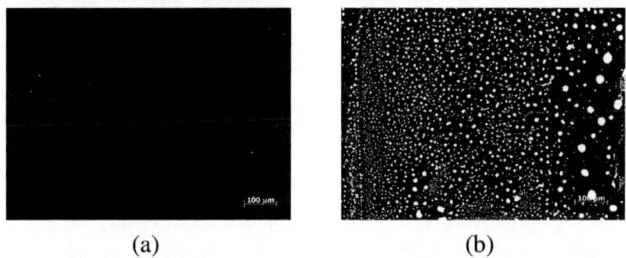

(a)   (b)

Figure 6. BSA adsorption on (a) hydrophilic and (b) hydrophobic PPy surface observed under green fluorescent light source.

Fibronectin adsorption was studied on wettability controlled PPy surfaces (-0.1 mA/cm^2 and +0.1 mA/cm^2) using AFM. The negatively reduced PPy surface in **Figure 7 (a)** exhibited few adhesion of fibronectin in **Figure 7 (c)**. Whereas the positively oxidized hydrophobic surface in **Figure 7 (b)** attracted proteins in larger degree, forming layers of fibronectin on nanostructures in **Figure 7 (d)**.

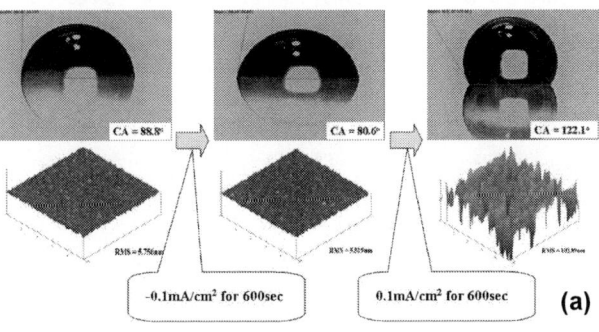 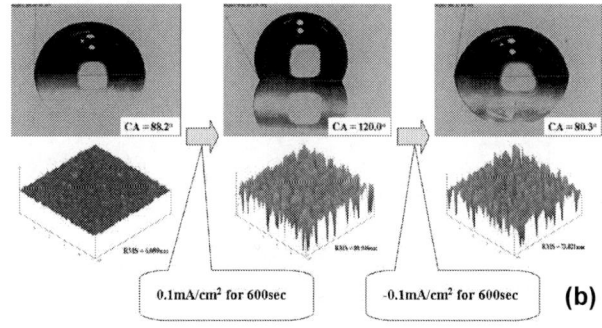

Figure 4. PPy wettability switching. (a) -0.1mA/cm^2 redox for 10min followed by +0.1mA//cm^2 redox for 10min, and (b)+0.1mA/ cm^2 redox for 10min followed by -0.1mA redox for 10min.

Our preliminary results from protein adhesion tests indicated that the wettability controlled PPy film can be used as a smart surface for promoting or preventing protein adsorption for implantable biomedical devices. It has been known that the initially adsorbed proteins act as an underlying layer for consequent adhesion of proteins and cells, and determine subsequent biological response to substrates. This unique property of the PPy indicates the bright potential for new biomedical device development.

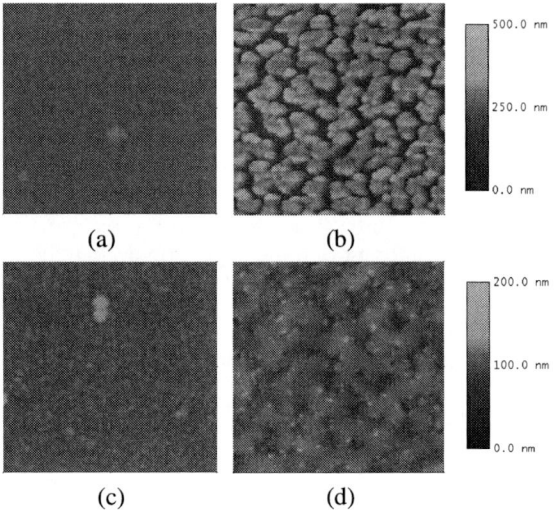

Figure 7. (a)(c) hydrophilic surface (RMS=5.125nm) and post fibronectin adsorption (RMS = 5.514nm). (b)(d) hydrophobic PPy surface (RMS = 72.922nm) and post fibronectin adsorption (RMS = 80.48nm).

CONCLUSION

We have demonstrated the switchable hydrophobic / hydrophilic surface by electrochemically induced polymer redox processes. The surface nanostructuring of biocompatible polymeric films was utilized controlling surface wettability, characterized by means of contact angle measurement and AFM. BSA and fibronectin adsorption characteristics on the wettability controlled surface were investigated. Both exhibited an affinity to the hydrophilic surface. The possibility of controlling protein adsorption via wettability switching indicates the great potential for rapid healing and novel implantable device development, since cell adhesion occurs on the surface protein layer.

ACKNOWLEDGEMENTS

This work is supported by the National Science Foundation under Award No ECCS 0802100. The authors thank the members from Semiconductor and Microsystems Fabrication Laboratory (SMFL) at R.I.T. for technical support and Dr. T. Smith for valuable discussion.

REFERENCES

[1] P. Roach, D. Farrar, and C. C. Perry, "Interpretation of protein adsorption: Surface-induced conformational changes," *Journal of the American Chemical Society,* vol. 127, pp. 8168-8173, 2005.

[2] Y. Arima and H. Iwata, "Effect of wettability and surface functional groups on protein adsorption and cell adhesion using well-defined mixed self-assembled monolayers," *Biomaterials,* vol. 28, pp. 3074-3082, Jul 2007.

[3] Y. Koc, A. J. de Mello, G. McHale, M. I. Newton, P. Roach, and N. J. Shirtcliffe, "Nano-scale superhydrophobicity: suppression of protein adsorption and promotion of flow-induced detachment," *Lab on a Chip,* vol. 8, pp. 582-586, 2008.

[4] N. Verplanck, Y. Coffinier, V. Thomy, and R. Boukherroub, "Wettability switching techniques on superhydrophobic surfaces," *Nanoscale Research Letters,* vol. 2, pp. 577-596, 2007.

[5] J. Deval, T. A. Umali, E. H. Lan, B. Dunn, and C. M. Ho, "Reconfigurable hydrophobic/hydrophilic surfaces in microelectromechanical systems (MEMS)," *Journal of Micromechanics and Microengineering,* vol. 14, pp. 91-95, Jan 2004.

[6] H. Y. Erbil, A. L. Demirel, Y. Avci, and O. Mert, "Transformation of a simple plastic into a superhydrophobic surface," *Science,* vol. 299, pp. 1377-1380, 2003.

[7] X. J. Feng, L. Feng, M. H. Jin, J. Zhai, L. Jiang, and D. B. Zhu, "Reversible super-hydrophobicity to super-hydrophilicity transition of aligned ZnO nanorod films," *Journal of the American Chemical Society,* vol. 126, pp. 62-63, 2004.

[8] J. Lahann, S. Mitragotri, T. N. Tran, H. Kaido, J. Sundaram, I. S. Choi, S. Hoffer, G. A. Somorjai, and R. Langer, "A reversibly switching surface," *Science,* vol. 299, pp. 371-374, 2003.

[9] Y. Takahashi, K.S. Teh, and Y.-W. Lu, "Fabrication and characterization of nano-structuring polymeric surface for MEMS applications," *The ASME International Mechanical Engineering Congress & Exposition*, Boston USA, 2008.

[10] A. B. D. Cassie and S. Baxter, "Wettability of porous surfaces," *Trans. Faraday Soc.*, vol. 40, pp. 546-51, 1944.

[11] S. Herminghaus, "Roughness-induced non-wetting," *Europhysics Letters,* vol. 52, pp. 165-170, 2000.

MICRO CORIOLIS MASS FLOW SENSOR WITH INTEGRATED CAPACITIVE READOUT

J. Haneveld, T.S.J. Lammerink, M.J. de Boer and R.J. Wiegerink.
MESA+ and IMPACT Research Institutes, University of Twente, The Netherlands.
e-mail: j.haneveld@utwente.nl

ABSTRACT

We have realized a micromachined micro Coriolis mass flow sensor with integrated capacitive readout to detect the extremely small Coriolis vibration of the sensor tube. A special comb-like detection electrode design eliminates the need for multiple metal layers and sacrificial layer etching methods. Using differential readout signals significantly reduces the influence of parasitic capacitances. In addition, the sensing electrodes have been realized on a suspended tube structure which will allow for tuning of the electrode separation by a DC current. First measurements using water, ethanol and white gas indicate that true mass flow is measured by the sensor and that sensor output is linear with mass flow. The measurement error is currently in the order of 2% of the full scale of 1.2 ml/hr for all measured liquids (which corresponds to 1.2 g/hr in the case of water).

INTRODUCTION

Integrated microfluidic systems have gained interest in recent years for many applications including (bio)chemical, medical, automotive, and industrial devices. A major reason is the need for accurate, reliable, and cost-effective liquid and gas handling systems with increasing complexity and reduced size. In these systems, flow sensors are generally one of the key components.

Most MEMS flow sensors are based on a thermal measurement principle. It has been demonstrated [1,2] that such sensors are capable of measuring liquid flow down to a few nl/min. These sensors require accurate measurement of very small flow-induced temperature changes. An important problem of thermal flow sensors is that the measurement is highly dependent on temperature and fluid properties like density and specific heat.

The sensor presented in this paper is based on the Coriolis force which acts on a fluid (mass) flowing in a vibrating channel. Coriolis flow meters [3-5] are mostly used for measuring large flow rates, since the relatively weak Coriolis forces are correspondingly harder to detect for small flows. In general, the signal-error ratio is very sensitive to fabrication and construction errors, as well as external influences of temperature and mechanical nature. However, an important advantage of this type of sensor is that the Coriolis force is directly proportional to the mass flow and independent of temperature, pressure, flow profile and fluid properties.

OPERATING PRINCIPLE

A Coriolis type flow sensor consists of a vibrating tube. Now consider a moving mass inside the tube. This mass is forced to change its velocity due to the externally imposed vibration. This results in Coriolis forces that can

be detected. Figure 1 shows a schematic drawing of a Coriolis sensor based on a rectangular tube shape. The tube is actuated in a "torsion mode" indicated by ω. A mass flow Φ_m inside the tube results in a Coriolis force F_c that can be expressed by:

$$\vec{F}_c = -2L\,\vec{\omega}\times\vec{\Phi}_m \qquad (1)$$

Where L is the length of the rectangular tube (see Figure 1). The Coriolis force induces a "flapping mode" vibration with an amplitude proportional to the mass flow.

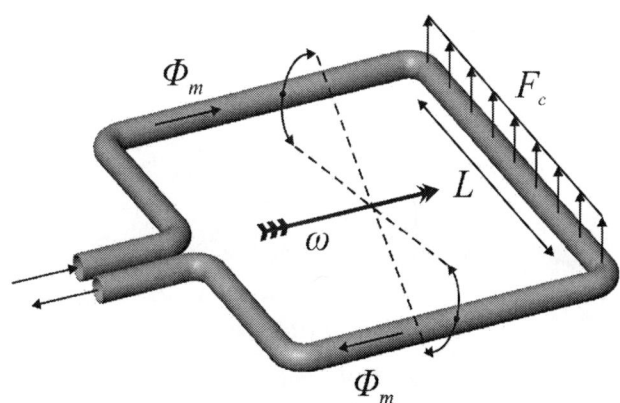

Figure 1: Rectangle-shaped Coriolis flow sensor (ω is the torsion mode actuation vector, F_c indicates the Coriolis force as a result of mass flow).

SENSOR DESIGN

In [6] we proposed to fabricate a micro Coriolis mass flow sensor using silicon nitride as the tube material. This resulted in a sensor with a very thin (1.2 μm) tube wall, so that the mass of the tube is small compared to the mass of the moving fluid. This was a significant improvement over [7] and [8], which use silicon as the tube material, leading to a relatively heavy and stiff tube. We demonstrated that our silicon nitride sensor could reach a resolution in the order of 10 milligram/hour [6], however at that time no readout structures were integrated on the chip, so a laser vibrometer was needed to measure the out-of-plane Coriolis motion of the tube.

The newly fabricated sensor incorporates capacitive readout structures, which are used to measure the very small Coriolis vibration. In addition, the fluid in- and outlets have been moved to the backside of the wafer, resulting in a more sturdy and leak-free fluidic connection. The Coriolis tube dimensions remain essentially the same: a low stress silicon-rich nitride (Si_xN_y) tube with an effective diameter of approximately

978-1-4244-2977-6/09 $25.00 © 2009 IEEE

40 µm and a wall thickness of approximately 1.2 µm. The Coriolis tube has a rectangular shape (Figure 1) with dimensions 2.5×4 mm. Opposite the Coriolis tube is a second tube, which contains the capacitive readout structures to detect the vibration of the Coriolis tube. The detection is achieved through a comb-like structure which functions as a parallel plate capacitor when the two halves of the comb are separated out-of-plane. If necessary the tube containing the sensing structures can be moved up and down in order to tune the sensitivity of the readout structure. This is done by means of Lorentz force actuation, using a DC current. Figure 2 shows a schematic layout of the sensor, and Figure 3 is a photograph of a finished device, mounted on a PCB.

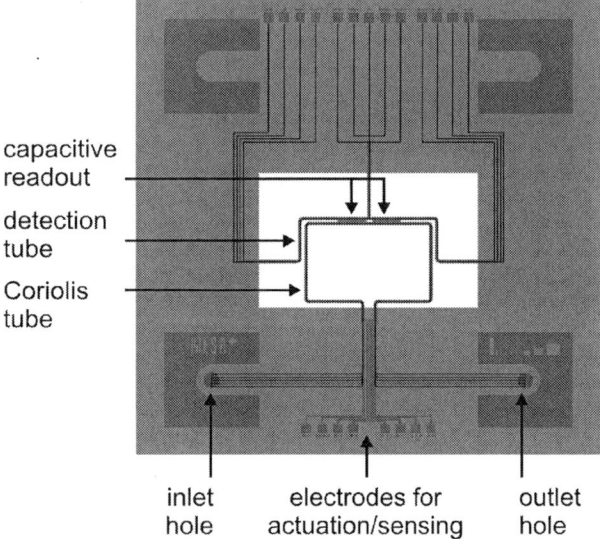

Figure 2: Layout of the sensor chip (size 15x15 mm).

Figure 3: Photograph of mounted chip with electrical connections and permanent magnets for Lorentz force actuation. Fluidic connections are located at the backside of the PCB.

FABRICATION

The sensor structures were fabricated using the surface channel technology described in [6,9,10] as a starting point. Figure 4 shows a summary of the fabrication process.

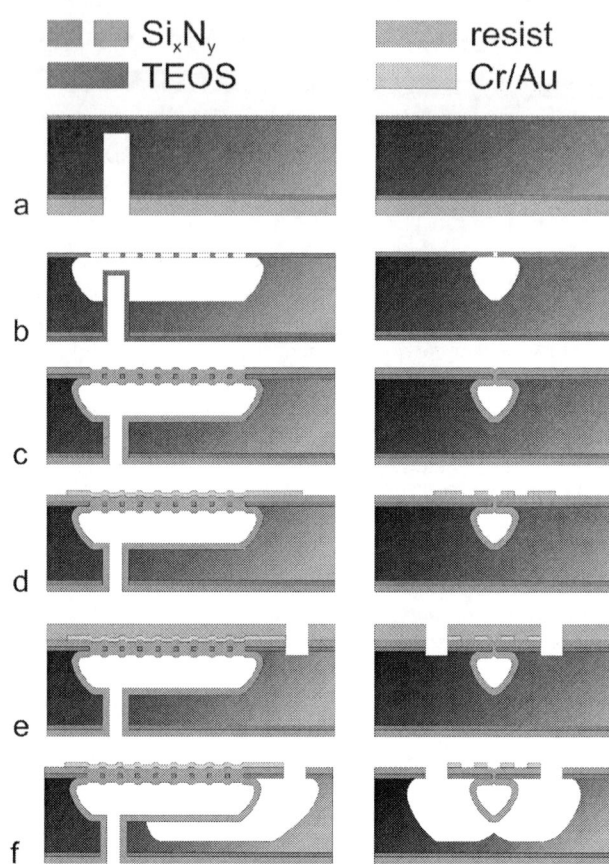

Figure 4: Outline of the fabrication process. Left column: cross-section along the length of the tube. Right column: cross-section perpendicular to the sensor tube.

Starting with a <100> p++ wafer, a 500 nm thick low stress LPCVD silicon-rich nitride (Si_xN_y) layer is deposited. Then fluidic access holes are etched (cryogenic pulsed CHF_3 plasma etching on an Adixen AMS 100 SE system) from the back side of the wafer. The endpoint of this etch step is aimed approximately 20 µm below the top surface of the wafer and has to be carefully monitored (Figure 4a). Next a 500 nm thick layer of TEOS oxide is deposited on the wafer. The oxide layer is removed from the front side of the wafer using BHF etching, while protecting the oxide on the back side with a layer of adhesive dicing foil. Then a 50 nm layer of chromium is sputtered on the front side of the substrate. This chromium layer is subsequently patterned using a mask containing arrays of 5×2 µm holes, spaced 3 µm apart. The pattern is then transferred to the nitride layer by means of reactive ion etching (RIE Electrotech PF 340). Then an isotropic plasma etch step is performed in an ADIXEN AMS 100 SE apparatus (Figure 4b): this defines the channel shape.

The TEOS layer and the chromium mask are then removed and another Si_xN_y layer is grown with a thickness of 1.8 µm; this forms the channel wall and seals the etch holes in the first nitride layer in one single step (Figure 4c). A 10/200 nm layer of chromium and gold is sputtered (chromium serving as the adhesion layer for the gold) and pre-patterned using wet etching in a KI/I_2 etch solution, followed by a short dip in chromium etchant (Figure 4d) to create the metal tracks on top of the tube which will facilitate Lorentz force actuation and

978-1-4244-2977-6/09 $25.00 © 2009 IEEE   464

capacitive sensing of the structure. Then a second lithography step is performed, in which the comb-like structures are defined, as well as the windows which are used to release the tube from the substrate. The Cr/Au layer is then patterned using ion beam etching, because wet etching makes it extremely hard to precisely control the width of the teeth of the comb structures. Directly after the metal patterning the release windows are opened by reactive ion etching of the Si_xN_y layer (Figure 4e). Then the structure is released by an isotropic silicon plasma etch step (using again the ADIXEN AMS 100 SE apparatus), followed by resist removal using a Tepla 300 oxygen plasma barrel stripper (Figure 4f)).

The entire resulting structure can be seen in Figure 5. Detailed views of the capacitive sensing structures can be seen in Figures 6 and 7.

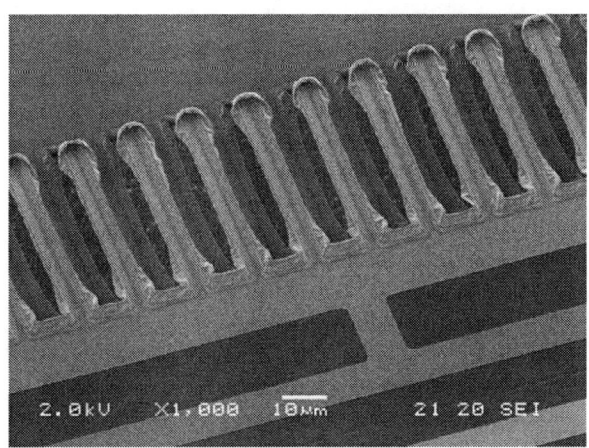

Figure 7: Detailed view of one of the comb-like capacitor structures. The lower structure is the Coriolis tube. In this case the other comb structure is attached to the bulk of the chip.

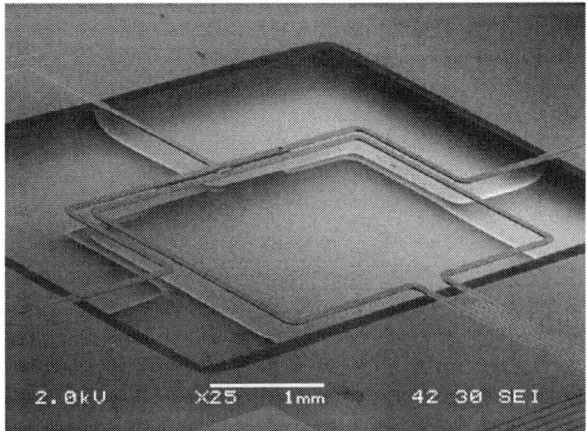

Figure 5: SEM photograph of the complete sensor chip.

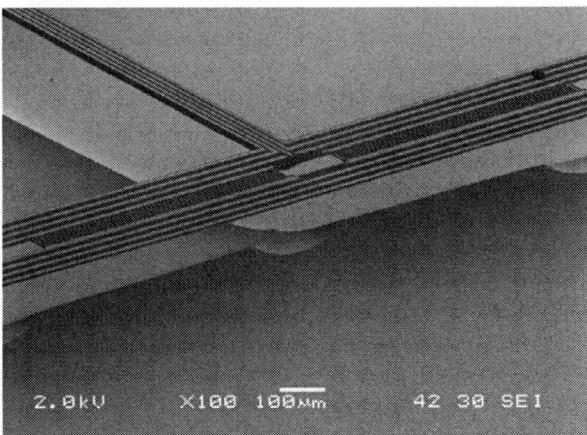

Figure 6: Close-up of the capacitive sensing structures with two comb structures to sense both the actuation and Coriolis amplitudes.

The finished chips are epoxy-glued to a PCB, which also contains pre-drilled fluidic access holes. A fluidic connector is then glued to the back of this PCB to allow for fluid/gas delivery. Permanent magnets are inserted in slots in the PCB, and permanently fixed using thermal glue. Electrical connections are made by wirebonding from the bondpads on the chip to the PCB.

MEASUREMENTS

Mass flow measurements were done using water, ethanol and white gas. The sensor was connected to electronics which deliver the actuation signal and process the detection signal. Figure 8a shows a schematic diagram of the interface electronics. The sensor was operated in resonance mode (1.77, 1.86 and 1.89 kHz for water, ethanol and white gas, respectively) by running an alternating current through a metal track running over the tube. At the same time two 1.4 MHz signals are fed to the parts of the combs which are attached to the sensor tube. These two signals are in counter-phase to eliminate the influence of parasitic capacitance to the substrate. The counter electrodes are kept at virtual ground by two charge amplifiers (see Figure 8b).

The output signals of the charge amplifiers are amplitude modulated signals of 1.4 MHz, where the amplitude is proportional to the sensor capacitance. The signals are demodulated using standard integrated analog multipliers (SA602A) and an op-amp based second-order low pass filter at 3 kHz. Summation of the two output signals gives a measure for the difference in capacitance $(C_1 - C_2)$, i.e. the actuation amplitude. The difference between the output signals is a measure for the common variation in the capacitors due to the Coriolis effect. A lock-in amplifier is used to extract the Coriolis signal from the differential output signal in order to eliminate noise and any remaining actuation signal due to nonlinearities.

A syringe pump (CMA Microdialysis CAD system) was used to generate water, ethanol and white gas volume flows between 0 and 1.2 cm^3/hr through the chip. During the measurement the sensor output signal is recorded, together with the pressure of the liquid using a separate pressure sensor. The pressure sensor signal is proportional to the volume flow rate. It is calibrated at the highest flow level generated by the syringe pump of 1.2 cm^3/hr.

Figure 10 shows the measured output signal as a function of the calculated volume flow (derived from the pressure sensor signal) for water, ethanol and white gas.

978-1-4244-2977-6/09 $25.00 © 2009 IEEE

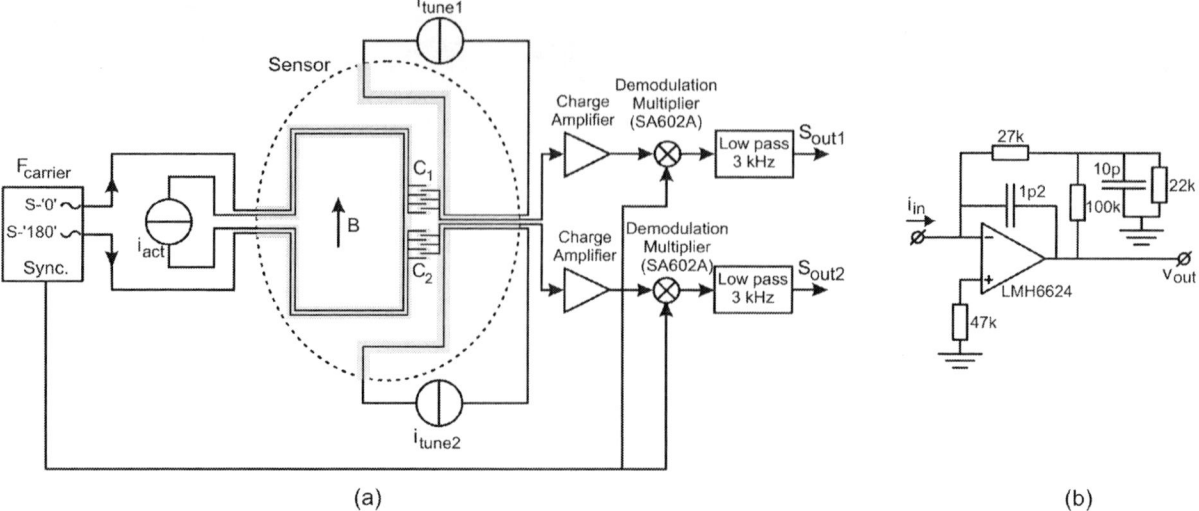

(a)   (b)

Figure 8: (a) Schematic diagram of the actuation and readout electronics. (b) Charge amplifier circuit that is used to keep the readout electrodes at virtual ground. The circuit is optimized for carrier frequencies between 1 and 5 MHz.

The difference in the slope of the lines is equal to the relative densities of the liquids, indicating that the device directly measures mass flow. Analysis of the measurement data shows that all deviations from the linear fit are less than 2% of the full scale output signal for each liquid.

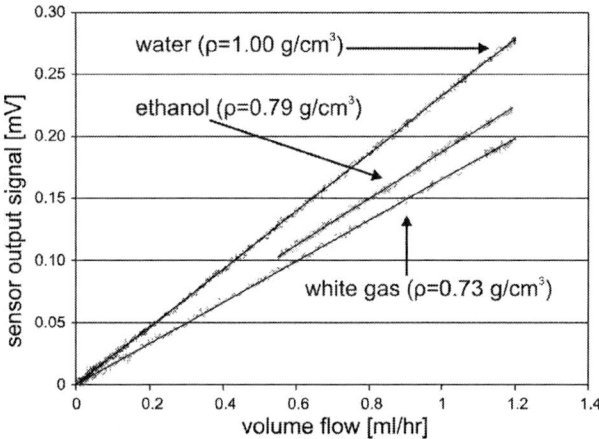

Figure 10: Measured output signal as a function of volume flow for water, ethanol and white gas.

CONCLUSIONS

We have successfully integrated a capacitive readout structure on a micro Coriolis mass flow sensor chip. The fluidic in- and outlet of the chip are on the back-side of the wafer, allowing for a more stable connection. The sensor shows excellent linear output and accurately measures mass flow in the flow range of 0-1.2 g/hr. The measurement error is currently in the order of 2% of the full scale for all measured liquids.

ACKNOWLEDGEMENT

This research was financed by the Dutch MicroNed program. The authors would like to thank the industrial partners in the project for many fruitful discussions.

REFERENCES

[1] Y. Mizuno, M. Liger, Y-C. Tai, "Nanofluidic Flowmeter Using Carbon Sensing Element", *Proc. MEMS 2004*, Maastricht, The Netherlands, Jan. 22-25, 2004, p. 322-325.

[2] M. Dijkstra, M.J. de Boer, J.W. Berenschot, T.S.J. Lammerink, R.J. Wiegerink, M. Elwenspoek, "Miniaturized Flow Sensor with Planar Integrated Sensor Structures on Semicircular Surface Channels", *Proc. MEMS 2007*, Kobe, Japan, Jan. 21-25, 2007, pp. 123-126.

[3] R.C. Baker, "Coriolis Flowmeters: Industrial Practice and Published Information", *Flow. Meas. Instrum.*, Vol. 5, No. 4, pp. 229-246, 1994.

[4] M. Anklin, W. Drahm, A. Rieder, "Coriolis Mass Flowmeters: Overview of the Current State of the Art and Latest Research", *Flow. Meas. Instrum.*, Vol. 17, pp. 317-323, 2006.

[5] A. Mehendale, P.P.L. Regtien, "System Design of Low Capacity Coriolis Mass-flow Meters", *3rd Int. Symp. Sens. Sci.*, Jülich, Germany, Jul. 18-21, 2005.

[6] J. Haneveld, T.S.J. Lammerink, M. Dijkstra, H. Droogendijk, M.J. de Boer, R.J. Wiegerink, "Highly Sensitive Micro Coriolis Mass Flow Sensor", *Proc. MEMS 2008*, Tucson, Az, Jan. 13-17, 2008, pp. 920-923.

[7] P. Enoksson G. Stemme, E. Stemme, "A Silicon Resonant Sensor Structure for Coriolis Mass-flow Measurements", *J. MEMS*, Vol. 6, pp. 119-125, 1997.

[8] D. Sparks, R. Smith, J. Cripe, R. Schneider, N. Najafi, "A Portable MEMS Coriolis Mass Flow Sensor", *Proc. IEEE Sensors 2003*, Toronto, Canada, Oct. 22-24, 2003, pp. 90-92.

[9] M. Dijkstra, M.J. de Boer, J.W. Berenschot, T.S.J. Lammerink, R.J. Wiegerink, M. Elwenspoek, "A Versatile Surface Channel Concept for Microfluidic Applications", *JMM*, Vol. 17, pp. 1971-1977, 2007.

[10] M.J. de Boer, R.W. Tjerkstra, J.W. Berenschot, H.V. Jansen, G.J. Burger, J.G.E. Gardeniers, M. Elwenspoek, A. van den Berg, "Micromachining of Buried Micro Channels in Silicon", *J. MEMS*, Vol. 9, No. 1, pp. 94-103, 2000.

978-1-4244-2977-6/09 $25.00 © 2009 IEEE

MEMS TACTILE DISPLAY
WITH HYDRAULIC DISPLACEMENT AMPLIFICATION MECHANISM
T. Ninomiya, K. Osawa, Y. Okayama, Y. Matsumoto and N. Miki
Keio University, Yokohama, Japan

Abstract

Tactile displays mechanically stimulate tactile receptors on a finger pad. They require a micro-actuator array capable of generating displacement greater than 100 μm. We propose a MEMS-based tactile display with hydraulic displacement amplification mechanisms (HDAM). HDAM contains micro-cavities that have different cross-sectional areas between at the contact and the drive parts and encapsulate incompressive glycerin with largely-deformable thin PDMS (polydimethylsiloxane) membranes. Displacement applied to the drive part is amplified at the contact part associated with the ratio of cross-sectional areas of the drive to the contact parts. HDAM was fabricated and achieved a seven-fold of amplification. We successfully demonstrated rewritable Braille codes.

INTRODUCTION

Background

Touch, or tactile sensing, has been widely utilized as a human interface. For example, iPhone (Apple Inc.) and Nintendo DS (Nintendo Inc.) exploit "Touch" and have become world wide hits. In addition, new products which introduce touch operation are going to be released for some years to come. This is because tactile perception is very intuitive and easy to recognize for users. Recently a variety of tactile displays have been developed to enhance user-friendliness of machines, immersiveness of games, and effectiveness of virtual training.

Tactile sense of human

Tactile information is detected by four tactile receptors that are densely distributed on a finger pad. The receptors are categorized into two types according to their frequency response. Merkel's disc and Ruffini ending are classified into SA (Slow adapting), while Meissner corpuscle and Pacinian corpuscle are classified into FA (Fast Adopting). SA detects the displacement of skins, which are translated to "rough" and "rugged" surfaces. On the other hand, FA detects the velocity and acceleration of skins, which are "smooth" and "textured" surfaces. These tactile receptors have detectable thresholds in depth direction with respect to frequencies of stimulus as shown in Figure 1 [1]. For example, the threshold of Pacinian corpuscle is 1 μm at 200 Hz. Merkel's disc has the threshold of 100 μm independent of the frequency. In planer direction, a finger tip has two-point discrimination threshold of 2-3 mm. Given these characteristics of tactile receptors, in order to display various textures, tactile displays require arrays of micro-actuators at less than 3 mm apart and capable of generating displacements greater than 100 μm and high-frequency operation at 200 Hz as shown in Figure 2, which is quite challenging.

Previously developed tactile displays

In previous work, two types of tactile displays have been developed associated with the two types of tactile receptors, SA and FA. SA type tactile displays generate displacements of finger skins by using actuators, such as SMA (Shape Memory Alloy) [2], multilayer PZT (Piezoelectric Zirconate Titanate) [3], PZT bimorph [4], pneumatic [5], RC servomotor [6] and thermopneumatic actuator [7]. They contain arrayed pins that are actuated in a vertical direction. Braille codes displays are classified into SA type tactile displays. However, it is difficult for SA type to be miniaturized because they require relatively large actuators to generate enough displacements. These actuators also have difficulty in generating as high-frequency vibration as 200 Hz. Meanwhile, FA type generates vibration using actuators like USM (UltraSonic Motor) [8], ICPF (Ionic Conductive Polymer Film) [9] and PZT [10][11]. Tactile display using electrical stimulation [12] is also proposed. They provide an illusion of textures to tactile receptors. However, it is difficult for FA type to generate as large displacement as 100 μm. Therefore, previously developed tactile displays cannot stimulate both SA and FA. The encountered challenge is development of micro-actuators that are capable of generating large displacement at a high frequency in an arrayed arrangement.

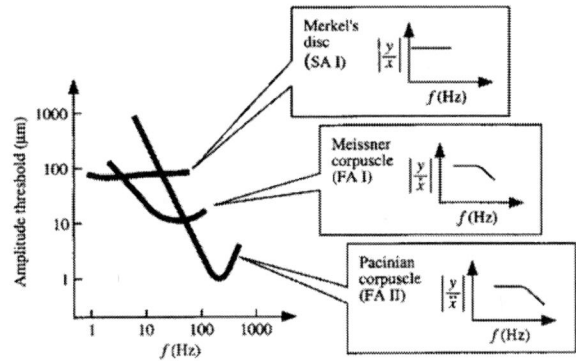

Figure 1: Thresholds of tactile receptors for vibratory stimulus and selective stimulation ranges [1].

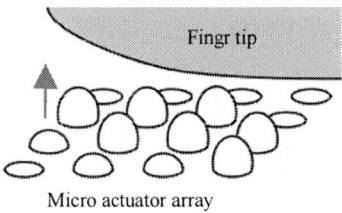

Figure 2: Image Schematic image of tactile display.

978-1-4244-2977-6/09 $25.00 © 2009 IEEE

In this research, we propose hydraulic displacement amplification mechanisms (HDAM). HDAM can amplify displacement of micro-actuators, such as USM and PZT actuators that are capable of high-frequency actuation but with only small displacement. HDAM will enable tactile displays to generate various tactile information by combining SA and FA stimulus.

CONCEPT

Tactile display contains arrayed micro-actuators as shown in Figure 2. The bumps represent contact parts to a finger pad. Figure 3 depicts a cross-sectional view of the HDAM. The top side of the chamber is the contact part while the bottom side is the drive part. Glycerin is encapsulated in the chamber. Because glycerin is an incompressible fluid, its volume extruded from drive part is transmitted to contact part without being decreased. Largely-deformable PDMS that can endure a stretch more than 450 % is synthesized to be elastic and bonded onto the contact and drive parts. Largely-deformable PDMS is detailed in the following chapter. Because the cross-sectional area of the drive part is larger than that of the contact part, the driven displacement by multilayer PZT actuator at the drive part is amplified at the contact part. PZT actuators excel in high-frequency actuation whereas they have difficulties in generating large displacements. HDAM overcomes this problem. The displacement generated by PZT actuators of tens of micrometers is amplified to be hundreds of micrometers by HDAM, which satisfies thresholds of all tactile receptors. In addition, HDAM isolates the contact part from actuators at the drive part that involve high voltage and heat, which is an excellent advantage from a safety perspective. Fabrication of HDAM is compatible with mass-producible MEMS process. Hence, the proposed mechanism enables the miniaturized densely-arrayed micro-actuators capable of stimulating both SA and FA.

DESIGN

The proposed HDAM was designed as follows; given the two point threshold of 2-3 mm, the interval of the micro-actuators was set to be 3 mm. Taking the highest threshold of Merkel's disc of 100 µm into consideration, each parameter was set as shown in Figure 4 assuming that deformation of the contact part was sphere, glycerin was an incompressible fluid and its volume was constant, and largely-deformable PDMS deformed elastically. Size of the drive part was fixed at 2.2 mm in diameter while some sizes of the contact part were prepared for comparison of amplified displacement. In order to make the contact part deform with less force, the largely-deformable PDMS sealing the contact part was fabricated more thinly than that at the drive part. Thicknesses of the PDMS membrane were 90 µm and 110 µm at the contact and drive parts, respectively.

FABRICATION
Largely-deformable PDMS

Largely-deformable PDMS polymer was made by mixing DC 3145 CLEAR and Thinner RTV (Dow Corning Toray Inc.). DC 3145 CLEAR is a PDMS elastomer that can stretch over 650 %. Thinner RTV was added to decrease the viscosity of the uncured largely-deformable PDMS, which enables spin-coating of the polymer with a thickness control capacity by spinning speed. Tensile test indicated Young's modulus of the Largely-deformable PDMS was 32 kPa and it could stretch by more than 450 %. The relation between spin coating revolution and thickness of the largely-deformable PDMS was investigated by ZYGO (ZYGO Inc.), which is optical 3D surface profiler.

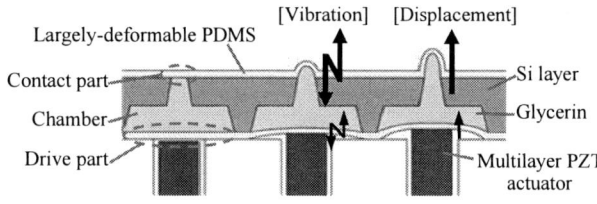

Figure 3: Schematic cross-sectional view of HDAM.

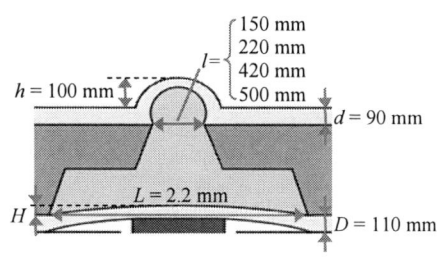

Figure 4: Model of HDAM.

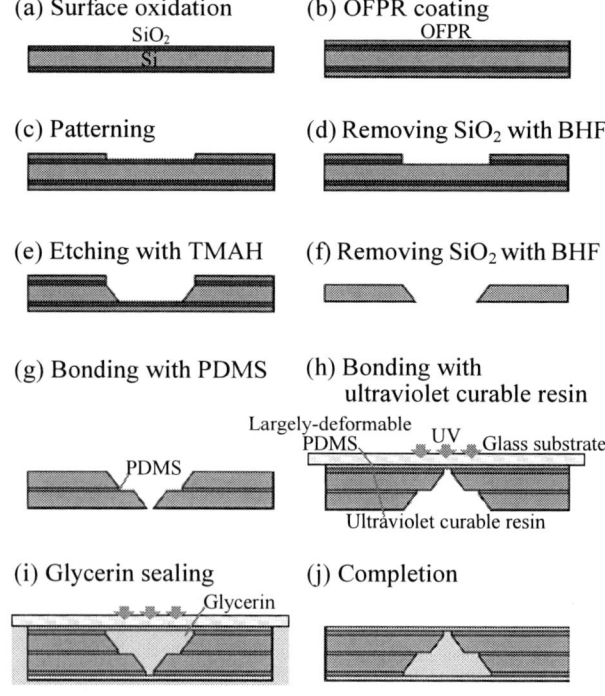

Figure 5: Fabrication processes of HDAM.

HDAM

Fabrication process is shown in Figure 5. Si wafer 380 μm in thickness was used for main body of HDAM. (a) SiO_2 film was thermally grown on a Si wafer at 1150 degrees C. (b) After OAP (Tokyo Ohka Kogyo Inc.) was coated on it as a primer, OFPR (Tokyo Ohka Kogyo Inc.) resist was patterned by spin coating. (c) SiO_2 film was etched by BHF (Buffered Hydrofluoric Acid) etching. (d) Si was anistropically etched by TMAH (tetramethylammonium hydroxide) with SiO_2 as a mask. OFPR was dissolved in the TMAH bath and removed. (e) SiO_2 was removed by BHF etching. (f) Si wafers forming the contact part and drive part were bonded via PDMS thin layer. (g) A largely-deformable PDMS thin layer was formed on a glass substrate and bonded to a Si structure via an ultraviolet curable resin. The ultraviolet curable resin is used as an adhesive that does not require heat or vacuum environment but ultraviolet to bond two layers. Largely-deformable PDMS was exfoliated from the glass substrate. (h) Glycerin was encapsulated in the chamber when a largely-deformable PDMS thin layer was bonded to the other side of the Si structure by ultraviolet curable resin in glycerin solution. Note that this process enables to encapsulate glycerin in the chamber without any bubbles. Although water is also an incompressive fluid, glycerin was advantageous over water due to the non-volatility considering the high permeability of PDMS membrane.

Figure 6 depicts the fabricated 3x3 HDAM. Figure 7 depicts side views of the deformation of the largely-deformable PDMS during actuations. This figure shows that the contact part is being deformed spherically.

RESULTS

Relation between driven and amplified displacement

Relation between the driven and amplified displacement was investigated. HDAM with the contact parts that have diameters of 150, 220, 420 and 500 μm and the drive parts 2.2 mm in diameter were tested. The amplified displacement was measured when the drive part was pushed up by a pin on the Z-axis stage. The amplified displacement was measured by image measurement of a microscope. Experimental result shown in Figure 8 indicated that displacement greater than 100 μm was achieved when 20μm-displacement was applied to 150μm-, 220μm- and 420μm-samples. The amplification of 500μm-sample was about two-fold, 420μm- four-fold, 220μm- six-fold and 150μm- seven-fold. The amplified displacement satisfied the thresholds of both SA and FA. The amplification became larger with the smaller the contact parts. It was indicated that further miniaturization of the contact parts enables further amplification. Further amplification enables miniaturization of the actuators in turn. As the contact part was fabricated smaller, however, largely-deformable PDMS were more likely to be exfoliated from the contact part. This is because the polymer membrane at the contact part experiences higher pressure with the smaller geometry. Augmentation of the amplification will be realized by miniaturizing HDAM with preventing largely-deformable PDMS from exfoliating.

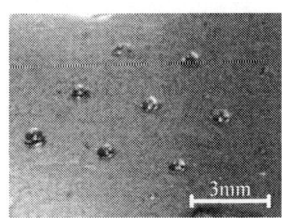

Figure 6: Fabricated 3x3 HDAM array.

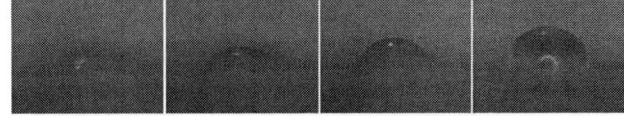

Figure 7: Deformation of contact part.

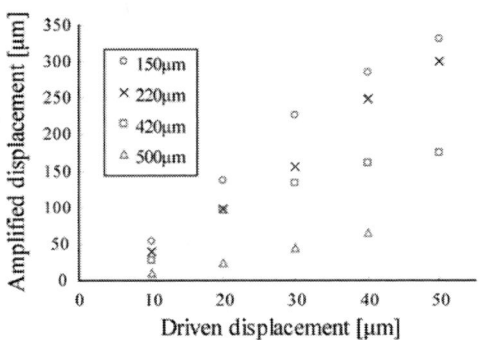

Figure 8: Relation between driven and amplified displacement.

Relation between frequency and amplification ratio

Relationship between actuation frequencies and amplification ratio was evaluated with 420μm-sample. The amplification ratio was calculated by dividing the amplified displacement by the driven displacement. The driven displacement of multilayer PZT actuators (NEC-TOKIN Inc.) and amplified displacement of the contact part were measured by LDV (Laser Doppler Vibrometer, GRAPHTEC Inc.). Because largely-deformable PDMS has optical transparency, LDV can not measure the displacements of the contact part. Therefore, the contact part was colored by white color spray to reflect laser well. The multilayer PZT actuators were driven by $100V_{p-p}$-sine-waves. The result is shown in Figure 9. Resonances at about 70 and 140 Hz were observed. The resonance frequencies depend on the geometry of HDAM. The required frequency of stimulus is mainly less than 200 Hz as shown in Figure 1. Effective amplification in the practical range will be realized by exploiting the resonances although the amplification ratio was lower when HDAM was generating vibration as shown in Figure 9 than it was generating displacement as shown in Figure 8.

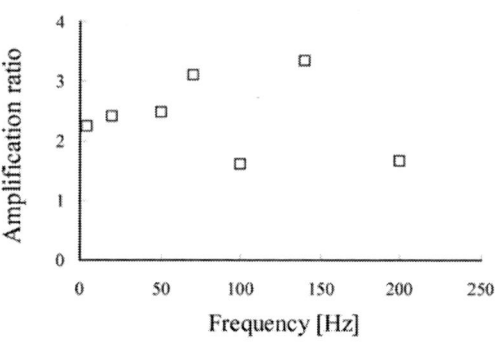

Figure 9: Relation between frequency and amplification ratio.

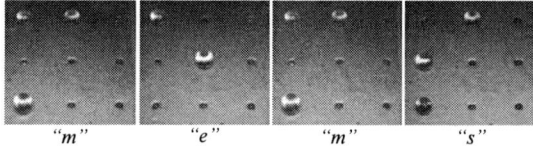

Figure 10: Demonstration of Braille code: mems.

Demonstration

We successfully demonstrated rewritable Braille codes with the developed tactile display with HDAM. Braille codes of "m" "e" "m" "s" were shown in Figure 10.

CONCLUSIONS

A hydraulic displacement amplification mechanism (HDAM) for a tactile display was proposed and fabricated. The experimental results showed that the driven displacement was amplified as greatly as seven-fold by HDAM. The amplified displacement without vibration satisfied the thresholds of both SA and FA. Rewritable Braille codes using the fabricated tactile display with HDAM were successfully demonstrated. In high-frequency actuation, HDAM satisfied the thresholds partly, which will be improved by exploiting the resonances of HDAM. The proposed MEMS tactile displays with HDAM have potentials to stimulate both SA and FA and convey various tactile information to users.

In future work, further amplification is realized by miniaturizing HDAM with preventing largely-deformable PDMS from exfoliating. This will enable tactile displays with HDAM to stimulate both SA and FA at the all range of frequency.

ACKNOWLEDGEMENTS

This work was partially supported by NEDO, Fine MEMS project.

REFERENCES

[1] T. Maeno, "Structure and Function of Finger Pad and Tactile Receptors," J. of the Robotics Society of Japan 2000, Vol. 18, No. 6, pp. 772-775.(in Japanese)

[2] X. Wu, S.H. Kim, C.H. Ji, and M.G. Allen, "A piezoelectrically-driven high flow rate axial polymer microvalve with solid hydraulic amplification," MEMS 2008, pp. 523-526.

[3] M. Ohka, H. Koga, Y. Mouri, T. Sugiura, T. Miyaoka and Y. Mitsuya, "Figure and texture presentation capabilities of a tactile mouse equipped with a display pad of stimulus pins," Robotica, Vol. 25, No. 4, pp. 451-460, Jul. 2007.

[4] G. Moy, C. Wagner, and R.S. Fearing, "A compliant tactile display for teletaction," Robotics and Automation, Proceedings. ICRA 00, Vol. 4, pp. 3409-3415, Apr. 2000.

[5] C.R. Wagner, S.J. Lederman, and R.D. Howe, "A tactile shape display using RC servomotors," HAPTICS 2002. Proceedings. 10th Symposium on, pp. 354 – 355, Mar. 2002.

[6] H.J. Kwon, S.W. Lee, and S.S. Lee, "Braille code display device with a PDMS membrane and thermopneumatic actuator," MEMS 2008, pp. 527-530.

[7] T. Maeno, K. Otokawa, and M. Konyo, "Tactile Display of Surface Texture by use of Amplitude Modulation of Ultrasonic Vibration," Ultrasonics Symposium, pp. 62-65, Oct. 2006.

[8] T. Tsuchiya, O. Tabata, J. Sakata, Y. Taga, "Specimen Size Effect on Tensile Strength of Surface Micromachined Polycrystalline Silicon Thin Films", J. Microelectromech. Syst., vol. 7, pp. 106-113, 1998.

[9] M. Konyo, S. Tadokoro, A. Yoshida, and N. Saiwaki, "A tactile synthesis method using multiple frequency vibrations for representing virtual touch," IROS 2005, pp. 3965-3971.

[10] M. Biet, G. Casiez, and F. Giraud, "Lemaire-Semail, B.Discrimination of Virtual Square Gratings by Dynamic Touch on Friction Based Tactile Displays," HAPTICS 2007, pp. 41-48.

[11] H. Kotani, M. Takasaki, and T. Mizuno, "Surface Acoustic Wave Tactile Display using a Large Size Glass Transducer," Mechatronics and Automation, 2007. ICMA 2007, pp. 198-203.

[12] H. Yokota, A. Yamamoto, H. Yamamoto, and T. Higuchi, "Producing Softness Sensation on an Electrostatic Texture Display for Rendering Diverse Tactile Feelings," World Haptics 2007. 2nd Joint, pp. 584-585.

HIGH-PURITY SEPARATION OF RARE SPECIES IN DROPLET MICROFLUIDICS USING DROPLET-CONDUIT STRUCTURES

Gaurav J. Shah and Chang-Jin "CJ" Kim
Mechanical and Aerospace Engr. Dept., University of California, Los Angeles (UCLA), CA, U.S.A.

ABSTRACT

We present a technique to achieve high-purity isolation of the target species in electrowetting-on-dielectric (EWOD)-based droplet microfluidics by forming a long and slender liquid path between droplets that acts as a "conduit" for actively transported target species, while minimizing non-specific transport due to diffusion and fluidic movement. By stabilizing the conduit chemically (e.g., using surfactants) and controlling the fluid by EWOD, we demonstrate, as an example, very high magnetic bead (MB) collection efficiency (> 99%) while eliminating ~97% non-magnetic beads (nonMBs) in just one separation step.

INTRODUCTION

Concentration/separation is critical in many biochemical assays, particularly where purity of isolated target species ("TS") (e.g. specific cells, proteins, DNA, either by themselves or bound to beads) is vital to their effectiveness. Unlike continuous microfluidics where TS are immobilized while wash-buffer is flowed through the channels to remove impurities, purification in droplet microfluidics (e.g. by EWOD) typically involves serial (i.e., repeated steps of) dilution of the non target species ("nonTS") [1],[2]. In each wash step, a buffer droplet is added to the sample. The TS is actively collected in one region of this combined ("parent") droplet using one or more differentiating properties, such as magnetic, electric, optical or dimensional (e.g., [3],[4]). The droplet is then split so that most of the TS are collected in one of the daughter droplets.

The distribution of the nonTS between the two daughter droplets, on the other hand, is governed by their original distribution in the parent droplet and non-specific phenomena such as diffusion and microfluidic movement that occur during the purification step. If the nonTS were uniformly distributed in the parent droplet, they would be distributed between the daughter droplets roughly in proportion to their volumes. Thus, even though many nonTS from the parent droplet will be removed in the form of the "depleted" droplet (from which TS have been depleted), this would still leave a significant proportion of nonTS in the "collected" droplet (wherein the TS has been collected). In order to improve the TS purity, therefore, TS is collected across the incoming buffer droplet [5], which is initially free from the nonTS (Fig. 1(1a-1b)). The droplet is then stretched (Fig. 1(1c)) and split (Fig. 1(1d)) into the collected (left) and depleted (right) droplets.

Even though the original distribution of nonTS in the parent droplet is favorable for high purity (Fig. 1(1b)), their non-specific transport and redistribution of the nonTS during the subsequent collection and droplet splitting (Fig. 1(1b-1c)) may lower purity of the TS in the collected droplet (Fig. 1(1d)), entailing multiple wash cycles (Fig. 1(2a-3a)). Two main mechanisms for the non-specific transport are: (a) diffusion and (b) microfluidic movement.

Fick's law relates the diffusion flux (J_{dif}) of a species across a fluidic section to its concentration gradient:

$$J_{dif} = -D\frac{\partial \phi}{\partial x}$$

where D is the diffusion coefficient determined by the particle radius, temperature and viscosity of the medium. For a given particle size, concentration gradient and viscosity of the medium, the rate of diffusion-driven transport across a fluidic section varies directly with its cross-sectional area and inversely with its length.

The second mechanism for contamination is fluidics-driven transport, i.e. species transported due to the viscous forces during fluidic movement. Since the nonTS are usually not actively manipulated, they are therefore highly susceptible to travel with the flow. Although appropriate choice of droplet actuation sequence can reduce the flow into the collected droplet, some flow is inevitable during neck creation and pinch-off required for droplet splitting. As the droplet is stretched, the ensuing flow drags the nonTS along with it. This is particularly pronounced along the droplet meniscus, where flow velocity is higher [6].

NonTS contamination into the collected droplet due to both the mechanisms described above could be reduced if a slender neck was created in the buffer droplet prior to sample introduction. While this neck could be a "conduit" for active TS transport, its slender (long and narrow) structure would make it an excellent diffusion barrier. Moreover, since little fluidic movement would be required from this pre-necked stage to split droplet formation, such a slender conduit would also help minimize the fluidics-driven nonTS transport during the splitting.

Making a narrow, physical channel or other such structures is one way to achieve high purity [7], but has disadvantages like complexity in fabrication and lack of re-configurability and control. Instead, we propose a purely fluidic conduit in this paper to overcome these issues. The challenge, however, is that for pure DI water or buffer media, such a slender liquid structure tends to be hydrodynamically unstable. The addition of surfactants, for example, can make the thin liquid column more stable and may therefore stabilize the "droplet-conduit structure", forming the basis of the proposed idea.

PROPOSED IDEA: DROPLET-CONDUIT

The use of surfactants on EWOD has recently received much attention [8],[9]. However, addition of surfactants to the solution tends to impede droplet splitting by stabilizing the neck. Although long necks are routinely found during the cutting steps, their dimensions and locations are not so controllable. Here, we report achieving the controllability using a combination of surfactants and electrode modification. Specifically, a surfactant was added in a moderate concentration to increase tendency for stable conduit formation, while a stabilizing electrode ("SE") was used to toggle between stability and breakage.

978-1-4244-2977-6/09 $25.00 © 2009 IEEE

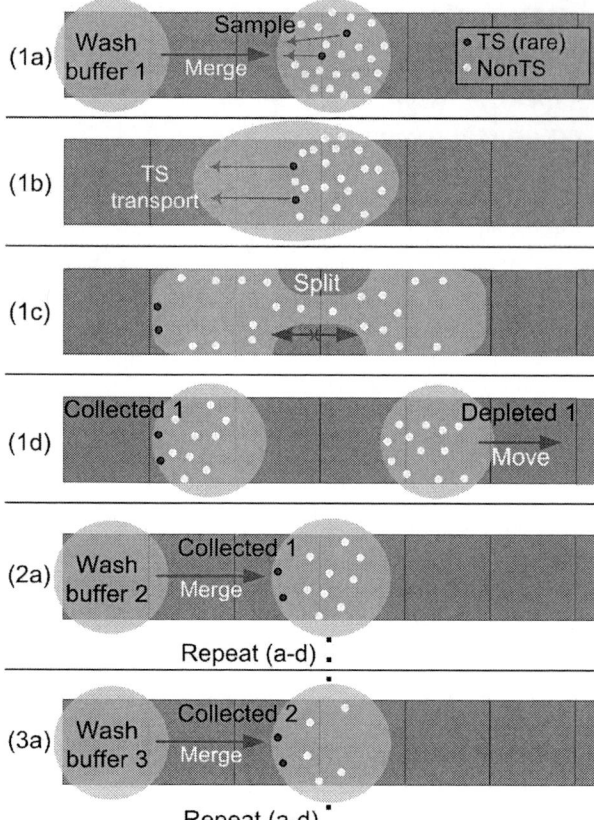

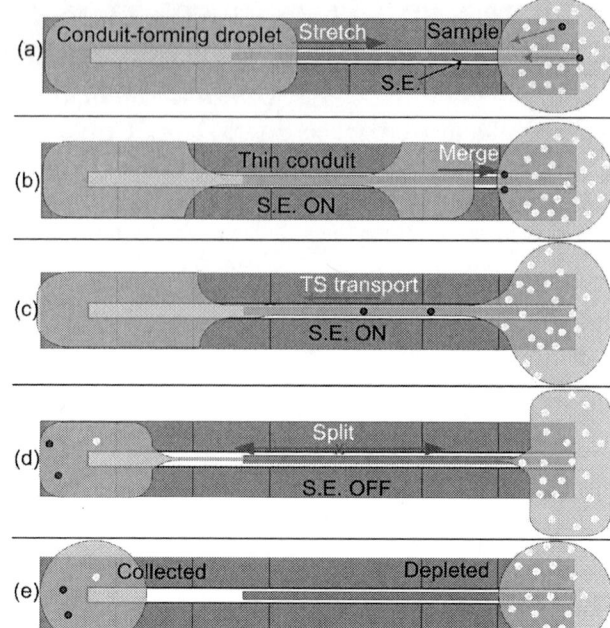

Fig. 1: Known technique of TS purification by serial dilution: (1a) Wash buffer is added to sample. (1b) TS (dark) is transported and collected in one region of the droplet. (1c) The droplet is then stretched and (1d) cut to form "collected" (wherein TS are collected) and depleted (depleted of TS) droplets. Many nonTS (bright) are removed with the depleted droplet, but some nonTS enter collected droplet due to diffusion, and/or fluidics-driven transport. (2a, 3a) More wash-buffer is added and the above steps are repeated to improve TS purity by serial dilution of the nonTS.

Fig. 2 illustrates the proposed idea. The typical EWOD electrode layout is modified to incorporate a slender line electrode (i.e. SE) through the center of the square EWOD electrodes. The conduit-forming droplet (left) is stretched towards the mixed sample containing TS and nonTS. Keeping the SE on, a slender droplet-conduit is created from the conduit-forming droplet (Fig. 2(a,b)), whose width is defined by the SE.

On merging with the sample, the TS are transported across the conduit using an active transport mechanism (Fig. 2(c)). However, very few nonTS can cross the diffusion barrier presented by the droplet-conduit. After TS transport, the SE is turned off, and the droplet is stretched further (Fig. 2(d)), breaking the conduit and completing the droplet split (Fig. 2(e)). It should be noted that much lesser fluidic movement is involved in this cutting operation as compared to Fig. 1, since the neck was already formed. As such, the fludically driven nonTS transport into the collected region is much reduced. The collected droplet (left) contains the TS and very few nonTS, most of which are left in the depleted droplet (right).

Fig. 2: Proposed technique for high purity rare TS separation using droplet conduit. (a) TS are transported to left edge of sample, while nonTS are randomly distributed in the droplet. With SE on, the conduit-forming droplet is stretched, (b) forming a slender "conduit". On merging with sample, (c) the conduit allows active TS transport. But the slender conduit restricts diffusion-driven nonTS transport. (d,e) When droplet is stretched with SE off, the droplet splits with minimal fluidically driven nonTS transport into the high purity "collected" droplet and the "depleted" droplet.

MATERIALS AND METHODS

The technique was demonstrated for high purity collection of magnetic beads (MBs) from a mixed population of MBs and non-magnetic beads (nonMBs) on an EWOD device. MBs of 4.5 µm diameter (Invitrogen) were used as the TS, and fluorescent nonMBs of 5.2 µm diameter (Molecular Probes) were used as the nonTS. Pluronic F68 surfactant of 0.15% w/v (Sigma-Aldrich) in PBS was used for both the sample and the buffer droplets.

A standard two-plate EWOD device (e.g. [10]) was used for the experiments (Fig. 3). For the bottom plate, EWOD electrodes were patterned into the 140 nm ITO layer on glass substrate. A 60 µm wide line going through the middle of the 1 mm x 1 mm EWOD electrodes defines the SE. A gap (1-1.5 mm) was left at the left end of the SE to ensure the breakage of the conduit during droplet splitting. Silicon nitride (1.1 µm thick) was deposited by plasma-enhanced chemical vapor deposition (PECVD) to form the dielectric layer, and Cytop® (~1 µm thick) was spun-coated to form the top hydrophobic layer. For the top plate, an unpatterned layer of ITO was coated with 100 nm silicon nitride and 50 nm Cytop®. Double-sided tape (~100 µm thick) was used as the spacer between the plates.

Droplet actuation was achieved by sequential application of voltage (70-80 V_{ac} @1 kHz) to the EWOD electrodes. Electronic control for the actuation sequence was controlled using LabVIEW (National Instruments) via a digital I/O device (DAQPad 6507, National

Instruments). For magnetic collection, an NdFeB permanent magnet (½" dia x ½" thick) was placed on top of the device to the left. Droplet actuation movies and low-magnification (for a larger field-of-view) images were captured by a video camera (Panasonic KR-222) mounted onto the microscope (Nikon TE 2000U), while better quality fluorescence images were taken using a cooled-CCD camera (Photometrics Coolsnap EZ) attachment.

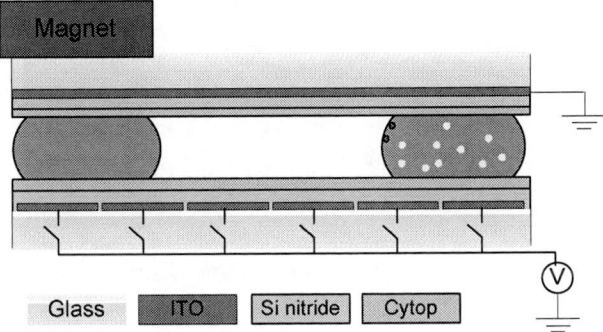

Fig. 3: Schematic cross-section of EWOD device used.

RESULTS AND DISCUSSION

Experiments were performed to show high purity separation of TS (i.e., MBs) from fluorescent nonTS (i.e., nonMB) using the slender droplet conduit. In order to demonstrate the utility of conduit for purification of rare species, a low (around 1:20) TS:nonTS ratio was chosen.

Fig. 4 shows the image sequence for the experiment performed. The sample droplet containing MBs and nonMBs, along with 0.15 %w/v pluronic surfactant F68 is placed on the right, while the conduit-forming (i.e., buffer) droplet, also containing the surfactant, is introduced from the left (Fig. 4(a)). The magnet is positioned at the left, so that the MBs collect at the left meniscus of the sample droplet (see Fig. 4(a) inset). A

stable, slender conduit is formed by stretching the conduit-forming droplet while keeping the 60 µm wide SE on (Fig. 4(b)). On merging with the sample (Fig. 4(c)), the MBs from the sample are actively transported across the conduit towards the magnet, to the leftmost edge of the combined droplet, while most of the nonMBs remain behind at the right (Fig. 4(d)). After all the MBs are transported, the SE is turned off and the droplet is stretched further (Fig. 4(e)) so as to split it into the collected (left) and depleted (right) droplets (Fig. 4(f)). Since the neck was already formed prior to the merging, the splitting operation involves very little fluidic movement.

To evaluate the purity of the separation, MBs (dark) and nonMBs (fluorescent) are counted in the collected (Fig. 5(a,b)) and depleted (Fig. 5(c,d)) droplets. Images were taken using both cameras under fluorescent excitation, but with some bright field illumination so as to visualize non-fluorescent features as well. The bright nonMBs can be easily distinguished from the background. To distinguish the MBs from the other dark features in the images, a magnet was introduced on one side of the droplet, and the magnetically responsive collected at the edge were counted. The original sample contained ~283 nonMBs and 16 MBs. Comparing Fig. 5 (a) and (b) it is clear that much fewer nonMBs (bright) can be seen in the collected droplet compared to the depleted droplet. The fluorescent nonMB count in the collected droplet is less than 10 while that in the depleted droplet is ~267. Thus, ~97% of nonMBs were kept out of the collected droplet.

Around 16 MBs were counted in the collected droplet (Fig. 5(a)), while no (0) MBs were seen in the depleted droplet (Fig. 5(b)), demonstrating that high purity magnetic separation was achieved while maintaining high collection efficiency (> 99%).

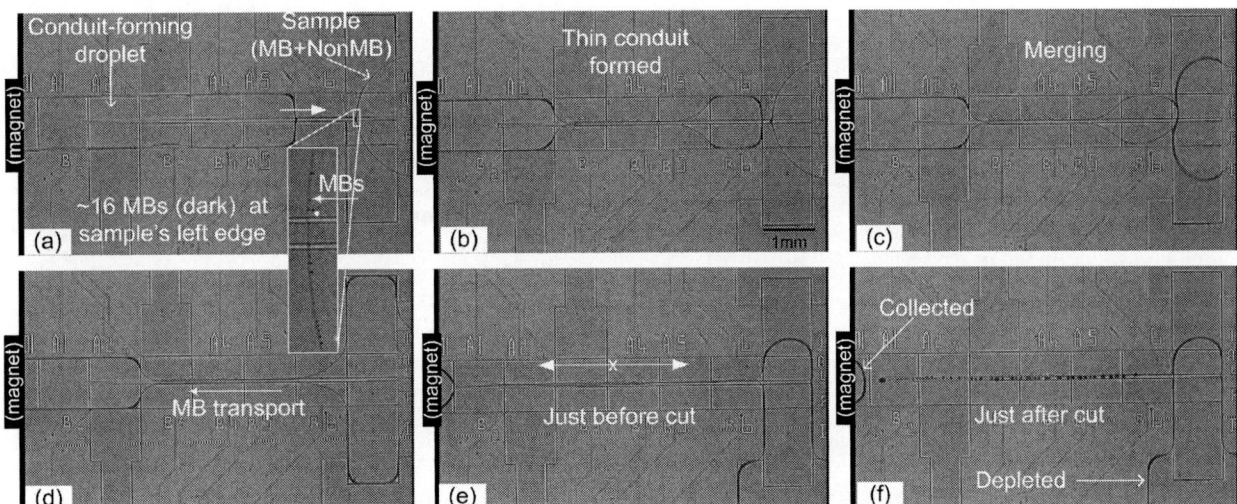

Fig. 4: Image sequence for high purity magnetic separation using droplet conduit structures: All droplets contain 0.15% pluronic F68 in PBS. (a) Magnet is positioned to the left of the sample and conduit-forming droplet, so that MBs (dark) are attracted to the left edge of sample (see inset). (b) Conduit is formed by stretching the droplet while the SE is on. (c) Sample is merged with the conduit-forming droplet, (d) allowing MBs to pass through. After transport, very little fluidic movement is involved as (e) the droplet further stretched with SE turned off, (f) cutting it into "collected" (MBs collected, very few nonMBs) and depleted droplets (depleted of MBs). Satellite droplets can be cleaned up by depleted droplet.

978-1-4244-2977-6/09 $25.00 © 2009 IEEE

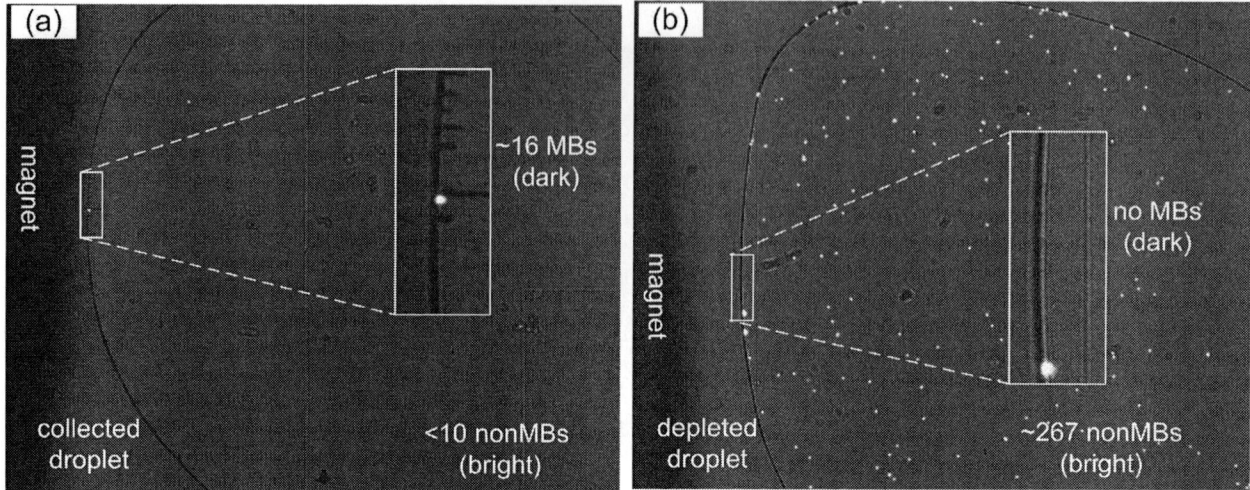

Fig. 5: High-purity separation is obtained using droplet-conduit structures, as estimated by counting the MBs (dark) and nonMBs (bright) in the collected and depleted droplets. Images were taken under fluorescent excitation but with some bright-field illumination so as to visualize non-fluorescent features as well. To count MBs (and not dirt etc.), a magnet was introduced right next to a droplet edge, and the magnetically responsive MBs attracted to it were counted. Original sample droplet had ~16 MBs and ~283 nonMBs. (a) All the MBs (~16) and very few nonMBs (<10) were counted in the "collected droplet". (b) No MBs (0) and a majority of nonMBs (~267) were seen in the "depleted droplet". The results correspond to target (MB) collection efficiency over 99% in the collected droplet, with nonMB concentration dropping to ~3% (or by more than 28 times) in just one step.

CONCLUSION

Purity of the TS in droplet based washing steps is adversely affected by the non-specific transport, mainly as a result of diffusion and fluidic movement. (For the present device geometry and particle size, it turns out that the latter is particularly dominant). Contamination due to both the factors is reduced using droplet-conduit structures, without sacrificing the TS collection efficiency.

The chemically stabilized slender droplet conduit structure controlled by EWOD not only creates a diffusion barrier, but also provides a pre-formed neck to minimize fluidic movement required for droplet splitting after TS collection, thus ensuring high purity as well as high TS collection efficiency.

ACKNOWLEDGEMENT

This work was supported by NASA through Institute for Cell Mimetic for Space Exploration (CMISE), NIH through Pacific Southwest RCE (grant AI065359), and Intramural Seed Grant of UCLA Department of Urology.

REFERENCES

[1] R. B. Fair, "Digital microfluidics: is a true lab-on-a-chip possible?," *Microfluidics and Nanofluidics,* vol. 3, pp. 245-281, 2007.

[2] J. Gong and C.-J. Kim, "All-electronic droplet generation on-chip with real-time feedback control for EWOD digital microfluidics," *Lab on a Chip,* vol. 8, pp. 898-906, 2008.

[3] S. K. Cho, Y. J. Zhao, and C.-J. Kim, "Concentration and binary separation of microparticles for droplet-based digital microfluidics," *Lab on a Chip,* vol. 7, pp. 490-498, 2007.

[4] G. J. Shah, J. L. Veale, Y. Korin, E. F. Reed, H. A. Gritsch, and C.-J. Kim, "Concentration of CD8+ lymphocytes on EWOD Platform for monitoring organ transplant rejection," in *Proc. Solid-State Sensors, Actuators and Microsystems Workshop,* Hilton Head Island, SC, USA, Jun. 2008, pp. 28-31.

[5] Y. Z. Wang, Y. Zhao, and S. K. Cho, "Efficient in-droplet separation of magnetic particles for digital microfluidics," *Journal of Micromechanics and Microengineering,* vol. 17, pp. 2148-2156, 2007.

[6] J. Fowler, H. J. Moon, and C.-J. Kim, "Enhancement of mixing by droplet-based microfluidics," in *Proc. IEEE Int. Conf. MEMS,* Las Vegas, NV, Jan. 2002, pp. 97-100.

[7] S. M. Kim, S. H. Lee, and K. Y. Suh, "Cell research with physically modified microfluidic channels: A review," *Lab on a Chip,* vol. 8, pp. 1015-1023, 2008.

[8] O. Raccurt, J. Berthier, P. Clementz, M. Borella, and M. Plissonnier, "On the influence of surfactants in electrowetting systems," *Journal of Micromechanics and Microengineering,* vol. 17, pp. 2217-2223, 2007.

[9] V. N. Luk, G. C. H. Mo, and A. R. Wheeler, "Pluronic additives: A solution to sticky problems in digital microfluidics," *Langmuir,* vol. 24, pp. 6382-6389, 2008.

[10] S. K. Cho, H. J. Moon, and C.-J. Kim, "Creating, transporting, cutting, and merging liquid droplets by electrowetting-based actuation for digital microfluidic circuits," *Journal of Microelectromechanical Systems,* vol. 12, pp. 70-80, 2003.

A HIGHLY FLEXIBLE SUPERHYDROPHOBIC MICROLENS ARRAY WITH SMALL CONTACT ANGLE HYSTERESIS FOR DROPLET-BASED MICROFLUIDICS

Maesoon Im[1], Dong-Haan Kim[1], Xing-Jiu Huang[2],
Joo-Hyung Lee[1], Jun-Bo Yoon[1], and Yang-Kyu Choi[1]
[1]Korea Advanced Institute of Science and Technology (KAIST), Daejeon, KOREA
[2]University of Oxford, Oxford, UK

ABSTRACT

This paper reports a highly flexible superhydrophobic and superhydrorepellent microlens array substrate with very low flow resistance. Even though the microlens array has no nanostructures, it shows hydrophobic property due solely to its geometrical effect. A contact angle of 165° and hysteresis of 3° are achieved on a flexible polydimethylsiloxane (PDMS) microlens array substrate with a Teflon (polytetrafluoroethylene) coating. Moreover, double-layered metals (Cr/Au) are sandwiched between the PDMS and Teflon layers for electrostatic or electrowetting-on-dielectric (EWOD) actuation. Due to its low flow resistance and superhydrophobicity, the array can be used as a microfluidic component that reduces external pressure and power consumption for mobility.

INTRODUCTION

Random [1-3] or ordered [4-7] nanostructures and microstructures can be used to realize superhydrophobic surfaces. To introduce nanoscaled rough surfaces, researchers have reported numerous approaches, including crystal growth [1], catalyzed growth [1], electrospraying [2], plasma treatment [3], dry etching [4], and replica molding with a porous anodic aluminum oxide template [5, 6]. Various materials have been used such as polydimethylsiloxane (PDMS) [3], silicon [4], carbon nanotube arrays [7, 8], and nanowires [9].

In addition to those previous works, we have reported a perfectly ordered microbowl array [10] that makes large-area superhydrophobic surfaces without nanostructures. A photoresist microbowl array fabricated by means of three-dimensional diffuser lithography [11] has extremely superhydrophobic features [10]. Recently, we fabricated a microbowl array and a microlens array on a flexible polymer substrate [12] with a soft lithography replica molding method.

A liquid droplet on a microbowl or microlens array follows the wetting behavior of a Cassie-Baxter model [13]. Because air is trapped among adjacent microlenses, the surface of the microlens array is more hydrophobic than a flat surface made of the same material. In Figure 1, a shape of the droplet on the microlens array is shown.

One helpful way of manipulating liquid droplets in microfluidic systems is to utilize electrostatic force or electrowetting-on-dielectric (EWOD) actuation; however, a few issues should be addressed in relation to the aforementioned structures before these methods can be used in flexible applications [14]. First, the low flow resistance is crucial. It is noticeable that the high contact angle does not guarantee small contact angle hysteresis [15]. Second, a thin metal film for the electrostatic or EWOD actuation should be conformal, uniform, and reliable on the substrate.

When the PDMS is used as a structural material, the microlens array has a smaller contact angle than the microbowl array [12], even though the microlens array is also hydrophobic and has a simpler fabrication process. Moreover, the high adhesive force of a PDMS microlens array is a fatal disadvantage for a microfluidic component because the adhesive force impedes the transportation of liquid samples to a designated location. By decreasing the adhesive force on the surface of the PDMS microlens array, we can ensure that the delivery of liquid samples with reduced power consumption is possible in a microfluidic system, especially as a microfluidic channel and as a form of droplet manipulation with electrostatic force or EWOD actuation.

In the case of nanostructured superhydrophobic surfaces, a process of filling gaps between nanostructures to form metal and dielectric layers may degrade the hydrophobic property that originates from nanoscaled geometrical shapes. Although metal electrodes can be integrated underneath nanostructures, higher operating voltages are needed on account of the thick nanostructured materials required for hydrophobicity [16]. On the surface of a microlens array, a metal layer can be deposited with conformal coverage due to its convex shape. While keeping up this advantage of the microlens array, we attain a lower flow resistance and a higher contact angle in this work.

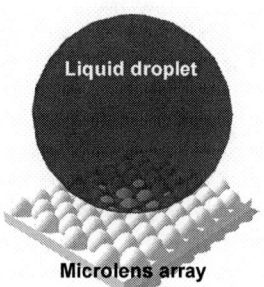

Figure 1: Schematic of a droplet shape on a microlens array in a Cassie-Baxter regime

FABRICATION PROCESS

The fabrication process of a microlens array with reduced flow resistance is shown in Figure 2. A thick positive-type photoresist (AZ9260, Clariant Co. Ltd.) is spin-coated on a silicon wafer. Three-dimensional diffuser lithography [11] is then applied to the photoresist to create microbowl patterns. As shown Figure 2(a), the direction of UV light is randomized by a sandblasted diffuser plate (F43-725, Edmund Optics Co. Ltd.) on a photomask, resulting in microlens-shape exposure profiles. The fabrication conditions of the photoresist microbowl array are well described in the literature [10].

978-1-4244-2977-6/09 $25.00 © 2009 IEEE

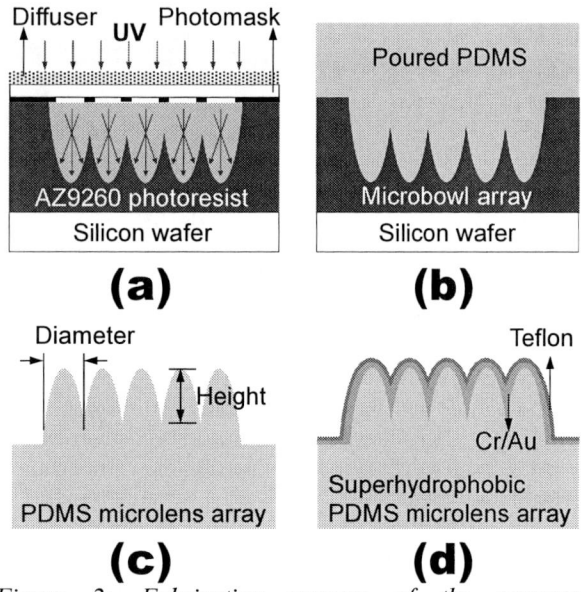

(a) **(b)**

(c) **(d)**

Figure 2: Fabrication process of the proposed hydrorepellent superhydrophobic PDMS microlens array (a) three-dimensional diffuser lithography; (b) the pouring of PDMS onto the fabricated microbowl array (photoresist mold); (c) the peeling of the PDMS from the photoresist mold; (d) Teflon (500 nm) coating on the sputtered Cr/Au (20 nm/300 nm)

Before the replica is formed, the surface of the photoresist microbowl array is passivated with the vapor phase of a silanizing agent (tridecafluoro-1,1,2,2-tetrahydrooctyl trichlorosilane) to help release the PDMS. A PDMS prepolymer is prepared by thorough mixing of the PDMS base and the curing agent (Sylgard 184 Silicone Elastomer Kit, Dow Corning, Midland, MI) in a 10:1 (base:agent) weight ratio. The prepolymer is then poured onto the mold of the photoresist microbowl array, and air bubbles created during the mixing and pouring process are removed in a low pressure chamber. After that, the PDMS prepolymer is solidified at room temperature for a day or at 80°C for an hour in a convection oven.

After peeling off the PDMS from the microbowl mold, we sputtered double-layered metals (Cr/Au) on it for electrostatic or EWOD actuation. Finally, a 500 nm Teflon layer is formed by the spin-coating of 2 wt% Teflon AF2400 (amorphous fluoropolymer, DuPont, Wilmington, DE) solution in FC-40 (perfluorocarbon, 3M, St. Paul, MN), which evaporates overnight at room temperature.

Figure 3 shows scanning electron microscope (SEM) images of the fabricated photoresist microbowl array and the PDMS microlens array, which is a replica of the microbowl array. As shown in Figures 3(a) to 3(d), the arrays are formed uniformly in a large area. The microlens has a diameter of 10 μm and a height of 13 μm. Figures 3(e) and 3(f) clearly show that metal layers are deposited over the microlens array with conformal coverage.

Photographs of the fabricated samples in Figure 4 demonstrate good flexibility and superhydrophobicity. The area of the PDMS microlens array sample is 1 cm^2, and it can be enlarged to wafer size or even further. On account of the flexible PDMS substrate and the very thin

metal layer, the superhydrophobic sample is highly flexible. Additionally, a spherical water droplet is sustained on the superhydrophobic surface, which assures a high contact angle.

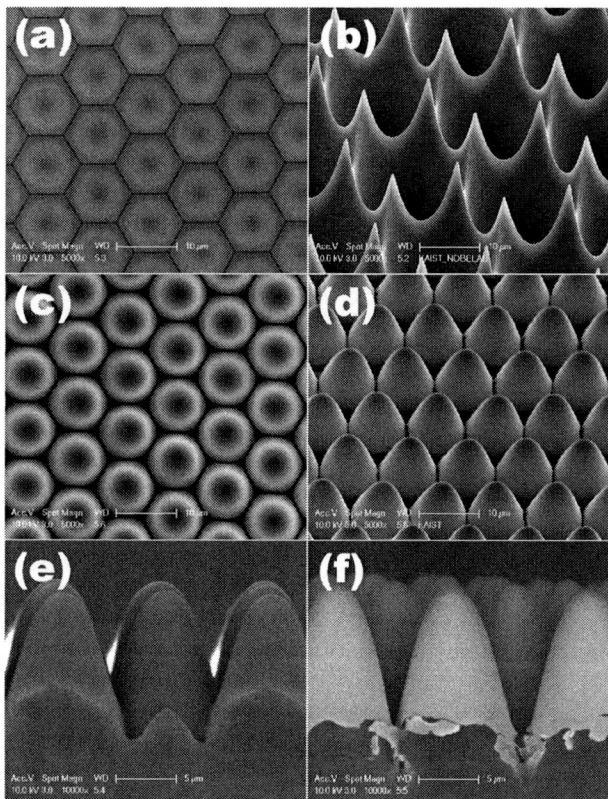

Figure 3: SEM images of fabricated sample (a) top view of a photoresist microbowl array mold fabricated by means of diffuser lithography; (b) tilted view of the photoresist microbowl array mold; (c) top view of a PDMS microlens array fabricated from a photoresist microbowl array mold; (d) tilted view of a PDMS microlens array; (e) cross section of a PDMS microlens array; (f) cross section of a PDMS microlens array after Cr/Au deposition

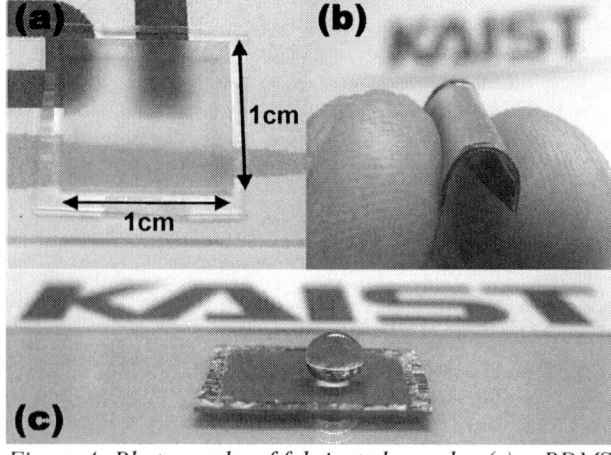

Figure 4: Photographs of fabricated samples (a) a PDMS microlens array; (b) flexibility test after Cr/Au deposition and Teflon coating on a PDMS microlens array; (c) a water droplet (approximately 10μl) on the fabricated superhydrophobic sample

978-1-4244-2977-6/09 $25.00 © 2009 IEEE 476

EXPERIMENTAL RESULTS

The wettability on the surface of the fabricated microlens array is in the Cassie-Baxter regime [13] because of the air trapped between the microstructures, as on the photoresist microbowl array [10]. With consideration given to the surface roughness, the contact angle (θ_{CB}) of the surfaces governed by the Cassie-Baxter model is expressed [1] as follows:

$$\cos \theta_{CB} = rf(\cos \theta_{FLAT}) + f - 1 \qquad (1)$$

where r is the roughness factor of the surface, f is the fraction of area that supports the liquid droplet, and θ_{FLAT} is the contact angle on a flat surface.

As shown in Figure 5, the contact angle of the flat Teflon surface ($\theta_{FLAT, Teflon}$) is slightly higher than that of the flat PDMS surface ($\theta_{FLAT, PDMS}$). Together with the geometrical effect (r and f) on hydrophobicity, the Teflon coating on the PDMS microlens array enhances the contact angle so that the angle is comparable to that of the microbowl array [10]. Figure 6 shows the contact angles before and after the Teflon coating. Note that the hydrophobic surface becomes superhydrophobic ($\theta_C > 150°$) when the Teflon layer is introduced.

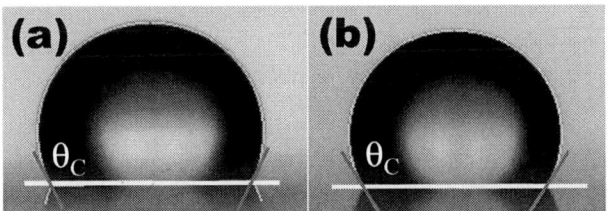

Figure 5: Contact angles of flat surfaces (a) PDMS on a silicon wafer ($\theta_{FLAT, PDMS} = 116°$) and (b) a Teflon-coated silicon wafer ($\theta_{FLAT, Teflon} = 122°$)

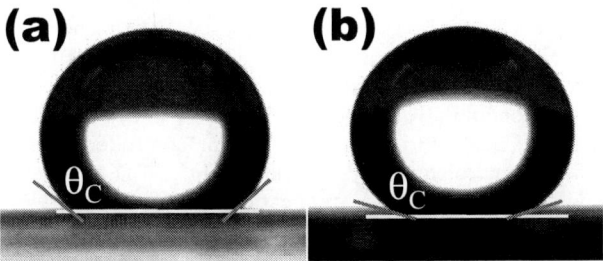

Figure 6: Contact angles of the microlens array (a) a PDMS-only microlens array ($\theta_C = 141°$) and (b) a Teflon-coated PDMS microlens array ($\theta_C = 165°$)

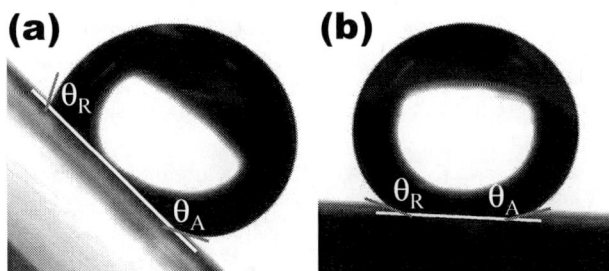

Figure 7: Contact angle hysteresis of the microlens array on a tilted plate (a) a PDMS microlens array ($\theta_{ADV} = 154°$, $\theta_{REC} = 117°$; $\theta_{HYS} = 37°$) and (b) a Teflon-coated PDMS microlens array ($\theta_{ADV} = 165°$, $\theta_{REC} = 162°$; $\theta_{HYS} = 3°$)

Reduction of the flow resistance on the Teflon-coated surfaces is confirmed by measurement of the contact angle hysteresis (θ_{HYS}), which is the difference between an advancing and receding contact angle. The tilting plate method is used to analyze the contact angle hysteresis, as shown in Figure 7. The hysteresis of the proposed structure (Teflon/Au/Cr/microlens) is reduced remarkably from 37° to 3°.

The small contact angle hysteresis means that a liquid droplet can roll off easily due to the negligible static friction force (F_f), which is calculated as follows [16]:

$$F_f = 2\gamma_{la}w(\cos \theta_{ADV} - \cos \theta_{REC}) \qquad (2)$$

where γ_{la} is the liquid-air interfacial energy, w is the width of the droplet, θ_{ADV} is the advancing contact angle, and θ_{REC} is the receding contact angle.

We can estimate the reduction of static friction force by the above equation (2). The friction force on the proposed structure is reduced enormously to just 3.3% of that on the initial PDMS microlens array. This level of reduction reveals that droplet manipulation is feasible on the proposed surface with low power consumption.

The contact angle and its hysteresis of the PDMS microlens array are strongly dependent on the aspect ratio of the microlens [12]. Therefore, by controlling the aspect ratio, we can adjust those parameters to satisfy the demands of end-users.

To demonstrate the superhydrorepellency of the fabricated sample, we recorded a video clip of a water droplet rolling off. Figure 8 shows the captured images with a time interval of 1/30 s. A deionized water droplet of 15 µl is dispensed by a micropipette on the sample, which is placed on a slide glass tilted about 6°. This demonstration clearly shows the exceedingly slippery surface characteristic, which is a very significant aspect of a self-cleaning application.

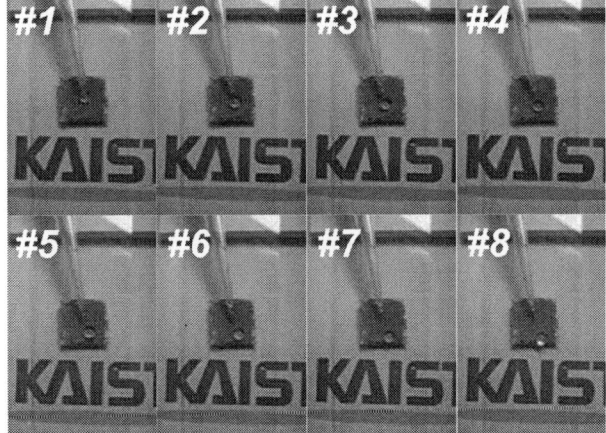

Figure 8: Captured images to show rolling off characteristics of a water droplet (approximately 15µl) on the fabricated sample with 6° tilted angle. The time interval between adjacent frames is 1/30 s.

An endurance test is carried out with a vortex mixer to give the cycled bending stress to the fabricated sample. It should be noted that the fabricated sample showed no significant degradation of the contact angle or electrical resistance (R_{AB}) of the metal layer after being bent more than 10^5 times as shown in Figure 9. This result is critical in applications involving liquid transportation on a flexible substrate for an arbitrarily curved shape [14]. The fact that the electrical connection is guaranteed after the repetitive bending highlights the potential use of the fabricated sample as a substrate for droplet movements by electrostatic force or EWOD actuation in droplet-based microfluidics.

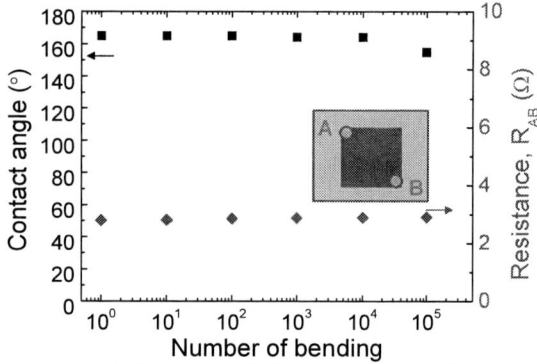

Figure 9: Endurance of the contact angle and the electrical resistance after the cycled bending stress

CONCLUSIONS

In this work, substrates with the characteristics of superhydrophobicity, superhydrorepellency, and flexibility were demonstrated with the aid of a Teflon-coated PDMS microlens array. The array consists of a unit microlens with a diameter of 10 μm and a height of 13 μm. The contact angle improvement from 141° to 165° ensures the attainment of a satisfactory level of hydrophobicity. In addition, the fact that the contact angle hysteresis is reduced from 37° to 3° ensures that a satisfactory level of hydrorepellency is achieved with the aid of Teflon coating and a three-dimensional microlens structure.

For potential droplet manipulation on the fabricated sample by electrostatic force or EWOD actuation, double-layered metals were integrated between the PDMS and the Teflon layer. The endurance of electrical continuity on the same flexible substrate was characterized in a cyclic bending test. The results confirm that the reduced small contact angle hysteresis can decrease any external pressure and power consumption in droplet manipulation for transportation of liquid samples with warranted reliability.

The proposed structure is expected to be utilized in applications for droplet-based microfluidics and for the self-cleaning of arbitrarily curved surfaces such as a swimsuit, goggles for swimmers, anti-fog glasses, and the windshield of a car.

ACKNOWLEDGEMENTS

This work was partially supported by a grant from the National Research Laboratory (NRL) program (No. R0A-2007-000-20028-0) of the Korea Science and Engineering Foundation (KOSEF), which is funded by the Korean Ministry of Education, Science and Technology (MEST). It was also partially supported by the National Research and Development Program (NRDP, 2005-01274) for the development of biomedical function monitoring biosensors; this program is also sponsored by the Korean Ministry of Education, Science and Technology.

REFERENCES

[1] P. Roach, N. J. Shirtcliffe, and M. I. Newton, "Progress in superhydrophobic surface development," *Soft Matter*, vol. 4, pp. 224-240, 2008.

[2] B. Burkarter, C. K. Saul, F. Thomazi, N. C. Cruz, L. S. Roman, and W. H. Schreiner, "Superhydrophobic electrosprayed PTFE," *Surf. Coat. Tech.*, vol. 202, pp. 194-198, 2007.

[3] A. D. Tserepi, M.-E. Vlachopoulou, and E. Gogolides, "Nanotexturing of poly(dimethylsiloxane) in plasmas for creating robust super-hydrophobic surfaces," *Nanotech.*, vol. 17, pp. 3977-3983, 2006.

[4] T. N. Krupenkin, J. A. Taylor, T. M. Schneider, and S. Yang, "From rolling ball to complete wetting: the dynamic tuning of liquids on nanostructured surfaces," *Langmuir*, vol. 20, pp. 3824-3827, 2004.

[5] L. Zhang, Z. Zhou, B. Cheng, J. M. DeSimone, and E. T. Samulski, "Superhydrophobic behavior of a perfluoropolyether lotus-leaf-like topography," *Langmuir*, vol. 22, pp. 8576-8580, 2006.

[6] M. Kim, K. Kim, N. Y. Lee, K. Shin, and Y. S. Kim, "A simple fabrication route to a highly transparent super-hydrophobic surface with a poly(dimethylsiloxane) coated flexible mold," *Chem. Commun.*, vol. 22, pp. 2237-2239, 2008.

[7] K. K. S. Lau, J. Bico, K. B. K. Teo, M. Chhowalla, G. A. J. Amaratunga, W. I. Milne, G. H. McKinley, and K. K. Gleason, "Superhydrophobic carbon nanotube forests," *Nano Lett.*, vol. 3, no. 12, pp.1701-1705, 2003.

[8] L. Ci, R. Vajtai, and P. M. Ajayan, "Vertically aligned large-diameter double-walled carbon nanotube arrays having ultralow density," *J. Phys. Chem. C*, vol. 111, pp. 9077-9080, 2007.

[9] J. Yuan, X. Liu, O. Akbulut, J. Hu, S. L. Suib, J. Kong, and F. Stellacci, "Superwetting nanowire membranes for selective absorption," *Nature Nanotech.*, vol. 3, pp. 332-336, 2008.

[10] X.-J. Huang, J.-H. Lee, J.-W. Lee, J.-B. Yoon, and Y.-K. Choi, "A one-step route to a perfectly ordered wafer-scale microbowl array for size-dependent superhydrophobicity," *small*, vol. 2, pp. 211-216, 2008.

[11] S.-I. Chang, and J.-B. Yoon, "Shape-controlled, high fill-factor microlens arrays fabricated by a 3D diffuser lithography and plastic replication method," *Optics Express*, vol. 12, no. 25, pp. 6366-6371, 2004.

[12] X.-J. Huang, D.-H. Kim, M. Im, J.-H. Lee, J.-B. Yoon, and Y.-K. Choi, "'Lock-and-key' geometry effect of patterned surfaces: the wettability and the switching of adhesive force," *small*, accepted to be published.

[13] A. B. D. Cassie, and S. Baxter, "Wettability of porous surfaces," *Trans. Faraday Soc.*, vol. 40, pp. 546-551, 1944.

[14] M. Abdelgawad, S. L. S. Freire, H. Yang, and A. R. Wheeler, "All-terrain droplet actuation," *Lab Chip*, vol. 8, pp. 672-677, 2008.

[15] J. Kim, and C.-J. Kim, "Nanostructured surfaces for dramatic reduction of flow resistance in droplet-based microfluidics," in *Digest Tech. Papers IEEE MEMS 2002 Conference*, Las Vegas, NV, Jan. 20-24, 2002, pp. 479-482.

[16] K.-S. Yun, and C.-J. Kim, "Low-voltage electrostatic actuation of droplet on thin superhydrophobic nanoturf," in *Digest Tech. Papers IEEE MEMS 2007 Conference*, Kobe, Japan, Jan. 21-25, 2007, pp. 139-142.

AMBIENT TEMPERATURE-GRADIENT COMPENSATED LOW-DRIFT THERMOPILE FLOW SENSOR

M. Dijkstra, T.S.J. Lammerink, M.J. de Boer, R.J. Wiegerink and M. Elwenspoek
MESA+ Institute for Nanotechnology, University of Twente, The Netherlands

ABSTRACT

A highly-sensitive thermal flow sensor for liquid flow with nl·min⁻¹ resolution has been realised. The sensor consists of freely-suspended silicon-rich silicon-nitride microchannels with integrated Al heater resistors and Al/poly-Si⁺⁺ thermopiles. The influence of drift in the thin-film metal resistors is effectively eliminated by using thermopiles combined with an adequate measurement method, where the power in the heater resistors is controlled, e.g. constant-power calorimetric method or temperature balancing method. The special meandering layout of the microchannels and the placement of thermopile junctions increases sensitivity by summing the thermopile voltages due to convection by fluid flow, whereas the influence of ambient temperature gradients is compensated for.

1. INTRODUCTION

The miniaturisation of microfluidic components asks for accurate and reliable measurement of tiny fluid flow rates in the order of nl·min⁻¹. Current micromechanical thermal flow sensors are capable of measuring down to nl·min⁻¹ resolution [1–4]. Important problems limiting the accuracy of these thermal flow sensors are: the drift in the electrical resistance of thin-film layers used for heating and temperature sensing, and the influence of external temperature gradients across the sensor chip. Thermistors used for temperature sensing can be avoided, by using thermopiles to measure a temperature difference, with inherent zero-offset [5–6]. Using thermopiles, the flow sensor can be made independent of resistance drift by applying power control on heater resistors. Additionally, a temperature-balancing feedback loop can be used to compensate for thermopile sensitivity drift [6].

This paper presents a low-drift calorimetric flow sensor for liquid flows with nl·min⁻¹ resolution. The flow sensor consists of freely-suspended microchannels with integrated thermopiles for temperature sensing, with power control applied on heater resistors used. Furthermore, a special meandering microchannel layout is utilised to compensate for the influence of external temperature gradients across the sensor chip.

2. COMPENSATION CONCEPT

Ambient temperature-gradients are compensated using a meandering microchannel layout, in which a fluid flow Q passes in alternating direction through several freely-suspended microchannels used for thermal flow sensing (Figure 1). Thermopiles are connected in series and positioned on the freely suspended microchannels in such a way that each thermopile measures a temperature difference ΔT between an upstream junction temperature T_{up} and a downstream junction temperature T_{down}.

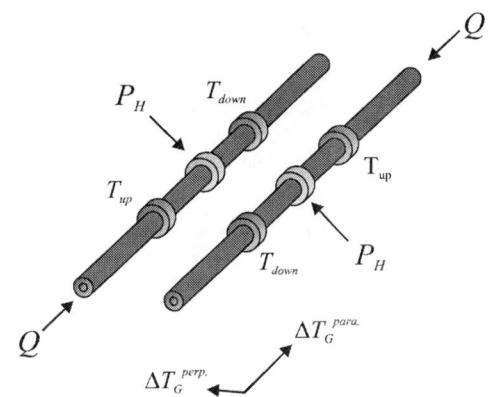

Figure 1: Ambient temperature-gradient compensated flow sensor concept.

Effectively the flow passes the same sensor configuration several times, while flowing through the meandering microchannel, giving a summation of multiple thermopile voltages due to convection by fluid flow.

External temperature gradients ΔT_G of first order can be decomposed in two directions. A temperature gradient parallel to the microchannels $\Delta T_G^{para.}$ changes T_{up} and T_{down} in opposite ways on alternating microchannels, not influencing the total measured ΔT. A gradient perpendicular to the microchannels $\Delta T_G^{perp.}$ changes the offset temperature of each microchannel, not influencing the difference between T_{up} and T_{down} on the same microchannel. First order temperature gradients can therefore be fully compensated. However, the heater resistors are slightly temperature dependent and therefore require control of the heating power P_H, in order for the compensation concept to work properly.

3. SENSOR FABRICATION

The fabricated thermopile flow sensor uses freely-suspended surface microchannels for on-chip transport of fluid. The microchannels are fabricated without the requirement for sacrifical layer etching and allow for the integration of sensor elements in close proximity to the fluid [7]. The technology has been applied in the fabrication of highly-sensitive flow sensors [4, 6, 8].

Figure 2 gives a schematic overview of the process scheme for the fabrication of the thermopile flow sensor. Surface microchannels are created by isotropic dry etching, using high-density SF₆ plasma with zero self-bias, through etch holes 2 μm in width, in a low-stress 500 nm silicon-rich silicon-nitride (SiRN) layer. The etch holes and inner surfaces of the microchannels are conformally coated by a second low-stress LPCVD deposited 1.3 μm SiRN layer, resulting in completely sealed microchannels, while leaving a planar substrate

978-1-4244-2977-6/09 $25.00 © 2009 IEEE

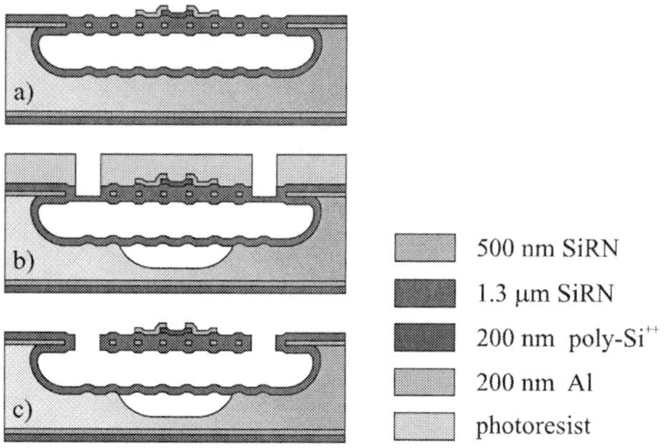

Figure 2: Process scheme for the fabrication of the temperature-gradient compensated thermopile flow sensor.

surface for the integration of Al/poly-Si^{++} thermopiles and Al heater resistors. The thermopiles are created by LPCVD deposition and boron doping, by solid source diffusion, of a 200 nm poly-Si^{++} layer and sputtering of a 200 nm Al layer (Figure 2a). The surface microchannels are released by SF_6 plasma etching for thermal isolation from the heat-conducting substrate, with the photoresist mask protecting the sensor elements during the release (Figure 2b). The photoresist is removed after fluidic entrance holes are etched through the SiRN layer (Figure 2c).

Figure 3 shows a micrograph of the fabricated temperature-gradient compensated flow sensor. The sensor contains three parallel 20 µm diameter microchannels, which cross a thermal-isolation cavity (1.2 × 2.0 mm, 250 µm deep) several times in alternating flow directions. Heater resistors are centered on the freely suspended microchannels and connected in series. Four-point contact is used for accurate power control.

Figure 4 shows a close-up of the surface microchannels with integrated Al/poly-Si^{++} thermopiles having eight junctions on either side, for measuring the down- and upstream temperature difference.

Figure 3: Microchannel temperature-gradient compensated thermopile flow sensor.

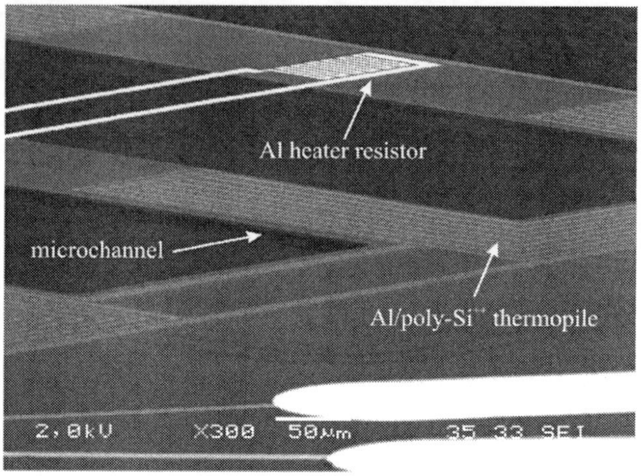

Figure 4: Close-up of the microchannels with integrated Al heater resistors and Al/poly-Si^{++} thermopiles, suspended over a thermal isolation cavity etched in the substrate.

4. SENSOR MODELLING

A thermal model of the flow sensor was constructed in COMSOL Multiphysics. The temperature field was solved including the thermal conduction in the flow sensor and surrounding air and forced convection by water flow through the microchannels. The flow inside the microchannels is approximated using a fixed velocity profile corresponding to a given flow rate, without affecting the accuracy of the thermal model. Additionally, deposited layers are modelled as highly conductive layers, including the thermopiles and heater resistors.

Figure 5 shows isothermal surfaces of the resulting temperature field at 350 nl·min^{-1} water flow through the microchannels, with 0.6 mW total heating power applied. Clearly, the effect of forced convection by the alternating direction of the flow through the microchannels can be observed at this relatively high flow rate. The high

Figure 5: Thermal FEM model of the flow sensor, showing the influence of convection by the alternating direction of 350 nl·min^{-1} water flow through the microchannels at 0.6 mW total heating power.

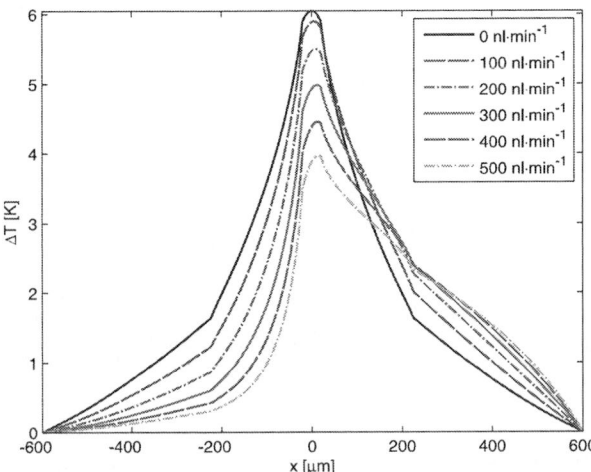

Figure 6: Temperature profiles for various water flow rates in the positive x-direction along a microchannel suspended at ±600 μm, with 0.6 mW total heating power applied.

thermal conduction of the Al leads connecting the heater resistors is also apparent. Although, there is no heat flow between adjacent microchannels for the temperature field by conduction alone.

Figure 6 shows temperature profiles along the microchannel at various water flow rates. A difference in the slope of the temperature profiles can be observed at ±225 μm from the centre, where the thermopile junctions are situated on the microchannel, with the thermopile leads causing much higher thermal conduction to the substrate.

5. EXPIREMENTAL RESULTS

Flow Rate Sensitivity

The fabricated thermopile flow sensors chips (12.5 × 12.5 mm) have bondpads and microchannel entrance holes on fixed positions for self alignment with a chip holder (Figure 7). The chip holder contains O-rings for fluidic interfacing and pogo-pins for electrical connections to the sensor chip. Water flow is applied by an elevation head Δh, giving a stable flow rate. The flow rate is calibrated by microbalance weighing, where evaporation from the balance, in the order of 100 nl·min⁻¹, has to be taken into consideration. A source measurement unit (Keithley 2420) was used for controlling a fixed total heating power P_H. The thermopile voltage V_{TC} was obtained by a nano-volt meter (HP 34420A).

Figure 8 shows the thermopile voltage V_{TC} dependency on water flow rate Q up to 550 nl·min⁻¹ at various applied heating powers. The sensor shows linear sensitivity up to flow rates of about 100 nl·min⁻¹. The heater temperature decreases at higher flow rates (Figure 6), causing the sensor output to decrease according to King's law. The dashed lines in figure 8 show fitted FEM results after scaling calculated temperature differences to measured voltages. Measurements show offsets in the thermopile voltage, which are dependent on the heating power and can be attributed to sensor asymmetries.

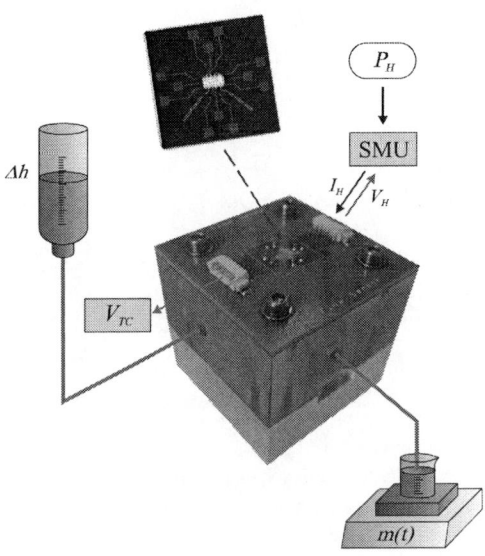

Figure 7: Overview of the flow measurement setup, with the chip holder establishing fluidic and electrical connections.

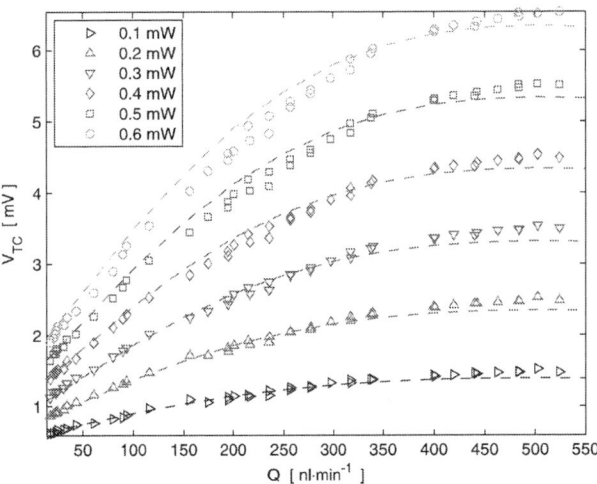

Figure 8: Flow sensor thermopile-voltage V_{TC} dependency on water flow rate Q at various applied total heating powers. Dashed lines show fitted FEM results.

Ambient Temperature-Gradient Sensitivity

A temperature-gradient measurement setup was constructed in order to determine the influence of external temperature gradients on the thermopile voltage (Figure 9). The setup consists of a 5 mm thick Cu plate, which is heated from one side by a resistor dissipating 25 W, while the other side is connected to a large aluminium plate acting as heat sink, thus creating a well defined temperature gradient. The temperature gradient was determined to be approximately 2.8 K·cm⁻¹, measured using two Pt-100 elements placed 10 cm apart. The sensor chip is placed in the centre of the Cu plate and probes are used to measure the thermopile voltage using a nano-volt meter (HP 34420A). This allows the sensor chip to be rotated and the influence of the external temperature gradient to be measured at any angle.

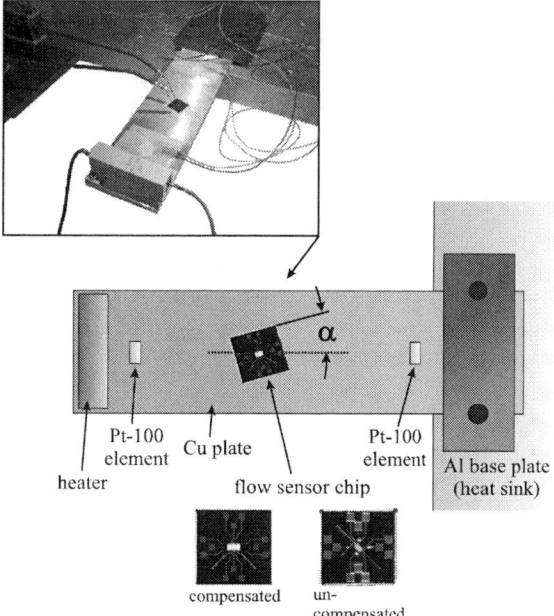

Figure 9: Temperature-gradient measurement setup for inducing a well defined temperature gradient across the sensor chip.

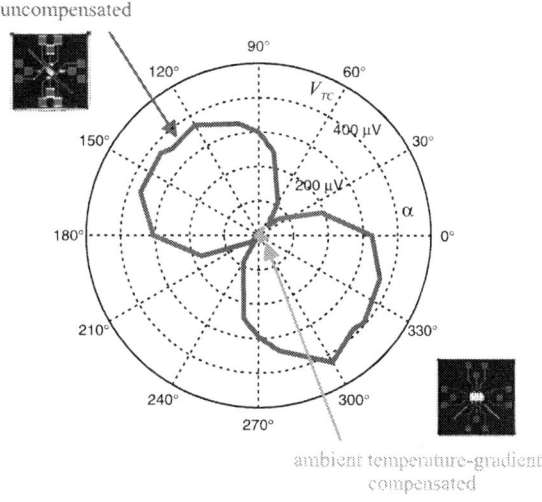

Figure 10: External temperature-gradient sensitivity of the thermopile voltage V_{TC} measured for a compensated sensor and a single-microchannel uncompensated flow sensor.

An ambient temperature-gradient compensated flow sensor and an uncompensated single-microchannel flow sensor, for comparison, were measured, with both sensors having nearly equal amounts of thermopile junctions. The ambient temperature-gradient compensated flow sensor shows to be completely compensated for external temperature gradients (Figure 10). The uncompensated flow sensor shows a figure-of-eight sensitivity pattern, because the flow sensor is only sensitive to temperature gradients parallel to the microchannel $\Delta T^{para.}$.

6. CONCLUSIONS

A calorimetric flow sensor measuring liquid flow down to nl·min⁻¹ resolution has been realised, using freely-suspended microchannels for thermal isolation form the silicon substrate. The flow sensor shows a linear response for water flow up to 100 nl·min⁻¹, which is accurately described by a thermal FEM model of the sensor.

Influence of drift in thin-film metal resistors is effectively eliminated by using power control, in conjunction with thermopiles for temperature sensing. It is demonstrated that this measurement method in combination with a special meandering microchannel layout can effectively compensate drift by ambient temperature-gradients across the sensor chip. Compared to a single microchannel uncompensated flow sensor a significant improvement is obtained in sensor accuracy.

ACKNOWLEDGEMENT

The authors would like to thank the Dutch Technology Foundation (STW) for their financial support through the low-drift micro-flowsensor project (TET.6634).

REFERENCES

[1] Y. Mizuno, M. Liger, Y-C. Tai, "Nanofluidic Flowmeter using Carbon Sensing Element", *Proc. IEEE MEMS*, 2004, pp. 322–325.

[2] S. Wu, Q. Lin, Y. Yuen, Y-C. Tai, "MEMS flow sensor for nano-fluidic applications", *Sensors and Actuators A*, vol. 89, pp. 152–158, 2001.

[3] H. Ernst, A. Jachimowicz, G.A. Urban, "High resolution flow characterization in Bio-MEMS", *Sensors and Actuators A*, vol. 100, pp. 54–62, 2002.

[4] M. Dijkstra, M.J. de Boer, J.W. Berenschot, T.S.J. Lammerink, R.J. Wiegerink, M. Elwenspoek, "Miniaturized Thermal Flow Sensor with Planar-Integrated Sensor Structures on Semicircular Surface Channels", *Sensors and Actuators A*, vol. 143, pp. 1–6, 2008.

[5] P. Brushi, A. Diligente, D. Navarrini, M. Piotto, "A Double Heater Integrated Gas Flow Sensor with Thermal Feedback", *Sensors and Actuators A*, vol. 123–124, pp. 210–215, 2005.

[6] M. Dijkstra, T.S.J. Lammerink, M.J. de Boer, J.W. Berenschot, R.J. Wiegerink, M. Elwenspoek, "Low-Drift Flow Sensor with Zero-Offset Thermopile-Based Power Feedback", *Proc. DTIP*, 2008.

[7] M. Dijkstra, M.J. de Boer, J.W. Berenschot, T.S.J. Lammerink, R.J. Wiegerink, M. Elwenspoek, "A Versatile Surface Channel Concept for Microfluidic Applications", *J. Micromech. Microeng.*, vol. 17, pp. 1971–1977, 2007.

[8] J. Haneveld, T.S.J. Lammerink, M. Dijkstra, H. Droogendijk, M.J. de Boer, R.J. Wiegerink, "Highly Sensitive Micro Coriolis Mass Flow Sensor", *Proc. IEEE MEMS*, 2008, pp. 920–923.

978-1-4244-2977-6/09 $25.00 © 2009 IEEE

AN ELECTRICALLY-DRIVEN, LARGE-DEFLECTION, HIGH-FORCE, MICRO PISTON HYDRAULIC ACTUATOR ARRAY FOR LARGE-SCALE MICROFLUIDIC SYSTEMS

Hanseup Kim and Khalil Najafi
Center for Wireless Integrated Microsystems
University of Michigan, Ann Arbor, MI 48109-2122, USA

ABSTRACT

This paper describes a proof-of-concept all-electrical individually-addressable micro piston actuator array that produces high displacement and force by utilizing hydraulic amplification and electrostatic control. This new class of actuator arrays can remove a critical hurdle, pneumatic control ports to lab-on-chips, and achieve fully-integrated, large-scale, and all-electrical microfluidic systems. The fabricated actuator consists of a 3×3 array of $2×2mm^2$ membranes and produces maximum deflections of 35, 23, and 11 μm when hydraulically-driven by piezoelectric actuation at 100, 80, 60V, respectively. The corresponding hydraulic amplification ratios are 3.2, 3, and 2.5, respectively. The array functions up to a frequency of 2Hz without failing, while allowing control over individual actuators by utilizing electrostatic latching at 100V. The active device part measures as $8.4×8.4×0.65mm^3$.

INTRODUCTION

Large-scale fluidic manipulation is crucial in emerging microfluidic applications, particularly for multi-functional, complex chemical and biological analysis. Despite significant progress, previous lab-on-chip devices still rely on external sources, such as highly-pressurized gas tanks and syringe pumps, to actuate numerous micro-valves and pumps for fluidic manipulation [1]. Such dependence inevitably requires pneumatic interconnections that often create various issues in space, integration, leakage, and control. Thus, total system miniaturization and portability have been prohibited.

Despite several problematic issues, external fluidic sources have continuously been used because they can provide high force and displacement in the micro-scale. Conventional micro-scale actuators have failed to provide such features despite excellent integration and control capability, and low power consumption.

To realize a high-pressure micro actuator, hydraulic amplification has been earlier introduced in the micro-domain [2-4]. Hydraulic systems have been widely used to amplify force and deflection in the macro world. In previous papers we reported the concept of electrostatic micro-hydraulic actuation that can achieve hydraulic amplification in force and deflection under electrostatic control [2-3].

In this paper we report the concept of all-electrical, programmable, 3-D, micro piston actuation arrays. The micro hydraulic actuator array consists of flexible

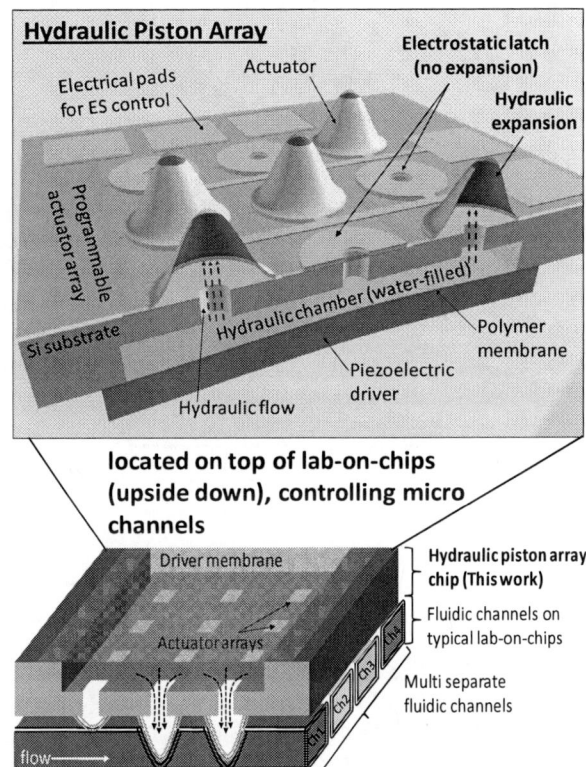

Figure 1: Illustration of (Top) a programmable, large-deflection, high-force hydraulic micro actuator arrays, (Bottom) the application: a no-fluidic-control-ports (no pneumatic connections), no-leakage, only-electric, large-deflection, high-force, and simple generic microfluidic manipulation platform.

membranes that can be electrically, individually, and selectively actuated to produce relatively large displacement and force in the direction vertical to the plane of the substrate. When placed on top of a typical microfluidic chip that contains flow channels and thin membranes at the intersection of these flow channels, the actuators can be used to close the channels thus allowing valving and pumping functions (Fig. 1). Thus, this actuator array obviates the need for conventional pneumatic control components, and their associated problems, such as large number of fluidic interconnections, gas leaks, complexity in control and structure, and the need for external components (gas tanks or syringe pumps).

Operation Principle

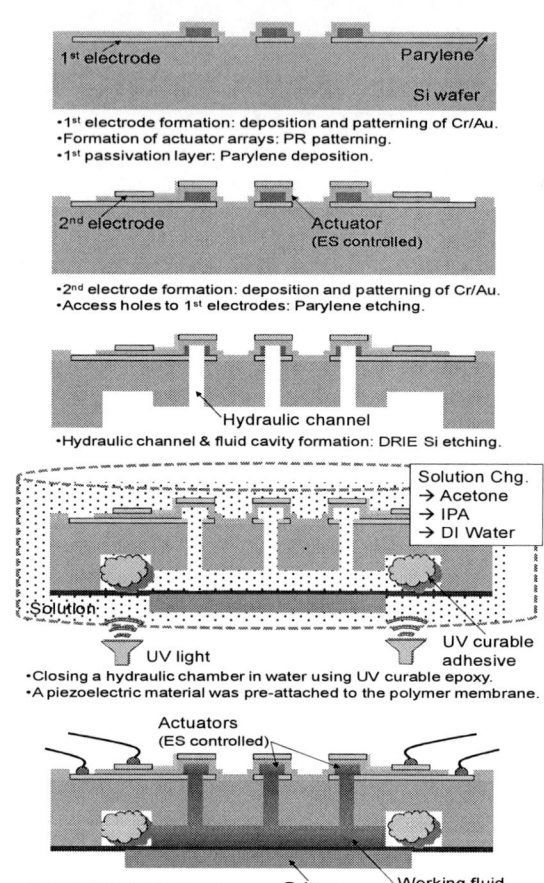

Figure 2: Operation principle: hydraulic amplification, only-electric actuation and programmability.

OPERATION PRINCIPLE

The micro piston actuator array consists of a matrix of flexible Parylene membranes formed on the front side of a silicon wafer, and a single driver membrane on the backside (Fig. 1). The front and back sides of the wafer are fluidically connected through a number of thru-wafer channels formed under each of the front-side membranes. The sealed cavity (hydraulic chamber) between the front and back sides is filled with a hydraulic fluid (e.g. de-ionized water). The matrix of flexible membranes on the front side is coated with a conductive metal layer for electrostatic control. The metal layer on the membrane forms a movable electrode against a counter-electrode that is formed on the substrate under each membrane. These metal layers are patterned in such a way, like a semiconductor memory, that individual membrane can be accessible.

The operation of the micro piston hydraulic array is illustrated in Figure 2. When the large driver membrane on the back side of the wafer is piezoelectrically deflected, the hydraulic fluid is compressed through the small channels to the other side, generating *large deflection* and *high force* through hydraulic amplification. Note that the area of the driver membrane on the backside is larger than those of actuator membranes in the front side. When the driver chamber is relaxed, the hydraulic fluid returns to the driving chamber deflating individual actuator membranes. Each actuator membrane is controlled by electrodes on the membrane and the substrate. When voltage is applied between these electrodes, the membrane is electrostatically latched to the substrate and does not expand, allowing electrical control (*programmability*) over the operation of

Figure 3: Microfabrication process.

each membrane. This device is fully functional by itself, and does not require any external (fluidic) components to create large deflection and force.

MICROFABRICATION

Fabrication consists of electrode formation, hydraulic chamber formation, and liquid encapsulation, as shown in Figure 3. First, the 1st metal electrode is patterned on an oxidized silicon wafer, and then photoresist and Parylene layers are deposited and patterned to form the matrix of actuator membranes. The photoresist layer defines the actuator cavity and will be removed later. Next, the 2nd metal electrode is deposited and patterned on top of the Parylene layer. The silicon substrate is then DRIE-etched from the backside to form hydraulic channels and the driving chamber. The device is immersed into Acetone, IPA, and DI water to dissolve the sacrificial layer, form the actuator chambers, and fill them with DI water. The chamber is sealed *in water* using UV curable adhesive and a polymer membrane with an attached piezoelectric material. Note that the sealing is performed in water to prevent air bubbles from being trapped inside. Final fabricated devices are shown in Figure 4. The figure describes the whole device (top left), a close-up view of the actuator membrane array on the front side of the wafer (top right), hydraulic channels to the actuator chamber from the backside (bottom

978-1-4244-2977-6/09 $25.00 © 2009 IEEE   484

Fabricated hydraulic actuator arrays

Figure 4: Photographs of a fabricated hydraulic actuator array.

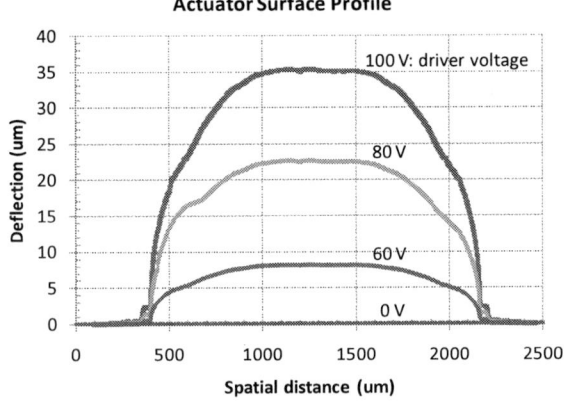

Figure 5: Measured hydraulic expansion of a single actuator at different voltages.

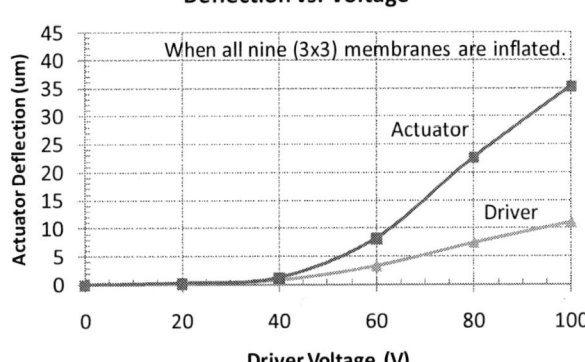

Figure 6: Measured deflections for both the driver and the actuators.

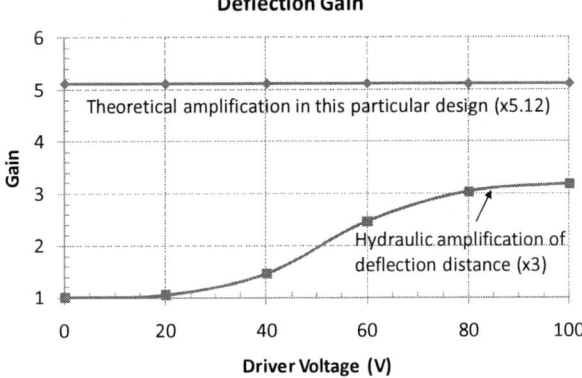

Figure 7: Measured hydraulic amplification at various voltages.

left), and SEM of the hydraulic channels (bottom right). The active area device part measures as $8.4 \times 8.4 \times 0.65 \text{mm}^3$. The actuator array (top right) is a metal-coated Parylene membrane.

TEST RESULTS

The fabricated 3×3 array (Fig. 4) was operated at different driving voltages (0-100 V), frequencies (0.25-2 Hz), and electrostatic latching conditions to examine hydraulic operation and programmability. The array was electrically actuated by bending a piezoelectric film attached to the driver membrane on the backside. Actuator membranes ($2 \times 2 \text{ mm}^2$) on the front side are selectively latched by appropriately applying DC voltages to the target membrane. The actuation was observed by utilizing optical, electrical (capacitance measurement), and mechanical (surface profile measurement) methods.

Control of a Single Actuator

The fabricated array was operated at various piezoelectric-driving DC voltages (0-100 V) to show different actuator deflection. The surface profile of an actuator membrane was measured by a Dektak surface profiler. The $2 \times 2 \text{ mm}^2$ actuator membrane is deflected by 35, 23, and 11 μm at driving voltages of 100, 80, and 60 V, respectively (Fig. 5). The resultant deflection distances are longer than those available from typical air-gap electrostatic actuators of the same sizes [4].

Hydraulic Amplification

While the fabricated array was operated at various piezoelectric-driving DC voltages (0-100 V), hydraulic amplification in deflection was measured between the driver membrane and an actuator membrane. During the measurement, all nine (3×3) membranes were actuated. The deflection distance was measured by utilizing a Dektak surface profiler.

Figure 6 shows the deflection distance vs. driving voltages for both the driver and an actuator. The $13.56 \times 13.56 \text{ mm}^2$ driver membrane deflected by 11.1, 7.48, and 3.35 μm at driving voltages of 100, 80, and 60 V, respectively. This corresponds to deflection amplification gains of 3.18, 3.03, and 2.46, respectively (Fig. 7). The ideal gain for this particular hydraulic amplification is 5.12. The deviation is possibly caused by partial inflation of the driver membrane due to its flexibility.

978-1-4244-2977-6/09 $25.00 © 2009 IEEE

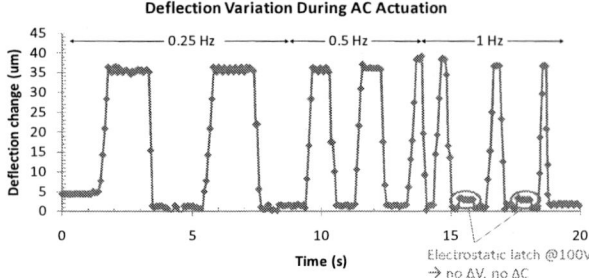

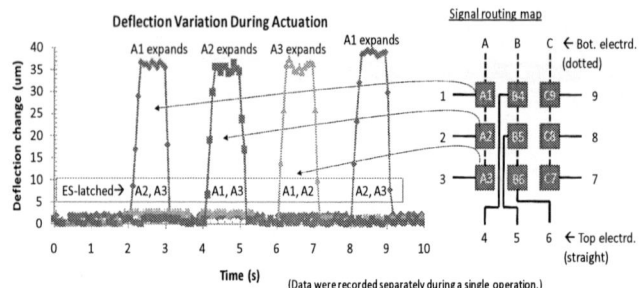

Figure 8: Deflection variation at a single hydraulic actuator operation at different frequencies and under electrostatic latching.

Figure 9: Deflection and control of multiple hydraulic actuators showing its electrostatic programmability.

Transient Characteristics

At a given voltage (100V), the actuator array was operated under different frequencies (0.25-2 Hz) with electrostatic latching to investigate maximum operation frequencies. The deflection was monitored by measuring capacitance variations utilizing a HP4208 LCR meter, and then was converted to deflection distance. Testing results show that the array functions up to a frequency of 2Hz without failing, while allowing control over individual actuators (Fig. 8). The actuator motion was in precise synchronization with the driver membrane movement up to a driving frequency of 2 Hz. At higher frequencies the motion of the actuator membrane did not show repeatable deflection over. Up to 1 Hz, the actuator movements were well controlled by electrostatic latching.

Electrostatic Programmability

Finally, the micro piston hydraulic array was operated under a given sequence, and selected membranes were monitored to investigate programmability. Both the piezoelectric-driving and electrostatic-latching voltages were 100 V. Electrostatic-latching was achieved by applying voltages to bottom electrode lines (A, B, and C) and to top individual electrodes (1-9). For example, when a high voltage is applied both to the line A and to the electrode 1, the A1 membrane is electrostatically latched. Figure 9 shows the testing results on actuator programmability. Actuator membranes in the same column were successfully latched alternatively (A1, A2, or A3 membranes are latched). Significant cross-talk was not observed.

CONCLUSIONS

We have successfully demonstrated a proof-of-concept all-electrical programmable micro piston actuator array to produce individually-addressable high displacement and force by utilizing hydraulic amplification and electrostatic control. The fabricated 3×3 array produces hydraulically-amplified (>3x) out-of-plane deflection of 35μm from each of the electrostatically-latchable 2×2mm² actuators at the operation voltages of 100, 80, 60V, respectively. The array functions up to a frequency of 2Hz without failing, while allowing control over individual actuators by utilizing electrostatic latching at 100V. The active device part measures as 8.4×8.4×0.65mm³.

ACKNOLEDGEMENT

The authors would like to thank Seunghyun Lee for his valuable advice and help during microfabrication, Mahdi Sadeghi for his help on testing devices, and Dr. Il-Joo Cho for his valuable advice on photography. This project is funded by the Engineering Research Centers Program of the National Science Foundation under Award Number EEC-9986866.

REFERENCES

[1] H. Kim, S. Lee, and K. Najafi, ""High-force liquid-gap electrostatic hydraulic micro actuators," *Proc. 11th Int. Conf. on Miniaturized Systems for Chemistry and Life Sciences (μTAS '07)*, Paris, France, Oct. 7-11, 2007, pp. 1735-1737.

[2] H. Kim and K. Najafi, "Electrostatic hydraulic three-way gas microvalve for high-pressure applications," *Proc. 12th Int. Conf. on Miniaturized Systems for Chemistry and Life Sciences (μTAS '08)*, San Diego, California, USA, Oct. 12-16, 2008, pp. ???.

[3] T. Thorson, S. Maerkl, and S. R. Quake, "Microfluidic large-scale integration," *Science,* v298, n5593, pp. 580-584, 2002.

[4] H. Kim, A. Astle, K. Najafi, L. Bernal, and P. Washabaugh, "A fully integrated high-efficiency peristaltic 18-stage gas micropump with active microvalves," *Proc. 20th IEEE Int. Conf. on Micro Electro Mechanical Systems (MEMS '07)*, Kobe, Japan, Jan. 21-25, 2007, pp. 131-134.

[4] X. Yang, A. Holke, S.A. Jacobson, J.H. Lang, M.A. Schmidt, and S.D. Umans, "An electrostatic, on/off microvalve designed for gas fuel delivery for the MIT microengine," *IEEE/ASME Journal of Microelectro-mechanical Systems (JMEMS),* v13, n4, pp. 676-687, 2004.

[5] D. Roberts, H. Li, J. Steyn, O. Yaglioglu, S. Spearing, M. Schmidt, and N. Hagood, "A piezoelectric microvalve for compact high-frequency, high-differential pressure hydraulic micropumping systems," *IEEE/ASME Journal of Microelectro-mechanical Systems (JMEMS)*, JMEMS, v12, n1, pp. 81-92, 2003.

[6] M. P. Chang and M.M. Maharbiz, "Electrostatically-actuated reconfigurable elastomer microfluidics," *Proc. 13th Solid-State Sensor, Actuator, and Microsystems Workshop (Hilton Head '08)*, Hilton Head Island, SC, June, 2008, pp. 122-125.

ELECTROHYDRODYNAMIC JET PRINTING CAPABLE OF REMOVING SUBSTRATE EFFECTS AND MODULATING PRINTING CHARACTERISTICS

Jun-Sung Lee[1], Young-Jae Kim[1,2], Byeong-Geun Kang[1], Sang-Yoon Kim[1], Jaehong Park[1], Jungho Hwang[1] and Yong-Jun Kim[1]

[1]School of Mechanical Engineering, Yonsei University, Republic of Korea
[2]Samsung Electro-Mechanics Co., Ltd., Republic of Korea

ABSTRACT

Electrohydrodynamic jet printing (EHDP) technique is widely used for the direct writing. However, in the existing EHDP method, the printing characteristics are affected by the printing substrate used, and the printed line width is determined by the geometry of the nozzle. We propose an EHDP method which is capable of (1) removing the effect from the substrate, and (2) controlling the line width through the ON/OFF control of the each nozzle in the nozzle array. Printing characteristics of our EHDP system were examined and successful ON/OFF control of the nozzle array were demonstrated. By using the proposed EHDP, it is expected that stable meniscus regardless of the effect from substrate and different line widths even using the same nozzle can be achieved.

1. INTRODUCTION

Recently, direct writing technologies have been used to form fine patterns that range from the micro to nanoscales [1]. In the direct writing approach, structures or patterns are obtained by directly depositing or dispensing ink onto a surface without using masking and etching processes. This method reduces cost, increases production speed and is an environment-friendly process [2]. A direct writing technology that has been widely used is inkjet printing. This technique works by generating one-by-one droplets that are smaller than 50 pL (typically 1-20 pL) [3]. However, in inkjet printing the droplet size is usually 1.89 times larger than the nozzle diameter [4]. In other words, to produce a 10μm droplet a 5.3μm sized nozzle is needed. Unfortunately, there are two major technical obstacles associated with using nozzles with very small diameters. First of all, there are severe blockage problems in narrow sized nozzles when nanocolloid ink, which consists of functional solid nanoparticle powder, liquid, dispersing agent, and surfactant, is used. Secondly, it is technically very difficult to produce and manufacture narrow sized nozzles. To overcome these problems, several studies have examined the potential use of electrohydrodynamic printing (EHDP), which can obtain the diameter of droplet under 10μm when using a nozzle that is larger than 100μm in diameter [5, 6]. Most of these studies have focused on making functional lines or structures through electrohydrodynamic printing using a single pin shaped nozzle. In addition, these studies have mainly focused on making a stable cone-jet and reducing the positioning error of the ejected jet. Two approaches have been used to solve these problems: using a pin shaped ground electrode positioned under the substrate or using a ring shaped electrode between the nozzle and the substrate [5, 6]. Researchers have also examined the DOP (drop on place) method to print place on demand rather than the whole place print. In this case, a pulse voltage is usually used to

eject the jet or droplet. When a pulse voltage is applied to the nozzle the meniscus height is extended and then the jet or droplet is ejected from the tip of the meniscus [7]. However, these methods, including using a continuous jet and DOP, have some limitation. One of these limitations is a low output due to the use of a pin shaped single nozzle. Another issue is the wetting phenomenon in the lateral direction of the nozzle. This occurs because the pin shaped nozzle obstructs the stable state of the meniscus. To solve this meniscus stability problem, an additional device, such as a ring shaped electrode, is needed. An additional limitation of these methods is a constant voltage must be applied to nozzle for meniscus controlling, which is done using a pulse voltage. However, when the pin shaped nozzle is arrayed, the required pulse voltage and onset voltage must be changed according to the distance of the operating nozzle due to interference and distortion in the electric field [8].

To overcome these limitations, we considered several potential alternative methods. First, to increase output and to stabilize the cone-jet mode, a multi-nozzle that consists of three single nozzles and three extractors were fabricated from a FR-4 substrate, which is a nonconductive material, using the commercial PCB process. If the nozzle covered with non-conductive plate, onset voltage lower than conductive plate case [9]. Second, to prevent the effect from substrate, the extractor, which alternative device of pin-shaped ground electrode, was used. Third, to achieve addressable jetting, which means jetting on demand, the applied voltage to extractor was controlled.

2. EHDP PRINCIPLE AND DESIGN

Figure 1 shows deformation of meniscus shape according to intensity of electric field. When an external electric field is applied to a conductive solution, ions in the solution will aggregate around the electrode of opposite polarity. Positive ions travel to the negatively charged electrode and negative ions travel toward the positive electrode. As a result, jet or droplet ejected from tip of meniscus. When a positive voltage is applied to the multi-nozzle, the ions in the solution of the same polarity will be forced to aggregate at the surface of meniscus and will cause the drop to deform into the shape of a cone. If the electric potential of the surface charge exceeds a critical value, the electrostatic forces will overcome the surface tension of the conductive solution. A thin jet of the conductive solution will be ejected from the surface of the cone and travel toward the nearest electrode of opposite polarity [10].

Figure 2 shows schematic of our EHDP system which consists of the three nozzles, spacer, extractor, and reservoir plate. In this system, ground electrode, which makes electric potential, was not needed. Because of the extractor alternate the role of ground electrode. If the ground electrode located under substrate, the electric field is changed according to material properties of the substrate. Therefore, effects from the substrate that is the major reason of the unstable meniscus can be removed.

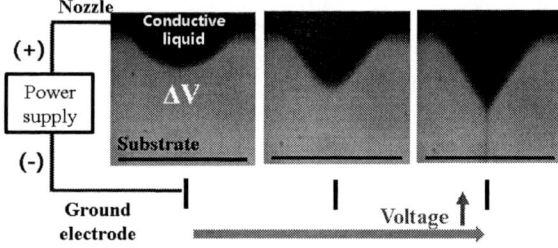

Figure 1: Jetting principle of EHDP

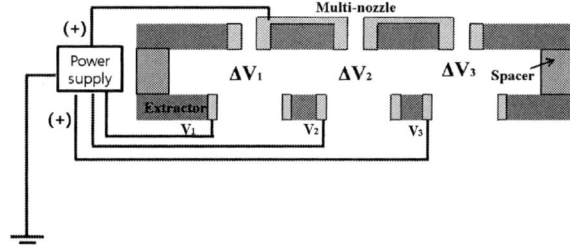

Figure 2: Scheme of addressable multi-nozzle

3. FABRICATION

Figure 3 shows the simplified fabrication steps of reservoir plate. And the multi-nozzle and extractor were fabricated using the conventional printed circuit board fabrication technology as shown in Fig. 4. Multi-nozzle fabrication processes were conducted using the following procedures. First, 150um-thick SU-8 layer was coated onto silicon substrate. Second, coated SU-8 was patterned using photolithography process. Third, PDMS was poured on the silicon substrate and cured at 100 °C for 1h. Fourth, nozzle part was coated with 4um-thick PDMS and then bottom layer was coated with Teflon (AF2400, Dufont Corp.). Fifth, molded PDMS and nozzle part are treated using RIE process at O2: 30sccm, pressure: 250torr, RF power: 25W and process time: 25sec. Finally, treated PDMS and nozzle part PCB were assembled by annealing with hot plate at 200 °C. After fabrication of the nozzle part, each part was then packaged as shown in Fig. 5. Fabricated nozzle part was assembled with extractor part. The spacer makes spacing between nozzle part and extractor part. The spacer is important that makes optimum distance that does not take place corona discharging.

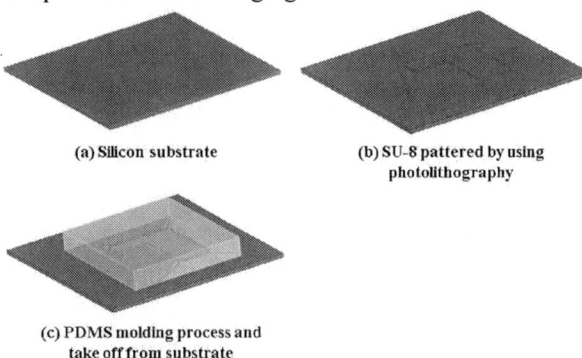

Figure 3: Simplified fabrication processes of the reservoir plate

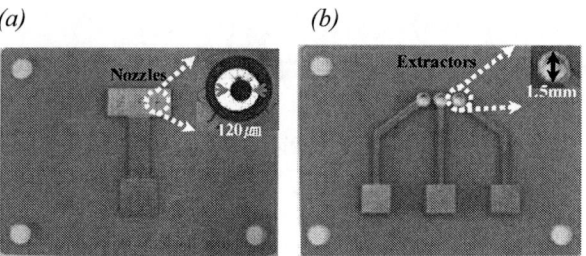

Figure 4: Optical photograph of the (a) multi-nozzle, (b) extractor

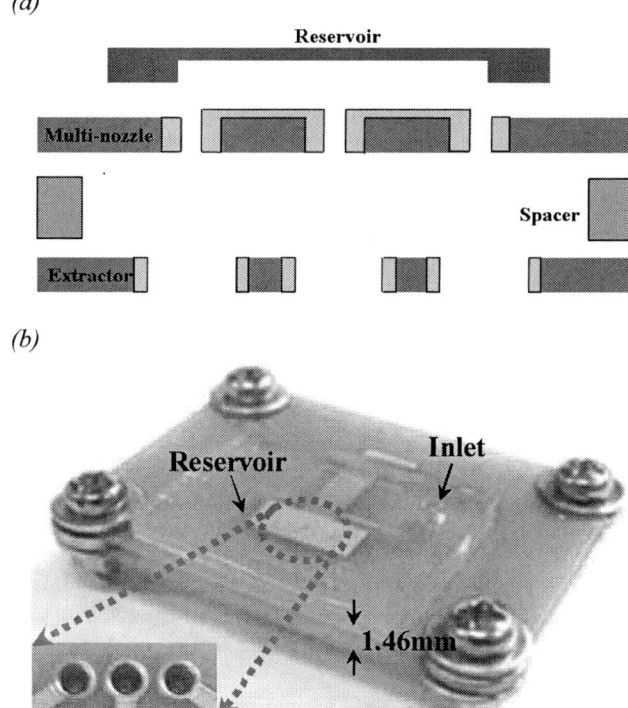

Figure 5: (a) packaging scheme and (b) the results of the addressable multi-nozzle.

4. EXPERIMENT AND RESULT

The EHDP system used in this study consisted of a multi-nozzle, power supply, and X-Y stage, as shown in Fig. 6. The multi-nozzle (diameter: 120 μm) was used to produce a jet containing silver nanoparticles (AG-IJ-100-S1, CABOT Corp.), which were uniformly supplied to the nozzle by a syringe pump (kds-100, KD Scientific Inc.). The nozzle was also used as anodes as well as the extractor (diameter: 1.5mm), which was located 1.46mm below the nozzle. The extractor with an inner hole reduced the chaotic motion of the jet and prevented the jet from digressing from the centerline [11]. Figure 7(a) is an image of the experimental results. The experiment was conducted using the following procedure. High voltage supply (HV-Rack, Ultravolt Inc.), which consists of ON-OFF controllable four channels, was used to applied voltage. Each channel was connected to nozzle part electrode and extractor. Using the electric signal produced by the ON-OFF controller, we could control whether the voltage was supplied to the extractor such as one nozzle ON, two nozzles ON or three nozzles ON or not. After 2ms that voltage was applied to nozzle and extractor because of

978-1-4244-2977-6/09 $25.00 © 2009 IEEE   488

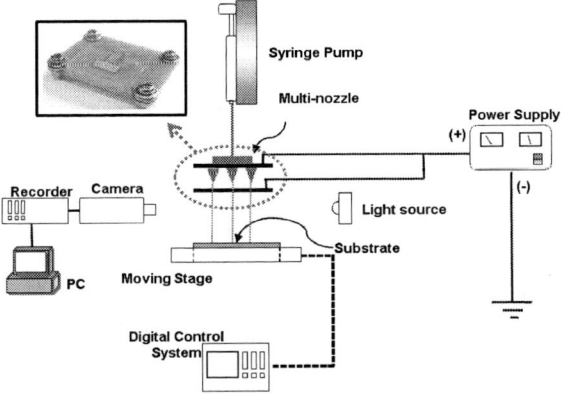

Figure 6: Schematic drawing of the experimental setup

ion in the solution move to surface, the meniscus became the cone-jet and eventually the cone-jet was ejected from the tip of meniscus. Figure 7(b) shows images acquired using the cone-jet mode with arrays of pin shaped nozzles [8]. Due to end effect in the pin shaped nozzle arrays, the cones are asymmetric except in the center nozzle. In contrast, every FR-4 based nozzle forms a symmetric cone-jet mode. Because of the potential line paralleled with the nozzle. Therefore, the end effect did not occur when the FR-4 based multi-nozzles were used. As a result, this novel technique simultaneously reduces positioning error and allows independent control over each nozzle. And also, during this experiment, coherent jet of 19μm in diameter was obtained when 3μl/min flow-rate and 1.8kV applied voltage. Although our proposed EHDP method used a nozzle (120μm) much larger than an ink-jet nozzle (about 20μm), it allowed the generation of a micron-sized jet.

(a)

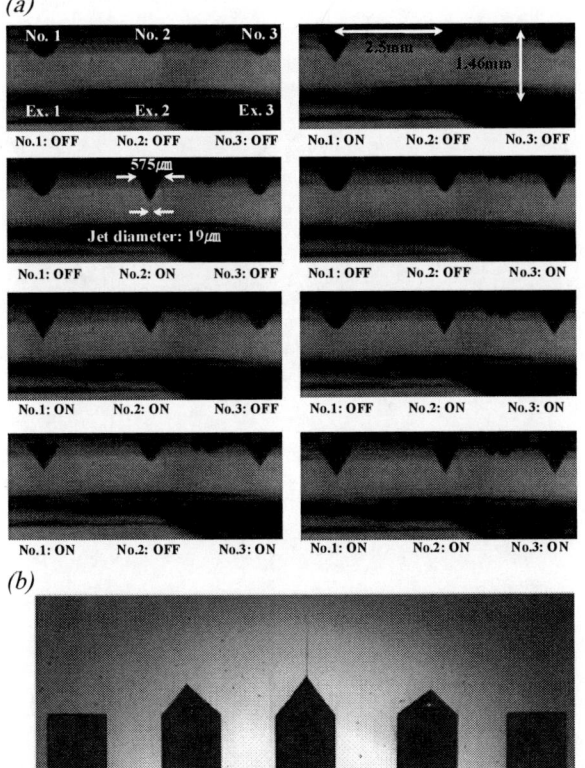

(b)

Figure 7: (a) results of addressable jetting by using an electric potential. (b) result of previous EHDP system using five nozzles.

Figure 8 shows the comparison results between proposed addressable EHDP and the existing EHDP method. In this result, the onset electric field range, which means forming cone-jet mode, was 1.2-2.0kV/mm. Generally, the pin ground type nozzle made narrow pattern line. In some range, however, narrow pattern line was printed by using proposed multi-nozzle. It proves that ground electrode is not certainly needed. And also, uniformity of printed pattern line which printed by using proposed EHDP method was better than the existing EHDP method.

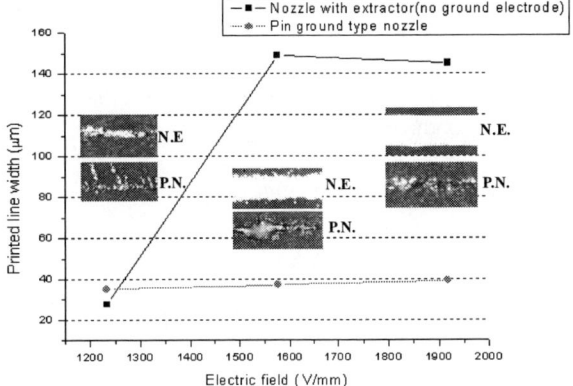

Figure 8: Printed line width according to applied electric potential. (N.E.: nozzle with extractor, P.N.: pin ground type nozzle)

Figure 9 shows the printed line width according to nozzle control. Figure9 (a) shows printed conductive line by using one nozzle. And also, (b) and (c) shows the printed conductive line by using two and three nozzles respectively. In this experiment, line thickness was changed according to the number of the used nozzle. Figure9 (d) and (e) shows the line width changing from 100μm to 197μm and from 100μm to 302μm respectively. Ejected jet diameter was 19μm. But printed line width by using one nozzle was 100μm. The problem was from the moving stage. The speed of the moving stage does not matched with the jet speed. As a result, it makes the accumulation phenomenon of the nanocolloid solution.

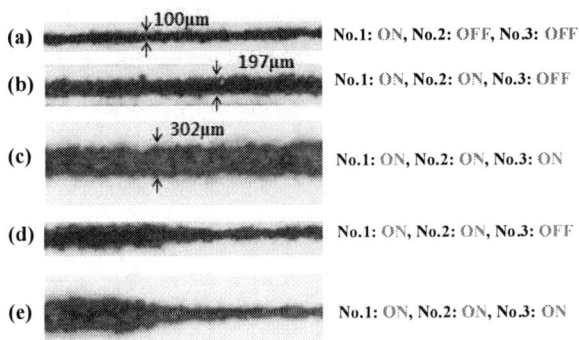

Figure 9: Printed line width according to number of operating nozzle: (a) one nozzle operating, (b) two nozzles operating, (c) three nozzles operating, (d) one nozzle operating after then two nozzles operating and (e) one nozzle operating after then three nozzles operating.

5. CONCLUSION

The EHDP multi-nozzle, which removing the effect from the substrate, has been demonstrated for making conductive line pattern using continuous jet and addressable jetting control. It developed to solve many problems associated with using arrays of pin shaped nozzles (previous EHDP method). The EHDP multi-nozzle was fabricated through simple photo-lithography process with conventional printed circuit board fabrication technology. The jetting characteristics of the proposed EHDP multi-nozzle was examined by controlling of applied voltage. In a continuous jetting, the optimum electric potential was successfully founded where the cone-jet mode gets the stable stage. Since the addressable jetting of the EHDP multi-nozzle can be independently operated without end-effect, which makes positioning error, the EHDP multi-nozzle can be applied for making accurate functional line such as conductive line, ceramic line. If the EHDP multi-nozzle was used, cost reduction and reducing contaminants from fabrication processes will be achieved.

ACKNOWLEDGEMENT

This work was supported by the Seoul R & BD program (GR070039).

REFERENCES

[1] A. Pique and D. B. Chrisey, "Direct Write Technologies for Rapid Prototyping Applications (Academic, San Diego)", 2002, preface.

[2] D. B. Chrisey, "MATERIAL PROCESSING: The Power of Direct Writing", Science, vol. 289, pp. 879 ~881, 2000.

[3] O. Yogi, T. Kawakami, M. yamauchi, J. Y. Ye, and M. Ishikawa, "On-Demand Droplet Spotter for Preparing Pico-to Femtoliter Droplets on Surfaces", Anal. Chem., vol. 73, pp. 1896~1902, 2001.

[4] H. F. Poon, "Electrohydrodynamic Printing" Ph.D. thesis, Princeton University, 2002.

[5] D. Y. Lee, Y. S. Shin, S. E. Park, T. U. Yu, and J. Hwang, "Electrohydrodynamic printing of silver nanoparticles by using a focused nanocolloid jet", J. Appl. Phys. Lett., vol. 90, 081905, 2007.

[6] J. H. YU, S. Y. Kim, and J. Hwang, "Effect of viscosity of silver nanoparticle suspension on conductive line patterned by electrohydrodynamic jet printing" Appl. Physics. A: Mater. Sci. Process., vol.89, pp. 157~ 159, 2007.

[7] J.L. Li, "EHD sprayings induced by the pulsed voltage superimposed to a bias voltage", J. Elec., Vol. 65, pp. 750~757, 2007.

[8] Bui Quang Tran Si, Doyoung Byun, and Sukhan Lee, "Experimental and theoretical study of a cone-jet for an electrospray microthruster considering the interference effect in an array of nozzles", J. Aerosol Sci. 38, pp. 924-934, 2007.

[9] H. Park, K. Kim, and S. Kim, "Effect of a guard plate on the characteristics of an electrospray in the cone-jet mode", J. Aerosol Sci., Vol. 35, pp. 1295~1312, 2004.

[10] Taylor, G. I., "Disintegration of water drops in an electric field", Proc. R. Soc., vol. A280, pp. 383-397, 1964.

[11] D. H. Reneker, A. L. Yarin, H. Fong, and S. Koombhongse, "Bending instability of electrically charged liquid jets of polymer solutions in electrospinning", J. Appl. Phys. Vol. 87, 4531, 2000.

HYBRID ON DEMAND JETTING SYSTEM FOR ULTRA FINE DROPLET BASED ON ELECTROHYDRODYNAMIC AND PIEZOELECTRIC ACTUATION

Young-Jae Kim[1,2], Jun-Sung Lee[1], Sang-Yoon Kim[1], Sung-Eun Park[1], Jungho Hwang[1], Yong-Jun Kim[1]
[1]Yonsei University, Seoul, KOREA
[2]Samsung Electro-Mechanics Co., Ltd., Suwon, KOREA

ABSTRACT

This paper demonstrates a hybrid jetting system (HJS) using both electrohydrodynamic (EHD) force and mechanical actuation to generate ultra fine droplets with drop on demand (DOD). In the proposed HJS, liquid meniscus was formed by a piezoelectric actuator and jets were generated by EHD force. Jetting characteristics were also examined for hydrophobicity and hydrophilicity of a surface of a capillary. With the HJS, on demand jetting of femto-liter was achieved with a frequency of 5 kHz.

INTRODUCTION

Recently, direct write techniques got much attentions to realize patterns and dots of micro or nano scale in their dimensions [1]. For manufacturing microstructures and fine patterns without the plating, etching and lithography process, direct write techniques have several advantages including low cost, high speed, noncontact and environmental friendly manufacturing capability to forming micro structures and fine patterns [2]. There are several direct writing techniques. Among those techniques, an inkjet DOD jetting is widely used for printed electronic devices such as DNA microarrays, ceramic printing and organic transistors [3]. The inkjet printing is capable of on demand jetting with a high operational frequency of 10-100 kHz. However, the volume of droplet, 10-100 pl in general, strongly depends on the size of nozzle. With such droplet of large volume, it is not easy to meet requirement of printed electronics having fine features, including thin film transistors and high-density printed circuit boards [4, 5].

A lot of research groups have studied EHD jetting techniques to produce drops on demand with tens of times smaller than those produced by inkjet printing. The EHD techniques can be mainly classified into two types, the pulsating jet mode [6, 7] and the pulsed-voltage cone-jet mode [8-13]. In the pulsating jet mode, droplets are jetted by periodical deformation of the meniscus with DC high voltage. Its complete cycle consists of liquid accumulation, cone formation, droplet ejection and relaxation. For this cycle, 3 to 10 ms are required at least, which corresponds to a jetting frequency of 100-300Hz. Also, its jetting process depends on the oscillation of the meniscus by the electric field. Therefore, it is difficult to control the jetting frequency and produce droplets with high frequency. The pulsed-voltage cone-jet mode uses pulse waves of high voltage to switch electrohydrodynamic force (on/off) instead of DC high voltage. Its jetting duration is determined by the width of the applied high voltage pulse. This method is also required delay time, 3.6 ms [8], to make a Taylor cone immediately after applying pulsed voltage. This delay time is originated from the high voltage used. Due to such delay time, it is theoretically impossible to realize jets of a frequency higher than 300 Hz and ultra fine droplets.

In this study, we propose a system that capable of jetting ultra fine droplets on demand with high jetting frequency up to 5 kHz and volume of a few femto-liter level. The proposed system is named a hybrid jetting systems (HJS) since a meniscus was controlled by piezoelectric actuation, and a Talyer cone was formed by the EHD force. In the following sections, details of operational principle of the HJS and results will be presented.

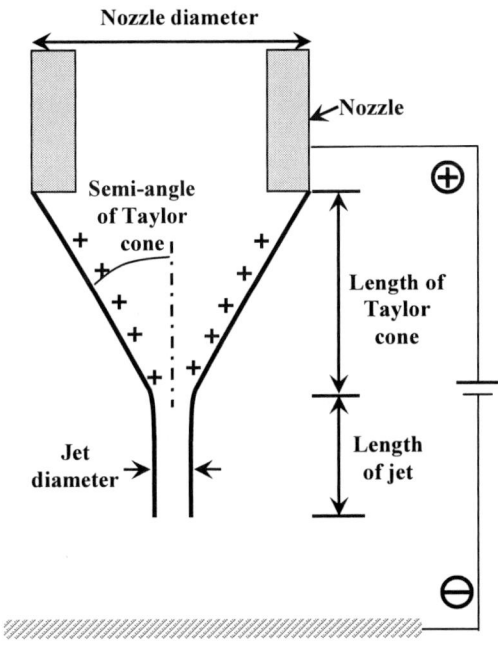

Figure 1: Principle of electrohydrodynamic jetting.

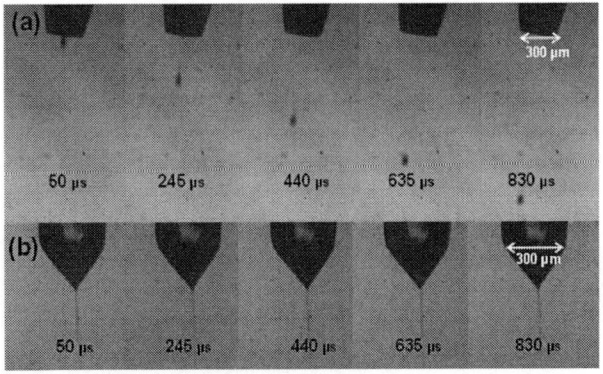

Figure 2 : General (a) piezoelectric DOD inkjet printing and (b) EHD continuous

EXPERIMENTS AND RESULTS

Figure 1 is a schematic showing changes of meniscus shape according to intensity of an electric field. When an external electric field is applied to a conductive solution, ions in the solution aggregate around the electrode of opposite polarity. Positive ions travel to the electrode charged negatively. The negative ions travel toward the positive electrode. When a positive voltage is applied to the capillary, ions in the solution of a same polarity will be moves to the surface of a meniscus. This deforms the shape of the meniscus into a cone. When the electric potential on the surface of the meniscus exceeds a critical value, the electrostatic forces overcome the surface tension of the conductive solution. In the event, a thin jet of the conductive solution is ejected from the surface of the cone and travels toward the nearest electrode of opposite polarity.

When a nozzle is operated solely by piezoelectric actuation (Fig. 2(a)), droplets can be jetted on demand However, the size of droplet depends on the size of orifice because these jetting phenomena are originated from the pressure variation inside the capillary. Therefore, in general, the size of droplet is similar to the size of orifice and its volume is large. When a nozzle is operated solely by an EHD force (Fig. 2(b)), very thin liquid stream can be generated even with a relatively large orifice. With this EHD method, we can extrude continuous liquid streams not droplets.

Figure 3 is a schematic diagram of the HJS. This system consists of the following components: (1) a glass capillary (0.06mm inner and 0.3mm outer diameters, respectively) pressurized by a piezoelectric actuator (Microfab, MJ-AT-01-50); (2) a driving waveform generator for piezoelectric actuation (Softmecha); (3) a high speed camera (Motion Pro HS-4, Redlake Inc.) for observation of droplet jetting in real time; (4) a power supply for generating a high voltage (HV-Rack, Ultravolt Inc., 4.7 kV) which is connected to a pin type electrode as an anode and a capillary holder as a cathode; and (5) a pneumatic device (EFD-2400) that kept the initial vertical position and the shape of meniscus uniform during all the experiments. Liquid used in this study was C_2H_5OH of 99.7% in its purity and its physical properties are listed in Table 1.

Figure 4 shows an operational principle of the HJS. In this system, a piezoelectric actuator is equipped in the capillary for on demand jetting. When a driving waveform is applied to the piezoelectric actuator, the wall of the capillary is deformed, and pressure inside the capillary increases. Such a higher pressure allows extrusion of liquid meniscus outside of the capillary. In here, it is important to optimize an applied driving waveform because large droplet is ejected by excessive displacement of the piezoelectric actuator.

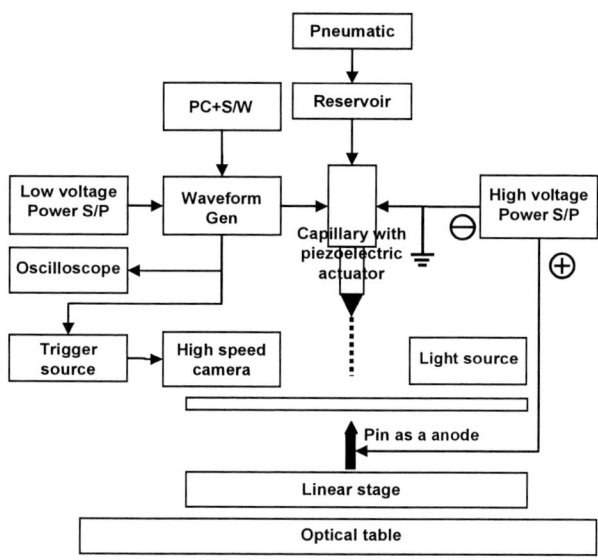

Figure 3. Schematic diagram of the experimental setup for the proposed HJS

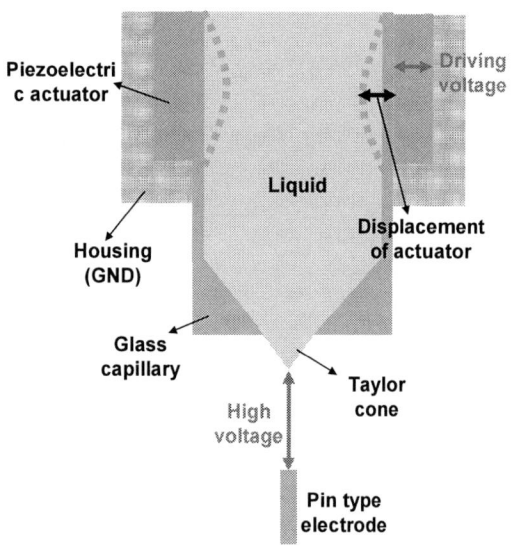

Figure 4. Principle of HJS.

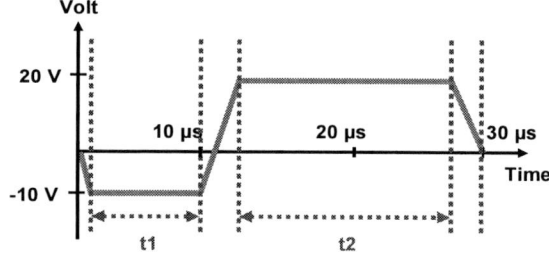

Figure 5. Driving waveform of piezoelectric actuator.

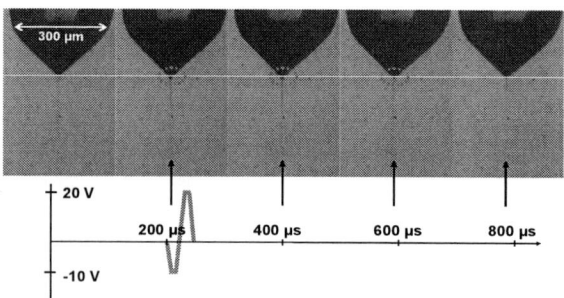

Figure 6. Hydrophobic capillary jetting.

When the liquid meniscus is deformed and extruded outside of capillary, EHD jetting is initiated because the electric potential on the surface of the meniscus exceeds a critical value due to the high voltage applied between capillary and bottom pin type electrode.

When the driving waveform is off and pressure inside the capillary is decreased, the meniscus of liquid returns to the initial position, and one cycle of the jetting is finished. Contrary to the pulsed-voltage cone-jet mode (operational voltage is several kV), the shape of liquid meniscus can be controlled by using a piezoelectric actuator which is operated by electric potential of tens of volts. Using this principle, ultra fine droplets could be generated with a high frequency up to 5 kHz. Furthermore, response characteristics of the HJS were better than those of pulsed-voltage cone-jet modes and pulsating jet modes, because delay time to make a Taylor cone is very small as much as it can be ignored.

Figure 5 shows representative waveform used to drive the piezoelectric actuator. When the section of *t1* is applied to the actuator, pressure inside the capillary is slightly decreased and liquid meniscus deforms into a concave shape. When the section of *t2* is applied to the actuator, pressure inside the capillary is increased. Eventually, liquid meniscus deforms into convex shape.

Figure 6 shows the HJS jetting images with a capillary that is hydrophilic. It was observed that the liquid meniscus has a similar size of outer diameter (300 μm) of glass capillary, and jetting through the deforming of cone shape could be achieved. The volume of Taylor cone was very large, and the volume change of liquid meniscus is quite smaller than the total volume of Taylor cone. This was originated from the hydrophilicity of the surface of glass capillary. Therefore, it was hard to control the meniscus precisely and also keep the meniscus stable. Furthermore, jetting was not stopped even after turning the driving wave form to the actuator off. The delay time, caused by a residual electric force and inertia of the liquid, was observed to be more than 400 μs.

To reduce effects resulted from hydrophilicity of the capillary as shown in Fig. 6 the glass capillary was treated with Teflon (AF2400, Dufont Corp.) to have hydrophobicity. As shown in Fig. 7, it was possible to produce a liquid meniscus with a similar size to an inner diameter of the glass capillary (60 μm) and form a Taylor cone of small volume of compared to Fig. 6. This was mainly due to hydrophobic treatment of the capillary. Also it was very important to precisely control the initial position of liquid meniscus by pneumatic controller.

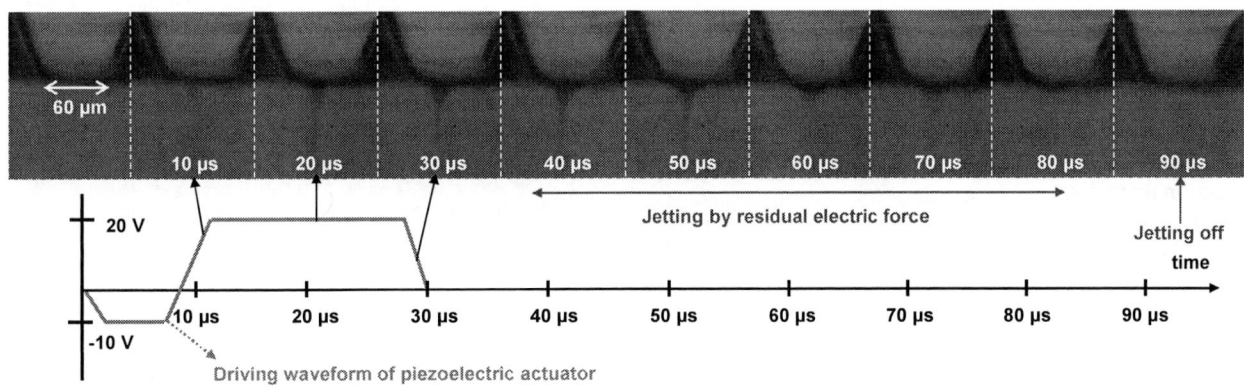

Figure 7. One cycle of the hybrid jetting using the proposed HJS.

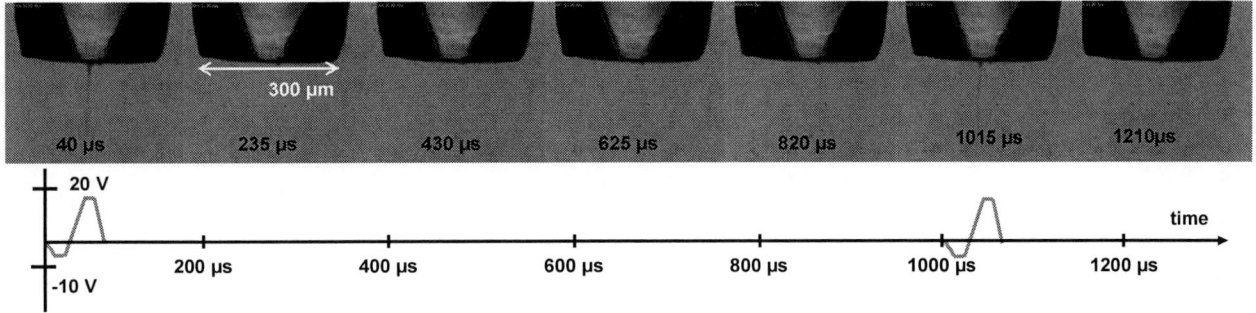

Figure 8. Detailed photograph which shows that the jetting are controlled at 1 kHz.

Table 1. Physical properties of the liquid used in this study

Viscosity (mPa s)	Electrical conductivity (Sm⁻¹)	Relative permittivity ($\varepsilon_i / \varepsilon_0$)	Surface tension coefficient γ (Nm⁻¹)	Density (kgm⁻³)	Charge relaxation time $\beta \varepsilon_0 / K$ (s)
1.16	3.0E-4	25	0.022	789	6.0E-7

At 10 μs, the displacement of actuator was produced and the inside pressure of capillary is increased when the driving waveform is applied to piezoelectric actuator. As a result, the liquid meniscus was extruded outside of the glass capillary. At 20 μs, extruded meniscus was transformed into a cone shape by an existing high voltage, and the cone jet ejection was begun. Although the driving waveform was turned off at 30 μs, jetting was not finished but continued till 80 μs because of the inertia of liquid and residual electric force. At 90 μs, the whole jetting process was fully finished and meniscus returned to the initial position. This means that one jetting period could be finished within 90 μs and theoretically jetting with frequency of about 11 kHz could be achieved.

Figure 8 shows an example of jetting on demand with operational frequency of 1 kHz. According to this result, it was ascertained that jetting frequency could be precisely controlled by our will. This may be a real DOD jetting.

DISCUSSION AND CONCLUSION

We have demonstrated a hybrid jetting system, termed HJS, for ultra fine droplets with a high jetting frequency. In the proposed HJS, the meniscus movement was controlled even with a voltage ranging 20 to 40 V. By virtue of this small driving voltage, the proposed HJS is capable of (1) generating ultra fine droplets by extremely short duration of a driving wave form (one droplet generation cycle: 90 μs) compared to the existing methods, (time for one droplet generation cycle of the pulsating jet mode: 3 to 10 ms [5] and delay time of the pulsed-voltage cone-jet mode: 3.6 ms [8]), (2) modulating the size of droplets by controlling duration of a driving wave form, and (3) high frequency operation up to 5 kHz.

In conclusion, the proposed HJS has unique features as follows: (1) we can achieve both ultra fine droplets and jetting on demand with high frequency simultaneously. This is a first unique feature of the proposed HJS compared to conventional ink jet printing and EHD jetting techniques, and (2) ultra fine droplets were generated by using mechanical actuation, not by using high voltage pulse that are widely used in EHD techniques.

ACKNOWLEDGMENT

This work was supported in part by the Information Technology Research Center support program (IITA-2005-C1090-0592-0012), and in part by the Seoul R&BD program (GR070039)

REFERENCES

[1] K. H. Church, C. Froe, T. Feeley, "Commercial applications and review for direct write technologies," *Proc. Material Research Sym.*, 2000, pp.3–8.

[2] D.B. Chrisey, "The power of direct writing," *Science*, vol. 289, pp. 1872–1875, 2000.

[3] P. Calvert, "Inkjet printing for materials and devices," *Chem. Mater.* vol. 13, pp. 3299–3305, 2001.

[4] O. Yogi, T. Kawakami, M. Yamauchi, J.Y. Ye, and M. Ishikawa, "On-demand droplet spotter for preparing pico- to femtoliter droplets on surfaces," *Anal. Chem.*, vol. 73, pp. 1896–1902, 2001.

[5] K. Hosokawa, T. Fujii, and I. Edo, "Handling of picoliter liquid samples in a poly

(dimethylsiloxane)-based microfluidic device," *Anal. Chem.*, vol. 71, pp. 4781–4785, 1999.

[6] J.-U. Park, M. Hardy, S.J. Kang, K. Barton, K. Adair, D.K. Mukhopadhyay, C.Y. Lee, M.S. Strano, A.G. Llleyne, J.G. Georgiadis, P. M. Ferreira and J.A. Rogers. "High-resolution electrohydrodynamic jet printing," *Nat. Mater.* vol. 6, pp.782–789, 2007.

[7] M.D. Paine, M.S. Alexander, K. L. Smith, M. Wang, and J.P.W. Stark, "Controlled electrospray pulsation for deposition of femtoliter fluid droplets onto surfaces," *J. Aerosol Sci.*, vol. 38, pp.315–324, 2007.

[8] J.L. Li, "On the meniscus deforming when the pulsed voltage is applied," *J. Electrostatics*, vol. 64, pp. 44–52, 2006.

[9] C.-H. Chen, D.A. Saville, and A. Aksay "Scaling laws for pulsed electrohydrodynamic drop formation," *Appl. Phys. Lett.*, vol. 89, 124103, 2006.

[10] J.L. Li, "EHD sprayings induced by the pulsed voltage superimposed to a bias voltage," *J. Electrostatics*, vol. 65, pp.750–757, 2007.

[11] J. Li, "On the stability of electrohydrodynamic spraying in the cone-jet mode," *J. Electrostatics*, vol. 65, pp. 251–255, 2007.

[12] B.S. Lee, H.-J. Cho, J.-G. Lee, N. Huh, J.-W. Choi, I.S. Kang, "Drop formationg via breakup of a liquid bridge in an AC electric field," *J. Colloid Int. Sci.*, vol. 302, pp. 294–307, 2006.

[13] W. Balachandran, W. Machowski, and C.N. Ahmad, "Electrostatic atomization of conducting liquids using AC imposed on DC fields," *IEEE Trans. Industry Applications*, vol. 30 , pp.850–855, 1994.

MICRO-MACHINED STENT-TYPE FLOW SENSOR FOR EVALUATION OF NASAL RESPIRATION

T. Yokota, J. Naito, M. Shikida, and K. Sato

Dept. of Micro-Nano Systems Engineering, Nagoya University, JAPAN

ABSTRACT

We present a thermal flow sensor that is suitable for measuring nasal respiration. To fix the flow sensor inside the nasal passage, we integrated it onto a stent structure, which is normally used as a medical device. The sensor was monolithically integrated on the Ti substrate by applying photolithography and wet etching processes. The developed fabrication has advantage that it is able to fabricate the both of the cavity for the thermal isolation and the stent structure during the same process. We mounted the flow sensor onto the inner surface of the silicone tube by inflating the balloon tube. The mechanical strength of the cylindrical stent under the compression condition was also studied, and it showed an elastic deformation when the force was less than 0.08 N. The developed flow sensor could detect the flow direction at the flow range of 0 - 2000 ccm. A response time of less than 260 msec was obtained by forming a cavity under the sensor.

INTRODUCTION

Thanks to the development of semiconductor technologies, many types of physical sensors used in industrial applications have been miniaturized and commercialized. Miniaturized sensors can be used to provide new type of measurement hardware that will be used with current medical diagnosis systems. Human beings breathe through both the mouth and the nose. The bacteria and viruses are trapped in the nose cavity, and therefore the cleaned inspired air only supplied into the inside of the lung. The inspirited air is also humidified here so that it is able to absorb oxygen effectively in the lung. Therefore, the flow characteristics of the air at the nose are closely related to health. We newly propose a stent-type thermal flow sensor for the evaluation of the nasal respiration in this paper.

STENT-TYPE FLOW SENSOR

There are two challenges to realize the flow sensor for the purpose of the nasal respiration.

(1) As the sensor structure itself disturbs the flow stream, it needs to be as small as possible.
(2) The flow sensor has to be fitted onto the inside surface of nasal passage, which has an irregular and non-constant cross-section.

The mechanism of hot-wire sensing detection is widely used in conventional thermal flow-rate measurements and involves the use of a sensor positioned at the center of a tube [1-4]. However, this positioning disturbs the flow stream. Therefore, this method is difficult to apply to meet our goal. To overcome these two problems, we propose a stent-type flow sensor (Fig. 1). The thermal type of the flow sensor is integrated on the stent structure, which is normally used as a medical device to expand and scaffold blood vessels. Because of the expandability of the stent structure, this sensor fits the inside surface of the nasal passage. Moreover, the structure of the flow sensor does not interfere flow stream because it is mounted on the inner wall surface.

FABRICATION

Planate structure fabrication

As shown in Fig. 2, we developed the fabrication process of the flow sensor by carrying out photolithography and wet etching. We first fabricated a sensor element on the metal substrate and then etched the substrate to form the stent structure and the cavity for the thermal isolation. The details of our developed fabrication method are as follows.

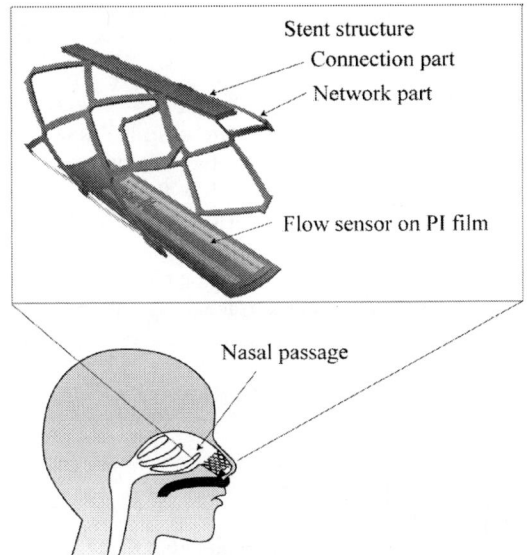

Figure 1: Schematic view of stent-type flow sensor and its application to evaluation of nasal respiration.

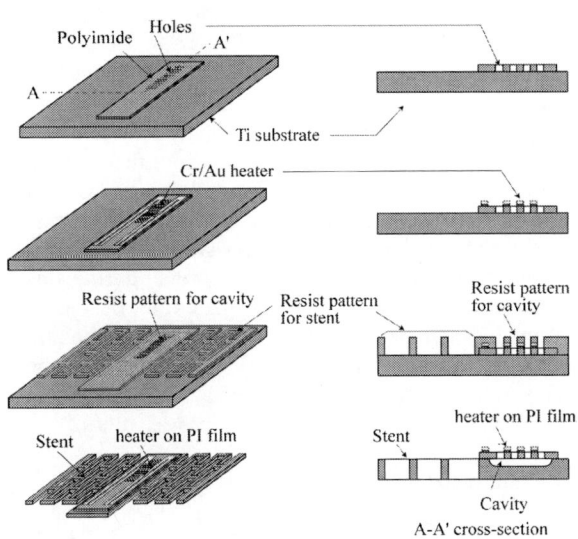

Figure 2: Fabrication process for stent-type flow sensor.

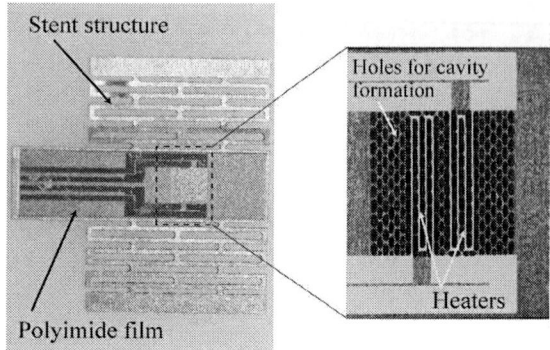

Figure 3: Fabricated planate flow sensor.

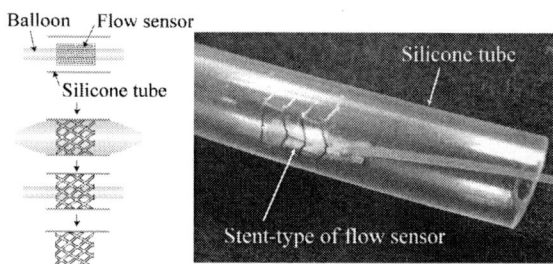

Figure 4: Mounted flow sensor onto inner surface of silicone tube by swelling balloon tube.

(a) An excellent biocompatible material, Ti, is widely used in implant products such as artificial cardiac valves and bones. We therefore used a Ti substrate with a thickness of 50 μm as the material of the stent. A 5.0-μm polyimide (PI) film as a substrate of the thermal flow sensor was deposited on the Ti substrate, and it was patterned by the photolithography for defining the outline and for forming the shapes of holes to form a cavity.

(b) An Au/Cr film (Cr was used as a membrane for adhesion) was deposited on the PI film by sputtering. The thickness of the Au and Cr was 300- and 50- nm, respectively. The metal film structures were patterned by applying the lift-off process. The metal film worked as a sensing heater, and we fabricated two parallel heaters that were used to detect the direction of the flow of aspirated and inspired air.

(c) We coated a photoresist layer onto the Ti substrate and it was patterned to define the both shapes of the stent and cavity for the thermal isolation. The cavity formed under the PI film is one of the important structures that shorten the response time.

(d) Isotropic wet etching was applied to the Ti substrate with the patterned PI and photoresist films. Both the stent and the cavity structures were formed during this process because the size of the hole pattern for the cavity was smaller than that for the stent (Isotropic wet etching was the diffusion-limited process, and therefore the etching rate decreases with the reduction of the hole size). The depth of the fabricated cavity was 30 μm.

The fabricated planate flow sensor is shown in Fig. 3.

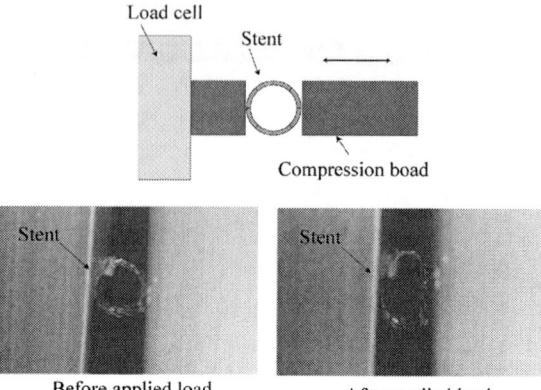

Figure 5: Compression test setup and stent cylindrically deformed

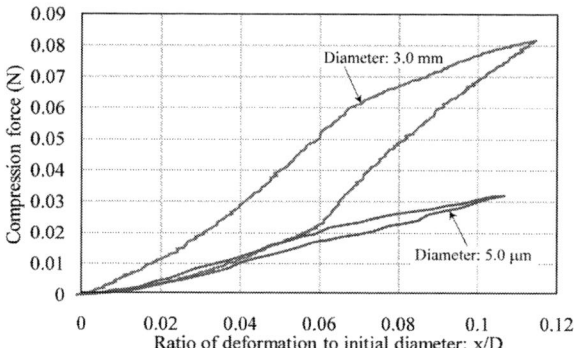

Figure 6: Elastic deformation of the cylindrical stent under compression condition.

Forming cylindrical shape

The planate flow sensor and a flexible printed circuit were bonded by an adhesive, and then the electrical pads on them were connected by conductive silver paste. The planate sensor was then tubulated by wrapping it by hand onto the Teflon rod with a diameter of 2.0 mm. A joint (two opposing sides) was connected by an adhesive, and then the rod was pull from the stent structure. Using Teflon material enabled the rod to be easily pulled from the tabulated sensor.

Mounting onto the inner surface of tube

To measure the flow characteristics, we fixed the sensor onto the inside surface of the silicone tube by inflating the balloon tube (Fig. 4). We first inserted the balloon into the inside of the cylindrical sensor, and then it was inflated to expand the stent structure. At this step, the flow sensor was extended by using the inflated balloon, and it was fixed onto the inner surface of the silicone tube. Finally, the balloon was deflated, and extracted from the silicone tube. We confirmed that the sensor was able to be fixed to the inside surface of a tube with an inner diameter of up to 5.0 mm.

EXPERIMENTAL RESULTS

Mechanical strength

We first measured the mechanical strength at the radial direction of the extended cylindrical stent. Two different sized stent structures with extended sizes of 3.0- and 5.0-

mm in diameter were prepared. The schematic view of the experiment was shown in Fig. 5. The compression force was applied to the stent by the movable load, and the applied force was detected by the using the load cell. The applied load and the deformation curves at the both stents are shown in Fig. 6. Extended stents showed an elastic deformation when the deformation was less than about 10 %. The maximum applied forces at 5.0 m and 3.0 in diameter were 0.03 N and 0.08 N, respectively.

Temperature coefficient of resistance (TCR)

We investigated the relationship between the sensor temperature and the electrical resistance change. The resistance linearly increased as the temperature increased, as shown in Fig. 7. From this graph, we obtained a TCR value of $0.0022K^{-1}$.

Relationship between sensor outputs versus flow rate

To evaluate the flow sensing performance, we fixed the sensor onto the inside surface of the silicone tube with an inner diameter of 3.0 mm.

As describe in "Fabrication", two sensing heaters were formed on a cavity structure to detect the flow direction. As shown in Fig. 8, to shorten the response time in the experiments, both heaters were operated by individual constant temperature circuit. The output signal from each circuit was put into the other final amplifier circuit. The output from each circuit was the same as when the sensor was in the no-flow condition, as a result, the output at the final circuit becomes zero. When the flow occurred, the output signals at each circuit changed because the warmed air by the upstream heater passes through the downstream heater. Therefore, we are able to detect the flow rate and the flow direction by measuring the magnitude and the polarity of the final output signal, respectively [3, 6].

The relationship between the sensor outputs and the flow-rate at both forward and reverse flow conditions is shown in Fig. 9. The applied flow rate ranged from 0 to 2000 ccm, which was enough to evaluate the inspired and expired air flows in a human nasal passage. We confirmed that the both output signals obtained at the forward- and the backward-direction of the flows obeyed the equation of King, which is known to show the typical signal curve of a thermal flow sensor.

Response time

To detect the inspired and expired air performances, the response time must be less than a few hundred of ms. Therefore, to evaluate the response time performance of the sensor, we used a three-port solenoid valve located between the sensor and the mass flow controller. We rapidly changed the flow rate by switching the tube connection and measured outputs of sensors with and without cavity, respectively. The response waveform at the various flow rate conditions is shown in Fig. 10. The fabricated sensor without the cavity needs a long time to steady its outputs because heat escapes to the polyimide film. In contrast, by forming the cavity under the sensing PI film, the sensor output initially increased rapidly and reached 90% of the maximum value within 260 msec.

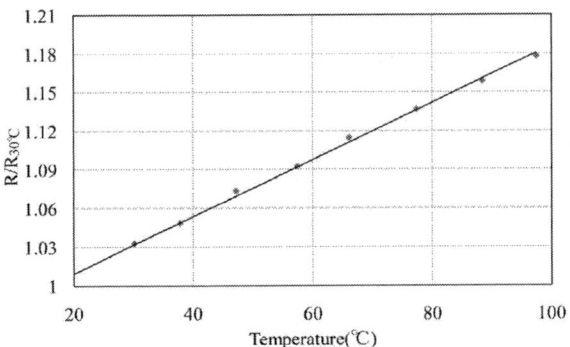

Figure 7: Change in resistance of thermal flow sensor versus temperature.

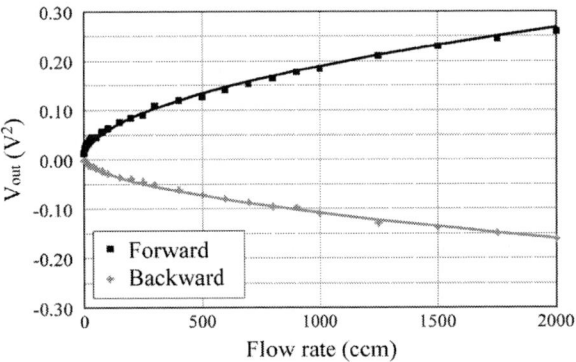

Figure 9: Relationship between flow rate and sensor output when flow was both forward and backward.

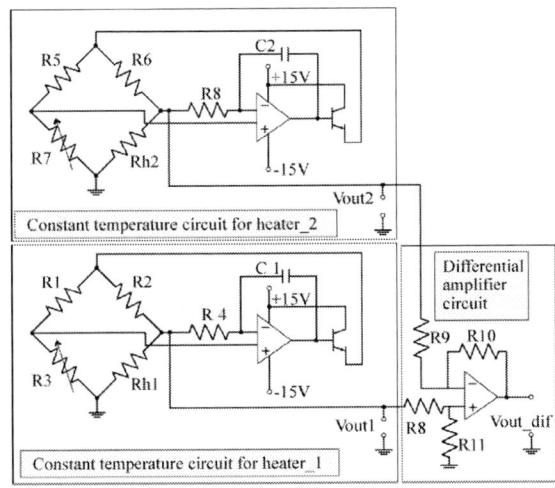

Figure 8: Constant temperature circuit.

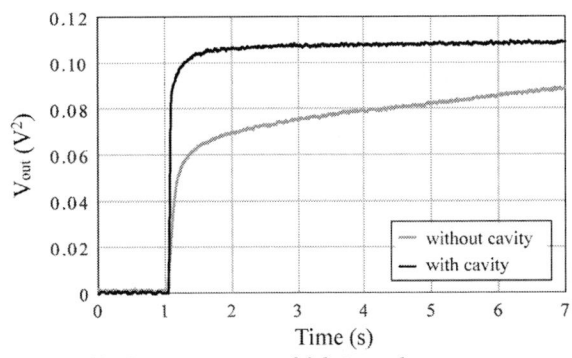

Figure 10: Response time of fabricated sensor.

CONCLUSION

We developed a stent-type flow sensor for evaluation nasal respiration. The sensor part was integrated on the stent structure, and it was fixed onto the inside surface of the silicone tube by using the expandability of the stent structure. The flow characteristics of the developed sensor **were as follows.**

(1) The developed sensor showed an elastic deformation when the force was less than 0.08 N.
(2) Relationship between the input power of the sensor and the flow-rate obeyed the equation of King under both forward and reverse flow conditions through 0 to 2000 ccm.
(3) Response time of less than 260 msec was obtained by forming cavity under the sensor.

These results suggest that we believe that this sensor will be able to evaluate flow characteristics in the nasal passage.

ACKNOWLEDGMENTS

I am deeply grateful to Prof. T. Kawabe and Dr. Y. Hayashi of Nagoya University for their kind advice.

REFERENCES

[1] B. W. van Oudheusden, "Silicon thermal flow sensors," Sensors and Actuators A, Vol. 30 (1992) pp. 5–26.

[2] H. Batles, O. Paul, and O. Brand, "Micromachined thermally based CMOS Microsensors," *Proc. of the IEEE*, Vol. 86, No. 8, pp. 1660–78, Aug. 1998.

[3] U. Buder, L. Henning, A. Neumann, and E. Obermeier, "AeroMEMS wall hot-wire sensor arrays on polyimide with through foil vias and bottom side electrical contacts," Tech. Dig. Transducers '07 (2007) pp. 2349–2352

[4] G. C. M. Meijer and A. W. Herwaarden, "Thermal Sensors," Institute of Physics Publishing Bristol and Philadelphia, 1994.

[5] K. Takahashi, I. Takayama, and R. Matsuhashi, "Effect of HNO_3 concentration in nitric-hydrofluoric acid solutions on surface morphology of titanium," Current advances in materials and processes: report of the ISIJ meeting, Vol. 9 (1996) pp.636

[6] J. Naito, M. Shikida, M. Hirota, Z.Y. Tan, and K. Sato, "Miniaturization of On-wall In-tube Flexible Thermal Flow Sensor Using Heat Shrinkable Tube," Proc. of MEMS 2008 pp. 927-930

EHD MICRO PUMP USING PYROLYZED POLYMER
3-D CARBON MESH ELECTRODES

Daisuke Wakui[1], Naoki Imai[1], Yasuhiro Nagaura[1], Hironobu Sato[1], Tetsushi Sekiguchi[1]
Satoshi Konishi[2], Shuichi Shoji[1] and Takayuki Homma[1]

[1]Major in Nano-science and Nano-engineering, Waseda University, Japan
[2]Department of Micro System Technology, Ritsumeikan University, Japan

ABSTRACT

This paper presents the EHD (electro-hydrodynamic) micro pump using 3-D carbon mesh electrodes. The carbon electrodes were fabricated by pyrolysis of SU-8 micro mesh structures [1]. Low temperature SU-8 bonding method under 90°C was developed to realize wafer level carbon structure packaging. The pumping behaviors were evaluated using fluorinert as a sample solution. The maximum pressure and volume flow rate are about 23Pa and 400nL/min under applied voltage of 500V.

INTRODUCTION

An EHD pump uses electric fields acting on the space charges embedded in the nonconductive fluid. The charges move in the direction of the electric field and drag the fluid. The typical injection type EHD pumps have two faced electrodes along with a micro channel and require no moving parts like impellers, bellows or valves. Fig.1 is the schematic drawing of an injection type EHD pump.

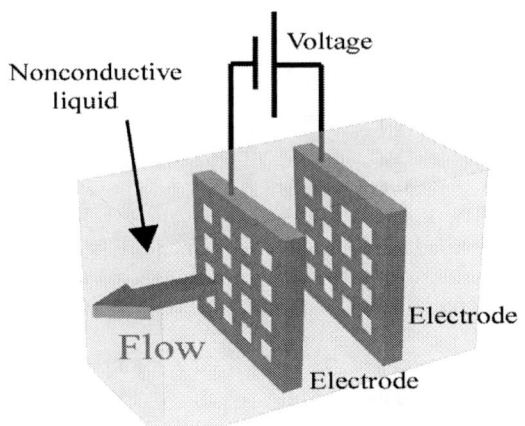

Figure 1: Schematic drawing of an injection type EHD pump

The injection type EHD micro pump using the silicon bulk micromachining and the wafer bonding technique was reported [2]. This kind of micro pump had a lumped configuration of grid electrode and was not convenient in the on-chip integration of micro-fluidic systems. Therefore, the several other types of EHD pumps with the 2D planar electrodes were realized [3]-[5]. They demonstrated the bi-direction pumping capability with much simpler fabrications than that of the first one.

In EHD micro pumps, electrochemical reaction on the surface of the metal electrodes and isolation between the electrodes were problems for practical uses [3]-[5].

Carbon materials have been used as electrodes in electrochemical applications. These materials have a very good electrochemical stability, biocompatibility and chemically inert surface. Pyrolyzed polymer is one of carbon materials and has been used as electrodes because of its characteristics [6]-[9].

In this paper, pyrolyzed polymer 3-D carbon mesh electrodes are utilized for injection type EHD micro pump. The 3-D electrodes were integrated on one chip. The wafer level packaging process (low temperature < 90°C) and direct SU-8 bonding technique were developed. The EHD micro pump was fabricated and its pumping properties were evaluated.

DEVICE DESIGN AND FABRICATION

The EHD micro pump design is shown in Fig.2. The gap between two electrodes is 90μm and the depth of the micro channel is about 100μm.

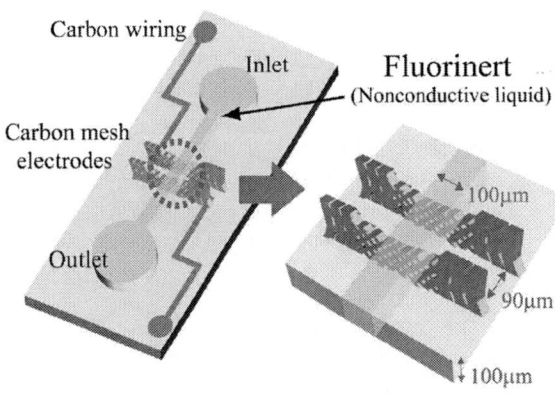

Figure 2: Schematic diagram of the 3-D carbon mesh EHD pump

The fabrication process of the EHD micro pump is shown in Fig.3. 1) Cr was evaporated on the glass substrate (Fig.3 (a)). 2) Cr was patterned and 3-D SU-8 mesh structures were fabricated by the backside UV exposure [10] (Fig.3 (b)). 3) SU-8 micro mesh structures were pyrolyzed in N2 atmosphere to form carbon electrodes (Fig.3 (c)). In pyrolysis, temperature was raised from room temperature to 300°C at a rate of 10°C/min, from 300°C to 850°C at a rate of 1.8°C/min and then, cooled down to room temperature at the same rate. 4) SU-8 was spin-coated to form micro channels and contact pad (Fig.3 (d). 5) LOR was spin-coated on the glass substrate (Fig.3 (e)). 6) SU-8 was spin-coated on the surface of the LOR layer (Fig.3 (f)). 7) The inlet and outlet were patterned by photolithography and, after that the SU-8 layer was lifted off by LOR layer etching (Fig.3 (g)). The SU-8 top cover was fabricated (Fig.3 (h)).

978-1-4244-2977-6/09 $25.00 © 2009 IEEE

This SU-8 structure was utilized as top ceiling of the device. 8) As an adhesion layer, SU-8 thin layer was spin-coated on the structure of Fig.3 (d). SU-8 top cover was put on it and bonded with the adhesion SU-8 layer [11] (Fig.3 (i)). 9) The residual SU-8 adhesive was removed by using SU-8 developer. The conductive wire was bonded to the contact pads (Fig.3 (j)). Fig.4 shows SU-8 mesh structure and pyrolyzed carbon structure. The pyrolyzed mesh structures were shrunk and distorted by the heat stress. But collapse of the structures was not observed. Fig.5 shows SEM photomicrograph of the carbon mesh electrodes formed in SU-8 straight channel. The carbon mesh electrodes were embedded in the SU-8 micro channel successfully.

Since many 3-D carbon mesh electrodes are integrated into the complex micro channels using this process, multi flow control micro systems can be fabricated. Fig.6 is the SEM image of the four 3-D carbon mesh electrodes integrated in the cross micro channel.

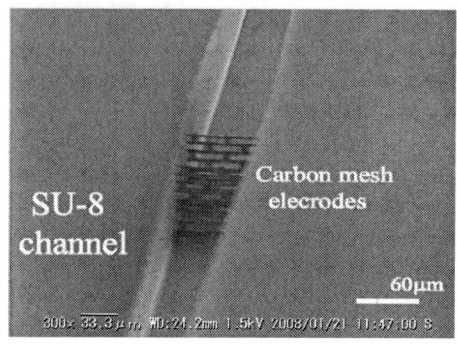

Figure 5: SEM images of the carbon mesh structures integrated in the straight micro channel

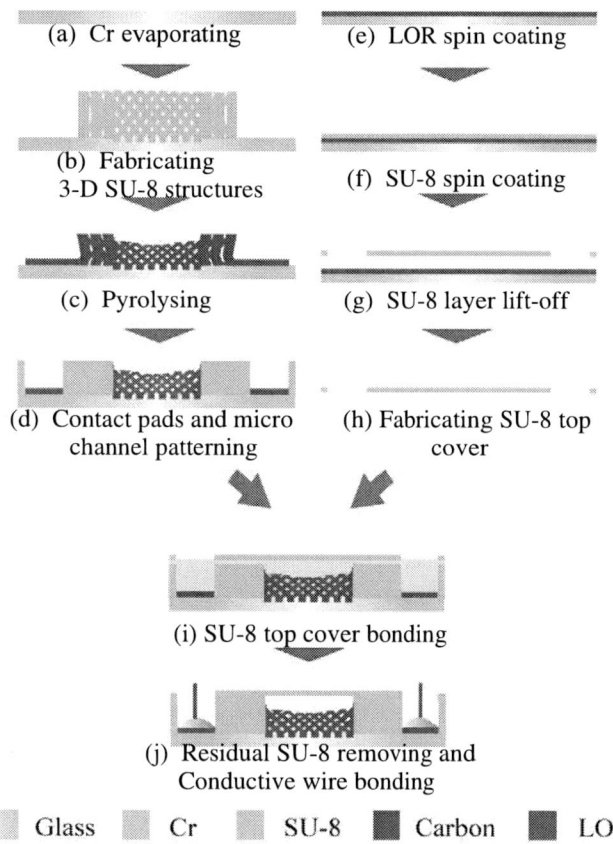

Figure 3: Fabrication process

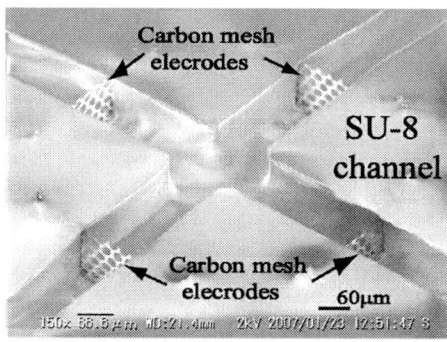

Figure 6: 3-D carbon mesh electrodes integrated in the cross micro channel.

EXPERIMENTS

In order to investigate EHD pump behaviors, two properties of the pump pressure and volume flow rate, as a function of applied DC voltage, was evaluated. The CCD camera was used to monitor the movement of the fluid-air interface. Fluorinert (FC-3283, 3M) was used as the non-conducting liquid. The applied voltage was from 100V DC to 500V DC.

The experimental setup of the pump pressure measurement is shown in Fig.7.

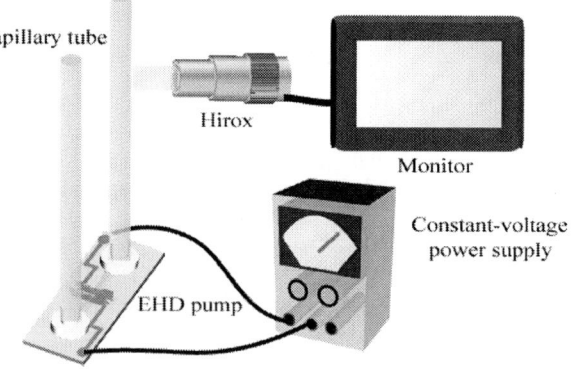

Figure 7: Setup of the pump pressure characteristic experiment

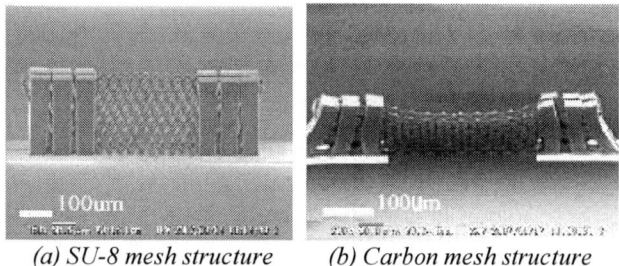

(a) SU-8 mesh structure *(b) Carbon mesh structure*

Figure 4: SEM images 3-D micro mesh structure

In this experiment, the liquid level difference was measured after 3min pumping. The pump pressure was obtained by using Eq. (1).

$$P = \rho g h \qquad (1)$$

P is the pressure of this pump, ρ is the density of fluorinert (= 1.83 kg/m^3), g is gravitational acceleration (= 9.81 m/s^2) and h is the liquid level difference.

Fig.8 shows the experimental setup of the flow rate measurement.

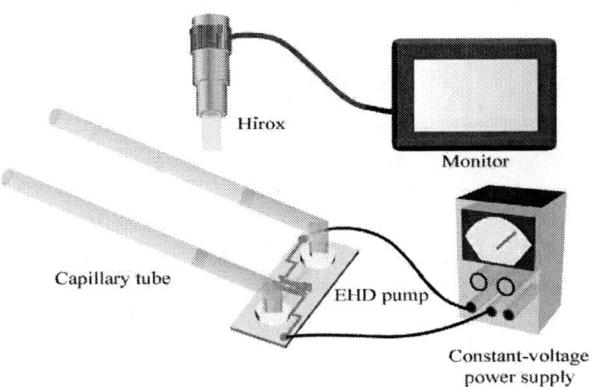

Figure 8: Setup of the volume flow rate experiment

In the flow rate characteristic experiment, the velocity of the liquid-air interface was measured. The volume flow rate was obtained by using Eq. (2).

$$U = D \times v \qquad (2)$$

U is the volume flow rate, D is the volume of capillary tube par 1mm (= 0.325nl/mm) and v is the velocity of liquid surface.

RESULTS AND DISCUSSION

Fig.9 is the image of liquid level change after 1min pumping at 500V. The liquid level change is about 1mm.

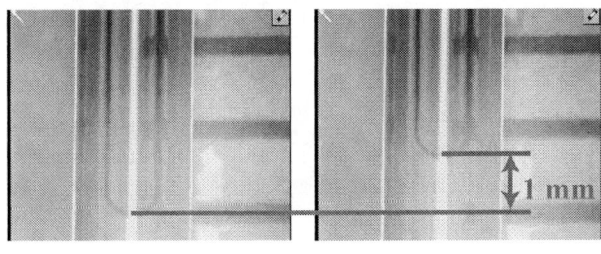

(a) 0min   *(b) After 1min pumping*

Figure 9: The liquid level change after 1min pumping at 500V

Figure 10 shows the relationship between applied voltage and pump pressure. The threshold voltage of this micro pump is about 200V. The pressure gradually increases with DC voltage. The controllable pressure of the EHD micro pump was from 0.8 Pa to 22.9 Pa.

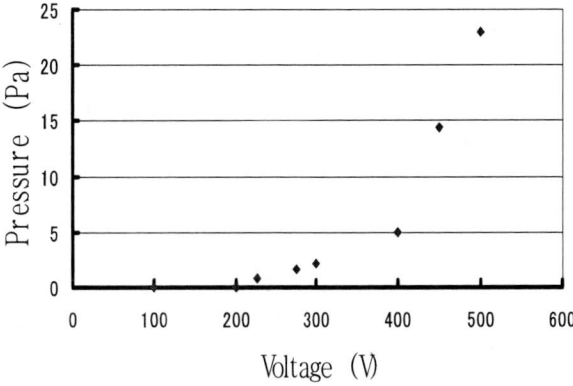

Figure 10: The relationship between applied voltage and pump pressure.

Figure 11 shows the relationship between applied voltage and volume flow rate. The volume flow rate gradually increases with DC voltage. The controllable volume flow rate of the EHD micro pump was from 48.1 nL/min to 398.4 nL/min.

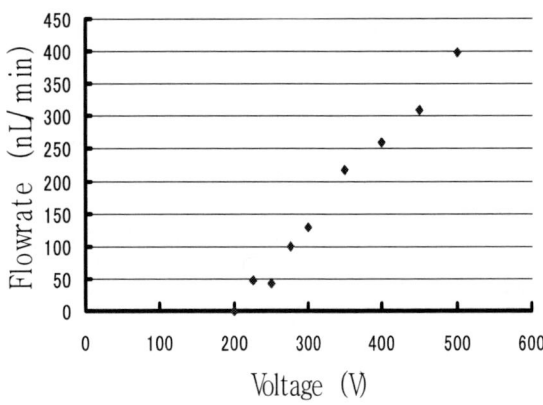

Figure11: The relationship between applied voltage and volume flow rate.

Increasing of the number of the electrode pair or shortening distance of the electrode is effective to improve the EHD micro pump performances. This EHD micro pump can be applied to the micro cooling devices, sample injectors and flow control micro systems.

CONCLUSIONS

This paper presents the novel EHD micro pump using pyrolyzed polymer 3-D carbon mesh electrodes. The wafer level SU-8 packaging method was developed. The two faced 3D carbon mesh electrodes were isolated perfectly. Since many 3D carbon mesh electrodes can be formed using this process, multiple flow control systems can be fabricated. Using Fluorinert as sample liquid, the maximum pressure of about 23Pa and maximum flow rate of 400nL/min were achieved. The micro pump still needs improvements of the pumping efficiency and the long-term stability. This novel EHD micro pump will be able to apply to many micro fluidic devices.

ACKNOWLEDGEMENTS

This work is partly supported by Japan Ministry of Education, Culture, Sports Science & Technology Grant-in-Aid for ASMeW, Global COE of Waseda University, Scientific Basic Research (A) No. 19206046, Grant-in-Aid for Specially promoted Research No.20002006. H.S. would like to thank the Research Fellowship of the Japan Society of the Promotion of Science for Young Scientists.

REFERENCES

[1] Kensaku Yamamoto, Keisuke Naka, Yasuhiro Nagaura, Hironobu Sato,Shuichi Shoji and Satoshi Konishi, "Pyrolyzed Polymer Mesh Electrode Integrated Into Fluidic Channel For Gate Type Sensor" the International Conference on Micro ElectroMechanical Systems 2007, 2007, pp. 271-274

[2] A. Richter and H. Sandmaier, "An EHD pump", the International Conference on Micro ElectroMechanical Systems 1990, 1990, pp.99-104

[3] S. Ahn and Y. Kim, "Fabrication and experiment of a planar micro ion drag pump", Sensors and Actuators A, 1998, Vol 70 (1-2), pp.1-5

[4] Yang, L. J.,Wang, J. M. and Huang, Y. L., "The Micro Ion Drag Pump Using Indium-Tin-Oxide (ITO) Electrodesto Resist Aging," Proc. IEEE MEMS'03, pp.
112_115 (2003), or *Sensors and Actuators A*, Vol. 111, 2004, pp. 118–122

[5] Jiun-Min Wang and Lung-Jieh Yang, "Electro-Hydro-Dynamic (EHD) Micropumps with electrode Protection by Parylene and Gelatin", Tamkang Journal of Science and Engineering, 2005 Vol.8, NO.3, pp.231-236

[6] K. Naka and S. Konishi, "Micro and Nano Structures of Carbonized Polymer through Pyrolytic Transform from Polymer Structure", IEE Micro and Nano Letters, pp. 79-82

[7] K.Naka, H.Hayashi, M.Senda, H Shiraishi and S. Konishi, "Eefect OF NANO STRIPE CARBONIZED-POLYMER ELECTRODE ON HIGH S/N RATIO IN ELECTRO CHEMICAL DETECTION", the International Conference on Micro Electro Mechanical Systems 2007, 2007, pp.195-198

[8] K. Naka, T. Hashishin, J. Tamaki and S. Konishi, "PYROLYZED POLYMER AS ABSORBENT OF NITROGENDIOXIDE SENSOR", the International Conference on Micro Electro Mechanical Systems, 2006, pp. 518-521

[9] R. Yokokawa, K. Yamamoto, S. Aoki, K. Naka and S. Konishi, "Characterization of Patterned Pyrolyzed Polymer for glucose sensor", Asia-Pacific Conference of Transducers and Micro-Nano Technology, 2006, pp. 287-290

[10] H.Sato, T.Kakinuma, J.S.Go, S.Shoji, "In-channel 3-D micromesh structures using maskless multi-angle exposure and their microfilter application", Sensors & Actuators A, Vol. 111, 2004, pp87-92

[11] H.Sato, H.Matsumura, S.Keino, S.Shoji, "An all SU-8 Microfluidic Chip with Built-in 3-D Fine Microstructures", Journal of Micromechanics and Microengineering, 16, 2006, pp.2318-2322.

978-1-4244-2977-6/09 $25.00 © 2009 IEEE

NON-LINEAR FLUIDIC INTEGRATED CIRCUITS REALIZED BY PNEUMATIC-FIELD EFFECT TRANSISTORS WITH CONTROLLABLE OUTPUT RESISTANCE

H. Takao, N. Tanaka, M. Sugiura, K. Sawada, M. Ishida
Toyohashi University of Technology, Toyohashi, Aichi, Japan

ABSTRACT

This paper reports 'non-linear' micro fluidic integrated circuit technology using newly developed vertical microvalve, Pneumatic-Field Effect Transistor (Pneumatic-FET) with controllability of its output resistance. In addition to the basic linear amplifiers of pressures obtained by positive output resistance of pneumatic-FET, the negative output resistance is attractive to realize non-linear functions in the smallest number of FETs. In this study, we have succeeded to fabricate 'non-linear' fluidic circuits with hysteresis or kinked transfer function utilizing negative output resistance of vertical pneumatic-FET for the first time. The novel device structure and the simple fabrication technology for controllable output resistance, and characteristics of the pneumatic integrated circuits are demonstrated.

INTRODUCTION

In microfluidic systems, pressure and flow-rate of fluid are basic parameters. A highly complicated microfluidic system includes many parameters to be controlled and monitored in the microfluidic network. In such systems, processing circuits of fluids to control and monitor the parameters will be effective for highly functional operation and precise flow control. For processing of large scale fluidic channels in microsystems, we have proposed and reported the original pneumatic analog circuit technology based on analogical relationship between pressure/flow circuits and voltage/current circuits. Pneumatic microvalve with MOSFET-like flow characteristics and linear pressure amplifier circuit were demonstrated at Transducers' 01 for the first time [1]. This MOSFET-like microvalve element corresponds to transistor in fluidic circuit controlled by pneumatic-field at movable diaphragm [2], and is called as 'pneumatic-field effect transistor (pneumatic-FET)'. The equivalent analysis model of the pressure amplifiers has been proposed [3] and demonstrated based on the MOSFET models analysis [4]. A differential acoustic amplifier has been reported recently [5], as an advanced version using the similar principle.

All the previous works of fluidic amplifiers are 'linear' analog circuit technology using 'positive' high output resistance of microvalve drain [1-5]. In this paper, non-linear fluidic integrated technology based on the newly developed 'vertical' pneumatic-FET device is presented. Unlike MOSFET, output resistance of microvalves can be controlled by geometry of the device structure. If negative output resistance is obtained and applied to circuit design of fluidic integrated circuit, 'non-linear' amplifiers with discontinuous or kinked transfer characteristics can be realized. In vertical pneumatic-FET, output flow resistance is controllable to positive or negative by changing the layout design geometry. Also, the occupied area of device is much reduced as compared to the previously reported planer

microvalves [1-5]. Utilizing its controllability of output resistance, microfluidic hysteresis inverter and differential pressure amplifier with kinked transfer curve have been realized for the first time.

THE VERTICAL PNEUMATIC-FET
Structure and Operation

In microfluidic circuits, 'pressure' and 'flow-rate' correspond to 'voltage' and 'current' in LSI, respectively. Pneumatic-FET works as active devices (transistors) in fluidic circuits similarly with MOSFETs in electronics. It realizes amplification of input 'pressure' by a large ratio of output conductance and transconductance [1]. Fig. 1 shows the newly developed vertical pneumatic-FET in this study. A very thin vertical diaphragm and narrow channel gap are formed with SOI active layer using Deep-RIE. Source-Drain conductance is modulated by gate pressure changing the gap. Top and bottom of the vertical diaphragm is released (i.e. not fixed) from the glass and handle wafer by making a small gap.

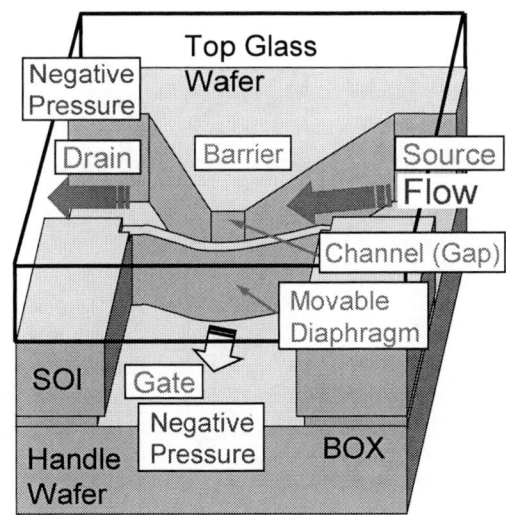

Figure 1: Newly developed pneumatic-FET structure with vertical diaphragm of SOI layer.

Vertical moving wall in microfluidics has been utilized in droplet size control, recently [6]. In this study, the vertical structure is used to control the precise device geometry. Also, it realizes remarkable device size-reduction to realize large scale integrated microfluidic circuits. In the previous research works, the movable diaphragm is formed by in-plane silicon or polyimide diaphragm, and occupied area of the device is so much larger than this vertical diaphragm. Forming the vertical diaphragm, device geometry is determined by single photomask, and device area is reduced to below 7% of the previous pneumatic-FETs with same flow capability at least. This simplicity is essentially advantageous for high density integration by device shrinking like "Moore's law

978-1-4244-2977-6/09 $25.00 © 2009 IEEE

in CMOS".

Fabrication of pneumatic-FET

The vertical pneumatic-FET is fabricated by very simple MEMS process with SOI wafers. Fig. 2 shows the outline of the fabrication process. The pneumatic-FET and its integrated circuits are fabricated only by 3-photomasks. Starting material of the fabricated device in this experiment was SOI wafer with 100μm-thick active layer.

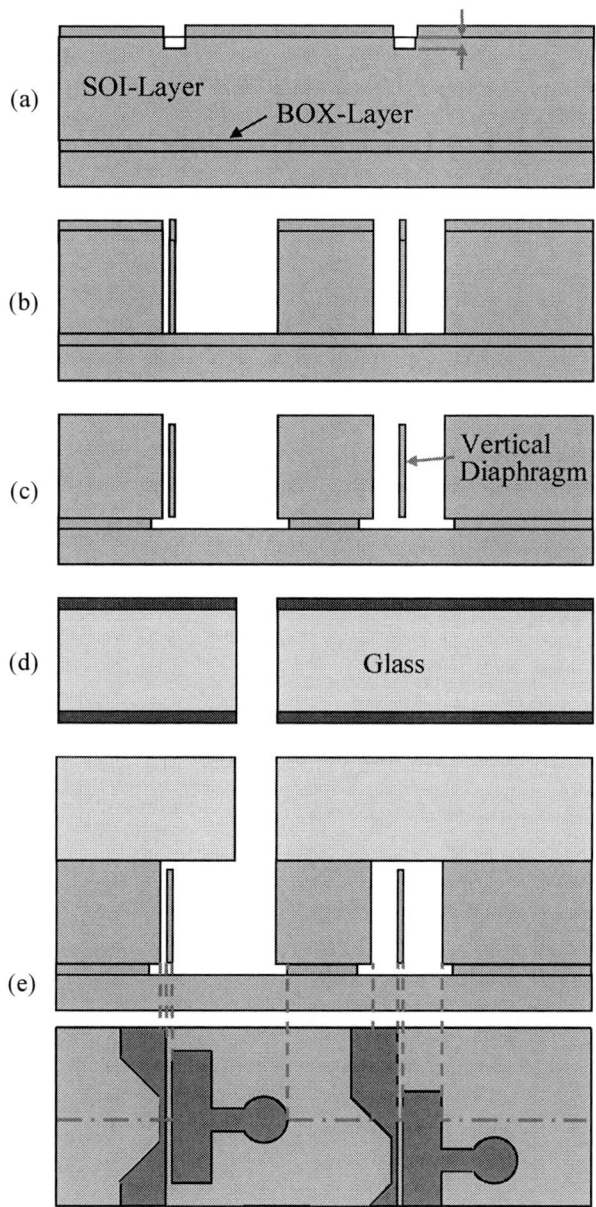

Figure. 2: Outline of the fabrication process.

(a); The first fabrication step is shallow etching (~0.5μm) of the movable diaphragm region to make a gap between the top glass wafer. (b); Vertical diaphragm structures and microvalve essential structures are formed by a Deep-RIE process at the same time. (c); SiO₂ layers are etched by BHF, and the vertical diaphragm is released. (d); Through holes are formed in top glass wafer by sandblast. (e); The SOI wafer and the glass wafer are bonded by anodic bonding process, and microfluidic integrated circuits with pneumatic-FETs are completed.

Fig. 3 shows an SEM photograph of a fabricated pneumatic-FET structure. The top glass wafer is not yet bonded. Narrow channel gap (~2μm) and thin diaphragm (~3μm) are formed by single Deep-RIE process, and symmetrical structure of Source/Gate/Drain is successfully made in a single step. Top bottom of the diaphragm is already released by sacrificial etching of BOX layer.

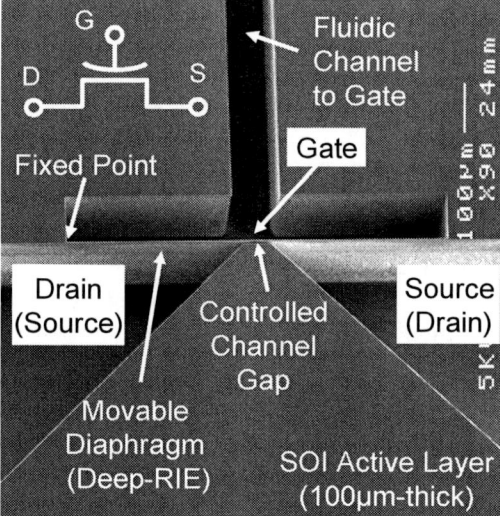

Figure. 3: SEM of a fabricated pneumatic-FET.

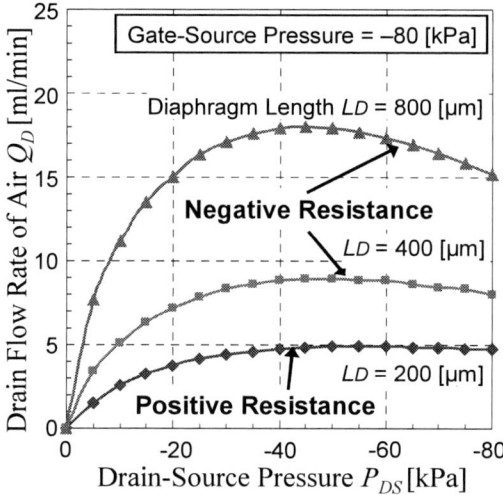

Figure. 4: Drain flow-rate (Q_D) vs. Drain pressure (P_{DS}) of pneumatic-FETs showing the controllability of the output resistance. Gate pressure (P_{GS}) is fixed at a -80kPa.

Fig. 4 shows measured flow characteristics of pneumatic-FETs with various diaphragm lengths (200, 400, 800μm). The relationship between Drain flow-rate (Q_D) and Drain-Source pressure (P_{DS}) for a constant Gate-source pressure (P_{GS}) is analogous with I_D-V_{DS} characteristics of a MOSFET. Similar to the MOSFET characteristics, drain flow-rate Q_D is saturated by P_{DS} due to the narrowed channel gap and increased flow resistance between the barrier and the movable diaphragm. The differential output resistance (dQ_D/dP_{DS}) becomes positive and negative by changing the length of the vertical diaphragm for the same barrier (gate) length. Since transfer characteristics of amplifier circuits strongly depend on the output resistance of active FETs, both conventional linear amplifiers and novel non-linear amplifiers can be realized with perfect freedom in fluidic circuit design. This is the most important feature of the vertical pneumatic-FET which can realize precise controllability of output resistance only by the layout design.

NON-LINEAR FLUIDIC CIRCUITS
Hysteresis Inverter

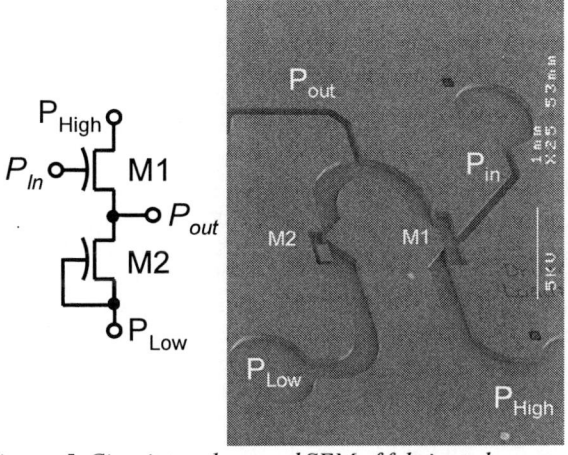

Figure. 5. Circuit topology and SEM of fabricated pressure inverting amplifier with integrated pneumatic-FETs. Depending on the output resistance, this circuit can realize different function.

Utilizing the controllable output resistance, non-linear function such as kinked and discontinuous transfer function can be realized in addition to the linear amplification functions realized in the previous studies. Fig. 5 shows the circuit topology and an SEM of fabricated pressure inverting amplifier with monolithically integrated two pneumatic-FETs. M1 is the drive FET with source terminal grounded to P$_{High}$, and M2 is the load FET with diode connection. So, this circuit works as common source inverting amplifier for input pressure. Choosing the polarity of the output resistance of FETs, inverting linear pressure-amplifier or non-linear hysteresis inverter can be designed and fabricated with the same technology. If the M1 has a positive output resistance, the gain of pressure amplification is almost determined by transconductance of M2, and becomes a finite negative value. The pressure gain of this inverting amplifier is expressed as the next equation;

$$G_{Pressure} = -\frac{g_{mM1}}{(g_{dsM1} + g_{mM2})} \qquad (1),$$

where g_{mM1} and g_{dsM1} are the transconductance and inverse of the output resistance of M1, respectively, and g_{mM2} is the transconductance of M2. If g_{dsM1} (i.e. inverse of the output resistance of M1) is a negative value equal to the positive g_{mM2}, the pressure gain becomes infinite value, and the operating point of the circuit will have discontinuous operating points. In this case, hysteresis operation is observed in the transfer characteristic.

Operating point analysis of the pressure inverting amplifier (Fig. 5) using measured flow characteristics of load/driver FETs is shown in Fig. 6. Since M1 has 800μm diaphragm length, the output resistance becomes a strong

negative value. Tracing the crossing point of the flow curve of M1 and M2, corresponding output pressure for each input pressure is obtained. Due to the snapbacked flow characteristic of M1, there are 'bi-state' crossing points on the same flow curve for an input pressure. This means that there are bi-state output pressures for an input pressure in the pressure transfer curve of the amplifier, and 'hysteresis' in the characteristic is obtained.

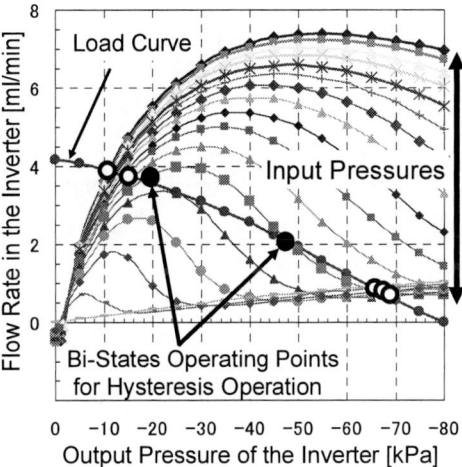

Figure. 6: Operating-point analysis of the pressure inverting amplifier (Fig. 5). The load curve corresponds to the flow characteristic of the diode-connected M2.

Fig. 7 shows a measured pressure transfer characteristic of the pressure inverting amplifier shown in Fig. 5. As expected in the operating point analysis using the flow characteristics of pneumatic-FETs shown in Fig. 6, hysteresis operation is clearly obtained in the pressure transfer curve. The window width of hysteresis is 7.2kPa, and it is determined by the shape of the flow characteristic of M1 with negative output resistance. As a result of detailed comparison between the operating point analysis, obtained transferred curve corresponds well with the analyzed characteristics in Fig. 6. This interesting bi-state operation is similar to FeRAM or Schmitt-trigger circuit in electronics, but realized only by two FETs.

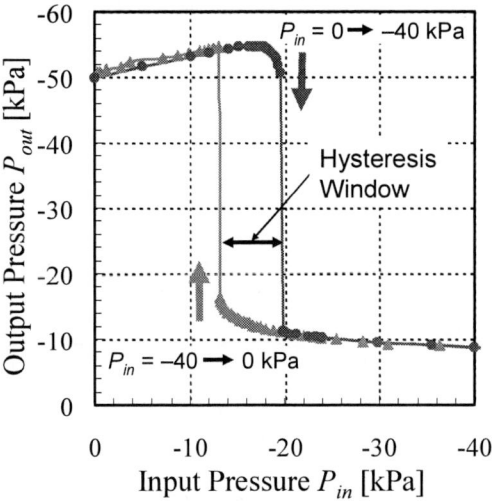

Figure. 7: Characteristic of the hysteresis inverter. The hysteresis window width is around 7.2kPa, and enough for bi-state operation of the circuit.

Differential Amplifier with Kinked Pressure Gain

Since the vertical pneumatic-FET is small-sized device and has very simple structure for large-scale integration, it is easy to form more complicated circuit topology. Fig. 8 shows a circuit topology of fully-differential pressure amplifier with gate-biased active loads, and SEM of a fabricated circuit integrating five pneumatic-FETs. This differential circuit has analogical circuit topology with a MOSFET differential amplifier. Both linear and non-linear differential amplifiers are realized by controlling the output resistance of the pneumatic-FETs (M2 and M3). Depending on the output resistance, the circuit becomes linear pressure differential amplifier or pressure comparator with input tolerance (low-gain region). In this design, the length of the diaphragm of drive FETs is 800μm, and non-linear transfer curve is obtained by the snapbacked (kinked) flow characteristics of M2 and M3.

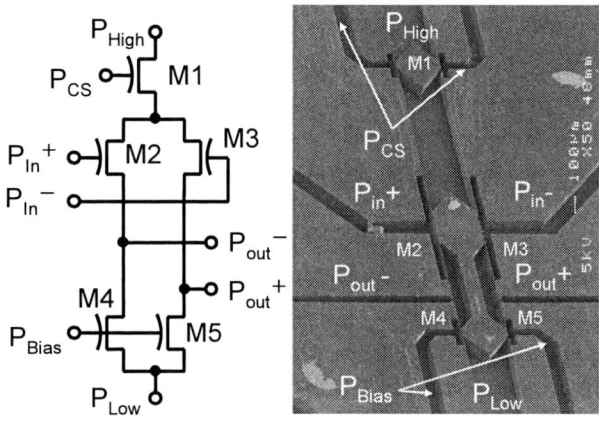

Figure. 8: Circuit topology and SEM of fully-differential pressure amplifier integrating five pneumatic-FETs.

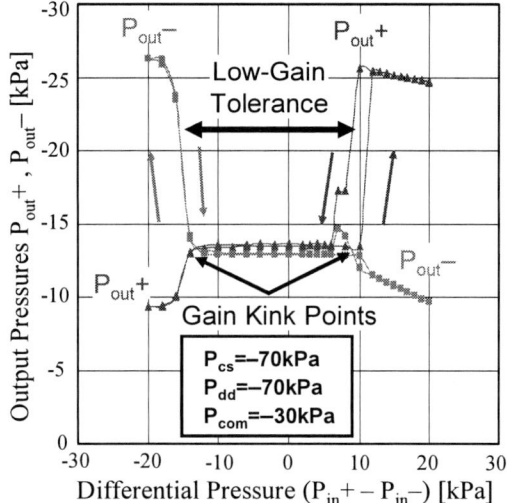

Figure. 9: Characteristic of the differential pressure amplifier with 'kinked' transfer function. Particular low gain window can be controlled and used as tolerance of pressure comparator.

Fig. 9 shows a measured pressure transfer characteristic of the fabricated non-linear differential pressure amplifier shown in Fig. 8. Symmetry 'gain-kinks' are observed particularly in the differential transfer curves due to the negative output resistance of M2 and M3. Around the neutral input point where the differential input pressure is 0kPa, the circuit has very small pressure gain. This low-gain region is suitable to make a tolerance of pressure comparator for the offset error caused by fabrication mismatch. The obtained window width of the low gain tolerance in the circuit is around 27kPa. The transfer characteristic of the amplifier is controllable by changing the flow characteristics of the drive pneumatic-FETs changing the layout design. Tolerance in pressure comparator can be applicable to robust feedback control of sample pressure/flow with switching margins.

CONCLUSIONS

In this paper, novel 'non-linear' micro fluidic integrated circuit technology using newly developed vertical pneumatic-FETs with controllability of the output resistance has been reported for the first time. 'Non-linear' fluidic circuits with hysteresis or kinked transfer functions utilizing negative output resistance were successfully fabricated, and their functionality has been confirmed. The non-linear functions in microfluidics are very attractive for various microfluidic systems .Unlike MOSFET integrated circuits with only positive output resistance, various and novel functions of integrated circuits can be realized utilizing this technology.

ACKNOWLEDGEMENTS

This work has been partially supported by The Ministry of Education, Science, Sports and Culture, Global COE Program 'Frontiers of Intelligent Sensing', and Japan Science and Technology Agency, CREST.

REFERENCES

[1] H. Takao, M. Ishida, K. Sawada, "A pneumatically-actuated silicon microvalve and its application to functional fluidic integrated circuits," *Tech. Digest Transducers'01*, pp. 946-949, 2001.

[2] H. Takao, M. Ishida, K. Sawada, "A Pneumatically Actuated Full In-Channel Microvalve with MOSFET-Like Function in Fluid Channel Networks," *J. Microelectromech. Syst.*, vol. 11, no. 5, pp. 421-426, 2002.

[3] H. Takao and M. Ishida, "Micro Fluidic Integrated Circuits for Signal Processing Using Analogous Relationship between Pneumatic Microvalve and MOSFET," *J. Microelectromech. Syst.*, vol. 12, no. 4, pp. 497-505, 2003.

[4] H. Takao and M. Ishida, *MEMS/NEMS Handbook: Techniques and Appl.*, Springer, Vol. 2. Chap. 4, "Novel MEMS Fluidic Integrated Circuit Technology with 'MOSFET-Like Microvalve Elements'," pp.111-140, 2006,.

[5] M. Tanaka, T. Sakashita, S. Konishi, "Acoustic Differential Amplifier Based on Acoustic FET," *Proc. IEEE MEMS 2005*, pp. 44-47, 2005.

[6] Y. H. Lin, C. H. Lee, G. B. Lee, "A New Droplet Formation Chip Utilizing Controllable Moving-Wall Structures for Druble Emulsion Applications," *Proc. IEEE MEMS 2008*, pp.22-25, 2008.

HIGH-RESOLUTION PIEZO INKJET PRINTHEAD FABRICATED BY THREE DIMENSIONAL ELECTRICAL CONNECTION METHOD USING THROUGH GLASS VIA

M. Murata, T. Kondoh, T. Yagi, N. Funatsu, K. Tanaka, H. Tsukuni, K. Ohno, H. Usami, R. Nayve, N. Inoue, S. Seto, and N. Morita

Fuji Xerox Co., Ltd.

ABSTRACT

High-resolution piezo inkjet (PIJ) printhead with 1200 nozzles per inch (npi) has been developed. A glass substrate with through holes is bonded to Si substrate with piezo actuators (PAs) placed in matrix array (32 x 16) via polyimide film planarized by grinding method. A thin PA composed of sputter-deposited Pb(Zr,Ti)O$_3$(PZT) film and Si diaphragm contributes to a shrink of PA size and then high dense integration. The through holes are electroplated with Ni, and Ti/Cu interconnect is formed by resist spray coating and wet etching method. The glass substrate also acts as an encapsulation plate to form a cavity space over the PA, which enables a desirable displacement of the PA. The PIJ printhead exhibited a successful ejection of 2pl ink drops by single jetting drive and 8pl ink drops by three multiple jetting drive.

INTRODUCTION

There are some kinds of inkjet printheads such as thermal inkjet (TIJ) and PIJ according to an actuator, which generates pressure and ejects ink drops. The PIJ printhead has some advantages such as wide usage of inks, corrosion resistance of an actuator, and high frequency ejection as compared with the TIJ printhead. However, it is difficult to array nozzles in high density, because an actuator is larger than that of the TIJ printhead, which 800npi in a row [1, 2] and 1200npi in two rows [3] have been reported. As for the nozzle resolution of the PIJ printhead, 360npi in a row utilizing a very narrow (71μm) pitch of PA has been reported [4]. And a printhead with 720npi in two rows was fabricated by using the PA.

In this work, PIJ printhead with 1200npi was developed by placing PA with matrix array as keeping a relatively wide pitch of PA. The primary subject for this approach is an electrical connection between the PA and driver IC. Since there is not sufficient area between PAs, it is difficult to draw interconnects on the same floor as that of the PA (two dimensional interconnection). For this reason, three dimensional electrical connection method was applied for the PIJ printhead.

DESIGN CONCEPT

Figure 1 and Figure 2 show components and cross-sectional illustration of the printhead, respectively. The printhead is constructed with fluid plate, PA plate, flexible printed circuit (FPC), and ink manifold. The fluid plate consists of mechanical components considering cost

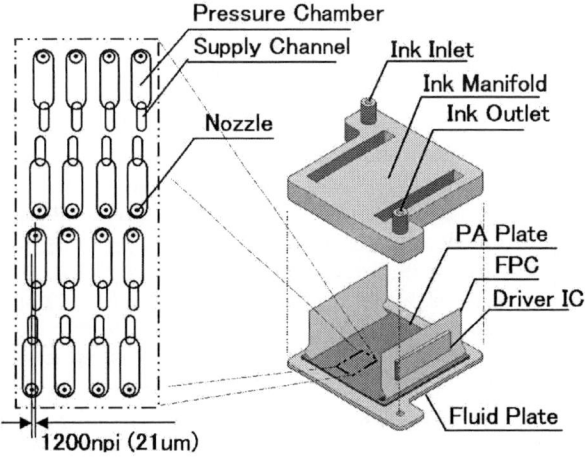

Figure 1: Components of the printhead.

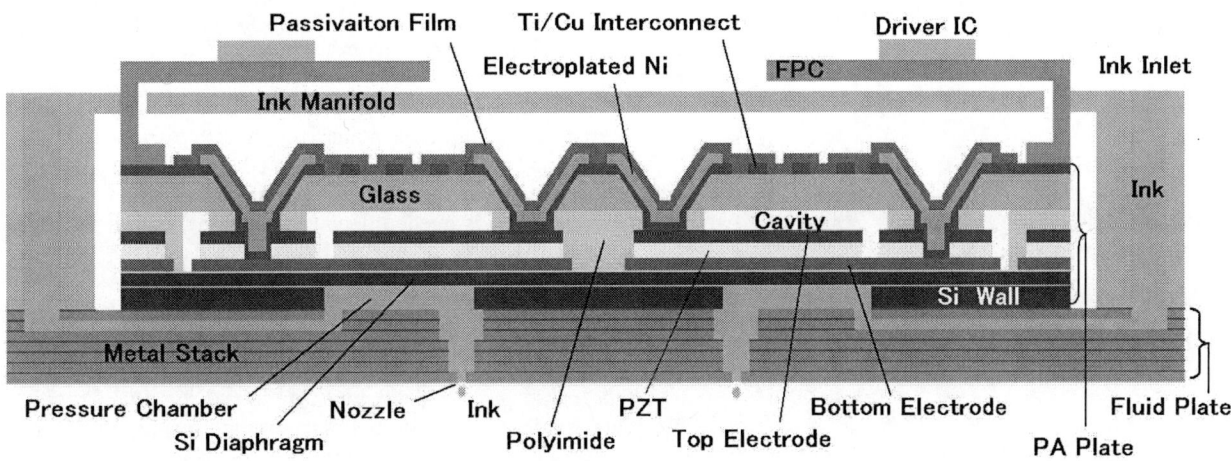

Figure 2: Cross-sectional illustration of the printhead.

merit, which are stacked thin metal plates and resin films, because a fluid channel has relatively large dimensions with hollow structures. The PA plate is fabricated using a semiconductor/MEMS process, because a fine design rule and a precise control are required.

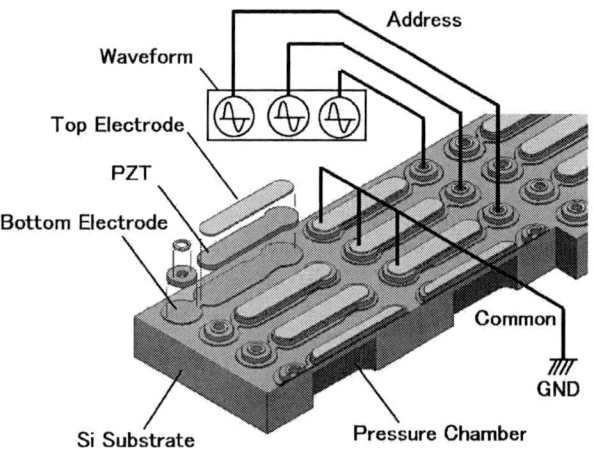

Figure 3: Interconnect structure of the printhead.

Figure 3 shows an interconnect structure of the printhead. The authors have reported that PA using a sputter-deposited PZT film exhibits an asymmetric D-E (Displacement–Electric field) hysteresis loop, and it is effective for high displacement efficiency and its stability to address a bottom electrode to a positive potential for drive, simultaneously keeping a top electrode to a ground potential [5]. Figure 4 shows displacements of PA with a poling treatment after heating at various temperatures. Two kinds of poling treatments were applied. "Normal poling" means that a higher potential is applied to bottom electrode. And "Reverse poling" means vice versa. A displacement of the PA with reverse poling was reduced after heating, while little change was observed for the PA with normal poling. Interconnect drawn from through glass VIA, which is placed on top and bottom electrodes of the PA, is connected to FPC equipped with driver IC at the peripheral region of the PA plate. Because any obstructive

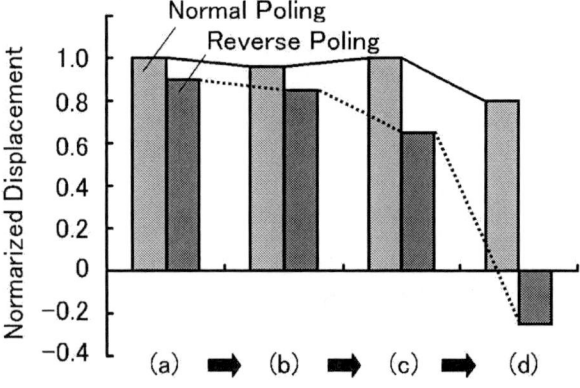

Figure 4: Normalized displacement of the PA after heating at various temperatures for 10min. (a) After poling. (b) After heating at 100°C. (c) After heating at 200°C. (d) After heating at 330°C.

components such as PA are not placed on the glass plate, a wider pitch of interconnection is possible as compared with two dimensional interconnection method.

Because the primary components, i.e., a stacked metal plate, Si substrate, and glass substrate, are heated to a few hundred degrees centigrade during some bonding processes, a coefficient of thermal expansion (CTE) of these materials is adjusted to that of Si substrate to minimize a warp problem.

FABRICATION
PA Plate

Figure 5 shows a fabrication process of the PA plate. Bottom electrode (Ti/Pt), piezoelectric element (PZT), and top electrode (Ir/Ti) are deposited by a sputtering method on thermally oxidized silicon on insulator (SOI) substrate and they are patterned in turn by using a reactive ion etching (RIE) method shown in Figure 5 (a) and Figure 6. Figure 7 shows an X-ray diffraction profile of the PZT film with top and bottom electrodes. A (100)–oriented PZT film is obtained for high displacement efficiency of the PA. The high efficient PA patterned by RIE contributes to a shrink of PA size and then a high dense integration. The PA is arrayed to 336μm pitch in a row.

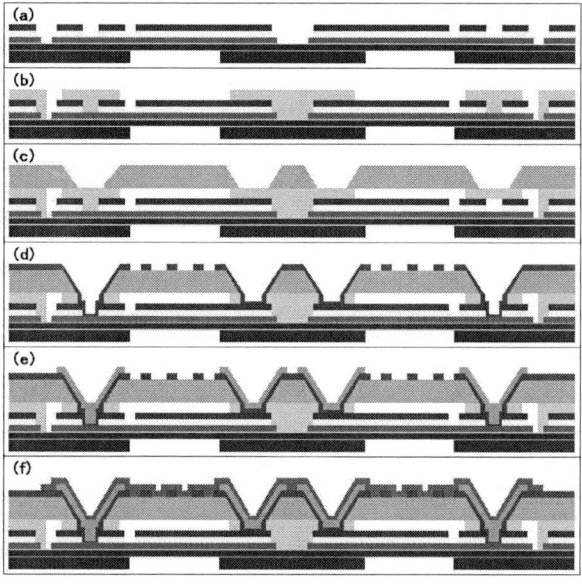

Figure 5: Fabrication process of the PA plate.

After that, a photo-sensitive polyimide film is patterned, cured, and grinded for planarization as shown in Figure 5 (b), and then bonded to blasted glass substrate by thermo-compression method as shown in Figure 5 (c). Here, curing is needed because a polyimide film without curing evolves out gas and forms voids at the bonded interface [6]. It was confirmed that a polyimide film planarized by a polishing method exhibited adhesion problems caused by a difference of polyimide thickness due to a pattern-dependent partial slurry flow. However, the grinding method did not cause such the non-uniformity, and exhibited superior bonding strength (share mode) and reliability even after temperature cycle stress (-20 to 60°C and 160 cycles).

978-1-4244-2977-6/09 $25.00 © 2009 IEEE   508

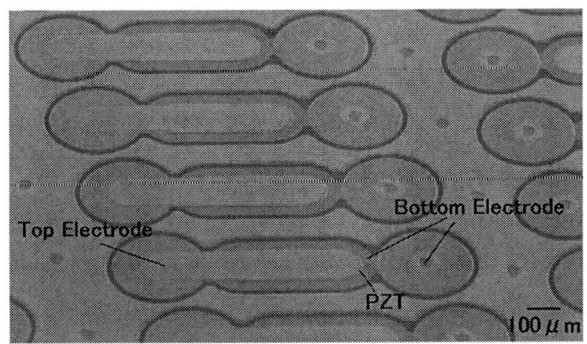

Figure 6: SEM photograph of the PA patterned by RIE.

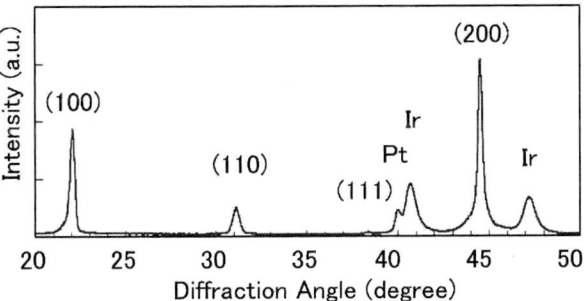

Figure 7: X-ray diffraction profile of PZT film with top and bottom electrodes.

Because the glass substrate is an insulator, a scratch or a microcrack formed on the glass surface does not cause a failure of leakage current observed in through silicon via (TSV) substrate. The glass plate also forms cavity space over the PA and acts as an encapsulation plate. The cavity is connected to air by using polyimide pattern to prevent a breakage of PA, which is caused by a pressure difference between the cavity and air during fabrication process, e.g., vacuum process and thermal process. Although the Si plate is very thin (less than 100μm) and easy to be broken alone, the bonded glass plate with 150μm thick plays an important role to keep the mechanical strength of the PA plate. The glass substrate is thinned from 340μm to 150μm by polishing performed after wafer bonding. The polishing process reduces a size of VIA top from 220μm to 150μm and then a length between VIA tops is increased and contributes to wide interconnect region.

Figure 5 (d) shows a sputtering process of Ti/Cu film that acts as seed layer for Ni electroplating and interconnects. Figure 5 (e) shows a Ni electroplating process and Ti/Cu patterning process. The Ni film is plated only at via hole by using a photo-sensitive dry film resist as a mask. Because a step around via hole is high, the Cu/Ti film is patterned using a resist spray coating method. Electrical connection between interconnect and electrode of the PA is possible even without the Ni plating. However, the sputter-deposited Ti/Cu film becomes extremely thin on a wall of the through hole, especially around the bottom region. Such a thin interconnect is easily broken by an increased current or a mechanical stress. Because the electroplated Ni film is thick sufficiently even at such region, the failure does not occur. Furthermore, this method enables to connect electrodes with different height

at a time. The polyimide film acts as an adjusting spacer.

As a last process, figure 5 (f) shows that the backside of the Si substrate is thinned to 70μm by grinding method and patterned by RIE to form a pressure chamber. Figure 8 shows SEM photographs of the completed PA plate. A slight bending of Si diaphragm observed in Figure 8 (b) was caused by the shrink of resin material, which was filled during SEM sample preparation.

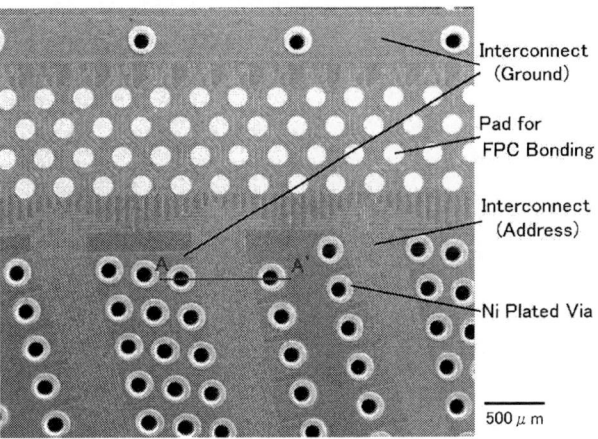

Figure 8: SEM photographs of the completed PA plate. (a) Plan view of glass side. (b) Cross-section of A-A' in (a). (c) Perspective view of Si side.

978-1-4244-2977-6/09 $25.00 © 2009 IEEE

Fluid Plate

The fluid plate is fabricated by laminating some metal plates and polyimide films. The material of the metal plate is a 42 alloy with CTE close to that of Si substrate to suppress a warp problem. The metal plate is patterned by wet etching method and has a poor patterning accuracy. On the other hand, the accuracy of metal thickness is superior because a sheet metal rolling process is utilized for the fabrication. This means that the metal stack method is useful for a high-precise fluid channel with large area. Furthermore, the polyimide film is useful for an acoustic damper to suppress a cross-talk among ejectors. Nozzle holes are opened in the polyimide film by laser processing.

Assembly

First, the PA plate is bonded to the fluid plate by thermo-compression method using an adhesive. Next, the FPC is bonded to the PA plate via bump formed on the FPC. Finally, the ink manifold is bonded to the fluid plate.

PRINTHEAD AND PERFORMANCE

Obtained specifications and jetting performance of the printhead are shown in Table 1 and Table 2, respectively. Ejection of 2pl ink at the velocity of 10m/sec., ejection of 8pl ink achieved by three multiple drive at 140kHz, and stable jetting at 40kHz were confirmed by the use of ink with the viscosity of 10mPa·s at a drive voltage of 21V.

Table 1: Specifications of the printhead.

Nozzle resolution	1200npi
Nozzle number	512 nozzles/head
Matrix array	32 x 16 (row x column)
Size of PA plate	25mm x 19mm
Nozzle density	0.4 nozzles/mm^2
PZT thickness	2μm
Diaphragm thickness	6μm

Table 2: Jetting performance of the printhead. Viscosity of Ink is 10mPa·s.

Minimum ink drop volume	2pl
Ink drop velocity	10m/sec.
Jetting frequency	140kHz (for multi-drive)
Drive voltage	21V

CONCLUSIONS

High-resolution PIJ printhead with 1200npi has been successfully fabricated by three dimensional electrical connection method using through glass VIA. The printhead exhibited superior jetting characteristics.

The electrical connection and encapsulation method are favorable for not only inkjet printhead, but also various kinds of sensors and actuators with matrix array as MEMS packaging technology.

REFERENCES

[1] M. Murata, M. Kataoka, R. Nayve, A. Fukugawa, Y. Ueda, T. Mihara, M. Fujii, and T. Iwamori, "High Resolution Long Array Thermal Ink Jet Printhead with On-Chip LSI Heater Plate and Micromachined Si Channel Plate," *IEICE Trans. ELECTRONICS*, vol. E-84-C, No.12, December, pp. 1792-1800, 2001.

[2] R. Nayve, A. Fukugawa, M. Fujii, and M. Murata, "HIGH RESOLUTION LONG ARRAY THERMAL INK JET PRINTHEAD FABRICATED BY ANISOTROPIC WET ETCHING AND DEEP SI RIE", in *Tech. Digest MEMS2003 Conference*, Kyoto, January 19-23, 2003, pp. 456-461.

[3] M. Kaneko, K. Nakajima, and H. Matsuda, "High Speed High Image Quality Printing on Plain Paper Using Symmetrically Arranged Color Bubble Jet Print Head", in *Tech. Digest NIP19*, New Orleans, September 29- October 4, 2003, pp. 354-358.

[4] M. Okumura and T. Takahashi, "Novel Micro Piezo Technology for Ink Jet Printhead", in *Tech. Digest NIP23*, Anchorage, September 16-11, 2007, pp. 314-318.

[5] S. Seto, H. Nakamura, M. Murata, and N. Morita, "Thin Film Piezo Inkjet Printhead having Matrix Nozzle Arrangement using MEMS Technology", in *Tech. Digest PPIC 2008*, Tokyo, June 25-27, 2008, pp. 52-55.

[6] X. X. Zhang and J. P. Raskin, "Low-Temperature Wafer Bonding: A Study of Void Formation and Influence on Bonding Strength", *J. Microelectromech. Syst.*, vol. 14, pp. 368-382, 2005.

SELF-RESONANT FLOW SENSOR USING RESONANCE FREQUENCY SHIFT BY FLOW-INDUCED VIBRATION

Kyoung Hwan Kim and Young Ho Seo
Mechatronics Division, Kangwon National University, Chuncheon, South Korea

ABSTRACT

A self-resonant flow sensor using resonance frequency shift by the flow-induced vibration is presented. All of MEMS-based flow sensors have measured thermal flux, mechanical strain, capacitance, and so on. In this paper, we used self-resonant vibration induced by a fluid flow and resonant frequency shift by a surface stress on the cantilever beam due to fluid drag force in order to measure mass flow. Vibration induced by air flow was measured by the piezoelectric material of PZT on the silicon cantilever beam. Theoretical resonant frequency of the cantilever beam (610 μm×300 μm×3.3 μm) was 12,416Hz. For the air flow velocities of 2.8m/sec and 9.7m/sec, shifted resonant frequencies were 12,810Hz and 15,602Hz, respectively. Sensitivity of the present self-resonant flow sensor was approximately 384Hz/(m/sec).

INTRODUCTION

MEMS-based flow sensor is one of the many researches with inertial sensor. MEMS-based flow sensors have measured thermal flux using mostly thermal resistance and temperature sensors or mechanical strain, capacitance and Coriolis force[1-4] with respect to the flow velocity or mass-flow. These MEMS-based flow sensors required the electrical power in order to operate thermal heaters, strain gage, capacitor and electromagnetic excitation.

In this work, we suggest a simple measurement method of the flow velocity using self-resonance phenomenon of a cantilever beam. When a cantilever beam is immersed in fluid flow, fluid flow induces mechanical oscillation of a cantilever beam. This phenomenon is known as the "flow-induced vibration" or "vortex-induced vibration". If, however, a cantilever beam is smaller than the boundary layer of the fluid flow, the cantilever beam is oscillated by turbulence-induced vibration. In the case of the turbulence-induced vibration, vibration is modulated with resonance frequency of the cantilever beam.[5] This resonant frequency modulated vibration is generated by just fluid flow without external excitation. We measured the vibration of the cantilever beam induced turbulence using piezoelectric effect.

THEORETICAL ANALYSIS
Flow-induced vibration

Generally, when the air-flow passes over a structure, periodic vortex streets are generated behind of the structure. Then, these periodic vortices shake the structure. Therefore, the vibration frequency is related with the vortex-shedding frequency. This kind of vibration is known as "vortex-induced vibration". Vortex-induced vibration is described by Strouhal number, St, velocity of fluid flow and dimension of the structure.[5]

$$f_s = St \frac{U}{D} \qquad (1)$$

where f_s, St, U, and D are vortex shedding frequency, Strouhal number, flow velocity, and characteristic length of a structure, respectively. Strouhal number is typically 0.15~0.25 for the arbitrary shape.[5] From Eq.(1), we found the vortex shedding frequency is proportional to the flow velocity. Moreover, Strouhal number is hardly defined for the microstructure, because periodic vortex streets could not generate at low Reynolds number condition (Re<40) and at turbulence flow condition.[5] It was famous research results that a microstructure in MEMS could generate turbulence.[1]

According to R.D. Blevins[5], a microstructure in the air-flow undergoes "turbulence-induced vibration". A structure exposed to turbulence will typically have a modulated response. The modulation is produced by the interaction of the components of turbulence that produce the resonant response. That is, vibration induced by turbulence flow is modulated with resonant frequency of the microstructure, while the vortex-induced vibration depends upon the vortex shedding frequency. Figure 1 shows the typical structural response in the turbulence-induced vibration.[5] Therefore, we could expect the output signal of self-resonant flow sensor generated by turbulence-induced vibration is modulated with its own natural frequency.

Resonant frequency shift by fluid drag force

Figure 2 shows the schematic illustration of self-resonant flow sensor using turbulence-induced vibration. Self-resonant flow sensor had a piezoelectric material of the PZT on the cantilever structure. When the cantilever beam with thin-film PZT is immersed in the fluid flow, the cantilever beam vibrates by turbulence-induced vibration as well as is bent by flow-drag force. The flow-drag force, F_{drag} increases with air velocity, and is well described by fluid dynamics.

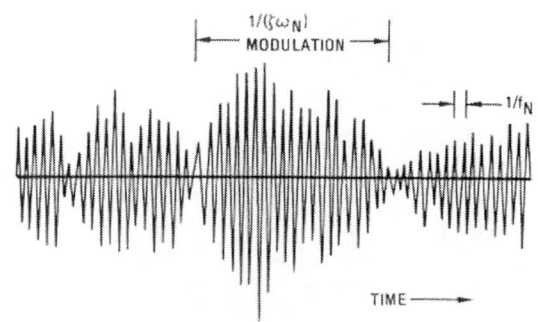

Figure 1: Typical time-history response of a lightly damped system to turbulence-induced vibration.[5]

The flow-drag force is depends on the velocity of the fluid flow, and the governing equation is changed at the low Reynolds number, Re, condition. Especially in case of Re<1, the flow-drag force is governed by Stoke's law. Flow-drag force by the fluid velocity is [6]:

$$F_{drag} = \frac{1}{2}C_D\rho V_\infty^2 A \qquad \text{for Re>}10^3 \qquad (2)$$

$$F_{drag}^s = C_D\mu V_\infty l \qquad \text{for Re<1,} \qquad (3)$$

where, C_D, ρ, V_∞, A, μ and l are coefficient of drag force, fluid density, flow velocity, base area, fluid viscosity and characteristic length, respectively. In case of cantilever to resist perpendicularly air velocity, if Re is higher than 10^3, the coefficient of drag force keep up a value of between 1.28~2.0. But, if Re is around 10^2, the coefficient of drag force shows a value about 2~4 in approach to Stoke's law.[6] Especially at Re<1, the coefficient of drag force increases up to 10, due to viscosity effect.

Magnitude of the flow-drag force depends on the flow velocity and Reynolds number. That is, when the fluid flow is increased, the flow-drag force on the cantilever beam is increased. The flow-drag force induces the mechanical strain on the cantilever beam, and then the mechanical strain increases the stiffness of the cantilever beam. Eventually, resonant frequency of the cantilever beam is increased by increasing the fluid flow.[7-9]

Theoretical equation of resonant frequency affected by surface stress on the cantilever beam was more complicate form rather than that of the both end supported beam. [9] Equation (4) was the theoretical resonant frequency of the both end supported beam with the uniform residual stress, σ.[8]

$$f_{shifted} = f_0\sqrt{\left(1 + 0.54\frac{\sigma l^2}{E t^2}\right)} \qquad (4)$$

f_0, l, E, and t are resonant frequency with residual stress, length, Young's modulus and thickness of the beam, respectively.

For, the cantilever beam without any axial force, i.e., residual stress, the resonant frequency is expressed by [10]

$$f_n = \frac{1}{2\pi}\beta_n^2\sqrt{\frac{EI}{\rho A}}, \qquad (5)$$

$$\cos\beta_n l \cosh\beta_n l = -1. \qquad (6)$$

Equation (6) is the characteristic equation. I and A are the inertia moment and cross-sectional area of the cantilever beam. In the case of the cantilever beam with an axial force, F, (residual stress, σ), the characteristic equation is complicated.[9]

$$\frac{q\sin ql + p\sinh pl}{q^2\cos ql + p^2\cosh pl} = -\frac{p^2\cos ql + q^2\cosh pl}{qp^2\sin ql - pq^2\sinh pl} \quad (7)$$

$$p = \sqrt{\frac{\sqrt{S^2 + 4\beta_n^4} + S}{2}}$$

$$q = \sqrt{\frac{\sqrt{S^2 + 4\beta_n^4} - S}{2}} \qquad (8)$$

$$S = \frac{F}{EI}$$

Equation (7) is the characteristic equation of the vibration mode of the cantilever beam with an axial force, F. The first eigenvalue of β_l indicates the primary vibration mode of the cantilever beam.

In summary, when the cantilever beam is immersed in the fluid flow of velocity, U_l, the cantilever beam vibrates by turbulence-induced vibration as well as is bent by flow-drag force. Vibration is modulated with resonant frequency, $f_{r,l}$. Next, velocity of fluid flow is increased to U_2 ($U_2>U_l$), the cantilever beam vibrates with resonant frequency of $f_{r,2}$ ($f_{r,2}>f_{r,l}$) as shown in Fig.3. Therefore, the velocity of the fluid flow or mass flow rate can be measured by tracking the resonant frequency change of the cantilever beam.

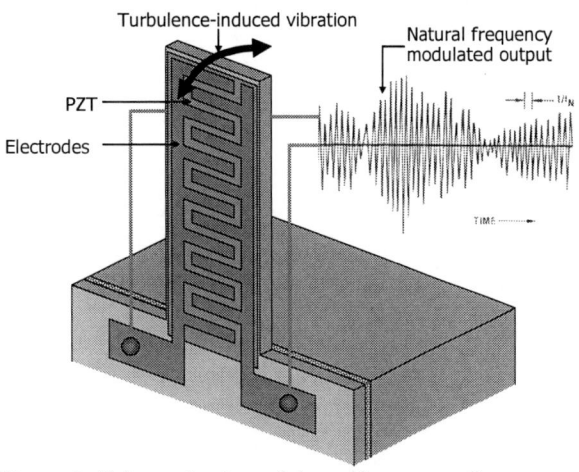

Figure 2: Schematic view of the self-resonant flow sensor.

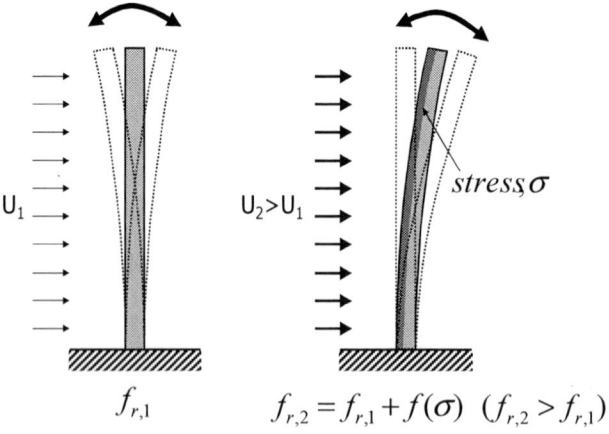

Figure 3: Working principle of the self-resonant flow sensor using turbulence-induced vibration of the fluid flow.

DESIGN AND FABRICATION

Size of the self-resonant flow sensor was designed as length of 610μm, width of 300μm and thickness of 3.3μm, respectively. Theoretical natural frequency of a cantilever beam without any axial force was estimated as 12,416Hz based on Eq.(5), where Young's modulus and density of silicon was considered as 170GPa and 2330kg/m³, respectively.

The self-resonant flow sensor was fabricated by silicon on insulator (SOI) wafer with 3μm-top layer and 1μm-buried oxide layer as shown in Fig.4. Lower electrode of Pt (0.2μm) was sputtered, and then PZT layer (0.4μm) was coated by spinning process of a PZT sol-gel. The PZT solution had a composition of Pb(Zr+Ti) along with a Zr/Ti ratio of 52/48. PZT was five times to fabricate 0.4μm-thick PZT layer. Upper electrode of Pt (0.2μm) was also sputtered, and the PZT layer was crystallized at 600 ℃. A photo-resist lift-off process was used to define the contact pad. Lower and upper electrodes of Pt and PZT layer were patterned by dry etching process. Finally, handle-silicon and buried thermal oxide were etched by DRIE process and HF solution (BOE) for a structure release. Fig.5 shows SEM image of the fabricated self-resonant flow sensor chip.

EXPERIMENTAL RESULT

In the experimental study, self-resonant flow sensor was fixed in the tube whose diameter of 5mm, and then air flow was applied. Air flow was measured by commercial air flow meter (Dwyer Instrument), and rate of the air flow was controlled by a regulator. Vibration signal was directly measured from PZT film using A/D converter(NI USB-6251, National Instrument) and LabVIEW.

As shown in Fig.6(upper), output signal was modulated with high frequency component. In order to find the relation between high frequency component in the output signal and the resonant frequency of the cantilever beam, time domain signal (upper graph in Figure 6) was converted to frequency domain signal (lower graph in Figure 6) by FFT(Fast Fourier Transform).

We could easily found the peak frequency from the FFT results which come from the vibration of the cantilever beam. The peak frequency at the air velocity of 2.8m/sec was 12,810Hz which was agree with theoretically estimated resonant frequency of the cantilever beam of 12,416Hz. Therefore, we concluded the vibration of the cantilever beam was modulated with its resonant frequency by turbulence-induced vibration. So, we called this phenomenon as "self-resonant" effect.

In Fig.7, two different FFT results measured at the velocities of the air flow of 2.8m/sec and 9.7m/sec were compared. The peak frequencies were 12,810Hz and 15,602Hz, respectively. We found the resonant frequency was changed by the air flow velocity. The sensitivity of the present self-resonant flow sensor was approximately 384Hz/(m/sec) as shown in Figure 8. From the experimental work, we verified the suggested self-resonant flow sensor could measure the fluid flow or mass-flow without external electrical power source.

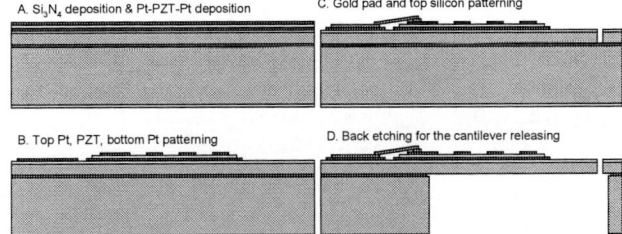

Figure 4: Fabrication Process of the self-resonant flow sensor.

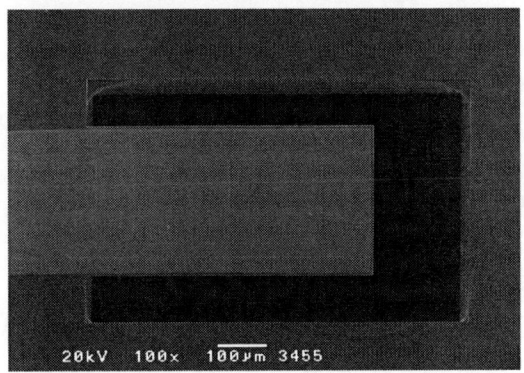

Figure 5: SEM image of the fabricated flow sensor which consists of the PZT film on the silicon cantilever beam.

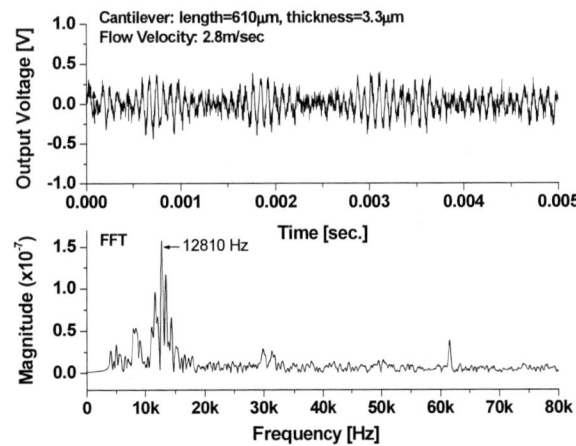

Figure 6: Output signal from the self-resonant flow sensor: (Upper) Time domain signal which was directly measured from PZT, (Lower) Frequency domain signal from FFT.

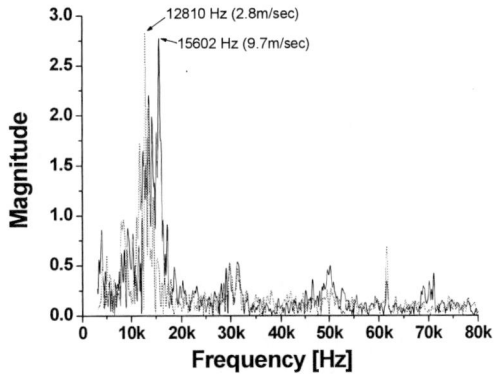

Figure 7: Comparison of two difference peak frequencies after FFT.

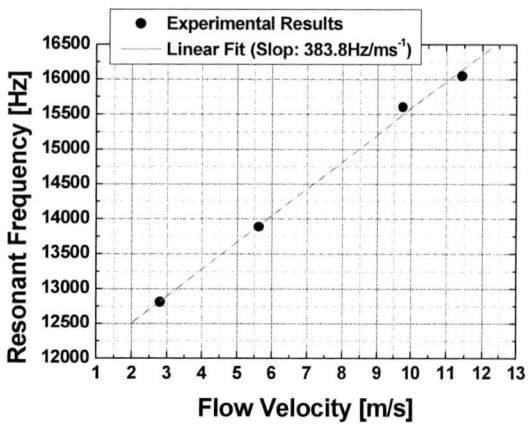

Figure 8: Sensitivity of the self-resonant flow sensor: Resonant frequency shift vs. air flow velocity.

CONCLUSIONS

We suggested self-resonant flow sensor using by turbulence-induced vibration. Response of the cantilever beam was modulated with its own resonant frequency. Flow-drag force induced the mechanical strain on the cantilever beam, and then modulated frequency was shift. The self-resonant flow sensor was fabricated to SOI wafer by PZT sol-gel method and DRIE process. From the experimental study, we found that

• Vibration response showed the vibration of the micro-cantilever was modulated with its own resonant frequency,
• Vibration mode of the micro-cantilever was turbulence-induced vibration, not vortex-induced vibration,
• Modulated resonant frequency increased as velocity of the air flow increased, and
• The sensitivity of the present self-resonant flow sensor was approximately 384Hz/(m/sec).

We experimentally verified the feasibility of the self-resonant flow sensor based on turbulence-induced vibration without external electrical power source.

ACKNOWDGEMENTS

This research was partially supported by a grant (08K1401-00521) from the Center for Nanoscale Mechatronics & Manufacturing, one of the 21st Century Frontier Research Programs, which are supported by Ministry of Science and Technology, Republic of Korea. This work was, also, partially supported by Ministry of Knowledge Economy, Republic of Korea. The authors would like to give special thank to SAMSUNG Electro-Mechanics Co., Ltd. for their financial supports.

REFERENCES

[1] Chih-Ming Ho and Yu-Chong Tai, "Micro-Electro-Mechanical- Systems (MEMS) and Fluid Flows," *Annu. Rev. Fluid Mech.*,vol. 30, pp. 579-612, 1998.
[2] Shuyun Wu, Qiao Lin, Yin Yuen, Yu-Chong Tai, "MEMS flow sensors for nano-fluidic applications," *Sensors and Actuators* A: Physical, 89 (1-2) pp. 152-158, 2001.
[3] Gijs Krijnen, Arjan Floris, Marcel Dijkstra, Theo Lammerink and Remco Wiegerink, "Biomimetic micromechanical adaptive flow-sensor arrays", *Proc. of SPIE* vol. 6592, pp. 6592-16, 2007.
[4] Peter Enoksson, Göran Stemme and Erik Stemme, "A silicon resonant sensor structure for Coriolis mass-flow measurements," *IEEE/ASME Journal of Microelectromechanical Systems*, vol.6, no.2, pp. 119-125, 1997.
[5] R.D. Blevins, *Flow-Induced Vibration*, Van Nostrand Reinhold Company, 1977.
[6] P. M. Gerhart, R. J. Gross and J. I. Hochstein, *Fundamentals of Fluid Mechanics*, Addison-Wesley Publishing Co., 1992.
[7] S. Timoshenko, D. H. Young, and W. Weaver, Jr., *Vibration Problems in Engineering*, John Wiley & Sons, Inc., 1974.
[8] R. M. Langdon, "Resonator sensors," *J. Phys. E: Sci. Instrum.*, vol.18, pp. 103-115. , 1985.
[9] P. Lu, F. Shen, S. J. O'Shea, K. H. Lee, and T. Y. Ng, "Analysis of surface effects on mechanical properties of micro cantilevers," *Mater. Phys. Mech,.* vol.4, pp.51-55, 2001.
[10] L. Meirovitch, *Principles and Techniques of Vibrations, Chap.7*, Prentice-Hall Inc. 1997.

MICROFLUIDIC-BASED DISPENSER FOR SUB-MICROLITER PIPETTING

Song-Bin Huang [1] and Gwo-Bin Lee[1,2]

[1]Department of Engineering Science, National Cheng Kung University, Tainan, Taiwan
[2]Medical Electronics and Device Technology Center, Industrial Technology Research Institute, Hsinchu, Taiwan

ABSTRACT

In the study, we report a microfluidic-based dispenser fabricated by standard soft lithography of PDMS (polydimethylsiloxane) for sub-microliter range precise pipetting. The key feature of the microfluidic-based dispenser is the incorporation of a pressure-generating unit providing the driving force for precise and quick aqueous liquid sampling and pipetting. The microfluidic-based dispenser features in the elegant control of the releasing time of the air pressure in the pneumatic chamber of the pressure-generating unit causing the underside membrane to provide suction, thus contributing to precise pipetting of aqueous liquid volumes ranging from 0.05 µl to 0.45 µl (the minimum unit is 0.05 µl) achieving the multi-volume dispensing capability. Therefore, the developed microfluidic-based dispenser can dispense multiple sub-microliter aqueous liquid volumes precisely and quickly and is promising for further micro-total-analysis-systems in the field of drug dosing or other biomedical applications.

1. INTRODUCTION

Precise pipetting of liquid samples and reagents is crucial for biochemical analysis, especially for liquid handling in the sub-micro-liter range. Typically, injection of a fixed amount samples can be easily achieved by the use of syringe pumps or peristaltic pumps for micro-flow injection analysis [1, 2]. Nevertheless, these external equipments are, to some extent, costly and bulky to operate for the requirement of integrated into a µ-TAS. Alternatively, with the rapid development of microfluidic technology, microscale devices for precise sample transportation methods have been extensively investigated. For example, micro-dosing systems actuated by electrokinetic forces [3, 4] only require a simple design and a fabrication process since it makes use of the electrokinetic phenomenon for pumping electrically conductive solutions. Nevertheless, such micro-dosing systems are restricted to sampling fluids with a certain level of electrical conductivity. Also, the high electric field intensity may influence the performance of the bio-sample. On the other hand, some silicon-based micro-dosing and pipetting systems [5, 6] which can sample the precise amount of sample liquids have been extensively used for chemical analysis. However, not only are the fabrication processes for these micro dosing devices highly complicated and relatively expensive, the silicon and metal based materials are not suitable for long-term biomedical applications. Recently, polydimethylsiloxane (PDMS) is popular in the flied of microfluidics. The advantages of it are fairly lower cost, simpler to fabricate and relatively more biocompatible, optically transparent, permeable to gases, and durable. Hence, the microfluidic chips fabricated by PDMS material with the ability of sub-microliter fluidic handling have been extensively explored. For example a PDMS constructed micro-dosing chips based on passive fluidic control were reported [7, 8]. However, the sample injection is controlled by the rotational motion of a hand-made apparatus, which is highly depended on manual controlling. In order to prevent the manual errors, a microfluidic device for automatic precise pipetting was proposed by our groups previously [9]. This micropipettor was composed of a series of pneumatic microvalves and a multi-width microchannel. The working principle of the micropipettor was based on controlling different pneumatic valves to dispense the specific liquid volume ratios of 1×, 5× and 30× and directly applying the external pneumatic pressure to pipette the dispensed liquid volume. However, the complex operation process and supplying systems with different compressed air pressures and five electro-magnetic-valves (EMV) hinder its practical applications for µ-TAS. Besides, the design of the hydrophobic based side-channel avoiding bubble formation complicates the chip fabrication process. Thus, to tackle these issues, a new microfluidic-based dispenser fabricated by standard soft lithography of PDMS (polydimethylsiloxane) is proposed in this study. The key feature of the microfluidic-based dispenser is the incorporation of a pressure-generating unit which provides a force for quick sampling and pipetting the precise variable liquid ranging from 0.05 µl to 0.45 µl. Also, Pluronic® F68 solution is used to modify the dispenser surface to maintian the hydrophilic situation [10] such that quick dispensing can be achieved. It also prevents the PDMS surface of the dispenser from adsorbing proteins or small hydrophobic molecules. Furthermore, the dispenser only requires a simple controller with only one air pressure and two EMV's. Therefore, the developed microfluidic-based dispenser can pipette multiple aqueous liquid volumes in the sub-microliter range precisely.

2. MATERIALS AND METHODS

Design of microfluidic-based dispenser

Figure 1 shows a schematic diagram of the microfluidic-based dispenser. It consists of reservoirs for samples and collecting, microvalves, a via hole for introducing the deionized water into the sample flow microchannel to clean the chip, and a sample flow microchannel (0.1mm deep, 0.5 mm wide, 10 mm long) connected with a pressure-generating unit. Structurally, the dispenser is composed of two layers of PDMS (layer A for the pneumatic chamber layer and layer B for the sample flow microchannel layer) and a bottom glass layer. They are permanently bonded after a plasma oxidation

978-1-4244-2977-6/09 $25.00 © 2009 IEEE

treatment to form a laminate structure encompassing channels and chambers, as shown in Fig. 2. Please note that three pillars are designed to prevent the PDMS membrane collapsing in the section of the pressure-generating unit, (see Fig. 1 and Fig. 2). The entire dispensing process is illustrated as Fig. 3. Briefly, the microvalves and pressure-generating unit are firstly pressurized pneumatically to deform the PDMS membranes sandwiched between the pneumatic chambers, microchannels and chambers so as to block the liquid flow microchannel (Fig. 3(a)). Then the pressurized microvalve B and the pressure-generating unit are released so that the membrane regains its original shape. A suction force is then generated to draw liquid into the microchannel while the microvalve A is still actuated to prevent the samples leaking into the collecting reservoir directly (Fig. 3(b)). Finally, the air pressure is introduced into the microvalve B and the pressure-generating unit again when the microvalve A is released at the same time. This leads to the deformation of the PDMS membrane in the pressure-generating unit again and thus produces a driving force pushing the liquid in the liquid flow microchannel into the collecting reservoirs. Please note that the microvalve B is actuated to prevent the sample liquids flowing back to the sample reservoir. It is worth mentioning that all the dispensing process is automatically controlled by a custom-made controller (Fig. 4). Hence, the microfluidic-based dispenser features in the elegant control of the release time of the air pressure in the pneumatic chamber of the pressure-generating unit causing the underside membrane to provide a suction (see Fig. 1 and Fig. 3), thus contributing to precise pipetting of liquid volumes ranging from 0.05 to 0.45 µl. Figure 5 indicates the relationship between the specific dispensed volumes and the releasing time of the air pressure released from pneumatic chamber of the pressure-generating unit. The maximum coefficient of variation of the dispensed liquid volume is measured to be 11.3 % for 0.05 µl dispensation.

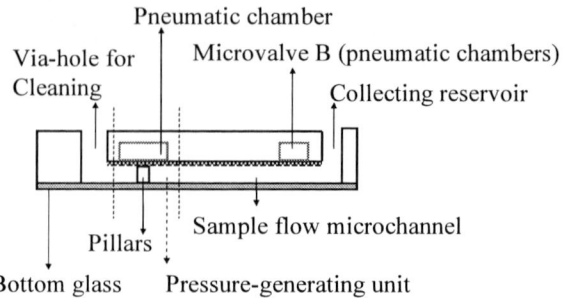

Figure 2: The cross-sectional view of the dispenser.

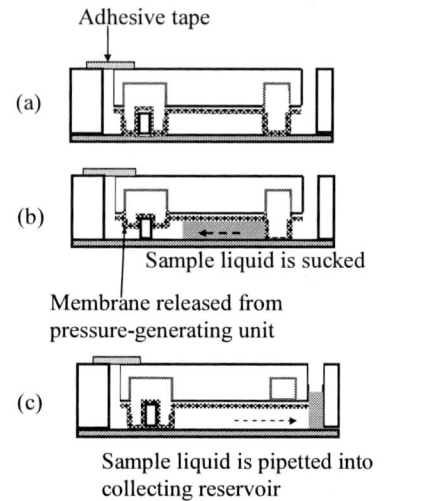

Figure 3: The illustration of the dispensing process.

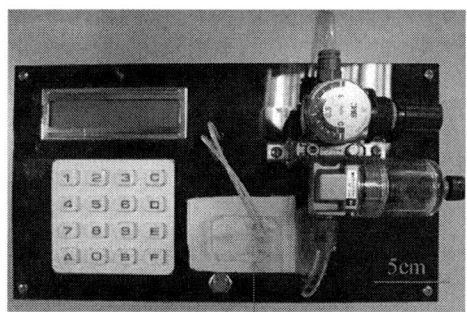

Microfluidic based dispenser platform

Figure 4: The photograph of the custom-made controller.

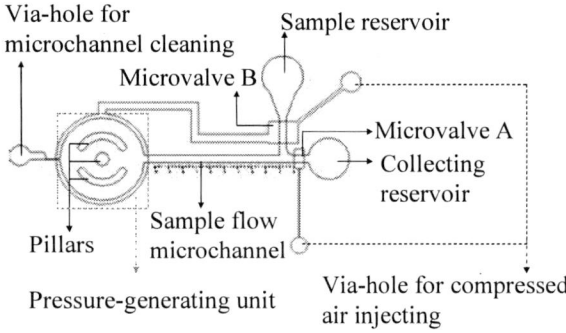

Figure 1: Top- view layout of the dispenser.

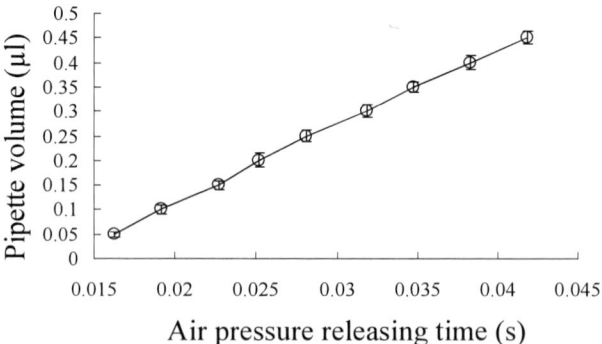

Figure 5: Relationship between the specific dispensing sample volume and air pressure releasing time from the pressure-generating cell.

Fabrication

The fabrication processes were based on SU-8 lithography and a PDMS (Sylgard® 184, Dow Corning, USA) replication processes. The basic process for SU-8 lithography is illustrated as Fig. 6. First, the master mold is fabricated by photolithography of SU-8 50 negative photoresist (MicroChem, USA) on a silicon wafer. The rotation speed of the spin-coating is 1000 rpm for 1 minute to achieve the required photoresist thickness of 100 μm, followed by a two-step soft bake initially at 65°C for 10 minutes and then at 95oC for 30 minutes. The baked wafer is then ultraviolet (UV) exposed with a total exposure dose of 500 mJ/cm². This is followed by another two-step post-exposure baking initially at 65°C for 3 minutes and then at 95°C for 10 minutes. The baked wafer is subsequently developed using the SU-8 developer (MicroChem., USA) for 2 minutes. The developed wafer is then rinsed with isopropyl alcohol and followed by a nitrogen blow-dry. In the following replica molding process, the PDMS polymer is prepared by thoroughly mixing the PDMS pre-polymer and the curing agent in a ratio of 10:1 by weight. The polymer is then de-aerated under vacuum to remove all air bubbles created during mixing. Then the mixture is poured onto the masters fabricated previously and cured at 70°C for 2.5 hours. The PDMS layer is then cautiously peeled in a mechanical de-molding process. The two layers of PDMS obtained from the replica molding and spin-coating processes and bottom glass layer are then bonded by an oxygen plasma treatment to form the micro-dispenser chip. The dimensions of the microfluidic chip are measured to be 3.5 cm x 1.5 cm, respectively. After the micro-dispenser chip is constructed, the 3% (w/v) of Pluronic® F68 (Sigma, Taiwan) surfactant solution is loaded into the sample flow microchannel for about 12 hours to modify the PDMS surface.

Experimental setup

The control system for automating the microfluidic-based dispenser is illustrated in Fig. 4. In the hand-held controller, there is an air compressor (MDR2-1A/11, Jun-Air Inc., Japan) which supplies air pressure to the pneumatic chambers of the microvalves. A programmable control circuit system with two EMV's (S070M-5BG-32, SMC Inc., Taiwan) to switch the air pressure on and off are built. The activation times and sequences of the microvalves and pressure-generating cell are programmed in advance to dispense the various aqueous liquid volumes. In addition, in order to observe the actions of the micro-dispenser to dispense varied various volumes, a high-speed CCD camera-coupled (MC1311, Mikrotron, Germany) microscope (TE300, Nikon, USA) is used. The frame rate of these images is set at 1000 frames/sec to distinctly capture the tiny differences during the movement of the membrane pulsations. Moreover, a balance (Mettler Toledo, AB54-S, Switzerland) is used to measure the weights of various volumes of the water during the weighting experiment.

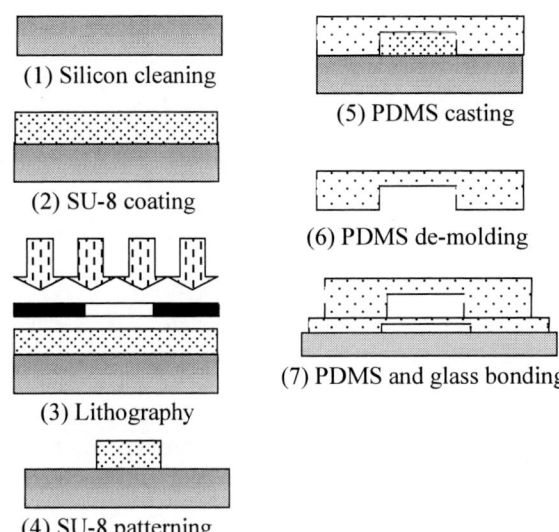

Figure 5: Schematic illustration of the fabrication process for the PDMS-based dispenser chip.

3. RESULTS AND DISCUSSION

Characterization of the dispenser

As mentioned previously, The key feature of the microfluidic-based dispenser is the incorporation of a pressure-generating unit providing a driving force for precise and quick aqueous liquid sampling and dispensing. By controlling the releasing time of the air pressure released from the pneumatic chamber of the pressure-generating unit causing the underside membrane to provide suction, the different liquid sample volumes ranging from 0.05 μl to 0.45 μl (the minimum unit is 0.05 μl) can be dispensed quickly. In order to confirm the relationship between every specific dispensed volume and the air pressure releasing time, high-speed CCD images are used to observe the needed time of the different liquid sample volume sucked into the sample flow microchannel when the air pressure is released from the pneumatic chamber of the pressure-generating unit. Since a marked line is fabricated below the sample flow microchannel as a reference (every interval of two long marked lines is set to be 1 mm and the width and depth of the sample flow microchannel is fabricated to be the 0.5 mm and 0.01mm respectively), the volume of the liquid sample sucked into the sample flow microchannel can be determined by the length. Hence, the releasing time of the air pressure released from the pneumatic chamber of the pressure-generating unit can be observed from the needed time of the liquid stopped at the specific location marked line by using the high-speed CCD images (see Fig. 7). Therefore, the relationship between every specific dispensed aqueous liquid sample volume and the air pressure releasing time can be characterized accordingly.

Volumes of 0.05 μl Volumes of 0.25 μl Volume of 0.45 μl

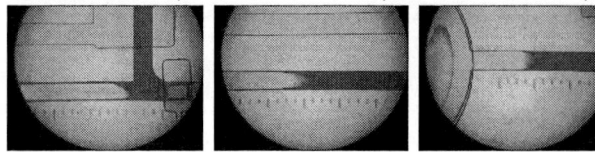

Figure 6: The CCD images of different dispensed volumes.

Dispensation tests

In order to confirm that the micro-dispenser has a reproducible performance, a weighting experiment is performed. Nine pieces of Eppendorf tubes are first marked and weighted. Then colored water is loaded into the sample reservoir of the micro-dispenser. The micro-dispenser injects specific volumes of liquids (0.05 µl to 0.45 µl) for twenty times into every marked Eppendorf tube, which are weighted afterwards. Then, the dispensed volumes ranging from 0.05 µl to 0.45 µl can be calculated. The weighting experiments are repeated at the 2nd, 4th, and 6th day after the micro-dispenser chip is prepared. From the experimental results (see Fig. 8), it reveals that the proposed micro-dispenser is able to perform aqueous liquid dispensing with high reproducibility. The maximum deviation of the dispensed liquid volume is about 0.008 µl, which is about 5 % of dispensed volume of 0.15 µl.

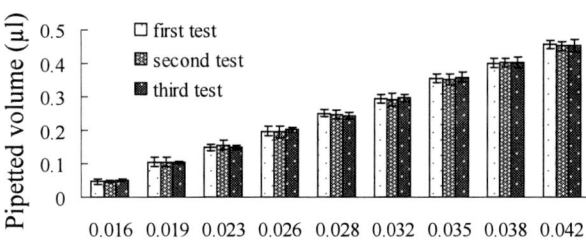

Figure 7: Dispensation experiments tested at the 2nd, 4th, and 6th day after the chip prepared.

4. CONCLUSIONS

In the study, we reported a microfluidic-based dispenser fabricated by standard soft lithography of PDMS for sub-microliter range precise and quick dispensing. It can be easily integrated with other microfluidic devices for precise dispensing. The key feature of the dispenser is the incorporation of a pressure-generating unit providing the driving force for precise and quick aqueous liquid volumes ranging from 0.05 µl to 0.45 µl (the minimum unit is 0.05 µl) sampling and pipetting. Experimental results reveal that the proposed dispenser is able to perform aqueous liquid dispensing ability with high reproducibility. Hence it is promising for further micro-total-analysis-systems in the field of drug dosing or other biomedical applications.

ACKNOWLEGEMENTS

The authors would like to thank financial support from the National Science Council in Taiwan (NSC 95-2221-E-006-012-MY3).

REFERENCES

[1] M. Hulsman, M. Bos, and W. E. van der Linden, "Automated injection of slurry samples in flow-injection analysis," Anal. Chim. Acta, vol. 324, pp. 13-19, 1996.

[2] M. D. Foster, M. A. Arnold, J. A. Nichols, and S. R. Bakalyar, "Performance of experimental sample injectors for high-performance liquid charomatography microcolumns," J. Chromatogr. A, vol. 869, pp. 231-241, 2000.

[3] C. K. Byun, X. Wang, Q. Pu, and S. Liu, "Electroosmosis-Based Nanopipettor," Anal. Chem., vol. 79, pp. 3862-3866, 2008.

[4] C. W. Huang, and G. B. Lee, "Microautosamplers for discrete sample injection and dispensation Electrophoresis," vol. 26, pp. 1807-1813, 2005.

[5] N. Szita, R. Sutter, J. Dual, and R. A. Buser, "A micropipettor with integrated sensors," Sensors and Actuators A, vol. 89, pp. 112-118, 2001.

[6] K. Handique, D. T. Burke, C. Mastrangelo, and M. A. Hand Burns, "On-Chip Thermopneumatic Pressure for Discrete Drop Pumping," Anal. Chem., vol. 73, pp. 1831-1838, 2001.

[7] S. H. Lee, C. S. Lee, B. G. Kim, and Y. K. Kim, "Quantitatively controlled nanoliter liquid manipulation using hydrophobic valving and control of surface wettability," J. Micromech. Microeng., vol. 13, pp. 89-97, 2003.

[8] S. H. Lee, C. S. Lee, B. G. Kim, and Y. K. Kim, "An integrated microfluidic chip for the analysis of biochemical reactions by MALDI mass spectrometry," Biomed Microdevices, vol. 10, pp. 1–9, 2008.

[9] C. W. Huang, S. B. Huang, and G. B. Lee, "A microfluidic device for precise pipetting," J. Micromech. Microeng., vol. 18, 035004 (7pp), 2008.

[10] M. H. Wu, J. P. G. Urban, Z. Cui, and Z. Cui, "Development of PDMS microbioreactor with well-defined and homogenous culture environment for chondrocyte 3-D culture," Biomedical Microdevices, vol. 8, pp. 331-340, 2006.

PRECISE TITRATION MICROFLUIDIC SYSTEM USING SUB-PICOLITER DROPLET INJECTION

Kentaro Kawai, Masaru Fujii, and Shuichi Shoji

Major in Nano-science and Nano-engineering, Waseda University, JAPAN

ABSTRACT

Precise titration using sub-picoliter droplet injection microfluidic system is proposed. This system can to measure less than 1 picoliter of reagent and to deliver to the micro chamber without diffusion which becomes major concern when the reagent volume is smaller. A prototype device can generate a droplet of 0.8 picoliter (800 femtoliter) in the metering channel of 40 μm × 5 μm × 4 μm, and step-by-step titration by precise reagent injection per two hundred thousand (2.0×10^{-5}) in volume of the micro chamber (157 nanoliter: 1 mm in diameter × 200 μm in height).

INTRODUCTION

In the field of bio-micro electro mechanical system (Bio-MEMS), reagent injection devices that measure, generate, and handle droplets have been developed [1-10]. The droplet based chemical/biochemical assay can reduce the total production and chemical/biochemical hazards of the harmful reactions. In addition, μTAS device has advantages in the reaction time, sample consumption, disposability, and so on.

However, droplet generation by microvalve actuation or electrowetting on dielectric (EWOD) has a problem in the minimum volume. The measurement volume of those devices [1-6] is the range of several microliter to tens of picoliter, which volume is as same as the reactor volume in microfluidic devices. It is difficult to mix the two samples at the difference in large volume ratio. On the other hand, continuous droplet generation can generate wide range of sample volume that is from hundreds picoliter to 1 picoliter [7-10]. However, continuous droplet generation systems generate from tens to hundreds of droplets per second, it is unsuitable for single droplet handling. In addition, the reason continuous droplet generation is carried out in two-phase flow is very small droplets has its large specific surface area comparing its volume. It causes the quick diffusion of the samples in the same phase solution. Using the two-phase flow such as water-in-oil or oil-in-water, the droplets are maintained their volume.

CONCEPT

To solve this sample volume and diffusion problems, we propose a sub-picolittler injection system as shown in Figure 1. This system can to measure less than 1 picoliter of the reagent. Since the volume of the micro chamber is about 157 nanoliter, the volume ratio between the droplet and the micro chamber is over two hundred thousand times. This volume ratio is same as conventional macro scale titration to about 60 liter to 80 liter.

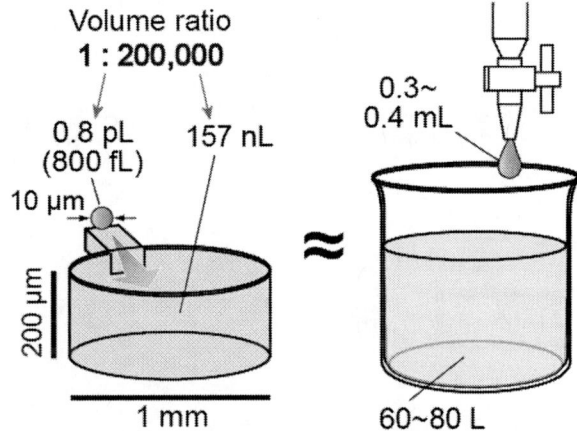

Figure 1 Schematic of the titration of sub-picoliter droplet injection system and the macro scale of same volume ratio.

The system is consisted of a micro chamber, and fluidic channels of droplet generation area and injection area as shown in Figure 2. The fluidic channels are consisted of the metering channel, the main channel, and the loading channel. The metering channel measures the volume of the reagent. The main channel cuts off the extra amount of reagent. The measured droplet is introduced to the micro chamber through the loading channel. To prevent the diffusion of the measured droplet, the droplet generation area is filled by an inert liquid for delivering fluid.

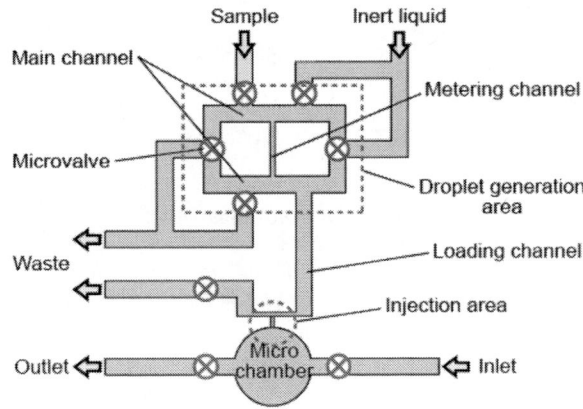

Figure 2 Schematic sub-picoliter injection system. Fluid control is carried out by integrated eight microvalves.

The generation and injection of the droplet are carried out by the four steps of the combinational actuation of eight microvalves as shown in Figures 3. Step 1 and step 2 are the droplet metering steps. Step 1 is the reagent introduction to the metering channel filled by the reagent

978-1-4244-2977-6/09 $25.00 © 2009 IEEE   519

through the main channel as shown in Figure 3(a). Step 2 is the reagent measurement in the metering channel by introducing the delivering fluid to the main channel as shown in Figure 3(b). Step 3 and step 4 are the droplet injection steps. Step 3 is the droplet generation and introducing to the injection area in the loading channel as shown in Figure 3(c). Step 4 is the reagent injection as shown in Figure 3(d). Since the droplet has a small volume and large specific surface area, the reagent diffuses immediately to the aqueous solution in the micro chamber as soon as the surfaces of the reagent droplet and the aqueous solution contact to each other at the injection area.

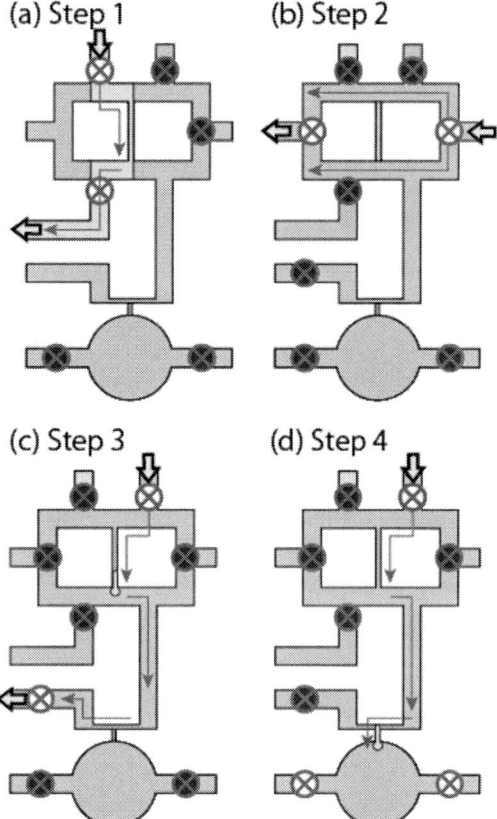

Figure 3 Four steps of reagent measurement and injection: (a)loading reagent to metering channel, (b)measuring reagent by introducing inert liquid to main channel, (c)loading reagent to injection area, (d)injecting reagent to micro chamber by diffusion.

DESIGN AND FABRICATION

Design

A prototype device with the metering channel of 40 μm × 5 μm × 4 μm (800 femtoliter) was fabricated as shown in Figure 4. The width and depth of the main channel is 40 μm and 4 μm respectively. The main channel where is before the loading channel is made into narrow so that the droplet becomes easy to flow into the loading channel. The width

and depth of the loading channel are 10 μm and 4 μm respectively. The injection channel to the micro chamber is 20 μm × 10 μm × 4 μm. The micro chamber is 1 mm in diameter and 200 μm in depth. The volume of the metering channel and the micro chamber is 800 femtoliter and 157 picoliter, respectively. The width and depth of the channels from reagent inlet to the droplet generation area is fabricated 40 μm and 50 μm respectively for low flow resistance. The microvalve area of top side is 400 μm × 200 × 10 μm. The diameter of the through hole is 40 μm.

Fabrication

The prototype device is consisted of Si, glass, and poly-dimethylsiloxane (PDMS, SILPOD 184, Dow Corning Toray, U.S.A.). The Si (100) 200 μm thick substrate was etched from both sides (Figure 4 (a-f)) to form the channels and through holes using deep inductive-coupled reactive ion etching (ICP-RIE; Surface Technology Systems, U.K.) and capacitive-coupled plasma ion etching (CCP-RIE; Samco, Japan), and the side of fluidic channels is sealed with Pyrex® glass plates (Corning, U.S.A.) by anodic bonding (Figure 4(g)). The pneumatic lines for the microvalves are fabricated by PDMS molding and bonded to the topside after O_2 plasma treatment (Aiplasma, Panasonic Electric Works, Japan) as shown in Figure 4(h). The PDMS mold is made from the patterned negative photoresist (SU-8 3020, Kayaku MicroChem, Japan) on the glass substrate. Tubing connector is bonded on the through hole as shown in Figure 4(i).

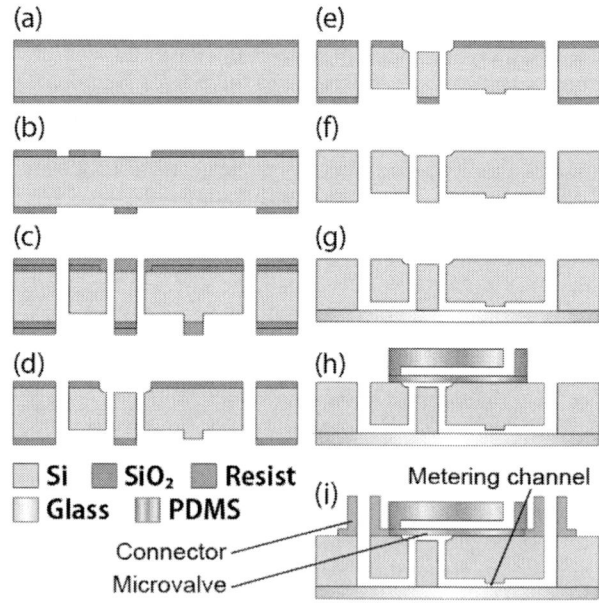

Figure 4 Fabrication process: (a-f) Si etching process, (g)anodic bonding, (h)PDMS bonding by plasma treatment, (i)bonding tubing connector.

Figures 5 show the scanning electro microscope (SEM) images of the fabricated fluidic channels. Figure 5(a) shows the overview of the bottom side. Figure 5(b) shows the droplet generation area with the main channel and the metering channel. Figure 5(c) shows the injection area with the loading channel and the micro chamber. Figure 5(d) shows the enlarged view of the metering channel. Figure 5(e) shows the enlarged view of injection area. Figure 5(f) is the microvalve area of the top layer. The PDMS membrane is pressed to this area by the pressure from the pneumatic line, and the through holes are closed. The advantage of this structure is no leak problem from the shape of the channel edge.

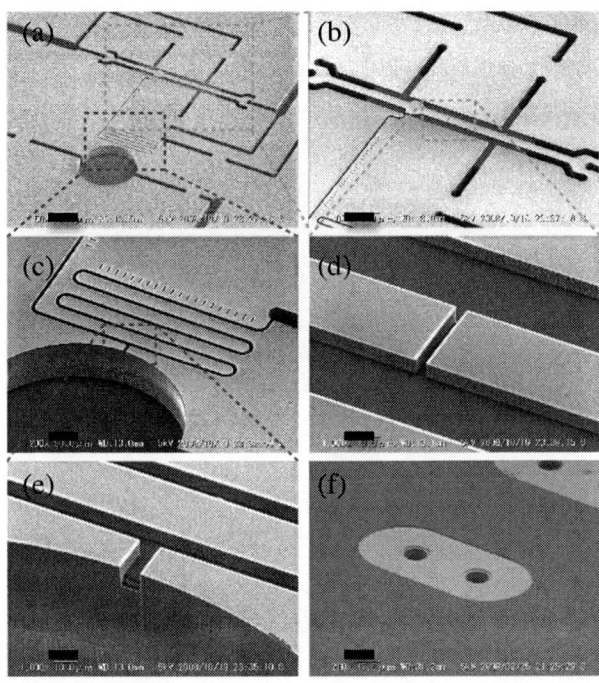

Figure 5 SEM image: (a) overview of the side of fluidic channels, (b) droplet generation area, (c) loading channel and micro chamber, (d) metering channel, (e) droplet injection area, (f) microvalve area. Scale bar: (a) 200 μm, (b) 100 μm, (c)(f) 50 μm, (d) (e) 10 μm.

EXPERIMENTAL SETUP

Figures 6 show the experimental set up for the reagent measurement and injection. The pneumatic pressure is generated by the pressure generator and regulated by the pressure regulator. The injection pressure of the reagents are 25 kPa and the actuation pressure of the microvalves are 180 kPa. The switching of the microvalve is controlled by the pnuematic lines from off-chip solenoid valves. The combination of the valve switching is controlled by signal from PC (LabView®, National Instruments, U.S.A.). The fluorescent images are observed using an inverted fluorescent microscope (IX71; Olympus, Japan). The images are captured using a 3 charge-coupled device (CCD) camera (JK-TU53H; Toshiba, Japan). The fluorescein solution is used as the reagent, and the inert liquid (Fluorinert™, 3M, U.S.A) is used as the delivering fluid.

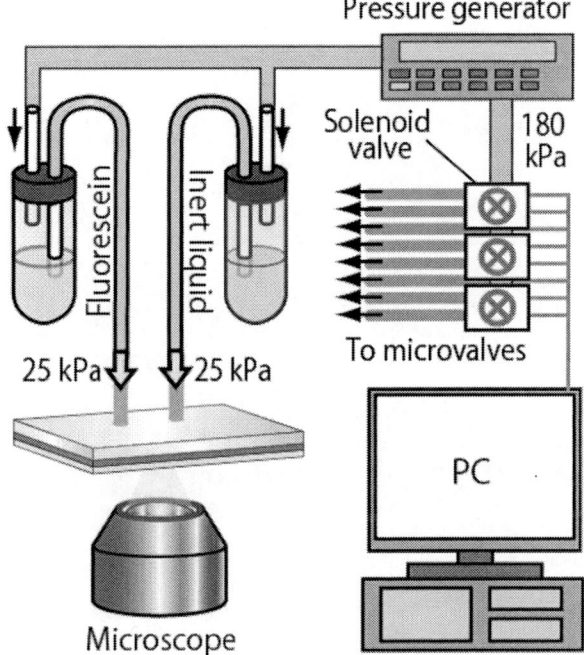

Figure 6 Schematic of the experimental setup.

RESULTS AND DISCUSSION

Figures 6 show the experimental result of the droplet generation and injection. First, hexamethyldisilazane (HMDS) was introduced for hydrophobic treatment to channel surface. The hydrophobic treatment prevents the droplet from adhesion to the channel surface. Then the droplet genalation area and the droplet injection area are filled with the inert liquid. The micro chamber is filled with water. Figure 6(a) shows the image of the introduction of the fluorescein solution to the metering channel. The fluorescein solution is introduced from the left-side inlet to the left-side waste outlet. Figures 6(b) and 6(c) show the reagent measurement by the inert liquid. The reagent in the metering channel is metered by the flow of the inert liquid in the main channel. Figure 6(d) shows the measured droplet of the fluorescein solution from the metering channel. The diameter of the droplet is 16.9 μm and the volume is about 896 femtoliter. The droplet is introduced to the loading channel by the flow of the inert liquid from right-side inlet to the right-side loading channel. Because the reagent waste side is opposite side from the loading channel, the unnecessary residue of the fluorescein solution does not be introduced to the loading channel. Figure 6(e) shows the image of introducing the droplet to the injecting area. The reagent droplet is diffused to the micro chamber immediately as soon as the droplet comes to the interface between the water in the micro chamber and the inert liquid. Figure 6(f) shows the image of introducing reagent to the micro chamber by diffusion. These results confirmed that sub-picoliter reagent injection is succeeded.

978-1-4244-2977-6/09 $25.00 © 2009 IEEE

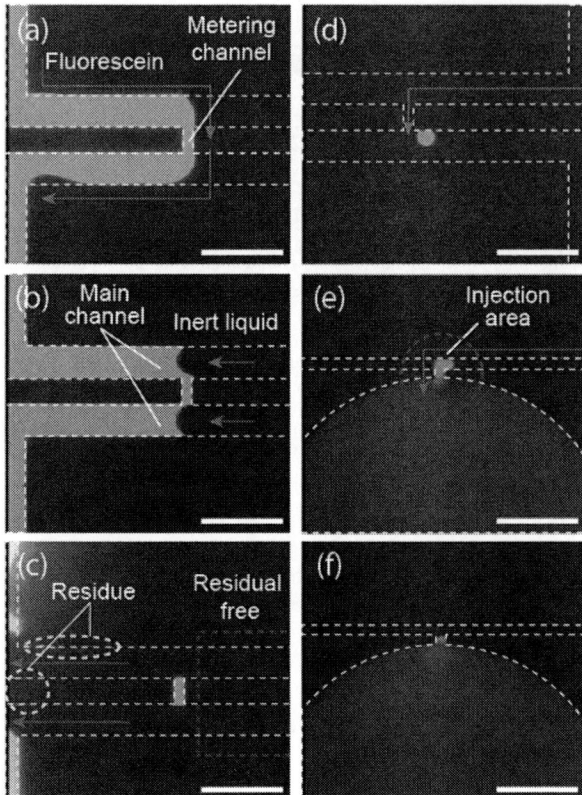

Figure 7 Measurement and injection of the reagent: (a-c) reagent measurement in metering channel by inert liquid, (d-f) reagent injection to micro chamber. Scale bar: 100 μm.

CONCLUSION

Sub-picoliter droplet injection microfluidic system is proposed. Using a prototype device with the metering channel of 40 μm × 5 μm × 4 μm (800 femtoliter), sub-picoliter volume of the reagent droplet is generated. The droplet is introduced to the micro chamber using the inert liquid for the delivering fluid. The diameter of the droplet is 16.9 μm in the channel depth of 4 μm, and the volume considering the droplet as a cylinder is about 896 femtoliter.

ACKNOWLEDGEMENT

This work is partly supported by Japan Ministry of Education, Culture, Sports Science & Technology Grant-in-Aid for Global COE of Waseda University, Scientific Basic Research (A) No. 19206046, and ASMeW.

REFERENCES

[1] M. Yamada and M. Seki, "Nanoliter-Sized Liquid Dispenser Array for Multiple Biochemical Analysis in Microfluidic Devices", Anal. Chem., 76 (2004), pp. 895-899

[2] M. Kanai, H. Abe, T. Munaka, Y. Fujiyama, D. Uchida, A. Yamayoshi, H. Nakanishi, A. Murakami and S. Shoji, "Micro chamber for cellar analysis integrated with negligible dead volume sample injector", Sens. Act. A, 114, pp. 129-134

[3] A. Hibara, M. Nonogi and T. Kitamori, "Microchip titration by utilizing Laplace valve", proc. μTAS2007 (2007), vol.1, pp. 601-603

[4] SK Cho, H. Moon and CJ Kim, "Creating, transporting, cutting, and merging liquid droplets by electrowetting-based actuation for digital microfluidic circuits", J.Microelectromech.Syst., vol.12, No.1 (2003), pp. 70-79

[5] O. D.Velev, B. G.Prevo and K. H. Bhatt, "On-chip manipulation of free droplets", Nature, vol.426 (2003), p.515-516

[6] P. Y. Chiou, H. Moon, H. Toshiyoshi, CJ Kim and MC Wu, "Light actuation of liquid by optoelectrowetting", Sens. Act. A, vol.104 (2003), p.222-228

[7] T. Nisisako, T. Torii and T. Higuchi, "Preparation of Picoliter-Sized Reaction / Analysis Chambers for Droplet-Based Chemical and Biochemical Systems", proc. μTAS2002 (2002), vol.1, pp.362-364

[8] V. Reddy, S. Yang and JD Zahn, "Organic/Aqueous Two Phase Microflow For Biological Sample Preparation", proc. μTAS200 3 (2003), vol.1, pp.437-440

[9] S. Abraham, E. H. Jeong, T. Arakawa, S. Shoji, K. C. Kim, I. Kim and J. S. Go, "Microfluidics assisted synthesis of well-difined spherical polymeric microcapsules and their utilization as potential encapsulants", *Lab on a Chip*, vol. 6 (2006), pp.752-756

[10] Y.C. Tan, V. Cristini and A.P. Lee, "Monodispersed microfluidic droplet generation by shear focusing microfluidic device", Sens. Act. B 114 (2006), pp. 350–356.

DROPLETS GENERATION METHOD FOR WATER-IN-OIL STATE IN THE POLYDIMETHYLSILOXANE MICROCHANNEL WITH GROOVES

Jihoon Kim[1], Doyoung Byun[1], Jongin Hong[2] and Andrew J. deMello[2]

[1]Dept. of Aerospace Information Engineering, Konkuk University, Seoul, Republic of Korea
[2]Dept. of Chemistry, Imperial College London, London, United Kingdom

ABSTRACT

We present a new method to generate droplets stored in cavity structures using microchannels containing grooves. We investigate the effects of flow rate and groove pitch microchannelondroplet size.

INTRODUCTION

Droplet-based microfluidic systems have been introduced as a fundamental experimental platform for high-throughput experimentation in biology and chemistry [1-4]. These systems enable the generation and manipulation of highly monodisperse liquid droplets in an immiscible carrier fluid. Such encapsulated droplets can be used to mimic artificial cells or isolated reaction vessels and then to probe kinetics and biology behind fundamental reactions. Compared to conventional single-phase microfluidic systems, localization of reagents within discrete and isolated compartments is an extremely effective way of enhancing reaction yields and narrowing residence time distributions [1,2,4]. Precise control of droplet size and polydispersity is of great importance for practical applications. Highly reproducible monodisperse droplets have been prepared using a variety of methods, including geometry-dominated break-up [5,6], cross-flow rupturing through microchannel arrays [7], capillary instability-mediated drop formation (in a flow focusing configuration) [8] and pressure-drop induced break-up (at a microfluidic T-shaped junction) [9,10]. These techniques can generate many droplets in a serial fashion, with their generation frequency dependent on the input fluid flow rates, the channel widths and the relative viscosity between the two phases. Additionally, microdroplets can be fused, subdivided, sorted, isolated or incubated to establish a multifunctional analytical device. Importantly, local storage and incubation of water-in-oil droplets are crucial for monitoring cell behavior and detecting *in vitro* expression. In this study, we present a simple and novel method to generate and isolate droplets simultaneously in a microfluidic system.

MECHANISM OF DROP GENERATION

Figure 1 illustrates a schematic the generation of droplets in a grooved microchannel. The wall is comprised of laterally alternating ribs and cavities running parallel to the flow direction with microscale dimensions ranging from 10 μm to 200 μm. If the pitch between these microribs is small enough and the surfaces are hydrophobic, the liquid will not wet or enter cavities (Figure 1a). In contrast, if the spacing between adjacent microribs is large enough, the liquid may fully wet grooves (Figure 1b). The groove

size including pitch (*p*) and height (*h*) necessary to determine wetting depends on the liquid-solid interfacial chemistry, the interfacial tension between the liquid and solid, and the pressure difference between the vapor and liquid phases. The former is beneficial to reduce total frictional resistance in a microchannel, whilst the latter is of no significance in single-phase flow microfluidics. We expect that cavities located on the side walls should play a role in both the formation and storage of microdroplets if one can use two immiscible fluids such as water and oil. A brief scenario is presented as follows. First, a microchannel is filled with water and followed by oil which flows smoothly into the microchannel. The oil replaces water in all regions except for a small portion inside groove features. Finally, because of equilibrium of interfacial tension, stationary water droplets are formed in the continuous oil phase.

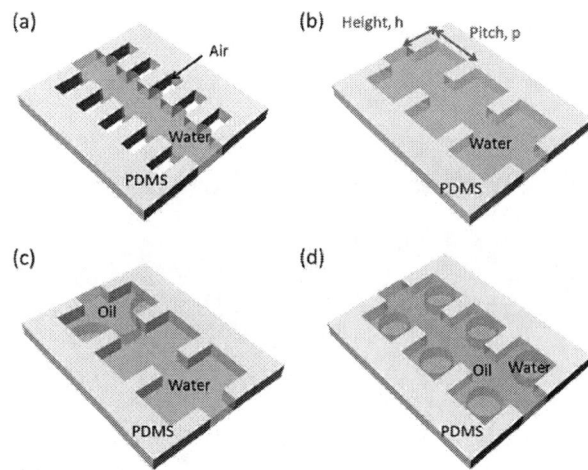

Figure 1: Schematics of droplet generation

EXPERIMENT SETUP

Fabrication of Microchannel

In order to verify the efficacy of this approach, microfludic devices have been fabricated in polydimethysiloxane (PDMS), a transparent elastomer, using soft lithographic techniques with Si masters (Figure 2). The Si masters are fabricated by using general photolithography with a deep reactive ion etching (DRIE) process. Initially, the polished Si substrate is cleaned with a Piranha solution (H_2SO_4:H_2O_2 = 3:1 by volume) and is then dehydrated for 10 minutes at 150°C on a hot plate. Then, AZ1512 photo-resist (PR) (AZ Electronic Materials) is spin-coated at 3000 rpm for 35 seconds. After the spin-coating, it is soft-baked at 110°C for 110 seconds on a hot plate. Lithography is performed using a mask aligner to define the desired region populated by microscale grooves.

978-1-4244-2977-6/09 $25.00 © 2009 IEEE

Development is accomplished in 90 s by immersing the exposed substrate into a developer solution (AZ300MIF, Clariant, Muttenz, Switzerland). The pattern is transferred into the Si substrate using DRIE. Subsequently, the PR is removed and cleaned with Piranha solution. A PDMS stamp with the grooved structures is cast using the Si mold. PDMS is cured for 30 minutes at 110°C and then peeled off the mold. The PDMS is bonded to a partially cured PDMS slab. Figure 3 shows a top view of the PDMS microchannel with microscale grooves. On both side-walls, a number of grooved structures are made to generate and isolate droplets.

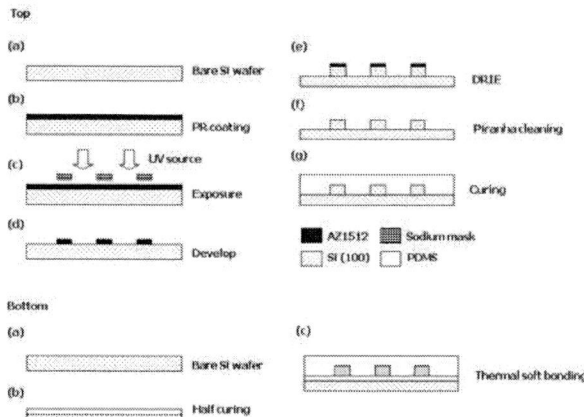

Figure 2: Fabrication process of grooved silicon molder

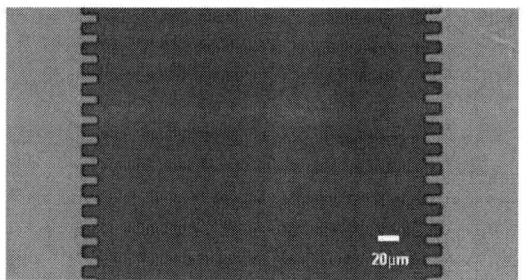

Figure 3: Picture of fabricated PDMS microchannel (top-view).

Materials and Experiment Equipments

An aqueous phase used for all experiments consists of a 78:22 (volume) mixture of water and glycerin. Olive oil (viscosity, 84 cP) is used for the oil phase. A precision syringe pump (11 Plus, Harvard Inc.) is used to deliver reagent solutions at flow rates ranging from 5 to 15 μl/min. A 1380 × 1040 resolution cooled CCD camera (Sensicam QE, PCO Inc.) is connected to an inverted microscope (Eclipse TE2000, Nikon Inc.).

RESULT AND DISCUSSION

Table 1 shows the dimensions of the grooved structures in the microchannel, where droplets can be isolated. In our study, the microchannel length and width are 30 mm and 200 μm, respectively. While the width, height, and depth of grooves are fixed at 15 μm, 45 μm and

20 μm, respectively, three different structures are fabricated with pitches of 100, 110, and 120 μm. Figure 4 shows a schematic of the grooved microchannel.

Table 1: Dimensions of microchannel with three kinds of grooves pitches.

Pitch (μm)	100	110	120
Depth (μm)		20	
Width (μm)		15	
Height (μm)		45	
Flow rate (μl/min)		5, 10, 15	

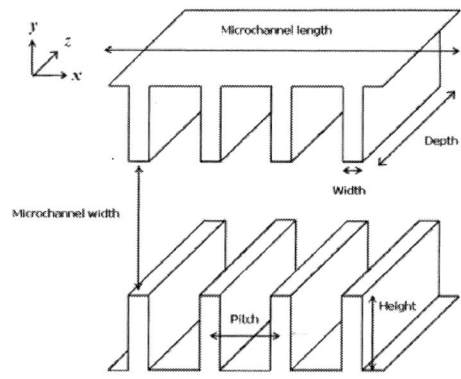

Figure 4: Schematics of grooved microchannel

Figure 5 illustrates the effect of pitch on droplets isolated in the cavities at different oil phase flow rates. The insets display sequential images of droplet generation at a flow rate of 10 μl/min with 110 μm pitch. When an interface of two immiscible liquids moves across two grooves and a cavity, the oil phase replaces the water except inside the cavity. The interface may isolate the water phase into a corner of the cavity forming a droplet. The isolated water volume directly affects the droplet size, which is determined by the speed of the interface, the geometry of the grooves, and physical properties of fluids, such as viscosity and surface tension.

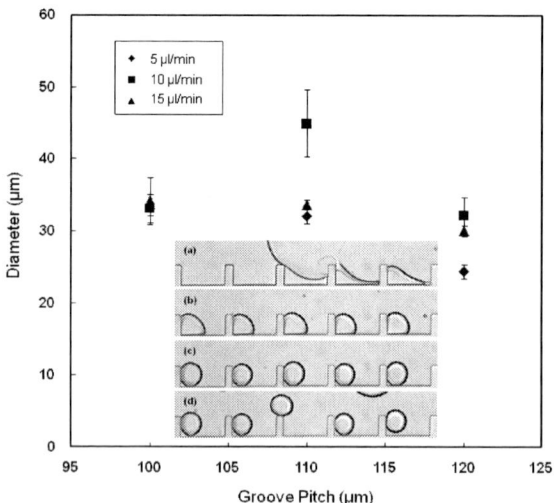

Figure 5: Comparison between grooves pitches and drop diameter depending on difference flow rate.

In our proposed geometries, the diameter of isolated droplets ranges from 25 μm to 46 μm depending on the oil phase flow rate and the pitch of the groove. Figure 5 depicts the existence of optimal conditions for stable generation of large droplets. The largest droplets were generated at 10 µl/min flow rate in grooves of 110 µm pitch. If the pitch is made higher or lower than 110 µm (at the flow rate of 10 µl/min) the droplet size decreases. At a fixed pitch of 110 µm, similar results may be observed for the effect of the flow rate. If the flow rate is faster or slower than 10 µl/min, the droplet size decreases. This means that droplets can be optimally formed when both the flow rate and pitch size are well matched. Interestingly, at the optimal condition, uniform droplets can be generated without observable daughter droplets in cavities.

Figure 6 describes the pattern of droplet generation as a function of flow rate. If the flow rate is too high or the pitch is too small, the interface between the two immiscible liquids moves too quickly through the cavity to generate stable droplets. Daughter droplets can also be generated in the corner of backward direction. As noted, both a flow rate of 10 µl/min and a pitch of 110 µm (Fig. 6 (b)) seem to be ideal for generating stable and uniform droplets in our system. When the flow rate is slower (Fig. 6 (a)) or faster (Fig. 6 (c)) than the optimal condition, daughter droplets are generated in the right corner of the cavities. It should be noted that there are two obvious factors when generating droplets. First, once the interface between the oil and aqueous phases moves into a cavity and across a groove, the interface should contact the top of the groove isolating the water in the left corner. This contact and the isolation of the aqueous phase determine the size of a droplet. Second, the speed of the two phase contact line on the wall determines the generation of the daughter droplet. In Fig. 6 (a), when the interface contacts the left groove top surface, the interface also contacts a bottom wall due to a low flow rate before the interface contact line on a vertical wall moves down to a bottom. Therefore, daughter droplets could be generated due to the isolation of the aqueous phase even in the right corner. When the flow rate is too fast, the interface may isolate almost the entire cavity region and then break into two droplets due to minimizing surface energies. Figure 6 (b) is a good example of optimal droplet generation due to simultaneous contact of the interface on the top of the left groove and on the right bottom of the cavity. Accordingly, the motion of the two-phase interface on the wall is of significant importance for determining the formation of daughter droplets. If the two-phase interface is able to touch earlier or later, the main droplet is formed in the forward corner of the cavity and the daughter droplet is formed in the backward corner of the cavity. It should be noted that the contact line movement on the wall is strongly related to the difference of viscosity and surface tension between the oil and aqueous phases. We are currently exploring the effect of both fluidic properties on this droplet generation technique .

Figure 7 shows the effect of pitch on droplet size at a flow rate of 5 µl/min. The groove pitch varies from 100 µm to 120 µm. When the oil phase moves from right to left within a cavity, the two-phase interface touches two points: one at the top of a left groove and one at any point on the bottom surface. As the groove pitch increases, the contacts are delayed and thus a small main droplet is generated. A daughter droplet is also generated in the cavity. Therefore,

if one wants to generate uniform droplets without daughter droplets, the synchronization of both contacts on the top of the forward groove and the bottom in the cavity must be considered. Further research is being currently directed on determining the effect of the shape and size of grooves on optimal droplet generation.

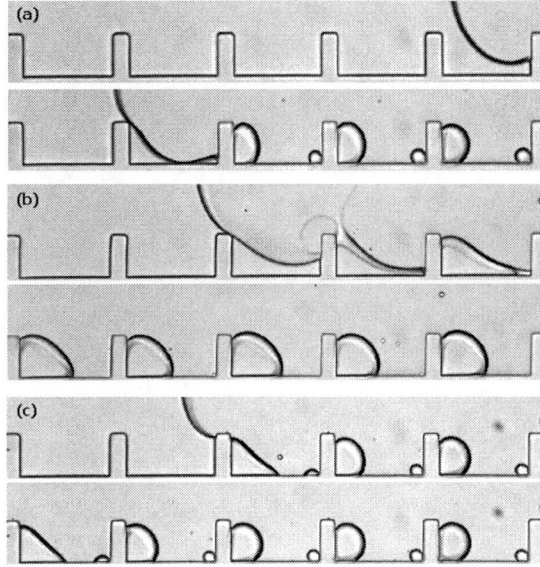

Figure 6: Effect of flow rate on droplet diameter.

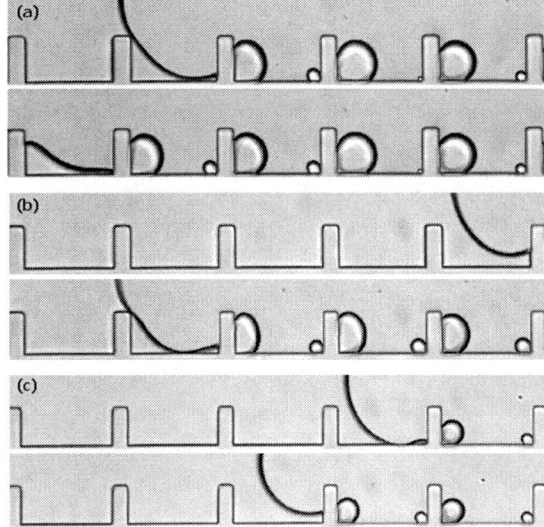

Figure 7: Effect of pitch of a groove on droplet diameter.

CONCLUSION

We have presented a new method to generate droplets stored in microfluidic cavity structures. We have designed and fabricated PDMS microchannels with various grooves on the walls. The mechanism of droplet generation in the cavity is proposed based on the observations of changing the flow rate of the oil phase and the pitch of the grooves. The two-phase interface plays a significant role in generating stable droplets. We expect that this technique will serve as a platform for monitoring cell behavior and *in vitro* expression.

REFERENCES

[1] H. Song, D.L. Chen and R.F. Ismagilov, Angew, "Reactions in Droplets in Microfluidic Channels", Chem. Int. Ed., vol. 45, pp. 7336–7356, 2006.

[2] A. Günther, K.F. Jensen, "Multiphase microfluidics: from flow characteristics to chemical and materials synthesis ", Lab Chip, vol. 6, pp. 1487-1503, 2006.

[3] S.Y. Teh, R. Lin, L.H. Hung and A.P. Lee, "Droplet microfluidics", Lab Chip, vol. 8, pp. 198-220, 2008.

[4] A. Huebner, S. Sharma, M. Srisa-Art, F. Hollfelder, J.B. Edel, A.J. deMello, "Microdroplets: A sea of applications?", Lab Chip, 2008, vol. 8, pp. 1244-1254,

[5] S. Sugiura, M. Nakajima, S. Iwamoto, S. Seki, "Interfacial Tension Driven Monodispersed Droplet Formation from Microfabricated Channel Array", Langmuir, vol. 17, pp. 5562-5566, 2001.

[6] D.R. Link, S.L. Anna, D.A. Weitz, H.A. Stone, "Geometrically Mediated Breakup of Drops in Microfluidic Devices", Phys. Rev. Lett., vol. 92, pp. 054503-054506, 2004.

[7] T. Kawakatsu, Y. Kikuchi, M. Nakajima, "Regular-sized cell creation in microchannel emulsification by visual microprocessing method", J. Am. Oil Chem. Soc., vol. 74, pp. 317-321, 1997.

[8] S.I. Anna, N. Bontoux, H.A. Stone, "Formation of dispersions using "flow focusing" in microchannels", Appl. Phys. Lett., vol. 82, pp. 364-366, 2003.

[9] T. Thorsen, R. Roberts, F. Arnold, S. Quake, "Dynamic Pattern Formation in a Vesicle-Generating Microfluidic Device", Phys. Rev. Lett., vol. 86, pp. 4163-4166, 2001.

[10] H. Song, R.F. Ismagilov, "Millisecond Kinetics on a Microfluidic Chip Using Nanoliters of Reagents ", J. Am. Chem. Soc., vol. 125, pp. 14613-14619, 2003.

A LOW-COST FLEXIBLE HOT-FILM SENSOR SYSTEM FOR FLOW SENSING AND ITS APPLICATION TO AIRCRAFT

R. Zhu[1], P. Liu[1], X.D. Liu[1], F.X. Zhang[1], and Z.Y. Zhou[1]

[1]State Key Laboratory of Precision Measurement Technology and Instruments, Department of Precision Instruments and Mechanology, Tsinghua University, Beijing, CHINA

ABSTRACT

This paper reports a simple, low-cost preparation method for fabricating a thin hot-film flow sensing system directly on a flexible polyimide Printed Circuit Board (PCB) by incorporating printed circuit technique with micromachining process. The sensor system takes merits of simple structure, low cost, easy fabrication and packaging, integrating sensors with processing circuits, good mechanical property and sensing capability. The paper also demonstrates a novel application of the sensor system to a micro aerial vehicle (MAV). The wing tunnel experiments are conducted to validate the effectiveness of the sensors and the MAV.

INTRODUCTION

The measurement of fluid mechanics is very important in various industrial fields, and flow sensors have been widely applied to execute accurate and timely measurements. Of these sensors, micromachined flow sensors have been developed for several decades. Silicon-based flow sensor was firstly demonstrated in 1974 [1] and then many application-oriented flow sensors have been developed based on various sensing principles, such as thermal anemometry (Perry 1982), Doppler frequency shift, and indirect inference from pressure differences (Rediniotis 1999; Richter et al. 1999) [2]. Among these sensors, thermal flow sensors possess merits of simple structure, easy use, and thus offering a practical solution for fluidics measurement applications [3].

Hot-wire/film anemometers are two commonly used flow sensor types. The fabrication and packaging process of conventional hot-wire/film flow sensor are delicate and often do not guarantee practical performances, also it is prohibitively difficult to form large array of sensors for measuring flow distribution. Although in recent years, many researchers have applied micromachining processes to realize micro hot-wire/film sensors on silicon or polyimide substrate [4-5], most processes are still complex, time consuming and expensive. These sensors are mostly applied to detect flow separation or vortex dynamics around aircraft wing surface [5-6]. Challenges still exist in the requirements of easy-fabrication, high-reliability, high-sensitivity and low-cost for real applications.

This paper reports a novel methodology for fabricating micro hot-film flow sensors directly on a flexible PCB, which serves as not only the substrate of the sensor array, but also the body of the signal conditional circuit. Comparing with previous technologies, the method takes merits of low cost, easy fabrication and packaging, integrating sensors with processing circuits, and providing good mechanical property for the sensor. Using the proposed hot-film flow sensor system, an integrative flexible wing MAV and a methodology for detecting flight parameters of the MAV, such as air speed, angle of attack, and angle of sideslip, are experimentally demonstrated.

FABRICATION OF SENSOR SYSTEM

The diagram of the fabrication process of the sensor system is shown in Figure 1. The flexible PCB with the thickness of 20μm is served as the sensor substrate, where electrical circuits (signal processing circuits of the sensor system) and wiring electrodes of sensors are prepared by using printed circuit technique. Afterwards, a thin composite metal film is deposited/sputtered and patterned to form a line shape with 3mm long and 300μm wide across two prepared electrodes. The nickel (Ni) thin film (150-200nm in thickness) is utilized as the sensitive material, which works as a thermal element that serves both as a Joule heater and temperature sensor [2]. A chromium (Cr) layer and a platinum (Pt) layer are used as an adhesion and a protection respectively. The way of directly sputtering/depositing metal thin film on the prepared flexible PCB benefits to simplify the packaging for the sensor, no complex solder bonding is needed. The electric connection between the sensor and the measurement circuit has been realized through the preprinted wiring. In addition, directly fabricating the thermal element on the FPCB also benefits to provide a smooth surface for a sensor to be exposed to the flow. The polyimide substrate also provides excellent thermal isolation. For stabilizing sensors, an annealing process is performed. Considering the heat-resistant capability of the polyimide substrate, we heat the prepared sensors at 200°C for 60 min in a vacuum environment. Then an electric annealing by electrifying the sensors at 30mA for 1 hour is performed subsequently. The fabricated sensor unit and array are shown in Figure 2.

1. Flexible PCB

2. Print circuits and electrodes

3. Sputter/Deposit Cr/Ni

4. Sputter/Deposit Pt

Figure 1: The fabrication process of the sensor system

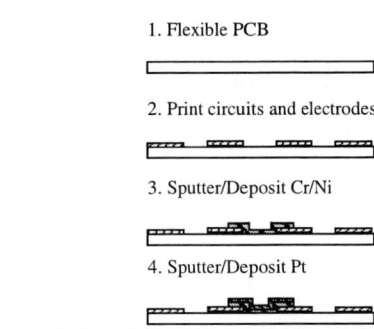

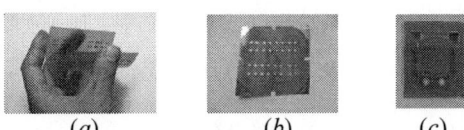

(a)   *(b)*   *(c)*

Figure 2: Fabricated sensor unit and array (a) Flexible PCB (b) Fabricated sensor array (c) Sensor unit

MEASUREMENTS OF SENSORS

The measurement of thermal flow sensors relies on the

detection of the convective heat transfer between an electrically heated resistive sensing element (i.e. hot-film) and surrounding fluid flow. Under a constant bias power, the thermal element assumes a steady-state temperature, which means the heat transfer system reaches equilibrium. If an external flow passes around the thermal element, the element experiences forced convective cooling. Accordingly, the temperature of the thermal element decreases, then the resistance of the element changes, and thus provides information on the flow that governs the cooling rate. As a result, the resistance of the sensor is the key parameters for detecting the flow velocity. A resistance measurement can be implemented by using a Wheatstone bridge circuit. Considering both functions of Joule heater and thermistor, either of constant current (CC), constant voltage (CV) and constant temperature (CT) mode can be utilized to operate the hot-film thermal sensors [7]. Amongst three modes, CT mode is with highest sensitivity and fastest response, which is realized by using a close-loop control to adjust the temperature of the hot-film element automatically. We establish a CT circuit shown in Figure 3 to work the flow sensors, where R_W refers to the resistance of the thermal sensor, R_1, R_2, R_3 are reference resistances in the bridge, the term R_2/R_1 is called the bridge ratio, and R_D is a resistance used to set the overheat ratio of the thermal sensor.

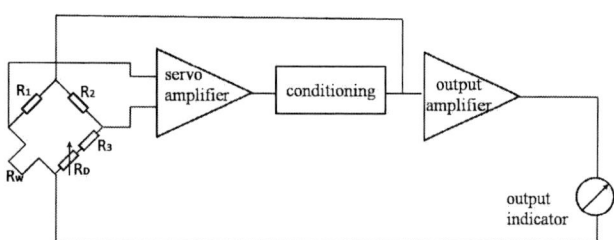

Figure 3: Schematic view of constant temperature circuit

A fabricated sensor system comprising of sensor units placed on the backside of the flexible PCB (Figure 4(b)) and the signal acquisition circuit (i.e. CT measurement circuit in Figure 3) distributed on the foreside of the flexible PCB (Figure 4(a)) is shown in Figure 4.1

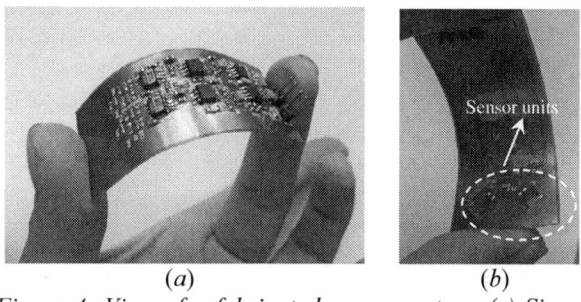

(a) (b)

Figure 4: View of a fabricated sensor system, (a) Signal acquisition circuit placed on the foreside of FPCB, (b) Sensor units on the backside of FPCB

Based on the heat transfer principle and CT measurement, the relationship between the readout of the measurement circuit (voltage V) and the measured airflow velocity (U) can be formularized as follows:

$$V^2 = A + BU^n \qquad (1)$$

where A and B are parameters determined by the material and geometric features of the thermal film, and n is the geometric factor. All these parameters can be determined through experimental calibrations.

The experimental tests to the performances of the flow sensors are carried out subsequently. The temperature coefficient of the resistance of the fabricated sensors is tested to be 0.001-0.006/°C. The sensor is also experimentally calibrated in a wind-tunnel. The steady state response results for airflow sensing under the CT mode with different overheat ratio (3% and 2%) are shown in Figure 5. The fit models shown as solid line are in accordance well with equation (1), where the geometric factor n is determined to be around 0.6. The results also indicate the overheat ratio r affects the sensitivity of the sensor, a higher overheat ratio implies a higher sensitivity, but unavoidably enlarges the power consumption. To mitigate the electrical consumption in some specific cases, such as application on a MAV, a tradeoff between the sensitivity and power of the sensor needs to be balanced. The time response of the sensor is roughly tested by manually rapidly exerting a wind flow on the sensor. Because generating an accurate pulsed or step flow velocity profile with sharp transient is difficult, the way of manually removing a baffle between a wind flow and the sensor is used. We do not use the conventional electric stimulation [2][4][7] for testing the time response because we expect to test a real response of the sensor to various flow speeds. The tested time response of the sensor is on the order of tens to hundreds of millisecond corresponding to the airflow speed from 15 to 3 m/s. Figure 6 gives a measurement result of the response to the flow at the speed of 10 m/s, where the time response is faster than 0.3s.

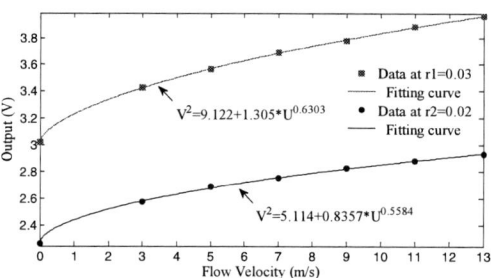

Figure 5: Testing results of the sensor operated under constant temperature mode. Bridge ratio of CT circuit is set as 20, and the overheat ratio r1 and r2 is set to be 0.03 and 0.02 respectively.

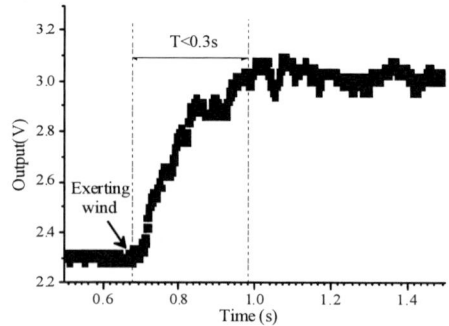

Figure 6: Time response of the sensor to an impact of a flow at a speed of 10 m/s.

SENSOR APPLICATION TO AIRCRAFT

Modern aircrafts require a vast amount of data coming from onboard sensors, either to provide information to the pilot, or to achieve automatic flight control as in unmanned air vehicle (UAVs). Specifically, air data is fundamental to infer many high-level flight parameters, including the free-stream air speed and the attitude of flight. Many researchers have taken efforts on sensor applications to aircrafts, such as flow separation and vortex dynamics detection, air flight parameters measurement. As for the measurement of flight parameters, these parameters could be indirectly inferred by measuring the pressure distribution on the wing of the aircraft [8][9]. The air speed and angle of attack affect the shear stress along the surface of the wing [10], in addition, the angle of sideslip could be inferred by the local angle of attack of the left and right wing of the air vehicle [11]. The requirements for the sensors utilized on an aircraft concern miniaturization, easily mounting arrays on non-planar surface, etc. Micromachined hot-wire and hot-film sensors are excellent candidates for implementing the applications. In this section, we will use the prepared hot-film flow sensors to infer a plurality of flight parameters for a MAV, including air speed, angle of attack, and angle of sideslip. First of all, a novel MAV with a flexible wing is constructed.

A flexible wing MAV integrating with the flow sensing system is developed and shown in Figure 7. A full piece of flexible PCB, on which the flow sensor array and its data acquisition and record circuits are distributed, is utilized to form the wing membrane as shown in Figure 7(a). Carbon fiber tape is cut into long narrow tacky strips with which the MAV's wing skeleton is constructed. The MAV is with a wingspan of 300mm and adopts the airfoil of S5010 and Zimmerman shape. The MAV is electrically powered, and we utilize an elevator to control longitudinal movement and a rudder to control lateral movement. Through experiments of real flights, the turning, climbing, maneuverability and wind resistant ability of the MAV are all tested. The flight tests show that the MAV with a flexible wing can fly with good stability and maneuverability.

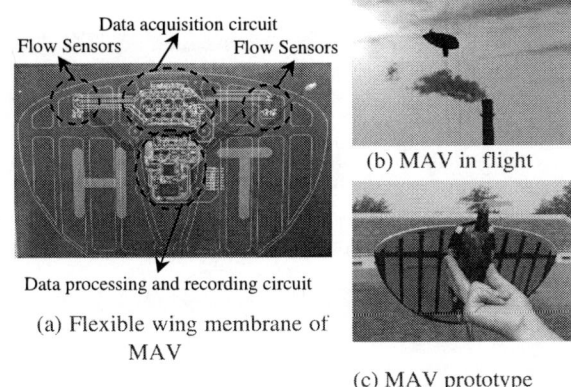

(a) Flexible wing membrane of MAV

(b) MAV in flight

(c) MAV prototype

Figure 7: A flexible membrane wing MAV integrating with flow sensing system

For a certain aircraft, the flow field around the wing surface is mainly determined by the free stream air speed, the angle of attack and the angle of sideslip. Therefore it is reasonable to detect these flight parameters by measuring the flow field around the wing surface. Also considering that on the front edge of the wing, the phenomena such as flow separation and vortex seldom occurs, in addition, the flow field there changes violently with the change of air speed and flight attitude angles, all of these imply that it is a good choice to place the sensors on the front edge of the wing. The prepared hot-film flow sensors are thus symmetrically placed on the front edges of the wing as shown in Figure 8. The data outputted from the sensors are processed and utilized to deduce the flight parameters of the MAV [12].

Figure 8: The collocation of hot-film flow sensors on the wing of a MAV

Suppose a set of n flow sensors are used, and denote their outputs in the vector form

$$U = [U_1 \quad U_2 \quad \cdots \quad U_n]^T \qquad (2)$$

A mathematic relation between the vector U and the three flight parameters (the free stream air speed U_∞, the angle of attack α and the angle of sideslip β) is expressed by (to simplify the analysis, the second order dependence of on other parameters such as the air density is not considered):

$$U = f(U_\infty, \alpha, \beta) \qquad (3)$$

where f represents the relationship model depending on aerodynamic characteristics of the air vehicle. Intuitively, the difference signal between the sensors on the top and bottom surfaces of the wing is mainly determined by the angle of attack, the difference signal between the sensors on the left and right sides of the wing is mainly determined by the angle of sideslip, and the average output of all sensors is mainly determined by the air speed. A primary experiment to justify this assumption is performed in a wind tunnel. The readings of four flow sensors (two in pair symmetrically placed on the upper and bottom surfaces of the front edge respectively) are acquired and converted into digital signals and transferred into an integrated microprocessor. The flight parameters are calculated and deduced in the processor and recorded in an on-board flash offering 32Mx8bit, NAND cell, which is an optimum solution for flight log requiring non-volatility. Figure 9 shows the primary experimental results for detecting the flight parameters. A systematic experiment is still in progress. In the future, we hope to utilize these inferred flight parameters to serve for an autonomous control.

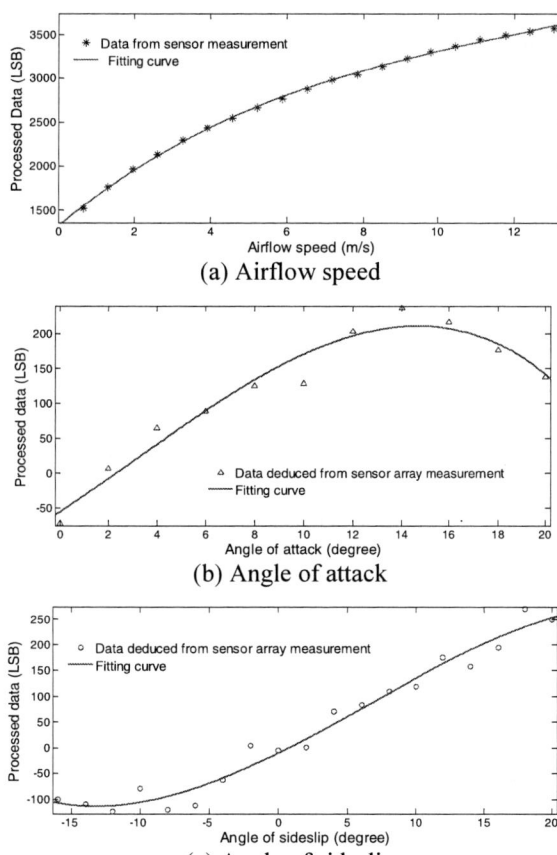

(a) Airflow speed

(b) Angle of attack

(c) Angle of sideslip

Figure 9: Experimental results of the flight parameters tested in a wind tunnel

Due to small size, array configuration, flexible capacity, high integration with signal processing circuit, the proposed hot-film flow sensor system exhibits a great potential application prospect. Other applications include surface flow dynamics tracking, laminar-separation determination, aerodynamic active control, etc.

CONCLUSION

In this paper, we propose a hot-film flow sensor and integrated system fabricated on a FPCB substrate. The fabrication for the sensor and system is relatively simple and economical by incorporating an industrial printed circuit technique with micromachining deposit/sputter process. The testing to the sensors validates the feasibility of practical uses. Afterwards, we father develop an electrically powered micro air vehicle with a flexible wing integrated with on-board electronics. The integrated onboard flow sensor array and system is used to determine a plurality of flight parameters. The experimental validations are also obtained. The proposed hot-film flow sensor system is unique owing to their high integration and reliability, and thus be compatible to practical applications.

ACKNOWLEDGMENT

This work is financially supported by the National High-tech Program "863" of China under the grants 2006AA04Z257.

REFERENCES

[1] A. F. P.Van, Putten, S. Middelhoek, "Integrated silicon anemometer", *Electron. Lett.*, vol. 10, pp. 425–426, 1974.

[2] J. Chen, Z. F. Fan, J. Zou, "Two-Dimensional micromachined flow sensor array for fluid Mechanics studies", *Journal of Aerospace Engineering*, vol. 16, no.2 pp.85-97, 2003.

[3] J.J. van Baar, R.W. Wiegerink, T.S.J. Lammerink, G.J.M. Krijnen, M. lwenspoek, "Micromachined structures for the thermal measurements of fluid and flow parameters", *J. Micromech. Microeng.*, vol. 11, pp. 311-318, 2001.

[4] C. Liu, J. B. Huang, Z. J. Zhu, F. K. Jiang, S. Tung, Y. C. Tai, C. M. Ho, "A Micromachined Flow Shear-Stress Sensor Based on Thermal Transfer Principles", *Journal of Microelectromechanical Systems*, vol. 8, pp. 90-99, 1999.

[5] Y. Xu, F. K. Jiang, S. Newbern, A. Huang, C. M. Ho and Y. C. Tai, "Flexible Shear-stress Sensor Skin and its Application to Unmanned Aerial Vehicle", *Sensors and Actuators A*, vol. 105, pp. 321-329, 2003.

[6] U. Buder, R. Petz, M. Kittel, W. Nitsche and E. Obermeier, "AeroMEMS Polyimide based Wall Double Hot-wire Sensors for Flow Separation Detection", *Sensors and Actuators A*, vol. 142, pp.130-137, 2008.

[7] H. H. Bruun, *Hot-Wire Anemometry Principles and Signal Analysis*, Oxford University Press, New York, 1995.

[8] S. Callegari, A. Talamelli, M. Zagnoni, A. Golfarelli, V. Rossi, M. Tartagni, E. Sangiorgi, "Aircraft Angle of Attack and Air Speed Detection by Redundant Strip Pressure Sensors", *Proceedings of IEEE*, vol. 3, 2004 pp.1526-1529.

[9] S. Callegari, M. Zagnoni, A. Golfarelli, M. Tartagni, A. Talamelli, P. Proli, A. Rossetti, "Experiments on aircraft flight parameter detection by on-skin sensors", *Sensors and Actuators A*, vol. 130–131, pp. 155–165, 2006.

[10] Y. Xu, Y. C. Tai, A. Huang and C. M. Ho, "IC-integrated Flexible Shear-stress Sensor Skin", *Journal of Microelectromechanical Systems*, vol. 12, pp. 740-747, 2003.

[11] J. Choisnet, L. Collot, N. Hanson, "Method for determining aerodynamic parameters and method for detecting failure of a probe used for determining aerodynamic parameters", US Patent Application 2004/0011124 A1.

[12] H. P. Fei, R. Zhu, Z. Y. Zhou and J. D. Wang, "Aircraft Flight Parameters Detection Based On Neural Network Using Multiple Hot-film Flow Speed Sensors", *Smart Mater. Struct.*, vol. 16, pp.1239-1245, 2007.

978-1-4244-2977-6/09 $25.00 © 2009 IEEE

INTEGRATING EWOD WITH SURFACE RATCHETS FOR ACTIVE DROPLET TRANSPORT AND SORTING

Todd A. Duncombe[1], Momoko Kumemura[2], Hiroyuki Fujita[2], Karl F. Böhringer[1]
[1]Electrical Engineering Department, University of Washington, USA
[2]CIRMM, Institute of Industrial Science, The University of Tokyo, JAPAN

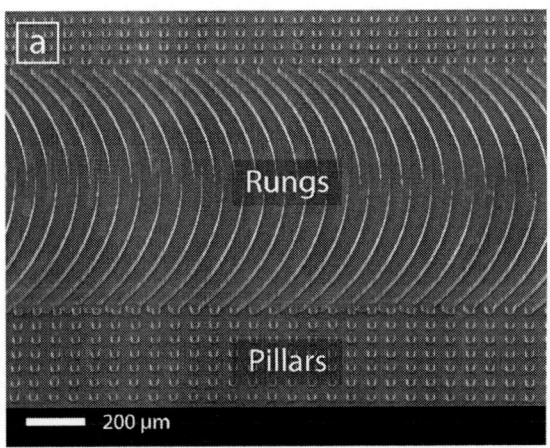

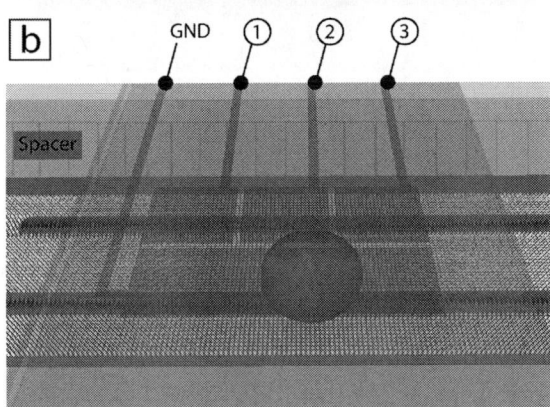

Figure 1: Experimental setup. (a) A SEM micrograph of a surface ratchet. (b) A schematic of the EWOD plate alignment above the surface ratchet, both surfaces contact the droplet. Electrodes 1, 2, 3 and Ground are labeled. (c) The closed system is attached to a stage mounted horizontally on an electromagnetic speaker, which delivers controlled vibrations to the system.

ABSTRACT

Combining surface ratchets and electrowetting on dielectric (EWOD) produces novel microfluidic systems that achieve passive droplet transport by vibration along microscopically-rough surfaces and active droplet sorting by electric signals. The super-hydrophobic surface ratchet and EWOD plate sandwich a droplet; when vibrated the device adopts passive droplet transportation via the surface ratchet. The EWOD technology is utilized at particular junctions to produce several droplet specific control functions, including the combination of a new ratchet design with an EWOD plate to develop a switch (serving the same purpose as a switch in train tracks) that sorts 10 µl droplets at a junction.

INTRODUCTION

Actuation methods for microfluidic devices designed for transporting discrete liquid volumes have traditionally fallen into two categories; passive droplet transport (global actuation) or active droplet transport (local actuation). There are a wide range of controls for local actuation including light, thermal and electric stimuli. Ichimura et al. [1] developed an active droplet transport device by designing a monolayer of photo-responsive azobenzene groups that altered their wetting properties when exposed to particular wavelengths of light. By applying blue light asymmetrically in intensity on an olive oil droplet, they actively controlled droplet transport. Their surface could then be 'reset' to its original wetting properties with the application of UV light, and droplet transport repeated. Darhuber et el. [2] built a thermocapillary microfluidic device, transporting water droplets by actively engaging micro-heaters in an array, moving a droplet to the adjacent colder region. Similarly to micro-heater arrays, Cho et al. [3] used an array of addressable electrodes for local actuation via the electric wetting properties of polar liquids. Electrowetting on dielectric (EWOD) devices function by reducing the solid-liquid surface tension, γ_{SL}, of the droplet-surface interface, and thus the droplet is pulled towards the region with high electric field.

Passive droplet transport refers to actuation via a global stimulus. A simple actuation scheme replaces the complex design required for droplet specific control. Physical configuration establishes a bias for motion that drives droplet transport. A classic example is Chaudhury and Whitesides [4] "How to make water run uphill," which presents a passive transport method by varying the chemical composition on a surface to move a droplet against gravity over a limited distance. Dos Santos et al. [5] reported self-propelling silane droplets on a glass or silicon surface. When transported, droplets alter the chemical composition of the surface thereby requiring extensive cleaning for re-

978-1-4244-2977-6/09 $25.00 © 2009 IEEE

peated transports. Bico & Quéré's recent demonstration of self-propelling bislugs [6] consists of ethylene glycol and silicone oil in a glass capillary. The discrepancy in the capillary pressure drives the motion. Similarly to bislugs, surface ratchets [7] are not limited in transport distance and are repeatable. Surface ratchets use the asymmetric surface contact of a droplet on a rough, hydrophobic surface to achieve transport through vibration. This global actuation transports numerous droplets simultaneously.

DESIGN AND FABRICATION

Ratchets have a rough, super-hydrophobic surface that creates a biphasic air-solid interface with water, the Fakir state, in which the droplet contacts only the top of the surface's asperities. A droplet in the Fakir state has high apparent contact angles and low resistance during movement [8]. The asymmetric portion of the ratchet is its track, 1.2 mm in width, which is delimited by a more hydrophobic, rougher region of sparsely spaced pillars (Figure 1A). Pillars have a diameter of 30 μm and are spaced in a 70 μm pitch. A track is constructed using periodic semicircular rungs (650 μm in curvature, 30 μm width) spaced by 75 μm. The rung's curvature establishes a discrepancy of surface contact between the two sides of the droplet while the low surface energy of the pillar region contains the droplet on the track. One side of a droplet is aligned with the rung's curvature, giving it uniform surface contact along the rung. The other side is not aligned with the rungs curvature, yielding non-uniform surface contact. When agitated by vibration this discrepancy in edge surface contact drives droplet transport [7].

The microscopically rough surface is fabricated by etching a silicon wafer to a depth of 70 μm. It is then coated with a monolayer of perfluoro-octyltrichlorosilane (FOTS), deposited from its vapor phase, producing a hydrophobic surface. Vibration is provided by a horizontal stage mounted on an electromagnetic speaker.

EWOD devices are typically designed as a two-plate closed system [3]. To integrate EWOD with surface ratchets we use a one-plate design [9], with grounded electrodes running parallel to control electrodes. The design consists of a 2 mm x 6 mm rectangular ground electrode running parallel to and spaced 4 μm from a series of three 2 mm x 2 mm control electrodes (control electrodes are separated by 4 μm) (Figure 1B). Each electrode is connected to a contact pad, allowing us to address each electrode individually. We use an AC power source to reduce the risk of electrolysis destroying the electrodes.

The device is constructed on a glass wafer. Fabrication begins with photolithography and a 50 nm ITO evaporation. A 500 nm layer of SiO_2 is coated using plasma deposition. In the final step the surface is spin-coated with a thin layer of CYTOP, a hydrophobic dielectric layer. During the process, tape is placed over the ITO contact pads during the deposition of SiO_2 and CYTOP, preserving an area where electrical connections can be made.

Combining these two technologies, we constructed a device that provides droplet specific control at discrete locations on a surface ratchet. Spacers are used to suspend the

EWOD plate above and face to face with the ratchet. The distance between the two surfaces is chosen such that the droplet contacts both. The surface ratchet, spacer and EWOD plate are placed on a stage mounted horizontally to an electromagnetic speaker, which provides vertical vibrations (Figure 1C). This is the first instance a surface ratchet has been shown functioning contained in a closed system and the first time a single EWOD plate was used to move large (>10 μl) droplets while inverted. The ratchets operating conditions require lower frequency and higher amplitude as compared to surface ratchets contacting only one surface.

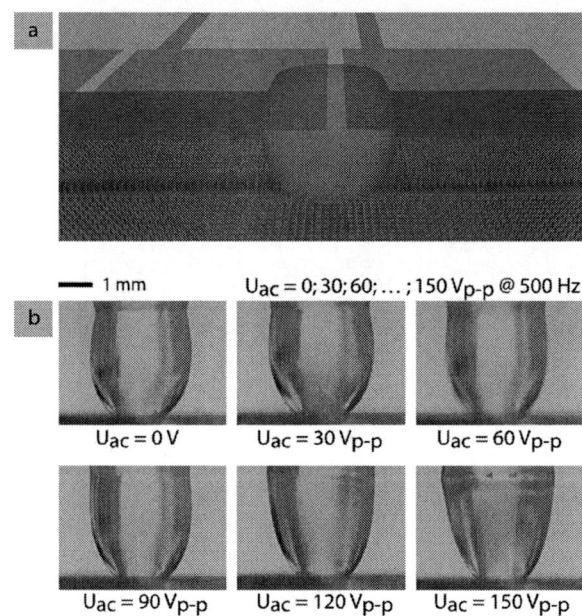

Figure 2: Droplet surface contact vs. applied voltage. (a) A 20 μL droplet is placed in between a surface ratchet and EWOD plate spaced by 3.5 mm. (b) The images are taken looking between the two surfaces. As voltage is increased the contact angles with the EWOD plate are reduced, lifting the droplet's center of gravity further from the ratchet.

RESULTS

When a droplet is under an activated EWOD device, the center of mass of the droplet is lifted up and away from the surface ratchet. The droplet's contact angle with the EWOD plate decreases as its area of contact expands (Figure 2). Increasing the applied voltage increases the electric field and further raises the droplet's center of mass.

Using this novel combination of microfluidic methods we demonstrate three important functions: a valve, a flow reversal device, and a switch for sorting droplets.

Valve

The most basic function of our system is a valve. Surface ratchet design and vibration cause the droplet to move in the direction of the track's asymmetrical bias. By activating electrodes placed at specific locations we are able to stop an advancing droplet and hold it in place as the system continues to vibrate. When the electrodes are deactivated the droplet resumes movement along the track.

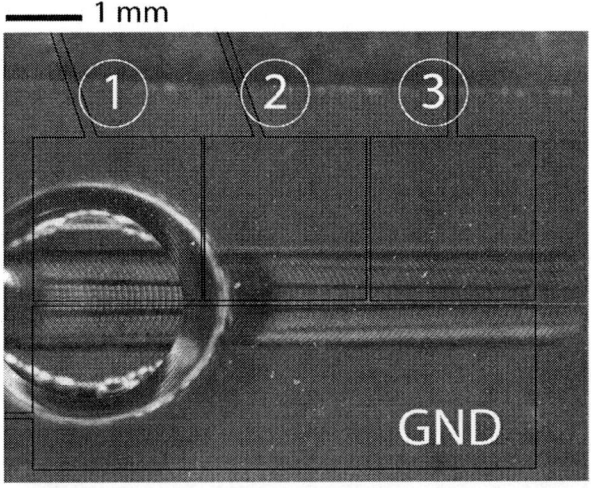

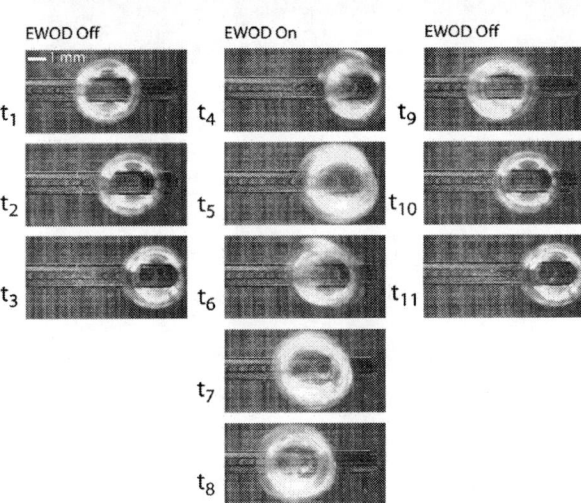

Figure 3: Flow Reversal Device. A 20 µL droplet is vibrated by the electromagnetic speaker at 4.2 V and 32 Hz; 150 V at 500 Hz is applied to the electrodes. In frames t_1 - t_3, EWOD is off and the droplet advances via the surface ratchet at an average pace of 3 mm/s. At t_4 electrode 3 is activated (1, 2 float). Droplets appear larger when affected by the EWOD device due to the increased contact with the EWOD plate. At t_5 electrode 2 is activated, stretching the droplet to the left. At t_6 electrode 3 is deactivated and the droplet is pulled to the left, centering itself on electrode 2. The process in t_4-t_6 is repeated in t_6-t_8 with electrodes 1 and 2, at t_8 the droplet has moved backwards returning to the original position in t_1. When EWOD is disengaged, in t_9-t_{11}, the droplet returns to its initial contact angles and resumes motion via the surface ratchet.

Figure 2A shows our implementation of the valve. The minimum voltage required to stop the droplet must be large enough to reduce the droplet's contact with the surface ratchet such that the adhesion force due to EWOD is stronger than the force from the track's bias. To stop and hold a 20 µl droplet advancing along the track the minimum applied voltage is 65 V at 500 Hz.

Flow Reversal Device

The second function we developed is a flow reversal device. The flow reversal device transports droplets in the opposite direction from the ratchet's asymmetrical bias while the stage continues to vibrate. Figure 3 outlines the electrode placement above the ratchet. One ground electrode and three control electrodes are placed over the surface ratchet. By sequentially activating and deactivating electrodes we are able to stop an advancing droplet and reverse its direction. We use a higher voltage (150 V at 500 Hz) in the flow reversal device than the valve to produce rapid and reliable transport.

The operation of the flow reversal device involves multiple stages. Electrode 3 in Figure 3 is activated while the other two electrodes are floating. Initially like a valve, the advancing droplet is stopped at electrode 3. At this point, electrode 2 is activated. The resulting electric field stretches the droplet against track's bias (as seen between t_4-t_5) thus aligning the droplet between electrodes 2 and 3. Electrode 3 is then deactivated and the droplet moves backwards (t_5-t_6) to center itself on electrode two. This sequence is then repeated to move a droplet backwards any desired distance.

A Switch (Sorting Droplets)

We demonstrate a switch function that enables droplet specific sorting using EWOD with a novel surface ratchet design (Figure 4). The new ratchet design is a four way intersection of tracks; one track's asymmetric bias is towards the junction while the other three are away from it. The EWOD switch at the junction transports droplets against the track's asymmetrical bias to a position over a new track oriented in a direction perpendicular to the incoming track. After the droplet has been moved, the electrodes are deactivated and the droplet resumes ratchet driven motion in the chosen direction.

Electrodes are controlled to turn a droplet left, right or let it pass straight through. As seen in Figure 4's electrode configuration, electrodes are above the right hand and left hand path. When one of these electrodes is activated, an advancing droplet will be pulled by the electric field as it reaches the junction against the track's bias. If the voltage level is sufficient, the droplet will center itself under the activated electrode. The track's bias under these electrodes favors movement along the new track's direction. Thus, when the electrode is deactivated the droplet begins movement via the surface ratchet in its new direction. If both electrodes are deactivated the droplet is unaffected by the junction and continues straight along the ratchet.

DISCUSSION

Active and passive droplet transport methods have their own independent applications, hinging on whether droplet specific control is required or not. A limitation of passive transportation is the difficulty to address and guide specific droplets. For surface ratchets, or any other passive droplet transportation method, to be useful in a wide range of microfluidic applications, a method of active, droplet-specific transport is essential.

We have implemented a device that integrates the two, taking advantage of surface ratchet's passive transport characteristics, and EWOD's droplet specific control. Using the integrated system, we have demonstrated three critical functions that could greatly expand the applications of surface ratchets. Now a device that functions primarily through surface ratchet's passive droplet transportation, moving several droplets over long periods of time, can be controlled at specific junctions, stopping droplet flow with a valve, moving droplets backwards with a flow reversal device, and sorting droplets at a switch.

ACKNOWLEDGEMENTS

This work was supported by an International Research and Education in Engineering (IREE) supplement to National Science Foundation grant ECCS-05-01628 (Rajinder Khosla, program director). The authors thank VLSI Design and Education (VDEC) for mask production, Laurent Jalabert, Cagatay Tarhan, Bernard Wee, Christophe Yamahata and all the members of CIRMM at the University of Tokyo for their generous help and support.

REFERENCES

[1] Kunihiro Ichimura, Sang-Keun Oh, Masaru Nakagawa. "Light-Driven Motion of Liquids on a Photoresponsive Surface." *Science* **288**(547):1624-1626, 2000.

[2] Anton A. Darhuber, Joseph P. Valentino, Sandra M. Troian, and Sigurd Wagner. "Thermocapillary Actuation of Droplets on Chemically Patterned Surfaces by Programmable Microheater Arrays." *Journal of Microelectromechanical Systems* **12**(6), 2003.

[3] S. K. Cho, H. Moon and C.-J. Kim, "Creating, Transporting, Cutting, and Merging Liquid Droplets by Electrowetting-Based Actuation for Digital Micro-fluidic Circuits," *Journal of Microelectromechanical Systems* **12**(1):70-80, 2003.

[4] Chaudhury, M.K. and G.M. Whitesides, "How to make water run uphill?" *Science* **256**(5063):1539-1541, 1992.

[5] Fabrice Domingues Dos Santos and Thierry Ondarcuhu. "Free-Running Droplets." *Physical Review Letters* **75**(16), 1995.

[6] J. Bico & D. Quéré, "Self-propelling slugs in a tube", *J. Fluid Mech.* **467**(201), 2002.

[7] A. Shastry, D. Taylor and K. F. Böhringer, "Micro-Structured Surface Ratchets for Droplet Transport," in *Transducers'07*, pp. 1353-1356, Lyon, France, 2007.

[8] Cottin-Bizonne, C., et al., "Low-friction flows of liquid at nanopatterned interfaces." *Nature Materials* **2**:237-240, 2003.

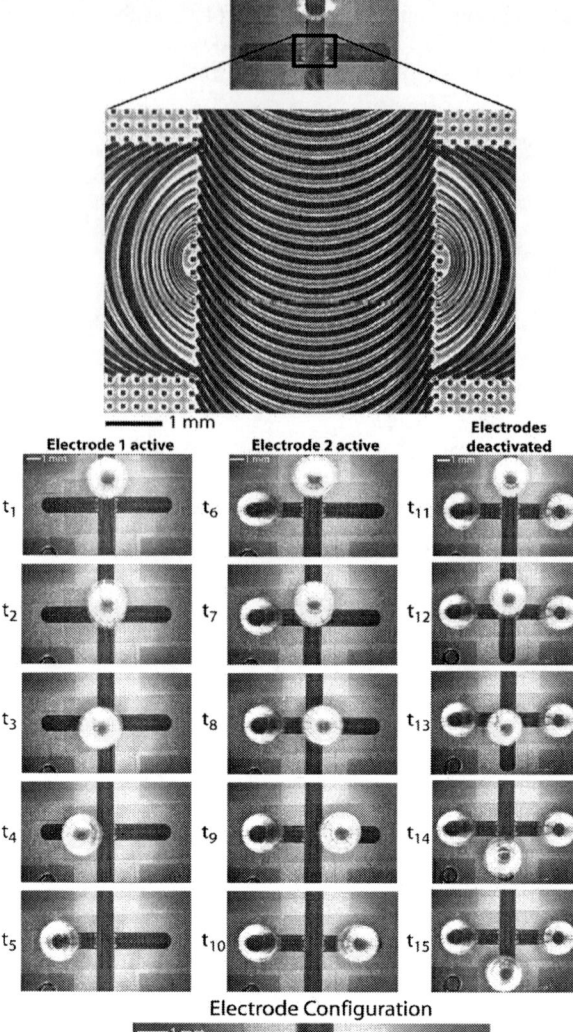

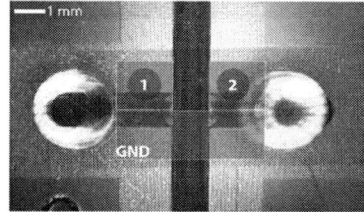

Figure 4: Sorting droplets. Depending whether electrode 1 or 2 is activated, the droplet turns to the left or right, respectively. A 10 μL droplet is vibrated at 8 V and 55 Hz, the applied voltage at the electrodes is 150 V at 500 Hz. In t_1-t_2 the droplet advances using the surface ratchet. In t_3-t_4 the droplet 'feels' the electric field and is pulled towards electrode 1. At t_5 the electrode is turned off and the droplet resumes its initial contact, and moves in the new direction. This procedure is repeated with electrode 2 in t_6-t_{10} to sort droplets in the alternate direction. In t_{11}–t_{15} the electrodes are deactivated; droplets are unaffected by the junction.

[9] U.-C. Yi and C.-J. Kim, "EWOD actuation with electrode-free cover plate," in *Transducers'05*, pp. 89- 92, Seoul, Korea, 2005.

THERMAL CHARACTERIZATION OF MICROLITER AMOUNTS OF LIQUIDS BY A MICROMACHINED CALORIMETRIC TRANSDUCER

G. Pärr[1], E. Santagata-Iervolino[1], A.W. van Herwaarden[1], W. Wien[2] and M.J. Vellekoop[3]
[1]Xensor Integration, Delfgauw, THE NETHERLANDS
[2]Delft University of Technology, Delft, THE NETHERLANDS
[3]Vienna University of Technology (ISAS), Vienna, AUSTRIA

ABSTRACT

We show a simple method of measuring thermal conductivity and thermal diffusivity of liquids. For this we used stacked liquid calorimeter chips fabricated with thin film membranes for optimum thermal isolation of the sample under test. DC joule heating allows measurement of conductivity, AC joule heating allows measurement of diffusivity. We measured the response for iso-propanol (IPA), methanol and water, and water with IPA, methanol and sugar, and the measurement results compare well with the conductivity values. Diffusivity could also be measured.

INTRODUCTION

Measuring thermal properties of liquids

In the literature various methods are described for measuring thermal properties of liquids. Optical techniques are based on, for example, photo-pyroelectric effects [1] or on interferometry [2]. These methods require, for example, an electromagnetic coupling between the optical actuation and the thermal property of the sample.

Zhang and Tadigadapa [3, 4] describe a method which uses micro-calorimetry chips [5]. They use a chip with a resistive heater and a thermopile on a single membrane. This allows very good measurement of the thermal conductivity. The proximity of both elements on the same membrane and the poor thermal isolation of the liquid cell hamper the measurement of the thermal diffusivity, though.

Our device consists of two calorimetric chips and two membranes, each with a resistive heater and a thermopile.

The twin chips are stacked on each other, and a liquid sample is isolated between the two membranes of the chips as a thin film. This results in a very good thermal isolation of the liquid sample.

A resistive heater on one of the membranes sends thermal waves through the sample. The resulting temperature oscillations are detected by thermopiles on both membranes. Attenuation and phase shift of the thermal waves are related to the conductivity and diffusivity of the liquid.

The closed fluidic structure allows in-line integration into other micro-fluidic systems as an autonomous unit.

DESIGN

Chip design

The heart of the thermal properties sensor is made up of two chips (XI-318 and XI-319) glued on top of each other, designated XI-318-9. Figure 1 gives a top view of the stacked chips, Figure 2 a schematic cross section (not to scale). The general idea is to create a liquid volume that is mainly contained between thin SiN membranes. In this way, a good thermal isolation towards the ambient is

obtained. Compared to devices made up of a single chip with a SiN membrane and a solid wall around the liquid volume [6, 7], or even single chips with silicon membranes [8], the thermal resistance in water is increased significantly from about 10 K/W for the NCM-9924 [8] up to nearly 1 kK/W for XI-318-9.

Figure 1: Photograph of the two chips (1) which form the device XI-318-9 with its thermopiles (2) and Joule heaters (3). The transparent membranes (4) are 3.8 x 0.8 and 2 x 0.8mm large.

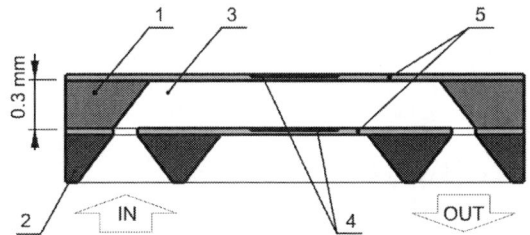

Figure 2: Lateral cross-section A-A from Figure 1 shows the stacked chips (1 = XI-318, 2 = XI-319), a 1µl sample volume (3), two thermopiles (4) and 2µm thin membranes (5). Not to scale.

Simultaneously, the volume of the sensor is also reduced, from 5-30 µl for the large liquid micro-calorimeter chips with silicon membranes (LCM-2506 and NCM-9924) to about 1 µl for the present design.

For both chips the thermopiles and heater resistances are designed to be identical, having a resistance of about 50 kOhm and 6.6 kΩ, respectively. The thermopiles consist of 36 p-type vs. n-type polysilicon thermocouples with an estimated sensitivity at room temperature of about 11 mV/K. With a thermal resistance of about 1 kK/W in non-flowing water this gives a theoretical transfer of the order of 11 V/W. The peak-to-peak noise in air is estimated to be around 0.23 µV in a 1 Hz bandwidth

978-1-4244-2977-6/09 $25.00 © 2009 IEEE

(thermal or Johnson noise). This gives a NEP_{pp} (Noise Equivalent Power) of the order of 20 nW and a $NETD_{pp}$ (Noise Equivalent Temperature Difference) of the order of 20 μK, all for a 1 Hz bandwidth.

FABRICATION
Chip fabrication
The chips are made using a thin-film bulk-micro-machining process. The starting material is a 300-μm thick <100> silicon wafer of 100 mm diameter, on which 600 nm LPCVD low-stress SiN and 300 nm LPCVD low-stress polysilicon are deposited. After implantation of the poly to make low-resistive n-type and p-type regions (50Ω/sq and 75 Ω/sq, respectively) the poly is patterned and covered by another 400 nm SiN. The contact openings to the poly are made, 0.6 μm aluminum is deposited and patterned, and a 0.8 μm PECVD oxide is deposited for scratch protection. Of course the wire bonding pads are opened. Then SiN membranes are etched at the front by anisotropic KOH-etching from the back side. After sawing the chips are ready for assembly.

Sensor build-up
The thermal properties sensor is made by gluing the XI-318 chip onto an aluminum substrate, and gluing the XI-319 chip on top of XI-318. The aluminum substrate has openings to allow liquids entering and exiting the access holes in the XI-318, a third hole is present to ventilate the cavity below the main membrane of XI-318. Fluid connections to the aluminum substrate are made by gluing tubes to the aluminum or by zero-dead-volume screw connections. Electrical connections to the outside are made with PCB or ceramic plates for wire bonding and soldering.

In general, the sensor element made up in this way is then (thermally) connected to a larger mass of aluminum and isolated with polystyrene to reduce thermal fluctuations and drift. In order to further minimize thermal fluctuation effects, twin sensors can be made consisting of two stacks of XI-318 + XI-319 chips, and measuring differential signals.

EXPERIMENTAL RESULTS
Sensor characterization
The measurements of the basic characteristics of the sensors agree well with the theoretical estimates. The transfer of the XI-318-9 sensor is 11 V/W when filled with non-flowing water. With flowing water the transfer decreases, but not even dramatically up to 500 μl/min when heating and measuring at the same membrane. For a 10 μl/min flow the decrease is 2.5%, at 500 μl/min the transfer is nearly halved.

A time constant for heat pulses by the 6.6 kΩ heater resistance on the membrane of the order of 0.5 sec has been measured in water.

The noise and interference on the signal have been observed by measuring the output signal using an Agilent 6-digit DMM, which in the normal mode filters with approximately 1 Hz band width and has a resolution of 0.1 μV. In air, the measured noise is of the order of 0.25-0.5 $μV_{pp}$, as theoretically expected for the Johnson noise.

In non-flowing water, the total noise and interference

is of the order of 2½ $μV_{pp}$, making the NEP and NETD 10 times worse. In flowing water the total noise and interference is larger still, the exact value depending upon the flow rate and measures taken to dampen the pressure pulses emitted by the liquid pumps. This makes the sensor resolution better in non-flowing conditions than in flowing conditions. It is therefore advantageous to (temporarily) stop the flow for accurate measurements.

Comparison with other liquid calorimeter chips
In the Table 1 below we have compared a few liquid calorimeter chips with the new chips XI-318-9.

Table 1: Comparison of liquid calorimeter chips.

Chip Parameters in water	NCM-9924 [8]	Köhler [6, 7]	XI-318-9
Transfer V/W	1.0	0.75	11
Volume μl	30	2	1
Volume transfer μl·V/W	30	1.5	11
Thermal resistance K/W	10	25	850
Output resistance kΩ	50	8	50
Time constant sec	0.7	5	0.5
Noise μV	2.5	2.5	2.5
NEP μW	2.5	3	0.2

The transfer in V/W of the new chip is much higher than previously reported chips, while the volume is generally lower. For use with experiments where the signal is generated by a volumetric reaction, the power produced in a much smaller volume is proportionally lower. This, however, is to some degree compensated by a much higher transfer, so that the sensitivity of the device for volumetric reactions, such as enzyme conversions, is comparable. Only, much less chemicals are needed, which can be an advantage when the chemicals are expensive or scarce. The smaller volume also has advantages in refreshing time and signal peak widening.

Measurements on gases
The device has first been characterized in air and helium, with thermal conductivities of 25 and 150 mW/Km, respectively. The transfer in air was found to be about 5.5 times as high as for water (59 V/W), when heating and measuring on the same membrane. The temperature rise in the opposite membrane, as measured by the thermopile, was very low, only 4.2 % of that (2.5 V/W). As we heated with 5 V into the heater resistance of 6.6 kΩ, this gave output voltages of about 224 mV and 9.6 mV.

Interesting is the situation when the air inside the device is changed for helium (outside is always air, also when liquids are inserted). While the helium cools the membrane where the heating takes place better than air (attenuation), it also transfers heat more efficiently to the opposite membrane. Consequently, the transfer for the membrane being heated decreases from 59 V/W to 44 V/W (165 mV), while the transfer of the opposite membrane rises from 2.6 V/W to 3.0 V/W or 7% of the membrane being heated (11.5 mV).

Water, that transfers heat even much more efficiently than helium, induces a transfer at the opposite membrane of about 11 % of the heated membrane.

978-1-4244-2977-6/09 $25.00 © 2009 IEEE

DC measurement results

The simplest measurements of the liquid thermal properties consist of DC heating using the heater resistance in the middle of the membrane, and then measuring the output voltage of the thermopile on the same and the opposite membrane. The thermopile on the same membrane is about 20 μm away. The thermopile on the opposite membrane is 300 μm away, the temperature increase of the heater being transferred but also attenuated by the liquid in between.

Figure 3 shows the output voltage of the sensor for a 3.8 mW input power for various liquids: water; methanol; IPA (iso-propanol); various sucrose-water solutions (0-100 %, 7-93 %, 14-86 %, 40-60 % by weight); mixture of water with methanol and mixture of water with IPA. The alcohol-water mixtures are 50%-50 vol%. On the x-axis we show the value of the thermal conductivity we found in the literature, the y-axis shows the thermopile voltage. The values of the thermal conductivity for the sucrose solutions and for the water-alcohol mixtures have been calculated using equation 4 of [9] and equation 1 of [10] respectively. In the calculations a temperature value of 20 °C has been considered.

At these output voltages, of the order of 100 mV, the temperature increase of the heater is about 10 K, opposite to the heater about 1 K.

As these measurements have been carried out in non-flowing conditions, the noise is of the order of 2.5 μV. With a change in output voltage of the order of 50 mV for a change in thermal conductivity from 0.1 to 0.6 W/Km, this would correspond to a sensitivity of about 100 mV/(W/Km). This is comparable to the 5.8 mV/(W/Km) that Zhang and Tadigadapa find [4], as their output signals are of the order of 5 mV. Compared to the 2.5 μV noise, a resolution in thermal conductivity value is obtained of the order of 0.25 mW/Km, or typically well below 1% of the thermal conductivity values of most liquids (usually around 200 mW/Km). Interesting is also to note that the resolution for the measurement of sucrose in water is of the order of 100 ppm in weight, or 0.3 mM (milli-mol per liter). This compares favorable with the resolution of the enzymatic measurement of glucose, which has a resolution of about 2 mM [11].

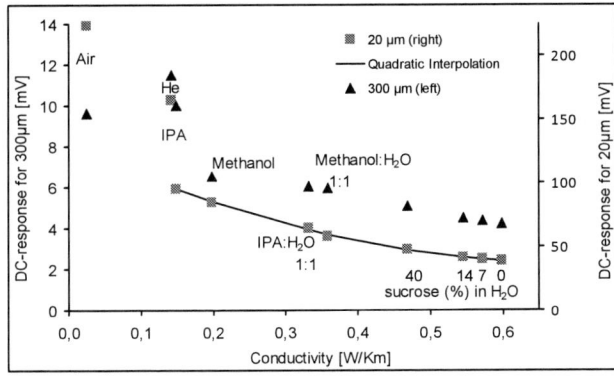

Figure 3: Output voltage of thermopile at 20 and 300 μm distance from the heater as function of thermal conductivity of various liquids. Input power of 3.8 mW.

But in contrast to enzymatic determinations, the measurement of the thermal conductivity does not have

any selectivity between sucrose and other substances.

In practice the stability of the sensor will also play a role in the resolution that can be obtained.

It is curious to see that the DC response for the gas helium is much higher than for the liquid IPA, although they have the same thermal conductivity. The cause for this is not yet known.

AC measurement results

When using a time-varying heating signal, other data can be distilled from the output signal. We performed measurements on water, methanol and IPA, using a sinusoidal heating signal of frequencies between 10 mHz and 1 Hz. In Figure 4 we show the AC signal amplitude and the phase shift as a function of heating voltage frequency for 300 μm passage through three different liquids.

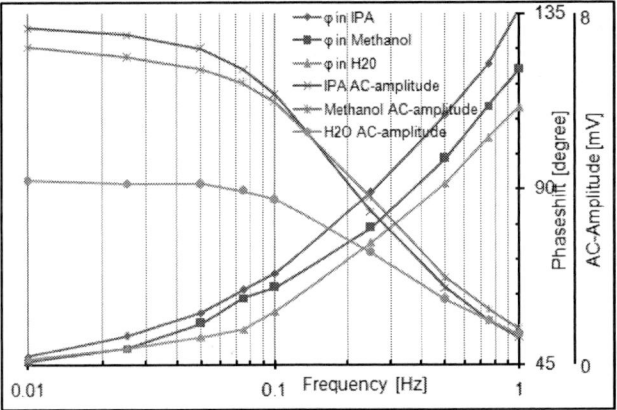

Figure 4: AC thermal measurements on liquids. Heating power $1mW_{p-p}$. Measured amplitudes in mV and phase shift in degrees after a passage through 300μm liquid.

For very low frequencies, the amplitude will approach the DC value, while the phase shift between input heating voltage and output voltage will approach the minimum value of 45 degrees.

In Figure 4 we can clearly see that the best resolution in amplitude is found for the smallest frequency, i.e., for DC values. The liquid acts as a low-pass filter that will attenuate the signal more as the frequency becomes higher. At 1 Hz, it is already very difficult to distinguish the water from the methanol or IPA, even though their thermal conductivity is very different.

More interesting is the behavior of the phase shift with frequency. The filtering action for thermal waves of the liquid is proportional to the thermal resistivity and the specific heat (per unit of volume) of the liquid, just as an RC time constant for transmission lines. The thermal diffusivity in m^2/sec is the parameter which combines these factors of thermal resistivity in Km/W, and specific heat in J/kgK and density in kg/m^3. As the dimensions show, the diffusivity is the inverse of the RC-time constant, expressing the capability of the liquid to quickly transport heat. And ten times further takes 100 times as long. Figure 5 shows the phase shift at 1 Hz heating signal for a distance between heater and thermopile of 300 μm and also at 20 μm, the thermopile directly next to the heater on the same membrane. The values are shown

978-1-4244-2977-6/09 $25.00 © 2009 IEEE   537

again as a function of the value of the diffusivity found in the literature.

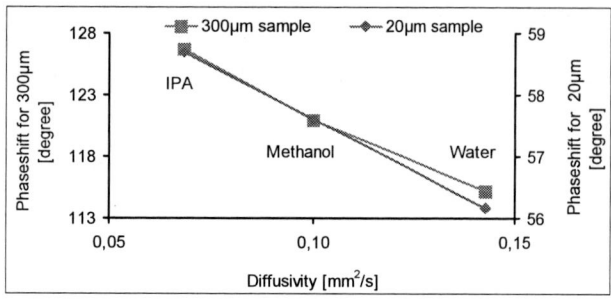

Figure 5: Phaseshift as a function of diffusivity for a 1 Hz heating signal.

Figure 5 shows that even for a thermopile very near the heater, distinguisable phase shifts can be observed. For a distance of 300 µm, the phase shift between water and IPA is approximately 13 degrees for a difference of 0.1 mm²/sec, three times the value at at 10-20 µm distance. On a total phase shift of about 80 degrees (when comparing to the 45 degrees minimum value), this is a 15% change, much less than for the amplitude, where a change of 100% is seen in the 300 µm distance measurement.

CONCLUSIONS

We have used liquid calorimeter chips to measure the thermal conductivity and thermal diffusivity of liquids. Detecting different liquids based on their thermal conductivity turns out to be rather easy to do, where liquids differing 1% in thermal conductivity can be distinguished. Because this is a DC measurement with constant heating and a convenient DC output signal, the method is rather easy to implement.

The thermal diffusivity of liquids can also be detected by liquid calorimeter chips, although the resolution for this parameter is less good. Moreover, the method requires AC heating and preferably lock-in amplification electronics, making this method more difficult to implement electronically. Nevertheless, liquids can also be distinguished on basis of their diffusivity.

OUTLOOK

Future work with these devices will focus on measurements at bio-chemically interesting liquids, where the good resolution will be exploited to the end. It is interesting to compare the measurements of almost identical liquids with this method, and also to compare with the measurement of the heating power of enzymatic reactions or living cells.

ACKNOWLEDGEMENTS

The Authors thank the European Commission for their financial support via the Marie Curie Research and Trainings Network project Cellcheck (www.cellcheck.sk).

They thank Jeroen Kool of the Vrije Universiteit of Amsterdam for the help with liquid measurement set up.

REFERENCES

[1] D. Paoletti et al, "A new method for measuring the diffusivity of liquid binary mixtures using DSPI", *Pure Appl. Opt.* vol. 2, pp. 489-498, 1993

[2] M. Marinelli et al, "High resolution simultaneous photothermal measurements of thermal parameters at a phase transition with the photopyroelectric technique", *J. Appl. Phys.*, vol. 72/3, 1992

[3] Y. Zhang and S. Tadigadapa, "Calorimetric biosensors with integrated microfluidic channels", *Biosensors & Bioelectronics*, vol. 19, pp.1733-1743, 2004

[4] Y. Zhang and S. Tadigadapa, "Thermal characterization of liquids and polymer films using a microcalorimeter", *Applied Physics Letters*, vol. 86, 034101, 2005

[5] A.W. van Herwaarden, "Overview of Calorimeter Chips for Various Applications", *Thermochimica Acta*, vol. 432, pp. 192-201, 2005

[6] J.M. Köhler and M. Zieren, "Micro Flow Calorimeter For Thermoelectrical Detection Of Heat Of Reaction In Small Volumes", *Fresenius J Anal Chem*, vol. 358, pp. 683-686, 1997

[7] M. Zieren and J.M. Köhler, "A micro-fluid channel calorimeter using BiSb/Sb thin film thermopiles", *Proc. Transducers '97"*, pp. 539-542, 1997

[8] Data sheet liquid nanocalorimeters, www.xensor.nl

[9] Y. Muramatsu, A. Tagawa and T. Kasai, "Thermal conductivity of several liquid foods", *Food Sci. Technol. Res.*, pp. 288-294, 2005

[10] M.J. Assael, E. Charitidou and W.A. Wakeham, "Absolute measurements of the thermal conductivity of mixtures of alcohols with water", *International Journal of Thermophysics,* vol. 10, pp. 793-803, 1989

[11] P. Bataillard, E. Steffgen, S. Haemmerli, A. Manz and H.M. Widmer, "An integrated silicon thermopile as biosensor for the thermal monitoring of glucose, urea and penicillin", *Biosensors & Bioelectronics*, vol. 8, pp. 89-98, 1993

978-1-4244-2977-6/09 $25.00 © 2009 IEEE

TIR-BASED DYNAMIC LIQUID-LEVEL AND FLOW-RATE SENSING AND ITS APPLICATION ON CENTRIFUGAL MICROFLUIDIC PLATFORMS

J. Hoffmann[1,‡], L. Riegger[2,‡], D. Mark[1], F. von Stetten[1], R. Zengerle[1,2], J. Ducrée[3]

[1] HSG-IMIT, Villingen-Schwenningen, GERMANY
[2] IMTEK - University of Freiburg, Freiburg, GERMANY
[3]BDI, Dublin City University, Dublin, IRELAND
‡ contributed equally

ABSTRACT

For the first time we present a technique for the spatio-temporally resolved localization of liquid-gas interfaces on centrifugal microfluidic platforms based on total internal reflection (TIR) at the channel wall. The simple setup consists of a line laser and a linear image sensor array mounted in a stationary instrument. Apart from identifying the presence of (usually unwanted) gas bubbles, the here described online meniscus detection allows to measure liquid volumes with a high precision of 1.9 %. Additionally, flow rates and viscosities (range: 1 – 10.7 mPa s) can be sensed even during rotation at frequencies up to 30 Hz with a precision of 4.7 % and 4.3 %, respectively.

INTRODUCTION

Microfluidic lab-on-a-chip systems make use of the rescaling of hydrodynamic force ratios in miniaturized fluidic networks. Towards the micron world, surface-to-volume ratios significantly increase, thus making surface mediated effects such as surface tension and viscous drag prevail over bulk forces such as inertia and gravity.

Lab-on-a-chip devices typically concatenate laboratory unit operations for liquid handling such as sample preparation, aliquoting, metering, and mixing on a single substrate [1,2] to perform bio-chemical assays in a process-integrated, automated, miniaturized and often parallel fashion. However, due to their strong dependency on often hard to control surface properties, lab-on-a-chip systems are intrinsically prone to inconsistencies such as the formation of gas bubbles or pockets during priming and operation. This tendency may lead to insufficient filling of microfluidic cavities with sample or reagents. In order to leverage a successful transfer of academic lab-on-a-chip technology to real-world applications, it is therefore key to introduce a quality control which monitors the course of liquid filling.

To this end, we recently presented a technique based on TIR on the channel wall to localize liquid-gas interfaces in microchannels at rest [3]. In this contribution, we present for the first time its implementation on a centrifugal microfluidic platform [4]. Our technique is based on linear illumination and detection elements allowing quasi-continuous meniscus detection with a spatial resolution of 50 μm at high speeds of rotation up to 30 Hz.

SETUP

The experimental setup consists of a rotatable lab-on-a-chip substrate (also referred to as lab-on-a-disk), and a device (called the "player") which accommodates the optical components and spins at well defined rotational frequencies and acceleration rates.

Disk Design and Fabrication

The disk contains four pairs of channels aligned at 90°-offset to each other according to the symmetry of the disk as illustrated in Figure 1.

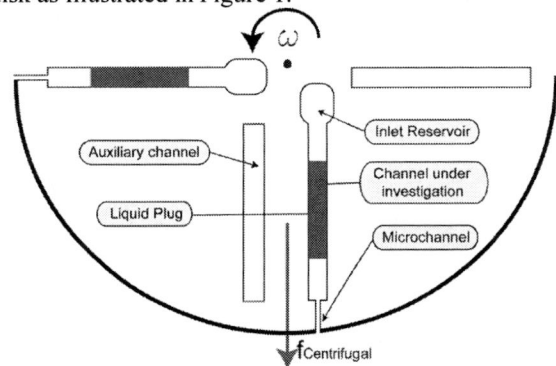

Figure 1: Schematic of the on-disk structure, showing all relevant channels and geometric features referred to in this section.

Each channels pair is laterally shifted with respect to a central radial line. The channel under investigation and the auxiliary channel possess a triangular cross-section, each possessing a wall inclination angle of 45° ("V-groove"). These channels are illuminated by a laser from the top, i.e. from a direction perpendicular to the plane of the disk. According to Snell's law TIR occurs for a perpendicular incidence in case the ratio of refractive indices of the bulk material of the disc and the medium inside the channel exceeds 1.41 at the location of the inclined channel wall. This condition is fulfilled for a gas filled channel segment, but not in the presence of a liquid.

Said channels have a depth of 1.5 mm and a width of 3 mm. The four inner endings of the channels near the center of rotation are flanked by an inlet reservoir. The radially outer side of the channel leads to a rectangular channel measuring 400 μm in width, 100 μm in depth, and ~ 4 mm in length, referred to as the "microchannel". The other four V-grooves are auxiliary channels which are permanently filled with air.

The 2-D fluidic design of the disk is generated by a CAD software. It provides the code for a micromilling machine which creates the corresponding 2.5-D design within a cyclo olefin colpolymer (COC) substrate (refracttive index $n = 1.49$). The total time from design to the fabricated disk takes approximately 4 hours (depending on the complexity of the fluidic design).

Optical and Read-out Components

The optical measurement components are a line laser and a linear image sensor array. The laser diode emits at the peak-wavelength $\lambda_{peak} = 650$ nm having a maximum optical power of 24 mW (Roithner RLLH 650-24-3). The housing contains this diode as well as a special line optics generating the probe line. The used CMOS linear image sensor (Hamamatsu S8377) is a compact sensor featuring a built-in timing generator as well as signal processing circuit. The active area has a length of 25.6 mm subdivided into 512 individual pixels (pixel pitch: 50 µm, pixel height: 500 µm).

An electrical controlling unit (Spectronics Devices) is used to address the operating mode of the sensor as well as to transfer optical data to a PC. The set comprises two small size printed circuit boards (PCB): a small 50 mm x 20 mm PCB holding the sensor incorporating a pre-amplifier and A/D converter electronics and a second PCB with the control electronics, a 1-MByte data storage memory for saving 1000 scans and a USB PC-interface. The processing time for one scan takes 2 µs, only, which allows the here required high-speed data acquisition.

OPERATING PRINCIPLE

Figure 1-A illustrates the functional principle for the TIR-based meniscus detection while the experimental setup is depicted in Figure 1-B:

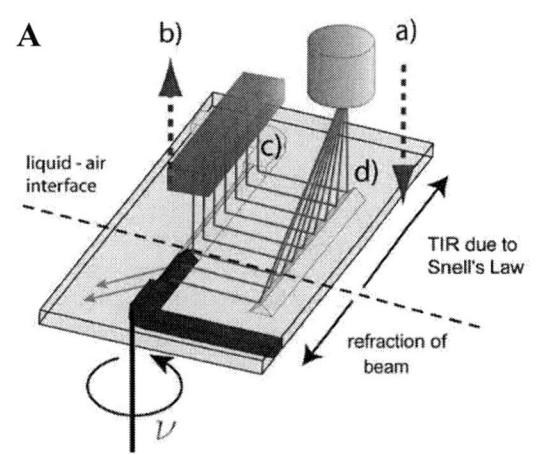

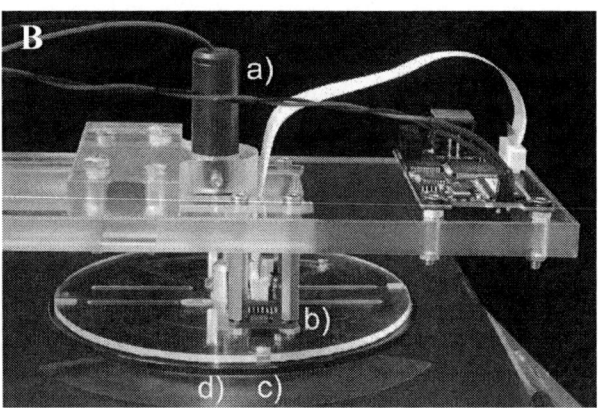

Figure 2: (A) Setup of TIR-based meniscus detection. (B) Photo of the experimental setup. The beam is issued from a line laser onto the rotating substrate and reflected back to a linear image sensor. Legend: a) line laser b) linear image sensor c) channel under investigation d) auxiliary channel.

Optical Pathway

The beam emitted by the line laser (a) impinges on an auxiliary, permanently air-filled channel (d). Due to the 45° inclination of its respective wall, the beam is deflected by 90° into the plane of the substrate. After a defined distance corresponding to the spacing between the laser and the detector, the beam impinges on the V-grooved segment of the channel under investigation (c). If a probed segment is filled with air, a second TIR redirects the laser beam by 90° towards the linear image sensor (b) placed perpendicular to the plane of the disk. The segments of the channel which are filled by liquid change the local refractive index and rule out TIR.

The emitting power of the line laser is set to a value (5 mW), thus always driving those pixels of the sensor receiving high intensity into saturation. This way a discrete, binary signal is obtained where 1 indicates an empty section, 0 a liquid-filled section of the channel (c). This works up to frequencies of 30 Hz. At higher frequencies, the received intensity is too low to push the sensor into saturation.

Measurements under Rotation

The azimuthal position of the channels on the disk and the sensor position are synchronized by using the TTL-signal provided by the player once per revolution as a command-signal for the sensor. The linear image sensor starts to acquire optical data each time the rising flank of the trigger signal occurs and stops after a certain integration time derived from the frequency of rotation. The integration time is set to a value that each revolution a disk-sector spanning over channels (c) and (d) is covered. Thereby, a consistent and stable measurement system is guaranteed requiring no alignment or signal strength calibration.

EXPERIMENTS
Bubble Detection

A representative plot for the pixel intensity along a channel entrapping an air bubble under static conditions is shown in Figure 2, revealing clearly discernible edges at the respective liquid-gas interfaces. Using proper curve fitting algorithms, the position of the meniscus can be pinpointed down to roughly 3 pixels, corresponding to 150 µm.

978-1-4244-2977-6/09 $25.00 © 2009 IEEE

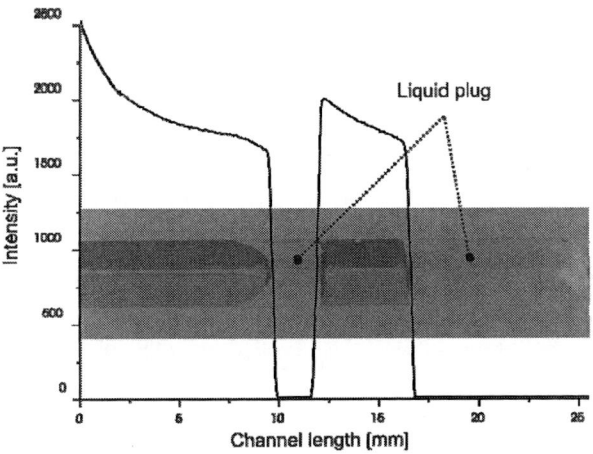

Figure 3: Measured pixel intensity plot of a bubble within a channel captured by the linear image sensor at rest. The transition between the liquid and the gas state is very distinct thus allowing a clear localization of the meniscus.

Volume Calibration

The detection of gas bubbles and liquid filling levels are essential to assure reproducible and quantitative bioassays on LoaC systems. This section describes the calibration of our TIR-based measurement system to meet these requirements. Figure 3 displays a calibration curve with five pre-metered volumes, featuring a high accuracy and a precision of 1.9 %.

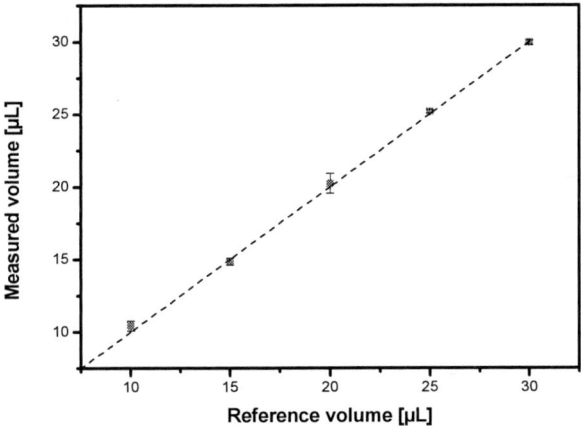

Figure 4: Calibration of the measured volumes with pre-metered volumes featuring a high precision of 1.9 %.

For a fixed microchannel geometry, the theoretical maximum filling level is calculated with respect to the length of the channel. In the next step, the active area of the linear image sensor is laterally shifted (positioned) in such a way that it provides output characteristics corresponding to the calculated liquid level. Once the system is calibrated, the measurements described in the following sections are performed. For the volume calibration, liquid-levels of five different volumes (10, 15, 20, 25 and 30 µL) are measured five times each. Each measurement is conducted under rotation at 10 Hz whereby the position of the liquid-gas interface is averaged over 10 revolutions with one measurement per revolution. Figure 3 shows the capability of this system to accurately measure and meter liquid volumes during rotation.

Flow Rates and Viscosities

Centrifugally driven discharge induced by a constant rotational frequency of 10 Hz is presented in Fig. 4. Tests are based on four different glycerol solutions with dynamic viscosities in the range of 1 – 10.7 mPa s.

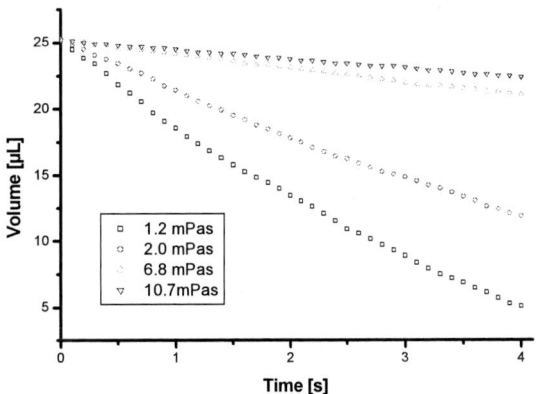

Figure 5: Flow rate and viscosity measurements derived from a time-resolved liquid-level determination with the calibrated system. The relative decrease of the flow rate correlates with increasing viscosity.

25 µL of each sample liquid are pipetted into the inlet reservoir. Fluids are driven through the inclined channel by centrifugal forces and introduced into the adjoining microchannel. The flow rate in the microchannel itself is indirectly measured via the movement of the liquid-gas interface in the channel under investigation using a curve-fitting algorithm.

Four different liquids with viscosities of 1.2, 2, 6.8, and 10.7 mPa s are investigated. The time resolved measurements for each moving liquid reveal the $1/\eta$ dependence for the flow rate during the outflow. Also, with the dynamically decreasing plug position, a multi-parameter dependent flow rate can be identified. The motion of the meniscus is affected by the spinning frequency, interfacial interactions, microchannel geometries, and the viscosity of the fluid.

Contact-free and on-line determination of the flow rate provides an insight into liquid guidance on lab-on-a-disk systems. Furthermore, the implementation of a feedback-loop can improve the liquid guidance on-disk and contribute to a better controllable fluidic system. Figure 5 demonstrates the capability of the system to measure the dynamic viscosity of low-viscosity liquids. The unprecedented dynamic recordings of the flow rates and viscosities during rotation are enabled by the linear optical elements in contrast to the static scanning principle of our previous work [3].

978-1-4244-2977-6/09 $25.00 © 2009 IEEE

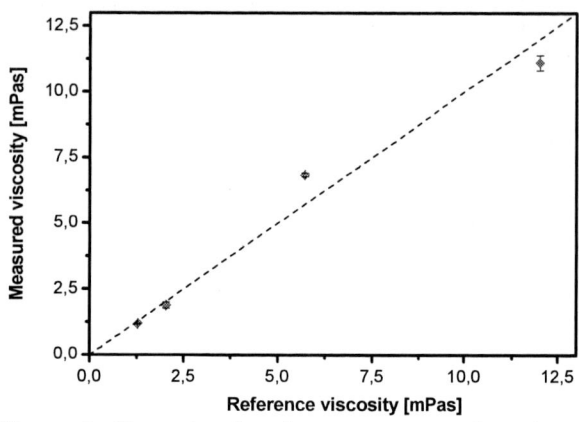

Figure 6: Dynamic viscosity measurement featuring a precision of 4.3 % for viscosities up to 10.7 mPa s.

The previously described glycerol solutions are used to highlight the correlation between the measured viscosities and the reference viscosities. First, defined volumes (25 µL) are introduced into the channel under investigation and centrifugally pumped at a constant frequency of $v = 10$ Hz through the connected microchannel. Each flow rate is measured five times.

Secondly, the viscosity is calculated taking the following physical parameters into account: flow rate and position of the passing liquid plug, temperature, angular frequency and the dimensions of the microchannel. A precision of 4.3 % is a good result compared to rheology measuring devices struggling with difficulties in the low viscosity regime. In addition to flow rates of unknown samples, our system reveals absolute viscosity data as a surplus.

CONCLUSION

This work investigates a novel TIR-based setup with linear optical elements for spatio-temporally resolved meniscus detection. The system is well suited for compact system integration and excels with its high-speed data acquisition and low cost compared to conventional image processing systems.

Due to the simple integration of the auxiliary channel structures in the microfabrication process of the substrate as well as the uncomplicated component upgrade of the disk-processing unit, the opto-fluidic technology can be easily incorporated into centrifugal microfluidic systems.

The system can be used for quality control, feedback loop and calibration purposes on centrifugal lab-on-a-disk platforms, e.g. for the on-line verification or liquid filling or outflow, for monitoring the flow of a liquid plug, for reagent metering or viscosity measurements of unknown patient samples. The contactless measuring principle does not require any pretreatment of the fluid, e.g. the introduction of particles, and prevents any impact on the fluidic function itself.

Implementation of our measurement system into real-world point-of-care devices could significantly improve the quality of assay results by the real-time monitoring of liquid levels, flow rates, viscosities and gas bubble formation in liquid handling operations.

REFERENCES

[1] P. S. Dittrich and A. Manz, "Lab-on-a-chip: microfluidics in drug discovery," *Nature Reviews Drug Discovery*, vol. 5, no. 3, pp. 210-218, Mar.2006.

[2] J. Y. Park and L. J. Kricka, "Prospects for nano- and microtechnologies in clinical point-of-care testing," *Lab Chip*, vol. 7, no. 5, pp. 547-549, 2007.

[3] F. Bundgaard, O. Geschke, R. Zengerle, and J. Ducrée, "A simple opto-fluidic switch detecting liquid filling in polymer-based microfluidic systems," in *Proceedings of the 14th International Conference on Solid-State Sensors, Actuators and Microsystems (Transducers & Eurosensors '07)* Lyon, France: 2007, pp. 759-762.

[4] J. Ducrée, S. Haeberle, S. Lutz, S. Pausch, F. von Stetten, and R. Zengerle, "The centrifugal microfluidic Bio-Disk platform," *J. Micromech. Microeng.*, vol. 17, p. S103-S115, 2007.

100 NANOMETER SCALE RESISTIVE HEATER-THERMOMETER ON A SILICON CANTILEVER

Zhenting Dai[1], Keunhan Park[2], and William P. King[1]

[1]Deptartment of Mechanical Science and Engineering, University of Illinois, Urbana, IL, USA
[2]Department of Mechanical Engineering and Applied Mechanics, University of Rhode Island,
Kingston, RI, USA

ABSTRACT

This paper reports a method to fabricate a 100 nm scale heater-thermometer into a silicon microcantilever based on contact photolithography and a controlled annealing process. The heater is formed during a photolithography process that can achieve a minimum feature size of about 1 μm, while careful control of doping and annealing parameters allows the heater size to be further decreased, to a width of 100 nm. The heater is fabricated onto the free end of a silicon cantilever suitable for scanning probe microscopy, and can be integrated into cantilevers with or without sharp tips. The fabricated heater has a maximum temperature of over 700 °C, and a heating time of 56 μsec to reach 500 °C.

INTRODUCTION

The ability to locally control the temperature of a nanometer scale hotspot on a surface with an atomic force microscope (AFM) tip has led to advances in data storage, nano-manufacturing, and materials characterization. For most of these applications, the heater is several micrometers on a side, while heat flows along the length of a tip to the substrate. The further miniaturization of the heating element would allow more local control of temperature and improved measurement of heat flow.

A few published articles have reported miniaturization of heaters or thermometers to the 100 nm scale. A metal-metal junction can be fabricated onto a scanning probe tip to form a scanning thermocouple element [1] and the junction can also be used as a nanometer scale heater element. One study used electron beam lithography to enable a resistive heater in silicon that had dimensions near 100 nm [2]. The nanoheater was capable of reaching temperatures exceeding 500 °C. While offering an ultrasmall heater, the electron beam lithography step adds cost and complexity. A few articles describe the development of an AFM probe with a nanometer-scale resistive element fabricated from doped silicon [3]. This fabrication process used photolithography to define a region of low-doped high resistivity silicon on a cantilever tip, which is connected to regions of higher doped low resistivity silicon on either side of the tip. The size of the highly resistive region was made smaller via lateral dopant diffusion during annealing. While this probe has some similarities with heated cantilevers for data storage [2], the probe was designed and used for imaging electrical fields at high spatial resolution and not for heating or thermometry.

This paper describes a simple batch fabrication process to form a 100 nm scale heater-thermometer on a silicon cantilever, both with a tip and in the absence of a tip. The heater is first fabricated onto a cantilever having no tip, and then integrated into a sharp tip. The electrical and thermal characteristics of the heater are thoroughly examined.

DESIGN AND FABRICATION

To form a heater-thermometer at the end of a silicon cantilever, the photolithographically defined high doping regions are first implanted with boron, and then high temperature annealing causes the dopants to diffuse in a way that shrinks the heater region to about 100 nm. Figure 1 shows the concept of this process. When the silicon is annealed at high temperature, the dopants diffuse within the silicon, narrowing the gap between the two high doping areas. Careful control of the annealing conditions allows the major carrier at the middle point between two high doping areas changes from N type to P type and the dopant concentration reaches 10^{17} atoms/cm^3.

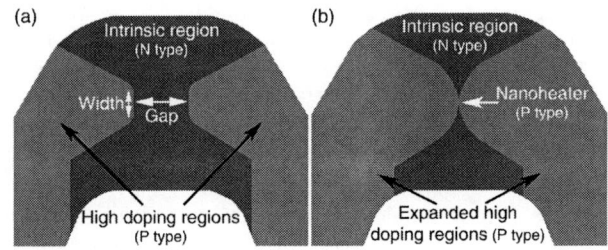

Figure 1: Concept illustration of how the anneal process changes the doping profile around nanoheater region. (a) Doping distribution before anneal process, (b) Doping distribution after anneal process. Careful control of the annealing conditions allows the doping concentration at nanoheater region to be about 10^{17} atoms/cm^3 with a gap width of 100 nm.

To understand how the annealing process affects the surface doping distribution, simulations were run in the process modeling tool TCAD 8.0 to explore the processing space and to optimize the annealing parameters. The dopant diffusion is a function of time, temperature, and dopant concentration. The initial dopant concentration is in turn a function of the lithographically defined doping mask as well as the implant parameters. The key challenge is to achieve a narrow region of high resistivity that is connected to regions of much lower resistivity, so that when current flows through the circuit the resistive heating will be highly localized. The target doping concentration for the heater is in the range 10^{17}-10^{18} atoms/cm^3, while the target concentration for the current-carrying legs is 10^{20} atoms/cm^3 or higher.

978-1-4244-2977-6/09 $25.00 © 2009 IEEE

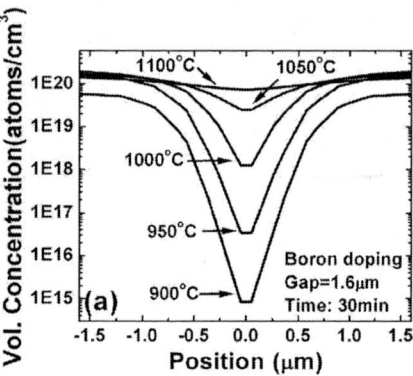

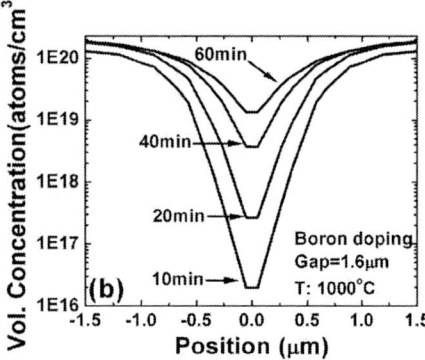

Figure 2: Simulation results of how anneal temperature and time affect the doping profile in the nanoheater area. (a) Doping profile of devices annealed at different temperature, (b) Doping profile of devices annealed at different times.

Figure 2 shows simulation results for various conditions. Many simulations were performed leading us to an optimal initial gap distance of 1.6 μm. The results of Figure 2 summarize the conditions around this parameter. The second assumption is that the silicon bordering the gap is doped to 10^{20} atoms/cm^3. Figure 2(a) shows predictions for the lateral doping concentration as a function of various anneal temperatures for a gap distance of 1.6 μm and anneal time at 30 minutes. Since the ideal doped level of the heater should fall in the 10^{17}-10^{18} atoms/cm^3 range, the optimized anneal temperature should be in the range 950-1000 °C. Figure 2(b) shows predictions for lateral dopant distribution for different anneal times when the anneal temperature is fixed at 1000 °C and the initial gap is 1.6 μm. In addition to the simulation results, a few practical limitations arise as well. The maximum operating temperature of our anneal tube is 1100 °C. The minimum feature that can be produced using our mask aligner is about 1 μm. Accounting for the simulation results and our experimental limitations, the optimized conditions are selected as 1000 °C anneal temperature, 20 minutes anneal time, and 1.6 μm initial gap.

The basic fabrication process is very similar to the fabrication process for silicon AFM cantilevers having integrated heaters [4], with the exception that the initial goal is to fabricate a tipless cantilever with a 100 nm heater-thermometer. A scanning electron microscope (SEM) image of final released device is shown in Figure 3. The nanoheater is located at the end of cantilever.

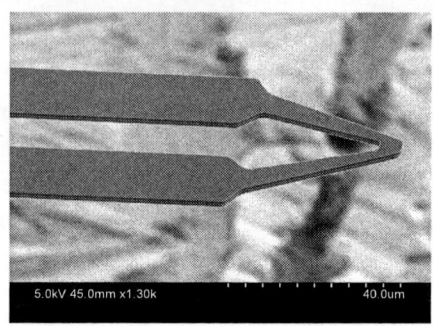

Figure 3: Scanning electron microscopy image of a tipless nanoheater cantilever. The heater is located at the free end of cantilever and the cantilever is 2 μm thick.

NANOHEATER CHARACTERIZATION

To verify the simulation result, the doping profile around the nanoheater region was measured using scanning capacitance microscopy (SCM). Figure 4 shows the SCM measurements of nanoheaters for different anneal times. The devices were annealed at 1000 °C and had an initial gap of 1.6 μm. From Figure 4(a) to 4(d), the devices were annealed at 0 minutes, 20 minutes, 40 minutes and 60 minutes, respectively. In these images, darker color corresponds to higher doping level. The gap between the two highly doped areas clearly decreases as the anneal time increases. When the anneal time was 20 minutes, the gap is only 100 nm and the width of the heater region is only 400 nm. If the anneal time is further increased, the two high doping areas start to merge. The doping level of the heater region increases and the width of the heater also increases. When the device was annealed at 60 minutes, the two highly doped areas merge, which is undesirable. Overall, the experimental results agree well with the simulations shown in Figure 2.

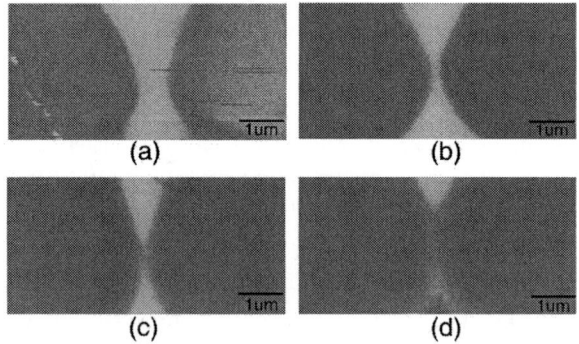

Figure 4: Surface doping level measurement by scanning capacitance microscopy (SCM) around the nanoheater region of tipless nanoheater cantilever. (a) Without annealing, (b) Anneal 20 min, (c) Anneal 40 min, (d) Anneal 60 min. All at 1000 °C.

The electrical characteristics of a heated silicon cantilever can offer insights into its operation [5]. The most important electrical measurement for a heated cantilever is the cantilever electrical resistance as a function of heating power, which is highly nonlinear [6]. Figure 5 shows this measurement for the nanoheater cantilevers annealed at different times, before and after release from the wafer. Measurements for the unreleased

978-1-4244-2977-6/09 $25.00 © 2009 IEEE

devices are shifted to higher power than for the released devices, since the thermal conductance into the wafer allows the heater to dissipate higher power than the cantilevers freely suspended in air. In general, the shape of the curve is what is expected for a doped silicon heater-thermometer. The electrical resistance slowly increases with heating power, as mobility decreases in the doped silicon. This increase is followed by a maximum and then drop in resistance with increased power, as thermally-generated intrinsic carriers outnumber the background carriers. The heater resistance decreases with increasing anneal time, which can be attributed to a decrease in the gap length of the nanoheater. The resistance measurements of devices annealed at 20 minutes and 40 minutes have a clear peak, which is similar to the standard heated cantilever having a micrometer-scale heater. But for the device annealed at 60 minutes, the resistance curve is nearly flat, which can be attributed to interdiffusion of the two highly doped areas on either side of the gap. In this case, the doping level at the targeted region is much higher than the optimal level for heater application.

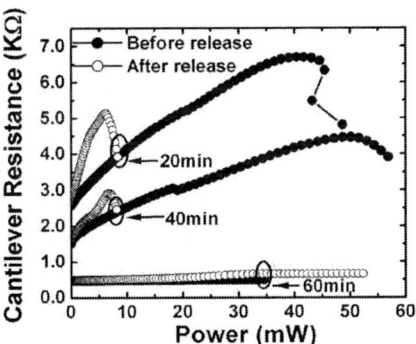

Figure 5: Resistance-power measurement of nanoheater cantilevers with different anneal times before and after release.

Raman spectroscopy can be used to determine temperature distribution on cantilevers with micrometer resolution [5]. A Renishaw InVia Raman Microscope with 180° backscattering geometry was used to measure the inelastic scattering from the heater. Based on the shift of Stokes peak, the average temperature around the nanoheater could be calculated. The nanoheater can reach temperatures above 700 °C based on the measurement results of Raman spectroscopy.

NANOHEATER INTEGRATION WITH A SHARP TIP

Having achieved a 100 nm scale resistive heater-thermometer in a silicon cantilever, the next challenge was to integrate such a heater onto a tip at the free end of a silicon cantilever. The fabrication process starts with making sharp tips on a silicon on insulator (SOI) wafer. The tips are made with a standard oxide sharpening process. After the sharp tips are made, the cantilever structures are defined by photolithography followed by dry etch in an inductively-coupled-plasma (ICP) system. The remainder of the fabrication process is similar to the fabrication process for silicon AFM cantilevers having integrated heaters [4]. To achieve best alignment of the

heater with the tip, a 20X lens is used in this photolithography step, and the alignment precision is around 500 nm. Since the base diameter (2.7 µm) of tips is much larger than alignment precision (500 nm), most of the heaters are located somewhere in the slope of the tips if not the center of tips. Figure 6 shows a SEM image of a released device, which clearly show the sharp tip and the edges of the implant mask.

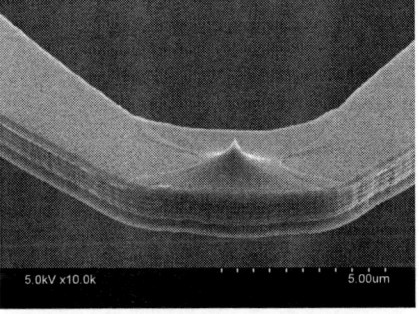

Figure 6: Scanning electron microscope image of a released nanoheater cantilever with a sharp tip. The cantilever is 1.5 µm thick.

The steady-state thermal behavior of the nanoheater cantilever having a tip is the same as Figure 5. In order to investigate the dynamic thermal behavior of the nanoheater cantilever, a pulse heating measurement and 3ω harmonic measurement were performed. In the pulse heating measurement, a voltage pulse was applied to the cantilever circuit with a 1 kΩ sense resistor in series. Upon the voltage pulse, the heater temperature increases to steady state under the sufficient pulse duration. After the pulse has elapsed, the cantilever temperature then drops to an initial value. Thus, the measurement of the cantilever resistance during a pulse heating reveals the transient thermal response of the cantilever. Figure 7(a) shows the transient change of the cantilever resistance. Overall transient features are quite similar to those of standard microheater cantilevers [5]. The peaks in the cantilever resistance curves begin to appear at high input voltages because these voltages heat the cantilever beyond its maximum resistance shown in Figure 5. The peak position shifts to a short duration time as the input voltage increases, indicating that the cantilever can be heated much faster with a higher voltage pulse. The heating time required to heat the cantilever to its maximum resistance (or ~ 500 °C) exponentially decreases and asymptotically approaches to 23 µsec. This value is slightly larger than that of a typical microheater cantilever, which is around 10-20 µsec, mainly due to the thickness of the nanoheater cantilever which was somewhat thicker than typical microheater cantilevers.

Figure 7(b) shows the out-of-phase 3ω voltage signal of the nanoheater cantilever as a function of the operation frequency. When the cantilever is driven by a sinusoidal current source at frequency ω, 3ω harmonics appear in the voltage signal due to the temperature oscillation at 2ω [7]. Since the 3ω harmonic signal is the consequence of the heat diffusion inside the cantilever, the 3ω signal provides the thermal diffusion time of the cantilever. The 3ω signal was measured under a differential scheme [8] where the input voltage of 10 mV-rms was applied to the circuit having sense resistors of 5 kΩ each, providing a 2 μA sinusoidal current to the cantilever. The imaginary 3ω signal of the cantilever has two peaks at different frequencies. The 3ω signal of the cantilever can be simply modeled as $V_{3\omega} = V_1/(1 + if\tau_1) + V_2/(1 + if\tau_2)$, where V is the asymptotic voltage at low frequency, f is the operational frequency, and τ is the time constant. Since the 3ω signal is the consequence of the thermal diffusion, τ represents the thermal diffusion time. By comparison of the measurement with the calculation, the two time constants are estimated to be 89 μsec and 3.7 msec, respectively. We speculate that the long time constant is required for heat to be transferred through the leg region while a short diffusion time is due to the thermal diffusion to the region around the nanoheater.

of 1.6 μm. The resulting structure was about 100 nm by 400 nm, visualized using SCM. The electrical characteristics closely matched those expected for a cantilever heater, and high temperature heating was verified using Raman spectroscopy. The heater was also integrated with a sharp tip, suitable for scanning probe microscopy.

REFERENCES

[1] L. Shi, O. Kwon, A. C. Miner, A. Majumdar, "Design and Batch Fabrication of Probes for Sub-100 nm Scanning Thermal Microscopy," *Journal of Microelectromechanical Systems,* vol. 10, pp. 370-378, 2001.

[2] U. Drechsler, N. Burer, M. Despont, U. Durig, B. Gotsmann, F. Robin, P. Vettiger, "Cantilevers with Nano-heaters for Thermomechanical Storage Application," *Microelectronic Engineering,* vol. 67-68, pp. 397-404, 2003.

[3] H. Park, J. Jung, D. K. Min, S. Kim, S. Hong, H. Shin, "Scanning Resistive Probe Microscopy: Imaging Ferroelectric Domains," *Applied Physics Letters,* vol. 84, pp. 1734-1736, 2004.

[4] B. W. Chui, T. D. Stowe, Y. S. Ju, K. E. Goodson, T. W. Kenny, H. J. Mamin, B. D. Terris, R. P. Ried, D. Ruger, "Low-Stiffness Silicon Cantilever with Integrated Heaters and Piezoresistive Sensors for High-Density Data Storage," *Journal of Microelectromechanical Systems,* vol. 7, pp. 69-78, 1998.

[5] J. Lee, T. Beechem, T. L. Wright, B. A. Nelson, S. Graham, W. P. King, "Electrical, Thermal, and Mechanical Characterization of Silicon Microcantilever Heaters," *Journal of Microelectromechanical Systems,* vol. 15, pp. 1644-1655, 2006.

[6] B. W. Chui, M. Asheghi, Y. S. Ju, K. E. Goodson, T. W. Kenny, H. J. Mamin, "Intrinsic-Carrier Thermal Runaway in Silicon Microcantilevers," *Microscale Thermophysical Engineering,* vol. 3, pp. 217-228, 1999.

[7] K. Park, J. Lee, Z. M. Zhang, W. P. King, "Frequency-Dependant Electrical and Thermal Response of Heated Atomic Force Microscope Cantilevers," *Journal of Microelectromechanical Systems,* vol. 16, pp. 213-222, 2007.

[8] D. G. Cahill, "Thermal Conductivity Measurement from 30 to 750 K: the 3ω Method," *Review of Scientific Instruments,* vol. 61, pp. 802-808, 1990.

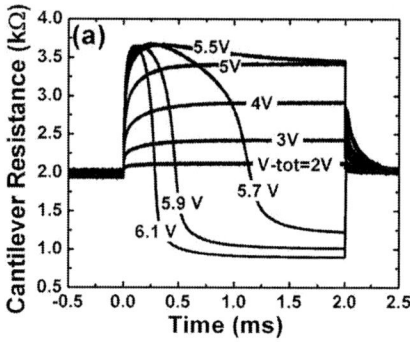

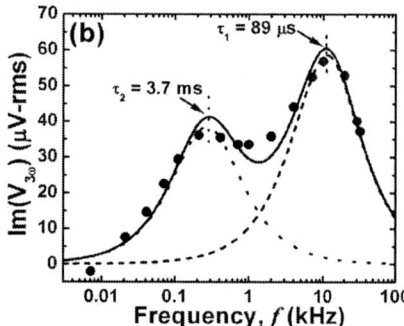

Figure 7: (a) Cantilever resistance change during pulse voltage input for 2 ms, (b) Out-of-phase 3ω voltage signals of the cantilever as a function of frequency.

CONCLUSION

This paper reports an approach to fabricate a 100 nm scale heater-thermometer using contact photolithography and controlled anneal conditions. A simulation first determined a window for optimal process parameters, which were 20 minutes at 1000 °C for an initial gap width

MEMS DIFFRACTIVE OPTICAL NANORULER TECHNOLOGY FOR TIP-BASED NANOFABRICATION AND METROLOGY

Norimasa Yoshimizu, Amit Lal, and Clifford R. Pollock
*Sonic*MEMS Laboratory, School of Electrical and Computer Engineering
Cornell University Ithaca NY 14853

ABSTRACT

This paper reports on a diffractive optical nanoruler used to guide tip-based nanofabrication. A precision of $\pm 3 \times 10^{-4}$ has been demonstrated across a 75 mm wafer; we can account for errors external to the system which reduce this figure to $\pm 1.5 \times 10^{-5}$. A microfabricated aluminum grating diffracts an external cavity laser beam stabilized to the rubidium D2 line (780nm). The resulting hexagonal lattice intensity pattern guides a PC-board assembly, consisting of a quadrature photodiode and an STM tip, on a flexural piezo stage. The STM tip is used to make indentations in resist spun on an Al film deposited on a silicon wafer. The precision is measured using electron microscopy by locating the indentations.

INTRODUCTION

There are a number of serial tip-based nanofabrication methods to exploit controlled solid-state properties only found at the nanoscale. The tip-based fabrication techniques include dip-pen nanolithography, SPM thermal surface modification, and seeded nanowire growth among other techniques [1,2]. However, their inclusion in large-scale commercial manufacturing will be practical only with enabling technologies which increase their throughput while preserving precision and repeatability over long stage travel. For example, 10 nm tip placement error over a 200 mm wafer is desired, corresponding to a 10^{-8} precision requirement.

Table 1: Comparison of stage metrology technologies and possibility for the optical ruler system.

	OTW	Optical encoder	Capacitive sensors
Travel	wafer	wafer	sub-mm
Frequency limit	high	low	medium
Short travel precision	sub-nm	10-100 nm	sub-nm
Long travel precision	nm	10-100 nm	N/A
Short travel accuracy	nm	low	nm
Long travel accuracy	nm	low	N/A

Currently, stage technology [3] for wafer-scale nanofabrication faces two challenges: stage motion position control and nanometer-scale resolution in tip motion, since the relative motion of the tip and stage results in errors in nanostructure location. Very high accuracy using feedback control on a high resolution flexural piezoelectric stage can be achieved using capacitive sensors, strain gauges, or compensation

modeling [4]. However, the motion of these stages is limited in travel to less than a millimeter due to sensor dynamic range and nonlinearities which overwhelm the control algorithms. On the other hand, long travel actuators such as stepper motors have poor resolution limited by the accuracy of the optical encoders and stick-slip errors. In addition, position control in wafer-scale stages with stepper motors are control bandwidth limited to hundreds of Hertz due to the need to count the optical encoder signals and motion time constants due to the large mass for a given actuator. Even smaller piezo stages are frequency limited below kHz due to mechanical instabilities.

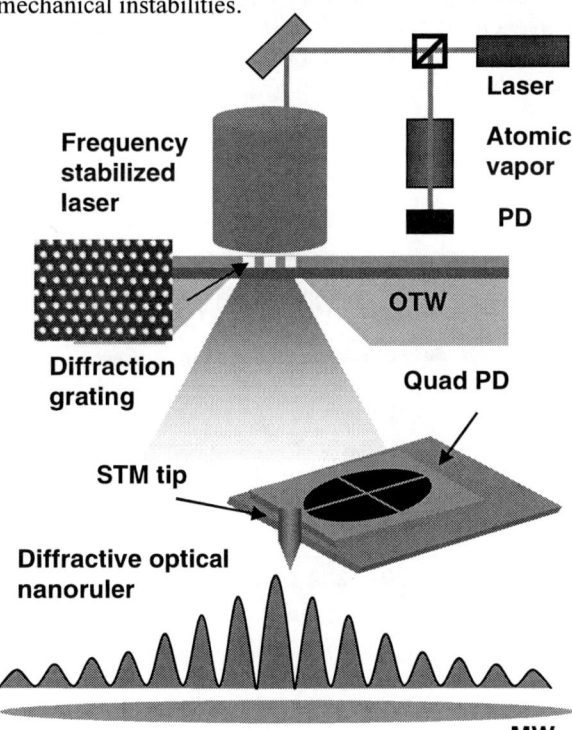

Figure 1: Schematic of the microfabricated diffractive optical nanoruler technology. A frequency stabilized laser illuminates a thin film pattern that diffracts the beam. The diffraction optical ruler is detected by a photosensitive device, here shown as a quadrature photodetector. A nanofabricating tip is rigidly attached to the assembly.

Our system circumvents the resolution and bandwidth problems by generating a diffraction optical grid over the manufacturing wafer (MW) to guide the fabrication process. The diffraction pattern, created by an atomically stabilized laser, sets up a three-dimensional ruler in the optical field to guide the manufacturing process. In this setup the manufacturing wafer does not move, eliminating the errors associated with stage motion. Because the metrology has been physically decoupled from the moving objects (the manufacturing tip), we can move a much smaller mass in the manufacturing tip and gain an advantage in the control bandwidth of the fabrication. As the diffraction pattern is, in general, anisotropic in all three dimensions, the tip placement can be potentially calculated independent of the path it takes, resulting in faster fabrication. We summarize the advantages of the method in Table 1.

PRINCIPLE OF OPERATION

Figure 1 shows a schematic of the MEMS diffractive optical nanoruler system. A laser is locked to an atomic resonance frequency. The laser illuminates the optical transfer wafer (OTW). The OTW has a reflective thin-film pattern which diffracts the incoming laser beam. The path length of the laser beam to the grating is sufficiently long so that it has diffracted slightly during its propagation and generously illuminates the whole area of the diffractive thin film.

The diffractive intensity pattern that is formed acts as an optical ruler to guide the nanofabrication tip. The field is translationally anisotropic in all three directions away from the diffractive thin film, and can have nearly 100% intensity variation across the ruler (see Figure 2). A photosensitive device measures the optical field and guides itself to the fabrication point on the MW.

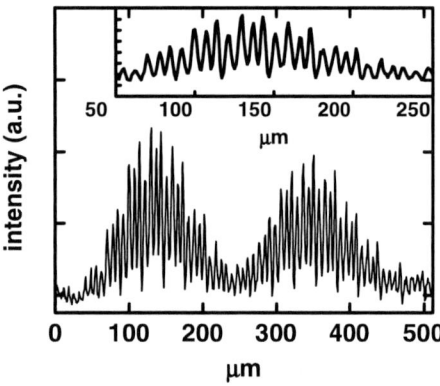

Figure 2: *Cross section of example optical diffraction pattern that serves as the ruler. Only a small portion of the total field is shown, and inset shows a closeup. Note nearly 100% variation in intensity.*

In addition to guiding the nanofabricating tip, the optical ruler provides a simple means of aligning the MW to the OTW as well (see Fig. 3). First, the angle between the MW and OTW can be fixed. The diffraction pattern expands in the xy-plane as it propagates along z toward the MW. When the MW is at some angle with respect to the OTW, the reflection of the diffraction pattern back to the MW will be compressed on one side (scaled smaller) and expanded on the other side (scaled larger). That is, assuming the diffraction pattern is symmetric across the alignment direction, the two halves should be symmetric if the wafers are parallel to each correction. The distance from the OTW to the MW can be fixed by the scale of the reflected diffraction image: for example, by aligning certain features to four points on the boundary of the

OTW.

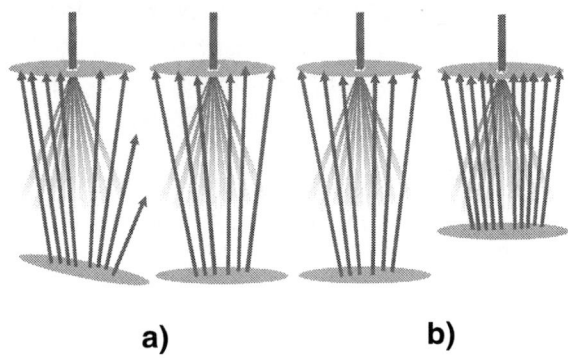

a)   **b)**

Figure 3: *Alignment and separation scheme between MW and OTW. a), the angular alignment scheme is achieved by symmetry of reflection of the diffractive optical ruler from the MW back to the OTW. b), the separation fixing scheme is achieved by the overall scaling of the image. This can be done by PN-junction photodiodes fabricated into the OTW, or by fabricating through-wafer holes in the bulk Si of the OTW that can be detected by a separate photodetector.*

EXPERIMENTAL RESULTS

In order to demonstrate the precision and function of the system, indentations were made on a MW. An external cavity single mode diode laser (Toptica DL100, <100 kHz linewidth, 3 mW at the OTW) is locked to the rubidium D2 line (780 nm). The frequency stabilized laser serves as the foundation of the precision and accuracy of the optical ruler. With such a small linewidth, such a laser can lower the initial wavelength precision to 10^{-11} and very easily start at 10^{-8}. The beam propagates approximately 2 meters until it reaches the OTW, and is chopped at 1.25 kHz. The laser beam, OTW, and MW are aligned using the scheme described in the previous section and by the direct reflection of the laser beam back on itself. The OTW can be aligned as well, as the reflection of the laser beam itself generates the optical ruler diffraction pattern due to its reflection from the thin film (i.e., Babinet's principle). The distance between the OTW and MW are fixed using the scheme above as well, in addition to measuring the gap directly.

The OTW is fabricated by depositing LPCVD low stress nitride on a bare Si wafer, followed by 400 nm of e-beam evaporated Al. The Al is patterned in a chlorine RIE. After patterning a window (1×1 mm^2) into the backside nitride, a backside KOH through-etch exposes the thin film through the bulk Si. The Al thin film is patterned with different patterns, for example a hexagonal lattice aperture grid with 10 μm separation between the 3 μm square holes. The resulting diffraction pattern is a hexagonal lattice of high intensity maxima, with separations of 9.3 mm at the MW. The large separation accommodates the large area (15 mm per side) quadrature photodetector. The process flow and an SEM of the resulting diffraction grating are shown in Figure 4. The MW is mounted on a diamond-lapped vacuum wafer chuck.

978-1-4244-2977-6/09 $25.00 © 2009 IEEE

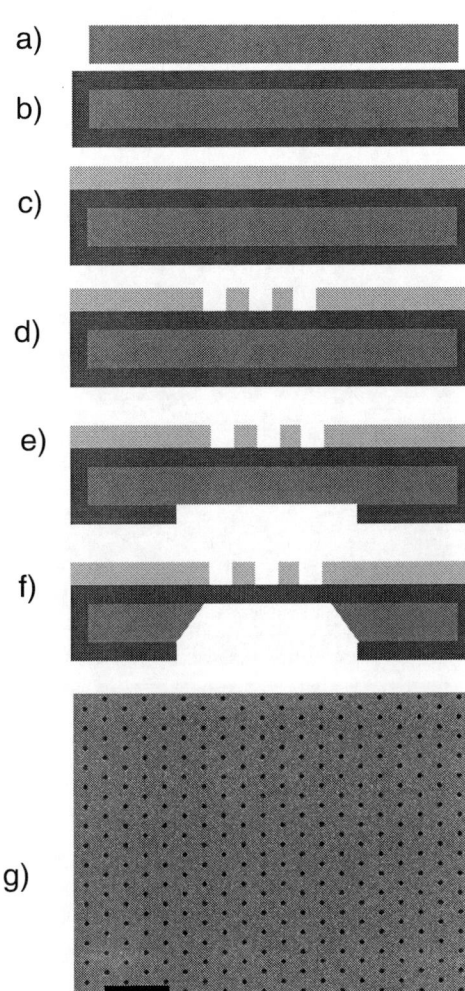

Figure 4: Process flow for OTW fabrication. a) Si wafer; b) 700 nm LPCVD low stress nitride; c) 400 nm Al; d) chlorine RIE patterning of Al film; e) backside patterning of nitride; f) through-wafer KOH etch for optical access; g) SEM of microfabricated diffraction grating: the bar shows 20 μm.

An STM tip and a quadrature photodetector (Pacific Silicon Sensors) are mounted on a PCB. The tip extends a millimeter below the PCB through a hole. The PCB is then mounted on a flexural piezoelectric stage (nPoint XYZ100B). The photodiode is reverse biased at 1.5 V and, after a low noise transimpedance amplifier, is fed to lock-in amplifiers (SRS SR810) with a time constant of 100 ms. The computer controls the piezo stage (100 μm × 100 μm × 10 μm travel range), as well as the long travel stepper motor stage that moves the fabricating tip across the whole wafer. Figure 5 shows the actual setup used.

The MW is a three inch wafer with 400 nm Al with 500 nm photoresist. The STM tip is lowered at 3 μm/sec, and a -10 V bias through the Al indicates a contact with the substrate, at which point the tip is retracted quickly. Figure 6 shows a typical indentation, and an SEM of a platinum / iridium STM tip. The PID control of the stage, running at a loop time constant of 30 ms, yields 190 nm standard deviation in error from its position in the optical field as indicated by the

quadrature photodiode. The limit of the lock-in with a quadrature photodiode sensor can be estimated by $\delta y = \partial y \, / \, \partial I \times$ intensity/SNR = 30 nm, where these parameters are empirical numbers from the experiment.

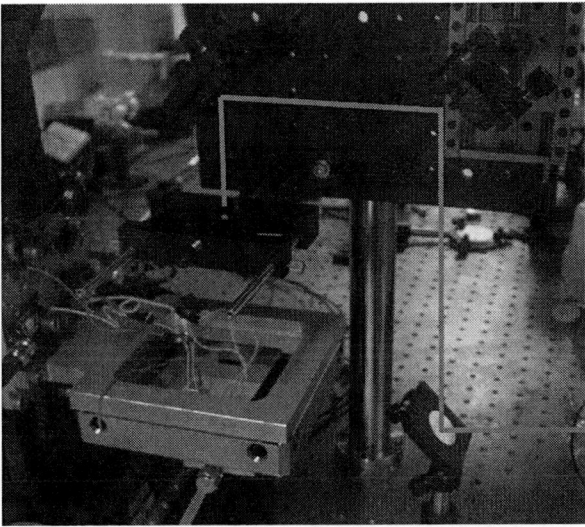

Figure 5: Setup of optical ruler technology. The laser beam enters from the right, then illuminates the diffraction grid on the OTW. An aluminum chuck holds the flexural piezo stage just above the MW (not visible). The chuck is attached to a long-travel, stepper motor stage. The MW sits on a diamond-lapped vacuum wafer chuck.

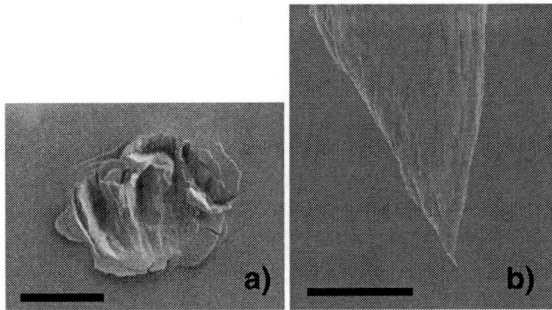

Figure 6: (a) Typical tip indentation about 400 nm in size; in this case, the tip made more than one indentation (see Table 2). Bar shows 1 μm. (b) Platinum/iridium STM tip used for fabrication. Bar shows 10 μm.

A set of 27 indentations made over an area spanning 66 mm by 28 mm were made, and the position of the points were measured in an SEM. Since the orientation of the wafer in the SEM is arbitrary, a calculated set of points is rotated and translated to align with the measured data set. The mean of the position error was 24 μm, or a precision of 3×10^{-4}.

With such high optical alignment precision – 2.7×10^{-6} – the error incurred during the transfer of precision from the photodiode to the MW is not clear. Some of the error may be attributed to the SEM stage which was not designed for long time- and length-scale precision. Two wafer alignment marks placed 80 mm apart on a 100 mm wafer were found to drift ±3 μm over a three

978-1-4244-2977-6/09 $25.00 © 2009 IEEE

hour period.   Due to the difficulty in finding indentations, it may take up to six or seven hours to find a complete set of indentations on a MW. Precision was also lost as the STM tip bent from wear and as it made an indentation in the stiff photoresist to make contact with the MW, and due to thermal expansion of the PC-Board. We estimate sources of error in Table 2.

Table 2: Estimates of sources of error in tip placement

Source of error	Error
Tip bend during indentation (10 mN, photoresist Young's modulus 3 GPa)	10 μm
Tip wear	5 μm
SEM metrology long term drift and hysteresis	5 μm
Thermal expansion of PC-Board (at 50 ppm)	1 μm

HIGH FILL-FACTOR WAFER-SCALE DIFFFRACTION PATTERNS

The data above used a hexagonal lattice optical field as the ruler with large separation so that the quadrature photodiode could lock to one maximum at specific points of the MW. When the optical ruler system needs to provide precise positioning at arbitrary places of the MW, the ruler will need to generate high signal across the entirety of the wafer. In addition, different diffraction patterns could be used to provide higher precision in certain regions of a wafer by generating high variation and intensity in those areas. We have demonstrated this ability by making diffraction grids that generate a strong optical ruler across 125 mm at the same distances from the OTW as the experimental results above. The results, show in Figure 7, show a rich variety of diffraction patterns that are possible.

CONCLUSION

We have demonstrated a diffractive optical ruler technology for tip-based nanofabrication by making indentations in 500 nm of photoresist with an STM tip. A precision of $\pm 3 \times 10^{-4}$ across a three inch wafer was demonstrated.   Given the sources of error discussed above, we believe we are actually closer to a precision of $\pm 1.5 \times 10^{-5}$. By better metrology of the indentations, the overall error may even be lower. We have also demonstrated that a rich variety of optical ruler patterns at the wafer-scale are possible. Future additions to the system will include continuous feedback control of the OTW/MW alignment, a spatially sampling CMOS array chip in place of the quadrature photodiode used here, and the addition of an in-plane alignment scheme to position fabrication with respect to the MW.

ACKNOWLEDGEMENTS

The authors thank DARPA/MTO's TBN program for generous funding, and the Cornell Nanoscale Science and Technology Facility (CNF) for fabrication and microscopy.

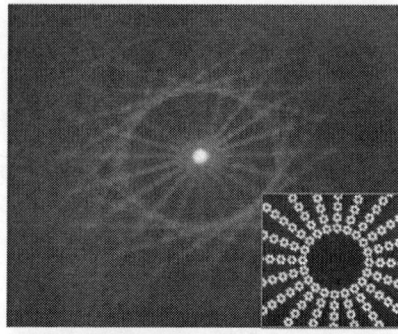

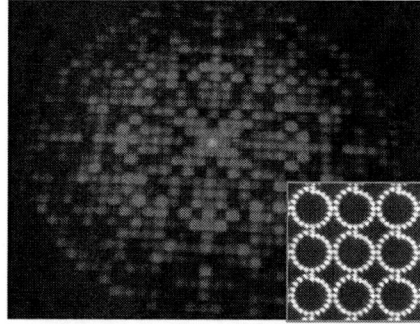

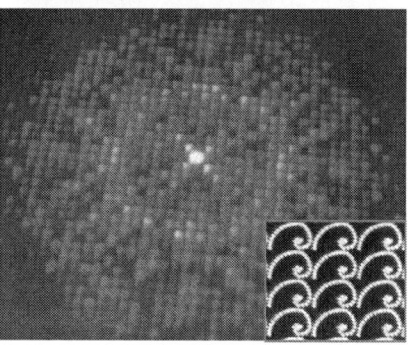

Figure 7: Wafer-scale, high fill-factor diffractive optical rulers. The diffraction pattern sizes are 125 mm in either dimension; insets show corresponding grating at 150 μm per side.   A HeNe laser was used for visibility.

REFERENCES

[1] K.-B. Lee, So-Jung Park, Chad A. Mirkin, Jennifer C. Smith, Milan Mrksich, "Protein Nanoarrays Generated By Dip-Pen Nanolithography", Science, vol. 295, pp. 1702-1705, 2002.

[2] Alexander A. Milner, Kaiyin Zhang, Yehiam Prior, "Floating Tip Nanolithography," Nano Letters, vol. 8, pp. 2017-2022, 2008.

[3] D. Croft, G. Shed, S. Devasia, "Creep, Hysteresis, and Vibration Compensation for Piezoactuators: Atomic Force Microscopy Application," ASME J. Dyn. Syst., Meas., Control, vol. 123, pp. 35-43, 2001.

[4] S. Salapaka, A Sebatsian, J. P. Cleveland, M. V. Salapaka, "High Bandwidth Nano-positioner: A Robust Control Approach," Rev. Sci. Inst., vol. 73, pp. 3232-3241, 2002.

INTEGRATION OF BRIDGING-STRUCTURAL SWNTS ON FLEXIBLE PDMS SHEET BY STAMPING TRANSFER

Yusuke Takei, Tetsuo Kan, Eiji Iwase, Kiyoshi Matsumoto, and Isao Shimoyama
The University of Tokyo, JAPAN

ABSTRACT

This paper describes an integration method of bridging-structural single-walled carbon nanotubes (SWNTs) onto the flexible PDMS sheet by stamping transfer. Silicon microstructures and the SWNTs which directly synthesized between the gaps of the microstructures were lifted off by PDMS stamp sheets with high yield (97.8 %) and accuracy (position error < 100 nm). From the SEM observation of our transferred structures, we confirmed that our method could transfer the bridging-structural SWNTs on to the flexible materials without any damage to the SWNTs' bridging.

(A) Press PDMS sheet on to the Si structures and bridging SWNTs.

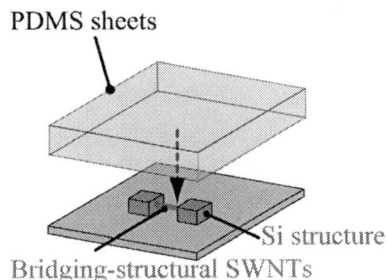

(B) Lift off the PDMS sheet.

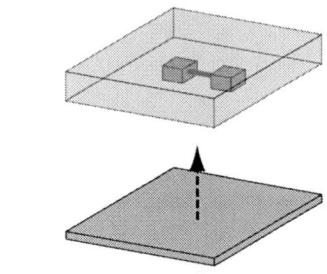

(C) When PDMS sheet stretched, SWNTs also stretch.

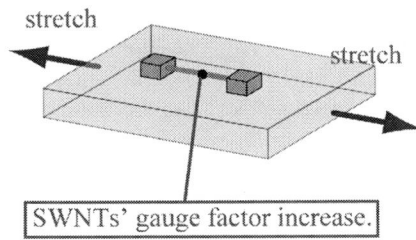

Figure 1: Concept diagram of our integration method of SWNTs+MEMS structures by liftoff and stamping transfer using PDMS sheets.

INTRODUCTION

SWNT is a material consisting of a network of six-membered rings of carbon atoms (called a graphene sheet) rolled into a single-walled tubular shape. Stretching a SWNT in axial direction causes changes in its electrical properties such as its band structure and electrical conductivity, and the resulting change in resistance is called the piezo-resistance effect [1-3].

In previous studies, Tombler et al. measured the changes in electrical conductivity when metal SWNTs were poked with an AFM probe [4]. Stampfer et al. measured the piezo-resistance of simulated bridging SWNTs [5][6]. Their researches indicated that gauge factor of the SWNTs got higher when the SWNTs stretched.

In order to build high gauge factor SWNTs into the MEMS devices, we need to fix SWNTs on the wafer with applying the tension. But if we use mechanism of micro-scale actuator (such as electrostatic actuator) to apply tension to the SWNTs physically, it will spoil the merit of the nano-scale SWNTs. One of the solutions for this problem is to fix the SWNTs on the flexible material, and apply tension to the material to deform both material and SWNTs. But we can't synthesize SWNTs on flexible materials because of the high temperature ($800°C$) of the CVD process.

Our proposed method is to use stamping transfer [7]. We synthesize SWNTs on the solid wafer and then transfer them onto the flexible sheet (Figure 1). This SWNTs-PDMS structure has an advantage that gauge factor of the SWNTs can be tuned by stretching the PDMS sheet. In addition, we use bridging-structural SWNTs grown between the silicon pillars as a transfer target (Figure 2(B)) Bridging-structural SWNTs are suitable for the sensors because we can easily control the number of SWNTs rather than synthesized on the substrate (Figure 2(A)).

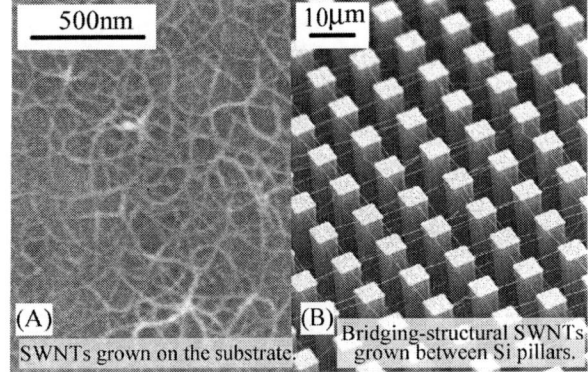

Figure 2: (A) SWNTs formed on a flat Si substrate, and (B) SWNTs formed on silicon pillar structures. On the flat Si substrate, the nanotubes take the form of tangled cobwebs, but on the silicon pillars, they produce bridging structures.

978-1-4244-2977-6/09 $25.00 © 2009 IEEE

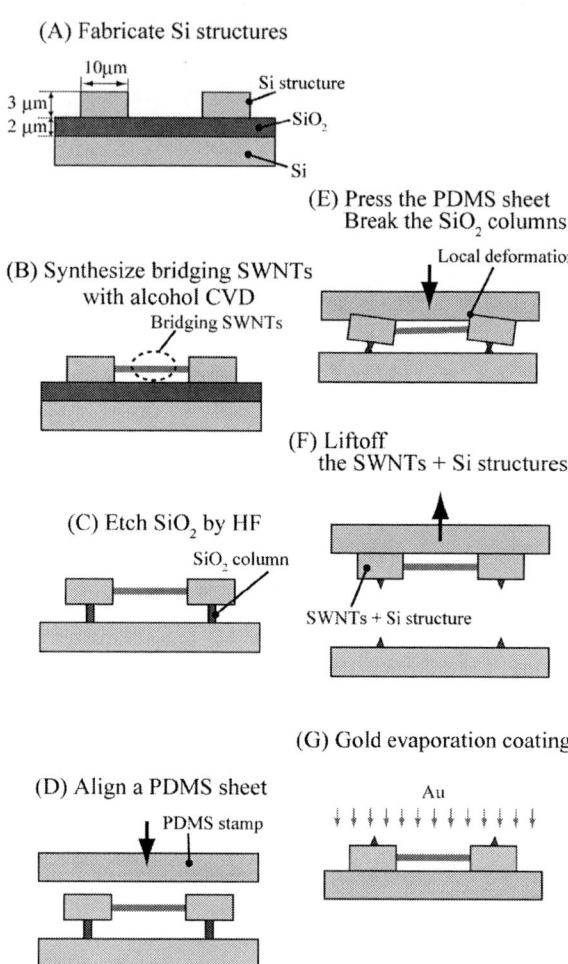

Figure 3: Process of integrating bridging-structural SWNTs on flexible PDMS sheet by stamping transfer.

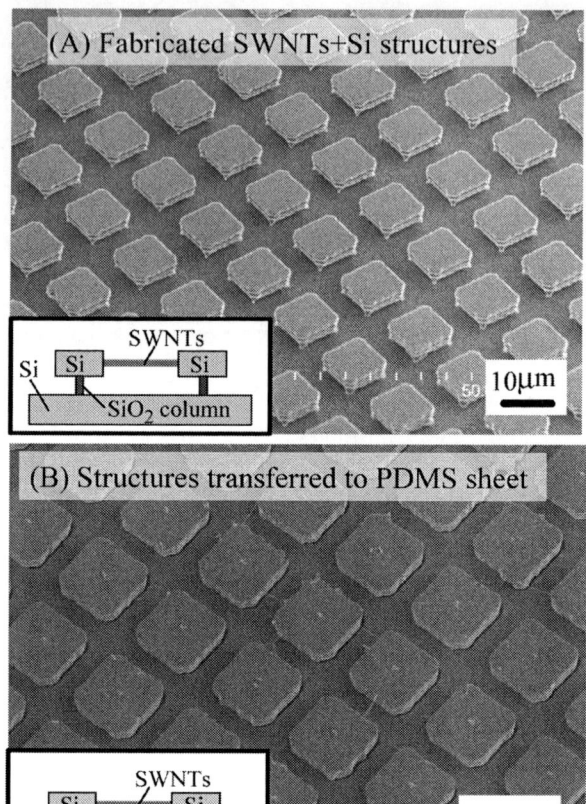

Figure 4: SEM images of SWNTs and Si structures before transfer (A), and after transferred to PDMS sheet (B). (A) Rectangular Si microstructures are supported by SiO₂ column of 1μm in diameter. And bridging-structural SWNTs are directly synthesized between the Si structures. (B) Si structures and bridging SWNTs are transferred on flexible PDMS sheet.

FABRICATION

Bridging-structural SWNTs

The process flow of our method is shown in Figure 3. We used rectangular silicon microstructures (10μm-square and 3μm in height) and bridging-structural SWNTs (directly synthesized between the silicon structures by CVD) as test structures for the stamping transfer. The transfer area was 300μm-by-300μm square (Figure 5). The silicon microstructures were fabricated by etching a top silicon layer of an SOI wafer (Figure 3(A)). Then bridging-structural SWNTs were formed by CVD between the silicon microstructures (Figure 3(B)).

We used alcohol CVD to synthesize the SWNTs used in this study, but CNTs produced by this method are usually oriented vertically. This phenomenon is strongly related to the type and density of the metal catalyst that provides the nuclei for CNT formation. As the density of the metal catalyst on the substrate increases, the resulting CNTs tend to grow along the surface of the substrate and interfere with neighboring CNTs with the result that they are formed oriented perpendicular to the substrate. Thus in order to form the CNTs so that they grow in the plane of the substrate surface and bridge across gaps in the substrate as in this study, it is necessary to keep the metal catalyst fixed to the substrate surface down to a low density. In this paper,

we used the dip-coating method to keep the density of the metal catalyst fixed to the silicon oxide film down to a low value [8].

Catalytic Loading

We used iron and cobalt as catalytic metals for the CVD. Particles of the catalytic metals were loaded on the SiO₂ surface. Iron and cobalt acetate were dissolved in ethanol to form a metal acetate solution, the weight concentration of the both metals were made equal (0.01 wt %). The solution was dip-coated on the SiO₂ surface at a constant pull-up speed 4 cm/min in the room temperature. After coating the catalysts on the wafer, the wafer was dried in a 400ºC oven for 2 minutes to remove organic residues and to form oxidized bimetallic particles on the SiO₂ surface.

Yield of Stamping Liftoff : 97.8%

70μm ■ : Missing Si structures

Figure 5: Top view of the transferred Si structures + SWNTs observed by SEM. The transferred area was 300μm-by-300μm square (20x20 Si structures in the area). The red squares draw in the image are void caused by the stamping failure. Yield of the stamping liftoff was 97.8%.

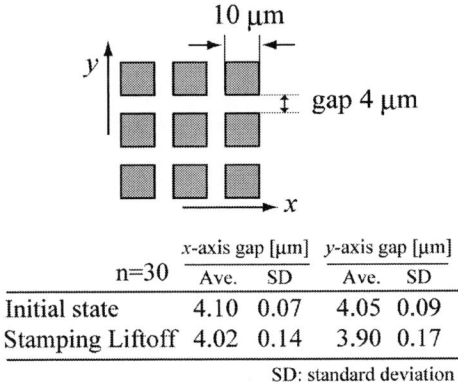

n=30	x-axis gap [μm]		y-axis gap [μm]	
	Ave.	SD	Ave.	SD
Initial state	4.10	0.07	4.05	0.09
Stamping Liftoff	4.02	0.14	3.90	0.17

SD: standard deviation

Figure 6: Evaluation of the accuracy of the stamping transfer. We measured the gap length between the Si structures.

Chemical Vapor Deposition (CVD)

The CVD process was carried out in a quartz tubes as follows. In the CVD process, a silicon wafer loaded on a quartz boat was placed in the center of the quartz tube. The center part of the quartz tube was heated up to 800°C with electric furnace, flowing 300 sccm of Ar gas containing 3% of hydrogen gas (Ar/H$_2$ gas) so that the pressure inside the quartz tube was kept at about 40 kPa. When reaction temperature (800°C) was achieved, the Ar/H$_2$ gas was stopped and the quartz tube was evacuated with a vacuum pump. Then ethanol vapor was introduced into the quartz tube, keeping the ethanol pressure at about 1.6 kPa. The reaction time for CVD was 30 minutes at that time. After the reaction, the ethanol gas was stopped and the electric furnace was turned off, cooling inside the tube to the room temperature with 100 sccm of Ar/H$_2$ gas flowing.

HF vapor Etching

In order to lift off the Si structures and bridging-structural SWNTs, we etched the SiO$_2$ layer of the SOI wafer by HF vapor etching after the SWNTs synthesis. We observed the wafer with infrared microscope and see how the HF vapor etching is going under the top Si structures. We stopped the etching just before the structures were completely released. If the SiO$_2$ columns were not enough slim, the following stamping transfer would fail. From the trial of stamping transfer with variable diameter of supporting SiO$_2$ columns, we found the diameter of less than 1μm was suitable for the stamping transfer. Figure 4(A) shows the fabricated Si structures and bridging-structural SWNTs. Si structures are supported by SiO$_2$ columns.

STAMPING TRANSFER

The transfer area was 300μm-by-300μm square (Figure 5). A PDMS stamp sheet was aligned to the silicon structures, and pressed vertically (Figure 3(D)). The SiO$_2$ columns supporting the structures were surely broken by the press as the PDMS stamp sheet locally deforming (Figure 3(E)). Then, the silicon structures were lifted off from the wafer by adhesion between the silicon structures and the PDMS stamp (Figure 3(F)). After the liftoff, we evaporated the gold to the PDMS sheet for the SEM observation (Figure 3(G), Figure 4(B)).

Yield of the PDMS liftoff was evaluated as shown in Figure 5. We examined the yield by counting the number of the transferred structures in 400 Si blocks. The accuracy was investigated by measuring the gap between the structures (Figure 6). We obtained high yields for the liftoff (97.8 %). As for the accuracy, the errors in the position of the structures were within 100nm. Figure 7 shows the SEM images of our transferred bridging-structural SWNTs and silicon microstructures on flexible PDMS sheet. On top of the silicon microstructures, remains of the SiO$_2$ column which caused by the breakage of the SiO$_2$ columns were observed. This structure indicates that SWNTs have high elasticity and this characteristic enables us to transfer the whole bridging structures on flexible PDMS sheet without any breakage on SWNTs' bridging.

CONCLUSION

In conclusion, we demonstrated integrating bridging-structural SWNTs on flexible PDMS sheet by the stamping transfer. Silicon microstructures and the SWNTs which directly synthesized between the gaps of the microstructures were lifted off by PDMS sheets with high yield (97.8 %) and accuracy (position error < 100 nm). From the SEM observation, we confirm that our method can transfer the bridging-structural SWNTs on to the flexible materials. We indicate that our process can be applied to various applications taking advantage of SWNT's high elasticity such as flexible wiring, and force sensor with high gauge factor SWNTs.

978-1-4244-2977-6/09 $25.00 © 2009 IEEE

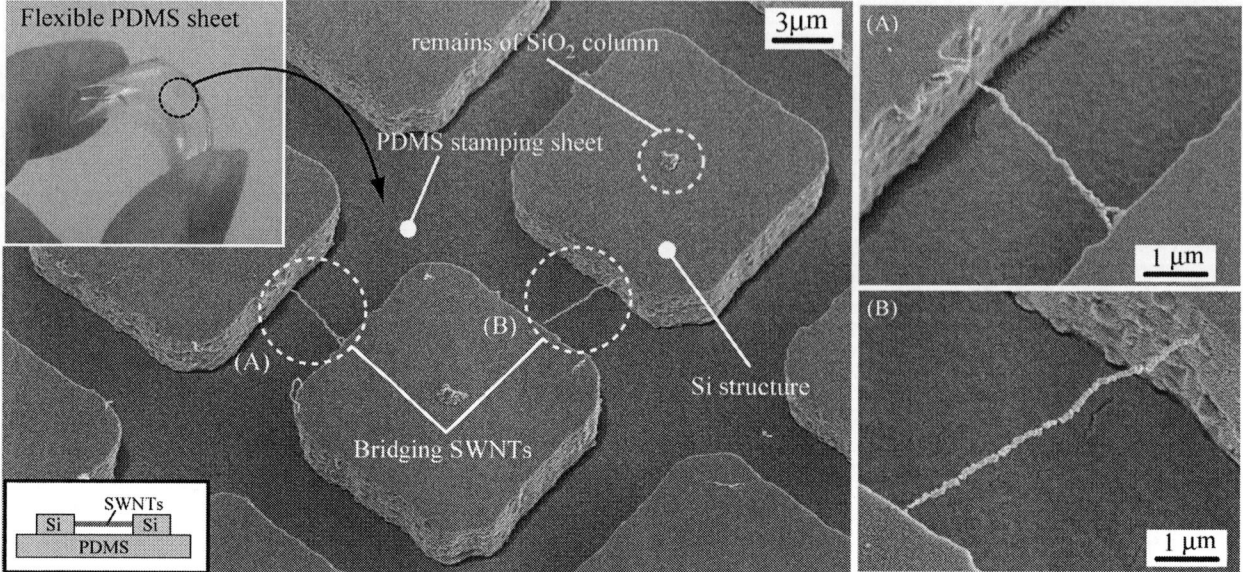

Figure 7: SEM images of our transferred bridging-structural SWNTs and silicon microstructures on flexible PDMS sheet. The SWNTs are thicken by gold evaporation. Remains of SiO$_2$ column on the Si blocks are caused by the breakage of the SiO$_2$ column at the PDMS sheet stamping process.

ACKNOWLEDGMENTS

This research is partly supported by Special Coordination Funds for Promoting Science and Technology, "IRT Foundation to Support Man and Aging Society". The photolithography masks were fabricated using the University of Tokyo VLSI Design and Education Center (VDEC)'s 8-inch EB writer F5112+VD01 donated by ADVANTEST Corporation. Alcatel ICP plasma etching system A601E (DRIE) was used to etch the silicon structures. VAN Partners Corporation is acknowledged by FESEM S-4700.

REFERENCES

[1] M. Ouyang, J. L. Huang, C. L. Cheung and C. M. Lieber, "Energy Gaps in "Metallic" Single- Walled Carbon Nanotubes," *Science*, Vol. 292, pp. 702-705, 2001.

[2] G. Zhang, P. Qi, X. Wang, Y. Lu, X. Li, R. Tu, S. Bangsaruntip, D. mann, L. Zhang and H. Dai, "Selective Etching of Metalllic Carbon Nanotubes by Gas-Phase Reaction," *Science*, Vol. 314, pp. 974-977, 2006.

[3] L. Liu, C. S. Jayanthi, M. Tang, S. Y. Wu, T. W. Tombler, C. Zhou, L. Alexseyev, J. kong and H. Dai, "Cotrollable Reversibility of an sp^2 to sp^3 Transition of a Single Wall Nanotube under the Manipulation of and AFM Tip: A Nanoscale Electromechanical Switch?," *Physical Review Letters*, Vol. 84, No. 21, pp. 4950-4953, 2000.

[4] T. W. Tombler, C. Zhou, L. Alexseyev, J. Kong, H. Dai, L. Liu, C. S. Jayanthi, M. Tang and S. Wu, "Reversible Electromechanical Characteristics of Carbon Nanotubes under Local-Probe Manipulation," *Nature*, Vol. 405, pp. 769-772, 2000.

[5] C. Stampfer, A. Jungen, R. Linderman, D. Obergfell, S. Roth and C. Hierold, "Nano-Electromechanical Displacement Sensing Based on Single-Walled Carbon Nanotubes," *Nano Letters*, vol. 6, No. 7, pp. 1449-1453, 2006.

[6] C. Stampfer, T. Helbling, D. Obergfell, B. Schoberle, M. K. Tripp, A. Jungen, S. Roth, V. M. Bright and C. Hierold, "Fabrication of Single Walled Carbon Nanotube-Based Pressure Sensors," *Nano Letters*, Vol. 6, No. 2, pp. 233-237, 2006.

[7] H. Onoe, E. Iwase, K. Matsumoto, I. Shimoyama, " Three Dimensional Integration of HeterogeneousSilicon Micro-Structures by Liftoff and StampingTransfer," *J. Micromech. Microeng.*, vol. 17, pp.1818-1827, 2007.

[8] Y. Murakami, S. Yamakita, T. Okubo and S. Maruyama, "Single-Walled Carbon Nanotubes Catalytically Grown from Mesoporous Silica Thin Film," *Chemical Physics Letters*, Vol. 375, pp. 393-398, 2003.

GIANT PIEZORESISTANCE OF NANO-THICK SILICON INDUCED BY INTERFACE ELECTRON TRAPING EFFECT

Yongliang Yang, Xinxin Li

State Key Lab of Transducer Technology, Shanghai Institute of Microsystem and Information Technology, Chinese Academy of Science 200050, China (Email: xxli@mail.sim.ac.cn)

ABSTRACT

Both n- and p-type nano-thick piezoresistors are fabricated on SOI (silicon on insulator) wafers using micro-fabrication processes. Giant piezoresistance is measured and theoretically explained for nano-thick silicon resistors. Compared to bulk silicon, one order of magnitude higher piezoresistive coefficients are, for the first time, tested with 13nm-thick n-type and 9nm p-type samples. Surpassing 2-D quantum effect, Si-SiO$_2$ interface electron trapping effect dominates the giant piezoresistance. Different from equivalent mobility change in conventional piezoresistance of bulk silicon, the giant piezoresistance come from carrier concentration change and have the same effect on the longitudinal and transverse piezoresistors.

1. INTRODUCTION

Piezoresistance effect has been extensively studied [1], [2] since the pioneering work of Smith in 1954 [3]. Due to the excellent electrical and mechanical properties of silicon and the availability of integrated circuit compatible fabrication processes [4], the piezoresistance effect of silicon has been used as a sensing principle for mechanical sensors, such as diaphragm type pressure sensors and micro-accelerometers [5], [6]. In recent year, nano-thick piezoresistor was theoretical studied and expected to obtain higher piezoresistance effect. Theoretical calculations considering 2-D quantum confinement effect predicted that piezoresistance in nano-thick p-type silicon can increase a couple of times compared with bulk silicon [7], [8]. All most all the published researches are concentrated on p-type nano-thick piezoresistance. In contrast, n-type nano-thick piezoresistance is seldom studied. Following the 2-D quantum confinement effect, our calculations show that n-type 2-D nano-thick piezoresistance will diminish to zero when the silicon layer thickness decreases to several nanometers.

In this work, both n- and p-type nano-thick piezoresistors are fabricated on SOI wafers. The experimental results show that in relative thicker specimen, 2-D quantum effect dominant the piezoresistance. While for relative thinner sample, surpassing 2-D quantum effect, Si-SiO$_2$ interface electron-trapping effect dominates the piezoresistance. Compared to bulk silicon, one order of magnitude higher piezoresistance were tested with 13nm-thick n-type and 9nm p-type samples. Different from equivalent mobility change in conventional piezoresistance of bulk silicon, the giant piezoresistance come from carrier concentration change and have the same effect on the longitudinal and transverse piezoresistors.

2. EXPERIMENTS

Fabrication

Both n- and p-type nano-thick piezoresistors are fabricated on SOI wafers using micro-fabrication processes

as schematically shown in Figure 1. With uni-axial stress applied on the chip, the piezoresistive coefficients in longitudinal (π_l) and transverse (π_t) direction can be measured simultaneously by longitudinal and transverse resistors. The process flow sketched in Figure 2. SOI (100) wafers with p-silicon active layer on top of buried oxide layer are used as starting material [Figure 2(a)]. The silicon active layer is thinned by several dry thermal oxidations. Every time we preserve the oxide on the terminal areas of the resistors and remove the oxide on the other areas. By this method, the terminal areas of the resistors are made thicker than other regions for Ohmic contact with metal wires. The resistors along <100> and <110> are both shaped by photolithography and TMAH etching. Then a SiO$_2$ layer is dry oxidized for insulation [Figure 2(b)]. Various resistor thicknesses are obtained by controlling the oxidation thickness. Phosphor for n-type or boron for p-type is doped by ion implantation and the wafers are annealed in N$_2$ at 1000 °C for 30 minutes. Both n-type and p-type resistors are formed with the doping level of 10^{19} cm^{-3} [Figure 2(c)]. Al interconnection is sputtered and sintered in N$_2$ and H$_2$ at 450 °C for 30 minutes [Figure 2(d)].

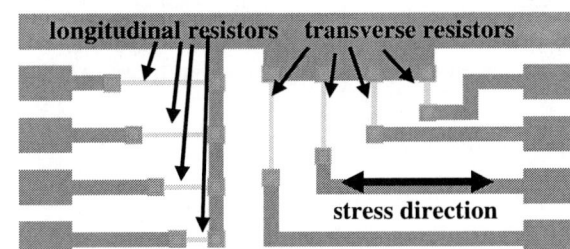

Figure 1: Layout of the nano-thick piezoresistor chip.

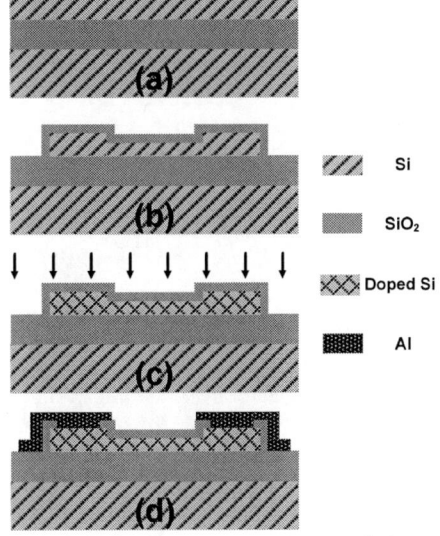

Figure 2: Fabrication process steps of the nano-thick piezoresistor.

978-1-4244-2977-6/09 $25.00 © 2009 IEEE

The finished chip is shown in Figure 3. The thicknesses of the silicon layer are measured by TEM.

Figure 3: Photograph of the fabricated chip.

Mechanical stress testing method

The mechanical stress is applied by a mental cantilever bending system. As shown in Figure 4(a), near the clamped end of the cantilever a trapezoid region was designed as working region to get uniformity strain. The chips are glued on the working region of the cantilever with <100>/<110> directions parallel to the longitudinal direction of the cantilever. Then the cantilever is fixed on the table and the weights are loaded on the free end of the cantilever [See Figure 4(b)]. The cantilever will bend according to the differential equation of the beam. The chips are deformed together with cantilever thus uni-axial tensile strain is applied on the chips. It is assumed that strains on the cantilever are completely transmitted into the chips. From this assumption, the strains on the nano-thick resistors are measured by the standard strain gauges glued near the chips on working region of the cantilever.

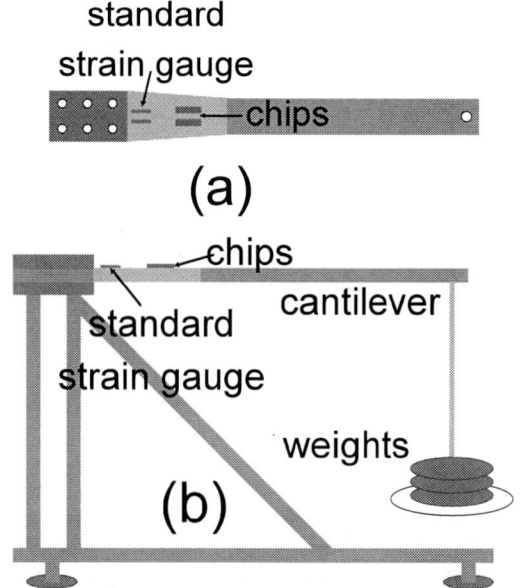

Figure 4: Schematic of the cantilever bending system, with the chip glued on the metal cantilever for stress loading.

Experimental results

The tested linear resistance change versus stress is shown in Figures 5. For n-type, the relatively thicker specimens show the same properties as bulk silicon [1].

The longitudinal resistance decreases with stress, while the transverse resistance increases, and the former is with the relative change value being about twice of the latter. In contrast, for relatively thinner specimens, both longitudinal and transverse resistances increase with stress and the change rate is almost the same for them [Figure 5(a)]. Similar phenomena are also observed for p-type resistors. The relatively thicker specimens show regular resistance change as reported in Ref [1]. The longitudinal resistance increases with stress, while the transverse resistance decreases, and the change of the former is a little large than the latter. While for the relatively thinner specimen, both longitudinal and transverse resistances decrease with stress and the change rate is almost the same for them [Figure 5(b)].

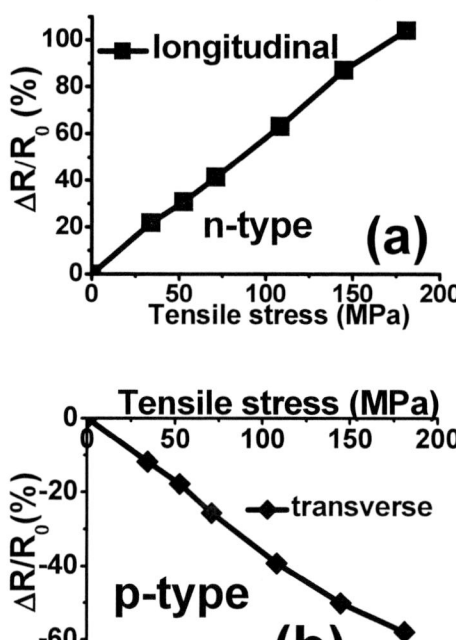

Figure 5: Resistance change versus stress of n-type (a) and (b) p-type nano-thick resistors.

Specimens with various active-layer thicknesses have been tested. The tested thickness-dependent piezoresistive coefficients are shown in Figure 6. For n-type silicon (See Figure 6), in the thickness range larger than about 30nm, both the longitudinal and transverse piezoresistive coefficients (denoted as π_l and π_t) gently decrease the values form $-37\times10^{-11}\text{Pa}^{-1}$ and $22\times10^{-11}\text{Pa}^{-1}$ (for bulk-Si) with thinning the Si-layer. For even thinner samples, π_l and π_t tend to positive values, and finally both π_l and π_t rapidly reach a giant value of about $460\times10^{-11}\text{Pa}^{-1}$ for 13nm-thickness. For p-type silicon, in the thickness range large than 25nm, both π_l and π_t increase the values from $37\times10^{-11}\text{Pa}^{-1}$ and $-31\times10^{-11}\text{Pa}^{-1}$ (bulk-Si). Then both tend to negative values with thinner the silicon layer and finally reach a giant value of about $-400\times10^{-11}\text{Pa}^{-1}$ for 9nm-thickness. Compared to the conventional piezoresistance for bulk silicon, herein the tested giant piezoresistive coefficients for nano-thick silicon exhibit an increase of higher than one order of magnitude.

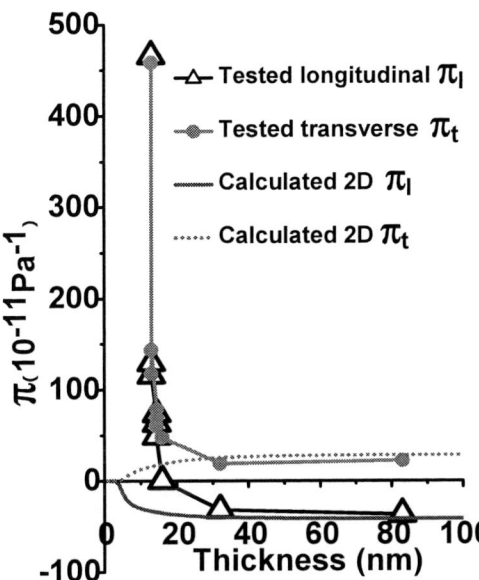

Figure 6: Tested and calculated piezoresistive coefficients in terms of varied thicknesses of n-type piezoresistors.

3. ANALYSIS AND DISCUSSION

2-D quantum confined piezoresistance

With the thickness of the piezoresistor decreasing, which comparable with the de Broglie wavelength of electron, the electrons are restricted in the space *(-d/2, d/2)* in the *z*-axis, where *d* is the thickness of the piezoresistor. The quantum size effect plays significant role and gives a quite different picture of the electron or hole dispersions by shift the energy structure of conduction band or valence band. Thus the 2-D quantum piezoresistance effect is different form the conventional bulk silicon piezoresistance effect.

For n-type nano silicon layer, following the quantum theory of a particle in a box, the energy valleys of conduction band will split according to

$$E_i^{l,t} = \frac{\pi^2 \hbar^2 i^2}{2m_{l,t}d^2} \; ; i = 1, 2, 3 \cdots \qquad (1)$$

where m_l and m_t are the longitudinal and transverse electron effective masses, respectively (See Figure 9). Since the longitudinal effective mass is about five times larger than the transversal effective mass. Under the quantum confinement, the confinement energies of longitudinal valleys (along *z*-axis) will be five times smaller than those of transverse valleys (along *x*- and *y*-axes) [Figure 7(a)]. According to Fermi distribution, most of the electrons will populate the valleys along *z*-axis. Following the deformation potential theory [9], the tensile mechanical stress lifts the energy minimums on the *x*-axis and lowers those on the *y*- and *z*-axes [See Figure 7(b)]. Thus part of the electrons populated in the *x*-axis valleys move out of the valleys and relocated in the other four valleys. From

$$1/\rho_x = \sigma_x = \left(\frac{n_x}{m_l} + \frac{n_y + n_x}{m_t} \right) e\tau \qquad (2)$$

the resistivity change under <100> direction stress is due to the change of electrons populated in *x*-axis valleys n_x. As

the thickness decreasing, more and more electrons are populated in *z*-axis valleys and the change of electrons in *x*-axis valleys under stress decrease to almost zero. Reduced piezoresistive coefficients for n-type quantum confined piezoresistor are obtained as the thickness decreasing. The calculations show that π_l and π_t will decrease down to zero when continually thinning the silicon-layer to several nanometers (See Figure 6).

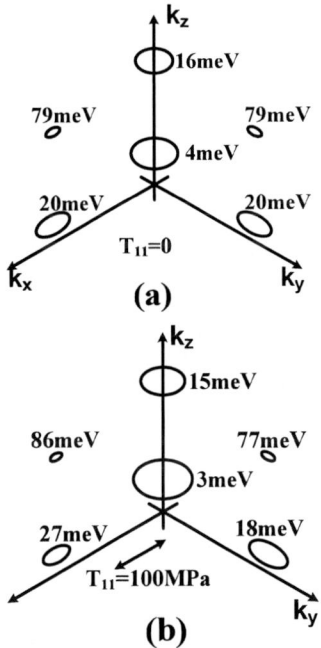

Figure 7: Calculated 2-D quantum confinement energies of valley for 10 nm-thick relaxed (a) and strained (b) silicon.

For p-type nano-thick silicon layer, theory studies considering 2-D quantum effect predicted that the piezoresistance is increase to several times with thinning the silicon layer.

In Figure 6, the 2-D quantum modeling results can only fit the experimental data for relatively thick silicon. However, it cannot be the real reason for the giant piezoresistance experimentally obtained for the thinner samples. Electron-trapping effect at Si-SiO$_2$ interface is found the dominant effect on the giant piezoresistance.

Interface electron trapping effect

For ultra-thin silicon, electron traps at the Si-SiO$_2$ interface can largely affect the conductivity of silicon by capturing or emitting electrons [10]. Researches have proved that the trapped electrons will increase under stress [11], [12], which means more movable electrons being trapped and the carrier concentration in silicon layer is changed. Only for thin silicon layer, the interface trapping effect caused carrier number change is comparable with the total carrier quantity. Thus the interface electron trapping effect will dominate the piezoresistance.

For n-type nano-thick resistors, under mechanical tensile stress, a portion of electrons in the conduction band will move to the interface band and. Thus the resistance is increased by decreasing the electron concentration in conduction band. As the electron concentration decrease is equal to longitudinal and transverse resistors, both π_l and π_t are positive with the almost same giant value in ultra-thin samples.

For p-type nano-thick resistors, under mechanical tensile stress, a portion of electrons in the valence band will move to the interface band and form holes in the valence band. Thus the resistance is decreased by increasing the hole concentration in valence band. Since the hole concentration increase is equal to longitudinal and transverse resistors, both π_l and π_t are negative with the almost same giant value in ultra-thin samples.

4. CONCLUSION & ACKNOWLEDGMENT

Giant piezoresistance effect was experimentally tested in both n- and p-type nano-thick piezoresistors for the first time. Different form mobility changes in conventional piezoresistance, the giant piezoresistance comes form carrier concentration change. The great enhanced piezoresistive coefficients are expected for ultra-sensitive applications. With brand-new properties, new design rules will be established for using new sensing effect in new micro/nano sensors.

This research is supported by the NSFC Project under Contract No. 60725414, the Chinese 973 Program (2006CB300405) and Chinese 863 Project (2006AA04Z365). The second author thanks to the support of the NSFC Project of No. 60721004.

REFERENCES

[1] Y. Kanda, "A graphical representation of the piezoresistance coefficients in silicon", *Electron Devices, IEEE Transactions on,* vol. 29, pp. 64-70, 1982.

[2] T. Toriyama, S. Sugiyama, "Analysis of piezoresistance in p-type silicon for mechanical sensors", *Microelectromechanical Systems, Journal of,* vol. 11, pp. 598-604, 2002.

[3] C. S. Smith, "Piezoresistance Effect in Germanium and Silicon", *Physical Review*, vol. 94, pp. 42-49, 1954.

[4] K. E. Petersen, "Silicon as a mechanical material", *Proceedings of the IEEE,* vol. 70, pp. 420-457, 1982.

[5] M. Bao, *Analysis and Design Principles of MEMS Devices*: Elsevier Science & Technology, 2005.

[6] S. S. Huang, Xinxin Li, Z. H. Song, Y. L. Wang, H. Yang, L. F. Che, and J. W. Jiao, "A high-performance micromachined piezoresistive accelerometer with axially stressed tiny beams", *Journal of Micromechanics and Microengineering,* vol. 15, pp. 993-1000, May 2005.

[7] T. Ivanov, T. Gotszalk, T. Sulzbach, I. W. Rangelow, "Quantum size aspects of the piezoresistive effect in ultra thin piezoresistors", *Ultramicroscopy,* vol. 97, pp. 377-384, 2003.

[8] J. Zhang, Q. a. Huang, H. Yu, S. Lei, "A Theoretical Study of the Piezoresistivity of a p-Type Silicon Nanoplate", *Chinese Journal of Semiconductors,* vol. 29, pp. 970-974, 2008.

[9] C. Herring, E. Vogt, "Transport and Deformation-Potential Theory for Many-Valley Semiconductors with Anisotropic Scattering", *Physical Review,* vol. 101, pp. 944-961, 1956.

[10] E. H. Nicollian, J. R. Brews, *MOS (Metal Oxide Semiconductor) Physics and Technology,* Wiley-Interscience, New York, 1982.

[11] J. Boch, D. M. Fleetwood, R. D. Schrimpf, R. R. Cizmarik, F. Saigne, "Impact of mechanical stress on total-dose effects in bipolar ICs", *IEEE Transactions on Nuclear Science,* vol. 50, pp. 2335-2340, Dec 2003

[12] A. Hamada, E. Takeda, "Hot-electron trapping activation energy in PMOSFET's under mechanical stress", *Electron Device Letters, IEEE,* vol. 15, pp. 31-32, 1994.

ELECTRICAL DISCHARGE BASED MICROFABRICATION ON ELECTROSPUN NANOFIBERS

H. Zeng and Y. Zhao
Department of Biomedical Engineering
The Ohio State University, Columbus, Ohio, USA

ABSTRACT

This paper reports the use of electrical discharge for fabricating microstructures on electrospun polymer nanofibers. Microchips containing an array of conductive microelectrodes are fabricated. Electrical discharges are induced by applying high electrical voltage to these microelectrodes. The thermal energy generated by the micropatterned discharge arcs elevates the temperature in localized regions and melts polymer nanofibers in the close vicinity. Microstructures with the minimum line width as small as 20 μm are demonstrated. This method provides a promising start point to interface biodegradable nanofibrous materials to microstructures, which is significant for a broad array of biomedical and industrial applications.

INTRODUCTION

Electrospun polymer nanofibers are composed of entangled fibrous structures with typical diameters from 5 nm to a few hundreds of nm, which form interconnected pores with controllable sizes and large surface-to-volume ratios. Due to the mechanical flexibility, biocompatibility, and unique morphology, such materials have recently gained rapid popularity in a broad range of biomedical and industrial applications, including high performance filtration membranes, drug delivery carriers, tissue engineering scaffolds and protective clothing [1-5].

In tissue engineering applications, multi-scale structures combining nanofibers and microstructures are often needed because they provide an ideal vehicle to interface subcellular objects with larger scaled cells and tissues. Specifically, the non-woven nanofibers closely mimic the network of nanometer-sized proteins and glycosaminoglycans in extracellular matrix (ECM) [6, 7], while the microstructures provide local engineered topography to guide cell growth and differentiation [8, 9]. For example, micropatterned carbon nanofibers can modulate the adhesion of osteoblasts and the deposition of calcium phosphate [10]. Microfabricated poly-caprolactone (PCL) scaffolds with nanopores can be used as the engineered counterparts of natural blood vessels by aligning vascular smooth muscle cells and enhancing nutrient diffusion [11]. However, despite the usefulness there are only a few technologies available for fabricating such combined micro/nanostructures in electrospun nanofibers.

In this work, we demonstrate a simple and inexpensive method for creating microstructures on electrospun polymer nanofiber mats by contacting polymer mats with a microelectrodes-patterned glass substrate. Electrical discharges are induced by applying high voltage bias. The thermal energy generated by the discharge arcs melts the nanofibers and create microstructures in defined regions.

EXPERIMENTAL CONCEPT

Figure 1 illustrates the experimental concept. Aluminum microelectrodes are patterned on a glass substrate using standard photolithography and lift-off processes. The tips of opposing microelectrodes are separated by a gap ranging from 50 μm to above 1 mm. Sharp tip asperities are designed to generate highly enhanced field for discharge emission and to guide the trajectory of the discharge arcs. Micropatterning is demonstrated using electrospun nanofibers made of polyester urethane urea (PEUU), which has a melting point above 180 °C.

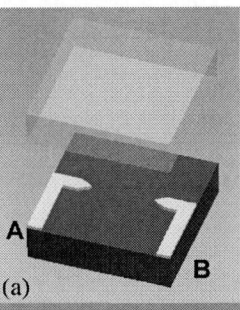

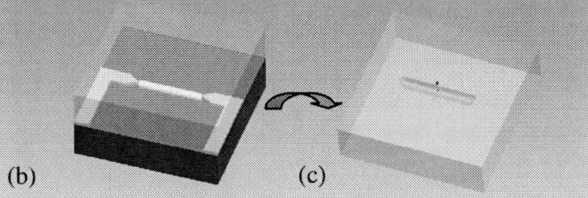

Figure 1: Schematics of the discharge-based microfabrication process. (a) Microelectrodes with opposing tips are patterned on a glass substrate; (b) Polymer nanofiber mat is covered on the glass substrate with a modest pressure. The discharge arcs are induced by applying a high frequency voltage bias between the microelectrodes; (c) Micropatterns are formed upon substrate removal.

After fabrication of the substrate, one of the microelectrodes is connected to the output A of a high frequency generator, while another microelectrode B is grounded (Figure 2). The power transformer T1 sets up a high voltage which causes a spark gap to break down at the rate of twice the line frequency (100-120 Hz). The spark gap charges capacitors C1 and C2 which are connected to the primary windings of the resonator coil T2 with an air core. Because of the inductance of the primary windings of T2 and the capacitors, an oscillating current of very high frequency is set up in the circuit. The spark gap is adjusted to reach the resonant frequency of

978-1-4244-2977-6/09 $25.00 © 2009 IEEE

the circuit about 3.8 MHz. High voltage is induced in the secondary windings of T2, and applied to the output A.

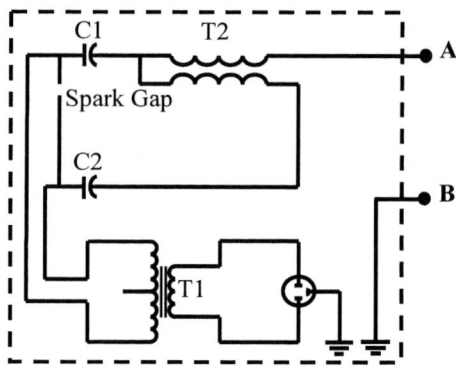

Figure 2: Schematic wiring diagram of the high frequency generator.

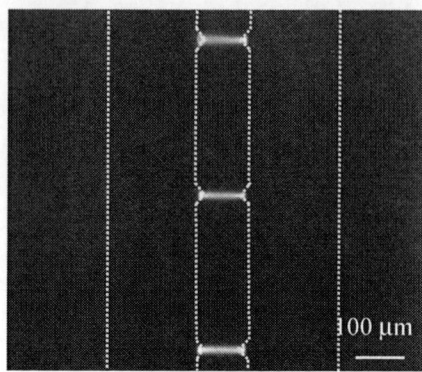

Figure 3: Discharge arcs are generated between the opposing tips once high frequency AC voltage is applied (Dark field image).

ON-CHIP ELECTRICAL DISCHARGE

According to Paschen's law, the static breakdown voltage for generating a self-sustained discharge arc can be calculated from the gap distance between the microelectrode tips and the air pressure in the gap [12]. Up to a certain critical frequency depending upon the gap distance, the breakdown voltage does not practically vary with frequency of the applied field. If the frequency f exceeds the critical value, the interval between any two consecutive onsets of electrical breakdown is small enough not to allow the occurrence of any significant deionization. As a result, positive ions are trapped in the gap and start oscillation, increasing the net positive charges in the gap and lowering the breakdown voltage from its DC value [13]:

$$V_b = \frac{G V_i f d}{\ln\left(\dfrac{G \cdot f \cdot d}{\ln(1 + 1/\gamma)}\right)}$$

(1)

where G is constant specific to the air; V_i is the effective ionizing potential of the air, γ is the Townsend's secondary ionization coefficient, and d is the gap distance between the electrode tips. Given the static breakdown field stress of air at normal atmospheric conditions and the geometries of the microelectrodes used in this work, the threshold breakdown voltage is estimated on the order of a few hundreds volts. The critical frequency is on the order of a few kHz. Since the applied electrical voltage is well beyond the breakdown voltage at 3.8 MHz, self-sustained discharge arcs are induced between the opposing tips of the microelectrodes once high frequency electrical voltage is applied (Figure 3).

HEAT TRANSFER OF ELECTRICAL DISCHARGE

Typically, the temperature at the center of the discharge arc is above 1000 °C by the onset of the arc current oscillation. The thermal energy is utilized to melt the polymer nanofibers therein. Heat transfer through porous nanofibers consists of conduction, convection, and radiation. Since the polymer nanofibers usually have a low fiber volume fraction (less than 10% of the total), heat conduction through the solid portion of the fibrous matrix

is negligible [14]. In addition, we consider radiation to be also a negligible factor. This is because the infrared radiation wavelengths are in the range of 750 nm to 1 mm. The polymer nanofibers with mean diameters of 500 nm would be too small to interact with the thermal radiation. Moreover, the convective heat transfer in nanofibers is also negligible because the very fine nanofibers tend to damp out all convection due to their huge surface area and entangled structures [15]. Therefore, the major mode of heat transfer is the conduction through the air portion of the polymer nanofibers and the conduction through the insulating glass substrate.

Although the complexity and variety of the phenomenon has prevented the development of a complete understanding of electrical discharge, the discharge based microfabrication can be understood using a simplified model by approximating the high-current discharge as a thermal band source sandwiched between the insulating glass substrate and the nanofiber mat. The width of the band source depends upon the arc current [16] and is approximated by the width of visible arc as observed. The lateral dimensions and thickness of the nanofiber mat and the glass substrate are orders greater than the width of the discharge arc.

The temperature distribution of the discharge arc is estimated using finite element method (COMSOL software). For analysis simplicity, we assume that the heat flux generated by the discharge arc is homogenous and keeps constant over time. It is further assumed that the polymer nanofibers are melted immediately once they are exposed to a temperature beyond the melting point. The latent heat during the melting is small and does not affect the local temperature. Therefore, the melting region can be described by an isothermal surface at the melting point of the polymer nanofibers. The nanofibers enclosed by the isothermal surface and the top surface of the glass substrate are subjected to a temperature above the melting point and are melted. The temperature outside the surface is below the melting point, where the nanofibers maintain their fibrous morphology. The temperature field at a given time point can thus be determined. By assuming that the melting point of PEUU is 500 K, the dimensions of the isothermal surface at the melting point can be obtained, as shown in Figure 4. It is seen that the isothermal surface varies with the discharge duration and the applied power. Particularly, the height of the melting region increases at the start of the discharge, and reaches the plateau as the discharge proceeds. The width of the melting region follows the similar trend (as shown in the subfigures). The melting region also depends on the heat flux generated by the discharge arc. With a higher heat flux, the dimensional change of the melting region at the start of the discharge is more drastic. It also takes longer for the thermal equilibrium to be reached. According to these results, the dimensions of the melting regions can be practically controlled by adjusting the discharge time and the applied power.

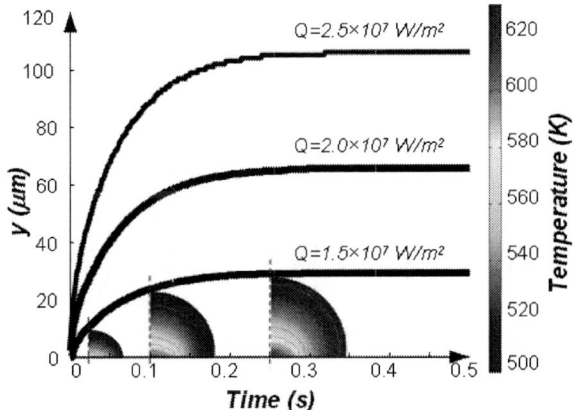

Figure 4: The discharge based microfabrication is understood using a finite element model which estimates the distribution of temperature field around the discharge arc. The dimensions of the melting regions vary with the heat flux generated by the discharge and the sustaining time. The subfigures show the melting regions at different time points at $Q=1.5\times10^7$ W/m²

EXPERIMENTAL VALIDATION

Nanofibers are electrospun using a custom-built electrospinning setup. The central components of the setup include a high voltage power supply, a precisely controlled syringe pump, a syringe with steel capillary, and a conductive collector. PEUU was synthesized as previously reported [17].

Prepolymer solution is driven to the steel capillary by the syringe pump, where the solution was charged by a positive electrical potential (e.g., 15 kV DC). Under such a potential, the polymer solution in the capillary is positively charged. Due to the electrostatic repulsion of the charged liquid and the electric field constructed between the steel capillary and the ground collector, a Taylor cone forms at the tip of the capillary. Once the force overcomes the surface tension, polymer jet is initiated to deposit nanofibers on the grounded aluminum foil. Typical diameter of a single fiber is around 500 nm.

After preparation of polymer nanofibers, experimental validation is performed by covering the aluminum microelectrodes with the polymer nanofiber mat and applying a modest pressure on the top to ensure an intimate contact. Short high-current AC voltage is applied to the microelectrodes with various gap distances. Microstructures with different geometries are successfully created (Figure 5). To create microstructures with small dimensions, the electrical voltage is set at its low limit and the sustaining time of the discharge arc is less than 1 second. The width of the structures is as small as about 20

μm. A closer examination shows that the bottom surface of the microgrooves has modest remelting, where the fibrous structures are not severely deteriorated. This is believed attributed to the high temperature at the center of the discharge arc which is enough to vaporize the polymer and redeposit within the polymer matrix. The limited damage on the nanofibrous structures is essential for cell/tissue engineering applications where the enhanced nutrient delivery through the nanofibrous structures can be achieved.

Furthermore, an array of microgrooves is created by applying the voltage bias to an interdigitated microelectrode array (Figure 6). The batch fabrication capacity in a relatively large area provides the promise of interfacing the combined micro/nanostructures with other microstructures fabricated at wafer-level, which is important for parallel assay and tissue engineering scaffolding.

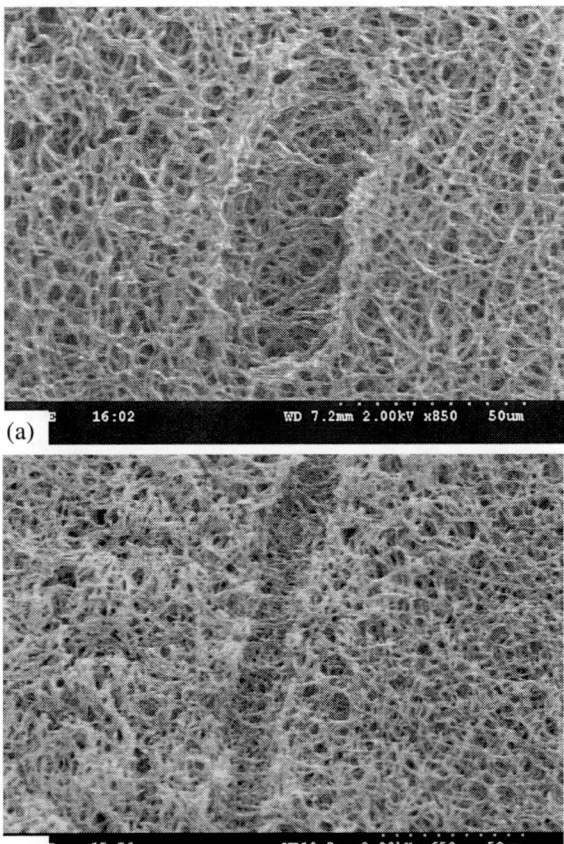

Figure 5: Micropatterns are created on PEUU nanofibers by electrical discharge arcs. (a) A "slotted eye" feature with dimensions 50μm×20μm. (b) A long microgroove with the width 20 μm and the length over 100 μm.

CONCLUSION

In summary, we demonstrate a method to fabricate micropatterns on nanofiber mats by utilizing the thermal energy of discharge arcs. By applying a sufficiently high electrical voltage bias to microelectrodes, self-sustained discharge arc is generated at the air gap. The thermal energy dissipated from the discharge arc melts the

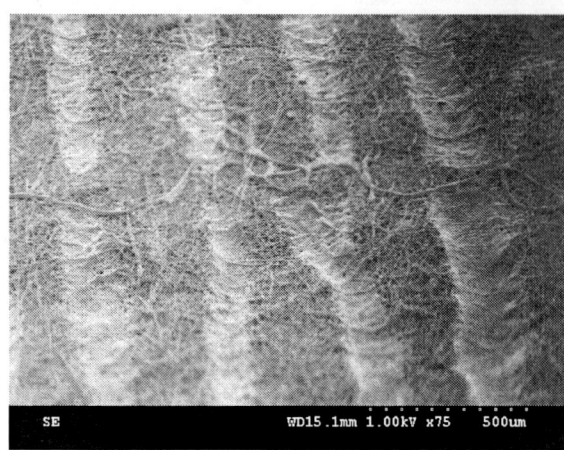

Figure 6: The batch fabrication capacity is demonstrated by creating an array of micropatterns using an interdigitated microelectrodes array.

surrounding nanofibers and creats microgrooves. By designing microelectrodes and controlling the applied voltage, micropatterns can be created. Different from electric discharge machining (EDM) where the electrode material is removed by the electrical arc, this method create microstructures in insulating polymer materials between the microelectrodes. For the best of our knowledge, this is the first effort that the thermal energy of discharge arcs is used as a microfabrication approach for creating microstructures in polymer nanofibers. This method provides a promising start point to interface nanofibers to the world, especially to microstructures, which is of significance for a broad range of applications including cell/tissue engineering, drug delivery and nanoelectronics.

ACKNOWLEDGEMENTS

The authors would like to thank Dr. Jianjun Guan and his laboratory at Ohio State University for their technical help on the preparation of PEUU nanofibers. The work is partially supported by the Institute of Material Research at OSU.

REFERENCES

[1] Q. P. Pham, U. Sharma, and A. G. Mikos, "Electrospinning of polymeric nanofibers for tissue engineering applications: a review," *Tissue Eng,* vol. 12, pp. 1197-1211, 2006.

[2] P. Gibson, H. Schreuder-Gibson, and D. Rivin, "Transport properties of porous membranes based on electrospun nanofibers," *Colloids and Surfaces A: Physicochemical and Engineering Aspects,* vol. 187/188, pp. 469-481, 2001.

[3] H. Schreuder-Gibson, P. Gibson, K. Senecal, M. Sennett, J. Walker, W. Yeomans, D. Ziegler, and P. P. Tsai, "Protective Textile Materials Based on Electrospun Nanofibers," *Advanced Materials,* vol. 34, pp. 44-55, 2002.

[4] R. S. Barhate and S. Ramakrishna, "Nanofibrous filtering media: Filtration problems and solutions from tiny materials," *Journal of Membrane Science,* vol. 296, pp. 1-8, 2007.

[5] S. Panseri, C. Cunha, J. Lowery, U. Del Carro, F. Taraballi, S. Amadio, A. Vescovi, and F. Gelain, "Electrospun micro- and nanofiber tubes for functional nervous regeneration in sciatic nerve transections," *BMC Biotechnol,* vol. 8, p. 39, 2008.

[6] M. Schindler, I. Ahmed, J. Kamal, E. K. A. Nur, T. H. Grafe, H. Young Chung, and S. Meiners, "A synthetic nanofibrillar matrix promotes in vivo-like organization and morphogenesis for cells in culture," *Biomaterials,* vol. 26, pp. 5624-5631, 2005.

[7] W. J. Li, K. G. Danielson, P. G. Alexander, and R. S. Tuan, "Biological response of chondrocytes cultured in three-dimensional nanofibrous poly(epsilon-caprolactone) scaffolds," *J Biomed Mater Res A,* vol. 67, pp. 1105-1114, 2003.

[8] J. L. Charest, A. J. Garcia, and W. P. King, "Myoblast alignment and differentiation on cell culture substrates with microscale topography and model chemistries," *Biomaterials,* vol. 28, pp. 2202-2010, 2007.

[9] Y. Zhao, H. Zeng, J. Nam, and S. Agarwal, "Fabrication of Skeletal Muscle Constructs by Topographic Activation of Cell Alignment," *Biotechnology and Bioengineering,* p. in press, 2008.

[10] D. Khang, M. Sato, R. Price, L,, A. E. Ribbe, and T. J. Webster, "Selective adhesion and mineral deposition by osteoblasts on carbon nanofiber patterns," *Int J Nanomedicine,* vol. 1, pp. 65-72, 2006.

[11] S. Sarkar, G. Y. Lee, J. Y. Wong, and T. A. Desai, "Development and characterization of a porous micro-patterned scaffold for vascular tissue engineering applications," *Biomaterials,* vol. 27, pp. 4775-82, 2006.

[12] G. G. Raju, *Gaseous Electronics: Theory and Practice.* Boca Raton, FL: CRC / Taylor & Francis, 2006.

[13] J. C. Biswas and V. Mitra, "High-Frequency Breakdown and Paschen Law," *Applied Physics,* vol. 19, pp. 377-381, 1979.

[14] A. Lefort, M. J. Parizet, S. Jellouli, and R. Bessege, "Heat transfer from an electric arc to insulating wall," *IEEE Transactions on Plasma Science,* vol. 21, pp. 593-597, 1993.

[15] P. W. Gibson, C. Lee, F. Ko, and D. Reneker, "Application of Nanofiber Technology to Nonwoven Thermal Insulation," *Journal of Engineered Fibers and Fabrics,* vol. 2, pp. 32-40, 2007.

[16] A. Tslaf, "Insulation as affected by immobile high-current discharge," *IEEE Transactions on Electrical Insulation,* vol. EI-14, pp. 51-57, 1979.

[17] J. Guan, M. S. Sacks, E. J. Beckman, and W. R. Wagner, "Synthesis, characterization, and cytocompatibility of elastomeric, biodegradable poly(ester-urethane)ureas based on poly(caprolactone) and putrescine," *J Biomed Mater Res,* vol. 61, pp. 493-503, 2002.

AUTHOR INDEX

Abbaspour-Tamijani, A. 935
Abe, T. 904
Acquaviva, D. 132
Agah, M. 288
Aïd, M. 619
Aiyar, A. R. 447
Akbarian, M. 160
Akgul, M. 19
Akin, T. 813, 817
Akiyama, S. 567
Aljasem, K. 999
Allen, M. G. 447
Alper, S. E. 813
Amaya, S. 709
Ambacher, O. 923
An, J. Y. 951
Ando, T. 661
Arai, F. 51, 375
Armbruster, S. 693
Arnold, D. P. 665
Autizi, E. 630
Averitt, R. D. 108
Ayazi, F. 749, 888, 931
Baca, A. I. 638
Baek, C.-W. 725
Baghoomian, E. 216
Bahl, G. 801, 1095
Bain, J. A. 1107
Baker, M. 841
Balkan, T. 1071
Barry, T. 356
Bartsch, U. 1039
Baskaran, R. 777
Baumann, W. 220
Baygents, J. C. 431
Bearda, T. 136

Beardslee, L. A. 284
Bedair, S. 268
Bedani, M. 860
Benito-Lopez, F. 439
Benzel, H. 693
Berenschot, E. 152
Berenschot, J. W. 324
Bergaud, C. 567
Berry, C. W. 216
Berry, H. 829
Bever, T. 880
Beyaz, M. I. 1091
Bhagavat, M. 88
Bhave, S. A. 23, 896
Bielen, J. A. 916
Binh-Khiem, N. 168, 963
Birkle, G. 43
Blattmann, R. 669
Boga, B. 817
Bogaerts, L. 136
Böhringer, K. F. 531
Bourouina, T. 975
Brama, R. 860
Brand, O. 284
Braun, S. 1063
Bright, V. M. 120, 638
Bronson, J. R. 1043
Brown, J. J. 638
Brueckner, K. 923
Brugger, J. 144
Bruinink, C. M. 152
Buchaillot, L. 868, 919
Burger, R. 443
Butler, D. P. 833
Byun, D. 523
Cai, H. 967, 975, 1023

AUTHOR INDEX

Candler, R. N. 801
Carlborg, C. F. 39
Carraro, C. 611
Casavant, B. P. 304
Casinovi, G. 931
Castaneda, J. 841
Cellere, G. 630
Ceyssens, F. 451
Chae, J. 96, 935
Chaitanya, C. R. Alla 701
Chandorkar, S. 657
Chandrahalim, H. 23, 896
Chang, C.-T. 296
Chang, C.-W. 200
Chang, J. 128
Chang, P.-Z. 204
Chang, Y. J. 120
Chang, Y.-C. 200
Chao, F.-S. 204
Chen, D.-S. 1031
Chen, H. 264
Chen, J.-S. 627
Chen, J.-W. P. 927
Chen, K. L. 801, 1067
Chen, K.-L. 23
Chen, K.-S. 1099
Chen, M.-H. 296, 455
Chen, P.-J. 196, 244
Chen, R. 200
Chen, T. 1015
Chen, Y. 360
Chen, Z. G. 399
Cheng, M.-Y. 92, 292
Cheng, Y. T. 567
Cheung, L. S. L. 379, 431
Chi, X. Z. 104, 809

Chia, B. T. 292
Chiang, A.-S. 100
Chiao, M. 995
Chicherin, D. 892
Chiravuri, S. 252
Chiu, N.-F. 204
Cho, J. 749
Cho, S. K. 1079
Cho, Y. H. 399
Cho, Y.-H. 64
Choi, C.-A. 793
Choi, D. 360
Choi, H.-C. 939
Choi, W. 160
Choi, Y.-K. 475
Chu, L.-A. 100
Chung, S. K. 1079
Cimalla, V. 923
Cismak, A. 220
Clausi, D. 1063
Collard, D. 607
Collins, S. D. 571
Cooper, J. R. 208
Cordero, M. 1075
Coronel, Ph. 132
Couderc, S. 642
Craighead, H. 646
Creemer, J. F. 76, 844
Cristman, P. 757
Cui, J. 104, 809
Currano, L. 591
Czarnecki, P. 852
Dai, Z. 543
Dainesi, P. 634
Dao, D. V. 709, 1007
Davidson, B. D. 120

AUTHOR INDEX

De Boer, M. J. 152, 463, 479

De Moor, P. 136

De Nooijer, C. 136

De Volder, M. 451

Defay, E. 619

Demartini, B. E. 1095

Demello, A. J. 523

Demirci, K. S. 284

Despont, M. 603

Dhull, R. K. 991

Diamond, D. 439

Dijkstra, M. 479

Dimov, I. K. 356

Ding, H. T. 809

Ding, Z. 407

Draghi, L. 713

Drechsler, U. 603

Drittenbass, S. 575

Ducree, J. 356, 439, 443, 539

Dueck, M. 304

Duncombe, T. A. 531

Durrer, L. 575

Dy, E. 391

Edamoto, M. 1055

Ekeom, D. 919

Ekkels, P. 852

Elata, D. 753, 927, 1111

Elwenspoek, M. C. 324, 479

Esashi, M. 705, 1027

Eskinazi, I. 665

Etzkorn, J. R. 713

Eun, Y. 164, 1035

Evans, A. T. 252

Fan, K. 108

Fan, L.-S. 627

Fang, L. 595

Fang, W. 100, 172, 200, 805, 1047

Fattaccioli, J. 567, 677

Fedder, G. K. 268, 1107

Fernández-Bolaños, M. 634

Fiorini, P. 623

Fischer, M. 693

Flood, A. H. 595

Fonseca, J. G. 443

Foulds, I. G. 583

Frazier, A. B. 399

French, P. J. 140, 844

Fujii, H. 1059

Fujii, M. 519

Fujii, T. 864

Fujita, H. 403, 531, 607, 697, 884, 1075, 1083

Fujita, M. 184

Fujiwara, N. 435

Fujiyoshi, M. 733

Fukuda, M. 260

Fukuoka, M. 124

Fuller, L. 991

Funatsu, N. 507

Fung, A. 360

Ganzerli, M. 860

Gao, X. 931

Garcia-Cordero, J. L. 356, 439

Garipov, G. 979

Garrell, R. L. 280

Gaspar, J. 7, 220, 599, 1039

Gehner, A. 1003

George, S. M. 120

Ghodssi, R. 1091

Gianchandani, Y. B. 212, 252, 837

Gieschke, P. 765

Gilmour, R. 336

Gnudi, A. 860

AUTHOR INDEX

Gojo, R. 256

Goodman, M. B. 188

Goryu, A. 224

Gradin, H. 1063

Graham, A. B. 23, 741, 801

Grasshoff, T. 947

Green, S. R. 212

Gruenberger, R. 880

Gudipaty, T. 379

Guo, T.-F. 415

Guo, Z. Y. 104, 809

Gurau, B. 35

Guzman, R. 431

Hagleitner, C. 603

Hagner, M. 220

Halder, S. 136

Han, A. 399

Han, J. 80

Han, K.-H. 308

Han, S.-I. 308

Haneveld, J. 463

Hao, Y. L. 673

Hartwel, P. 741

Hartwell, P. G. 1095

Hasegawa, Y. 769

Hashimoto, S. 60

Haspeslagh, L. 136

Hassanin, H. S. 717

Hatakeyama, Y. 705

Heilmann, A. 220

Heimark, R. L. 431

Hein, M. A. 923

Helbling, T. 575

Held, J. 7, 220, 599

Helveg, S. 76

Heo, Y. J. 68

Herfst, R. W. 916

Herrmann, M. 765

Herwik, S. 232

Hibert, C. 144

Hida, H. 403, 908

Hierold, C. H. 300, 575, 669

Higgs, G. 407

Higo, A. 697, 1083

Hill, T. A. 841

Hilton, J. P. 344

Hirshberg, A. 753

Ho, C. M. 391

Ho, D. 360

Hoffmann, J. 539

Homma, T. 499

Honda, H. 908

Hong, C.-T. 100

Hong, J. 523

Hoveling, G. H. 76

Howe, R. T. 23, 741, 927

Hsiao, S.-Y. 172

Hsieh, C.-Y. 627

Hsieh, J. 927

Hsu, H. Y. 579

Hsu, H.-Y. 411

Hsu, T.-H. 455

Huang, R. 745

Huang, S.-B. 515

Huang, T. J. 332, 595

Huang, W.-L. 19

Huang, X. 352

Huang, X.-J. 475

Humayun, M. 196

Humayun, M. S. 244

Hwang, G. 793

Hwang, J. 387, 487, 491

AUTHOR INDEX

Hwang, Y.-S 725
Hyeon, I.-J. 725
Iannacci, J. 860
Ihara, K. 864
Ikeda, A. 677
Ikedo, A. 224, 721
Ikeuchi, M. 124, 348
Ikuta, K. 11, 124, 348
Ilic, B. 646
Im, M. 475
Imai, N. 499
Inoue, N. 507
Inoue, S. 653, 1051, 1059
Ionescu, A. M. 132, 634
Irieda, T. 864
Ishida, M. 192, 224, 503, 721
Ishihara, H. 367
Ito, Y. 419
Itoh, T. 797
Iwai, K. 371
Iwase, E. 68, 176, 551, 943, 963
Izadi, H. 583
Jaganatharaja, R. K. 152
Jakobs, F. 773
Jalabert, L. 607
Jamshidi, A. 411, 579
Jang, W. G. 959
Jansen, H. V. 324
Je, S.-S. 96
Jee, H. 729
Jeon, J. A. 979
Jeong, B. 164
Jeong, D.-H. 793
Jiang, K. 717
Jiang, L. 208, 379
Jiang, Y. Q. 587

Jin, J. Y. 979
Jo, K. W. 959
Johnson, A. T. 320
Jones, D. 841
Jones, T. S. 320
Judy, D. C. 591, 896
Judy, M. 88
Juluri, B. K. 595
Jun, M. H. 951
Jung, H.-K. 725
Jung, K.-D. 156
Junge, S. 773
Kaanta, B. C. 264
Kamezawa, C. 615
Kamiya, S. 599
Kan, T. 551, 963
Kaneko, H. 192
Kang, B.-G. 487
Kanzaki, R. 180
Karim, K. S. 583
Karle, M. 276
Kasagi, N. 60, 615, 1055
Kashiwagi, K. 1055
Kashiwagi, S. 653
Katahira, K. 419
Kato, D. 260
Kato-Negishi, M. 423
Kaul, R. 896
Kawai, K. 519
Kawai, Y. 1027
Kawano, K. 955
Kawano, T. 192, 224, 721
Kawasaki, S. 884
Kawashima, T. 192, 721
Ke, F. 856, 872
Keller, M. 232

AUTHOR INDEX

Kenny, T. W. 23, 657, 741, 801, 1067

Khamis, S. M. 320

Khrenov, B. 979

Khuri-Yakub, B. T. 757

Kim, B. 19, 741

Kim, B. J. 567, 642, 677

Kim, C.-H. 156

Kim, C.-J. 160, 280, 471, 900

Kim, D.-H. 475

Kim, D.-S. 939

Kim, E. S. 316, 761

Kim, H. 483

Kim, H. S. 399

Kim, J. 96, 164, 523, 789, 1035

Kim, J. E. 979

Kim, J. G. 677

Kim, J. K. 729

Kim, J.-K. 84

Kim, J.-M. 27

Kim, K. H. 511

Kim, M. 979

Kim, M. G. 951, 959

Kim, S. 563

Kim, S. G. 563

Kim, S. H. 447

Kim, S.-Y. 487, 491

Kim, W. 156

Kim, Y. K. 979

Kim, Y.-H. 387

Kim, Y.-J. 272, 387, 487, 487, 491, 491

Kim, Y.-K. 725

King, W. P. 543, 848

Kinoshita, Y. 260

Kishi, H. 904

Kitajima, T. 419

Klose, T. 947

Knese, K. 693

Kobayashi, K. 11

Kobayashi, T. 260, 797

Koester, P. J. 220

Koga, H. 260

Kolster, M. L. 152

Koltay, P. 43, 1087

Komatsubara, M. 653, 1051

Kondoh, T. 507

Konishi, S. 260, 435, 499, 737

Kooyman, P. J. 76

Korvink, J. G. 1015

Kotake, N. 256

Kozicki, M. N. 96

Kratt, K. 777

Krausse, S. 1015

Krijnen, G. J. M. 152

Kroninger, C. M. 1043

Krumbholz, S. 1087

Krylov, S. 646

Kuhl, M. 765

Kühne, S. 669

Külah, H. 817, 1071

Kumar, A. S. 595

Kumar, K. 1011

Kumar, S. 833

Kumemura, M. 531, 607

Kuo, P.-L. 340

Kuo, S.-M. 983

Kupnik, M. 757

Kuribayashi, K. 367

Kuypers, F. A. 64

Kwon, S. 387

Kwon, T. H. 785

Lackey, C. 841

Lai, R.-J. 1031

AUTHOR INDEX

Lai, Y.-T. 92
Lal, A. 336, 547
Lambertus, G. 264
Lammerink, T. S. J. 152, 463, 479
Lang, W. 773
Lapisa, M. 1003
Larcher, L. 860
Lauria, J. 136
Lazarus, N. 268
Le Rhun, G. 619
Le, N. C. H. 1007
Lee, B.-K. 148, 785
Lee, C.-C. 172
Lee, C.-I. 272
Lee, D.-W. 84
Lee, G.-B. 47, 340, 415, 515
Lee, H. 308
Lee, H. J. 757
Lee, H. K. 657, 801
Lee, H. S. 785
Lee, H. W. 563
Lee, J. 848, 979
Lee, J. H. 951, 959
Lee, J.-H. 475, 793
Lee, J.-I. 164
Lee, J.-S. 487, 491
Lee, L. P. 304
Lee, Luke P. 55
Lee, M.-L. 793
Lee, S. K. 951
Lee, S.-H. 749
Lee, S.-W. 749
Lee, T.-M. 939
Lee, W. C. 64
Lee, Y. K. 395
Lee, Y.-T. 200

Lei, U. 204
Lemke, B. 777
Leung, M. 360
Leus, V. 1111
Li, C.-M. 312
Li, S. 352
Li, W. 248
Li, W.-C. 19
Li, X. 328, 555
Liao, C. Y. 689
Liao, H.-H. 204
Lien, K.-Y. 340
Lin, A. 316
Lin, C.-H. 983
Lin, C.-L. 292
Lin, C.-W. 200, 204, 292
Lin, J. C.-H. 196
Lin, L. 31, 72, 128, 304, 587, 864
Lin, L. T. 104, 809
Lin, Q. 344, 352
Lin, W.-Y. 47
Lin, Y. 19
Lin, Y.-H. 47, 415
Liu, A. Q. 967, 975, 1023
Liu, B. 967, 975, 1023
Liu, C.-J. 340
Liu, F. 611
Liu, J. 1107
Liu, L. 1103
Liu, L. J. 619
Liu, M. C. 427
Liu, P. 527
Liu, P.-C. 204
Liu, X. D. 527
Liu, X. S. 809
Liu, Y. 595

AUTHOR INDEX

Liu, Z. 1015
Liu, Z. B. 395
Lo, C. 268
Lo, R. 236
Loo, J. A. 280
Lopez, G. 184
Lord, S. F. 657
Löw, P. 567
Lu, S.-S. 204
Lu, Y. W. 459, 991
Luzinova, Y. 284
Maboudian, R. 611
Mabuchi, K. 256
Mader, D. 987
Maeda, M. 184
Mahajerin, A. 304
Maharbiz, M. M. 216
Mao, H. Y. 673
Margesin, B. 860
Mark, D. 539
Maruyama, H. 375
Masel, R. I. 35
Massieu, J. L. 947
Masuda, T. 797
Matsumoto, K. 68, 168, 176, 551,
    781, 821, 943, 963, 971, 1019
Matsumoto, Y. 467
McCarthy, M. 1091
McCormick, D. T. 864
Melamud, R. 657, 741, 1067
Meneghesso, G. 630
Meng, E. 236
Messana, M. 741
Metz, M. 693
Metz, T. 43
Miao, J. 312, 856, 872
Miki, N. 467

Miki, T. 1051
Miller, M. 619
Millet, O. 868
Misawa, N. 180
Mita, M. 697
Mitsuno, H. 180
Miwa, J. 60, 276
Mizaikoff, B. 284
Mogulkoc, B. 324
Molenbroek, A. M. 76
Momose, T. 681
Moon, H. 360
Morimoto, K. 260
Morimoto, Y. 56
Morita, N. 507
Morizawa, Y. 1055
Mukerjee, E. V. 848
Müller, C. 1087
Munt, S. 1087
Murase, H. 419
Murata, M. 507
Murmann, B. 657
Na, G. W. 979
Na, H. 1035
Nagamori, E. 403
Nagasawa, H. 1051
Nagaura, Y. 499
Naito, J. 495
Najafi, K. 483, 749
Naka, N. 653
Nakada, M. 697, 1083
Nakai, A. 176
Nakamura, K. 864
Nam, S. 979
Namazu, T. 653, 1051, 1059
Nawaz, M. 880

AUTHOR INDEX

Nayve, R. ...507
Neale, S. ...411, 579
Nelson, W. ...280
Neves, H. ...232
Nguyen, C. T.-C. ...19
Nguyen, T. ...344
Niebelschuetz, F. ...923
Niklaus, F. ...1003
Ninomiya, T. ...467
Nishimura, T. ...60
Noda, K. ...781
Noge, H. ...955
Noh, J.-H. ...939
Noman, M. ...1107
Nonomura, Y. ...733
Norman, J. J. ...407
Nurcahyo, Y. ...232, 765
Oba, M. ...1055
Oberhammer, J. ...15, 856, 892
Obinata, K. ...403
Ocak, I. E. ...817
O'Grady, J. ...356
Ogura, D. ...769
Ohkubo, T. ...681
Ohmori, H. ...419
Ohno, K. ...507
Ohta, A. T. ...411, 579
Ohtsuki, K. ...653
Okada, H. ...797
Okandan, M. ...841
Okayama, Y. ...467
Okitsu, T. ...423
Okochi, M. ...908
Okuda, Y. ...348
Okumura, R. ...943
Olsson, R. ...841

Onda, K. ...51
Ono, T. ...1027
Onoe, H. ...176
Oralkan, Ö. ...757
Osawa, K. ...467
Osten, W. ...7
Ota, K. ...864
Ou, K.-S. ...1099
Ou, Y.-C. ...204
Paccagnella, A. ...630
Pacheco, S. ...619
Padilla, W. J. ...108
Pakula, L. S. ...140
Palanivelu, R. ...208
Parameswaran, M. ...583
Park, D. ...387
Park, H. G. ...363
Park, I. H. ...979
Park, J. ...487
Park, J. H. ...979
Park, K. ...543
Park, K. H. ...363
Park, K. K. ...757
Park, S. ...164
Park, S. E. ...491
Park, S.-J. ...188
Park, S.-W. ...685
Park, Y. S. ...959
Park, Y.-S. ...979
Pärr, G. ...535
Parra, E. ...31
Parviz, B. A. ...713
Paul, O. ...7, 220, 232, 599, 765, 777, 1039
Paust, N. ...1087
Pearson, S. ...841
Pedrini, G. ...7

AUTHOR INDEX

Peeri, Y. 216
Pei, R. 344
Peirs, J. 1063
Pellegrin, S. 829
Peng, I. 280
Peretti, V. 630
Perlin, G. E. 228
Perruchot, F. 619
Petzold, B. 188
Pham, H. T. M. 140, 383
Piazza, G. 320, 912
Pillans, B. 876
Pisano, A. P. 64, 685
Polcawich, R. G. 591, 896, 1043
Pollock, C. R. 547
Presser, N. 360
Provine, J. 741
Pruitt, B. L. 188, 407
Puchades, I. 991
Puers, R. 451, 852
Pulskamp, J. S. 591, 896, 1043
Qin, Y. 208
Qiu, X. 1103
Qu, Y. Q. 657
Quévy, E. P. 927
Raberg, W. 880
Räisenen, A. V. 892
Rajaraman, V. 140
Rebeiz, G. M. 27, 876
Reines, I. 876
Reis, N. 443
Ren, Z. 19
Resnick, P. 841
Rey, P. 619
Reynaerts, D. 451, 1063
Ricco, A. J. 356, 439

Riegger, L. 539
Rinaldi, M. 320, 912
Robin, R. 868
Rodger, D. C. 248
Roman, C. 575
Roper, C. 611
Roth, G. 276
Rothuizen, H. 603
Rottenberg, X. 852
Rubtsov, V. 160
Ruther, P. 220, 232, 765
Ryu, K. 1079
Saati, S. 244
Saeedi, E. 713
Sahin, K. 813
Saif, T. 80
Saito, T. 681
Sakuma, S. 51, 375
Salvia, J. 657, 801
Samarao, A. K. 888
Sammoura, F. 88
Samukawa, Seiji 112
Sandner, T. 947
Santagata, F. 844
Santagata-Iervolino, E. 535
Sarajlic, E. 607, 1075
Sari, I. 1071
Sarro, P. M. 76, 140, 383, 844
Sasaki, H. 403
Sasao, T. 348
Sato, H. 216, 499
Sato, K. 495, 661, 769, 908
Sato, R. 184
Sawada, K. 192, 224, 503, 721
Sawano, S. 435
Sawyer, W. 88

AUTHOR INDEX

Schenk, H. 947

Schlatmann, B. 136

Schlesinger, T. E. 1107

Schmid, S. 300

Schmidt, M. E. 599

Schoen, F. 880

Schroeder, J. 431

Schuck, P. J. 128

Schultz, J. 352

Schüttler, M. 232

Sebastian, A. 603

Sedaghat-Pisheh, H. 27

Seghete, D. 120

Segueni, K. 868

Seidel, H. 693

Seidl, K. 232

Seifert, A. 987, 999

Seita, H. 884

Seki, T. 1055

Sekiguchi, T. 499

Sen, P. 900

Senda, H. 733

Senesky, D. G. 685

Seo, J. H. 284

Seo, Y. H. 511

Seto, S. 507

Shah, G. J. 471

Shannon, M. A. 35

Shao, C. Y. 1027

Shen, C. 383

Shen, C. J. 336

Shetye, S. B. 665

Shi, J. 332

Shih, W.-P. 204

Shikida, M. 403, 495, 769, 908

Shimizu, K. 403, 737

Shimogaki, Y. 681

Shimoyama, I. 68, 168, 176, 551, 781, 821, 943, 963, 971, 1019

Shin, D.-M. 399

Shoji, S. 499, 519

Simard, K. 583

Singh, P. 304

Siu, C. P. B. 995

Skotnicki, T. 132

Smith, R. L. 571

Sochol, R. D. 304

Someya, T. 642

Somjit, N. 15

Song, C. 447

Song, E.-S. 725

Song, Y.-H. 148

Sosnowchik, B. D. 128

Spadaccini, C. M. 848

Sparks, A. 88

Spinney, P. S. 571

Stamm, M. T. 379

Steinecker, W. H. 264

Stemme, G. 15, 39, 892, 1003, 1063

Stephan, R. 923

Sterner, M. 892

Stieglitz, T. 232

Stoddart, J. F. 595

Stojanovic, M. 344

Stopa, A. 431

Strikwerda, A. 108

Stulemeijer, J. 916

Stupian, G. 360

Su, H.-T. 1047

Su, Y. C. 689

Sugano, K. 184

Sugiura, M. 503

Sugiyama, M. 681

AUTHOR INDEX

Sugiyama, S. 709, 1007

Suhm, A. 619

Sun, C.-M. 805

Sun, W. 884

Suzuki, T. 256

Suzuki, Y. 60, 615, 769, 1055

Swaminathan, V. V. 35

Tabata, O. 184

Tachibana, H. 955

Tai, Y. C. 248, 427, 745

Tai, Y.-C. 196, 244

Takahashi, H. 821

Takahashi, K. 697, 1083

Takahashi, T. 1083

Takahashi, Y. 459

Takahata, K. 701

Takahata, T. 971

Takama, N. 567, 677

Takamatsu, S. 943, 1019

Takao, H. 503

Takei, A. 963

Takei, K. 192, 224

Takei, Y. 551

Takeuchi, S. 56, 180, 256, 363, 367, 371, 423

Takumi, T. 661

Tamil, J. 967, 975, 1023

Tan, C. W. 872

Tanaka, K. 507

Tanaka, N. 503

Tanaka, S. 705

Tane, R. 124

Tanemura, T. 184

Tang, T.-L. 1047

Tang, Z. 825

Tao, H. 108

Tautorat, C. 220

Taylor, R. 407

Tazzoli, A. 630

Teh, K. S. 459

Tilmans, H. A. C. 136, 852

Tonisch, K. 923

Tonomura, W. 737

Torfs, T. 232

Toshiyoshi, H. 697, 884, 1083

Trott, W. 841

Truax, S. 284

Tsai, J.-C. 1031

Tsai, M.-H. 805

Tsai, Y.-C. 204

Tsamados, D. 132, 634

Tsang, S.-H. 583

Tsao, C.-M. 92

Tseng, F.-G. 296, 455

Tseng, Y.-T. 296

Tsou, W.-A. 627

Tsuchiya, T. 184

Tsuda, Y. 56, 423

Tsukuni, H. 507

Tsurui, T. 1051

Turner, K. L. 1095

Uchic, M. 80

Ueda, H. 955

Uejima, T. 681

Ullmann, S. 76

Umezu, S. 419

Unnikrishnan, S. 324

Ur, S. C. 761

Usami, H. 507

Valley, J. K. 411, 579

Van Beek, J. T. M. 650

Van Bommel, M. 136

Van Der Avoort, C. 650

AUTHOR INDEX

Van Der Wijngaart, W. 39, 1063

Van Drieenhuizen, B. 136

Van Herwaarden, A. W. 535

Van Hoof, C. 623

Van Wingerden, J. 650

Vanneer, R. 136

Varma, R. 244

Vazquez-Mena, O. 144

Vellekoop, M. J. 535

Vigna, Benedetto 1

Villanueva, G. 144

Villareal, G. 360

Von Stetten, F. 276, 539

Wägli, P. 300

Waguespack, R. P. 829

Waibel, P. 987

Waits, C. M. 1091

Wakui, D. 499

Wallrabe, U. 1015

Walmsley, R. G. 1095

Wang, H. 995

Wang, J. 825

Wang, Q. 352

Wang, S. 1067

Wang, W. 415

Wang, X. 1103

Wang, Z. 312, 623

Watamura, K. 348

Weber, H. 693

Weber, W. 880

Wei, P. 407

Weigel, R. 880

Weiss, P. S. 595

Wells, J. 1007

Wen, K.-A. 627

Wetzel, E. D. 1043

Wiegerink, R. J. 152, 463, 479

Wien, W. 535

Wiesmann, D. 603

Wilson, C. G. 829

Winkler, B. 880

Wise, K. D. 228

Witvrouw, A. 136

Wright, S. A. 837

Wu, D. 673

Wu, M. C. 411, 579

Wu, W. G. 673

Xia, X. 328

Xu, B. J. 395

Xu, J. 673

Xu, T. 312

Xu, W. 935

Yagi, T. 507

Yama, G. 801

Yamahata, C. 607, 1075

Yamane, D. 697, 884

Yamanishi, Y. 51, 375

Yan, G. Z. 104, 809

Yang, C. K. 844

Yang, K. 88

Yang, L.-J. 204

Yang, M. 395

Yang, S. 951

Yang, S.-A. 292

Yang, Y. 328, 555

Yang, Y. W. 595

Yang, Y.-J. 92, 204, 292

Yang, Z. C. 104, 809

Ye, T. 595

Yeh, Y.-T. 627

Yi, S. H. 761

Yin, C.-Y. 1031

AUTHOR INDEX

Yokokawa, R. 1007
Yokota, T. 495
Yokoyama, T. 1055
Yoneoka, S. 741, 801
Yoo, B. W. 979
Yoo, K. 789
Yoon, J.-B. 148, 475
Yoon, S.-I. 272
Yoon, Y. K. 729
Yoshida, H. 260
Yoshihata, Y. 963
Yoshimizu, N. 547
Yu, B. 196
Yu, J. J. 395
Yu, L. 312
Yun, S.-S. 793
Yun, T.-S. 729
Zaman, M. F. 749
Zandbergen, H. W. 76
Zappe, H. 987, 999
Zareian-Jahromi, M. A. 288
Zeng, H. 240, 559, 995
Zengerle, R. 43, 276, 539, 1087
Zhang, F. X. 527
Zhang, H. X. 673
Zhang, Q. 1103
Zhang, Q. X. 967, 1023
Zhang, W. 975, 1023
Zhang, X. 108, 264, 935
Zhang, X. J. 1011
Zhang, X. M. 967, 975, 1023
Zhang, Y. 84
Zhao, Q. C. 104, 809
Zhao, Y. 240, 559
Zhdaneev, O. 264
Zheng, X. J. 431

Zheng, Y. B. 332
Zhou, Q. 72, 587
Zhou, Y. 1103
Zhou, Z. Y. 527
Zhu, L. 35
Zhu, R. 527
Zhu, W. M. 975, 1023
Zhu, Y. 1103
Ziaei-Moayyed, M. 927
Ziaie, B. 407
Ziegler, C. 1087
Zimmer, F. 1003
Zohar, Y. 208, 379, 431
Zuniga, C. 320, 912

9781424429776

2009 IEEE 22nd International Conference on Micro Electro Mechanical Systems (MEMS)

Sorrento, Italy
25-29 January 2009

IEEE Catalog Number: CFP09MEM-POD
ISBN: 978-1-42442-977-6

2009 IEEE 22nd International Conference on Micro Electro Mechanical Systems

(MEMS)

Sorrento, Italy
25-29 January 2009

Pages 563-1114

IEEE Catalog Number: CFP09MEM-PRT
ISBN: 978-1-4244-2977-6

Copyright © 2009 by the Institute of Electrical and Electronic Engineers, Inc
All Rights Reserved

Copyright and Reprint Permissions: Abstracting is permitted with credit to the source. Libraries are permitted to photocopy beyond the limit of U.S. copyright law for private use of patrons those articles in this volume that carry a code at the bottom of the first page, provided the per-copy fee indicated in the code is paid through Copyright Clearance Center, 222 Rosewood Drive, Danvers, MA 01923.

For other copying, reprint or republication permission, write to IEEE Copyrights Manager, IEEE Service Center, 445 Hoes Lane, Piscataway, NJ 08854. All rights reserved.

***This publication is a representation of what appears in the IEEE Digital Libraries. Some format issues inherent in the e-media version may also appear in this print version.**

IEEE Catalog Number:	CFP09MEM-PRT
ISBN 13:	978-1-4244-2977-6
ISSN:	1084-6999

Additional Copies of This Publication Are Available From:

Curran Associates, Inc
57 Morehouse Lane
Red Hook, NY 12571 USA
Phone: (845) 758-0400
Fax: (845) 758-2633
E-mail: curran@proceedings.com

TABLE OF CONTENTS

INVITED SPEAKER I

MEMS EPIPHANY 1
Benedetto Vigna

SESSION I – MEMS ACTUATORS

DEVELOPMENT OF CALIBRATION STANDARDS FOR THE OPTICAL MEASUREMENT OF IN-PLANE DISPLACEMENTS OF MICROMECHANICAL COMPONENTS 7
J. Gaspar, J. Held, G. Pedrini, W. Osten, and O. Paul

3D MAGNETIC MICROACTUATOR MADE OF NEWLY DEVELOPED MAGNETICALLY MODIFIED PHOTOCURABLE POLYMER AND APPLICATION TO SWIMMING MICROMACHINE AND MICROSCREWPUMP...11
K. Kobayashi and K. Ikuta

SESSION II – RF MEMS

NOVEL CONCEPT OF MICROWAVE MEMS RECONFIGURABLE 7X45º MULTI-STAGE DIELECTRIC-BLOCK PHASE SHIFTER...15
N. Somjit, G. Stemme, and J. Oberhammer

MICROMECHANICAL RESONANT DISPLACEMENT GAIN STAGES 19
B. Kim, Y. Lin, W.-L. Huang, M. Akgul, W.-C. Li, Z. Ren, and C.T.-C. Nguyen

EPITAXIAL SILICON MICROSHELL VACUUM-ENCAPSULATED CMOS-COMPATIBLE 200 MHz BULK-MODE RESONATOR..23
K.-L. Chen, H. Chandrahalim, A.B. Graham, S.A. Bhave, R.T. Howe, and T.W. Kenny

A NOVEL STRESS-GRADIENT-ROBUST METAL-CONTACT SWITCH 27
H. Sedaghat-Pisheh, J.-M. Kim, and G.M. Rebeiz

SESSION III – FUEL CELL POWER MEMS

MICROBIAL FUEL CELL BASED ON ELECTRODE-EXOELECTROGENIC BACTERIA INTERFACE 31
E. Parra and L. Lin

A MICRO HYDROGEN GENERATOR WITH A MICROFLUIDIC SELF-REGULATING VALVE FOR SENSORS AND FUEL CELLS 35
V.V. Swaminathan, L. Zhu, B. Gurau, R.I. Masel, and M.A. Shannon

SESSION IV- MEMS IN MICROFLUIDICS

MICROCHANNELS WITH SUBSTANTIAL FRICTION REDUCTION AT LARGE PRESSURE AND LARGE FLOW 39
C.F. Carlborg, G. Stemme, and W. van der Wijngaart

STARJET: PNEUMATIC DISPENSING OF NANO- TO PICOLITER DROPLETS OF LIQUID METAL 43
T. Metz, G. Birkle, R. Zengerle, and P. Koltay

CONTINUOUS MICRO-PARTICLE SEPARATION USING OPTICALLY-INDUCED DIELECTROPHORETIC FORCES 47
W.-Y. Lin, Y.-H. Lin, and G.-B. Lee

POWERFUL ACTUATION OF MAGNETIZED MICROTOOL BY FOCUSED MAGNETIC FIELD ON A DISPOSABLE MICROFLUIDIC CHIP 51
F. Arai, S. Sakuma, Y. Yamanishi, and K. Onda

INVITED SPEAKER II

NANOBIOPHOTONICS AND BIOASICS FOR BIOMEDICAL INNOVATIONS 55
Luke P. Lee

SESSION V- CELLS AND PARTICLES

RECONSTRUCTION OF 3D HIERARCHIC MICRO-TISSUES USING MONODISPERSE COLLAGEN MICROBEADS 56
Y. Morimoto, Y. Tsuda, and S. Takeuchi

LABEL-FREE CONTINUOUS MICRO CELL SORTER WITH ANTIBODY-IMMOBILIZED OBLIQUE GROOVES 60
S. Hashimoto, T. Nishimura, J. Miwa, Y. Suzuki, and N. Kasagi

FLOW-LYSOMETRY AND ITS BIOMEDICAL APPLICATION: CYTOSOLIC ANALYSIS OF SINGLE CELLS IN LARGE POPULATIONS 64
W.C. Lee, F.A. Kuypers, Y.-H. Cho, and A.P. Pisano

STRETCHABLE SUBSTRATES FOR THE MEASUREMENT OF INTRACELLULAR CALCIUM ION CONCENTRATION RESPONDING TO MECHANICAL STRESS 68
Y.J. Heo, E. Iwase, K. Matsumoto, and I. Shimoyama

SESSION VI – MICRO PROCESSING AND CHARACTERIZATION

MICRO CHEMICAL VAPOR DEPOSTION SYSTEM: DESIGN AND VERIFICATION 72
Q. Zhou and L. Lin

MEMS NANOREACTOR FOR ATOMIC-RESOLUTION MICROSCOPY OF NANOMATERIALS IN THEIR WORKING STATE 76
J.F. Creemer, S. Helveg, G.H. Hoveling, S. Ullmann, P.J. Kooyman, A.M. Molenbroek, H.W. Zandbergen, and P.M. Sarro

NOVEL MEMS APPARATUS FOR IN SITU THERMO-MECHANICAL TENSILE TESTING OF MATERIALS AT THE MICRO- AND NANO-SCALE 80
J. Han, M. Uchic, and T. Saif

A NEW MICRO-FOUR-POINT PROBE DESIGN FOR VARIOUS APPLICATIONS 84
J.-K. Kim, Y. Zhang, and D.-W. Lee

STICTION IN LOW HUMIDITY ENVIRONMENT 88
F. Sammoura, A. Sparks, W. Sawyer, M. Bhagavat, M. Judy, and K. Yang

SESSION VII – PHYSICAL MEMS

A NOVEL HIGHLY-TWISTABLE TACTILE SENSING ARRAY USING EXTENDABLE SPIRAL ELECTRODES...92
  M.-Y. Cheng, C.-M. Tsao, Y.-T. Lai, and Y.-J. Yang

A DIRECTIONAL CAPACITIVE MEMS MICROPHONE USING NANO-ELECTRODEPOSITS 96
  S.-S. Je, J. Kim, M.N. Kozicki, and J. Chae

NANOTEXTURE ELECTRODE ON NANOPOROUS AAO DIELECTRIC FOR MICRO TACTILE SENSING DEVICES...100
  C.-T. Hong, L.-A. Chu, A.-S. Chiang, and W. Fang

AN ELECTRICALLY DECOUPLED LATERAL-AXIS TUNING FORK GYROSCOPE OPERATING AT ATMOSPHERIC PRESSURE 104
  Z.Y. Guo, L.T. Lin, Q.C. Zhao, J. Cui, X.Z. Chi, Z.C. Yang, and G.Z. Yan

TERAHERTZ METAMATERIALS WITH SIMULTANEOUSLY NEGATIVE ELECTRIC AND MAGNETIC RESONANCE RESPONSES BASED ON BIMATERIAL POP UP STRUCTURES...108
  H. Tao, A. Strikwerda, K. Fan, W.J. Padilla, R.D. Averitt, and X. Zhang

INVITED SPEAKER III

DAMAGE-FREE PLASMA ETCHING PROCESSES FOR FUTURE NANOSCALE DEVICES...112
  Seiji Samukawa

SESSION VIII – NANOPROCESSING AND NEMS

ATOMIC LAYER DEPOSITION (ALD) TUNGSTEN NEMS DEVISES VIA A NOVEL TOP-DOWN APPROACH...120
  B.D. Davidson, Y.J. Chang, D. Seghete, S.M. George, and V.M. Bright

NANOFIBROUS SURFACE PATTERNING USING NANO-MESHED MICROCAPSULES INDUCED BY PHASE SEPARATION ASSISTED ELECTROSPRAY 124
  M. Ikeuchi, R. Tane, M. Fukuoka, and K. Ikuta

TUNABLE OPTICAL ENHANCEMENT FROM A MEMS-INTEGRATED TIO$_2$ NANOSWORD PLASMONIC ANTENNA 128
  B.D. Sosnowchik, P.J. Schuck, J. Chang, and L. Lin

MICROELECTROMECHANICAL METAL-AIR-INSULATOR-SEMICONDUCTOR (MEM-MAIS) DIODE: A NOVEL HYBRID DEVICE FOR ESD PROTECTION.......................132
  D. Acquaviva, D. Tsamados, Ph. Coronel, T. Skotnicki, and A.M. Ionescu

SESSION IX- FABRICATION AND PACKAGING

PACKAGING OF 11 MPIXEL CMOS-INTEGRATED SIGE MICRO-MIRROR ARRAYS 136
  A. Witvrouw, H.A.C. Tilmans, L. Bogaerts, P. De Moor, T. Bearda, S. Halder, L. Haspeslagh, B. Schlatmann, M. van Bommel, C. de Nooijer, J. Lauria, R. Vanneer, and B. van Drieenhuizen

ROBUST WAFER-LEVEL THIN-FILM ENCAPSULATION OF MICROSTRUCTURES USING LOW STRESS PECVD SILICON CARBIDE..140
  V. Rajaraman, L.S. Pakula, H.T.M. Pham, P.M. Sarro, and P.J. French

DIRECT ETCHING OF HIGH ASPECT RATIO STRUCTURES THROUGH A STENCIL 144
G. Villanueva, O. Vazquez-Mena, C. Hibert, and J. Brugger

INDIUM TIN OXIDE (ITO) TRANSPARENT MEMS SWITCHES 148
B.-K. Lee, Y.-H. Song, and J.-B. Yoon

ADVANCEMENTS IN TECHNOLOGY AND DESIGN OF BIOMIMETIC FLOW-SENSOR ARRAYS 152
C.M. Bruinink, R.K. Jaganatharaja, M.J. de Boer, E. Berenschot, M.L. Kolster, T.S.J. Lammerink, R.J. Wiegerink, and G.J.M. Krijnen

SESSION X- OPTICAL MEMS

A WAFER-LEVEL MICRO MECHANICAL GLOBAL SHUTTER FOR A MICRO CAMERA 156
C.-H. Kim, K.-D. Jung, and W. Kim

MICROFABRICATED FLIPPING GLASS DISC FOR STEREO IMAGING IN ENDOSCOPIC VISUAL INSPECTION 160
W. Choi, M. Akbarian, V. Rubtsov, and C.-J. Kim

RESONANT FREQUENCY TUNING OF TORSIONAL MICROSCANNER BY MECHANICAL RESTRICTION USING MEMS ACTUATOR 164
J.-I. Lee, S. Park, Y. Eun, B. Jeong, and J. Kim

REFLECTIVE DISPLAY USING IONIC LIQUID 168
N. Binh-Khiem, K. Matsumoto, and I. Shimoyama

FORMATION AND INTEGRATION OF A BALL LENS UTILIZING TWO PHASES LIQUID TECHNOLOGY 172
C.-C. Lee, S.-Y. Hsiao, and W. Fang

TEMPERATURE-CONTROLLED TRANSFER AND SELF-WIRING FOR MULTI-COLOR LED DISPLAY ON A FLEXIBLE SUBSTRATE 176
E. Iwase, H. Onoe, A. Nakai, K. Matsumoto, and I. Shimoyama

SESSION XI – BIOMEMS

MICROFLUIDIC ODORANT SENSOR WITH FROG EGGS EXPRESSING OLFACTORY RECEPTORS 180
N. Misawa, H. Mitsuno, R. Kanzaki, and S. Takeuchi

SEQUENTIAL AND SELECTIVE SELF-ASSEMBLY OF MICRO COMPONENTS BY DNA GRAFTED POLYMER 184
T. Tanemura, G. Lopez, R. Sato, K. Sugano, T. Tsuchiya, O. Tabata, M. Fujita, and M. Maeda

PIEZORESISTIVE CANTILEVER-BASED FORCE-CLAMP SYSTEM FOR THE STUDY OF MECHANOTRANSDUCTION IN C. ELEGANS 188
S.-J. Park, B. Petzold, M.B. Goodman, and B.L. Pruitt

SUB-5μM DIAMETER SILICON MICROPROBE AND SILICON DIOXIDE MICROTUBE ARRAYS FOR LOW-INVASIVE ANIMAL EXPERIMENTS 192
K. Takei, T. Kawano, T. Kawashima, K. Sawada, H. Kaneko, and M. Ishida

MINIMALLY INVASIVE PARYLENE DUAL-VALVED FLOW DRAINAGE SHUNT FOR GLAUCOMA IMPLANT 196
J.C.-H. Lin, P.-J. Chen, B. Yu, M. Humayun, and Y.-C. Tai

GLASS MICROPROBE WITH EMBEDDED SILICON VIAS FOR 3D INTEGRATION 200
C.-W. Lin, C.-W. Chang, Y.-T. Lee, R. Chen, Y.-C. Chang, and W. Fang

POSTER/ORAL PRESENTATIONS

BIOMEDICAL APPLICATIONS

FABRICATION PROCESSES OF INTEGRATED MULTI-ANALYTE BIOCHIP SYSTEM FOR IMPLANTABLE APPLICATIONS 204
Y.-C. Tsai, N.-F. Chiu, P.-C. Liu, Y.-C. Ou, H.-H. Liao, Y.-J. Yang, L.-J. Yang, U. Lei, F.-S. Chao, S.-S. Lu, C.-W. Lin, P.-Z. Chang, and W.-P. Shih

MICROSYSTEM-BASED STUDY OF POLLEN-TUBE ATTRACTANTS SECRETED BY OVULES 208
J.R. Cooper, Y. Qin, L. Jiang, R. Palanivelu, and Y. Zohar

A BATCH-PATTERNED SELF-EXPANDING BILIARY STENT WITH CONFORMAL MAGNETIC PDMS LAYER AND TOPOLOGICALLY-MATCHED WIRELESS MAGNETOELASTIC SENSOR 212
S.R. Green and Y.B. Gianchandani

RADIO-CONTROLLED CYBORG BEETLES: A RADIO-FREQUENCY SYSTEM FOR INSECT NEURAL FLIGHT CONTROL 216
H. Sato, Y. Peeri, E. Baghoomian, C.W. Berry, and M.M. Maharbiz

HOLLOW MICRONEEDLE ELECTRODE ARRAYS FOR INTRACELLULAR RECORDING APPLICATIONS 220
J. Held, J. Gaspar, P.J. Koester, C. Tautorat, M. Hagner, A. Cismak, A. Heilmann, W. Baumann, P. Ruther, and O. Paul

BATCH FABRICATION OF OUT-OF-PLANE, IC-COMPATIBLE, NANOSCALE-TIP SILICON NEUROPROBE ARRAYS 224
A. Goryu, A. Ikedo, K. Takei, K. Sawada, T. Kawano, and M. Ishida

ULTRA-COMPACT INTEGRATION FOR FULLY-IMPLANTABLE NEURAL MICROSYSTEMS 228
G.E. Perlin and K.D. Wise

CMOS-BASED HIGH-DENSITY SILICON MICROPROBE ARRAY FOR ELECTRONIC DEPTH CONTROL IN NEURAL RECORDING 232
K. Seidl, S. Herwik, Y. Nurcahyo, T. Torfs, M. Keller, M. Schüttler, H. Neves, T. Stieglitz, O. Paul, and P. Ruther

IN-PLANE BANDPASS REGULATION CHECK VALVE IN HEAT-SHRINK PACKAGING FOR DRUG DELIVERY 236
R. Lo and E. Meng

ON-CHIP BLOOD VISCOMETER TOWARDS POINT-OF-CARE HEMATOLOGICAL DIAGNOSIS 240
H. Zeng and Y. Zhao

IMPLANTABLE FLEXIBLE-COILED WIRELESS INTRAOCULAR PRESSURE SENSOR 244
P.-J. Chen, S. Saati, R. Varma, M.S. Humayun, and Y.-C. Tai

INTEGRATED WIRELESS NEUROSTIMULATOR 248
W. Li, D.C. Rodger, and Y.C. Tai

TRANSDERMAL POWER TRANSFER FOR IMPLANTED DRUG DELIVERY DEVICES USING A SMART NEEDLE AND REFILL PORT 252
A.T. Evans, S. Chiravuri, and Y.B. Gianchandani

A FLEXIBLE REGENERATION MICROELECTRODE WITH CELL-GROWTH GUIDANCES 256
R. Gojo, N. Kotake, T. Suzuki, K. Mabuchi, and S. Takeuchi

ADAPTIVE BLOOD CELL SEPARATION USING COOPERATION OF SERIALLY CONNECTED MEMBRANE FILTERS 260
T. Kobayashi, D. Kato, H. Koga, K. Morimoto, M. Fukuda, Y. Kinoshita, H. Yoshida, and S. Konishi

BIOCHEMICAL SENSORS

HIGH SENSITIVITY MICRO-THERMAL CONDUCTIVITY DETECTOR FOR GAS CHROMATOGRAPHY 264
B.C. Kaanta, H. Chen, G. Lambertus, W.H. Steinecker, O. Zhdaneev, and X. Zhang

CMOS-MEMS CAPACITIVE HUMIDITY SENSOR 268
N. Lazarus, S. Bedair, C. Lo, and G.K. Fedder

A THERMOELECTRIC GAS SENSOR BASED ON AN EMBEDDED TIN OXIDE CATALYST FOR DETECTING HYDROGEN AND NOX GASES 272
S.-I. Yoon, C.-i. Lee, and Y.-J. Kim

A NOVEL MICROFLUIDIC PLATFORM FOR CONTINUOUS DNA EXTRACTION AND PURIFICATION USING LAMINAR FLOW MAGNETOPHORESIS 276
M. Karle, J. Miwa, G. Roth, R. Zengerle, and F. von Stetten

AN EWOD DROPLET MICROFLUIDIC CHIP WITH INTEGRATED LOCALTEMPERATURE CONTROL FOR MULTIPLEX PROTEOMICS 280
W. Nelson, I. Peng, J.A. Loo, R.L. Garrell, and C.-J. Kim

FREQUENCY DRIFT COMPENSATION IN MASS-SENSITIVE CHEMICALSENSORS BASED ON PERIODIC STIFFNESS MODULATION 284
K.S. Demirci, J.H. Seo, S. Truax, L.A. Beardslee, Y. Luzinova, B. Mizaikoff, and O. Brand

μGC MULTICAPILLARY COLUMNS WITH MONO-LAYER PROTECTED GOLD AS A STATIONARY PHASE 288
M.A. Zareian-Jahromi, and M. Agah

A SELF-CONTAINED MINIATURIZED PCR SYSTEM USING ELECTROMAGNETIC ACTUATORS 292
B.T. Chia, S.-A. Yang, M.-Y. Cheng, C.-L. Lin, C.-W. Lin, and Y.-J. Yang

NANOSTRUCTURE-ENHANCED FIBER-OPTIC INTERFEROMETRY FOR LABEL-FREE IMMUNE SENSING 296
Y.-T. Tseng, C.-T. Chang, M.-H. Chen, and F.-G. Tseng

BIOSENSORS BASED ON ALL-POLYMER RESONANT MICROBEAMS 300
S. Schmid, P. Wägli, and C.H. Hierold

BEAD-IMMOBILIZED MOLECULAR BEACONS FOR HIGH THROUGHPUT SNP GENOTYPING VIA A MICROFLUIDIC SYSTEM 304
R.D. Sochol, A. Mahajerin, B.P. Casavant, P. Singh, M. Dueck, L.P. Lee, and L. Lin

A FULLY AUTOMATED MICRO-SOLID PHASE EXTRACTION CHIP FOR GENETIC SAMPLE PREPARATION SYSTEM 308
S.-I. Han, H. Lee, and K.-H. Han

SELF-POLARIZED PIEZOELECTRIC BIOSENSOR ARRAY FOR MULTIPLE IMMUNOASSYS APPLICATIONS 312
T. Xu, J. Miao, Z. Wang, L. Yu, and C.-M. Li

SELECTIVITY AND LONG-TERM RELIABILITY OF RESONANT EXPLOSIVE-VAPOR-TRACE DETECTION BASED ON ANTIGEN-ANTIBODY BINDING 316
A. Lin and E.S. Kim

DNA-DECORATED CARBON NANOTUBES AS SENSITIVE LAYER FOR AlN CONTOUR-MODE RESONANT-MEMS GRAVIMETRIC SENSOR 320
C. Zuniga, M. Rinaldi, S.M. Khamis, T.S. Jones, A.T. Johnson, and G. Piazza

MICROFLUIDICS WITHIN A SWAGELOK®: A MEMS-ON-TUBE ASSEMBLY 324
S. Unnikrishnan, H.V. Jansen, J.W. Berenschot, B. Mogulkoc, and M.C. Elwenspoek

NANO-THICKNESS INTEGRATED RESONANT CANTILEVERS WITH SURFACE-STIFFENING SCHEME FOR ULTRA-SENSITIVE DETECTION OF TRACE CHEMICALS 328
X. Xia, Y. Yang, and X. Li

SURFACE ACOUSTIC WAVE-DRIVEN ACTIVE PLASMONICS BASED ON DYNAMIC PATTERNING OF NANOPARTICLES IN MICROFLUIDIC CHANNELS 332
J. Shi, Y.B. Zheng, and T.J. Huang

PUPAL-STAGE INSERTED SILICON MICROPROBE NEURAL INTERFACE IN INSECTS FOR GAS SENSING 336
C.J. Shen, R. Gilmour, and A. Lal

BEAD-BASED MICROFLUIDIC PLATFORM INTEGRATED WITH OPTICAL DETECTION DEVICES FOR RAPID DETECTION OF GENETIC DELETION FROM SALIVA 340
K.-Y. Lien, C.-J. Liu, P.-L. Kuo, and G.-B. Lee

A MICROFLUIDIC AFFINITY COCAINE SENSOR 344
J.P. Hilton, T. Nguyen, R. Pei, M. Stojanovic, and Q. Lin

DEVELOPMENT OF NEW BIOCHEMICAL IC CHIP-SET FOR REAL-TIME PCR 348
K. Ikuta, T. Sasao, Y. Okuda, K. Watamura, and M. Ikeuchi

A MEMS SENSOR FOR CONTINUOUS MONITORING OF GLUCOSE IN SUBCUTANEOUS TISSUE 352
X. Huang, S. Li, J. Schultz, Q. Wang, and Q. Lin

MONOLITHIC CENTRIFUGAL MICROFLUIDIC PLATFORM FOR BACTERIA CAPTURE AND CONCENTRATION, LYSIS, NUCLEIC-ACID AMPLIFICATION, AND REAL-TIME DETECTION 356
J.L. Garcia-Cordero, I.K. Dimov, J. O'Grady, J. Ducree, T. Barry, and A.J. Ricco

CELLS AND PARTICLES

IN VITRO DETECTION OF NEURAL ACTIVITY WITH VERTICALLY GROWN SINGLE PLATINUM NANOWIRE 360
D. Choi, A. Fung, H. Moon, G. Villareal, Y. Chen, D. Ho, N. Presser, G. Stupian, and M. Leung

GENERATION AND SELECTIVE RETRIEVAL OF MICRO DROPLETS IN AN ARRAY FOR MICRO-PCR USING A GLASS-SILICON-GLASS CHIP 363
K.H. Park, H.G. Park, and S. Takeuchi

ARRAYING SINGLE ADHERENT CELLS BY MICROPLATE SELF-ASSEMBLY 367
H. Ishihara, K. Kuribayashi, and S. Takeuchi

A DYNAMIC MICROARRAY WITH PNEUMATIC VALVES FOR SELECTIVE TRAPPING AND RELEASING OF MICROBEADS 371
K. Iwai and S. Takeuchi

SIZE-DEPENDENT PARTICLE FILTERATION USING MAGNETICALLY DRIVEN MICROTOOL AND CENTRIFUGAL FORCE IN MICROCHIP 375
H. Maruyama, S. Sakuma, Y. Yamanishi, and F. Arai

CLUSTER DYNAMICS IN FLOW OF SUSPENDED PARTICLES IN MICROCHANNELS 379
T. Gudipaty, M.T. Stamm, L.S.L. Cheung, L. Jiang, and Y. Zohar

A MULTIFUNCTIONAL VERTICAL MICROSIEVE FOR MICRO AND NANO PARTICLES SEPARATION 383
C. Shen, H.T.M. Pham, and P.M. Sarro

CLASSIFICATION AND CONDENSATION OF NANO-SIZED AIRBORNE PARTICLES BY ELECTRICALLY TUNNING COLLECTION SIZE 387
Y.-H. Kim, S. Kwon, D. Park, J. Hwang, and Y.-J. Kim

DEVELOPMENT OF A CYTOMIC FORCE TRANSDUCER FOR EXPERIMENTAL MECHANOBIOLOGY 391
E. Dy and C.M. Ho

A PDMS PLANAR PATCH-CLAMP ARRAY CHIP WITH POLY (ETHYLENE GLYCOL)/SU-8 BASED CELL-PATCH INTERFACE 395
B.J. Xu, Z.B. Liu, J.J. Yu, Y.K. Lee, and M. Yang

WHOLE CELL IMPEDANCE ANALYSIS OF METASTATIC AND NON-METASTATIC CANCER CELLS 399
H.S. Kim, Y.H. Cho, A.B. Frazier, Z.G. Chen, D.-M. Shin, and A. Han

INTEGRATION OF SKELETAL MUSCLE CELL ONTO SI-MEMS AND ITS GENERATIVE FORCE MEASUREMENT 403
K. Shimizu, H. Sasaki, H. Hida, H. Fujita, K. Obinata, M. Shikida, and E. Nagamori

A STRETCHABLE CELL CULTURE PLATFORM WITH EMBEDDED ELECTRODE ARRAY 407
P. Wei, R. Taylor, Z. Ding, G. Higgs, J.J. Norman, B.L. Pruitt, and B. Ziaie

ASSESSMENT OF SINGLE CELL VIABILITY FOLLOWING LIGHT-INDUCED ELECTROPORATION THROUGH USE OF ON-CHIP MICROFLUIDICS 411
J.K. Valley, H.-Y. Hsu, S. Neale, A.T. Ohta, A. Jamshidi, and M.C. Wu

MANIPULATION OF BIOSAMPLES AND MICROPARTICLES USING OPTICAL IMAGES ON POLYMER DEVICES 415
W. Wang, Y.-H. Lin, T.-F. Guo, and G.-B. Lee

FABRICATION OF LIVING CELL STRUCTURE UTILIZING ELECTROSTATIC INKJET PHENOMENA 419
S. Umezu, T. Kitajima, H. Murase, H. Ohmori, K. Katahira, and Y. Ito

SIZE-CONTROLLED ISLET-CELL SPHEROIDS FOR GEOMETRIC ANALYSIS OF INSULIN SECRETION 423
Y. Tsuda, M. Kato-Negishi, T. Okitsu, and S. Takeuchi

A 3-D MICROFLUIDIC COMBINATORIAL CELL CULTURE ARRAY 427
M.C. Liu and Y.C. Tai

FLOW ACCELERATION EFFECT ON CANCER CELL DEFORMATION AND DETACHMENT......431

L.S.L. Cheung, X.J. Zheng, A. Stopa, J. Schroeder, R.L. Heimark, J.C. Baygents, R. Guzman, and Y. Zohar

MICROFLUIDICS AND FLOW SENSORS

LINEAR EXPANSION AND CONTRACTION OF PAIRED PNEUMATIC BALOON BENDING ACTUATORS TOWARD TELESCOPIC MOTION......435

N. Fujiwara, S. Sawano, and S. Konishi

LOW-COST MICROFLUIDIC SINGLE-USE VALVES AND ON-BOARD REAGENT STORAGE USING LASER-PRINTER TECHNOLOGY......439

J.L. Garcia-Cordero, F. Benito-Lopez, D. Diamond, J. Ducree, and A.J. Ricco

DROPLET MIXER BASED ON SIPHON-INDUCED FLOW DISCRETIZATION AND PHASE SHIFTING......443

R. Burger, N. Reis, J.G. Fonseca, and J. Ducreé

AN ALL-POLYMER AIR-FLOW SENSOR ARRAY USING A PIEZORESISTIVE COMPOSITE ELASTOMER......447

A.R. Aiyar, C. Song, S.H. Kim, and M.G. Allen

DESIGN AND CHARACTERISATION OF A HYDRAULIC MICROACTUATOR FABRICATED BY LITHOGRAPHY......451

M. De Volder, F. Ceyssens, D. Reynaerts, and R. Puers

BIOMIMIC LOW HYSTERESIS SURFACE PREPARED BY HIERARCHICAL MICRO/NANO TEXTURES......455

M.-H. Chen1, T.-H. Hsu1, and F.-G. Tseng1,2

WETTABILITY SWITCHING TECHNIQUE OF A BIOCOMPATIBLE POLYMER......459

Y. Takahashi, K.S. Teh, and Y.W. Lu

MICRO CORIOLIS MASS FLOW SENSOR WITH INTEGRATED CAPACITIVE READOUT......463

J. Haneveld, T.S.J. Lammerink, M.J. de Boer, and R.J. Wiegerink

MEMS TACTILE DISPLAY WITH HYDRAULIC DISPLACEMENT AMPLIFICATION MECHANISM......467

T. Ninomiya, K. Osawa, Y. Okayama, Y. Matsumoto, and N. Miki

HIGH PURITY SEPARATION OF RARE SPECIES IN DROPLET MICROFLUIDICS USING DROPLET-CONDUIT STRUCTURES......471

G.J. Shah and C.-J. Kim

A HIGHLY FLEXIBLE SUPERHYDROPHOBIC MICROLENS ARRAY WITH SMALL CONTACT ANGLE HYSTERESIS FOR DROPLET-BASED MICROFLUIDICS......475

M. Im, D.-H. Kim, X.-J. Huang, J.-H. Lee, J.-B. Yoon, and Y.-K. Choi

AMBIENT TEMPERATURE-GRADIENT COMPENSATED LOW-DRIFT THERMOPILE FLOW SENSOR......479

M. Dijkstra, T.S.J. Lammerink, M.J. de Boer, R.J. Wiegerink, and M.C. Elwenspoek

AN ELECTRICALLY-DRIVEN, LARGE-DEFLECTION, HIGH-FORCE, MICRO PISTON HYDRAULIC ACTUATOR ARRAY FOR LARGE-SCALE MICROFLUIDIC SYSTEMS......483

H. Kim and K. Najafi

ELECTROHYDRODYNAMIC JET PRINTING CAPABLE OF REMOVING SUBSTRATE EFFECTS AND MODULATING PRINTING CHARACTERISTICS 487
J.-S. Lee, Y.-J. Kim, B.-G. Kang, S.-Y. Kim, J. Park, J. Hwang, and Y.-J. Kim

HYBRID ON DEMAND JETTING SYSTEM FOR ULTRA FINE DROPLET BASED ON ELECTROHYDRODYNAMIC AND PIEZOELECTRIC ACTUATION 491
Y.-J. Kim1,2, J.-S. Lee1, S.-Y. Kim1, S.E. Park1, J. Hwang1, and Y.-J. Kim1

MICRO-MACHINED STENT-TYPE FLOW SENSOR FOR EVALUATION OF NASAL RESPIRATION 495
T. Yokota, J. Naito, M. Shikida, and K. Sato

EHD MICRO PUMP USING PYROLYZED POLYMER 3-D CARBON MESH ELECTRODES 499
D. Wakui, N. Imai, Y. Nagaura, H. Sato, T. Sekiguchi, S. Konishi, S. Shoji, and T. Homma

NON-LINEAR FLUIDIC INTEGRATED CIRCUITS REALIZED BY PNEUMATIC-FIELD EFFECT TRANSISTORS WITH CONTROLLABLE OUTPUT RESISTANCE 503
H. Takao, N. Tanaka, M. Sugiura, K. Sawada, and M. Ishida

HIGH-RESOLUTION PIEZO INKJET PRINTHEAD FABRICATED BY THREE DIMENSIONAL ELECTRICAL CONNECTION METHOD USING THROUGH GLASS VIA 507
M. Murata, T. Kondoh, T. Yagi, N. Funatsu, K. Tanaka, H. Tsukuni, K. Ohno, H. Usami, R. Nayve, N. Inoue, S. Seto, and N. Morita

SELF-RESONANT FLOW SENSOR USING RESONANCE FREQUENCY SHIFT BY FLOW-INDUCED VIBRATION 511
K.H. Kim and Y.H. Seo

MICROFLUIDIC-BASED DISPENSERS FOR SUB-MICROLITER PIPETTING 515
S.-B. Huang and G.-B. Lee

PRECISE TITRATION MICROFLUIDIC SYSTEM USING SUB-PICOLITER DROPLET INJECTION 519
K. Kawai, M. Fujii, and S. Shoji

DROPLETS GENERATION METHOD FOR WATER–IN–OIL STATE IN THE POLYDIMETHYLSILOXANE MICROCHANNEL WITH GROOVES 523
J. Kim, D. Byun, J. Hong, and A.J. deMello

A LOW-COST FLEXIBLE HOT-FILM SENSOR SYSTEM FOR FLOW SENSING AND ITS APPLICATION TO AIRCRAFT 527
R. Zhu, P. Liu, X.D. Liu, F.X. Zhang, and Z.Y. Zhou

INTEGRATING EWOD WITH SURFACE RATCHETS FOR ACTIVE DROPLET TRANSPORT AND SORTING 531
T.A. Duncombe, M. Kumemura, H. Fujita, and K.F. Böhringer

THERMAL CHARACTERIZATION OF MICROLITER AMOUNTS OF LIQUIDS BY A MICROMACHINED CALORIMETRIC TRANSDUCER 535
G. Pärr, E. Santagata-Iervolino, A.W. van Herwaarden, W. Wien, and M.J. Vellekoop

TIR-BASED DYNAMIC LIQUID-LEVEL AND FLOW-RATE SENSING AND ITS APPLICATION ON CENTRIFUGAL MICROFLUIDIC PLATFORMS 539
J. Hoffmann, L. Riegger, D. Mark, F. von Stetten, R. Zengerle, and J. Ducreé

NANO DEVICES AND CARBON NANOTUBES

100 NANOMETER SCALE RESISTIVE HEATER-THERMOMETER ON A SILICON CANTILEVER 543
Z. Dai, K. Park, and W.P. King

MEMS DIFFRACTIVE OPTICAL NANORULER TECHNOLOGY FOR TIP-BASED NANOFABRICATION AND METROLOGY 547
N. Yoshimizu, A. Lal, and C.R. Pollock

INTEGRATION OF BRIDGING-STRUCTURAL SWNTS ON FLEXIBLE PDMS SHEET BY STAMPING TRANSFER 551
Y. Takei, T. Kan, E. Iwase, K. Matsumoto, and I. Shimoyama

GIANT PIEZORESISTANCE OF NANO-THICK SILICON INDUCED BY INTERFACE ELECTRON TRAPPING EFFECT 555
Y. Yang and X. Li

ELECTRICAL DISCHARGE BASED MICROFABRICATION ON ELECTROSPUN NANOFIBERS 559
H. Zeng and Y. Zhao

INTEGRATION OF CARBON NANOTUBES INTO MEMS DEVICES VIA DETERMINISTIC TRANSPLANTING ASSEMBLY 563
S. Kim, H.W. Lee, and S.G. Kim

SURFACE-TEMPERATURE CONTROL OF SILICON NANOWIRES IN DRY AND LIQUID CONDITIONS 567
S. Akiyama, Y.T. Cheng, J. Fattaccioli, N. Takama, P. Löw, C. Bergaud, and B.J. Kim

ELECTRON BEAM STIMULATED OXIDATION OF CARBON (EBSOC) 571
P.S. Spinney, S.D. Collins, and R.L. Smith

ULTRA SMALL SINGLE WALLED CARBON NANOTUBE PRESSURE SENSORS 575
T. Helbling, S. Drittenbass, L. Durrer, C. Roman, and C.H. Hierold

METALLIC NANOPARTICLE MANIPULATION USING OPTOELECTRONIC TWEEZERS 579
A. Jamshidi, H.Y. Hsu, J.K. Valley, A.T. Ohta, S. Neale, and M.C. Wu

THIN FILM TRANSISTOR (TFT) SENSING ELEMENTS FABRICATED IN SURFACE MICROMACHINED POLYMERMEMS FOR A DIFFERENTIAL CALORIMETRIC FLOW SENSOR 583
S.-H. Tsang, K. Simard, I.G. Foulds, H. Izadi, K.S. Karim, and M. Parameswaran

PLANAR MEMS SUPERCAPACITOR USING CARBON NANOTUBE FORESTS 587
Y.Q. Jiang, Q. Zhou, and L. Lin

PIEZOELECTRIC NANOSWITCH 591
D.C. Judy, J.S. Pulskamp, R.G. Polcawich, and L. Currano

A MOLECULAR MUSCLE BASED NANO-ELECTRO-MECHANICAL-SYSTEMS (NEMS) 595
B.K. Juluri, A.S. Kumar, Y. Liu, T. Ye, Y.W. Yang, A.H. Flood, L. Fang, J.F. Stoddart, P.S. Weiss, and T.J. Huang

MATERIALS AND DEVICE-CHARACTERIZATION

HIGHLY EFFICIENT EXTRACTION OF MECHANICAL AND LINEAR AND QUADRATIC PIEZORESISTIVE PROPERTIES OF POLY-SI FILMS USING WAFER-SCALE MICROTENSILE TESTING 599
M.E. Schmidt, J. Gaspar, J. Held, S. Kamiya, and O. Paul

DESIGN OF POWER-OPTIMIZED THERMAL CANTILEVERS FOR SCANNING PROBE TOPOGRAPHY SENSING 603
H. Rothuizen, M. Despont, U. Drechsler, C. Hagleitner, A. Sebastian, and D. Wiesmann

MECHANICAL CHARACTERIZATION OF BIOMOLECULES IN LIQUID USING SILICON TWEEZERS WITH SUBNANONEWTON RESOLUTION 607
C. Yamahata, E. Sarajlic, L. Jalabert, M. Kumemura, D. Collard, and H. Fujita

EFFECT OF ELECTRODE GEOMETRY AND SURFACE PASSIVATION ON CORROSION OF POLYCRYSTALLINE SILICON UNDER HIGH RELATIVE HUMIDITY AND BIAS 611
F. Liu, C. Roper, C. Carraro, and R. Maboudian

MECHANICAL RESPONSE EVALUATION OF HIGH-THERMALLY-STABLE-GRADE PARYLENE SPRING 615
C. Kamezawa, Y. Suzuki, and N. Kasagi

PIEZOELECTRIC PZT VERY THIN FILMS IN THE 100NM RANGE: A SOLUTION FOR ACTUATORS EMBEDDED IN LOW VOLTAGE DEVICES 619
E. Defay, G. Le Rhun, F. Perruchot, P. Rey, A. Suhm, M. Aïd, L.J. Liu, S. Pacheco, and M. Miller

CMOS-COMPATIBLE SURFACE-MICROMACHINED TEST STRUCTURE FOR DETERMINATION OF THERMAL CONDUCTIVITY OF THIN FILM MATERIALS BASED ON SEEBECK EFFECT 623
Z. Wang, P. Fiorini, and C. van Hoof

A BIOCOMPATIBLE AND FLEXIBLE RF CMOS TECHNOLOGY AND THE CHARACTERIZATION OF THE FLEXIBLE MOS TRANSISTORS UNDER BENDING STRESSES 627
C.-Y. Hsieh, J.-S. Chen, W.-A. Tsou, Y.-T. Yeh, K.-A. Wen, and L.-S. Fan

RADIATION SENSITIVITY OF OHMIC RF-MEMS SWITCHES FOR SPATIAL APPLICATIONS 630
A. Tazzoli, G. Cellere, E. Autizi, V. Peretti, A. Paccagnella, and G. Meneghesso

RELIABILITY OF RF MEMS CAPACITIVE SWITCHES AND DISTRIBUTED MEMS PHASE SHIFTERS USING ALN DIELECTRIC 634
M. Fernández-Bolaños, D. Tsamados, P. Dainesi, and A.M. Ionescu

TENSILE MEASUREMENT OF A SINGLE CRYSTAL GALLIUM NITRIDE NANOWIRE 638
J.J. Brown, A.I. Baca, and V.M. Bright

CELLULOSE-BASED COMPOSITE AS A RAW MATERIAL FOR FLEXIBLE AND ULTRA-LIGHTWEIGHT MECHANICAL SWITCH DEVICE 642
S. Couderc, B.J. Kim, and T. Someya

DETERMINATION OF DENSITY AND YOUNG'S MODULUS OF ATOMIC LAYER DEPOSITED THIN FILMS BY RESONANT FREQUENCY MEASUREMENTS OF OPTICALLY EXCITED NANOCANTILEVERS 646
B. Ilic, S. Krylov, and H. Craighead

THE EFFECTS OF THERMAL OXIDATION OF MEMS RESONATOR ON TEMPERATURE DRIFT AND ABSOLUTE FREQUENCY 650
C. van der Avoort, J. van Wingerden, and J.T.M. van Beek

NON-DESTRUCTIVE QUANTITATIVE MEASUREMENT METHOD FOR NORMAL AND SHEAR STRESSES ON SINGLE-CRYSTALLINE SILICON STRUCTURES FOR RELIABILITY OF SILICON-MEMS 653
M. Komatsubara, T. Namazu, N. Naka, S. Kashiwagi, K. Ohtsuki, and S. Inoue

PHASE LOCK LOOP BASED TEMPERATURE COMPENSATION FOR MEMS OSCILLATORS 657
J. Salvia, R. Melamud, S. Chandorkar, H.K. Lee, Y.Q. Qu, S.F. Lord, B. Murmann, and T.W. Kenny

DEGRADATION OF MECHANICAL STRENGTH AT SI/SIO$_2$ INTERFACE ON SOI WAFERS UNDER CYCLIC LOADING 661
T. Ando, T. Takumi, and K. Sato

FABRICATION PACKAGING ENCAPSULATION

PART-TO-PART AND PART-TO-SUBSTRATE MAGNETIC SELF-ASSEMBLY OF MILLIMETER SCALE COMPONENTS WITH ANGULAR ORIENTATION 665
S.B. Shetye, I. Eskinazi, and D.P. Arnold

LOW TEMPERATURE FABRICATION PROCESS FOR HIGH-ASPECT-RATIO AND MULTI-COMPLIANT MEMS 669
S. Kühne, R. Blattmann, and C.H. Hierold

FABRICATION OF NANOPILLARS BASED ON SILICON OXIDE NANOPATTERNS SYNTHESIZED IN OXYGEN PLASMA REMOVAL OF PHOTORESIST 673
H.Y. Mao, D. Wu, W.G. Wu, J. Xu, H.X. Zhang, and Y.L. Hao

VISUAL OBSERVATION OF PDMS TIP IN LIQUID MICROCONTACT PRINTING 677
J. Fattaccioli, A. Ikeda, J.G. Kim, N. Takama, and B.J. Kim

SUPERCRITICAL FLUID DEPOSITION (SCFD) TECHNIQUE AS A NOVEL TOOL FOR MEMS FABRICATION 681
T. Momose, T. Ohkubo, T. Uejima, T. Saito, M. Sugiyama, and Y. Shimogaki

ELECTRODEPOSITION OF PERMALLOY IN DEEP SILICON TRENCHES WITHOUT EDGE-OVERGROWTH UTILIZING DRY FILM PHOTORESIST 685
S.-W. Park, D.G. Senesky, and A.P. Pisano

FABRICATION OF BIODEGRADABLE MAGNETIC MICROCAPSULES UTILIZING 3D, SURFACE MODIFIED PDMS MICROFLUIDIC DEVICES 689
C.Y. Liao and Y.C. Su

NOVEL TECHNOLOGY FOR CAPACITIVE PRESSURE SENSORS WITH MONOCRYSTALLINE SILICON MEMBRANES 693
K. Knese, S. Armbruster, H. Weber, M. Fischer, H. Benzel, M. Metz, and H. Seidel

DEVELOPMENT OF MULTI-USER MULTI-CHIP SOI CMOS-MEMS PROCESSES 697
K. Takahashi, M. Mita, M. Nakada, D. Yamane, A. Higo, H. Fujita, and H. Toshiyoshi

MEMS-BASED BATCH-MODE MICRO-ELECTRO-DISCHARGE MACHINING USING MICROELECTRODE ARRAYS ACTUATED BY HYDRODYNAMIC FORCE 701
C.R. Alla Chaitanya, and K. Takahata

SILICON CARBIDE SURFACE MICROMACHINING TECHNOLOGY BY TETRAMETHYLSILANE-BASED ATMOSPHERIC VAPOR DEPOSITION 705
Y. Hatakeyama, M. Esashi, and S. Tanaka

DEVELOPMENT OF A MONOLITHIC PMMA COMB-DRIVE MICRO ACTUATOR UTILIZING HOT EMBOSSING AND ULTRA-PRECISION MACHINING 709
S. Amaya, D.V. Dao, and S. Sugiyama

OPTICALLY PROGRAMMABLE SELF-ASSEMBLY OF HETEROGENEOUS MICRO-COMPONENTS ON UNCONVENTIONAL SUBSTRATES 713
E. Saeedi, J.R. Etzkorn, L. Draghi, and B.A. Parviz

FABRICATION OF NEAR NET SHAPE ALUMINA NICKEL COMPOSITE MICRO PARTS USING AQUEOUS SUSPENSION 717
H.S. Hassanin and K. Jiang

VERTICALLY ALIGNED VARIOUS LENGTHS DOPED-SILICON MICROWIRE ARRAYS BY REPEATED SELECTIVE VAPOR-LIQUID-SOLID GROWTH 721
T. Kawano, A. Ikedo, T. Kawashima, K. Sawada, and M. Ishida

FABRICATION OF ELECTROSTATICALLY-ACTUATED, IN-PLANE FUSED QUARTZ RESONATORS USING SILICON-ON-QUARTZ (SOQ) BONDING AND QUARTZ DRIE 725
Y.-S. Hwang, H.-K. Jung, E.-S. Song, I.-J. Hyeon, Y.-K. Kim, and C.-W. Baek

ADJUSTABLE REFRACTIVE INDEX METHOD FOR COMPLEX MICROSTRUCTURE BY AUTOMATED DYNAMIC MODE MULTIDIRECTIONAL UV LITHOGRAPHY 729
J.K. Kim, T.-S. Yun, H. Jee, and Y.K. Yoon

A NEW HIGH-SENSITIVITY PACKAGE-LEAK TESTING METHOD FOR MEMS SENSORS 733
M. Fujiyoshi, Y. Nonomura, and H. Senda

SPATIALLY ARRANGED MICROELECTRODES USING WIRE BONDING TECHNOLOGY FOR SPATIALLY DISTRIBUTED CHEMICAL INFORMATION ACQUISION 737
W. Tonomura, K. Shimizu, and S. Konishi

WAFER SCALE ENCAPSULATION OF LARGE LATERAL DEFLECTION MEMS STRUCTURES 741
A.B. Graham, M. Messana, P. Hartwel, J. Provine, S. Yoneoka, B. Kim, R. Melamud, R.T. Howe, and T.W. Kenny

PARYLENE-POCKET CHIP INTEGRATION 745
R. Huang and Y.C. Tai

A LOW-POWER OVEN-CONTROLLED VACUUM PACKAGE TECHNOLOGY FOR HIGH-PERFORMANCE MEMS 749
S.-H. Lee, J. Cho, S.-W. Lee, M.F. Zaman, F. Ayazi, and K. Najafi

DIRECT WIRE-BONDING OF SILICON DEVICES WITHOUT METAL PADS 753
A. Hirshberg and D. Elata

PHYSICAL SENSORS

A LOW-NOISE OSCILLATOR BASED ON A MULTI-MEMBRANE CMUT FOR HIGH SENSITIVITY RESONANT CHEMICAL SENSOR 757
H.J. Lee, K.K. Park, P. Cristman, Ö. Oralkan, M. Kupnik, and B.T. Khuri-Yakub

PERFORMANCE OF PACKAGED PIEZOELECTRIC MICROSPEAKERS DEPENDING ON THE MATERIAL PROPERTIES 761
S.H. Yi, S.C. Ur, and E.S. Kim

CMOS INTEGRATED STRESS MAPPING CHIPS WITH 32 N-TYPE OR P-TYPE PIEZORESISTIVE FIELD EFFECT TRANSISTORS 765
P. Gieschke, Y. Nurcahyo, M. Herrmann, M. Kuhl, P. Ruther, and O. Paul

DEVELOPMENT OF FABRIC-TYPE OF TENSIONAL FORCE SENSOR 769
Y. Suzuki, D. Ogura, M. Shikida, Y. Hasegawa, and K. Sato

MEASUREMENT RESULTS OF THE FIRST TWO CHIP SILICON MICROPHONE WITH LOW STRESS NICKEL MEMBRANE COVERING FULL AUDIO RANGE 773
S. Junge, F. Jakobs, and W. Lang

TOWARDS PIEZORESISTIVE CMOS SENSORS FOR OUT-OF-PLANE STRESS 777
B. Lemke, K. Kratt, R. Baskaran, and O. Paul

FLEXIBLE TACTILE SENSOR SHEET WITH LIQUID FILTER FOR SHEAR FORCE DETECTION 781
K. Noda, K. Matsumoto, and I. Shimoyama

BIOMIMETIC DESIGN AND FABRICATION OF COMPONENTS FOR ARTIFICIAL HAIR CELL SENSOR 785
H.S. Lee, B.-K. Lee, and T.H. Kwon

A NOVEL CONFIGURABLE MEMS INERTIAL SWITCH USING MICROSCALE LIQUID-METAL DROPLET 789
K. Yoo and J. Kim

NOVEL MICRO CAPACITIVE INCLINOMETER WITH OBLIQUE COMB ELECTRODE AND SUSPENSION SPRING ALIGNED PARALLEL TO {111} VERTICAL PLANES OF (110) SILICON 793
D.-H. Jeong, S.-S. Yun, M.-L. Lee, G. Hwang, C.-A. Choi, and J.-H. Lee

A DIGITAL OUTPUT PIEZOELECTRIC ACCELEROMETER USING PATTERNED PB(ZR,TI)O3 THIN FILMS ELECTRICALLY CONNECTED IN SERIES 797
T. Kobayashi, H. Okada, T. Masuda, and T. Itoh

ACCELERATION COMPENSATION OF MEMS RESONATORS USING ELECTROSTATIC TUNING 801
S. Yoneoka, G. Bahl, J. Salvia, K.L. Chen, A.B. Graham, H.K. Lee, G. Yama, R.N. Candler, and T.W. Kenny

DESIGN AND IMPLEMENTATION OF A NOVEL CMOS-MEMS SINGLE PROOF-MASS TRI-AXIS ACCELEROMETER 805
C.-M. Sun, M.-H. Tsai, and W. Fang

A LATCHING ACCELERATION SWITCH WITH MULTI-CONTACTS INDEPENDENT TO THE PROOF-MASS 809
Z.Y. Guo, Z.C. Yang, L.T. Lin, Q.C. Zhao, H.T. Ding, X.S. Liu, X.Z. Chi, J. Cui, and G.Z. Yan

AN ANALYSIS TO IMPROVE STABILITY OF DRIVE-MODE OSCILLATIONS IN CAPACITIVE VIBRATORY MEMS GYROSCOPES 813
S.E. Alper, K. Sahin, and T. Akin

MODELING OF A CAPACITIVE Σ-Δ ACCELEROMETER SYSTEM INCLUDING THE NOISE COMPONENTS AND VERIFICATION WITH TEST RESULTS 817
B. Boga, I.E. Ocak, H. Külah, and T. Akin

AIR PRESSURE SENSOR ON AN INSECT WING 821
H. Takahashi, K. Matsumoto, and I. Shimoyama

A FULLY CMOS-COMPATIBLE MICRO-PIRANI GAUGE BASED ON A CONSTANT CURRENT CIRCUIT 825
J. Wang and Z. Tang

GLASS AND QUARTZ MICROSCINTILLATORS FOR CMOS COMPATIBLE MULTI-SPECIES RADIATION DETECTION 829
R.P. Waguespack, H. Berry, S. Pellegrin, and C.G. Wilson

SEMICONDUCTING Y-BA-CU-O INFRARED MICROBOLOMETER ARRAY FABRICATED AND CHARACTERIZED WITH CCBDI ROIC 833
S. Kumar1 and D.P. Butler2

A MICROMACHINED QUARTZ AND STEEL PRESSURE SENSOR OPERATING UP TO 1000°C AND 2000 TORR 837
S.A. Wright and Y.B. Gianchandani

HIGH SPEED (GHZ), ULTRA-HIGH PRESSURE (GPA) SENSOR ARRAY FABRICATED IN INTEGRATED CMOS+MEMS PROCESS 841
M. Okandan, R. Olsson, M. Baker, P. Resnick, T.A. Hill, C. Lackey, S. Pearson, J. Castaneda, W. Trott, and D. Jones

SINGLE WAFER SURFACE MICROMACHINED FIELD EMISSION ELECTRON SOURCE 844
F. Santagata, C.K. Yang, J.F. Creemer, P.J. French, and P.M. Sarro

SUSPENDED MEMBRANE SINGLE CRYSTAL SILICON MICRO HOTPLATE FOR DIFFERENTIAL SCANNING CALORIMETRY 848
J. Lee1, C.M. Spadaccini2, E.V. Mukerjee2, and W.P. King1

RF MEMS AND RESONATORS

SIMPLE AND ROBUST AIR GAP-BASED MEMS SWITCH TECHNOLOGY FOR RF-APPLICATIONS 852
P. Ekkels, X. Rottenberg, P. Czarnecki, R. Puers, and H.A.C. Tilmans

RUTHENIUM/GOLD HARD-SURFACE/LOW-RESISTIVITY CONTACT METALLIZATION FOR POLYMER-ENCAPSULATED MICROSWITCH WITH STRESS-REDUCED CORRUGATED SIN/SIO$_2$ DIAPHRAGM 856
F. Ke, J. Miao, and J. Oberhammer

A MEMS RECONFIGURABLE QUAD-BAND CLASS-E POWER AMPLIFIER FOR GSM STANDARD 860
L. Larcher, R. Brama, M. Ganzerli, J. Iannacci, B. Margesin, M. Bedani, and A. Gnudi

A RF-MEMS BASED MONOLITHICALLY INTEGRATED WIDE RANGE TUNABLE LOW PASS FILTER FOR RF FRONT END APPLICATION 864
T. Fujii, T. Irieda, K. Nakamura, K. Ota, K. Ihara, D.T. McCormick, and L. Lin

LOW ACTUATION VOLTAGE SPDT RF MEMS K BAND SWITCH USING A SINGLE GOLD MEMBRANE 868
R. Robin, O. Millet, K. Segueni, and L. Buchaillot

PERFORMANCE ENHANNCEMENT BY SUBSTRATE PERFORATION FOR A WAFER-LEVEL ENCAPSULATED RF MEMS DC SHUNT SWITCH 872
F. Ke, J. Miao, and C.W. Tan

A STRESS TOLERANT TEMPERATURE-STABLE RF MEMS SWITCHED CAPACITOR 876
I. Reines1, B. Pillans2, and G.M. Rebeiz1

TEMPERATURE COMPENSATION IN SILICON-BASED MICRO-ELECTROMECHANICAL RESONATORS 880
F. Schoen, M. Nawaz, T. Bever, R. Gruenberger, W. Raberg, W. Weber, B. Winkler, and R. Weigel

A 12-GHZ DPDT RF-MEMS SWITCH WITH LAYER-WISE WAVEGUIDE/ACTUATOR DESIGN TECHNIQUE 884
D. Yamane, H. Seita, W. Sun, S. Kawasaki, H. Fujita, and H. Toshiyoshi

POST-FABRICATION ELECTRICAL TRIMMING OF SILICON BULK ACOUSTIC RESONATORS USING JOULE HEATING 888
A.K. Samarao and F. Ayazi

RF MEMS HIGH-IMPEDANCE TUNEABLE METAMATERIALS FOR MILLIMETER-WAVE BEAM STEERING 892
M. Sterner, D. Chicherin, A.V. Räisenen, G. Stemme, and J. Oberhammer

MONOLITHICALLY INTEGRATED PIEZOMEMS SP2T SWITCH AND CONTOUR-MODE FILTERS 896
J.S. Pulskamp, D.C. Judy, R.G. Polcawich, R. Kaul, H. Chandrahalim, and S.A. Bhave

A LIQUID-METAL RF MEMS SWITCH WITH DC-TO-40 GHZ PERFORMANCE 900
P. Sen and C.-J. Kim

DESIGN AND FABRICATION OF A GAUSSIAN-SHAPED AT-CUT QUARTZ CRYSTAL RESONATOR 904
T. Abe and H. Kishi

GAS-LIQUID SEPARATED RESONATOR FOR BIO-CHEMICAL APLICATION 908
H. Hida, M. Shikida, M. Okochi, H. Honda, and K. Sato

5 - 10 GHZ ALN CONTOUR-MODE NANOELECTROMECHANICAL RESONATORS 912
M. Rinaldi, C. Zuniga, and G. Piazza

CHAOS IN ELECTROSTATICALLY ACTUATED RF-MEMS MEASURED AND MODELED 916
J. Stulemeijer, R.W. Herfst, and J.A. Bielen

THERMOELASTIC FEM-BEM MODEL FOR MEMS RESONATOR LOSS SIMULATION 919
D. Ekeom and L. Buchaillot

RESONANT PIEZOELECTRIC ALGAN/GAN MEMS SENSORS IN LONGITUDINAL MODE OPERATION 923
K. Brueckner, F. Niebelschuetz, K. Tonisch, R. Stephan, V. Cimalla, O. Ambacher, and M.A. Hein

FULLY DIFFERENTIAL INTERNAL ELECTROSTATIC TRANSDUCTION OF A LAMÉ-MODE RESONATOR 927
M. Ziaei-Moayyed, D. Elata, J. Hsieh, J.-W.P. Chen, E.P. Quévy, and R.T. Howe

ANALYTICAL MODELING AND NUMERICAL SIMULATION OF CAPACITIVE SILICON BULK ACOUSTIC RESONATORS 931
G. Casinovi, X. Gao, and F. Ayazi

THERMAL ANALYSIS AND CHARACTERIZATION OF HIGH Q FILM BULK ACOUSTIC RESONATOR (FBAR) AS BIOSENSERS IN LIQUIDS 935
X. Zhang, W. Xu, A. Abbaspour-Tamijani, and J. Chae

OPTICAL MEMS

EL DISPLAY PRINTED ON CURVED SURFACE 939
T.-M. Lee, H.-C. Choi, J.-H. Noh, and D.-S. Kim

SOLUTION ELECTROCHEMILUMINESCENT MICROFLUIDIC CELL FOR FLEXIBLE AND STRETCHABLE DISPLAY 943
R. Okumura, S. Takamatsu, E. Iwase, K. Matsumoto, and I. Shimoyama

MEMS BASED LASER IMAGER WITH DIAGONAL PROGRESSIVE SCANNING 947
T. Sandner, T. Grasshoff, T. Klose, H. Schenk, and J.L. Massieu

POLYMERIC (SU-8) OPTICAL MICROSCANNER DRIVEN BY ELECTROSTATIC ACTUATION 951
S.K. Lee, M.G. Kim, J.Y. An, M.H. Jun, S. Yang, and J.H. Lee

VACUUM WAFER LEVEL PACKAGED TWO-DIMENSIONAL OPTICAL SCANNER BY ANODIC BONDING 955
H. Tachibana, K. Kawano, H. Ueda, and H. Noge

ACCURACY ENHANCEMENT IN THE IN-PLANE DYNAMIC MEASUREMENT OF MEMS ACTUATORS USING A LASER DOPPLER VIBROMETER WITH A 45°-ANGLED OPTICAL FIBER 959
M.G. Kim1, K.W. Jo1, Y.S. Park2, W.G. Jang, and J.H. Lee1

MICRO LIQUID PRISM 963
Y. Yoshihata, A. Takei, N. Binh-Khiem, T. Kan, E. Iwase, K. Matsumoto, and I. Shimoyama

MEMS OPTICAL LOGIC NOR GATE USING INTEGRATED TUNABLE LASERS 967
B. Liu, H. Cai, X.M. Zhang, J. Tamil, Q.X. Zhang, and A.Q. Liu

TILTED PARABOLOIDAL REFLECTIVE LENS FOR FAR INFRARED FABRICATED BY MASK WITH RECTANGULAR OPENINGS 971
T. Takahata, K. Matsumoto, and I. Shimoyama

MEMS LASER WITH TUNABLE WAVELENGTH AND POLARIZATION USING OPTICAL TUNNELING EFFECT 975
W.M. Zhu, X.M. Zhang, H. Cai, J. Tamil, W. Zhang, B. Liu, T. Bourouina, and A.Q. Liu

A NOVEL TELESCOPE WITH MICROMIRROR FOR OBSERVATION OF TRANSIENT LUMINOUS EVENTS FROM SPACE 979
J.H. Park, S. Nam, G. Garipov, J.A. Jeon, J.Y. Jin, B. Khrenov, J.E. Kim, M. Kim, Y.K. Kim, J. Lee, G.W. Na, I.H. Park, Y.-S. Park, and B.W. Yoo

NON-SPHERICAL SU-8 MICROLENS ARRAY FABRICATED UTILIZING A NOVEL STAMPING PROCESS AND AN ELECTRO-STATIC PULLING METHOD 983
S.-M. Kuo and C.-H. Lin

MULTICHAMBER TUNABLE LIQUID MICROLENSES WITH ACTIVE ABERRATION CORRECTION 987
D. Mader, P. Waibel, A. Seifert, and H. Zappe

OPTICAL MICROMIRROR ACTUATION USING THERMOCAPILLARY EFFECT IN MICRODROPLETS 991
R.K. Dhull, I. Puchades, L. Fuller, and Y.W. Lu

DUAL-AXES CONFOCAL MICROLENS FOR RAMAN SPECTROSCOPY 995
C.P.B. Siu, H. Wang, H. Zeng, and M. Chiao

TUNABLE SCANNING FIBER OPTIC MEMS-PROBE FOR ENDOSCOPIC OPTICAL COHERENCE TOMOGRAPHY 999
K. Aljasem, A. Seifert, and H. Zappe

CMOS-INTEGRABLE PISTON-TYPE MICRO-MIRROR ARRAY FOR ADAPTIVE OPTICS MADE OF MONO-CRYSTALLINE SILICON USING 3-D INTEGRATION 1003
M. Lapisa, F. Zimmer, F. Niklaus, A. Gehner, and G. Stemme

MONOLITHIC MULTICOLOR TOTAL INTERNAL REFLECTION (TIR)-BASED CHIP FOR HIGHLY-SENSITIVE MULTIFLUORESCENCE DETECTION AND IMAGING 1007
N.C.H. Le, D.V. Dao, R. Yokokawa, J. Wells, and S. Sugiyama

CMOS-COMPATIBLE 2-AXIS SELF-ALIGNED VERTICAL COMB-DRIVEN MICROMIRROR FOR LARGE FIELD-OF-VIEW MICROENDOSCOPES 1011
K. Kumar and X.J. Zhang

TOPOLOGY OPTIMIZATION FOR MICRO ROTATIONAL MIRROR DESIGN AND SAFE MANUFACTURING 1015
T. Chen, Z. Liu, J.G. Korvink, S. Krausse, and U. Wallrabe

STRETCHABLE YARN OF DISPLAY ELEMENTS 1019
S. Takamatsu, K. Matsumoto, and I. Shimoyama

A MICROMACHINED THERMO-OPTIC TUNABLE LASER 1023
H. Cai, B. Liu, X.M. Zhang, W.M. Zhu, J. Tamil, W. Zhang, Q.X. Zhang, and A.Q. Liu

MEMS ACTUATORS & POWER MEMS

MASS-ANALYSIS SCANNING FORCE MICROSCOPY WITH ELECTROSTATIC SWITCHING MECHANISM 1027
C.Y. Shao, Y. Kawai, T. Ono, and M. Esashi

A MULTIPLE DEGREES OF FREEDOM ELECTROTHERMAL ACTUATOR FOR A VERSATILE MEMS GRIPPER 1031
D.-S. Chen, C.-Y. Yin, R.-J. Lai, and J.-C. Tsai

BIDIRECTIONAL ELECTROTHERMAL ELECTROMAGNETICTORSIONAL MICROACTUATORS 1035
Y. Eun, H. Na, and J. Kim

A 2D ELECTRET-BASED RESONANT MICRO ENERGY HARVESTER 1039
U. Bartsch, J. Gaspar, and O. Paul

PZT MEMS ACTUATED FLAPPING WINGS FOR INSECT-INSPIRED ROBOTICS 1043
J.R. Bronson, J.S. Pulskamp, R.G. Polcawich, C.M. Kroninger, and E.D. Wetzel

A NOVEL UNDERWATER ACTUATOR DRIVEN BY MAGNETIZATION REPULSION/ATTRACTION 1047
H.-T. Su, T.-L. Tang, and W. Fang

DEVELOPMENT OF THE FORWARD-LOOKING ACTIVE MICRO-CATHETER ACTUATED BY TI-NI SHAPE MEMORY ALLOY SPRINGS 1051
M. Komatsubara, T. Namazu, H. Nagasawa, T. Miki, T. Tsurui, and S. Inoue

LOW-RESONANT-FREQUENCY MICRO ELECTRET GENERATOR FOR ENERGY HARVESTING APPLICATION 1055
M. Edamoto, Y. Suzuki, N. Kasagi, K. Kashiwagi, Y. Morizawa, T. Yokoyama, T. Seki, and M. Oba

DEVELOPMENT OF THE NOVEL ELONGATION-MEASUREMENT DEVICE WITH IN-PLANE BIMORPH ACTUATOR FOR THE TENSILE TEST 1059
H. Fujii, T. Namazu, and S. Inoue

MICROACTUATION UTILIZING WAFER-LEVEL INTEGRATED SMA WIRES 1063
D. Clausi, H. Gradin, S. Braun, J. Peirs, G. Stemme, D. Reynaerts, and W. van der Wijngaart

AN INTEGRATED SOLUTION FOR WAFER-LEVEL PACKAGING AND ELECTROSTATIC ACTUATION OF OUT-OF-PLANE DEVICES 1067
K.L. Chen, R. Melamud, S. Wang, and T.W. Kenny

AN ELECTROMAGNETIC MICRO POWER GENERATOR FOR LOW FREQUENCY ENVIRONMENTAL VIBRATIONS BASED ON THE FREQUENCY UP-CONVERSION TECHNIQUE 1071
I. Sari, T. Balkan, and H. Külah

ELECTROSTATIC ROTARY STEPPER MICROMOTOR FOR SKEW ANGLE COMPENSATION IN HARD DISK DRIVES...1075
E. Sarajlic, C. Yamahata, M. Cordero, and H. Fujita

A SURFACE-TENSION-DRIVEN PROPULSION AND ROTATION PRINCIPLE FOR WATER-FLOATING MINI/MICRO ROBOTS...1079
S.K. Chung, K. Ryu, and S.K. Cho

DEVELOPMENT OF SKEWED DRIE PROCESS AND ITS APPLICATION TO ELECTROSTATIC TILT MIRROR...1083
M. Nakada, K. Takahashi, T. Takahashi, A. Higo, H. Fujita, and H. Toshiyoshi

DESIGN OF A PASSIVE AND PORTABLE DMFC OPERATING IN ALL ORIENTATIONS...1087
N. Paust, S. Krumbholz, S. Munt, C. Müller, R. Zengerle, C. Ziegler, and P. Koltay

A ROTARY MICROACTUATOR SUPPORTED ON ENCAPSULATED MICROBALL BEARINGS USING AN ELECTRO-PNEUMATIC THRUST BALANCE...1091
M. McCarthy, C.M. Waits, M.I. Beyaz, and R. Ghodssi

PASSIVATED ELECTRODE ACTUATOR WITH STABLE RESONANCE AMPLITUDE...1095
G. Bahl, R.G. Walmsley, B.E. DeMartini, K.L. Turner, and P.G. Hartwell

FAST POSITIONING AND IMPACT MINIMIZING OF MEMS DEVICES BY SUPPRESSION MOTION-INDUCED VIBRATION BY COMMAND SHAPING METHOD...1099
K.-S. Chen and K.-S. Ou

A MICRO DIRECT METHANOL FUEL CELL INTEGRATED WITH A TEMPERATURE CONTROL SYSTEM FOR EXTREME ENVIRONMENTS...1103
Q. Zhang, X. Wang, Y. Zhu, Y. Zhou, X. Qiu, and L. Liu

LEVER-BASED CMOS-MEMS PROBES FOR RECONFIGURABLE RF IC'S...1107
J. Liu, M. Noman, J.A. Bain, T.E. Schlesinger, and G.K. Fedder

A NEW EFFICIENT METHOD FOR SIMULATING THE DYNAMIC RESPONSE OF ELECTROSTATIC SWITCHES...1111
V. Leus and D. Elata

Author Index

INTEGRATION OF CARBON NANOTUBES INTO MEMS DEVICES VIA DETERMINISTIC TRANSPLANTING ASSEMBLY

S. Kim, H.W. Lee, and S.G. Kim
Massachusetts Institute of Technology, Cambridge, Massachusetts, USA

ABSTRACT

This paper presents a new method to assemble individual nanostructures such as a single strand carbon nanotube (CNT) at a deterministic location of MEMS devices. An array of vertically aligned CNTs grown on nickel nano-dots was encapsulated into MEMS scale polymer blocks that were then transplanted to MEMS cantilevers to make CNT-tipped atomic force microscope (AFM) probes. Even manual transfer of a CNT-bearing polymer block to the target location leads to reliable and reproducible fabrication of a CNT-based device such as the CNT-tipped AFM probe. The scanning experimental results on AFM standard gratings and biological samples confirm the superior characteristic of CNTs (high aspect ratio and toughness) and justify this assembly method can be useful to mass produce nanostructured MEMS devices.

INTRODUCTION

Assembly of individual nanostructures is an essential step in their integration into MEMS devices while preserving their novel properties at nanometer scales. However, handling and assembling individual nanostructures has been a challenge with no adequate solutions yet [1-3].

This paper reports a new method of handling and locating individual CNTs via transplanting process. The key idea of transplanting assembly is to grow the vertically aligned individual CNTs on a substrate at optimal growth conditions, to encapsulate them to become transplantable and to transplant embedded CNT strands to the target locations (Figure 1). The major technical issues we solved are: (1) how to grow vertically aligned single strand CNTs at predefined locations, (2) how to preserve/control the orientation/length of an individual CNT during transplanting processes, and (3) how to release/locate an individual CNT at the target location.

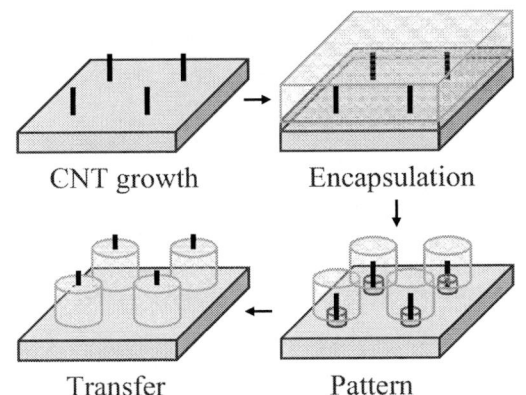

Figure 1: A schematic showing the procedure of transplanting assembly of the individual CNTs.

This assembly transforms the scale of tools necessary for CNT assembly from nano-scale to micro-scale, which enables automated, parallel or even manual assembly of individual CNTs to MEMS in a deterministic and reproducible way. Previously, authors reported a bundle of CNTs could be transplanted from where they were grown to where they could function [4]. This paper presents that a single strand CNT can be transplanted to make a CNT-tipped AFM probes which includes detailed design and fabrication process of integrating individual CNTs into MEMS devices and quantitative characterization of physicochemical interactions of the individual CNTs with encapsulating polymers and etchants during the transplanting process.

DESIGN OF CNT CARRIER

The key issue in assembling individual CNTs is how to control the number, location, shape (the diameter and the length), and orientation of CNTs before transplanting them. We show in this paper that they can be controlled by the shape and location of individual catalytic dots as well as CNT growth parameters.

The diameter and length of vertically grown individual CNTs are governed by the size and thickness of catalytic dots. The geometries (the diameter, thickness, and the location) of the Ni catalytic dots are defined by electron-beam lithography in combination with thin metal deposition. The patterns written by electron beam lithography are in an array (21 by 21) of circular dots. Various size dots are made which ranges between 100 nm and 200 nm in diameter and can be well controlled by the path and exposure dose of electron beams at the electron-beam writing facility available at MIT. The thickness of the Ni catalytic dots has been varied 10~40 nm, which was then optimized through the CNT growth experiments. The spacing between dots ranges from 5~40 μm.

The next step is to freeze each CNT into micro-scale polymer blocks, which is transplantable. In encapsulating the individual CNTs into MEMS carriers, SU8-2015 (MicroChem Corp., Newton, Massachusetts) is selected as a carrier body for individual CNs, while the sacrificial bottom layer (polymethylglutarimide (PMGI) SF 11, MicroChem, Newton, Massachusetts) defines the exposed length of the CN tips after assembly as well as it enables controlled release of the CN-bearing SU8 blocks from the substrate. Both are chemically compatible with CNTs and easy to pattern with the thickness of 1~30 μm.

The key idea in the double polymeric layer encapsulation design for MEMS carriers is to provide a substantial undercut in the bottom layer as illustrated in Figure 1 using the different etch selectivity of them. After the CNT growth step, the two layers of the polymers are coated using spin-coating processes. The top layer (SU8 2015) forms a MEMS carrier through patterning processes using photolithography. The key role of the bottom layer is to define the length of the CNT tips and to release the CNT bearing carrier easily from the substrate when the thin bottom layer is sheared off gently.

Among many candidate materials for the bottom layer,

978-1-4244-2977-6/09 $25.00 © 2009 IEEE

polymethylglutarimide (PMGI) resist is chosen primarily because of its good selectivity to SU8-2015. PMGI SF series is an organic polymer solution, and its primary application has been to promote the adhesion of SU8 to a substrate or to lift-off the SU8 layer. PMGI-based resists are positive-toned and can be patterned using deep-UV (DUV) radiation. PMGI has been patterned using electron-beam and proton beam exposure. Its insolubility to the casting solvent used by most novolac photoresist formulations enables the application for lift-off processes. Based on the chemical composition, PMGI SF series show a wide range of achievable thicknesses according to the viscosity. Among them, PMGI SF 11 is chosen for its achievable thickness range (1~3µm) and low viscosity (1/100~1/50 of SU8-2015).

FABRICATION

A 33 by 33 array of vertically aligned single strand CNTs was grown from the nickel (Ni) nano-dots, which were defined on Si/TiN substrates using electron-beam lithography followed by metal deposition and lift-off processes (Figure 2).

A 100 nm thick TiN layer is deposited on a Si wafer through the metal sputtering process. This TiN layer promotes adhesion of the catalytic (Ni) layer to the substrate, and prevents the formation of Ni silicide ($NiSi_X$) during the CNT growth step. If Ni catalysts are deposited directly on a Si substrate, Ni diffuses into Si to form $NiSi_X$ at temperatures above 450°C, resulting in poor growth yield. A 75~100 nm thick polymethylmethacrylate (PMMA) layer is spin-coated on top of the TiN layer. Patterns of small holes are formed by using a scanning electron-beam lithography system, Raith-150, followed by PMMA development with MIBK (methyl isobutyl ketone). After patterning holes on the PMMA layer, a 30 nm thick Ni layer is deposited using electron beam metal evaporator. The optimal thickness of Ni catalytic dots, 30 nm, in combination with the diameter of 150 nm, was determined experimentally to achieve straighter CNTs with length longer than 5 µm. The Ni layer is then patterned by lift-offing the PMMA layer with NMP (1-methyl-2-pyrrolidinone) etchant.

A home-built plasma enhanced chemical vapor deposition (PECVD) machine was used to grow individual CNTs vertically. PECVD is widely used to obtain well-aligned multi-walled CNTs at temperatures below 700°C. The essential component used in a PECVD system is its electric field (plasma). The electric fields across the electrodes induce the dipole moment in metal catalysts guiding the alignment of each CNT during the growth process [6].

Among various plasma sources for CNT growth available, we use a DC plasma method. The dc plasma reactor consists of a pair of electrodes in a grounded chamber: one electrode is grounded while the other is connected to a power supply. The negative dc bias voltage applied to one electrode on the sample side (cathode) dissociates the feedstock gas (C_2H_2) and generates many carbon-bearing radicals for carbon nanotube growth.

The diameter of CNTs is 100~200 nm depending on the size of Ni dots, and the length is 5~10 µm. Figures 3 and 4 show vertically grown CNTs which match well with the previous report [5], but show more uniform cylindrical

CNTs. The process parameters such as ratio of NH_3 and C_2H_2 have been optimized to result in the uniform cylindrical shape.

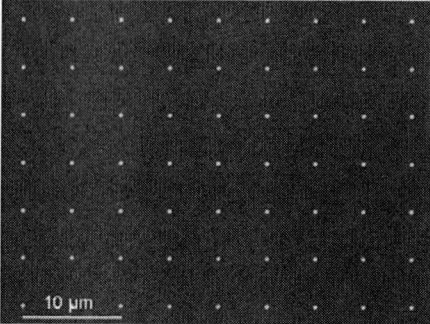

Figure 2: An array of catalytic dots with 5 µm spacing between each dot.

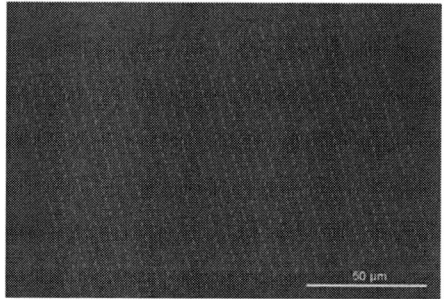

Figure 3: A 21 x 21 array of vertically aligned CNTs.

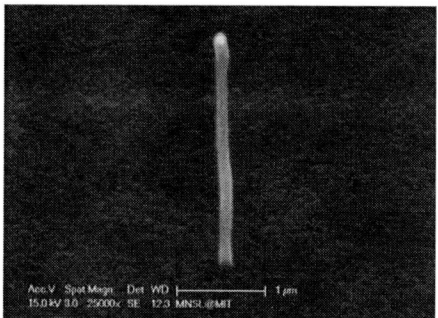

Figure 4: Individual CNTs with a uniform cylindrical shape. (30° tilted image from the top)

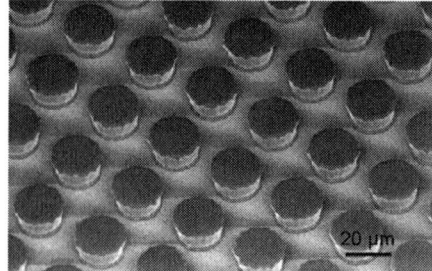

Figure 5: An array of SU8 pellets which are used as a MEMS carrier during nano structure assembly.

Each grown CNT strands were embedded into a 15 µm-tall MEMS polymer block which serves as a CNT carrier (Figure 5). We used a double polymeric layer encapsulation process with SU8 (top) and polymethylglutarimide (PMGI) (bottom): the top SU-8 forms the body of the carrier while the bottom PMGI layer holds the body until the release of the carrier from the

978-1-4244-2977-6/09 $25.00 © 2009 IEEE

substrate and then is going to be removed to expose the CNT.

5 ml of PMGI SF 11 solution is statically dispensed over the array of CNTs on the substrate. The spinning at 3000 rpm for 45 sec results in the 1.5μm thick bottom layer. Then the sample is baked on a hot plate at 200°C for 5 min. After the bottom layer is coated, 10ml of SU8-2015 solution is statically dispensed over the bottom layer and the array of CNTs. The sample is spin coated at 3000 rpm for 30 sec in order to reach the thickness of 20 μm, and is baked on a hot plate at 95 °C for 4 min. Then the sample is cured at the room temperature for 24 hours before the edge removal process. The top, SU8-2015, layer is patterned to form an array of cylinders with 20 μm in diameter and each contain only one CNT inside.

The bottom (PMGI SF 11) layer was etched using PMGI 101 developer (MicroChem, Newton, MA) with the top patterned SU8 pellets as etch masks. The PMGI 101 developer is a tetraethylammonium hydroxide (TEAH)-based positive radiation resist developer, which exhibits etch selectivity to PMGI over novolac imaging resists such as SU8. This etch selectivity enables development of PMGI films independent of an overlying resist layer, and can be used for enhanced undercut in bi-layer lift-off processes.

For easier assembly, the bottom (PMGI) layer was etched further until the adhesion between the top (SU8) and the bottom (PMGI) layers are smaller than the adhesion between the top (SU8) and the Si cantilever while the orientation of the SU pellet with the single CNT remains vertical, so the orientation of the CNT can be preserved during the assembly step.

An analytic model is developed to estimate flow-induced deflections of the individual CNTs during the spin casting of polymers over the array of CNTs, which predicts negligible deflection of vertically standing CNTs. Shearing off the bottom layer releases the CNT bearing SU8-2015 carrier from the substrate. The exposed CNT tip, located at the center of the MEMS carrier, remains parallel to the axis of the MEMS carrier, and the CNT tip is about 1.5μm long, which matches the thickness of bottom layers. The transplanted CNT remain parallel to the pellet axis, which also matches the calculated results of less than 1 degree deflection from the vertical axis (Figure 6).

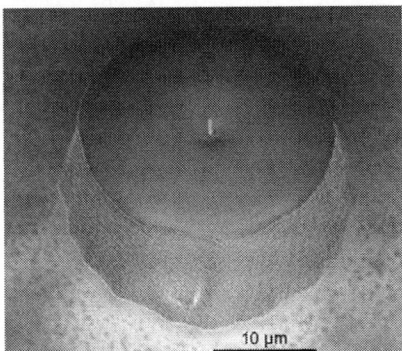

Figure 6: A released SU-8 block with a single CNT.

Picking and placing the individual nanostructures is the essential step in a deterministic assembly of the nanostructures. The bottom layer is etched further with PMGI 101 remover until a small amount of the bottom

layer remains to protect the CNTs from buckling. The bonding force between the bottom layer and the substrate will be small enough to break the bonding when the bottom layer is sheared off mechanically. A CNT tipped AFM probe is made by attaching a CNT bearing polymeric carrier at the end of a MEMS cantilever with humid curing adhesive.

A tipless AFM cantilever (NCS12 of MIKROMASCH, San Jose, California) is mounted on a 3 axis motion stage under a 60X (6X optical and 10X digital) optical microscope. The tip which already picked up a small droplet of a humidity curing adhesive (LOCTITE® 408™) is approached to a target carrier using a micro motion stage while being monitored with the optical microscope. The LOCTITE® 408™ is chosen because of its low viscosity (~ 5 cSt) which enables us to pick up a minimal amount of the adhesive at the end of the cantilever. The adhesive provides both tensile and shear bonding strength larger than 5 MPa once it is cured.

The adhesive is cured in 5~20 minutes depending on the undercut of the bottom (PMGI SF 11) layer. 120 minute's etching remains the bottom layer with 7 μm in diameter, which requires 5 minute's curing for a strong bonding between the MEMS carrier and the Si cantilever. After the adhesive is cured, the bottom PMGI SF 11 layer is sheared mechanically in parallel to the substrate using the micro motion stage under an optical microscope. The bonding of the bottom PMGI layer and CNT roots to the substrate is broken by mechanical shearing, and the individual CNTs with MEMS carriers are released from the substrate. The existence of a CNT core through the top SU8 layer to the bottom PMGI layer may enforce the bonding between the top and bottom layers. The entire block with a single CNT is released from the substrate by being picked up with the micro motion stage. After the probe is released from the substrate, the whole probe is put inside of the PMGI 101 developer until the remaining bottom layer is fully removed. Finally, a CNT tipped AFM probe is fabricated with the exposed CNT tip. The CNT tip is 1.5 μm long, which corresponds to the thickness of the bottom PMGI layer, and normal to the sample surfaces (Figure 7). Even manual assembly of a polymer block to the target location leads to reliable and reproducible fabrication of a CNT-based device such as the CNT-tipped AFM probe.

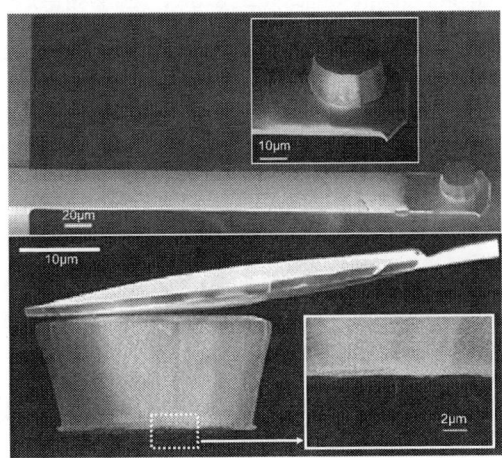

Figure 7: The CNT-tipped AFM probes which was made by transplanting assembly technique.

SCANNING EXPERIMENTS

The scanning experiments on standard AFM nanogratings and biological samples confirm that the superior characteristic of CNTs (high aspect ratio scanning and minimal wear for long life) can be widely utilized by assembling individual CNT into a MEMS device deterministically (Figure 8).

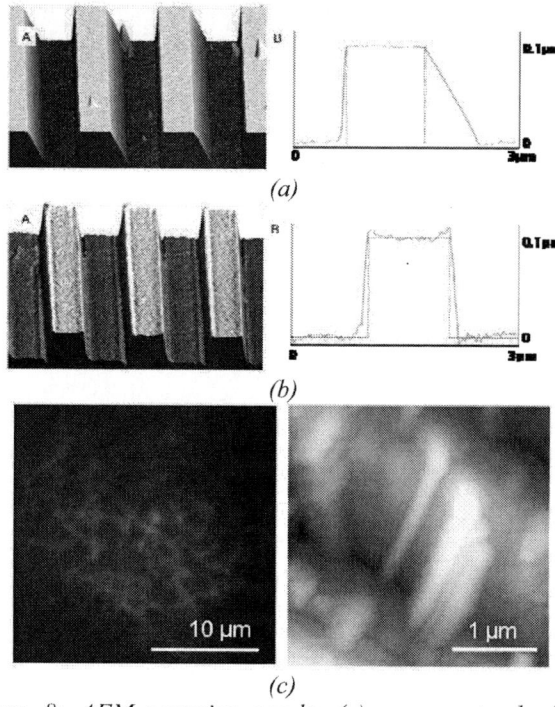

(a)

(b)

(c)

Figure 8: AFM scanning results (a) over a standard grating with standard Si AFM tip, (b) with CNT-tipped AFM tip made by transplanting assembly technique, and (c) fluorescence microscopy image and AFM scanning image of a biological sample.

Figures 8 (a) and (b) were obtained using contact mode scanning at the speed of 5 μm/sec. It is obvious that the CNT tipped AFM probe scan the vertical trenches closer to its original features than the standard pyramidal tip Si probes do. The little overshoot at both edges results from non-optimal AFM scanning parameters such as the proportion and integration control gains (Figure 8 (b)). The scanning capability of the CNT-tipped nanoprobe over biological samples was checked by scanning a filament actin (F-actin) sample (Figure 8 (c)). 64μl/2.3μM of F-actins was prepared in the 50 ml buffer solution (5mM Tris-HCl, 0.5mM ATP, 0.2mM DTT, 100mM KCl, 0.2mM $CaCl_2$, and 2mM $MgCl_2$) and stored at 4 °C before scanning. Fluorescent microscopy (left) shows bundles of F-actins in the buffer solution. 10 ml of the solution was dropped on a glass slide, and AFM scanning was performed right after the buffer solution is dried. Figure 8 (c, right) shows the contact mode AFM image with our CNT-tipped nanoprobe. It is noted that the sample has large surface roughness which may results from the clustered bundles of actins, remnants from the buffer solution, and contaminants on the substrate.

CONCLUSION

Deterministic transplanting assembly technique provides reliable and reproducible assembly method in the integration of carbon nanotubes into MEMS devices. Each vertically grown CNT is encapsulated in to a polymeric MEMS carrier and is transplanted to the end of an AFM cantilever. Through the use of double polymeric layer encapsulation with different etch selectivity, the length of the exposed carbon nanotube can be controlled. SU-8 block was used as a CNT bearing carrier during the assembly process.

We also showed that transplanting assembled CN-tipped AFM probes could be used for scanning and imaging nano-bio-objects. Scanning experiments using the fabricated CNT-tipped AFM probes confirm that the original properties of the CNT tips are preserved through deterministic assembly of the individual CNTs on a MEMS device and provide high quality scanning and imaging of nano- and bio-objects.

Massive parallel assembly of nanostructure to MEMS becomes feasible with our method. We demonstrated even manual assembly of a CNT-tipped AFM probe was possible by transplanting assembly. We believe that this assembly method can be scaled up to make a massive parallel array of CNTs on MEMS devices for high throughput applications.

ACKNOLWLEDGEMENT

This work is supported by Intelligent Microsystem Center in Korea.

REFERENCES

[1] H. Dai, J. H. Hafner, A. G. Rinzler, D. T. Colbert, R. E. Smalley, "Nanotubes as nanoprobes in scanning probe microscopy", *Nature*, vol. 384, no. 14, pp. 147-150, 1996.

[2] J. Tang, G. Yang, Q. Zhang, A. Parhat, B. Maynor, J. Liu, L. Qin, O. Zhou, "Rapid and reproducible fabrication of carbon nanotube AFM probes by dielectrophoresis", *Nano Lett.*, vol. 5, no. 1, pp. 11-14, 2005.

[3] J. H. Hafner, C. Cheung, T. H. Oosterkamp, C. H. Lieber, "High-yield assembly of individual single-walled carbon nanotube tips for scanning probe microscopies", *J. Phys. Chem. B*, vol. 105, pp. 743-746, 2001.

[4] T. A. El-Aguizy, J. H. Jeong, Y. B. Jeon, W. Z. Li, Z. F. Ren, S. G. Kim, "Transplanting Carbon Nanotubes", *Appl. Phys. Lett.*, vol. 85, pp. 5995-5997, 2004.

[5] Z. F. Ren, Z. P. Huang, J. W. Xu, J. H. Wang, P. M. Bush, P. Siegal, P. N. Provencio, "Synthesis of large arrays of well-aligned carbon nanotubes on glass", *Science*, vol. 282, pp. 1105-1107, 1998.

[6] Z. F. Ren, Z. P. Huang, D. Z. Wang, J. G. Wen, J. W. Xu, J. H. Wang, L. E. Calvet, J. Chen, J. F. Klemic, M. A. Reed, "Growth of a single freestanding multiwall carbon nanotube on each nanonickel dot", *Appl. Phys. Lett.*, vol. 75, pp. 1086-1088, 1999.

SURFACE-TEMPERATURE CONTROL OF SILICON NANOWIRES IN DRY AND LIQUID CONDITIONS

S. Akiyama[1], Y. T. Cheng[1], J. Fattaccioli[2], N. Takama[1], P. Löw[3], C. Bergaud[3] and B. J. Kim[1]
[1]Institute of Industrial Science, The University of Tokyo, Tokyo, JAPAN
[2]LIMMS/CNRS-IIS (UMI2820), Institute of Industrial Science, Tokyo, JAPAN
[3]LAAS-CNRS, Toulouse, FRANCE

ABSTRACT

In this paper we present the results of surface temperature control of silicon nanowires by using fluorescent thermometry at the nanometer scale. Rhodamine B is one of the stable fluorescent molecules, which rely on the characteristic of temperature-dependent change in fluorescent intensity, and it was used for nano-scale surface temperature sensing interface. The resistive heating on Si nanowires was carried out with applying voltage potential of 6 ~ 12 V. Surface-temperature measurement was performed by converting the changes in fluorescent intensity with calibration curve of Rhodamine B. The temperature at the central line along nanowires increasing from 30 degrees to 35~70 degrees was observed.

INTRODUCTION

Silicon nanowires have attracted much attention for many years because they are most attractive from the point of view of high sensitivity and surface-modifying for various sensors. Recently, silicon nanowires are widely applied for biosensors [1]. But a few attempts [2] were made so far to control their temperature although this parameter has been shown to be crucial when dealing with biochemical reactions [3]. There are several methods which measure temperature under the micrometer scale, for example Raman spectroscopy and Infrared thermometry. But considering the spatial resolution and cost of setup, fluorescent thermometry has considerable promise as thermometer at the sub-micro scale. It relies on the measurement of the photoluminescent light emitted by a probe in order to determine the temperature. Since the probe used in fluorescent thermometry transmit information via light, the method is useful as a non-contact approach in the characterization of temperature.

Using a fluorescent probe, Rhodamine B, which shows fluorescent intensity dependence as a function of the temperature [4], Löw *et al.* demonstrated the possibility to monitor the temperature of nickel nanowires by fluorescence thermometry in dry condition [5]. However, this work showed several issues, such as a high thermal inertia. Besides, nickel is difficult to functionalize and its reflectivity makes difficult the microscopic studies of low fluorescence signals.

From these backgrounds, we think that it is important to control surface-temperature of silicon nanowires in dry as well as in liquid conditions. The main advantage of silicon nanowire heating device is the capability for fast temperature cycling due to the lower thermal mass and heat capacitance.

2. MATERIALS AND METHODS
Fabrication of the nanowires

We fabricated the silicon nanowires from Silicon-on-Insulator (SOI) wafer with conventional microfabrication processes only [6]: photolithography, anisotropic wet etching, local thermal oxidation of silicon (LOCOS), and removal of silicon dioxide. The basic principles and processes are shown in figure 2. First, wafer cleaning and silicon nitride deposition were carried out (fig. 2a and 2b). Approximately 100 nm-thick silicon nitride (Si_3N_4) film was deposited on the surface of SOI substrate by using Low-Pressure Chemical Vapor Deposition (LPCVD). This silicon nitride layer was used as a masking layer in the silicon anisotropic KOH solution etching (Fig. 2c) and local oxidation of silicon process (Fig. 2d). Next, in order to remove silicon nitride film on the pre-wire structure, photolithography and Reactive Ion Etching (RIE) process were carried out (Fig. 2e and 2f). After that, anisotropic KOH solution etching was performed again to remove silicon partially in the pre-wire structure (Fig. 2g). Finally remaining Si_3N_4 film was removed by RIE process (Fig. 2h). As additional process, sputtering process was carried out to change the surface characteristic. The surface of silicon nanowires was changed from silicon to silicon dioxide to attach the biological molecules.

The most important point of this process is that the width and thickness of fabricated silicon nanowires would be determined by the thickness of top silicon layer of SOI substrates only. The final length, width and thickness of the nanowires are 20 μm, 360 nm and 240 nm, respectively. Due to the anisotropic etching process dependant on Si crystalline, the shape of cross sectional nanowire is a triangle. The SEM images of silicon nanowires are shown in figure 1.

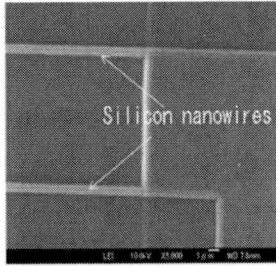

Figure 1: SEM images the nanowires. Their final length, width and thickness are 20 μm, 360 nm and 240 nm, respectively.

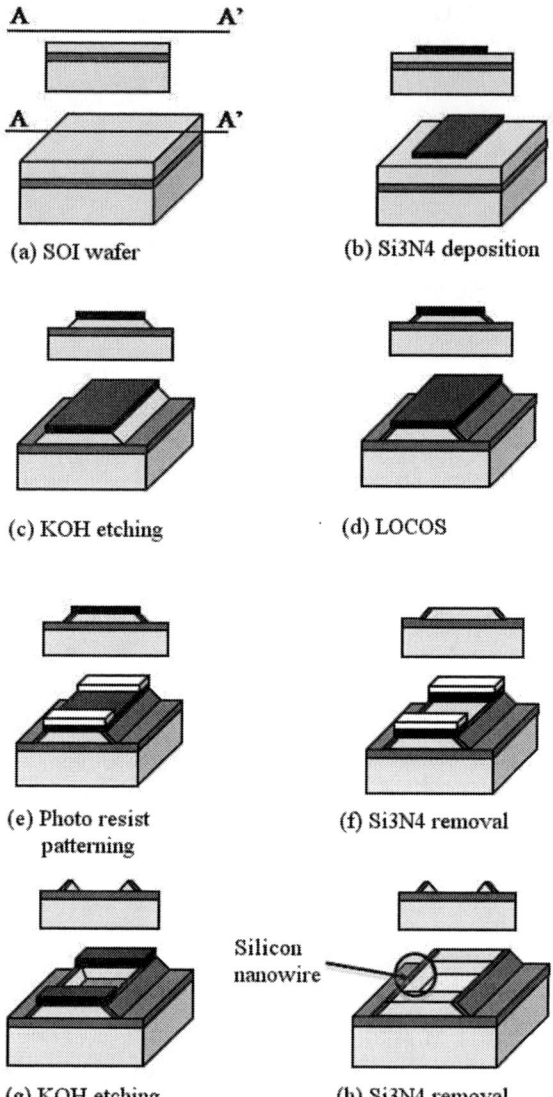

(a) SOI wafer

(b) Si3N4 deposition

(c) KOH etching

(d) LOCOS

(e) Photo resist patterning

(f) Si3N4 removal

Silicon nanowire

(g) KOH etching

(h) Si3N4 removal

Figure 2: Fabrication process of the nanowires:
(a) Preparation of the SOI substrates; (b) Silicon nitride deposition; (c) Top silicon layer etching by KOH; (d) Top silicon layer oxidation (LOCOS); (e) Lithography; (f) Si_3N_4 film removal; (g) Nanowire formation by KOH etching; (h) Removal of remaining silicon nitride.

Finally, we decrease the resistivity of the nanowires by a p-type ion implantation of silicon (ULVAC UP-150, Japan, 3×10^{15} ion/cm^2, 15 μA) in order to improve the performances of the nanowires toward resistive heating.

Functionnalization of the Nanowires by Rhodamin-B

For the experiment in air (dry condition), the Rhodamine B molecules (Molecular Probes, Invitrogen) are dissolved in deionized water (Millipore) and simply dried on the surface of the nanowires before carrying out the measurements.

However, in order to perform the measurement with the nanowires immersed in an aqueous buffer (*liquid condition*), rhodamine has to be firmly attached to surface to avoid its spontaneous desorption enhanced by the high temperature [7]. Instead of using bare rhodamine, we use a functionnalization route involving the biotin-streptavidin complex since it is easy to handle [8] and it has been shown to be thermally stable until a temperature of 95 degrees, which is sufficient for our application [9].

We first adsorb a positive polyelectrolyte (poly-L-lysine, *Sigma-Aldrich*) to which we attach a biotin derivative Biotin-$(C_2H_4O)_3$-NH_2 ($C_{16}H_{30}N_4SO_4$, Pierce) and finally adsorb a protein, rhodamin-streptavidin (Molecular Probes, Invitrogen), thanks to its high chemical affinity with biotin [10].

Fluorescent Thermometry Setup

Fluorescent measurements were made using a BX-51 upright fluorescence microscope and a conventional mercury lamp from Olympus, Japan. In order to capture the fluorescence for digital analysis, we use an EM-CCD (electron-multiplying CCD) camera (Cascade II 512, Photometrics, USA) connected to a computer. During the experiments we use a thermal plate (MATS-1002RO, TOKAI HIP Co., Ltd., Japan) to set the temperature of the silicon support of the nanowires to 30 degrees as a reference. The schematic diagram of fluorescence thermometry is shown in figure 3.

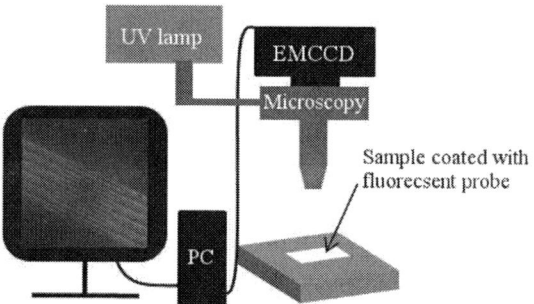

Figure 3: Schematic representation of the fluorescent thermometry setup under the microscope we use to perform the temperature mapping of the nanowires.

The Al-electrodes were patterned on the substrate by evaporation and lithography. Next, the silicon nanowire tip was connected to a commercial PCB board by using a wire bonder (MB-2100, Nippon Avionics Co., Ltd., Japan). The resistive heating on nanowires was carried out with applying voltage potential by a DC power supply (R6240A DC Voltage current source/monitor, Advantest, Japan). In dry condition, silicon nanowire tip was covered by a coverslip, which fixed with tapes. On the other hand in liquid condition, the silicon nanowire tip was sealed with commercial nail lacquer after this process to avoid evaporation of the aqueous buffer.

3. RESULTS

Calibration of the Fluorescence Intensity Measurement

978-1-4244-2977-6/09 $25.00 © 2009 IEEE

To calibrate the measurement, we measure the fluorescence intensity of a functionalized and oxidized silicon surface in *dry* and *liquid* condition when varying its temperature with an accurate hot-plate mounted on the microscope. Since the intensity of fluorescence is decreasing with the increasing temperature, we calculate the ratio

$$r = \frac{I_T}{I_{30}}$$

of the intensity I_{30} at 30 degrees to the intensity I_T at higher temperatures. On figure 4, we plot the intensity ratios as a function of the hot-plate temperature. In air and in liquid, the intensity decrease is similar and scales linearly with the temperature. In the following experiments we use these data to calibrate the temperature of the nanowires.

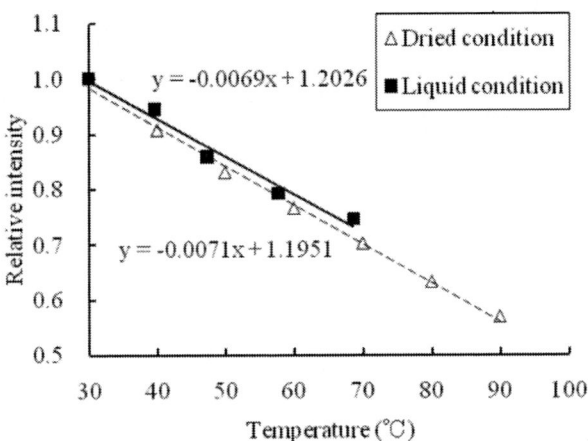

Figure 4: Calibration curve of the fluorescent intensity of Rhodamine B molecules as a function of temperature in dry (△) and liquid (■) conditions. In both cases, the relative intensity scales linearly with the temperature.

Temperature Mapping of the Nanowires

In dry condition, we measured the intensity of fluorescence of silicon nanowires coated with Rhodamine B when a voltage ranging from 0 to 13 V is applied (Fig. 5). We can see silicon nanowires clearly with no voltage in the left image. But in the right image, because of applied voltage, surface temperature of silicon nanowires was increased. As a result the white parts of silicon nanowires changed black.

From the calibration data, we are able to make the temperature mapping of the silicon nanowires when the tension is increased. First, the intensity of background was subtracted from the averaged image. Next, the intensity of every pixel on the applied voltage images was divided by the no voltage images to discover the change of intensity, which means the ratio of intensity. At last, the changes in intensity were converted to changes in temperature by utilizing calibration curve of fluorescent intensity of Rhodamine B.

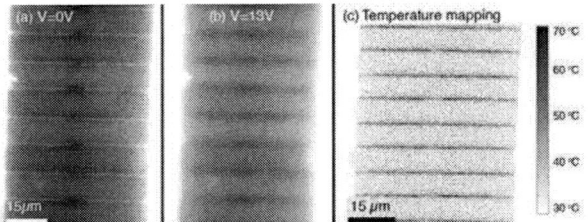

Figure 5: Fluorescence microscopy images of silicon nanowires (Length=50 μm) coated with Rhodamine B molecules in dry condition with no voltage (a) and 13 V (b). The fluorescence intensity decreases with the increase of temperature.

The distribution of surface temperature along and cross the silicon nanowires is shown on figure 6. The temperature of the wires ranges from 35 to almost 70°C.

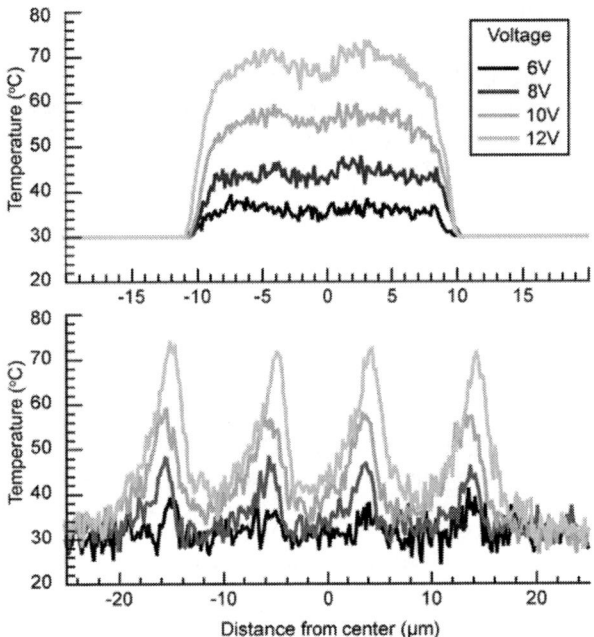

Figure 6: (a) The surface temperature distribution along single silicon nanowires as a function of the applied voltage. The temperature is almost constant over the full length of the nanowire and ranges between 35 and 70 degrees. (b) Comparison of the temperature distribution over four adjacent nanowires on the same silicon substrate as a function of the applied voltage. The nanowires have the same temperature.

On figure 7, Mean temperature of the nanowires plotted as a function of the electrical tension applied. The temperature increases linearly over the whole tension range. Based on these results, it is clear to understand that the surface temperature of silicon nanowires increased corresponding to the increased applied voltage potential.

978-1-4244-2977-6/09 $25.00 © 2009 IEEE

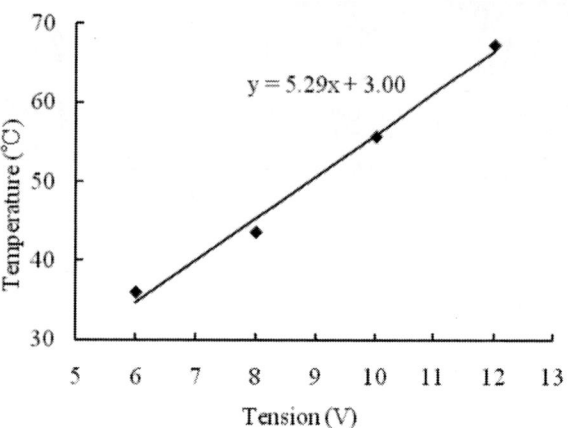

Figure 7: *Mean temperature of a nanowire as a function of the electrical tension. The temperature increases linearly over the whole tension range.*

4. CONCLUSION

In this article, we measure the temperature of silicon nanowires as a function of the applied voltage by fluorescence thermometry. After a calibration step of the measurement in *dry* and *liquid* condition, we are able to make the spatial mapping of the temperature of the nanowires, and we show that this temperature ranges linearly until 70 degrees with the applied voltage in *dry* condition. The next step will be to perform the same measurement in *liquid* condition in order to use the devices for biomolecular studies.

5. ACKNOWLEDGMENT

This research is partially supported by Sharp Corporation, and the Japanese Society of Promotion of Science (JSPS, Fellowship P06733). We also acknowledge the laboratories of Prof. H. Fujita and Prof. H. Toshiyoshi for their support regarding experimental equipment and cleanroom facilities. The authors would like to acknowledge Professor Hiroyuki Fujita for his fruitful advice of the fabrication of Si nanowires.

6. REFERENCES

[1] Y. Cui, Q. Wei, H. Park, C. M. Lieber, "Nanowire Crossbar Arrays as Address Decoders for Integrated Nanosystems", *Science*, 293, pp. 1289, 2001.

[2] M.S-C Lu, D.H. Liu, L.S. Zheng and S.H. Tseng, "CMOS micromachined structures using transistors in subthreshold region for thermal sensing", *J. Micromech. Microeng.*, vol.16, pp.1734–1739, 2006.

[3] H.F. Arata, Y. Rondelez, H. Noji, and H. Fujita, "Temperature Alternation by an On-Chip Microheater To Reveal Enzymatic Activity of β-Galactosidase at High Temperatures", *Anal Chem.*, vol.77, pp. 4810-4814, 2005.

[4] S.Wang, S.Westcott, and W.Chen, "Nanoparticle Luminescence Thermometry", *J. Phys. Chem. B*, 106, pp. 11203-11209, 2002.

[5] P. Löw, B.J. Kim, N. Takama, and C. Bergaud, "High spatial resolution surface temperature mapping using fluorescent thermometry", *Small*, 4, No.7, pp. 908-914, 2008.

[6] K. Kakushima, T. Watanabe, K. Shimamoto, T. Gouda, M. Ataka, H. Mimura, Y. Isono, G. Hashiguchi, Y. Mihara, H. Fujita, "Atomic Force Microscope Cantilever Array for Parallel Lithography of Quantum Devices", *Japanese Journal of Applied Physics,* vol.43, pp. 4041-4044, 2004.

[7] G. T. Hermanson, *Bioconjugate Techniques*, Academic Press, North Carolina, 1996

[8] N. M. Green, "Avidin and streptavidin", *Methods in Enzymology*, Vol. 184, 51-67, 1990.

[9] M. Gonzales, C.E. Aragana, G.D. Fidelio, "Extremely high thermal stability of streptavidin and avidin upon biotin binding", *Biomolecular Engineering*, vol.16, pp. 67-72, 1999.

[10] J. Fattaccioli, J. Baudry, N. Henry, F. Brochard-Wyart and J. Bibette, "Specific Wetting Probed with Biomimetic Droplets", *Soft Matter*, 2008, DOI: 10.1039/B806635C.

ELECTRON BEAM STIMULATED OXIDATION OF CARBON (EBSOC)

P. S. Spinney, S. D. Collins, R. L. Smith

Micro Instruments and Systems Laboratory, University of Maine, Orono, Maine, USA

ABSTRACT

The patterning of carbon nanostructures by electron beam stimulated oxidation is described. Sputter deposited carbon thin films and carbon nanotubes are locally oxidized in a scanning electron microscope using injected water vapor. The resulting structures are examined with scanning electron microscopy (SEM) and transmission electron microscopy (TEM). Electrical resistance obtained post- processing is comparable to as-deposited values. Etched linewidths on the order of 20nm are demonstrated along with sub-2nm wide carbon wires.

INTRODUCTION

The burgeoning field of nanotechnology has spawned a sophisticated control over the detailed sculpting of metals[1,2], semiconductors[3,4] and insulators[5.6] producing an impressive inventory of nanostructures. With the advent of carbon nanotubes[7], graphene[8], and "bio" nanotechnology[9], carbon has gained increased interest as a promising material for nanoscale device fabrication. Carbon has many advantageous properties that make it an attractive candidate for nanostructures, including: high chemical stability, excellent mechanical properties, electrical behavior ranging from insulating to conducting, and a relative ease of fabrication. Current applications of carbon include direct patterning of carbon nanotubes and graphene films for molecular electronics[10,11], chemical etch mask fabrication[12], low-damage x-ray mask repair[13] and micro-electrodes for electrochemistry[14]. In this report, electron beam stimulated oxidation of carbon (EBSOC) by water vapor is described, demonstrating a facile and versatile method for nano-sculpting of carbon films.

The etching of carbon in an electron beam dates back to the early 1950's when biologists first noticed that their cryogenically prepared samples showed electron beam damage during SEM and/or TEM investigation. The damage resulted from e-beam oxidation of the carbonaceous material when water vapor, produced from the local heating of the ice, reacted with the sample. The phenomenon was considered mostly a nuisance for the next 50 years. With the advent of Focused Ion Beams (FIB) came a resurgent interest in gas phase etching of materials with both FIB and Focused Electron Beams (FEB).

Electron beam stimulated gas etching was reported by Coburn and Winters in 1979[15]. It was found that in the presence of an electron beam, XeF_2 decomposes and locally etches SiO_2, Si_3N_4 and SiC. The technique was extended by Fujioka et al.[16] to include Cl_2 and ClF gases as etchants. Randolph et al. [17] performed detailed studies to further the understanding of the etchant/substrate combination of XeF_2/SiO_2. Since then, the FEB and/or FIB gas phase etching of a multitude of gas phase/substrate combinations has been reported. An excellent review on the subject is given in Utke et al.[18]. Interestingly, despite the intense fervor for gas phase FEB/FIB etching, little attention has been given to the substrate/etchant combination that introduced the field, i.e. carbon/water. In 1996, Taniguchi et al.[19] reported the used of oxygen combined with electron beam stimulation to etch single crystal diamond. More recently, Yuzvinsky et al.[20] demonstrated that electron beam stimulated gas etching could be used to cut multi-walled carbon nanotubes using O_2, N_2, H_2 and H_2O. The précis of this report is to further the understanding of the etching process and demonstrate the potential of FEB etching for generating carbon nanostructures.

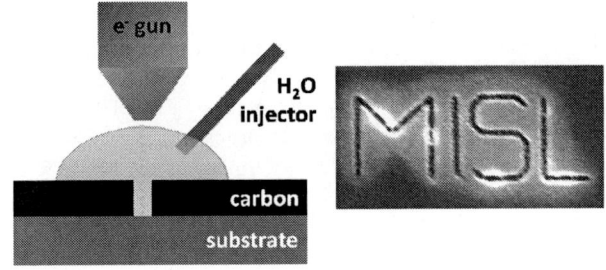

Figure 1: (left)Schematic diagram of electron beam stimulated oxidation of carbon. Water vapor is locally injected into the SEM chamber and the electron beam is rastered over the area of carbon to be etched. (right) An example of EBSOC is shown; etched linewidth is 20-30 nm.

The technique is depicted in Figure 1. High energy electrons from a standard SEM electron column are focused onto a thin carbon film. A gas injection nozzle delivers a local environment of water vapor to the carbon surface, oxidizing it to volatile products, CO_x, that are subsequently removed by the vacuum pump. Etching is potentially a combination of two factors: a) direct carbon activation with subsequent oxidation by water vapor, and b) activation/ionization of the water vapor above the substrate with subsequent reaction of the water radicals/ions with carbon.

EXPERIMENTAL

In this report a Zeiss N-Vision 40 FIB/SEM with an imaging resolution of 1.25nm at 15kV, and 2.4nm at 500V was used as the electron source. In most experiments, a Raith Elphy Quantum lithography system controlled the e-beam deflection and dose. The carbon film (40 nm) was deposited by pulse DC sputtering from a graphite target onto a dielectric coated silicon wafer. The dielectric coating consisted of 100nm of thermally grown SiO_2 and a 160nm film of low-pressure chemical vapor deposited (LP-CVD) Si_3N_4. The carbon film was patterned for electrical characterization into arrays of 0.5-2.5μm wide, 10μm long carbon wires using standard photolithography and reactive oxygen ion etching. Aluminum was deposited and patterned by lift-off at the ends of the carbon wires for electrical contacts. Both patterned and unpatterned samples were loaded in the SEM and water vapor injected across the carbon using a

978-1-4244-2977-6/09 $25.00 © 2009 IEEE

Seiko SII gas injection system. The pressure in the vacuum chamber was allowed to stabilize at approximately 2.5x10⁻⁵ Torr before the electron beam was rastered across the carbon area to be etched. Control experiments showed no appreciable carbon etching when either the e-beam or water vapor was absent, i.e. the e-beam alone and water vapor alone did not etch carbon. It was only in the presence of both that the carbon film was etched.

RESULTS & DISCUSSION

The efficiency of the carbon etching process was characterized according to: 1) The electron beam energy and 2) electron fluence. Figure 2 graphs the etch efficiency of carbon as a function of the electron energy. Etch efficiency is measured in cubic microns per coulomb. It is found by measuring the time need to etch a fixed volume (1μm x 1 μm x 40 nm) and measuring the beam current using a faraday cap. Etch efficiency follows a clear power law, decreasing with increasing acceleration voltage. This result suggests that carbon etching results primarily from a gas-phase ionization mechanism and not lattice activation of carbon through radiolysis. This is consistent with previous reports using XeF_2[15-17] in which a decrease in etch efficiency was observed with increasing acceleration voltage.

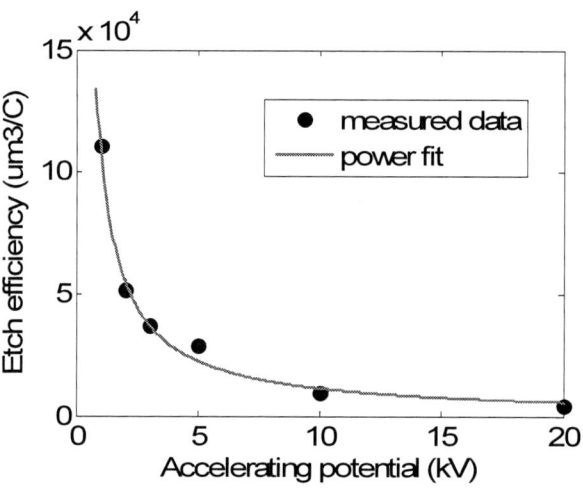

Figure 2. Plot of EBSOC electron etch efficiency for sputtered carbon versus electron beam accelerating potential. The solid line shows a power fit to the measured data, etch efficiency = 109500 · acceleration potential$^{-0.9805}$ where accelerating potential is in kV and etch efficiency is in $\mu m^3/C$, with a correlation coefficient of 0.992.

Figure 3 shows the carbon etch rate (volume/time) versus electron fluence at a fixed acceleration voltage (5kV). As expected, the carbon etch rate increases linearly with increasing current. This indicates that the water vapor is not being locally depleted through the current range tested.

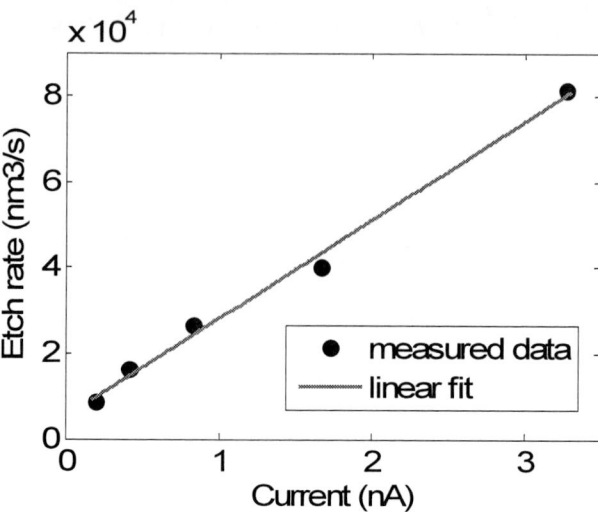

Figure 3. Plot of EBSOC etch rate for a 1μm x 1μm x 40nm volume of sputtered carbon versus current at 5kV electron beam acceleration potential. The solid line shows a linear fit to the measured data points, etch rate = 22930*current + 5044 where etch rate is in nm^3/s and current is in nA, with correlation coefficient (R-squared) of 0.994.

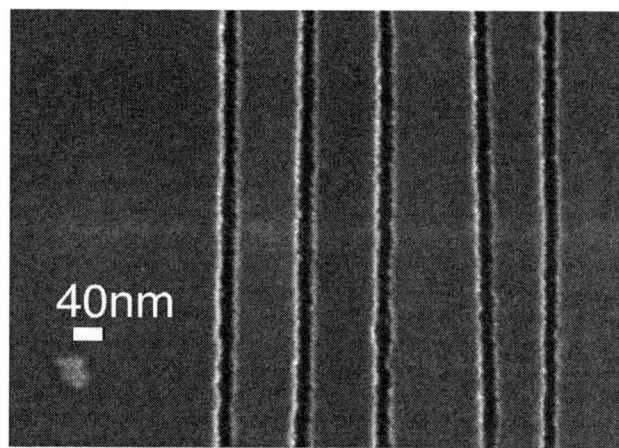

Figure 4. SEM image of lines etched into conductive carbon thin film. Average etched linewidth is 20nm. Carbon film is 40nm thick on a 160nm thick insulating layer of LP-CVD Si_3N_4. Electron beam current was 200pA with an accelerating voltage of 5kV.

Figure 4 shows a series of lines etched into a 40nm carbon film using an electron beam current of 200pA at an acceleration voltage of 5kV. The lines were fabricated using a simple SEM line scan with a step size of 1.6nm and a dwell time of 30ms. An estimated dosage for the etching is 9.36 C/cm². The etched linewidth is 20nm with an peak to peak edge roughness ≤ 5 nm, presumably due to inhomogeneties in the carbon film both in composition and thickness. Charging of the ionized water immediately above the sample was not found to be a significant factor in beam deflection. Results to date suggest that the resolution is limited by vapor diffusion since the linewidths achieved are an order of magnitude wider than the electron beam width. However, the process is stable and controllable, allowing narrower features to be created by etching away material on either side, as shown in Figure 5.

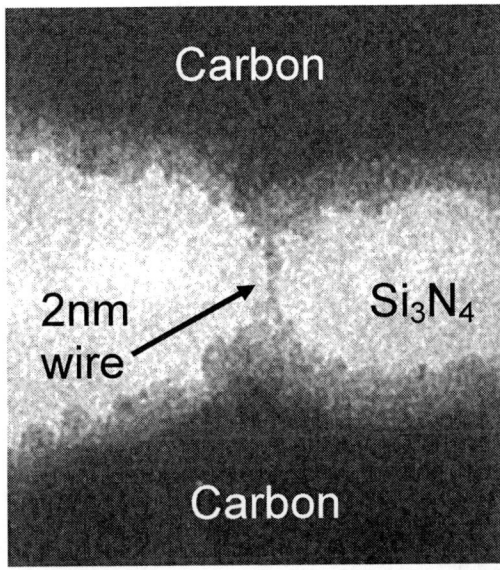

Figure 5. TEM image of a 2nm carbon nanowire formed using water vapor etching of a 40nm thick carbon film on a 50nm thermal Si_3N_4 membrane. Dark area is carbon, light area is Si_3N_4 membrane.

Figures 5 and 6 are micrographs for a 2nm and 16nm carbon wire fabricated by the EBSOC method. An accelerating potential of 5kV was selected to maximize resolution while maintaining a reasonable etch rate. At lower accelerating potentials, insertion of the gas injection needle resulted in a loss of resolution. Rough etching (> 50nm feature sizes) was carried out with a 120 μm aperture and beam currents of 3.4nA while fine etching (<50nm) was carried out using a 30 μm aperture and a beam current of 200pA. The wires were made by rastering the e-beam on either side of the wire under an injected water vapor blanket until visual inspection indicated that all the carbon film was removed. Both images show a significant thinning of the wires around the edges which is more evident in the TEM photograph (Figure 5) where the change in contrast indicates that the nanowire has been thinned. It is estimated that the remaining wire thickness is approximately 20 nm. This is indicative of the diffuse etching distribution inherent in gas phase activation/ionization processes. Current work is underway to delineate the conditions promoting carbon lattice activation (radiolysis) and improved carbon etching resolution.

The SEM image in Figure 6 shows a carbon nanowire formed by EBSOC etching of a 2μm wide conductive carbon wire. The nanowire is 170nm long with an average width of 16nm. The measured thickness after etching is 25nm in the center, measured using a Park Systems XE-100 AFM in contact mode. Figure 7 is an i/V plot of the carbon wire before and after EBSOC. The carbon nanowire remains continuous and conductive. The increase in resistance is proportional to the increase in length to width ratio of the nanowire (sheet resistance = 2.4MΩ/□) indicating that the electrical properties of the carbon are not affected by EBSOC.

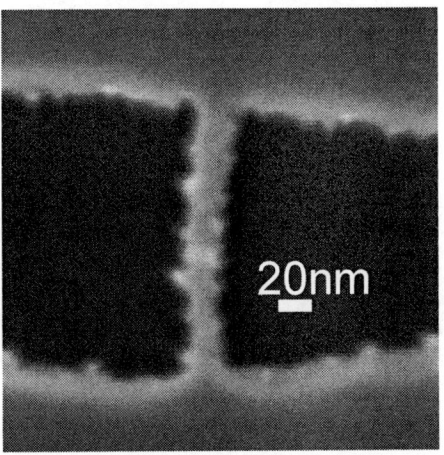

Figure 6. SEM image of 16nm wide 170nm long and 40nm thick carbon nanowire formed using EBSOC. Black area is Si_3N_4, lighter areas are carbon.

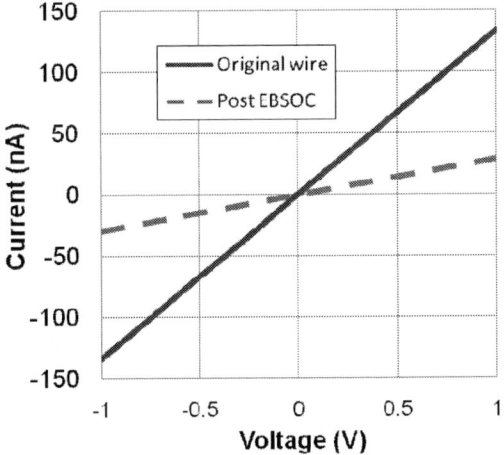

Figure 7. Plot showing current versus voltage relationship of carbon wires before and after etching using EBSOC. The solid line shows the conductivity of the initial 2μm wide carbon line. The dashed line shows the conductivity of the 16nm x 170nm nanowire shown in Figure 6 formed using EBSOC.

Figure 8 shows the application of EBSOC to machine a multi-wall carbon nanotube (≈ 25 nm O.D.). The carbon nanotubes were suspended in methanol using ultrasonic agitation. Silicon substrates were prepared by solvent cleaning, followed by a de-ionized water rinse, and a dilute HF (1:50) dip. The nanotube solution was poured onto the silicon substrate, which was then spun at 1000 rpm for 1 minute. The prepared samples were immediately transferred to the SEM for EBSOC and imaging. A beam energy of 5kV was used in the presence of water vapor to carve a 10nm gap.

978-1-4244-2977-6/09 $25.00 © 2009 IEEE   573

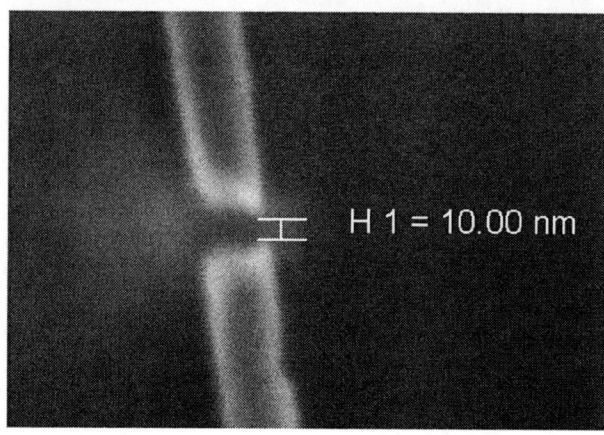

Figure 8. SEM image of multi-wall carbon nanotube with 10nm gap created using EBSOC with a 5kV accelerating potential. Carbon nanotube appears bright on the dark silicon substrate. Image captured with in-lens secondary electron detector.

CONCLUSIONS

EBSOC provides a technique for nanomachining carbon fullerenes and thin films without the potential contamination and damage associated with FIB machining. Applications include direct patterning of carbon nanotubes and graphene films for electronic devices, chemical etch mask fabrication, low-damage x-ray mask repair and fabrication of carbon nanoelectrodes for single-molecule studies.

ACKNOWLEDGEMENTS

This work was supported by the David and Lucile Packard Foundation (2002-22776A) and the National Science Foundation (NSG-IGERT-O504494).

REFERENCES

[1] G. C. Gazzadi, E. Angeli, P. Facci, S. Frabboni, "Electrical characterization and Auger depth profiling of nanogap electrodes fabricated by I_2-assisted focused ion beam", *Appl. Phys. Lett.*, vol. 89, pp. 173112, 2006.

[2] B. C. Gierhart et al. "Nanopore with transverse nanoelectrodes for electrical characterization and sequencing of DNA", *Sensors & Actuators B: Chemical*, vol. 132(2), pp. 593-600, 2008.

[3] F. Yang et al. "5nm-Gate Nanowire FinFET", *Symp. On VLSI Tech.*, pp. 196-197,2004.

[4] C. M. Bruinink et al., "Nanoimprint Lithography for Nanophotonics in Silicon," *NANO Lett.*, vol 8(9), pp. 2872-2877, 2008.

[5] C. Srinivasan, J. N. Hohman, M. E. Anderson, P. S. Weiss, M. W. Hornb, "Sub-30-nm patterning on quartz for imprint lithography templates," *App. Phys. Lett.*, vol. 93, pp. 083123, 2008.

[6] A. Campo, E. Arzt, "Fabrication Approaches for Generating Complex Micro- and Nanopatterns on Polymeric Surfaces", *Chem. Rev.*, vol. 108, pp. 911–945, 2008.

[7] S. Iijima, "Helical microtubules of graphitic carbon," *Nature*, vol. 354, pp. 56-58, 1991.

[8] K. S. Novoselov, A. K. Geim, S. V. Morozov, D. Jiang, Y. Zhang, S. V. Dubonos, I. V. Grigorieva, A. A. Firsovl, "Electric Field Effect in Atomically Thin Carbon Films", *Science*, vol. 308, pp. 666, 2004.

[9] C. S. S. R. Kumar, J. Hormes, C. Leuschner, *Nanofabrication Towards Biomedical Applications: Techniques, Tools, Applications, and Impact*, Wiley-VCH, 2005.

[10] C. Berger, Z. Song, X. Li *et. al.*, "Electronic confinement and coherence in patterned epitaxial grapheme", *Science*, vol. 312, pp. 1191-1196 , 2006.

[11] G. Gu, S. Nie, R. M. Feenstra, R. P. Devaty, W. J. Choyke, W. K. Chan, M. G. Kane, "Field effect in epitaxial graphene on a silicon carbide substrate", *Appl. Phys. Lett.*, vol. 90, pp. 253507, 2007.

[12] T. Muhl, H. Bruckl, D. Kraut, J. Kretz, I. Monch, G. Reiss, "Nanolithography of metal films using scanning force microscope patterned carbon masks", *J. Vac. Sci. Technol. B*, vol. 16, pp. 3379-3992, 1998.

[13] Mohamed Gad-el-Hak, *The MEMS Handbook, 2nd ed.* CRC Press, pp. 5-6, 1995.

[14] G. Fiaccabrino, M. Koudelkahep, Y. Hsueh, S. Collins, R. Smith, "Electrochemiluminescence of Tris(2,2'-bipyridine)ruthenium in Water at Carbon Microelectrodes", *Anal. Chem.*, vol. 70, pp. 4157-4161, 1998.

[15] J. W. Coburn,H. F. Winters, "Ion- and Electron-assisted gas-surface chemistry—An important effect in plasma etching", *J. Appl. Phys.*, vol. 50, pp. 3189, 1979.

[16] H. Fujioka, K. Nakamae, M. Hirota, K. Ura, N. Tamura, and T. Takagi, J., "Measurements of the energy dependence of electron beam assisted etching of, and deposition on, silica", *J. Phys. D: Appl. Phys.*, vol. 23, pp. 266-268, 1990.

[17] S. J. Randolph, J. D. Fowlkes, P. D. Rack, "Focused electron-beam-induced etching of silicon dioxide", *J. Appl. Phys.*, vol. 98, pp. 034902, 2005.

[18] I. Utke, P. Hoffmann, J. Melngailis, "Gas-assisted focused electron beam and ion beam processing and fabrication", *J. of Vac. Sci. & Tech. B*, vol. 26, pp. 1197-1276, 2008.

[19] J. Taniguchi, I. Miyamoto, N. Ohno, S. Honda, "Electron Beam Assisted Chemical Etching of Single Crystal Diamond Substrates", *Japanese J. of Appl. Phys.*, vol. 35, pp. 6574-6578, 1996.

[20] T. D. Yuzvinsky, A. M. Fennimore, W. Mickelson, C. Esquivias, A. Zettl, "Precision cutting of nanotubes with a low-energy electron beam", *Appl. Phys. Lett.*, vol. 86, pp. 053109, 2005.

[21] D.G. Howitt, S.J. Chen, B. Gierhart, R.L. Smith, S.D. Collins, "The electron beam hole drilling of silicon nitride thin films", *J. of Appl. Phys.*, vol. 103, pp. 024310, 2008.

[22] M. T. Yin, M. L. Cohen, "Structural theory of graphite and graphitic silicon", *Phys. Rev. B.*, vol. 29(12), pp. 6996-6998, 1984.

[23] D. G. Howitt, "Radiation Effects Encountered by Inorganic materials in Analytical Electron Microscopy", *Principles of Analytical Electron Microscopy Eds.* Joy, D.C., Romig, A. D and Goldstein, J. I. Plenum, 1986.

978-1-4244-2977-6/09 $25.00 © 2009 IEEE

ULTRA SMALL SINGLE WALLED CARBON NANOTUBE PRESSURE SENSORS

T. Helbling, S. Drittenbass, L. Durrer, C. Roman, and C. Hierold
Micro and Nanosystems, Department of Mechanical and Process Engineering, ETH Zurich,
SWITZERLAND

ABSTRACT

This paper reports on the fabrication and characterization of ultra small pressure sensors with individual single-walled carbon nanotube (SWNT) field effect transistors (CNFETs) as strain gauges. The smallest piezoresistive pressure sensor with a membrane diameter of d~40µm is fabricated and characterized. This miniaturization is made possible due to the nanoscaled size, electronic properties, and high piezoresistive gauge factors (GF) of SWNTs and is currently limited by the membrane fabrication capabilities using a 200µm thick Si wafer. In summary the sensor performance is: Gauge factor: ~450 to 700 (strain dependent), sensitivity: -54pA/mbar (V_{ds}=200mV), resolution: 15mbar, power consumption: ~100nW.

INTRODUCTION

Microelectromechanical systems (MEMS) in general and piezoresistive MEMS pressure sensors in particular have become important products over the last few decades due to their small size, robustness, low production costs and ease of fabrication. Further miniaturization towards nano-electromechanical systems (NEMS) pressure sensors employing novel nanoscaled functional elements shows high potential with regard to reduced power consumption, and increased sensitivity. Carbon nanotubes and especially SWNTs, discovered in 1993 [1], show outstanding properties to be used as novel functional material in NEMS devices. Their electronic, mechanical and electromechanical properties have already been demonstrated in carbon nanotube based inertial balances [2], resonators [3] and pressure sensors [4,5]. Electronically, SWNTs are either metallic, small-bandgap semiconducting (SGS) or semiconducting depending on their chiral angle and diameter. Investigations on the piezoresistivity of CNFETs were carried out in a number of studies [4-7]. For freestanding CNTs strains up to 3%, and gauge factors (GF) up to 2900 were assessed in pre-strained SWNTs [7].

In this paper we present the fabrication and characterization of pressure sensors with carbon nanotubes as the functional material. For the first time, the advantage of the nanoscaled size of the SWNT is exploited in downscaling piezoresistive pressure sensors to diameters of 40µm. These sensors show, in contrast to state-of-the art pressure sensors, besides the reduced size also very low power consumption and increased sensitivity. A schematic of the SWNT based sensor is depicted in Fig. 1. The sensors consist of a double layer membrane made of SiO_2 and atomic layer deposited (ALD) alumina (Al_2O_3). The SWNT strain gauges are embedded between the bottom SiO_2 and the top Al_2O_3 layer and contacted in a field effect transistor configuration. The top and bottom oxide layers are used to encapsulate the SWNTs to protect the SWNTs from the environment, as examined in a previous study [8]. The working principle is analogous to MEMS pressure sensors. The pressure sensor membrane is strained due to an applied differential pressure. This strain in the membrane is transferred to the SWNT embedded in the membrane which results in a change in the SWNT resistance [9].

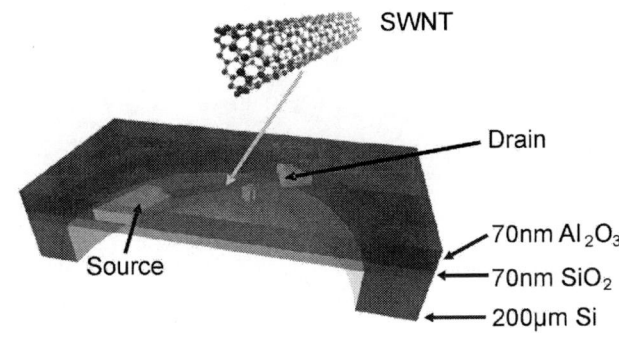

Figure 1: Schematic of a CNT based pressure sensor. The sensor is made of a double layer membrane of 70nm SiO_2 and 70nm of Al_2O_3. The SWNT is embedded between the two layers and contacted by a source and a drain metal electrode. A nearby gate electrode (not shown) is used to bias the SWNT in a transistor configuration.

FABRICATION

The fabrication process is divided into two main phases. First, the fabrication and integration of CNTs into CNFETs which is described in detailed in Refs. [10,11] and second, the bulk micromachining of the sensor membrane. The fabrication starts with a 200µm thick highly doped Si substrate where 70nm of SiO_2 is grown at 1000°C by dry oxidation. Individual SWNTs are grown on Ferritin-based Fe catalyst nanoparticles using chemical vapor deposition (CVD) with methane as feedstock. The SWNT growth took place at 850°C under a H_2/CH_4 atmosphere (50/150mbar) resulting in SWNT mean diameters d=1.9+1.2/-1 nm [11]. Then the outer ring of electrodes and wire-bonding contact pads are structured by standard UV lithography (see Fig. 2a), physical vapor deposition (PVD) and lift-off. In order to obtain the location and orientation of the randomly distributed CNTs, alignment markers are used. These markers are structured by electron beam (e-beam) lithography, PVD of 2nm Cr and 40nm Au, followed by lift-off. The CNT location relative to the alignment markers was recorded by atomic force microscopy (AFM). This mapping is used to design individual contact electrodes to the CNTs, which are patterned by a second e-beam structuring, PVD (2nm Cr and 40nm Au) and lift-off. The separation of the source-drain electrodes is nominally d_{sd}= 1.5µm. Immediately after lift-off, the samples were transferred

978-1-4244-2977-6/09 $25.00 © 2009 IEEE

into the preheated (T=150°C) Picosun Sunale R-150B ALD reactor where the conformal growth of 70nm Al_2O_3 is carried out using N_2 as a carrier gas. An Al_2O_3 thickness of 70nm is enough to fully cover and encapsulate the SWNTs [12].

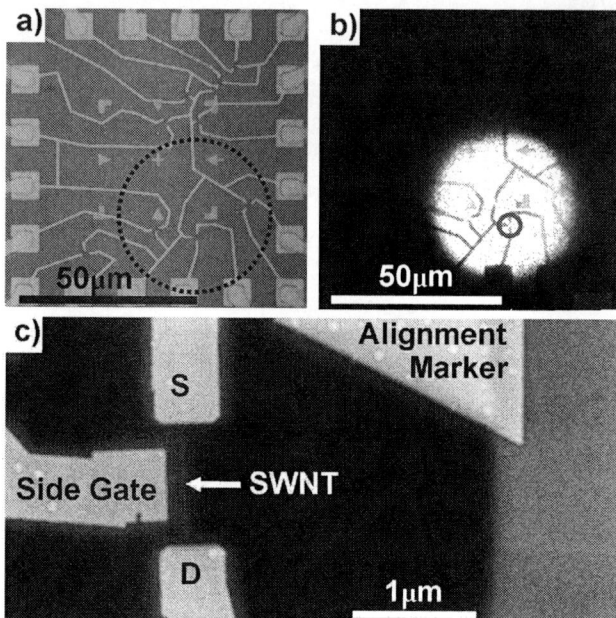

Figure 2: Processing images of CNT pressure sensors. a) 9 contacted CNFETs after ALD encapsulation. Each CNFET is made of source (S), drain (D), and sidegate electrodes. The dashed circle indicates the location where the membrane is later fabricated. b) Microscope image of a final sensor illuminated from the backside through the transparent 140nm thick SiO_2/Al_2O_3 membrane. c) SEM image of one CNFET contacted on the sensor membrane. The edge of the membrane is clearly visible as a color change from black to grey.

The pressure sensor fabrication is proceeded with UV lithography of the membrane opening at the backside of the sample using infrared alignment. After hydrofluoric acid etching of the backside hardmask, the membrane was released by the BOSCH® dry etching process with 3 of 9 CNFETs located on the sensor membrane (see Fig. 2b). Fig. 2a shows 9 contacted CNFETs. The location where the membrane was later fabricated is shown by the dashed circle. Fig. 2b shows a backlight illuminated microscope image through the transparent sensor membrane (d~40μm). In Fig. 2c a close up SEM image of one integrated CNFET from a similar sensor with a side gate and an Au alignment marker is shown.

RESULTS
Electrical CNFET Characterization

Electrical CNFET characterization of the source-drain current I_{ds} by changing the nearby sidegate voltage V_g (see the sidegate in Fig. 2c and in the inset of Fig. 3) is used to determine the type of the SWNT.

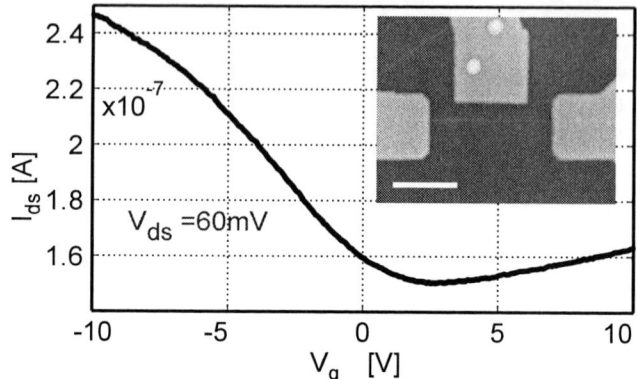

Figure 3: Small-bandgap gate characteristics of a CNFET on the membrane from Fig 2b. Due to a bandgap of a few meV, thermally activated transport is apparent in the off-state of the transistor at V_g=2.9V. The inset shows a typical CNFET with source-drain and side gate (scale bar 1μm).

Figure 3 shows the gate characteristic of the CNFET in the center of the sensor membrane (See Fig. 2b, red circle). The characteristic is typical for a small-bandgap SWNT. Here the bandgap is in the order of a few meV [13]. From Fig. 3 it is observed that the coupling of the electric field from the sidegate to the CNFET channel is strong enough to modulate the current by a factor ~1.7 but not strong enough to fully open the CNT transistor at large negative gate voltages. The minimum resistance at V_g = -10V is ~240kΩ, which is above the minimum theoretical resistance of 6.5kΩ. It is believed that barriers at the Cr/Au-nanotube contacts and defects in the SWNT account for this. Due to different injection barriers for hole and electrons, the gate characteristic shows asymmetric behavior [14].

Electromechanical sensor characterization

For the electromechanical characterization, a bulge test setup is used to automatically apply differential pressure to the sensor membrane while I_{ds} and the applied pressure p are measured. Measurements on the pressure sensor include 1) CNFET source-drain current $I_{ds}(t)$ at fixed side gate voltage V_g=4V with varying pressure $p(t)$ ($I_{ds}(t,V_g$=4V,$p(t)$)), and 2) measurements of the CNFET gate characteristics $I_{ds}(p,V_g)$ at different pressure levels. Figure 4 shows pressure dependent measurements for a pressure shape $p(t)$ which is depicted in the upper (black) curve while the acquired current I_{ds} biased at V_{ds}= 200mV and V_g = 4V is shown in the lower (red) curve. While the pressure is linearly increased from p_0 to p_{max} a monotonic decrease of I_{ds} is observed. Upon releasing the pressure back to p_0, the current I_{ds} changes reversibly and monotonically to its initial value. The sensitivity of the sensor for fixed electrical CNFET bias values is measured by applying linear pressure ramps to the sensor and measuring I_{ds}, as shown in Fig. 5. The black triangles indicate the measured points for the ramp up and the red

978-1-4244-2977-6/09 $25.00 © 2009 IEEE

squares for the pressure ramp down. Empirically it is found that the sensor transfer function is linear. By linearly fitting the data points, an input to output sensitivity of S=-54pA/mbar is obtained. Hysteresis is not visible for the speed at which the pressure ramp was applied (~25s up/down). The resolution of the sensor is 15mbar which is calculated by taking twice the standard deviation ( ) from the transfer function of the sensor. The I_{ds} response of the sensor to a pressure peak of 50mbar, shown in Fig. 4 (the first triangle in the trace), verifies that the resolution of 15mbar is reasonable and achievable.

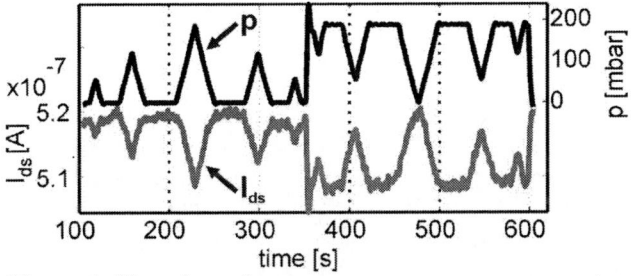

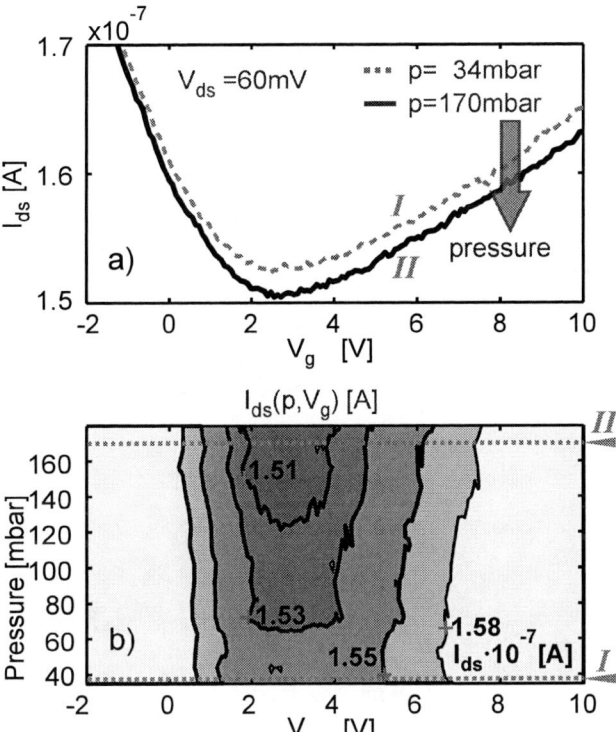

Figure 4: Time dependent pressure sensor measurements $I_{ds}(t,V_g=4V,p(t))$. The upper black curve shows the measured pressure applied to the membrane and the lower (red) curve shows the measured I_{ds} at V_{ds}=200mV.

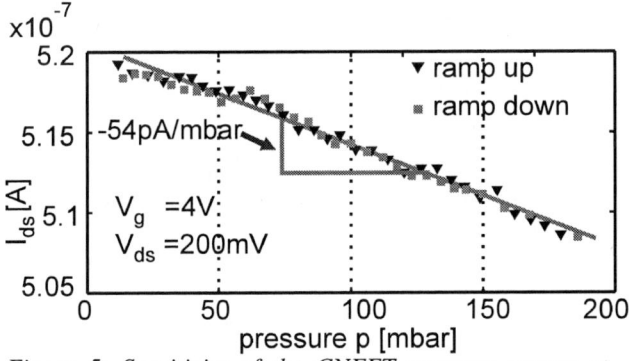

Figure 5: Sensitivity of the CNFET pressure sensor at V_g=4V for a linear pressure ramp up (triangles) and down (squares). The slope of the linear fit (solid line) is -54pA/mbar at a bias voltage V_{ds}=200mV. Hysteresis is not visible for the speed at which the pressure ramp was applied (~25s up/down).

Sensor characterization: CNFET gate characteristic

In order to obtain more in depth information on the piezoresistive behavior of CNFETs, the gate characteristics are measured for a number of differential pressures across the sensor membrane. In Fig. 6a the gate characteristics $I_{ds}(V_g,p)$ at p=34mbar (red dashed curve) and at p=170mbar (black solid curve) are depicted. Both curves show ambipolar behavior. A decrease of I_{ds} is observed for all V_g at increased pressure. A more detailed measurement of the gate characteristics was made at 15 different pressure values and the result is shown in Fig. 6b in a contour plot. It shows that the current is monotonically decreased for all values of V_g over the measured pressure range $p \in [0,200]$mbar. From Fig. 6a and from the different slopes of the contour lines in 6b it is obvious that the change in I_{ds} due to pressure is not constant over V_g, and thus the sensor sensitivity is dependent on V_g.

Figure 6: CNFET gate characteristics versus pressure $I_{ds}(p,V_g)$. a) CNFET gate characteristic $I_{ds}(V_g)$ for p=34mbar (**I**) and p=170mbar (**II**). b) Contour plot of $I_{ds}(p,V_g)$ with 15 pressure values to highlight the monotonicity of the sensor response for all values of V_g. The blue dashed lines (**I, II**) in b) correspond to $I_{ds}(p,V_g)$ of the two curves shown in a). The highest sensitivity is at $V_g \sim 3V$.

The result of the sensitivity dependency on V_g for one CNFET is shown in Fig. 7. The values of the red solid line are found by linearly fitting $I_{ds}(p)$ for $V_g \in [-10,10]$V (dashed black curve is smoothed). It is observed that the maximum sensitivity is $|S|_{max}$=20pA/mbar at V_g=2.9V, where the current in the gate characteristic is in the off state. Towards more positive gate voltage, the sensitivity is decreasing to a value around $|S| \sim 15$pA/mbar, while for more negative V_g the value reaches $|S| \sim 7$pA/mbar. Note the dependency of the sensitivity on V_{ds}. The sensitivity of $|S|$=54pA/mbar at V_{ds}=200mV and V_g=4V (Fig. 5) is in good agreement with the sensitivity of $|S|$=10pA/mbar at V_{ds}=60mV and V_g=4V from Fig. 7.

These results can be understood by the theoretical prediction on the piezoresistance of CNTs due to a strain dependent change in the bandgap. By assuming thermally activated transport the current in the SWNT strain gauge is qualitatively most sensitive to the bandgap change when the Fermi level is in the middle of the bandgap. This is the case when $I_{ds} = I_{min}$. Influences of the metal contacts are however not taken into account and may also play an important role.

978-1-4244-2977-6/09 $25.00 © 2009 IEEE

CNFET gauge factors

In this section the intrinsic SWNT strain sensitivity (GF) is separated from the sensor transfer function. By means of mechanical bulge test the pressure deflection behavior of the membrane is measured. Then the strain in the center of the membrane, where the SWNT is located, is approximated by analytical models (see Ref. [5] for details). The piezoresistive gauge factor is defined as $GF=(\Delta R/R_0)^{-1}$, where $\Delta R/R_0$ is the relative change in the SWNT resistance and ε is the strain at the SWNT. The GF is strain dependent and values between $450<GF<700$ are found when the CNFET is biased with $V_g=2.9V$. This result is in agreement with previously measured GF in which GF between -400 and 850 (210) were reported in Ref. [4] (Ref. [5]).

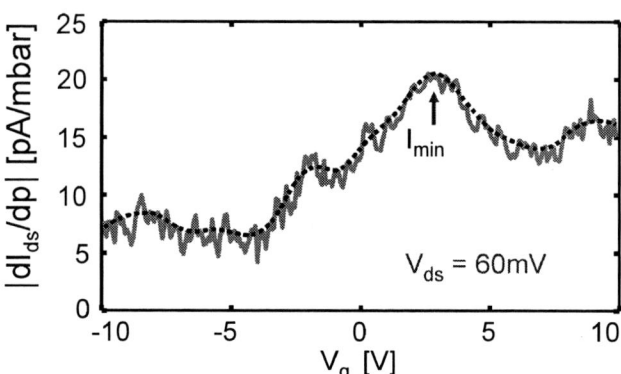

Figure 7: Sensor sensitivity versus gate voltage. The maximum sensitivity $|S|=20pA/mbar$ ($V_{ds}=60mV$) is at $V_g=2.9V$.

CONCLUSIONS

In conclusion, the feasibility of miniaturizing piezoresistive pressure sensors by using carbon nanotube strain gauges was demonstrated. CNT based pressure sensors with membrane diameters as small as 40µm were fabricated and characterized. The sensors have linear transfer characteristics, and the presented performance of the small pressure sensor with a resolution of 15mbar is achieved by the high piezoresistance of the SWNT strain gauge with strain dependent GF between 450 to 700. Furthermore, the power consumption of ~100nW is exceptionally low for piezoresistive sensors. It is shown that the sensitivity can be tuned between 7 and 20pA/mbar ($V_{ds}=60mV$) by adjusting V_g, allowing to easily calibrate SWNT pressure sensors for future applications.

ACKNOWLEDGEMENTS

The authors wish to thank Moritz Mattmann, Matthias Muoth, Shih-Wei Lee, Christoph Stampfer, Otte Homan, Ronald Grundbacher and Pascal Behr for support and discussions. Support by the ETH FIRST Lab and financial support by the ETH Zurich (TH-13/05-3) are gratefully acknowledged.

REFERENCES

[1] S. Iijima and T. Ichihashi, "Single-shell carbon nanotubes of 1-nm diameter", *Nature*, vol. 363, pp. 603-605, 1993

[2] K. Jensen, K. Kim and A. Zettl, "An atomic-resolution nanomechanical mass sensor", *Nat. Nanotechnol.*, vol. 3, pp. 533-537, 2008

[3] V. Sazonova, Y. Yaish, H. Üstünel, D. Roundy, T. A. Arias and P. L. McEuen, "A tunable carbon nanotube electromechanical oscillator", *Nature*, vol. 431, pp. 284-287, 2004

[4] R. Grow, Q. Wang, J. Cao, D. Wang and H. Dai, "Piezoresistance of carbon nanotubes on deformable thin-film membranes", *Appl. Phys. Lett.*, vol. 86, pp. 093104, 2005

[5] C. Stampfer, T. Helbling, D. Obergfell, B. Schoberle, M. K. Tripp, A. Jungen, S. Roth, V. M. Bright and C. Hierold, "Fabrication of single-walled carbon-nanotube-based pressure sensors", *Nano Lett.*, vol. 6, pp. 233-237, 2006

[6] T. W. Tombler, C. W. Zhou, L. Alexseyev, J. Kong, H. J. Dai, L. Lei, C. S. Jayanthi, M. J. Tang and S. Y. Wu, "Reversible electromechanical characteristics of carbon nanotubes under local-probe manipulation", *Nature*, vol. 405, pp. 769-772, 2000

[7] C. Stampfer, A. Jungen, R. Linderman, D. Obergfell, S. Roth and C. Hierold, "Nano-Electromechanical Displacement Sensing Based on Single-Walled Carbon Nanotubes", *Nano Lett.*, vol. 6, pp. 1449-1453, 2006

[8] T. Helbling, L. Durrer, C. Roman, V. M. Bright and C. Hierold, "Zero level packaging and long term stability of carbon nanotube based transistors in pressure sensors", in *Digest Tech. Papers Eurosensors XXII Conference*, Dresden, September 7-10, 2008 pp. 652-655

[9] L. Yang and J. Han, "Electronic structure of deformed carbon nanotubes", *Phys. Rev. Lett.*, vol. 85, pp. 154-157, 2000

[10] C. Stampfer, A. Jungen and C. Hierold, "Fabrication of discrete nanoscaled force sensors based on single-walled carbon nanotubes", *IEEE Sens. J.*, vol. 6, pp. 613-617, 2006

[11] L. Durrer, T. Helbling, C. Zenger, A. Jungen, C. Stampfer and C. Hierold, "SWNT growth by CVD on Ferritin-based iron catalyst nanoparticles towards CNT sensors", *Sens. Actuators, B*, vol. 132, pp. 485-490, 2008

[12] M. D. Groner, S. M. George, R. S. McLean and P. F. Carcia, "Gas diffusion barriers on polymers using Al2O3 atomic layer deposition", *Appl. Phys. Lett.*, vol. 88, pp. 051907, 2006

[13] C. W. Zhou, J. Kong and H. J. Dai, "Intrinsic electrical properties of individual single-walled carbon nanotubes with small band gaps", *Phys. Rev. Lett.*, vol. 84, pp. 5604-5607, 2000

[14] J. Appenzeller, "Carbon nanotubes for high-performance electronics - Progress and prospect", *Proc. IEEE*, vol. 96, pp. 201-211, 2008

METALLIC NANOPARTICLE MANIPULATION USING OPTOELECTRONIC TWEEZERS

Arash Jamshidi, Hsan-Yin Hsu, Justin K. Valley, Aaron T. Ohta, Steven Neale, and Ming C. Wu

Berkeley Sensor & Actuator Center (BSAC) and Department of Electrical Engineering and
Computer Sciences, University of California, Berkeley, California 94720, USA

ABSTRACT

We report on trapping of single and multiple spherical gold nanoparticles with 60 to 250 nm diameters using optoelectronic tweezers (OET). Thanks to the low optical intensities required for stable trapping (20 μW over 1.7 μm spot), we estimate the temperature increase in OET-trapped nanoparticles due to absorption to be ΔT < 0.1°C, making OET-trapped nanoparticles suitable for biological imaging and sensing applications. In addition, we observe translational velocities of 68 μm/s and demonstrate trapping of both single and multiple nanoparticles in a single trap.

INTRODUCTION

In recent years, there has been much interest in metallic nanoparticles as biological nano-sensors due to their interesting optical properties [1]. However, a persistent challenge has been to find techniques for interaction with and manipulation of these nanoparticles. Optical tweezers have been used previously to trap metallic nanoparticles of different sizes [2, 3]; however, the high optical power intensities required for stable trapping (~ 10^7 W/cm^2) result in excessive heating in metallic nanoparticles (ΔT > 55°C) [4], hampering the application of optical tweezer-trapped particles in biological environments. Dielectrophoresis (DEP) can trap nanoparticles using fixed electrodes [5]; however, since the trapping positions are lithographically defined, fixed-electrode DEP lacks the capability to dynamically scan and manipulate the trapped particles. Trapping of single molecules has also been achieved using an Anti-Brownian Electrokinetic (ABEL) trap [6] which provides extensive information about the particle dynamics. However, this technique requires the molecules to be fluorescent.

In contrast, OET is an optical manipulation technique capable of dynamically manipulating a large number of micro and nanoparticles or cells over large areas using optical intensities 5 orders of magnitude smaller than optical tweezers [7]. Previously, the smallest particles that OET could trap were limited to nanowires of diameters below 100 nm and approximately 5 μm length [8]. In this paper, we report, for the first time, trapping of metallic spherical nanoparticles with 60 to 250 nm diameter using optoelectronic tweezers (OET).

THEORETICAL BACKGROUND

OET Device Operation Principles

Figure 1a shows the optoelectronic tweezers (OET) device structure. The OET device consists of a top and a bottom indium-tin-oxide (ITO) coated glass electrode with an AC voltage applied between the two electrodes. A 1-μm-thick layer of photoconductive material (hydrogenated amorphous silicon) is deposited on the bottom ITO substrate. The liquid solution containing the dispersed metallic nanoparticles is sandwiched between the top ITO electrode and the photoconductor substrate. A 635-nm diode laser is used to interact with the photoconductive layer and trap the metallic nanoparticles.

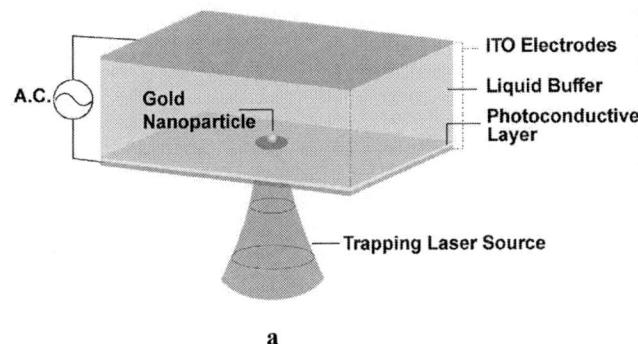

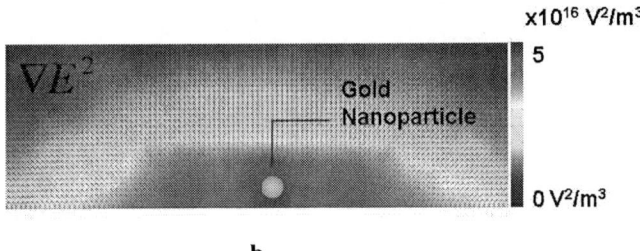

Figure 1: (a) Optoelectronic tweezers (OET) device structure for manipulation of nanoparticles. OET works based on the principle of optically-induced dielectrophoresis (DEP), where optically defined virtual electrodes create non-uniform electric fields to polarize objects in the vicinity of the fields. The objects are then attracted to or repelled from areas of high electric field intensity gradient depending on their effective induced polarization relative to the medium. The metallic nanoparticles experience an attractive (positive) DEP force due to their high polarizability relative to the liquid medium. (b) The simulated gradient of electric field intensity is shown near the OET surface. The nanoparticles are immersed and trapped in the high field gradient region near the OET surface.

978-1-4244-2977-6/09 $25.00 © 2009 IEEE

When there is no laser light present, the impedance of the photoconductive layer is higher than that of the liquid layer and the majority of the applied AC voltage is dropped across the photoconductive layer. However, once the laser light is introduced, it generates electron-hole pairs in the photoconductive layer, reducing the impedance of the photoconductor layer below that of the liquid layer. Therefore, the majority of voltage is switched from the photoconductive layer to the liquid layer in the area that laser is present. Since the voltage switch occurs only in the area that laser source is present, the electric field in the liquid will have a non-uniform profile. This non-uniform field polarizes the metallic nanoparticles in its vicinity, attracting them to areas of high electric field intensity gradient according to the DEP force principle.

Figure 1b shows the finite-element simulation of the gradient of the non-uniform electric field intensity for an applied bias of 20 Vpp at 100 kHz.

Dielectrophoresis Force

The non-uniform electric field present in the liquid layer induces a dipole moment (p) in the metallic nanoparticles. The induced dipole interacts with the electric field, resulting in a dielectrophoretic force, $F = (p \cdot \nabla)E$, which attracts the nanoparticles to areas of highest field intensity gradient [9].

The DEP force expression for a spherical particle is given by [9]:

$$F_{DEP} = 2\pi r^3 \varepsilon_m \, \text{Re}\{K\} \nabla(E^2) \quad (1)$$

where, r is the radius of the particle, ε_m is the permittivity of the liquid medium, $\text{Re}\{K\}$ is the real part of the Clausius-Mossotti (CM) factor given by, $\text{Re}\{K\} = \text{Re}\{(\varepsilon_p^* - \varepsilon_m^*)/(\varepsilon_p^* + 2\varepsilon_m^*)\}$, where $\varepsilon^* = \varepsilon - j\sigma/\omega$, with p and m subscripts referring to the particle and the liquid medium, respectively.

As shown in Figure 1b, the gradient of field intensity is strongest near the OET surface and falls off sharply as we move away from the surface. Due to the nanoparticles small size, they are immersed in the high-∇E^2 region near the OET surface. The gradient of field intensity can be simulated using COMSOL finite-element modeling and is estimated to be $10^{16} - 10^{17}$ V^2/m^3 near the OET surface. Using this value, we can estimate the strength of the DEP force for a 100 nm diameter nanoparticle to be approximately 0.1 pN.

To estimate the velocity of the nanoparticles due to this DEP force, we can use the drag force acting on the spherical nanoparticles [10],

$$F_{Drag} = 6\pi\eta r v_{drag} \quad (2)$$

where, r is the particle's radius, η is dynamic viscosity of water, and v_{Drag} is the drag velocity. Equating this force to the DEP force ($F_{DEP} = F_{Drag}$), we can achieve v_{Drag} close to 100 μm/s.

Temperature Analysis

Using a similar analysis to ref. [4], we can estimate the temperature increase in the OET-trapped nanoparticles as $\Delta T = P_{abs}/(4\pi r C)$, where C is the thermal conductivity of water (0.6 W/K.m), r is the radial distance from the nanoparticle's center, and P_{abs} is the absorbed power in the nanoparticle given by $P_{abs} = \sigma_{abs} I^2$, where I is the laser intensity and σ_{abs} is the absorption cross section of the nanoparticle given by $\sigma_{abs} = (2\pi n_m / \lambda) \times \text{Im}[3V(\varepsilon_p^* - \varepsilon_m^*)/(\varepsilon_p^* + 2\varepsilon_m^*)]$, where $\varepsilon_m^* \approx 1.77$ ($n_m \approx 1.33$) and $\varepsilon_p^* \approx -10.66 + i1.37$ at $\lambda = 635$ nm and V is the volume of the nanoparticle. For a 20 μW trapping laser source with 1.7 μm (FWHM) spot size, we estimate the temperature increase at the surface of 60 to 250 nm diameter gold nanoparticles due to absorption to be less than 0.1°C.

It is important to note that this calculation does not take into account the temperature increase due to the joule heating in the liquid layer. The joule heating effect can be roughly estimated as [11]: $\Delta T_{joule} = \sigma_{liquid} V^2 /(2C)$, where σ_{liquid} is the liquid conductivity, V is the applied voltage, and C is the thermal conductivity of water. Using the typical experimental values for nanoparticle trapping (σ_{liquid} = 1-10 mS/m, V = 10-20 Vpp), we can estimate the temperature increase due to joule heating to be of orders of a few °C which is about an order of magnitude larger than the temperature increase due to absorption in metallic nanoparticles. Therefore, joule heating would be the dominant effect in calculating the total temperature increase in the trapping environment.

EXPERIMENTAL RESULTS
Experimental Setup

Figure 2 shows the experimental setup used for the metallic nanoparticle manipulation using OET. Gold nanoparticles with 60 to 250 nm diameters with an approximately 10^{10} particles/ml density were diluted in a 2.6 mS/m conductivity solution of DI water and KCl. 4 μL of the sample was introduced into the OET device. Majority of the nanoparticles showed strong Brownian while a portion of the particles adhered to the surface. A 635 nm diode laser with 20 μW power and 1.7 μm (FWHM) optical spot size at the OET surface was used to trap the nanoparticles. AC voltages of 10-20 Vpp at 50-100 kHz frequency were applied to the OET device. Dark field microscopy using a BX51M Olympus microscope was used to visualize the nanoparticles and images were captured using a CCD camera.

978-1-4244-2977-6/09 $25.00 © 2009 IEEE

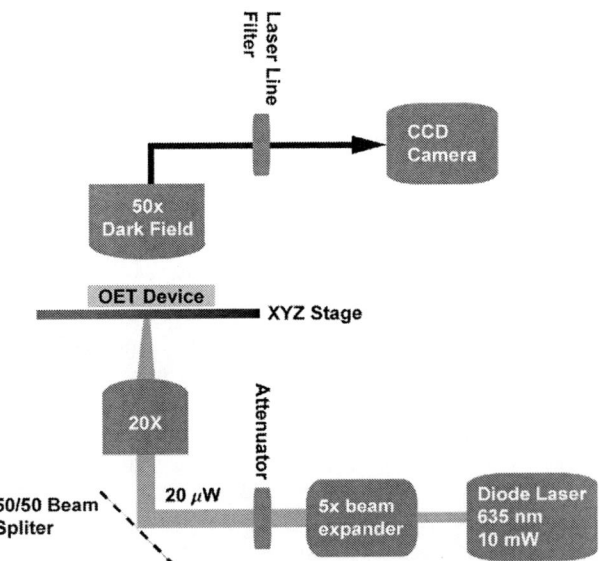

Figure 2: Experimental setup for manipulation of metallic nanoparticles. A 10-mW,635nm diode laser was expanded 5×, attenuated to 20 μW, and focused onto the OET chip with 1.7 μm (FWHM) using a 20× objective lens. The nanoparticles were visualized using dark field microscopy and a CCD camera.

DEP Manipulation of Metallic Nanoparticles

Figure 3 shows trapping of a single 100 nm gold nanoparticle using OET. The nanoparticle experiences a positive DEP force and is attracted to the laser trapping source. By manually adjusting the laser trap position, the nanoparticle is transported over an approximately 200 μm² area in 12 seconds.

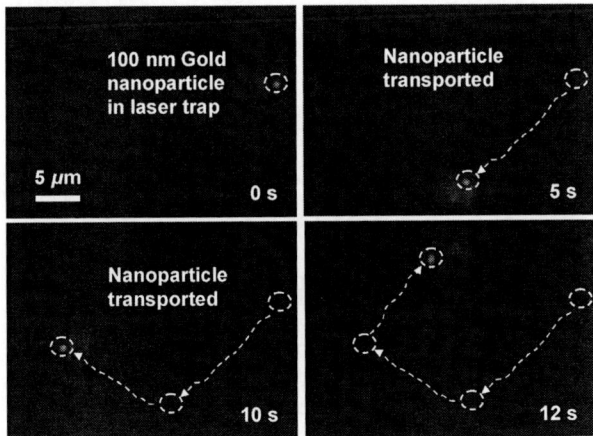

Figure 3: Trapping and transport process of a single 100 nm gold nanoparticle using OET. The nanoparticle is transported over an approximately 200 μm² area in 12 seconds.

Trap Characterization

To quantify the maximum trapping speed of the metallic gold nanoparticles, we used an ESP-300 Newport motorized actuator controller and a LTA-HL motorized actuator to move the microscope stage relative to the optical pattern. Figure 4 shows the maximum translational speeds of 100 nm gold nanoparticles as a function of the applied AC voltage. The experimental data follows a quadratic trend (black fitted line) which is expected since the DEP force is proportional to the gradient of the field intensity.

A 68 μm/s maximum translation speed is measured for an applied AC voltage of 20 Vpp. This measured translational speed is close to the calculated speeds for metallic nanoparticles. In addition, a maximum trapping radius of approximately 28 μm is measured at 20 Vpp.

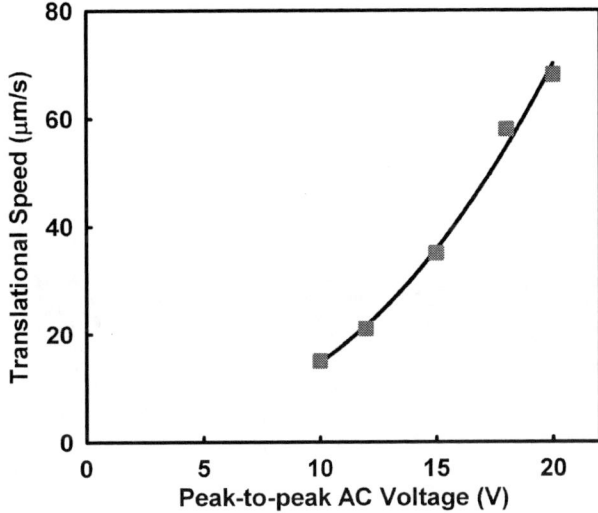

Figure 4: OET-trapped nanoparticles' translational speed as a function of the applied voltage. A maximum translational speed of 68 μm/s at 20 Vpp is achieved. The experimental data follows a quadratic trend (black fitted curve). This is expected since the DEP force is proportional to ∇E^2.

When more than one nanoparticle is trapped in the laser, the particles experience a mutual repulsive force due to two effects. First, nanoparticles carry a negative surface charge with a zeta potential of ξ = -70 mV resulting in a mutual columbic repulsion. Second, the dipoles induced in the nanoparticles interact with each other resulting in a dipole-dipole repulsive force which is a function of the applied voltage. By measuring the translation speeds of the particles after removing the laser trap, we can calculate the net repulsive force between two 100 nm gold nanoparticles to be approximately 23 fN at 20 Vpp (Figure 5a).

We can also observe this repulsive force acting on more than two particles as shown in Figure 5b for three nanoparticles. In the beginning the three particles are trapped in the laser source (filtered out), once the laser trap is removed, the mutual repulsion between the particles repels them from each other.

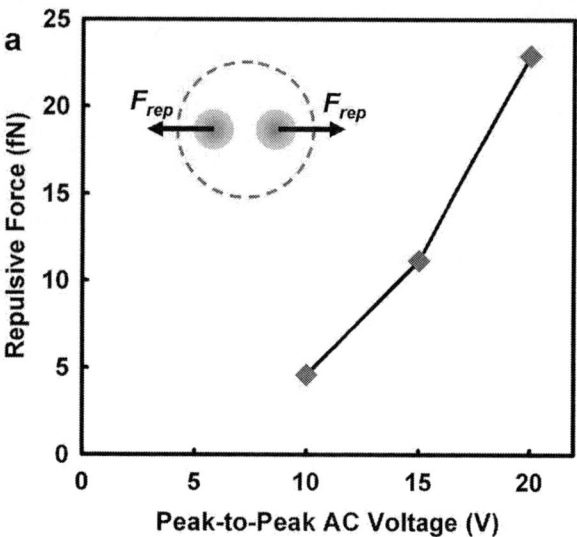

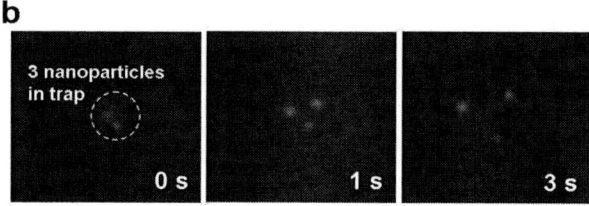

concentrated in the laser trap. (b) Nanoparticles are transported to a new location by manually adjusting the laser position. (c) Laser trap is removed and nanoparticles undergo Brownian motion. (d) Five gold nanoparticles are distinctly observable after removal of the trap.

Figure 5: (a) The repulsive force (F_{rep}) between two nanoparticles (due to dipole-dipole interaction and columbic repulsion) as a function of applied voltage. (b) The repulsive force interaction is also observed for three nanoparticles. In the beginning, the particles are trapped using OET (laser filtered out), once the trap is removed, the nanoparticles repel each other.

CONCLUSION

In conclusion, we report on trapping single and multiple spherical gold nanoparticles with 60 to 250 nm diameters using optoelectronic tweezers. Due to low optical intensities required for stable trapping we estimate the temperature increase in OET-trapped nanoparticles due to absorption to be $\Delta T < 0.1°C$. In addition, we observe translation speeds of 68 μm/s for 100 nm gold nanoparticles at 20 Vpp applied voltage.

ACKNOWLEDGEMENT

This work was supported in part by the National Institutes of Health through the NIH Roadmap for Medical Research (Grant #PN2 EY018228) and DARPA. We also thank Lawrence Berkeley National Laboratory (LBNL) Molecular Foundry and UC Berkeley Microfabrication Laboratory.

REFERENCES

[1] Anker, J.N., et al., Nature Materials 7, pp. 442 - 453 (2008).

[2] Svoboda, K. and Block, S. M. Optics Letters 19, pp. 930 - 932, (1994).

[3] Hansen, P. M., et al., Nano Letters 5, pp. 2429 - 2431, (2005).

[4] Seol, Y., Carpenter, A. E., Perkins, T. T., Opt. Lett. 31, pp. 1937 - 1942, (2006).

[5] Zheng, L., Li, S., Brody, J. P., Burke, P. J., Langmuir 20, pp. 8612-8619, (2004).

[6] Cohen, A. E., Moerner, W. E., Optics Express 10, pp. 6941-6956, (2008).

[7] Chiou, P.Y., Ohta, A. T., Wu, M. C., Nature, 436, pp. 370-372, (2005).

[8] Jamshidi, A., et al., Nature Photonics 2, pp. 85 - 89, (2008).

[9] T. B. Jones, Electromechanics of Particles (Cambridge: Cambridge University Press), 1995.

[10] H. Morgan and N. Green, AC Electrokinetics: colloids and nanoparticles (Research Studies Press Ltd.), 2003.

[11] Ramos, A., et al., J. Phys. D: Appl. Phys. 31, pp. 2338 - 2353 (1998).

Trapping of Multiple Nanoparticles

Figure 6a-d shows trapping and transport of five 250 nm gold nanoparticles using OET. Nanoparticles are concentrated in the OET trap and can be transported by scanning the laser trap manually, once the laser trap is removed, nanoparticles undergo Brownian motion and the five gold nanoparticles are distinctly observable. The ability to concentrate the nanoparticles in a single spot is important to enhance the sensitivity of the dynamic hot-spots for imaging and sensing applications.

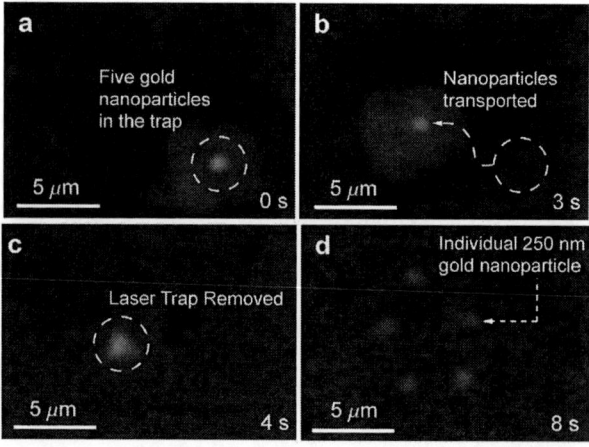

Figure 6: Trapping and transport of five 250 nm gold nanoparticles. (a) Nanoparticles are trapped and

978-1-4244-2977-6/09 $25.00 © 2009 IEEE   582

THIN FILM TRANSISTOR (TFT) SENSING ELEMENTS FABRICATED IN SURFACE MICROMACHINED POLYMERMEMS FOR A DIFFERENTIAL CALORIMETRIC FLOW SENSOR

S.-H. Tsang[1], K. Simard[2], I. G. Foulds[2], H. Izadi[3], K. S. Karim[3], M. Parameswaran[2]

[1]Analog Devices Inc., Cambridge, MA, USA
[2]Simon Fraser University, Burnaby, BC, Canada
[3]University of Waterloo, Waterloo, ON, Canada

ABSTRACT

This paper presents the first process to allow the embedding of amorphous silicon thin film transistors in surface micromachined polymerMEMS. The presented polymerMEMs fabrication technique embeds hydrogenated amorphous silicon thin film transistors (α-Si:H TFT) into polyimide for active sensing. An out-of-plane differential calorimetric flow sensor has been fabricated using this method to demonstrate the feasibility of this fabrication process. The calorimetric flow sensor uses the α-Si:H TFTs as active sensing elements and has a linear unamplified sensitivity of 2.2mV/(cm/s). This paper provides details for the fabrication process and reports on the design and functional results of calorimetric flow sensor fabricated by embedding α-Si:H TFTs into polymerMEMS.

INTRODUCTION

The development of fabrication processes that use polymers for surface micromachining have increased the flexibility and capabilities for MEMS especially in the biomedical and microfluidic fields. However, one major drawback to typical polymerMEMS processes is the lack of electrical signal conditioning or active sensing capabilities. PolymerMEMS devices currently rely on external circuits for signal conditioning during actuation or sensing, and typical processes can only fabricate passive structures without any transduction capabilities. Although embedded organic thin film transistor (TFT) circuits in polymerMEMS have been demonstrated [1], the fabrication technique requires microstereolithography which does not scale well for mass production as compared to the well established amorphous silicon TFT technology [2] currently used for flat panel displays.

The polymerMEMS process presented here allows for the embedding of hydrogenated amorphous silicon thin film transistors (α-Si:H TFT) inside the polymer microstructures for active sensing. A calorimetric flow sensor [3], shown in Figure1, is fabricated to demonstrate the feasibility of this process.

THIN FILM TRANSSITOR POLYMERMEMS PROCESS FLOW

The following describes the process steps. This process uses 8 masks: ANCHOR, DIMPLE, DSMETAL, ISLAND, GMETAL, HEATER, BONDBAD, and STRUCT. Figure 2 illustrates the key process steps as outlined below:

a) A 1 µm thick spin-on-glass (SOG) layer is applied to a 500µm thick silicon substrate which has been RCA cleaned followed by a dip in 10:1 HF (48%wt.) solution and primed by a 30 second spin with HMDS at 3000 rpm. The SOG layer is spun at 900 rpm for 30 seconds and is cured by ramping the wafer from room temperature to 350°C at 450°C/hour and is held for 10 minutes and cooled naturally. A second 1µm thick SOG layer is also spun on also using the same steps and cured. The second layer is cured and ramped back down to room temperature at a controlled 300°C/hour to reduce the thermal stress between the two SOG layers during cooling and eliminate cracking between the layers. HDMS is spun on followed by Shipley SC1813 photoresist spun at 4000 rpm for 30 seconds. This spin speed results in a 1.2µm thick layer of SC1813. The photoresist is softbaked at 103°C for 2 minutes. All subsequent photoresist application steps are identical and will be omitted to reduce redundant procedures. The photoresist is exposed to the ANCHOR mask using a Quintel 2001 aligner and developed in MF-319 to completion. The typical dosage used for the photoresist is102mJ/cm^2 and the development time is 50 seconds. These values will vary depending on the processing equipment used and re-optimization is required for different process conditions. The optical masks used for the exposure are emulsion on Mylar film and required overexposure to create sharp features. The edges of the optical mask are diffuse and unless the photoresist is overexposed, the feature edges do not fully develop.

b) The dual thickness SOG layer is etched using an Axic Benchmark 800 Reactive Ion Etch (RIE) at 150 mTorr pressure, 50 SCCM CF$_4$, at 250W for 20 minutes with a circular electrode of 23cm in diameter and a one inch electrode gap.

c) The photoresist masking layer is stripped using an oxygen plasma RIE without breaking vacuum. The oxygen RIE parameters are 400 mTorr pressure, 100 SCCM O$_2$, at 300W for 5 minutes. Following the etch, a second photoresist layer is applied using the procedure as described in step b) and patterned using the DIMPLE mask with the same conditions as step a).

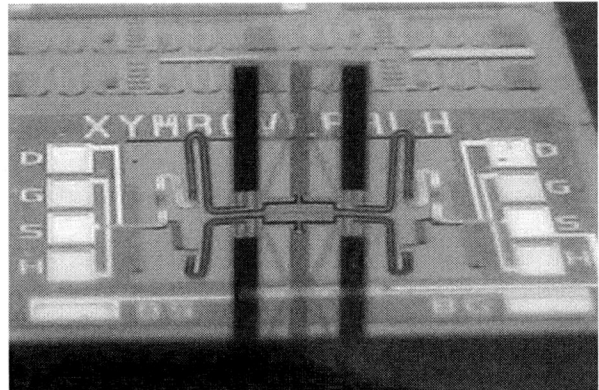

Figure 1: Fabricated and assembled differential calorimetric flow sensor with active TFT elements

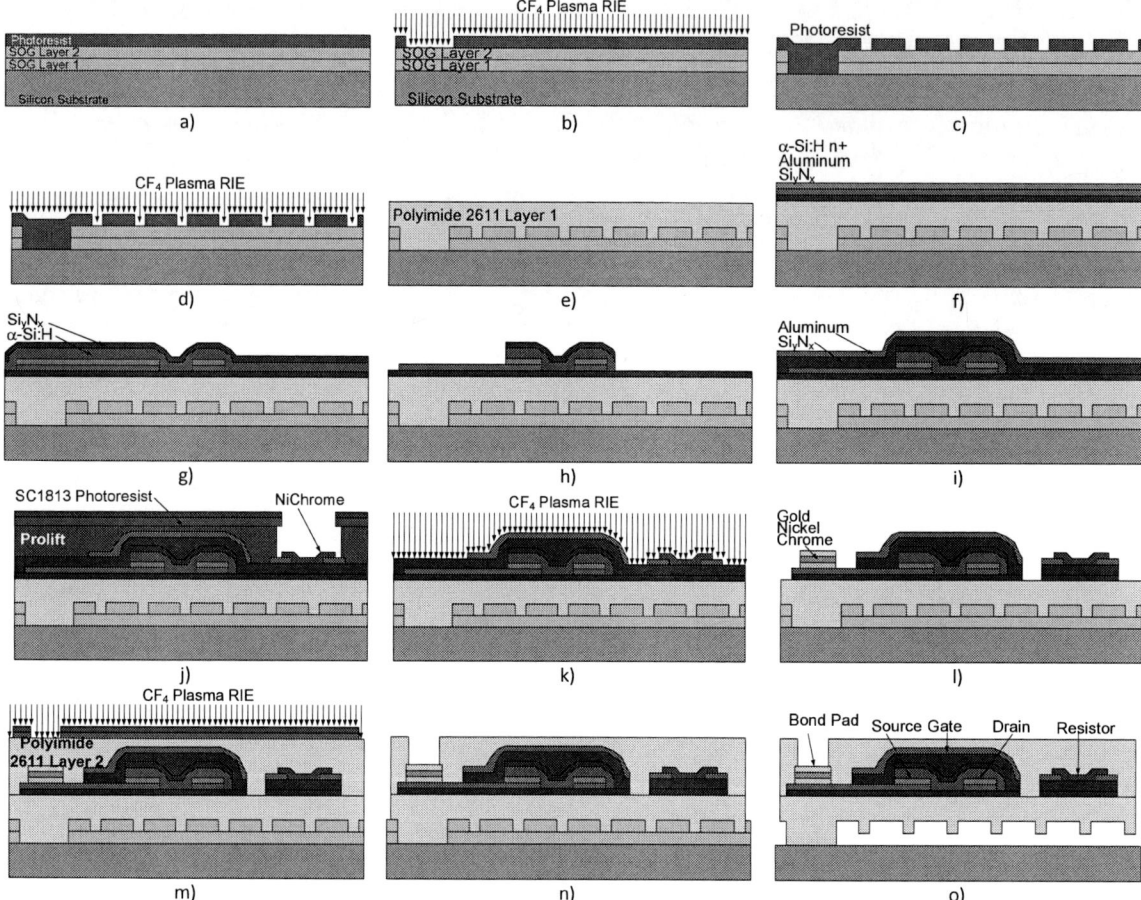

Figure 2: Embedded TFT Polymer MEMS Process Flow

d) The SOG dimple layer is etched with similar parameters as the anchor layer except for the reduction in time (total of 8 minutes) which partially etches the SOG to a 1 μm depth. The photoresist masking layer is again removed using oxygen plasma. During ashing a thin layer of SiO_2 forms on the anchor openings which aids in the adhesion for the subsequent polyimide layer.

e) Prior to spinning on the first polyimide layer, an adhesion promoter, supplied as VM-651 (3-aminopropyltriethoxysilane) is spun at 3000 rpm for 30 seconds and baked in a convection oven at 120°C. The polyimide layer, consisting of PI-2611 (S-Biphenyl-dianhydride/p-Phenylenediamine in N-Methyl-2-Pyrro-lidone casting solvent), is spun at 500 rpm and held for 5 seconds and ramped to 5000 rpm over 20 seconds and held at speed for 30 seconds. The polyimide layer is soft-baked at 80°C, held for 3mins, and is immediately transferred to another hotplate at 150° C and held for another 3mins. The polyimide is cured by ramping the wafer to 285°C at 240°C/hour and held for two hours and cooled back to room temperature at a natural rate. Once cured, the polyimide layer shrinks in the vertical direction and results in a 3μm thick film.

f) Following the curing of the polyimide layer, the polyimide is surface activated to increase adhesion to subsequent layers by subjecting the layer to low power oxygen plasma. The plasma treatment is performed using RIE at 100mTorr pressure, 30 SCCM O_2, at 50W for 1 minute. After the surface activation, a three layer stack of silicon nitride (Si_yN_x), aluminum, and n+ doped hydrogenated amorphous silicon (α-Si:H n+) is deposited. These layers are deposited in a cluster tool consisting of an aluminum DC sputter chamber, a Plasma Enhanced Chemical Vapor Deposition (PECVD) chamber with a 13.56MHz RF supply and an interlocked transfer zone (ITZ). The three layers are deposited without breaking vacuum to reduce contamination from chemical interaction with air. The wafer is held at 250°C in the PECVD chamber with a hydrogen flow of 40 SCCM at a pressure of 1.9 Torr for 30 minutes to reach temperature equilibrium prior to deposition. Silicon nitride is then deposited at 1.5mTorr, 200 SCCM H_2, 16 SCCM NH_3, 4 SCCM SiH_4, at 20W for 15 minutes. The wafer is transferred into the sputter chamber and flushed with argon and pumped down for 30 minutes and flushed for 15 minutes before reaching a deposition pressure of 5mTorr. Aluminum is deposited at 50W for 8 minutes for a thickness of 60nm. After the aluminum deposition, the wafer is returned to the PECVD chamber and held for another 30 minutes with hydrogen flow to ensure temperature equilibrium. α-Si:H n+ is then deposited at 1900 mTorr pressure, 1 SCCM SiH_4, 1 SCCM PH_3, 200 SCCM H_2, at 20W for 9 minutes resulting in a 70nm thick layer. After the deposition, the wafer is returned to the ITZ and left to cool for 1 hour before breaking the vacuum. The cooling step in vacuum reduces the interaction of the α-Si:H n+ layer with air.

g) The wafer is then masked using the standard photoresist steps as described previously using the DSMETAL mask which defines the drain source

978-1-4244-2977-6/09 $25.00 © 2009 IEEE   584

connections of the TFTs. The α-Si:H n+ is etched using RIE at 100 mTorr pressure, 50 SCCM CF_4, 5 SSCM O_2, at 100W for 2 minutes. The aluminum layer is then etched using standard Type-A aluminum etchant, held at 50°C, until completion. Following the aluminum etch, the photoresist is stripped using an acetone bath. After stripping the photoresist, any remaining residue is ashed in oxygen RIE at 100 mTorr pressure, 30 SCCM O_2, at 50W for 30 seconds. The wafer is then dipped in a weak solution of 50:1 HF (48% wt.) to remove the surface oxide that may have formed during the ashing step. The wafer is returned to the PECVD and reheated using the same equilibrium procedure as before. The surface of the α-Si:H n+ is then exposed to a hydrogen plasma to remove dangling bonds using the following parameters 1900 mTorr pressure, 30 SCCM H_2, at 20W for 2 minutes. After the plasma treatment, α-Si:H is deposited at 500 mTorr pressure, 20 SCCM H_2, 3 SCCM SiH_2, at 1W for 13 minutes resulting in a thickness of 80nm. The wafer is held at 1900 Torr pressure, 40 SCCM H_2 flow for 10 minutes to return to temperature equilibrium. Si_yN_x is then deposited at 1500 mTorr pressure, 200 SCCM H_2, 16 SCCM NH_3, 4 SCCM SiH_4, at 20W for 6 minutes. The wafer is again left in the ITZ at vacuum to cool for 1 hour to reduce interaction with air.

h) The wafer is removed from PECVD and photoresist is reapplied and patterned using the ISLAND mask. The Si_yN_x and α-Si:H is etched consecutively using RIE at 100mTorr pressure, 50 SCCM CF_4, 5 SCCM O_2, at 100W for 3minutes and 30 seconds. The etch does not penetrate the aluminum layer, and only partially removes the underlying Si_yN_x layer. The photoresist is removed in an acetone bath and the wafer is ashed again in oxygen RIE.

i) A second layer of Si_yN_x is deposited using the same conditions for 13 minutes followed by another aluminum layer with a deposition time of 10 minutes. The second nitride layer is used to completely encapsulate the active areas of the α-Si:H to protect it against contamination during subsequent processing steps. The aluminum layer is patterned using the GMETAL mask to define the TFT gates and resistor connection lines.

j) Once the aluminum is patterned, Prolift 100-24 (a polyimide lift-off resist supplied by Brewer Science) is spun on at 500 rpm and ramped to 6000 rpm over 20 seconds and held for 90 seconds resulting in a 2μm thick layer. The Prolift is soft baked at 150°C on a hotplate for 3 minutes and then placed on a 275°C hotplate for 5 minutes to completely drive out the solvent and to partially cure the Prolift. Shipley SC1813 is spun on top of the Prolift layer using the same spin on and soft-bake procedure as before and exposed using the HEATER mask but is developed for 1 minute and 30 seconds. Development in MF-319 co-develops the SC1813 and undercuts Prolift for a reentrant profile for metal lift-off. 100nm of NiChrome is then sputtered on to the wafer. After the sputtering step, the photoresist is stripped in an acetone bath using ultrasonics. Prolift is highly resistant to acetone which protects encapsulated layers. Prolift is dissolved using Prolift remover as sold by the manufacturer. The wafer is then ashed for 1 minute to remove any remaining residue.

k) The exposed Si_yN_x is etched away using the metal as a self aligned mask at 100mTorr pressure, 50 SCCM CF_4, 5 SCCM O_2, at 100W for 4 minutes. The removal of the exposed Si_yN_x provides access to the buried aluminum lines and exposes the lower polyimide layer for bonding to the encapsulating polyimide layer in subsequent steps.

l) A three metal stack of Chrome, Nickel, and Gold (50nm/50nm/300nm) is then patterned with the BONDPAD mask using Prolift. Following Prolift strip wafer is again ashed in oxygen plasma to remove any residue. The ashing step in this case has two uses. One of which is to remove any remaining resist material and to activate the surface of the exposed polyimide to increase adhesion to the encapsulating polyimide layer.

m) The second layer of polyimide is then spun on using the same conditions as the first layer. This layer is also soft-baked, and cured using the same temperatures as the first layer. After curing, the nominal total thickness of the two polyimide layers is 6μm. An aluminum hard mask (200nm) is sputtered and patterned using the STRUCT mask. The photoresist used to pattern the aluminum is not stripped and is used to minimize the ion sputtering of the aluminum which can cause micro-masking effects on to the wafer during the etching of the polyimide. The STRUCT mask is used to pattern the polyimide structures as well open vias to the bond pads.

n) The polyimide is etched in RIE at 300 mTorr pressure, 50 SCCM O_2, 10 SCCM CF_4, at 150W for 30 minutes which also completely strips away the photoresist layer. The aluminum hard mask is stripped using etchant.

o) The structures are released in a bath of 4:1 HF (48%wt.) for 7 minutes and dipped into a bath of 2:1 ammonium hydroxide (28%wt.) for 1 minute to neutralize the HF solution. The wafer is rinsed in isopropyl alcohol and then dipped into an acetone bath. The wafer is then put into contact with a hotplate at 120°C to drive off the acetone. This effect provides an upward force on the structures as the liquid is driven off to reduce stiction. The completed device contains encapsulated hydrogenated amorphous silicon transistors, as well as resistor heating elements which can be assembled out-of-plane as a thermally isolated sensor.

FLOW SENSOR DESIGN AND RESULTS

The calorimetric flow sensor consists of a differential pair and biasing transistors as shown in Figure 3. The M1 and M2 TFTs are designed for active sensing and are assembled out-of-plane during operation to provide

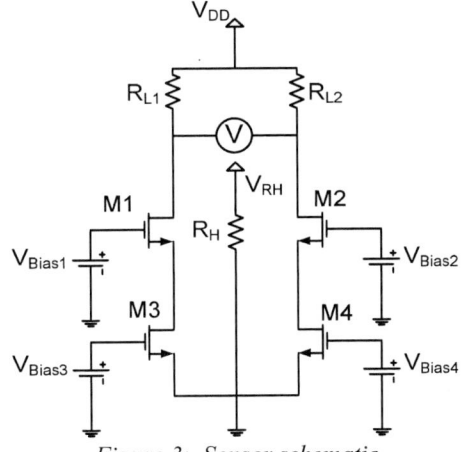

Figure 3: Sensor schematic

978-1-4244-2977-6/09 $25.00 © 2009 IEEE

thermal isolation from the substrate. M3 and M4 can be selectively connected as biasing transistors to provide additional gain, or to operate the circuit in a differential configuration. Resistor R_H is a central heater that provides a temperature gradient for flow sensing. R_{L1} and R_{L2} are external resisters used to bias the TFT circuit. Figure 4 illustrates the layout of the calorimetric flow sensor and a photomicrograph of the active region of a transistor. The active regions are designed with interdigitated drain and source fingers. As fabricated, the channel length is 12.2μm with a channel width of 3550μm resulting in a W/L ratio of approximately 291. The resistance of the embedded resistor R_H is 3kΩ.

The TFTs were characterized over 63 samples using a semiconductor parameter analyzer (SPA) with an average standard deviation of 10.16μA for I_D. The temperature dependence of I_D [4] is also characterized by taking SPA measurements of the device at temperatures between -30°C to 70°C. The temperature was controlled using an environmental chamber where each SPA measurement is performed after the equilibrium temperature is reached. Figure 5 illustrates the V_{ds} vs I_D curve as well as the temperature dependence of the drain current for the TFTs.

Incorporating these parameters, the calorimetric flow sensor was fabricated using the aforementioned embedded TFT polymerMEMS process. The sensor as shown in Figure 1 is packaged in a 16-pin ceramic dual in-line package, wirebonded and assembled out-of-plane using the technique as described by Tsang et al [5]. The sensor is placed into a flow shroud with a 1cm wide channel and a commercial gas flow meter is attached to the channel and used as a reference. Compressed dry air from a cleanroom source is then used to provide laminar flow into the channel. The biasing parameters are set at V_{Bias1} = V_{Bias2} = 8.5V, R_{L1} = R_{L2} = 15kΩ. M3/M4 was disconnected by grounding the source of M1/M2. Figure 6 shows the output of the calorimetric flow sensor. The fabricated device has a sensitivity of 2.2mV/(cm/s) for linear flow velocity, which can be calibrated to provide volumetric measurements in the L/min or mL/min ranges by adjusting the shroud size.

CONCLUSION

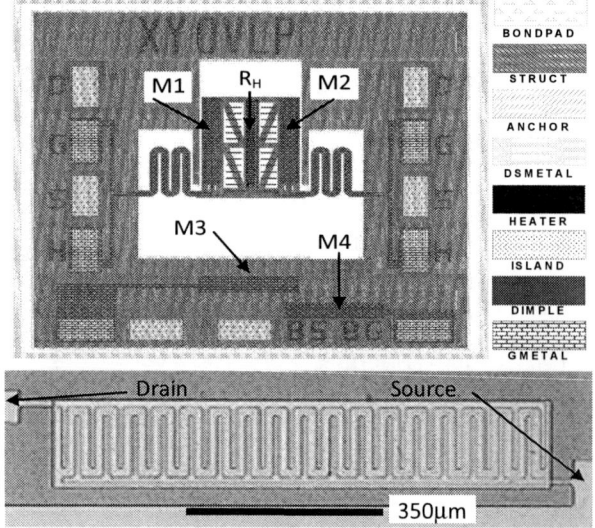

Figure 4: Layout of the sensor and photomicrograph of the fabricated active region of the TFT

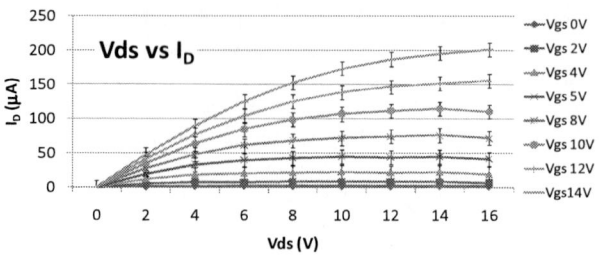

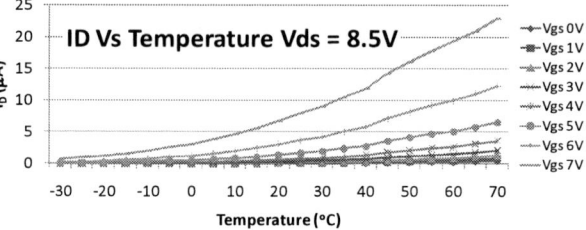

Figure 5: TFT sensor parameters and characteristics

We have developed an embedded TFT polymer-MEMS process that allows for the integration of active sensing elements into polymerMEMS. A prototype calorimetric flow sensor has been fabricated to validate the process feasibility.

REFERENCES

[1] V. K. Varadan, "Three dimensional polymer MEMS with functionalized carbon nanotubes and modified organic electronics," presented at Nanotechnology, 2003. IEEE-NANO 2003. 2003 Third IEEE Conference on, 2003.

[2] R. A. Street, Ed. , "Technology and Applications of Amorphous Silicon", New York, Springer Series, 2000

[3] M. Elwenspoek and R. J. Wiegerink, "Mechanical Microsensors," Springer, 2001.

[4] B. Chen, W. Wu, J. Chen, and C. Hong, "Temperature dependence of hydrogenated amorphous silicon thin-film transistors," *Journal of Materials Science*, vol. 30, pp. 2254-2256, 1995.

[5] S. H. Tsang, D. Sameoto, I. G. Foulds, R. W. Johnstone, and M. Parameswaran, "Automated assembly of hingeless 90°; out-of-plane microstructures," Journal of Micromechanics and Microengineering, vol. 17, pp. 1314-1325, 2007.

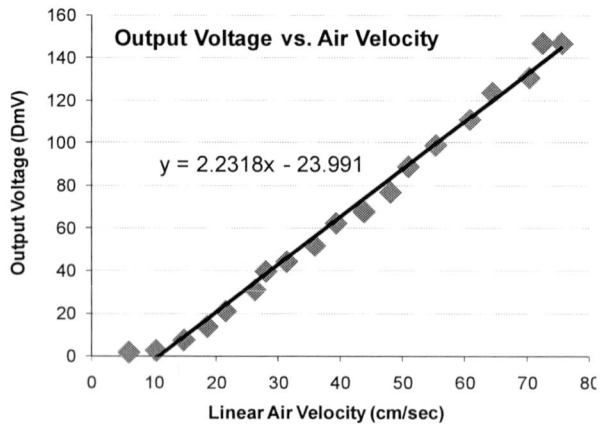

Figure 6: Unamplified calorimetric flow output versus air velocity

PLANAR MEMS SUPERCAPACITOR USING CARBON NANOTUBE FORESTS

Y.Q. Jiang, Q. Zhou, and L. Lin

Mechanical Engineering Department, Berkeley Sensor and Actuator Center

University of California, Berkeley CA 94720, USA

ABSTRACT

Planar micro supercapacitors utilizing vertically aligned carbon nanotube (CNT) forests with the design of interdigital electrodes have been demonstrated. Conductive substrates using Mo/Al/Fe metal stack shows best CNT forest synthesis results with high electrical conductivity and dense CNT structures. Prototype devices show measured specific capacitances about 1000 times higher than those with plain metal electrodes without CNT forests. Furthermore, charging/discharging experiments show over 92% efficiency and very robust cycling stability. As such, we believe these planar MEMS supercapacitors could be applicable in various systems including energy harvesters, pulse-power supplies and advanced microelectronics as on-chip capacitors.

INTRODUCTION

A supercapacitor is a charge storage device composed of two electrodes with electrolyte in between. There are two fundamental differences between a supercapacitor and a regular capacitor: First, ion-conducting electrolyte is used in the supercapacitor instead of dielectrics. As a result, under a biased voltage, the positive and negative ions in electrolyte will separate, migrate and accumulate around the electrode surface, forming a nm-thin layer called electro-chemical double layer (EDL). Therefore, there are essentially two EDL capacitors connected in serial within a single supercapacitor. Second, extremely porous electrode materials are needed in supercapacitors to enhance the interfacial capacitance. Porous electrodes maximize the EDL phenomenon and give supercapacitors an astonishingly high value of specific capacitance (capacitance per unit area). Furthermore, unlike fuel cells and Li-ion batteries, supercapacitors require no chemical reaction during its charging/ discharging process to rapidly store/release energy, making it a popular candidate in applications like vehicle regenerative braking and camera flash systems.

In the field of MEMS, supercapacitors could find very promising applications. For example, as a power source without involving chemical reactions, supercapacitors are ideally suited for the energy suppliers in pulse-power applications such as solid-state sensors. Supercapacitors have simple configuration, stable performance (millions cycles *vs.* ~1000 cycles of Li-ion batteries [1]), and less temperature dependency, making them suitable as the main power sources for short-term, high power-density usages in many applications. Moreover, one might envision that supercapacitor can temporarily store energy from energy harvester and power the system for sensing and wireless communication needs. We also believe that the demonstration of the capability of "planar" structure for supercapacitors could inspire new architecture in circuit designs such as on-chip supercapacitors for advanced microelectronics.

The concept of "planar supercapacitor" has been investigated sparsely previously. For example, Sung et al. have used conducting polymer-coated metal layer [2] as planar electrodes to make supercapacitors. However, their electrodes have rather flat metal surface as essentially 2-D structure without the highly porous characteristics. In another work, In et al used KOH etching to increase the surface area [3]. Carbon nanotubes (CNT) are well-known for its high surface area-to-volume ratio and good conductivity such that several groups have used them in large-scale supercapacitors. For example, An et al. have deposited random CNT network as electrodes [4], in which contact resistance among tubes as well as contact to the substrate are major technical challenges. Pushparaj et al. have proposed and demonstrated a complicated process by transferring CNT arrays to a conductive substrate after the synthesis process while no CNT-based planar architecture has been proposed [5].

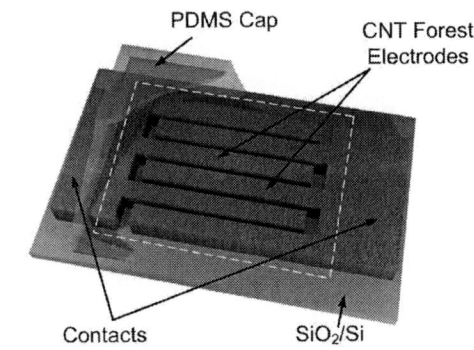

Figure 1: Schematic of the planar micro supercapacitor using aligned, 3D CNT forest electrodes.

Here we present a simple, yet versatile planar MEMS supercapacitor using porous CNT forests as electrodes with low contact resistance without using the top/bottom electrode architecture. The goal of our research is to demonstrate on-chip supercapacitor that can be integrated with MEMS devices and CMOS in a single chip. Figure 1 shows the schematic diagram of a planar supercapacitor with a top cap to contain electrolyte. Unlike the conventional sandwich structure, the two electrodes are made on the same plane. The major advantages of this design include size reduction and simplification of fabrication process such as sophisticated bonding. Furthermore, the travelling distance of ions in the electrolyte, a major performance factor in supercapacitors, can be well controlled and shortened while eliminating the necessity of a separator, which is indispensible in the sandwich-type supercapacitors to prevent electrical short.

CNTS ON CONDUCTIVE SUBSTRATE

Activated carbon (AC) has been widely used as the

electrode materials for supercapacitors currently in use. It could have a very large surface area (~1200 m²/g [6]) but the major drawback of AC is its pore distribution. There are three classes of pores in AC: micropore (D<2nm), mesopore (2nm<D<50nm), and macropore (50nm<D) [1, 6]. Micropores can significantly increase surface area but fail to produce the effect of double-layer capacitance due to the difficulties of ion diffusion and ion-sieving effect [1, 6]. AC also has high internal resistance caused by its low conductivity (~2.5S/cm) and numerous contacts between AC particles. CNTs, on the other hand, have a moderate surface area (100-500m²/g [7]), however, most of them are mesopores having double-layer contribution [8] with excellent conductivity (100S/cm [9]). Furthermore, each tube is individually connected to the substrate to lower the internal/contact resistance. The construction of CNT forests can be binder-free thanks to van der Waals force among them. Compared to powered AC, the much better structured pores help facilitating the transfer of electrolyte ions, resulting in rapid rate capability [8].

One major challenge in CNT-based MEMS supercapacitor is the choice of substrate as it plays two important roles: 1) a back layer suitable for the synthesis of CNTs and 2) a low-resistance current collector. Considering the fact that CNTs grow predominantly on non-conducting substrates [10], we carried out extensive experiments to investigate the quality of CNT and their contact resistance to the substrate. Thermally oxidized silicon wafers were used as the common insulating substrate and iron as the common catalyst with results summarized in Table 1. Resistances were measured with a pair of probes on the samples of same size with a constant distance. Some trends are observed based on limited results. We observed two opposite extremes: aluminum (Al) and molybdenum (Mo) are representative examples. Al is a well-known metal suitable for CNT growth. We were able to grow thick and dense aligned CNT array on the Al layer. However, although the Al layer initially exhibited low resistance of about 10Ω, the resistance jumped to 16kΩ at the end of the synthesis process. One possible explanation is that, under high temperature synthesis, Al may react with the carbon source, forming an insulating AlC layer. Contrarily, the resistance of Mo layer reduced from ~200Ω to 40Ω after the synthesis process, probably due to thermal annealing process during the CNT synthesis. Unfortunately, the problem comes that Mo is not adequate for the growth of CNTs since we hardly found CNT growth using the Mo layer under Scanning electron microscope (SEM). Other metal layers showed more or less intermediate characteristics, either close to Al or to Mo, but none showed the combined

advantages of the two extremes: thriving CNTs and low contact resistance. We proposed and, after numerous tests, experimentally verified the concept of metal stack method to solve such dilemma. It is found the combination of Mo/Al layer gives very satisfactory results. We hypothesize that the insertion of Al layer prohibits the alloying of Mo and Fe, which would otherwise make Fe lose its catalytic activities on the formation of CNTs. Meanwhile, the alloying of Mo and Al somehow avoid the formation of AlC, leaving native Al_2O_3 a well-known material for the growth of CNT on the top. It should be pointed out that we also got similar success on the recipe of W/Al stack layer, demonstrating that this methodology can be a general methodology. Further study should be carried out to investigate whether Al or native Al_2O_3 layer plays the major role in this methodology and why Al_2O_3 or AlC does not affect the contact resistance as people intuitively would expect.

Figure 2 shows vertically aligned multiwalled CNTs synthesized on the lithography-patterned Mo/Al substrate. The width/gap of the porous CNT electrodes was 40μm/20μm in (a) and (b) and 60μm/30μm in (c) and (d) all with a uniform height of about 80μm. A close view in Figure 2(d) shows that, while each individual CNT is directly connected to the substrate, and minimizing the contact resistance, the "dense" forests are actually highly porous for the diffusion of electrolyte, forming a real 3-D conductive network.

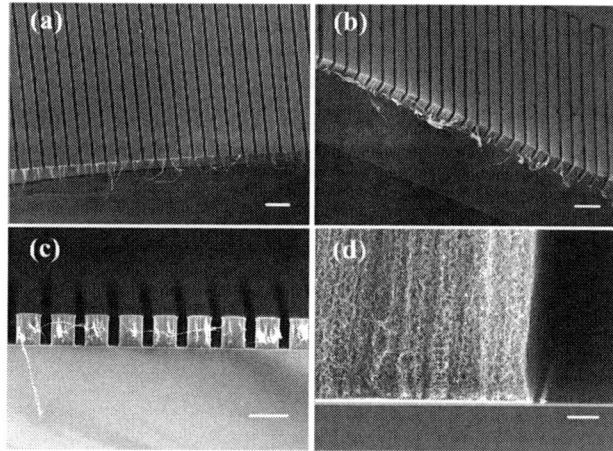

Figure 2: SEM pictures of interdigital CNT electrodes grown on the Mo/Al/Fe stack layer. (a) top view; (b) oblique view; (c) cross-section view; and (d) close view of the contact region. All the bars except (d) are 100μm. The bar in (d) is 4μm.

Metal type	Ti (50)	Cr (50)	Ni (50)	Al (50)	Mo (50)	Mo/Al (50nm/10nm)
Resistance before growth, R1	145	140	21	7.5	233	216
Resistance after growth, R2	11.7k	49.6k	373	16.7k	40	26
Ratio of R2/R1	>80	>354	>17	>15560	~0.17	~0.12
CNT profile	Short and sparse	Long but sparse	Hardly any	Thick and dense	No at all	Thick and dense
SEM pictures						

Table 1: CNT syntheses on different metal layers. The number in the brackets is the thickness (unit: nm). The resistance unit is ohm. The synthesis parameters were exactly the same. The bar in the SEM picture of Ti is 2μm and applies to all the other pictures.

FABRICATION

The fabrication process began with thermal oxidation of a bare silicon wafer to isolate the substrate from the electrodes. Then lithography was made to pattern two comb-like interdigital electrodes. After that, Mo, Al, and Fe were evaporated onto the substrate with a thickness of 50nm, 10nm, and 5nm, respectively. Fe acts as the common catalyst of CNT growth in this paper. A photoresist lift-off followed to remove the metals on unwanted areas. Thermal chemical vapor deposition (CVD) furnace (Lindberg/Blue M® three-zone tube furnace, Thermo Electron Corp., Asheville, NC) was used to grow the CNTs. The furnace was firstly purged with hydrogen and then heated up to 720°C in hydrogen environment. Subsequently, the mixture of ethylene and hydrogen with a proportion of about 1:3 was flowed through the quartz tube for 10 minutes. Finally the tube was left cooling down to room temperature with the help of a fan. Leading wires were bonded to the CNT electrodes using silver epoxy paste. 1-Butyl-3-methylimidazolium tetrafluoro-borate ([BMIM][BF4]) ionic liquid (Sigma-Aldrich inc., MO) was used as the electrolyte with a PDMS cap to the chip and limit its flow. The main reason that we chose ionic liquid rather than aqueous electrolyte (such as KOH solution) is its high breakdown voltage (~3V vs. ~1V for aqueous electrolyte).

Figure 3 shows the assembled prototype. The CNT electrode area was 5mm×7mm with about 30 comb fingers on a single electrode. Note that the area can be easily scaled down by simply changing the lithography mask.

Figure 3: Assembled prototype. A PDMS cap was used to limit the electrolyte within a fixed 5mm×7mm CNT region.

RESULTS AND DISCUSSION

The characterization of the CNT supercapacitor was conducted by the CHI 750A electrochemical workstation (CH Instruments, Inc. USA). Figure 4 shows the cyclic voltammetry curves of supercapacitors using CNT electrodes. As can been seen, our device presents a very symmetric rectangular shape (the ideal supercapacitor should have a perfect rectangular shape). The specific capacitance of our CNT-based supercapacitor can be calculated by:

$$C_{sp} = \frac{I}{dV/dt} \cdot \frac{1}{A} \qquad (1)$$

Here we represent I with the current at V=0; dV/dt was set manually on the equipment as50mV/sec in Figure 4. A is the effective CNT forest area exposed to electrolyte, 26.7mm² (this value is less than the visible 5mm×7mm area because we exclude the artificial gap area between two electrodes.). Therefore, the capacitance of a single

device was calculated as 149.7μF and specific capacitance as 428μF/cm². As a comparison, we also show the supercapacitor using the identical Mo/Al/Fe electrode but without CNT forests. It is observed that the CV curve with no CNT forests was squeezed into a "straight line". An enlarged view is shown on the upper left corner. Comparing the current level, the CNT supercapacitor has roughly three orders of magnitude higher than those with only plain metal electrodes. It is noteworthy that this ratio was quite consistent among dozens of samples we tested, firmly contributing this dramatic increase to the benefit of 3D CNT forests. The porosity of 3D CNT forest was also verified by the apagoge that, if the CNT forest were solid (no electrolyte could diffuse into the CNT forests), the capacitance should have been increased by only a factor of three due to increases in the sidewall area of the CNT finger electrodes.

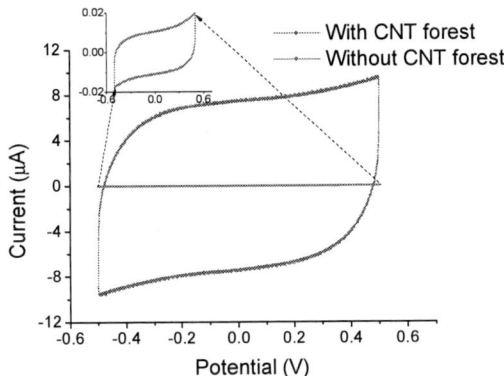

Figure 4: Cyclic voltammetry curves of the samples with (green) and without (red) CNT covering. Scanning rate is 50mV/sec. The inset shows an enlarged view of the curve of bare-metal sample. The units are the same in both coordinates.

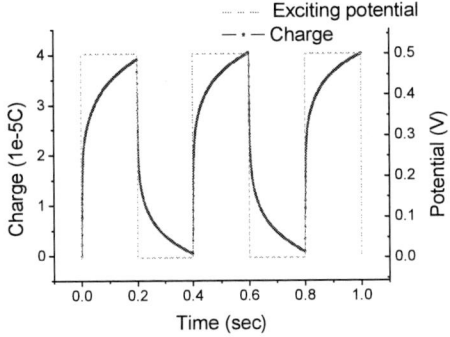

Figure 5: Chronocoulometry curve of the CNT supercapacitor. The device was charged and discharged under a square-wave voltage toggling between 0.5V and 0V at a period of 0.4s.

Figure 5 shows the chronocoulometry curve, presenting the charge/discharge capacity of our device. Defining the charge efficiency as $\Delta Q_{DISCHARGE} / \Delta Q_{CHARGE}$ per cycle, we got an impressively high efficiency of over 92%. This is partly owed to little energy dissipation because of low internal resistance. Under these conditions, an average power density of 0.28mW/cm² was calculated using the equation.

$$\bar{P} = \frac{\Delta Q V_0}{\Delta t} \cdot \frac{1}{A} \qquad (2)$$

where ΔQ is the charge accumulated during the charge session, Δt is the charge time. V_0 is the constant charge voltage, and again A is the effective carbon forest area. Figure 6 shows the charge efficiency for 10 continuous cycles. As expected, little performance degradation was observed, demonstrating a robust operation, a fundamental superiority of supercapacitors over those chemical reaction-based micro power sources.

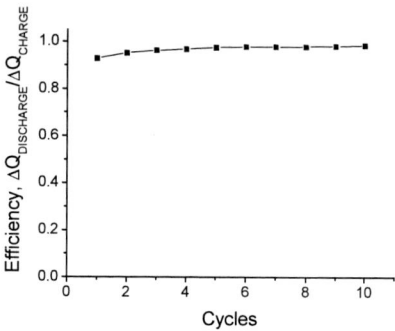

Figure 6: Efficiency variation during 10 continuous charge/discharge cycles. An excellent efficiency of over 92% was observed demonstrating a robust device.

To investigate how well the ionic liquid interacts with CNT forest, we evaluate the theoretical maximum capacitance for our current device. The supercapacitor can be expressed as

$$C = \frac{\varepsilon_0 \varepsilon_r A}{d} \qquad (3)$$

where ε_0 is the permittivity of the vacuum, 8.854×10^{-12}F/m. ε_r is the relative permittivity of the electrolyte, 11.7 for [BMIM][BF4] [11]; d is the double-layer thickness, also electrolyte-dependent parameter, 0.69nm here [12], and A is the overall surface area of electrodes. Because our CNT forests are made of multiwalled carbon nanotubes with a diameter of about 30nm, we choose $110\text{m}^2/\text{g}$ accordingly as an approximation of surface-to-mass ratio [7]. We measured the mass of CNT forests on a device at about 3×10^{-5}g, which then gives a total surface area of $3.3 \times 10^{-3}\text{m}^2$. Noticing this value should be divided equally between two electrodes, such that the area in Eq. (3) in the equation should be $1.65 \times 10^{-3}\text{m}^2$. Substituting these values into the equation, we get the theoretical peak capacitance for a single device as 248μF. Compared to our measured value 149.7μF, the utilization of surface area is about 60%. The reason that the CNT forests are not fully employed may be because that there is some air bubble trapped within the interior of CNT forests, blocking electrolyte from accessing part of CNT surface. Slowing down the filling speed of electrolyte to let the air vent and filling the electrolyte within low pressure environment may help to solve this problem.

CONCLUSIONS

We introduced the supercapacitor into micro scale by designing a novel planar MEMS supercapacitor using vertically aligned CNT forest as the electrodes. Unique metal stack layers are proposed as the conductive substrate, which generated both highly ordered CNT forest and low contact resistance. The results show our CNT-covered supercapacitor has generated a specific capacitance of 428μF/cm², three orders higher than those merely using metal electrodes. The samples show an excellent charge efficiency of 92% and a robust cycling stability. We also analyzed the difference between theoretical and measured values and proposed directions in future investigations. We believe MEMS supercapacitors will have promising applications in Microsystem including energy storage, pulse-power supply and on-chip capacitive components. Besides, the methodology of directly synthesize aligned and patterned CNTs using metal stack layers offers a general tool extend the utilization of CNTs in diverse MEMS applications.

ACKNOWLEDGEMENTS

The authors thank Professor Luke P. Lee and John Waldeisen in the department of Bioengineering at UC Berkeley, for generous help in sample measurements. This project is supported in part by the DARPA N/MEMS Fundamentals Program and Siemens Inc..

REFERENCES

[1] B. E. Conway, *Electrochemical Supercapacitors: Scientific Fundamentals and Technological Applications*, Kluwer Academic/Plenum Publishers, New York, 1999.

[2] J.H. Sung, S.J. Kim, and K.H. Lee, "Fabrication of microcapacitors using conducting polymer micro-electrodes", *J. Power Sources*, vol. 124, pp. 343-350, 2003

[3] H.J. In, S. Kumar, Y. Shao-Horn, and G. Barbastathis, "Nanostructured Origami™ 3D Fabrication and Assembly of Electrochemical Energy Storage Devices", *Proc 2005 IEEE on Nanotechnology*, Nagoya, Japan, July 2005 pp. 374-377.

[4] K.H. An, W.S. Kim, Y.S. Park, Y.C. Choi, S.M. Lee, D.C. Chung, D.J. Bae, S.C. Lim, and Y.H. Lee, "Supercapacitors Using Single-Walled Carbon Nanotube Electrodes", *Adv. Mater.*, vol. 13, pp. 497-500, 2001

[5] V.L. Pushparaj, M.M. Shaijumon, A. Kumar, S. Murugesan, L.J. Ci, R. Vajtai, R.J. Linhardt, O. Nalamasu, and P.M. Ajayan, "Flexible energy storage devices based on nanocomposite paper", *PNAS*, vol. 104, pp. 13574-13577, 2007.

[6] A.G. Pandolfo and A.F. Hollenkamp, "Carbon properties and their role in supercapacitors", *J. Power Sources*, vol. 157, pp.11-27, 2006

[7] http://www.cheaptubes.com/MWNTs.htm

[8] H. Zhang, G. Cao, and Y. Yang, "Electrochemical properties of ultra-long, aligned, carbon nanotube array electrode in organic electrolyte", *J. Power Sources*, vol. 172, pp.476-480, 2007

[9] A.E. Aliev, C. Guthy, M. Zhang, S. Fang, A.A. Zakhidov, J.E. Fischer, and R.H. Baughman, "Thermal transport in MWCNT sheets and yarns", *Carbon*, vol. 45, pp.2880-2888, 2007

[10] S. Talapatra, S. Kar, S.K. Pal, R. Vajtai, L. Ci, P. Victor, M.M. Shaijumon, S. Kaur, O. Nalamasu and P.M. Ajayan, "Direct growth of aligned carbon nanotubes on bulk metals", *Nature nanotechnology*, vol. 1, pp. 112-116, 2006

[11] C. Wakai, A. Oleinikova, M. Ott, and H. Weingartner, "How Polar Are Ionic Liquids? Determination of the Static Dielectric Constant of an Imidazolium-based Ionic Liquid by Microwave Dielectric Spectroscopy", J. Phys. Chem. B, vol. 109, pp.17028-17030, 2005

[12] http://www.utexas.edu/research/chemed/lagowski/WSSP /houston_2006_paper_maass_katie.pdf

PIEZOELECTRIC NANOSWITCH

D.C. Judy, J.S. Pulskamp, R.G. Polcawich, and L. Currano
US Army Research Laboratory, Adelphi, MD, USA

ABSTRACT

This paper details the design, fabrication, and measurement of a nano-scale mechanical switch. The switch uses lead zirconate titanate (PZT) actuators fabricated on silicon wafers and is intended to realize very low leakage complimentary mechanical logic elements. The switches were fabricated using an eleven step electron beam process using a novel dual electron beam photo resist technique to create gold air bridges. A normally closed switch was successfully demonstrated with switching times as fast as 17ns, operating as low as 6 volt actuation.

INTRODUCTION

As MEMS device complexity increases, there is an interest in integrating control elements with MEMS actuators. Potential applications include control elements within the ground plane of RF MEMS phase shifters and integrated microcontrollers for small scale robotic actuators. The primary method of implementing control elements with MEMS devices is to integrate complementary metal-oxide semiconductor (CMOS) electronics with MEMS [1]. An alternative approach leverages a MEMS fabrication process to create logic and memory based on mechanical switches [2, 3].

The first attempts at creating MEMS logic used the Polymumps process [4, 5] with electrostatic actuation for the switching elements. More recently, Judy et. al. demonstrated piezoelectrically actuated logic elements including an inverter, NAND gate, and ring oscillator [6]. One advantage of the piezoelectric approach is low voltage actuation and low leakage which enables low dynamic and static power drain. Additionally, the fabrication process used to create the piezoelectric MEMS logic is identical to the process used to demonstrate low voltage piezoelectric RF switches and phase shifters [7].

A limitation of the piezoelectric MEMS logic is the operational speed of the logic elements. With dimensions of 125 um in length, switching speeds are limited to tens of microseconds. To achieve speeds on the order of a few to tens of nanoseconds, this research examines the scaling of the piezoelectric switch to create a piezoelectric nanoswitch.

DESIGN

To achieve the logic switching speeds faster than one nanosecond, a normally open PZT nanoswitch was modeled as a 3-dimensionally scaled version of the PZT RF MEMS switch reported in Reference 1. Both finite element modeling using ANSYS and analytical models were used to examine the design space of PZT actuator length, PZT thickness, and contact gap. Modeling the switch as a composite cantilever experiencing a piezoelectric actuation moment acting at the tip of the top electrode, the free displacement of the contact location, y_{cf}, can be expressed as:

$$y_{cf} = \frac{e_{31}(E)VwhL_{act}(2L_{sw} - L_{act})}{2(YI_{comp})} \quad (1)$$

where $e_{31}(E)$ is the field dependant effective piezoelectric stress constant, V is the voltage, w is the width of the top electrode, h is distance between the structure's bending neutral axis and the midplane of the PZT layer, L_{sw} is the distance from the anchor to the contact, L_{act} is the length of the top electrode, and YI_{eff} is the flexural rigidity of the composite structure.

The maximum possible switching speed t_{max}, without the presence of contact bounce, is approximately equal to a third of the resonant period of the switch and can be expressed as:

$$t_{max} \approx \frac{1}{3f_o} = \frac{2\pi L_{sw}^2}{3\lambda_i^2 \sqrt{\frac{(YI)_{comp}}{\mu}}} \quad (2)$$

where f_o is the fundamental length flexural resonance frequency, λ_i is the mode number, and μ is the mass per unit length of the structure. The results of these models predicted PZT normally open nanoswitches could achieve better than a one nanosecond operation by using PZT thin films less than 100 nm thick, actuators less than 600nm long, and with contact gaps less than 10 nm (see Figure 1).

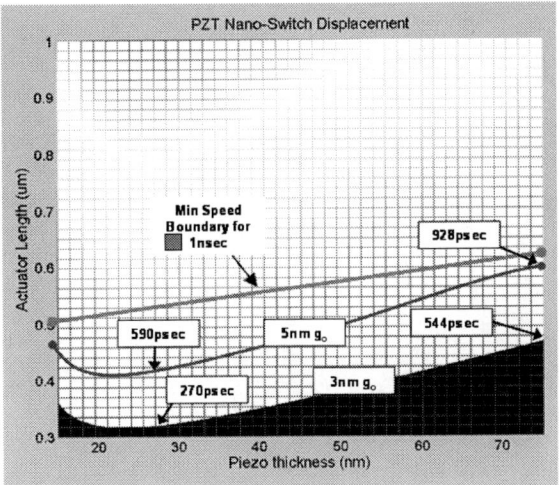

Figure 1. Theoretical predictions of a normally open PZT nanoswitch designed to operate at better than one nanosecond.

Significant assumptions with these predictions included zero change of the PZT material properties as a function of thickness (5000 nm to 5 nm) and zero residual stress gradients both of which are difficult to account for in the theoretical predictions. To accommodate variations in residual stress switches with various lengths were fabricated. In addition, normally closed switches were added to enable the creation of the nanoscale inverters and

NAND gates necessary for mechanical logic.

The normally closed switches are similar to the normally open switch described in [8] except that the gold contact cantilever (see Figure 2) is attached to the actuator instead of the silicon wafer and the bottom gold contact is attached to the silicon wafer. Thus, when the actuator is released by a XeF$_2$ etch of the underlying silicon, the residual stress in the actuator causes the actuator to bend down into the wafer until the contact cantilever hits the input and output electrodes preventing further deflection and completing the electrical connection with zero applied voltage. By applying an actuation voltage across the PZT, the piezoelectric strain creates a moment relative to the neutral axis of the actuator structure, such that the actuator

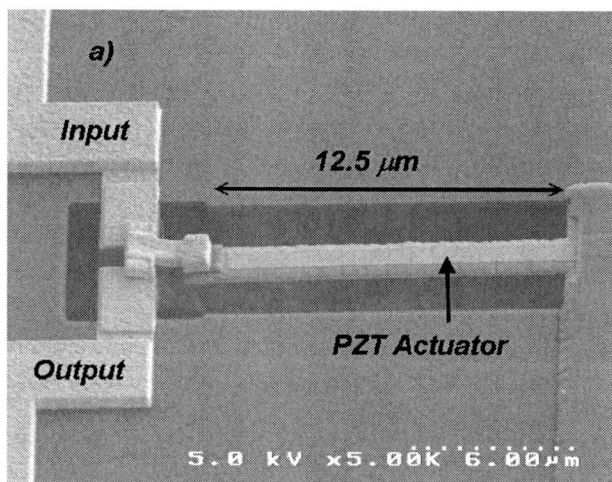

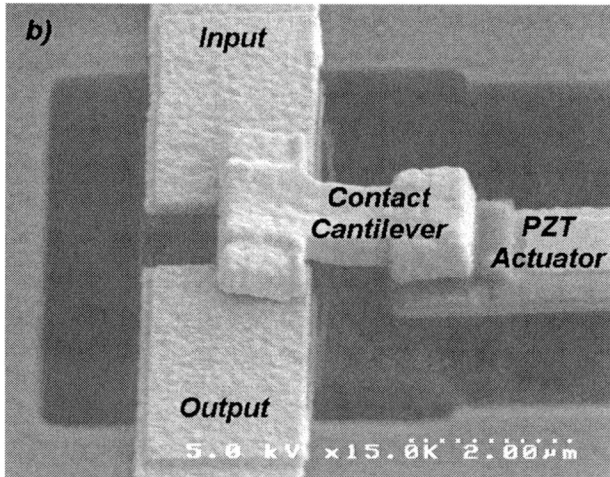

Figure 2: SEM of a normally closed PZT nanoswitch. a) overview of the normally closed switch showing the entire 12.5 μm long actuator. b) zoomed in view of the switch contact region.

bends up breaking the electrical contact. When the actuation voltage is removed the residual stress forces the actuator downward until the top contact hits the bottom contact completing the electrical contact.

FABRICATION

Device fabrication was done at the Specialty Electronic Materials and Sensors Cleanroom Facility, U.S. Army Research Laboratory, Adelphi, MD. Using an eleven step, aligned electron beam lithography process,

PZT nanoswitches were fabricated using an actuator stack comprised of SiO$_2$/Ti/Pt/PZT/Pt. The nanoscale switches were designed using a minimum feature size of 500nm along with a 50nm alignment tolerance. The conservative alignment tolerances used for the multiple lithography process steps in the fabricated devices created restrictions limiting the actuator dimensions to 1000 nm wide and lengths varying from 700 nm to 15 μm. With careful control, even smaller devices can be realized in the same process. The individual process steps used to create the nanoscale switches were nearly identical to the process used to create the MEMS switches in Ref. 8. The process includes two ion-milling steps to pattern the top platinum electrode and subsequently pattern the PZT and base Ti/Pt electrode. A wet etch is used to open access to the bottom Ti/Pt electrode. A reactive ion etch is used to define the actuator and open access to the silicon substrate. An additional thin silicon dioxide layer is deposited and patterned to provide electrical isolation on the sidewall of the PZT and Ti/Pt layer. This is followed by metal evaporation processes including Ti/Au for the co-planar waveguide transmission line for testing and Au/Pt for contact dimples. An evaporated silicon sacrificial layer is then used to create the <50nm gap necessary for switch operation. The silicon was deposited with a thickness of either 10nm or 30nm on different samples. The next step of the fabrication process involved the use of a novel sacrificial electron beam resist process to create suspended gold air bridges and cantilever structures (see Figure 3). The final step after the metal lift-off patterning of the 1000nm thick

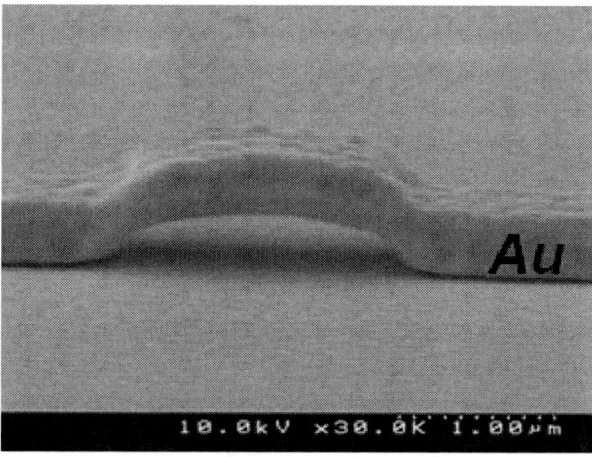

Figure 3: SEM of a gold air bridge created using a novel dual electron beam photoresist process.

gold was the release etches consisting of an oxygen plasma for the sacrificial resist and XeF$_2$ to undercut the actuator, input and output traces, and remove the evaporated silicon sacrificial layer.

MEASUREMENT

All of the switches were designed into 50Ω co-planar waveguide (CPW) transmission lines (see Figure 4). The CPW geometry was used to reduce the effects of parasitics on switch risetime and thus facilitate switching speed measurements. The actuation pulse is provided by an arbitrary waveform generator with a risetime of 10ns and variable pulse amplitude with a 10 volt maximum.

The input transmission line was connected to a power supply set at 100mV (V_{input}) using a 200um pitch coplanar wafer probe. The actuation pulse is recorded on channel 1 of the oscilloscope (1 MΩ) before being connected to the switch actuation electrodes using a pair of DC needle probes. The output CPW was connected to channel 2 (50Ω) of the scope via a second coplanar probe. This configuration eliminates the effect of mismatch reflections that would make interpretation of the switching speed difficult as well as allowing the determination of the contact resistance. The switch contact resistance (R_c) and the scope internal load resistor (R_l) form a voltage divider that allows the contact resistance to be determined by

$$R_c = \frac{R_l (V_{input} - V_{scope})}{V_{scope}} \qquad (3)$$

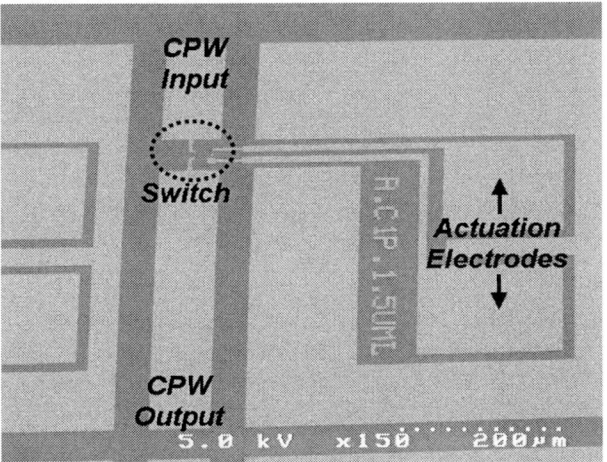

Figure 4: SEM image of the CPW configuration used to measure the PZT nanoswitch.

Figure 5 shows the switching response of the normally closed switch. From the switching response curve it can be seen that V_{scope} prior to switching is 45 mV. Using equation (3) the contact resistance is calculated to be 61 ohms. This would be poor contact resistance for RF switches, but is more than acceptable for use in mechanical logic circuits. The cable delay was measured by applying the actuation pulse directly to the center conductor of the output CPW near the switch using the DC needle probe. Figure 6 shows the measurement of the cable delay, which is observed to be 17ns.

After studying the switches in an SEM, it was determined that the longer, normally closed switches had the best chance of working because of the larger than anticipated residual stress deformation. This rendered the normally open switches unusable, because the gap between the actuator and the contact cantilever was too large to close with the PZT actuators. Normally open switches and switches with shorter lengths can be realized with additional control of the residual stress gradient and thickness choice of the evaporated silicon sacrificial layer.

Measurements in this work concentrated on the longer, normally closed switches that the SEM images suggested had closed contacts. The switch reported here had an actuator length of 12.5 µm as seen in Figure 2. Referring to Figure 5b, it can be seen that the measured actuation time of the on-off (unactuated-actuated)

switching event was observed to be 33.8ns, or 16.8ns after accounting for cable delay. Referring to Figure 5c, the observed off-on (actuated-unactuated) switching time was 50 ns for first contact, or 33 ns after accounting for cable delay. There was a significant bounce time in this particular case as seen in Figure 5c. Switch bounce was observed to slow the off-on rise time to nearly 500 ns. However, the switch dynamics are still being studied as zero bounce was also observed during some switch cycle tests (see Figure 7). These measurements were repeated for several cycles and switching times (minus cable delay) were observed to vary from 17 – 33 ns for the on-off switching event and off-on switching time varying from 33 – 43 ns. The observed actuation voltage for this device was 6 volts, but this is higher than expected due mainly to the same excessive residual stress that rendered the

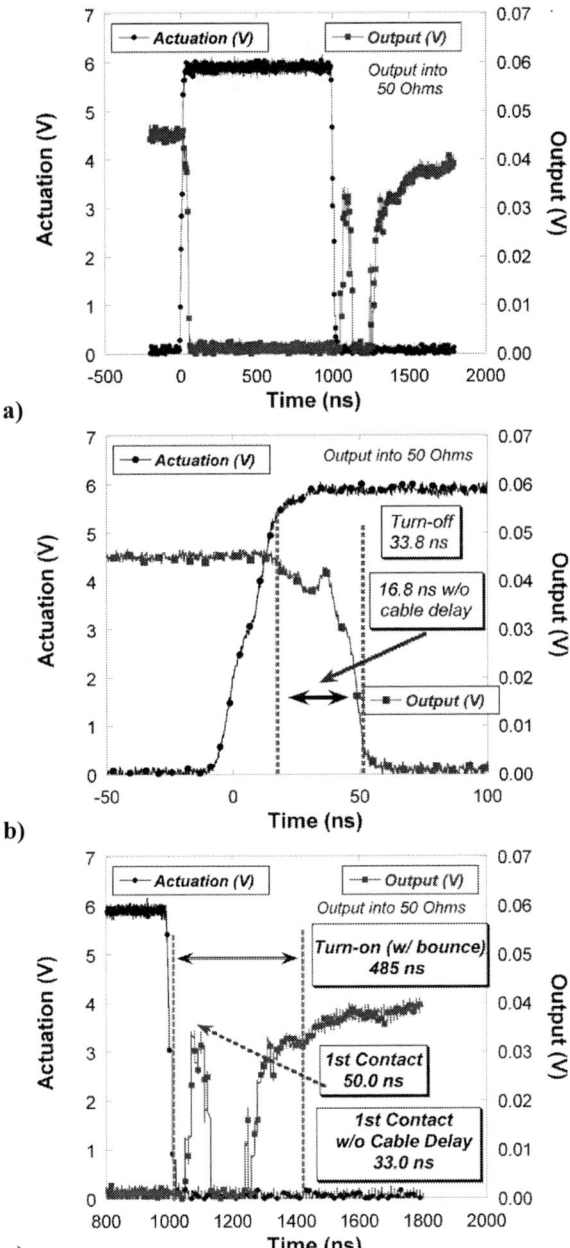

Figure 5: Switching response for a normally closed PZT nanoswitch: a) full cycle, b) off-state (6V), and c) on-state (0V).

normally open switches nonfunctional. The extra residual stress adds to the contact force that has to be overcome by the PZT strain, leading to a larger than expected voltage required to operate the device.

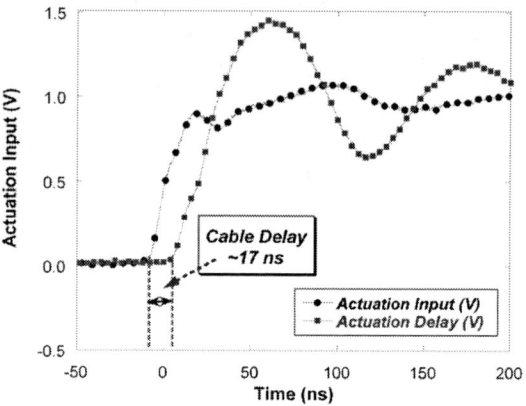

Figure 6: Experimental results for the cable delay in the measurement setup.

CONCLUSIONS

A normally closed nanoscale switch using piezoelectric actuators has been demonstrated. The switch showed switching times as low as 17 ns with an actuation voltage of 6 volts. Residual stress prevented the demonstration of normally open switches and complementary inverters. The excessive residual stress contributed to the unexpectedly high actuation voltage of 6 volts. These results demonstrate the feasibility of creating single nanosecond and faster nanoswitches using PZT thin films and electron beam lithography for mechanical logic elements. Future efforts will be devoted to improving the excessive residual stress, inceasing yield of the devices, and demonstrating nanoscale logic elements such as inverters and nand gates.

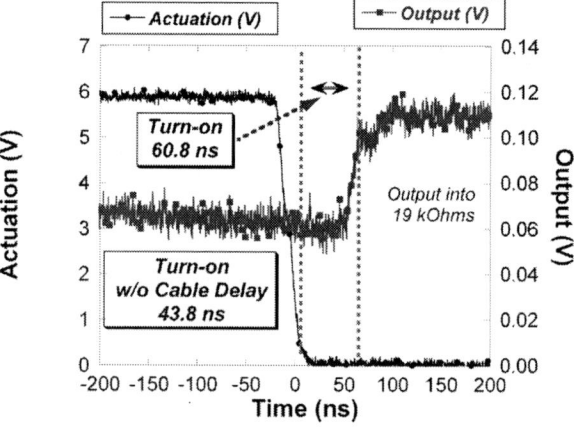

Figure 7: PZT nanoswitch operation without contact bounce.

ACKNOWLEDGEMENTS

The authors would like to thank Joel Martin and Richard Piekarz of the US Army Research Laboratory and Brian Power of General Technical Services for their roles in the fabrication of the PZT nanoswitch.

REFERENCES

[1] P. J. Gilgunn, J. Liu, N. Sarkar, and G. Fedder, IEEE Journal of Microelectromechanical Systems, vol. 17, no. 1, february 2008, pp. 103-114

[2] W. Jang, J. Lee,1 J. Yoon, M. Kim, J. Lee, S. Kim, K. Cho, D.Kim, D. Park, and W. Lee, Applied Physics letters 92, 103110 2008, pp.1-3

[3] M. Kim, W. Jang, J. Lee, S. Kim, E. Yun, K. Cho, S. Lee, I. Choi, Yong, J. Yoon, D. Kim, and D. Park, ISDRS 2007, December 12-14, 2007, College Park, MD, USA

[4] S.W. Lee, R. W. Johnstone, and A. M. Parameswaran, IEEE CCECE/CCGEI, Saskatoon, May 2005, pp. 1513 – 1516.

[5] S.W. Lee, R. W. Johnstone, and A. M. Parameswaran, Proc. SPIE, 6037, pp. 1A1-1A10, 2005.

[6] D. Judy, R. G. Polcawich, J. Pulskamp, *2008 Solid State Sensor, Actuator and Microsystems Workshop* (Hilton Head 2008), Hilton Head Island, South Carolina, June 1-5, pp. 328 – 331, 2008.

[7] R.G. Polcawich, D. Judy, J. Pulskamp, *2007 International Microwave Symposium*, Honolulu, Hawaii, June 3-8, pp. 2083-2086, 2007.

[8] R.G. Polcawich, et. al., *IEEE Trans. Microw. Theory Techn.*, vol 55, pp. 2642 – 2654, 2007.

MOLECULAR MUSCLE BASED NANO-ELECTRO-MECHANICAL-SYSTEMS (NEMS)

Bala Krishna Juluri,[1] Ajeet S. Kumar,[1] Yi Liu,[2] Tao Ye,[1] Ying-Wei Yang,[2] Amar H. Flood,[2] Lei Fang,[3] J. Fraser Stoddart,[3] Paul S. Weiss,[1] and Tony Jun Huang[1]

[1]The Pennsylvania State University, University Park, PA
[2]University of California, Los Angeles, CA
[3]Northwestern University, Evanston, IL

ABSTRACT

Created by a bottom-up approach based on self-assembly and molecular recognition, molecular motors and muscles such as bistable rotaxanes are recognized as promising actuation materials for Micro/Nano Electro-Mechanical-Systems (MEMS/NEMS). In this work, we report a hybrid NEMS actuation device based on *electrochemical activation* of surface-bound bistable rotaxanes (molecular muscles). This NEMS actuator consists of a micro-cantilever coated with a self-assembled monolayer of redox-active palindromic bistable [3]rotaxane R^{8+} molecules, and when subjected to oxidizing and reducing electrochemical potential undergoes controllable and reversible bending. This work represents an important step towards the practical realization of artificial molecular muscle based actuators for various NEMS/MEMS applications.

INTRODUCTION

The past two decades have witnessed huge research interest from both academia and industry to develop numerous micro/nano electro-mechanical systems (MEMS/NEMS). Many of these systems rely on some form of actuation mechanisms and traditionally electrostatic and piezoelectric materials are used in these devices.[1] However these conventional actuation materials require high driving voltages, and have to be fabricated using photolithography and other top-down manufacturing techniques. These fabrication methods limit the ease of making feature sizes below 100 nm. A more recent approach to overcome this problem has led researchers to fabricate hybrid NEMS/MEMS actuation devices that are based on an integrated approach of using both top-down and bottom-up fabrications methods. This integrated approach is more favorable due to advantages of bottom-up approach which enables the elegant employment of molecular machines (either biological or artificial) to work coherently and perform macroscopic work at much larger scales that begins at nanoscale.

A model hybrid NEMS actuation device consists of a top-down fabricated micro-cantilever with large surface area and ultralow stiffness constant coated on one side with bottom-up assembled layer of molecular machines. Various molecular machines have been used for hybrid NEMS actuators in the recent past [2-6]. Our work towards building artificial molecular machinery for hybrid NEMS actuator applications was inspired by the classiness with which skeleton muscles operate in nature. In skeleton muscles, chemical energy is converted to relative mechanical sliding of numerous very well organized and interlocked myosin molecules between actin filaments.[7] Coherent and cooperative

conformation changes in this system are efficiently harnessed to obtain macroscopic muscle contraction and expansion that starts from single molecule. Inspired and in an attempt to imitate this type of energy conversion, palindromic bistable [3]rotaxane R^{8+} molecule was designed to undergo controlled conformational changes in the presence of external stimulus. As shown in Figure 1(a), a palindromic bistable [3]rotaxane R^{8+} molecule synthesized by template-directed procedure contains two spacer separated pairs of redox active tetrathiafulvalene (TTF) and competing naphthalene (NP) stations. Two tetracationic cyclophanic, cyclobis (paraquat-para-phenylene) (CBPQT4+) rings are recognized by each TTF site through electron donor and acceptor interactions. Under the absence of any perturbations, the cyclobis rings are designed to station at TTF sites (ground state) and the removal of one or two electrons (oxidation of TTF) from each TTF's station causes the rings to immediately move towards NP stations due to columbic repulsion between the ring and the TTF station (metastable state). Reducing back the TTF stations with a supply of electrons (reduction of TTF) causes the rings to shuttle back to TTF's stations (ground state). Disulphide groups are tethered to each ring of the molecule so that these molecules can form a SAM layer on gold and collective ring motions can be harnessed efficiently to perform mechanical work on micro-cantilever.

There are several advantages of palindromic bistable [3]rotaxane R^{8+} molecule for usage as actuation materials in hybrid NEMS devices. First of all, rotaxanes can generate large strains up to 42%, while the strains generated by the gold-standard actuation materials – piezoelectric materials – are typically 0.1–0.2% [1]. Second, they have high force density, e.g., a bistable rotaxane generates 100 pN force, while a kinesin biomotor, which is much larger than a bistable rotaxane, can only generate 6 pN. Third, they can undergo controlled mechanical motion for a variety of external stimuli, while traditional actuation materials and biomotors must both rely on a single stimulus. Owing to these advantages, in our earlier work, we utilized bistable rotaxanes to build a hybrid NEMS actuator. We showed that by chemically controlling the redox state of individual surface bound [3] rotaxane molecules attached on the gold coated micro-cantilever, it is possible to perform mechanical work on a micro-cantilever [8, 9]. In principle, the removal or supply of electrons to change the molecular conformation of bistable [3] rotaxane R^{8+} molecule can be obtained by various methods which include chemical, electrochemical and photochemical methods. In contrast to chemical method, electrochemical and

978-1-4244-2977-6/09 $25.00 © 2009 IEEE

photochemical methods have advantages of robustness and fast operating speeds. In addition they provide a way to both "read" and "write" the state of the molecular conformations and therefore helpful in controlling and monitoring the system. Using UV-VIS spectroscopy, we have previously shown that the position of rings in liquid dispersed palindromic bistable [3]rotaxane R^{8+} molecules can be controlled by electrochemical method. However the question whether electrochemical method can be used to activate artificial molecular muscles attached to solid surfaces still remains unanswered. Herein we report our investigation made on electrochemical redox control of solid bound bistable [3]rotaxane R^{8+} molecules and on the possibility of realizing a hybrid NEMS actuator based on electrochemical activation of artificial molecular muscles.

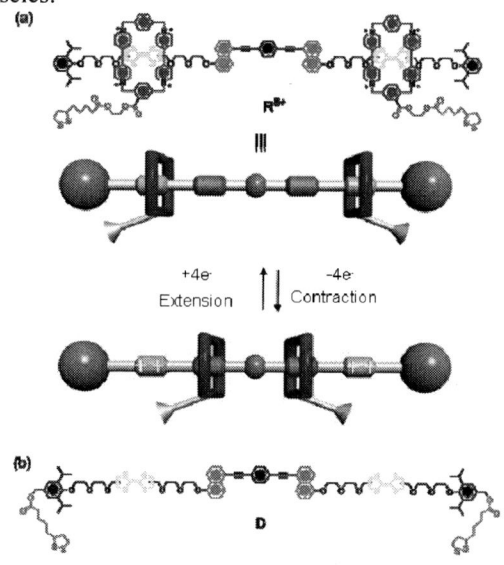

Fig. 1: *Molecular structures of (a) palindromic bistable [3] rotaxane R^{8+} and (b) disulfide-tethered dumbbell D (control compound related to R^{8+}).*

EXPERIMENTAL

Working Principle behind the hybrid NEMS actuator based on Electrochemical Activation of R^{8+} molecules

With artificial molecular muscles attached to gold surface of micro-cantilever, it is expected that the application of sufficient oxidizing potential will remove electrons from that TTF station and make it positively charged. The change in the charge causes the substrate attached rings to move towards DNP station due to columbic repulsion between the rings and TTF stations. This in turn would build up bending moment on the supporting micro-cantilever and bend the cantilever downward. Applying a reducing potential consequently reduces the TTF station causing the rings to move back to TTF station from DNP station and hence relax the collective bending moment built up in the previous step and cause the cantilever reach to its original position. Therefore by applying electrochemical potentials, it is possible to obtain control over collective buildup or relaxation of bending moment on a micro-cantilever.

Sample Preparation and Experimental Setup

Rectangular Si micro-cantilevers of length 500 μm, width 100 μm and thickness 1 μm were used in all the experiments and were obtained from commercial manufacturer (NanoAndMore Inc, Lady's Island, SC). These micro-cantilevers were pre-coated with a 20 nm thick gold layer. Before functionalization, each micro-cantilever was thoroughly cleaned for 5 minutes with UV/Ozone process and later washed with DI water. The palindromic bistable [3] rotaxanes R^{8+} molecules and control molecules were synthesized using the method reported earlier [9]. Thoroughly cleaned micro-cantilevers were then immersed in solutions containing target molecules for 48 hours enabling self assembly on the gold side of the micro-cantilever. In order to detect the mechanical work exerted by surface bound bistable [3] rotaxane R^{8+} molecules on micro-cantilever during electrochemical activation, we combined atomic force microscopy based optical deflection measurement technique with in-situ electrochemistry setup (Pico SPM 2500, Molecular Imaging) as shown in Figure 2. Gold coated micro-cantilever with self assembled monolayer's of rotaxane molecules were used as working electrode, Ag and Pt wires were used as reference and counter electrode respectively. All these electrodes were then immersed in a teflon cell filled with 0.1 M $NaClO_4$ electrolyte. The redox state of the molecules is changed by applying a desired potential between a working electrode and counter electrode using a three electrode potentiostat. Beam reflected by the uncoated side of the micro-cantilever is collected by a four quadrant position sensitive photo-diode. Any deflections in the micro-cantilever due to build-up and relaxation of bending moment during electrochemical activation are readout from the variations in reflective laser beam position on the photodiode. Sensitivities of each micro-cantilever to convert deflection signal from photodiode in volts to nanometers were calculated by fitting the slopes of force curves.

RESULTS AND DISCUSSION

Figure 3 (a) shows the response of hybrid NEMS actuator based on molecular muscles to a triangular potential scan. During the anodic sweep, the initial deflection went upward and after reaching a potential of 350 mV, the micro-cantilever started to deflect downwards. The initial upward deflection could be correlated to specific adsorption of perchlorate anions.[10] The latter downward deflection could be correlated to the collective action of molecular muscles acting against the micro-cantilever's restoring force and the compressive stresses originating due to specific adsorption. Consequent application of cathodic sweep caused the micro-cantilever to restore to its neutral position. The downward deflection

978-1-4244-2977-6/09 $25.00 © 2009 IEEE   596

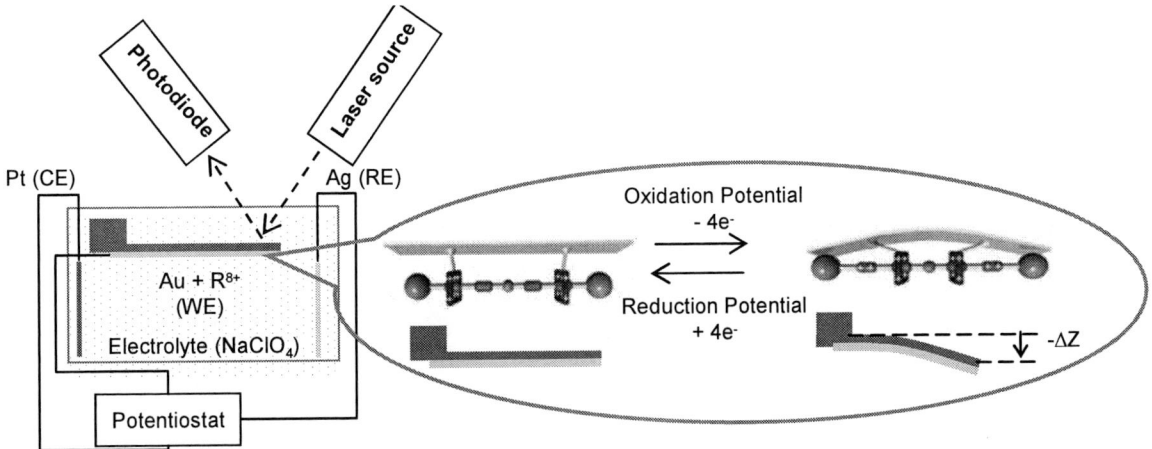

Fig. 2: Schematic of the experimental setup used for in situ electrochemical activation of the palindromic bistable [3] rotaxanes molecules. The inset shows the reversible electrochemical oxidation and reduction of R^{8+} to produce the micro-cantilever deflection. (WE: working electrode, CE: counter electrode, RE: reference electrode).

as the potential went below 280mV indicates desorption of perchlorate anions from the micro-cantilever. To verify whether other affects may cause the observed deflection signature, we conducted control experiment with the control compound, **D**. As shown in Figure 1(b), **D** molecule is equivalent to bistable [3] rotaxane but with no cyclobis rings and with two disulphide tethers on either ends of the molecule. This control compound is devoid of moving elements and is expected not to perform any mechanical work on micro-cantilever. The deflection signature of a triangular sweep on this control compound attached to a micro-cantilever is shown in figure 3(b). It can be seen that there is a slight upward and downward deflection in the micro-cantilevers coated with control compounds and can be attributed to the perchlorate adsorption and desorption on the gold side of micro-cantilever. The differences between the direction and magnitude of deflection in R^{8+} coated micro-cantilever and that of micro-cantilever coated with control compound provides clear evidence

about the ability of electrochemical method to activate solid bound molecular muscles.

We further extended our studies to dynamic performance by applying square shaped potential steps. The results with a step width of 20sec are presented in Figure 4. It can be observed that at higher potential (enough to oxidize the R^{8+} molecules), the micro-cantilever bends downward and at lower potential, the micro-cantilever deflect upwards towards its neutral position over a series of potential steps. The alternative bending down and bending up of the micro-cantilever can be correlated to the alternative buildup and relaxation of collective bending moments originating from molecular muscles. Two effects should be noted in the dynamic performance: 1) after the completion of the first oxidation step, there is a sudden decrease in downward deflection amplitude in the following oxidation steps and 2) there is a gradual downward deflection of the micro-cantilever as the number of oxidation and reduction cycles increase.

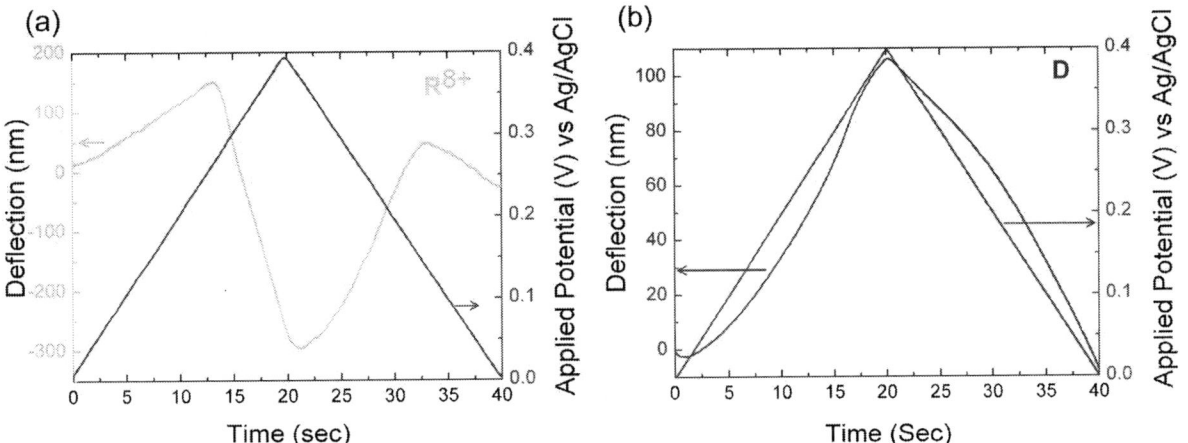

*Fig. 3: Deflection and applied potential vs time for micro-cantilevers coated with monolayer films of (a) bistable [3] rotaxane R^{8+} and (b) the dumbbell **D** molecules.*

978-1-4244-2977-6/09 $25.00 © 2009 IEEE

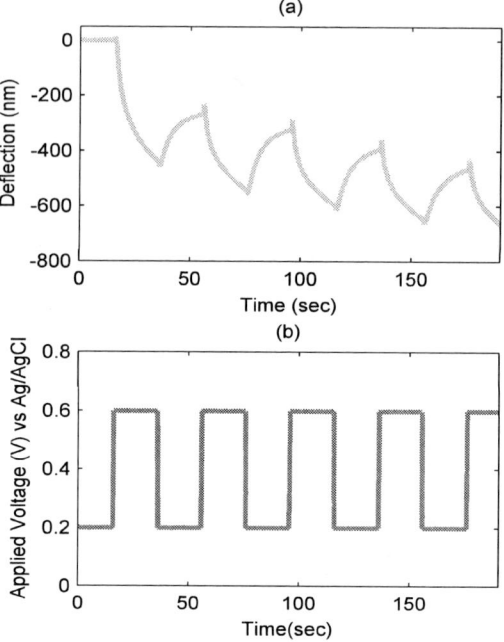

Fig. 4: Time-dependent operation of micro cantilevers coated with the palindromic bistable [3] rotaxane R^{8+}, (a) Measured deflection and (b) Applied Voltage vs Ag/Agcl

The sudden decrease in the downward deflection amplitude can be explained by the incomplete recovery of metastable state rotaxane molecules to ground state in the first reduction cycle. The gradual downward deflection of micro-cantilever or creep has been observed in our experiments with control compound **D** and in other electrochemical actuator systems [11-12]. We believe that a creep mechanism arising from a reorganization of the muscle molecules within the monolayer film plays an important role during the electrochemical perturbation.

CONCLUSIONS

In summary, we have demonstrated that surface-bound artificial muscle molecules, when electrochemically activated, cause a micro-cantilever to bend up and down. Conversely, micro-cantilever beams that are coated with redox-active but mechanically inert control compounds do not display the same bending characteristics. Compared with their chemically-driven counterparts [8,9], the electrochemically-driven molecular-muscle-based actuators can be operated much faster, more conveniently, and with larger responses. These results constitute a key step towards engineering applications of NEMS based on artificial molecular muscles.

ACKNOWLEDGEMENTS

This research was supported in part by the Grace Woodward Grants for Collaborative Research in Engineering and Medicine, and the NSF NIRT grant (ECCS-0609128). Components of this work were conducted at the Penn State node of the NSF-funded National Nanotechnology Infrastructure Network.

REFERENCES

[1] Liu, C. *Foundations of MEMS*, Pearson Education: New Jersey, 2006.

[2] Raguse, B.; Muller, K. H.; Wieczorek, L. Nanoparticle Actuators. *Adv. Mater.* vol. 15, pp. 922–926, 2003.

[3] Bay, L.; West, K.; Sommer-Larsen, P.; Skaarup, S.; Benslimane, M. A Conducting Polymer Artificial Muscle with 12% Linear Strain. *Adv. Mater.*, vol.15, pp. 310–313, 2003.

[4] Shu, W. M.; Liu, D. S.; Watari, M.; Riener, C. K.; Strunz, T.; Welland, M. E.; Balasubramanian, S.; McKendry, R. A. DNA Molecular Motor Driven Micromechanical Cantilever Arrays. *J. Am. Chem. Soc.*, vol. 127, pp. 17054–17060, 2005.

[5] Ren, Q.; Zhao, Y. P.; Han, L.; Zhao, H. B. A Nanomechanical Device Based on Light-Driven Proton Pumps. *Nanotechnology*, vol. 17, pp. 1778–1785, 2006.

[6] Ji, H. F.; Feng, Y.; Xu, X. H.; Purushotham, V.; Thundat, T.; Brown, G. M. Photon-Driven Nanomechanical Cyclic Motion. *Chem. Commun.*, vol. 22, pp. 2532–2533, 2007.

[7] Goodsell, D. S.: *Our Molecular Nature: The Body's Motors, Machines, and Messages*, Copernicus, New York, 1996.

[8] Huang, T. J.; Brough, B.; Ho, C. M.; Liu, Y.; Flood, A. H.; Bonvallet, P. A.; Tseng, H.-R.; Stoddart, J. F.; Baller, M.; Magonov, S. A Nanomechanical Device Based on Linear Molecular Motors. *Appl. Phys. Lett.*, vol. 85, pp 5391–5393, 2004.

[9] Liu, Y.; Flood, A. H.; Bonvallett, P. A.; Vignon, S. A.; Northrop, B. H.; Tseng, H. -R.; Jeppesen, J. O.; Huang, T. J.; Brough, B.; Baller, M.; Magonov, S.; Solares, S. D.; Goddard, W. A.; Ho, C. M.; Stoddart, J. F. Linear Artificial Molecular Muscles. *J. Am. Chem. Soc.*, vol. 127, pp. 9745–9759, 2005.

[10] Ibach, H.; Bach, C. E.; Giesen, M.; Grossmann, A. Potential-Induced Stress in the Solid-Liquid Interface: Au(111) and Au(100) in an HClO4 Electrolyte. *Surf. Sci.*, vol. 375, pp. 107–119, 1997.

[11] Mirfakhrai, T.; Oh, J.; Kozlov, M.; Fok, E. C. W.; Zhang, M.; Fang, S.; Baughman, R. H.; Madden, J. D. W. Electrochemical Actuation of Carbon Nanotube Yarns. *Smart Mater. Struct.*, vol. 16, pp. S243–S249, 2007.

[12] Smela, E.; Lu, W.; Mattes, B. R. Polyaniline Actuators: Part 1. PANI(AMPS) in HCl. *Synth. Met.*, vol. 151, pp. 25–42, 2005.

978-1-4244-2977-6/09 $25.00 © 2009 IEEE

HIGHLY EFFICIENT EXTRACTION OF MECHANICAL AND LINEAR AND QUADRATIC PIEZORESISTIVE PROPERTIES OF POLY-SI FILMS USING WAFER-SCALE MICROTENSILE TESTING

M. E. Schmidt[1], J. Gaspar[1], J. Held[1], S. Kamiya[2], and O. Paul[1]

[1]Department of Microsystems Engineering (IMTEK), University of Freiburg, GERMANY
[2]Department of Mechanical Engineering, Nagoya Institute of Technology, JAPAN

ABSTRACT

This paper reports on the extension of the wafer-scale microtensile technique to the piezoresistive characterization of thin-films, demonstrated for in-situ n-doped poly-Si layers. In addition to the reliable extraction of mechanical properties, this extended high-throughput method enables the acquisition of linear and, for the first time, nonlinear piezoresistive coefficients, namely the first and second order longitudinal parameters, $\pi_{l,1}$ and $\pi_{l,2}$, respectively, with statistical significance. In contrast to previous studies, the data presented here are extracted for strains up to the fracture value, leading to meaningful linear and quadratic piezoresistive parameters. Along with the mechanical properties, these values are especially important for sensing applications where poly-Si films are subjected to significant levels of stress and strain.

INTRODUCTION

Poly-Si films are widely used in integrated circuits (ICs) and microelectromechanical systems (MEMS) devices [1]. In addition, the piezoresistivity of poly-Si is useful in sensor applications, such as pressure, stress, force, and torque sensors [2],[3]. Accurate knowledge of the piezoresistive coefficients is crucial for sensor design and performance prediction.

The resistivity of materials changes under strain variations, which is known as the piezoresistance effect. The geometrical piezoresistance effect describes the relative change of resistivity due to the change in the cross section and length of a body under mechanical stress. It is present in all materials. Semiconductors exhibit an additional, stronger effect, caused by the deformation of the charge carrier band structure [4]. In the case of monocrystalline silicon (c-Si), its three piezoresistive coefficients have been reported in the literature [5]. However, the theoretical prediction of the piezoresistive parameters of poly-Si is more delicate, since this material consists of small monocrystalline grains separated by grain boundaries [6]. In addition, the size and orientation of the grains depends strongly on deposition and annealing conditions [7], which leads to a wide range of gauge factors. One way of determining such parameters is to inject an electrical current through a microtensile specimen while loading it mechanically and simultaneously monitoring its relative resistance change.

Tensile testing is a fundamental, simple and well-established mechanical characterization method. One of its advantages is that uniaxially loaded specimens experience a uniform stress distribution, allowing a straightforward analysis of the load-displacement curves and extraction of mechanical parameters [8]. Due to the deviation of mechanical and piezoresistive material behavior from the macro to the micro scale, there is a need to characterize samples of similar dimensions as those in ICs

and MEMS structures. For these reasons, the previously developed wafer-scale microtensile testing method [9] is extended here in order to allow additional resistance measurements of poly-Si microtensile specimens.

In contrast with previous studies where just the linear response was characterized or nonlinear piezoresistive coefficients were acquired for strains only up to 0.1-0.2 % [6],[10]-[12], the data presented here are extracted for strains up to the fracture point at a strain of about 1%. This leads to more accurate and credible values of higher order piezoresistive parameters. The following two sections describe the design and fabrication of the test structures. The automated setup, capable of measuring simultaneously force and electrical resistance as a function of the displacement imposed to the specimens, and the extraction of mechanical and piezoresistive material parameters of poly-Si films annealed under different conditions are then reported in the susbsequent sections.

TEST STRUCTURE DESIGN

An individual test structure is schematically shown in Fig. 1. It consists of a fixed c-Si outer frame and a movable inner frame, connected by parallel springs defined by deep reactive ion etching (DRIE). In comparison with the previous process [9], the deep etch mask used here was improved by adding pronounced rounded corners that lead to a constant etch rate at the locations defining the mechanical springs. A 400-μm-long, 50-μm-wide and 1-μm-thick poly-Si specimen, bridging the gap between the movable inner and fixed outer c-Si frames, is uniaxially deformed when the inner part is displaced by an actuator inserted into its opening. Additional aluminum (Al) lines and 18 contact pads enable electrical measurements. Electrical signals are transmitted to the inner movable frame by metal lines patterned on top of the parallel springs.

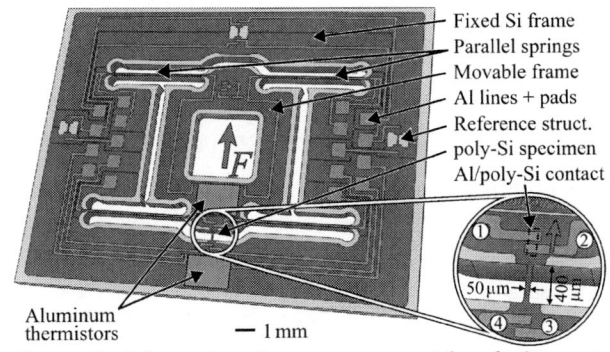

Figure 1: Schematics of test structure with poly-Si specimen bridging the gap between the movable and fixed frame. Upon applied force, the specimen deforms uniaxially. Its resistance is measured by injecting a fixed current between contacts ① and ③ and monitoring the voltage drop between ② and ④.

978-1-4244-2977-6/09 $25.00 © 2009 IEEE

Possible configurations of the electrical measurements include four-point resistance measurement of the microtensile specimen and an identical reference structure, the ability to configure a Wheatstone bridge configuration using the microtensile specimen together with three reference structures, and four-point resistance measurements of two Al thermistors. The thermistors allow the temperature at the two anchor points of the specimen to be monitored.

The four-point resistance measurement of the specimen for extraction of longitudinal piezoresistive coefficients is performed with currents below 1 mA, small enough to keep the resistive heating of the specimen at a negligible level. The measurement configuration is schematically shown in Fig. 1. Homogeneous flow of the complete injected current has to be achieved for accurate measurements. This is done by defining electrically insulated poly-Si areas that are contacted using the metal layer only at the designated contact areas through openings in the oxide.

Additional test structures are fabricated in different locations on the wafer in order to evaluate the fabrication process and its homogeneity. These include Kelvin structures for Al/poly-Si contact resistance measurements [13] with $10 \times 10 \ \mu m^2$ contact area, van der Pauw Greek cross structures for the measurement of the poly-Si sheet resistance [14] and calibration structures for the probe card.

FABRICATION

A total of 26 devices are fabricated per 4-inch silicon wafer. The fabrication process of test structures with piezoresistive microtensile specimens is summarized in Fig. 2: (a) a 1-μm-thick in-situ n-doped poly-Si film, deposited by low-pressure chemical vapor deposition (LPCVD) at 580°C using 59.6 sccm of silane (SiH_4) and 0.4 sccm of phosphine (PH_3) and annealed at 1050°C for three different times (1, 5, 10 h), is patterned on top of a previously deposited low-temperature oxide (LTO), and subsequently covered with a thin plasma-enhanced chemical vapor deposition (PECVD) oxide. After opening the electrical contact area in the oxide by reactive ion etching (RIE), 1-μm-thick Al lines are sputtered and patterned using wet etching. An annealing step performed at 450°C during 1 hour in an N_2 atmosphere ensures a low metal-poly contact resistance. (b) A previously thickened rear oxide is now opened by RIE and serves as an etch mask

for (c) a through-wafer deep reactive ion etch (DRIE), which stops on the LTO layer at the front of the silicon substrate. (d) The specimens are finally released by removing the LTO membranes using a highly concentrated solution of 73%-hydrofluoric acid (HF), providing high selectivity between the oxide etching with respect to the poly-Si and Al layers [15]. Optical and scanning electron microscope (SEM) graphs of fabricated devices are shown in Figs. 3 (a) and (b), respectively.

The measured unstressed resistance of the specimens, $R_{spec} = 61.60 \pm 3.26 \ \Omega$, is compatible with the poly-Si sheet resistance, $R_{sq} = 6.37 \pm 0.30 \ \Omega$, extracted from van der Pauw structures. This corresponds to a resistivity ρ of $6.4 \times 10^{-4} \ \Omega cm$ and an estimated doping concentration of $3 \times 10^{20} \ cm^{-3}$ [6]. From the measured Al/poly-Si contact resistivity, $\rho_c = (2.17 \pm 0.19) \times 10^{-4} \ \Omega cm^2$, obtained from Kelvin structures, a contact resistance of $R_c = 1.95 \pm 0.17 \ \Omega$ is estimated for the $90 \times 100 \ \mu m^2$ specimen contacts, far below the resistance of the specimen.

MEASUREMENT SETUP

The wafer-scale microtensile setup is shown in Fig. 4 (a), with a close-up of the actuation tip, probe card and the device under test in Fig. 4 (b). The wafer is mounted on a thermal chuck fixed on a motorized $xy\theta$ table. This allows the positioning and alignment of each of the 26 test structures with respect to the loading tip and probe card. After alignment, the actuation tip and probe card are lowered for actuation and electrical contacting purposes, respectively, and the automated force-displacement-resistance measurement is performed. In this way, the 26 test structures are consecutively measured without the need of realignment or user interaction. The tip is mounted on a force sensor with a resolution of 0.1 mN, connected to a PZT actuator operated in closed-loop displacement mode for sample actuation. Displacements are measured with a laser-deflection position sensor with a resolution of 10 nm. Specimen resistance measurement resolution is 0.05 Ω. The achieved stress σ and strain ε resolutions for typical specimen dimensions are 10 MPa and 0.0025 %, respectively.

RESULTS AND DISCUSSION

Load-displacement curves

An example of the load-displacement curves obtained from test structures with poly-Si specimens is shown in

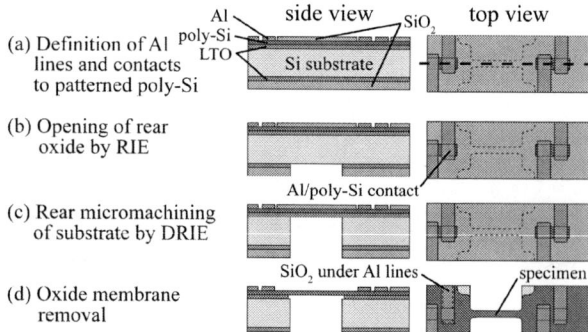

Figure 2: Fabrication of poly-Si specimens: (a) patterned 1-μm-thick poly-Si films are passivated with oxide (SiO$_2$, LTO) and contacted by Al; (b), (c) through-wafer DRIE; (d) LTO oxide removal.

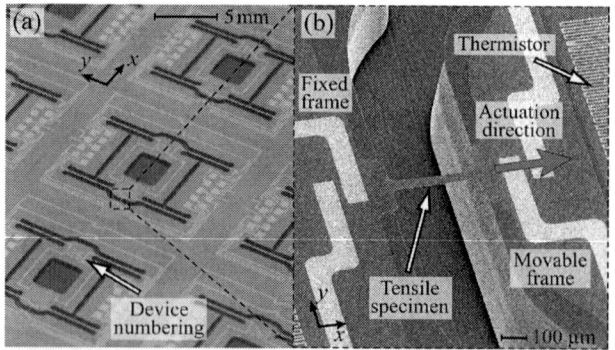

Figure 3: (a) Photograph of fabricated test structures processed on a 4-inch silicon wafer containing 26 devices, and (b) SEM graph emphasizing one Al-contacted poly-Si tensile and thermistor for temperature measurements.

978-1-4244-2977-6/09 $25.00 © 2009 IEEE

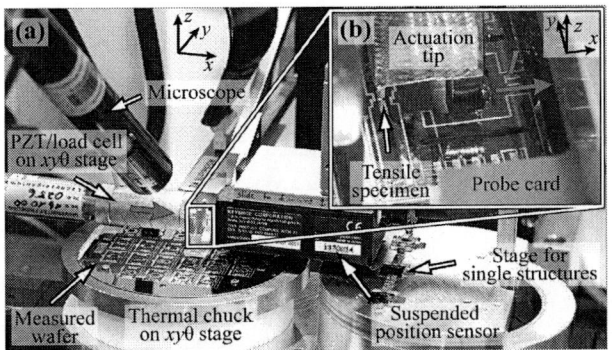

Figure 4: (a) Automated wafer-scale microtensile setup for sequential measurement of all test structures on a Si wafer. Force, displacement and resistance are monitored simultaneously. (b) Measured test structure with inserted actuation tip and touching probe card.

Fig. 5 (a). Upon loading and after the initial contact, the slope of the curve is the sum of the spring constants of the specimen, k_1, and of the four parallel springs, k_2: the force response increases due to deformation of the specimen and the springs. After failure of the specimen, the force drops abruptly, leaving only the response of the parallel springs. The resulting engineering stress-strain curve shown in Fig. 5 (b) is obtained by normalizing the difference of the two load-displacement curves (load and unload), corresponding to the test structure with and without microtensile specimen. With the specimen length L, width w, and thickness t, the engineering stress is $\sigma = F/wt$ and the engineering strain becomes $\varepsilon = d/L$, where F and d denote the applied force and displacement, respectively. The extraction of mechanical parameters such as Young's modulus E ($\sigma = E\varepsilon$) and ultimate tensile strength σ_u (maximum stress experienced by the specimen) from these curves is straightforward.

Load-resistance curves

Figure 5 (c) shows a measured specimen resistance as a function of the applied force, which becomes infinite at fracture. The relative resistivity change $\Delta\rho/\rho$, obtained after removal of the geometrical piezoresistive effect [16], is plotted as a function of σ in Fig. 5 (d), showing a decrease of 12.59 % over the entire stress range, which corresponds to a strain of about 0.92 %. The nonlinear piezoresistive response is modeled with the fit function [5]

$$\Delta\rho/\rho = \pi_{l,1}\sigma + \pi_{l,2}\sigma^2 \tag{1}$$

defining the linear (or first-order) and quadratic (or second-order) piezoresistive coefficients $\pi_{l,1}$, and $\pi_{l,2}$, respectively. As the quality of the fit suggests, coefficients up to second order model the observed piezoresistive reponse satisfactorily.

Elastic and piezoresistive constants

The wafer distribution of the mechanical parameters (Young's modulus and fracture strength), obtained for poly-Si films annealed for 1 h is shown in Fig. 6 (a). The average value of $E = 153.4\pm6.4$ GPa represented in the histogram of Fig. 6 (b) is well within the range of literature data [9],[17], as are the mean tensile strength $\mu = 1.41$ GPa and Weibull modulus $m = 15.3$, obtained from the Weibull fits displayed in Fig. 6 (c) [18],[19].

The extracted linear and second-order piezoresistive coefficients from one wafer, plotted as a function of device number, are shown in Figs. 7 (a) and (b), respectively. The average longitudinal piezoresistive coefficients are $\pi_{l,1} = -(1.36\pm0.04)\times10^{-10}$ Pa^{-1} and $\pi_{l,2} = (3.27\pm0.21)\times10^{-20}$ Pa^{-2}, as shown in Figs. 7 (c) and (d), respectively. Along with the mechanical parameters, the wafer distribution of the piezoresistive coefficients shows relatively low scattering.

The dependence of the mechanical and piezoresistive properties on annealing time is shown in Fig. 8. Both Young's modulus E and the strength μ slightly decrease by 11 % and 5.5 %, respectively, for longer times. Within the experimental uncertainty, no variation of the piezoresistive coefficients is noticed.

CONCLUSIONS

The experimental extraction of first and second-order piezoresistive coefficients of poly-Si thin films, along with mechanical properties, for strains up to fracture is demonstrated for the first time. In combination with the wafer-scale microtensile technique, a tool for extraction of process specific mechanical and piezoresistive properties, with statistically relevance, is established, further increasing the portfolio of reliable testing methods and allowing improved device design.

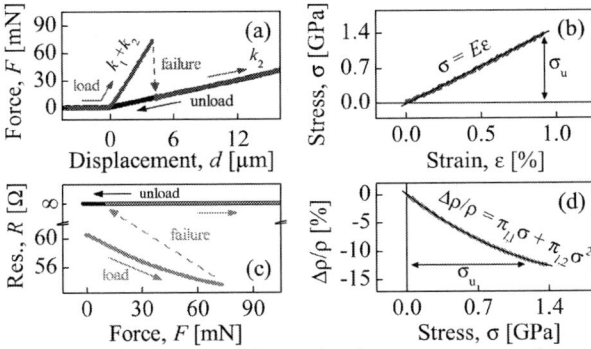

Figure 5: (a) Measured force-displacement curve of a test structure, and (b) resulting stress-strain curve. (c) Resistance of the poly-Si specimen measured as a function of the applied force and (d) resulting relative resistivity change with stress.

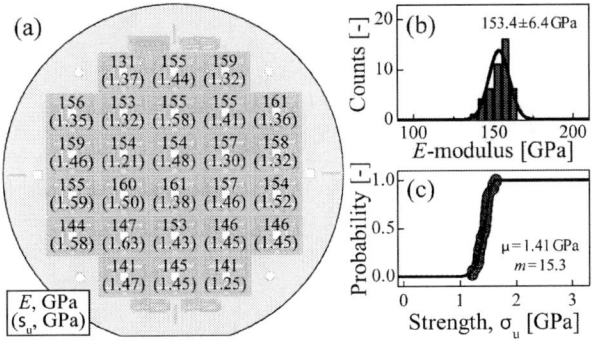

Figure 6: (a) Wafer distribution of the extracted Young's modulus E and strength σ_u for poly-Si films annealed at 1050°C for 1 h, (b) resulting histogram of E and (c) Weibull-plot of the σ_u, from which the mean fracture strength μ and Weibull modulus m are obtained.

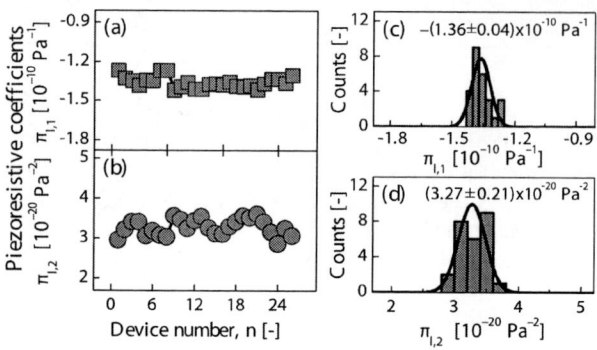

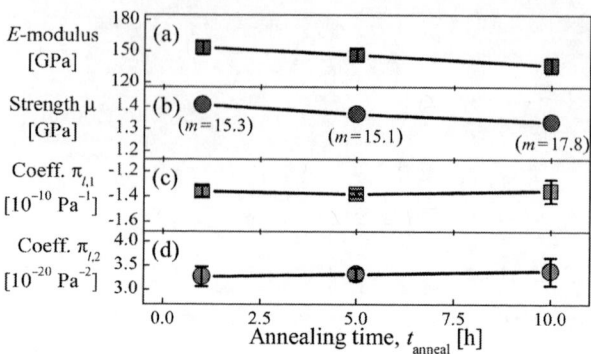

Figure 7: Extracted longitudinal piezoresistive coefficients of (a) first and (b) second order for the 26 devices of a wafer; (c) histograms of $\pi_{l,1}$; (d) histogram of $\pi_{l,2}$.

Figure 8: Dependence of the extracted mechanical and piezoresistive properties of poly-Si films on annealing time 1050°C: (a) Young's modulus, (b) mean strength, and (c) linear and (d) quadratic piezoresistive coefficients.

ACKNOWLEDGEMENTS

The authors gratefully acknowledge A. Baur, M. Reichel and F. Dieterle (IMTEK Cleanroom Service Center) for fruitful discussions concerning process development and professional support of device fabrication.

REFERENCES

[1] H. Baltes, O. Brand, G. K. Fedder, C. Hierold, J. G. Korvink, and O. Tabata, *CMOS-MEMS: Advanced Micro and Nanosystems*, Wiley-VCH, 2005.

[2] V. Mosser, J. Suski, and J. Goss, "Piezoresistive Pressure Sensors Based on Polycrystalline Silicon," *Sens. Actuators*, vol. A28, no. 2, pp. 113-132, 1991.

[3] M. Doelle, J. Held, P. Ruther, and O. Paul, "Simultaneous and Independent Measurement of Stress and Temperature Using a Single Field-Effect Transistor Structure," *IEEE J. Microelectromech. Syst.*, vol. 16, no. 5, pp. 1232-1242, 2007.

[4] S. M. Sze, *Semiconductor Sensors*, Wiley-Interscience, 1994.

[5] Y. Kanda, "A Graphical Representation of the Piezoresistance Coefficients in Silicon," *IEEE Trans. Electr. Dev.*, vol. ED-29, no. 1, pp. 64-70, 1982.

[6] P. J. French, "Polysilicon: a Versatile Material for Microsystems," *Sens. Actuators*, vol. A99, no. 1-2, pp. 3-12, 2002.

[7] T. I. Kamins, M. M. Mandurah, and K. C. Saraswat, "Structure and Stability of Low Pressure Chemically Vapor-Deposited Silicon Films," *J. Electrochem. Soc.*, vol. 125, no. 6, pp. 927-932, 1978.

[8] See, for example, J. R. Davis, *Tensile Testing*, Ohio: ASM International, 2004.

[9] J. Gaspar, M. Schmidt, J. Held, and O. Paul, "High-Throughput Wafer-Scale Microtensile Testing of Thin Films," in *Tech. Dig. 21st IEEE MEMS Conf.*, Tucson, AZ, Jan.13-17, 2008, pp. 439-442.

[10] E. Luder, "Polycrystalline Silicon-Based Sensors," *Sens. Actuators*, vol. 10, pp. 9-23, 1986.

[11] J. Y. Seto, "Piezoresistive Properties of Polycrystalline Silicon," *J. Appl. Phys.*, vol. 47, no. 11, pp. 4780-4783, 1976.

[12] V. A. Gridchin, V. M. Lubimsby, and M. P. Sarina, "Piezoresistive Properties of Polysilicon Films," *Sens. Actuators*, vol. A49, no. 1-2, pp. 67-72, 1995.

[13] J. Santander, M. Lozano, A. Collado, M. Ullán, and E. Cabruja, "Accurate Contact Resistivity Extraction on Kelvin Structures with Upper and Lower Resistive Layers," *IEEE Trans. Electr. Dev.*, vol. 47, no. 7, pp. 1431-1439, 2000.

[14] M. G. Buehler and W. R. Thurber, "An Experimental Study of Various Cross Sheet Resistor Test Structures," *J. Electrochem. Soc.*, vol. 125, no. 4, pp. 645–650, 1978.

[15] P. T. J. Gennissen, and P. J. French, "Sacrificial Oxide Etching Compatible with Aluminum Metallization," in *Tech. Dig. Tranducers'97*, Chicago, June 16-19, 1997, pp. 225-228.

[16] K. Matsuda, K. Suzuki, and Y. Kanda, "Nonlinear Piezoresistance Effects in Silicon," *J. Appl. Phys*, vol. 73, no. 4, pp. 1838-1847, 1993.

[17] M. A. Haque, and M. T. Saif, "In-situ tensile testing of nano-scale specimens in SEM and TEM," *Exp. Mech.*, vol. 42, no. 1, pp. 123-128, 2002.

[18] O. M. Jadaan, N. N. Nemeth, J. Bagdahn, and W. N. Sharpe, "Probabilistic Weibull behavior and mechanical properties of MEMS brittle materials," *J. Mat. Sci.*, vol. 38, no. 20, pp. 4087-4113, 2003.

[19] See, for example, W. Weibull, "A statistical function of wide applicability", *J. Appl. Mech.*, vol. 18, pp. 293-297, 1951.

DESIGN OF POWER-OPTIMIZED THERMAL CANTILEVERS FOR SCANNING PROBE TOPOGRAPHY SENSING

H. Rothuizen, M. Despont, U. Drechsler, C. Hagleitner, A. Sebastian, and D. Wiesmann

IBM Research, Zurich Research Laboratory, 8803 Rüschlikon, Switzerland

ABSTRACT

This paper reports on design optimization of cantilever-based thermal topography sensors using finite-element simulations, their fabrication and characterization. Measurements demonstrate vertical distance sensitivities $\Delta R/R$ of 2×10^{-4} nm^{-1}, yielding sub-angstrom resolution at bandwidths of several tens of kHz and at powers on the order of 1 mW.

INTRODUCTION

Microfabricated silicon cantilevers with integrated heating elements are powerful tools for manipulation and interrogation at the nanoscale. They can be used for highly localized heating of surfaces in probe-based data storage [1-3] and nanoscale manufacturing [4] as well as for thermo-resistive topography sensing [1-3,5].

The performance of cantilever-based silicon thermal topography sensors is characterized by sensitivity magnitude and bandwidth as well as power consumption. The dependence of these performance attributes on geometry and materials parameters has been studied by numerical simulations [3,6] and analytically [7], without however attempting to optimize resolution and bandwidth under a constrained power budget, as would be required for applications using large arrays of cantilevers.

Experimentally, a topography sensitivity $\Delta R/R$ of $\sim 1 \times 10^{-4}$ nm^{-1} has been demonstrated [5] in the useable regime below thermal runaway [8] and requiring ~ 4 mW of dissipated power. Although there are no reports on the measured sensing bandwidth, the thermal dynamics of silicon microheaters integrated in atomic force microscopy (AFM) cantilevers has been investigated previously. This earlier work reports two time constants dominating

the frequency response [1,9]. The faster one (~ 1 μs) is attributed to exclusive heating of the slightly constricted heater region; the slower one (~ 100 μs) corresponds to heating/cooling of the entire cantilever. While it is possible to take advantage of the fast time constant in pulsed or periodic heating configurations, the incomplete isolation of the heater region from the cantilever body has two significant drawbacks when using the heater as a topography sensor. First, the multi-pole transfer function limits its use to characteristic times below the slower thermal time constant. Second, the heat flux to the larger-volume cantilever body decreases the achievable sensitivity and increases power consumption. In [10,11], we reported on a frequency response with a single thermal time constant below 10 μs, achieved through careful isolation of the heater region.

In this paper, we report on a further design optimization. Through the introduction of reinforcement straps (Fig. 1), which allow the mechanical stiffness of the device to be maintained, we have been able to minimize the cross-section and maximize the length of the beams holding and isolating the heater region. Thus, we optimize the sensitivity-power relationship and limit the thermal mass that governs the sensing bandwidth to the volume of the heater and the isolation beams.

Second, we have comprehensively studied the tradeoff between increasing bandwidth and decreasing resolution when reducing said volume. The decrease in resolution is induced by the volume-dependent 1/f electrical noise. Finite element (FE) modeling, in combination with an experimentally verified model for the electrical noise, was applied to identify an optimal design based on the cantilever design template shown in Fig. 1.

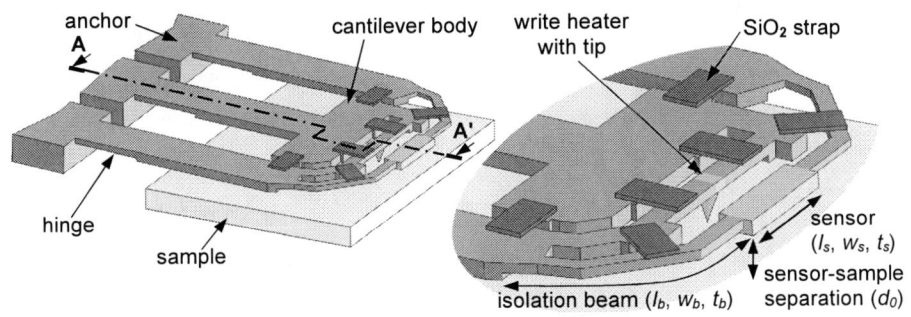

Isolation Beam:
Length: $l_b \sim 20$ μm
Width: 1.5 μm $< w_b < 2.0$ μm
Thickness: $t_b \sim 0.15$ μm

Sensor:
Length: 2.0 μm $< l_s < 10$ μm
Width: 1.5 μm $< w_s < 3.0$ μm
Thickness: 0.15 μm $< t_s < 0.5$ μm

Sensor–Sample Separation:
0.125 μm $< d_0 < 1.0$ μm

Figure 1: Cantilever schematic comprising geometric features subject to optimization as discussed in text (isolation beams, sensor resistor), and list of typical dimensions. The distance sensor is operated by applying a voltage across the two outermost cantilever legs, resulting in joule heating in the low-doped sensor region; distance is monitored via the current. While the figure also displays features related to other functionalities (e.g. a write heater), these are not discussed further in this paper, which focuses on distance sensing only.

978-1-4244-2977-6/09 $25.00 © 2009 IEEE

FABRICATION

The fabrication process developed to implement the power-optimized cantilever design outlined in Fig. 1 is adapted from a previous generation of thermomechanical levers [12] and is compatible with heterogeneous integration onto CMOS circuitry [13].

To fabricate the stiffening SiO_2 straps, we utilize the buried oxide (BOX) of the starting silicon-on-insulator (SOI) wafer material and employ a technique that allows patterning of the back-side of the cantilever. This technique consists of a temporary bonding of the cantilever wafer to a glass carrier, removal of the SOI handle layer, enabling access to the back-side of structured features, and a second, permanent transfer to a wafer from which a chip handle is bulk micromachined. The overall process is outlined in Fig. 2. The starting substrate is a 4″ SOI wafer with a 2-μm-thick, epitaxially grown silicon membrane and a 0.4-μm-thick BOX. The membrane is n-doped with phosphorous at a concentration of 2.0×10^{17} at/cm^3, which is the desired doping level for the sensor. The first step consists of thermally growing a 500-nm layer of oxide, which is used as mask material when etching thick anchor legs and the tip. CHF_3/O_2-based reactive ion etching (RIE) is used to open this oxide, followed by an isotropic SF_6/Ar RIE process to etch an hourglass-shaped precursor of the tip into the membrane (Fig. 2a). After removing the oxide mask, an oxidation sharpening technique is used to finalize the tip shape (Fig. 2b), and the oxide is removed using BHF wet etching. This technique reliably yields tips with apex radii < 5 nm.

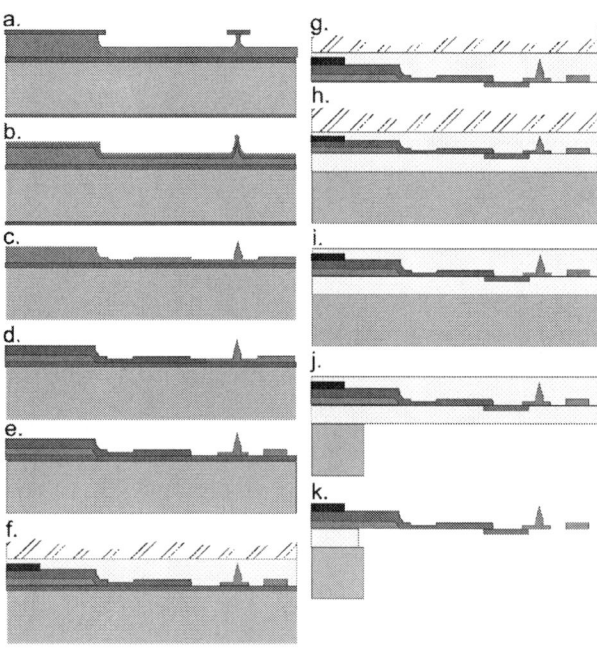

Figure 2: Process flow. The cross section shown corresponds to cut AA′ in Fig. 1

Then, regions of the remaining Si membrane corresponding to the hinges and isolation beams are thinned using conventional lithography and timed SF_6-based RIE (Fig. 2c). The next fabrication step confers high electrical conductivity to regions of the levers that function as electrical leads (i.e., all but the sensor/heaters) by a sequence of capping (30-nm thermal oxide), ion implan-

tation (90-keV, 1×10^{16} ions/cm^2, phosphorous), and activation of the dopants (furnace anneal at 1000°C, 30 min). After removal of the capping oxide (Fig. 2d), the cantilever shape is transferred into the remaining Si membrane using RIE and the buried SOI oxide acts as etch stop (Fig. 2e). The front-side processing is finalized by the deposition and lift-off of 500-nm-thick aluminum contact pads.

Once the cantilevers have been fabricated, a 5-μm-thick polyimide layer is spun and cured, followed by lamination of the wafer to a glass substrate with a PTFE (Teflon) foil in between (Fig. 2f). Lamination is typically done under isostatic pressure at 6 bar and 380°C for 15 min. The SOI handle wafer is then removed by grinding down most of its thickness and switching to SF_6-based plasma etching to remove the remaining part using the BOX as an etch stop. The cantilever back-side processing consists of defining the oxide straps and, in some design variants, a passivation layer overlapping the sensor, using CHF_3/O_2-based RIE to remove a complementary pattern in the remaining BOX (Fig. 2g, passivation not shown). Next, the new handle wafer is coated with an adhesive polyimide and bonding is performed using the same conditions as for bonding to the glass wafer (Fig. 2h). Finally, the chip body itself is structured using deep RIE and the cantilever is released from the polyimide using an O_2 plasma etch (Fig. 2k). An example of a finished cantilever is shown in Fig. 3.

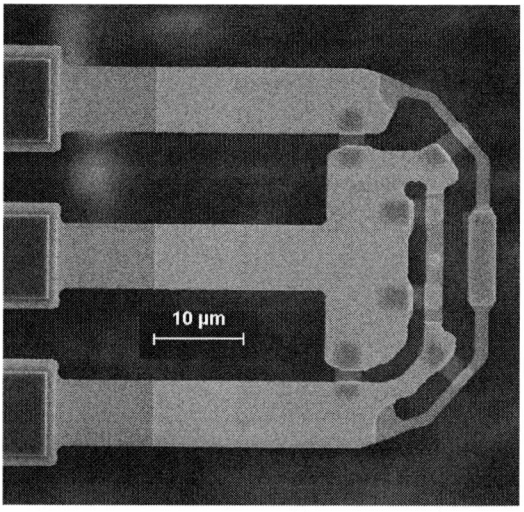

Figure 3. SEM of a fabricated cantilever. The low-doped front sensor "paddle" has a volume of 8.2 μm × 3.0 μm × 0.46 μm, and the slender suspension/isolation beams have a length of 23 μm and a cross section of 1.5 μm × 0.18 μm to minimize thermal conductance. These dimensions constitute a standard set throughout this paper unless stated otherwise. The SiO_2 straps micro-machined from the starting SOI material's BOX are 0.12 μm thick.

FE MODEL

Working under the assumption that certain lateral heat conduction paths (e.g. through the SiO_2 straps) can be neglected, we use a geometrically simplified FE model that includes only the sensor portion of the cantilever and its immediately relevant thermal environment, i.e., the isolation beams, a small portion of the lever body that emulates a suitable heat sink, and a volume of air sur-

978-1-4244-2977-6/09 $25.00 © 2009 IEEE

rounding the lever and sample. The model accounts for the coupling between thermal and electrical domains, and for the dependence of thermal and electrical conductivities on temperature, doping and thickness (for Si regions) via precalculated tabular data based on semi-empirical models described in [7,14]. Because of their sensitivity to the actual materials present, a small number of scattering coefficients used within these material models (electron–phonon scattering exponent, phonon-boundary scattering strength, and solid-gas interfacial thermal resistance) are the object of a one-time fit to obtain agreement with experimental DC resistance and sensitivity values. Figure 4 shows a typical solution temperature field.

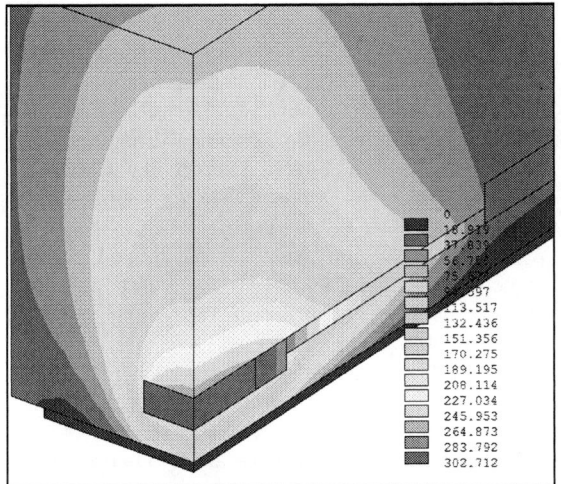

Figure 4: Example FE-calculated temperature field in a ¼ symmetry model of the distance sensor. The temperature scale is in °C relative to the heat sink.

Figures 5 and 6 show the calculated dependence of low-frequency sensitivity ($\Delta R/R$) on power for varying sensor-to-sample distance and isolation beam length, respectively. The experimental data were obtained by scanning a watermark sample with 30-nm step height; its overlay with the corresponding calculation is due to the aforementioned one-time fit of scattering coefficients.

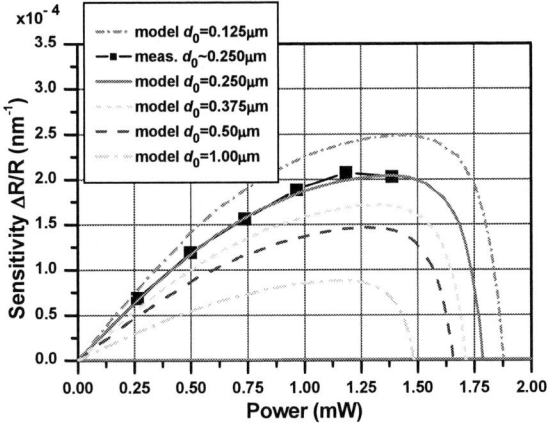

Figure 5: Calculated sensitivity vs. power, varying the sensor-to-sample distance d_0. Other dimensions as in the device of Fig. 3.

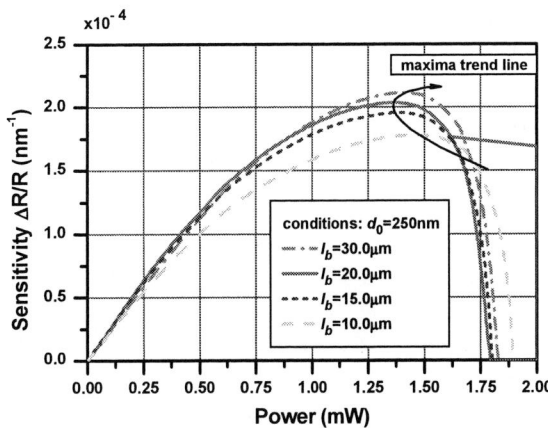

Figure 6: Calculated sensitivity vs. power, varying the length of the isolation beams l_b

In Fig. 6, the sensitivity at realistic operating power levels (i.e., 0.25–1 mW) is seen to increase with the isolation beam length, and then saturate when l_b exceeds approx. 20 μm. The trajectory of the sensitivity maximum also reverses around this value. From this, one can predict that resistive losses in the beams begin to offset the benefits of increased thermal isolation at this length, which is therefore optimal for the doping level (2×10^{20} cm^{-3}) and beam cross-section considered.

RESOLUTION

The key characteristic for applications such as metrology and data storage is the frequency-dependent resolution, which requires optimization not only for the sensitivity magnitude (as described above) but also for the bandwidth and noise figures, both of which depend critically on the sensor volume.

To predict and compare the resolution for a range of sensor dimensions, we first simulate the dynamic sensing response of the system by applying a small-signal sine modulation to d_0 in the FE model described. Figure 7 shows the sensing transfer function $\mathbf{T}_{\Delta R/R d, d0}$ relating the relative resistance fluctuations $\Delta R/R$ to the sensor-sample distance fluctuations d around the nominal distance d_0, for the standard geometric dimensions (see Fig. 3).

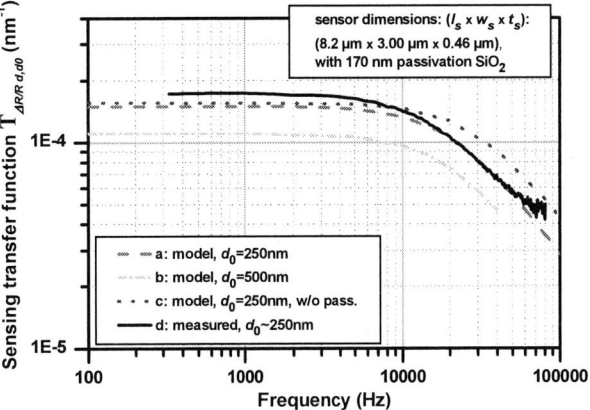

Figure 7: Sensing transfer function. Curves a, b: for two values of d_0; c: without the mass of the passivation oxide, and d: as measured.

To verify the simulations, we measured the thermal sensing response of the corresponding cantilever. To this effect we capacitively excited a vertical sensor motion, which we monitored optically while at the same time measuring the thermoresistive sensing response. Figure 7 shows good agreement between simulation and experiment.

The two predominant noise sources in thermoresistive silicon cantilever sensors are Johnson thermal noise and $1/f$ noise [11,14,]. The noise power spectral density S_{nR} including these two components is

$$ S_{nR} = \frac{4k_B T_0 R_0^2}{P_0} + \frac{\alpha R_0^2}{f\,N} \,, \qquad (1) $$

where k_b is the Boltzmann constant, T_0 and R_0 are the temperature and electrical resistance at the DC operating point (d_0, P_0) selected, respectively. The Hooge factor is assumed to be $\alpha = 1.0 \times 10^{-5}$, based on previous experiments [14], and N is the number of charge carriers in the read resistor. Convolving the simulation results with the semi-empirical model for the noise spectral density leads to the predicted achievable resolution shown in Fig. 8. The three different sets of sensor dimensions (and volumes) clearly show the trade-off between resolution and bandwidth.

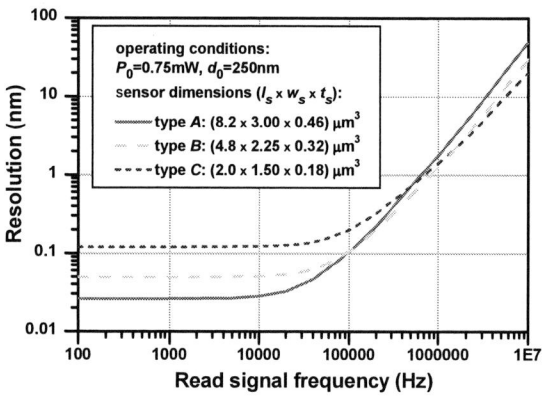

Figure 8: *Predicted resolution for three sets of the sensor geometry*

CONCLUSION

In summary, we optimized geometry considering all characteristics of sensitivity and bandwidth, noise, and power consumption, yielding designs with sensitivities of 2×10^{-4} nm^{-1}, comparable to the best published [5], but with more than one order of magnitude higher bandwidths and at significantly lower power.

ACKNOWLEDGEMENTS

We thank members of the IBM Probe Storage Group for their support and fruitful discussions. This work is supported in part by the European Research Council under the project ProTeM. The information here represents the authors' views and does not constitute a liability for IBM Corporation or its subsidiaries.

REFERENCES

[1] B.W. Chui *et al.*, "Low-stiffness Silicon Cantilevers with Integrated Heaters and Piezoresistive Sensors for High-Density AFM Thermomechanical Data Storage," *J. Microelectromech. Syst.*, vol. 7(1), pp. 69-78, 1998.

[2] P. Vettiger *et al.*, "The 'Millipede'— Nanotechnology Entering Data Storage," *IEEE Trans. Nanotechnol.*, vol. 1(1), pp. 39-55, Mar. 2002.

[3] W.P. King *et al.*, "Design of Atomic Force Microscope Cantilevers for Combined Thermomechanical Writing and Thermal Reading in Array Operation", *J. Microelectromech. Syst.*, vol. 11(6), pp. 765-774, 2002.

[4] P.E. Sheehan *et al.*, "Nanoscale Deposition of Solid Inks via Thermal Dip Pen Nanolithography," *Appl. Phys. Lett.*, vol. 85(9), pp. 1589-1591, 2004.

[5] K.J. Kim *et al.*, "Nanotopographical Imaging Using a Heated Atomic Force Microscope Cantilever Probe", *Sensors & Actuators A*, vol. 136, pp. 95-103, 2007.

[6] W.P. King, "Design Analysis of Heated Atomic Force Microscope Cantilevers for Nanotopography Measurements", *J. Micromech. Microeng.*, vol. 15, pp. 2441-2448, 2005.

[7] U. Dürig, "Fundamentals of Micromechanical Thermoelectric Sensors", *J. Appl. Phys.*, vol. 98, ref. 044906, 2005.

[8] B.W. Chui *et al.*, "Intrinsic-carrier Thermal Runaway in Silicon Microcantilevers", *Microscale Thermophys. Eng.*, vol. 3, pp. 217-228, 1999.

[9] J. Lee *et al.*, "Electrical, Thermal, and Mechanical Characterization of Silicon Microcantilever Heaters", *J. Microelectromech. Syst.*, vol. 15(6), pp. 1644-1655, 2006.

[10] D. Wiesmann *et al.*, "Dynamics of Silicon Microheaters: Modeling and Experimental Verification", *Proc. IEEE MEMS Conference*, Istanbul, Turkey, January 2006, pp. 182-185/

[11] A. Sebastian *et al.*, "Modeling and Experimental Identification of Silicon Micro-heater Dynamics: A Systems Approach", *J. Microelectromech. Syst.*, vol. 17(4), pp. 911-920, 2008.

[12] M. Despont *et al.* "VLSI-NEMS Chip for Parallel AFM Data Storage", *Sensors & Actuators A*, vol. 80, pp. 100-107, 2000.

[13] M. Despont *et al.*, "Wafer-scale Microdevice Transfer/Interconnect: Its Application in an AFM-based Data Storage System", *J. Microelectromech. Syst.*, vol. 13(6), pp. 895-901, 2004.

[14] C. Hagleitner *et al.*, "Modeling, Design, and Verification for the Analog Front-End of a MEMS-Based Parallel Scanning-Probe Storage System", *IEEE J. Solid State Circuits*, vol. 42, pp. 1779-1789, 2007.

978-1-4244-2977-6/09 $25.00 © 2009 IEEE

MECHANICAL CHARACTERIZATION OF BIOMOLECULES IN LIQUID USING SILICON TWEEZERS WITH SUBNANONEWTON RESOLUTION

C. Yamahata[1,2], E. Sarajlic[1], L. Jalabert[3], M. Kumemura[1], D. Collard[3] and H. Fujita[1]

[1]CIRMM, Institute of Industrial Science, The University of Tokyo, JAPAN
[2]Present address: Microsystems Laboratory, EPFL, SWITZERLAND
[3]LIMMS/CNRS-IIS, Institute of Industrial Science, The University of Tokyo, JAPAN

ABSTRACT

Molecular biophysicists seek to understand how biological systems work through mechanical or electrical characterizations performed at the molecular scale. From this perspective, we have devised a silicon-based micromechanical tool for stress-strain measurements of molecular fibers and demonstrated micromanipulation and biomechanical characterization of DNA bundles in a liquid solution. By combining this instrument with a microscopic displacement measurement technique based on Fourier transform image processing, we could achieve a force resolution of 25 pN – a level which is within the single-molecule sensing range – and validate a novel approach for stress-strain measurements.

INTRODUCTION

A survey of the state-of-the-art in cell and molecular force probing shows that Atomic Force Microscopes (AFM), Optical Tweezers (OT) and Magnetic Tweezers (MT) are the most commonly used biomechanical instruments [1]. In the AFM approach, micromechanical probes with sharp tips are employed to perform direct "contact" probing [2]. AFMs are used either for stretching assays on biomolecules or for indentation on cells. On the contrary, OT and MT are "non-contact" techniques in which microscopic-sized particles – including cells and bacteria – can be non-intrusively handled by means of a focused light or a magnetic field, respectively. The sensitivities of the latter instruments are generally higher than those of AFMs. In return, experiments with OT and MT are also more time-consuming.

In our laboratory, we have been working on a distinct method based on micromachined silicon tweezers that is complementary to the AFM stress-strain measurement method. We have recently demonstrated the dielectrophoretic (DEP) trapping of DNA bundles using silicon nanotweezers [3] and developed a tool for the micromanipulation of filamentary biomolecules [4]. With such a tool, we could stretch molecules and simultaneously measure their electrical properties [5].

Here, we present a new design that, in addition to the previous capabilities, enables molecular stress-strain measurements to be performed in liquids. Our system combines: (i) DEP trapping capability, (ii) strain actuation and sensing, and (iii) enables straightforward force measurements. The novelty of our approach is in the force measurement method which is based on Fast Fourier Transform (FFT) image analysis. In this paper, we will mainly focus on the latter point.

WORKING PRINCIPLE

Design

The design of the tweezers is a modified version of a previous prototype whose working principle and microfabrication details can be found in earlier publications [4, 5]. Figure 1 shows a 3-D schematic of the new device. It consists of two sharp silicon tips that act as electrodes for DEP trapping (sinusoidal voltage V_{DEP}). One tip acts as a *strain probe*, while the other one is the force *sensing probe*. The strain probe is a rigid probe that can be moved in both ways in the x-direction by means of a series of electrostatic comb-drive actuators (actuation

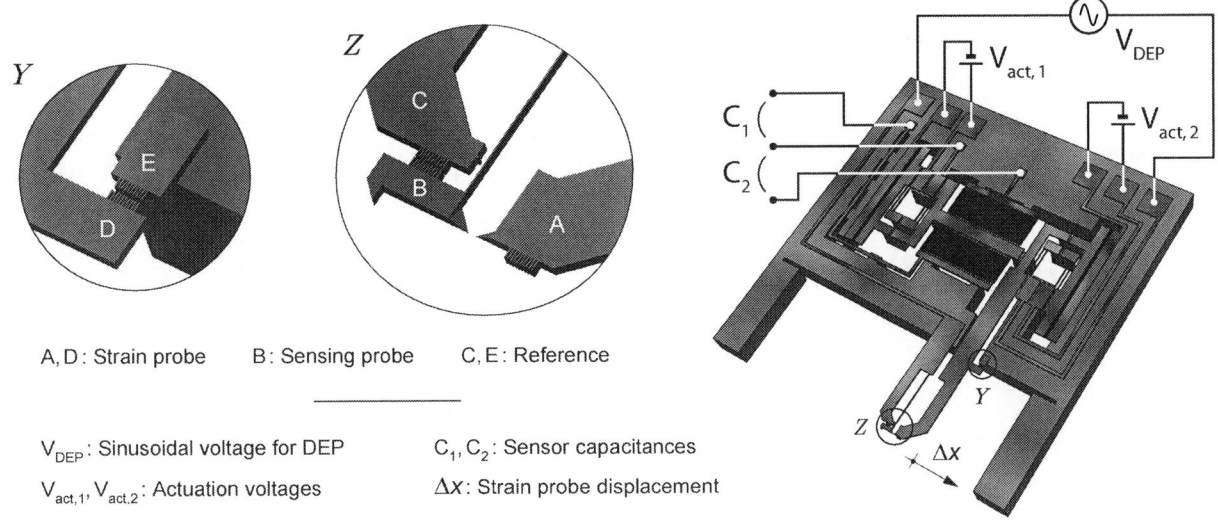

A, D : Strain probe   B : Sensing probe   C, E : Reference

V_{DEP} : Sinusoidal voltage for DEP C_1, C_2 : Sensor capacitances

$V_{act,1}, V_{act,2}$: Actuation voltages Δx : Strain probe displacement

Figure 1: 3-D Schematic of the silicon nanotweezers with close-up views on the patterns used for Fourier transform image processing. The strain probe (A, D) is electrostatically actuated in the x-direction (+/−). Its displacement is measured by differential capacitive sensing and calibration is performed by FFT image processing. The displacement of the sensing probe (B) is measured by FFT method.

978-1-4244-2977-6/09 $25.00 © 2009 IEEE   607

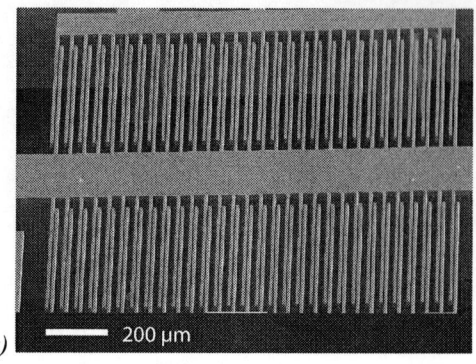 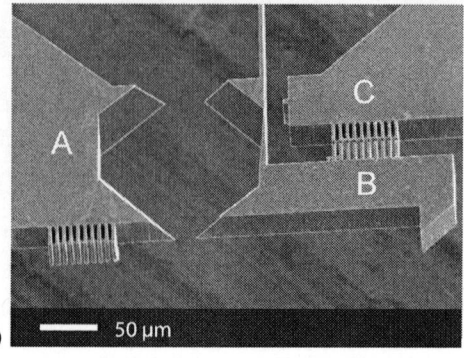

*Figure 2: SEM micrograph showing close-up views of **(a)** the differential capacitive sensor and **(b)** the sharp tips. The gap between the sharp tips is 16 µm. The repeating patterns used for FFT measurements have a spatial period t = 6 µm.*

voltages $V_{act,1}$, $V_{act,2}$). Its displacement, Δx, is measured with a differential capacitor ($\Delta C = C_1 - C_2$). The force sensing probe is an AFM-like compliant beam with integrated periodic patterns used for Fourier Transform image analysis. Figure 2 shows SEM micrographs of the device fabricated out of a Silicon-On-Insulator (SOI) wafer. Figure 2a is a close-up view of the differential capacitor. Figure 2b is a close-up view of the sharp probes (the letters refer to Figure 1).

Fourier Transform image analysis

Intelligent vision systems use image processing techniques to extract information on the orientation, position, and displacement of moving targets observed with a static camera. Sandoz *et al.* have recently applied phase measurement techniques to perform in-plane orientation and position measurements on microstructured patterns [6, 7]. Microelectromechanical systems appear to be perfect candidates, as demonstrated in References [8, 9]. Periodic patterns required for these measurements already exist (e.g. comb-drive electrostatic actuators) or can be introduced at no cost on MEMS devices.

Here, we have applied such image processing technique to one-dimensional translation measurements in order to perform force measurements. As depicted in Figure 3, the displacement of the force sensing probe, $\Delta x'$, is measured as the phase difference in FFT analysis of the periodic patterns B and C drawn on the sensing tip and on

the reference structure, respectively. Identical patterns (D and E, close-up view Y) were also drawn on the strain probe. In this way, we could compare the measurements obtained using FFT method with those obtained from the differential capacitive sensor.

Using calibration techniques commonly used to estimate AFM cantilever spring constants [10, 11], the force could be directly evaluated from the phase measurements.

EXPERIMENTAL SETUP

Figures 4a and 4b are 3-D views showing the tweezers introduced into a microfluidic device and observed under a microscope. The experimental setup shown in Figure 4c consists of a Keyence digital microscope VHX-500 placed on an anti-vibration stage, a lock-in amplifier electronic circuit, and a LabVIEW® interface that controls the tweezers and communicates with the lock-in amplifier. AVI videos were recorded with the microscope (lens VH-Z500, magnification ×5000) and

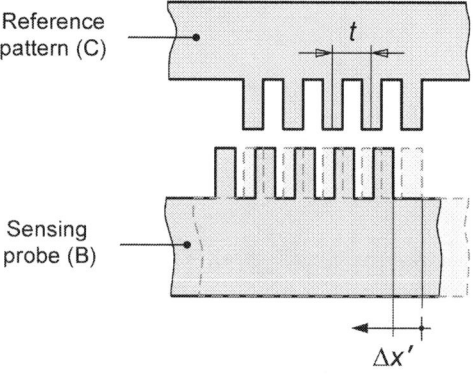

Figure 3: Principle of the displacement measurement method based on Fourier transform image processing. The displacement of the sensing probe (B) is compared with the reference (C) by FFT analysis of the phase shift.

Table 1: Main parameters of the experimental setup.

SOI wafer	(100)-oriented Si layer: 25 ± 1 µm; buried oxide layer: 1000 ± 50 nm; base wafer: 300 ± 25 µm; Young's modulus: $E = 160$ GPa (Si)
Differential capacitive sensor	number of comb units: $N = 30$; plates length: $L = 450$ µm; height: $w = 27$ µm; initial gap: $d_0 = 5$ µm; distance between comb units: $d_1 = 20$ µm
Keyence digital microscope	model VHX-500; VH-Z500 lens (× 5000 magnification); monitoring range: 46 µm × 61 µm; AVI: 600 × 800 pixels @ 15 frames/sec. (actual image size: 480 × 800 pixels)
AFM probe	'Type A' – length: $l = 850$ µm; width: $w = 25$ µm; thickness: $b = 5$ µm; stiffness $K_A = 200$ pN/nm; 'Type B' – length: $l = 1$ mm; width: $w = 27$ µm; thickness: $b = 3.5$ µm; stiffness $K_B = 46$ pN/nm

978-1-4244-2977-6/09 $25.00 © 2009 IEEE

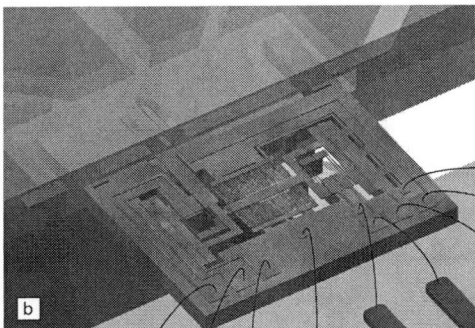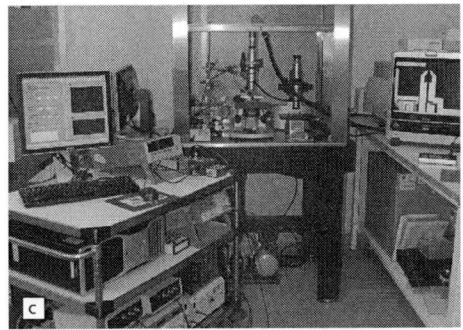

Figure 4: **(a, b)** 3-D view of the experimental setup used for biomechanical assays in liquids (we have represented an inverted microscope for ease of understanding). **(c)** Photograph of the experimental setup comprising a Keyence digital microscope VHX-500, an anti-vibration stage and a LabVIEW® 8 interface for the control of the tweezers and communication with the lock-in amplifier (NF, model LI 5640).

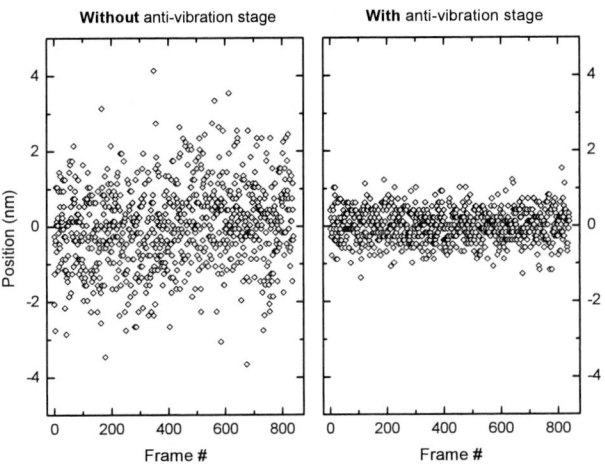

Figure 5: Observation of the effect of the anti-vibration stage on noise reduction. The data were extracted from phase measurements on patterns B and C recorded from 'Type A' prototype. The standard deviation could be reduced by a factor 2.

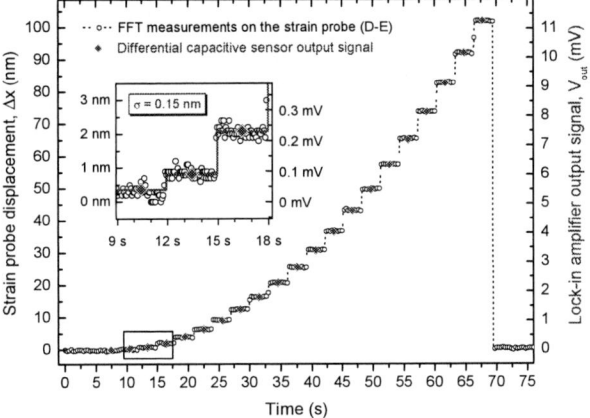

Figure 6: Electrostatic actuation of the strain probe ($V_{act,1} = 0\ V;\ 0.5\ V;\ 1\ V;\ ...;\ 10\ V$) and measurement of its displacement by Fourier transform image processing (patterns D, E in Figure 1) and differential capacitive sensing methods. The standard deviation achieved with the FFT method is $\sigma = 0.15$ nm, which enables a resolution of 0.5 nm.

analyzed in Matlab® (v.2007b, Image Processing Toolbox). Table 1 summarizes the main parameters used for this computation and for other calculations.

RESULTS

Effect of the anti-vibration stage

Our first experiment consisted of observing the effect of the anti-vibration stage on measurements errors. This was done by recording a video of the patterns B and C. Typical measurements are shown in Figure 5. From the Gaussian distribution of these data, we could demonstrate that the standard deviation, σ, was reduced by a factor 2 due to vibration reduction.

Device characterization and calibration

In Figure 6, we have actuated the strain probe (actuation voltage $V_{act,1}$) and recorded the movement of the patterns integrated on this probe relative to the fixed reference (patterns D and E, respectively). The graph compares the data obtained by FFT analysis with the output signal of the differential capacitive sensor obtained with the lock-in amplifier circuit [4]. The standard deviation for the FFT measurements was 0.15 nm. This

corresponds to an achievable resolution of 0.5 nm ($\sim 3\sigma$). In addition to the verification of the method, these results show that FFT analysis is also a straightforward method to calibrate the capacitive sensor. FFT measurements were also performed on the sensing probe and showed comparable standard deviation levels. Given the achievable spatial resolution and the stiffness of the most sensitive probe fabricated ($K_B = 46$ pN/nm, 'Type B' prototype), we could achieve a force resolution of about 25 pN.

Biomechanical experiments

Using the experimental setup shown in Figure 4, we have performed biomechanical experiments on DNA filaments. In this experiment, the tweezers arms were immersed in a diluted λ-DNA solution (Takara Bio Inc.) that was introduced in the slit formed between glass microslides. After dielectrophoresis ($V_{DEP} = 40\ V_{pk\text{-}pk}$ @ 1 MHz), we applied a triangular wave to the comb-drive ($V_{act,2}$) and recorded a video of the patterns B and C in order to determine the deformation (bending) of the

978-1-4244-2977-6/09 $25.00 © 2009 IEEE

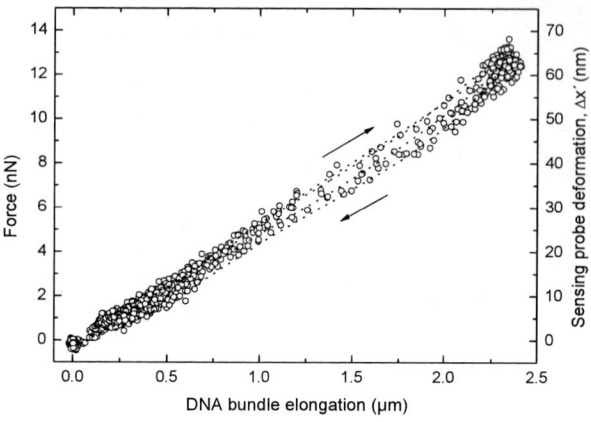

Figure 7: Stress-strain assay on a DNA bundle in solution. In this experiment, we have used the 'Type A' prototype. From the stiffness of a single DNA strand reported in the literature [1], we can estimate that this bundle consists of approximately 200 DNA strands.

strain probe. Figure 7 shows the force working on the DNA bundle versus its elongation. The stiffness of the DNA bundle can be determined from the slope of the curve. By comparing this stiffness with the initial stiffness of a single DNA strand reported in the literature ($\sim 3 \times 10^{-5}$ N/m) [1], we can estimate that this bundle consisted of about 200 DNA strands.

CONCLUSION

We have proposed a new approach for force measurements at the molecular scale and demonstrated successful mechanical measurements on DNA molecules in solution. Using Fourier transform image analysis, we achieved sub-pixel resolution (better than 7/1000 pixel) and a subnanonewton force resolution of 25 pN suitable for molecular biomechanical experiments.

ACKNOWLEDGEMENTS

This work was supported by the Japan Society for the Promotion of Science (JSPS, fellowship No. P06373) and the Japan Science and Technology Corporation (JST).

The photolithography masks were fabricated with the 8-in EB writer F5112+VD01 donated by the Advantest Corporation to the VLSI Design and Education Center (VDEC, The University of Tokyo).

REFERENCES

[1] D. Collard, C. Yamahata, B. Legrand, T. Takekawa, M. Kumemura, N. Sakaki, G. Hashiguchi, H. Fujita, "Towards mechanical characterization of biomolecules by MNEMS tools", *IEEJ Trans. Electr. Electron. Eng.*, vol. 2 (3), pp. 262-271, 2007.

[2] G. Kada, F. Kienberger, P. Hinterdorfer, "Atomic force microscopy in bionanotechnology", *Nano today*, vol. 3 (1-2), pp. 12-19, 2008.

[3] G. Hashiguchi, T. Goda, M. Hosogi, K. Hirano, N. Kaji, Y. Baba, K. Kakushima, H. Fujita "DNA Manipulation and Retrieval from an Aqueous Solution with Micromachined Nanotweezers", *Anal. Chem.*, vol. 75 (17), pp. 4347-4350, 2003.

[4] C. Yamahata, D. Collard, B. Legrand, T. Takekawa, M. Kumemura, G. Hashiguchi, H. Fujita, "Silicon nanotweezers with subnanometer resolution for the micromanipulation of biomolecules", *J. Microelectromech. Syst.*, vol. 17 (3), pp. 623-631, 2008.

[5] C. Yamahata, D. Collard, T. Takekawa, M. Kumemura, G. Hashiguchi, H. Fujita, "Humidity dependence of charge transport through DNA revealed by silicon-based nanotweezers manipulation", *Biophys. J.*, vol. 94 (1), pp. 63-70, 2008.

[6] P. Sandoz, J. C. Ravassard, S. Dembélé, A. Janex, "Phase-sensitive vision technique for high accuracy position measurement of moving targets", *IEEE Trans. Instr. Meas.*, vol. 49 (4), pp. 867-872, 2000.

[7] P. Sandoz, V. Bonnans, T. Gharbi, "High-accuracy position and orientation measurement of extended two-dimensional surfaces by a phase-sensitive vision method", *Appl. Opt.*, vol. 41 (26), pp. 5503-5511, 2002.

[8] S. Vanapalli, *Techniques for characterzation of in-plane displacement for microactuators*, M.Sc. Thesis, University of Twente, The Netherlands, 2004.

[9] E. Sarajlic, C. Yamahata, H. Fujita, "Electrostatic 3-Phase Linear Stepper Motor Fabricated by Vertical Trench Isolation Technology", in *Proc. 19th MicroMechanics Europe Workshop (MME 2008)*, Aachen, Germany, September 28-30, 2008.

[10] J. P. Cleveland, S. Manne, D. Bocek, P. K. Hansma, "A nondestructive method for determining the spring constant of cantilevers for scanning force microscopy", *Rev. Sci. Instrum.*, vol. 64 (2), pp. 403-405, 1993.

[11] B. Ohler, "Cantilever spring constant calibration using Doppler vibrometry", *Rev. Sci. Instrum.*, vol. 78, 066701/1-5, 2007.

EFFECT OF ELECTRODE GEOMETRY AND SURFACE PASSIVATION ON CORROSION OF POLYCRYSTALLINE SILICON UNDER HIGH RELATIVE HUMIDITY AND BIAS

F. Liu[1,2], C. S. Roper[1,3], C. Carraro[1] and R. Maboudian[1]
[1]University of California, Berkeley, CA 94720, USA
[2]University of Science and Technology of China, Hefei, Anhui 230027, China
[3]HRL Laboratories, LLC, Malibu, CA 90256, USA

ABSTRACT

The combination of high electric fields and moisture from the environment can lead to electrochemical reactions causing corrosion, which can severely impact micro-electromechanical systems (MEMS) device reliability and lifetime. In this paper, we present a detailed study of the corrosion behavior and surface passivation methods for enhancing the corrosion resistance of surface micromachined polycrystalline Si (poly-silicon) electrodes at high relative humidity and applied voltage. The results indicate the occurrence of anodic oxidation under positive bias and absence of cathodic protection under negative bias. Additionally, surface passivation chemistry, in particular employing self-assembled monolayers (SAM), is found to effectively enhance the corrosion resistance of the electrodes.

INTRODUCTION

Recent developments in the MEMS field have created a growing interest in the reliability of these miniaturized devices [1]. Due to their reduced dimensions, MEMS devices often operate under high fields ($>10^7$ V/m) even for moderate applied voltages. The combination of high electric fields and moisture from the environment can lead to electrochemical reactions leading to corrosion, which can severely impact device lifetime.

We have previously reported damage to a negatively biased electrode (cathode) in poly-silicon cantilever beams. The effect has been reproduced in test structures consisting of two geometrically symmetric, electrically isolated electrodes, where one electrode was contacted to the substrate through a breach in the silicon nitride isolation layer [2]. Systematic studies of corrosion in MEMS with varying geometry electrodes have not yet been reported. Separate efforts to address the stiction-related reliability issues in MEMS devices have resulted in a number of surface passivation treatments [3]. However, surface treatments to enhance the corrosion-resistance of poly-silicon MEMS electrodes have not been reported.

The objective of this paper is to investigate the corrosion mechanism of poly-silicon electrodes with different geometries and study the effect of surface chemistry on the corrosion behavior under high relative humidity environment. Poly-silicon electrodes with different anode to cathode area ratios are fabricated and tested. Since most MEMS electrodes are fabricated on an isolation layer, the designs presented in this work are more representative of MEMS devices. Cathodic corrosion is found to occur when the anode to cathode area ratio is higher than unity. A model is proposed to explain these results. Furthermore, the effects of hydrogen-termination and $CF_3(CF_2)_7(CH_2)_2SiCl_3$ (FDTS) SAM coating are investigated. The results yield valuable information on corrosion mechanisms of MEMS devices and the enhancement of corrosion-resistance by surface chemistry treatments.

EXPERIMENTAL

Fabrication process

The test structures consist of two rectangular electrodes which are 500 nm thick and separated by a 2 μm gap. The electrodes are isolated from the substrate by a silicon nitride layer. One electrode has a length of 300 μm and a width of 80 μm, while the other has the same length but variable width ranging from 80 to 1500 μm. A one-mask microfabrication process is executed in the Berkeley Microfabriction Laboratory. The process begins with n-type (100) silicon wafers of 10-20 Ω·cm resistivity. The wafer surfaces are first heavily doped with phosphorus in a standard diffusion furnace using a phosphosilicate glass (PSG) sacrificial layer as the dopant source. Next, the PSG film is removed by HF, and a 600 nm low-stress silicon nitride layer is grown by low pressure chemical vapor deposition on the wafers as an electrical isolation layer. Afterwards, a 500 nm phosphorus-doped poly-silicon film is deposited on the wafers using LPCVD employing SiH_4 and PH_3 gases at 615°C. The resistivity of the poly-silicon film, obtained by means of a Tencor RS35C 4-point probe, is about 6.2×10^{-3} Ω·cm. The poly-silicon layer is patterned by photolithography and a transformer-coupled plasma etch using Cl_2 and HBr. After surface treatments, which will be discussed in detail later, the electrodes are electrically connected to a standard 16-pin package via Al wire bonding.

Surface treatments

For chips undergoing the hydrogen terminating treatment, the samples are soaked for 2 min in a solution of 1:1 (v/v) concentrated HF (49 wt%) and concentrated HCl (36 wt%) which has a good etch selectivity to silicon oxide over silicon nitride. The etch solution is displaced with isopropanol (IPA) to limit the oxidation of hydrogen-terminated silicon [4]. For chips undergoing self-assembled monolayer coating, FDTS precursor (PCR Chemicals, 98% purity) from the vapor phase was applied following the sequence described elsewhere [5] after UV ozone (Jelight Mode 42) treatment for 20 min.

Testing and characterization methods

It has been reported that the corrosion rate of poly-silicon electrodes increases with increased relative humidity and applied voltage, as well as decreased electrode separation [2]. In order to accelerate possible failure mechanisms, the test chips are operated in environments with high relative humidity of 89%, at high anode voltage (100 V), and with the cathode grounded.

978-1-4244-2977-6/09 $25.00 © 2009 IEEE

Humidity during testing is controlled using a potassium chloride saturated solution. X-ray photoelectron spectroscopy confirms no trace of potassium or chlorine on the specimen surface, indicating that only water vapor is released from the salt solution during testing. Water contact angle measurements are used to determine the degree of surface hydrophobicity before and after each surface treatment. Contact angle data are taken with DI water (resistivity 18 MΩ. cm) according to the sessile droplet method with a Rame Hart goniometer. The droplet size is approximately 4 μL.

Test specimens are monitored throughout the measurements using a Leica DMLM microscope equipped with a Sony CCD camera. Chemical composition is evaluated using energy dispersive X-ray spectroscopy (EDXS). Surface morphology is analyzed with a Leo 1550 Schottky field emission scanning electron microscope (SEM). Atomic force microscopy (AFM, Digital Instruments, Multimode Nanoscope IIIa with Extender Electronics Module) is operated in tapping mode to image the electrode surfaces and quantify roughness.

RESULTS AND DISCUSSIONS
Effect of electrode geometry

Figure 1 shows the corrosion behavior of unpassivated (i. e., covered with native oxide) poly-silicon electrodes after 20 hrs exposure to 89% RH under 100 V applied bias. As seen in all cases, the left electrode (positively biased anode) shows volume expansion across the gap and delamination as a result of anodic oxidation. EDXS analysis confirms the oxidation of the anode, by showing clear Si and O peaks. From the elemental maps of oxygen, the pixel intensity for oxygen increases by more than 25× on the delaminated edges of the anode and by about 3× on the discolored regions for the symmetric case. Shown in Figure 2, the top of the corroded anode is as high as 2.2μm above the isolation layer. This is attributed to the corrosion induced stress leading to the detachment and the upward curling of the anode edge. After HF etching of the oxide, the anode disappears almost entirely (Fig. 2c).

While anodic oxidation is observed in all cases, damage to the cathode is observed only when the anode to cathode area ratio is higher than unity. EDXS analysis of the discolored regions of the cathode indicates increased oxygen content, hence the cathode oxidation.

In this study, the poly-silicon coated with native oxide and the silicon nitride isolation layer both have hydrophilic surfaces as seen by the low water contact angles, tabulated in Table 1. In the presence of water, oxidation of silicon is the predominant reaction at the anode where holes are supplied:

$$H_2O \rightarrow H^+ + OH^- \tag{1}$$

$$Si + 2OH^- + 2h^+ \rightarrow SiO_2 + H_2 \tag{2}$$

At the cathode where electrons are supplied, reduction of water and dissolved oxygen are proposed:

$$2H_2O + 2e^- \rightarrow H_2 + 2OH^- \tag{3}$$

$$O_2 + 2H_2O + 4e^- \rightarrow 4OH^- \tag{4}$$

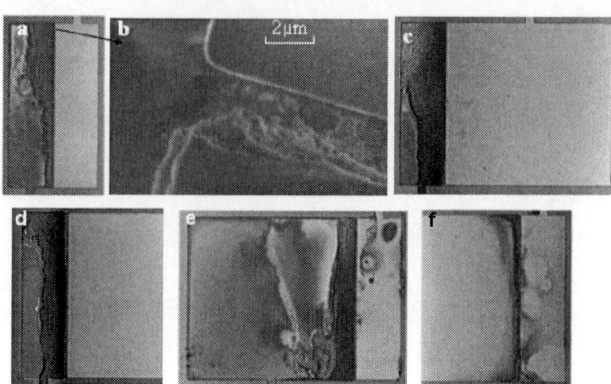

Figure 1: Poly-silicon corrosion test electrodes initially covered with native oxide after 20 hrs exposure to 89% RH Anode=left electrode at +100 V; Cathode=right electrode, electrically grounded. a) Both electrodes are 80 μm wide-symmetric electrodes. b) SEM of the symmetric electrodes; c) Anode is 80 μm wide while cathode is 300 μm wide. d) Anode is 80 μm wide while cathode is 1500 μm wide. e) Anode is 300 μm wide while cathode is 80 μm wide. f) Anode is1500 μm wide while cathode is 80 μm wide.

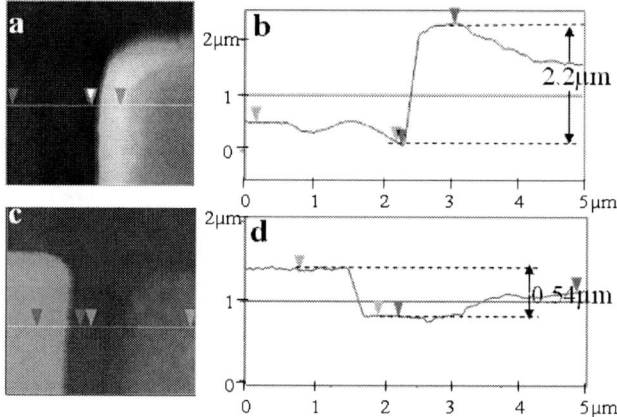

Figure 2: AFM analysis of symmetric poly-silicon corrosion test electrodes after 20 hrs exposure to 89% RH at +100 V anodic potential. Scan size: 5μm×5μm, z scale: 2.5μm. Anode=right electrode; Cathode=left electrode. a) AFM image and b) the cross sectional analysis; c) AFM image after HF etching to remove the oxidized portion of the electrodes; and d) the cross sectional analysis.

Table 1: Water contact angle data on the surfaces examined.

Surface treatment	Water contact angle
Poly-silicon with native oxide	0-20°
H-terminated poly-silicon	75°
FDTS coated poly-silicon	115°
As fabricated silicon nitride	0-25°
FDTS coated silicon nitride	87°

Factors such as ohmic loss (or IR drop) across the electrolyte, electrode geometry, mass transport and reaction kinetics can influence the state of the electrochemical system. Our results are consistent with the earlier hypothesis, proposed in Ref. 2, that the damage observed at the cathode may be associated with the accumulation of OH- and large ohmic loss. Lack of bulk electrolyte limits diffusion of redox species at the electrode surfaces. In the electrochemical system, reaction at the poly-silicon cathode occurs when its potential is insufficient for cathodic protection. The protection potential of the electrode depends on pH, temperature, pressure, composition and electrical state. Cathodic protection is difficult to achieve under highly alkaline conditions [6]. This is illustrated in the Pourbaix diagram of the silicon-water system in the high pH region. Theoretically, a region of cathodic protection still exists, but at very high pH achieving cathodic protection may require such a large cathodic voltage that it is impractical to implement.

In a typical electrochemical cell where electrodes are immersed in a conductive electrolyte, IR drop, or the ohmic loss, across the electrolyte is usually insignificant. The small ohmic loss allows for greater surface potential at the electrode interface and cathodic protection can occur at lower applied voltages. However, in systems with only surface electrolyte, the large ohmic loss across the electrolyte results in significant reduction in surface potential at the electrodes. Voltage drop across the anode compared to cathode is another source of ohmic loss. The resistance of an electrode is inversely proportional to its width. When the width of the anode is greater than the width of the cathode (area ratio of $A_{anode}/A_{cathode}>1$ in this test system), larger voltage drop across the anode results in further potential drop at the cathode. Consequently, together with the high pH environment, the effective potential may be insufficient for cathodic protection to take place. Reaction at the poly-silicon cathode occurs when its potential is insufficient for cathodic protection. On the other hand, when $A_{anode}/A_{cathode} \leq 1$, the voltage drop across the anode is small compared to that of the cathode, hence the cathode may maintain the potential required for cathodic protection and remain unaffected.

Effect of surface termination

Table 1 shows the hydrophobicity of the device surfaces with different surface treatments. As shown in Fig. 3, H-termination helps to reduce corrosion of the anode but corrosion is eventually observed after 20 hrs at 89% relative humidity. On the other hand, no corrosion is observed on SAM-coated electrodes. Different samples with the same surface treatments are tested several times to duplicate and verify this behavior. AFM studies of FDTS coated surface before and after the corrosion test (Figure 4) show no changes in the electrode morphology.

Electrochemical reactions occur when there is a finite surface leakage current between neighboring poly-silicon electrodes on the surface of the insulating isolation layer in presence of moisture. The number of holes consumed at the anode must be the same to the number of electrons consumed at the cathode. For the same electrode geometry, the reaction rate of both electrodes should be proportional to the leakage current between the two electrodes, and thus the net reaction observed on the anode should be proportional to the integrated current, or total charge transferred [8]. From the current integration results tabulated in Table 2, it can be

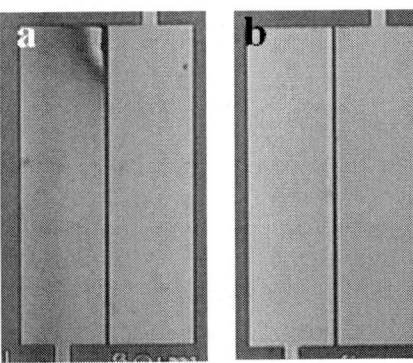

Figure 3: Poly-silicon symmetric corrosion test structures with surface pretreatments after 20 hrs exposure to 89% RH at +100 V anodic potential. Anode=left electrode; Cathode=right electrode. a) H-terminated electrodes; b) FDTS coated electrodes.

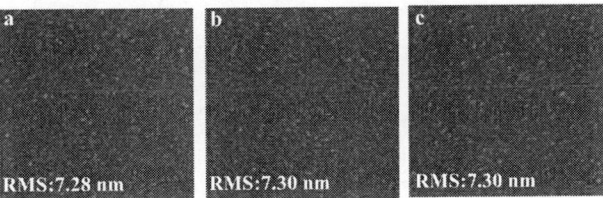

Figure 4: Surface topography of electrodes. Scan size: 8 μm×8 μm, z scale: 200 nm. a) poly-silicon electrode as fabricated. b) FDTS coated electrode before testing. c) FDTS coated anode after 20hrs exposure at 89% RH and +100 V anodic potential.

seen that the total charge transferred during the testing of the H-terminated and FDTS coated electrodes is dramatically lower than that of poly-silicon electrodes initially covered with native oxide, consistent with the observed corrosion behavior.

Table 2: Total charge transferred during 20 hr corrosion testing of symmetric electrodes with different surface pretreatments at 89% RH and +100 V anodic potential.

Surface treatment	Oxide terminated	H terminated	FDTS coated
Total charge (mC)	643. 8	31. 9	5. 4

The contact angle data for FDTS-coated poly-silicon as given in Table 1 confirms that well-packed monolayers are formed [1]. While HF treatment would not appreciably affect the silicon nitride isolation layer, FDTS treatment leads to the formation of a monolayer on silicon nitride and affects its hydrophobicity. Contact-angle measurements probe the interfacial phenomena, which are inherently related to the surface free energies of the two contacting phases. Tencer, et al. [7] have reported that in constant relative humidity, the number of water layers present on a surface is inversely proportional to surface energy. Thus, water is not only the reactant on the electrode surfaces but also the ionic carrier across the inter-electrode gap. Lowering the surface energy by surface pretreatments reduces the number of water layers absorbed on the microstructures, thus reducing the corrosion. For the H-terminating treatment, the surface energy of poly-silicon electrodes decreases, while the surface energy of the silicon nitride is not significantly altered, thus the water layer across the inter-electrode gap remains. In this case, corrosion is still observed after 2 hrs of testing although at much reduced rates. Furthermore, only the edge is corroded, which may be due to electric field enhancement at the corners. For FDTS coated structures, the surface energies of both the electrodes and the silicon nitride present in the inter-electrode gap are decreased, hence water adsorption is hindered. This barrier to the formation of water layers effectively reduces corrosion at both electrodes.

CONCLUSIONS

In summary, the significant impact of humidity on the electrical reliability of surface micromachined polysilicon electrodes is presented. In particular, the corrosion behavior of poly-silicon electrodes with different geometries is investigated. It is found that cathodic oxidation occurs when the ratios of the anode to cathode areas are greater than unity. Anodic oxidation is observed under all geometries. We report for the first time the effect of surface passivation pretreatments on corrosion phenomena in MEMS. Hydrophobic coatings that can reduce water adsorption on the surfaces of the electrodes and the isolation layers are found quite effective at hindering the corrosion. This work highlights that once the combined effect of surface chemistry and electrode geometry is understood, one can choose surface treatments and layouts to prevent corrosion of MEMS electrodes.

ACKNOWLEDGEMENTS

The authors wish to acknowledge DARPA Science & Technology Center, CIEMS for financial support. One author (Fang Liu) thanks China Scholarship Council for additional support.

REFERENCES

[1] J. Frechette, R. Maboudian and C. Carraro, "Effect of Temperature on In-use Stiction of Cantilever Beams Coated with Perfluorinated Alkysiloxane Monolayers", *Journal of Microelectro-mechanical Systems,* vol. 15, pp. 737-744, 2006.

[2] M. Hon, F. W. DelRio, J. T. White, M. Kendig, C. Carraro, and R. Maboudian, "Cathodic corrosion of polycrystalline silicon MEMS", *Sensors and Actuators A*, vol. 145–146, pp. 323–329, 2008.

[3] R. Maboudian and C. Carraro, "Surface chemistry and tribology of MEMS", *Annu. Rev. Phys. Chem.* vol. 55, 35–54M, 2004.

[4] M.R. Houston, R. Maboudian, and R.T. Howe, "Effect of hydrogen termination on the work of adhesion between rough polycrystalline silicon surfaces", *Journal of Applied Physics.* vol. 81, pp.3474-3483, 1997.

[5] W. R. Ashurst, C. Carraro, R. Maboudian, and W. Frey, "Wafer level anti-stiction coatings for MEMS," *Sens. Actuators A, Phys.* vol. 104, pp. 213–221, 2003.

[6] John Newman, *Electrochemical Systems*, John Wiley& Sons, Inc. New Jersey, 2004.

[7] M. Tencer and J. S. Moss, "Humidity Management of Outdoor Electronic Equipment: Methods, Pitfalls, and Recommendations", IEEE Transactions on Components and Packaging Technologies, 25 (2002).

[8] H. Shea, A. Gasparyan, H. B. Chan, S. Arney, R. E. Frahm, D. Lopez, S. Jin, and R. P. McConnell, "Effects of electrical leakage currents on MEMS reliability and performance", *IEEE Trans. Dev. Mat. Rel.* vol. 4, pp. 198-207, 2004.

MECHANICAL RESPONSE EVALUATION OF HIGH-THERMALLY-STABLE-GRADE PARYLENE SPRING

Chihiro Kamezawa, Yuji Suzuki and Nobuhide Kasagi
Department of Mechanical Engineering, The University of Tokyo, JAPAN

ABSTRACT

Mechanical response of high-thermally-stable-grade parylene (diX-SR/HR) is investigated for flexible spring applications. Pendulum structures with a high-aspect-ratio beam are microfabricated using different parylene materials, and their amplitude and stress at resonant oscillation are measured. Based on fatigue tests and measurements of the temperature coefficient of Young's modulus, MEMS structures with diX-SR/HR are proved to have better thermal tolerance than parylene-C, while they have almost the same mechanical properties as parylene-C.

INTRODUCTION

Parylene (poly-para-xylylene) is a MEMS-compatible polymer with favorable physical properties, and used for various applications [1-5]. Especially, parylene-C has good mechanical properties such as low Young's modulus and large yield strain [6, 7]. Suzuki & Tai [8] develop a microfabrication technology for high-aspect-ratio parylene-C beams, and realize spring structures with low resonant frequency and large oscillation amplitude.

However, the glass transition temperature T_g of parylene-C is around 50 °C [9], and parylene-C degrades in oxygen environment for temperature above 100 °C. Thus, the continuous service temperature of parylene-C is 80 °C, which is lower than the requirement for applications such as automotive ones. Rapid thermal cycling beyond T_g may also causes reliability issues due to tensile stress build-up [10]. Fluorinated parylene, i.e., parylene-AF4 or parylene-HT [11] has excellent thermal stability, but its elongation to break is much smaller than that of parylene-C [6].

Recently, diX-SR/HR, high-thermally-stable grades of parylene, are released by KISCO. It is reported that diX-SR/HR bulk films have similar mechanical properties with parylene-C, while their tensile strength is almost unchanged even at 140 °C for 400 hours.

Fatigue is also important issue to be considered for MEMS springs/structural members. For instance, springs used for vibration-driven energy harvesting devices [12] will experience up to 10^{10} cycle oscillations during its lifetime. Unlike metals, fatigue process of polymer is complicated; their degradation mechanisms include scission of molecular chains and partial re-crystallization [13-15]. Thus, experimental evaluation for the fatigue characteristics is also necessary.

In the present study, MEMS-compatibility of diX-SR/HR is investigated. In addition, the mechanical properties of MEMS springs including long-term stability and thermal coefficient of Young's modulus (TCY) are examined in a series of experiments.

DESIGN AND MICROFABRICATION

In order to evaluate diX-SR/HR, we adopt micro pendulum structures supported with a high-aspect-ratio parylene spring [8]. Figure 1 shows the schematic of the present measurement method. Micromachined pendulums with a parylene beam are fixed on an electromagnetic shaker, where sinusoidal oscillation is applied at a given frequency. Strain in the parylene beams is imposed by vibration of the pendulum. Amplitude of each pendulum is measured with photointerrupters, and the strain is estimated on-line. Any mechanical degradation results in changes of the amplitude. Since four pendulums with different beam lengths are integrated on a single chip, mechanical response tests with four different values of strain can be made simultaneously.

Figure 2 shows the fabrication process. The process starts with DRIE of 350-μm-deep trenches using a 1.5-μm-thick SiO_2 mask (Figs. 2a-c). The trenches also define the boundaries of the Si islands to be left. Next, the trenches are filled with parylene (Fig. 2d). The deposition condition of diX-SR/HR is the same as that of parylene-C; the pyrolysis temperature and the deposition pressure are respectively 690 °C and 20 mTorr. Deposition rate of diX-SR/HR with this setting is the same as that of parlene-C. After the parylene deposition, the metal mask is patterned (Fig. 2e) followed by O_2 plasma etch of the parylene film (Fig. 2f). After stripping the metal mask (Fig. 2g), Si substrate surrounding the beams is etched away with XeF_2 to release the pendulum structures (Fig. 2h).

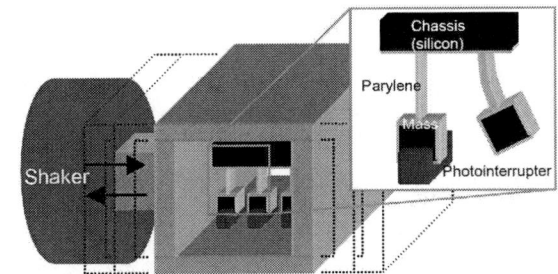

Figure 1: Measurement set-up with on-line strain monitoring.

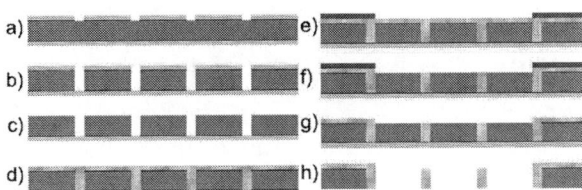

Figure 2: Process flow of parylene high-aspect-ratio springs.

RESULT AND DISCUSSION

Figure 3 shows photos of the microfabricated parylene pendulums. Beam width and length of the pendulums are respectively 30 μm and 3.85~4.0 mm. Designed value of the resonant frequency is about 30Hz for 3 mm$_{p-p}$ oscillation. The maximum stress of the parylene beam under deflection is located at the fixed end of the beam.

Firstly, thermal tolerance of parylene-C and diX-HR/SR is examined. Figure 4 shows the parylene pendulum structures after thermal annealing at 140 °C for 5 hours. As clearly seen, little deformation is found in the diX-HR beams, whereas the parylene-C beams are significantly deformed due to thermal softening. Thus, it is confirmed that diX-HR MEMS structures have higher thermal tolerance than parylene-C.

Figure 5 shows the experimental setup for oscillation tests. A test fixture is mounted on an electromagnetic shaker. This system allows simultaneous examination of four test chips including 16 pendulum structures. Air temperature inside the fixture is measured with a thermocouple, and kept constant with a heater.

Figures 6a and 6b show a long-time exposure photo and a snap shot of the parylene pendulums at the resonance. The transit time Δt, in which the mass goes through the photointerrupter, is monitored with a digital oscilloscope, and stored onto a PC (Fig. 6c).

With the oscillation frequency imposed and the measurement data of transit time, the amplitude of pendulums is computed. However, under the present experimental conditions, the beam deflection is very large, so that the conventional linear beam model cannot be used to estimate the strain from the amplitude. We employ a non-linear model [16, 17] including axial elongation of the beam. Figure 7 shows the maximum strain thus obtained, which is located at the fixed end of the beam, in comparison with the stain computed from high-speed camera images of the pendulums. Since the maximum strain estimated from the transit time measurement is in good agreement with the results of the optical measurement, accuracy of the amplitude measurement using the photointerupter is confirmed. Note that, as also shown in Fig. 7, the linear

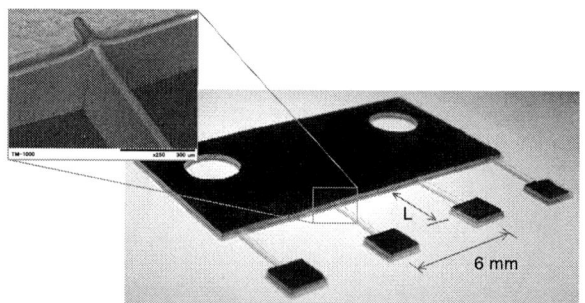

Figure 3: Test chip with 4 pendulums. Parylene beam width is 30 μm. Beam lengths (L) from left to right are 4.0, 3.95, 3.9 and 3.85 mm. Dimension of the mass is 2.0 mm × 2.0 mm. The inset is a SEM image of the beam junction.

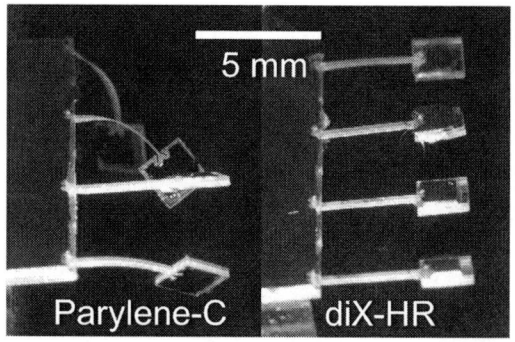

Figure 4: Free-standing parylene-C (left) and diX-HR (right) beams after thermal annealing at 140 °C for 5 hours. Beam width and height are respectively 30 μm and 350 μm.

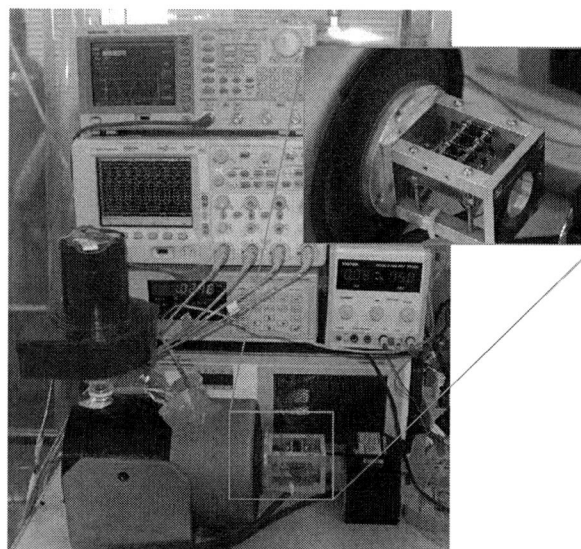

Figure 5: Experimental setup for oscillation tests.

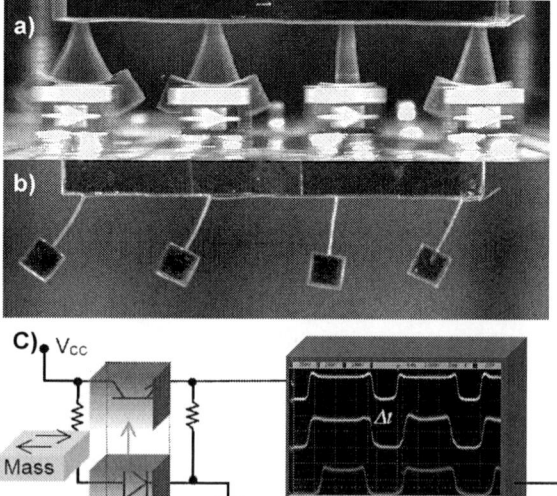

Figure 6: Parylene pendulum structures in resonance at 60 Hz. a)Long-exposure photo, b) Snap shot, c) Read-out circuit with the photointerrupters and output signals taken with a digital oscilloscope.

978-1-4244-2977-6/09 $25.00 © 2009 IEEE

model considerably underestimates the strain for small transit time, where the amplitude and the maximum strain are both large.

Fatigue characteristics of the parylene beams at room temperature are examined through cyclic load tests, in which the inertia force of the pendulums is kept constant. The strain is monitored by the measurement of transit time as explained above. Since typical strain expected in actual device such as seismic generators [12] is as large as 1%, the maximum strain is set to about 0.4 and 1 %. Since the yield strain of the parylenes is 3 % [6], the maximum strain imposed to the fixed end of the beam is about one third of the yield strain. Figure 8 shows the strain versus the number of cycles. The strain for parylene-C and diX-HR is almost constant without any sudden change up to 10^7 cycles, indicating no mechanical degradation occurs during the test. Therefore, both parylene materials have comparable mechanical durability, although the number of cycles tested is much smaller than the requirement in real applications. Note that the strain slightly varies in time due to the temperature change of specimen.

As mentioned earlier, diX-HR has higher annealing tolerance if compared with parylene-C. We also examine their mechanical properties under different temperature conditions. Figure 9 shows the transit time versus the oscillation frequency for different temperature. External

amplitude is kept constant. The valley of each curve corresponds to the resonant frequency at the temperature. Shift in the resonant frequency is due to thermal softening of parylenes, which is represented with the temperature coefficient of Young's modulus (TCY).

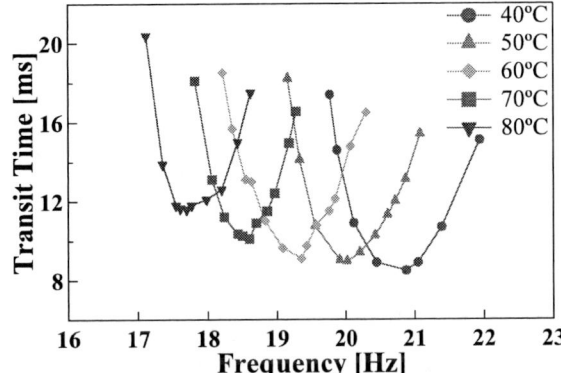

Figure 9: Frequency response of parylene-C pendulums for different temperatures. The minimum transit time of each curve corresponds to the resonant frequency.

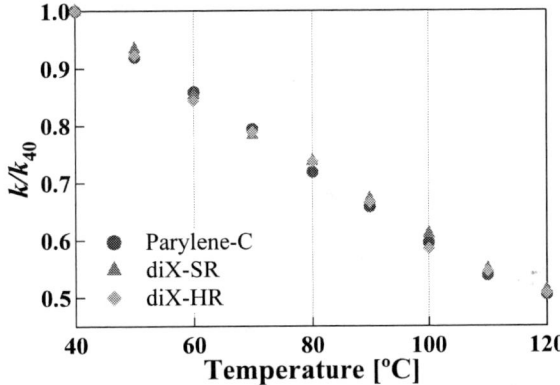

Figure 10: Spring constant normalized with the value at 40 °C. Three different parylene materials have the same TCY of 0.6 %/K.

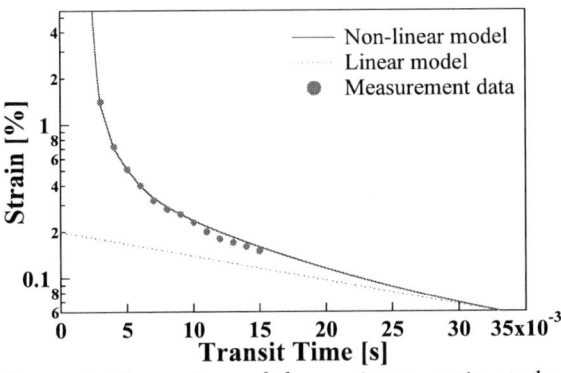

Figure 7: Comparison of the maximum strain at the point of maximum stress on deflecting beams. Beam length is 4 mm and oscillation frequency is 30 Hz.

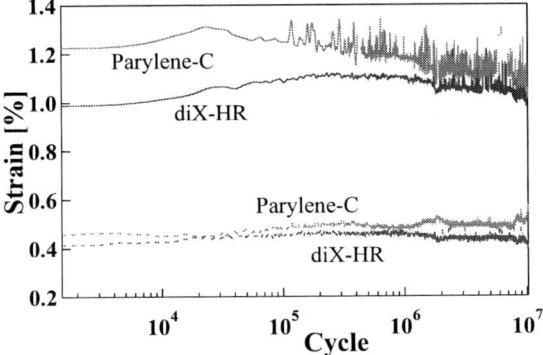

Figure 8: Experimental results for the cyclic load tests of parylene-C/diX-HR beams.

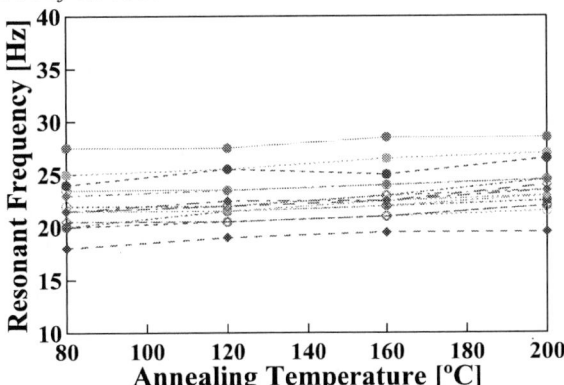

Figure 11: Effect of short-term thermal annealing on the resonant frequency: solid lines, diX-HR pendulums; dashed lines, parylene-C pendulums. Data for 4 pendulums with different beam lengths are plotted for both parylene-C and diX-HR.

Figure 10 shows spring constant of the parylene-C and diX-SR/HR beams. The spring constant is normalized with the value at 40 °C. The temperature coefficient of Young's modulus (TCY) of parylene-C thus obtained is 0.6 %/K, which is in good agreement with our previous data using leaf-spring structures [8]. It is found that TCY of diX-SR/HR is the same as that of parylene-C. Note that, although the glass transition temperature of parylenes is around 50 °C [9], no hysteresis of the spring constant is observed up to 120 °C.

Finally, the effect of short-time annealing, which is expected to occur during fabrication processes or reflow soldering, is examined. Figure 11 shows resonant frequencies after annealing at different temperatures for 5 mins. It is found that the resonant frequency of parylene beams remains almost unchanged after the short-term annealing up to 200 °C, although it is slightly increased by 1-2 Hz from 80 to 200 °C.

CONCLUSION

We have developed a cyclic-load test bench for micromachined high-aspect-ratio parylene springs. The mechanical properties of high-thermally-stable-grade parylenes (diX-HR/SR) are examined in a series of experiments. Compatibility of diX-SR/HR to MEMS processes is confirmed. It is also confirmed from an annealing test that diX-HR/SR MEMS structures have higher thermal tolerance than parylene-C. On the other hand, mechanical properties of diX-HR/SR and parylene-C are almost the same at lower temperatures; no mechanical degradation of the beams is observed up to 10^7 cycles at the maximum strain as large as 1%. The temperature coefficient of diX-SR/HR is 0.6 %/K, which is also the same as that of parylene-C. Therefore, since the diX-SR/HR has better thermal stability than parylene-C, diX-SR/HR can replace parylene-C as a MEMS structural material for higher temperature applications up to 120-140 °C.

This work is supported through New Energy and Industrial Technology Development Organization (NEDO). Photomasks are made using the University of Tokyo VLSI Design and Education Center (VDEC)'s 8-inch EB writer F5112+VD01 donated by ADVANTEST Corporation.

REFERENCES

[1] L. Licklider, X. Q. Wang, A. Desai, Y.-C. Tai, and T. D. Lee, "A micromachined chip-based electrospray source for mass spectrometry," *Anal. Chem.*, Vol. 72, 2000, pp. 367-375.

[2] T. N. Pornsin-sirirak, Y.-C. Tai, H. Nassef, and C.-M. Ho, "Titanium-alloy MEMS wing technology for a micro aerial vehicle application," *Sensors Actuators, A*, Vol. 89, 2001, pp. 95-103.

[3] H. S. Noh, K. S. Moon, A. Cannon, P. J. Hesketh, and C. P. Wong, "Wafer bonding using microwave heating of parylene intermediate layers," *J. Micromech. Microeng.*, Vol. 14, 2004, pp. 625-631.

[4] S. Takeuchi, D. Ziegler, Y. Yoshida, K. Mabuchi, and T. Suzuki, "Parylene flexible neural probes integrated with microfluidic channels," *Lab Chip*, Vol. 5, 2005, pp. 519-523.

[5] P. J. Chen, D. C. Rogers, E. M. Meng, M. S. Humayun, and Y.-C. Tai, "Surface-micromachined parylene dual valves for on-chip unpowered microflow regulation," *J. Microelectromech. Syst.*, Vol. 16, 2007, pp. 223-231.

[6] http://www.kisoco-net.co.jp/dix/index.html

[7] C. Y. Shih, T. A. Harder, and Y.-C. Tai, "Yield strength of Parylene-C," *Microsyst. Tech.*, Vol. 10, 2004, pp. 407-411.

[8] Y. Suzuki, and Y.-C. Tai, "Micromachined high-sapect-ratio parylene spring and its application to low-frequency accelerometers," *J. Microelectromech. Syst.*, Vol. 15, 2006, pp. 1364-1370.

[9] J. J. Senkevich, and S. B. Desu, "Morphology of poly(chloro-*p*-xylylene) CVD thin films," *Polymer*, Vol. 40, 1999, pp. 5751-5759.

[10] S. Dabral, J. Van Etten, X. Zhang, C. Apblett, G.-R. Yang, P. Ficalora, and J. F. McDonald, "Stress in thermally annealed parylene films," *J. Elec. Mater.*, Vol. 21, 1992, pp. 989-995.

[11] W. R. Dolbier Jr., and W. F. Beach, "Parylene-AF4: a polymer with exceptional dielectric and thermal properties," *J. Fluorine. Chem.*, Vol. 122, 2003, pp. 97-104.

[12] M. Edamoto, Y. Suzuki, N. Kasagi, K. Kashiwagi, Y. Morizawa, T. Yokoyama, T. Seki, and M. Oba, "Low-resonant-frequency micro electret generator for energy harvesting application," Proc. 22nd IEEE Int. Conf. MEMS, Sorrento, 2009, to be presented.

[13] J. A. Sauer, and G. C. Richardson, "Fatigue of polymers," *Int. J. Fracture*, Vol. 16, 1980, pp. 499-532.

[14] J. H. Golden, B. L. Hammant, and E. A. Hazell, "Degradation of polycarbonates. 4. Effect of molecular weight on flexural properties," *J. Polym. Sci. A*, Vol. 2, 1964, pp. 4787-4794.

[15] A. N. Gent, and S. Madan, "Plastic yielding of partially crystalline polymers," *J. Polym. Sci.*, Vol. 27, 1989, pp. 1529-1542.

[16] R. Frisch-Fay, *Flexible Bars*, Butter Worths, 1962.

[17] C. González, and J. LLoca, "Stiffness of a curved beam subjected to axial load and large displacements," *Int. J. Solid Struc.*, Vol. 42, 2005, pp. 1537-1545.

PIEZOELECTRIC PZT THIN FILMS IN THE 100NM RANGE: A SOLUTION FOR ACTUATORS EMBEDDED IN LOW VOLTAGE DEVICES

E. Defay[1], G. Le Rhun[1], F. Perruchot[1], P. Rey[1], A. Suhm[1], M. Aïd[1], L.J. Liu[2], S. Pacheco[2], M. Miller[2]

[1]CEA, LETI, Minatec, F38054 Grenoble, FRANCE
[2]Freescale Semiconductor, Tempe, Arizona, USA

ABSTRACT

This paper reports the piezoelectric properties of sputtered and sol gel PZT thin films investigated in a low thickness range (100-250nm). Piezoelectric-elastic bimorphs including very thin PZT films were realized and the maximum reachable deflection at 5V was characterized. The depoling effect experienced by the PZT versus the maximum post process temperature is also discussed which leads to the motivation of developing low thickness range PZT thin films. Elastic-piezoelectric bimorphs were realized and exhibited deflection higher than 5µm at 5V. Moreover, these bimorphs showed only a 10% decrease of the deflection after 5 billions cycles.

INTRODUCTION

PZT thin films have attracted much attention for transduction applications as they exhibit outstanding piezoelectric properties. Several examples of devices using PZT thin films have already been reported in literature: micro-mirrors [1], micro-switches [2] or acoustic transducers [3]. The thickness range reported in these work is greater than 0.5µm and often in the 1-2µm range [4]. The main reason of thicknesses over 500nm is that piezoelectric coefficients decrease when thickness decreases as reported by Haccart et al. [5]. However, the desired integration of these films in low voltage devices induces issues that can be overcome by decreasing the thickness to the 100-200nm range.

For most integrated devices, it is mandatory to withstand a temperature around 260°C for the final board assembly. This temperature can result in the removal of the remnant polarization of PZT and piezoelectricity as will be detailed in the following section of the paper. To avoid a high voltage repoling process, it is highly desirable to repole the PZT by using the low voltage available in the operating system. For example, this low voltage can be provided by the battery in mobile applications, provided that the electric field induced by that voltage is significantly higher than the coercive field E_C of PZT films. Typically, the optimized poling field for PZT films is around 200kV/cm. At 3V or 5V, the thickness should be around 150nm or 250nm, respectively. This approach also has the added advantages that 1) re-poling during each operation improves the stability and reliability of the device; and 2) lower PZT thickness increases piezoelectric moment. In this paper, we detail the development of $Pb(Zr_{0.52}Ti_{0.48})O_3$ thin films in the 100-200nm range. Both sol gel and sputtering method were applied. A specific piezoelectric characterization set-up was used based on the direct piezoelectric effect. Correlations with structural characterization (X-Ray diffraction) are discussed. Thereafter, we describe how we realized elastic-piezoelectric bimorphs. The main steps of the analytical model are discussed allowing the extraction of the piezoelectric coefficient e_{31} based on the reverse piezoelectric effect. Results are compared to those observed in the literatures. The final part of the paper is dedicated to a mechanical reliability study of the realized bimorph.

CHARACTERIZATION OF PZT VERY THIN FILMS

Technology

The first goal of this study is to characterize the piezoelectric properties of PZT thin films in the 100-250nm range. These films were prepared by sol gel or sputtering methods on the following standard stack: Si substrate 100mm diameter /SiO$_2$ 500nm insulation layer/TiO$_2$ 20nm adhesion layer/Pt 100nm bottom electrode. Sputtered films were deposited from a ceramic target in a RF magnetron system using pure argon as sputtering gas. The composition of the target was $Pb_{1.1}(Zr_{0.52}Ti_{0.48})O_3$ (10% Pb excess). The sputtering power and pressure were respectively 2W/cm^2 and 2.10^{-2}mbar. The films were not heated during deposition. A final annealing under air at 700°C for 30 minutes allowed the perovskite crystallisation of the PZT thin films.

Sol gel films were prepared from a commercial source, also containing 10% excess of the lead precursor. Each spin coating gives at the end of the process a PZT thickness between 50 and 60nm. The final thickness is then obtained by adjusting the number of spin coating. The standard process is spin coating, calcination at 400°C under air during 15min and a final rapid thermal annealing at 700°C under oxygen for 1min.

X-Ray diffraction

It is possible to obtain randomly or (100) or (111) oriented layers following the preparation of the PZT layers. The bottom electrode has a huge influence on the final orientation of the PZT layer. Figure 1 shows an X-Ray Diffraction (XRD) diagram of a 220nm thick sputtered PZT film with (100) orientation.

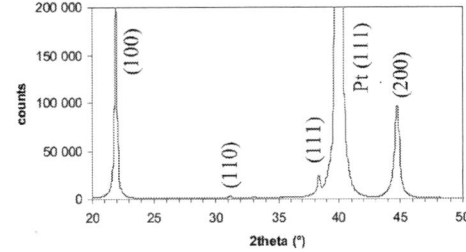

Figure 1: X-Ray diffraction diagram obtained from a sputtered PZT film showing (100) orientation

978-1-4244-2977-6/09 $25.00 © 2009 IEEE

Figure 2 is an XRD pattern of a 260nm thick PZT film randomly oriented. The main difference between this and that shown in Fig. 1 is the deposition temperature of the bottom electrode.

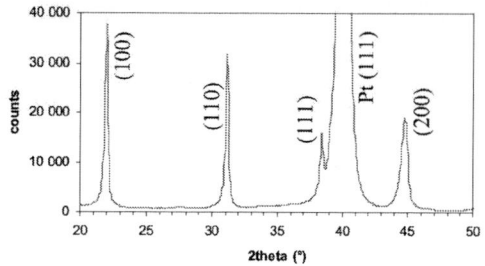

Figure 2: X-Ray diffraction diagram obtained from a sputtered PZT film showing random orientation

All the sol gel films prepared for the results in the following sections are (111) oriented.

e_{31} set up: direct piezoelectric effect

We used a set up developed in-house for the e_{31} characterization of the PZT thin films, so-called the vibrating beam method. The set-up diagram is shown in figure 3. The general idea of this characterization is to vibrate a macroscopic Si beam (4cm length) in the fundamental flexure resonant mode. The whole piezoelectric stack was deposited on this Si beam and electrodes are connected to an oscilloscope in order to capture the voltage obtained from the direct piezoelectric effect once the beam is released from a known initial deflection.

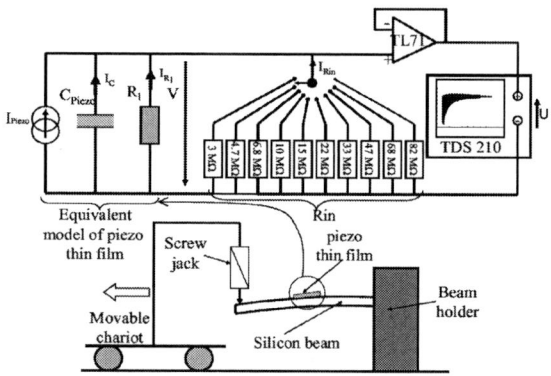

Figure 3: The set-up used for the piezoelectric characterization of PZT thin films

The input impedance of the oscilloscope is controlled by choosing a resistance R_{in} from a dedicated circuit. This allows varying the characteristic time of the discharge. Thereafter, several measurements by changing the value of the resistance enhance the accuracy of the e_{31} characterization. Further details are given elsewhere [6]. Figure 4 shows an example behavior of a PZT thin film, typical for both sol gel and sputtered films. This specific case is from a 260nm-thick PZT film after poling at 8V for 20min.

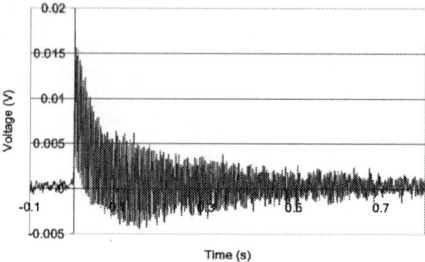

Figure 4: typical piezoelectric voltage measured from a sputtered PZT film using the detailed set-up after the release of the macroscopic beam

The e_{31} extraction is obtained from the slope of the reverse time integral of the voltage versus the reverse of the input resistance. The model gives the following relationship:

$$\frac{1}{\int_0^\infty U dt} = -\frac{1}{A e_{31eff} k \delta(0)} \left(\frac{1}{R_l} + \frac{1}{R_{in}} \right) \quad (1)$$

With $k = \left(1 + \frac{s_{21Si}}{s_{11Si}} \right) \frac{3 d_{Si} (L - l_m)}{2 L^3}$ (2)

s_{ijSi} are the compliance coefficients of Si, d_{Si} is the Si thickness, L is the length of the vibrating beam, l_m is the location of the electrode measured from the clamping, A is the surface of the PZT capacitor, $\delta(0)$ is the initial deflection of the beam and e_{31} is the effective transverse piezoelectric coefficient. This coefficient is effective as the film is clamped on the substrate. The relationship between this coefficient and the true piezoelectric e_{31} and e_{33} coefficients is given by:

$$e_{31eff} = e_{31} - e_{33} \frac{c_{13}^E}{c_{33}^E} \quad (3)$$

With c_{ij}^E the stiffness coefficients of PZT and e_{33} the longitudinal piezoelectric coefficient.

Sputtered films

The reported e_{31} values for 3 different sputtered PZT thicknesses (130nm-180nm-260nm) after poling at 5V for 5s are shown in Table 1. They are comparable to most of the values reported in literature for sputtered films [4,5]. The lower the thickness, the lower the e_{31}. The main explanation given for this behavior is based on the decrease of the domain boundaries mobility when thickness decreases.

Table 1. Thickness and e31 of sputtered PZT films poled at 5V for 5s

Thickness (nm)	130	180	260
e_{31} (C/m²)	-2.5	-3.5	-5.3

Finally, we did not notice any correlation between crystalline orientation and e_{31} values, whereas it is known that (100) orientation enhances the piezoelectric properties for thicker films with the same composition (morphotropic phase). The low thickness should play a role here.

Sol gel films

Comparatively, a sol gel film 160nm-thick exhibited a higher e_{31} value after a 5V poling: -5.8C/m². This is a general trend on all the films we prepared: sol gel piezoelectric properties are stronger than sputtered ones. In literature, Ledermann et al. reported a e_{31} around -7 C/m² for 1 µm thick (111) PZT film. However, much higher values (-18C/m²) have been reported for optimized 2µm thick sol gel films [7]. The comparison with literature is therefore difficult as all the data were given for thicknesses in the µm range.

Heating effect on the piezoelectric effect of PZT films

As mentioned in the introduction, the poling of PZT is mandatory for PZT in order to be piezoelectric. The PZT remnant polarization can be dramatically influenced by the process temperature experienced by the final device. The board assembly process for all integrated devices is done around 260°C. The PZT sample characterized in figure 4 was heated at 300°C during 1min in order to simulate the temperature of this final assembly. The piezoelectric voltage obtained thereafter is shown in figure 5. PZT films annealed at 260°C showed very similar results.

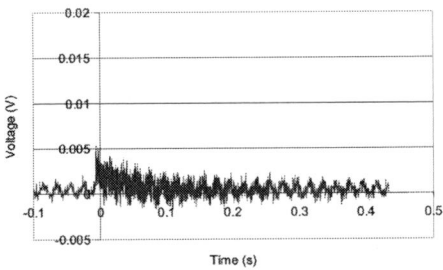

Figure 5: piezoelectric voltage obtained on a PZT thin film after heating at 300°C (or 260°C) during 1 min

The extracted effective e_{31} value is only -0.2C/m², meaning that the poling is almost completely removed by the heating process. This experiment proves that this assembly reflow process can be detrimental for the integrated devices with PZT thin films. As already discussed, the solution proposed is the use of PZT films with very low thickness.

ELASTIC-PIEZOELECTRIC BIMORPH
Technology

The elastic material used in the stack is silicon nitride (SiN). The main reason is that it gives the opportunity to control the residual stress in the released beam. Thereafter, the final shape of the beam profile can be flat. All the layers from the piezoelectric stack Pt/PZT/Pt are under tensile stress after the PZT annealing. The residual stresses are 1.1GPa for Pt and 200MPa for PZT. The only way to balance this stress is to use an elastic layer under tensile stress. Since the final released is done using XeF₂ process, Si cannot be used as elastic layer. In this study, the Si rich-

SiN exhibits a 400MPa-tensile stress. The simplest stack used for the realization of bimorphs is Si substrate/SiN 1 µm elastic layer/TiO₂ 2nm adhesion layer/Pt 50nm bottom electrode / PZT of various thickness / Pt 25nm top electrode. It is shown schematically in figure 6.

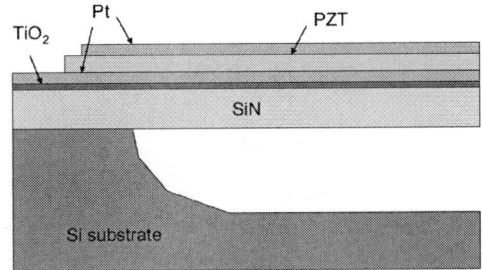

Figure 6: Profile of the realized stack

Analytical model

The model describing the deflection of a free-clamped beam versus a constant voltage is well described in the literature. See for example Smits [8]. The model gives the opportunity to take into account the influence of the four layers. However, in our case, the most important mechanical layer is the SiN film as its thickness is around 10 times higher than other individual layers. Moreover, its Young modulus is the highest among the materials involved. For a coarse estimation, it is by far easier to use the 2-layers model, allowing a better understanding of the whole behavior of the bimorph. The equation (4) gives the deflection z of the bimorph free-end versus the voltage applied V.

$$z = \frac{3E_2 t_1 t_2 (t_1 + t_2) L^2}{E_1^2 t_1^4 + E_2^2 t_2^4 + 2E_1 E_2 t_1 t_2 (2t_1^2 + 3t_1 t_2 + 2t_2^2)} \left(e_{31} \frac{V}{t_1} \right) \quad (4)$$

Here, 1 and 2 indexes refer respectively to the piezoelectric and the SiN layer. E_i is the Young modulus of the layer i, The Young modulus of both SiN and PZT are respectively 220GPa and 120GPa. t_i is the thickness of the layer i and L is the length of the bimorph. Since $t2 >> t1$, equation (4) can be simplified to:

$$z \approx \frac{3e_{31} V}{E_2 t_2^2} \quad (5)$$

This relationship is quite simple and gives an estimation of the e_{31} extracted from an experiment based on the reverse piezoelectric effect. The results obtained from this simplified equation are very similar to the one obtained from the complete model which takes into account all the layers.

Deflection characterization

The deflection is measured by using a white light interferometer (Wyko). A DC voltage is applied between the two electrodes and the deflection is measured at the free end of the bimorph as shown in Figure 7. This specific bimorph

978-1-4244-2977-6/09 $25.00 © 2009 IEEE

is made with a 260nm-thick sputtered PZT film. The ferroelectric behavior of PZT films is revealed by the hysteretic curve. The maximum differential deflection measured is 7.5μm at 5.4V, which is similar to literature for comparable cantilevers [2], but with 3 times thinner PZT films in our case.

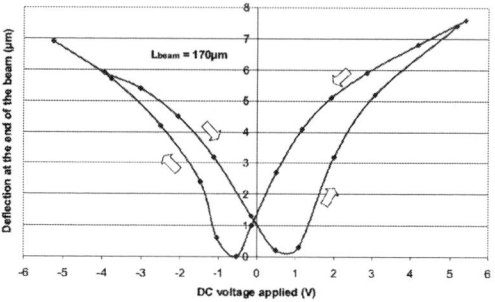

Figure 7: Deflection of a SiN/Pt/PZT/Pt cantilever versus the DC voltage applied measured with a white light interferometer (5s duration for each point – no pre-poling)

Figure 8 is an illustration of the bimorph shape following the DC voltage applied.

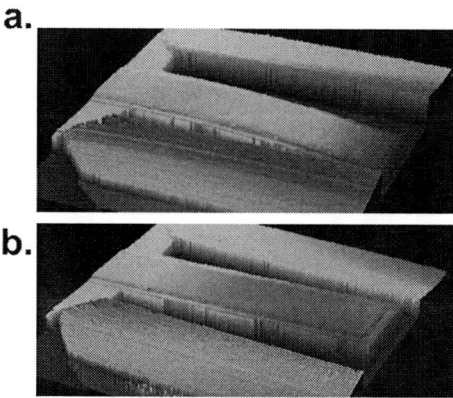

Figure 8: Deflection of the cantilever (L=170μm) a. at 0V; b. at 5V

One can notice that the deflection in this case is always upwards at high voltage, whatever the polarity. Indeed, as the voltage applied induces an electric field stronger than the coercive field, polarization and electric field are in the same direction. In this case, the PZT film always experiences a positive out-of-plane strain and then a negative in-plane strain, resulting in the upwards deflection of the bimorph.

Piezoelectric reverse effect characterization

The e_{31} extracted from the deflection with the linear analytical model previously discussed is -5.5C/m², close to the results obtained from the direct measurement method. This reverse characterization gives a high confidence of the e_{31} values obtained from the direct piezoelectric measurement.

RELIABILITY - AGING

Figure 9 shows the deflection measured on a bimorph realized with a 160nm-thick sol gel film versus the number of unipolar cycles (0V-3V). The bimorph exhibits a 10% decrease in its deflection at 3V after 5 billions cycles. This proves that the very thin PZT films can be very reliable for low voltage applications.

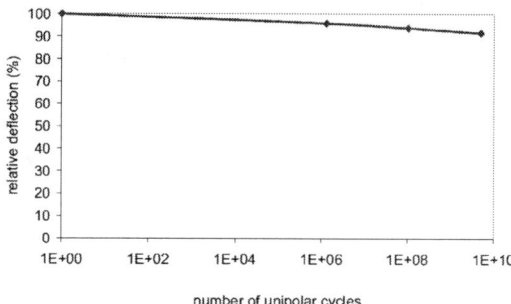

Figure 9: Relative deflection of a bimorph measured after unipolar cycling

CONCLUSION

This study demonstrated the potential of very thin PZT films for low voltage applications. In the 100-250nm range, both sputtered and sol gel exhibit high e_{31} values reaching -5.8C/m² for sol gel films. However, the influence of the thickness is obvious and a compromise should be done depending on applications. Bimorphs were realized exhibiting deflections higher than 5μm at 5V. Moreover, these bimorphs showed only a 10% decrease in their deflection after 5 billions unipolar cycles.

REFERENCES

[1] Filhol F; Defaÿ E; Divoux C; Zinck C; Delaye M-T ; Proceedings IEEE-LEOS Conference Optical MEMS 2004 (Takamatsu, Japan); p. 190-191.

[2] Jae-Hyoung Park, Youg-Dae Kim, Youg-Hee Park, Hee-Chul Lee, Hyouk Kwon, Hyo-Jin Nam and Jong-Uk Bu, MEMS 2007 (Kobe, Japan), pp. 163-166.

[3] Muralt-P; Ledermann-N; Paborowski-J; Barzegar-A; Gentil-S; Belgacem-B; Petitgrand-S; Bosseboeuf-A; Setter-N, IEEE Transactions on UFFC 2005; 52(12): 2276-88

[4] Ledermann-N; Muralt-P; Baborowski-J; Gentil-S; Mukati-K; Cantoni-M; Seifert-A; Setter-N, Sensors and Actuators A; A105(2) (2003),pp. 162-170

[5] Haccart T., Cattan E. and Rèmiens D., Semicond. Phys. – Quantum Electronics and Optoelectronics, vol. 77 (2002), pp. 78-88

[6] Defaÿ E; Zinck C; Malhaire C, Baboux N, Barbier D; Review of Scientific Instrument; vol. 77 (2006), pp. 103903-7

[7] Calame-F; Muralt-P, Applied Physics Letters, 90(6) (2007), pp. 62907-1-3

[8] Smits-JG; Dalke-SI; Cooney-TK, Sensors-and-Actuators-A, 28(1), 41 (1991)

CMOS-COMPATIBLE SURFACE-MICROMACHINED TEST STRUCTURE FOR DETERMINATION OF THERMAL CONDUCTIVITY OF THIN FILM MATERIALS BASED ON SEEBECK EFFECT

Z. Wang[1,2], P. Fiorini[3], and C. Van Hoof[3,2]
[1]Holst Center/IMEC, Eindhoven, THE NETHERLANDS
[2]Catholic University of Leuven, Leuven, BELGIUM
[3]IMEC vzw, Heverlee, BELGIUM

ABSTRACT

This paper reports the design, modeling, fabrication and measurement of a CMOS-compatible surface-micromachined test structure for the determination of the thermal conductivity of thin films based on the Seebeck effect. The Seebeck effect-based temperature sensing is more advantageous for thin film materials with a relatively large Seebeck coefficient, such as lightly doped poly-Si and poly-SiGe. In this paper, the conceptual design is first analyzed and then verified with finite element modeling. The test structure is fabricated with poly-Si$_{70\%}$Ge$_{30\%}$. Its functionality is demonstrated from experimental results. The sources of the measurement error are discussed and the solutions to minimize the measurement error are proposed.

INTRODUCTION

The thermal conductivity of thin film materials, such as polycrystalline silicon (poly-Si) and polycrystalline silicon germanium (poly-SiGe), is an essential material property for both modern very large scale integration (VLSI) circuits and microelectromechanical system (MEMS) devices. The thermal conductivity of thin film sample can be largely different from that of bulk sample [1]. Moreover, the thermal conductivity is strongly dependent on the exact processing condition, because the heat transport is to a large extent determined by various process-related factors, such as the stoichiometry, the crystalline structure and the grain size. These facts highlight the need to precisely measure the thermal conductivity on thin film basis.

To measure the thermal conductivity, the temperature difference, induced by a certain heat flow, across the sample material needs to be measured. Most of the traditional micromachined test structures take advantage of the fact that the electrical resistance of some metals and semiconductors, such as aluminum and poly-Si, changes with temperature [2-5]. The percentage of resistance change per unit temperature difference is denoted as temperature coefficient of resistance (TCR). However, the TCR-based measurement can be complicated by the variable temperature profile in the sample material and by the temperature-dependent parasitic resistance. Moreover, KOH bulk etching is used in the fabrication process of most of the traditional test structures, which prevents or complicates their monolithic integration.

Besides the TCR-based principle, the Seebeck effect [6] can be employed for temperature sensing as well. Because the Seebeck voltage is determined only by the net temperature difference, it is independent from the particular temperature profile. This feature gives an edge to the Seebeck effect-based temperature measurement in terms of accuracy.

This paper introduces a new surface-micromachined test structure based on the Seebeck effect. Its use is especially advantageous for materials with a relatively large Seebeck coefficient α, such as lightly doped poly-Si and poly-SiGe. For these materials, the test structure can be made relatively simple by merging together the heater, the temperature sensor and the sample material. For materials having a relatively low α, such as heavily doped poly-Si and poly-SiGe, the test structure can be employed as well (for details please refer to the DESIGN section). The CMOS-compatible fabrication process of this test structure makes it possible to monolithically integrate it with VLSI circuits or MEMS devices.

DESIGN

The test structure is schematically shown in Figure 1. It essentially consists of two suspended thermocouples, which are connected by a heater, anchored on Si$_3$N$_4$-covered Si substrate. As illustrated in Figure 1, the n-type thermocouple legs, devised in meandering shape, are longer and narrower than the p-type thermocouple legs. Metal pads are made at the thermocouple junctions ("A" and "B") for electrical interconnect and on top of the anchors for proper measurement ("C", "D", "E" and "F").

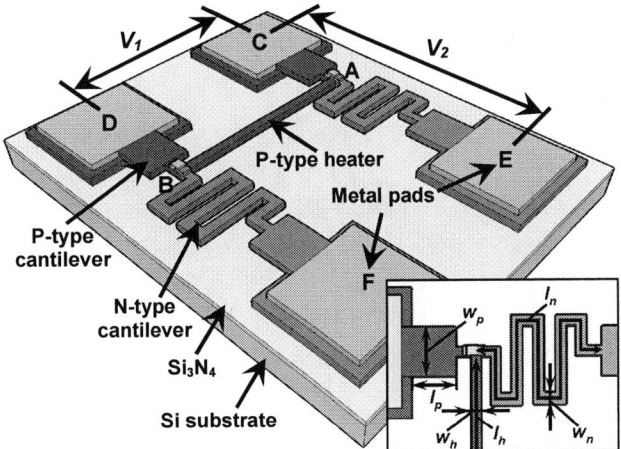

Figure 1: Schematic design of the Seebeck effect-based test structure (not drawn to scale); inset defines the geometries.

To measure the thermal conductivity λ_p of p-type material, an AC square voltage V_1 is first applied between "C" and "D". Due to Joule heating, a heat Q is generated mainly between "A" and "B", because the portion between "A" and "B" is much narrower and longer than the two cantilevers. The value of Q can be roughly calculated from the applied AC voltage V_1 and the measured electrical resistance R between "C" and "D".

978-1-4244-2977-6/09 $25.00 © 2009 IEEE

The heat flows into the Si substrate, causing a temperature difference ΔT between points "A" or "B" and the Si substrate. This ΔT can be determined by measuring the established Seebeck voltage V_2 either between "C" and "E" or between "D" and "F", provided that the Seebeck coefficients α_p and α_n for p-type and n-type materials are known. V_2 as a DC voltage can be measured accurately despite the partial superimposition of the AC voltage V_1 onto it. On the other hand, due to the large difference in thermal resistances of p-type and n-type cantilevers, most of the generated heat Q flows through p-type cantilevers into the Si substrate. Thus the thermal resistance R_{th} of p-type cantilevers is expressed as

$$R_{th} = \frac{\Delta T}{Q} = \frac{V_2 R}{V_1^2 (\alpha_p + \alpha_n)} \qquad (1).$$

R_{th} is composed of two components: the one-dimensional thermal resistance R_{th_1D} and the spreading thermal resistance R_{th_s}. R_{th_1D} is expressed as

$$R_{th\_1D} = \frac{l_p}{\lambda_p w_p t} \qquad (2)$$

where l_p, w_p and t denote the length, the width and the thickness of p-type cantilevers, respectively, as shown in the inset of Figure 1. The R_{th_s} arises from the different cross-sectional areas of the heater and the cantilevers. Our case can be simplified as a heat source of isoflux strip attached to an isotropic rectangular flux channel [7]. R_{th_s} is expressed then as

$$R_{th\_s} = \frac{1}{\lambda_p \pi^3 \left(\dfrac{w_h}{w_p}\right)^2} \sum_{n=1}^{\infty} \frac{\sin^2\left(n\pi \dfrac{w_h}{w_p}\right)}{n^3} \qquad (3)$$

where w_h denotes the width of the p-type heater. From Equation (2) and (3), it can be observed that R_{th} is inversely proportional to λ_p through a constant c, which is determined only by the geometries of the test structure. Thus, R_{th} is rewritten as

$$R_{th} = R_{th\_1D} + R_{th\_s} = \frac{c}{\lambda_p} \qquad (4),$$

$$c = \frac{l_p}{w_p t} + \frac{1}{\pi^3 \left(\dfrac{w_h}{w_p}\right)^2} \sum_{n=1}^{\infty} \frac{\sin^2\left(n\pi \dfrac{w_h}{w_p}\right)}{n^3} \qquad (5).$$

Hence, λ_p is obtained as

$$\lambda_p = \frac{c V_1^2 (\alpha_p + \alpha_n)}{V_2 R} \qquad (6).$$

The thermal conductivity of n-type material can be measured in the same way on a test structure wherein the p-type and n-type materials are switched over as opposed to the scheme shown in Figure 1. For sample materials having a relatively low α, such as heavily doped poly-Si and poly-SiGe, this test structure can be built in such a way that the short cantilevers and the heater are made with sample material while the long meandering cantilevers are made with another material having a large α. In this case, the heating is implemented in the sample material while the temperature sensing is implemented mainly through the material having a large α.

The rest of this paper is focused on the detailed analysis of a test structure made of poly-Si$_{70\%}$Ge$_{30\%}$ deposited by Low Temperature Chemical Vapor Deposition (LPCVD) due to out interest in its thermoelectric properties [8].

MODELING

Finite element modeling (FEM) is implemented to verify the analytical modeling with the software MSC.Marc [9]. The finite element model is shown in Figure 2. Thanks to the structural symmetry, it is enough to model only half of the test structure. An electrical-thermal coupled analysis is performed to simulate the function of the test structure. A current is injected into the middle cross section ("A" in Figure 2) of the p-type heater. At the bottom surface ("B" in Figure 2) of the anchor, the electrical potential is grounded and the temperature is fixed.

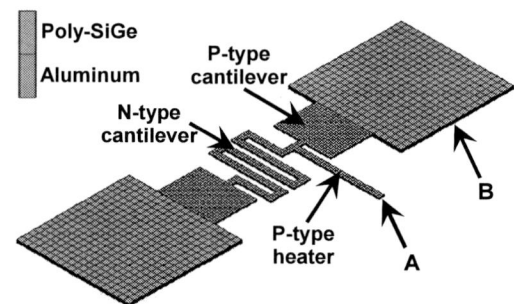

Figure 2: Finite element model for the designed test structure built with MSC.Marc.

The simulated temperature distribution within the p-type cantilever due to Joule heating with the mentioned boundary conditions is shown in Figure 3.

Figure 3: Simulated temperature distribution within the p-type cantilever due to Joule heating. The spreading of heat flow is clearly observed.

Based on FEM result, the thermal resistance R_{th} of the p-type cantilever can be calculated. It is 4.39×10^5 K/W with the assumption that the generated heat completely flows through the p-type cantilever. Calculated with the same set of parameters analytically with Equation (4) and (5), the thermal resistance R_{th} of the p-type cantilever is 4.58×10^5 K/W, which is 4.3% larger than the value obtained from FEM. If the heat loss through the n-type cantilever is considered, the gap between the two values narrows down further to 3.6%. This consistency confirms the correctness of the analytical expressions for the thermal resistance of the cantilever and hence of the value of the thermal conductivity derived from it.

FABRICATION

The designed test structure is fabricated by using surface micromachining technology, as schematically depicted in Figure 4. A layer of 1.2-μm-thick sacrificial SiO_2 is first deposited with Plasma Enhance Chemical Vapor Deposition (PECVD) and patterned with BHF on Si_3N_4-covered Si substrate (Figure 4(a)). Then a layer of 1.0-μm-thick p-type poly-$Si_{70\%}Ge_{30\%}$ is deposited with LPCVD and patterned by Reactive Ion Etching (RIE) (Figure 4(b)). So is the n-type poly-$Si_{70\%}Ge_{30\%}$ layer (Figure 4(c)). To electrically interconnect the p-type and n-type poly-$Si_{70\%}Ge_{30\%}$, a layer of 1.0-μm-thick aluminum is sputtered and then patterned with wet etch (Figure 4(d)). Finally, the sacrificial SiO_2 is released in diluted BHF and then dried in critical point dryer (Figure 4(e)).

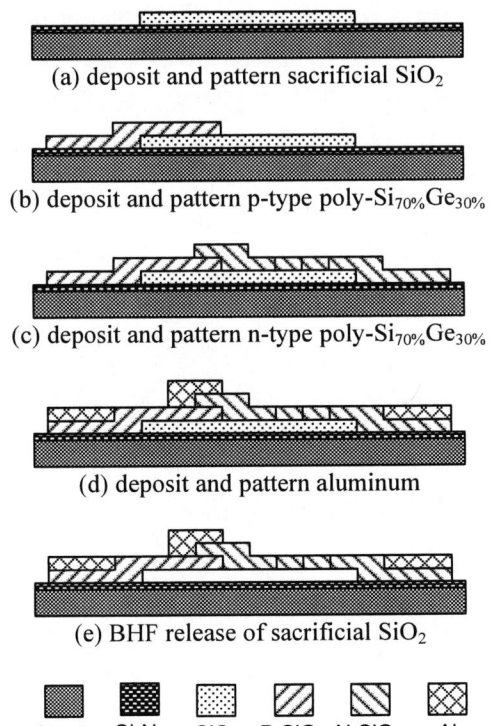

(a) deposit and pattern sacrificial SiO_2

(b) deposit and pattern p-type poly-$Si_{70\%}Ge_{30\%}$

(c) deposit and pattern n-type poly-$Si_{70\%}Ge_{30\%}$

(d) deposit and pattern aluminum

(e) BHF release of sacrificial SiO_2

Si sub Si_3N_4 SiO_2 P-SiGe   N-SiGe   Al

Figure 4: Schematic process flow for the designed test structure (not drawn to scale).

Because no exotic process steps are involved in this process flow, it is completely CMOS-compatible. The fabricated test structures are show in Figure 5. Each test structure occupies a footprint area of around 200 μm × 300 μm. The width of the p-type heater and the n-type

cantilever is variable from 4 μm to 10 μm. From Figure 5(b), it can be clearly observed that the release is complete and moreover, no stiction occurs.

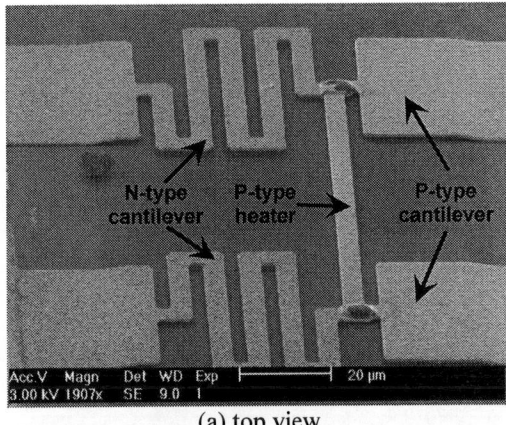

(a) top view

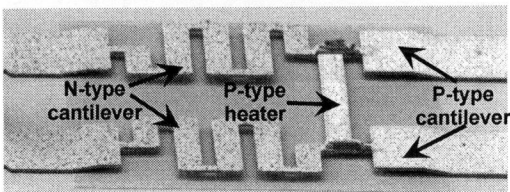

(b) side view

Figure 5: SEM micrographs of the fabricated test structure: (a) top view; (b) side view.

MEASUREMENT

The application of the designed test structure requires *a priori* knowledge of the Seebeck coefficients for the p-type and n-type poly-$Si_{70\%}Ge_{30\%}$. This issue is addressed by measuring macroscopic poly-$Si_{70\%}Ge_{30\%}$ samples on a custom designed set-up. As shown in Figure 6, the Seebeck coefficients measured at variable temperature differences are about 40 μV/K and 190 μV/K, respectively.

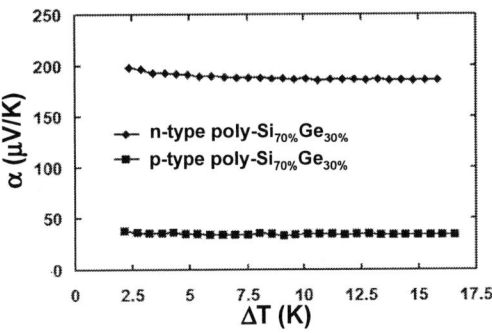

Figure 6: Measured Seebeck coefficients for p-type and n-type poly-$Si_{70\%}Ge_{30\%}$ at variable temperature differences.

The thermal measurement is performed in a manual vacuum prober system PMV150 from SüssMicrotec to minimize the thermal loss through air convection and conduction. The experimental vacuum is below 10^{-4} mBar. A function generator is used to apply the AC square voltage V_1 and meanwhile, a Keithley digital multimeter is used to measure the generated Seebeck voltage V_2. In the experiment, V_1 is increased stepwise from 50 mV to 400 mV. The measured V_2 on a fabricated

978-1-4244-2977-6/09 $25.00 © 2009 IEEE

test structure having a 7.0-μm-wide heater is plotted against the square of V_1, as shown in Figure 7. As anticipated from Equation (1), V_2 is linearly proportional with V_1^2.

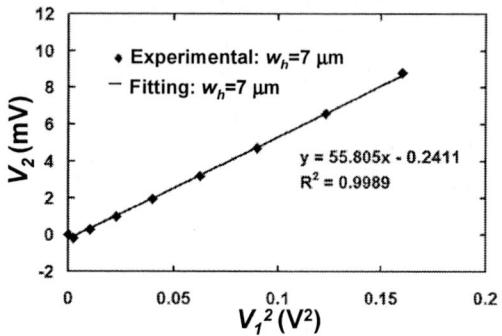

Figure 7: Measured Seebeck voltage V_2 as a function of applied AC voltage V_1.

In the analytical modeling, R denotes the total electrical resistance between "C" and "D" in Figure 1. However, heat is generated mainly along the heater between "A" and "B" in Figure 1. Therefore, λ_p needs to be corrected by multiplying it with

$$c_1 = \frac{2l_p/w_p + l_h/w_h}{l_h/w_h} \quad (7).$$

Another measurement error arises from the heat loss through the n-type cantilevers. Thus, λ_p can be further corrected by multiplying it with

$$c_2 = \frac{l_p/w_p + l_n/w_n}{l_n/w_n} \quad (8).$$

Hence after corrections, the measured λ_p of the p-type poly-Si$_{70\%}$Ge$_{30\%}$ is about 3.3 W/m/K. A variety of test structures having different feature sizes are measured in the same way. The results are shown in Figure 8.

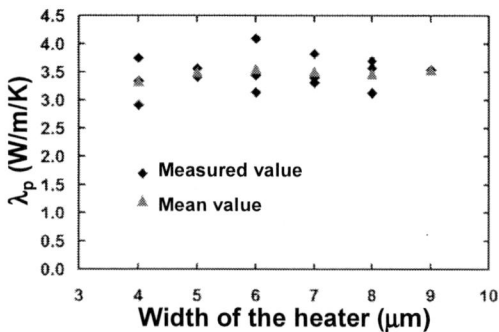

Figure 8: Measured thermal conductivity of p-type poly-Si$_{70\%}$Ge$_{30\%}$ on a variety of test structures.

DISCUSSION

As shown in Figure 8, the mean values ("▲") of λ_p measured on various test structures tend to converge towards about 3.5 W/m/K, which is consistent with values reported in literature [10]. This convergence demonstrates the functionality of the designed test structure and the absence of systematic errors related to structural dimensions. Although some scattering is observed, it can be largely minimized. The scattering is firstly attributed to the non-uniformity associated with the fabrication process. The variation in the thicknesses of various layers is estimated to be about 5~10% within an 8-inch Si wafer. Similarly, the Seebeck coefficients of poly-Si$_{70\%}$Ge$_{30\%}$ layers also slightly vary due to processing non-uniformity. The influence of process non-uniformity can be minimized by optimizing processing parameters and performing individual thickness measurement for each test structure. Moreover, the electrical resistance of the test structure is actually coupled to the temperature as well, especially for materials having a large TCR. This issue can be addressed by a more comprehensive data analysis, in which the change of the electrical resistance due to the Joule heating is taken into consideration.

CONCLUSIONS

A new test structure for determination of thermal conductivity of thin film materials is designed, modeled, fabricated and measured. Based on the Seebeck effect, the temperature sensing is more accurate than the TCR-based principle. The analytical modeling is verified by FEM modeling. Fabricated with surface micromachining technology, the test structure can be monolithically integrated with CMOS circuits and MEMS devices. Measured with this test structure, the thermal conductivity of the p-type poly-Si$_{70\%}$Ge$_{30\%}$ thin film is about 3.5 W/m/K. The sources of measurement error associated with this test structure are discussed. And the solutions to minimize the measurement error are proposed.

REFERENCES

[1] M. von Arx, O. Paul, H. Baltes, *J. Microelectromech Syst.*, vol. 9, pp.136-145, 2000.

[2] O. Paul, P. Ruther, L. Plattner, H. Baltes, *IEEE Trans. Semi. Manu.*, vol. 13, pp.159-166, 2000.

[3] N. Stojanovic, J. Yun, E. B. K. Washington, J. M. Berg, M. W. Holtz, H. Temkin, *J. Microelectromech Syst.*, vol.16, pp.1269-1275, 2007.

[4] L. La Spina, A. W. van Herwaarden, H. Schellevis, W. H. A. Wien, N. Nenadovic, L. K. Nanver, *J. Microelectromech Syst.*, vol.16, pp.675-683, 2007.

[5] A. McConnell, S. Uma, K. E. Goodson, *J. Microelectromech Syst.*, vol. 10, pp.360-369, 2001.

[6] D. M. Rowe, *CRC Handbook of Thermoelectrics*, CRC Press, Inc., New York, 1995.

[7] M. M. Yovanovich, Y. S. Muzychka, J. R. Culham, *Journal of Thermophysics and Heat Transfer*, Vol. 13, pp.495-500, 1999.

[8] Z. Wang, V. Leonov, P. Fiorini, C. Van Hoof, *Proc. Eurosensors '08*, Dresden, Sep. 7-10, 2008, pp.1420-1423.

[9] *MSC.Marc User's Guide Version 2000*, MSC.Software Corporation, Santa Ana, CA, 2000.

[10] D. D. L. Wijngaards, R. F. Wolffenbuttel, *IEEE Trans. Elec. Dev.*, vol. 52, pp. 1014-1025, 2005.

A BIOCOMPATIBLE AND FLEXIBLE RF CMOS TECHNOLOGY AND THE CHARACTERIZATION OF THE FLEXIBLE MOS TRANSISTORS UNDER BENDING STRESSES

C-Y Hsieh[1], J-S Chen[2], W-A Tsou[2], Y-T Yeh[1], K-A Wen[2], and L-S Fan[1]*

[1]Institute of NanoEngineering and MicroSystems, National Tsing Hua University, Taiwan
[2]Department of Electronics Engineering, National Chiao Tung University, Taiwan
*Correspondence: lsfan@ieee.org

ABSTRACT

To enable medical implants such as artificial retina, smart stent microsensors and other implantable wireless sensors, we developed a biocompatible and flexible RF CMOS technology based on a 0.18 μm CMOS process on 8-inch SOI (silicon on insulator) wafers. The silicon substrate for MOS transistors is 1μm thick and sandwiched between two parylene layers. Since the potential implantable microsystems are intended to operate under external mechanical stresses, the effects of bending stresses (between -100 MPa to 100 MPa) on the flexible electronic devices are characterized. The piezo-coefficients for the flexible MOS transistors are extracted from measured I-V characteristics. While carrier mobility is linearly related to the stresses in both longitudinal and transverse directions, the threshold voltage is relatively insensitive to stresses. The experiment results can be used for pre-compensations in circuits design based on this technology.

INTRODUCTION

With the progress of technology, implantable microsystems will become more and more important for future medical applications. There are bottlenecks of the development of implantable microsystems including the biocompatibility, reliability, package methods, energy source and communications issues. We started an effort of integration of RF transmitter and low noise amplifier circuit in a biocompatible, flexible CMOS MEMS process. To enable medical implants such as artificial retina, smart stent microsensors and other implantable wireless sensors, we developed a biocompatible and flexible RF CMOS technology based on a 0.18 μm CMOS technology on 8-inch SOI (silicon on insulator) wafers. The silicon substrate for MOS transistors is around 1μm in thickness and sandwiched between two layers of parylene material. Since the potential implantable microsystems are intended to operate under external mechanical stresses, the effects of bending stresses (between -100 MPa to 100 MPa) on the flexible electronic devices are characterized. The details of the flexible CMOS technology and the piezo characteristics of transistors are described in the following two sections. The RF microsystems characteristics will be reported elsewhere.

FLEXIBLE CMOS TECHNOLOGY

Figure 1 shows the die photo of a 0.18μm RF CMOS MEMS die integrated with 400MHz MICS transceiver and pressure sensors. Flexible IC was fabricated following a 0.18μm 1P6M CMOS process on 8" SOI wafers and a set of post-CMOS processing steps. The CMOS wafer was first coated with a parylene layer and the front side is bonded to a handle wafer, the silicon substrate was mechanically thinned down to 50 μm and the remaining substrate with the mechanical damage layer is removed by a silicon etch process. Finally, the backside was coated with another parylene layer and optionally the contact pads were opened by a masked RIE etching. The finished CMOS silicon substrate is 1μm thick and sandwiched between two 10μm-thick parylene passivation layers, and the final total thickness is 30 μm that light can pass through it (Fig.1(b)). Although the silicon substrate has an energy band gap of 1.1 eV, colors of shorter visible wavelength can still pass through this thin silicon layer. A 1.2cmX1.2cm strip was diced from the finished thin RF CMOS substrate, curled into a cylindrical shape and inserted into a glass capillary tube as shown in Figure 1(a). The 0.18um CMOS process is used so RF transceivers as shown in Figure 1(c) can be implemented. The I-V characteristics of the MOS transistor before and after the post processing are shown in Figure 2 and Table 1. The small degradation of drain current after the post-CMOS process may be due to the changed residual stresses and/or mechanical defects induced during the post processing.

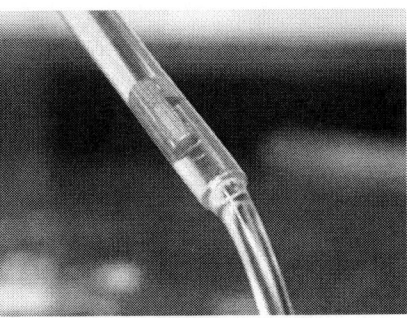

Fig. 1 (a)

Fig. 1 (b)

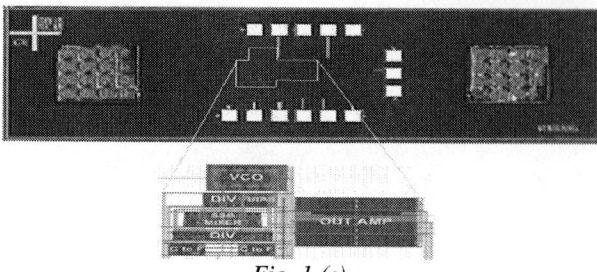

Fig. 1 (c)

Fig. 1(a) A flexible RF CMOS MEMS chip inserted into a capillary tube for flow measurements. (b) The flexible CMOS wafer become transparent after the post-CMOS process (5cmX4cm) (c) Photo of a 0.18um RF CMOS MEMS die integrated with 400MHz MICS transceiver and pressure sensors.

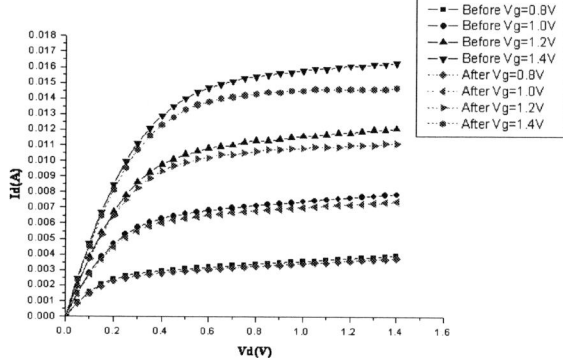

Fig. 2 The effect of post-processing on the measured I-V characteristics of an MOS transistor with W/L=50/0.18μm.

Table 1 Post-processing effects on MOS transistor

W \ Vt(Volts)	Before Process	After Process
25μm	0.567	0.560
50μm	0.549	0.551
75μm	0.546	0.543

TRANSISTOR CHARACTERIZATION

Since the implantable & flexible microsystems are intended to operate under external mechanical stresses, the effects of bending stresses on the MOS transistors are characterized. When a transistor is under mechanical stresses, the energy band structure will change, thus the effective mass of carriers and the associate density of states etc. will vary accordingly. A four-point-bending (4PB) test setup as shown in the schematics of Fig.3 is used to apply external mechanical stress on the flexible chips.

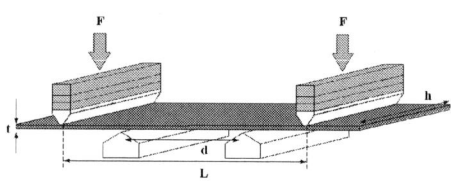

Fig. 3 The schematics of the 4PB fixture used to generate

uniform stress within the inner props.

A uniform uni-axial stress between the middle two props is established on the surface and it can be shown to be

$$\sigma = \frac{3(L-d)F}{ht^2}$$

(1)

when the beam deformation is small and the dimensions of t and h are small in comparison to d and L, where t denotes the thickness of the plate, h denotes the width of the plate, F denotes the applied force, L denotes the distance between the two upper props and d denotes the distance between the two lower props. The applied stress can be either tensile or compressive depends on whether $d < L$ or $d > L$. The mobility of electrons in the channel of the NMOS transistors and the threshold voltages are extracted from its I-V characteristics when the channel and immediate surrounding region is under tensile or compressive stress (between -100 MPa to 100 MPa) in both longitudinal and transverse directions in reference to the channel length direction. NMOS transistors with channel width of W= 45μm and 55μm, (channel length L = 0.18 μm and the gate oxide thickness is 42Å) were used for stress characterization. A piece of the flexible IC containing these individual MOS transistors was bonded onto a rectangular silicon strip 500μm in thickness. We kept the applied mechanical stress within +-100 MPa to avoid cracking the Si strip initiated from edge defects which happened when the stress was approaching 200MPa. The calibration of 4PB fixture was carried out by using a strain gauge (Vishay C061117-C) and Keithly 2410. The calibrated four-point bending jig is adapted to a shielded probe station to apply controlled uniform stresses (tensile or compressive) on the sample and the I-V characteristics are measured.

The current to voltage characteristics are measured by semiconductor parameter analyzer HP4156C. The measurements of transistor threshold voltage were all performed at V_{DS}=0.1V with linear extrapolation method. From the standard drain current expression of n-channel MOSFET in the linear region

$$I_d = \left(\mu_n C_{ox}\right)\left(\frac{W}{L}\right)\left[(V_{GS} - V_t)V_{DS} - \frac{1}{2}V_{DS}^2\right],$$

(2)

where μ_n is the carrier mobility in the channel, and C_{ox} is the oxide capacitance, V_{GS} is the gate to source voltage, V_t is the transistor threshold voltage and V_{DS} is the drain to source voltage. From the x-intercept of the I_d vs. V_{DS} curves, we get the threshold voltage as $V_t = V_{GS} - \frac{V_{DS}}{2}$ and from the slope $Slope = \mu_n C_{ox}(W/L)V_{DS}$ we can obtain the electron mobility.

The variation of threshold voltage and mobility under stress can be measured from the slope and intercept of transfer characteristics of the transistor the linear region as a function of applied stress. The normalized mobility change versus stresses is shown in Fig.4. The normalized mobility variation $\frac{\Delta\mu}{\mu}$ of the flexible MOS transistor under stresses is shown to be

$$\frac{\Delta\mu}{\mu} = \Pi_{T(orL)}\sigma \quad , \tag{3}$$

where Π_L is the longitudinal piezo-coefficient, and Π_T is the transversal piezo-coefficient. Table 2 summarizes the piezo-coefficients of the flexible transistors. The threshold voltage is found to only have small sensitivity to stress as shown in Fig.5. These parameters will be used for pre-compensations in circuits design based on this technology.

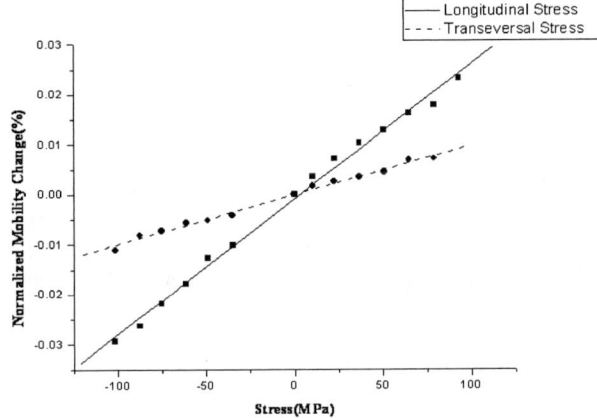

Fig. 4 Normalized mobility change versus applied stress of MOS transistors under tensile and compressive stresses in both the longitudinal direction and transverse directions relative to the transistor channel direction.

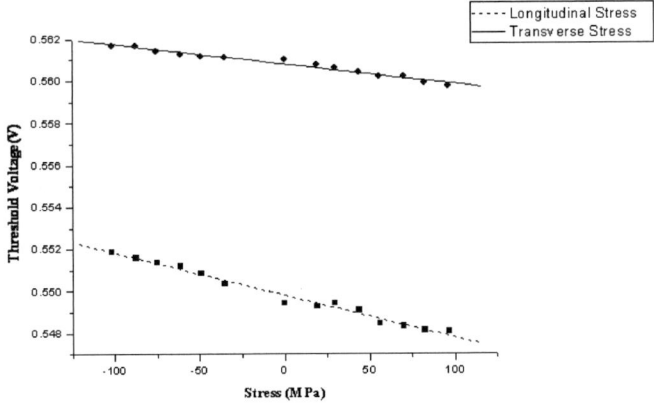

Fig. 5 Measured MOS transistor threshold voltage versus applied stress both tensile and compressive in longitudinal and transverse directions.

Table 2 Stress effects on a 0.18μm MOS transistor

Piezo Parameters	NMOS Devices($x10^{-12}Pa^{-1}$)
Π_L	266
Π_T	131

CONCLUSION

We have characterized the piezo-coefficients of the biocompatible flexible MOS transistors. A uniaxial mechanical stress has been applied to the test devices. The flexible MOS transistors function under either tensile or compressive strain. It has been shown that the threshold voltages of these MOS transistors are relatively insensitive to the applied stress and the mobility variation is linearly related to the external stresses within our measurement range. This work provides the needed MOS piezo-characteristics for circuit designers of the flexible integrated circuits.

REFERENCES

[1] C.Y. Hsieh, W.A. Tsou, Y.T. Yeh, K.A. Wen, L.S. Fan, *Proc. APCOT 2008*, pp. 746-748.

[2] R. Dekker, M.H.W.A. van Deurzen, W.van der. Einden, H.G.R. Mass, and A.G. Wagemans, "A low-cost substrate transfer technology for fully integrated transceivers," *in Proc-IEEE Bipolar/BiCMOS Circuits and Technology Meeting (BCTM)*, 1998, pp. 132-135.

RADIATION SENSITIVITY OF OHMIC RF-MEMS SWITCHES FOR SPATIAL APPLICATIONS

A. Tazzoli, G. Cellere, E. Autizi, V. Peretti, A. Paccagnella, and G. Meneghesso
Department of Information Engineering, University of Padova, Italy

ABSTRACT

The impact of 2MeV protons and 10keV x-rays radiation stresses on electrostatically actuated ohmic RF-MEMS switches has been analyzed at increasing radiation dose and during subsequent annealing at room temperature. Small variations of electrical parameters (actuation and release voltages) have been identified, accompanied by a strong rf-performances degradation. Monte Carlo TRIM simulations have been carried out to understand the mechanisms responsible of such degradations, finding that both NIEL and ionizing damages appear to play an important role.

INTRODUCTION

Radio Frequency Micro-Electro-Mechanical Systems (RF-MEMS) have been proved to be interesting candidates to overcome the limits of actual state-of-art solid state devices. Good rf performances, optimum linearity, low power consumption, and low cost production are just some peculiarities of RF-MEMS switches very useful for today wired and wireless terrestrial applications [1]. Furthermore, small dimensions and light weight make such devices very appealing for spatial applications. Despite these positive aspects, all the above and further benefits go along with a series of present-day shortcomings; these shortcomings are mostly related to the maturity of today's still-evolving design methodologies, lack of fabrication processes standardization, limited reliability database, and poor knowledge of ageing mechanisms and reliable design practices. These limits are even more important for the performance-hungry but reliability-critical satellite applications, since in the harsh space environment radiation induced damage is one of the main causes of failure [2]. In particular, the behavior after exposure to ionizing radiation of devices having mechanical motion governed by electric fields across insulators has been seldom studied, mainly in MEMS sensors [3] or capacitor-like structures [4]. Just few authors have analyzed complete RF-MEMS devices [5], [6].

This paper begins to fill the gap, presenting results on the reliability of ohmic RF-MEMS switches irradiated with 2MeV protons (AN2000 accelerator at INFN, Legnaro, Italy), and 10keV x-rays (INFN, Legnaro, Italy).

RADIATION STRESS

It is well known that space can be a harsh environment for traditional solid state devices. Ionizing radiation comes in form of protons or ions trapped in the van Allen belts, of solar wind (with its periodical variations), and of protons and ions of galactic origin [7]. Since several years, radiation and EOS/ESD phenomena have been addressed as the main causes of mission failure, as reported by Koons et al. [2]. Most works were carried out considering traditional solid state electronic components, but recent studies have indicated the possibility that also micro-mechanical structures can be hampered by radiation damages. The standard satellite radiation shielding has been proved to be a good solution to guarantee the electrical device lifetime during space missions, however, since the potential killer application of RF-MEMS devices would be on nano-satellites, the efficiency of such a protection on small and light satellites is yet to be verified. Further, even if a relatively thick shielding helps reducing the electrostatic charging and the total dose delivered to the electronic devices, it cannot totally suppress it.

The effects of ionizing radiation on microelectronic devices can be broadly divided into three categories. **Single Event Effects** are the macroscopic manifestation of single ions, such as the bit flip in a SRAM. To date, this should not be a problem with MEMS. **Total Ionizing Dose** effects are the progressive buildup of defects, mainly in the dielectric layers, due to the energy loss via ionization. Electrically active defects can be of different kinds, including charge trapping (E' and P_b centers), Si/SiO_2 interface defects, and the generation of mobile charge [8]. These sort of effects need to be considered for RF-MEMS since charge trapping is universally recognized as one of the most impairing problems for the reliability of electrostatically actuated switches. Finally, even the most ionizing particles lead to a certain amount of **Non Ionizing Energy Loss** (NIEL), that is, an energy loss due to interactions with atomic nuclei and not to the direct generation of electron/hole pairs. These interactions typically result in the generation of point defects which can, for example, degrade the gain of bipolar transistors [9]. In the same way, NIEL can lead to a modification of the Young module of adopted materials, going to modify the mechanical properties of the MEMS structural parts, but also making the device more vulnerable to creeping and fracture.

DEVICE DESCRIPTION

Focus of this work are several typologies of ohmic RF-MEMS switches built by FBK-IRST (TN, Italy). Devices have been designed in both shunt (normally closed) and series (normally open) configurations, with two kind of suspension structures: meanders based, with a low spring constant (see Figure 1a), and with straight beams (see Figure 1b) with an higher spring constant leading to an higher restoring force. Another design variation regards the way in which the suspended membrane gets in touch with the bottom metal layer when actuated. The contact is achieved by the impact on dimples (see Figure 1c), or simply with flat surfaces. A detailed description of the switches design, adopting an interdigitated topology in order to reduce charge trapping phenomena can be found in [10].

The technology utilized for the fabrication of ohmic RF-MEMS switches uses a surface micromachining process based on electrodeposited suspended gold for the membrane layer (see the cross-section in Figure 1d).

978-1-4244-2977-6/09 $25.00 © 2009 IEEE

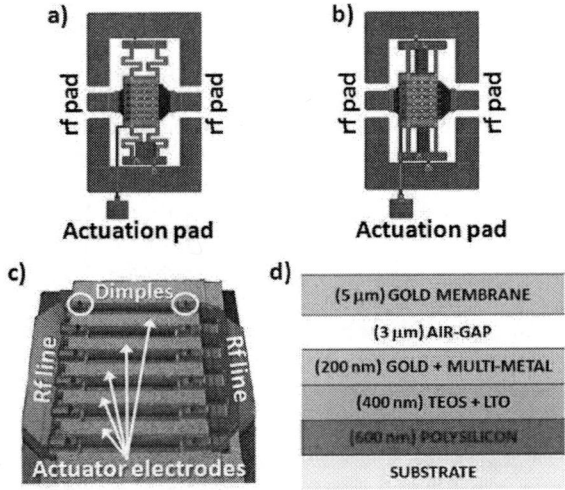

Figure 1: Tested devices layout: meanders based suspensions (a), and straight beam suspensions (b); optical profilometer image of a series switch (c) after suspended membrane removal; schematic process description (d).

The same gold layer is also available as low-loss RF signal path, whereas an high resistivity polysilicon layer is used for actuation electrodes, and a further Al-Ti-TiN multilayer for RF-signal underpasses. A thin evaporated gold layer is deposited over the signal underpass electrodes in order to achieve a gold-gold ohmic contact upon actuation. Direct contact is allowed only between the plate and the signal electrodes, by raising the signal underpass metal above the level of the poly electrodes, through the placement of poly dummy rectangular bricks. A 5μm thick electroplated gold layer is used for the plate to improve its rigidity, while a thinner (1.5μm) gold membrane implements the four suspending beam springs. Due to requirements for the release etch step, the suspended membrane is realized as a perforated plate structure with 20x20μm holes with 20μm separation. More details on the FBK-IRST process can be found in [11].

EXPERIMENTAL RESULTS

In this work we have submitted RF-MEMS switches to protons and x-rays radiation sources analyzing the impact of such a stress on the rf and electrical parameters of the devices with the increase of the radiation dose, and during the successive days of annealing. The setup adopted to characterize the switches is based on a Vector Network Analyzer (Agilent HP8753), for the measurement of the rf-performances, and a Source Meter (Keithley 2612) to bias the device and measuring the current drained by the actuator structure. The measurement starts from 0V up to $+V_{MAX}$, goes back to 0V, decreases down to $-V_{MAX}$, and finally comes back to 0V. This procedure leads to the traditional hysteresis-like diagram reported in Figure 2 (consider the "fresh device" curve), and it is possible to extract the actuation voltage ($|V_{ACT}|$) the release voltage ($|V_{REL}|$), and the S_{11} and S_{21} parameters. Furthermore, considering the graphs symmetry or translation, it is possible to study the presence of charge trapping or redistribution phenomena. More details on the adopted setups are reported in [12].

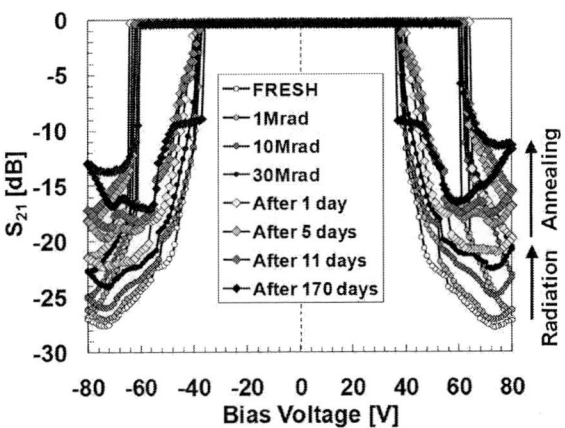

Figure 2: Degradation of S_{21} (V_{BIAS}) of a shunt switch (straight beams) during the protons radiation stress and the successive days of storage (test conditions: f_{rf} = 6GHz, P_{rf} = 0dBm).

An excerpt of the complete characterization carried out on a shunt switch after growing doses of 1, 10, and 30Mrad(SiO₂) protons stresses, and during the subsequent room temperature anneal, is reported in Figure 2.

The most notable result is the heavy degradation of the insertion losses, see Figure 2, with just a small variation of the actuation voltage, see Figure 3. The relatively small variations of the actuation voltage with the increasing dose appears to exclude large charge trapping as the only cause for insertion losses degradation.

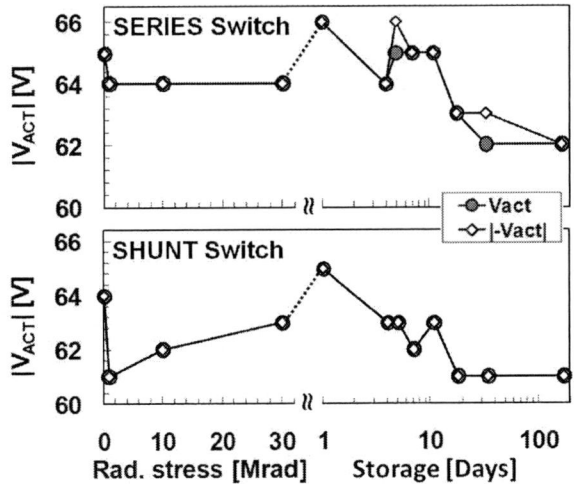

Figure 3: Evolution of V_{ACT} and $-V_{ACT}$ for a series switch (up) and a shunt one (bottom, from Figure 2) during the protons radiation stress and the successive days of storage.

In fact, positive charge trapping would have changed both the actuation and the release voltages toward lower values. Then charge de-trapping would have moved $|V_{ACT}|$ and $|V_{REL}|$ toward their pre-irradiation values. On the opposite, both follow a complex behavior with a fast degradation, a partial recovery during irradiation, and an apparent worsening during post-irradiation storage. Such a complex behavior suggests that the phenomena underlying the degradation of devices should have a complex nature, probably resulting from the superimposition of different effects having different temporal evolutions. This is confirmed by Figure 4, where we show the evolution of the S-parameters, which severely degrade during the storage (anneal).

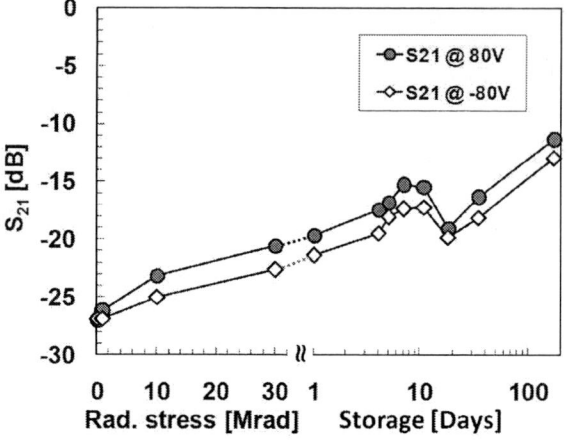

Figure 4: Evolution of the S_{21} parameter @ 80V and -80V extracted from curves of Figure 2.

It is believed that the degradation is not due to an increase of the coplanar waveguides (CPW) or substrate losses, since repeating the measurements at different frequency from 100kHz to 6GHz (also on simple CPW irradiated test structures) it does not show any frequency-related dependence, as shown in Figure 5.

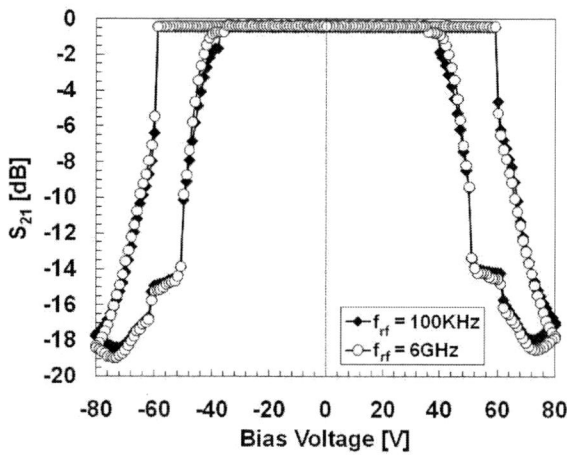

Figure 5: Comparison of S_{21} (V_{BIAS}) of a 30Mrad protons stressed shunt switch measured applying an rf signal of 100kHz, and 6GHz ($P_{rf} = 0dBm$). The curves are almost identical.

Furthermore, stressed devices have shown a faster degradation rate during cycling stresses (see Figure 6), and a possible explanation could be found in the degradation of the metal-to-metal contact (increased series resistance). In principle, the damage induced by 2MeV protons could be due to either displacement damage (lattice defects directly induced by proton-nucleus interactions), or to ionizing damage (protons generate columns of charges in the device that recombine and/or

move generating the actual damage) [13].

Monte Carlo simulations with TRIM code [14] (see Figure 7) show that (i) the range of 2MeV protons is more than enough to cross all the device active area.

This confirms that protons are crossing the whole device, and that all parts (bridge, actuators, substrate) may in principle be the origin for the observed degradation; (ii) the displacement damage is mainly located in the bulk of the silicon. Since this part of the device has no role in the electrical performances of the switch, the easiest conclusion one can draw is that performance degradation should be linked someway to ionizing damage; (iii) however, displacement damage in the gold layer is much higher than that in the surface silicon, due to the larger mass of the former. Hence, a contribution of NIEL cannot be neglected.

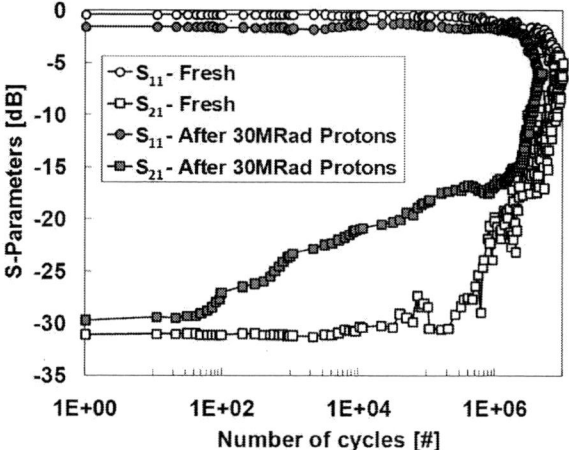

Figure 6: Comparison of the cycling robustness of an untreated series switch with a 30Mrad protons stressed one ($V_{bias} = 80V$, $f_{rf} = 6GHz$, $P_{rf} = 0dBm$).

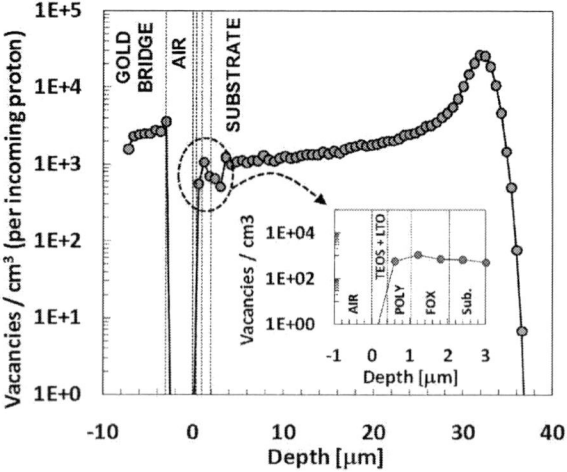

Figure 7: Monte Carlo TRIM simulation of displacement damage for 2MeV protons in our devices. The inset highlights the vacancies/cm^3 just below the air-gap.

The fluences used (up to 1.7×10^{13} for the higher dose) are actually compatible with literature results on displacement damage in silicon devices [15]. To shed a better light on this, we moved to a radiation source where displacement damage is in first approximation negligible, such as 10keV x-rays. 70 devices were tested with x-rays up to 1Mrad(SiO$_2$), and the summary of results are reported in Table 1.

Table 1: X-ray radiation stress induced failures. Number of tested devices: 30 meanders based, 40 straight beam.

Type of damage	Meanders based suspensions failure [%]	Straight beams suspensions failure [%]
Stiction	50	6
S-parameters degradation	42	24
Negligible variations	8	66
Actuation line damage	0	4

An interesting result is reported in Figure 8 (series switch), in which the radiation stress has degraded the S-parameters (S$_{21}$ in on state), with almost no changes in both V$_{ACT}$ and V$_{REL}$, and showing a good recovery after 1 month.

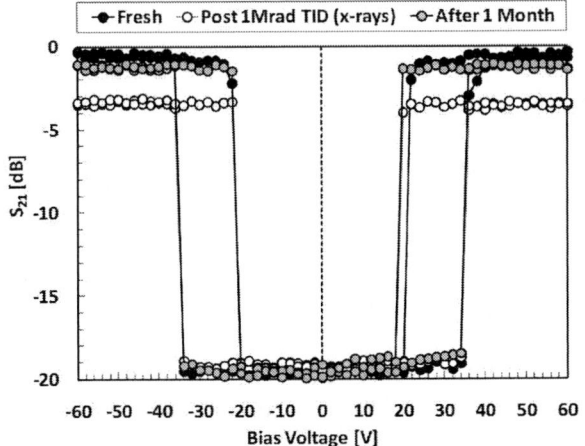

Figure 8: Comparison of S$_{21}$(V$_{BIAS}$) of a fresh shunt switch, after 1Mrad x-rays stress, and after 1 month of storage (showing the recovery).

This more pronounced recovery found after x-rays irradiation, if compared to protons, suggests that displacement damage may actually play a role in the degradation of switches performance. On the other side, recombination kinetics is not the same after irradiation with x-rays and protons (especially at relatively low energies such as those used here [13]), and phenomena purely linked to ionization damage cannot be excluded.

CONCLUSIONS

We have shown that RF-MEMS switches performances can be impacted by ionizing radiation. In particular, both the actuation voltage and the S-parameters are impaired.

The relatively small shift in actuation and release voltages do not appear as a limiting factor for the device reliability during operations, but the S-parameters degradations are potentially more critical, in particular as a consequence of the complex time evolution which complicate the development of a predictive model. Moreover, both NIEL and ionizing damage appear to play a role. Which one of the two is prevailing and why is still an open question which we will address with future works.

ACKWNOLEDGEMENT

Authors would like to thank Benno Margesin, Flavio Giacomozzi, and Jacopo Iannacci from FBK-IRST (Italy), and Roberto Gaddi (Cavendish Kinetics) for fruitful discussions and devices design and manufacturing. This work was partially supported by the Italian project PRIN 2005 entitled "Reliability of RF-MEMS for high frequency applications".

REFERENCES

[1] G. Rebeiz, et al., "RF MEMS Switches and Switch Circuits", IEEE Microw. Mag., Dec. 2001, pp. 59-71.

[2] H. C. Koons, et al., "The impact of the space environment on space systems", TR-99(1670)-1, 1999, pp. 7-10.

[3] A. R. Knudsen et al., "The effects of radiation on MEMS accelerometers", IEEE Trans. Nucl. Sci., 1996, vol. 43, pp. 3122.

[4] M. Exarchos et al., "Charging of Radiation Induced Defects in RF MEMS Dielectric Films", doi:10.1016/j.microrel.2006.07.045

[5] S. S. McClure, et al., "Radiation Effects in Micro-Electro-Mechanical Systems (MEMS): RF Relays", IEEE Trans. Nucl. Sci. 2002, vol. 49, pp. 3197.

[6] J. Ruan, et al., "Alpha particle radiation effects in RF MEMS capacitive switches", Micr. Rel. 48 (2008), pp. 1241-1244.

[7] E. Stassinopoulos, et al., "The space radiation environment for electronics", Proceedings of the IEEE, 76 (11), Nov. 1988, pp. 1423 – 1442.

[8] J. Schwank, "Total dose effects in MOS devices", in IEEE NSREC Short Course, 2002.

[9] D. Ball, R. Schrimpf, H. Barnaby, "Separation of ionization and displacement damage using gate-controlled lateral PNP bipolar transistors", IEEE Trans. on Nuclear Science, 49(6), 2002, pp. 3185-3190.

[10] R. Gaddi et. al, "Interdigitated Low-loss Ohmic RF-MEMS Switches", NSTI 2004, Vol. 2, pp. 327-330.

[11] F. Giacomozzi, et al., "Electromechanical Aspects in the Optimization of the Transmission Characteristics of Series Ohmic RF-Switches", MEMSWAVE 2004, pp. C25-C28.

[12] A. Tazzoli, et al., "Electrostatic Discharge and Cycling Effects on Ohmic and Capacitive RF-MEMS Switches", IEEE Trans. on Device and Materials Reliability, vol. 7, no. 3, Sept. 2007, pp. 429-437.

[13] T. P. Ma, et al., "Ionizing radiation effects in MOS devices and circuits", Wiley, New York, 1989.

[14] J. Ziegler, SRIM/TRIM computer code, available on line at www.srim.org.

[15] S. Messenger, et al., "Non-ionizing energy loss (NIEL) for heavy ions", IEEE Trans. on Nuclear Science, 1999, pp. 1595-1602.

RELIABILITY OF RF MEMS CAPACITIVE SWITCHES AND DISTRIBUTED MEMS PHASE SHIFTERS USING ALN DIELECTRIC

M. Fernández-Bolaños, D. Tsamadós, P. Dainesi, A. M. Ionescu

Nanolab, École Polytechnique Fédérale de Lausanne (EPFL), Lausanne, Switzerland

ABSTRACT

The reliability and charging/discharging dynamics of wideband (1.5-14GHz) phase shifters made of MEMS capacitive switches using Aluminum Nitride (AlN) as dielectric are originally reported. Phase shifter lifetimes exceeding 10^9 cycles are achieved in hot-cycling (+5dBm RF power). Dynamic tests were done for the first time under ambient conditions (50% humidity) over 2×10^9 cycles with no major degradation of *individual switches* performances. It is demonstrated that the phase shifter is very robust (no permanent failure or stiction) and can withstand environmental effects as well as high temperature variations, without the need of expensive hermetical packaging. The excellent reliability is attributed to the *slow dielectric charging* (a square-root time law) and *fast discharging mechanism* of AlN (an exponential time law proposed and validated in our work). We extend the validity of charging and discharging models from single device to arrays of parallel MEMS capacitors.

INTRODUCTION

Radio-frequency Micro-Electro-Mechanical Systems (RF MEMS) capacitive switches are very attractive for use in wireless applications such in phase shifters due to their excellent RF properties, potential low cost and low power consumption [1]. However, the reliability in RF MEMS capacitive switches is still the object of many publications and efforts [2-7] because the success of MEMS switches appears to be finally conditioned by their reliability and an adapted (low-cost) package. The reliability is still an issue and dielectric charging, the main responsible mechanism for failures in capacitive switches, is only partially understood. Moreover, very little results have been published on the discharging mechanism [5] and/or on the reliability of full RF-MEMS system [6], as the distributed MEMS transmission line (DMTL) *phase shifter*, which consists of *arrays of tens of MEMS capacitors* loaded in parallel on a coplanar waveguide (CPW). On the other hand, Aluminum Nitride (AlN) is found to be an excellent medium-dielectric-constant ($\varepsilon_r = 8.9$) insulating material and has many others outstanding physical properties [8], such as high thermal conductivity, and elevated breakdown dielectric strength, which could contribute to enhance device lifetimes.

The paper investigates the reliability of the digital DMTL phase shifter [7], which consists in 13 MEMS shunt capacitors (Figure 1). The novelty of this paper is the study on the charge's effects of AlN in MEMS switches, validating existing charging model [2] and developing an exponential discharging experimental law, valid also for arrays of parallel MEMS capacitors. For the first time excellent reliability has been measured under ambient conditions (50% of relative humidity). The devices are very robust and can withstand environmental effects as well as high temperature variations ($\Delta T = 180°$) [7] without permanent failures or stiction.

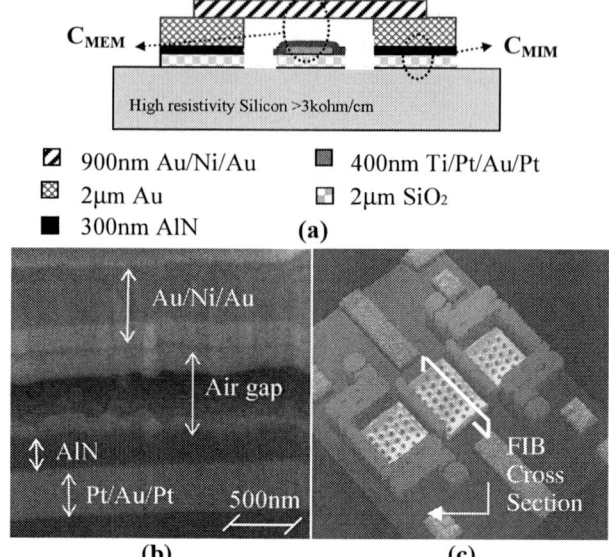

Figure 1: RF MEMS capacitive switch (a) Cross section schematic view illustrating the different material and thicknesses. C_{MEM} and C_{MIM} denoted the MEMS and the metal-insulator-metal capacitors. (b) Focused Ion Beam (FIB) cross section of the fabricated device. (c) Profilometer optic image of the device.

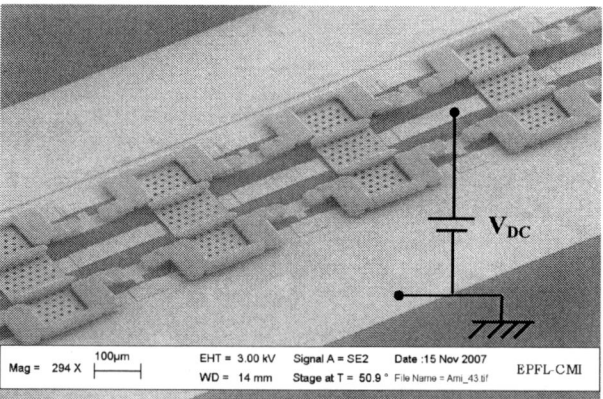

Figure 2: SEM image of the fabricated distributed MEMS digital phase shifter using 13-loaded RF MEMS capacitive switches which corresponds to the one of Figure 1 (only 4 switches are shown in this figure).

DEVICE CHARACTERIZATION

The fabrication and RF characterization of a single digital capacitor and the DMTL phase shifter is reported elsewhere [7]. A digital MEMS capacitor is defined as a standard MEMS capacitive switch in series with a metal-insulator-metal (MIM) fix capacitance which limits the value of the down state MEMS capacitance. A schematic representation of the capacitive switch with a SEM cross section view of a fabricated device is shown in Figure 1.

The RF MEMS shunt capacitors are electrostatically actuated from the bottom electrode applying a DC voltage

978-1-4244-2977-6/09 $25.00 © 2009 IEEE

inline with the RF signal. The suspended bridge (top electrode of the capacitor) is anchored to the ground plane of the CPW (Figure 2). When a voltage is applied across the electrodes, the electrostatic force pulls the suspended plate towards the bottom electrode, increasing quasi-linearly the value of the capacitance. However, at a certain voltage (pull-in, V_{PI}) the bridge becomes unstable and snaps down, suddenly increasing the value of the capacitance. Once the actuation voltage is decrease again, as the electrostatic force is larger to the restoring force of the bridge, it will state in the down state up to a pull-out voltage, (V_{PO}) lower to the V_{PI} (Figure 3).

The polarity of the applied voltage has no influence on the actuation since the electrostatic force is proportional to the voltage squared. Thus, the C-V curve should be symmetrical around V=0, although due to mobile and inhomogeneous charge in the dielectric, this is not always the case.

Figure 3 depicts the typical C-V curve of both, a single digital capacitor and a full array of capacitors loaded on a CPW, namely the DMTL phase shifter, which are *simultaneously operated*; One can observe a quasi-identical pull-in voltage (V_{PI}) for all the switches (~22V) and a symmetrical curve around V=0, which indicates either that no mobile residual charges exits or that they are compensated. The pull-out is slightly higher than 10V.

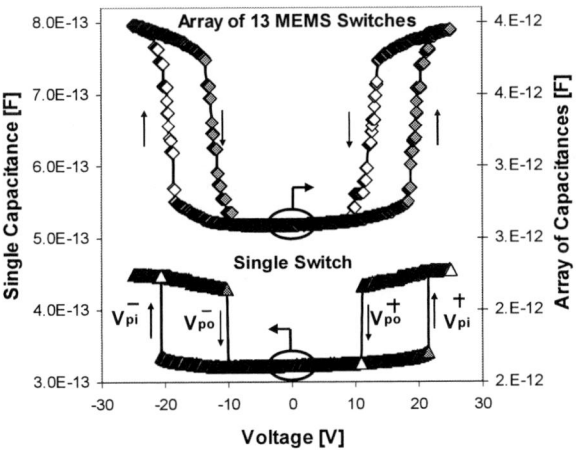

Figure 3: Measured C-V curves of a single digital capacitor and the array of parallel loaded 13 MEMS capacitors forming the phase shifter.

The RF performance of the *fresh phase shifter* has been measured (Figure 4) achieving a constant time delay of 25ps over a wide range of frequency (from 5GHz to 14GHz), with high reproducibility and accuracy (within few %). A phase shifter lifetime larger than 1×10^9 cycles and over 2×10^9 for individual switches have been demonstrated for "hot switching" conditions with a continuous RF power of +5dBm and a relative humidity of 50%.

These promising results suggest that the fabricated devices could work quite reliable without the need of an expensive hermetical packaging (mechanical protection is however, needed). The full system is actuated with a switching time lower than 22μs; the return loss is better than 10dB for both states along the 10^9 cycles and the insertion loss is in average lower than 2.5dB for a 5mm-long phase shifter (inset of Figure 4); both the speed and

the RF characteristics, are acceptable performance values for most phase shifter applications.

A slight degradation of the phase shifter performance is observed along with the first 10^8 switching cycles. However it seems to appear a fast trend toward saturation with no further degradation (Figure 4). The phase delay is reduced with the ageing tests in hot-switching conditions (+5dBm). After 10^9 cycles the phase shifter has reduced in 5ps its time delay which corresponds to an error of 2° at 1GHz (Figure 4). In the inset of Figure 4 it can be observed how the return loss (S_{11} and S_{22}) in the up state shifts with the cycling as well as the insertion loss slightly increases (presumably due to the increase in the up state capacitance). Nevertheless, the down-state performance does not apparently change most probably thanks to the MIM fix series capacitance [7].

After a number of cycles we presume that the suspended membranes do not recover the exactly original position due to inhomogeneous dielectric charging and/or mechanical degradation of the springs [9] and as a consequence the up-state capacitance (C_{UP}) slightly increases with the ageing (Figure 5). Thus, the maximum up state capacitance degradation is of the order of 2% after 10^3s of accumulated stress (A constant of 30s relaxation time is applied after each stress step).

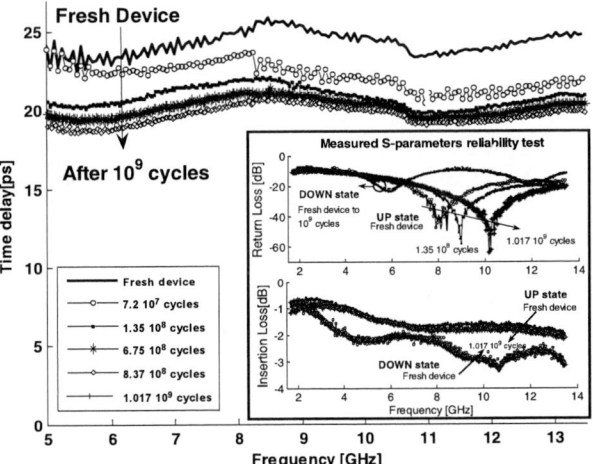

Figure 4: Time delay degradation versus frequency over the cycles. Inset: Evolution of S-parameters measurements of the DMTL phase shifter during/after 10^9 cycles.

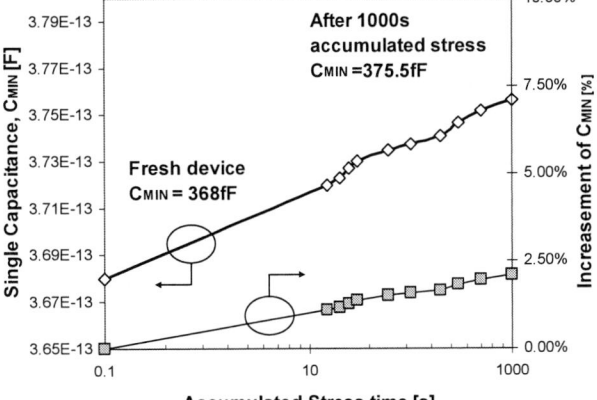

Figure 5: Evolution of the up-state capacitance, C_{UP}, of a single capacitor versus the accumulating stressing time and the increase of the C_{UP} in percentage with ageing.

978-1-4244-2977-6/09 $25.00 © 2009 IEEE

ALN CHARGING MECHANISM

In order to explain the reasons of the excellent reliability performances of the fabricated devices the charging and discharging mechanism of the AlN dielectric are investigated

To analyze the AlN dielectric charging, the switches are stressed at a given voltage (V_{STRESS}) for different stress times, t_{STRESS}. We measure periodically the shift in the up state of the C-V curve, (in order to not charge the dielectric when measuring [4]), V_{Shift}, as well as the minimum capacitance, C_{UP} (Figure 6). Two effects are observed: (i) the C-V curve shifts to the right, due to injection of positive charges in the dielectric, and (ii) the up-state capacitance increases. Interestingly, V_{Shift} and C_{UP} can be modeled by a similar analytical dependence, being proportional to the square root of the stress time (Figure 7):

$$V_{Shift} = \alpha \cdot \sqrt{t_{STRESS}} \cdot e^{(\beta_1 \cdot V_{STRESS})}, \qquad (1)$$

$$C_{UP} = C_0 \cdot \sqrt{t_{STRESS}} \cdot e^{(\beta_2 \cdot V_{STRESS})}, \qquad (2)$$

Being C_0 the initial C_{UP} for the fresh device and α, β_1 and β_2 experimental fitting constants.

The investigation of the C_{UP} is shown to provide useful complementary information, especially for the phase shifter where key performance depends on the capacitance. The AlN behaves as the silicon nitride (SiN) dielectric charging following a square-root time law proposed by Herfst [2].

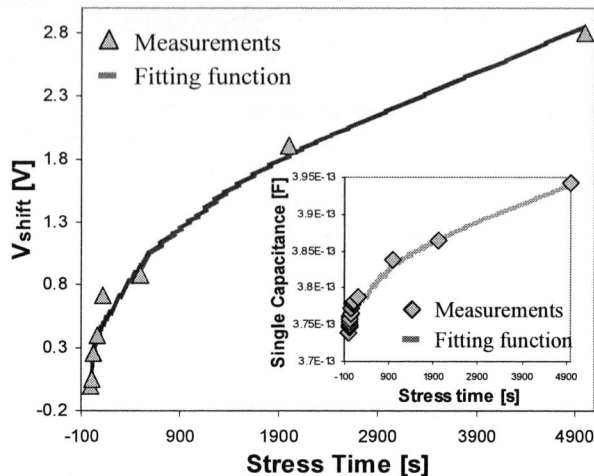

Figure 7: C-V curve shift versus the stressing time; a dependence of V_{Shift} with $t^{1/2}$ is confirmed. Inset: Minimum capacitive value versus the stressing time.

In Figure 8, it can be observed how C_{UP} decreases exponentially with the relaxation time at the same time that the C-V curve shifts to the left. Measuring the return loss of S-parameters, it is possible to observe how the main discharging process takes place among the first 60sec after releasing the stressing voltage (Figure 9).

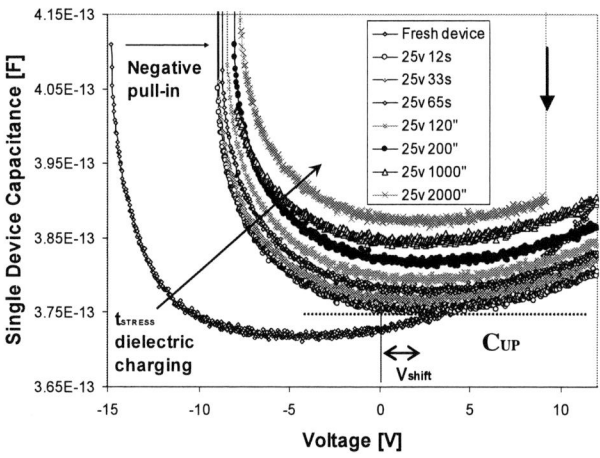

Figure 6: Evolution of the up-state capacitance, C_{UP} of a single capacitor versus the voltage in the open state for different stressing times and V_{STRESS} of +25V.

ALN DISCHARGING MECHANISM

Finally, discharging investigations are carried out in order to conclude about dielectric reversible or irreversible degradation and the related relaxation times of injected charges. Once that the dielectric is charged during t_{STRESS}, the voltage bias is switched to zero and the induced charge in the AlN dielectric need a certain time to recombine (by diffusing towards the bottom electrode). The MEMS capacitive switch recovers approximately the initial air gap when most of the injected in the dielectric have been neutralized. The charged AlN needs less than 90sec to recombine all the injected charges recovering the original C-V curve and their initial pull-in/pull-out voltages.

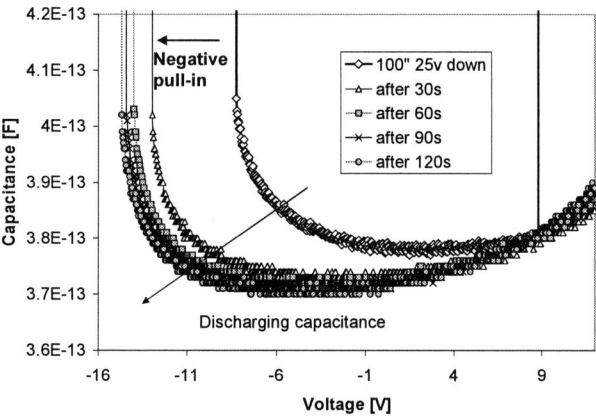

Figure 8: Discharging capacitance versus voltage in the open state region. When releasing the applied voltage, the membrane takes a relaxation time to recover the original position.

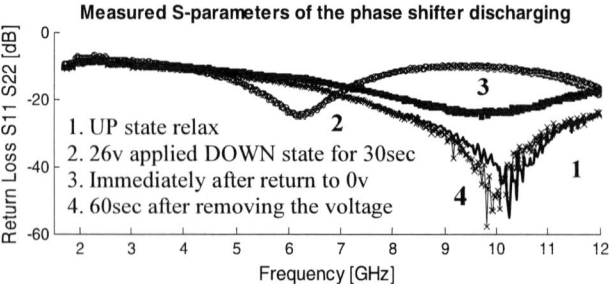

Figure 9: The relaxation of the membrane can be also monitored with the S-parameter measurement; full relaxation corresponds to the time to pass from state '3' to '4' (='1').

Figure 10 compares the relaxation time of an individual capacitor with the one of the full phase shifter; both following an exponential function. The shift of the C-V curve, proportional to the V_{PI}, follows the same law as the discharging capacitance (inset Figure 10).

$$C_{RELAX} = A_1 \cdot e^{\left(-t/\tau_1\right)} + C_0 , \qquad (3)$$

$$V_{PI}^{-} = A_2 \cdot e^{\left(-t/\tau_2\right)} + V_0 , \qquad (4)$$

Being V_0 the pull-in voltage for the fresh device and A_1, A_2, τ_1, τ_2 the experimental fitting constants.

From the capacitance values C_{UP}, during the recombination of the residual charges and the electromechanical characteristics of the suspended bridge (spring constant $k \sim = 9N/m$), it is possible to extract the residual charge at each moment ($t=0s$, $Q_{res}=6.65 \times 10^{-13}C$) until it is fully recombined ($t=90s$, $Q\sim 0$). No permanent stiction was observed in our devices, even when keeping the 13-MEMS-capacitor phase shifter in down position for more than two hours; none of the 13 MEMS capacitors had permanent failure after 1×10^9 cycles.

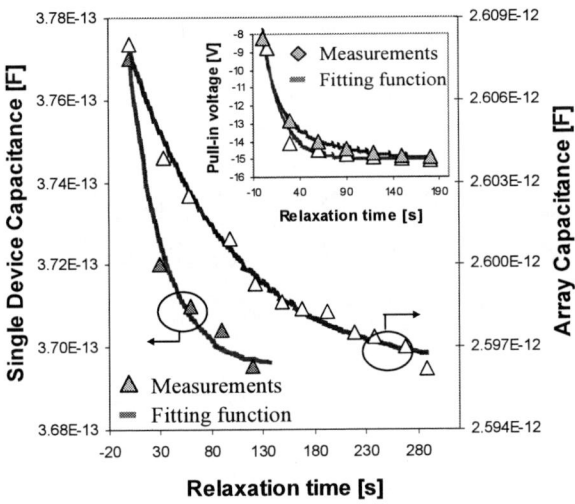

Figure 10: Comparison between discharging time dependence of a single capacitor and of an array of 13 MEMS capacitors, for 100s stressing time at 25V. Inset: the negative pull-in, V_{PI}^{-}, versus the relaxation time.

CONCLUSIONS

Our investigation identifies and validates the charging/discharging analytical laws in AlN dielectric, *slow charging* ($\propto \sqrt{t_{STRESS.}}$) and *fast discharging* ($\propto \exp(-t_{STRESS})$) mechanisms. Shifts of C-V curve and of the up-state capacitances, induced by DC stress, follow square root time laws. The charging of AlN is found to be easily reversible after the applied stress conditions, with a fast relaxation time, following an exponential time law for both individual devices and arrayed capacitive switches, namely the phase shifter.

It is demonstrated that wideband (1.5-14GHz) RF MEMS phase shifters and individual devices are very robust and reliable and can withstand more than 10^9 cycles in hot switching (+5dBm RF power) and for the first time under ambient condition (50% of relative humidity) without permanent failure or stiction.

These promising results suggest firstly, that a low-cost no hermetical packaging could be appropriate and secondly, that the good results seem to be responsible for the good reliability of AlN.

AlN seems to be a good medium-dielectric-constant insulator candidate for MEMS devices and their future physical understanding is key for progress in MEM suited reliability.

ACKNOLEDGEMENTS

We thank the Fraunhofer ISiT for the assistance and fabrication support. This work was supported by the European Commission FP6 IST Integrated Project e-CUBES.

REFERENCES

[1] Gabriel .M. Rebeiz, *RF MEMS – Theory, Design and Technology*, John Wiley & Sons, Inc., New Jersey, 2003.

[2] R.W. Hertst, P.G. Steeneken, J. Schmitz, "Time and voltage dependence of dielectric charging in RF MEMS capacitive switches" in *45th Annual International Reliability Phy. Symp.*, Phoenix, April 15-19, 2007, pp. 417-421.

[3] W.M. van Spengen, R. Puers, R. Mertens, I. De Wolf, "A comprehensive model to predict the charging and reliability of capacitive RF MEMS switches", *J. of Micromech. Microeng.*, vol. 14, pp. 514-521, 2004

[4] R.W. Herfst, P.G. Steeneken, H.G.A. (Bert) Huizing, J. Schmitz, "Center-Shift Method for Characterization of Dielectric Charging in RF MEMS Capacitive Switches", *IEEE Semic. Manuf.*, vol. 21, pp. 148-153, 2008.

[5] J. C. M. Hwang, "Reliability of Electrostatically Actuated RF MEMS Switches", *Radio-Frequency Integration Technology, 2007 IEEE International Workshop on*, Singapore, Dec. 9-11, 2007, pp. 168-171.

[6] J.G. Teti, F.P. Darreff, "MEMS 2-bit phase shifter failure mode ande reliability consideration for large X-band arrays", *IEEE Trans. Microwave Theory Tech*, vol. 52, no. 2, pp. 693-701.

[7] M. Fernandez-Bolaños, T. Lisec, P. Dainesi, A.M. Ionescu, "Thermally Stable Distributed MEMS Phase Shifter for Airborne and Space Applications", to be presented at *38th European Microwave Conference*, Amsterdam, Oct. 27-31, 2008.

[8] J. P. Kar, G. Bose, "DC stress effect on charge distribution in sputtered AlN films", *J. Electron. Devices*, vol. 2, pp. 57-61, 2003.

[9] R.W Herfst, P.G. Steeneken, J. Schmitz, "Identifying degradation mechanisms in RF MEMS capacitive switches", *MEMS 2008*, Tuscson, Jan. 13-17, 2008, pp. 168-171.

TENSILE MEASUREMENT OF A SINGLE CRYSTAL GALLIUM NITRIDE NANOWIRE

J.J. Brown[1], A.I. Baca[1], and V.M. Bright[1]

[1]Department of Mechanical Engineering, University of Colorado, Boulder, Colorado, USA

ABSTRACT

This paper reports the first direct tensile test on a nearly defect free, n-type (Si-doped) GaN nanowire single crystal. Here, for the first time, nanowires have been integrated with actuated, active microelectromechanical (MEMS) structures using dielectrophoresis-driven self-assembly and Pt-C clamps created using a gallium focused ion beam. The nanowire modulus of elasticity is measured to be 201 GPa, and the nanowire demonstrated more than 4% elongation before one of the clamps failed. Failure of the test sample occurred at the interface of one of the Pt-C clamps with the fixed MEMS stage, rather than in the nanowire itself.

INTRODUCTION

Recent developments in the synthesis of gallium nitride materials have raised the possibility that forms of this material with few defects could be used in numerous applications. It is a direct bandgap semiconductor with good thermal conductivity and notable piezoresistive and strain-dependent optical properties. [1 – 4] It may also have a high quality factor that makes it appealing in mechanical resonator designs. [1] Integration of this material with active MEMS structures may allow development of a new class of tunable mechanical resonators, LEDs, lasers, and switches, among other possible new sensor and transducer technologies. In earlier work, limited mechanical testing of this nanowire material used mechanical resonance of a nanowire bonded to a fixed substrate. [1, 2] Development of comprehensive mechanical data on gallium nitride nanowires (GaN NWs) enables the contemplation of this material in new devices.

The use of microfabricated structures enables mechanical testing of materials in small quantities or for which bulk sample preparation is difficult. Additionally the integration of nanomaterial specimens to a MEMS mechanical tester serves as a test case for chip and wafer-scale integration of microtechnologies with a nanoscale material that has specific synthesis requirements.

EXPERIMENT

MEMS Structure

The MEMS mechanical test structure (Figure 1) is a simplified version of the one reported at Hilton Head 2008 [5, 6], consisting of a fixed stage electrically isolated from a moving stage that is laterally stabilized and actuated using a buckling beam thermal actuator. This test structure was fabricated using the PolyMUMPS service. [7]

Motion of the moving stage is measured directly from scanning electron microscope (SEM) micrographs. The force applied by the stage can be calibrated using the approach identified in [5, 6], which is described here. The actual stage displacement d is compared to an expected displacement d_0. The force applied F is calculated according to Equation 1 using the discrepancy between these displacements multiplied by the spring constant k of the system.

$$F = k (d_0 - d) \qquad (1)$$

The system spring constant is calculated assuming small displacements to a system of bending beams, and was found using mathematical analysis to be $k = 170.8$ $\mu N/\mu m$. Finite element simulations in CoventorWare software supported this value for the spring constant.

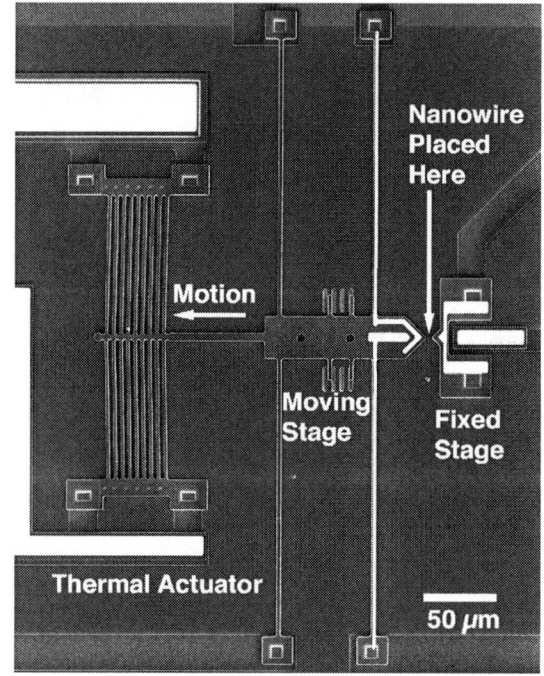

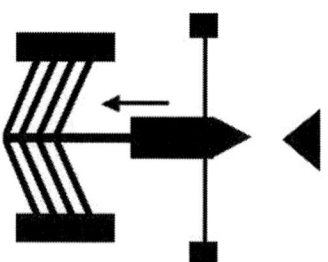

Figure 1: (Top) Microfabricated tensile test structure consisting of electrically isolated moving and fixed stages. During a tensile test the moving stage moves away from the fixed stage. Thermal actuator beams are angled 89° from the direction of motion. (Bottom) Simplified schematic of the device seen above. (Reproduced from [5,6].) The thermal actuator consists of an array of angled beams, which expand as current flows through them. The actuator pulls the moving stage, which is laterally stabilized by pairs of opposing beams.

The expected displacement d_0 is calculated by using a fit to a curve of displacement versus input power P

978-1-4244-2977-6/09 $25.00 © 2009 IEEE

measured from a test device. For a freely moving stage, displacement depends on thermal strain, which in turn derives from temperature change and thermal expansion coefficients in the actuator beams. Due to the thermal resistance of the beams, the temperature profile of the beams depends upon the power dissipated within them, which can be measured by multiplying the current and voltage supplied to the actuator. The d_0 vs. P curve as observed for freely moving test systems is usually linear or slightly parabolic. Calibrating from measured experimental curves allows the fit parameter to incorporate the nonuniform actuator temperature profile, the nonlinear thermal expansion and temperature-dependent changes in the electrical and thermal conductivities.

For the force measurements used here, a linear fit between d_0 and P was used, $d_0 = BP$, and the fit parameter B is calculated from a regression analysis of d_0 and P data measured as output and input, respectively, of a freely moving test structure.

Nanowire Placement

Dielectrophoresis has been previously suggested [3] as a means of integrating nanowires to microfabricated structures in a wafer-level assembly technique, rather than the time consuming approach of micromanipulator placement. We have demonstrated that this approach to self-assembly can be extended to complete MEMS devices.

The dielectrophoretic nanowire placement is performed using a modified probe station. Two electrical probes are placed in contact with electrical pads that allow an alternating (AC) electrical field to be placed across the fixed and moving stages. The stages taper to opposing points, where the nanowire is desired. Consequentially, there is a gradient in the electric field between these stages, with the maximum field at the location where the stages are in closest proximity. The nanowires polarize in the presence of an electric field, causing them to align parallel to the field lines. In the presence of the electric field gradient, nanowires are swept towards the field maximum, bridging the fixed and moving stages.

The C-axis oriented, Si-doped (n-type) GaN nanowires were provided by the National Institute of Standards and Technology, Boulder, CO, USA, (NIST-Boulder), having been synthesized by molecular beam epitaxy in the same manner as those described in [1, 2]. A nanowire suspension was formed by briefly (2 – 5 minutes) sonicating a substrate with nanowires in several milliliters of isopropanol. The suspension was drawn into a syringe, and the tip of the syringe was positioned over the MEMS tester using a probe station micromanipulator. For the dielectrophoretic nanowire placement, a signal of 10 V peak-to-peak amplitude was provided at 70 – 75 kHz to the stage. While the field was applied, about 0.1 µL of the nanowire suspension was dispensed onto the stage. As the solvent evaporated, surface tension and dielectrophoretic electrical forces caused the self-assembly of one or more nanowires into position on the stage. The dielectrophoresis parameters were derived through experimentation with signal amplitude and frequency, and were found to be most effective for the given stage and NW samples at the values above. This dielectrophoretic self-assembly process functions for both unreleased and released MEMS structures. Released structures do not exhibit stiction due to solvent evaporation.

Nanowire Clamping

The nanowire specimens were clamped (Figure 2) to the MEMS test structures using Pt-C deposits formed in a FEI NOVA 600i dual beam focused ion beam (FIB) system using ion-beam induced deposition (IBID), similar to the approach taken in prior work. [8 - 10] After deposition of the contacts, the test structure was released in aqueous HF and dried using supercritical CO_2. GaN survives HF etching without significant damage. It was observed that HF release after clamping nanowires to structures using IBID can leave a membrane visible on the substrate of the test device (visible in Figure 3). This membrane can also form in instances where devices are subjected to SEM imaging followed by HF release. These membranes are not seen in cases where HF release is performed before imaging, indicating they may be thin carbon or Pt-C deposits formed as byproducts of the imaging process and of IBID.

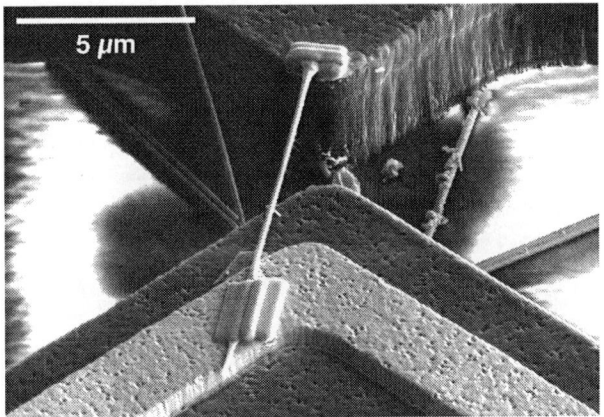

Figure 2: The mounted GaN NW specimen #1, 200 nm in diameter, was bonded to the unreleased tensile stage using Pt-C deposits formed using a focused ion beam system.

Measurements

Measurement of nanowire strain was taken directly from SEM images of the nanowire using ImageJ software, measuring changes in the nanowire length between the clamped regions. The measurement of elongation directly from electron micrographs removes the effect of clamp deformation from the strain data. Current and voltage supplied to the MEMS actuator were measured using Hewlett-Packard 34401A multimeters or a Hewlett-Packard 3425A Universal Source. Stress is calculated by dividing the force applied by the cross-sectional area of the nanowire. The cross section area is found by measuring the nanowire diameter from SEM micrographs.

RESULTS

Tensile tests were performed on several nanowire specimens. These results are summarized in Table 1, below. From this table it should be noted that the nanowire specimens can withstand significant strain.

Table 1: Tensile test data for three nanowire specimens.

Specimen	1	2	3
Maximum Engineering Strain	4%	3.1%	~1%
Modulus	201 GPa	--	--
Failure Mode	Clamp Failure	C-plane Fracture	Insufficient MEMS actuator force

Specimen 1 stands out in both the measured result and the accuracy of this measurement because its relatively long 10.1 μm length and narrow 200 nm diameter enabled it to stretch significantly under the forces provided by the MEMS test system. Measurement and regression fit error was typically less than 8%. For a first analysis, loading in the nanowire was assumed to be uniaxial, although in reality the nanowire was angled 5° off of the tensile axis. The normal stress applied to the nanowire when the clamp failed was about 7 GPa. The failure shear stress of the clamp-microdevice interface was about 76 MPa.

Figure 3: Specimen 1, GaN nanowire under tensile load after failure of the bottom clamp at the interface between the clamp and the fixed stage. The area of the failed platinum/carbon clamp is approximately 3 μm².

The expected displacement used in force calibration for this test device was derived from motion of the stage after the specimen clamp failed (Figure 4). This is a new development in tester calibration beyond the approach reported in [5, 6], in that the fitting parameter between d_0 and P is derived from the device which was used to perform a mechanical test.

Specimen 2 did not appear to fracture during the tensile test, but subsequent high-resolution SEM examination and FIB cross-sectioning showed that the nanowire developed two fractures, one near each clamp. Nonuniform stress states due to non-uniaxial loading may have contributed to these fractures. The nanowire was oriented about 10° off the tensile axis.

In contrast, specimen 3 was oriented nearly parallel with the tensile axis, but it was a twinned crystal with a large cross-sectional area of about 480 x 890 nm (0.46 μm²) and the force available from the actuator (about 600 μN maximum) was insufficient to cause failure of this

fiber. The FIB was used to reduce the cross-section of this nanofiber to about 250 x 290 nm, at which point the nanowire spontaneously buckled. When tensile stress was applied, failure occurred with about 2.4 GPa stress applied to the fiber. This stress was estimated from the fiber cross-section and the assumption of uniform loading in the fiber. The failure stress was likely higher because the specimen remained bowed as it failed, indicating stress nonuniformity in the fiber cross section.

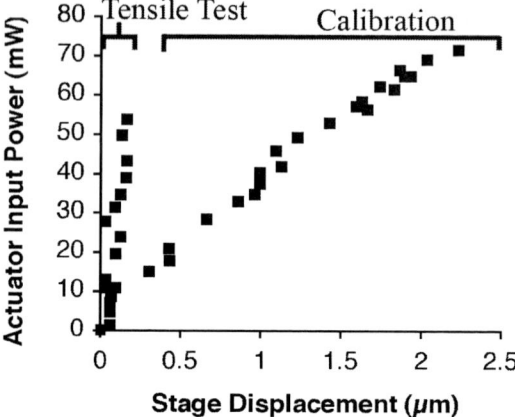

Figure 4: Device input power and output displacement during the initial tensile test and then after the failure seen in Figure 3.

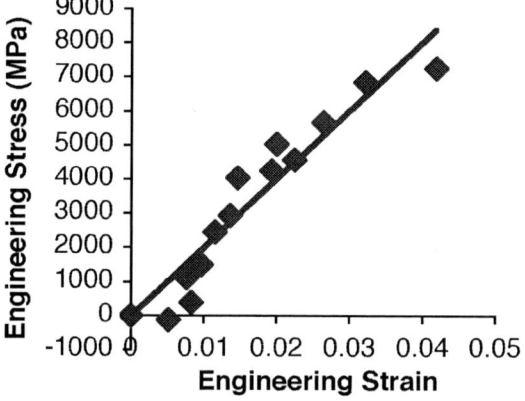

Figure 5: Tensile test data for specimen 1 derived from the experiment reported above. The regression line corresponds with an elastic modulus of 201 GPa.

DISCUSSION

The robustness demonstrated by specimen 1 in withstanding a 4% elongation is perhaps the most notable result of this work. Image superposition confirms the strain measurement for this specimen. Observation of the specimen after the clamp failure indicates that it did not exhibit failure, nor did it experience plastic deformation. The estimated maximum stress applied to this fiber, 7 GPa appears to be a reasonable value in that the ultimate failure of specimen 3 occurred for a stress value of a similar order of magnitude.

The ability to withstand 4% elongation, as reported here, reflects an exceptional strength for what might otherwise be considered a brittle material, but it is not entirely surprising given that this measurement was performed on a single crystal of a covalent material with a

978-1-4244-2977-6/09 $25.00 © 2009 IEEE

hexagonal lattice. All of these are factors known to increase mechanical strength.

Measurement in the SEM used to observe the tensile experiments is limited to the resolution of images obtained from this instrument. In practice, it appears that +/- 50 nm is about the best accuracy obtained from this instrument in the measurements reported here, so large deformations and large specimen gauge lengths will be more accurately recorded with this technique.

The MEMS test structures exhibit repeatable behavior as long as the input power to the actuator remains below about 60 mW. Structures that experience inputs above this threshold appear to undergo several effects: the actuator resistance varies over time, SEM imagery shows discoloration in the vicinity of the hottest part of the actuator, and upon release of the input power, the moving stage returns to a point closer to the fixed stage than when it began. These observations appear to indicate that the highest input powers also lead to plastic deformation of the thermal actuator.

CONCLUSION

This work has demonstrated that dielectrophoretic self-assembly of GaN nanowires can effectively place nanowires on active MEMS devices without manipulation of individual nanowires. IBID platinum-carbon clamps can secure individual nanowires to a polysilicon surface for mechanical loading of the nanowires, but the interface between these clamps and the surface can fail with a shear stress on the order of 76 MPa, therefore clamps must be deposited over a sufficient area to withstand the force applied to the nanowire specimen. Measurement of specimen strain from SEM images removes the effect of clamp deformation from the tensile curve, but the clarity of tensile data will be limited by the resolution of the SEM. Single crystal gallium nitride appears capable of withstanding uniaxial strain at least 1% and as much as 4% strain and 7 GPa normal stress without failure. SEM and IBID processes give secondary deposition of material coatings on the imaged and nearby surfaces, and these coatings survive HF release. For this reason, future nanowire specimen preparation should be performed on released devices. Continued work should aim to increase the number of specimens measured and to better quantify and reduce the error of the measurements obtained.

ACKNOWLEDGEMENTS

This research was supported by DARPA Award #HR0011-06-1-0048 and the DARPA Focus Center for Integrated Micro/NanoMechanical Transducers (iMINT Center) at the University of Colorado, Boulder. This material is based upon work supported under a National Science Foundation Graduate Research Fellowship, which supports J. J. Brown. Special thanks to Dr. K. Bertness and her colleagues at NIST-Boulder for the nanowire samples.

REFERENCES

[1] S. M. Tanner, J. M. Gray, C. T. Rogers, K. A. Bertness, N. A. Sanford, "High-Q GaN nanowire resonators and oscillators", *App. Phys. Lett.,* vol. 91, p. 203117, 2007.

[2] J. B. Schlager, K. A. Bertness, P. T. Blanchard, L. H. Robins, A. Roshko, N. A. Sanford, "Steady-state and time-resolved photoluminescence from relaxed and strained GaN nanowires grown by catalyst-free molecular-beam epitaxy", *J. App. Phys.,* vol. 103, p. 124309, 2008.

[3] Motayed, M. He, A. V. Davydov, J. Melngailis, S. N. Mohammed, "Realization of reliable GaN nanowire transistors utilizing dielectrophoretic alignment technique", *J. App. Phys.,* vol. 100, p. 114310, 2006.

[4] H. W. Seo, S. Y. Bae, J. Park, H. Yang, K. S. Park S. Kim, "Strained gallium nitride nanowires", *J. Chem. Phys.,* vol. 116, pp. 9492-9499, 2002.

[5] J. J. Brown, J. W. Suk, G. Singh, D. A. Dikin, R. S. Ruoff, and V. M. Bright, "Microsystem for electromechanical measurements of carbon nanofiber loading and failure", Proc. Solid-State Sensors, Actuators, and Microsystems Workshop 2008, Transducer Research Foundation, Hilton Head, SC, USA, p. 135-139, 2008.

[6] J. J. Brown, J. W. Suk, G. Singh, A. I. Baca, D. A. Dikin, R. S. Ruoff, V. M. Bright, "Microsystem for Nanofiber Electromechanical Measurements", to be published in *Sensors and Actuators A.*

[7] MEMSCAP, Inc., Research Triangle Park, NC, USA.

[8] C. Y. Nam, D. Tham, J. E. Fischer, "Disorder effects in focused-ion-beam-deposited Pt contacts on GaN nanowires", vol. 5, pp. 2029-2033, 2005.

[9] C. Y. Nam, P. Jaroenapibal, D. Tham, D. E. Luzzi, S. Evoy, J. E. Fischer, "Diameter-dependent electromechanical properties of GaN nanowires", *Nano Lett.,* vol. 6, pp. 153-158, 2006.

[10] D. Tham, C. Y. Nam, J. E. Fischer, "Microstructure and composition of focused-ion-beam-deposited Pt contacts to GaN nanowires", *Adv. Mater.,* vol. 18, pp. 290-294, 2006.

CELLULOSE-BASED COMPOSITE AS A RAW MATERIAL FOR FLEXIBLE AND ULTRA-LIGHTWEIGHT MECHANICAL SWITCH DEVICES

S. Couderc[1], B.J. Kim[2], and T. Someya[3]

[1]LIMMS, CNRS-IIS (UMI 2820), The University of Tokyo, Tokyo, JAPAN
[2]CIRMM, Institute of Industrial Sciences, The University of Tokyo, Tokyo, JAPAN
[3]School of Engineering, The University of Tokyo, Tokyo, JAPAN

ABSTRACT

This article covers investigations on an original, almost fully biodegradable and flexible mechanical switch device fabricated from cellulose composite. The cellulose films, mainly composed of microfibrils, revealed a high surface roughness and poor dielectric properties making them unsuitable substrates for electronic applications. By coating the cellulose film with a specific polyimide, these drawbacks were overcome. The mechanical switch was operated through the deflection of an electrostatically actuated cantilever beam. Its displacement, induced by the electrostatic force, was validated and the switch ON state detection was performed for an actuation voltage of 55 V and for a beam-substrate distance around 30 µm.

INTRODUCTION

The development of fully or partially biodegradable electronic components is a great challenge for environmental considerations. In this context, few studies were carried out on the fabrication of electronic devices based on cellulose composites such as the so-called Electro-Active Paper (EAPaps) actuators [1-3] or more recently, a transparent flexible substrate reinforced with bacterial cellulose for Light-Emitting Diodes applications [4]. The interest of using cellulose in electronics is that cellulose is a fully biodegradable and biocompatible material, which exists in abundance on earth at low-cost. In the framework of "green research", this article reports innovative results on an environmentally friendly, flexible and ultra-lightweight mechanical switch device made of cellulose composite. The switch operation is based on an electrostatically actuated cantilever beam deflection.

FILM CHARACTERIZATIONS

Film elaboration

The cellulose films are elaborated from a wood-pulp paste provided by CELISH - DAICEL Chemical Industries LTD. The wood-pulp paste is dissolved into water and the homogeneity of the solution is improved with magnetic stirrer and ultrasonicator. The solution is then poured within a Petri dish and the water is evaporated by oven annealing. In order to improve the film flatness, a press with heater system is implemented. The microstructural and the dielectric properties of ~15 µm-thick cellulose films are first investigated. This film thickness is similar to the one used in the switch fabrication.

Surface morphology

The surface of such cellulose films are characterised by a Scanning Electron Microscopy (SEM) and by an Atomic Force Microscope (AFM Nanoscope IIIa, Digital Instruments). The SEM analysis reveals that the cellulose film is composed of a tangle of microfibrils with a diameter and length around 100 nm and 20 µm respectively and with few fibers bundles (~1 µm – ~10 µm in diameter) resulting from residual inhomogeneities (Fig. 1). The cellulose film also presents a porous microstructure and the holes diameter ranges from ~100 nm up to ~1 µm (Fig. 1). The AFM scanning profile of the surface shows a very high surface roughness due to the fibrous-like morphology, with a Root-Mean Square (RMS) value around 1 µm (Fig. 2).

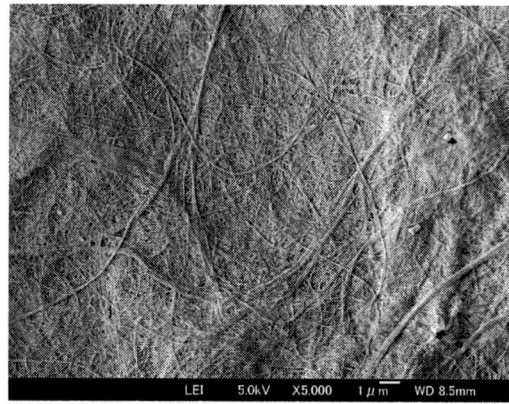

Figure 1: SEM micrograph of the surface of a ~15 µm-thick cellulose film. The scale bar is 1 µm.

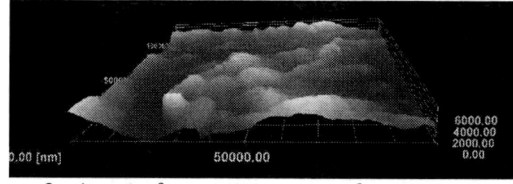

Figure 2: Atomic force microscope observation on a ~15 µm-thick cellulose film surface - the scan size is 100 µm x 100 µm. The average extrapolated RMS value is ~1.2 µm.

Dielectric properties

The dielectric properties of a ~15 µm-thick cellulose film are investigated by measuring the leakage currents. All measurements are performed in a 35% humidity ratio ambient environment that may influence the results. Indeed, the water molecules directly act on hydrogen bonds linking the cellulose chains together and cause crinkling and the formation of holes. Many studies have already been carried out on dielectric characterizations of hydrophilic materials [5-6] to understand the moisture effect. The leakage currents were performed using Agilent 4155C Semiconductor Parameter Analyzer and by applying a voltage ranging from 0 V up to 100 V between two Au electrodes sandwiching the cellulose film. The leakage current density as a function of the electric field exhibits very high values, from 10^2 times up to 10^4 times higher than values usually obtained for polymer films such as polyimide [7-8] (Fig. 3). This may result from the porous microstructure and the hydrophilic properties.

978-1-4244-2977-6/09 $25.00 © 2009 IEEE

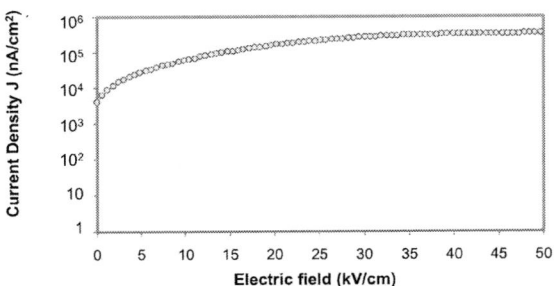

Figure 3: Current density as a function of the electric field for a ~15 μm-thick cellulose film.

Effect of polymer coating

The high surface roughness and poor dielectric behaviour are the main drawbacks of cellulose films as substrates for electronic applications. To overcome these issues, the cellulose film surface is covered by a spin-coated and thermo-cross-linkage polyimide precursor (Kemitite CT4112A, Kyocera Chemical, Kawasaki, Japan). The spin-coater operating speed is fixed at 1000 tr/mn and a ~2 μm-thick polyimide layer forms after solvent evaporation at 90°C for 10 mn and cross-linking at 180°C for 1 h. The high viscosity of the polyimide precursor masks the cellulose microfibrils as shown on the SEM image (Fig. 4), which greatly improves the surface smoothness. Indeed, the RMS value extracted from AFM measurements is around 200 nm for the polyimide-coated cellulose film (Fig. 5), while the pure cellulose film exhibits a five times higher roughness value (~ 1 μm) (Fig. 2).

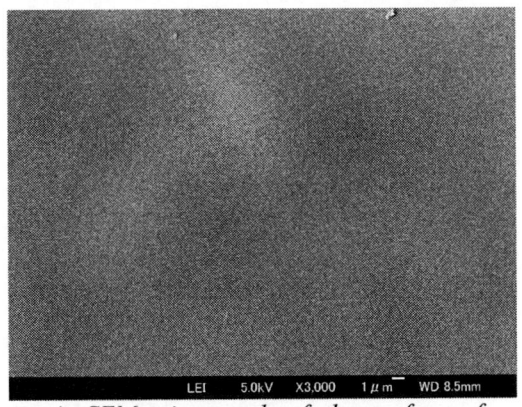

Figure 4: SEM micrograph of the surface of a ~15 μm-thick cellulose film coated on both sides by ~2 μm-thick polyimide. The scale bar is 1 μm.

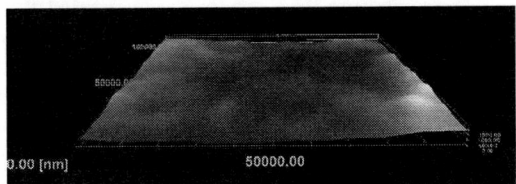

Figure 5: The surface of ~20 μm-thick polyimide-coated cellulose film characterized by an atomic force microscope - the scan size is 100 μm x 100 μm. The average extrapolated RMS value is ~220 nm.

Moreover, with polyimide coating, cellulose films present a hydrophobic behaviour as attested by the water

droplet test, as shown in Fig. 6 (a). The contact angle when the droplet is deposited is greater than 90° for the cellulose film coated with polyimide indicating a non-wetting surface (Fig. 6(a)), while it is ~ 36° for the pure cellulose film (Fig. 6(b)).

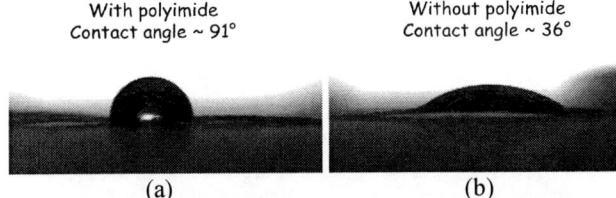

(a)   (b)

Figure 6: Water droplet test on a polyimide-coated cellulose film (a) and compared to the same cellulose film without polyimide (b).

The dielectric properties of a ~20 μm-thick polyimide-coated cellulose film are investigated in the same conditions as those for pure cellulose films. With polyimide coating, the leakage current density can be reduced down to a factor 10^3 compared to one for pure cellulose film and are comparable to the values obtained for 1 μm-thick polyimide film given in ref. [8] (Fig. 7). As a result, it is a great advantage to use polyimide coating thanks to its hydrophobic properties that completely remove the cellulose film moisture sensitivity. Moreover, its high viscosity masks the microfibrils and fills the holes making the surface smoother and therefore improving the dielectric properties.

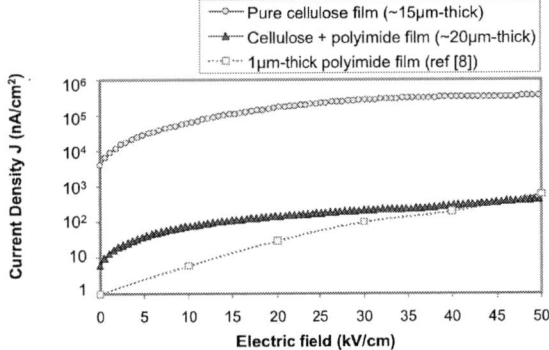

Figure 7: Leakage current density as a function of the electric field for a ~20 μm-thick polyimide-coated cellulose film and compared to a ~15 μm-thick pure cellulose film. The traces are also compared to the current density for a 1 μm-thick polyimide film (the data are extracted from the ref. [8]).

CELLULOSE MECHANICAL SWITCH
Design

The mechanical switch is composed of a cantilever beam with a proof mass at the free end and is relatively large size (~4 mm x 4 mm) (see on Fig. 8). To flip from the OFF state to the ON state, the switch operation is based on the electrostatically actuated beam deflection. The electrostatic force is created by applying an actuation voltage between the electrodes on the beam and on the substrate (called "Actuation top" and "Actuation bottom" respectively on the Fig. 8). The ON state detection is performed through electrical contact when the beam is in contact with the substrate thanks to the additional face-to-face electrodes (called "Top contact" and "Bottom contact" on Fig. 8). The distance between the beam and the substrate (called gap hereafter) is created by the addition of a cellulose interlayer. All the dimensions are specified on the schematic below (Fig. 8).

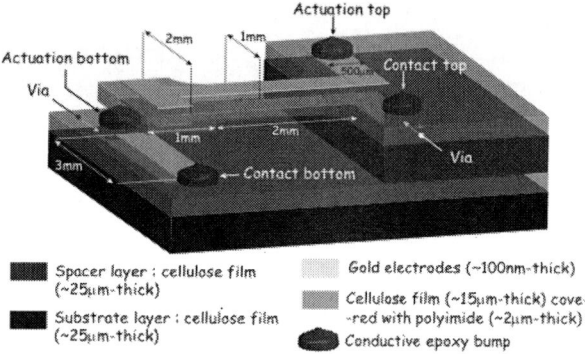

Figure 8: Schematic of the mechanical switch.

Fabrication

The mechanical switch device consists in a ~15 μm-thick pure cellulose film coated on both sides with ~2 μm-thick polyimide, using the same process as mentioned before. Firstly, a high precision punching machine draws the cantilever beam shape by mechanically drilling of successive holes into the film. Then, the 100 nm-thick Au electrodes are vacuum-evaporated through shadow masks on both sides of the films. To allow electrical connexion between the bottom electrodes and the device topside, a CO_2 laser marker forms 150 μm-diameter via holes through the cellulose at the edge of the electrodes. Then, holes are filled up with conductive epoxy glue (Chemtronics®), and dried at ambient temperature for 8 h to allow the electrical contact implementation. Afterwards, a ~25 μm-thick pure cellulose film is stuck on the substrate bottom side to consolidate it. Finally, a ~25 μm-thick pure cellulose interlayer is sandwiched between the two cellulose films to create the gap and is stuck with glue to form the final device (Fig. 9). The gap ranges from 30 μm up to 100 μm depending on the cellulose film flatness. The biodegradability level for this device, calculated from the ratio between the volume of fully biodegradable material (cellulose) and the device total volume, is evaluated around 85%.

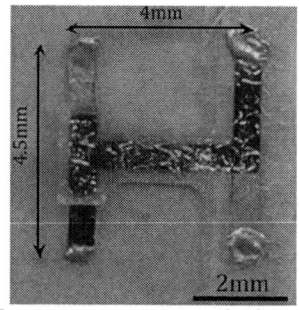

Figure 9: Top view picture of the cellulose-based mechanical switch device. The scale bar is 2 mm.

Cantilever beam deflection and contact detection

To validate the operation of the electrostatic actuation, an electric voltage is applied between the actuation electrodes of the beam and the substrate. The beam deflection is measured simultaneously with a laser Doppler vibrometer. By increasing the actuation voltage from 5 V up to 100 V, the electrostatic force induces a beam tip displacement ranging from 20 nm up to 10 μm when the switch gap height is approximately 80 μm (Fig. 10).

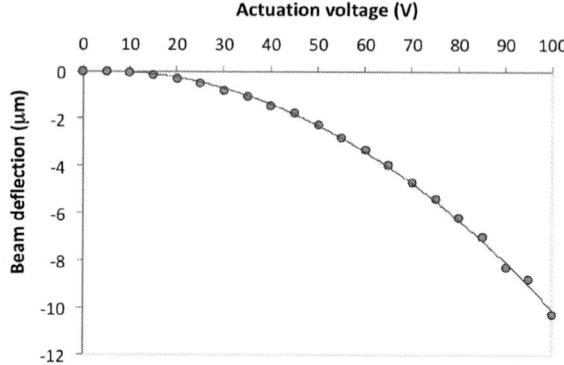

Figure 10: Measured cantilever beam tip displacement as a function of the actuation voltage for a ~80 μm-high gap.

The ON state is obtained after reducing the gap down to ~30 μm and applying an actuation voltage of 55 V. The contact detection is achieved by measuring the voltage variations through a 220 kΩ resistor in series with the resistance of the face-to-face electrodes (see electrical circuit schematic on Fig. 11).

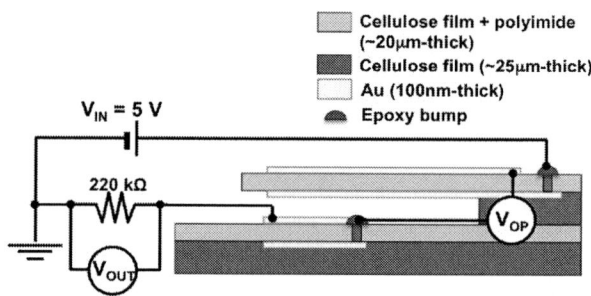

Figure 11: Schematic of the electrical circuit for the switch operation test.

Figure 12 shows the switch response through the output voltage measurement when the device is subjected to a 10 Hz square wave signal. The device operates as expected. The switching time is measured around 1.8 ms while the release time is 3 ms (Fig. 13). These delays are relatively high compared to silicon-based MEMS (MicroElectro Mechanical Systems) switch performances [9]. Nevertheless they are about the same order of magnitude as those typically measured for polymer-based mechanical switches [10].

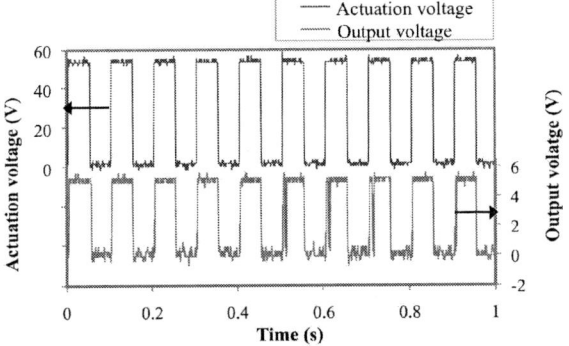

Figure 12: The switch response through the output voltage measurement when the device is subjected to a 10 Hz square wave actuation voltage with amplitude of 55 V. The switch gap height is ~30 μm.

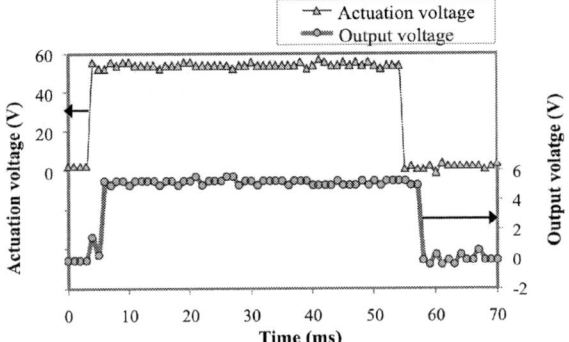

Figure 13: The switching and the release times of the mechanical switch response when the device is subjected to a 10 Hz square wave actuation voltage with amplitude of 55 V. The switch gap height is ~30 μm.

CONCLUSION

We demonstrated the fabrication of an electrostatically actuated cellulose-based mechanical switch device. The cellulose films elaborated from a wood-pulp paste exhibit a fibrous-like morphology leading to a high surface roughness. Moreover, the hydrophilic behavior of cellulose fibrils and the porous microstructure strongly degrade the dielectric properties. Therefore, polyimide polymer was spin-coated on the cellulose film, making its surface completely hydrophobic and reducing both the roughness by a factor five and the leakage currents up to a factor 10^3. The cellulose switch was fabricated by implementing punched-draw pattern defining, metal evaporation through shadow mask and glue assembling. The biodegradability level of this device is ~85%. The principle of cantilever beam deflection caused by electrostatic force was validated. After reducing the gap height down to ~30 μm, the electric contact was performed for an actuation voltage of 55 V. The results are conclusive and the device operates as expected. Further investigations could be done on the device size reduction using micro-fabrication techniques. It could also be interesting to replace the polyimide by a biodegradable material in order to obtain a fully biodegradable device.

ACKNOWLEDGEMENTS

We acknowledge JSPS (Japan Society Promotion for Science - P07737) for their financial supports. The authors also thank Olivier Ducloux, Laurent Jalabert and Nobuyuki Takama for their assistance and the laboratories of Professor Fujita and Professor Toshiyoshi for the support regarding equipment and clean room facilities.

REFERENCES

[1] J. Kim, S. Yun, Z. Ounaies, "Discovery of Cellulose as a Smart Material", *Macromolecules*, vol. 39, pp. 4202-4206, 2006.

[2] J. Kim, N. Wang, Y. Chen, G.-Y. Yun, "An Electro-Active Paper Actuator Made with Lithium Chloride/ Cellulose Films: Effects of Glycerol Content and Film Thickness", *Smart Mater. Struct.*, vol. 16, pp. 1564-1569, 2007.

[3] J. Kim, Y.B. Seo, "Electro-Active Paper Actuators", *Smart Mater. Struct.*, vol. 11, pp. 355-360, 2002.

[4] M. Nogi, H. Yano, "Transparent Nanocomposites Based on Cellulose Produced by Bacteria Offer Potential Innovation in the Electronics Device Industry", *Adv. Mat.*, vol. 20, pp. 1849-1852, 2008.

[5] S. Jang, P. Basappa, J. Kim, "Study of Dielectric Properties Electro-Active Paper", *Annual Report on Electrical Insulation and Dielectric Phenomena*, pp. 675-678, 2007.

[6] A.M.A Nada, M. El-Sakhawy, S. Kamel, M.A.M Eid, A.M. Adel, "Mechanical and Electrical Properties of Paper Sheets Treated with Chitosan and its Derivatives", *Carbohydrate Polymers*, vol. 63, pp. 113-121, 2006.

[7] A. Dubey, D.L. Lile, "Electrical Properties of Polyimides for Interlevel Isolation and Active Device Gate Isolation", *VLSI Multilevel Interconnection Conference*, June 12-13, 1989, pp. 390–396.

[8] Y. Zhang, I.W. Boyd, "UV Light-Induced Deposition of Low Dielectric Constant Organic Polymer for Interlayer Dielectrics", *Optical Materials*, vol. 9, pp. 251–254, 1998.

[9] Z.J. Guo, N.E. McGruer, G.G. Adams, "Modeling, simulation and measurement of the dynamic performance of an ohmic contact, electrostatically actuated RF MEMS switch", *J. Micromech. Microeng.*, vol. 17, pp. 1899-1909, 2007.

[10] S. Nakano, T. Sekitani, T. Yokota, T. Someya, "Low Operation Voltage of Inkjet-Printed Plastic Sheet-type Micromechanical Switches", *Appl. Phys, Lett.*, vol. 92, pp. 053302, 2008.

DETERMINATION OF DENSITY AND YOUNG'S MODULUS OF ATOMIC LAYER DEPOSITED THIN FILMS BY RESONANT FREQUENCY MEASUREMENTS OF OPTICALLY EXCITED NANOCANTILEVERS

B. Ilic[1], S. Krylov[2], and H. Craighead[1]
[1]Cornell University and Cornell Nanoscale Facility, USA
[2]Tel Aviv University, ISRAEL

ABSTRACT

We report on a methodology for simultaneous determination of the Young's modulus and density of ultrathin films from a resonance experiment. The approach is based on an interferrometric detection of the in-plane and out-of-plane resonant responses of an optically excited single crystal Si nano cantilever prior and after Atomic Layer Deposition (ALD) of a thin film. The frequencies shifts were measured at the same structure reducing sensitivity to scattering in geometric parameters and clamping compliances. Experimental results obtained for Al_2O_3 (alumina) and HfO_2 (hafnia) were consistent with the model predictions and the data available in literature.

INTRODUCTION

Resonant method has been widely exploited for determination of elastic constants of materials used in micro-[1-4] and nano-devices [5] due to high accuracy, non-destructive character and simplicity of resonant experiments suitable for *in-situ* measurements. While determination of Young's modulus, shear modulus and Poisson's ratio using excitation of a fundamental bending mode [1, 2], combination of bending and torsion [3] or higher [4] modes of free-standing [1-3] or deposited on a substrate [4] thin films were reported, the mass density of the tested material, which has to be known for the implementation of the resonant method, was measured separately prior to experiment [3] or estimated based on bulk values available in literature [2] (comparative discussion of different experimental techniques for extraction of material properties can be found in [6]) . However, unabated decrease in the film thicknesses and overall dimensions of devices combined with the development of new deposition methods, such as Atomic Layer Deposition (ALD) affecting material properties precludes the use of the bulk values, complicates density measurement and stimulates development of new approaches for material characterization.

In this work we present an approach allowing simultaneous determination of the Young's modulus and density of ultrathin films from a resonance experiment. The approach is based on a non-invasive optical detection of the in-plane and out-of-plane resonant responses of a nano cantilever fabricated from a material with known mechanical properties (in this work single crystal Si was used) prior and after deposition of a thin ALD film. The shifts of in-plane and out-of-plane frequencies were measured at the same structure therefore reducing sensitivity to scattering in geometric parameters and clamping compliances.

EXPERIMENT

Our cantilevers of varying length (6-8μm) and width (45nm-1μm) serving as a carrier structure and made of single crystal Si were fabricated from silicon-on-insulator (SOI) (100) wafers with 250nm thick structural layer. The corresponding process flow along with the scanning electron micrographs of the devices is shown in Fig. 1.

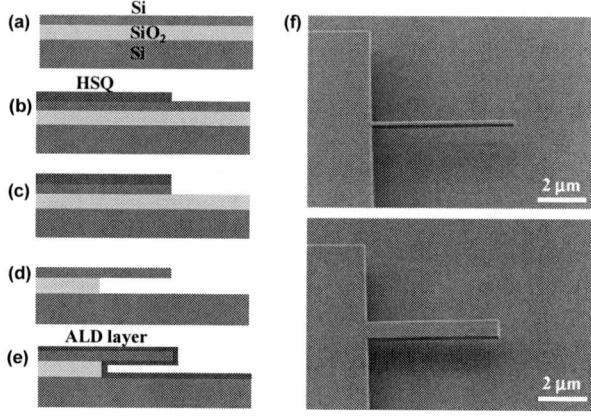

Figure 1: Fabrication of nano cantilevers: (a) SOI wafer (b) HSQ patterning using a 100keV JEOL9300FS E-beam lithography system (c) Reactive Ion Etching (RIE) of the device layer (d) release of the HSQ and 1μm thick sacrificial SiO_2 layer by hydrofluoric acid HF (e) atomic layer deposition of a 212-215 Å thick conformal layer. (f) Scanning electron micrographs of the fabricated devices of different width.

Devices were operated within a vacuum chamber that was evacuated to a pressure of 2×10^{-7} Torr. Vibrational excitation and motion detection were accomplished using an optical setup [7, 8]. An amplitude modulated 415nm diode laser in conjunction with a Fabry-Perot interferometric system incorporating HeNe laser focused at the free end of the cantilever were used to excite and detect the motion of the nanomechanical oscillator. A single cell photodetector and a spectrum analyzer were used to collect the output signal and apply the excitation to the electro-optic modulator (EOM) (Fig. 2), respectively. The energy of the incident light couples into the device layer through localized temperature variations. The frequency of oscillation were determined by fitting the frequency spectra to Lorentzian functions while observed resonances, corresponding to different modes of vibration, were identified by comparison to the values predicted by the model [8].

978-1-4244-2977-6/09 $25.00 © 2009 IEEE

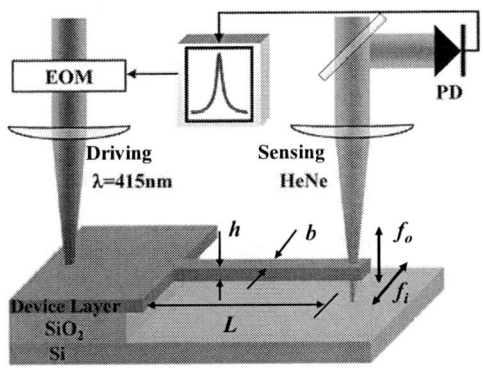

Figure 2: Schematic of the optical actuation and detection setup. Arrows indicate transverse out-of-plane and in-plane motion with the frequencies f_o and f_i, respectively.

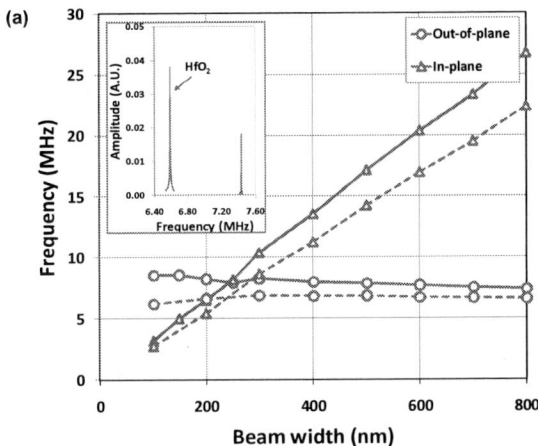

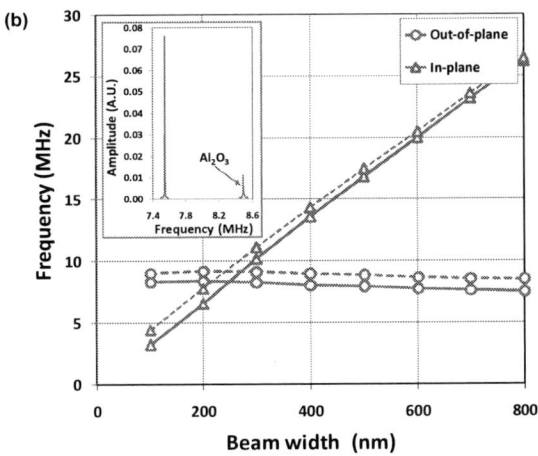

Figure 3: Measured resonant out-of-plane and in-plane frequencies of 6 μm long cantilevers of differing width. Solid lines correspond to the baseline measurements, dashed lines correspond to the cantilevers with ALD deposited (a) 215nm thick HfO₂ (hafnia) layer and (b) 212 nm thick Al₂O₃ (alumina) layer. Inserts represent the out-of-plane response spectra for the 800nm wide beam before and after the layer depositions.

By analyzing resonant responses of the beams of different dimensions it was observed that the in-plane resonant frequencies increased linearly with width and exceeded the nearly constant out of plane frequency for the beams with h/b<1, Figs. 3, (solid lines). Note that a slight decrease of the fundamental frequency of the out-of-plane vibrations was observed and was attributed to the non-ideality of the clamping conditions originated in the compliance of the undercut produced by the wet release process.

The measurements were repeated after conformal ALD of 212-215Å thick alumina or hafnia layers. Depending on the material parameters of the film, the ALD resulted in an increase or decrease of the resonant frequency (Figs. 3, dashed lines). The deposition of the hafnia layer resulted in the decrease of the frequency while deposition of the alumina layer resulted in the increase in the resonant frequency. Relative large frequency shifts, up to 25%, were observed, especially for the wider beams, Fig. 3(a).

MODEL

We consider a nanomechanical cantilever of length L, width b and thickness h serving as a carrier structure with known (baseline) values of Young's modulus E^{bl} and mass density ρ^{bl} and conformally covered by a thin layer of the thickness t with the Young's modulus and density E and ρ, respectively. The beam is allowed to vibrate in the out-of-plane (y) and in-plane (z) directions.

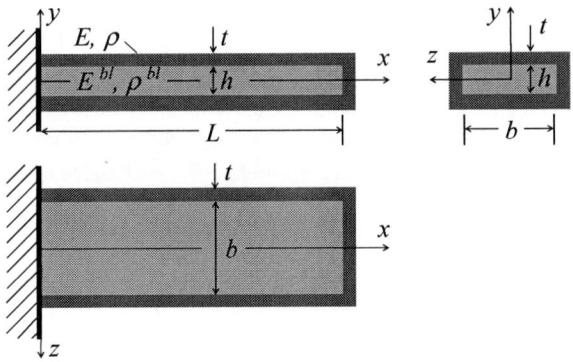

Figure 4: Illustration of the device and notations used in the development.

In the framework of the Euler-Bernoulli theory, the fundamental frequencies of the out-of-plane and in-plane vibrations of the carrier beam without the thin layer (baseline values) are given by the expressions

$$f_o^{bl} = \frac{\lambda_o^2}{2\pi}\sqrt{\frac{B_o^{bl}}{m^{bl}L^4}} \qquad f_i^{bl} = \frac{\lambda_i^2}{2\pi}\sqrt{\frac{B_i^{bl}}{m^{bl}L^4}} \qquad (1)$$

Here $B_o^{bl} = E^{bl}I_{zz}^{bl}$, $B_i^{bl} = E^{bl}I_{yy}^{bl}$ are out-of-plane and in-plane bending stiffness, $m^{bl} = \rho^{bl}A^{bl}$ is the mass per unit length of the beam, $I_{yy}^{bl} = hb^3/12$, $I_{zz}^{bl} = bh^3/12$ are second moments of area of the crossection. Additionally $A^{bl} = bh$ is the crossectional area and λ_o, λ_i are the fundamental eigenvalues of the out-of-plane and in-plane vibrations, both which depend on boundary conditions ($\lambda_o = \lambda_i = 1.875$ for the case of ideal clamped boundary condition). Conformal deposition of the thin layer is accompanied by the change in the distributed stiffness and mass properties of the cantilever that consequently results in the change of the natural frequencies

$$f_o = \frac{\lambda_o^2}{2\pi}\sqrt{\frac{B_o^{eff}}{m^{eff}L^4}} \qquad f_i = \frac{\lambda_i^2}{2\pi}\sqrt{\frac{B_i^{eff}}{m^{eff}L^4}} \qquad (2)$$

By measuring the change in the out-of-plane and in-plane natural frequencies and taking into consideration that the effective values associated with the bending stiffness B_o^{eff}, B_i^{eff} and mass properties M^{eff} are influenced by the geometry of the carrier beam and of the layer as well as by the material properties of both materials, the Young's modulus E and density ρ of the thin layer can be extracted from the following two equations

$$R_0 = \sqrt{\frac{B_o^{eff}}{B_o^{bl}}\frac{m^{bl}}{m^{eff}}} \qquad R_i = \sqrt{\frac{B_i^{eff}}{B_i^{bl}}\frac{m^{bl}}{m^{eff}}} \qquad (3)$$

where $R_o = f_o / f_o^{bl}, R_i = f_i / f_i^{bl}$ are the ratios between the out-of-plan and in-plane natural frequencies after and before ALD. In these equations, the left hand sides incorporate measured values while the right hand side expressions are provided by the model. The natural frequencies before and after the thin layer deposition are measured on the same structure and separately for the out-of-plane and in-plane vibrations, and under the assumption that the deposition of the thin layer does not affect boundary conditions. The eigenvalues which depend on highly uncertain clamping conditions do not appear in eqs. (3), thereby reducing the influence of the scattering in the geometric properties of the beam on the results accuracy.

The effective stiffness and mass properties dependence on the material properties of the thin film were built within the framework of the Euler-Bernoulli beam theory. Assumptions here were such that (i) the sections are planar and perpendicular to the axis of the beam before and after deformation and (ii) the stresses in the width and thickness direction are zero, $\sigma_{yy} = \sigma_{zz} = 0$, and that the beam is under plane stress conditions. In the considered case, the width of the thin film is typically much larger than its thickness and the film's material is under plane strain rather than plane stress conditions (as in the case of the mismatch between the Poisson's ratios of the carrier beam and the film materials). As a result, the extracted value of the Young's modulus is higher than the actual value and is bounded from above by the effective value $\tilde{E} = E / (1 - \nu^2)$ (where ν is the Poisson's ratio of the film material) corresponding to the case of zero Poisson's ratio of the carrier beam material. Our numerical results suggest that the influence of this mismatch is relatively minor for typical values of Poisson's ratio of the materials under consideration.

The effective out-of-plane and in-plane bending stiffness of the composite beam as well as effective mass per unit length are given by the expressions

$$B_o^{eff} = B_o^{bl}[1 + 2\alpha\varepsilon\mathcal{F}_o(\gamma, \varepsilon)]$$
$$B_i^{eff} = B_i^{bl}[1 + 2\alpha\varepsilon\mathcal{F}_i(\gamma, \varepsilon)] \qquad (4)$$
$$m^{eff} = m^{bl}[1 + 2\beta\varepsilon\mathcal{F}_m(\gamma, \xi, \varepsilon)]$$

where the non-dimensional functions

$$\mathcal{F}_o(\gamma, \varepsilon) = 3 + \gamma + 6(1+\gamma)\varepsilon + 4(1+3\gamma)\varepsilon^2 + 8\gamma\varepsilon^3$$
$$\mathcal{F}_i(\gamma, \varepsilon) = 1 + 3\gamma + 6(1+\gamma)\gamma\varepsilon + 4(3+\gamma)\gamma^2\varepsilon^2 + 8\gamma^3\varepsilon^3 \qquad (5)$$
$$F_m(\gamma, \xi, \varepsilon) = 1 + \gamma + 2\gamma\varepsilon + 2\xi(1+2\gamma\varepsilon)(1+2\varepsilon)$$

depend on the beam geometric parameters and

$$\alpha = \frac{E}{E^{bl}} \qquad \beta = \frac{\rho}{\rho^{bl}} \qquad \gamma = \frac{h}{b} \qquad \varepsilon = \frac{t}{h} \qquad \xi = \frac{h}{L} \qquad (6)$$

are non-dimensional quantities introduced for convenience. In view of eqs. (4)-(6), eq.(3) is a system of two algebraic equations in terms of the stiffness and density parameters α and β. These equations can be represented as a system of two linear equations

$$2\varepsilon\mathcal{F}_o\alpha - 2\varepsilon R_o^2\mathcal{F}_m\beta = R_0^2 - 1$$
$$2\varepsilon\mathcal{F}_i\alpha - 2\varepsilon R_i^2\mathcal{F}_m\beta = R_i^2 - 1 \qquad (7)$$

whose solution is

$$\alpha = \frac{R_o^2 - R_i^2}{2\varepsilon(\mathcal{F}_o R_i^2 - \mathcal{F}_i R_o^2)}$$
$$\beta = \frac{\mathcal{F}_o - \mathcal{F}_i - (\mathcal{F}_o R_i^2 - \mathcal{F}_i R_o^2)}{2\varepsilon\mathcal{F}_m(\mathcal{F}_o R_i^2 - \mathcal{F}_i R_o^2)} \qquad (8)$$

and allows to find the Young's modulus and mass density of the layer material using notations of eq. (6) and known material properties of the carrier beam material. Note that in the degenerated case of a square crossection, we have $b=h$, $\gamma=1$ and $\mathcal{F}_o = \mathcal{F}_i$ implying that the out-of-plane and in-plane effective bending stiffnesses are identical (see eqs. (4), (5)). In this case the matrix of the system (7) is singular. We conclude therefore that beams with the square cross-sections are not suitable for the simultaneous extraction of E and ρ.

RESULTS AND DISCUSSION

In order to verify the model as well as to investigate the influence of various parameters on the accuracy of the results, we conducted a set of numerical experiments. The three-dimensional finite elements (FE) model was built using a commercially available package ADINA; three-dimensional 27 node solid elements with quadratic interpolating function in each direction were used. The values of the Young's modulus, Poisson's ratio and mass density of the single crystal Si adopted in our calculations were E^{bl} = 169MPa, ν^{bl}= 0.28 and ρ^{bl}= 2300kg/m3 respectively. First, the out-of-plane and in-plane natural frequencies of the beam with ideal clamping conditions (and dimensions consistent with those used in experiments) were calculated before and after "deposition" of a thin layer with known a priori (prescribed) properties. The Young's modulus and the density were then extracted using eq. (8) and very good agreement (to within 0.5%) with the prescribed values was obtained. Next, the beams of realistic geometries were analyzed. Three-dimensional finite element analysis showed that non-ideal clamping conditions arising due to the undercut of the sacrificial layer as well as residual stress and mismatch in the Poisson's ratio between the deposited film and the cantilever materials, while affecting frequency itself, have minor influence on the determined values. This is as a consequence of the fact that the eigenvalues do not appear in eq. (8). In particular, in the case of largest possible mismatch in the Poisson's ratio the relative error in the Young's modulus was 3.8%.

The values of E and ρ of two materials - Al_2O_3 (alumina) and HfO_2 (hafnia) obtained using measured values of the out-of-plane and in-plane frequencies (see Fig. 3) and eq. (8) are presented in Table 1. Our theoretical and experimental results which are consistent with numerical data and values available in literature (see [9] where the influence of porosity on E and ρ of alumina was investigated) indicate that the suggested approach can be efficiently used for the non-destructive *in situ* material parameters extraction of very thin films incorporated in nano-scale oscillators as well as for combined stiffness-density based material identification.

Table 1: Measured frequencies and extracted values of E and ρ.

b (nm)	f_o^{BL} (MHz)	f_i^{BL} (MHz)	f_o (MHz)	f_i (MHz)	E (GPa)	ρ (kg/m³)
HfO₂ (Hafnia)						
500	7.865	17.134	6.820	14.253	73.4	6053.3
600	7.686	20.366	6.735	16.984	75.7	6120.4
700	7.540	23.356	6.644	19.502	74.9	6112.3
800	7.433	26.718	6.591	22.393	73.0	5990.5
Average					74.2±1.2	6069.1±60.4
Al₂fO₃ (Alumina)						
500	7.909	16.833	8.831	17.462	163.3	2984.8
600	7.759	19.992	8.690	20.494	169.2	3122.1
700	7.641	23.191	8.580	23.577	173.4	3207.5
800	7.549	26.158	8.490	26.428	174.5	3227.7
Average					170.1±5.1	3135.5±110.4

REFERENCES

[1] K.E. Petersen and C.R. Guarnieri, "Young's Modulus Measurements of Thin Films using Micromechanics", *J. Appl. Phys.*, vol. 50, pp. 6761-6766, 1979.

[2] L. Buchaillot, E. Farnault, M. Hoummady, and H. Fujita, "Silicon Nitride Thin Films Young's Modulus Determination by an Optical non Destructive Method", *Jpn. J. Appl. Phys.*, vol. 36, pp. L794-L797, 1997.

[3] S. Nakano, R. Maeda, and K. Yamanaka, "Evaluation of the Elastic Properties of a Cantilever using Resonant Frequencies", *Jpn. J. Appl. Phys.*, vol. 36, pp. 3265-3266, 1997.

[4] H. Guo and A. Lal, "Die-Level Characterization of Silicon-Nitride Membrane/Silicon Structures using Resonant Ultrasonic Spectroscopy", *J. Microelectromech. Syst.*, vol. 12, pp. 53–63, 2003.

[5] P. A. Yuya, Y. Wen, J. A. Turner, Y. A. Dzenis, and Z. Li, "Determination of Young's Modulus of Individual Electrospun Nanofibers by Microcantilever Vibration Method", *Appl. Phys. Lett.*, vol. 90, 111909, 2007.

[6] M. Radovic, E. Lara-Curzio, L. Riester, "Comparison of Different Experimental Techniques for Determination of Elastic Properties of Solids", *Mat. Science Eng.* Vol. A368, pp. 56–70, 2004.

[7] B. Ilic, S. Krylov, K. Aubin, R. Reichenbach, H. G. Craighead, "Optical Excitation of Nanoelectromechanical Oscillators", *Appl. Phys. Lett.*, vol. 86, 193114, 2005.

[8] B. Ilic, S. Krylov, M. Kondratovich, and H. G. Craighead, "Selective Vibrational Detachment of Microspheres using Optically Excited In-plane Motion of Nanomechanical Beams", *Nano Letters*, vol. 7, pp. 2171-2177, 2007.

[9] S. Puchegger, F. Dose, D. Loidl, K. Kromp, R. Janssen, D. Brandhuber, N. Hüsing, and H. Peterlik, "The Dependence of the Elastic Moduli of Reaction Bonded Alumina on Porosity", *J. Europ. Ceramic Soc.*, vol. 27, pp. 35–39, 2007.

THE EFFECTS OF THERMAL OXIDATION OF A MEMS RESONATOR ON TEMPERATURE DRIFT AND ABSOLUTE FREQUENCY

C. van der Avoort, J. van Wingerden and J.T.M. van Beek
NXP Semiconductors Research, Eindhoven, THE NETHERLANDS

ABSTRACT

The dependency of the resonance frequency on temperature, in short temperature drift, of Si resonators is the major contributor to frequency inaccuracy of MEMS based oscillators. The temperature dependency can be reduced through thermal oxidation of the resonator [1]. However, in this paper it is concluded that thickness control of the oxidation layer has a major influence on the absolute frequency, besides on the temperature drift, and as such dominates the frequency spread of a MEMS resonator after production including oxidation. This is a result of the large difference in Young's modulus of Si and SiO2 and the added mass due to oxidation. The overall frequency range in operation—including both thermal drift range and spread in absolute frequency—can even *deteriorate*, rather than improve, through thermal oxidation, depending on the accuracy at which the oxide thickness is controlled.

INTRODUCTION

For most applications of timing or frequency references, a small frequency budget of typically less than 0.1%, or 1000 ppm (parts per million), is required. MEMS-based oscillators are near to ready to replace the quartz-based oscillator market [2]. At the core of a MEMS-based oscillator lies a purely mechanical MEMS resonator [3]. For low-power applications it is necessary to fulfill the frequency budget requirement passively, i.e. without consuming power. Hence, oven-control for operation of the MEMS resonator at elevated temperature or compensating IC electronics [2] are not allowed in meeting the specified budget. For a low-cost solution, trimming of produced resonators is not allowed either. The cost of trimming is identified as a major "gap to commercialization" [4]. Without trimming or active compensation, the temperature coefficient (TC) of the Young's modulus of silicon constitutes a frequency range

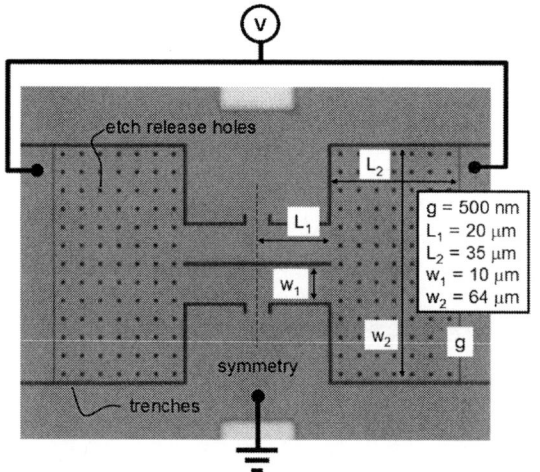

Figure 1: Optical micrograph of the oxidized extensional MEMS resonator under study. Dimensions before oxidation are indicated. $f_0 \approx 20$ MHz.

of thousands of ppms and this thermal range alone is already a showstopper for most applications.

The thermal spread can be reduced effectively by covering the silicon resonator with SiO2 [1]. The Young's moduli of these two materials [5] is here expressed in GPa as

$$
\begin{aligned}
E_{Si} &= 131 \cdot (1 - 60 \cdot 10^{-6} \Delta T) \\
E_{SiO_2} &= 73 \cdot (1 + 196 \cdot 10^{-6} \Delta T)
\end{aligned}
, \tag{1}
$$

where ΔT is the change in temperature. The temperature dependencies have opposite signs. Incorporating the proper amount of SiO2, depending on the geometry and mode of operation of the resonator, the resulting effective temperature dependency can be brought to zero. The temperature-dependent effective modulus of elasticity of an extensional resonator can be denoted as

$$
E_{eff}(T) = \frac{E_{Si} A_{Si} + E_{SiO2} A_{SiO2}}{A_{Si} + A_{SiO2}}, \tag{2}
$$

where the two areas A correspond to the cross-sectional areas of silicon and silicon-dioxide, respectively. The effective modulus is inserted into the equation for the frequency of this resonator

$$
f(T) = \frac{\beta^2}{2\pi} \sqrt{\frac{E_{eff}(T)}{\rho_{eff}}}, \tag{3}
$$

where the constant β^2 is a function of the dimensions w_i and L_i indicated in Figure 1. The effective mass density ρ_{eff} is determined analogue to Eq. (2). The linear TC of the frequency of the resonator (hereafter, simply TC) is the derivative of $f(T)$ with respect to temperature T. For a silicon-only resonator the linear TC is -30 ppm/K.

MEASUREMENTS

Four identical SOI wafers are produced with elongational mode resonators, as depicted in Figure 1. Actuation as well as measurement of the vibration is performed electrostatically. The actuation gap dimension is chosen such that oxidation up to several hundreds of nm is possible. Thermal oxidation of the four wafers with a 1.5 μm thick SOI layer leads to average oxide skins of 102 nm, 186 nm, 234 nm and 334 nm thickness, respectively. The found thicknesses differ from the targeted values and were found by ellipsometry on dummy wafers.

The temperature coefficients (TCs) of the resonators on these wafers are assessed experimentally. Figure 2 shows frequency versus temperature measurements on the four wafers, expressed in ppm change in frequency with respect to the frequency at room temperature. Table 1a summarizes the linear TCs (also depicted in Figure 4) and quadratic TCs that are found by a least-squares polynomial fit. The quadratic component found in our experiments (Figure 5) is in agreement with literature [5] and is unaltered by the addition of silicon-dioxide. The quadratic component hence renders a ~100 ppm spread

978-1-4244-2977-6/09 $25.00 © 2009 IEEE

Table 1. Measured linear and quadratic temperature coefficients (TCs) on four different wafers and analysis of the variation in oxide thickness needed to account for the measured spread in frequencies. Wafer 11* has such a thick oxide skin that the 500 nm gap is nearly filled and not many working devices are found.

Wafer	Thickness	Temperature dependence (1a)		Frequency spread (1b)			
		TClin [ppm/°C]	TCquad [ppb/°C²]	f_{avg} [MHz]	spread [ppm]	t_{ox} var. [nm](pct)	direct [nm]
1	102 nm	-20.8 ± 0.2	-23 ± 3	19.98	1450	6 (6%)	(6)
2	186 nm	-12.3 ± 0.4	-24 ± 9	19.36	2300	7.5 (4%)	(6.7)
3	234 nm	-8.6 ± 1.2	-24 ± 1	19.03	4280	14 (6%)	-
11*	336 nm	+3.0 ± 0.6	-22 ± 6				

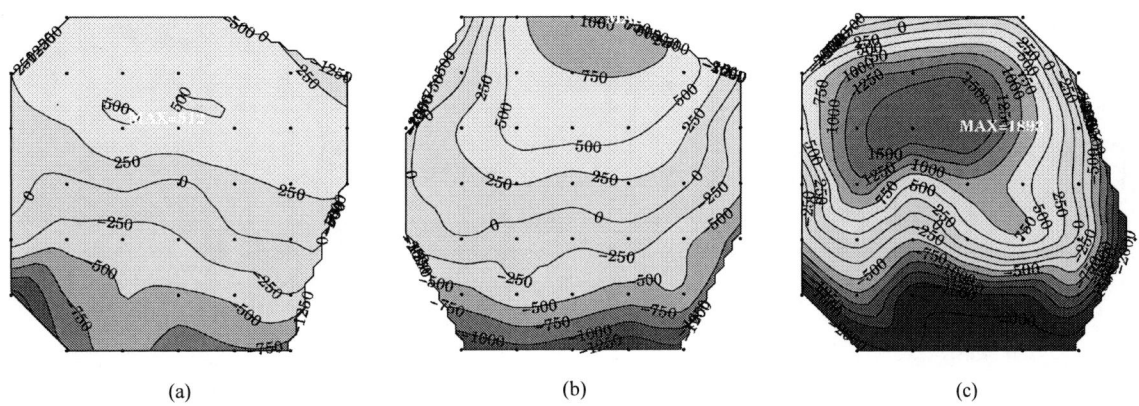

   (a)   (b)   (c)

Figure 3: Direct measurements of the frequency spread on three wafers. Figures a, b and c correspond to wafers 1, 2 and 3, respectively. Color coding is equal in all figures. Iso-levels indicate the local frequency relative to the average of the measurements, in ppm. Black dots indicate the locations of working devices on the wafer.

over a 100°C window that cannot be compensated for by the thermal oxide layer.

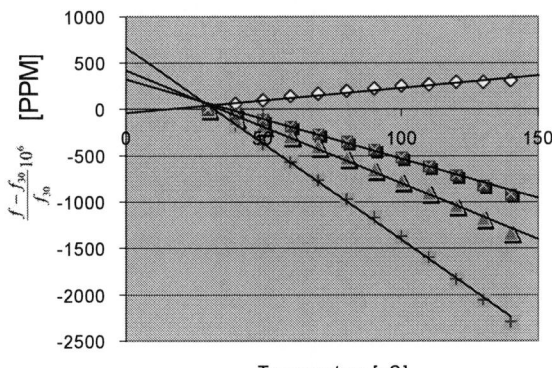

Figure 2: Frequency versus temperature measurement, expressed in ppm change relative to the frequency @ 30°C.

The spread of resonance frequencies over these wafers is measured. Figure 3 shows the measured frequencies per wafer, relative to the average frequency. Table 1b shows the resulting spread for which a minimum-to-maximum range is taken. The oxide thickness variation of 5% (on average) deduced from a simple linear model (Figure 4) for frequency versus oxide thickness is in agreement with direct measurements of the variation of the thickness of the oxide. For wafers 1 and 2, dummy wafers were available, allowing a 9-point oxide layer thickness measurement. On wafer 11, due to the thick oxide layer compared to the gap width, not enough

working devices could be found in order to make a map. The maps in Figure 3 show a clear trend of increasing frequencies from bottom to top. This result was expected from fabrication. As a result of the clamping and the oven used, all wafers were prone to varying oxidation speeds as a result of a temperature gradient in the oven.

On wafer 2 two frequency maps were generated, one at 25 °C and one at 140 °C. Using these maps, a crude estimate of the spread in linear TC can be deduced. The spread in the linear TCs is again a result of the 5% relative thickness error in the oxide layer.

RESULTS

The results of the experiments are summarized in figure 6a, where frequency spreads of mainly two origins for oxidized resonators are collected into one graph. For a low-cost production process, one requires any of the resonators to be deployable in the total temperature range. Therefore, a range of 100 °C is taken, to be multiplied by the linear TC in ppm/K. The total frequency spread is composed of a part due to temperature variation and a part due to spread in the oxide thickness. The horizontal axis on the graph poses the *intended* oxide thickness. The thermal drift is at first hand diminishing when we start to grow oxide. The range for absolute frequency is only increasing. This is because the frequency *variation* is a function of the relative range of 5% on the oxide thickness. The two straight lines in Figure 4 are hence used to construct an *overall* frequency spread. The temperature-dependent and variation-related spreads are summed. In Figure 6a, measurements A, B and C stem from the directly performed spread measurements.

978-1-4244-2977-6/09 $25.00 © 2009 IEEE

Measurements D, E, F and G are found from the accurate frequency-versus-temperature measurements. The value J is the 3100 ppm spread intrinsic for silicon. Value H is the actually measured spread in linear TC across wafer 2. The found variation is two times larger than what could be expected from a 5% oxide thickness tolerance projected on the linear TC-versus-thickness line. However, the uncertainty in actual TC as a result of the 5% error on oxide thickness (red line in Figure 6a) constitutes a negligible contribution to the overall frequency spread.

In Figure 6b the overall spread is again calculated using the straight lines from figure 4, now considering a 1% relative accuracy on oxide layer thickness. In this case, a reduction of the overall frequency spread can be expected when approaching the oxide thickness for which TC = 0.

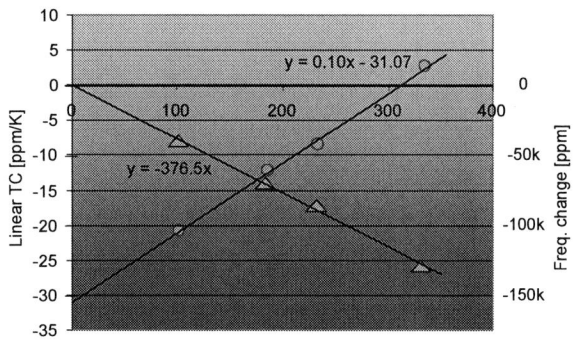

Figure 4: Measured linear TCs on four wafers (circles, left axis) and change in absolute frequency (triangles, right).

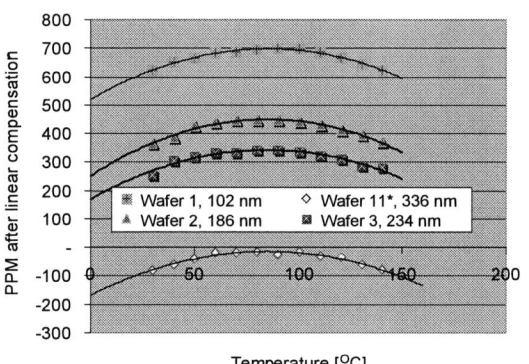

Figure 5: Residual ppms after subtraction of the linear fit in Figure 2. The vertical axis abscissae have no meaning. The second-order TCs are equal.

CONCLUSIONS

From the conducted experiments it is concluded that instead of decreasing the *overall* frequency spread of a MEMS resonator by oxidizing it, it is increased. At a 3% tolerance the spread will remain at ~3000 ppm and Figure 6b shows that reduction of the oxidation inaccuracy to a hypothetic 1% will lead to an improved overall spread. One is left with the major consumer of the overall frequency budget being the spread on absolute frequency after oxidation. It is concluded that uniform thermal oxidation of Si resonators *alone* does not lead to the level of accuracy in resonance frequency that is typically required for reference oscillators and timing applications.

Tighter control of oxide layer thickness is not realistic. Trimming is not desired with respect to the costs. As such, it is the slope of frequency versus oxide layer thickness (now -377 ppm/nm) that has to be lowered to reduce the overall frequency spread.

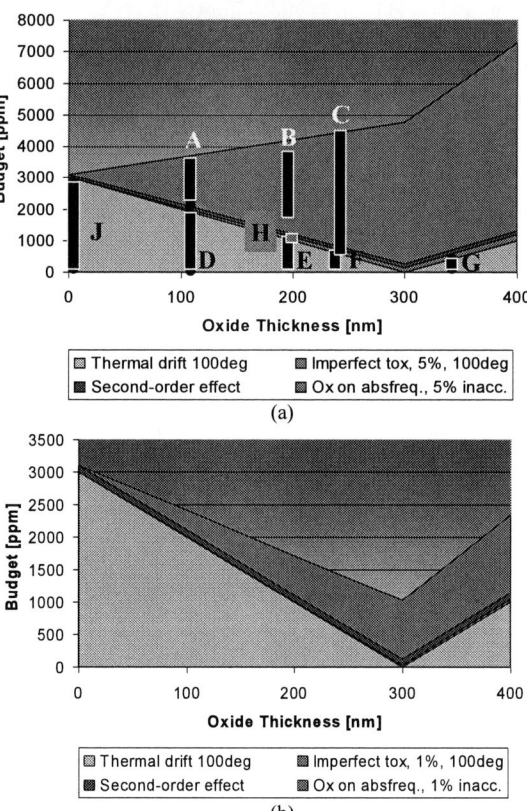

Figure 6: (a) Skin thickness-dependent frequency spread for the case of 5% oxide accuracy (including measured values) and (b) for the hypothetical case of 1% variation. Description of measurements A to J in the main text.

ACKNOWLEDGEMENT

The measured data presented in this paper were to a large extent collected by Diego Villar.

REFERENCES

[1] R. Melamud et al., "Composite flexural-mode resonator with controllable turnover temperature", in *IEEE international conference on micro electro mechanical systems*, pp. 199–202 (2007)

[2] W.T. Hsu, "Vibrating RF MEMS for timing and frequency references", in *IEEE MTT-S International Microwave Symposium Digest*, pp. 672–675 (2006)

[3] J.T.M. van Beek et al., "Scalable 1.1 GHz fundamental mode piezo-resistive silicon MEMS resonator", in *IEDM 2007*, pp. 411–414 (2007)

[4] W.T. Hsu and A.R. Brown, "Frequency trimming for MEMS resonator oscillators", in *IEEE International Frequency Control Symposium 2007*, pp. 1088–1091 (2007)

[5] C. Bourgeois et al., "Design of resonators for the determination of the temperature coefficients of elastic constants of monocrystalline silicon", in *Proc. of the 1997 IEEE International Frequency Control Symposium*, pp. 791–799 (1997)

NON-DESTRUCTIVE QUANTITATIVE MEASUREMENT METHOD FOR NORMAL AND SHEAR STRESSES ON SINGLE-CRYSTALLINE SILICON STRUCTURES FOR RELIABILITY OF SILICON-MEMS

M. Komatsubara[1], T. Namazu[1], N. Naka[2], S. Kashiwagi[2], K. Ohtsuki[2], and S. Inoue[1]

[1]Department of Mechanical and Systems Engineering, University of Hyogo, JAPAN
[2]Semiconductor Systems R&D Department, HORIBA Ltd., JAPAN
E-mails: namazu@eng.u-hyogo.ac.jp (T. Namazu); nobuyuki.naka@horiba.jp (N. Naka)

ABSTRACT

This paper describes an experimental analysis method for evaluating surface stress distribution in single-crystalline silicon (SCS) microstructures using laser Raman spectroscope. A biaxial tensile tester designed for film specimens was employed to apply uni/biaxial stresses to SCS specimen having 270-nm-high, 4-μm-square SCS convex structures in the gauge section. As reported in Transducers 2007 [1], two-curve fitting of Raman spectrum was useful for analyzing stress magnitude at the edge of convex structures. In this study, the partial least-square (PLS) method was adopted for the obtained Raman spectra at the convex edge in order to determine stress components as well as their magnitudes. By using the PLS method, the shear stress component was able to be measured in addition to the normal stress components. The stress magnitude in respective stress components was in very good agreement with that estimated by finite element analysis (FEA).

1. INTRODUCTION

Reliable stress evaluation at the micro- or nanoscale has been in high demand for micro devices, such as semiconductor products and microelectromechanical systems (MEMS). Micro-Raman spectroscopy is a powerful tool for measuring the stress and strain in single-crystalline silicon (SCS) that is the most common structural material in MEMS [1]-[4]. Micro-Raman spectroscopy offers the following advantages over other stress or strain evaluation methods, such as X-ray diffraction and converged beam electron diffraction: (1) nondestructive measurement, (2) short measurement time, and (3) a high spatial resolution of less than 1 μm. However, the pattern size continually shrinks as a result of the rapid development in semiconductor fabrication technologies, so that it is difficult to apply micro-Raman spectroscopy to submicron-scale structures like a channel or shallow trench isolation (STI) because of the technique's spatial resolution limit. In addition, the stress conditions around the edge of such structures are sure to be complex: tensile, compressive, and shear stresses will coexist. Therefore, the evaluation of the magnitude and component of stress or strain in micro- and submicron-scale structures with Raman spectroscopy becomes a challenging technical requirement.

The objective of this study is to develop an evaluation method for measuring the magnitude and component of surface stress in SCS microstructure by using micro-Raman spectroscopy. A specially developed biaxial tensile tester [5] was employed to apply uniaxial and biaxial tensile stresses on SCS specimens with "convex" structures indicating

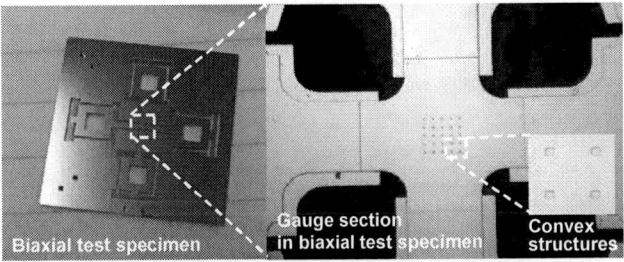

Fig. 1 The produced SCS specimen for biaxial tensile test.

Fig. 2 Biaxial tensile tester under Raman spectroscope.

step-like and square-shaped SCS structures. The test setup enabled us to produce a multi-axial stress state around the convex structures for micro-Raman spectroscopy measurement. To specify both the stress component and magnitude, the analytical calibration curves were newly constructed by multivariate analysis using the PLS method.

2. EXPERIMENTAL PROCEDURE

SCS specimen for biaxial tensile test

Fig. 1 shows the produced SCS specimen for biaxial tensile test. The specimen was fabricated through conventional micromachining techniques. The tensile specimen consists of a gauge section with 5×5 SCS convex structures, a fillet section for reducing stress concentration during tensile loading, four etched holes for fastening the specimen, SCS springs for supporting the gauge section, and a frame. The specimen gauge section has a 300×300 μm² square shape, and is oriented along the [110] directions in the (001) plane.

Raman spectroscopic stress measurement

When a tensile stress is applied to a specimen, the atomic distance along the tensile direction expands. Consequently, the peak position of the Raman spectrum shifts to lower wavenumbers. The amount of shift in the Raman spectrum peak is known to be linearly proportional to the applied stress

978-1-4244-2977-6/09 $25.00 © 2009 IEEE

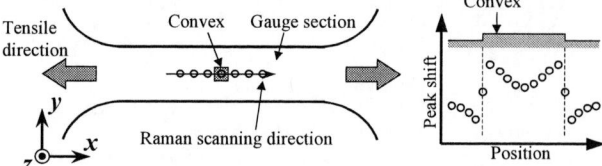

Fig. 3 Schematic of line scan measurement.

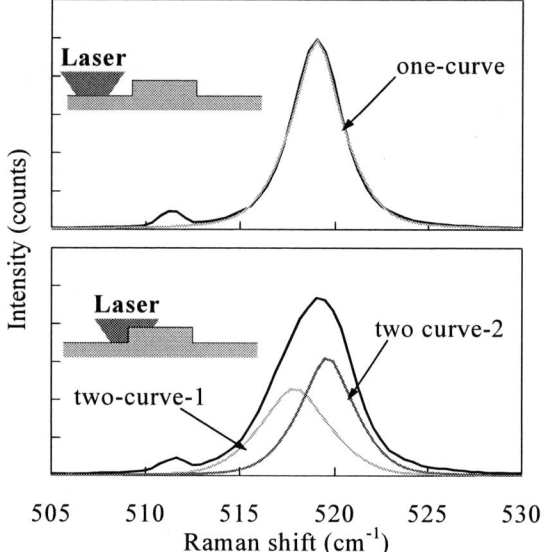

Figs. 4 Typical Raman spectrum at the flat and edge sections.

[1]. For a biaxial stress state, the relation between the peak position and the stress is expressed as:

$$\sigma_{xx} + \sigma_{yy} = a \times \Delta v \qquad (1)$$

The coefficient a is estimated to be in the 430-500 MPa/cm^{-1} range.

Fig. 2 shows a photograph of the biaxial tensile tester set under the commercial Raman spectrometer (HORIBA Jobin Yvon, LabRAM HR-800). The tensile tester consists of four sets of a piezoelectric actuator, a linear variable differential transformer (LVDT), a load cell, and specimen holders [5]. The tester was set onto the Raman spectroscope stage driven by a piezoelectric actuator with a minimum moving step of 5 nm. A water-cooled argon ion laser with a wavelength of 363.8 nm was utilized as the excitation light for Raman spectroscopy. The incident depth of the light into SCS is estimated to be less than 10 nm, indicating that only the Raman spectrum near the top surface of SCS is obtained. The spot size of the focused laser through an objective lens is approximately 1 μm in diameter. All the spectral parameters, such as peak position, peak intensity, and full width at half maximum (FWHM), were derived from a Gaussian/Lorentzian curve fitted to the Raman spectrum obtained.

We conducted two types of experiments:

(1) Raman measurement on a flat section of specimen to determine the relations between the spectrum peak shift and the applied uniaxial and biaxial tensile stresses.

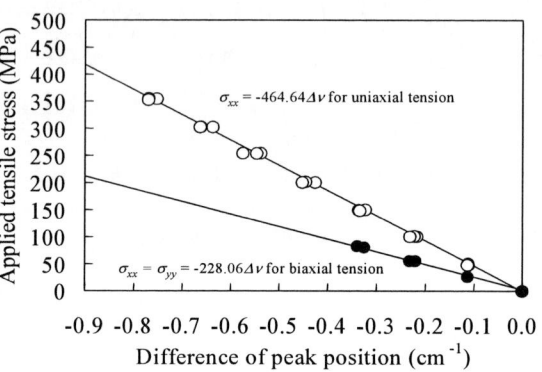

Fig. 5 Difference of peak position vs. applied tensile stress.

(2) Raman measurement around a SCS convex structure to determine the surface stress distribution. As shown in Fig. 3, measurements were conducted over a convex structure at 50 nm intervals along the tensile direction.

3. RESULTS AND DISCUSSIONS
Calibration curve

Fig. 4(a) shows the Raman spectrum around 520 cm^{-1} for the flat section of a specimen under 500 MPa. The spectrum has two peaks, indicating that the sharp and weak peaks are related to SCS and the laser. The SCS peak has a symmetrical shape, so that the peak can be fitted by one curve. As can be seen in Fig. 4(b), the Raman spectrum for the edge of a convex structure at 500 MPa exhibits a broader peak than that for the flat section, and is asymmetric. This is because the spectrum obtained from the vicinity of the edge includes information of the sidewall, the top and bottom corners of a convex structure, or there may be several peaks due to the lifting of degeneracy.

Fig. 5 depicts the relationships between the peak position of the spectrum and the uniaxial and biaxial tensile stresses. The results are derived from the shifts in the strongest peak position located around 520 cm^{-1} obtained at a flat section of specimen. The amount of peak shift in the flat section clearly increases linearly with increasing applied tensile stress regardless of uniaxial or biaxial stress state. The slopes are calculated to be -464.64 and -228.06 MPa/cm^{-1} under uniaxial and biaxial stress states, respectively. We used the slope (-464.64 MPa/cm^{-1}) as coefficient a in Eq. (1) in order to estimate the stress on the specimen from the peak position data of Raman spectra.

Line scan measurement result

In order to assess the credibility of the stress distribution from Raman measurements, finite element analysis (FEA) was conducted. Figs. 6(a)-(d) show the distributions of σ_{xx}, σ_{yy}, σ_{zz}, and σ_{xz} derived from FEA calculations. The specimen model, in particular, the convex shape, was built on the basis of the AFM and SEM results for a specimen that was tested. All the stress distributions were obtained under a tensile stress of 500 MPa applied to the specimen model longitudinally. Since σ_{xy} and σ_{yz} were almost 0 MPa for the entire specimen model, these stress components were not considered. As for σ_{xx} in Fig. 6(a), the maximum tensile

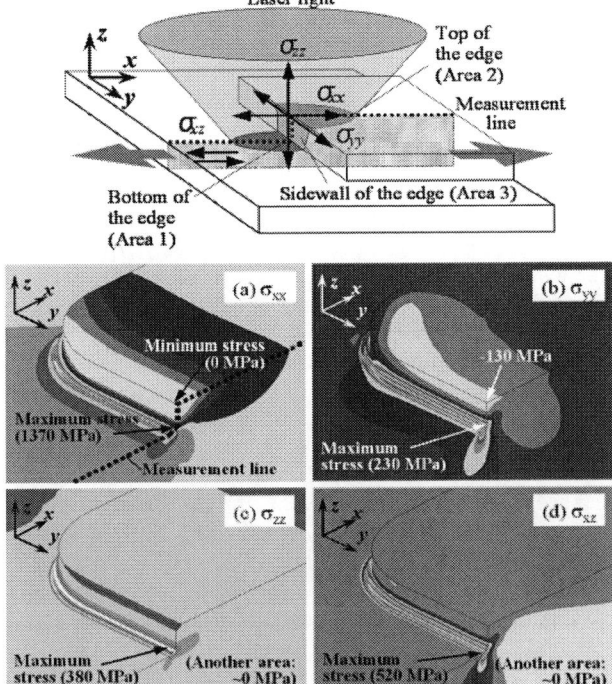

Figs. 6 FEA stress distributions in the edge section.

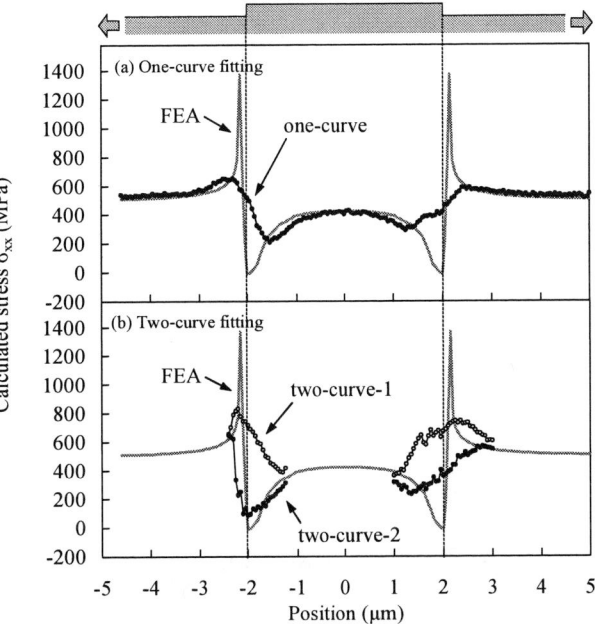

Figs. 7 Stress distributions derived from one- and two-curve fittings for Raman spectra obtained at the convex region.

stress occurred at an extremely small area in the bottom corner of the convex edge. The maximum value of σ_{xx} is found to be 1370 MPa, which is approximately 2.7 times the applied tensile stress. The stress decreases along the convex edge from the bottom to the top corner. At the top corner, the stress is found to be almost 0 MPa. On the other hand, in Fig. 6(b), the transverse stress, σ_{yy}, at the top corner is not 0 MPa, but rather, a few hundred MPa. At the bottom corner, a stress was generated in the opposite direction along the y-axis. For σ_{zz} shown in Fig. 6(c), the maximum stress occurred at the bottom corner, whereas the minimum stress, i.e., almost 0 MPa, was produced at the top corner, as with σ_{xx}. In addition to these normal stresses, a shear stress, σ_{xz}, was clearly produced around the convex structure despite the uniaxial stress state of the flat section. In Fig. 6(d), the distribution of σ_{xz} is observed around the bottom edge where other stress components were produced. The area where these stresses were generated was located within approximately 200 nm from the bottom corner of the convex structure, which can produce multi-axial stress components on a portion of the SCS specimen when a uniaxial tensile stress is applied to the entire specimen.

Under the assumption that the Raman spectrum from around the edge only contains information on the top and bottom corners, and no information on the side wall, we conducted two-curve fittings of the spectra obtained near the edge in order to separate the data for the top and bottom corners. Figs. 7(a) and (b) show the distributions of the tensile stress σ_{xx} derived from one- and two-curve fittings of the peak position distribution, respectively. These stress distributions were obtained when a tensile stress of 500 MPa was applied to the flat section of the specimen. In the one-curve fitting results shown in Fig. 7(a), the stress distribution over the flat section is in very good agreement with that obtained by FEA. In the flat section, the error margin between the stress distributions obtained by one-curve fitting and FEA is within ±40 MPa. Around the edge, however, the one-curve fitting result differs completely from the FEA result. This indicates that it is difficult to estimate the stress at the edge of the convex structure precisely by one-curve fitting of the peak position data alone. As seen in Fig. 7(b), at the top and bottom corners of the convex structure, the stress distribution obtained by two-curve fitting exhibits a better approximation for the FEA stress distribution than that obtained by one-curve fitting. For example, at the top and bottom corners of the convex edge, the differences in the stress values between the Raman measurements and FEA results were 540 MPa and 97 MPa, respectively, which are approximately 25% and 56% lower than the values obtained by one-curve fitting. This indicates that, for the evaluation of σ_{xx}, two approximation curves produced by the two-curve fitting process can independently provide stress information for the top and bottom corners of the convex structure. From the viewpoint of the ease of the procedure, the two-curve fitting method would be effective for the evaluation of uniaxial plane stress distribution on a microscale SCS structure, though there are slight differences in the stress magnitude at the edge between the fitting method and FEA.

Raman spectrum analysis with the PLS method

During the tensile test, a multi-axial stress is sure to be produced around the edge of convex structures on the specimen. Even if Raman spectroscopy is conducted around this structure, it is difficult to directly obtain the stress components as well as the magnitude of the stress from the

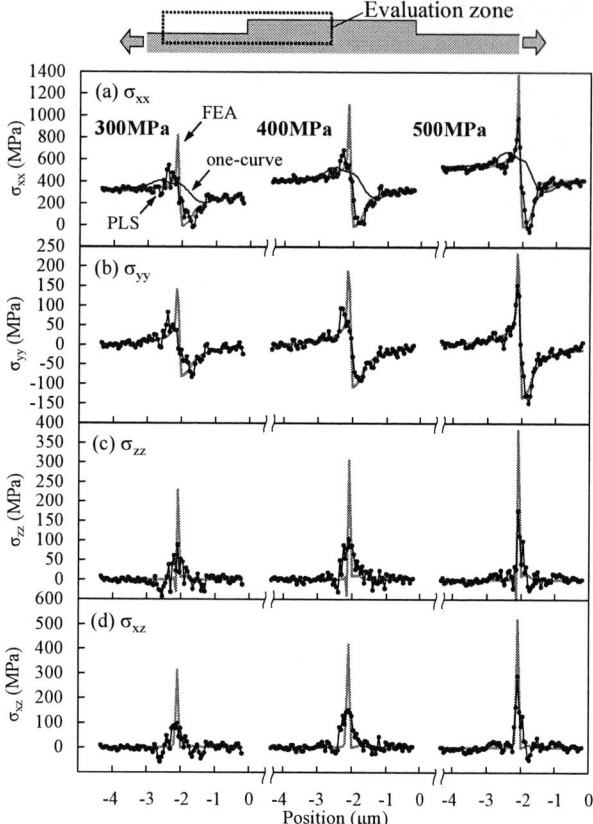

Figs. 8 Stress distributions derived from the multivariate analyses with the PLS method.

the peak positions in two-curve fitting, and c_{ij} denotes the calibration coefficients obtained from PLS analyses. In actual fact, PLS analyses were carried out by using the data of 9 Raman measurement points for each FEA calculation point.

Figs. 8(a)-(d) show the PLS analysis results, which indicate the stress distributions for σ_{xx}, σ_{yy}, σ_{zz}, and σ_{xz} around the edge of the convex structure. In Fig. 8(a), the stress distributions, σ_{xx}, obtained by PLS analysis show a closer agreement with those obtained by FEA than those obtained by the one-curve fitting method. For instance, regardless of the magnitude of the applied stress, the differences in the minimum and maximum stresses around the edge between PLS and FEA are almost 0% and 35-50%, respectively, which are smaller than those between one- and two-curve fittings and FEA. The differences in the minimum stresses at the bottom corner of the edge are relatively larger than those in the maximum stresses at the top corner. In Fig. 8(b), the stress distributions σ_{yy} obtained by PLS analysis also agree with those determined by FEA, as do σ_{xx}. For all stress conditions, the regression errors at the top corner are approximately 0 MPa, whereas those at the bottom corner are 82, 93, and 57 MPa, a 30-40% error margin relative to FEA. In Figs. 8(c) and (d), stress distributions σ_{zz} and σ_{xz} derived by FEA show a dramatic change in an extremely small area. The PLS method can be used to determine these distributions, but its accuracy for estimating σ_{zz} and σ_{xz} is not beyond those for σ_{xx} and σ_{yy}. If better spatial resolution of micro-Raman spectroscopy is achieved, stress distributions obtained by the PLS method would have a better accuracy even in the case of extremely small areas.

4. CONCLUSIONS

We evaluated local surface stress distributions on SCS microstructures by Raman spectroscopy using a biaxial tensile tester. The two-curve fitting to the Raman spectra obtained near the edge exhibited a better approximation for FEA than the one-curve fitting. In the multivariate analysis with the PLS method, the stress distributions calculated from Raman spectral parameters coincided almost exactly with the FEA stress distributions. The PLS method was able to express shear stress as well as normal stress on a SCS microstructure.

ACKNOWLEDGEMENTS

The authors would like to express their gratitude to members of SyncMEMS Laboratory in Kagawa University, Japan, for specimen fabrication. This study is partly supported by Grant-in-Aid 20760075 for Scientific Research from Japan Society for the Promotion of Science (JSPS).

REFERENCES

[1] T. Namazu, *et al.*, *Transducers 2007*, p. 627, 2007.
[2] M. Komatsubara, *et al.*, *SSDM 2008*, p. 378, 2008.
[3] V. Senez, *et al.*, *J. Appl. Phys.*, vol. 94, p. 5574, 2003.
[4] M. Kodera, *et al.*, *Jpn. J. Appl. Phys.*, vol. 47, p. 2506, 2008.
[5] Y. Nagai, *et al.*, *IEEE MEMS 2008*, p. 443, 2008.

measured Raman spectrum alone (i.e., these data cannot be obtained from one curve fitting method). To estimate the stress distribution around the convex structures, we conducted multivariate analyses with the PLS method, using Raman spectral parameters that were obtained around the edge of the convex structure. In these analyses, the explanatory variables are Raman spectral parameters, such as peak shift, peak intensity, FWHM for one-curve fitting, and peak shift for two-curve fitting, and the response variables are stress distributions calculated by FEA. The PLS method employs three principal stress components (σ_{xx}, σ_{yy}, σ_{zz}) and one shear stress component (σ_{xy}). The other two shear stress components (σ_{zy}, σ_{zx}) have not been considered because of they are almost zero at the edge of the convex structure. By comparing the Raman spectral parameters with the FEA stress data at one measurement point, the stress components derived from the PLS method can be expressed as follows:

$$\begin{pmatrix} \sigma_{xx} \\ \sigma_{yy} \\ \sigma_{zz} \\ \sigma_{xy} \end{pmatrix} = \begin{pmatrix} c_{11} & c_{12} & c_{13} & c_{14} & c_{15} \\ c_{21} & c_{22} & c_{23} & c_{24} & c_{25} \\ c_{31} & c_{32} & c_{33} & c_{34} & c_{35} \\ c_{41} & c_{42} & c_{43} & c_{44} & c_{45} \end{pmatrix} \begin{pmatrix} \nu_1 \\ A_1 \\ \omega_1 \\ \nu_{21} \\ \nu_{22} \end{pmatrix} \quad (2)$$

where ν_1, A_1, and ω_1 are the peak position, peak intensity, and FWHM in one-curve fitting, respectively. ν_{21} and ν_{22} are

PHASE LOCK LOOP BASED TEMPERATURE COMPENSATION FOR MEMS OSCILLATORS

J. Salvia, R. Melamud, S. Chandorkar, H.K. Lee, Y.Q. Qu, S.F. Lord, B. Murmann, and T.W. Kenny

Stanford University, Stanford, CA, USA

ABSTRACT

We present a new temperature compensation system for microresonator based frequency references. It consists of a phase lock loop whose inputs are derived from two microresonators with different temperature coefficients of frequency. The resonators are suspended within an encapsulated cavity and are heated to constant temperature by the phase lock loop controller, thereby achieving active temperature compensation. We show repeated real-time measurements of two prototypes which achieve frequency stability of better than ±1 ppm from -20 °C to +80 °C without calibration look-up tables and ±0.05 ppm with calibration.

INTRODUCTION

Largely because of their reliability, small size, and system integration, Microelectromechanical Systems (MEMS) oscillators have begun to replace the current state of the art—quartz crystal oscillators—in segments of the frequency reference market [1]. However, the temperature stability of MEMS oscillators is inferior to that of compensated quartz crystal oscillators, and this is one of several shortcomings that have precluded the use of MEMS references in high precision applications like navigation or radar. In this paper we present a temperature compensation scheme that significantly improves the temperature stability of MEMS oscillators and may thereby facilitate the use of low power, space saving MEMS oscillators in high precision applications.

Many compensation schemes have been proposed in the past to improve the temperature stability of MEMS resonators, including using electrostatic forces, alternative materials, stresses, or variable gaps to correct for temperature effects, but few achieve frequency stability below 1 ppm [2-5]. Our group previously described a

temperature compensation scheme that could achieve sub-ppm temperature stability using quality factor as a feedback signal in an active control loop [6]. However, a practical implementation of this scheme revealed that even minor variations in feedback components could severely degrade system performance [7]. In contrast, the compensation scheme presented in this work relies upon the dual resonator temperature sensing method presented in [8], the thermal isolation structure from [9], and a new phase lock loop based controller to achieve sub-ppm stability even when the entire system is implemented on a breadboard and exposed to temperature variations alongside the oscillator.

DEVICE

Our microstructure is comprised of two composite Si/SiO$_2$ tuning fork resonators with resonant frequencies near 1.2 MHz (Fig. 1). The two resonators have uniform coatings of SiO$_2$, but their silicon cores have different widths. Consequently, the resonant frequencies of resonators 1 and 2, called f_1 and f_2, have different temperature dependences (Fig. 2) [3]. The entire structure is hermetically vacuum encapsulated at the wafer level using a high temperature epitaxial silicon sealing process [10]. The resonators are anchored to the substrate by a folded beam suspension that provides both thermal isolation and Joule heating resistance. In combination with the vacuum encapsulation, the folded beam suspension forms a low power 'micro-oven' [9]. DC current flowed through the structure can be used to maintain the resonators' temperature at a constant elevated value with an efficiency of 6.7 K/mW. The micro-oven also ensures that there is negligible temperature difference between the two resonators.

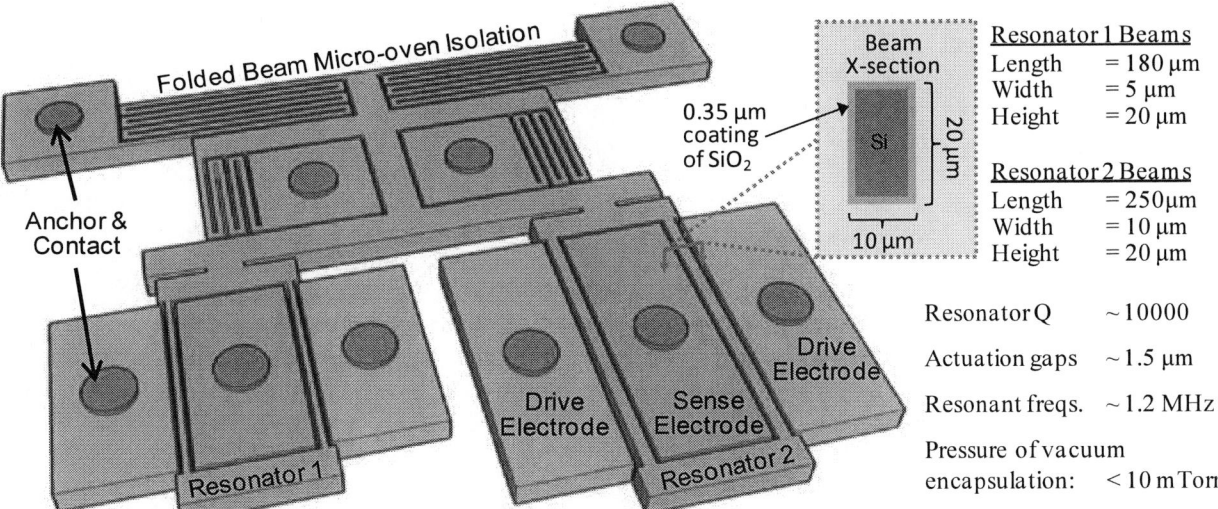

Figure 1: Double-ended-tuning-fork resonators suspended using micro-oven isolation. They are fabricated in the single crystal silicon device layer of a silicon-on-insulator wafer, uniformly coated in a 0.35 µm layer of SiO$_2$, and vacuum packaged at the wafer level using a low-pressure high-temperature epitaxial silicon thin film encapsulation process [10].

978-1-4244-2977-6/09 $25.00 © 2009 IEEE

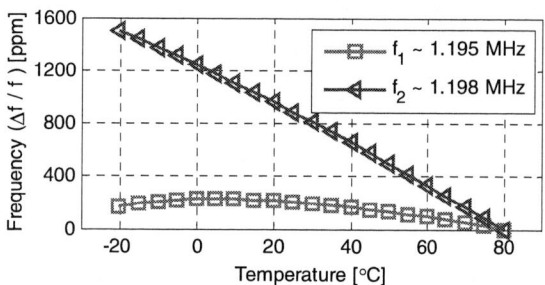

Figure 2: Uncompensated resonant frequency vs. temperature characteristics of the two oxide-coated silicon resonators.

COMPENSATION SYSTEM

Fig. 3 shows a block diagram of our temperature compensation system. Separate transimpedance amplifiers (TIAs) electrostatically actuate resonators 1 and 2 to steady state oscillations at f_1 and f_2, respectively. The outputs of these oscillators are mixed and filtered to form a signal at the difference frequency f_2-f_1. The output of the first oscillator is also passed through an integer frequency divider to form a signal at the reference frequency f_1/N. Because f_1 and f_2 have different temperature dependences, the difference frequency f_2-f_1 depends much more strongly on temperature than the reference frequency f_1/N (-15 Hz/°C compared to less than -0.02 Hz/°C). We selected N such that f_2-f_1 equals f_1/N when the resonators are heated to 90 °C—just above the ambient temperature range of interest (Fig. 4). Our compensation system applies power to the micro-oven to maintain f_2-f_1 and f_1/N in phase lock, thereby ensuring that the resonators in the micro-oven are held at 90 °C despite changes in ambient temperature. Either f_1 or f_2 can then be used as the system's output.

An optional calibration step and digital lookup table can be added to the system to remove residual errors caused by the temperature sensitivity of the TIA oscillator electronics and the thermal expansion of the micro-oven. This compensation is achieved by slightly adjusting a resonator's drive level depending on the ambient temperature, which can be inferred from the voltage input to the micro-oven (see dashed path in Fig. 3). In this way,

the system can take advantage of the nonlinear A-f effect [11] to correct for residual errors left over by the main phase lock loop controller. In our experiments, the resonators, oscillator electronics, and phase lock loop controller (i.e. everything drawn with solid lines in Fig. 3) were implemented at the board level and experienced the same temperature variations as the MEMS resonator (Fig. 5). The optional look-up table was implemented with bench top equipment.

MEASUREMENT RESULTS

We built two prototypes with lithographically identical MEMS devices. Because of process variations, these devices exhibited different resonant frequencies, as shown in Table 1. Table 2 shows the power consumption of our compensation system. The transient performance of our prototypes without the optional calibration and lookup table is shown in Fig. 6. The data demonstrate that our compensation system reduces frequency deviations to less than ±1 ppm in a robust and repeatable manner even for transient temperature ramps of 5 °C/min. Fig. 7 shows the time-averaged performance of the systems when subjected to stationary ambient temperatures. This figure indicates that the optional calibration and lookup table can reduce residual frequency errors to ±0.05 ppm.

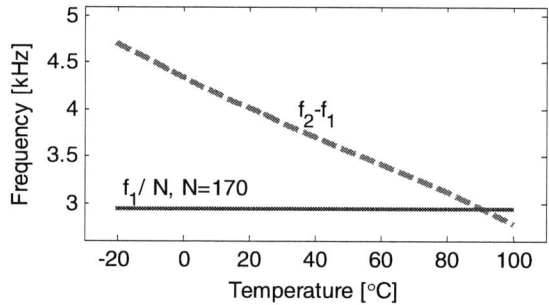

Figure 4: Measurement of the temperature dependence of the reference frequency and the difference frequency. N is chosen so that the two curves intersect around 90 °C.

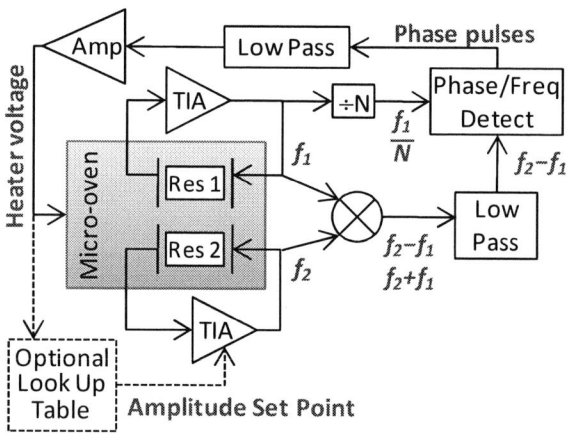

Figure 3: Block diagram of prototype temperature compensation system. Negative feedback holds f_2-f_1 in phase lock with f_1/N (see Fig. 4), ensuring constant micro-oven temperature despite changes in ambient temperature.

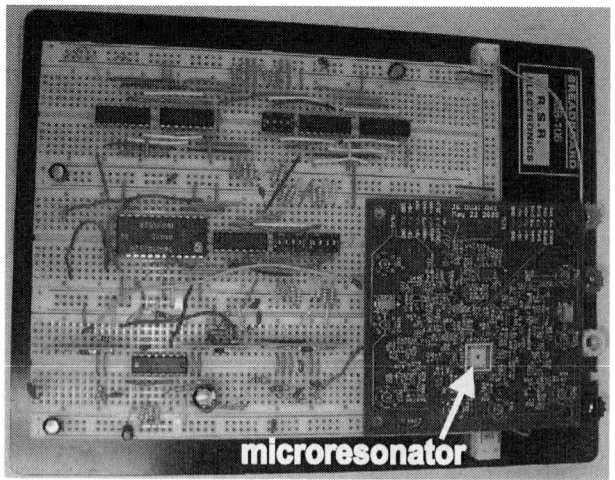

Figure 5: Picture of the board level implementation of our temperature compensated MEMS oscillator.

Table 1: Comparison of the two prototypes.

	Prototype A	**Prototype B**
f_1	1.186 MHz	1.195 MHz
f_2	1.193 MHz	1.198 MHz
f_2-f_1	6.98 kHz	2.95 kHz
N	170	405

Table 2: Prototype B power consumption over the temperature range -20 °C to +80 °C

TIA #1	46 to 47 mW
TIA #2	46 to 47 mW
Micro-oven (Joule heating power)	2 to 17 mW
Micro-oven Amplifier	10 to 13 mW
Phase Detector, Filters, & Frequency Divider	10 to 13 mW
Total	**114 to 137 mW**

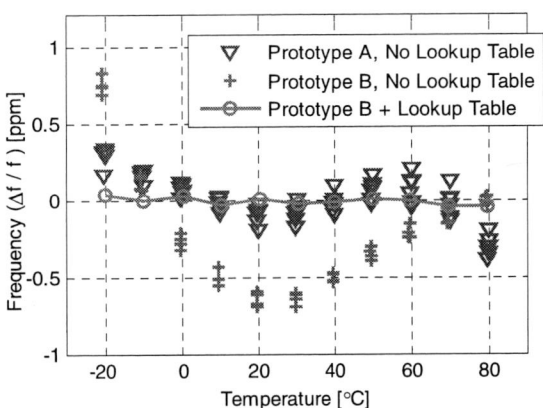

Figure 7: Multiple measurements of steady state frequency deviation vs. ambient temperature for two similar prototypes.

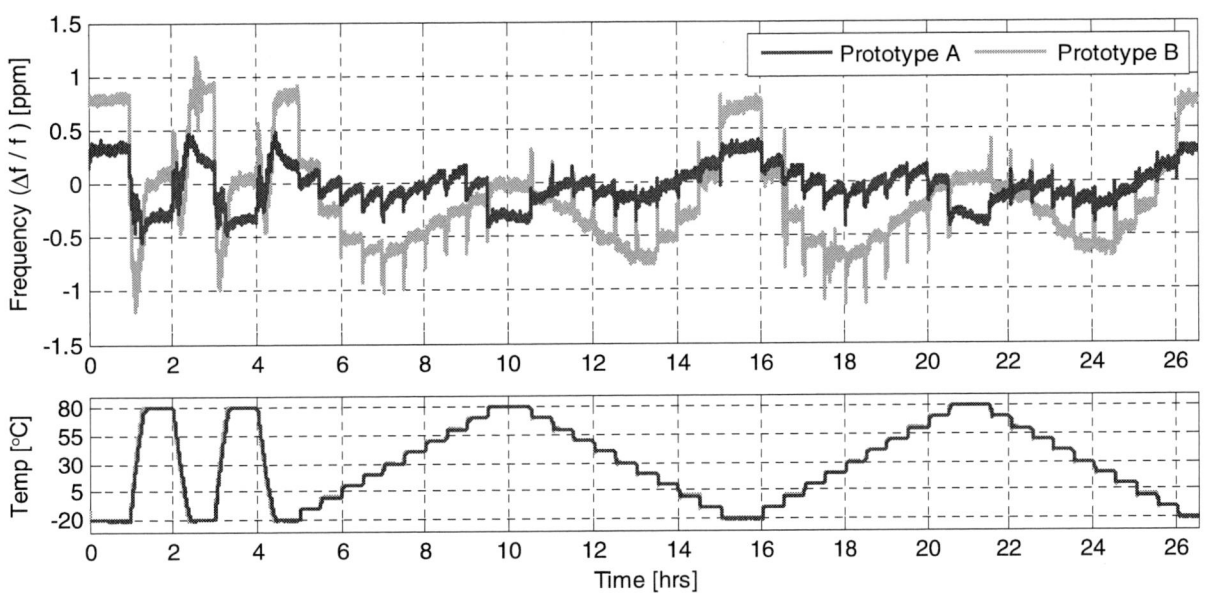

Figure 6: Real-time frequency measurements of two prototypes subjected to 5 °C/min ambient temperature ramps. In this experiment, the optional digital calibration and lookup table were not used.

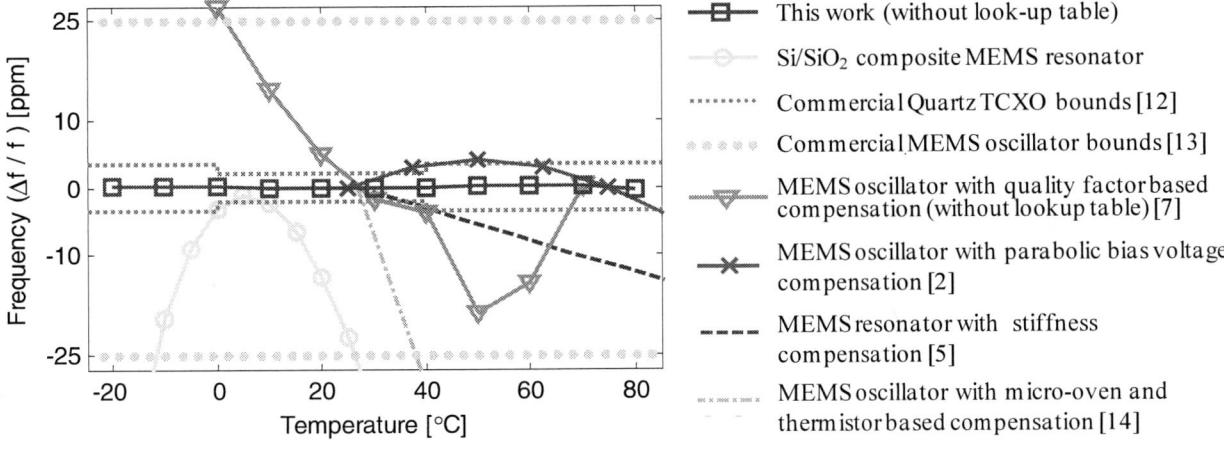

Figure 8: Frequency vs. temperature curves for a variety of academic and commercial resonators and oscillators.

978-1-4244-2977-6/09 $25.00 © 2009 IEEE

DISCUSSION

Fig. 8 compares the temperature sensitivity of our prototypes to alternative systems. Even without calibration, our system outperforms some of the best temperature compensated MEMS oscillators to date, and the temperature performance falls well within bounds for Temperature Compensated Quartz Crystal Oscillators (TCXOs). In fact, with a digital calibration and lookup table our prototype can compete with Oven Compensated Quartz Crystal Oscillators (OCXOs), which offer temperature stability in the range of ppb but require watts of power and cubic centimeters of volume [15]. Our resonators and micro-oven occupy less than 1 mm^3.

The power consumption listed in Table 2 should be interpreted with care, as our board level prototype was limited to commercially available parts. An integrated circuit (IC) designed specifically for this purpose would consume far less power. For example, the IC MEMS oscillator in [16], which includes a TIA comparable to those in our system, consumes only 1.8 mW. Thus, we expect the power budget of an IC implementation of our system to be dominated by the 17 mW of micro-oven heating power and not by the oscillator electronics.

Some significant challenges still remain in the development of a high precision MEMS oscillator. Temperature stability is only one of many important specifications (e.g. phase noise, aging, process and supply variation, acceleration sensitivity), and our prototypes are still far from meeting all of them. One disadvantage of our system is the coupling between the two separate oscillator loops, which can generate unwanted harmonics and mixing products in the output spectrum. Another potential disadvantage is our use of Si/SiO$_2$ composite resonators, whose long term stability is still uncertain [17]. Future design iterations could alternatively use the dual mode temperature sensing scheme from [18] instead of composite resonators to avoid this issue.

CONCLUSIONS

We have demonstrated a new method of temperature compensation for MEMS frequency references. It relies on a phase lock loop controller to regulate the temperature of a low power micro-oven that contains two microresonators. Our board level prototypes achieve temperature stability that is competitive with compensated quartz crystal oscillators—a substantial improvement over previous temperature compensation schemes for microresonators.

ACKNOWLEDGEMENTS

This work was supported by DARPA HERMIT (ONRN66001-03-1-8942), an NDSEG fellowship, the Robert Bosch Corporation Palo Alto RTC, a CIS Seed Grant, and the National Nanofabrication Users Network facilities funded by the National Science Foundation under award ECS-9731294. Matt Hopcroft, Bongang Kim, and John Vig also helped make this work possible.

REFERENCES

[1] M. Lutz et al., "MEMS oscillators for high volume commercial applications," in *Digest Tech Papers Transducers '07 Conference*, Lyon, France, Jun. 10-14, 2007, pp. 49-52.

[2] G.K. Ho, K. Sundaresan, S. Pourkamali, F. Ayazi, "Temperature compensated IBAR reference oscillators," *Proc. 19th IEEE MEMS 2006*, Istanbul, Turkey, Jan. 22-26, 2006, pp. 910-913.

[3] R. Melamud et al., "Temperature-compensated high-stability silicon resonators," *Applied Physics Letters*, vol. 90, 244107, Jun. 2007.

[4] W.-T. Hsu, C.T.-C. Nguyen, "Geometric stress compensation for enhanced thermal stability in micromechanical resonators," *Proc. IEEE Itnl. Ultrasonic Symp.*, Sendai, Japan, Oct. 5-8, 1998, pp. 945-948.

[5] W.-T. Hsu, C.T.-C. Nguyen, "Stiffness compensated temperature insensitive micromechanical resonators," *Proc. 15th IEEE MEMS 2002*, Las Vegas, NV, USA, Jan. 20-24, 2002, pp. 731-734.

[6] M. A. Hopcroft et al., "A high stability MEMS frequency reference," in *Digest Tech Papers Transducers '07 Conference*, Lyon, France, Jun. 10-14, 2007, pp. 1307-1309.

[7] J. Salvia et al., "Exploring the limits and practicality of Q-based temperature compensation for silicon resonators" to be presented at *IEDM 2008*, San Francisco, CA, USA, Dec. 15-17, 2008.

[8] C. M. Jha et al., "High resolution microresonator-based digital temperature sensor," *Applied Physics Letters*, vol. 91, 074101-3, Aug. 2007.

[9] C. M. Jha et al., "Thermal isolation of encapsulated MEMS resonators," *J. of Microelectromech. Syst.*, vol. 17, no. 1, pp. 175-184, Feb. 2008.

[10] A. Partridge, A.E. Rice, T.W. Kenny, M. Lutz, "New thin film epitaxial polysilicon encapsulation for piezoresistive accelerometers," *Proc. 14th IEEE MEMS 2001*, Interlaken, Switzerland, Jan. 21-25, 2001, pp. 54-59.

[11] M. Agarwal et al., "Nonlinear characterization of electrostatic MEMS resonators," *Proc. IEEE Frequency Control Symp., 2006*, Miami, FL, USA, Jun. 4-7, 2006, pp. 209-121.

[12] "Timekeeping accuracy, automatic and affordable," Application Note 3566, Maxim Integrated Products, Aug. 17, 2005.

[13] SiT8102 Datasheet Rev 1.04, SiTime Corporation, Aug. 12, 2008.

[14] C.T.-C. Nguyen, R.T. Howe, "Microresonator frequency control and stabilization using an integrated micro oven," *Digest Tech Papers Transducers '93 Conference*, Yokohama, Japan, Jun. 7-10, 1993, pp. 1040-1043.

[15] C4600 OCXO Datasheet, Vectron International, v2006-11-10.

[16] K. Sundaresan, G.K. Ho, S. Pourkamali, F. Ayazi, "A two-chip, 4-MHz, microelectromechanical reference oscillator," *Proc. Itnl. Symp. Circ. Sys. 2005*, Kobe, Japan, May 23-26, 2005, pp. 5461-5464.

[17] G. Bahl et al., "Observations of fixed and mobile charge in composite MEMS resonators," *Proc. Hilton Head Workshop*, Jun. 1-5, 2008, pp. 102-105.

[18] M. Koskenvuori, V. Kaajakari, T. Mattila, I. Tittonen, "Temperature measurement and compensation based on two vibrating modes of a bulk acoustic mode microresonator," *Proc. 21nd IEEE MEMS 2008*, Tucson, AZ, USA, Jan. 13-17, 2008, pp. 78-81.

DEGRADATION OF MECHANICAL STRENGTH AT SI/SIO₂ INTERFACE ON SOI WAFERS UNDER CYCLIC LOADING

T. Ando[1], T. Takumi[2], and K. Sato[1]

[1]Dept. of Micro-Nano System Engineering, Nagoya University, JAPAN
[2] Dept. of Mechanical Engineering, Nagoya University, JAPAN

ABSTRACT

Fatigue tests of silicon stepped cantilevers fabricated from silicon on insulator (SOI) wafers were conducted under the bending mode to evaluate the effect of cyclic loading on fractures occurring in silicon and Si/SiO₂ interfaces. The specimen in the quasi-static mode fractured at the stress concentration site on the silicon specimens. However, during the fatigue tests the cantilever broke after 10^4 cycles with stress amplitude of nearly half of the bending strength at the fixed end comprising the Si/SiO₂ interface. The results demonstrated that the cyclic stress durability in the Si/SiO₂ interface is significantly lower than that of the silicon body.

1. INTRODUCTION

The use of silicon on insulator (SOI) wafers is expanding the possibilities for miniaturization, improved performance, and increased accuracy of microelectromechanical systems (MEMS) devices. Silicon dioxide (SiO₂) used under the top silicon layer performs a number of functions. In the fabrication process, it serves as an etch stop and as a sacrificial layer. In the device itself, it serves as an insulating layer and as a supporting plate of the floating structure from the substrate. In the fabrication process, the use of SiO₂ makes it easy to adopt conventional etching processes such as surface micromachining and bulk micromachining to the top silicon layer and the substrate, respectively. Because of the advantages they provide, MEMS devices on SOI wafers are being manufactured in increasing quantities despite their higher cost compared to that of conventional bulk silicon wafers [1]. For example, the microactuators in SOI-MEMS devices have a high-aspect ratio structure that achieves high driving force with low consumption energy [2]. AFM probes are typical examples of SOI-MEMS devices in which the cantilever or the tip is fabricated on the SOI layer [3, 4].

In a microstructure consisted of two or more layers, the interface is one of the most destructive site that includes the physical misalignment of each layer and has a high probability of stress concentration. It is well known that the silica is subjected to stress corrosion and that this causes delayed failures [5]. This phenomenon is strongly correlated to failures that occur in SiO₂ sandwiched between single crystal silicon layers in SOI wafers. In commercial devices, a great concern is their reliability and durability for a long term. To date, however, few studies have been performed on the occurrence of mechanical failures in SiO₂ or in Si/SiO₂ interfaces [6, 7].

A subject of major concern is how the forming of an SiO₂ layer over a silicon structure affects device durability. This paper reports experimental results of bending tests carried out under quasi-static and cyclic loading mode by using a newly developed testing method.

2. EXPERIMENTAL

2.1 Specimen

Figure 1 shows the schematic illustration of the test specimen used in this study. The cantilever beam specimen having a step change in width is 90 μm and 30 μm width for the larger and smaller parts, respectively. The thickness is 10 μm throughout the specimen. The external load is applied at the point 50 μm apart from the stepped site in width. When the length of the wider part of the stepped cantilever is enough shorter than that of smaller width part, the stress concentration occurs at the corner of the step change. The step corner was designed to be 5 μm in length from the fixed end. The silicon cantilever specimen is fabricated in the SOI layer and fixed to the SiO₂ intermediate layer at the end by chemical bonds.

2.2 Bending Test

Figure 2 shows our newly developed bending test device for micro-sized cyclic tests. The test device has a mass structure supported on a rigid frame by four parallel beams, shown in Figure 2 (a). The cantilever specimens were fixed to the frame towards the mass that has banks underlying the free ends of the cantilever. The bending test was carried out by driving the mass perpendicular to the device as shown in Figure 2 (b). When the frame and mass moved relatively up and down in the vertical direction, respectively, the cantilever specimen was pushed up by the

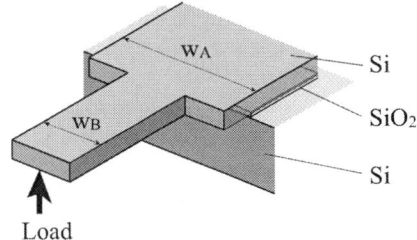

Figure 1: Schematic illustration of stepped cantilever specimen fabricated from SOI wafer.

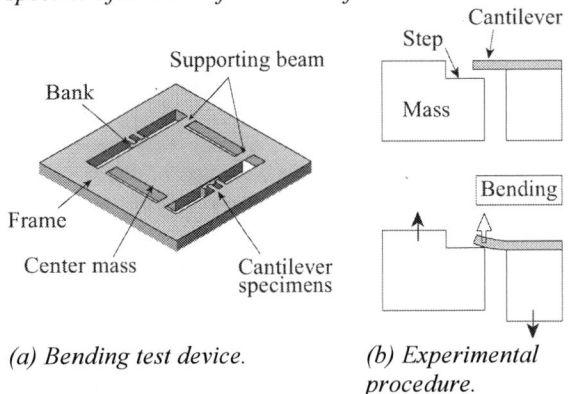

(a) Bending test device.   *(b) Experimental procedure.*

Figure 2: Bending test device for quasi-static and cyclic loading test.

978-1-4244-2977-6/09 $25.00 © 2009 IEEE

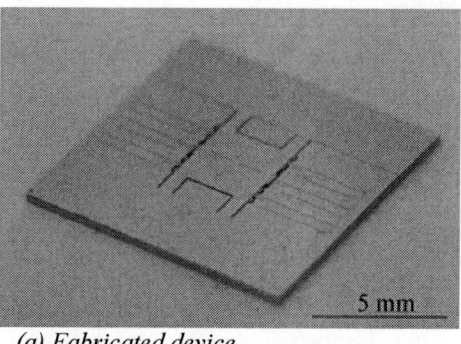

(a) Fabricated device

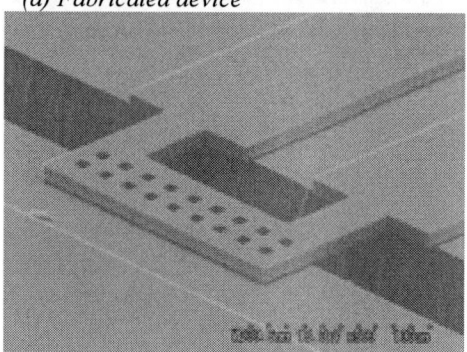

(b) SEM photograph of specimen.
Figure 3: SEM photographs of fabricated testing device.

bank.

The testing device was fabricated from SOI wafer by using MEMS processes. A (100) CZ-silicon layer with a thickness of 10 μm are bonded on the (100) silicon substrate through the thermally oxidized SiO_2 layer with a thickness of 1 μm. The top silicon layer is etched by deep reactive ion etching (DRIE) to form the cantilever specimen and parallel beams. The longitudinal direction of the cantilever is aligned to <110>. The 350 μm-thick silicon substrate is etched away from the back side by DRIE to attain the SiO_2 intermediate layer. The specimen is released to make it freestanding by immersing it in a buffered hydrofluoric acid (BHF) solution to etch the SiO_2 layer. Figure 3 shows the fabricated testing device (Figure 3 (a)) having four cantilever specimens (Figure 3 (b)) in total.

2.3 Measurement System

Figure 4 shows a schematic illustration and photograph of the measurement system. By appropriate control, a monotonic and cyclic loads were applied to the specimen by using a voice coil motor (VCM) connected to the stage. The displacement of the stage was detected and recorded by using a laser displacement sensor. The applied load was measured by a load cell connecting a quartz rod that vertically restrains the center mass of the testing device. The fatigue tests were carried out under a load-controlled mode with a sinusoidal waveform applied at a frequency of 200 Hz. All of the tests were conducted at room temperature within negligible change.

3. RESULTS AND DISCUSSION
3.1 Quasi-static Bending Test

Figure 5 shows the measured displacement and load during a quasi-static bending test. The temporal trajectories of displacement and load are almost same at first. It is

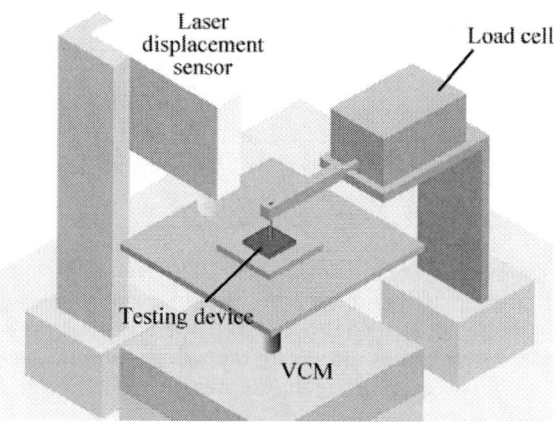

(a) Schematic of measurement system.

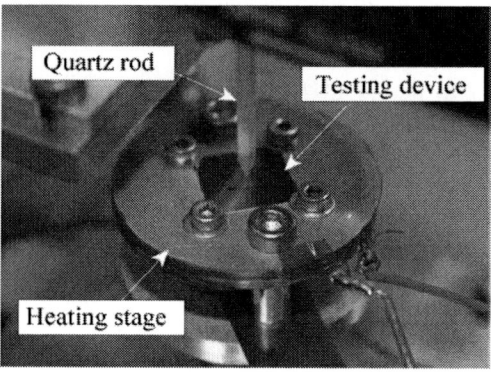

(b) Photograph of magnified view of testing device.
Figure 4: Measurement system for quasi-static and fatigue test.

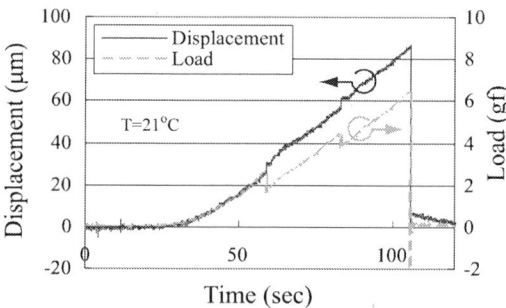

Figure 5: Measured displacement and load in quasi-static bending test.

noted that the elastic deformation is recognized in this field that the load linearly increases against the displacement. The load increases in proportion to time until one specimen of four fractured at the load drop point in Figure 5, while the displacement monotonically increases during the test. After the first fracture of the specimen was confirmed, the VCM was stopped even if another three specimens were not broken because the lack of the specimen produced the loss of balance in vertical drive of the frame.

Fracture morphology of the specimen after the bending test was observed by SEM, all specimens failed in brittle manner. A SEM photograph of the fractured specimen is shown in Figure 6. For all specimens the fracture surface appeared at the stepped area in cantilever and inclined with respect to the longitudinal axis. Based on

978-1-4244-2977-6/09 $25.00 © 2009 IEEE   662

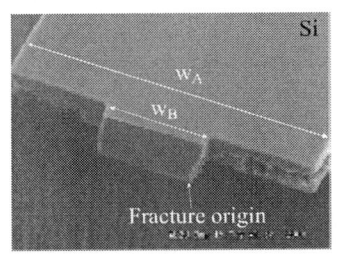

(a) SEM photograph of fracture specimen.

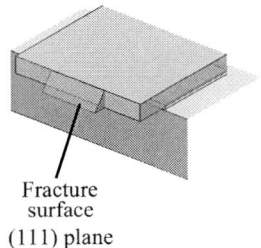

(b) Schematic of fractured surface.

Figure 6: Fractured specimen after quasi-static bending test.

this figure, it can be said that fracture occurred from the edge and propagate across the width on (111) cleavage plane, as shown in Figure 6 (b). The stress was concentrated on the lower half of cantilever where damaged area was formed due to the DRIE process. Fracture originated at the site with a massive roughness on it.

The bending strength in this study is simply defined as the normal stress at the step, instead of troublesome analysis for stepped cantilever with inflection point. The bending strength was estimated from the amount of the drop of the load at fracture shown in Figure 5 by simply applying the formula in material mechanics to calculate the bending stress in the longitudinal direction of the cantilever.

$$\sigma = \frac{Pl}{I}\frac{h}{2},$$

where σ is maximum normal bending stress, P is applied load, l, h, I is length, thickness, moment of inertia of the specimen, respectively. The measured bending strength was 3.20 ± 0.48 GPa.

3.2 Fatigue Test

Fatigue tests were conducted under the cyclic load in fatigue tests at a frequency of 200 Hz with various maximum stress levels. The stress ratio R was 0.1. The measured loads and displacements before and after the fracture are shown in Figure 7. The fracture of the specimen was confirmed from the change of the amplitude of the load, while the amplitude of the displacement didn't change. The change of the load amplitude enables us to evaluate the fatigue life of the specimen. The specimens fractured after over 10^4 cycles and one of them did not break even after 10^7 cycles. The relation between the bending stress and fatigue cycle for some samples is shown in Figure 8. As the stress level decreased, the fatigue life tends to increase although there is a wide range of experimental data. The specimens broke in fatigue tests at a stress around half that of the static strength. In this graph, the stress value means the amplitude of the stress at the stepped corner, instead of the bending strength calculated at the fracture point for convenience.

Fractured specimen subjected to cyclic bending load was observed under a SEM to investigate the fracture behavior. All of the specimens failed in brittle manner, same as the quasi-static test. Figure 9 shows SEM photograph of the typical fractured specimen. The fracture surface shows inclined plane that consists of (111) and

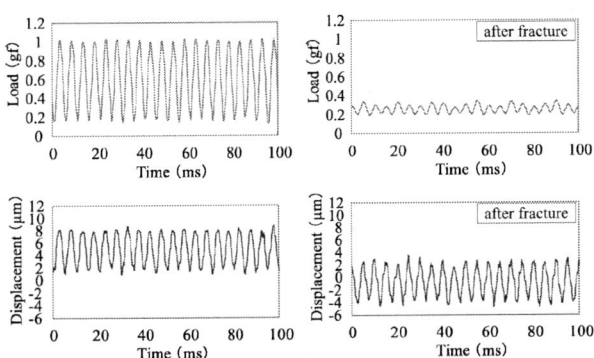

Figure 7: Obtained data of load and displacement during fatigue test.

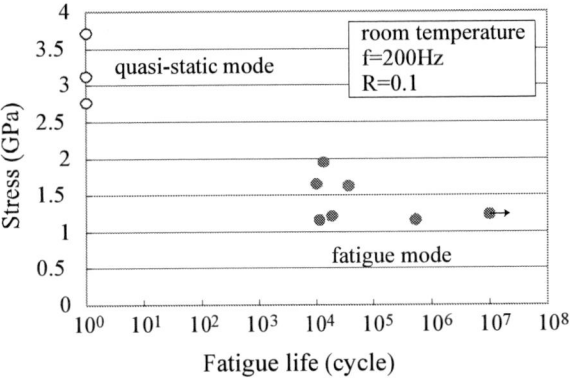

Figure 8: Relationship between fatigue life and applied maximum stress.

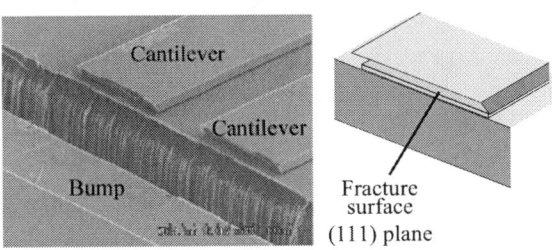

(a) SEM photograph of fracture specimen.

(b) Schematic of fatigue-fractured specimen.

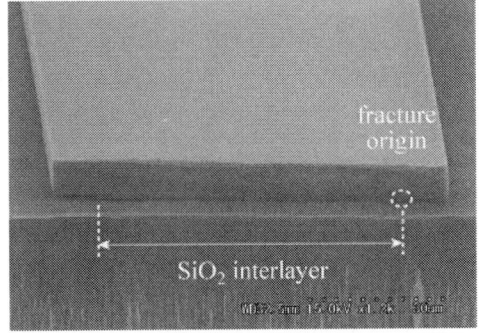

(c) Magnified view of fracture surface.

Figure 9: Fractured specimen after fatigue test.

crack initiation site. Unlike the specimen under the quasi-static load, in the case of fatigue tests fracture occurred at the fixed end on the edge of the interface of Si/SiO$_2$. The bending stress generated at the edge of the interface is intrinsically smaller than that at stepped area. It is clear that the crack initiated from the end of the interface of Si/SiO$_2$ by breaking a Si-O bond at the lower stress level in cyclic test. The initiated crack propagated to the silicon inside and formed the (111) plane.

As for bi-material, the interface between two different solid is weak for load applied from various directions. The vertical load applied to the thin film might cause delamination of the film from the substrate. The structure of the specimen fabricated on SOI wafer used in this study is similar to the experimental setup for delamination test of bi-layer material [8]. However, the crack propagated in silicon body instead of the interface of Si/SiO$_2$ and SiO$_2$layer. The crack initiation due to cyclic load is more likely to occur at the edge of the Si/SiO$_2$ interface, while the silicon is weak to propagation of the existing crack.

4. CONCLUSION

We estimated the strengths of silicon cantilevers fabricated from SOI wafers under quasi-static and cyclic loading conditions The cantilever specimens were designed to have two steps in width to enable the local stress to be concentrated in the silicon layer. In quasi-static bending tests, the specimens fractured at the stress concentration point with an average strength of 3.20±0.48 GPa. In fatigue tests with cyclic loading, the fracture origin shifted to the fixed end where the edge of the intersection of Si/SiO$_2$ and the applied stress amplitude was nearly half of the bending strength. The difference in the fractures in the static and fatigue modes indicates that the cyclic loading significantly affected Si/SiO2 interfaces in terms of the generation and propagation of cracks.

REFERENCES

[1] S. Renard, "Industrial MEMS on SOI", J. Micromech. Microeng., 10, (2000), pp.245-249

[2] B.-H. Kim and K. Chun, "Fabrication of an electrostatic track-following micro actuator for hard disk drives using SOI wafer", J. Micromech. Microeng., 11, 2001, pp.1-6

[3] R.P. Ried, H.J. Mamin, B.D. Terris, L.-S. Fan, and D. Rugar, "6-MHz 2-N/m piezoresistive atomic-force-microscope cantilevers with INCISIVE tips", Journal of Microelectromech. Syst., vol. 6, 1997, pp.294-302

[4] J. Thaysen, A. Boisen, O. Hansen, and S. Bouwstra, "Atomic force microscopy probe with piezoresistive read-out and a highly symmetrical Wheatstone bridge arrangement", Sensors and Actuators:A, vol. 83, 2000, pp.47-53

[5] T. A. Michalske and S. W. Freiman, "A Molecular Interpretation of Stress Corrosion in Silica", Nature, vol. 295, 1982, pp.511-512

[6] K. Miyamoto, K. Sugano, T. Tsuchiya, and O. Tabata, "Effect of Surface Oxidation on Mechanical Properties of Single Crystalline Silicon", Proc. of the 24th Sensor Symposium, 2007, pp.165-168

[7] R. J. Jaccodine and W. A. Schlegel "Measurement of Strains at Si-SiO2 Interface", Journal of Applied Physics, vol. 37, no. 6, 1966, pp. 2429-2434.

[8] T. Kitamura, H. Hirakata, D.V. Truong, "Initiation of interface crack at free edge between thin films with weak stress singularity", Thin Solid Films, vol. 515, 2007, pp. 3005-3010.

PART-TO-PART AND PART-TO-SUBSTRATE MAGNETIC SELF-ASSEMBLY OF MILLIMETER SCALE COMPONENTS WITH ANGULAR ORIENTATION

S.B. Shetye, I. Eskinazi, and D.P. Arnold

Interdisciplinary Microsystems Group, University of Florida, Gainesville, FL, 32611

ABSTRACT

This paper presents multifunctional self-assembly of millimeter scale components using magnetic forces between permanent micromagnets integrated on the component surfaces. Part-to-part assembly is demonstrated by batch assembly of free-floating 1mm x 1mm x 0.5mm silicon parts in a liquid environment with the assembly yield varying from 88% to 90%. Part-to-substrate assembly is demonstrated by assembling an ordered array on a substrate in a dry environment with the assembly yield varying from 87% to 98%. In both cases, diverse magnetic shapes/patterns are used to control the alignment and angular orientation of the components and assembly times range from 15 – 240 s.

INTRODUCTION

The demand for low-cost, highly functional miniaturized electronic systems places significant technologic and economic challenges on system manufacturers, especially consumer electronics. Complex systems such as mobile phones are currently made by board-level integration of various components and subsystems, each made using different materials and fabrication technologies. The individual subcomponents are generally manufactured using batch processes, but then assembled in serial fashion using conventional robotic assembly techniques. To sustain in the market with ever increasing demands of multifunctional end user products, manufacturers and various research groups are exploring different self-assembly techniques to automate the assembly and packaging processes [1].

In the past, researchers have explored gravity, fluidic forces, capillary forces, and electrostatic and magnetic fields to drive the self-assembly process [2-6]. Most of these methods do not match the full functionality, cost, throughput, or accuracy offered by robotic and manual assembly. Hence, improved self-assembly techniques are required to replace the robotic and manual assembly.

Our approach, magnetic self-assembly (MSA), uses magnetic forces between permanent micromagnets integrated on the component surfaces. Assembly occurs when magnets with opposite polarities bond to one another, resulting in energy minima. The magnetic forces and torques—controlled by the size, shape, material, and magnetization direction of the magnets—compel the components to rotate and align. Thus, with proper design, MSA offers the potential for advanced self-assembly features such as angular orientation (assembly is restricted to one physical orientation), inter-part bonding (assembly of free-floating components to one another, rather than assembly of free-floating components to a substrate), and bonding specificity (assembly restricted to one type of component when multiple components may be present).

In a prior publication, the feasibility of the MSA approach was validated by demonstrating part-to-substrate

assembly using simple square magnets with a maximum yield of 98.7 % in 7 s [7]. Here, we present part-to-part MSA in a liquid environment and part-to-substrate assembly in dry environment, both with angular orientation. Different magnetic patterns such as squares, stripes, ovals, triangles, and arrow-heads are explored, covering 4 - 75 % of the bonding surface area. Analytical [8] and finite element methods are used to estimate the bonding forces.

FABRICATION

The components used for this study are 1 mm x 1 mm x 0.5 mm silicon blocks with micromagnets integrated on the bonding surface. As shown in Figure 1, the magnets are batch-fabricated at the wafer level using two different previously developed magnetic powder processes [9-10]. Starting with 100 mm diameter, 500 μm thick Si wafers, small cavities are etched in the silicon. For the polyimide-capped magnets, the cavities are then packed with magnetic powder, and a thin layer of polyimide is spun and cured on the wafer to lock the powder in place. For the wax bonded magnets, instead of using the polyimide cap, a wax powder is used to bond the magnetic powder in the trenches.

For the experiments here, polyimide capping was used for SmCo magnets, while wax bonding was used for NdFeB and ferrite magnets. Typical properties of the fabricated magnets measured using a vibrating sample magnetometer (VSM) are shown in Table 1, showing the NdFeB magnets posses the highest magnetic properties. Additionally, the magnets (SmCo only) used for the part-to-part assembly are 60 μm thick, and the magnets (SmCo, NdFeB, and ferrite) used for the part-to-substrate assembly are 150 μm thick.

After pulse magnetization to pole the magnets, the wafer is diced into 1 mm x 1 mm individual components. Figure 2 shows a top view and cross-section of a single component with polyimide-capped embedded magnet. A

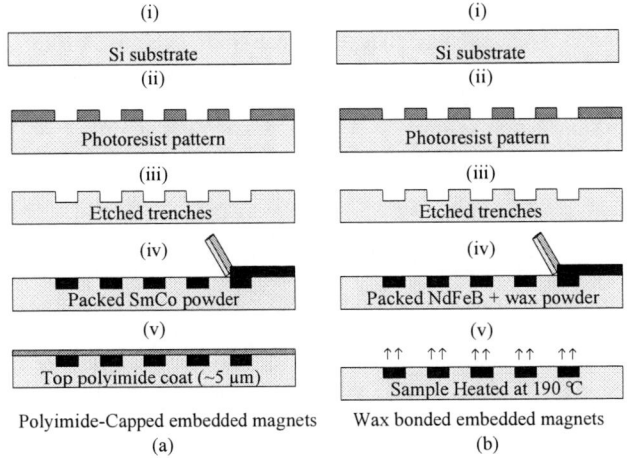

Figure 1: Fabrication processes for embedded magnets.

Table 1: Magnetic properties of embedded micromagnets.

Material (size)	Fabrication Method	H_c (kA/m)	B_r (T)	BH_{max} (kJ/m³)
SmCo (10 µm)	Polyimide-capped	103	0.11	1.25
NdFeB (50 µm)	Wax-bonded	727.2	0.36	21.5
Ferrite (2 µm)	Wax-bonded	312.8	0.1	1.7

Figure 2: Top view and cross-section of component with embedded polyimide-capped SmCo magnet.

similar process is used for fabrication of substrates for the part-to-substrate assembly. The substrates possess a 5 x 5 array of "receptor sites" with center-to-center spacing of 1.4 mm.

PART-TO-PART ASSEMBLY

Experiment

The experimental setup for demonstration of part-to-part self-assembly is shown in Figure 3. For each set of experiments, eight components to be assembled are magnetized with one polarity, and an excess of components (~32) are magnetized with opposite polarity. The parts are mixed together in a rotating conical tube with approximately 1 ml of methanol. The methanol enables the components to randomly mix and tumble over each other. It also prevents the components from sticking to the tube surface and mitigates air bubble formation. The tube is inclined at an angle of 35° - 45° and rotated at ~60 rpm. Because of the magnetization direction, like parts do not bond with each other since they have the same polarity (like poles repel), thus avoiding agglomeration.

60 µm-thick polyimide-capped SmCo magnets are used to demonstrate the part-to-part self-assembly process. Figure 4 shows the various magnetic patterns used. The square magnetic patterns exhibit 4-fold rotational symmetry resulting in four possible bonding orientations; the corner squares, stripe, and oval patterns exhibit 2-fold symmetry permitting assembly in only two orientations. Fig. 5 shows the components mixed in the conical tube and the assembled pairs.

Results

The number of component pairs assembled after 20 s is recorded. Table 2 and Table 3 summarize the results for the magnetic patterns with 4-fold and 2-fold symmetry respectively. The percentage yields are averaged over 15 runs and values presented are with the 95 % confidence intervals. The percentage yield for the components with 4-fold symmetry varies from 89 % to 97 %. The results in Table 2 show an increase in yield with decreasing bonding force and surface area. This is because of excessive

bonding forces that cause multiple components to stick to one another in an uncontrollable manner, observed predominantly in the Square 4 case where the estimated bond force was 0.2 mN. The yield for the components with 2-fold symmetry varies from 88 % to 90 %, slightly lower than the 4-fold case. One general trend found for the part-to-part experiments is that better yields are achieved with smaller forces.

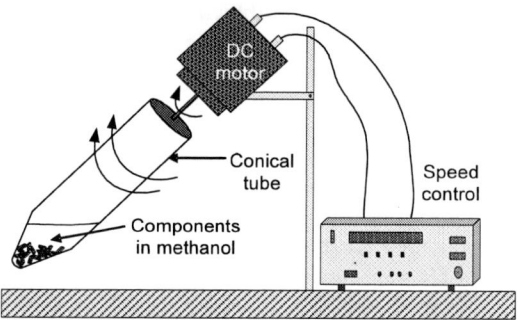

Figure 3: Experimental set-up for part-to-part magnetic self-assembly.

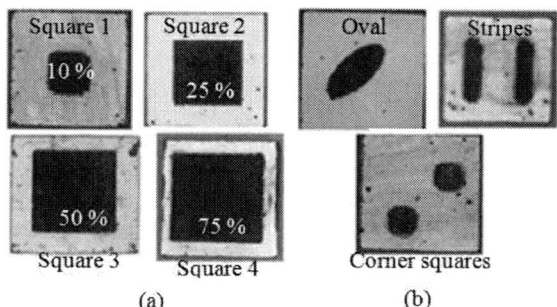

(a)　　　　　　　　(b)

Figure 4: Various magnetic patterns used for magnetic self-assembly (a) 4-fold symmetric (b) 2-fold symmetric patterns.

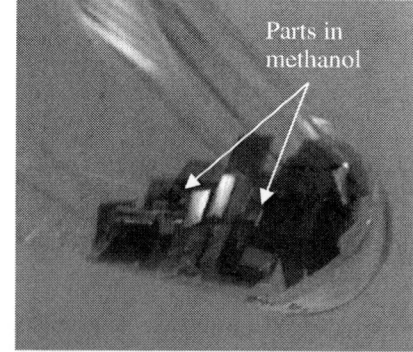

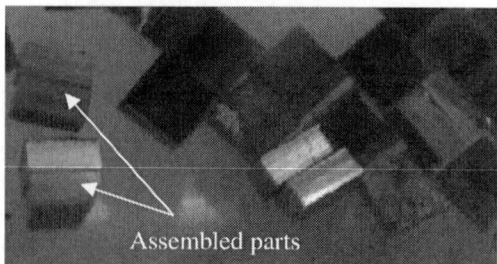

Figure 5: Part-to-part magnetic self-assembly of free floating parts.

978-1-4244-2977-6/09 $25.00 © 2009 IEEE

Table 2: Part-to-part magnetic self-assembly results for magnetic patterns with 4-fold symmetry.

Component Type	Magnet Area	Force (FEM)	% Yield (20 s)
Square 1	10 %	0.08 mN	97 ± 2.6
Square 2	25 %	0.12 mN	97 ± 2.4
Square 3	50 %	0.17 mN	93 ± 4.2
Square 4	75 %	0.20 mN	89 ± 4.3

Table 3: Part-to-part magnetic self-assembly results for magnetic patterns with 2-fold symmetry.

Component Type	Magnet Area	Force (FEM)	% Yield (20 s)
Oval	16 %	0.05 mN	90 ± 4.4
Stripes	12 %	0.17 mN	89 ± 4.8
Corner squares	10 %	0.10 mN	88 ± 4.2

PART-TO-SUBSTRATE ASSEMBLY

Experiment

Part-to-substrate assembly is demonstrated by batch assembly of components into an ordered array on a planar substrate. The substrate used is a 5 x 5 array of receptor sites, and ~6x redundant parts (150) are used for the experiment. As shown in Figure 6, the experimental setup extends the setup used in [7]. The components are agitated using an electromechanical shaker and assembled onto an inverted substrate. A 3 Hz, 2 V_{p-p}, square wave is fed from a signal generator through an amplifier to the shaker in order to bounce the components ~10 mm. The substrate is attached to a rod that is connected to a vibrating piezoelectric plate. The piezoelectric plate provides a secondary vibration that helps align the components and prevent the stacking of components (an extra component sticking to the back of the assembled component). The frequency and amplitude of the piezo vibrator are tuned by trial and error, but depend on the force between the magnets on the components being assembled. A digital camera is used to capture and record the data into a computer for image processing.

In addition to the magnetic patterns shown in Figure 4, three asymmetrical patterns shown in Figure 7 are used for the part-to-substrate experiments. The arrow-head, single triangle, and two triangle patterns restrict assembly to only one physical orientation. The components are fabricated with either polyimide-capped SmCo magnets or wax-bonded NdFeB or ferrite magnets. The magnets cover about 4 % to 75 % of the component surface area and the thickness is about 150 μm. Figure 8 shows an example of the assembled components on the inverted substrate.

Results

Table 4 summarizes the results for the components with 4-fold, 2-fold and asymmetric patterns. The recorded percentage yield is averaged over 15 test runs and reported with 95 % confidence intervals.

The first experiments examined the 4-fold symmetry cases using small ("Square 1") and large ("Square 4") square patterns with either SmCo or NdFeB. The percentage yield for 4-fold symmetry varies from 88 % to 98 %. For the "Square 1" case, higher yields and faster assembly rates were achieved with NdFeB, attributed to

the higher bonding force as compared to the SmCo. Here we note that higher forces appear to increase yield, as compared to the part-to-part assembly where the opposite trend was observed. However, there is a limit. The "Square 4" components with NdFeB magnets caused stacking (a free component sticking onto the back of another component) due excessively strong bond forces, thus hindering the assembly process. This indicates there is an optimum for the magnetic force where yield is maximized. Another observation is that high yields and fast assembly times can also be achieved with lower bonding forces if the magnet surface area is increased, as in the case of the SmCo "Square 4."

For the 2-fold symmetry experiments, to avoid stacking, different types of magnets were used for the substrate and components—the substrate being the stronger NdFeB, and the components being the weaker SmCo. The yield for the 2-fold symmetry experiments varies from 85 % to 97 %. The oval magnet components assemble much faster with higher yield compared to the stripes and corner squares patterns, despite having the lowest FEM-predicted bonding force. The oval possessed the largest surface area, suggesting the assembly rate and yield improves with increasing magnet surface area.

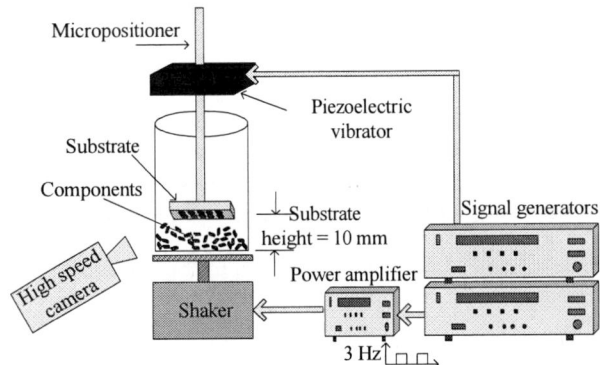

Figure 6: Experimental set-up for part-to-substrate magnetic self-assembly

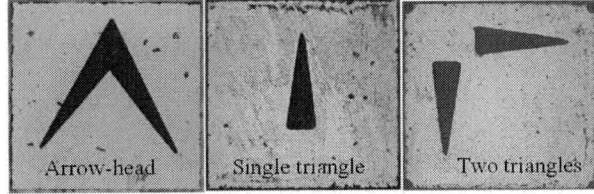

Figure 7: Asymmetric magnetic patterns for MSA with orientation uniqueness.

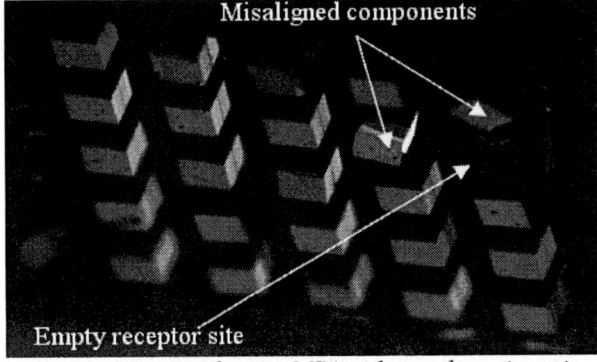

Figure 8: Part-to-substrate MSA with angular orientation

Table 4: Part-to-substrate magnetic self-assembly results.

Component Type	Symmetry	Magnet Area	Substrate Magnets	Part Magnets	Force (FEM)	Time (s)	% Yield
Square 1	4-fold	10 %	SmCo	SmCo	0.13 mN	58	88.3 ± 2.0
Square 1	4-fold	10 %	NdFeB	NdFeB	2.2 mN	33	97.6 ± 1.3
Square 4	4-fold	75 %	SmCo	SmCo	0.37 mN	33	97.7 ± 1.1
Square 4	4-fold	75 %	NdFeB	NdFeB	6.6 mN	Stacking	
Oval	2-fold	16 %	NdFeB	SmCo	0.28 mN	15	96.8 ± 2.1
Stripes	2-fold	12 %	NdFeB	SmCo	0.86 mN	22	94.7 ± 2.5
Corner squares	2-fold	10 %	NdFeB	SmCo	0.54 mN	33	85.6 ± 2.8
Arrow-head	Asymmetric	11 %	NdFeB	Ferrite	0.48 mN	33	87.2 ± 2.8
Single triangle	Asymmetric	4.3 %	NdFeB	Ferrite	0.22 mN	120	88.5 ± 3.2
Two triangles	Asymmetric	8.5 %	NdFeB	Ferrite	0.45 mN	240	82.0 ± 6.5

For the experiments with asymmetric magnetic patterns, the substrate magnets are again the stronger NdFeB and the parts this time are ferrite. The ferrite magnets are chosen for the same reason as the SmCo magnets—that is for their weaker bonding force. The smaller physical diameter ferrite powder (~2 μm) also permits the fabrication of sharp magnet apexes, important for the asymmetric patterns. The yield in this case varies from 82 % to 89 %, the highest yields being for the arrow-head and single triangle patterns. The highest assembly rate is achieved for the arrow head pattern, which also has the highest surface area, again indicating that rate and yield correlate with surface area. However, the two triangle rate is less than the single triangle rate, even though the two triangle total magnet surface area is twice that of the single triangle. This suggests that higher rates can be achieved with higher surface area provided that the magnet surface area is continuous, i.e. a single magnet is better compared to multiple magnets.

CONCLUSION

The feasibility of magnetic self-assembly for part-to-part and part-to-substrate assembly with angular orientation was validated here. Both types of assembly were shown to produce reasonably high yields (up to 98 %) with fast assembly times (15 – 240 s).

Part-to-part assembly was achieved using free-floating parts in a liquid environment using a tumbler apparatus. In the case of part-to-substrate assembly, controlled assembly of parts was achieved using a shaker setup in conjunction with active vibration of the substrate. The use of stronger magnets on the substrate and weaker magnets for the components mitigated stacking issues and permitted an additional level of process control.

The experiments showed that higher bonding forces are required for part-to-substrate assembly as compared to part-to-part assembly, presumably because stronger forces are necessary to capture the part during the more "dynamic" assembly part-to-substrate process. In both types of assemblies, the assembly rate and yield increases with the magnet symmetry. This indicates that symmetry, which increases the number of possible bonding orientations, directly increases the probability of a individual bonding event. The interesting results shown here prompt further experiments to correlate the desired outcomes (yield, rate, alignment, etc.) with the assembly parameters (magnet properties, assembly process, etc.).

ACKNOWLEDGEMENTS

Financial support for this project was provided by NSF grant DMI-0556056.

REFERENCES

[1] C. J. Morris, et al., "Self-assembly for microscale and nanoscale packaging: steps toward self-packaging," *IEEE transactions on Advanced Packaging*, vol. 28, no. 4, pp. 600- 611, November 2005.

[2] H.-J. J. Yeh and J. S. Smith, "Fluidic self-assembly for the integration of GaAs light-emitting diodes on Si substrates," *IEEE Photonics Tech. Letters*, vol. 6, no. 6, pp. 706-708, 1994.

[3] U. Srinivasan, et al., "Microstructure to substrate self-assembly using capillary forces," *J. Microelectromech .Syst.*, vol. 10, no. 1, pp. 17–24, March 2001.

[4] C. F. Edman, et al., "Electric field directed assembly of an InGaAs LED onto silicon circuitry," *IEEE Photonics Lett.*, vol. 12, no. 9, pp. 1198-1200, 2000.

[5] Y. Murakami, et al., "Application of micromachine techniques to biotechnological research," *Mat. Sci. & Eng. C*, vol. 12, pp. 67-70, 2000.

[6] K. F. Böhringer, et al., "Parallel microassembly using electrostatic force fields," *IEEE Int. Conf. Robotics and Automation (ICRA)*, Leuven, Belgium, May 1998, pp. 1204–1211.

[7] Shetye, S., et al., "Self-assembly of millimeter-scale components using integrated micromagnets," *IEEE Trans. Magn*, vol. 44, no. 11, November 2008.

[8] J. Agashe and D. P. Arnold, "A study of scaling and geometry effects on the forces between cuboidal and cylindrical magnets using analytical force solutions," *J. Phys. D: Appl. Phys.*, vol. 41, pp. 105001-1-9, May 2008.

[9] B. Bowers, et al., "A method to form bonded micromagnets embedded in silicon," in *Proc. 14th Int. Conf. Solid-State Sensors, Actuators, and Microsystems (Transducers 07)*, 2007, pp. 1581-1584.

[10] N. Wang, et al., "Wax-bonded NdFeB micromagnets for microelectromechanical systems applications," *J. Appl. Phys.*, vol. 103, pp. 07E109-1-3, 2008.

LOW TEMPERATURE FABRICATION PROCESS FOR HIGH-ASPECT-RATIO AND MULTI-COMPLIANT MEMS

S. Kühne, R. Blattmann, and Ch. Hierold

Micro and Nanosystems, Department of Mechanical and Process Engineering, ETH Zurich,
SWITZERLAND

ABSTRACT

The presented process technology enables the fabrication of 3-dimensional high-aspect-ratio microstructures featuring the integration of multiple materials. The technology offers the possibility to benefit from unique material properties for optimized MEMS functionality and performance. The fabrication process relies on wafer stacking by low temperature plasma activated direct bonding, SOI layer transfer and SU-8 structuring. The proposed process flow is validated by the fabrication of micro mirrors featuring a soft polymeric suspension and a high-aspect-ratio vertical comb-drive actuator. The devices are characterized and confirm the expected performance with dynamic resonant optical deflection angles of 59° at 50V.

INTRODUCTION

There is a growing field of applications for high-aspect-ratio 3-dimensional MEMS as well as an interest in enabling the integration of different materials for optimal device performance [1]. However established wafer stacking technologies based on fusion or anodic bonding are not compatible with the integration of polymers as a functional material. The fabrication process requires similar or even tougher low temperature compatibility as for monolithic integration of CMOS-MEMS [2]. The integration of functional polymer structures, e.g. SU-8, restricts the processing temperature to below 300°C.

In this paper we present a fabrication process exploiting the advantages of low temperature hydrophilic wafer bonding of plasma activated silicon and oxide surfaces [3, 4]. The plasma activation and low temperature bonding enables the integration of polymer as functional material in complex MEMS structures. The prevailing bonding mechanism is the formation of reactive hydroxyl functional groups by plasma activation as well as their polymerization, which form strong covalent siloxane bonds [5]. This reaction can take place at temperatures as low as room temperature, provided the silanol groups are in close enough contact and hydrogen bonded.

The presented technology enables the fabrication of 3-dimensional high-aspect-ratio MEMS structures through wafer stacking as well as the integration of multiple functional materials for an optimized device performance. To demonstrate and characterize the proposed process flow we use an improved design of earlier presented scanning micro mirrors featuring a multi-compliant structure [6].

Design

The developed and fabricated devices are scanning micro mirrors, which feature a soft polymeric suspension and a high-aspect-ratio vertical comb-drive actuator. The device set-up is schematically shown in Figure 1. It consists of two silicon layers and the polymer suspension layer. The lower substrate is 250μm thick and forms the fixed electrodes and electrical connection pads of the vertical comb-drive. The upper substrate constitutes the mirror plate and the movable combs. Its thickness strongly influences the system performance by defining the system inertia and the maximum tilt angle. Therefore the device layer thickness and the cross-section of the polymeric torsionbar suspensions are important design parameters. Device layers of 50μm and 100μm, and polymer suspensions of 20μm and 30μm were used for different mirror designs. The soft polymeric suspension is designed for improved separation of system resonance frequencies for specific modal excitation and operation. The vertical comb-drive actuators and the suspensions are situated on both sides of the mirror. The symmetric arrangement allows tilt and piston operation of the device. Further the mirror actuation is based on charge separation, since the movable structure is not electrically connected [6].

Thus the structure is optimized for large tilting angles at low actuation voltages. Furthermore the mirror plate distortion during operation is minimized due to the multi-compliant structure. The presented scanning mirrors with different designs of actuator and suspension feature system resonant frequencies ranging from 500 Hz to 3 kHz.

a)

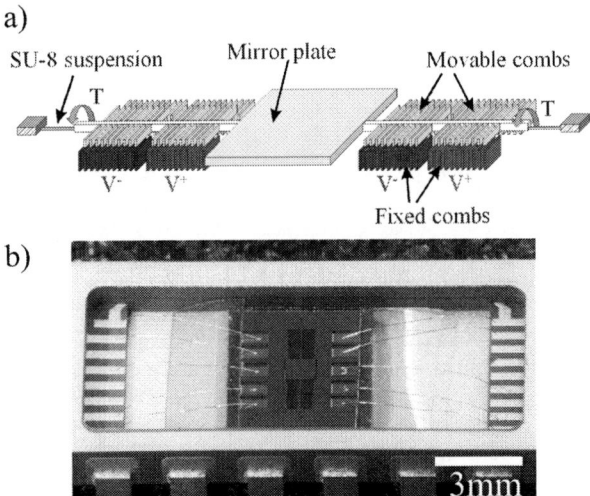

b)

Figure 1: a) Schematic of scanning micro mirror design with mirror plate, symmetric high-aspect-ratio comb-drive actuator, and polymeric suspension, b) photograph of fabricated and packaged device.

978-1-4244-2977-6/09 $25.00 © 2009 IEEE

FABRICATION

The micro mirrors are built from three bonded substrates, a silicon support wafer, a silicon wafer forming the fixed comb-drive electrodes and an SOI device layer featuring the mirror, the movable combs and the anchors for the polymer suspension (Figure 1, 2e).

The developed process flow is depicted in Figure 2. The substrate forming the fixed electrodes and contact pads of the vertical comb-drive is a 250 µm thick double side polished silicon wafer. A 500nm thick thermal oxide is grown on the substrate and is forming the electrical insulation layer, which prevents shorts between fixed comb-drive electrodes as well as through the support or device wafers. The wafer is then bonded onto a silicon support wafer by low temperature direct bonding (Figure 2a). The support wafer features alignment marks for the accurate etch mask photolithography and bond alignment. The use of the same marks for all critical alignment steps ensures a minimal misalignment of individual mirror components, e.g. SU-8 springs, fixed and movable combs.

The O_2-plasma activation prior to the hydrophilic low temperature direct bonding is performed in an ICP system. The ICP allows for precise plasma control by coil and platen power tuning. The wafers are exposed to activation plasma for 60s with a coil and platen power of 900W and 30W, respectively. The activation is followed by a DI water rinse, spin dry and bringing the wafers into contact. After the room temperature bonding, the wafer stack is annealed in vacuum at a temperature of 250°C for five hours.

This substrate is then structured by an RIE/DRIE step (Figure 2b). The fixed combs, the electrodes and contact pads, isolation trenches as well as the cavities for the mirror and polymer suspension are formed in one etch step by taking advantage of aspect ratio dependent etching [7].

The SOI substrate preparation starts with etching alignment marks on the bulk layer. SOI wafers with device layer thickness of 50µm and 100µm, respectively, were used. The SU-8 springs are structured onto the SOI device layer by photolithography (Figure 2c). SU-8 suspensions layers of 20±1µm and 30±1µm were structured. The SOI wafer is then bonded onto the prepared fixed electrodes substrate by the same low temperature hydrophilic direct bonding process with an O_2-plasma activation as described above. An additional bond alignment before contacting the wafers is necessary for countersinking the SU-8 into the etched suspension cavities of the fixed electrodes substrate (Figure 2d). The SOI device layer is transferred by removing the bulk silicon layer in an isotropic dry etching step. The release of the movable combs, mirror plate and polymer suspension is again carried out by an RIE/DRIE process (Figure 2e). The electrical connection pads of the comb-drive electrodes are also released in this etch step, since they are located between the support and device substrates (Figure 2b).

After dicing the chips are packaged in DIL ceramic packages for characterization and testing of the micro mirror devices.

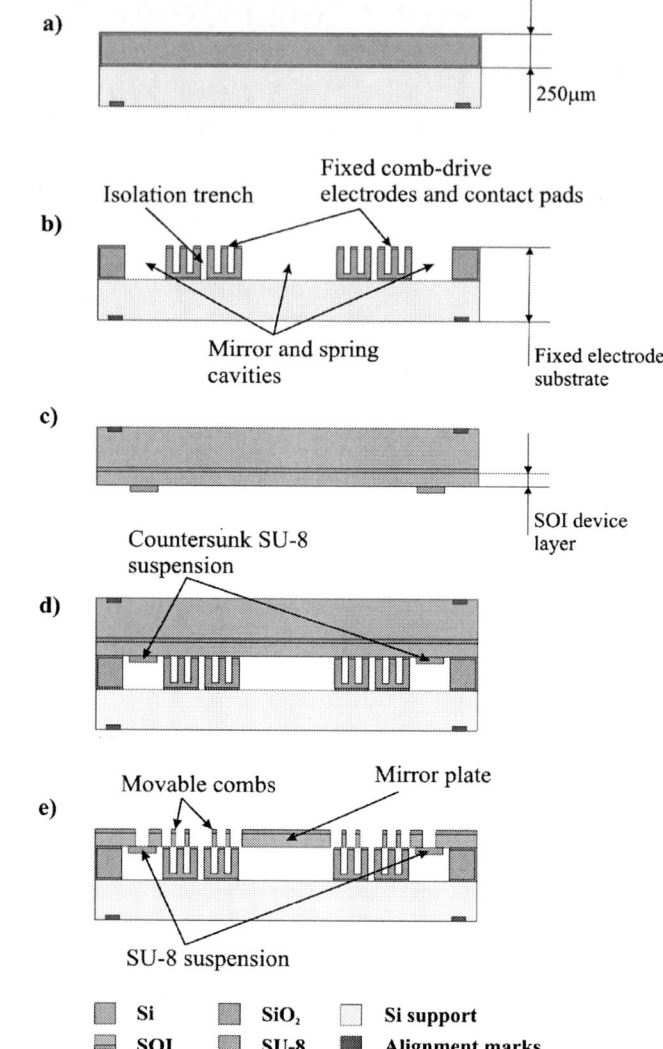

Figure 2: Low temperature fabrication process flow for multiple material integration in high-aspect-ratio MEMS.

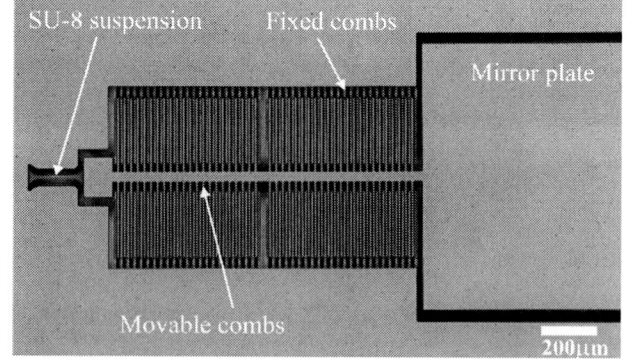

Figure 3: Microscope image of released micro mirror device showing the SU-8 suspension, fixed and movable combs and the mirror plate.

RESULTS

Fabrication

The identified three major factors influencing the achievable bond strength at low temperature are the plasma surface activation, the present bond interface materials and their surface roughness.

The prevailing mechanism for achieving high bond strength at low annealing temperatures is the formation of reactive oxide surfaces with hydroxyl functional groups and their density [1, 3, 5]. The reactive hydrophilic oxide surface is formed by O2-plasma treatment and DI water rinse. The silanol groups (Si-OH) present on the surfaces lead to hydrogen bonds forming when the activated surfaces are brought into contact at room temperature. The bond mechanism during the subsequent annealing is considerably depending on interaction distance and material properties. The silanol is known to polymerize and form strong covalent siloxane (Si-O-Si) bonds under synthesis of water (1), which again may diffuse through the surrounding oxide and react with Si to form SiO_2 and hydrogen (2):

$$Si\text{-}OH + HO\text{-}Si = Si\text{-}O\text{-}Si + H_2O \qquad (1)$$
$$Si + 2H_2O = SiO_2 + 2H_2. \qquad (2)$$

Bond yield and strength are strongly influenced by the absorption and diffusion capabilities of water and hydrogen molecules into the surrounding material or cavities [4]. Remaining water and hydrogen can lead to bubbles at the interface and therefore weaken the bond. Throughout the performed experiments with Si-SiO2 bonds, no bubble formation was observed.

In the presented fabrication process the bonding interfaces are Si and wet thermally grown SiO_2. All bond strength experiments were performed with double side polished silicon wafers and corresponding ones with 1μm oxide. Bond strength was determined by means of blade insertion tests [8]. An IR transmission pictures of bonded wafers under test are shown in Figure 4.

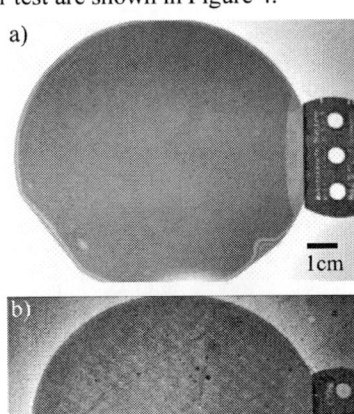

Figure 4: IR transmission images of bonded wafers under blade insertion test: a) bare Si-SiO₂ interface and b) bond featuring etched combs and cavities on the oxidized wafer.

The plasma surface activation is a key issue for hydrophilic low temperature wafer bonding and was therefore optimized. The influence of different plasma exposures of the silicon and SiO2 interfaces were investigated. The varied parameters are the plasma duration and the ICP platen power, which influences the intensity of the ion bombardment. However activated surfaces show an increased surface roughness with increased plasma exposure duration and ion bombardment, which leads to severe bond strength degradation. Figure 5 shows AFM surface roughness measurements, which are related to the resulting bond energies presented in Figure 6. Plasma exposures of 60s and 100s with platen powers of 30W and 45W, respectively, lead to a significant rms surface roughness increase of about 100nm. The achieved bonding energies and the process parameters are summarized in Figure 6 and Table 1. Bond energies of 2.74±0.1 J/m² were achieved by O2-plasma activations with 900W coil power, 30W platen power and 60s exposure duration. The annealing was always carried out in vacuum at 250°C for 5 hours.

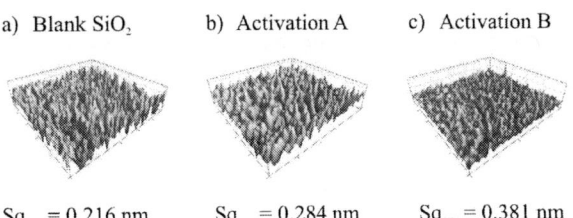

Figure 5: RMS surface roughness Sqrms of 1μm² AFM scans: a) thermal oxide, b) oxide after plasma activation A in Fig.6, c) oxide after plasma activation B in Fig.6.

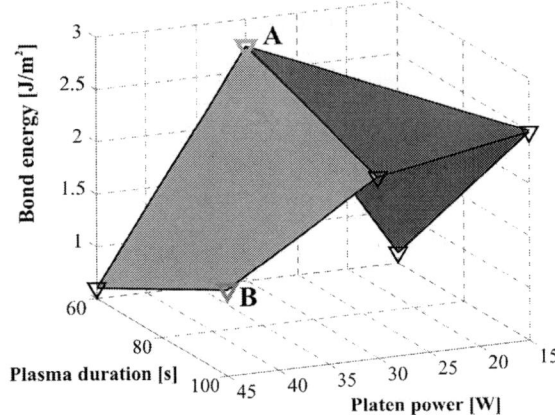

Figure 6: Achieved bond energies vs. O₂-plasma duration and ICP platen power.

Table 1: ICP O₂-plasma activation parameters with resulting bond energies determined by blade insertion tests.

Activation Parameters	O₂ - Plasma			
Coil power [W]	900	900	900	900
Platen power [W]	15	30	30	45
Duration [s]	100	100	60	100
Bond energy [J/m²]	2.49	2.23	2.74	1.34

Mirror Characterization

The fabricated devices are characterized and tested by means of Laser Doppler Vibrometer and laser beam deflection measurements. The mirror devices are excited by a periodic chirp for the frequency spectrum measurements (inset Figure 7). The device resonance frequencies and system Q-factors are determined in ambient air and vacuum. Table 2 summarizes results for different device designs, device layers and SU-8 thicknesses.

The laser beam deflection measurements were performed in ambient air and with resonant mirror excitation. Dynamic resonant optical deflection angles of 59° where achieved at 50V (Figure 7). The results are compared to two analytical models, which regard the comb-drive torque as a function of the actuation voltage ($T_A(V_A)$) or additionally of the deflection angle ($T_A(V_A,\theta)$). The refined model accounts for the actuation force of the comb-drive being deflection angle dependant ($T_A(V_A,\theta)$), since the actuation force component contributing to the torque scales with the cosine of the mechanical mirror deflection angle θ. The system dynamics are described as a 2nd order system with damping and excitation moment T_e

$$\ddot{\varphi} + 2\zeta\omega_0\dot{\varphi} + \omega_0^2\varphi = \frac{T_e}{I_\varphi} \qquad (3)$$

where I_φ is the momentum of inertia, ω_0 the resonance frequency and ζ the damping ratio.

The mirror performance agrees well with the model for scan angles below 40°. The larger deviation for very large tilting angles is assumed to be due to increased damping effects in this operation regime. The air damping is assumed to increase with increasing mirror deflection angles due to squeeze film damping acting on the mirror plate as well as viscous damping in the comb-drive actuator.

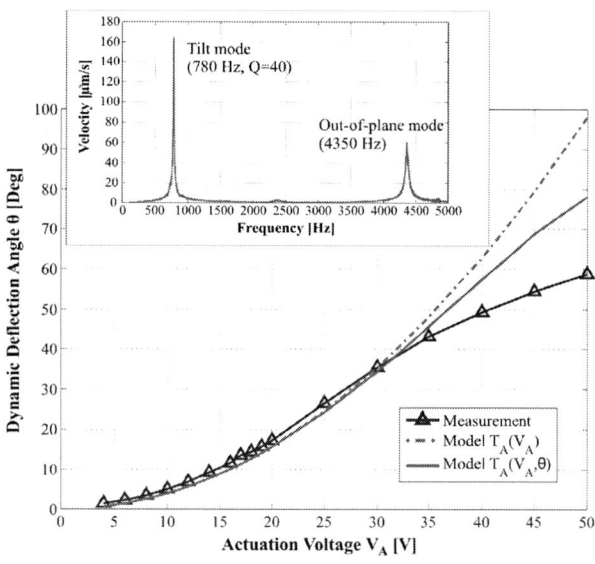

Figure 7: Dynamic optical scan angle in air vs. actuation voltage at device resonance frequency. The inset shows a measured device frequency spectrum in vacuum.

Table 2: Comparison of micro mirror designs: 1)-2) with 100μm device layer and 20μm SU-8, 3) 50μm device layer and 30μm SU-8.

Parameters	Micro mirrors				
Design	**1**		**2**		**3**
Ambient air / vacuum < 10⁻⁵mbar	**air**	**vac**	**air**	**vac**	**air**
Tilt mode [kHz]	0.5	0.5	0.8	0.8	1.5
Damping ratio ζ [ ]	0.05	0.01	0.04	0.01	0.04
Quality factor Q [ ]	10	50	12	40	12
Out-of-plane mode [kHz]	3.2	3.6	3.9	4.3	5.8

CONCLUSIONS

The proposed fabrication process flow was applied for torsional micro mirror devices. The optimized O_2-plasma activation parameters yielded high bond energies of 2.74 ± 0.1 J/m² at the low annealing temperature of 250°C.

Further the fabricated micro mirror devices were characterized. The high-aspect-ratio structures with soft polymeric suspension achieved high optical deflection angles of 59° at 50V resonant actuation. Thus the measurement results confirm the expected device performance as well as the potential of low temperature multiple material integration in 3-dimensional high-aspect-ratio MEMS.

REFERENCES

[1] Q.-Y. Tong and U. M. Gosele, "Wafer bonding and layer splitting for microsystems," *Adv. Materials,* vol. 11, pp. 1409-1425, 12 1999.

[2] H. Baltes and O. Brand, "CMOS-based microsensors and packaging," *Sensors and Actuators A: Physical,* vol. 92, pp. 1-9, 2001.

[3] S. N. Farrens, J. R. Dekker, J. K. Smith, and B. E. Roberds, "Chemical Free Room-Temperature Wafer to Wafer Direct Bonding," *J. Electrochem. Soc.,* vol. 142, pp. 3949-3955, Nov 1995.

[4] T. Suni, K. Henttinen, I. Suni, and J. Makinen, "Effects of Plasma Activation on Hydrophilic Bonding of Si and SiO_2," *J. Electrochem. Soc.,* vol. 149, pp. G348-G351, 2002.

[5] T. Qin-Yi, T. Qin-Yi, G. Cha, R. Gafiteanu, and U. A. G. U. Gosele, "Low temperature wafer direct bonding," *J. Microelectromech. Syst.,* vol. 3, pp. 29-35, 1994.

[6] D. Bachmann, S. Kuhne, and C. Hierold, "MEMS scanning mirror supported by soft polymeric springs and actuated by electrostatic charge separation," in *Proc. MEMS 2007,* pp. 723-726.

[7] J. Yeom, Y. Wu, J. C. Selby, and M. A. Shannon, "Maximum achievable aspect ratio in deep reactive ion etching of silicon due to aspect ratio dependent transport and the microloading effect," *J. Vac. Sci. Technol. B,* vol. 23, pp. 2319-2329, Nov-Dec 2005.

[8] W. P. Maszara, G. Goetz, A. Caviglia, and J. B. McKitterick, "Bonding of silicon wafers for silicon-on-insulator," *Journal of Applied Physics,* vol. 64, pp. 4943-4950, 1988.

978-1-4244-2977-6/09 $25.00 © 2009 IEEE

FABRICATION OF NANOPILLARS BASED ON SILICON OXIDE NANOPATTERNS SYNTHESIZED IN OXYGEN PLASMA REMOVAL OF PHOTORESIST

H.Y. Mao[1], D. Wu[1], W.G. Wu[1]*, J. Xu[2], H.X. Zhang[1] and Y.L. Hao[1]

[1]National Key Laboratory of Micro/Nano Fabrication Technology, Institute of Microelectronics, Peking University, Beijing 100871, P. R. China

[2]Electron Microscopy Laboratory, Peking University, Beijing 100871, P. R. China

*Email: wuwg@ime.pku.edu.cn

ABSTRACT

We report for the first time a facile lithography-free approach for fabricating nanopillars over large areas or in patterns. The key technique of this approach is that randomly-distributed nanoscale SiO_2 patterns can be synthesized on substrates simply by removing photoresist with oxygen plasma bombardment. Those SiO_2 nanopatterns may further function as masks in the following etching process for nanopillars. Based on this approach, a variety of microstructures containing nanopillars with diameters of 30~200 nm, which include surface micro channels, micro-cantilever probes and nanofences, have been fabricated. This approach can be applied both to silicon and metal substrates compatible with conventional micro-electromechanical systems (MEMS) fabrication.

INTRODUCTION

The increasing interest in nanopillars has boomed over the past decade due to their diverse applications in biosensors [1], DNA molecule separation devices [2], photonic crystal devices [3], field electron emitters [4], catalyst supports in liquid-fed microfuel cells [5] and nanoimprint lithography stamps [6]. Therefore, a wide range of research activities have been triggered, aimed at finding simpler and faster fabrication processes for such nanostructures.

Standard lithography methods for fabricating nanopillars include conventional photolithography, electron beam (EB) lithography [7], and focused ion beam (FIB) lithography [8], etc. Optical lithography methods using photomasks allow parallel fabrication processes of accurately defined and localized pillars down to a size of ~100 nm, but the cost is exorbitant. On the other hand, the EB and FIB methods allow accurate patterning of sub-100 nm features, but the serial nature of these technologies makes them both time- and labor-consuming, and the costs of special apparatuses are also considerably high, even though no photomasks are required in these two methods.

Due to the limitations of those standard lithography methods, several novel technologies aimed at convenient and cheap nanofabrication have been investigated. The most commonly used technologies are nanosphere lithography [9] and thin-film deposition lithography [10]. Nanospheres or particles can be prepared on substrates by spin coating or thin film deposition. Nanospheres can also be linked on to wafers using self-assembling proteins [11], but the self-assembly process may take a considerable amount of time. After the spheres or particles are introduced, they will act as etching masks for creating nanopillars. However, to fabricate nanopillar in patterns, which could be useful in many biological or chemical applications, both of these approaches require some specially-designed complicated techniques [9, 10].

In this paper, an easy and fast lithography-free method for fabricating nanopillar in patterns as well as over large areas is presented. The core technique of this method is that SiO_2 nanopatterns can be synthesized on the wafer only by spinning and removing photoresist, in which no lithography process is required. The nanopatterns are then used as etching masks in the following deep reactive ion etching (DRIE) process for producing nanopillars. Based on this approach, we have successfully fabricated a variety of microstructures containing nanopillars in diameters of 30~200 nm. Compared with other nanopillar fabrication methods, this approach is both time- and labor-saving. Moreover, the approach we have developed is also suitable for different substrate materials, including single-crystal silicon, polysilicon, copper and gold, etc.

FORMATION OF NANOPATTERNS

Oxygen plasma bombardment is usually utilized to remove used photoresist masks or residual photoresists in conventional micromachining processes. After a long enough period of bombardment, as taken for granted, the surface of the substrate should be very clean. However, through scanning electron microscope (SEM), large areas of nanoscale residues were observed on the surface of silicon and some metal substrates, which were previously covered by photoresist before the oxygen plasma treatment. We also found that the nanoscale residues were able to withstand long-time oxygen plasma bombardment afterwards.

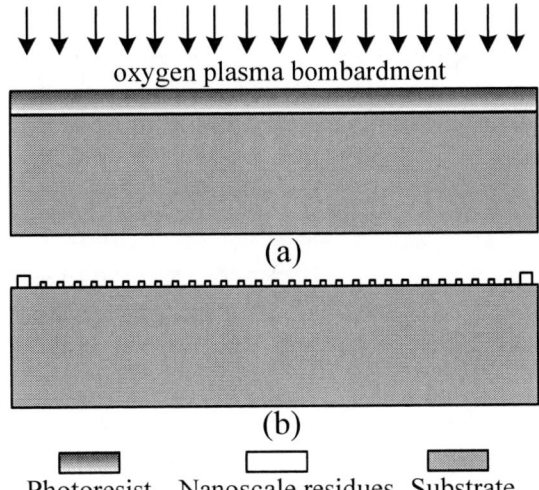

Figure 1: Schematic process for generating nanoscale residues.

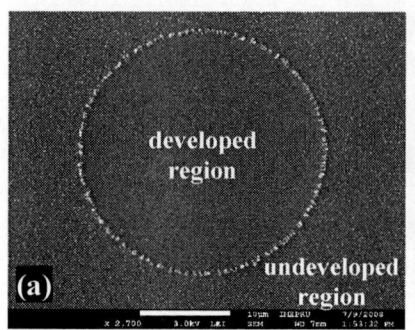 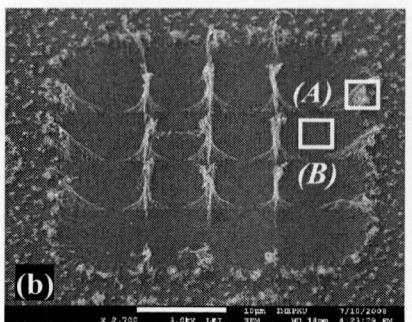 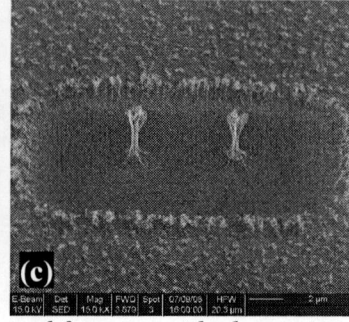

Figure 2: SEM images of the patterned nanoresidues. The undeveloped regions stand for areas with photoresist after development, while the developed regions stand for areas without photoresist after development.

Figure 1 schematically depicts the process for generating the nanoscale residues. Some obtained SEM images of the residues are shown in Figure 2. For a better view of the residues, the photoresists which we used to generate the residues were patterned before being removed. No residues were observed in areas where the photoresists had been exposed and developed, while nanoscale residues randomly distributed in areas where unexposed photoresist had been removed by oxygen plasma bombardment. Experimental results also show that the size of the nanoscale residues is at the level of tens to hundreds nanometers, varying with the thickness of photoresist. It is noteworthy that, due to some special edge-effects of the photoresist patterns, the residues reach maximum size as well as their distribution becomes ordered to some extent near all the edges.

To find out what kind of materials the nanoresidues are, two measurements were carried out. An FEI Strata DB 235 FIB/SEM dual beam system was utilized to measure the energy spectra of the residues and, meanwhile, to capture their SEM images. For a chemical-component comparison between the residues and the silicon substrate, the energy spectra of the substrate were measured together with those of the residues on the substrate. Figure 3 (a) shows an energy spectrum of the residues on silicon substrate, corresponding to position *A* in Figure 2(b). Figure 3 (b) (the inset) shows an energy spectrum of the silicon substrate, corresponding to position *B* (not far away from position *A*) in Figure 2(b). Comparing the two spectra, we find that the major difference between the chemical components of the generated residues and silicon

substrate is the element of oxygen, which strongly suggests that the residues are probably to be silicon oxide.

Subsequently, RENISHAW Raman spectrometer was used to further confirm the chemical component of the residues. Figure 4 gives both the Raman spectrum of standard SiO_2 and that of the residues at position *A* in Figure 2 (b). As shown in Figure 4, the peaks of these two spectra appear at the same Raman shift, which is ~1117 cm⁻¹, indicating that the residues are made of SiO_2. The formation of SiO_2 might be due to some promoters used to promote the adhesion between photoresist and the substrate.

FABRICATION OF NANOPILLARS

Nanopillar forests and even nanopillar patterns can be fabricated by employing these residual SiO_2 nanopatterns as etching masks and combining these nanopatterns with conventional patterning methods. Figure 5 shows the schematic process of fabricating silicon nanopillars on chosen areas. After the optical lithography process, micro photoresist patterns are defined on the substrate (Figure 5 (a)). The substrate covered with patterned photoresist is then bombarded by oxygen plasma. As a result, nanopatterns of SiO_2 are synthesized within the areas where photoresist exists before the bombardment process (Figure 5 (b)). Taking advantage of the SiO_2 nanopatterns as etching masks, the first DRIE step is adopted to etch silicon (Figure 5 (c)). In this way, silicon nanopillars with a diameter of ~30 nm can be achieved, and the SEM image of the nanopillars is illustrated in Figure 6.

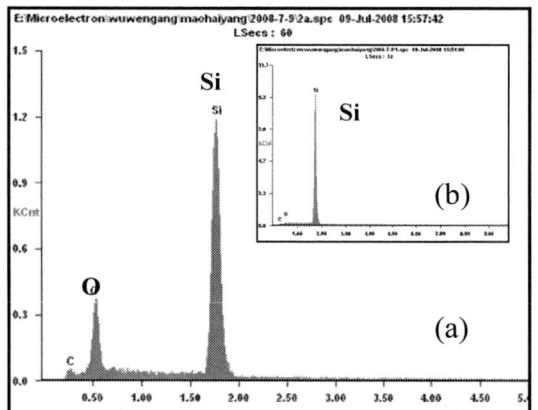

Figure 3: Energy spectra of nanoresidues and silicon substrate.

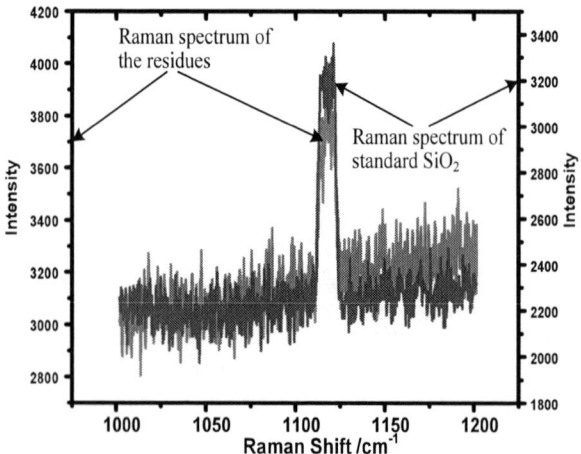

Figure 4: Raman spectra of the residues and standard SiO_2.

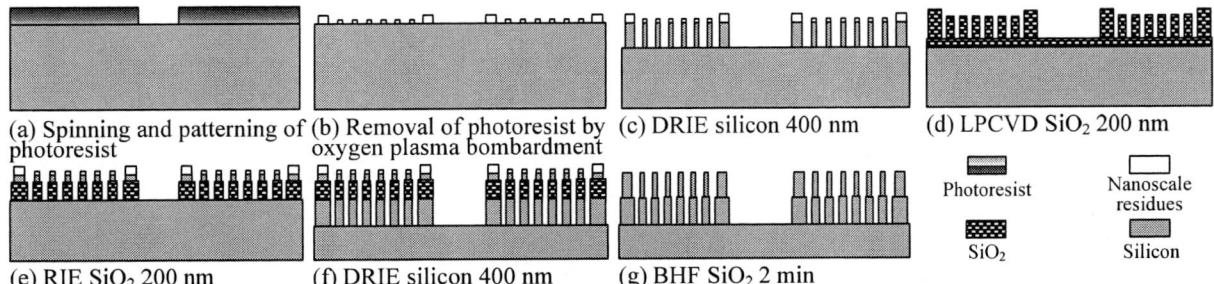

(a) Spinning and patterning of photoresist

(b) Removal of photoresist by oxygen plasma bombardment

(c) DRIE silicon 400 nm

(d) LPCVD SiO$_2$ 200 nm

Photoresist | Nanoscale residues

SiO$_2$ | Silicon

Figure 5: Fabrication process of silicon nanopillars.

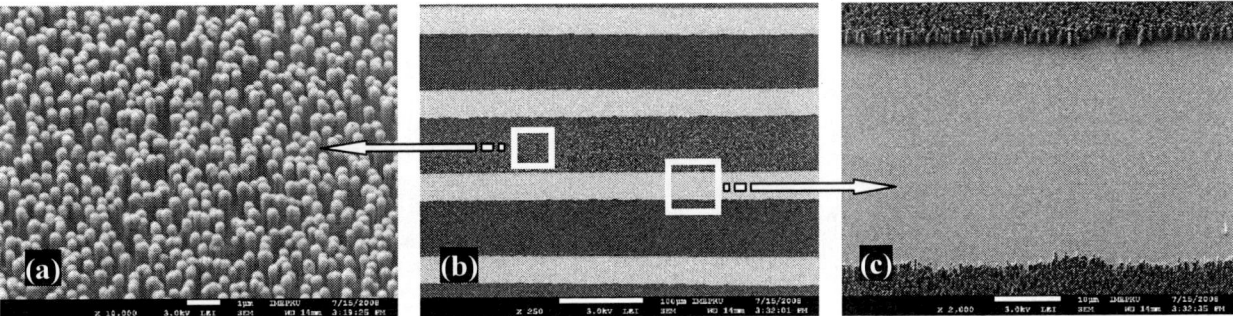

(e) RIE SiO$_2$ 200 nm

(f) DRIE silicon 400 nm

(g) BHF SiO$_2$ 2 min

Figure 7: SEM images of surface micro channels defined by nanopillar forests.

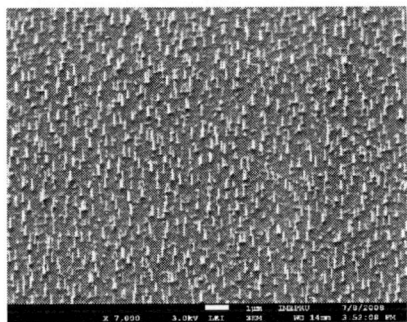

Figure 6: A large area of silicon pillars with a diameter of ~30 nm.

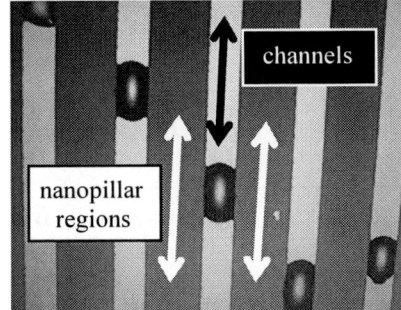

Figure 8: Water droplets distribute in micro silicon channels, indicating the nanopillar forests being more hydrophobic.

Figure 9: Water droplet on silicon nanopillars, with a contact angle of 151°.

To further modulate the parameters of the nanopillars, such as increasing the height-to-diameter ratio (noted that the larger the diameters, the higher the heights can be obtained in etching), spacer technology is introduced to form SiO$_2$ sidewalls around the nanopillars, functioning as the second etching masks for higher nanopillars. Firstly, 200 nm SiO$_2$ is conformally deposited on the initial nanopillars by low pressure chemical vapor deposition (LPCVD) (Figure 5(d)), and anisotropic reactive ion etching (RIE) is then employed to etch the SiO$_2$ film at the base of those pillars (Figure 5(e)). The obtained sidewalls serve as etching masks in the following DRIE process (Figure 5(f)). Finally, the SiO$_2$ sidewalls are removed by buffered HF (BHF) solution, leaving more uniform nanopillars on the substrate (Figure 5(g)).

Based on the fabrication process described above, a variety of microstructures containing nanopillars were produced. Figure 7 shows surface micro channels fabricated by patterning the nanopillar forests. It can be seen that the boundaries between the micro channels and the nanopillar forest areas are very clear. As is shown in Figure 8, tiny water droplets are more likely to occupy micro channels, rather than the nanopillar forest areas, indicating that the nanopillar forest areas are more hydrophobic than the silicon channels. Particularly, when a ~2 μL water droplet was dropped on the nanopillar-wholly-covered region, super-hydrophobicity with a contact angle of 151° was observed (Figure 9).

Moreover, micro-cantilever probes with nanopillars selectively created on defined areas were fabricated, shown in Figure 10(a). The nanopillars can be made in dual-lines, and the distance between the lines can reach just a few nanometers. Nanofences around defined patterns were also obtained, shown in Figure 10(b) and 10(c).

978-1-4244-2977-6/09 $25.00 © 2009 IEEE

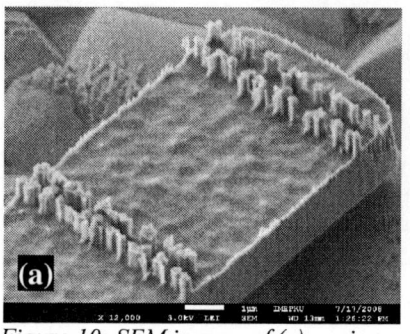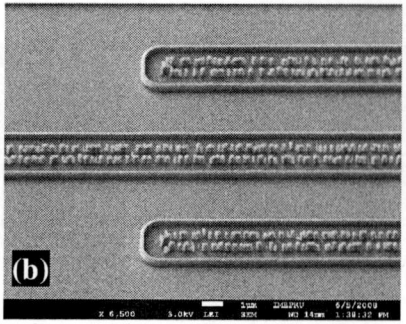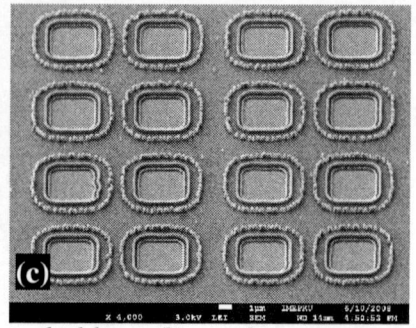

Figure 10: SEM images of (a) a micro-cantilever probe structure with nanopillars in dual-lines; (b) nanofences composed of nanopillars between nanowires; (c) nanofences around parallel nanowires. The nanowires in (b) and (c) are fabricated using spacer technology, with a typical width of <100 nm.

CONCLUSIONS

An easy and fast lithography-free approach for fabricating nanopillars over large areas or in patterns has been demonstrated. In the approach, silicon oxide nanopatterns synthesized during oxygen plasma removal of photoresist are used as etching masks for producing nanopillars. The approach might be applied to semiconductor, metal and other materials compatible with micro/nano-electromechanical systems (MEMS/NEMS) fabrication. By using the approach, silicon nanopillars with 30~200 nm in diameters are successfully fabricated. The nanopillars themselves and the microstructures with the nanopillars are of great importance in many MEMS/NEMS devices. They can play the roles as protein carriers in bio-devices, as nanoscale emitter tips in filed emission devices and as catalyst supports in liquid-fed microfuel cells.

ACKNOWLEDGMENT

This work is supported in part by the National Basic Research Program of China (973 Program) under Grant No. 2009CB320300.

REFERENCES

[1] Y.-K. Choi, J. S. Lee, J. Zhu, G. A. Somorjai, L. P. Lee, J. Bokor, "Sublithographic Nanofabrication Technology for Nanocatalysts and DNA Chips", *J. Vac. Sci. Technol. B*, vol. 21, pp. 2951-2955, 2003.

[2] Y. C. Chan, Y.-K. Lee, Y. Zohar, "High-throughput Design and Fabrication of An Integrated Microsystem with High Aspect-ratio Sub-micron Pillar Arrays for Free-solution Micro Capillary Electrophoresis", *J. Micromech.Microeng.*, vol. 16, pp. 699-707, 2006.

[3] V. Poborchii, T. Tada, T. Kanayama, "Si Pillar Photonic Crystal Slab with Linear Defects: Transmittance and Waveguide Properties", *Opt. Commun.*, vol. 210, pp. 285-290, 2002.

[4] M. Terauchi, N. Shigyo, A. Nitayama, F. Horiguchi, "'Depletion Isolation Effect' of Surrounding Gate Transistors", *IEEE Trans. Electron Devices*, vol. 44, pp. 2303-2305, 1997.

[5] C.H. Feng, Z.Y. Xiao, P.C.H. Chan and I.M. Hsing, "Lithography-free Silicon Micro-pillars as Catalyst Supports for Microfabricated Fuel Cell Applications", *Electrochem. Commun.*, vol. 8, pp. 1235-1238, 2006.

[6] I. Maximov, E.-L. Sarwe, M. Beck, K. Deppert, M. Graczyk, M. H. Magnusson, L. Montelius, "Fabrication of Si-based Nanoimprint Stamps with Sub-20nm Features", *Microelectron. Eng.*, vol. 61/62, pp. 449-454, 2002.

[7] W. Chen, H. Ahmed, "Fabrication of 5-7 nm Wide Etched Lines in Silicon Using 100 kev Electron-beam Lithography and PolymethylmethaCrylate Resist", *Appl. Phys. Lett.*, vol. 62, pp. 1499-1501, 1993.

[8] L. Xia; W. G. Wu; J. Xu; Y. L. Hao; Y. Y. Wang, "3D Nanohelix Fabrication and 3D Nanometer Assembly by Focused Ion Beam Stress-introducing Technique", in *Digest Tech. Papers MEMS'03 Conference*, Istanbul, Jan. 22-26, 2006, pp. 118-121.

[9] C. L. Haynes, R. P. Van Duyne, "Nanosphere Lithography: A Versatile Nanofabrication Tool for Studies of Size-Dependent Nanoparticle Optics", *J. Phys. Chem. B*, vol. 105, pp. 5599-5611, 2001.

[10] V. Ovchinnikov, A. Malinin, S. Novikov, C. Tuovinen, "Fabrication of Silicon Nanopillars Using Self-organized Gold–chromium Mask", *Mater. Sci. Eng. B*, vol. 69-70, pp. 459-463, 2000.

[11] S. S. Mark, M. Bergkvist, P. Bhatnagar, C. Welch, A. L. Goodyear, X. Yang, E. R. Angert, C. A. Batt, "Thin Film Processing Using S-layer Proteins: Biotemplated Assembly of Colloidal Gold Etch Masks for Fabrication of Silicon Nanopillar Arrays", *Colloid. Surface. B*, vol. 57, pp. 161-173, 2007.

VISUAL OBSERVATION OF PDMS TIP IN LIQUID MICROCONTACT PRINTING

J. Fattaccioli[1], A. Ikeda[2], J.G. Kim[2], N. Takama[2] and B.J. Kim[2]
[1]LIMMS, CNRS-IIS (UMI 2820), The University of Tokyo, Tokyo, JAPAN
[2]CIRMM, Institute of Industrial Science, The University of Tokyo, Tokyo, JAPAN

ABSTRACT

Microcontact printing has been shown to be a viable lithographic technique for the fabrication of microstructures, through the deposition of molecules by conformal contact between a surface and an elastomeric stamp. However, the diffusion of the molecules on the substrate and the deformation of the stamp during the contact are severe drawbacks when considering the resolution of the technique. In this paper, we show the effect of the diffusion of the molecules on the size of gold patterns on silicon and we observe *in-situ* the deformation of the PDMS stamp in liquid and air environment using an interferential microscopy technique.

1. INTRODUCTION

Microcontact printing [1] is a versatile technology used to create molecular micropatterns through the conformal contact of a microstructured elastomeric stamp with a wide range of substrates [2]. Low-cost and straightforward, this technique is used in several engineering domains, from chips microfabrication [3] to biophysical studies [2].

An elastomeric material, commonly PDMS, is molded in a hard microfabricated template in order to create a to a high precision soft replica of the master. After detaching the stamp from the mold, it can be *inked* with the chemical material that is to be printed. The stamp is then pushed toward the substrate with a controlled pressure and time in order to transfer the *ink* molecules to the substrate and then removed. The molecules are located where the conformal contact between the stamp and the surface takes place with a submicrometer resolution.

Although the PDMS stamp can be fabricated with high precision [4], the limitation of the resolution of the microcontact printing technique is a consequence of two phenomena occurring during the stamping [2] : the deformation of the stamp under pressure and the diffusion of the *ink* molecules on the substrate during the contact [5,6]. The effect of the diffusion can be avoided with a decrease of the contact time, a decrease of the *ink* concentration or a different solvent [7]. On the other hand, if it is easy to avoid large scale deformation by modifying the design of the stamp [8], it is difficult to quantify in real time the deformations of the surface of the stamp during the contact.

In this article, we apply Reflection Interference Contrast Microscopy (RICM) to visualize *in-situ* the deformation of a PDMS stamp during the inking process on a glass surface. RICM, developed in 1964 by Curtis *et al.* [9] to study the adhesion patterns of biological cells, allows imaging the Newton rings created by an object in close vicinity of a transparent surface and then reconstructing its three-dimensional profile [10]. RICM was recently used to study the adhesion [11] and detachment [12] of elastomeric colloidal particles on glass substrates.

In this article, we introduce the RICM and we apply it to the study of the deformation of the tips with the applied pressure and we show we can reconstruct the tip profile from the analysis of the interference patterns. Then, we compare these results with the ones obtained by performing a microcontact printing experiment on a gold surface and we conclude on the role of the deformation on the pattern area in air and water environment.

2. PDMS STAMP FABRICATION

In this study, we use a millimetric PDMS stamp having microscopic pyramidal features, in order to make easier the study of the deformation during the stamping process (fig. 1). We fabricate the silicon master with pyramidal grooves by photolithography and anisotropic KOH etching. A fine control of the etching time allows controlling the final surface of the pyramidal tip, which is set to $3.5 \times 3.5 \ \mu m^2$ in our experiments. Then we pour a PDMS/Curing agent (9:1wt/wt, Dow Corning Sylgard 184) mixture onto the master and we cure it at 180°C for three hours. Finally, we detach the stamp from the master and we attach it on a glass slide for easy handling. Each single PDMS tip molded from the master has a pyramidal structure with $3.5 \times 3.5 \ \mu m^2$ square surface on the top. The stamp itself has a square shape and a total surface of $3 \times 3 \ mm^2$.

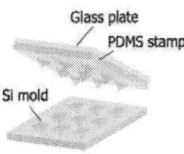 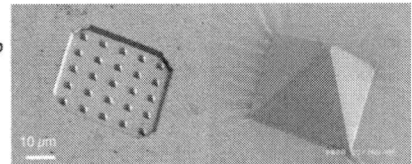

Figure 1: Schematic view of the molding process of a PDMS stamp and SEM image of the stamp itself. On the center image, we see clearly the supporting area around the tips. The tips have a pyramidal shape ending up with a flat surface of $3.5 \times 3.5 \ \mu m^2$.

3. RICM IMAGING OF CONTACT AREAS
Reflection Interference Contrast Microscopy

The principle of Reflection Interference Contrast Microscopy (RICM) is depicted on fig. 2. An object, in this study a PDMS stamp, in contact with a glass substrate is observed by reflection microscopy with epi-illumination with monochromatic light. The incident beam is partly reflected at the glass interface, the object reflects the transmitted part and interference fringes, also called Newton rings, are created on the glass surface. The RICM allows imaging patterns smaller than the micron and allows reconstructing the vertical profile of the object over several microns far from the glass substrate.

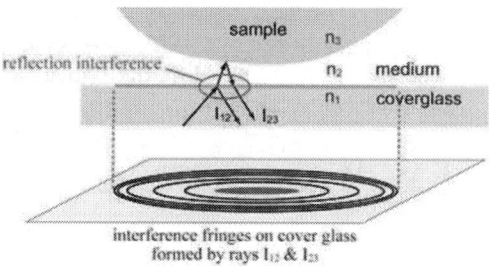

Figure 2: Schematic representation of an RICM experiment. The proximity between the stamp and the glass surface creates interference patterns under monochromatic illumination. The interference fringes inter-distance depends on the vertical profile of the stamp.

Optical Setup

We observe the RICM interference patterns on an inverted light microscope Olympus IX 51. We create a monochromatic illumination with an epifluorescence mercury lamp and a set of filters allowing only a single wavelength (λ=570 nm) going through the optical setup. We perform the image acquisition with a QImaging Retiga EXi 12 bit-camera and the image analysis with the NIH ImageJ software.

Glass substrates

The RICM study of the stamp deformation is done on commercial glass coverslips used for microscopy. Prior to the experiment the coverslips are rinsed with ethanol and deionized water. The experiments in liquid are done in cell culture chamber having a coverslips at the bottom.

In-situ Observation of the Stamping Process

We press a PDMS stamp without ink molecules on a glass slide with a stamping force ranging from 1 to 20 mN and controlled by calibrated weights. By RICM, we take a picture of each tip and we measure numerically the contact area in air (a) and (b) and we report the results on the figure 3.

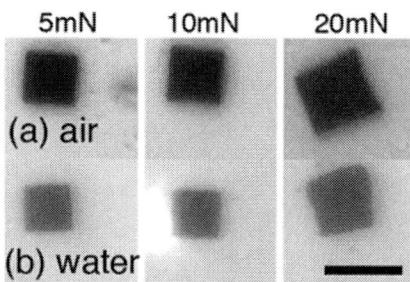

Figure 3: RICM images of the PDMS tip in contact with the glass slide for different stamping pressures in air and water. The difference of contrast is due to the different refractive index of air and water. (Scalebar: 5 µm)

The difference of contrast between the pictures in air (a) and in water (b) relies in the difference of refractive index between air and water.

We see that the contact area increases with the applied force. Moreover, as the increase of the area is more important in air condition, the shape of the stamp is modified for an applied force of 20 mN as the structure collapses.

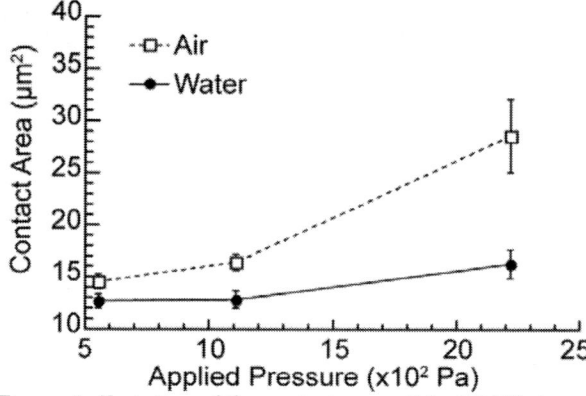

Figure 4: Variation of the contact area of the PDMS tips as a function of the stamping pressure.

On figure 4, we report the evolution of contact area of the tip as a function of the experimental environment and the applied pressure. Neglecting the buoyancy force, the applied pressure is expressed as the applied weight divided by the total area of the stamp. In air, the deformation of the stamp leads to an important increase of the stamping area with the applied pressure. In water, this deformation doesn't occur since liquid is entrapped within the stamp thanks to the PDMS support.

Reconstruction of the stamp profile

The RICM technique allows bringing more quantitative information than the contact area solely. Indeed, we can also observe on a tip image the interference fringes and analyze their spacing since it depends on the vertical profile of the tip itself, as shown on figure 5(inset).

On figure 5(top) we plot the intensity profile of the interference pattern along one direction of the tip image starting from the center of the contact area, for an experiment in water with an applied force of 20 mN.

The figure 5(bottom) shows the numerical reconstruction of the profile of the stamp over several hundred of nanometers far from the glass substrate based on the image analysis of the interference fringes inter-distance shown on figure 5(top): each local minimum of the graph corresponds to a height difference on the profile equal to half of the illumination wavelength [10].

As inferred from the asymmetric RICM image, the profile of the stamp itself is not symmetric, and hence the applied force is not uniform on the stamp.

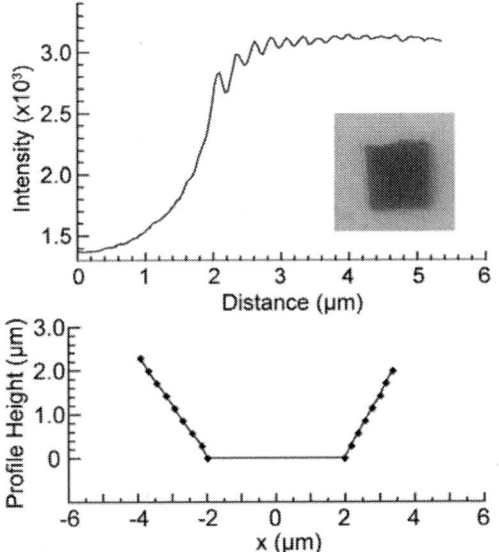

Figure 5: (Top) Intensity profile of the interference fringes along one direction of the stamp RICM image for a stamping in water with a force of 20 mN. (Inset) RICM image of the tip in contact with the glass slide: the center part corresponds to the contact area with the glass. (Bottom) Stamp profile reconstructed from the interference pattern analysis.

4. GOLD PATTERNS ON SILICON

Silicon substrates

In order to study the effect of the pressure and stamping time on the shape of gold patterns by microcontact printing, we thermally evaporate a 50 nm thick gold layer with a 10 nm thick chrome adhesion layer (Nilaco, Japan) on silicon wafers cleaned by a Piranha mixture (H_2SO_4/H_2O_2 1:1 vol.).

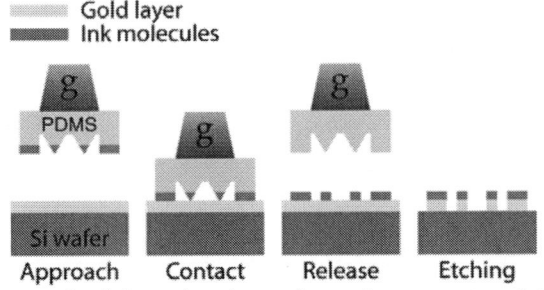

Figure 6: Schematic view of a microcontact printing experiment: (a) approach, (b) contact and transfer, (c) release and (d) etching of the gold layer. The pressure on the stamp is controlled by calibration weights. Gold is solely etched where no molecules are bound to it.

Microcontact Printing Experiment

We cover the PDMS stamp with a SAM molecule, hexadecanethiol (HDT, 100 mM in ethanol), which has the ability to self-assemble on the gold surface [1]. Then we apply the stamp on the gold surface with a controlled pressure and time in air or deionized water environment (Figure 6). We detach the stamp from the sample and finally

we immerse the sample into a gold etching solution (KOH 1M, $Na_2S_2O_3$ 1M, $K_3Fe(CN)_6$ 0.01M, $K_4Fe(CN)_6$ 0.001M) for 150 seconds until complete etching of the gold layer where no SAM molecules are adsorbed.

Gold patterns on silicon

On Figure 7, we show the gold patterns obtained with an applied force ranging between 5 and 20 mN during 5 seconds in air (a) and water (b) and observed with a JEOL JSM 740 field-effect scanning electron microscope. The patterns are larger in air than in water environment and the molecular diffusion of the molecules tends to round the patterns.

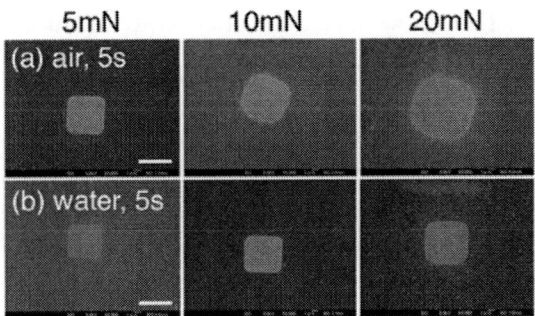

Figure 7: SEM images of gold patterns fabricated by microcontact printing of SAM molecules in air and liquid environment. The contact time, the applied pressure and the environment have an influence on the pattern size and its shape. (scalebar : 5 µm)

On figure 8, we report the evolution of the gold pattern area as a function of the experimental environment, the stamping time and the applied pressure. In air, the pattern area increases by a factor 2 over the applied pressure range and we can see a strong influence of the molecular diffusion since the area increases with the stamping time. When the stamping process is performed in water, nor the applied pressure, neither the molecular diffusion has a strong influence on the pattern area.

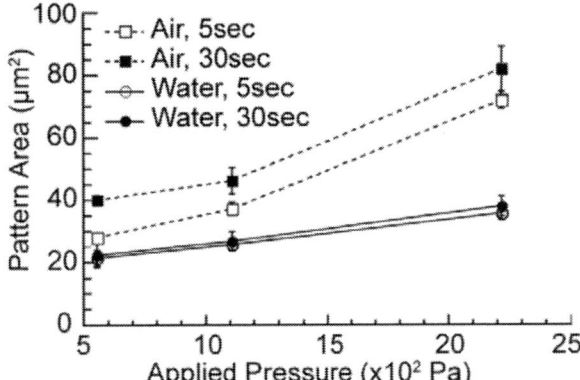

Figure 8: Gold pattern area as a function of the applied pressure. The data are averaged over the 25 tips and the error bar represents the standard deviation of the area.

978-1-4244-2977-6/09 $25.00 © 2009 IEEE

5. DISCUSSION

Performed in air or in water, the contact area of the stamp and the gold pattern area have the same trend toward the applied pressure. In our experimental configuration, the PDMS tips collapse in air environment, leading to a change of the shape of the tip and a broadening of the contact area. However, in water, since the liquid is trapped between the stamp and the substrate, the shape of the tip is less deformed and doesn't collapse. Compared to the experiment on gold surface, the diffusion accounts for less than the stamp deformation itself on the final surface area of the patterns, especially in air.

6. CONCLUSION

In this article, we have shown that the RICM technique can be used to analyze the tip deformation of a stamp in microcontact printing experiments. The analysis of the contact area showed that depending on the experimental conditions, the pressure can deform and make collapse the tip, with dramatic effect on the patterns surface area. We confirmed this with the analysis of the surface area of gold patterns in the same conditions.

RICM is a powerful technique allowing the reconstruction of the tridimensional profile of the stamp in the close vicinity of the substrate. Used as a monitoring instrument, and not as an analysis technique, it can be really helpful during microcontact printing experiments in order to control the stamping conditions *in-situ*.

ACKNOWLEDGMENT

The work was partially supported by the Japanese Society of Promotion of Science (JSPS, Fellowship P06733). We also acknowledge the laboratories of Prof. H. Fujita and Prof. H. Toshiyoshi for their support regarding experimental equipment and cleanroom facilities.

REFERENCES

[1] J.L. Wilbur, A. Kumar, E. Kim, and G. Whitesides, "Microfabrication by microcontact printing of self-assembled monolayers", *Advanced Materials*, vol. 6, pp. 600-604, 1994.

[2] S.A. Ruiz and C.S. Chen, "Microcontact Printing: A tool to pattern", *Soft Matter*, vol. 3, pp. 168-177, 2007.

[3] D. Burdinski, M. Saalmink,and J.P.W.G Berg, and C. Marel. "Universal Ink for Microcontact Printing", *Angew. Chem. Int. Ed.*, vol.45, pp.4355-4358, 2006.

[4] H. Li, B. Muir, G. Fichet, and W. Huck, "Nanocontact printing: A route to sub-50-nm-scale chemical and biological patterning", *Langmuir*, vol. 19, pp. 1963-1965, 2003.

[5] C. Hui, A. Jagota, Y. Lin, and E. Kramer, "Constraints on microcontact printing imposed by stamp deformation", *Langmuir*, vol. 18, pp. 1394-1407, 2002.

[6] E. Delamarche, H. Schmid, A. Bietsch, N. Larsen, H. Rothuizen, B. Michel, and H. Biebuyck, "Transport mechanisms of alkanethiols during microcontact printing on gold", *J. Phys. Chem. B*, vol. 102, pp. 3324-3334, 1998.

[7] L. Libioulle, A. Bietsch, H. Schmid, B. Michel, and E. Delamarche, "Contact-inking stamps for microcontact printing of alkanethiols on gold", *Langmuir*, vol. 15, pp. 300-304, 1999.

[8] H. Schmid and B. Michel, "Siloxane polymers for high resolution, high accuracy softlithography", *Macromolecules*, vol. 33, pp. 3042-3049, 2000.

[9] A. S. G. Curtis, "The mechanism of adhesion of cells to glass. A study by interference reflection microscopy", *J. Cell Biol.*, vol.20, pp. 199, 1964.

[10] J. Rädler and E. Sackmann, "Imaging optical thicknesses and separation distances of phospholipid-vesicles at solid-surfaces", *Journal De Physique II*, vol. 3, pp. 727-748, 1993.

[11] S. Schmidt, M. Nolte and A. Fery, "Single-colloidal particle microcontact printing", *Phys. Chem. Chem. Phys.*, vol. 9, pp. 4967-4969, 2007

[12] H. Gérardin, A. Burdeau, A. Buguin and F. Brochard-Wyart, "Force detachment of immersed elastic rubber beads", *Langmuir*, vol. 23, pp. 9704-9712, 2007

SUPERCRITICAL FLUID DEPOSITION (SCFD) TECHNIQUE AS A NOVEL TOOL FOR MEMS FABRICATION

T. Momose, T. Ohkubo, T. Uejima, T. Saito, M. Sugiyama, and Y. Shimogaki

The University of Tokyo, JAPAN

ABSTRACT

In the present paper, supercritical fluid deposition (SCFD) process is proposed for making functional films and coatings on the surface of MEMS devices. SCFD can provide superior step coverage on to high aspect ratio features at relatively low temperature. It also provides possibility to deposit wide variety of materials, including metals, oxides, and organic compounds, with controllability of substrate selectivity. For example, pure Cu was conformally dposited onto high aspect ratio vias for ULSI interconnects with 50nm in diameter and 1μm in depth at 220 °C. The Cu-SCFD process is a selective process, which can deposit Cu film only on conductive surface, however, non-selective deposition of Cu is also possible by using CuMnOx as a buffer layer between Cu and insulative substrate. SiO_2 film was conformally fabricated in high aspect ratio trenches with 500 nm in width and 5 μm in depth at 200 °C. Biocompatible poly ethylene glycol monomethacrylate (PEGMA) was grafted onto the surface of micro-channel with aspect ratio of 2500 at 80 °C, and proved effective protection of protein adsorption onto SiO_2 surface.

INTRODUCTION

In MEMS structures, functional films, for example, metal electrode, passivation film, and biocompatible coating, have to be made on the high-aspect-ratio features, even on over-hung structures to create functional devices. It is also important to lower the process temperature not to damage the temperature sensitive substrate such as organic parylene. Control of selective/non-selective deposition is also required for composite materials structures. Conventional deposition method of physical vapor deposition (PVD) is, however, unacceptable for high-aspect-ratio features due to its poor step coverage. Chemical vapor deposition (CVD) is a well-known technology for conformal deposition for high-aspect-ratio features, however, suitable precursor with high volatility should be developed to enable CVD process as the device fabrication steps. For example, Cu-CVD was intensively studied for ULSI interconnects, however, poor adhesion was caused by fluorine which is incorporated in the Cu precursor. [1-3].

On the other hand, we have been investigating supercritical fluid deposition (SCFD) as nanostructure fabrication method for various materials, and presented its excellent step coverage and superior gap-filling in nano-scale dimensions [4-9]. This method involves reduction/oxidation of organic compounds in supercritical carbon dioxide ($scCO_2$) to make thin films. The benefits of SCFD over conventional deposition techniques are a consequence of the unique properties of supercritical fluid (SCF), which is a material heated and compressed beyond its critical temperature and pressure [10, 11]. High density

of SCF enables dissolution of a number of metal precursors and deposition at relatively high precursor concentrations. In contrast, CVD has a limitation in choice of reactants because of their insufficient volatility [12]. High precursor concentration and diffusivity of SCF enables excellent step coverage in sub-micrometer features. Besides, in Cu-SCFD, Langmuir-Hinshellwood-type (LH-type) nonlinear surface reaction kinetics was reported [13, 14], which also contributes to conformal deposition. Solvation effect in SCF accomplishes low process temperature suitable for the deposition onto temperature sensitive materials [8, 10, 11]. Solubility of SCF is also effective to achieve high purity of films by removing contaminations from the film surface during deposition. Adding small amount of organic compounds into $scCO_2$ is effective to enhance the solubility and reactivity of the precursors, and thus extends the applicability of SCFD process [4]. Owing to these characters, SCFD is expected to be a powerful solution for nano-processing. This paper introduces application of this technique to MEMS fabrication, focusing on the uniformity of step coverage and substrate selectivity in Cu-SCFD, SiO_2-SCFD, and poly ethylene glycol monomethacrylate (PEGMA)-grafting.

FABRICATION

All the deposition runs were carried out using high pressure resistant batch reactor with inner volume of 9 ml. Depositions were initiated by heating the substrate from 40 - 60 °C to a given temperature. First, the reactor was opened to air, and substrate was placed inside the reactor. Then, precursor and/or reducing/oxidizing agent were loaded into the reactor, which was then pressurized by CO_2 to initial condition shown below. Finally, the reactor was sealed, and the substrate and/or the reactor were heated up to a given temperature to enhance the deposition reaction. After the deposition runs, the surface morphology and the film thickness of the deposited films were observed using field emission scanning electron microscope (FE-SEM; JEOL JSM6340F). The chemical composition and stoichiometry of the deposited films were analyzed using X-ray photoelectron spectroscopy (XPS; ULVAC-Φ XPS Model 1600C).

Cu/CuMnOx-SCFD

Deposition was carried out using a cold wall type reactor with inner volume of 9 ml. Initial condition of $scCO_2$ was 9.5 MPa at 40 °C in case of Cu deposition, 11 MPa at 60 °C in case of CuMnOx deposition. The substrate was heated to 200 °C when the precursor concentration was 1.7×10^{-4} mol/L, or to 220 °C when the precursor concentration was 1.7×10^{-3} mol/L in case of Cu deposition, and to 230 °C in case of CuMnOx deposition.

The precursors used here were a Cu organic compound, bis(2,2,6,6-tetramethyl-3,5-heptanedionato) copper(II) ($Cu(tmhd)_2$), which does not includes F atoms and thus improves adhesion of the Cu film to a metallic underlayer [8], and bis(pentamethylcyclo-pentadienyl) manganese ((pmcp)$_2$Mn). H$_2$ was used as a reducing agent. Via patterned substrate with 20nm-thick ALD Ru was used for evaluating the conformality and gap-filling property of Cu deposition. A test element group (TEG) with overhung structure consisting of poly-Si, boron phosphorus silicon glass (BPSG), and tetraethyl ortho-silicate (TEOS) SiO$_2$ was used for evaluating conformality of CuMnOx deposition. Parylene-F evaporated on Si substrate was also used. These substrates were used without any pre-treatment.

To investigate deposition profile and gap-filling in ultra narrow vias, deposited via patterned substrate was polished with angle of 0.5°, and then the plain-view of round-sliced via holes at continuous depth from the top to the bottom was observed by FE-SEM.

SiO$_2$-SCFD

Deposition was carried out using a cold wall type reactor with inner volume of 9 ml. Initial condition of scCO$_2$ was 12 MPa at 50 °C. Before deposition, the reactor was evacuated by rotary pump to eliminate contamination. The substrate was heated to 200 °C. After the substrate temperature reached 200 °C, scCO$_2$ in the reactor was purged by scCO$_2$ to prevent the precipitation of SiO$_2$ particle derived from the remaining precursors in scCO$_2$. The precursor used here was TEOS. Trench patterned Si substrate was used without any pre-treatment. After deposition, infrared absorption spectrum was measured by fourier transform infrared spectrometer (FTIR; Nicolet Impact 410).

PEGMA-Grafting

Grafting of PEGMA was carried out using a hot wall type reactor with inner volume of 9 ml.

Procedure consists of three steps, silanol treatment, silanization, and PEGMA grafting. First, a surface silanol group on the SiO$_2$ substrate was increased by silanol treatment using HF. Then, silanization of SiO$_2$ was performed using allyltriethoxysilane (ATES), where ATES was immobilized at the surface of SiO$_2$ by the reaction between the silanol group and ethoxyl group (silanization). Finally, PEGMA monomer was grafted at the surface of SiO$_2$ by in situ graft polymerization.

A glass based micro-chip (ICC-SY500, Institute of Microchemical Technology, Japan) was used as the test element for the micro-channel modification. This micro-chip was used without any pre-treatment.

Silanol treatment was performed at 40 °C under 12 MPa scCO$_2$ dissolving 1.6 ml HF for 40 min. Silanization was performed at 80 °C under 23 MPa scCO$_2$ dissolving 1.0 ml ATES for 21 hours. PEGMA was grafted at 80 °C under 25 MPa scCO$_2$ dissolving 3.3ml PEGMA and 20mg 2,2'-azobisisobutyronitrile (AIBN) for 24 hours. After each step, the micro-chip was taken out from the reactor, and the substrate was washed with water and then ethanol (three times each), and then dried in vacuo at room temperature.

After grafted, protein-adsorption-prevention effect on the surface of a micro-channel was evaluated by monitoring the adsorbed fluorescein isothiocyanate conjugate bovine serum albumin (FITC-BSA) at the surface by using a confocal fluorescence microscope (Axioskop 2 plus, Carl Zeiss Inc., Germany) integrated with a CCD camera (VB7010, Keyence Co., Japan).

RESULTS

Cu/CuMnOx-SCFD

To investigate the potential of conformal step-coverage and gap-filling property in Cu-SCFD, ultra thin Cu film was fabricated on patterned substrate, and the gap was filled. Two different precursor concentrations were introduced into the batch reactor to control the film thickness, namely, 1.7×10^{-4} and 1.7×10^{-3} mol/L, which were equivalent to deposition of 10nm-thick film and 100nm-thick film, respectively. Left hand side of Fig.1 shows cross-sectional SEM images of the deposited vias after filling (50nm in diameter, 1μm in depth). The pictures at the center of Fig.1 show plain-view SEM images of the round-sliced vias after angled polishing. The vias were deposited with precursor concentration of 1.7×10^{-4} mol/L, and the right hand side shows that with precursor concentration of 1.7×10^{-3} mol/L. 10nm-thick smooth and continuous Cu film was conformally deposited onto via patterns. Vias were completely filled without any seems and voids as shown in Fig.1. As we mentioned, the potential reason of this excellent step coverage was LH-type nonlinear surface reaction kinetics of Cu-SCFD cased by strong adsorption and surface saturation of the precursor. Although the precursor concentration at the bottom of the vias was expected to be lower than that at the top, conformal coverage was achieved, because the precursor concentration had little effect on Cu growth rate in LH-type kinetics.

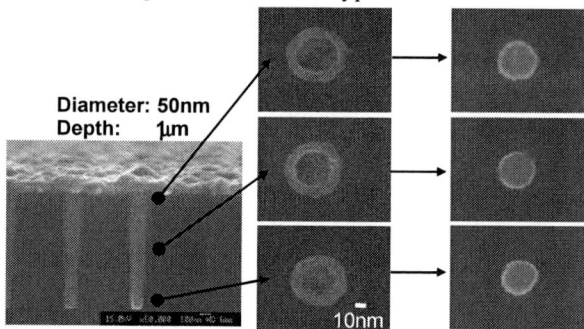

Figure 1: Cross-sectional SEM image of patterned substrate after Cu-SCFD and plain-view SEM images after angled polishing. Center pictures are after making 10-nm-thick continuous Cu film, left and rightpictures are after filling. Depth from top was a) 100nm, b) 500nm, and c) 900nm. Cu(tmhd)$_2$ was used as precurso, and H$_2$ was used as reducing agent.

In addition, Cu deposition onto micro-scale high aspect ratio Si trenches, which were fabricated by BOSCH etching, was examined. The deposition condition was same with the previous experiments. Cu was conformally deposited from top to bottom with thickness of 100 nm on patterned Si substrate as shown in Fig.2. In metal deposition by SCFD, metal film was selectively deposited on conductive substrate, because hydrogen molecule in scCO$_2$ is stable, and hydrogen atoms generated by dissociative adsorption on the conductive substrate react with precursor to make Cu film. In Si substrate, by removing the native oxide and terminating H atom by HF solution, metal film can be fabricated on Si surface.

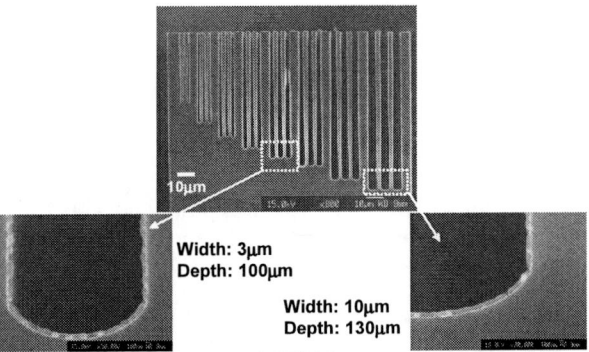

Figure 2: Cross-sectional SEM images of patterned Si substrate after Cu-SCFD. Cu(tmhd)₂ was used as precursor and H₂ was used as reducing agent.

Figure 2: Cross-sectional SEM images of patterned Si substrate after Cu-SCFD. $Cu(tmhd)_2$ was used as precursor and H_2 was used as reducing agent.

A trial on one-step fabrication of Cu films on insulators by SCFD was made, because conventional SCFD cannot directly deposit Cu on insulators. In our one-step method, $CuMnO_x$ composite layer was used as a buffer layer for Cu deposition. This method was successfully demonstrated by depositing Cu on overhung TEG consists of Poly-Si, TEOS-SiO₂, BPSG as shown in Fig.3. It was deposited until the substrate was heated up to 230 °C under 3.8×10^{-3} mol/L $Cu(tmhd)_2$ and 1.7×10^{-3} mol/L $(pmcp)_2Mn$. The purity of the Cu film was evaluated by XPS analysis, and found contaminations of C, O was about 2 atomic% each. This method also successfully applied on Cu film (not shown here). These results indicated that CuMnOx composite layer deposits non-selectively on various kinds of substrate and then Cu film can deposit on top of CuMnOx layer.

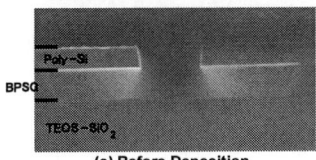 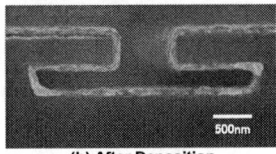

(a) Before Deposition   (b) After Deposition

Figure 3: Cross-sectional SEM images of over-hung substrate a) before and b) after one-step direct Cu-SCFD on insulator. Substrate consists of poly-Si, BPSG, and TEOS-SiO₂. $Cu(tmhd)_2$ was used for Cu deposition, and $(pmcp)_2Mn$ was used for buffer layer deposition

Further investigation of our one-step Cu-SCFD revealed that, during heating the substrate, CuMnOx buffer layer was deposited at 160-180 °C, and Cu deposition started at 180 °C (not shown here). It suggests this method can be applied at low temperature of 200 °C or less, which extends the applicability of this method. We successfully demonstrated Cu fabrication on parylene at 200 °C by our one-step method as shown in Fig.4.

In summary, SCFD can make high purity Cu film onto high aspect ratio structures at low temperature of 200 °C. Surface selectivity of Cu-SCFD could be controlled by our new method, namely, selective deposition on conductor and non-selective deposition using CuMnOx as the buffer layer.

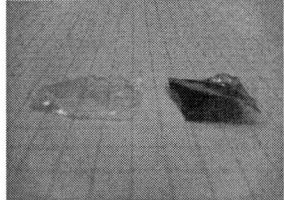 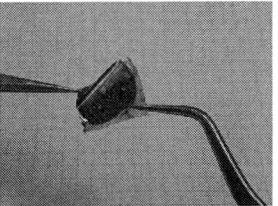

(a) Left; before deposition,   (b) Flexibility of the sample
Right; after deposition

Figure 4: a) Outline of parylene film before and after Cu-SCFD using CuMnOx deposition as buffer layer for Cu deposition. b) Outline of Cu deposited parylene film bended to show flexibility of it.

SiO₂-SCFD

The potential of step-coverage by SiO₂-SCFD was investigated. Figure 5 shows SiO₂ film fabricated on patterned Si substrate (500 nm in width, and 5 µm in depth) with thickness of 100 nm. This film was deposited until the substrate was heated up to 200 °C. Precursor (TEOS) concentration was 0.13 mol/L. The purity of this film was evaluated by XPS analysis, which could not detect any C contamination (data is not shown), while deposited sample at 100 °C include several percentage of C contamination. With the increase of the maximum temperature during our batch process, C contamination was decreased and disappeared at 200 °C. Considering that high purity SiO₂ is typically deposited at 500-800 °C without any oxidizing agent in LPCVD, the low deposition temperature of 200 °C achieved in SCFD was quit low temperature. The infrared absorption spectrum of SiO₂ deposited by SCFD was evaluated by using FTIR (not shown here). Peak of Si-O-Si stretching mode was 1072 cm^{-1}, and FWHM was 80 cm^{-1}, which was similar quality with that achieved by PECVD. About surface selectivity, different from Cu-SCFD, SiO₂-SCFD does not involve catalytic effect of substrate surface, resulting in non-selective deposition. Unfortunately, it also suggests that TEOS is possible to precipitate in $scCO_2$. Adding small amount of O₂ or water to SiO₂-SCFD promoted the nucleation reaction in SCF, namely, resulted in particle generation in SCF (not shown here). O in the reactor should be carefully eliminated in SiO₂-SCFD. Consequently, conformal and high quality SiO₂ was fabricated by SCFD on high aspect ratio features at low temperature of 200 °C.

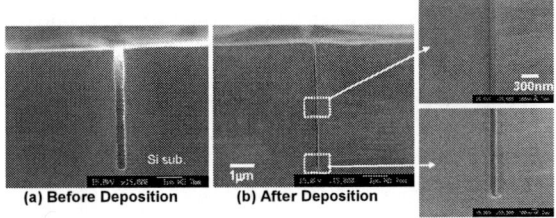

(a) Before Deposition   (b) After Deposition

Figure 5: Cross-sectional SEM images of patterned substrate a) before and b) after SiO₂-SCFD on Si trenches. TEOS was used as precursor.

PEGMA-Grafting

SCFD can also make functional organic polymers onto glass substrate used for biocompatible MEMS. For example, PEGMA is able to protect BSA adsorption onto SiO₂ surface. Figure 5 shows a demonstration of PEGMA

978-1-4244-2977-6/09 $25.00 © 2009 IEEE

graft onto high aspect ratio micro-channel fabricated on glass substrate at 80 °C. The micro-channel was 98.5 μm in width, 40.0 μm in depth, and 500 mm in length. PEGMA graft at the center of the micro-channel was evaluated by measuring the protection capability of BSA adsorption using a confocal fluorescence microscope, and was successfully confirmed as shown in Fig.6 (b)-(d). On the other hand, BSA adsorption test on blanket SiO_2 shows the effect by PEGMA grafted in $scCO_2$ was as high as that grafted in water (not shown here). Consequently, PEGMA grafting in $scCO_2$ exhibited excellent step coverage with having similar BSA adsorption protection effect with that grafted by conventional wet method. This is quite important for the fabrication of micro-TAS devices.

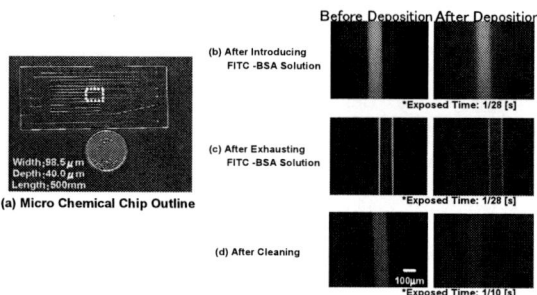

Figure 6: a) Outline of micro chemical chip. Fluorescence microscope images of adsorbed FITC-albumin on SiO_2 micro channel b) after introducing FITC-BSA, c) after exhausting FITC-BSA, and d) after cleaning.

CONCLUSIONS

SCFD was proposed for making functional films on the surface of MEMS due to its superior step coverage, wide variety of deposition materials, low temperature, and controllability of substrate selectivity. Pure Cu was conformally fabricated on high aspect ratio vias with 50nm in diameter and 1μm in depth at 220 °C, and surface selectivity (selective deposition on conductor/non-selective deposition) of Cu-SCFD was controlled by using CuMnOx as a buffer layer for Cu deposition. SiO_2 was conformally fabricated on high aspect ratio trenches with 500 nm in width and 5 μm in depth at 200 °C. Biocompatible PEGMA was grafted on the micro-channel with aspect ratio of 2500 at 80 °C, and proved effective protection of protein adsorption onto SiO_2 surface.

REFERENCES

[1] Y. S. Kim and Y. Shimogaki, "X-ray photoelectron spectroscopic characterization of the adhesion behavior of chemical vapor deposited copper films", *J. Vac. Sci. Tech.*, vol. A19, 2642-2651, 2001.

[2] Y. S. Kim and Y.Shimogaki, "Physical and chemical contributions of interfacial impurities to film adhesion", *Jpn. J. Appl. Phys.*, vol. 40, L1380-1383, 2001.

[3] Y. S. Kim, H. Hamamura, and Y. Shimogaki, "Adhesion Characteristics between Chemical Vapor Deposited Cu and TiN Films: Aspects of

Process Integration", *Jpn. J. Appl. Phys.*, vol. 41, 1500-1506, 2002.

[4] T. Momose, T. Ohkubo, M. Sugiyama, and Y. Shimogaki, "Effect of liquid additives in supercritical fluid deposition of copper for enhancing deposition chemistry", *Thin Solid Films*, in press.

[5] T. Momose, M. Sugiyama, and Y. Shimogaki, "In situ observation of initial nucleation and growth processes in supercritical fluid deposition of copper", *Jpn. J. Appl. Phys.*, vol. 47, 885-890, 2008.

[6] B. Zhao, T. Momose, T. Ohkubo, and Y. Shimogaki, *J. Microelectronic Eng.*, vol. 85, 675-681, 2008.

[7] B. Zhao, T. Momose, and Y. Shimogaki, "Deposition of Cu-Ag alloy film by supercritical fluid deposition", *Jpn. J. Appl. Phys.*, vol. 45, L1296-1299. 2006.

[8] T. Momose, M. Sugiyama, and Y. Shimogaki, "Precursor Evaluation for Cu-Supercritical Fluid Deposition Based on Adhesion Properties and Surface Morphology", *Jpn. J. Appl. Phys.*, vol. 44, L1199-1202, 2005.

[9] T. Momose, M. Sugiyama, and Y. Shimogaki, "In situ observation of initial nucleation and growth processes in supercritical fluid deposition of copper", *Appl. Phys. Express*, vol. 1, 097002, 2008.

[10] J. M. Blackburn, D. P. Long, A. Cabanas, and J. J. Watkins, "Deposition of Conformal Copper and Nickel Films from Supercritical Carbon Dioxide", *SCIENCE*, vol. 294,141-145, 2001.

[11] E. Kondoh and H. Kato, "Characteristics of copper deposition in a supercritical CO_2 fluid", *Microelect. Eng.*, vol. 64, 495-499, 2002.

[12] T. T. Kodas and M. J. Hampden-Smith, *The Chemistry of Metal CVD*, VCH Publishers, New York, 1994.

[13] Y. Zong and J. J. Watkins, "Deposition of Copper by the H_2-Assisted Reduction of $Cu(tmod)_2$ in Supercritical Carbon Dioxide: Kinetics and Reaction Mechanism", *Chem. Mater.*, vol. 17, 560-565, 2005.

[14] E. Kondoh and J. Fukuda. "Deposition kinetics and narrow-gap-filling in Cu thin film growth from supercritical carbon dioxide fluids", *J. Supercritical Fluids*, vol. 44, 466-474, 2008.

ELECTRODEPOSITION OF PERMALLOY IN DEEP SILICON TRENCHES WITHOUT EDGE-OVERGROWTH UTILIZING DRY FILM PHOTORESIST

Sang-Won Park, Debbie G. Senesky, and Albert P. Pisano
Berkeley Sensor and Actuator Center (BSAC)
University of California, Berkeley, California 94720, USA

ABSTRACT

An electrodeposition process for embedding NiFe alloys in deep silicon trenches (100 μm) without edge-overgrowth was developed by utilizing dry film photoresist as a sacrificial trench top. The dry film photoresist prevented high concentration of current flux at the trench edges during electrodeposition leading to planar deposition topographies. In addition, the effect of the applied current density on material composition of the electrodeposited NiFe film was investigated, and the composition for permalloy ($Ni_{80}Fe_{20}$) was obtained with a current density of $100 \, mA/cm^2$. Furthermore, the B-H response of the electrodeposited permalloy films exhibited a saturation magnetic flux density and relative permeability of 1.23 Tesla and 82, respectively. The electrodeposition technique developed in this work was utilized to fabricate a free-standing MEMS electromagnetic linear actuator composed of silicon and permalloy structures.

INTRODUCTION

Ferromagnetic materials embedded in silicon have been utilized to develop Microelectromechanical System (MEMS) devices [1]. For example, MEMS researchers have utilized ferromagnets to develop microactuators, sensors and micromotors because of the ability to produce high force and large stroke electromagnetic actuation [2-4]. The most commonly used ferromagnetic material is permalloy ($Ni_{80}Fe_{20}$) due to its superior magnetic properties and manufacturability. Permalloy is known for its high permeability, high magnetic saturation, low coactivity, good corrosion resistance and near-zero magnetostriction [5]. In addition, low stress deposition processes have been developed to obtain film thicknesses between a few nanometers and millimeters [6, 7].

Electrodeposition is a fabrication technique that is often used to deposit permalloy. This technique has the advantage of obtaining high growth rate while maintaining low film stress; unlike evaporation and sputtering techniques [8]. In addition, selective growth can be obtained to create isolated features that can be embedded or released to create microstructures. Furthermore, the magnetic and physical properties of electrodeposited films can be altered by adjusting deposition parameters such as current density, temperature and pH of electrolyte [1, 7, 9].

The drawback of the electrodeposition process is non-uniform deposition topography in trenches which requires planarization techniques. The deposition profile of

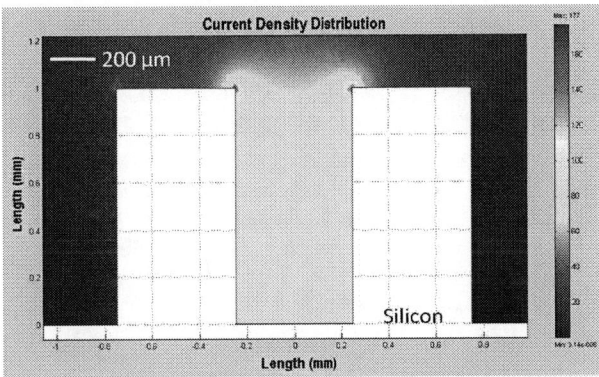

Figure 1: The FEM analysis was performed to simulate the current flux density distribution inside of an isolated trench during electrodeposition. The simulation demonstrates high flux concentration at the top edges of the trench, which causes edge-overgrowth.

electrodeposited material within a trench is affected by the geometrical configuration. More specifically, the large area of the anode in comparison to the cathode (deposition surface) causes crowding of the electric field at the top perimeter of the trench. In Figure 1, the results of a finite element modeling (FEM) analysis are shown to observe the current flux density distribution during electrodeposition inside of an isolated silicon trench. The simulation demonstrates a high concentration of current flux at the top trench edge surface. This high concentration of current flux causes an increased growth rate at the edge of the trench, leading to edge-overgrowth. In many fabrication sequences, a planar topography is necessary for lithography steps, and etchback methods such as chemical mechanical polishing (CMP), mechanical lapping or reverse electroplating are often utilized [10, 11]. However, these techniques reduce the fabrication throughput leading to higher manufacturing costs.

This paper presents fabrication technology for the electrodeposition of thick (above 100 μm) material without edge-overgrowth through the utilization of dry film photoresist as a sacrificial trench top. The technology can be used for applications that utilize electrodeposition processes such as electrical interconnects or MEMS structures. The advantage of this process is an elimination of the etchback process, which is typically employed to remove the edge-overgrowth material. In addition to the deposition topography, the material properties such as composition, relative permeability and saturation flux density of electrodeposited permalloy films are investigated.

978-1-4244-2977-6/09 $25.00 © 2009 IEEE

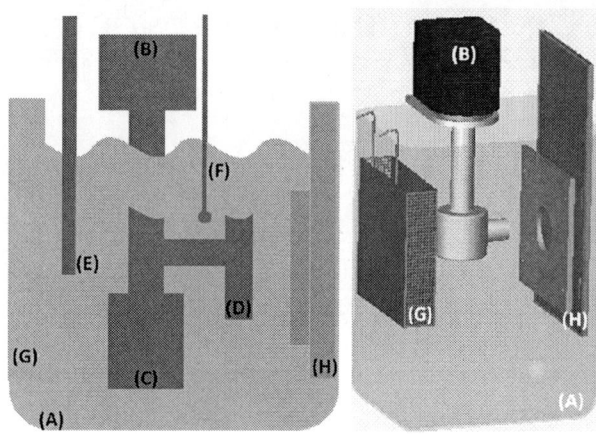

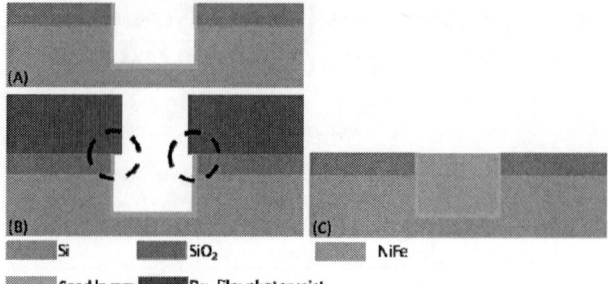

Figure 2: The schematic of NiFe alloy electrodeposition apparatus. (A) 10 liter chromatography jar, (B) Pump, (C) Filter, (D) Multi-holes nozzle, (E) Immersion heater, (F) Temperature probe, (G) Ni basket (Anode) and (H) Sample holder (Cathode).

EXPERIMENTAL
Electrodeposition Apparatus

A citrate-complexed NiFe electrolyte was utilized for electrodeposition as it has demonstrated deposition of thin films with superior magnetic properties during long electrodeposition times (up to approximately 25 hours) and insensitivity to deviations in deposition parameters (current density, pH and temperature) [7, 9]. The electrolyte is composed of 112 g/l of nickel sulfate ($NiSO_4 \cdot 6H2O$), 5 g/l of iron sulfate ($FeSO_4 \cdot 7H2O$), 75 g/l of sodium citrate ($Na_3C_6H_5O_7 \cdot 2H_2O$), 1.5 g/l of potassium sulfate (K_2SO_4), 0.2 g/l of sodium n-dodecyl sulfate ($C_{12}H_{15}SO_4Na$) and 1 g/l of saccharin ($C_7H_5NO_3S$). Sodium n-dodecyl sulfate is an anionic surfactant compound which promotes wetting of the cathode surface, and saccharin is utilized as a stress reducing agent [7]. A schematic of electroplating apparatus is shown in Figure 2. A 10 liter chromatography jar (A) holds electrolyte while a pump (B) agitates the electrolyte through a filter (C). Direct impingement of the electrolyte into the surface of the sample is achieved by a nozzle with multiple holes (D). A constant electrolyte temperature of 45°C is obtained by a temperature controller, which is coupled with an immersion heater (E) and a temperature probe (F). A titanium anode basket (G) holds electrolytic nickel rounds and is covered by an anode bag filter, which prevents nickel particulates from depositing on the sample surface. A cathode sample holder (H) fixtures the sample inside the electrodeposition cell, providing a consistent sample location. A power supply with current control provides a current flow between the anode and the cathode and initiates electron transfer. The current density has been shown to affect the film deposition rate and composition [7], and a constant current source, therefore, was utilized to aid in obtaining a uniform composition through the depth of the deposit.

Figure 3: Fabrication process implemented to prevent edge-overgrowth utilizing dry film photoresist as a sacrificial trench top. (A) Conductive seed layers (chromium/copper) were evaporated onto a 100 μm deep trench, (B) Dry film photoresist was deposited and patterned and (C) Electrodeposition of NiFe film was performed followed by lift-off of the dry film photoresist layer.

Fabrication Process

The characterization of electrodeposited NiFe alloys during trench filling was performed by utilizing trenches, which were etched into a silicon substrate (Figure 3). Thermal silicon dioxide film (1 μm thick) was grown on top of a 500 μm thick silicon substrate, and trenches (100 μm deep and 500 μm wide) were anisotropically etched with a Deep Reactive Ion Etching (DRIE) process. Thin films of chromium and copper were deposited sequentially with an E-beam evaporator to obtain thicknesses of 50 nm and 200 nm, respectively (Figure 3, (A)). The chromium film serves as an adhesion layer between the silicon substrate and the copper seed layer. In order to prevent edge-overgrowth, a 40 μm thick dry film photoresist (Riston FX-940) was deposited on top of the substrate (Figure 3, (B)). Deposition of the dry film photoresist serves as a sacrificial trench top for the prevention of edge-overgrowth. The dry film photoresist was then patterned with photolithography. A 15 μm overhang of dry film photoresist was utilized to accommodate misalignment errors during the photolithography process and ensured complete coverage of the trench sidewalls. An additional benefit of employing of the dry film photoresist is the selective electrodeposition across the wafer; the dry film photoresist masks the selected regions of silicon surface preventing electrodeposition. In addition, unlike liquid photoresist, dry film photoresist has non-conformal wafer coverage and enables lithography above pre-existing features. Upon patterning of the dry film photoresist, timed electrodeposition of NiFe alloy into the trench is followed (Figure 3, (C)). Finally, lift-off of the sacrificial dry film photoresist was performed with an acetone etch to reveal surface of the silicon substrate.

RESULTS
Deposition Topography

To characterize the topography of the electrodeposited NiFe structures, the electroplating was performed in silicon trenches with and without deposition of sacrificial dry film photoresist.

Figure 4: Microscope and SEM images of the top and cross sectional views for the topography of electrodeposited NiFe structures in silicon trenches. (A)-(B) topography of electrodeposition without dry film photoresist demonstrating edge-overgrowth and (C)-(D) topography of electrodeposition that utilized the sacrificial dry film photoresist demonstrating prevention of edge-overgrowth.

Microscope and Scanning Electron Microscope (SEM) images of the deposition topographies and cross sections are shown in Figure 4. The deposition profile in the trench without the dry film photoresist (Figure 4, (A)-(B)) shows edge-overgrowth as was predicted in the FEM analysis. In contrast, the deposition profile in the trench that utilized the dry film photoresist (Figure 4, (C)-(D)) demonstrates no edge-overgrowth and a planar topography. In both cases, good trench filling was demonstrated; no voids or pinholes were observed. The thickness of the embedded NiFe structure was measured to be approximately 100 μm. In addition, the electroplated trench structure utilizing the dry film photoresist demonstrated a lower deposition rate compared with the trench deposition without the dry film photoresist even with identical deposition conditions.

Material Properties

The material composition of electrodeposited NiFe films for various applied current densities was measured (Figure 5). Previous work has shown that the composition of electrodeposited NiFe alloys can be controlled with the applied current density [7, 12]. In this work, Energy Dispersive X-ray (EDX) spectrum analysis was performed to obtain the nickel and iron concentrations of the electrodeposited NiFe films. As shown in Figure 5, a decrease of iron weight percentage was observed with respect to an increase in the applied current density. Furthermore, the figure demonstrates that the composition of permalloy ($Ni_{80}Fe_{20}$) was achieved with an applied current density of approximately 100 mA/cm^2.

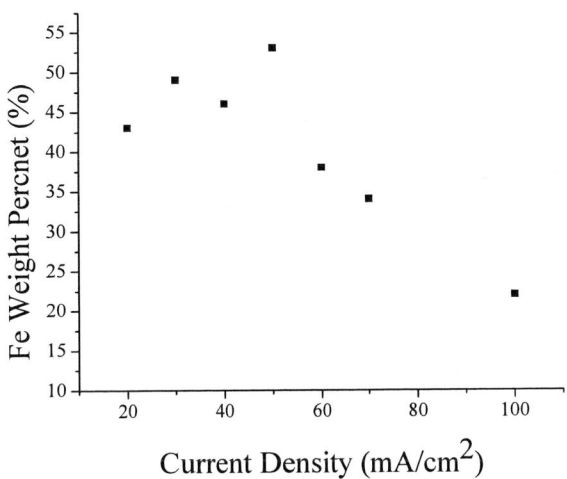

Figure 5: Weight percent of iron for various NiFe films with respect to the applied current density during electrodeposition. Permalloy composition ($Ni_{80}Fe_{20}$) was achieved with applied current density of 100 mA/cm^2.

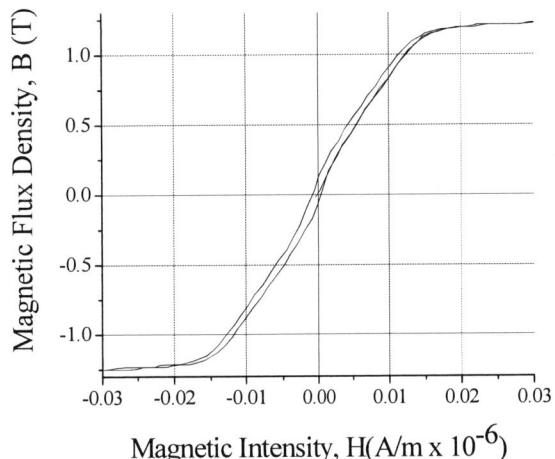

Figure 6: The B-H response of 100 μm thick electrodeposited permalloy film showing a saturation magnetic flux density of 1.23 Tesla and relative permeability as 82 of B = 0.4 Tesla.

In addition to the film composition, the magnetic properties of the 100 μm thick electrodeposited permalloy structure were characterized. The B-H response of the sample was measured and is shown in Figure 6. The saturation magnetic flux density (B_s) of the permalloy sample was measured to be approximately 1.23 Tesla. In addition, the relative permeability (μ_r) was determined as approximately 82 at magnetic flux density of 0.4 Tesla. It should be noted that the measured saturation flux density agrees with previously reported values of 1.18 Tesla for permalloy films [13].

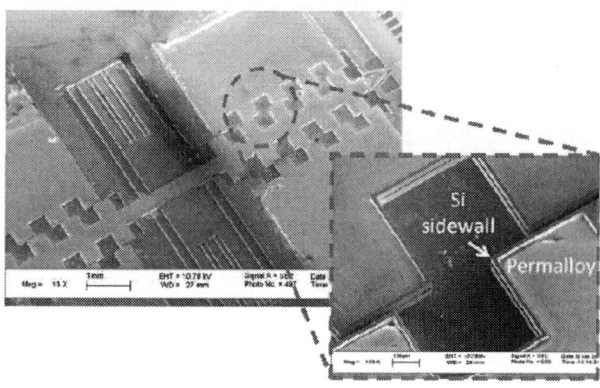

Figure 7: The electrodeposition process with the utilization of dry film photoresist as a sacrificial trench top was utilized to fabricate a MEMS electromagnetic linear actuator composed of silicon and permalloy structures. The permalloy structure was electrodeposited inside of silicon trenches without edge-overgrowth.

However, the relative permeability obtained in this work is higher than the previously reported work, which demonstrated the minimum relative permeability of 50 with permalloy films thickness of 70 µm. [13]. The contrasting values are affected by the variation in the reference flux used during the B-H measurement and the stress properties of the films.

Device Fabrication

The electrodeposition process described above was utilized to fabricate a MEMS electromagnetic linear actuator (Figure 7). The integration of post processing steps that include photolithography and DRIE were utilized to define features in the silicon mold. More specifically, the mechanical flexures, armature and stator of the MEMS electromagnetic linear actuator were defined. In addition, a releasing step was used to free the composite silicon and permalloy device. The deposition profile of the permalloy structure with a thickness of 100 µm showed no edge-overgrowth leading to a planar topography that enables photolithography without post-etchback steps. In addition, the release of closely-packed features was accomplished to create an actuatable, free-standing electromagnetic device.

CONCLUSION

A unique electrodeposition method to embed NiFe alloy structures in silicon was developed through the utilization of dry film photoresist as a sacrificial trench top. The dry film photoresist served as a temporary mask, which prevents high deposition rate at the trench edges. This process demonstrated the prevention of edge-overgrowth; eliminating the need for post planarization or etchback steps. The material composition of NiFe alloys with respect to applied current density was investigated, and permalloy composition was obtained with a current density of 100 mA/cm^2. In addition, the B-H response of permalloy structures (100 µm thick) was measured and showed a saturation magnetic flux density and relative permeability of 1.23 Tesla and 82, respectively. Furthermore, the feasibility of this process to create free-standing MEMS structures was demonstrated through the

fabrication of a composite MEMS electromagnetic linear actuator. The prevention of edge-overgrowth enabled planar topographies, which are necessary for photolithography used to define structural features.

REFERENCES

[1] N.V. Myung, D.-Y. Park, B.-Y. Yoo, and T.A. Paulo, "Development of electroplated magnetic materials for MEMS," *Magnetism and Magnetic Materials*, vol. 265, 2003, pp. 189-198.

[2] Z. Cui, X. Wang, Y. Li, and G.Y. Tian, "Fabrication of micromirror-based magnetic sensor," *J. Phys.: Conf. Ser*, vol. 76, 2007.

[3] C.T. Pan, and S.C. Shen, "Magnetically actuated bi-directional microactuators with permalloy and Fe/Pt hard magnet," *J. Magnetism and Magnetic Materials*, vol. 285, no. 3, 2005, pp. 422-432.

[4] E.J. O'Sullivan, C. E.I., L.T. Romankiw, K.T. Kwietniak, P.L. Trouilloud, J. JHorkans, C.V. Jahnes, I.V. Babich, S. Krongelb, H. S.G., J.A. Tornello, N.C. LaBianca, J.M. Cotte, and T.J. Chainer, "Integrated variable-reluctance magnetic minimotor," *IBM J. Research and Development*, vol. 42, no. 5, 1998.

[5] R.M. Bozorth, *Ferromagnetism*, Wiley-IEEE Press 1993.

[6] J.-M. Quemper, S. Nicolas, J.P. Gilles, J.P. Grandchamp, A. Bosseboeuf, T. Bourouina, and E. Dufour-Gergam, "Permalloy electroplating through photoresist molds," *Sensors and Actuators A: Physical*, vol. 74, no. 1-3, 1999, pp. 1-4.

[7] D.G. Jones, and A.P. Pisano, "Fabrication of ultra thick ferromagnetic structures in silicon," *ASME International Mechanical Engineering Congress and Exposition*, Anaheim, November 13-19, 2004.

[8] J.W. Judy, "Batch-fabricated ferromagnetic microactuators with silicon flexures," Mechanical Engineering, Univerisity of California, Berkeley, 1996.

[9] H.V. Venkatasetty, "Electrodeposition of thin magnetic permalloy films," *J. Electrochemical Society*, vol. 117, no. 3, 1970, pp. 403-407.

[10] X. Li, T. Abe, Y. Liu, and M. Esashi, "Fabrication of high-density electrical feed-throughs by deep-reactive-ion etching of pyrex glass," *J. Microelectromech. Syst.*, vol. 11, no. 6, 2002, pp. 625-630.

[11] T. Abe, X. Lia, and M. Esashi, "Endpoint detectable plating through femtosecond laser drilled glass wafers for electrical interconnections," *Sensors & Actuators A-Physical*, vol. 108, no. 1-3, 2003, pp. 234-238.

[12] S.D. Leith, and D.T. Schwartz, "In-situ fabrication of sacrificial layers in electrodeposited NiFe microstructures," *J. Micromech. Microeng.*, vol. 9, 1999, pp. 97-104.

[13] M.C. Wurz, D. Dinulovic, and H.H. Gatzen, "Investigation of permeability on electroplated and sputtered permalloy," *206ndMeeting of The Electrochemical Society*, Honolulu, October 3-8, 2004.

978-1-4244-2977-6/09 $25.00 © 2009 IEEE

FABRICATION OF BIODEGRADABLE MICROCAPSULES UTILIZING 3D, SURFACE MODIFIED PDMS MICROFLUIDIC DEVICES

C.Y. Liao and Y.C. Su

Department of Engineering and System Science

National Tsing Hua University, Taiwan

ABSTRACT

We have successfully demonstrated the fabrication of biodegradable and magnetic microcapsules utilizing PDMS double emulsification devices. Specially designed 3D microfluidic channels with surface modified by a self-aligned photo-grafting process are employed to shape immiscible fluids into monodisperse W/O/W emulsions. By varying the outer and inner fluid flow-rates, the overall and core sizes of the resulting double emulsions can be adjusted accordingly. Biodegradable materials and surface-treated γ-Fe_2O_3 nanoparticles are dispersed uniformly in the middle organic phase, and solidified into magnetic microcapsules once the solvent is extracted. In the prototype demonstration, microcapsules made up of poly(L-lactic acid), trilaurin and phosphocholine (DOPC) are successfully fabricated. As such, the proposed PDMS micro-devices could potentially serve as versatile encapsulation tools, which are desired for a variety of biological and pharmaceutical applications.

INTRODUCTION

Double emulsions are structured fluids that consist of emulsion drops with smaller droplets inside. With the employment of a middle fluid, the inner fluid is completely separated from the outer continuous-phase fluid. Compared to simple emulsions, double emulsions are inherently more difficult to form and stabilize. Conventionally, two-step emulsification processes, in which the inner droplets are emulsified in the middle fluid first and the mixture is then emulsified in the outer continuous-phase fluid, are employed for the massive production of double emulsions. However, the results are usually ill-controlled in both size and structure, owing to the polydispersity resulting from the processes. Monodisperse emulsions are highly desirable for a variety of applications. Researchers have already demonstrated double emulsification in various microfluidic devices utilizing two-step droplet breakup [1, 2] and 3D flow focusing [3, 4]. The two-step breakup approaches employ surface modification to definite the required hydrophobic and hydrophilic patterns, while the flow focusing schemes utilize manual assembly to construct the desired 3D structures. In general, emulsions with <10% size variation can be readily produced by these approaches.

Because of the great application potential, considerable attention has been attracted to the development of reliable and controllable fabrication schemes for biodegradable microcapsules [5]. This paper presents an easy-to-fabricate PDMS micro-device that employs specially patterned 3D microfluidic channels to generate W/O/W emulsions. Biodegradable materials and surface-treated γ-Fe_2O_3 nanoparticles are dispersed uniformly in the middle organic phase, and solidified into magnetic microcapsules once the solvent is extracted. As such, the need of improved applicability and process compatibility is addressed, and biodegradable microcapsules can be formed in a reliable and controllable manner.

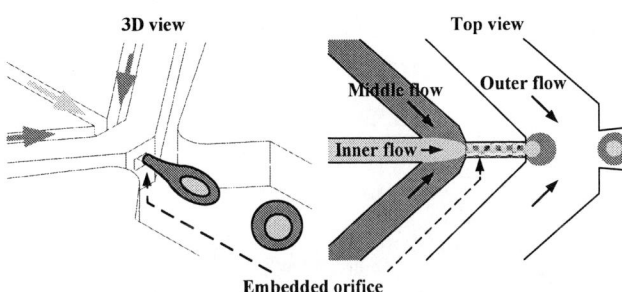

Figure 1: Schematic illustration of the proposed 3D microfluidic channels for double emulsification.

OPERATING PRINCIPLE

Figure 1 illustrates the proposed emulsification device, which is made up of two adjacent focusing junctions with one embedded orifice connecting in between. The employed inner, middle and outer fluids are driven by 3 independently controlled syringe pumps operating at varying flow rates. Before entering the embedded orifice, the inner water-phase flow is intersected and focused by the middle organic-phase flow at the upstream junction. While flowing inside the embedded orifice, the inner water-phase flow is either shaped into a thread or broken into a series of droplets by the surrounding organic-phase flow, depending mainly on the flow rates and viscosities of the fluids involved. Once coming out of the embedded orifice, the water-in-organic coaxial flow is intersected and focused by the outer water-phase flow, without touching the downstream channel surfaces. Ideally, with the assistance of viscous and/or inertial forces that act against interfacial tension, the water-in-organic coaxial flow could be broken into drops with water-in-organic-in-water structure.

It is crucial to include the embedded orifice as part of the 3D focusing mechanism, which could significantly reduce the undesired surface distraction. For hydrophobic PDMS channels, it is most likely that the middle organic-phase fluid, whose interfacial tension with PDMS is much lower than that of the outer water-phase fluid, would attach itself to the channel surfaces. The embedded orifice (located in the sidewall separating the two adjacent focusing junctions) is utilized to restrict the water-in-organic coaxial flow away from the surfaces of the downstream channel. This could potentially prevent the attachment, but it is quit possible that some unexpected disturbance (especially during the starting period) might still lead to the undesired attachment and halt the emulsification process immediately. To further reduce the undesired surface distraction, the downstream channel surfaces are selectively modified into hydrophobic. In addition, surfactants and osmotic agents are employed to

978-1-4244-2977-6/09 $25.00 © 2009 IEEE

facilitate the breakup process and stabilize the resulting emulsion structures. Meanwhile, biodegradable materials and surface-treated γ-Fe$_2$O$_3$ nanoparticles are dispersed uniformly in the middle organic phase, and solidified into magnetic microcapsules once the solvent is extracted. As such, biodegradable and magnetic microcapsules can be fabricated in a reliable and controllable manner.

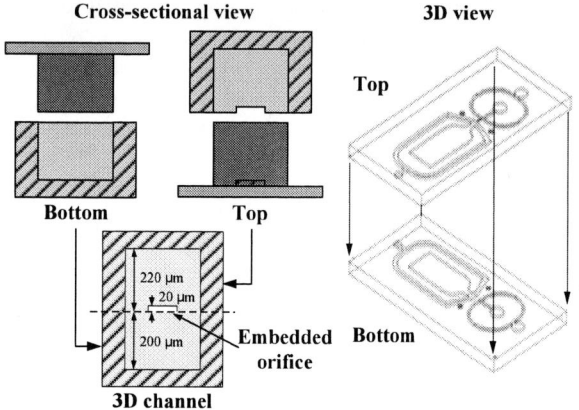

Figure 2: Fabrication process of the 3D micro-channels.

FABRICATION PROCESSES

In the prototype demonstration, two-level photo-lithography, PDMS molding and irreversible bonding processes are employed to fabricate the proposed 3D microchannels, as illustrated in Figure 2. First of all, a layer of 200 μm thick negative photo-resist (SU-8, Micro-Chem) is spin-coated and patterned on top of a clean silicon wafer to fabricate the mold used for duplicating the bottom layer of PDMS microchannels. Meanwhile, the mold used for duplicating the top PDMS layer, which has microchannels of two different depths on it, is fabricated by coating and patterning a 20 μm thick SU-8 layer first, followed by a second SU-8 layer of 200 μm in thickness. A mixture of 10:1 PDMS pre-polymer and curing agent (Sylgard 184, Dow-Corning) is stirred thoroughly and then degassed under vacuum to remove trapped air bubbles. The pre-polymer and curing agent mixture is then poured onto the molds, degassed, and cured for 2 hours at 85°C. After thoroughly cured, the PDMS replicas are peeled off from the molds. The top PDMS layer is punched through with a sharp metal-tube array to fabricate the connecting holes for 3 inlets and 1 outlet, and ultrasonically cleaned in an ethanol bath to remove residual debris from its surface. The surfaces of the two PDMS layers are then treated with a hand-held corona treater (BD-20AC, Electro-Technic Products), which ionizes surrounding air and creates localized plasma to activate the surfaces for irreversible bonding. The intensity of the corona is set at a relatively low level in order to produce a stable but soft corona with minimal crackling and sparking. The corona-treated surfaces are then pressed together and left undisturbed for at least one hour at 85°C for the bonding to take effect. At the end, PTFE tubes are inserted into the punched holes to build the necessary interconnection for sample injection and discharge.

Once the 3D PDMS micro-channels are fabricated, the downstream channel surfaces are selectively modified into hydrophobic utilizing a self-aligned photo-grafting process.

As illustrated in Figure 3(a), ethyl acetate (mixed with oleic acid to adjust its viscosity) is injected into the upstream channel, while its front automatically stopped at the exit of the embedded orifice by surface tension. In this scheme, ethyl acetate functions as a mask to protect the upstream channel surfaces from photo-grafting reaction. Afterward, a grafting solution containing 0.34 g of benzophenone (photo-initiator), 0.32 g of Pluronic F-68 (to improve the solubility of benzophenone in aqueous phase), 15 g of acrylic acid (monomer), 27.6 g of tert-butanol (solvent) and 43.6 g of water is prepared and injected into the downstream channel. Since the grafting solution has a higher interfacial energy with PDMS than ethyl acetate does, it is not flowing further into the embedded orifice. The 3D PDMS channels are then exposed with a UV lamp ($\lambda = 253.7$ nm) for about 10 minutes to activate the photo-grafting reaction. As a result, the downstream channel surfaces (e.g. the area with a darker color in Figure 3(c)) are converted into hydrophilic. After the photo-reaction, the channels are washed with ethanol and then DI-water to remove the residue ethyl acetate and the grafting solution.

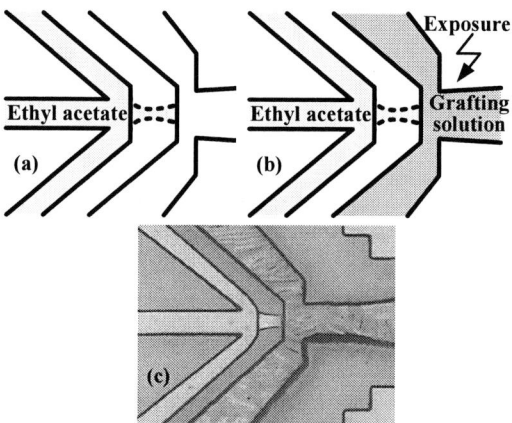

Figure 3(a) and (b): Schematic of the self-aligned photo-grafting process and (c): Photo of a modified PDMS device.

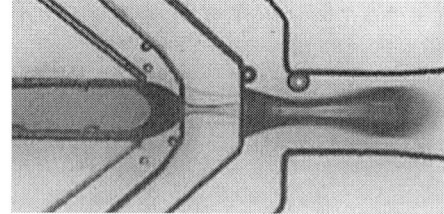

Figure 4: A captured one-step double emulsification sequence.

EXPERIMENTAL AND RESULTS

Water-in-organic-in-water emulsions are prepared using ethyl acetate with 5 wt% of Span 85 as the middle organic-phase fluid. In addition, less than 10 wt% of biodegradable and magnetic materials are dispersed in the organic-phase. Meanwhile, de-ionized water is used as the inner and outer water-phase fluid, with 30 wt% of glycerol and 5 wt% of Tween 20 dissolved within, respectively. The outer water-phase fluid is injected into the device first, followed by the injection of the middle organic-phase fluid.

978-1-4244-2977-6/09 $25.00 © 2009 IEEE

Meanwhile, the flow-rates of the two fluids are increased slowly, until stable organic-in-water emulsification is reached. Finally, the inner water-phase fluid is injected into the device, with a slowly increasing flow-rate. As such, stable emulsification can be reached eventually and the flow-rates are adjusted afterward to characterize their effects on emulsification behavior.

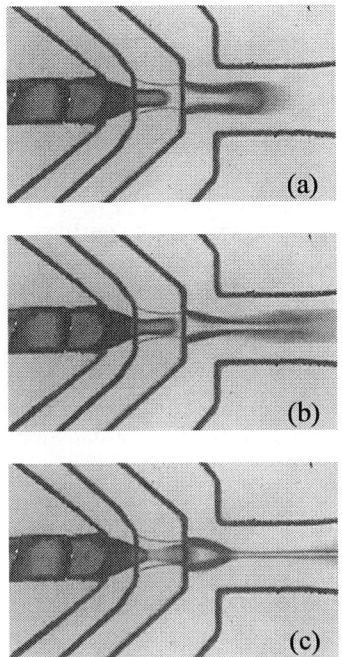

Figure 5: A captured two-step double emulsification sequence.

In the emulsification trials, dripping is intentionally induced to produce monodisperse double emulsions with either one-step (Figure 4) or two-step (Figure 5) breakup. For low viscosity organic solutions (e.g. using ethyl acetate as the solvent), the breakup is more likely to be two-step, in which the inner fluid is broken inside the embedded orifice and the middle fluid (with a droplet in it) is broken around the exit of the orifice. It is found that by varying the flow-rates of the three involved fluids (Q_i, Q_m, and Q_o), the overall (D_m) and core (D_i) diameters of the resulting double emulsions can be adjusted. In general, the overall diameter (D_m) is mainly determined by the outer fluid flow-rate (Q_o). As illustrated in Figure 6(a), the overall diameter (D_m) decreases in case that a higher outer fluid flow-rate (Q_o) is employed, while the core diameter (D_i) and breakup period decrease as well. While the outer fluid flow-rate (Q_o) is varied from 0 to 4 μl/min, it results in overall diameter (D_m) of 350 to 200 μm. Further increase in the outer fluid flow-rate (Q_o) can eventually switch the breakup into jetting mode, which usually leads to polydisperse results and is beyond the scope of this work. Meanwhile, the core diameter (D_i) increases when the inner fluid flow-rate (Q_o) rises, as illustrated in Figure 6(b). Even though the break-up period decreases when the inner fluid flow-rate (Q_i) rises, the core diameter (D_i) rises from 90 to 120 μm. Further increase in the inner fluid flow-rate (Q_i) will result in the failure of double emulsification, while a high-speed water jet bursts the middle organic-phase flow. Therefore, it is recommended that the inner fluid flow-rate (Q_i) should be adjusted from low to high at the last, after stable

organic-in-water emulsification with the desired drop size (D_m) is roughly achieved.

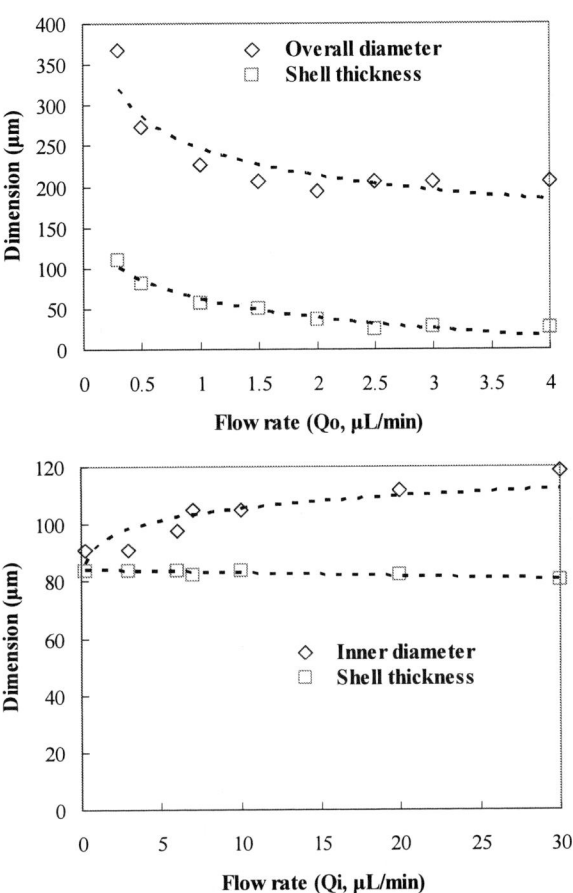

Figure 6: (a) Relationship between emulsion sizes and outer fluid flow-rate and (b) Relationship between emulsion sizes and inner fluid flow-rate.

At the end, ethyl acetate (organic solvent) is extracted from the organic phase and the dissolved biodegradable and magnetic materials therefore solidify into a solid capsule. The solubility of ethyl acetate in water is about 8 wt%, and the extraction typically takes about minutes to complete. In the prototype demonstration, microcapsules made up of poly(L-lactic acid), trilaurin and phospho-choline (DOPC) are successfully fabricated. Figure 7 shows the captured solidification sequences of poly(L-lactic acid) capsules with and without aqueous droplet encapsulated inside. The overall drop size decreases over time while the color turns dark due to the change in optical properties. Roughly 85% (or more) of the volume is lost because of the extraction of solvent. In case that a magnetic field is applied externally, the fabricated capsule is attracted to the sidewall and rotated with its axis aligned to the field as shown in Figure 8. Figure 9(a) shows the SEM micrograph of a fabricated poly(L-lactic acid) micro-capsule with embedded magnetic nanoparticles clearly observed on the surface. The embedded γ-Fe$_2$O$_3$ nano-particles, which have been widely used experimentally for numerous biomedical applications, can potentially function as targeting and heating elements. For example, the magnetic capsules can be targeted to some specific sites with the guidance of an external field, and remotely energized by the field to melt and accelerate the release from the capsules. In addition to poly(L-lactic acid), phosphocholine (DOPC) is also tested in the trials to form lipid microvesicles as shown in Figure 9(b).

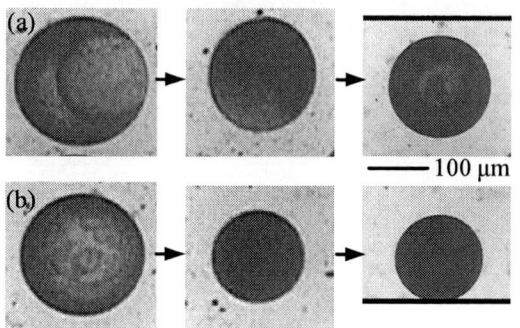

Figure 7: Captured solidification sequences (a) with aqueous droplet encapsulated and (b) without drop encapsulated.

Figure 8: Dynamic response of a magnetic microcapsule to an external magnetic field.

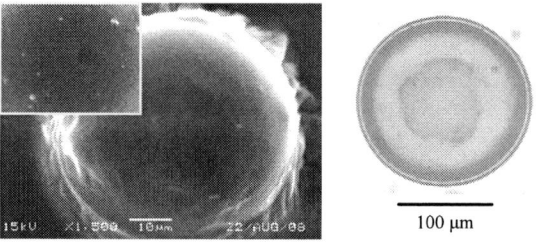

Figure 9: (a) SEM micrograph of a magnetic PLA microcapsule and (b) photograph of a DOPC micro-vesicle.

CONCLUSION

The paper presents the fabrication of biodegradable and magnetic microcapsules utilizing PDMS emulsification devices. Specially designed 3D microfluidic channels with surface modified by a self-aligned photografting process are employed to shape immiscible fluids into monodisperse W/O/W emulsions. By varying the outer and inner fluid flow-rates, the overall and core sizes of the resulting double emulsions can be adjusted accordingly. Biodegradable materials and surface-treated γ-Fe$_2$O$_3$ nanoparticles are dispersed uniformly in the middle organic phase, and solidified once the solvent is extracted to form magnetic microcapsules. The embedment of γ-Fe$_2$O$_3$ nanoparticles can potentially enable the functions of targeting and controlled release. In the prototype demonstration, microcapsules made up of poly(L-lactic acid), trilaurin, and phosphocholine (DOPC) are successfully fabricated. As such, the proposed PDMS microfluidic devices can serve as low-cost and versatile encapsulation tools, which are desired for a variety of biological and pharmaceutical applications.

ACKNOWLEDGEMENTS

This work was supported in part by the National Science Council of Taiwan under Contract No. NSC 96-2221-E-007-116-MY3. The demonstrated systems were fabricated in the ESS Microfabrication Lab. at National Tsing Hua University, Taiwan.

REFERENCES

[1] S. Okushima, T. Nisisako, T. Torii and T. Higuchi, "Controlled Production of Monodisperse Double Emulsions by Two-Step Droplet Breakup in Microfluidic Devices", *Langmuir,* vol. 20, pp. 9905-9908, 2004.

[2] M. Seo, C. Paquet, Z. Nie, S. Xu and E. Kumacheva, "Microfluidic Consecutive Flow-Focusing Droplet Generators", *Soft Matter*, vol. 3, pp. 986-992, 2007

[3] R. Bocanegra, J. Sampedro, A. Ganan-Calvo and M. Marquez, "Monodisperse Structured Multi-Vesicle Microencapsulation Using Flow-Focusing and Controlled Disturbance", *J. Microencapsul.*, vol. 22, pp. 745-759, 2005

[4] A. Utada, E. Lorenceau, D. Link, P. Kaplan, H. Stone and D. Weitz, "Monodisperse Double Emulsions Generated from a Microcapillary Device", *Science*, vol. 308, pp. 537-41, 2005

[5] C. Berkland, E. Pollauf, D. Pack and K. Kim, "Uniform Double-Walled Polymer Microspheres of Controllable Shell Thickness", *J. Control Release*, vol. 96, pp. 101-11, 2004

NOVEL TECHNOLOGY FOR CAPACITIVE PRESSURE SENSORS WITH MONOCRYSTALLINE SILICON MEMBRANES

K. Knese[1], S. Armbruster[1], H. Weber[1], M. Fischer[1], H. Benzel[1], M. Metz[1] and H. Seidel[2]

[1]Robert Bosch GmbH – Engineering-Sensor Technology Center, Reutlingen, GERMANY
[2]Chair of Micromechanics, Microfluidics/Microactuators, Universität des Saarlandes, GERMANY

ABSTRACT

We report on a novel surface micromachining technology for the fabrication of capacitive absolute pressure sensors. The pressure sensitive membrane is formed by single crystal silicon enabling excellent long term stability. The membrane formation is based on the Advanced Porous Silicon Membrane (APSM) process [1], which is currently applied to piezoresistive transducers. Expanding this technology to capacitive transduction allows for a greater flexibility in tailoring the sensor properties to specific applications [2]. This expansion is implemented by adding a poly-Si counter electrode layer on top of the membrane in a surface micromachining step. Since only front side processing on standard silicon substrates is used, this method is very cost-efficient and fully CMOS-compatible, enabling monolithic integration of circuitry.

INTRODUCTION

A widely-used approach for the fabrication of capacitive MEMS pressure sensors is surface micromachining (SMM) [3]. Typically, on top of a silicon substrate a sacrificial silicon dioxide film is etched from underneath a polysilicon layer creating a cavity. SMM processes allow for fully CMOS compatible, low cost capacitive pressure sensors, since only front side processing is used. On the other hand, the formation of stress-free polysilicon is difficult [4]. Furthermore, the necessary subsequent sealing of the cavity typically done with dielectric layers is a possible weak spot regarding long term stability and leak tightness.

A second bulk micromachining approach uses continuous single crystal silicon membranes as pressure sensitive element leading to higher mechanical stability. The diaphragm is formed by a shallow KOH etch on the front surface and a deep etch from the backside. The surface side is then attached to a glass plate by electrostatic bonding. The latter is provided with a thin metal film serving as counter electrode [3]. Due to the backside processing, large required space and additional need of a glass wafer bulk micromachined pressure sensors are quite cost-intensive.

Recently, a concept for a capacitive pressure sensor with a boss membrane using the APSM technology was reported [5]. The APSM process enables fabrication of self-sealing single crystal membranes above a vacuum cavity. The novel concept presented in this paper combines the benefits of both approaches mentioned above. First, it provides a continuous single crystal pressure sensitive membrane taking advantage of its unique properties like high strength, high reliability and no mechanical hysteresis. Second, this membrane as well as the polycrystalline counter electrode added on top of it is formed by only using front side and fully CMOS-compatible process steps. Third, since the dimensions of the electrodes are confined by an etched trench instead of a *pn*-junction and

the electrodes are isolated by a dielectric layer, the sensor is designed for stable temperature behaviour.

SENSOR CONCEPT

A schematic view of the sensor concept is displayed in Figure 1a). The sensing element consists of a single crystal silicon membrane covering a vacuum cavity which serves as pressure sensitive lower electrode. To realize the capacitive measuring principle a perforated polysilicon counter electrode is added on top of the membrane. A thin silicon oxide layer separates and electrically isolates the electrodes. To define the lateral confinement of the counter electrode and thereby reduce the parasitic capacitance of the sensor, an additional trench is etched around the membrane region. The vent-holes allow the applied pressure to penetrate the polysilicon electrode and deflect the single crystal membrane. With a proper design of the vent holes the perforated polysilicon electrode is not affected by the pressure. Therefore the deflection results in a variation of the electrode gap. This can be detected electrically as a change of capacitance ΔC.

To reduce temperature dependence and parasitic capacitances, an additional reference element is used [6]. This is identical to the sensing element except for the cavity (Figure 1b)). Hence, in case of the reference element

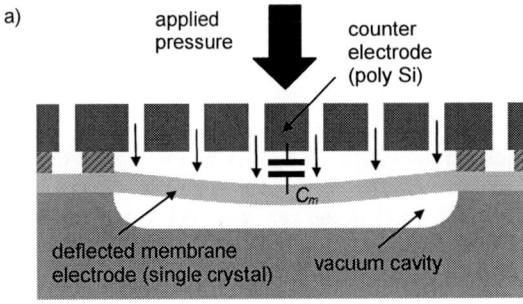

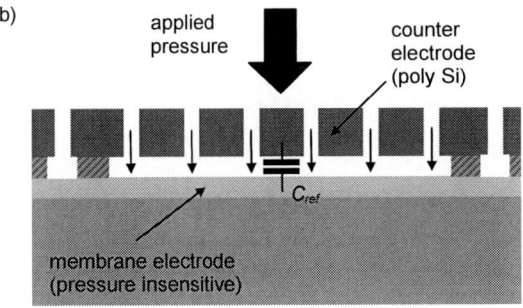

Figure 1: Functional principle of capacitive transducer: a) sensing element: applied pressure deflects single crystal membrane covering a vacuum cavity, b) reference element without cavity providing a pressure insensitive membrane electrode.

978-1-4244-2977-6/09 $25.00 © 2009 IEEE

the single crystal lower electrode is pressure insensitive. Therefore, by taking the difference between the two capacitances, most temperature and parasitic effects can be eliminated.

FABRICATION

The fabrication process of the capacitive sensor is described by means of schematic cross-sectional illustrations (Figure 2). It consists of the APSM process [1] for the single crystal membrane formation (Figure 2a)-c)) and some further SMM process steps for the implementation of the counter electrode (Figure 2d)-f)).

Process flow

The core of the APSM process is the so-called anodization, during which the silicon substrate is locally porosified by electrochemical etching in concentrated hydrofluoric acid (HF). Starting material is a moderately p-doped silicon substrate with (100)-orientation.

First, a deep n^+-doping and a shallow p^+-doping are implanted to the substrate. The n^+-doping, which is not porosified during the anodization [7], serves as lateral confinement of the membrane region into the bulk substrate. It is designed as quadratic frame enclosing a shallowly p^+-doped area (cf. Figure 2a). The p^+-doping subsequently serves as single crystal seed layer for the epitaxial growth of silicon. Additionally, a silicon nitride layer is deposited onto the wafer to avoid porous etching of the substrate surface. The subsequent anodic etching in concentrated HF is performed in a two step process. First, the p^+-Si layer is etched using a low current density resulting in a mesoporous silicon layer. Second, the subjacent p-Si is anodized with an increased current density leading to a nanoporous silicon layer buried under the mesoporous layer (Figure 2a). Afterwards, the silicon nitride layer is removed in HF solution. In a CVD reactor the substrate is then exposed to a hydrogen atmosphere. At temperatures between 900°C and 1100°C the hydrogen desorbs native oxide residuals from the porous silicon surface. As soon as the oxide is removed, both porous silicon layers start to rearrange driven by the minimization of the surface energy: the surface pores in the mesoporous layer are sealed, while the buried nanoprous silicon starts to merge into one single cavity (Figure 2b).

In the same CVD reactor, an epitaxial Si-layer is grown, using the sealed mesoporous silicon as a seed layer. Simultaneously, any porous silicon has completely rearranged leaving one single cavity (Figure 2c). The thickness of the epitaxial layer (~10 µm) defines the membrane thickness. The hydrogen enclosed inside the cavity diffuses through the silicon during subsequent high temperature processes leaving a vacuum.

This membrane constitutes the pressure sensitive first electrode of the capacitive transducer. The pressure insensitive counter electrode is implemented as follows.

First, a thin (~1 µm) silicon oxide film is deposited onto the Si-epitaxy. It serves as sacrificial layer and defines the electrode gap. In the contact area of the membrane the oxide is removed by wet etchig (Figure 2d). In a CVD reactor, a second silicon layer of ~10 µm is subsequently deposited. In the etched area the Si-layer grows as a single crystal onto the substrate preparing the membrane contact. At the same time, the deposition on the sacrificial

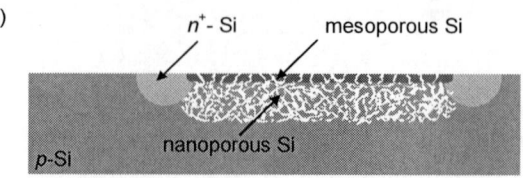

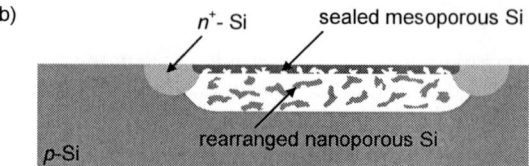

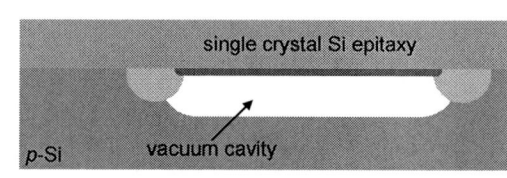

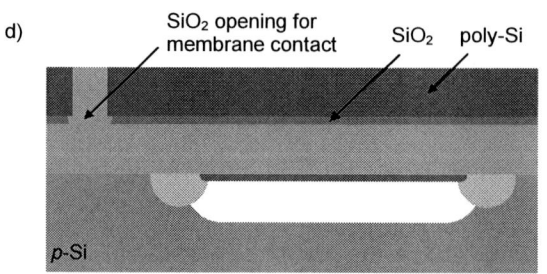

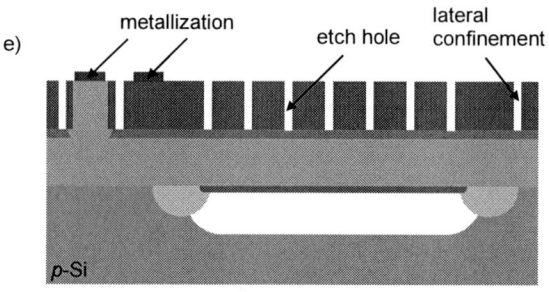

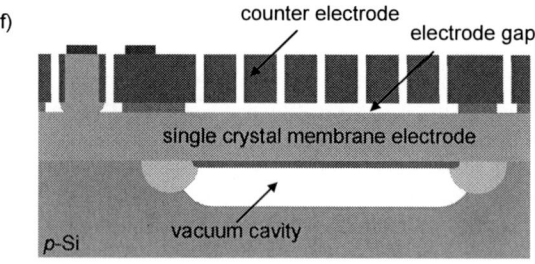

Figure 2: Schematic process flow: a) porous etching of mesoporous seed layer and nanoporous cavity layer, b) thermal rearrangement of porous silicon during hydrogen prebake, c) sealing of cavity by lateral overgrowth during Si epitaxy, d) deposition and structuring of sacrificial layer (SiO₂) and deposition of poly-Si (counter electrode), e) metallization and perforation of poly-Si by DRIE, f) removal of sacrificial layer by HF vapor etch and thereby releasing membrane from counter electrode.

oxide results in a polysilicon layer, which constitutes the pressure insensitive counter electrode (Figure 2d)).

After depositing and structuring an Al metallization layer for the electrical connection of the electrodes, the membrane region of the polysilicon layer is provided with access holes by DRIE (Deep Reactive Ion Etch) (Figure 2e)). The dimensions of the holes (2-3 μm diameter and 15-19 μm distance) result from the trade-off between maximizing the sensing capacitor area and minimizing the duration of the sacrificial layer etching. With the same DRIE step the trenches for separating the membrane contact from the poly-Si layer and confining the counter electrode are etched. By means of HF vapor etching through the vent holes the sacrificial silicon oxide is removed in the membrane region (Figure 2f)) and the electrodes are released.

Results

Figure 3 pictures cross-sectional SEM micrographs of the cavity and membrane region at certain process steps. Figure 3a) shows the cavity region after anodization. In the whole membrane region a double porous stack with an upper mesoporous layer and a buried nanoporous layer were generated.

Figure 3b) displays the cavity region after hydrogen prebake. Driven by the minimization of the free surface energy, both porous silicon layers start to rearrange. The buried nanoporous silicon shows a coarsened morphology, while the upper mesoporous layer is sealed providing the seed layer for the epitaxial growth. Since the rearrangement of porous silicon is solely thermally activated, it continues with each high temperature process, even in the sealed cavity, until a single cavity is formed.

Figure 3c) shows the membrane and cavity region after the deposition of an approximately 10 μm thick Si-membrane. The porous silicon has now completely rearranged creating a single vacuum cavity. The Si epitaxy is grown onto the sealed mesoporous silicon. Since the mesoporous silicon layer and thereby the Si growth is perfectly single crystalline, a high quality single crystal silicon layer is formed. The mesoporous layer is part of the membrane and has been completely smoothed by thermal rearrangement.

Figure 3d) shows the final membrane stack above a vacuum cavity after HF vapor etching. The silicon dioxide layer has been fully undercut and removed, resulting in two freestanding membranes: a continuous lower silicon membrane and a perforated upper membrane. The latter shows the typical polycrystalline structure, in contrast to the perfectly single crystal lower membrane. The cut through the vent-hole shows the characteristic periodic profile of the DRIE. The gap between the membranes corresponds to the sacrificial oxide thickness of 1 μm.

Top views of the sensing element are given in Figures 4. Figure 4a) shows the single crystal lower membrane before the sacrificial layer deposition. Membrane electrode and contact area are connected by an n^+-doped conductive path. A p^+-guard ring surrounds the whole membrane and contact region confining the lower electrode by a pn-junction. Due to the vacuum inside the cavity the single crystal membrane is slightly bent under atmospheric pressure. This is visualised as contrast given by a DIC (Differential Interference Contrast) microscope,

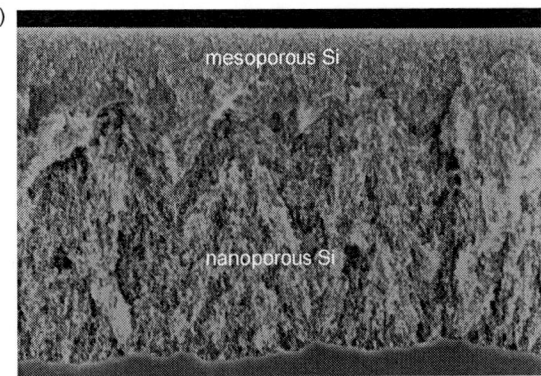

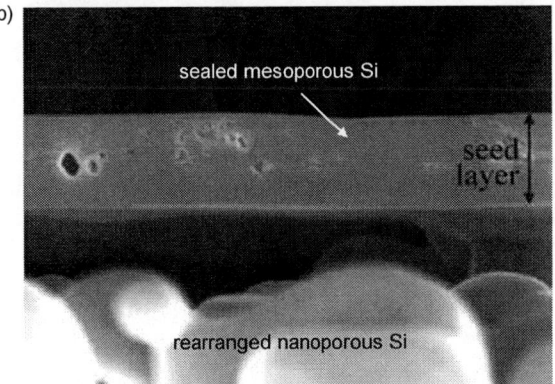

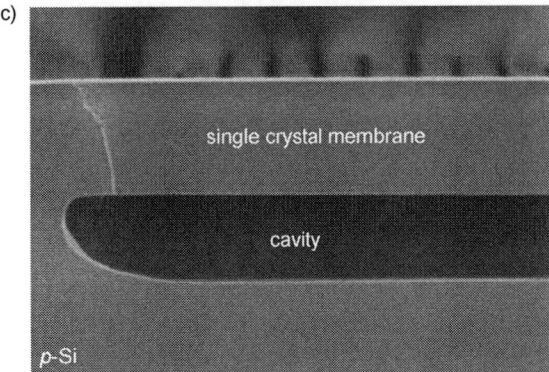

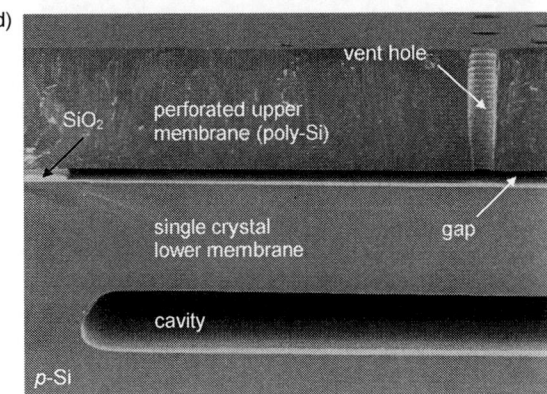

Figure 3: Cross-sectional SEM micrographs of the cavity and membrane region a) nanoporous and mesoporous layer after anodization, b) rearranged porous silicon layers after hydrogen prebake, c) single crystal membrane and cavity after epitaxial deposition, d) full membrane stack after etching the sacrificial layer.

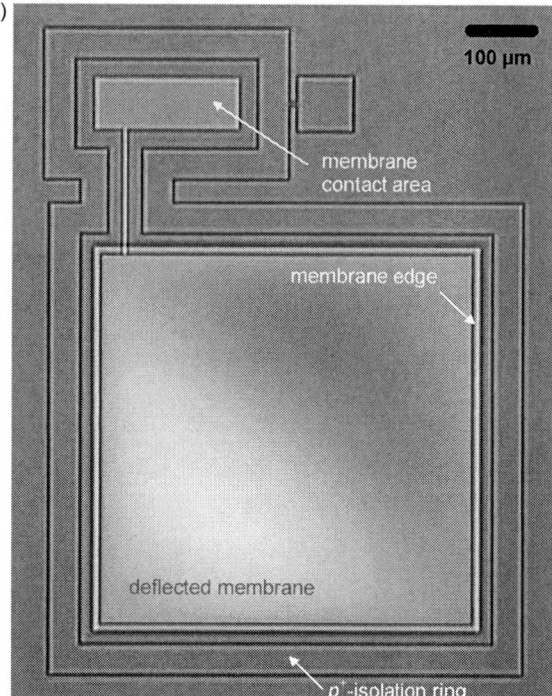

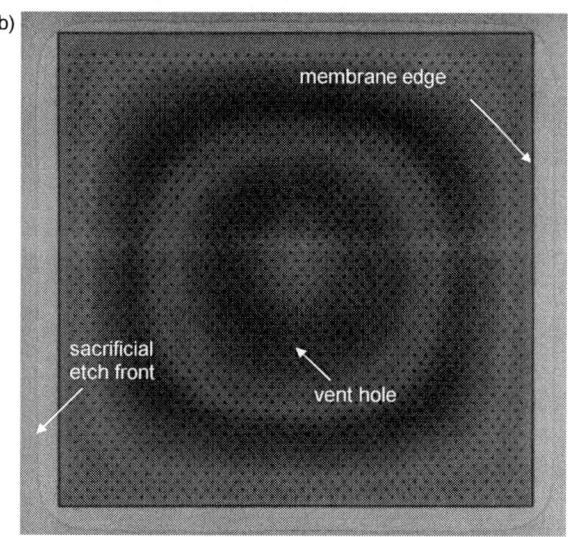

Figure 4: Top view of the sensing element: a) optical microscope view before sacrificial layer deposition, b) infrared microscope view after sacrificial layer etching.

which detects the optical path difference generated by the deflected membrane (Figure 4a)).

Figure 4b) shows an infrared microscope view of the membrane region after the full process demonstrating the relative position of the membrane stack. The polysilicon perforation is centered above the cavity region (indicated by the membrane edges). The etch front of the sacrificial oxide layer reaches the membrane edge releasing the whole single crystal membrane. The apparent Newton's Rings indicate that the capacitor plates are not coplanar.

This is attributed to the deflection of the single crystal lower membrane.

CONCLUSIONS

This paper introduces a novel technology for the fabrication of capacitive absolute pressure sensors. Single crystal silicon serves as material for the pressure sensitive membrane electrode and a perforated polysilicon layer as pressure insensitive counter electrode. The single crystal membrane is formed using the APSM process. The main process steps are: (i) local anodic etching of porous silicon in concentrated hydrofluoric acid; (ii) sintering of the porosified areas in hydrogen atmosphere at elevated temperature; (iii) epitaxial growth of the monocrystalline membrane onto rearranged porous silicon [1]. The counter electrode is added on top by deposition of a thin oxide and a thick polysilicon layer. The sacrificial oxide layer separates and electrically isolates both membranes. The vent holes allow the applied pressure to reach and deflect the single crystal membrane.

By using a self-sealing continuous single crystal membrane as pressure sensitive electrode, the sensor elements exhibit high long term stability and leak tightness. Since the thickness of the sacrificial oxide layer defines the electrode gap, a considerably smaller and well-defined gap than previously [5] can be achieved leading to higher capacitance and sensitivity. Finally, by using only CMOS compatible, front side process steps on standard silicon substrates this method is very cost-efficient and allows for the integration of circuitry.

REFERENCES

[1] S. Armbruster, F. Schäfer, G. Lammel, H. Artmann, C. Schelling, H. Benzel, S. Finkbeiner, F. Lärmer, P Ruhter, O. Paul, "A novel micromachining process for the fabrication of monocrystalline Si-membranes using porous silicon" *Proc. Transducers '03*, pp. 246-249, 2003.

[2] M.J. Madou, „*Fundamentals of Microfabrication: the science of miniaturization*", 2nd edition, CRC Press, Boca Raton, 2002.

[3] R. Puers, "Capacitive Sensors: when and how to use them", in *Sensors and Actuators A, 37-38*, 1993, pp. 93-108.

[4] Min-Hang Bao, "Handbook of Sensors and Actuators 8: Micro Mechanical Transducers", Elvesier, 2004.

[5] K. Knese, S. Armbruster, H. Benzel, H. Seidel, "Microsystems Technology in Germany 2008", in *Editor Trias Consult*, Berlin, 2008, pp. 46-47.

[6] R. Puers, E. Peeters, A. van den Bossche, W. Sansen, „A capacitive Pressure Sensor with Low Impedance Output and Active Suppression of Parasitiv Effects", in *Sensors and Actuators A, 21-23*, 1990, pp. 108-114.

[7] M.I.J. Beale, J.D. Benjamin, M.J. Uren, N.G. Jew, A.J. Cullis, "An experimental and theoretical study of the formation and microstructure of porous silicon", in *J. Cryst. Growth*, Vol. 73, 1985, pp. 622-636.

DEVELOPMENT OF MULTI-USER MULTI-CHIP SOI CMOS-MEMS PROCESSES

K. Takahashi[1], M. Mita[2], M. Nakada[1], D. Yamane[1], A. Higo[3], H. Fujita[1], and H. Toshiyoshi[1]

[1]Institute of Industrial Science, the University of Tokyo, Tokyo, JAPAN

[2]Japan Aerospace Exploration Agency, Kanagawa, JAPAN

[3]Research Center for Advanced Science and Technology, the University of Tokyo, Tokyo, JAPAN

ABSTRACT

This paper presents a new method of integrating multiple MEMS designs with 40V class CMOS driver circuits in a multi-user-multi-chip manner. The multi-chip multi-user CMOS-MEMS process was done at 35 mm x 35 mm SOI chip. More than six different designs of SOI-bulk micromachined actuators including the pitch-tunable gratings were monolithically integrated onto the pre-fabricated high-voltage level-shifter circuits. We measured electro mechanical characteristics of the grating light valve integrated with high-voltage level-shifter and successfully demonstrated 1MHz operation.

INTRODUCTION

Monolithic integration of CMOS circuitry and MEMS (CMOS-MEMS) technology improves the performance of micro sensors and actuators, programmability, miniaturization and simplicity of total device packaging. CMOS-MEMS is also expected to be an enabling micro fabrication technology that delivers a more-than-Moore solution to the existing silicon VLSI. There are sophisticated examples of integrated MEMS reported, including the Texas Instruments DMD [1] and Analog Devices ADXL accelerometers [2]; nevertheless, integration of MEMS is not an easy task for most MEMS designers due to costly fabrication process equipment as well as the set of photomasks or reticules. In a similar manner as the MCNC MUMPs [3], we believe that multi-user multi-chip CMOS-MEMS processes are needed to encourage more trail and error for exploring the new phase of integrated MEMS.

Our version of multi-user multi-chip CMOS-process is characterized by the high-voltage capability upward of 40 V [4]; high voltage is beneficial to improve the microactuator's performance such as larger displacement under given voltage and higher mechanical stiffness. To attain versatility for MEMS designers, we adopted a CMOS-first MEMS-last approach of integration. In addition, SOI bulk-micromachining enables us to make high mechanical property, accuracy, and reliability of MEMS sensors and actuators. High aspect ratio structures can be constructed by silicon deep etching without any high temperature deposition processes, which is appreciated to the CMOS-first integration. We firstly prepared a set of 40V CMOS circuits of 8-channel level-shifters and allocate multiple MEMS designs in the identical photomask shot, as illustrated in Figure 1. In this paper, we demonstrate the multi-user multi-chip CMOS-MEMS process with a CMOS-first MEMS-last manner.

MULTI-USER CHIP DESIGN

As shown in Figure 2, we designed MEMS actuator reticules to fit a 2.9 mm x 2.9 mm die site in the SOI CMOS wafer of an 8-μm-thick device layer, a 1-μm-thick buried oxide, and a 150-μm-thick handle layer. Our multi-chip integration process had total four masks; two masks were fixed design for SiO2 etching and interconnection between CMOS and MEMS, and another two were left to MEMS designers for their own structures made by the double side DRIE processes.

The photograph of the pre-fabricated IC is shown in Figure 3. The IC and blank field for MEMS area are 2.9 mm each, and a MEMS user can use 6 x 6 areas including 13

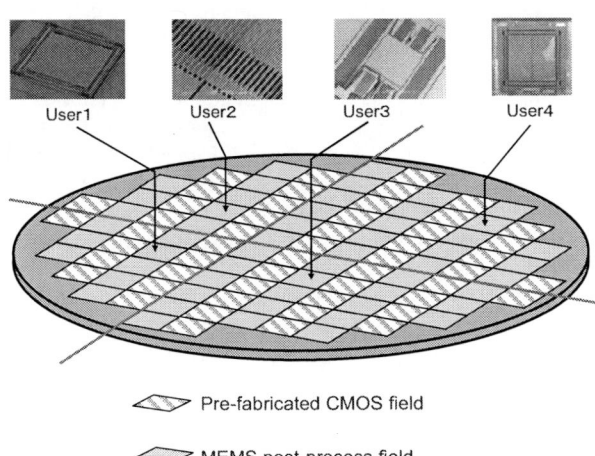

Figure 1 Schematic view of multi-user multi-chip SOI CMOS-MEMS processes.

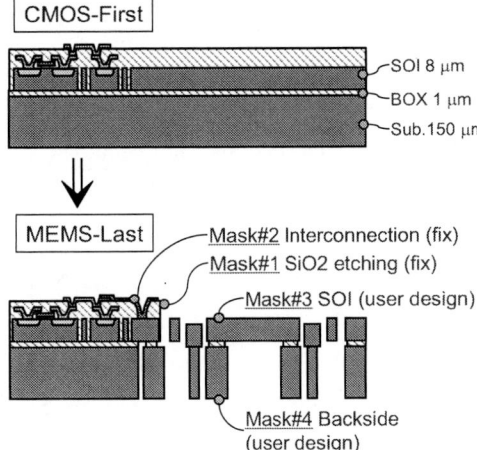

Figure 2 Layer thickness of SOI CMOS wafer and configuration of photomasks for multi-chip process.

MEMS fields. We arranged 8-channel high-voltage level-shifters per chip, which were electrically interconnected to the MEMS actuators sitting next to the field, by using thin aluminous layer.

Table I lists the design rules of the multi-chip process. The minimum gap was standardized to 2 microns in the SOI layer due to the surface roughness of the initial chip condition; and that in the substrate layer was 10 microns for deep trench etching. We decided maximum release width to 20 microns as fast as we can finish the sacrificial etching. Effective pattern area is 2.8 mm each due to the dummy pattern as dicing line at the edges of the chip. Double side alignment capability was defined to 2 microns by a mask aligner resolution.

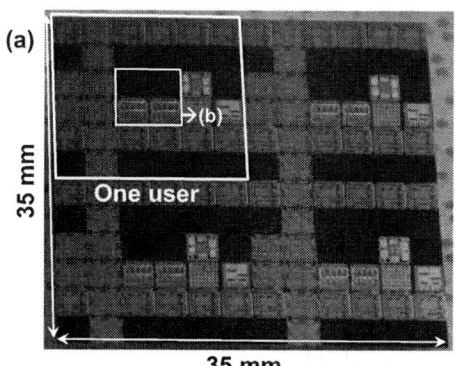

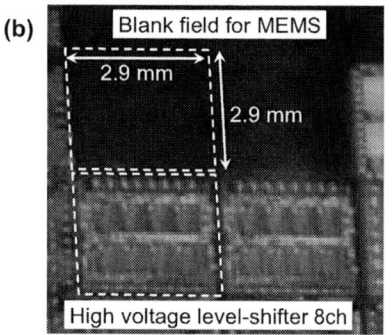

Figure 3 Chip layout of the pre-fabricated IC and blank field for MEMS

Table I Design rules of multi-chip CMOS-MEMS process

	Parameter	Value (μm)
SOI	Min. line	3.0
	Min. gap	2.0
	Max. release line	20
	Anchor size	60 x 60
	Usable area per die	2800 x 2800
Backside	Min. line	5.0
	Min. gap	10
	Dicing line	30
	Alignment capability	2.0

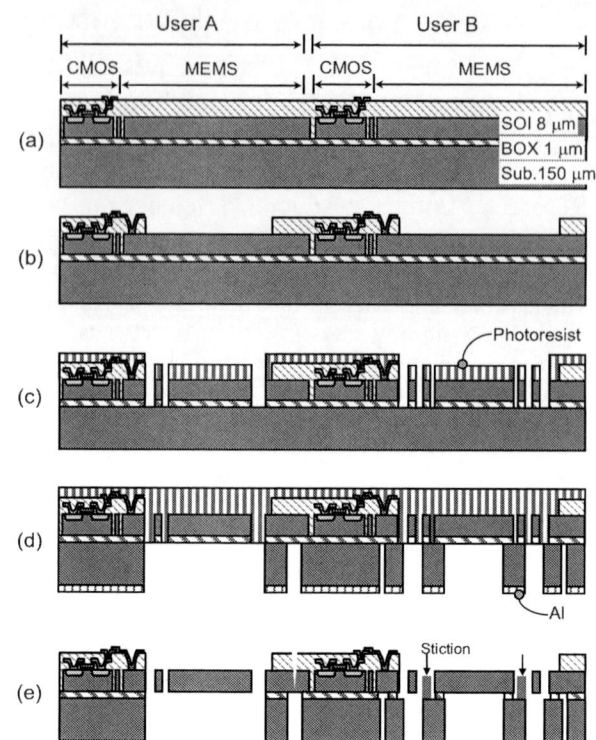

Figure 4 Fabrication process for multi-user multi-chip SOI CMOS-MEMS process

FABRICATION

The multi-user multi-chip SOI CMOS-MEMS process outline is shown in Figure 4. (a) MEMS actuators were post-processed after the TOSHIBA 0.6 micron LDMOS (Laterally Diffused MOS) [5]. The pre-fabricated IC chip had been covered with an interlayer insulation TEOS. (b) The interlayer insulation dioxide on the MEMS-filed was removed in buffered HF. Aluminum layer was deposited by vacuum evaporation and patterned onto the level-shifter's output ports. (c) At this stage, we implemented the MEMS actuator designs into the 8-micron-thick SOI layer by using the DRIE process with a photoresist mask. The BOX layer was partially etched by the reactive ion etching of CHF3 gas to relieve the residual stress in the membrane; this helped significantly to make fine structures in the double sides of an SOI chip. (d) After the top surface was passivated with a photoresist, the backside of the silicon substrate was etched by the DRIE with aluminum mask; some designs utilize both the SOI and the substrate layers for more complex micro structures. It is also possible to have electrical intra-connection between the SOI and the substrate layers by using the stiction bar reported elsewhere [6]. (e) The sacrificial layer was removed by buffered HF, and the passivation photoresist was removed by oxygen ashing. Chips ware separated by the back-side DRIE trenches, and hence no dicing was needed.

In Figure 5, we show an example of such multi-user

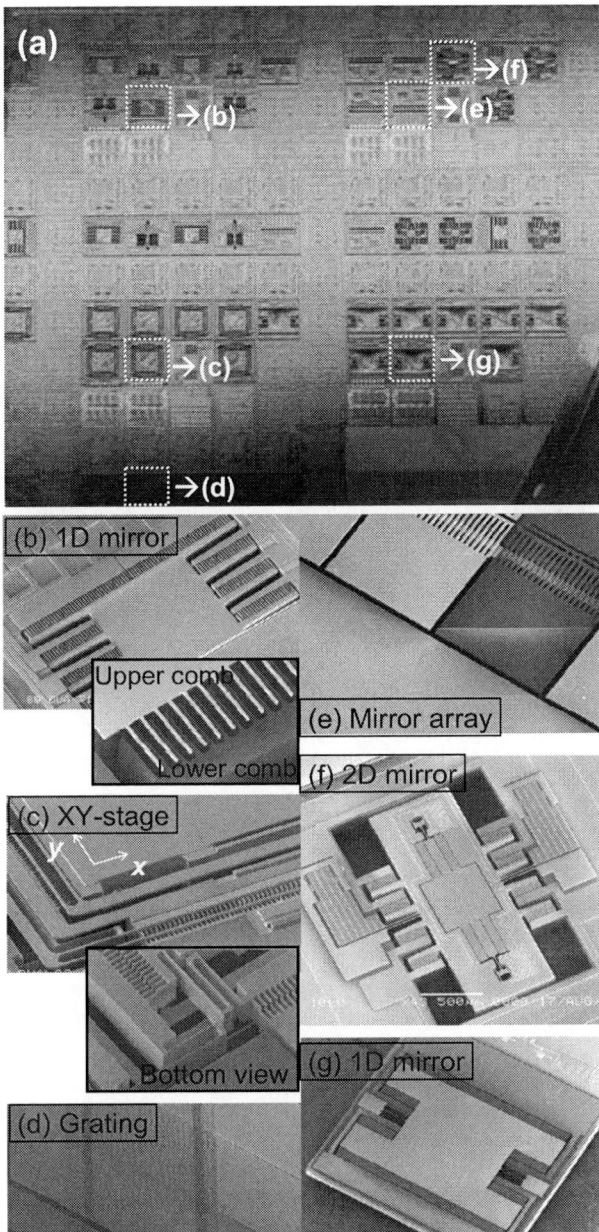

Figure 5 Photograph of multi-user multi-chip wafer that carries different designs such as 1D mirror scanner with vertical-combs, XY-stage, electrostatically tunable gratings, digital micro mirror array, 2D optical scanner, and 1D optical scanner for OCT endoscope.

multi-chip wafer that carries different designs such as (b) 1D mirror scanner with vertical-combs using double side alignment, (c) double-deck XY-stage using interlayer connection [6], (d) electrostatically tunable gratings [4], (e) digital micro mirror array, (f) 2D optical scanner, and (g)1D optical scanner for OCT (optical coherence tomography) endoscope [7]; they have been developed for their own purposes and were put together this time to demonstrate the

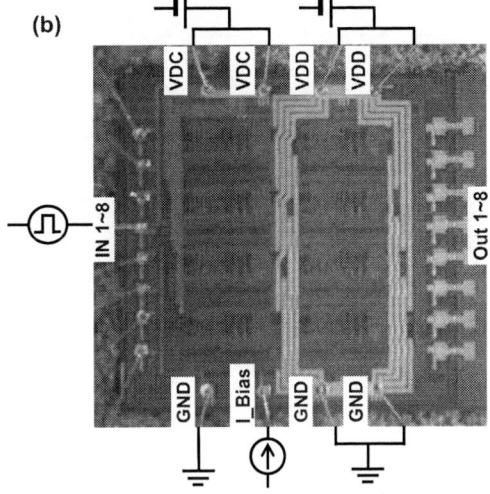

Figure 6 (a) Photograph of wire-bonded CMOS-MEMS device with digital level-shifters. (b) Microphotograph of pad layout for 8-channel digital level-shifters.

multi-chip processing capability. Careful actuator design was needed not to cause vertical or lateral stiction [8]. Due to the limited circuit/footprint resource on the current pre-fabricated CMOS wafer, only a fraction of such designs (20%) were electrically integrated with the driver circuits, and the rest remained to be a TEG (test element group).

EXPERIMENTAL RESULT

Figure 6 shows a micrograph of wire-bonded CMOS-MEMS chip with digital level shifters and tunable gratings on the left- and right-hand side, respectively. This voltage level shifter consisted of two voltage supplies, namely, a high voltage state (VDD) of 40V and a low voltage state (VDC) of 5V. At the input stage, a 5V input pulse voltage (IN) was given to the inverter circuit. For high frequency operation over 1MHz, the voltage level shifter used a current bias stage (I_bias) which effectively reduced the on-resistance of the DMOS transistors.

As shown in Figure 7, a square wave voltage of 5 V was given to control the 40 V output; the output ports were individually connected to the MEMS actuator electrodes. Table II summaries drive condition of the voltage level-shifter and measured results at such conditions. Output rise time and fall time was measured to be 13.69ns and

9.76ns, respectively, at 1MHz operation, which was thought to be fast enough for most of our MEMS actuator designs.

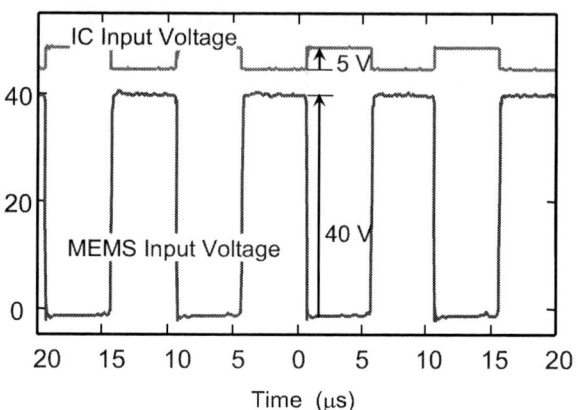

Figure 7 Input and output voltage of high voltage level-shifter.

Table II Drive condition for digital level-shifter and response time measured at such conditions.

	Parameter	Value	Units
Input	VDD	40	Volts
	VDC	5.0	Volts
	I_Bias	500	μA
	Input signal	5.0	Volts
	Input frequency	1.0	MHz
	Offset voltage	2.5	Volts
Output	Output rise time	13.69	ns
	Output fall time	9.76	ns

CONCLUSION

A multi-user multi-chip process of integrated MEMS has been demonstrated with a CMOS-first MEMS-last manner. We measured electro mechanical characteristics of grating light valves integrated with the 40V compatible high-voltage level-shifter and successfully demonstrated 1MHz operation. The newly developed CMOS-first type MEMS integration exhibited versatility and adaptation to various types of SOI bulk micromachined structures and actuators, which is highly appreciated for the MEMS R&D today where multi-products in small quantity are dominant.

ACKNOWLEDGMENTS

The authors would like to thank K. Suzuki, H. Funaki, and K. Itaya for their technical assistance for developing the SOI CMOS chip. The photomasks used in this work were made using the 8-inch EB writer F5112+VD01 donated by ADVANTEST Corporation to The University of Tokyo VLSI Design & Education Center (VDEC). This research has been supported in part by the G-COE Program of the

Electrical Engineering and Electronics, The University of Tokyo.

REFERENCES

[1] P. F. Van Kessel, L. J. Hornbeck, R. E. Meier, and M. R. Douglass, "A MEMS-Based Projection Display," *Proceedings of the IEEE*, vol. 86, no. 8, pp.1687-1704, 1998.

[2] W. Kuehnel and S. Sherman, "A surface micromachined silicon accelerometer with on-chip detection circuitry," *Sensors and Actuators A*, vol. 45, no. 1, pp. 7-16, 1994.

[3] MUMPs (multi-user MEMS processes) is currently provided by MEMSCAP. Available: http://www.memsrus.com/nc-mumps.html

[4] Kazuhiro Takahashi, Hiroyuki Fujita, Hiroshi Toshiyoshi, Kazuhiro Suzuki, Hideyuki Funaki, and Kazuhiko Itaya, "Tunable Light Grating Integrated with High-voltage Driver IC for Image Projection Display," *Proc. 20th IEEE Int. Conf. Micro Electro Mechanical Systems (MEMS'07)*, Kobe, Japan, Jan. 21-25, 2007, pp. 147-150.

[5] H. Funaki, Y. Yamaguchi, Y. Kawaguchi, Y. Terazaki, H. Mochizuki, and A. Nakagawa, "High Voltage BiCDMOS Technology on Bonded 2 μm SOI Integrating Vertical npn pnp, 60V-LDMOS and MPU, Capable of 200 °C Operation," *Proc. IEEE International Electron Devices Meeting (IEDM '95)*, Washington, USA, Dec. 10-13, 1995, pp. 967-970.

[6] K. Takahashi, M. Mita, H. Fujita, and H. Toshiyoshi, "An SOI-MEMS compatible designing technique of reversed double-deck actuators with interlayer electrical connection," *Proc. 4th Asia Pacific Conference on Transducers and Micro-Nano Technology (APCOT 2008)*, Tainan, Taiwan, Jun. 22-25, 2008, 1A2-6.

[7] M. Nakada, C. Chong, K. Isamoto, H. Fujita, and H. Toshiyoshi, "Design and Fabrication of Optical MEMS Scanners for Optically Modulated Fiber Endoscopes," *Proc. IEEE/LEOS Int. Conf. on Optical MEMS and Their Applications (Optical MEMS 2006)*, Montana, USA, Aug. 21-24, 2006, pp. 34-35.

[8] K. Isamoto, T. Makino, A. Morosawa, C. Chong, H. Fujita, and H. Toshiyoshi, "Self-Assembly Technique for MEMS Vertical Comb Electrostatic Actuators," *IEICE Electronics Express*, vol. 2, no. 9, pp.311-315, May 2005.

MEMS-BASED BATCH-MODE MICRO-ELECTRO-DISCHARGE MACHINING USING MICROELECTRODE ARRAYS ACTUATED BY HYDRODYNAMIC FORCE

Chakravarty Reddy Alla Chaitanya and Kenichi Takahata[*]
The University of British Columbia, Vancouver, Canada

ABSTRACT

This paper reports a batch-mode micro-electro-discharge machining technique that is enabled by the actuation of the suspended microelectrode arrays fabricated on the workpiece using the flow of the machining fluid. The electrode devices are microfabricated to be the double-layer construction that has 25-μm-thick copper electrodes with custom patterns on the bottom of the 18-μm-thick planar structures suspended above the workpiece surfaces. The built-in capacitance of the electrode device is utilized to construct a resistance-capacitance pulse generation/timing circuit. The suspended planar electrodes are advanced into the workpiece material with the hydrodynamic force while sustaining high-frequency discharge pulses, removing the workpiece material. Arrays of microstructures with 26-μm depth were machined in stainless steel using the machining voltage of 90 V. The dynamic behavior of the built-in capacitance of the double-layer electrode devices with the actuation was experimentally evaluated.

INTRODUCTION

The use of a broad range of functional engineering materials for the fabrication of microelectromechanical systems (MEMS) and device is a promising approach to extend their functionalities and performance. There is also high demand for micromachining of replication tools for the manufacturing of large-scale products such as flat panel displays and functional films with micro/nanostructured surfaces. Micro-electro-discharge machining (μEDM) is an attractive micromachining technique for these applications as it can be used for any type of electrically conductive materials including all kinds of metals and alloys regardless of its hardness or brittleness. It involves an electrothermal process that removes workpiece material by localized intense heat induced by microscopic sparks generated between an electrode and a workpiece in a dielectric fluid. Microstructures with minimum feature size of 5 μm and aspect ratios of > 20 can be produced by the technique [1]. μEDM has been leveraged for prototyping various MEMS devices [2, 3] as well for commercial applications such as inkjet nozzles [4] and magnetic heads [5].

One of the major constraints in traditional μEDM is its low throughput associated with the use of a single electrode tip along with the numerical control (NC) of the tip and the workpiece for the machining, producing structures individually. This issue has been addressed by the batch mode machining approach, where arrays of microelectrodes are sunk into the workpiece using the NC stage in a μEDM apparatus [6, 7]. This implementation is still limited in the machinable area due to the available size of the substrate that holds the arrays as well as the use of the mechanical stages. A new μEDM method, M³EDM (MEMS-based micro-electro-discharge machining) has recently been reported, which uses micromachined actuators with movable planar electrodes that are fabricated directly on the workpiece material using standard lithography techniques to implement the machining [8, 9]. The electrodes are actuated by either the electrostatic force generated by the machining voltage applied between the electrode and the workpiece or an external force produced by downflow of the dielectric EDM fluid. This approach offers an opportunity to eliminate the need of NC machines from the process, achieving high scalability of the process to very large areas for high-throughput micromanufacturing as well as a drastic reduction of the equipment cost associated with the process. This paper reports an advanced batch-mode M³EDM process based on the use of double-layer planar electrodes that have microstructure arrays with custom patterns on the backside of the electrodes and are actuated with the hydrodynamic force to perform the machining.

PRINCIPLE AND DEVICE DESIGN

Figure 1 shows the machining method of the M³EDM implemented by the double-layer electrode device. The planar electrodes are suspended above the surfaces of the conductive workpiece material through the tether-anchor structures with a gap separation in a dielectric fluid so that they can be actuated vertically. The pulse generation is implemented with a resistance-capacitance (*RC*) circuit with a DC voltage source [10] similar to those used in the previous efforts, where the electrode device itself serves as the capacitive element with its parasitic/built-in capacitance (shown in Figure 1) in the circuit [8, 9]. The voltage applied between the electrode and the workpiece electrostatically pulls the electrode towards the workpiece.

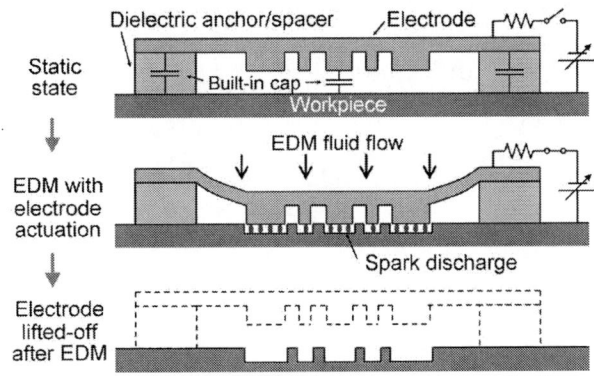

Figure 1: Cross sectional view of the M³EDM method and its process steps.

---

[*]*Corresponding author*: 2332 Main Mall Vancouver, B.C. V6T 1Z4 Canada; Tel: +1-604-827-4241; Fax: +1-604-822-5949; E-mail: takahata@ece.ubc.ca

This effort utilizes the electrodes with double-layer structures design to form microstructures on the bottom surface of the planar electrodes with custom patterns to be machined on the work material. Compared to the single-layer electrodes used in [8], the electrode plates need to be located above the workpiece surfaces with a larger distance due to the presence of the custom features on the bottom of the plate. This can result in a smaller electrostatic force, leading to a failure of pulling the electrodes to initiate the breakdown and discharge pulses. In addition, the electrostatic force can vary depending on the geometry of the custom features.

The use of external force for the electrode actuation would minimize such application-associated dependences in the operations of the M³EDM process. Another major advantage of this actuation method over the electrostatic principle is that it offers a larger actuation range for deeper machining. It also permits having a large initial gap space between the electrode and the workpiece that promotes easier flushing of the removed particles from the gap spacing when the electrodes are in resting state. The external force can be effectively applied to the electrodes for their vertical actuation using downflow of the EDM fluid [9]. The flow rate, i.e. the hydrodynamic pressure can be controlled so that the electrodes are actuated towards the workpiece, implementing the pull-in and inducing the repeated cycle of the pulse generation and charging of the built-in capacitor to remove the workpiece material. The fluid flow also enhances the removal of the byproducts during the μEDM process.

The electrode plates of the double-layer devices were designed to be similar to the single-layer electrode devices, which have 30×30-μm² through-hole arrays in the plate structures for a sacrificial etching purpose. The electrode-tether structures and the micro features underneath the electrodes were constructed by copper that has been used as an electrode material for μEDM [6, 8, 9]. Test electrodes were designed to have various structures such as crab-legged, fixed-fixed, and cantilever configurations with varying sizes. A sample layout of the double-layer electrode structure with the crab-leg tethers is shown in Figure 2.

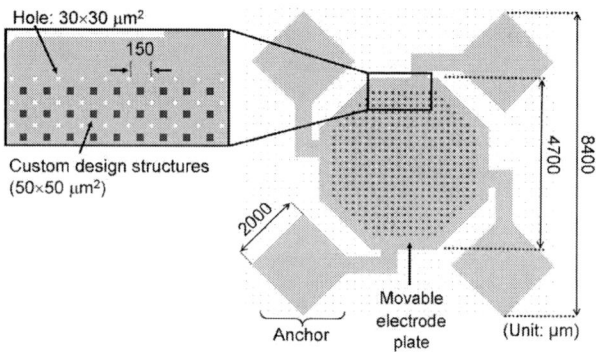

Figure 2: A sample design layout of the μEDM device that has 445 microstructures in array.

FABRICATION

Figure 3 shows the fabrication process flow for the double-layer devices that uses 18-μm-thick copper foil for the planar electrode fabrication [9]. The

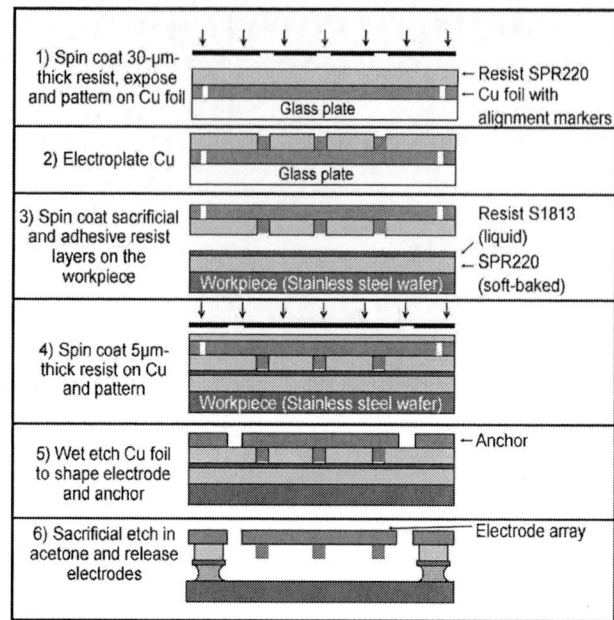

Figure 3: Sample fabrication process for double-layer copper electrode devices.

microstructures to be formed on the bottom of the electrode are fabricated by copper electroplating on the copper foil. The stainless steel was used as the workpiece material and served as the substrate (in the form of 3" wafers) for the fabrication. All the patterning steps in this process were implemented with UV lithography using a standard mask aligner with mylar masks.

First, a photoresist (SPR220, Rohm and Hass Co.) is spin coated on the copper foil to obtain a thick layer of the resist that is used to create molds for the subsequent electroplating, which is followed by a soft baking process on a hot plate 90 °C for 30 min. In this effort, the resist are double coated to obtain 30-μm-thick layer. Prior to the resist coating, through-hole markers are created in the copper foil for the alignment of the patterns to be created on the top and bottom of the copper foil. The resist is then exposed and patterned to create the molds (step 1). Next, copper is electroplated in the resist mold, forming the 25-μm-thick microstructures on the copper foil (step 2). Figure 4 shows an example of the electroplated structures obtained after this step (the resist mold was removed to show the electroplated structures). As a separate process, 25-μm-thick SPR220 resist is coated on the stainless-steel

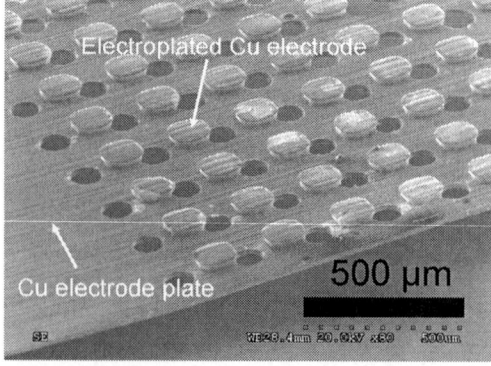

Figure 4: An SEM image of the electroplated copper on the copper supporting electrode.

978-1-4244-2977-6/09 $25.00 © 2009 IEEE

workpiece to form the sacrificial layer, followed by a soft baking process. Next, a 1-μm-thick resist (S1813, Rohm and Haas Co.) is coated on the sacrificial layer to serve as an adhesive layer for the bonding of the copper foil with microstructures underneath. The bonding is completed by a soft baking process to solidify the adhesive layer of the S1813 resist (step 3). Then, a 5-μm-thick SPR220 is coated on the top of the copper foil and patterned (step 4), which serves as a mask for wet etching of copper to define the overall shapes of the devices (step 5). Finally, the timed sacrificial etching is performed in acetone to obtain suspended electrode structures (step 6). As discussed earlier, the perforations created in the electrodes promote undercutting during the sacrificial etching process. Figure 5 shows the arrayed devices with the crab-leg design shown in Figure 2 fabricated by the developed process. The built-in capacitance of the single device in Figure 2 and the arrayed device in Figure 5a was measured in air to be 6.4 pF and 68.4 pF, respectively.

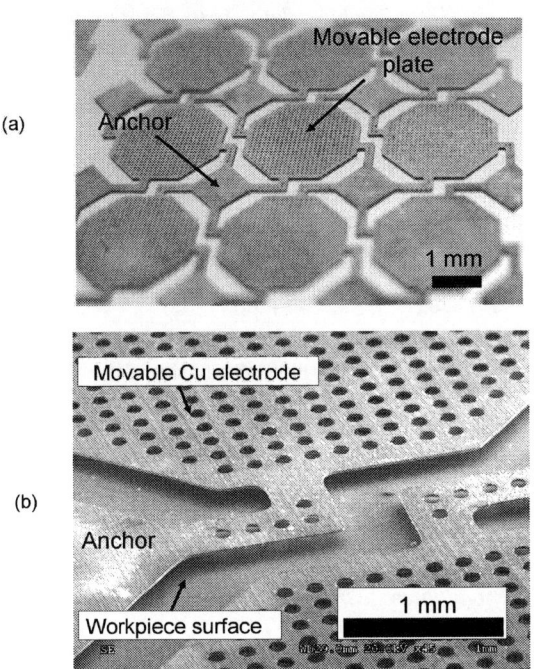

Figure 5: (a) An optical image of the fabricated device with the design in Figure 2; (b) an SEM image of the close-up.

EXPERIMENTAL RESULTS

Figure 6 shows a set-up used for μEDM tests as well as the characterization of the electrode devices. The stainless-steel wafer along with the fabricated device was placed in an ultrasonic bath filled with the dielectric EDM oil (EDM 185, Commonwealth Oil Co.). The device was connected to a DC voltage source through a 20-KΩ charging resistor as shown in Figure 6. A continuous flow pump was used to inject the EDM fluid from a nozzle with the inner diameter of 1.3 mm, at a controlled rate to generate the pressure high enough to actuate the electrode and sustain the μEDM process. The ultrasonic waves were applied during the machining to disperse the byproducts generated in the machining process.

The pressure generated with the oil injection was measured using a pressure transducer with the full scale of 1 psi (PX309-001GV, Omega Engineering, Inc.) and

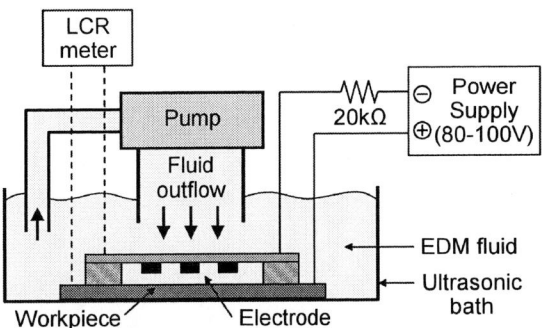

Figure 6: Characterization of the electrode actuation and μEDM tests.

plotted in Figure 7a. The nozzle was kept at a distance of 19 mm centered above the diaphragm of the transducer. This distance was commonly used in the characterization and machining tests. The result shows the pressure increase at an approximate rate of 147 Pa/(m/s). Figure 7b shows the measured built-in capacitance of a single crab-leg device with the design in Figure 2 as a function of flow velocity of the injected EDM oil. The measurement was performed with an HP 4275A LCR meter as shown in Figure 6 in both increasing and decreasing directions while applying no voltage to the electrode. The result indicates consistent increase of the built-in capacitance, i.e. displacement of the electrode towards the substrate due to the hydrodynamic pressure. The plot in Figure 7b also suggests the highly elastic behavior of the electrode device during the actuation.

With the application of machining voltage of 90V along with the injection of the fluid to the fabricated electrodes, the sequential pulses of micro spark discharge were continuously generated with the devices. The fluid flow velocity that was required to initiate and sustain the discharge pulses was 2.5-3 m/s for the devices with the crab-legged configuration. Arrays of microstructures were successfully machined in the stainless-steel workpiece

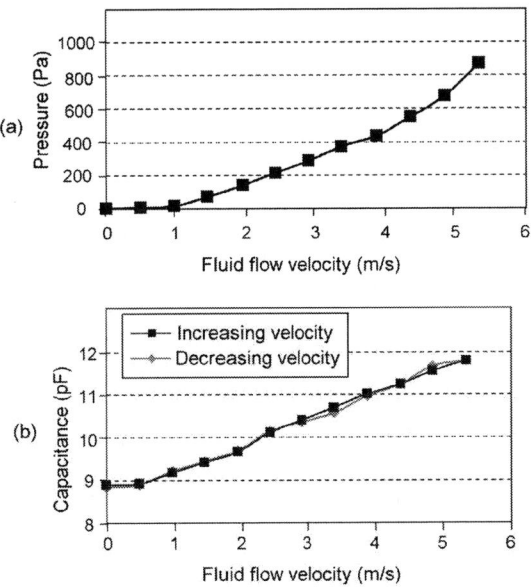

Figure 7: (a) Measured gauge pressure vs. flow velocity of the injected EDM oil; (b) Built-in capacitance vs. flow velocity for the crab-legged electrode device.

978-1-4244-2977-6/09 $25.00 © 2009 IEEE

using the electrode device in Figure 2 at the machining voltage of 90V for ~30 min (Figure 8). The array of the holes created in the workpiece material corresponds to the pattern of the electroplated structures underneath the supporting electrode. The machined structures were characterized using a Wyko™ NT110 optical profiler and were measured to have the removal depth of 26 μm (Figure 9).

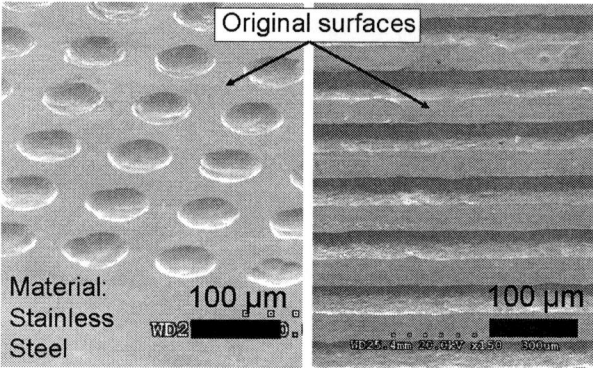

Figure 8: An SEM image of microstructures machined using different electrode patterns in stainless steel (electrodes removed after the machining).

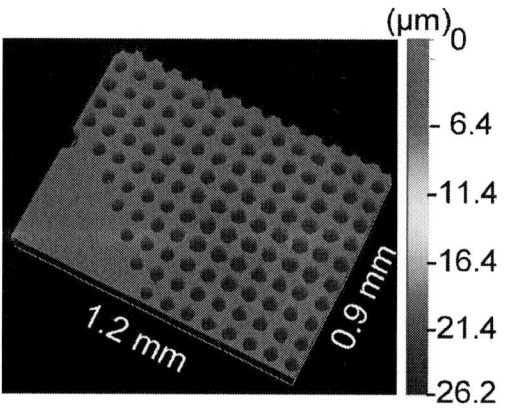

Figure 9: An optically measured geometry of the microstructure array machined with the device in Figure 2.

CONCLUSION

A batch-mode M³EDM method that actuates double-layer planar electrodes fabricated on the workpiece surfaces for micromachining of custom patterns has been explored. The developed electrode construction allows one to incorporate arbitrary designs of the electrode structures for μEDM without the need of NC systems for the processing. A microfabrication process was developed to obtain the movable copper electrode devices with electroplated micro features that were formed on the bottom of the planar electrode structures shaped from 18-μm-thick stock copper foil. The actuation was achieved by the externally applied force produced by downflow of the EDM fluid towards the electrode devices. The built-in capacitance of the electrode devices, which was used to form the *RC* pulse generation/timing circuit, was experimentally measured to characterize the dynamic behavior of the electrode with the hydrodynamic force. Micromachining of stainless steel with the removal depth of 26 μm was demonstrated using the fabricated devices

with the machining voltage of 90 V and the flow velocity of 2.5-3 m/s, successfully transferring the patterns of electroplated micro features of the electrodes. Future work will encompass optimization of the design and fabrication for the double-layer electrode devices as well as the μEDM process for finer and deeper machining.

ACKNOWLEDGMENT

We would like to thank NSERC for their financial support to this research. We also thank Ms. Vijayalakshmi Sridhar and Mr. Greg Wong for assisting in the fabrication process.

REFERENCES

[1] T. Masaki, K. Kawata, T. Masuzawa, "Micro Electro-Discharge Machining and its Applications," *Proc. IEEE MEMS*, 1990, pp. 21-26.

[2] L.L. Chu, K. Takahata, P. Selvaganapathy, Y.B. Gianchandani, J.L. Shohet, "A Micromachined Kelvin Probe with Integrated Actuator for Microfluidic and Solid-State Applications," *J. Microelectromech. Syst.*, vol. 14, 2005, pp. 691-698.

[3] K. Takahata, Y.B. Gianchandani, "Bulk-Metal-Based MEMS Fabricated by Micro-Electro-Discharge Machining," *Proc. IEEE Canadian Conf. Electr. Comput. Eng. (CCECE)*, 2007, pp. 1-4.

[4] D. M. Allen, A. Lecheheb, "Micro-Electro-Discharge Machining of Ink Jet Nozzles: Optimum Selection of Material and Machining Parameters", *J. Mater. Process. Tech.*, vol. 58, 1996, pp. 53-66.

[5] Y. Honma, K. Takahashi, M. Muro, "Micro-Machining of Magnetic Metal Film using Electro-Discharge Technique", *Adv. Inform. Stor. Syst.*, vol. 10, 1999, pp. 383-399.

[6] K. Takahata, Y.B. Gianchandani, "Batch Mode Micro-Electro-Discharge Machining," *J. Microelectromech. Syst.*, vol. 11, 2002, pp.102-110.

[7] Y. M. Sang, P. S. Min, L. S. Young, C. N. Chong, "Fabrication of Stainless Steel Shadow Mask using Batch Mode micro-EDM", *J. Microsyst. Technol*, vol. 14, 2007, pp. 411-417.

[8] C. R. Alla Chaitanya, K. Takahata, "Micro-Electro-Discharge Machining by MEMS Actuators with Planar Electrodes Microfabricated on the Work Surfaces", *Proc. IEEE MEMS*, 2008, pp. 375-378.

[9] C. R. Alla Chaitanya, K. Takahata, "M³EDM: MEMS-Enabled Micro-Electro-Discharge Machining", *J. Microelectromech. Syst.*, vol. 18, 105009, 2008, 7pp.

[10] T. Masuzawa, T. Sata, "The occurring mechanism of the continuous arc in micro-energy EDM by RC circuit", *J. Elec. Machining*, vol. 5, pp. 35-52.

SILICON CARBIDE SURFACE MICROMACHINING TECHNOLOGY BY TETRAMETHYLSILANE-BASED ATMOSPHERIC VAPOR DEPOSITION

Y. Hatakeyama[1], M. Esashi[2] and S. Tanaka[1]
[1]Graduate School of Engineering, Tohoku University
[2]World Premier Advanced Institute for Materials Research, Tohoku University

ABSTRACT

Systematic study on selective silicon carbide (SiC) deposition on SiN, SiO_2 and Si was performed. Poly-SiC was deposited by atmospheric pressure vapor deposition (APCVD) using tetramethylsilane (TMS), which allows easy handling and low temperature deposition. We found a process condition enabling the selective SiC deposition as well as a design guideline for SiC surface micromachining based on the selective SiC deposition. If both process condition and design guideline are used, continuous fine-grained SiC films are deposited on SiN and Si, while SiC islands are deposited on SiO_2. SiC films on SiN are used as the microstructures, and SiC islands on SiO_2 are lifted off by SiO_2 sacrificial etching. Finally, movable SiC microstructures were fabricated by the developed surface micromachining technology.

INTRODUCTION

Silicon carbide (SiC) is important not only for high power and high temperature semiconductor devices but also for MEMS used in harsh environments owing to its excellent high temperature mechanical property and chemical inertness [1]. However, SiC is difficult to microfabricate due to its chemical inertness. For instance, reactive ion etching (RIE) can be applied to SiC, but the use of common masking materials, e.g. photoresist, and etch stop on Si, SiN and SiO_2 are practically difficult due to low etching selectivity.

Instead of RIE, selective deposition can be used to pattern poly-SiC on a Si or SiC wafer. It is known that the deposition of poly-SiC is considerably affected by the surface of substrates. In general, a fine-grained SiC film is deposited on Si and SiN, while a rough-grained SiC film or SiC islands are deposited on SiO_2 [2]. This phenomenon can be exploited for the surface micromachining of SiC, i.e. a discontinuous SiC film or SiC islands on SiO_2 can be lifted off by etching the underlying SiO_2.

In previous works, silane and hydrocarbon were used as source gases for atmospheric pressure vapor deposition (APCVD) of poly-SiC [3]. However, a liquid source of tetramethylsilane (TMS) was used in this study, because TMS is easy to handle and allows low temperature deposition in comparison with silane and hydrocarbon. We investigated the dependence of SiC deposition on the surface of substrates and process conditions. On the substrates, various patterns of SiN, SiO_2 and Si were formed to investigate pattern dependence as well as material dependence. Finally, new poly-SiC surface micromachining technology by TMS-based selective APCVD was demonstrated.

SiC DEPOSITION METHOD

A laboratory-made, inductively-heated, vertical type, cold wall APCVD reactor with an inner diameter of 80 mm is used to deposit poly-SiC. TMS is introduced to the APCVD reactor using H_2 carrier gas. Deposition process is composed of three steps: preheating, deposition and annealing. In the preheating and annealing step, only H_2 is introduced into the APCVD reactor. Throughout the deposition process, deposition temperature and H_2 flow rate are kept constant.

Substrates used for SiC deposition are n-type (100) Si substrates with SiN and SiO_2 thin film patterns. SiN and SiO_2 are deposited by APCVD, and etched by hot H_3PO_4 and buffered HF, respectively. The thickness of SiN and SiO_2 are 100 nm and 200 nm, respectively.

DEPENDECE ON SURFACE MATERIALS AND PATTERNS

Experimental method

The deposition of poly-SiC is considerably affected by the surface of substrates. Therefore, Si substrates with SiN and SiO_2 thin films patterns shown in Figure 1 are used to investigate the dependence of SiC deposition on surface materials and patterns. a, b and c are parameters, and each pattern with different parameters is prepared on a Si substrate. The deposition temperature is 1020 °C, and the flow rate of H_2 and TMS are 1.7 ℓ/min and 0.5 sccm, respectively. Preheating, deposition and annealing are performed for 25 min, 30 min and 3 min, respectively. Before setting the substrate in the APCVD reactor, the substrate is cleaned by piranha etch (H_2SO_4 + H_2O_2), but not treated by diluted HF to remove surface natural oxide.

Results

Figures 2, 3 and 4 show typical deposition results on the substrates with SiN, SiO_2 and Si areas. In general, SiC islands with diameter of 10~20 μm were deposited on SiO_2, while continuous SiC films were deposited on SiN and Si. Difference in the surface roughness of SiC films on SiN and Si is found as the brightness of the films

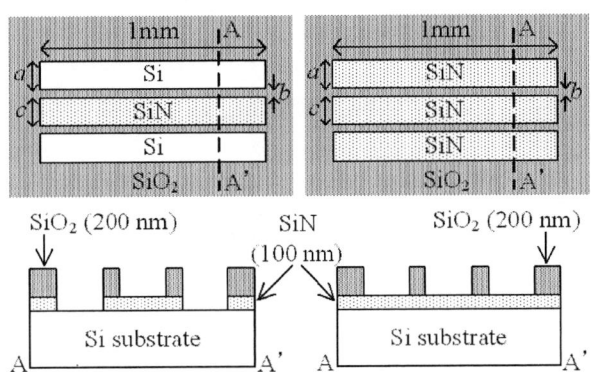

Figure 1: Two type substrates with SiN and SiO_2 thin film patterns used for SiC deposition experiments: a, b, and c are parameters.

indicates in the figures. For instance, the surface roughness of the dark and bright areas on SiN in Figures 4 and 5 is R_a 20~40 nm and ca. R_a 10 nm, respectively.

All SiN and Si areas in Figure 2 show dark color. In Figures 3 and 4, the circumferences of the SiN and Si areas adjoining the outer SiO$_2$ area show dark color. In addition, the center of the SiN area in Figures 3 and 4 has a gradation of darkness from the left to the right. By comparing Figures 2 and 3, it is found that the width of the silicon area, a, affects the surface roughness of the SiN area as well as the Si areas. By comparing Figures 3 and 4, it is found that the SiO$_2$ gap between the SiN and Si areas, b, also affects the surface roughness of the SiN area.

Figure 5 shows a typical deposition result on a substrate with SiN and SiO$_2$ areas, i.e. no Si area. Similarly, SiC islands with diameter of 5~15 μm were deposited on SiO$_2$. SiC films on SiN are rougher (ca. R_a 37 nm) in Figure 5 in comparison with Figure 3. This suggests the necessity of Si areas to deposit fine-grained SiC films on SiN.

Discussion

The obtained results suggest that SiC seeds on SiO$_2$ migrate and cluster together at high temperature until they are trapped on SiC islands on SiO$_2$ or settle on SiN or Si to form rough-grained films. This is supported by the facts that SiC is rougher at the circumferences of the SiN and Si areas within 5~10 μm from the border with the SiO$_2$ areas, and that there are less SiC islands at the circumferences of the SiO$_2$ areas adjoining the SiN or Si areas. This may be because the "wettability" of SiO$_2$ to SiC is lower than that of SiN and Si.

Comparison between Figures 3 and 5 suggests that Si has an effect of making SiC films on SiN smooth. As shown in Figures 3 and 4, at the left and right (not shown) ends of the SiN areas, rough (dark) SiC areas extend into a distance of ca. 100 μm (Figure 3) or longer (Figure 4).

However, such a phenomenon is not found in Figure 5, i.e. on the substrate without Si areas, also suggesting that Si (or Si in conjunction with SiO$_2$ as discussed later) influences poly-SiC deposition on adjoining SiN areas. The physical and chemical mechanism of this effect is under investigation.

An important finding from the viewpoint of microprocess is that SiN areas should be surrounded by Si areas larger than a certain size. From deposition results for different a (10~200 μm), the minimum required width of the Si area is 80 μm, which is also confirmed by the comparison between Figures 3 and 4. This is an important guideline for the design of SiC MEMS fabricated by TMS-based selective SiC APCVD.

DEPENDENCE ON DEPOSITION PROCESS

Experimental method

In this experiment, we investigated the dependence of poly-SiC morphology on whether a substrate is treated by diluted HF or not and preheating time. Diluted HF treatment just before SiC deposition removes natural surface oxide. Preheating also removes natural surface oxide, because the substrate is exposed to H$_2$ at high temperature. Therefore, we can know the influence of natural surface oxide to poly-SiC deposition in this experiment.

The deposition temperature is 1020 °C, and the flow rate of H$_2$ and TMS are 1.7 ℓ/min and 0.1 sccm, respectively. Deposition and annealing are performed for 30 min and 3 min, respectively. The Si substrates have SiN, SiO$_2$ and Si patterns shown in Figure 6. An energy dispersive X-ray fluorescence spectrometer (EDX) and a scanning electron microscope (SEM) are used to analyze the surface of the substrates after SiC deposition.

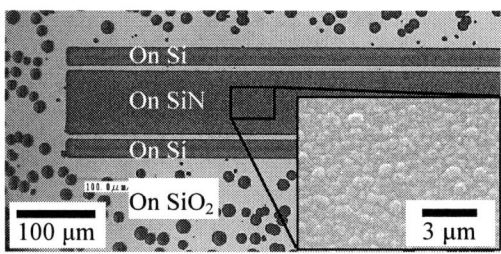

Figure 2: SiC deposition result for a = 20 μm, b = 10 μm and c = 80 μm

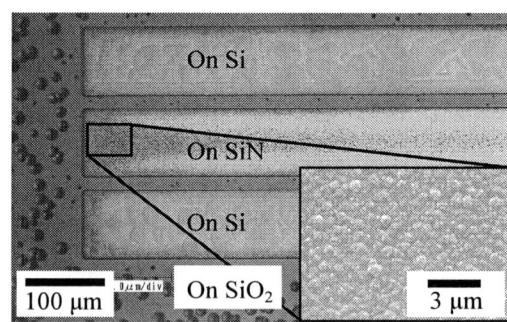

Figure 4: SiC deposition result for a = 80 μm, b = 20 μm and c = 80 μm

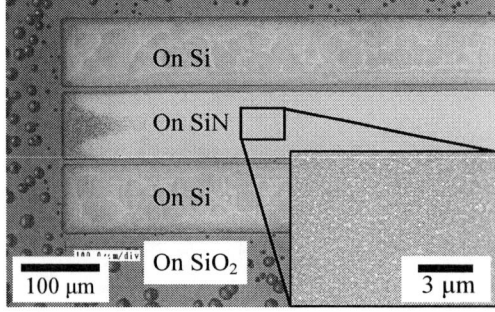

Figure 3: SiC deposition result for a = 80 μm, b = 10 μm and c = 80 μm

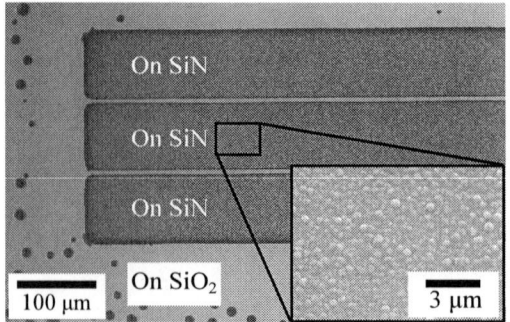

Figure 5: Rough-grained SiC on a substrate without Si areas: a = 80 μm, b = 10 μm and c = 80 μm

Results

Table 1 summarizes representative results. The SiC deposition result for process No. 3 is consistent with the results presented in the previous section, because the process conditions except for a flow rate of TMS are identical. Fine-grained SiC films were deposited on SiN and Si, and SiC islands were deposited on SiO₂. By comparing processes No. 2 and 3, the influence of diluted HF treatment just before SiC deposition is made clear. If the substrate is treated by diluted HF, SiC islands are deposited on Si, and continuous SiC films are deposited only at the edges adjoining the SiO₂ areas. Also, the SiC film on SiN is rougher than that for process No. 3.

In processes No. 1 and 4, preheating time is shortened to 10 min and prolonged to 55 min from 25 min in process No. 3, respectively, and the other process conditions are identical. For process No. 1 with a shorter preheating time of 10 min, SiC deposition was not confirmed on the Si areas by EDX and SEM, and the SiC film on the SiN area is rougher than that for process No. 3. For process No. 4 with a longer preheating time of 55 min, on the other hand, the deposition on Si result is similar to the process No. 2, except for the size of SiC islands (20 μm in process No. 2 and 30~40 μm in process No. 4).

Discussion

Natural surface oxide might not exist on the substrates before the deposition step in processes No. 2 and 4, because natural surface oxide was removed by diluted HF treatment and preheating, respectively. In these processes, a continuous SiC film was not obtained in the Si areas except for the edges adjoining to the SiO₂ areas. On the other hand, natural surface oxide might cover the whole Si areas throughout the deposition step in process No. 1, because the substrate was not treated by diluted HF and preheating time was short. As a result, SiC was not deposited on the Si areas.

In Figures 3 and 4 and the photograph for process No. 3 in Table 1, interference fringes are observed in the Si areas. We can infer the mechanism of the generation of the interference fringes from the above assumption as follows: The substrates were not treated by diluted HF, and thus a natural surface oxide remained on the substrates before the deposition process. SiC deposition started before the natural surface oxide was completely etched in preheating, and the partly remaining natural surface oxide "catalyzed" SiC deposition on Si. The interference fringes might be the traces of the natural surface oxide just being etched, i.e. the history of SiC growth edges. This "catalysis" effect is also seen at the circumferences of the Si areas adjoining SiO₂ in the photograph for processes No. 2 and 4 in Table 1.

In conclusion, to deposit continuous fine-grained SiC films on SiN for poly-SiC surface micromachining (described later), the following two conditions are required: (1) Natural surface oxide must remains on Si

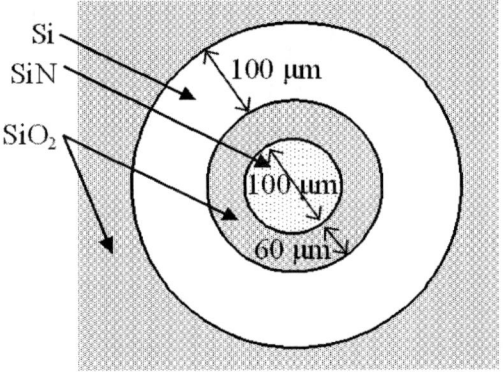

Figure 6: SiN, SiO₂ and Si patterns on a substrate: The cross section is identical to Figure 2.

Table 1: Summary of SiC deposition results for different process conditions

Process number		No. 1	No. 2	No. 3	No. 4
Diluted HF treatment just before SiC deposition		No	Done	No	No
Preheating time		10 min	25 min	25 min	55 min
Images of samples after SiC deposition					
Morphology of SiC	On Si	No deposition (Confirmed by EDX)	Small islands Continuous films at SiO₂ edges	Smooth film with optical interference fringes	Large islands Continuous films at SiO₂ edges
	On SiN	Fine-grained film	Medium-grained film	Fine-grained glossy film	Large-grained film
	On SiO₂	Little deposition	Small islands (dia. ca. 5 μm)	Medium islands (dia. 10 ~20 μm)	Large islands (dia. 10~30 μm)

before the SiC deposition process. (2) Preheating must be done for an appropriate time to produce the coexistence of SiO_2 and Si during SiC deposition.

SiC SURFACE MICROMACHINING BY SELECTIVE SIC DEPOSITION

Poly-SiC surface micromachining technology by selective SiC deposition has been developed based on the above findings. An important point for design is that SiN areas, on which SiC MEMS is fabricated, should be surrounded by Si areas with more than 80 μm width to obtain fine-grained SiC films. Figure 7 shows the developed fabrication process. SiO_2 and SiN are deposited on a Si substrate by APCVD, and patterned by wet etching and $CF_4 + O_2$ plasma etching, respectively (Figure 7 (a)). The SiN layer is the basement of SiC structures and also works as an insulation layer. The SiO_2 layer works as a sacrificial layer. The thickness of SiO_2 and SiN are 1 μm and 0.1 μm, respectively. Poly-SiC is deposited on the conditions of process No. 1 in Table 1. Fine-grained SiC films are deposited on SiN and Si, and SiC islands are deposited on SiO_2 (Figure 7 (b)). The thickness of the SiC film is 1 μm. Finally, the substrate is dipped in HF to etch the sacrificial SiO_2 layer and lift off the SiC islands on SiO_2 (Figure 7 (c)).

Figure 8 shows a fabricated movable poly-SiC microstructure. Insets (b) and (c) show the magnified views before and after the final HF etching, respectively. The poly-SiC microstructure was successfully released without damage and bending.

CONCLUSION

We systematically investigated selective SiC deposition on SiN, SiO_2 and Si. Poly-SiC was deposited by atmospheric pressure vapor deposition (APCVD) using tetramethylsilane (TMS), which is easy to handle and allows low temperature deposition. In general, fine-grained SiC films are deposited on SiN and Si, while

SiC islands are deposited on SiO_2. However, this is the case only when the following two conditions are satisfied: (1) SiN areas must be surrounded by Si areas, otherwise SiC films on SiN become too rough to be used for MEMS. (2) The substrate should not be treated by diluted HF just before SiC deposition process, and also preheating, in which the substrate is exposed to H_2 at high temperature, should be done for an appropriate time.

Based on the above findings, we have developed poly-SiC surface micromachining technology by selective SiC deposition. SiC MEMS are fabricated on SiN areas, which are surrounded by Si areas to obtain fine-grained continuous SiC films. A SiO_2 layer is used as a sacrificial layer, on which discontinuous SiC islands are deposited and then lifted off by SiO_2 etching. Finally, movable SiC microstructures were successfully fabricated.

ACKNOWLEDGEMENT

This study was performed in R&D Center of Excellence for Integrated Microsystems, Tohoku University under the program "Formation of Innovation Center for Fusion of Advanced Technologies" supported by Special Coordination Funds for Promoting Science and Technology.

REFERENCES

[1] Mehran Mehregany *et al.*, "Silicon Carbide MEMS for Harsh Environments", *Proc. IEEE*, 86, 8 (1998) pp. 1594–1610.

[2] Chien Hung Wu *et al.*, "Growth of Polycrystalline SiC Film on SiO_2 and Si_3N_4 by APCVD", *Thin Solid Films*, 355-356 (1999) pp. 179–183.

[3] Robert F. Wiser *et al.*, "Polycrystalline Silicon-Carbide Surface-Micromachined Vertical Resonators – Part 1: Growth Study and Device Fabrication", *J. Microelectromech. Syst.*, 14, 3 (2005) pp. 567–578.

(a) Before SiC deposition: SiO_2 and SiN are patterned on a Si substrate. SiN areas are surrounded by Si areas.

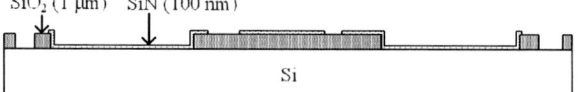

(b) After SiC deposition: Fine-grained SiC films are deposited on SiN and Si, while SiC islands are deposited on SiO_2.

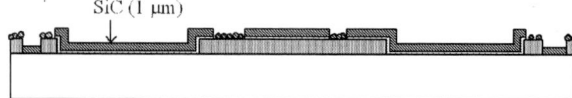

(c) After sacrificial SiO_2 etching: SiC islands on SiO_2 are lifted off, and SiC microstructures are released.

Figure 7: SiC surface micromachining process by selective SiC deposition

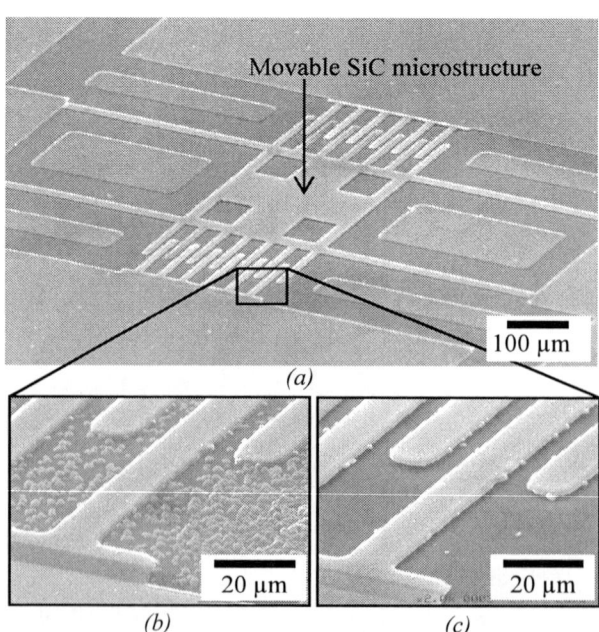

Figure 8: Movable SiC microstructure: (a) After SiO_2 etching, (b) Before SiO_2 etching, (c) After SiO_2 etching

DEVELOPMENT OF A MONOLITHIC PMMA COMB-DRIVE MICRO ACTUATOR UTILIZING HOT EMBOSSING AND ULTRA-PRECISION MACHINING

Satoshi Amaya[1], Dzung Viet Dao[2] and Susumu Sugiyama[2]
[1]TOWA Corporation, Kyoto, JAPAN
[2]Ritsumeikan University, Shiga, JAPAN

ABSTRACT

This paper presents a novel fabrication process of monolithic PMMA comb-drive actuators utilizing hot embossing and ultra-precision machining. The robustness and capability of the method are demonstrated through the fabrication of sophisticated PMMA freestanding micro structures, such as lateral and vertical comb actuators, and a torsion micro mirror. Firstly, a silicon mold is fabricated by bulk micromachining technology. Next, comb-drive micro actuator structures are formed on PMMA plate by hot embossing. Then, the PMMA layer remained after hot embossing is removed by ultra-precision cutting to release the movable parts. Finally, the device is coated with a gold layer for the electrode. A monolithic PMMA lateral and vertical comb actuators with finger's width, thickness, and gap between two fingers are 5μm, 60μm and 5μm, respectively, have been fabricated and tested successfully.

INTRODUCTION

Recently, polymer micro devices become one of the attractive research topics in MEMS because of several advantages against traditional silicon material, such as lower material and fabrication costs, softer, transparent, biocompatible and environmental friendly properties, etc. Soft material electrostatic actuator requires less power to drive, and therefore it can save energy; transparent property of PMMA is suitable for research in biomedical, micro fluidic devices and optical devices [1, 2].

In this work, we propose a new fabrication process of monolithic PMMA comb-drive actuators utilizing hot embossing and ultra-precision machining. Unlike the method reported in [3], where only photoresist-based polymers were possible, and the device was not all-polymer since a glass base must be used, the proposed method can be used for varieties of polymers and monolithic devices are straightforward. Lateral PMMA comb actuator by hot embossing was reported recently [4], however, O_2 plasma was used to remove a thin PMMA layer remained after hot embossing process; therefore, surface quality was strongly affected. In this paper, the fragile microstructures are protected by sacrificial material casting, and then, ultra-precision machining is used to remove the PMMA layer; therefore, freestanding microstructures were released successfully and nanometer order of surface roughness was obtained.

STRUCTURES OF MICRO ACTUATORS

In order to demonstrate the robustness of the proposed fabrication method for polymer microstructures, lateral, vertical comb-drive actuators and torsion mirror actuator device are introduced. Configuration of the PMMA lateral comb actuator is shown in figure 1. Dimensions of a finger are $50\times5\times60\mu m^3$ (L×W×T), gap between two fingers is 5 μm, and dimensions of the suspended beam are $600\times5\times60\mu m^3$ (L×W×T). The gap between device layer and the substrate is about 40 μm. Figure 2 shows the structure and dimensions of the vertical comb micro actuator. This device consists of a mirror suspended by two torsion bars. The mirror can rotate reciprocally around the torsion bars by using the vertical comb-drive actuators located at both sides of the mirror. To create the vertical comb actuator, the fixed fingers are designed so that it can be pushed downward easily to make a vertical offset between movable fingers and fixed fingers (as shown in figure 2). Dimensions of a finger are $50\times5\times60\mu m^3$ (L×W×T), gap between the two fingers is 5μm, and the dimensions of the torsion bar are $200\times10\times60\mu m^3$ (L×W×T). The size of the mirror is 1mm×1mm. The gap between device layer and the substrate is about 30 μm. This is also the vertical offset between the movable and fixed electrodes.

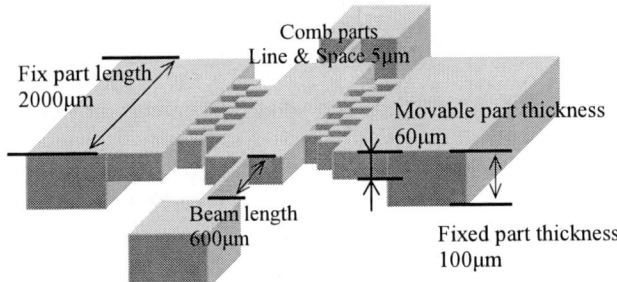

Figure 1: Structure of the PMMA lateral comb micro actuator.

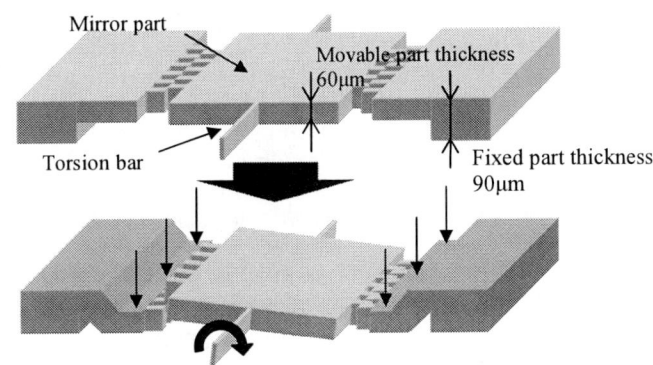

Figure 2: Structure of the electrostatic torsional micromirror device. The upper figure shows the mirror device before the vertical offset is formed.

978-1-4244-2977-6/09 $25.00 © 2009 IEEE

FABRICATION PROCESS OF MICRO ACTUATORS

The fabrication process of the PMMA micro actuator is shown schematically in figure 3. Firstly, two-stepped Si mold was fabricated by ICP-RIE (figure 3a), and then the device structures were formed by hot embossing (figure 3b). Then sample was bonded to a PMMA substrate by surface activation method to create a firm bonding without using intermediate material (figure 3c). A thin PMMA layer is always remained after hot embossing process. Removing this layer is a challenging step in creating freestanding structures. In this paper, a sacrificial material was cast into the embossed PMMA structures to protect the microstructures from fracture in releasing step. Next, the remained PMMA layer with a thickness of about 30µm to 50µm was removed by using the ultra-precision mechanically cutting (figure 3d). Then, the sacrificial material is removed by self-evaporation in vacuum, and the freestanding PMMA microstructures, such as fingers, beams and mirror, are successfully released. The surface roughness of the mirror was measured to be less than 87nm (Rz). Finally, a gold layer was sputtered to the surfaces of the PMMA comb fingers to create the electrodes for the actuators (figure 3e).

In order to assure the micro actuator operate, it is necessary to make the movable part release from the substrate. Usually, to produce floating structure, the moving parts and fixed parts are produced separately, and then bonded together. However, this method requires alignment step before bonding; therefore, it strongly affects the accuracy and fabrication time of the device. In this work, we propose a method, in which the fixed part and movable part are fabricated from a single molding process, therefore, the accuracy is high and process time is short. To apply this method, the mold structure must have two steps as shown in figure 3a.

Details of each fabrication step are described in the following sections.

Fabrication of two-step Si mold

The fabrication process of the two-step Si mold is shown schematically in figure 4 [5]. Firstly, SiO_2 layer is formed on the surface of a Si wafer by thermal oxidation process, and then movable parts and fixed parts are patterned by photolithography and SiO_2 etching in BHF acid (figure 4a). Next, the fixed parts are patterned by photolithography to create a photoresist mask for ICP-RIE process (figure 4b). Then, ICP-RIE process is performed to a depth of 30 µm (figure 4c). Next, photoresist is removed and the second ICP-RIE process is performed (with the etching mask is SiO_2) to a depth of 70 µm, and the patterns of the movable part are created (figure 4d). ICP-RIE conditions was investigated carefully to guarantee the positive taper of the side wall of the mold [6]. Finally lubricating agent for easier de-molding is applied to the Si mold before performing embossing process. Figures 5 and 6 show the Si mold structure.

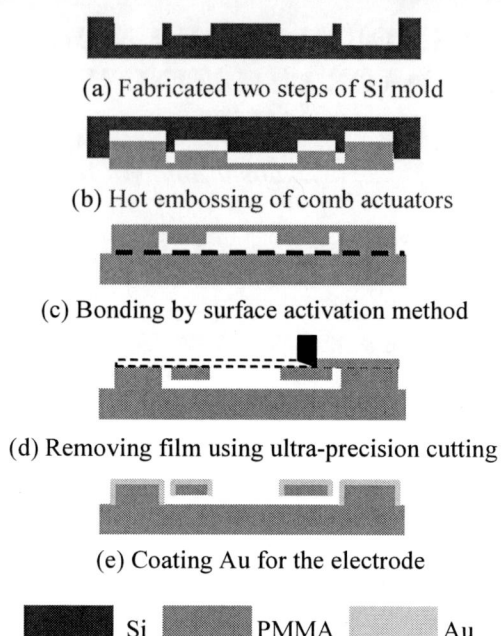

(a) Fabricated two steps of Si mold

(b) Hot embossing of comb actuators

(c) Bonding by surface activation method

(d) Removing film using ultra-precision cutting

(e) Coating Au for the electrode

Si   PMMA   Au

Figure 3: Fabrication process of the PMMA micro actuator devices.

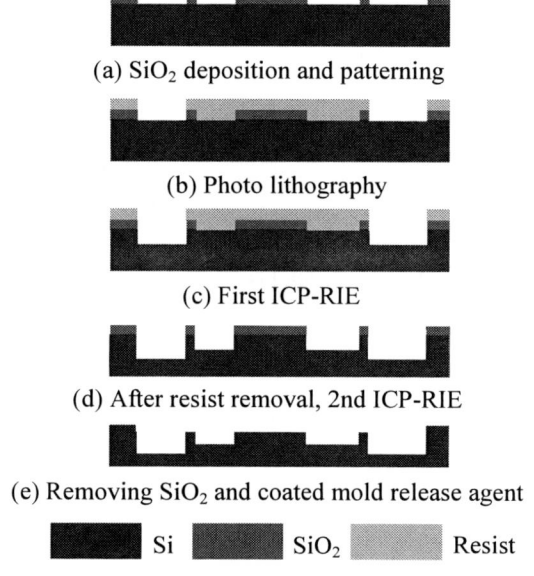

(a) SiO_2 deposition and patterning

(b) Photo lithography

(c) First ICP-RIE

(d) After resist removal, 2nd ICP-RIE

(e) Removing SiO_2 and coated mold release agent

Si SiO_2 Resist

Figure 4: Fabrication process of two-stepped Si mold.

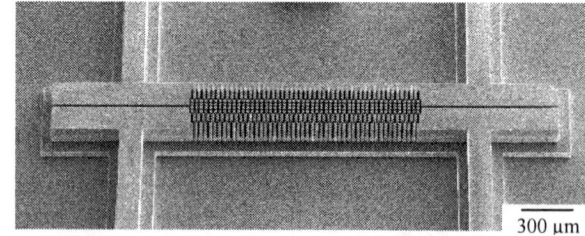

300 µm

Figure 5: SEM image of the two-stepped Si mold for lateral comb micro actuator.

978-1-4244-2977-6/09 $25.00 © 2009 IEEE

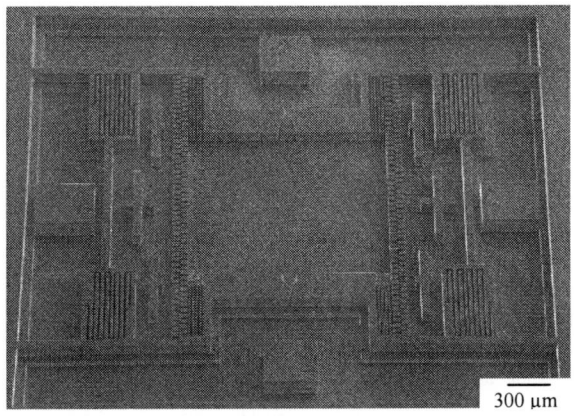

Figure 5: SEM image of the two-stepped Si mold for vertical comb actuator and micromirror device.

Hot Embossing and Bonding

The fabricated silicon molds are used for hot embossing process. Micro actuator structures were hot embossed on PMMA material. The entire hot embossing procedure and condition parameters are as below:

(1) Set up a silicon mold and PMMA sheet of 1mm into a chamber and evacuate to a pressure under 40 Torr. This is important condition to obtain sharp edges of PMMA structure.

(2) Heat the silicon mold and PMMA at the same time to a temperature of 180 °C and maintain this temperature for 15 minutes.

(3) Insert the mold into PMMA and increase the molding force to 2500 N with the molding rate of 0. 1mm/min. Maintain the molding force of 2500 N for 15 minutes.

(4) Cool the mold and PMMA to 80 °C.

(5) De-molding rate is 0.01 mm/min at 80 °C.

(6) Vent chamber, remove silicon mold and hot embossed PMMA sheet.

Figures 6 and 7 show the PMMA micro structure fabricated by hot embossing using two-stepped silicon mold. The silicon mold was not broken after de-molding, and can be re-used for the next hot embossing experiments.

Thickness of hot embossed PMMA structure are measured by alpha-step machine. In the case of lateral comb actuator, movable parts have the thickness of about 73 μm and the thickness of the fixed parts is 108 μm. In the case of vertical comb actuator, movable parts have the thickness of about 69 μm and the thickness of the fixed parts is 101 μm. The PMMA layer, which is unintentionally remained after the hot embossing, has a thickness of 30μm to 50 μm. This layer later will be removed by ultra-precision cutting to release the movable parts.

Figure 6: Laser scanning microscope image of the PMMA lateral comb actuator after hot embossing process.

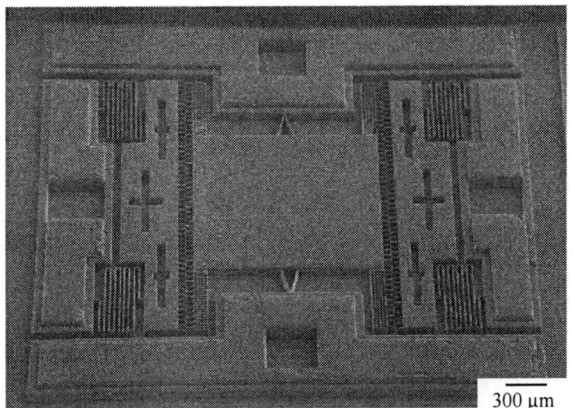

Figure 7: SEM image of the PMMA vertical comb actuator and micromirror device after hot embossing process.

After hot embossing, the top surface of PMMA structure was bonded to a PMMA plate by surface activation method (figure 3c) [7].

Releasing and metallization

After the hot embossed PMMA micro structure was bonded to a PMMA substrate, the next step is to release the movable parts of the device by removing the remained PMMA thin film (figure 3d). To protect the fragile structures during removing, a reinforcement agent was filled with the structure. Next, the remained PMMA thin film is removed by using ultra-precision cutting method [8]. The cutting amount of one cutting cycle was adjusted to 10-50 μm/cycle. Finally, the reinforce agent is automatically removed in vacuum at room temperature. Freestanding PMMA micro structures, such as comb fingers and beams, were successfully released (figures 8, 9). The surface roughness of the mirror was measured to be 87 nm (Rz). After the structures were fully released, a thin Au layer was deposited by sputtering to create electrode for electrostatic actuator. In the case of vartical actuator, the fixed fingers are pushed downward to create a vertical distance offset between movable fingers and fixed finger (figure 10). The offset was about 25 μm measured by laser scaning microscope.

Accordingly, the PMMA micro actuator with 100 comb fingers has been fabricated successfully.

Figure 8: SEM image of the PMMA lateral comb actuator after released.

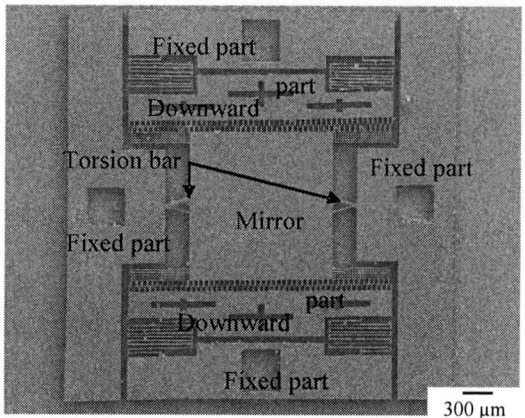

Figure 9: SEM image of PMMA vertical comb actuator after released.

Figure 10: SEM image of PMMA vertical comb fingers after the fixed comb part was pushed downward.

OPERATIONAL TEST

Now the device is ready for operational test. The comb actuator is placed under a microscope with a digital camera. A voltage supplier is connected to the movable part and fixed part of the actuator. The relation between applied voltage and displacement of lateral comb actuator is shown in Fig.11. The displacement is about 20 times larger than that of Si counterpart. In the case of vertical actuator and the micro mirror, the vertical comb actuator was operated well and the mirror plate could rotate about 4.4 degrees at 100V@100Hz applied consecutively to the left and right actuators at two edges of the mirror (figure 2).

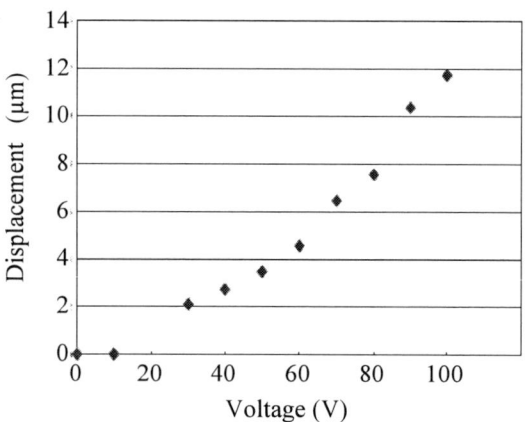

Figure 11: Displacement vs. applied voltage of the PMMA lateral comb actuator.

CONCLUSIONS

A monolithic PMMA lateral and vertical micro actuators have been successfully fabricated on PMMA by hot embossing and ultra-precision cutting. The operational test results showed the device worked properly, although the experiment and calculation values were still different. Several advantages of this technique in our experiments can be summarized as follows: (i) whole process is simple and low cost; (ii) reproduction is easy; (iii) the PMMA structure has higher flexibility compared with the counterpart made of silicon; (iv) the driving voltage is also much lower compared with silicon-based devices because PMMA has a lower Young's modulus; (v) device surface is very smooth.

Because this work is still at early stage of our research, there are several problems need to be solved to improve the properties of polymer-based micro actuators, such as reducing the actuation potential, increasing the stroke distance, and adjusting the natural frequency

REFERENCES

[1] Chao-Heng Chien And Zhi-Peng Chen, *Tech. Digest APCOT2006, B-16.*

[2] T. Suzuki, F. Kitagawa, H. Shinohara, J. Mizuno, K. Otsuka, and S. Shoji, *Tech. Digest Transducers'07, pp. 1617-1620.*

[3] Junwei Chung, Yuande Huang, and Wensyang Hsu *Proceedings of MEMS 2008, pp 475- 478*

[4] Y Zhao, T Cui, *Journal of Micromechanics and Microengineering, 13, 2003, 430-435*

[5] M. Mita, Y. Mita, H. Toshiyoshi, H. Fujita," Multiple-height Microstructures Fabricated by ICP-RIE and Embedded Masking Layers", *T.IEE Japan, Vol. 120-E, No.11, 2000, pp.493-497*

[6] P. H. Pham et al, "Fabrication and Characterization of Smooth Si Mold for Hot Embossing Process", *J. Micromech. Microeng. Vol. 127 (2007), No. 3 pp.187-191*

[7] T. Suzuki, F. Kitagawa, H. Shinohara, J. Mizuno, K. Otsuka and S. Shoji, "Polymer Microchip for Electrophoresis-Mass Spectrometry Fabricated by Hot Embossing and Low Temperature Direct Bonding", *Tech. Digest Transducers'07, pp. 1617-1620.*

[8] Suzuki, N. Nakamura, A. Shamoto, E. Harada, K. Matsuo, M. Osada, M." Ultraprecision micromachining of hardened steel by applying ultrasonic elliptical vibration cutting " *Proceedings of MHS 2003, pp 221- 226*

OPTICALLY PROGRAMMABLE SELF-ASSEMBLY OF HETEROGENEOUS MICRO-COMPONENTS ON UNCONVENTIONAL SUBSTRATES

Ehsan Saeedi[*], James R. Etzkorn, Louis Draghi, Babak A. Parviz
Dept. of Electrical Engineering, University of Washington, WA USA

ABSTRACT

Self-assembly is a promising technique for fast and cost-effective integration of microcomponents especially at smaller scales. Methods are needed to program the self-assembly process so that we can assemble heterogeneous components necessary to build a complex system. Here we present a new method of programming the self-assembly process which is based on optically removing blocking polymer from designated receptor sites. In order to perform the self-assembly process we need to fabricate free-standing microcomponents and templates. Templates were fabricated on both plastic and glass. We have shown successful assembly of four types of silicon microcomponents on plastic. The blocking AZ4620 resist was removed from the designated receptor sites immediately prior to the assembly of the desired microcomponents by using optical masks, UV exposure, and resist developer. 98% yield of proper positioning of microcomponents on a plastic template was achieved within ~10min of pipetting the components onto a template in a fluidic medium.

INTRODUCTION

There is a constant demand in the market for more portable light-weight devices that are capable of doing multiple functions. Due to this trend in the market, the construction of microsystems increasingly requires the integration of multiple devices and components such as electronic circuitry parts, MEMS, and sensors which are continuously shrinking in size. Often, these devices and microcomponents are made using incompatible fabrication processes, and therefore fast and cost-effective methods are needed to integrate a large variety of dissimilar parts to form complex systems. Currently, robotic assembly is dominantly used for integration of millimeter scale (or even smaller than one millimeter) components. But the assembly speed and cost are the restricting factors when we go to even smaller size scales.

We have previously shown that self-assembly is a strong candidate for performing system integration, particularly at the micro-scale [1, 2, 3]. Using such methodology, functional micro-components are independently microfabricated and released from a carrier wafer to form a collection of free-standing parts suspended within a fluid. These components are subsequently passed over a template containing receptor sites that provide locations for the micro-components to be assembled and complete the system (Figure 1).

Contact author: Ehsan Saeedi, ehsan@u.washington.edu

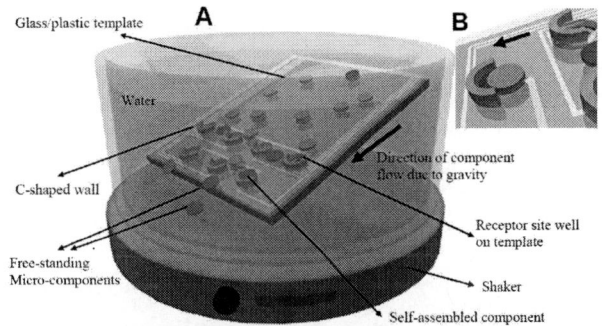

Figure 1. The self-assembly process. (A) Microcomponents are introduced over a template submerged in water and moved toward the bottom of the template due to external agitation and gravity. Self-assembly occurs as microcomponents first hit the C-shaped traps and then fall into receptor site wells. (B) Magnified view of a component entering a c-shape trap on the template.

To achieve heterogeneous integration, methods are needed to "program" the self-assembly process on the template surface. Capillary based programmable self-assembly methods have been previously shown by changing the hydrophobicity of the contacts on the template [4], or by using solder alloys with different melting temperatures [5]. Chung et.al. used micro-heaters to locally melt the solder alloy and assemble different color LEDs on designated locations on a flexible template [6]. Also, single particle/cell assembly with programmable patterns on photoconductive surfaces has been shown, using optical image driven dielectrophoresis technique [7]. We have already shown that shape recognition between the microcomponents and receptor site wells on template can be used to program the self-assembly method [1]. This programming method is limited by the number of simple geometric shapes that don't overlap, and also the assembly yield varies depending on the shape that is used.

Here we present a method to optically program the micro-scale self-assembly process which is compatible with unconventional substrates such as plastic or glass. In this method, the proper binding site was "activated" upon light exposure immediately prior to the introduction of the specified component type. An available, or open, binding site accepted components as they were trapped by the C-shaped structures (Figure 1B). After the completion of the first step, the components were fixed in their location by spin-coating a thin polymer film, and then a new set of traps was made available by selectively exposing resist through an optical mask and removing the blocking photoresist once again. The newly available sites were populated with a new type of component in a manner similar to that explained above. As this cycle was

978-1-4244-2977-6/09 $25.00 © 2009 IEEE

repeated, various microcomponents were guided and placed onto the desired locations of the template to complete a heterogeneous system (Figure 2).

To demonstrate the described programmable self-assembly process, we fabricated four types of circular micro-components as shown in Figure 3.

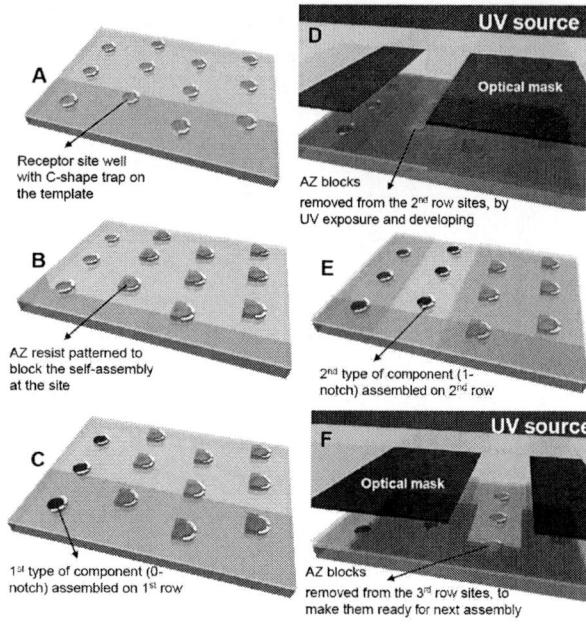

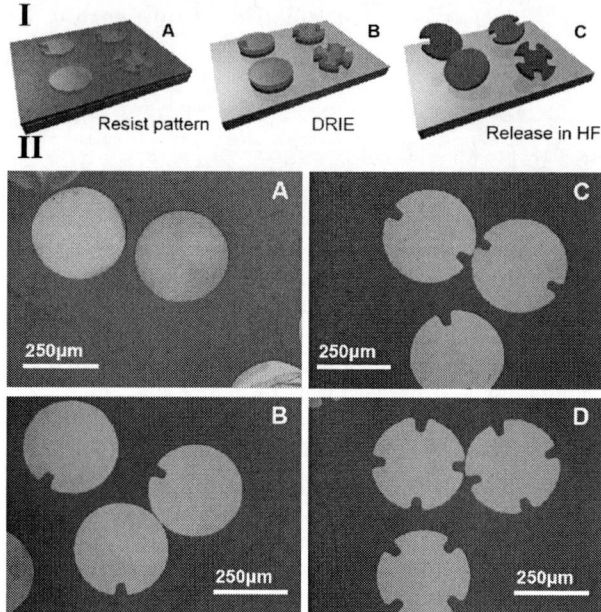

Figure 2. Optical programming of the self-assembly process. AZ4620 resist was patterned on to the template to cover the receptor sites and block the self-assembly of components. The optical mask and UV exposure were used to remove the AZ blocks from the designated areas before each step of the self-assembly process. (A) Fabricated template with receptor site wells and C-shaped traps. (B) AZ resist was spin-coated and patterned to cover all the receptor sites, except the sites on the 1st row. (C) 1st type of microcomponents (circular 0-notch) was assembled on 1st row. (D) Thin layer of AZ resist was spin-coated on the whole template to protect the assembled components. Then an optical mask, UV exposure, and photoresist developer were used to remove the AZ blocks from the 2nd row sites on template. (E) 2nd type of components (circular 1-notch) was assembled on 2nd row. (F) Similar to part D, the AZ blocks were removed from the 3rd row to make it ready for assembly of the 3rd type of components. By repeating these steps, we can assemble many types of components on a common template.

In the following sections the fabrication methods used to make components and templates for the self-assembly process are described. In addition the self-assembly results and parameters affecting the yield are discussed.

FABRICATION

Conventional fabrication processes were used to make Si microcomponents and glass templates. The process flow for fabrication of plastic templates was similar to glass ones, but to make sure we stay below the glass transition temperature of PET some changes were applied to the processes, specifically to the baking temperatures.

Figure 3. Si micro-component fabrication. (I) Process flow (A) AZ4620 resist spin-coated and patterned on SOI wafer to define micro-component shape. (B) DRIE down to reach sacrificial oxide layer. (C) Release components in HF by etching the oxide. (II) Released micro-components on a glass slide. (A) 0-notch components (B) 1-notch components (C) 2-notch components (D) 4-notch components

Components were fabricated on a SOI (Silicon On Insulator) wafer, using lithography and DRIE steps to define the shape of each type of component. All components had a diameter of 320μm with a different number of notches (0, 1, 2 and 4). The 20μm thick components were released from the carrier silicon wafer in HF to form a powder like collection of microcomponents. The components were cleaned with acetone and HCl and were stored in DI water to be used in the self-assembly process.

The template for self-assembly was microfabricated on glass or plastic by first pattering metal interconnects and then defining binding sites with SU8 (Figure 4). Each binding site consisted of a circular well (16.1 μm deep) under a C-shaped trap (17.3 μm tall). Lastly, all the binding sites were "blocked" by photolithographically patterning a thick layer of AZ4620 photoresist (23.2 μm). A completed flexible plastic template and microscope images of three receptor sites (with no blocking resist) are shown in Figure 4 (B) and (C) respectively.

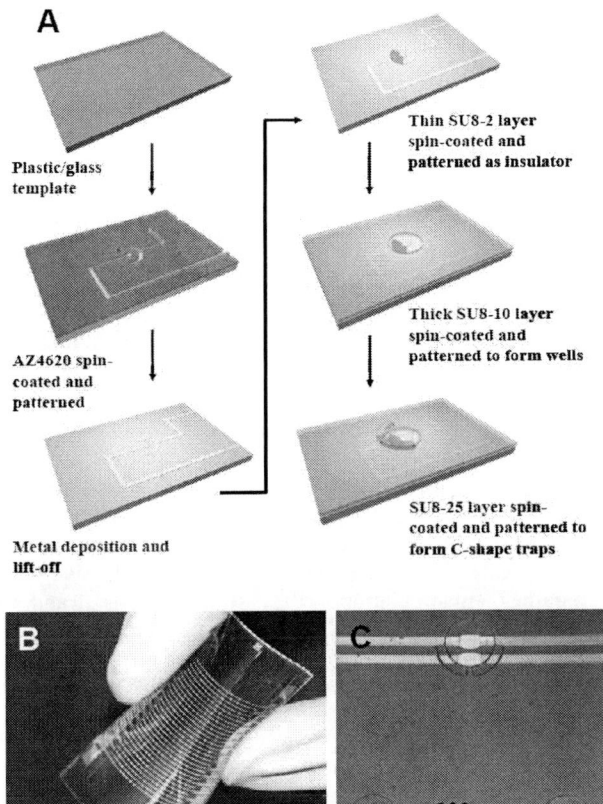

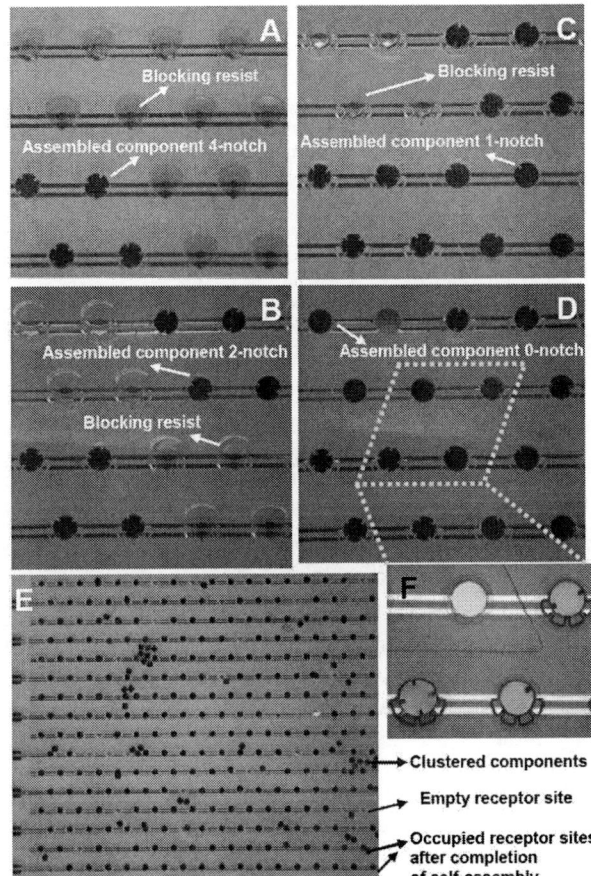

Figure 4. (A) Process flow for plastic or glass template fabrication. (B) A complete flexible plastic template. (C) Microscope image of three circular receptor site wells with C-shape traps on a plastic template.

Figure 5. Self-assembly results for 4 types of components on a chip with 320 receptor sites. A parallelogram pattern was used to optically program the self-assembly. Before assembling each type of component, AZ blocks were removed from designated sites using a parallelogram pattern that opened a 2×2 receptor site area. (A) The 1st type of components (circular 4-notch) was assembled within the 2×2 parallelogram pattern, shown on the bottom left corner of the image while the other receptor sites were blocked by AZ resist. (B) The 2nd type of components (circular 2-notch) was assembled. After assembling the 1st type of components, a thin layer of AZ was patterned on the whole chip to protect the assembled components. Then the blocking resist was removed from the receptor sites designated for the 2nd type of component. (C) The 3rd type of components (circular 1-notch) were assembled by repeating the steps explained in part B. (D) The 4th and final type of components (circular 0-notch) were assembled. (E) Zoomed-out view of the completed chip after assembly of all 4 component types. (F) Magnified view of dashed area shown in part D that has all 4 types of components assembled in the proper location.

SELF-ASSEMBLY RESULTS

To perform the self-assembly process, the template was kept at a ~30° angle and submerged in water while components were introduced and flowed over its surface. Gravity, fluidic forces, and external agitation helped the components move around randomly on the template until they were trapped in a receptor site well. During the first step all the receptor sites except the ones that were designated for assembly of 4-notch components were blocked by AZ resist. After the completion of the first self-assembly step, 4-notch components were populated in proper receptor site wells with a high yield (Figure 5A). Then a very thin layer of AZ1512 resist (1.8 µm) was spin-coated over the template to keep the assembled components in place, and also protect the remaining blocking resist from etching and degradation during the upcoming developing steps.

Before performing the second self-assembly step, the template was exposed to UV under the optical mask and then developed to remove the blocking resist from the receptor sites designated for the second type of components (2- notch). Next, the second assembly step was performed to integrate 2-notch components on the template (Figure 5B). We repeated these steps two more times to assemble all four component types.

We have performed a large number of successful self-assembly experiments to quantify the yield of the process (Figure 5). Typical yield results are shown in Table 1. As shown in the table we can get very high yield of positioning the components in proper receptor sites on the template (~98% yield) and this is achieved after about 10min of pipetting the components on to the template with ~3 times the number of components compared to the number of available receptor sites. The yield was reduced to 85% later mainly during the polymer coating step. This can be resolved in future by using molten alloy or other adhesives on the template to keep the component in place during the polymer coating step.

row no.	no. of empty receptor sites from total of 20 sites per row on the chip			
	after SA when chip is still in water	after taking chip out of water	after spin coating AZ on the chip	After developing
1	1	1	2	2
2	0	2	4	4
3	0	0	3	3
4	0	0	2	2
5	0	0	0	0
6	1	1	3	3
7	0	2	4	4
8	1	2	6	6
yield	98.125	95	85	85

Table 1. Typical yield table for all the steps needed to assemble 1 type of component. We stopped the self-assembly after ~10min, which resulted in ~98% yield. The assembly yield reduced to ~85% after manual wafer handling and spin-coating the AZ sealing layer. This reduction in yield can be avoided by using an automated system.

A common defect mode is the adhesion of the micro-components onto unwanted locations on the template due to surface forces. Surface forces become much stronger at small scales, increasing the chance of component sticking to each other or template. Strong fluid flow can remove most the sticking components, but some of them still remained in place as shown in Figure 5 (E).

CONCLUSION
Optically programmable micron-scale self-assembly offers an excellent method for the integration of complex microsystems made from incompatible microfabricated components. The integration of this method with digital micro-mirror exposure promises to provide a versatile and flexible manufacturing method for microsystems of the future.

We have shown that we can assemble heterogeneous microcomponents with very high yield on plastic which is the necessary step to build many kinds of complex electronic systems such as displays. In future we will use molten alloy to keep the components in place after the assembly, which will prevent the yield reduction, and will also enable the electrical connection between the template and components.

ACKNOWLEDGEMENTS
We acknowledge financial support from the National Institute of Health through NHGRI. We also acknowledge support from the UW Technology Gap Innovation Fund and Washington Research Foundation. The staff of the Microfabrication Laboratory at the Washington Technology Center has provided crucial help in cleanroom microfabrication processes.

REFERENCES
[1] S.A. Stauth, B.A. Parviz, "Self-assembled single-crystal silicon circuits on plastic", *PNAS,* Vol. 103, 2006, pp. 13922.

[2] E. Saeedi, S. Kim, B.A. Parviz, "Self-assembled crystalline semiconductor optoelectronics on glass and plastic", *J. Micromech. Microeng.,* Vol. 18 No 7, 2008, pp. 075019.

[3] C.J. Morris, B.A Parviz, "Micro-scale metal contacts for capillary force-driven self-assembly", *J. Micromech. Microeng.,* Vol. 18 , 2008, pp. 015022.

[4] X. Xiong, Y. Hanein, J. Fang, Y. Wang, W. Wang, D.T. Schwartz, K.F. Böhringer, "Controlled Multibatch Self-Assembly of Microdevices", *JMEMS,* Vol. 12, No 2, 2003, pp117.

[5] M. Liu, W.M. Lau, J. Yang, "On-demand multi-batch self-assembly of hybrid MEMS by patterning solders of different melting points", *J. Micromech. Microeng.,* Vol. 17, 2007, pp.2163.

[6] J. Chung, W. Zheng, T. J. Hatch, H.O. Jacobs, "Programmable Reconfigurable Self-Assembly: Parallel Heterogeneous Integration of Chip-Scale Components on Planar and Nonplanar Surfaces", *JMEMS,* Vol. 15, No 3, 2006, pp457.

[7] P. Y. Chiou, A. T. Ohta, M. C. Wu, "Massively parallel manipulation of single cells and microparticles using optical images", Nature, Vol. 436, 2005, pp.370.

FABRICATION OF NEAR NET SHAPE ALUMINA NICKEL COMPOSITE MICRO PARTS USING AQUEOUS SUSPENSION

H. Hassanin, and K. Jiang
School of Mechanical Engineering, University of Birmingham, UK

ABSTRACT

Ultra thick alumina-nickel composite micro parts were successfully fabricated using a combination of micromoulding soft lithography and colloidal powder processing of Al_2O_3-Ni suspension. The softlithographic method involves a deep UV lithography of SU-8, followed by the replication of Polydimethylsiloxane (PDMS), shaping the green by filling PDMS moulds with the composite paste. Next, drying, demoulding and finally sintering using TiO_2 as a sintering aid. The rheological properties have been characterized for the preparation of dispersed slurry. The optimal parameters have been achieved at pH=2; concentration of dispersant= 0.004 g/ml; solid loading= 65 wt%. After sintering, dense and free standing Al_2O_3-Ni composite micro parts have been achieved. Well dispersed nickel particles are found in the composite components. This approach provides a useful way to produce MEMS ceramic composites with tuneable properties by changing the amount of nickel in the matrix.

INTRODUCTION

Much attention has been focused on the development of ceramic micro parts in recent years. They are highly appropriate for MEMS applications due to their outstanding characteristics. Their chemical stability property makes them beneficial in biocompatible applications [1]. Their corrosion resistance at high temperatures makes them suitable for chemical processes and microsensor applications [2]. Their outstanding thermal and mechanical properties against very high temperatures of up to 1500 °C make them applicable for high temperature systems, such as micro engine and micro scale gas turbine for small power generation [3-5]. Their hardness and wearing resistant properties make them vital in overcoming problems of friction and stiction in micro motorized systems [6].

The main constraint to their MEMS applications is the low fracture toughness, which provides challenges and opportunities to further explore promising materials and fabrication techniques in order to broaden the application ranges of ceramic MEMS devices.

Ceramic composites are a type of promising ceramics with improved property. For example, incorporation of SiC and Si_3N_4 particles into an alumina matrix has been found to bring benefits in their mechanical properties [7]. The results encourage researchers to develop novel microfabrication techniques using these materials. Hassanin et al used preceramic polymers as both a binder and as SiC additive to fabricate ceramic composites micro components in a non-aqueous suspension [8]. Wei et al, fabricated Ni–Al_2O_3 composite micro component with enhanced properties by electroforming technique. This method is limited by a low addition of alumina in the nickel matrix [9]. Another approach to improve their fracture properties is by addition of metallic inclusions. The addition of metallic particles such as Ni (about 5 %) inside the alumina matrix can reinforce it due to microcrack toughening, crack deflection and crack bridging by a ductile material. Residual stresses will be developed due to the differences in thermal expansion between alumina and nickel. This stress will force the crack to divert away from the nickel grains [10-13].

The research work reported in this paper was set up to introduce a combination of soft lithography and colloidal powder processing technologies in order to develop a fabrication process of producing 3D free standing Al_2O_3-Ni composite near net shape micro parts. The soft lithography process is used to develop high quality micromoulds while colloidal powder processing is used to shape micro parts with the help of TiO_2 in promoting sintering. In the preparation of high performance aqueous Al_2O_3-Ni suspension, zeta potential and sedimentation, characteristics were investigated to optimize pH and dispersant concentration.

EXPERIMENTAL

Fabrication Layout

In this fabrication approach, complex shape ultra thick SU-8 micro master moulds are fabricated using UV lithography followed by producing polydimethylsiloxane (PDMS) soft moulds from the master moulds. Afterwards, the soft moulds are filled with well prepared Al_2O_3-Ni slurry. Finally, the green patterns are demoulded and sintered to produce the micro parts. The fabrication procedure is illustrated in Figure 1.

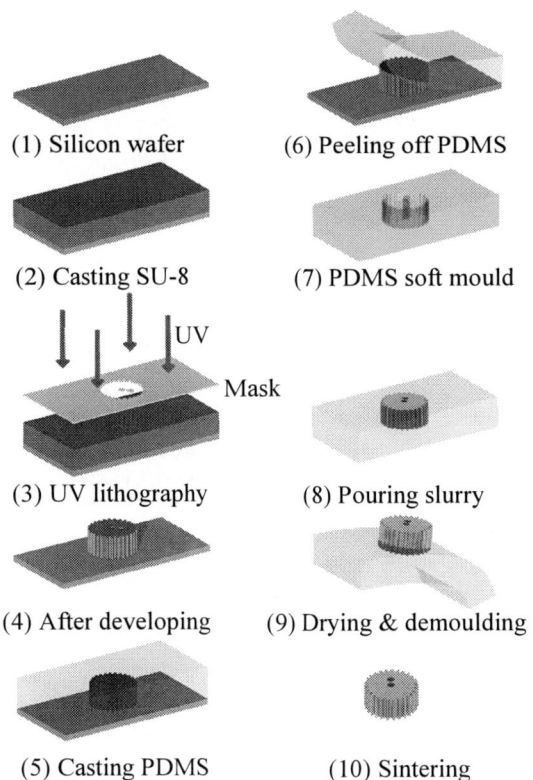

(1) Silicon wafer   (6) Peeling off PDMS

(2) Casting SU-8   (7) PDMS soft mould

(3) UV lithography   (8) Pouring slurry

(4) After developing   (9) Drying & demoulding

(5) Casting PDMS   (10) Sintering

Figure 1: Schematic diagram of the fabrication process.

Master Moulds

High quality SU-8 master moulds were fabricated using UV lithographic process. The moulds have a thickness of 1000 μm, aspect ratio of 10 and feature size less than 100 μm. SU-8 is a negative photoresist and widely used for fabrication of thick microstructures [14,15]. SU-8 has a low UV absorption characteristics. This property allows a uniformed exposure for higher thickness range compared with other photoresists. The perfect vertical sidewall profile is achieved based on the low UV absorption property. However, the transparency of SU-8 degrades as the thickness gets higher and it becomes clearer for thickness over 500 μm.

The demanding aspect ratio and heights of the required moulds caused many problems at beginning, such as non-straight and non-vertical sidewalls. Therefore, an optimized process was used. 4.9 ml SU-8 2050 (Microchem, USA) was casted to a 100 mm Silicon wafer to produce a 1000 μm thick resist layer and baking it on a levelled hot plate. UV exposure was carried on using Canon PLA-501 UV-mask aligner and the exposure energy density was 2.5 J cm^{-2}. The post exposure bake was performed first at 65 °C for 15 minutes and then at 95 °C for 25 minutes. Afterwards, the wafer was developed in ultrasonic bath SU-8 developer (Microchem, USA) for 1 hour, further details are explained in reference [15]. SEM image of the fabricated micro gear is shown in Figure 2.

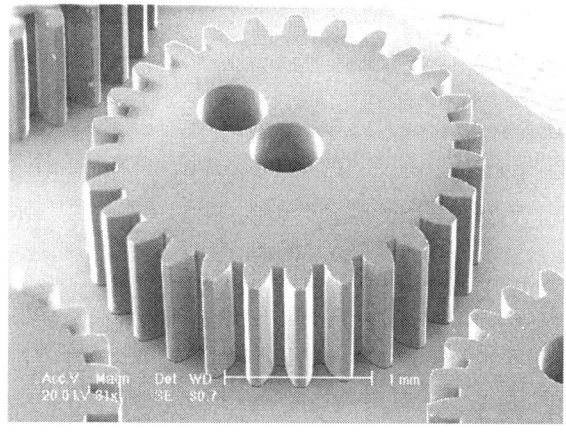

Figure 2: An SEM image of 1000 μm thick SU-8 micro gear.

Soft Moulds

Soft moulds were produced using PDMS from replication of SU-8 master moulds. In this process, the silicon wafer holding the master moulds was placed in a container. PDMS moulds were prepared by first mixing the Sylgard 184 precursors (Dow Corning Corp.) in a glass beaker in a weight ratio of 10:1 in base to curing agent ratio and degassing the mixture under vacuum to remove bubbles formed during the mixing. An approximately 2 mm thick layer of the PDMS mixture was then poured onto the patterned silicon wafer and the wafer was placed in a vacuum again until all residual bubbles were removed before it was cured at 65 °C for 4 hours. After cooling down, the cured PDMS was gently peeled off from the SU-8 master mould template, as seen in Figure 3.

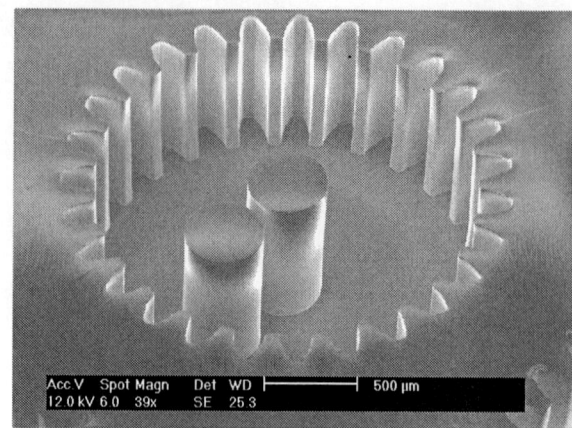

Figure 3: An SEM image of PDMS micro gear soft mould.

Suspension Composition

The composition of the suspension consists of Al_2O_3 powder with an average particle size of $d_{50}=0.4$ μm (ABSCO Materials, UK), Ni powder $d_{50}= 1.2$ μm (Vale Inco, UK) and TiO_2 powder which was prepared by hydrothermal synthesis of titanium tetrachloride $TiCl_4$ [16]. D-3005 (Duramax, Rohm and Haas, USA) was used as the dispersant, B-1000 and B-1007 (Duramax, Rohm and Haas, USA) were used as the binder, HCl and NaOH diluted solutions were used to adjust the pH value of the suspensions in processing, and distilled water was used as the solvent in all experiments.

The mixing process consists of the following steps: (1) 0–0.06 g/ml Duramax D-3005 was added to the distilled water under constant magnetic stirring for 30 minutes. Next, alumina and nickel powders in 95:5 ratio were slowly added into the solution followed by the addition of titania to the mixture (0-5) wt.% and then stirred for 4 hours; (2) 0.4 wt% B-1000 + 0.6 wt% B-1007 were added to the mixture for binding, with adjusted pH value and then stirred for 1 hour; (3) The mixed slurry was poured into a beaker and degassed under vacuum until all residual bubbles were removed.

Filling, Demoulding and Sintering

PDMS moulds were filled up with the high solid loading aqueous composite suspension by pressurized filling using stainless steel spatula. The pressure induced from the spatula was able to push the air bubble out from the PDMS and to fill the suspension in every corner of the moulds. Transparency of PDMS soft moulds enables to check the filling characteristics. Afterwards, residual of the composite suspension on the top surface of the mould was removed by means of stainless steel blade.

It is important to avoid cracks in the green structures by controlling the residual stress during the drying process. Therefore, a low drying rate was used by placing the moulds in a closed container, to minimize evaporation rate of the aqueous contents. Once the moulds were dry, the green was achieved by peeling off the soft PDMS moulds. The green components were then ready for the sintering process.

The free standing green micro parts were placed in a tube furnace with a flowing of Nitrogen. The binder and dispersant in the component should be decomposed in around 600 °C. Therefore, the heating rate of the furnace to 600 °C was first adjusted at a low ramping rate of 100 °C/hr and then set at 1350 °C at a high ramping rate of 200 °C/hr. The temperature remained at 1350 °C for 2 hours before the furnace was cooled down naturally to room temperature.

Measuring

Zeta potential values were measured by the use of a zeta potential analyzer (Zetamaster, Malvern instruments, U.K.). The suspension was diluted to 2 wt% by distilled water and the pH value was adjusted with NaOH and HCl.

Sedimentation test was used to optimize the dispersant amount. The powder was mixed with distilled water in solid loading of 10 wt.% and stirred thoroughly for 1 hour, then transferred to a graduated cylinder and held in a vertical position. The final settled heights of alumina (h) were recorded after 48 hours. Final height ratio (h/ho) of the suspension was studied as a function of D-3005 concentration in the mediums in a range of (0-0.06) g/ml.

SEM images were analyzed with a scanning electron microscope (SEM, Philips Xl-30, UK), while the density of the resultant micro parts was measured using Archimedes method.

RESULTS

Suspension Optimization

Zeta potential is a very important in determining the degree of stability of ceramic slurry. High zeta potential values will result in the higher the repulsive energy which makes slurries more stable. Figure 4 shows the relationship between the pH and zeta potential. The isoelectric point at which the zeta potential is equal to zero was around pH=8 for both alumina and nickel. The figure shows that both alumina and nickel particles are highly and negatively charged when the pH value is high, and positively charged when the pH is low. When pH is 2, the absolute zeta potential is more than 40 mV, which indicates that repulsion forces between dispersed particles are larger than van der Waals forces, usually about 30 mV. This value is able to retain well-dispersed slurry.

The gravity settling behaviour of the alumina suspensions at different D3005 concentrations is shown in Figure 5. The final settling heights decrease sharply as D3005 concentrations increases until it reaches to minimum value at D3005= 0.004 g/ml, and then gradually and slowly increases as D3005 concentrations increases (up to 0.06 g/ml).

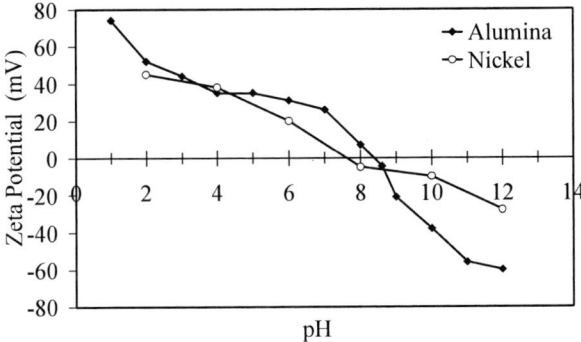

Figure 4: Zeta potential of Alumina and nickel particles as a function of pH.

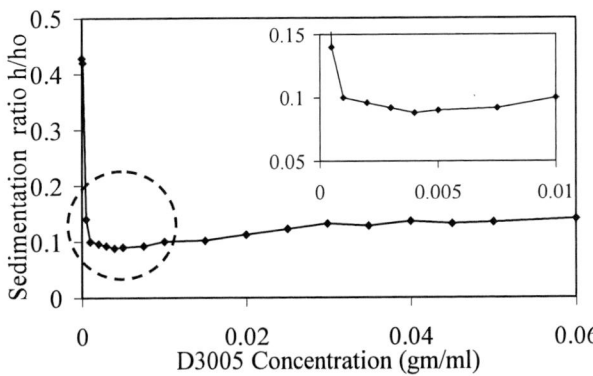

Figure 5: Sedimentation behaviour of aqueous Al_2O_3 suspensions of different D3005 concentration.

TiO_2 Effect

The influence on the density of Al_2O_3-Ni by adding TiO_2 was characterised by measuring the density of sintered samples at lower temperature. When TiO_2 was added to Al_2O_3-Ni composite, a high densification up to 94% of the theoretical density was attained at temperature as low as 1350 °C for 2 hours. Figure 6 illustrates the effect of TiO_2 content on the density of sintered Al_2O_3-Ni. The density increases from about 78 to 94% of theoretical density (T.D) by the addition of about 4% TiO_2, which is almost equal to the density of monolithic alumina fabricated by the same method and sintered at 1650 °C for 2 hours.

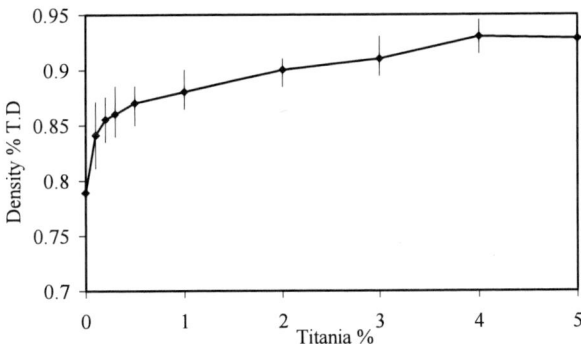

Figure 6: Effect of TiO_2 on sintered density of Al_2O_3-Ni

Properties of the Micro Parts

For fabrication of Al_2O_3-Ni composite micro parts with high aspect ratio and deep thickness, the decisive process steps were the high quality moulds, the well dispersed slurry and high solid loading. High solid loading is necessary to improve a uniform dispersion of nickel particles in the alumina suspension during filling and drying steps. This is due to the difference in densities between the two materials. Whereas the filling of the moulds can be achieved by an appropriate dispersed suspension and the high solid loading, demoulding is another crucial step. With increasing the thickness, the friction during demoulding will enlarge and can cause damage to the micro parts. Demoulding result is considerably affected by the strength of the green parts and the surface quality of the mould, and is noticeably facilitated by the use of soft moulds like PDMS and the optimization of the suspension composition that provide a

978-1-4244-2977-6/09 $25.00 © 2009 IEEE

smooth dried surface.

The sintered gear was examined under SEM as shown in Figure 7. For complex structures, such as gears with precious teeth profile, the image shows uniform teeth and vertical sidewalls. Moreover, the micro gear exhibits no visible cracks and the overall linear shrinkage was found to be about 18%. The composite Al_2O_3-Ni fabrication process enabled the fabrication of free standing 3D micro parts and preserved the shape retention of the gear even the sharp edges of its teeth.

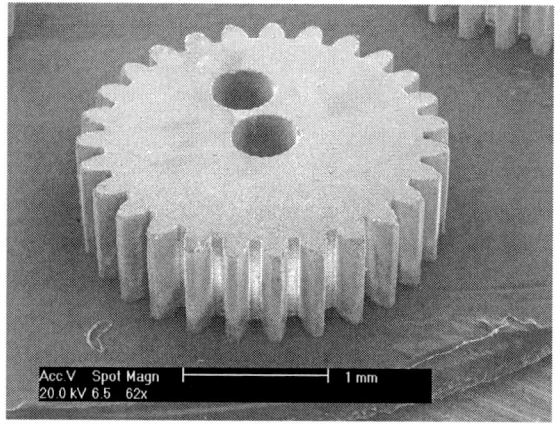

Figure 7: An SEM image of a free standing Al_2O_3-Ni micro gear.

CONCLUSIONS

This paper introduces an optimized process to fabricate ultra thick Al_2O_3-Ni composite micro parts. It provides the potential to be used in high temperature, high wearing with improved toughness MEMS applications. The results show excellent shape retention of the sintered composite micro parts, keeping the sidewall angle, dimensions and geometry of the master moulds very well. The sintering temperature used in this work is considered lower than the melting point of nickel. Furthermore, the density of the sintered structures is improved with the optimization of TiO_2 contents in the composite at a temperature of 1350 °C. This approach enables the control of the well dispersed Al_2O_3-Ni contents by using of optimal colloidal powder processing. Moreover, it is suitable for mass production of composite micro parts.

ACKNOWLEDGMENT

The authors would like to acknowledge Sandvik Osprey Powder Group and Vale Inco Europe Ltd for providing free nickel powder samples.

REFERENCES

[1] J. Liu, N. I. Nemchuk, D. G. Ast, and J. G.-. Couillard, "Etch rate and surface morphology of plasma etched glass and glass-ceramic substrates," *Journal of Non-Crystalline Solids,* vol. 342, pp. 110-115, 2004.

[2] M. Heule and L. J. Gauckler, "Gas sensors fabricated from ceramic suspensions by micromolding in capillaries," *Advanced Materials,* vol. 13, pp. 1790-1793, 2001.

[3] T. Wien, H. C. Liu, S. Kang, F. B. Prinz, and J. Stampfl, "Fabrication of ceramic components for micro gas turbine engines," Cocoa Beach, FL, United States, 2002, pp. 43-50.

[4] P. Baldus, M. Jansen, and D. Sporn, "Ceramic fibers for matrix composites in high-temperature engine applications," *Science,* vol. 285, pp. 699-703, 1999.

[5] A. H. Epstein, "Millimeter-scale, MEMS gas turbine engines," Atlanta, United States, 2003, pp. 669-696.

[6] M. Mehregany, C. A. Zorman, N. Rajan, and C. H. Wu, "Silicon carbide MEMS for harsh environments," *Proceedings of the IEEE,* vol. 86, pp. 1594-1610, 1998.

[7] K. Niihara and A. Nakahira, "Particulate strengthened oxide ceramics-nanocomposites," in *7th CIMTEC - World Ceramics Congress*, Montecatini Terme, Italy, 1991, pp. 637-44.

[8] H. Hassanin and K. Jiang, "Alumina composite suspension preparation for softlithography microfabrication," *proceeding of MNE 2008,* 2008.

[9] X. Y. Wei, Z. G. Zhu, P. D. Prewett, and K. Jiang, "Fabrication of Ni-Al_2O_3 composite microcomponent by electroforming," *Microelectronic Engineering,* vol. 84, pp. 1256-9, 2007.

[10] A. J. Sanchez-Herencia, N. Hernandez, and R. Moreno, "Fracture behaviour of pressureless sintered nickel-reinforced alumina composites," *Key Engineering Materials,* vol. 290, pp. 324-327, 2005.

[11] R. Z. Chen and W. H. Tuan, "Pressureless sintering of Al_2O_3/Ni nanocomposites," *Journal of the European Ceramic Society,* vol. 19, pp. 463-8, 1999.

[12] J. Husheng, L. Xin, L. Tianbao, Y. Hong, L. Xuguang, and X. Bingshe, "Nickel and zirconia toughened alumina prepared by hydrothermal processing," *Journal of Materials Science,* vol. 42, pp. 4707-11, 2007.

[13] A. J. Sanchez-Herencia, N. Hernandez, and R. Moreno, "Rheological behavior and slip casting of Al_2O_3-Ni aqueous suspensions," *Journal of the American Ceramic Society,* vol. 89, pp. 1890-6, 2006.

[14] K. Jiang, M. J. Lancaster, I. Llamas-Garro, and P. Jin, "SU-8 ka-band filter and its microfabrication," *Journal of Micromechanics and Microengineering,* vol. 15, pp. 1522-6, 2005.

[15] P. Jin, K. Jiang and N Sun, "Ultra-thick SU-8 Fabrication for Micro Reciprocating Engines", *Journal of Microlithography, Microfabrication, and Microsystems,* Vol 3, No 4, pp 569-573, 2004

[16] N. Manaswita, B. Pratyay, and S. V. Manorama, "Low-temperature hydrothermal synthesis of phase-pure rutile titania nanocrystals: time temperature tuning of morphology and photocatalytic activity," *Materials Research Bulletin,* pp. 1691-704, 2007.

VERTICALLY ALIGNED VARIOUS LENGTHS DOPED-SILICON MICROWIRE ARRAYS BY REPEATED SELECTIVE VAPOR-LIQUID-SOLID GROWTH

Takeshi Kawano, Akihito Ikedo, Takahiro Kawashima, Kazuaki Sawada and Makoto Ishida
Department of Electrical and Electronic Engineering, Toyohashi University of Technology
1-1 Hibarigaoka Tempaku-cho, Toyohashi, Aichi, 441-8580 JAPAN
Telephone: +81(532)-44-6746 Fax: +81(532)-44-6757, E-mail: kawano@eee.tut.ac.jp

ABSTRACT

We have proposed a growth technique of various lengths, 2-4μm diameter, conductive-silicon micowire arrays, by repeated vapor-liquid-solid (VLS) growth of n-type silicon, using Au as the growth catalyst and a mixture gas of 1% PH_3 with 100% Si_2H_6 as the silicon gas source. We obtained a longer 100μm-length silicon wire by both the first growth of 50μm-length wire and an additional growth of 50μm-length wire over the first wire, while a shorter 50μm-length silicon wire had simultaneously been grown from the substrate by the additional growth. We investigated the junction existing at the interface between the first and the second n-type silicon wire bodies. Current *I*-voltage *V* measurements on a two-step grown n-type/n-type silicon wire exhibit linear behavior with the overall resistance of 850Ω, confirming no electrical barrier at the interface junction. Several bending tests on the wires with the junction confirmed no significant change in the mechanical properties of the wire. We developed the microwire arrays for a potential application to investigations of multiple cell layers in brain cortex or retina (Fig.1). We also believe that the proposed technique becomes new approach to construct three-dimensional devices in MEMS fields.

INTRODUCTION

Vapor-liquid-solid growth method [1] has commonly been used in synthesis of one-dimensional semiconducting nanowires, metallic nanowires, carbon nanotubes as well as microwires for various device applications including optical devices, bio/chemical sensors and NEMS/MEMS devices. Further requirements of the VLS growth method are controlled size (diameter) and position of the wire, that can be realized by lithographic patterning of catalytic-particles, followed by the VLS growth [2,3]. In addition, the length of the wire can be controlled by a constant growth rate, which is depending on the growth temperature and the pressure of the gas source [4]. We have already demonstrated the Au-catalyzed VLS growth of silicon microwire arrays, with the controlled diameter of 2-4μm and the controlled length of 10-300μm, for use in particular applications

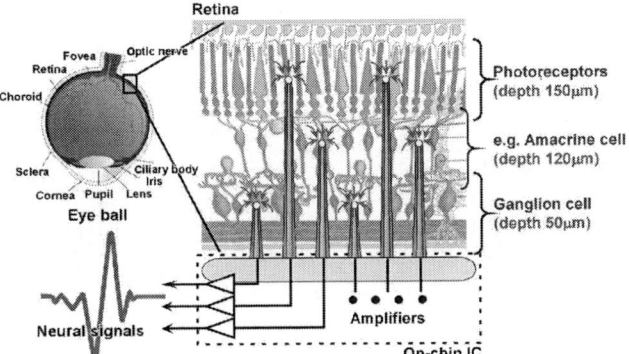

Figure 1: Different lengths microwire array device for an application to investigations of multiple cell layers in retina/brain cortex.

that the microwire arrays are inserted into a neuronal tissue in order to detect signals from a large number of neurons.

Although the above VLS growth method provides wire arrays with a same wire length due to same parameters of the VLS growth (catalyst, growth temperature, growth pressure and growth time), individually controlled length of wire in the array is still problematic. To realize the different lengths of wires, here we demonstrated a repeated selective VLS growth of silicon microwires. We first grow several silicon wires from catalytic-Au particles in the first VLS growth, while other Au particles in the array are covered with a layer of silicon dioxide (SiO_2), in order to prevent the Au particles from the reaction of the first VLS growth. After the exposing the Au particles underneath the dioxide layer, the second VLS growth is carried out to grow wires from both Au particles at the tips of the first wires (for longer wires) and Au particles at the substrate (for shorter wires), resulting in different wire lengths in the same array. In fact, the fabricated longer wire by both the first and the second VLS growths shows a change in the wire diameter in the middle portion of the silicon wire body, indicating the interface junction between the tip of the first silicon wire body and the bottom of the second wire body. Also, electrical and mechanical characterizations of the junction are discussed.

978-1-4244-2977-6/09 $25.00 © 2009 IEEE

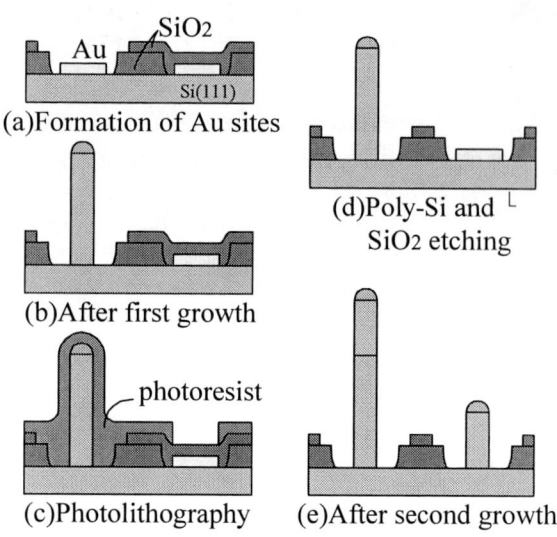

Figure 2: *Process sequence of the repeated (two-step) selective VLS growth of n-type silicon, using catalytic-Au dot and a mixture gas of PH_3 with Si_2H_6.*

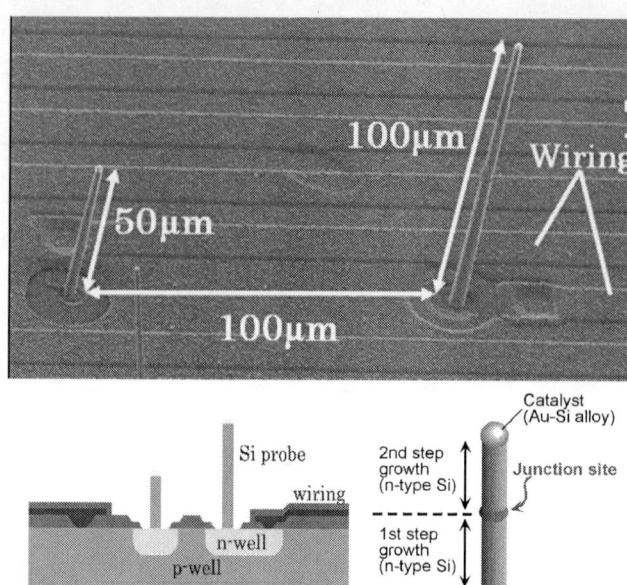

Figure 3: *Fabricated wires with 50μm and 100μm different lengths, integrated with on-chip interconnections: SEM (top), schematic image of the wire array (lower left), and image of the junction existing in the 100-μm-length wire by two-step VLS growth (lower right).*

REPEATED SELECTIVE VLS GROWTH

Figure 2 schematically illustrates the process flow of the repeated (two-step) selective VLS growth to realize the different lengths of silicon microwires in the array. The selective VLS growth can be carried out after Integrated circuits processes [5,6]. We first start with a silicon (111) substrate, to achieve the perpendicular silicon wires to the silicon substrate by the VLS growth. Heavily doped n-type region (resistivity of 10^{-3} Ω·cm) is formed by phosphors diffusion, where silicon wire is grown in the selective VLS growth. Each microwire region is connected with individual interconnection of 400nm-thick WSi/TiN/Ti [6], for use the silicon wire arrays in electrical measurements (Fig. 1). After the on-chip interconnection processes, a 160-nm-thick Au film is formed by evaporation and patterned by lift off, resulting in 4-μm-diamter catalytic-Au dots array. To grow silicon wires in the first VLS growth, we use a Si_2H_6 gas, while another Au dot for the growth of the shorter wire is selectively covered with a 1-μm-thick SiO_2 in order to prevent the Au dot from the gas reaction of the first VLS growth (Fig. 2a). After the first VLS growth (Fig. 2b), the Au dot underneath the SiO_2 layer is exposed by photolithography and chemical etching of the SiO_2 (Fig.2 c-d) and the second VLS growth is carried out, resulting in the growth of silicon wires from both the Au dot at the tip of the firstly grown wire and the another Au dot at the substrate (Fig. 2e). A technique of spray coating of photoresist is used in the photolithography, in order to completely cover the three-dimensional silicon wires with an uniform-thick photoresist (>2 μm-thick). Figure 3 shows scanning electron microscope (SEM) image of a 100-μm-length silicon wire and a 50-μm-length silicon wire integrated with on-chip interconnections of WSi/TiN/Ti. Also we can observe the junction site between the first and the second grown silicon wire bodies, due to the change in the size/diameter of the catalyst during the two-step growth process (Fig. 3 lower right). Here we used a mixture gas of 1% PH_3 (diluted in 99% hydrogen) with 100% Si_2H_6 as the silicon gas source, to obtain conductive n-type silicon wires with the impurity concentration of the order of 10^{-2} Ω·cm (resistivity 10^{-3}-10^{-2} Ω·cm) [7]. All of VLS growths in this work were carried out at the gas pressure of 0.6 Pa, at the growth temperature of 680-700°C.

ELECTRICAL CHARACTERISTICS

Figure 4 shows the current *I*–voltage *V* characteristics of both a single n-type/n-type silicon wire grown by the two-step growth and a single n-type silicon wire by only one-step growth, at the PH_3 and Si_2H_6 ratio of 8000ppm, at 680°C. *I-V* curves were taken by using a 5μm tip diameter tungsten needle contacting with the tip of the silicon wire. Due to lateral growth of silicon during the VLS growth, the measured wire exhibits a circular-cone shape with the tip diameter of 2.2μm, the bottom diameter of 3.1μm and the total wire length of 50μm. *I-V* curve of the two-step grown n-type/n-type silicon wire (Fig. 4a) exhibits the electrical liner behavior under the voltage of ±1.0V with the overall resistance of 850Ω, indicating no significant electrical barrier at the interface between the tip of the first silicon wire and the bottom of the second silicon wire. Actually, the interface junction might play as an electrical discontinuous/barrier region in the silicon wire. For example, the change in the impurity concentration of phosphorous at the

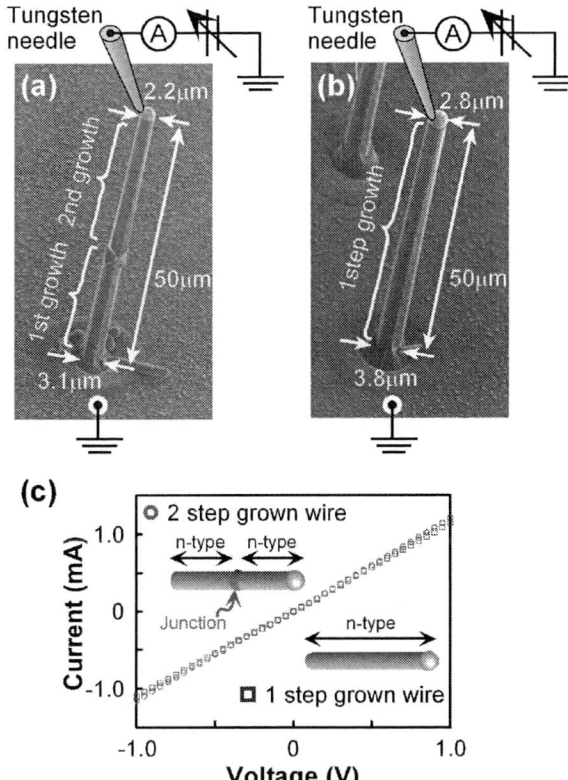

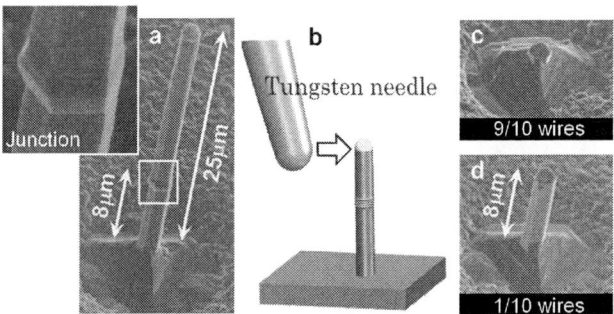

Figure 5: *Bending test on two-step grown silicon wire with a junction. (a) tested wire (inset image is the junction), (b) image of the bending test, (c) broken at the bottom and (d) at the junction site after the test.*

Figure 4: Current (*I*)-voltage (*V*) *curves of two-step grown (a) and one-step grown (b) microwires. All wires have the same length of 50μm.*

interface results in the change in its electrical resistance of the whole silicon wire. When we assume that used *in-situ* doping of the VLS growth (PH_3 and Si_2H_6 ratio of 8000ppm) can provide a constant impurity concentration of phosphorous in the whole silicon wire, the electrical resistivity of the two-step VLS grown silicon wirer, ρ, can be calculated with both the resistance of the measured wire, R, and the volume of the circular-cone-shaped wire (tip diameter a, bottom diameter b, and length l), as $\rho = \pi abR/4l$. The formula gives the resistivity of $9.1 \times 10^{-3} \Omega \cdot cm$ (impurity concentration of $5 \times 10^{18} cm^{-3}$) and this value is similar to the resistivity (or impurity concentration) of the one-step grown silicon wire of Fig. 4b without junction ($1.5 \times 10^{-2} \Omega \cdot cm$, $2 \times 10^{18} cm^{-3}$). Diameters of the measured one-step grown silicon wire are 2.8μm for the tip, 3.8μm for the bottom, and the same length of 50μm (wire resistance is 890Ω). Based on liner *I-V* curves shown in Fig. 4c and the comparison of resistivity/impurity concentration between the one-step and the two-step grown wires, the two-step VLS grown silicon wire consisting of n-type/n-type system has no significant change in the electrical properties in the wirer body [8].

BENDING TEST

We expected that mechanical properties of the silicon wire by the two-step growth is same as that of the silicon wire grown by the one-step VLS growth, because the whole wire body must be single-crystalline silicon structure, even though the

wire consists of the junction site in the body. Figure 5b schematically shows the image of the bending test on a single two-step grown silicon wire, using the same 5μm tip diameter tungsten needle. Figure 5a shows the SEM of the tested typical column-shaped silicon wire, with the diameter of 3μm and the total length of 25μm (8μm by the first growth and 17μm by the second growth). Also we predicted that the maximum stress associated with the bending force applied at the wire tip must be at near the base of the wire body. A bending force via the tungsten needle was applied at the wire tip until the breakdown of the silicon wire. Consequently, we observed that 9 of 10 wires were broken with the breakdown portion at the base of the silicon wire, as shown in Fig 5c. Only one of 10 wires indicated that the breakdown portion was at the junction site located at 8μm height from the substrate, as shown in Fig. 5d. This breakdown behavior is probably due to the nick shape appeared at the junction, which was formed by the change in the diameter of the catalyst at the beginning point of the second VLS growth.

DISCUSSION

Typically we have grown three types of silicon wires, as schematically shown in Fig. 6, following vertically aligned single silicon wire (top image in Fig. 6b and Fig. 6c), silicon wire/s grown with different growth direction (middle image) and randomly grown silicon nanowires over firstly grown silicon microwire (bottom image). In VLS growth, the liquid Au catalyst consists of the certain amount of silicon atoms, for example the amount of the silicon in the liquid Au at the first growth temperature of 700°C can be expected to be 32 at. % (atomic percent) by the phase diagram of Au-Si system (Fig. 6a) [9]. After the first VLS growth, the growth temperature is decreased to room temperature, and the precipitation of silicon atoms occurs at the interface between the catalyst and the tip of the silicon wire [10], resulting in the reduced 19 at. % silicon in the catalyst (this value given by the eutectic point of 363°C). In the second VLS growth, while the sample is heated up to the same growth temperature (700°C), the liquid Au

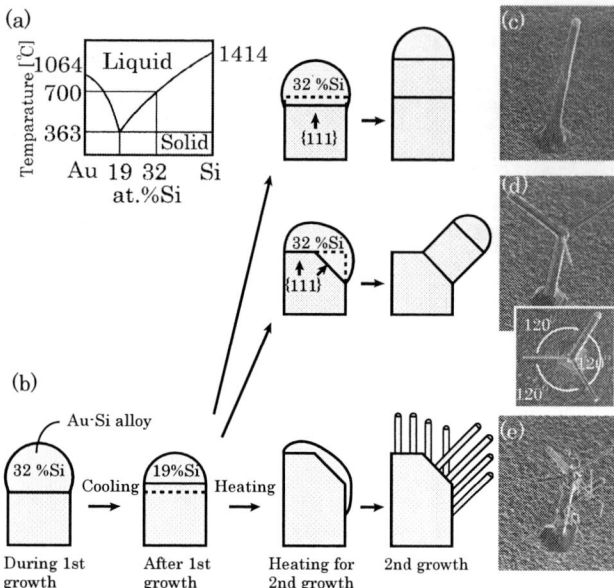

Figure 6: *Different morphologies of silicon wire observed after the second VLS growth. (a) phase diagram of Au-Si system, (b-c) vertically aligned single wire (top), spread wires with other <111> direction (middle), and massive nanowires (bottom). The morphologies are caused by different types of catalyst formation before the second growth.*

catalyst requires the same amount of 32 at. % silicon, which can be provided from the tip of the firstly grown silicon wire. However, we may have mainly three different types of formations of the liquid catalyst at this point, depending on the pathway of silicon atoms dissolved in the catalyst, such as from the exactly tip of the wire with (111) surface plane same as the silicon substrate (top image in Fig. 6b), from the edge of the tip of the silicon wire with the other (111) surface plane (middle image) and from the larger area due to the catalyst spill from the tip region (bottom image). As the results, we obtained not only single vertically aligned silicon wire, but also silicon wires with other growth direction as well as bunch of nanowires, as confirmed by SEM observations (Fig. 6c-e) [11]. Although we have used same growth parameters (temperature, pressure, catalyst-diameter), the different growth phenomena of repeated VLS grown wire have been observed and the further investigation will be required for high yield of the single vertically aligned silicon wire, as shown in Fig. 6c. Unfortunately, we used the temperature reduction (from the growth temperature to room temperature) between the first and the second growths in order for the additional photolithography process for the second VLS growth, and the yield of the single vertically aligned wire could be improved by using the continuous growth temperature from the first to the second VLS growths.

CONCLUSION

In summary, various length silicon wires had been fabricated in the same array by selective repeated VLS growth of silicon. Although a junction is found at the interface between the silicon wire bodies, we conclude that there are no significant changes in electrical and mechanical properties of the wire. Based on the capability of the repeated VLS growth of silicon wire, a large number of various length wires for electrical recording of neurons would be possible. This technique may also enable vertically aligned wires consisting of p-n junction for electrical/optical device applications, by using p-type and n-type gas sources (e.g. PH_3-, B_2H_6-Si_2H_6 mixtures). We have demonstrated vertically aligned various lengths silicon wire arrays for only micro-scale device fabrications, but this technique will be used in nano-scale device fabrications.

ACKNOWLEDGEMENTS

The authors would like to thank Kuniharu Takei for discussion and Mitsuaki Ashiki for his assistance with fabrication processes. All devices in this work were fabricated at the Electron Device Research Center (EDRC) at the Toyohashi University of Technology. This work was supported by a Grant-in-Aid for Scientific Research (S), by the global COE Program "Frontiers of Intelligent Sensing" and by a research grant for Brain-Machine Interface (subject B) from the Ministry of Education, Culture, Sports, Science and Technology of Japan.

REFERENCES

[1] R. S. Wagner and W. C. Ellis, Applied Physics Letters, vol. 4, no. 5, pp. 89-90 (1964)

[2] B. M. Kayes, *et al.*, Applied Physics Letters, vol. 91, no. 10, 103110 (2007)

[3] T. Sato, *et al.*, Applied Physics Letters, vol. 66, no. 2, pp. 159-161 (1995)

[4] T. Kawano, *et al.*, Sensors and Actuators A, vol. 97-98, pp. 709-715 (2002)

[5] T. Kawano, *et al.*, IEEE Transactions on Electron Devices, vol. 51, no. 3, pp. 415-420, (2004)

[6] K. Takei, *et al.*, Journal of Micromechanics and Microengineering, vol. 18, no. 3, 035033, (2008)

[7] Md. S. Islam, *et al.*, Jpn. Journal Applied Physics, vol. 44, pp. 2161-2165(2005)

[8] M.S. Gudiksen, *et al.*, Nature (London) vol. 415, pp. 617-620 (2002)

[9] T. B. Massalski, *et al.*, "Binary Alloy Phase Diagrams Second Edition", pp. 428 (1990)

[10] R. S. Wagner, Journal Applied Physics, vol. 38, no. 4, 1554 (1967)

[11] R. S. Wagner and C. J. Ooherty, J. Electrochem. Soc. Vol. 115, no. 1, pp. 93-99 (1968)

978-1-4244-2977-6/09 $25.00 © 2009 IEEE

FABRICATION OF ELECTROSTATICALLY-ACTUATED, IN-PLANE FUSED QUARTZ RESONATORS USING SILICON-ON-QUARTZ (SOQ) BONDING AND QUARTZ DRIE

Young-Suk Hwang[1], Hyoung-Kyoon Jung[1], Eun-Seok Song[1], Ik-Jae Hyeon[2], Yong-Kweon Kim[1], and Chang-Wook Baek[2,*]

[1]Seoul National University, Seoul, KOREA
[2]Chung-Ang University, Seoul, KOREA
*Corresponding author: cwbaek@cau.ac.kr

ABSTRACT

This paper reports a novel process to fabricate *electrostatically-actuated*, in-plane micromechanical resonators made of *fused quartz* for high-Q microsensor applications. Two key processes – low temperature plasma-assisted Silicon-on-Quartz (SoQ) direct bonding and quartz DRIE using C_4F_8/He plasma – have been used in combination with thin metallization to fabricate fused quartz resonators driven by electrostatic force. The proposed method enables wafer-level fabrication of fused quartz resonators readily mounted on the substrate, which is advantageous over the conventional fabrication method of quartz crystal resonators. By using the proposed process, 40-μm-thick laterally-driven fused quartz cantilever resonators have been successfully fabricated. The measured Q-values of the metal-coated fused quartz cantilevers are 21,700~48,900 according to the length of the cantilever.

INTRODUCTION

Quartz material has been applied to resonators in time keepers, oscillators and high-frequency filters, as well as to microsensors such as piezoelectric micro gyroscopes and quartz crystal microbalances (QCM), because of high quality factors, low aging rates and good temperature stability of quartz [1-3]. In these classical applications, quartz devices are usually manufactured from a single crystal quartz wafer of proper cuts such as X, Z, or AT. Orientation-dependent anisotropic etching property of single crystal quartz in an ammonium bifluoride solution has been used to make quartz microstructures [4]. Also, piezoelectricity of single crystal quartz is suitable for driving/sensing of quartz resonators and sensors.

On the other hand, fused quartz is also a very good resonator material which has a high quality factor, and has been used for bulk hemispherical resonator gyroscopes (HRG). Especially, fused quartz is known to have a very low thermoelastic damping, which is a dominant loss mechanism in microscale, compared to silicon or single-crystal quartz [5]. However, fused quartz has not been frequently used for micromechanical resonators because high aspect ratio, precise anisotropic etching is difficult since wet etching technique does not work anymore, and piezoelectric properties for actuation/sensing are lost.

Recent advances in SiO_2 DRIE technology make it possible to fabricate high aspect ratio, thick glass or quartz microstructures with good sidewall profiles [6, 7]. Therefore, DRIE technique is a useful tool for micromachining of fused quartz, which has no specific crystal orientations. In addition, low-temperature wafer bonding between dissimilar materials such as silicon, SiO_2,

or quartz have been reported [8, 9]. If quartz DRIE is properly combined with wafer bonding techniques, fused quartz resonant microstructures can be fabricated in a similar way to produce silicon resonators using the anodically-bonded SiOG wafer, as reported in our previous result [10]. In this case, by conformal coating of fused quartz with a very thin metal film, quartz resonators *actuated by electrostatic force* with high Q-factors due to the excellent properties of fused quartz can be realized.

In this paper, a novel process to fabricate laterally-driven electrostatic fused quartz resonators is proposed by using Silicon-on-Quartz (SoQ) bonding and quartz DRIE. A low temperature, plasma-assisted SoQ bonding is developed, and the bonded silicon wafer is used for both a quartz DRIE mask and a handling wafer of quartz devices. Fused quartz is used as the device layer for resonators, and patterned by DRIE using a C_4F_8/He gas mixture with the bonded silicon mask. By bonding a trench-formed silicon wafer with a fused quartz wafer, and coating a thin metal film on the fused quartz, wafer-level fabrication of electrostatically-actuated fused quartz resonators is possible. In order to demonstrate the usefulness of the propose method for high-Q resonant quartz devices, fused quartz cantilever resonators have been fabricated and their resonant frequencies and Q-factors have been evaluated.

FABRICATION PROCESS DETAILS
SoQ (Silicon-on -Quartz) direct bonding

In this study, SoQ direct bonding process has been developed for two purposes. A silicon wafer bonded with a quartz wafer can be used as an excellent etch mask for quartz DRIE. In DRIE of glass or quartz, metal masks (e.g. aluminum or nickel) are general due to their superior etch selectivity (~10) over resist masks. However, in order to etch quartz more than tens of microns, a thick metal layer is needed, which in turn requires an additional process like electroplating [6]. The wafer-bonded silicon is a good alternative for an etch mask of quartz: it has a high etch selectivity (>15), the thickness of silicon can be easily adjusted within a full range of wafer thickness, and silicon is precisely patterned by DRIE. In addition, the wafer bonded silicon can also be used as a handling substrate for the fabricated quartz devices. In contrast to the conventional fabrication process of quartz crystal resonators, in which each resonator is separately etched from quartz crystal and mounted to the package individually, wafer bonding approach enables to fabricate quartz resonators self-mounted on the handing substrate in wafer level. Also, it can provide an opportunity for wafer-level packaging of quartz devices.

In this work, an oxygen plasma-assisted bonding

process is used to keep the bonding temperature as low as possible. Low temperature bonding is important since thermal expansion coefficients of silicon and quartz are quite different. The process details are summarized in Fig. 1. First, 4-inch, 500-μm-thick double-side polished silicon and fused quartz (VIOSIL-SX, ShinEtsu) wafers are cleaned by standard RCA1 solution. RCA1 cleaning removes surface contaminants and makes the surface hydrophilic with OH⁻ terminations in order for spontaneous bonding to occur [8]. Then the wafers are exposed to oxygen plasma in RIE mode to activate the wafer surfaces. It is known that the physical/chemical effects of plasma treatment increase surface energies, which help to obtain higher bonding strength at a relatively low annealing temperature [9]. After plasma treatment, the wafers are rinsed in DI water and brought into contact at room temperature for the spontaneous bonding to be occurred. Finally, the wafers are annealed at a temperature of 300 °C for 8 hours to increase the bonding strength.

Bonding experiments under different plasma treatment conditions have been performed using the proposed bonding procedures. As a typical result, photograph of the SoQ wafer, where a quartz wafer is bonded onto a silicon wafer with 5-μm-deep trenches, is shown in Fig. 2. Shear strength of the bonded wafer for each condition has been measured by System 522 universal bond tester. The highest bonding strength of 10 MPa is obtained at a minimal oxygen plasma treatment time of 15 seconds. The measured strength is about half of the nominal strength of anodic bonding (~20 MPa), but is sufficient to sustain subsequent chemical thinning or mechanical polishing processes.

Quartz DRIE using wafer-bonded silicon mask

Quartz DRIE process has been studied using an AOE (Advanced Oxide Etcher, STS) system with a gas mixture of C_4F_8 and He due to its excellent selectivity for silicon mask. A single crystal silicon wafer is bonded with the fused quartz wafer using the developed process, thinned down to 30 μm by lapping and CMP, and patterned by silicon DRIE. Using this silicon mask, DRIE of fused quartz has been performed. Process conditions are summarized in Table 1.

Table 1: Process conditions for quartz DRIE.

Process parameters	Values
Gas flow	C_4F_8 = 18 sccm He = 25 sccm
Chamber pressure	4 mTorr
Antenna power	1300 W
Bias power	500 W

The SEM image of the etched 50-μm-thick fused quartz microstructures is shown in Fig. 3. Almost vertical profile angle is achieved by using the above process conditions. Etch rate and selectivity are measured to be 0.38 μm/min and 15:1, respectively. Effect of He flow rate on the etch rate of quartz is negligible, but etch selectivity is improved as He flow rate is increased. However, too much increase of He flow is not effective to eliminate so-called a cusping effect, which is the distortion of etch profile at the bottom of the structure [10]. Since the thickness of silicon mask is sufficiently large to overcome the decrease of etch selectivity, He flow rate was reduced in our experiment. Deep etching of quartz more than 50 μm will be possible using this bonded silicon mask.

Fabrication of electrostatic fused quartz resonators

Based on the developed SoQ bonding and quartz DRIE techniques, a simple two-mask, wafer-level process to fabricate electrostatic fused quartz resonators has been designed. The fabrication process flow is illustrated in Fig. 4. It starts with forming 5-μm-deep trenches by DRIE in a single-crystal silicon wafer. This silicon wafer becomes a supporting substrate of quartz devices. The trenches are

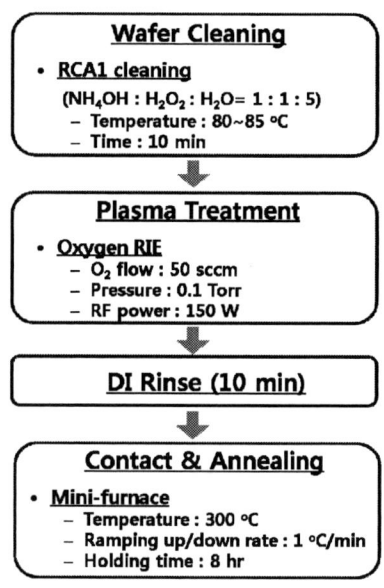

Figure 1: Process flow of SoQ bonding.

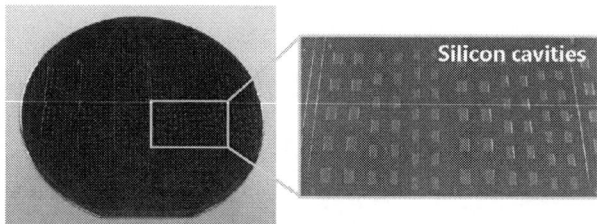

Figure 2: Photograph of the bonded SoQ wafer. Fused quartz wafer is bonded onto the silicon wafer with 5-μm-deep trenches.

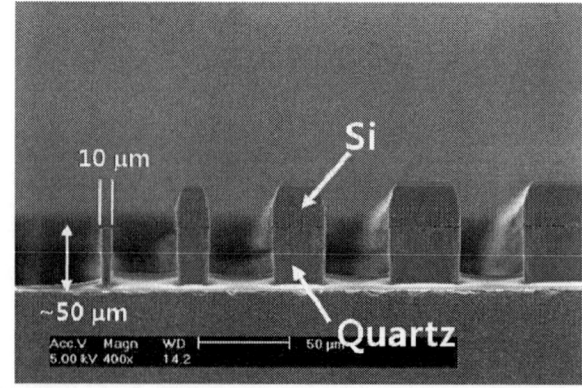

Figure 3: SEM image of 50-μm-thick quartz microstructures etched by DRIE using the bonded silicon mask layer.

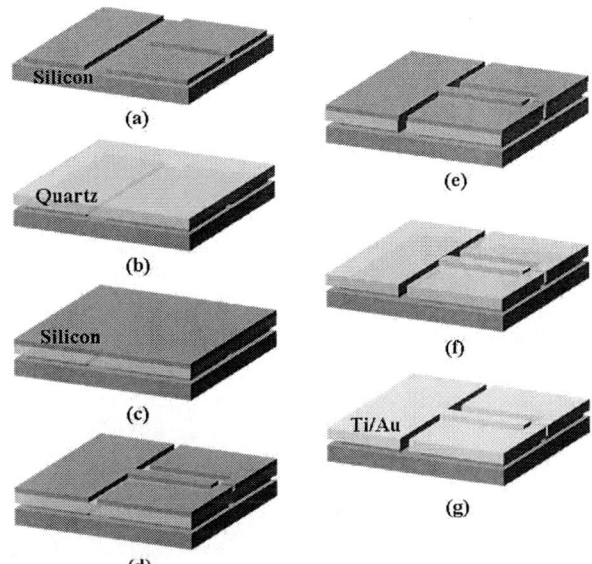

Figure 4: Fabrication process of electrostatic fused quartz resonators: (a) trench formation by Si DRIE, (b) 1st SoQ bonding and quartz thinning, (c) 2nd SoQ bonding and Si thinning, (d) Si mask formation by Si DRIE, (e) Quartz DRIE, (f) Si mask removal, and (g) sputtering of thin metal films.

formed to electrically isolate driving electrodes from the resonator through the silicon substrate, as well as to release resonators automatically at the moment quartz DRIE is finished. This trench-formed silicon substrate is bonded to a fused quartz wafer. After reducing the thickness of the quartz wafer down to the target value of 40 μm by lapping and CMP, another silicon wafer is bonded onto the surface of quartz. The second silicon wafer is also thinned down to 30 μm and patterned by DRIE to form silicon etch masks for quartz DRIE. After quartz DRIE using C_4F_8/He plasma, the silicon mask is removed by DRIE. Finally, the exposed quartz surface is metallized with a conformal coating of Ti/Au (5/50 nm) by sputtering to make the quartz layer conductive for electrostatic actuation.

In order to show the feasibility of the process, cantilever resonators having beam lengths of 900~1200 μm have been fabricated. Photographs of the fabricated fused quartz resonator wafer and the die attached on the PCB board are shown in Fig. 5. Due to the pre-etched trenches, individual cantilevers and two neighboring electrodes are kept electrically disconnected after sputtering. SEM images of the fabricated fused quartz cantilever are shown in Fig. 6. The width of the fabricated beam is 12.8 μm,

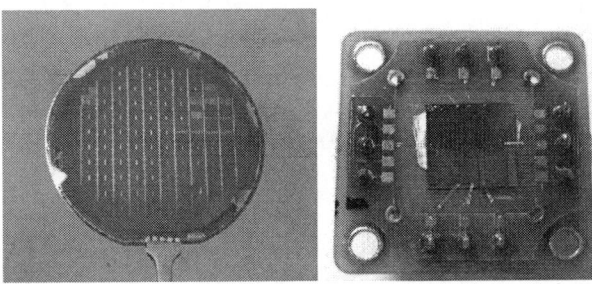

Figure 5: Photograph of the wafer-level fabricated fused quartz resonator and the resonator die attached on the PCB board.

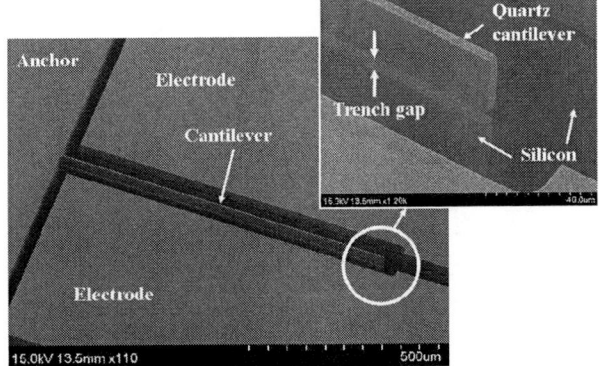

Figure 6: SEM images of the fabricated fused quartz cantilever resonator.

which is reduced from the designed value of 20 μm due to the lateral overetching during the process.

EXPERIMETAL RESULTS

Resonant frequencies and quality factors of the fabricated fused quartz cantilever resonators have been measured using Micro System Analyzer (MSA-400, Polytec) under a vacuum level of < 2 mTorr. The resonator wafer is placed in the SUSS vacuum chamber, and the DC bias voltage of V_P = 7 V and ac driving voltage of v_{pk-pk} = 1 V are applied between one of the electrodes and the DC bias voltage of V_P = 7 V and ac driving voltage of v_{pk-pk} = 1 V are applied between one of the electrodes and the cantilever. The fundamental mode resonant frequency is determined first using the out-of-plane laser Doppler vibrometry. Fast, broadband sweeping is possible in this mode because of the extremely high sensitivity of the system. Then the in-plane motion of interest is analyzed using stroboscopic video microscopy by narrowband sweeping around the resonant frequency.

A typical measured frequency response curve of the cantilever resonator is shown in Fig. 7. Quality factors are evaluated from the frequency response using the equation $Q = f/\Delta f$, where f is the resonant frequency and Δf is the 3-dB bandwidth. The measured resonant frequencies and quality factors are summarized in Table 2. The resonator

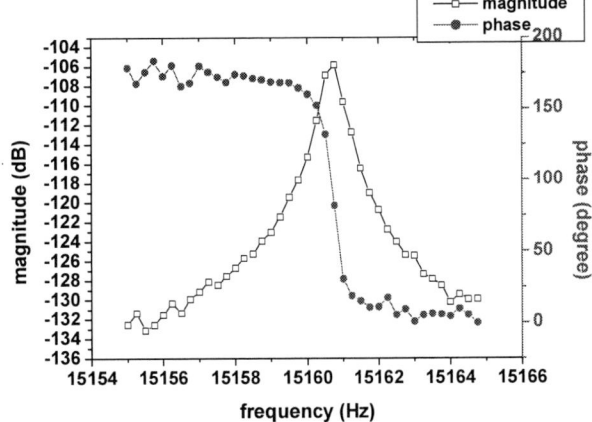

Figure 7: Typical measured frequency response of the fabricated fused quartz cantilever (length of the beam L = 900 μm, width of the beam w = 12.8 μm, thickness of the beam t = 40 μm). Measurement is performed at a vacuum of 2 mTorr.

Table 2: The fundamental resonant frequencies and quality factors of the fabricated fused quartz cantilever resonators.

Beam length [μm]	Measured resonant frequency [kHz]	Quality factors			
		Measured Q	$Q_{support}$ (calculated)	Q_{TED} (calculated)	$Q_{support+TED}$ (calculated)
900	15.16	33,100	753,300	907,300	411,600
1000	12.16	38,100	1,033,000	830,800	460,500
1100	10.11	48,900	1,375,000	793,500	503,200
1200	8.33	21,700	1,786,000	784,800	545,200

Q-values from 21,700 to 48,900 are obtained according to the beam length of the cantilever. Support quality factor ($Q_{support}$) and thermoelastic damping quality factor (Q_{TED}) of the quartz cantilever calculated from the analytic models given in [11] are also included in Table 2 for comparisons. The measured quality factors are lower than those from the analytic models by an order of magnitude. Degradation of quality factors might be attributed partially to the surface loss due to the defects and absorbed contaminants at the quartz surface during and after quartz DRIE process. Another possible reason is internal energy loss in the thin metal film deposited on the quartz surface for resonator actuation. According to our FEM simulation results, the resonant frequency of the cantilever is not significantly changed by coating of very thin metal film. However, it is reported that even very thin metal coating of 100 nm may reduce quality factors by more than an order of magnitude [12]. It is expected that further reduction of the metal thickness or partial removal of metal unnecessary for actuation will increase quality factors of fused quartz resonators further.

CONCLUSIONS

In this work, a novel process combining SoQ direct bonding and quartz DRIE has been proposed to demonstrate electrostatically-actuated fused quartz resonators for high-Q microsensor applications. Two processes of plasma-assisted SoQ bonding and quartz DRIE using wafer-bonded silicon masks are utilized, and these approaches enable us to fabricate resonator arrays supported on the handling substrate in wafer level. Fused quartz cantilever resonators with a thin film metal coating have been fabricated and their resonant frequencies and quality factors have been measured. The measured quality factors are from 21,700 to 48,900. The proposed technique will be very useful to develop quartz-based high-Q microsensors, taking advantages of extremely low thermoelastic damping of fused quartz and simple electrostatic actuation mechanism.

ACKNOWLEDGEMENTS

This research was financially supported by a grant to MEMS Research Center for National Defense funded by Defense Acquisition Program Administration.

REFERENCES

[1] E. P. Eernisse, R. W. Ward, and R. B. Wiggins, "Survey of quartz bulk resonator sensor technologies," *IEEE Trans. Ultrason., Ferroelect., Freq. Contr.,* vol. 35, no. 3, pp. 323-330, 1988.

[2] A. Madni and R. Geddes, "A micromachined quartz angular rate sensor for automotive & advanced inertial applications," *IEEE Sensors,* vol. 16, no. 8, pp. 26-34, 1999.

[3] L. Li, T. Abe, and M. Esashi, "Fabrication of miniaturized bi-convex quartz crystal microbalance using reactive ion etching and melting photoresist," *Sens. Actuators A,* vol. 114, pp. 496-500, 2004.

[4] C. Hedlund, U. Lindberg, U. Bucht, and J. Söderkvist, "Anisotropic etching of Z-cut quartz," *J. Micromech. Microeng.,* vol. 3, pp. 65-73, 1993.

[5] R. L. Kubena and D. T. Chang, US Patent 0017287, 2007.

[6] L. Li, T. Abe, and M. Esashi, "Smooth surface glass etching by deep reactive ion etching with SF6 and Xe gases," *J. Vac. Sci. Technol. B,* vol. 21, no. 6, pp. 2545-2549, 2003.

[7] M. Pavius, C. Hilbert, Ph. Flückieger, Ph. Renaud, L Rolland, and M. Puech, "Profile angle control in SiO2 deep anisotropic dry etching for MEMS fabrication," in *Tech. Dig. IEEE MEMS 2004,* Maastricht, The Netherlands, Jan. 25-29, 2004, pp. 669-672.

[8] Q. –Y. Tong, U. Gösele, T. Martini, and M. Reiche, "Ultrathin single-crystalline silicon on quartz (SOQ) by 150 °C wafer bonding," *Sens. Actuators A,* vol. 48, pp. 117-123, 1995.

[9] C. –H. Chang, C. –J. Peng, G. –L. Lu and T. –S. Lin, "Single-crystalline silicon on quartz (SOQ) wafer by ultra-low temperature (100 °C) wafer bonding and thinning approaches," *Tamkang J. Sci. Eng.,* vol. 8, no. 3, pp. 207-210, 2005.

[10] H. –K. Jung, Y. –S. Hwang, I. –J. Hyeon, Y. –K. Kim, and C. –W. Baek, "Silicon/quartz bonding and quartz DRIE for the fabrication of quartz resonator structure," in *Proc. IEEE NEMS 2008,* Sanya, China, Jan. 6-9, 2008, pp. 1172-1176.

[11] Z. Hao, A. Erbil and F. Ayazi, "An analytical model for support loss in micromachined beam resonators with in-plane flexural vibrations," *Sens. Actuators A,* vol. 109, pp. 156-164, 2003.

[12] R. Sandberg, K. Mølhave, A. Boisen, and W. Svendsen, "Effect of gold coating on the Q-factor of a resonant cantilever," *J. Micromech. Microeng.,* vol. 15, pp. 2249-2253, 2005.

ADJUSTABLE REFRACTIVE INDEX METHOD FOR COMPLEX MICROSTRUCTURES BY AUTOMATED DYNAMIC MODE MULTIDIRECTIONAL UV LITHOGRAPHY

Jungkwun 'JK' Kim, Tae-Soon Yun, Hongsub Jee, and Yong Kyu 'YK' Yoon
Department of Electrical Engineering, University at Buffalo, the State University of New York,
Buffalo, NY 14260, USA

ABSTRACT

A method to use a liquid-state refractive index matching medium is introduced to overcome the limit of the inclined angle for the three-dimensional (3D) microstructures by dynamic mode multidirectional ultraviolet (UV) lithography. The proposed approach uses an isolated container for an index matching medium without direct contact between the index matching medium and the sample, reducing the chance of contamination, simplifying the fabrication process, and broadening the selection of an index matching medium. In addition, the liquid container is designed to allow *in-situ* adjustable refractive index matching performance during a dynamic mode operation. A refracted angle of 58.5° for an incident angle of 67.5° has been obtained for an SU-8 structure using glycerol as an index matching medium. UV lithography using water, solvent, acid, oil, and starch syrup as an index matching medium has been demonstrated. Various microstructures with large inclined or flare angles such as a chevron shape, an ellipsoidal horn, and a chained wind vane are successfully demonstrated with dynamic mode multidirectional UV lithography.

Index Terms- inclined exposure, multidirectional UV lithography, dynamic mode UV lithography, rotational exposure, *in situ* refractive index matching medium, adjustable refractive index.

INTRODUCTION

Recently, an inclined ultraviolet (UV) exposure scheme has been demonstrated to fabricate complex three dimensional (3-D) microstructures such as vertical screen filters, mixers, horn, and nozzles [1-4]. This process has been further advanced using an automated multidirectional scheme, where a collimated UV source is incorporated with a movable stage equipped with two computer controlled motors and a microcontroller for more complex 3-D microstructures such as a vertical triangular slab, a quadruple triangular slab, a cardiac horn, screwed wind vane shapes [5]. However, those previous exercises have been performed in air environment and thus are limited in the achievable flare or inclined angle of the fabricated structures due to the high refractive index difference between air and SU-8, photopatternable photoresist with a refractive index of 1.69 [1]. Note the refractive index of air is unity. Thus the maximum achievable angle of the inclined structure is less than 35° from the vertical line to the substrate surface.

To overcome the limit of the inclined angle, the index matching approach using a liquid medium has been demonstrated [6]. To reduce the difference of refractive indices between air and SU-8, glycerol with a refractive index of 1.56 has been employed between air and the substrate. However, in this approach the static tilting stage

and the sample substrate are submerged all together in glycerol, which potentially increases the chance of contamination during the process and may not be directly applicable to dynamic operation if possible.

In this work, an adjustable refractive index method is demonstrated using various index matching liquids in a separate container, by which the chance of contamination during the process is reduced, the fabrication process is simplified not needing to submerge the whole system into liquid, and the selection of index matching materials is broad. Also, the container for refractive index materials is devised to be flexible to accommodate dynamic mode stage movement while preserving index matching performance. Also, UV lithography experiments using various index matching materials are performed and 3-D complex microstructures with enlarged inclined or flare angles are demonstrated such as a chevron shape pillar array, an ellipsoidal horn array, and the chained vane array using automated dynamic mode multidirectional UV lithography.

ADJUSTABLE REFRACTIVE INDEX MEDIUM

To increase the achievable inclined angle of the tilted 3-D microstructures from the multidirectional UV exposure scheme, an index matching medium between the light source surrounded by air and photoresist such as SU-8 with a high refractive index of 1.69 for an i-line light source needs to be introduced.

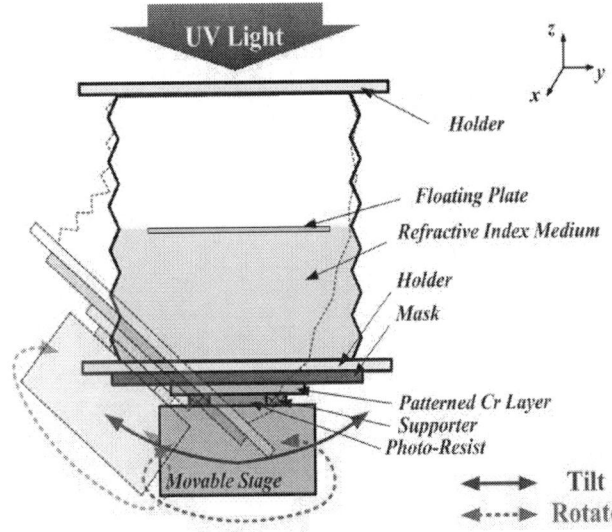

Figure 1: Refractive index matching container with a dynamic mode multidirectional UV lithography stage.

Moreover, for the dynamic mode multidirectional exposure scheme using a fixed collimated light source and a movable stage holding a polymer substrate and photomask, the continuous adjustment of the geometry of an index matching medium is necessary, where the interface between air and the index matching material should be perpendicular to the incident light all the time while the tilting angle of the interface between the index matching material and photoresist is varying. To satisfy these requirements, an approach to use a liquid index medium in a deformable container is introduced.

Figure 1 shows the schematic diagram of the system, where the container filled with a liquid index matching medium is attached to the dynamic stage subject to tilting and rotational movement. The cylindrical side wall of the container with a pleated membrane is flexibly expanded and contracted responding to the movement of the attached stage. The liquid medium is reflowing to fit to the deformed container. To maintain the interface horizontally during dynamic operation, the stage is driven in a slow motion. To help the incident light to be perpendicular to the air-liquid interface and minimize the vibration or perturbation effect during the dynamic mode operation, a UV transparent acrylic floating disc covering the liquid surface is placed in the air-liquid interface. Because of the inertia of the floating disc, the small liquid fluctuation during the dynamic operation is greatly suppressed. Also, a UV transparent acrylic plate is used for the bottom of the container, which is interfaced with the photomask and an SU-8 coated substrate, attached to the movable stage which is equipped with two stepper motors (MD-2A, Arrick Robotics, Inc.) and is controlled by computer. To secure contact between the photomask and the substrate, an elastomeric polydimethylsiloxane (PDMS) layer can be optionally coated on the photomask. The sample supporter can be placed near to the sample to prevent the deformation of photoresist due to the weight of the flexible container. The holder is maintained horizontally during the dynamic mode operation.

Figure 2 shows the system setup. The pleated side wall has been implemented using a bellows shape

aluminum membrane. The rigid aluminum side wall helps prevent the rotational deformation of the container during rotational movement.

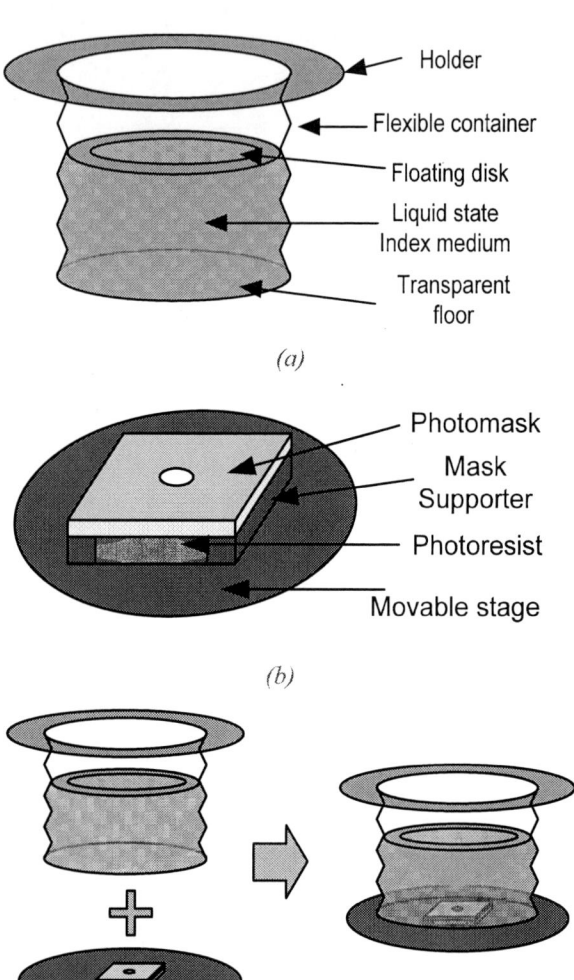

(a)

(b)

(c)

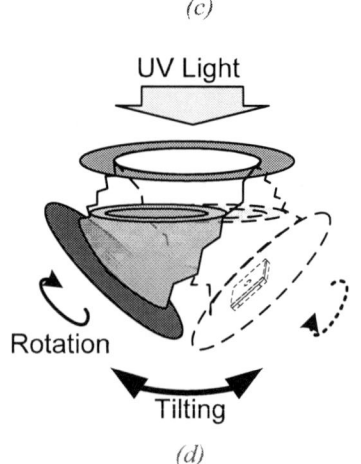

(d)

Figure 3. Fabrication process. (a) Index medium container. (b) Sample preparation. (c) Assembly of the container with the index matching medium and the sample with the movable stage. (d) Dynamic mode UV exposure.

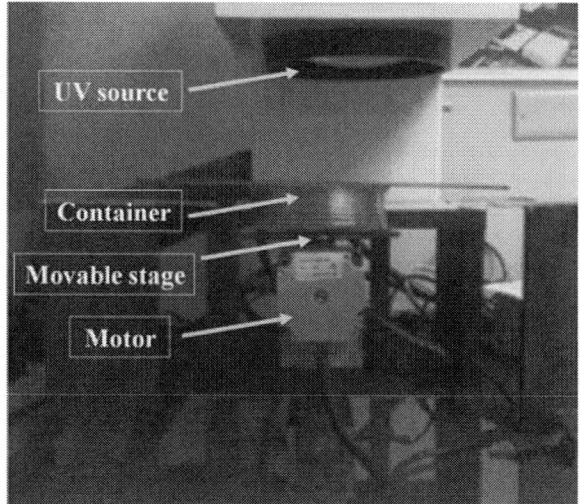

Figure 2: System apparatus.

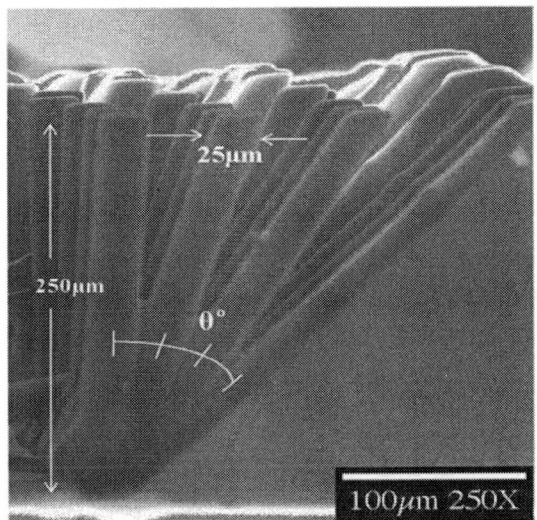

Figure 4: Tilted pillars by refracted UV light.

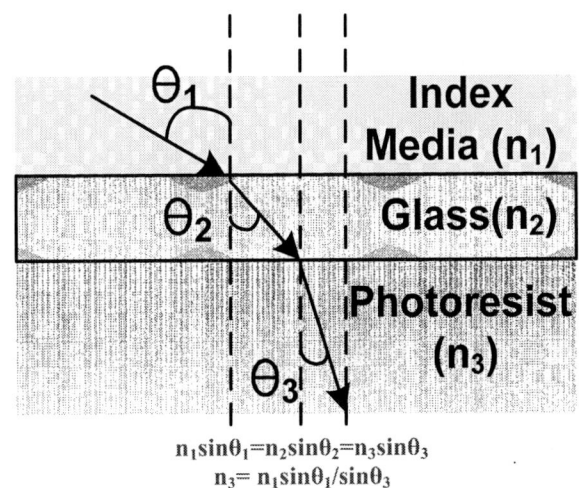

$$n_1\sin\theta_1=n_2\sin\theta_2=n_3\sin\theta_3$$
$$n_3=n_1\sin\theta_1/\sin\theta_3$$

Figure 6. Light refracted at the interfaces due to different refractive indices of materials.

FABRICATION PROCESS

Figure 3 shows an operational process for the dynamic mode multidirectional exposure scheme with the adjustable refractive index medium. The flexible container is filled with an index matching medium. The floating disk is placed on the surface of the liquid in the flexible container (3a). A pre-patterned chromium-coated glass plate is used as a substrate as well as a photomask. SU-8 is coated on the substrate to a thickness which will ultimately define the height of the structure (3b). After the SU-8 is baked, the substrate is turned over and attached to the bottom of the flexible container (3c). The container with the loaded sample is exposed at different angles where the tilting and rotational angles are dynamically changing during UV exposure as the user programmed (3d).

REFRACTIVE INDEX

Since the index matching liquid does not directly contact with the sample substrate, the selection of the index matching material is very broad. Multiple exposures with different incident angles through the clear window photomask have been performed using various refractive index media. A fabricated SU-8 structure from an open window with a diameter of 25 μm and a thickness of 250 μm is shown in Figure 4.

Various incident angles and refracted angles for water, phosphoric acid, starch syrup, and glycerol media have been measured for SU-8 structures. The refractive indices of the materials are experimentally determined for the i-line light source (LS 30, OAI Inc.). Glycerol shows the largest achievable refracted angle among them with a refractive index glycerol of 1.56 which is the closest to that of SU-8 (n_{SU-8} = 1.69) as shown in Figure 5. While a refracted angle of 45°, which is an important angle for many optical applications such as prism and mirror, could not be achieved with an air environment multidirectional scheme [1], it has been obtained with an incident angle of 50° and 51.4° for an index matching medium of glycerol and starch syrup, respectively. Other refracted angles for incident angles of 0°, 22.5°, 45°, and 67.5° are summarized in Table 1, where the refractive index is calculated using measured angles and the Snell's law as shown in Figure 6. The n_1, n_2, and n_3 are the refractive indices of air, glass, and photoresist, respectively, and θ_1, θ_2, and θ_3 are the incident angle, the refracted angle in glass, and the refracted angle in photoresist, respectively.

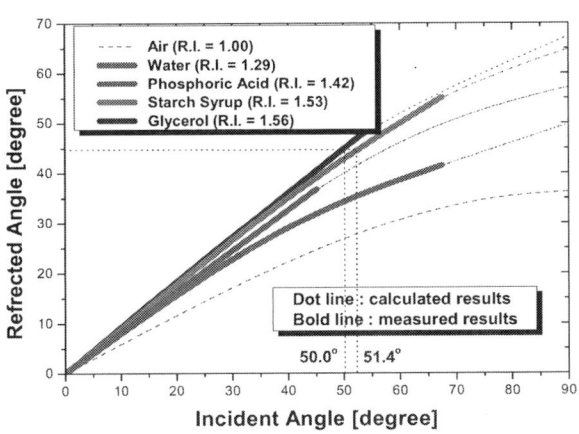

Figure 5: Incident angle versus refracted angle.

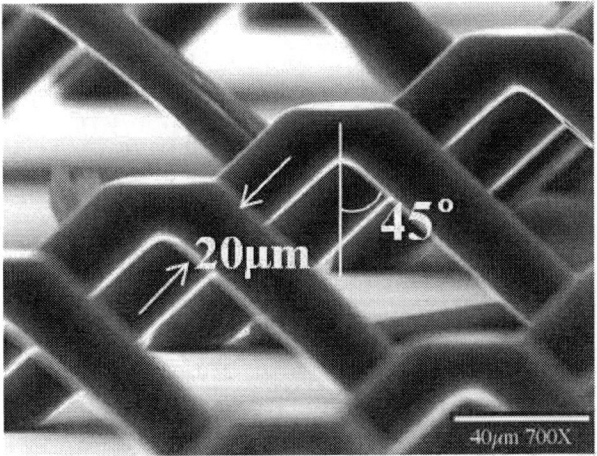

Figure 7: Inclined and reflected pillar array.

978-1-4244-2977-6/09 $25.00 © 2009 IEEE

Table 1. Various liquid state media for refractive index matching.

Incident Angle	Water	IPA	Acetone	Methanol	Nitric Acid	Phosphoric Acid	Starch Syrup	Glycerol
22.5°	18.09°	17.78°	18.43°	20.40°	18.00°	18.44°	20.22°	20.82°
45.0°	32.85°	34.70°	36.00°	40.19°	34.21°	36.74°	39.72°	40.12°
67.5°	44.00°	-	-	-	-	-	-	58.52°
Refractive index	1.29	1.36	1.39	1.54	1.36	1.42	1.53	1.56

RESULT

Since the refractive indices of glycerol and starch syrup are relatively closer to that of the SU-8 than other tested liquids, they are utilized as index matching media in the fabrication of structures with a tilting angle of 45° or greater.

Figure 7 shows an inclined pillar array with a refracted angle of 45° from bottom exposure. The top surface has been interfaced with a reflecting surface and the reflected pillar and the incident pillar form a chevron shape array with an angle of 90° between two branches.

Figure 8(a) shows an array of ellipsoidal horns fabricated using a dynamic mode operation, where a flare angle of 90° is successfully implemented. In Figure 8(b), another dynamic mode operation with glycerol as a matching medium shows a chained wind vane with a flare angle of 85° compared to the isolated wind vane performed in an air environment with a flare angle of 45° [5].

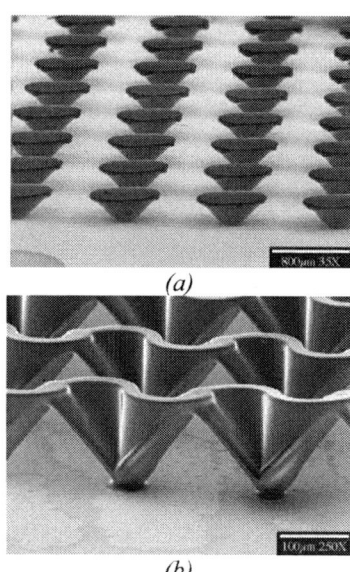

(a)

(b)

Figure 8: Array pattern. (a) Ellipsoidal horn. (b) The chained vane.

CONCLUSION

An adjustable refractive index method with liquid state media in a preferentially deformable isolated container has been demonstrated for complex microstructure fabrication using dynamic mode multidirectional UV lithography. The container compliant to the tilting movement and rigid to the rotational movement has been successfully implemented. Appropriately selected index matching materials allow to tailor the achievable tilting angle of the fabricated structures. Various refractive index media including water, acid, solvent, glycerol, and starch syrup have been tested using the implemented adjustable refractive index medium system. Experimentally fabricated microstructures with glycerol as a refractive index medium show the largest flare angles among them. With the expanded refracted angle capability, more complex 3-D microstructures are available in dynamic mode multidirectional UV lithography, which can find useful applications in optics, RF, and bio fields.

ACKNOWLEDGEMENTS

This work is funded by National Science Foundation CAREER-ECCS 0748153, CMMI 0826434, and University at Buffalo Interdisciplinary Research Development Funding (IRDF 1064615-1-39297).

REFERENCES

[1] Y.K. Yoon, J.-H. Park, and M.G. Allen, "Multidirectional UV lithography for complex 3-D MEMS Structures," *Journal of MEMS*, vol. 15, no. 5. pp. 1121-1130, 2006.

[2] H. Sato, T. Kakinuma, J. S. Go, and S. Shoji, "In-channel 3-D Micromesh Structures using Maskless Multi-Angle Exposures and their Microfilter Application", *Sensors and Actuators* A, 111 (2004), pp. 87-92.

[3] M. Han, W. Lee, S.-K. Lee, and S.S. Lee, "Fabrication of 3D microstructures with inclined/rotated UV lithography," *Proceedings of IEEE Micro Electro Mechanical Systems*, 2003, pp. 554-557.

[4] Y.K. Yoon and M.G. Allen, "Proximity Mode Inclined UV Lithography," *Solid-State Sensor, Actuator, and Microsystems Workshop*, Hilton Head Island, SC, June 4-8, 2006, pp. 98 – 99.

[5] J.K. Kim, M.G. Allen, Y.K. Yoon, "Automated dynamic mode multidirectional UV lithography for complex 3-D microstructures," *Proceedings of IEEE Micro Electro Mechanical Systems*, Jan. 13-17, 2008, Tucson, AZ, pp. 399 – 402.

[6] K. Y. Hung, H. T. Hu, F. G. Tseng, "A novel fabrication technology for smooth 3D inclined polymer microstructures with adjustable angles," Proceedings of Solid State Sensors Actuators, and Microsystems (Transducers '03), Boston, MA, 2003, pp. 821-824.

A NEW HIGH-SENSITIVITY PACKAGE-LEAK TESTING METHOD FOR MEMS SENSORS

M. Fujiyoshi[1], Y. Nonomura[1], and H. Senda[2]
[1]Toyota Central R&D Labs., Inc., Aichi, JAPAN
[2]Toyota Motor Corporation, Aichi, JAPAN

ABSTRACT

A new high-sensitivity package-leak testing method has been developed for micro electro mechanical system (MEMS) sensors such as a vibrating angular-rate sensor housed in a vacuum package. Procedures of the method to obtain high leak-rate resolution are as follows. (1) A package filled with helium gas is kept in a small and closed chamber to accumulate helium gas leaking out of the package. (2) The chamber is chilled with liquid-N_2 after the accumulation to reduce background gases except the helium gas. (3) The helium gas is transported to a mass spectrometer in a short time and recorded as a transient waveform. The leak-rate resolution of 1×10^{-17} Pa·m^3/s was obtained. This resolution was 10^5 times superior to that of the conventional method.

INTRODUCTION

Vibrating sensors, such as angular-rate sensors, fabricated by MEMS technology are usually housed in vacuum packages, because the sensitivity of the sensors is strongly influenced by the air-viscous effect that occurs in small gaps of a comb drive and between a mass portion and a substrate. If a tiny leak is present in a package, the sensitivity of a sensor will be degraded after a long time. Thus, to ensure the sensitivity of the sensors for long time, both a package-sealing technique for reduction of the leak and a package-leak-testing method with high sensitivity are required. These requests have recently become stronger with decreasing size of sensors.

Detectable ranges of various leak-testing methods [1]-[3] are shown in Fig. 1. The helium vacuum integration method has the highest sensitivity among conventional leak-testing methods. However, the sensitivity of the method is only as low as 10^{-12} Pa·m^3/s, a level which is not enough to ensure the sensitivity of vibration sensors for more than ten years. In this method, a package filled with helium gas is kept in a large vacuum chamber, and the leak-rate is obtained by monitoring the concentration of the helium gas in the chamber with a mass spectrometer for a long time (a few hours). Because of the large chamber and the long measurement time, the contribution of background gases desorbing from the vacuum apparatus to the mass spectrometer signal is large. As another method for leak-tests, micro pirani gauge had been applied to monitoring the pressure in small packages [4],[5], but the origin of the pressure variation can not be identified, e.g. background gases from the inside of the package or leakage from out of the package.

In this paper, we have developed a high-sensitivity package-leak testing method using a high-resolution vacuum integration method (the HR method). The leak rate resolution of 1×10^{-15} Pa·m^3/s was achieved by the HR method. Furthermore, we have improved the HR method by using a liquid-N_2 chilled vacuum integration chamber (the chilling HR method). The leak rate resolution of 1×10^{-17} Pa·m^3/s was attained by the chilling HR method.

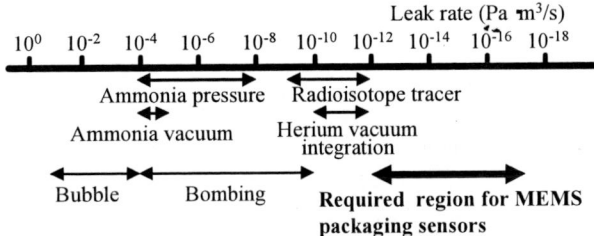

Figure 1: Detectable ranges of various leak-testing methods.

HR METHOD

Principle of the basic HR method

Configuration of the basic HR method measurement system is shown in Fig. 2. A sample was small package for sensors. The package filled with helium gas was set in a small chamber (Fig. 3), and the chamber was evacuated and shut by a valve connecting to a mass spectrometer. Then, the helium gas leaking out of the package was accumulated in the chamber. After the accumulation time t_s, the helium gas was transported to the mass spectrometer by opening the valve.

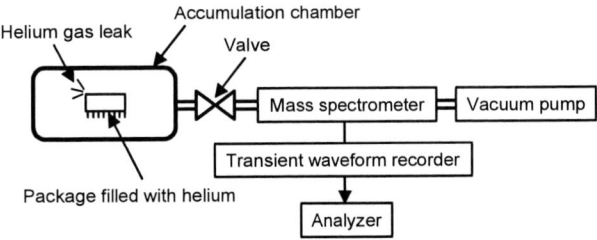

Figure 2: Configuration of basic HR method.

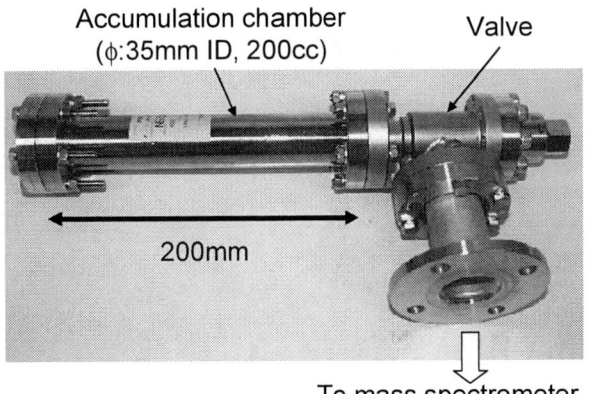

Figure 3: Photograph of package-leak testing apparatus.

978-1-4244-2977-6/09 $25.00 © 2009 IEEE

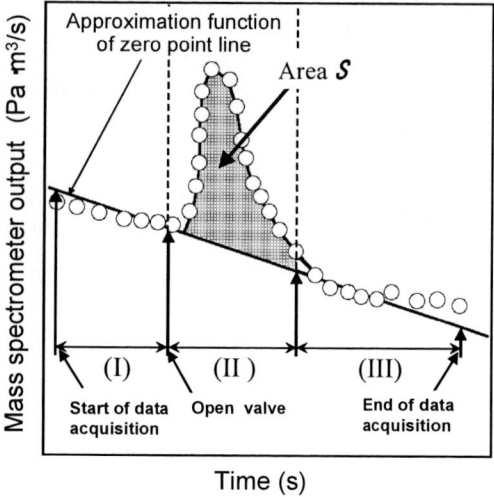

Figure 4: Waveform of mass spectrometer output at mass number 4 during the helium gas transportation from accumulation chamber to mass spectrometer. The area S, integration of waveform, corresponds to total amount of the helium gas accumulated in accumulation chamber.

$$Q = \frac{S}{t_s \cdot \left(\frac{P}{P_0} \right)} \qquad (1)$$

Whereas;
Q : Leak-rate ($Pa \cdot m^3/s$)
t_s : Accumulation time (s)
S : Total helium gas quantity ($Pa \cdot m^3$)
P : Helium gas pressure sealed in package (Pa)
P_0 : Standard pressure (Pa), P_0 : 1×10^5 (Pa).

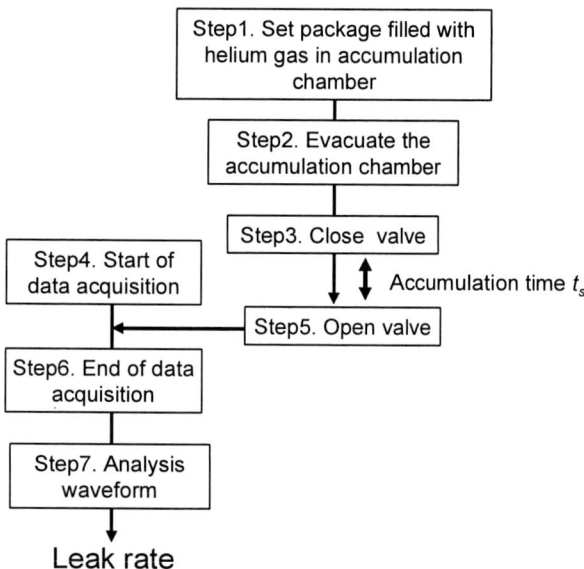

Figure 5: Measuring procedures.

The amount of the helium gas flux from the chamber was monitored with the spectrometer and recorded as a transient waveform shown in Fig. 4. Usually, the transient

measurement finished within 30s. The total amount of the helium gas accumulated in the chamber was obtained as an area S between the waveform and the approximation function of zero point line. The leak-rate Q of the package was obtained by dividing the total amount of the helium gas S by the accumulation time t_s as shown in Eq. 1. These measuring procedures are summarized in Fig. 5.

The basic HR method has the next three merits for leak-tests.

(1) Long term accumulation increases the partial pressure of the helium gas leaking from the package.

(2) Since the volume of the accumulation chamber is limited to small one and the mass spectrometer is out of the chamber, the influence of background gases is small, compared with the conventional method. These ideas are effective to improve signal-noise ratio.

(3) Measurement is carried out in a short time. The peak profile of transient waveform becomes keen at the mass spectrometer, and the measurement resolution goes up.

Chilling HR method

During long term accumulation, background gases such as water, hydrocarbon, and carbon dioxide are desorbed from the surfaces of the accumulation chamber. The fragmentation of the hydrocarbon during ionized at the mass spectrometer also arose. The background gases and the fragmentation from the background gases were misunderstood as helium gas signal. Therefore, the resolution of the leak-tests was restricted in the basic HR method.

We reduced background gases by chilling the chamber. At the end of the accumulation of the helium gas leaking out of the package, the chamber was chilled with liquid-N_2. Since the boiling point of N_2 (77K) is higher than that of the helium (4.2K) and lower than those of background gases such as water and hydrocarbon, the background gases must be selectively adsorbed on the chamber wall by liquid-N_2 chilling. Also, the background gases were further reduced by prebaking the chamber and increasing the evacuation time before the accumulation.

A configuration of the chilling HR method is shown in Fig. 6. The chilling HR method has a liquid-N_2 chiller on the accumulation chamber in addition to the configuration of the basic HR method in Fig. 2.

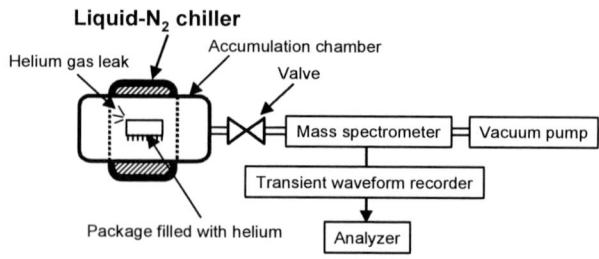

Figure 6: Configuration of chilling HR method.

EXPERIMENTS

To determine the leak-rate sensitivity of the basic HR method and the chilling HR method, the amount of background gases accumulated in the chamber was measured in the same manner without any packages. In this case, the mass spectrometer output at mass number 4 does not originate from the helium but from the background gases.

Chilling effect on background gas reduction

Experiments were carried out for two patterns (A) and (B) as follows.

(A) Basic HR method: without chilling; evacuation time t_p; 10 minutes and accumulation time t_s; from 76 min to 2 days.

(B) Chilling HR method: with chilling; evacuation time t_p; 10 minutes and accumulation time t_s; from 6 days to 8 days.

A photograph of the chilling HR method apparatus is shown in Fig. 7. The liquid-N_2 was supplied from the inlet and kept in the chiller made of thermal insulators. The liquid-N_2 chilled the accumulation chamber outer wall directly. As a measuring procedure, at the end of the accumulation time t_s, the accumulation chamber was chilled by injecting liquid-N_2 through the inlet. The chamber was kept chilled for the chilling time t_{N2}. The chilling time t_{N2} was set to 2 minutes after injecting the liquid-N_2. The chilling time t_{N2} was derived from the experimental results of the pressure measurement in the accumulation chamber with N_2 chilling.

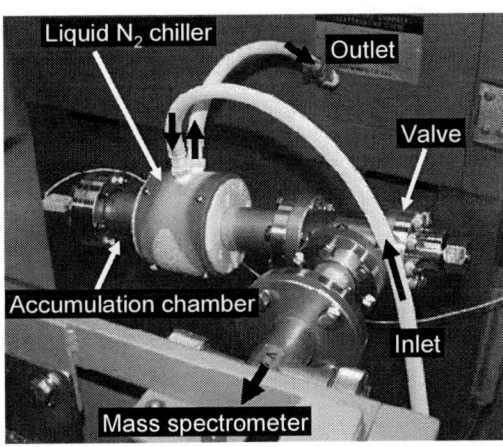

Figure 7: Photograph of leak-testing apparatus with liquid-N_2 chiller.

Baking and evacuation effect on background gas reduction

Effects of baking and evacuating the chamber were evaluated. Experiment was carried out for pattern (C) as follows and the results were compared with those of the pattern (B).

(C) Chilling HR method : with baking, long evacuation, and chilling: baking time; 120 minutes at 150°C, evacuation for latter half of the baking time, evacuation time t_p; 60 minutes, and accumulation time t_s; from 6 days to 13 days.

Accuracy verification

To verify accuracy of the basic HR method and the chilling HR one, the methods were calibrated with a standard helium gas leak source, which leak-rate was 2.1×10^{-10} Pa·m³/s (in terms of Air). The standard leak source was connected to the end of the accumulation

chamber as shown in Fig. 8. The verification was done by measurement of the total helium quantity. The total helium quantity was realized by multiplying the time during the source valve open and the leak-rate of the standard leak source. In this experiment, the time during the source valve open was equivalent to the accumulation time t_s in the HR method.

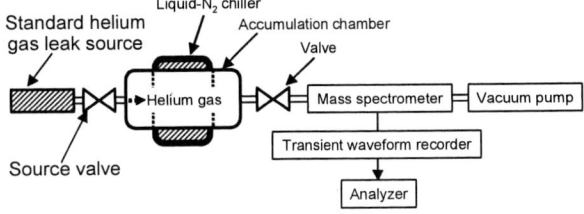

Figure 8: Accuracy verification method.

RESULTS

Chilling effect on background gas reduction

Transient waveforms of two results, the pattern (A) (accumulation time t_s: 73 hours) and the pattern (B) (accumulation time t_s: 64 hours), are shown in Fig. 9 with the timing of opening the valve. Each waveform had a peak just after opening the valve. There was no helium gas in the accumulation chamber of each the pattern (A) and (B). Therefore, the shadowed areas were the integrated mass spectrometer outputs due to the background gases. The shadowed area of the (B) was significantly few compared with that of the (A). This result means chilling the accumulation chamber is effective to reduce the background gases.

Figure 10 shows the integrated mass spectrometer outputs as a function of the accumulation time for the different conditions. The data for each condition are roughly on a straight line, whose gradient corresponds to the leak-rate.

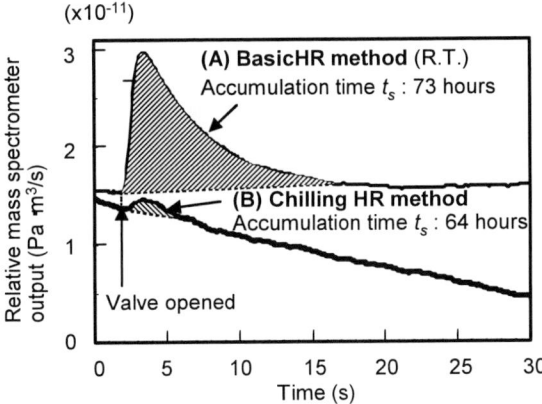

Figure 9: Transitional waveform of the mass spectrometer.

In these experiments there were no packages in the chamber. Therefore, the obtained leak-rates were equal to the leak-rate resolutions of the methods. The leak-rate resolution of the (A) was 1×10^{-15} Pa·m³/s and that of the (B) was 4×10^{-17} Pa·m³/s. The leak-rates resolution of the (A) was 10^3 times superior to the conventional vacuum integration method. The leak-rates resolution of the (B)

978-1-4244-2977-6/09 $25.00 © 2009 IEEE

with the chilling was enhanced for twenty five.

Baking and evacuation effect on background gas reduction

Baking to 150°C and long time evacuation were also effective to reduce the background gases. Comparing the pattern (B) and the pattern (C) in Fig. 10, the integrated mass spectrometer output was reduced to one fourth. The chilling HR method of the pattern (C) achieved the leak-rate resolution of 1×10^{-17} Pa·m³/s. This resolution was 10^5 times superior to the conventional vacuum integration method. The leak-rate resolutions under different conditions are summarized in Table 1.

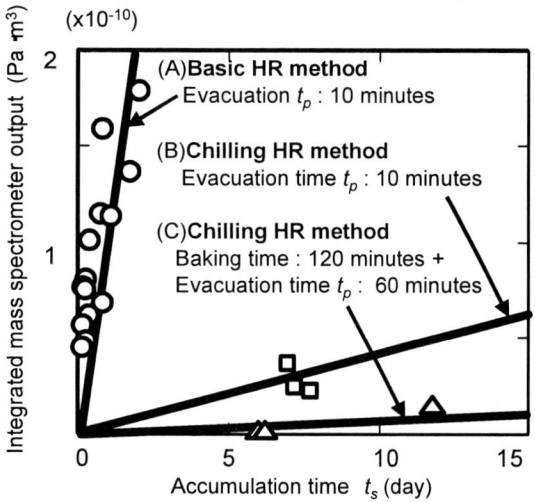

Figure 10: Integrated mass spectrometer output as a function of accumulation time ts. The measurement was done without any package to evaluate the background effect.

Table 1. Leak-rate resolution under different conditions for the three testing methods.

Conditions	Baking time (min)	Evacuation time t_p (min)	Resolution (Pa·m³/s)
(A) Basic HR	—	10	1×10^{-15}
(B) Chilling HR	—	10	4×10^{-17}
(C) Chilling HR with baking	120 (at 150°C)	60	1×10^{-17}

Accuracy verification

The measuring system was calibrated with the standard helium gas leak source. The experimental results of the basic HR method, the chilling HR method and the calculation result were compared as shown in Fig. 11. The difference between the results of the basic HR method and the results of the chilling HR method was within 0.5%. The differences between the measured results and the calculation one were within 8%. Chilling the chamber with liquid-N_2 reduced the background gases effectively and no absorption for helium was observed.

CONCLUSION

We have developed the high-sensitivity package-leak testing method of the HR method. By the basic HR method, the leak-rate resolution of 1×10^{-15} Pa·m³/s was obtained. By chilling the accumulation chamber by liquid-N_2, the background gases except the helium were efficiently reduced. The leak-rate resolution of 1×10^{-17} Pa·m³/s was obtained by the chilling HR method with baking, with baking time: 120 minutes at 150°C, evacuation time: 60 minutes, and accumulation time: from 6 days to 13 days. The result was 10^5 times superior to the conventional vacuum integration method. The accuracy of the measuring methods was verified with a standard helium gas leak source and was proved within 8%.

The HR method was applied to development of package-sealing technique for automotive angular-rate sensors. The HR method is also useful for development pressure sensor, accelerometer sensor and inspection of leak check for ultrahigh vacuum apparatus.

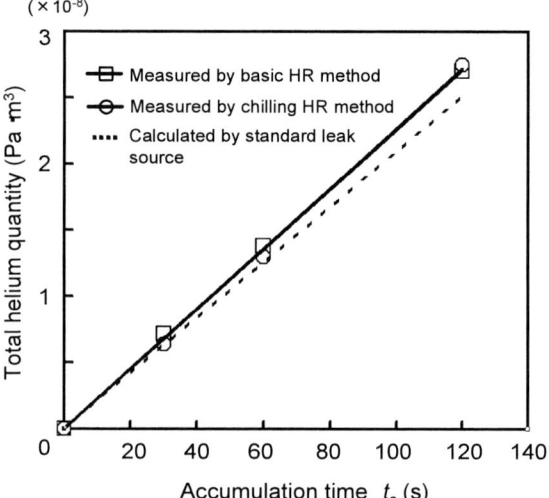

Figure 11: Results of accuracy verification.

REFERENCES

[1] JIS (Japanese Industrial Standards), Z 2331, 2332, 2333, C0026, C6701

[2] ASTM E499-95 Standard Test Methods for Leaks, 2000

[3] A. Igawa, "Helium leak detector", J of Vacuum, vol. 34, No.11, 1991

[4] B.H. Stark, J. Chae, A. Kuo, A. Oliver, K. Najafi, "A high-performance surface-micromachined pirani gauge in SUMMIT V", Proc. IEEE MEMS'05, Miami, USA, pp.295-298, 2005

[5] J. Chae, JM. Giachino, K. Najafi, "Wafer-level vacuum package with vertical feedthroughs", Proc. IEEE MEMS'05, Miami, USA, pp.548-551, 2005

SPATIALLY ARRANGED MICROELECTRODES USING WIRE BONDING TECHNOLOGY FOR SPATIALLY DISTRIBUTED CHEMICAL INFORMATION ACQUISION

W. Tonomura[1], K. Shimizu[1], and S. Konishi[2]

[1]Graduate School of Science and Engineering, Ritsumeikan University, Kusatsu, Shiga, JAPAN

[2]Department of Micro System Technology, Ritsumeikan University, Kusatsu, Shiga, JAPAN

ABSTRACT

This paper presents spatially arranged microelectrodes to allow real-time monitoring of behavior of spatially distributed chemical. Out-of-plane microelectrodes standing on a substrate with gradation in height are developed for the purpose. Wire-bonding-based probe technology [1] makes it possible to provide spatially arranged microelectrodes. The wire-bonding-based probe technology combines wire bonding and laser machining. Bonded metal wires are converted to probe arrays by cutting the bridge. Spatially arranged microelectrodes can detect spatially distributed chemical in real time would understand various phenomena caused by spatially distributed chemical. This paper demonstrates spatially arranged microelectrodes could be used as the working electrodes of spatially electrochemical sensors in chronoamperometric measurements using $K_3Fe(CN)_6$.

INTRODUCTION

Recently owing to the advances of microfabrication techniques, microfluidic devices integrated with electrodes have been extensively exploited to facilitate the electrochemical detection in a micro-TAS platform.

We presented a gate type electrochemical sensor integrated three-dimensional mesh electrode into a microfluidic channel as shown in Fig. 1 (a) [2]. A planar electrode which is used as a sensor in the microfluidic channel has difficulty in sensing characteristics of non-uniform flow such as laminar flow. The gate type sensor can be expected to overcome this limitation of a planar electrode. Insulation structures of polymer materials can be transformed into conductive structures through pyrolysis. The three-dimensional SU-8 original structure was micromachined by multi-angle inclined lithography. Three-dimensional structures of pyrolyzed SU-8, which have meshes of 10 μm × 20 μm in dimension, could be obtained by pyrolysis in N_2 atmosphere. Furthermore, the structures were integrated into the SU-8 fluidic channel with the 100 μm in height and 200 μm in width by a post-pyrolysis process.

This paper presents spatially arranged microelectrodes which is isolated each other for real-time monitoring of spatially distributed chemical as shown in Fig. 1 (b). Spatially arranged microelectrodes are composed of metal wires and microelectrode arrays with a suction hole. This study focuses on spatially electrochemical sensors while it is also expected to detect spatially oxygen consumption of embryo [3-5] using a suction hole.

Dr. Shiku et al. reported a study quantifying the oxygen consumption of each single embryo by scanning electrochemical microscopy (SECM), a technique in which the tip of a microelectrode is scanned to monitor the local distribution of electroactive species near the sample surface [3]. The embryo was manipulated and held at the center of the optical microscopic view with a manual micropositioner and a manual microinjector. The microelectrode tip approached the embryo by scanning in the X-, Y-, and Z- directions using a motor-driven XYZ stage located on the microscope stage. For the X- and Y- directions, the tip scanned a 500 μm step. For the Z- direction, the tip scanned a 160 μm step. However, skillful manipulations using a microinjector and the SECM take much time to practice and the rate of successful manipulations is low.

Proposed spatially arranged microelectrodes would be able to overcome the above drawbacks of manual manipulation. Spatially arranged microelectrodes are composed of metal wires and microelectrode arrays with a suction hole for clamping of a bio particle. The measurement system allows self-alignment between a clamped bio particle and each spatially arranged microelectrode. The self-alignment measurement system can reduce difficulties in manipulating a microinjector and scanning a microelectrode in conventional methods.

Here, the design and fabrication of spatially arranged microelectrodes for real-time spatial monitoring is described. Successful chronoamperometric measurement results as spatially electrochemical sensors will be also reported.

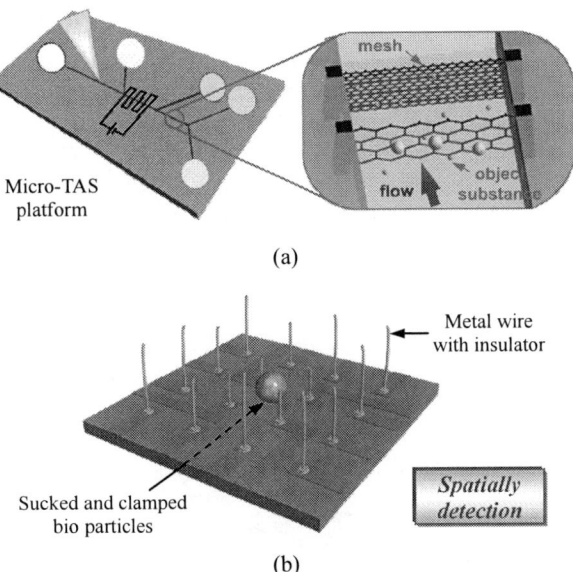

(a)

(b)

Fig. 1: Electrochemical sensor for spatially distributed chemical information acquision. (a) Three-dimensional mesh electrode integrated into flow channel and (b) Spatially arranged microelectrodes.

ISOLATED SPATIALLY ARRANGED MICROELECTRODES

Structure and Specification

The proposed spatially arranged microelectrodes has 4 × 4 arrayed metal wires bonded on the tip of each microelectrode by using fabrication technologies of wire bonding and laser machining. Compared with previous gate type electrochemical sensor integrated three-dimensional mesh electrode into a microfluidic channel, spatially arranged microelectrodes has advantages as follows: i) Isolated spatially arranged microelectrodes for real-time monitoring of spatially distributed chemical; ii) Various heights of microelectrodes on a substrate.

A cross sectional view of spatially arranged microelectrodes composed of upper and lower structures is shown in Fig. 2. The upper structure consists of a suction hole, a suction channel and microelectrode arrays. On the other hand, the lower structure consists of a microelectrode under the suction hole. After the upper and lower structures are bonded, spatially arranged microelectrodes are formed by using wire bonding and laser machining.

The specification of spatially arranged microelectrodes is designed for real-time monitoring of spatially distributed chemical. An array of 4 × 4 independent metal wires bonded on the tip of each microelectrode is arranged for spatially multipoint recording. The diameters of the metal wire and suction hole are Φ 25 μm and Φ 30 μm, respectively. The distance between the tips of each microelectrode is designed for 200 μm by taking account of previous reports on measurement of oxygen consumption of embryo [3-5] as an example of spatially distributed chemical.

Measurement Principle

Isolated 16 metal wires bonded on the tip of each microelectrode are developed as the spatially arranged microelectrodes for working electrodes in electrochemical detection. The tip of metal wire coated by a thin polymer as an insulator is exposed by the laser cutting. Spatially arranged microelectrodes allow self-aligned measurements between a bio particle clamped in the center of chip and spatially arranged microelectrodes. The self-alignment measurement system can solve difficulties and low throughput in manipulating a microinjector and scanning a microelectrode in conventional methods.

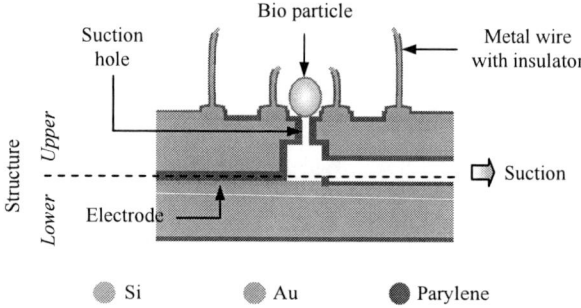

Fig. 2: A cross sectional view of spatially arranged microelectrodes composed of upper and lower structures.

FABRICATION OF ISOLATED SPATIALLY ARRANGED MICROELECTRODES

Fabrication Process

Figure 3 explains a proposed fabrication method for the probe based on wire bonding technology. At first, an upper structure is fabricated by using 2-step DRIE process, parylene deposition (10 μm) and Cr / Au (50 nm / 200 nm) electrode patterning (see Fig. 3 (a)). A lower structure is fabricated by Cr / Au (50 nm / 200 nm) electrode patterning, followed by parylene deposition (10 μm) (see Fig. 3 (b)). After that, upper and lower structures are bonded by parylene-parylene thermo compression bonding (see Fig. 3 (c-1)). Finally, spatially arranged microelectrodes are fabricated. The bridge of Au wire between each microelectrode and dummy substrates (250 μm or 500 μm) coated Cr / Au (50 nm / 200 nm) is provided by a manual wire bonding machine (see Fig. 3 (c-2)). Au wires coated parylene (1 μm) as an insulator are converted to spatially arranged microelectrodes by cutting the bridge of wire using an excimer laser machining (see Fig. 3 (c-3) and (c-4)).

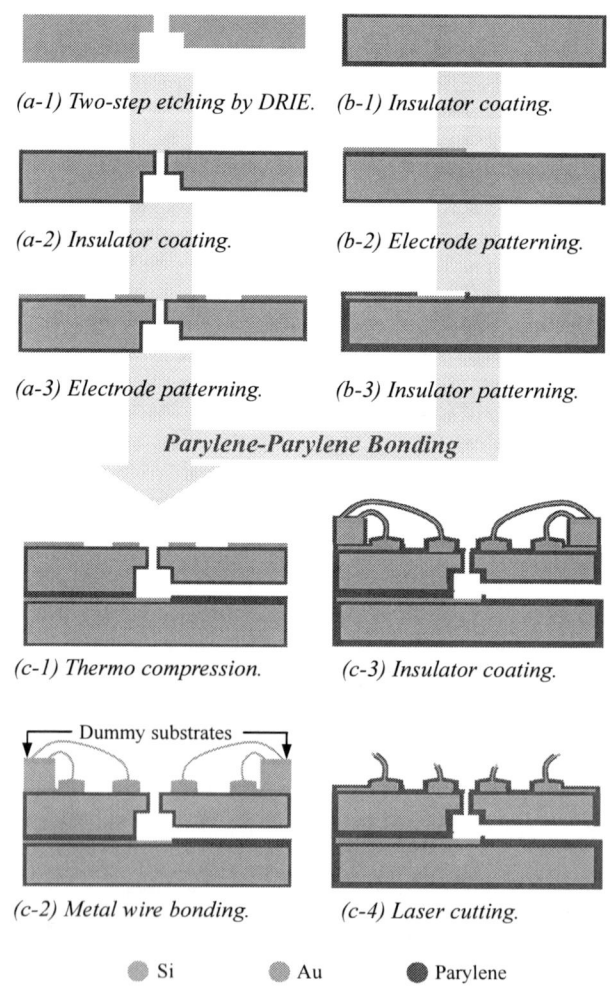

Fig. 3: Fabrication process of spatially arranged microelectrodes. (a) Upper structure, (b) Lower structure and (c) Combination of wire bonding and laser machining after bonding of upper and lower structures.

Fabricated Spatially Arranged Microelectrodes

Figure 4 shows fabricated spatially arranged microelectrodes. It consists of upper and lower structures. Whole size is 20 mm × 20 mm (see Fig. 4 (a)). A SEM image of spatially arranged microelectrodes is shown in Fig. 4 (b). 4 × 4 isolated microelectrodes forested successfully on the planar microelectrode arrays. Microelectrodes having different heights are spatially as designed. The height of standing electrode is limited by the height difference between planar microelectrode arrays and the top surface of dummy substrates. The height of each standing electrode is finally decided by laser cutting within the above limitation.

Magnified view of the tip of a spatial microelectrode is shown in Fig. 4 (c). The tip of Au wire (Φ 25 μm) coated by parylene as an insulator is exposed by the laser cutting.

THREE-DIMENSIONAL POSITION COORDINATES OF SPATIALLY ARRANGED MICROELECTRODES

Measurement System

Fabricated spatially arranged microelectrodes have different heights so as to execute real-time monitoring of distributed chemical. It is important to correspond monitored data to three-dimensional position of each microelectrode tip for spatial sensing. Three-dimensional position of each microelectrode tip was measured by using a digital microscope (KH-7700, HiROX Corporation). The digital microscope was equipped with a three-dimensional surface analysis soft (SurfLab/KH, Mitani Corporation) and XYZ-electric driven stage (SHX-100, HiROX Corporation).

It becomes also important to apply the position measurement system to a calibration against individual microelectrode chips. Figure 5 (a) shows three-dimensional position coordinates of the tips of standing microelectrodes. The suction hole in the center of chip was defined as the origin of coordinates. Each numerical value of X-, Y- and Z-position represents the distance from the suction hole. The height of each standing electrode is limited by the thickness of the dummy substrates as explained ($Z \leq t_D + h_W$). The thickness of dummy substrates was 250 μm in this work. Figure 5 (b) shows a three-dimensional graphical representation based on the measured position coordinates of microelectrode tips. The highest and lowest heights of demonstrated microelectrodes were 298 μm and 161 μm, respectively.

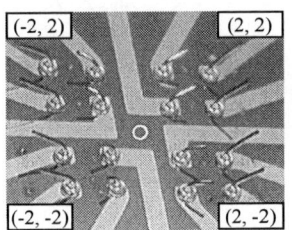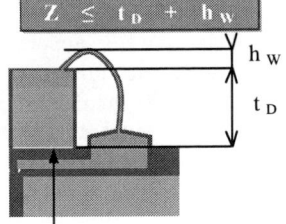

(⊚ : origin of coordinate)   dummy substrate

	(-2, 2)	(-1, 2)	(1, 2)	(2, 2)
X	-871	-469	217	622
Y	575	546	503	465
Z	285	163	171	286
	(-2, 1)	**(-1, 1)**	**(1, 1)**	**(-2, 1)**
X	-912	-486	198	592
Y	275	278	199	127
Z	280	169	162	292
	(-2, -1)	**(-1, -1)**	**(1, -1)**	**(-2, -1)**
X	-976	-529	159	546
Y	-184	-210	-297	-260
Z	283	163	161	298
	(-2, -2)	**(-1, -2)**	**(1, -2)**	**(2, -2)**
X	-999	-543	117	532
Y	-473	-506	-560	-544
Z	286	166	181	287

(μm)

(a)

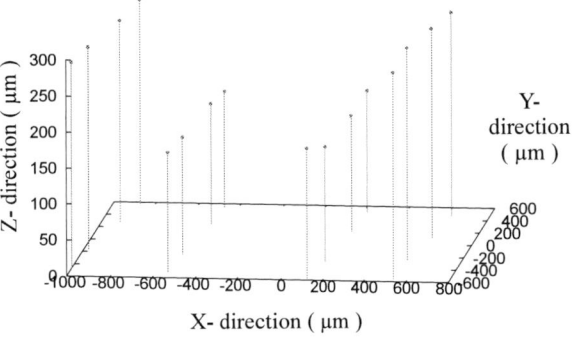

(b)

Fig. 5: Three-dimensional position coordinates of spatially arranged microelectrodes. (a) Three-dimensional position coordinates of the tips of standing microelectrodes and the definition of height of each standing microelectrode and (b) Three-dimensional graphical representation.

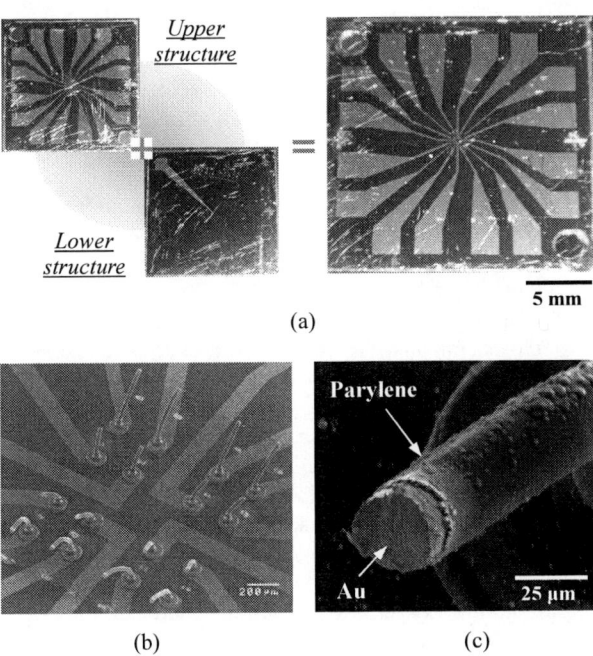

(b)   (c)

Fig. 4: Fabricated spatially arranged microelectrodes. (a) Whole view bonded upper and lower structures, (b) A SEM image of spatially arranged microelectrodes and (c) A SEM image of the tip of one microelectrode.

ELECTROCHEMICAL SENSING OF ISOLATED SPATIALLY ARRANGED MICROELECTRODES

Electrochemical characteristics of the spatially arranged microelectrodes were evaluated by chronoamperometric measurements. Electrochemical measurements were conducted using standard three-electrode configuration. The fabricated spatially arranged microelectrodes were employed as working electrodes. A Pt wire was employed as a counter electrode. A silver / silver chloride (Ag / AgCl) was served as a reference electrode. Electrochemical analyzer (Model 701C, ALS / CH Instrument) was used to conduct electrochemical measurements including a cyclic voltammetry and a potential-step chronoamperometry.

$K_3Fe (CN)_6$ (Kanto Chemical Co., Inc.) and KNO_3 (Wako Pure Chemical Industries, Ltd.) are employed for an electrolyte and a supporting electrolyte, respectively. All solutions are prepared with deionized water of resistivity not less than 18.2 MΩcm.

Figure 6 shows the current response of the dual spatial microelectrodes to drop 2 mM $K_3Fe (CN)_6$ droplet into PBS as the supporting electrolyte. The tip positions of employed two spatial microelectrodes are (- 546, - 260, 298) and (159, - 297, 161), respectively. As a result, more quick response to the droplet of $K_3Fe (CN)_6$ could be observed at (- 546, - 260, 298) near the drop point. These successful results proved that the spatially arranged microelectrodes could detect the behavior of spatially distributed chemical in real time.

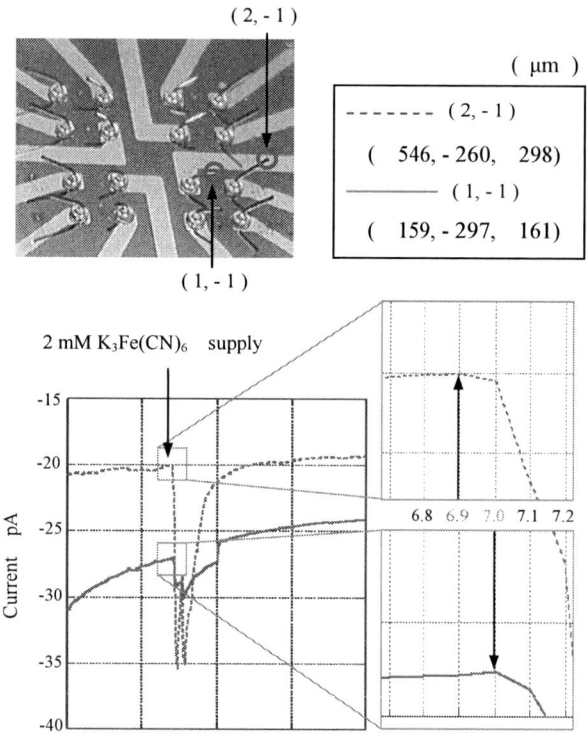

Fig. 6: Electrochemical real-time responses of the dual spatial microelectrodes to drop electrolyte droplet into PBS as the supporting electrolyte.

CONCLUSION

Novel spatially 16 arranged microelectrodes to allow real-time monitoring of behavior of spatially distributed chemical is described in this paper. Wire-bonding-based probe technology makes it possible to provide out-of-plane arranged microelectrodes. The tip of each metal wire (Φ 25 μm) coated parylene as an insulator is exposed by the laser cutting. The highest and lowest height of fabricated microelectrodes could be obtained 298 μm and 161 μm, respectively. The difference of dummy substrates thickness and laser cutting position provide various heights of structures for a spatial microelectrode.

In electrochemical spatial detection, spatially arranged microelectrodes could be successfully used as the working electrodes of spatially electrochemical sensors, because more quick response to an electrolyte droplet could be observed at an isolated spatial microelectrode near the drop point.

Given the results described here, further studies of spatially distributed chemical information acquision using developed spatially arranged microelectrodes are underway in order to high-throughput analysis of cellular respiration toward fertility treatment.

ACKNOWLEDGMENT

This work was supported by Grant-in-Aid for Scientific Research on Priority Areas "Lifesurveyor" from the Ministry of Education, Culture, Sports, Science and Technology of Japan, and was also supported by Research Fellowships of the Japan Society for the Promotion of Science (JSPS) for Young Scientists.

REFERENCES

[1] K. Shimizu, M. Nakanishi, M. Makikawa, S. Asajima and S. Konishi, "Combination of Wire Bonding and Laser Machining for Flexible Probes as Low Damage Electrical Interface", *Proc. of APCOT 2008 Conference*, Tainan, Taiwan, June 22-25, 2008.

[2] K. Yamamoto, K. Naka, Y. Nagaura, H. Sato, S. Shoji and S. Konishi, "Pyrolyzed Polymer Mesh Electrode Integrated into Fluidic Channel for Gate Type Sensor", *Proc. of MEMS 2007 Conference*, Kobe, Japan, Jan. 21-25, 2007, pp. 271-274.

[3] H. Shiku, T. Shiraishi, H. Ohya, T. Matsue, H. Abe, H. Hoshi and M. Kobayashi, "Oxygen Consumption of Single Bovine Embryos Probed by Scanning Electrochemical Microscopy", *Analytical Chemistry*, Vol. 73, No. 15, pp. 3751-3758, 2001.

[4] H. Abe, "A Non-Invasive and Sensitive Method for Measuring Cellular Respiration with a Scanning Electrochemical Microscopy to Evaluate Embryo Quality", *J. Mamm. Ova Res.*, Vol. 24, pp. 70-78.

[5] C.C. Wu, T. Saito, T Yasukawa, H. Shiku, H. Abe, H. Hoshi and T. Matsue "Microfluidic Chip Integrated with Amperometric Detector Array for in-situ Estimating Oxygen Consumption Characteristics of Single Bovine Embryos", *Sensors and Actuators B*, Vol. 125, pp. 680-687, 2007.

WAFER SCALE ENCAPSULATION OF LARGE LATERAL DEFLECTION MEMS STRUCTURES

A.B. Graham[1], M.Messana[1], P. Hartwell[2], J. Provine[1], S. Yoneoka[1],
B. Kim[3], R. Melamud[1], R.T. Howe[1], and T.W. Kenny[1]

[1]Stanford University, Stanford, California, USA
[2]Hewlett-Packard Laboratories, Palo Alto, California, USA
[3]University of California at Berkeley, Berkeley, California, USA

ABSTRACT

Packaging of microelectromechanical systems (MEMS) is a critical step in the transition from product development to production. This paper presents a robust, hermetically-sealed encapsulation method that can accommodate many traditional MEMS devices by allowing large lateral deflection structures within a clean environment. Using the new technology described in this paper, trench widths ranging from 1μm to 100μm were successfully encapsulated at the wafer level while maintaining devices as thick as 20μm. Devices produced with this method have proven durable enough to withstand harsh post-processing such as dicing and wire bonding. Two different types of MEMS resonators are also discussed, demonstrating the use of both large and small trench widths within the encapsulation.

INTRODUCTION

Microelectromechanical systems (MEMS) are increasingly moving from expensive prototypes to commercial products. A critical step in this process is the packaging of such devices. Not only does the packaging serve as protection from the external environment, but it must provide a controlled environment in which to operate the device. A detailed description of the challenges involved in the packaging and assembly of MEMS is provided by *Najafi* [1]. Packaging methods can be broadly categorized into two categories: device level and wafer level packaging. Because device level packaging often involves handling each individual part (typically decreasing yield and increasing cost), wafer level packaging has experienced increasing attention.

The most commonly used method of encapsulating at the wafer level is bonding a separate substrate (typically glass or silicon) onto completed structures [2]. While this method has been successful, thin-film encapsulation (depositing a structural cap layer on top of a sacrificial layer) has several advantages. By encapsulating as part of the fabrication process, the overall package footprint only needs to be slightly larger than the device itself, leaving just enough material to maintain structural integrity. This method also allows for vertical electrical interconnects, which has the potential to further reduce die area. Alternatively, bonded packages require sufficient bonding area to fully enclose the structure, and they typically increase the overall package thickness due to the bonding cap. For these reasons, we have chosen the method of encapsulating by means of a deposited cap layer.

In work preceding that discussed here, *Candler et al.* demonstrated the encapsulation of piezoresistive and capacitive structures by depositing a thin film epitaxial polysilicon cap, etching vent holes in the cap to provide access to the sacrificial oxide, etching the sacrificial material, and closing the access holes using a low pressure chemical vapor deposition (LPCVD) silicon dioxide [3]. In addition, the sealing step was performed directly after the release etch in a cleanroom environment, with no intermediate processing that could cause additional contamination. This process had greater than 90% yield for capacitively actuated, double-ended tuning fork resonators. Similar resonators were later used by *Kim et al.* to demonstrate the stability of an 'epi-seal' encapsulation in which the etch access holes were sealed using a second epitaxial silicon deposition rather than LPCVD silicon dioxide [4]. Their work showed that the resonant frequency stayed within the measurement uncertainty for a period of over one year.

While the 'epi-seal' encapsulation technique developed jointly by Stanford and Bosch has produced devices in a very stable and robust environment, many common MEMS structures cannot be encapsulated using such a technology due to its design constraints. For example, the current rules for designing in such a process limit maximum trench widths to 2μm, preventing the incorporation of even basic comb-drive structures. *Ayanoor-Vitikkate et al.* attempted to address this using a timed thermal oxidation of sacrificial beams to fill in trenches as wide as 10-20μm, but working devices have yet to be demonstrated using this method [5].

In this paper, we describe a clean, robust, hermetically-sealed encapsulation method that can accommodate more traditional MEMS devices by allowing large lateral deflection structures. The process addresses the previous trench width limitations by filling in wider trenches via the deposition of a very thick plasma enhanced chemical vapor deposition (PECVD) silicon dioxide layer (up to 23μm). Devices produced with this encapsulation method have proven durable enough to withstand harsh post-processing such as dicing and wire bonding. We also present two different types of functioning MEMS resonators fabricated within this encapsulation.

FABRICATION

The fabrication process begins with silicon-on-insulator (SOI) wafers having a 2μm thick buried oxide and device layer thicknesses that varied among 5μm, 10μm, and 20μm. As shown in Fig. 1(a), the first step is to define the functional layer using a deep reactive ion etch (DRIE). Because this step includes both high and low aspect ratio trenches, an etch recipe was developed that eliminates "grassing" in wide trenches while maintaining feature critical dimensions in narrow trenches. This recipe included adding decreasing C_4F_8 passivation gas during the first few minutes of the etching, similar to the process outlined by *Liu* [6].

A thick silicon dioxide is then deposited via PECVD

978-1-4244-2977-6/09 $25.00 © 2009 IEEE

using a silane and nitrous oxide chemistry, as shown in Fig. 1(b). PECVD was chosen over the two LPCVD (low pressure chemical vapor deposition) methods available to us for its rapid deposition rate, good conformality and low stress. While the standard 'epi-seal' process only needs 2-3μm of oxide to seal over the narrow trenches and provide the upper sacrificial layer, as much as 23μm of oxide is deposited here to fill in the wide trenches found on the 20μm device layer wafers. This ensures that the minimum height of the oxide is above the top of the device layer. The film stress of the PECVD oxide was measured to be -160MPa.

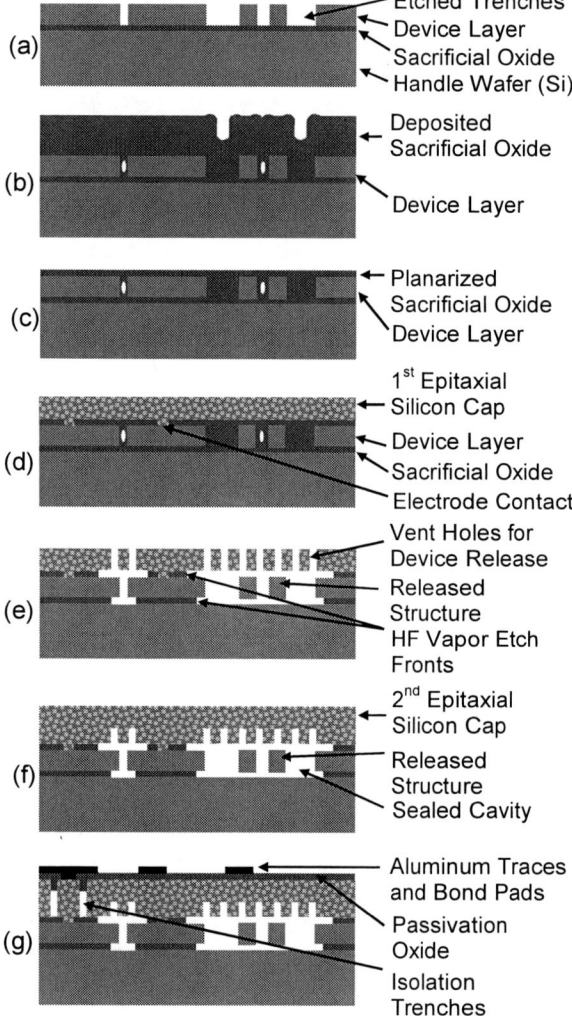

Figure 1: Wafer-scale encapsulation fabrication steps (a) SOI wafer after deep reactive ion etching (DRIE). (b) Thick silicon dioxide deposition. (c) Plasma planarization. (d) First epitaxial silicon deposition. (e) DRIE vent etch and HF vapor etch to release devices. (f) Epitaxial silicon deposition to seal devices. (g) Electrical isolation and metallization for bond pads and metal traces.

A combination of plasma planarization and chemical-mechanical polishing (CMP) is then used to planarize this sacrificial layer. After coating with photoresist, the plasma planarization is performed using CHF_3, CF_4, Ar, and O_2 as etch gases, followed by a short CMP using silicon dioxide specific slurry. This results in a surface topology that is suitable for subsequent lithography steps while resulting in oxide-filled trenches, shown in Fig. 1(c). Electrical contact vias are then etched in the silicon dioxide to allow contact with various device layer electrodes.

The 20μm thick epitaxial silicon encapsulation layer, shown in Fig. 1(d), is deposited at 1080°C using Dichlorosilane, hydrogen, and Phosphine as a dopant gas. Small trenches (approximately 0.7μm by 8μm) are etched through the cap to provide access to the sacrificial oxide for the subsequent release etch. To prevent stiction problems, devices are released using a hydrofluoric acid vapor etch, seen in Fig. 1(e). As the HF vapor etches the sacrificial and buried silicon dioxide layers, etch fronts are monitored using an infrared microscope to determine when the release etch is complete.

Wafers are then immediately sealed using a second epitaxial silicon deposition, shown in Fig. 1(f). Performed at 1080°C and 40mbar, this deposition includes gaseous hydrochloric acid to achieve a recipe that is selective between the silicon and oxide. This prevents silicon deposition on the silicon dioxide layers separating the handle wafer, device layer, and encapsulation layer, which would create electrical shorts. An important feature of the sealing process is that it results in a single crystal silicon device layer free of native oxide. As the deposition chamber is being brought up to temperature, the released and unsealed wafer is only exposed to hydrogen gas. By reacting hydrogen with the thin native oxide at temperatures in excess of 1000°C, the native oxide is consumed leaving a pure silicon surface [7]. The primary gas remaining in the encapsulated cavity following the seal is hydrogen.

Several straightforward steps are then taken to complete the fabrication, as shown in Fig. 1(g). An isolation trench is etched to isolate electrical contacts, followed by the deposition of a silicon dioxide passivation layer. After opening contacts in the oxide, aluminum is deposited, patterned, and etched to create the wire traces and bond pads necessary for testing.

To reduce the pressure inside the sealed cavity, the wafers are placed in a nitrogen furnace at 400°C. As discussed by *Candler et al.*, although nitrogen will begin to slowly diffuse into the cavity, the much smaller hydrogen molecules diffuse out at a much higher rate, reducing the pressure inside the encapsulation [8]. A high temperature vacuum anneal would allow a further reduction in the amount of hydrogen, however a tool capable of this was not available at the time of processing.

RESULTS AND DISCUSSION

Using the process described, numerous functioning devices were successfully encapsulated. Among these are structures that include design features not possible in any previous iteration of the 'epi-seal' fabrication process, most notably trenches ranging from 1μm to 100μm wide. An important outcome of the increase in allowable trench widths is the inclusion of comb-drive structures. As examples, Fig. 2 shows a fully released comb-drive structure prior to the sealing step and Fig. 3 shows the interdigitated comb-drive fingers of a sealed device. Note that the structure in Fig. 2 broke as part of the preparation of the sample for the SEM.

Figure 2: SEM cross-section showing vents etched into the first epitaxial silicon cap layer (top-most layer) over part of a released comb drive resonator.

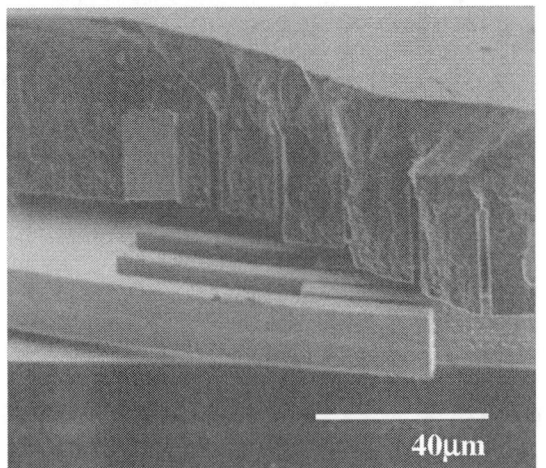

Figure 3: SEM cross-section of released and sealed interdigitated comb-drive fingers.

Though the intention of this new method was to allow the encapsulation of large deflection devices, it is still quite suitable for encapsulating the narrow trench devices found in other fabrication processes. Double-ended tuning fork resonators, similar to those previously encapsulated [4], were fabricated. Because the quality factor is limited by damping in such resonators, these structures were used to monitor pressure changes within the encapsulation by measuring their quality factor at various times during the hydrogen diffusion anneal, as shown in Fig. 4. The pressure is estimated to be well below 10mbar at room temperature based on the sealing pressure and temperature. *Candler et al* showed that the pressure can be reduced to less than 1mbar for a similar encapsulation process and demonstrated a method for determining pressure from quality factor [8].

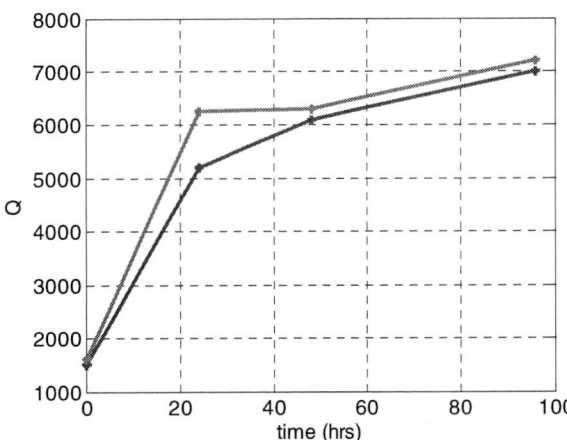

Figure 4: Plot of resonator quality factor, Q, increasing with anneal time in a 400°C nitrogen ambient, corresponding to a decrease in the cavity pressure. Plot shows data for two different devices. Cavity pressure is estimated to be below 10mbar.

As it is the driving force of this work, large lateral deflection devices were also successfully encapsulated. Among these is a comb-driven, resonating fan structure for studying fatigue, shown in Fig. 5, that has been electrostatically actuated to achieve a displacement of 8.6μm and a stress of 1.6 GPa (calculated). The frequency response of this test structure is shown in Fig. 6. As frequency is sensitive to many environmental factors, the constant temperature frequency stability of ±3ppm over 700 hours of operation (at low amplitude) demonstrates the robustness of the encapsulation method, as seen in Fig. 7. More details on the testing of this structure are available elsewhere in these proceedings [9].

Figure 5: CAD drawing of a functioning comb-drive resonator encapsulated in this process.

It is often the case that wafer singulation and electrical contact to external circuitry are areas of concern for MEMS devices. The hermetic epitaxial silicon encapsulation presented here, however, is very robust and allows for dicing using a wafer saw and electrical contact via wire bonding without affecting the device integrity. For working devices, there was no loss in yield in going from completed wafers (tested using a probe station) to individual devices that were diced in a wafer saw, epoxied into packages, and wire bonded to those packages.

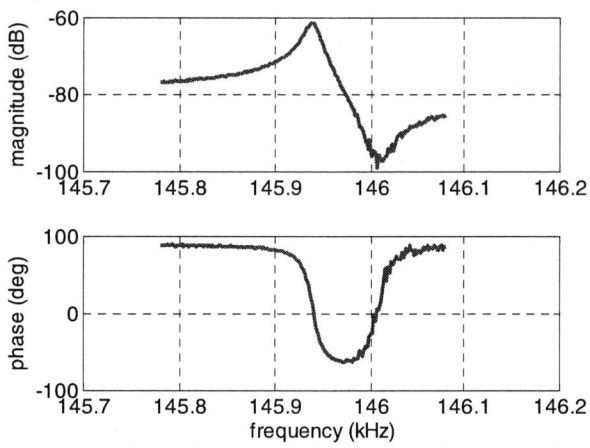

Figure 6: Plots of gain and phase versus frequency for a large deflection comb-drive resonator.

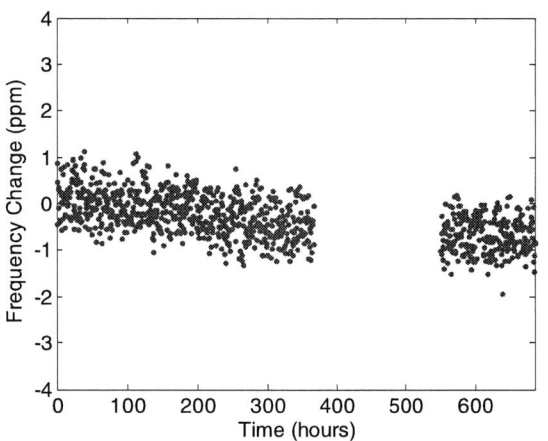

Figure 7: Plot of frequency change versus time for a large deflection comb-drive resonator operating at a nominal frequency of 137kHz. Temperature was maintained at 29°C±0.1°C. The loss of data between 400-600 hours is a result of a memory error during data acquisition, but the device was continuously operating during this time.

CONCLUSIONS

With this fabrication platform, devices with large lateral displacements can now be packaged at the wafer-scale into a stable, hermetic environment. The method of encapsulation has proven durable enough to withstand standard backend processing such as dicing and wire bonding, with no loss of performance. This has the potential to improve the reliability, performance and yield of common devices such as gyroscopes, accelerometers and comb-drive structures. In addition, the well-controlled environment can be used for fundamental studies of surface phenomenon such as fatigue, adhesion, and friction. Future work will focus on characterization of devices fabricated using this technique.

ACKNOWLEDGEMENTS

This work was performed under the Center on Interfacial Engineering for Microelectromechanical Systems, funded by DARPA grant HR0011-06-0049 and managed by Dr. D. L. Polla. Additional support was provided by the National Nanofabrication Users Network facilities funded by the National Science Foundation under award ECS-9731294, and the National Science Foundation Graduate Research Fellowship support for A.B. Graham. The authors would also like to thank Uija Yoon[1] and Gary Yama[2] for their assistance, without whom this work would not have been possible ([1]Hewlett-Packard Laboratories; [2]Robert Bosch LLC Research and Technology Center).

REFERENCES

[1] K. Najafi, "Micropackaging technologies for integrated microsystems: Applications to MEMS and MOEMS," *Proc. SPIE - Micromachining and Microfabrication Process Technology VIII*, pp. 1-19, 2003.

[2] H. Henmi, S. Shoji, Y. Shoji, K. Yosimi, and M. Esashi, "Vacuum package for microresonators by glass-silicon anodic bonding," *Proc. 7th Int. Conf. Solid-State Sens. and Actuators (TRANSDUCERS '93)*, pp. 584-587, 1993.

[3] R.N. Candler, W.-T. Park, H. Li, G. Yama, A. Partridge, M. Lutz, and T.W. Kenny, "Single wafer encapsulation of MEMS devices," *IEEE Trans. Adv. Packag.*, vol. 26, no. 3, pp. 227-232, 2003.

[4] B. Kim, R.N. Candler, M. Hopcroft, M. Agarwal, W.-T. Park, and T.W. Kenny, "Frequency stability of wafer-scale film encapsulated silicon based MEMS resonators," *Sensors and Actuators, A: Physical*, vol. 136, pp. 125-131, 2007.

[5] V. Ayanoor-Vitikkate, K.-L. Chen, W.-T. Park, G. Yama, and T.W. Kenny, "Wafer scale encapsulation of wide gaps using oxidation of sacrificial beams," *IEEE IEMT*, pp. 300-306, 2006.

[6] X. Liu, C. Wang, Y. Zhu, and G Yan, "Vertical Profiles and CD Loss Control in Deep RIE Technology," *Proc. 7th Int. Conf. Solid-State and Integrated Circuits Technology*, vol.3, pp. 1848-1851, Oct. 2004.

[7] S.T. Liu, L. Chan, and J.O. Borland, "Reaction Kinetics of SiO_2/Si (100) Interface in H_2 Ambient in A Reduced Pressure Epitaxial Reactor," *Proc. 10th Int. Conf. Chemical Vapor Deposition. Electrochem. Soc.*, pp. 428-434, 1987.

[8] R.N. Candler, W.-T. Park, M. Hopcroft, B. Kim, and T.W. Kenny, "Hydrogen diffusion and pressure control of encapsulated MEMS resonators," *Proc. 13th Int. Conf. Solid-State Sens. and Actuators (TRANSDUCERS '05)*, pp. 920–923, 2005.

[9] S. Yoneoka, Y.Q. Qu, S. Wang, M.W. Messana, A.B. Graham, J. Salvia, B. Kim, R. Melamud, G. Bahl, and T.W. Kenny, "Fatigue of Single Crystal MEMS Resonators in a Vacuum Sealed Environment." *Proc. 22nd IEEE Int. Conf. Micro Electro Mech. Syst. (MEMS'09)*, 2009.

978-1-4244-2977-6/09 $25.00 © 2009 IEEE

PARYLENE-POCKET CHIP INTEGRATION

R. Huang, and Y.C. Tai

Micromachining Laboratory, California Institute of Technology, Pasadena, CA, USA

ABSTRACT

In this paper, we present a novel packaging technique that utilizes a simple, flexible parylene (chip) pocket on silicon substrate with metal pads. This pocket can house an IC chip or a discrete component inside and provide electrical connections to it. On the other hand, recent achievement in silicon probes implantation in the parietal cortex enables technological advances in neural prosthesis research. However, most of these technologies suffer from high signal-to-noise ratio and expensive integration scheme with IC chips or lack thereof. As a demonstration, this work uses this technique to produce an 8-shank silicon probe array integrated with a fully functional 16-channel amplifier CMOS chip.

INTRODUCTION

An important goal in neural prosthesis is to be able to decode the movement intention in the parietal cortex from neurons by implanting neuroprobes [1]. While 3-D integrated silicon probes have been successfully manufactured [2], the degradation of the signal to noise ratio (SNR) is still a major challenge because electronics are too far away from the recording site. Additionally, recent development in bioimplantable devices such as retinal, cochlear and cortical prosthesis implants also increases the demand for totally implanted technologies. Therefore, a totally biocompatible packaging/integration solution to embed amplifiers near the recording sites must be developed.

However, regardless of the technology used to achieve this IC chip integration, an ideal well-packaged device for implantation would 1) cause no infections and inflammatory responses and 2) be able to withstand the harsh physiological environment of human body and 3) be inexpensive and easy to make. Current state technologies have yet to balance the trade-offs between these criteria. A few have well developed CMOS/MEMS integrated process but is very expensive and painstaking to make. On the other hand, some have no integrated circuits at all. [2–5]

To overcome these challenges, we develop here a new packaging technique that utilizes a flexible parylene (chip) pocket on silicon substrate with metal pads. This pocket can both house an IC chip inside and provide electrical connections to it, and can be totally integrated with silicon probes. This way, one can make the chips and probes separately and later package them together, which is a huge advantage over IC/probe integration. The pocket size and the configuration of the electrodes can be modified to suit different chips and applications and is not sensitive to dry etching flatness [6]. Figure 1 shows the drawing of a wireless parylene pocket integration scheme. The whole bonding structure is conformally coated and sealed with parylene-C (poly-para-xylylene-C), and with medical grade epoxy to achieve total encapsulation for biocompatibility.

As a result, we are able to fabricate IC-cabled packages that are highly customizable, fully biocompatible, and easy to mass fabricate, which will facilitate future research in developing more complex bioimplantable system.

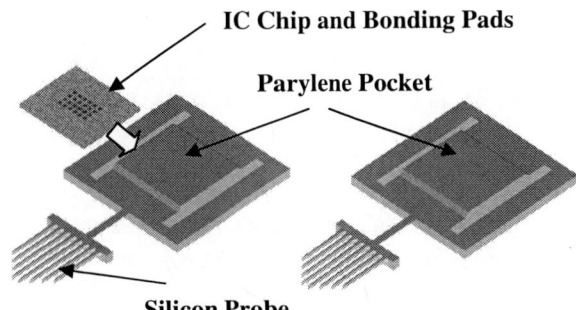

Figure 1: The concept illustration of a complete wireless parylene pocket with signal processing chip. The chip may be powered through RF coil (not shown). Left) IC chip being inserted into the pocket. Right) complete packaged pocket structure.

DESIGN

The parylene pockets presented in this paper were designed to fit a 300 µm thick, 0.5 mm x 0.5 mm sized commercial IC chip. The parylene C cable and pocket thickness is 12 µm with gold metal connection traces embedded. The two openings on two sides of this pocket were designed to provide stress relief during the insertion. The edge of the pocket has a 1 mm wide melted parylene adhesion layer section to provide the adhesion between the parylene and the silicon substrate.

This structure also consists of a parylene pocket integrated with a 2-D 32 channel flexible cabled electrode array device, which can be expanded to 3-D 32xN channel structures by probe stacking (figure 2). The silicon base of the pocket is 0.8 mm x 0.8 mm in size and is 500 µm thick. The parylene pocket is connected to a 7 cm cable with a 60° Y shape pattern at the end (figure 3). These are circular platinum rings arranged to be electrically bonded to commercial available connectors with conductive epoxy on a circular PC board. The silicon shank is 150 µm thick and has length of 5.1 mm, 4.6 mm, 4.1 mm and 3.6 mm from longest to shortest, respectively (figure 4).

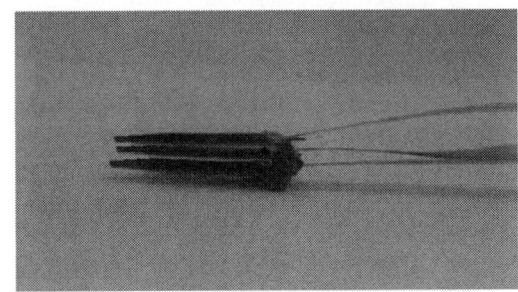

Figure 2: Probe stacking capability of the silicon probe structure. Figure shows three 32 channel probes stacked together to form a 96 channel 3-D structure. The spacing between the probes can be modified.

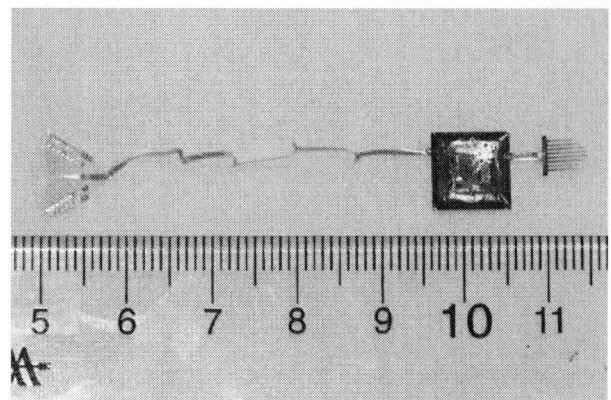

Figure 3: Silicon probe integrated with parylene pocket. A conduction chip has been inserted to demonstrate the functionality of the integrated pocket structure. This device has been totally coated with parylene and sealed with epoxy.

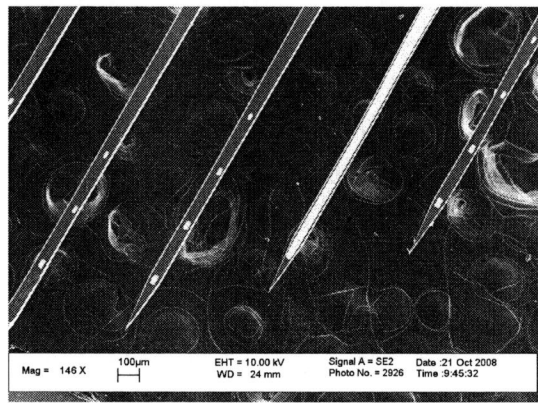

Figure 4: SEM picture of the electrodes on the silicon shank with electrodes.

FABRICATION

The parylene pocket structures are fabricated on double side polished wafers with DRIE (Deep Reactive Ion Etching) technology. Figure 5 shows the fabrication process.

First, 0.5 μm of parylene-C is deposited by room temperature CVD (chemical vapor deposition) and is melted in an oven with N_2 backflow at 350 °C. This film is then patterned by oxygen plasma to leave areas for adhesion enhancement between the parylene/silicon interfaces. A sacrificial photoresist layer is then spin-coated for pocket releasing. A bottom layer of parylene-C (6 μm) is then deposited, followed by Cr/Au (0.05/0.2 μm) lift-off process with electron beam evaporation to provide electrical connection. The top layer of parylene-C (6 μm) is deposited to complete the parylene-metal-parylene sandwich structure. Electrode sites and the device definition are then opened by a two-step RIE with O_2 Plasma (Reactive Ion Etching) process. The outline of the parylene pocket structure is subsequently etched by DRIE from both sides of the wafer. In the final step, the devices are released in photoresist stripper and dried. An IC chip is then inserted into the parylene pocket (figure 6), aligned, bonded with conductive epoxy (figure 7 and 8) and totally coated with parylene-C again for complete encapsulation and to ensure biocompatibility. Parylene-C was selected due to its

insulating capability, flexibility, biocompatibility, and its prior use in medical applications.

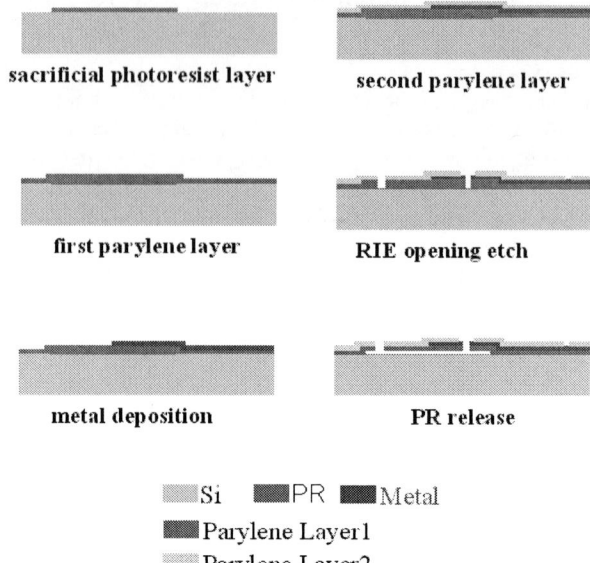

Figure 5: Fabrication process steps. These steps can be totally integrated with the parylene-cabled silicon prosthesis probes [2].

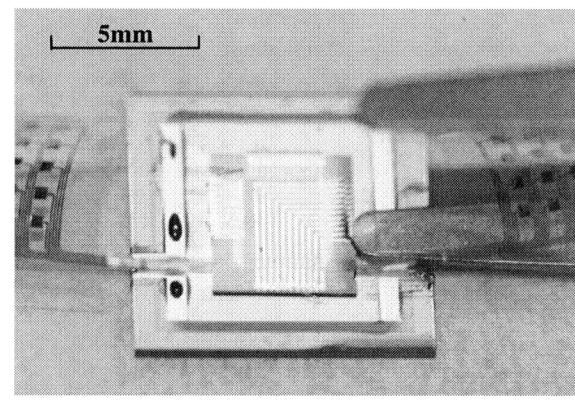

Figure 6: Insertion of the IC chip into the parylene pocket. Tweezers are used to align the bonding pads to the metal pads on the chip under the microscope.

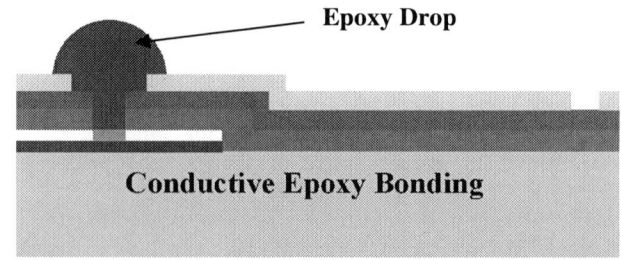

Figure 7: Bonding scheme for the parylene pocket. A drop of conductive epoxy is applied on the bonding hole, which exposes the metal pads on the IC chip underneath. The drop of the conductive epoxy can be controlled to 200 μm by hand application.

978-1-4244-2977-6/09 $25.00 © 2009 IEEE

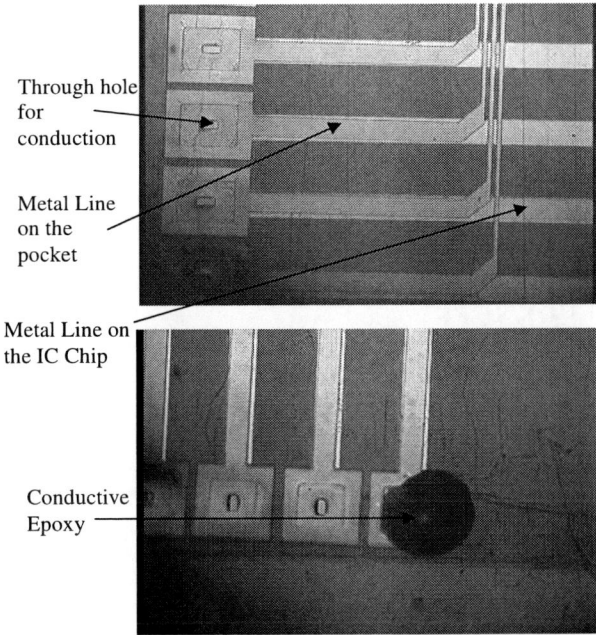

Through hole for conduction

Metal Line on the pocket

Metal Line on the IC Chip

Conductive Epoxy

Figure 8: Top) Alignment of the pads on the parylene pocket and the metal lines on the inserted chip. The alignment offset was on the order of 10 to 20 μm. Bottom) A drop of biocompatible conductive epoxy is applied over the metal pads to provide electrical conduction. The size of the drop is on the order of 150 to 200 μm.

EXPERIMENT AND RESULT

Device testing was performed in two phases: chip integration testing for both conduction and amplifier chips (figure 9), and accelerated life testing in saline environment.

The embedded CMOS amplifier chip [7] is a scalable 16-channel pre-amplifier and buffer chip with an in-band gain of 35.5 dB. This chip was tested using a function generator on a pocket only structure (figure 10). Both sine waves and square waves with frequency of 0.5 kHz, 1 kHz, 2 kHz and 5 kHz and amplitude of 5 mV were passed into the chip. The amplified output (figure 11) from the oscilloscope concludes a successful functionality testing of our packaging technology. The conduction chip was bonded in a silicon probe integrated pocket and the impedances of the electrodes were measured.

An on-going accelerated soaking life-time test was also done on the parylene pocket structure to determine the mean time to failure. Testing shows the pocket structure with the embedded amplifier chip sealed by 1.5 mm of biocompatible epoxy is able to function after soaking in 90°C saline for more than 30 days, which roughly translates a lifetime of years in saline at 37 °C. A study is ongoing to investigate the optimal size, geometry and shape of the epoxy that complies with surgical protocols.

In addition to the totally integrated silicon probe array, a second design was developed. It consists of a probe only device with 36 electrodes and 2 reference electrodes separated from the parylene pocket only structure (figure 12). The probe only device has identical geometry as that of the totally integrated device mentioned above. The user

would be able to test the pocket structure with the integrated IC chip before connecting it to the electrodes on the silicon probe, resulting in a big advantage in terms of yielding and reliable testing. The assembled device is shown in figure 13. We use conductive epoxy as the medium for bonding.

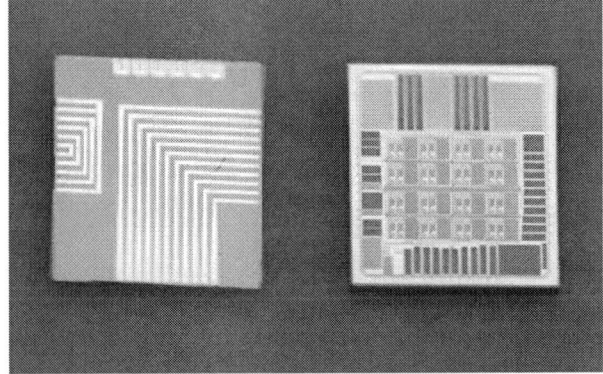

Figure 9: The conduction chip and the amplifier chip. The metal pads on the conduction dummy chip were made to mimic the metal pads on the amplifier chip.

Figure 10: Parylene pocket only structure. This device is used to test the amplifier chip after bonding.

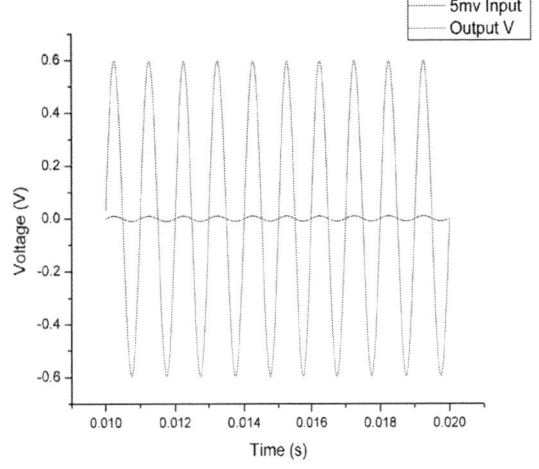

Figure 11: Input of 5mV (10 mV peak to peak) sine wave of 1 kHz is passed into the amplifier (gain = 60) and its oscilloscope output is recorded. Signals of 0.5 kHz, 2 kHz, and 5 kHz were also tested with the chip.

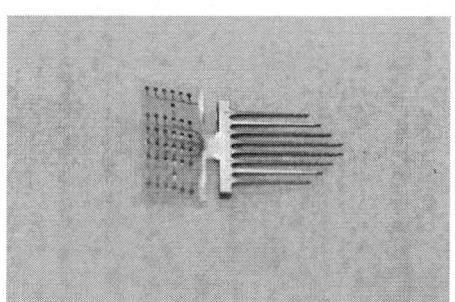

Figure 12: Silicon probe only structure for testing purpose. The parylene sheet extension provides metal pads for electrical connection to the parylene pocket only structure.

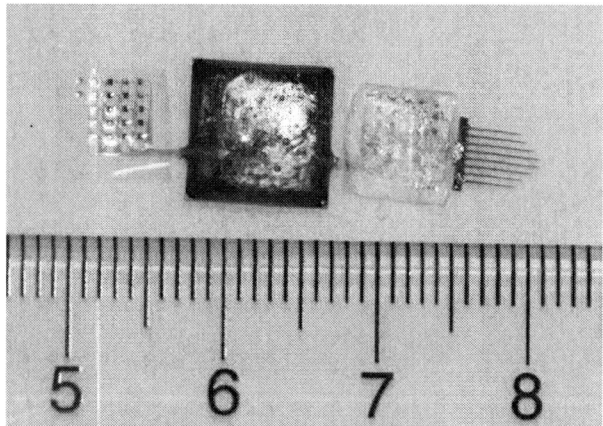

Figure 13: Bonded and packaged structure. In this device, the conduction chip was used. The impedance of the electrode was measured to be 600 kΩ.

CONCLUSIONS

We have developed and tested a new parylene-pocket packaging technology to fabricate a silicon probe array with an IC chip close to its recording sites. The packaging schemes and the fabrication details are presented. A commercial amplifier chip was used and its full functionality after packaging was verified. High-temperature accelerated life-time saline soaking testing shows satisfactory performance in both the pocket structure and the overall system. From the testing results, this new packaging technique shows promise for other more complex integrated biomedical implants needed for neural prosthesis research.

Future work is underway to optimize the space in the pocket and the pitch of the pads for connection. Gold electrodeless plating is also under investigation as an alternative pad-to-pad interconnection technique for our next generation device.

ACKNOWLEDGMENT

This work is supported by Defense Advanced Research Projects Agency (DARPA) (Award # 908165) and by National Institute of Health (NIH) moveable probe project (Award # R01EY015545). The authors also would like to thank Trevor Roper, Agnes Tong, Christine Garske, other members of the Caltech Micromachining Laboratory and members of the Andersen lab at Caltech for their advice and help.

REFERENCES

[1] S. Musallam, B. D. Corneil, B. Greger, H. Scherberger, and R. A. Andersen, "Cognitive Control Signals for Neural Prosthetics," *Science* 305 (5681) pp. 258 – 262, 2004.

[2] R. Huang, C. Pang, Y.C. Tai, K. Emken, C. Ustun, R. Andersen, J. Burdick, "Integrated Parylene-Cabled Silicon Probes for Neural Prosthetics," *22nd IEEE International Conference on Micro Electro Mechanical Systems (MEMS)* , pp. 240-243, 2008

[3] K.D. Wise, D.J. Andersen, J.F Hetke, D.R. Kipke, K. Najafi, "Wireless implantable Microsystems: high-density electronic interfaces to the nervous system," *Proceedings of the IEEE*, pp. 76-97, Vol. 92, Issue 1, Jan 2004

[4] A.E. Ayoub, B. Gosselin, and M. Sawan, "A Microsystem Integration Platform Dedicated to Build Multi-Chip Neural Interfaces," *Proceedings of the 29th Annual International Conference of the IEEE EMBS*, pp. 6604-6607, 2007

[5] M. Töpper, M. Klein, K. Buschick, V. Glaw, K. Orth, O. Ehrmann, M. Hutter, H. Oppermann, K.-F. Becker, T. Braun, F. Ebling, H. Reichl, "Biocompatible Hybrid Flip Chip Microsystem Integration for Next Generation Wireless Neural Interfaces," *Electronic Components and Technology Conference*, pp. 705-708, 2006.

[6] W. Li, D. Rodger, Y.C. Tai, "Implantable RF-Coiled Chip Packaging," *22nd IEEE International Conference on Micro Electro Mechanical Systems (MEMS)*, pp. 108-111, 2008

[7] W. Liu, M. Sivaprakasam, G. Wang, and M. Chae, "A Multi-Channel Neural Recording System for Monitoring Shark Behaviour," *IEEE International Symposium on Circuits and Systems*, May 2006.

A LOW-POWER OVEN-CONTROLLED VACUUM PACKAGE TECHNOLOGY FOR HIGH-PERFORMANCE MEMS

S. -H. Lee[1], J. Cho[1], S. W. Lee[1], M. F. Zaman[2], F. Ayazi[2], and K. Najafi[1]
[1]The University of Michigan, Ann Arbor, MI 48109 USA
[2]Georgia Institute of Technology, Atlanta, GA 30332 USA

ABSTRACT

This paper presents a generic vacuum packaging technology for environment-resistant MEMS devices. This packaging approach simultaneously provides low-power oven-controlled thermal environment and vibration isolation using an isolation platform. The oven-controlled structure is thermally isolated from the environment by crab-leg suspensions made out of a 100 µm-thick glass wafer, an anti-radiation shield, and vacuum encapsulation. Performance is evaluated by packaging Pirani gauges and mode-matched tuning fork gyroscopes (M^2-TFGs). The package has maintained vacuum pressure of ~6 mTorr for ~1 year. A packaged M^2-TFG shows a high-Q mode-matched operation (Q~65,000) at a constant temperature of −5 °C. Allan variance analysis displays an estimated angle random walk (ARW) of 0.012 °/√hr and a bias instability value of 0.55 °/hr at a constant −5 °C. Drive frequency stability of 0.22 ppm/°C is obtained using a compensated oven-control approach. Low power consumption of 33 mW for oven-control at 80 °C is demonstrated when the environment temperature is −30 °C.

1. INTRODUCTION

Increasingly, precision instruments are required for many emerging applications for gas and chemical analysis, environmental sensing, and health monitoring. To achieve high performance, these instruments often require a well-controlled environment (i.e. vacuum/hermetic), temperature isolation, and mechanical isolation. Environmental parameters, especially temperature, can easily compromise the output of a MEMS device, and can induce long-term undesirable effects. One approach for thermal stabilization is to package the device in an oven-controllable micromachined dewar. In order to achieve low power consumption, a high thermal isolation is needed [1,2].

Packaging cost is another issue because it is more than half of the total manufacturing cost, and increases as the structure becomes more complex [1]. MEMS devices have a variety of designs and applications, and they often utilize a customized packaging technology, which is only suitable for a given application. Therefore, a generic packaging approach that can suit a number of different devices and applications is desirable to reduce design/fabrication lead-time and costs. A wafer-level packaging approach can help reduce cost, and helps to protect devices during post-processes such as cleaning and dicing.

This paper reports a new technology for packaging and isolating MEMS devices from adverse environmental conditions. It reports a vacuum package with a new technology for forming vertical feedthroughs, and provides a robust isolation platform to support the MEMS device and thermally and mechanically isolate it from the external environment.

2. PACKAGE DESIGN

Figure 1 shows the package structure, which has three major components; (i) a platform substrate which provides thermal and mechanical isolation using suspensions made out of 100 µm-thick glass, (ii) a MEMS device supported and flip-chip attached on the platform, and (iii) a package cap incorporating vertical signal feedthroughs and providing final vacuum encapsulation. Thermal stabilization is provided by oven-controlling the device at a temperature higher than the maximum environment temperature utilizing a heater and a temperature sensor located on the isolation platform. The heated structure is thermally isolated from the environment by the glass isolation suspensions, anti-radiation shield, and vacuum encapsulation to minimize power dissipation. The suspensions are designed with sufficient stiffness for mechanical support, and flexibility for rejecting environmental vibrations. A wafer-level package cap provides vacuum and vertical feedthroughs. These vertical feedthroughs allow direct 0/1-level packaging. Shock absorption layers [3], and a getter layer for achieving and maintaining high vacuum [4] are formed inside the package.

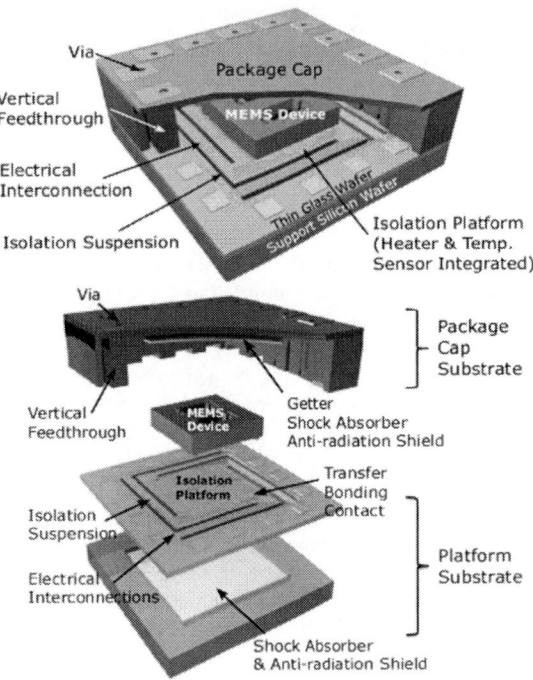

Figure 1: Schematic illustrations of the package.

3. FABRICATION

Figure 2 shows the fabrication process sequence. A recess of ~10 μm is first formed in a silicon wafer and is coated with a shock absorption layer (e.g., gold) (Figure 2 (a1)). The wafer is anodically bonded to a 100 μm-thick Pyrex glass wafer (Figure 2 (a2)). Metal interconnection lines, heater and temperature sensor are then defined using Ti/Pt/Au layers on the glass wafer (Figure 2 (a3)). Next, isolation platform/suspensions are patterned by wet etching the glass using 49% HF solution (Figure 2 (a4)). MEMS chips are batch transferred onto the isolation platform using In-Au Transient Liquid Phase (TLP) bonding technique (Figure 2 (a5)). The batch transfer technique is described in [2]. It is noted that any kind of MEMS devices can be assembled since the chips are packaged after they are fabricated using any given process.

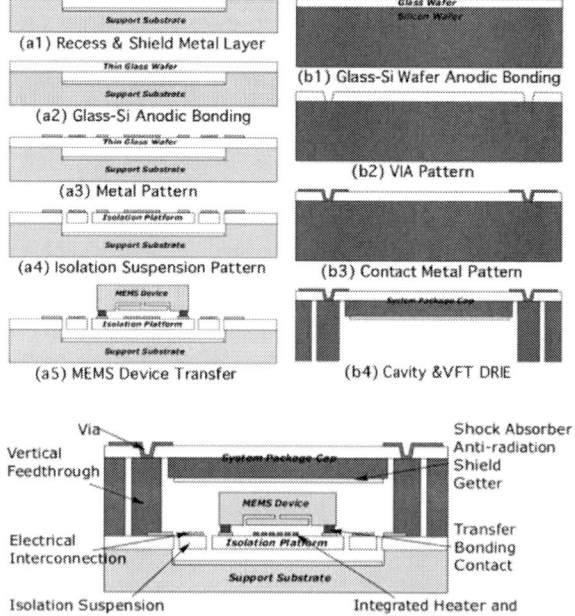

Figure 2: Fabrication process sequence.

The cap substrate fabrication starts with a highly-doped silicon wafer that is anodically bonded to a second 100 μm-thick glass wafer (Figure 2 (b1)). Via holes are formed by wet etching the glass using 49% HF solution (Figure 2 (b2)). Contact metal of aluminum is then patterned over the via holes (Figure 2 (b3)). The vertical feedthroughs and the package cavity are then defined by silicon DRIE (Figure 2 (b4)). A getter layer, anti-radiation shield layer, and shock absorption layer are deposited inside the cavity.

Finally, the platform substrate containing the MEMS chip and the package cap substrate are anodically bonded at the wafer-level. Figure 3 shows wafer- and die-level views of the prepared platform, cap, and completed package. The package die size is 1.2×1.2×0.17 cm³.

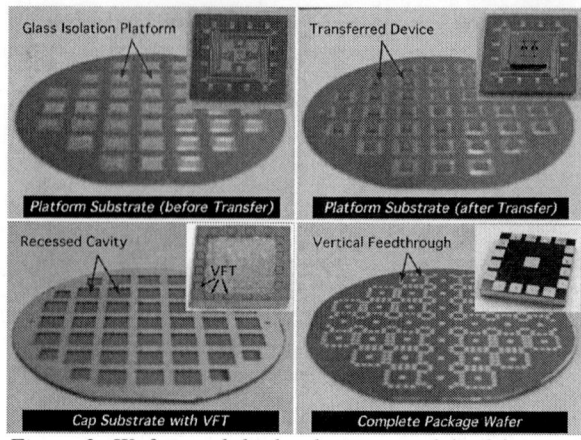

Figure 3: Wafer- and die-level pictures of the fabricated substrates.

Figure 4 shows photographs of the isolation platform and suspensions. The isolation platform is suspended over the substrate by 10 μm, and includes the heater and temperature sensor for oven control.

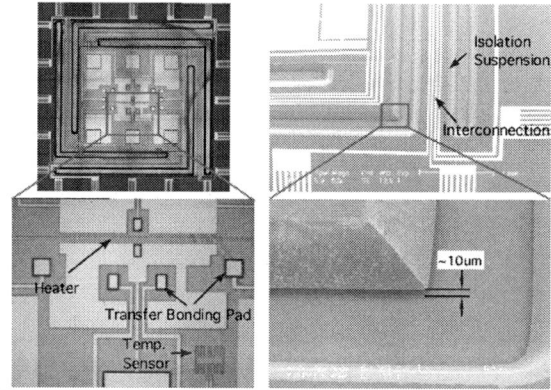

Figure 4: Photographs of the fabricated isolation platform and suspensions.

Figure 5 shows the complete package. The package die is torn apart or diced for visual inspection. The package is mechanically robust to survive dicing.

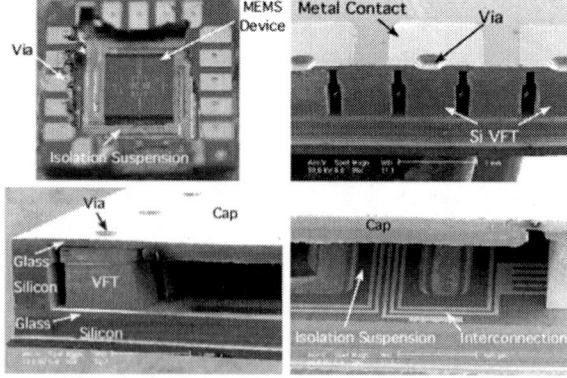

Figure 5: Pictures of the complete package.

4. RESULTS

Performance has been evaluated by packaging Pirani gauges and mode-matched tuning fork gyroscopes (M^2-TFGs) [5] (Figure 6). The dimension of these device chips is $4.5 \times 4.5 \times 0.5$ mm^3.

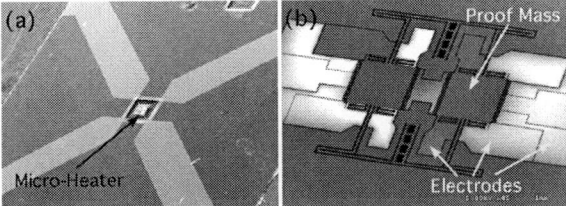

Figure 6: SEM views of (a) the Pirani gauge, and (b) the mode-matched tuning fork gyroscope (M^2-TFG).

4.1 Vacuum Measurement

The vacuum levels inside the package cavity have been directly measured using the Pirani gauge. At the center of the Pirani gauge chip, a thin film Pt heater is suspended by thin dielectric bridges (Figure 6 (a)). By measuring the power consumption for heating the Pt heater, pressure levels inside the package can be extracted. Figure 7 shows the long-term pressure measurement data from the packages after bonding. The pressures inside the cavities range from 6 mTorr to 23 mTorr, and have remained stable for about 1 year without any obvious leaks.

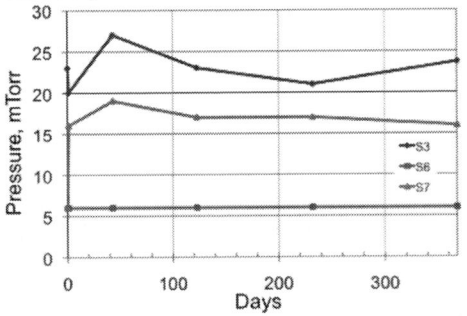

Figure 7: Long-term vacuum level data directly measured using the vacuum packaged Pirani gauge chip.

4.2 Packaged Gyroscope Characterization

The 1.2 mm × 1.2 mm M^2-TFG prototypes have been fabricated on a 100 μm-thick single crystalline silicon substrate [5]. The packaged M^2-TFG prototype demonstrated high-Q resonant operating modes. The measured Q-factor for the drive and sense modes were 82,000 and 60,000 respectively at room temperature. These high Q-factors indicate the gyroscope is operating under high vacuum. Through a combination of quadrature-signal reduction and electrostatic spring softening, the device is operated under high-Q mode-matched resonance. In an effort to enhance device sensitivity and overall resolution, the device has been operated at −5 °C (to reduce thermoelastic damping). Figure 8 shows the mode-matched Q_{EFF} at −5 °C is 65,000. Figure 9 illustrates the response of the device to a 0.01 deg/s (or 36 deg/hr) sinusoidal rotation at 0 °C. In this set-up, the device undergoes ±1.6 degrees of

sinusoidal rotation at 1 mHz (time-period = 1000 s). It is evident from Figure 9 that rotation rates below earth-rate (~12-15 deg/hr) can be measured using the vacuum packaged M^2-TFG, potentially making it suitable for navigation-grade applications such as gyro-compassing.

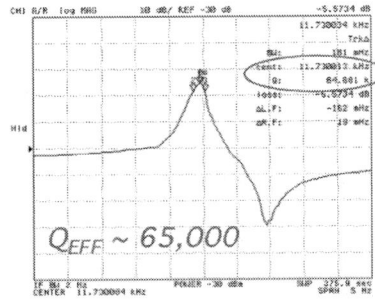

Figure 8: High-Q mode-matched operation of the vacuum packaged M^2-TFG prototype at $T = -5$ °C.

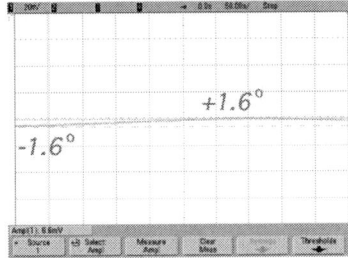

Figure 9: Response to 0.01 deg/s (36 deg/hr) sinusoidal input rotation signal at $T = 0$ °C.

Allan variance analysis has been performed with ZRO data collected over a period of 6 hours at two constant temperature settings. Figure 10 shows the measured ARW and bias instability values of 0.025 °/√hr and 0.72 °/hr, respectively, when operated at constant 25 °C. The ARW and bias instability values are improved at −5 °C to 0.012 °/√hr and 0.55 °/hr, respectively.

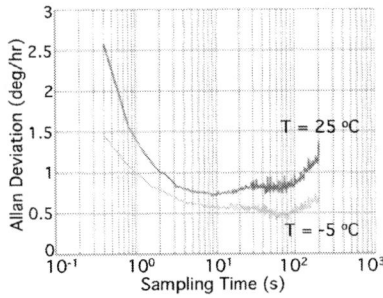

Figure 10: Root Allan variance plot of the vacuum packaged M^2-TFG prototype at $T = 25$ °C and −5 °C.

4.3 Thermal Characterization of the Package

The vacuum packaged M^2-TFG has been tested in a temperature chamber for its thermal performance evaluation. Without oven-control, the drive frequency varies at 17.3 ppm/°C over a temperature range of 100 °C, but stays on average <1 ppm at the same temperature

978-1-4244-2977-6/09 $25.00 © 2009 IEEE

without any hysteresis during thermal cycling as shown in Figure 11. The drive Q-factor reaches 124,000 at −25 °C, and 35,000 at 75 °C.

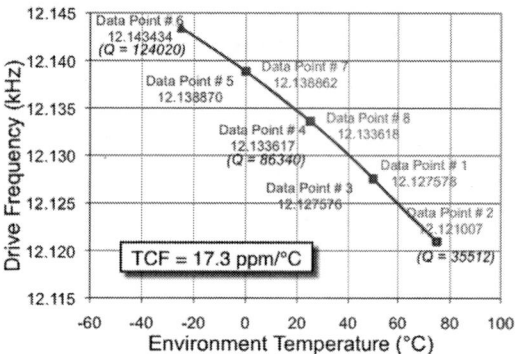

Figure 11: Drive frequency variation with environment temperature cycling from −25 °C to 75 °C.

The drive frequency variation can be reduced by oven-controlling the gyroscope. With oven-control, the isolation platform and the gyroscope are maintained at a temperature higher than the maximum environment temperature utilizing the integrated heater and temperature sensor. The heater is on/off controlled by a comparator chip and feedback signal from the integrated temperature sensor on the isolation platform. The package is tested in a temperature oven from −30 °C to 70 °C. As shown in Figure 12, the drive frequency stays within 0.96 ppm/°C when the oven-control set-temperature is fixed at 80 °C. This frequency change may come from a non-uniform temperature distribution across the gyroscope, because the heater is located on the platform and heat is transferred only through the bonding contact pads and radiation.

Higher frequency stability of 0.22 ppm/°C is obtained using a compensated oven-control (Figure 12), where the set-temperature is modified based on the actual environment temperature. The compensated set-temperature is obtained from the temperature dependency data of the drive frequency operating without oven-control.

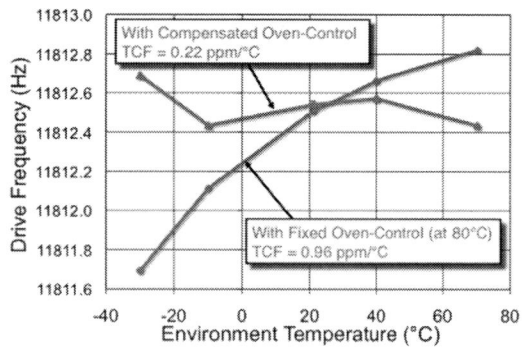

Figure 12: Drive frequency variation of the packaged M^2-TFG with oven control.

Power consumption for the oven-control has been measured (Figure 13) to be 33 mW when heating the device to 80 °C with an environment temperature of −30 °C; it decreases as the environment temperature increases. This

corresponds to a thermal resistance of 3,300 K/W.

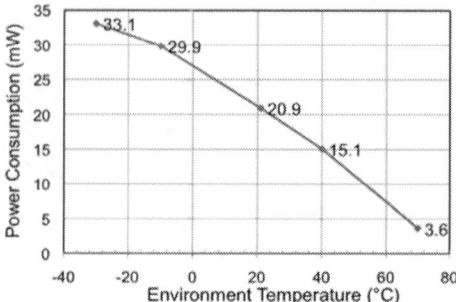

Figure 13: Measured power consumption for oven control at 80 °C.

CONCLUSION

We have developed and demonstrated a low power oven-controllable vacuum package technology for high performance microinstruments. The package provides high thermal and mechanical isolation utilizing glass isolation suspensions. This wafer-level packaging technology allows for handling and packaging a wide variety of MEMS devices. This package is capable of long-term stable vacuum encapsulation, and provides robust vertical feedthroughs. By packaging a high performance MEMS gyroscope, the feasibility of this technology has been demonstrated. With the low-power oven-control, the thermal stability of the packaged gyroscope has been enhanced by 80 times. This approach is suitable for many different high performance MEMS devices including resonators, accelerometers, infrared imagers, or any application requiring low power oven control, vibration isolation, and hermetic/vacuum packaging for stable operation.

ACKNOWLEDGMENTS

This work was supported in part by DARPA's HERMIT program under contract number W31P4Q-04-1-R001.

REFERENCES

[1] K. Najafi, "Micropackaging Technologies for Integrated Microsystems: Applications to MEMS and MOEMS," *Proceedings of SPIE*, vol. 4979, pp. 1-19, 2003.

[2] S. -H. Lee, S. W. Lee, and K. Najafi, "A Generic Environment-Resistant Packaging Technology for MEMS," in *Solid-State Sensors, Actuators and Microsystems Conference, 2007. TRANSDUCERS 2007*, Jun 2007, pp. 335-338.

[3] S. W. Yoon, S. Lee, N. C. Perkins, and K. Najafi, "Shock Protection Using Soft Coating as Shock Stops," in *Solid-State Sensors, Actuators, and Microsystems Workshop*, Hilton Head, South Carolina, Jun 2006, pp. 396-399.

[4] J. Mitchell, G. R. Lahiji, and K. Najafi, "Long-term Reliability, Burn-in and Analysis of Outgassing in Au-Si Eutectic Wafer-level Vacuum Packages," in *Solid-State Sensors, Actuators, and Microsystems Workshop*, Hilton Head, South Carolina, Jun 2006, pp. 376-379.

[5] A. Sharma, M. F. Zaman, M. Zucher, and F. Ayazi, "A 0.1°/hr Bias Drift Electrically Matched Tuning Fork Microgyroscope," in *IEEE 21st International Conference on Micro Electro Mechanical Systems, 2008. MEMS 2008*, Jan 2008, pp. 6-9.

DIRECT WIRE-BONDING OF SILICON DEVICES WITHOUT METAL PADS

A. Hirshberg and D. Elata

Technion - Israel Institute of Technology

ABSTRACT

We demonstrate a new method for direct wire-bonding of silicon devices, which does not require any metal bond-pads. A wire bond-ball is pressed into a hole etched in the silicon device layer, and is wedged in the hole by plastic deformation. Experimental measurements show that strength and conductivity of direct-bonds are comparable to those of standard wire-bonds on metal pads. The relevance of direct wire-bonding is that by eliminating metal bond-pads, constraints on high temperature processing steps and limitations on sacrificial release steps, are alleviated.

INTRODUCTION

Wire-bonding is a prevalent method for interconnecting silicon and MEMS devices [1, 2]. Since metal wires do not naturally adhere to smooth, clean silicon surfaces, a metal bond-pad is usually necessary. In standard wire-bonding, a metal wire is welded to the metal pad by applying pressure, heat and ultrasonic energy (Fig. 1a). The metal pad provides both mechanical adhesion and electrical connectivity between the wire and the conductive silicon substrate.

It is common practice to pattern metallic layers which include bond-pads on smooth silicon surface, at the beginning of the fabrication before MEMS structures are etched. Once metallic layers are deposited on the device, subsequent fabrication steps are often limited to temperatures below *350°C*. If however, following high-temperature fabrication processes are necessary, metal pads can be deposited on micromachined devices using shadow-masks, but this technology is limited in resolution and precision.

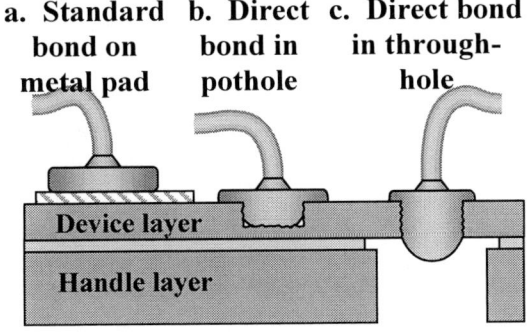

Figure 1. Three methods for wire bonding on a silicon die: a) standard wire bond on metal pad, b) bond-ball in a pothole, and c) bond-ball in a through-hole.

In this work, we propose and demonstrate a method for direct wire-bonding onto a silicon die, without metal bond-pads. While fabricating MEMS devices from SOI wafers, we utilize DRIE steps to etch potholes in the device layer of the substrate (Fig. 1b). If our process flow includes back-side etching, we may etch through-holes in the device layer with a back side opening in the handle below (Fig. 1c).

Following the completion of the micromachining steps, our devices are singulated and packaged. Then, a standard wire bonding machine is used to plastically deform wire bond-balls into the potholes or through-holes.

EXPERIMENTAL DEVICES

To measure mechanical adhesion and electrical conductance of the proposed direct bonding, dedicated MEMS devices were designed and fabricated (Fig. 2). They are built from P-type SOI (*50:2:400*) wafers and patterned by DRIE. Figure 2 shows a SEM photograph of a typical device.

The test devices are designed to emulate a wire bond interconnecting a silicon device to its package. In the lower section of the test device (Fig. 2), a direct bond in the island on the left (electrically isolated region) is connected by a wedge bond to a metal bond-pad in the island on the right. The left island with the direct bond section also includes a metal bond-pad for probing. In the top, the left island (Fig. 2C) includes two direct bonds each connected by a wedge bond to bond-pad on an isolated island on the right (*A* and *B*).

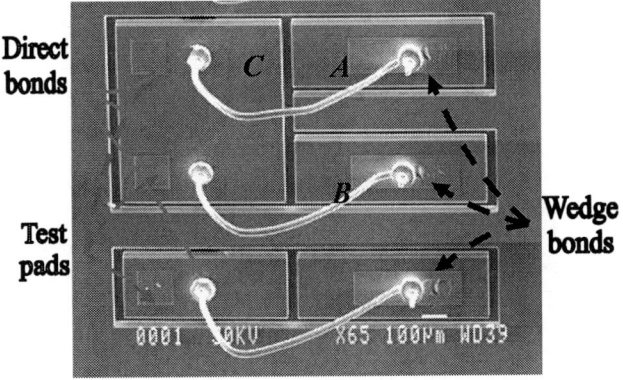

Figure 2: Test device for measuring the resistance and the strength of direct-bonds. Two direct-bonds in island C are connected to metal-pads A and B.

ADHESION

In standard wire-ball-bonding the adhesion is easily achieved by metal to metal migration [3, 4]. In the proposed method a gold wire is directly ball-bonded to silicon. Gold eutectically bonds with silicon at a temperature above *363°C* [5-8]. Due to the fact that wire-bonding is usually done during packaging, temperature-sensitive materials can be damaged or degraded if such high temperature is applied. Therefore, only localized induced heating (e.g. Joule heating) or heating by standard

---

978-1-4244-2977-6/09 $25.00 © 2009 IEEE   753

wire-ball-bonding process (e.g. ultrasonic friction heating) is permissible. Furthermore, bare silicon is covered with native silicon-dioxide which limits Gold diffusion, especially at room temperature [9, 10].

The solution proposed in this work is a mechanical connection achieved by plastically deforming a bond-ball into a silicon pothole.

To demonstrate the concept, several potholes and through-holes with different geometries were designed and fabricated. The primary consideration in the designs was a large and rough interface with the silicon. Specific features were introduced into the designs to increase plastic flow in horizontal direction to increase pressure on the pothole walls (Fig. 1b).

Several different test devices were wire bonded with a standard K&S manual ball bonder, loaded with 1mil Gold wire. The wire-bonds I-V characteristics were measured to determine their electrical conductance. Subsequently, the wire bonds were subjected to standard pull-tests [11, 12].

Figure 3 present some of the pull-tests results. The results show that most of these bonds failed by wire-fracture at an average of *7.56 grams* of pull force. This value is well above the minimum *3 grams* bond strength required by standards [11].

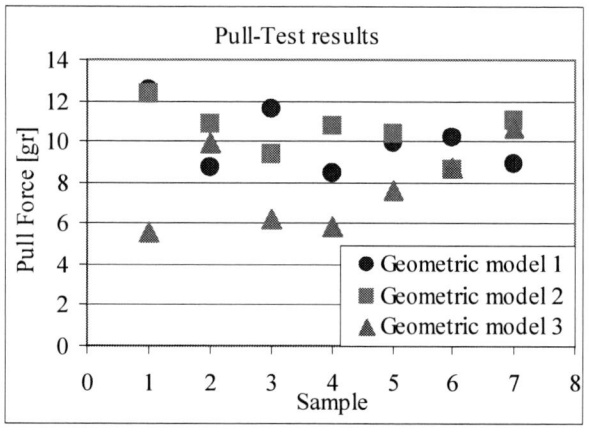

Figure 3: Pull-test results of three different hole geometries. 7 samples of each geometry produced, on average, strength of 10.05±1.53g, 10.47±1.19g, 7.78±2.04g, for geometries 1, 2, and 3, respectively

Figure 4 presents SEM photos of some direct-bonding holes and ball-bonds. Figure 4a shows a through-hole before direct-bonding. The circumferential fins are intended to increase plastic deformation and interface area. Figure 4b shows a ball-bond directly bonded to the pothole in the silicon substrate without use of a metal-pad. Its gold wire tore at over *10 grams* in a pull-test. Figure 4c shows a pothole after pull-test. The pothole geometry is created by dense parallel silicon fins. The fins have two purposes, to elevate the aspect ratio of the DRIE etching thus creating a pothole instead of through-hole, and to increase plastic deformation. In this case the fins broke during bonding and pull-test. Figure 4d shows fractured fins which remained fixed to the bond-ball after the pull-test. In this case bond failed by detaching from the pothole.

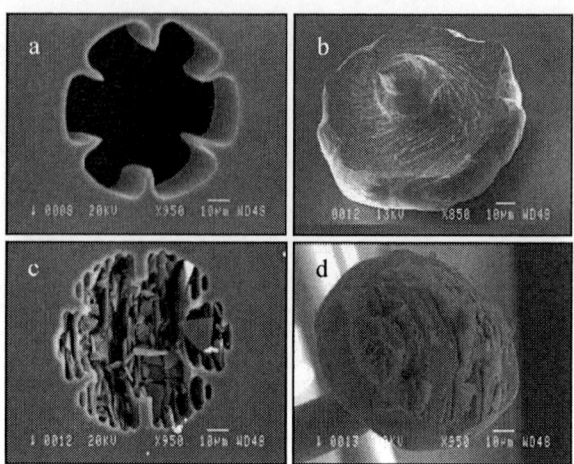

Figure 4. a. Through-hole before direct-bonding. b. ball-bond directly bonded to the pothole in the silicon substrate (no metal-pads!). Wire tore at over 10 grams in a pull-test. c. pothole after pull-test. Parallel fins increase plastic deformation and interface area. Fins broke in pull-test. d. Fins fractured in pull-test and remained fixed to the bond-ball. In this case bond failed by detaching from hole.

CONDUCTION

Metal-Semiconductor contacts are present in every semiconductor device. They can behave either as a Schottky barrier or as an ohmic contact, depending on the characteristics of the interface. The Schottky potential barrier results from work function difference between the interfacing materials. Usually the metal is used as an electrical interface to the semiconductor and as such we wish to lower the barrier as much possible (*i.e.* ohmic contact). The lowest barrier height for Gold is achieved for P-type silicon (*i.e.* Boron doping). An ohmic P-type silicon contact is achieved when the work function of the metal is higher then that of the semiconductor [13, 14]. Another way to improve the contact is by heavily doping the contact area. The high doping increases the barrier height but narrows it down until the electrons can tunnel across the barrier (i.e. tunneling).

In this work the interface is achieved by a direct contact between Gold and silicon, where the average work function of gold is *5.1V* [13, 15]. We can determine the work function of the silicon by controlling the doping concentration. In P-type silicon work function decreases if lower doping concentration is used. Hence an ohmic contact can be designed. We have measured several different wafers with high resistivity (HR) and low resistivity (LR) to obtain the ohmic contact.

All measurements are made between two metal pads A and B (Fig. 2). The electric path includes two direct-bonds back to back which will measure as two Schottky diodes, back to back, in series. The electric path also includes the bulk silicon between the direct-bonds and the negligible resistance of the bond-wires. For reference we measure the conduction between the two metal pads on island C (Fig. 2), which create a back to back Schottky contact as well.

The direct-bond originally has a high contact resistivity in order of tens of kΩ in addition to a Schottky behavior. During measurements we notice that the conductance is

978-1-4244-2977-6/09 $25.00 © 2009 IEEE

permanently improving with increase of applied voltage. In order to study this phenomenon we have developed a custom I-V voltage sweep which comprises of an increasing square wave form (Fig. 5). The aim is to monitor conductance improvement during measurement.

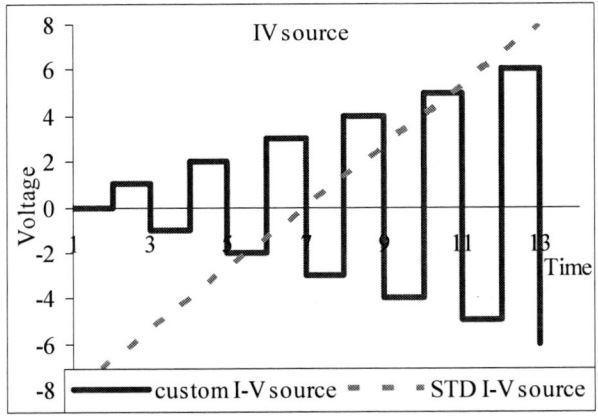

Figure 5: A standard I-V sweep source in dash line. A custom increasing square curve used for monitoring the contact priming

Figure 6 presents typical I-V measure with the custom source between metal pads A and B (Fig. 2). The first measure has a distinct back to back Schottky behavior as expected. Moreover, the conductance is markedly improved for applied voltages above 3V. The electrical resistance drops from 220Ω at 3V to 70Ω at 4V. Following repetitive measurements shows constant results, with the improved Schottky behavior. We assume the contact is priming during measurement by diffusion which is induced by localized Joule heating.

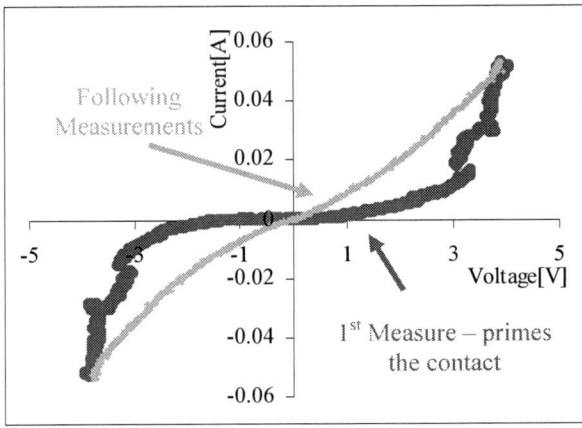

Figure 6: A typical I-V curve of a back to back direct-bond. The first measure shows conductance improvement from 220Ω at 3V to 70Ω at 4V. Following measurements show improved conductance.

Figure 7 shows test results for several wafers with different doping concentrations. The doping levels are designed to match the work-function of the gold/direct-bond thus creating a linear ohmic contact [13, 15]. The doping concentration is measured by a 4-point probe measurement for sheet resistivity. The higher doped silicon (LR wafer) had a resistivity of 0.012Ωcm and the lower doped silicon (HR wafer) had resistivity of 8Ωcm. These compare to a work function of 5.13eV and 4.91eV respectively. Since the gold work function has an average of 5.1eV, the LR wafer was expected to show a distinct Schottky behavior while the HR wafer should show an ohmic contact behavior.

The HR wafer measurement shows a distinct linear curve which implies work function matching and ohmic contact (Fig. 7 "HR wafer"). It has a high resistivity of ~5000Ω due to the bulk silicon.

Two different measures are made on the LR wafer – a reference measurement and a direct-bond measurement. The reference measurement has a distinct Schottky behavior and low resistance of 30Ω at 5V (Fig. 7 "LR wafer reference"). The direct-bond curve show improved Schottky behavior but higher resistance of 55Ω at 7V (Fig. 7 "LR wafer"). Another attempt for improving the contact was made by extra doping the LR wafer to create a tunneling contact (Fig. 7 "LR wafer + doping"). The measurement show excellent results with linear ohmic behavior and low resistivity of 12Ω at 2V.

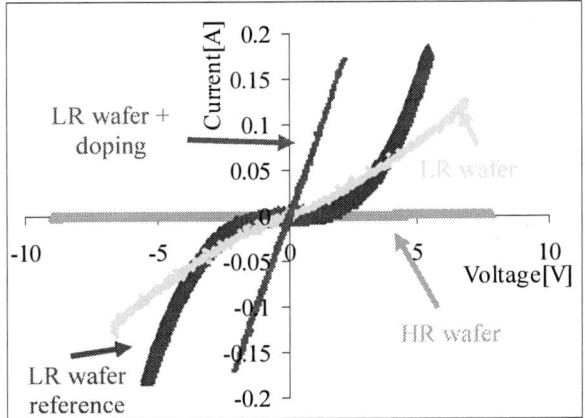

Figure 7: I-V curves for direct Gold bonds onto high and low-resistivity silicon devices. Direct-bonds on LR wafer resulted in Schottky contacts whereas direct-bonds on HR wafer and LR wafers with extra doping resulted in linear ohmic contact.

CONCLUSION

A direct wire bonding to silicon is presented. Pull-tests proved the strength of the direct bonds to be compatible with current standards for wire-bonds [11]. In our search for improving the direct-bonding electrical conductance, we have successfully controlled the contact interface levels by wafer doping while thriving toward linear ohmic contact. Moreover we have shown a better conductance contact then a standard contact of metal pads.

Future research aims to optimize the priming process, and to expand the research to Aluminum wedge bonding.

ACKNOWLEDGMENTS

The support of I.M.H. Ltd. in test-device fabrication it gratefully acknowledged.

978-1-4244-2977-6/09 $25.00 © 2009 IEEE

REFERENCES

[1] G. G. Harman, *Wire bonding in microelectronics*, 2nd ed. McGraw-Hill, 1997.

[2] R. R. Tummala and E. J. Rymaszewski, *Microelectronics packaging handbook*, 2nd ed. Chapman & Hall, 1997.

[3] G. Harman and J. Albers, "*The Ultrasonic Welding Mechanism as Applied to Aluminum-and Gold-Wire Bonding in Microelectronics*," Parts, Hybrids, and Packaging, IEEE Transactions on, vol. 13, pp. 406-412, 1977.

[4] G. E. S. a. S. D. Brandenburg, "*Wire Bonding - A Closer Look*," presented at ISTFA, Los Angeles, California, USA, 1991.

[5] T. J. Harpster and K. Najafi, "*Field-assisted bonding of glass to Si-Au eutectic solder for packaging applications*," MEMS-03 Kyoto. IEEE

[6] A. L. Tiensuu at.el, "*Assembling 3-Dimensional Microstructures Using Gold Silicon Eutectic Bonding*," Sensors and Actuators a-Physical, vol. 45, pp. 227-236, 1994.

[7] L. W. Lin, Y. T. Cheng, and K. Najafi, "*Formation of Silicon-gold eutectic bond using localized heating method*," Japanese Journal of Applied Physics Part 2-Letters, vol. 37, pp. L1412-L1414, 1998.

[8] Y. T. Cheng, L. Lin, and K. Najafi, "*Localized Silicon fusion and eutectic bonding for MEMS fabrication and packaging,*" Microelectromechanical Systems, Journal of, vol. 9, pp. 3-8, 2000.

[9] M. Morita, T. Ohmi, E. Hasegawa, M. Kawakami, and M. Ohwada, "*Growth of Native Oxide on a Silicon Surface*," Journal of Applied Physics, vol. 68, pp. 1272-1281, 1990.

[10] J. C. Anderson, "*Applications of thin films in microelectronics*," Thin Solid Films, vol. 12, pp. 1-15, 1972.

[11] MIL-STD-883G, "*Bond strength (destructive bond pull test), Method 2011.7*," 2006.

[12] G. Harman and C. Cannon, "*The Microelectronic Wire Bond Pull Test-How to use It, How to Abuse It,*" Components, Hybrids, and Manufacturing Technology, IEEE Transactions, vol. 1, pp. 203-210, 1978.

[13] E. H. Rhoderick and R. H. Williams, *Metal-semiconductor contacts*, 2nd ed Claredon Press; Oxford University Press, 1988.

[14] J. Singh, *Semiconductor devices : basic principles.* New York: Wiley, 2000.

[15] S. M. Sze and K. K. Ng, *Physics of semiconductor devices*, 3rd ed. Hoboken, N.J.: Wiley-Interscience, 2007.

A LOW-NOISE OSCILLATOR BASED ON A MULTI-MEMBRANE CMUT FOR HIGH SENSITIVITY RESONANT CHEMICAL SENSORS

H. J. Lee, K. K. Park, P. Cristman, Ö. Oralkan, M. Kupnik and B. T. Khuri-Yakub
Edward. L. Ginzton Laboratory, Stanford University, Stanford, California, USA

ABSTRACT

We present 17.5-MHz and 42.7-MHz low-noise Colpitts oscillators employing capacitive micromachined ultrasonic transducers (CMUTs), each composed of a thousand resonator cells electrically connected in parallel. The massive parallelism lowers the motional impedance, and thus, reduces frequency noise and provides better matching to low-noise oscillator topologies. The 42.7-MHz oscillator achieved a phase noise of -105 dBc/Hz and -148 dBc/Hz at offset frequencies of 1 kHz and 1 MHz, respectively in air. The performance is comparable to MEMS oscillators based on resonators with high Q in vacuum. The lowest Allan deviation of the oscillator was measured to be 4.7×10^{-9} implying a mass resolution of 0.96 attogram per membrane. In addition, using the 17.5-MHz CMUT resonator, the performance of the Colpitts topology is compared to that of the amplifier based oscillator topology.

INTRODUCTION

Resonant sensing has been widely applied to micromechanical systems to measure a variety of measurands, such as pressure, acceleration and chemical or biological agents. These sensor systems benefit from direct frequency output, high sensitivity, low noise and large dynamic range [1]. The operation principle of resonant sensing is the change of resonant characteristics in response to the effect of measurands on the spring constant or the mass of the resonant structure.

One important figure of merit for the resonant sensor system is the noise performance of the oscillator, which limits the minimum detectable signal of the sensor (i.e. resolution). Various methods have been explored in resonator design [2, 3] as well as in circuit design to improve the frequency noise. One method to further reduce the noise of an optimized single resonator is to use multiple in series or parallel to make a resonator with a lower motional resistance (R_x). Driscoll [4] demonstrated a reduction in phase noise of the oscillator by connecting multiple Quartz resonators in series. Recently, Demirci [5] and Lin *et al.* [6] reported reduction in motional resistance and phase noise by mechanically-coupling

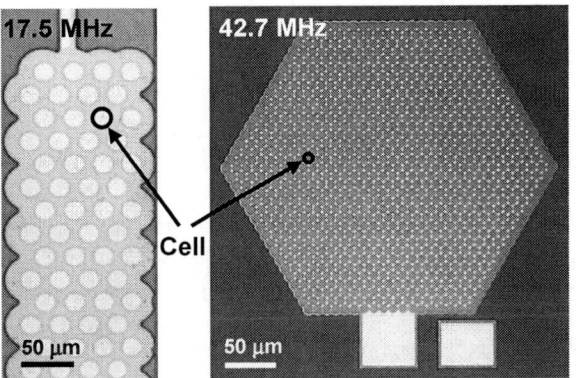

Figure 2: *Photographs of the 17.5-MHz and 42.7-MHz CMUT devices with thousand resonator cells.*

multiple microresonators.

Connecting multiple microresonators electrically has been avoided due to the reduced Q resulting from the resonator-to-resonator non-uniformities. If the non-uniformities can be tightly controlled or the Q of a single resonator is relatively low, electrically connecting multiple microresonators is a simpler approach than that of mechanically coupling in terms of the resonator design.

We demonstrate in this work that we achieve a low phase noise by electrically connecting multiple microresonators. Motivated by the goal to improve the resolution of a chemical sensor based on the capacitive micromachined ultrasonic transducer (CMUT) technology [7, 8], we fabricated CMUTs composed of approximately 1000 resonator cells electrically connected in parallel. We first introduce design and characterization of CMUT resonators and discuss the implication of parallelism on oscillator design. Further, we demonstrate the noise characteristics of two implemented oscillators and discuss its implication on the resolution of the CMUT chemical sensor.

CMUT RESONATOR DESIGN
Operation and Structure

A CMUT with one membrane is a capacitor (Figure 1). The top and bottom electrodes are composed of a single crystal silicon membrane and a conductive substrate, respectively. These electrodes are separated by a thin vacuum gap and an insulation layer and are circularly anchored by an oxide post. When the capacitor is actuated by electrostatic (DC and AC) force, it resonates at a resonant frequency determined by the material properties and dimensions of the circular membrane resonator (1),

$$ w_0 = \frac{2.98t}{r^2} \sqrt{\frac{E}{(1-v^2)\rho}} \ , \tag{1} $$

where t, r, E, v and ρ are the thickness, radius, Young's modulus, Poisson's ratio and density.

The 17.5-MHz CMUT resonator used for this work is

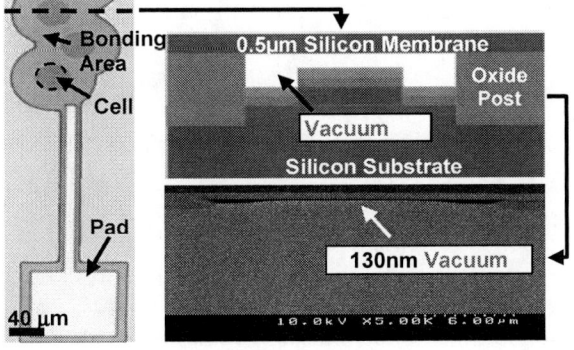

Figure 1: *Schematic diagram and cross-sectional SEM of a single cell from the multi-cell CMUT resonator.*

978-1-4244-2977-6/09 $25.00 © 2009 IEEE

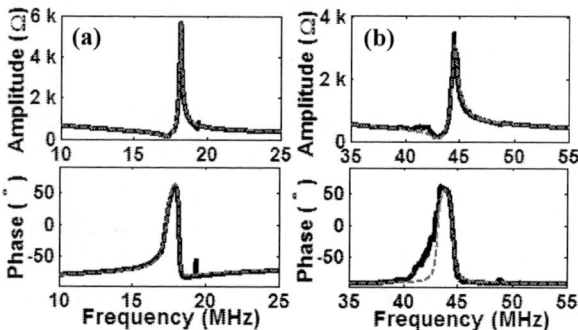

Figure 3: Measured and fitted impedance characteristics of (a) the 17.5-MHz CMUT, (b) the 42.7-MHz CMUT, in air. (Solid line: measurement, dotted lines: fitted).

composed of 1000 circular cells with 500-nm thick silicon membranes and radii of 9 μm (Figure 1). The membranes are supported by a 0.9-μm-thick oxide post and separated from the bottom electrodes by a 130-nm vacuum gap. The 42.7-MHz CMUT is composed of 1027 circular cells with radii of 5.3 μm (Figure 2). The vacuum gap height is 50 nm and the membranes are supported by a 1-μm-thick oxide post. The details of the fabrication processes of the CMUT based on direct wafer-bonding and LOCOS techniques are described in [8].

Input Impedance Characteristics

The resonant characteristics of two CMUT resonators were measured using an impedance analyzer (Agilent Technologies, Model 4294A, Palo Alto, CA). One was biased at 41 V with a parallel resonance frequency (f_p) at 17.5 MHz and one biased at 66 V with f_p at 42.7 MHz (Figure 3). The input impedance measurements were fitted to a 6-element equivalent circuit (Figure 4), which includes parasitic effects of the substrate and the electrode contacts in addition to the conventional 4-element RLC van Dyke model [9]. The standard van Dyke circuit consists of four real circuit parameters, R_X, L_X, C_X, and C_0, which physically represent loss, mass, stiffness and electric capacitance of the resonant structure, respectively. Due to the massive parallelism, the motional resistance dropped below 100 Ω for both resonators (Figure 4).

One disadvantage of massive parallelism is the degradation of Q due to process variations. Our analysis shows that Q at the parallel resonance (Q_p) is degraded less than Q at the series resonance (Q_s) from the resonator-to-resonator non-uniformity (Figure 5). The effect of process variation on the quality factor is investigated by computing total impedance of 169 CMUT resonators placed in parallel. The radius and thickness of each resonator is given a normal distribution with a mean value of 5.3 μm and 0.5 μm, respectively. For various standard deviation values, we computed the quality factors of the overall impedance.

Q_s and Q_p are almost identical when the device variations are negligible (Figure 5). However, as the variation increases, Q_s degrades more than Q_p. The input impedance measurement also verifies that Q_s is smaller than Q_p for various numbers of microresonators connected in parallel across the operating bias voltage [10]. Therefore, instead of oscillating at the series resonance, the multi-membrane CMUT microresonators can be

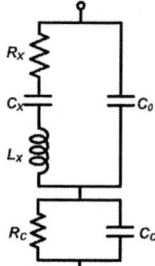

	17.5-MHz	42.5-MHz
V_{DC}	41 V	66 V
R_X	50 Ω	75 Ω
C_X	1.5 pF	0.61 pF
L_X	54 μH	24 μH
C_0	21 pF	7.4 pF
R_C	154 Ω	1.6 Ω
C_C	8.8 pF	0.46 pF

Figure 4: 6-element equivalent circuit model used to fit the input impedance of the CMUT. The values for each components for the two CMUT reosonators are shown.

oscillated in its inductive region with a reasonable Q and still provide the advantages of parallelism.

CMUT OSCILLATOR DESIGN

The reduction in motional impedance has two implications in the oscillator design: better impedance matching to various oscillator circuit topologies and lower mechanical noise.

Oscillator Design for CMUT

MEMS resonator based oscillators can be largely divided into two types: transresistance amplifier based oscillators and single-stage tuned oscillators, such as Pierce and Colpitts. The open-loop gain of a transresistance amplifier based oscillator is determined by the resistive elements while that of a single-stage tuned oscillator is determined by the capacitive elements offering superior frequency stability. Further, for the CMUT resonator, the Colpitts topology is more suitable than the Pierce topology for two reasons: the constraint to ground one electrode of the CMUTs (Figure 6) and higher Q at parallel resonance than at series resonance (Figure 5). Colpitts oscillators operate at the inductive region between the series and parallel resonances while Pierce oscillators operate at the series resonance.

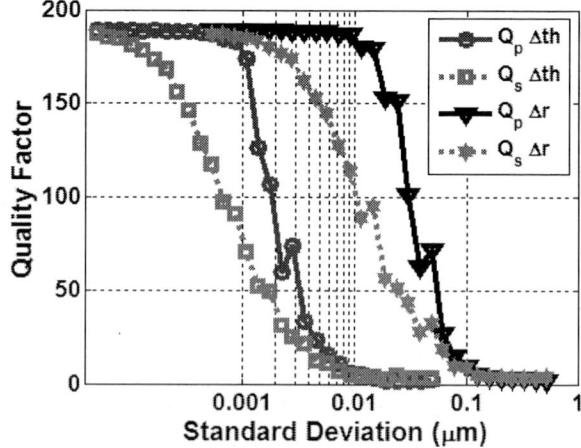

Figure 5: Simulation results showing the effect of process variations on Q_s and Q_p assuming the process variation is a normal distribution. 169 CMUT resonators are placed in parallel with mean radius and thickness of 5.3 μm and 0.5 μm, respectively. The standard deviation of the thickness (Δth) and the radius (Δr) of the membrane are varied.

(a)

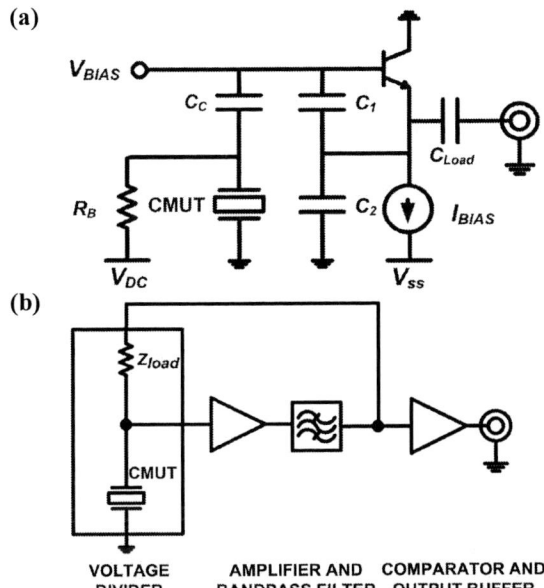

(b)

VOLTAGE DIVIDER   AMPLIFIER AND BANDPASS FILTER   COMPARATOR AND OUTPUT BUFFER

Figure 6: (a) Circuit diagram of the Colpitts oscillator with the bias circuitry of the CMUT resonator shown. (b) Block diagram of the amplifier based oscillator.

Therefore, the Colpitts topology was chosen to implement an oscillator based on the CMUT resonator.

The CMUT resonators are first modeled as a lossy inductor ($R_e + jL_e$) in the narrow frequency region between the series and the parallel resonance (Figure 6 (a)). The design values for the circuit parameters, C_1 and C_2, and the bias point for the bipolar transistor (BJT) are determined according to the start-up criteria (2) and the desired oscillation frequency (3),

$$\frac{g_m}{w_0^2 R_e C_1 C_2} \geq 1 \qquad (2)$$

$$w_0 = \frac{1}{\sqrt{L_e C_T}}, \qquad (3)$$

where g_m is the transconductance, w_0 is the resonant frequency, R_e and L_e are the real and imaginary part of the modeled lossy inductor, and C_T is the total capacitance in parallel with the resonator [11]. In addition, the values of the biasing resistors must be large enough to avoid significant loading of the emitter-follower amplifier. Following this design method, the 17.5-MHz and 42.7-MHz CMUT Colpitts oscillators are implemented on PCBs, on which the CMUT resonators are directly wire-bonded to eliminate parasitic effects of a chip carrier.

Noise Characteristics

The stability of the oscillator is characterized in the frequency domain using a signal source analyzer (Agilent Technologies, Model E5052B, Palo Alto, CA). For the 42.7-MHz oscillator, we achieved a phase noise of -105 dBc/Hz and -148 dBc/Hz at offset frequencies of 1 kHz and 1 MHz, respectively (Figure 7). The 17.5-MHz oscillator achieved a phase noise of -97 dBc/Hz and -155 dBc/Hz at offset frequencies of 1 kHz and 1 MHz, respectively. (Figure 7). The 42.7-MHz oscillator exhibits a better close-in phase noise but a worse

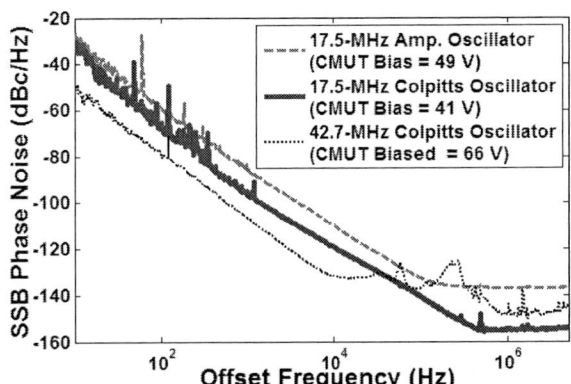

Figure 7: Single side band (SSB) phase noise of the 17.5-MHz amplifier based oscillator, the 17.5-MHz Colpitts oscillator, and the 42.7-MHz Colpitts oscillator, measured using a signal source analyzer (Agilent Technologies, Model E5052B, Palo Alto, CA).

broadband phase noise than the 17.5-MHz oscillator because 42.7-MHz resonator has a higher Q_p than the 17.5-MHz resonator.

The performance of 42.7-MHz oscillator is comparable to state of the art MEMS oscillators based on microresonators with high Q in vacuum (Table 1). For the comparison, the phase noise is converted according to (4) to incorporate the differences in the oscillation frequencies.

$$L(f)\big|_{f_0=X} = L(f)\big|_{f_0=Y} - 20\log\left(\frac{Y}{X}\right) \qquad (4)$$

where X is the frequency to convert into and Y is the actual oscillation frequency.

For the 17.5-MHz resonator, we compared the performance of the Colpitts topology to that of the high-gain amplifier based oscillator topology (Figure 6 (b)). Figure 7 shows that the noise floor of the 17.5-MHz Colpitts oscillator is 19 dB smaller than that of the amplifier based oscillator. In addition, the total power consumption has decreased from 80 mW to 6 mW (Table 1). Therefore, the single-stage tuned oscillator topology is advantageous in two aspects; the gain elements are no longer noisy resistors and the required component counts are much smaller, resulting in a smaller thermal noise and potentially less power consumption.

Table 1: Comparison of the 42.7-MHz CMUT oscillator to low phase noise MEMS oscillators.

	This work	Hsu, 2004 [2]	Lin, 2004 [6]	Lavasani, 2008 [3]
f_0 **(MHz)**	42.7	70	60	467
Q	~ 300	16000	~115000	1850
Medium	Air	Vacuum	Vacuum	Air
L(1 kHz) (dBc/Hz)	-105	-117	-123	-90
L(1 kHz) (dBc/Hz) (f_0 converted to 42.7MHz)	-105	-121	-126	-110
Phase noise floor (dBc/Hz)	-148	-130	-136	-147
Integrated	No	No	Yes	Yes

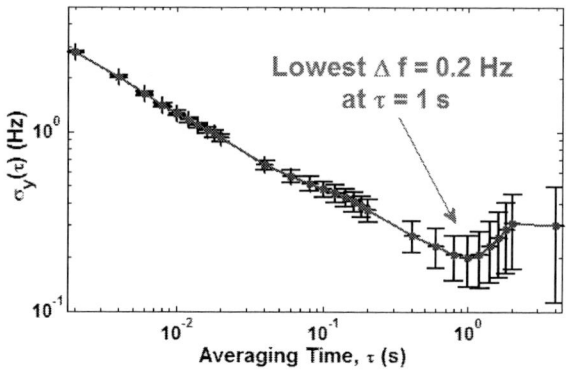

Figure 8: Overlapped Allan deviation calculated from the frequency counter data with a gate time of 2 ms (Stanford Research Systems, Model SRS620, Sunnyvale, CA). The error bars indicate the 1-sigma confidence level.

MASS SENSOR IMPLICATION

The detection limit of the mass sensor, Δm, is determined by short-term frequency stability of an oscillator (Δf). Thus, we characterized the oscillators in time domain using a frequency counter (Stanford Research Systems, Model SRS 620, Sunnyvale, CA). The 42.7-MHz oscillator exhibited the lowest Δf. The Allan deviation (σ_y) was measured as 4.7×10^{-9} at an averaging time of 1 s (Figure 8). Using the mass sensitivity equation, $\Delta m = -2m \cdot (\Delta f / f)$, the mass resolution was calculated to be 0.96 attogram per membrane (Table 2). For the resonant chemical sensor application, the massive parallelism has other benefits such as robustness and large sensing area required for efficient chemical detection.

CONCLUSION

By electrically connecting large number of microresonators, we demonstrated a 42.7-MHz CMUT-based oscillator with excellent phase noise. In theory, the sensor system is capable of detecting masses as low as 0.96 attogram. Using the high-gain amplifier based oscillator, we have previously demonstrated a 6-MHz CMUT chemical sensor array [12] and an 18-MHz CMUT chemical sensor with a volume sensitivity of 37 ppbv/Hz to DMMP in air [7]. Thus, with new oscillators with much smaller frequency noise, we expect a significant improvement in the volume sensitivity. In addition, we have submitted an application-specific integrated circuit

Table 2: Summary of Colpitts oscillator performance.

	17.5-MHz Oscillator	42.7-MHz Oscillator
CMUT Bias (V)	41	66
R_x **(Ω)**	50.2	73.7
Q	127	160
Power (mW)	6	16
Output Power (dBm)	4.6	-15.7
Allan Deviation	2.8×10^{-8}	4.7×10^{-9}
Mass Sensitivity (g/Hz)	27×10^{-18}	4.8×10^{-18}
Mass Resolution (ag)	13.3	0.96
Mass Resolution per Unit Area (ag/μm^2)	0.067	0.011

chip containing oscillator circuits for future studies.

ACKNOWLEDGEMENT

This work is funded by DARPA, Microsystems Technology Office under grant N66001-06-1-2030. We would like to thank National Semiconductor, Santa Clara, CA, for providing the phase noise test setup.

REFERENCES

[1] A. A. Seshia, M. Palaniapan, T. A. Roessig, R. T. Howe, R. W. Gooch, T. R. Schimert, and S. Montague, "A Vacuum Packaged Surface Micromachined Resonant Accelerometer", *IEEE J. Microelectromech. Sys.*, vol. 11, pp. 784-793, 2002.

[2] W. –T. Hsu, K. Cioffi, "Low Phase Noise 70 MHz Micromechanical Reference Oscillators," in *Micro. Symp. Dig., 2004 IEEE MTT-S Int.*, vol. 3, pp. 1927-1930, Jun., 2004.

[3] H. M. Lavasani, R. Abdolvand, and F. Ayazi, "Low Phase-Noise UHF Thin-Film Piezoelectric-On-Substrate LBAR Oscillators", in *Proc. IEEE MEMS Conference*, Tucson, pp. 1012-1015, Jan., 2008.

[4] M. M. Driscoll, "Reduction of Quartz Crystal Oscillator Flicker-of-Frequency and White Phase Noise (Floor) Levels and Acceleration Sensitivity via Use of Multiple Resonators," in *Proc. IEEE Frequ. Contr. Symp.*, pp. 334-339, May, 1992.

[5] M. U. Demirci and C. Nguyen, "Mechanically Corner-Coupled Square Microresonator Array for Reduced Series Motional Resistance," *IEEE J. Microelectromech. Sys.*, vol. 15, no. 6, pp. 1419-1436, Dec. 2006.

[6] Y. Lin, S. Lee, S. Li, Y. Xie, Z. Ren and C. Nguyen, "60-MHz Wine-Glass Micromechanical-Disk Reference Oscillator," in *Digest of Tech. Papers, IEEE ISSCC*, San Francisco, pp. 322-331, Feb., 2004.

[7] H. J. Lee, K. K. Park, Ö. Oralkan, M. Kupnik and B. T. Khuri-Yakub, "CMUT as a Chemical Sensor for DMMP Detection," in *Proc. IEEE Intern. Frequ. Contr. Symp.*, pp. 434-439, May, 2008.

[8] K. K. Park, H. J. Lee, M. Kupnik, Ö. Oralkan, and B. T. Khuri-Yakub, "Fabricating Capacitive Micromachined Ultrasonic Transducers with Direct Wafer-Bonding and LOCOS Technology," in *Proc. IEEE MEMS Conference*, Tucson, USA, pp. 339-342, 2008.

[9] S. Sherrit, H. D. Wiederick, B. K. Mukherjee, "Accurate Equivalent Circuits for Unloaded Piezoelectric Resonators," in *Proc. IEEE Ultrason. Symp.*, pp. 931-935, vol. 2, 1997.

[10] H. J. Lee, K. K. Park, P. Cristman, Ö. Oralkan, M. Kupnik and B. T. Khuri-Yakub, "The Effect of Parallelism of CMUT Cells on Phase Noise for Chem/Bio Sensor Applications," accepted for *Proc. IEEE Ultrason. Symp.*, 2008.

[11] G. Gonzalez, *Foundations of Oscillator Circuit Design*. Norwood, MA: Artech House, Inc., 2006.

[12] K. K. Park, H. J. Lee, G. G. Yaralioglu, Ö. Oralkan, M. Kupnik, C. F. Quate, B. T. Khuri-Yakub, T. Braun, H. P. Lang, M. Hegner, C. Gerber, and J. Gimzewski, "Capacitive Micromachined Ultrasonic Transducers for Chemical Detection in Nitrogen," *Applied Physics Letters*, Lett. 91, 094102, 2007.

PERFORMANCE OF PACKAGED PIEZOELECTRIC MICROSPEAKERS DEPENDING ON THE MATERIAL PROPERTIES

SeungHwan Yi[1], Soon Chul Ur[2], and Eun Sok Kim[3]

[1]Dept. of Mechanical Eng., Chungju National University/RIC-ReSEM, Chungju, Chungbuk, KOREA
[2]Dept. of Material & Sci., Chungju National University/RIC-ReSEM, Chungju, Chungbuk, KOREA
[3]Dept. of EE-Electrophysics, University of Southern California, Los Angeles, CA, USA

ABSTRACT

This paper presents the improved piezoelectric microspeakers that implement a packaging structure with AlN film and are made of composite diaphragms of different residual stresses. The output Sound Pressure Level (SPL) is enhanced more than 10 dB with a higher compressively stressed composite diaphragm (having -20 MPa silicon nitride film as a supporting layer) from 100 Hz to 3 kHz. The best microspeaker produces more than 100 dB SPL into open field around 3 kHz (when measured 10 mm away from the speaker that is driven with 20 $V_{peak-to-peak}$ sinusoidal signal). A microspeaker with a square electrode pattern produces more uniform frequency response above 4 kHz than that with a circular electrode pattern.

INTRODUCTION

Personal Communication System (PCS) including mobile phones demands compact size and lightweight, as the components used in PCS are getting increasingly smaller and lighter. In case of audio components for mobile phone application, MEMS microphones are rapidly replacing the conventional electret microphone that cannot withstand a solder-reflow temperature (250 °C), and have been packaged into a mobile phone through an expensive manual assembly. But the speakers used in PCS are still all electrodynamic-type that is bulky and thick, as it is larger than 10 mm in diameter and thicker than 2 mm.

The early versions of the piezoelectric microspeakers attempted to increase their sound output by minimizing the tensile residual stress of the composite diaphragm [1, 2] or making the diaphragm compliant through employing polymer layer to make a cantilever-like diaphragm [3]. However, the potential of a piezoelectric microspeaker as a loud speaker was not demonstrated until silicon nitride diaphragm was compressively stressed so heavily as to make the whole diaphragm wrinkled except the piezoelectrically active area [4]. Its refined version with piezoelectric ZnO film that improves the sound output will be presented in [5].

Though the piezoelectric coefficient d_{31} of ZnO film is twice that of AlN film, we have explored AlN film as the piezoelectric transducing material, since AlN has successfully been commercialized for RF-front-end filter based on film bulk acoustic resonator. In this paper, we have intentionally varied the residual stress of the supporting diaphragm (composed of silicon nitride) and the shape of the electrode pattern on the piezoelectrically active area, and characterized their effects on the performance of the piezoelectric microspeakers based on piezoelectric AlN film. Also, the packaging structure has been developed to compare the sound output of the microspeakers with and without the packaging.

FABRICATION & EXPERIMENT

Device Fabrication

Five masks were used in the fabrication processes of the piezoelectric microspeaker shown in Fig. 1.

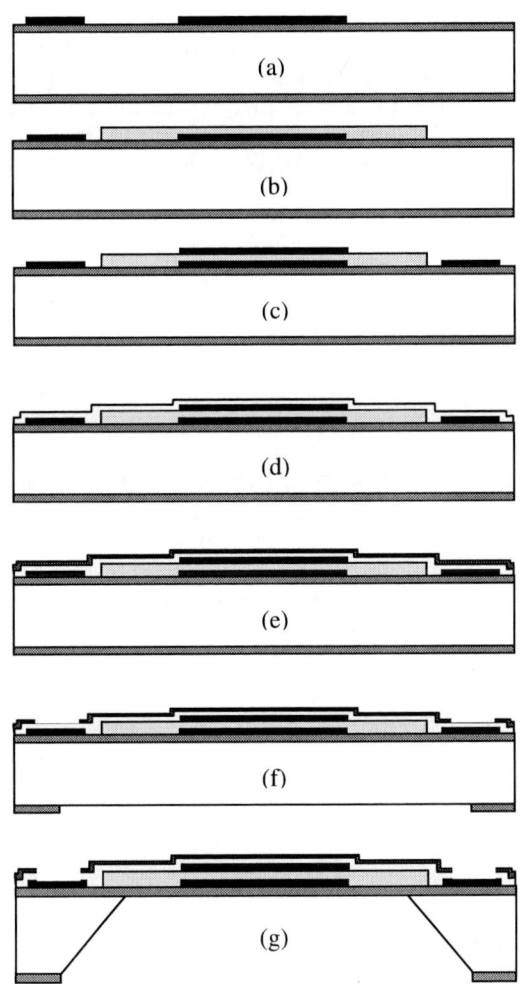

Figure 1: Fabrication process steps of the piezoelectric microspeakers (circular electrode type).

First, on double-polished 4-inch silicon wafers, 1.0 μm thick silicon nitride film was deposited in a low-pressure chemical vapor deposition (LPCVD) system. In this process, we deposited two kinds of silicon nitride films by varying the ratio of the reactant Si_2H_6 and NH_3 gases at 830 °C; one for a mildly tensile stressed film and the other for slightly compressive stressed film. In both cases, the stress level was measured after deposition of each film. Then 0.2

978-1-4244-2977-6/09 $25.00 © 2009 IEEE

μm thick Mo film along with 0.02 μm thick Ti adhesion layer for Mo film are sputter-deposited at room temperature on the front side of the wafers for the floating electrode of some microspeakers' configuration. After patterning the Mo/Ti film (Fig. 1 (a)), 0.5 μm thick piezoelectric AlN film was deposited in a RF magnetron sputtering system with 300 °C substrate temperature, 1 mTorr chamber pressure, 3 kWatt RF power and Ar/N$_2$ gas ratio set at 1/4. Then some parts of the piezoelectric AlN film were patterned by ICP (Inductively Coupled Plasma) RIE (Reactive Ion Etch) system (Fig. 1 (b)). After the AlN film patterning, Mo/Ti film was deposited for the top electrodes and contact pads, and was etched in ICP RIE system again (Fig. 1 (c)). In order to increase the mechanical sturdiness of the released diaphragms during the sawing process, we deposited Parylene-C film only onto the front side of the silicon wafer (Fig. 1 (d)). In this case, the Parylene-C film covered all area of the pre-fabricated structures, and needed to be patterned to expose the contact pads. In order to pattern the Parylene, we deposited Al film as a masking layer of Parylene-C film during an oxygen plasma ashing process. With the fourth mask, the Al masking layer was patterned, followed by removal of the exposed Parylene-C film in an oxygen plasma process (Fig. 1 (e) and (f)).

Then, after removing the aluminum etch mask in a wet etchant composed of KOH: K$_3$ Fe(CN)$_6$: H$_2$O = 1 g : 10 g : 100 ml, we patterned the backside silicon nitride film with a double-side alignment, etching it in an RIE system with CF$_4$ gas (Fig. 1 (f)). The silicon wafer was then etched from its backside in KOH in order to release the diaphragm (Fig. 1 (g)). During the backside etching process, the front-side of the silicon wafer was protected from KOH solution by spin-coating photo-resist on the front side and specially designed protection layers (composed of polyethylene film, quartz substrate and silicone rubber). After the backside etching process, the completed wafer was cleaned in D.I. water, and was dried in a convection oven for 30 min.

Figure 2 shows the fabricated microspeakers on a 4-inch silicon wafer. Even though there were 24 different designs for the microspeaker, only two microspeakers (the square and the circular electrode types) will be discussed in this paper. After completing the process on the wafer, we diced the wafer into individual chips, each of which was attached to a printed circuit board (PCB) as shown in Figure 3 in order to secure solid electrical connection for testing the packaged microspeaker. Then the PCB with the microspeaker was inserted into a package made of acryl, of which the total volume including the PCB was about 4 cc (Figure 3).

Experiment

Since the quality of the piezoelectric film is directly related to the performance of the piezoelectric microspeaker, we investigated the film quality with SEM and X-ray diffraction (XRD). The cross-sectional SEM photos were taken in order to confirm the columnar, polycrystalline structure of the deposited AlN film, while XRD measurements were made to quantify the c-axis orientation of the film and the uniformity of the c-axis among different grains. As will be reported in [5], the measured product (Bd$_{31}$) between biaxial modulus (B) and

piezoelectric coefficient (d$_{31}$) is strongly related to XRD measurements. We measured the curvatures of the wafer and the film thickness after the deposition of each film in order to measure the residual stress of each film, using Stoney's equation [6]

$$\sigma = \frac{E_s \cdot t_s^{\,2}}{6(1-v_s)t_f}\left(\frac{1}{R_s}-\frac{1}{R_f}\right) \qquad (1)$$

where E_s=1.301×10^5 MPa and v =0.279, while t_s, t_f, R_s and R_f are the wafer thickness, film thickness, wafer curvature without the film and wafer curvature with the film, respectively.

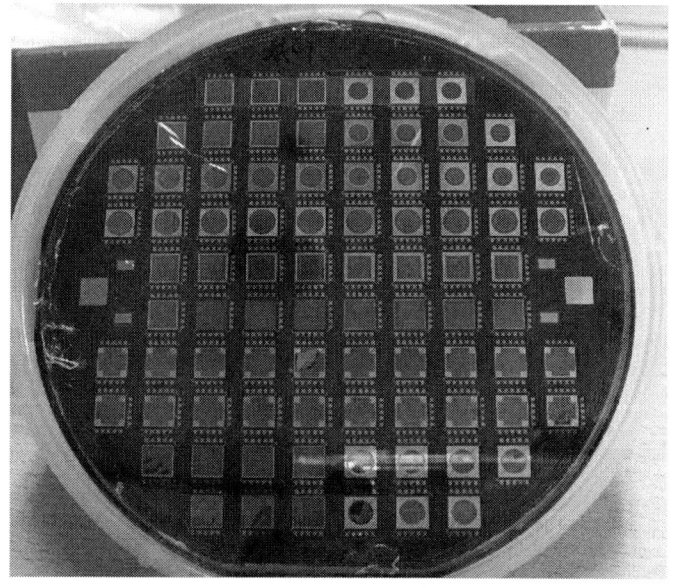

Figure 2: Photo of fabricated microspeakers on a 4-inch silicon wafer.

Figure 3: Photo of a packaged microspeaker attached on a PCB.

Figure 4 shows the experimental setup for measuring sound output pressure level of the fabricated microspeakers. The frequency-modulated sinusoidal voltage (1.25 V$_{peak-to-peak}$) from NI Sound & Vibration Board of a computer was fed to B&K power amplifier, which amplified it to 20 $V_{peak-to-peak}$. The amplified sinusoidal signal was applied to a fabricated microspeaker (and also to the Data Acquisition board of the computer in order to record the frequency and amplitude of the applied signal). The sound output from the microspeaker was measured with a reference microphone (B&K 4192L that has an amplified sensitivity of 1 V/Pa), and was recorded in the computer through the Data Acquisition board, as we varied the frequency from 100 Hz to 15 kHz.

The reference microphone was kept 10 mm away from the microspeaker in both cases of the microspeaker alone and the packaged microspeaker. The SPL of the fabricated microspeaker was obtained through the following equation: $SPL = 20 \times \log(P_e / P_r)$ [dB], where P_e and P_r are the output pressure measured with the reference microphone and a reference sound pressure (2×10^{-6} Pa), respectively.

In order to obtain reliable SPLs of the fabricated microspeakers, we tested the microspeakers more than ten times and averaged the measured data.

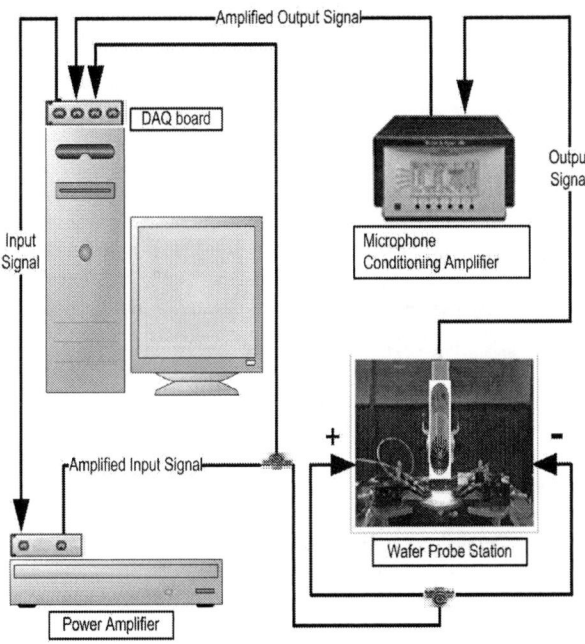

Figure 4: Block diagram of the measurement set-up.

RESULTS AND DISCUSSION

Figure 5 shows a SEM photo of AlN thin film deposited onto Mo/Ti bilayers. The Mo/Ti films (that were used as an electrode) show a well-oriented columnar structure over the supporting silicon nitride film. The piezoelectric AlN film also has a columnar structure with fine grain size and high density. Even though measured XRD data are not shown in this paper, the XRD spectrum shows only one peak for AlN (002) that indicates a c-axis oriented perpendicular to the silicon substrate.

Figure 6 describes the fabricated microspeaker chips and their packaged structures. As can be seen in Fig. 6 (a) and (c), the fabricated microspeaker chips are wrinkled since the supporting diaphragm and the piezoelectric AlN are compressively stressed. (The residual stress of the AlN film was measured to be about -100 MPa.) The contact

pads on the microspeaker chips were wire-bonded onto several of the contact pads on the PCB, which were connected to solder pads as shown in Fig. (b) and (d).

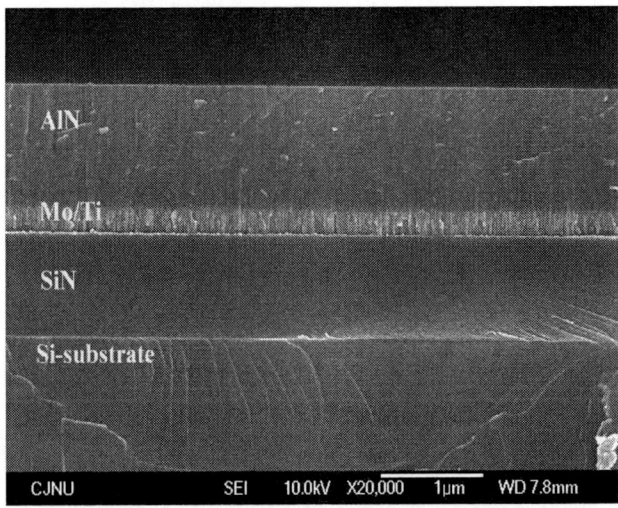

Figure 5: Cross-sectional SEM photo of the AlN/MoTi/SiN/Si layers.

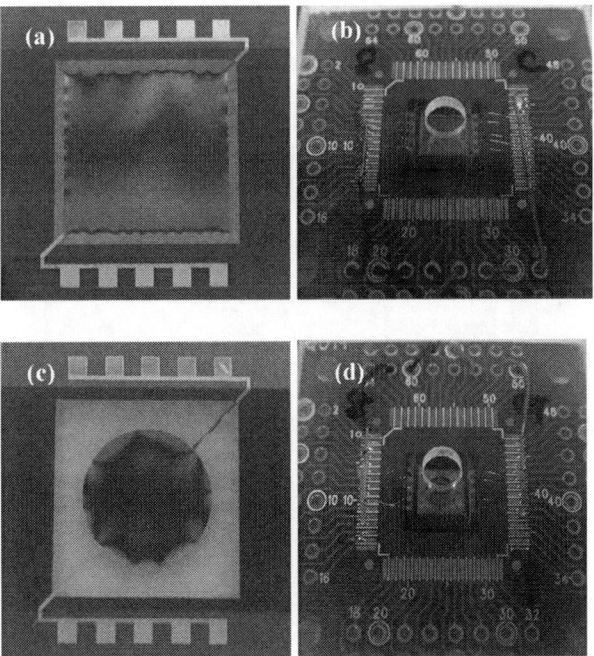

Figure 6: Photos of tested microspeaker chips and packaged structures: (a) and (b) are a microspeaker with square electrodes and its packaged structure, respectively, while (c) and (d) are a microspeaker with circular electrodes and its packaged structure, respectively.

Figure 7 shows the measured frequency response of the fabricated microspeaker as a function of the residual stress in the composite diaphragm of the microspeaker. The measured data confirms that as the residual stress becomes more compressive, the microspeaker produces a higher sound output, especially at low frequency region where the difference is more than 10 dB SPL.

978-1-4244-2977-6/09 $25.00 © 2009 IEEE   763

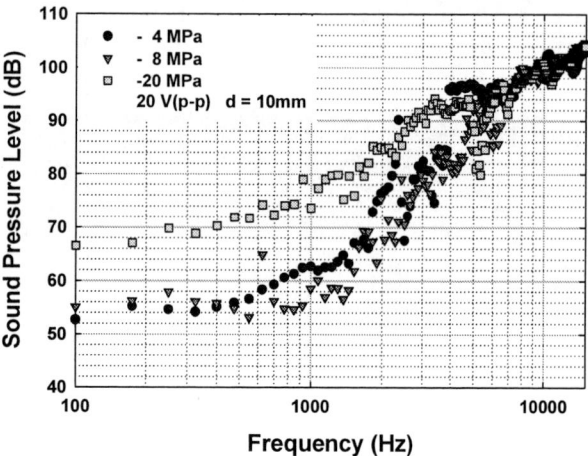

Figure 7: Measured frequency response of the fabricated microspeakers between 100 Hz and 15 kHz as a function of the residual stresses in the composite diaphragm.

Measured SPLs of a microspeaker are compared in Figure 8 before and after packaging it as in Figure 3. A pre-packaged microspeaker (which means a microspeaker chip alone) showed a relatively low sound pressure level at low frequency, but produced a much higher output pressure as the frequency was increased. When the microspeaker was packaged as shown in Figure 3, the packaged speaker produced a significantly higher sound output pressure from 100 Hz to 4 kHz with its peak of more than 100 dB SPL near 3 kHz. However, the packaging structure was not good enough to improve the sound output from 4 kHz to 15 kHz.

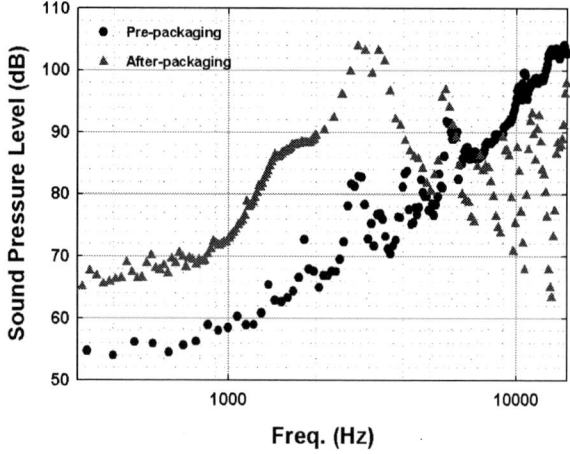

Figure 8: Comparison of the measured sound pressure levels of a fabricated microspeaker before and after packaging it as shown in Figure 3.

Figure 9 compares two different types of the fabricated microspeakers that were packaged into similar packages. The microspeaker consisting of circular electrodes (8G1) showed a higher SPL in the low frequency region than that with square electrodes, but showed a very non-uniform response beyond 4 kHz. On the other hand, the microspeaker consisting of square electrodes (8C1) produced a little lower SPL below 4 kHz, but a higher and

more uniform SPL between 4 kHz and 15 kHz.

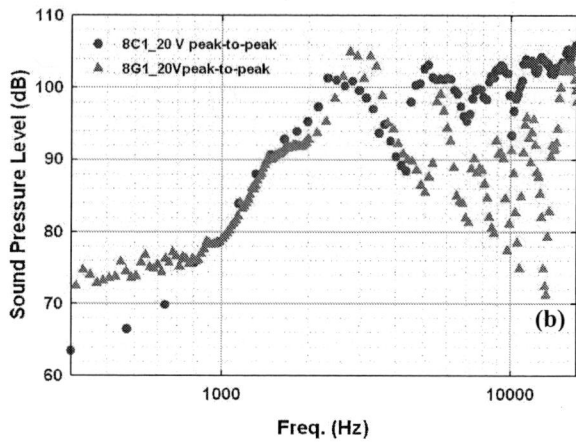

Figure 9: Comparison of the measured sound pressure levels produced by the microspeakers consisting of two different electrode shapes: square (8C1) and circle (8G1).

CONCLUSION

The fabricated microspeakers as well as the packaged microspeakers produced audible sound in open air. From the experimental results described in this paper, the performance of the piezoelectric microspeakers can be enhanced with a compressively-stressed supporting diaphragm and AlN thin film, with greater improvement for higher compressive stress. Since the residual stress of LPCVD silicon nitride can be controlled over a very wide range, a LPCVD silicon nitride would be a good diaphragm material to support a piezoelectric film and electrodes for microspeaker application. Acoustically well-designed packaging structure can improve the frequency response of the microspeakers.

REFERENCES

[1] C.H. Han and E.S. Kim, "Parylene-Diaphragm Piezoelectric Acoustic Transducers", *Digest Tech. papers IEEE MEMS 2000 Conference*, Miyazaki, January 23-27, 2000, pp. 148-152.

[2] K.W. Seo, J.S. Park, H.J. Kim, D.K. Kim, S.C. Ur, and S.H. Yi, "Micromachined Piezoelectric Microspeakers Fabricated with High Quality AlN Thin Film", *J. Integrated Ferroelectrics*, vol. 95, pp.74-82, 2007.

[3] C.H. Han and E.S. Kim, "Fabrication of Piezoelectric Acoustic Transducers", *Digest Tech. papers IEEE MEMS 2001 Conference*, Interlaken, January 21-25, 2001, pp. 110-113.

[4] S.H. Yi and E.S. Kim, "Micromachined Piezoelectric Microspeakers", *Japanese J. Appl. Phys.*, vol. 44, pp. 3836-3841, 2005.

[5] S.H. Yi, M.S. Yoon, and S.C. Ur, "Piezoelectric Microspeakers with High Compressive ZnO Film and Floating Electrode", *J. Electroceramics*, will be published in 2008.

[6] A.K. Sinha and T.T. Sheng, "The Temperature Dependence of Stresses in Aluminum Film on Oxidized Silicon Substrate", *Thin Solid Films*, vol. 48, pp.117-126, 1978.

CMOS INTEGRATED STRESS MAPPING CHIPS WITH 32 N-TYPE OR P-TYPE PIEZORESISTIVE FIELD EFFECT TRANSISTORS

P. Gieschke, Y. Nurcahyo, M. Herrmann, M. Kuhl, P. Ruther, and O. Paul
Department of Microsystems Engineering (IMTEK), University of Freiburg, Germany

ABSTRACT

This paper reports a novel generation of CMOS stress mapping chips comprising 32 square field effect transistors (FET) with four source/drain contacts (piezo-FETs) exploiting the shear piezoresistive effect in n-type (NMOS) or p-type (PMOS) inversion layers. The sensor chips with a total die area of 2.5×2 mm^2 are integrated with analog circuitry and digital logic. When exposed to homogenous shear or normal stress, all 32 integrated stress sensors show a linear response in excellent agreement with theoretical predictions and exhibit identical stress sensitivities. Piezo-FETs fabricated as separate devices are characterized with respect to stress sensitivity, intrinsic offset, and noise behavior. Stress sensitivities are enhanced by incorporating a central hole into the piezo-FETs. Sensitivities of -448 µV/(V MPa) and 477 µV/(V MPa) were measured for NMOS and PMOS devices, respectively.

INTRODUCTION

Mechanical stress induced during fabrication and packaging leads to parameter shifts influencing the behavior of analog and digital circuits. Therefore, reliable packaging of integrated circuit (IC) chips requires low, constant stresses acting on the packaged devices. The co-integration of digital circuits with analog components and CMOS-compatible sensors increases the need for the evaluation of these stresses. Piezoresistive stress sensors, typically in Wheatstone bridge configurations, have been used since the early 1980s for the experimental evaluation of package induced stresses [1]. In order to increase the resolution of die stress maps, CMOS chips with integrated circuitry have been designed, e.g. using 32 octagonal n-well resistors [2] and 512 current mirror type piezoresistive metal oxide semiconductor field effect transistors (MOSFETs) [3] as stress sensors. To integrate larger numbers of stress sensors in mixed-signal CMOS chips, piezo-FETs are superior to diffused piezoresistors since the gate contact serves as an intrinsic switch. A chip with 32 n-type piezo-FETs [4] and a method for accurately extracting the temperature with piezo-FETs [5] have been reported.

In addition, external loads acting on a sensor package can be determined by measuring the stress distribution induced into the package. As an example, smart orthodontic brackets using piezoresistive stress sensor chips exploit this principle to measure forces and moments acting on individual teeth [6].

The sensor chips introduced here represent the next generation of piezo-FET based stress mapping chips fully integrated with analog and digital CMOS circuitry. They contain either 32 NMOS or 32 PMOS piezo-FETs. Furthermore, the integrated piezo-FETs were individually characterized with respect to sensitivity, noise, intrinsic offset, and their scatter.

PIEZO-FET STRESS SENSORS
Design and Fabrication

The piezo-FETs reported in this paper are square MOSFETs with four source/drain contacts. They exploit the shear piezoresistive effect in their inversion layer. Two different types of these sensors exist, with either an n-type (NMOS) or a p-type (PMOS) inversion layer.

Figure 1 shows optical micrographs of both types of piezo-FET stress sensors used in this work. The four source/drain contacts at the corners of the transistors are clearly visible. For symmetry reasons, the polysilicon gate is contacted on all four sides by a square metal ring around the sensors. A p$^+$-guard ring is placed around the NMOS sensors shown in Figs. 1 (a) and (b) for setting the bulk-potential to a well-defined value. The p-channel sensors shown in Figs. 1 (c) and (d) are placed inside an n-well, surrounded by an n$^+$-contact ring. In contrast to the standard sensors in Figs. 1 (a) and (c), those in Figs. 1 (b)

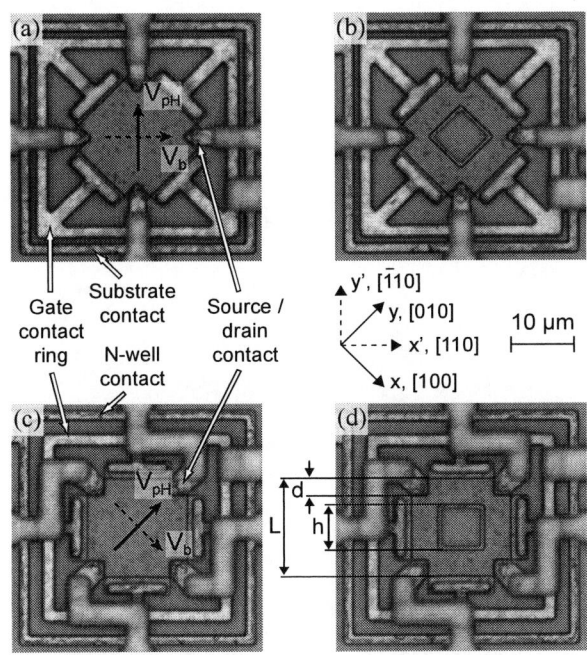

Figure 1: Optical micrographs of pseudo-Hall piezo-FETs: (a) Standard NMOS, (b) NMOS with central hole, (c) standard PMOS, and (d) PMOS with central hole. The unprimed crystallographic and primed chip coordinate systems are defined as shown.

Table 1: Geometrical parameters of the implemented n-channel and p-channel piezo-FETs.

Geometrical parameter (µm)	NMOS	PMOS
Transistor side length L	17.1	16.6
Contact side length d	3.3	3
Central hole side length h	8	8
Effective channel length L_{eff}	15	15

978-1-4244-2977-6/09 $25.00 © 2009 IEEE

and (d) have non-conducting central holes enhancing their sensitivity [2, 7]. The most important geometrical parameters are listed in Table 1.

All devices described in this work were fabricated in the commercial 0.6 μm CMOS process XC06 of *X-FAB Semiconductor Foundries AG* (Erfurt, Germany). This is a double-poly, triple-metal, 5 V process on 6-inch (100)-oriented p-type wafers.

Theory and Operation Mode

The shear piezoresistance effect, also termed pseudo-Hall effect, describes the occurrence of a voltage V_{pH} perpendicular to the bias voltage V_b applied between two opposite contacts, due to the anisotropy of the conductivity tensor caused by in-plane stresses. The sensitivity of MOSFET inversion layers is described by three coefficients Π_{11}, Π_{12}, and Π_{44}, corresponding to the π-coefficients of conventional piezoresistors. The general pseudo-Hall response of a four-contact device to mechanical in-plane stresses defined in the primed chip coordinate system as illustrated in Fig. 1 is given by [7, 8]

$$V_{pH} = \left[\left(\Pi_{11} - \Pi_{12} \right) \sigma_{x'y'} \cos\left(2\phi\right) - 0.5\,\Pi_{44} \left(\sigma_{x'x'} - \sigma_{y'y'} \right) \sin\left(2\phi\right) \right] V_b G \quad , \tag{1}$$

where ϕ denotes the angle between the [110]-axis and the direction of the bias current. G is a geometrical correction factor.

In order to maximize the sensitivity, n-type sensors are used to measure the in-plane shear stress $\sigma_{x'y'}$ with a bias current in in-plane ⟨110⟩-directions ($\phi = n\times\pi/2$), whereas p-type sensors are used to measure differences of in-plane normal stresses ($\sigma_{x'x'} - \sigma_{y'y'}$) with a bias current in ⟨100⟩-directions ($\phi = \pi/4 + n\times\pi/2$) [7, 9]. The resulting voltage related relative sensitivities S_V are:

$$S_V^{NMOS} = \frac{1}{V_b} \frac{dV_{pH}}{d\left(\sigma_{x'y'}\right)} = \left(\Pi_{11} - \Pi_{12} \right) G \tag{2}$$

$$S_V^{PMOS} = \frac{1}{V_b} \frac{dV_{pH}}{d\left(\sigma_{x'x'} - \sigma_{y'y'}\right)} = \frac{1}{2}\Pi_{44} G \tag{3}$$

In order to reduce non-mechanical offset contributions to V_{pH}, the sensors are operated in a four-fold orthogonal current switching method [7]. The pseudo-Hall voltage is calculated from four measurements where V_b and V_{pH} are rotated in steps of 90° with the first direction ($V_{b,1}$, $V_{pH,1}$) defined as shown in Fig. 1 as

$$V_{pH} = \frac{V_{pH,1} - V_{pH,2} + V_{pH,3} - V_{pH,4}}{4} \quad . \tag{4}$$

Experimental Setups

To characterize the stress sensitivity of the fabricated devices, well-defined stress states are applied. For this purpose, the 6-inch CMOS wafers of thickness $t = 650$ μm with n-channel and p-channel piezo-FET test structures and integrated stress sensor chips were cut into strips of width $w = 9$ mm. Electrical connection to a custom-made printed circuit board (PCB) is established using standard ball-wedge bonding.

The strips connected to a PCB are inserted into a four-point bending bridge (4PBB) for application of normal stresses $\sigma_{x'x'}$ and a torsional bridge (TB) for application of in-plane shear stresses $\sigma_{x'y'}$. Both setups are schematically shown in Fig. 2. During the measurements, the strips are released from the PCB; the remaining load due to the bond wires is negligible.

The normal stress $\sigma_{x'x'}$ generated by the 4PBB as a function of the displacement Δd of the inner supports is given by [4, 10]

$$\sigma_{x'x'}(\Delta d) = \frac{3\,\Delta d\, E_{Si}\, t}{2\,L_o{}^2 + 6\,L_o\,L_i} \quad , \tag{5}$$

where E_{Si}, t, L_o, and L_i denote Young's modulus of silicon, the thickness of the beam, and the distances between inner and outer supports and between the center and the inner supports, respectively. Figure 3 shows a silicon strip containing the test structures mounted in the 4PBB and the connections to external circuitry.

In order to exert in-plane shear stress $\sigma_{x'y'}$ on the same silicon strips, a novel TB was used [11]. The setup consists of two clamps precisely aligned on a rotational axis. While one clamp is fixed, the other clamp is mounted on a torque sensor attached to a motor-driven precision rotation stage. It was shown by finite-element simulations that the TB applies a sufficiently constant shear stress $\sigma_{x'y'}$ across the surface of the silicon strip

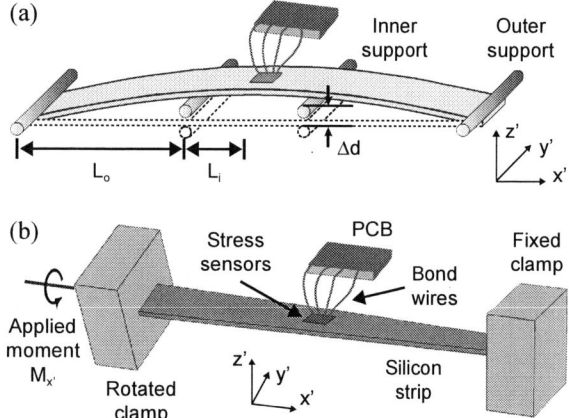

Figure 2: Schematic illustrations of (a) bending and (b) torsional bridges for the application of well-defined in-plane normal and shear stresses, respectively.

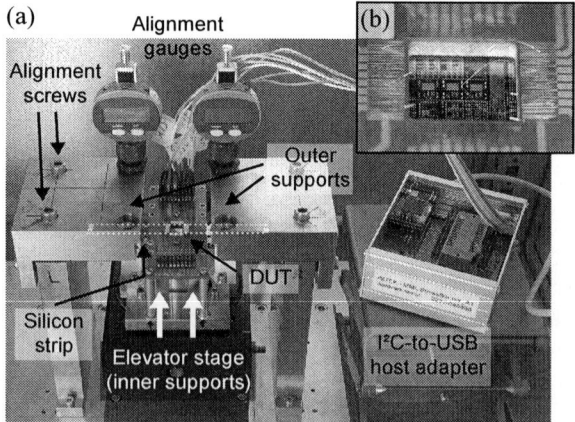

Figure 3: (a) Four-point bending bridge with silicon strip mounted. (b) Close-up of device under test (DUT) wire bonded to a printed circuit board.

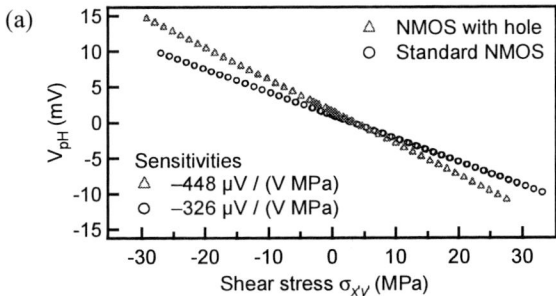

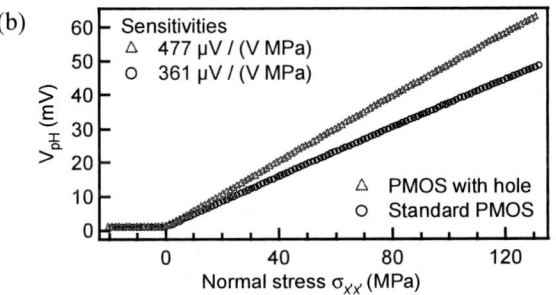

Figure 4: Pseudo-Hall response V_{pH} of (a) n-type piezo-FETs subjected to homogenous shear stress $\sigma_{x'y'}$ and (b) p-type piezo-FETs subjected to homogenous normal stress $\sigma_{x'x'}$ operated at $|V_{GS}| = 5\ V$ and $|V_{DS}| = 1\ V$.

except for a 1.5 mm-wide area along the edges. The applied shear stress as a function of the applied moment $M_{x'}$ is given analytically by [11]

$$\sigma_{x'y'}(M_{x'}) = \frac{M_{x'}}{\eta\,w\,t^2}\ , \qquad (6)$$

where w, t, and η denote the beam width, thickness and a geometrical correction factor which is 0.32 for the used strips with $w/t \approx 14$.

Characterization Results

NMOS and PMOS piezo-FETs were fabricated as separate devices without amplifier and digital circuitry for characterization of electrical parameters needed for their integration into mixed-signal CMOS chips. Figure 4

Table 2: Measured parameters of standard NMOS and PMOS piezo-FETs at a gate voltage of $|V_{GS}| = 5\ V$ and a bias voltage of $|V_{DS}| = 1\ V$. For sensitivities, also see Fig. 4.

Parameter	NMOS [Fig. 1 (a)]	PMOS [Fig. 1 (c)]
Sensitivity [μV / (V MPa)]	−326 ($\sigma_{x'y'}$)	361 ($\sigma_{x'x'} - \sigma_{y'y'}$)
Mean offset (mV) (single direction)	−0.01 ± 0.43	0.01 ± 1.70
Mean offset (mV) (four-fold orthogonal switching)	0.26 ± 0.35	1.29 ± 1.12
Thermal noise (nV/√Hz)	10	18
1/f noise corner frequency (Hz)	1000	400

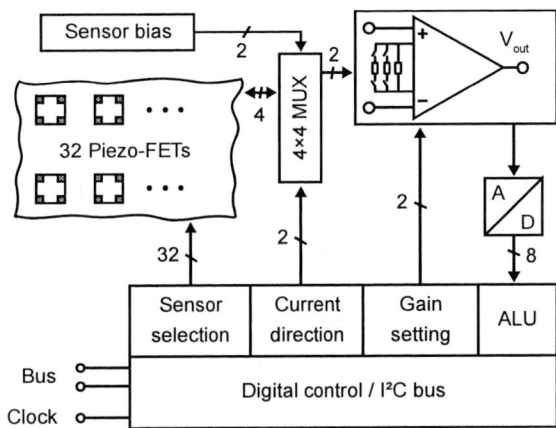

Figure 5: Schematic of system components, including 32 stress sensors, amplifier, ADC, multiplexer for current switching, and digital control, as well as the signal flow of the integrated stress sensor chips.

shows measurement signals extracted from NMOS and PMOS piezo-FETs according to Eqn. (4) when subjected to defined mechanical loads as described. The response is linear for all devices with a sensitivity increase due to the central non-conducting hole of 37% and 32% for n-type and p-type devices, respectively. Performance data of standard sensors are summarized in Table 2.

CMOS INTEGRATION

For many applications, a lateral resolution and thus interfacing to a multitude of stress sensors is desirable. This is achieved by an integrated system.

Concept and Design

A schematic drawing of all components of the CMOS integrated stress sensor system including the signal flow is illustrated in Fig. 5. Currently, 32 piezo-FETs are placed at desired positions on the chip surface. The sensor output is amplified using an instrumentation amplifier with selectable gain between 100 and 700. Furthermore, an 8-bit analog-to-digital converter (ADC) and digital logic are integrated on-chip. The digital part successively addresses all sensors, extracts the signal proportional to the mechanical stress, performs a digital low-pass filtering, and transmits the data via an I²C-bus. In addition, a stabilized voltage of $V_{DS} = 1\ V$ is generated for biasing the stress sensors and a set of transmission gates is included

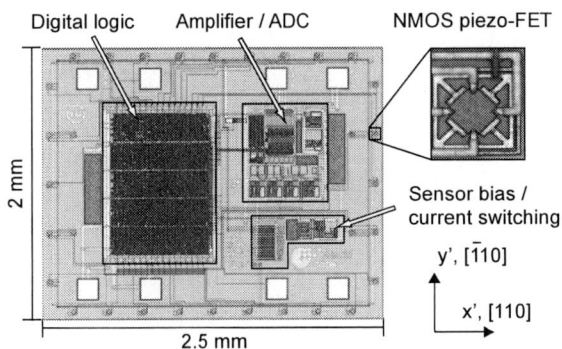

Figure 6: Micrograph of the complete stress sensor chip with integrated electronics and 32 n-type stress sensors distributed along the chip edges.

978-1-4244-2977-6/09 $25.00 © 2009 IEEE   767

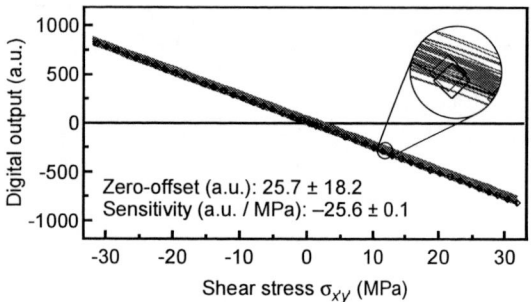

Figure 7: Measured digital signals of 32 NMOS piezo-FETs (blue lines) of a sensor chip with shear stress $\sigma_{x'y'}$ applied to the silicon strip. Measurement points of one sensor are indicated by symbols.

for switching the current direction. Complete stress maps can be obtained with a maximum frequency of 30 Hz.

Integrated stress sensor chips with a size of 2.5×2 mm^2 containing analog and digital circuitry and either 32 NMOS or 32 PMOS piezo-FETs were designed and fabricated. The chip shown in Fig. 6 contains 32 NMOS piezo-FETs distributed along its edges. The sensors are operated in a switched current mode and exhibit a pseudo-Hall response when exposed to an in-plane shear stress $\sigma_{x'y'}$. In contrast, chips comprising 32 PMOS piezo-FETs are sensitive to the difference of in-plane normal stresses ($\sigma_{x'x'} - \sigma_{y'y'}$).

Measurement Results

For characterization, the integrated sensor chips were tested under well-defined mechanical loads as described in the previous section.

Figure 7 shows the digital signal of all 32 NMOS sensors with a homogeneous shear stress applied to a silicon strip containing the stress sensor chip. All sensors show a linear response and nearly identical sensitivity. The corresponding result for a homogeneous normal stress applied to a sensor chip containing 32 PMOS piezo-FETs is shown in Fig. 8. The variation of the zero offset indicates the presence of an intrinsic normal stress distribution over the chip.

CONCLUSIONS

This paper presented CMOS integrated stress mapping chips with significantly improved performance. On an area of 2.5×2 mm^2, the chips contain either 32 n-type or 32 p-type piezo-FETs, highly sensitive to in-plane shear or normal stress, respectively. Important operating parameters of both types of sensors, viz., sensitivity, intrinsic offset, and noise were reported. The CMOS chips are useful for the characterization of packaging stresses and for force/torque sensors that evaluate characteristic stress distributions to monitor the applied load.

ACKNOWLEDGEMENTS

The authors would like to thank the German Research Foundation (DFG) for financial support through project PA 792/5-1.

REFERENCES

[1] J.C. Suhling and R.C. Jaeger, "Silicon Piezoresistive Stress Sensors and Their Application in Electronic Packaging," *IEEE Sens. J*, vol. 1, no 1, pp. 14-30, 2001.

[2] P. Ruther, J. Bartholomeyczik, S. Kibbel, T. Schelb, P. Gieschke, O. Paul, "Integrated CMOS-Based Sensor Array for Mechanical Stress Mapping," in *Proc. IEEE Sensors 2006*, Daegu, Korea, pp. 1131-1134, 2006.

[3] Y. Chen, R.C. Jaeger, J.C. Suhling, "Multiplexed CMOS Sensor Arrays for Die Stress Mapping," in *Proc. European Solid-State Circuits Conf.*, Montreux, Switzerland, pp. 424-427, 2006.

[4] P. Gieschke, Y. Nurcahyo, M. Herrmann, M. Kuhl, P. Ruther, O. Paul, "Integrated Stress Mapping Chip with 32 Piezoresistive Field Effect Transistors," in *Proc. Eurosensors XXII*, Dresden, Germany, pp. 292-295, 2008.

[5] M. Doelle, J. Held, P. Ruther, O. Paul, "Simultaneous and Independent Measurement of Stress and Temperature Using a Single Field-Effect Transistor Structure," *J. Microelectromech. Syst.*, vol. 16, no. 5, pp. 1232-1242, 2007.

[6] P. Gieschke, B. Lapatki, J. Bartholomeyczik, O. Paul, "First 1:1 Scale Smart Orthodontic Bracket," in *Proc. Smart Systems Integration 2007*, Paris, France, pp. 207-214, 2007.

[7] M. Doelle, *Field Effect Transistor Based CMOS Stress Sensors*, Ph.D. thesis, IMTEK, University of Freiburg, Germany, 2006.

[8] J. Bartholomeyczik, *Advanced CMOS-based stress sensing*, Ph.D. thesis, IMTEK, University of Freiburg, Germany, 2006.

[9] A.T. Bradley, R.C. Jaeger, J.C. Suhling, Y. Zou, "Test Chips for Die Stress Characterization Using Arrays of CMOS Sensors," in *Proc. 1999 IEEE Custom Integrated Circuits Conf.*, San Diego, USA, pp. 147-150, 1999.

[10] S.A. Gee, V.R. Akylas, W.F. van den Bogert, "The Design and Calibration of a Semiconductor Strain Gauge Array," in *IEEE Proc. Microelectronic Test Structures*, Kanazawa, Japan, vol. 1, no. 1, pp. 185-191, 1988.

[11] M. Herrmann, P. Gieschke, Z. Liu, J. Korvink, P. Ruther, O. Paul, "Design and Characterization of In-Plane Silicon Stress Sensors with Isotropic Sensitivity," in *Proc. IEEE Sensors 2008*, Lecce, Italy, in press, 2008.

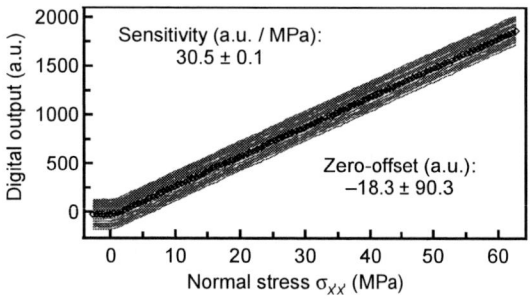

Figure 8: Measured digital signals of 32 PMOS piezo-FETs (blue lines) of a sensor chip with normal stress $\sigma_{x'x'}$ applied to the silicon strip. Measurement points of one sensor are indicated by symbols.

978-1-4244-2977-6/09 $25.00 © 2009 IEEE

DEVELOPMENT OF FABRIC-TYPE OF TENSIONAL FORCE SENSOR

Y. Suzuki, D. Ogura, M. Shikida, Y. Hasegawa, and K. Sato
Department of Micro-Nano Systems Engineering, Nagoya University, Japan

ABSTRACT

We developed a tensional force sensor by applying artificial tetrafluoroethylene-perfluoroalkylvinyl ether copolymer (PFA) hollow fibers. The fibers were fabricated by depositing thin metal and an insulation layer on the PFA tube, and the fabric sensor was made by weaving the decorated PFA tube and the conventional cotton yarns together. The sensor was 57.0 x 4.0 mm. The sensor output linearly increased with the increase of the tension. We confirmed that the sensor output increased with applied tension at a rate of 2.1 %/N.

1. INTRODUCTION

Many physical sensors, for example pressure, acceleration, gyro, and tactile ones used in micro Electro Mechanical Systems (MEMS), have been developed for applications in the automotive and medical industries [1-4]. These sensors now come into usage for recording human behavior. They are attached to the surface of clothes . To be fitted in the surface of the clothes, these sensors need to have a flexible structure. Furthermore, they have to be enlarged to be used in applications used by people and those designed for robots. Therefore, flexible resin materials, such as PDMS and polyimide films, used as a substrate for their structures [5, 6]. As film-based sensors can bend along one axis, they can be attached to applications that are cylindrical. However, they are difficult to fit on any 3-dimensional arbitrary surfaces. To overcome these problems, J. Chen et al. proposed the use of a novel fabrication process that can produce an arbitrary bendable thin film structures that are similar to a fabric [7]. As this process involves the use of photolithography, it cannot be easily used to produce large (i.e., meter-sized) fabric structures. The sensor has to be enlarged to recognize human behaviors, as shown in Fig. 1.

We therefore propose an artificial hollow fiber structure as a new material for MEMS and applied this fiber to a fabric tactile sensor that detects a normal load [8]. This fiber was then made into a tensional force sensor, a sensor that has rarely been presented at a MEMS conference, with the goal of expanding the number of applications of fabric sensing devices and developing this field.

2. DESIGN AND PRINCIPLES

A schematic view of the fabric tensional force sensor and its detection mechanism are shown in Fig. 2(a). The surface of the hollow fiber is covered with thin metal and insulation layers. The fabric tactile sensor is formed by weaving the hollow fibers as shown in Fig. 2(a). The contact area at the intersection between warp and weft fibers is the part that senses force. As shown in Fig. 2(b), when a tensional force is applied to the end of the fabric tactile sensor, both the warp and weft fibers deform. As a result, the capacitance between them increases with applied force. The sensor can therefore be used to detect

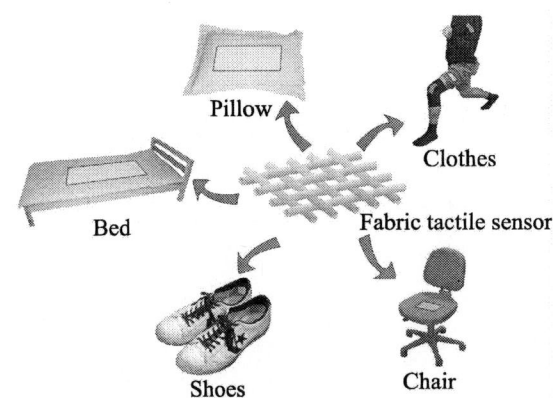

Fig. 1 Applications of a fabric tactile sensor.

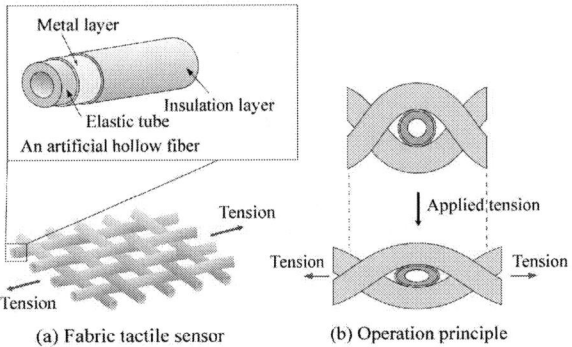

(a) Fabric tactile sensor   (b) Operation principle

Fig. 2. Fabric sensor and its mechanism of detecting tension.

the tensional force by measuring the change of capacitance between the fibers.

3. FABRICATION

The fabrication of the sensor is shown in Fig. 3. We used commercially available tetrafluoroethylene-perfluoroalkylvinyl ether copolymer (PFA) tube as the elastic hollow fiber (Fig. 3(a)). Its external diameter and thickness were 0.3 and 0.1 mm, respectively, and it had a Young's modules of 28 MPa. To deposit Au film on the tube surface, we developed a rotating mechanism by which the PFA tube was rotated in a vacuum chamber during sputtering. The thickness of the deposited Au was 250 nm (Fig. 3(b)). The electrical wirings were then attached at the one end of the fiber by using silver paste, as shown in Fig. 3(c). We then evaporated 1.0-μm-thick parylene-C film on the surface as an insulating layer. Finally, the rectangular shaped fabric tension sensor was fabricated by weaving the hollow fibers and conventional cotton yarn (0.4 mm in diameter) manually(Fig. 3(d)). The total size was 57.0 x 4.0 mm, and the pitch of the warp was 3.5 mm. A single pair of the sensing fiber was knitted in the fabric sensor to evaluate the performance of the tensional force detection. A schematic

978-1-4244-2977-6/09 $25.00 © 2009 IEEE

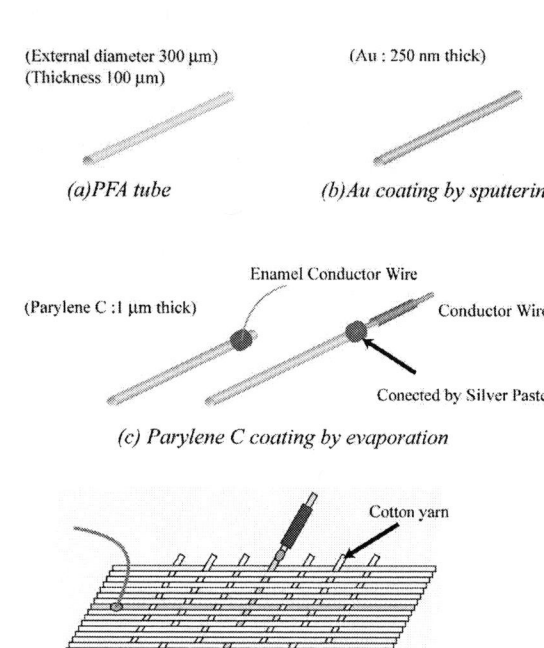

(External diameter 300 μm)
(Thickness 100 μm)

(a)PFA tube

(Au : 250 nm thick)

(b)Au coating by sputtering

(Parylene C :1 μm thick)

Enamel Conductor Wire

Conductor Wire

Conected by Silver Paste

(c) Parylene C coating by evaporation

Cotton yarn

(d)Weaving tubes to produce fabric tactile sensor

Fig. 3. Fabrication of fabric tactile sensor.

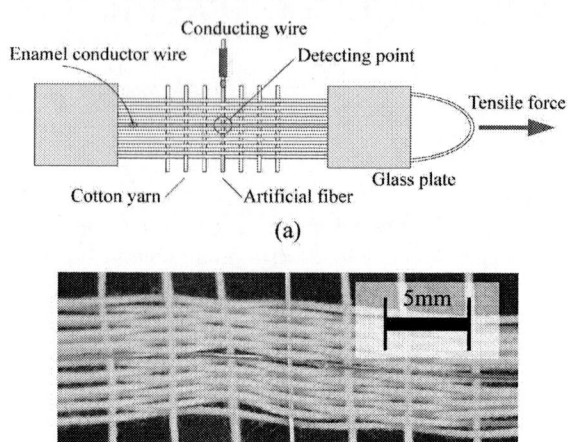

(a)

(b)

Fig. 4. Experimental device (a) Schematic (b) Photograph

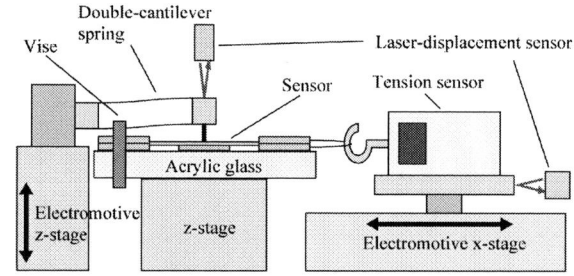

Fig. 5. Experimental setup for evaluating normal load, tension, and sensor output.

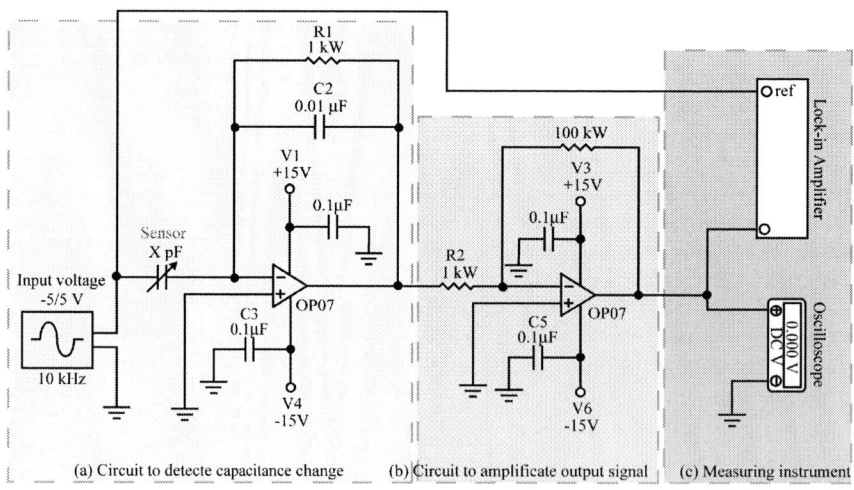

(a) Circuit to detecte capacitance change (b) Circuit to amplificate output signal (c) Measuring instrument

Fig. 6. Conventional CV circuit for evaluating sensor output.

view and a photograph of the fabricated fabric sensor are shown in Fig. 4. A 25-mm-square glass plate was attached at both end of the fabric to apply the tension.

4. RESULTS

The experimental setup and a detection circuit are shown in Figs. 5 and 6, respectively. The fabric sensor was placed on an acrylic plate in z-state, and one end of it was fixed on the plate. The other end was fixed on the head of

the force sensor placed on the x-stage, and the tensional force was applied to the fabric sensor by automatically actuating the stage. The velocity of the movement of the x-stage was 0.30-1.35 mm/s. First, the capacitance value at the intersection between the warp and weft fibers was detected by the conventional CV circuit, and then it was synchronously-demodulated by a lock-in amplifier (Fig. 6). The signal of 10 kHz was used as the capacitance detection.

978-1-4244-2977-6/09 $25.00 © 2009 IEEE

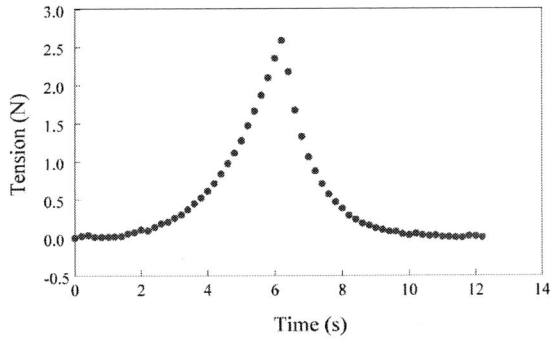

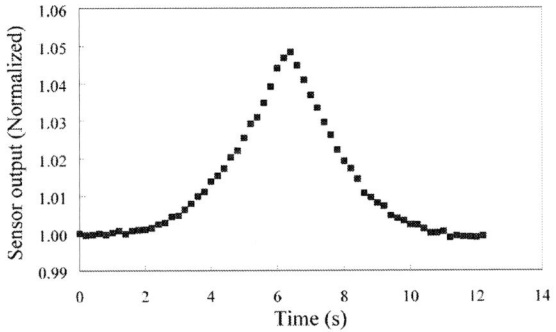

Fig. 7. Change of applied tension and sensor output with movement of x-stage.

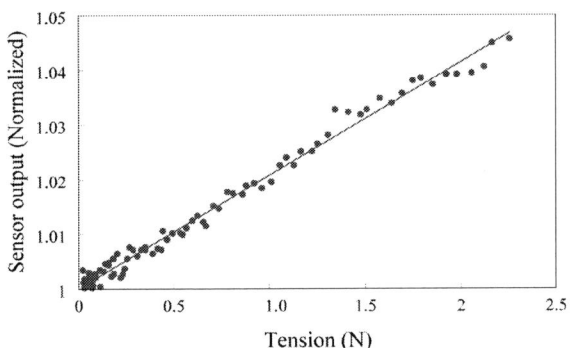

Fig. 8. Relationship between sensor output and applied tension.

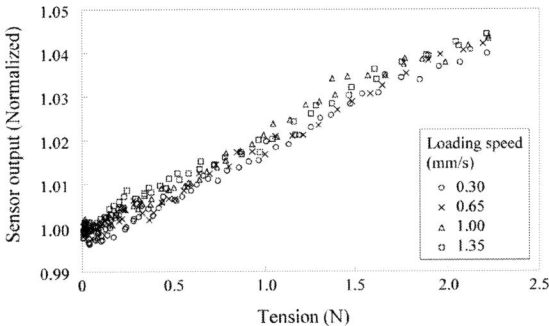

Fig. 9. of pulling speed on detecting tension.

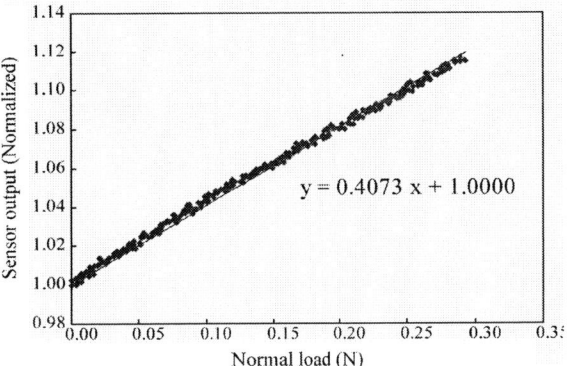

$$y = 0.4073\ x + 1.0000$$

Fig. 10. Relationship between sensor output and applied normal load.

Performance as Tensional Force Sensor

We first measured the changes of the applied tensional force and sensor output with the movement of x-stage, and the result was shown in Fig. 7. Both values were similarly changed with the back-and-forth motion of the x-stage. Therefore, we re-plotted the graph clearly showing the relationship between the applied tension and the sensor output (Fig. 8). The sensor output linearly increased with the tension. The increasing rate of the sensor output against the applied tension was 2.1 %/N, and the output changed 4.2 % when the applied tension was 2.0 N.

Influence of Pulling velocity

We also studied the influence of pulling velocity on the performance of the tension detection, and confirmed that it did not depend on the velocity in the range of 0.30-1.35 mm/s (Fig. 9).

5. DISCUSSION

The fabric sensor stated above also has a high sensitivity to the normal force because, as described in Ref. 8, the fibers at the crossing point between the warp and the weft also deforms by the normal force. We therefore investigated the relationship between the applied normal force and the sensor output, and compared the sensitivity to both forces. The sensor output linearly increased with the increase of the normal load at the no-tensional force conditions (Fig. 10). The increasing rate of the sensor output against the applied normal load was 40.7 %/N. This result means that the sensitivity of the normal force was 20 times higher than that of the tensional force.

Therefore the fabric sensor has to be used for detecting the two forces as follows.

(1) Tensional force detection: As the sensor has a much high sensitivity to normal force, adding normal force should be avoided. The sensor detects tensional force with the sensitivity of 2.1 %/N.

(2) Normal force detection: We are able to ignore the small amount of the tensional force when we detected normal force. The sensor detects the normal force with the sensitivity of 40.7 %/N.

6. CONCLUSION

We developed a tensional force sensor by applying artificial PFA hollow fibers. We made the artificial fibers by depositing the thin metal and the insulation layer on the PFA tube, and the fabric sensor was obtained by weaving the tube and the conventional cotton yarns together. The total size was 57 x 4.0 mm.

The following results were obtained.

(1) The sensor output linearly increased with the increase of the tension. The increasing rate of the sensor output against the applied tension was 2.1 %/N.

(2) The increasing rate of the sensor output against the applied normal load was 40.7 %/N. This result means that the sensitivity of the normal force was 20 times higher than that of the tensional force.

REFERENCES

[1] B. J. Kane and G. T. A. Kovacs. A CMOS compatible traction stress sensing element for use in high resolution tactile image. Proceedings of the 8th International Conference on Solid-State Sensors and Actuators. 1995. p. 648-651.

[2] B. L. Gray and R.S. Fearing. A surface micromachined microtactile sensor array. Proceedings of IEEE Robotics and Automation conference. 1996. p. 1-6.

[3] J. Engel, J. Chen, Z. Fan, and C. Liu, Polymer micromachined multimodal tactile sensors. Sensors and Actuators A: Physical. 2005. A117. pp. 50-61.

[4] Y Hasegawa, M Shikida, H Sasaki, K Itoigawa, and K Sato, An active tactile sensor for detecting mechanical characteristics of contacted objects. Journal of Micromechnics.and Microengineering. 2006. 16(No 8), p. 1625-1632.

[5] E-S. Hwang, J-H. Seo, and Y-J. Kim. A polymer based flexible tactile sensor for normal and shear load detection. Proceedings of IEEE MEMS'06 Conference. 2006. p. 714-717.

[6] H-K. Lee, S-I. Chang, and E. Yoon. A capacitive proximity sensor in dual implementation with tactile imaging capability on a single flexible platform for robot assistant applications. Proceedings of IEEE MEMS'06 Conference. 2006. p. 606-609.

[7] N. Chen, J. Engel, S. Panadya, and C. Liu. Flexible skin with two-axis bending capability made using weaving by-lithography fabrication method. Proceedings of IEEE MEMS'06 Conference. 2006. p. 330-333.

[8] Y. Hasegawa, M. Shikida, D. Ogura, and K. Sato, "Novel type of fabric tactile sensor made from artificial hollow fiber", Proc. of IEEE MEMS'07, 2007, pp. 603–606.

MEASUREMENT RESULTS OF THE FIRST TWO CHIP SILICON MICROPHONE WITH LOW STRESS NICKEL MEMBRANE COVERING FULL AUDIO RANGE

S. Junge[1,2], F. Jakobs[1,2], and W. Lang[1]

[1]Institut for Mikrosensors, -actuators and -systems (IMSAS), University of Bremen,
Otto-Hahn-Allee, 28359 Bremen, Germany
[2]Friedrich Wilhelm Bessel Institute Research Society GmbH (FWBI),
P.O. Box 106364, 28063 Bremen, Germany

ABSTRACT

This paper provides an overview of the development, fabrication and measurement results of a new type of silicon microphone. It mainly consists of a low stress nickel membrane with a nonlinear frequency response and a gold backplate electrode. The microphone can be used as condenser or electret microphone.

A sensitivity of -64dBV/Pa at 52V bias was obtained with an air gap of 20μm and membrane resonance at 20kHz.

INTRODUCTION

Several silicon microphone developments over the last decades had the goal to reach a flat frequency response within the audio range. But common high end microphones for studio recordings do not always have a flat response. The lower frequencies are recorded with a lower sensitivity and the resonance is typically in the higher audio region around 14kHz up to 20kHz. The reason is that the bass frequencies have a low absorbability contrary to the higher audio frequencies. This nonlinear response assists to record a better natural vocal tone.

Typically, silicon microphones use very small air gaps. Single chip microphones have an air gap in the range of 0.5μm to 2μm [1-2], two chip microphones typically around 10μm [3]. In contrast to that, the new microphone has an air gap of 20μm; this allows the use of a very high polarization voltage in combination with a membrane with a very high compliance. The polarization can either be applied externally as DC voltage or by integrating a local PTFE electret in the future. Figure 1 shows a simplified drawing of the developed microphone system.

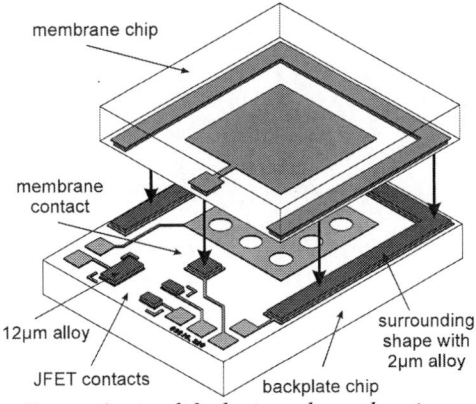

Figure 1: simplified microphone drawing

FABRICATION

The fabrication describes all main process steps for the realization of the membrane and backplate chip as well as a brief overview of the assembling technology to complete the two chip acoustic sensor.

The device wafer for the membrane is 380μm thick. Figure 2 shows cross-sections of all main process steps. At first, the backplate is anisotropically prepared. A 2μm thermal oxide and 100nm LPCVD silicon nitride were etched on the backside by using reactive iron etching (RIE) followed by a 300μm anisotropic potassium hydroxide (KOH) etch step. Phosphoric acid is used to remove the nitride layer.

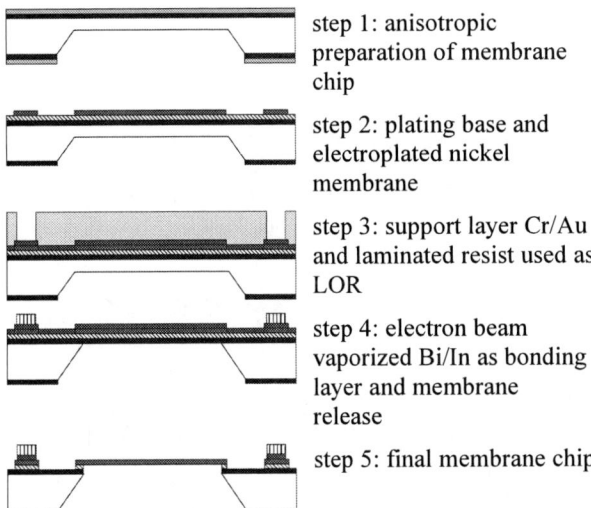

step 1: anisotropic preparation of membrane chip

step 2: plating base and electroplated nickel membrane

step 3: support layer Cr/Au and laminated resist used as LOR

step 4: electron beam vaporized Bi/In as bonding layer and membrane release

step 5: final membrane chip

Figure 2: main process steps for the membrane chip

The second step is the electroplating of the nickel membrane. A plating base layer stack of 15nm chromium and 80nm copper was sputtered. Thereafter a 1μm thin nickel membrane with an edge length of 1,6mm was electroplated using a sulfamate electrolyte. The membrane has a tensile stress of 25MPa and a centered 22μm hole for atmospheric pressure compensation. The low stress was achieved with a low standard deviation of 3MPa across full 4in wafer scale. To reach those nickel characteristics it was necessary to operate it outside recommendations given by the manufacturer. The complete development of this electrolyte was previously published elsewhere [4]. The measured Young's modules was between 170GPa and 190GPa and can be influenced by changing the concentration of a specific component of the used electrolyte. The tensile stress was characterized by thirty mechanical stress indicators distributed over the wafer.

978-1-4244-2977-6/09 $25.00 © 2009 IEEE 773

Thereafter an adhesion layer of 15nm Cr and 80nm Au was sputtered. This undercoating is required to increase the adhesion of the eutectic layer because bismuth and indium have a insufficient adhesion directly on nickel. As liftoff resist a laminated 38μm thick dry resist is used. Figure 3 shows pictures of the developed dry resist process depending on the time of exposition.

Subsequently indium and bismuth are deposited as eutectic bonding layer to attach the membrane, backplate chip and a JFET amplification stage [5] by electron beam vaporization. After the deposition the membrane was released by etching the backside in TMAH stopping on the thermal oxide below the membrane. The anisotropic etching of the substrate has to be done in a wafer holder with single side protection. The protected front side must also flushed be constantly with nitrogen because otherwise the TMAH vapors will etch the copper of the plating base through the venting hole of the wafer holder. The liftoff was performed by using acetone.

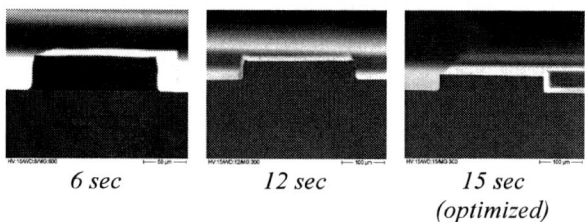

6 sec *12 sec* *15 sec (optimized)*

Figure 3: SEM pictures of exposed and developed resist

The final step is to remove all unnecessary layers, e.g. the plating base for the electroplated nickel membrane and the adhesion layer for the eutectic bonding layer using basic etching solutions. Figure 4 shows a SEM picture of the membrane chip from both sides.

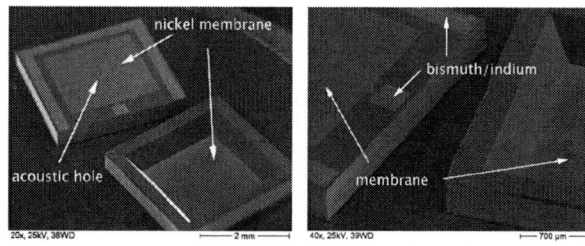

Figure 4: SEM pictures of the realized membrane chip

Starting from now the process steps for the backplate chip will be explained. All main steps are depicted in Figure 5. For the backplate chip the same kind of device wafer was used. For the first anisotropic etch a masking layer stack of 100nm thermal oxide and 100nm LPCVD silicon nitride was used. The stack was RIE etched on the backside and followed by an anisotropic KOH etch step to realize a back volume below the chip. Thereafter the same was done on the front side to etch the later air gap. At this stage both sides are etched simultaneously.

After reaching the final depth (22μm) for the air gab including depth compensation for the thickness of the bonding layer all masking layers will be stripped followed by and a new thermal oxidation of 2μm SiO2.

Afterwards a layer stack of 15nm chromium and 150nm gold was sputtered and structured by wet chemical etching. This layer contains the electrode with round acoustic holes, a contact pad for the membrane, contact pads for the amplification stage and the bond pads for the electrical connections to the characterization board.

The fourth step is the DRIE etch of the acoustic holes through the complete silicon substrate below the gold electrode. Before this a 2μm aluminum layer is sputtered on the backside to prevent a backside colling leakage which tends to destabilize the etching plasma. The masking layer for the DRIE process is a 13μm photo resist. It must be spun on in two steps, each 6.5μm thick, because of the bulk micromachined substrate.

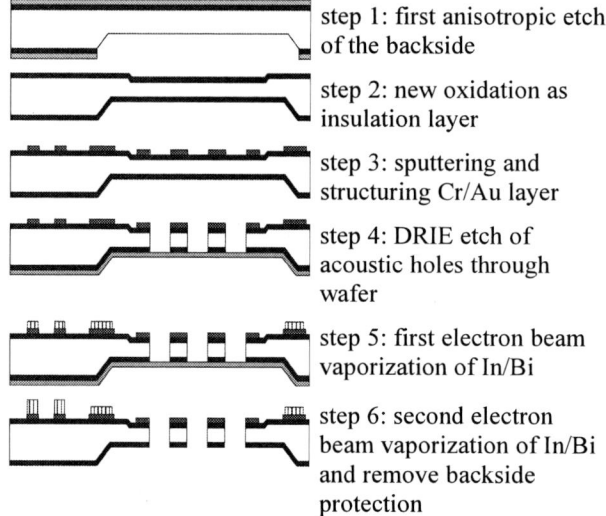

step 1: first anisotropic etch of the backside

step 2: new oxidation as insulation layer

step 3: sputtering and structuring Cr/Au layer

step 4: DRIE etch of acoustic holes through wafer

step 5: first electron beam vaporization of In/Bi

step 6: second electron beam vaporization of In/Bi and remove backside protection

Figure 5: main process steps of the backplate chip

For the assembling of both chips the same 2μm indium bismuth layer as eutectic soft solder is needed on the backplate chip and additionally a 12μm thick layer for the die attach of the amplification JFET. Therefore the deposition of the soft solder is split in two parts. The first one is the implementation of the 2μm layer to stack both chips. The second is the deposition of the 12μm thick soft solder above the gold contacts for drain, source and gate contacts to attach the JFET die. It is the same deposition technology that was used for the membrane chip and previously published in detail [5].

The last step is the removal of the backside aluminum and the final backplate ship is shown in Figure 6.

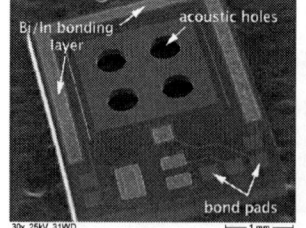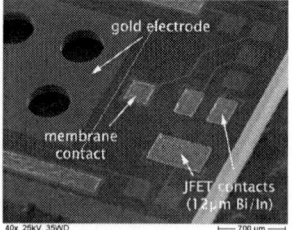

Figure 6: SEM pictures of the realized backplate chip

The assembling of both chips was developed for low temperatures because the microphone design is capable to include also a 7μm to 9μm thick Teflon electret which is quite temperature sensitive. However the measurements presented here are currently without an electret but this will increase the sensitivity drastically. Figure 7 is

978-1-4244-2977-6/09 $25.00 © 2009 IEEE

showing the assembling technology and a mounted silicon microphone system prepared for the measurements.

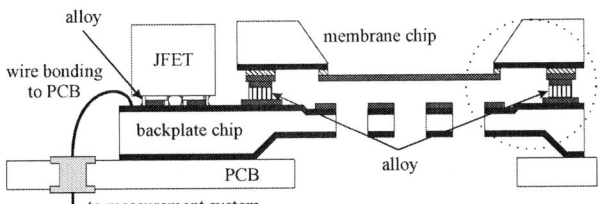

a) cross section of the assembling technology

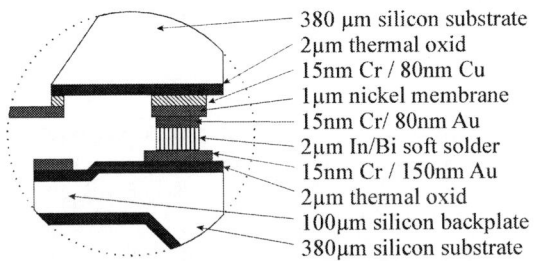

b) overview of used layer thicknesses (see circle from picture above)

c) SEM picture of assembled silicon chips including JFET

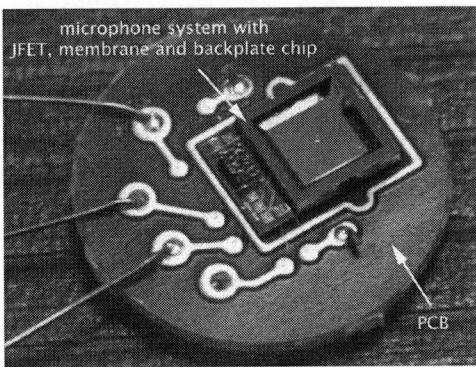

d) microphone mounted on the PCB

Figure 7: assembling overview from the silicon microphone

Figure 7a shows a cross section of the silicon chips and the PCB for the characterization. The JFET as well as both chips are bonded by the low melting bismuth/indium soft solder. Before this step it is mandatory to remove a thin oxide layer on top of the soft solder in diluted hydrochloric acid. The chip stack is fixed on the PCB using an adhesive and the electrical connections are done by wire bonding.

Figure 7b provides an overview of the layer thicknesses of the mounted microphone system and illustrates the circled part of the previous Figure 7a.

Figure 7c shows a SEM picture of the assembled silicon chips. The membrane chip was mounted to the backplate chip as well as the amplification JFET. It also shows the bond pads to realize the electrical connection the the PCB which is depicted in Figure 7d.

MEASUREMENT

For the first measurement the housing of the device is not yet completed, so the hole in the PCB was closed manually, similar to the later housing. Therefore the air between the membrane and the PCB is slightly less, than in the final housing. To measure the acoustic response the microphone was placed in an acoustically field free room. The membrane was grounded and a 52V bias was applied at the backplate electrode. In a certain distance to the microphone an acoustic source is placed. This way a well defined acoustic field which is generated.

This acoustic source generates an acoustic signal with given frequencies. For each frequency the corresponding output signal is measured.

RESULTS

The frequency response of our new type of microphone is shown in Figure 8 for the range of 40 Hz to 20 kHz. It is nearly flat between 300Hz to 8kHz and the sensitivity is increasing to -50dBV at resonance. The frequencies lower than 300 Hz will be recorded with a lower sensitivity as desired.

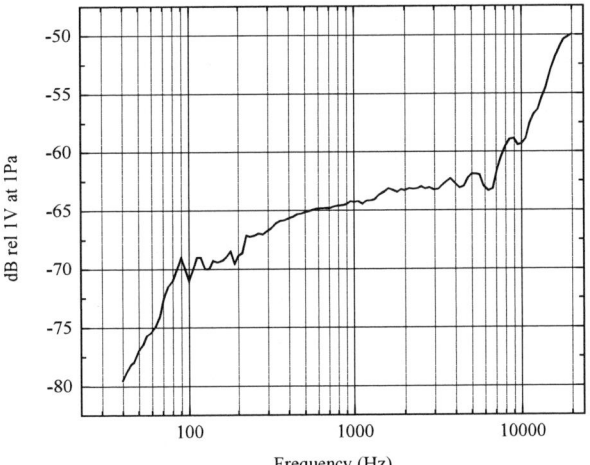

Figure 8: first measured frequency response

The slope at lower frequencies can be adjusted and modified by changing the diameter of the centered hole in the membrane. The resonance can mainly be shifted by changing the tensile stress but due to the high density of nickel it has a lower effect as common silicon membrane materials like poly silicon or silicon nitride.

This response characteristic is well known and is used in hand held high end microphones. Due to the wanted non-flat response with a resonance in high audio range a replayed voice sounds more natural. This is an improvement with very good prospect to develop a miniaturized silicon microphone for high end applications.

978-1-4244-2977-6/09 $25.00 © 2009 IEEE

FUTURE WORK

The next goal in this research progress is to include a localized Teflon layer in the process flow as shown in Figure 9. The SEM pictures depict the first fabricated backplate chips with a local 10μm Teflon layer. This Teflon shall be charged by a corona process and therefore replace the now used DC bias. The result is an increased sensitivity of the microphone.

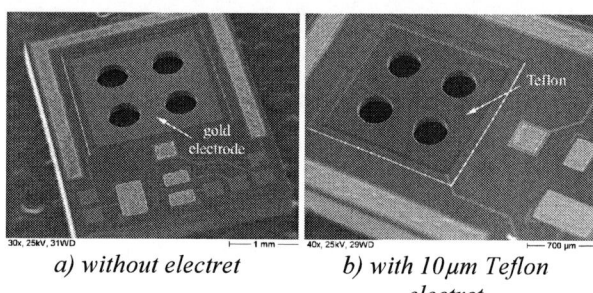

a) without electret   b) with 10μm Teflon electret

Figure 9: integration of a Teflon electret for self biasing

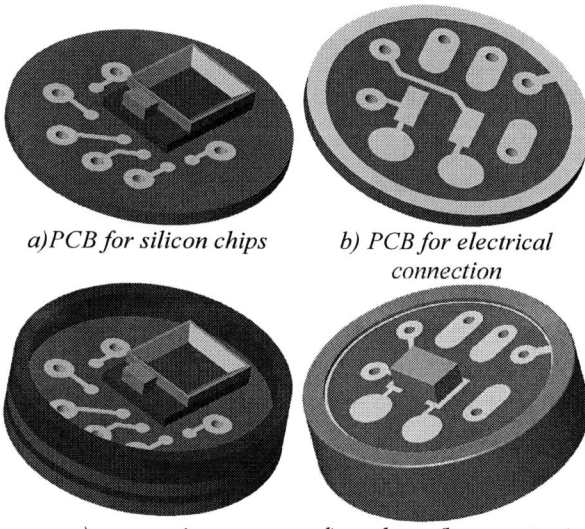

a)PCB for silicon chips   b) PCB for electrical connection

c) spacer rings   d) package (bottom view) with matching capacitor

Figure 10: projected round packaging for the microphone

Further work has also to be done regarding the assembling process and the housing of the microphone. There are two PCBs, the first one is the carrier for the silicon microphone (see Figure 10a) and the second is the outer PCB for the electrical connection of the round packaging and an additional SMD capacitor. The housing has a diameter of 9,2mm and a total height of 3,2mm.

Both PCBs are 0,5mm thick and connected by vias and soldered vertical wires. Between both PCBs a 0,5mm thick spacer ring is placed to increase the back volume below the backplate. This will avoid a membrane resonance frequency shift and a decreased sensitivity by air damping. On top of the PCB with the microphone a 1,0mm spacer ring provide a specific distance from the membrane to the sound inlet on the front side.

CONCLUSTION

We have developed a very new type of a two chip silicon microphone with a wanted nonlinear frequency response like common high end microphones. We have obtained a sensitivity of -64dBV/Pa at 52V bias with an air gap of 20μm and a membrane resonance at 20kHz.

The future work is the integration of a charged Teflon electret for self biasing to increase the sensitivity and the placement into a round package.

ACKNOWLEDGEMENTS

We acknowledge the Deutsche Forschungsvereinigung für Meß-, Regelungs- und Systemtechnik e.V. (DFMRS) for their financial support of this research which was part of project (AiF-No. 14382N/1), granted from the Federal Ministry of Economics and Work (BMWA) via the Association of Industrial Research Organisation (Arbeitsgemeinschaft industrieller Forschungsvereinigungen "Otto-von-Guericke" e.V., AiF).

REFERENCES

[1] P. Rombach, F. Carsten, U. Klein, „Chip scale packaged digital silicon microphone, technology and applications", *Proc. Congress on Acoustics (ICA), 2007*

[2] A. Dehé, „Silicon microphone development and application", *Sensors and Actuators A Physical 133 (2007)*, pp. 283–287

[3] M. Goto, Y. Iguchi, K. Ono, A. Ando, F. Takeshi, S. Matsunaga, Y. Yasuno, K. Tanioka, T. Tajima, „High-performance condenser microphone with single-crystalline silicon diaphragm and backplate", *IEEE Sensors Journal 7 (2007)*, pp. 4–10

[4] S. Junge, F. Jakobs, W. Benecke, W. Lang, „Ultra low tensile stress electroplated nickel layers: a new application as membrane material for acoustic devices", *Proc. Transducers 2007*

[5] S. Junge, F. Jakobs, W. Benecke, W. Lang, „Indium-Bismuth: lead free alloy for electrical and mechanical die attach at low temperatures", *Proc. Eurosensors 2008*

TOWARDS PIEZORESISTIVE CMOS SENSORS FOR OUT-OF-PLANE STRESS

B. Lemke[1], K. Kratt[1], R. Baskaran[2], and O. Paul[1]

[1]Department of Microsystems Engineering (IMTEK), University of Freiburg, Germany
[2]Components Research, Intel Corporation, USA

ABSTRACT

This paper presents a novel piezoresistive sensor detecting a temperature compensated sum of the three mechanical normal stress components including the out-of-plane stress σ_{zz}. The sensor is based on CMOS-compatible diffusions designed to exploit vertical currents. We show that the temperature compensated stress sum can be extracted via two different sensor designs and even with a single device exploiting the strong influence of the junction field effect in the sensor. For this purpose, an experimental setup that induces homogenous vertical normal stress σ_{zz} by applying a vertical force via a gold bump to the sensor is presented. For the characterization, the resulting mechanical stress at the sensor location was evaluated using finite element simulations. Sensitivities, depending on gold bump shape and bias current, were experimentally determined and vertical forces were successfully extracted between 10°C and 60°C, independent of temperature.

INTRODUCTION

Piezoresistive silicon sensors are an established tool for the detection of mechanical stress in die packages [1] and stress mapping with high lateral resolution [2]. New lead-free solder alloys, low-k materials, and a further reduction of the die thickness pose new challenges concerning the reliability of interconnects [3]. There, knowledge of the stress state inside the silicon chips is of crucial importance to optimize their stability. Furthermore, experimental data may help to verify simulations of complex material compounds [4].

Up to now, CMOS-compatible piezoresistors feature only in-plane current components limiting the accessible stress information to the temperature compensated in-plane shear stresses σ_{xy} and $\sigma_{xx} - \sigma_{yy}$ [2], [5] and a

temperature uncompensated linear combination of $\sigma_{xx} + \sigma_{yy}$ and σ_{zz} including the piezoresistance coefficients. The shear stresses can be measured accurately using various techniques [2], [5], [6]. Using n- and p-type piezoresistors, theoretically, $\sigma_{xx} + \sigma_{yy}$ and σ_{zz} can be separated [7], though in a temperature uncompensated way. In practice, however, unsuitable values of the piezoresistance coefficients of p-type silicon [8] prohibit a meaningful separation.

A novel CMOS silicon stress sensor featuring vertical current components is presented in this paper. The piezoresistance effect on these vertical current components provides an additional linearly independent measurement of mechanical stress and enables the extraction of the stress state including σ_{zz}. As a first step towards out-of-plane stress detection, it allows extracting of a sum of the three normal stress components with inherent temperature compensation.

DESIGN AND FABRICATION

The novel stress sensors were fabricated using the 0.6 µm high voltage, double poly, triple metal CMOS process *XC06* of X-FAB (Erfurt, Germany). An optical micrograph of a sensor structure is shown in Figure 1. The large bond pad (150×150 µm²) centered on top of the sensitive area (~30×30 µm²) offers the possibility to place gold bumps on top of the sensor for characterization purposes.

The symmetric sensor shown in Figure 1 (a) is based on a square n-well comprising a central contact C_C surrounded by a ring contact C_R separated by a p-well ring. The ring is reverse-biased to form a variable insulating depletion zone at the pn-junction. Bias currents applied between C_C and C_R result in large vertical current densities in the sensor center. They depend on the depth of the n-well and on the width w_{eff} defining the diameter of the vertical channel in the center of the structure. Below the central channel, the current density decreases due to the radially spreading current. The piezoresistive change of the potential between the contacts recorded in a 4-point measurement is exploited in stress detection. In this paper, three realized sensors are presented including standard and deep n-wells with different w_{eff}.

THEORY

The total resistance of the sensor $R = R_{V1} + R_{V2} + R_H$ varies with mechanical stress σ due to the piezoresistance effect, with the voltage V through the junction field effect, and with temperature T.

Fundamental piezoresistance theory describes relative resistance changes for vertical currents in [001] direction as

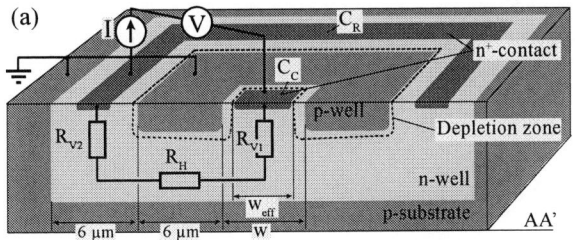

Sensor identifier	n-well type	w (µm)
⊙ #1	deep	5.0
▣ #2	deep	6.0
△ #3	standard	5.2

Figure 1: (a) Schematic and (b) optical top view of the novel stress sensor design fabricated in a 0.6 µm CMOS process of X-FAB, Germany. The table indicates geometry parameters of three investigated sensors.

$$\frac{\Delta\rho_V}{\rho_V} = \pi_{12}\left(\sigma_{xx} + \sigma_{yy}\right) + \pi_{11}\sigma_{zz} \qquad (1)$$

and for an equally distributed, radial, horizontal current in the (001) plane, as may be approximated in the sensor below the p-wells, as

$$\frac{\Delta \rho_H}{\rho_H} = \frac{\pi_{11} + \pi_{12}}{2}(\sigma_{xx} + \sigma_{yy}) + \pi_{12}\,\sigma_{zz}. \qquad (2)$$

The relative change of the total resistance R due to an applied stress depends on the ratio of vertical to horizontal current density and thus is a combination of (1) and (2). The influence of each component on the total resistance change is described by a geometry factor Γ $(0 \le \Gamma \le 1)$ defined as the ratio of vertical resistance $R_V = R_{V1} + R_{V2}$ to the total resistance R. It mainly depends on the effective width w_{eff} of the inner current channel affected by the junction field effect and on the n-well depth. Thus, the overall relative resistance change of the sensor including thermal resistance changes is

$$\frac{\Delta R}{R} = \left[\pi_{11} + \pi_{12} - \Gamma(\pi_{11} - \pi_{12})\right]\frac{\sigma_{xx} + \sigma_{yy}}{2},$$
$$+ \left[\pi_{12} + \Gamma(\pi_{11} - \pi_{12})\right]\sigma_{zz} + \alpha\,\Delta T \qquad (3)$$

where π_{ij}, σ_{ij}, and α are piezoresistance coefficients, stress components, and the thermal coefficient of resistance, respectively. Two sensors with different geometry factors Γ execute linearly independent measurements of the temperature and the combination $\sigma_\Sigma = (\sigma_{xx} + \sigma_{yy})/2 - \sigma_{zz}$ of the three normal stress components. Combining their outputs makes it possible to extract σ_Σ in a temperature-compensated manner with sensitivity ΔS_V

$$\frac{\Delta R_1}{R_1} - \frac{\alpha_1}{\alpha_2}\frac{\Delta R_2}{R_2} = -(\Gamma_1 - \frac{\alpha_1}{\alpha_2}\Gamma_2)(\pi_{11} - \pi_{12})\,\sigma_\Sigma \qquad (4)$$
$$= \Delta S_V\,\sigma_\Sigma$$

For a high sensitivity of the measurement, sensors with a large difference in Γ are required. Different factors Γ are achieved varying the sensor design or, using a single sensor device, by operating the sensor at different bias currents. These different bias conditions change the depletion zones and consequently the current densities and the ratio Γ.

EXPERIMENTAL SETUP

Mechanical stress is applied via accurately positioned, highly reproducible gold bumps realized using a commercial wire bonder. They act as force transmitters between a pressure tool and the sensors, homogenize the stress distribution in the sensor surface, and avoid local stress concentration.

The experimental setup, shown in Fig. 2, applies vertical forces F_V to the gold bumps in a controlled way. This is achieved by measuring the force using a load cell

Table 1: Measured bump contact diameters d_c and heights h realized at various bond forces F_b. The values are extracted from cross sectional cuts.

F_b (N)	0.5	0.75	1	1.3
d_c (μm)	69	90	98	107
h (μm)	53	43	38	32

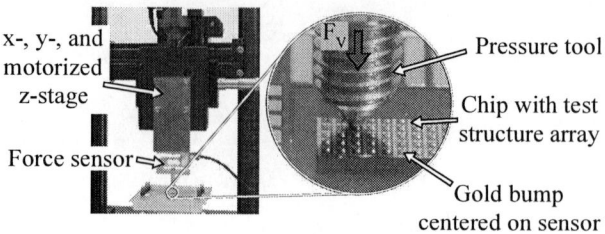

Figure 2: Experimental setup used for the application of a vertical force F_V to the stress sensor via a gold bump using a pressure tool.

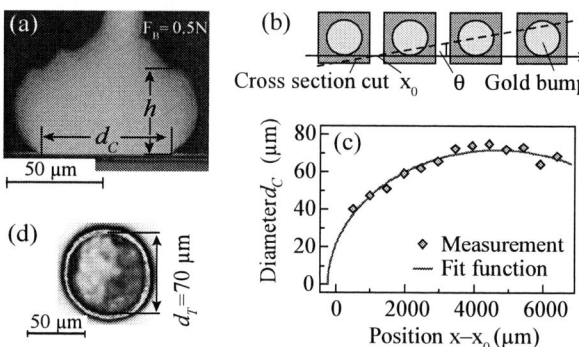

Figure 3: (a) Cross sectional view of a gold bump, cut as schematically shown in (b) and subsequently polished. (c) Corresponding fit function for contact diameters d_c, x_0, and θ. (d) shows an optical top view of a bump after force application in a measurement, where the remaining bond tail is compressed into a cylinder with diameter $d_T = 70$ μm.

and controlling the displacement of a pressure tool using a motorized precision stage.

The pressure tool features a circular area parallel to the chip surface with a diameter of 300 μm to contact the gold bump. For precise simulations of the stress distribution and elaborate characterizations of the sensors, accurate geometry data of the bump shapes are essential. These data were identified for different bond forces from cross sectional cuts through bumps placed in a straight line as shown in Figure 3 (a,b). The contact diameter d_C was determined by a fit function based on 13 cut bumps shown in Figure 3 (c). The bond-force-dependent contact diameter d_c and the height h of the bond capillary imprint are given in Table 1 for different bond forces F_B.

The wire bonder places stud bumps by tearing off the bond wire leaving behind a wire tail protruding over the height h visible in Figure 3 (a). This tail is compressed to a cylindrical shape of 70 μm diameter and 10 μm height during the application of characterization forces F_V between 0.5 N and 1 N as shown in the optical micrograph in Figure 3 (d).

FINITE ELEMENT SIMULATIONS

Finite element simulations were performed using COMSOL Multiphysics 3.4 to determine the stress state in the sensor. The simulation model consists of a silicon block $(300 \times 300 \times 300 \; \mu m^3)$ with anisotropic mechanical properties and two gold quarter cylinders (isotropic

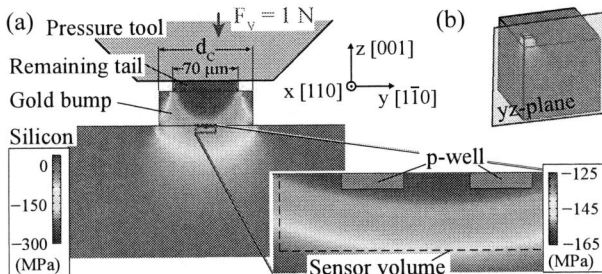

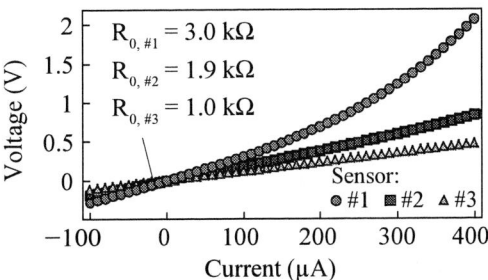

Figure 4: (a) Close-up and (b) overview of the numerical simulation model. The graphics show simulated σ_{zz}- stress distribution within the stress sensor volume for an applied vertical force $F_V = 1$ N.

Figure 5: Non-linear current-voltage characteristics measured for the three different sensor geometries. Resistances for small currents are given.

Young's modulus $E = 70$ GPa, Poisson's ratio $\nu = 0.44$). The radius of the lower cylinder is adapted to the contact diameter d_C measured for a specific bond force F_B, while the diameter of the upper cylinder of 70 μm represents the remaining tail of the bond wire shown in Figure 3 (d). The lower silicon surface is fixed and the movement of the upper gold surface is restricted in x- and y-direction reflecting its constraint by the pressure tool. An area load F_V/A is applied to the upper surface. The mesh is highly refined in the sensor volume centered directly below the gold cylinders.

A force F_V applied to the setup induces a large vertical normal stress increasing linearly with F_V. The average value of σ_{zz}/F_V inside the sensor volume is −143 MPa/N for a gold bump fabricated with $F_B = 1$ N. Due to the lateral expansion of the materials, in-plane normal stresses (−13 MPa/N) and out-of-plane shear stresses (±9 MPa/N) occur as well. The in-plane shear stress of ±0.5 MPa/N is negligible. Figure 4 shows the distribution of the vertical normal stress in a cross section in a (110)-plane.

MEASUREMENT RESULTS

Sensor Characterization

The resistance of the sensor varies with the applied bias current due to the junction field effect. A decreased effective width w_{eff} amplifies this non-linearity in the current-voltage characteristic since the relative change of the conductive diameter in the vertical channel increases. This effect is demonstrated in Figure 5 for sensors #1 and #2 both incorporating deep n-wells designed with different width w ($w_{#1} = 5$ μm, $w_{#2} = 6$ μm). The resistance R_0 for small currents also reflects the strong influence of the width w. The horizontal current and thus R_H dominates the resistance of device #3 based on a standard n-well. Its resistance change due to the junction field effect is negligible.

The measured sensitivity S_V of the sensors defined as $\Delta R/R\ F$ is given in Figure 6 as a function of the bias current for gold bumps realized with a bond force $F_B = 0.75$ N. Similar to the resistance measurements, sensor #3 has a constant sensitivity independent of the bias current, since the horizontal current component dominates the device under all bias conditions. The increasing sensitivity of device #2 is due the contraction of the central vertical channel, where R_{V1} and thus the ratio Γ increases significantly. The increase in the

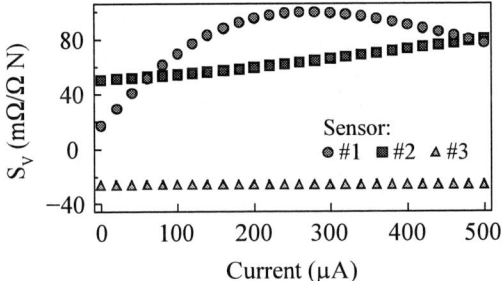

Figure 6: Measured sensitivities S_V to a vertical force F_V vs. bias current, using bumps fabricated at a bond force $F_B = 0.75$ N.

sensitivity of #1 up to $I_{bias} = 250$ μA has the same reason. The origin of the decreasing sensitivity at higher bias currents might be a pinch off of the channel at the surface. The expanding depletion zone forms a conical shape of the conductive path that possibly favors horizontal current components even in the vertical channel. As a consequence Γ decreases for $I_{bias} > 250$ μA.

Applying the simulated stress distribution considering bonds fabricated at $F_B = 0.75$ N to (1) and (2) and assuming the well known piezoresistance coefficients for low doped silicon [8], the expected sensitivity to a vertical force, is $-79 \times 10^{-3}/F_V$ in case of solely horizontal currents ($\Gamma = 0$) and $128 \times 10^{-3}/F_V$ for exclusively vertical currents ($\Gamma = 1$). The experimentally extracted sensitivity S_V of a sensor lies between these limits, thus making it possible to extract its geometry factor Γ. Therefore, measurements are carried out under isothermal conditions and $\Delta T = 0$ is assumed in Equation (3). From the gold bump dependent

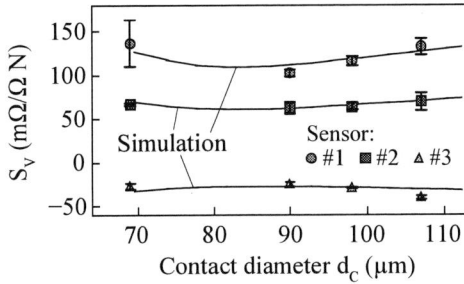

Figure 7: Comparison of simulated and measured sensitivities S_V for different contact diameters d_C at a bias current $I_{bias} = 250$ μA.

978-1-4244-2977-6/09 $25.00 © 2009 IEEE

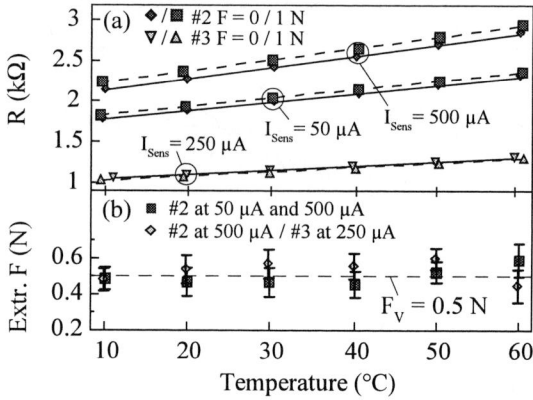

Figure 8: (a) Measured resistance of sensors #2 and #3 with and without applied force F_V for different temperatures (b) temperature compensated force values extracted using one single sensor and two different sensors, compared to the known applied force F_V.

stress simulation and the measured relative resistance changes Γ is finally calculated. Averaged Γ values determined with four different gold bump shapes ($F_B = 0.5$ N / 0.75 N / 1 N / 1.3 N) are $\Gamma_{\#1} = 0.86 \pm 0.014$, $\Gamma_{\#2} = 0.64 \pm 0.020$, and $\Gamma_{\#3} = 0.22 \pm 0.023$ for those sensor types. The measured sensitivities for bumps with different contact diameter d_C are shown in Figure 7 with simulated sensitivities. The simulation incorporates fit functions for the bump height h gained in the cross section measurements shown in Figure 3 as well as the ratio Γ.

Temperature Compensated Force Measurements

To verify the temperature compensation principle, resistance measurements were carried out between 10°C and 60°C with and without applying a vertical force F_V to the gold bump ($F_B = 0.75$ N). Sensor #2 was operated at bias currents $I_{bias} = 50$ μA and $I_{bias} = 500$ μA, sensor #3 at $I_{bias} = 250$ μA to obtain independent measurements with differing geometry factors Γ. Figure 8 (a) clearly shows the temperature influence on the sensor resistance. Two methods to extract the force independent of the temperature influence were implemented. First, the signal of sensors #2, biased with 500 μA and #3 were subtracted as described in Equation (4). This yielded a measurement with large sensitivity, as the geometry factors of the devices showed a large difference. The extraction using only sensor #2 operated in different biasing modes allows a highly localized measurement with a smaller sensitivity due to smaller Γ variations. Extracted force values are shown in Figure 8 (b). The determined temperature compensated values differ from the applied force of 0.5 N by less than 15 % for both methods. One reason for the deviation is the temperature dependence of the piezoresistance coefficients that cannot be compensated.

CONCLUSION

A novel piezoresistive stress sensor featuring vertical currents was described. It is based on CMOS compatible diffusions and enables the temperature compensated detection of a sum of the three normal stresses including the vertical component σ_{zz}. We presented an experimental

setup for the controlled application of a vertical force, inducing a large vertical normal stress in the sensor together with other stress components one order of magnitude smaller. After successful characterization, the functionality of the device was demonstrated in a temperature compensated measurement of a vertical force.

The presented sensors are a first step towards the detection of the complete local stress state. The presented device is attractive due to its low-cost fabrication and the possibility to integrate it into packages in combination with established in-plane sensors and other MEMS components.

ACKNOWLEDGEMENT

The authors gratefully acknowledge Shankar Ganapathysubramanian (Intel Corporation) for inspiring discussions. The authors also thank Andreas Frank for his assistance in the simulation and measurements and Georgi Todorov for the introduction to polishing the cross section cuts.

This work is funded by Intel Corporation.

REFERENCES

[1] M.K. Rahim, J. Roberts, J.C. Suhling, R.C. Jaeger, P. Lall, "Continuous In-Situ Die Stress Measurements During Thermal Cycling Accelerated Life Testing, " in *Proc. Electronic Components and Technology Conference*, pp. 1478-1489, 2007.

[2] M. Doelle, C. Peters, P. Ruther, O. Paul, "Piezo-FET Stress Sensor Arrays for Wire Bonding Characterization," *IEEE Journal of Microelectromechanical Systems.*, vol. 15, no. 1, pp. 24-31, 2005.

[3] S. Agraharam, N. Deshpande, J. Jackson, R. Mahajan, R. Manepalli, M. Pang, N. Patel, P. Stover, R. Tanikella, P. Tiwari, V. Wakharkar, "Flip-Chip Packaging Technology for Enabling 45nm Products," *Intel Technology Journal*, vol. 12, no. 2, pp. 145-156, 2008.

[4] R.R. Dimagiba, S. Ganapathysubramanian, M. Modi, "A Review of First Level Interconnect Modeling Methodology," in *Proc. Electronics Manufacturing and Technology*, Petaling Jaya, Malaysia, pp. 529-533, 2007.

[5] R.C. Jaeger, J.C. Suhling, R. Ramani, "Errors Associated with the Design, Calibration and Application of Piezoresistive Stress Sensors in (100) Silicon," *IEEE Transaction on Components, Packaging, and Manufacturing Technology – Part B*, vol. 17, no. 1, pp. 97-107, 1994.

[6] J. Bartholomeyczik, S. Brugger, P. Ruther, O. Paul, "Multidimensional CMOS In-Plane Stress Sensor," *IEEE Sensors Journal*, vol. 5, no. 5, pp. 872-882, 2005.

[7] D.A. Bittle, J.C. Suhling, R.E. Beaty, R.C. Jaeger, R.W. Johnson, "Piezoresistive Stress Sensors for Structural Analysis of Electronic Packages," *Journal of Electronic Packaging*, vol. 113, no. 3, pp. 203-215, 1991.

[8] C.S. Smith, "Piezoresistance Effect in Germanium and Silicon," *Physical Review*, vol. 94, no. 1, pp. 42-49, 1954.

FLEXIBLE TACTILE SENSOR SHEET
WITH LIQUID FILTER FOR SHEAR FORCE DETECTION
K. Noda[1], K. Matsumoto[1], and I. Shimoyama[1]
[1]The University of Tokyo, Tokyo, Japan

ABSTRACT

We propose a tactile sensor with standing cantilevers embedded in viscous liquid. The viscous liquid was covered with an elastic body. Since the micro cantilevers were kept free in the liquid, they can detect large shear forces without damaging. When shear forces were applied to the sensor surface, they deform the elastic body and the viscous liquid. Since the standing cantilevers follow deformations of viscous liquid, shear forces can be detected with resistance changes of the cantilevers. If the shear force was kept constant, the cantilevers in liquid return to their initial posture. From this mechanism, the proposed sensor can detect the change of shear forces without damaging.

Since the standing cantilever in single liquid filter detect the force applied to its surface. It becomes difficult to detect the contact position with the proposed sensor. Therefore, we arranged channel in the liquid filter to control the liquid flow. By arranging standing cantilevers into the channel, the cantilever's resistance changes with the distance from the contact position. In this paper, we measured the relationship between the contact position and the resistance changes of the cantilevers arranged in channel.

INTRODUCTION

In the field of robotics, manipulation of objects is one of the major studies. To control grasping force in minimum, it becomes important to detect a slippage between the object and the robot-hand surface. Since the slippage causes shear deformation to the sensor surface, tactile sensors to detect shear forces are required.

Recently, several kinds of MEMS tactile sensors to detect 3-axis forces are reported. One major approach for the force measurement is to measure the deformation of the elastic body with strain gauges, for example the gauges using piezoresistor or thin metal layer [1-4]. Although, these sensors are useful to detect the applied forces, the micro sensing structures composed of thin metal layers are fixed to the elastic body to detect its deformation directly. Since the thin metal material cannot be stretched in large range, the micro structures were easily damaged with the large deformation of the elastic body. To resolve this problem, it is necessary to measure the deformation without fixing to the elastic body.

In [5] and [6], force sensors which measure the deformation of elastic body by using a vision were reported. These sensors detect the movements of markers designed on the surface elastic body with image sensors. So that it can detects the force without damaging the structures. However the vision system is too large so that it is difficult to cover the curved small area as the finger tips.

In this paper, we propose a tactile sensor with standing piezoresistive cantilever embedded in viscous liquid as shown in Figure 1. The viscous liquid is covered with an elastic body.

When a shear force is applied to the sensor surface, it

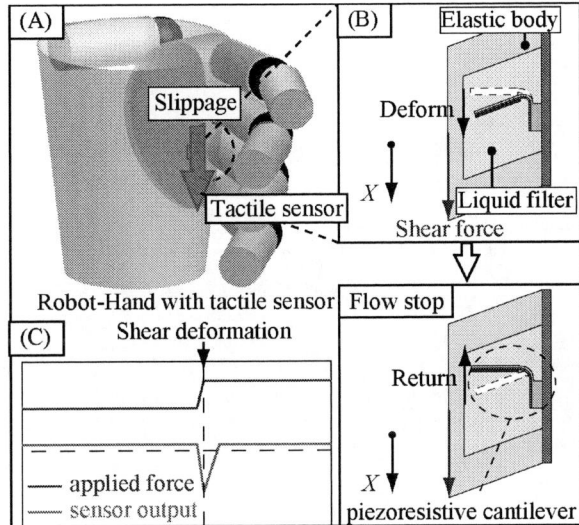

Figure 1. Concept sketch of shear force measurement.
(A) is the image of the sensor attached on a robot-hand.
(B) is the deformation of the standing piezoresistive cantilever in liquid.
(C) is the image of resistance change of the standing cantilever during its deformation shown in (B).

deforms the elastic body and the viscous liquid in the shear direction. Since the standing cantilever follows the deformation of viscous liquid, it is deformed in the x-axis direction by the shear force as show in Figure 1(B). The piezoresistor is formed on the hinge of the standing cantilever, so that its deformation in x-axis direction can be measured as a resistance change. Therefore, the shear force applied to the sensor surface can be measured as a resistance change of the standing cantilever.

When the shear force was kept constant, cantilever returns to its initial position and also its resistance returns to the initial value (Figure 1(C)). From this mechanism, the proposed sensor can detect the changes of applied shear forces.

Additionally the cantilever is unfixed to the elastic body, so that it is not damaged even though the over range deformation for the cantilevers were occurred to the elastic body. Because of this characteristics, wide range shear force can be measured with the proposed sensor.

PROTOTYPE LIQUID FILTER SENSOR
Fabrication

Fabrication steps are shown in Figure 2.

First, we fabricated a piezoresistive cantilever on an SOI wafer. The piezoresistor was composed to the surface of SOI with thermal diffusion. It was covered with 50nm thick Au and 300nm-thick Ni layers vaporized on the SOI surface. By patterning the Ni / Au / Si layers, the piezoresistive cantilever was fabricated. The cantilever was released by removing the part of backside Si and SiO$_2$ layer. The detail of this process is reported by us on [7].

978-1-4244-2977-6/09 $25.00 © 2009 IEEE

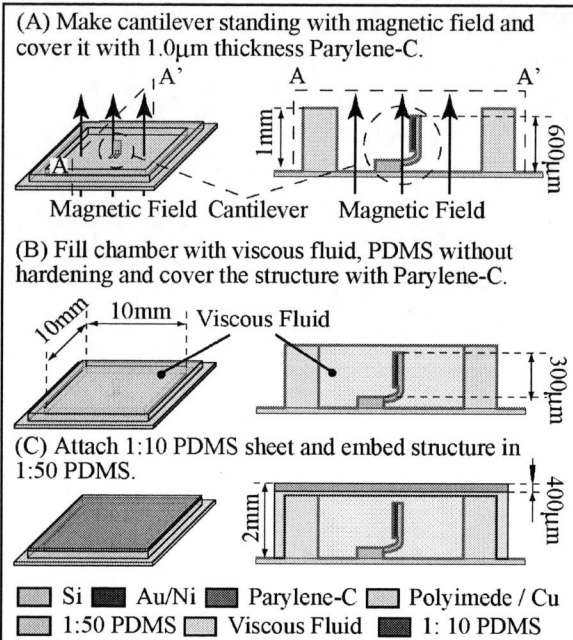

(A) Make cantilever standing with magnetic field and cover it with 1.0μm thickness Parylene-C.

(B) Fill chamber with viscous fluid, PDMS without hardening and cover the structure with Parylene-C.

(C) Attach 1:10 PDMS sheet and embed structure in 1:50 PDMS.

■ Si   ■ Au/Ni   ■ Parylene-C   □ Polyimede / Cu
□ 1:50 PDMS   □ Viscous Fluid   ■ 1: 10 PDMS

Figure 2. Fabrication process of the tactile sensor with liquid filter.

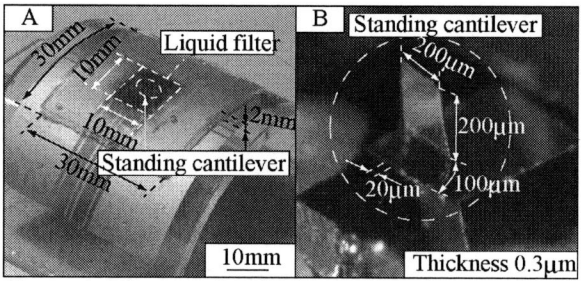

Figure 3. Photographs of fabricated tactile sensor.
(A) is the whole structure of the sensor with viscous liquid.
(B) is the standing piezoresistive cantilever.

The cantilever was attached and wired onto a Cu/Polyimide film. A liquid chamber was composed to the film with 1mm high spacers on the film. The spacer was composed with the PDMS. The mixture rate of the PDMS and its hardening was 50:1. The cantilever was made to stand by applying magnetic field and covered with 1μm thick Parylene-C (Figure 2(A)).

Then the liquid chamber was filled with viscous liquid and covered it with 1μm thick Parylene-C (Figure 2(B)). We used PDMS without hardening as a viscous liquid. The liquid was colored in blue by applying ink. Since the PDMS has quite high vapor pressure, Parylene-C can be vaporized to keep the shape of liquid.

Finally, a 400μm-thick 10:1 mixture rate PDMS sheet was attached to the sensor surface and whole structure was embedded into 50:1 mixture rate PDMS (Figure 2(C)). By using soft 50:1 mixture rate PDMS, the sensor become flexible enough to attach to curved surfaces.

The photographs of the prototype sensor were shown in Figure 3. The surface area of the sensor was 30×30mm² and its thickness was 2mm. The blue area in the Figure 3(A) is the viscous liquid. As shown in the photo, the fabricated sensor can be attached onto a curved surface

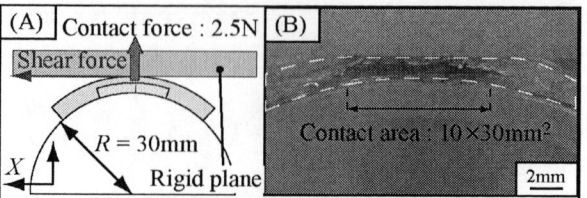

Figure 4. Experimental setup to apply shear force.
(A) is the concept sketch of the shear force supply.
(B) is the photograph of the side view of contact between the sensor and the rigid plane.

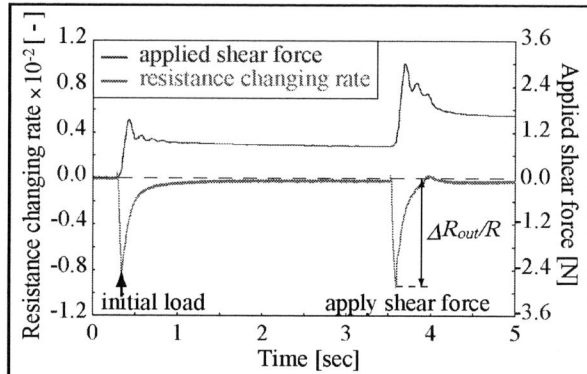

Figure 5. Sensing characteristics of the fabricated sensor to detect the changes of shear force. 2.0N shear force was applied to the sensor surface after 1.5N initial force.

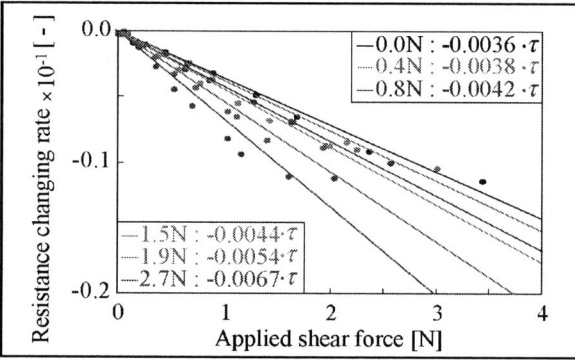

Figure 6. Relationship between the applied shear force and the resistance changing rate of the fabricated sensor. Influence of the initial shear force to the sensitivity was measured.

with 30mm in radius. The initial resistance of the piezoresistive cantilever was 2.0kΩ in average.

Shear force detection

The sensing characteristic of the fabricated sensor was measured with the setup shown in Figure 4. The sensor was attached to a 30mm radius half-rod and it was contacted to a rigid plane with 2.5N force. Shear forces were applied to the sensor surface by moving the sensor in X-axis direction with 10mm/sec speed. The applied normal and shear forces were measured with the 6-axis force sensor attached on the bottom of the half-rod. As shown in Figure 4(B), the contact area between the sensor and the rigid plane was 10×30mm² at the 2.5N contact force.

Figure 5 shows the magnitudes of shear forces measured with 6-axis force sensor and the resistance

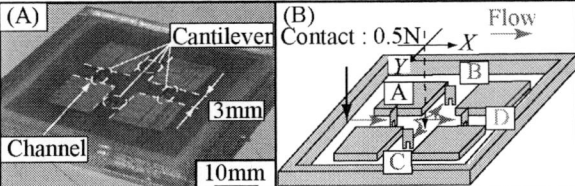

Figure 7. Sensor with channel for contact detection.
(A) is the fabricated sensor with channel.
(B) is the flow direction occurred in the channel.

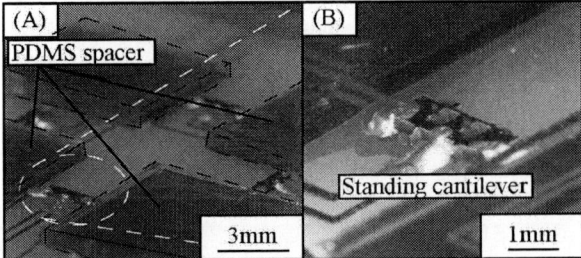

Figure 8. Channel made with PDMS spacer and the cantilever chips arranged in the channel.

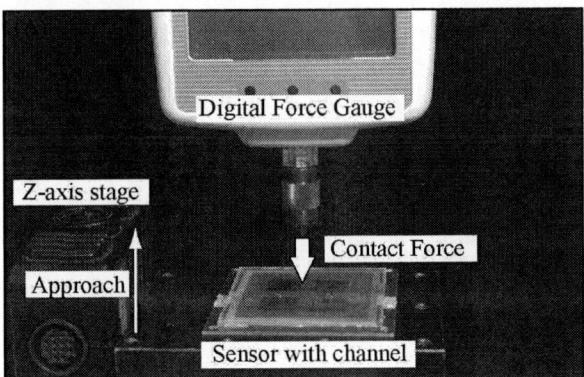

Figure 9. Experimental setup to apply contact force to the sensor surface.

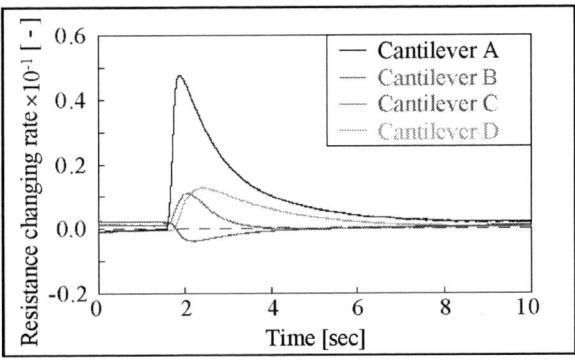

Figure 10. Resistance change of the cantilevers to the contact at X=-15mm, Y=0mm position.

change of fabricated sensor to the applied shear forces. In this experiment, 1.5N initial shear force and the 2.0N shear force was applied to the sensor. As shown in the result, the cantilever's resistance changed with the changes of the magnitude of applied shear forces. The response speed to the applied force was 45msec in average. After the deformation, the resistance changes of the fabricated sensor went back to the initial value after 1.5sec in average. Since these response speeds are strongly influenced with the viscosity of the liquid, the response speeds can be increased by using low viscosity liquid as silicone oil.

The relationships between the peaks of applied shear forces, and the resistance changing rate of the fabricated sensor are shown in Figure 6. In this experiment, we changed the initial shear force from 0 to 2.7N to confirm the influence of the initial shear force to the sensing characteristics of the fabricated sensor. As shown in the result, the correlation coefficient R^2 was 0.97 in average, this shows that the resistances of the sensor changes linearly to the magnitude of applied shear forces. The minimum sensing range of the sensor was 0.05N and the maximum range was 3.0N.

According to Figure 6, the absolute value of the sensitivity increased from 0.0036 to 0.0067N^{-1} when the initial shear forces increased. The shear deformation of the

liquid chamber caused the increase of the sensitivity. When the shear forces were applied to the sensor surface, it reduces the height of liquid chamber around the standing cantilever. Thus the elastic body surface layer comes close to the standing cantilever and influents to the deformation of standing cantilever.

The reduction of the height of the liquid chamber will be caused with the initial shear force and the contact forces. Therefore, it becomes necessary to detect the contacting force to measure the sensor posture and to calibrate its sensitivity for the shear force detection.

CONTACT DETECTION WITH CHANNEL

If the standing cantilever was embedded into a single large liquid chamber as Figure 3, the liquid deforms free. Since the whole liquid will be deformed at once with the contact, it becomes difficult to detect the contact positions by measuring the resistance changes of the cantilevers. To detect contact position, we arranged 4 cantilever chips in a 3mm width cross type channels as shown in Figure 7. The surface area of the sensor was 30×30mm^2 and its thickness was 1.5mm. The average resistances of the cantilevers were 2.0kΩ.

When the object contacted to the sensor surface, the liquid deformed in particular directions as shown in Figure 7(A). Therefore the responses of each cantilever changes with the distance from the contact positions even though the cantilevers were arranged in near position with 10mm distances.

The brief fabrication process was the same as the Figure 2. The sensor was fabricated on a Cu substrate. The channel was fabricated by attaching 1:10 mixture rate PDMS spacers with 1mm height to the substrate as shown in Figure 8. After covering the channel with viscous liquid and Parylene-C as the Figure 2(B), the whole structure was embedded in 1:10 mixture rate PDMS.

The contact positions were measured with the setup shown in Figure 9. In this setup, the sensor surface was touched with a corn type tip which was attached to the force gauge. 0.5N force was applied around the channel with 5mm pitch. Figure 10 shows the resistance changes of the 4 cantilevers when the contact force was applied to the X=15mm and Y=0mm position where the arrow in Figure 7(B) shows. Since the cantilever A was most close to the contact position and the distance between other cantilevers and the contact position was the same, only the resistance change of cantilever A was higher than others. The relationships between the contact positions and the

978-1-4244-2977-6/09 $25.00 © 2009 IEEE

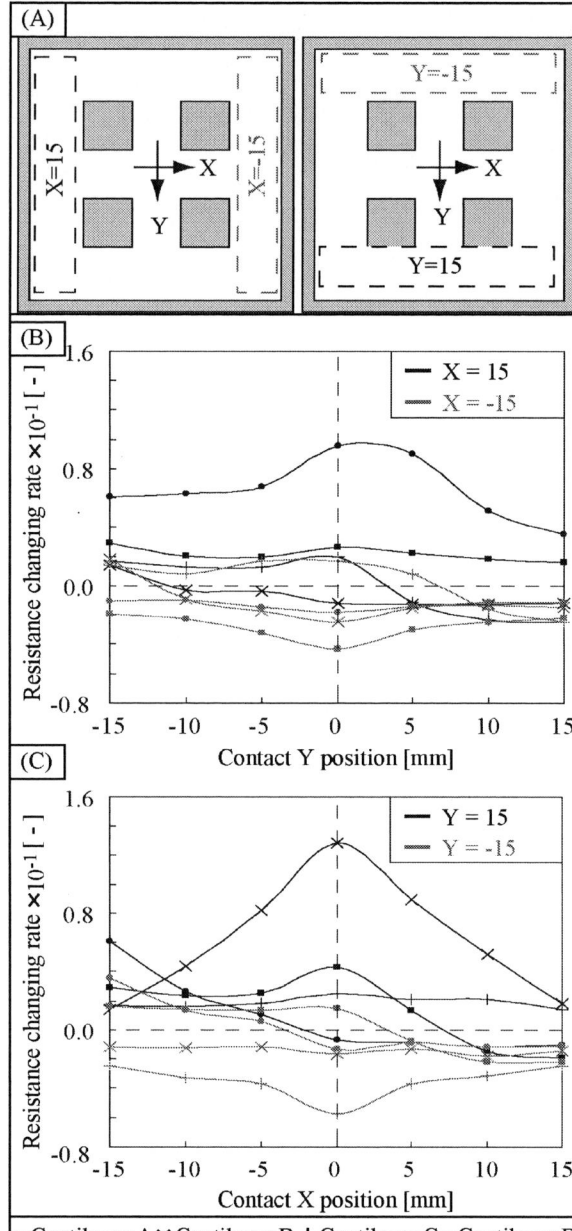

Figure 11. Relationships between contact position and the resistance changes of the cantilevers.

(A) is the contact positions in the experiments.

(B) is the relationship between the contact positions and the resistance changes of the cantilevers when X-axis position was -15mm or 15mm.

(C) is the relationship between the contact positions and the resistance changes of the cantilevers when Y-axis position was -15mm or 15mm.

resistance change of each cantilever is shown in Figure 11.

Figure 11(B) shows the resistance changes of the cantilevers when the X-axis positions were kept in -15 or 15mm and the y-axis positions were changed from -15 to 15mm. Figure 11(C) shows the resistance changes of cantilevers when Y-axis positions were kept in -15 or

15mm and x-axis positions were changed. As same as Figure 10, the resistance changing rates of the cantilevers changed by the distance from the contact positions. Therefore, the contact position can be detected by comparing the magnitudes of resistance changes of the cantilevers.

CONCLUISON

In conclusion, we proposed and fabricated a tactile sensor with a liquid chamber. Since the standing cantilever can deform free in the liquid, changes of the large shear force can be detected without damaging the micro cantilever. By using the prototype sensor we measured 0.05N change of shear force in minimum.

The contact positions were detected by arranging 4 cantilevers in 3mm width channel. The channel determines the direction of the liquid deformation, so that the resistance changes of each cantilever will be changed by the distance from the contact position. By using this sensor, contact position of 0.5N force was measured with 5mm pitch.

These results confirm that our proposed sensor mechanism is efficient to detect the contact position and the shear forces. The mechanism which reduces the damage to the sensor structure enables to detect the large force for grasping objects with robot-hands.

ACKNOWLEDGEMENTS

Using mask writer: Photomasks were made using theUniversity of Tokyo VLSI Design and Education Center (VDEC)'s 8-inch EB writer F5112+VD01 donated by ADVANTEST Corporation.

References

[1] K. Kim, *et al.,* "A silicon-based flexible tactile sensor for ubiquitous robot companion applications," *Journal of Physics,* Vol. 34, pp. 399-403, 2006.

[2] E. S. Hwang, *et al.,* "A Polymer-based Flexible Tactile Sensor for Normal and Shear Load Detection," *16th IEEE Conference of Micro Electro Mechanical Systems,* Istanbul, Turkey, pp. 714-717, 2006.

[3] J. Engel, et al., "Development of polyimide flexible tactile senor skin," *Journal of Micromechanics and Micro engineering,* Vol. 13, pp. 359-366, 2003.

[4] Y. Tanaka, et al., "TRIAXIAL TACTILE SENSOR CHPS WITH PIEZORESISTIVE CANTILEVERS MOUNTABLE ON CURVED SURFACE," 4th APCOT Conference, TAINAN, TAIWAN, pp. 29-32, 2008.

[5] M. Ohka, *et al.,* "An Experimental Optical Three-axis Tactile Sensor for Micro-Robots," *Robotica,* Vol. 23, pp. 457-465, 2005.

[6] M.Ohka, *et al.,* "Sensing Characteristics of an optical three-axis tactile sensor mounted on a multi-fingered robotic hand," *Proceedings of 2005 IEEE/RSJ International Conference of Intelligent Robots System,* pp. 1959-1964, 2005.

[7] K. Noda, *et al.,* "A shear stress sensor for tactile sensing with the piezoresistive cantilever standing in elastic material," *Sensors and Actuators A,* Vol. 127, pp. 295-301, 2006.

BIOMIMETIC DESIGN AND FABRICATION OF COMPONENTS FOR ARTIFICIAL HAIR CELL SENSOR

H.S. Lee[1,2], B.-K. Lee,[1,2] and T.H. Kwon[1,2]

[1]Department of Mechanical Engineering, POSTECH, Korea
[2]MEMS Research Center for National Defense, POSTECH, Korea

ABSTRACT

As a part of effort to develop an artificial hair cell (AHC) sensor, we designed and fabricated a high-aspect-ratio/single cilium structure and a mechanoreceptor since they are the most important components for AHC sensor. The high-aspect-ratio single/cilium structure was successfully replicated by means of a hot embossing process with a help of a double-sided mold system. Especially for the high-aspect-ratio microstructure, we have proposed a new concept of a separated micro mold system utilizing LIGA process. A multi-wall carbon nanotube polydimethylsiloxane (MWNT-PDMS) composite was used as a force sensitive resistive of a mechanoreceptor. The top-bottom electrodes type mechanoreceptor was designed since it is more effective than in-plane electrodes type. The performance of the mechanoreceptor was characterized by a nano indentation system.

INTRODUCTION

A biological system has evolved for a long time so that living organism becomes highly optimized and efficient. The present work is focused on a biomimetic application of biological hair cells to the development of an artificial hair cell sensor [1-3]. An artificial hair cell sensor basically consists of three components, namely a cilium structure, a mechanoreceptor and a signal processor, as indicated in Figure 1. The cilium is usually a high-aspect-ratio single structure that can be effectively bent by the external flow in its surrounding medium, which in turn generates a mechanical quantity as an input to the senor system. The mechanoreceptor transforms the mechanical quantity to a measurable electrical quantity. These two components play the most important roles in the artificial hair cell sensor. In the present study, we successfully designed and fabricated a high-aspect-ratio single cilium structure and a multi-wall carbon nanotube polydimethylsiloxane (MWNT-PDMS) composite based mechanoreceptor.

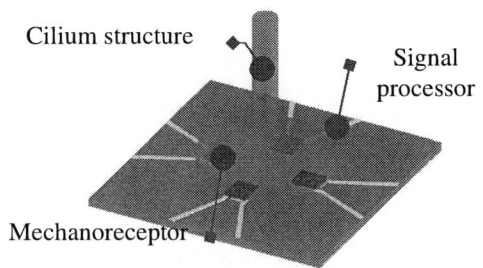

Figure 1: components of artificial hair cell sensor

CILIUM STRUCTURE

Mold system for single micro structure

Microstructures replicated by mass production technology such as injection molding, hot embossing and UV embossing usually has macro scale ground structure which can support these microstructures. However, the desired cilium structure in this study should be a single part which can be bent freely by the external flow. In this regard, we propose a new method to fabricate the single microstructures by hot embossing process without the ground structure.

The proposed design is a double-sided mold system which has the upper mold and the lower mold as schematically illustrated in Figure 2. This mold system has the high-aspect-ratio cavity and low-aspect-ratio cavity which will be the cilium structure and the supporting disc shape after molding, respectively. During embossing process, the heated up polymer will fill the cavity part of upper and lower molds and the residual layer will be getting thinner with high pressure (Figure 2b). It can be thin (less than several µm) enough to easily separate the molded cilium structure from the remaining polymer (Figure 2c).

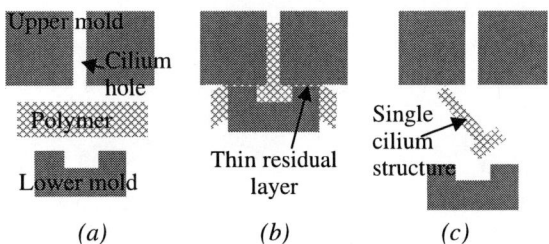

Figure 2: Double-sided mold system having upper mold and lower mold: (a) before embossing, (b) embossing and (c) demolding and final product

Mold system for high-aspect-ratio structure

The separated micro mold system, which is the upper mold of the double-sided mold system, was realized by the LIGA process which consists of three steps, i.e., deep X-ray lithography, electroforming, and plastic molding. The deep X-ray lithography enables us to make the high-aspect-ratio structure (usually above aspect ratio 50) and good sidewall quality (less than Ra 30nm).

The X-ray mask was fabricated on 300 µm thick Si-wafer by SU-8 lithography and Au electroplating. The electroplated 20µm thick gold absorbed X-ray during the X-ray exposure. The 2 mm thick PMMA was attached to the polished titanium plate using liquid PMMA. The deep X-ray lithography was performed at PLS (Pohang

Light Source). The exposure time was 500 minute and the exposed energy at bottom level was 3.6 kJ/m³. The exposed PMMA structure was immersed in GG-developer (ethanol 99% 180ml, Morpholine 60ml, 2-Aminoetanol 15ml and DI water 45ml) at 35°C for 24 hour. Nickel was electroformed over 2mm thick from the master of the developed PMMA structure. After removing the remained PMMA, the nickel cilium mold inserts were obtained.

The separated micro mold inserts are divided into two mirrored shaped mold inserts. As each part has a half cylindrical cilium cavity, an assembly of two parts becomes to have full cylindrical cilium cavity. Also each part has positive or negative aligning structures for assembling. This design gives us two dominant advantages: one is easy electroforming and the other is easy demolding. During electroforming process, the small area and high aspect-ratio of the developed PMMA structure cause poor adhesion between PMMA structure and substrate which is one of serious problems in LIGA. However, the area of PMMA structure can be enlarged in the proposed design so that nickel can be easily electroformed. As usually the case, demolding without damaging the replicated structure is a critical issue in the case of high-aspect-ratio structures. The higher aspect ratio structure, the more serious demolding problem is. However, the molded high-aspect-ratio structure can be easily demolded by separating two mold inserts during demolding. Figure 3 shows mold inserts of the separated micro mold system fabricated by LIGA process.

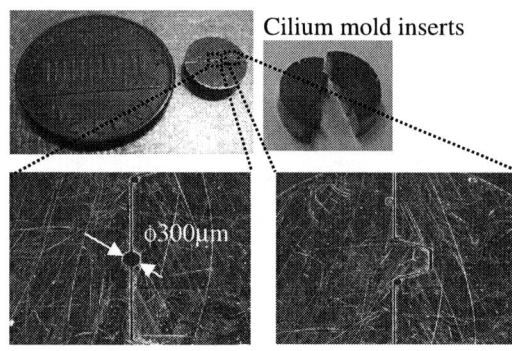

Figure 3: Mold inserts of the separated micro mold system fabricated by LIGA process

Fabrication

The hot embossing process was performed with double-sided mold system and the separated micro mold system. At the first embossing process with embossing temperature 150℃ and embossing pressure 91 MPa in the case of polycarbonate (PC), the cilium structure was not fully filled and the residual layer was not thin enough to separate the cilium structure from the remaining polymer. In order to overcome these problems such as an unfilled polymer and thick residual layer, we increased the embossing temperature and the embossing pressure. In the case of increasing pressure, mold inserts were damaged. So we decide to increase only temperature (up to 170℃) and to decrease the pressure (down to 42 MPa). The PC cilium structure was successfully replicated and the residual layer was thin enough. Also it

was successfully replicated in case of polymethyl methacrylate (PMMA) with the processing conditions of temperature 150℃ and pressure 42 MPa. These processing conditions were 30℃ higher and 38 MPa lower than the usual processing conditions in hot embossing process. The high-aspect-ratio single cilium structure which has 300μm diameter and 2mm height was successfully replicated with the proposed mold systems as shown in Figure 4.

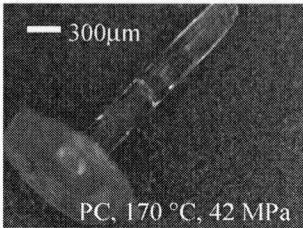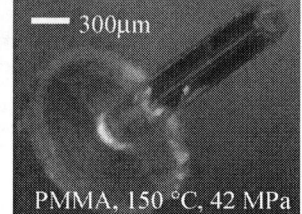

Figure 4: Replicated high-aspect-ratio single cilium structure

MECHANORECEPTOR

Materials

A multi-walled carbon nanotube polydimethylsiloxane (MWNT-PDMS) composite was used as a force sensitive resistive of the mechanoreceptor since its electrical resistance decreases as an applied pressure is increased. In order to uniformly disperse MWNT inside PDMS (Sylgard-184 A), we dissolved PDMS in tetrahydrofuran (THF) solvent which lowered the viscosity of PDMS and easily evaporated (boiling point 66°C). The MWNT and PDMS-THF solution was mixed by an ultrasonic wave. Before pattering the MWNT-PDMS composite, a curing reagent (Sylgard-184 B, 10:1) is added to the prepared MWNT-PDMS composite.

Design and fabrication

A new design for top-bottom electrode type mechanoreceptor was proposed in this study. Figure 5 shows the fabrication sequence of the proposed design. At first, a top stainless steel electrode and a stainless steel stencil mask was fabricated by the photolithography. The thickness of the top electrode was thin (30μm) enough to be bent and to be handled easily. A MWNT-PDMS composite was patterned on four sensing sites of the top electrode using a stencil method as shown in Figure 5a. A copper electrode on PCB was used as a bottom electrode. An adhesive was patterned on outside of PCB (Figure 5b). The top and bottom electrode parts were integrated (Figure 5c). The integrated part was in vacuum for 1 hour to get rid of the entrapped air. And then, compressive pressure of 1MPa was applied for bonding two part and temperature of 90°C was maintained for curing the MWNT-PDMS composite for 10 hour via hot embossing machine. The patterned adhesive made the gap of two electrodes so that the top electrode works like a cantilever in final mechanoreceptor. The gap thickness varied from 96.1μm to 116μm depending on the thickness of the adhesive, the bonding pressure and the size of the

978-1-4244-2977-6/09 $25.00 © 2009 IEEE

patterned composite. The size of each sensing site was about 100μm×250μm and the overall diameter of four sensing sites was 1mm.

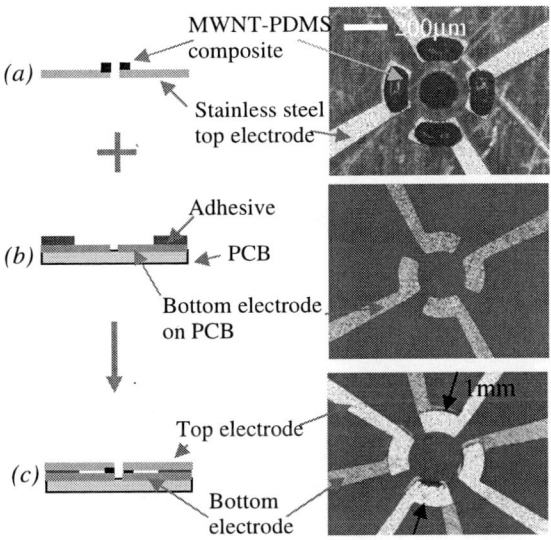

Figure 5: Fabrication sequence for top-bottom electrodes mechanoreceptor: (a) pattering MWNT-PDMS composite on stainless steel top electrode using stencil method, (b) pattering adhesive on outside of PCB and (c) maintaining compression of 1MPa for bonding and temperature of 90 °C for curing MWNT-PDMS composite via hot embossing machine for 10 hour

Performance test

A computer-controlled nanoindenter (Nanoindenter XP, MTS) equipped with a spherical sapphire indenter tip was used to characterize the performance of mechanoreceptors by driving indenter into the specimen surface and dynamically collecting the applied force and displacement data. The large size of indenter tip, φ200μm, was utilized for our own purpose.

An electrical resistance (R) can be expressed as Equation 1.

$$R = \rho \frac{L}{A} \tag{1}$$

where ρ is the resistivity, L is the interdistance and A is the area of the material. Assuming that the resistivity ρ is constant and the change of area (ΔA) is much smaller than the change of interdistance (ΔL) in Equation 1, the change of resistance (ΔR) can be expressed as below:

$$\Delta R = \rho \frac{\Delta L}{A} \tag{2}$$

From Equation 1 and Equation 2, a following equation is obtained.

$$\frac{\Delta R}{R} = \frac{\Delta L}{L} = \varepsilon \tag{3}$$

where ε is strain of the composite. From Equation 3, the normalized electrical resistance $\Delta R/R$ is a linear function of strain ε. The results of the nanoindentation shows experimentally the linear tendency of $\Delta R/R$ and ε indicated in Figure 6a although the initial electrical resistance of mechanoreceptors varies from several kilo ohms to several mega ohms. As shown in Figure 6b, the strain-stress relation is almost linear and the Young's modulus of the MWNT-PDMS composite is about 3.3 MPa comparing with the Young's modulus 1 MPa of a pure PDMS. Finally, the normalized resistance was found to be $\Delta R/R \cong -0.2355\sigma$ (MPa) where σ is an applied stress.

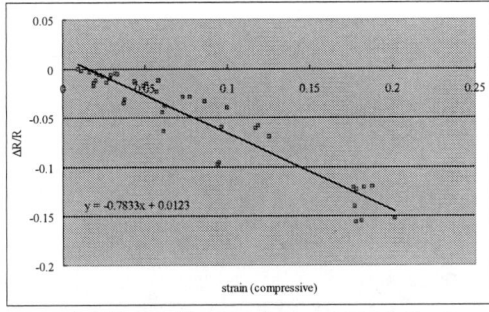

(a) Normalized resistance vs. strain

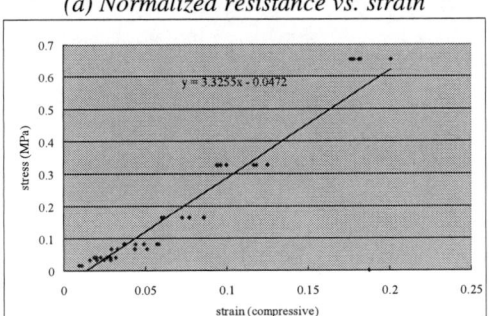

(b) Stress vs. strain
Figure 6: Plots of the characteristics of the proposed mechanoreceptor investigated by nanoindention system

CONCLUDING REMARKS

In the present study, we designed and fabricated a high-aspect-ratio/single cilium structure and a mechanoreceptor which play the most important roles in an artificial hair cell (AHC) sensor.

The high-aspect-ratio/single cilium structure was replicated by means of a hot embossing process. A double-sided mold system was designed and fabricated for the single cilium structure. This mold system consists of the upper and the lower mold. During the embossing process, heated-up polymer will fill the cavity part of the upper and lower molds, yielding a residual layer between the two molds. The higher the applied pressure, the thinner the residual layer is. The thin residual layer enables an easy separation from the molded cilium structure. It may be noted that for the high-aspect-ratio microstructure we proposed a new concept of a separated micro mold system making use of LIGA process: the mold inserts are divided into two mirror-shaped mold inserts, each having a half cylindrical cilium cavity; thus an assembly of two parts makes a full cylindrical cilium cavity. This separated micro mold system provides us with two marvelous advantages: i) an easy electroplating in LIGA process and ii) an easy demolding during the hot embossing process. The high-aspect-ratio/single cilium structure with the diameter of 300μm and the height of 2mm was successfully replicated in this manner.

As for the mechanoreceptor a multi-walled carbon

nanotube polydimethylsiloxane (MWNT-PDMS) composite was used as a force sensitive resistive material since it has a feature of decreasing resistance with an increasing applied pressure. In the developed mechanoreceptor we have designed and fabricated the top-bottom electrodes type since it is more effective than in-plane electrodes type. The performance characteristic of the mechanoreceptor proposed in the current study was investigated by a nano indentation system: the normalized resistance is found to be $\Delta R/R = -0.2355\sigma$ (MPa)

ACKNOWLEDGEMENT

This work was supported by Defense Acquisition Program Administration and Agency for Defense Development under the contract UD060049AD.

REFERENCES

[1] Y. Yang, N. Chen, C. Tucker, J. Engel, S. Pandya and C.
Liu, *Proc. MEMS 2007*, pp. 577-580.

[2] J. M. Engel, J. Chen , C. Liu and D. Bullen, *J. Microelectromech. Syst.,* 15(2006), pp.729-736

[3] M. Dijkstra, J. J. van Baar, R. J. Wiegerink, T. S. J. Lammerink, J. H. de Boer and G. J. M Krijnen, *J. Micromech. Microeng.,* 15 (2005), pp.132–138

[4] Patent applied in Korea

A NOVEL CONFIGURABLE MEMS INERTIAL SWITCH USING MICROSCALE LIQUID-METAL DROPLET

Kwanghyun Yoo and Joonwon Kim

Department of Mechanical Engineering of Science and Technology (POSTECH), KOREA

ABSTRACT

This paper presents a novel MEMS inertial switch using a microscale liquid-metal droplet as an active component. Our inertial switch is realized using the advantageous properties of the liquid–metal droplet (i.e., high surface tension, high density and electrical conduction) and the selectively modified contact surface. The modified contact surface contains microstructures which provide enhancement of non-wetting behavior as well as minimization of the contact angle hysteresis (CAH) of the liquid-metal droplet on the surface. Integrating the liquid-metal droplet, which is used as an inertial proof mass, inside the configured channel containing the microstructures on the surface, a dramatically simplified yet effective mechanism of inertial switch is realized. Various channel configurations can be designed for different threshold-levels of acceleration.

INTRODUCTION

Recently, many different inertial switches have been developed to detect an applied acceleration [1-4] for various applications. The detected acceleration can be used in the inertial device to trigger a safety mechanism to prevent major damage from a sudden impact during the device operation when the acceleration is more than the threshold-level of the device [5-6]. Conventionally, in order to detect the applied acceleration, a solid-type microscale inertial mass and beams are required to sense the acceleration which call for complex fabrication processes. Also, the switching contact is based on solid-to-solid contact which can cause reliability problems such as signal bouncing and contact wear during switching motion.

Unlike the others, our device uses the advantageous properties of liquid-metal (e.g., mercury) droplet to solve the problems of the conventional inertial switches. Also, the configured channel shape with the modified contact surface is used to control the movement of the droplet. The modified contact surface provides the easy movement of the liquid-metal droplet inside the channel by reducing the CAH of the droplet on the surface [7, 8]. The CAH is measured under the sliding condition of the droplet on the surface [9, 10].

Various channel configurations can be designed for different threshold-levels of acceleration. The relationship between the threshold-level and the channel dimension (i.e., the neck dimension) is predicted theoretically and verified experimentally. The concept and systematically performed experiments that bring about the novel inertial switch is reported.

DEVICE CONCEPT AND DESIGN

Configuration and operation principle

Fig.1 shows schematics of the operation concept of the inertial switch. Fig.1a shows an overview of the

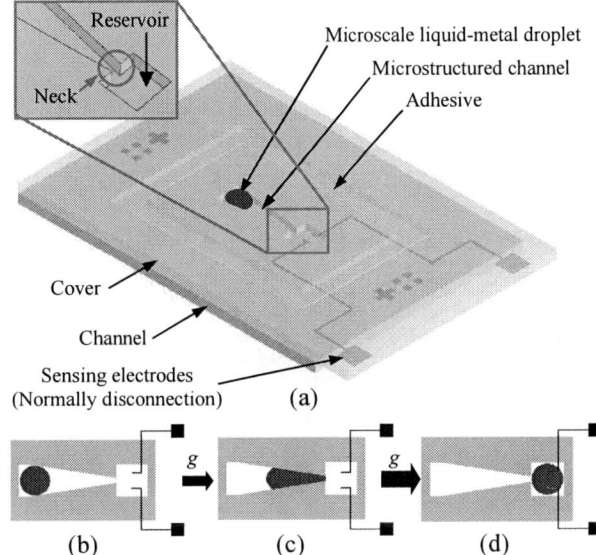

Figure 1: Schematic of the operation concept (a) Overview of the inertial switch, (b) Initial position, (c) Less than threshold-level, and (d) Over threshold-level

inertial switch. A liquid-metal droplet is placed in the selectively modified channel as an initial off-state (Fig.1b). When the acceleration is less than the threshold-level, the droplet can not pass the neck of the channel (Fig.1c). However, once the acceleration is more than the threshold-level, the droplet passes through the neck and switch on the sensing electrodes (Fig.1d). The threshold-level can be configured based on the geometry of the channel and the wetting property of the channel surface with a given volume of the droplet.

Theoretical approach

In order to estimate the relationship between the threshold-level for switching and the dimension of the channel, various parameters are considered. Fig.2 shows the notation of the design parameters.

According to previous researches regarding the liquid flow in a configured channel shape, which contains

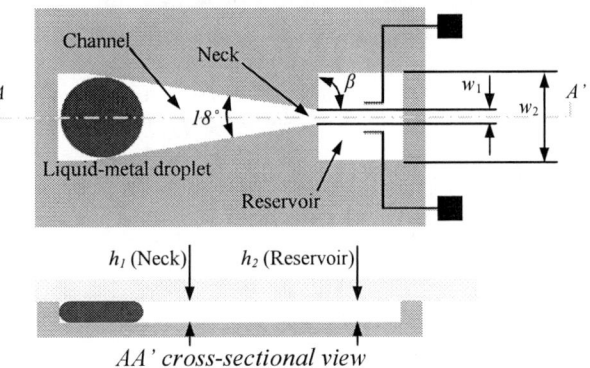

Figure 2: Notation of our design parameters

978-1-4244-2977-6/09 $25.00 © 2009 IEEE   789

different channel sizes in the flow path, the difference in capillary forces are computed to estimate the required pressure to move the liquid in the channel [11, 12]. However, since our device uses the modified surface which provides less CAH than the flat surface, an additional characteristic constant (i.e., C_c) is introduced to represent the surface property of the modified surface.

Eq. (1) shows the relationship between the threshold-level (i.e., expressed as number of g-value) and the channel dimension for the device switching. It is derived and modified from the previous works [11, 12].

$$T_g = \left(\frac{2C_c\gamma_{LV}}{g}\right)\left(\frac{w_2h_2 - w_1h_1}{\rho V}\right)\left[\left(\frac{\cos\theta_a}{w_2} + \frac{\cos\theta_a}{h_2}\right) - \left(\frac{\cos(\theta_a + \beta)}{w_1} + \frac{\cos\theta_a}{h_1}\right)\right] \quad (1)$$

T_g: Threshold g-value	C_c: Characteristic constant
γ_{LV}: Surface tension	g: Gravitational acceleration
ρ: Density of droplet	V: Droplet volume
θ_a: Advancing contact angle	β: Expansion angle
w_1: Width of neck	w_2: Width of reservoir
h_1: Height of neck	h_2: Height of reservoir

The value of C_c depends mainly upon the CAH of the droplet on the surface which indicates the property of the modified surface that we use in our device. More detailed information about the parameters in Eq. (1) can be obtained from [11, 12].

Measurements of the CAH

In order to obtain the CAH value of our modified surface, the advancing contact angle (θ_a) and receding contact angle (θr) are measured using the drop-shape analyzer (DSA-100, Krüss Co.). The measurement was taken when the droplet starts to move on the surface with a given sliding angle.

Fig.3 shows the CAH comparison of the flat and the microstructured surfaces. The microstructures provide reduction of the droplet flow resistance causing easy movement of the droplet on the surfaces, thus a discrete condition of on and off-state during switching motion is possible.

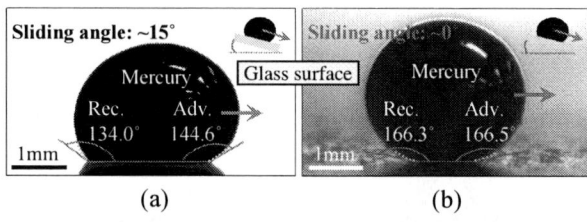

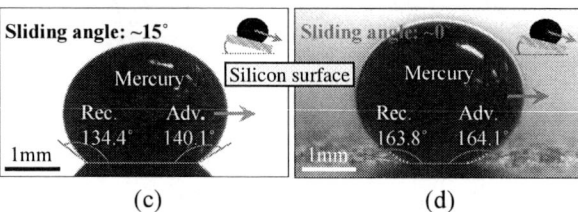

Figure3: Comparison of contact angle hysteresis (CAH)
(a) On flat glass (CAH=10.6°), (b) On microstructured
glass (CAH=0.2°), (c) On flat silicon (CAH=9.7°), and
(d) On microstructured silicon (CAH=0.3°)

FABRICATION

Two different substrates are used in our device. The channel is patterned on a silicon substrate using microfabrication techniques and a glass substrate is used for cover to visualize the droplet movement.

Fig.4 shows the fabrication process with corresponding images of the device. The channel (Fig. 4a) is patterned using DRIE and the selective surface modification is done sequentially using a sandblaster to form the microstructures inside the channel including the sidewall. Then, the cover (Fig.4b) is prepared having the patterned Ni/Cr (2000Å/500Å) sensing electrodes on a glass substrate (Pyrex 7740). A protective layer (e.g., PR) is then patterned on the substrate for the selective surface modification. Again, the sandblaster is used for the microstructure formation. Then, the device assembly (Fig.4c) is performed by placing the liquid-metal droplet (180nl) inside the microstructured channel and bonded using an UV curable adhesive (Polytech-3016).

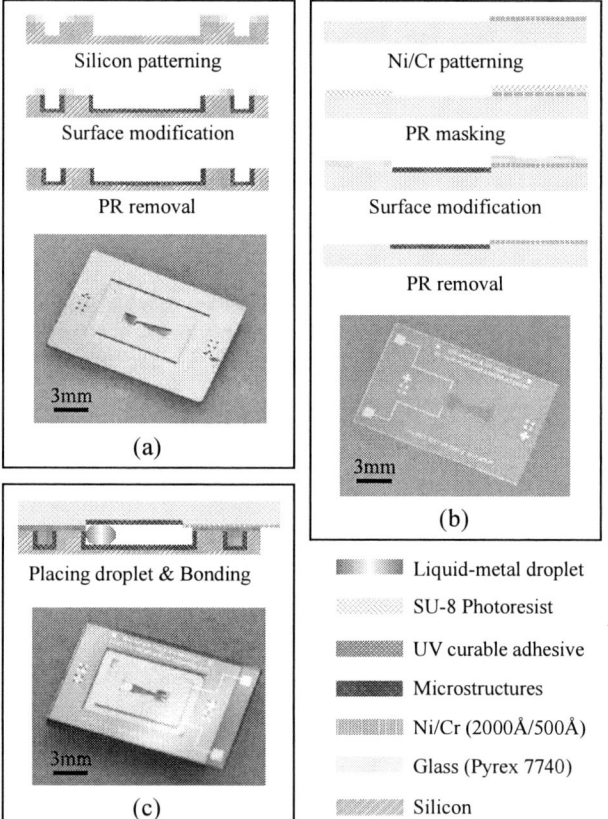

Figure 4: Fabrication process with corresponding
images of the device
(a) Channel, (b) Cover, and (c) Device assembly

EXPERIMENT

After the device fabrication, switching experiments are conducted using a custom-built testing stage (Fig.5). A reference accelerometer (PCB- 353B01), a signal circuit, a pushing-device and other electronics including NI-PXI 4472 are used to measure the given acceleration and the position of the droplet is taken by an optical camera.

Our current testing devices are fabricated based on the parameter values in Table 1. Four different widths of the neck area are considered, while other parameters are fixed,

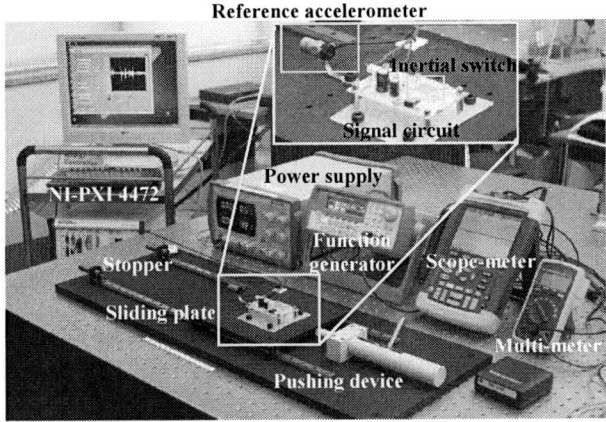

Figure 5: Custom-built testing setup

Table 1: Parameter values for testing devices

Parameters	values
Volume of droplet	$V = 180\text{nl}$
Surface tension	$\gamma_{LV} = 0.484$ N/m
Density of droplet	$\rho = 13546$ kg/m^3
Width of neck	$w_1 = 200, 300, 400, 500\mu\text{m}$
Width of reservoir	$w_2 = 1200\mu\text{m}$
Height of neck & reservoir	$h_1 = h_2 = 300\mu\text{m}$
Advancing contact angle	$\theta_a = 164.1°$
Expansion angle	$\beta = 90°$

to verify the relationship between the threshold-level and the neck dimension.

During device testing, the fabricated inertial switch is placed on the sliding plate and the sensing electrodes on the device are connected to the copper wires from signal circuit using a conductive silver paste. Once the device is mounted, the sliding plate is accelerated by the pushing-device. When the inertial switch is collided with stopper, the liquid-metal droplet in the channel moves to on-state and the sensing electrodes are closed sending the signal to the PXI. Then, the corresponding acceleration is measured and recorded using PXI. The detailed testing process is summarized in Fig. 6.

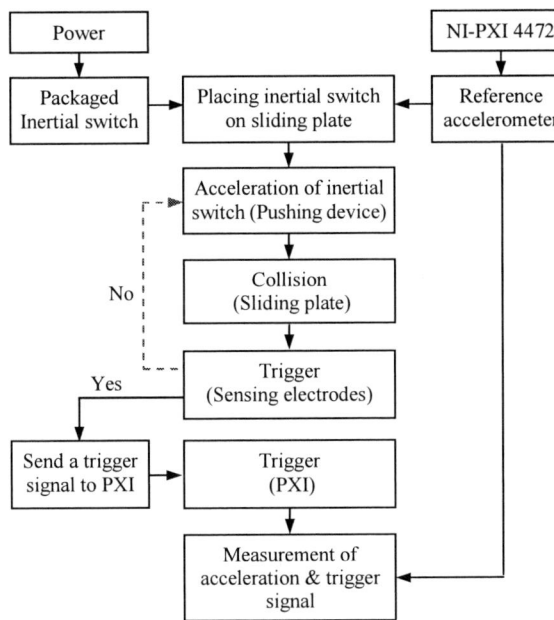

Figure 6: Testing process diagram for inertial switch

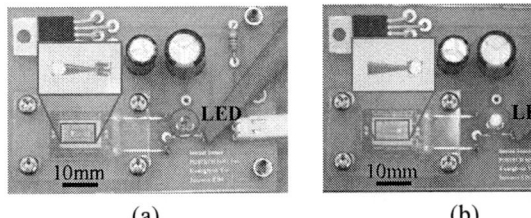

Figure 7: Optical images of the droplet positions (a) Off-state and (b) On-state

Fig. 7 shows the positions of the liquid-metal droplet before and after the switching test. Fig.7a shows the initial off-state and Fig.7b shows on-state under a given threshold acceleration (i.e., 8g).

RESULT AND DISCUSSION

After we obtained the corresponding threshold-level of accelerations for four different inertial switches, a plot showing the relationship between the threshold g-value and the neck dimension is realized. Fig. 8 shows the experimental results (dotted points) and the theoretical prediction of the threshold g-value with respect to the neck dimension. The theoretical curve was completed by obtaining the value of Cc from the experimental data. According to the calculation using the experiment data, Cc was ~0.3 and this value along with Eq. (1) provides well fitted curve to the experimental data points as shown in Fig. 8. Therefore, using this curve, we can predict the threshold g-value for various neck dimensions which can be used in many different applications with a specific threshold g-value requirement.

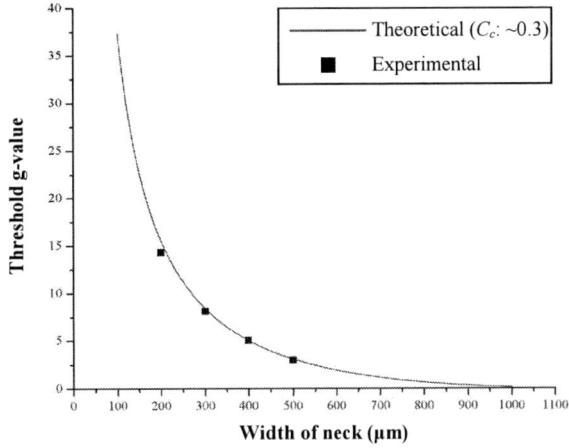

Figure 8: Threshold g-value vs. width of neck (Theoretical and experimental result)

CONCLUSION

A simple yet effective mechanism of the MEMS inertial switch has been introduced. Using the advantageous properties of the liquid-metal combining with the selective surface modification inside the channel, the movement of the liquid-metal droplet is enhanced to work as an active component in the inertial switch. Also, an equation was carried out to provide the relationship between the threshold g-value and various neck dimensions. Especially, the characteristic constant C_c was introduced to represent the surface property of the inertial switch which tells us how the droplet can move easily in the channel. According to our current design configuration, the value of C_c showed ~0.3. The variation in the magnitude of C_c indicates more or less the threshold g-values with a given neck dimension as shown in Fig. 9.

We are currently working on further experiments for the detailed characterization of our device and improving device design as well as the fabrication process to optimize the device performance.

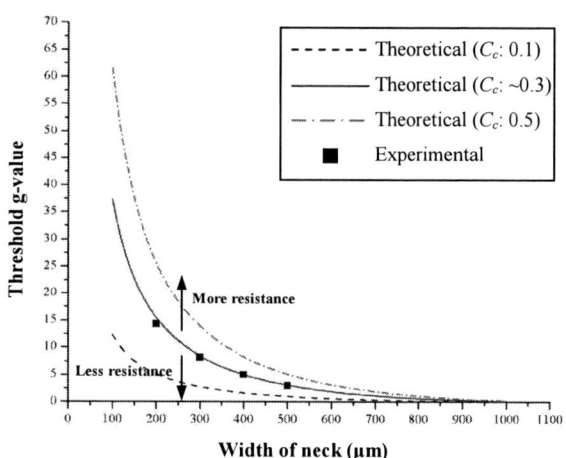

Figure 9: Threshold g-value vs. width of neck
(with various values of C_c)

ACKNOWLEDGEMENTS

This work was partially supported by "a grant to MEMS Research Center for National Defense funded by Defense Acquisition Program Administration," and "the Nano R&D Program grant (No. R01-2006-000-10585-0) from the Korea Science & Engineering Foundation (KOSEF)".

REFERENCES

[1] D. Frobenius, A. Zeitman, G. Hamel, "Microminiature Ganged Threshold Accelerometers Compatible with Integrated Circuit Technology", *IEEE Transaction on Electron Devices*, vol.19, pp. 37-40, 1972.

[2] S. Michaelis, H. Timme, M. Wycisk, J. Binder, "Additive electroplating technology as a post-CMOS process for the production of MEMS acceleration-threshold switches for transfortation applications", *J. Micromech. Microeng.* vol.10, pp. 120-123, 2000.

[3] J.Zhao, J. Jia, G. Chen, "A Novel MEMS Parallel-beam Acceleration Switch", in *Digest Tech. Papers Proceedings' 2nd Conference,* August, 2006, pp. 1-5.

[4] J. Mengjun, L. Xinxin, Y. Heng, "Micro-cantilever shocking-acceleration switches with threshold adjusting and 'on'-state latching functions", *J. Micromech. Microeng.* vol.17, pp. 567-575, 2007.

[5] C. Chan, "A Treatise on Crash Sensing for Automotive Air Bag Systems", in *Digest Tech. Papers Transaction on Mechatronics,* June, vol. 7, 2002, pp. 220-234.

[6] T. Matsunaga, M. Esashi, "Acceleration switch with extended holding time using squeeze film effect for airbag systems", *Sensors and actuators* vol.A100, pp. 10-17, 2002.

[7] J. Kim, W. Shen, L. Latorre, and C.-J. Kim "A Micromechanical Switch with Electrostatically Driven Liquid-Metal Droplet", *Sensors and actuator*, Vol. A97-98, pp. 672-679, 2002.

[8] C. H. Chen, D. Peroulis, "Liquid Metal Droplets for RF MEMS Switches", *in Digest Tech. Papers Silicon Monolithic Integrated Circuits in RF Systems'06 Conference.* January 18-20, 2006, pp. 4-7.

[9] J. Kim, C. J. Kim, "Nanostructured surfaces for dramatic reduction of flow resistance in droplet-based microfluidics", *in Digest Tech. Papers IEEE MEMS'02 Conference*, Las Vegas, January 20-24, 2002, pp. 479-482.

[10] H. Y. Kim, H. J. Lee, B. H. Kang, "Sliding of Liquid Drops Down an Inclined Solid Surface", *J. Colloid and Interface Science* vol. 247, pp. 372–380, 2002.

[11] H. Cho, H. Y. Kim, J. Y. Kang, T. S. Kim, "How the capillary burst microvalve works", *J. Colloid and Interface Science* vol. 306, pp. 379-386, 2007.

[12] M. R. McNeely, M. K. Spute, N. A. Tusneem, A. R. Oliphant, " Hydrophobic Microfluidics", *Proceeding of SPIE-Microfluidic Devices and Systems II*, vol. 3877, pp.210-220, 1999.

NOVEL MICRO CAPACITIVE INCLINOMETER WITH OBLIQUE COMB ELECTRODE AND SUSPENSION SPRING ALIGNED PARALLEL TO {111} VERTICAL PLANES OF (110) SILICON

Dae-Hun Jeong[1], Sung-Sik Yun[1], Myung-Lae Lee[2], Gunn Hwang[2], Chang-Auk Choi[2], and Jong-Hyun Lee[1]

[1]Shool of Information and Mechatronics, Gwangju Institute of Science and Technology (GIST), Korea (ROK)
[2]Next-generation I-MEMS team, Electronics and Telecommunications Research Institute (ETRI), Korea (ROK)

ABSTRACT

A novel high resolution micro capacitive inclinometer has been developed using (110) silicon. KOH crystalline wet etching was employed after silicon deep reactive ion etching (DRIE) to reduce morphologic defects on the sidewalls of oblique comb electrodes aligned parallel to vertical {111} plane. Suspension springs are also parallel to other vertical {111} plane to secure the width during KOH wet etching. The sensitivity (pF/°) was increased because the oblique comb electrodes change both the overlapped area and gap during operation. The capacitance changed from -0.8 to 0.8 pF for -90°–90° and resolution was estimated at 0.18° or less for ±80°.

INTRODUCTION

For many years, Microelectromechanical systems (MEMS) sensors have been widely used to detect physical properties, such as pressure, acceleration, angular velocity, position for mechatronics applications like automobiles, navigation systems, medical devices, robot control systems, portable electronic devices, etc [1-2]. Recently, a micro capacitive inclinometer based on inertia sensors has been paid attention as an important component of the position sensing systems [3]. Comparing to conventional tilt sensors using the detection methods, such as heat convection or electrolytes [4-5], the capacitive detection method has some merits due to its low power consumption, long life time, simple fabrication, and insusceptibility to surrounding temperature.

The silicon deep reactive ion etching (DRIE) process has been conventionally employed to fabricate capacitive micro devices that require a high aspect ratio [6]. However, the vertical sidewalls fabricated using the silicon DRIE process show serious morphologic defects, such as footing, scalloping, verticality error, etc. These defects on sidewalls of the comb electrodes might incur a signal noise, which should degrade the stability or resolution of the capacitive sensors.

In this paper, we propose a novel micro capacitive inclinometer in which the morphologic defects are drastically removed. In order to reduce the morphologic defects, the sidewalls of the sensing electrodes, which are aligned parallel to vertical {111} plane of (110) silicon, are slightly etched in KOH solution after silicon DRIE process [7]. Suspension springs are also aligned parallel to another vertical {111} plane of (110) silicon to secure the width during a crystalline wet etching in KOH solution. To prevent inversion of the capacitance change for the negative (-) inclination angle, oblique comb electrodes are designed with the optimized gaps between movable and stationary electrodes. The fabricated inclinometer is experimentally investigated in terms of capacitance change and resolution is discussed based on temporal variation of the capacitance.

DESIGN

A micro capacitive inclinometer consists of proof mass, oblique comb electrodes, and suspension springs. Fig. 1 shows the schematic of the proposed inclinometer. Because oblique comb electrodes and suspension springs are aligned parallel to two vertical {111} planes of (110) silicon that show negligible etch rate, they can be etched in KOH solution with very small reduction of width.

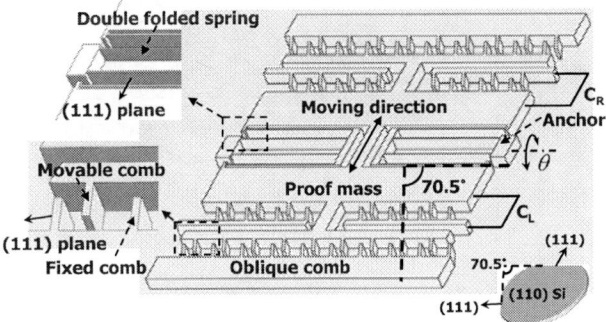

Figure 1: Schematic of the proposed micro capacitive inclinometer with oblique comb electrodes and suspension springs parallel to two vertical {111} planes of (110) silicon.

Operational principles

Operational principles of the proposed inclinometer are shown in Fig. 2. The gravitational force acting on the proof mass can be decomposed into the in-plane and out-of-plane components. When the sensor is inclined, the in-plane component of the gravitational force induces the in-plane motion of the proof mass with respect to inclination angle as shown in Fig. 2(a). The displacement of the proof mass is detected by capacitance change between movable and stationary comb electrodes.

Because oblique comb electrodes are slanted to the suspension springs by 70.5°(φ), capacitance changes depending on both overlapped area and gaps between movable and stationary comb electrodes as shown in Fig. 2(b). This will lead to the higher sensitivity (pF/°) compared with that of normal comb electrodes ($\varphi = 90°$).

Design of oblique comb electrodes

Because the oblique comb electrodes are slanted to the suspension springs, the mechanism for the capacitance change is different from that of normal comb electrodes. For instance, the capacitance increases for the overlapped areas between movable and stationary comb electrodes (x_0) both on the left and right sides when the inclinometer is inclined to the positive (+) inclination angle. On the contrary, the capacitance on left gap decreases (increase of d_{0L}) but the capacitance on right gap increases (decrease of d_{0R}) on the same inclination angle. This mechanism will increase capacitance change for a higher sensitivity.

978-1-4244-2977-6/09 $25.00 © 2009 IEEE

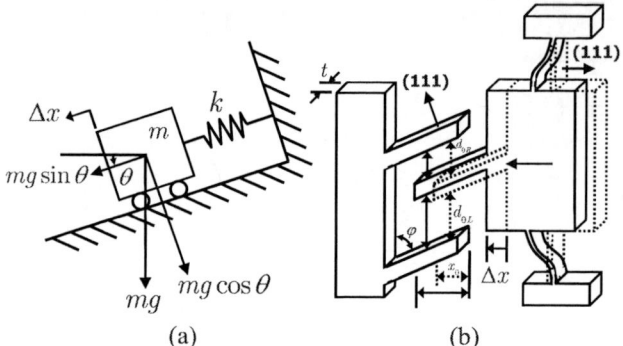

(a)   (b)

Figure 2: Schematics of operational principles for the proposed micro capacitive inclinometer: (a) components of the gravitational force acting on the proof mass, and (b) variations of the overlapped area and gaps between movable and stationary oblique comb electrodes.

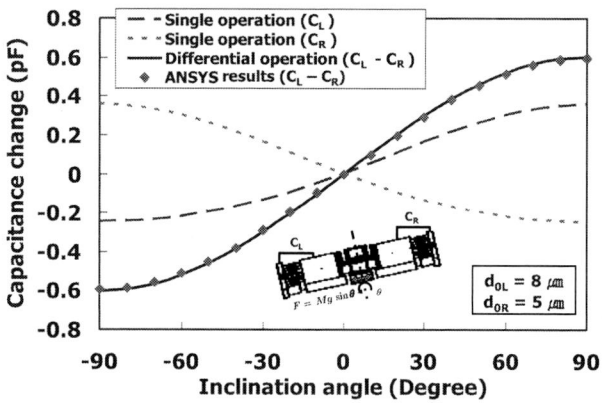

Figure 3: Theoretical and simulation results of capacitance change with respect to inclination angles in single and differential operations.

Oblique comb electrodes with different gaps on the left and right sides are employed not only to prevent inversion of the capacitance change for the negative (-) inclination angle but also to increase the sensitivity (pF/°) for the full range of the inclination angle.

If the oblique comb electrodes have the same gap for left and right gaps between movable and stationary comb electrodes (for example, $d_{0L} = d_{0R} = 5$ μm), an inversion can be seen in capacitance change for the negative (-) inclination angle. It is because the capacitance change in the oblique comb electrodes is very sensitive to gap closing (decrease of d_{0L}). On the contrary, there is no inversion in capacitance change when the left gap is larger than right gap ($d_{0L} > d_{0R}$). Particularly, when the right and left gaps are 5 μm and 8 μm, respectively, the inversion disappears in capacitance change but the sensitivity is saturated for the full range of the inclination angle.

Meanwhile, the proposed micro capacitive inclinometer can be operated in differential mode (C_L - C_R) because it has the independent comb electrode pair for each side (C_L and C_R). The ANSYS simulation indicates that the differential mode shows a higher sensitivity than single operation (C_L or C_R) as shown in Fig. 3.

FABRICATION

The proposed micro capacitive inclinometer was fabricated using (110) silicon as a structural active layer on a patterned (100) silicon substrate as shown in Fig. 4. The fan shape pattern was employed to find exact position of vertical {111} planes of (110) silicon. This will allow crystalline wet etching to be used for obtaining footing-free and scalloping-free sidewall in the oblique comb electrodes and suspension springs, which can be frequently noticeable in the presence of insulating layer beneath structural layer after silicon DRIE process.

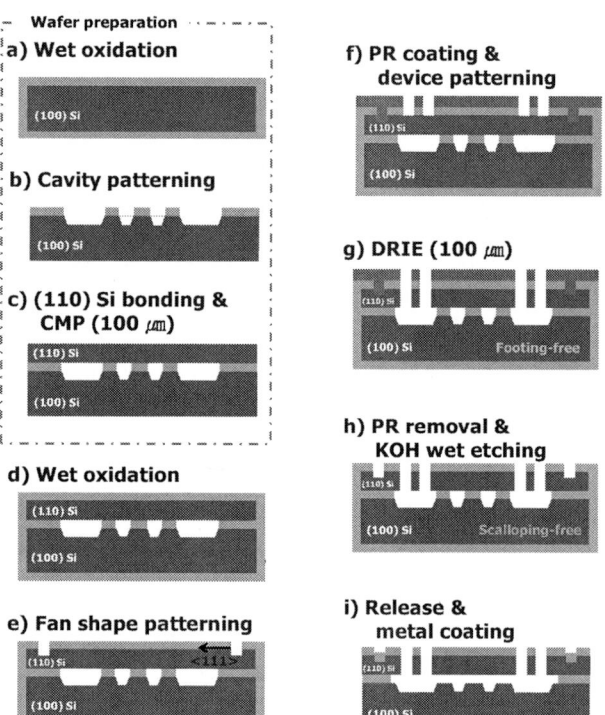

Figure 4: Fabrication sequence of the proposed micro inclinometer using (110) silicon.

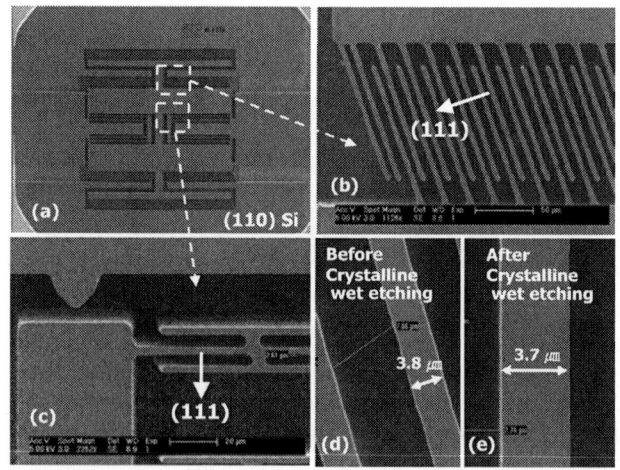

Figure 5: SEM images of the fabricated inclinometer: (a) top view, (b) oblique comb electrodes with different gaps on left and right sides, (c) suspension spring, (d) width of oblique comb finger before crystalline wet etching, and (e) width of oblique comb after crystalline wet etching.

978-1-4244-2977-6/09 $25.00 © 2009 IEEE

Fig. 5 shows the fabricated micro capacitive inclinometer before a crystalline wet etching treatment. The oblique comb electrodes and suspension springs were aligned parallel to two vertical {111} planes of (110) silicon. So a crystalline wet etching process can be done with a negligible reduction of the width of structures, the spring in particular, because it shows the slowest etch rate in <111> directions, as shown in Fig. 5(d) and (e).

Crystalline wet etching

In order to reduce the morphologic defects, which might incur a signal noise in capacitance measurements, on the sidewalls of the oblique comb electrodes, additional crystalline wet etching was employed using a KOH solution after silicon DRIE process. After KOH crystalline wet etching at 75°C for 10 minutes, the surface roughness of the oblique comb electrodes are remarkably reduced down to 20 nm (R_{RMS}) and the verticality was also improved up to 89.7° as shown in Fig. 6.

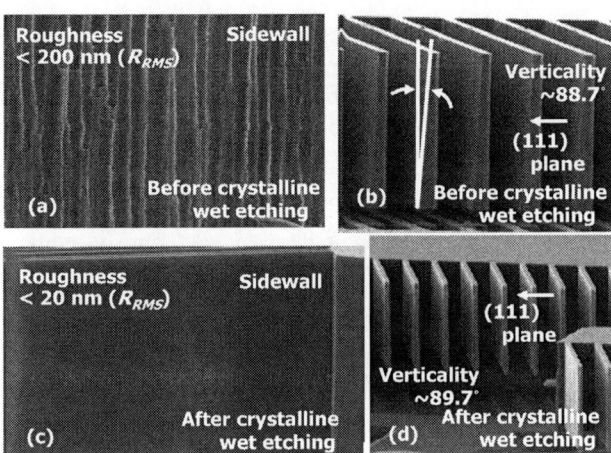

Figure 6: SEM images of the fabricated oblique comb electrodes: (a) scalloping on sidewall before crystalline wet etching, (b) verticality of sidewalls before crystalline wet etching, (c) smooth plane on sidewall after crystalline wet etching, and (d) verticality of sidewalls after crystalline wet etching.

EXPERIMENTS

To investigate performance of the fabricated inclinometer, the experimental setup consisting of 360° rotation stage, and LCR meter (Agilent 4284A precision LCR meter) was constituted as shown in Fig. 7. The fabricated inclinometer mounted on PCB with Au wiring was connected to LCR meter using soldering. Then PCB was attached to rotation stage, and LCR meter was connected to personal computer to record capacitance change with respect to inclination angle.

The capacitance change of the fabricated inclinometer was experimentally evaluated with respect to inclination angle, and the corresponding resolution is estimated. Capacitance was measured at every inclination angle of 10° for the full range (-90° – 90°).

The capacitance of the fabricated inclinometer varies from -0.8 pF to 0.8 pF for the full range of inclination angle (-90° – 90°) in a differential mode as shown in Fig. 8(a). The inclinometer shows larger capacitance change than designed value because the width of suspension springs (w_s) smaller than designed one after silicon DRIE process.

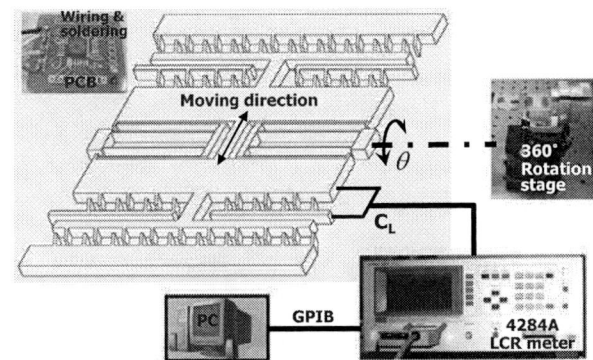

Figure 7: Schematic of the experimental setup for the capacitance measurements.

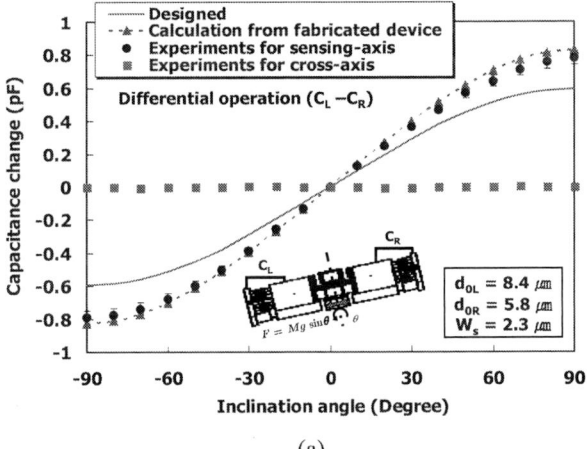

(a)

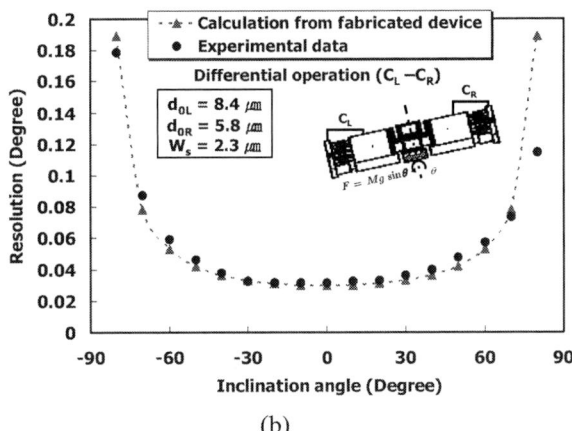

(b)

Figure 8: Experimental results of the fabricated inclinometer: (a) capacitance change in differential operation, and (b) estimated resolution with respect to inclination angle.

Meanwhile, the morphologic defects are very small in the sidewall of the oblique comb electrodes and suspension springs. This can explain the fact that experimental result matches very well with theoretical calculation based on dimensions of fabricated device measured by confocal microscopy (NanoFocus™).

In addition, cross-axis effect was measured at 5 % or smaller so that it is very acceptable value comparing to commercialized micro inclinometer. The resolution of the fabricated inclinometer is estimated based on signal-to-noise ratio (SNR) using measured capacitance change and temporal variation of the capacitance. Temporal variation of the measured capacitance was as low as 0.4 fF during experiments leading to a resolution of 0.18° or less for ±80° as shown in Fig. 8(b). The maximum and minimum resolutions of the fabricated inclinometer were shown at 0° and ±80°, respectively.

CONCLUSION

A novel micro capacitive inclinometer with oblique comb electrodes and suspension springs parallel to two vertical {111} planes of (110) silicon was proposed in this paper. In order to reduce morphologic defects on the sidewalls of oblique comb electrodes, a crystalline wet etching was employed with robust suspension springs.

Two kinds of structures were aligned parallel to two vertical {111} planes of (110) silicon, against crystalline wet etching due to the slowest etch rate in <111> direction. The inclinometer was designed and fabricated with optimal dimensions on left and right gaps between movable and stationary oblique comb electrodes to prevent inversion of capacitance change for the negative (-) inclination angle and to increase sensitivity of the inclinometer.

The fabricated inclinometer was evaluated in terms of capacitance change and resolution. The inclinometer shows high sensitivity using oblique comb electrodes and high resolution through a crystalline wet etching process. Specifications of the fabricated inclinometer are summarized in Table 1. The proposed oblique comb electrode and suspension spring structures can be applied to other micro device, such as optical filter, micro sensors, and micro actuators.

Table 1: Specifications of the fabricated inclinometer.

Parameters	Value
Gap between the movable and stationary oblique comb electrodes	8.4 μm (d_{0L}) 5.8 μm (d_{0R})
Spring constant	0.97 *N/m*
Mass	372.07 μg
Measurement range	-90° − 90°
Capacitance change	-0.793 − 0.783 *pF*
Resolution	0.18° or less for ±80°

ACKNOWLEDGEMENT

This work was supported by the IT R&D program of MIC/IITA. [2006-S054-03, Development of CMOS based MEMS processed multi-functional sensor for ubiquitous environment]

REFERENCES

[1] S. Łuczak, W. Oleksiuk, and M. Bodnicki, "Sensing Tilt With MEMS Accelerometers", *IEEE Sensors Journal*, vol. 6, no. 6, pp.1669–1675, 2004.

[2] R. H. Dixon, and J. Bouchaud, "Markets and applications for MEMS inertia sensors", *Proc. of SPIE*, vol. 6113, pp. 611306-1–611306-10, 2006.

[3] J. Bergeron, and M. Looney, "Making MEMS accelerometers work in motion control", *Electronic Engineering Times*, Issue. 1487, pp. 35–36, 2007.

[4] S. Billat, H. Glosch, M. Kunze, F. Hedrich, J. Frech, J. Auber, H. Sandmaier, W. Wimmer, and W. Lang, "Micromachined inclinometer with high sensitivity and very good stability", *Sensors and Actuators A*, vol. 97–98, pp. 125–130, 2002.

[5] H. Jung, C. J. Kim, and S. H. Kong, "An optimized MEMS-based electrolytic tilt sensor", *Sensors and Actuators A*, vol. 139, pp. 23–30, 2007.

[6] J. Bhardwaj, and H. Ashraf, "Advanced silicon etching using high-density plasmas", *Proc, of SPIE*, vol. 2639, pp. 224-233, 1995.

[7] S.-S. Yun, S.-K. You, and J.-H. Lee, "Fabrication of vertical optical plane using DRIE and KOH crystalline etching of (110) silicon wafer", *Sensors and Actuators A*, vol. 128, pp. 387–394, 2006.

978-1-4244-2977-6/09 $25.00 © 2009 IEEE

A DIGITAL OUTPUT PIEZOELECTRIC ACCELEROMETER USING PATTENRNED PB(ZR,TI)O₃ THIN FILMS ELECTRICALLY CONNECTED IN SERIES

T. Kobayashi[1,3], H. Okada[1,3], T. Masuda[2,3], and T. Itoh[1,3]

[1]National Institute of Advanced Industrial Science and Technology (AIST), Tsukuba, Japan
[2]The University of Tokyo, Tokyo, Japan
[3]CREST-JST, Tokyo, Japan

ABSTRACT

We propose a digital output piezoelectric accelerometer for chicken health monitoring. The accelerometer has patterned Pb(Zr,Ti)O₃ (PZT) thin films electrically acc connected in series, which accompany CMOS switches. The elerometer converts acceleration into number of on-state CMOS switches, which can be called as "digital output". Since the accelerometer are based on piezoelectric effect, does not need voltage amplifier and A/D converter, ultra low power consumption is expected. We have fabricated the digital output piezoelectric accelerometer through sol-gel deposition of PZT thin films and MEMS micro fabrication process. The output voltage has increased with the number of connected PZT thin films, which indicate the potential to realize the proposed digital output piezoelectric accelerometer.

INTRODUCTION

Recently, there has been a growing interest in sensor network using MEMS-based wireless sensor nodes. The previous studies reported the MEMS-based sensor network integrated into textile-reinforced composite [1] and that for environmental sensor system [2].

As a new application of sensor network, we have been developing a chicken health monitoring system for global avian influenza [3]. The system monitors the body temperature and movement of chickens by using temperature sensors and accelerometers mounted on wireless sensor nodes. For stress-free monitoring for chickens, the wireless sensor nodes should weigh as low as possible, which means that buttery capacity is limited. Thus, the temperature sensors and accelerometers of ultra low power consumption are indispensable for the longer operation time of the wireless sensor node.

At present, various kinds of capacitive or piezoresistive accelerometers are commercially available. Such accelerometers, however, consume power even under stand-by state. On the other hand, piezoelectric accelerometers are principally zero power consumption. MEMS-based piezoelectric accelerometers using piezoelectric thin films have been reported to date [4-6]. However, the piezoelectric accelerometers usually need voltage or charge amplifier for practical application, which consume power. Moreover, the analog output from the accelerometers should be digitalized by A/D converter for wireless data transmission, which also consumes power. Then, we propose a "digital output" piezoelectric accelerometer which enables ultra low power consumption.

CONCEPT

Figure 1 illustrates the concept of the digital output piezoelectric accelerometer. The piezoelectric accelerometer has patterned PZT thin films ("PZT plate" hereafter) arrayed in parallel and electrically connected in series to amplify an output voltage. CMOS switches are connected at each end of some of the PZT plate. The CMOS switches turn on when the output voltage of the PZT plate are higher than the CMOS threshold voltage V_{th}. In Fig. 1, 9 pieces of PZT plates (PZT-1 to 9) are electrically connected in series and CMOS switches are connected to PZT-3, 6 and 9. Now let us assume that the sensitivity of the PZT plate and V_{th} are 0.07/G and 0.3V, respectively, and the output voltage of PZT plate is amplified linearly with the number of connected PZT plate. When 1 G of acceleration is applied, the output voltages at PZT-3, 6 and 9 are 0.21, 0.42, and 0.63 V, respectively. Then, CMOS switches 2 and 3 become on-state.

Table 1 summarizes the state of CMOS switches for various accelerations. If the on and off state of CMOS switches are defied as 0 and 1, acceleration is directly converted into digital output (e.g. 0.5 G: 001, 1 G: 011) without using A/D converter. So, we can call it as "digital output piezoelectric accelerometer". The digital output accelerometer cannot distinguish the small difference in acceleration; the digital output for 0.5 G acceleration is same as that for 0.6 G acceleration. But it can distinguish the state of chickens, which is sufficient for our application. Since the digital output piezoelectric accelerometer does not need stand-by power consumption, amplifier to amplify output voltage, and A/D converter to transmit data, the accelerometers of ultra low power consumption is expected.

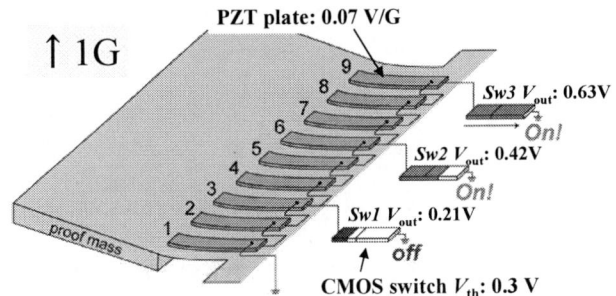

Figure 1: Concept of "digital output" piezoelectric accelerometer. PZT plate are arrayed in parallel and electrically connected in series to amplify output voltage. CMOS switches are connected at each end of some of the PZT plate.

978-1-4244-2977-6/09 $25.00 © 2009 IEEE

Table 1: State of CMOS switches at various accelerations.

Accel./G	V@Sw1	V@Sw2	V@Sw3	digital output	state
0	0/off	0/off	0/off	000	death
0.2	0.042/off	0.084/off	0.126/off	000	
0.5	0.105/off	0.21/off	0.315/On	001	small motion
0.6	0.126/off	0.252/off	0.378/On	001	
1	0.21/off	0.42/On	0.63/On	011	large motion
2	0.42/On	0.84/On	1.26/On	111	influenza

DESIGN

The digital output piezoelectric accelerometer is based on voltage amplification with the connected PZT plates. Thus, we have designed the accelerometer to investigate the voltage amplification. Figure 2 shows the design of the digital piezoelectric accelerometer. 10 pieces of PZT plate are arrayed in parallel and electrically connected in series. The cantilever is 3 mm long x 5 mm wide x 5 μm thick and the proof mass is 5 mm long x 5 mm wide x 0.4 mm thick, respectively. The PZT plates are 3 mm long x 0.24 mm wide and 3 μm thick. If acceleration a is applied to the gravity center of the proof mass which weigh m_{mass}, the output voltage V of each of the PZT plate is expressed as

$$V = \frac{\frac{d_{31}E_{PZT}m_{mass}a}{K_f}w_{TE}\left\{y_n - \left(h_{SiO_2} + h_{TE} + \frac{h_{PZT}}{2}\right)\right\}\left\{\frac{l_{TE}^2}{2} + \alpha l_{TE}\right\}}{\varepsilon_{PZT}\frac{w_{TE}l_{TE}}{h_{PZT}} + C_0} \quad (1),$$

where d_{31}, E_{pzt}, ε_{pzt} and h_{pzt} are transverse piezoelectric constant, Young's modulus, dielectric constant, and thickness of PZT thin films. K_f is the bending rigidity of the cantilever. y_n is the distance to the neutral axis of the cantilever, α is the distance between tip and the gravity center of the cantilever, C_0 is the capacitance of a CMOS switch, w_{TE}, h_{TE} and l_{TE} are width, thickness and length of the top electrode, h_{SiO2} is the thickness of SiO$_2$ as insulation layer. Equation (1) estimates the output voltage of each of the PZT plate under sinusoidal acceleration application to be 61.6 mV$_{pp}$/G. In order to turn on CMOS switch with 0.3 V of V_{th}, at least 10 pieces of PZT plate should be connected in series.

DEVICE FABRICATION

Multilayer deposition

The designed piezoelectric accelerometers were fabricated from the multilayers of Pt/Ti/PZT/Pt/Ti/SiO$_2$ deposited on SOI wafers (structural Si: 5 μm, buried SiO$_2$: 1 μm, substrate Si: 400 μm) through MEMS microfabrication process [7]. Figure 3 shows the schematic cross-section of the fabrication process. The deposition of the multilayers started from thermal oxidation of the SOI wafers followed by Pt/Ti bottom electrodes sputtering. Then, 3 μm thick (100)-oriented PZT thin films were deposited by sol-gel process [8]. The cross-section SEM image and XRD pattern is shown in Fig. 4(a) and (b). Finally, Ti/Pt/Ti top electrodes were sputtered.

Accelerometer fabrication

After multilayer deposition, Ti/Pt/Ti, PZT thin films and Pt/Ti were etched by Ar-ion etching (Ti/Pt/Ti and Pt/Ti) and wet etching (PZT) through mask 1-3. Then, 0.8 μm thick SiO$_2$ thin films as insulation layer were deposited by RF-magnetron sputtering at room temperature, which was followed by contact hole etching by reactive ion etching (RIE) with CHF$_3$ gas (mask 4). After that, wiring Pt (1 μm) with Ti adhesion layer was deposited by RF-magnetron sputtering at room temperature and etched by Ar-ion to connect the PZT plate in series (mask 5). Figure 4(c) shows the SEM image around the contact hole. We can see that the wiring Pt well covers the sidewall of the PZT thin films. Next, thermal SiO$_2$, structural Si and BOX were etched by RIE with CHF$_3$ gas (SiO$_2$) and SF$_6$ gas (Si) through mask 6. Finally, substrate Si and BOX were etched from backside to release cantilever and proof mass.

Packaging

After the fabrication process, the obtained piezoelectric accelerometers were annealed at 450 °C in air to recover the process-derived degradation [7]. Then, the accelerometers were bonded onto a metal package, and the pads and pins were connected by Au wire as shown in Fig. 4(d).

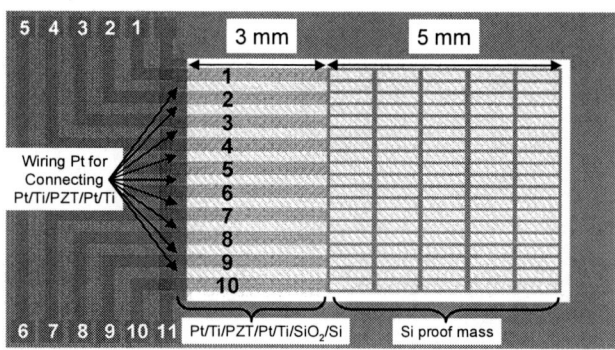

Figure 2: Design of digital output piezoelectric accelerometer. 10 pieces of PZT plate are arrayed in parallel and electrically connected in series.

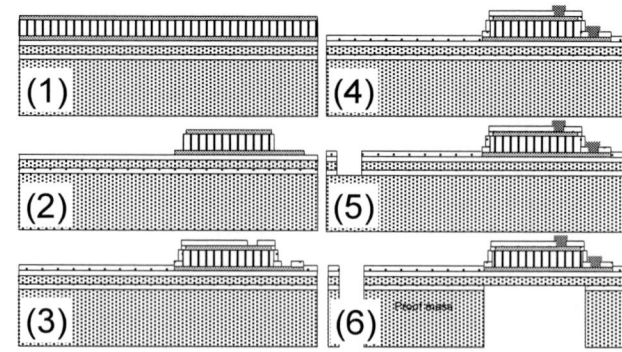

Figure 3: Schematic of the fabrication process. (1)Ti/Pt/Ti/PZT/Pt/Ti/SiO$_2$ deposition, (2)Ti/Pt/Ti/PZT/Pt/Ti etching (Mask 1-3), (3)SiO$_2$ sputtering and contact hall formation (Mask 4), (4)Wiring Pt sputtering and patterning (Mask 5), (5)Cantilever patterning (Mask 6), (6) Substrate etching from backside to release proof mass and cantilever (Mask 7).

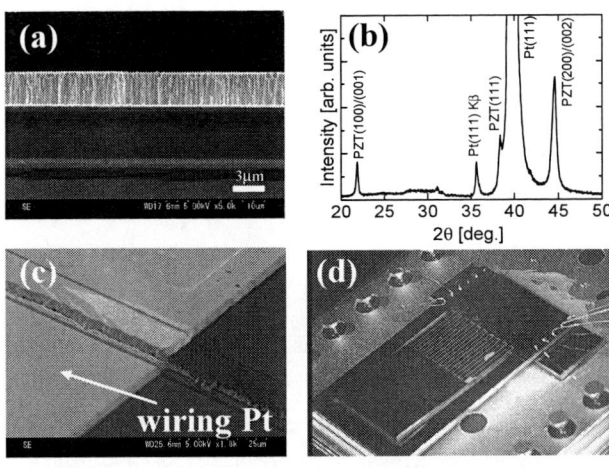

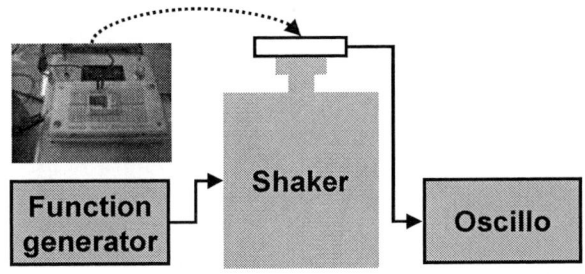

Figure 5: Measurement setup for characterization.

Figure 4: (a)Cross sectional SEM and (b) XRD pattern for PZT thin films. (c) SEM of fabricated accelerometer around the contact hole. (d) Overview of packaged accelerometer.

TEST AND RESULTS

We characterized the fabricated accelerometers by using the measurement setup shown in Fig. 5. The packaged accelerometer was inserted into a breadboard fixed on the stage of a shaker (Model-8100, Showa Sokki, Japan). The shaker was controlled by a function generator, and the output voltage of the accelerometer was directly measured by an oscilloscope. Note that the each of the PZT plates was biased at 25 V for 5min. as a poling treatment before measurement. The poling direction was from bottom electrode to top electrode.

Figure 6 shows the frequency spectra for PZT-10 in Fig. 2 measured at the acceleration of 0.051G. The resonant frequency of the accelerometer was measured to be 33 Hz. Since the output voltage under non-resonant condition is utilized for the practical application in the chicken health monitoring system, we measured the output voltage at 90 Hz, which is apart from the resonant frequency.

Figure 7 shows the peak to peak value of output voltage for PZT-10 in Fig. 2 as a function of acceleration. The accelerometer exhibits good linearity toward acceleration. Figure 8(a) shows the peak to peak values of output voltage for PZT-1 to PZT-10. The measured output voltage is only 10-20 % of the calculated value (61.6 mV_{pp}/G). The high resistivity of wiring Pt thin films and leak current in the PZT thin films may cause the loss in the generated charge. In order to reduce the loss, wiring metal with lower resistivity such as Al should be introduced and deposition condition of the PZT thin films should be improved.

The output voltage of the PZT plate at the edge of the cantilever (PZT-1, 2, 9, 10) is about twice as large as that at the center of the cantilever (PZT-3 to -8). The PZT plates at the edge of the cantilever were found to be compressed. The compression may increase the piezoelectric constant of the PZT plate at the edge of the cantilever leading to the larger output voltage.

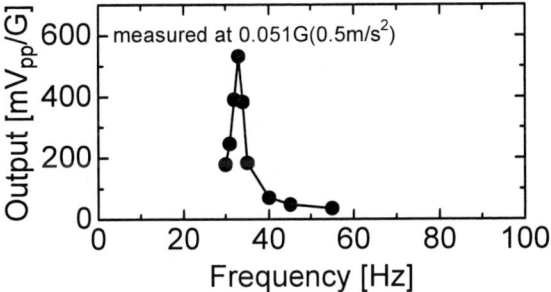

Figure 6: Frequency spectra for PZT-10 in Fig. 2 measured at acceleration of 0.051G. Resonant frequency of fabricated accelerometer was measured to be 33 Hz.

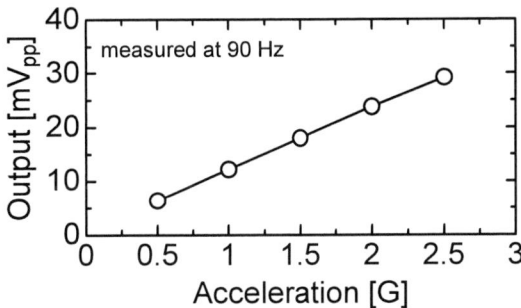

Figure 7: Peak to peak value of output voltage for PZT-10 in Fig. 2 as a function of acceleration.

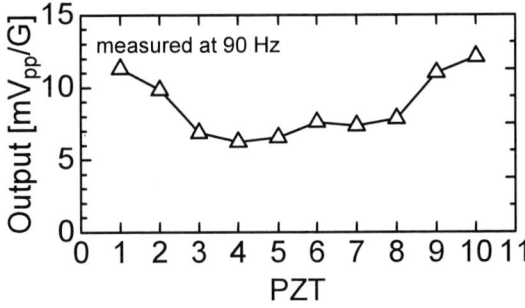

Figure 8: Peak to peak values of output voltage for PZT-1 to PZT-10.

978-1-4244-2977-6/09 $25.00 © 2009 IEEE

We have investigated the voltage amplification with the PZT plates electrically connected in series. Figure 9(a) shows the waveforms of the output voltage for 1, 2 and 10 connected PZT plates. The peak to peak output voltage as a function of the number of the connected PZT plate is shown in Fig. 9(b). The output voltage increases with the number of the connected PZT plates. However, the increase in the output voltage tends to saturate. Additionally, the phase of the output voltage delays with the number of the connected PZT plates. It is probable that the PZT and wiring Pt thin films form RC delayed circuit leading to the saturation and phase delay. This should be improved for the practical application.

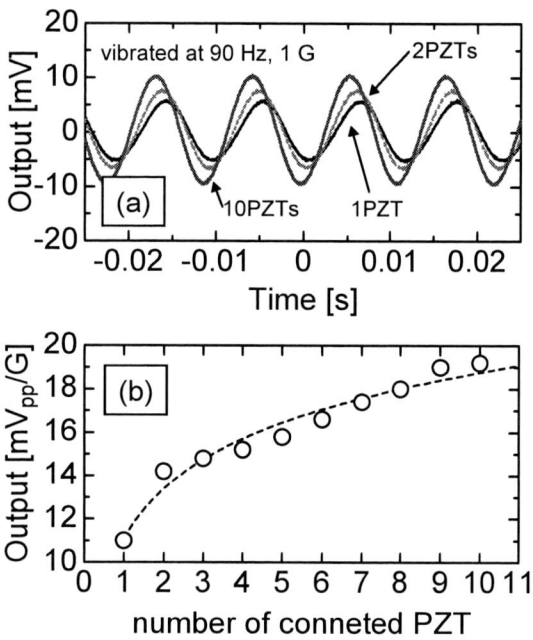

Figure 9(a): Waveforms of output voltage for 1, 2 and 10 connected PZT plates. (b) Peak to peak output voltage as a function of the number of connected PZT plate.

CONCLUSION

We have proposed a digital output piezoelectric accelerometer for the chicken health monitoring in the present study. The accelerometer has PZT plates electrically connected in series, which accompany CMOS switches. The accelerometer converts acceleration into number of on-state CMOS switches, which can be called as digital output. The digital output piezoelectric accelerometer was fabricated through sol-gel deposition of PZT thin films and MEMS micro fabrication process. The output voltage has increased with the number of the connected PZT plates. Although the increase in the output voltage tends to saturate with the number of the connected PZT plates, the results indicate the potential to realize the proposed digital output piezoelectric accelerometer.

ACKNOWLDGEMENT

This research is supported by fund of the CREST (Core Research for Evolutional Science and Technology) of JST (Japan Science and Technology Agency). The authors would like to thank Mr. Yushita for his contributions to the device fabrication during his master's course in the University of Tokyo.

REFERENCES

[1] K. U. Roscher, W. J. Fischer, J. Landgraf, G. Pfeifer, and E. Starke, "Sensor Networks for Integration into Textile-reinforced Composite", in *Digest Tech. Papers Transducers'07 Conference*, Lyon, June 10-14, 2007, pp. U803-U803.

[2] E. Yoon and K. S. Yun, "Development of a Wireless Environmental Sensor System and MEMS-based RF Circuit Components", in *Digest Tech. Papers Transducers'05 Conference*, Seoul, June 05-09, 2005, pp. 1981-1985.

[3] Y. Zhang, T. Ikehara, J. Lu, T. Kobayashi, M. Ichiki, T. Itoh, and R. Maeda, "Novel MEMS-based Thermometer with Low Power-consumption for Health-monitoring Network Application", *Proc. of SPIE*, vol. 6800, art. no. 68001V, 2008.

[4] Q. Zou, W. Tan, E. S. Kim, and G. E. Loeb, "Single- and Triaxis Piezoelectric-Bimorph Accelerometers", *J. Microelectromech. Syst.*, vol. 17, pp. 45-57, 2008.

[5] H. G. Yu, L. Zou, K. Deng, R. Wolf, S. Tadigadapa, and S. Trolier-McKinstry, "Lead Zirconate Titanate MEMS Accelerometer Using Interdigitated Electrodes", *Sensors and Actutators A*, vol. 107, pp. 26-35, 2006.

[6] K. Kunz, P. Enoksson, and G. Stemme, "Highly Sensitive Triaxial Silicon Accelerometer with Integrated PZT thin film detectors", *Sensors and Actuators A*, vol. 92, pp. 156-160, 2001.

[7] T. Kobayashi, M. Ichiki, R. Kondou, K. Nakamura, and R. Maeda, "Degradation in the ferroelectric and piezoelectric properties of $Pb(Zr,Ti)O_3$ thin films derived form a MEMS microfabrication process", *J. Micromech. Microeng.*, vol. 17, pp. 1238-41, 2007.

[8] T. Kobayashi, M. Ichiki, J. Tsaur, and R. Maeda, "Effect of multi-coating process on the orientation and microstructure of lead zirconate titanate (PZT) thin films derived by chemical solution deposition", *Thin Solid Films*, vol. 489, pp. 74-78, 2005.

ACCELERATION COMPENSATION OF MEMS RESONATORS USING ELECTROSTATIC TUNING

S. Yoneoka[1], G. Bahl[1], J. Salvia[1], K. L. Chen[1], A. B. Graham[1], H. K. Lee[1],
G. Yama[2], R. N. Candler[3], and T. W. Kenny[1]

[1]Stanford University, California, USA
[2]Robert Bosch LLC Research and Technology Center, California, USA
[3]University of California, Los Angeles, California, USA

ABSTRACT

A method of using electrostatic tuning to compensate the phase noise due to external vibrations in a MEMS oscillator is presented in this paper. An accelerometer measures the acceleration applied to a resonator, and a compensation signal generated by the accelerometer is added to the bias voltage. We achieve 91% reduction of the acceleration sensitivity for sinusoidal accelerations from 100 Hz to 300 Hz using a double-ended tuning fork resonator. This is the first demonstration of active acceleration compensation for MEMS resonators.

INTRODUCTION

MEMS resonators are being studied for use as timing references in a wide variety of electronic systems and are considered as replacements for quartz crystal resonators because of their small size and low cost. In many applications, frequency stability is the key feature, and our group has developed methods for encapsulation and temperature compensation that result in variations below 1 ppm over temperature and time [1-6]. For many applications, short-term variations in frequency, often referred to as phase noise, are the most challenging problem, and this is especially true for applications where vibrations are present, such as power tools, automobiles, helicopters, and many others. Inertial forces from vibrations can cause stresses inside resonant structures, which lead to frequency changes in quartz crystal resonators and MEMS resonators [7-10].

Several approaches to compensate the phase noise due to acceleration are proposed for quartz crystal resonators, including passive and active methods [11,12]. However, there is no study to demonstrate the acceleration compensation of MEMS resonators.

In this paper, we explore one additional advantage of MEMS – the ability to construct an accelerometer alongside a resonator and use the accelerometer signal to generate compensation for external vibrations to the resonator.

DESIGN AND FABRICATION

For this study we use a single anchored double-ended tuning fork resonator fabricated from single crystalline silicon and encapsulated at the wafer level (Fig. 1) [1]. The dimensions of the resonator beams are 200 μm long, 8 μm width, and 20 μm thickness. A bias voltage V_{bias} is applied to the resonator beams, which are electrostatically actuated by applying an AC signal v_{AC} to the input electrode. The output signal of the resonator is measured by the capacitive change between the beam and the output electrode. The resonator is oscillated by a closed loop oscillator circuit. The resonance frequency of the device is 1.5 MHz, and the quality factor is about 5000.

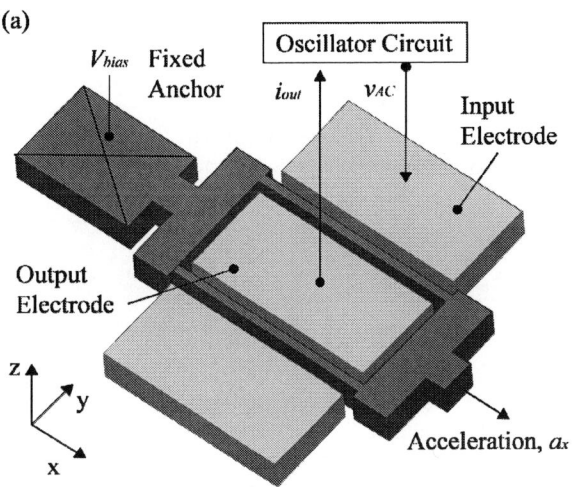

(a)

(b)

Figure 1: (a) Schematic of a single anchored double-ended tuning fork resonator. Acceleration is applied in the x-direction during experiments. (b) IR image of tested device. The resonator is sealed by an oxide-seal encapsulation technology [1].

METHOD

The resonant frequency of a resonator is shifted when an external force is applied to the resonant structure. If accelerations are applied to the resonator, an internal force is generated by the mass of the resonator itself, and the resonant frequency is shifted. In the case of our double-ended tuning fork resonator, the mass of the coupling bar that connects the beams generates axial stress in the resonant beams and causes frequency change. The effect of acceleration on the resonant frequency is measured by the acceleration sensitivity Γ_{accel}. The frequency change due to acceleration is described as

$$f = f_0\left(1 + \vec{\Gamma}_{accel} \cdot \vec{a}\right) \qquad (1)$$

where f_0 is a resonant frequency with no acceleration and

$\vec{a}$ is the acceleration that the resonator experiences.

The resonance frequency of an electrostatically actuated MEMS resonator can also be changed using the spring softening effect by varying the applied bias voltage [13]. The frequency change due to this electrostatic tuning is measured by the bias sensitivity Γ_{bias} [10]. The frequency change due to the bias voltage can be linearized

$$f \approx f_0(1 - \Gamma_{bias}\Delta V_{bias}) \qquad (2)$$

where ΔV_{bias} is a change of bias voltage.

Our double-ended tuning fork resonators have the largest acceleration sensitivity Γ_{accel} in the direction parallel to the resonant beams [9]. Therefore, we compensate the effect of acceleration in the x direction a_x in this work (see Fig. 1(a)). The change of frequency due to an acceleration a_x can be compensated by controlling the bias voltage as following.

$$\Delta V_{bias} = \frac{\Gamma_{accel}}{\Gamma_{bias}} a_x \qquad (3)$$

EXPERIMENT
Measurement of acceleration sensitivity

We used the dynamic measurement method described in Fig. 2 to measure an acceleration sensitivity Γ_{accel} and demonstrate the acceleration compensation [7-10]. A MEMS resonator and a commercial accelerometer (Analog Devices: ADXL321) were soldered on a printed circuit board (PCB), which was mounted to a shaker. The resonator was subjected to a sinusoidal acceleration in the x direction at varying frequencies f_v. The oscillator circuit was not mounted to the shaker, and steps were taken to

ensure that cables between the resonator and the oscillator circuit didn't affect the measurement. The acceleration of the resonator was measured by a laser Doppler vibrometer (LDV). We measured the output signal from the oscillator circuit using a spectrum analyzer and extracted the relative strength of the sideband peak L with respect to the strength of main peak to calculate acceleration sensitivity Γ_{accel}.

Characterization

To compensate the phase noise due to sinusoidal accelerations, an AC voltage v_c is added to a DC voltage V_{DC} from a voltage supply and applied to the resonator beam as a bias voltage. Since we generate the AC voltage v_c from the output signal of an accelerometer, it is necessary to adjust the amplitude and phase of v_c with respect to the acceleration. To determine an optimal amplitude and phase, acceleration and bias sensitivity need to be measured for each resonator. Phase differences between the acceleration on the resonator and the output signal from the accelerometer also need to be characterized.

Bias and acceleration sensitivity are measured using the experimental setup in Fig. 2 (a). The bias sensitivity of our device is 111 ppm/V for a bias voltage of 42 V, and the acceleration sensitivity is 24.7 ppb/m/s^2. This acceleration sensitivity is larger than is expected from analytical calculations. We suspect that this is due to fabrication error, but we are still working to determine exactly why the acceleration sensitivity is larger than predicted by theory.

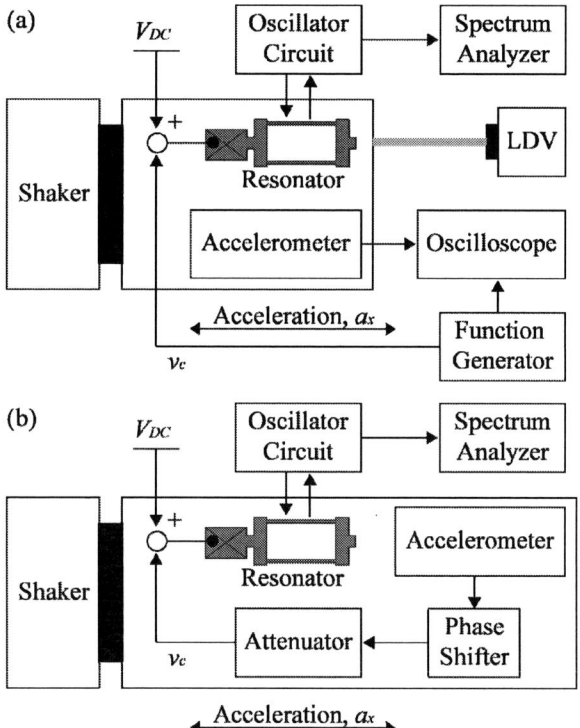

Figure 2: Schematic of the experimental setup. (a) Function generator and (b) accelerometer generates the compensation signal v_c.

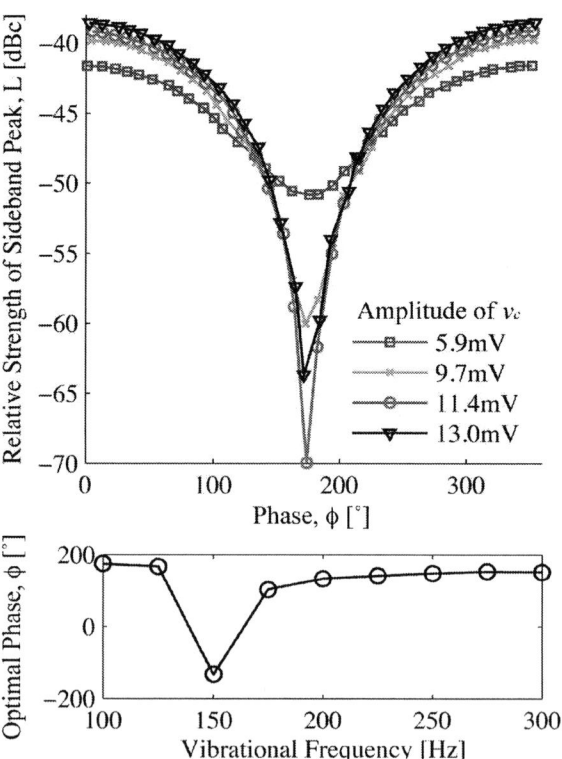

Figure 3: (a) Relative strength of sideband peak L of the compensated resonator with sinusoidal acceleration. L changes based on the phase and amplitude of the compensation signal. (b) Phase ϕ to achieve the maximum compensation for different vibration frequencies.

Phase differences between the optimal v_c and the acceleration on the resonator are determined by an exhaustive search. We use a function generator to create compensation signals v_c and measure the acceleration sensitivity of the compensated resonator as a function of phase ϕ, which is the phase difference between v_c and the output signal from the accelerometer (Fig. 2(a)). Phase ϕ is measured by an oscilloscope. We also repeat the same experiment with different amplitudes of v_c. Fig. 3 (a) shows the relative strength of sideband peak L of the compensated resonator as a function of the amplitude and the phase ϕ of compensation signal v_c. We observe that the relative strength of the sideband peak L varies with different amplitudes and phases. The optimal phase ϕ for the compensation signal v_c is different for different vibration frequencies (Fig. 3(b)), and we expect that those differences stem from the flexing PCB.

Acceleration compensation

Fig. 2(b) shows the experimental setup to prove our proposed method. In this configuration, we use the output signal from the accelerometer to generate the compensation signal v_c. The output signal of the accelerometer goes through a signal conditioning circuit on the same PCB to adjust desired amplitude and phase ϕ of compensation signal. Those values are set manually based on previous experimental result using a function generator. For the electronics, we implement the phase shifter with an all-pass filter and the attenuator with voltage divider. Fig. 4 shows a picture of our compensation system. The accelerometer is aligned with the resonator.

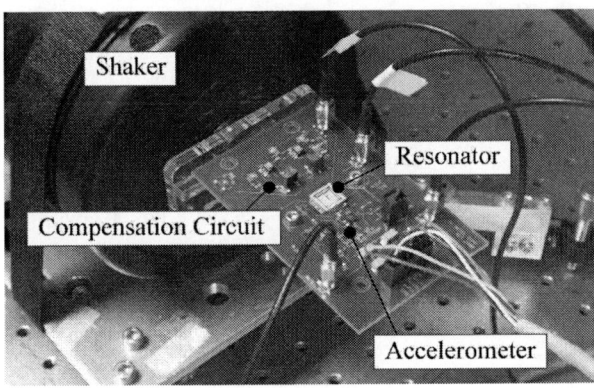

Figure 4: Picture of experimental setup. A resonator, an accelerometer, and signal conditioning circuits are implemented on a PCB board.

RESULTS AND DISCUSSION

Our experiment for acceleration compensation is performed using the output signal from an accelerometer (Fig. 2 (b)). The transfer function of the signal conditioning circuit is set manually to achieve the optimal amplitude and phase of v_c based on the previous experiment (Fig. 2(a)). Fig. 5 shows a comparison of the output spectrums from an uncompensated resonator and a compensated resonator for a 100 Hz sinusoidal acceleration with 39.5 m/s² amplitude. We observe a 23 dB reduction of the sideband peak using our compensation method. Fig. 6 shows a comparison of the acceleration sensitivity of the resonator with and without compensation

for accelerations with different vibration frequencies ($f_v = 100 - 300$ Hz, 39 m/s² amplitude). The acceleration sensitivity is reduced from 24.7 ppb/m/s² to 2.13 ppb/m/s². This is a 91 % reduction of acceleration sensitivity. More precise control of attenuation and phase are needed to achieve further reduction of acceleration sensitivity.

This method can be applied to other types of electrostatically actuated MEMS resonator and, although not demonstrated in this work, this method would also work for static accelerations. Furthermore, MEMS resonators and MEMS accelerometers can be fabricated in the same die using the same materials and technology. This level of integration gives our method an advantage in terms of size, packaging, and complexity compared with similar vibration compensation systems for quartz oscillators.

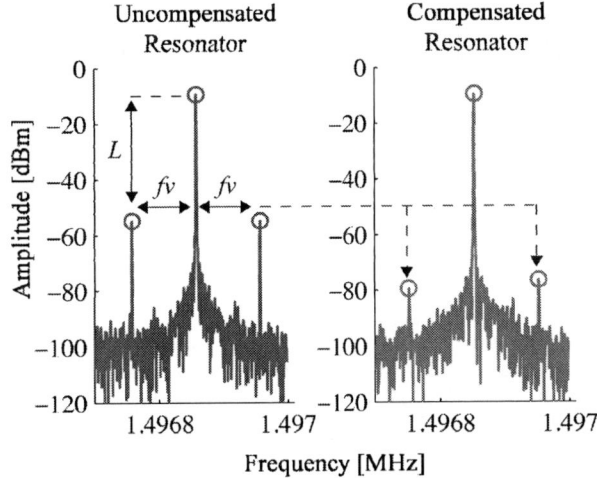

Figure 5: Output spectrum of uncompensated and compensated resonator with a sinusoidal acceleration ($f_v = 100$ Hz, 39.5 m/s² amplitude). The amplitude of the side band is reduced about 23 dB by our method.

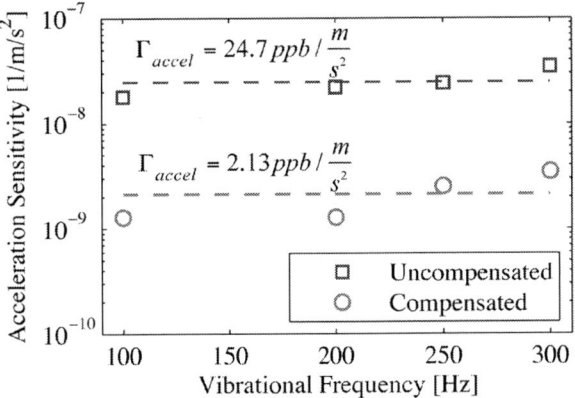

Figure 6: Comparison of acceleration sensitivity between uncompensated and compensated resonator for a sinusoidal acceleration with 39 m/s² amplitude.

CONCLUSION

In this paper, we demonstrate an active compensation method to reduce the phase noise of a MEMS resonator due to external vibrations with electrostatic tuning. Using the

978-1-4244-2977-6/09 $25.00 © 2009 IEEE

output signal from an accelerometer to generate a compensation signal, we achieve a 91% reduction of acceleration sensitivity for sinusoidal accelerations from 100 to 300 Hz. Future work will focus on the integration of a resonator and an accelerometer in the same encapsulation technology.

ACKNOWLEDGEMENTS

This work was supported by DARPA HERMIT (ONR N66001-03-1-8942), the Bosch Palo Alto Research and Technology Center, the National Users Network facilities funded by the National Science Foundation under award ECS-9731294, and NDSEG fellowship support for J. Salvia. Authors would like to thank Dr. B. Kim, Dr. M. A. Hopcroft, Prof. C. T.-C. Nguyen, Dr. J. R. Vig, Dr. A. Lal, Dr. A. Partridge and Prof. R. T. Howe for their supports.

REFERENCES

[1] A. Partridge, A. E. Ross, T. W. Kenny, M. Lutz, "New thin film epitaxial polysilicon encapsulation for piezoresistive accelerometers", *Proc. 14th IEEE MEMS 2001,* Interlaken, Switzerland, Jan. 21-25, 2001, pp. 54-59.

[2] R. N. Candler, W-T Park, G. Yama, A. Partridge, M. Lutz, T. W. Kenny, "Single wafer encapsulation of MEMS devices", *IEEE Trans. Adv. Packag.,* vol. 26, pp. 227-232, 2003.

[3] B. Kim, R. N. Candler, M. A. Hopcroft, M. Agarwal, W-T Park, T. W. Kenny, "Frequency stability of wafer-scale film encapsulated silicon based MEMS resonators", *Sens. Actuators, A,* vol. 136, pp. 125-131, 2007.

[4] J. Salvia, R. Melamud, S. Chandorkar, H. K. Lee, Y. Q. Qu, S. F. Lord, B. Murmann, T. W. Kenny, "Phase lock loop based temperature compensation for MEMS oscillators", *Proc. 22nd IEEE MEMS 2009,* Sorrento, Italy, Jan. 25-29, 2009.

[5] M. A. Hopcroft, M. Agarwal, K. K. Park, B. Kim, C. M. Jha, R. N. Candler, G. Yama, B. Murmann, T. W. Kenny, "Temperature compensation of MEMS resonator using quality factor as a thermometer",

Proc. 19th IEEE MEMS 2006, Istanbul, Turkey, Jan. 22-26, 2006, pp. 222-225.

[6] H. K. Lee, M. A. Hopcroft, B. Kim, J. Salvia, S. Chandorkar, T. W. Kenny, "Electrostatic–tuning of hermetically encapsulated composite resonator", *Proc. Hilton Head Workshop,* Jun. 1-5, 2008, pp. 48-51.

[7] R. L. Filler, "The acceleration sensitivity of quartz crystal oscillators: a review", *IEEE. Trans. Ultrason. Ferroelectr. Freq. Control,* vol. 35, pp. 297-305, 1988.

[8] J. R. Vig, C. Audoin, L. S. Cutler, M. M. Driscoll, E. P. EerNisse, R. L. Filler, R. M. Garvey, W. J. Riley, R. C. Smythe, R. D. Weglein, "Acceleration, vibration and shock effects-IEEE standards project P1193", *46th Ann. Freq. Control Symposium,* May 27-29, 1992, pp. 763-781.

[9] M. Agarwal, K. K. Park, S. A. Chandorkar, R. N. Candler, B. Kim, M. A. Hopcroft, R. Melamud, T. W. Kenny, B. Murmann, "Acceleration sensitivity in beam-type electrostatic microresonators", *Appl. Phys. Lett.,* vol. 90, pp. 4103-05, 2007.

[10] M. Agarwal, K. K. Park, M. Hopcroft, S. Chandorkar, R. N. Candler, B. Kim, R. Melamud, G. Yama, B. Murmann, T. W. Kenny, "Effects of mechanical vibrations and bias voltage noise on phase noise of MEMS resonator based oscillators", *Proc. 19th IEEE MEMS 2006,* Istanbul, Turkey, Jan. 22-26, 2006, pp. 154-157.

[11] V. R. Rosati, R. L. Filler, "Reduction of the effects of vibration of SC-cut quartz crystal oscillator", *Proc. 35th Ann. Freq. Control Symposium,* May 27-29, 1981, pp. 117 – 121.

[12] J. M. Przyjemski, "Improvement in system performance using a crystal oscillator compensated for acceleration sensitivity", *32nd Ann. Freq. Control Symposium,* May 31, 1978, pp.426-431.

[13] G. K. Ho, K. Sundaresan, S. Pourkamali, F. Ayazi, "Temperature compensated IBAR reference oscillators", *Proc. 19th IEEE MEMS 2006,* Istanbul, Turkey, Jan. 22-26, 2006, pp. 910-913.

978-1-4244-2977-6/09 $25.00 © 2009 IEEE

DESIGN AND IMPLEMENTATION OF A NOVEL CMOS-MEMS SINGLE PROOF-MASS TRI-AXIS ACCELEROMETER

Chih-Ming Sun[1], Ming-Han Tsai[1], and Weileun Fang[1,2]
[1]Nanoengineering and Microsystem Institute, [2]Power Mech. Eng. Dept,
National Tsing Hua University, Hsinchu, Taiwan

ABSTRACT

This study presents a novel single proof-mass tri-axis capacitive CMOS MEMS accelerometer to reduce the footprint of chip. A serpentine out-of-plane (Z-axis) spring is designed to reduce cross-axis error. A magnetic actuation for Z-axis self-test is also presented. The tri-axis accelerometer has been successfully implemented using TSMC 2P4M process and our in-house post-process. Measurement results show that sensitivities (non-linearity) of etch direction are 0.53mV/G (2.64%) of X-axis, 0.28mV/G (3.15%) of Y-axis, and 0.2mV/G (3.36%) of Z-axis. The cross-axis sensitivities range from 1% to 8.3%, and the measurement range is between 0.8~6G, respectively.

INTRODUCTION

The integration of smaller MEMS inertial sensor can find many applications in automobile industry, consumer electronics (e.g. digital camera, mobile phone, notebook, and video games), etc. Thus, the need of multi-axes (dual-axis or tri-axis) accelerometer is increased. Various batch fabrication technologies, such as surface micromachining [1], bulk micromachining [2], SOI process [3], and CMOS technology [4-6], have been reported to realize the MEMS accelerometers. As to the device design, the tri-axis MEMS accelerometer has been extensively investigated in [2-5]. In 2006, the integration of tri-axis MEMS accelerometer [7,8] has been exploited for video game motion-controller, named "Wii Remote" [9]. The multi proof-mass designs for in-plane and out-of-plane sensing are adopted in most reported monolithic multi-axis accelerometers [2-4]. The multi proof-mass accelerometer has the advantages of less cross-talk between sensing axes and relatively easier to design, however, larger chip size is required.

It is possible to reduce the size as well as the cost by implementing the tri-axes MEMS accelerometer using the single proof-mass design. For instance, as reported in [4,10], near 50% chip size can be reduced by the single proof-mass tri-axis accelerometer design. However, the complicated mechanical structures and sensing electrodes of the single proof-mass tri-axis accelerometer are critical design considerations. It is even challenge to design the single proof-mass tri-axis accelerometer using the existing CMOS process. This study presents a novel single proof-mass tri-axis capacitive CMOS MEMS accelerometer to reduce the footprint of chip. A serpentine out-of-plane (Z-axis) spring and sensing finger structures are designed to reduce the cross-axis sensitivity. In application, the accelerometer has been implemented using TSMC 0.35μm 2P4M CMOS process plus the present post-release technique.

DESIGN CONCEPT

Fig.1a shows the present tri-axis accelerometer consisting of one proof-mass, two supporting frames, and three sets of springs (named X-, Y-, and Z-spring) and sensing electrodes. These three sets of springs are designed only flexible in one axis. The exploded drawing in Fig.1b shows the proof-mass supported by Z-spring to form the Z-axis sensing element. Such Z-axis sensing element connects to a frame supported by Y-spring to form the Y-axis sensing element. Similarly, the Y-axis sensing element connects to a frame supported by X-spring to form the X-axis sensing element. These three parts form the three orthogonal sensing elements of the tri-axis accelerometer. The detail information of the tri-axis accelerometer design, including (1) Z-spring design, (2) sensing electrode for cross-axis decoupling design, and (3) out-of-plane magnetic actuation self-test design, is discussed as follows,

(a)

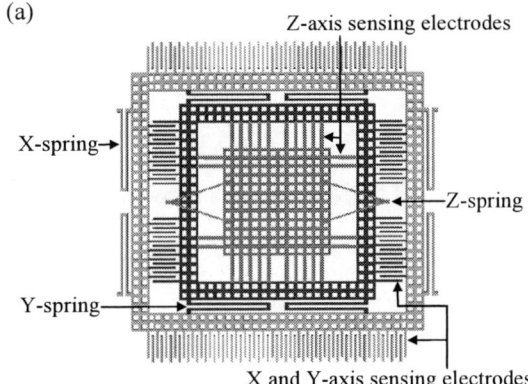

(b)

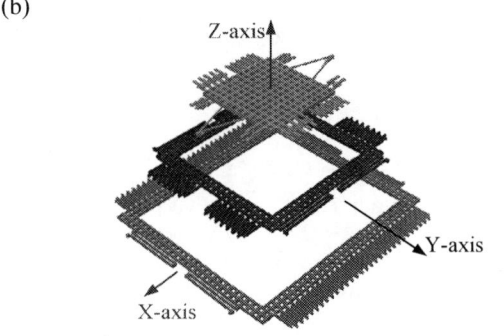

Figure 1 Design concept of single proof-mass tri-axis accelerometer, (a) top view, (b) exploded drawing of the sensing units for three different axes.

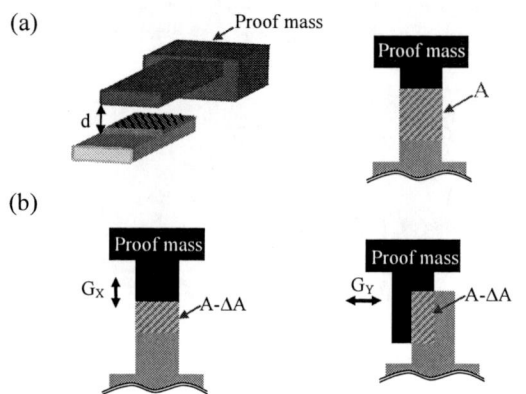

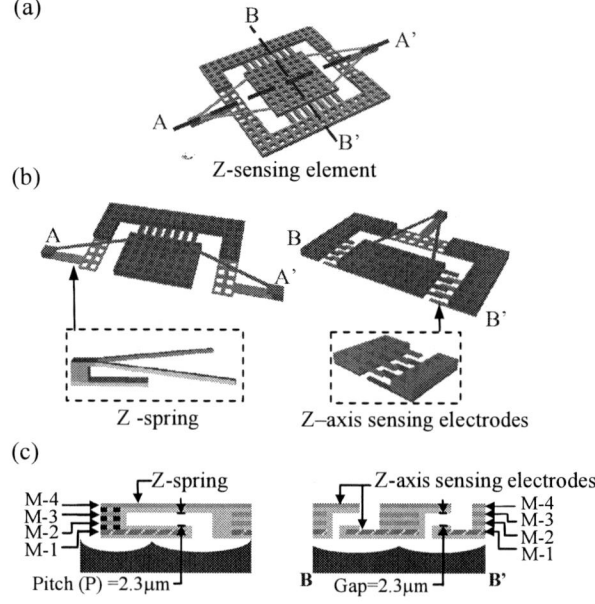

Figure 2 The overlap area of sensing electrodes (in dashed line) for Z-axis comb-finger, (a) ideal case without in-plane motions, and (b) real case with in-plane motions.

Figure 3 Design concept of Z-sensing element, (a) top view, (b) cross-section view of Z- spring and Z-axis sensing electrodes, and (c) structure layer of Z-spring and Z-axis sensing electrodes.

- **Z-spring design**

The cross-axis single coupling is the primary design consideration for the presented tri-axis accelerometer. The capacitance C between two parallel sensing electrodes varies with A/d, where A is the overlap area and d is the gap of these two electrodes. For instance, consider an ideal Z-axis sensing element in Fig.2a; the gap-closing capacitive sensing electrodes have a constant overlap area A (in dashed line). Thus, the capacitance change of Z-axis sensing electrodes is resulted from the variation of gap d caused by the inertial force in Z-axis. However, in the real application, the acceleration in X- or Y-axis will lead to the in-plane motion of the sensing electrodes, and further cause the change of

overlap area ΔA as shown in Fig.2b. As a result, the Z-axis capacitance change will couple with the in-plane accelerations.

The cross-axis signal coupling can be reduced by spring design. It is straightforward to design and fabricate desirable planar X- and Y-springs using micro fabrication processes. On the other hand, it is challenge to realize a promising Z-spring for sensing decoupling. This study presents a novel Z-spring design to reduce the cross-axis signal coupling for Z-axis sensing element, as shown in Fig.3. Figure 3a shows the Z-axis sensing element. The illustrations in Fig.3b respectively show the cross-section views of Z-spring (AA'-section in Fig.3a) and Z-axis sensing fingers (BB' -section in Fig.3a). The novel serpentine CMOS MEMS Z-spring is a structure folded in the out-of-plane direction, and designed to increase the stiffness ratio of in-plane/out-of-plane. As shown in Fig.3c, the pitch (P=2.3μm) of the folded structure is realized by the sacrificial metal etching. As indicated in Fig.3b-c, the top structure (M-4) of folded Z-spring connects to proof-mass, and the bottom structure (M-1) of folded Z-spring connects to Y-supporting frame. The joint of top and bottom structures is formed by metal layers (M-3 and M-2) and vias. In this study, the out-of-plane stiffness of Z-spring is two orders smaller than its in-plane stiffness. The coupling of sensing signals between Z-axis and the in-plane axes is significantly suppressed. Moreover, the folded Z-spring also has the advantages of wide linear loading range, and the release of thin film residual stress of CMOS layers.

- **Sensing electrode design for cross-axis decoupling**

In addition to spring design, this study also presents the sensing electrode design to reduce the cross-axis signal coupling. As shown in Fig.4a, this study proposed the design of Z-axis sensing electrodes with different in-plane dimensions. The edge length of the bottom electrode (dashed line) is 1μm smaller than the top electrode. Thus, the in-plane motion of the electrodes induced by the X- or Y-axis acceleration will not lead to the change of overlap area (the 0.5μm edge length can tolerate in-plane acceleration), as indicated in Fig.4b. Only the Z-acceleration will lead capacitance change. Similarly, the in-plane sensing electrode also employs the same design concept to reduce the cross-axis coupling of X-axis and Y-axis sensing signals.

- **Out-of-plane magnetic actuation self-test**

A novel magnetic actuation design in Fig.5a was implemented for self-test of Z-axis sensing element. As shown in Fig. 5b, the wire coils formed by two aluminum layers (M-2 and M-3) and the permanent magnet (outside of the ceramic housing) were exploited to produce a magnetic force to drive the Z-axis accelerometer. As compare with the self-test by electrostatic actuation, the magnetic actuation has the following advantages: (1) no complicated electrical routing required, (2) smaller driving voltage, and (3) no actuation electrodes required.

(a)

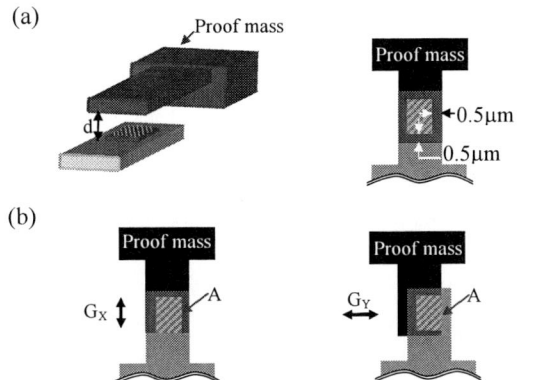

(b)

Figure 4 Overlap of the sensing electrodes(in dashed line) on comb-finger, (a) new Z-axis sensing electrodes design with different areas, and (b) the overlap sensing area A is not changed by the in-plane motions .

(a)   (b)

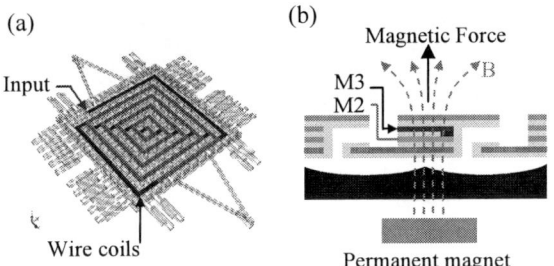

Figure 5 The Z-axis self-test design by magnetic actuation, (a) the magnetic coils prepared using two CMOS metal layers (M2 and M3 layers), and (b) magnetic force produced by magnetic coils and permanent magnet.

CMOS MEMS POST PORCESS

Figure 6 shows the present fabrication steps to realize a monolithic integrated single proof-mass tri-axis CMOS accelerometer. The two cross-section views respectively indicate the process steps for Z-spring (AA' cross section in Fig.3) and out-of-plane sensing comb-fingers (BB' cross section in Fig.3). The fabrication begins with the TSMC foundry service 0.35µm CMOS 2P4M process, as shown in Fig.6a. After that, the post CMOS processes consist of metal wet-etching, dielectric dry-etching, and bulk Si etching processes from the front-side of substrate are in Fig.6b-d.

As shown in Fig.6a, the pattern and stacking designs of metal films and tungsten vias are clearly observed. These metal films were exploited as sacrificial layers and sensing electrodes. The passivation layer was patterned to define the region for the following wet etching. As shown in Fig.6b, the H_2SO_4 and H_2O_2 solution was used to etch through the metal and tungsten vias [11, 12]. The passivation layer acted as a protection film during wet etching. In addition, the dielectric film distributed along the path of wet etching was also employed as the protection sidewall in this process.

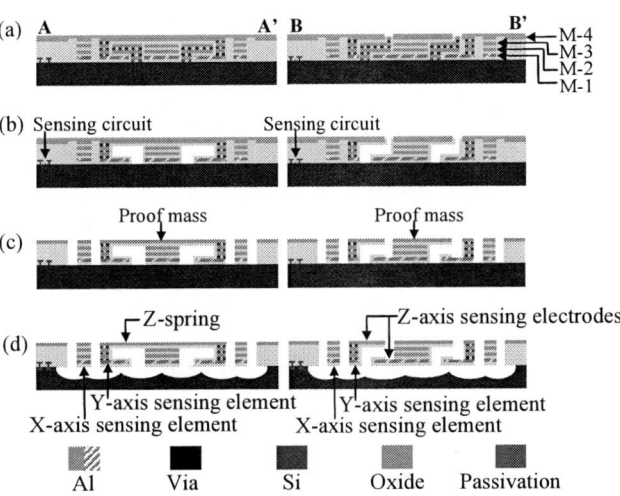

Figure 6 The fabrication steps of post-CMOS processes.

However, the dielectric blocks surrounded by sacrificial metal layers were fully released and then removed from substrate after the metal layers were etched away. Thus, (1) thickness and pitch of out-of-plane springs, and (2) the gap between top and bottom electrodes, were defined. After removing the top passivation layer, as illustrated in Fig.6c, the RIE was employed to etch dielectric SiO_2 layers (top metal layer M4 acted as etching mask) to define: (1) the shape of accelerometer, and (2) the gap between the stationary and the moving comb-fingers of in-plane accelerometers. Finally, the dry silicon isotropic etching using XeF_2 was used to release the accelerometers, as shown in Fig.6d.

EXPERIMENT AND RESULTS

This study has successfully demonstrated the tri-axis accelerometer using the process in Fig.6. The SEM micrograph in Fig.7 is a typical fabricated CMOS-MEMS single proof-mass tri-axis accelerometer (770µm×770µm). Near 50% chip footprint is reduced. The out-of-plane serpentine spring is clearly observed in zoom-in micrographs, as shown in Fig.7. The chip also contains CMOS circuits to form a standard two-stage cascode differential amplifier [13], which is employed to act as the capacitance readout IC. Fig.8 further shows the accelerometer chip after wire bonding and packaging. To calibrate the performance of present accelerometer, a commercial accelerometer was also packaged on the test board.

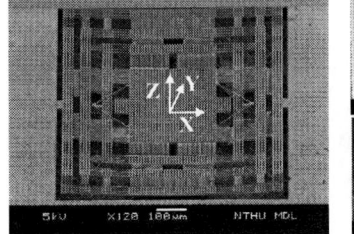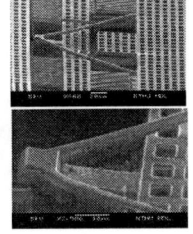

Figure 7 The SEM micrographs of single proof-mass tri-axis accelerometer, and the zoom-in of serpentine Z-spring.

978-1-4244-2977-6/09 $25.00 © 2009 IEEE

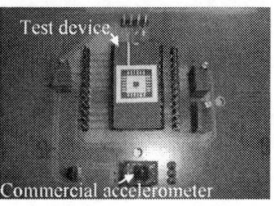

Figure 8 The accelerometer chip, (a) after wire bonding and packaged in ceramic housing, and (b) assemble on test board.

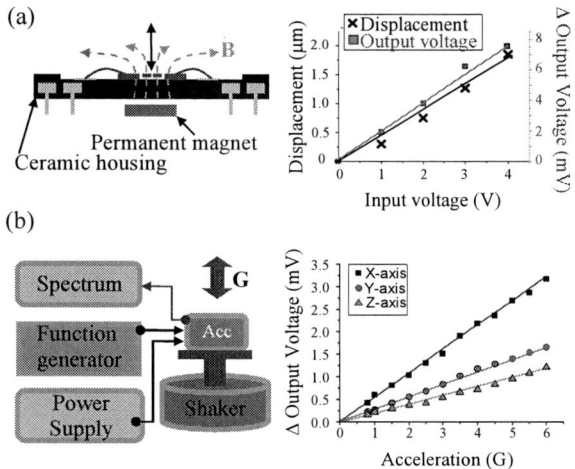

Figure 9 (a) The Z-axis self-test and the typical measured input voltage vs. static displacement and associated output sensing signal, and (b) the test setup to characterize the tri-axis accelerometer and the typical measured sensitivities.

Fig.9a shows experiment setup for the Z-axis self-test by magnetic actuation. The measurement results show the driving voltages (<5V) versus out-of-plane displacements as well as the associated output sensing-signal. The results demonstrated the feasibility of the magnetic actuation for Z-axis self-test. The test setup in Fig.9b was established to characterize the performance of accelerometer. The accelerometer was excited by the shaker. The misalignment of the shaker and accelerometer is used to measure the cross talk of the sensor. The excitation range provided by shaker is from 0.8G to 6G. Fig.9b shows the measured sensitivities of accelerometer are 0.53mV/G (X-axis), 0.28mV/G (Y-axis), and 0.2mV/G (Z-axis). The cross-axis sensitivity and non-linearity of accelerometer are respectively less than 8.3% and 4%. Table 1 summarizes the characteristics of accelerometer. Measurement results demonstrate the feasibility of the single proof-mass tri-axis accelerometer. Moreover, the serpentine Z-spring designs can reduce the cross-axis sensitivity.

CONCLUSIONS

This study presents a novel design of single proof-mass tri-axis capacitive CMOS MEMS accelerometer. To reduce the cross-axis sensitivity, the serpentine out-of-plane (Z-axis) spring and the special in-plane and out-of-plane sensing electrodes are designed. In addition, a magnetic actuation for

Z-axis self-test is also presented. The tri-axis accelerometer has been successfully implemented using TSMC 2P4M process and our in-house post-process. Measurement results show that sensitivities (non-linearity) in etch direction are: 0.53mV/G (2.64%) of X-axis, 0.28mV/G (3.15%) of Y-axis, and 0.2mV/G (3.36%) of Z-axis. The cross-axis sensitivity ranges from 1% to 8.3%, and the measurement range is between 0.8~6G, respectively. In summary, the presented single proof-mass tri-axis CMOS MEMS accelerometer offers reasonable sensitivities and acceptable cross-axis sensitivity.

Table 1 Characterizations of the tri-axis accelerometer.

Sensing-axis	Tri-axis accelerometer		
	X-axis	*Y-axis*	*Z-axis*
Sensitivity (mV/G)	0.53	0.28	0.2
Non-linearity (%)	2.64	3.15	3.36
Resonant Freq. (KHz)	7.89	16.15	9.65
Cross-axis Sensitivity X(%)		< 7.46	< 8.05
Cross-axis Sensitivity Y(%)	< 1		< 2.88
Cross-axis Sensitivity Z (%)	< 1	< 8.33	

ACKNOWLEDGEMENTS

This project was (partially) supported by the NSC of Taiwan under grants 96-2628-E-007-008-MY3 and 95-2221-E-007-068-MY3. The authors would like to thank TSMC Ltd., and National Chip Implementation Center (CIC), Taiwan, for supporting the IC Manufacturing. The authors would also like to thank National Tsing Hua Universuty., and National Chiao Tung University. for providing the fabrication facilities.

REFERENCE

[1] K.H.L. Chau, S.R. Lewis, Y. Zhao, R.T. Howe, S.F. Bart, and R.G. Marcheselli, *Sensor and Actuator A*, **54**, pp 472-476, 1996.

[2] K. Kwon, and S. Park, *Sensors and Actuators A*, **66**, pp.250-255, 1998

[3] J. Chae, H. Kulah, and K. Najafi, *J. of.MEMS*, **14**, pp 235-241, 2005.

[4] M.A. Lemkin, M.A. Ortiz, N. Wongkomet, B.E. Boser, and J.H. Smith, *44th ISSCC 1997*, pp.202-203, 457.

[5] H. Qu, D. Fang, and H. Xie, *IEEE Sensors J.*, **8**, 2008, pp.1511-1518.

[6] H. Luo, G. Zhang, L.R. Carley, and G.K Fedder, *J. of. MEMS*, **11**, 2002, pp.188-195.

[7] Analog Device Inc., ADXL 330, http://www.analog.com

[8] *ST* Microelectronics LIS3L02, http://eu.st.com

[9] Nintendo Inc., http://www.nintendo.com

[10] M.A. Lemkin, B.E. Boser, D. Auslander, and J.H. Smith, *Transducer '97*, pp.1185-1188. 1997.

[11] C. Wang, M.-H. Tsai, C.-M. Sun, and W. Fang, *J. Micromech. Microeng* , **17**, 2007, pp. 1275-1280.

[12] O. Paul, and H. Baltes, *Sensors and Actuators A*, **46/47**, no.1-3, 1995, pp. 143-146.

[13] J.M. Tsai, and G.K. Fedder, *IEEE MEMS'05*, 2005, pp. 630-633.

A LATCHING ACCELERATION SWITCH
WITH MULTI-CONTACTS INDEPENDENT TO THE PROOF-MASS

Z.Y. Guo, Z.C. Yang, L.T. Lin, Q.C. Zhao, H.T. Ding , X.S. Liu, X.Z. Chi, J. Cui, G.Z. Yan

Institute of Microelectronics, Peking University, Beijing, 100871 China

National Key Laboratory of Micro/Nano Fabrication Technology

ABSTRACT

An acceleration latching switch with independent multi-contacts is presented in this paper. All the contacts and their beams are independent to the proof-mass so as to prevent the contacts from the impact resulting from the rebound or vibration of the proof mass once the switch is latched. Moreover, multiple contacts are used in order to get high reliable contact, to lower the contact resistance and to increase the maximum allowable current. The switch was fabricated by low-cost process and tested. The latching shock is 4500G and the response time is less than 0.1ms. The contact resistance is no more than 5 ohms while the isolation resistance is more than 200M ohms and the maximum allowable current is up to 100mA.

INTRODUCTION

High reliability, high performance and low costs are very important for acceleration threshold switches which can be widely used in airbags, transportation systems and crash recorders [1] [2] [3]. Microelectromechanical systems (MEMS) technology makes it possible to realize low-cost acceleration switches with small size; many different acceleration MEMS switches have been reported [1-9]. Pull-in or snap through behavior of electrostatic force can be used to realize the needed acceleration thresholds [4]. However, it not only has caused failures of MEMS devices frequently but also has a risk of mis-operation by an electromagnetic noise [7] [8].

Mechanical threshold acceleration switches are immune to outside disturbances such as electro-magnetic noises and shocks of short duration [2][3][8]. Bistable beams are commonly used as latching mechanism but increase the difficulty of design [3] [9]. A mechanically

latching switch has been reported [2]. However, the proof mass and the contacts are incorporated, so the contacts are inevitably impacted by the rebound or vibration of the proof mass once the switch is latched. In reference [2], the contacts will be detached when the acceleration exceeds the threshold. This paper presents a low-cost acceleration latching switch with high performance and reliability. In our design, multi-contacts are used and at least two contacts of different electrodes keep contacted when the acceleration equals or exceeds the threshold. Furthermore, the contacts are independent to the proof mass, so the impact due to the rebound and vibration of the proof mass can be avoided and the contact reliability can be improved further.

DEVICE DESIGN
Structure and Working Principle

The schematic diagram of the switch is shown in Figure 1. It uses four groups of independent beams and contacts to realize multi-contacts independent to the proof mass. When the substrate get a downward acceleration along the substrate, the proof mass moves upwards and pushes contact 4 move in the same direction and contact 4 pushes contact 2 and contact 3 aside. Once the acceleration is equal to or more than the latching threshold, contact 4 will pass over contact 2 and contact 3 which will move back under the elasticity of their beams and contact 4 will be latched. The proof mass will be apart from contact 4 once the acceleration decreased. Contact 4 is designed to contact with contact 1, 2 and 3 at the same time so as to improve the contact reliability, to increase the maximum allowable current and to decrease the contact resistance.

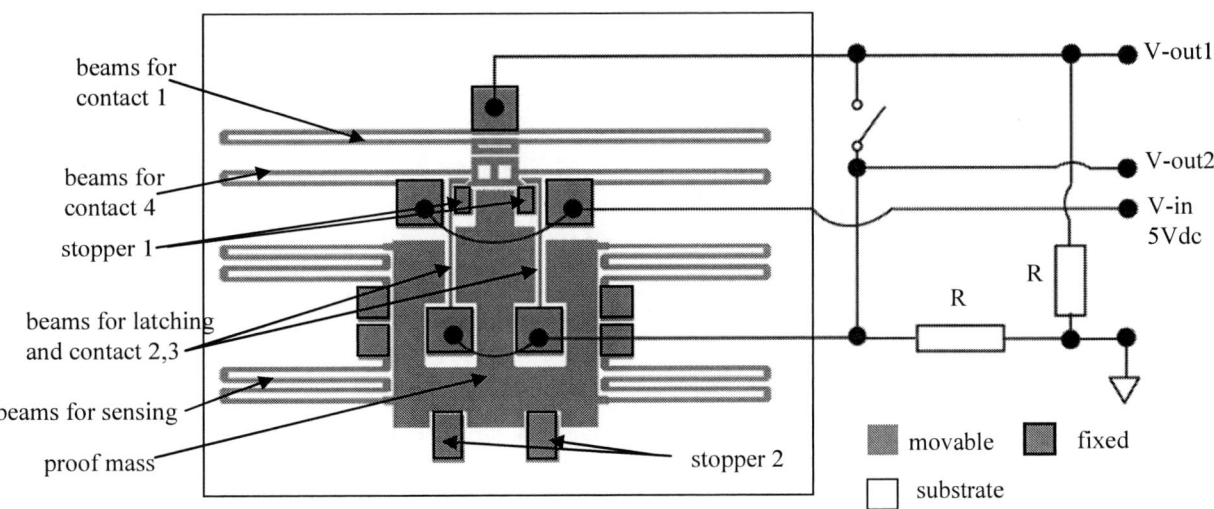

Figure 1: schematic of the switch and test system.

978-1-4244-2977-6/09 $25.00 © 2009 IEEE

Latching Mechanism

In order to insure the switch to be latched reliably, the latching mechanism must be designed to latch easily but difficult to release after latching. In this design, two L-shape latching beams symmetrically located at two sides of contact 4, as shown in Figure 1. The analysis of the deformation of one latching beam under two typical applied forces is illustrated in Figure 2. The deformation of the short part of the latching beam is ignored here. Figure 2(a) shows the schematic diagram of the beam's deformation at the instant of latching. The force, F_1, from contact 4 forces the latching beams to open. However, when the switch is in the state of being latched, the latching beams have the tendency of closing under the force, F_2, from contact 4, as that shown in Figure 2(b).

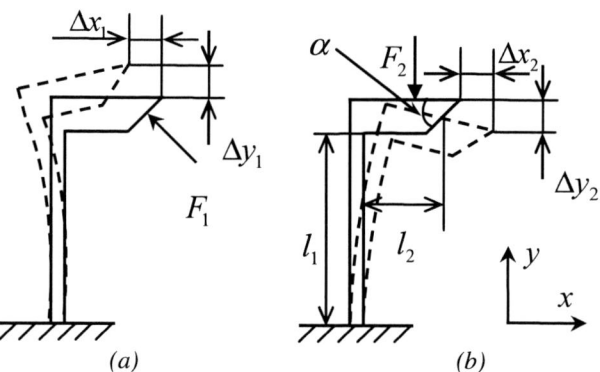

(a) *(b)*

Figure 2: the schematic diagram of the latching beam's deformation under two typical applied forces.

Furthermore, the displacements, Δx and Δy ,along x and y axes in the two states shown in Figure 2 must have large ratios for the purpose of easy opening and difficult releasing. The ratio $\Delta x / \Delta y$ in Figure 2(a) can be written as

$$R_{\Delta a} = \frac{3k_l + 2k_l^2 \tan \alpha}{3k_l \tan \alpha + 6} \tag{1}$$

where $k_l = l_1 / l_2$.

And the ratio $\Delta x / \Delta y$ in Figure 2(b) can be described as

$$R_{\Delta b} = \frac{1}{2} k_l \tag{2}$$

So $R_{\Delta a}$ and $R_{\Delta b}$ increase with both k_l and α . However, too large k_l and α will result in not only large size of the whole device but also reduce the natural frequency of the latching beams, witch can decrease the response speed of the latching mechanism and in turn lower the reliability of latching. Here $\alpha = \frac{1}{8}\pi$ and $k_l = 8$, so $R_{\Delta a}$ and $R_{\Delta b}$ are 5 and 4 respectively.

Displacement Limitation

In order to improve the reliability of the switch especially under large acceleration or shocks, measure must be taken to limit the displacement of the proof mass. The function is performed by the stoppers and anchors in this design.

The two anchors connected with the beam of contact 4 are adopted to limit the open displacements of the latching beams so as to shorten the time of their repositioning. At the same time, the vibration of contact 4 and contact 1 under large shocks can bring adverse effect to the contact reliability though they have been perforated to reduce their masses. In order to reduce the effect, stopper 1 is used to limit the displacements of the latching beams along y axis, as shown in Figure 1.

The in-plane rotational movement of the proof mass must be limited so as to meet the needed acceleration threshold precisely. Though the center of the proof mass is designed to be coinciding with its support center, more measure is necessary because imperfect fabrication will result in the misalignment of the two centers. So any in-plane acceleration or shock will result in the in-plane rotational movement of the proof mass. In this work, stoppers and anchors are designed to have the function of limiting the in-plane rotational movement of the proof mass.

Moreover, all of the stoppers and anchors except the one connected with contact 1 will limit the displacements of the proof mass along x and y axes in order to protect the switch from transverse impact or over impact under large shocks. Specially, stopper 2 are used to limit the displacement of the proof mass in anti-sensing direction along sensing axis for the purpose of protecting the switch from large acceleration or shocks.

Response Time Evaluation

The response time of the switch is mainly determined by the beam-mass system combined by the sensing beams and proof mass, which is typically a two order system. So the response time can be evaluated by the rise time of the two order system,

$$t_r = \frac{\pi - \beta}{\omega_d} \tag{3}$$

where $\beta = \arccos(\zeta)$, and $\omega_d = \omega_n \sqrt{1 - \zeta^2}$. And ζ is the damping ratio which can be taken as zero here; ω_n is the natural frequency of the system. Considering $\omega_n = 2\pi f_n$, the response time can be written as

$$t_r = \frac{1}{4 f_n} \tag{4}$$

If the response time is to be less than 0.1 ms, the natural frequency should be more than 2.5 kHz. For this switch, $f_n = 3.4$ kHz, so the response time is about 0.07ms.

FABRICATION

The switch was fabricated by two-mask silicon on glass process using silicon/glass wafer bonding, DRIE and maskless gold sputtering, as shown in Figure 3. Figure 4 shows the SEM photo of the fabricated switch and Figure 5 shows the latching state of the contacts. The close-up views of the latched contacts are shown in Figure 6.

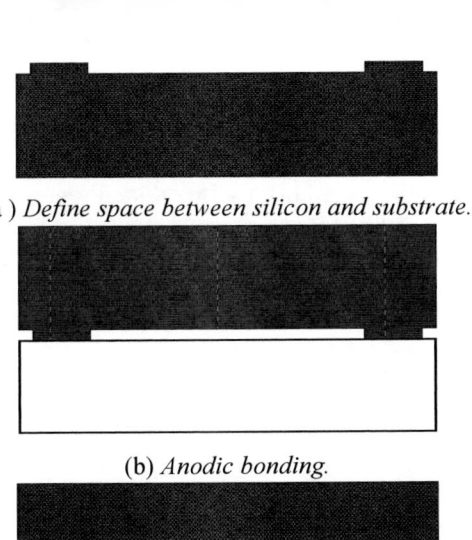

(a) *Define space between silicon and substrate.*

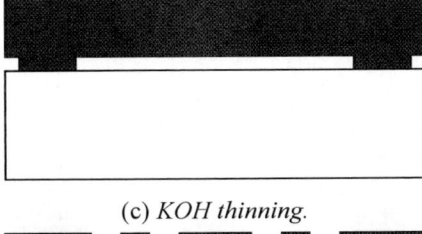

(b) *Anodic bonding.*

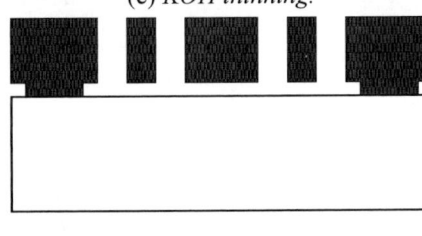

(c) *KOH thinning.*

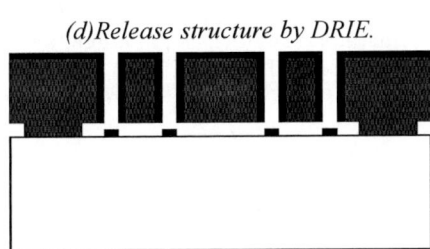

(d)*Release structure by DRIE.*

(e) Maskless gold sputtering.

■ *Si* □ *glass* ■ *Au*

Figure 3: *Fabrication process of the switch.*

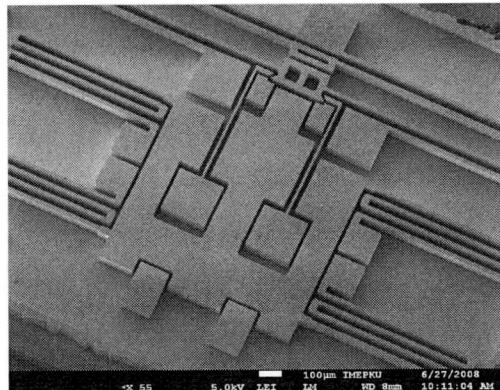

Figure 4: *SEM photo of the switch.*

CHARACTERIZATION

The switch was tested on a drop table as shown in Figure 7. Figure 1 shows the schematic of the measurement system. In order to test the contact characteristics, contact 2

and 3 are connected but they do not connect with contact 1. Figure 8 shows the contacts contacted under 4400G's shock but the switch are not latched. Figure 9, 10 show the outputs of the switch under 5800G's shock. The contacts contacted at about the same time and the response time is less than 0.1ms. As Figure 10 shows, there are at least two contacts of different electrodes keeping contacted when the acceleration is equal to or more than the threshold. The contact resistance is no more than 5 ohms while the isolation resistance is more than 200M ohms and the maximum allowable current is up to 100mA.

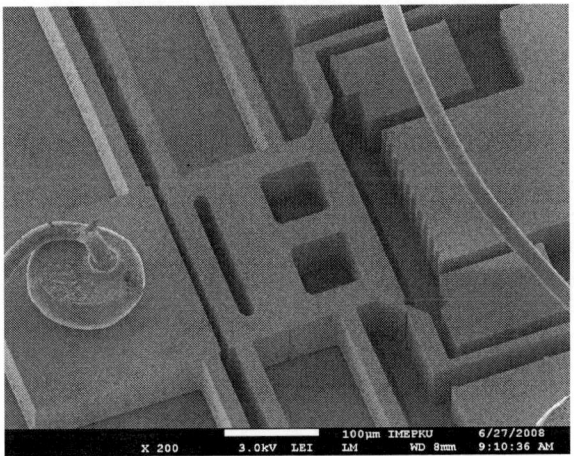

Figure 5: *SEM photo of the latched contacts.*

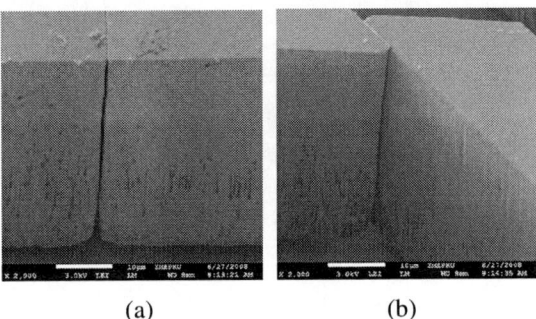

(a)　　　　　　　　　(b)

Figure 6: *SEM photo of close-up view of the latched contacts: (a) contact1 and contact 4;(b) contact 4 and contact 2 or 3.*

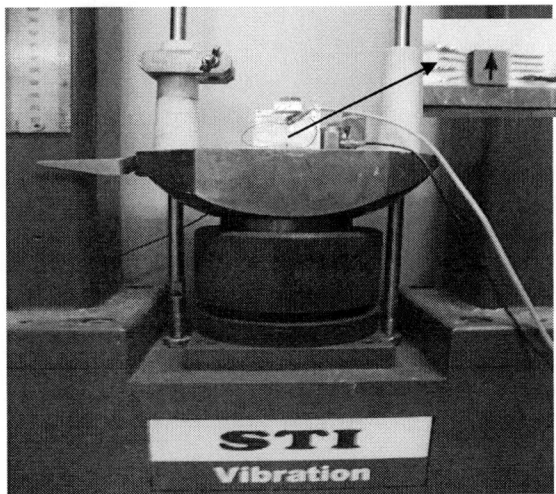

Figure 7: *Photo of the packaged switch mounted on the drop table.*

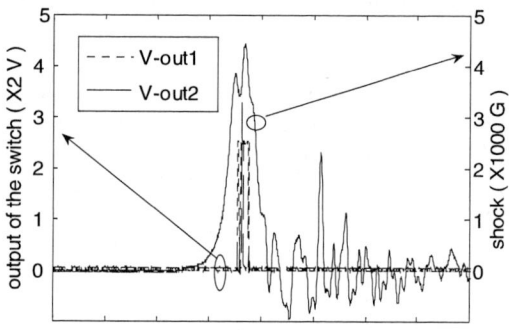

Figure 8: Output under 4400G shock: contacted but not latched.

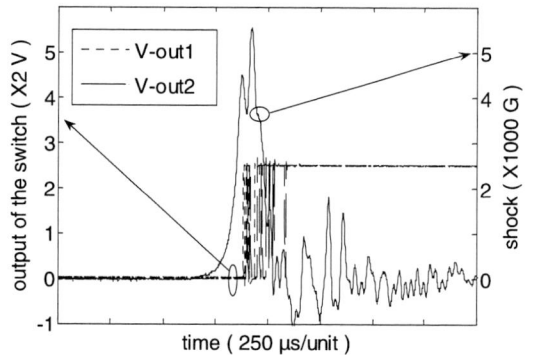

Figure 9: Output of the switch under 5800G shock.

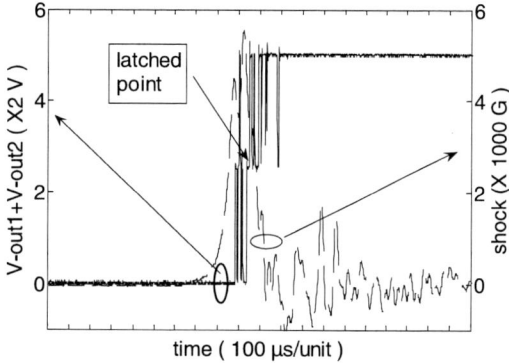

Figure 10: Output of the switch under 5800G shock: V-out1+V-out2

CONCLUSION

An acceleration latching switch with independent multi-contacts was designed and implemented successfully. All the contacts and their beams are independent to the proof-mass so as to prevent the contacts from the impact resulting from the rebound or vibration of the proof mass once the switch is latched. It has highly reliable contact, low contact resistance and high maximum allowable current. The latching shock is 4500G and the response time is less than 0.1ms. The contact resistance is no more than 5 ohms while the isolation resistance is more than 200M ohms and the maximum allowable current is up to 100mA.

ACKNOWLEDGEMENTS

The authors would like to thank the technical staffs in National Key Laboratory of Micro/Nano Fabrication Technology for the support on device fabrication.

REFERENCES

[1] Sven Michaelis , Hans-Joerg Timme , Michael Wycisk , Josef Binder, "Acceleration threshold switches from an additive electroplating MEMS", *Sensors and Actuators A,* vol. 85, pp.418–423, 2000.

[2] Dino R Ciarlo, "A latching accelerometer fabricated by the anisotropic etching of (110) oriented silicon wafers," *J. Micromech Microeng,* vol. 2, pp.10-13, 1992.

[3] Jian Zhao, Jianyuan Jia, and Guangyan Chen, "A Novel MEMS Parallel-beam Acceleration Switch," in *Proceedings of the 2nd IEEE/ASME International Conference.* Beijing, Aug. 2006, pp.1-5.

[4] Jeung Sang Go, Young-Ho Cho, Byung Man Kwak, Kwanhum Park, "Snapping microswitches with adjustable acceleration threshold," *Sensors and Actuatars A,* vol.*54,* pp.579-583,1996.

[5] Sven Michaelis, Hans-Jorg Timme, Michael Wycisk and Josef Binder, "Additive electroplating technology as a post-CMOS process for the production of MEMS acceleration-threshold switches for transportation applications," *J. Micromech. Microeng.,* vol.10, pp.120–123, 2000.

[6] Wei Ma , Gang Li , Yitshak Zohar , Man Wong , "Fabrication and packaging of inertia micro-switch using low-temperature photo-resist molded metal-electroplating technology," *Sensors and Actuators A,* vol. 111, pp.63–70, 2004.

[7] Zhuoqing Yang, Guifu Ding,Weiqiang Chen, Shi Fu, Xiaofeng Sun and Xiaolin Zhao, "Design, simulation and characterizationof an inertia micro-switch fabricated by non-silicon surface micromachining," *J. Micromech. Microeng.,* vol.17 , pp.1598–1604, 2007.

[8] Tadao Matsunaga, Masayoshi Esashi, "Acceleration switch with extended holding time using squeeze film effect for side airbag systems," *Sensors and Actuators A,* vol. 100, pp. 10–17, 2002.

[9] B J Hansen, C J Carron, B D Jensen, A R Hawkins and S M Schultz, "Plastic latching accelerometer based on bistable compliant mechanisms," *Smart Mater. Struct.,* Vol.16, pp.1967–1972, 2007.

978-1-4244-2977-6/09 $25.00 © 2009 IEEE   812

AN ANALYSIS TO IMPROVE STABILITY OF DRIVE-MODE OSCILLATIONS IN CAPACITIVE VIBRATORY MEMS GYROSCOPES

S. E. Alper[1], K. Sahin[1], and T. Akin[1,2]

[1]Middle East Technical University, Dept. of Electrical and Electronics Eng., Ankara, TURKEY
[2]METU-MEMS Research and Application Center, Ankara, TURKEY

ABSTRACT

This paper presents an analysis showing that it is possible to determine the factors that limit the stability of drive-mode oscillations of a vibratory MEMS gyroscope with precise modeling. Specifically, unwanted electrical resonance characteristics of the gyroscope drive-mode are analyzed together with analytical expressions, whereas the results of the theoretical analyses are verified both by simulations and measurements, which agree very well. It is demonstrated that the stability of the drive-mode oscillations significantly improves by using a compensation capacitor across resistive preamplifiers of drive-mode oscillator output when the oscillator is driven by square-wave input signal generated by conventional automatic gain control (AGC) loops. The model can be used to identify and eliminate other factors that limit the bias stability.

INTRODUCTION

The performance of MEMS gyroscopes has been improved by a factor of ten every two years [1], from early 1990s till 2000s. However, the progress has slowed down within the past few years after reaching to bias stabilities of about 1deg/hr. In order to further improve gyroscope performances, it is necessary to identify and eliminate the factors that limit the bias stability. One of the most important factors that limit the bias stability is the stability of drive-mode oscillations, since the rate output of a vibratory gyroscope is directly proportional to the velocity of these oscillations. Stabilizing the drive-mode oscillation velocity requires the stabilization of both the oscillation amplitude and frequency, which is achieved by AGC loops [2-6]. These loops find the correct oscillation frequency and keep the oscillation amplitude constant against variations in sensor resonance frequency, quality-factor, etc. However, this approach assumes that the preamplifiers at the drive-mode output generate stable voltages in response to physical vibrations. This assumption fails when improper excitation waveforms and high-gain preamplifiers are used, and when the effects of stray/parasitic capacitances associated with the drive-mode are neglected.

This paper first time analyzes these non-ideal behaviors in detail by developing an extensive model including the effects of stray and parasitic capacitances as well as preamplifier parameters, in order to improve stability of drive mode oscillations in capacitive vibratory MEMS gyroscopes. The analysis is made especially for the case of a common-practice [2, 4-6], i.e., when the drive-mode of the gyroscope is excited by a square-wave signal. Both the simulations and measurements show that a square-wave excitation generates ringing distortion on the drive-mode output voltage, which causes instability in physical drive-mode oscillations since the AGC loop regulates the oscillator's output voltage rather than the physical vibrations of the oscillator.

DRIVE-MODE OSCILLATOR MODEL

Stabilization of drive-mode oscillations is experimentally studied on one of the capacitive vibratory MEMS gyroscopes that is fabricated with an in-house silicon-on-glass micromachining process [7]. Figure 1 shows the fabricated gyroscope integrated with preamplifiers on a hybrid platform package, prior to vacuum sealing. Transresistance amplifier with 1Mohm feedback resistance is used as the preamplifier that buffers the output of the drive-mode oscillator. Identical preamplifiers are also integrated inside the package for buffering the differential sense outputs of the gyroscope.

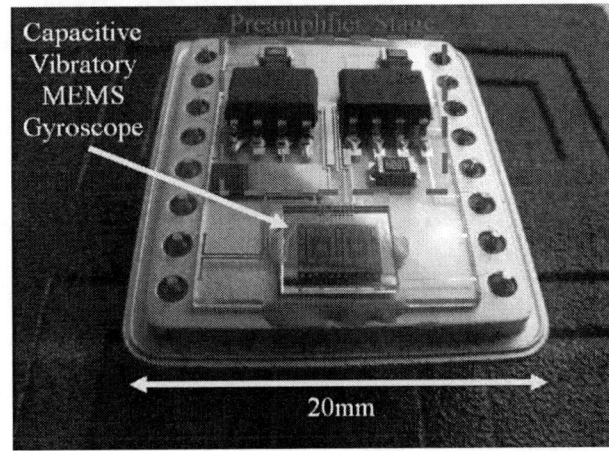

Figure 1: Fabricated MEMS gyroscope and preamplifiers integrated on a hybrid platform package, prior to vacuum sealing.

The transfer characteristics of the drive-mode of the gyroscope are obtained by using a dynamic signal analyzer and applying a DC polarization voltage to the proof mass of the gyroscope. The DC polarization voltage is switched on and off for obtaining the transfer characteristics of the drive-mode oscillator including and excluding the mechanical resonance, respectively. Switching off the DC polarization voltage yields the transfer characteristics only due to stray capacitances from the input to the output. On the other hand, switching on the polarization voltage provides the transfer characteristics due to both stray capacitances and mechanical resonance of the drive-mode oscillator.

The drive-mode mechanical oscillator is then modeled as the combination of a 2nd order mechanical resonator and a 1st order electrical feedthrough (stray) capacitance, yielding a 3rd-order analytical model. Figure 2 shows the measured and simulated drive-mode resonance characteristics of the fabricated gyroscope and the fitted model, respectively, which agree well around 10.7kHz mechanical resonance peak. The total effective stray capacitance from the differential drive inputs to the drive-mode output of the gyroscope is extracted to be only 18fF from the fitted model.

978-1-4244-2977-6/09 $25.00 © 2009 IEEE

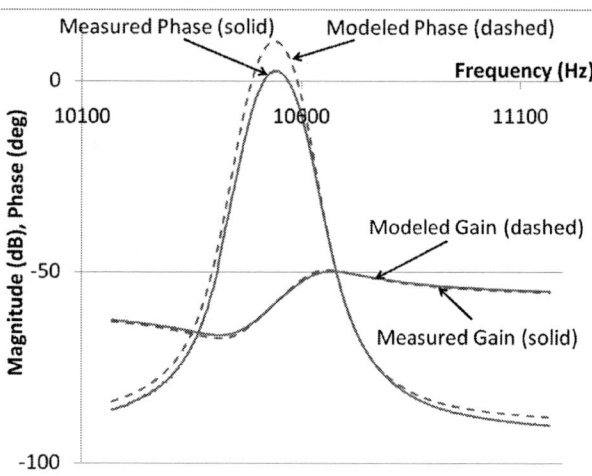

Figure 2: Measured and modeled resonance characteristics of the drive-mode oscillator, including effects of stray capacitances.

COMPLETE ELECTROMECHANICAL MODEL OF THE DRIVE-MODE

Figure 3 shows the complete electromechanical model of the analyzed system constructed with OrCad, including the mechanical drive-mode oscillator with stray capacitances from the input to the output, and the preamplifier structure with parasitic and compensation capacitances. Note that, a 100fF compensation capacitance shown in the model is due to unavoidable stray capacitance across the feedback path of the hybrid preamplifier stage. Moreover, a 15pF input differential capacitance is used for LF353 FET op amp, which is used as the preamplifier.

Figure 4 shows the simulated frequency response of the system of Figure 3. The system has an electrical resonance at a frequency about 200kHz, whose magnitude is greater than that of the mechanical resonance. This unwanted electrical resonance peak may result in unstable oscillations at the preamplifier output. Successive simulations performed by OrCad verify that the gain and frequency of electrical resonance can be analytically expressed in terms of the stray and parasitic capacitances as well as other preamplifier parameters as follows:

$$A_{electrical} \approx \frac{C_{s\_effective}}{C_c + C_{s\_effective}} \tag{1}$$

$$F_{electrical} \approx \frac{1}{2\pi} \cdot \sqrt{\frac{\omega_{cut-off}}{R_{preamp} \cdot (C_p + C_{DP+} + C_c + C_{s\_effective})}} \tag{2}$$

where $A_{electrical}$ and $F_{electrical}$ are the gain and frequency of the electrical resonance peak, respectively, $C_{s\_effective}$ is the effective stray capacitance obtained from Figure 2, and $\omega_{cut-off}$ is the cut-off frequency of the preamplifier op amp. From Equation 1 it is clear that the gain of electrical resonance peak can be suppressed by increasing the amount of compensation capacitance. However, the required compensation capacitance value gets larger for large stray capacitances. Therefore, it is highly desirable to integrate the preamplifiers and the gyroscope over the same substrate if possible.

On the other hand, observing Equation 2 shows the fact that the frequency of the electrical resonance can be shifted to higher frequencies by using an op amp with a higher cut-off frequency and a smaller preamplifier resistance.

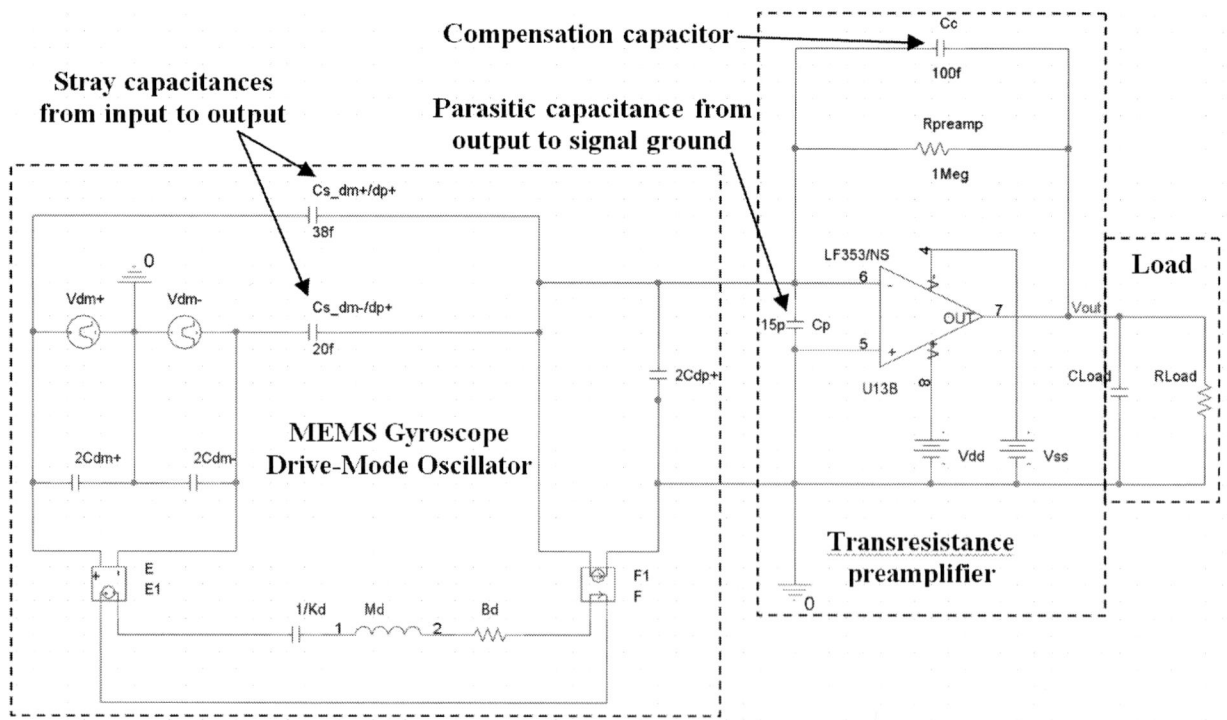

Figure 3: Complete electromechanical model of the analyzed MEMS oscillator and preamplifier.

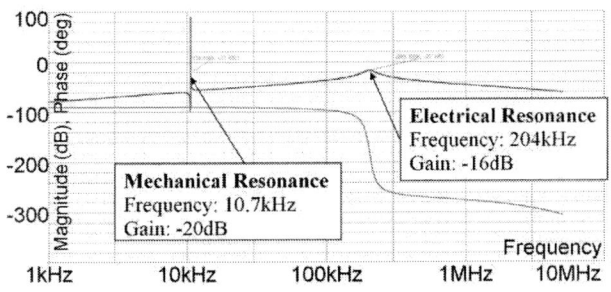

Figure 4: Simulated frequency response of the analyzed system without a compensation capacitor, having an electrical resonance peak at a frequency about 200kHz, whose magnitude is greater than that of the mechanical resonance.

OBSERVED OUTPUT INSTABILITY FOR DRIVE-MODE OSCILLATIONS

Figure 5 shows the simulated and measured input-output waveforms for the drive-mode oscillator when excited by a 500mV$_{peak}$ square-wave input signal at 10.7kHz. No compensation capacitor is used in this test except the inherent 100fF across preamplifier stage as shown in Figure 3. Both simulated and measured waveforms show a ringing oscillation at the preamplifier output at the frequency of simulated electrical resonance. Even though this ringing decay before the end of a single period of drive-mode oscillations, it still causes unstable phase and amplitude variations at the output signal, and has the potential of disturbing phase-sensitive demodulation used in the constructed AGC loop. The source of ringing oscillations is clearly due to the use of square-wave input signal, whose higher-order harmonics cause non-negligible excitation of the electrical resonance shown in Figure 4.

Fortunately, this unwanted resonance oscillation decays in time thanks to the AGC loop, whose phase response does not allow amplification of the electrical resonance signal. This is simply because of the fact that the phase angles of mechanical and electrical resonance signals have 180° phase difference, and phase of AGC loop is designed for locking on the mechanical resonance only.

PROPOSED METHODS AND RESULTS FOR STABILIZING THE DRIVE-MODE OSCILLATOR OUTPUT

One solution to prevent ringing oscillations is to use a sinusoidal input signal, which is free from high-frequency harmonics. Figure 6 shows the simulated and measured input-output waveforms of the same drive-mode oscillator when excited by a sinusoidal input signal, which almost completely eliminates the output distortion observed in Figure 5.

The problem with generating a sinusoidal input signal, on the other hand, is the requirement to design a totally different AGC loop compared to the conventional loops designed to generate square-wave drive signals. In classical AGC loops, the oscillator output signal is demodulated to the baseband and the oscillator input signal is regenerated by modulating an amplitude-controlled DC signal generated inside the loop. This technique inherently produces a square-wave input signal for driving the oscillator, unless the input signal is filtered using a low-pass filter, which is generally not desirable due to phase-shift concerns of the closed-loop system. Alternative solution is to design a new AGC loop with a variable-gain amplifier (VGA), for which the sinusoidal output signal of the oscillator is not demodulated and only its amplitude is controlled by a VGA, generating a pure sinusoidal input signal as in Figure 6.

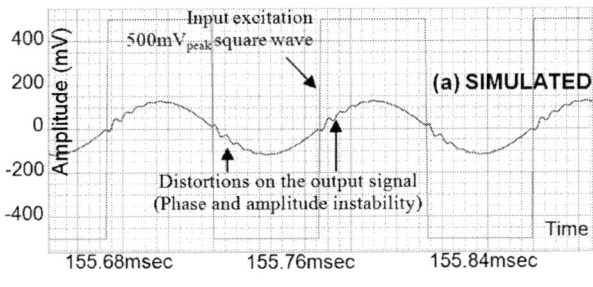

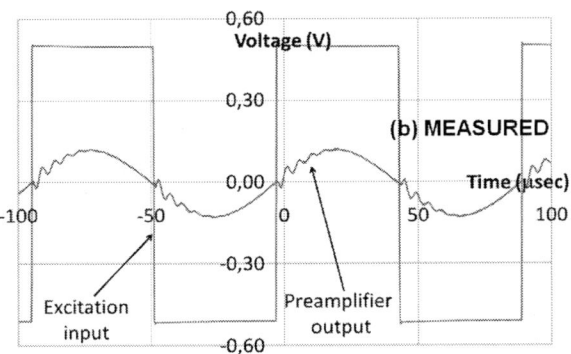

Figure 5: (a) Simulated and (b) measured input-output waveforms for the drive-mode oscillator, when excited by a 500mVpeak square-wave input signal at mechanical resonance frequency of the MEMS oscillator.

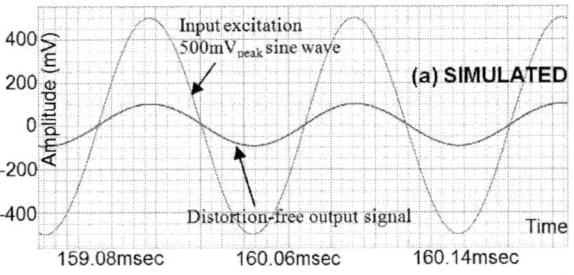

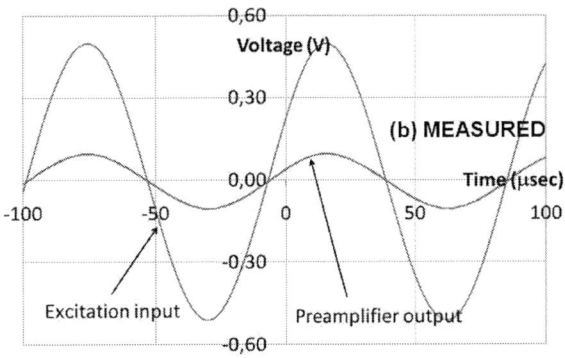

Figure 6: (a) Simulated and (b) measured input-output waveforms for the drive-mode oscillator, when excited by a 500mVpeak sinusoidal input signal, eliminating the distortion in Figure 5.

Yet a more convenient solution for stabilizing the drive-mode oscillator output is proposed, which does not require a new AGC loop design. Observing Equations 1 and 2 and results of analyses in Figures 4 to 6, it is possible to conclude that using a proper compensation capacitor across the preamplifier feedback path can remove the ringing distortion observed in Figure 5. Figure 7 presents the OrCad simulation results verifying this fact for a compensation capacitor of 1pF for the case of a square-wave input excitation applied to the oscillator.

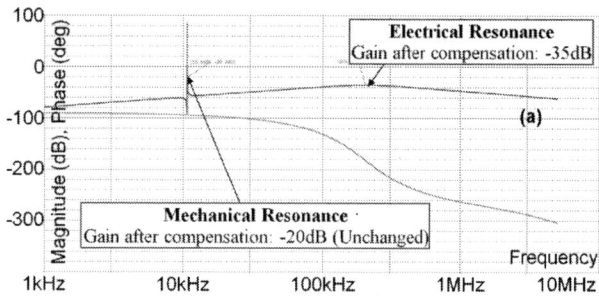

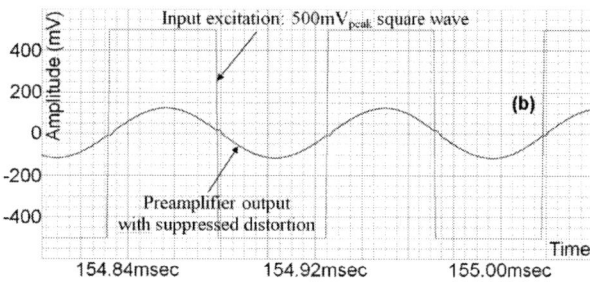

Figure 7: Simulated (a) wideband frequency response and (b) input-output waveforms for the drive-mode oscillator, when a compensation capacitor is used across the preamplifier.

The value of compensation capacitor must be selected carefully, as it also causes a deviation in the preamplifier transfer characteristics from resistive to capacitive. In other words, a high compensation capacitor will cause a significant phase shift for the signal at the preamplifier output, as referred to the oscillator input signal. Such a phase shift will result in the AGC loop lock to a frequency slightly shifted from the pure mechanical resonance frequency of the drive-mode oscillator, resulting in a lower mechanical gain for the oscillator at the operation point. OrCad simulations show that the normalized mechanical gain of the oscillator drops from 0,997 to 0,685 in closed-loop operation for compensation capacitors increased from 780fF up to 10pF.

Figure 8 shows the photograph of the AGC module integrated over a hybrid platform package. This module is connected to the sensor module of Figure 1, in order to test the efficiency of compensation capacitor during closed-loop operation. Figure 9 shows the input and output waveforms obtained from the drive-mode oscillator in closed-loop operation, showing that the output waveform is free of ringing with a compensation capacitor of 1pF. Characterization of the effects of drive-mode oscillation stability on the overall bias stability of the gyroscope is still under inspection by ongoing research.

Figure 8: Photograph of the AGC loop module integrated over a hybrid platform package.

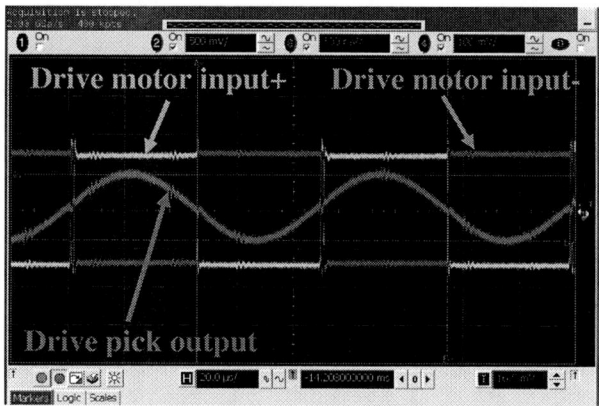

Figure 9: Input and output waveforms obtained from the drive-mode oscillator that use a using a compensation capacitor in closed-loop operation, showing that the output waveform is free of ringing.

REFERENCES

[1] N. Yazdi, F. Ayazi, and K. Najafi, "Micromachined Inertial Sensors", Proceedings of the IEEE, vol. 86, no. 8, pp. 1640-1659, 1998.

[2] J.A.Geen, S.J.Sherman, J.F.Chang, and S.R.Lewis, "Single-Chip Surface Micromachined Integrated Gyroscope With 50 °/h Allan Deviation", J.Solid State Cct., vol.37, no.12, pp.1860-1866, 2002.

[3] M. Palaniapan, R. T. Howe, and J. Yasaitis, "Performance Comparison of Integrated Z-Axis Frame Microgyroscopes", in Digest Tech. Papers MEMS 2003 Conf., Kyoto, January 19-23, 2003, pp. 482-485.

[4] A. Sharma, M. F. Zaman, F. Ayazi, "A 0.2Deg/Hr Micro-Gyroscope with Automatic CMOS Mode Matching", ISSCC2007, San Fransisco, February 11-15, 2007, pp. 386-387.

[5] T. Tsuchiya, Y. Kageyama, H. Funabashi, and J. Sakata, "Vibrating Gyroscope Consisting of Three Layers of Polysilicon Thin Films", Sensors Actuators A, vol. 82, pp. 114-119, 2000.

[6] F. Ayazi and K. Najafi, "A HARPSS Polysilicon Vibrating Ring Gyroscope", J. Microelectromech. Syst., vol. 10, no. 2, pp. 169-179, 2001.

[7] S. E. Alper, K. M. Silay, and T. Akin, "Tactical-Grade Silicon-on-Glass Gyroscope with Very Small Quadrature Coupling," in Digest Tech. Papers Eurosensors XX, Goteborg, September 17-20, 2006, Vol.1, pp. 56-57.

MODELING OF A CAPACITIVE Σ-Δ MEMS ACCELEROMETER SYSTEM INCLUDING THE NOISE COMPONENTS AND VERIFICATION WITH TEST RESULTS

Biter Boğa[1,2], Ilker Ender Ocak[1], Haluk Külah[1,3], and Tayfun Akın[1,3]

[1]Middle East Technical University, Dept. of Electrical and Electronics Eng., Ankara, TURKEY
[2]TÜBİTAK-SAGE, Ankara, TURKEY
[3]METU-MEMS Research and Application Center, Ankara, TURKEY

ABSTRACT

This paper presents a detailed SIMULINK model for a conventional capacitive Σ-Δ accelerometer system consisting of a MEMS accelerometer, closed-loop readout electronics, and signal processing units (e.g. decimation filters). By using this model, it is possible to estimate the performance of the full system, including the effect of individual noise components, operation range, scale factor, etc. The model has been verified through test results using a lateral accelerometer, full-custom designed 2nd-order Σ-Δ readout electronics, and a signal processing unit implemented on PIC. The implemented system operates at 500 kHz sampling rate and has 0.53 V/g open-loop sensitivity, 58.7 μg/√Hz resolution, ±12g operation range, and 0.97e-6 g/(output units) scale factor, where these numbers are in close agreement with the estimated results.

INTRODUCTION

Micromachined accelerometers are extensively used in different areas such as automotive, inertial navigation, guidance, industry, space applications etc. because of low cost, small size, low power, and high reliability. Among various sensing schemes of accelerometers, capacitive sensing is generally preferred since it provides low temperature dependency, high voltage sensitivity, low noise floor, and low drift. Capacitive accelerometers require special readout electronics to sense the capacitance change and to operate in force-feedback for increased operation range and linearity. With the force-feedback circuit, the overall system becomes complicated because of having both mechanical and electrical components defining the overall performance.

There are various studies aiming to model electromechanical Σ-Δ accelerometer systems in the literature [1-3]. Most of these studies consider specific parts of the system (e.g. the accelerometer itself or the Σ-Δ modulator) and they do not take into account secondary effects, such as noise components of the individual parts, which have a significant role on the overall system performance. Besides, most of the developed models lack experimental verification through detailed system level tests. This paper presents a detailed model for conventional capacitive Σ-Δ accelerometer systems with experimental verification. In the next section, modeling of each individual component is explained. Then the overall system model with simulations is presented. Finally, experimental verification and test results are given.

MODELING

Figure 1 shows a typical capacitive accelerometer system operating in closed-loop [2, 3]. The readout

electronics form a 2nd order electromechanical Σ-Δ modulator together with the conventional lateral accelerometer, and provides a single-bit PDM (Pulse Density Modulated) output, which is decimated and filtered by a signal processing unit. This system has both mechanical and electrical components and the overall performance is dependent on each of these individual parts. Figure 3 shows the overall system SIMULINK model developed in this study.

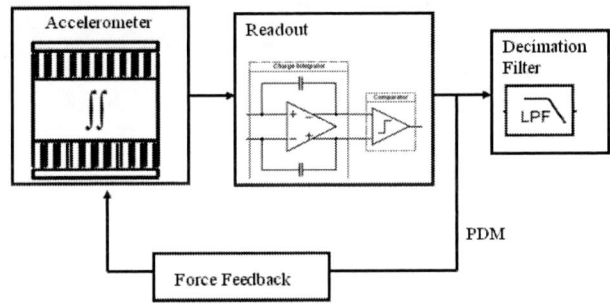

Figure 1: Typical capacitive accelerometer system

MEMS Accelerometer Model

A typical capacitive MEMS accelerometer is composed of capacitors formed between the proof mass and a fixed conductive electrode as shown in Figure 2.

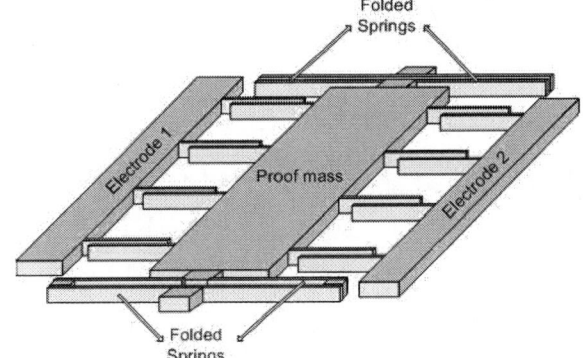

Figure 2: Typical capacitive MEMS accelerometer structure

The sensor is generally modeled with a mass-spring-damper system [2-6], where the relation between the applied acceleration and the displacement can be expressed as:

$$\frac{X(s)}{a_{ext}(s)} = \frac{1}{s^2 + \frac{B}{m}s + \frac{K}{m}} \qquad (1)$$

where B is damping, m is mass and k is spring constant. This displacement results in capacitance change across the electrodes and the proof mass, which can be derived with the equation given in Eq(2).

978-1-4244-2977-6/09 $25.00 © 2009 IEEE

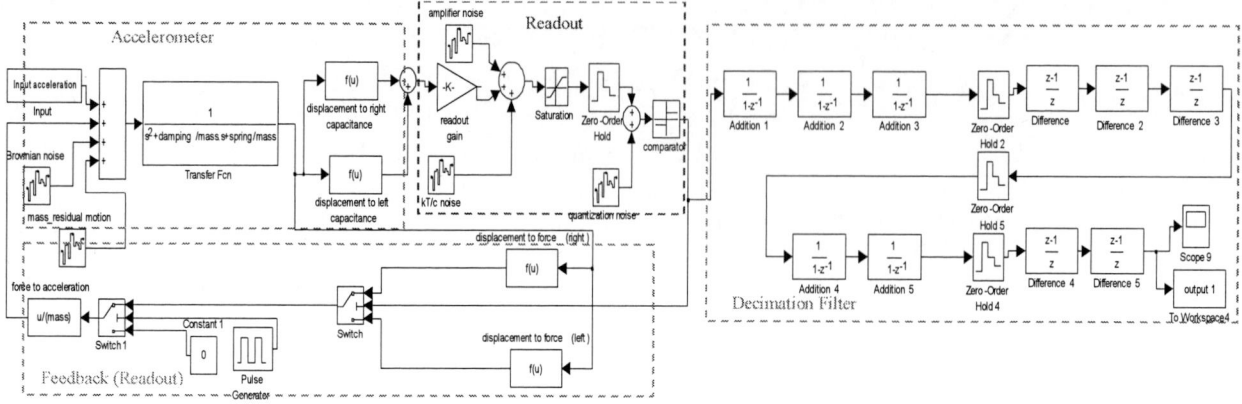

Figure 3: Capacitive - MEMS accelerometer system SIMULINK model

$$C_{1,2} = \frac{N\varepsilon_0 h L_{finoverlap}}{d_1 \mp x} + \frac{(N-1)\varepsilon_0 h L_{finoverlap}}{d_2 \pm x} \quad (2)$$

where N is number of fingers per side, h is structural thickness, $L_{finoverlap}$ is the overlap length of fingers, d_1 and d_2 are the gap and antigap spacings, and x is the proof mass displacement. The accelerometer part of the full system model is represented using the described transfer function in Eq(1) and Eq(2) as shown in Figure 3.

Σ-Δ Readout Electronics Model

Accelerometer part provides differential capacitance change which is usually in the range of tens of atto-farads, and this change should be sensed by a special electronic circuitry. Among various techniques for sensing such small capacitance difference, 2nd order Σ-Δ modulation is generally preferred because of its force feedback structure and inherent analog-to-digital conversion providing linearity and large operating range. Figure 4 shows a typical architecture for such an interface electronics [4].

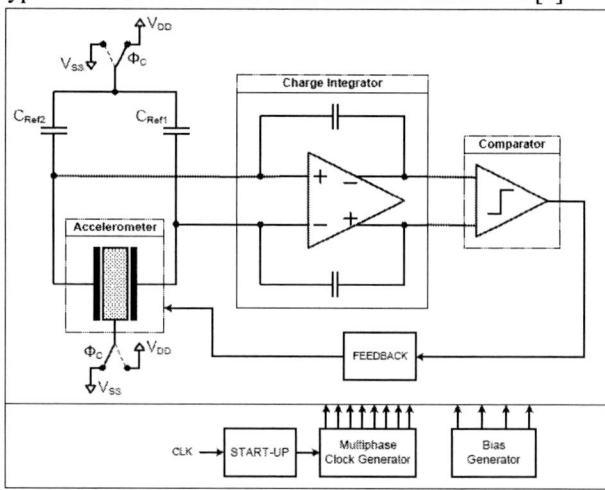

Figure 4: Σ-Δ readout electronics structure

Readout circuit is composed of three main blocks: charge integrator, comparator, and feedback. This circuit operates in two phases (sense and feedback) in a sampling period. In the sense phase, the capacitance change is converted into differential voltage and then single-bit output is generated by means of a comparator. In the feedback phase, this single-bit output is applied as a

electrostatic feedback to the proof mass to realize force-feedback operation. The circuit utilizes oversampling in order to increase the resolution, to decrease the overall in-band noise by shifting the quantization noise to high frequencies [7], and to provide linear operation by keeping the mass displacement very small. In the SIMULINK model, the charge integrator and the comparator are modeled by a gain, saturation, zero-order-hold, and sign blocks [2, 6]. Feedback part is represented as a function block which takes the proof mass displacement and the comparator output as the input and applies feedback force as shown in Figure 3.

Decimation Filter Model

The PDM output of the readout circuit has oversampled and redundant data, and it must be processed to extract the incoming acceleration information. This signal processing unit may affect the performance of the system and it should be included in the overall model.

The oversampled and noise shaped output of the circuit is low-pass filtered to get rid of the redundant data by decreasing the data rate, and to filter out the high frequency noise. For this purpose, generally sinc filters are used. The structure of a Sinc^M filter (M^{th} order) is shown in Figure 5 [7]. The degree of the sinc filter should be at least 1 greater than that of the closed-loop system [7]. Here, cascaded Sinc^3 and Sinc^2 filters are used to satisfy this condition, and they are modeled with the transfer functions of addition and subtraction blocks as shown in Figure 3. Decimation order is determined according to the required signal bandwidth, which is f_s/N, where f_s is the sampling frequency and N is the decimation order. The switch used for decimation is realized with zero-order-hold block (Figure 3).

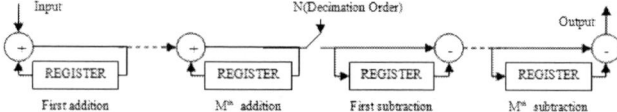

Figure 5: Sinc^M decimation filter structure [7]

Noise Sources Model

The accelerometer system has different mechanical and electrical noise components affecting the resolution of the system. Each of these noise sources shows different behaviors in closed-loop operation. The developed model takes these individual noise sources and their closed-loop

978-1-4244-2977-6/09 $25.00 © 2009 IEEE

behaviors into account and estimates the resolution in a more accurate manner.

Basic noise components for accelerometer system can be listed as Brownian noise, amplifier noise, kT/C noise, quantization noise, and mass residual motion which are described below in detail [5]. Table 1 shows the expressions for these noise sources.

Brownian Noise: Brownian noise is due to the thermal motion of the proof mass, and it is represented as a white noise input in the system model (Figure 3) since f_s is much greater than the accelerometer bandwidth [8].

Amplifier Noise: Amplifier noise is related with the thermal and flicker noise of the main amplifier utilized in the front-end readout. In general, flicker noise is cancelled out by using correlated double sampling (CDS), and therefore this noise source is represented as a band-limited white noise at the output of readout gain block (Figure 3).

kT/C Noise: kT/C noise is a thermal noise due to the switched-capacitor nature of the readout electronics, and mostly depends on the integration capacitance and the sampling frequency. This noise source is represented at the output of the gain block (Figure 3).

Quantization Noise: Quantization noise is effective under closed loop operation, and it is one of the dominant noise sources in the system. This noise source is generated during the analog-to-digital conversion and dominantly depends on the oversampling ratio. The quantization error is assumed to be input independent, uniformly distributed, and independent identically distributed and therefore modeled as a white noise source [7].

Mass Residual Motion Noise: Mass residual motion is one of the most dominant noise sources especially at low oversampling ratios. In closed loop operation, the pulse train output of the Σ-Δ readout is given as feedback to the proof mass, and this results in the oscillation of it around its equilibrium position. This oscillation exists even at zero input acceleration, and it is taken into account as a noise source and added to the model as shown in Figure 3.

Table 1: Noise source expressions

Noise	Expression
Brownian Noise	$a_n^2 = \dfrac{4k_b T b}{9.81^2 m^2}$
Amplifier Noise	$Vout\_thermal = \sqrt{\dfrac{16}{3}\dfrac{Cs+Cp}{Cint}\dfrac{k_bT}{Cout}\dfrac{1}{fs}}$
kT/c Noise	$Vout\_kT/C = \sqrt{\dfrac{4k_bT}{fsCint}}$
Quantization Noise	$erms \dfrac{\Pi^n}{M^{n+0,5}\sqrt{2n+1}}$
Mass Residual Motion	$N_{rm} = 4\dfrac{f_{BW}}{f_s}\left(\dfrac{K}{M}\right)\dfrac{a_{fb}}{(2\pi f_s/4)^2}$

k_b :Boltzmann Constant, T: Temperature (Kelvin), m: mass, C_s: sense capacitance, C_{int}: integration capacitance, f_s: sampling frequency, M: oversampling ratio, K: spring constant, a_{fb}: feedback acceleration, n: order of sigma-delta

SIMULATIONS AND TEST RESULTS

In order to verify the developed model; an accelerometer system is constructed including the capacitive MEMS accelerometer, full-custom CMOS readout electronics, and the decimation filter implemented on PIC. The MEMS accelerometer is fabricated using Dissolved Wafer Process and the readout circuit is implemented using XFab 0.6 μm CMOS process [4]. The

fabricated MEMS accelerometer and readout circuit are shown in Figure 6(a) and Figure 6(b), respectively. Table 2 shows the combined system parameters used in the modeling.

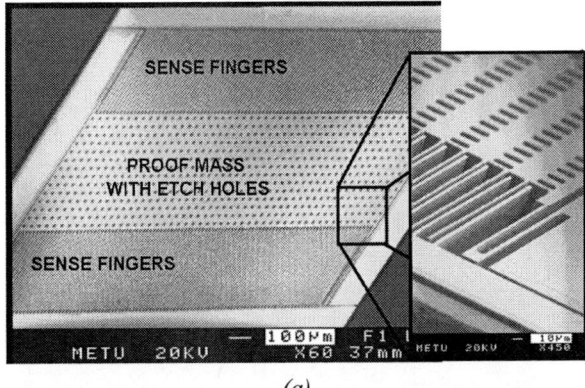

(a)

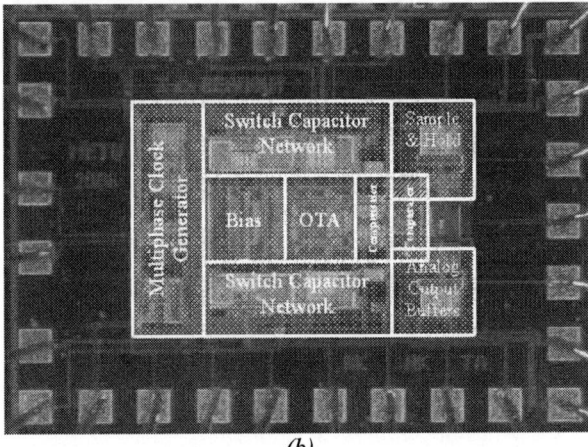

(b)

Figure 6: (a) MEMS accelerometer, (b) CMOS readout

Table 2: Overall system parameters used in modeling

System Parameters	
Proof Mass	1.67e-7 kg
Damping	0.0027
Spring Constant	25.36 N/m
Sensor Sensitivity	9.8 x10⁻⁶ F/m
Readout Gain	1.25x10¹²
Sampling Frequency	500 kHz
Integration Capacitance	4 pF
Output Data Rate	800 Hz

Accelerometer system is simulated using the parameters given in Table 2 and the model shown in Figure 3. Figure 7 shows both simulation and test results for 4-position (0g, -1g, 0g, +1g) acceleration tests. As the figure shows, results are in agreement with simulations, and the system has 11 mg peak-to-peak noise at 800 Hz data rate, corresponding to 58μg/√Hz resolution.

The accelerometer has nearly 15 μm thickness resulting in a relatively small proof mass. Therefore, mass residual motion is expected to be the most dominant noise and in order to verify this expectation and to observe the effect of mass residual motion, f_s of the system is increased from 500 kHz to 750 kHz with 50 kHz increments and overall noise is investigated. It has been observed in Figure 8 that the overall noise is decreased from 58.7 μg/√Hz to 30.5 μg/√Hz, which is in consistence with the simulations.

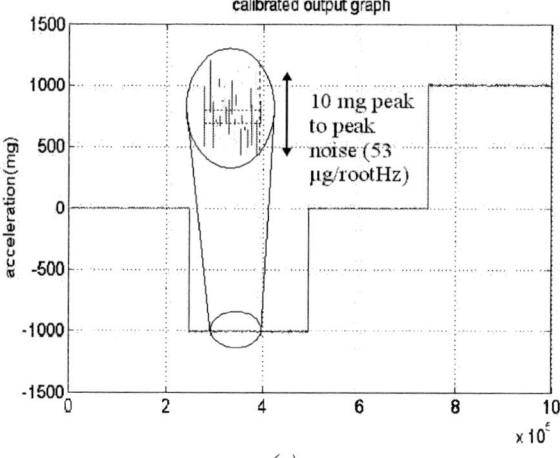

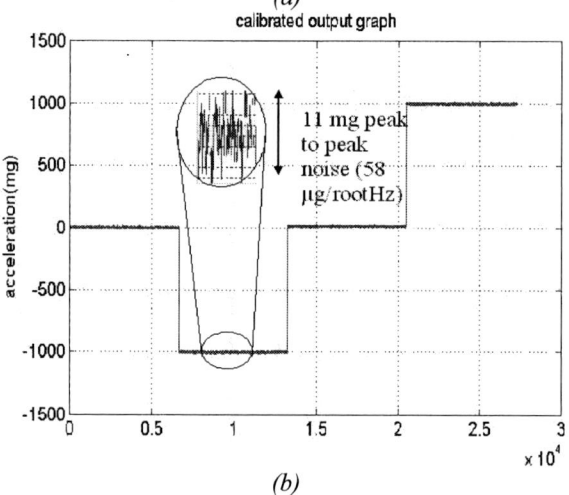

Figure 7: 4-position data (a) simulation (b) test

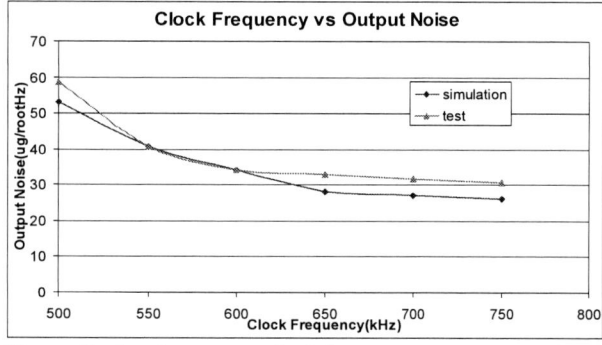

Figure 8: Accelerometer clock frequency vs output noise

As Figure 8 shows, the system noise is dominated by mass residual motion noise for sampling frequencies smaller than 700 kHz. When the sampling frequency is increased further, mass residual motion becomes insignificant compared to other noise sources. In this regime, quantization noise becomes dominant.

Table 3 shows the comparison of hand-calculated, simulated, and tested values for individual noise sources. The difference between the hand-calculated and simulated values comes from the closed loop effect of the system.

The model also provides estimates of other system performance parameters such as scale factor, operational range, and open loop sensitivity. These performance parameters are investigated both with simulations and

tests and compared in Table 4, indicating that the estimated values are consistent with test results. The difference in the operational range is because of the sensing fingers buckling due to mechanical stress.

Table 3: Accelerometer system individual noise sources

Noise	Expected ($\mu g/\bullet Hz$)		Simulation ($\mu g/\bullet Hz$)	Test ($\mu g/\bullet Hz$)
Mass Residual Motion Noise	27.1		46.4	50.1
Other Noises	Brownian Noise	0.14	26.1	30.5
	Amplifier Noise	3.41		
	kT/C Noise	1.09		
	Quantization Noise	20.9		
Total Noise	34.4		53.3	58.7

Table 4: Accelerometer system overall performance

Performance Parameter	Simulation	Test
Scale Factor (g/Output units)	0.97e-6	1.17e-6
Operational Range	±17g	±12g
Open Loop Sensitivity	0.35 V/g	0.48 V/g

CONCLUSIONS

A system level model of a capacitive Σ-Δ MEMS accelerometer including individual noise components is constructed in this study to analyze the full system behavior. This model gives information about the output noise, range, scale factor, and open loop sensitivity of the accelerometer. A prototype accelerometer is developed to justify this model through test results. The developed system has 58.7 $\mu g/\sqrt{Hz}$ resolution, 1.17e-6 scale factor, ±12g operational range, and 0.48 V/g open loop sensitivity; where these values are in agreement with developed model simulations.

REFERENCES

[1] M. A. Lemkin et al., "A Three-axis micromachined accelerometer with a CMOS position-sense interface and digital offset-trim electronics," *IEEE JSSC*, pp. 456-468, 1999

[2] V. P. Petkov et al, "A fourth-order $\Sigma\Delta$ interface for micromachined inertial sensors," *IEEE JSSC*, vol. 40, pp. 1602-1609, 2005

[3] N. Yazdi et al., "Micromachined inertial sensors," *Proceedings of the IEEE*, vol. 86, pp. 1640-1659, 1998

[4] R. Kepenek et al., "A μg Resolution Micro Accelerometer System with A Second-Order Σ-Δ Readout Circuitry," *Research in Microelectronics and Electronics PRIME 2008*, pp. 41-44, 2008

[5] H. Kulah et al., "Noise Analysis and Characterization of A $\Sigma\Delta$ Capacitive μ-accelerometer," *IEEE JSSC*, vol. 41, pp. 352-361, 2006

[6] B. V. Amini and F. Ayazi, "Micro-gravity capacitive silicon-on-insulator accelerometers," *JMM*, vol. 15, pp. 2113-2120, 2005

[7] S. R. Norsworthy, R. Schreier, and G. C. Temes, *Delta-Sigma Data Converters Theory, Design, and Simulation*. IEEE Press, 1996

[8] T. B. Gabrielson, "Mechanical-Thermal Noise in μmachined Acoustic and Vibration Sensors," *IEEE Transactions on Electron Devices*, vol. 40, no.5, 1993

AIR PRESSURE SENSOR FOR AN INSECT WING

Hidetoshi TAKAHASHI, Kiyoshi MATSUMOTO, and Isao SHIMOYAMA
The University of Tokyo, Tokyo, JAPAN.

ABSTRACT

This paper reports on an air pressure sensor for the measurement of the aerodynamic force on an insect wing. A piezoresistive cantilever was fabricated as a pressure sensor. The size of the cantilever was 250 μm × 200 μm × 0.3 μm. The sensitivity to air pressure was 0.29×10^{-3} Pa^{-1} below 2.0 kHz. By attaching the sensors on the prototype of an insect wing, pressure distribution over the surface were measured when the flow was both steady and unsteady.

INTRODUCTION

Insects fly with a variety of maneuvers such as hovering and quick-turning. These maneuvers are made possible by the aerodynamic forces that act on their wings [1]. These forces are caused by the distribution of pressure difference between the upper face and lower face of the wing. Therefore, it is important to analyze the pressure distribution in order to understand insect flight mechanism.

In the literature, the total aerodynamic force on the wing was measured by detecting the strain on an artificial insect wing root [2-3]. However, this method can not provide data on the distribution of pressure differences, especially local pressure differences. In the study of airplane wings, local air pressure was measured by pressure sensors that are directly attached onto the wings' surfaces. The pressure distribution is evaluate from the local pressure measured using multiple sensors.

In this study, we propose a micro pressure sensor that can be attached on an artificial insect wing in order to measure the local pressure difference. By using an array of such sensors, we can evaluate the pressure distribution of an insect wing. Our sensor uses a piezoresistive cantilever to measure the local pressure difference between the upper face and lower face of the wing. The concept of the measurement system is shown in Figure 1. Piezoresistive cantilever type sensors have the advantage that they can measure from static pressure to high frequency pressure compared with other type pressure sensors [4]. Moreover, since the three edges of the cantilever are freed up, the sensor has a higher sensitivity than a traditional diaphragm sensor [5].

We measured the sensitivity and the response speed of the fabricated sensor. By attaching the sensors on an insect wing model, we demonstrated that the sensors detected the pressure distribution of the wing.

DESIGN AND FABRICATION

Our wing is modeled based on large-size insects such as a butterfly or a dragonfly, whose wing span is 50 to 100 mm long. These insects flap at 0 to 100 Hz. Thus, the pressure difference is estimated to be on the order to 1 Pa. Therefore, it is necessary for the sensor to have sufficient sensitivity to detect 1 Pa with response speed higher than 100 Hz. Moreover, the sensor needs to be small enough to have no influence on airflow around the wing.

A design of the cantilever is shown in detail in Figure

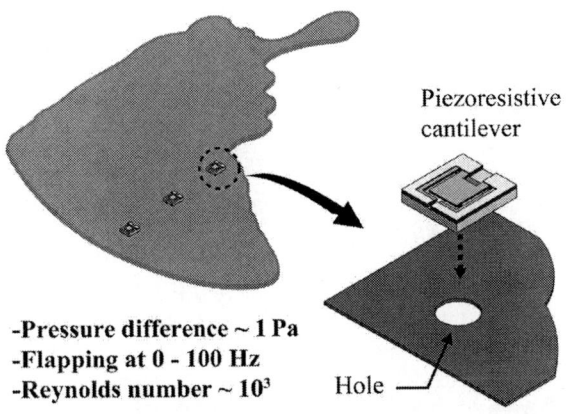

-Pressure difference ~ 1 Pa
-Flapping at 0 - 100 Hz
-Reynolds number ~ 10^3   Hole

Figure 1: Concept of the measurement of the air pressure distribution on an insect wing. Air pressure sensors are attached on the prototype of the insect wing.

2. The sensor chip and the cantilever were 1.5 mm × 1.5 mm × 0.3 mm and 250 μm × 200 μm × 0.3 μm, respectively. Piezoresistor is formed on the surface of the cantilever. With the design shown in Figure 1, air can flow through the gap between the cantilever and its frame. We fabricated devices with gap widths of 10, 20, 40 μm, and did experiments to evaluate the influence of the airflow through the gap on the sensor performance. The cantilever is designed to have a sensitivity of 22 μstrain/Pa and resonant frequency of 5.9 kHz. The sensitivity is sufficiently high to detect the pressure difference on the wings. The resonant frequency is higher than the flapping frequency.

The fabrication process of the sensor is shown in Figure 3(a). In the process, we used a 0.3 μm / 0.4 μm / 300 μm thick SOI (silicon on insulator) wafer. Firstly, N-type resistor was formed on the SOI wafer by the rapid thermal diffusion (Figure 3(a)(i)) [6]. Secondly, a gold layer was deposited, and patterned. Using this gold layer as a mask, the topside silicon layer was etched by DRIE (Figure 3(a)(ii)). Then the gold layer was etched again to remove the metal layer from the surface of the piezoresister (Figure 3(a)(iii)). The bottom-side silicon layer was etched by DRIE from the backside, and the cantilevers were released by etching the glass layer in HF vapor (Figure 3(a)(iv)).

The photographs of the cantilever are shown in Figure 3(b). The initial resistance of the cantilevers ranged from 1.2 to 1.5 kΩ.

SENSING CHARACTERISTICS
Sensitivity

We measured the fractional resistance change of the sensor against static pressure difference. The cantilever was mounted on a wall which separated two chambers as shown in Figure 4(a). We used syringes as the chambers. By applying air pressure to one chamber, we provided the constant pressure difference between the upper and lower surface of the cantilever. The piezoresistor on the cantilever was used as one of the four resistors in a

978-1-4244-2977-6/09 $25.00 © 2009 IEEE

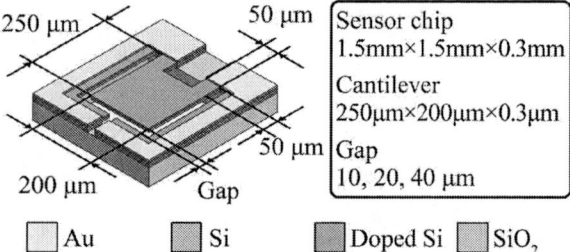

Sensor chip
1.5mm×1.5mm×0.3mm
Cantilever
250μm×200μm×0.3μm
Gap
10, 20, 40 μm

■ Au ■ Si ■ Doped Si ■ SiO₂

Figure 2: A schematic design of the cantilever. The size of a sensor unit was 1.5 mm × 1.5 mm × 0.3 mm. The cantilever was 250 μm × 200 μm × 0.3 μm. The gaps between the cantilever and the surround were 10, 20, 40 μm.

(a) Fabrication process flow of the cantilever

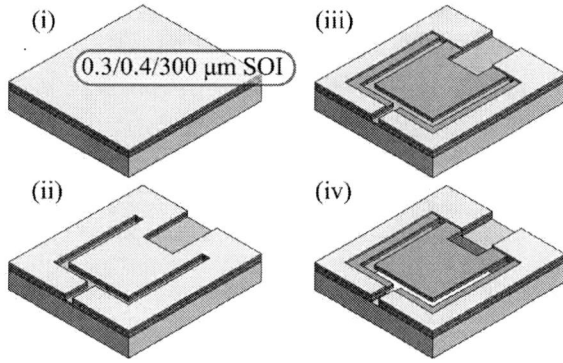

(b) Photographs of the fabricated sensor

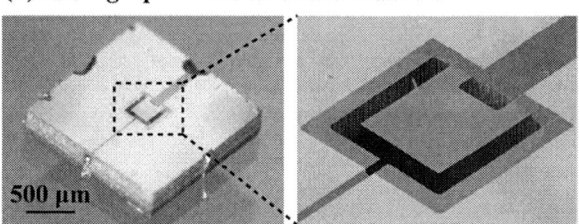

Figure 3: (a) Fabrication process of the cantilever. (b) Photographs of the pressure sensor unit. The initial resistance of the cantilevers was 1.2 - 1.5 kΩ.

Wheatstone-bridge circuit. The bridge circuit was connected to an oscilloscope through an amplifier circuit. We applied pressure difference from -20 to 20 Pa in this experiment. We defined the pressure difference as positive when the cantilever resistance becomes larger than the initial value. On the other hand, the pressure difference was negative when the resistance becomes smaller.

Figure 4(b) shows the relationship between the resistance change and the static pressure difference. The width of the gap did not affect the response. There was a good linearity between the magnitude of the pressure difference and the fractional resistance change. The fitted line, $\Delta R/R = 0.29 \times 10^{-3} \times p$, was obtained by least-square fit. The coefficient of determination (R^2) was 0.994.

Response speed

The frequency response of the sensor was measured to estimate the response speed. To apply dynamic air pressure, we emitted sound wave to the sensor from a speaker as shown in Figure 5(a). A bridge circuit was

(a) Experimental setup

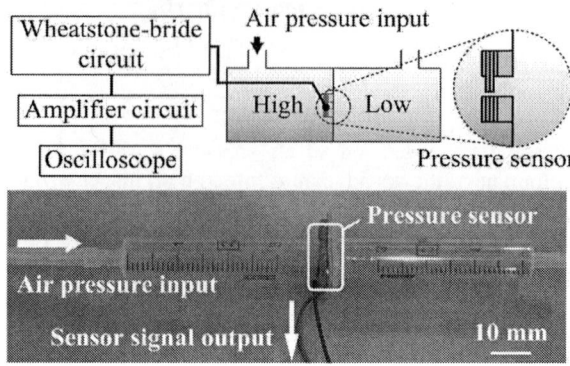

(b) Response to the static air pressure

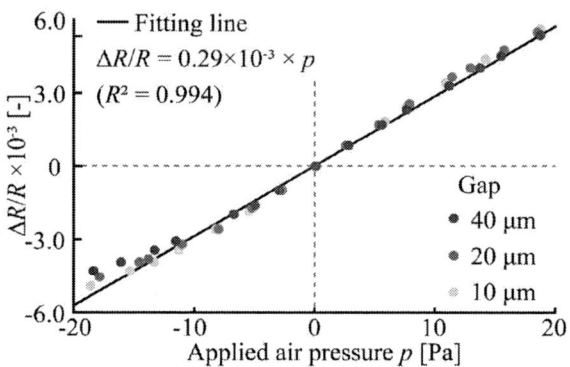

Figure 4: (a) The schematic and photograph of the experimental setup to measure the response to pressure difference. Air pressure was applied to one chamber to produce pressure difference. (b) Response to static pressure difference. The response had little influence over the gap width around the cantilever. The sensitivity was 0.29×10^{-3} Pa^{-1}.

connected to a network analyzer through an amplifier circuit. We measured the magnitude of the air pressure by a microphone at the same position of the sensor. In this experiment, we applied sinusoidal wave from 10 Hz to 10 kHz every 50 Hz.

The relationship between the frequency and the fractional resistance change is shown in Figure 5(b). The measured fractional resistance change was divided by the magnitude of air pressure. The response was largest at 4.1 kHz. Thus, this frequency was regarded as the first resonant frequency of the cantilever. The response remained stable at 0.29×10^{-3} Pa^{-1} below 2.0 kHz.

These results showed that the sensitivity of the sensor remained unchanged at 0.29×10^{-3} Pa^{-1} from static to 2.0 kHz. It was suggested that the sensor had a high sensitivity and high response speed enough to measure the pressure difference.

EXPERIMENT AND RESULT

We measured the pressure difference on the prototype of an insect wing. Figure 6(a) shows the prototype wing with the sensors. The wing was modeled after a large butterfly wing [7] and made of a 1.0 mm thick substrate. The wing length was 60 mm. We made holes of 1.0 mm in diameter on the wing and attached the sensors above the

978-1-4244-2977-6/09 $25.00 © 2009 IEEE

(a) Experimental setup

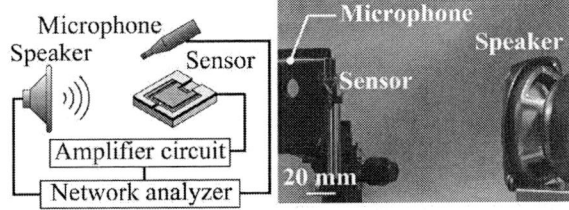

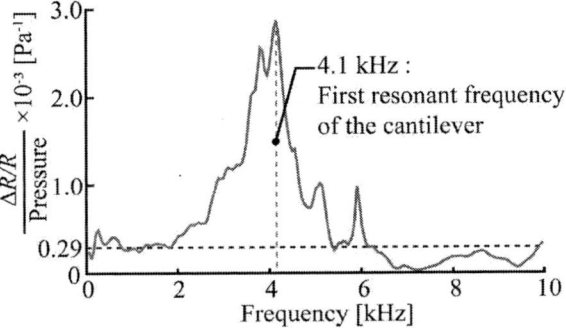

(b) Response to the pressure frequency

4.1 kHz :
First resonant frequency
of the cantilever

Figure 5: (a) The schematic and photograph of the experimental setup to measure the response to pressure frequency. Sound wave was applied to the sensor from 10 Hz to 10 kHz. (b) The response to the frequency of the sound wave. The response remained stable at 0.29×10^{-3} Pa^{-1} below 2.0 kHz.

holes as shown in Figure 6(b). The sensors were attached on the leading edge, the center, and the trailing edge of the wing, respectively. We measured both the pressure difference with the sensors and the total aerodynamic force with a load cell. The response of the road cell was divided by the wing area. This response was regarded as the mean pressure difference.

Steady flow

Figure 7 shows the measurement of the pressure difference when the flow was steady. The wing was mounted on a linear actuator to provide constant airflow as shown in Figure 7(a). The flow velocity was 2.0 m/s in every measurement. In this experiment, the angle of attack of the wing was changed from -30 to 30 degrees every 5 degrees.

Figure 7(b) shows the relationship between the angle of attack and the pressure difference at each point. The fitted line shown in Figure 4(b) provides the pressure difference from the fractional resistance change. The pressure distribution on the wing ranged from -6.0 to 6.0 Pa. As the angle of attack became larger, the pressure difference became larger at each point. When the angle of attack was 0 degrees, the pressure differences detected by the sensors and the load cell were almost 0 Pa. And, the responses were almost origin symmetry. The pressure difference was largest at the leading edge and smallest at the trailing edge at each angle. It was suggested that the mean pressure difference detected by load cell were in agreement with the pressure difference between the leading edge and the center at each angle of attack. These results coincided with the theoretical pressure distribution.

(a) Photograph of the wing

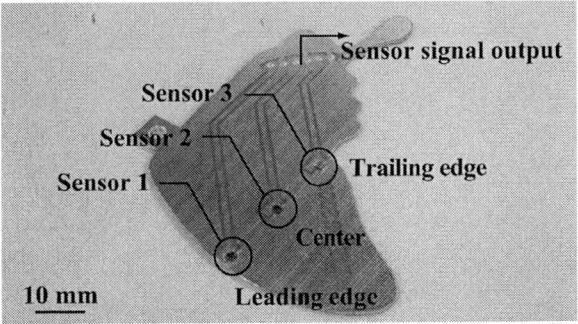

(b) Attachment of the sensor

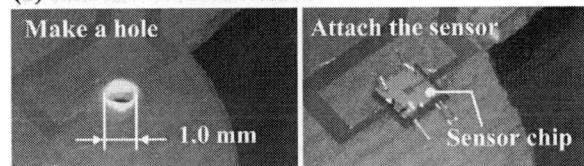

Figure 6: (a) Photographs of the prototype of an insect wing with the pressure sensors. (b) The sensors were attached above the holes which were made on the wing.

(a) Experimental setup

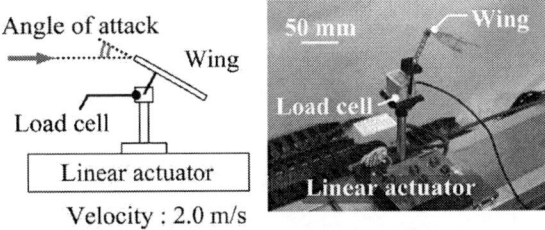

Velocity : 2.0 m/s

(b) Response to the pressure frequency

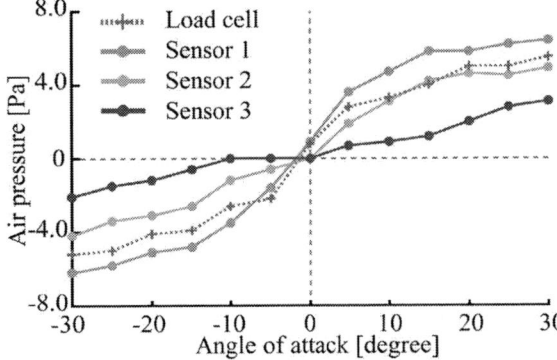

Figure 7: (a) The schematic and photograph of the experimental setup to measure the pressure difference on the wing when the flow was steady. (b) The relationship between the angle of attack and the pressure difference. At each angle, the pressure difference was largest and smallest at the leading edge and at the trailing edge, respectively.

Unsteady flow

Figure 8 shows the measurement of the pressure difference when the flow was unsteady. The wing was fixed on and a spring as shown in Figure 8(a). The prototype wing was pulled down with 15 mm and released. Then wing vibrated up-and-down at 3 Hz sinusoidal wave.

Figure 8(b) shows the pressure difference changes while the wing was vibrating. The pressure difference at

(a) Experimental setup

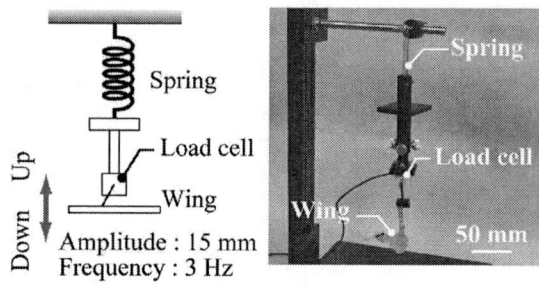

(b) Response to the pressure frequency

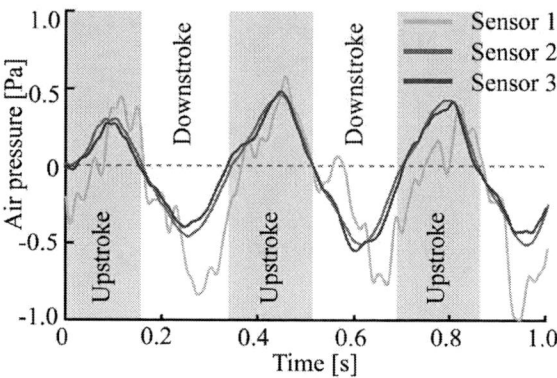

Figure 8: (a) The schematic and photograph of the experimental setup to measure the pressure difference on the wing when the wing was vibrating. (b) Pressure difference change at each point. The pressure changed from -0.5 to 0.5 Pa caused by the wing motion.

each point changed from -0.5 to 0.5 Pa symmetrically caused by the wing motion. These results were in agreement with the theoretical pressure difference. On the other hand, the output of the load cell was calculated to range from -6 to 6 Pa, because the output contained not only the pressure difference but also the initial force. These results suggested that micro sensors were suitable to measure the pressure distribution on the flapping wing.

CONCLUSION

In conclusion, we proposed the piezoresistive cantilever to detect the pressure difference on the artificial wing. The response of the sensor was not influenced by the gap width around the cantilever. The sensitivity was 0.29×10^{-3} Pa^{-1} and the response remained stable below 2.0

kHz. We attached the sensors on the prototype of the wing 60 mm long. In 2.0 m/s airflow, we were able to detect the pressure distribution changing caused by the angle of attack. While the wing was vibrating at 3 Hz sinusoidal wave, the sensors detected the pressure differences ranging from -0.5 to 0.5 Pa symmetrically in line with the wing motion. By attaching the pressure sensors on a flapping insect wing, the pressure distribution over the surface of the wing will be revealed.

ACKNOWLEDGMENTS

The photolithography masks were made using the University of Tokyo VLSI Design and Education Center (VDEC)'s 8 inch EB writer F5112 + VD01 donated by ADVANTEST Corporation. This research was partly supported by MEXT, Grant-in-Aid for Scientific Research (S) (18100002)

REFERENCES

[1] C. P. Ellington , V. D. B. Coen, A. P. Willmott, and A. L. R. Tomas, "Leading-edge Vortices in Insect Flight," *Nature*, vol. 384, pp. 626-630, 1996.

[2] J. M. Birch, W. B. Dickson, and M. H. Dickinson, "Force Production and Flow Structure of the Reading Edge Vortex on Flapping Wings at High and Row Reynolds Numbers," *J. Experimental Biology*, vol. 207, No. 7, pp. 1063-1072, 2004.

[3] W. J. Maybury and F-O. Lehmann, "The Fluid Dynamics of Flight Control by Kinematic Phase Lag Variation between Two Robotic Insect Wings," *J. Experimental Biology*, vol. 207, pp. 4707-4726, 2004.

[4] P. R. Scheeper, A. G. H. van der Donk, W. Olthuis, and P. Bergveld, "A Review of Silicon Microphone," *Sens. Actuator A-Phys.*, vol. 44, pp. 1-11, 1994.

[5] Lin-Tao Zhang, Tian-Ling Ren, Jian-She Liu, Li-Tian Liu, and Zhi-Jian Li, "Fabrication of a Cantilever Structure for Piezoelectric Microphone," *Jpn. J. Appl. Phys.*, vol. 41, pp.7158-7159, 2002.

[6] M. Gel and I. Shimoyama, "Force Sensing Submicrometer Thick Cantilevers with Ultra-Thin Piezoresistors by Rapid Thermal Diffusion," *J. Micromech. Microeng.*, vol. 14, pp. 423-428, 2004.

[7] H. Tanaka, K. Matsumoto, and I. Shimoyama, "Fabrication of a Three-dimensional Insect-wing Model by Micromolding of Thermosetting Resin with a Thin Elastmeric Mold," *J. Micromech. Microeng.*, vol. 17, pp. 2485-2490, 2007.

A FULLY CMOS-COMPATIBLE MICRO-PIRANI GAUGE BASED ON A CONSTANT CURRENT

Jiaqi Wang, and Zhenan Tang

Dalian University of Technology, Department of Electronic Engineering, Dalian, China

ABSTRACT

The paper presents a CMOS-compatible micro-Pirani sensor which consists of the tungsten micro-hotplate and the constant current circuit based on the operational amplifier. The advantage of the constant current circuit for the micro-Pirani sensor is that it is more sensitive than the constant temperature one in the low gas pressure. The micro-Pirani system is implemented in an industrial 0.5-μm CMOS process. The measurement results show that the designed micro-Pirani sensor has a good response to the gas pressure, especially in the low gas pressure ranging from 1Pa to 100Pa, the sensitivity of the sensor is 0.23mV/Pa and the linearity is 4.95%.

INTRODUCTION

The Pirani gauge is a thermal-conductivity-type vacuum sensor which has been widely used in vacuum testing [1]. Its working principle is based on the fact that the heat loss of a hotplate to its ambient through gas conduction is proportional to the molecular density of gas in the vacuum system [2]. The surface micromachined micro-hotplate is usually employed as the sensory component for the micro-Pirani gauge [3] [4]. The major bias circuits for the micro-hotplate to form a micro-Pirani sensor consist of the constant temperature circuits and the constant current circuit. C. H. Mastrangelo introduced a constant temperature circuit for the micro-Pirani sensor based on the microbridge [5]. E. H. Klaassen reported a temperature control system for the micro-Pirani sensor [6]. The more complex temperature control system was introduced by D. Barrettino for the micro-hotplate in the gas detection system [7]. Some researchers focused on the constant current circuit for the micro-hotplate. M. Y. Afridi studied a micro-hotplate working under the constant current circuit for the gas detection and he obtained a good temperature distribution of the micro-hotplate [8]. Zhang introduced a micro-Pirani based on the micro-hotplate technology in the constant current circuit [9].

In our study, we implement a micro-Pirani gauge under constant current circuit based on the surface micromachined micro-hotplate. The advantage of the constant current circuit is that the sensor has more sensitivity in the low gas pressure than that in the constant temperature circuit, as shown in Figure 1. This is because in the low gas pressure, the micro-hotplate temperature in the constant current circuit is much higher than that in the constant temperature circuit. The drawback of the constant current circuit is that the micro-hotplate may be destroyed due to the high temperature in the low gas pressure. Thus, we design a tungsten micro-hotplate which has the high melting point. Also, the tungsten is usually used as the plug material in a standard CMOS process, thus the designed micro-Pirani gauge is fully CMOS compatible.

The measurement results show that the micro-Pirani pressure sensor with the tungsten micro-hotplate in the constant current circuit has a good response to the gas pressure from 1Pa to 10^5Pa, especially in the range from 1Pa to 100Pa, the sensitivity of the sensor is 0.23mV/Pa and the linearity is 4.95%.

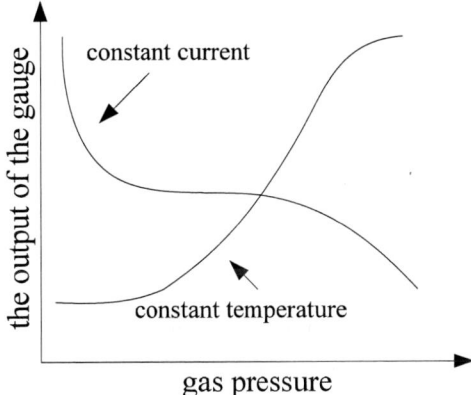

Figure 1: The schematic of the different output range between the constant current circuit and constant temperature circuit for the same micro-Pirani gauge.

THE FABRACATION OF TUNGSTEN MICRO-HOTPLATE

Tungsten has been traditionally used as a plug material to form via pathways between various metal layers and the silicon substrate because of its ability to uniformly fill the high-aspect ratio vias when deposited by chemical vapor deposition (CVD) methods. It also serves as a barrier to inhibit diffusion and reaction between silicon and the first metal layer [10].

In our design, the tungsten micro-hotplate is fabricated in an industrial 0.5-μm CMOS process that has two polysilicon layers (Poly1 and Poly2) and three metal layers (Metal1, Metal2 and Metal3). The metal plug between Metal1 and Metal2 is tungsten and the one between Metal2 and Metal3 is aluminum. In the design of tungsten micro-hotplate, tungsten is employed as the heater in the form of serpentine resistor instead of via plug. The anchors of the tungsten resistor are connected to Metal2, leaving the Metal1 unconnected. The 0.34μm thick Poly2 is used as a sacrificial layer below the tungsten micro-hotplate. The etch-windows of the tungsten micro-hotplate are opened during the bonding pad patterning in a standard CMOS process, as shown in Figure 2(a).

In Figure 2(b), the etch-windows and the bonding pads are etched simultaneously during the bonding pad etching process and the Poly2 is exposed. When we get the die of the micro-Pirani sensor from the integrated circuit (IC) foundry, the post-CMOS fabrication is in process in our laboratory, as shown in Figure 2(c). The Poly2 is etched to suspend the tungsten micro-hotplate by an improved tetramethyl ammonium hydroxide (TMAH) solution which doesn't etch the exposed aluminum pads [11]. Before starting the process, the new aluminum layer is sputtered at the backside of the die to prevent etching from the backside. There are about 8 hours required to remove the sacrificial layer in TMAH solution.

978-1-4244-2977-6/09 $25.00 © 2009 IEEE   825

Figure 3 shows the microscope picture of tungsten micro-hotplate. The tungsten micro-hotplate with the square area of 40μm×40μm is suspended by four beams which have the length of 30μm and the width of 15μm. The width of the tungsten resistor is 0.8μm. The suspended gap between the tungsten micro-hotplate and the substrate is 0.34μm.

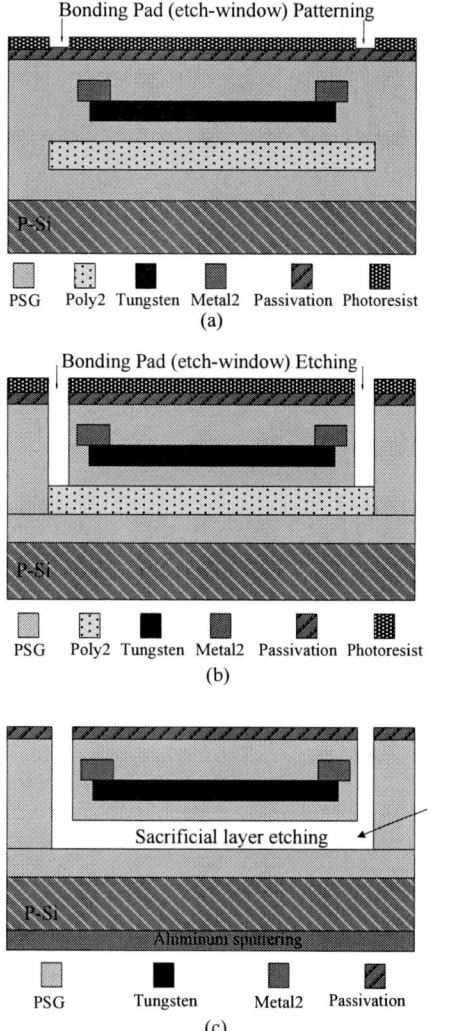

Figure 2: The tungsten micro-hotplate fabrication process.

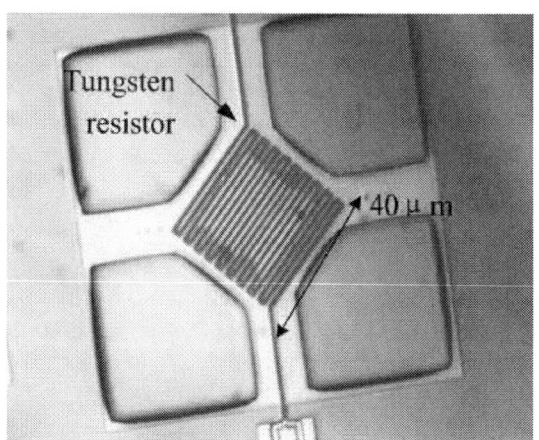

Figure 3: the microscope picture of the tungsten micro-hotplate.

Tungsten micro-hotplates have been encapsulated in the standard IC ceramic package and exposed to the temperatures ranging from 35°C to 150°C to measure the temperature coefficient. The result is that the temperature coefficient of tungsten is about $1.5 \times 10^{-3}/°C$.

A current source (Keithley 2400) forces a constant current to the resistor of the tungsten micro-hotplate to test its thermal impedance. Heating power can be calculated from the current and voltage across the tungsten resistor. Temperature is determined from the measured resistance with the temperature coefficient of resistance. Then a linear curve fit is applied to the power versus temperature data to extract thermal impedance that is the slope of the linear curve [12]. The value of the thermal impedance of the tungsten micro-hotplate is about 17°C/mW, seen in Figure 4.

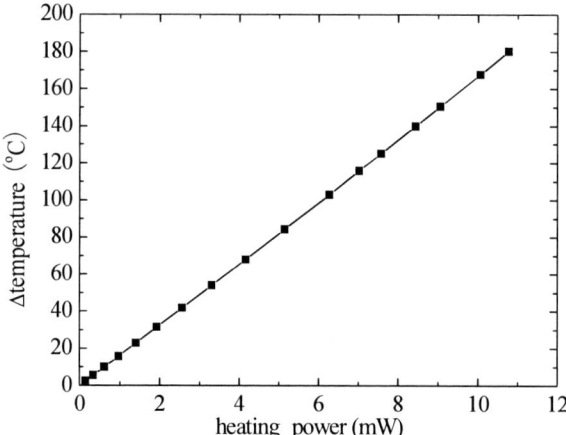

Figure 4: The thermal impedance (slope) of the tungsten micro-hotplate.

IMPLEMENTATION OF MICRO-PIRANI GAUGE IN THE CONSTANT CURRENT CIRCUIT

The micro-Pirani pressure sensor is driven by a constant current circuit which has a good response to the low gas pressure. Due to the large temperature coefficient of tungsten, the output of the constant current circuit can provide a wide measurement range.

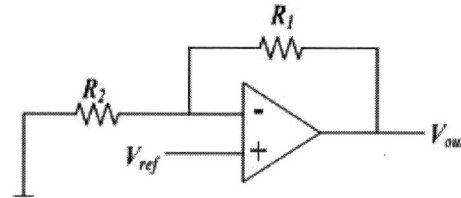

Figure 5: The constant current circuit for the sensor.

Figure 5 shows the schematic of the constant current circuit. R_1 is the tungsten resistor in the micro-hotplate and R_2 is the external reference resistor. V_{ref} is the external reference voltage. The current through R_1 is constant when the V_{ref} is a constant. V_{out} is the output voltage.

$$V_{out} = V_{ref}(1 + \frac{R_1}{R_2}) \qquad (1)$$

The output of the amplifier reflects the resistance

978-1-4244-2977-6/09 $25.00 © 2009 IEEE

variation of R_1. The resistance of R_1 increases as the gas pressure decreases. Thus, the output voltage increases as the gas pressure decreases.

The monolithic operational amplifier is designed and configured the constant current circuit. Figure 6 shows a microscope picture of micro-Pirani sensor including the tungsten micro-hotplate and the operational amplifier in the same chip.

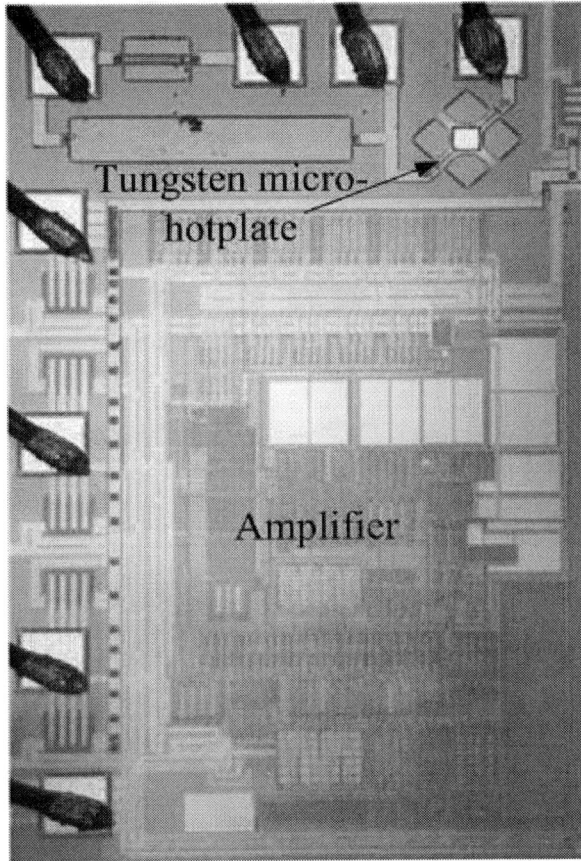

Figure 6: The microscope picture of the micro-Pirani gauge with tungsten micro-hotplate and constant current circuit.

MEASUREMENT

The value of R_1 is 230Ω and V_{ref} is 1.25V in the measurement. By adjusting R_2, we can obtain the proper current which not only has a good performance of pressure sensor, but also cannot destroy the micro-hotplate in the low gas pressure. The bias current is 7mA in the measurement. The pressure range of the vacuum system used for the measurement is from 0.1Pa to atmosphere. The sensor is fixed in the cylindrical chamber and the measurement procedure can be divided into two steps. First, the gas pressure is decreased from atmosphere to 0.1Pa. Second, when the gas pressure of the vacuum system is below 1Pa, the pump is isolated, then the gas slowly bleeds into the cylindrical chamber to the desired pressure, and the real-time output voltage of the constant current circuit is recorded with a personal computer using the A/D card. The results are shown in Figure 7 and Figure 8. In Figure 7, as the gas pressure changes from 1Pa to 10^5Pa, the output voltage changes from 2900mV to 3543mV. In Figure 8, when the gas pressure is from 1Pa to 100Pa, the full range of output voltage is from 3520mV to

3543mV, the sensitivity of the sensor is 0.23mV/Pa and the linearity is 4.95%. The designed micro-Pirani pressure sensor can be applied to the medium vacuum measurement.

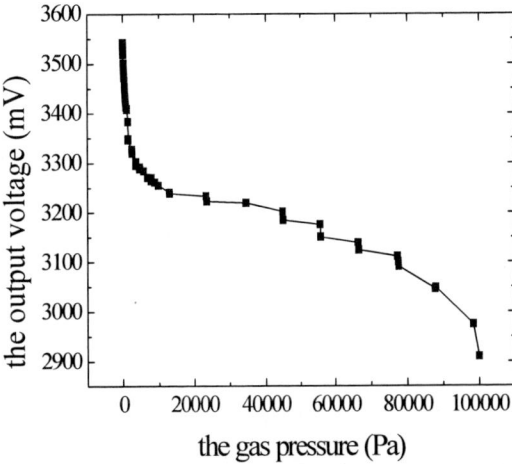

Figure 7: The measurement results of the micro-Pirani gauge, the gas pressure ranging from 1Pa to 10^5Pa.

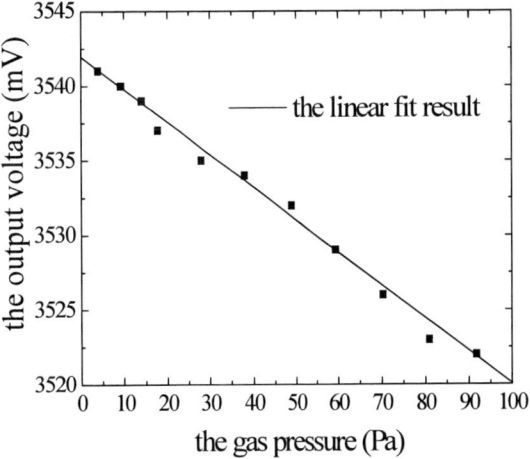

Figure 8: The measurement results of the micro-Pirani gauge, the gas pressure ranging from 1Pa to 100Pa.

CONCLUSIONS

This paper presents the micro-Pirani gauge working under the constant current circuit which is sensitive to the low gas pressure. In order to prevent the damage of the sensor in the low gas pressure due to the high temperature, a tungsten micro-hotplate is fabricated as the sensory component. The temperature coefficient of tungsten is $1.5 \times 10^{-3}/°C$. The thermal impedance of the designed tungsten micro-hotplate is $17°C/mW$. The micro-Pirani system is implemented in an industrial 0.5μm-CMOS process.

The measurement results show that the designed micro-Pirani sensor has a good response to the gas pressure, especially in the low gas pressure ranging from 1Pa to 100Pa, the sensitivity of the sensor is 0.23mV/Pa and the linearity is 4.95%. The micro-Pirani pressure sensor in our study can be applied to the medium vacuum measurement.

Because tungsten in a standard CMOS process has the large temperature coefficient and high melting point, a new series of thermal-based sensor products compatible with the standard CMOS process can be developed.

978-1-4244-2977-6/09 $25.00 © 2009 IEEE

ACKNOWLEDGEMENTS

This research was supported by the National Natural Science Foundation of China under Grant no. 90607003.

REFERENCES

[1] K. Khosraviani, A. M Leung, "The nanogap pirani - a pressure sensor with superior linearity in atmospheric pressure range", *IEEE 21st international conference on Micro Electro Mechanical Systems*, Tucson, Jan 13-17, 2008, pp. 900-903.

[2] P. K. Weng and J. S. Shie, "Micro Pirani vacuum gauge", *Rev. Sci. Instrum*, vol. 65, no. 2, pp. 492-499, 1994.

[3] Stéphane Leclerc, Robert Antaki and John F. Currie, "Novel simple and complementary metal oxide semiconductor compatible membrane release design and process for thermal sensors", *J. Vac. Sci. Technol. A*, vol. 16, no. 2, pp. 876-880, 1998.

[4] F. Völklein and A. Meier, "Microstructured vacuum gauges and their future perspectives", *Vacuum.*, vol. 82, no. 4, pp. 420-430, 2007.

[5] C. H. Mastrangelo and R. S. Muller, "Microfabricated thermal absolutepressure sensor with on-chip digital front-end processor", *IEEE J. Solid-State Circuits*, vol. 26, no. 1, pp. 1998-2007, 1991.

[6] Erno H.Klaassen, Gregory T. A. Kovacs, "Integrated thermal-conductivity vacuum sensor", *Sens. Actuators A, phys.*, vol. 58, no. 1, pp. 37-42, 1997.

[7] D. Barrettino, M. Graf and W. H. Song, "Hotplate Based Monolithic CMOS Microsystems for Gas Detection and Material Characterization for Operating Temperatures up to 500 °C ", *IEEE J. Solid-State Circuits*, vol. 39, no. 7, pp. 1202-1207, 2004.

[8] M. Y. Afridi, J. S. Suehle, M. E. Zaghoul, "A monolithic CMOS microhotplate-Based gas sensor system", *IEEE Sensors Journal*, vol. 2, no. 6, pp. 645-655, 2002.

[9] F. T. Zhang, Z. Tang, J. Yu, R. C. Jin, "A micro-Pirani vacuum gauge based on micro-hotplate technology", *Sens. Actuators A, phys.*, vol. 126, no. 2, pp. 300-305, 2006.

[10] M. Quirk and J. Serda, *Semiconductor Manufacturing Technology*, Prentice Hall., New Jersey, 2001.

[11] G. Yan, P. Chan, M. Hsing, R. Sharma, J. Sin and Y. Wang, "An improved TMAH Si-etching solution without attacking exposed aluminum", *Sens. Actuators A, phys.*, vol. 89, no. 1/2, pp. 135-141, 2001.

[12] J. Chae, B. H. Stark and K. Najafi, "A micromachined Pirani gauge with dual heat sinks", *IEEE Trans. Adv. Packag.*, vol. 28, no. 4, pp. 619-625, 2005.

GLASS AND QUARTZ MICROSCINTILLATORS FOR CMOS COMPATIBLE MULTI-SPECIES RADIATION DETECTION

Randy Waguespack, Heath Berry, Scott Pellegrin, and Chester G. Wilson

Department of Electrical Engineering/Institute for Micromanufacturing, Louisiana Tech University

ABSTRACT

This paper reports small scale radiation detectors sensitive to alpha, beta, gamma, and neutron radiation using glass and quartz doped with ^{10}B nanoparticles utilizing CMOS fabrication techniques. Two microscintillators have been fabricated and tested; one relies on sintered glass frit doped with boron nanoparticles; the other relies on diffused boron. Radiation impinging the scintillation matrix produces varying optical pulses which are differentiated by on-chip pulse height spectroscopy. The quartz substrates are more transparent to the wavelength of the created optical pulses, resulting in higher count rates when compared to glass.

INTRODUCTION

Micro-scale radiation detectors are needed to combat illicit nuclear proliferation. Currently Geiger counters are the most commonly used device for measuring radiation in the field [1]; these detectors are typically the size of a cell phone. There is a MicroGeiger counter that has been developed, but it only can detect the presence of beta radiation and requires high voltage to operate [2]. When high electric fields are present in Geiger-Muller mode of operation, single radiation particles will produce exponentially more pulses making the detector very efficient [3]. Geiger counters can be used to detect all types of nuclear radiation but are unable to differentiate radiation species and are unable to detect neutrons. Scintillators remain one of the most useful methods available for detecting a wide range of radiation species [4]. Scintillators are used to convert radiation particles into optical light which can be detected and processed [5]. Optical light is processed through the use of photodiodes and photomultiplier (PM) tubes which convert the pulse into a measurable electrical signal [6]. There have been several reports of multichannel micro-radiation detectors

which use polymer scintillating resins [7]. Polymers can absorb moisture so they exhibit long term drift and have a relatively high thickness; glass and quartz have a higher density than polymer allowing further miniaturization.

An integrated photodiode allows for the detector to be processed using CMOS fabrication techniques. Other multi-species radiation scintillator detectors have been developed but they require the use of an external PM tube which increases packaging size [8]. Photodiodes offer the advantage of higher quantum efficiency allowing for potentially better energy resolution, low power consumption, and improved ruggedness compared with PM tubes that are used for scintillation counting. Because of the relatively small distance which charge must move in the diode, the response time is comparable to that of PM tubes. Photodiodes also have the advantage that they are good for use with timing applications [4].

The CMOS compatible detector is a compact package consisting of on chip detection as well as pulse high spectroscopy of radiation particles. This device detects and discriminates neutron, gamma, beta and alpha radiation. The use of quartz and a photodiode allows for the device to be fabricated using standard CMOS techniques.

CMOS COMPATIBLE DETECTOR

On-chip CMOS interfacing is key to the production of small integrated radiation detection packages that are cheaper, more reliable, and easier to produce than assembled devices that use commercial off-the-shelf parts. The CMOS compatible detector reported in this work uses an on chip photodiode to detect radiation entering the device window. Two types of microscintillator have been fabricated for use in the device. The glass scintillator uses sintered glass frit doped with ^{10}B nanoparticles; the glass

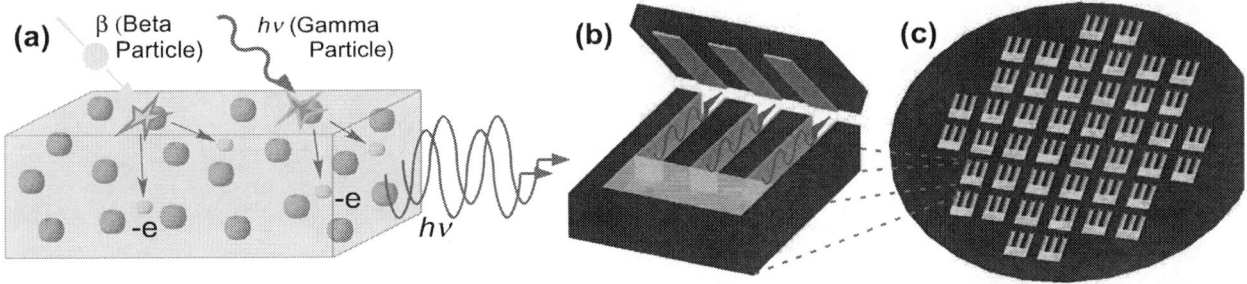

Figure 1: Glass substrate doped with boron nanoparticles converts impinging radiation particles into optical pulses which exit the scintillator (a). Optical pulses exiting the scintillator are directed away from the scintillating window by gold coated waveguides (b). The microdevice is patterned on chip for a completely integrated radiation detection package (c).

is anodically bonded to the device. Quartz scintillator relies on boron nanoparticles defused into a SiO_2 layer grown onto the photodiode. Radiation impinging the scintillation matrix produces varying optical pulses which exit the material. Gold coated waveguides embedded in the device direct the created optical pulses to the photodiode, converting the light pulse to electrical current (Figure 1). The detector is capable of detecting all types of emitted radiation particles (Figure 2).

Neutron Detection: The detection of neutrons is difficult due to their limited reaction with other particles. Using [10]B nanoparticles allows for the detection of these neutral charges. Neutrons interact with the [10]B nanoparticles releasing alpha particles. The alpha particles then interact with the quartz creating light. [10]B has a high thermal neutron cross section ensuring that the maximum number of impinging neutrons can be absorb, converted and detected.

Alpha Detection: The detection of alpha particles is accomplished using the glass and quartz substrates. Alphas transfer their energy to valence electrons due to columbic forces. If the amount of energy transferred from the alpha particle to an electron is enough to strip the valence electron, the electron scintillates the substrate.

Beta Detection: The detection of beta particles occurs when the particles are scintillated through the substrate. Fast electrons, beta particles, have a light output that is ten times greater than the output of alpha particles [4]. The higher energy allows for the better beta detection when using translucent substrates.

Gamma/X-ray Detection: The [10]B nanoparticles allow for the detection of gamma and X-ray particles. The gamma and X-ray particles interact with the nanoparticles resulting in the emission of secondary electrons. These secondary electrons are emitted through photoelectric absorption, pair production and Compton scattering [4].

Photons that correspond to scintillating light carry about 3-4 eV of energy which is sufficient to create electron-hole pairs. Unlike conventional a photocathode the conversion is not limited by the need for charge carriers to escape from a surface. This allows for a higher

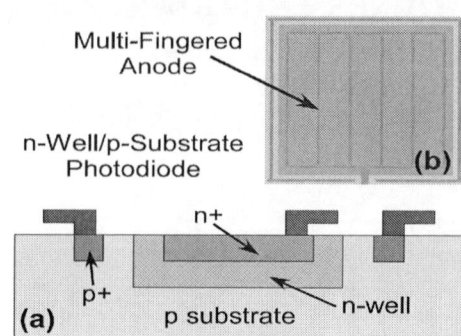

Figure 3: N/P CMOS photodiode detects scintillation light pulses from radiation particles impinging on scintillation matrix (a). On-chip layout of the photodiode detector (b).

primary charge to be created by the light from the scintillator [4]. Light particles leaving the detector induce a charge which is converted into electrical current pulses (Figure 3). The current information is processed through on chip circuitry. Each current pulse is then amplified and processed. Different particles give off different levels of energy which allows for discrimination between all four types of radiation through pulse height spectroscopy.

FABRICATION

The initial device is fabricated using borosilicate glass. The glass was chosen because of the ease of processing for initial device testing. The cavities of the device were etched using micro sandblasting techniques (Figure 4a,c). Gold is then deposited into the etched channels and on to the cap that covers the waveguides (Figure 4b). Glass and quartz scintillator are then patterned into the device window (Figure 4d). The glass cap is then bonded to the device to enclose the waveguides. The outside of the device is then covered with an opaque powder coating to cover the exposed glass to prevent particles from scintillating through the walls of the device (Figure 5).

Signal discrimination is done with on-chip CMOS circuitry and a photodiode. The photodiode design is an n-

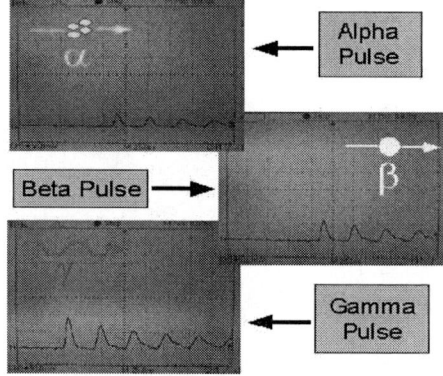

Figure 2: The boron doped scintillators produces varying output voltages each specific to one species of radiation. This allows for discrimination of particles using pulse height spectroscopy.

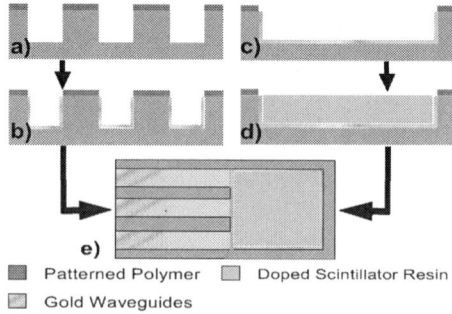

Figure 4: Waveguide channels etched into borosilicate glass wafer (a). Gold is deposited into waveguide channels (b). Scintillating window area is deposited with gold (c). Scintillator is patterned into the detector window (d). Completed device (e).

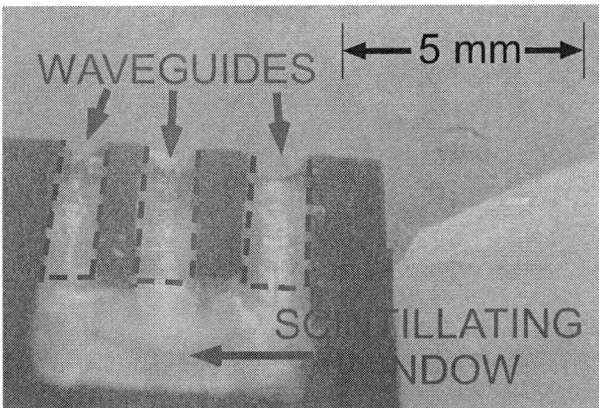

Figure 5: Photo of scintillating micro device with patterned scintillating quartz material inside of the etched cavity and gold coated waveguides.

well/p-substrate layout which is compatible with conventional 0.35 μm CMOS fabrication techniques and utilizes a multi-fingered anode for charge collection on the detector surface. The photons produced by the interaction of the radiation species and the nanoparticle doped scintillating resin are converted into electron-hole pairs by the photodiode collector surface, producing a detectable current. The produced signal is then amplified and the size of the voltage pulse produced determines the species of radiation (Figure 6). The CMOS detector is being fabricated by the MOSIS foundry.

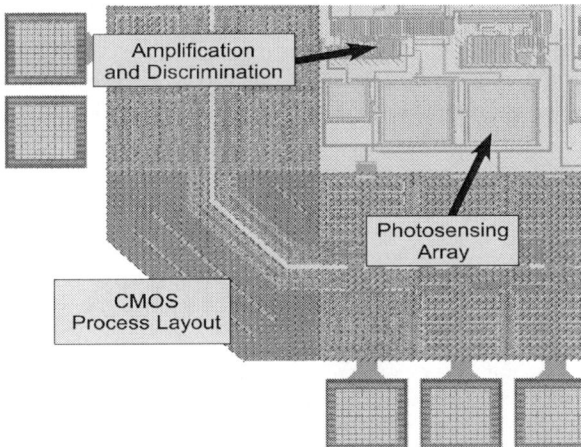

Figure 6: On-chip circuitry provides discrimination for all species of radiation through pulse height analysis of optical signals detected by the photodiode.

EXPERIMENTAL RESULTS

The detector was tested using a PM tube connected to an oscilloscope to find the performance of the detector with beta and gamma/X-ray radiation (Figure 7). As the ^{90}Sr beta source is moved away from the detector the count rate decreases. The count rate per minute of the beta particles decrease rapidly because the low energy beta particles are absorbed by the air gap between the source and the detector; beta detection count rates versus source distance were found similar to more expensive detectors (Figure 8). At larger detector source distance the counts per minute starts to fall off more exponentially.

The count rate versus distance is increased using the ^{60}Co source; the gamma particles interact with the boron causing lower energy particles to be created (Figure 9). The attenuation for the gamma source also falls off exponentially over greater distances.

A quartz scintillator with diffused boron has also been tested with this device. This scintillator was tested with both the ^{90}Sr for beta detection and the ^{60}Co for gamma particle detection. The quartz shows similar energy spectra to the glass scintillator, allowing for the microdevice to be constructed with glass or quartz scintillators (Figure 10). The quartz scintillator is able to detect neutron radiation particles, which being neutral, are undetected with traditional solid state radiation detectors (Figure 11).

The CMOS integrated circuit for the signal discrimination was also fully simulated and tested before being fabricated. In order to conserve space on the layout, the circuit employs an amplifier utilizing switched capacitor feedback to provide the amount of amplification necessary while keeping the area used minimal.

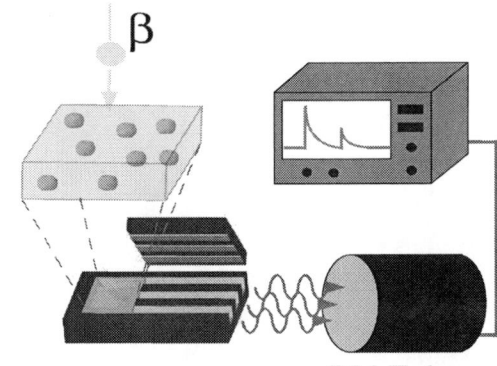

Figure 7: Glass and quartz scintillators were tested with different radiation species, a PM tube connected to an oscilloscope was used to detect the optical pulses.

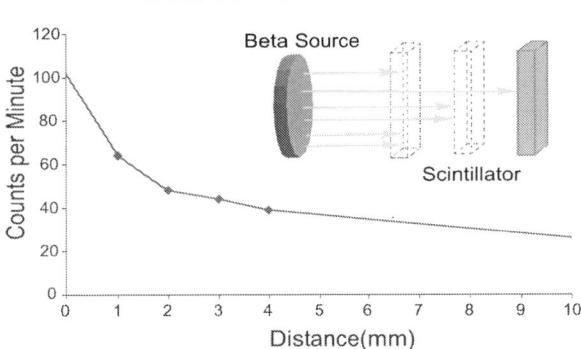

Figure 8: Testing the beta detector with a beta source for a function of counts per minute over varying distances. As distance increases the counts per minute reduce because of absorption of low energy betas by the air gap.

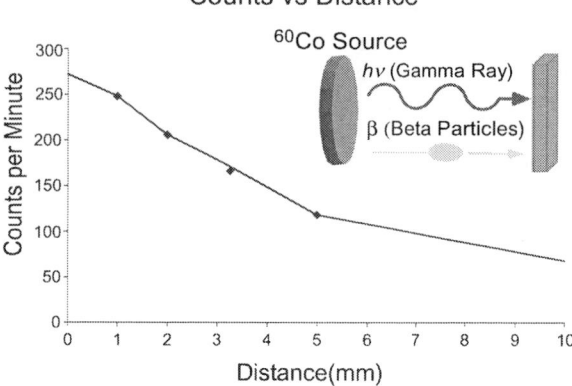

Figure 9: ^{60}Co emits both gammas and betas having a larger radiation flux with higher energy particles.

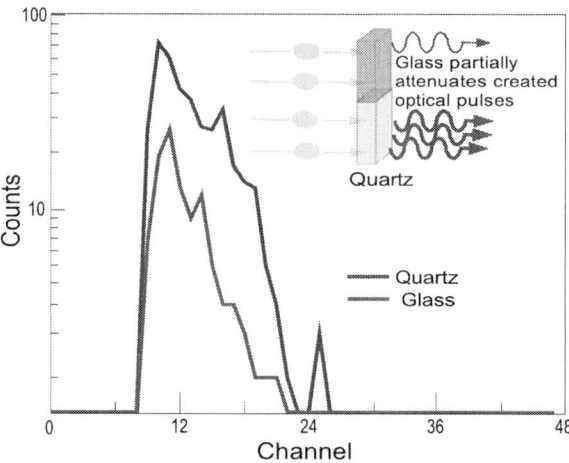

Figure 10: Glass and quartz have similar energy spectra allowing for glass to be anodically bonded or quartz grown onto the silicon wafer. The quartz spectra is higher because it is more transparent to the wavelength of created optical pulses.

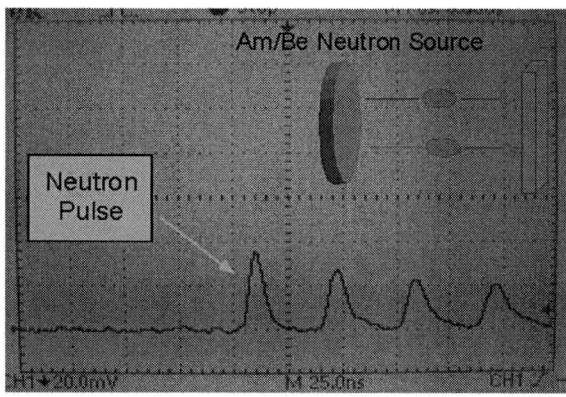

Figure 11: Quartz scintillator has the ability to detect neutrons allowing for traditional fabrication techniques to be utilized in creating a CMOS device.

CONCLUSION

The CMOS compatible detector is able to detect and discriminate all four types of radiation species; alpha, beta, gamma, and neutron radiation particles. The glass and quartz scintillators use ^{10}B nanoparticles to interact with all species of radiation. The energy is converted in to charged particles that are absorbed thought the photodiode. Quartz scintillator showed similar energy spectra to the glass scintillator which allows for the entire device to be fabricated along with the device's detection circuitry. Using quartz doped with boron scintillator allows for the detection of neutrons to be possible. The charged particles released after interacting with different types of radiation contain a specific amount of energy that can be discriminated using pulse height spectroscopy. Therefore, the type and energy of the impinging radiation can be determined.

ACKNOWLEDGEMENTS

The authors would like to thank the department of energy for partial funding. Funding was also provided by the Louisiana Board of Regents Research Competitiveness Subprogram.

REFERENCES

[1] Yogesh Gianchandandi, "From Antenna Stents to Wireless Geiger Counters: The Promise of Electrical Micro-discharges in the Fabrication and Operation of Microsensors", IEEE International Symposium on Micro-NanoMechatronics and Human Science, 2006.

[2] C. Wilson, C. Eun, Y. Gianchandani, "D-MicroGeiger: A Microfabricated Beta-Particle Detector with Dual Cavities for Energy Spectroscopy", *IEEE MEMS 2005*, Jan. 2005, pp. 622-625.

[3] K. Kudo, N. Takeda, A. Fukunda, "Measurement of Gamma Ray Dose in a Thermal Neutron Field by Using a ^{3}He-filtered GM Counter", IEEE Nuclear Symposium and Medical Imaging Conference Record, 1995, pp. 819-822.

[4] G. K. Knoll, *Radiation Detection and Measurement 3 ed.*, John Wiley & Sons, Inc., Danvers, MA, 2000.

[5] W. Price, *Nuclear Radiation Detection 2ed.*, McGraw-Hill, New York, 1964.

[6] J. B. Birks, *The Theory and Practice of Scintillation Counting*, Pergamon Press, Oxford, 1964.

[7] S. Pellegrin, C. Whitney, C.Wilson, "A multichannel nanoparticle scintillation microdevice with integrated waveguides for alpha, beta, gamma, x-ray, and neutron detection", *IEEE International Conference on Micro Electro Mechanical Systems (MEMS)*, v 2006, pp 682-685.

[8] R. Dasaka, S. Pellegrin, M. Kamavaram, C.Wilson, "Micromachined Scintillation Devices with Charge Conversion Nanoparticles for Neutron and Beta Particle Detection", Proceeding-MicroTas 2005, 1, Oct 2005, pp. 472-475.

SEMICONDUCTING Y-BA-CU-O INFRARED MICROBOLOMETER ARRAY FABRICATED AND CHARACTERIZED WITH CCBDI ROIC

S. Kumar[1], and D.P. Butler[2]

[1,2]The University of Texas at Arlington, Texas, USA

[1]Semiconductor Group at Texas Instruments, Texas, USA

ABSTRACT

Semiconducting Yttrium Barium Copper Oxide (Y-Ba-Cu-O) microbolometers based on self-supporting structure are fabricated and characterized with on-chip constant current buffered direct injection (CCBDI) readout integrated circuit (ROIC) in AMI 1.5μm double-poly-double-metal n-well 2.5V complementary metal-oxide-semiconductor (CMOS) technology. Self-supporting Y-Ba-Cu-O microbolometers exhibit low thermal mass, hence low thermal time constant, and are integrated with a silicon readout circuit for first time. The CCBDI readout circuit is designed to work both with a traditional frame rate of 30Hz as well as a novel, improved frame rate of 200Hz for a 640x480 array for faster thermal imaging in commercial, military and biomedical applications.

INTRODUCTION

The self-supporting Y-Ba-Cu-O microbolometers in the semiconducting phase are fabricated and characterized with the CCBDI readout circuit. Infrared (IR) detectors built using low-resistivity semiconducting Y-Ba-Cu-O are post-CMOS compatible and thus enabling the IR sensitive pixels to be built on top of the silicon-based readout circuitry

The thermal isolation of the microbolometers is varied by designing two different electrode arm geometries. Figure 1 shows cross-sectional view of the fabricated Y-Ba-Cu-O microbolometer where the first geometry is designed to obtain a relatively fast 200Hz frame rate (Figure 2a) while maintaining moderate detectivity and the second arm geometry is designed to achieve a traditional 30Hz frame rate (Figure 2b) with high detectivity microbolometers. The 18-μm-long electrode arm geometry capitalizes on the low thermal mass of self-supporting Y-Ba-Cu-O microbolometer to provide a higher frame rate of 200Hz. The 75-μm-long electrode arm geometry is designed to achieve a traditional 30Hz frame rate.

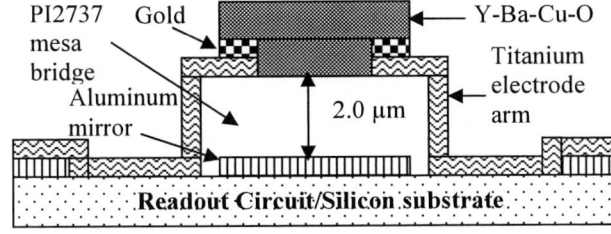

Figure 1: Cross-sectional view of a Y-Ba-Cu-O microbolometer.

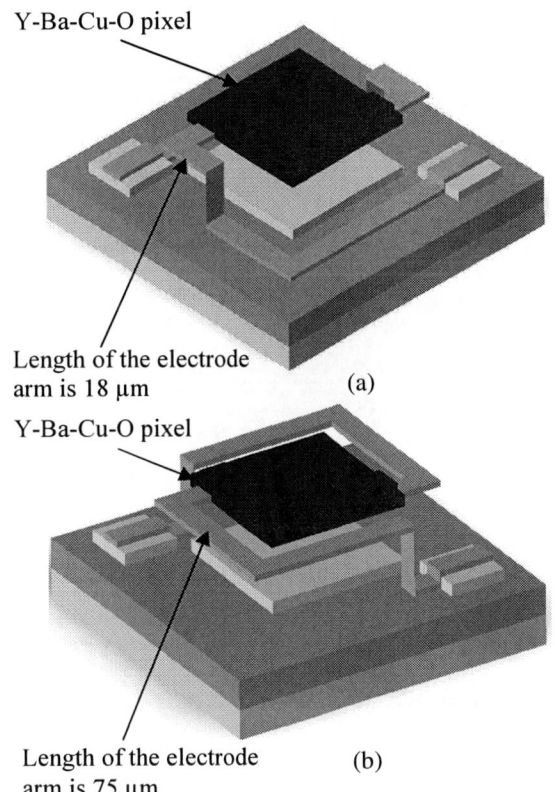

Figure 2: CoventorWare® model of the micromachined microbolometers showing a) the 18-μm-long electrode arm geometry for 200Hz frame rate and b) the 75-μm-long electrode arm geometry for 30Hz frame rate.

Y-Ba-Cu-O microbolometers possess a relatively high temperature coefficient of resistance (TCR) of -3 to -3.5 %/K and can be deposited by rf magnetron sputtering at ambient temperature from a single composite target [1]. The low thermal mass of the microbolometers enables a faster response time with the same degree of thermal isolation which can be utilized for imaging applications where higher frame infrared cameras are desired. By not requiring cryogenic cooling, uncooled microbolometers have found an attractive application for thermal detection with substantial reduction in system cost, size, weight, and power consumption as compared to cooled photon detectors. A number of other material and systems have been investigated for their use in uncooled infrared detection [2].

The CCBDI readout circuit is designed to provide an extrapolated frame rate of 200Hz for a 640x480 focal plane array. This will help improve the performance of uncooled IR cameras by faster thermal imaging in industrial, military, and biomedical applications. The CCBDI readout circuit is designed to detect the resistance

978-1-4244-2977-6/09 $25.00 © 2009 IEEE

change caused by the incident IR radiation on the microbolometer. The CCBDI circuit offers advantages of high linearity and uniformity, low offset error with maximum sensitivity for a bias current and great potential in the application of large uncooled bolometric focal plane arrays (FPAs) [3]. CCBDI readout circuit senses the signal from Y-Ba-Cu-O pixel array and multiplex the readout to a common output using row and column shift registers.

DC analysis (Figure 3) of 4x4 array CCBDI readout circuit plotting chip output voltage (readout voltage) versus bolometer resistance confirms the readout voltage of -174.45mV at Y-Ba-Cu-O microbolometer resistance of 1.58MΩ with a 1MΩ Y-Ba-Cu-O reference resistance.

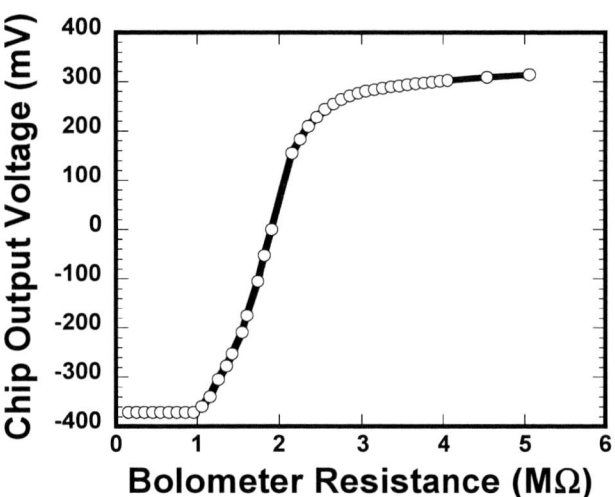

Figure 3: DC analysis of the 4x4 array with CCBDI readout circuit showing the change in output voltage with bolometer resistance.

For a 640x480 CCBDI array (extrapolated from the 4x4 array CCBDI readout circuit), the readout time for 307200 Y-Ba-Cu-O pixels is 4.79 ms, and this translates to a frame rate of 208Hz. For a 640x480 CCBDI array, the power dissipation calculated for Y-Ba-Cu-O microbolometers and reference resistors is 1.92 mW, for the 3.5 V and 1.5 V biasing circuits is 576 mW, and for CCBDI amplifier is 1.25 mW. The overall power dissipation calculated for the 640x480 CCBDI array is 579.17 mW.

INFRARED MICROBOLOMETER ARRAY FABRICATION

The self-supporting semiconducting Y-Ba-Cu-O microbolometers with the cell size of 31x31 μm^2 are fabricated on top of CCBDI readout circuitry with two different electrode arm geometries to provide the thermal isolation from the substrate. This is achieved by changing only one mask, which is used to pattern the sacrificial polyimide PI2737.

The MOSIS foundry service is used to fabricate the CCBDI readout circuit. The Y-Ba-Cu-O microbolometers are fabricated on the CMOS die with the CCBDI readout circuitry to form FPAs using surface micromachining.

Surface micromachining enables the fabrication of microbolometers by stacking and patterning thin films on the die integrated with CCBDI readout circuitry and wiring. The fabrication technique also allows for control over microbolometer response time and detectivity. Higher sensitivity is obtained by further increasing the thermal isolation, resulting in increased responsivity.

The CMOS readout circuit is fabricated with 8x8 μm^2 glass cuts to aid the electrical connection between the readout circuitry, the Y-Ba-Cu-O reference resistors, and the Y-Ba-Cu-O microbolometers. The microbolometer fabrication process started with bonding the CMOS readout die onto a silicon carrier wafer (Figure 4a) using polyimide PI2555.

At first, the aluminum (Al) mirror (Figure 4b) is deposited by thermal evaporation to obtain 400 nm of thickness and then patterned by liftoff, to form an optically resonant cavity with the suspended Y-Ba-Cu-O thermometer. The cavity height is designed to be 2 μm, at approximately $\lambda/4$ wavelength, to increase absorption in the 8-14 μm atmospheric transmission window. Next, the sacrificial polyimide layer (HD Microsystems PI2737) is spin coated and patterned using negative lithography (Figure 4c). The polyimide is cured at 275^{0}C for up to 4 hours. The PI2737 polyimide layer thickness, around 2 μm, forms the polyimide mesa bridge to determine the distance between the Y-Ba-Cu-O and aluminum mirror. The 100-nm-thick or 200-nm-thick titanium is sputtered onto the die and patterned by liftoff to form electrode arms that support the Y-Ba-Cu-O thermometer above the substrate. One of two electrode arm geometries investigated is selected by the sacrificial polyimide pattern. The 200-nm-thick and 18-μm-long electrode arm is designed to achieve a relatively fast 200Hz frame rate. Similarly, the 100-nm-thick and 75-μm-long electrode arm is designed to achieve a traditional 30Hz frame rate. The titanium (Figure 4d) is deposited by rf magnetron sputtering at 150W in pure argon environment at 10 mTorr pressure. The titanium electrode arms hold the Y-Ba-Cu-O thermometer above the substrate to achieve thermal isolation. The titanium has low thermal conductivity compared to other materials such as Au or Al. The low thermal conductance of the electrode arms minimizes the heat transfer from the Y-Ba-Cu-O thermometer to the substrate and results in high responsivity.

Next, a 40-nm-thick gold film is deposited and patterned by liftoff to form an Ohmic contact between the Ti arms and the Y-Ba-Cu-O thermometer (Figure 5a). The Au is sputtered at 100W in pure argon environment at 10 mTorr pressure. Au forms a good electrical contact with semiconducting Y-Ba-Cu-O. Then, the 400-nm-thick Y-Ba-Cu-O thermometer (Figure 5b) is deposited by RF magnetron sputtering at 75W in pure argon environment at 10 mTorr pressure. The Y-Ba-Cu-O thermometer is patterned by liftoff. Finally, the sacrificial polyimide layer (Figure 5c) is removed by ashing in oxygen plasma for approximately 10 hours. After ashing the sacrificial polyimide, the self-supporting   Y-Ba-Cu-O pixel is suspended by titanium arms providing thermal isolation from the substrate.

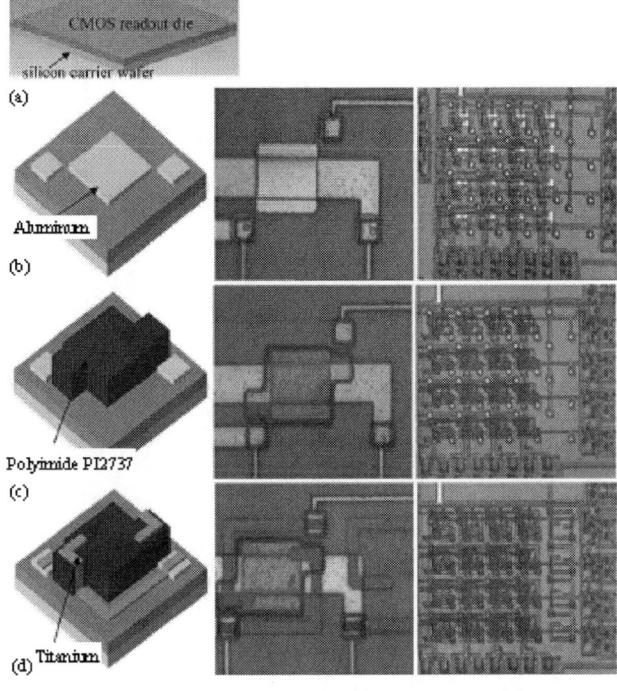

Figure 4: (a) Fabrication starts with the CMOS readout die bonded to silicon carrier wafer. (b) Aluminum deposition. Shown in a CoventorWare® schematic, a Unitcell, and a 4x4 array integrated with the CCBDI readout circuitry. (c) Sacrificial polyimide PI2737 coating and patterning for a 200Hz frame rate shown for the same series of photos. (d) Titanium deposition and patterning to form the 18-μm-long electrode arms with the same series of photos.

Figure 5: Microbolometer fabrication continued using a Coventorware® schematic, image of a Unitcell, and a 4x4 array integrated with the CCBDI readout circuitry. (a) Gold contact deposition and patterning. (b) Y-Ba-Cu-O thermometer deposition and patterning. (c) Sacrificial polyimide PI2737 ashing to complete the microbolometer

fabrication process.

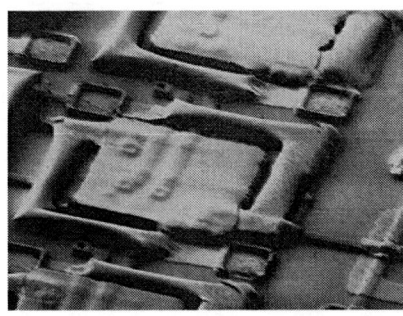

Figure 6: SEM micrograph of a portion of the fabricated Y-Ba-Cu-O 4x4 pixel microbolometer array integrated with the CCBDI readout integrated circuit.

Figure 7: Bonded packaged die of 0.5x0.5 cm² comprising self-supporting Y-Ba-Cu-O microbolometers fabricated and characterized with CCBDI readout integrated circuit.

A scanning electron microscope image of the self-supporting-Y-Ba-Cu-O microbolometer array fabricated and integrated with CCBDI readout circuit is shown in Figure 6. An optical photograph of the packaged die (0.5x0.5 cm²) comprised of the self-supporting Y-Ba-Cu-O microbolometer array integrated with the CCBDI readout circuitry is shown in Figure 7.

MICROBOLOMETER CHARACTERIZATION WITH CCBDI ROIC

The optical characterization involved the responsivity and detectivity measurements of the self-supporting Y-Ba-Cu-O microbolometer detector array with CCBDI readout integrated circuitry using an IR blackbody as the broadband source. The 4x4 pixel microbolometer array mounted inside the cryostat was illuminated using ZnSe window and a 2.5 μm long pass filter. The frequency chopper was used to modulate the infrared incident on the microbolometer array. The measured temperature coefficient of resistance (TCR) is -3.16 %/K at 295K and measured value of thermal conductance is 1.01×10^{-7} W/K for a 30Hz frame rate. The maximum responsivity of 1.62×10^{5} V/W and maximum detectivity of 3.51×10^{7} cmHz$^{1/2}$/W are measured of a single Y-Ba-Cu-O microbolometer pixel (pixel size of 31x31 μm²) from the 4x4 array CCBDI readout circuit for a 200Hz frame rate (18-μm-long electrode arm geometry) as shown in Figure 8.

978-1-4244-2977-6/09 $25.00 © 2009 IEEE

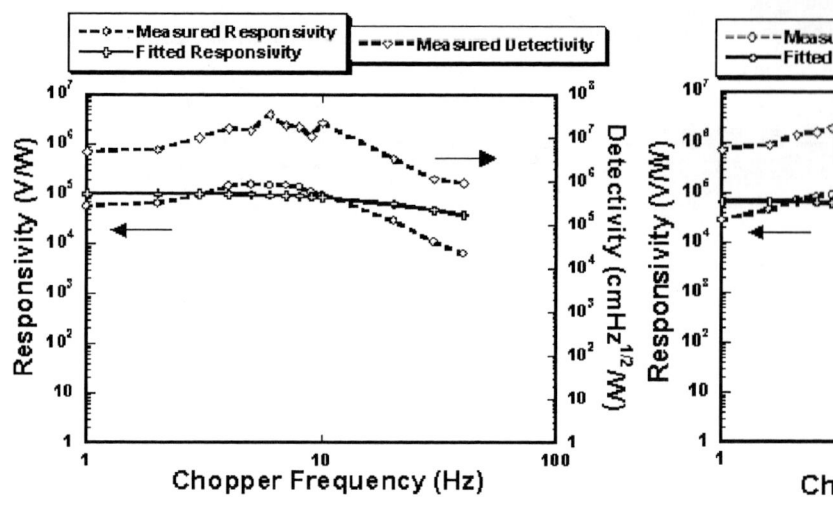

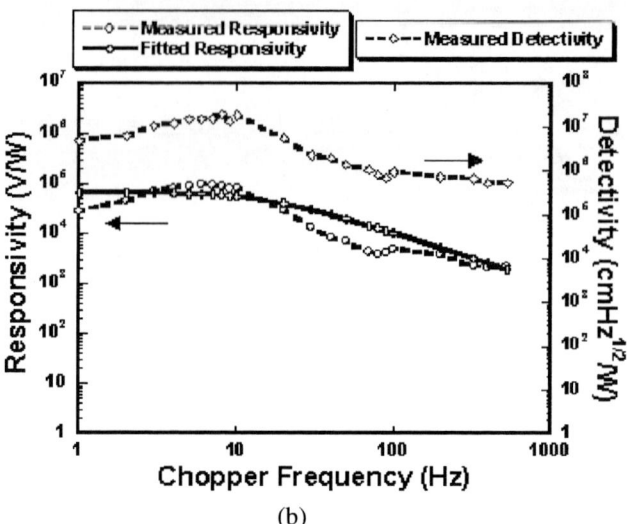

(b)

Figure 8: Responsivity and Detectivity versus chopper frequency of a single Y-Ba-Cu-O microbolometer pixel (pixel size of 31x31 μm^2) from the 4x4 array CCBDI readout circuit with 18-μm-long electrode arm geometry.

For the 30Hz frame rate (75-μm-long electrode arm geometry), the maximum responsivity of 1.24×10^5 V/W and maximum detectivity of 1.03×10^7 cmHz$^{1/2}$/W (Figure 9a), and the maximum responsivity of 9.48×10^4 V/W maximum detectivity of 1.98×10^7 cmHz$^{1/2}$/W (Figure 9b) were measured of a single Y-Ba-Cu-O microbolometer pixel (pixel size of 31x31 μm^2) from the 4x4 array CCBDI readout circuit.

These higher measured values of responsivity of a single Y-Ba-Cu-O microbolometer pixel from the 4x4 array CCBDI readout circuit for a faster frame rate of 200Hz confirms higher sensitivity as a novel approach over that of a traditional 30Hz frame rate.

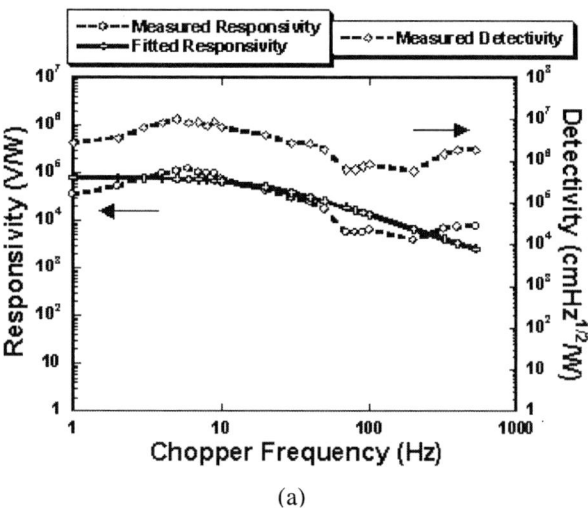

(a)

Figure 9a,b): Responsivity and Detectivity versus chopper frequency of a single Y-Ba-Cu-O microbolometer pixel (pixel size of 31x31 μm^2) from the 4x4 array CCBDI readout circuit with 75-μm-long electrode arm geometry.

CONCLUSIONS

This work has demonstrated the integration of self-supporting Y-Ba-Cu-O microbolometers with a CMOS CCBDI readout circuit.

ACKNOWLEDGEMENT

This material is based in part upon work funded by the Air Force Office of Scientific Research under grant F49620-03-1-0042. The authors would like to thank MOSIS for the fabrication of the readout circuits. The microbolometers were fabricated using the University of Texas Arlington NanoFab Teaching and Research Facility.

REFERENCES

[1] A. Jahanzeb, C. M. Travers, D. P. Butler, Z. Celik-Butler, S. G. Tan "A Semiconductor YBaCuO Microbolometer for Room Temperature IR Imaging" IEEE Transactions on Electron Devices, vol. 44, pp. 1795-1801, October 1997.

[2] P. Neuzil, Y. Liu, H.-H. Feng, and W. Zeng, "Micromachined bolometer with single-crystal silicon diode as temperature sensor," IEEE Electron Device Letters, Vol. 26, No.5, pp. 320-322, May 2005

[3] T.-H. Yu, C.-Y. Wu, Y.-C. Chin, P.-Y. Chen, F.-W. Chi, J.-J. Luo, C.D. Chiang, and Y.-T. Cherng "A New CMOS Readout Circuit for Uncooled Bolometric Infrared Focal Plane Arrays", ISCAS 2000
IEEE International Symposium on Circuits and Systems, May 28-31,2000, II-493, Geneva, Switzerland.

978-1-4244-2977-6/09 $25.00 © 2009 IEEE   836

A MICROMACHINED QUARTZ AND STEEL PRESSURE SENSOR OPERATING UPTO 1000°C AND 2000 TORR

Scott A. Wright and Yogesh B. Gianchandani

Engineering Research Center for Wireless Integrated MicroSystems (WIMS)

University of Michigan, Ann Arbor, MI, USA

ABSTRACT

This paper describes microdischarge-based pressure sensors which operate by measuring the change, with pressure, in the spatial current distribution of pulsed DC microdischarges. These devices are well-suited for high temperature operation because of the inherently high temperatures of the ions and electrons in the microdischarges, and are designed to allow for unequal expansion of electrodes and substrate during high temperature operation. These sensors use three-dimensional arrays of horizontal bulk metal electrodes embedded in quartz substrates with electrode diameters of 1-2 mm and 50-100 μm inter-electrode spacing. The sensors were operated in nitrogen over a range of 10-2,000 Torr, at temperatures as high as 1,000°C. The maximum measured sensitivity was 5,420 ppm/Torr at the low end of the dynamic range and 500 ppm/Torr at the high end, while the temperature coefficient of sensitivity ranged from -925 ppm/K to -550 ppm/K.

INTRODUCTION

High temperature pressure sensors have uses in numerous industrial sectors, and have been used in gas turbine engines, coal boilers, furnaces, and machinery for oil/gas exploration. A variety of microscale solutions have been explored. Fabry-Perot, and other interferometers, have been used on the ends of fiber optic cables and have been operated at temperatures of 800°C using sapphire membranes [1]. Another sensing technology uses Bragg gratings, which are photo inscribed into fibers, and used to trace wavelength shifts caused by pressure and temperature changes at temperatures exceeding 350°C, and potentially over 1,500°C [2]. Piezoresisitve pressure sensors with diaphragms made from silicon carbide [3], and more recently even Si [4], have been reported to operate at 600°C. Sapphire membranes have also been used in this context [5].

Microdischarges, or microplasmas, are miniature plasmas created in gases between electrodes and are used for on-chip chemical sensing and other applications. Devices utilizing microdischarges are well suited for high temperature operation as the electrons have average thermal energies exceeding 3 eV (34,815 K) [6] away from the cathode. Ions have thermal energies exceeding 0.03 eV (621 K) in a 23°C (296 K) ambient environment.

These microdischarge-based pressure sensors operate by measuring the change in spatial current distribution of microdischarges with pressure. The targeted pressure range is 10-2,000 Torr, as might be encountered in a variety of manufacturing applications. As gas pressure increases, the mean free path of ionized molecules is reduced and consequently, the breakdown and discharge characteristics are altered. Microdischarge-based pressure sensors are fundamentally different than ion gauges, which are not effective at atmospheric pressure because

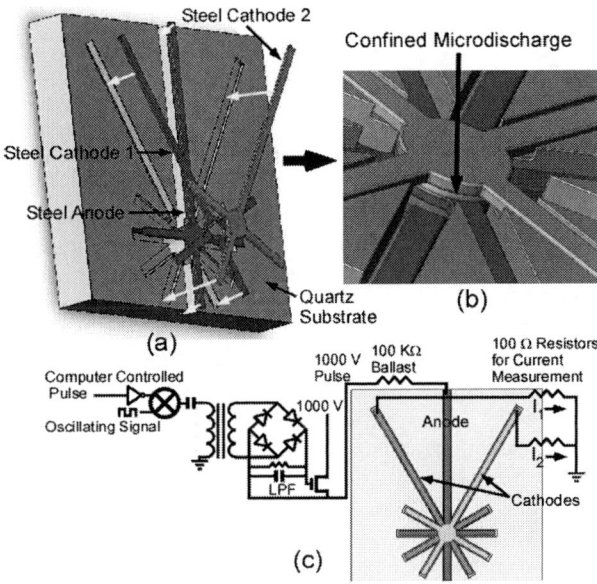

Figure 1: *Schematic of (a) stainless steel electrodes above quartz chip, illustrating electrode placement, (b) microdischarge chamber during operation, and (c) pulse generating, isolation, and readout circuitry used for sensor operation.*

the small mean free path of the created ions, 20-65 nm, makes them difficult to detect at the collector [7].

A planar thin-film version of the microdischarge-based pressure sensor was first reported in [8], and operated at temperatures upto 200°C, while this paper expands upon [9].

This paper describes a microdischarge-based microscale pressure sensor geometry that uses bulk foils in a two-cathode stack in a quartz substrate. In this effort, we explore the use of multiple cathodes. Multiple anodes may also be used; however, anode current shows excessive pressure dependence because of the high mobility of electrons that dominate it [10]. This high sensitivity results in relatively small dynamic ranges, thereby limiting the utility of multi-anode configurations. While this effort focuses on the performance of these devices in a nitrogen ambient, with appropriate encapsulation, they may be used in corrosive or liquid ambients.

DEVICE DESIGN AND OPERATION

The pressure sensor structure consists of several electrodes suspended over a cavity in a quartz chip (Fig. 1). Each electrode has a single lead for electrical contact and between one and three additional supports, which maintain the suspended position of the electrode. A microdischarge chamber exists in the center of the chip, in a through-hole, as shown in Fig. 1(b). A single disk-shaped anode electrode serves as the bottom of the

978-1-4244-2977-6/09 $25.00 © 2009 IEEE

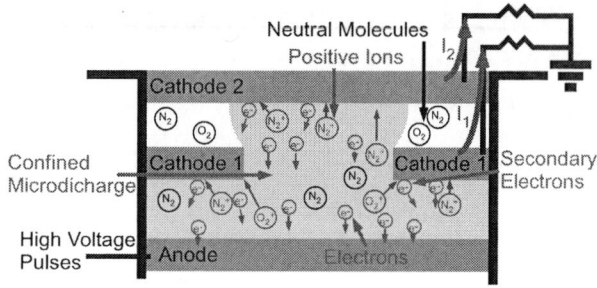

Figure 2: *Theoretical diagram of a microdischarge between a single anode and two cathodes in a microdischarge chamber. Changes in the ion distribution vary the currents in the cathodes.*

chamber while the center electrode is torus-shaped, allowing the discharges to exist between the bottom anode and both cathodes. The top cathode is disk-shaped as well, confining the discharges.

To determine the pressure, it is first necessary to separately determine the current in two of the cathodes. These current components are denoted as I_1 in the proximal cathode (cathode 1) and I_2 in the distal cathode (cathode 2). The differential current, expressed as a fraction of the total peak current, $(I_1 - I_2) / (I_1 + I_2)$, is treated as the sensor output. At low pressures, current favors the farthest cathodes, while at high pressures, the opposite occurs. An important benefit of using a differential output that is expressed as a fraction of the total is that the exact magnitudes are less important than fractional changes.

To control power consumption and parasitic heating in the pressure sensors, pulsed DC microdischarges are used, as opposed to constant DC discharges. A computer controlled, single ended, transformer coupled, gate drive circuit creates the pulses (Fig. 1(c)). A current limiting ballast resistor is used in series with the anode, and 100-Ω resistors are used in series with each cathode to measure current.

The basic operation of a DC microdischarge in a sensor is illustrated in Fig. 2, indicating electron and ion transport. The electrons are drawn towards the anode, whereas the positive ions are drawn to the two separate cathodes forming positively charged sheaths around them. Upon cathode impact, the energetic ions eject high energy secondary electrons from the cathodes, which sustain the microdischarges by ionizing additional neutral molecules and continuing the breakdown process. The current in each cathode is composed of a combination of positive ions impacting the cathodes from the microdischarges and secondary electrons ejected from the cathodes upon ion impact. Further away from the cathodes, the current is carried primarily by the faster moving electrons, which cannot reach the cathodes because of surrounding sheaths.

Sensor characteristics such as the sensitivity, pressure dynamic range, and temperature dynamic range depend on a variety of dimensional parameters, including inter-electrode spacing, electrode diameter, and the cathode thickness. (Cathode thickness effects sheath sizes as well as electrode spacing). The anode/cathode spacing in these sensors is set to produce measurable results up to 1,000°C. (The sensors are designed to function with an applied voltage of 1,000 V; altering the voltage results in different sensitivities.) A typical sensor design with electrodes

spaced 50 µm apart, 1-2 mm in diameter, and 125 µm thick is shown in Fig. 3. These sensors have active areas of 0.8-3 mm² and are fabricated on 1 cm² chips.

Figure 3: *Pressure sensor with electrodes spaced 50 µm apart and 125 µm in thickness on a U.S. penny.*

FABRICATION

A fabrication method that accommodates the expansion mismatch between electrodes and substrate is used (Fig. 4). For the electrodes, #302 stainless steel is used for several reasons. Primarily, it is robust, inexpensive, easily machinable by micro-electro-discharge machining and photochemical etching, and has a sufficient secondary emission coefficient (i.e. 0.02 secondary electrons per incident ion in a nitrogen ambient). Additionally, it is oxidation-resistant at high temperatures and can be heated to 1,420°C before melting. Alternate refractory metals such as tungsten, molybdenum, and niobium oxidize at high temperatures in air, making them less desirable. Platinum, iridium, and platinum-rhodium are attractive options but are significantly more expensive than stainless steel.

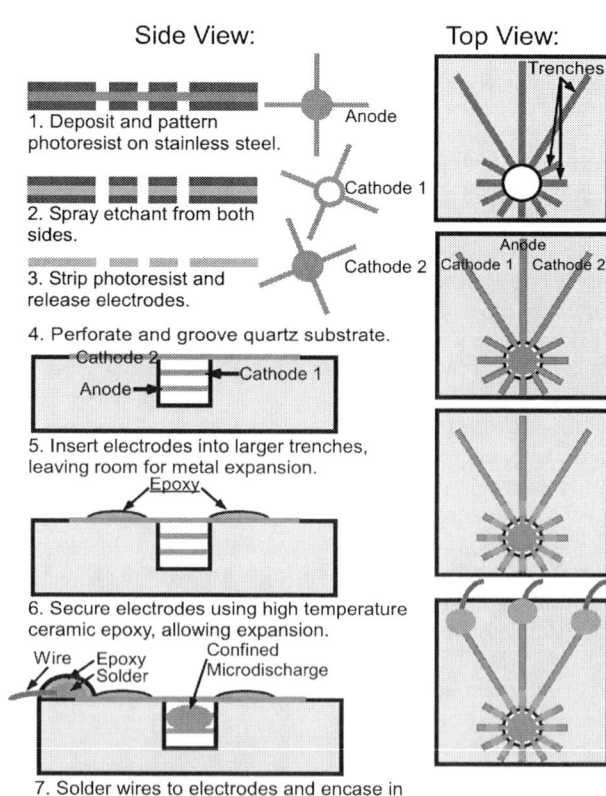

Figure 4: *Top and side view of the fabrication process.*

The electrodes are lithographically patterned and etched from stainless steel foil, using photochemical machining. This process involves coating a thin sheet of metal with photoresist, exposing the resist, and spraying the sheet with a chemical etchant to dissolve the exposed metal. The exposed metal is completely removed, leaving through-holes in the sheet, and the resist is stripped (Fotofab, Chicago IL).

Next, the electrodes are integrated into the substrate. An arrangement that accommodates the expansion mismatch between electrodes and substrate is necessary. Trenches of specified depths and a through-hole in the center are cut into a quartz chip. Both mechanical and wet-etch processes can be used for this purpose. The electrodes are assembled into the trenches, and the circular electrodes are placed in the through-hole. The different depths between the various trenches specify the discharge gap spacing, as the electrodes lie flush with the bottom of the trenches. Ceramic epoxy holds the electrode lead and support arms in place, without adhering to the stainless steel. This allows the leads and supports to expand separately from the quartz chip and the ceramic epoxy without buckling. High temperature-compatible wires are soldered to the electrodes and the solder is encased in ceramic. This ceramic keeps the solder in position so it maintains electrical contact, even at high temperatures.

RESULTS AND DISCUSSION

Pressure sensors were fabricated and tested at various temperatures upto 1,000°C, measuring pressures between 10 Torr and 2,000 Torr. Pulses, 1-20 ms in duration, were applied at a rate of 2-10 Hz to the anodes of the sensors with positive voltages between 700 V and 1,000 V. The pulses consumed between 168 µJ and 6 mJ each.

The applied voltage pulses resulted in current pulses through each cathode. The transient current peaks were approximately 50-200 ns in duration, with amplitudes of 1.3 mA to 2.85 A, varying with temperature and pressure. The sum of the measured cathode current peaks for a device is presented in Fig. 5(a). At each temperature, this sum conforms to the equation:

$$I_{pk1} + I_{pk2} = C_1 \cdot \ln(p) - C_2 \qquad (1)$$

The terms C_1 and C_2, determined by a least-squares fit to the measured data.

The sensors were tested in a nitrogen chamber with temperature and pressure control. Figure 5(b) shows fractional cathode currents, at 1,000°C, for a sensor with electrodes that were spaced 50 µm apart, 1 mm in diameter, and 125 µm thick. As noted, the output of the sensor is the differential peak current between two cathodes, expressed as a fraction of the total peak current.

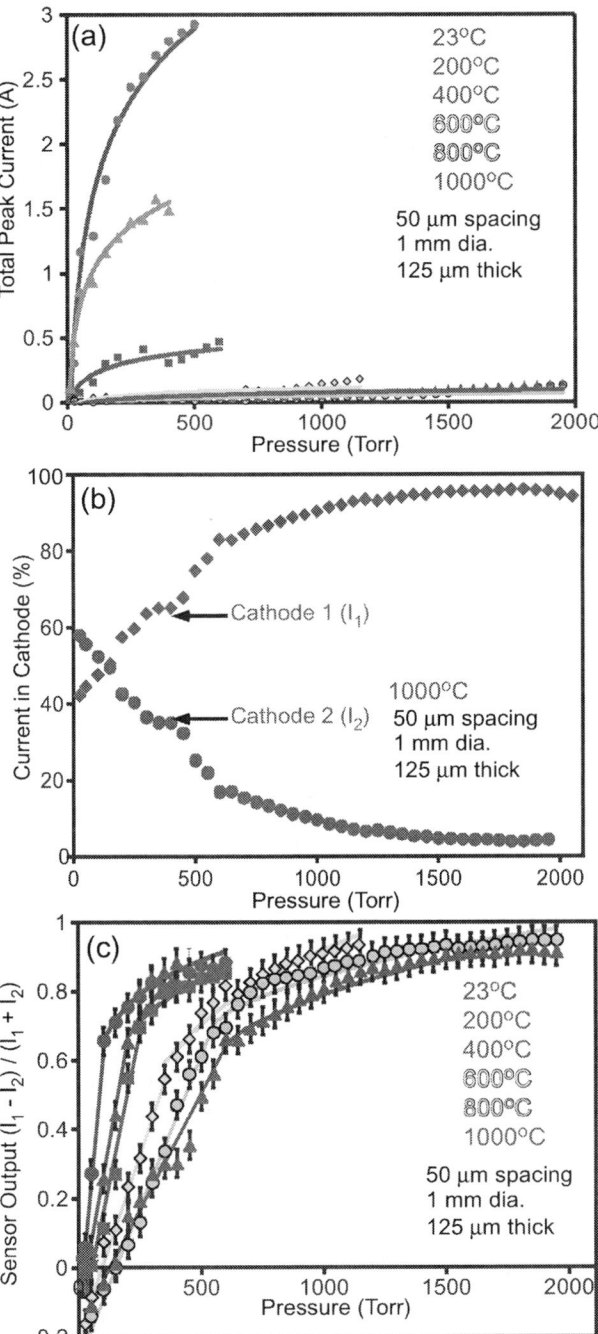

Figure 5: *Sensor output as a function of pressure in sensors with electrodes spaced 50 µm apart, 1 mm in diameter, and 125 µm thick. (a) Sum of the pulse currents in two cathodes as a function of pressure and temperature. The empirical curves for each temperature are indicated by the solid lines. (b) Percentage of total current in the cathodes at 1,000°C. (c) Differential current output determined from the percentage of total current. Each data point is the average of 100 measurements. The two empirical curves per temperature are indicated by the solid lines.*

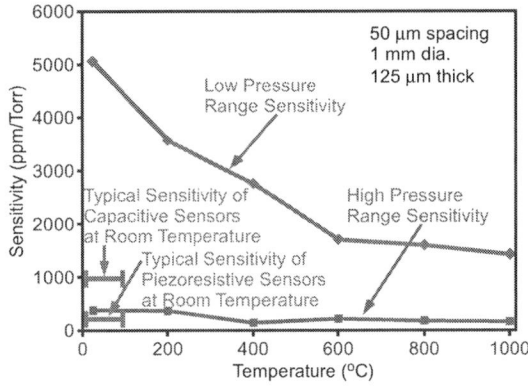

Figure 6: *The average sensitivities in both the low and high pressure ranges for the sensor in Fig. 5 as functions of temperature.*

978-1-4244-2977-6/09 $25.00 © 2009 IEEE   839

This is shown in Figure 5(c). The sensors demonstrate two regions of sensitivity (similar in some sense to touch-mode capacitive pressure sensors [11]). At low pressures the response is highly linear, whereas at high pressures it conforms to equation 1. The transition between these two regions rises from about 100 Torr at room temperature to about 500 Torr at 1,000°C.

The average sensitivities in the low pressure and high pressure operating regions of this sensor are shown in Fig. 6 as functions of temperature. One sensor design typically demonstrated the maximum lower pressure sensitivity, 5,420 ppm/Torr, as well as the maximum higher pressure sensitivity, 500 ppm/Torr. Other design variations were also explored and the typical results are listed in Table 1. The minimum average temperature coefficient of sensitivity was -550 ppm/K.

Table 1: *Typical performance of three different sensor designs with the highest performance in each category in bold.*

Sensor Parameters	Max Low Pressure Sensitivity (ppm/Torr)	Max High Pressure Sensitivity (ppm/Torr)	Dyn. Range (Torr)	Temp. Coeff. of Sensitivity (ppm/K)
D=1 mm W=125 μm G=50 μm	5,060	380	**2,000**	-650
D=1 mm W=125 μm G=100 μm	2,170	220	1,150	**-550**
D=2 mm W=125 μm G=100 μm	**5,420**	**500**	900	-925

Sensitivity, dynamic range, and the temperature coefficient of sensitivity are metrics used to compare microdischarge-based pressure sensor designs to one another and other pressure sensors. The performance of three different sensor designs (data from two are not presented in this paper) are shown in Table 1, with the highest performance in each category in bold. The data in the table represents typical operation of each sensor design. For comparison, typical piezoresistive and capacitive pressure sensors have sensitivities of 100 ppm/Torr and 1,000 ppm/Torr respectively, and temperature coefficients of sensitivity of ±1,000 ppm/K to ± 5,000 ppm/K [12, 13].

CONCLUSIONS

The microdischarge-based pressure sensors have demonstrated an ability to operate from room temperature to 1,000°C over the pressure range of 10-2,000 Torr. It is expected these sensors can potentially operate at temperatures below room temperature and over larger dynamic pressure ranges. They provide an electrical readout, avoiding an intermediate transduction step, which can be convenient in some cases. The active areas for these devices are small enough to permit hybrid or monolithic integration with other components that constitute functional microsystems. The sensitivity achieved compares favorably with conventional piezoresistive and capacitive pressure sensors of comparable size. The absence of a diaphragm, which is commonly used in piezoresistive and capacitive pressure sensors, also provides natural tolerance for overpressure

and consequently, mechanical robustness. However, encapsulating the devices within a sealed cavity with a flexible diaphragm would permit them to be operated in a broad range of environments.

ACKNOWLEDGMENTS

This work was supported primarily by the Engineering Research Center Program of the National Science Foundation under Award No. EEC-9986866. Y. Gianchandani acknowledges support through the IR/D program while working at the National Science Foundation. The findings do not necessarily reflect the views of the NSF.

REFERENCES

[1] R. Fielder, K. Stingson-Bagby, M. Palmer, "State of the art in high-temperature fiber optic sensors," *SPIE*, 5589(1), pp. 60-69, 2004.

[2] T. Li, Z. Wang, Q. Wang, X. Wei, B. Xu, W. Hao, F. Meng, S. Dong, "High pressure and temperature sensing for the downhole applications," *SPIE*, 6757(1), pp. 1-7, 2007.

[3] A. Ned, R. Okojie, A. Kurtz, "6H-SiC pressure sensor operation at 600°C," *HITEC*, Albuquerque, NM, pp. 257-260, 1998.

[4] S. Guo, H. Eriksen, K. Childress, A. Fink, M. Hoffman, "High temperature high accuracy piezoresistive pressure sensor based on smart-cut SOI," *MEMS*, Tucson, AR, pp. 892-895, 2008.

[5] S. Fricke, A. Friedberg, T. Ziemann, E. Rose, G. Muller, D. Telitschkin, S. Ziegenhagen, H. Seidel, U. Schmidt, "High temperature (800°C) MEMS pressure sensor development including reusable packaging for rocket engine applications," *MNT for Aerospace Applications, CANEUS*, pp. 5p, 2006.

[6] M. Kushner, "Modeling of microdischarge devices: Plasma and gas dynamics," *J. Phys. D: App. Phys.*, 38(11), pp. 1633-1643, 2005.

[7] C. Edelmann, "Measurement of high pressures in the vacuum range with the help of hot filament ionization gauges," *Vacuum*, 41(7-9), pp. 2006-2008, 1990.

[8] S. Wright, Y. Gianchandani, "A harsh environment, multi- plasma microsystem with pressure sensor, gas purifier, and chemical detector, *MEMS*, Kobe, Japan, pp. 115-118, 2007.

[9] S. Wright, Y. Gianchandani, "Microdischarge-based pressure sensors for operation at 1000°C," *Solid-State Sensors and Actuators and Microsystems Workshop*, Hilton Head, SC, pp. 332-335, 2008.

[10] C. Wilson, Y. Gianchandani, R. Arslanbekov, V. Kolobov, A. Wendt, "Profiling and modeling of dc nitrogen microplasmas," *J. Appl. Phys.*, 94(5), pp. 2845-2851, 2003.

[11] S. Cho, K. Najafi, C. Lowman, K. Wise, "An ultrasensitive silicon pressure-based microflow sensor," *IEEE Trans. on Electron Devices*, 39(X) pp. 825-835, 1992.

[12] Y. Zhang, K. Wise, "Performance of nonplanar silicon diaphragms under large deflections," *J. Microelectromech.Sys.*, 3(2) pp. 59-68, 1994.

[13] M. Gad-el-Hak, *The MEMS Handbook, Second Edition*, Boca Raton, FL, CRC Press, 2006.

HIGH SPEED (GHZ), ULTRA-HIGH PRESSURE (GPA) SENSOR ARRAY FABRICATED IN INTEGRATED CMOS+MEMS PROCESS

Murat Okandan, Roy Olsson, Michael Baker, Paul Resnick, Thomas A. Hill, Chad Lackey,
Sean Pearson, Jaime Castaneda, Wayne Trott, David Jones
Sandia National Laboratories, Albuquerque, NM 87185 USA

ABSTRACT

In order to observe and quantify pressure levels generated during testing of energetic materials, a sensor array with high temporal resolution (~1 ns) and extremely high pressure range (> 1 GPa) is needed. We have developed such a sensor array which utilizes a novel integrated high performance CMOS+MEMS process.

INTRODUCTION

Observation and quantification of pressure levels during testing of energetic materials is necessary for basic physical characterization of these materials. To achieve the desired temporal and pressure resolution, we have designed, fabricated and tested novel pressure sensors in an array format with integrated CMOS preamplification circuitry on the same sensor chip.

DESIGN

To utilize the large design space available in our technology, we used two different sensing methodologies (piezoelectric and capacitive), four preamplifier designs and four different sensor sizes. These 32 design combinations were fabricated in a CMOS first process, followed with aluminum nitride process split forming the piezoelectric sensing elements and molded tungsten process split for the capacitive sensors (Figure 1).

In our array designs we included both buffered and unbuffered sensors to compare the response of the raw sensor and the amplified signals. Sensors are laid out in a linear array to observe the evolving pressure wave as the energetic material is reacting. The different sizes allow us to access different dynamic ranges for the sensors and also provide the possibility for a higher density sensor array in later iterations. In the current iteration, sensors are spaced by 500 um from each other, with 8 sensors in each array; the total array length is 4 mm.

The four different CMOS buffer circuits (Figure 2) were designed to have slightly different gain and bandwidth characteristics and flexibility in terms of external biasing. The differential pair was the most flexible but also required the largest number of interconnects to the external biasing circuitry, other designs were more constrained but had simpler biasing requirements.

Modeling studies were also carried out to determine the expected output ranges from the aluminum nitride and capacitive sensors. These values were used in the design of the preamp circuits to provide the best possible match between sensor output and the amplifiers.

FABRICATION

CMOS and MEMS fabrication processes were carried out in the MESA fabrication facility at Sandia National Laboratories. For the capacitive sensors, CMOS process stopped at the tungsten plug level and the remaining mechanical structures were fabricated in the molded tungsten process [1]. For the piezoelectric sensors, the CMOS process stopped at metal1 level and the sensors were fabricated using the aluminum nitride process [2].

Molded Tungsten Process

In this process, tungsten layers are formed using a damascene process. Silicon dioxide is used as the sacrificial material in which trenches are etched in the shape of the desired structures. 1-2 um thick tungsten is deposited in these features and excess material is removed using a chemical-mechanical polishing process.

In our process flow, we utilized a thin silicon nitride layer for CMOS passivation. The first tungsten layer serves as an interconnect film and bottom electrodes. The second tungsten layer provides a gap between electrodes and the third tungsten layer forms the deformable top

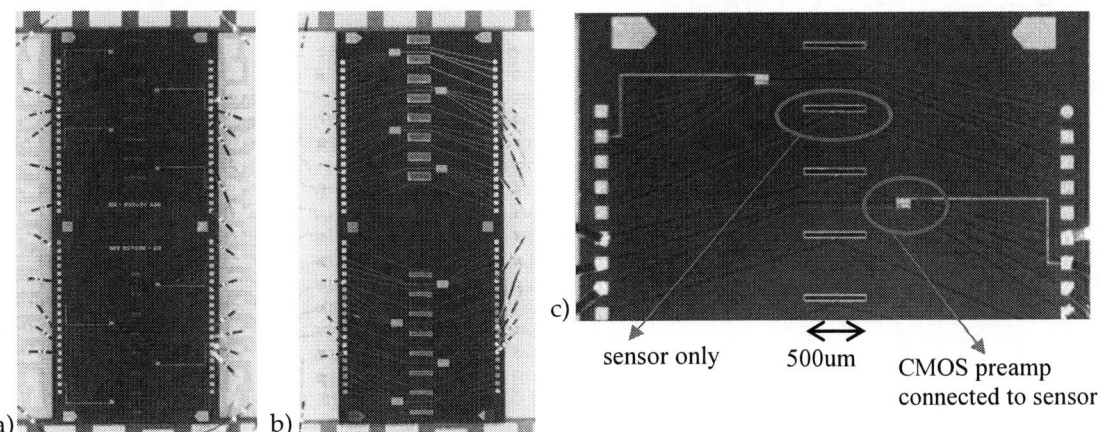

Figure 1. a) Aluminum nitride and b) molded tungsten sensors, c) close-up of the sensor and preamp.

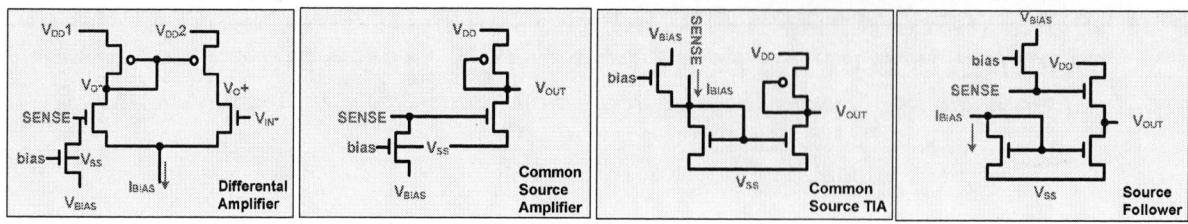

Figure 2. Four CMOS pre-amp configurations.

electrode of the capacitive sensors. At the end of the process, oxide was etched out, releasing the final structure. To withstand the high pressure levels during testing, the upper electrode layer was designed to be as stiff as possible. We also deposited parylene, filling in the space between upper and lower electrodes to form a more rigid structure.

Aluminum Nitride Process

Aluminum nitride sensors were formed as a rigid capacitor, with aluminum electrodes above and below the aluminum nitride sensing element. After the completion of metal1 layer in CMOS process, an oxide layer was deposited to isolate underlying metal lines. Contact holes were etched and tungsten plugs were formed to provide connections between sensors and the CMOS circuitry. Next, a thin aluminum layer was deposited and patterned which served as the bottom electrode for the sensors. Aluminum nitride was deposited and etched into sensor shapes. This layer also covered all of the bottom electrode sections. Top aluminum layer was then deposited and patterned which completed the process.

CHARACTERIZATION

CMOS Circuitry

In order to test the functionality of the CMOS components, identical pre-amp test circuits which were fabricated in a separate section of the chip were independently biased and characterized. Expected electrical parameters (gain, frequency response) were observed.

Pressure Sensors

Pressure sensor responses are being examined using two calibrated approaches – laser driven flyers and gas

gun loading. Both of these methods are routinely used to characterize the behavior of materials at high pressures [3] and were adapted for this purpose. In a laser driven flyer, a high energy laser pulse is coupled into an optical fiber which has an aluminum coating at the end. The interaction of the laser with this aluminum layer creates a plasma which launches a thin (~10 um) aluminum disk from the tip of the fiber. Using optical techniques such as VISAR (<u>V</u>elocity <u>I</u>nterferometer <u>S</u>ystem for <u>A</u>ny <u>R</u>eflector), resulting pressure levels in the 1-10 GPa range can be observed and calibrated [4].

Another benefit of this technique is that the test sites can be localized (~500um diameter spot) for the high pressure impacts and the resulting impact site can be visualized after testing. One such image is shown in Figure 4 where the impact site is clearly visible. This also allows multiple tests on a single chip and simultaneous monitoring of non-impact site sensors for investigating effects of coupling through the sensor substrate.

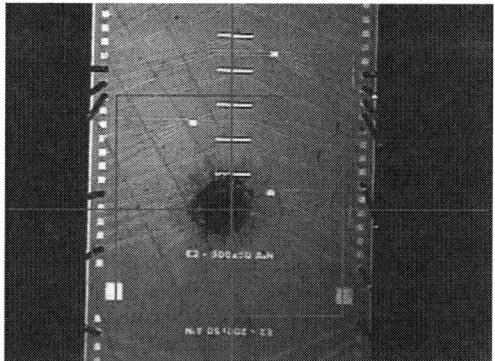

Figure 4. Image of the sensor chip after laser driven flyer loading of a single sensor.

Results of two laser driven flyer tests on aluminum nitride (sensor only) test structures are shown in Figure 5. These sensors show a high speed (1-10 ns scale) damped oscillatory response to the impact loading in the expected time scale when laser-drive pulse timing and flyer transit times are taken into account. The scaling of the output response with sensor area, consistent waveform shape and response duration provide a good indication of the repeatable device-to-device operation of these sensors under such extreme loading conditions.

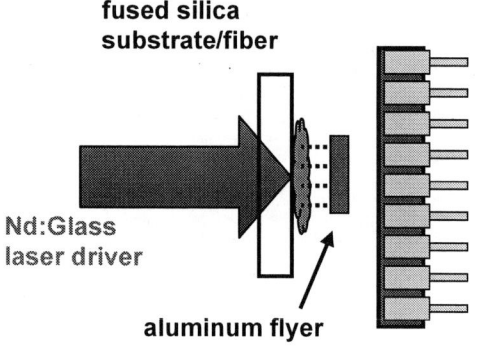

Figure 3. Schematic of laser driven flyers used in calibration of pressure sensors.

978-1-4244-2977-6/09 $25.00 © 2009 IEEE   842

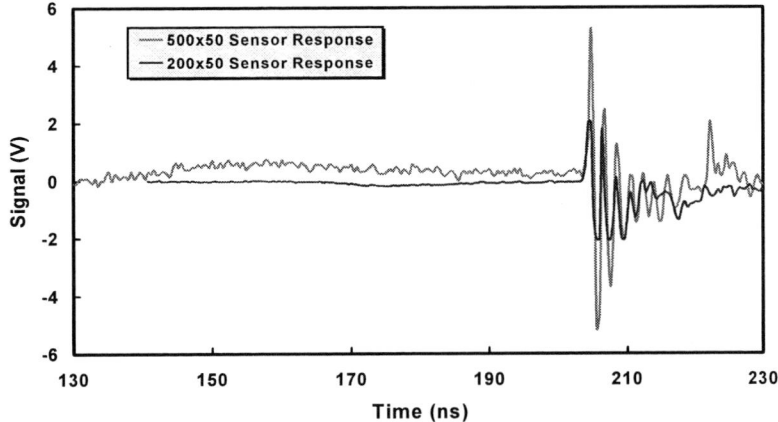

Figure 5. Output waveforms from aluminum nitride sensors (unbuffered) in laser driven flyer tests.

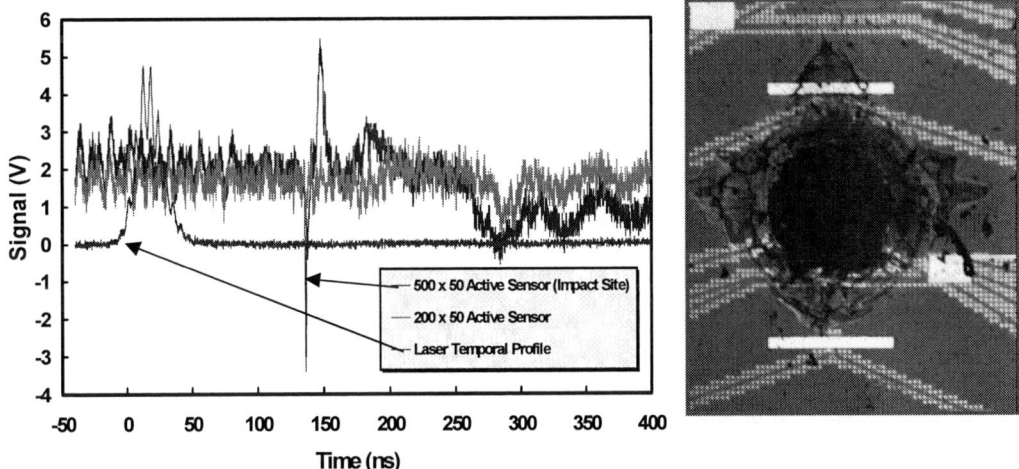

Figure 6. Laser driven flyer tests with a buffered aluminum nitride sensor – the second active sensor is on the same chip but further away from the impact site which does not show any coupling during the impact. Image on the right is the impact site.

These waveforms are recorded using a high speed oscilloscope and active probes to eliminate possible loading effects and reduce coupling of external noise from signal lines.

We have also conducted tests with the buffered (active) aluminum nitride sensors which can be seen in Figure 6. The laser output is monitored on a separate channel to provide an indication of the timescales involved. Duration between the laser pulse and the impact time correlates well with the expected flight time and speed of the flyer which was calibrated at 1.4 km/s. The second sensor which was also on the same chip shows no coupling during the impact time.

Gas gun tests are also being setup to characterize these sensors. In the gas gun, a projectile is launched at the target inside which the sensors are mounted. Again using the optical methods mentioned earlier, these tests are characterized and repeated in the GPa loading range. One key difference between gas gun tests and the laser driven flyers is the timescales and loading rates involved. Gas-gun loading can be adjusted to run up to the GPa pressure range in longer timescales (hundreds of nanoseconds to microseconds) than the laser driven flyers, which will allow us to test and characterize these sensors under varying conditions. Testing of the capacitive tungsten structures is currently underway.

CONCLUSIONS

We have designed, fabricated and tested ultra-high pressure (GPa) high temporal resolution (ns) pressure sensor arrays in a novel CMOS + MEMS process for characterization of energetic materials. Further tests are planned with gas-gun tests for detailed calibration of the sensors and coupling to energetic materials.

This work was supported by the Laboratory Directed Research and Development Program at Sandia National Laboratories. Sandia is a multiprogram laboratory operated by Sandia Corporation, a Lockheed Martin Company, for the United States Department of Energy's NNSA under contract DE-AC04-94AL85000.

REFERENCES

[1] S. Y. Lin, J. Moreno, and J. G. Fleming, *Appl. Phys. Lett.* **83**, 380 (2003)

[2] R. H. Olsson III, J. G. Fleming, K. E. Wojciechowski, M. S. Baker and M. R. Tuck, "Post-CMOS Compatible Aluminum Nitride MEMS Filters and Resonant Sensors," *Proc. IEEE Frequency Control Symposium*, (2007), pp. 412-419.

[3] W. M. Trott, R. E. Setchell, and A. V. Farnsworth, Jr., in *Shock Compression of Condensed Matter—2001* (AIP Press, Melville, NY, 2002), p. 1347.

[4] L. M. Barker and R. E. Hollenbach, *J. Appl. Phys.* **43**, 4669 (1972).

978-1-4244-2977-6/09 $25.00 © 2009 IEEE

SINGLE WAFER SURFACE MICROMACHINED FIELD EMISSION ELECTRON SOURCE

F. Santagata[1], C.K. Yang[2], J.F. Creemer[1], P. J. French[2], P.M. Sarro[1]

[1] Laboratory of Electronic Components, Technology & Materials (ECTM), DIMES, Delft University of Technology, P.O. Box 5053, Feldmannweg 17, 2600 GB Delft, The Netherlands.
[2] Laboratory for Electronic Instrumentation, T.U. Delft, Mekelweg 4, 2628 CD Delft, The Netherlands.

ABSTRACT

This paper reports on the design, fabrication and characterization of a MEMS electron source for low voltage electron emission (10 µA below 100 V) entirely built within a single silicon wafer. This device consists of a hermetically sealed microcavity that encapsulates an array of field emission silicon tips and a reaction chamber to which the emitted electrons are accelerated. An electron transparent membrane (20 nm-thick SiN) is positioned in between the two cavities. SiN circular pillars are designed to support the microcavities, control the tip-to-membrane distance and isolate the electrodes. The device is fabricated using a surface micromachining process that reduces the distance between the emitters and the electrodes leading to low voltage field emission.

INTRODUCTION

Free electron sources are needed in many applications like e-beam based material analysis [1], e-beam lithography [2], displays, but also sensing devices like vacuum gauges [3] and flow sensors [4], etc. However, for handling stable electron emissions, high vacuum conditions are needed, which require complex (and thus expensive) systems. Fabricating electron sources using MEMS leads to strongly miniaturized devices with performance and cost benefits. Miniaturization strongly reduced the volumes. Consequently, the requirements for the vacuum are significantly relaxed. This means that the required high vacuum can be reached easily using a micropump or a simple one stage pumping system. Another advantage of miniaturization are the high electric fields that can be generated by low voltages, thus reducing the power consumption and allowing the use of standard integrated circuits to drive the device.

Furthermore, a range of applications need emitted electrons traveling towards a region under controlled pressure and temperature conditions. Therefore, an electron source needs a vacuum chamber in which electrons are emitted, a reaction chamber to which the electrons are accelerated and an electron transparent membrane in between them [5]. A schematic configuration of such a micromachined electron source is showed in Figure 1.

In [1] a device without reaction chamber has been presented. The cavity was fabricated by wafer bonding. In the novel device reported here a sealed microcavity used as vacuum chamber, and a microfluidic device used as reaction chamber, have been integrated on a single substrate using an IC-compatible surface micromachining process. In comparison to the bonding technique [1] the thin film approach presented here allows to achieve field emission at voltages that are two

orders of magnitude lower because of the reduced distance between emitters and electrodes. In addition, our solution eliminates the need of the second substrate, thereby avoiding issues in the bonding of processed wafers and decreasing the footprint of the devices. Finally, the surface micromachining approach makes this device more IC compatible and thus more functions could be integrated.

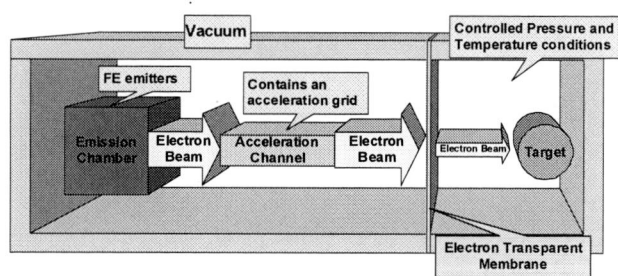

Figure 1: Schematic configuration of a micromachined free electron source consisting of a vacuum chamber containing the emitters and a reaction chamber to which the electrons are accelerated. The chambers are divided by an electron transparent membrane.

DESIGN

The device developed mainly consists of two parts; a hermetically sealed microcavity that encapsulates an array of field emission silicon tips and a reaction chamber to which the emitted electrons are accelerated. An electron transparent SiN membrane is positioned in between the two cavities. Both the sealed microcavity and the reaction chamber are fabricated following a surface micromachining process.

A schematic configuration of our device is shown in Figure 2. The main elements of the device are the tip, the electron transparent membrane, the electrodes and the insulating pillars between them. Each tip is 3 µm high and has a base radius of 4 µm.

Concerning the membrane, it has been shown, that electrons with relatively small energies of 2–3 keV generated in vacuum can be transferred through thin membranes of SiN (20-100 nm) to ambient conditions and are able to traverse a few hundred microns in air before they hit the target material [6]. Therefore SiN has been selected as material for the membrane. The chosen membrane thickness is 20 nm, the diameter is 2 µm.

SiN pillars with a diameter of 3 µm have been designed to support the microcavity structures, to control the distance between each tip and its corresponding membrane and to isolate the two electrodes. A thickness of 1 µm for the SiN capping layer and a distance of 20 µm between the pillars have been chosen. This thickness is enough to neglect the current leakage through the

978-1-4244-2977-6/09 $25.00 © 2009 IEEE

nitride layer. In order to obtain a surface micromachined cavity, access holes for sacrificial etching and sealing plugs are required. In the Figure 2 it is also reported the schematic top view of the device, showing the distribution of electron emitters, plugged access holes and pillars.

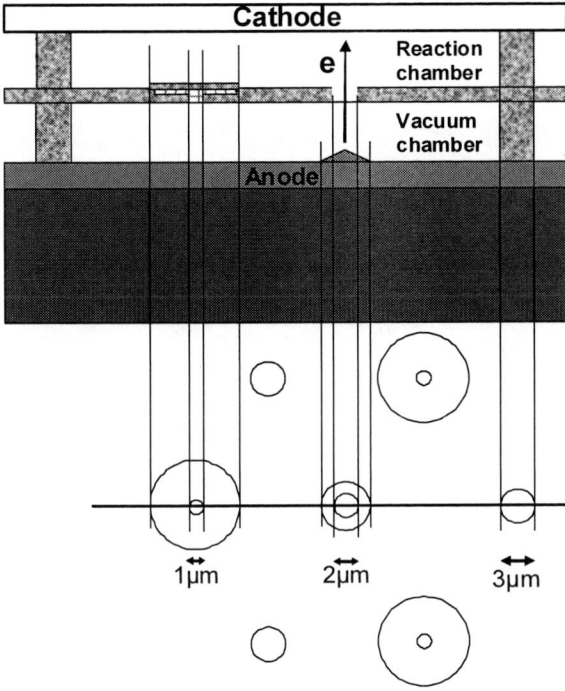

Figure 2. Schematic configuration (cross section and top view) of the device showing the electron emitter, the plugged access holes, and the pillars as circles.

FABRICATION

The starting material is a silicon wafers with a 10 μm high- phosphor-doped epitaxial layer representing the anode layer.

The fabrication process starts with the tips definition (Figure 3.a-c). The doped layer is patterned with silicon dioxide circular pads and isotropically dry etched by using SF_6 into a narrow neck column. The narrow-neck is then sent to thermal oxidation until silicon tip is formed under the grown layer. This extra oxidation step is used in order to sharpen the silicon tip [7].

In order to form the vacuum chamber (Figure 3.d), thick PECVD BPSG oxide is deposited on the oxidized narrow-neck structure and allowed to reflow. The previous oxidation step forms also a barrier to the diffusion of the dopants in the following high temperature steps. The BPSG deposition and reflow can be repeated in order to achieve a good planarization result. Using this technique a height difference of 5 μm has been reduced down to 200 nm. The next step is the deposition of a thin LPCVD low-stress silicon nitride layer followed by a PECVD silicon dioxide layer. The SiN layer will serve as the electron transparent membrane and the SiO_2 layer will act as etch stop layer during the further membrane etching step.

In order to fabricate the pillars (Figure 3.e-f), the stack of layers (SiO_2 + SiN + BPSG) is dry etched down to the silicon. Then low-stressed SiN to form the pillars and the capping layer has been deposited. During the SiN deposition, cracks of the BPSG layer are observed. A good reflow/annealing step before the SiN deposition reduces the cracking issue suggesting that the cause of the cracks is a large amount of gases trapped into the BPSG layer.

This thick SiN layer is then locally etched to form 1 μm diameter access holes for further sacrificial etching, reaching the underlying BPSG (Figure 3.g). The device is then submerged in HF solution for sacrificial etching of the BPSG layer thus forming the vacuum chamber. During the drying step, no stiction has been observed thanks to the support of the pillars distributed all around the chamber.

The released chamber is loaded in a PECVD oxide deposition system for performing the plugging step (Figure 3.h). The deposition recipe is tuned to specifically seal the 1 μm access hole gap, while minimize its deposition inside the chamber. Cross-section inspections in an SEM confirmed that this sealing process did not affect the sharpness of the tip.

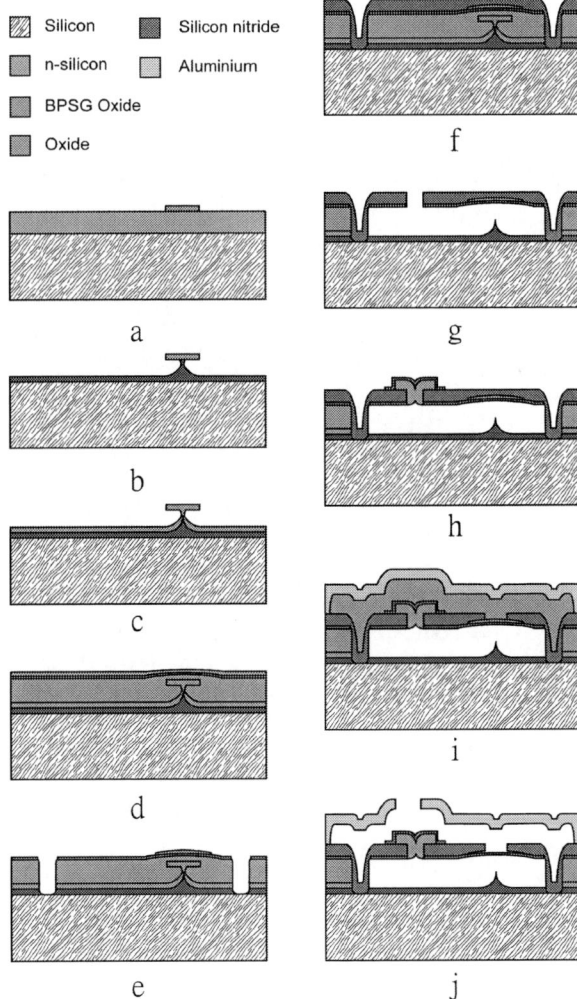

Figure 3. Main steps of the device process flow.

The PECVD oxide is etched away everywhere leaving the plug structure and a further LPCVD SiN layer is deposited on top of the oxide plug in order to improve the hermetic sealing.

After the sealing step the membrane etching step is then performed (Figure 3.i). A window pattern aligned to the top of the tip is dry etched through the stack of sealing and capping SiN layers landing on the buried PECVD silicon dioxide layer. Then a further wet etch step in HF to expose the SiN membrane has been performed. Both the sacrificial etching and the final HF etching inevitably reduces the thickness of the SiN membrane. Therefore a careful calculation of the etching rate and time is needed to decide the initial SiN deposition thickness.

The microfluidic chamber (Figure 3.l) has been fabricated using PECVD silicon dioxide as sacrificial layer. An array of pillars has been designed also for the reaction chamber in order to support the structure.

Aluminum is used as anode for accelerating the emitted electrons from the vacuum chamber. HF 73% has been used to etch the sacrificial silicon dioxide because it is very selective towards aluminum. The Figure 4 shows a SEM cross section of an encapsulated tip. The two chambers with the pillars and the tips distribution is clearly visible.

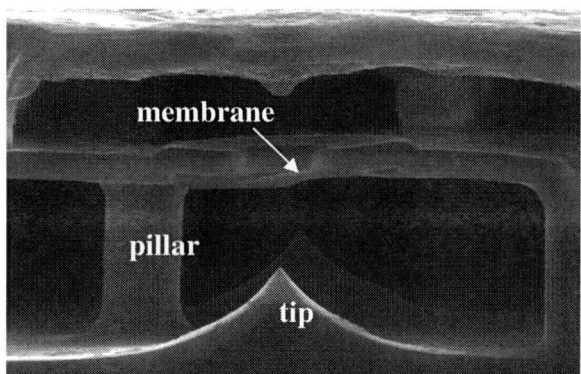

Figure 4. SEM cross section of encapsulated tips. The membrane window above the tip is in this case 50 nm thick here in order to withstand the cross section dicing process and to appear visibly in SEM imaging.

MEASUREMENTS AND DISCUSSION

In order to fully characterize the device, the emission characteristics of the electron source and the transmittance characteristics of the electron permeable membrane have to be studied.

In this paper only the electrical characterization of the tips emission has been reported. For this purpose test structures containing arrays with different number of tips, without permeable membranes, have been considered.

The experimental results are obtained using a Cascade probing station with an HP 4156C parameter analyzer. Field emission current versus anode-cathode voltage for two test structures containing a single tip and 400 tips respectively is showed in the Figs. 5 and 6.

For the single tip test structure, the cathode-anode voltage was varied from 0 to 70 V. The measured

emission current at 60 V is about 0.1 μA. The difference in the emission current between the single tip device and the one containing the tips array is about 100 times, and the array has overall less fluctuation due to the effect of averaging.

The Fowler-Nordheim plots showed in the insets of the figures 5 and 6 demonstrate that the measured currents follow the Fowler-Nordheim tunneling model, proving that the encapsulation process did not negatively affect the emission properties of the tips. Fig. 7 shows an optical image of the working test structure containing the array of 400 tips.

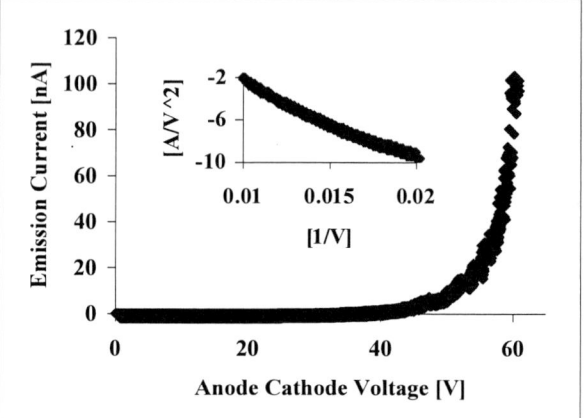

Figure 5. Field emission current versus anode-cathode voltage for a test structure containing one single tip. The inset shows the Fowler Nordheim plot.

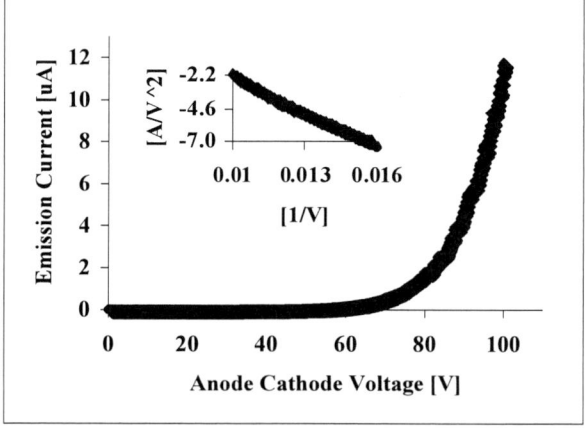

Figure 6. Field emission current versus anode-cathode voltage for a test structure containing an array of 400 tips. Inset: Fowler Nordheim plot.

Long-duration measurements have been performed in order to evaluate the stability of the field emission current; see Figure 8. An average emission current of 2.2 nA for 1 hour at 45 V cathode-anode voltage with a variation of 0.4 nA/hour has been measured. In order to stabilize the current, preprocess such as baking could be performed to remove contaminants on the emitters.

It was also observed that when a single tip device is driven with a current larger than 100 nA it explodes; this is illustrated in the optical image of an array test structure where some of the tips have blown up during the measurement (Fig.9). An SEM top view of an

978-1-4244-2977-6/09 $25.00 © 2009 IEEE   846

exploded tip is shown in Fig.10.

Figure 7. Optical image of the tested device (the tip array area is 260μm*370μm).

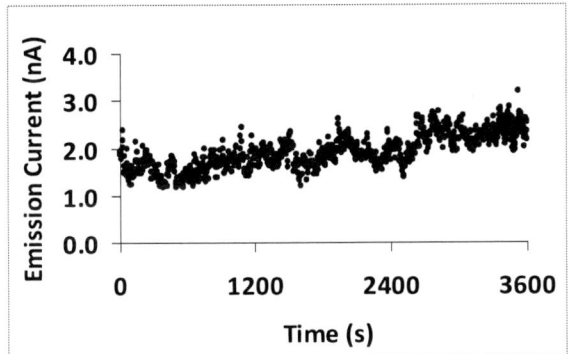

Figure 8. Stability measurement of a single tip test structure. The measured current is 2.2 ± 0.2 nA.

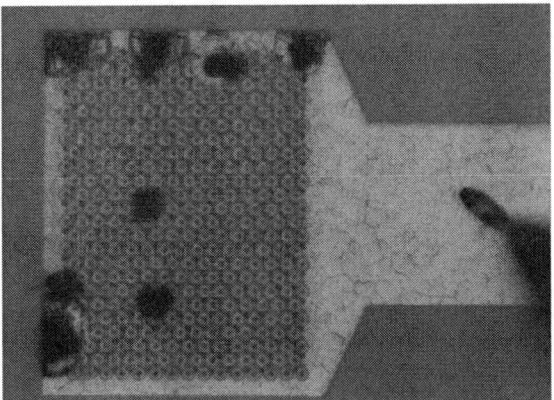

Figure 9. Optical image of array test structure where some of the tips have blown up during the measurement.

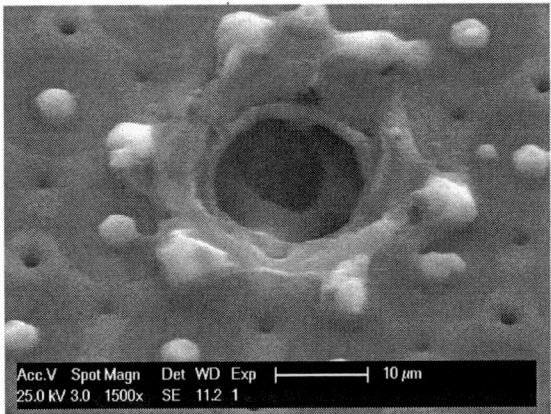

Figure 10. SEM top view of an exploded tip.

CONCLUSIONS

In this paper design, fabrication and characterization of a MEMS electron source for low voltage electron emission (10 μA below 100 V) entirely built within a single silicon wafer has been presented.

This device consists of a hermetically sealed microcavity that encapsulates an array of field emission silicon tips. The surface micromachining process enables field emission at low voltages.

Particular emphasis has been put to the fabrication issues related to the encapsulation steps, which did not negatively affect the tips sharpness.

Moreover the field emission behavior has been proven. Measurements of the emission current as a function of the applied voltage show a logarithmic behavior corresponding well to the Fowler-Nordheim theory.

ACKNOWLEDGEMENTS

The authors wish to acknowledge the DIMES IC Process Group for their assistance with device fabrication. This work is partly supported by the Point One project MEMSLand (www.memsland.nl).

REFERENCES

[1] J.E. Feldman, J.Z. Wilcox, T. George, D. Barsic, A. Scherer, Elemental surface analysis at ambient pressure by electron-induced X-ray fluorescence, *Review of ScientificInstruments* (2003), pp. 1251-1254.

[2] Wonie Cho, T. Ono, M. Esashi, Micro proximity electron source with apertured electron window for nanolithography in atmosphere, *Proceedings of Transducers & Eurosensors* (2007), pp. 1581-1584.

[3] In-Mook Choi, Sam-Yong Woo, Han-Wook Song, Improved metrological characteristics of a carbon-nanotube-based ionization gauge, *Applied physics letters* (2007), pp. 90-93.

[4] Bahram Ghodsian, M. Parameswaran, M. Syrzycki, Gas detector with low-cost micromachined field ionization tips, *IEEE Electron Device Letters 19*, N. 7 (1998).

[5] F. Haase, P. Detemple, S. Schmitt, *et al.*, Electron permeable membranes for MEMS electron sources, *Sensors and Actuators A* 132 (2006), pp. 98–103.

[6] J. Feldman, J.Z. Wilcox, T. George, D.N. Barsic, T. Doll, A. Scherer, Atmospheric electron X-Ray spectrometer, *Proceedings of the 46th Symposium Am. Vac. Soc,* (1999).

[7] M.A.R. Alves, D.F. Takeuti and E.S. Braga, Fabrication of sharp silicon tips employing anisotropic wet etching and reactive ion etching, *Microelectronics Journal 36* (2005), pp. 51–54.

SUSPENDED MEMBRANE SINGLE CRYSTAL SILICON MICRO HOTPLATE FOR DIFFERENTIAL SCANNING CALORIMETRY

J. Lee[1], C.M. Spadaccini[2], E.V. Mukerjee[2], and W.P. King[1]
[1]University of Illinois at Urbana-Champaign, Illinois, USA
[2]Lawrence Livermore National Laboratory, Livermore, California, USA

ABSTRACT

This paper introduces an array of single crystal silicon micro hotplates for differential scanning calorimetry. Based on heat transfer analysis considering tradeoffs between response time, temperature uniformity, and measurement sensitivity, suspended membrane micro hotplates with full backside release were found to be optimal designs. Due to the requirements of routine sample loading, the size of the heater is 100 or 200 μm while the size of the backside membrane cavity is 400 μm. Our design achieves a combination of time constant, temperature sensitivity, and heating efficiency that are comparable or superior to previously reported microcalorimeters.

INTRODUCTION

A calorimeter is a widely used analytical instrument that measures heat release or absorption of a material sample undergoing temperature change. A differential scanning calorimeter (DSC) is one of the most routinely used instruments for characterizing metals or polymers. These systems are capable of scanning over a wide temperature range while rejecting common-mode signals, thus enhancing sensitivity. Since DSC is a differential metrology tool, the reference and sample must be thermally decoupled. While both heaters are operated with a fixed temperature ramp, the measured differential power provides valuable information regarding thermal phase transitions and reactions.

Microelectromechanical systems (MEMS) fabrication technologies have aided the development of micro-scale DSC (microDSC) devices [1], [2]. Because of its small size, microDSC requires orders of magnitude smaller sample volumes, as well as offering ultra-fast operation and improved sensitivity. Several different configurations have been offered for MEMS-based DSC. Thin-film DSC

consisting of a metal heater and thermometer on a thin dielectric film has been used to measure the size-dependent depression of a melting point [3], [4] and the reduction of enthalpy of fusion [3], [4] in thin metallic films as well as the glass transition of thin polymeric films [5]. Suspended bridge, small-scale DSC with heater and temperature sensors connected to a bulk chip via tethers was reported to achieve improved temperature uniformity [6]. A rectangular micro hotplate DSC was employed for combustible gas sensing [7]. Although previous work on microDSC focused on reducing thermal capacity of the device, there is a lack of detailed thermal design analysis and experimental validation regarding heating efficiency and response time.

This paper reports a single crystal silicon micro hotplate for differential scanning calorimetry to pursue high temperature operation and long-term stability. We first introduce lumped parameter heat transfer models for micro hotplates with different geometries. Based on this modeling, the geometry with the best thermal efficiency was chosen. Suspended micro hotplates were fabricated and characterized. Using the fabricated silicon micro hotplates, calorimetric measurements of a paraffin wax were performed and thermodynamic properties measured.

DESIGN AND FABRICATION

Figure 1 shows the design of our microfabricated calorimeter. A square membrane is linked to the bulk silicon chip via four tethers. Two serpentine tracks of doped silicon run across the membrane to provide heating. For optimization, three possible types of micro hotplates are considered. Type A is a suspended membrane micro hotplate released from the top side, type B is a micro hotplate on a dielectric closed membrane released from the bottom side, and type C is a suspended membrane micro hotplate released from the bottom side. Lumped parameter

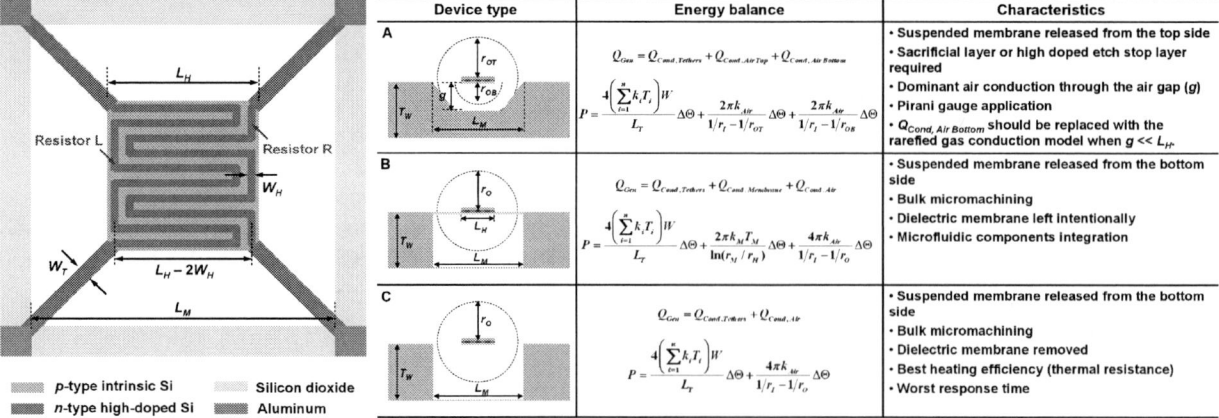

Figure 1: Design of the microfabricated hotplate calorimeter. A squared island at the center is linked to the bulk silicon chip with four tethers. Two serpentine heater-thermometer tracks run through the island and will serve as a heater platform. Each doped silicon trace can be used either as a heater or a resistive thermometer. Table shows lumped parameter heat transfer model and characteristics of three micro hotplate types.

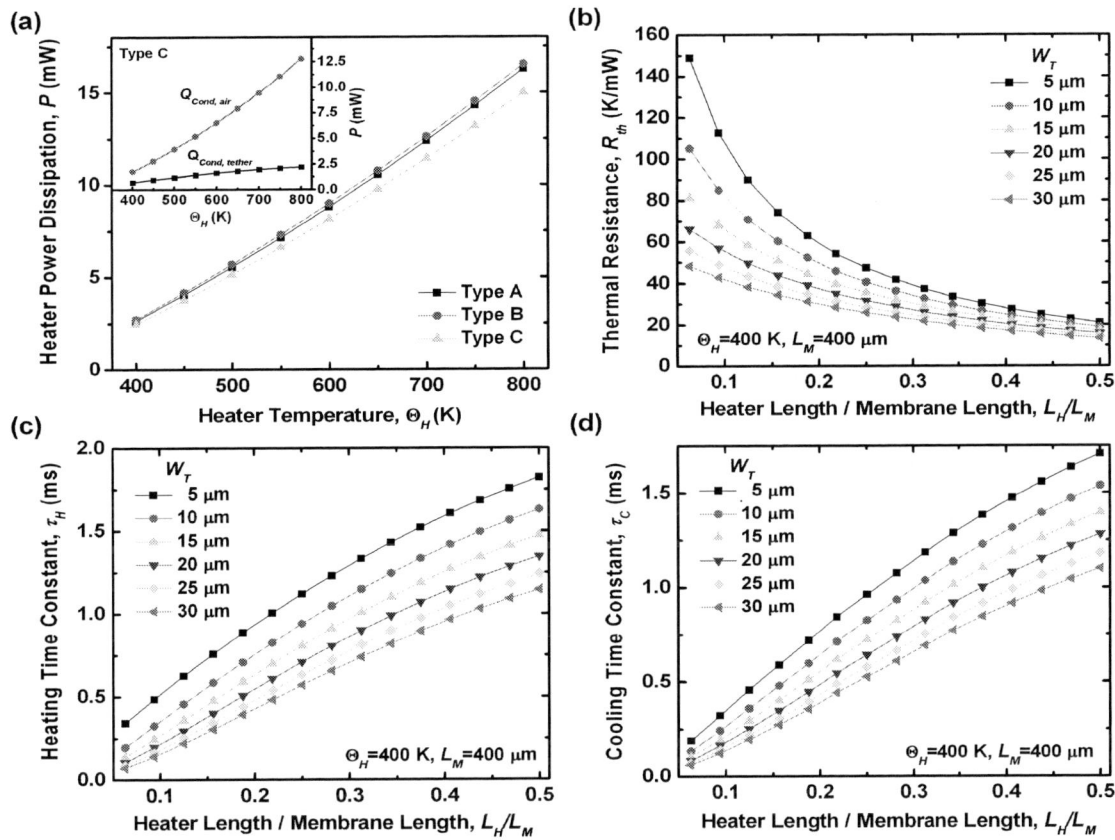

Figure 2: Lumped parameter heat transfer analysis results. (a) Heater power as a function of the heater type (b) Thermal resistance, (c) heating, and (d) cooling time constant as a function of the heater length-to-membrane cavity length ratio, L_H / L_M, respectively. Inset in (a) shows the two major heat transfer mechanisms of the type C micro hotplate.

heat transfer analysis on each type predicts that the type C hotplate is the most efficient and was chosen for further design consideration. Figure 2 shows a small heater can improve both heating efficiency and response time but the tether width offers a trade-off between the heating efficiency and the time constant. Based on the heat transfer analysis and consideration of practical sample loading, geometries for our micro hotplate were determined for fabrication. The square cavity size was chosen to be 400 μm on a side and two different square heater (L_H) sizes are used (100 and 200 μm, respectively). The widths of tethers for 100 and 200 μm heaters are determined to be 5 and 10 μm, respectively. To guarantee low noise, rapid heating, and high temperature sensitivity, doping concentration above $10^{20}/cm^3$ is recommended based on a temperature dependent resistivity model.

The fabrication process started with a p-type silicon-on-insulator wafer of orientation <100>, with a silicon device layer of 340 nm, a buried oxide layer of 400 nm, and a silicon handle layer of 450 μm (see Fig. 3). Device layer resistivity was 14 ~ 22 Ω-cm. The tethered membrane structures were defined using an inductively coupled plasma (ICP) etcher. The serpentine heater and tethers patterned with a photoresist were implanted with 2.51×10^{16} cm^{-2} of phosphorous at 180 keV. The photoresist masking window was removed with acetone and hot (120 °C) Piranha (70% H_2SO_4: 30% H_2O_2 in volume ratio) solution. Then, the wafer was annealed in a tube furnace with a 200 nm thick silicon dioxide layer deposited on the entire wafer using plasma enhanced chemical vapor deposition (PECVD) to prevent dopants

from diffusing back to the ambient. After the heat treatment, the PECVD oxide was removed with a buffered oxide etch to expose the doped silicon for metallization. Electron beam evaporation in conjunction with a lift-off process defined aluminum-doped silicon contacts and a 30-minute annealing step at 400 °C was performed to allow inter-diffusion of doped silicon and aluminum. Then, the backside of the wafer was patterned with a thick

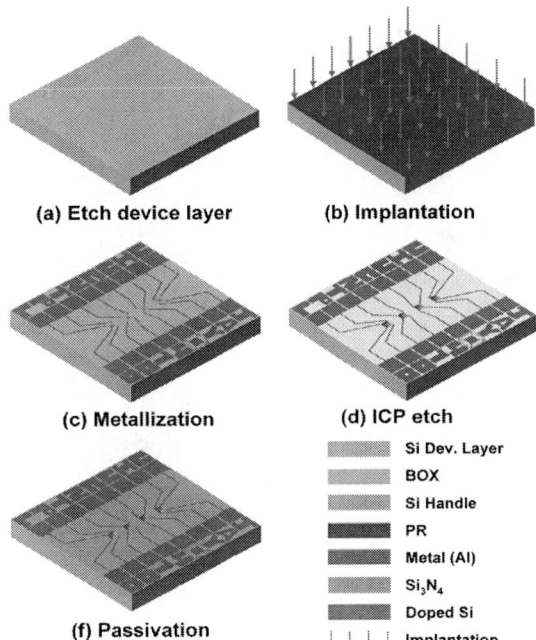

Figure 3: Five major fabrication steps.

978-1-4244-2977-6/09 $25.00 © 2009 IEEE

photoresist and etched using ICP. Finally, 400 nm of PECVD nitride was deposited on the front side for passivation. Figures 4(a) and (b) show scanning electron micrographs of the fabricated micro hotplates having a heater length of 100 μm. In a microDSC unit cell, there are four identical micro hotplates so that any two of them can perform differential measurements. Each unit cell was wire-bonded to and packaged in a 28 pin dual-in-line package as shown in Fig. 4(c).

RESULTS

Figure 5 shows electrical characterization and temperature calibration for two different operating configurations with an 8 $k\Omega$ sense resistor. We used a micro hotplate having a heater length of 100 μm and the heater temperature was measured at the center with laser Raman thermometry while the micro hotplate was resistively heated [8] . When the resistor L was operated alone (configuration L), temperature coefficient of resistance (TCR) was 1311 ppm/K in the temperature range of 300-500 K. When both resistors were operated (configuration L+R), TCR was 1254 ppm/K in the same temperature range. Configuration L+R is expected to have better temperature uniformity than configuration L since Joule heating would be more evenly distributed. Figure 5(b) shows the heater temperature at the center as a function of the heater power dissipation. Both configurations have a similar thermal resistance of 36.7 K/mW in the temperature range of 300-400 K. This is in agreement with the value of 40.6 K/mW obtained from the lumped parameter heat transfer model. Two micro hotplates from the same unit die were used and found to have nearly identical properties confirming their suitability for differential metrology.

Following characterization, DSC measurements were performed on an analyte of paraffin wax. Figure 6(a) shows the heater temperature at the center as a function of time with and without the paraffin sample, where a slow linear voltage ramp of 0.2 V/s is applied. This differential measurement rejected common-mode signals. Paraffin

melting transition was observed around $t = 15$ s and optical microscopy confirmed the transition. The heater temperature of the paraffin loaded calorimeter at $t = 15$ s is about 55 °C which is at the melting point of the bulk paraffin sample (Fisher Scientific, P31-500), which ranges 53–57 °C. Figure 6(b) shows heater power dissipation as a function of time with a fast linear voltage ramp of 8 V/s where only the resistor L is powered and voltage and current in both resistors L and R are monitored. Loaded paraffin shown in the inset was completely consumed in a single heating cycle of 1 s duration. At approximately 0.9 s, sudden changes in the heater power dissipation are observed. Since the heater temperature was increased above thermal runaway of the intrinsic silicon, electrical crosstalk between the heater (Resistor L) and sensor (Resistor R) existed. The structural membrane which has very low doping levels became less resistive due to the intrinsic carrier generation at temperatures higher than 500-550 K. By subtracting the total power dissipation in the paraffin-loaded hotplate from that in the reference hotplate, a thermal energy of 0.317 mJ was extracted using

$$E = \int_0^t \sum_{L,R} \left(P_{\text{Reference}} - P_{\text{Paraffin}} \right) dt .$$

Several measurements were repeated with varying initial paraffin mass on the micro hotplate and the extracted energy was generally proportional to the initial mass of the paraffin loaded, although the mass estimate was

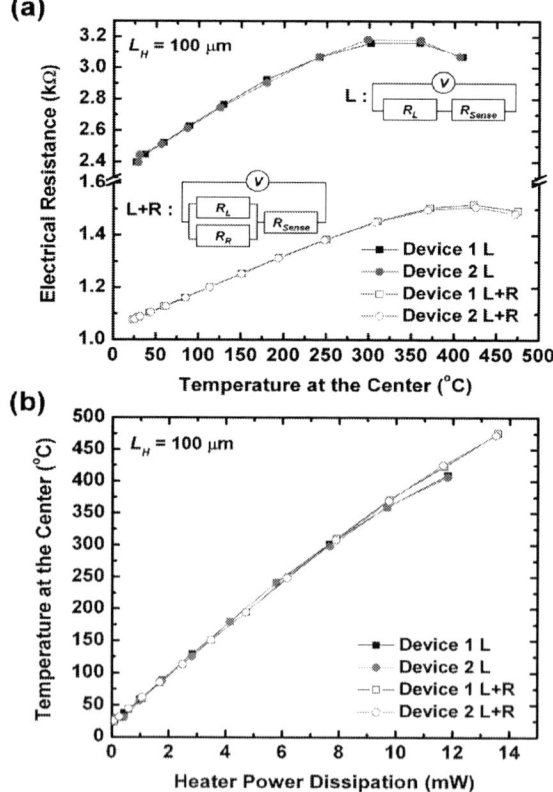

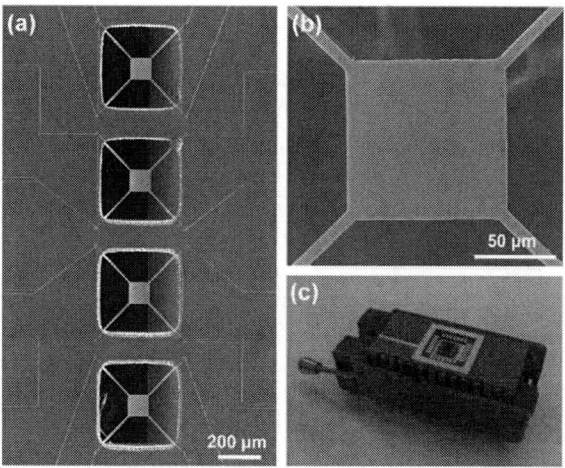

Figure 4: Scanning electron micrographs of the fabricated microcalorimeter. (a) A differential scanning calorimeter (DSC) unit die having four identical micro hotplates in a row. (b) A close up of one device. (c) A picture showing a microDSC unit attached to and wire-bonded to a dual-inline package.

Figure 5: (a) Electrical resistance of resistor L and the parallel resistor network of both resistors as a function of the temperature at the center of the heater. (b) Temperature at the center of the heater as a function of heater power dissipation for two circuit configurations using both resistors as heaters.

978-1-4244-2977-6/09 $25.00 © 2009 IEEE

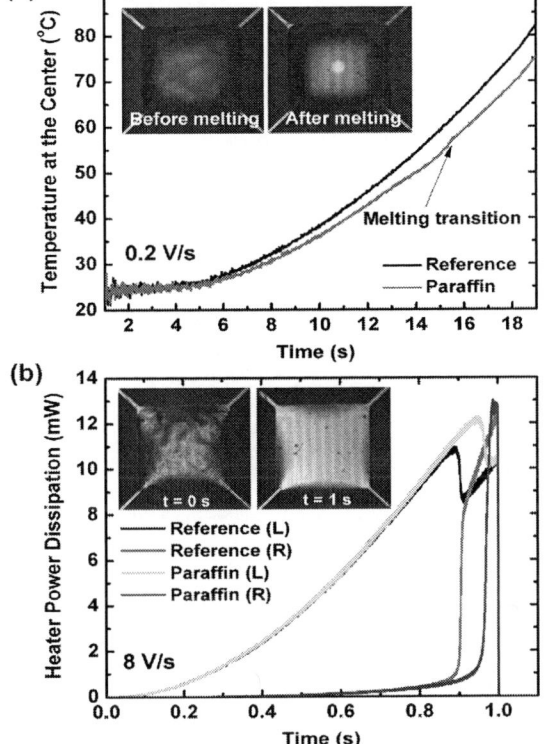

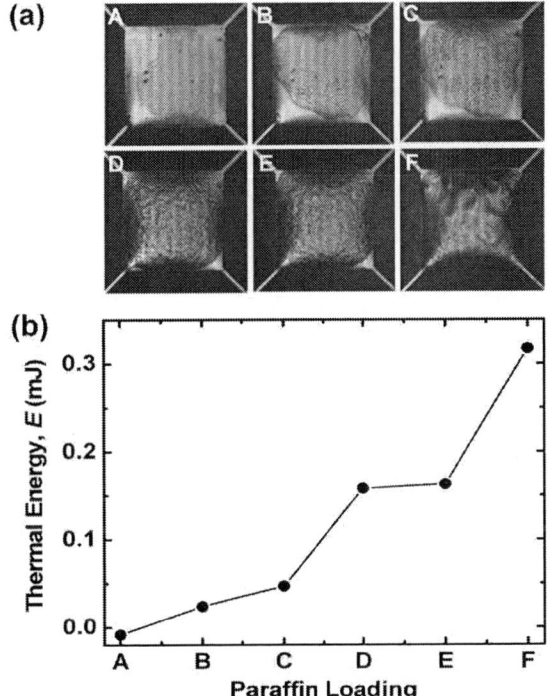

Figure 6: (a) Heater temperature and (b) power of reference and paraffin-loaded micro hotplates. Paraffin melting transition occurs approximately at 55 °C and melting is confirmed with optical micrographs.

Figure 7: (a) Paraffin sample loading with different initial mass (increasing order from A to F) (b) Energy associated with a given heating cycle having the same temperature range and ramp. Each paraffin sample loaded on the hotplate completely desorbs after the heating cycle.

qualitatively approximate as it was based on optical microscopy (see Fig. 7). The energy was attributed to the combination of sensible and latent heats since the micro hotplate underwent a wide range of temperature change.

CONCLUSIONS

This paper describes the design, fabrication, and characterization of a single crystal silicon micro hotplate for differential scanning calorimetry. Lumped parameter heat transfer analysis suggested a suspended membrane micro hotplate ensures the best heating efficiency. Two different sized suspended membrane micro hotplates were fabricated and their electrical and thermal behaviors agreed with model predictions. Not only were their heating efficiency and response time better than those of most micro hotplates previously reported, but the temperature uniformity was improved due to the relatively high thermal conductivity of silicon. By eliminating metals, the silicon microcalorimeters described here would offer extended temperature ranges creating new opportunities for microscale thermal analysis or other sensing applications.

ACKNOWLEDGEMENTS

Portions of this work were performed under the auspices of the U.S. Department of Energy by Lawrence Livermore National Laboratory under Contract DE-AC52-07NA27344 LLNL-PROC-407768.

REFERENCES

[1] D. W. Denlinger, E. N. Abarra, K. Allen, P. W. Rooney, M. T. Messer, S. K. Watson, and F. Hellman, "Thin-film microcalorimeter for heat-capacity

measurements from 1.5 K to 800 K," *Rev. Sci. Instrum.*, vol. 65, pp. 946-958, 1994.

[2] S. L. Lai, G. Ramanath, L. H. Allen, P. Infante, and Z. Ma, "High-speed (10^4 °C/S) scanning microcalorimetry with monolayer sensitivity (J/m^2)," *Appl. Phys. Lett.*, vol. 67, pp. 1229-1231, 1995.

[3] S. L. Lai, J. Y. Guo, V. Petrova, G. Ramanath, and L. H. Allen, "Size-dependent melting properties of small tin particles: Nanocalorimetric measurements," *Phys. Rev. Lett.*, vol. 77, pp. 99-102, 1996.

[4] M. Zhang, M. Y. Efremov, F. Schiettekatte, E. A. Olson, A. T. Kwan, S. L. Lai, T. Wisleder, J. E. Greene, and L. H. Allen, "Size-dependent melting point depression of nanostructures: Nanocalorimetric measurements," *Phys. Rev. B*, vol. 62, pp. 10548-10557, 2000.

[5] M. Y. Efremov, E. A. Olson, M. Zhang, Z. Zhang, and L. H. Allen, "Glass transition in ultrathin polymer films: Calorimetric study," *Phys. Rev. Lett.*, vol. 91, p. 085703, 2003.

[6] S. Zhang, Y. Rabin, Y. Yang, and M. Asheghi, "Nanoscale calorimetry using a suspended bridge configuration," *J. Microelectromech. Syst.*, vol. 16, pp. 861-871, 2007.

[7] R. E. Cavicchi, G. E. Poirier, N. H. Tea, M. Afridi, D. Berning, A. Hefner, I. Suehle, M. Gaitan, S. Semancik, and C. Montgomery, "Micro-differential scanning calorimeter for combustible gas sensing," *Sens. Actuators B, Chem.*, vol. 97, pp. 22-30, 2004.

[8] J. Lee, T. Beechem, T. L. Wright, B. A. Nelson, S. Graham, and W. P. King, "Electrical, thermal, and mechanical characterization of silicon microcantilever heaters," *J. Microelectromech. Syst.*, vol. 15, pp. 1644-1655, 2006.

978-1-4244-2977-6/09 $25.00 © 2009 IEEE

SIMPLE AND ROBUST AIR GAP-BASED MEMS SWITCH TECHNOLOGY FOR RF-APPLICATIONS

P. Ekkels[1,2], X. Rottenberg[1,2], P. Czarnecki[1,2], R. Puers[2] and H.A.C. Tilmans[1]

[1]IMEC, Kapeldreef 75, 3001 Heverlee, Belgium

[2]K.U.Leuven, Kasteelpark Arenberg 10, 3001 Heverlee, Belgium

ABSTRACT

This paper presents a simple and robust process for fabrication of functional electrostatic RF-MEMS switching devices with lifetimes easily exceeding 10^8 cycles with unipolar actuation at 100Hz. The device implements a switchable air gap capacitor and is therefore not limited in lifetime by dielectric charging as opposed to contact-type capacitive switches implementing high-k dielectrics. It is shown how these switched capacitors, even though having a capacitance ratio of only 2.8, can still form adequate switching devices and RF-circuits by proper design and combining of these devices with high-Q inductors and transmission lines. The novelty of the proposed process is that it combines a sacrificial layer consisting of a single layer with a single dry etching step for the dimples which define the air gap in the down-state. This airgap is switched by electrostatic actuation of a thick electroplated Nickel bridge-structure. The device is realized in a 4-lithographic steps process with low complexity and high robustness.

INTRODUCTION

The main failure mechanisms in RF-MEMS switching technologies are related to either dielectric charging for capacitive switches or contact degradation for ohmic switches [1]. To overcome these problems an increasing demand on the packaging and of the process complexity can be observed. By removing the dielectric from the active area of capacitive switches, many of the original failure mechanisms are removed. However, substrate charging may still remain an issue, as demonstrated in [2] [3]. At the same time, removing the typically thin high-k dielectric from the capacitive device reduces the capacitance ratio, which is an important figure of merit of an RF-MEMS capacitive switch. Nevertheless, several groups have demonstrated capacitive switching devices without dielectric that show a good performance [4][5]. But, even though capacitance ratios in the order of 5 to 10 are achieved, the used processes are relatively complex, using multiple sacrificial layers or dimples that result from sacrificial layer planarization effects. Furthermore, through proper design, even low capacitance ratio technologies combined with high-Q inductors and transmission lines still offer the possibility to design adequate switching devices as well as RF-circuits. In this paper we present a simple process that allows the realization of reliable RF-MEMS switching devices with a high functionality and a long lifetime.

BASIC CONCEPT

The switchable capacitor we propose consists of an electrostatically actuated bridge structure that switches an air gap capacitor. The bridge consists of a thick membrane which is suspended by four relatively weak springs. Dimples on the bottom side of the bridge define the air gap in the down state (after pull-in) as illustrated in Figure

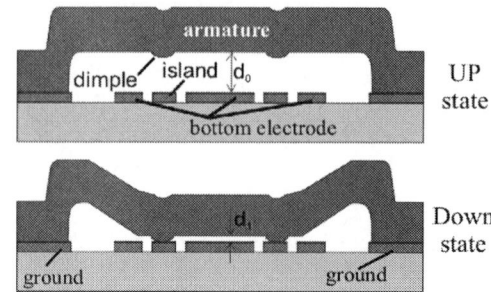

Figure 1: Sketch of the basic working principle of the switchable capacitor.

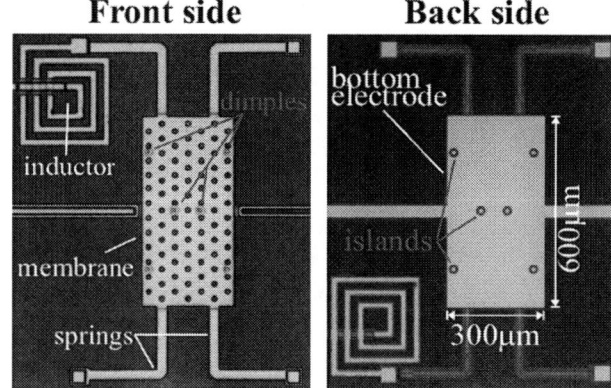

Figure 2: microphotograph of an air gap switch seen from front side and back sides.

1. The membrane has a much higher (bending) stiffness than the springs. This high stiffness together with a strategic placing of the dimples below the membrane prevents the membrane to collapse on the bottom electrode in the down-state. The 500nm high dimples land after pull-in on circular electrically isolated islands within the bottom electrode as shown in Figure 2. Since the capacitance of a parallel plate capacitor is directly related to the distance between the plates, the capacitance ratio C_{down}/C_{up} is defined by the ratio of the gap height in the up-state (d_0) and in down-state (d_1). The capacitance ratio can be improved by either increasing d_0, which would result in a higher pull-in voltage, or reducing d_1, by decreasing the height of the dimples. The scaling of both parameters is limited by the electric field in the air gap in the down state, which should remain below the breakdown field ($E_{breakdown}$) of the operating atmosphere:

$$\frac{V_{pull-in}}{d_1} = \sqrt{\frac{8kd_0^3}{27\varepsilon_0 A}} \cdot \frac{1}{d_1} \leq E_{breakdown}$$

where k the spring constant for the 4 springs combined, A the area of the membrane and ε_0 the permittivity of free space. The breakdown field is strongly dependent on the operation atmosphere and the air gap. The presented structure has a pull-in voltage of around 50 Volt, which result in an electric field close to 10^8V/m for a 500nm gap.

978-1-4244-2977-6/09 $25.00 © 2009 IEEE

FABRICATION PROCESS

The process consists of 4 lithographic steps. Figure 3 shows how the process starts with patterning a 1µm thick aluminum alloy (AlCu$_{0.5\%}$) bottom metallization on a 200mm quartz wafer (1). On top of this layer a 3µm thick PW-1530 polyimide sacrificial layer is spin-coated and photolithographically patterned. After the 250°C cross-link bake, 500nm deep pits which define the dimples below the membrane are dry etched in the polyimide (2). Next a 10 µm thick nickel film is electroplated through a photoresist mold (3). This layer serves both as the armature material and as signal line metallization as Figure 4 illustrates. Finally, the sacrificial layer is removed in an O$_2$-plasma (4).

The dimples below the membrane land on islands fabricated in the same bottom metallization as the electrode. A good planarization of the slots between the islands and the electrodes in the bottom metallization is needed to guarantee that the polyimide has the same thickness above the islands and the electrode. This way only the depth of the etched pits will determine the down-state air gap in the final device as illustrated in Figure 5. By reducing the slot width to 5µm the polyimide film only follows the bottom metallization topology above the slot, but remains the same thickness above the islands and the electrode. This results in a rim around the dimple on the membrane backside as can be observed in Figure 6.

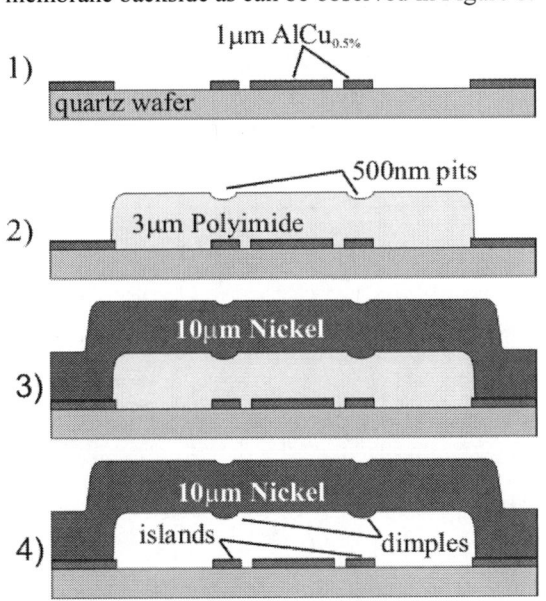

Figure 3: Sketch of fabrication process.

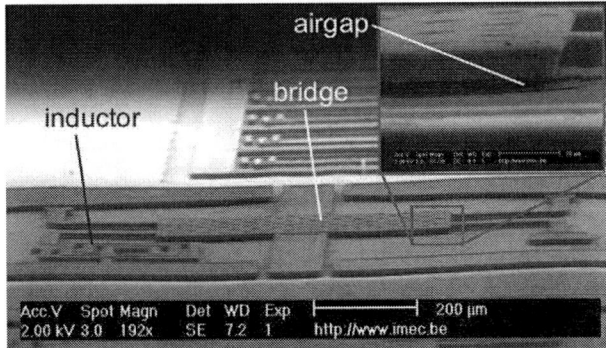

Figure 4: SEM image of topview of the device.

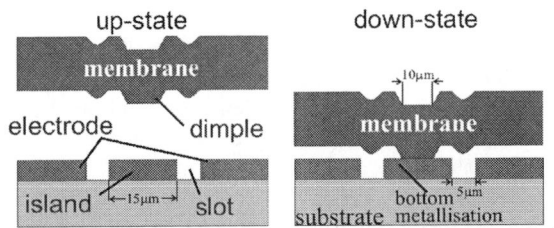

Figure 5: Sketch of membrane at the location of the dimples.

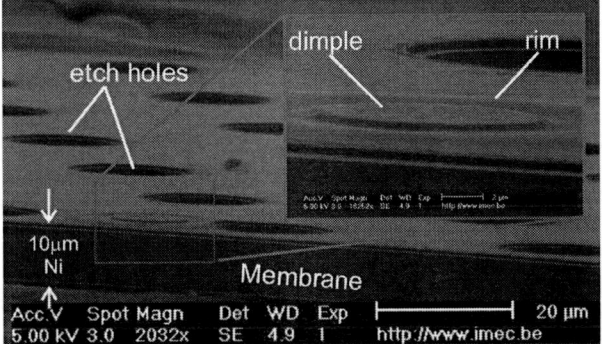

Figure 6: SEM image of the membrane backside.

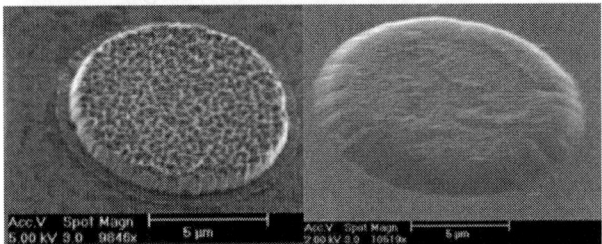

Figure 7: SEM picture of a 1µm high dimple etched in a parallel plate reactive ion etching system.

Figure 8: SEM picture of a 1µm high dimple etched in a downstream resist asher system.

For the etching of the pits a comparison has been made between parallel plate reactive ion etching (RIE) and etching in a downflow reactor system. Tips that result from pit etching in the RIE system as shown in Figure 7, present a much higher roughness than tips resulting from pit etching in a Matrix Bobcat downflow reactor system as shown in Figure 8. The latter also have a more isotropic profile more favorable for step coverage by the seed layer. Since the etching of the pits is done through a hard mask with a small number of openings, the loading during the plasma etch is relatively low. This results in a high uniformity of the depth of the etched pits for both processes (etch-depth variation of ~2% on 200mm wafer). The armature is realized by electroplating nickel through a photoresist mold with a Ti(30nm)/Cu(500nm) seed layer. The relatively thick seed layer guarantees a good step coverage at the edges of the sacrificial layer which is important to achieve a uniform distribution of the plating current-density over the wafer. This uniformity is required for a high plating uniformity. A high density of nickel structures further helps to increase the plating uniformity. Nevertheless a variation in thickness of 3% within a structure and of ~6-8% on a 200mm wafer is observed. The stress of the nickel film after plating is ~50MPa compressive, but after an anneal at 180°C the stress becomes 150-200MPa tensile.

MEASUREMENTS & DISCUSSION

For the realized devices, the stiffness of the 400μm long, 30μm wide springs by which the membranes are suspended is relatively low compared to the stiffness of the membrane itself. Nevertheless a voltage of 45V is required for pull-in. The C-V measurement, depicted in Figure 9, does not show the sharp pull-in as commonly observed with capacitive switches but show an initially more graduate change in capacitance. The low capacitance ratio of 1:1.8 is expected to result from parasitic capacitances and the measurement equipment.

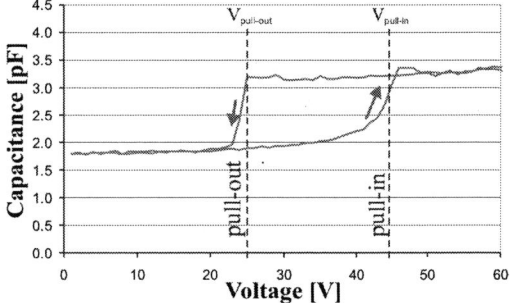

Figure 9: C-V measurement of the switchable capacitor.

The device with these basic characteristics enables the realization of several RF-circuits, two of which are presented in this paper. The first is the narrow band RF-switch shown in Figure 10 that functions as a tunable notch band filter while the second is the tunable stop band filter depicted in Figure 13. For the RF-switch the signal transmission in the up-state, characterized by the S_{21} parameter, is optimized by matching the up-state capacitance of the switchable capacitor and the inductive

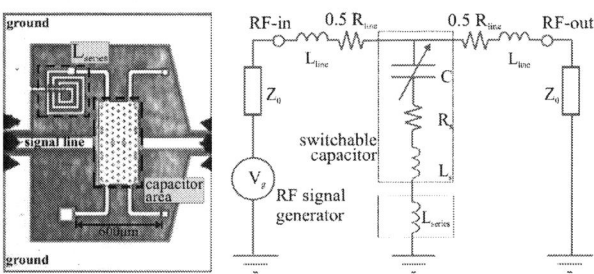

Figure 10: Micro-photograph of a narrow band RF-switch.

Figure 11: Circuit representation of a narrow band RF-switch.

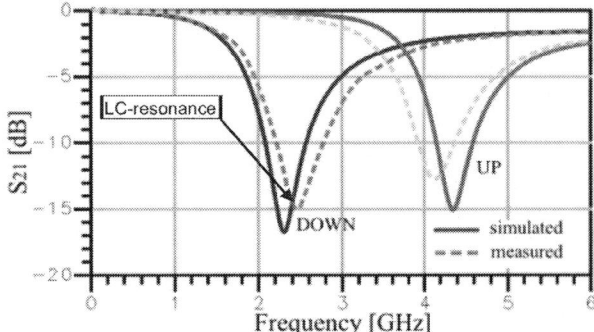

Figure 12: Simulation results and measured characteristics of the capacitive air gap RF-MEMS-switch of Figure 10 and 11, illustrating the narrow band behavior of the switch.

feeding lines. In the down- state, S_{21} is strongly influenced by the resonance of the lumped LC-circuit between the MEMS and the ground that is created by adding series inductor L_{series}. This high-Q inductor is realized in the thick Ni layer and placed in the space created by the matched feeding lines. This way a compact low frequency LC tank is defined of which the circuit representation is shown in Figure 11. The device shows a shift in LC-resonance from 4.2 GHz in the UP-state to 2.5 GHz in the DOWN-state as depicted in Figure 12. This results in a transmission loss of 0.3dB and an isolation of -15dB at 2.5GHz. The shift in resonance frequency matches a capacitance ratio of 2.8, which is lower than the capacitance ratio of ~6 expected from the change in air gap. The cause for this deviation is currently being investigated. We can further improve the switching characteristics of this circuit by implementing a more complex 2-stage circuit shown in Figure 13 and Figure 14. Here two RF switches in shunt configuration are coupled by an inductor (L_{couple}). This way a tunable stopband filter is defined which has a steep cut off and a wide rejection

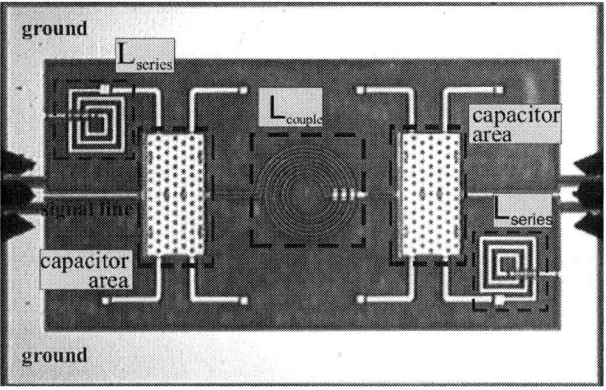

Figure 13: Microphotograph of a tunable stopband filter.

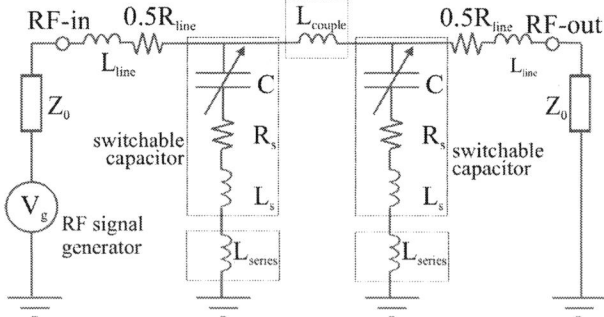

Figure 14: Circuit representation of the tunable stopband filter.

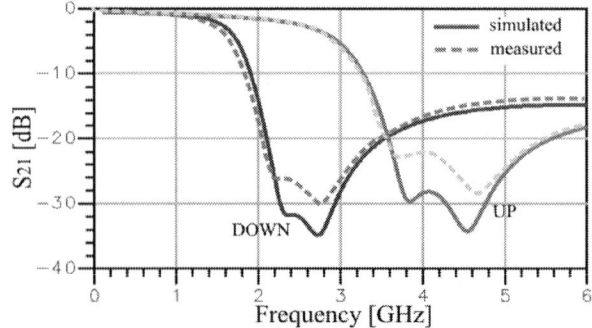

Figure 15: Simulation results and measured characteristics of the tunable stopband filter of Figure 12.

band as shown in Figure 15. This device shows a shift in LC-resonance from 3.8GHz in the UP-state to 2.2GHz in the DOWN-state, and a transmission loss of 1.4dB with an isolation of -26dB at 2.2GHz. Improvements of the design should significantly reduce the transmission loss. The main cause of failure for capacitive switching devices is stiction due to interposer dielectric charging of the high-k dielectric between the bridge and bottom electrode. Since there is no interposer dielectric in the device presented here, the only remaining source of charging is substrate charging [2]. Lifetime measurements with 50V unipolar actuation, 100Hz and 50% duty cycle in N_2 as depicted in Figure 16 showed no failure during 10^8 cycles, after which the measurement was stopped. The only significant shifts in the up and down capacitance are the result of thermal drift in the equipment during the measurement. In order to obtain more information on the charging behavior both the pull-in and pull-out voltage were monitored during the lifetime measurements as shown in Figure 17. Only a mild shift of the overall C-V characteristics is visible while both pull-in and pull-out window clearly narrow, testifying of non-uniform charging [6] [7]. The narrowing is most significant for the pull-out window is defined as $d\Delta V_{po}=(V_{po}+)-(V_{po}-)$. Figure 18 shows how the

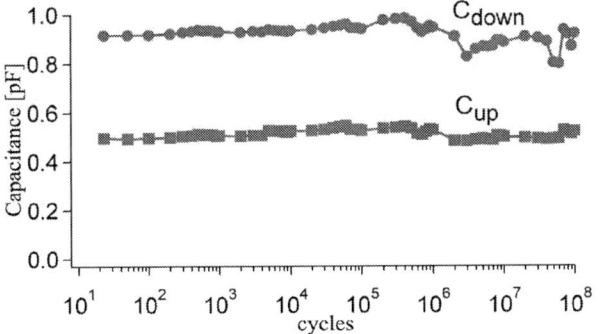

Figure 16: Lifetime of an air gap switch using 50V unipolar actuation, 100 Hz and 50% duty cycle in N_2 environment.

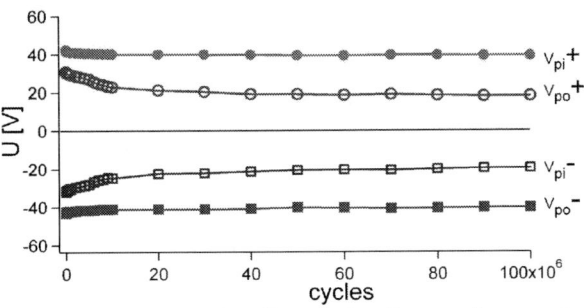

Figure 17: Monitored pull-in and pull-out voltages of air gap switch using 50V unipolar actuation, 100 Hz and 50% duty cycle in N_2 environment.

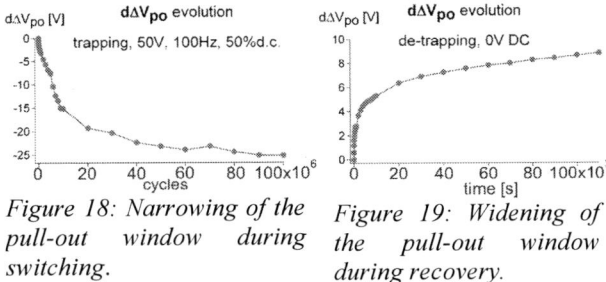

Figure 18: Narrowing of the pull-out window during switching.

Figure 19: Widening of the pull-out window during recovery.

narrowing stabilizes after $\sim 4 \cdot 10^7$ cycles due to saturation of the substrate charges. It is observed that after the loading ended the charges quickly disappear, which results in a widening of the pull-out window as shown by the de-trapping experiment of Figure 19. The lifetime of these devices is thus not limited by the substrate charging and significant longer lifetimes can be realized.

CONCLUSION

In this paper we have presented a simple and robust technology for realization of RF-MEMS circuits with a high functionality and long lifetimes The technology consists of a 4 lithographic steps process which is used to create a MEMS switchable capacitor. It consists of a 10µm thick Ni membrane which is used to switch the height of an air gap dielectric. Even though the resulting capacitance ratio of of 2.8 is relatively small the thick metallization can be used to create compact low frequency LC-tanks which allows the creation of several interesting RF-circuits. Due to the absence of an interposer dielectric the only source of charging is substrate charging of the quartz substrate. Stable lifetimes of at least 10^8 cycles with unipolar actuation have been demonstrated but a much longer lifetime is expected.

REFERENCES

[1] G. M. Rebeiz, "RF MEMS in theory, design and technology", Hoboken, NJ, USA, Wiley, 2003.

[2] P. Czarnecki, X. Rottenberg, P. Soussan, P. Ekkels, P. Muller, P. Nolmans, W. De Raedt, H. A. C. Tilmans, R. Puers, L. Marchand and I. De Wolf, "Influence of the substrate on the lifetime of capacitive RF MEMS switches", proc. of MEMS 2008, pp 172-175, 2008.

[3] X. Rottenberg, P. Nolmans, P. Ekkels, P. Czarnecki, R. P. Mertens, B. Nauwelaers, I. De Wolf, W. De Raedt and H. A. C. Tilmans, "Electrostatic fringing-field actuator (EFFA): application towards a low-complexity RF-MEMS technology, proc. of MEMS 2007, pp 203-206, 2007.

[4] Qin Shen, N. S. Barker, "Distributed MEMS Tunable Matching Network Using Minimal-Contact RF-MEMS Varactors", IEEE transactions on microwave theory and techniques, Vol 54, no 6, pp2646-2658, 2006.

[5] P. Blondy, A. Crunteanu, C. Champeaux, A. Catherinot, P. Tristant, O. Vendier, J. L. Caw and L. Marchand, "Dielectric Less Capacitive MEMS Switches", MTT-S Microwave Symposium Digest, Vol. 2, june 2004, pp. 573-576.

[6] X. Rottenberg, I. De Wolf, B. K. J. C. Nauwelaers, W. De Raedt and H. A. C. Tilmans, "Analytical model of the DC actuation of electrostatic MEMS devices with distributed dielectric charging and non-planar electrodes", JMEMS, vol. 16(5), pp. 1243-1253, October 2007.

[7] P. Czarnecki, X. Rottenberg, P. Soussan, P. Nolmans, P. Ekkels, P. Muller, H. A. C. Tilmans, W. De Raedt, R. Puers, L. Marchand, I. De Wolf, "Experimental evidence of non-uniform dielectric charging in capacitive RF-MEMS switches", 18th Workshop on MicroMechanics Europe, pp 227-230, 2007.

RUTHENIUM/GOLD HARD-SURFACE/LOW-RESISTIVITY CONTACT METALLIZATION FOR POLYMER-ENCAPSULATED MICROSWITCH WITH STRESS-REDUCED CORRUGATED SIN/SIO2 DIAPHRAGM

F. Ke[1], J. Miao[1], and J. Oberhammer[2]
[1]Nanyang Technological University, SINGAPORE
[2]KTH-Royal Institute of Technology, Stockholm, SWEDEN

ABSTRACT

This paper presents a RF MEMS switch with a new ruthenium/gold multi-layer contact metallization scheme, which combines the advantages of a hard ruthenium contact surface for high contact reliability and of a low, total contact resistance as typical for gold alloys. The performance of the new concept has been analyzed theoretically and was experimentally verified by contact resistance and life-time characterization of fabricated MEMS switches with conventional Au-Au and with the novel Au/Ru-Ru/Au contact metallization scheme. The switches are based on a low-stress SiN/SiO2 diaphragm which is polymer transfer-bonded and equipped with corrugations for reducing the stiffness and for lowering the stress. The reduced stiffness allows for early encapsulation by clamping the membrane all around its circumference, by maintaining medium actuation voltages.

INTRODUCTION

Metal-contact microrelays are suitable for a large range of applications since they are characterized by near-ideal signal switching behavior from DC to many tens of GHz, offering very low insertion loss, high isolation, and high linearity [1]. The metallization of the switch contacts in MEMS switches are still an issue of ongoing investigations, because the contact/opening force profile of microactuators is different from macroscopic relays. Soft metals such as gold and gold alloys offer low contact resistance even at low contact forces but suffer limited life time because of increased contact wear [2]. Furthermore, they need large opening forces to overcome the large adhesion force between closed soft-metal contacts, which is especially difficult to be achieved in conventional switch designs with deflected cantilever/membrane structures characterized by active closing but passive opening [3]. Hard materials such as rhodium have been utilized in MEMS switches and have proven good reliability but suffer large contact resistance. Ruthenium, even harder than rhodium, has so far only been employed in macro-relays as a contact material [4] because of its large contact resistance for the low contact forces in MEMS switches.

The present paper utilizes a gold/ruthenium multilayer contact metallization scheme which is designed to address contact reliability issues by having a hard contact surface but still keeping the total contact resistance in the same order as gold alloy contacts. The contact performance in terms of contact resistance and switch life time for different signal currents is compared for switches of the same device design with conventional and with the proposed contact metallization scheme.

MULTI-LAYER CONTACT DESIGN

The total contact resistance R_{CR} is composed by a constriction term R_C, describing the resistance of the contact asperity, and a film resistance term R_f, describing the resistance of the actual contact surface film [5]:

$$R_{CR} = R_c + R_f = \frac{\rho}{2a} + \frac{\rho_f d}{\pi a^2} \tag{1}$$

(R_{CR} ... total contact resistance; R_C, ... constriction resistance; R_f... film resistance; ρ ... asperity resistivity; ρ_f ... film resistivity; a ... asperity radius; d ... surface film thickness).

The effective contact area of the asperity in this model is given by $A_c = \pi a^2$, and the relationship between the effective contact area A_c and the contact force F is given via the material hardness H by $F = A_c H$ [6]. Thus, for equation (1), the (equivalent) asperity radius a of the contact point(s), which is impossible to determine in reality, can be expressed and substituted by $a = \sqrt{\frac{F}{\pi H}}$, resulting in:

$$R_{CR} = R_c + R_f = \sqrt{\frac{\rho^2 \pi H}{4F}} + \frac{\rho_f d H}{F} \tag{2}$$

In equation (2), the total contact resistance is expressed by the surface film thickness as the only geometrical parameter, besides the contact force and material parameters (asperity and surface material resistivity, surface hardness), and is therefore very suitable for estimating the total contact resistance.

By using different materials for the surface film and the contact asperity, the surface properties and the total contact resistance can be controlled with reduced interdependence. Figure 1 shows how the principal concept of the proposed contact metallization scheme: a thin surface coating with a hard and thus reliable ruthenium layer is expected only to slightly increases the total contact resistance, since only the film resistance term but not the constriction term is determined by the higher resistivity of ruthenium as compared to gold.

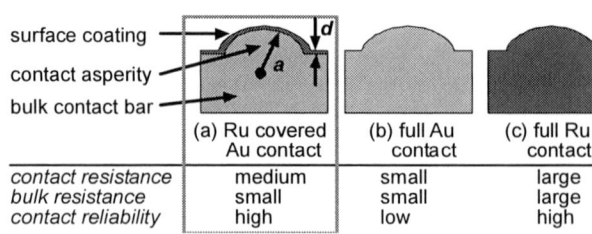

Figure 1: Novel Ru/Au contact metallization scheme as compared to conventional Au and Ru contacts.

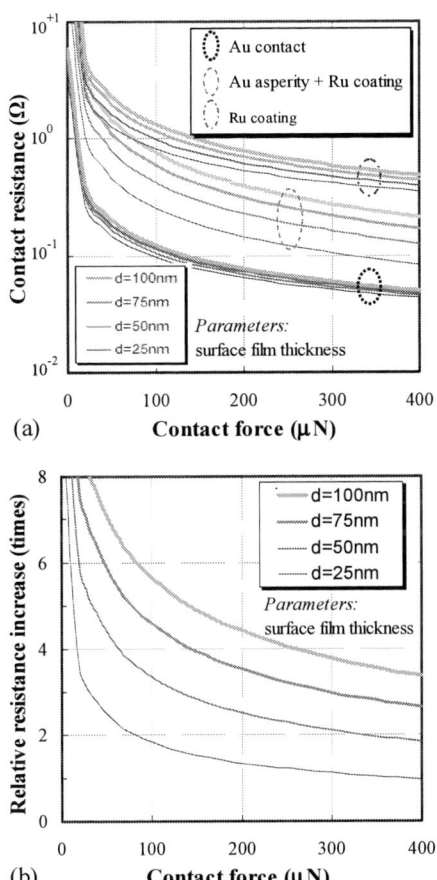

(a)

(b)

Figure 2: Simulated total contact resistance: (a) for conventional Au-Au, Ru-Ru and the novel Au/Ru-Ru/Au contact metallization, depending on the contact force; (b) relative contact resistance increase of the Au/Ru-Ru/Au contacts as compared to soft Au-Au contacts.

Figure 2 shows calculation results of the total contact resistance R_{CR} of the novel Au/Ru-Ru/Au contacts, which is about 3 times smaller than full Ru-Ru contacts, and only about 2-3 times larger than soft gold contacts for a film thickness of 50 nm and for contact forces as typical in MEMS switches (100-400 μN), despite offering the surface hardness of full ruthenium contacts and therefore expected similar contact reliability. Thus, for the fabrication, a surface film thickness of 50 nm was chosen since it is a good compromise between not too large relative resistance increase (as compared to gold) and stable and reliable contact coating.

CORRUGATED MEMBRANE SWITCH DESIGN AND FABRICATION

Figure 3 shows the fabrication procedure. The low-stress 1200nm LPCVD SiN/100nm thermally-grown SiO_2 diaphragm, fabricated on a silicon wafer, is clamped all around its circumference, which results in a fully (even though non-hermetically) encapsulated switch after the CYTOP polymer bonding process. This concept protects the precious metal contacts from contamination as early as possible during the fabrication. A completely clamped membrane, even if millimeter-sized, is very stiff and

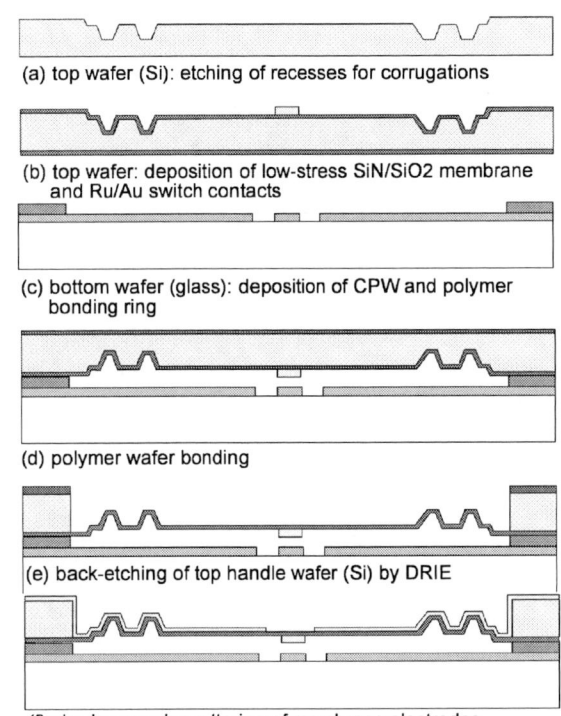

(a) top wafer (Si): etching of recesses for corrugations

(b) top wafer: deposition of low-stress SiN/SiO2 membrane and Ru/Au switch contacts

(c) bottom wafer (glass): deposition of CPW and polymer bonding ring

(d) polymer wafer bonding

(e) back-etching of top handle wafer (Si) by DRIE

(f) shadow-mask sputtering of membrane electrodes

Figure 3: Simplified fabrication sequence.

would require large actuation voltage. Thus, to reduce the stiffness, the diaphragm is endowed with specially designed corrugations [7], which were fabricated on different wafers either by deep reactive ion etching or by KOH etching (see Figure 4). The corrugations reduce the intrinsic stress from 56 MPa to 11 MPa (measured). The measured and simulated stress reduction factor is shown in Figure 4, for different corrugation depths both for KOH and DRIEtched recesses. Measurements of switches with and without DRIEtched corrugations, as specified in Figure 4, have confirmed the expected stiffness reduction by an actuation voltage decrease from 61V to 36V. All switch contacts are sputter-deposited.

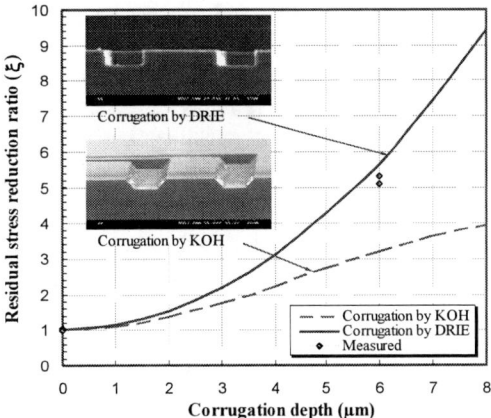

Figure 4: KOH and DRIEtched recesses in top wafer for fabricating diaphragm corrugations, with simulated and measured effective stress reduction factor in the membrane (two corrugation rings in 1400 μm × 1200 μm large, 1200nm SiN/100nm SiO₂ membrane).

978-1-4244-2977-6/09 $25.00 © 2009 IEEE   857

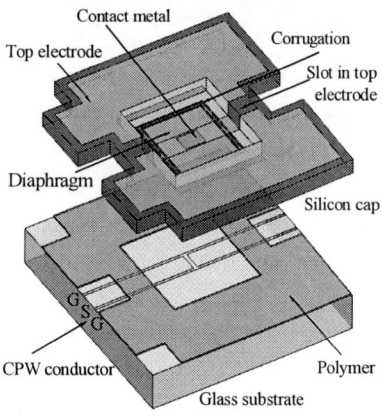

Figure 5: Schematic drawing of the transfer-bonded switch.

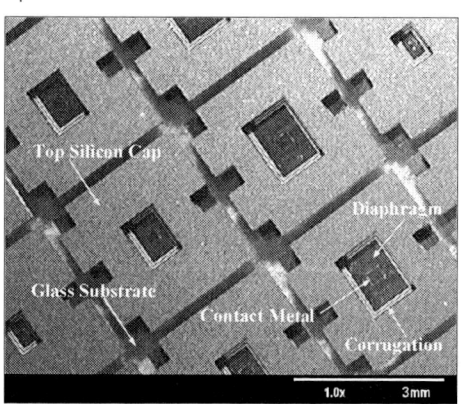

Figure 6: Photograph of wafer-scale packaged RF MEMS switches with a 1400 µm × 1200 µm × 1.7 µm corrugated diaphragm, before the wafer dicing. The overall size of a single switch after dicing is 3mm × 2.8mm × 0.95mm.

Figure 5 shows a schematic overview of both device parts before the bonding, and Figure 6 shows a microscope photograph of the fabricated switches before the final dicing of the wafer.

CONTACT PERFORMANCE

Contact resistance measurements were carried out for switches with the new contact metallization scheme and for switches with conventional, pure gold contacts. The total contact resistance of the Au/Ru-Ru/Au contacts is only about a factor of 2.5 times larger than soft Au-Au contacts, which confirms the theoretically predicted relatively small increase of the contact resistance. The measurement results are shown in Figure 7 for different actuation voltages resulting in different contact forces, for Au-Au and Au/Ru-Ru/Au switches. The contact resistance for Au-Au switches is between 250 and 450 mΩ for 90 to 50 V, respectively, and for the proposed Au/Ru-Ru/Au contacts between 700 and 1100 mΩ for 80 to 50 V, respectively. The figure also shows the burn-in behavior of newly fabricated switch contacts, which requires more cycles to reach stable values for soft gold than for ruthenium surfaces, indicating that more surface adaptation happens for gold surfaces (signal current of

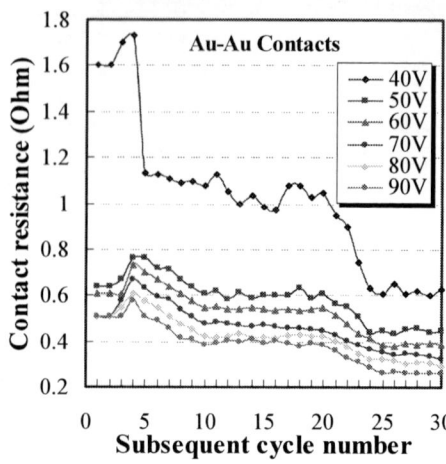

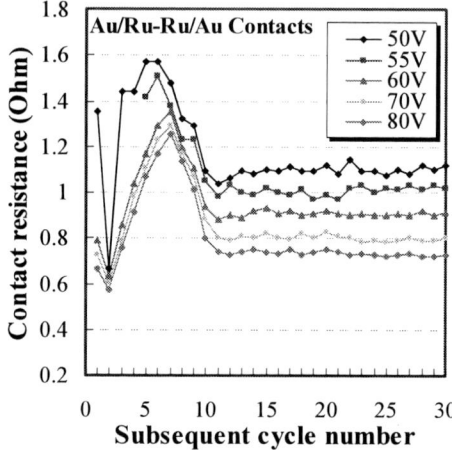

Figure 7: Contact resistance over initial switching cycles (burn-in behavior) at different actuation voltage levels for conventional soft Au-Au contacts and for the proposed Au/Ru-Ru/Au metallization scheme, with actuation voltage (i.e. contact force) as parameter.

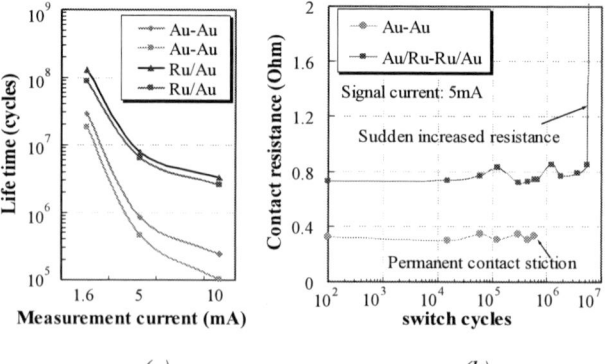

(a)   (b)

Figure 8: Hot-switching life cycle tests for conventional soft Au-Au contacts compared to the novel Au/Ru-Ru/Au metallization scheme: (a) life time for signal current of 1.6, 5, and 10 mA; (b) total contact resistance plot over life time for 5 mA hot-switched devices.

978-1-4244-2977-6/09 $25.00 © 2009 IEEE   858

5 mA, applied also during the opening of the contacts).

Figure 8 shows hot-switching life-time tests to compare the contact reliability of the different contact metallization schemes. For signal currents of 10 mA, the life-time of the Au/Ru-Ru/Au contacts is more than 10 times longer than the life-time of the Au contacts. The lifetime benefit is increasing for larger signal current. The contact resistance for both materials is stable within 20% throughout its life-time. Interestingly, it was found that Ru-surface switches die by sudden contact resistance increase, whereas the soft gold switches die in short-circuit by permanent contact stiction, which is an important criterion for fail-safe circuit design. These life time tests were carried out on (non-hermetically) encapsulated devices as described in the fabrication section. The switching frequency for the tests was 1 kHz, with 20% duty cycle, and the measurement signal is applied all the time, i.e. the signal current is applied during opening of the switch contacts.

RF CHARACTERIZATION

The measured and simulated RF performance is shown in Figure 9. The isolation is 30 dB at 4 GHz and 21 dB at 15 GHz. The measured insertion loss of the wafer-scale encapsulated Au/Ru-Ru/Au switches including a 3 mm long transmission line is 0.7 and 2.8 dB at 4 and 15 GHz, respectively, which does not match the simulated insertion loss of 0.2 and 0.9 dB at 4 and 15 GHz, respectively. The main reason for that, beside material losses, was identified to be the inaccuracy of the patterning of the top electrodes on the membrane: the metal electrodes are fabricated after the bonding by sputtering gold with a shadow mask placed at a distance of 200 µm above the membrane (Figure 3f). The nominal slot width and the pattern in the shadow mask is 320 µm wide. However, because of the distance to the membrane, the actual slot width of the sputtered material on the membrane is around 280 µm, which creates a mismatch to the 50 Ohm line, reflecting about 25% of the incoming

power at 15 GHz (measured). Simulations with 280 µm gap in the top electrode have verified this mismatch, and a correct gap in the top electrode results in a simulated insertion loss of 0.6 dB and 2.4 dB at 4 GHz and 15 GHz, respectively (Figure 9).

The switching time was measured to 176µs, mainly attributed to squeeze-film damping of the 1200x1400 µm^2 large diaphragm.

CONCLUSIONS

This paper presents a metal-contact switch design which uses, for the first time, ruthenium on the contact surface. The total contact resistance, as compared to pure ruthenium contacts for MEMS switches with low contact forces, is drastically reduced since a special, novel, multi-layer gold-ruthenium contact meatallization scheme is used. The contact resistance and life time of switches with the novel metallization scheme has been evaluated and compared to pure gold contact switches, demonstrating a life time enhancement of over 10 times, measured even in non hermetic environment, and an increase of total contact resistance of only 2.5 as compared to pure, soft gold contacts, matching very well the theoretical prediction. The switches were successfully fabricated on a stress-compensated SiN/SiO$_2$ membrane with corrugations for further stress reduction and for lowering the stiffness of the membrane, which allows the membrane to be clamped all over its circumference, resulting in a wafer-scale encapsulation already early during the fabrication.

REFERENCES

[1] G. M. Rebeiz, *RF MEMS Theory, Design, and Technology*, John Willey & Sons Publication, 2003.

[2] J. Schimkat, "Contact measurement providing basic design data for microrelay actuators," *Sensors and Actuators A: Physical*, volume 73, issue 1–2, pp. 138–143, March 1999.

[3] J. Oberhammer and G. Stemme, "Active opening force and passive contact force electrostatic switches for soft metal contact materials", *Journal of Microelectromechanical Systems*, vol. 15, no. 5, October 2006, pp. 1235-1242.

[4] W. Johler, "Precious metal-reduced contact materials in telecom and signal relays," *Electrical Contacts, 2001, Proceedings of the 47th IEEE Holm Conference on*, 10-12 September 2001, pp. 104-116.

[5] P. G. Slade, *Electrical Contacts Principles and Applications*, Cutler-Hammer, Horseheads, New York, 1999.

[6] F. P. Bowden, D Tabor, *Friction and Lubrication of Solids, Vol. II*, Oxford University Press, 1964.

[7] J. Miao, R. Lin, L. Chen, Q. Zou, S. Y. Lim, and S. H. Seah, "Design considerations in micromachined silicon microphones," *Microelectronics Journal*, volume 33, issue 1, January 2002, pp. 21-28.

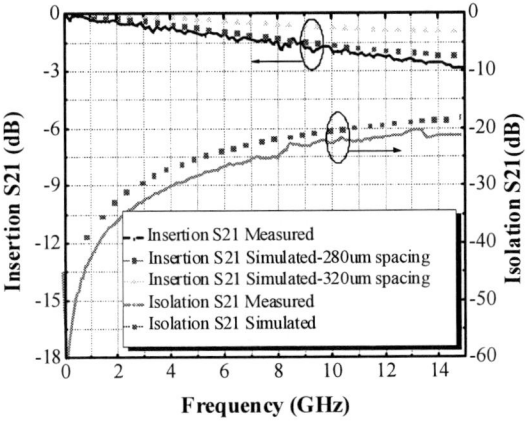

Figure 9: Simulated and measured RF isolation and insertion loss.

A MEMS RECONFIGURABLE QUAD-BAND CLASS-E POWER AMPLIFIER FOR GSM STANDARD

L. Larcher[1], R. Brama[1], M. Ganzerli[1], J. Iannacci[2], B. Margesin[2], M. Bedani[3] and A. Gnudi[3]

[1]DISMI Università degli Studi di Modena e Reggio Emilia, Reggio Emilia, Italy
[2]Fondazione Bruno Kessler (FBK), Povo di Trento, Italy
[3]DEIS-ARCES Università degli Studi di Bologna, Bologna, Italy

ABSTRACT

In this work we present a reconfigurable mid-power Class-E Power Amplifier (PA) [1,2] operating at ~900MHz and ~1800MHz (GSM standard [3]) realized hybridizing one chip manufactured in AMS 0.35μm CMOS technology and one MEMS sub-network. The CMOS chip realizes the active part of the circuit, whereas the MEMS block (realized in FBK technology) implements a reconfigurable impedance Matching Network (MN) that transforms the 50Ω antenna load to the 12Ω impedance required by the PA in order to deliver 20dBm output power in both the GSM operating frequency bands. The prototype of the MEMS/CMOS PA we realized delivers 20dBm with 38% and 26% drain efficiencies at 900MHz and 1800MHz, respectively, demonstrating to be a feasible option compared to standard commercial solutions.

INTRODUCTION

Micro-Electro-Mechanical-Systems (MEMS) technology for Radio Frequency (RF) applications has been proven in recent years to enable the manufacturing of high performance, low loss and low cost passives like high Q inductors and wide-range variable capacitors. Therefore, much interest is nowadays being drawn by the possibility of employing MEMS components to entirely synthesize low loss networks for an efficient signal processing.

Telecommunication standards impose the need for mobile phone transceivers to work on different frequency bands (GSM protocol, as an example, employs four uplink bands), hence PAs have to deal with a wide frequency range. Designing PAs with bandwidth exceeding 100MHz is not an issue [1-3], and commercial solutions for GSM employ usually two PAs, whose operation frequencies are centered at ~900MHz and ~1800MHz to allow signal amplification at lower or upper bands, respectively [3]. Despite this solution is effective, it imposes duplicating circuitry while having one PA in idle mode. On the contrary a reconfigurable PA would allow covering both lower and upper bands enabling the reduction of chip area.

In this scenario, we propose a reconfigurable mid-power Class-E PA operating at ~900MHz and ~1800MHz, realized connecting two chips manufactured using AMS 0.35μm CMOS and MEMS technologies, respectively. The CMOS chip realizes the active part of the circuit, whereas MEMS switches and passives implement the reconfigurable PA output network working at lower and upper bands. The adoption of a Class-E topology allows mitigating the circuit sensitivity against components variation. Indeed, in spite of the very high performances, RF-MEMS in FBK technology present a few parasitic effects due to critical process step

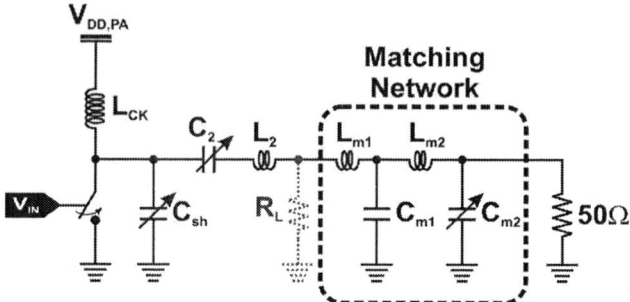

Figure 1: Schematic of a reconfigurable Class-E PA. The Matching Network down transforms the 50Ω antenna to R_L optimum load.

that will be further discussed [4].

The paper is organized as follows. Section 2 describes theoretical aspects and circuit design of a reconfigurable Class-E PA and gives details on the MEMS MN. In Section 3 we present and discuss experimental results. Conclusions follow.

RECONFIGURABLE CLASS-E PA DESIGN

Class-E power amplifiers are switched mode amplifiers invented by Sokal in 1975 [5]. Their components, shown in fig. 1, are sized according to simple design equations [6].

$$ R_L = \frac{\alpha V_{DD}^2}{P_{OUT}} ; C_{sh} = \frac{\beta}{\omega_0 R_L} ; C_2 = \frac{\gamma}{\omega_0 R_L} ; L_2 = \frac{Q_{out} R_L}{\omega_0} \quad (1) $$

α, β, γ are parameters that depend on the output network quality factor Q_{out}; P_{OUT} is the output power with ideal 100% efficiency; V_{DD} is the voltage supply; ω$_0$ is the working frequency; R$_L$ is the load resistance.

V_{DD} cannot exceed the maximum limit ~3V due to reliability constraints [7], and therefore increasing P_{OUT} needs to lower R_L. For a maximum output power ~21dBm at both lower and upper bands, we calculated $R_L = 12\Omega$, assuming a real ~40% efficiency, a $Q_{out} = 2.5$, and $V_{DD} = 3V$. In order to obtain this load impedance from a 50Ω antenna resistance at both lower and upper bands, we need a reconfigurable matching network. In our design this MN is completely realized by an external MEMS network. The matching network is comprised of two LC ladders which down transforms the antenna resistance at ~900MHz and ~1800MHz using a minimum number of reconfigurable components, namely C_2 and C_{m2}, and avoiding to switch inductors to reduce power losses.

In order to switch between upper and lower bands while

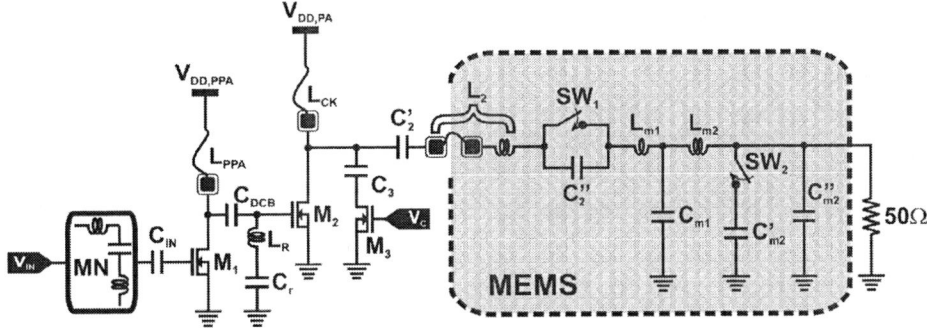

Figure 2: Schematic of the proposed reconfigurable Class-E PA. The MEMS based matching network (circled by the dashed line) implements the LC series filter and the matching network, composed by two LC sections.

keeping a high operation efficiency, output network components (C_{sh}, C_2 and L_2) have to be reconfigured according to (1). Since switch losses are one of the most significant power loss contributions, we reduced the number of switches used to reconfigure the network, keeping L_2 constant regardless operation frequency. Thus, the output network quality factor Q_{out} doubles at *1800MHz* as shown in (1). Selecting a low Q_{out} is crucial to reduce output network power losses [1], hence we choose Q_{out}=2.5 at *900MHz*. This way, capacitance values at low (L) and high (H) bands are given by $C_{sh,L}$=3.2pF and $C_{2,L}$=14.9pF, and $C_{sh,H}$=1.6pF and $C_{2,H}$=2pF, respectively.

In fig. 2 we show the circuit schematic of the reconfigurable power amplifier, including the external MEMS network. The PA is comprised of two common-source stages. To avoid gate oxide breakdown, we used MOSFETs with high breakdown voltage (*7.5V*) and limited the supply voltage of the PA driving stage to $V_{DD,PPA}$=1.25V, thus limiting the maximum voltage across the gate oxide stress below the DC breakdown limit [7].

In the implementation of the reconfigurable shunt capacitance C_{sh} we employed the switch M_3 in order to ground C_3 while using M_2 ($C_{dd,M2}$) and M_3 ($C_{dd,M3}$) drain parasitic capacitances, see fig. 2. At lower bands, ~*900MHz*, M_3 is switched ON in order to increase the shunt capacitance implementing $C_{sh,L}$. On the contrary, when working at ~*1800MHz*, M_3 is switched OFF to obtain $C_{sh,H}$. In order to find values for drain MOSFET capacitances and C_3 we solved the following equation system:

$$\begin{cases} C_{sh,L} = C_{dd,M2} + C_3 \\ C_{sh,H} = C_{dd,M2} + \dfrac{C_{dd,M3}C_3}{C_{dd,M3} + C_3} \end{cases} \quad (2)$$

Assuming that C_3 is k times the M_3 drain parasitic capacitance ($C_3 = k \cdot C_{dd,M3}$) and equating $C_{sh,L}$=2·$C_{sh,H}$ allows calculating capacitance values after some mathematical manipulations

$$\begin{cases} C_{dd,M2} = \dfrac{k-1}{2k} C_{sh,L} \\ C_{dd,M3} = \dfrac{k+1}{2k^2} C_{sh,L} \\ C_3 = \dfrac{k+1}{2k} C_{sh,L} \end{cases} \quad (3)$$

$k > 1$ has to be selected to maximize the efficiency. MOSFET power loss ($P_{L,MOS}$) depends inversely on transistor width (W), hence a large W is desirable to minimize on-state resistance [1]. Unfortunately, this increases the drain capacitance, which results to be proportional to $1/P_{L,MOS}$. Thus, minimizing MOSFET power losses requires to increase both $C_{dd,M2}$ and $C_{dd,M3}$, but this cannot be done acting on k, as their dependence on k is opposite. The optimum k=3.56 is found imposing $C_{dd,M2}$=2·$C_{dd,M3}$, that corresponds to equalize the power loss due to M_2 to the half of the M_3 one, as M_3 is switched on only at the lower frequency band, i.e. half of the time. Substituting k=3.56 into (3) allows calculating the optimum M_2 (W=6000μm) and M_3 (W=4500μm) widths and C_3=9.5pF. The finite RF Choke inductor L_{CK}=2.1nH is realized by a bondwire inductor.

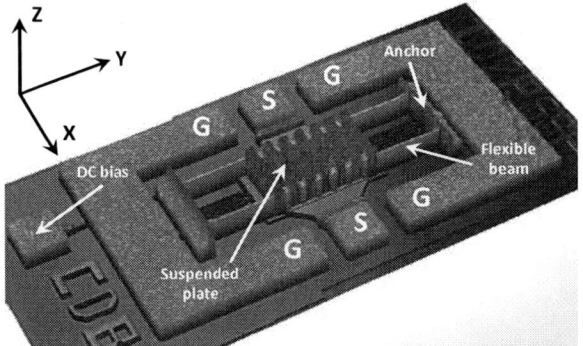

Figure 3: 3D image of the MEMS ohmic series switch, needed to reconfigure C_2, obtained with an optical profiling system. The color scale refers to the height along the Z-axis.

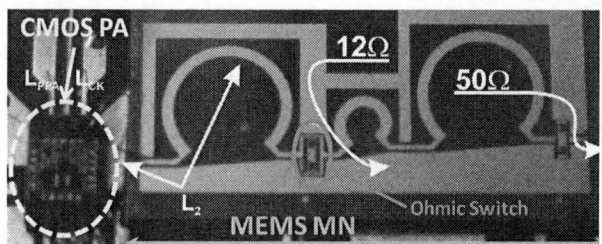

Figure 4: Photograph of the PA prototype. CMOS and MEMS dies, implementing the active part of the circuit and the output network respectively, are wire-bonded onto a DC testing board.

To drive efficiently the M_2 gate, we used a double resonant matching network synthesizing high impedance at both *900MHz* and *1800MHz*, which is comprised of the bondwire inductor $L_{PPA}=1.8nH$, the planar inductor $L_R=3.1nH$, $C_r=4.2pF$ and the M_2 gate-source capacitance.

The MEMS network realizing the series L_2-C_2 filter and the matching network is shown in fig. 2. L_2, realized by the series of a bondwire and a planar inductor, is kept constant over both frequency bands. C_2 is reconfigured by an ohmic series switch SW_1 shorting C_2''. Thus, $C_{2,H}$ is realized as the series between C'_2 and C_2'' while $C_{2,L}$ is implemented only by C'_2.

The matching network is comprised of two cascaded LC sections, down transforming the antenna impedance to *12Ω* at both frequency bands. At *900MHz* the ohmic shunt MEMS switch SW_2 is actuated, and $C_{m2}=C'_{m2}+C''_{m2}$.

The close up of the switch SW_1 used to reconfigure C_2 is depicted in fig. 3 by means of an optical profiling system based on the interferometry of a standalone device fabricated purposely. SW_1 is an ohmic series switch [8] based on a central rigid gold plate (*5μm* thick) with four straight suspending beams. The switch is electrostatically actuated and the bias DC line is realized by an additional conductive layer i.e. separated by the input/output RF signal lines [8]. The switch is open when the central plate is not biased while it connects the input/output lines when the membrane reaches the pull-in (closed switch). SW_1 can be modeled as the parallel of the two MIM (metal-insulator-metal) capacitors when the plate is not actuated. On the contrary, when the bridge is actuated, the two capacitors are shorted.

SW_2 is an ohmic shunt switch, whose behavior is opposite to the SW_1 one. SW_2 is open when the plate is not actuated and shorted when the plate reaches the pull in. L_{m1} and L_{m2}, shown in fig. 2 and fig.4, are 3/4-a-circle gold inductors.

MEASUREMENTS RESULTS

The microphotograph of the PA circuit prototype we realized is shown in fig. 4. Two dies manufactured in CMOS and FBK RF-MEMS technologies have been glued on top of a DC biasing board on FR4 substrate. Dies have been connected by means of bondwires in a System in a Package (SiP) fashion (fig. 4). Critical bondwires inductors are L_{CK}, L_{PPA} and the output one that synthesizes part of L_2

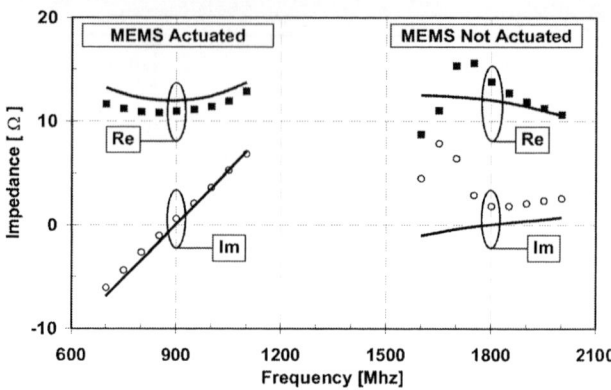

Figure 5: Simulated (solid lines) vs. measured (symbols) input impedance of the MEMS matching network at both lower and upper frequency bands.

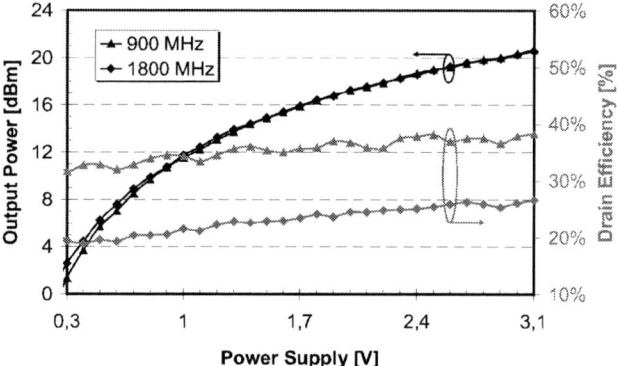

Figure 6: Output power and drain efficiency measured at 900MHz and 1.8GHz.

(see fig.2). The chip realized using AMS *0.35μm* technology implements the active part of the PA circuit. Grounding is provided by means of 10 bondwires while two dedicated ground bonding are used to connect the MEMS output network to ground. The MEMS network implementing series filter and matching network has been manufactured using fully compatible surface micromachining process available at the Bruno Kessler Foundation (FBK, Trento, Italy) [4,8]. The CMOS die area is *1.44mm²* including pads while the MEMS network area occupation is *29.88mm²*. The MEMS MN input/output and the PA RF input/output are configured for GSG (Ground Signal Ground) probing, enabling also the characterization of the matching network alone.

The MEMS MN has firstly been measured, without connecting it to the CMOS active part, with a Vector Network Analyzer (VNA) with probes in order to determine the effective input and output impedances. As the probes are matched to *50Ω*, a *12Ω* input impedance is expected for the MEMS MN. Measurements of the MN input impedance depicted in fig. 5 show that both real and imaginary part are slightly different from target values at both frequency bands. This issue is mainly due to parasitics of MEMS switches, not properly taken into account when designing the network. In particular, vias connecting the gold layer to the buried multi-metal layer realizing the RF lines underneath the

978-1-4244-2977-6/09 $25.00 © 2009 IEEE   862

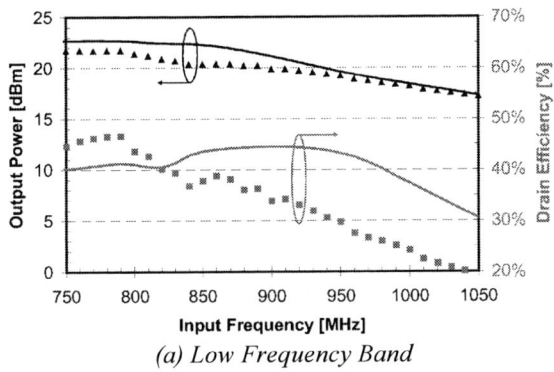

(a) Low Frequency Band

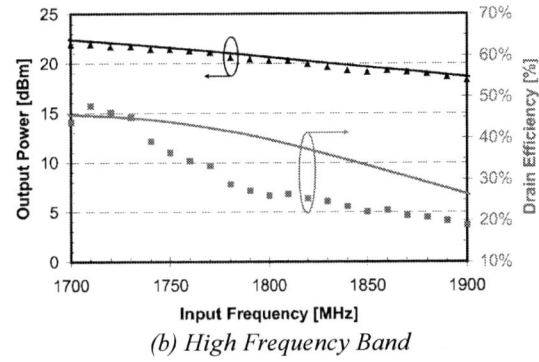

(b) High Frequency Band

Figure 7: Output power (triangles) and drain efficiency (squares) measured at both lower and upper frequency bands. Solid lines depict circuit simulations performed considering real behavior of output network obtained from the MEMS network characterization (see Fig. 5).

suspended plate (see fig. 3) introduce a series capacitance on the input/output lines explaining the difference from the expected RF behavior. However, such issue has been mitigated in recent technology runs thanks to a high temperature annealing step [4].

The complete PA has been measured using microprobes to connect input and output GSG pads. In order to make the PA working at lower frequencies, V_C=3V (see fig. 2) is applied to the M_3 gate while MEMS switches are actuated supplying 30V DC biasing. The input matching network MN is realized using external components.

Fig. 6 shows the output power and the drain efficiency of the PA prototype measured at 900MHz and 1.8GHz. With the maximum $V_{DD,PA}$=3V supply, the PA delivers 20dBm with 38% and 26% drain efficiencies at 900MHz and 1.8GHz, respectively. The difference with respect to post-layout simulations (44% and 40% at 900MHz and 1800MHz, respectively) is mainly caused by the behavior of the MEMS matching network that forces the power amplifier to operate with a mistuned output network, thus degrading circuit efficiency and performances, as shown in fig. 7. In addition, the unavoidable uncertainty on L_{PPA} and L_{CK} bondwire inductances contributes to reduce PA performances, explaining the not negligible difference between simulations and measurements, see fig. 7. Here output power and drain efficiency are plotted versus frequency at both lower and upper bands.

Despite such technological issues, the low output network quality factor we selected allows reducing circuit sensitivity against component variations, allowing the PA delivering more than 18dBm at both lower and upper bands, see fig. 7.

CONCLUSIONS

In this paper we presented a reconfigurable Class-E power amplifier able to operate in the up-link bands of the GSM standard. Prototypes realized assembling in a SiP fashion two dies manufactured using CMOS and MEMS technologies deliver 20dBm with 38% and 26% drain efficiencies at both 900MHz and 1800MHz, respectively. Even if PA performance are below the ones expected from post-layout simulations due to technological issues (switch losses and parasitics caused by a few technology issues of the FBK RF-MEMS process that, however, have recently been mitigated), we believe that MEMS technology can play a key role for future RF applications.

REFERENCES

[1] Mazzanti, L. Larcher, R. Brama, and F. Svelto, "Analysis of Reliability and Power Efficiency in Cascode Class E PAs," *IEEE J. Solid-State Circuits*, vol. 41, no. 5, pp. 1222–1229, 2006.

[2] R. Brama, L. Larcher, A. Mazzanti, and F. Svelto, "A 1.7-GHz 31dBm differential CMOS Class-E Power Amplifier with 58% PAE," in *Proc. 2007 IEEE Custom Integrated Circuits Conf.*, Sept 2007, pp. 551–554.

[3] Aoki, S. Kee, R. Magoon, R. Aparicio, F. Bohn, J. Zachan, G. Hatcher, D. McClymont, and A. Hajimiri, "A Fully Integrated Quad-Band GSM/GPRS CMOS Power Amplifier," in *IEEE 2008 International Solid-State Circuits Conf. Digest of Technical Papers*, Feb 2008.

[4] J. Iannacci, A. Gnudi, B. Margesin, F. Giacomozzi, L. Larcher and R. Brama, "Reconfigurable RF-MEMS Based Impedance Matching Network of a CMOS Power Amplifier". *Proc. 9th International Symp. on RF-MEMS and RF Microsystems MEMSWAVE 2008*, Heraklion, Greece, 30 June – 3 July 2008.

[5] N. O. Sokal and A. D. Sokal. "Class E-A New Class of High-Efficiency Tuned Single-Ended Switching Power Amplifier". *IEEE J. Solid-State Circuits*, 10(3):168–176, Jun 1975.

[6] N. O. Sokal. "Class-E Power Amplifiers". *QEX/Comm. Quarterly*, pages 9–20, Jan/Feb 2001.

[7] L. Larcher, D. Sanzogni, R. Brama, A. Mazzanti, and F. Svelto. "Oxide Breakdown after RF stress: Experimental Analysis and Effects on Power Amplifier Operation". *In Proc. of the International Reliability and Physics Symp. 2006*, pages 283–288, March 2006

[8] R. Gaddi, M. Bellei, A. Gnudi, B. Margesin, and F. Giacomozzi. "Interdigitated Low-Loss Ohmic RF MEMS Switches". In *Proc. of NSTI 2004 Nanotechnology Conf. and Trade Show, Nanotech 2004*, volume 2, pages 327–330, Mar. 2004.

978-1-4244-2977-6/09 $25.00 © 2009 IEEE

A RF-MEMS BASED MONOLITHICALLY INTEGRATED WIDE RANGE TUNABLE LOW PASS FILTER FOR RF FRONT END APPLICATIONS

Tomonori Fujii[1], Taisei Irieda[1], Kentaro Nakamura[1], Kenichi Ota[1], Kiyoyuki Ihara[2],
Daniel T. McCormick[3] and Liwie Lin[3].

[1] Taiyo Yuden Co., Ltd. 5607-2 Nakamuroda, Haruna-machi, Takasaki , Gunma, Japan, Zip 370-3347,
Tel +81-27-360-8303, Fax +81-27-360-8323, E-mail: fujiit@jty.yuden.co.jp
[2] TRDA Inc, 10180 Telesis Court, Suite 120,San Diego, California 92121, U.S.A.
[3] Berkeley Sensor and Actuator Center, University of California, Berkeley. Berkeley, CA 94720-1774

ABSTRACT

Monolithically integrated, wide tuning range low pass filters have been demonstrated based on electrothermally driven MEMS switches. The integrated system is comprised of copper spiral inductors, polysilicon capacitors and electrothermal MEMS actuators on a single 4.3 x 2.4 mm^2 chip; the design is optimized for the selection of three commonly used commercial cut-off frequencies: (f_c) at 0.9GHz, 1.9GHz and 2.5GHz. The driving voltage is only 3.7V and out-of-band attenuation is -39dB at 3.5GHz, which is comparable to the state-of-art commercial ceramic filters. As such, this tunable filter and related fabrication technologies are applicable for various RF front end applications.

INTRODUCTION

Since the late 1980s major advances have been seen in wireless communication technologies, today numerous standards such as CDMA and GSM exist. Voice, data and broadband technologies utilize various frequencies and network architectures. However, the continuing adoption of stringent requirements demanded by advanced wireless communication systems, including software defined radios (SDRs) and cognitive radio systems, require the development and adoption of emerging technologies such as RF-MEMS devices [1]. One of the primary premises of these emerging technologies is the ability to access multiple frequency bands and enable the utilization of several architectures via a single radio.

In order to realize the next generation of wireless communication systems an enabling technology, providing the ability to select frequency bands in the 0.7GHz to 2.5GHz range, is required. To date the majority of published works have targeted frequencies less than several hundred MHz or higher than 10GHz [2]-[3]. There are very few reported results applicable in the 0.9GHz to 2.5GHz range, where major wireless communication systems are currently deployed.

DESIGN

The integrated, wide tuning range, low-pass filter architecture is comprised of several sub-systems, including: MEMS actuators and contact switches, RF signal lines, spiral inductors and parallel plate capacitors. The mechanical properties of the MEMS actuators, RF properties of the switches, transmission lines, capacitors and inductors as well as the monolithic fabrication of the structures must all be considered during the design process. In the following sections the design of the MEMS actuators and RF components are discussed.

MEMS Actuator Design

The design of electro-thermal actuator was based on results reported by Wang [4] and Enkov [5]. Following prototype characterization runs the designs were optimized via FEM analysis.

As reported in [5], given the configuration shown in Figure 1 the lateral displacement (δ) can be defined as a function of the beam's geometry (width (w), length (l) and thickness (t)) and displacement angle (θ), as shown in equation (1). In equation (2) the axial buckling force (P_{Axial}) definition is provided.

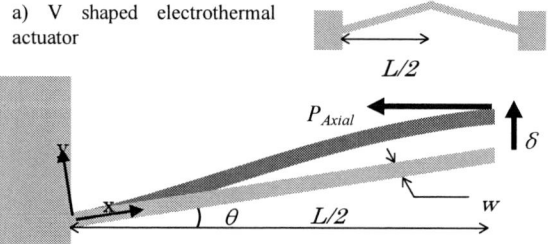

a) V shaped electrothermal actuator

b) Deformation of actuator and definition of geometrical parameters.

Figure 1: Schematic diagram showing the geometry and deformation of the V shaped electro-thermal actuator

$$\delta = \tan\theta \frac{2\tan\dfrac{kL}{2} - kL}{k\cos\theta} \qquad (1)$$

$$P_{Axial} = k^2 E I_{in} \qquad (2)$$

Employing the Euler defined buckling condition, given in equation (3), the maximum displacement of the V-shaped thermal actuator without buckling (δ_{max}) is calculated for a given geometry, (4) .

$$P_{critical} = 4\frac{\pi E I_{out}}{(2L)^2} \qquad (3)$$

$$\delta_{max} = \tan\theta \cdot \left[\frac{2\tan\left(\dfrac{\pi}{2}\dfrac{t}{w}\right) - \pi\dfrac{t}{w}}{\pi\dfrac{t}{w}\cos\theta} \right] \cdot L \qquad (4)$$

During the process of optimizing the design of the V shaped electrothermal actuators, power consumption as well as spring stiffness must be considered. Power consumption is one of the important design parameters for most applications, especially in the case of mobile radios.

The rigidity of the suspension is also an important design consideration as it relates to the contact force and contact resistance of the RF switch.

A graph presenting the design window as a function of actuator beam width and angle is shown in Figure 2. The "buckling area" is defined by (4). Power consumptions is estimated using FEM analysis. The actuator is designed such that power consumption will be less than 100mW. The design specification for spring stiffness is taken to be 20uN/m or greater.

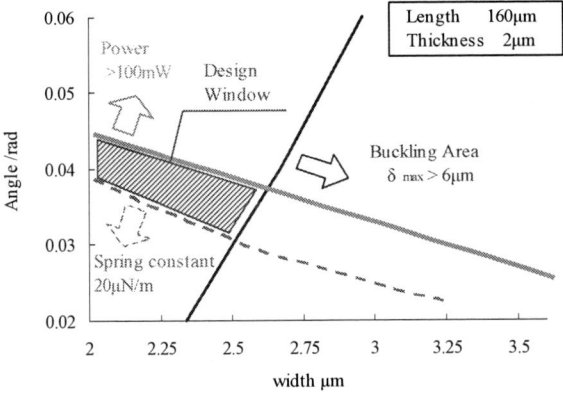

Figure 2: Design window for the electrothemal actuator. The hatched area meets the design criteria imposed by buckling of the actuator, power consumption and stiffness.

The final actuator design dimensions for the devices presented in this work are provided in Table 1.

Table 1: Dimensions of the presented thermal actuators .

Beam Length	L	160 µm
Beam Width	w	2.0 µm
Thickness	t	2.6 µm
Angle	θ	0.04 rad

RF Design

The fabricated LPFs are based on LC filter topologies. In order to realize both wide range tunability (between 0.9GHz to 2.5GHz) and minimal chip size, the inductance is selected via RF-MEMS switching.

Optimization of the inductor, capacitor and ground layer structures was achieved through HFSS analysis and the finalized element layout is depicted in Figure 3. In this design L1 and L4 are both 7.0nH while L2 and L3 have a designed value of 1.3nH. Capacitors C1, C2 and C3 are each 1.5pF and C4 and C5 have a design capacitance of 0.4pF.

Four RF-Switches are embedded in a single filter element to select the designed f_c's of 0.9GHz, 1.9GHz and 2.5GHz. Table 2 presents the switch configurations required to achieve the designed cut-off frequencies.

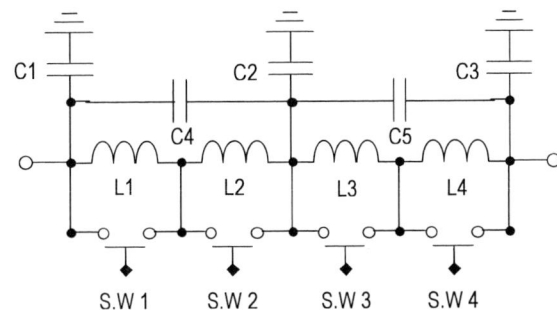

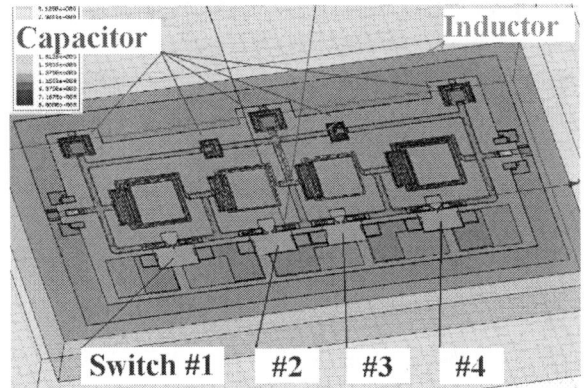

Figure 3: Tunable low pass filter system configuration: (top) schematic diagram of the principle lamped circuit, (bottom) 3D model showing the optimized device layout.

Table 2: Designed cut-off frequencies and switch matrix.

f_c	SW1	SW2	SW3	SW4
0.9 GHz	OFF	OFF	OFF	OFF
1.9 GHz	ON	ON	OFF	OFF
2.5 GHz	ON	ON	ON	ON

FABRICATION

Figure 4 provides an overview of the fabrication process. High resistivity wafers are employed to minimize the effect of parasitic capacitance from the substrate. First the electrothermal actuators are fabricated. After growing 0.7µm of thermal oxide on the wafer, a 300nm thick silicon nitride was deposited. This nitride layer serves as an etch stop layer for release etching later in the process. A 2.0µm, sacrificial layer of oxide was deposited via LP-CVD, followed by 1.0µm of silicon nitride. This nitride layer was patterned and etched to form a thermal and electrical insulating structure between the switch contact tip and the thermal actuator. *In-situ* doped polysilicon, 2.6µm thick, was also deposited via LP-CVD. Due to the high conductivity of the polysilicon both lowering of the actuation voltage and reducing the ESR of capacitor was achieved. The resistivity of the deposited polysilicon is approximately 5×10^{-6} ohm-meter. The V-shaped electrothermal actuators, lower electrodes of MIS capacitors and under-paths of the inductors were formed simultaneously by reactive ion etching of the poly silicon. Next un-doped LTO is deposited and patterned to create a dielectric layer for the MIS capacitors. Metallization is achieved by gold evaporation onto the contact area of the switch heads, GSG RF probing pads and DC contact pads of the thermal actuators. The inductors and upper electrode of the MIS capacitors are fabricated using a semi-additive copper plating process. During the release process, the inductors and capacitors are protected by photoresist as removal of the LTO sacrifice layer is not desired in these regions. Finally, the photoresist is stripped and the devices are dried using a carbon-dioxide super-critical point drying system.

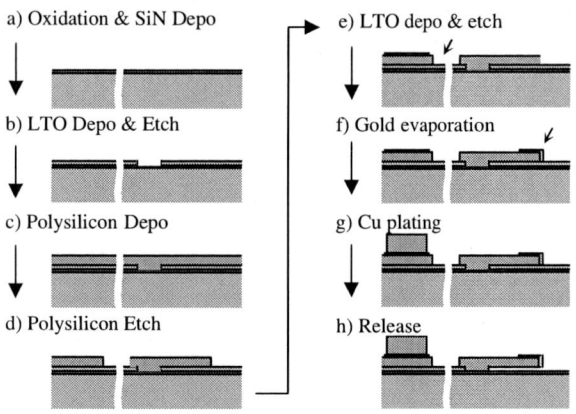

a) Oxidation & SiN Depo

b) LTO Depo & Etch

c) Polysilicon Depo

d) Polysilicon Etch

e) LTO depo & etch

f) Gold evaporation

g) Cu plating

h) Release

Figure 4: Fabrication process overview.

CHARACTERIZATION
RF MEMS SWITCH

The RF properties of the fabricated RF-MEMS switches were measured using a network analyzer (Anristu 37393). In Figure 5 plots of isolation and insertion loss from 100MHz to 10GHz are shown for a typical switch. Isolation is better than -38dB up to 5GHz. Insertion loss is dependent on the contact force of the switch head and therefore actuation current. In the range of 29mA to 33mA the insertion loss is decreased with increasing current; however, at 34mA insertion loss is increased with increasing current, this is due to bulking of the actuator beams. Nominal insertion loss is approximately -0.5dB at 5GHz and -0.8dB at 10GHz.

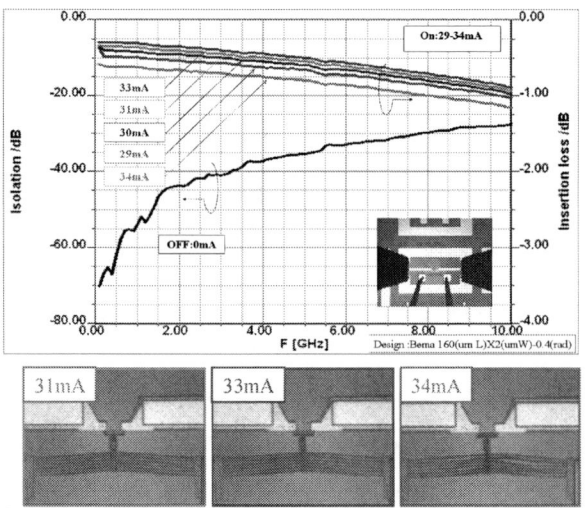

Figure 5:Measured RF switch properties. Isolation is better than -35dB below 5GHz. Insertion loss is -0.6dB at 5GHz.

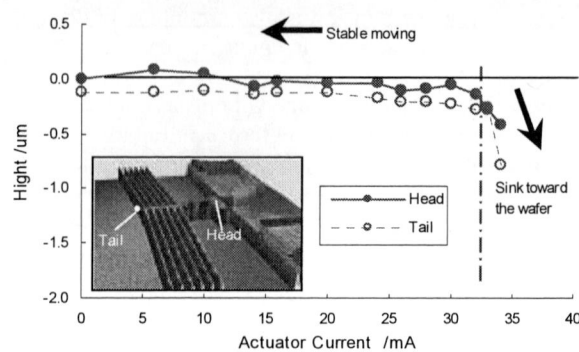

Figure 6: Measured Actuator movement via Wyko.

Actuator movement is observed using a wyko white-light interferometer as shown in Figure 6. After the actuator current is increased to approximately 33mA the actuator begins to buckle and sink towards the wafer, this agrees well with the fem modeling results in Figure 7. FEM analysis shows that the contact force begins decreasing after out of plane buckling occurs. Thus 31mA is selected as the actuation current of the RF MEMS switches. Typical insertion loss is -0.8dB at 31mA.

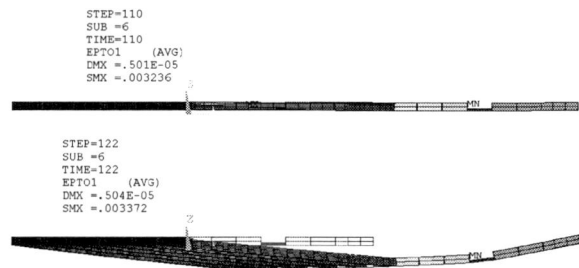

Figure 7: FEM analysis of V shaped actuator deflection.

RF MEMS SWITCHING TIMES

The switching time of our fabricated devices is measured by applying a step "turn-on" voltage to the actuator, and measuring the signal current through the switched signal line. The switching delay for an actuator at rest device is 154μsec, as shown in Figure 8.

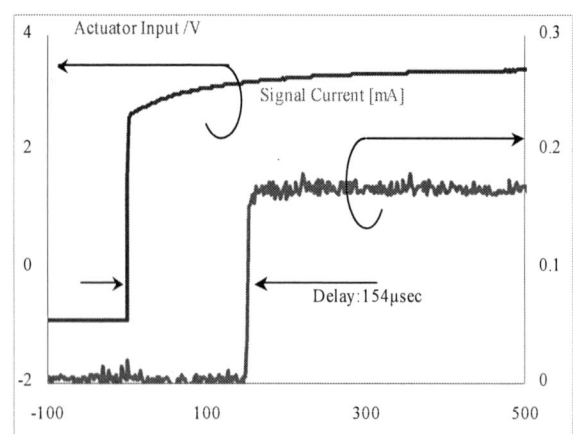

Figure 8: Switching characteristics of the fabricated RF-MEMS Switch. The time scale is in microseconds.

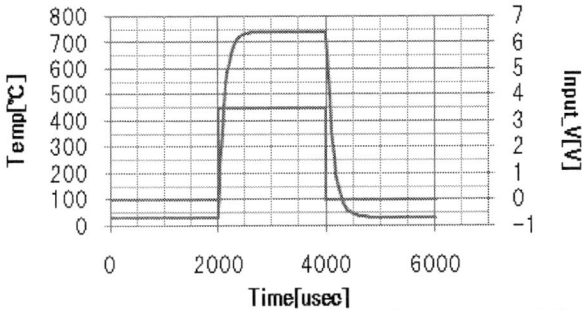

Figure 9: FEM simulation of thermal transient of the polysilicon actuator beams.

In addition, FEM calculations estimate the thermal transient time constant (τ) to be 146µsec in Fig. 9. On the other hand the mechanical resonance frequency of the in plane vibrating mode is 410 kHz. Thus the switching speed is determined by the thermal time constant of the system.

TUNABLE FILTER

The footprint of the fabricated tunable, low-pass filter is $4.3 \times 2.4 mm^2$. RF characterization of the fabricated system is achieved by actuating the individual switches while measuring the S-paramters (S11, S12, S21 and S22) of the device using GSG RF wafer probes and a network analyzer. An image of the fabricated device with is shown in Figure 10a, along with detailed images of a L,C pair (Figure 10b) and the MEMS switch (Figure 10c,d). Measured RF properties of the fabricated LPF in each of the three designed configurations are provided in Figure 11. With all the switches in the off state, the f_c is 0.9GHz, and out-of-band attenuation is -30dB. In the case of both switch 1, and switch 2 actuated, the f_c is moved to 1.9GHz. Finally, in the configuration with all switches are activated the device functions as a 2.4GHz LPF.

Typical out band attenuation in the 1.9GHz filter configuration is -39dB at 3.5GHz it's observed that the insertion loss of the realized RF switch does not have significant impact on the insertion loss of the filter.

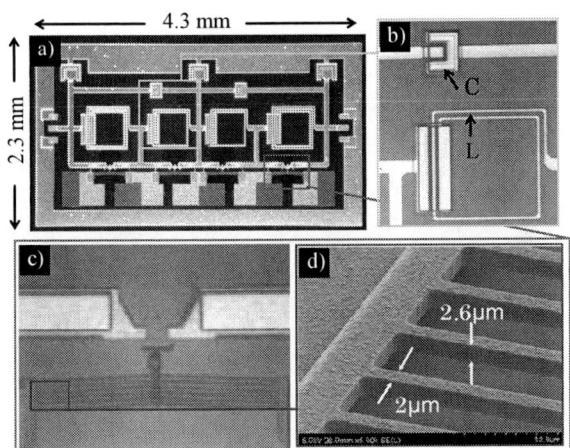

Figure 10: Overview of fabricated tunable filter. (a) Chip size of $2.3 \times 4.6 mm^2$. (b) Inductor and capacitor. (c) MEMS switch with electrothermal actuator. Beam length is 160µm with a 6-beam design. (d) SEM of the electrothermal actuator. Beam width is 2.0µm and thickness is 2.6µm.

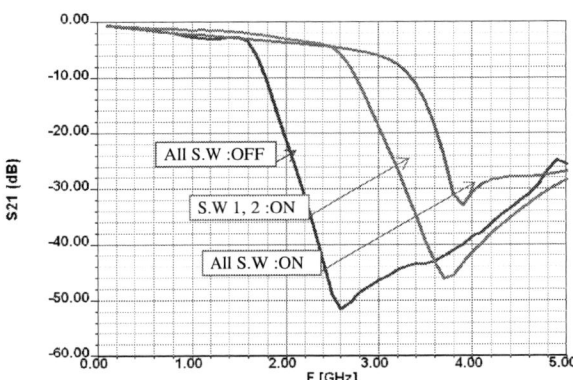

Figure 11. Measured RF transmission properties of the fabricated reconfigurable LPF for the three designed fc's. Out of band attenuation is better than -25dB below 5.0GHz.

CONCLUSIONS

Monolithically integrated, wide tuning range, low pass filters have been successfully demonstrated based on electrothermally driven MEMS actuators. The integrated system is comprised of copper spiral inductors, polysilicon capacitors and electrothermal MEMS actuators on a single $4.3 \times 2.4 mm^2$ chip. The design is optimized for the selection of three commonly used commercial cut-off frequencies: (f_c) at 0.9GHz, 1.9GHz and 2.5GHz. The tuning range of the filter is greater than a factor of two, due to the availability of a switched array of inductors. The driving voltage is only 3.7V and out-of-band attenuation is -39dB at 3.5GHz, which is comparable to the state-of-art commercial ceramic filters. As such, this tunable filter and related fabrication technologies are applicable to a wide range of RF front end applications.

REFERENCES

[1] De Los Santos, H.J. Fischer, H.A.C Timans, J.T van Beek,"RF MEMS for ubiquitous wireless connectivity. PartII. Application", *Mirowave Magazine*, VOL. 5 NO.4, December 2004.

[2] A. Pothier, J-C Orianges, Guizhen Zheng, C. Champeaux, A. Catherinot, D. Cros, P.Blondy, J. Papapolymerou, "Low-loss 2-bit tunable bandpass filters using MEMS DC contact switches", *IEEE Transactions on MTT*, VOL. 53 NO. 1, January 2005.

[3] Kamran Entesari, Gabriel M. Rebeiz, "A 12-18GHz Three-Pole RF MEMS Tunable Filter", *IEEE Transactions on MTT*, VOL. 53 NO. 8, August 2005.

[4] Ye Wang, Zhihong Li, Daniel T. McCormick, Norman C. Tien, "A Low-Voltage Lateral MEMS Switch With High RF Performance", *J. Microelectromech Syst.*, VOL. 13, NO. 6, DECEMBER 2004.

[5] E.T.Enikov, S.S Kedar, K.V Lazarov, "Analytical Model for Analysis and Design of V-shaped Thermal Microactuators", *J. Microelectromech Syst.*, VOL. 14, NO. 4, August 2005.

LOW ACTUATION VOLTAGE SPDT RF MEMS K BAND SWITCH USING A SINGLE GOLD MEMBRANE

R. Robin[1,2], O. Millet[1], K. Segueni[1], and L. Buchaillot[2]

[1]DelfMEMS SAS, Villeneuve D'Ascq, FRANCE

[2]IEMN, ISEN Dpt, Nano And Microsystems group, Villeneuve D'Ascq, FRANCE

ABSTRACT

This paper presents a SPDT (Single Pole Double Throw) system for K band applications using RF MEMS (Radio-Frequency MicroElectroMechanical Systems) switch. The switch is based on a single anchorless gold membrane which is simply supported. It achieves large deflections of 4µm at a low actuaction voltage (7.5V simulated and 12V demonstrated) and with a high reliability. New mechanical testing results, showing good agreements with simulations, and RF measurements of the structure, using a wide band coplanar waveguides (CPW) to microstrip (MS) transition, will be shown.

INTRODUCTION

Reliability still remains one of the major problem of RF MEMS switches. Phenomenoma such as dielectric charging, auto-commutation and also the bad temperature handling, weakens the potentiality of MEMS technology. A SPDT system, employing a new geometry approach which provides more reliability, is proposed [1]. It is based on a free flexible membrane instead of cantilevers or clamped-clamped beams.

The SPDT system is composed by two microstrip lines, the switch and three CPW to MS transitions (one for input and one for each outputs) enabling easier RF measurements. The SPDT was designed for K band applications requiring a double capacitive shunt configuration.

The switch is based on a gold membrane simply supported over three pillars [2]. Four electrodes, one on each extremity and one on each side of the central pillar, allow to create four different states as displayed in figure 1.

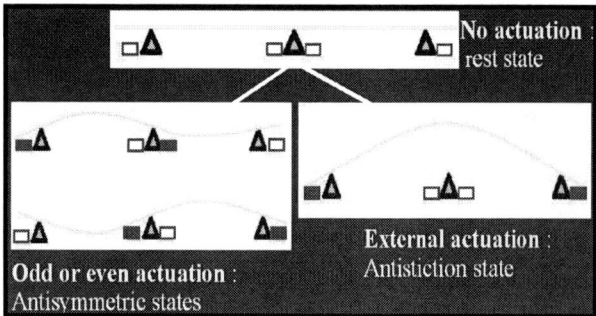

Figure 1: Illustration of the four states of the structure. Triangles stands for pillars and rectangles for electrodes. The filled rectangles correspond to the actuated ones.

Odd and even states are used to transmit the signal to one or another output. The external actuation state creates a lever effect and act as an anti-stiction system [3]. In case of odd or even actuation, a push-pull effect is created which grants a high reliability to the structure : both improved contact forces and large deflections are achieved for a low actuation voltage .

MECHANICAL DESIGN

The key point of this design was to provide large deflections over one MS line while the membrane is slapped to the second MS line on the other side (Figure 2). To make it possible, coupled-physics FEM simulations have been done using Ansys software.

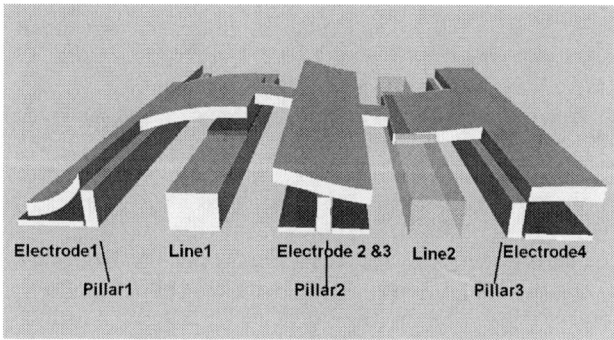

Figure 2: Schematic cross section of the structure

The structure has been designed for a very low actuation voltage. The final design presents several specificities. First of all, large areas have been designed on both membrane and electrodes to maximize the actuation force which has for consequence a huge reduction of the actuation voltage, resulting in a double H shape for the membrane.

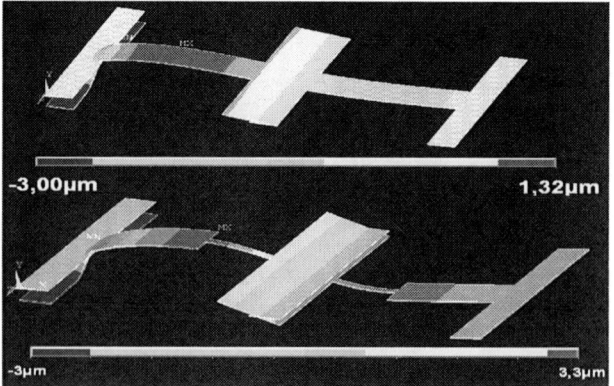

Figure 3: Simulated deflections with and without locally reduced sections using FEM analysis. The actuation voltage is 7.5V .

978-1-4244-2977-6/09 $25.00 © 2009 IEEE

More over, the section of the membrane between external and center pillars is locally reduced. As a benefit, the correlation between the two sides is thus lower and the deflection on the actuated side is larger without changing the mechanical behavior of the other side (Figure 3). It also reduces the actuation voltage because the stiffness is lower.

In addition to this, wings of balancing have been added on each internal corner of the membrane. They have several purposes : stress level reduction, zipping effect limitation and harmonic spurious modes suppresion [4]. Finally, mechanical stops units have been used to maintain the structure without limiting the movements and the deflexions when actuated. They are processed by auto-alignment in order to ensure that x-y-z translations are limited (Figure 4). Deflections of 4µm for a pull-in voltage of 7.5V have been simulated using the design previously describe.

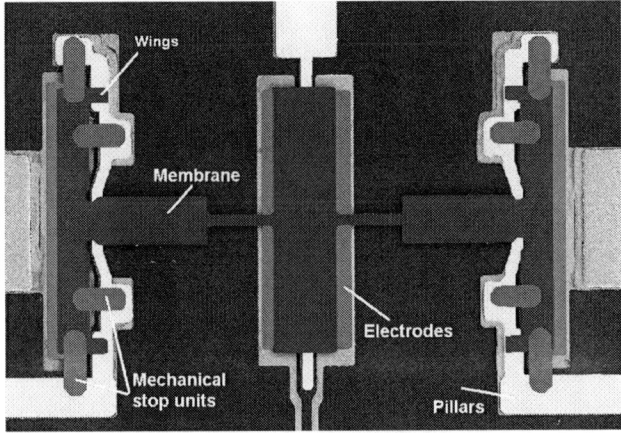

Figure 4: Upper view of the different parts of the structure (Colored SEM picture) .

PROCESS FLOW

A low temperature process (<200°C) is used, based on surface micromachining, and combined with three sacrificial layers of chromium, titanium-tungsten (TiW) alloy and silicon dioxide. The TiW is used as a barrier to avoid diffusion of the chromium into the gold. Eight masks are needed for this process. The first step consists in patterning 200nm actuation gold electrodes and 200nm PECVD (Plasma Enhanced Chemical Vapor Deposition) Si_3N_4 layer for the dielectric insulation. Then a 200nm chromium layer is sputtered in order to protect the Si_3N_4 during releasing step. This chromium layer is also used for electrical interconnection for the next gold electroplating. A 2.8µm SiO_2 is then deposited, patterned and etched by plasma; this layer is used as mold for the pillars electroplating. A 150nm thick chromium layer is sputtered and locally opened to package the two capacitive contacts. Then a 100nm thick Si_3N_4 is deposited and patterned to form the capacitive contact. Another 150nm thick chromium layer is sputtered for completing the package, in order to prevent etching of the contact dielectric during the releasing. It also prevents

from contact between the membrane and pillars. The structural membrane is performed by a 1.5µm thick electroplated gold. A third 500nm chromium layer is deposited and the mechanical stop units are patterned by gold electroplating in a positive photoresist mold. The releasing consists in wet etching of the silicon dioxide in hydrofluoric acid (HF), chromium in chromium etchant and TiW alloy in buffered H_2O_2 solution. Finally, the structures are dried using super critical CO_2 dryer.

MECHANICAL TESTING
Results

Simplified structures using only portions of MS lines instead of the complete circuit have been used as mechanical test structures (figure 5). A unipolar rectangular 2Hz electrical signal has been applied between the membrane and the electrodes under a probe station.

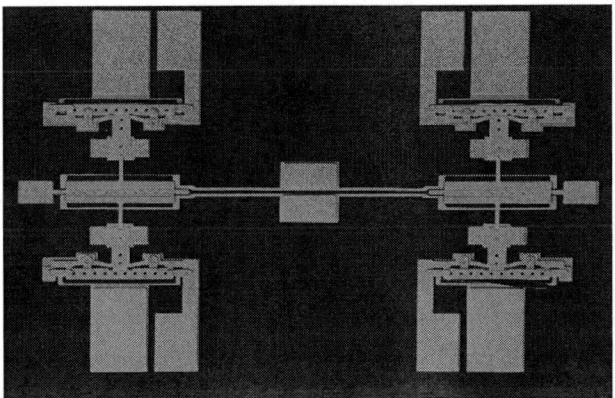

Figure 5: SEM View of two mechanical test structures.

Previous batches, processed without TiW alloy, have illustrated that the average pull-in voltage is about 26V [4]. For this run, pull-in voltage measurements show an average value of 12V for 14 structures and a minimal value of 8V which agree well with FEM simulations. Similar large deflections have been noticed for all the batches (figure 6). The actuation voltage values in odd and even states are quite the same as in anti-stiction state.

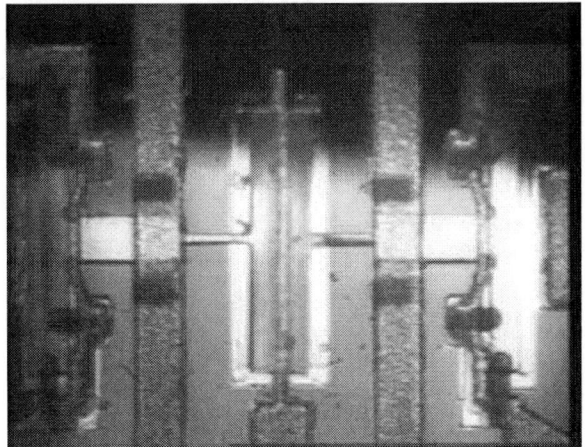

Figure 6: Optical capture of the deflected membrane in

fourth state.

Analysis

The decrease of the pull-in voltage is due to three major reasons. First, the membranes of the new batch are thinner. It contributes to the decrease but it does not explain a so high difference. The second reason is a better control of the electrolysis, which prevents from unwanted electroplating that occured during the previous batches. Finally, the most important reason is a drastically reduction of the structures bending once released. It is due to the use of TiW alloy which avoid diffusion of chromium into gold. A study of the shape of the membranes, using an interferrometer microscope, illustrates they were flat (figure 7). The intrinsic stress in the membranes without diffusion is not enough to deform the structures. More over, small buckling can occur as a consequence of gold-gold adhesion forces during the releasing step. They can be easily countered by a running in anti-stiction state.

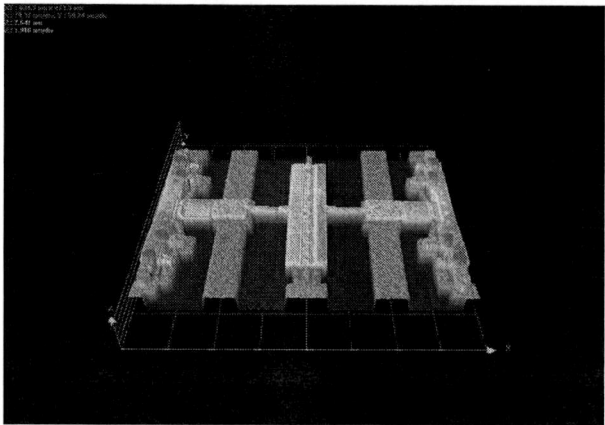

Figure 7: 3D view of the structure by interferrometrical microscope.

RF DESIGN AND RESULTS

Coplanar waveguides (CPW) to microstrip (MS) transition

CPW to MS transition are used in order to ease RF measurements. Their second purpose is to transform the 50Ω charateristic impedance of the measurement setup into 90Ω, which is the characteristic impedance of the MS lines. In fact, the geometry of the structure requires narrow MS lines (60μm), inducing higher characteristic impedance (90Ω in K band on silicon wafer).

The transition is composed of three parts (figure 8). A 50Ω CPW RF pads is followed by a 67Ω quarter-wavelength CPW. Then, quarter-wavelength open stubs suppress reflective modes [5] while the central conductor of the CPW is progressively enlarged in order to fit to the MS line's width.

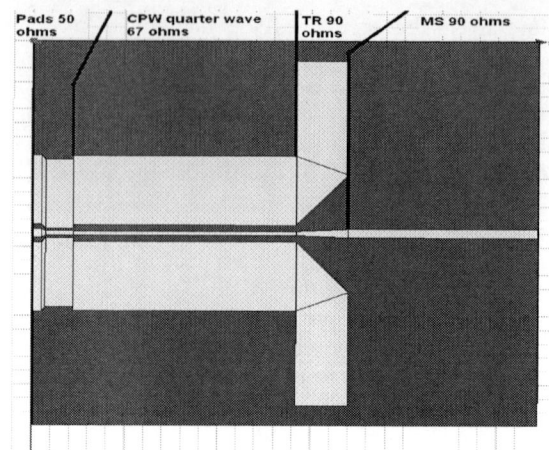

Figure 8: Top view of the CPW to MS transition.

Several transitions have been characterized, showing a large bandwidth behavior. The adaptation is at least -20dB between 15GHz and 30GHz. The normalized insertion losses are 0.63dB/mm at 24GHz on intrinsic silicon wafer. Good agreement between simulations and measurements are noticed (Figure 9).

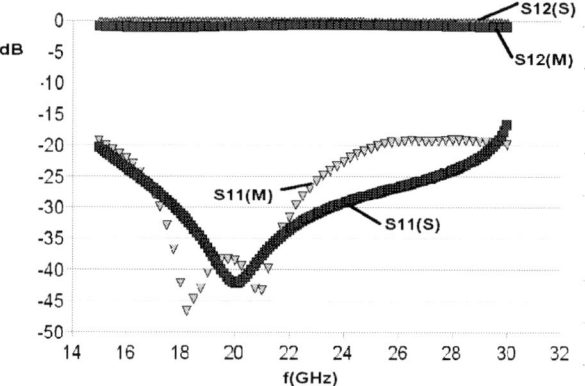

Figure 9: Measured (M) and simulated (S) S-parameters of the CPW to MS transition.

RF design and measurements of the SPDT

The SPDT have been processed on intrinsic silicon wafers. The dielectric of the capacitive contact is PECVD silicon nitride. Radial stubs are designed to create a virtual RF ground. Their choice rather than vias has been done to have an easier fabrication process even though radial stubs have a smaller bandwidth. Coupling between stubs and transmission lines is used to improve the impedance matching. It explains the proximity between MS lines and stubs. A quarter-wavelength line is placed between the capacitive contact and the microstrip separation to transform the short circuit in open circuit and thereby presents an infinite impedance for the blocked line (figure 10).

978-1-4244-2977-6/09 $25.00 © 2009 IEEE

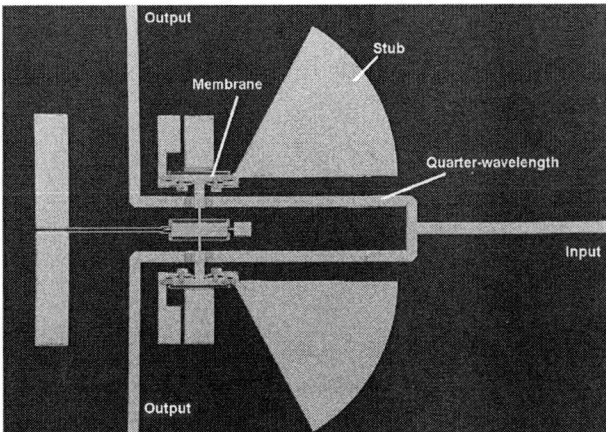

Figure 10: SEM view of the SPDT circuit.

Blocked and transmitted sides are measured separately. For the blocked side, the isolation is more than 40dB at 20GHz. Insertion loss of 0.36dB/mm at 20GHz is measured (figure 11). The adaptation was more than -20dB from 17GHz up to 30GHz for both the input and the output of the transmitted side. The maximal isolation is supposed to be obtain for 24GHz but unexpected issues during processing have shifted the frequency of the isolation peak.

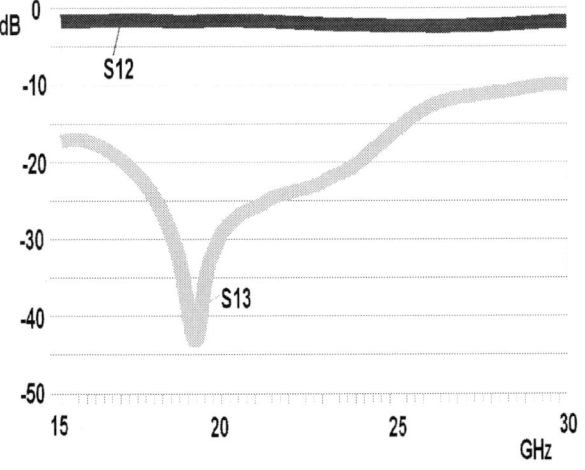

Figure 11: S parameters measurements : isolation and insertion losses of the SPDT.

CONCLUSION

Whole SPDT system has been proposed using a new kind of MEMS switch based on an anchorless membrane. The mechanical design has been optimized to obtain large deflexions at low actuation voltage with high reliability thanks to the degree of freedom of the structure and the push pull effect. First, mechanical structures have been tested once the process flow was well optimized. The average pull-in voltage is only 12V and demonstrates a good agreement with FEM simulations. In order to ease RF measurements, a CPW to MS transition have been designed. Its insertion loss is 0.63dB/mm at 24GHz and its return loss is below -20dB between 15GHz and 30GHz, which demonstrates a wide band operation. Finally, S-parameters of the SPDT have been measured.

They show an isolation greater than 40dB at 20GHz and insertion losses of 0.36dB/mm at 20GHz. Return loss is below -20dB from 17GHz up to 30GHz. Studies are still in progress in order to improve the handling of the capacitive contact during the releasing step and some others process aspects.

REFERENCES

[1] K.Segueni, A.-S. Rollier, L. Le Garrec, R. Robin, S. Touati, et al, "A totally free flexible membrane : a design for low electrostatic actuation MEMS", Transducers'07 Conference, pp. 461-464, June 10-14 2007, Lyon, France.

[2] R. Robin, L. Le Garrec, K.Segueni, O. Millet, L. Buchaillot, "A Design For 24GHz SPDT Using A Single RF MEMS Switch Based On A Totally Free Flexible Membrane", in *Proc.* of MEMSWave 2007 conference, pp. 193-195, June 26-29 2007, Barcelona, Spain.

[3] S. Touati, N. Lorphelin, A. Kanciurzewski, R. Robin, A.-S. Rollier, et al, "Low actuation voltage totally free flexible RF MEMS switch with antistiction system", in *Proc.* of Design Test Integration and Packging for MEMS and MOEMS 2008 conference, DTIP 2008, pp.66-70, April 9-11 2008, Nice, France.

[4] R. Robin, S. Touati, K. Segueni, O. Millet and L. Buchaillot, "A new four states high deflection low actuation voltage electrostatic MEMS switch for RF applications", in *Proc.* of Design Test Integration and Packging for MEMS and MOEMS 2008 conference, DTIP2008, pp 56-59, April 8-11 2008, Nice, France.

[5] J.-P. Raskin G. Gauthier, L. P. Katehi, and G. M. Rebeiz, "Mode Conversion at GCPW-to-Microstrip-Line Transitions", IEEE Transactions On Microwave Theory And Techniques, vol. 48, no. 1, pp 158-161, January 2000.

978-1-4244-2977-6/09 $25.00 © 2009 IEEE

PERFORMANCE ENHANCEMENT BY SUBSTRATE PERFORATION FOR A WAFER-LEVEL ENCAPSULATED RF MEMS DC SHUNT SWITCH

Feixiang Ke, Jianmin Miao, and Chee Wee Tan
Nanyang Technological University, SINGAPORE

ABSTRACT

A wafer-level encapsulated RF MEMS shunt switch with perforated base substrate and corrugated diaphragm was developed. For the first time, the perforated base substrate by meticulous design is introduced to tremendously reduce squeeze-film damping and thus significantly increase the switching speed. The characterization of the fabricated and wafer-level encapsulated switch in terms of the switching time and RF performance is presented in this paper as well.

INTRODUCTION

A MEMS shunt switch based on metal-metal contact can replace RF MEMS series switch for operation between DC-20 GHz [1]. However, packaging is an important aspect with respect to metal-metal contact switch. Exposure to air is an issue in terms of reliability and lifetime of the device. Therefore, there is a need to address the encapsulation of such MEMS switch, especially for the metal contact part. As a solution to the challenge, this paper proposes a micro-structure employing stacked-wafer concept to achieve a DC contact RF MEMS shunt switch which is encapsulated at wafer-level. The encapsulation of switch-contact is conducted at an early stage in the fabrication, providing a means of protection for switch-contact against contamination, and also an ease of handling for the subsequent process steps.

On the other hand, the switching speed of MEMS switch still remains one of the limitations of MEMS switches due to their mechanical movements of moving-parts. A method for increasing switching speed is to perforate the moving part, such as bridge or diaphragm, in order to reduce the damping due to the air flow in a narrow air gap. Contrary to perforating in bridge/diaphragm adopted by typical MEMS switch design [2, 3], our approach to reduce the squeeze-film damping applies perforation in the base substrate instead of the diaphragm due to the fact that the diaphragm is part of the encapsulated micro-chamber formed between the diaphragm and the base substrate. The proposed concept fulfills the performance enhancement by wafer-scale encapsulation integrated with diaphragm and by perforation to reduce the damping at the same time.

DESIGN

The Switch Configuration

The proposed DC contact shunt switch involves three wafers stacked together by wafer-bonding process, enabling the encapsulation of the contact metals in a micro-chamber filled with nitrogen gas. The switch consists of a moving diaphragm on a top silicon cap together with a contact metal; coplanar waveguide (CPW) and perforation in a silicon base substrate; and a cavity in a silicon backplate as shown in Fig.1. The core of the switch is the thin and flexible silicon nitride diaphragm suspended above the CPW, moving up and down under electrostatic actuation for switching purpose. The corrugation technique was also introduced to the diaphragm to alleviate the possible high residual stress in the diaphragm [4]. To demonstrate the switching speed increase by the perforation in the base substrate, the switches were made purposely with a large diaphragm of 1750 µm in radius and an air gap of 8 µm.

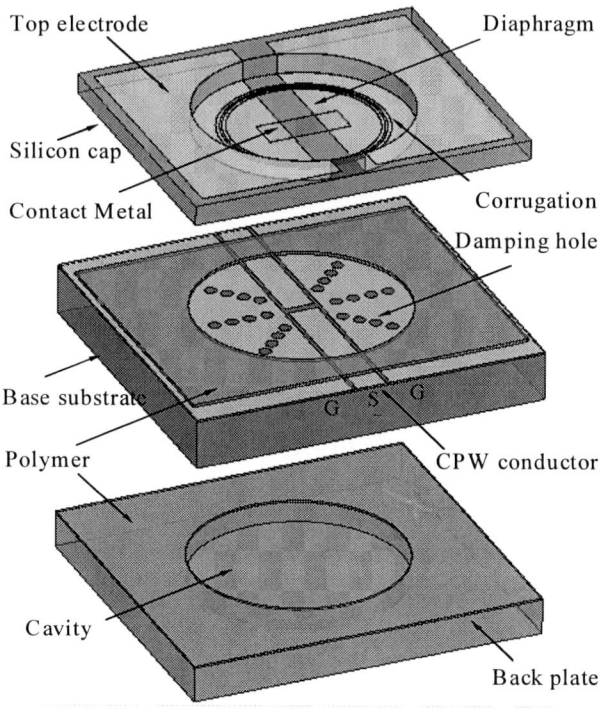

Figure 1: Schematic drawing of three parts before wafer bonding.

Damping Effect with Perforated Base Substrate

The damping coefficient (B_a), also termed as air resistance, of a diaphragm due to the viscous damping losses in the air gap and the damping holes is governed by following equation [5-7]

$$B_a = \frac{8\pi T \, \text{Im}(D)}{\omega} \tag{1}$$

where T is the diaphragm tension, ω is the angular velocity, and D is the air layer damping term (an dimensionless quantity) and is given by,

$$D = \frac{j4\omega\rho_0 J_0(Ka)}{\pi} \sum_{m=0}^{\infty} \left[-j\omega\left(\frac{z_m}{z_0}\right) + \sum_{s=1}^{q}\left(\frac{f_s}{z_0}\right)S_s\Gamma_m J_0(\xi_m a_s) \right]$$
$$\times \frac{1}{T_m J_0(\xi_m a)} \left\{ \frac{J_2(Ka)\delta(m)}{4K^2 a^2} + \frac{J_1(Ka)[1-\delta(m)]}{2Ka\left(K^2 a^2 - \xi_m^2 a^2\right)} \right\} \tag{2}$$

where ρ_0 is the static air density (1.205 kg m^{-3}), a is the radius of the diaphragm, m is the number of summation terms, z_m is the diaphragm displacement function [7], z_0 is the interim unknown diaphragm displacement in a trial expression [7], q is the number of holes, f_s is the air velocity at the kth opening in the back-support, S_s is the area of the kth opening in the back-support, Γ_m and T_m are numerical constants [7], ξ_m is the symmetric radical component of the scalar wave number, a_s is the location of the kth opening in the back-support, $\delta(m)$ is a conditional function ($\delta(m) = 1$ when $m = 0$; $\delta(m) = 0$ otherwise), $J_0(Ka)$, $J_1(Ka)$ and $J_2(Ka)$ are the zero, first-order and second-order Bessel functions of Ka, respectively. Here K is the wave number of sound in the diaphragm and can be expressed by

$$ K = \omega \sqrt{\left(\frac{\sigma_M}{T}\right)} \tag{3} $$

where σ_M is the mass surface density of diaphragm. One method of increasing switching speed of the diaphragm is to reduce the damping resistance through the meticulous design of base substrate (with damping holes) and air gap. When holes are perforated in the substrate, the damping can be reduced significantly as the air has an alternative path to escape through the holes rather than being squeezed out from the edge of the substrate. Obviously from Equation (1) and Equation (2), the damping coefficient is dependent on the location of holes in the base substrate. It should be noted that the diaphragm deflection is non-uniform in the radial direction and location dependent, and the f_s is also dependent on the hole location due to the non-uniform distributed pressure on the base substrate. Therefore, the physical location of holes in the base-substrate does play an important role on the damping characteristics.

Table 1: Parameters of the proposed MEMS switch with corrugated diaphragm and perforated base substrate.

Parameters	Symbol	Values
Diaphragm radius	a	1750 μm
Diaphragm thickness	d	1.75 μm
Mass surface density of diaphragm	σ_m	$3290 \times 1.75 \times 10^{-6}$ kg/m^2
diaphragm tension	T	87.5 N/m
Substrate radius	$b = a$	1750 μm
Unpolarized air gap	g	8 μm
Back chamber volume	V_{om}	$\times$ (1750 μm)^2 $\times$ 150 μm
Number of holes per ring	$b1$	10
Location of radius ring	$a1$	450, 620, 790, 960, 1130, 1300, 1470 μm
Hole radius	$r1$	50 μm
Hole depth	$l1$	500 μm
Polarization voltage	V	60 V

A simulation of damping coefficients of the diaphragm of the proposed MEMS switch was performed, based on the parameters listed in Table 1. For comparison, the damping coefficients of diaphragms with

different air gaps are also computed as shown in Fig. 2. It is seen that at switching speeds below a few thousands of cycles/second, there is not much change of damping coefficient with the increase of switching speed. The damping coefficient keeps, roughly, constant at the low switching speed range of 50 to around 5000 cycles/second. However, the damping coefficient is dependent strongly on the opening in the base substrate and air gap. The damping coefficient is reduced significantly with the implementation of damping holes in the base substrate. The reduction of the damping coefficient for each air gap is as high as 13 times. This implies that with the perforated substrate, the air finds an alternative path and takes much less time to escape from the gap. Furthermore, one can imagine that the air takes less time to escape from the air gap if it is enlarged. Therefore, the damping coefficient should be reduced with the increase of air gap. And this is true, which is clearly shown in Fig. 2. For example, the damping coefficient of the proposed diaphragm will be reduced from 8.03 Ns/m to 0.219 Ns/m if the air gap increases from 3 to 10 μm, for the case when the base substrate is not perforated.

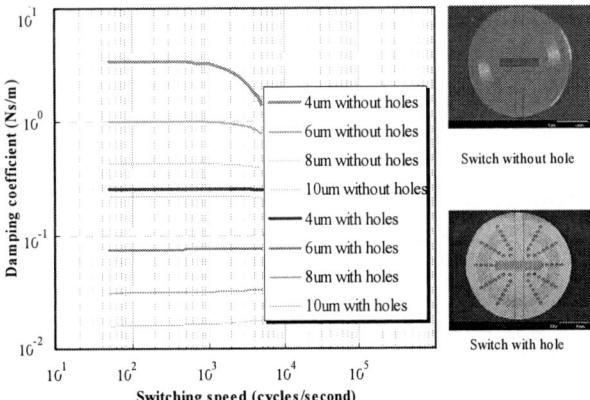

Figure 2: Simulation results of damping coefficient of diaphragm with different base substrate and air gap.

Switching Time

The switching time can be estimated by the following equations derived by Steenken, et. al. [2]. For the pull-in motion from the up-position ($\bar{z}_1$) to the down-position ($\bar{z}_0$), where the actuation voltage (V) is larger than pull-in voltage (V_{pi}), the switching time is,

$$ t_{pull-in}(\bar{z})\Big|_{\bar{z}_1}^{\bar{z}_0} = \frac{-8B_a g^3}{A\varepsilon_0 V_{pi}^2 f_1(-\gamma) f_2(-\gamma) f_3(2\delta)} \times [f_3(2\delta)\ln(\bar{z} + \gamma) $$
$$ - \frac{f_1(-\gamma)}{3}\ln(f_2(\bar{z})) + \frac{f_1(-\gamma) - 3f_3(2\delta)}{6}\ln(f_1(\bar{z})) $$
$$ - \frac{f_2(-\gamma)(f_2(-\delta\gamma) - 3\delta)}{\sqrt{3f_3(\delta)}} \arctan\frac{f_2(\bar{z}) - 3\delta}{\sqrt{3f_3(\delta)}}\Big]_{\bar{z}_1}^{\bar{z}_0} \tag{4} $$

where the normalized position $\bar{z} = z/g$, and γ is the slip-flow constant and is given by $\gamma = 9\lambda/g$. Here, λ is the mean free path of gas molecules that is about 70 nm under atmospheric pressure at room temperature.

$$\delta = \cosh\left[\frac{1}{3}\operatorname{arccos}h\left(\frac{2V_{eff}^2}{V_{pi}^2}-1\right)\right] \qquad (5)$$

$$f_1(x) = f_2(x)[f_2(x)-6\delta]+3f_3(2\delta) \qquad (6)$$

$$f_2(x) = 3x+2\delta-1 \qquad (7)$$

$$f_3(x) = x^2-1 \qquad (8)$$

For the release motion ($V = 0$), from the down-position ($\bar{z}_0$) to the up-position ($\bar{z}_1$), the switching time is [2],

$$t_{release}(\bar{z})\Big|_{\bar{z}_0}^{\bar{z}_1} = \frac{-8B_a g^3}{27A\varepsilon_0 V_{pi}^2(1+\gamma)}$$

$$\times\left[\frac{1+\gamma}{\gamma\bar{z}}+\ln\left(\frac{1}{\bar{z}}-1\right)-\frac{1}{\gamma^2}\ln\left(1+\frac{\gamma}{\bar{z}}\right)\right]_{\bar{z}_0}^{\bar{z}_1} \qquad (9)$$

It is seen that the switching time depends strongly on the damping coefficient B_a, pull-in voltage V_{pi} and ratio of V/V_{pi} (for pull-in motion). The switching time can be reduced by decreasing B_a by perforating the base substrate, increasing the pull-in voltage V_{pi} and actuation voltage V, since smaller B_a and larger voltages mean weaker damping force and stronger electrostatic force.

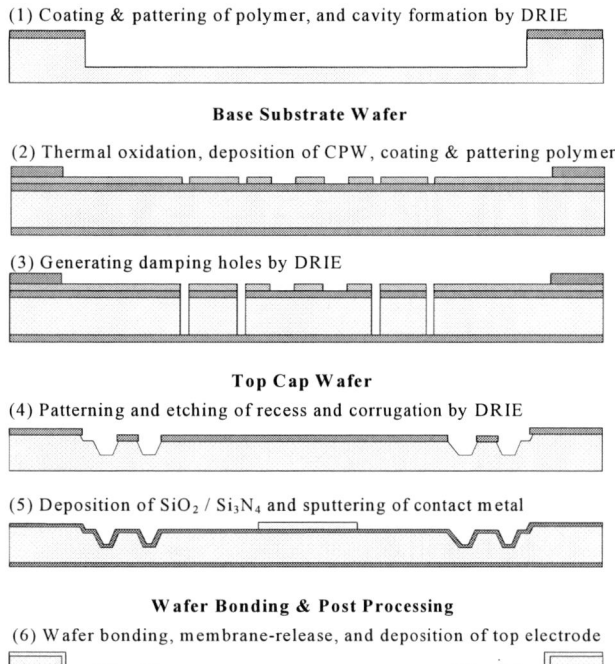

Figure 3: Fabrication process flow of the MEMS switch with perforated base substrate.

FABRICATION

The switch fabrication involves three-wafer stacking processes as illustrated in Fig. 3. The supporting backplate silicon (300 μm thick) is firstly spin-coated and patterned with a *CYTOP* bonding layer. A cavity with a depth of 150 μm is then generated by deep reactive ion etching (DRIE). A 550 μm thick and high resistive (~5000 Ω.cm) silicon as the switch base substrate is coated with 1.8 μm thick thermal SiO_2. CPW and actuation electrodes are subsequently fabricated by sputtering and lift-off process. Afterwards, *CYTOP* is spin-coated, followed by DRIE for through-silicon holes of 50 μm in radius. The cap silicon wafer is firstly etched to form 6 μm deep recesses, followed by DRIE for corrugation grooves. A film of 100 nm thermal SiO_2/1200 nm low stress LPCVD Si_3N_4 is deposited on the grooved surface for the corrugated diaphragm, on which the contact metal is sputtered. The three wafers are joined together by wafer bonding using *CYTOP* polymer. After that, the diaphragm is released by DRIE. Finally, the gold membrane electrode is deposited by sputtering through a shadow mask. Fig. 4 shows a SEM photo of a fabricated MEMS switch with perforated base substrate.

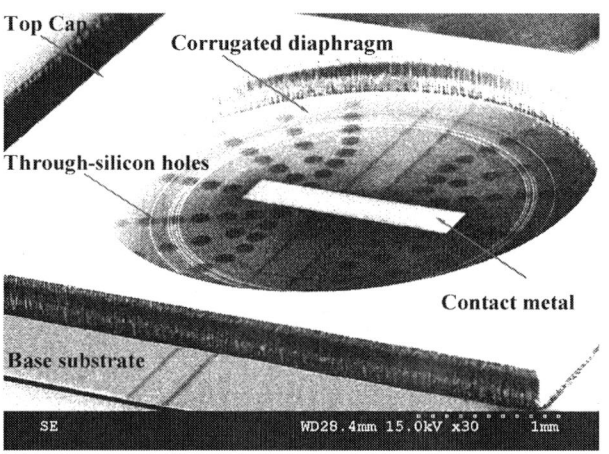

Figure 4: SEM photo of the fabricated RF MEMS shunt switch with perforated base substrate and a circular corrugated diaphragm (1750 μm in radius).

CHARACTERIZATION

Both pull-in and release time were measured for the fabricated switches by applying a square-wave with a duty-cycle of 20 % and an amplitude of 60 V for the actuation, which is 10 % higher than the pull-in voltage. The measurement results are tabulated in Table 2 and also shown in Fig. 5. For comparison, the calculated values by Equation (4) and Equation (9) are also given. With through-silicon holes in the base substrate, the measured pull-in time is enormously reduced from 5.4 ms to 0.435 ms, 12 times in reduction. At the same time, the release time is also significantly reduced from 40.6 ms to 3.2 ms, about 13 times in reduction, referring to Fig. 5.

It should be noted that the switching speeds of the current switch design, with a diaphragm of 1750 μm in radius, are quite slow even with a perforated base substrate. The main reason is that a large diaphragm is used. The proposed switch as a demonstrator was developed for proof of concept, focusing on the possible improvement in the performance of MEMS switch by employing closured diaphragm and perforated base substrate.

978-1-4244-2977-6/09 $25.00 © 2009 IEEE

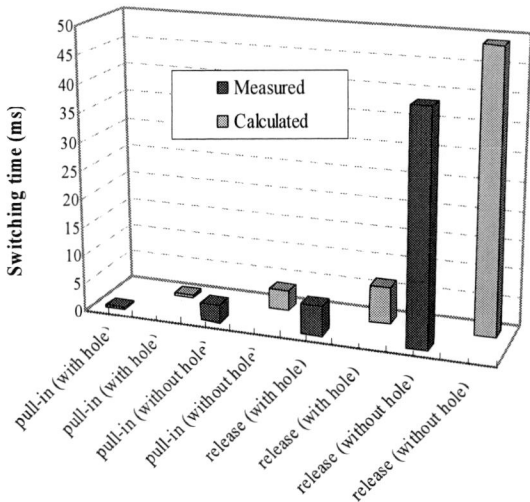

Figure 5: Measured switching times of MEMS switches in comparison with calculated values

Table 2: Switching times by measurement and calculation using Equation (4) and Equation (9)

Description of time	Calculat.	Meas.
Pull-in time with holes (ms)	0.473	0.435
Pull-in time without holes (ms)	6.46	5.4
Release time with holes (ms)	3.59	3.2
Release time without holes (ms)	49.0	40.6

The S-parameters of the fabricated switches are simulated by *HFSS* simulator and measured as shown in Fig. 6. The isolations are below −40 dB up to 14 GHz, and −35 dB at 25 GHz. The measured insertion losses are −0.7 dB up to 4 GHz, and −2.1 dB at 25 GHz for an encapsulated switch with a 5 mm long transmission line.

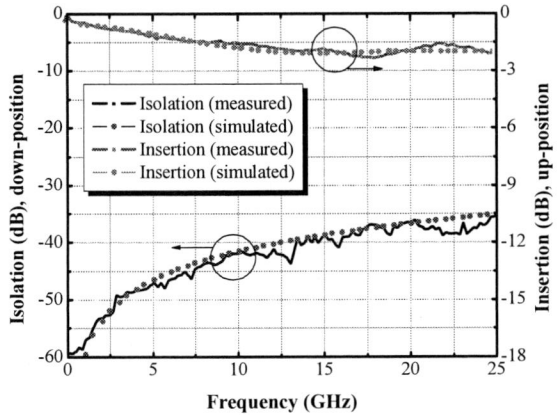

Figure 6: Measured and simulated S-parameters for a prototype switch.

CONCLUSION

The design, fabrication and characterization aspects of a DC contact RF MEMS shunt switch with a perforated base substrate are presented in this paper. An approach for the wafer-level encapsulation was proposed, which applies, for the first time for MEMS switch, the perforation in the base substrate to provide an effective way for reducing the damping coefficient and thus increasing the switching speed of the switch.

The switching times were measured for the fabricated switch with an air gap of 8 μm by applying a square-wave actuation voltage with an amplitude of 10 % higher than the pull-in voltage. The pull-in time is greatly reduced from 5.4 ms to 0.435 ms, 12 times in reduction, with the implementation of holes in the base substrate. Similarly, the release time is also significantly reduced from 40.6 ms to 3.2 ms, about 13 times in reduction, by means of the artificially creation of holes in the base substrate. The RF performance of the fabricated shunt switch was obtained by measurement as well. The isolation is below −40 dB up to 14 GHz, and about −35 dB at 25 GHz with an air gap of 8 μm. The measured insertion loss is about −0.7 dB up to 4 GHz, and −2.1 dB at 25 GHz for the encapsulated shunt switch with a 5 mm long CPW transmission line.

REFERENCES

[1] G. M. Rebeiz, *RF MEMS Theory, Design, and Technology*, John Willey & Sons Publication, 2003.

[2] P. G. Steeneken, et.al., "Dynamics and Squeeze Film Gas Damping of a Capacitive RF MEMS Switch," *Journal of Micromechanics and Microengineering*, 15 (2005), pp. 176-184.

[3] T. Veijola, et .al., "Gas Damping Model for a RF MEMS Switch and its Dynamic Characteristics", *Microwave Symposium Digest, 2002 IEEE MTT-S International*, volume 2, 2002, pp. 1213 – 1216.

[4] J. Miao, R. Lin, L. Chen, Q. Zou, S. Y. Lim, and S. H. Seah, "Design Considerations in Micromachined Silicon Microphones," *Microelectronics Journal*, vol. 33, iss. 1, January 2002, pp. 21-28.

[5] C. W. Tan and J. M. Miao, "Analytical Modeling of Bulk-micromachined Condenser Microphones", *Journal of the Acoustical Society of America* 120, No. 2, 2006, pp. 750-761.

[6] C. W. Tan, Z. H. Wang, J. M. Miao, and X.F. Chen, "A study on viscous damping effect for diaphragm-based acoustic MEMS applications," *Journal of Micromechanics and Microengineering*, 17 (2007), pp. 2253-2263.

[7] A. J. Zuckerwar, *AIP Handbook of Condenser Microphones: Principles of Operation of Condenser Microphones* (Edited by G. S. K. Wong and T. F. W. Embleton.), New York, American Institute of Physics, 1995, pp. 37-69.

A STRESS-TOLERANT TEMPERATURE-STABLE RF MEMS SWITCHED CAPACITOR

I. Reines[1], B. Pillans[2], and G.M. Rebeiz[1]

[1]University of California San Diego, La Jolla, California, USA
[2]Raytheon Systems Co. Dallas, Texas, USA

ABSTRACT

We present the design, fabrication, and measurement of an RF MEMS (Radio Frequency Micro-Electromechanical System) switched capacitor that exhibits reduced sensitivity to residual stress and temperature. The device is based on a circularly symmetric geometry with arc-type springs placed between the anchors and suspended beam. This design compensates for the effects of the residual biaxial stress in the beam, resulting in a pull-in voltage slope versus temperature of only -50 mV/ °C from -5 °C to 95 °C. Reducing the device sensitivity to residual stress improves the performance uniformity on a wafer-scale, and from wafer-to-wafer lots.

1. INTRODUCTION

RF MEMS switched capacitors offer key advantages in terms of linearity, loss, and power consumption compared to conventional solid-state switches [1]. To date, they have been used extensively in a wide range of RF switching and reconfigurable networks with excellent performance [2-3]. One drawback of commonly used fixed-fixed beam capacitive RF MEMS switches is their sensitivity to residual bi-axial stress. This limits the operating temperature, leads to excessively varying actuation voltage and up-state capacitance versus temperature, and reduces the power handling and device uniformity [4-6]. The device performance varies with temperature because a large portion of the spring constant is dependent on the residual stress, which changes with temperature due to the coefficient of thermal expansion (CTE) mismatch between the substrate and bridge. H. Nieminen previously reported a temperature-stable RF MEM capacitor whose design necessitated 4-10 μm thick spring metal to reduce initial beam displacement from a stress gradient [7].

The circular beam geometry presented here can be designed with the appropriate stiffness within a thin beam metal process ($t_{beam} < 1$ μm) while maintaining excellent electrical and mechanical performance over a wide temperature range. The symmetry of the design also reduces the sensitivity to vertical stress gradients and therefore minimizes the initial displacement of the suspended beam.

2. SWITCH DESIGN

Mechanical Properties

The spring constant of fixed-fixed beams is given by $k_{total} = k_1 + k_2$, where the k_1 portion is due to the stiffness of the bridge which is determined by material characteristics such as Young's modulus and the moment of inertia. The k_2 portion is due to residual biaxial stress in the beam and is controlled by the fabrication process. A perfectly symmetric circular geometry has been investigated to try and minimize the contribution of k_2 relative to k_1. This

will decrease the variation in beam stiffness versus residual stress and temperature. In general this is easier for thicker beam metals since $k_1 \sim t^3$ and $k_2 \sim t$.

The top view and cross section of the CPW implemented shunt switched capacitor are presented in Fig.1. The suspended plate is anchored symmetrically in four locations with arc-angled cutouts placed next to each anchor that define the springs that function to reduce k_2. The device is actuated by supplying a DC voltage between the ground and signal electrodes.

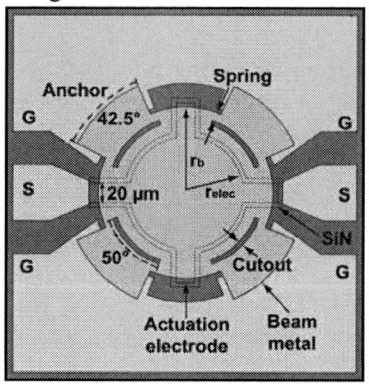

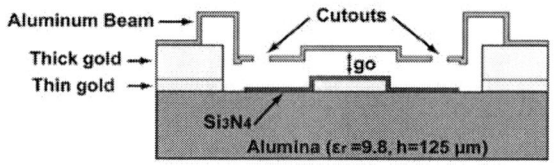

Figure 1: Circular MEMS switched capacitor, (top) layout, (bottom) cross-section.

The switched capacitor was designed to result in an up-state capacitance of 55 fF and a down-state capacitance of 1.5 pF ($C_r = 27$), given fixed process parameters. Electro-mechanical simulations of this structure were performed using CoventorWare [8], and the mechanical parameters are given in Table 1. The spring constant is defined for a displacement in the center of the beam given a distributed force applied to the top of the membrane over the actuation area.

Table 1: Mechanical parameters of the switched capacitor.

Parameter	Value
Effective area: A (μm^2)	13923
Beam radius: r_b (μm)	97
Electrode radius: r_{elec} (μm)	60
Spring width: w_s (μm)	10
Cutout width: w_c (μm)	5
Beam height: g_o (μm)	3.5
Spring constant[1]: k (N/m)	8.46
Pull-down voltage: V_p (V)	25.9
Resonant frequency[1]: f_o (kHz)	123

[1]simulated using CoventorWare assuming $\sigma_{res} = 60$ MPa

In-Plane Stress and Temperature

Mechanical simulations were performed on the circular switched capacitor to investigate the resulting change in the beam stiffness versus average in-plane tensile stress. As shown in Fig. 2, the spring constant increases by only 1.44 N/m for a residual stress of 180 MPa, which results in a maximum k_2/k_1 ratio of 18 %. For comparison, an ideal rectangular beam with dimensions width = 80 µm, length = 145 µm, and t_{beam}= 0.5 µm results in a k_1 portion equal to the circular design (k_1 = 8 N/m) as calculated from [1]. However, a beam stress of 180 MPa will result in a k_2/k_1 ratio of 2064%, demonstrating its sensitivity to in-plane residual stress.

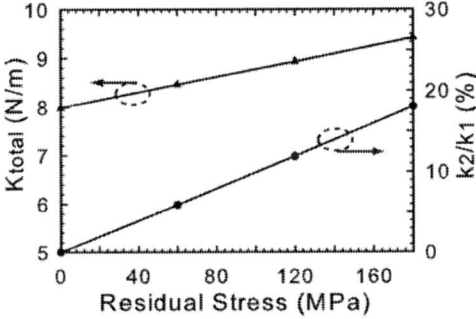

Figure 2: Simulated circular beam stiffness and k_2/k_1 for a varying in-plane residual stress.

When the switch is subjected to a temperature increase ΔT, the fixed anchors will exert a force on the suspended beam which induces a compressive stress, and this reduces the initial residual stress by:

$$\sigma_{total} = \sigma_{residual} - \alpha' E \Delta T \qquad (1)$$

$$\alpha' = \alpha_{beam} - \alpha_{substrate} \qquad (2)$$

where E is Young's modulus of the beam material, ΔT is the temperature difference, and α' is the CTE mismatch between the substrate and the beam (10-20 ppm/°C).

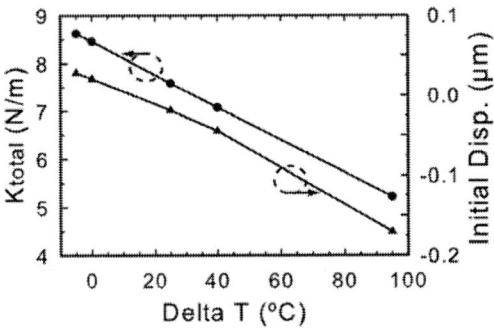

Figure 3: Simulated circular beam stiffness and maximum initial beam displacement (σ_{res} = 60 MPa, $\alpha_{substrate}$= 0 ppm/°C).

The total spring constant and displacement of the beam were simulated versus ΔT for a residual in–plane tensile stress of 60 MPa, assuming no substrate expansion. As shown in Fig. 3 the spring constant decreases linearly from 8.62 N/m to 5.22 N/m across the 100 °C range. The displacement is maximum in the center of the plate which deflects downwards by 0.17 µm for ΔT = +95 °C. This is due to the topography in the circular plate from the underlying bottom electrode and dielectric layers which creates bending moments in the beam that increase displacement versus initial residual stress and temperature, and should therefore be minimized if possible.

Stress Gradients and Temperature

The switched capacitor was also simulated versus varying vertical stress gradients. To approximate a stress gradient in Coventorware, the beam was divided into two equal-thickness layers that were assigned different average in-plane stress values while keeping the total average beam stress fixed at 60 MPa. As shown in Fig. 4, a vertical stress gradient increases both the beam stiffness and ratio of k_2/k_1 compared to an in-plane stress of 60 MPa. Simulations show that a varying stress gradient of 10-50 MPa/µm will result in an initial upward deflection from 0.1 - 0.34 µm.

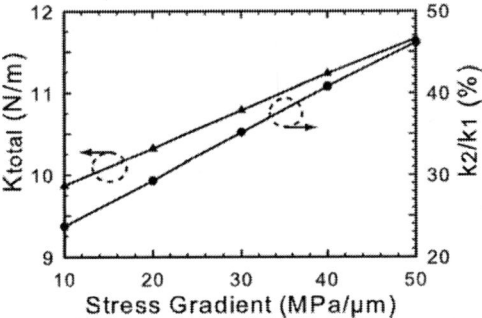

Figure 4: Simulated circular beam stiffness and ratio of k_2/k_1 for a varying vertical stress gradient.

The effects of a 20 MPa/µm stress gradient on beam stiffness and maximum initial displacement were simulated versus ΔT as shown in Fig. 5. The spring constant decreases linearly from 10.46 N/m to 6.48 N/m, while the center of the beam deflects downwards by 0.23 um across this 100 °C temperature range. Simulations of the circular beam geometry indicate that there is a tradeoff between reduced sensitivity to in-plane stress and initial beam deflection due to a stress gradient. For example, if the cutout angle in the beam is increased, the ratio of k_2/k_1 is reduced for an in-plane stress at the expense of more initial displacement from a stress gradient.

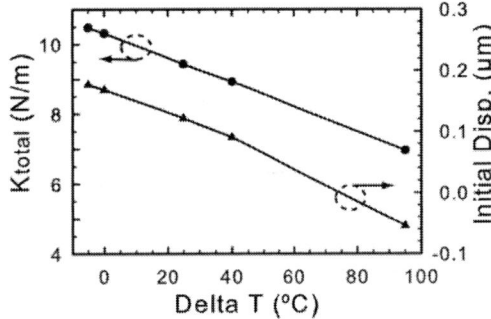

Figure 5: Simulated circular beam stiffness and maximum initial displacement ($\Delta\sigma_{grad}$ = 20 MPa/µm, $\alpha_{substrate}$= 0 ppm/°C).

3. FABRICATION

The switched capacitor is fabricated on a 125 µm alumina substrate (ε_r = 9.8, tan δ = 0.0001) using an RF MEMS capacitive switch process developed at Raytheon Co. A thin gold layer is first deposited that creates the bottom electrode of the capacitor. A layer of Si_3N_4 is then deposited on top of the bottom electrode to form the dielectric. A thick gold layer is used as support posts for the suspended beam and to reduce conductor losses in the transmission lines. Next a thin aluminum beam metal is deposited on top of a sacrificial layer which is later removed with surface micromachining to release the suspended membrane. To accommodate the release process, small circular holes are etched in the beam metal. It should be noted that these holes were not included in the electro-mechanical simulations. Test structures were also not included on this first fabrication run to de-embed the resulting biaxial stress in the fabricated devices but the in-plane stress is typically 60 MPa. A micrograph and 3-D white light interferometer image of the fabricated device are shown in Fig. 6.

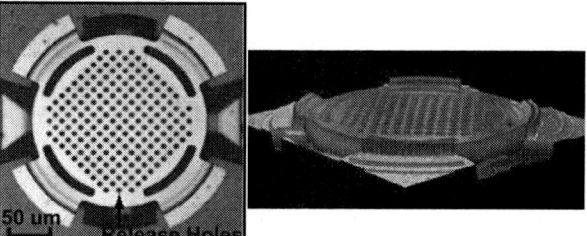

Figure 6: The fabricated switched capacitor, (left) micrograph, (right) 3-D white light interferometer image.

4. MEASUREMENTS

Pull-in Voltage vs. Temperature

The actuation voltage for both circular and standard [5] switched capacitors from two separate wafer lots were measured across a temperature range from -5 °C to 95 °C as shown in Fig. 8. The average pull-in voltage slope versus temperature for the two wafers is -50 mV/ °C for the circular design, compared to -340 mV/ °C for the standard switch. The spring constants were calculated using the measured actuation voltage and as-released gap heights for the devices from each wafer. The circular switches from wafers 1 and 2 have a k_{total} = 5.05 N/m and 7.84 N/m respectively. After temperature cycling, the circular devices were retested at room temperature and showed no change in actuation voltage or up-state capacitance from initial measurements.

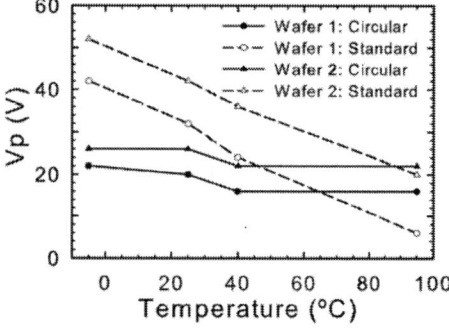

Figure 8: Measured pull-in voltage vs. temperature for both circular and standard switched capacitors from 2 wafer lots.

RF Characterization

The CPW implemented shunt switched capacitor was measured from 0.4 to 30 GHz in both the up-and-down states and versus temperature as shown in Figs.9-10. The up and down-state capacitances were extracted by fitting

the measured S-parameters to the lumped element model (Fig. 11). At room temperature, the switch has less than 0.51 dB insertion loss with better than 10 dB return loss up to 30 GHz (the insertion loss is dominated by the reflection coefficient, $1-|\Gamma|^2$). The equivalent up and down-state capacitances are C_u = 64 fF and C_d = 0.89 pF (C_r = 14). The down-state capacitance was reduced due to a rougher-than-normal dielectric. The up-state capacitance from the bridge increases from 53 fF to 58 fF from -5 °C - 95 °C as the average gap height decreases by 0.26 μm. The down-state capacitance did not change across the 100 °C temperature range. The insertion loss of the switch can be reduced in future designs by further optimization of the transitions in the CPW line surrounding the switch and resulting in a reduced reflection coefficient up to 30 GHz.

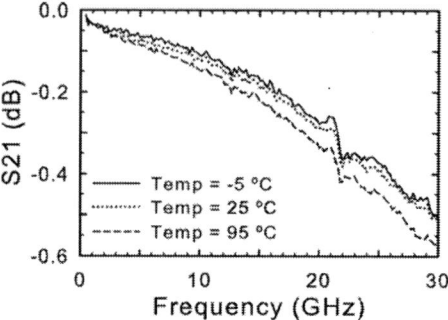

Figure 9: Up-state insertion loss of the switched capacitor from -5 °C to 95 °C (C_{bridge}=55.5 fF +/- 2.5 fF).

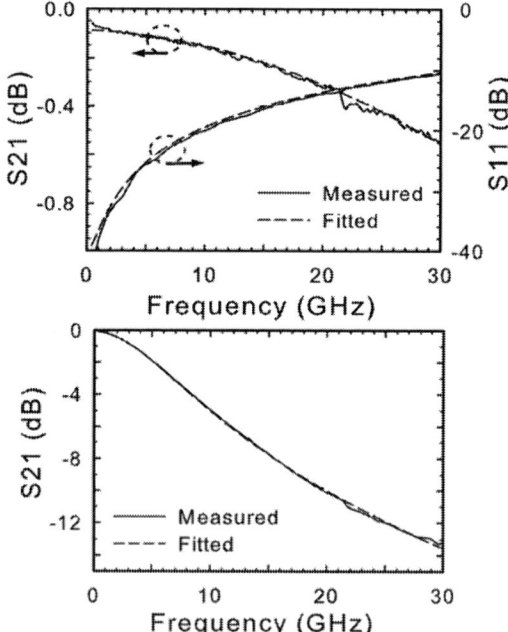

Figure 10: Measured and fitted S-parameters, (top) up-state, (bottom) down-state.

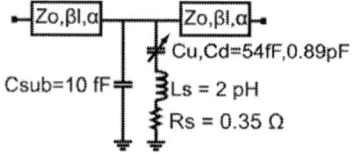

Figure 11: Equivalent lumped- element circuit model

The RF power handling of the switched capacitor was tested under a continuous power at 14 GHz, and the release voltage was measured versus incident power (Fig.

978-1-4244-2977-6/09 $25.00 © 2009 IEEE

12). The switched circular capacitors from both wafer lots handled 2W of continuous RF power before latching occurred.

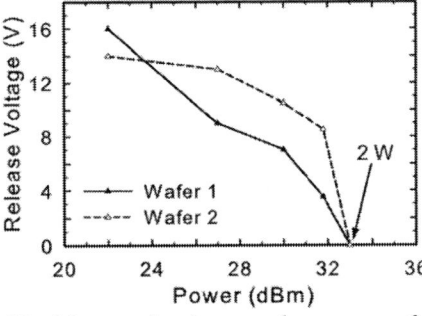

Figure 12: Measured release voltage as a function of continuous RF power at 14 GHz.

Switching Speed & Resonance Frequency

The switching speed measurement setup is depicted in Fig. 13. The switch is actuated at a rate of 100 Hz with a unipolar voltage from 0-40 V, and injected with 10 dBm of incident power at 10 GHz. The modulated output power is monitored with a power detector. The back-to-back waveguide to coax transitions act as a low frequency block to prevent leakage from the stepped voltage into the detector. A switching speed of 8 µs from up-to-down was measured as shown if Fig.13.

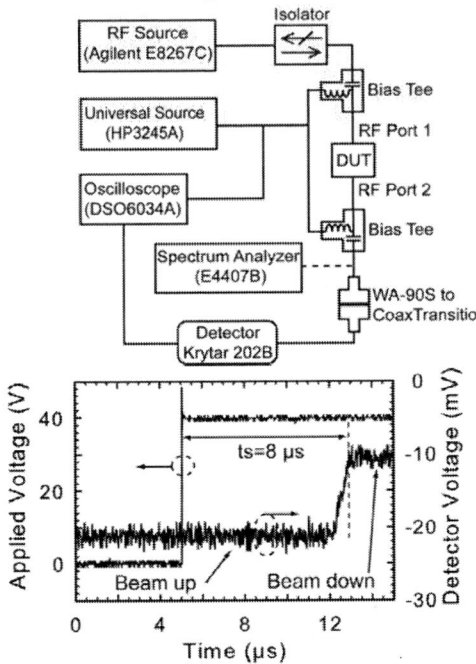

Figure 13: Switching speed and resonance frequency measurement setup (top), and measured switching speed from up-to-down state (bottom).

The mechanical resonance frequency was measured with the setup depicted in Fig. 12. For this measurement the switch is subject to a voltage given by:

$$V_{TOTAL} = V_{DC} + V_{LF} \cos(\omega_{LF}t) + V_{RF} \cos(\omega_{RF}t) \quad (3)$$

where there first two terms are applied from the universal source through the bias tees with $V_{DC} = 10$ V, $V_{LF} = 10$ V, and $f_{LF} = 10 - 200$ kHz. An RF power of 10 dBm at 10

GHz is applied and the modulated output signal from the switch is connected directly to a spectrum analyzer. The resonance frequency is measured by plotting the amplitude of the intermodulation products at $f_{RF} +/- f_{LF}$ while sweeping f_{LF}. The mechanical resonance frequency was measured at 102 kHz with a fitted quality factor of 1.95 as shown in Fig. 14.

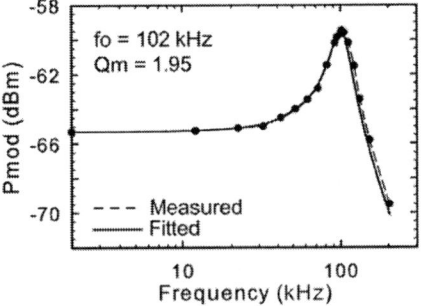

Figure 14: Measured mechanical resonance frequency of the circular switched capacitor.

5. CONCLUSIONS

An RF MEMS switched capacitor has been designed fabricated and tested showing reduced sensitivity to residual bi-axial stress and temperature. This device has a measured pull-in voltage slope versus temperature of only -50 mV/ °C from -5°C to 95°C compared to -340 mV / °C for standard rectangular devices. Across this temperature range the up-state capacitance varied by only 7.6 % or +/- 2.5 fF. This design improves the uniformity on a wafer-scale and from wafer-to-wafer lots increasing yield and manufacturability. RF measurements verify that this design results in good electrical performance up to 30 GHz over a wide temperature range. This work was supported under the DARPA (N/MEMS) Science and Technology Center on RF MEMS.

REFERENCES

[1] G. M. Rebeiz, RF MEMS Theory, Design, and Technology. John Wiley, 2003

[2] C. Nordquist et al. "A DC to 10-GHz 6-b RF MEMS Time Delay Circuit", IEEE Microwave and Wireless Component Lett., vol. 16, no. 5, pp. 305-307, May 2006.

[3] S. Park et al. "Low-Loss 4-6 GHz Tunable Filter With 3-Bit High-Q Orthogonal Bias RF-MEMS Capacitance Network", IEEE Trans. on Microwave. Theory and Tech., vol. 56, no. 10, pp. 2348-2355, October, 2008.

[4] C. Goldsmith et al. "Temperature Variation of Actuation Voltage in Capacitive MEMS Switches", IEEE Microwave. and Wireless Component Lett., vol. 15, no. 10, pp. 718-720, October, 2005.

[5] B. Pillans, "RF MEMS reliability at Raytheon", in Proc. IEEE MTT-S Workshop WFF: Reliability Testing & Reliability Enhancement in RF MEMS Switches, June, 2004.

[6] J. Reid et al. "RF Actuation of Capacitive MEMS Switches", IEEE MTT-S, pp. 1919-1922, June 2003.

[7] H. Nieminen et al. "Design of a Temperature-Stable RF MEM Capacitor", J. MEMS, vol.13, no.5, pp. 705-714, October, 2004.

[8] CoventorWare[tm], http:// www.coventor.com

TEMPERATURE COMPENSATION IN SILICON-BASED MICRO-ELECTROMECHANICAL RESONATORS

F. Schoen[1,2], M. Nawaz[1,2], T. Bever[1], R. Gruenberger[1], W. Raberg[1], W. Weber[1], B. Winkler[1] and R. Weigel[2]

[1]Infineon Technologies AG, Munich, GERMANY
[2]University of Erlangen-Nuremberg, Erlangen, GERMANY

ABSTRACT

In this paper, we present passive temperature compensation by means of silicon dioxide. Using an oxide refilling technique it avoids gap distance reduction and therefore prevents degradation of electromechanical coupling and motional resistance of the micro-electromechanical resonator. Samples are fabricated and electrically characterized to demonstrate the feasibility of the process concept. A constant quality factor (Q) and only a slight increase of the series resistance value (R_m) are achieved, while the frequency inaccuracy due to temperature variation is reduced. ANSYS simulations are carried out to evaluate the potential of the technique, resulting in a remaining inaccuracy of less than 40ppm.

INTRODUCTION

Micro-electromechanical resonators gain more and more interest as an adequate replacement for quartz crystals in timing and frequency reference applications. However the compensation of the temperature drift of the resonant frequency is still an important problem. Frequency shifts as high as -45ppm/K have been reported [1]. This decrease of resonant frequency can be essentially attributed to a decrease in Young's modulus of silicon with temperature, leading to a softening of the structure [2].

For low and medium performance applications, active compensation techniques (i.e. the use of a Delta-Sigma PLL [3], bias voltage trimming [4] or active heating concepts [2]) can be used. However, for high performance low power systems, active compensation techniques can pose significant problems due to their additional power consumption. Additionally, they degrade the phase noise performance and start-up time for the oscillator system. Therefore it is desirable to achieve temperature compensation by passive techniques. A very promising approach is to combine silicon with a material such as silicon dioxide which is known to get stiffer with increasing temperature [5]. By surrounding a silicon resonator with oxide Kim et al [6] achieved good temperature compensation. However this technique degrades both the electromechanical coupling (η), and the motional resistance (R_m) since it increases the electrically active gap d (Figure 1). The electromechanical coupling is given by,

$$\eta = u_{Bias} \frac{\varepsilon A}{d^2} \qquad (1)$$

where u_{bias} is the applied bias voltage, ε is the free space permittivity and A is the coupling area of the resonator.

The motional resistance is function of the electromechanical coupling, the stiffness k and mass m of the resonator and the Quality factor Q

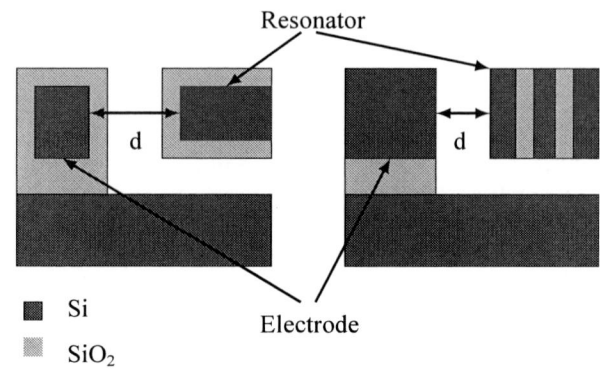

Resonator

d   d

■ Si
▨ SiO₂

Electrode

Figure 1: Temperature compensation by SiO₂ surrounding technique [6], which increases the distance between the electrodes at a given gap distance (left) whereas oxide filling of deep trenches allows to keep the same gap distance (right)

$$R_m = \frac{\sqrt{km}}{Q\eta^2} \qquad (2)$$

One can see from equations (1) and (2) that the impact of an increased electrical gap on these parameters is considerable.

PROCESS CONCEPT

To overcome the aforementioned problems, we have developed a new process concept (patent application ongoing), which allows to combine silicon and silicon dioxide for temperature compensation without degrading the electrical performance.

The processing starts with a silicon-on-insulator (SOI) wafer. Highly anisotropic trenches are formed by deep reactive ion etching using an oxide hard mask. A poly-spacer is formed to reduce the width of the electrical active gap from 800 to 400nm (Figure 2a). After oxide refill of the trenches and CMP with stop on silicon, a poly-silicon layer is deposited and structured to cover the trenches at locations, where the oxide will be maintained for temperature compensation (Figure 2b). Besides protection for the underlying oxide during later resonator release etch, the polysilicon layer acts as electrical connection between the parts of the resonator. Through three successive deposition, lithography and etch steps a local offset is created, which serves as etch channel during release etch, as well as protection during the cavity sealing (Figure 2c). A wet-chemical etch is performed to release the resonators before the device is hermetically sealed by a low pressure CVD of oxide. The local offset prevents the deposited material from reaching the

978-1-4244-2977-6/09 $25.00 © 2009 IEEE

resonator. The duration of the wet-chemical etch is optimized to release the resonator, without damaging the oxide filled trenches. To complete the sealing process an additional protection layer is deposited and patterned. A thick oxide layer is deposited to protect the resonator. Finally metal through vias are created and metallization forms contacts to electrodes and resonator (Figure 2d).

a)

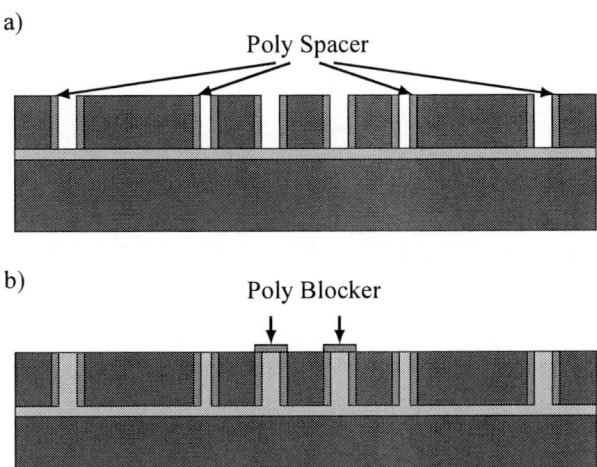

b)

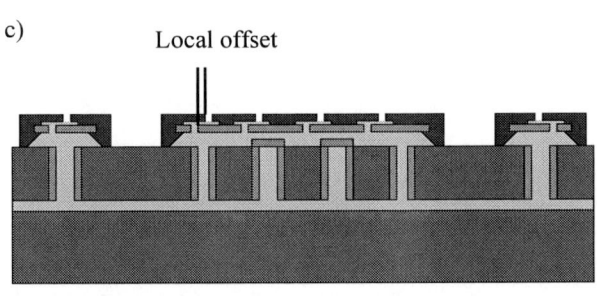

c)

d)

Figure 2: Schematic diagram of the device after trench etch and poly-spacer formation (a), after poly-silicon formation to cover the trenches where the oxide will be maintained for temperature compensation (b), after formation of a local offset (c) and after wet-chemical release etch. Hermetic sealing is achieved using the offset. Metallization forms contacts to electrodes and resonator (d).

DESIGN

To demonstrate process feasibility, 13MHz wheel resonators with and without compensation trenches are designed. The wheel resonator is anchored to the substrate at its center position, while the other parts of the structure (support beams and outer mass) are released. Electrostatic actuation is performed through a ring electrode surrounding the whole outer mass of the resonator. As depicted in Figure 3, oxide trenches (drawn with black color) are only placed within the outer mass, where the main movement of the structure is known to take place. Seven trenches, each with a final width of 1.35μm are formed.

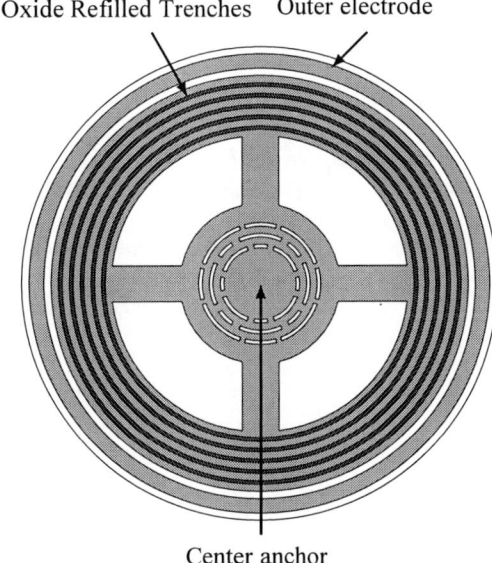

Figure 3 Schematic Layout of the 13MHz wheel resonator device with silicon dioxide fill structures (drawn with black color) in the outer mass for temperature drift compensation

RESULTS

Measurement setup

The fabricated devices are evaluated electrically through on-wafer measurements using an Agilent 4294A Impedance Analyzer. On-wafer measurement is chosen to avoid any influence of package stress on the temperature coefficient, as reported by Melamud [7]. The needed bias voltage is applied directly through an internal voltage source of the Impedance Analyzer. No additional devices or circuitry are used in order to prevent any distortion of the measurement data. The devices are contacted using Kelvin Probes to reduce the effect of parasitics. Temperature control is done using a heatable / coolable ceramic chuck.

Measurement results

Measurements are performed for a temperature range from 15°C to 95°C in 10°C steps, and a bias voltage of 40V. Figure 4 shows the measured resonant frequency as a function of temperature for a compensated and an uncompensated device respectively. For better comparison the measured frequencies are normalized to the measured values at 15°C. Temperature coefficients of resonant frequency are extracted from the measurement using a linear model, neglecting any higher order terms.

While the device without compensation has a temperature coefficient of the resonant frequency around -22ppm/K, the temperature induced variation is reduced to -9.9ppm for the oxide refilled device. The impedance in series resonance increases slightly from 12.6kΩ to 13.9kΩ, which can be attributed to small gap width variations. The quality factor decreases by less than 1%.

978-1-4244-2977-6/09 $25.00 © 2009 IEEE

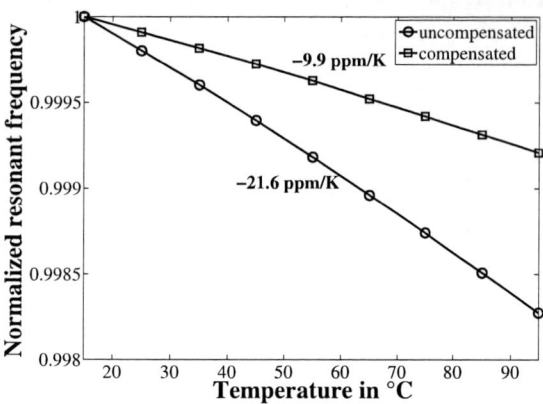

Figure 4: Measured resonant frequency dependence as a function of temperature for a compensated and an uncompensated 13MHz wheel resonator. The measured frequencies are normalized to the frequency at 15°C. The used u_{bias} is 40V.

SIMULATION

Finite element simulations using ANSYS are carried out for 13MHz and 27MHz wheel devices with different oxide trench width, to evaluate the potential of our technique. For 13MHz devices the number of trenches is varied additionally. The material data of silicon used for the simulation is taken from literature [5]. To account for deposition dependant mechanical parameters, the data for silicon dioxide are extracted from the performed measurements.

As can be seen in Figure 5 good compensation is achieved for the 13MHz device, while even overcompensation is possible for the 27MHz device.

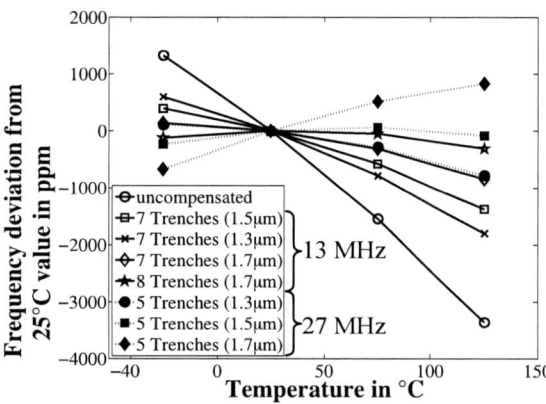

Figure 5: ANSYS simulated temperature compensation in 13MHz and 27MHz devices for different oxide trench widths and different number of oxide trenches. The frequency offsets are with reference to the value at 25°C

From the performed simulations first (α), second (β) and third order (γ) temperature coefficients of the resonant frequency are extracted. The dependence of the resonant frequency f on the temperature offset ΔT with reference to the initial frequency f_0 can then be written as,

$$f(T) = f_0(1 + \alpha\Delta T + \beta\Delta T^2 + \gamma\Delta T^3) \qquad (3)$$

An overview of the extracted data is given in table 1. As

expected, the first order temperature coefficient decreases with increasing oxide gap width. An overcompensation of the first order term for a 13MHz Wheel device can be achieved by using eight trenches, each with a width of 1.7μm. In addition to the first order term, the second order term also decreases with increasing oxide mass. The third order term is seen to be three orders of magnitude smaller than the second order term, and thus can be neglected for most applications. The potential of the technique becomes even more visible with the 27MHz wheel device. For reduced resonator dimensions the resonant frequency increases and the oxide mass needed for temperature compensation is reduced. A decrease of the first order term down to 2.09ppm/K can already be seen with five trenches, each with a width of 1.3μm. Increasing the trench width up to 1.7μm strong overcompensation (+13.49 ppm/K) is achieved. Both for the 13 and 27MHz devices the second order term is increased with increased oxide to silicon ratio.

Table 1: Extracted first (α), second (β) and third order (γ) temperature coefficients of resonant frequencies for 13MHz and 27MHz wheel resonators with oxide compensation extracted from ANSYS Simulations for different numbers of trenches and different trench widths. The devices are denoted (Frequency / Number of trenches / width of a single trench)

Device	α ppm/K	β ppm/K^2	γ ppm/K^3
13MHz/ 7T/ (1,3μm)	-11.88	-0.034	-5.33e-5
13MHz/ 7T/ (1,5μm)	-7.82	-0.033	-5.20e-5
13MHz/ 7T/ (1,7μm)	-2.72	-0.031	-5.46e-5
13MHz/ 8T/ (1,7μm)	2.41	-0.029	-5.33e-5
27MHz/ 5T/ (1,3μm)	-2.09	-0.031	-5.60e-5
27MHz/ 5T/ (1,5μm)	4.66	-0.029	-5.86e-5
27MHz/ 5T/ (1,7μm)	13.49	-0.026	-6.26e-5

For further evaluation a full first order compensated (α=0) 27MHz device is simulated. In Figure 6 the remaining temperature dependence of the resonant frequency is depicted. A nonlinear second order effect leads to a variation of 40ppm. This value complies with the typical specifications for the automotive temperature range (-40°C – 135°C).

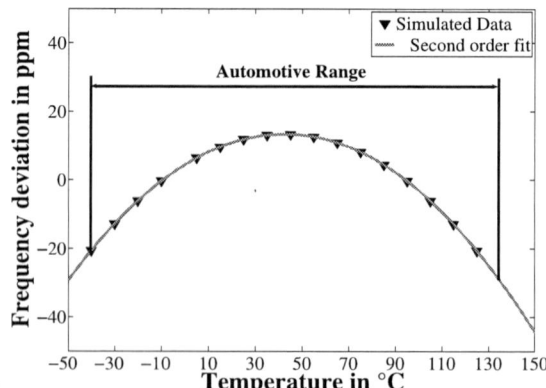

Figure 6: Simulated remaining temperature dependence of a fully SiO₂ compensated resonator (ANSYS simulation)

In fact, the performed simulations also predict that the

quality factor of the resonator can be improved using the oxide fill process. From simulations and from the measured temperature dependence of the quality factor it can be concluded that thermoelastic damping is the dominant loss mechanism of the resonator. This seems reasonable, as air damping is avoided through a vacuum sealing process and anchor loss is largely reduced through the center anchored design. As described by Candler et al. [8] the losses due to thermoelastic damping can be influenced by modifying the coupling between the mechanical and thermal eigenmodes of the resonator. This was achieved by placing slots within the resonator structure. In fact skillful placement of oxide within the structure leads to a comparable effect, therefore decreases thermoelastic losses and increases the total quality factor.

CONCLUSION

A novel oxide filling process flow is presented which offers the ability to passively compensate for temperature inaccuracy of silicon based microelectromechanical resonator. The process is designed in a way, that it does not degrade electrical parameters like electromechanical coupling or motional resistance.

Devices have been fabricated using the oxide fill process, showing the feasibility of the process concept and leading to a reduction of the temperature coefficient of frequency from -22.6ppm/K to -9.9ppm/K.

ANSYS simulations are carried out to evaluate the further potential of this technique. It is shown, that a full first order compensation can be achieved. The remaining temperature inaccuracy is as small as 40ppm. Additionally, simulations predict that the thermoelastic losses can be minimized and the quality factor of the resonator increased. To validate these results further compensated devices will be designed according to the results of the simulations.

ACKNOWLEDGEMENT

The authors would like to thank Richard Stepp, Waldemar Marsetz and Mihail Sararoiu at Infineon Technologies for their invaluable help.

REFERENCES

[1] J. T. M. Van Beek, H.-P. Loebl, F. W. M. Vanhelmont, "A MEMS RESONATOR, A METHOD OF MANUFACTORING THEREOF, AND A MEMS OSCILLTOR," *International Patent Application* WO 2007/072409 A2, June, 28, 2007.

[2] M. A. Hopcroft, "Temperature-Stabilized silicon resonators for frequency references," *Ph.D. dissertation,* Dept. Mech. Eng., Stanford Univ., Stanford, CA, 2007.

[3] Wan-Thai Hsu, A.R Brown., "Frequency Trimming for MEMS Resonator Oscillators," *in IEEE Int. Frequency Control Symp. Joint with 21st European Frequency and Time Forum*, Geneva, May 29 – June 1, 2007 pp.1088-1091

[4] G.K. Ho, K. Sundaresan, S. Pourkamali, F. Ayazi, "Temperature Compensated IBAR Reference Oscillators," *19th IEEE Int. Conf. Micro Electro Mechanical Systems*, Istanbul, January 22-26, 2006, pp.910-913

[5] H. J. McSkimin, "Measurement of Elastic Constants at Low Temperatures by Means of Ultrasonic Waves – Data for Silicon and Germanium Single Crystals, and for Fused Silica", *J. Appl. Phys.* , vol. 24, pp. 988-993, 1953.

[6] K .Bongsang, R. Melamud, M. A. Hopcroft, S. A. Chandorkar, G. Bahl, M. Messana, R. N. Candler, G. Yama, T. Kenny, "Si-SiO2 Composite MEMS Resonators in CMOS Compatible Wafer-scale Thin-Film Encapsulation," *in IEEE Int. Frequency Control Symp. Joint with 21st European Frequency and Time Forum*, Geneva, May 29 – June 1, 2007, pp.1214-1219

[7] R. Melamud, M. A. Hopcroft, C. Jha, B. Kim, S. Chandorkar, R. N. Candler, R. T. W. Kenny, "Effects of stress on the temperature coefficient of frequency in double clamped resonators", *Digest Tech. Papers Transducers'05 Conference*, Seoul, June 5-9, 2005, pp. 392-395

[8] R. N. Candler, A. Duwel, M. Varghese, S. A. Chandorkar, M. A. Hopcroft, Woo-Tae Park; B. Kim, G. Yama, A. Partridge, M. Lutz, R. T. W. Kenny, "Impact of geometry on thermoelastic dissipation in micromechanical resonant beams," *J. Microelectromech. Syst*, vol. 15, pp. 927-934, 2006

A 12-GHZ DPDT RF-MEMS SWITCH WITH LAYER-WISE WAVEGUIDE/ACTUATOR DESIGN TECHNIQUE

D. Yamane[1], H. Seita[2], W. Sun[1], S. Kawasaki[2,3], H. Fujita[1], and H. Toshiyoshi[1]

[1]Institute of Industrial Science, The University of Tokyo, JAPAN
[2]Kyoto University, JAPAN
[3]Japan Aerospace Exploration Agency (JAXA), JAPAN

ABSTRACT

A novel design of double-pole double-throw (DPDT) RF-MEMS switch for 12-GHz phase shifter has been developed to minimize the electrical crosstalk between the signal waveguide and electrostatic actuator by allocating them in separate layers of an SOI wafer. Compared with the previous other reports, our design can use relatively large area on the chip to accommodate more electrostatic actuator and to have more electrical ground planes. With this newly developed method, silicon RF-MEMS devices will be able to overcome the drawback of solid-state devices in terms of performance, device size, and cost. This paper reports driving voltage of 4V with the switching speed of 12 microseconds, insertion loss of 3 dB, return loss of 12 dB, isolation of 30dB at 12 GHz.

INTRODUCTION

Active phased array antenna (APAA) system is expected to be widely used for the mobile vehicle satellite communication in the near future thanks to the compactness and high scanning performance. Every APAA has a large number of phase shifters [1], where RF MEMS switches [2] are expected to deliver superior performances in terms of low insertion loss and high isolation compared to those of compound semiconductor devices.

Previously reported RF-MEMS switches include a surface-micromachined electroplated-type [3, 4] and a bulk-micromachined SOI-type [5, 6]. Ideally speaking, the electromechanical structures should be separated from the waveguides to reduce the electromagnetic coupling.

Therefore we newly developed a switch design, in which the actuators and the waveguides were made on the different surfaces of an SOI wafer. In addition to that layer-wise designing, we also used an SOI wafer of two different resistivities; the handle layer had high resistivity to minimize the microwave loss through the waveguides, and the SOI device layer had low resistivity for the electrical interconnection of electrostatic actuators.

By using the newly developed designing technique, we fabricated a DPDT RF-MEMS switch for line-switching phase shifter for the APAA application.

DESIGN

DPDT Switch for Line-Switching Phase Shifter

Figure 1 illustrates the RF waveguide-side of the DPDT switch with schematic view of the delay lines of a phase shifter. The switch is flip-chip bonded onto a low temperature co-fired ceramics (LTCC) [7, 8] substrate of microwave delay lines, where either longer or shorter pass is chosen to alter the phase delay. Two pieces of SPDT are needed to compose 1-bit of phase shifter, which leads to difficulty in mounting large number of chips onto the LTCC substrate. To cut the device packaging cost to nearly the half, we developed a DPDT RF-MEMS switch. In our design one DPDT switch occupies less device area than two SPDT ones; it even releases more space to accommodate larger electrostatic actuator that drives the switch in more reliable manner.

Figure 2 shows the actuator side of the DPDT switch, where electrostatic comb-drive actuators and suspensions are arranged. Thanks to the new design of arranging the waveguide on the other surface of substrate, the actuator used nearly the entire area for generating larger force.

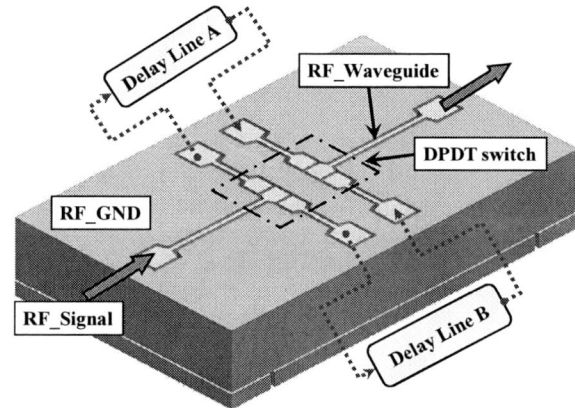

Figure 1: Schematic of the RF waveguide-side of the DPDT switch

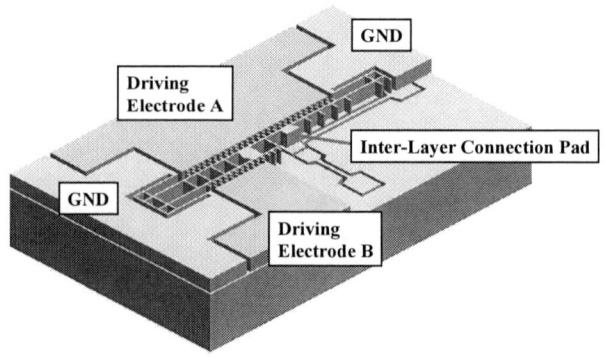

Figure 2: Schematic of the actuator-side of the DPDT switch

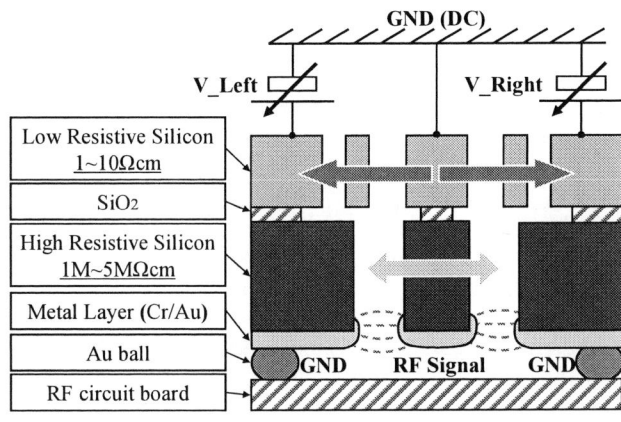

Figure 3: Schematic of the cross section of waveguide (substrate) and actuator (SOI) of the DPDT switch

It is also beneficial for the RF-waveguides because larger area can be assigned to make RF ground plane. Driving voltages are applied to the fixed parts of the comb-drive actuator, and the movable structures are always grounded. The actuators are shared for both input and output ports to save area.

The cross section image of the DPDT and its material properties are presented in Figure 3. Low resistive silicon (1~10Ωcm) layer was used for electrostatic actuator and high resistive silicon (10,000~50,000Ωcm) layer for the RF waveguide structures to minimize dielectric losses. The design of RF waveguide was an ideal coplanar waveguide (CPW) because the signal line was separated from the ground lines by air, which was the material of a lowest electric permittivity. Thanks to the geometrical and electrical separation between the waveguides and the actuators, the signal line was electrically decoupled from the actuators' drive electrodes. By the lateral motion of the actuator, the hammer tips of the movable waveguides are electromechanically brought in contact with the input and output ports to reroute the signals.

From an implementation point of view, flip-chip bonding is supposed to be rather desirable than wire bonding to lower the effect of wire inductance that hampers the impedance matching. Since the presented design of DPDT switch has layer-wise functions, the RF-waveguide surface can be flip-chip bonded onto an LTCC phase shifter substrate and the actuator surface is interconnected by wire bonding. It means that bonding areas for RF signals and electrostatic actuation can be well separated to minimize undesired electrical coupling.

FABRICATION
Double Side DRIE Process on SOI Wafer

We used silicon bulk micro-machining technique of deep reactive ion etch (DRIE) process to make trenches of high-aspect ratio. We also used the double-side DRIE processes to make fine three-dimensional silicon structures on both sides of the SOI substrate. Figure 4 shows the process flow of the DPDT switch. It took only two photolithography steps, while conventional surface micro-machined MEMS switches need more than four

photomasks [9, 10]. Original thickness of the handle layer on our SOI wafer in Figure 4(i) was 400 μm and the layer was grinded and then polished by chemical mechanical polishing (CMP), as shown in Figure 4(ii). The device layer with low resistive SOI silicon was 30 μm thick, and the handle layer with high resistive silicon was 100 um thick. The thickness of silicon dioxide layer was 2 μm. The first DRIE process shown in Figure 4(iii) defined the structure of comb-drive actuators and suspensions, followed by the surface passivation. Then second DRIE process was done to have the MEMS co-planar waveguides (Figure 4(iv)). After the double side DRIE processes, the silicon dioxide layer was wet-etched by using the buffered hydrofluoric acid (BHF) for releasing movable parts. During this process, the connection part of the silicon dioxide between the actuator and the waveguide was intentionally left connected by controlling the etching time in BHF, as shown in Figure 4(v); this part was designed to be larger than the other released areas for the easiness of etching control. Finally, on the top of the MEMS CPW was furnished with sputtered metals of 20 nm chromium and 800 nm gold.

SEM pictures of the DPDT switch are in Figure 5. Figure 5(a) and Figure 5(b) shows the actuator-side of the DPDT switch. The chip size is 2 mm by 4 mm in area, and the comb-drive electrodes have a 2 μm gap. There are four suspensions in the actuator layer, and the each one of them is 1.5 μm wide and 530 μm long.

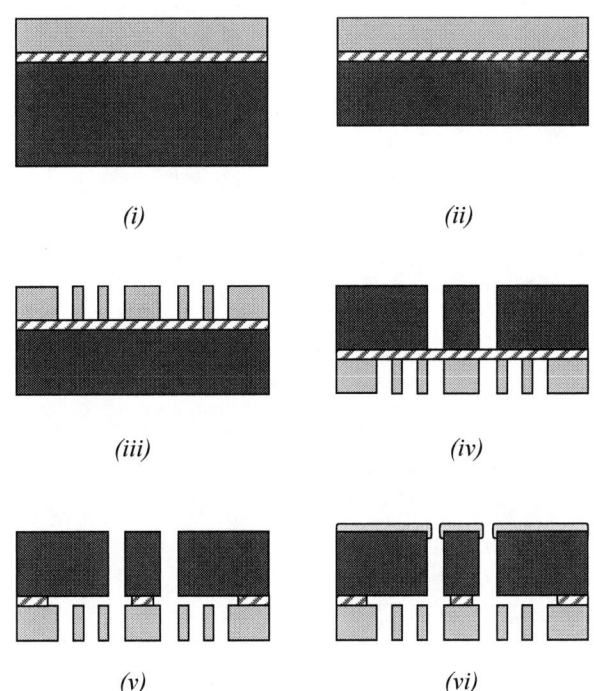

Figure 4: Process chart of the DPDT switch :
(i) Initial state, (ii) After grinding and CMP on handle layer,
(iii) First DRIE process on device layer, (iv) Second DRIE,
(v) BHF wet-etching for silicon dioxide layer
(vi) Metallization on the top of MEMS CPW

978-1-4244-2977-6/09 $25.00 © 2009 IEEE

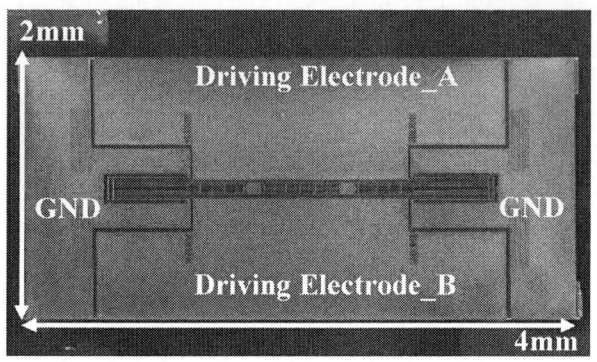

(a)

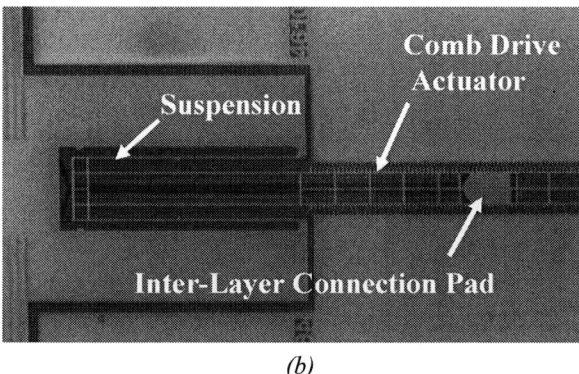

(b)

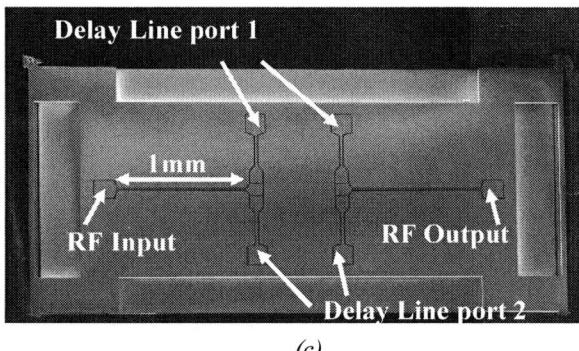

(c)

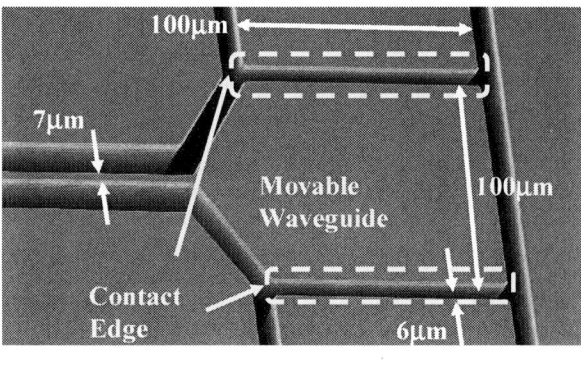

(d)

Figure 5: SEM pictures of developed DPDT RF-MEMS switch: (a) Actuator entire view, (b) Close-up image of suspension and comb drive actuator, (c) Waveguide entire view, (d) Close-up view of the movable waveguide hammer head

Figure 5(c) and Figure 5(d) are the images of the RF waveguide-side of the DPDT switch. The movable signal lines are 1 mm long and 7μm wide. The contact edge of the movable-waveguide tip is 100μm long and the initial contact gap is 6μm wide.

RESULTS

Mechanical Characteristics

Electrical switching operation was confirmed at the DC level signal as shown in Figure 6(a) and Figure 6(b), which correspond to the switching state of ON and OFF, respectively. The actuator was driven with square voltage of 3.7 V of upwards of 5 kHz. Switch response (rise and fall) with respect to the voltage was found to be 12 microseconds, which was fast enough compared with the switching speed needed for the APAA system, 1 milliseconds.

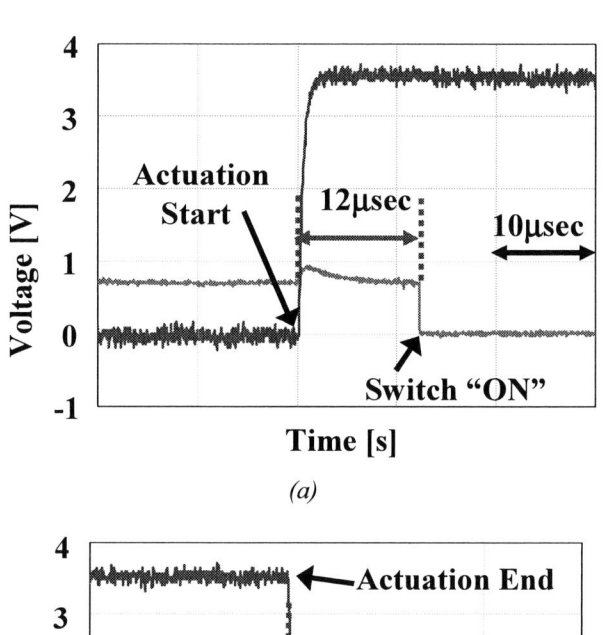

(a)

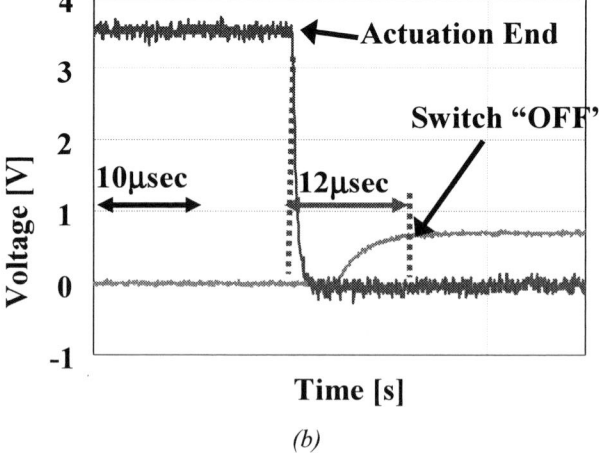

(b)

Figure 6: Switching operation at 5 kHz; Blue line shows the applied driving voltage and red line shows the constantly applied voltage between the input port and the output port. If the switch contact begins, the red line falls to zero potential for its short circuit. :
(a) ON, (b) OFF

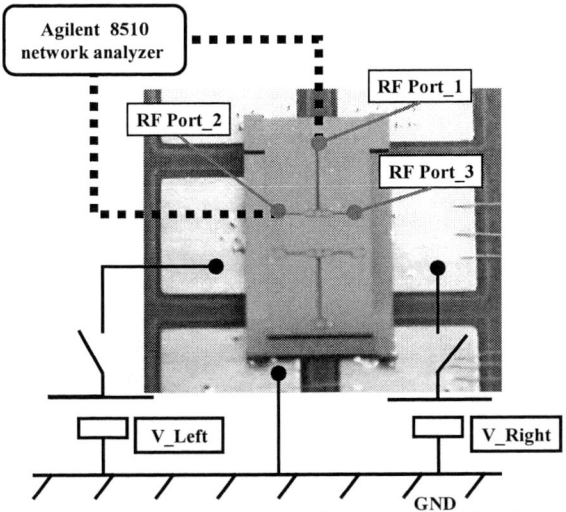

Figure 7: DPDT RF-MEMS switch on a measuring board with actuator side down

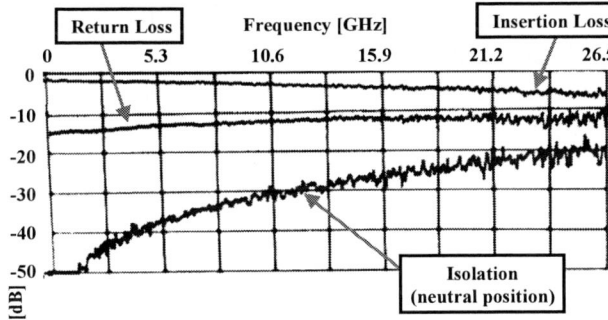

Figure 8: RF characteristics

RF Characteristics

Microwave performance as a bare MEMS switch (excluding the LTCC phase shifter) was measured with an Agilent 8510 network analyzer under the condition of 10mW RF signal. Figure 7 shows a test bench of the measurements.

The switch was bonded to a printed circuit board (PCB) with gold balls and silver paste putting the waveguide side up for the packaging convenient. For applying the driving voltage, wire bonding was performed to the metal pads on the underlying PCB from which the actuator was energized. The plot in Figure 8 shows the measured characteristics of the switch. At the target frequency of 12 GHz, insertion loss of 3 dB and return loss of 12 dB were obtained. Isolation was found to be 30 dB between the movable signal line at the neutral rest position.

CONCLUSIONS

Layer-wise waveguide/actuator design technique of RF-MEMS switch was successfully demonstrated as a DPDT switch developed on an SOI wafer. This technique provides us with a higher degree of freedom to create novel silicon MEMS devices not only for RF-MEMS field but also for the whole silicon MEMS field, because it effectively increases the device footprint within a give chip area.

In our study on the DPDT switch, the switch contact

was observed at 4V operation and switching speed was 12 microseconds. From a single switch point of view, the insertion loss of 3dB, the return loss of 12 dB and the isolation of 30dB were measured at 12 GHz. Insertion loss and reliability of switching are under improvement.

ACKNOWLEDGEMENTS

This work was founded by the *Radio-wave research and development for prospect of radio wave usage in the medium and long term foundation* conducted by the Japan Ministry of Internal Affairs and Communication.

REFERENCES

[1] G.-L. T. Gabriel M. Rebeiz, Joseph S. Hayden, "RF MEMS Phase Shifters: Design and Applications," *IEEE Microwave Magazine*, pp.72-81, 2002.

[2] J. B. M. Gabriel M. Rebeiz, "RF MEMS Switches and Switch Circuits," *IEEE Microwave Magazine*, pp.59-71, 2001.

[3] W. B. Zheng, Q. A. Huang, X. P. Liao, And F. X. Li, "MEMS Membrane Switches on GaAs Substrates For X-Band Applications," *Journal of Microelectromechanical Systems*, vol. 14, pp. 464-471, Jun 2005.

[4] G. David A, Richard D. Nelson and John S. Mckillop, "Design of A 20GHz Low Loss Ohmic Contact RF MEMS Switch," pp. 371-374, 2007.

[5] J. Oberhammer, M. Tang, A. Q. Liu, and G. Stemme, "Mechanically Tri-Stable, True Single-Pole-Double-Throw (SPDT) Switches," *Journal of Micromechanics and Microengineering*, vol. 16, pp. 2251-2258, Nov 2006.

[6] A.-Q. L. M.Tang, A. Agarwal, Z.-S. Liu, C. Lu, "A Single-Pole Double-Throw (SPDT) Circuit Using Lateral Metal-Contact Micromachined Switches," *Sensors and Actuators*, pp.187-196, 2004.

[7] K. I. Kim, J. M. Kim, J. M. Kim, G. C. Hwang, C. W. Baek, and Y. K. Kim, "Packaging for RF MEMS Devices Using LTCC Substrate and BCB Adhesive Layer," *Journal of Micromechanics and Microengineering*, vol. 16, pp. 150-156, Jan 2006.

[8] J.-M. K. Yong-Seung Bang, Yongsung Kim, Jung-Mu Kim, and Yong-Kweon Kim, "Fabrication and Characterization of RF MEMS Packagebased on LTCC Lid Substrate and Gold-Tin Eutectic Bonding," *Proc. Solid-State Sensors, Actuators and Microsystems Conference 2007.*, pp.2115-2118, 2007.

[9] Y. Wang, Z. H. Li, D. T. Mccormick, and N. C. Tien, "A Low-Voltage Lateral MEMS Switch with High RF Performance," *Journal of Microelectromechanical Systems*, vol. 13, pp. 902-911, Dec 2004.

[10] S. K. B. Devarajan Bahaman, Farrokh Ayazi, John Papapolymerou, "Low-Cost Low Actuation Voltage Copper RF MEMS Switches," *Microwave Symposium Digest, 2002 IEEE MTT-S International*, vol. 2, pp.1225-1228, 2002.

POST-FABRICATION ELECTRICAL TRIMMING OF SILICON BULK ACOUSTIC RESONATORS USING JOULE HEATING

A.K. Samarao and F. Ayazi
Georgia Institute of Technology, Atlanta, Georgia, USA

ABSTRACT

This paper presents a new method to electrically trim the resonance frequency of a Silicon Bulk Acoustic Resonator (SiBAR) post fabrication. Width-extensional mode silicon resonators are heated by passing a current through their resonating elements. This causes a mass loading gold pattern to diffuse into the bulk of the resonator. Upon cooling, the gold diffusion increases the stiffness of the resonating structure slightly, which reflects as an upward shift in resonance frequency. Thus, silicon resonators can be permanently trimmed to a desired frequency value by an electrical calibration step. As a proof of concept, an upward frequency shift of 240 kHz is demonstrated for a 40% mass loaded 100 MHz SiBAR after one hour of Joule heating with 30 mA of DC current.

INTRODUCTION

Silicon micromechanical resonators have been gaining importance in recent years owing to their small form factor, ease of integration and high $f.Q$ products. High-frequency and high-Q width-extensional mode SiBARs fabricated using the HARPSS process have shown atmospheric Q factors in excess of 10'000 at or above 100 MHz, with moderate motional resistances (~1000 Ω) [1,2]. The resonance frequency of silicon micromechanical resonators is dependent on the physical dimensions of the resonating structure. This causes the frequency of the micromachined resonator to deviate from a designed target value due to variations in photolithography, etching and film thickness. It can be shown that 2 μm variations in the thickness of an optimized 100 MHz width-extensional-mode SiBAR can cause 0.5% variation in its center frequency [3], while lithographic variations of ±0.1 μm in the width of the resonator can cause additional 0.5% frequency variations.

Our group had previously reported on the tuning of SiBAR frequency using electrical signals [1, 4]. Electrostatic voltage tuning of high frequency SiBARs is inefficient due to the large stiffness of the device [1] and heat-induced continuous current tuning of the device consumes large power [4]. In addition, these tuning techniques are limited in their tuning range and cannot be used to adjust the resonance frequency in case of large offsets (~1 %), which are quite typical in microfabrication. It is also well known that the frequency of mechanical resonators can be shifted downwards by deposition of a mass loading layer such as a metal on the surface of the resonating structure [5]. However, the thickness of the mass loading layer cannot be accurately controlled, and are subjected to the limitations of the metal evaporation systems. Though laser trimming [6] has been shown to shift the resonance frequency of silicon resonators downwards or upwards, the trimming is not precise as it is difficult to control the amount of material deposited or removed by the laser. This calls for an efficient post-fabrication trimming technique which can adjust the resonance frequency precisely to compensate for all possible inaccuracies that stem from the microfabrication processes.

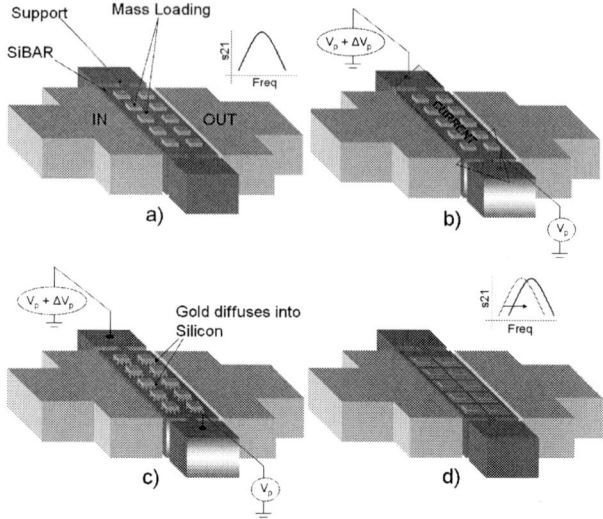

Figure 1: Schematic of electrical trimming of SiBAR using Joule heating; (a) Mass-loaded SiBAR; (b) Joule heating by passing a current through the body of the resonator; (c) Diffusion of gold into silicon; (d) Gold diffused SiBAR at room temperature shows an upward shift in frequency.

ELECTRICAL TRIMMING PRINCIPLE

In this work, a thin-film layer of gold is evaporated and patterned on the surface of the SiBAR during the fabrication process of the device. After the resonator is packaged, the SiBAR is heated up by passing a relatively large current through its resonating body during an electrical calibration step, as illustrated in Figure 1. High current densities due to the small cross-section area of the SiBAR create enough Joule heating to enable the diffusion of gold into the bulk of the silicon resonator. The advantage of gold over other metals is that gold diffuses into silicon at a much lower eutectic temperature of the silicon-gold binary system (360°C), which is very low compared to the individual melting points of gold (1064°C) and silicon (1414°C).

To calculate the temperature of a 100 MHz SiBAR for various durations of Joule heating with a given cross section area (41.5 μm × 20 μm) and resistivity (0.01 Ω-cm), the electro-thermal model based on the conservation of energy [7, 8] was used. As discussed later, a silicon beam of such dimensions can be heated to the eutectic temperature in less than five minutes by using currents of 600 mA or more. However, the maximum value of the current in the case of a SiBAR is limited by the small cross-section area of its two narrow supports (illustrated in Figure 1(a)). These supports are designed to be as narrow as possible to reduce acoustic loss and achieve high-Q, which causes increased current densities at the support regions leading to higher temperatures that can melt the supports. For the proof of concept, an optimum

978-1-4244-2977-6/09 $25.00 © 2009 IEEE   888

current of 30 mA was found to create the required Joule heating for gold diffusion without affecting the performance of the SiBAR under test and within reasonable lengths of time. Figure 2 shows the calculated temperature of the SiBAR for various durations of Joule heating at 30 mA. The glow color due to Joule heating can be seen by placing the SiBAR under an optical microscope, as shown in Figure 3. It can be seen that the glow is maximum at the support regions indicating that they heat up the most as expected. The supports can be made wider to pass higher current to reduce the duration of Joule heating, but at the cost of reducing the Q factor.

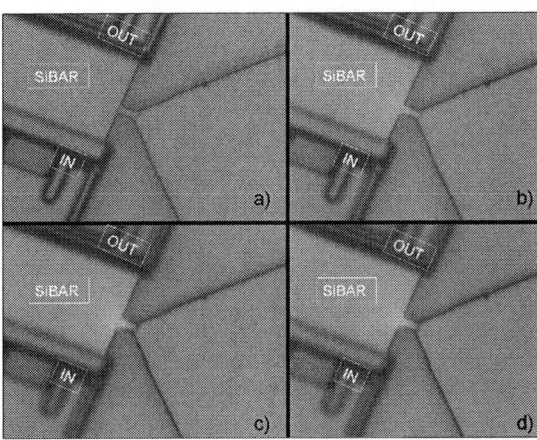

Figure 3: Optical images of Joule heating of the SiBAR at 30 mA after (a) 1 hour (b) 2 hours (c) 3 hours (d) 4 hours.

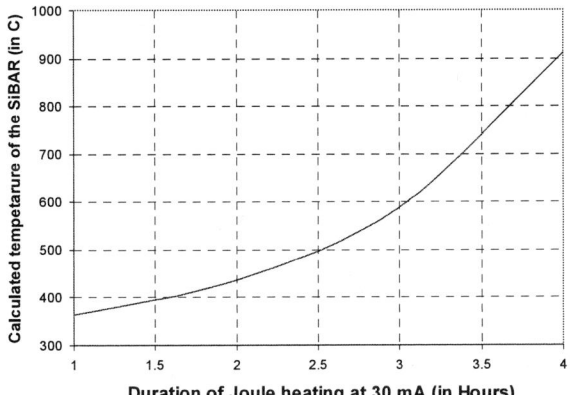

Figure 2: Calculated temperature of the 100 MHz SiBAR for various durations of Joule heating at 30 mA.

About one hour of Joule heating heats up the SiBAR to 363°C which will favor the formation of the silicon-gold eutectic. The thin mass-loading gold layer will diffuse into the bulk of the silicon at this temperature to form a eutectic alloy which has 19% silicon by atomic weight [9]. As gold atoms diffuse into silicon, they initially form a metastable gold-silicide [10, 11], wherein gold gets into the silicon interstitials, breaking Si-Si bonds and creating voids owing to its relatively larger atomic size. Upon further heating, supersaturation occurs followed by decomposition of the gold-silicide to a more stable polysilicon with intermediate voids and Au-Si bonds. The Au-Si bonds are stronger than the Si-Si bonds they replace [12]. As a result, the gold diffusion increases the stiffness (E) of the resonating silicon structure upon cooling. However, the voids introduced in silicon due to gold diffusion reduce its density (ρ). These collectively increase the acoustic velocity of the resonating structure which corresponds to a higher resonance frequency. Thus, the SiBAR can be permanently trimmed to a desired value with mass loading by gold followed by an electrical calibration step. With further Joule heating, the structure stabilizes more until no metastable structure exists.

FABRICATION

The fabrication process of the SiBAR is similar to the one reported in [1]. Trenches are etched into the device layer of a 20 μm thick SOI using an oxide mask (Figure 4(a)). These trenches define the dimensions of the SiBAR and its supports. The 100 MHz SiBARs are 41.5 μm wide and 415 μm long. The supports are 3 μm wide and 6 μm long. The capacitive gap of the SiBAR is defined by the thickness of the grown thermal oxide. In this work, a capacitive gap of 100 nm is achieved.

The trenches are refilled with doped LPCVD polysilicon and are etched back to the surface. The oxide on the surface is patterned (Figure 4(b)) to define the shape of the polarization voltage (V_p) pads. Input/Output pads are patterned through a second doped LPCVD polysilicon layer (Figure 4(c)), and the remaining silicon is etched back to the BOX layer of the SOI to isolate input/output and V_p pads. Gold is evaporated and patterned (Figure 4(d)) on the SiBAR surface using a lift-off process. Gold is deposited on silicon without any adhesive layer as that will hinder the gold diffusion into the SiBAR. Finally, the device is released in hydrofluoric acid.

ANSYS predicted a 1 MHz downshift in frequency for a blanket deposition of 150 nm thick gold layer on the entire surface of the SiBAR, referred to in this work as 100% mass loading. Every 10% mass loading results in a 100 kHz downward shift in frequency. The SEM images of the 40% and 80% mass-loaded SiBARs are shown in Figures 5 and 6. The different pattern densities of gold can be clearly seen.

RESULTS

Figure 7 shows the measured frequencies of various mass-loaded SiBARs before and after trimming via Joule heating. All devices were tested in vacuum and a DC current of 30mA was used for post-fabrication trimming of the resonators.

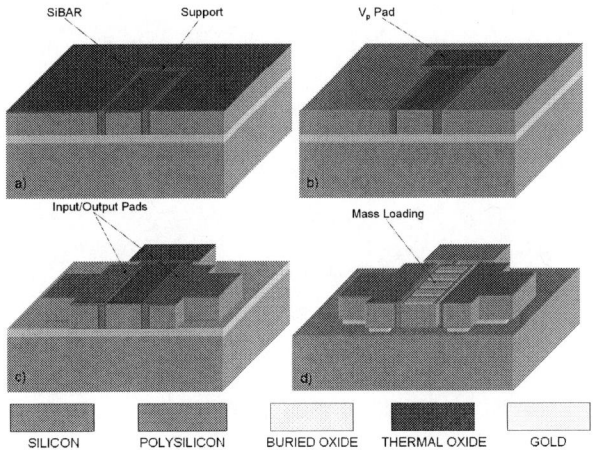

Figure 4: Fabrication process flow of mass-loaded SiBAR.

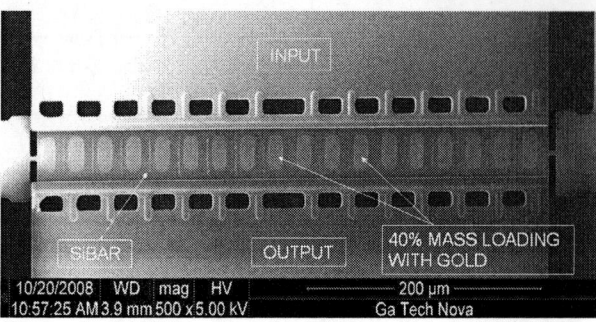

Figure 5: A SEM image of a 40% mass-loaded SiBAR.

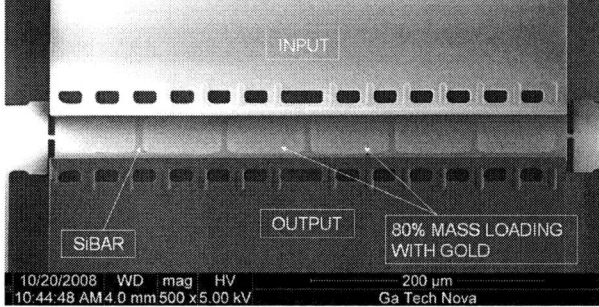

Figure 6: A SEM image of a 80% mass-loaded SiBAR.

In Figure 7, the curve labeled as 'Pre-Joule-Heating', shows the downward shift in frequency due to the mass-loading with various pattern densities of 150 nm thick gold. The 100% mass-loading offers a downward shift of 996.2 kHz in resonance frequency which is in very good agreement with ANSYS simulations. At a polarization voltage of 15 V, the unloaded SiBAR has a Q of 48'000. The mass loading lowers the Q to 23'000 in 40% and to 18'000 in 100% mass-loaded devices.

One hour of Joule heating at 30 mA shifts up the 40% mass loaded SiBAR with smaller islands of gold by 240 kHz and 80% mass loaded SiBAR with larger islands of gold by 17 kHz. This suggests that the localized heating of the SiBAR diffuses smaller islands of gold more readily than larger islands thereby showing larger frequency shifts than the later.

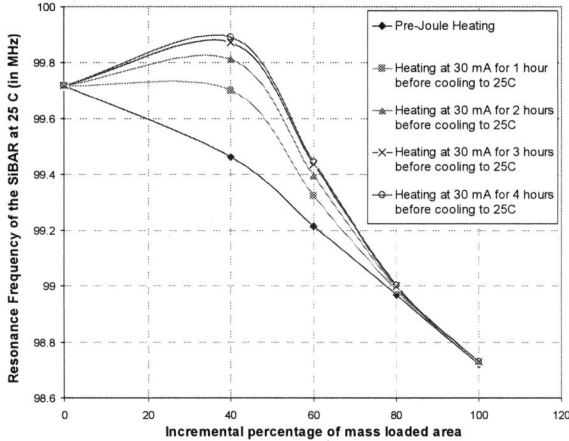

Figure 7: Measured post-fabrication electrical trimming of the SiBAR using Joule heating.

It can also be seen that, for a given mass-loaded SiBAR, the percentage increase in resonance frequency decreases with increasing durations of Joule heating. As explained earlier, the diffusion mainly occurs at the first hour of

Joule heating when the eutectic temperature is reached. Subsequent heating leads to a more stable resonating structure which will correspond to smaller frequency shifts. Hence, four hours of Joule heating shifts up the 40% and 80% mass loaded SiBAR by only 430 kHz and 35 kHz respectively. A shift of 430 kHz over four hours corresponds to a trimming rate of approximately 2 kHz per minute, which makes very precise and controlled electrical trimming possible.

The 40% mass loaded SiBAR is designed to give a resonance frequency of 99.6 MHz (i.e., a downshift of 400 kHz from 100 MHz). But it can be seen that variations in the SiBAR fabrication and also in the thickness of the deposited gold offsets the resonance frequency to 99.46 MHz. The electrical trimming time needed to shift up this frequency to the designed 99.6 MHz can be calculated to be 35 minutes from Figure 7. Thus, all variations of the SiBAR fabrication can be compensated successfully. From Figure 7, it can also be seen that the 40% mass loaded SiBAR exceeds the resonance frequency of unloaded SiBAR with longer hours of Joule heating. This suggests the formation of a structure with stronger Au-Si bonds and less dense packing with voids, to provide a higher acoustic velocity than crystalline silicon.

After electrical trimming, these devices were taken through temperature cycling by heating them in an oven to 85°C for 6 hours and back to room temperature. No frequency hysteresis was observed, confirming the temperature stability of the trimmed resonator. Although the mass loading reduces the Q of a SiBAR from its unloaded pure-silicon value, Figure 8 shows that the Q increases slightly with longer hours of Joule heating.

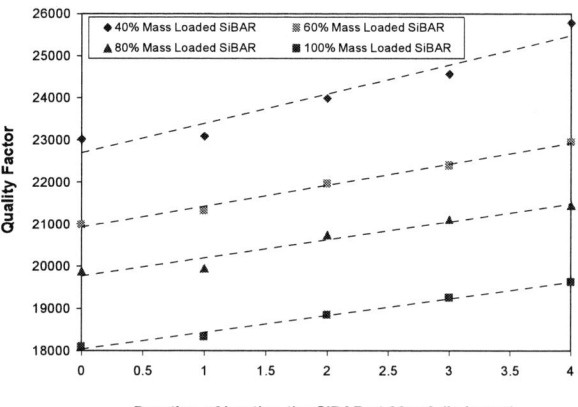

Figure 8: Measured increase in the Q of the mass loaded SiBAR for increasing duration of Joule heating.

DISCUSSIONS

In applications where the silicon resonators are designed with wider supports, the Joule heating can be performed with a higher current, and thereby reducing the frequency trimming time. If the silicon resonator is assumed to have to have a uniform cross section area (41.5 μm × 20 μm) and resistivity (0.01 Ω-cm) as presented in this work, but without any narrow support regions, then the duration of Joule heating for such a silicon resonator at various high currents and the corresponding temperature of the silicon beam [7, 8] is shown in Figure 9.

978-1-4244-2977-6/09 $25.00 © 2009 IEEE   890

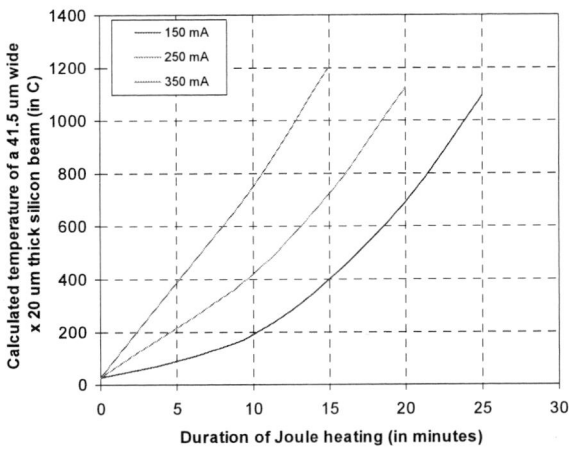

Figure 9: Calculated temperature of a 41.5 μm wide and 20 μm thick silicon resonator for various duration of Joule heating at different currents.

Using such higher currents, the eutectic temperature can be reached within 5 to 15 minutes of Joule heating. It can be seen from Figure 10 that the glow color of such a silicon beam due to Joule heating at 600 mA can be seen within one minute. The beam glows bright red after 5 minutes of heating, after which it melts. These suggest the possibility of performing electrical trimming of silicon micromechanical resonators within few minutes for resonators with wider support elements.

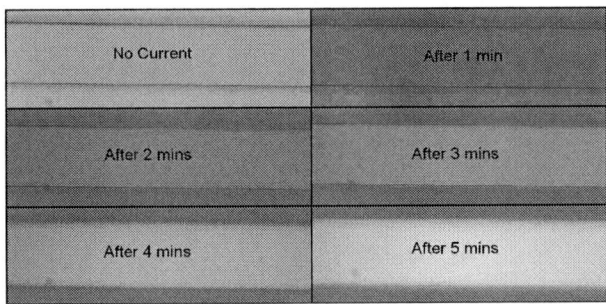

Figure 10: Optical images of Joule heating of a 41.5 μm wide and 20 μm thick silicon beam for various durations of Joule heating at 600 mA.

CONCLUSION

Post-fabrication electrical trimming of SiBARs has been demonstrated. The SiBARs were mass loaded with different pattern densities of gold to achieve different downward shift in the resonance frequency. Using Joule heating, the gold patterns were diffused into the resonating silicon structure, which upon cooling exhibits an increased stiffness. This reflects as an upward shift in the resonance frequency. An upward shift of up to 240 kHz at 99.45 MHz has been demonstrated with one hour of Joule heating at 30 mA. This electrical trimming technique offers the possibility to compensate for all the variations in the microfabrication processes without compromising on the performance of the resonator. The possibilities of reducing the electrical trimming durations down to few minutes have also been discussed.

ACKNOWLEDGEMENTS

Authors would like to thank the staff at the Georgia Tech Microelectronics Research Center (MiRC) for their assistance. This work was supported by DARPA under the Analog Spectral Processors (ASP) Program.

REFERENCES

[1] S. Pourkamali, G. K. Ho and F. Ayazi, "Low-Impedance VHF and UHF Capacitive Silicon Bulk Acoustic Wave Resonators—Part I: Concept and Fabrication", *IEEE Trans. Electron Devices,* pp. 2017-2023, 2007.

[2] H. M. Lavasani, A. K. Samarao, G. Casinovi and F. Ayazi, "A 145 MHz Low Phase-Noise Capacitive Silicon Micromechanical Oscillator", *IEEE Intl. Electron Devices Meeting 2008.*

[3] G. Casinovi, X. Gao and F. Ayazi, "Analytical Modeling and Numerical Simulation of Capacitive Silicon Bulk Acoustic Resonators", *IEEE Intl. Conf. on Microelectromech. Syst. 2009.*

[4] K. Sundaresan, G. K. Ho, S. Pourkamali and F. Ayazi, "Electronically Temperature Compensated Silicon Bulk Acoustic Resonator Reference Oscillators", *J. Solid State Circuits*, pp. 1425-1434, 2007.

[5] C. G. Courcimault and M. G. Allen, "High-Q Mechanical Tuning of MEMS Resonators using a Metal Deposition - Annealing technique", *Transducers 2005*, pp. 875-878.

[6] W- T. Hsu, A. R. Brown, "Frequency Trimming of MEMS Resonator Oscillators", *IEEE Intl. Freq. Control Sym. 2007*, pp 1088-1091

[7] L. Lin, "Selective encapsulations of MEMS: Micro channels, needles, resonators, and electromechanical filters", Ph.D. dissertation, Dept. Mech. Eng., Univ. California at Berkeley, Berkeley, CA, 1993.

[8] Y. T. Cheng, L. Lin and K. Najafi, "Localized Silicon Fusion and Eutectic Bonding for MEMS Fabrication and Packaging", *J. Microelectromech. Syst.*, vol. 9, pp. 3-8, 2000.

[9] H. Okamoto and T. B. Massalski: *Binary Phase Diagrams* (1986), pp. 312.

[10] B. Bokhonov and M. Korchagin, "In Situ Investigation of Stage of the Formation of Eutectic Alloys in Si-Au and Si-Al Systems", *J. Alloys and Compounds*, vol. 312, pp. 238-250, 2000.

[11] S. Takeda, H. Fujii, Y. Kawakita, S. Tahara, S. Nakashima, S. Kohara and M. Itou, "Structure of eutectic alloys of Au with Si and Ge", *J. Alloys and Compounds*, vol. 452, pp. 149-153, 2008.

[12] L. Lin, Y. T. Cheng and K. Najafi, "Formation of Silicon-Gold Eutectic Bond Using Localized Heating Method", *Japanese J. Applied Physics*, vol. 37, pp. 1412-1414, 1998.

RF MEMS HIGH-IMPEDANCE TUNEABLE METAMATERIALS FOR MILLIMETER-WAVE BEAM STEERING

M. Sterner[1], D. Chicherin[2], A.V. Räisenen[2], G. Stemme[1], J. Oberhammer[1]

[1]KTH – Royal Institute of Technology, Stockholm, Sweden

[2]TKK – Helsinki University of Technology, Helsinki, Finland

ABSTRACT

This paper presents the design, fabrication and evaluation of RF MEMS analog tuneable metamaterial high-impedance surfaces (HIS). These miniaturized structures are designed for W-band beam steering applications and are intended to replace a large multi-component subsystem by a single chip. Furthermore, the MEMS tuneable microwave metamaterials of this paper present a new class of microsystems interacting with microwaves, by uniquely combining the functionality of the microwave structures with the tuning MEMS actuators in one and the same distributed surface elements. A high-impedance surface array with 200×52 elements and a pitch of 350 µm has been successfully fabricated and evaluated. The device features monocrystalline silicon membranes which are transfer-bonded on a multi-wafer silicon-glass substrate. The measured pull-in voltage is 15.9 V. Microwave measurements from 70 GHz to 114 GHz confirm the frequency selective nature of the surface. The fabricated devices showed a resonance frequency of 111.3 GHz to 111.8 GHz with losses ranging from -18 dB to -23 dB at the resonance and from -5 dB to -7 dB outside the resonance, which is worse than theoretically predicted but mainly attributed to imperfections in the design and fabrication of the first prototypes.

INTRODUCTION

Millimeter-wave signals, having for decades been used in radio astronomy [1], are currently of increasing interest in high-capacity communication links, spectroscopy, medical diagnostics and radar, for different reasons such as bandwidth and resolution demands. The size of micro-machined MEMS structures is very suitable for designing new types of millimeter-wave devices, because the typical micro-to-millimeter lateral and vertical dimensions in microfabrication are in the same order of magnitude as the signal wavelength.

Electronic beam steering is typically achieved by creating a phase gradient over an emitting or reflecting surface by an array of phase shifters [2]. At millimeter wavelength, conventional phase shifters are unsuitable for such phased arrays due to large losses (e.g. solid-state diodes) or too large size (e.g. switched delay lines).

Metamaterial surfaces are artificial periodic patterns with surface properties not available in nature. High-impedance metamaterial surfaces (Fig. 1), which exhibit unnaturally high surface impedance at their resonance frequency, have attracted attention because of their promising applications in improvement of antenna radiation patterns, suppression of surface waves [3] and phase shifting [4]. The use of diode varactor tuned high-impedance surfaces has been demonstrated for electronic beam steering [5]. However, such beam steering would not work at millimeter wavelength due to high losses in the solid-state components and since the small wavelengths make it difficult to integrate discrete components.

RF MEMS has been successfully used to electromechanically reconfigure conventional RF structures with low losses [6]. Such reconfigurability is commonly implemented by actuators reconfiguring separated RF structures, for instance using discrete switches to select different signal paths [7].

In this paper we use RF MEMS to tune high-impedance surfaces designed for millimeter-wave beam steering. The design, fabrication and evaluation of microwave MEMS analog tuneable metamaterial high-impedance surfaces (HIS) is presented. The W-band beam steering applications of these structures include $76 - 81$ GHz automotive radar and 94 GHz to beyond-100 GHz reconfigurable point-to-point communication links. Using a high-impedance surface array potentially enables the replacement of a multi-component beam steering subsystem—encompassing for example power-splitters, switches, phase shifter line arrays, antenna arrays—by a single reflective chip.

Furthermore, the structures of this paper uniquely unify electromechanical tunability and microwave functionality in one and the same distributed high-impedance surface elements, rather than conventional reconfiguration of separated elements, thus presenting a new way of microsystems interacting with microwaves.

CONCEPT AND DESIGN OF TUNEABLE HIGH-IMPEDANCE SURFACES

At their resonance frequency, high-impedance surfaces (Fig. 1) have an effective surface impedance which is approaching $\pm j\infty$ and which is thus very large as compared to the free space impedance. This results in the phase of the reflection coefficient having a steep transition between $+180°$ and $-180°$, whereas its amplitude ideally remains constant. For a given signal frequency, this can be utilized to create a tuneable phase-shift, if the resonance frequency of the high-impedance surface is tuneable, as shown in (Fig. 2). The authors have previously proposed such tuning of multilayer high-impedance surfaces for direct beam-

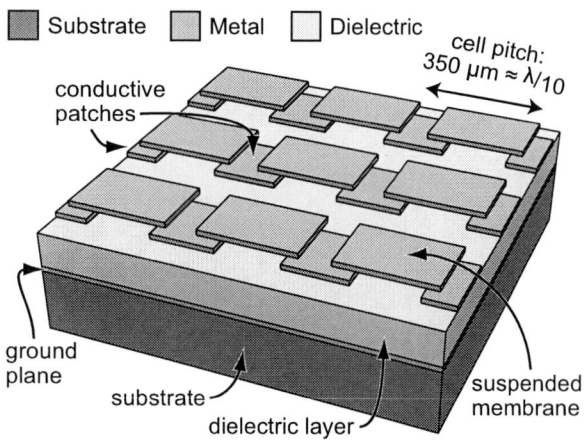

Figure 1: Illustration of 3×3 elements of a periodic high-impedance metamaterial surface.

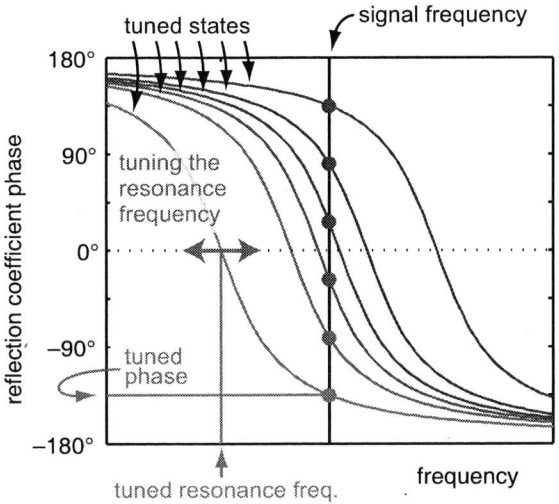

Figure 2: The phase of the reflection coefficient of the high-impedance surface is tuned by tuning the resonance frequency, translating the reflection phase curve to achieve a certain tuned phase at the signal frequency.

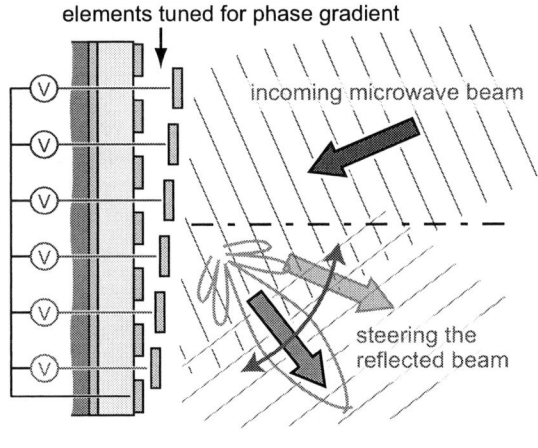

Figure 3: Beam steering by tuning of individual elements to create a gradient in the reflection phase along the surface. By controlling the gradient, the angle of the reflected beam is determined.

steering by creating a gradient in the phase of the reflection coefficient along the surface (Fig. 3) [8].

The high-impedance surfaces presented in this paper are composed of an array of electrostatically tuneable elements. Each element consists of a vertically moveable conductive membrane and conductive patches on the surface of a ground-backed 100 μm thick low-loss glass substrate (Fig. 4). A monocrystalline silicon core is used for the membrane to provide mechanical robustness and for optimizing the membrane flatness. The membrane is connected by meander shaped springs to supporting metal posts. The initial distance between the membrane and the patches is designed to 2 μm and the total stiffness is 8.3 N/m which gives a theoretical tuning distance of 0.6 μm and a pull-in voltage of 15.4 V.

The resonance frequency of the high-impedance surface is determined by the element's geometry and by the capacitive coupling between the membrane and the patches. The designed resonance frequency is 99 GHz. Fig. 5 shows HFSS simulation data on tuning the phase of the reflection coefficient by tuning the membrane gap.

The conductive patches and membranes are used both as actuation electrodes and as part of the microwave resonance circuit, thus combining the microwave and actuator functionality in the same elements. The distance between the two electrodes of a tuneable electrostatic actuator fits very well to the capacitive distance of a high-impedance surface element in the W-band.

FABRICATION

The fabrication is shown in Fig. 6. The 100 μm thin microwave glass substrate is bonded to a metal covered silicon wafer by Benzocyclobutene (BCB) adhesive wafer bonding at 250°C and 4 bar for 1 h. Since a low-loss microwave material (Schott AF45) with a coefficient of thermal expansion (CTE) non-matching to silicon must be used, a second glass wafer must be counter-bonded below the silicon substrate for stress compensation. After fabricating the metal patches on top of the wafer stack (Fig. 6c), the 1 μm thick monocrystalline-silicon device layer of a SOI wafer, which is covered by a 500 nm thick gold metallization layer, is polymer transfer-bonded to the wafer stack using a sacrificial polymer (MR-I 9150XP) at 200°C and 4 bar for 1 h [9]. The handle wafer of the SOI wafer is subsequently removed by Reactive-Ion-Etching (Fig. 6f). After adding another 500 nm thick gold metallization layer, the multilayer membrane is structured by plasma and wet etching steps. Mechanical support is created by electroplating of posts, and the membranes are finally released by etching the sacrificial bonding polymer in O_2 plasma (Fig. 6i). The dicing of the wafer is done before the final release-etch.

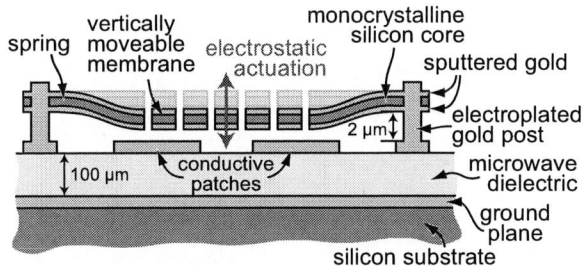

Figure 4: A single tuneable high-impedance element consisting of a conductive membrane, patches and a 100 μm thick, ground-backed dielectric layer.

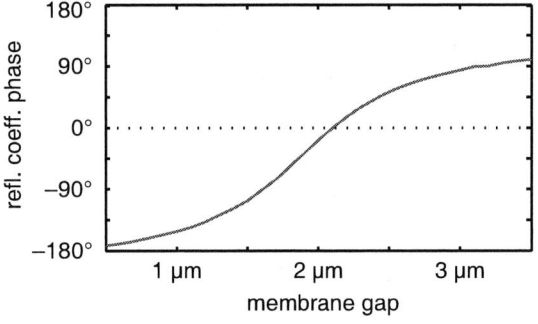

Figure 5: Electromagnetic simulation (HFSS) of the phase of the reflection coefficient depending on the gap between the electrodes and the vertically moving membrane.

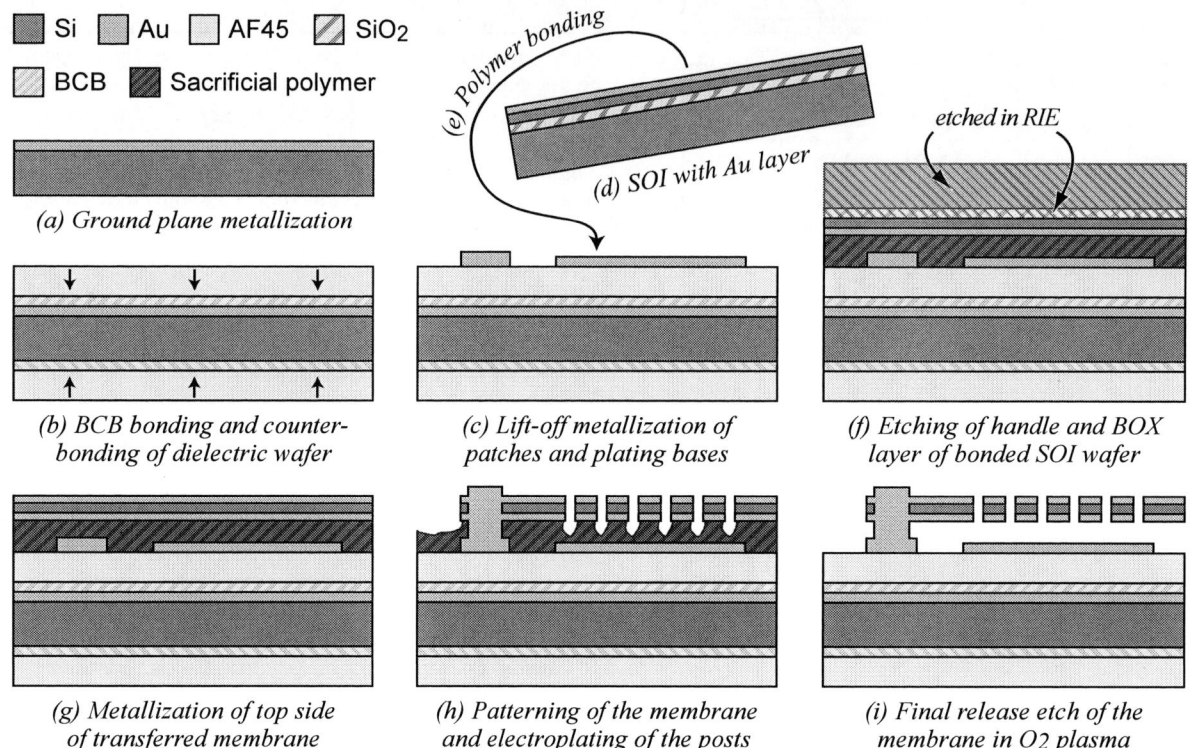

(a) Ground plane metallization

(b) BCB bonding and counter-bonding of dielectric wafer

(c) Lift-off metallization of patches and plating bases

(d) SOI with Au layer

(e) Polymer bonding

(f) Etching of handle and BOX layer of bonded SOI wafer

etched in RIE

(g) Metallization of top side of transferred membrane

(h) Patterning of the membrane and electroplating of the posts

(i) Final release etch of the membrane in O2 plasma

Si   Au   AF45   SiO2
BCB   Sacrificial polymer

Figure 6: Process flow showing the wafer-scale multi-wafer fabrication process for the high-impedance surface. The transferred monocrystalline membrane layer provides mechanical robustness and flatness. Counter-bonding of a second dielectric substrate (b) compensates for CTE mismatch. The sacrificial polymer layer defines the initial membrane gap after the release etch. The electroplated posts provide mechanical support.

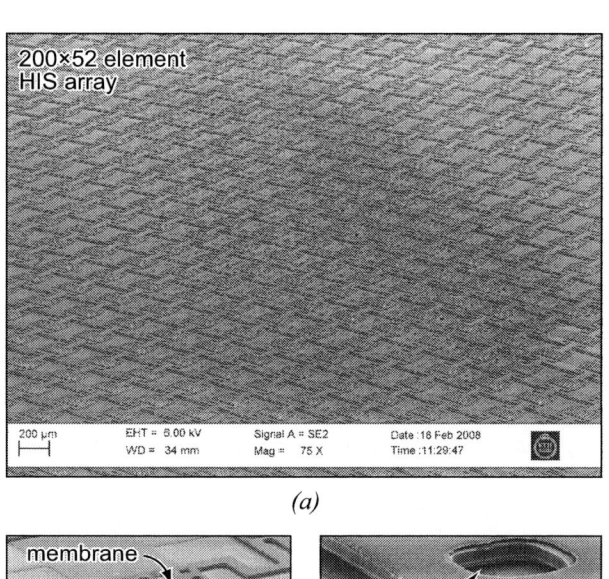

Figure 7: SEM pictures of a fabricated high-impedance surface array – (a) a partial view of the large-scale array, (b) the basic high-impedance surface element, (c) a closeup view of the membrane.

SEM pictures of a large-scale high-impedance surface with 200×52 array elements with a pitch of 350 µm resulting in a total size of 70×18 mm^2 are shown in Fig. 7.

MEASUREMENT RESULTS

Actuator characterization

The mechanical tuning range of the membranes has been determined by a WYKO NT1100 whitelight-interferometric profilometer to 21 % of the full deflection (Fig. 8a). The measured pull-in voltage of 15.9 V corresponds well to the 15.4 V determined by 3D electromechanical FEM simulations in COMSOL Multiphysics. The membrane non-uniformity is only 84 nm on a membrane length of 210 µm. The slight curvature is believed to be attributed by imperfections in the transfer bonding and asymmetric irregularities in the patterning of the metal coating, and is currently being improved by optimising the fabrication. The roughness (R_a) of the sputter metal coating is less than 3 nm.

Microwave characterization

The microwave properties of the fabricated high-impedance surfaces are evaluated from 70 GHz to 114 GHz with a HP 8510 VNA by placing the device as a back-short termination of a rectangular WR-10 waveguide. The phase and amplitude of the measured reflection coefficient are shown for three structures in Fig. 9. The resonance frequencies are between 111.3 GHz and 111.8 GHz and the characteristic steep phase transition of over 245° clearly confirms the frequency-selective high-impedance surface behavior of the device. The losses of the first prototypes

978-1-4244-2977-6/09 $25.00 © 2009 IEEE

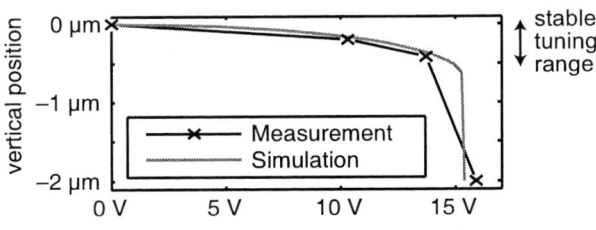

(a) Actuation characterization graph

(b) Height map of a high-impedance element

Figure 8: *Interferometric measurements of membrane deflection show pull-in behavior, which corresponds well to FEM simulations. The double-side metal coated membrane has very low roughness (< 3 nm).*

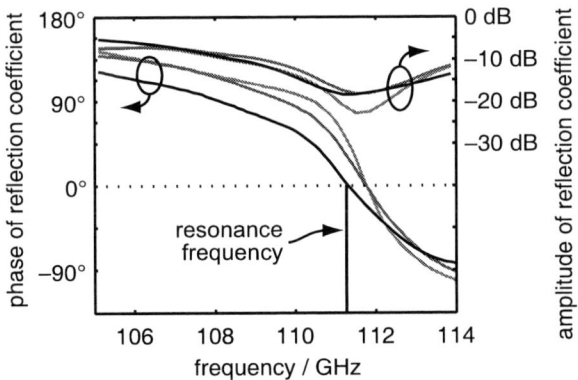

Figure 9: *Backshort measurements (70 − 114 GHz) of the reflection coefficient show the characteristic steep phase transition around the resonance frequency.*

range from −18 dB to −23 dB at the resonance and from −5 dB to −7 dB outside the resonance, which is worse than theoretically predicted and is attributed to imperfections in the current design and the prototype fabrication, along with the difference between the design and the measured resonance frequency.

CONCLUSIONS

We have successfully designed and fabricated a large-scale microwave MEMS analog tuneable high-impedance surface array and characterized its actuation and microwave performance. The measurements clearly show the phase transition at the resonance frequency as predicted for the micromachined high-impedance surface, the tuneability of the actuator and thus its suitability for tuning the phase over the array.

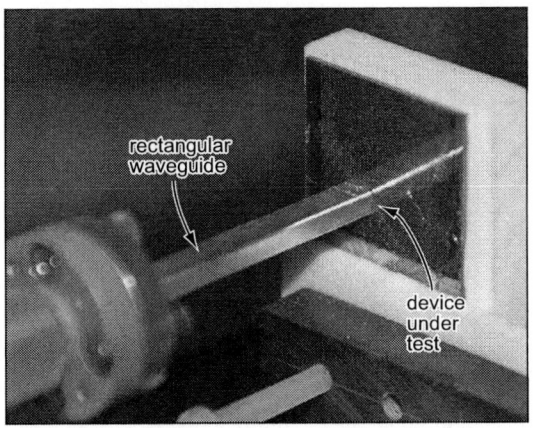

Figure 10: *Photograph of the measurement setup with the high-impedance surface terminating a rectangular waveguide in back-short configuration.*

REFERENCES

[1] R. Blundell and C.-y. Tong, "Submillimeter receivers for radio astronomy," *Proceedings of the IEEE*, vol. 80, no. 11, pp. 1702–1720, Nov 1992.

[2] C. Chekroun, D. Herrick, Y. M. Michel, R. Pauchard, and P. Vidal, "Radant - New method of electronic scanning," *L'Onde Electrique*, vol. 59, pp. 89–94, February 1981.

[3] D. Sievenpiper, "High-impedance electromagnetic surfaces," Ph.D. dissertation, Dept. Elect. Eng., Univ. of California, Los Angeles, 1999.

[4] J. Higgins, H. Xin, A. Sailer, and M. Rosker, "*Ka*-band waveguide phase shifter using tunable electromagnetic crystal sidewalls," *IEEE Trans. Microw. Theory and Techniques*, vol. 51, no. 4, pp. 1281–1288, April 2003.

[5] D. Sievenpiper, J. Schaffner, B. Loo, G. Tangonan, R. Harold, J. Pikulski, and R. Garcia, "Electronic beam steering using a varactor-tuned impedance surface," *IEEE Antennas and Propagation Society International Symposium*, vol. 1, pp. 174–177, 2001.

[6] L. E. Larson, R. H. Hackett, M. A. Melendes, and R. F. Lohr, "Micromachined microwave actuator (MIMAC) technology-a new tuning approach for microwave integrated circuits," in *IEEE Microwave and Millimeter-Wave Monolithic Circuits Symposium*, 1991, pp. 27–30.

[7] B. Schoenlinner, A. Abbaspour-Tamijani, L. Kempel, and G. Rebeiz, "Switchable low-loss RF MEMS Ka-band frequency-selective surface," *IEEE Transactions on Microwave Theory and Techniques*, vol. 52, no. 11, pp. 2474–2480, November 2004.

[8] D. Chicherin, S. Dudorov, D. Lioubtchenko, V. Ovchinnikov, and A. Räisänen, "Millimetre wave phase shifters based on a metal waveguide with a MEMS-based high-impedance surface," in *Proceedings of the 36th European Microwave Conference*, September 2006, pp. 372–375.

[9] M. Populin, A. Decharat, F. Niklaus, and G. Stemme, "Thermosetting nano-imprint resists: novel materials for adhesive wafer bonding," in *Proc. IEEE 20th International Conference on Micro Electro Mechanical Systems (MEMS)*, 2007.

978-1-4244-2977-6/09 $25.00 © 2009 IEEE

MONOLITHICALLY INTEGRATED PIEZOMEMS SP2T SWITCH AND CONTOUR-MODE FILTERS

J.S. Pulskamp[1], D.C. Judy[1], R.G. Polcawich[1], R. Kaul[1], H. Chandrahalim[2], and S.A. Bhave[2]
[1]US Army Research Laboratory, Adelphi, MD, USA
[2]Cornell University, Ithaca, NY, USA

ABSTRACT

This paper provides the first experimental demonstration of monolithically integrated piezoelectric MEMS RF switches with contour mode filters. Lead zirconate titanate (PZT) thin films are utilized to enable both low-voltage switch operation and filter tunability. This research leverages previous work using PZT actuators for low-voltage, wide-band switches and PZT transduced silicon resonators. The two device technologies are combined using a hybrid fabrication process that combines the key components of each device fabrication into a single unified process using silicon-on-insulator (SOI) substrates. The voltage tunable and switchable PiezoMEMS filter array provides a drop-in solution for frequency-agile channel selectivity.

INTRODUCTION

RF MEMS has been a significant area of research for well over a decade due to the promise of improved performance and integration potential in commercial and military wireless communication and radar systems. RF MEMS switches have demonstrated superior performance in terms of insertion loss, isolation, power consumption, and linearity [1]. RF MEMS filter technologies can provide insertion loss, percent bandwidth, and rejection performance similar to off-chip crystal filters and surface acoustic wave (SAW) devices with the compact integration of multiple frequencies on the same chip [2-4]. The integration of these two technologies has long been a goal of researchers and will enable not only more compact and lower cost systems but previously unachievable signal processing functions [5].

A number of transduction approaches have been utilized for both MEMS switch and filter operation, including electrostatic, electromagnetic, electro-thermal, and piezoelectric. Piezoelectric and or ferroelectric transduction can provide superior electro-mechanical coupling and mechanical energy densities; metrics applicable to both switch and filter operation. Piezoelectric AlN is a popular filter material due to the high acoustic velocity, high mechanical quality factor, and the ease of post-CMOS integration [3]. AlN is generally the preferred material for direct piezoelectric effect transduction due to its favorable ratio of stress constant to dielectric constant.

Ferroelectric lead zirconate titanate (PZT) has an effective piezoelectric stress constant that is an order of magnitude larger than AlN and is generally the preferred material for indirect piezoelectric effect transduction [6]. Thin films of ferroelectric PZT also permit tuning of the piezoelectric coefficients, some elastic constants, permittivity, and hence the electro-mechanical coupling factor of the material with a modest DC bias. Low-voltage (<10 V) ohmic contact series switches with high isolation and good insertion loss characteristics up

through 65 GHz have been demonstrated with surface micro-machined PZT MEMS [7]. Recently, PZT and PZT-on-SOI resonators and filters have shown low motional impedance (<50Ω) and good quality factor (2000-5100) with superior frequency tunability (5.1% and 0.2% respectively) [8]. The integration of PZT thin films with high mechanical quality factor single crystal silicon features allows a tradeoff between steep-walled narrow-bandwidth filters, low-Q wide-bandwidth filters, linearity and center frequency agility. Single pole dual throw (SP2T) PZT MEMS switches have also been previously demonstrated with better than 50 dB of isolation and less than 0.4 dB of insertion loss at 2 GHz [9]. Unlike AlN, integration of high quality piezoelectric PZT films with CMOS remains a challenge for MEMS applications because of the high process temperatures. However, PZT thin films with high quality ferroelectric properties have been successfully integrated with CMOS for ferroelectric random access memory (FeRAM) [10].

Recently, Sinha et. al. successfully co-fabricated aluminum nitride (AlN) switches and resonators and reported cascaded S-parameter data illustrating how a switchable resonator should function [11]. In this paper we report on the integration of a PZT RF MEMS single pole dual throw (SP2T) series switch with contour mode mechanically coupled PZT-on-SOI high quality factor filters and the first experimental demonstration of monolithically integrated piezoelectric MEMS RF switches and filters.

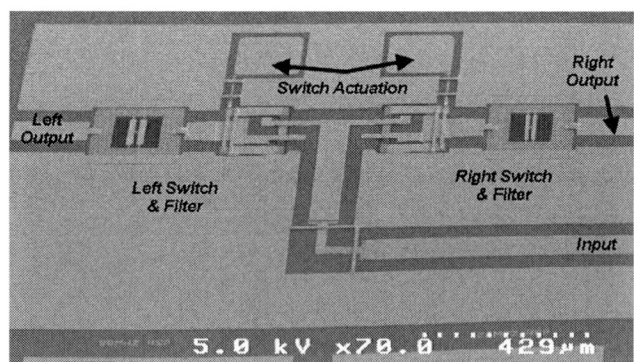

Figure 1: SEM of an integrated single pole dual throw PZT MEMS switch and two PZT contour mode filters.

DESIGN

Using a single pole dual throw (SP2T) architecture similar to that reported previously [8], two normally-open, ohmic contact, series switches were used to select between two contour mode mechanically coupled PZT-on- SOI high quality factor filters (see Fig. 1). As seen in Fig. 1, a conductive pad and contacts are located on the dielectric structure mechanically coupling the two cantilevered PZT unimorph actuators. The switch resides in the gaps of the co-planar waveguide (CPW)

978-1-4244-2977-6/09 $25.00 © 2009 IEEE

transmission line. Unlike the cantilevered designs in [7,9], the RF gold air bridge contact structures above the switch contact pad were redesigned as clamped-clamped structures to mitigate deformation of these structures during and after fabrication. Two bias line air bridges located at the anchors of the actuators electrically connect the top and bottom electrodes of the two actuators. To enable switch biasing with a single trace, the top electrode bias line air bridge anchors to one of the CPW ground planes and the bottom electrode bias line air bridge anchors to a single bias line. The typical switch actuator composite stress states, composition, and thicknesses are designed to provide static negative curvature in the switch to dictate the preferred initial contact gap. The application of voltage to the top and bottom electrodes provides d_{31} mode bending actuation to the switch and raises the contact pad into contact with the two clamped-clamped RF gold air bridges. These devices were designed with analytical models, ANSYS, ADS Momentum, and HFSS. The Ohmic series switches and SP2T design are similar to the designs presented in [7,9].

The filter designs in this work are based on the work presented in [8]. The designs in this work feature significantly reduced parasitic shunt capacitances by reducing the top electrode contact area, not associated with the active transducer, through the use of air bridge structures available with the switch process steps (See Fig. 2b). The constituent resonators are fundamental width-extensional contour mode designs. Two of these resonators are coupled via an acoustic quarter-wave coupling spring (not simulated) to create a two-pole mechanically coupled filter (See Fig. 3). The filters presented in this paper were designed as 209MHz and 313MHz. The fundamental resonant frequency of the width extensional mode is given by:

$$f_o = \frac{n}{2W}\sqrt{\frac{Y_{eff}}{\rho_{eff}}} \qquad (1)$$

where W is the width of the resonator, Y_{eff} and ρ_{eff} are elastic modulus and mass density of the composite resonator respectively, and n is the harmonic order. The bandwidth (BW) of such a mechanically-coupled filter is given by:

$$BW = \frac{f_o}{k_{ij}}\frac{k_s}{k_r} \qquad (2)$$

where f_o is the resonant frequency, k_s and k_r are the spring stiffness of the coupling spring and resonator respectively and k_{ij} is the filter coefficient [12].

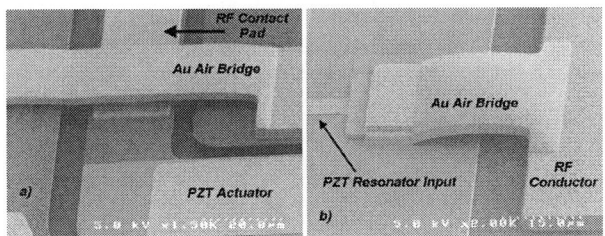

Figure 2: (a) SEM image of the air bridges used for switch operation and (b) reduced parasitics in the resonators.

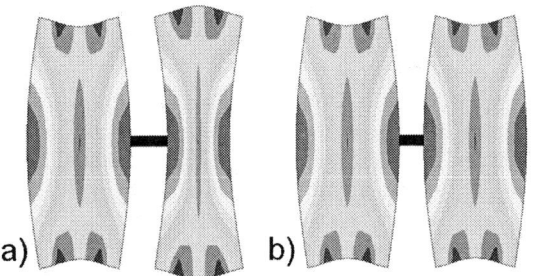

Figure 3: ANSYS mode shape of a fundamental width-extensional mode filter. (a) Anti-symmetric mode, (b)Symmetric mode

Relatively few modifications were required to integrate the two device types as most of the steps of the two individual processes were initially common. The substrate was chosen as SOI to accommodate the filters. The switch sacrificial layer features were altered to ensure proper release timing in the presence of the added passivation layer necessary to protect the device silicon features of the filter during the XeF$_2$ release and the release sequence was reversed to accommodate the integration.

FABRICATION

Device fabrication was done at the Specialty Electronic Materials and Sensors Cleanroom Facility, U.S. Army Research Laboratory, Adelphi, MD. The fabrication process for the monolithically integrated piezoelectric MEMS RF switches and filters utilized a silicon-on-insulator substrate with a 5 μm thick device layer with nominal resistivity of 30 Ohm-cm. A 5000 Å silicon dioxide thin-film was deposited by plasma-enhanced chemical vapor deposition (PECVD) and was followed by a sputtered Ti/Pt bottom electrode for the PZT deposition. The 5185 Å PZT thin films were prepared via a chemical solution derived deposition process modified from that outline in [13]. After the final PZT anneal, a 1050 Å platinum thin film was sputter deposited directly onto the PZT surface at 300°C.

The switch actuator and filter drive and sense electrode were patterned with the argon ion-milling of the top platinum layer and was followed by an additional ion-milling of the PZT and bottom electrode features. A wet etch then opened up contact vias to the local bottom electrodes of the switches and filters. The switch structure was then further defined by patterning the silicon dioxide layer with a reactive ion etch to provide access to the silicon device layer for the eventual release etch. A titanium / gold bi-layer was then deposited with electron beam evaporation and patterned via liftoff to define the CPW transmission line, contact structures for the switch, and anchor features required for gold air bridges. A lift-off process was then used to pattern the switch contact material. The filters were defined using a single photomask by an ion-milling of the PZT and bottom electrode, an RIE of the silicon dioxide layer, a DRIE of the device silicon layer, followed by an RIE of the buried oxide layer. A photo-resist sacrificial layer was then patterned and cured and was followed by the deposition and lift-off of 2 μm gold air-bridge features necessary for both the switches and filters. A thick photo-resist layer was then patterned to encase the silicon resonator features in resist and the buried oxide during release. The sidewalls of the switches and regions near the filters were not coated to permit a timed XeF_2 etch of the silicon device layer directly beneath the switch and the silicon handle layer beneath the filter. The release process reversed the typical release sequence utilized in [6], with the XeF_2 silicon etch preceding the oxygen plasma release of the gold air-bridge structures.

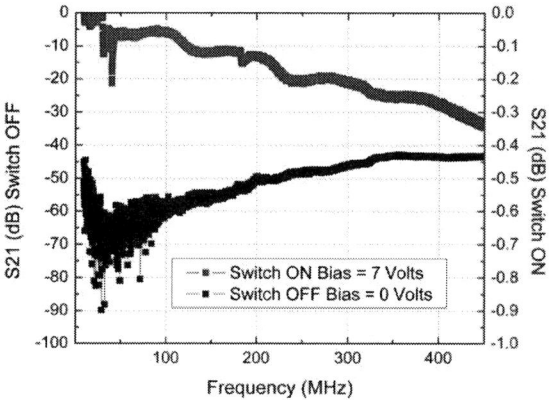

Figure 4: S_{21} data at 0 and 7 volts for the PZT MEMS switch co-fabricated with the PZT resonators.

EXPERIMENTAL RESULTS

All switch and filter measurements were completed with a network analyzer using a 50 Ohm termination. Individual switches exhibited isolation better than -40 dB from DC through 500 MHz (see Fig. 4). Switch actuation was achieved with the application of 7 volts resulting in an insertion loss less than -0.4 dB from DC to 500 MHz. Processing these devices on a relatively low resistivity SOI wafer resulted in the insertion loss being higher than the individual switches processed on silicon substrates with a resistivity greater than 10 kOhm-cm in previously reported research [6].

With both switches in the offstate (i.e. 0 volts applied), the output from both the left or right filters exhibit a flat S21 response with a magnitude of -60 dB (see Fig. 5a). This is somewhat surprising since a simple linear cascade of S parameters of the switch and resonator should show the filter response shifted down by the individual switch isolation as in [7]. Although this curiosity is still under investigation, we believe it is most likely due to current shunting around the open switch effectively reducing its isolation. The exclusion of air bridges at the SP2T junctions is a likely source of this reduced isolation.

As each switch was actuated at 7 volts, the associated filter response was detected at the relevant output port for each filter (see Fig. 5b and 5c). For both filters, an improvement in the insertion loss was observed with the application of a DC voltage applied to the RF signal through the bias tees of the network analyzer. The impedance tuning agrees with the previous observation of Chandrahalim et al. [7]. The filters exhibited out-of-band rejection of nearly -30 dB and a 50Ω terminated insertion loss of -17 dB and -24 dB (as measured from the minimum insertion loss) with the application of a 10 volt DC bias (see Fig. 5b and 5c). Similar filter responses were observed in the integrated (add 0.4 dB insertion loss) as in the standalone filters fabricated on the same wafer (See Fig 6). The similar performance between individual filters and filters integrated with SP2T switches suggest obtaining improved device performance is a matter of updating the filter design, which is currently in progress.

A series of time domain measurements were used to examine the switching and filter characteristics as a function of switch cycling. As shown in Fig. 7, the output response from the integrated switch and filter is ringing up with each switching pulse. It is unclear whether this is due to ring up of the filter or is influenced by the electrical time constant due to the switch contact resistance. Similar to previous PZT switches, the switch cycle lifetime is expected to last in excess of tens of millions of switch cycles and improve with more suitable contact materials, switch optimization, and integrated packaging.

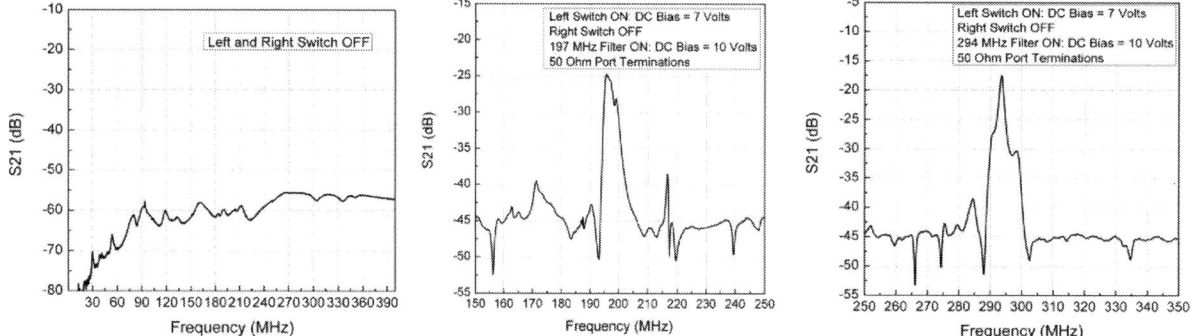

Figure 5: (a) S_{21} response for an integrated SP2T switch and filter with the switch in the off-state (0 V), (b) S_{21} response for the left switch and filter with the switch on (7 V) and with 10 V_{DC} applied to the filter, and (c) S_{21} response for the right switch and filter

978-1-4244-2977-6/09 $25.00 © 2009 IEEE

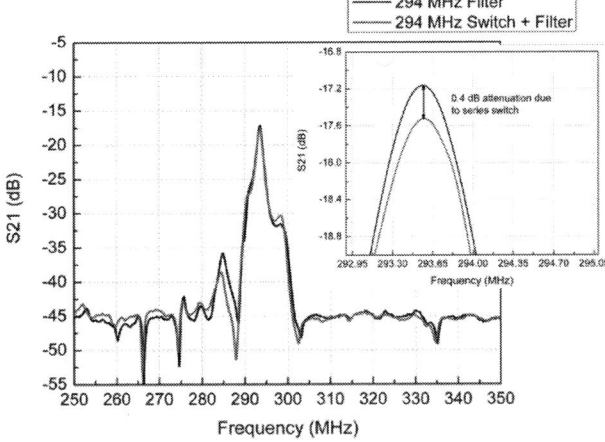

Figure 6: S_{21} response for a stand-alone PZT-on-SOI filter and an integrated PZT-on-SOI switch+filter with the DC bias at 10 volts for both devices (a). The switch increases insertion loss by 0.4dB (b).

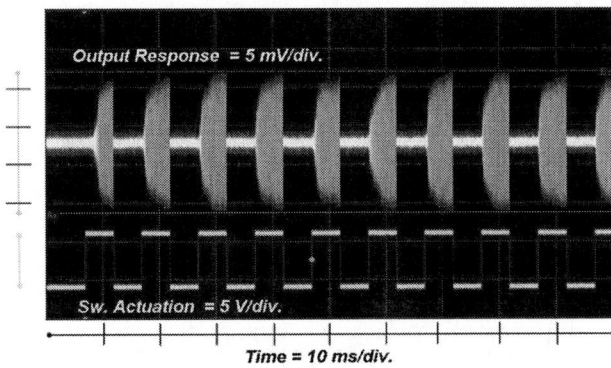

Figure 7: Time domain measurements of the switch+filter highlighting the 7V switching actuation pulses and the ring up response of the filter with the switch in the on state.

CONCLUSION

For the first time, piezoelectric MEMS RF switches and contour mode filters have been monolithically integrated and demonstrated as a switchable filter array. Lead zirconate titanate (PZT) thin films were utilized to enable both low voltage switch operation and filter tunability. On-going research is focusing on improving the loss performance of the filters, suppressing spurious modes, addressing switch lifetime, and incorporating wafer-level packaging. The low voltage switches and voltage tunable PiezoMEMS filter array provides a drop-in solution for frequency-agile channel selectivity.

ACKNOWLEGEMENTS

The authors wish to acknowledge the assistance and support of Joel Martin and Brian Power of General Technical Services and Richard Piekarz from the ARL for their hard work with device fabrication.

REFERENCES

[1] G. Rebeiz, *RF MEMS Theory, Design, and Technology*. Hoboken, NJ: Wiley, 2003.

[2] F. D. Bannon , J. R. Clark and C.T.-C. Nguyen "High-Q HF microelectromechanical filters," *IEEE J. Solid-State Circuits*, vol. 35, pp. 512, 2000.

[3] G. Piazza, et. al., "Single-Chip Multiple-Frequency AlN MEMS Filters Based on Contour-Mode Piezoelectric Resonators", *J. Microelectromech. Syst.*, vol. 16, pp. 319-328, 2007.

[4] H. Chandrahalim, D. Weinstein, L.F. Cheow, and S. A. Bhave, "Channel-select micromechanical filters using high-K dielectrically transduced MEMS resonators," MEMS'06, Istanbul, Turkey, pp. 894-897, 2006.

[5] Nguyen, C.T.-C, "Integrated Micromechanical Radio Front-Ends", VLSI Technology, Systems and Applications, 2008. VLSI-TSA 2008, pp. 3-4, 2008.

[6] P. Muralt, "PZT Thin Films for Microsensors and Actuators: Where Do We Stand?", IEEE Trans. Ultras. Ferro. Freq. Control, Vol. 47, pp. 903 – 915, 2000.

[7] R. G. Polcawich, et. al., "Surface micromachined microelectromechanical ohmic series switch using thin-film piezoelectric actuators," *IEEE Transactions Microwave Theory and Techniques*, Vol. 55, No. 12, pp. 2642 – 2654, 2007.

[8] H. Chandrahalim, et. al., "Influence of silicon on quality factor, motional impedance, and tuning range of PZT-transduced resonators," *2008 Solid State Sensor, Actuator and Microsystems Workshop*, Hilton Head Island, SC, pp. 360-363, 2008.

[9] D. J. Chung, et. al., "A SP2T and a SP4T Switch using Low Loss Piezoelectric MEMS," *IEEE MTT-S Microwave Symposium*, June 2008.

[10] Y. Arimoto and H. Ishiwara, "Current State of Ferroelectric Random-Access Memory", *MRS Bulletin, Nov. 2004, pp. 823-828.*

[11] N. Sinha, et. al., "Dual-beam actuation of piezoelectric AlN RF MEMS switches monolithically integrated with AlN contour-mode resonators," *2008 Solid State Sensor, Actuator and Microsystems Workshop*, Hilton Head Island, SC, pp. 22 - 25, 2008.

[12] A. I. Zverev, Handbook of Filter Synthesis. New York: Wiley, 1967.

[13] S. Dey, K. Budd, and D. Payne, "Thin-film ferroelectrics of PZT of sol-gel processing," TUFFC 35 (1) 1988, pp. 80-81.

A LIQUID-METAL RF MEMS SWITCH WITH DC-TO-40 GHZ PERFORMANCE

Prosenjit Sen and Chang-Jin "CJ" Kim

University of California, Los Angeles (UCLA), Los Angeles, California, U.S.A.

ABSTRACT

We report a low-loss shunt RF-MEMS switch based on solid-to-liquid contact. This switch uses electrowetting-on-dielectric (EWOD) actuation of a liquid-metal (LM) droplet confined by a microframe, allowing bounce-free operation and fast switching (60 µs). Carrying the inherent solution LM provides for the nagging contact reliability problem, this paper introduces a two-droplet design that allows a low-loss RF performance. We further describe the microfabrication process, and report the test results. The switch demonstrates better than 0.3 dB insertion loss and 20 dB isolation up to 40 GHz, achieving significant improvements over previous demonstrations.

INTRODUCTION

Reliability of solid-solid contacts becomes more pronounced for MEMS microswitches as surface plays an important role at microscale. The contact degradation [1] due to arcing, welding and material transfer, which is aggravated by contact bounce, limits the operational life of these devices [2]. In order to solve these problems and enhance reliability, liquid-solid contact has been used by several microswitching technologies. One approach in development of a LM based microswitch involves moving the LM droplet to achieve switching by actuation mechanisms such as electrothermal [3, 4], electrostatic [5, 6] and EWOD [7, 8]. These switches have demonstrated no contact bounce [7], low switch-on latency (60 µs) [7], fast signal rise/fall time (5 µs) [7], low contact resistance [3], long-life [3] and capability to handle large currents (1 A) [3]. Though some of these devices have been tested for RF performance, they have been mostly limited to 20 GHz and suffer from slow switching in the order of 1 ms.

First known implementation of a LM based microswitch reporting RF performance was measured to be 40 dB isolation and 0.1 dB insertion loss up to 2 GHz [4]. In this implementation a LM droplet immersed in water was actuated by thermal bubble generation. However, presence of water in the off state would have severely degraded the isolation at higher frequencies. Another thermally actuated switch used thermal expansion of air to break and move LM droplets [3]. The use of air instead of water as a working fluid improved their RF performance significantly. Better than 1 dB insertion loss and 20 dB isolation were reported up to 18 GHz along with ~0.92 ms switching latency and 10 µJ power requirement. Recently electrostatically actuated LM capacitive switch was reported with 0.6 dB insertion loss up to 20 GHz [9]. The isolation, however, degraded to 10 dB at 15 GHz and was attributed to the slotline mode arising due to the asymmetrical switch design. The switch-on time was reported to be ~1 ms. Possibility of using LM alloys for reflective switch using a LM alloy Galinstan in Teflon solution [10] and reflective and absorptive switch using water have been reported [11]. In these reports, however, the LM alloy or water was manually pumped to achieve actuation. Less than 1.3 dB insertion loss and 20 dB isolation were demonstrated up to 100 GHz. Integration of an actuation mechanism and development of a legitimate microswitch still remained to be demonstrated. Pumping of fluid also limited the expected switching latency to 10 ms [11].

The previously reported EWOD-driven LM switch [7], demonstrating bounce-free operation and low latency in comparison to other LM microswitches, was for DC applications only. Based on this actuation mechanism, this paper presents design, fabrication and testing of a low-loss, LM-based RF MEMS switch with DC-to-40 GHz performance.

ACTUATION MECHANISM

Liquid-metal actuation based devices have suffered due to their slow actuation speeds, limited by the accuracy with which the droplets have been formed (i.e., volume) and deposited (i.e., position). We have previously reported an actuation mechanism capable of high-speed operation [7], illustrated in Figure 1. A microframe structure, in this case made of SU-8, was used to hold the droplet in position. High surface tension of the LM droplet ensured that the interface position is defined accurately. A larger opening at the back (i.e., between the two left-most SU-8 posts in Figure 1) compared to the front (i.e., between the two right-most SU-8 posts, where the droplet is actuated to make contact with the signal electrode) absorbed any variation in the deposited LM volume, thus maintaining the interface position at the front. These features allowed switch design with very small switching gaps, e.g., 10 µm for 600 µm diameter droplet.

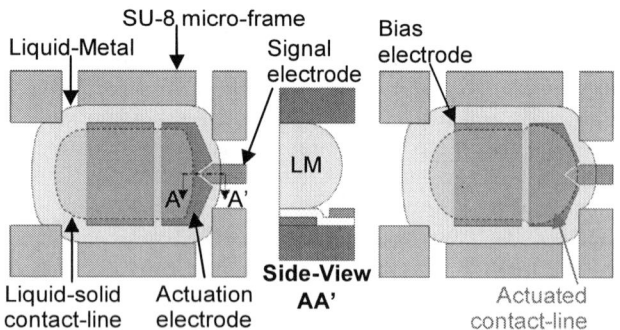

Figure 1: (Left & center) LM based EWOD switch concept at no actuation. (Right) EWOD causes the contact line to spread and make contact with the signal electrode.

The bias electrode grounded the LM droplet. When a potential (100 V) was applied to the actuation electrode, the droplet interface spread and made contact with the signal electrode. Using the initial regime of fast (0.5 m/s) contact line motion of EWOD, 60 µs of switch-on latency was demonstrated. Signal rise- and fall-time was better than 5 µs with no contact bounce. The switch required 10 nJ of energy per cycle. It was also demonstrated that dielectric charging has little ill effect on the actuation mechanism for 10^5 cycles. We use this actuation mechanism to design and demonstrate a low loss, LM RF switch.

DEVICE DESIGN

The schematic of the design is shown in Figure 2. The design is based on 20-180-20 µm coplanar waveguide (CPW) on quartz. The design uses two mirror-imaged microframe to accurately position one LM droplet on each of the two ground planes of the CPW, biasing the droplets at ground. When a potential is applied to the actuation electrode, the actuated droplets short the signal lines to the ground planes. Actuation electrode is capacitively coupled to the LM droplet, which means that the actuation electrode and the bias lines are grounded at RF frequencies.

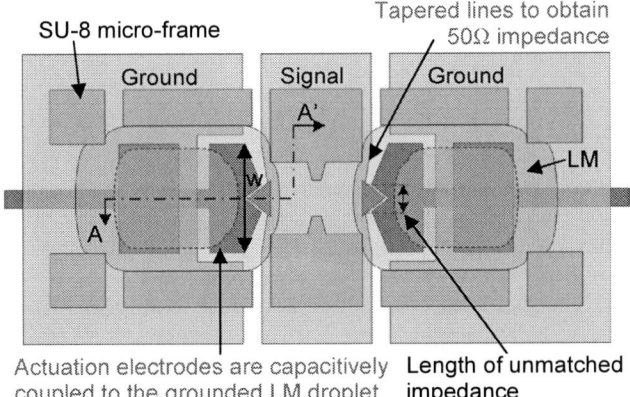

Figure 2: Schematic of a LM based direct-contact shunt switch. Two droplets are enclosed in mirror-imaged microframes. When actuated, the contact-line spreads and shunts the signal line to the ground planes.

Insertion loss of a switch is due to variation of the switch impedance from the characteristic impedance. Presence of actuation electrode requires a change in the gap between the signal line and the ground plane. At first sight it may seem that this discontinuity will lead to large insertion loss. However, in our design the actuation electrode, which is capacitively coupled to the ground plane through the LM droplet, acts as an extension of the ground plane at RF frequencies. To reduce losses the signal line of the CPW is tapered in accord with the actuation electrode to maintain 50 Ω impedance. This minimizes the impedance mismatched section to the pointed signal line extension where the actuated LM droplet makes contact, as seen in Figure 2. Thus the insertion loss in this design is from the capacitance due to the impedance mismatched contact region and LM droplet. The impedance mismatched contact region is 100 µm long, and the gap between the actuation electrode and the signal line in this region is 5 µm. The minimum gap between the signal line and the LM interface is 15 µm.

Though the LM surface area is larger than the contact region, most of the LM surface is significantly farther away from the signal line. Thus the signal electrode is responsible for most of the switch capacitance, and a smaller contribution is expected from the LM droplets. To minimize the effect of SU-8 microframe structures on insertion loss, they were positioned away from the gaps of the CPW.

The design of the microframe and its positioning with respect to the contact region have been reported before [7]. In the current device the opening at the back is 400 µm, and the opening at the front is 250 µm. SU-8 microframe is designed to be 400 µm high. The actuation electrode length 'w' is 250 µm. At the ends of the actuation electrode, the gap between the actuation electrode and the signal line is 20 µm, forming a 20-180-20 µm CPW. Prior to the contact region the gap between the actuation electrode and the signal line is 17 µm, and the signal line width is tapered to 146 µm to obtain a 50 Ω 17-146-17 µm CPW.

DEVICE FABRICATION

Device fabrication as shown in Figure 3 starts with a 700 µm thick fused silica substrate. 1000 Å chromium is evaporated on the substrate and patterned lithographically using a wet etchant. 8000 Å oxide, which isolates the bias lines from the CPW, is deposited using plasma enhanced chemical vapor deposition (PECVD) and etched using RIE. CPW is formed by lift-off of 8000 Å thick gold using 200 Å Cr as adhesion layer. Isotropic etching of LOR-20B from MicroChem is used with AZ5214 to obtain a clean lift-off of the thick metal. For the actuation dielectric, 3500 Å nitride is deposited using PECVD and etched by RIE. Since LM's attack most of the metals, a protective layer is required at the contact regions. A layer of 2000 Å Cr/Ni is deposited as the protective layer at the contact regions using lift-off. To reduce hysteresis (static friction which restricts contact line motion) and have a reasonable actuation voltage, a thin layer of hydrophobic coating of Teflon is used. Teflon is spin coated on the wafer to obtain a 2000 Å film, which is then baked at 320 °C for 3 hours. Further processing of Teflon-coated wafer is difficult because adhesion of any film is poor on Teflon that has a low surface energy. To successfully coat photoresist (PR) we add surfactant to the PR. The Teflon layer is patterned lithographically and etched in oxygen plasma. After the PR is removed in acetone, the patterned Teflon layer is baked again at 320 °C for 3 hours. To allow building the microframe in subsequent steps, Teflon covered area is minimized. For microframes, 400 µm thick SU-8 is obtained in a single spin using SU-8 2150 from MicroChem. SU-8 is soft-baked at 95 °C for 3 hours. Temperature is always ramped up or down with a rate of 60 °C / hr from 50 °C. A 300 s step exposure with each step consisting of 30 s exposure and 20 s delay is used, helping reduce surface hardening due to heat. After 45 min post exposure bake, the features are developed with agitation. The Teflon layer is again baked in a nitrogen environment at 200 °C for 3 hours. A lower temperature is used to prevent SU-8 burning. Finally a ~400 µm diameter LM droplet is placed manually.

978-1-4244-2977-6/09 $25.00 © 2009 IEEE

Cr is evaporated and patterned lithographically

Deposit 8000 Å PECVD oxide and etch in RIE

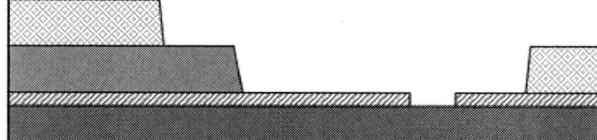

Lift-off 8000 Å Au using lift-off resist

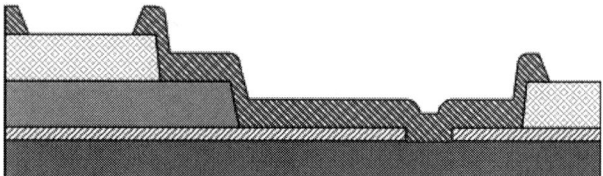

Deposit 3500 Å nitride and etch in RIE

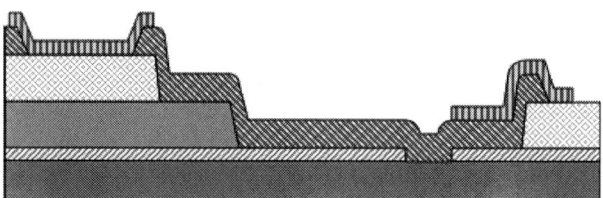

Lift-off Cr/Ni to form contact regions

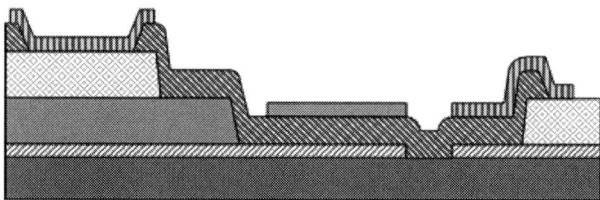

Teflon is spun coated and patterned using RIE

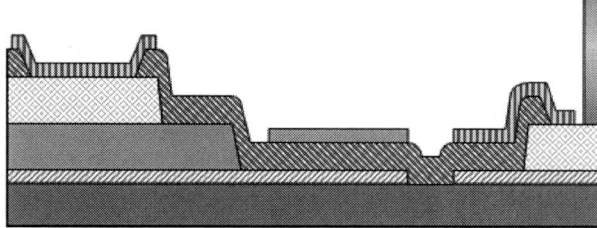

Lithographically define 400 µm SU-8 microframe

Figure 3: Process flow for device fabrication. The steps show section AA' from Figure 2.

It is important to note that this process deviates from conventional process where Teflon layer is spin-coated and patterned last. Conventionally this is done to protect the Teflon layer from any further chemical processing which may degrade its quality. For our case, however, this

is not possible due to the presence of the tall SU-8 microstructures. These microstructures destabilize the surfactant mixed PR film while spin coating, leading to dewetting of the PR. Furthermore, capillary force causes PR accumulation in the small spaces between the SU-8 microstructures. Teflon film is baked after every step to recover from any degradation in the film quality during processing. Figure 4 shows the final two steps where the Teflon is patterned before the SU-8 microstructures are fabricated.

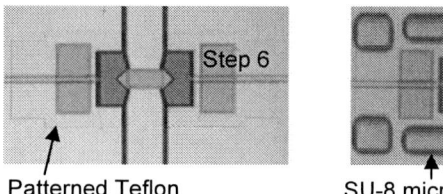

Patterned Teflon   SU-8 microframe

Figure 4: Teflon hydrophobic layer is coated and patterned (shown left) before fabricating SU-8 microframe structures (shown right) to solve the issues related to PR coating and lithography in presence of tall microframe structures.

RESULTS

HP 8510C or Agilent E8361A network analyzer is used to measure the device performance. A DC signal form a National Instrument multifunctional DAQ, amplified using a Trek amplifier, is used to actuate the switch. Ground-signal-ground (GSG) probe tips from Picoprobe are used to contact the CPW. To calibrate the setup on wafer, thru-reflect-line (TRL) calibration was performed.

The measured insertion loss is better than 0.3 dB up to 40 GHz as seen in Figure 5. A good match is obtained with the simulation result. The return loss is given by

$$|S_{11}| = \frac{\omega C_u Z_0}{2} \qquad (1)$$

where C_u is the switch capacitance and Z_0 is the characteristic impedance. The switch capacitance is calculated to be 14 fF by curve fitting the return loss.

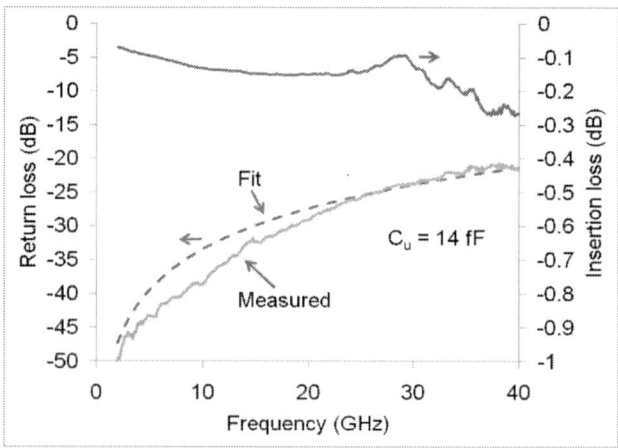

Figure 5: Measured insertion loss and return loss. Switch capacitance is calculated by curve fitting return loss.

The switch is actuated using 100 V, and isolation is measured. The switch isolation is given by

$$|S_{21}| = \frac{\sqrt{R^2 + \omega^2 L^2}}{\sqrt{(R + 0.5Z_0)^2 + \omega^2 L^2}} \qquad (2)$$

where R and L are the switch resistance and impedance, respectively. The isolation measured is better than 20 dB up to 40 GHz (see Figure 6). The fitted value for the inductance is 6.2 pH. The switch resistance (of two contacts in parallel) of 1.32Ω extracted from curve fitting is in good agreement with previously reported value of 2.35Ω for a single contact [7].

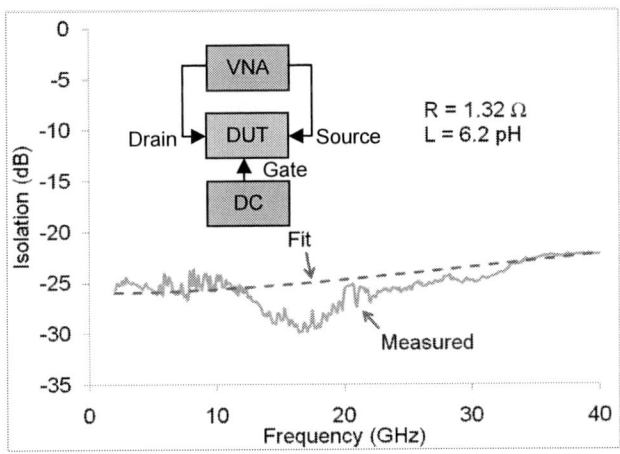

Figure 6: Measured isolation of the switch. Fitting a RL model allows extraction of switch resistance and inductance. (Inset) Schematic of the RF test setup.

SUMMARY

In this paper a contact LM droplet based RF switch has been presented. The fast actuation using EWOD combined with the accuracy provided by the microframe led to a switch with bounce-less operation and less than 5 μs signal rise/fall time, as reported previously. To implement for RF switches, a unique two-droplet design has been developed to allow for a fast switch with low loss. The design was optimized to minimize insertion loss. The measured insertion loss was better than 0.3 dB and the isolation better than 20 dB up to 40 GHz. These values are a significant improvement over previously reported values [3, 9].

ACKNOWLEDGEMENTS

We would like to thank Mr. James Jenkins and Mr. Tao Wu for their valuable discussions about the project. We would like to thank the staff at The Center for High Frequency Electronics at UCLA for their help with the RF experimental setup and the staff at UCLA Nanoelectronics Research Facility (Nanolab) for their help with device fabrication. This work was supported by DARPA HERMIT program.

REFERENCES

[1] D. Hyman and M. Mehregany, "Contact Physics of Gold Microcontacts for MEMS Switches," *IEEE Transactions on Components and Packaging Technology*, vol. 22(3), pp. 357 - 364, Sept. 1993.

[2] G.M. Rebeiz, "RF MEMS Theory, Design, and Technology," Weily Interscience, Hoboken, New Jersey, 1st. ed., 2003.

[3] Y. Kondoh, T. Takenaka, T. Hidaka, G. Tejima, Y. Kaneko, and M. Saitoh, "High-reliability, high-performance RF micromachined switch using liquid metal," *Journal of Microelectromechanical Systems*, vol. 14(2), pp. 214-220, 2005.

[4] J. Simon, S. Saffer, and C.-J. Kim, "A Liquid-Filled Microrelay with a Moving Mercury Microdrop," *Journal of Microelectromechanical Systems*, vol. 6(3), pp. 208-216, 1997.

[5] J. Kim, W. Shen, L. Latorre, and C.-J. Kim, "A micromechanical switch with electrostatically driven liquid-metal droplet," *Sensors and Actuators A*, vol. 97-98, pp. 672-679, 2002.

[6] W. Shen, R.T. Edwards, and C.-J. Kim, "Electrostatically Actuated Metal-Droplet Microswitches Integrated on CMOS Chip," *Journal of Microelectromechanical Systems*, vol. 15(4), pp. 879-889, 2006.

[7] P. Sen and C.-J. Kim, "A Fast Liquid-Metal Droplet Microswitch Using EWOD-Driven Contact-Line Sliding," *Journal of Microelectromechanical Systems*" (accepted for publication).

[8] P. Sen and C.-J. Kim, "Electrostatic Fringe-Field Actuation for Liquid-Metal Droplets," *Proceedings of The 13th International Conference on Solid-State Sensors, Actuators and Microsystems*, Seoul, Korea, pp. 705-708, Jun. 2005.

[9] C.-H. Chen and D. Peroulis, "Electrostatic Liquid-Metal Capacitive Shunt MEMS Switch," *Microwave Symposium Digest, IEEE MTT-S International*, pp. 263-266, June 2006.

[10] C.-H. Chen, J. Whalen, and D. Peroulis, "Non-Toxic Liquid-Metal 2-100 GHz MEMS Switch," *IEEE MTT-S International Microwave Symposium Digest*, pp. 363-366, Jun 2007.

[11] C.-H. Chen and D. Peroulis, "Liquid RF MEMS Wideband Reflective and Absorptive Switches," *IEEE Transactions on Microwave Theory and Techniques*, vol. 55(2), pp. 2919-2929, 2007.

DESIGN AND FABRICATION OF
A GAUSSIAN-SHAPED AT-CUT QUARTZ CRYSTAL RESONATOR

T. Abe, H. Kishi

Graduate school of Engineering, Tohoku University, Sendai, JAPAN

ABSTRACT

This paper reports design and fabrication of 3D shaped AT-cut quartz crystal resonator. The thickness distribution of the resonator is designed following a Gaussian-type function. The distribution was postulated from the analysis of thickness shearing vibration calculated by a finite element analysis (FEA) for a flat-shape resonator. The vibration analysis showed improvement of the energy trapping effect, and the vibration at the wafer edge was negligible. For the experimental verification, we attempted to fabricate a Gaussian-shaped resonator using MEMS technologies. The resonant characteristics were improved well by the shaping.

INTRODUCTION

A quartz-crystal microbalance (QCM) is one kind of the acoustic wave sensor, which is called bulk acoustic wave oscillators operating in the thickness shear mode (TSM). Almost all the quartz resonators in today's QCM applications use AT-cut quartz crystal resonator because of its high frequency stability and low temperature coefficient.

The QCM is shown to be a useful tool for fast and simply viscosity monitoring and viscous Newtonian liquids. Moreover, broadly used in the detection of chemicals by using a chemically sensitive layer on the surface of QCM to make it more specific [1].

In almost all QCM fabrication, high Q-factor, high sensitivity, high stability and the reduction of unwanted spurious responses are paramount importance. The parameter to indicate high resolution and high accuracy of the measurement is known as Q factor. The Q factor is defined as the ratio of the electrical power stored in the resonator to the electrical energy dissipated per cycle. It can be considered a mechanical parameter, which is related to the sharpness of the resonance. The maximum achievable Q times frequency is a constant, e.g., 16 million for AT-cut resonators, when f is in MHz [2]. The Q factor of quartz resonator is influenced by factors like impurities, mobility of ratio and parallelity of the quartz surfaces [3]. Due to the additional loses, the value calculated from the equation is not reached in practice [4]. The other important factor is resonator size. When small-diameter electrodes have been deposited on the surface of a flat crystal, mode coupling occurs between the fundamental thickness shear mode and the other vibration modes [5]. The suppression of the mode coupling is important to achieve high Q factor. A properly shaped quartz resonator can reduce the coupling. Since the first advent of circular plates with beveled edges by Florisson in 1926, there are many reports on the theoretical and experimental analysis of the shaping of the resonator [6]. Unfortunately, up to now beveled edge QCM is very common as commercial representative QCM fabricated by polishing technology. The main problem is that it is difficult to etch quartz crystal. Recently, our groups overcome it by using reactive ion etching (RIE) process [7]. Now we can fabricate 3D shaped quartz crystal resonator.

In this study, the shape of the quartz resonator was designed again based on the vibration mechanism. First, the finite element analysis (FEA) method was used to optimize the 3D shape of the quartz crystal resonator. Next, the resonator was fabricated by using newly developed RIE process. Finally, the performance was experimentally verified by the impedance measurements.

DESIGN

In this study, a single QCM with the resonant frequency of 5.6 MHz was constructed by ANSYS package. The coupled-fields element provided by the package was used for simulating the AT-cut element and the electrodes on the surfaces. The simulation method is same as that reported by F. Lu et al. [8]. A quartz resonator with circular plates was used for our simulation. The dimensions of quartz plate and electrode used for the simulation were shown in Table 1.

It is well known that the thickness vibration mode decays exponentially at the outside of the electrode. The distribution of TSM vibration follows a Gaussian like function. Ideally, the mass distribution of quartz plate must be followed the distribution of TSM vibration. Because the excess mass induces a stress inside the crystal. The Q factor will be influenced by the friction caused by the stress. To prove the hypothesis, a Gaussian shaped AT cut quartz crystal resonator was constructed by the ANSYS package. Figure 1 shows the models of the resonators analyzed by the package.

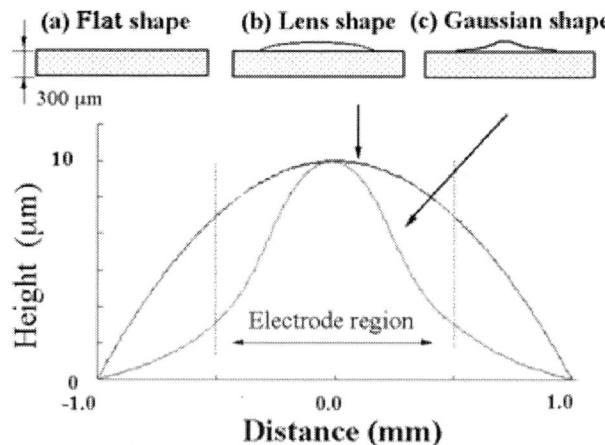

Figure 1: Models of the resonators analyzed using ANSYS®.

Table 1: Dimensions of quartz wafer and gold electrode.

Dimensions	Value
Thickness (quartz)	0.3 mm
Diameter (quartz)	10 mm
Thickness (gold electrode)	100 nm
Diameter (gold electrode)	1 mm

FABRICATION

Figure 3 shows the process flow. The synthetic AT-cut quartz crystals used in this study are provided for mass production (class C). The wafers were 0.1 mm thick with polished faces. The fundamental frequency was 16.6 MHz with a variation of 0.2 MHz. At first, the photoresist profile on a quartz wafer was adjusted to be a lens shape by the photoresist reflow (a). The method was shown in Ref. 7. The diameter of the patterned photoresist was 2 mm. Then, the shape was transferred into the quartz wafer using a RIE (b-d). By varying the etch selectivity in RIE, the shape can be controlled to be the Gaussian type shape. Finally, the electrode patterns were formed by rf sputtering a 10 nm chromium adhesion layer, followed by a 100-nm gold layer(e). The electrode diameters were 1 mm.

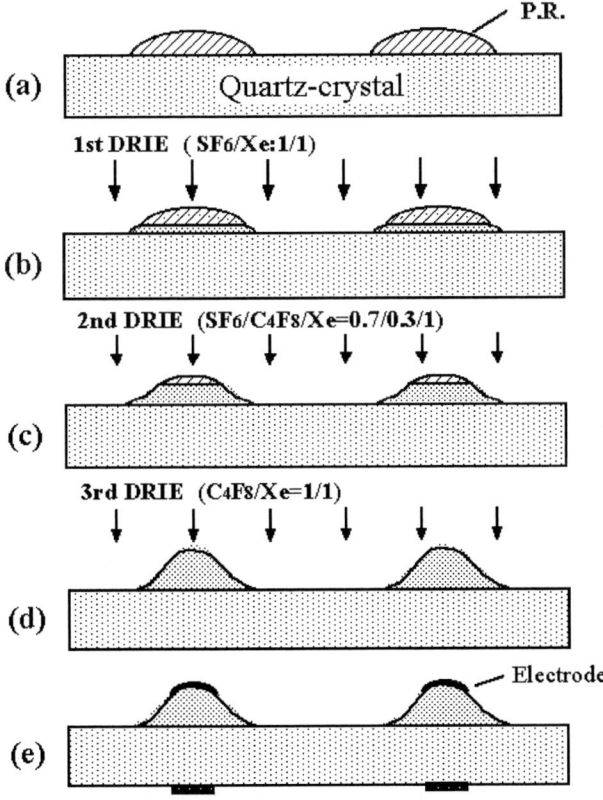

Figure 2: Process flow for the fabrication of the Gaussian shaped AT-cut quartz crystal resonators.

Reactive ion etching system was also developed for varying the etch selectivity. The etch selectivity control for fabricating 3D microstructure is already reported in case of silicon micromachining [10]. Our report is a first report as to quartz micromachining. The parameters of the system used to control the etch selectivity include input gases and their ratio. The etching was performed with SF_6, C_4F_8 and Xe gases. Figure 2 shows the relative etch rates of quartz and photoresist (etch selectivity). The total pressure of SF_6 and C_4F_8 was 0.1 Pa and the pressure ratio of SF_6/C_4F_8 was varied for the etch selectivity control. Addition to these gases, Xe was added for obtaining smooth etched surface[9]. The pressure of Xe was controlled to be 0.1 Pa. Therefore, the total pressure was 0.2 Pa. The other etch parameters are electrode power and

temperature. The electrode power was controlled to be 130 W. The cooling temperature was 293 K. The profiles of the photoresist and the etched quartz were measured by Tencor P-10.

To make impedance measurements, the resonator was mounted in a gold rf test fixture. An Agilent rf impedance analyzer (E4991A) measured conductance spectra. All the measurements were carried out at room temperature.

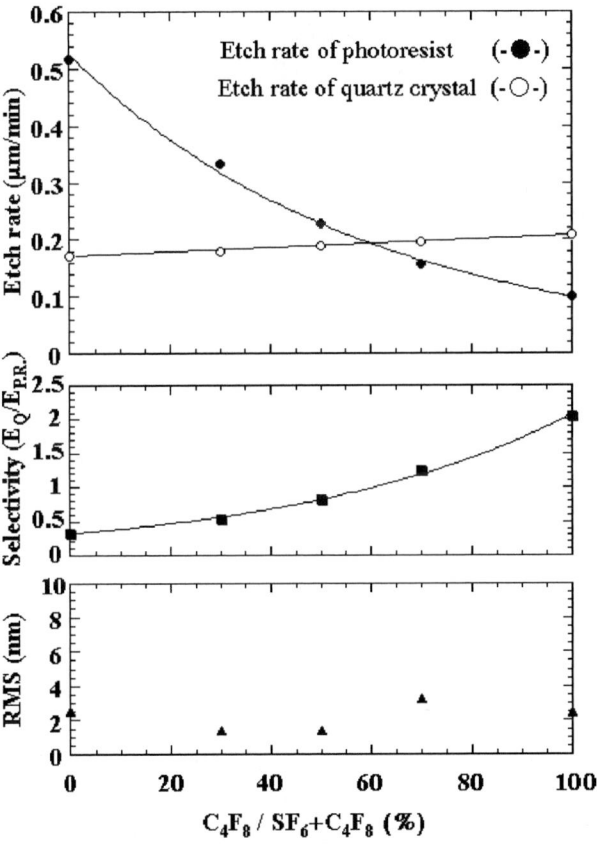

Figure 3: Etch selectivity, etch rate and surface roughness of reactive ion etching of quartz.

RESULTS AND DISCUSSION
Theoretical verification

Figure 4 shows the vibration shape of thickness shearing mode for the three models. These models were constructed using FEA method with a commercially available software package (ANSYS®) as shown in the previous section. The amplitude of the TSM vibration was shown by light and shade. The white circle means the shaped region. As shown in figure 4(a), the thickness shearing vibration was propagated to the wafer edge in case of a flat-shaped resonator. On the other hand, the vibration at the wafer edge was negligible in case of a Gaussian-shaped resonator (Fig.4(c)). The thickness shearing vibration decays quickly within the Gaussian-shaped region. The TSM vibration shape of lens-shaped resonator was also evaluated. The suppression was not perfect as shown in figure 4 (b). These simulation results support out hypothesis.

that of class A. The Gaussian-shaped resonator has two times higher Q factor than the Q factor of a flat shape resonator (Q = c.a. 44,000), and the spurious responses are reduced as expected from the FEA method.

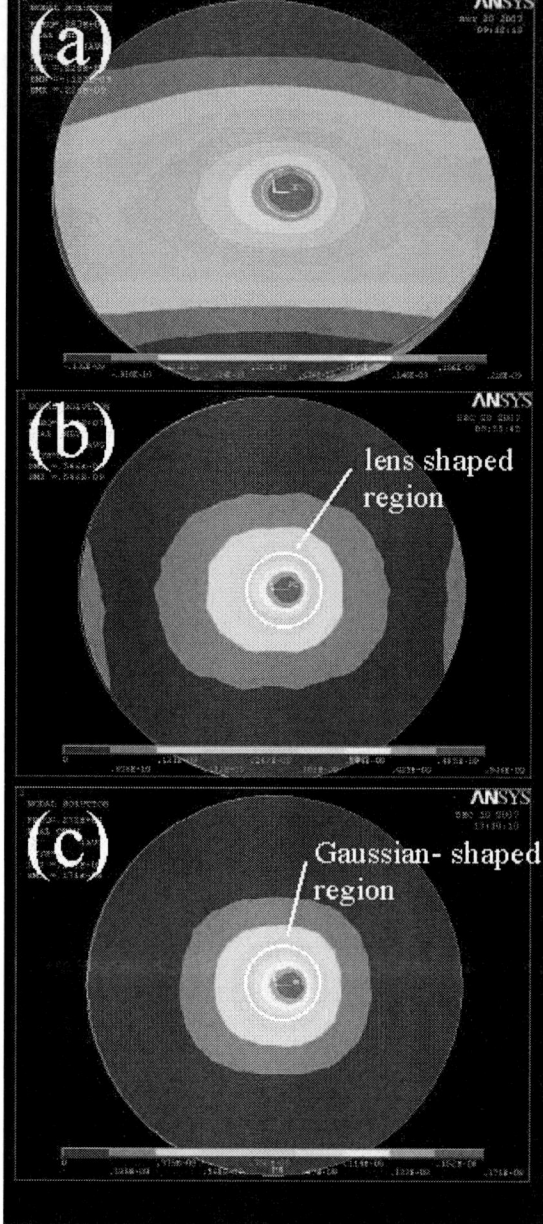

Figure 4: Vibration shapes for the designed resonators. (a) Flat shape, (b) Lens-shape, (c) Gaussian-shape

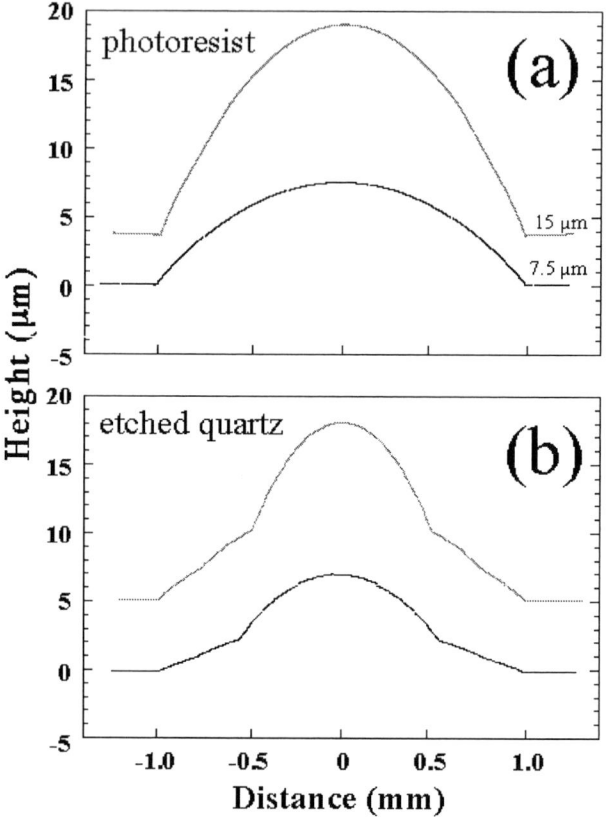

Figure 5: Examples of the pattern transfer by varying the etch selectivity. (a) Photoresist shape before the RIE. (b) Transferred pattern into the quartz wafer.

Experimental verification

Figure 5 shows the examples of the pattern transfer by varying the etch selectivity in the RIE. Figure 5 (a) shows the photoresist shape before the RIE. The shape was adjusted to be a spherical (lens) shape. The process recipe was selected to obtain a desired shape. In this study, the shape of the etched quartz was adjusted to be a Gaussian shape by using three steps RIE as shown in figure 2. The examples of the adjusted shapes are shown in figure 5(b). The photograph of the fabricated Gaussian-shaped resonator is shown in figure 6.

Figure 7 shows the conductance spectra. The Q factors were calculated from the spectra. It should be noted that the relative change in the Q factor is important because the quartz wafer used here is provided for mass production (class C). Ordinary, the Q factor is a half of

Figure 6: Photograph of the fabricated Gaussian shaped resonator. The diameter of the electrode is 1 mm.

978-1-4244-2977-6/09 $25.00 © 2009 IEEE

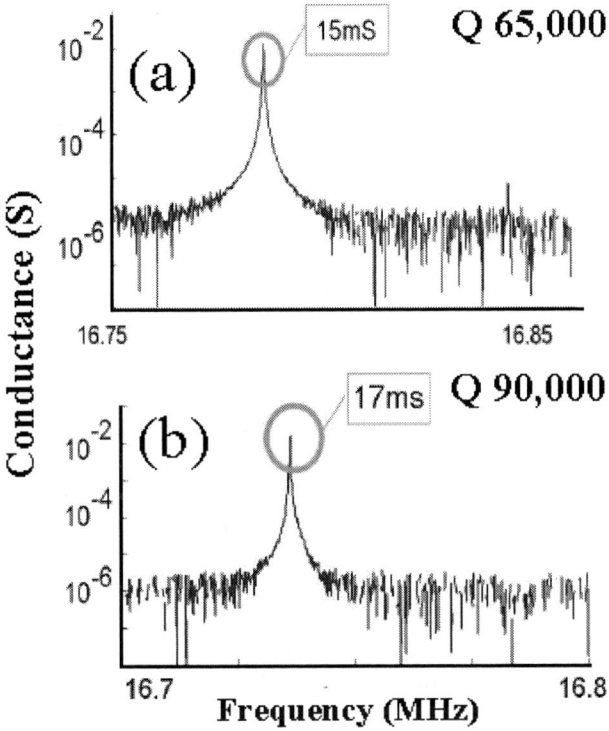

Figure 7: Conductance spectrum for lens–shaped resonator (a) and for Gaussian-shaped resonator (b).

CONCLUSIONS

In this paper, we proposed a Gaussian shaped AT-cut quartz crystal resonator as an ideal resonator. The resonant characteristics were improved well by the shaping. However, the obtained Q factor was still low value in relative to the expected maximum value. One of the reasons is that the fabricated resonator is very small in size. To obtain maximum value of Q factor, we have to develop a MEMS technology suitable for large size resonator because the Q factor depends on the resonator size. The development of the process recipe to form the shape accurately is also under investigation.

ACKNOWLEDGEMENTS

This work was supported in part by a Grant-in-Aid for Scientific Research from the Ministry of Education, Science, Sports and Culture of Japan and by a Grant-in-Aid Industrial Technology Research Grant Program in 2006 from NEDO of Japan.

REFERENCES

[1] A. Arnau, "Piezoelectric Transducers and Applications", Springer-Verlag, Berlin, 2008.

[2] J. R. Vig, "Quartz crystal resonators and oscillators for frequency control and timing applications", A Tutorial, SLCET-TR-88-1 (rev. 8.5.1.7), 2003.

[3] A. Ballato, J.G. Gualtieri, "Advances in high-Q piezoelectric resonator materials and devices", *IEEE Trans. of UFFC*, 41, pp.834, 1994.

[4] B. Zimmermann, R. Lucklum, P. Hauptmann, J. Rabe, and S. BuEttgenbach, "Electrical characterization of high-frequency thickness-shear-mode resonators by impedance analysis", *Sens. & Actuators B*, 76, pp.47, 2001.

[5] S. Goka, H. Sekimoto, Y. Watanabe and K. Sato, "Effect of stepped bi-mesa structures on spurious vibrations of AT-cut quartz plates", *Jpn.J.Appl.Phys.*, 39, pp.3054, 2000.

[6] M.C.Florisson, patent.No.667387 (1928) (in French)

[7] L. Li, M. Esashi, and T. Abe, "A miniaturized bi-convex quartz-crystal microbalance with large-radius spherical thickness distribution" *Appl. Phys. Lett,*, 85, pp.2652-2654 , 2004.

[8] F.Lu, H.P. Lee, P. Lu, S.P. Lim, "Finite element analysis of interference for laterally coupled quartz crystal microbalances", *Sens & Actuators A*, 119, pp90-99, 2005.

[9] L. Li, T.Abe, M. Esashi, "Smooth surface glass etching by deep reactive ion etching with SF_6 and Xe gases", *J. Vac.Sci.Technol.*, B21, pp2545-2549, 2003.

[10] C.M. Waits, B. Morgan, M. Kastantin, R, Ghodssi, "Microfabrication of 3D silicon MEMS structures using gray-scale lithography and deep reactive ion etching", *Sens & Actuators A*, 119, pp.245-253, 2005.

GAS-LIQUID SEPARATED RESONATOR FOR BIO-CHEMICAL APPLICATION

H. Hida[1], M. Shikida[1], M. Okochi[2], H. Honda[2], and K. Sato[1]

[1]Dept. of Micro-Nano Systems Engineering, Nagoya University, JAPAN
[2]Dept. of Biotechnology, Nagoya University, JAPAN

ABSTRACT

We have developed a novel type of gas-liquid separated micro-resonator for measuring the multifunctional characteristics of biological cells. The resonator consists of an area for the resonator/sensor (gas-phase) and one for the biochemical substance (liquid-phase), with a water-shedding wall separating them. The specimen support extends from the sensor body and passes through the water-shedding wall to the inside of the sample chamber We fabricated the comb-drive resonator on a silicon-on-insulator (SOI) wafer by using a deep reactive ion etching process and resonated it at around 2.5 kHz. Experimental results confirmed that the water-shedding wall was able to prevent solution leakage at the gap between the wall and the support.

INTRODUCTION

Recently, various types of biochemical analysis devices have been developed by using microelectromechanical system (MEMS) technology. Two analysis methods are usually applied for evaluating the characteristics of living biological cells. One is a static measurement method, in which a micro-cantilever is used as a tool for measuring generative force [1-2]. In this method, the cells are placed on the end of the cantilever, and the generative force is detected by measuring the cantilever's elastic deformation value. This method is advantageous in that it is quite simple in terms of structure and principle. The static mode, however, generally makes it difficult to detect the generative force value with high resolution. For this reason we used the other usual method, i.e., a dynamic method in which a micro-resonator is applied as a detection tool, to take our measurements. The most famous resonators used in biochemical analysis are quartz crystal microbalance (QCM) resonators [3], in which the mass change caused by a biochemical reaction is detected by measuring the changes in resonance characteristics. This method is able to detect mass changes with high resolution; however, it is difficult to use it to evaluate generative force measurements. Another disadvantage to this method is that with it a cell manipulation function cannot be incorporated into the device chip.

With most MEMS resonators, e.g., electrostatic comb-drive actuators, it is possible to integrate both a multifunctional detection mechanism and handling tools on the same chip. However, there are still two problems involved in applying them as tools for biochemical analysis.

The first concerns high salt buffer solutions: With such solutions, the living cells in them must be suspended to maintain their activity. Therefore, it is difficult to directly immerse the resonator into a buffer solution due to the solution's high electrical conductivity.

The second involves liquid-phase conditions in general. As stated above, living cells must be suspended in liquid-phase conditions. Consequently, the solution's high damping force causes the Q-factor at the resonator to reduce. This makes it difficult to directly apply a comb-drive actuator used as a sensing tool in cases where solutions are involved.

To overcome these problems, we have developed a

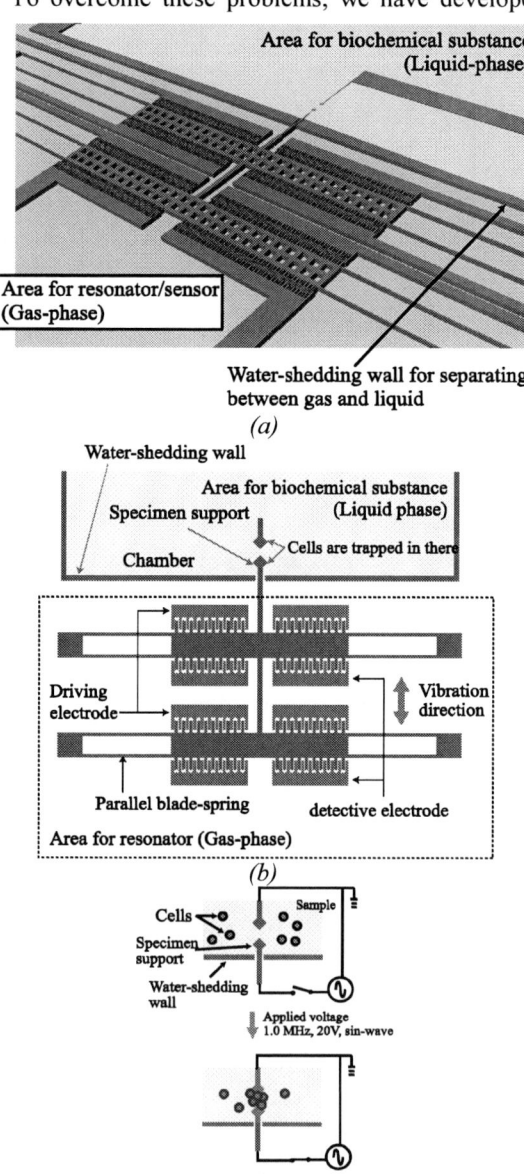

Figure 1: Schematic view of gas-liquid separated resonator and cell positioning mechanism onto specimen support. (a) Schematic view of gas-liquid separated resonator. (b) Detailed view of gas-liquid separated resonator; and (c) Cell positioning on support by dielectrophoresis force.

978-1-4244-2977-6/09 $25.00 © 2009 IEEE

novel gas-liquid separated resonator in which a water-shedding wall separates the areas for the sensor itself and for the biochemical substance.

GAS-LIQUID SEPARATED RESONATOR

Principle

The gas-liquid separated resonator we have developed is shown in Fig. 1. The device comprises an area for the resonator /sensor (gas-phase) and one for the biochemical substance (liquid-phase). The comb-drive resonator has excitation and detection functions and is used as the sensing element (Fig. 1(b)). The resonator's moving parts are suspended by four parallel blade springs. The specimen support extends from the sensor body and passes through the water-shedding wall to the inside of the sample chamber. The buffer solution including the biochemical substances is placed inside the chamber. Applying a water-shedding surface onto the wall prevents leakage at the gap between the wall and the support. The bio-chemical substances, e.g., cells, are positioned at the end of the support by using a dielectrophoresis force, as shown in Fig. 1(c).

The mass change and the generative force in the cells positioned on the support by the bio-chemical reactions are detected by measuring changes in the resonance characteristics. The resonator's mechanical resonant frequency f_0, is given by

$$f_0 = \frac{1}{2\pi}\sqrt{\frac{k}{m}} \qquad (1),$$

where k is the spring constant and m is the resonator's effective mass. When the cells are placed in the resonator, the resonant frequency is decreased by mass change (Δm).

$$f_{mass} = \frac{1}{2\pi}\sqrt{\frac{k}{m + \Delta m}} \qquad (2)$$

Additionally, when the generative force in the cell is applied to the resonator, the changed resonant frequency f_{force}, is given by

$$f_{force} = \frac{1}{2\pi}\sqrt{\frac{k + \Delta k}{m}} \qquad (3)$$

where Δk is an equivalent spring constant of the force applied by the cells. Using (1), (2) and (3), we can measure the mass change and the generative force in the cells by detecting the frequency difference. The resonant frequency change caused by the generative force is given by

$$\Delta f_{force} = f_{force} - f_0 = \frac{\sqrt{k + \Delta k} - \sqrt{k}}{2\pi\sqrt{m}} \qquad (4)$$

This equation shows that reducing the spring constant

Table 1: Mechanical properties of the designed resonator.

Length l [μm]	Spring constant k [N/m]	Resonant frequency f_0 [kHz]
250	234.4	17.012
500	97.8	5.938
750	62.5	3.089
1000	45.7	1.966

can increase the frequency difference Δf_{force}, thus improving the resonator sensitivity.

Mechanical properties

In designing the micro-resonator, we first determined its mechanical properties (i.e., spring constant and resonant frequency). The parallel blade springs were designed to have four values of length l, i.e., 250, 500, 750, and 1000 μm. We set their width w and thickness b at 4 μm and 20 μm respectively. The spring constants and resonant frequencies were calculated by using Coventorware finite element modeling (FEM) analysis. Table 1 shows the relationship between the parallel beam springs' mechanical properties and length.

Electrostatic driving force

We then designed the electrostatic comb-drive resonator and calculated its excitation/detection properties. When a DC voltage V is applied to the resonator, the generative electrostatic force F_{drive}, and the force displacement are given by

$$F_{drive} = \frac{2\varepsilon v V^2}{g - h}N = k\Delta x \qquad (5)$$

where ε (= 8.86×10^{-12} F/m) is the permittivity of the air, b is the device thickness, g is the gap between the electrodes, h is the electrodes' thickness, N is the number of the comb structure's fundamental units, and Δx is the resonator displacement. We determined that the values of b, g, h, and N were 20 μm, 8 μm, 4 μm, and 200 respectively. The relationship between the applied voltage and the resonator's generative electrostatic force is given by

$$F_{drive} = 1.77 \times 10^{-8}V^2 \qquad (6)$$

Sensitivity

The resonator's vibration behavior can be measured by detecting the capacitance. The resonator displacement can be related to the capacitance change ΔC by the following equation:

$$\Delta C = \frac{4\varepsilon b\Delta x}{g - h}N \qquad (7)$$

Using the dimensional values given above, the relationship between the capacitance change and the resonator displacement is given by

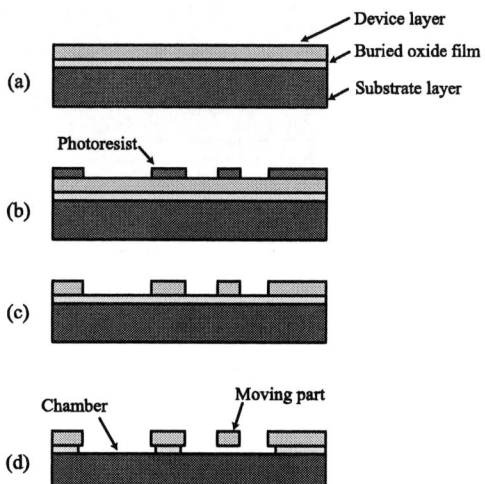

Figure 2: Fabrication process of the gas-liquid separated resonator. (a) Silicon on insulator water. (b) Patterning of photoresist. (c) Si etching by D-RIE; and (d) Removing the oxide film by wet etching process.

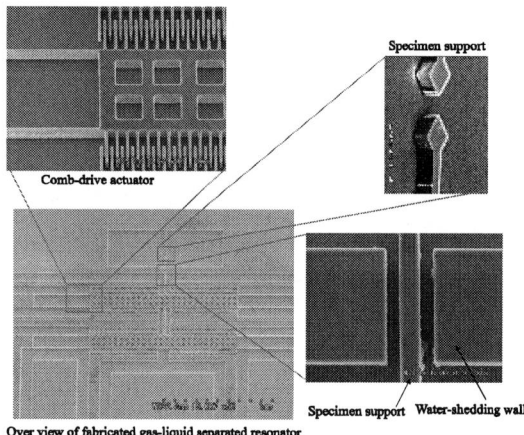

Figure 3: Fabricated gas-liquid separated resonator.

$$\Delta C = 3.54 \times 10^{-8} \Delta x \qquad (8)$$

Water shedding wall

We also designed a water-shedding wall to separate the excitation/detection and biochemical substance areas. The wall thickness and that of the gap between the wall and the support were set at 50 μm and 5 μm respectively. We believe that the wall is able to suppress leakage not only due to its high aspect ratio but also due to the hydrophobic character of its silicon surface.

Fabrication

We fabricated the resonator on a silicon-on-insulator (SOI) wafer as shown in Fig. 2. First, a photoresist was applied to the wafer's device layer and patterned by using an exposure system. The resonator and chamber structures were then formed by the deep reactive ion etching (D-RIE) process. After the process was completed, the wafer was cut into dices. The resonator's moving parts were then

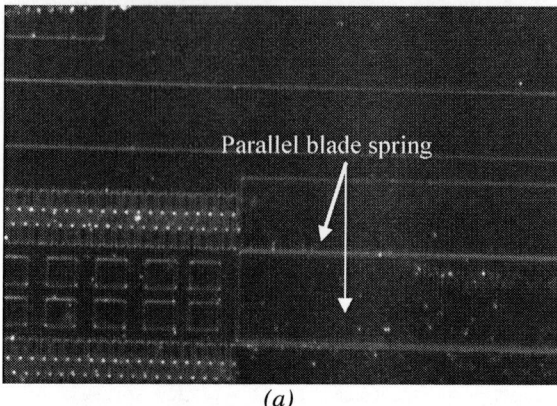

(a)

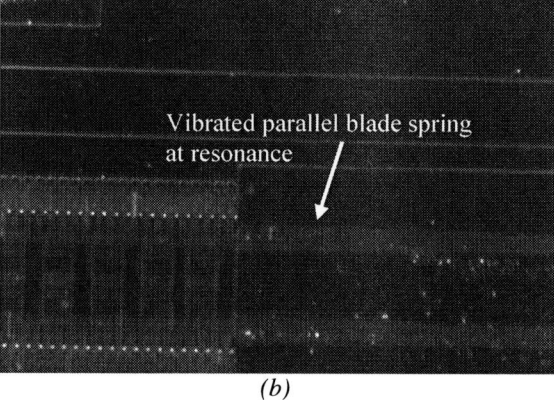

(b)

Figure 4: Resonated com-drive actuator at around 2.5 kHz. (a)Before applied voltage; and (b) after applied voltage.

released to selectively remove the buried silicon dioxide layer through an HF etching process. The device was then immersed in isopropyl alcohol and dried in the air to decrease the surface tension acting between the moving parts and the silicon substrate. Figure 3 shows an overview of the fabricated resonator.

RESONATOR PERFORMANCE

Frequency response

We evaluated the resonator's resonant frequency in air. First, the AC voltage required for excitation was amplified with a power amplifier and then applied to the fixed electrodes. The moving parts and substrate were grounded. At this time, we observed the resonator's vibration behavior with an optical microscope. We applied AC voltage of 50 V amplitude with various frequencies. The parallel springs' length and the theoretical resonant frequency were 750 μm and 3.09 kHz respectively. Figure 4 shows the resonator excited at around 2.5 kHz resonant frequency. The maximum displacement in air at this frequency was 20 μm. We found that the experimentally obtained resonance frequency coincided with the designed value.

Sample positioning in liquid environment

We also attempted to manipulate sample beads in a liquid environment. First, we selectively placed a liquid solution inside the chamber by using a pipette and tested

978-1-4244-2977-6/09 $25.00 © 2009 IEEE

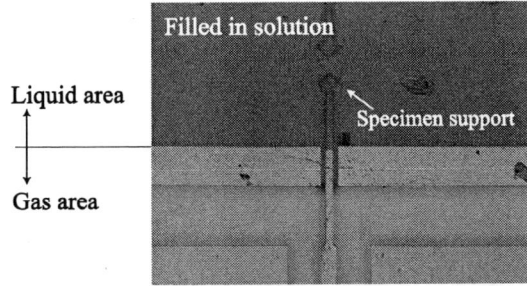

Figure 5: Gas-liquid separation by water- shedding wall.

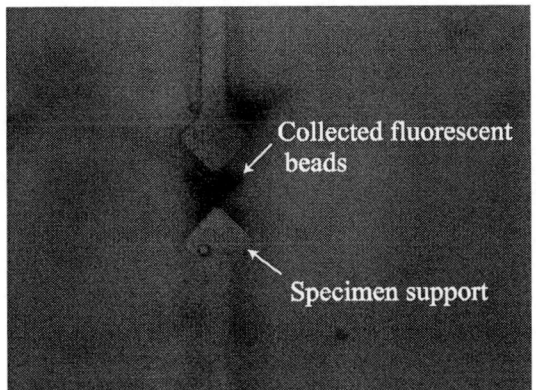

Figure 6: Beads positioning by dielectrophoresis force.

the resonator for leakage. As Fig. 5 shows, we confirmed that the water-shedding wall was successfully able to prevent solution leakage at the wall-support gap.

Finally, as a first cell-positioning trial, we used fluorescent beads 4.5 μm in diameter in place of a cell, and suspended them in water. A high-frequency voltage (1.0 MHz, 20V) was applied between the specimen support and the opposite pole to apply dielectrophoresis force to the beads. We found that the force successfully manipulated the beads so that they could be collected at the gap, as shown in Fig. 6.

CONCLUSION

We designed and fabricated a gas-liquid separated resonator for evaluating the multifunctional characteristics of biological cells. The micro-resonator was fabricated on an SOI wafer through the use of an D-RIE process. The resonator was resonated in air at around 2.5 kHz. In a cell-positioning trial, we also manipulated sample beads as cell substitutes by applying dielectrophoresis force to them in water and obtained very positive results. In the future, we will attempt to refine our gas-liquid separated resonator as follows:

First, by evaluating its resonant frequency in detail: We will try to evaluate the resonator's resonant frequency by using a chamber filled with liquid. In addition, we will also construct a circuit that can detect capacitance change in it.

Second, by determining the position and culture of actual living cells: We have succeeded in selectively positioning cell-substitute beads in water. The next challenge will be to ascertain the position and culture of living cells in buffered solutions to measure the cell-bonding force between them.

REFERENCES

[1]   U. Dammer, O. Popescu, P. Wagner, D. Anselmetti, H.-J. Güntherodt, G. N. Misevic, "Bending Strength Between cell Adhesion Proteoglycans Measured by Atomic Force Microscopy", *Science*, vol. 267, pp. 1173-1174, 1995.

[2]   J. Fritz, M. K. Baller, H. P. Lang, H. Rothuizen, P. Vettiger, E. Meyer, H. J. Güntherodt, Ch. Geber, J. K. Gimzewski, "Translating Biomolecular Recognition into Nanomechanics", *Science*, vol. 288, pp. 316-318, 2000.

[3]   P. Hauptmann, R. Lucklum, J. Hartmann, J. Auge, B. Alder, "Using the Quartz Microbalance Principle for Sensing Mass Changes and Damping Properties", *Sensors and Actuators A: Physical*, vol. 37-38, 1993.

5-10 GHZ ALN CONTOUR-MODE NANOELECTROMECHANICAL RESONATORS

M. Rinaldi, C. Zuniga and G. Piazza

Department of Electrical and Systems Engineering, University of Pennsylvania, Philadelphia, USA

ABSTRACT

This paper reports on the design and experimental verification of Super High Frequency (SHF) laterally vibrating NanoElctroMechanical (NEMS) resonators. For the first time, AlN piezoelectric nanoresonators with multiple frequencies of operation ranging between 5 and 10 GHz have been fabricated on the same chip and attained the highest *f-Q* product (~ 4.6· 10^{12} Hz) ever reported in AlN contour-mode devices. These piezoelectric NEMS resonators are the first of their class to demonstrate on-chip sensing and actuation of nanostructures without the need of cumbersome or power consuming excitation and readout systems. Effective piezoelectric activity has been demonstrated in thin AlN films having vertical and lateral features in the range of 250 nm.

I. INTRODUCTION

In recent years the use of Nanoelectromechanical systems (NEMS) for a large number of applications spanning from semiconductor based technology to fundamental science [1] has been extensively explored. In particular, NEMS resonators have been exploited as transducers suitable for the realization of extremely sensitive gravimetric sensors. Sub-attogram mass resolution has been demonstrated in NEMS cantilevers [2] thanks to the minuscule mass and the relatively high quality factor (*Q*).

A fundamental figure of merit for all resonant sensors is the *f-Q* product. In fact, the mass sensitivity of the resonator scales with the square of its frequency of operation while the minimum detectable frequency shift is intrinsically related to the frequency stability of the device, hence to the inverse of its quality factor [3]. High *Q* values have been measured for nanoresonators [4], but frequencies of operation have been limited to the UHF range. In order to take full advantage of scaling laws of miniaturization and achieve higher values of sensitivity and figure of merit than corresponding macroscale counterparts (*i.e.* quartz crystal-based sensors), NEMS resonators should operate in the SHF band (3 – 30 GHz).

The greatly reduced dimensions of the NEMS devices demonstrated to date make them very sensitive to added mass, but also render their transduction very hard to implement [5]. In particular, the most common sensing and actuation techniques (piezoelectric and electrostatic) employed for MEMS devices have not been directly applied to the NEMS domain, due to the increase of parasitic effects associated with both the scaled dimensions of the devices and their higher frequency of operation.

In this paper, fundamental transduction problems in NEMS resonators are solved by using on chip piezoelectric actuation and sensing of Aluminum Nitride (AlN) nanostructures. In addition, previously unexplored frequencies of operation for NEMS resonators in the 5-10 GHz range are demonstrated on the same chip and high quality factors between 400 and 700 are attained in ambient conditions. This work experimentally demonstrates the highest *f-Q* product ever achieved in NEMS resonators (~ 4.6· 10^{12} Hz).

These results have been made possible by exploiting the excellent scaling capability of the AlN contour-mode technology. Resonance frequencies in the SHF band have been obtained by piezoelectric excitation of nano-strips (500-1000 nm wide) of AlN in their contour–extensional mode of vibration (contrary to flexural vibrations generally employed in the NEMS devices demonstrated to date). At the same time, the increase of parasitic effects associated with the reduced dimensions of these AlN nano-strips has been overcome by mechanically coupling a large number (49-99) of these devices (Fig. 1). In this way all the nanoresonators comprised in the resulting array can be directly actuated and sensed piezoelectrically and are mechanically forced to simultaneously vibrate at the same frequency. Finally, a high *Q* factor in air has been demonstrated by taking advantage of the intrinsic high quality of the AlN nano-film (250 nm thick) grown directly on top of a silicon substrate.

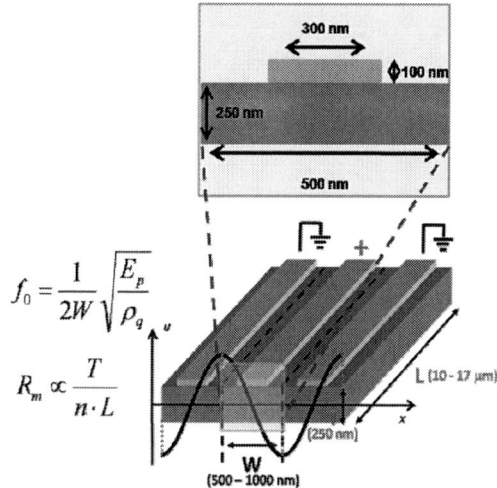

Figure 1: *Schematic representation of three mechanically coupled AlN nanoresonators: the resonance frequency of the overall system is set by the width (W) of the single nanoresonator. The frequency determining dimension is effectively uncoupled from the overall dimensions of the AlN nano-plate; therefore, the number (n) of nano-strips and their length (L) can be arbitrarily adjusted to set the motional impedance (Rₘ) of the resonator.*

A similar range of frequencies and quality factors have been demonstrated in thickness-extensional film bulk acoustic resonators (FBARs) [6]. Differently from, FBARs, laterally vibrating contour-mode NEMS resonators can easily offer multiple frequencies of operation on the same silicon chip, and, from a fundamental physics perspective, have shown, for the first

978-1-4244-2977-6/09 $25.00 © 2009 IEEE

time, piezoelectric excitation of nanoscale lateral features. The ability to define multi-frequency devices on the same die constitutes a breakthrough in the demonstration of multi-band communication devices and sensing platforms with extended sensitivity range.

II. NEMS RESONATOR DESIGN

High performance AlN contur-mode MEMS resonators in the frequency range of 20-1700 MHz and quality factor between 1000 and 3000 have been previously demonstrated [7,8]. Piezoelectric sensing and actuation of these devices have been successfully employed in the MEMS domain. Resonators with motional impedance as low as 50 Ω have been demonstrated, therefore making the interface of these devices with RF electronics relatively easy to implement.

According to the working principle of AlN contour-mode resonators, the application of an electric field across the thickness of the AlN film causes an in plane deformation of the structure through the equivalent d_{31} piezoelectric coefficient and excites the resonator in a contour-extensional mode of vibration. The resonance frequency of these devices can be lithographically set as expressed in equation (1)

$$f_0 = \frac{1}{2W}\sqrt{\frac{E_0}{\rho_0}} \qquad (1)$$

where W is the width of the resonator and E_0 and ρ_0 are respectively the equivalent Young's modulus and density of the AlN. In this way, in a rectangular plate, one device dimension (W) sets the resonance frequency whereas the other two (length, L, and thickness, T) can be used to define the equivalent motional impedance of the resonator [8]:

$$R_m \propto \frac{T}{L} \qquad (2)$$

From a scaling perspective, this is an important advantage of the AlN contour-mode technology over electrostatically-transduced resonators for which frequency scaling is generally associated with an increase in device impedance [9]. Even if improved device impedance with higher frequency of operation has been demonstrated using internal dielectric transduction [10], the achieved value of impedance is still considerably high.

In this work, the device dimensions are scaled to the nano realm both in the lateral (W) and vertical (T) direction (Fig. 1) in order to increase the frequency of operation while keeping relatively small values of motional resistance. The sub-micron width of the AlN nano-strip (Fig. 1) sets the resonance frequency in the SHF band, while an extremely thin AlN film (250 nm) increases the intrinsic electrical capacitance associated with the device to values above the substrate parasitics. Considering the reduced thickness of the AlN nano-strip, its length (L) is proportionally scaled in order to maintain a T/L ratio close to the one employed in MEMS devices. Even if a longer device would effectively yield a smaller motional resistance (eq. 2), the increase in length is associated with a higher electrical resistance in the electrodes, which directly affects the device Q factor. Therefore, the length of the device is set taking into account the trade-off between decrease in motional resistance and increase in electrical resistance associated

with the metal electrode.

In order to further reduce the motional resistance of the resonator and increase its electrical capacitance, which are both essential conditions to enable the device on chip actuation and sensing, a large number of AlN nano-strips are mechanically coupled into an array. According to this strategy, the final NEMS device is formed by an array (varying from 49-99) of mechanically coupled nano-strips of AlN (width of 500-1000 nm). The overall motional resistance proportionally decreases by the number (n) of mechanically coupled nano-strips, while the electrical capacitance increases by the same factor.

Differently from any previously demonstrated AlN contour-mode device, the very thin film of AlN (250 nm) is directly grown on top of a Si substrate. No bottom electrode is employed in order to maintain extremely high quality AlN in each of the nano-strips (rocking curve value of 2.1° was obtained and it is comparable to micron-thick films). Consequently, a scheme based on lateral field excitation (LFE), which requires solely a top electrode, is adopted to set contour-extensional vibrations in each of the nanostrips. An alternating electric field is applied across the thickness of each AlN nano-strip by coplanar signal and ground electrodes patterned on top of the AlN nano-plate (Fig. 2). Since all the AlN nano-strips are mechanically coupled, a lamb-wave like mode is excited in the nano-plate (Fig. 2).

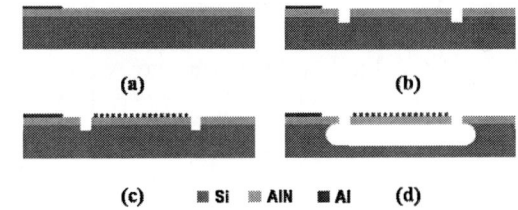

Figure 2: *Results of FEM analysis of the mode of vibration of a 8.5 GHz resonator composed of 89 mechanically coupled nano-strips of AlN. The figure shows only few of the sub-resonators as representatives of the overall device behavior.*

III. FABRICATION PROCESS

The AlN NEMS resonators are fabricated combining optical and electron-beam lithography techniques. A simple 3 mask, potentially post-CMOS compatible ($T_{max} < 400$ °C), fabrication process has been employed (Fig. 3). The thin AlN film (250 nm) is directly grown on top of a high resistivity Silicon wafer. Optical lithography is first performed for the definition of both the device contact pads and the alignment markers for the subsequent electron-beam lithography step. Then, the in-plane dimensions of the AlN nano-plate are defined by ICP etching in a Cl_2-based chemistry using photoresit as a mask. Afterwards, electron-beam lithography is performed to define the sub-micron electrodes necessary to excite each nano-strip of AlN. Finally, the resonator is released from the Si substrate by isotropic dry etching in XeF_2.

Figure 3: *Basic steps for fabrication of the AlN NEMS resonators. (a) Optical lithography and lift-off of Al (100* *nm thick). (b) Dry etching of the AlN in Cl_2. (c) Electron-beam lithography and lift-off of the sub-micron Al electrodes (100 nm thick). (d) Nanoresonator release in XeF_2.*

It is important to note that, although electron-beam lithography was employed, the minimum feature size (300 nm) can also be defined by state-of-the-art optical lithography conventionally used in CMOS foundries, therefore making this device amenable to large scale manufacturing.

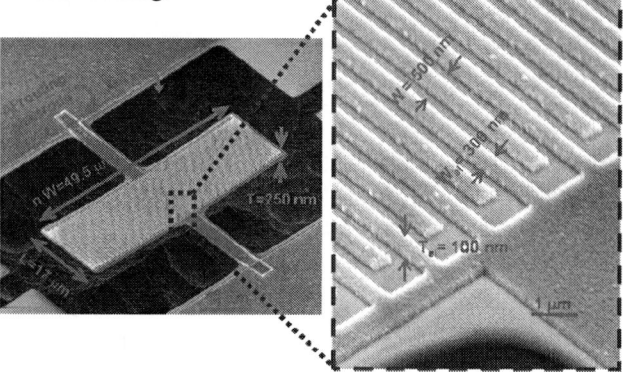

Figure 4: *SEM picture of a fabricated 9.9 GHz NEMS resonator with zoomed view of the nano-strip array (nano-strip width is 500 nm; electrode width is 300 nm).*

An SEM picture of one of the fabricated devices is shown in Figure 4. In this case, the overall dimensions of the mechanically coupled AlN nano-strip array ($n = 99$) are approximately 17 x 49.5 µm.

IV. RESULTS AND DISCUSSIONS

The fabricated AlN NEMS resonators were tested in an RF probe station and the admittance curves of the devices were measured by an Agilent® N5230A Network Analyzer after performing a short-open-load (SOL) calibration on a reference substrate. Furthermore, in order to access the true electrical response of the device a de-embedding procedure to eliminate pad parasitics was performed [11]. The value of parasitic capacitance associated with the substrate was found to be approximately 16 fF.

The admittance curve of an 8.5 GHz resonator after de-embedding is shown in Figures 5 and 6. As in all other tested devices a single sharp resonance is attained at the designed mode of vibration between 2.5 and 12 GHz.

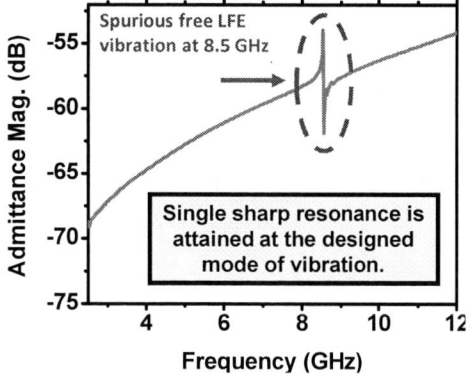

Figure 5: *Electrical response of a fabricated 8.5 GHz nanoresonator.*

The measured responses of the fabricated devices were fitted to the Butterworth van Dyke (BVD) model and the equivalent electrical parameters were extracted (Fig. 6). Moreover, 2D Finite Element Method (FEM) analysis using COMSOL Multiphysics was performed in order to evaluate the experimental performances of these nanoresonators and verify that all the properties of AlN film were preserved when scaled to the nano realm. In the FEM analysis, the quality factors of the devices were forced to match the experimental values, since they cannot be predicted a priori trough FEM simulation.

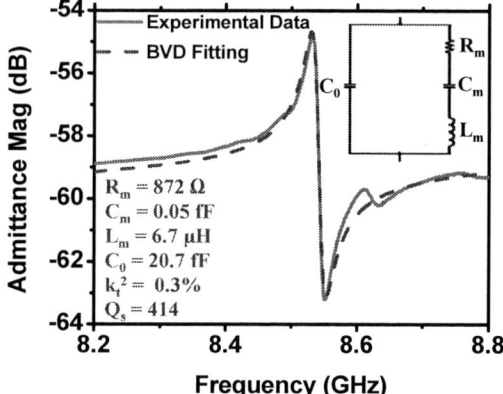

Figure 6: *Butterworth van Dyke (BVD) model fitting and parameters extraction for a fabricated 8.5 GHz nanoresoantor.*

The electromechanical parameters extracted by BVD fitting both FEM analysis and experimental response of a 8.5 GHz resonator are compared in Table 1. Considering the very good agreement between FEM analysis and actual experimental response it is possible to conclude that the AlN physical and piezoelectric properties are conserved in the fabricated nanoresonators.

Table 1: *Electromechanical parameters extracted from FEM analysis and experimental response of a 8.5 GHz nanoresonator.*

Extracted Parameters	FEM Simulations	Experimental Data
f_0	8.544 GHz	8.526 GHz
C_0	23 fF	20.7 fF
R_m	900 Ω	872 Ω
k_t^2	0.31%	0.3%

Comsol simulations also closely predict the frequencies of vibration of the AlN NEMS resonators. A discrepancy of less than 0.25% was found between FEM analysis and experimental data, further proving that the actual behavior of this new class of nanodevices can be described by conventional models valid for larger structures.

The measured resonance frequencies and the respective values of quality factor (Q_s) in air for all the fabricated nano devices are reported in Table 2. The best $f \cdot Q$ value that was achieved ($4.58 \cdot 10^{12}$ Hz) is approximately 2.5 times higher than what has ever been reported for AlN contour-mode MEMS resonators [8,12] and it is the highest ever measured for nanoelectromechanical devices.

The same electrical measurements were also

978-1-4244-2977-6/09 $25.00 © 2009 IEEE

performed in vacuum and showed a slight improvement in Q (approximately 10%). Although the improvement is not significant, it is still a sign that surface effects need to be taken into account in the design of these NEMS resonators.

Table 2: Dimensions and measured performances of the fabricated devices.

Width (W) and number (n) of AlN nano-strips	f [GHz]	Q_s	$f \cdot Q_s$ [Hz]
W=1000 nm, n=49	5.2	740	$3.85 \cdot 10^{12}$
W=700 nm, n=69	7.3	627	$4.58 \cdot 10^{12}$
W=600 nm, n=89	8.5	414	$3.52 \cdot 10^{12}$
W=500 nm, n=99	9.9	202	$4.16 \cdot 10^{12}$

The temperature coefficient of frequency (TCF) of the fabricated nano resonators was also measured and found to be - 35.8 ppm/°C. This value is higher than what is encountered in equivalent MEMS devices [13]. This is due to the effect of the Al electrodes, whose thickness (100 nm) becomes a considerable fraction of the AlN (250 nm) film in the NEMS implementation. Indeed, accounting for the temperature coefficient of Young's modulus of Al (about 10 X that of AlN), the analytical model fits the experimental data (Figure 7). The TCF of these NEMS resonators can be reduced by decreasing the thickness of the metal electrodes or by employing metals with temperature coefficient of Young's modulus lower than the one of Al (*i.e.* Pt). In addition, total temperature compensation can be achieved by depositing on the top surface of the NEMS resonators a material with positive TCF (such as SiO_2) in the same way it has been previously demonstrated for MEMS AlN resonators [13].

Finally, the sensitivity to mass per unit area of the fabricated devices was estimated accordingly to [14]. Device sensitivity as high as 6 MHz·μm^2/fg at 9.9 GHz was extracted. This value is approximately 10 X higher than what has been recently reported for nano cantilever devices [2]. This shows the great potential of the nanoresonators presented in this paper for gas sensing applications. The combination of high frequency of operation, high quality factor in air, large surface area available for sensing and on-chip transduction make these NEMS AlN technology the best candidate for the development of volatile organic chemical sensors.

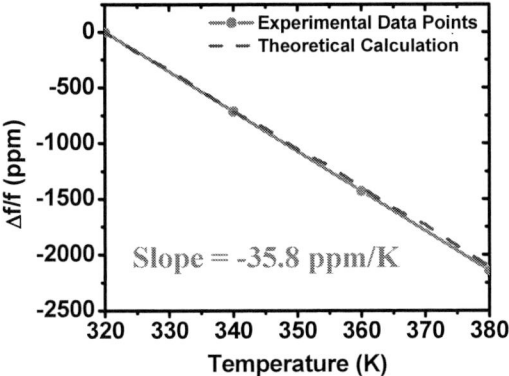

Figure 7: Experimental measurement of the TCF of the fabricated 8.5 GHz NEMS resonator compared with its

analytical prediction. The coefficient of Young's modulus of Al is about 10 X that of AlN and perfectly accounts for the recorded experimental data.

V. CONCLUSIONS

Design fabrication and testing of novel AlN piezoelectric NEMS resonators operating in the SHF band between 5 and 10 GHz have been demonstrated. For the first time on-chip piezoelectric sensing and actuation of nanostructures have been achieved by mechanically coupling a large number (49-99) of AlN vibrating nano-strips. AlN films have maintained high polycrystalline quality and orientation even in nanoscale devices, for which the highest $f \cdot Q$ product ($4.58 \cdot 10^{12}$ Hz) ever reported in NEMS resonators has been achieved. The super high frequencies of operation combined with on-chip sensing and actuation make the technology demonstrated herein amenable to the development of miniaturized components for wireless communications (WiMax, Satellite Radios and Radar), ultra sensitive chemical sensing and mechanical computing.

ACKNOWLEDGEMENT

This work was supported by NCMR/NSF grant no. IIS-07-15024 and NSF grant no. ECCS-08-22968. The authors wish to thank the staff of the Wolf Nanofabrication Facility (WNF) at The University of Pennsylvania and Chengjie Zuo and Dr. Marcelo Pisani for precious discussions.

REFERENCES

[1] M.L. Roukes, *Physics World*, 14, 25, 2001.

[2] M. Li, H. X. Tang, M.L. Roukes, *Nature Nanotechnology*, vol. 2, pp. 114-120, 2007.

[3] J. R. Vig, and F. L. Walls, *Proceedings IEEE Int. Frequency Control Symposium'00*, pp. 30-33, 2000.

[4] S. S. Verbridge et al, *Nano Letters*, vol. 7, no. 6, pp. 1728-1735, 2007.

[5] K. L. Ekinci, M.L. Roukes, *Review of Scientific Instruments*, 76, 061101, 2005.

[6] M. Hara et al, *Proceedings IEEE Ultrasonic Symposium'07*, pp. 1152-1155, 2007.

[7] P. J. Stephanou, J. P. Black, A. L. Benjamin, *Proceedings IEEE Radio Frequency Integrated Circuits Symposium'08*, pp. 171-174, 2008.

[8] G. Piazza, P.J. Stephanou, A.P. Pisano, *Journal of MicroElectroMechanical Systems,* vol. 15, no.6, pp. 1406-1418, December 2006.

[9] D. L. DeVoe, *Sensors and Actuators A,* 88, pp. 263-272, 2001.

[10] D. Weinstein, S. A. Bhave, *Proceedings IEEE International Electron Devices Meeting'07*, pp. 415-418, 2007.

[11] M. H. Cho et al, *Proceedings IEEE MTT-S Microwave Symposium'04*, pp. 1237-1240, 2004.

[12] P. J. Stephanou, A. P. Pisano, *Proceedings IEEE Ultrasonics Symposium'06*, pp. 2401-2404, 2006.

[13] R. H. Olsson III et al, *Proceedings IEEE Int. Frequency Control Symposium'08*, pp. 634-639, 2008.

[14] M. Rinaldi, C. Zuniga et al, *Proceedings IEEE Int. Frequency Control Symposium'08*, pp. 443-448, 2008.

CHAOS IN ELECTROSTATICALLY ACTUATED RF-MEMS MEASURED AND MODELED

J. Stulemeijer, R.W. Herfst, and J.A. Bielen
Epcos SAW, Nijmegen, The Netherlands

ABSTRACT

Observation of period doubling bifurcations and chaos in electrostatically actuated RF-MEMS are reported in this paper for the first time. Period doubling, quadrupling and chaotic motion can clearly be observed in the measurement data. A 1D nonlinear device model is fitted to the capacitance versus voltage (C-V) curves and mechanical resonance curve data. The features of the observed bifurcation diagram have been reproduced with the 1D model.

INTRODUCTION

Autonomous impact resonators were observed to exhibit chaotic motion [1]. Modeling work [2] predicted the existence of chaotic motion in electrostatic MEMS. This paper presents the first observations of chaotic motion in electrostatic RF-MEMS. Regardless of their excellent RF linearity, RF-MEMS can react in a strongly nonlinear manner to low frequency input stimuli close to the dynamic pull-in. These non-linearities can lead to chaotic motion. The chaotic motion was observed using a single frequency high bandwidth custom built 1-port network analyzer with improved resolution over the one described in [3], which is based on [4].

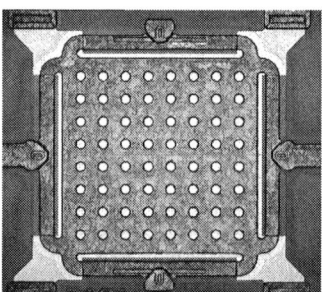

Figure 1: Geometry of RF-MEMS device used in all measurements.

The capacitive RF-MEMS switch used as test device is shown in Figure 1 with its capacitance versus voltage (CV) curve in Figure 2. The device consists of an aluminum plate suspended by beams, which is separated by an air gap from a dielectric layer on top of a lower electrode, see [5].

1D MODEL

A model [5] including inertia, nonlinear squeeze film damping, linear mechanical spring, mechanical contact, with exponential contact force displacement relation, and nonlinear electrostatic driving force was used:

$$m_{eff}\,\ddot{g}(t) + \frac{b_{eff}}{(g_o - g(t))^3}\dot{g}(t) + k_{eff}\,g(t) + c_1\,e^{c_2(g(t) - g_o + g_c)} = \frac{\varepsilon_0\,A_{eff}\,V^2(t)}{2(g_o - g(t))^2},$$

$$C(t) = C_0 + \frac{\varepsilon_0\,A_{eff}}{g_o - g(t)},$$

(1)

with $\varepsilon_0 = 8.85$ pF/m, open state air gap g_o, the spring constant k_{eff}, the plate area A_{eff}, the parasitic capacitance C_0, the contact parameters c_1 and c_2, the mass m_{eff}, and the damping constant b_{eff}.

The parameter g_o was designed at 3.2 µm. With white light interferometry it was confirmed that the actual gap did not deviate significantly from the designed gap.

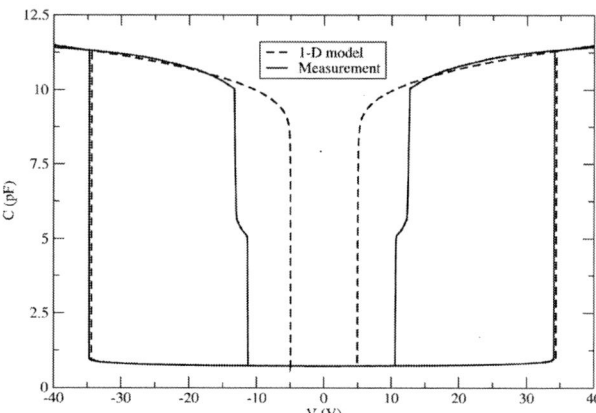

Figure 2: Modeled and measured C-V curve

The parameters k_{eff}, A_{eff}, C_0, c_1, and c_2 were obtained by fitting the modeled C-V curve to measurements as shown in Figure 2. The release voltage is not modeled well, because e.g. the plate starts to unzip in the corners when the voltage in the closed state is lowered towards the release voltage which is not captured in the 1D model.

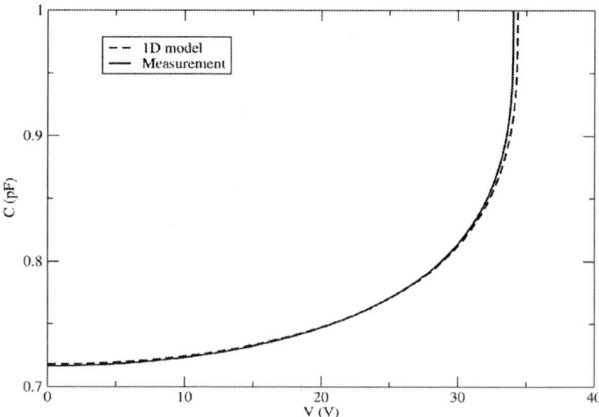

Figure 3: Modeled and measured C-V curve for Device in open state.

Figure 3 illustrates that the obtained fit in the open state is very good.

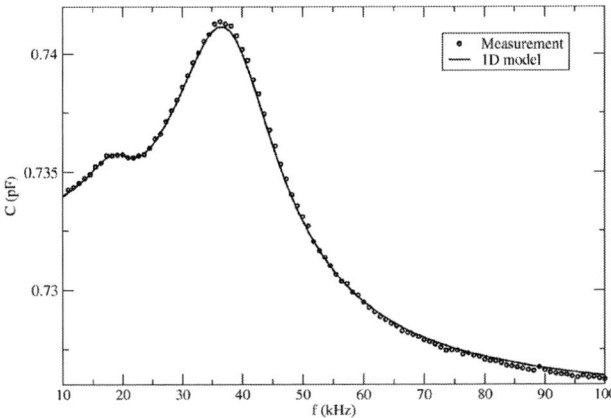

Figure 4: Measured and modeled frequency sweep response for V_{dc}=10 V and V_{ac}=5 V.

In Figure 4 a linear resonance curve of the RF-MEMS device is shown. The peak at half the resonance frequency is excited because the squaring of the voltage in the electrostatic force generates force components at the doubled excitation frequency. The 1D model parameters k_{eff} and m_{eff} were fitted to the resonance curve.

Table 1: Parameter values in fitted 1D model.

Parameter	value	unit	Parameter	value	unit
g_0	3.2	μm	c_1	45	kN
A_{eff}	177000	μm^2	c_2	150	μm^{-1}
k_{eff}	190	N/m	m_{eff}	3.08	ng
C_0	228	fF	b_{eff}	11.4	g μm^3/s

MEASUREMENT RESULTS

The capacitance was measured as function of time with V_{DC}=32 V at f=18 kHz, while V_{AC} was swept from 6.6 to 7.25 V. This resulted in motions with a single period (V_{AC} =6.6 V Figure 5), doubled period (V_{AC} =6.96 V Figure 6), quadrupled period (V_{AC} =7.06 V Figure 7), and a chaotic motion (V_{AC} =7.20 V Figure 8). For each of the motions the measured capacitance versus time, the spectral content of motion, and the Poincaré section is shown. A Poincaré section is obtained by sampling the velocity versus displacement curve at an integer times the driving period. The period doubled motion shows an additional peak at half the driving frequency and the quadrupled motion shows a peak at a quarter of the driving frequency. The chaotic motion shows increased power on all frequencies. In a chaotic motion each cycle will be unique, thus, as observed in Figure 8c, a cloud of points is formed in the Poincaré section.

When a projection of the Poincaré section to the displacement axis is plotted on the y-axis versus the swept V_{AC} on the x-axis a bifurcation diagram is formed. Figure 9 shows the simulated bifurcation diagram with V_{DC} =30 V. When we compare the simulated results with the measured results we see a small discrepancy in the V_{DC} at which the motion becomes chaotic. For V_{DC} =32 V no chaos was found. This small change in V_{DC} resulted in a large shift in the V_{AC} range of the chaotic region. Apparently the model is not completely describing all nonlinearities. Nonlinear mechanical spring behavior and squeeze film compression are expected to be needed for a better quantitative fit. This work has increased insight into the role of nonlinearities in RF-MEMS behavior.

CONCLUSIONS

The predicted chaotic behavior has been observed in RF-MEMS. A very simple 1D model suffices to get qualitative agreement between modeling and measurement. More comprehensive models are needed to get quantitative agreement. All measurements presented are taken at one frequency. Future work might look at the existence of chaotic regions over wider frequency and bias region.

ACKNOWLEDGEMENTS

The authors gratefully acknowledge the financial support by the Dutch Ministry of Economic Affairs in the framework of the Point One project MEMSLand [www.hitechprojects.com/euprojects/memsland/].

REFERENCES

[1] J. Bienstman, R. Puers, and J. Vandewalle, "Periodic and chaotic behaviour of the autonomous impact resonator," in *Micro Electro Mechanical Systems, 1998. MEMS 98. Proceedings., The Eleventh Annual International Workshop on*, 1998, pp. 562-567.

[2] S. K. De and N. R. Aluru, "Complex nonlinear oscillations in electrostatically actuated microstructures," *Microelectromechanical Systems, Journal of,* vol. 15, pp. 355-369, 2006.

[3] R. W. Herfst, P. G. Steeneken, and J. Schmitz, "Time and voltage dependence of dielectric charging in RF MEMS capacitive switches," in *Reliability physics symposium, 2007. proceedings. 45th annual. ieee international*, 2007, pp. 417-421.

[4] H. Nieminen, J. Hyyryläinen, T. Veijola, T. Ryhänen, and V. Ermolov, "Transient capacitance measurement of MEM capacitor," *Sensors & Actuators: A. Physical,* vol. 117, pp. 267-272, 2005.

[5] P. G. Steeneken, T. G. S. M. Rijks, J. T. M. v. Beek, M. J. E. Ulenaers, J. D. Coster, and R. Puers, "Dynamics and squeeze film gas damping of a capacitive RF MEMS switch," *Journal of Micromechanics and Microengineering,* vol. 15, pp. 176-184, 2005.

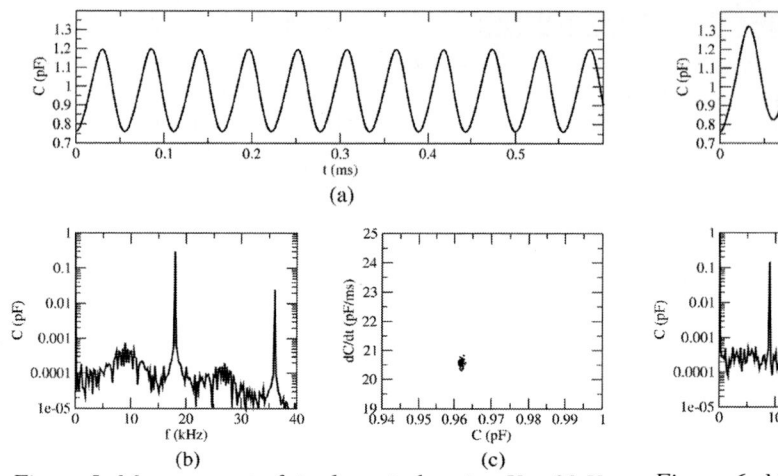

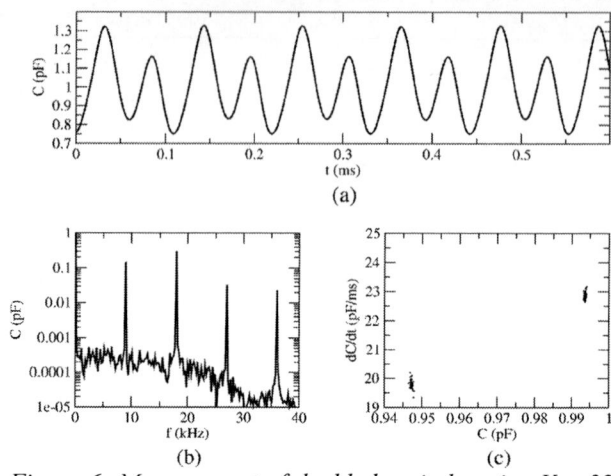

Figure 5: Measurement of single period motion V_{dc}=32 V, V_{ac}=6.6 V, and f=18 kHz, a) capacitance versus time, b) frequency response, and c) Poincaré section.

Figure 6: Measurement of doubled period motion V_{dc}=32 V, V_{ac}=6.96 V, and f=18 kHz, a) capacitance versus time, b) frequency response, and c) Poincaré section.

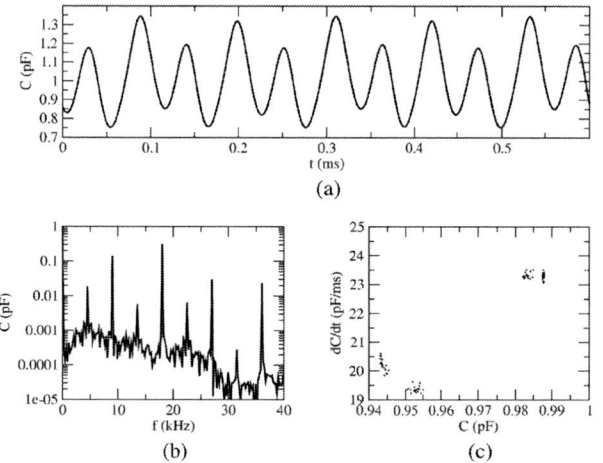

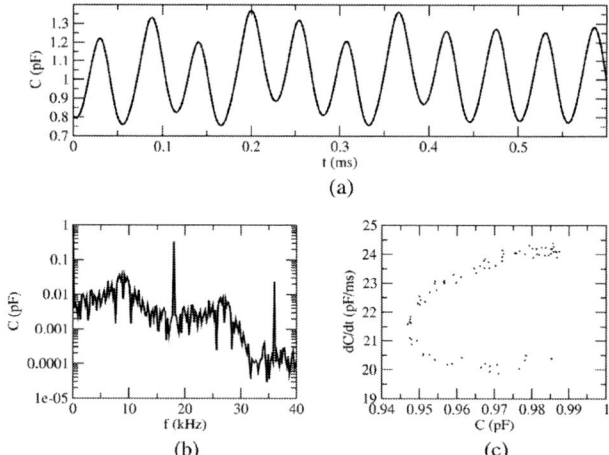

Figure 7: Measurement of quadrupled period motion V_{dc}=32 V, V_{ac}=7.06 V, and f=18 kHz, a) capacitance versus time, b) frequency response, and c) Poincaré section.

Figure 8: Measurement of chaotic motion V_{dc}=32 V, V_{ac}=7.2 V, and f=18 kHz, a) capacitance versus time, b) frequency response, and c) Poincaré section.

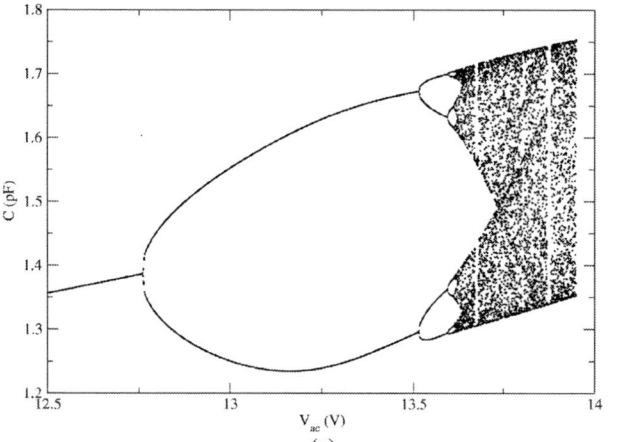

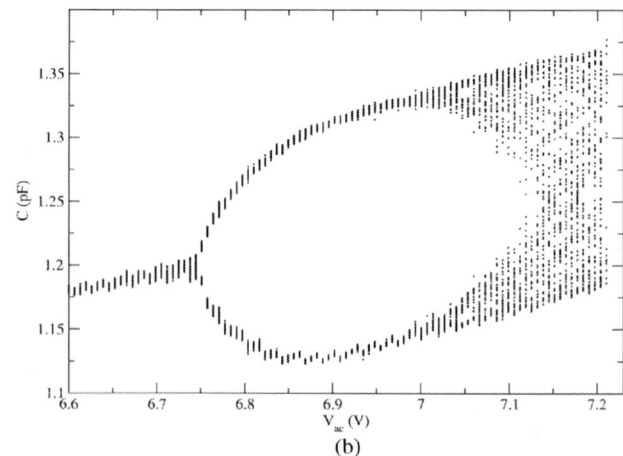

Figure 9: a) Modeled bifurcation diagram V_{dc}=30 V and f=18 kHz, and b) Measured bifurcation diagram V_{dc}=32 V and f=18 kHz.

THERMOELASTIC FEM-BEM MODEL FOR MEMS RESONATOR SIMULATION

D. Ekeom[1] and L. Buchaillot[2]

[1]Microsonics, Saint-Avertin, FRANCE

[2]IEMN, département ISEN, UMR CNRS 8520 Lille, FRANCE

ABSTRACT

Minimization of dissipative losses is a major goal in MEMS resonator design [1]-[2]. For an accurate simulation of a MEMS resonator vibrating in vacuum, thermoelastic damping phenomenon related to the irreversible heat dissipation induced by the coupling between heat transfer and strain rate during the resonator vibration and acoustic radiation into the substrate have to be taken into account. The finite element method (FEM) is suitable for structural simulation, especially for thermoelastic damped structures vibrating in vacuum. When the vibrating structure is deposited on an unbounded elastic medium, radiating conditions have to be taken into account. The boundary element method (BEM) is suitable for unbounded medium, because the radiating conditions are exact. It is often used to complete of the finite element method. The goal of our work is to develop and to validate a thermal-electromechanical FEM-BEM tool, which is helpful to predict and understand MEMS energy loss dissipation.

INTRODUCTION

MEMS resonators are important signal processing elements in communication systems. Over the past decade, there has been substantial progress in developing new types of miniaturized MEMS resonators using microfabrication processes [3]-[4]. For these MEMS to find application, it is essential they have high quality factors (Q-factor). FEM can be used to determine resonance frequency, but not to determine Q-factor. The reason for this is that it is difficult to model the relevant loss mechanisms. In high frequency resonators, the acoustic radiation into the substrate (anchors loss radiation) and thermoelastic dissipation play crucial roles in determining energy losses from a resonator.

D. Sherman [5] has presented some initial studies of microresonator anchors loss radiation. Results are mainly qualitative and applicable only to bending modes. More recently J. Judge [6] and al. has presented analytical estimation of Q-factor due to radiation through the substrate, for beam and substrate made from same material. The results shown that anchor Q-factor is mainly depending of the beam aspect ratios. David S. Bindel *et al.* [7] proposed a numerical method based on perfectly matched layer, recently extended time harmonic elastodynamics problems. The PML is coupled to FEM. Y.-H. Park *et al.* [1]-[2] has presented semi-analytical approach of wave propagation in the multilayered-substrate. This model can capture wave motions of submicron scale and is weakly coupled to FEM. Thermoelastic damping is taken into account with constant structural Q-factor. Zener [8]-[9] and R. Lifshitz *et al.* [10] have developed analytical solution to evaluated thermoelastic Q-factor versus of frequency of clamped-clamped beam resonator. Numerical tool for thermoelastic

damping can commonly be found in commercial FEM packages [11].

This work deals with three-dimensional thermal-electromechanical FEM-BEM strongly coupled simulation. To take into account energy loss by anchor loss and thermoelastic dissipation, the resonator is modelled with thermal-electromechanical FEM, while the substrate is modelled with an integral representation based on a half-space isotropic Green function (i.e. to BEM). The model is analytically and experimentally validated.

NUMERICAL MODEL

The resonator constitutes a close solid domain $_s$. This domain is in contact with an elastic homogeneous isotropic half-space by surface ( ). All the variables exhibit an implicit $e^{+j\omega\tau}$ dependence where ω denotes the angular frequency and τ the time. In $_s$, the physical quantities of interest are the displacement vector **u** and the electric potential ϕ. The nodal values of **u** and ϕ are the unknowns of the following set of finite element equations [11].

$$\begin{bmatrix} [K]-[\nabla_u F]-\omega^2[M] & [K_{ut}] & -[\nabla_\Phi F]^T \\ -j\omega T_0[K_{ut}]^T & [K_t]+j\omega[C_t] & [0] \\ [\nabla_\Phi F] & [0] & [\nabla_\Phi Q] \end{bmatrix} \begin{Bmatrix} U \\ T \\ \Phi \end{Bmatrix} = \begin{Bmatrix} F \\ Q_t \\ Q_e \end{Bmatrix} \quad (1)$$

U (resp. **F**) is the vector of the nodal values of the displacement (resp. force), **T** (resp. Q_t) is the vector of the nodal values of temperature (resp. heat generation). Φ (resp. Q_e) is the vector of the nodal values of the electric potential (resp. electrical charge). [K], [K_{ut}], [K_p] and [K_d] are respectively the mechanical and thermoelastic stiffness matrices. [K_t] is the thermal conductivity matrix and [C_t] specific heat matrix. [M] is the consistent mass matrix. T_0 is the absolute reference temperature. [$\nabla_u F$], [$\nabla_\Phi F$] and [$\nabla_\Phi Q$] are respectively, the nodal electric forces derivatives matrix with respect to nodal displacements, the nodal electric forces derivatives matrix with respect to nodal charge and the nodal electrostatic charges derivatives matrix with respect to nodal potentials. This last matrix is equivalent to dielectric stiffness matrix. The three matrices are limited to the nodes located on the electrodes boundaries. As energy is provided by the electrical voltage generator, $F = 0$, $Q_t = 0$ and $Q_e = 0$ (no applied external load) and $\Phi = \Phi_0$ on hot electrode and $\Phi = 0$ on mass electrode, equation (1) leads to:

$$\begin{bmatrix} [K]-[\nabla_u F]-\omega^2[M] & [K_{ut}] \\ -j\omega T_0[K_{ut}]^T & [K_t]+j\omega[C_t] \end{bmatrix} \begin{Bmatrix} U \\ T \end{Bmatrix} = \begin{Bmatrix} F+[\nabla_\Phi F]^T\Phi_0 \\ 0 \end{Bmatrix} \quad (2)$$

978-1-4244-2977-6/09 $25.00 © 2009 IEEE

For a three-dimensional substrate, the displacement **u** on (Γ) is given by the integral representation [12].

$$\mathbf{u}(\underline{r}) = \int_{\Gamma} \left[G(\underline{r} - \underline{r}') \right] \mathbf{t}(\underline{r}') d\Gamma_{r'} \qquad (3)$$

t is the vector of the surface force density applied on (Γ), $\underline{r}$ (resp. $\underline{r}'$) is the vector position of the observation (source) point and [G] is the Green tensor of the isotropic half-space with stress-free surface. For each finite element e of (Γ), the spatial discretization of **t** is performed by using the classical FEM interpolation functions [N^e] [6]

$$\mathbf{t}^e(\underline{r}) = \left[N^e \right] \mathbf{T}_{\Gamma}^e, \qquad (4)$$

where $\mathbf{T}_{\Gamma}^e$ is the elementary nodal vector of surface force density. The subscript applies to nodes located on (Γ). Eq. (5) is combined with equations (2) and (3) for each node on (Γ) to lead after assembling to the matrix equation

$$\mathbf{U}_{\Gamma} = \left[X(\omega) \right] \mathbf{T}_{\Gamma}, \qquad (5)$$

With,

$$\left[X(\omega) \right] = \sum_e \int_{\Gamma^e} \left[G(\underline{r} - \underline{r}') \right] \left[N^e(\underline{r}') \right] d\Gamma_{\underline{r}'}^e . \qquad (6)$$

In the considered problem, nodal forces **F** (sub-vector of **F**, restricted to nodes located on (Γ)) appearing in eq. (1) and (7) are reaction forces of the substrate on $_s$ expressed as [13]

$$\mathbf{F}_{\Gamma} = -\sum_e \int_{\Gamma^e} \left[N^e(\underline{r}) \right]^T \mathbf{t}(\underline{r}) d\Gamma_{\underline{r}}^e$$

$$= -\left(\sum_e \int_{\Gamma^e} \left[N^e(\underline{r}) \right]^T \left[N^e(\underline{r}) \right] d\Gamma_{\underline{r}}^e \right) \mathbf{T}_{\Gamma}. \qquad (7)$$

This last equation is equivalent to a projection on Galerkin based function. Eq. (7) combined with eq. (5) to leads the matrix the following equation

$$\mathbf{F}_{\Gamma} = -\left[Z(\omega) \right] \mathbf{U}_{\Gamma}, \qquad (8)$$

Where, [Z(ω)] is an impedance matrix. It is a complex non-symmetric and fully populated. The final set of equations is obtained by combining equations (2) and (8)

$$\begin{bmatrix} [K] - [\nabla_u F] - \omega^2 [M] + [Z(\omega)] & [K_{ut}] \\ -j\omega T_0 [K_{ut}]^T & [K_t] + j\omega [C_t] \end{bmatrix} \begin{Bmatrix} U \\ T \end{Bmatrix} = \begin{Bmatrix} [\nabla_\Phi F]^T \Phi_0 \\ 0 \end{Bmatrix}. \qquad (9)$$

This formulation has been implemented by coupling strongly the FEM software (ANSYS) to the integral representation software (Microsonics). To take into account the bias voltage a pre-stressed nonlinear static analysis is run before the dynamic analysis.

THEORETICAL VALIDATION

A clamped-free beam polysilicon resonator is used to validate anchor loss modelled by the electromechanical FEM-BEM coupling. For the fundamental mode, with resonator and substrate made from the same material, the analytical approximation is given by J. Judge *et al.* [6].

$$Q_{ana}^{-1} = A \frac{W_r}{L_r} \left(\frac{h}{L_r} \right)^4, \qquad (10)$$

Where, L_r and W_r are respectively beam length and width, thickness h is equal 1 μm and the coefficient A is equal to 2.9. Numerical and analytical results are shown in table 1. Q_{num} and F_r are respectively, Q-factor and resonance frequency obtained by electromechanical FEM-BEM computation. Q_{ana} is the analytical Q-factor obtained with eq. (10).

Table 1: Analytical validation summary results

L_r (μm)	W_r (μm)	F_r (MHz)	Q_{num}	Q_{ana}
20	4	11.548	263778	275863
20	5	11.597	208288	220690
15	3	19.914	89402	87284
15	5	20.076	52276	52371

A thin polysilicon clamped-clamped beam of length 400 μm (x-direction), width = 0.5 μm (y-direction) and high 5 μm (z-direction) vibrating transversely under a uniform pressure 1 MPa applied in the z-direction, is used to validate the thermal-mechanical FEM model. The beam absolute reference temperature is T_0 = 25°C. The analytical solution was proposed by Zener [8]-[9]. Results are display in log scale on Figure 1.

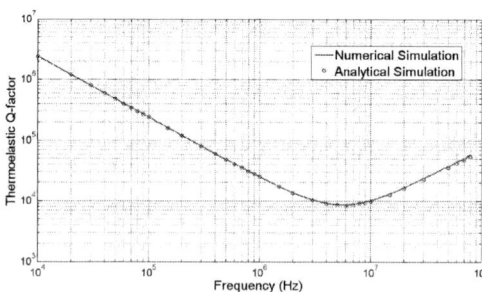

Figure 1: Thermoelastic Q-factor of a CC-beam resonator

Beam resonator material data used in this paper is given bellow: ρ = 2300 kg/m³, E = 150 Gpa, ν = 0.30, C_p = 712 J/kg, k = 148 W/m °K and α = 2.96 1/°K. Substrate material data is given bellow: ρ = 2300 kg/m³, E = 190 Gpa and ν = 0.17. Anchor and thermoelastic losses are analytically validated separately. In both cases, a good agreement is obtained between numerical and analytical Q-factors. It can be notice that Q-factors are frequency and geometrically dependent.

EXPERIMENTAL VALIDATION

The developed model is validated by correlating published experimental results [4] for two clamped-clamped beam resonators (Figure 2.) vibrating at 54.2 MHz and 71.8 Mhz, and two free-free beam resonators (Figure 3.) vibrating at 31.52 MHz and 92.25 MHz.

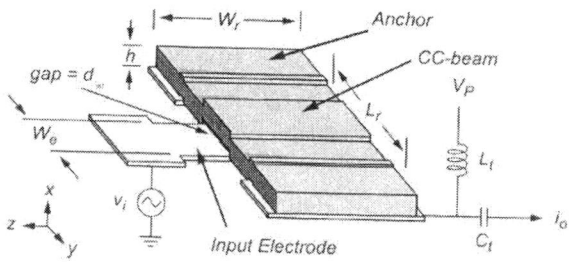

Figure 2 : Clamped-Clamped resonator shematic view

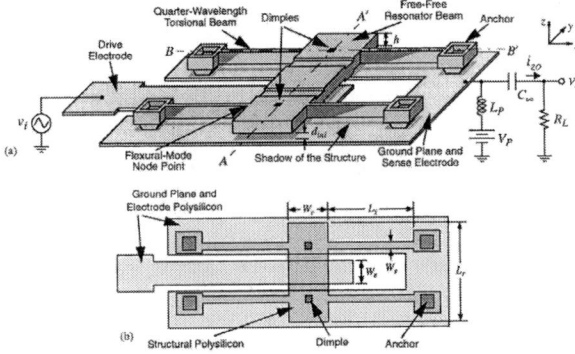

Figure 3 : Free-Free resonator shematic view

Resonator geometrical data are presented in table 2, beam lengths were adjusted to match experimental frequency and material data were defined previously.

Table 2: Resonators geometrical data

Parameter	Target frequency			
	30 MHz	50 MHz	70 MHz	90 MHz
	ff-beam	cc-beam	cc-beam	ff-beam
L_r (µm)	23.2 [1]	16.0 [2]	14.0 [3]	13.1 [4]
W_r (µm)	10.0	8.0	6.0	6.00
W_e (µm)	7.40	8.0	7.0	2.80
L_s (µm)	30.6	x	x	10.3
W_s (µm)	1	x	x	1
d_{ini} (µm)	0.16	0.30	0.30	0.16
h (µm)	2.05	2.10	2.10	2.05

Values use for simulation:
[1] L_r = 23.2 -0.230 µm; [2] L_r = 16.0 -1.070 µm
[3] L_r = 14.0 -1.560 µm; [4] L_r = 13.1 -0.105 µm

The experimental resonance frequencies F_{exp} and Q-factors at -3 dB, are directly measured and the unloaded resonator Q-factor Q_{exp} is deduced. A turbo-molecular pump was used to evacuate the chamber pressure on the order of 50 µtorr, to remove the viscous gas damping. Thus, the measured Q-factor is due to thermoelastic and acoustic radiation through anchor. Three kind of three-dimensional simulations are conducted. The first is a thermal-electromechanical FEM simulation, with clamped resonator-substrate interface. Resonance frequency is named F_{ted} and Q-factor Q_{ted} due to thermoelastic dissipation is evaluated. Secondly an electromechanical FEM simulation, with boundary element attached at the resonator-substrate interface. The multilayered substrate is assumed to be a semi-infinite silicon substrate, because the two first layers are very small compared to the problem wavelengths. Resonance frequency is named F_{anc} and Q-factor Q_{anc} due to radiation through anchor is evaluated. Finally, a full simulation with thermal-electromechanical finite element, and boundary element attached at the resonator-substrate interface is performed. Resonance frequency is named F_{num} and total Q-factor Q_{num} due to radiation through anchor and thermoelastic dissipation is evaluated. All numerical Q-factor are computed with resonance peak and frequency range at -3 dB, of the motional admittance. Contributions of thermoelastic loss (τ_{ted}) and anchor loss (τ_{anc}) to the global damping ratios of the coupled substrate resonator are estimated. Experimental and numerical results are presented table 3. Finally, simulated electrical admittance is display (Fig. 4 and Fig. 5).

Table 3: Experimental validation summary results

	Target frequency			
	30 MHz	50 MHz	70 MHz	90 MHz
	ff-beam	cc-beam	cc-beam	ff-beam
F_{ted}	31.51	57.23	76.51	92.37
F_{anc}	31.51	54.27	71.80	92.26
F_{num}	**31.51**	**54.27**	**71.80**	**92.26**
F_{exp}	**31.51**	**54.20**	**71.80**	**92.25**
Q_{ted}	8262	8628	10636	11739
Q_{anc}	7122673	513	338	20514
Q_{num}	**8258**	**487**	**329**	**7422**
Q_{exp}	**8214**	**857**	**301**	**7468**
τ_{ted} (%)	99.88	5.64	3.08	63.60
τ_{anc} (%)	0.12	94.39	96.92	36.40

The experimental resonance frequencies (F_{exp}) and Q-factors (Q_{exp}) matched well with the simulated resonance frequencies (F_{num}) and Q-factors (Q_{num}). These results show clearly that for an accurate simulation of MEMS resonator, radiation into the substrate and themoelastic effects have to be taken into account. Acoustic radiation into the substrate and themoelastic effects associated physical mechanisms discussed hereafter:

The augmented mass effect due to substrate modelling has an influence on resonance frequency estimation, in particular for clamped-clamped beam. The augmented mass effect is associated to the effective vibrating substrate volume. It can be characterized by $F_{ted} - F_{num}$.

The formula $1/Q_{num} = 1/Q_{anc} + 1/Q_{ted}$ is satisfied. Thus, contributions of anchor loss and thermoelastic loss to the global damping ratio of MEMS resonator are estimated.

Q_{anc}, Q-factor due to acoustic radiation through anchor is frequency and geometrically dependent. It increases drastically with the frequency. Acoustic radiation into the substrate is major source of energy dissipation for clamped-clamped resonator vibrating at 54.27 MHz and 71.80 MHz. Its contribution to global energy dissipation represents about 95 % of the total damping ration. While it represents less than 1% (resp. 36 %) for a free-free resonator vibrating at 31.51 MHz (resp. 92.25 MHz). Clearly, there is a correlation between augmented mass effect and anchor radiation: Q_{anc} is low when $F_{ted} - F_{num}$ is high.

Thermoelastic Q-factor Q_{ted} is also frequency and geometrically dependent, and not constant as often use in FEM simulation.

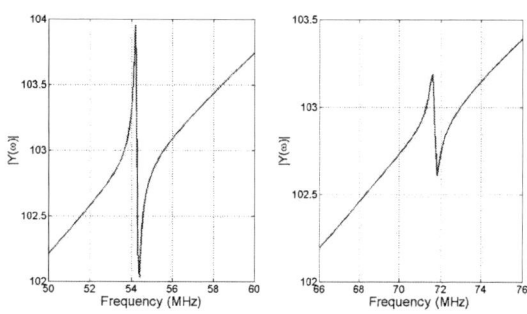

Figure 4 : CC-beams module of electrical admittance

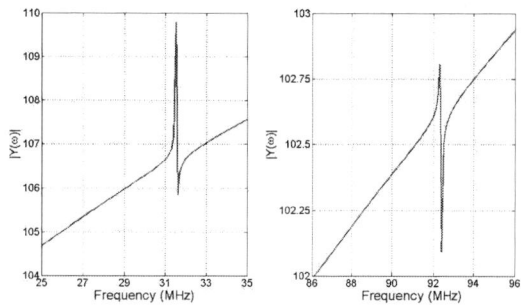

Figure 5 : FF-beams module of electrical admittance

CONCLUSION

To simulate MEMS Resonators, a three-dimensional numerical model, based on the thermal-electromechanical FEM representation of the resonator and the BEM description of the substrate, is developed. Elastodynamic energy radiation into the substrate and the thermoelastic damping are included in the model. Thus, it is possible to evaluate resonance frequency and Q-factor of MEMS

resonators. Analytical validation was successfully conducted. Numerical results are presented and compared to experimental results, for two clamped-clamped beam resonators vibrating at 54.2 MHz and 71.8 Mhz, and two free-free beam resonators vibrating at 31.52 MHz and 92.25 MHz. Thermoelastic and acoustic radiation effects are shown, their contributions in the resonator Q-factor value are clearly identified.

ACKNOWLEDGEMENTS

This work was supported by Oséo, European social funds, MITI Incubateur Nord Pas de Calais.

REFERENCES

[1] Y.-H. Park and K. C. Park," High-Fidelity Modeling of MEMS Resonators—Part I: Anchor Loss Mechanisms Through Substrate" *J. of Microelectromechanical syst.*, vol. 13, no. 2, (2004), pp 238-247.

[2] Y.-H. Park and K. C. Park," High-Fidelity Modeling of MEMS Resonators—Part II: Coupled Beam-Substrate Dynamics and Validation"*J. of Microelectromechanical syst.*, vol. 13, no. 2, (2004), pp 248-257.

[3] C. Nguyen, "Transceiver front-end architectures using vibrating micromechanical signal processors," in *Dig. of Papers, Topical Meeting on Silicon Monolithic Integrated Circuits in RF Systems*, pp. 23-32, (2001).

[4] K. Wang, A.-C. Wong, and C. Nguyen," VHF Free–Free Beam High-Q Micromechanical Resonators"*J. of Microelectromechanical syst.*, vol. 9, no. 3, (2000), pp 347-360.

[5] D. Sherman "An investigation of MEMS anchor design for optimal stiffness and damping," University of California, Departement of Mechanical Engineering, Tech. Rep., M.S. Report, (1996).

[6] J. Judge, D. Photiadis, J. Vignola, B. Houston and J. Jarzynski, *J. of Applied Phys.*, vol. 101, no. 013521, (2007), pp 1-11.

[7] D. S. Bindel, E. Quévy, T. Koyama, S. Govindjee, J. W. Demmel and R. T. Howe, "Anchor Loss Simulation in Resonators" *18th IEEE Conference on MEMS*, (2005), pp 133-136.

[8] C. Zener, "Internal Friction in Solids", *Phys. Rev.*, Vol. Vol. 52, (1937), p 230-235.

[9] C. Zener, "Internal Friction in Solids", *Phys. Rev.*, Vol. 53, (1938), pp 90-99.

[10] R. Lifshitz, and M. L. Roukes,"Thermoelastic Damping in Micro- and Nanomechanical Systems," Phys. Rev. B, 61, pp. 5600–5609, (2000).

[11] ANSYS version 11.0, user's guide, ANSYS Inc., Pittsburgh, (2007).

[12] G.F Miller and H. Pursey,"The field and radiation impedance of mechanical radiators on the free surface of a semi-infinite isotropic solid". *Proc R. Soc.* Vol. 223, (1954), pp 521-542.

[13] O.C. Zienkiewicz and R.L. Taylor, "The Finite Element Method", 4th ed., McGraw-Hill New Yokk, (1989).

RESONANT PIEZOELECTRIC ALGAN/GAN MEMS SENSORS IN LONGITUDINAL MODE OPERATION

K. Brueckner[1], F. Niebelschuetz[1], K. Tonisch[1], R. Stephan[1], V. Cimalla[2],
O. Ambacher[2], and M.A. Hein[1]

[1]Institute of Micro- and Nanotechnologies, Ilmenau University of Technology, Ilmenau, GERMANY
[2]Fraunhofer Institute for Applied Solid-State Physics IAF, Freiburg, GERMANY

ABSTRACT

Free-standing piezoelectric AlGaN/GaN beam resonators have been prepared on silicon substrates using a semiconductor fabrication process. To realize the back electrode for the piezoelectric active layer, the two-dimensional electron gas at the interface of the III/V heterostructure was employed. Longitudinal acoustic resonances have been excited and detected electrically. The fundamental and higher order vibration modes were analyzed in the frequency domain. The dependences of the measured resonant frequencies between 3.8 and 63.0 MHz are related to geometrical and material parameters. The sensitivity of the resonant response to environmental parameters is demonstrated exemplarily by investigating its dependence on ambient pressure.

INTRODUCTION

Resonant micro-electromechanical systems (MEMS) are progressively being adopted for the development of advantageous radio frequency (RF) devices [1], e.g., in terms of filters and oscillators. In addition, they are also suitable for the realization of novel sensor concepts, e.g., for the detection of chemical or biological species [2], or for the measurement of selected environmental parameters [3]. Among other types of actuation schemes like electrostatic or magnetomotive excitation [4], the piezoelectric scheme is favorable for many applications due to the potential for reducing the device dimensions while maintaining the electromechanical coupling strength [5].

Piezoelectric MEMS resonators have been realized previously using different materials, e.g., ZnO, AlN, AlGaAs, and PbZrTiO₃ (PZT) [5,6]. The free-standing structures feature piezoelectric active layers that are enclosed between two metallic layers which serve as top and back electrodes. In many cases, the geometry is enhanced by additional structural piezoelectric inactive layers, in order to define the elastic properties and hence the resonant behavior of the devices.

Here, AlGaN/GaN heterostructures are presented that combine the pyroelectric and piezoelectric properties of the group III nitrides [7]. The structural quality and the intrinsic properties of the material are suitable for size reduction, to achieve high frequencies at low intrinsic losses. At the AlGaN/GaN interface, a two-dimensional electron gas (2DEG) is confined, for which Hall measurements revealed an electron density and mobility of 10^{13} cm^{-2} and 1000 cm^2/V s, respectively. Thus, the 2DEG provides sufficient conductivity to form the back electrode of a piezoelectric transducer configuration without metallic interlayers. This has been demonstrated recently with interdigital transducers for a surface acoustic wave device [8] as well as for the piezoelectric actuation of a mesa heterostructure [9]. Whereas the piezoelectricity enables the realization of an integrated coupling mechanism, the

2DEG additionally delivers a pronounced sensitivity to mechanical stress and other environmental parameters, which opens the potential to exploit novel sensor principles based on pyroelectric effects [7].

Cantilevers and doubly-clamped piezoelectric beam resonators have been reported, to operate in mechanical vibration modes of flexural bending [5,10]. Another group of resonators is based on contour-mode vibrations, e.g., longitudinal width or length extensional, of suspended geometric bodies on flexible supports [6,11]. These modes can be described by standing bulk acoustic waves which occur at frequencies where integer multiples of the half-wavelength coincide with the geometric dimensions of the resonator. Here, we demonstrate the electrical excitation and detection of longitudinal acoustic vibration modes within piezoelectric doubly-clamped beam resonators, i.e. at fixed-fixed boundary conditions. This kind of miniaturized flexural plate wave device can also be referred to as Lamb wave resonator [12]. The acoustic resonances have been excited and detected electrically in the fundamental and higher order modes at frequencies up to 63 MHz.

DESIGN AND FABRICATION

The free-standing doubly-clamped beam resonators, as the ones shown in the scanning electron microscope images of Fig. 1, consist of a AlGaN/GaN heterostructure of a total thickness of 680 nm. The lower GaN and the upper $Al_{0.31}Ga_{0.69}N$ layer are 650-nm and 30-nm thick, respectively. Whereas the thickness t was kept constant for all fabricated beam resonators, they vary in length l from 10 to 1000 µm and width w from 2 to 10 µm.

Epitaxial layers of the heterostructure have been deposited on Si substrates by metal-organic chemical vapor deposition [13]. The lateral dimensions of the beam resonators and of the electrical contacts were defined by optical lithography. To release the structures, a special combination of dry anisotropic and isotropic etching was applied that conserves the 2DEG along the entire beam

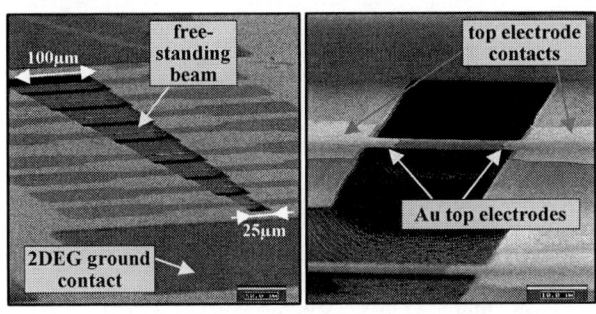

Figure 1: Scanning electron microscope images showing an array of AlGaN/GaN beams of lengths l between 25 and 100 µm (left) and one of the shortest realized beam resonators of 20 µm in length (right).

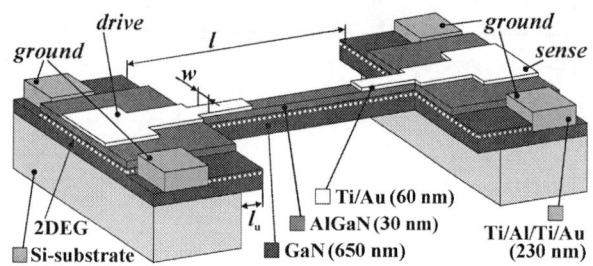

Figure 2: Scheme of the doubly-clamped beam resonators analyzed that vary in length l, undercut length l_u, and width w. The layer thicknesses are given in brackets. The drive, sense, and ground contact areas are used for electrical excitation and detection.

length [14]. 50-nm thick Au top electrodes with a thin adhesive Ti layer underneath were deposited at both ends of the beams, each extending along the beams by one quarter of the total length. The electrical contact to the 2DEG back electrode was provided by the alloyed Ti/Al/Ti/Au ground contact areas.

Fig. 2 sketches the basic resonator structure. If RF signals are applied to the drive contacts, a vertically oriented electric field evolves between the driving top electrode and the conductive 2DEG sheet charge. Consequently, longitudinal in-plane mechanical oscillations of the suspended microbridges can be excited utilizing the inverse piezoelectric effect (piezoelectric modulus d_{31} ≈ 2 pm/V). The signal detection is achieved at the sense electrode contacts through the charges generated by the direct piezoelectric effect.

The resonant frequencies f_n of the nth harmonic resonant longitudinal mode are given by the relation

$$f_n = c_1/\lambda = n \cdot c_1/(2l) . \qquad (1)$$

Here, c_1 denotes the phase velocity and λ the wavelength of the corresponding standing acoustic wave; c_1 depends strongly on the excited wave mode. As discussed below, for the geometry considered here, c_1 can be estimated as

$$c_1 \approx \sqrt{E_{GaN}/\rho_{GaN}} , \qquad (2)$$

where E_{GaN} and ρ_{GaN} are the Young's modulus and the mass density of the dominating beam material GaN, respectively.

To evaluate the strength of the electromechanical signal transduction, the structure was investigated numerically by finite element models using the commercial computer code ANSYS®. The electromechanical coupling factor η, which describes the force generated per applied input voltage or equivalently the output charge per displacement, was estimated to be about five times higher for the longitudinal modes as for the modes of flexural bending which can also be excited in these resonator structures [15]. The numerical analysis included the effects originating from the high residual tensile layer stress σ_l inherent to the devices ($\sigma_l \approx 800$ MPa). Whereas tensile stressed flexural resonators exhibit notably higher f_n than their relaxed equivalents, the influence on the longitudinal modes was found to be only weak. The resulting tensile stress σ_b of the etched individual resonator beams is additionally influenced by the undercut length l_u, (cf. Fig. 2) which results from the etching process. As the anchor regions can relax during the release process, $\sigma_b = \sigma_b(l)$ increases to values slightly above the value σ_l of the layers.

EXPERIMENTAL CHARACTERIZATION

The piezoelectric beam resonators have been contacted electrically inside a vacuum wafer probe station SUSS PMV150 by two 650-µm pitch coplanar probes, one at each port, cf. Fig. 2. The probe station allows the operation of the resonators in ambient atmosphere, as performed solely in this study, or in gaseous nitrogen. The ambient pressure p can be varied between a high vacuum of 10^{-3} Pa and normal ambient pressure of 10^5 Pa.

A vector network analyzer Anritsu MS4630B was used to characterize the beam resonators by recording the frequency-dependent scattering parameters S_{21}. The driving power could be varied between -40 and +10 dBm, where the lower limit was set by the noise level of the receiving port and the upper limit by the onset of nonlinear distortion of the response curves.

In Fig. 3, the magnitude of the transmission coefficient $|S_{21}|$ of the first acoustic resonant frequency f_1 is visualized for the longest realized beam resonator ($l = 1000$ µm) at four different pressure levels p. At constant driving power, the magnitude of the resonant peak under vacuum is about three times higher, and shifted to higher frequencies by about 1 kHz, compared to ambient conditions, revealing pressure-dependent quality factors Q and resonant frequencies f_n.

The measured frequency responses could be simulated using the equivalent electrical circuit model of Fig. 4. It includes a series resonant circuit (R_m, L_m, C_m) which is the transformed model of the mechanical resonator. The capacitances of the drive and sense contact areas are denoted as C_{pad}. The transformer T_{sense} accounts for the 180° phase shift between the drive and sense port signals in the fundamental mode. R_{bg}, C_{bg}, and L_{bg} represent direct ohmic, capacitive, and inductive couplings, respectively, that contribute to the background signal superimposed on the response of the mechanical resonator. In our case, the ohmic background caused by the limited conductance of the 2DEG back electrode was found to be dominating.

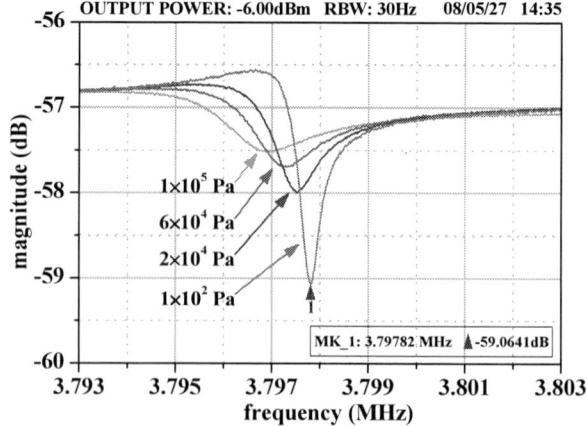

Figure 3: Pressure dependence of the transmission coefficient S_{21} of a 1000-µm long and 5-µm wide resonator bridge. Resonant frequency and quality factor shift downwards as the pressure is increased.

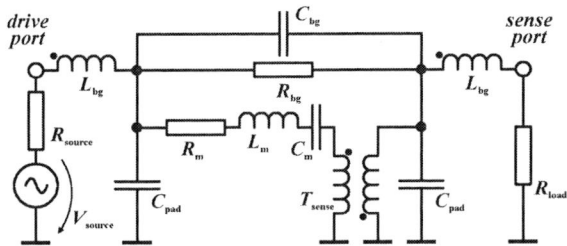

Figure 4: Simplified equivalent electrical circuit model of the longitudinal acoustic beam resonators. The mechanical components of mass m, stiffness k, and damping coefficient c are transformed to inductance L_m, capacitance C_m, and ohmic resistance R_m, respectively.

The quality factors were determined by curve fitting assuming a Lorentzian frequency response superimposed by a frequency-independent background signal. For the resonator analyzed in Fig. 3, the Q-factors were determined to decrease from 5950 under vacuum to 1725 at normal ambient conditions.

RESULTS AND DISCUSSION

In the experimental study of our first samples, the resonant frequencies under vacuum have been measured on a chip containing 5-μm wide doubly-clamped resonator bridges of lengths l between 75 and 1000 μm. The results are compiled in Fig. 5, where double-logarithmic scaling was chosen to emphasize the power-law dependence of the resonant frequency on the beam length l: $f_n \sim l^b$ with a characteristic exponent b. According to Fig. 5, the slopes of the $f_n(l)$ curves fell slightly above -1, e.g., $b_1 = -0.935$.

The observed oscillatory mode can be described as 0th-order symmetric Lamb wave [16]. The sound velocity c_l defined by Eq. (2) is representative of such Lamb wave modes in the case of very thin beams, i.e. with a thickness t much smaller than the lateral acoustic wavelength λ. The observed small deviation from the inverse proportionality expected by Eq. (1) is caused by the dispersion relation of the wave mode, implying decreasing phase velocities for increasing frequencies. For the limit of thick beams, c_l is approximately reduced by a factor of 0.6 and approaches the value of a Rayleigh surface wave. The values $c_l \approx$

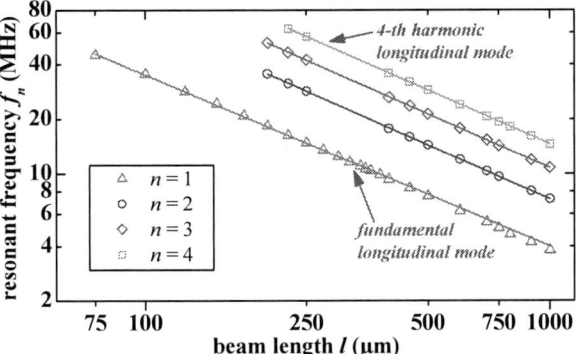

Figure 5: Resonant frequency f_n of the nth longitudinal vibration mode, for $1 \leq n \leq 4$, versus beam length l on double-logarithmic scales. The measurement was carried out for a set of 5-μm wide doubly-clamped beams with lengths l between 75 and 1000 μm. The lines represent power-law fits. For all curves, the slope is close to -1.

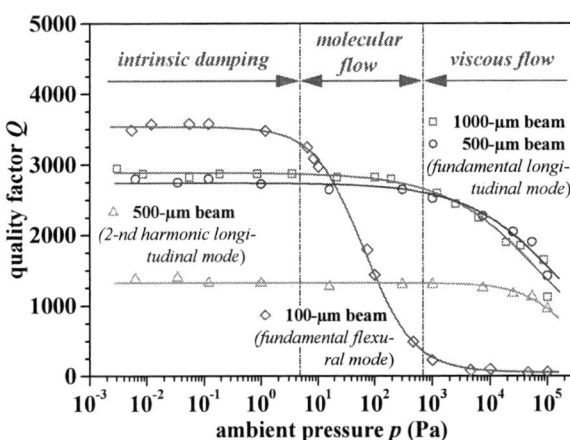

Figure 6: Pressure dependence of the quality factor Q of the fundamental longitudinal vibration mode measured for two 5-μm wide resonator beams having lengths of 500 μm and 1000 μm. For the 500-μm long beam, Q(p) is also shown for the 2nd harmonic. The lines are fits according to the analytical models described in the text.

7500 m/s obtained for the AlGaN/GaN resonators agree very well with results from literature which were obtained from measurements of AlGaN layers for a slightly lower level of tensile stress [17].

In addition to the curves shown in Fig. 3, the pressure dependence of the doubly-clamped longitudinal mode beam resonators is further detailed in Fig. 6 for the fundamental mode of two resonators of lengths $l = 500$ and 1000 μm and, for the shorter beam, also for the 2nd harmonic mode. The resonant frequencies f_1 were measured as 3.794 MHz and 7.536 MHz for the 1000-μm and the 500-μm long beam, respectively. The second harmonic of the 500-μm long beam was found at $f_2 = 14.425$ MHz. For comparison, the Q(p)-curve of a 100-μm long doubly-clamped flexural bending beam resonator ($f_1 = 559$ kHz) is included in the diagram [18].

The total Q-factor can be described as $Q^{-1}(p) = Q_{vac}^{-1} + Q_{fluid}^{-1}(p)$, where Q_{vac} and $Q_{fluid}(p)$ represent the intrinsic and the p-dependent fluid damping losses, respectively. At pressures below about 5 Pa, the vibrations are unaffected by the environment and the Q-factor is dominated by intrinsic and clamping losses. In the region of molecular damping, a pronounced Q-drop can be observed for the flexural bending mode ($Q_{fluid} \sim p^{-1}$). In the continuum of viscous flow, for $p \geq 1$ kPa, the Q-factors of the longitudinal modes also decrease but with a lower sensitivity $\partial \log(Q)/\partial(\log_{10}(p))$ compared to the flexural mode beam in the molecular region. In the viscous region, one finds $Q_{fluid} \sim p^{-1/2}$ for the flexural mode, which seems to hold for the longitudinal modes, too. This overall behavior is expected as the mechanical deflections of the acoustic modes are much smaller than those of the flexural bending modes. Similar results for Q(p) have been obtained from measurements for contour-mode MEMS resonators [19,20]. The Q-factor for ambient conditions was found to be of the same order of magnitude as the vacuum value Q_{vac}. Hence, intrinsic Q_{vac} and external Q_{fluid} are of comparable magnitude. For the 2nd harmonic longitudinal mode of the 500-μm long beam, the intrinsic losses were found to exceed the viscous losses under ambient conditions leading to a reduced sensitivity of Q(p).

Despite the lower sensitivity on p, longitudinal mode resonators can be advantageous in sensor applications where the high Q maintained in gaseous and presumably as well in liquid environments simplifies the detection than for the low-Q flexural mode devices. Compared to the reported Q-factors of several hundred thousands for the flexible supported contour-mode devices, the values of the doubly clamped longitudinal mode MEMS resonators of this study are by far lower. It is tempting to speculate that the coupling of mechanical energy into the substrate at the clamping points is a possible source of losses, which has to be studied in more detail. Additionally, parasitic capacitances, mechanical, galvanic and capacitive couplings will be minimized in the next design step, by applying bridge measurement methods, using either external [11,18] or on-chip [6] compensation elements.

CONCLUSIONS

We have demonstrated piezoelectrically actuated doubly-clamped AlGaN/GaN MEMS resonators where a two-dimensional electron gas forms the back electrode of the piezoelectric transducer. Resonant longitudinal acoustic vibration modes have been electrically excited and detected. Facilitating all-electrical transduction, the electromechanical coupling was found to be much higher for the longitudinal modes than for the flexural bending modes. The measured resonant frequencies varied between 3.8 and 63.0 MHz and scaled inversely with the beam length. The Q-factors of several thousands under vacuum decreased to values around one thousand under normal ambient pressure. The highest sensitivity was observed in the region of viscous damping which may be utilized in future sensor applications for the determination of the properties of gaseous or even liquid fluids. Improvements of the actuation and read-out schemes including the reduction of the background signal will be faced next.

ACKNOWLEDGEMENT

This work has been funded by the German Research Foundation (DFG), Priority Program 1157 'Integrated electroceramic functional structures' (Grants HE3642/2 and AM105/2).

REFERENCES

[1] C.T.-C. Nguyen, "MEMS technology for timing and frequency control", *IEEE Trans. Ultrason. Ferroelectr. Freq. Control*, vol. 54, pp. 251-270, 2007.

[2] N.V. Lavrik, M.J. Sepaniak, and P.G. Datskos, "Cantilever transducers as a platform for chemical and biological sensors", *Rev. Sci. Instrum.*, vol. 75, pp. 2229-2253, 2004.

[3] K. Khosraviani and A.M. Leung, "The nanogap pirani – a pressure sensor with superior linearity in atmospheric pressure range", in *Proc. 21st IEEE Int. Conf. on Micro Electro Mechanical Systems (MEMS 2008)*, Tucson, Arizona, Jan. 13-17, 2008, pp. 900-903.

[4] K.L. Ekinci, "Electromechanical transducers at the nanoscale: actuation and sensing of motion in nano-electromechanical systems (NEMS)", *Small*, vol. 1, pp. 786-797, 2005.

[5] D.L. DeVoe, "Piezoelectric thin film micromechanical beam resonators", *Sens. Actuators A*, vol. 88, pp. 263-272, 2001.

[6] L. Li, P. Kumar, L. Calhoun, and D.L. DeVoe, "Piezoelectric $Al_{0.3}Ga_{0.7}As$ longitudinal mode bar resonators", *J. Microelectromech. Syst.*, vol. 15, pp. 465-470, 2006.

[7] V. Cimalla, J. Pezoldt, and O. Ambacher, "Group III nitride and SiC based MEMS and NEMS: materials properties, technology and applications", *J. Phys. D*, vol. 40, pp. 6386-6434, 2007.

[8] K.-Y. Wong, W. Tang, K.M. Lau, and K.J. Chen, "Surface acoustic wave device on AlGaN/GaN heterostructure using two-dimensional electron gas interdigital transducers", *Appl. Phys. Lett.*, vol. 90, art. no. 213506, 3 pp., 2007.

[9] K. Tonisch, C. Buchheim, F. Niebelschütz, A. Schober, G. Gobsch, V. Cimalla, O. Ambacher, and R. Goldhahn, "Piezoelectric actuation of (GaN/)AlGaN/ GaN heterostructures", *J. Appl. Phys.*, vol. 104, no. 8, art. no. 084516, 8 pp., 2008.

[10] D.L. DeVoe and A.P. Pisano, "Modeling and optimal design of piezoelectric cantilever microactuators", *J. Microelectromech. Syst.*, vol. 6, pp. 266-270, 1997.

[11] K. Sundaresan, G.K. Ho, S. Pourkamali, and F. Ayazi, "Electronically temperature compensated silicon bulk acoustic resonator reference oscillators", *IEEE J. Solid-State Circuits*, vol. 42, pp. 1425-1434, 2007.

[12] J. Bjurström, I. Katardjiev, and V.Yantchev, "Lateral-field-excited thin-film Lamb wave resonator", *Appl. Phys. Lett.*, vol. 86, art. no. 154103, 3 pp., 2005.

[13] A. Dadgar, F. Schulze, M. Wienecke, A. Gadanecz, J. Bläsing, P. Veit, T. Hempel, A. Diez, J. Christen, and A. Krost, "Epitaxy of GaN on silicon-impact of symmetry and surface reconstruction", *New J. Phys.*, vol. 9, art. no. 389, 10 pp., 2007.

[14] F. Niebelschütz, V. Cimalla, K. Tonisch, Ch. Haupt, K. Brückner, R. Stephan, M. Hein, and O. Ambacher, "AlGaN/GaN-based MEMS with two-dimensional electron gas for novel sensor applications", *phys. stat. sol. (c)*, vol. 5, pp. 1914-1916, 2008.

[15] K. Brueckner, F. Niebelschuetz, K. Tonisch, S. Michael, A. Dadgar, A. Krost, V. Cimalla, O. Ambacher, R. Stephan, and M.A. Hein, "Two-dimensional electron gas based actuation of piezoelectric AlGaN/ GaN microelectromechanical resonators", *Appl. Phys. Lett*, vol. 93, no. 17, art. no. 173504, 3 pp., 2008.

[16] D.S. Ballantine, R.M. White, S.J. Martin, A.J. Ricco, E.T. Zellers, G.C. Frye, H. Wohltjen, *Acoustic wave sensors*, Academic Press, 1st edn., San Diego, 1997, pp. 115-119.

[17] R. J. Jiménez Riobóo, E. Rodríguez-Cañas, M. Vila, C. Prieto, F. Calle, T. Palacios, M.A. Sánchez, F. Omnès, O. Ambacher, B. Assouar, O. Elmazria, "Hypersonic characterization of sound propagation velocity in $Al_xGa_{1-x}N$ thin films", *J. Appl. Phys.*, vol. 92, pp. 6868-6874, 2002.

[18] K. Brueckner, V. Cimalla, F. Niebelschütz, R. Stephan, K. Tonisch, O. Ambacher, and M.A. Hein, "Strain- and pressure-dependent RF response of microelectromechanical resonators for sensing applications", *J. Micromech. Microeng.*, vol. 17, pp. 2016-2023, 2007.

[19] J. Wang, Z. Ren, and C.T.-C. Nguyen, "1.156-GHz self-aligned vibrating micromechanical disk resonator", *IEEE Trans. Ultrason. Ferroelectr. Freq. Control*, vol. 51, pp. 1607-1628, 2004.

[20] J.E.-Y. Lee, Y. Zhu, and A.A. Seshia, "A bulk acoustic mode single-crystal silicon microresonator with a high-quality factor", *J. Micromech. Microeng.*, vol. 18, art. no. 064001, 6 pp., 2008.

FULLY DIFFERENTIAL INTERNAL ELECTROSTATIC TRANSDUCTION OF A LAMÉ-MODE RESONATOR

M. Ziaei-Moayyed[1], D. Elata[2], J. Hsieh[1], J.-W. P. Chen[1], E. P. Quévy[3], and R. T. Howe[1]

[1]Stanford University, Stanford, California, USA
[2]Technion – Israel Institute of Technology, Haifa, Israel
[3]Silicon Clocks Inc., Fremont, California, USA

ABSTRACT

This paper reports the parallel internal electrostatic transduction of a laterally driven Lamé-mode polysilicon resonator. This resonator is fabricated using a manufacturable double nanogap process that provides ultrathin high-aspect ratio lateral gaps. The transduction electrodes are optimally placed and oriented to maximize electromechanical transduction efficiency for the fundamental Lamé mode. A 128.15 MHz Lamé-mode resonator is driven and sensed differentially with a 20 V DC polarization voltage: the motional resistance is about 30 kΩ and the quality factor $Q > 12000$ in air.

INTRODUCTION

The promise to integrate fully wireless transceiver systems has focused research efforts on developing on-chip high quality factor micro-electro-mechanical systems (MEMS) resonators. Electrostatic air- or vacuum-gap transduction has been used to drive and sense micromechanical resonators at frequencies above 1 GHz, with quality factors exceeding 10,000 [1]. However, one of the challenges of conventional electrostatic transduction is their relatively low efficiency. Previous efforts to minimize this impedance have been focused on reducing the gap, increasing the transduction area by using geometries such as rings [2] or by coupling arrays of resonators [3]. Internal electrostatic transduction, where a dielectric material replaces the air gap, offers an alternative route to reducing the motional impedance and also offers greater reliability by eliminating air gaps [4][5]. By optimally placing the dielectric within the resonator one can achieve higher frequencies and quality factors [6].

Another challenge facing electrostatic transduction of MEMS resonators is the presence of capacitive feedthrough, which makes direct measurement of resonators especially difficult at RF frequencies and motivates different measurement techniques [7]. Differential sense and excitation of electrostatic resonators allow for cancellation of capacitive feedthrough and offers extended linearity and better power handling [8]. The Lamé-mode resonator is ideal for differential measurements since the stresses in the perpendicular principle orientations in the plate are 180 degrees out of phase during resonance. Quality factors exceeding 1,000,000 have also been reported in vacuum [9]. For the Lamé-mode resonator reported in this paper (Figure 2), the four arrays of internal electrodes are optimally placed in the quadrants of the resonator such that adjacent electrodes are subjected to opposite stresses. Two signals that are 180 degrees out of phase are applied to the drive electrodes (1) and (2) and the currents out of sense electrodes (3) and (4) are combined for the total motional current (Figure 2). The net current from the resonator body (gray area in Figure 2) can approach zero, if the Lamé mode is differentially excited, due to the matched drive electrodes. This operating mode reduces ohmic loss through the anchor, thereby increasing Q.

DESIGN AND SIMULATION

The Lamé-mode resonator is a square plate that is anchored at its four corners. The Lamé mode shape is isochoric, which minimizes energy loss due to thermoelastic dissipation (TED), resulting in high quality factors. Figure 1 shows the results for a Lamé-mode resonator simulated in COMSOL Multiphysics, showing that the adjacent edges vibrate 180° out of phase.

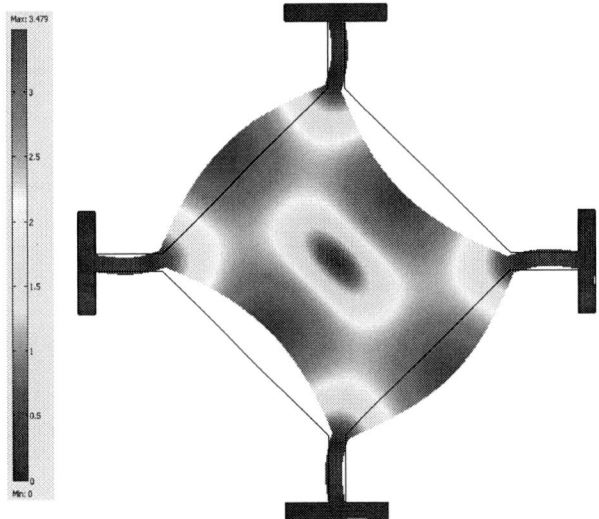

Figure 1: COMSOL simulation of a Lamé-mode resonator.

Figure 2 is a schematic depiction of the Lamé-mode resonator, where the transduction electrodes are integrated within the vibrating plate. The four arrays of internal electrodes are optimally placed in the quadrants on the resonator such that adjacent electrodes experience opposite stresses. Electrodes (1) and (3) go under compressive stresses and electrodes (2) and (4) experience tensile stresses. The presence of multiple dielectric gaps increase transduction efficiency and coupling to the fundamental Lamé mode. The resonant frequency of the Lamé-mode resonator is given approximately by [10]:

$$f_o \cong \frac{1}{\sqrt{2} \cdot L_r} \sqrt{\frac{G}{\rho}}$$

where L_r is the length of the resonating plate, ρ and G are the mass density and shear modulus of polysilicon respectively. For the resonator shown in Figure 2, since the silicon nitride dielectric is acoustically well matched to silicon and very thin (few nanometers), we assume that the presence of the dielectric does not affect the resonance

978-1-4244-2977-6/09 $25.00 © 2009 IEEE

frequency or the mode shape significantly. The overall performance of the resonator, including motional impedance, also depends on the thickness of the silicon nitride "dielectric gap", and the size and spacing of the internal electrodes.

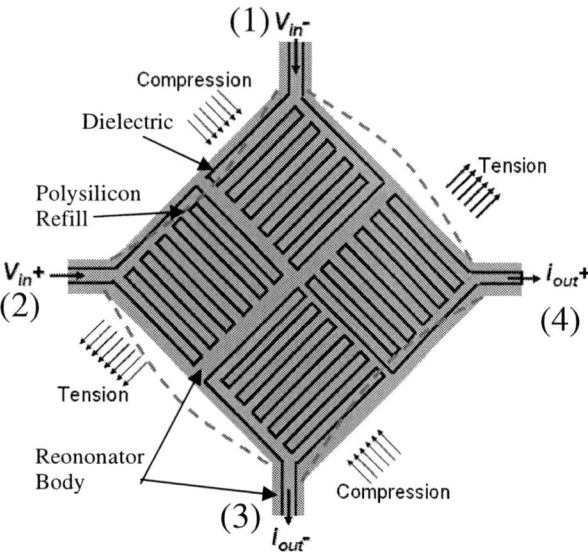

Figure 2: Overview of the Lamé-mode resonator with the integrated internal transduction electrodes (1-4). The DC bias voltage is applied to the resonator body. The electrodes are isolated from each other by dielectric routing (black).

FABRICATION

The resonators were fabricated using a manufacturable double nanogap process shown in detail in Figure 3. First a 1μm layer of thermal silicon oxide is grown on a conductive silicon substrate, followed by 200nm layer of LPVD silicon nitride. This stack provides the necessary isolation and also acts as a etch stop during the release. (1) A 500nm layer of phosphorus (n+) doped polysilicon is deposited and annealed. This layer, which acts as a routing layer for the resonators, is then patterned using an ASML stepper and etched in a RIE step. (2) A 1μm-thick sacrificial silicon oxide layer (LTO) is then deposited. (3) This oxide is then patterned and etched to define the anchors. (4) A 2.5um-thick n+ doped LPCVD structural polysilicon layer is deposited followed by a 0.25μm layer of oxide hard mask. (5) The 600nm-wide trenches are defined lithographically and etched into the hard mask and structural layer using a two-step RIE etch. (6) A 50nm layer of conformal stoichiometric silicon nitride defines the transducing dielectric. (7) The remainder of the trench is filled by a conformal layer of doped LPCVD polysilicon. This polysilicon layer defines the electrodes for this resonator. (8) A chemical mechanical polishing (CMP) step is used to planarize the surface and remove the polysilicon on top of the structural layer to isolate the electrodes from the resonator body. (9) The resonating plate and contact pads are defined and etched using an RIE step. (10) The resonator is then released in a buffered HF solution in a timed etch step. This is followed by a critical point drying (CPD) step to prevent stiction.

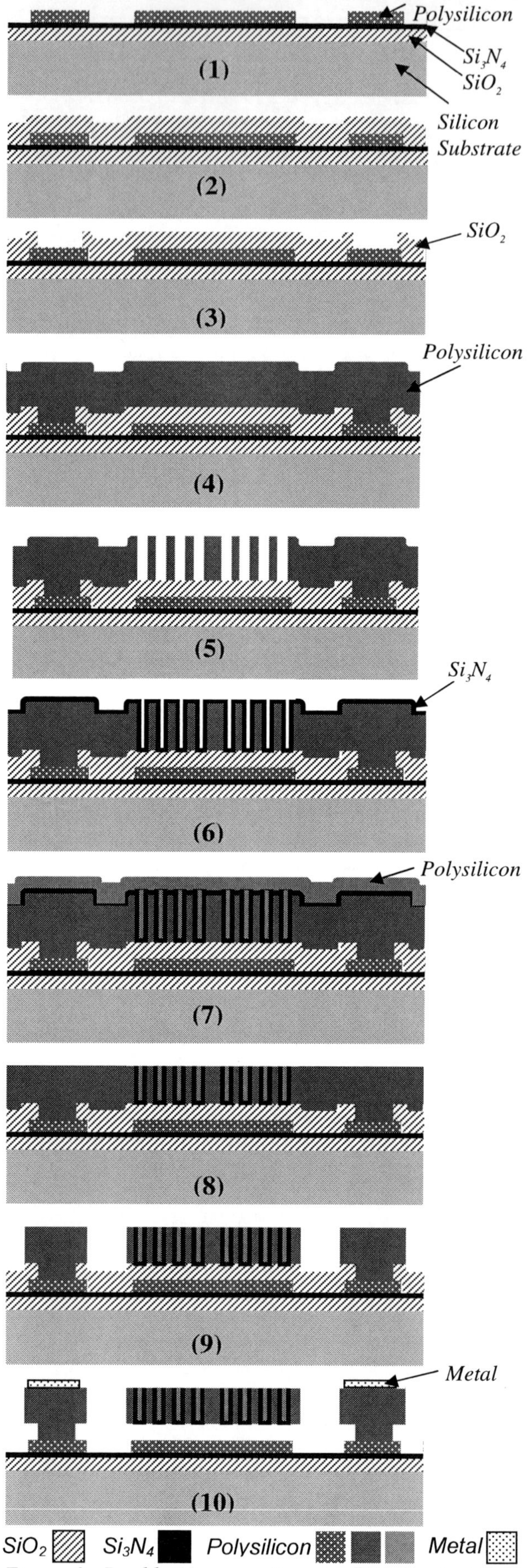

Figure 3: Double Nanogap Process flow for the Lamé-mode resonator.

One of the main features of this process is that nanoscale high-aspect ratio gaps can be fabricated using optical lithography. This is not possible using the conventional fabrication processes, where the gaps are defined by direct etching or a spacer process. Figure 4 shows SEMs of the Lamé-more resonator including the routing lines and contacts pads. The four corner pads are connected to the resonator body and are used for biasing the resonator. The other four electrodes are used for drive and sense. Isolation between pads is done by properly arranging the dielectric routing.

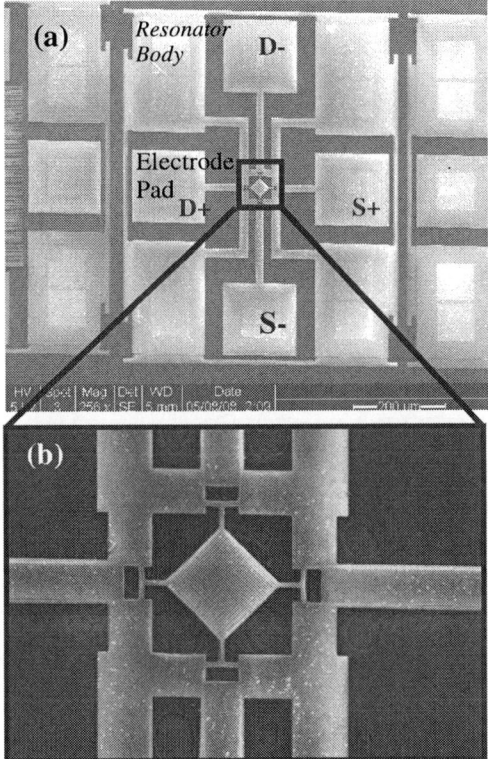

Figure 4: SEMs of the Lamé-mode resonator (a) overall view showing Drive (D+/D-) and Sense (S+/S-) electrodes. The four corner electrodes are connected to the resonator body (b) close up of the released resonating structure.

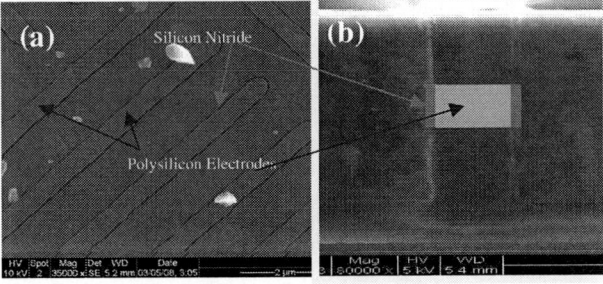

Figure 5: SEMs of the Lamé-mode resonator. (a) Close-up view of the polysilicon electrodes and silicon nitride gaps (b) cross-sectional view showing the trenches filled with silicon nitride/polysilicon.

EXPERIMENTAL RESULTS

Electrostatic resonators are often driven using a two port drive/sense configuration. This scheme limits the feedthrough capacitance to the substrate capacitance plus fringing field capacitance above the resonator. Even though this method has improved the ability to characterize many resonating structures, it deems insufficient to detect the resonance for some structures. As we move towards higher frequencies, the current due to feedthrough capacitance increases and eventually completely masks the motional current of the resonator. In order to cancel out feedthrough, different solutions have been proposed, such as the use of mixing techniques [7]. Differential measurements using a dummy resonator to null out feedthrough has been demonstrated as an effective method. This method is not desirable for on-chip applications because of the added area and complexity of having a dummy resonator next to the actual device [11].

In the single-ended drive and sense configuration, the polarization DC bias voltage is applied to the body of the resonator through the corner electrodes in Figure 4 (a). One of the electrodes in adjacent quadrants is used for drive and sense, respectively. The two other electrodes are grounded. In this measurement scheme, the ohmic resistances of the anchors are in series with the motional impedance and can significantly increase the damping in the resonator. Also, the effective capacitive feedthrough is much higher, which results in a much higher noise floor.

Single-Ended Drive and Differential Sense

Figure 7 shows the measurement setup for a single-ended drive and differential sense scheme. The input voltage is applied to electrode (1), and the signals from the two other adjacent electrodes (3) and (4) were added together using an off-chip power combiner [12]. The fourth electrode, (2) was grounded. The DC bias was applied to the resonator body through the anchors. By adding the sense signals that are 180° out of phase the capacitive feedthrough is reduced to ~ -80 dB, making it easier to detect a resonant peak. The resistive losses from the anchors greatly degrade the quality factor to ~ 2500.

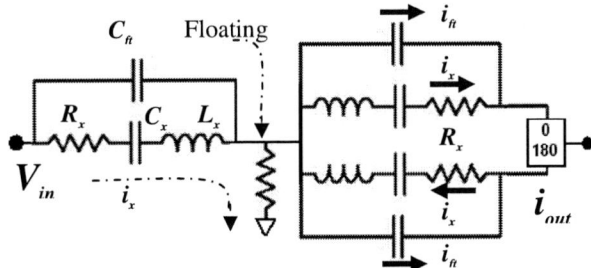

Figure 7: Equivalent circuit model for Single-ended drive, differential sense configuration.

Differential Drive and Sense

A fully differential drive and sense scheme takes advantage of the matched internal electrodes to suppress feedthrough. Figure 8 shows the test setup for this configuration. The input signal from port 1 of a network analyzer (Agilent PNA E8361A) was divided into two signals that are 180° out of phase using a power splitter [12]. These two signals were applied to the two adjacent electrodes (1) and (2) in Figure 2. The signals from the two other electrodes (3) and (4) were again combined together using a second power combiner [12]. This signal was then input to the second port of the network analyzer. In this setup, the electrostatic charges are internal between the electrodes and there is no current that passes through the anchors. The "resonator body" node is therefore a

978-1-4244-2977-6/09 $25.00 © 2009 IEEE

virtual ground and capacitive feedthrough is significantly reduced. This measurement scheme also allows for much higher quality factor since the anchor losses are minimized. Using this fully differential drive and sense method, a quality factor of 12050 in air has been demonstrated at a resonant frequency of 128.15MHz. The measured frequency is close to the simulated value of 125.9MHz. The effective motional impedance for this Lamé-mode resonator is calculated to be 31KΩ, including an interconnect resistance of about 8kΩ. The motional impedance can be further reduced by decreasing the thickness of the silicon nitride dielectric, and by further optimization of the spacing and width of the electrodes.

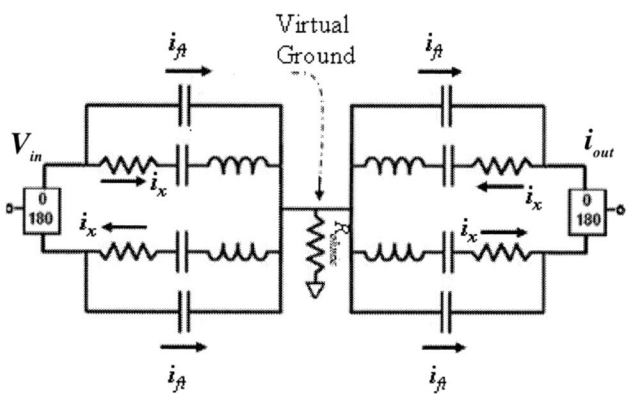

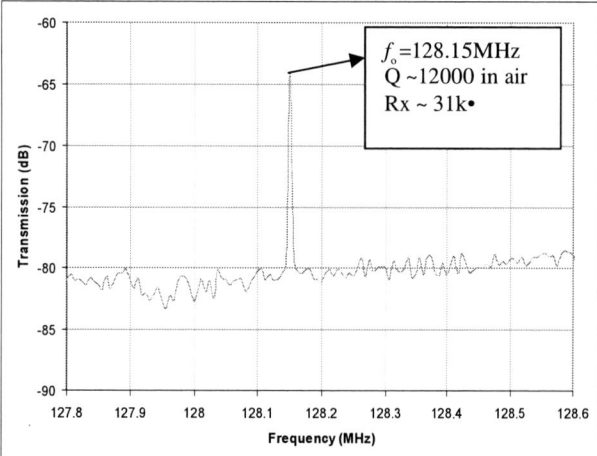

Figure 8: Fully differential configuration (a) Equivalent circuit model and (b) Measured frequency response with a Q of 12050 at 128.15MHz. $P_{in} = 0dBm$, $V_{DC} = 20V$.

CONCLUSION

This paper, reports a polysilicon Lamé-mode resonator that is driven and sensed by internal capacitive electrodes at 128.15MHz. By optimally placing and orienting the electrodes, the fundamental Lamé mode is efficiently excited and a fully differential drive and sense interface can be implemented within the resonator structure. Using this interface, ohmic losses through the anchors are minimized and a quality factor exceeding 12000 is demonstrated in air.

ACKNOWLEDEMENT

This work was supported by Intel Corp. (Drs. Vijay K. Nair and John M. Heck), with partial support through the DARPA Center on Interfacial Engineering for Microelectromechanical Systems, funded by DARPA grant HR0011-0600049 and managed by Dr. Dennis L. Polla. The fabrication work was performed at the Stanford Nanofabrication Facility (a member of the National Nanotechnology Infrastructure Network) which is supported by the National Science Foundation under Grant ECS-9731293.

REFERENCES

[1] C. T.-C. Nguyen, "Vibrating RF MEMS for next generation wireless applications," Proceeding IEEE Custom Integrated Circuits Conf., Orlando, Fl. October 2004, pp. 257-264

[2] B. Birchumshaw, G. Liu, H. Takeuchi, T.-J. King, R. T. Howe, O. O'Reily, and A. Pisano, "The Radial Bulk Annular Resonator: Towards a 50Ω RF MEMS Filter," Transducers 2003, Boston, June 2003. pp. 875-878

[3] M. U. Demirci and C. Nguyen, "Micromechanically Corner-Coupled Square Microresonator Arrays for Reduced Series Motional Resistance," IEEE/ASME J. Microelectromechanical Systems, vol.15, no. 6, pp. 1419-1436. December 2006

[4] S. A. Bhave and R. T. Howe, "Silicon Nitride-on-Silicon Bar Resonator Using Internal Electrostatic Transduction," Transducers 2005, Seoul, Korea, June 2005, pp. 2139-2142.

[5] L. Hung, C. Nguyen, Y. Xie, Y. Lin, S. Li, and Z. Ren, "UHF micromechanical compound-(2,4) mode ring resonators with solid-gap transducers," in Proc. 2007 IEEE Intl. Frequency Control Symposium 2007, pp. 1370-1375.

[6] D. Weinstein and S. A. Bhave, "Internal Dielectric Transduction of a 4.5GHz Silicon Bar Resonator," IEEE International Electron Devices Meeting, Washington, DC, December 2007, pp. 415-418.

[7] J.R. Clark, W.-T. Hsu, and C. T.-C. Nguyen, "Measurement Techniques for Capacitively-Transduced VHF-to-UHF Micromechanical Resonators," Transducers 2001, Munich, Germany, June 2001, pp. 1118-21

[8] S. A. Bhave, D. Gao, R. Maboudian, and R. T. Howe, "Fully-Differential Poly-SiC Lamé-mode Resonator and Checkerboard filter," IEEE Intl. Conference of Micro Electro Mechanical Systems, Miami, Florida, January 30 - February 3, 2005, pp. 223-226

[9] L. Kline, M. Pilaniapan, and W.-K. Wong, "6 MHz bulk-mode resonator with Q values exceeding one million," Transducers 2007, Lyon, France, June 2007, pp. 2445-48

[10] H. Majjad, J.-R. Coudevylle, S. Basrour, and M. Labachelerie, "Modeling and Characerization of Lamé-mode microresonators realized by UV-LIGA," Transducer 2001, Munich, Germany, June 2001, pp. 300-303

[11] P.Rantakari, J. Kiihamaki, M. Koskenvuori, T. Lemminmaki, and I. Tittonen, "Reducing the effect of parasitic capacitance on MEMS measurements," Transducers 2001, Munich, Germany, June 2001, pp. 1156-59

[12] Minicircuits Inc., 2 way-180° Power Combiner and Splitter: http://www.minicircuits.com/pdfs/ZFSCJ-2-4.pdf

ANALYTICAL MODELING AND NUMERICAL SIMULATION OF CAPACITIVE SILICON BULK ACOUSTIC RESONATORS

G. Casinovi, X. Gao, and F. Ayazi

School of Electrical and Computer Engineering

Georgia Institute of Technology, Atlanta, Georgia, USA

ABSTRACT

This paper introduces two newly developed models of capacitive silicon bulk acoustic resonators (SiBARs). The first model is analytical and is obtained from an approximate solution of the linear elastodynamics equations for the SiBAR geometry. The second is numerical and is based on finite-element, multi-physics simulation of both acoustic wave propagation in the resonator and electromechanical transduction in the capacitive gaps of the device. This latter model makes it possible to compute SiBAR performance parameters that cannot be obtained from the analytical model, e.g. the relationship between transduction area and insertion loss. Comparisons with measurements taken on a set of silicon resonators fabricated using electron-beam lithography show that both models can predict the resonant frequencies of SiBARs with a relative error smaller than 1%.

INTRODUCTION

Much research activity in recent years has been directed at the development of bulk acoustic resonators that are compatible with standard integrated circuit technologies. In this respect, capacitive resonators [1–3] offer a particularly attractive option, since they can be made entirely of materials that are used routinely in IC fabrication processes, resulting in significant advantages in terms of ease of integration and cost savings.

Disk resonators were among the first examples of devices of this type [1], but more recently width-extensional-mode resonators based on an alternative, rectangular-bar geometry were demonstrated [2], [3], which are referred to as silicon bulk acoustic resonators, or SiBARs. The basic structure of a SiBAR is schematically shown in Figure 1: the resonating bar element is placed between two electrodes, supported by two thin tethers. A DC polarization voltage applied between the resonator and the electrodes generates an electrostatic field in the capacitive gaps. When an AC voltage is applied to the drive electrode, the electrostatic force applied to the corresponding face of the resonator induces an acoustic wave that propagates through the bar.

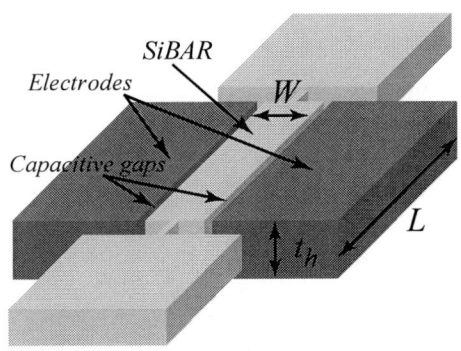

Figure 1: Structure of a SiBAR.

Small changes in the size of the capacitive gap on the other side of the device induce a voltage on the sense electrode, whose amplitude peaks near the mechanical resonant frequencies of the bar.

SiBARs offer several potential advantages over their disk-shaped counterparts, the most important of which is that the electrostatic transduction area can be increased without changing the main frequency-setting dimension, resulting in significantly lower motional resistance while maintaining high Q values [3].

While the behavior of SiBARs is well understood in broad, qualitative terms, a major obstacle to the design of high performance devices is a lack of sufficiently accurate analytical or numerical models. For example, to the authors' best knowledge there is currently no quantitative analysis of how exactly the dimensions of a SiBAR affect its insertion loss. Even the computation of the resonant frequencies is based on approximate formulae that, as will be shown in this paper, can be quite inaccurate, especially in the case of an anisotropic material such as single-crystal silicon.

This paper presents two newly developed SiBAR models, one analytical and the other numerical, that are significantly more accurate than currently available models. The analytical model is obtained from an approximate solution to the linear elastodynamics equations that satisfies the boundary conditions imposed by the SiBAR geometry. The second model is based on numerical multi-physics simulations of the device, performed in ANSYS. Comparisons with measurements taken on a set of devices fabricated using electron-beam lithography show that both models can predict the resonant frequencies of SiBARs of varying dimensions with a relative error of less than 1%.

ANALYTICAL MODEL

This section presents an analysis of acoustic wave propagation in a SiBAR, modeled as a bar of rectangular cross-section and made of a single, homogeneous, orthotropic material. The objective of this analysis is to obtain a quantitative model of the mechanical behavior of the resonator, which can then be used to compute the values of the resonant frequencies based on the resonator's dimensions and material properties.

From a mathematical point of view, an acoustic wave propagating through the resonator is represented by a vector $\mathbf{u} = [u_x, u_y, u_z]^T$, describing the displacement of a generic point in the device with respect to its position in the undeformed structure. In the analysis that follows, the x axis corresponds to the direction of the resonator's length (L), the y axis to the direction of its thickness (t_h), and the z axis to the direction of its width (W).

To simplify the analysis, the resonator length will be assumed to be infinite. Numerical simulation results and experimental measurements, presented later in this paper, show that this assumption does not result in a significant

978-1-4244-2977-6/09 $25.00 © 2009 IEEE   931

loss of accuracy, provided that the resonator's length is sufficiently large compared to the two other dimensions.

Symmetry considerations following from this assumption dictate that the solutions of interest are those that are independent of x and have no displacement component in the x direction (i.e. $u_x = 0$). Limiting the analysis to sinusoidal steady-state, such solutions can be expressed as

$$\mathbf{u}(y,z,t) = \begin{bmatrix} u_{y0} \\ u_{z0} \end{bmatrix} e^{j(\omega t - k_y y - k_z z)} \quad (1)$$

The expression above is a solution of the linear elastodynamics equations if and only if the following equation — generally known as the Christoffel equation [4] — is satisfied

$$\begin{bmatrix} c_{22}k_y^2 + c_{44}k_z^2 - \rho\omega^2 & (c_{23}+c_{44})k_y k_z \\ (c_{23}+c_{44})k_y k_z & c_{44}k_y^2 + c_{33}k_z^2 - \rho\omega^2 \end{bmatrix} \begin{bmatrix} u_{y0} \\ u_{z0} \end{bmatrix} = 0 \quad (2)$$

where ρ is the resonator mass density, and c_{22}, c_{23}, c_{33} and c_{44} are stiffness matrix coefficients.

Equation (2) has non-trivial solutions if and only if the determinant of the coefficient matrix is equal to zero, which, after some algebraic manipulation, leads to the following equation

$$\rho^2 v^4 - (c_{33}+c_{44})\rho v^2 + c_{33}c_{44}$$
$$+ [c_{44}^2 + c_{22}c_{33} - (c_{23}+c_{44})^2 - \rho(c_{22}+c_{44})v^2]\zeta \quad (3)$$
$$+ c_{22}c_{44}\zeta^2 = 0$$

where $\zeta = (k_y/k_z)^2$ and $v = \omega/k_z$. For a given value of v, there are in general two solutions to (3), regarded as an equation in ζ, hence four values of the ratio k_y/k_z for which (2) has non-trivial solutions.

It is readily observed that the constant term in (3) — that is, the term that is independent of ζ — can also be written as $(\rho v^2 - c_{33})(\rho v^2 - c_{44})$. Let $v_u = \sqrt{c_{33}/\rho}$ and $v_l = \sqrt{c_{44}/\rho}$; note that $v_l \le v_u$ because $c_{44} \le c_{33}$. Consequently, if $v_l \le v \le v_u$ the constant term in (3) is negative, which means that in this case (3) has two real solutions, one positive and one negative. Letting $\zeta_1 = -\alpha^2$ and $\zeta_2 = \beta^2$, the possible values for k_y are $\pm j\alpha k_z$ and $\pm \beta k_z$. For each value of k_y, the corresponding values of u_{y0} and u_{z0} can be obtained from (2).

By superposition, a generic wave propagating through the resonator is a linear combination of four waves of the type given in (1), one for each possible value of k_y/k_z. This leads to the following expression for such wave

$$\mathbf{u}(y,z,t) = A_0 \begin{bmatrix} j\alpha \sinh(\alpha k_z y) \\ u_{z0} \cosh(\alpha k_z y) \end{bmatrix} e^{j(\omega t - k_z z)}$$
$$+ A_1 \begin{bmatrix} ju_{y1} \sin(\beta k_z y) \\ \beta \cos(\beta k_z y) \end{bmatrix} e^{j(\omega t - k_z z)} \quad (4)$$

where

$$u_{z0} = \frac{\rho v^2 + c_{22}\alpha^2 - c_{44}}{c_{23} + c_{44}}$$

and

$$u_{y1} = \frac{c_{44}\beta^2 + c_{33} - \rho v^2}{c_{23} + c_{44}}$$

Finally, coefficients A_0 and A_1 in (4) must be chosen so that the wave satisfies traction-free boundary conditions on the top and bottom faces of the resonator, that is

$$\sigma_{xy} = \sigma_{yy} = \sigma_{yz} = 0, \qquad y = \pm t_h/2$$

where σ_{xy}, σ_{yy} and σ_{yz} are components of the stress tensor. After some algebraic manipulation, those conditions lead to a homogeneous system of two linear equations in the two unknowns A_0 and A_1, which can have non-trivial solutions only if the determinant of the coefficient matrix is equal to zero. Making the substitution $k_z = 2\pi/\lambda_z$, where λ_z is the wavelength in the z direction, the condition on the determinant yields the following equation

$$(c_{22}\alpha^2 - c_{23}u_{z0})(u_{y1} - \beta^2)\cosh(\pi\alpha\xi)\sin(\pi\beta\xi)$$
$$- \alpha\beta(1 + u_{z0})(c_{22}u_{y1} - c_{23})\sinh(\pi\alpha\xi)\cos(\pi\beta\xi) \quad (5)$$
$$= 0$$

where $\xi = t_h/\lambda_z$. For fixed α and β, this is an equation in ξ which has infinitely many solutions, because of the periodicity of the sine and cosine terms. The expression for $\mathbf{u}$ in (4) and the relation between v and ξ derived from (5) are valid only for $v_l \le v \le v_u$, but the procedure outlined above requires only minor modifications to handle other ranges of values for v, e.g. $v \ge v_u$.

Since α and β depend on v through (3), equations (3) and (5), taken together, define a relation between v and ξ. Since (5) has multiple solutions, this relation defines a function $v = v(\xi)$ that has multiple branches. To each point (ξ, v) that lies on a branch of this function there corresponds an acoustic wave that propagates across the resonator.

Figure 2 shows the graph of $v(\xi)$ for single-crystal silicon, when the x and z reference axes are aligned with the and $\langle 0\overline{1}1 \rangle$ and $\langle 011 \rangle$ crystallographic directions, and the y axis with the $\langle 100 \rangle$ direction. Only the first four of infinitely many branches of $v(\xi)$, or dispersion curves, are

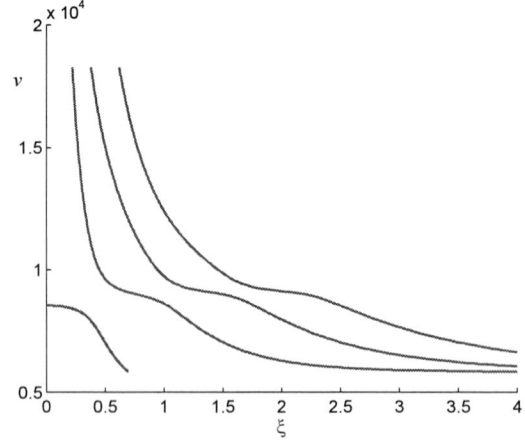

Figure 2: Propagation velocity of acoustic waves along the $\langle 011 \rangle$ crystallographic direction in a (100) silicon wafer ($c_{22} = 165.7$, $c_{33} = 194.4$, $c_{23} = 63.9$, $c_{44} = 79.6$ GPa).

shown in the figure. It should be pointed out that those curves depend only on the properties of the material, and not on the geometry of the resonator.

Once $v(\xi)$ has been obtained, it is straightforward to relate it to the resonant frequencies of a resonator of given dimensions. By definition, $v = \omega/k_z$ and $k_z = 2\pi/\lambda_z$, and from these two equalities it follows that $f\lambda_z = v$. Moreover, at resonance λ_z must be an integer submultiple of $2W$, that is $\lambda_z = 2W/n_z$, hence

$$f = \frac{1}{\lambda_z} v(\xi) = \frac{n_z}{2W} v(n_z t_h / 2W) \qquad (6)$$

Therefore the relationship between the resonator dimensions and its resonant frequencies can be obtained from $v(\xi)$ simply by changing the scales on the v and ξ axes.

ANSYS MODEL

To complement to the analytical model described in the previous section, we have also developed a SiBAR model for the ANSYS simulator that accounts for both the finite length of the resonator and the electromechanical transduction in the capacitive gaps, which is an integral part of the device behavior.

The orthotropic SOLID95 model is used for the resonating bar. Electromechanical transduction is modeled with two arrays of TRANS126 elements generated by the EMTGEN macro after the bar has been meshed. A number of resistors and capacitors model the test setup used for resonator testing and characterization [3]. The equivalent schematic diagram of the complete ANSYS model is shown in Figure 3: C_s and C_d model the gap capacitances, C_{ps} and C_{pd} the parasitic pad capacitances, and R_S and R_L the internal resistances of the test instruments.

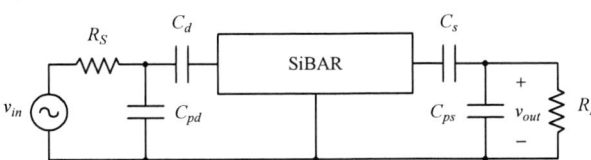

Figure 3: Equivalent circuit of the ANSYS model, including test setup.

Each simulation of the ANSYS model consists of a static analysis, which accounts for the effect of the DC polarization voltage, followed by a harmonic analysis. The simulation results include the values of all the node voltages, which makes it possible to generate plots of the voltage gain $A_v = v_{out} / v_{in}$ over the given range of frequencies, as shown in Figure 4. Many parameters related to the resonator performance can then be evaluated based on the location and magnitude of the peaks in the graph of $|A_v|$, including the effects of the resonator dimensions, the polarization voltage and the magnitude of the capacitive gaps not only on the resonant frequency, but also on the insertion loss.

The selection of the damping ratio used by ANSYS in its harmonic analysis (DMPRAT) merits some comment. The value of DMPRAT was chosen so that the simulated insertion loss of the resonator matched previously measured insertion losses of similar devices in the frequency range of interest. Consequently, this ANSYS model cannot provide reliable a-priori estimates of a resonator's insertion loss. On the other hand, once the value of DMPRAT has been selected in this way, the model can be expected to provide reasonably accurate information about how changes in the resonator dimensions affect the overall voltage gain A_v, provided that the resonant frequency does not change much. This is based on the assumption that the total rate of energy losses remains relatively constant within a relatively narrow frequency range.

To illustrate the model's capabilities, we present the results of the simulations of a set of SiBARs having the same length (400 μm) and width (40 μm), but varying thickness. The thickness values were chosen so that the main resonant peak would fall on the first dispersion curve of Figure 2. Figure 5 compares the values of the resonant frequency obtained from the ANSYS simulations with those predicted by the analytical model: it can be seen that the two models are in excellent agreement.

The plot of the simulated values of $|A_v|$ for the same set of SiBARs is shown in Figure 6. As can be seen in the figure, at first the magnitude of the voltage gain increases with the thickness of the device, because of the corresponding increase in the capacitive transduction area. Beyond a certain point, however, further increases in the thickness actually cause the voltage gain to decrease. This

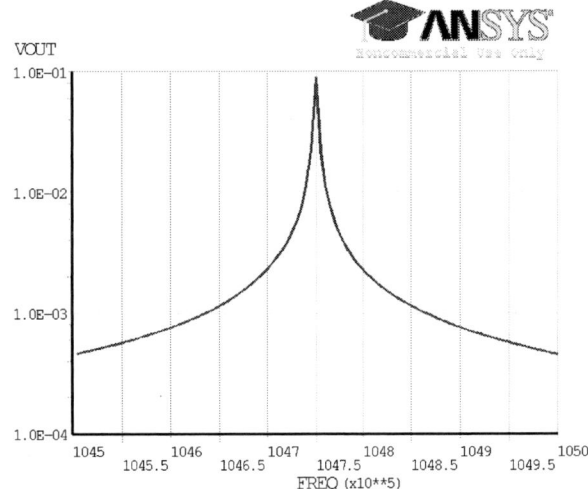

Figure 4: Simulated frequency response of a $400\,\mu m \times 40\,\mu m \times 20\,\mu m$ SiBAR.

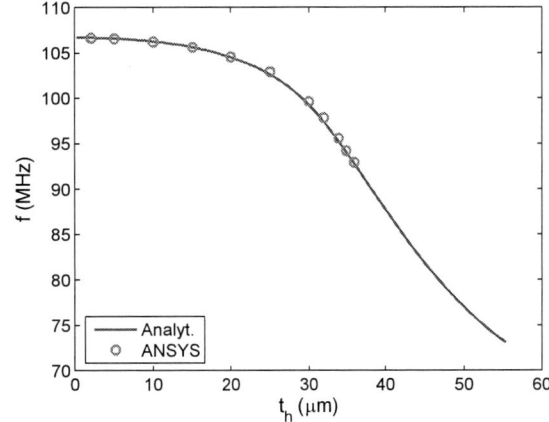

Figure 5: Resonant frequency vs. SiBAR thickness.

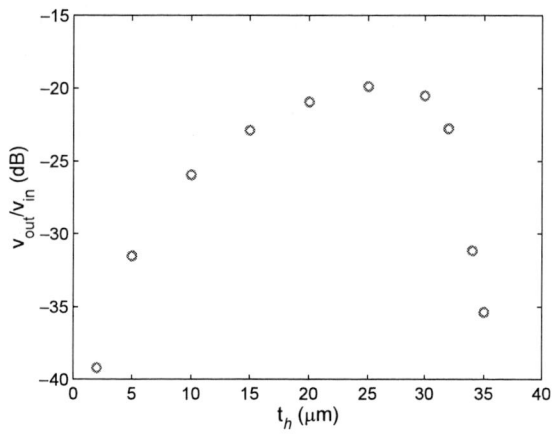

Figure 6: Plot of voltage gain vs. SiBAR thickness obtained from ANSYS simulations.

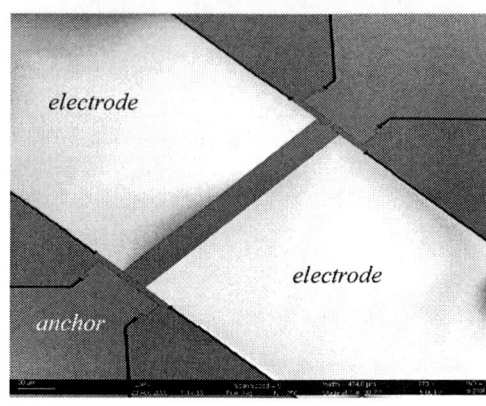

Figure 7: SEM view of a SiBAR fabricated in 10μm thick SOI with a capacitive gap of 300 nm.

phenomenon can be explained, at least in part, by a decrease in the efficiency of the electrostatic transduction in the capacitive gaps due to deterioration of the mode shape [5].

EXPERIMENTAL VERIFICATION

Both models were validated against measurements taken on SiBARs of various dimensions fabricated in 10 μm thick SOI. Device definition includes sub-micron trench formation using DRIE, followed by 3 μm wide peripheral trench etching, and HF release. A SEM micrograph of a sample device fabricated with this process is shown in Figure 7.

Table 1 compares the values of the resonant frequencies predicted by both models with those obtained from device measurements. The table shows also the relative error incurred when the approximate formula

$$f = \frac{n_z}{2W}\sqrt{\frac{E}{\rho}} \qquad (7)$$

is used to estimate the resonant frequency of the SiBAR [3]. In all models, the value of W was set equal to the measured width of the device, so that the comparison would not be affected by process variations. In (7), the value of E was set to 169 GPa.

It is readily seen that the values of the resonant frequencies predicted by the models described in this paper are substantially more accurate than those given by (7), and that the error associated with the latter formula grows as the ratio $\xi = t_h/\lambda_z = n_z t_h/2W$ increases. This is consistent with the relationship between resonant frequency and SiBAR thickness predicted by both models

(Figures 2 and 5), which show a gradual decrease in the resonant frequency of the device as its thickness increases.

CONCLUSIONS

For the first time, the models described in this paper make it possible to perform an accurate quantitative analysis of SiBARs, and to predict with a certain degree of confidence key parameters such as resonant frequency, frequency response and insertion loss. Both models can be effective aids in the design of high performance SiBARs.

ACKNOWLEDGEMENTS

This work was supported by DARPA under the Analog Spectral Processors program.

REFERENCES

[1] J. Wang, Z. Ren and C. T.-C. Nguyen, "Self-aligned 1.14 GHz vibrating radial-mode disk resonators," in *Proc. TRANSDUCERS 2003*, vol. 2, pp. 947–950.

[2] S. Pourkamali, G.K. Ho and F. Ayazi, "Vertical capacitive SiBARs," in *Proc. MEMS 2005*, pp. 211–214.

[3] S. Pourkamali, G. K. Ho and F. Ayazi, "Low-impedance VHF and UHF capacitive silicon bulk acoustic wave resonators – Parts I and II," *IEEE Trans. Electron Devices*, vol. 54, no. 8, pp. 2017–2030, Aug. 2007.

[4] J. F. Rosenbaum, *Bulk Acoustic Wave Theory and Devices*, Artech House, Boston, 1988.

[5] G. K. Ho, *"Design and Characterization of Silicon Micromechanical Resonators,"* Ph. D. Thesis, Georgia Institute of Technology, 2008.

Table 1: Comparison with measured data

Length (μm)	Width (drawn) (μm)	Width (meas.) (μm)	Mode order (n_z)	Res. freq. (meas.) (MHz)	Analytical model		ANSYS model		Equation (7)	
					Freq. (MHz)	Error (%)	Freq. (MHz)	Error (%)	Freq. (MHz)	Error (%)
310	40	39.97	1	106.308	106.34	0.03	106.23	-0.07	106.5	0.18
			3	299.082	297.92	-0.40	297.85	-0.41	319.61	6.86
400	40	39.86	1	106.5	106.6	0.09	106.59	0.08	106.8	0.28
216	27	26.94	1	157.064	156.84	-0.14	156.86	-0.13	158.1	0.66
270	27	26.96	1	157.062	156.73	-0.21	156.74	-0.20	157.9	0.53
240	24	23.91	1	176.452	176.12	-0.19	176.17	-0.16	178.1	0.93
200	20	19.90	1	210.628	209.97	-0.31	210.06	-0.27	214	1.6

THERMAL ANALYSIS AND CHARACTERIZATION OF A HIGH Q FILM BULK ACOUSTIC RESONATOR (FBAR) AS BIOSENSERS IN LIQUIDS

Xu Zhang, Wencheng Xu, Abbas Abbaspour-Tamijani, and Junseok Chae
Department of Electrical Engineering, Arizona State University, Tempe, AZ 85287

ABSTRACT

This paper presents the thermal analysis and characterization of a zinc oxide (ZnO) based film bulk acoustic resonator (FBAR) having a high quality factor (Q) in liquid environments for biosensing applications. Q of up to 120, an improvement of at least 8× greater than state-of-the-art devices in liquids, are achieved by integrating microfluidic channels with heights comparable to the acoustic wavelength in FBAR. In order to achieve temperature stability in the highly sensitive FBAR sensor, we analyze sources of thermal effects and characterize FBAR in a Pierce oscillator. Measurements show a temperature coefficient of oscillation frequency (TCF) of -112 ppm/K for the uncompensated circuit. We show that this thermal drift can be reduced to less than 1ppm/K by applying a properly chosen bias to the oscillator, which suggests the possibility of a feedback approach to achieve thermal stability.

INTRODUCTION

Offering the common benefits of micromachining technologies, FBAR is an attractive candidate for implementing miniature biomolecule sensors due to its non-labeling process, high sensitivity and excellent resolution [1]. However, previous research reveals a significant resolution compromise caused by Q degradation when FBARs are exposed to liquid environments; ~10 ng/cm^2 with Q of 15 in water vs. 0.5-1 ng/cm^2 with Q of 250 in the air [2, 3]. Besides the Q degradation in liquid, FBAR suffers from thermal instability. Temperature drifts in the resonance frequency of FBAR can be as high as -25 ppm/K to -60 ppm/K [4], and if are not compensated or otherwise corrected for can lead to false-positive/negative responses. In addition, characterization of FBAR requires bulky test equipments and includes time-consuming calibration process, which poses problems in biomolecule sensing applications.

This paper presents approaches to enhance the Q of FBAR for liquid applications and to minimize the thermal sensitivity of the FBAR-based oscillator that facilitates the characterization of FBAR.

APPROACH

Typical MEMS FBAR's have high Q in the air/vacuum ranging from a few hundreds to a few thousands. This is because the acoustic wave generated in the FBAR is well confined by the very large acoustic impedance mismatch between the solid materials and air [5]. However, when immersed in liquid, the solid-liquid interface become leaky for acoustic wave as the impedance mismatch is small, resulting in the degradation of Q of FBAR. To improve Q of FBAR in liquid, we confine the liquid to a height comparable to the acoustic wavelength by forming microfluidic channels on top of FBAR (Figure 1). This minimizes the dissipation in the liquid and hence improves the resonator Q.

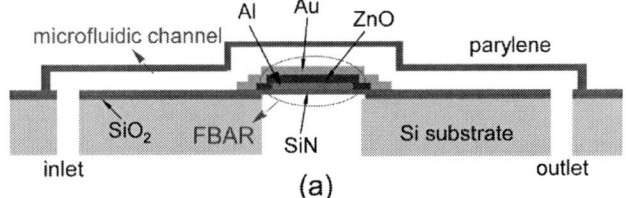

(a)

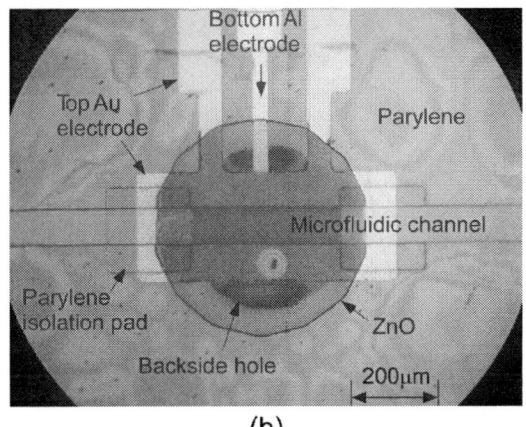

(b)

Figure 1: (a) Schematic of the FBAR sensor integrated with a microfluidic channel and (b) Top view of the fabricated FBAR sensor; a microfluidic channel run across the FBAR sensor.

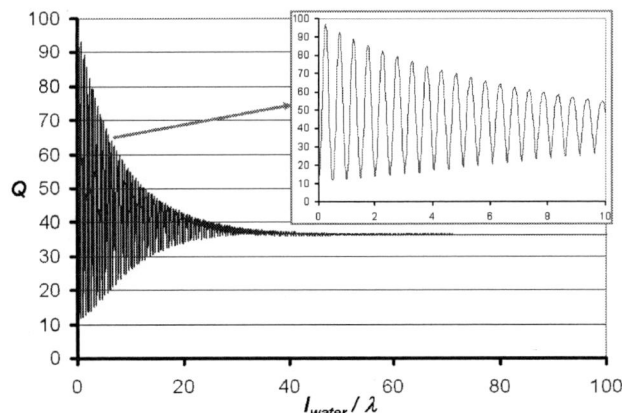

Figure 2: Theoretically predicted Q versus the ratio of fluidic channel height to the acoustic wavelength in water.

A transmission-line model [5, 6] with Mason's in-line equivalent circuit of piezoelectric composite is used to study the Q behavior of the device. Q as a function of the ratio of channel height and acoustic wavelength is plotted in Figure 2. Q shows an oscillatory behavior at the small channel height regime where the height of the liquid layer is comparable to the acoustic wavelength of the FBAR. It peaks when the height becomes an odd integer multiple of the quarter wavelength.

However, Q is not constant over a wide temperature range and usually deteriorates at higher temperatures. Temperature alters the ratio of the liquid layer thickness to

978-1-4244-2977-6/09 $25.00 © 2009 IEEE

the acoustic wavelength, l_{water}/λ (x-axis in Figure 2), in several ways: i) CTE (Coefficient of Thermal Expansion) mismatch, ii) acoustic velocity change, iii) viscosity change, and iv) volume expansion. Figure 3 illustrates the temperature effects on the FBAR. Temperature change causes the expansion of SiN membrane, electrodes, ZnO film, liquid, and parylene enclosure as their CTEs, resulting in a geometrical deformation in the channel (a). The temperature effects on Q (b) and the overall performance deterioration (c) are plotted as well. CTE mismatch between FBAR and the channel can be identified as the dominant factor.

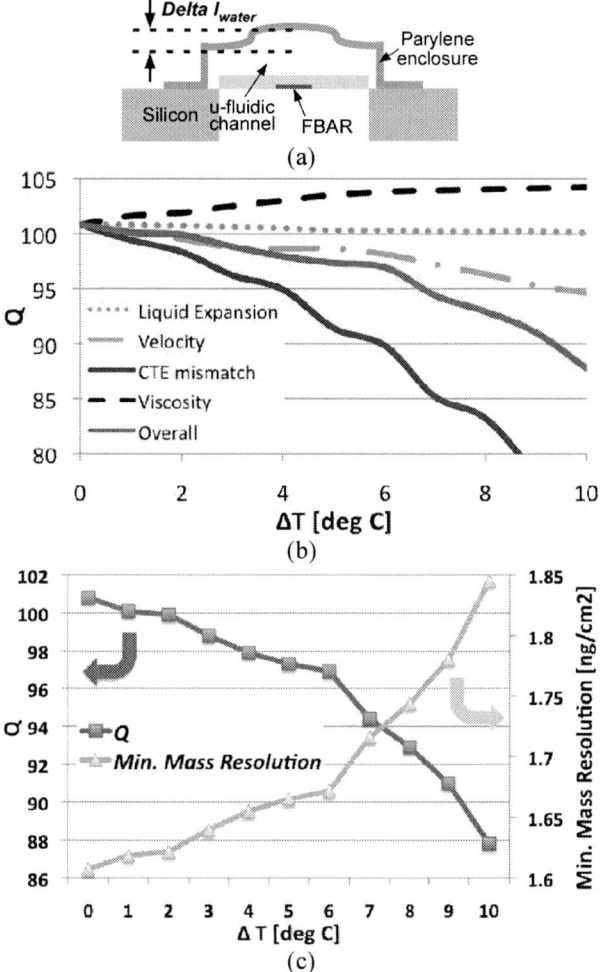

Figure 3: Temperature effects on the sensor performance; (a) cross sectional view of the FBAR sensor with a channel showing temperature change causes Δl_{water} that change the ratio of l_{water}/λ, (b) Q variations of different sources of thermal effects, (c) Q degradation and minimum detectable mass resolution over a temperature range of 10 °C.

Besides the Q variation, temperature has a significant impact on the resonance frequency of FBAR. As a result, when FBAR is characterized in a Pierce oscillator, where it is modeled by Modified Butterworth-Van Dyke (BVD) circuit [7], oscillation frequency varies with temperature fluctuations. To compensate the thermal effect, Supply voltage of the oscillator is tuned to change DC and AC voltage across the FBAR. The DC voltage change induces a positive change in the stiffness of the piezoelectric material (electrical stiffening) [8], which changes

resonance frequency of FBAR and in turn changes the oscillation frequency. On the other hand, the AC voltage regulates the temperature of FBAR as AC voltage generates heat within the resonator that also causes oscillation frequency change as mentioned before. The temperature compensation is achieved when the net frequency drift caused by the two effects sums to be zero, which is expressed by

$$t_v \Delta V + TCf_{osc} \Delta T = 0 \qquad (1)$$

where TCf_{osc} is the temperature coefficient of oscillation frequency, t_v is the relative frequency tunability of supply voltage, ΔT is the temperature change and ΔV is the tuning voltage needed for compensation.

FABRICATION AND ASSEMBLY

The FBAR sensor is fabricated using standard silicon micromachining technologies. The fabrication process is illustrated in Figure 4.

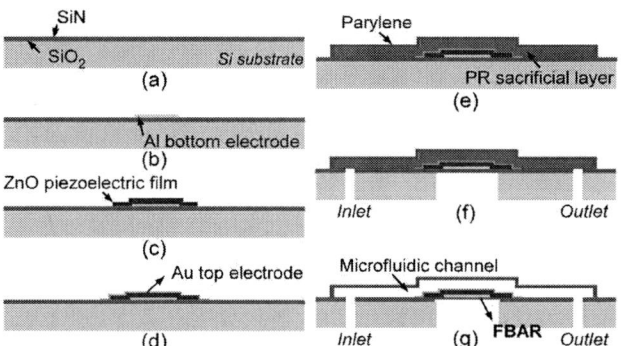

Figure 4: Fabrication process of the FBAR integrated with a microfluidic channel

Fabrication starts with a <100> silicon wafer; 2000Å thick oxide is first thermally grown on the wafer, followed by a deposition of 6000Å thick low-stress LPCVD SiN layer. 1600Å thick Al is evaporated and patterned on the SiN layer as the bottom electrode, which determines the effective area of the FBAR. ZnO is then sputtered with thicknesses from 0.55 to 4.2 µm. 300Å/1500Å thick Cr/Au is sputtered and patterned by lift-off over the ZnO layer as a top electrode. Au is used as the top electrode due to its excellent conductivity and good affinity for biomolecular binding. 3µm thick photoresist is patterned as a sacrificial layer on the top electrode, coated by Parylene C film. The parylene C thin film is 3 µm thick which provides enough mechanical strength and forms the microfluidic channel enclosure. The wafer is etched through from the backside by Deep RIE (Reactive Ion Etch) to release the SiN membrane and to form the channel inlets/outlets. At the end, the sacrificial photoresist layer is removed by soaking devices in acetone and the microfluidic channels are dried by supercritical CO_2 process. The SiO_2 layer underneath SiN is removed by buffered oxide etchant to achieve a smooth SiN supporting membrane.

A fabricated FBAR is glued on a PCB board and wire bonded to the PCB. The PCB board is fabricated by a milling machine that patterns the plated copper on both sides of PCB board. At the center of the board, locates a BJT transistor, NEC NE68119. Standard SMD components with case 0402 are soldered around on the board. A 50 Ω connector is fixed at the edge to connect the oscillator output to a spectrum analyzer via 50 Ω cable. Two wires are soldered on the board to connect the board to a DC power supply.

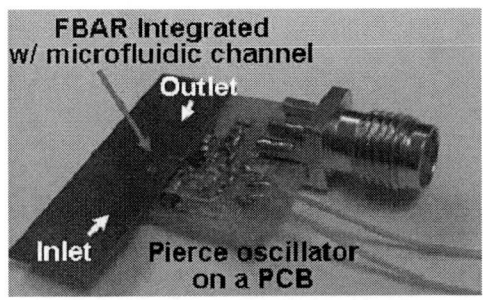

Figure 5: PCB implementation of the FBAR in Pierce oscillator; the FBAR is wire-bonded to connect to the oscillator components.

RESULTS AND DISCUSSIONS

Q performance of the FBAR is first examined. FBAR chips are mounted on a probe station and characterized using HP 8510C network analyzer, 85105A millimeter-wave controller and 8517B S-Parameter test set. S_{11} and the impedance of the sensor are extracted from the S_{11} spectrum. Then, the Q of resonator is calculated by

$$Q = \frac{f}{2} \cdot \frac{d\varphi}{df} \qquad (2)$$

where f is frequency and $d\varphi/df$ is the slope of impedance phase vs. frequency curve.

Table 1: Measured Q for different damping conditions.

ZnO thickness (μm)	0.55	2.1	4.1
Resonant freq. (GHz)	2.2	1.05	0.6
Q in the air (l_{water}=0)	220	240	430
Q in the water (l_{water}~∞)	12	7	15
Q w/ microchannels (l_{water}=3 μm)	35~40	65~120	85~110

FBAR exhibits high Q (~250) in the air, but the Q decreases drastically (~10) while contacting with large depth water. By flowing water in the microfluidic channels, significant Q improvement (up to 120) is obtained (Table 1). This data shows 8 to 10 times higher Q than any previously reported state-of-the-art longitudinal FBARs in liquid.

To evaluate the frequency drift caused by temperature, the series and parallel resonance frequencies of FBAR and oscillation frequency of the oscillator are measured at different temperatures. A heater is used to control ambient temperature and a thermocouple is fixed to read the temperature. The measurement data are plotted in Figure 6; the slopes of lines representing the temperature coefficients. The measured temperature coefficients of series and parallel resonance frequencies are – 46 ppm/K and – 41 ppm/K, respectively. The oscillation frequency shows a temperature coefficient of -112 ppm/K, which is larger than the temperature coefficients of the resonance frequencies. The larger temperature coefficient is because the oscillation frequency is not only decided by resonance of FBAR, but also associated with the energy loss of FBAR,

transistor parameters, SMD capacitors in parallel with FBAR and parasitics from the interconnects, all of which contribute to the temperature coefficient of oscillation frequency.

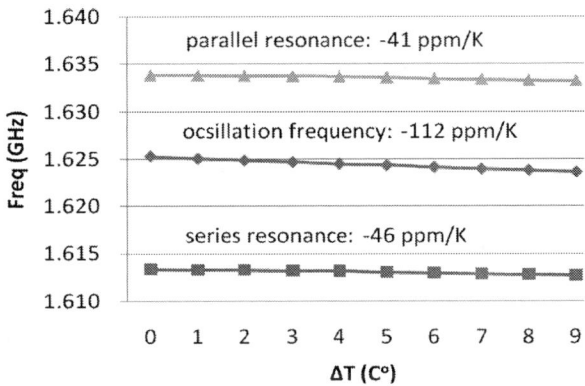

Figure 6: Temperature coefficient of series resonance, parallel resonance and oscillation frequency.

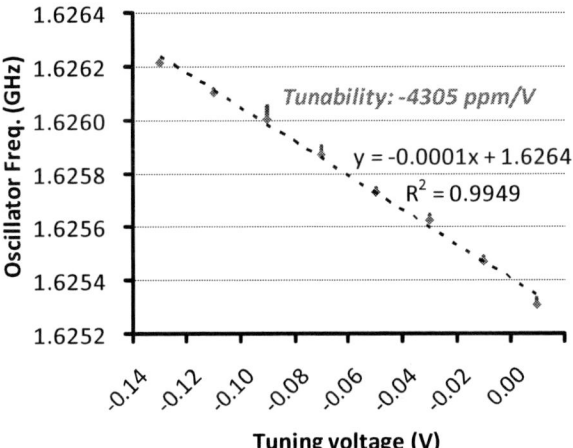

Figure 7: Voltage tuning on the oscillation frequency (Measurement error bars are shown)

The frequency tunability of the oscillator is characterized by measuring oscillation frequency vs. tuning voltage. While oscillation frequency is measured by spectrum analyzer, the tuning voltage is read and calculated from the DC power supply. At the same time, a thermocouple is used to exclude the influence of ambient temperature. The tunability measurement starts at 25°C with a supply voltage of 3V. Measured oscillation frequency and the tuning voltage are plotted in Figure 7. Oscillation frequency shows very linear with the tuning voltage, and it demonstrates a large tunability of -4305 ppm/V.

By knowing frequency tunability of supply voltage and temperature coefficient of the oscillator, the tuning voltage for compensation is calculated from equation (1). In Figure 8(a), to compensate a temperature change of 5°C, the supply voltage only needs to be tuned by 0.14 V. This is because of the large tunability of the oscillator frequency. Figure 8(b) demonstrates uncompensated and compensated oscillator frequencies for 5°C change in temperature. Frequency fluctuation exists, which is caused by, we believe, the inaccuracy of DC power supply. Yet the temperature coefficient is greatly reduced from -112 ppm/K to, on an average, of less than 1 ppm/K, upon the compensation.

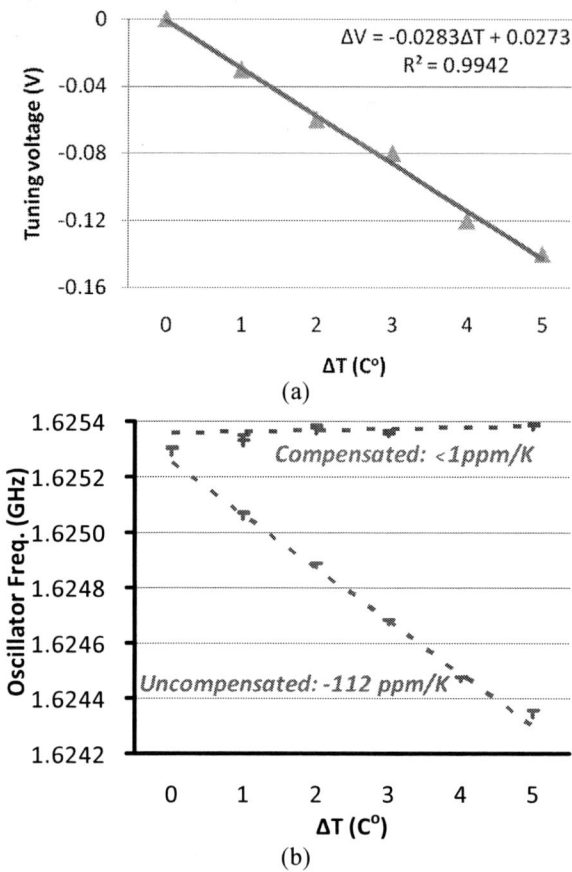

(a)

(b)

Figure 8: (a) Tuning voltage over temperature variation of 5°C and (b) Temperature compensation on oscillation frequency (Measurement error bars are shown); the compensated curve has less than 1 ppm/K over temperature variation of 5°C.

CONCLUSION

A high Q, up to about 120, FBAR for detection in liquid environment, is demonstrated. Q is improved by integrating a microfluidic channel on FBAR, which confines the liquid to a height comparable to the acoustic wavelength. However, the high Q FBAR has large temperature sensitivity. Increasing temperature degrades Q, resulting in degradation of resolution. Also,

temperature changes the resonant frequency which needs to be compensated to minimize false-positive/negative responses of FBAR. To characterize the temperature effect, FBAR is implemented in a Pierce oscillator. Resonance frequency of FBAR is characterized by oscillation frequency. As a result, oscillation frequency of the Pierce oscillator varies with temperature and has a negative temperature coefficient of -112 ppm/K. The oscillation frequency could be tuned by supply voltage of the oscillator and the tunability of the frequency is measured as -4305 ppm/V. By using the voltage tuning technique the output frequency of the oscillator is kept within 1ppm/K over the temperature change of 5°C.

REFERENCES

[1] H. Zhang, E. S. Kim, "Micromachined acoustic resonant mass sensor", *J. Microelectromech. Syst.*, vol. 14, pp. 699-706, 2005.

[2] H. Zhang, M. S. Marma, E. S. Kim, C.E. McKenna, M. E. Thompson, "A film bulk acoustic resonator in liquid environments", *J. Micromech. Microeng.*, vol. 15, pp. 1911-1916, 2005.

[3] J. Weber, W. M. Albers, J. Tuppurainen, M. Link, R. GABL, W. Wersing, M. Schreiter, "Shear mode FBARs as highly sensitive liquid biosensors", *Sens. Actuators A*, vol. 128, pp.84-88, 2006.

[4] J. Bjurstrom, G. Wingqvist, V. Yantchev, I. Katardjiev, "Temperature compensation of liquid FBAR sensors", *J. Micromech. Microeng.*, vol. 17, pp. 651–658, 2007.

[5] W. P. Mason, R. N. Thurston, *Physical Acoustics*, Academic Press, Inc., New York, 1972.

[6] Y. Zhang, Z. Wang, J.D.N. Cheeke, "Resonant spectrum method to characterize piezoelectric films in composite resonators", *IEEE Trans. Ultrason. Ferroelectr. Freq. Control*, vol. 50, pp. 321-333. 2003.

[7] W. Pang, H. Zhang, E. S. Kim, "Micromachined acoustic wave resonator isolated from substrate", *IEEE Trans. Ultrason. Ferroelectr. Freq. Control*, vol. 50, pp. 1239-1246, 2005.

[8] W. Pang, H. Zhang, H. Yu, C. Y. Lee, E. S. Kim, "Electrical Frequency Tuning of Film Bulk Acoustic Resonator", *J. Microelectromech. Syst.*, vol. 16, no. 6, pp. 1303-1313, 2007.

EL DISPLAY PRINTED ON CURVED SURFACE

Taik-Min Lee[1†], Hyun-Cheol Choi[1], Jae-Ho Noh[1], and Dong-Soo Kim[1]

[1]Nano Machinery Research Division, Korea Institute of Machinery and Materials, Daejeon, Korea

ABSTRACT

This paper presents the electro-luminescence (EL) display lamp which is patterned on a curved surface by the pad printing method. The EL display lamp consists of 5 layers: Bottom electrode; Dielectric layer; Phosphor; Transparent electrode; Bus electrode. For proper pad printing, the ink of each layer was formulated and the pad printing condition was controlled. A PEN film was used first in order to realize the pad printing process condition of each layer. Finally, the EL display lamp was printed on a dish, which has a radius of curvature 80mm.

INTRODUCTION

Recently, various micro-printing technologies have begun to hit the spotlight. This is due to the superior price competitiveness of these technologies as compared to existing lithographic processes for manufacturing some parts used in the manufacture of display units, electronic papers, RF-ID information devices, and so on. The semiconductor manufacturing process requires a long manufacturing time and expensive equipment [1,2]. The photolithography and electro plating methods require acid washes and a substantially larger amount of material. Therefore, in spite of worse resolution and limited application areas, the printing technology has come into the spotlight for its fast manufacturing time, mass production ability, low cost, and so on. The word, 'PEMS' stands for "Printed Electro-Mechanical System" which is fabricated by means of various printing technologies. Passive and active components in 2D or 3D such as conducting lines, resistors, capacitors, inductors and TFT, which are printed with functional materials, can be classified in this category. PEMS products also include assemblies of printed passive and active components, such as RFID tags, E-paper displays, solar-cell devices and printed sensors. Such PEMS products have already been commercially available in the display area. The Flexible electronics products at a low cost, such as RFID tags, E-paper displays, solar-cell devices and etc., would have the market size of $20B per annum in 5 years and $100B per annum in 10 years, according to the predictions of international market researches, such as DARPA in 2004[3].

For the exemplary development cases of PEMS processes and systems, the corporate R&D center of GE and ECD Ovonic Corp. which has gained the pronounced reputation with solar-cells launched a joint project for the development of the printing system which enables the production of polymeric devices in September of 2003 [4]. The system to develop is announced that the world first Roll-to-Roll based printing system in combination of gravure and screen printing technologies for the mass production of printed electronics products. GE will take a part of the design of printed electronics products and Ovonic will develop the printing system in size of 23 m by 4 m.

Fraunhofer IZM institute in Germany belongs to one of leading groups in the manufacture of flexible electronics, which made the prototype for the Roll-to-Roll based lithography system for the production of PEMS devices. It devised numerous Roll-to-Roll based systems for etch-rinse-cleansing, electroplating and screen printing too. These systems are equipped with the precise alignment mechanism, enabling the multi-layer printing of different materials on top of each other. Diverse PEMS passive and active components such as ring oscillator, invertor, RFID tag and OTFT were produced at the precision of 10 μm[5].

Man Roland Corp. in Germany announced the concepts of its next generation printing/packaging systems and it puts a main focus on the research and development for the patterning method of PEMS components such as RFID and smart packaging. The total packaging concept is, (1) printing electronics such as circuits, antennas, low end active and passive components, (2) repeating the same procedure for the next printing layers, and (3) completing the final product by placing and bonding a chip on the printed components. This will be the ultimate goal for the production of PEMS components and assemblies [6].

PolyApply is the name of the consortium which ties up with 20 members of companies, research institutes and universities in Europe. It is noteworthy that this consortium excludes the production of RFID tags based on silicon technologies. Its main aims are the development of in-line processes and protocols for RFID tags with all polymeric materials at an ultra-low cost.

The printing methods for PEMS devices, including the gravure [7], screen [8], flexo, inkjet, and pad printing, have an advantage of one-step direct patterning. However, in general, the printing, except pad printing method, cannot be applied for patterning on a curved surface. This paper presents the precision of micro pad printing process, the characteristics of the formulated inks, pad printing process for the electro-luminescence (EL) display lamp, and the EL display lamp pad-printed on a curved surface.

PRECISION OF MICRO PAD PRINTING

The pad printing is the ink transferring process, in which the ink moves from an engraved plate to a substrate via a pad, which is made by silicon rubber. The pad printing is the combination technology between the pad, engraved plate, substrate, and ink. The figure 1 shows the simple schematic diagram of pad printing process. The engraved plate is doctor-bladed, the silicon pad proceeds toward the plate, the ink is transferred onto the pad, the pad proceeds toward the curved surface, and the ink on the pad is finally transferred onto the surface[9]. Since the pad is soft, the pad printing process can be applied to the curved substrate. However, in order to utilize the pad printing process for micro patterning, the minimal pattern size and pattern distortion, which is caused by the use of pad, should be clarified.

978-1-4244-2977-6/09 $25.00 © 2009 IEEE

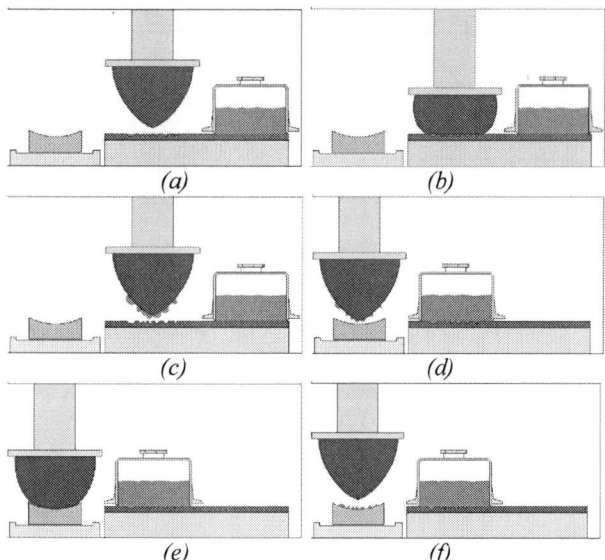

Figure 1: Schematic diagram of pad printing process: (a) Doctor blading; (b) Pad approaching to the engraved pattern; (c) Ink transferring onto the pad; (d) Moving toward the substrate; (e) Pad approaching to the substrate; (f) Ink transferring onto the substrate.

Although there is the conceptual possibility of several microns patterning in the pad printing process, from our pad printing experiments using the engraved plate with various pattern widths, it is found that the minimal pattern width of pad printing is several tens of microns. The figure 2 shows the pad-printed pattern which is 35 μm wide, which was the minimal size in our experiment. The pattern thickness was 2.4 μm.

In order to inspect the pattern distortion of pad printing caused by the deformation of silicon pad, concentric circles which have the diameters increased by 2mm was pad-printed on a flat surface. The interval of the patterns which is supposed to be 2mm was measured by using an optical microscope. The figure 3 shows the measured pattern distortion errors. The pattern distortion is less than 5 μm on a flat surface. If the substrate is a curved surface, then the pattern distortion will depend on the curvature of the surface.

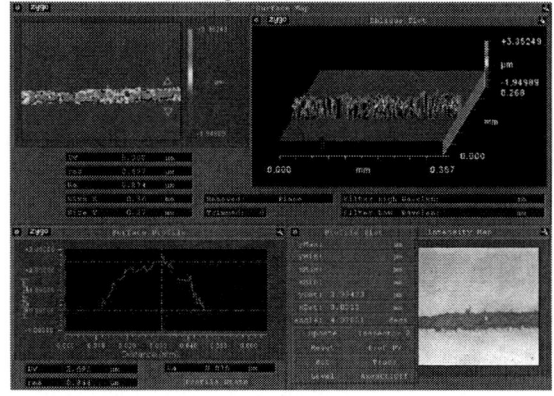

Figure 2: Pad printed line pattern of 35 μm width and 2.4 μm depth.

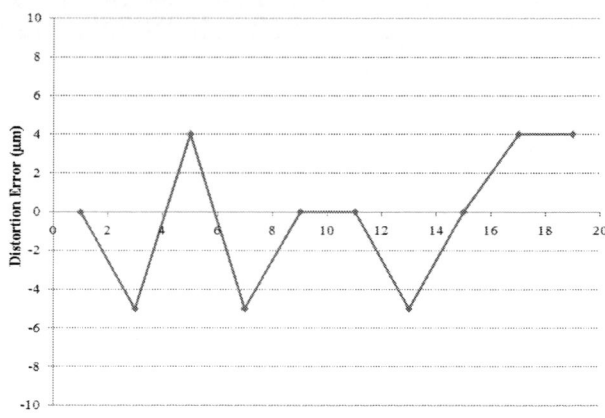

Figure 3: Pattern distortion error caused by the deformation of silicon pad (measured at every 2mm distances from the center of pad).

EL DISPLAY LAMP PAD-PRINTED ON CURVED SURFACE

The EL display lamp consists of 5 layers as shown in the figure 4: Bottom electrode; Dielectric layer; Phosphor; Transparent electrode; Bus electrode [10]. For proper pad printing, the ink of each layer should be formulated and the pad printing condition should be well controlled. In order to fill ink to the engraved pattern plate, the viscosity of the ink should be low. After the ink is transferred to the pad, in view point of the ink transferring from the pad to a substrate, the viscosity of the ink on the pad should become higher. Namely, the ink should contain a volatile solvent which induces viscosity change. In this experiment, we used several kinds of solvents like Ethyl acetate, BK, and so on. The table 1 shows the initial viscosity of ink for each layer. The viscosities were measured by using a Brookfield DV-II+ Pro.

Table 1: Viscosity of ink for each layer.

Ink	Viscosity (cPs)	Remarks
Silver	7,000 – 7,500	screen print ink
Carbon	34,000	screen print ink
Dielectric	4,139	BK treated
Phosphor	2,076	screen print ink
Transparent	4,511	BK treated

As the bottom electrode, silver ink and carbon ink were used. Since the resistance of silver is low enough for electrodes, 1 time printing results in the sufficient conductivity, even though the printed thickness is as thin as 1.4μm. However, in case of carbon, we printed 5 times for the sufficient conductivity. The pattern height of carbon was 5.77μm. Since the dielectric layer should be thick in order to prevent a short circuit at a high voltage, we printed this layer 5 times. The phosphor and transparent layer was also printed 5 times for sufficient layer thickness. The bus electrode is same as the bottom electrode. However, since the bus electrode is printed on the transparent electrode, the adherence between the layers is critical. The adherence of the carbon ink is almost 5 times better than the adherence of the silver ink. We used the carbon ink for the bus electrode. The printing process condition for each layer is described in the Table 2.

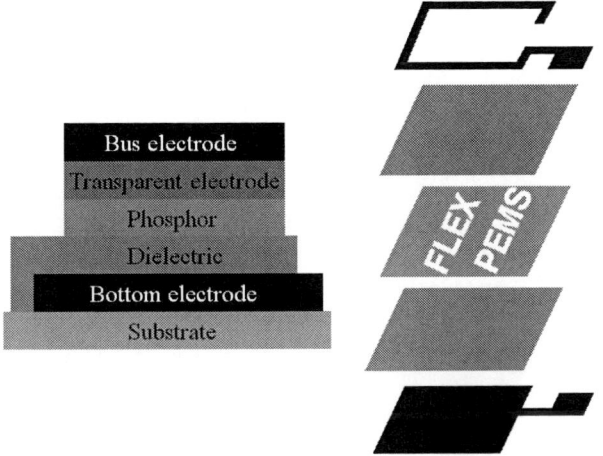

Figure 4: Design and fabrication process of the EL display lamp printed on a curved surface.

Table 2: Pad printing process condition for each layer.

Layer	Printing Condition	Pattern Height	Remarks
Bottom electrode	1 time printing 4 minutes curing at 120 ℃	1.4μm	1.1 Ω/cm
Dielectric	5 times printing 2 minutes drying at 120 ℃	4.6μm	
Phosphor	5 times printing 2 minutes drying at 120 ℃	2.2μm	
Transparent electrode	5 times printing 3 minutes drying at 120 ℃	3.5μm	8~20 kΩ/cm
Bus electrode	5 times printing 4 minutes curing at 120 ℃	5.77μm	2~6 kΩ/cm

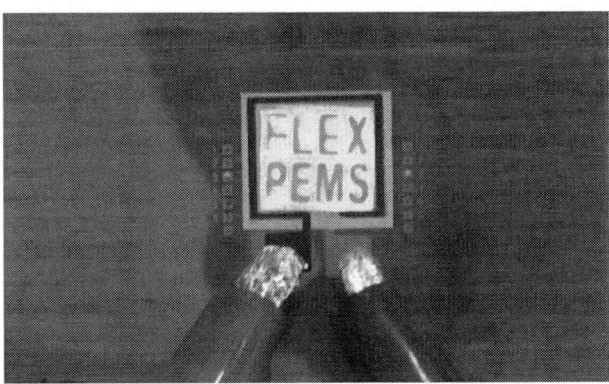

Figure 5: EL display lamp pad-printed on a PEN film.

The figure 5 and figure 6 show the fabrication results of the pad printed EL display lamp. The PEN film was used first in order to realize the pad printing process condition of each layer. The pad printed EL display lamp on PEN film is flexible as much as a radius of curvature 5 mm. Finally, we printed EL display lamp on a dish as shown in Fig.6. The dish has a radius of curvature 80mm. The EL display lamp was driven at AC 200V of 1kHz. The difference between the PEN film and the dish is surface adherence. The adherence of dish is so bad that the carbon ink was used for the bottom electrode. Consequently, we can conclude that EL display lamp was fabricated on a curved surface by using pad printing method, which has a great potential as the micro fabrication technique onto various shapes.

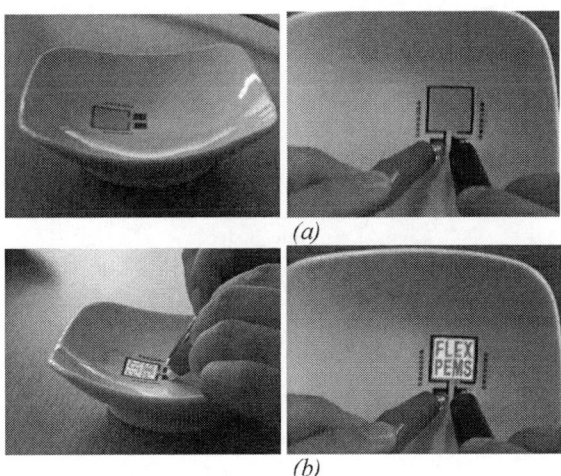

Figure 6: EL display lamp pad-printed on a dish having radius of curvature 80mm: (a) Off state; (b) On state (AC 200V, 1kHz).

CONCLUSION

In this paper, we presents the electro-luminescence (EL) display lamp which is patterned on a curved surface by the pad printing method. In the pad printing process, the engraved pattern plate is doctor-bladed, the silicon pad proceeds toward the plate, the ink is transferred onto the pad, the pad proceeds toward the curved surface, and the ink on the pad is finally transferred onto the surface. The minimal pattern of pad printing was 35 μm wide and 2.4 μm thick. The pattern distortion of pad printing caused by the deformation of silicon pad was measured at every 2mm distances from the center of pad. The pattern distortion was less than 5 μm on a flat surface. It means, as one of the micro patterning process, the pad printing method can be utilized for 35 μm wide precise patterns having 5μm location error.

The EL display lamp consists of 5 layers: Bottom electrode; Dielectric layer; Phosphor; Transparent electrode; Bus electrode. For proper pad printing, the ink of each layer was formulated and the pad printing condition was controlled. A PEN film used first in order to realize the pad printing process condition of each layer. Finally, the EL display lamp was printed on a dish, which has a radius of curvature 80mm. The EL display lamp was driven at AC 200V of 1kHz. Consequently, we can conclude that EL display lamp was fabricated on a curved surface by using pad printing method, which has a great potential as the micro fabrication technique onto various shapes.

ACKNOWLEDGEMENT

This work was supported by the Korea Research Council for Industrial Science & Technology (KOCI) through Center of Excellent Program and the Korea Foundation for International Cooperation of Science & Technology (KICOS) through the Global Partnership Program in K20602000002.

REFERENCES

[1] T.-M. Lee, T. G. Kang, J. S. Yang, J. Jo, K.-Y. Kim, B.-O. Choi, D.-S. Kim, "Drop-on-demand Solder Droplet Jetting System for Fabricating Micro Structure," *IEEE Transactions on Electronics Packaging Manufacturing*, vol.31, no.3, pp.202-210, 2008.

[2] T.-M. Lee, T. G. Kang, J. S. Yang, J. Jo, K.-Y. Kim, B.-O. Choi, D.-S. Kim, "Gap Adjustable Molten Metal DoD Inkjet System with Cone-Shaped Piston Head," *Journal of Manufacturing Science and Engineering – Transactions of the ASME,* vol.130, no.3, 031113, 2008.

[3] R. Reuss, "Large-Area, Integrated, Distributed Electronics," in *IMAPS 3Rd Advanced Technology Workshop on Printing an Intelligent Future*, Annapolis, USA, Sep. 2004

[4] M. Schen, "Accelerating Innovative Manufacturing Technologies and Positioning Flexible Electronics," *IMAPS 3Rd Advanced Technology Workshop on Printing an Intelligent Future*, Annapolis, USA, Sep. 2004

[5] K. Bock, "Reel to Reel Manufacturing of Printed Electronics and Systems," in *Printed Electronics Europe 2005*, Cambridge, UK, Apr. 2005.

[6] Reinhard R. Baumann, Robert Weiss, "Printing Beyond Color," in *Printed Electronics Europe 2005*, Cambridge, UK, Apr. 2005.

[7] W.-X. Huang, S.-H. Lee, H. W. Kang, H. J. Sung, T.-M. Lee, D.-S. Kim, "Simulation of liquid transfer between separating walls for modeling micro-gravure-offset printing," *International Journal of Heat and Fluid Flow,* vol.29, pp.1436-1446, 2008.

[8] T.-M. Lee, Y.-J. Choi, S.-Y. Nam, C.-W. You, D.-Y. Na, H.-C. Choi, D.-Y. Shin, K.-Y. Kim, K.-I. Jung, "Color Filter Patterned by Screen Printing," *Thin Solid Films*, vol.516, iss.21, pp.7875-7880, 2008.

[9] L. Mooring, N. G. Karousosa, C. Livingstonea, J. Davisa, G. G. Wildgooseb, S. J. Wilkinsc, R. G. Compton, "Evaluation of a novel pad printing technique for the fabrication of disposable electrode assemblies," *Sensors and Actuators B,* vol.107, pp.491-496, 2005.

[10] T.-M. Lee, D.-Y. Shin, Y.-S. Kim, C.-H. Kim, J. Jo, B.-O. Choi, D.-S. Kim, "Flexible EL Display Printed on a Paper," in *IMID*, Daegoo, Korea, 2007 Aug.

SOLUTION ELECTROCHEMILUMINESCENT MICROFLUIDIC CELL FOR FLEXIBLE AND STRETCHABLE DISPLAY

R. Okumura, S. Takamatsu, E. Iwase, K. Matsumoto and I. Shimoyama

The University of Tokyo, Tokyo, JAPAN

ABSTRACT

We developed liquid luminescent flexible displays using electrochemiluminescent solution as deformable luminescent material. The solution consisted of ruthenium complexes as light emitting material and ionic liquid as nonvolatile solvent. We injected the solution, to pattern and seal it, into microfluidic cells between two flexible films of polyethylene terephthalate substrates with indium tin oxide electrodes. The light emitting area of one cell was 2×2 mm^2. The gap between the electrodes was 100 μm. We measured the luminance of the cell during application of 50Hz/4.0Vpp (peak-to-peak value) rectangular waves of voltage in bent condition. The device was bent along the convex curvature with radii of 60, 30 and 20 mm. The luminance was reduced by only 30 % when the radius of curvature reached 20 mm. We conclude the electrochemiluminescent cells are suitable for flexible displays because they can tolerate large deformation of bending.

INTRODUCTION

Recently, the rapid advance in material engineering has promoted active research of flexible displays for mobile applications [1]. One of the most important issues on the flexible displays is how to achieve higher mechanical robustness of their components because the devices are largely deformed when used [2]. Therefore, deformable luminescent material is required. Solid luminescent material like organic light emitting diodes is widely used for conventional flexible displays [3]. Solid material is, however, cracked and cut down by bending strains [4]. On the other hand, liquid material can avoid mechanical destruction because they can change its shape freely and repetitively without degradation. Thus, liquid luminescent flexible displays offer the higher flexibility and mechanical robustness (Figure 1).

Electrochemiluminescent solution is the most promising liquid luminescent material because it can be controlled by voltage of transistor-transistor logic and is self-luminous material [5]. In addition to that, abundance of luminescent species provides the electrochemiluminescence (ECL) with variety of luminescent colors [6]. Moreover, the electrochemiluminescent cells have simple structures of the electrochemiluminescent solution between the electrodes. Yet the lack of packaging technique remains its drawback. Encapsulation method for the electrochemiluminescent solution is desired to be developed.

This paper reports on the electrochemiluminescent flexible displays. The electrochemiluminescent solution consists of ruthenium complexes as light emitting material and ionic liquid as nonvolatile solvent. We have developed patterning and sealing method for the solution. The emission of light was observed during application of voltage to the electrodes of the electrochemiluminescent cells. The luminance measurement of the electrochemiluminescent cells in bent condition confirmed their mechanical robustness.

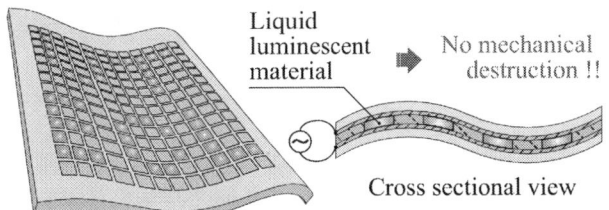

Figure 1: Concept of a liquid luminescent flexible display. The liquid luminescent material can avoid mechanical destruction if the device is largely deformed.

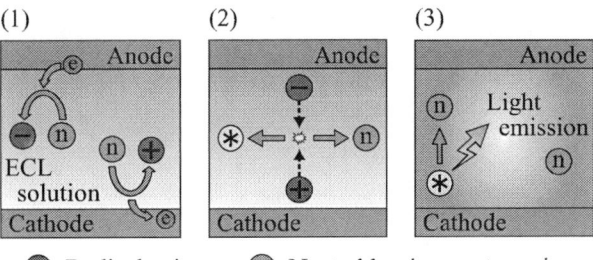

Figure 2: Principle of the ECL. (1) Production of radical anion and cation by redox. (2) Excitation of luminescent species by collision of the anion and cation. (3) Light emission during radiative deactivation.

PRINCIPLE

The electrochemiluminescence is produced during electrochemical reactions in the solution of luminescent species (polycyclic aromatic hydrocarbons or metal complexes) in aprotic organic solvent or ionic liquid. It is generated by radiative deactivation of excited luminescent species in the solution (Figure 2). First, redox reactions on the electrodes produce radical anion and cation of luminescent species in the electrochemiluminescent solution. Second, collision of the anion and cation excites the luminescent species. Third, the excited luminescent species emit light during radiative deactivation. These reactions are cyclic.

DESIGN AND FABRICATION

We have prepared three compositions of the ECL solution (Figure 3). Rubrene and 9,10-diphenylanthracene (Wako Pure Chemical Industries) are kinds of polycyclic aromatic hydrocarbons [7]. Tris(2,2'-bipyridyl)ruthenium (II) chloride (Alfa Aesar), Ru(bpy)$_3$Cl$_2$, is known as a prototypical example of ion transfer metal complexes (iTMCs) [6]. 1-ethyl-3-methylimidazolium bis(trifluoromethylsulfonyl) imide (Wako Pure Chemical Industries), EMITFSI, is an ionic liquid. Figure 3 shows the luminescence of each solution. The luminescent color was yellow (Solution A), blue (Solution B) and orange (Solution C). Each solution was sandwiched between indium tin oxide (ITO) sputtered glass substrates.

978-1-4244-2977-6/09 $25.00 © 2009 IEEE

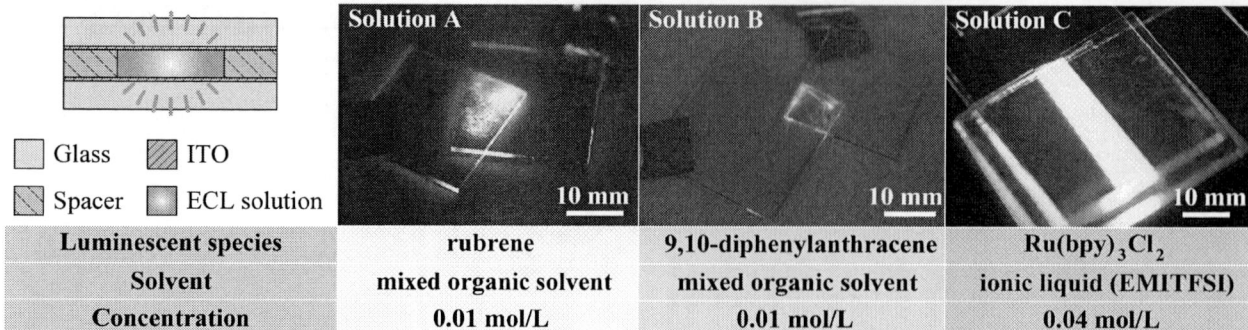

Luminescent species	rubrene	9,10-diphenylanthracene	Ru(bpy)$_3$Cl$_2$
Solvent	mixed organic solvent	mixed organic solvent	ionic liquid (EMITFSI)
Concentration	0.01 mol/L	0.01 mol/L	0.04 mol/L

Figure 3: The electrochemiluminescence of rubrene (Solution A), 9,10-diphenylanthracene (Solution B) and Ru(bpy)$_3$Cl$_2$ (Solution C). The voltage of 8.0 V was applied to Solution A and B, and the 50Hz/4.0Vpp (peak-to-peak value) rectangular waves of voltage to Solution C. The mixed solvent consisted of o-dichlorobenzene and acetonitrile (volume ratio was 2:1). Ru(bpy)$_3$Cl$_2$ is the formula of tris(2,2'-bipyridyl) ruthenium(II) chloride. EMITFSI is the abbreviation of 1-ethyl-3-methylimidazolium bis(trifluoromethylsulfonyl)imide.

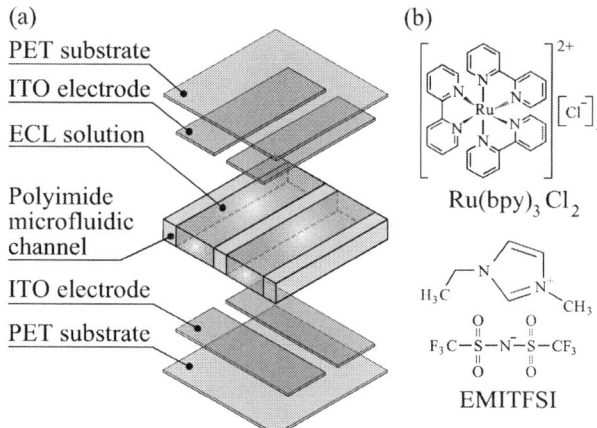

Figure 4: (a) Configurations of the flexible ECL cells. Each cell is addressed by a passive matrix. (b) Molecular structures of Ru(bpy)$_3$Cl$_2$ (luminescent species) and EMITFSI (ionic liquid).

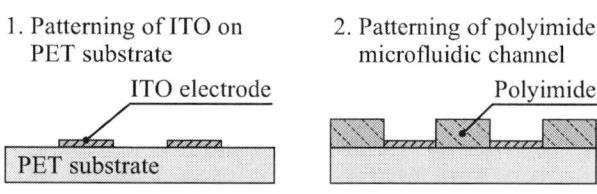

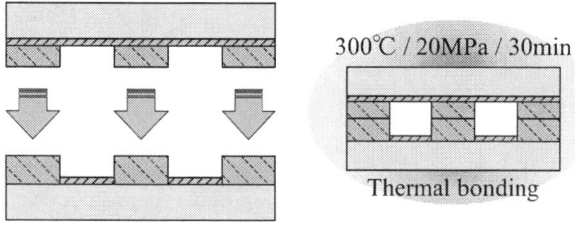

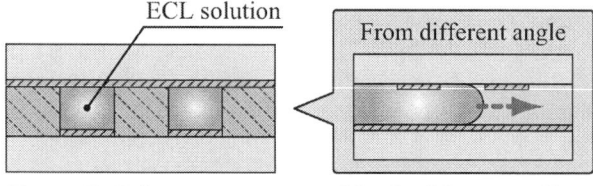

Figure 5: Fabrication process of the flexible ECL cells.

The flexible ECL cells consisted of the ECL solution sandwiched between two flexible films of polyethylene terephthalate (PET) substrates with ITO electrodes (figure 4 (a)). Each cell was addressed by a passive matrix. The emitting area of one cell was 2×2 mm^2. The distance between the ITO electrodes was 100 μm. The thickness of the PET substrates was 125 μm. The surface resistance of the ITO electrodes was 10 Ω/□. The Solution C was employed as the ECL solution. Molecular structures of the Ru(bpy)$_3$Cl$_2$ and EMITFSI (ionic liquid) are shown in Figure 4 (b). The extremely low vapor pressure of the EMITFSI ensures the stability of the solution. The vapor pressure of the EMITFSI is 6.2×10^{-6} kPa at temperature of 442 K [8]. The applying voltage should not exceed the electrochemical window of the EMITFSI (from -2.1 to +2.4 V) [9].

Figure 5 shows fabrication process of the flexible ECL cells. To pattern and seal the ECL solution between the flexible ITO-PET films, we used injection process of the solution. First, the ITO layer on the PET film was etched. Second, polyimide was patterned on the substrates. The thickness of the polyimide was 50 μm. Third, microfluidic channels were formed between the ITO-PET films by polyimide-polyimide thermal bonding at 300 degree C and 20 MPa for 30 minutes. Fourth, the ECL solution was injected into the microfluidic channels. The solution flowed into the channels by capillary force without external pumps. Bubble did not remain within the channels because the capillary force was quite large. Finally, microfluidic channels were sealed with epoxy resin.

We fabricated 2×2 flexible ECL cells. The four cells emitted light, being bent along a convex curvature with radius of 30 mm (Figure 6). They were driven by 50Hz/4.0Vpp rectangular waves of voltage.

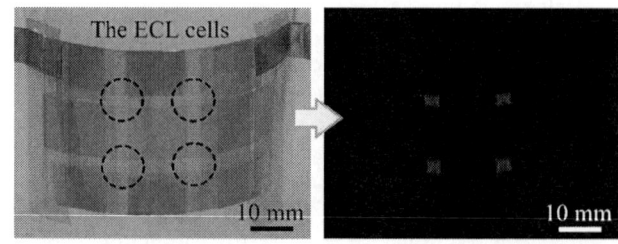

Figure 6: Emission of the flexible ECL cells. The emitting area of one cell is 2×2 mm^2. The device was bent along a convex curvature of an acrylic pipe with radius of 30 mm.

EXPERIMENTAL

We have evaluated the basic luminescent characteristics of the Solution C and mechanical robustness of the fabricated flexible ECL cell. The luminance was measured by an optical power meter (Advantest Corporation Q82214). The light receiving element of the power meter was 8 mm in diameter. The measurement was performed in a dark room. The ECL cell was driven by rectangular waves of voltage from function generator.

Basic luminescent characteristics of the Solution C

The basic luminescent characteristics of the solution C were measured (Figure 7). Figure 7 (a) shows the experimental setup for the luminance measurement of the Solution C. The solution was sandwiched between ITO sputtered glass substrates. 100 μm thick polyimide tapes were inserted between the substrates as spacers. The emitting are of the cell was 10×10 mm^2. The distance between the cell and the power meter was 10 mm.

Figure 7 (b) shows the luminance of the cell as a function of the peak-to-peak voltage of the rectangular waves. The frequency of the applied voltage was 50 Hz. The peak-to-peak voltage was changed from 0.5 to 4.0 Vpp. The luminance at the voltage of 4.0 Vpp was normalized to 1.0 (arbitrary unit). The luminance was almost zero until the voltage reached 2.5 Vpp, and then increased rapidly. This result is explained by the fact that the reduction of the $Ru(bpy)_3^{2+}$ occurs at -1.09 V and the oxidation of it occurs at +1.63 V [10]. In order to obtain both reduction and oxidation, at least 3.0 Vpp is needed. One can control the luminance of the solution by changing the voltage within 3.0 to 4.0 Vpp. Higher voltage than 4.5 Vpp should not be applied to the solution because the electrochemical window of the EMITFSI ranges from -2.1 to +2.4 V.

Figure 7 (c) shows the luminance of the cell as a function of the frequency of the applied voltage. The peak-to-peak voltage of the rectangular waves was 2.5 Vpp. The frequency of the applied voltage was changed from 10 to 200 Hz. The luminance at the frequency of 10 Hz was normalized to 1.0 (arbitrary unit). The luminance showed rapid decline from 20 to 60 Hz, followed by a slow approach to zero.

Figure 7 (d) shows the luminance versus time during the application of the rectangular waves of voltage (50 Hz / 4.0 Vpp). The temporal average luminance was normalized to 1.0 (arbitrary unit). Two peaks were observed in 0.02 seconds (one cycle of the applied rectangular waves of voltage). It suggests that light emission occurs following both the negative and positive sweeps of the rectangular waves of voltage [11].

Mechanical robustness of the flexible ECL cell

In order to verify the mechanical robustness of the flexible ECL cell, we measured the luminance of it in bent condition (Figure 8). Figure 8 (a) shows the experimental setup for the luminance measurement of the flexible ECL cell. The Solution C was encapsulated between ITO-PET films with the injection method. The gap between the ITO electrodes was 100 μm. The emitting area of the cell was 2×2 mm^2. The distance between the cell and the power meter was 5 mm. The cell was driven by 50Hz/4.0Vpp rectangular waves of voltage. It was bent along convex curvature of acrylic pipes with radii of 60, 30 and 20 mm.

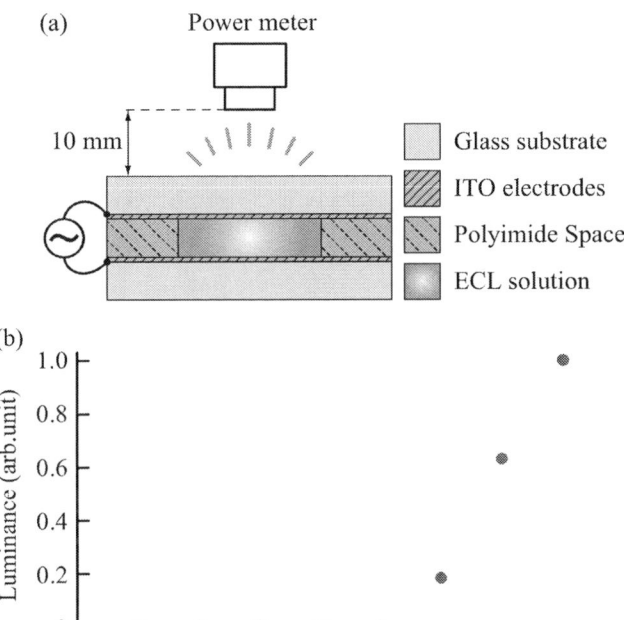

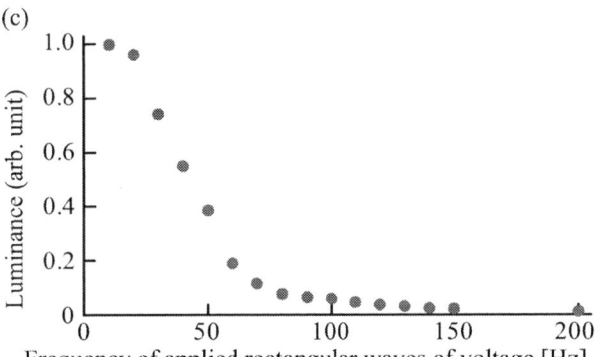

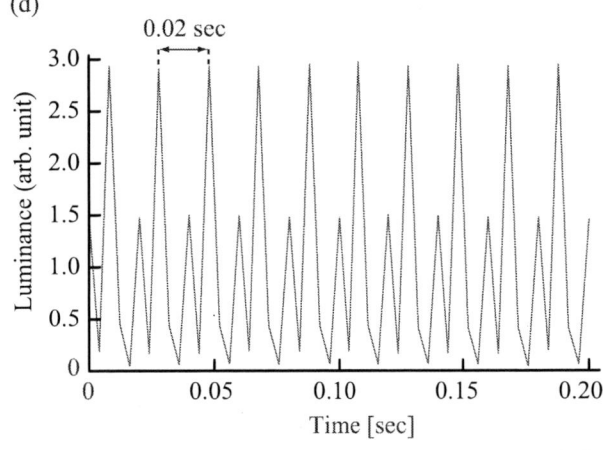

Figure 7: The basic Luminescent characteristics of the Solution C. The luminance is expressed by arbitrary unit. (a) The experimental setup for the luminance measurement of the Solution C. (b) The luminance as a function of the peak-to-peak voltage of the applied rectangular waves. (c) The luminance as a function of the frequency of the applied rectangular waves of voltage. (d) The luminance versus time during the application of the 50Hz/4.0Vpp rectangular waves of voltage.

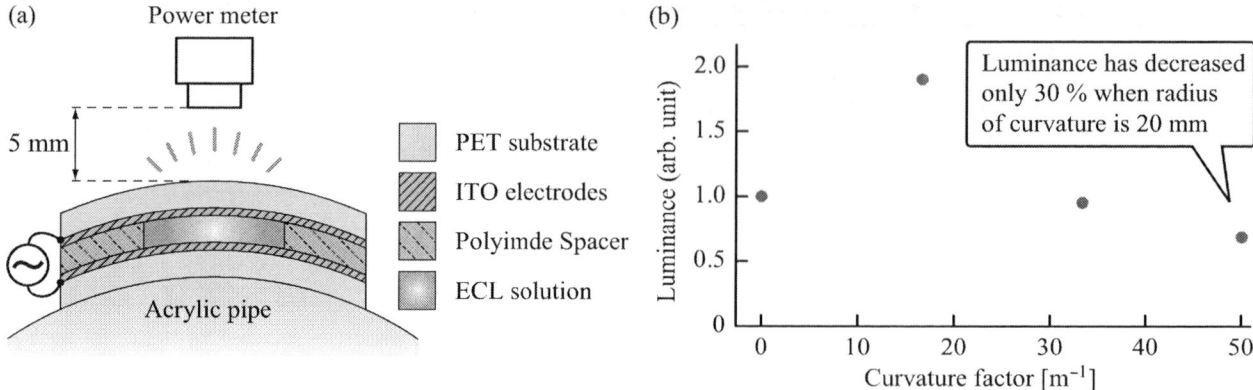

Figure 8: The luminescent characteristics of the flexible ECL cell in bent condition. The luminance is expressed by arbitrary unit. (a) The experimental setup for the luminance measurement of the flexible ECL cell. (b) The luminance of the flexible ECL cell as a function of the curvature factor of the cell.

Figure 8 (b) shows the relationship between the luminance and the curvature factor (inverse of the radius) of the cell. The luminance was reduced by only 30 % when the curvature factor reached 50 m^{-1} (the radius of the cell was 20 mm). The result demonstrates the mechanical robustness of the fabricated flexible ECL cell.

CONCLUSION

The electrochemiluminescent flexible displays are proposed. We have fabricated the flexible ECL cells with injection method of the ECL solution. The ECL solution consists of Ru(bpy)$_3$Cl$_2$ in EMITFSI (ionic liquid). The emission of light was generated from the fabricated ECL cells in bent condition during the application of the rectangular waves. The basic characteristics of the ECL solution were evaluated. The luminance was almost zero when the applied voltage was below 2.5 Vpp, and then increased rapidly with the voltage from 2.5 to 4.0 Vpp. The luminance decreased monotonically with the frequency of the applied rectangular waves of voltage. Two peaks of the luminance were observed in one cycle of the applied rectangular waves. Then, the mechanical robustness of the flexible ECL cell was evaluated. The luminance of the cell was reduced by only 30 % when the radius of the cell reached 20 mm. This result confirms the mechanical robustness of the flexible ECL cells. Most significantly, this work reveals the possibilities to fabricate the flexible electrochemiluminescent displays which tolerate large deformation of bending or stretching.

ACKNOWLEDGMENTS

Photomasks were made using the University of Tokyo VLSI Design and Education Center (VDEC)'s 8-inch EB writer F5112+VD01 donated by Advantest Corporation.

REFERENCES

[1] L. Zhou, A. Wanga, S.C. Wu, J. Sun, S. Park, and T.N. Jackson, "All-organic active matrix flexible display," *Appl. Phys. Lett.*, vol. 88, pp. 083502, 2006.

[2] T. Sekitani, Y. Noguchi, K. Hata, T. Fukushima, T. Aida and T. Someya, "A Rubberlike Stretchable Active Matrix Using Elastic Conductors," *Science*, vol. 321, pp. 1468-1472, 2008.

[3] H. Y. Low and S. J. Chua, "Mechanical properties of organic light-emitting thin films deposited on polymer-based barrier substrate: potential for flexible organic light-emitting displays," *Mater. Lett.*, vol. 53, No. 7, pp. 227-232, 2002.

[4] R. Bhattacharya, S. Wagner, Y. J. Tung, J. R. Esler and M. hack, "Organic LED Pixel Array on a Dome," *Proc. of the IEEE*, vol. 93, No. 7, pp. 1273-1280, 2005.

[5] Radha Pyati and Mark M. Richter, "ECL — Electrochemical luminescence," *Annu. Rep. Prog. Chem.*, Sect. C, vol. 103, pp. 12-78, 2007.

[6] J.D. Slinker, J. Rivnay, J.S. Moskowitz, J.B. Parker, S. Bernhard, H.D. Abruna and George G. Malliaras, "Electroluminescent devices from ionic transition metal complexes," *J. Mater. Chem.*, vol. 17, pp. 2976-2988, 2007.

[7] K. Nishimura, Y. Hamada, T. Tsujioka, S. Matsuta, K. Shibata and T. Fuyuki, "Solution Electro-chemiluminescent Cell with a High Luminance Using an Ion Conductive Assistant Dopant," *Jpn. J. Appl. Phys.*, vol. 40, pp. L1323-L1326, 2001.

[8] D.H. Zaitsau, G.J. Kabo, A.A. Strechan, Y.U. Paulechka, A. Tschersich, S.P. Verevkin and A. Heintz, "Experimental Vapor Pressures of 1-Alkyl-3-methylimidazolium Bis(trifluoromethyl-sulfonyl)imides and a Correlation Scheme for Estimation of Vaporization Enthalpies of Ionic Liquids," *J. Phys. Chem. A*, vol. 110, pp. 7303-7306, 2006.

[9] M. Holzapfel, C. Jost, A. Prodi-Schwab, F. Krumeich, A. Wursig, H. Buqa and P. Novak, "Stabilisation of lithiated graphite in an electrolyte based on ionic liquids: an electrochemical and scanning electron microscopy study," *Carbon*, vol. 43, pp. 1488-1498, 2005.

[10] T. Kado, M. Takenouchi, S. Okamoto, W. Takashima, K. Kaneto and S. Hayase, "Enhanced Electrochemiluminescence by Use of Nanoporous TiO$_2$ Electrodes: Electrochemiluminescence Devices Operated with Alternating Current," *Jpn. J. Appl. Phys.*, vol. 44, No. 11, pp. 8161-8164, 2005.

[11] J. D. Slinker, J. Rivnay, J. A. DeFranco, D. A. Bernards, A. A. Gorodetsky, S. T. Parker, M. P. Cox, R. Rohl, G. G. Malliarasa, S. Flores-Torres and H. D. Abruña, "Direct 120 V, 60 Hz operation of an organic light emitting device," *J. Appl. Phys.*, vol. 99, pp.074502, 2006.

978-1-4244-2977-6/09 $25.00 © 2009 IEEE

MEMS BASED LASER IMAGER
WITH DIAGONAL PROGRESSIVE SCANNING

T. Sandner[1], T. Grasshoff, T. Klose[1], H. Schenk[1] and J.L. Massieu[2]
[1] Fraunhofer Institute of Photonic Microsystems, GERMANY
[2] Intermec Inc., Toulouse, FRANCE

ABSTRACT

This paper presents a novel concept of a laser scanning imager with diagonal progressive scanning using electrostatic resonant 2D-gimbal mirrors. Typically a horizontal progressive raster scanning cannot be realized for resonant 2D-scanners due to a lower frequency limit. Alternative to raster scanning we introduce the novel concept of diagonal progressive scanning using a high frequency resonant 2D-scanner of specific constant frequency ratio. The focus of this work is on the design of the novel 2D-scanner, where an increased frequency bandwidth was realized by using multiple degressive spring suspensions and more efficient comb geometries. Simulated and experimental results of the 2D-scanner and VGA laser imaging system are shown.

INTRODUCTION

Micro scanning mirrors are essential components in miniaturized and highly accurate scanning systems [1]. At Fraunhofer IPMS an electrostatic in-plane vertical comb drive was developed for micro scanning mirrors (MSM) firstly reported in [2]. The driving principle of MSM with in-plane-vertical comb-drive as well as the timing between mirror oscillation and driving voltage is shown in Figure 1.

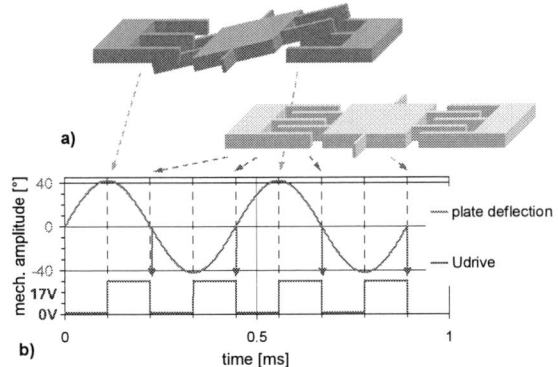

Figure 1: Driving principle of resonant MSM of IPMS

Using an appropriate spring design allows for either rotational [3] or translational [6] operation mode, needed for 1D/2D light beam deflection or optical path modulation, respectively. This resonate MSM are promising for laser scanning imaging applications, e.g. 2D-data capture or medical endoscopes [5], due to their larger scan angles of up to 140° [5], low power consumption and qualified SOI technology [3] suitable for mass fabrication [1]. Their shock and vibration insensitivity make them attractive for portable applications like 1D/2D-barcode readers [1], ultracompact laser projection displays [4] or miniaturized FTIR spectrometers [6]. In contrast to this the dynamic response is characterized by a limited frequency bandwidth, hysteresis and an instable resonance point for fixed

driving frequency [3]. By synchronization of driving voltage to phase of mirrors oscillation a stable operation at resonance point is possible, but the resonance frequency is affected by fabrication tolerances.

Today, reading of 2D-codes is realized with CCD-imagers. It is expected that 2D laser scanning based data code reading has several advantages like large depth of focus and improved signal to noise ratio. This article presents the novel concept of a laser scanning imager with diagonal progressive scanning using electrostatic resonant 2D-gimbal mirrors. Typically the scanning of resonant comb driven MSM is limited to a complex Lissajou figure due to a lower limit of resonance frequency [4]. However, a horizontal progressive raster scanning, where the vertical scan frequency is equal to the frame rate, is advantageous to prevent image artefacts of a moving target [1].

Laser imaging with diagonal progressive scanning

In contrast to raster scanning we introduce the novel concept of staggered diagonal progressive scanning using a high frequency resonant 2D-scanner of a specific frequency ratio defined by:

$$f_v = k \cdot f_h \text{ with } k = (N+1)/N \qquad (1)$$

where f_v and f_h are the vertical and horizontal resonant frequencies, respectively. To guarantee an angular grid pitch smaller or equal FOV/n the parameter $N=\pi/2 \cdot n/0.9$ is defined by the number of pixels n increased the by factors of $\pi/2$ to compensate the sinusoidal velocity of the scanned laser spot, and by 1/0.9 to compensate the unusable border at the periphery of the field of view (FOV). Using a 2D-scanner of a frequency ratio equal to (1) and a data sampling rate of $f_c= (N+1)f_h$ we get a squared grid of data points captured by a diagonal progressive scanning (see figure 2).

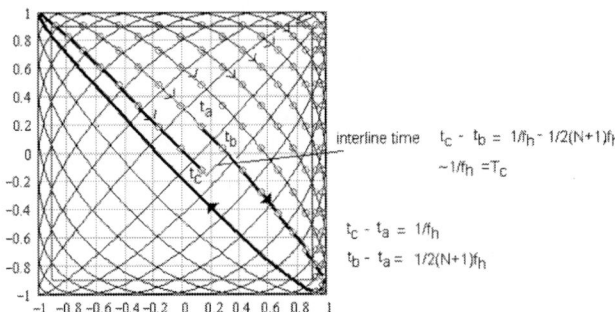

Figure 2: Schematic of laser imaging with diagonal progressive scanning

In this novel approach of a laser imager with diagonal progressive scanning the parasitic sensitivity of a common Lissajou scanning imager to motion artifacts can be eliminated.

From figure 3 it can be seen that a moving target causes only a slight geometrical deformation of the captured image whereas for a Lissajou laser imager significant uncorrelated changes of the captured image occurs.

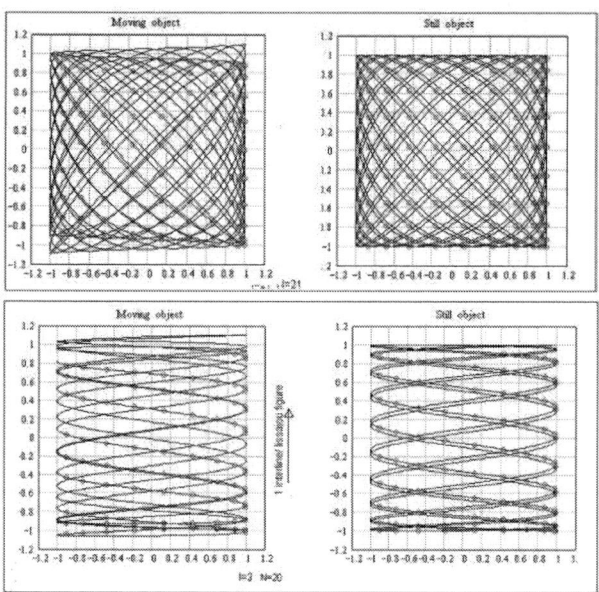

Figure 3: Comparison of laser imaging with diagonal progressive scanning (upper) and Lissajou scanning (lower) for moving object (left) and sill object (right).

This work targets on a laser imager of VGA resolution (640x640 pixels) with a frame rate of 5 frames /s and an optical field of view adjustable between 5° to 44° enabling a zoom function. This results in a fixed frequency ratio of f_v/f_h=5585Hz/5580Hz according to (1). The specifications of the required 2D-scanner are summarized in table 1.

SCANNER DESIGN

It is obvious that a resonant driven 2D-scanner with fixed frequency ratio of f_v/f_h=5580Hz/5585Hz and variable scan angle of MSA=1.25°...11° is a challenge for the MEMS design. To guarantee a fixed frequency ratio the novel 2D-scanner has to operate in open loop due to frequency fabrication tolerances. Therefore a fault tolerant scanner design with suitable frequency band width is required to adjust frequency ratio.

Table 1: Parameters of 2D-scanner designed for laser scanning imager with progressive diagonal scanning

Parameter	Inner Mirror	Movable Frame
Resonance frequency	5580 Hz	5585 Hz
Optical FOV (adjustable)	5° - 44°	5° - 44°
Mirror diameter	1.2…1.5mm	-
Frequency bandwidth @ common scanner design	7 Hz	3 Hz
Frequency bandwidth @ optimized scanner design	70 Hz	34 Hz

With a conventional scanner design, presented in [3], only a small frequency bandwidth limited to 3Hz(MF) and

7Hz(M) could be realized. To compensate the fabrication tolerances of typically $\sigma(\Delta f) = 15...20$Hz an increased frequency bandwidth of at least 30Hz is required.

Hence, the main challenge of the 2D-scanner design was to increase the frequency bandwidth required to adjust the fixed frequency ratio in open loop operation and to meat an acceptable fabrication yield. For that reason we investigated several design options to increase bandwidth, e.g by optimized spring characteristics and comb drive geometries, or to compensate fabrications tolerances, e.g. by compensation structures or active frequency tuning. In total we have designed, fabricated and experimentally investigated twelve different scanner variants including an integrated 2D- piezo-resistive position sensor [7]. In this paper we are focusing on 2D-scanner designs with increased frequency bandwidth suitable for diagonal progressive scanning.

Methodology of scanner design

The scanner designs have been developed within an iterative design process using two simulation techniques:

- FEA simulations using ANSYS™ multiphysics
- Reduced Order Model (ROM) simulations of dynamic frequency response behavior using MATLAB™

In figure 4 the schematic design workflow of the scanner design is illustrated. After model initialization the nonlinear angular characteristics of all relevant physical domains, (1) structural mechanics (e.g. spring characteristics), (2) electrostatics (e.g. capacitance characteristics) and (c) fluid-mechanics (e.g. deflection dependent viscous air flow / damping), are calculated by FEA (see figure 4b).

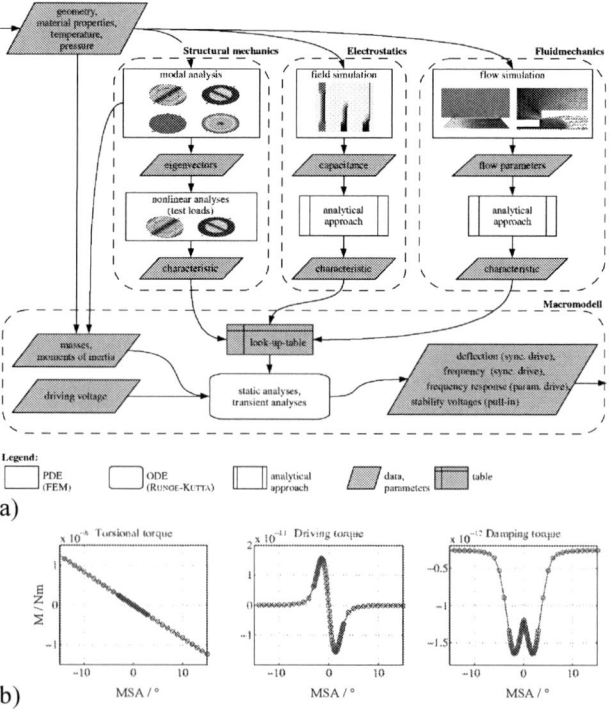

Figure 4: Simulation methodology of scanner design: (a) simulation work flow with order reduction, (b) simulation of nonlinear influence quantities (structural nonlinearities, electrostatic and damping torque) required for simulation of dynamic frequency response

The nonlinear angular characteristics are stored in lock-up-tables required for reduced order modeling of dynamic scanner behavior. Then the transient and frequency response behavior is simulated based on an analytical / numerical solution of the fundamental nonlinear MATHIEU equitation of scanner oscillation by means of nonlinear dynamics using our MATLAB™ based toolkit IMTK developed at IPMS. This results in a reliable estimation of open loop frequency response, bandwidth and required driving voltage.

Simulation results

With a common 2D-scanner design using simple torsional beam suspensions with progressive characteristic and low efficient comb drives located at frame/mirror plate a poor bandwidth of only 3Hz / 7Hz (MF/M) can be realized (see figure 5a) in comparison to 15…30Hz frequency variation due to DRIE fabrication tolerances. The bandwidth is defined as frequency range at $MSA=11°$ adjustable by driving voltage $U \leq 200V$ limited by electrostatic stability. To increase bandwidth & range of operation (for frequency independent amplitude control) we investigated the following main design concepts:

- optimization of in-plane vertical comb drives by using additional comb drives located near to torsional flexures enabling a higher driving efficiency due to a broader angular capacitance characteristic,
- increase of bandwidth using multiple beam suspensions of up to nine parallel springs resulting in degressive spring characteristics reducing the slope of frequency response curve at larger deflection angles.

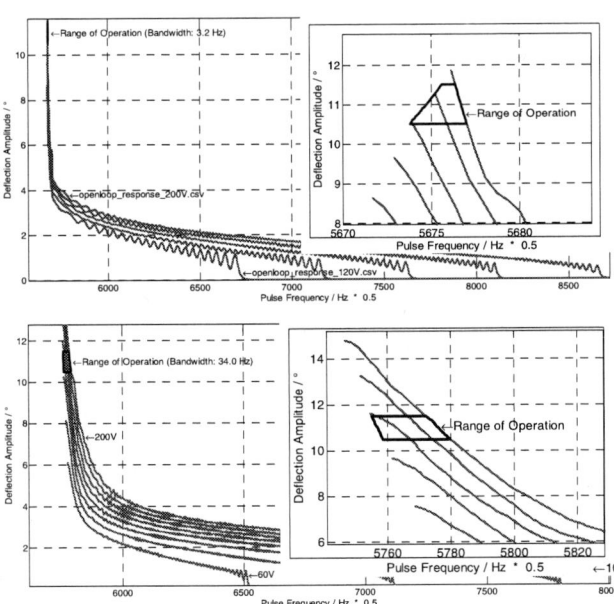

Figure 5: Simulated frequency response curves of movable frame for common scanner design, and (b) optimized design

Using additional comb drives at conventional flexures the bandwidth could be increased to 19Hz / 37Hz (MF/M). But the combination of optimized spring and comb drive geometry using degressive multiple spring suspensions and more efficient comb drives located near to flexures was found most promising for diagonal progressive scanning. For this 2D-scanner with degressive springs and additional comb drives at flexures and increased bandwidth of 34Hz / 70Hz (MF/M) could be realized (see figure 5b).

EXPERIMENTAL RESULTS

Finally twelve different design variants of 2D-MSM were designed and fabricated based on CMOS compatible 30µm thick SOI-MEMS technology presented in [3].

In this article we can present only results of the two preferred design variants both using additional comb drives located near to torsional flexures. In figure 6 photographs of a 2D-scanner chip with degressive suspensions of multiple parallel beams in both axes are shown. From the measured frequency response curves shown in figure 7 it is visible that the required fixed frequency ratio of f_v/f_h=5580Hz/5585Hz and a symmetric FOV could be meat only for MSA =11°..9° and MSA≤.4.2° due to a lower slope of mirrors frequency response curves compared to frequency curves of frame.

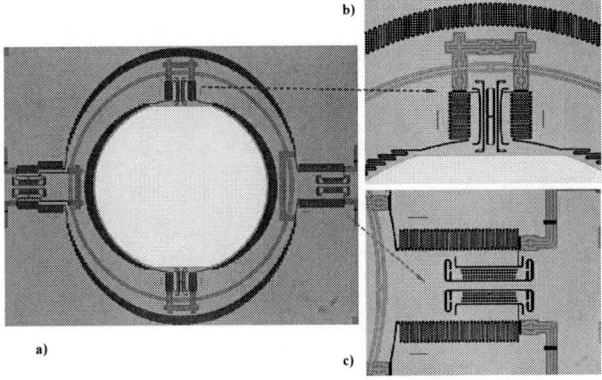

Figure 6: 2D-scanner chip (a) with degressive multiple springs suspensions of mirror (b) and frame (c).

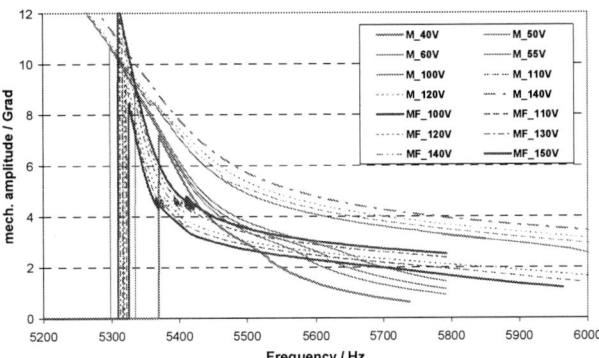

Figure 7: Measured frequency response of 2D-MSM shown in figure 6 with degressive spring suspensions in both axes.

In figure 8 the preferred 2D-scanner design is shown. Here, a degressive spring suspension of frame was combined with a slightly progressive torsional spring of mirror to achieve an equal slope of M/MF-frequency response enabling also comb drives with higher efficiency for better amplitude control. From the measured frequency response curves (see figure 9) it is obvious that the fixed frequency ratio can be guaranteed for the complete range of symmetric FOV adjustable between 11° and 1.25°.

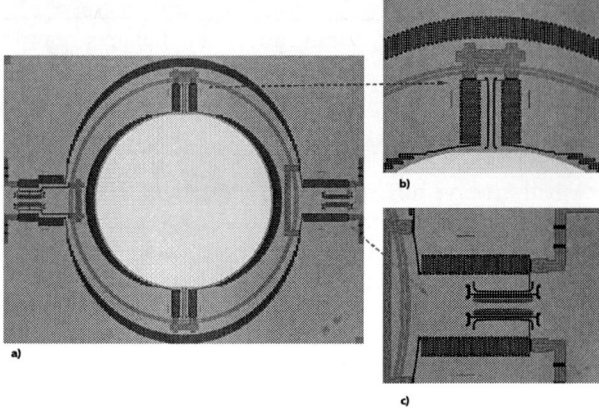

Figure 8: Photograph of scanner chip optimized for diagonal progressive scanning, a) 2D-gimbal scanner, b) detail of mirror suspension with highly efficient comb drives located near to torsional axis, c) degressive spring of frame.

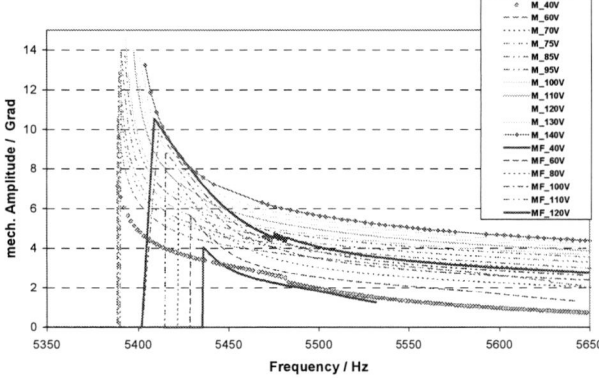

Figure 9: Measured frequency response curves of preferred scanner design shown in figure 8.

System tests of laser imager

The preferred 2D-scanner design shown in figure 7 was integrated into a first system demonstrator to verify the concept of diagonal progressive laser scanning experimentally. This demonstrator system is based on FPGA electronics for 2D-scanner driving and data capture enabling 2D-laser imaging of VGA resolution and data capture of currently 5 frames /s. In principle the frame rate can be increased by a factor of 4 using the same 2D-scanner. In figure 10 two experimental test images of a stationary and moving test pattern are shown.

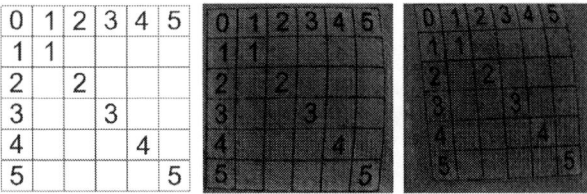

Figure 10: system test results of VGA laser scanning imager @ MSA = ±6°: test pattern (left), image of a stationary target (middle), image of moving target (right)

The image of the moving target verifies the concept of diagonal progressive scanning, because only geometrical distortions similar to raster scanning but no

stochastic motion artifacts of a Lissajou imager are visible. The image distortions of the stationary target are due to deterministic deviations of the real scanner oscillation from an ideal sinusoidal function assumed for the current system demonstrator and can be compensated in principle.

CONCLUSIONS

This article presents the novel concept of a VGA laser scanning imager with diagonal progressive scanning using electrostatic resonant 2D-gimbal mirrors, which requires a fixed frequency ratio of f_v/f_h=5585Hz/5580Hz not possible with a common scanner design due to limited frequency bandwidth. We developed a specific 2D-MSM suitable for diagonal progressive scanning by using multiple parallel spring suspensions with degressive spring characteristics, especially suitable for the movable frame, combined with optimized comb drives located nearby the torsional springs for improved efficiency. This results in a broader frequency bandwidth and enables an independed adjustment of frequency and amplitude. Hence, the constant frequency ratio can be guaranteed for the complete range of symmetric FOV adjustable between 44° and 5°. In addition the concept of progressive diagonal laser scanning was verified successfully by a VGA system demonstrator. Finally our simulation methodology is from general interest for design of resonate comb driven micro mirrors, because it includes all relevant nonlinearities of structural mechanics, electrostatics and fluid mechanics and enables a precise simulation of dynamic scanner behavior using advanced reduced order models.

REFERENCES

[1] A. Wolter, S. Hsu, H. Schenk, H. Lakner, "Applications and requirements for MEMS scanner mirrors", Proc. SPIE vol. 5719, pp. 64-75, 2005.

[2] H. Schenk, P. Dürr, D. Kunze, H. Kück, "A new driving principle for micromechanical torsional actuators", MEMS vol. 1, pp. 333-338, 1999.

[3] H. Schenk, P. Dürr, T. Haase, D. Kunze, U. Sobe, H. Lakner, H. Kück, "Large deflection micromechanical scanning mirrors for linear scans and pattern generation", IEEE J. Selected Topics of Quantum Electronics 6 (5), pp. 715- 722, 2000.

[4] M. Scholles, A. Bräuer, K. Frommhagen, C. Gerwig, H. Lakner, H. Schenk, M. Schwarzenberg, "Ultracompact laser projection systems based on two-dimensional resonat microscanning mirrors", SPIE J. Micro/Nanolith. MEMS MOEMS 7 (2), pp. 021001: 1-11, 2008.

[5] C. Drabe, R. James, T. Klose, A. Wolter, H. Schenk, H. Lakner "A New Micro Laser Camera", Proc. SPIE vol. 6466, pp. 64660I: 1-8, 2007.

[6] T. Sandner, C. Drabe, H. Schenk, A. Kenda, W. Scherf, "Translatory MEMS actuators for optical path length modulation in miniaturized Fourier-transform IR spectrometers", SPIE J. Micro/ Nanolith. MEMS MOEMS 7(2), pp. 021006:1-12, 2008.

[7] T. Sandner, H. Conrad, T. Klose, H. Schenk, "Integrated Piezo-resistive Positionssensor for Microscanning Mirrors", IEEE LEOS OMEMS 2007, pp. 195-196, 2007.

978-1-4244-2977-6/09 $25.00 © 2009 IEEE

POLYMERIC (SU-8) OPTICAL MICROSCANNER DRIVEN BY ELECTROSTATIC ACTUATION

S. K. Lee[1], M. G. Kim[1], J. Y. An[1], M. H. Jun[1], S Yang[1,2], and J. H. Lee[1,2]

[1]School of Information and Mechatronics, [2]School of Medical System Engineering,
Gwangju Institute of Science and Technology (GIST), Korea, South

ABSTRACT

A torsional micromechanical scanner was fabricated using photosensitive polymer (SU-8). The scanner consists of two parts; top layer (micro mirror) and bottom layer (an anchor and electrodes). The SU-8 scanner is actuated by electrostatic force generated from the gap-closing electrodes. For the fabricated optical scanner with the mirror area of $3\times3mm^2$, the experimentally determined scanning angles were 0.85° and 3.1° for 60Hz (non-resonant) and 1.13 kHz (resonant) driving frequencies, respectively, at an input voltage of 160V. A polymer-based actuator offers advantages including a low-cost fabrication process and low frequency operation.

INTRODUCTION

A MEMS scanner is one of the key elements for the applications of laser display, optical communication, bar code reading, medical devices (optical coherence tomography: OCT) and so on [1, 2, 3]. The key issues for the MEMS scanners are the actuation mechanism and the materials for the structures. There are four kinds of the actuation methods to operate MEMS scanners; electrostatic, electromagnetic, piezoelectric [4] and thermal actuators [5]. Electromagnetic actuators make use of a Lorentz force generated by magnetic fields. They are, however, bulky due to the use of permanent magnet although they provide a large driving force. Piezoelectric actuators feature fast response, but require relatively complicate fabrication process to deposit piezoelectric materials on the actuation regions to minimize the size of the device. Thermal actuators can be used for the applications requiring large force and deformation, but they still need improvement in operation speed and power consumption. Electrostatic actuators (gap-closing or combs) utilize an electric potential difference between two electrodes. In micro systems, electrostatic actuators have been preferred owing to low power consumption and high speed operation, although they have several drawbacks such as relatively high driving voltage and small displacement.

For the materials of microstructures, silicon has been widely used because of its good mechanical characteristics and batch fabrication using bulk and/or surface micromachining processes. However, the silicon process is usually complicated and expensive. Recently, new polymer materials such as photo resist (SU-8), polycarbonate (PC), polydimethylsiloxane (PDMS), cyclic olefin copolymer (COC), and fire resistant 4 (FR4) [6, 7], have been introduced to take advantages of simple process and low cost fabrication.

As for the application of polymer to micromachined optical scanners, a torsional scanner was fabricated using fire resistant 4 (FR4) to realize a low-cost device [8]. Another example of the polymer used for scanner fabrication is RenShape SL5195, with which pattern was formed by mechanically machined aluminum mold [9]. PDMS and SU-8 were also used to fabricate the spring structure with a higher pattern resolution so that the reduced stiffness of the spring could afford a large scanning angle or a low-frequency operation [10, 11]. These will be a low-cost alternative to silicon MEMS scanners, but have not used to fabricate the whole device structure.

In this paper, we propose a polymeric one-dimensional torsional scanner made of SU-8 as a whole structure to exploit a high patterning resolution, low cost fabrication and low frequency operation. An anisotropic wet etching of (100) Si wafer is used to fabricate a silicon mold for a micro mirror and springs. The bottom layer including anchor and electrodes is separately made using another (100) Si substrate. The micromirror and springs are patterned using a standard photo lithography process, and are assembled with another substrate prepared anchors and electrodes in advance. The roughness of the fabricated micromirror is investigated, and the scanner is characterized in terms of frequency response and scanning angle with respect to driving frequency.

DEVICE DESIGN

In general, the thickness of the top structural layer including the mirror should be increased to reduce the dynamic deformation as the mirror size is increased. Thereby, scanner operation requires, however, a higher driving voltage due to the increased spring stiffness as the thicknesses of the mirror and spring are increased. Because the mirror size of the SU-8 scanner in this paper is as large as $3\times3mm^2$, the spring thickness should be reduced to keep the driving voltage as low as possible.

Figure 1 shows the cross-sectional and top views of the proposed SU-8 scanner consisting of two substrates. As shown in Figure 1a, a silicon mold was designed to have V-grooves at both sides of the mirror, which will be used for the fabrication of the torsional spring. The cross-sectional shape of the torsional spring is pentagonal as shown in Figure 1b. Figure 1c represents the completed SU-8 scanner where the gap closing actuator is realized by assembling the top and bottom layers with metal electrodes.

978-1-4244-2977-6/09 $25.00 © 2009 IEEE

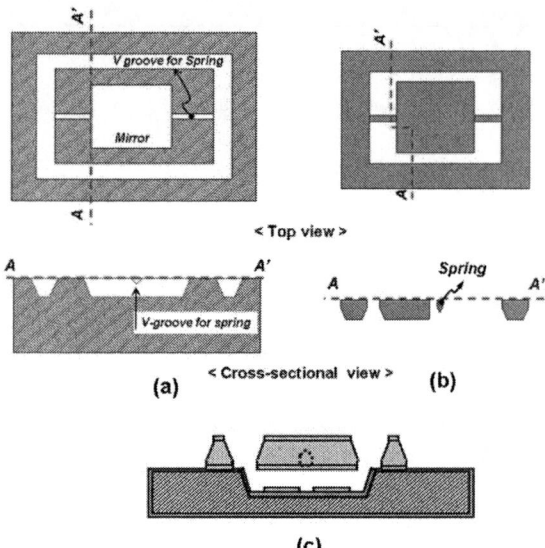

Figure 1: Schematics of polymeric (SU-8) torsional scanner: (a) a silicon mold for polymeric scanner, (b) top layer of the polymeric scanner, and (c) the completed SU-8 scanner assembled.

FABRICATION

The SU-8 scanner is actuated by an electrostatic force generated between two electrodes (backside of the mirror and bottom electrodes). To fabricate two parts, the standard photo lithography and anisotropic wet etching of (100) Si substrate were adopted as shown in Figure 2. Two parts, SU-8 top layer (Figure 2.a-1 to 2.a-7) and bottom (Figure 2.b-1 to 2.b-5) layer with metal electrodes are individually prepared.

For a SU-8 top layer, silicon nitride was deposited on the silicon substrate and was patterned by reactive ion etching (RIE) (Figure 2.a-1 to 2.a-2). Then, the anisotropic wet etching of (100) Si substrate was conducted in KOH solution to obtain V-grooves for pentagonal torsional springs (Figure 2.a-3). SU-8 was coated on the silicon mold by spin coating followed by photolithography process to form a SU-8 top scanner layer (Figure 2.a-4 to 2.a-5). Additional anisotropic wet etching of the silicon mold in KOH solution was conducted to strip the SU-8 layer off from the silicon mold (Figure 2.a-6). In order to avoid thermal deformation of SU-8 layer, which might be occurred during second anisotropic wet etching process, the second wet etching process was conducted at low temperature ($< 50\ ^{\circ}C$). To prepare the mirror surface and metal electrodes, both sides of a SU-8 top layer were coated with Cr/Au by sputtering (Figure 2.a-7) to secure a good reflectivity in visual region.

For the fabrication of the bottom layer, anisotropic wet etching of (100) Si substrate was also used to control the gap between two electrodes by timely controlled silicon etching with KOH solution (Figure 2.b-1 to 2.b-3). Then, silicon nitride was deposited on top of the silicon mold to secure electrical isolation between two electrodes (Figure 2.b-4). The bottom electrodes (Cr/Au) were patterned by an etch back process with PR mask (Figure 2.b-5).

Finally, two parts were assembled by an epoxy bonding to complete the fabrication of the SU-8 scanner, as shown in Figure 3a. The spring thickness at center region was reduced due to the V-groove on the silicon mold, and

the cross sectional shape of the spring was pentagon, as shown in Figure 3b. As a result, the thickness of pentagonal torsional spring was thinner than that of the mirror plate, which allows a low frequency operation with a large scanning angle.

After fabrication of the bottom layer (Figure 3c), two parts were assembled (Figure 3d). The area of the fabricated mirror is as large as $3\times3mm^2$, and its thickness is $130\mu m$. The width of the spring is $100\mu m$ and the air gap between electrodes is $50\mu um$.

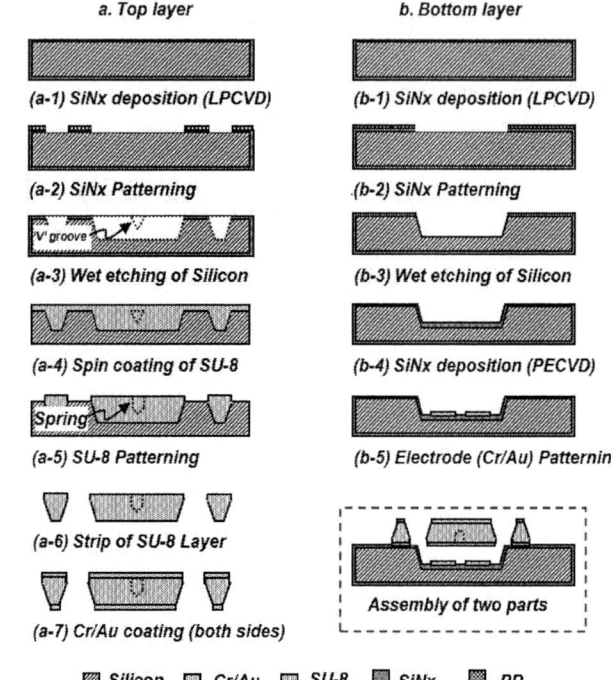

Figure 2: Process flow for the fabrication of the SU-8 scanner actuated by an electrostatic force.

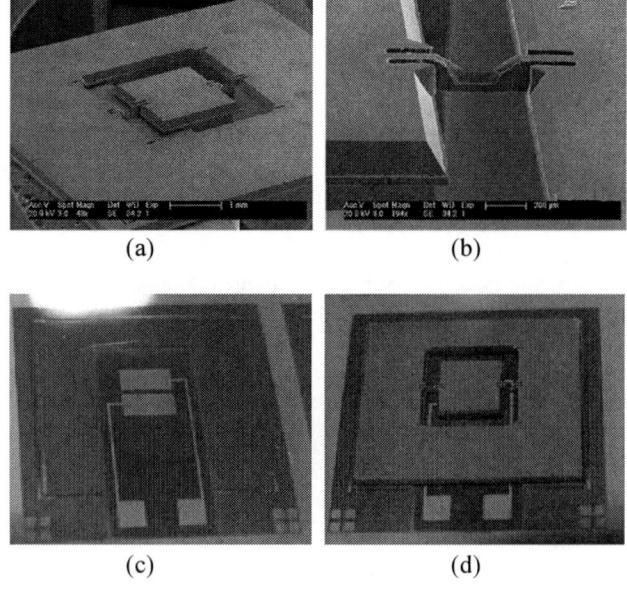

Figure 3: Images of the SU-8 scanner: (a) top layer, (b) spring with a pentagonal cross section, whose thickness is smaller than that of the mirror plate, (c) bottom electrodes, and (d) assembled scanner.

978-1-4244-2977-6/09 $25.00 © 2009 IEEE

The roughness of the fabricated mirror surface, which is one of the major players of the optical scattering loss, is measured by 3D profiler (Nano Focus™) as shown in Figure 4. The measured roughness was 54nm in r.m.s. (root mean square) so that scattering effect is negligible because the roughness of mirror surface is smaller than λ/10. In addition, the surface roughness can be further reduced by optimizing anisotropic wet etching process.

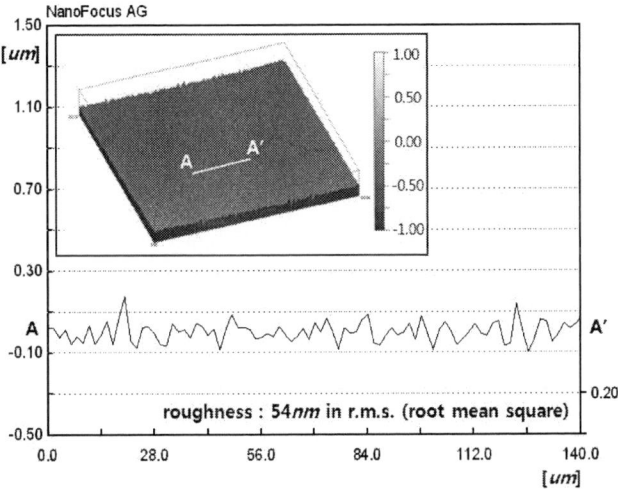

Figure 4: Surface profile of the SU-8 mirror surface. The measured roughness was 54nm in r.m.s. (root mean square)

EXPERIMENTAL RESULTS

The fabricated SU-8 scanner was evaluated in terms of driving frequency and scanning angle. The frequency response was measured by a conventional Laser Doppler Vibrometer (LDV) as shown in Figure 5. A 1.13 kHz of the 1st resonant frequency was obtained, which is high enough for low frequency applications such as OCT and barcode reading.

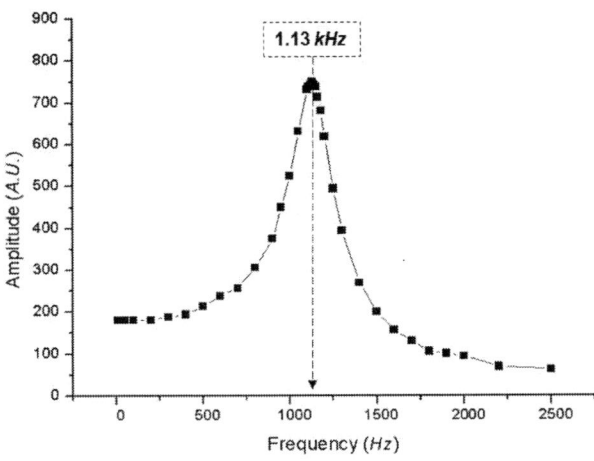

Figure 5: Frequency response of the SU-8 scanner (1st resonant frequency: 1.13 kHz)

The schematic view of the experimental setup for the measurement of the optical scanning angle is shown in Figure 6. A laser diode (LD), optical fibers and GRIN

(GRadient INdex) lens were used for the measurement of the scanning angle. After aligning each component, a laser light emanated from the GRIN lens was redirected toward the screen using the SU-8 scanner. To measure the optical scanning angle, the SU-8 scanner was oscillated with a sinusoidal voltage (peak to peak: 0-170V) at driving frequencies of 60 Hz (non-resonant) and 1.13 kHz (resonant), respectively. Then, we obtained scanned beam images from the screen to calculate the scanning angle, as shown in the inset of Figure 7.

The scanning angle can be calculated by geometrical optics as shown in Figure 6a, and the equation for scanning angle was obtained as in the equation (1).

$$\theta = \tan^{-1}\left(\frac{l_d \sin\theta_0}{l - l_d \cos\theta_0}\right) \quad (1)$$

θ : Optical Scanning Angle
θ_0 : Incidence angle
l : Distance from the scanner to the screen
l_d : Scan range on the screen

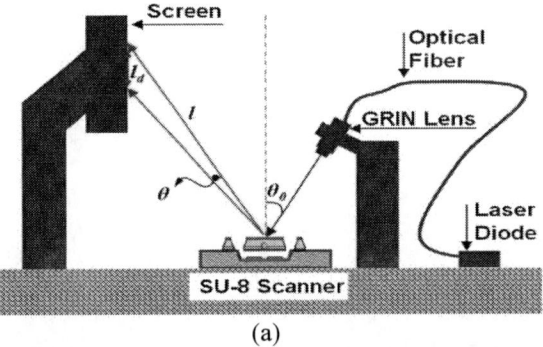

(a)

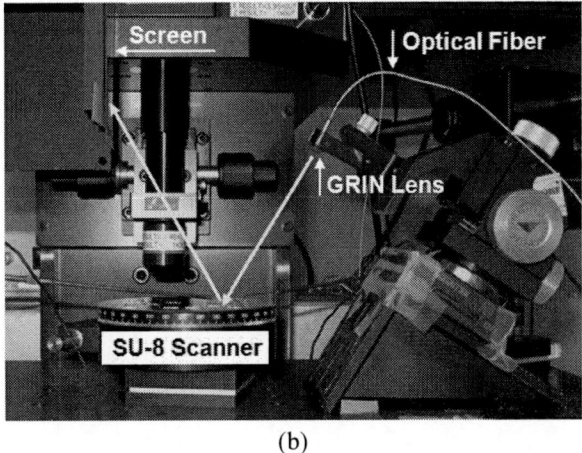

(b)

Figure 6: Experimental Setup for the measurement of the optical scanning angle: (a) schematic view and (b) photograph of the experiment setup.

The scanned beam images are shown in the inset of the Figure 7. The optical scanning angle was also estimated by applying the equation (1) and plotted as shown in Figure 7. The maximum optical scanning angles were 3.1° and 0.85° for resonant (1.13 kHz) and non-resonant (60 Hz) driving frequencies at the driving voltage of 160V. The scanning

978-1-4244-2977-6/09 $25.00 © 2009 IEEE

angle can be increased by enlarging the gap and the thinner spring. The detailed methodologies to increase scanning angle are under the investigation and the reliability issue will be also addressed in bear future.

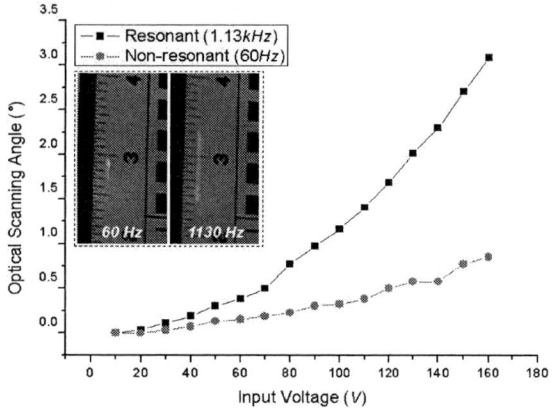

Figure 7: Experimental result: for the optical scanning angle with respect to an applied voltage. The inset represents the photographs of scanned beam at each frequency.

CONCLUSIONS

A polymeric scanner was successfully fabricated by a standard photolithography process using photosensitive SU-8 and anisotropic wet etching of (100) Si substrate.

A spring thickness was controlled by the etch depth of a silicon mold to reduce spring stiffness for operating in low frequencies with large scanning angles. The maximum optical scanning angles were 3.1° and 0.85° for resonant (1.13 *kHz*) and non-resonant (60 *Hz*) driving frequencies at the driving voltage of 160V. As a conclusion, a polymer-based actuator offers advantages including a low-cost fabrication process and low frequency operation.

ACKNOWLEDGMENT

The research was supported by a grant from the institute of Medical System Engineering (iMSE) in the GIST, Korea.

REFERENCES

[1] Arda D. Yalcinkaya, H. Urey, D. Brown, T. Montague, and R. Sprague, "Two-Axis Electromagnetic Microscanner for High Resolution Display," *J. Microelectromech. Syst.*, Vol. 15, No. 4, pp. 786-794, 2006

[2] H. N. Kwon, J-H Lee, K. Takahashi and H. Toshiyoshi, "MicroXY Stages with Spider-leg Actuators for Two-dimensional Optical Scanning," *Sensors and Actuators A*, Vol. 130-131, pp. 468–477, 2006

[3] J. T. W. Yeow, V. X. D. Yang, A. Chahwan, M. L. Gordon, B. Qi, I. A. Vitkin, B. C. Wilson and A. A. Goldenberg, "Micromachined 2-D Scanner for 3-D Optical Coherence Tomography," *Sensors and Actuators A*, Vol. 117, pp. 331-340, 2005

[4] J. Tsaur, L. Zhang, R. Maeda and S. Matsumoto, " 2D Micro Scanner Actuated by Sol-gel Derived Double Layered PZT," *Microelectromechanical systems Conference*, Las Vegas, January 20-24, 2002, pp. 548-551

[5] A. Jain, H. Qu, S. Todd and H. Xie, "A thermal bimorph with large bi-directional and vertical actuation," *Sensors and Actuators A*, Vol. 122, pp. 9-15, 2005

[6] C. Y. Chang, S. Y. Yang, L. S. Huang and J. H. Chang, "Fabrication of plastic microlens array using gas-assisted micro-hot-embossing with a silicon," *Infrared Physics & Technology*, Vol. 48, pp. 163-173. 2006

[7] T. Pan, S. J. McDonald, E. M. Kai, and B. Ziaie, "A magnetically driven PDMS micropump with ball check-valves," *J. Micromech. Microeng.*, Vol. 15, pp. 1021-1026, 2005

[8] H. Urey, S. Holmstrom and A. D. Yalcinkaya, "Electromagnetically Actuated FR4 Scanners," *IEEE Photonics Tech. Lett.*, Vol. 20, No. 1, pp. 30-32, January, 2008

[9] A. D. Yalcinkaya, O. Ergeneman and H. Urey, "Polymer magnetic scanner for barcode applications," *Sensors and Actuators A*, Vol. 135, pp. 236-243, 2007

[10] E. Leclerc, et al., "Silicon based optical scanner using PDMS as torsion springs," *IEEE/LEOS Int. Conf. on Optical MEMS and Nanophotonics(2003)*, pp. 95-96

[11] T. Fujita, K. Maenaka and Y. Takayama, "Dual-axis MEMS mirror for large deflection-angle using SU-8 soft torsion beam," *Sensors and Actuators A*, Vo. 121, pp. 16-21, 2005

VACUUM WAFER LEVEL PACKAGED
TWO-DIMENSIONAL OPTICAL SCANNER BY ANODIC BONDING

H. Tachibana, K. Kawano, H. Ueda, and H. Noge
Panasonic Electric Works Co., Ltd.

ABSTRACT

Vacuum wafer-level packaged two-dimensional optical scanners actuated by vertical electrostatic combs have been developed. A 1 mm-diameter gimbal mirror with high frequency deflection and a surrounding 2.5 × 3 mm movable frame with low frequency deflection are fabricated in a SOI wafer. Wafer-level packaging with anodically bonded borosilicate glass substrates on both sides together with non- evaporable getters fixed on the lower glass substrate allows hermetic sealing in high vacuum. The gimbal mirror and the movable frame are both deflected mechanically ±12 deg. at a resonant frequency of 25.1 kHz with a driving voltage of 18 V, and at a resonant frequency of 78.6 Hz with a driving voltage of 10 V, respectively. The internal pressure of the device is 20 Pa.

1. INTRODUCTION

Optical micro-mirror scanners based on MEMS technology have attractive features in their small footprint and high frequency vibration. They are expected to be used in devices such as compact projectors [1], head-up displays, high-speed bar code readers, and endoscopes [2]. In particular, scanners having vertical comb electrostatic actuators are especially promising due to their small size, low power consumption, and excellent controllability [3].

However, a serious drawback of such scanners is that extremely high voltage is necessary to obtain a large mirror deflection angle at high resonant frequencies using electrostatic comb drives operated in air [4]. In order to decrease the driving voltage of the scanner, the damping of torsional vibrations due to air resistance must be reduced. For this purpose, effective hermetic sealing in vacuum is necessary. Wafer level packaging (WLP) is promising not only for vacuum sealing but also for high-yield and low-cost production [5].

Glass substrates are best for WLP of optical scanners because they have good transparency to an incoming light beam, and from this point of view, a glass-silicon anodic bonding process which has been proven in other MEMS devices, for example, micro pressure sensors [5] was adopted for WLP of optical scanners. In the anodic bonding process, sufficient fracture toughness for hermetic sealing can be realized because a Si wafer is bonded with a borosilicate glass wafer at the atomic level by applying a suitable voltage at high temperature [6]. However, gas generation during the anodic bonding process induces loss of internal vacuum even when the process is carried out in a vacuum chamber. In order to obtain a high-vacuum sealed MEMS device, residual gasses and gasses leaking into the cavity have to be eliminated by means such as a non-evaporable getter activated at adequate temperature [5, 7].

In this study, two-dimensional optical scanners having electrostatic combs are hermetically sealed in vacuum by WLP. The top and bottom surfaces of the SOI (Silicon on Insulator) scanner wafer have been bonded anodically with borosilicate glass substrates together with non-evaporable getters placed on the lower glass substrate. Using a vacuum sealing process, large two-dimensional deflection of a gimbal mirror excited with low driving voltage has been realized.

2. BASIC STRUCTURE AND FABRICATION PROCESSES

Basic Structure

Figure 1 shows a schematic of our resonant optical scanner. A 1 mm diameter, 35 µm-thick gimbal mirror located in the center, a surrounding 2.5 × 3 mm movable frame, and an outermost fixed frame are fabricated in the top silicon layer of an SOI wafer. The mirror and the movable frame are supported by coupled orthogonal hinges, i.e. mirror and frame hinges, and rotate each other around the corresponding hinges. Electrostatic vertical comb actuators are positioned on the edges of the mirror and the frame to give them acceleration torques. High-resolution raster scanning of a laser beam can be realized by increasing the ratio of the resonant frequency of the mirror to that of the movable frame. In order to provide sufficient inertia to maintain a low resonant frequency for the movable frame, the bottom Si layer was left intact to act as a weight under the isolation trenches in the movable frame [8]. The bottom Si layer also prevents the inner and outer parts of the trenches of the movable frame from separating from each other.

As it is difficult to obtain a large deflection angle of the mirror in ambient pressure at high resonant frequencies, we performed vacuum WLP of the SOI scanner wafer with borosilicate glass substrates by anodic bonding to its top and bottom surfaces. The top glass substrate has 300 µm-deep cavities and through holes for electrical contact, and the bottom one has 700 µm-deep cavities on which non evaporable getters were attached to absorb residual gasses in the cavities.

Fabrication Processes

Figure 2 shows the fabrication processes of the optical scanner. The scanners are fabricated in an SOI wafer with 35 µm-thick top active-layer / 1 µm-thick buried oxide (BOX) layer/ 400 µm-thick bottom handle wafer. Initially, the top and the bottom surfaces of the SOI wafer were thermally oxidized in a furnace (Fig. 2 (a)). The silicon oxide film on the top active layer was patterned by photolithography and RIE (Reactive Ion Etching) processes, which served as a hard mask for the subsequent deep etching process to form fine patterns of combs and hinges (Fig. 2 (b)). Aluminum was deposited on the mirror to increase its reflectivity, as shown in Fig. 2 (c). The active layer Si was removed to reach the BOX layer by ICP (Inductively coupled plasma) etching (Fig. 2 (d)). The next step was backside processing, in which the thermally oxidized Si film on the handle wafer was patterned by

978-1-4244-2977-6/09 $25.00 © 2009 IEEE

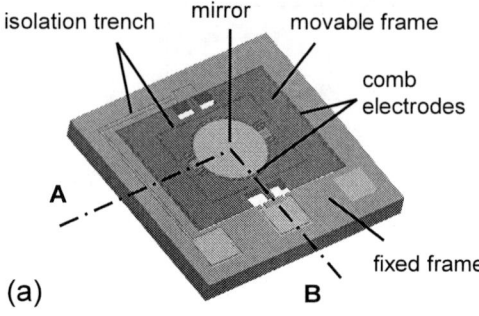

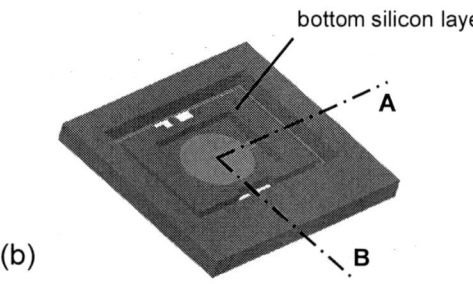

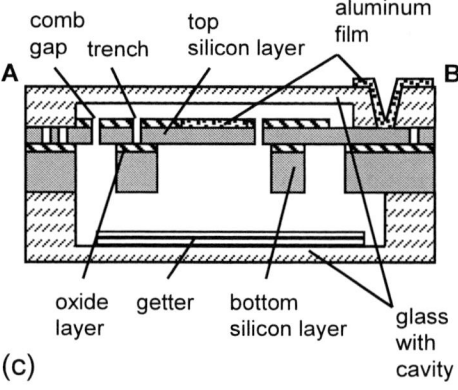

Figure 1: Schematic structure of the developed optical scanner. (a) top view, (b) bottom view before WLP, and (c) cross-sectional view after WLP along the line A-B.

photolithograpy and RIE, and then the entire 400 µm-thick handle wafer was removed by ICP etching to reveal the BOX layer (Fig. 2 (e)).

After MEMS structures were fabricated in the SOI wafer, WLP processes were carried out for both the top and bottom surfaces as follows. A borosilicate glass substrate in which 300 µm-deep cavities and through holes were formed was anodically bonded to the top Si layer of the SOI scanner wafer at 400 °C for 30 min with applied voltage of 600 V under a pressure of 0.4 Pa (Fig. 2 (f)). After removing the BOX and bottom SiO$_2$ film from the handle wafer, the bottom surface of the SOI scanner wafer was bonded to a borosilicate glass substrate by anodic bonding. 700 µm-deep cavities were formed in the glass substrate by sandblasting, in which non-evaporable getters were attached. After the getters were annealed inside the bonding apparatus to desorb gasses from their surfaces [4], anodic bonding was performed at 400 °C for 30 min with applied voltage of 600 V under a pressure of 0.4 Pa. The getters inside were activated during the bonding process to absorb residual gasses in the cavities (Fig. 2 (g)).

(a) Thermal oxidization of SOI

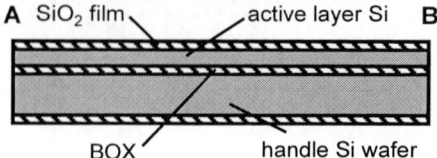

(b) Patterning of SiO$_2$ film (for hard mask)

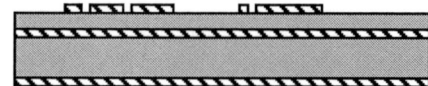

(c) Deposition of aluminum film

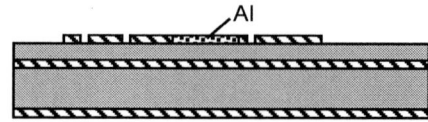

(d) ICP etching of the active layer

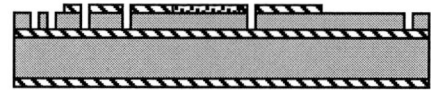

(e) Removing handle Si wafer

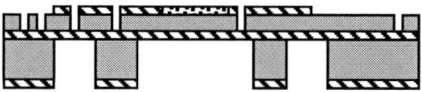

(f) Anodic bonding of top surface with glass, removing BOX layer

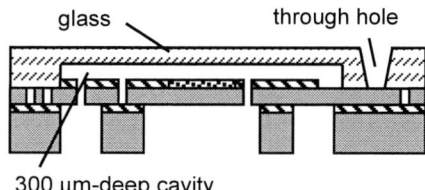

(g) Anodic bonding of bottom surface with glass, depositing aluminum film for electrode

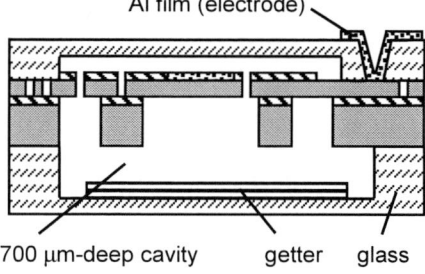

Figure 2: Schematics of the fabrication process of the wafer-level packaged MEMS optical scanner.

978-1-4244-2977-6/09 $25.00 © 2009 IEEE

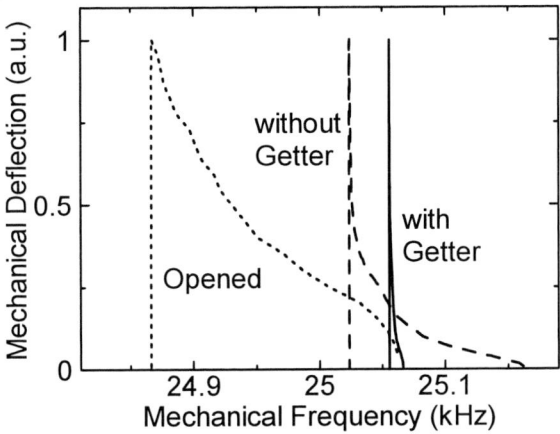

Figure 3: Mirror deflection angle vs. vibration frequency curves. Solid, dashed, and dotted lines correspond to sealed-with-getter, sealed-without-getter, and open-package samples, respectively.

3. CHARACTERIZATION OF THE SCANNER

Torsional vibration of the mirror and the movable frame is excited by applying periodic voltage pulses between the corresponding inner and outer combs. Their mechanical frequency is half of the electrical one since the actuators are symmetrically located about the rotational axis on the same plane. The rotational motion of the mirror and the frame is sinusoidal and symmetric with respect to their equilibrium positions. Here we define their mechanical deflection angle as the maximum of the sinusoidal motion relative to the equilibrium. A laser beam deflected by a fabricated optical scanner was projected to a screen. The mechanical deflection angles of the mirror and movable frame are half the optical deflection angles obtained from the lengths of the scanned lines projected on the screen.

The fabricated scanners were evaluated as follows. For precise evaluation of the WLP and getters, two additional scanners, one sealed in vacuum without a getter and one opened to ambient pressure after packaging, were prepared for comparison. The mirror deflection angle versus vibration frequency curves for these three scanners in which the electrical excitation frequencies were swept downwards are shown in Fig. 3. Note that all the curves are normalized. They are highly asymmetric around the resonant peaks, typical of vertical comb drive scanners with large out-of-plane motion [2]. The full-width at half-maxima (FWHM) of these curves, which are believed to be related to the Q-factors of the mirrors, were 0.60 Hz, 2.4 Hz, and 64 Hz for the sealed sample with getter, that without getter, and the opened sample, respectively, at the resonant frequency of about 25 kHz. Since the Q-factor is closely related to the air damping, a narrower FWHM means the inner pressure of the cavities is lower, i.e. the vacuum WLP process was obviously effective. Data for the movable frame showed the same improvement.

Figure 4 shows the applied pulse voltage dependence of deflection angles of the mirror (a) and the movable frame (b). The deflection angles become larger with increasing voltage for both the mirror and the movable frame because the acceleration torques increase with

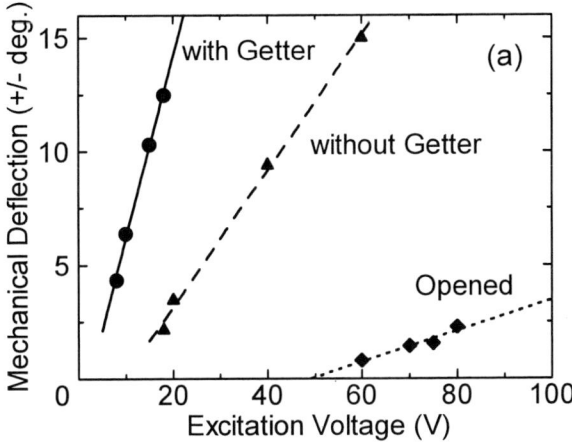

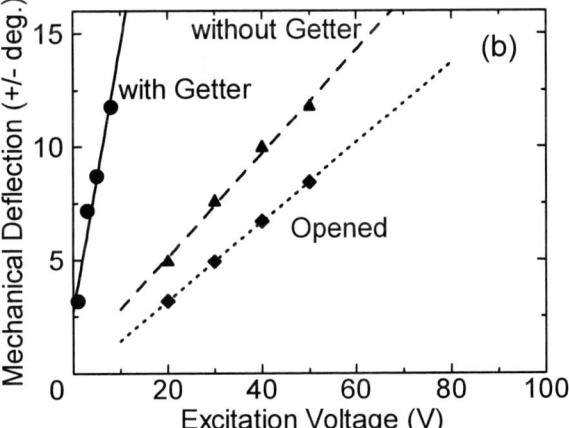

Figure 4: Deflection angle vs. excitation voltage relationships for (a) mirror and (b) movable frame. Circles, triangles, and diamonds correspond to sealed-with-getter, sealed-without-getter, and open-package samples, respectively.

applied voltage. They all showed nearly linear characteristics. In the opened sample, the mechanical deflection of the mirror was only ±2.5 deg. at the resonant frequency even with excitation voltage as high as 80 V. In contrast, a mechanical deflection of ±12deg. was obtained for the mirror in the sealed sample with getter at the resonant frequency of 25.1 kHz. Only 10 V of excitation voltage was sufficient to mechanically deflect the movable frame of this device ±12 deg. at the resonant frequency of 78.6 Hz. The mechanical deflection/excitation voltage slopes of both the mirror and movable frame are largest for the sealed sample with getter, intermediate for sealed sample without getter, and smallest for the opened sample. As the slopes are closely related to the internal pressures of the cavities, it is possible to estimate the internal pressures of the cavities from the slope values.

In order to evaluate the pressure inside the scanners, the corresponding measurements of the applied pulse voltage dependence of the deflection angles of the mirror and the movable frame in a vacuum chamber were both performed at various pressures using an open-package device. The deflection angle/voltage slopes of the mirror (a) and the movable frame (b) as a function of pressure are shown in Fig. 5. Both slopes increase with decreasing

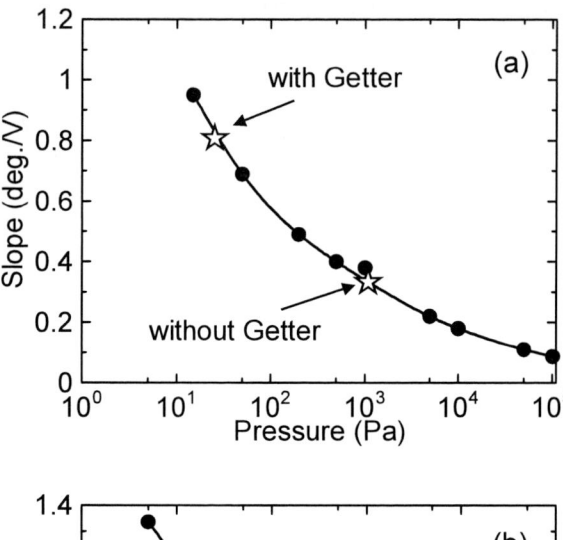

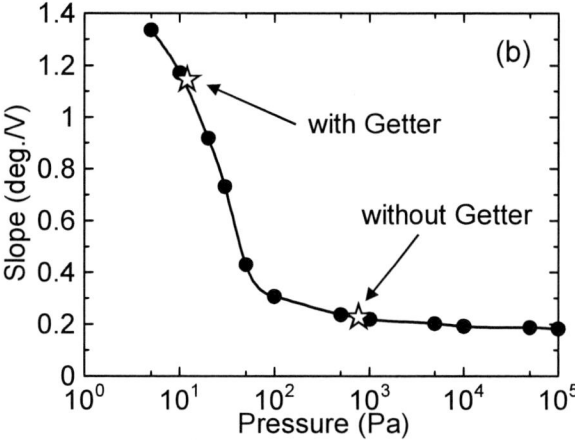

Figure 5: Deflection angle/excitation voltage slopes of (a) mirror and (b) movable frame as a function of pressure for open-package sample in a vacuum chamber (solid circles). Corresponding data for sealed samples are also shown (stars).

pressure and appear to saturate below 10 Pa, as the internal friction of the torsional motion dominates over air damping at low pressures. The slopes of the mirror and the movable frame for the sealed scanners with and without getters obtained from Fig. 4 are also plotted in Fig. 5 as starred data points. The comparison in Fig. 5 indicates that the resultant internal pressures of the sealed scanners with and without getters were approximately 1000 Pa and 20 Pa, respectively.

4. CONCLUSIONS

We have developed vacuum wafer-level packaged two-dimensional MEMS optical scanners sealed with non-evaporable getters. Electrostatic combs were fabricated to actuate the gimbal mirror and surrounding movable frame, and anodic bonding was used to seal the packages. The mechanical deflection angle of the mirror was ±12 deg. at a resonant frequency of 25.1 kHz with a driving voltage of 18 V. For the movable frame, these values were 78.6 Hz and 10 V. The estimated internal pressure of the device was 20 Pa.

This work demonstrates substantial progress in development of MEMS optical scanners with vertical electrostatic comb drives. Novel applications taking advantages of the small size, low cost and low driving voltage of these scanners will likely be realized in the near future.

5. REFERENCES

[1] W.O. Davis, R. Sprague, and J. Miller, "MEMS-Based Pico Projector Display", in *Digest Tech. Papers IEEE/LEOS Int. Conf. on Optical MEMS and Nanophotonics 2008*, Freiburg, Aug. 11-14, 2008, pp. 31-32.

[2] H. Miyajima, K. Murakami, and M. Katashiro, "MEMS Optical Scanners for Microscopes", *IEEE J. Select. Top. Quantum Electron.*, vol. 10, pp. 514-527, 2004.

[3] H. Shenk, P. Dürr, T. Haase, D. Kunze, U. Sobe, H. Lakner, and H. Kück, "Large Deflection Micromechanical Scanning Mirrors for Linear Scans and Pattern Generation", *IEEE J. Select. Top. Quantum Electron.*, vol. 6, pp.715-722, 2000.

[4] S. Hsu, T. Klose, C. Drabe, H. Schenk, "Two dimensional microscanners with large horizontal-vertical scanning frequency ratio for high resolution laser projectors", *Proc. of SPIE*, vol. 6887, pp. 688703-688703-13, 2008.

[5] H. Henmi, S. Shoji, Y. Shoji, K. Yoshimi, and M. Esashi, "Vacuum packaging for microsensors by glass-sillicon anodic bonding", *Sens. and Act. A*, vol. 43, pp. 243-248, 1994.

[6] G.Wallis, D.I. Pomerantz, "Field assisted glass-metal sealing", *J. Appl. Phys.* vol. 40, pp. 3946-3049, 1969.

[7] G. Longoni, A. Conte, M. Moraja, "Stagle and Reliable Q-Factor in Resonant MEMS with Getter Film", in *Digest Tech. Papers IEEE 44th Ann. Int. Rel. Physics Symp.*, San Jose, 2006, pp. 416-420.

[8] H. Noge, Y. Hagihara, K. Kawano, H. Ueda, T. Yoshihara, "Electrostatically actuated two-dimensional optical scanner having a high resonant frequency ratio of fast/slow axes", *IEICE Trans. Electron.*, vol. E91-C, pp. 1611-1615, 2008.

ACCURACY ENHANCEMENT IN THE IN-PLANE DYNAMIC MEASUREMENT OF MEMS ACTUATORS USING A LASER DOPPLER VIBROMETER WITH A 45°-ANGLED OPTICAL FIBER

M. G. Kim[1], K. W. Jo[1], Y. S. Park[2], W.G. Jang[2], and J. H. Lee[1]

[1]MEMS Lab., Gwangju Institute of Science and Technology (GIST), KOREA, SOUTH
[2]Micro Optics Team, Korea Photonics Technology Institute (KOPTI), KOREA, SOUTH

ABSTRACT

On-axis LDV (Laser Doppler Vibrometer), where the direction of incident beam is the same as that of the movement, intrinsically offers an accurate dynamic measurement, but it is not presently applicable to the MEMS actuator with in-plane motion due to no accessibility of the sidewall. The only available measurement method is off-axis LDV, which still shows a serious measurement error depending on the device pattern of the target surface. The on-axis FLDV (Fiber-optic LDV) with a 45°-angled optical fiber is proposed to accurately measure the in-plane motion of MEMS actuator. The performance of the proposed FLDV is evaluated in terms of signal stability and the measurable range.

INTRODUCTION

There are conventionally two types of dynamic measurement method for the moving object. One is an on-axis measurement where the direction of incident beam is the same as that of the movement. This method is intrinsically accurate because the interference signal can be readily obtained as the gap between the optical head and the object becomes smaller or larger. It is, however, not presently applicable to the MEMS actuator with in-plane motion due to no accessibility of the sidewall, as shown in Figure 1(a). The other one is off-axis measurement, which is frequently used when it is impossible for the objective lens of the measurement system to be positioned in the moving direction of the object, as shown in Figure 1(b).

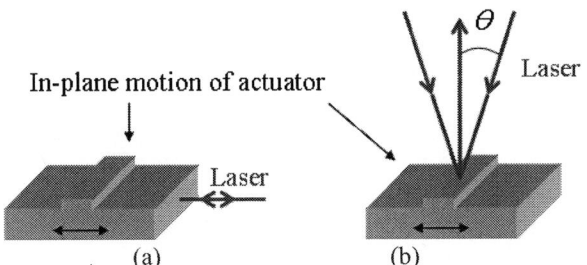

Figure 1: Dynamic measurement methods for the in-plane motion of moving object; (a) on-axis measurement and (b) off-axis measurement

The off-axis measurement is designed to measure the velocity of a surface moving in a plane perpendicular to the bisector of the two converging laser beams [1]. These beams exit the optical head, converge and overlap with a known angle, θ, on the moving surface to be measured. Diffusely scattered light from the surface is collected by a lens, and focused onto a detector.

The off-axis differential LDV has, however, some demerits in the dynamic characterization of in-plane motion of the object. Firstly, the consequence of speckle noise is represented by momentarily signal drop-out (spike) typical in a LDV output [2]. Secondly, a measurement point makes shift by the amount of displacement of the actuator because the incident laser is perpendicular to the direction of actuation. This will cause the serious instability in the measured signal. Thus, there are especially limitations to the off-axis measurement for the in-plane motion of MEMS actuator, which will be shown in the experiment.

The demerits of off-axis differential LDV can be overcome by the on-axis FLDV with optical fiber where the incident laser vertically impinges on the sidewall of moving object, as shown in Figure 1(a). This will allow the measuring position to be stationary during actuation because the incident angle of the measuring beam coincides with the direction of actuation.

The measuring beam using a flat-ended optical fiber in the U-groove on the substrate can be easily aligned to the sidewall of MEMS actuator, as shown in Figure 2(a). In this case, the thickness of the actuator (or height of the sidewall) should be large enough to receive the incident beam; the optimum thickness is around 80 μm. However, the sidewall is not accessible to the measuring beam when the thickness of the actuator is too small and/or U-groove cannot be engraved onto the substrate. In this case it is possible for measuring beam to access the sidewall by bending the beam by 90° without a U-groove.

In this paper, we proposed a novel dynamic measuring method using a 45°-angled optical fiber irrespective of the actuator thickness and U-groove. The proposed method does not change the position of measuring beam during the actuation, because the incident beam is perpendicular to the sidewall by changing the beam direction by 90°, as shown in Figure 2 (b). The accuracy can be further enhanced by employing the microlens fabricated on the side surface of the 45°-angled optical fiber [3].

978-1-4244-2977-6/09 $25.00 © 2009 IEEE

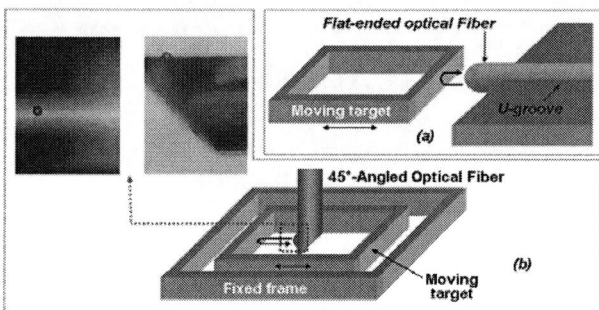

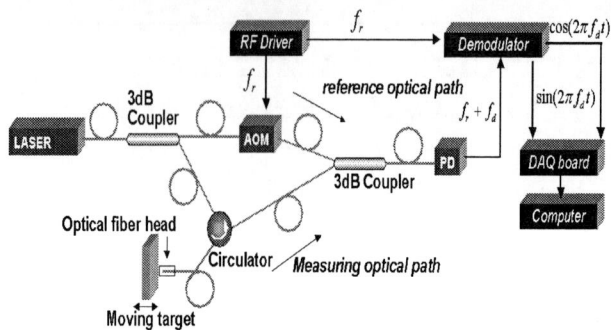

Figure 2: The alignments of the optical fiber head for the enhancement of the measuring accuracy in in-plane motion; (a) with a flat-ended optical fiber with lens, (b) with a 45°-angled optical fiber with the self-aligned microlens.

Figure 3: Optical configuration of an optical fiber based laser Doppler Vibrometer (FLDV); fr: reference frequency from AOM driver, fd: Doppler frequency induced by moving actuator.

SYSTEM CONFIGURATION

As shown in Figure 3, the laser beam of 1.55 μm in wavelength is launched into an optical fiber, and split into two beams by the 1×2 fiber coupler. One of the two beams, measuring beam, is focused onto the moving target through an optical fiber head. The other beam, reference beam, is modulated at 80MHz by an AOM (acousto-optic modulator) connected with pig-tailed fiber. Both measuring beam and reference beam are traveled into the other 1×2 fiber coupler where the mixed beam is guided into photo detectors. The interference signal of the mixed beam is expressed as in eq. (1):

$$I_m + I_r + 2\sqrt{I_r I_m} \cos\left\{2\pi(f_m + f_d)t + \phi\right\} \qquad (1)$$

ϕ : phase difference
I_r : intensity at the reference optical path
I_m : intensity at the measuring optical path
f_d : Doppler frequency
f_m : modulation frequency

Two quadrature Doppler signals, which have a phase difference of 90° each other, are extracted by the demodulation. Then, the phase change, induced by the moving object, is estimated from quadrature signals by an arctangent method [4]. The displacement, $d(t)$, is finally calculated as the amount of phase change in Eq. (2). Note that the displacement is directly obtained by quadrature signals not by integrating the velocity calculated using Doppler frequency.

$$d(t) = \sum_{i=0}^{i=N} \left(\frac{\lambda}{2} \frac{1}{2\pi} \tan^{-1} \left(\frac{\sin\left(2\pi f_d i \Delta t\right)}{\cos\left(2\pi f_d i \Delta t\right)} \right) \right) \qquad (2)$$

λ : wavelength
Δt : sampling time
I : the order of sampling
N : total numbers of sampling

EXPERIMENT
Measurement by the off-axis LDV

Conventional off-axis LDV shows a limitation to the measurement of the in-plane motion of the MEMS actuator. The limitation comes from the fact that the measured signal is not stable because the measurement position changes during actuation in the off-axis measurement. Thus, the displacement data measured by off-axis LDV generally shows a considerable variation; in case of sinusoidal input of the actuator, the output amplitude seriously changes depending on the morphology of the surface to be measured, as shown in Figure 4. Even worse, it is apparently impossible to measure the in-plane displacement of the MEMS actuator when the top layer of the MEMS actuator has mirror-like surface. It is because the LDV cannot recognize any changes from the laser beam reflected from the mirror surface though the surface is really moving horizontally.

However, when the actuator surface has some patterns such as etch holes or device patterns, part of incident beam can be scattered with a Doppler signal. This is often the case in the most of Si MEMS actuator, opposed to the case of randomly distributed roughness which is proven to offer a sufficiently accurate signal in the measurement of in-plane motion.

In order to obtain the best output signal, the measuring beam should be initially aimed at the position near the edge of the device pattern in the moving shuttle so that the incident beam can be effectively scattered back with a displacement-related signal to the detector. However, the experimental displacement measured by the off-axis LDV was considerably fluctuated with several hundred-percent variance even though the initial position of the laser beam slightly changes near the edge of the etch holes. The best position to obtain the highest signal was chosen by adjusting the initial position of measuring beam around the edge of etch holes.

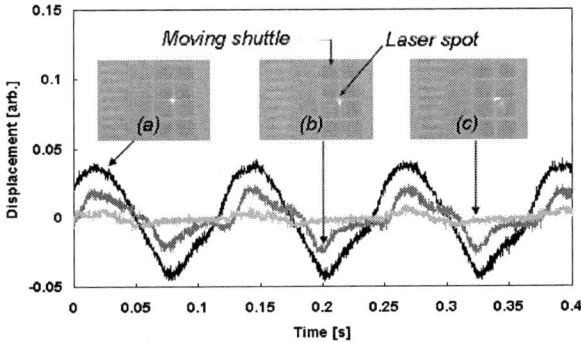

Figure 4: Displacement output experimentally obtained using the off-axis LDV. When a sinusoidal voltage was applied to the MEMS actuator with an in-plane motion, the displacement data seriously varies depending on the initial position of measuring beam.

Measurement by the proposed on-axis FLDV

The measuring beam using the flat-ended optical fiber with a microlens cannot access the sidewall of MEMS actuator when the thickness of the actuator is much less than 80 μm and/or it is difficult to engrave U-groove for fiber alignment in the substrate. To facilitate the laser beam access to the sidewall of the MEMS actuator, we proposed an on-axis FLDV using a 45°-angled optical fiber with a self-aligned micro lens, as shown in Figure 5. The measuring beam, bent by 90° at the 45°-angled facet, was readily aligned perpendicular to the sidewall of the actuator. The opening area requires only about 150×15 μm² to receive the 45°-angled optical fiber with a reasonable tolerance considering that the diameter of the optical fiber is 125 μm.

Figure 6 (a) and (b) explain the procedure for parallel alignment to the sidewall of the actuator. Firstly, the 6-axis stage holding the 45°-angled optical fiber roughly moves to the opening area for the parallel alignment. Then, the fiber is precisely aligned in x-y plane to the target position keeping the enough gap between the fiber and opening boundary through monitoring the CCD image. Finally, the fiber comes down until the reflected optical power reaches to its maximum value, which indicates the best alignment.

Figure 7 shows the measured Doppler signal by the on-axis FLDV using a 45°-angled optical fiber. Figure 7(a), (b) and (c) represents two quadrature Doppler signals with a phase difference of 90° each other, phase extracted from the quadrature signals and displacement calculated from the phase, respectively. A positive slope of the phase means upward displacement and a negative one means a downward displacement. The displacement versus time can be obtained by unwrapping the phase change.

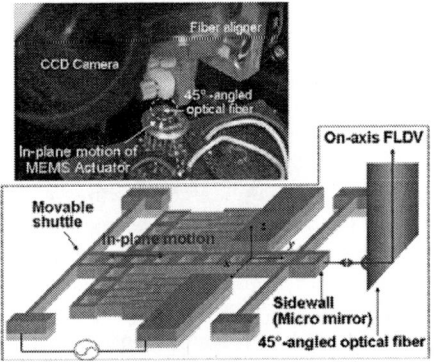

Figure 5: Photograph and schematic of the experimental setup using on-axis LDV with a 45°-angled optical fiber.

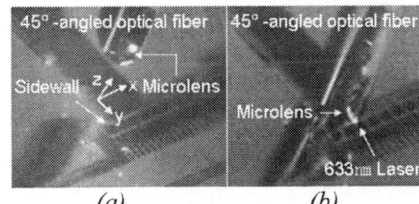

Figure 6: Procedure for parallel alignment to the sidewall of the MEMS actuator using on-axis LDV with a 45°-angled optical fiber; (a) pre-alignment to the position of MEMS actuator, and (b) final alignment to the sidewall of MEMS actuator in x-y plane. (The required opening area for parallel alignment of the optical fiber is only about 150×150 μm2.)

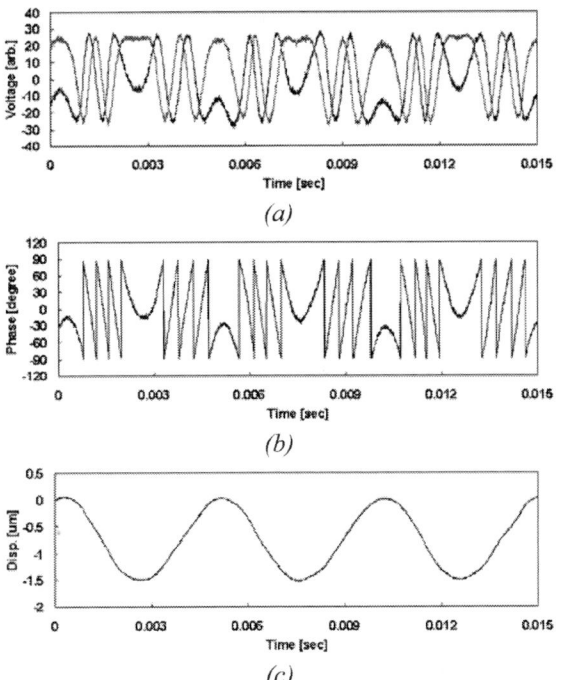

Figure 7: The measured Doppler signal by the on-axis FLDV using a 45°-angled optical fiber; (a) quadrature signals of I and Q, (b) phase extracted from the quadrature signals, and (c) displacement calculated from the phase

Measurable range of the proposed on-axis FLDV

Figure 8 shows measurable range in terms of operation frequency and displacement on the condition that the displacement of MEMS actuator driven by the sinusoidal input is measured by the on-axis FLDV with a 45°-angled optical fiber.

The minimum velocity and maximum velocity demonstrated for examples in this paper were 81.6 μm/s (▲) and 19.3 mm/s (■), respectively. The corresponding operation frequency for measuring the displacement of 1 μm and *100 μm* is 13Hz and 30 *Hz,* respectively. The circles(○) indicate the measured points within the theoretically measurable range that showed a sufficiently accuracy in the displacement measurement.

The measurable range can be easily extended to lower and higher Doppler frequency by employing a laser with a higher power and DAQ board with a higher sampling frequency, respectively. The proposed system is applicable to the dynamic measurement of MEMS actuator considering the displacement of MEMS actuator mostly ranges from 1 μm to 100 μm.

It is confirmed that the displacement of the MEMS actuator can be measured using the proposed FLDV with a 45°-angled optical fiber even when the actuator, surrounded by an outer frame allowing no optical access, is in in-plane motion.

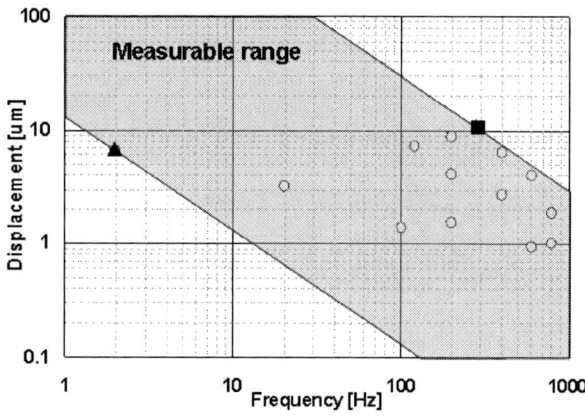

Figure 8: Measurable range (shaded area) of the proposed FLDV with a 45°-angled optical fiber. The triangle (▲) and square (■) represent 105.3Hz and 24.9 kHz in Doppler frequency, respectively. The circles(○) indicate the measured points within the theoretically measurable range that showed a sufficiently accuracy in the displacement measurement.

CONCLUSION

The displacement data measured by an off-axis differential LDV showed a considerable variation; in case of sinusoidal input, the amplitude of the displacement output for the MEMS actuator show a serious variation depending on the initial position of measuring beam. To overcome the problem of an off-axis differential laser Doppler vibrometer, we proposed FLDV using a 45°-angled optical fiber head to measure the dynamic characteristics of in-plane motion of MEMS actuator. Displacement of the MEMS actuator was measured using the proposed FLDV even when the actuator is surrounded by an outer frame allowing no optical access.

The minimum velocity and maximum velocity demonstrated for examples in this paper were 81.6 μm/s (▲) and 19.3 mm/s (■), respectively. The corresponding operation frequency for measuring the displacement of 1 μm and *100 μm* is 13Hz and 30 *Hz,* respectively. The measurable range can be easily improved by employing a laser with a higher power and DAQ board with a higher sampling frequency. The proposed FLDV with optical fiber head will be a good candidate to measure the displacement of in-plane motion with a high accuracy. Therefore the FLDV using a 45°-angled optical fiber is applicable to the dynamic measurement of MEMS actuator considering the displacement of MEMS actuator mostly ranges from 1 μm to 100 μm.

ACKNOWLEDGEMENT

The research was supported by a grant from the institute of Medical System Engineering (iMSE) in the GIST, Korea.

REFERENCES

[1] Polytec, Germany, Available: http://www.polytec.com/

[2] E. P. Tomasini, G. M. Revel, and P. Castellini, "Encyclopedia of Vibration," D. J. Ewins and S. S. Rao, vol. 2, chap. L, pp. 699 (Academic, London, 2001).

[3] Kyoung Woo Jo, Myun Sik Kim, Jong-Hyun Lee, Eun-Kyung Kim, Kang-Ho Park, "Optical Characteristics of a Self-aligned Microlens Fabricated on the Sidewall of 45°-angled Optical Fiber," IEEE Photonics Technology Letters, vol. 16, no. 1 pp. 138-140 (2004).

[4] Tae Bong Eom, Tae Young Choi, Keon Hee, Lee, Hyun Seung, Choi and Sun Kyu, Lee, "A simple method for the compensation of the nonlinearity in the heterodyne interferometer," Measurement Science and Technology, vol. 13, Issue 2, pp. 222-225 (2002).

MICRO LIQUID PRISM

Y. Yoshihata, A. Takei, N. Binh-Khiem, T. Kan, E. Iwase,
K. Matsumoto, and I. Shimoyama

Department of Mechano-Informatics, Graduate School of Information Science and Technology,
The University of Tokyo,
7-3-1 Hongo, Bunkyo-ku, Tokyo, 113-8656, Japan.
E-mail: yoshihata@leopard.t.u-tokyo.ac.jp

ABSTRACT

This paper presents a micro liquid prism. Two flat transparent plates float on a liquid droplet and these plates serve as prism faces. The two plates are positioned automatically by surface tension, just by putting the plates on the ellipsoidal droplet. Therefore the proposed prism can be fabricated accurately in micro scale without difficulty. As the liquid and the plates are encapsulated by Parylene, the prism can retain its shape. The prism which faces were 400μm in diameter and whole size was smaller than 1mm^3 was fabricated. The adequate function of the fabricated prism for Surface Plasmon Resonance (SPR) measurement was verified.

INTRODUCTION

Prisms have been playing essential roles in various optical systems such as cameras, microscopy, projectors, survey instruments, optical measurement instruments, and Surface Plasomon Resonance (SPR) sensors. Recently, miniaturization of optical systems is demanded and prisms are not exception. However, it is difficult to fabricate micro prisms with high accuracy. In this paper, we propose a micro liquid prism which is made of a liquid droplet and two plates (Figure 1(a)). The prism faces which are made of the plates can be made smooth and flat easily. As surface tension of the liquid automatically positions the two plates, the prism shape can be formed with high accuracy. The liquid and the plates are encapsulated by the thin transparent film, Parylene [1], which retains the shape of the prism.

In micro scale, surface tension is dominant and useful for self-assembly and shape formation. Many types of two dimensional (2-D) and three dimensional (3-D) self-assembly of micro components using surface tension of the liquid were reported [2,3]. Surface tension were also used to form the unique 3-D shape of the liquid which devices were made by solidifying or encapsulating the liquid [4,5]. Meanwhile, liquid is suitable for optical devices because of its transparency. For example, the lenses [6] and the wedge prism [7] were fabricated by the liquid.

In this research, we fabricate the prism using such as the characteristics of the liquid. The prism can be fabricated easily with high accuracy in micro scale. The prism works as a triangular prism and is useful for SPR measurement.

POSITIONNING BY SURFACE TENSION

When two plates are put on an ellipsoidal droplet, the plates move in the direction of height and rotation by internal hydraulic pressure and surface tension. First, we explain the phenomenon of the plate moving to the top of the droplet by considering the case one plate floating on the

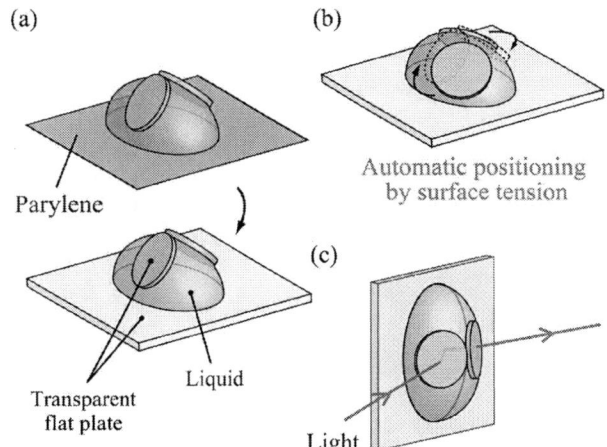

Figure 1: Concept of the micro liquid prism. (a) Two plates which serve as prism faces float on a liquid droplet, and all of them are encapsulated by Parylene. (b) The position of the plates is automatically adjusted by surface tension. (c) The micro liquid prism works as a triangular prism.

droplet. Next, we explain the principle of the positioning of the two plates.

In the case of one plate

When a plate is put on a spherical liquid droplet, the plate floats and is stabilized at the top of the droplet [8]. This phenomenon is caused by the difference of the internal hydraulic pressure and surface tension as shown in Figure 2(a). The internal hydraulic pressure of the liquid droplet is different depending on the height due to the gravity. Therefore, when the floating plate is inclined, the moment is generated and keeps the plate in a horizontal position. The plate is also moved by surface tension. If the plate is laterally displaced, the liquid shape gets distorted and the contact angles at the left and right edges of the plate become different. Consequently, the surface tension components parallel to the plate at the left and right edges become different. This generates the force which moves the plate toward the center of the droplet. By these two actions, the plate is moved and is stabilized at the top of the droplet.

As shown in Figure 2(b), the plate is always stabilized perpendicular to the direction of the gravity even if the substrate is tilted. This verifies that the phenomenon that the plate floats on the droplet is related to the gravity.

The sequential photographs of the plate moving to the top of the droplet are shown in Figure 2(c). The silicon plate with diameter of 1.5mm was put on a droplet with a base diameter of 2.4mm. The liquid was silicone oil. It took about 9 seconds for the plate to move from the initial position to the stable position. This speed is relatively slow

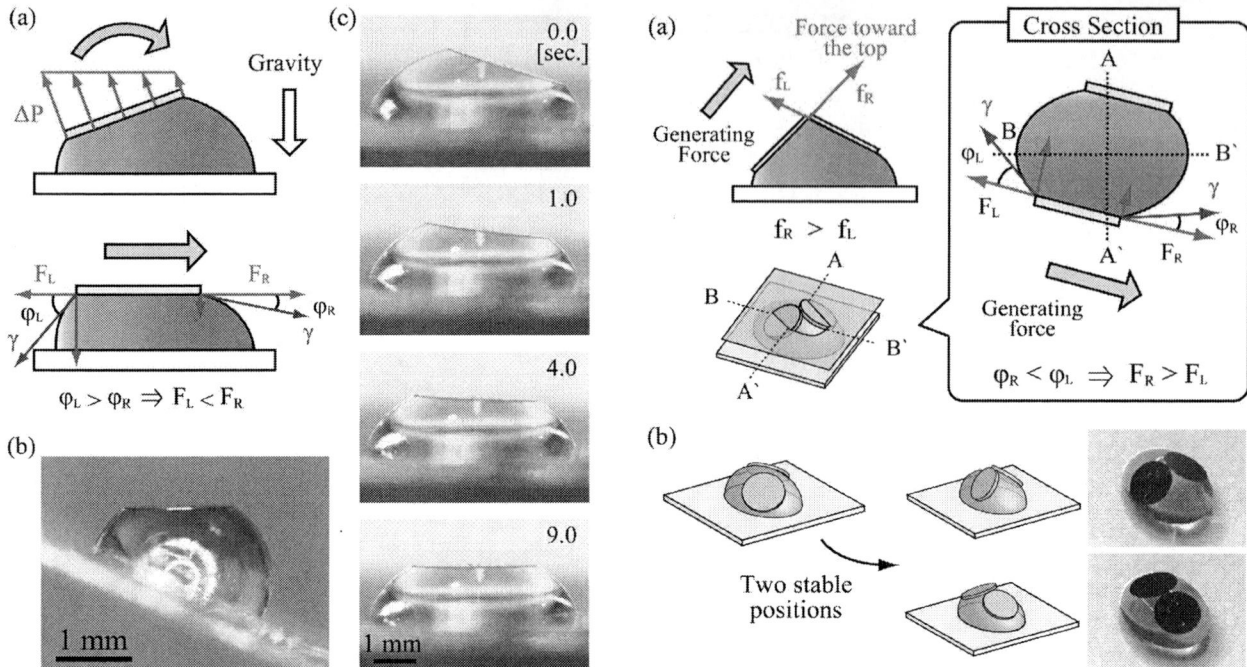

Figure 2: Principle of the plates floating and being stabilized at the top of the droplet. (a) The plate is moved by the difference of the internal hydraulic pressure and surface tension. (b) The photograph of the tilted substrate. (c) Time sequence photographs of the plate moving to the top.

Figure 3: Principle of the automatic positioning of the two plates by surface tension. (a) When the two plates float on the ellipsoidal droplet, the lower position plate moves toward the top, and the plates rotate to reduce the difference of the contact angles at the left and right edge of the plate. (b) The plates on the droplet are stabilized at the two positions.

Figure 4: Time sequence photographs of the plates moving to the stable position.

because the silicone oil has high viscosity and low surface tension. If the liquid which has lower viscosity and high surface tension such as water is used, the moving speed of the plate becomes faster.

In the case of two plates

When two plates are put on the spherical droplet, both plates attempt to move toward the top on the same principle as in the case of one plate. If the height of the two plates is different, the force which moves the lower position plate toward the top is generated (see Figure 3(a)). Additionally, if the droplet's shape is not spherical but ellipsoidal, the plates are also stabilized against rotation about the axis perpendicular to the substrate. This is similarly caused by the difference of the contact angles at the left and right edges of the plate as shown in Figure 3(a). The plates are driven in the direction that the difference of the contact angles is decreasing and are stabilized at the point that the two angles are same. There are only two stable positions: parallel to the major and minor axes (Figure 3(b)). Therefore, the two plates are automatically positioned at

either one of them, just by putting the plates on the droplet. These two positions can be switched easily by stirring the liquid by such as tweezers.

The sequential photographs during the plates moving to the stable position automatically are shown in Figure 4. The silicon plates with diameter of 1.5mm were put on a droplet with an elliptical base of 2.5mm × 3.5mm. The liquid was ionic liquid. The plates moved quickly upward to the top of the droplet until the two plates collided. Then, the plates rotated slowly and were stabilized at the intended position.

FABRICATION

The composition and the process flow of the fabricated prism are shown in Figure 5. An ionic liquid (1-Butyl-3-methylimidazolium hexafluorophosphate) is used because it has both large contact angle on perfluoro resin, Cytop (about 90 degree) and low vapor pressure. The former characteristic leads to small angle between the two prism faces, and the latter one is needed for Parylene

978-1-4244-2977-6/09 $25.00 © 2009 IEEE   964

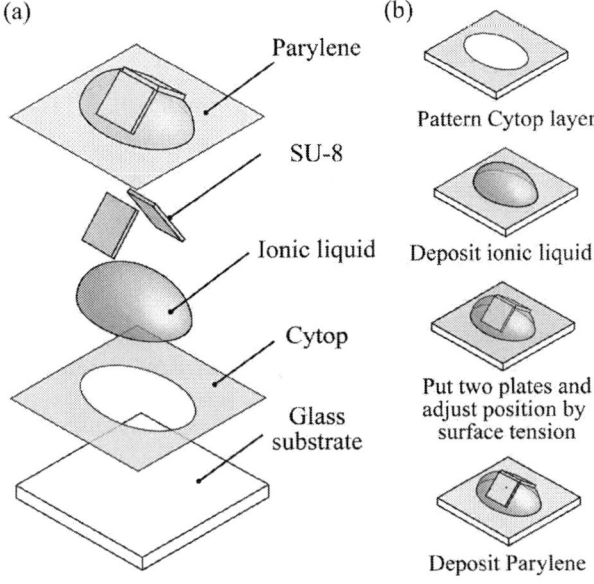

Figure 5: The composition and the fabrication process of the micro liquid prism.

deposition. The refraction index of the ionic liquid is 1.41, which is higher than that of water. The flat plates which serve as prism faces are made of 50μm thick SU-8. The plate shape is formed by exposing and developing the SU-8 film (XP FILM TRIAL-50). On the glass substrate, a Cytop layer is spincoated and an elliptical hydrophilic domain is patterned by oxygen plasma etching. Then, the ionic liquid is dropped on the hydrophilic domain on the glass substrate. The bottom shape of the liquid is defined by this hydrophilic domain and its shape becomes ellipsoidal-like geometry. Then, the two plates are put on the droplet with tweezers. After positioning the plates automatically by surface tension, all components are encapsulated by Parylene [1]. The vapor pressure of the ionic liquid is sufficiently low for Parylene deposition in vacuum [9].

We fabricated two types of the micro liquid prisms. One which has round plate is shown in Figure 6(a). The prism faces were 400μm in diameter, and the bottom of the liquid was 800μm × 1000μm elliptical. The other which has square plates is shown in Figure 6(b). The prism faces were 500μm on a side, and the bottom was 1000μm × 1200μm elliptical. Both of them were smaller than 1mm³. The angle between the two plates of the both prisms was 116 degree. From the photographs, it can be seen that the plates were symmetrically tilted and positioned accurately. The prism faces were enough smooth to see the view of the opposite side.

EXPERIMENT

We verified that the fabricated prism can be used for SPR measurement. A schematic view of SPR measurement is shown in Figure 7(a). The polarized laser light passes through the prism and is totally reflected at the bottom of the glass substrate covered by a thin metal (gold) layer. When the incident angle of the light matched the SPR condition, surface plasmons resonate with the light. The resonation results in absorption of the light, and a SPR dip is measured. The SPR curve is obtained by measuring the intensity of the reflected light with the prism rotating.

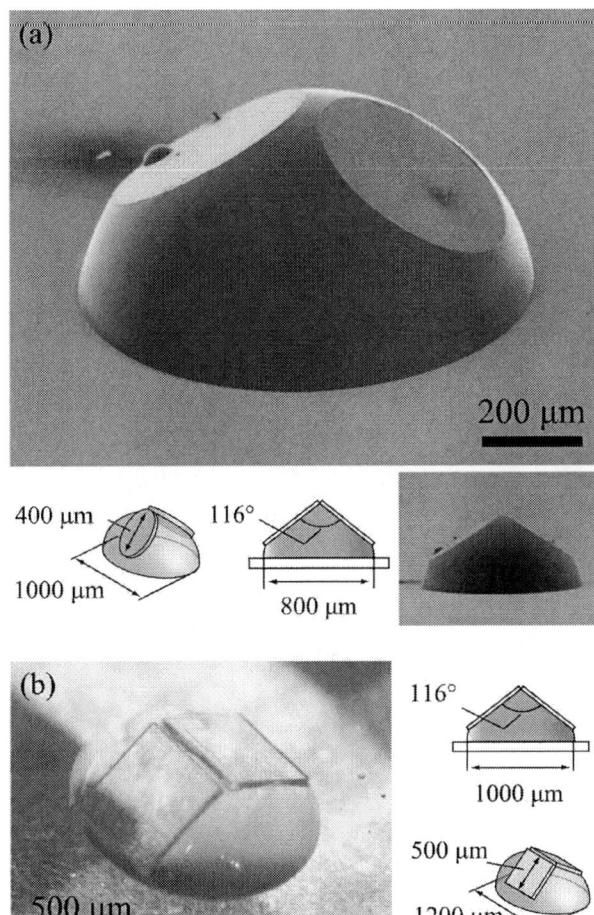

Figure 6: The fabricated micro liquid prisms. (a) The SEM photographs and the size of the prism with round plates. (b) The photograph and the size of the prism with square plates.

The geometry and optical properties of the prism used in the measurement are shown in Figure 7(b). The prism faces were square 1.3mm on a side, the bottom of the liquid was 2.5μm × 3.5mm elliptical, and the angle between the two prism faces was 118.5 degree. The refraction index of the ionic liquid and the glass are 1.410 and 1.525, respectively.

We compared the SPR curves for air measured by the fabricated prism and a conventional glass prism (Figure 7(c)). The smooth SPR curve and sharp dip were obtained by the fabricated prism. In addition, the SPR dips measured by the fabricated prism and the conventional prism were appeared at the angle. The similar shape of the SPR curves was obtained. This proves that the fabricated prism had sufficient performance for the SPR measurement.

CONCLUSION

A micro liquid prism composed of a liquid droplet and two flat plates, and encapsulated by Parylene has been proposed. We demonstrated that the position of the plates on the droplet was automatically adjusted by surface tension. The smallest prism we fabricated had the faces of 400μm in diameter and whole size of it was smaller than 1mm³. A SPR curve similar to that measured by a conventional glass prism was measured by the fabricated

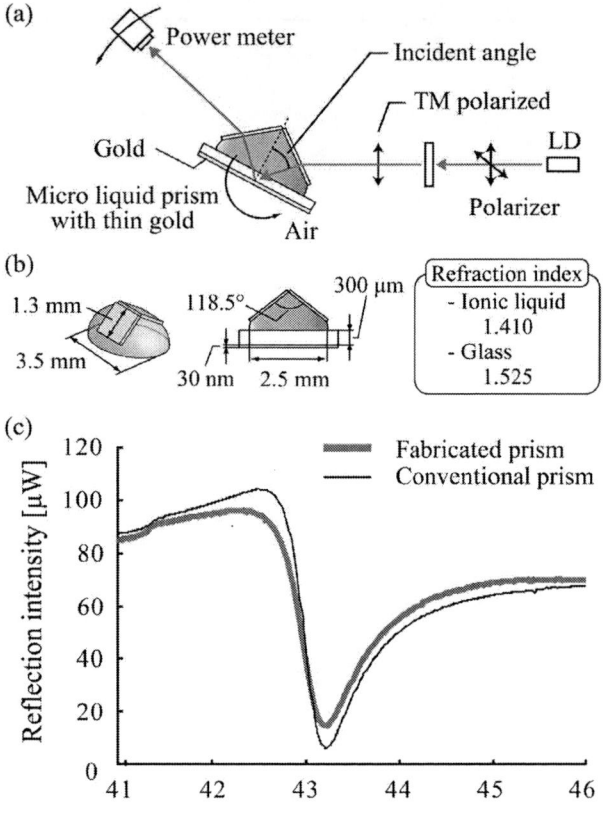

Figure 7: SPR measurement using the fabricated prism. (a) Schematic view of the SPR measurement. (b) The properties of the prism used in the measurement. (c) SPR curves measuring by the fabricated prism and the conventional prism.

prism. This proves that the prism was fabricated accurately and can achieve adequate performance for SPR measurement.

ACNOWLEDGMENTS

The photolithography masks were made using the University of Tokyo VLSI Design and Education Center (VDEC)'s 8 inch EB writer F5112 + VD01 donated by ADVANTEST Corporation. This research is supported by Special Coordination Funds for Promoting Science and Technology, "IRT Foundation to Support Man and Aging Society".

REFERENCES

[1] N. Binh-Khiem, K. Matsumoto, I. Shimoyama, "Polymer thin film deposited on liquid for varifocal encapsulated Liquid Lenses," *Applied Physics Leteers,* Vol. 93, pp. 124101, 2008.

[2] K. Hosokawa, I. Shimoyama, H. Miura, "Two-dimensional micro-self-assembly using the surface tension of water", *Sensors and Actuators A*, Vol. 57, No. 6, pp. 1329-1338, 2005.

[3] R. R. A. Syms, E. M. Yeatman, V. M. Bright, G. M. Whitesides, "Surface Tension-Powered Self-Assembly of Microstructures—The State-of-the-Art", *J. Microelectromech. Syst.*, vol. 12, No. 4, pp. 387-416, 2003.

[4] Y. Yoshihata, N. Binh-Khiem, A. Takei, E. Iwase, K. Matsumoto, I. Shimoyama, "Scanning Micromirror Using Deformation of A Parylene-Encapsulated Liquid Structure", in *Digest Tech. Papers MEMS'08 Conference*, Tucson, January 13-17, 2008, pp. 770-773.

[5] J. M. Bauer, T. A. Saif, D. J. Beebe, "Surface Tension Driven Formation of Microstructures", *J. of Microelectromech. Syst.*, Vol. 13, No. 4, pp. 553-558, 2004.

[6] S. Kuiper, B. H. Hendriks, "Variable-focus liquid lens for miniature camera", *Applied Physics Letters*, Vol. 85, No. 7, pp. 1128-1130, 2004.

[7] A. Takei, E. Iwase, K. Hoshino, K. Matsumoto, I. Shimoyama, "Angle-tunable liquid wedge prism driven by electrowetting", *J. of Microelectromech. Syst.*, Vol. 16, No. 6, pp. 1537-1542, 2007.

[8] A. Takei, N. Binh-Khiem, E. Iwase, K. Matsumoto, I. Shimoyama, "Liquid Motor Driven By Electrowetting", in *Digest Tech. Papers MEMS'08 Conference*, Tucson, January 13-17, 2008, pp. 42-45.

[9] E. F. Borra, O. Seddiki, R. Angel, D. Eisentein, P. Hickson, K. R. Seddon, S. P. Worden, "Deposition of metal films on a ionic liquid as a basis for a lunar telescope", *nature*, Vol. 447, No. 7147, pp. 979-981, 2007.

MEMS OPTICAL LOGIC NOR GATE USING INTEGRATED TUNABLE LASERS

B. Liu[1], H. Cai[1], X. M. Zhang[2] J. Tamil[1], Q. X. Zhang[3] and A. Q. Liu[1†]

[1] School of Electrical & Electronic Engineering,
Nanyang Technological University, Singapore 639798
[2] Department of Mechanical Engineering, University of Maryland, USA 20742
[3] Institute of Microelectronics, Singapore 117685
([†]Corresponding author: A.Q. Liu; Email: eaqliu@ntu.edu.sg; Tel: +65-6790 4336)

ABSTRACT

This paper presents an optical logic NOR gate constructed by two MEMS tunable lasers and a Fabry-Pérot (FP) laser chip. The key idea is that the FP chip is directly controlled by two tunable optical signals via the physical mechanism of injection locking. Compared with other fiber-based gating mechanisms, this logic operation only depends on the optical power without need for accurate phase matching or polarization control. In experiment, the logic NOR function is demonstrated successfully at 100 Mb/s with an extinction ratio of more than 20 dB, and it has a potential to reach higher speed of 10 Gb/s level. The work demonstrates a MEMS single-chip solution for optical logic gates with the advantages of compact footprint, simple optical alignment and easy integration, and may find potential applications in high-speed all-optical communication networks.

1. INTRODUCTION

The capacity of fiber-optic transmission has increased dramatically since the invention of wavelength-division multiplexing (WDM) technology. In vision of the future high-speed optical networks, all-optical signal processing has long been a big attraction to tackle the speed bottleneck of electronic signal processing and to increase the bit rate of optical links. All-optical logic gates are key functional elements in various all-optical signal processing devices, such as all-optical demultiplexers, switches, buffers, and regenerator. Consequently, intensive research efforts have been devoted to pursue optical signal processing and computing, aiming to develop low-cost technologies for high performance all-optical logic operations for ultrafast signal processing. Another natural motivation is from the attempt of replacing electrons by photons, a highly anticipated approach due to the fundamental differences between the behavior of light and electrons. Indeed, not only the operation processes through photons can be carried on in parallel and almost instantaneously, but also the energy dissipation can be reduced significantly.

Several approaches have been proposed in the literature to realize the all optical gates based on the optical nonlinearities in glass (e.g. optical fiber) and semiconductor material components (e.g. semiconductor optical amplifiers, namely SOA) [1-4]. They are based on various working mechanisms, such as four-wave mixing (FWM) [5], cross gain modulation (XGM) [6-9], cross-phase modulation (XPM) and cross-absorption modulation [10], cross-polarization modulation (XPolM) [11], or their combinations. Among these, the FWM method is limited by the conversion efficiency, while the XPM has the need of strict phase control. The XPolM requires matching both the polarization and wavelengths of

the input beams, which poses a technical challenge for practical applications. Consequently, the XGM method that usually employs more than one SOA presents to be easier to implement as it circumvents many problems of the other methods. In terms of the nonlinear media for optical signal processing, the optical fibers have demonstrated advantage of ultrafast processing speed by use of the femtosecond response of Kerr effect in silica [12]. However, the fiber-based configuration tends to be large in size as it usually has to involve on the cascading of several stages. In contrast, the semiconductor components (e.g. SOAs) feature many benefits such as small footprint, easy implement, low power consumption and easy integration to other systems for cost effective on-chip solutions.

The progress of advanced fabrication and packaging technology of microelectromechanical systems (MEMS) has facilitated the development of complicated optical and photonic systems by hybrid integration of MEMS parts with external function components such as laser chips, optical fibers and microlenses. This capability has been well demonstrated in the recently presented micro-scale laser systems such as MEMS tunable lasers [13], injection-locked lasers [14] and coupled cavity lasers [15]. These studies have paved the way to the development of more complicated devices that utilize a few laser chips for optical logic gate systems. With this vision, this paper focuses on the experimental demonstration of a MEMS-based optical logic NOR gate.

2. DESIGN OF MEMS LOGIC GATE

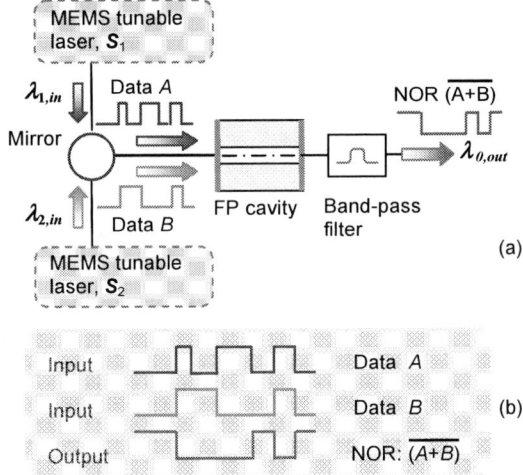

Figure 1. (a)Schematic diagram of the MEMS optical logic NOR gate, and (b) principle of logic gate process.

The schematic diagram of the proposed optical NOR gate is shown in Fig. 1. It consists of two MEMS external cavity tunable lasers (ECTLs) [13] (S_1 and S_2), a Fabry-Pérot (FP) chip and a λ_0 band-pass filter. During the operation, the two input signals (A and B), emitted from two individual MEMS ECTLs, are coupled into the FP chip through the coupler. As a result, the FP chip is optically locked by the signals A and B due to its nonlinearities. The band-pass filter defines the transfer window and selects the wavelength, resulting in the NOR signal at $\lambda_{0,out}$. Although signals A and B may carry different data information at different wavelengths (e.g., $\lambda_{1,in}$ and $\lambda_{2,in}$), such scheme allows simple and efficient implementation because the injection locking depends on only the optical power rather than the accurate phase matching or polarization control.

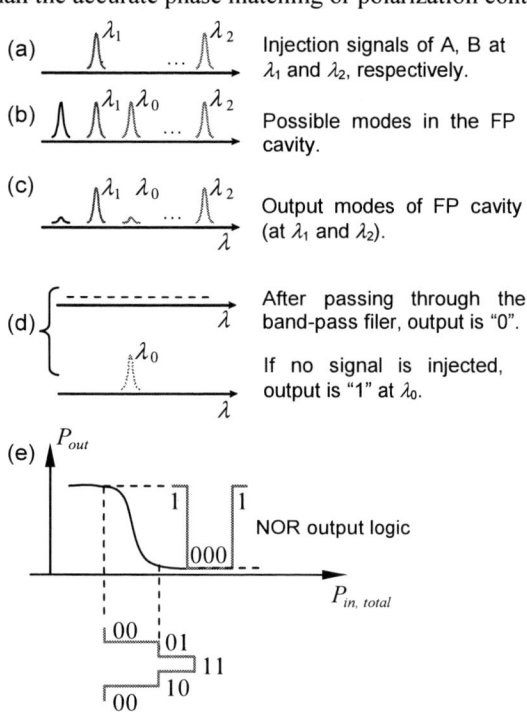

Figure 2. Working principle of the logic NOR function based on the MEMS tunable lasers injection locking the FP chip. (a) – (d) NOR logic relation as explained in the spectrum domain, and (e) relation in optical power.

Based on the above configuration, the detailed explanation of the logic NOR function can be illustrated as shown in Fig. 2. Data A (at λ_1) and B (at λ_2) from S_1 and S_2 are injected into the FP chip simultaneously (Fig. 2a). The FP chip basically has a multi-mode/wavelength output (Fig. 2b). With the presence of either signal λ_1, λ_2 or both, the output of the FP chip can be locked at the corresponding wavelengths (Fig. 2c). The band-pass filter only allows the λ_0 signal to pass through. Therefore, the mode of λ_0 is suppressed to low power level (corresponding to bit "0") if one of the two injection signals is at high power level (corresponding to bit "1", which includes the possible combinations "01", "10" and "11"). Only when A and B are both "0", the final output turns to bit "1" at λ_0 (Fig. 2d) since the carrier density of the FP chip recovers all modes to a high level. Therefore, the logic NOR operation is realized.

It should be noted that the optical logic NOR gate is implemented by injection-locked FP chip, which also control the output power of the signal. Generally, as long as the FP chip remains locked, the output intensity is insensitive to the large variation in the injected signal power. Moreover, as addressed by Liu *et al.* [14], the signal wavelength does not need to match exactly the targeted FP mode wavelength. There is a lockable wavelength range for the injected beam. Thus, if the wavelength detuning (i.e., the difference between the input and the target FP mode) falls into the lockable range, the injected signal is strongly amplified and all the other inherent modes of the FP chip are suppressed to very low level. An output with higher power is obtained. In contrast, when the input wavelength goes beyond the locking range, the output power is thereby kept at low level even if the input power is high enough. Generally, higher input power tolerates a larger wavelength detuning and makes it easier to lock the FP chip.

3. FABRICTION AND DEVICE

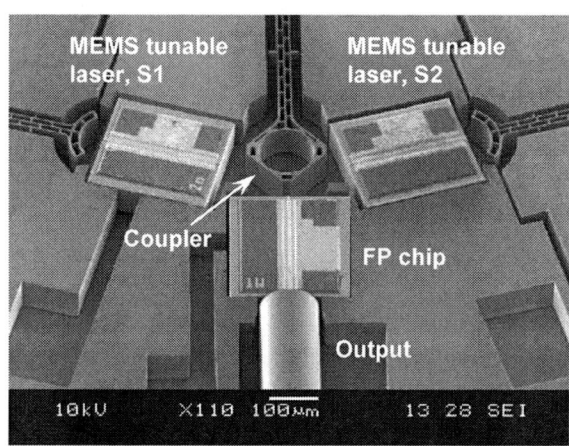

Figure 3. Scanning electron micrograph of the integrated MEMS logic NOR gate.

The overview of an assembled MEMS logic NOR gate is shown in Fig. 3 by the scanning electron micrograph. Two identical MEMS ECTLs (S_1 and S_2), each formed by a FP chip, a curved mirror and comb-drive actuator are used to generate input data singles. The coupler is also driven by another comb-drive actuator to adjust its position for optimal coupling. All the MEMS structures including the adjustable coupler and the movable mirror are fabricated on an SOI wafer by using deep-reactive-ion-etching (DRIE) process, with structure layer of 100 μm thick. The three FP chips (two for MEMS ECTLs, one for logic gate output) share the same dimension of 300 μm × 250 μm × 100 μm. They are assembled into the etched laser trenches on the MEMS substrate. Finally, the output fiber is inserted and fixed to the etched groove for output detection. Due to the use of fiber grooves and the laser trenches, the optical alignment is guaranteed by the accuracy of photo lithography patterning and is no more a big problem. The curved mirrors of the MEMS ECTLs are designed identically, with the radius of 66 μm. To further improve the mirror reflectivity, a 0.2-μm aluminum layer is evaporated onto the mirror surfaces.

4. RESULTS AND DISCUSSIONS

In the experiments, the operation of the injection-locking is first characterized as shown in Fig. 4. Various output spectra of the FP chip at different states have been measured. The driving current of the FP chip is $1.5\,I_{th}$, where I_{th} (~12 mA) is the threshold current. Initially, the single FP chip has multi-mode output, with an emitting band of ~ 30 nm (Fig. 4a). After the FP chip is assembled in the MEMS ECTLs configuration, its output becomes almost single mode (λ_0 = 1553.14 nm, Fig. 4b), though there still has some other side modes. After injection locking, the output of the FP chip becomes pure single-mode emission. As shown in Fig. 4c, the output spectra is locked to the single mode at λ_0 = 1553.34 nm, with a side mode suppression ratio (SMSR) of 26.9 dB. Thanks to the technique of injection lock, the wavelength purity of the input data signal is not that strict in the proposed logic NOR scheme. More importantly, the output spectrum has high wavelength purity. This property is important for construction of all-optical cross connects, in which other components-like amplifiers may degrade the spectral quality.

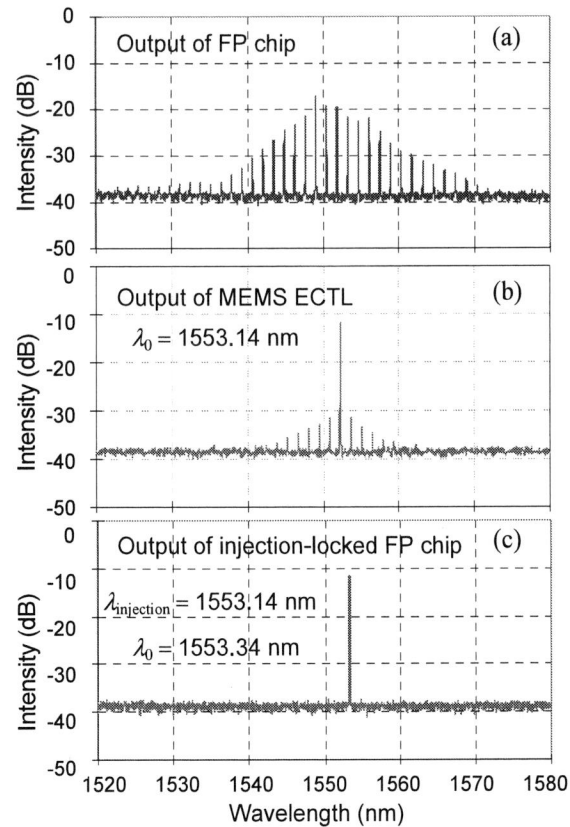

Figure 4. Output spectra of the FP chip in different states (with a driving current of $1.5I_{th}$). (a) Original multi-mode output of the single chip, (b) output spectrum of the MEMS ECTL, with center emission at λ_0 = 1553.15 nm, SMSR = 19.7 dB, and (c) single mode output spectrum of the injection-locked FP chip, with λ_0 = 1553.34 nm , SMSR = 26.9 dB.

It is also observed that there is a red shift (e.g. $\Delta\lambda$ = 0.2 nm in this case) in the output wavelength after the FP chip is locked. The reason might be the optical injection to the FP chip modifies its carrier density and refractive index, leading to the FP mode shift. As a result, the wavelength detuning appears and is usually within the lockable wavelength range.

The static characteristics are measured for the logic NOR gate configuration. It is done by using different input wavelengths (λ_{in}) injected to the FP chip. The output power evolution (at the target wavelength λ_0) versus the injection power is presented at different injected wavelengths as shown in Fig. 5. For any wavelength injection (λ_{in}), the increase of its injection power causes the decrease of the output power at the target wavelength (λ_0), excluding the case of when λ_{in} is close to λ_0. For instance, an increase of the injection power (at λ_{in} = 1559.44 nm) from -15 to 0 dBm produces a large decrease (~ -20.8 dBm) at the FP output power (initial power is -27.4 dBm). Meanwhile, it is observed that for a proper logic NOR gate operation, the injection power level of input data signal should be either high or low enough, which can guarantee the logic NOR operation (e.g. output at "1" or "0" state). Otherwise, a middle-level power injection may result in another type of logic operation, which is beyond the focus of this paper.

For the required input power, experiment finds that the FP chip can be injection-locked at lower injection power if the injection signal is carried at the wavelength (λ_{in}), which has smaller wavelength detune ($\Delta\lambda = \lambda_{in} - \lambda_0$), from the target wavelength λ_0. Particularly, when the injection power is higher than -30.7 dBm, the FP chip is locked sufficiently and sets the logic NOR gate at high logic level "1". In contrast, the FP chip is fully released from the locking state when the power is below -46.0 dBm. In such case, the output at λ_0 is at low logic level "0".

Additionally, the variation of the injection wavelength can strongly affect the operation of the logic NOR gate. Signals with smaller wavelength detuning requires less injection power for the switch of the logic state (e.g. from "1" to "0" state).

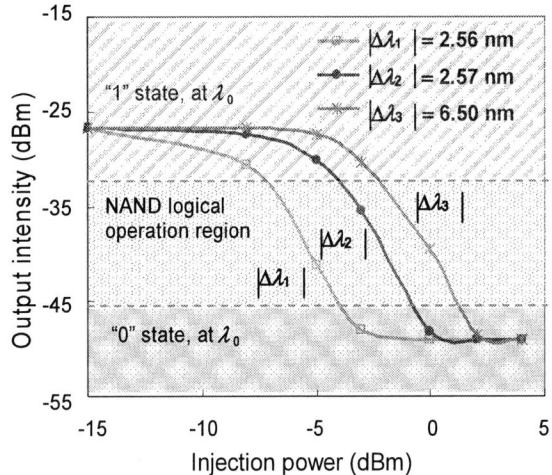

Figure 5. Output power of the logic NOR gate versus external injection power under different injection wavelengths.

The measured waveforms of different signals are presented in Fig. 6, in which the input and output bit rates are at 100 Mb/s. The output power is -30.7 dBm for bit "1" and -46.0 dBm for bit "0", the corresponding extinction ratio is 15.3 dB. The waveform in Fig. 6a represents the bit streams of the injection signal A, with a period of 001000110011101. The waveform in Fig. 6b represents the bit streams of signal B, with a period of 000110011001010. For both two input signals, the optical power is above -28 dBm, which is sufficiently high for optical injection locking. The waveform in Fig. 6c represents the output waveform C at $\lambda_0 = 1553.34$ nm. It has the bit streams of 100001000100000. A comparison between the measured output signal and the NOR truth table (as shown in Table1), shows that it corresponding exactly to the Boolean NOR between signal A and signal B. As a result, the logic NOR function is demonstrated.

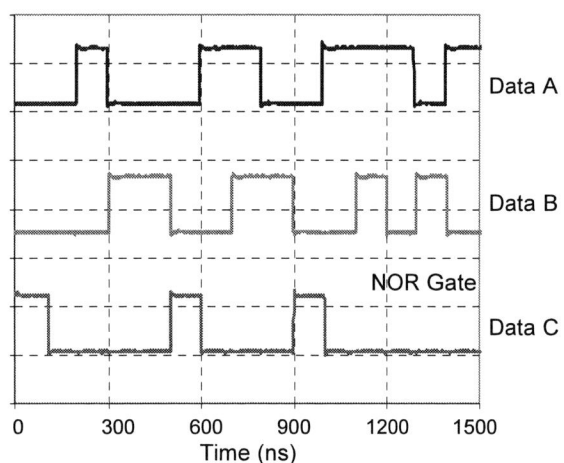

Figure 6. Waveforms of the input and output signals of MEMS logic NOR gate operation.

Table 1: Truth table for logic NOR operation.

Input	A	0	0	1	1
	B	0	1	0	1
Output	$C = \overline{A+B}$	1	0	0	0

5. CONCLUSIONS

A micromachined optical logic NOR gate using MEMS tunable lasers is designed and demonstrated at 100Mb/s. The logic gate operation is realized by two MEMS ECTLs, a FP chip and a band-pass filter. NOR operation with the output extinction ratio higher than 20 dB is obtained experimentally. This is the first attempt to realize a single-chip solution for the optical logic gate using the MEMS technology, and may inspire the development of various all-optical signal process devices for next generation fiber-optic communication, optical computation, etc.

ACKNOWLEDGMENTS

The authors gratefully acknowledge the Agency for Science, Technology and Research (A*STAR) of Singapore for the financial support through the grant No. 042 108 0097.

REFERENCES

[1] Q. Wang, G. Zhu, H. Chen, J. Jaques, J. Leuthold, A. B. Piccirilli, N. K. Dutta, "Study of all-optical XOR using Mach-Zehnder interferometer and differential scheme," *J. Quantum Electron.* vol. 40, pp. 703-710, 2004.

[2] J. M. Jeong and M. E. Marhic, "All-otical logic gates based on cross-phase modulation in a nonlinear fiber interferometer," *Opt. Commun.* 85, pp.430-436, 1991.

[3] A. Bogoni, L. Poti, R. Proietti, G. Meloni, F. Ponzini, and P. Ghelfi, "Regenerative and Reconfigurable all-optical logic gates for ultra-fast applications," *Electron. Lett.* vol. 41, pp. 435-436, 2005.

[4] Z. Li and G. Li, "Ultrahigh-Speed Reconfigurable Logic Gates based on Four-wave Mixing in a Semiconductor Optical Amplifier," *IEEE Photonics Technology Lett.* vol. 18, pp. 1341-1343, 2006.

[5] K. Chan, C. K. Chan, L. K. Chen, F. Tong, "Demonstration of 20-Gb/s all-optical XOR gate by four-wave mixing in semiconductor optical amplifier with RZ-DPSK modulated inputs," *IEEE Photon. Technol. Lett.* vol. 16, pp. 897-899, 2004.

[6] J. H. Kim, Y. M. Jhon, Y. T. Byun, S. Lee, D. H. Woo, S. H. Kim, "All-optical XOR gate using semiconductor optical amplifiers without additional input beam," *IEEE Photon. Technol. Lett.* vol. 14. pp. 1436-1438, 2002.

[7] A. Sharaiha, H. W. Li, F. Marchese, J. Le Bihan, "All-optical logic NOR gate using a semiconductor laser amplifier," *Electron. Lett.* vol. 3, pp. 323-325, 1997.

[8] A. Hamie, A. Sharaiha, M. Guegan, B. Pucel, "All-optical logic NOR gate using two-cascaded semiconductor optical amplifiers," *IEEE Photon. Technol. Lett.* vol. 14, pp. 1439-1441, 2002.

[9] X. Zhang, Y. Wang, J. Sun, D. Liu, D. Huang, "All-optical AND gate at 10 Gbit/ s based on cascaded single-port-coupled SOAs" *Opt. Exp.* vol. 12 pp.361-366 2004.

[10] R. P. Webb, R. J. Manning, G. D. Maxwell, A. J. Pousite, "40 Gbit/s all-optical XOR gate based on hybrid-integrated Mach-Zehnder interferometer" *Electron. Lett.* vol. 39, pp. 79-81, 2003.

[11] H. Soto, C. A. Diaz, J. Topomondzo, D. Erasme, L. Schares, G. Guekos, "All-optical AND gate implementation using cross-polarization modulation in a semiconductor optical amplifier," *IEEE Photon. Technol. Lett.* vol. 14, pp. 498-500, 2002.

[12] G. P. Agrawal, *Nonlinear fiber optics*, 4th ed., (Academic Press, 2006).

[13] X. M. Zhang, A. Q. Liu, D. Y. Tang, and C. Lu, "Discrete wavelength tunable laser using microelectromechanical systems technology," *Appl. Phys. Lett.* vol. 84, pp.329–331, 2004

[14] A. Q. Liu , X. M. Zhang, H. Cai, D. Y. Tang and C. Lu, "Miniaturized injection-locked laser using microelectromechanical systems technology," *Applied Physics Letters*, vol. 87, 101101, 2005.

[15] H. Cai, A. Q. Liu, X. M. Zhang, J. Tamil, Q. X. Zhang, D. Y. Tang, C. Lu, "A miniature tunable coupled-cavity laser constructed by micromachining technology," *Appl. Phys. Lett.* vol. 92, (03), 031105, 2008.

TILTED PARABOLOIDAL REFLECTIVE LENS FOR FAR INFRARED SENSOR FABRICATED BY MASK WITH RECTANGULAR OPENINGS

Tomoyuki Takahata, Kiyoshi Matsumoto, and Isao Shimoyama
The University of Tokyo, Tokyo, JAPAN

ABSTRACT

A tilted paraboloidal reflective lens for focusing far infrared (FIR) rays has been devised. Tilted paraboloids with smooth surface and vertical walls were fabricated by two-step etching using microloading effect with a mask which had rectangular openings of various sizes. FIR light was focused by a 30-degree-tilted paraboloid lens. The size of a real image of an FIR source was 110 µm, which agrees with a lens formula.

INTRODUCTION

FIR sensors (i.e. temperature sensors) are widely used for many applications, such as detecting human in security systems, measuring temperature of a food in a kitchen utensil. In general, a thermography, a commonly used FIR imager, has been expensive because lenses for FIR have been made of material like germanium that is difficult to be machined. Reflective lenses can reduce the cost of FIR imagers, because any material can be used as a lens as long as it is coated some metal (such as gold and aluminum), which have high-reflectivity for FIR.

A paraboloid can focus rays that are parallel to the central axis of the paraboloid (Figure 1A). But the paraboloid can not focus rays from a certain angle (Figure 1B). When the central axis of a paraboloid is tilted and parallel to an incident angle of FIR, the rays are focused at the focal point of the paraboloid (Figure 1D).

The tilted paraboloids must have (1) depth varying in two independent directions, (2) smooth surface, and (3) vertical walls. Vertical walls are needed to densely arrange lenses. By assembling the lens array to some FIR sensor array, a high-sensitive and directional FIR sensor can be achieved.

In a past research [1], two-step silicon etching was employed to fabricate a paraboloid lens array; the first step was anisotropic etching and the second was isotropic etching. However, tilted paraboloids can not been achieved by that method, because one mask opening corresponds one paraboloid. Arbitrary structures, whose depth varies in two directions, can be fabricated using microloading effect [2, 3] or grayscale mask [4, 5]. In the grayscale mask method, three dimensional resist profiles are transferred to silicon by a reactive ion etching. A disadvantage of that method is rough surface, because it is difficult to control the etching selectivity of silicon and the resist. Microloading effect is used in this paper.

A three-dimensional structure that was fabricated by two-step isotropic etching using a mask with circular openings arranged at even intervals had an arbitrary shape and a smooth surface [2]. However vertical walls can not be fabricated by that way. In [3], vertical walls were available using ICP-RIE at first etching. The width of the mask was constant for a smooth surface, but the depth

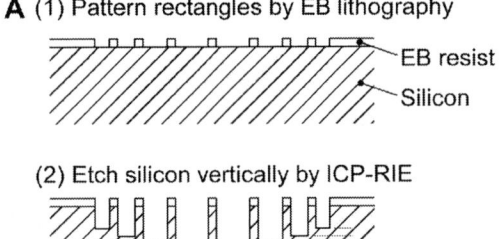

A (1) Pattern rectangles by EB lithography

— EB resist
— Silicon

(2) Etch silicon vertically by ICP-RIE

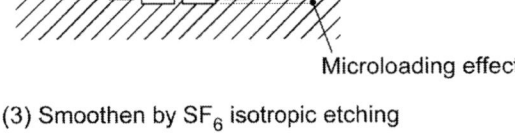

Microloading effect

(3) Smoothen by SF$_6$ isotropic etching

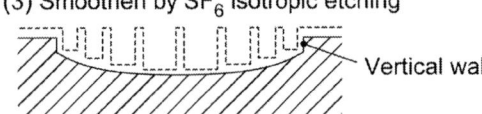

Vertical wall

B

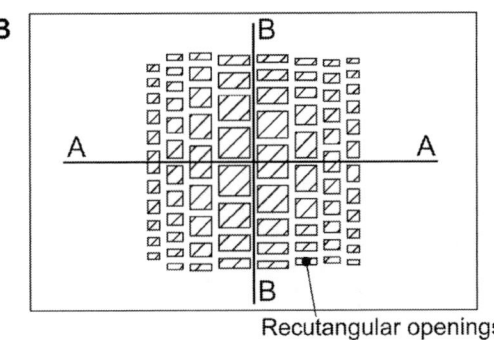

Recutangular openings

Figure 2. Fabrication of the lens. (A) A schema to fabricate an arbitrary shape smooth structure with vertical walls using rectangular openings. (B) An example of mask for a paraboloid lens.

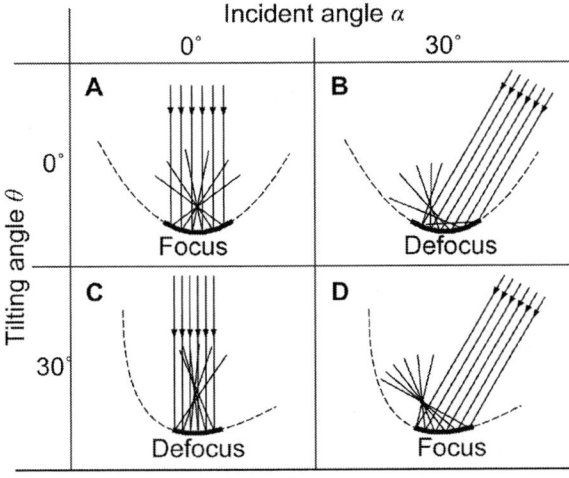

Figure 1. A schematic drawing of tilted paraboloidal reflective lenses to gather far infrared (FIR) rays. The spot size is smallest when the tilting angle equals the incident angle of the ray (A and D).

978-1-4244-2977-6/09 $25.00 © 2009 IEEE

varied only in one direction because the mask had only square openings.

In this paper, two-step etching that consists of ICP-RIE, for which a mask with rectangular openings and constant width was used, and another smoothing etching satisfied abovementioned three requirements.

FABRICATION

The lenses were fabricated by following steps (Figure 2A). An electron beam (EB) resist layer on a silicon wafer was patterned by EB lithography. Then, the silicon wafer was vertically etched by ICP-RIE to depths according to the areas and the shapes of mask openings. Trenches with different depths were formed. Finally, the surface was smoothed by removing sidewalls of trenches using isotropic RIE.

The etched depth depends on the area of openings (Figure 3(a)). The thinner the mask width between the openings is, the smoother the resulting surface is. The mask width was fixed to 200 nm. Adopting rectangular openings enabled an arbitrary shape though the mask width was fixed.

The mask design procedure is as follows. Firstly, a width of rectangles on a certain line (along B-B direction in Figure 2(b)) is determined corresponding to the average depth on the line according to the relationship shown in Figure 3(a). Then, the other side length of the rectangle is calculated from depth of that point. The eventual etched depth varies with the aspect ratios of the rectangles, which is the ratio of the short side to the long side (Figure 3(b)).

A lens array that has 5 tilted paraboloid lenses was fabricated (Figure 4). The tilting angle was from 0° to 60° by 15°. The lenses ware 200 μm in width and placed at 220 μm intervals. The shapes of the fabricated lenses were measured with a laser microscope (Figure 5). The profile along A-A agrees with the designed one. The profile along B-B is slightly different at edges, because the aspect ratios of the rectangles at the edges are so small that the etched surface got shallow due to the effect shown in Figure 3(b). The arithmetic surface roughness was about 100 nm, which was smooth enough for FIR lens, because it was 1% of the wavelength of FIR (~10 μm).

EXPERIMENTS

Since silicon is transparent in FIR, the surface of the lenses should be coated by a metal film to enhance reflectance. The reflectance as a function of thickness of a metal film was measured using an FIR source (Helioworks Inc., KE-8523) and an FIR camera (Wuhan Guide Infrared Co., Ltd., GUIDIR IR106) (Figure 6). The incident angle

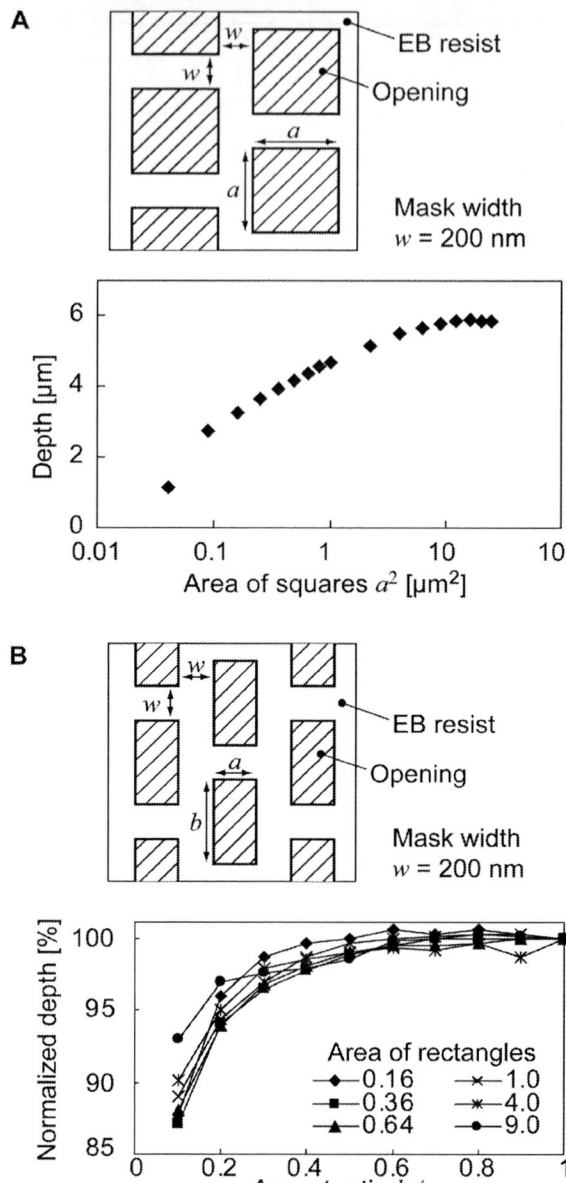

Figure 3. Relationship between etched depth and shape of openings. (a) Etched depth as a function of area of square openings. As the area gets larger, the depth increases as the area increases due to a microloading effect. (b) Effect of aspect ratio on the etched depth, which was normalized by the depth of a square opening (i.e. the depth when the aspect ratio was 1).

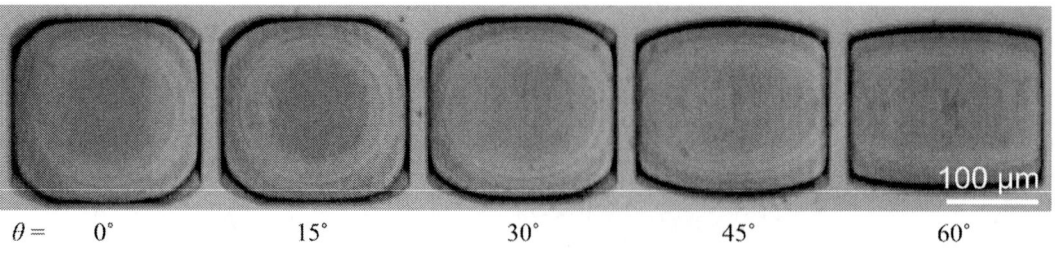

Figure 4. A photograph of lenses. The tilted angle was from 0° to 60° by 15°.

from an FIR source was 45°. Reflected power was measured with an FIR camera. The reflectance of aluminum was larger than that of gold when the film was thicker. The reflectance of 40-nm-thick aluminum was 80%.

A real image of an FIR source focused by a lens tilted by 30° and coated by 50-nm-thick aluminum was observed with the FIR source and the FIR camera (Figure 7 and 8). The incident angle was 30°, which is same as the tilting angle of the lens under test. The size of a focal spot was 110 μm. The power at the center of the spot was 8 times larger than that at the background. The size of the source was 7 mm, the distance between the source and the lens was 85 mm, and the distance between the lens and the spot was 1.2 mm, the spot size and the focal length calculated from a lens formulas were 100 μm and 1.18 mm, respectively. These results show that the lens was useful for focusing FIR.

CONCLUSION

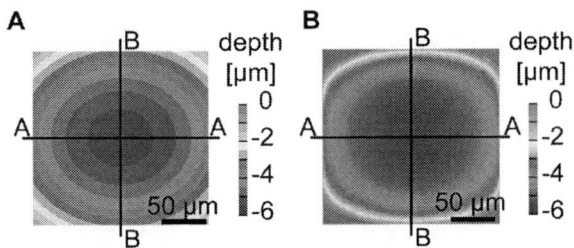

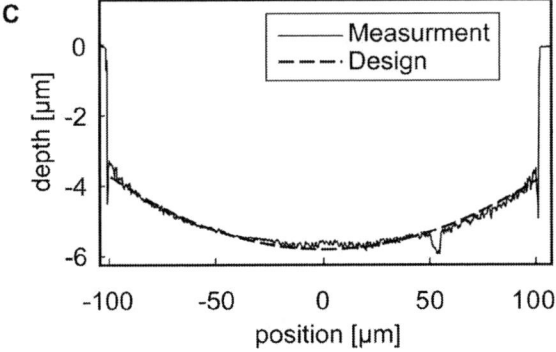

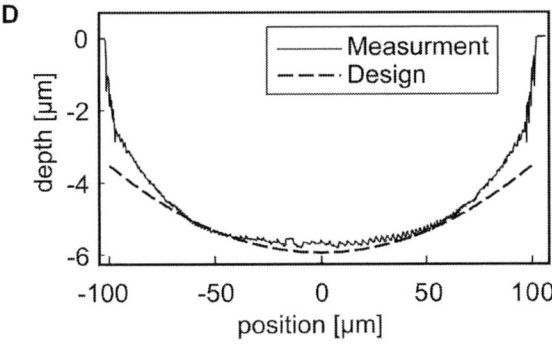

Figure 5. Shapes of a 30°-tilted lens measured with a laser microscope. (a) A contour plot of designed shape. (b) A contour plot of measured shape. (c) A-A profile of a lens that is tilted by 30° in A-A direction. (d) B-B profile of the same lens.

Tilted paraboloidal reflective lenses have been devised. The lenses were fabricated by two-step etching using microloading effect. The mask had rectangular openings and a constant width. An arithmetic surface roughness of the fabricated lens was about 100 nm, which was smooth enough for FIR. FIR was focused by a 30-degree-tilted lens. The focal length of the lens and the size of a real image of an FIR source agree with theoretical predictions. The power at the spot center was 8 times larger than that at the background.

ACKNOWLEDGEMENTS

The lenses were fabricated with the University of Tokyo VDEC's EB writer F5112+VD01 donated by ADVANTEST Corporation.

REFERENCES

[1] Kerwin Wang, Karl F. Böhringer, "Time-Multiplexed-Plasma-Etching of High Numerical Aperture Paraboloidal Micromirror Arrays," in *Proc. of the 5th Pacific Rim Conference on Lasers and Electro-Optics*, Taipei, Dec. 15–19, 2003, pp. 317–319.

[2] T. Bourouina, T. Masuzawa, and H. Fujita, "The MEMSNAS process: microloading effect for micromachining 3-D structures of nearly all shapes," *Journal of Microelectromechanical Systems*, vol. 13, no. 2, pp. 190-199, 2004.

[3] M. P. Rao, M. F. Aimi, and N. C. MacDonald,

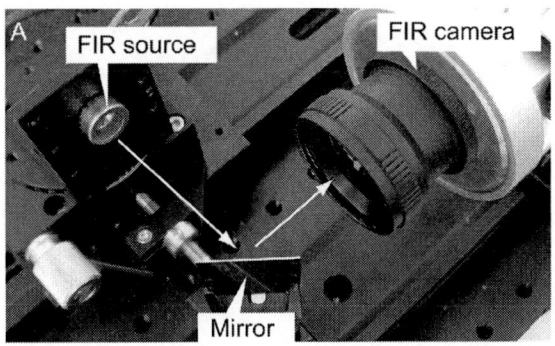

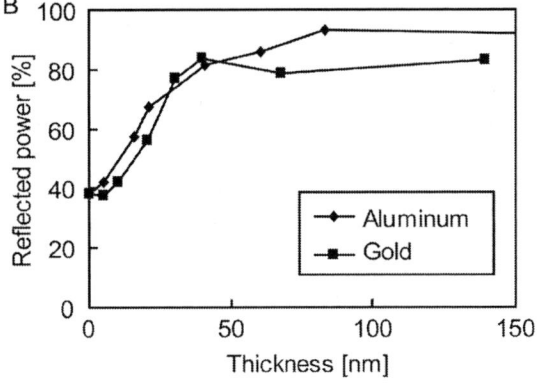

Figure 6. FIR reflectance as a function of thickness of metal films deposited on a silicon wafer. The incident angle was 45°. The power of reflected FIR rays was measured by FIR camera.

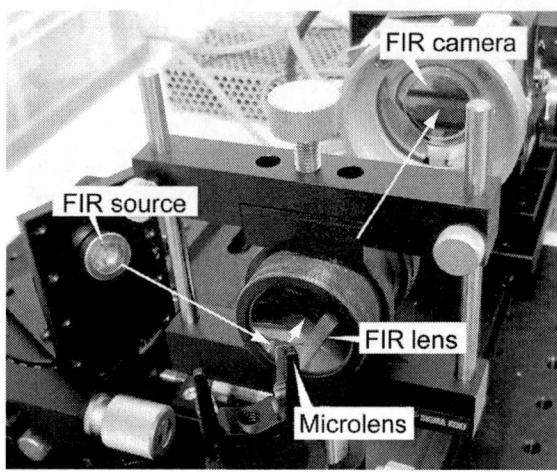

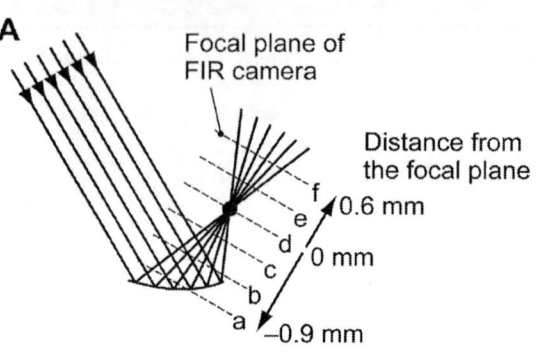

Figure 7. A photograph of experimental setup. The incident angle was 30°.

"Single-mask, three dimensional microfabrication of high aspect-ratio structures in bulk silicon using reactive ion etching lag and sacrificial oxidation," *Applied Physics Letters*, vol. 85, no. 25, pp. 6281-6283, 2004.

[4] A. Sure, T. Dillon, J. Murakowski, C. Lin, D. Pustai, and D. W. Prather, "Fabrication and characterization of three-dimensional silicon tapers," *Optics Express*, vol. 11, no. 26, pp. 3555-3561, 2006.

[5] C. M. Waits, B. Morgan, M. Kastantin, and R. Ghodssi, "Microfabrication of 3D silicon MEMS structures using gray-scale lithography and deep reactive ion etching," *Sensors and Actuators A: Physical*, vol. 119, no. 1, pp. 245-253, 2005.

[6] R. A. Gottscho, C. W. Jurgensen, and D. J. Vitkavage, "Microscopic uniformity in plasma etching," *Journal of Vacuum Science & Technology B*, vol. 10, no. 5, pp. 2133-2147, 1992.

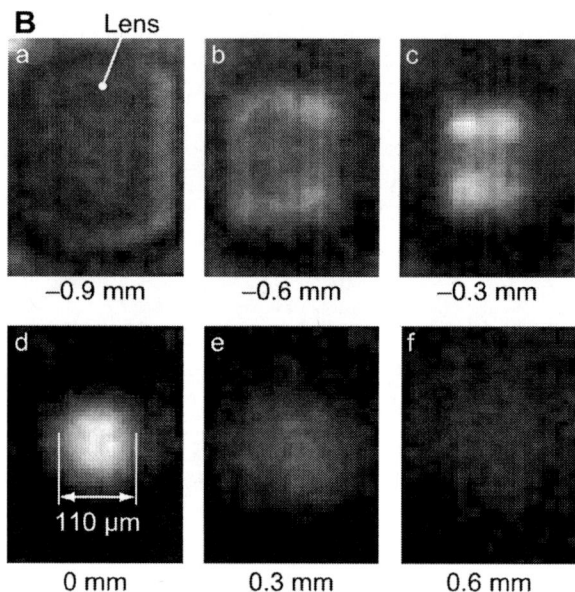

Figure 8. FIR images of a spot by the lens. The diameter of the spot was about 90 µm. The lens can be seen in (a), and a real image of the FIR source by the lens can be seen in (d).

MEMS LASER WITH TUNABLE WAVELENGTH AND POLARIZATION USING OPTICAL TUNNELING EFFECT

W. M. Zhu[1, 3], X. M. Zhang[2], H. Cai[1], J. Tamil[1], W. Zhang[1], B. Liu[1], T. Bourouina[3] and A. Q. Liu[1†]

[1] School of Electrical & Electronic Engineering,
Nanyang Technological University, Singapore 639798

[2] Department of Mechanical Engineering,University of Maryland College Park, MD 20742, USA

[3] Ecole Supérieur des Ingénieurs en Electronique et Electrotechnique (ESIEE),
University of Paris East, France, 93162

(† Corresponding author: A.Q. Liu; Email: eaqliu@ntu.edu.sg; Tel: +65-6790 4336)

ABSTRACT

This paper presents a micromachined tunable laser that utilizes a silicon optical tunneling structure to tune both the polarization state and the wavelength of the laser output. The device is fabricated on silicon-on-insulator wafer using deep reactive ion etching. In experiment, the wavelength and the polarization are tuned by heating up the silicon optical tunneling structure via the thermo-optic effect. A 90° change of polarization direction is obtained using a heating current of 61.2 mA, and a wavelength tuning of 2 nm is also demonstrated. The output spectrum shows a high suppress ratio above 30 dB. Compared with the previous MEMS tunable lasers that have a random or fixed polarization state, this device provides a special capability in tuning the polarization state in addition to the wavelength, and would find niche applications in biomedical research, interferometry, coherent communications, instrumentations and sensors.

1. INTRODUCTION

Tunable diode lasers are enabling components to the next generation of optical networks, the research in atomic and molecular physics, and the development of many sensor and instrumentation systems [1-4] thanks to their superior performance such as wide tuning range, excellent wavelength accuracy, narrow line width and high reliability. For these reasons, the tunable diode lasers have long been the research focus. The tendency of today's tunable lasers is to have small device scale, wide tuning range and good single mode suppression. A prominent progress is the miniaturized tunable lasers that have been developed recently using the advanced fabrication and integration technologies of microelectromechanical systems (MEMS). Compared with the conventional tunable lasers, the MEMS lasers bear more benefits such as low cost, low power consumption and feasibility of integration in addition to the inherent features of compact footprint and fast tuning speed. On the other hand, many emerging applications require the tuning capability not only in the wavelength but also in the polarization state of the output laser light. This is particular important to those applications that rely on the coherence and polarization state of the light, such as coherent communications, biomedical instrumentations, sensor systems and so on.

However, the developed MEMS tunable lasers [5-7] have limited space to put in any external components for polarization control due to their micro-scaled size. For example, it is difficult to integrate a small crystal with controllable optical nonlinearity or a micro-scaled polarizer into the external cavity of a MEMS tunable laser system. As a result, the majority of MEMS lasers have no control of the polarization state. The output has either randomly polarization state or a fixed state as determined by the diode laser chip. Therefore, a polarization tunable laser surely has considerable impact on both scientific research and engineering applications.

In this paper, a MEMS polarization tunable laser is designed and demonstrated by integrating tunable optical tunneling structure with the semiconductor laser diode. The wavelength and the polarization state can both be tuned by the tunable optical tunneling structure via thermo-optical effect. The experimental results show that the output laser can be switched between two perpendicular polarization states while maintaining the linear wavelength tunablity between 1548 nm and 1550 nm. The heating current is within 100 mA due to the highly sensitivity of the optical tunneling structure.

This paper is organized as follows. The design and theoretical study of the proposed tunable laser will be elaborated in section 2. And the experimental results will be presented in section 3, along with some in-depth discussions. Finally, some conclusions will be given in section 4.

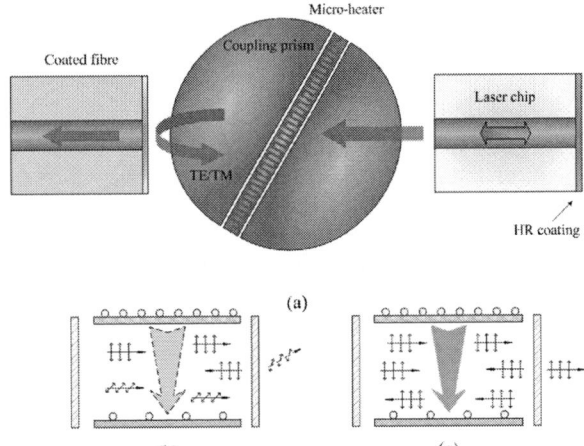

Figure 1: Schematic and working principle of the polarization tunable laser. (a) Configuration of the tunable lasers; (b) randomly polarized light by the stimulated radiation; and (c) the laser emits light with a polarization state as selected the optical tunneling structure.

2. DESIGN AND THEORY

2.1 Design of MEMS polarization tunable laser

The schematic of the MEMS polarization tunable laser is shown in Fig. 1(a). It consists of a laser diode, an optical tunneling structure and a coated fiber. The laser diode is functioned as the gain media when pumped electrically. If

978-1-4244-2977-6/09 $25.00 © 2009 IEEE

there is no polarization control, it generates randomly polarized light via transition of the stimulated electrons as shown in Fig. 1(b). The left side of the laser diode is coated with high reflective (HR) film which prevents light from leaking. The other side of the laser diode is anti-reflection (AR) coated, with a reflectivity of approximately 0.2%. The fiber end is coated to improve its reflectivity from its intrinsic value of 4% to about 50% to make it semi-transparent. It collects the output laser light for detection and also acts as a flat mirror. As a result, the coated fiber end and the optical tunneling structure build up the external cavity for the laser system. The output wavelengths of the laser must satisfy the resonance condition of the external cavity and the gain region of the laser diode. Instead of tuning the wavelength by changing the distance of between the two mirrors as most MEMS tunable external cavity diode lasers rely on, the optical tunneling structure is introduced for both wavelength and polarization state tuning.

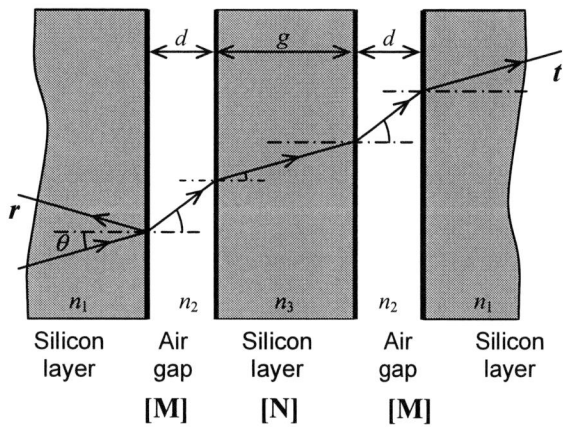

Figure 2: Theoretical model of the optical polarization selective structure as presented in a multi-layer model.

2.2 Working principle

This optical tunneling structure consists of five layers, including two air gaps and three silicon layers as shown in Fig. 2. Photons with proper polarization state and wavelength can transmit through this structure while other photons are reflected and lost totally. In another word, the optical tunneling structure is transparent for photons with certain wavelength and polarization states [8]. Therefore, only those photons that can resonate within the laser cavity are amplified to form the laser output. In this sense, the optical tunneling structure acts as an intracavity filter to select certain wavelength and polarization state. Further more, as the tunneling property can be adjusted by the refractive index change of the central silicon layer, both the wavelength and polarization can be tuned simultaneously by simply heating up the central silicon slab. As the silicon material has a strong thermo-optic effect, the temperature change would vary the refractive index of the central silicon layer effectively. For this purpose, a micro-heater can be patterned on top of the central silicon layer as illustrated in Fig. 1.

2.3 Theoretical analysis

The transfer matrix method has been well developed for dealing with the multi-layer films. As the optical tunneling structure can be treated as a multi-layer structure as illustrated in Fig. 2. Here the transfer matrix method is employed for the theoretical analysis of the optical tunneling structure. For single air gap, the reflection coefficient R can be expressed as [9],

$$R = \frac{r_{10} + r_{02}e^{i\delta}}{1 + r_{10}r_{02}e^{i\delta}} \tag{1}$$

The term δ is the phase delay of the reflected light as given by,

$$\delta = (4\pi d/\lambda)\sqrt{1 - n^2 \sin^2 \theta} \tag{2}$$

where d is the width of the air gap, λ is the wavelength, θ is the incident angle, and r_{mn} (m, $n = 0$, 1 or 2) represent the reflection coefficients at different interfaces given by the Fresnel equations [10].

It is worth highlighting that the incident angle has to be above the critical angle. Under this condition, the light passes through the air gaps by the optical tunneling effect. Otherwise, the single air gap is just a simple Fabry-Pérot cavity for propagation lightwave. Due to the optical tunneling effect, the phase shift of the light traveling through the air gap is imaginary. This results in the attenuation of the light amplitude. The transmission coefficient T can be written as,

$$T = \frac{t_{10}t_{02}e^{i\Delta}}{1 + r_{02}r_{10}e^{i\delta}} \tag{3}$$

where t_{mn} (m, $n = 0$, 1 or 2) are the transmission coefficients at different interfaces, and Δ is the phase delay of the light transmitted through the air gap.

The optical tunneling structure can be treated as the cascading of two air gaps separated by a silicon slab (corresponding to the central silicon layer). Therefore, the transfer matrix of the optical tunneling structure can be expressed by R and T as given by

$$M_{double} = MNM = \begin{pmatrix} m_{11} & m_{12} \\ m_{21} & m_{22} \end{pmatrix} \tag{4}$$

$$M = \begin{bmatrix} \dfrac{T^2 - R^2}{T} & \dfrac{R}{T} \\ -\dfrac{R}{T} & \dfrac{1}{T} \end{bmatrix} \tag{5}$$

$$N = \begin{bmatrix} e^{i\phi} & 0 \\ 0 & e^{-i\phi} \end{bmatrix} \tag{6}$$

where M and N are the transfer matrices of the air gap and the central silicon layer, respectively, and ϕ is the phase delay of the central silicon layer. Here it is assumed that the two air gaps are identical.

The contour map of the output laser power versus wavelength and refractive index change is shown in Fig 3(a). The bright color represents high transmission while the dark one stands for low transmission. It can be seen from the simulation results that both the transverse electric (TE) polarization and the transverse magnetic (TM) polarization have a bright stripe in the region of interest. The two stripes are separated but share the same inclination. The separation implies that it is not possible to get high transmission for both the TE and TM with a given combination of the refractive index change and the wavelength. In another word, the transmission is highly selective to the polarization state. The inclination suggests that the wavelength of the laser output is red shifted with the increase of the refractive index of the central silicon. This acts as the fundamental mechanism for the wavelength tuning even though the cavity length between the fiber end and the HR laser facet does not change. It can be observed from Fig. 3(a) that the TE light have narrower stripe width than the TM light because the TM light is easier to penetrate the optical tunneling structure than the TE light due to the in-plane electric field.

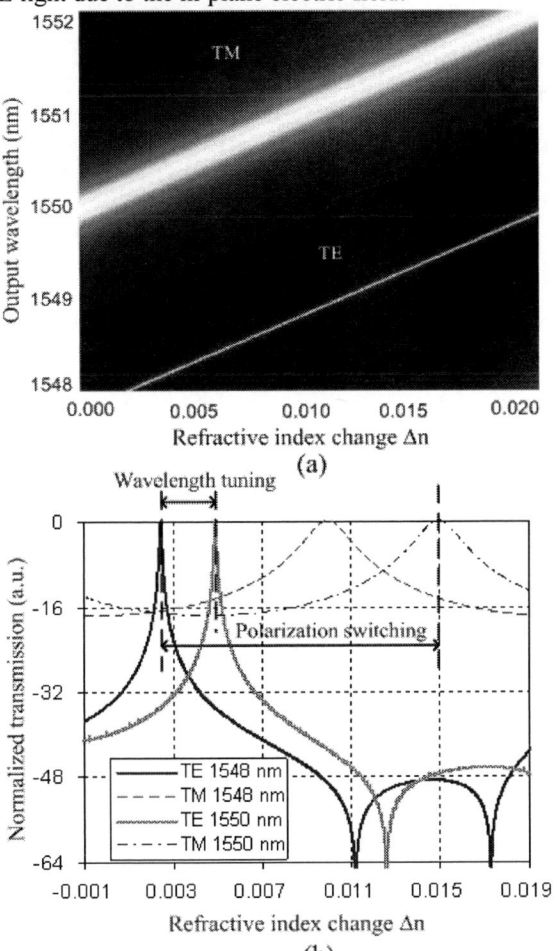

(a)

(b)

Figure 3: Simulation results of the output of the proposed polarization tunable laser. (a) The contour map of the transmission at different wavelength and refractive index change, and (b) the normalized transmission as a function of the refractive index change at the output wavelength 1548 nm and 1550 nm, respectively.

The numerical results are shown in Fig. 3(b) for the transmission of the optical tunneling structure versus the refractive index change of the central silicon layer. According to the simulation results, the polarization state remained unchanged when the refractive index change Δn is less than 0.01. Under this condition, the refractive index change only affects the wavelength of the peak transmission. However, when the Δn is larger than 0.01, the polarization of the transmitted light is switched to its

perpendicular position. It should be pointed out that the output laser should satisfy two conditions to suppress other spontaneous radiation and to form a steady laser output. One is to satisfy the tunneling condition of the optical tunneling structure, the other condition is the resonance condition of the laser cavity. Those two conditions will guarantee the single mode output of the tunable laser, similar to the Vernier effect of the tunable external cavity lasers. However, the free spectrum range (FPR) of the optical tunneling spectrum can reach 30 nm [8], much larger than the typical values of 1 nm in the conventional external-cavity tunable lasers. This helps avoid the mode hopping problem.

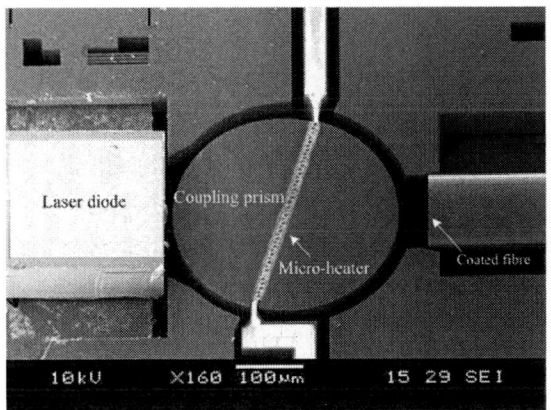

Figure 4: SEM of the polarization tunable laser.

3. RESULTS AND DISCUSSIONS

The overview of an integrated MEMS polarization tunable laser is shown in Fig. 4 by a scanning electron micrograph (SEM). The whole structure is fabricated on a SOI wafer with a silicon structure layer of 75 μm thick. The coated fiber is aligned by the fiber groove. The semi-cylinder shaped prism is fabricated to have a radius of 265 μm for the optimal coupling of the light into the fiber. The air gap between the central silicon layer and the prism is 0.9 μm and the central silicon layer is 30 μm wide. The incident angle of the light is 18°, larger than the critical angle of 17° for the silicon/air interface. The gain chip is a multiple quantum well InGaAsP/InP device having dimensions of 250 μm × 300 μm × 100 μm (length × width × height). The divergent angles in the vertical and horizontal directions are both 42°. It is aligned by an alignment marker and placed upside down on the silicon substrate. This is for the vertical alignment of the optical axis of the laser chip to the prism and the fiber because the output aperture of the laser diode is close to the upper surface of the laser diode. The silicon substrate underneath the gain chip is coated with an aluminum layer of 0.3 μm. In this way, the laser diode can be pumped by applying a current between the coated silicon substrate and the back surface of the laser diode. The coated fiber is connected to an optical spectrum analyzer (OSA) to monitor the alignment of the device. The position of the fiber is adjusted by the fiber stage until a single mode output with relatively strong power is achieved, which means the output laser satisfies both the tunneling condition and the resonance condition of the laser cavity.

The micro-heater is connected to a signal measurement unit (SMU). The output at different heating currents is measured by the OSA and the polarization analyzer (POD-101A) so as to get the information of both the output wavelength and polarization states. The laser output at different heating current is summarized in Fig. 5. The refractive index change Δn increases with higher heating current. And it can be seen from the Fig. 5 that the output spectrum undergoes a red shift as the refractive index change is increased, which agrees well with the simulation results in Fig. 3(a). When the heating current goes higher than 47.2 mA, the polarization state is switched from the TE to the TM.

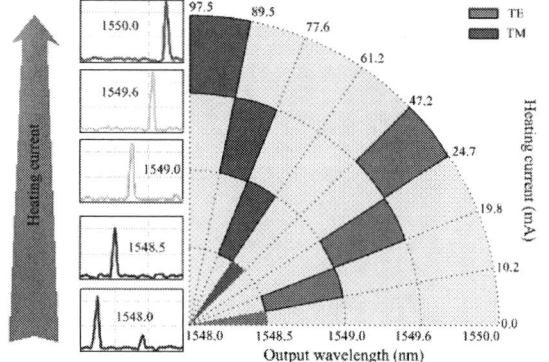

Figure 5: Measured polarization states and wavelength of the output laser under different heating current. The insects exemplify the spectra of different output wavelengths.

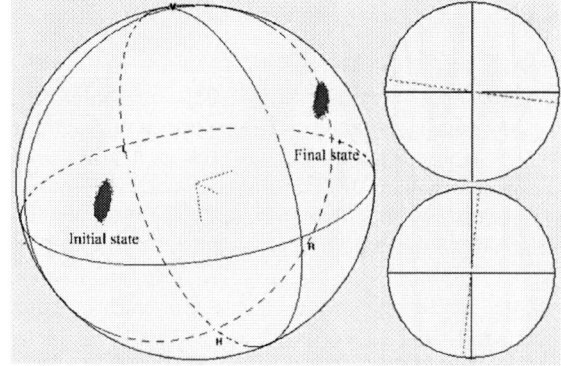

Figure 6: The Poincaré sphere diagram of the output laser with different polarization state. The heating power is 0 mA and 61.2 mA for the initial state and final state respectively.

For a detailed study of the polarization tuning, the SMU is initially switched off so that there is no heating current injected to the micro-heater and the refractive index change is equal to zero. It should be noted that the inevitable twist of the fiber might cause the change of the polarization state before the output light goes to the polarization analyzer. In the experiment, the output fiber is fixed to make sure that the output light from the tunable laser undergoes the same change before they goes to the polarization analyzer. Therefore, the difference of the polarization states between the initial and final states can be monitored by the polarization analyzer.

The measurement results of the polarization analyzer are shown in Fig. 6. The left side is the Poincaré sphere diagram of the tunable output. The upper right shows the initial polarization state. The lower right shows the tuned polarization state when the micro-heater is pumped with

61.2 mA. The initial and the finial polarization state are perpendicular to each other while the output wavelength remains unchanged, which demonstrates the capability of independent tuning of the polarization states.

4. CONCLUSIONS

In this paper, a MEMS polarization tunable laser has been designed and demonstrated using a micromachined silicon optical tunneling structure. The wavelength and the polarization states can be tuned by the thermo-optic effect without need for any mechanical movement and any external polarizer. In experiment, the laser output measures a 2-nm wavelength tuning range and maintains single mode with a side mode suppression ration of more than 30 dB. Polarization tuning between two perpendicular states has also been demonstrated without changing the laser wavelength. The unique capability of both wavelength tuning and polarization tuning makes this laser potential for many applications such as the research of atomic and molecular physics, biomedical monitoring and coherent communications.

5. REFERENCES

[1] H. Ukita, Y. Uenishi, and H. Tanaka, "A Photomicrodynamic System with a Mechanical Resonator Monolithically Integrated with Laser Diodes on Gallium Arsenide", *Science*, vol. 260, pp. 786-789, 1993.

[2] C. J. Myatt, N. R. Newbury and C. E. Wieman, "Simplified atom trap by using direct microwave modulation of a diode lasers", *Opt. Lett.*, vol. 18, pp. 649-651, 1993.

[3] K. Toyada, A. Miura, S. Urabe, K. Hayasaka and M. Watanabe, "Laser cooling of calcium ions by use of ultraviolet laser diodes: significant induction of electron-shelving transition", *Opt. Lett.*, vol. 26, pp. 1897-1899, 2001.

[4] J. Kitching, H. G. Robinson, L. Hollberg, S. Knappe and R. Wynands, "Optical-pumping noise in laser-pumped, all-optical microwave frequency references", *J. Opt. Soc. Am. B*, vol. 18, pp. 1676-1683, 2001.

[5] H. Cai, A. Q. Liu, X. M. Zhang, J. Tamil, Q. X. Zhang, D. Y. Tang, C. Lu, "A miniature tunable coupled-cavity laser constructed by micromachining technology", *Appl. Phys. Lett.*, vol. 92, 031105, 2008.

[6] X. M. Zhang, A. Q. Liu, C. Lu and D. Y. Tang, "A real pivot structure for MEMS tunable lasers", *IEEE JMEMS*, vol. 16, pp. 269-278, 2007.

[7] H. Cai, A. Q. Liu, X. M. Zhang, J. Tamil, D. Y. Tang, J. Wu and Q. X. Zhang, "Tunable dual-wavelength laser constructed by silicon micromachining", *Appl. Phys. Lett.*, vol. 92, 051113, 2008.

[8] W. M. Zhu, X. M. Zhang, A. Q. Liu, H. Cai, J. Tamil, And T. Buorouina, "A micromachined optical double well for thermo-optic switching via resonant tunneling effect", *Appl. Phys. Lett.*, vol. 92, 251101, 2008.

[9] S. Zhu, A.W. Yu, D. Hawley and R. Roy, "Frustrated total internal reflection: A demonstration and review," *J. Opt. Soc. Am. B*, vol. 54, pp. 601-607, 1986.

[10] M. Born and E. Wolf, *Principle of optics*, Pergamon, Oxford, 1975.

A NOVEL TELESCOPE WITH MICROMIRROR FOR OBSERVATION OF TRANSIENT LUMINOUS EVENTS FROM SPACE

J. H. Park[1], S. Nam[1], G. Garipov[2], J. A. Jeon[1], J. Y. Jin[3], B. Khrenov[2], J. E. Kim[1], M. Kim[3], Y. K. Kim[3], J. Lee[1], G. W. Na[1], I. H. Park[1], Y. –S. Park[3], and B. W. Yoo[3]

[1]Ewha Womans University, Seoul, Korea
[2]Moscow State University, Moscow, Russia
[3]Seoul National University, Seoul Korea

ABSTRACT

A novel type of telescope, a pinhole-like camera with a micromirror array, is introduced for space observation of Transient Luminous Events (TLEs) like gigantic lightning occurring at upper atmosphere currently under question or investigation. The presented telescope has a unique feature of wide field of view (FOV) of surveillance, fast zoom-in and tracking. A high-fill factor, two-axis rotational micromirror array is proposed for the key component of the telescope. To obtain a large static deflection of a mirror, the self-aligned vertical comb-drive actuator is used. Multi-layered structures of a micromirror are fabricated using bulk micromachining techniques, such as deep reactive ion etching and wafer bonding on the full wafer-scale.

INTRODUCTION

In recent years, observation of TLEs has raised several questions of the understanding of global electrical phenomena in the atmosphere and the influence of those phenomena on atmospheric properties [1-3]. Ground observation of TLE does not give a global picture of the phenomenon as it could be done only at a limited number of places on the Earth. Optical observations from space are more promising due to their capability to scan the entire atmosphere. Moreover, these observations can be combined with the data on normal lightning and other electrical discharges collected in the global network of radiometers, which provides information about atmosphere electric activity in almost all geographical regions of the Earth. TLEs have been reported in the form of Blue jets, Sprites, Elves, Gigantic jets occurring between stratosphere and ionosphere with fast development time (1~100 ms) over large dimensions. The scale of a TLE is 10-70 km in height and 50-100 km in width. Observation of such extremely fast phenomena from space requires a special type of instrument that allows wide FOV to find the event and fast follow-up of the triggered event for more detailed study.

MEMS (Micro-Electro-Mechanical Systems) technology has been widely used for many optical applications. One of the most commonly used optical MEMS devices is a micromirror due to very fast tilting speed and high reliability compared to large conventional mirror rotated by electromagnetic motor [4-10]. MEMS-based micromirror has very small size and low power consumption, which can provide accurate and robust operation for optical applications. In this paper, a novel type of telescope based on the optical MEMS technology has been proposed for the observation from space of transient events like TLEs. Compared to the previously reported space telescope using a pinhole [11],

presented in this paper is a simple telescope in which the pinhole is substituted for a micromirror array. Fast rotatable micromirror allows important functions like fast zoom-in and tracking of a given object. The micromirror directs lights quickly to photo-detectors, and thus to follow the moving light source, which is impossible in pinhole camera. The payload of the presented telescope will be carried into orbit by a Russian microsatellite Tatyana II rocket, of which the primary aim is to observe TLEs over a time period lasting at least one year.

PRINCIPLE AND DESIGN OF THE TELESCOPE

A schematic view of the telescope is shown in Figure 1. The telescope is composed of the trigger camera and the zoom-in camera. The trigger mirror positioned at distance L_1 from the focal plane detector is used for locating the object in the wide FOV. The zoom-in mirror, installed at distance $L_2 >> L_1$, is used to detect an object image with higher lateral resolution which provides the zoom-in effect. The focal length of trigger mirror is 90 mm, making the FOV of 11.3°. The FOV in the zoom mirror regime is 2.9° with the focal length of 360 mm. When the telescope uses the trigger mirror for observation from a satellite at a height 800 km above the Earth's surface, it views an area in the atmosphere of 160×160 km^2. When using the zoom mirror, the telescope views an area 16 times smaller than it does with the trigger mirror, meaning that an object can be observed in greater detail via the zoom-in mirror.

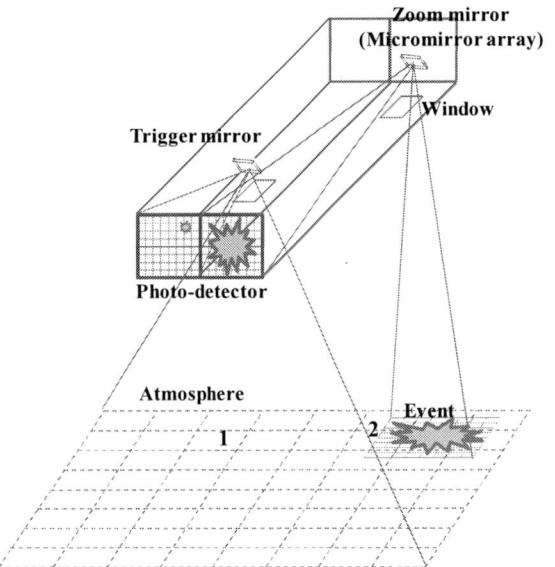

Figure 1: Schematic view of the proposed telescope

978-1-4244-2977-6/09 $25.00 © 2009 IEEE

At the beginning stage, namely in the trigger mode, trigger mirror is reflecting lights from watching area to the photo-detector. Photo-detector's signals of the light are quickly analyzed in order to determine the direction of the light source in wide FOV (area 1 in Figure 1). Once the direction is obtained, zoom mirror is rotated in correspondence to the trigger mirror information so that the image (shown in area 2 in Figure 1) is aligned to the center of the photo-detector. With the above described method, the features of zoom-in and tracking are obtained by the zoom mirror while the event is selected in wide FOV by the trigger mirror. It is important that the zoom mirror should be fast enough in rotating its viewing angle to observe the first instance of TLE immediately after the trigger.

Figure 2 shows the fabricated telescope. A plane mirror with the size of 3 mm × 3 mm is used for the trigger mirror and micromirror array is installed for the zoom mirror. A multi-anode photomultiplier tube (MAPMT) is used as a focal plane detector to meet the requirements in fast response of the presented telescope. The square MAPMT (HAMAMATSU H7546A) is sub-divided into 8×8 square pixels, each 2 mm × 2 mm. The size of the telescope window is determined by the required FOV. Baffle walls inside the telescope are implemented to minimize the background of scattered light in the telescope produced by light coming through the aperture.

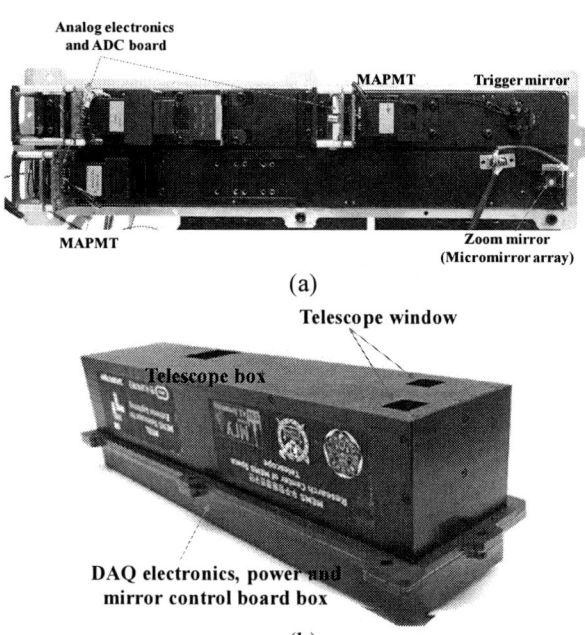

(a)

(b)

Figure 2: Fabricated telescope (a) top view of the telescope without covering a box, (b) telescope covered with a box

MICROMIRROR

The micromirror array is composed of 8×8 square pixels with the size of 340×340 μm². The micromirror utilizes two-axis electrostatic vertical comb actuators that allow one to continuously change the viewing angle of the mirror plate biaxially. Single crystalline silicon is selected as a structural material for the micromirror due to its properties such as a negligible residual stress, high yield strength, high temperature resistance, and a flat surface. The tilting angle of all reflectors in an array is consistent

when the same voltage signal is applied to individual mirrors.

A micromirror consists of a mirror plate, an actuator with top and bottom comb electrodes, and a glass substrate with addressing lines. A schematic view of the micromirror is shown in Figure 3, which has two rotational axes. The moving part of the actuator has a gimbal-like frame. A bias voltage at the comb electrodes attached to the frame provides electrical torque for tilting the frame itself, while the DC bias at the comb electrodes located inside the frame generates torque to tilt the inner plate. Two orthogonal pairs of springs allow the mirror plate to be tilted in independently in two orthogonal directions. By combining rotations along the two axes, the tilting angle of the micromirror can be controlled in any direction according to the data from the control circuit, indicating the direction to the useful event.

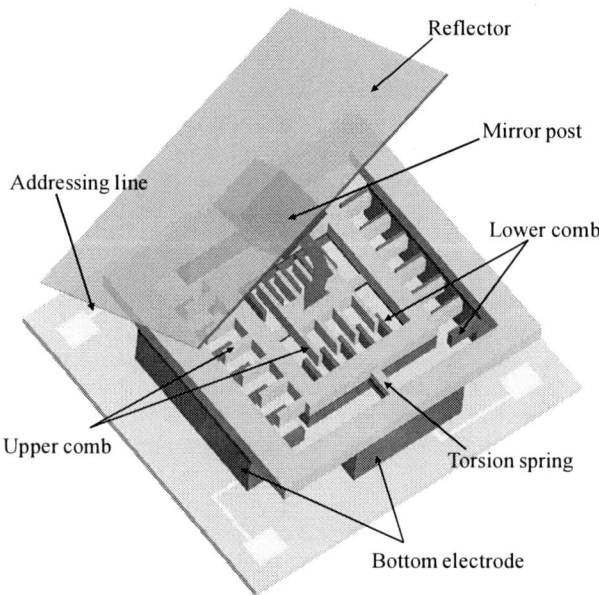

Figure 3: Schematic view of the micromirror

Figure 4 illustrates the fabrication process of the mirror. Self-aligned vertical comb structures are fabricated on a SOI (Silicon-On-Insulator) wafer to construct the actuation part of the mirror. The bottom silicon layer of SOI wafer is patterned to form comb electrodes to which the actuation voltage is applied and the top silicon layer is used as the ground electrode. The mirror plate is formed at the inner plate of the actuator using a wafer bonding process and coated with an Al layer to make a highly reflective surface. The actuator part and the glass substrate with electrical lines are bonded together to make electrical contacts between comb electrodes and addressing lines. The actuator and addressing lines can be hidden behind the mirror plate, resulting in a high fill-in factor of 84%.

Figure 5 shows a Scanning Electron Microscope (SEM) image of the fabricated micromirror array with one of the mirror plate removed to reveal the actuator. The actuator and addressing lines are located underneath the mirror plate. The measured average roughness (Ra) is 4 nm and the curvature radius is 0.9 m, which shows the mirror plate has a surface profile sufficient for the proposed telescope.

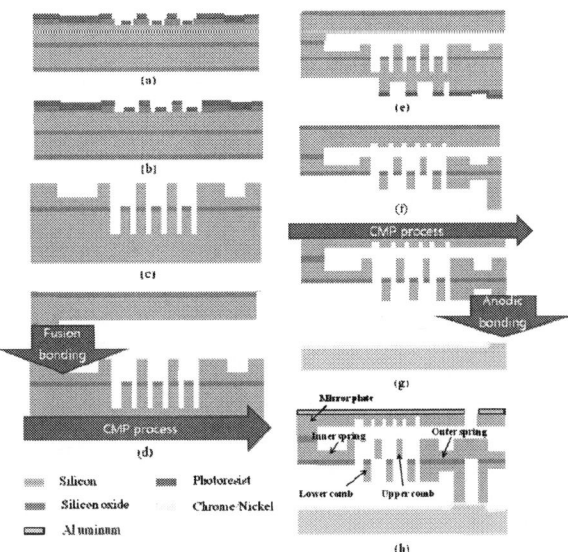

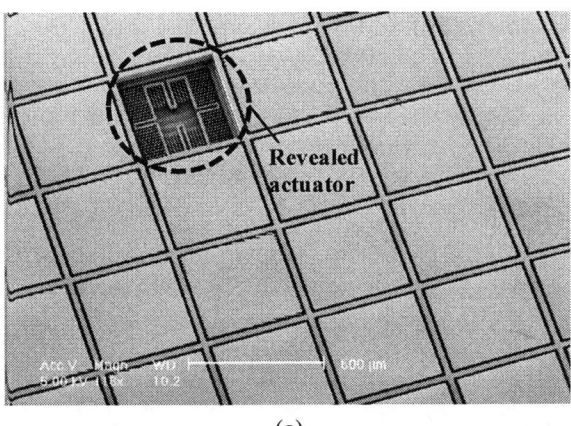

Figure 4: Fabrication process of the micromirror

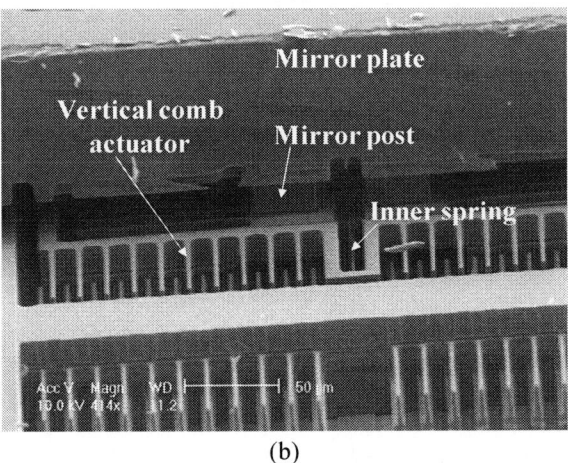

(a)

(b)

Figure 5: SEM images of the fabricated micromirror

The static tilting angle of the mirror is measured using a helium-neon laser source and a Position Sensing Detector (PSD). Figure 6 shows the measured optical tilting angles of a micromirror with respect to the applied voltage. The maximum optical tilting angle measured is 2.2° and 6.2°, for the two rotational axes, at the voltage of 45 V and 75 V, respectively. The dynamic responses of the micromirror

were measured using a non-contact laser displacement meter and dynamic signal analyzer. The signal analyzer produces an input voltage signal and its frequency varies within specific range in real-time. Meanwhile, the laser displacement meter provides a voltage signal which is in proportion to the deflection angle of the micromirror. The torsion resonance of the structure was detected at the frequency of 2 kHz and 1 kHz, respectively, which shows that the micromirror is capable of making observations of fast-moving objects such as TLEs when used as a component of the proposed telescope.

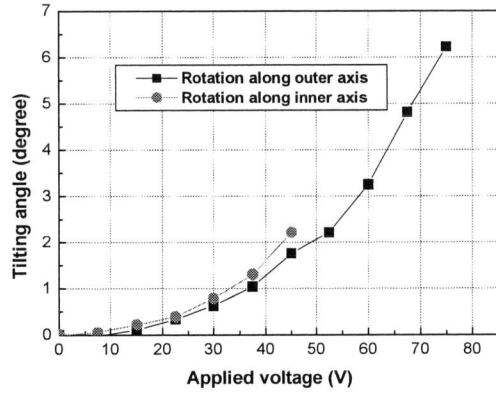

Figure 6: Measured static responses of the micromirror

TEST RESULTS OF THE TELESCOPE

The performance of the fabricated telescope has been tested using UV LED as a light source, which has peak intensity at a wavelength of 400 nm with FWHM of 20 nm. The distance between telescope and the light source was 6 m and the detector images from trigger and zoom mirror were recorded with the control of micromirror tilting angle for the different positions of light source. The diameter of the UV LED light source is equivalent to a 20 km source observed from an altitude of 800 km.

Figure 7 shows the test results of the proposed telescope. The responses of 64 pixels of MAPMT from the trigger and the zoom mirror are shown and z-axis denotes the digitized light intensity detected by each pixel. A control circuit receives the digitalized photon detector signal and the data are then analyzed to make a trigger decision. In parallel with this, the position of the brightest point of event is stored to determine the tilting angle of zoom mirror. When the trigger condition is satisfied, the tilting angle information is converted to the actual bias values to be applied to the micromirrors. Next, the micromirror driving circuit applies the necessary voltage to change the orientation angle of micromirrors. Figure 7 (a) represents the recorded images at the beginning stage and 7 (b) shows the images after the zoom mirror points the light source and zooms into the center of the source. Figure 7 (c) shows the tracking of the light source taken by the zoom mirror, when the light source moved to the other position. If the position of the light source changes, the control circuit calculates the information and adjusts the bias voltage for the micromirror to align the tilting angle to the center the source.

978-1-4244-2977-6/09 $25.00 © 2009 IEEE

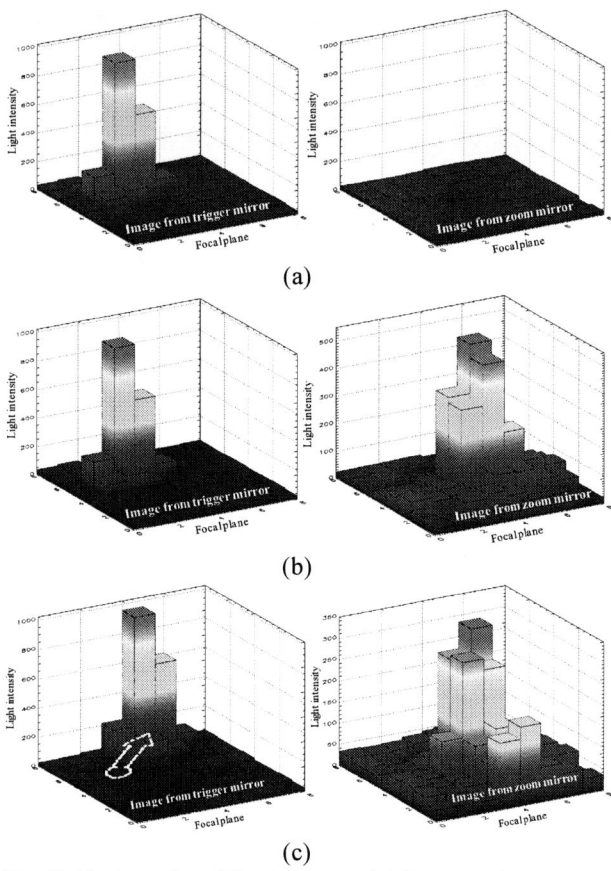

(a)

(b)

(c)

Fig. 7: Test results of the telescope (a) Detector images at the beginning stage (b) Images after the zoom mirror tracks the light source (c) Images, when the light source moved from the point shown in (b)

The reliability of the micromirror was evaluated by actuating the micromirror repeatedly to a tilting angle of 3.1° at atmospheric environments. Even after 10^8 cycles repeated actuation, there is no change in the resonant frequency and no degradation in static responses, which means that the micromirror has lifetime of over 100 millions actuation cycles. In addition, there was no significant degradation in the telescope after the space qualification tests such as shock, vibration, humid, vacuum, radiation and thermal tests. For example, shock and vibration test of the telescope was performed in three mutually perpendicular directions on the basis of Russian satellite test procedure. The qualification level random vibration tests ranged from 20 Hz to 2000 Hz. The 40g shock at the orbital stage was the maximum operational g-load to qualify the test. After all the shock and vibration tests, not only no damage was observed, but also resonance frequency of the micromirror was not changed.

CONCLUSION

A new type of telescope with rotatable micromirror array is proposed for the observation of fast-evolving events like TLEs in the upper atmosphere. The advantage of the presented telescope over conventional one is the capability of searching for TLEs in a wide field of view and to immediately zoom in for investigation of the detailed structure of TLEs. It also allows the tracking of space-time development of an event by placing the interesting part of an image onto the center of the photo-detector array. For the key component of the telescope, the two-axis micromirror array using the self-aligned vertical comb-drive actuator with high fill factor has been designed, fabricated and tested.

ACKNOWLEDGMENTS

This work was supported by Creative Research Initiatives (RCMST) of MOST/KOSEF.

REFERENCES

[1] V. P. Pasko, "Electric jets," *Nature,* vol. 423, pp. 927-929, 2003.

[2] R. A. Roussel-Dupre, A. V. Gurevich, "On runaway breakdown and upward propagating discharges," *J. Geophys. Res.*, vol. 101, pp. 2297, 1996.

[3] Y. Yair, C. Price, Z. Levin, J. Joseph, P. Israelevitch, A. Devir, M. Moalem, B. Ziv, M. Asfur, "Sprite observations from the space shuttle during the Mediterranean Israeli dust experiment (MEIDEX)," *J. Atmos. Terr. Phys.*, vol. 65, pp. 635–642, 2003.

[4] P. F. Kessel, L. J. Hornbeck, R. E. Meier, M. R. Douglass, "A mems-based projection display," *Proc. IEEE*, vol. 86, pp. 1687-1704, 1998.

[5] I. W. Jung, U. Krishnamoorthy, O. Solgaard, "High fill-factor two-axis gimbaled tip-tilt-piston micromirror array actuated by self-aligned vertical electrostatic comb drives," *J. Microelectromech. Syst.*, vol. 15, pp. 563-571, 2006.

[6] D. Hah, C. A. Choi, C. K. Kim, C. H. Jun, "A self-aligned vertical comb-drive actuator on an SOI wafer for a 2D scanning micro mirror," *J. Micromech. Microeng.*, vol. 14, pp. 1148-1156, 2004.

[7] Y. Hishinuma, E. H. Yang, "Piezoelectric unimorph micro actuator arrays for single-crystal silicon continuous-membrane deformable mirror," *J. Microelectromech. Syst.*, vol. 15, pp. 370-379, 2006.

[8] J. Kim, D. Christensen, L. Lin, "Monolithic 2-D scanning mirror using self-aligned angular vertical comb drives," *IEEE Photon. Tech. Lett.,* vol. 17, pp. 2307-2309, 2005.

[9] U. Krishnamoorthy, D. Lee, and O. Solgaard, "Self-aligned vertical electrostatic combdrives for micromirror actuation," *J. Microelectromech. Syst.*, vol. 12, pp. 458-464, 2003.

[10] V. Milanović, G. A. Matus and D. T. McCormick, "Gimbal-less monolithic silicon actuators for tip–tilt–piston micromirror applications," *IEEE J. Select. Topics Quantum Electron.*, vol. 10, no. 3, pp. 462-471, 2004.

[11] G. K. Garipov, B. A. Khrenov, M. I. Panasyuk, V. I. Tulupov, A. V. Shirokov, I. V. Yashin, H. Salazar, "UV radiation from the atmosphere: results of the MSU "Tatiana" satellite measurements," *Astropart. Phys.*, vol. 24, pp. 400-408, 2005.

NON-SPHERICAL SU-8 MICROLENS ARRAY FABRICATED UTILIZING A NOVEL STAMPING PROCESS AND AN ELECTRO-STATIC PULLING METHOD

Shu-Ming Kuo[1] and Che-Hsin Lin[1,2]

[1]Department of Mechanical and Electro-mechanical Engineering,
National Sun Yat-sen University, Kaohsiung, 804, Taiwan
[2]Advanced Crystal Opto-electronics Research Center,
National Sun Yat-sen University, Kaohsiung, 804, Taiwan

ABSTRACT

This paper presents a simple and novel method to fabricate non-spherical SU-8 microlens array utilizing a soft SU-8 stamping approach and an electro-static pulling process. A SU-8 based stamp composed of micro-nozzle arrays and a reservoir structure is first fabricated on a glass substrate using a process of dosage control exposure. The unexposed SU-8 encapsulated in the cross-linked SU-8 shell is used as the "ink" for the stamping process. The presented SU-8 microlens array is then formed by stamping the formed SU-8 structure on an ITO glass substrate at a temperature higher than the glass transition temperature of unexposed SU-8 PR. The formed SU-8 array is then electro-statically pulled in a uniform electric field to generate non-spherical microlenses with various radii of curvature. Microlens arrays with diameters ranging from 20 to 500 µm are successfully fabricated with the presented method. Measured surface roughness (Ra) of 3.84 nm and good optical uniformity confirm the optical quality of the microlens arrays fabricated with this innovative fabrication technique.

INTRODUCTION

SU-8 PR is an excellent material for microlens array fabrications duo to its high optical transmittance from visible to near IR wavelength and high refractive index (~1.6). Furthermore, SU-8 PR also has better mechanical strength and chemical resistance than some other common-use polymers, such as polycarbonate (PC) or PMMA. In addition, microlens array formation is one of the hottest topics in the optical MEMS field. Microlens arrays have various applications, such as optical communication [1], flat panel display module [2], optical storage [3], and scanning micro-optical systems. In this regards, a number of fabrication techniques for producing microlens arrays have been developed including the photoresist or sol-gel glass reflowing technique [4], hot embossing, replica molding [5, 6], ink-jet printing [7, 8], gray scale photolithography [9], and excimer laser ablation [10]. Of these methods of microlens array fabrication, direct photoresist reflow process is the most popular method since the process is simple and controllable. However, some drawbacks might come with the reflow process for microlens array fabrication, including limited lens shape, high processing temperature and long processing time.

Alternatively, hot embossing or thermal molding to replicate PC, PMMA or even PDMS based polymers are often used to mass-produce microlens arrays without using the reflow process [11]. These materials have good mechanical and optical transparent properties for optical applications. A variety of lens shapes can also be formed

with this replication process. A precise metal mold is required prior to the molding procedure, however, and such fabrication is challenging for micromachining technology. In addition, the surface roughness of the micro-molds fabricated with these delicate approaches was also problematic for optical applications.

Therefore, in order to meet the requirement of high surface quality and various radii of curvature for practical uses, a variety of methods, including electro-wetting[12], electrostatic force modulation [13], hydraulic extruding [14] and etc, had been reported to modify the surface profile of the fabricated microlenses. The electro-wetting scheme usually relies on a well-defined electrode array and a low-viscous lens material to change the radius of curvature of the lenses. The hydraulic extruding method required an external pumping system to provide the extruding force and the system is comparable not easy to control.

This paper reports a novel method to fabricate microlens arrays utilizing the SU-8 negative PR. A SU-8 stamp and cartridge structure is first fabricated by a novel dose-control exposure process. The unexposed SU-8 is then injected through the via-hole in the SU-8 stamp at a temperature higher than the glass transition temperature (T_g) of the unexposed SU-8. A constant volume of SU-8 adheres on the substrate to form the microlens structure. The shape and the radius of the curvature of the microlens can be easily controlled by the designed radius of the via-hole in the stamp and the temperature of the stamping process. Furthermore, the surface profile of the stamped microlenses can be easily modified utilizing the electro-static pulling effect to produce the radii of curvature of the microlenses on-demand.

WORKING PRINCIPLE

The glass temperature of SU-8 is around 55°C before the exposure procedure and it is highly sticky at its liquid phase. On contrast, the T_g of SU-8 after exposure is up to 225°C and it has a low surface energy such that the surface is not sticky. Therefore, the basic principle of the fabrication method is to manipulate the physical properties of SU-8 PR to meet different fabrication purposes. Figure 1 presents the concept of the developed method for fabricating non-spherical microlens array. A SU-8 stamp structure composed of a hard shell and a micro-nozzle array as the via-holes is fabricated using a dose-control exposure process. The un-exposed SU-8 PR is encapsulated in the exposed SU-8 shell. This single exposure process can simultaneously produce a stamp structure (exposed SU-8) including the cartridge (unexposed SU-8 as the ink) (Fig.1 (A)). The exposed SU-8 structure is then used as a stamp to define the pattern of the microlens array and as reservoir to store the SU-8 ink. The un-exposed SU-8 ink can be squeezed out to the stamp surface and adheres to the substrate by applying a small pressure on the SU-8 stamp at the temperature higher than the T_g of un-exposed SU-8 PR. The adhered SU-8 forms a ball shape droplet array due to the force

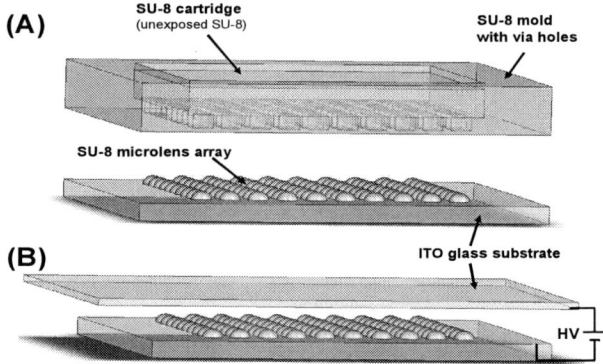

Figure 1: Schematic of the concept for fabricating non-spherical SU-8 microlens array (A) SU-8 lens molding and (B) electrostatic pulling.

FABRICATION PROCESS

Fig.2 shows the simplified fabrication process of the proposed technique and the fabrication details of the proposed method are described below.

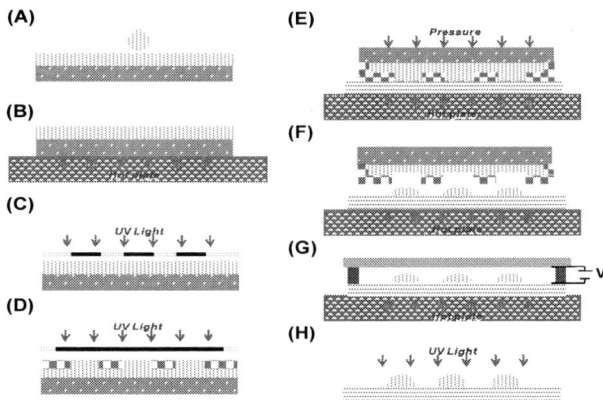

Figure 2: Illustration of the process for fabricating the microlens array. (A) SU-8 coating (B) Soft-baking (C)Controlled dose exposure (D) Well formation (E) SU-8 mold stamping (F) Demolding (G) Electro-static pulling (H) UV-curing.

In order to coat a thick SU-8 PR in a single coating process, we employed the "constant-volume-injection" method [15]. The injection volume for forming an SU-8 layer of about 500 um was 0.38 ml. Then glass with a thick SU-8 layer is placed on a 95°C for at least 1 hour to perform the soft-baking process.(Fig 2(A)) During this process, the force of gravity and surface tension allows coated SU-8 PR to self-planarize naturally since the unexposed SU-8 is in liquid phase at the soft-bake temperature.(Fig 2(B)) An underexposure process (60% of the commonly sufficient dose) was employed to provide minor cross-linking of the SU-8 photoresist.(Fig 2(C)). A second flux exposure process was then performed to define the reservoir structure area. (Fig 2(D)) The SU-8 stamp was then directly contact to an ITO glass at a temperature higher than the T_g of non-crosslinked SU-8 PR. The melted

balance of surface tension and gravity force. In addition, SU-8 molecule is with polarity and is capable of being attracted and deformed under external electric field. Non-spherical microlens array can be successfully fabricated with this simple approach. (Fig.1 (B))

SU-8 PR was squeezed out from the via-holes and adhered onto the ITO glass substrate under a small applied pressure.(Fig 2(E)) The stamp structure was then vertically stripped from the glass substrate and the microlens array was formed.(Fig 2(F)). The formed microlens array was then again placed on a 65°C hot plate and in an uniform electric field between 0 – 33000 V/m for electro-static pulling process.(Fig 2(G)) Once the designed radius of curvature was obtained, an UV-lamp was then used to cure the SU-8 structure and fix the shape of the microlenses.(Fig 2(H))

RESULTS AND DISCUSSION

Figure 3 shows the SEM image of a hexagonal SU-8 microlens array fabricated using the presented method. Microlens arrays with various shapes and lens distribution are easy to define using this novel stamping approach. On contrast, only ball-shape lens structure can be produced using the ink-jet printing method. The formed microlens array shows good shape uniformity and good surface quality. In addition, the calculated filling ratio of the microlens array can be up to 92.5% with a proper mask design.

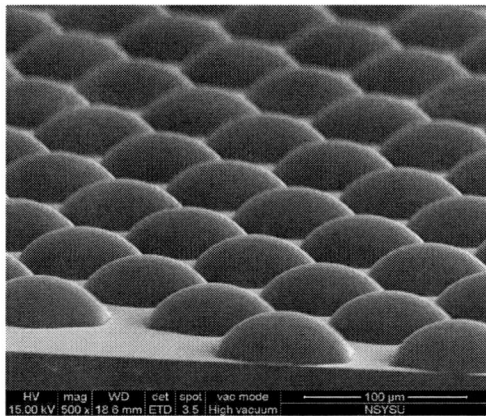

Figure 3: SEM image of the fabricated microlens array.

Figure 4 presents the measurement result of lens surface protruding height while manufactured under different diameter and temperature conditions. These data were averaged from measuring 10 individual microlenses from the captured SEM image. The T_g of SU-8 PR was about 55°C such that the fabricated temperatures ranged from 60°C to 100°C. The surface tension and viscosity of SU-8 PR was sensitive to processing temperature, higher temperature induced a lower surface tension and viscosity. From the results, the lens height reduced with rising temperature and the reduced ratio was nearly linear with the processing temperature. The lens height produced in this manufacturing method can reach to about 40% of designed diameter. It is very clear that we could manipulate the droplet shape by this approach.

978-1-4244-2977-6/09 $25.00 © 2009 IEEE

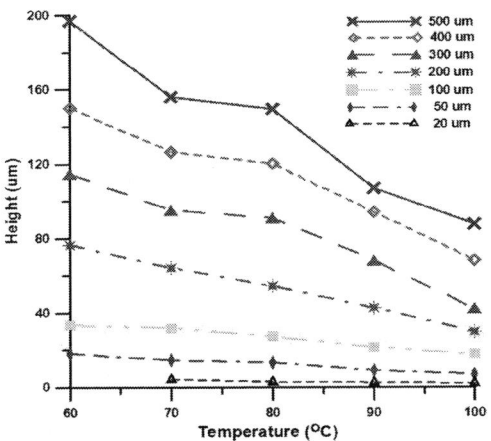

Figure 4: Measured lens heights of the micro-lenses fabricated with different stamping temperature and lens diameters.

Figure 5 shows the SEM images of the microlens arrays fabricated without (Fig. 5(A)) and with the electro-static pulling process (Fig. 5(B)). It is clear that a cone shape microlens array was produced after the electro-static pulling process. This is caused by the force balance between surface tension of SU-8 PR and the applied electro-static pulling force. Note that the top of the microlens fabricated with this approach is always near to spherical shape. The radius of curvature was significantly reduced after the electro field treating. Microlens array with non-spherical shapes of various radii of curvature can be fabricated by controlling the applied electric field strength.

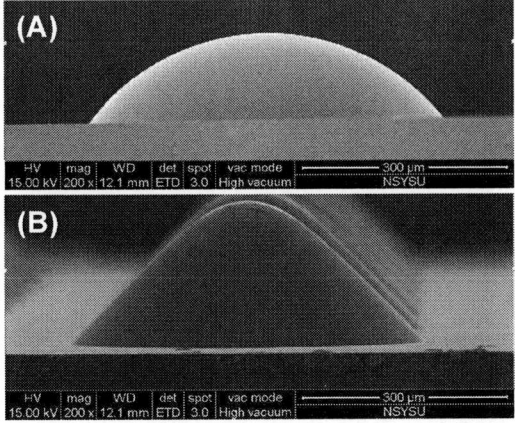

Figure 5: SEM images of the microlens fabricated (A) before and (B) after electro-static pulling at 2500 V.

Figure 6 presents the OM images showing the optical behavior for the microlens fabricated under different electro-static pulling voltages. These results show that the higher electric field strength produced microlenses with smaller radius of curvature. From basic optical formulae, the imaging results indicate that the higher electric pulling force produced the microlens array with a shorter focal length. It also suggests the fabricated lenses had a smaller radius of curvature and higher protruding height.

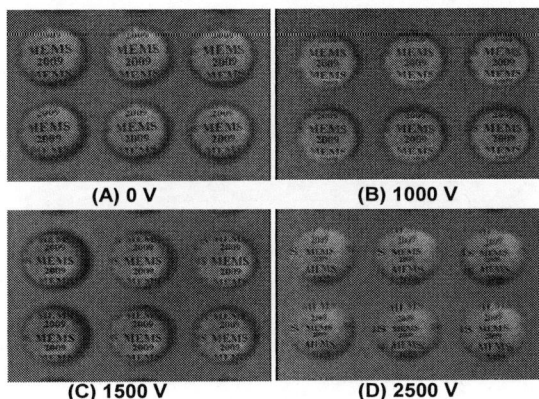

Figure 6: OM images showing the optical behavior for the microlens fabricated under different electro-static pulling voltages.

Figure 7 shows the relationship between the radii of curvature of the lenses (with a base diameter of 500 μm for the lens) fabricated under different applied electric fields. The gap between two ITO-glass electrodes was 770 μm and the applied electric field was from 0 up to about 32500 V/cm. Results indicate the measured radius of curvature of the microlenses reduced with the increasing applied electric field. The smallest radius of curvature fabricated with this approached was around 105 μm for the lens with a base diameter of 500 μm. However, the minimum radius of curvature of the microlens can be as small as 10 μm if the base diameter was 50 μm. Note that higher applied electric field might cause electro-spinning effect or electrical discharge between the electrodes and fail the experiment.

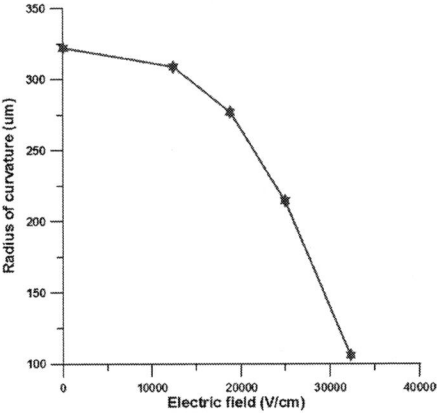

Figure 7: Relationship between the radius of curvature of the lens fabricated with different applied electrical field.

In order to evaluate the optical preference of the fabricated microlens array, we also set up a testing platform. At the focal plane of the light spot, where the spot had the minimum size and the maximum intensity .Figure 8 shows the captured image of the emitted lights projecting through the microlenses and the calculated light intensity profiles of the emitted light spots. Microlens arrays with excellent optical properties and uniformities can be fabricated with this novel stamping process and electro-static pulling method.

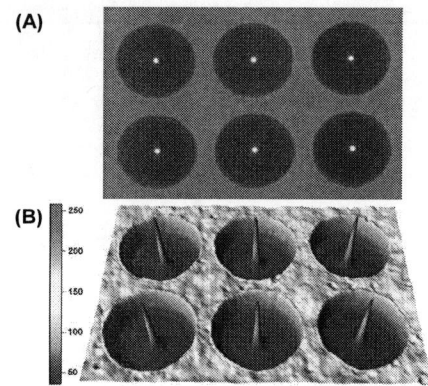

Figure 8: (A) Photograph of the emitted light from the microlenses (B) Measured intensity profiles.

CONCLUSION

This paper reports a simple and rapid method to successfully fabricate non-spherical SU-8 microlens arrays utilizing a novel stamping and electro-static pulling approach. An SU-8 stamp structure with an integrated SU-8 cartridge was first fabricated using the exposure process. The SU-8 microlens array was then precisely fabricated by printing the stamp under different temperatures and further modified utilizing the static pulling scheme. Excellent uniformity and optical properties of the fabricated microlens arrays were experimentally confirmed. Microlenses with diameters from 20 to 500 μm were successfully fabricated. A variety of lenses shape and different of optical properties was fabricated by controlling the mask design, the process temperature and the applied electro-field. The proposed method also shows its potential for a substantial impact on the mass fabrication of microlens arrays.

ACKNOWLEDGEMENTS

The author would like to thank partial financial supports from National Science Council (NSC 97-2221-E-110-018-MY3) and Minister of Education (5Y50B project).

REFERENCES

[1] M. He, X. C. Yuan, N. Q. Ngo, and S. H. Tao, "Single-step fabrication of a microlens array in sol-gel material by direct laser writing and its application in optical coupling," *Journal of Optics a-Pure and Applied Optics*, vol. 6, pp. 94-97, 2004.

[2] H. Urey and K. D. Powell, "Microlens-array-based exit-pupil expander for full-color displays," *Applied Optics*, vol. 44, pp. 4930-4936, 2005.

[3] K. Kurihara, I. D. Nikolov, S. Mitsugi, K. Nanri, and K. Goto, "Design and fabrication of microlens array for near-field vertical cavity surface emitting laser parallel optical head," *Optical Review*, vol. 10, pp. 89-95, 2003.

[4] M. He, X. C. Yuan, N. Q. Ngo, J. Bu, and V. Kudryashov, "Simple reflow technique for fabrication of a microlens array in solgel glass," *Optics Letters*, vol. 28, pp. 731-733, 2003.

[5] J. R. Ho, T. K. Shih, J. W. J. Cheng, C. K. Sung, and C. F. Chen, "A novel method for fabrication of self-aligned double microlens arrays," *Sensors and Actuators a-Physical*, vol. 135, pp. 465-471, 2007.

[6] J. N. Kuo, C. C. Hsieh, S. Y. Yang, and G. B. Lee, "An SU-8 microlens array fabricated by soft replica molding for cell counting applications," *Journal of Micromechanics and Microengineering*, vol. 17, pp. 693-699, 2007.

[7] D. L. Macfarlane, V. Narayan, J. A. Tatum, W. R. Cox, T. Chen, and D. J. Hayes, "Microjet Fabrication of Microlens Arrays," *Ieee Photonics Technology Letters*, vol. 6, pp. 1112-1114, 1994.

[8] V. C. Fakhfouri, N.; Mermoud, G.; Kim, J.Y.; Boiko, D.; Charbon, E.; Martinoli, A.; Brugger, J., "Inkjet printing of SU-8 for polymer-based MEMS a case study for microlenses," presented at Micro Electro Mechanical Systems, 2008., Tucson, USA, 2008.

[9] W. X. Yu and X. C. Yuan, "UV induced controllable volume growth in hybrid sol-gel glass for fabrication of a refractive microlens by use of a grayscale mask," *Optics Express*, vol. 11, pp. 2253-2258, 2003.

[10] S. Mihailov and S. Lazare, "Fabrication of Refractive Microlens Arrays by Excimer-Laser Ablation of Amorphous Teflon," *Applied Optics*, vol. 32, pp. 6211-6218, 1993.

[11] T. K. Shih, C. F. Chen, J. R. Ho, and F. T. Chuang, "Fabrication of PDMS (polydimethylsiloxane) microlens and diffuser using replica molding," *Microelectronic Engineering*, vol. 83, pp. 2499-2503, 2006.

[12] B. Berge and J. Peseux, "Variable focal lens controlled by an external voltage: An application of electrowetting," *European Physical Journal E*, vol. 3, pp. 159-163, 2000.

[13] K. Y. Hung, F. G. Tseng, and T. H. Liao, "Electrostatic-force-modulated microaspherical lens for optical pickup head," *Journal of Microelectromechanical Systems*, vol. 17, pp. 370-380, 2008.

[14] H. W. Ren, D. Fox, P. A. Anderson, B. Wu, and S. T. Wu, "Tunable-focus liquid lens controlled using a servo motor," *Optics Express*, vol. 14, pp. 8031-8036, 2006.

[15] C. H. Lin, G. B. Lee, B. W. Chang, and G. L. Chang, "A new fabrication process for ultra-thick microfluidic microstructures utilizing SU-8 photoresist," *Journal of Micromechanics and Microengineering*, vol. 12, pp. 590-597, 2002.

MULTICHAMBER TUNABLE LIQUID MICROLENSES WITH ACTIVE ABERRATION CORRECTION

D. Mader, P. Waibel, A. Seifert, and H. Zappe
Department of Microsystems Engineering – IMTEK, University of Freiburg, Germany

ABSTRACT

A design approach and new manufacturing technique for a novel type of stacked fluidic multi-chamber tunable lenses is presented. The design offers flexibility and extensibility, leading to fully functional miniature tunable optical lens systems with the ability for low order aberration control.

INTRODUCTION

Pressure-actuated liquid-filled microlenses (Figure 1) may be tuned in focal length from several millimeters to infinity and are thus suitable for use in tunable optical systems. We present here a novel fabrication method for pneumatic multi-chamber lens systems which provide a wide tuning range combined with the ability to correct imaging errors, such as chromatic and spherical aberrations. The approach is a highly flexible, accurate and inexpensive way to implement lens combinations on the micro-scale which are similar to those used in conventional macroscopic systems and offer similar options for optical design.

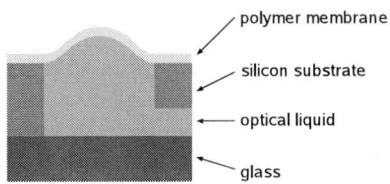

Figure 1: The concept of liquid-filled membrane lenses: Changes in pressure of a fluid in the cavity cause variable distension of the membrane and thus a change in focal length of the lens.

Using membrane lens technology, we can fabricate a tunable MEMS-based implementation of the conventional achromatic Fraunhofer glass doublet; the system contains optical liquids with appropriate refractive indices and dispersion characteristics, separated by thin flexible polymer membranes (Figure 2). The selected membrane material (DowCorning Sylgard 184) shows very low optical absorption in the visible spectrum (< 1 %, see Figure 3). Additionally, it offers an exceptionally high bond strength after oxygen plasma activation treatment to itself, glass and silicon. As a result, a promising assembly technology for the multi-chamber lens system was developed, applicable to both wafer-level as well as single device assembly.

LENS DESIGN LAYOUT

Fundamental theory

The fundamental design principals considered for the achromatic lens design will be briefly explained using the paraxial approximation [1].

The refractive power ϕ_k of a single lens k is given by its refractive index n_k and the lens bending c_k which is determined by the two radii R_{ki},

$$\phi_k = (n_k - 1)\left(\frac{1}{R_{k1}} - \frac{1}{R_{k2}}\right) = (n_k - 1)c_k. \quad (1)$$

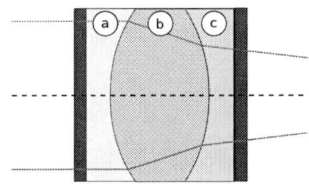

Figure 2: Schematic drawing of the proposed multi-chamber lens: a) flint-like liquid, b) crown-like liquid, and c) air. The individual chambers are separated by thin flexible polymer membranes.

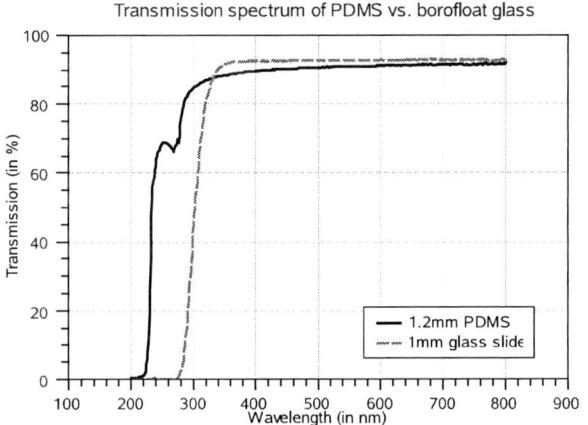

Figure 3: Measured transmission spectrum of Sylgard 184 compared to glass. Most of the loss is due to reflection at the interfaces, not due to absorption.

For a combination of lenses, the resulting lens power is then simply the sum of the individual lens powers ϕ_k, namely

$$\phi = \sum_{k=1}^{l} \phi_k. \quad (2)$$

Each lens has a dispersion expressed by the Abbe number v_k, defined by a ratio of the refractive indices at certain Fraunhofer lines,

$$v_k = \frac{n_d - 1}{\Delta n} = \frac{n_d - 1}{n_F - n_C}. \quad (3)$$

To correct chromatic aberrations, the sum of the individual dispersion characteristics v_k must add to zero, thus

$$\sum_{k=1}^{l} \frac{\phi_k}{v_k} = 0. \quad (4)$$

For the three-chamber lens system as described above (Figure 2), Equation 2 and 4 lead to

$$\phi = c_a(n_a - 1) + c_b(n_b - 1) + c_c(n_c - 1) \quad (5)$$
$$0 = c_a\Delta n_a + c_b\Delta n_b + c_c\Delta n_c. \quad (6)$$

Since the first and last refractive interface are flat cover glasses, the respective curvatures are $c_0 = c_3 = \infty$. This simplifies the equations to an equation system with two variables for the two refractive interfaces, c_1 and c_2,

$$\phi = c_1(n_b - n_a) + c_2(n_c - n_b) \tag{7}$$
$$0 = c_1(\Delta n_b - \Delta n_a) + c_2(\Delta n_c - \Delta n_b). \tag{8}$$

The described configuration allows correction of chromatic aberration, but leaves no further degree of freedom for the active correction of other imaging errors.

With an additional fourth lens chamber and thus an additional refractive interface, one more degree of freedom in the resulting under-determined system of linear equations would allow the correction of a second imaging error, such as spherical aberration. The manufacturing technology and design layout we employ here allows up to four fluidic chambers with three separating membranes and thus enables the realization of such a setup.

Since the primary spherical aberration expressed as a function of lens curvature is not a linear problem, it can only be solved numerically. In order to continue with an analytical predesign of a spherical corrected achromatic lens system, R. Kingslake has proposed the use of a *Gsum* ansatz [2].

For a tunable 3-chamber lens system, the above analytical predesign and subsequent numerical calculation using ZEMAX raytracing software suggest a certain arrangement of the optical liquids in order to minimize the spherical aberration for a tunable achromatic lens. In contrast to the classical Fraunhofer achromatic doublet setup with crown-like material (low refractive index, high dispersion) at first and flint-like material (high refractive index, low dispersion) behind, it could be shown that a reverse arrangement of the liquids is favorable [3].

Fluidic layout

To tune the lens power and simultaneously correct both mono- and polychromatic aberrations, the hydrostatic pressures within the individual chambers must be adjusted independently. Currently this is done by means of external pressure controlling; however, the on-chip microfluidic structures have been designed to accommodate a later integration of miniaturized pressure sensors and actuator components.

Figure 4 shows a schematic drawing of a 4-chamber stack wherein each of the fluidic chambers can be pressurized individually using channels accessed from the top side by vias positioned around the lens aperture. The lens stack can thus be combined with pressure transducers either integrated into each individual lens element or externally attached onto the topside.

PROCESS TECHNOLOGY

The fluidic channels and the lens cavity itself are defined by two-step silicon micro-machining. Figure 5 shows both RIE-only and combined KOH-RIE processes. The DRIE variant offers more layout options while the process with a KOH pre-etch is cheaper.

Though technically feasible [4], the spin-coating of PDMS on the substrate and using the PDMS membranes as etch-stop layers for the definition of the apertures is

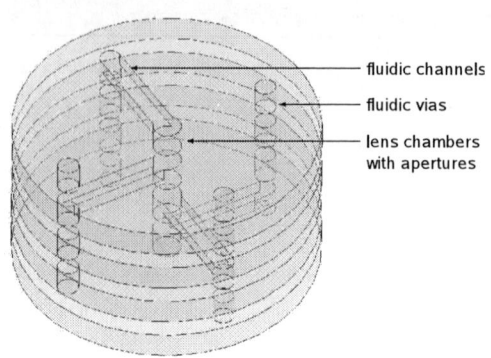

Figure 4: Schematic explosion stack layout for the multi-chamber system. Each of the four vias around the centered lens aperture connects to an individual chamber for actuation.

not advisable since the selected polymer is very sensitive to plasma damage. To circumvent this, the PDMS membrane is bonded to the silicon lens frame after the etch process by means of plasma-activated bonding [5, 6]. Proper plasma treatment parameters for the chosen silicone have been developed so that strong covalent Si–O–Si bonds can be formed. The membranes are formed in a separate spin-coating process on substrates which allow for easy membrane liftoff: Teflon (PTFE) coated glass and plastics substrates (PP, PC, and PMMA) have been shown to be suitable. Soaking in a mild organic solvent leads to an easy detachment of the membranes due to the reversible swelling of the polymer. The single lens cavities with the bonded membranes are then tempered to assure mechanical stability of the PDMS [7].

The subsequent stack assembly is realized in the same glueless and reliable plasma bonding technology which is schematically illustrated in Figure 5. A resulting 3-chamber prototype is shown in Figure 6 from the top and the bottom [8], while Figure 7 shows a cross-sectional view of the devices.

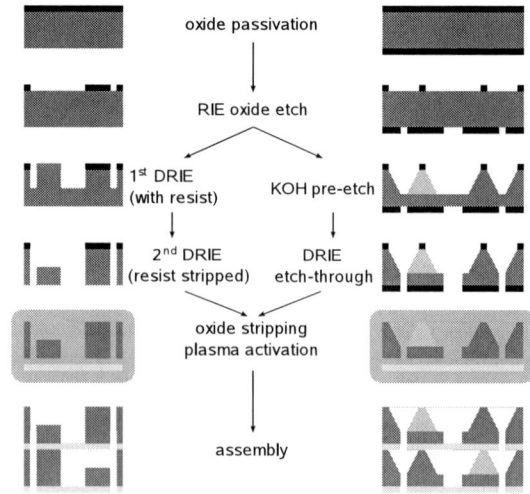

Figure 5: Schematic of the two process variants: Two step deep reactive ion etching (DRIE) micro-machining on the left vs. combined KOH-DRIE micro-machining on the right.

Figure 6: Photograph showing both sides of first prototypes of the silicon micro-machined devices. The KOH etch grooves can be clearly seen on the left.

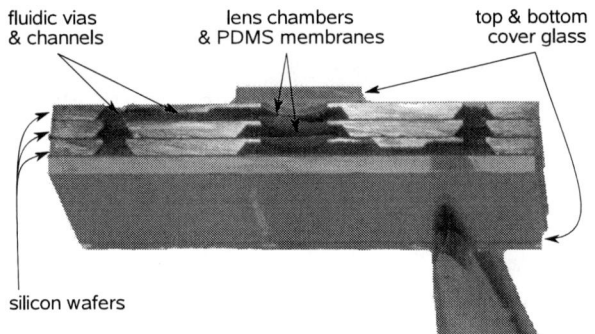

Figure 7: Photograph showing a diced 3-chamber lens system. The fluidic connections through the whole stack can be clearly seen for the top and bottom chamber.

RAYTRACING SIMULATIONS

Due to the flexible design and process technology employed here, an aperture stop can also be placed inside the lens chambers, thereby allowing a reduction of distortion. Distortion implies that structures in the object and image planes are no longer geometrically similar. Figure 8 shows the distortion for different positions of the aperture stop, as simulated by raytracing.

For a solid glass lens, the aperture stop can only be placed in front of or behind the lens, leading to significant distortion in both cases. For this stacked, membrane-based lens system, this distortion can be minimized by placing the aperture stop inside the lens. Optimum position and size of the stop, and hence optimized optical performance, can easily be found by ray-tracing methods. Such an arrangement is not possible in conventional optical systems and should enable new design options for imaging systems on the micro-scale (Figure 11).

Raytracing simulations show an optimum position of the aperture stop inside the lens system with minimized distortion (Figure 8). The lens system will unfortunately become thicker by adding this aperture stop using the current technology unmodified, which somewhat counteracts the benefit of it. However, wafer thinning processes and

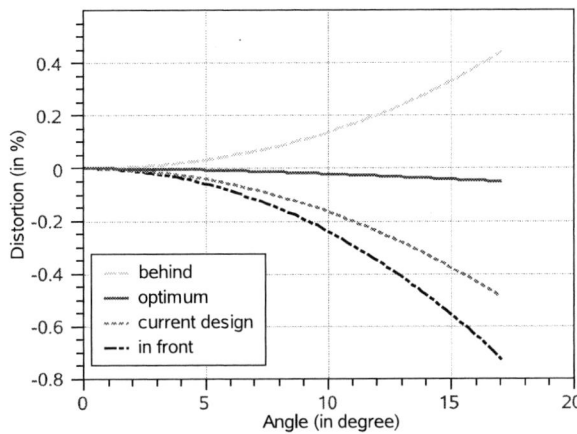

Figure 8: Calculated amount of distortion for different aperture stop positions. The angle includes the optical axis and the image ray.

an appropriate design of the aperture element make such a realization possible.

MEASUREMENT TECHNIQUES

The monochromatic transmission wavefront errors of the stacked membrane lens system are measured using a Mach-Zehnder interferometer at a wavelength of 633 nm (Figure 9). The quality of the lens stacks is characterized quantitatively by Zernike polynomial analysis. Polychromatic wavefront errors may subsequently be measured using an additional wavelength of 441 nm. The longitudinal chromatic aberration can be extracted from the defocus terms of the wavefront errors at these wavelengths. Figure 10 shows first measurements of a 3-chamber lens stack. Low-order aberrations such as coma and astigmatism can be minimized by improving the fabrication process and stack assembly.

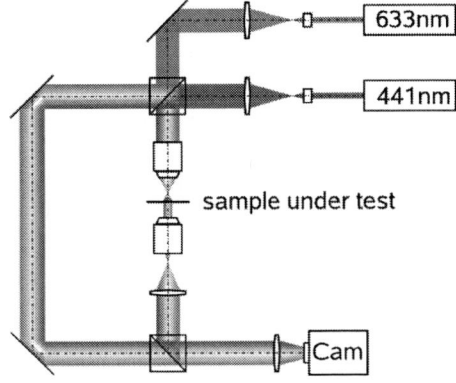

Figure 9: Schematic of the dual-wavelength interferometer used for lens testing.

CONCLUSION AND OUTLOOK

An oxygen plasma bonding technique has been successfully combined with silicon micro-machining for the fabrication and assembly of liquid multi-chamber microlenses. No adhesive or glue is required, the bond is irreversible, and the cavities can be aligned accurately prior to the bonding.

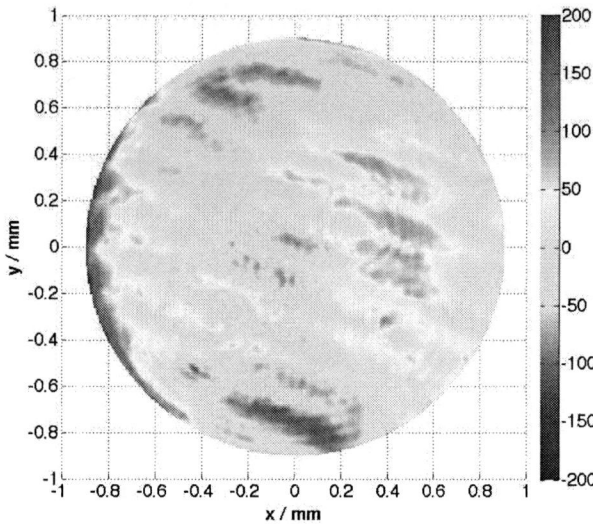

Figure 10: First measurements of the residual wavefront error of a 3-chamber system after subtraction of coma, astigmatism and spherical aberrations. The residual stripes are artefacts due to a non-optimized interferometric setup.

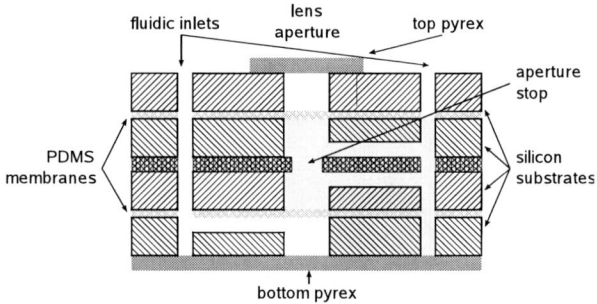

Figure 11: Schematic cross section of a 3-chamber lens system with an aperture stop included in the central lens element.

The resulting devices withstand fluid pressures up to 10^5 Pa before any delamination occurs.

The approach allows a flexible combination of optical parts such as aperture stops together with lenses in a single device. It also enables the integration of actuation principles such as piezo displacement elements or thermopneumatics [9] within the encapsulated assembly in order to control the optical performance by a complete and integrated system.

ACKNOWLEDGMENTS

The authors gratefully acknowledge financial support from the Landesstiftung Baden-Württemberg within the framework of the project MEMLENS. Fundamental PDMS ideas and research were contributed by Dr. Armin Werber. Additional thanks to Bernd Aatz, the IMTEK clean room service team, and the Laboratory for Biomedical Microtechnology for their invaluable assistance and equipment.

REFERENCES

[1] Reichelt, Stephan and Hans Zappe: *Design of spherically corrected, achromatic variable-focus liquid lenses.* Optics Express, 15(21):14146–14154, 2007.

[2] Kingslake, Rudolf: *Lens Design Fundamentals.* Academic Press, 1978.

[3] Waibel, P., D. Mader, D. Lämmle, A. Seifert, and H. Zappe: *Simulation and characterization of tunable achromatic micro-lenses.* In *Proc. IEEE-LEOS Optical MEMS*, W2.4, pages 94–95, 2008.

[4] Werber, Armin and Hans Zappe: *Tunable microfluidic microlenses.* Applied Optics, 44(16):3238–3245, June 2005.

[5] Jo, Byung Ho, Linda M. Van Lerberghe, Kathleen M. Motsegood, and David J. Beebe: *Three-Dimensional Micro-Channel Fabrication in Polydimethylsiloxane (PDMS) Elastomer.* Journal of Microelectromechanical Systems, 9(1):76–81, March 2000, ISSN 1057-7157.

[6] Agarwal, M., R. A. Gunasekaran, P. Coane, and K. Varahramyan: *Polymer-based variable focal length microlens system.* Journal of Micromechanics and Microengineering, 14:1665–1673, December 2004.

[7] F. Schneider, T. Fellner, J. Wilde and U. Wallrabe: *Mechanical properties of silicones for MEMS.* JOURNAL OF MICROMECHANICS AND MICROENGINEERING, 18(6):1–9, April 2008.

[8] Mader, D., A. Seifert, and H. Zappe: *Fabrication of aberration-corrected tunable micro-lenses.* In *Proc. IEEE-LEOS Optical MEMS*, Tu3.2, pages 60–61, 2008.

[9] Werber, Armin and Hans Zappe: *Thermopneumatically actuated, membrane-based micromirror devices.* Journal of Micromechanics and Microengineering, 16(12):2524–2531, 2006.

OPTICAL MICROMIRROR ACTUATION USING THERMOCAPILLARY EFFECT IN MICRODROPLETS

R. K. Dhull, I. Puchades, L. Fuller and Y. W. Lu
Rochester Institute of Technology, New York, USA

ABSTRACT

This paper presents a simple means that utilizes surface tension gradient to cause droplet deformation, and to tilt micro-objects. Thermocapillary or Marangoni effect, and contact angle hysteresis are employed to control the droplet shape and position. The device consists of a microplate placed onto a microdroplet, and can produce a 6.5° tilting angle when actuated at 30 V. It shows the potential applications in scanning micromirror and display technology.

INTRODUCTION

The emergence of MEMS technology has enabled pathbreaking applications in the fields of science and engineering. For example, various micromirrors, made by MEMS fabrication, have been demonstrated. These mirrors are mostly actuated by using thermo-pneumatic [1], electrostatic [2] and temperature gradient [3] techniques. Many of them need complicated microfabrication processes, and contain beam structures, which may suffer degradation due to torsion or contact. To address these issues, micromirror actuation, in conjunction with microfluidics, has gained commendable interest recently. Several mechanisms, such as electrowetting-on-dielectric (or EWOD) and electrostatic forces, have been exploited to control the droplet shapes so that the micromirrors on the top can be moved or tilted [4-6]. However, they often require a high actuation voltage, or suffer from the water droplet evaporation problems.

Meanwhile, the thermocapillary forces due to the low actuation voltages have drawn great attention in the microfluidic applications [7-8]. Our device has exploited the thermocapillary effect to change the droplet shape and tilt the micromirror on the top. Contact angle hysteresis is utilized to increase the stability of the device. The device fabrication and operation is relatively simple while the actuation voltage is also low.

THERMOCAPILLARY EFFECT OR MARANGONI EFFECT

Thermocapillary effect or Marangoni effect can be explained by the variation of interfacial tension at a fluid-fluid interface caused by temperature gradients. Since the surface tension of a liquid is a linear function of temperature, it can be expressed by:

$$\sigma_{lv}(T) = \sigma_{lv0} - \varepsilon(T - T_0) \qquad (1)$$

where σ_{lv0} is the liquid-vapor surface tension at room temperature T_0 and ε is the temperature coefficient of

surface tension. Therefore, when a temperature gradient is presented, the surface tension in the warmer region decreases, as most of the liquid remains at a constant temperature. The liquid in the warmer region is pulled towards the colder region, where the surface tension is higher, to minimize the total surface energy. Such a mechanism can change the contact angle of the liquid in advancing and receding ends, making the droplet to move or changing its shape.

Taking advantage of the thermocapillary effect, our device in Figure 1 utilizes Marangoni flow, and contact angle hysteresis to change the droplet shape and to maintain the droplet position.

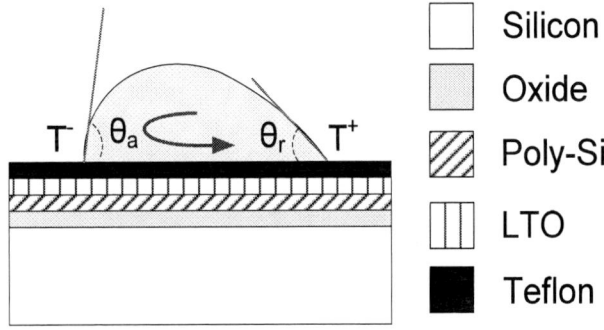

Figure 1: Schematic of the thermocapillary actuation or Marangoni effect. Liquid moves from the hot (T^+) to the cold (T) side on a hydrophobic substrate.

As shown in Figure 1, while the right side of the liquid is heated, the Marangoni flow tends to actuate the droplet to the left. Polysilicon microheaters are used to create sufficient temperature and surface tension gradient.

The difference between the advancing angle (θ_a) and the receding angle (θ_r) of the droplet determine the contact angle hysteresis (H):

$$H = [\max(\theta_a) - \min(\theta_r)] \qquad (2)$$

This hysteresis is the measure of the droplet's reluctance to any motion. The larger the contact angle hysteresis, greater is the force required to move the liquid meniscus, and the droplet. Therefore, the choice and realization of the liquid-substrate configuration are the key points of the experiment. Table 1 lists our liquid selections and their material properties. Contact angle hysteresis measurements for different liquids are made using VCA Optima contact angle analysis equipment. Mustard oil is chosen for better device stability and performance because of its high viscosity and large hysteresis. Although DI water shows high hysteresis, it has evaporation problems due to the heat, which prevents its use in our device. Moreover, the oil droplet has a lower vapor pressure that eliminates the evaporation problem as

978-1-4244-2977-6/09 $25.00 © 2009 IEEE

seen otherwise in aqueous-solution based actuation mechanisms.

Table 1: Liquid choices for the thermocapillary device and their properties (from measurement).

Liquid	Viscosity (Pa·s @ 25°C)	Contact angle on Teflon©	Contact angle hysteresis (H)
Mustard oil	70	82.4°	4.6°
Silicone oil	9.3×10^{-3}	51.4°	3.2°
Glycerin	1.5	108.5°	4.5°
DI water	8.9×10^{-4}	120°	5.5°

DEVICE DESIGN AND SIMULATION

The layout of the device is designed using Mentor Graphics software. It requires a total of four mask layers as shown in Figure 2 (a), namely: poly, contact, metal and Teflon. Polysilicon microheaters are designed in a circular shape to completely cover the droplet circumference. Each microheater (430 Ω) thus heats up one-fourth of the droplet volume. Figure 2 (b) and (c) show the top view of the droplet actuation without any micromirror on the top. A 15 V actuation voltage was applied to the first microheater causing Marangoni flow in the south-west direction.

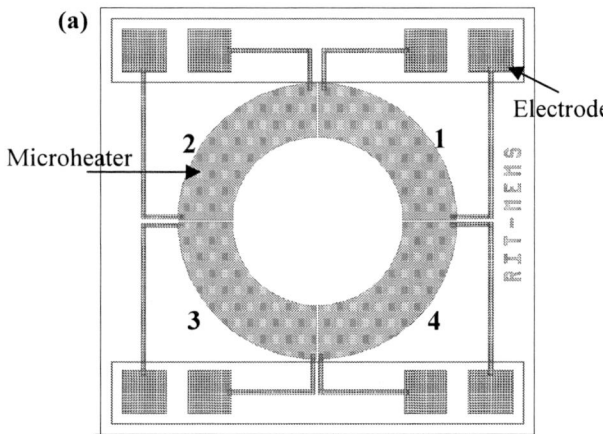

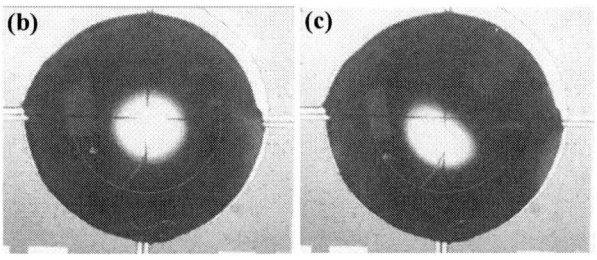

Figure 2: (a) Layout of the device: microheaters (1, 2, 3 and 4) and metal electrodes. Teflon AF 1600 is coated on top of the chip and etched away from the top of the electrodes. (b) and (c) show the top view of an oil droplet sitting on the device without microplate: (b) before actuation, and (c) when 15 V is applied to the microheater 1.

Numerical calculations were first conducted to evaluate the device performance and to realize the relationship between the applied voltages and the surface temperature on the chip. The simulation was performed using electrothermal simulator in Coventor software suite. As shown in Figure 3, a temperature difference of ~2 K was obtained on the chip surface, when a 30 V actuation voltage was applied, which was sufficient for actuating the droplet [9]. An infrared thermal camera was also used to validate the results.

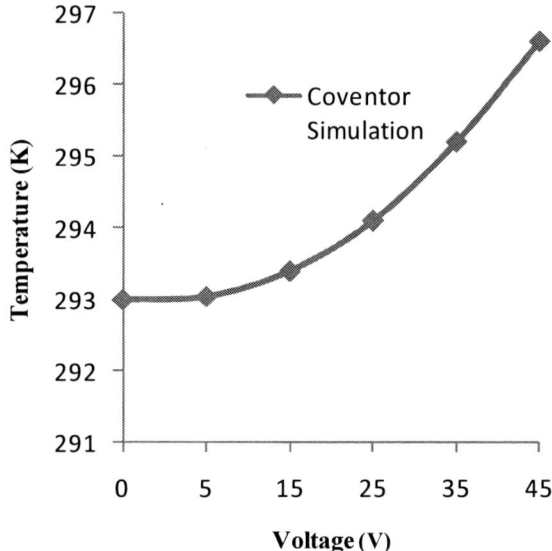

Figure 3: The temperature on the surface of the chip with the applied voltages. Simulation results were obtained using Coventor software, and validated by an infrared thermal camera.

FABRICATION

Our process flow is based on the RIT CMOS/MEMS processes, and it is illustrated in Figure 4. A silicon wafer was first deposited with a 5000 Å thick thermal oxide and 6000 Å thick LPCVD polysilicon layer. Spin-on-glass (N-250) was coated and followed by thermal diffusion to increase the conductivity of the polysilicon layer. The first mask (Poly) was used to define the polysilicon microheaters using photolithography, and RIE dry etching. An 8000 Å thick LTO layer was then deposited using LPCVD. Subsequently, the second mask was employed to define the contact cuts. A 1 μm aluminum layer was sputtered and patterned to create electrodes. A Teflon layer of 2000 Å thickness was spin-coated and patterned to increase the hydrophobicity of the surface, which increased the contact angle of the oil droplet.

The wafer was then diced into 4 x 4 mm^2 chips, and subsequently an oil droplet of 1.2 μL, equivalent to a 2 mm base diameter, was placed in the center of the chip using a computer controlled syringe. Finally, for our preliminary test, a 1 x 1 x 0.1 mm^3 Polypropylene (PP) microplate was placed onto the droplet using a micromanipulator.

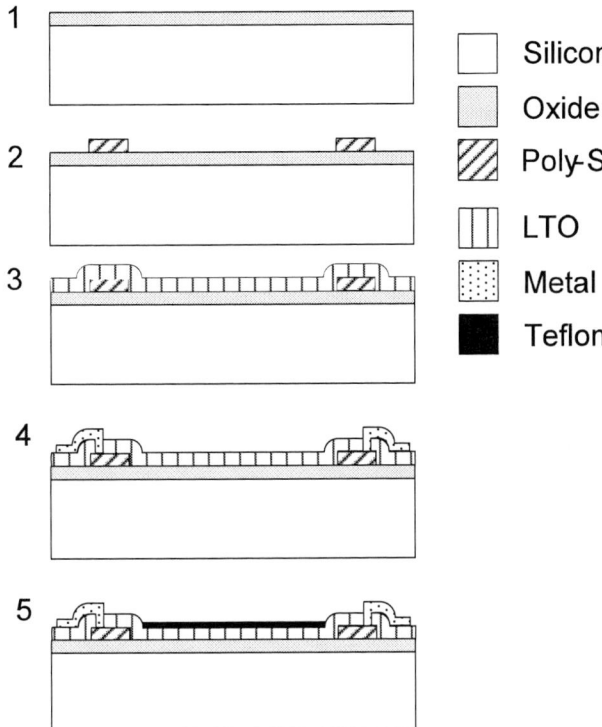

Figure 6 shows the measurement result. The tilting angle of the microplate was found to be linearly dependent on the applied voltage. As the voltage was increased, the advancing (θ_a) angle of the droplet increased, effectively increasing the microplate tilt angle. The receding angle (θ_r) was found to be increasing, instead of decreasing, due to the presence of the microplate. The advancing angle of the droplet increased from 79° to 86° whereas the receding angle increased from 78.9° to 81.8°.

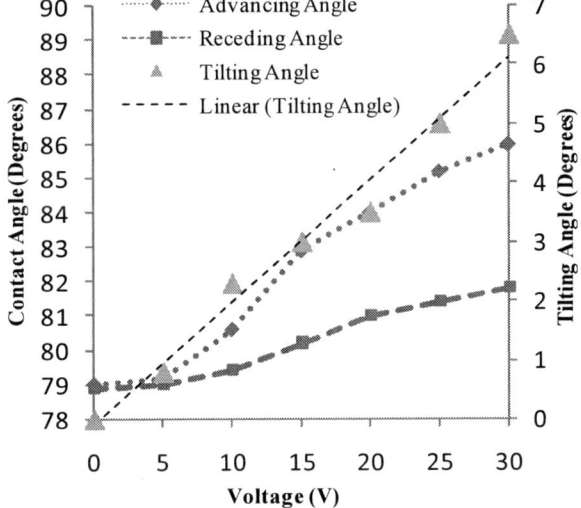

Figure 4: Process flow: (1) Thermal oxide on top of silicon substrate. (2) Poly layer is patterned to create microheaters. (3) LTO deposition. (4) LTO is etched for creating contact cuts. Then aluminum is sputtered and patterned for defining electrodes. (5) Finally, a Teflon layer is spin-coated and patterned.

Figure 6: The dependence of the advancing angle and receding angle of the droplet, and the tilting angle of the microplate at different applied voltages. The linear curve is drawn to fit the experimental result.

RESULTS AND DISCUSSION

To find the relationship between the actuation voltage, the contact angles of the droplet, and the tilting angles of the plate, the measurement setup in Figure 5 was utilized. A CCD camera was used to capture video images of the droplet and the microplate. The pulsed DC voltage was generated by employing a transistor switch and a function generator (Agilent 33250A). The contact angles of the microdroplet were calculated with an accuracy of ± 0.1 degree using a goniometer (VCA Optima).

The snap shots in Figure 7 depict the operation of the actual device. Figure 7 (a) shows the normal state of the device when no voltage is applied, whereas Figure 7 (b) shows a micromirror tilt angle of 6.5° at 30 V, 7 Hz frequency with 50% duty cycle supply voltage. Once the micromirror was actuated, lesser voltage was needed to maintain the micromirror in the tilted state thus lowering the power requirements of the device. The frequency response of the device can be improved by selecting a liquid with a higher surface tension, and by decreasing the droplet diameter [10]. Furthermore, our device can work as a scanning micromirror with two degrees-of-freedom (2-DOF) motions by individually addressing each of the microheaters (1, 2, 3, and 4) in an appropriate order.

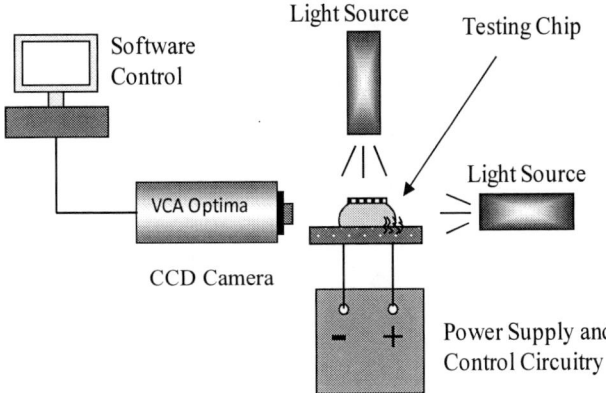

Figure 5: The schematic of the device test setup used for measuring the droplet advancing, receding and the micromirror tilt angle.

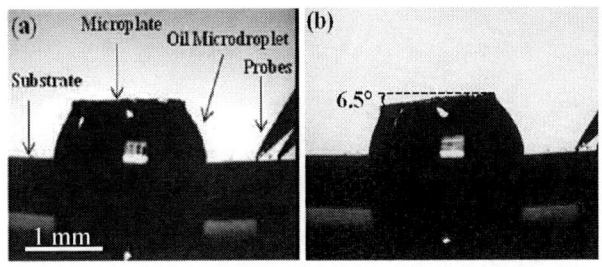

Figure 7: (a) Microplate on top of the oil droplet @ 0V. (b) Microplate tilted 6.5° when a 30 V actuation voltage was applied to the microheater at right.

CONCLUSIONS

The successful design, fabrication and operation of the device have been presented, showing its potential in the micromirror applications. The device is made by our existing RIT CMOS/MEMS processes, and actuated at a lower voltage with a comparable tilting angle to the other mechanisms. A maximum of 6.5° micromirror tilt is obtained at 30 V, 7 Hz frequency. Thus, it provides an opportunity of a complete microsystem on a single chip. The microplate in the current configuration can be coated with a reflective material (*e.g.* Aluminum), and replaced by a thinner plate (e.g. Silicon) for further improvement.

ACKNOWLEDGEMENTS

This research work is supported by Texas Instruments/Harvey Award. The authors would like to thank the members from the Semiconductor and Microsystems Fabrication Laboratory (SMFL) for their technical support, and Mr. Pow from the Center of Imaging Science at RIT for IR camera measurement.

REFERENCES

[1] A. Werber and H. Zappe, "Thermo-pneumatically actuated, membrane-based micro-mirror devices", *Journal of Micromechanics and Microengineering*, Oct 2006, pp. 2524-2531.

[2] Y. Yoshihata, N. B. Khiem, A. Takei, E. Iwase, K. Matsumoto and I. Shimoyama, "Scanning micromirror using deformation of a Parylene-encapsulated liquid structure", in *Proc. IEEE MEMS Conference*, Tucson, AZ, USA, Jan 2008, pp. 770-773.

[3] D. Elata and R. Mahameed, "A temperature-gradient driven micromirror with large angles and high frequencies", in *Proc. IEEE MEMS Conference, Istanbul*, Turkey, Jan 2006, pp. 850-853.

[4] H. Kang and J. Kim, "EWOD (Electrowetting-on-dielectric) actuated optical micromirror", in *Proc. IEEE MEMS Conference*, Istanbul, Turkey, Jan 2006, pp. 742-745.

[5] J. Gong, G. Cha, Y. S. Ju and C. J. Kim, "Thermal switches based on coplanar EWOD for satellite thermal control", in *Proc. IEEE MEMS Conference*, Tucson, Jan 2008, pp. 848-851.

[6] A. Takei, *N. B-Khiem*, E. Iwase, K. Matsumoto and I. Shimoyama, "Liquid motor driven by electrowetting", in *Proc. IEEE MEMS Conference*, Jan 2008, pp. 42-45.

[7] A. A. Darhuber, J. P. Valentino, S. M. Troian and S. Wagner, "Thermocapillary actuation of droplets on chemically patterned surfaces by programmable microheater arrays", *Journal of MEMS,* Dec 2003, vol. 12, pp. 873-879.

[8] Z. Jiao, X. Huang, N. T. Nguyen and P. Abgrall, "Thermocapillary actuation of droplet in a planar microchannel", in *Journal of Microfluidics and Nanofluidics,* Aug 2008, vol. 5, no. 2, pp. 205-214.

[9] J. Z. Chen, S. M. Troian, A. A. Darhuber and S. Wagner, "Effect of contact angle hysteresis on thermocapillary droplet actuation", *Journal of Applied Physics*, Dec 2004, vol. 97, no. 1, pp. 014906.

[10] Z. Wan, H. Zeng and A. Feinerman, "Reversible electrowetting of liquid-metal droplet", *Transactions of the ASME*, April 2007, vol. 129, pp. 388-394.

DUAL-AXES CONFOCAL MICROLENS FOR RAMAN SPECTROSCOPY

C.P.B. Siu[1], H. Wang[2], H. Zeng[2], and M. Chiao[1]

[1]The University of British Columbia, CANADA
[2]British Columbia Cancer Research Centre, CANADA

ABSTRACT

This paper presents a novel dual-axes scanning microlens for *in vivo* near infrared (NIR) Raman spectroscopy. *In vivo* Raman spectra of fibrosarcoma tumor skin and normal mouse skin measured from the microlens are reported. Out-of-plane (Z-axis) and transverse (X-axis) actuations are demonstrated using electrostatic and electromagnetic forces, respectively. The microlens achieves 120 μm scan in Z-axis at 286 Hz, and 163 μm scan in X-axis at 480 Hz.

INTRODUCTION

Raman spectroscopy is a powerful spectroscopic technique in biochemistry, virology and pathology diagnostics [1]-[4]. It provides an accurate, rapid and non-invasive diagnostic scheme down to molecular scale [1]-[5]. The principle of Raman spectroscopy is based on the quantum effect that results from energy transition of the scattering molecule to a virtual excited state. The scattering molecule is then retuned to a higher or lower vibration state with the emission of a photon in altered frequency from the incident photon [4]. This inelastic scattering is named as Raman Effect to recognize the merit of experimental discovery by C. V. Raman in 1928 [2]. Figure 1 shows a schematic energy diagram of Raman scattering and Rayleigh scattering. Unlike the Rayleigh scattering that the elastic scattered photon has the same frequency as the incident photon, the frequency difference between incident photon and the scattering photon in the Raman scattering gives a Raman shift of the molecular structure from the specimen [4]. The Raman shift is defined as

$$R_s = \nu_i - \nu_s \qquad (1)$$

where R_s is the Raman shift, ν_i is the frequency of incident photon, and ν_s is the frequency of the emitted photon. Since Raman scattering is independent of the frequency of incident photon, the spectrum of Raman shift provides a unique fingerprint spectrum for molecular identification [1]. Statistically, the quantity of Raman scattered photon and incident photon are in the ratio of 10^{-10} [6]. Therefore, to utilize the Raman spectroscopy, the incident photon beam need to be converged on the target specimen in order to generate and achieve enough Raman scattered photons for spectroscopic analysis [2]-[4].

Biological application of Raman spectroscopy for cellular analysis was developed since 1986 [7]. *In vivo* confocal NIR Raman microscopy enables assessment of pathological tissue at precancer stage without taking biopsy samples. Malignant cellular and nuclear structure change can be identified through Raman spectrum inspection [2], [8], [9]. However, the existing scanning lens systems are bulky, expensive and inconvenient for routine clinical use. Microfabricated scanning microlens is promising but has never been demonstrated in Raman spectroscopy. Recent confocal microlens demonstrated multi-axes scanning by stacking 3 uniaxial electrostatic microlens scanners together, and reported 55 μm and 100 μm, out-of-plan and transverse scan, respectively [10]. However, Specific combination of lenses is required to reduce the accumulated aberrations in stacking lens system. Recently, we developed a magnetic scanning 1-D microlens with 125 μm transverse scan [11], [12]. In this paper, we demonstrate a new architecture of dual-axes confocal microlens using a single lens with large scanning range for Raman spectroscopy.

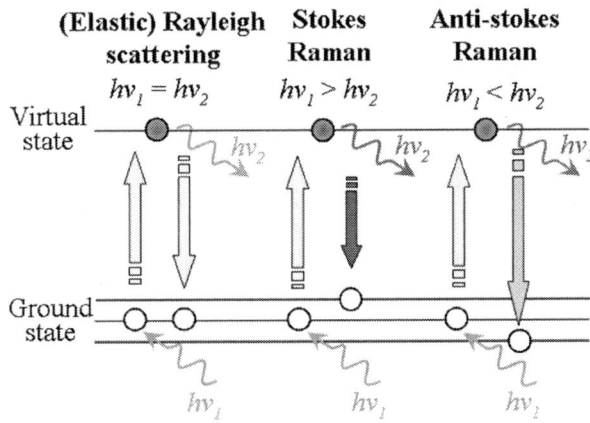

Figure 1: Schematic energy diagram of scattered photons: (i) Rayleigh (elastic) scattering, (ii) Raman stokes scattering (molecule absorbs energy from incident photon), and (iii) Raman anti-stokes scatting (molecule release extra energy to the scattered photon).

DESIGN

Figure 2 shows the microlens fabricated using a foundry service, METAL MUMPS®. A nickel structure, composes of gimbaled electrode, microlens platform, V-shaped springs, and serpentine springs, is formed by electroplating on a silicon substrate. A buried polysilicon electrode is encapsulated by two silicon nitride layers which provide over 400 V short-circuit free protection [13]. A Polydimethylsiloxane (PDMS) lens of 1.25 mm in diameter and 4.2 mm in effective focal length was added on the nickel platform after sacrificial release.

Figure 3a shows the working principle of the microlens. The Z-axis scan utilizes electrostatic potential between the gimbaled nickel electrode and a buried polysilicon electrode underneath the substrate surface.

978-1-4244-2977-6/09 $25.00 © 2009 IEEE

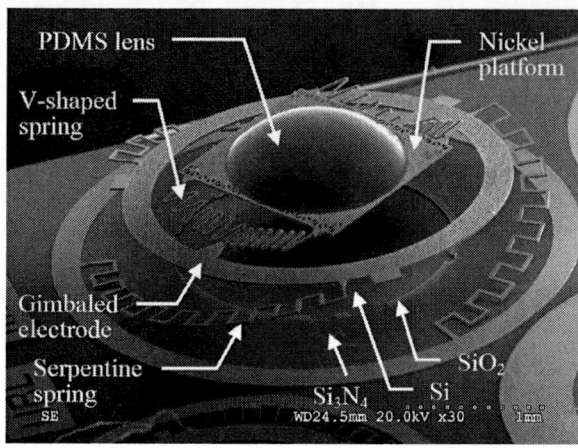

Figure 2: SEM of dual-axes scanning microlens.

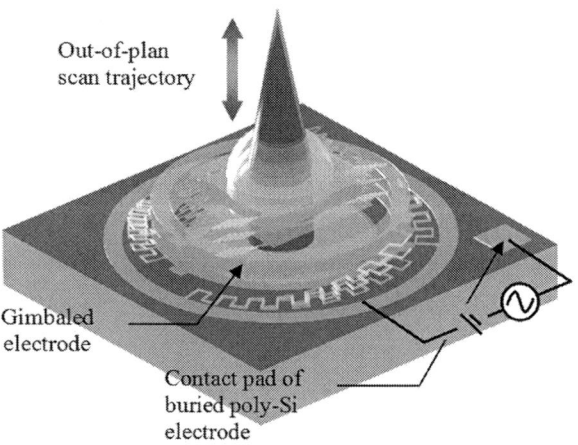

(a) Out-of-plane scanning

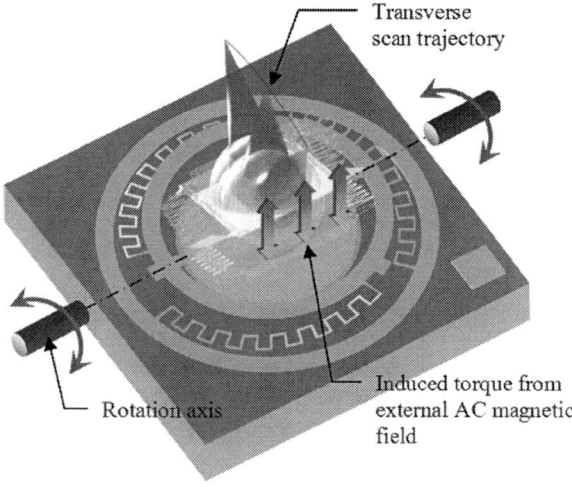

(b) Transverse scanning

Figure 3: Dual-axes scanning principle of the microlens.

The X-axis scan is achieved by an induced magnetic torque on the nickel platform from an external AC magnetic field (Figure 3b). This magnetic torque is a functional integral of the area of the nickel platform and the angle between the normal area to the applied ac magnetic field [14].

FABRICATION

Figure 4 shows the schematic diagram of the fabrication process. The nickel structure is fabricated on a double side polished n-type (100) silicon wafer. First, a 2 μm-thick silicon dioxide (SiO$_2$) is grown on a silicon wafer followed by a 200 nm phosphosilicate glass (PSG). The SiO$_2$ serves as an insulation layer and provides residue stress compensation for the later silicon nitride depositions. The PSG layer defined the optical entrance size for the optical bulk silicon etch. Then a 0.35 μm-thick low-stress silicon nitride (Si$_3$N$_4$) is deposited (Figure 4(a)). Afterwards, a 0.7 μm-thick polysilicon (poly-Si) layer is deposited on the Si$_3$N$_4$ and is patterned with reactive ion etching to form the buried electrode. Another layer of 0.35 μm Si$_3$N$_4$ is deposited. The two Si$_3$N$_4$ layers form the poly-Si electrode encapsulation, whereas the Si$_3$N$_4$ on the contact pad area are removed (Figure 4(b)). A 1.1 μm-thick PSG layer is deposited and annealed at 1050°C for 1 hour. It is then patterned for the sacrificial metal release. Anchoring metals consist of 10 nm-thick Cr and 25 nm-thick Pt layers are deposited on the polysilicon pads by a lift-off process (not showing in the figure). The wafer surface is then coated and patterned with a thick photoresist as the electroplating stencil. A 20 μm-thick Nickel and 0.5 μm-thick gold are then electroplated and filled inside the stencils (Figure 4(c)). The thick photoresist, the two PSG sacrificial layers and the silicon dioxide underneath the silicon nitride pupil are released by solvents and 49% hydrofluoric acid respectively. Post-fabrication steps include an optional bulk silicon etch to remove the silicon under the aperture, manufacturing of PDMS microlens, nickel platform elevation, and assembling (Figure 4(d)).

PERFORMANCE

Figure 5 shows the frequency response of the microlens at Z-axis and X-axis scans measured by a laser Doppler vibrometer (Polytec OFC-5000, Displacement decoder DD-200). The maximum range of Z-axis scan is 120 μm at 286 Hz using a 28 V$_{DC}$ plus 56 V$_{AC}$ sinusoidal input. The maximum range of the X-axis scan is 163 μm at 480 Hz driven by a 34.5 mT external magnetic field strength is shown in Figure 6. This external magnetic field is generated by a solenoid (S-17-85-39QH, Magnetic Sensor system) with the centre core replaced by a 1 mm diameter soft iron bar in order to generate a small scale local magnetic field on the microlens. The tip of the extruded iron core from the solenoid is orientated 60° and 1 mm above the midpoint of one side of the nickel platform edge. The magnetic field is measured by a Gauss / Teslameter (F.W. BELL model 6010, Sypris Test & Measurement) and sensor of hall probe is placed at the same position as the microlens platform in the static state. Through triggering the electrical inputs for the electrostatic and electromagnetic forces, coupling biaxial scan of the microlens can be achieved. A 2 Hz interlaced scan is achieved with the X-scan operated at resonant mode and the Z-scan manipulated at 2 Hz. A 2 Hz interlaced scan is achieved with the X-scan operated at resonant mode and the Z-scan manipulated at 240 Hz. To increase the scan range of the Z-scan at non-resonant state, a higher applied voltage can be provided between gimbaled electrode and the polysilicon electrode. When the microlens is actuated at resonance in Z-axis, the X-axis scan can be toned close to the 3rd harmonic frequency at 1430 Hz for faster interlaced scan performance.

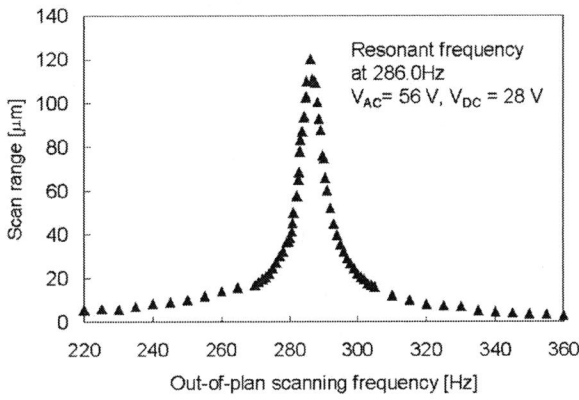

Figure 5: Out-of-plan scanning frequency response

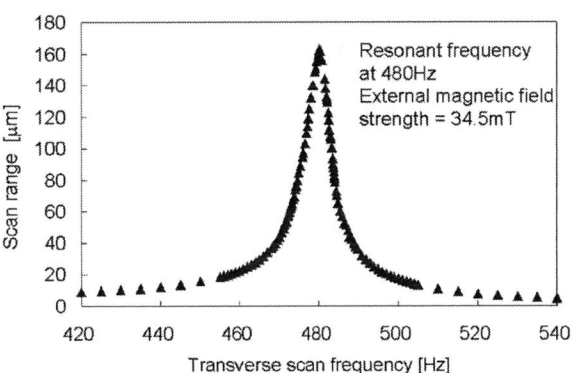

Figure 6: Transverse scanning frequency response.

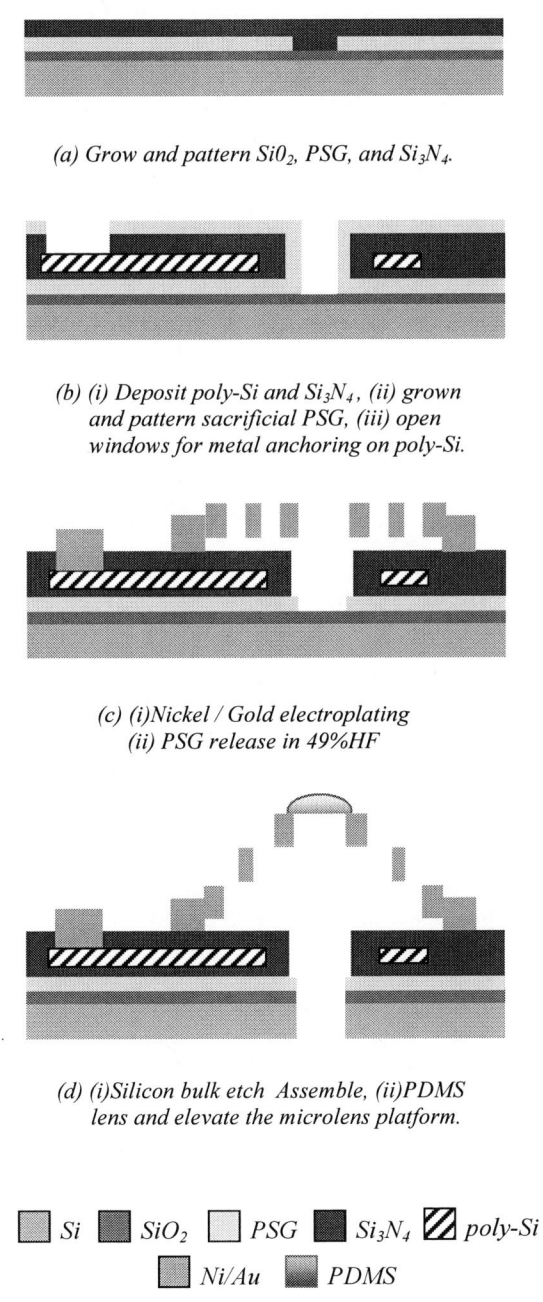

(a) Grow and pattern SiO_2, PSG, and Si_3N_4.

(b) (i) Deposit poly-Si and Si_3N_4, (ii) grown and pattern sacrificial PSG, (iii) open windows for metal anchoring on poly-Si.

(c) (i)Nickel / Gold electroplating (ii) PSG release in 49%HF

(d) (i)Silicon bulk etch Assemble, (ii)PDMS lens and elevate the microlens platform.

Si SiO_2 PSG Si_3N_4 poly-Si
Ni/Au   PDMS

Figure 4: Schematic fabrication process of the microlens

Figure 7 shows Raman spectra measured using a 785 nm NIR Raman spectroscopic setup integrated with the microlens. The intensity of the focused laser spot after the microlens is 35 mW measured by a laser power meter (PM100, Thorlabs). The incident laser (785nm, SDL XC30) is guided through a series of band-pass filters, NIR reflection mirrors, dichroic mirror, lens before entering to the microlens. The Raman scatterings from the sample are received via the microlens synchronously. A dichroic mirror split out the Raman scatterings to an optical fiber (100 μm, Ocean Optics) and transmits to a spectrometer (HoloSpec-f/2.2 NIR, resolution: 8cm^{-1}, Kaiser optical system) for analysis. A murine tumor model is analyzed in the Raman spectroscopic measurement and compared with the normal skin tissue. The BALB/c mice (Simons laboratory) were implanted with MCA 205 fibrosarcoma cells into the subcutaneous region of the lower back. The acquisition time for the normal skin and the tumor skin measurement are 100 s and 150 s, respectively.

The Raman bands observed in both fibrosarcoma tumor and the normal skin at 1209 cm^{-1}, 1270 cm^{-1}, 1307 cm^{-1}, 1558 cm^{-1}, 1655 cm^{-1} are presumably attributed to the ν(C-C$_6$H$_5$) phenylalanine, (ν(CN)) and (δ(NH)) amide III, δ(CH$_2$) twisting/wagging of lipid/collagen, ν(CN) and δ(NH) amide II, and ν(C=O) amide I [15]. The increase of the intensities in the 1209 cm^{-1}, 1270 cm^{-1}, 1307 cm^{-1}, and 1558 cm^{-1} bands as well as the decrease in the 1655 cm^{-1}

band reflect the cellular changes associated with the fibrosarcoma transformation from the mouse normal skin tissue.

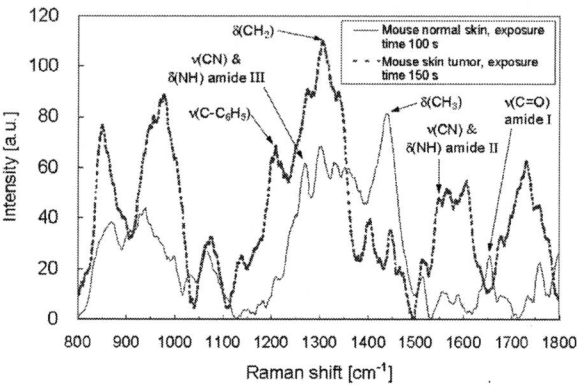

Figure 7: In vivo Raman spectra acquired from microlens integrated Raman system with mouse skin samples: (i) normal skin, and (ii) fibrosarcoma tumor skin.

CONCLUSION

In this paper, a dual-axes scanning microlens for *in vivo* near infrared (NIR) Raman spectroscopy is reported for the first time. The microlens is fabricated from the standardized foundry MetalMUMPs process with a thermal setting PDMS microlens. It has both out-of-plane (Z-axis) and transverse (X-axis) scanning capabilities designed on a single device without lens stacking. The actuations are driven by electrostatic and electromagnetic forces for the Z-axis and X-axis scan, respectively. The microlens achieves 120 μm scan in Z-axis at 286 Hz, and 163 μm scan in X-axis at 480 Hz. A murine fibrosarcoma tumor *vs* the normal skin from a mouse model are investigated. Differential protein/amide group intensities are detected from the Raman spectra.

ACKNOWLEDGEMENT

The Metal MUMPs microfabrication process is provided by the CMC Microsystems, a Canadian federally incorporated non-profit corporation. This work is supported by the Faculty of Applied Science, UBC and NSERC Discovery Grant to MC. MC is supported by Canada Research Chairs Tier 2.

REFERENCES

[1] Y. W. C. Cao, R. Jin, and C. A. Mirkin, "Nanoparticles with Raman Spectroscopic Fingerprints for DNA and RNA Detection," *Science*, vol. 297, pp.1536-1540, 2002.

[2] E. B. Hanlon, R. Manoharan, T. W. Koo, K. E. Shafer, J. T. Motz, M. Fitzmaurice, J. R. Kramer, I. Itzkan, R. R. Dasari, and M. S. Feld, "Prospects for in vivo Raman spectroscopy," *Phys. Med. Biol.*, vol. 45, pp. R1-59, 2000.

[3] P. J. Lambert, A. G. Whitman, O. F. Dyson, and S. M. Akula, "Raman spectroscopy: the gateway into tomorrow's virology," *Virol. J.*, vol. 3, pp. 1-8, 2006.

[4] A. Mahadevan-Jansen, "Raman spectroscopy: from Benchtop to bedside," in *Biomedical Photonics Handbook*, T. Vo-Dinh, Ed. Boca Raton, Fla,: CRC Press, 2003, ch. 30.

[5] C. A. Lieber, S. K. Majumder, D. Billheimer, D. L. Ellis, and A. Mahadevan-Jansen, "Raman microspectroscopy for skin cancer detection in vitro," *J. Biomed. Opt.*, vol. 13, DOI:10.1117/1.2899155, 2008.

[6] J. F. McGilp, "Epioptics: linear and nonlinear optical spectroscopy of surfaces and interfaces", *J. Phys. Condensed Matter*, vol. 2, pp. 7985-8006, 1990.

[7] D. A. Butterfield, "Spectroscopic methods in degenerative neurological disease," *CRC Crit. Rev. Clin. Neurobiol.*, vol. 2, pp.169-240, 1986.

[8] Z. Huang, H. Lui, X. K. Chen, A. Alajlan, D. I. McLean H. Zeng, "Raman spectroscopy of in vivo cutaneous melanin," *J. Biomed. Opt.* vol. 9, pp.1198-1205, 2004.

[9] H. Zeng, A. Weiss, R. Cline, and C. E. MacAulay, "Real-time endoscopic fluorescence imaging for early cancer detection in the gastrointestinal tract," *Bioimaging*, vol. 6, pp.151-165, 1998.

[10] S. Kwon, V. Milanović, and L. P. Lee, "Vertical microlens scanner for 3D imaging," in *Tech. Digest of the Solid-state Sensor & Actuator 2002*, Hilton Head Isl., SC, June 2-6 2002, pp.227-230.

[11] C. P. B. Siu, H. Zeng, and M. Chiao, "*Magnetically actuated scanning microlens for NIR Raman spectroscopy*," in *Proc. IEEE MEMS 2007*, Kobe Japan, January 21-25, 2007 pp.735-738.

[12] C. P. B. Siu, H. Zeng, and M. Chiao, "*Magnetically actuated MEMS microlens scanner for in vivo medical imaging*," Opt. Express, vol. 15, pp.11154-11166, 2007.

[13] H. C. Cheng, H. W. Liu, H. P. Su, and G. Hong, "*Superior Low-Pressure-Oxidized Si3N4 films on rapid-thermal-nitride poly-Si for high-density DRAM's*" IEEE Electr. Device L., vol. 16, pp. 509-511, 1995.

[14] J. W. Judy, R. S. Muller, "Magnetically actuated, addressable microstructures," *J. Microelectroemch. Syst.*, vol. 6, pp.249-256, 1997.

[15] Z. Huang, H. Lui, D. I. McLean, M. Korbelik, and H. Zeng, "Raman spectroscopy in combination with background near-infared autofluorescence enhances the in vivo assessment of malignant tissues," *Photochem. Photobiol.* vol. 81, pp.1219-122, 2005.

TUNABLE SCANNING FIBER OPTIC MEMS-PROBE FOR ENDOSCOPIC OPTICAL COHERENCE TOMOGRAPHY

K. Aljasem, A. Seifert, and H. Zappe

Department of Microsystems Engineering – IMTEK, University of Freiburg, Germany

ABSTRACT

A novel 3D probe for endoscopic optical coherence tomography (OCT) based on tunable and movable MEMS components is presented. A tunable micro-lens and a 2D scanning micro-mirror are integrated into a probe to enable two-dimensional movement with simultaneous dynamic focusing of a beam onto a target. The tunable system is based on a pneumatically actuated micro-lens for the axial movement of the focus position concomitantly with the depth scan of the OCT, whereas an electrostatically actuated micro-mirror is integrated to obtain the 2D lateral scan of the beam. High resolution imaging at high scan rates is expected for the entire scan depth using this concept. Probe design, assembly, and integration into an OCT system are discussed.

MOTIVATION

OCT as a high resolution optical technique is used to provide cross-sectional imaging of biological tissue [1]. Current OCT systems are able to provide real-time sub-cellular imaging and is thus a promising emerging technique in biomedical and clinical diagnostics. Similarly to other conventional optical imaging systems, such as confocal, multiphoton, and fluorescence microscopy, OCT can penetrate only a few millimeters below the surface.

Endoscopic OCT will expand the range of potential clinical applications. Numerous research groups have developed different endoscopic OCT systems, almost all of which were devoted to the integration of MEMS scanners into endoscopic probes to obtain 2D or 3D endoscopic OCT images [2]. Although the advantage of tunability of the beam focus has previously been shown [3], only few tunable devices have been presented in an OCT configuration; but no endoscopic solution with scanning optics has emerged from these ideas. The miniaturization of endoscopically integratable tunable systems is challenging but microsystems fabrication techniques and advanced assembly technologies together with high precision mechanics can offer a convenient means to achieve a sufficiently miniaturized multifunctional endoscopic probe.

INTRODUCTION

OCT can provide 1D, 2D and 3D optical imaging, depending on the system configuration. As depicted in figure 1, OCT is based on white light interferometry using a broadband light source with a small coherence length.

Axial and lateral scanning

3D OCT imaging is performed by an axial scan (z-direction) and two lateral scan directions (x, y). Depth information, below the surface of the tissue, is derived from the z-scan (the interferometric signal) and when combined with the two-dimensional x- and y-scans, a three-dimensional volume image of the sample is generated.

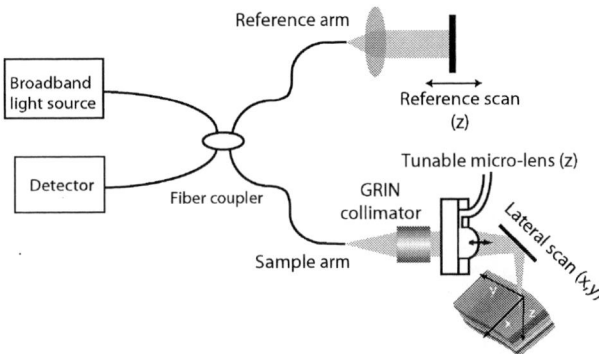

Figure 1: Illustration of a basic 3D OCT setup. In the sample arm, a tunable micro-lens and a lateral scanner enable 3D movement of the focal point on the target under test.

Fast scanning rate, small size, and low power consumption are the main parameters required for optimum endoscopic OCT integration. A MEMS-based scanning mirror together with a tunable micro-lens are the core elements of the endoscopic probe for providing enhanced 3D OCT images presented here.

Dynamic focusing for OCT imaging

Conventional OCT systems use fixed-focal length lenses in the sample arm. Due to the properties of Gaussian beams, high lateral resolution can only be obtained at the smallest beam waist such that resolution decreases along the scan (z) depth. Moreover, the number of backscattered photons from the tissue also depends on the focus or beam waist position and it is smaller for structures located far from the smallest beam waist, hence decreasing the sensitivity. To overcome this problem, the smallest beam waist (i.e., the focus) should move into the tissue synchronously with the reference scan. As moving the entire focusing element is not practicable in endoscopes, a scan can be achieved by tuning the focal point using a tunable micro-lens. We use a pneumatically actuated tunable micro-lens integrated in the endoscopic probe to accomplish this here.

PROBE COMPONENTS

The probe consists of two components: the scanning micro-mirror and the tunable micro-lens. Each component and its requirements for endoscopic integration will be presented separately.

Scanning micro-mirror

A miniaturized parallel plate electrostatically actuated scanning micro-mirror is implemented into the probe. The scanner is based on a silicon substrate bonded on a glass substrate, on which the electrodes for x and y scanning directions were defined. The Si mirror is illustrated in figure 2. The mirror is fabricated using KOH etch processes applied on the backside of the Si substrate [4]. The mirror has a 1 mm² active surface and is suspended by 6 µm wide springs. The size of the active surface mirror is larger than $\sqrt{2}$ times the beam diameter. A process for thinning the substrate, reduces the distance between electrodes and mirror down to 170 µm, and hence reduces the required actuation voltage. The resonant frequencies of the mirror are roughly 640 and 280 Hz for x and y directions, respectively. These scan rates are, however, sufficient to obtain OCT images at video-rates. A 4° tilting angle could be obtained for both directions for applied biases of less than 100V.

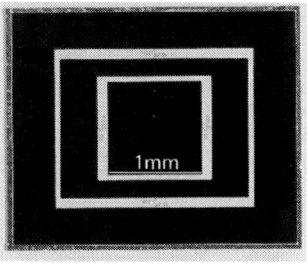 

a- Top view of the 1 mm active surface mirror

b- Top view of the electrodes

Figure 2: Micro-mirror system of the 3D probe.

Tunable micro-lens

Miniaturization and integration of a tunable lens in a scanning probe are the most challenging issues in this work. Based on previously published technologies [5], the lens is fabricated using standard silicon (Si) micromachining combined with polydimethylsiloxane (PDMS). The PDMS is spun on the front side of the Si substrate and two deep-reactive ion etch (DRIE) processes were implemented on the backside to define the cavities of the lens, as shown in the backside view of Figure 3. The 1 mm diameter circular opening defines the aperture of the lens. Two micro-channels enable the liquid to pass through the cavities of the lens and the arrangement of cavities has been optimized to obtain smallest lens size. The lens of 1 mm aperture has an overall diameter of 3 mm, the smallest pneumatically actuated lens demonstrated to date.

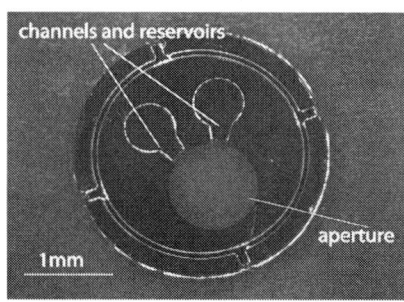

Figure 3: Backside of the tunable lens substrate, which is integrated in the probe showing the fluidic channels and reservoirs.

PROBE DESIGN

Arrangement of lens and mirror

The absolute focal length, the tuning range as well as the optical quality of the entire probe determine the design and integration strategy. These parameters are not only dependent on the lens and mirror surface profiles, but also on the arrangement of mirror and lens.

Figure 4 illustrates two possible arrangements of these micro devices and their simulated spot diagrams, indicating their optical quality. The initial position of the mirror is 45° with respect to the incident beam, and the maximum tilting angle of the mirror is ±5°.

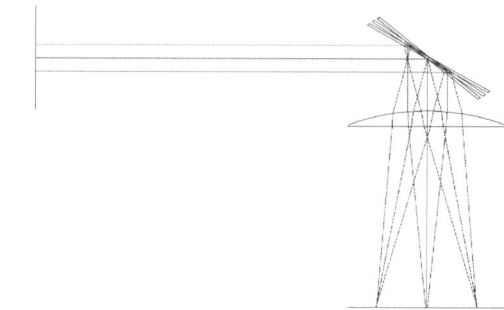

a- Mirror-lens (M-L) arrangement

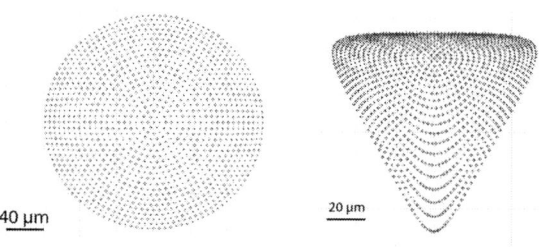

b- M-L Spot diagram for (0°:left and 5°:right)

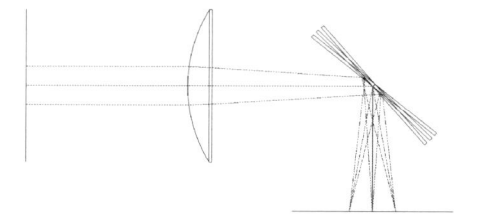

c- Lens-Mirror (L-M) arrangement

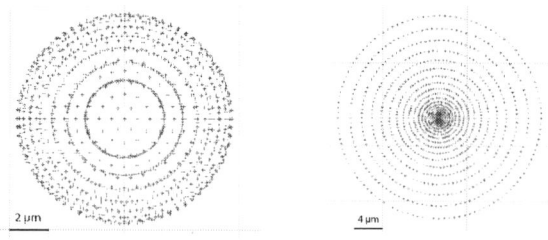

d- L-M Spot diagram for (0°:left and 5°:right)

Figure 4: Arrangements of mirror and lens and corresponding spot shapes. The mirror tilt is ±5° with respect to its initial 45° position.

A comparison of the optical quality achievable for each situation can be obtained from the spot diagrams. As shown in figure 4, it is clearly seen that the lens-mirror arrangement (Figure 4c) gives a better spot profile, predicting higher resolution, than that in the mirror-lens arrangement (a,b). But one drawback of this arrangement (Figure 4c) is a shorter effective tuning depth obtained from the system. The lens integrated in the probe has long tuning range, sufficient to cover the entire scan depth of OCT.

Assembly and packaging

The assembly of the probe is based on the lens-mirror arrangement given in figure 4c leading to higher lateral resolution.

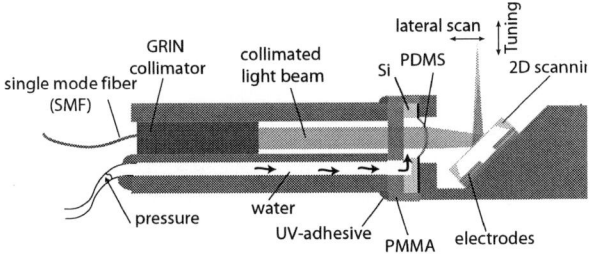

Figure 5: Design concept of the assembled probe.

The optical probe is designed and assembled as shown in figure 5: A mechanical aluminum body fabricated by high precision machining has three circular openings defining two pressure channels and an opening for the GRIN lens collimator. The GRIN collimator has a diameter of 1.25 mm and gives a 0.8 mm wide collimated light beam. A high optical quality polymethylmethacrylate (PMMA) disc is bonded between the mechanical cylinder and the lens. Two cylindrical openings matching the pressure openings of the cylinder were machined into the PMMA substrate to enable the pressure activation of the lens cavities. The lens is bonded on the PMMA substrate after aligning it to the optical path. UV-curable adhesives were used to fix the devices.

For mounting and adjustment of the scanning mirror, a 45° aluminum body, also fabricated by precision machining, was bonded onto the cylinder. The mirror chip containing the electrode and Si substrate was aligned and bonded onto the mechanical substrate. Figure 6 shows a photo of the assembled probe. The entire diameter of the probe with an optical aperture of 1 mm is 4 mm.

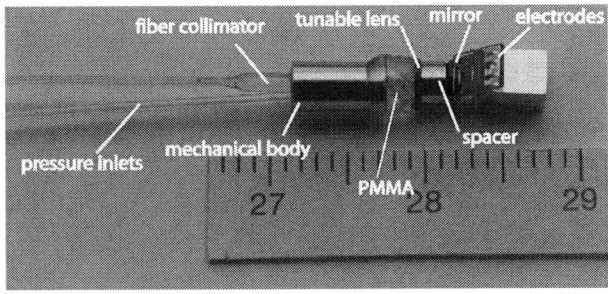

Figure 6: Photo of the assembled probe

PROBE CHARACTERIZATION

One of the most important properties of the probe is the tuning range of the effective back focal length. Figure 7 illustrates the back focal length (BFL) of the lens as a function of applied pressure. As the optical path between the lens and the side-output of the lens is 5 mm long, more than 5 mm effective focal length change is still available. The inverse relation between pressure and focal length allows the probe to be tuned with low pressure actuation.

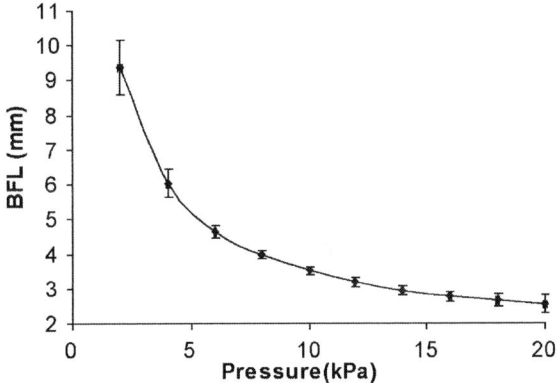

Figure 7: Back focal lens as function of applied pressure

To test the durability of the lens, it has been actuated with up to 40 kPa; no delamination has been observed even at these pressures. Figure 8 shows a photo of the probe under high pressure.

Figure 8: The tunable part of the probe under high pressure (exceeding the range required for application). The needle of a syringe shows an approximate scaling of the probe.

OCT MEASURMENTS

The performance of lateral scans using different scanning micro-mirrors [2] has previously been presented. In this paper the effect of tunability on the lateral resolution is discussed, as an innovation for OCT functionality.

To demonstrate the efficiency of tunability for OCT imaging, a 20 μm periodic binary grating made of glass and aluminum was used to test the resolution of the lens for different pressure values resulting in different focal lengths. The position of the grating was adapted to each active focal length, in order to have the smallest beam waist in the plane of the reflective grating. This demonstrates the quality of the lateral resolution for different depths by synchronizing the reference arm with the tuning of the focal length.

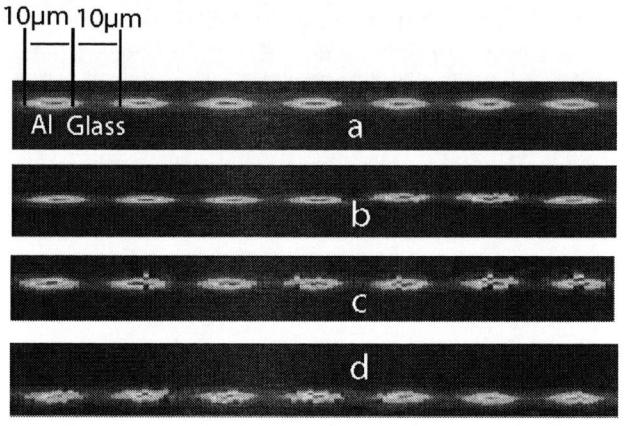

Figure 9: Four OCT images of an aluminum (Al)-Glass binary grating using a tunable lens at different pressures: a) 2kPa, b) 5kPa, c) 10kPa, and d) 15kPa. The images show that 10 μm is still resolved over a wide tuning range.

Figure 9 shows that structures of 10 μm are still resolved over a wide tuning range. This clearly indicates that a high lateral resolution is maintained over a wide tuning range and thus for a correspondingly large scanning depth of the OCT. These images can not be obtained using a conventional lens with a fixed focal length.

CONCLUSION

A novel optical and micro-mechanical design of a MEMS probe offering a 3D manipulation of the focal point for endoscopic OCT has been presented. A 2D scanning micro-mirror and a pneumatically actuated tunable micro-lens are integrated into the probe. Design, assembly, and integration of the entire probe in an OCT setup was realized. The high lateral resolution, a wide tuning range and a fast lateral scanning rate are promising properties for implementing this system in a clinical environment, particularly for high resolution and real time in-vivo optical biopsies. On the one hand side this type of probe can be implemented into OCT configurations, and on the other hand also into all other modern optical imaging systems requiring a 3D manipulation of the focal point like confocal scanning microscopy.

ACKNOWLEDGEMENT

The authors are grateful to Dr. Armin Werber, Dr. David Kallweit, and Dipl. Ing. Andreas Fischer for their contributions in this work.

REFERENCES

[1] A. F. Fercher, W. Drexler, C. K Hitzenberger, T. Lasser, "Optical coherence tomography -- principles and applications", *Reports on Progress in Physics*, vol. 66, pp. 239-303, 2003.

[2] Z. Yaqoob, J. Wu, J.E. J. McDowell, X. Heng, C. Yang, "Methods and application areas of endoscopic optical coherence tomography", *Journal of Biomedical Optics, SPIE,* vol. 11, pp. 063001, 2006.

[3] K. Aljasem, A. Werber, A. Seifert, H. Zappe, "Fiber optic tunable probe for endoscopic optical coherence tomography", *Journal of Optics A: Pure and Applied Optics*, vol. 10, pp 044012 (8pp), 2008.

[4] D. Kallweit and H. Zappe, "Fabrication of bulk-Si micromirrors with an integrated tilt sensing mechanism", *Journal of Micromechanics and Microengineering*, Vol. 16, pp. 463-469, 2006.

[5] A. Werber, and H. Zappe, "Tunable microfluidic microlenses", *Applied Optics*, vol. 44, pp. 3238-3245, 2005.

CMOS-INTEGRABLE PISTON-TYPE MICRO-MIRROR ARRAY FOR ADAPTIVE OPTICS MADE OF MONO-CRYSTALLINE SILICON USING 3-D INTEGRATION

Martin Lapisa[1], Fabian Zimmer[2], Frank Niklaus[1], Andreas Gehner[2], Göran Stemme[1]
[1]Microsystem Technology Laboratory, School of Electrical Engineering
Royal Institute of Technology (KTH), Stockholm, Sweden
[2]Fraunhofer Institut für Photonische Mikrosysteme (IPMS), Dresden, Germany

ABSTRACT

This paper presents a novel CMOS-compatible fabrication process and evaluations of a micro mirror array (MMA) made of mono-crystalline silicon (m-Si) for adaptive optic (AO) applications. The m-Si mirror layer is transfer bonded from a silicon-on-insulator (SOI) donor wafer with adhesive wafer bonding towards an intermediate patterned polymer spacer layer and clamped with metal plating. We present a CMOS compatible, bond alignment-free fabrication scheme offering the potential for high air gap distances between substrate and mirrors and we show first measurements of the fabricated mirrors.

INTRODUCTION

Micro-mirrors have found a wide range of applications in the past decades, e.g. in projection systems [1], optical scanners [2], optical switches [3] and maskless lithography systems [4] to mention a few examples.

General requirements especially in high precision applications are to achieve high mirror planarity and high deflection reproducibility without any mechanical drift. Usually, such deflection drifts are accompanied with plastic deformations or imprinting originating from material creep and internal stress relaxation processes [5]. However, for many applications a stable analog deflection over a long period is vital. Therefore the use of mechanically more stable materials has been investigated for CMOS integration [6].

One of them is mono-crystalline silicon (m-Si) that can be integrated by utilizing wafer-bonding. M-Si shows a moderate reflectivity in the ultraviolet (UV) spectral range but the reflectance drops in the visible and infra-red region [7]. Despite that m-Si shows superior mechanical properties as it does not contain any internal stress that might relax and it is fully elastic deformable. Furthermore, m-Si is supplied with a very low surface roughness without the need of additional surface treatments. As a result, m-Si has the potential to allow for drift-free, highly planar micro-mirror devices [5].

Adaptive optics (AO) for correction of phase aberrations in the optical-path requires 2-dimensional arrays of vertically movable micro-mirrors with μm-stroke capabilities, high fill factor, high surface flatness, nanometer range deflection precision and short response times. Especially in those cases, where a high number of actuator elements are required, the driving electronic has to be incorporated by means of a monolithically integrated CMOS address circuitry within the same chip. CMOS compatibility arises constrains to allowed fabrication methods and processes regarding the thermal budget, radiation and applicable electric field strength.

For integration of m-Si micro-mirror arrays (MMA) with CMOS driving circuits the assembly after processing

of MEMS and electronics on separate wafers in a flip-chip process was reported [8], [9]. In that way processing restrictions can be bypassed for arrays with large mirrors that offer the space for flip-chip interconnects. However, small pixel sizes with high fill-factors cannot be realized with this technique as the scalability of flip-chip bonding is limited.

The wafer-level integration of micro-mirrors on CMOS driving circuits requires particle tolerant processes that allow for high yield. Adhesive wafer bonding that utilizes a particle insensitive adhesion layer is a well suited integration technique [10], [11]. It was successfully used in the fabrication of tilting mirror-arrays for maskless deep-UV lithography [12].

In our previously reported work [13] small arrays of tilting m-Si micro-mirrors with small air gaps between mirror and electrode was presented. The current work focuses on the need of adaptive optic applications for phase aberration correction. We present a process that allows for large air gaps by utilizing a thick patterned polymer spacer in combination with adhesive wafer bonding of m-Si. The CMOS compatible process is wafer-alignment free and the fabrication complexity is relatively low. No surface finishing processes like chemical-mechanical polishing (CMP) to achieve high mirror flatness are required.

DESIGN & FABRICATION

Design

The MMA consist of 96 x 96 piston-type mirrors with 40 μm pitches, a fill factor of 73% and a 2.2 μm air gap in between mirror plate and electrode. The mirrors are deflected in a parallel plate actuator manner.

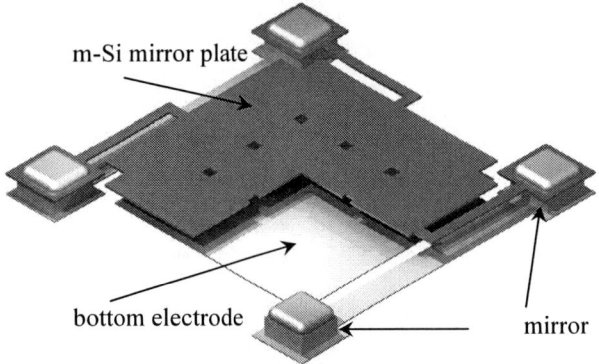

Figure 1: Sketch of single mirror, mirror plate sectioned

To save time and costs for first evaluations of the process technology the MMA has been implemented on CMOS dummy devices employing fixed wired address electrodes. The surface of these devises has not been planarized and shows a peak-to-valley topography of 300

978-1-4244-2977-6/09 $25.00 © 2009 IEEE

nm in between the electrodes and wires. Eight columns of mirrors are interconnected and actuated simultaneously.

Figure 1 shows the design of a single MMA pixel. The m-Si mirror plate is suspended on four flexures which are connected to metal posts. The metal posts define the air gap in between mirror plate and electrode and connect the mirrors electrically to common ground potential. When actuated, an electrostatic field between mirror plate and bottom electrode creates an attractive force resulting in a piston-like downward movement of the mirror.

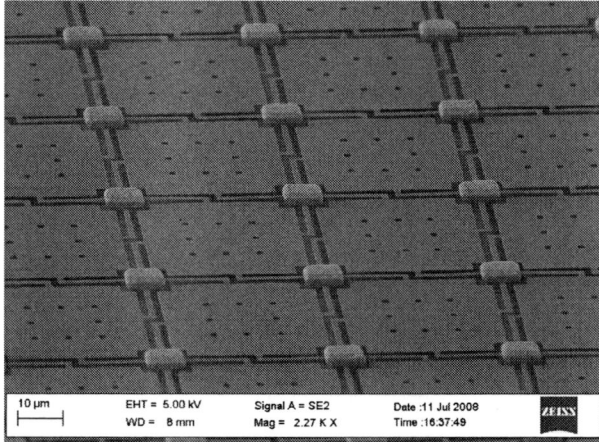

Figure 2: SEM picture of a section of the fabricated micro-mirror array with 96x96 mirrors. The mirror layer is made from 340 nm thick mono-crystalline Silicon.

Figure 2 shows a scanning-electron-microscope (SEM) picture of a section of the fabricated mirror-array. All mirrors are interconnected by sharing the post with the neighboring mirror.

Fabrication

The fabrication process of the passive MMA device is depicted in Figure 3. The process starts with fabrication of the bottom substrate by sputter deposition of aluminum (AlSi2Cu0.5) onto an oxidized silicon wafer. Electrodes and wires are etched with reactive-ion-etching (RIE) into the metal layer. Hereafter, a thin layer of nickel is deposited onto the aluminum in a lift-off process at the positions of later electroless nickel plated posts (Fig. Figure 3a). This layer is required for the auto-catalytic chemical reaction of electroless plating to start.

The negative photoresist AZ nLOF 2070 is used as a sacrificial layer and is spun onto the substrate in a standard photoresist spinner. Trenches in between electrodes and wires on the substrate cause surface topography in the applied sacrificial polymer. To lower the topography the resist spin-on is done in two passes. In the first pass the photoresist is diluted to a viscosity convenient to provide resist layer thickness of 200 nm. The resist is spun-on for 5 s at 3000 rpm and left on the chuck for 2 minutes to allow a reflow and the trenches to be filled. After subsequent soft-baking non-diluted photo-resist with a thickness of 2 μm is applied in the second pass. A total layer thickness of 2.2 μm defines the mirror air gap. At the positions of the plating bases openings are photo-lithographically patterned into the resist-stack (Fig. Figure 3b). These openings act as plating moulds later in the process. The negative resist is thereafter fully cross-

linked by flood exposure in a mask aligner and hardened by hard-baking in an oven for 120 minutes at 205°C.

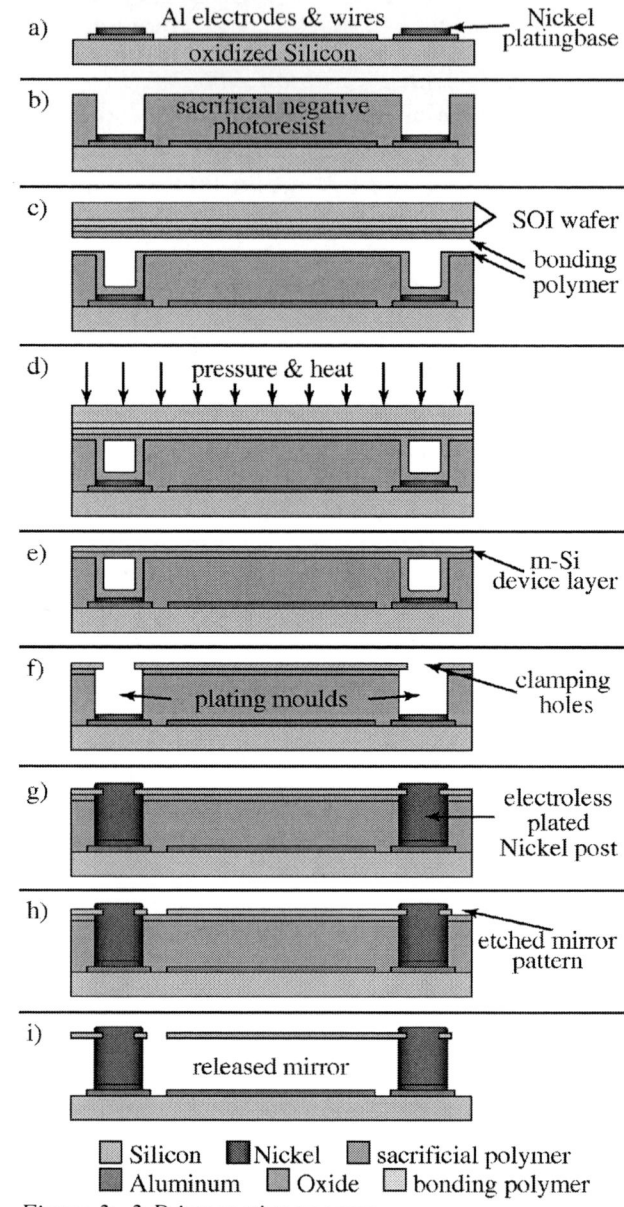

Figure 3: 3-D integration process

For the m-Si layer transfer from the silicon-on-insulator (SOI) donor wafer an adhesive wafer-bonding process is used that we reported earlier [14]. The bond-polymer is spun-on and pre-cured on both the device layer of the SOI-wafer and the patterned sacrificial polymer of the substrate (Fig. Figure 3c). The wafers are bonded in a wafer-bonder under high vacuum by applying pressure and ramping up the temperature of the wafer stack (Fig. Figure 3d). Bond alignment is not needed for the assembly since the SOI-wafer does not contain any pattern at this point. After bonding, the handle wafer is etched away isotropically with SF6 plasma in an inductively coupled plasma etcher (ICP). The 400 nm thick buried oxide of the SOI wafer is an etch-stop layer with this etch-chemistry and protects the m-Si device layer from being attacked. Thereafter the oxide is stripped in buffered HF (BHF), leaving only the thin device layer of the SOI-wafer behind (Fig. Figure 3e).

978-1-4244-2977-6/09 $25.00 © 2009 IEEE   1004

The wafer alignment marks on the bottom substrate are visible to the stepper optics through the thin m-Si device layer and can be used for consecutive lithography steps. Holes are etched into the m-Si layer by RIE to access the plating moulds. The moulds are cleared from bonding polymer in an anisotropic oxygen plasma etch to make the underlying nickel accessible (Fig. Figure 3f). Because the bond polymer is etched 3 times faster than the highly cross-linked sacrificial negative resist, the latter is almost not attacked at all in this step.

To clamp the m-Si device layer mechanically and electrically, nickel posts are grown in the plating moulds. Nickel deposition starts spontaneously as the nickel platingbase gets in contact with the nickel hypophosphite ($Ni[H_2PO_2]_2$) plating solution at elevated temperatures. The wafer is immersed into the electrolyte until nickel is plated slightly above the edge of the m-Si device layer (Fig. Figure 3g).

With the last lithography mask the mirror pattern is RIE etched into the m-Si layer (Fig. Figure 3h). The wafer is thereafter coated with a thick photoresist to protect the surface from particles, dirt and mechanical impact that could appear during die separation in the dicing saw.

After dicing, the mirror structures are released in isotropic oxygen plasma in which the sacrificial resist spacer is etched away selectively (Fig. Figure 3i).

MEASUREMENTS

All micro-mirror measurements are done with a white-light interferometer from Wyko. The MMA devices were housed in a ceramic pin grid array (PGA 68), which can be plugged into an external driving setup.

In order to keep the number of external interconnects limited, sub-regions of 8 mirror columns have been electrically combined to one input and thus can be actuated only simultaneously. Figure 4 shows a 3-D deflection profile of a MMA section. The left part of the array is actuated at 30 V while the right part is set to 0 V. In this measurement the metal posts which support the micro-mirror flexures are masked and not shown for better illustration.

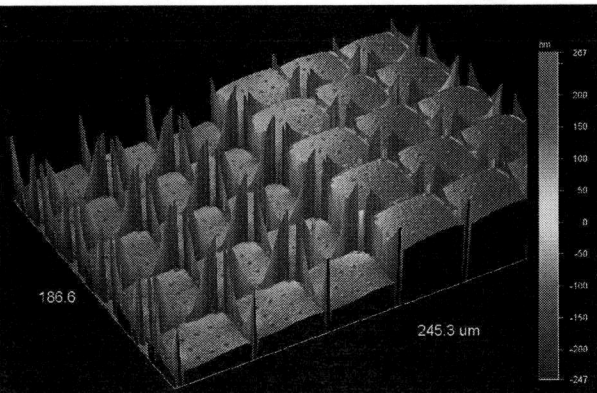

Figure 4: Measurement with white light interferometer.

A line-scan of this measurement is shown in Figure 5. The scan is taken along a line in the middle of the third row, as indicated by the green line drawn in the inset.

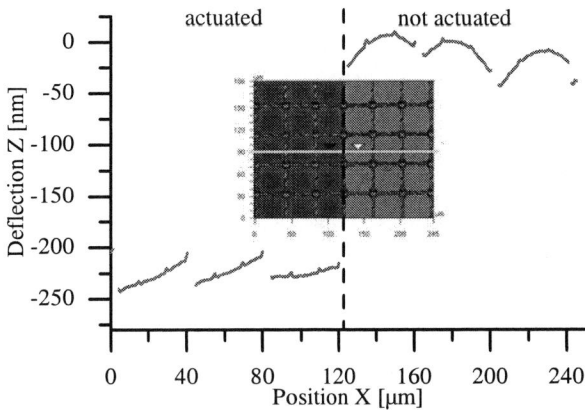

Figure 5: Line-scan along green marked line (inset).

Figure 6 depicts the same line scan taken from a measurement with all mirror non-deflected.

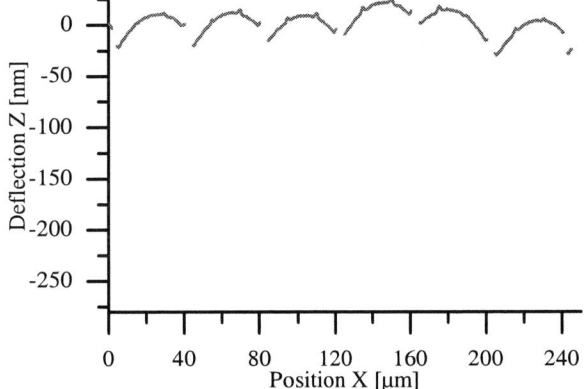

Figure 6: Zero-deflection line-scan

The mirrors have a slight convex bow in the non-deflected initial state. Apart from that a slight non-uniformity in height is observed together with a certain tilt of the mirrors.

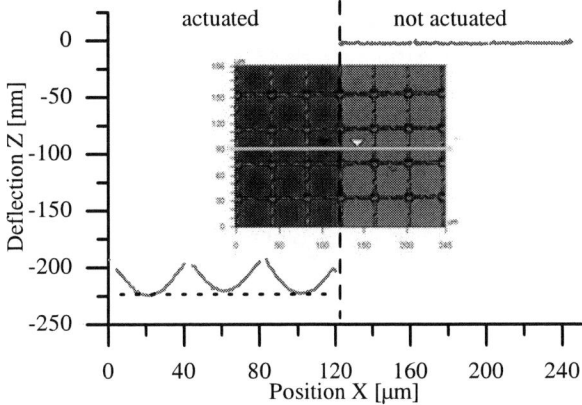

Figure 7: Deflection profile along the green marked line (inset), with zero-voltage surface profile subtracted.

The deflection profile in Figure 7 shows the subtracted profiles of Figure 5 and Figure 6, thus eliminating any pre-deformation effects like initial curvature or tilt. As can be seen, upon actuation the mirrors undergo a warpage downwards in the center.

DISCUSSION

In this first evaluation of the process technology non-planarized, CMOS-compatible test devices with a surface topography of 300 nm are used to build up MMA devices. With a 2-pass spin-on process the topography of the sacrificial layer was lowered by 40% compared to a single pass process and ended up at 150 nm prior bonding. In spite of this remaining non-planarity, the micro-mirrors show a 1σ-height deviation of only about 9 nm in the non-deflected state. This confirms the excellent leveling capability of the adhesive wafer-bonding process. A more uniform height distribution can be expected on planarized substrates such as CMOS-wafers.

In the non-deflected zero-voltage state the mirrors show a convex curvature corresponding to a pixel RMS of about 10 nm or a peak-to-valley value of about 25 nm. Upon actuation, the mirrors change from their initial convex shape to a slight concave shape. This is due to the electrical field forces acting on the mirror plate and thus introducing a bending towards the pixel center. The center-deflection of the measured MMA section showed a uniform displacement in Z-direction of 220 nm ± 5 nm at 30 V.

Due to process variations, the mirrors have a statistical spread pre-tilt of up to 50 nm. With the pre-tilted mirrors, a pull-in deflection of about 400 nm was reached. The RMS mirror planarity and height deviations obtained within this first fabrication run are quite promising to meet the requirements for most AO applications and the maximal deflection is sufficient for 2π-phase modulation in the visible light spectrum. With an optimized processing scheme, the observed variations can possibly be further reduced, allowing for a reduced pre-tilt of the mirrors.

CONCLUSION

A novel CMOS compatible fabrication technique utilizing a patterned polymer spacer in combination with adhesive wafer-bonding is presented. Piston-type micro-mirror arrays consisting of 96x96 mirrors made from mono-crystalline silicon with an air-gap of 2.2 μm were successfully fabricated and first evaluations were presented.

ACKNOWLEGEMENT

This work is part of the European project Q2M and sponsored by the EU through its 6th Framework Program.

REFERENCES

[1] P. F. van Kessel, L. J. Hornbeck, R. E. Meier, M. R. Douglass, „A MEMS-Based Projection Display", *Proc. of the IEEE*, Vol. 86, No. 8, pp.1687-1704, 1998

[2] R. A. Miller, G. W. Burr, Y. C. Tai, D. Psaltis, "Magnetically actuated MEMS scanning mirror", *Proc. of SPIE*, Vol. 2687, pp. 47-52, 1996

[3] M. F. Dautartas, A. M. Benzoni, Y. C. Chen, G. E. Blonder, B. H. Johnson, C. R. Paola, E. Rice, Y. H. Wong, "A silicon-based moving-mirror optical switch", *J. of Lightwave Technology*, Vol. 8, pp. 1078-1085, 1992

[4] N. Choksi, D. S. Pickard, M. McCord, R. F. D. Pease, Y. Shroff, Y. Chen, W. Oldham, D. Markle, "Maskless extreme ultraviolet lithography", *J. of Vacuum Science and Technology*, Vol. 17, pp. 3047-3051, 1999

[5] J. U. Schmidt, M. Friedrichs, T. Bakke, B. Voelker, D. Rudloff, H. Lakner, "Technology development for micromirror arrays with high optical fill factor and stable analogue deflection integrated on CMOS substrates", *Proc. of SPIE*, Vol. 6993, pp. 69930D-1 – 69930D-7, 2008

[6] J.U. Schmidt, J. Knobbe, A. Gehner, H. Lakner, „CMOS integrable micro mirrors with highly improved drift-stability", *Proc. of SPIE*, Vol. 6467, pp. 64670R-1 – 64670R-11, 2007

[7] H. Ehrenreich, H. R. Philipp, B. Segall, "Optical Properties of Aluminum", *J. Physical Review*, Vol. 132, pp. 1918-1928, 1963

[8] I. W. Jung, U. Krishnamoorthy, O. Solgaard, "High fill-factor two-axis gimbaled tip-tilt-piston micromirror array actuated by self-Aligned vertical electrostatic combdrives", *J. of Microelectro-mechanical Systems*, Vol. 15, pp. 563-571, 2006

[9] M. A. Michalicek, V. M. Bright, "Flip-chip fabrication of advanced micromirror arrays", *Sensors and Actuators*, Vol. 95, pp. 152-167, 2002

[10] F. Niklaus, G. Stemme, J.Q. Lu, R.J. Gutmann, "Adhesive wafer bonding", *J. of Applied Physics*, Vol. 99, pp. 031101-1 - 031101-28, 2006

[11] S. van der Groeny, M. Rosmeulen, K. Baert, P. Jansen, L. Deferm, "Substrate bonding techniques for CMOS processed wafers", *J. of Micromechanics and Microengineering*, Vol. 7, pp. 108-110, 1997

[12] T. Bakke, M. Friedrichs, B. Voelker, M. Reiche, L. Leonardsson, H. Schenk, H. K Lakner, „Spatial light modulators with monocrystalline silicon micro-mirrors made by wafer bonding", *Proc. SPIE*, Vol. 5715, pp. 69-79, 2005

[13] F. Niklaus, S. Haasl, G. Stemme, "Arrays of mono-crystalline silicon micromirrors fabricated using CMOS compatible transfer bonding", *J. of Micro-electromechanical Systems*, Vol. 12, pp. 465-469 2003

[14] M. Populin, A. Decharat, F. Niklaus, G. Stemme, "Thermosetting nano-imprint resists: novel materials for adhesive wafer bonding", *Proc. of MEMS*, pp. 239-242, 2007

MONOLITHIC MULTICOLOR TOTAL INTERNAL REFLECTION (TIR)-BASED CHIP FOR HIGHLY-SENSITIVE MULTIFLUORESCENCE DETECTION AND IMAGING

Nam Cao Hoai Le[1], Dzung Viet Dao[1], Ryuji Yokokawa[1, 2], John Wells[1] and Susumu Sugiyama[1]

[1]Ritsumeikan University, Shiga, JAPAN
[2]CREST/PRESTO, JST, Saitama, JAPAN

ABSTRACT

We present a multicolor Total Internal Reflection (TIR)-based chip capable of generating up to four overlapping evanescent fields for highly-sensitive multifluorescence detection and imaging. The monolithic chip was fabricated using Si bulk micromachining and PDMS casting. Our proposed method integrates all miniaturized components, including prism, cylindrical microlenses, fiber stopper and alignment grooves, into one single PDMS chip; thus assembly is unnecessary, and misalignment is eliminated. The chip capabilities were demonstrated by detecting simultaneously two fluorescent dyes, namely Tetramethylrhodamine (TMR) and Fluorescein (Fl), and then imaging a mixture of Nile-red (NR) and Dragon-green (DG) microbeads in two overlapping evanescent illumination modes. Such a device could be a key component in highly-sensitive multifluorescence molecular detection in μ-TAS.

1. INTRODUCTION

Multifluorescence microscopy is a well established technique widely used to visualize and discriminate among several structural and functional elements in biological samples ranging from cells to whole organisms [1]. Multifluorescence microscopy is typically implemented into two fluorescent microscopy configurations: fluorescent confocal microscope [2] and epi-fluorescent microscope [3]. Although these two configurations have been popular tools in multicolor fluorescence-based biological imaging, they still contain several drawbacks. For instance, fluorescent confocal microscopes are limited by their high cost due to the indispensable needs of high numerical aperture, high quality objective lenses and good vibration isolation. Furthermore, they suffer from slow scanning speed and time-consuming image reconstruction [4]. As for epi-fluorescent microscope, since it uses volumetric illumination which leads to the acquisition of emission not only in the depth of focus but also in a region well beyond that, the out-of-focus fluorophores in these regions naturally reduce the signal to noise ratio (SNR) and limit its resolution [5].

On the other hand, evanescent illumination by Total Internal Reflection (TIR) with its narrow, optically defined excitation depth (on the order of 100-300 nm) is considered as a very effective way to overcome the background noise problem which is often the biggest problem in fluorescent detection/imaging. Accordingly, the SNR and resolution of this microscope technique are significantly improved [6]. Therefore, this technique is increasingly applied in single molecule detection in biological systems [7] or characterizing of micro/nanofluidic flows near a solid/liquid

interface [8]. Recently, some groups have proposed dual-color Total Internal Reflection Fluorescent Microscopy (dual-color TIRFM) for simultaneous detection of two fluorescent bio-molecules [9-12]. These dual-color TIRFM systems were an indispensable tool to study the interaction, binding or colocalization of two differently fluorescently-labeled molecules either inside a cell [9-10] or on a glass substrate [11-12]. Although these systems are powerful tools, they are expensive, bulky, elaborate and time-consuming in assembly, and more importantly they can only generate at most two overlapping evanescent fields.

We propose here the first integrated and miniaturized multicolor TIR-based chip capable of generating up to four overlapping evanescent fields. To examine the feasibility of our chip for multicolor fluorescent detection, we used coaxial and perpendicular overlapping TIR modes with two optical fibers providing two excitation wavelengths. However, three or four excitation wavelengths can be straightforwardly introduced by adding third or fourth fiber based on the design concept of the chip. Our miniaturized, integrated device could be an alternative to the conventional dual-color TIRFM systems. It could also be a useful component of a μ-TAS for highly-sensitive multifluorescence detection and imaging.

2. DESIGN AND WORKING PRINCIPLE

Fig. 1(a) shows the A-A cross section view of the multicolor TIR-based chip from Fig. 1(b) with the optical path for analysis. The chip consists of a truncated pyramid-shaped prism in the center, four cylindrical microlenses, four fiber stoppers and four fiber alignment grooves (Fig. 1(b)). All miniaturized optical components namely prism, lenses etc., are supported underneath by a common base. The chip is attached to a glass flow-cell which serves as a sample delivery platform. Since the chip is monolithic in PDMS, several sample delivery platforms, namely flow-cells, microchannels etc., can be attached to the chip by van-der-Waals contact between PDMS and an intermediate glass slide. Furthermore, by simply turning the device up side-down, it is possible to use the chip with both upright and inverted microscope [13].

Fig. 1(a) also explains the working principle of the multicolor TIR-based chip in the coaxial overlapping TIR mode. In this mode, we use to two single mode optical fibers in opposite direction (i.e. either fiber 1 – fiber 3 or fiber 2 – fiber 4) to deliver two excitation beams to generate two TIR spots (Fig. 1(b)). Blue laser (wavelength 473 nm) from left optical fiber, which is aligned and stopped by fiber alignment groove and fiber stopper on the left, is collimated by a cylindrical microlens, refracted twice at air/PDMS and PDMS/glass interfaces, respectively, and finally TIR at

978-1-4244-2977-6/09 $25.00 © 2009 IEEE

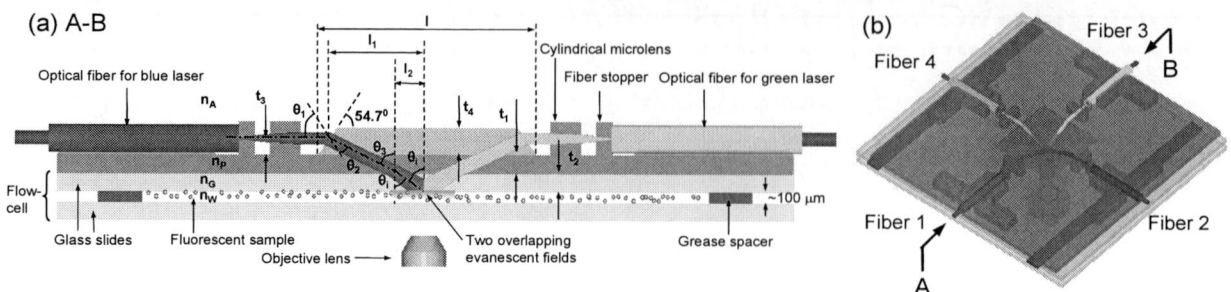

Fig. 1(a) Cross sectional view (A-B) (not to scale) from (b) of the multicolor TIR-based chip with glass flow-cell attached, the incident angle $\theta_i = 69.5^0$ is larger than the critical angle $\theta_C = 61^0$ (glass/water interface) satisfying the TIR condition for four excitation lasers (b) 3-D model of the chip.

glass/water interface. Similarly, green laser (wavelength 543 nm) from right optical fiber will also cause another TIR at the same glass/water interface. When the chip is in perpendicular overlapping TIR mode, two perpendicular fibers (i.e. either fiber 1 – fiber 2 or fiber 2 – fiber 3 or fiber 3 – fiber 4 or fiber 4 – fiber 1) will be used to generate two TIR spots (Fig. 1(b)). By choosing the appropriate device design parameters based on the optical analysis in the following, we can guarantee the overlap of those two TIR spots in the center of the prism projected on the flow-cell.

For simplicity, the optical analysis only considers the central ray of the beam and assumes the bottom surface of the fiber touch the chip base. Using Snell's law and geometric relationships from Fig. 1(a), the following equations can be easily derived.

$$\theta_1 = 90^o - 54.7^o = 35.3^o \qquad (1)$$

$$\theta_2 = \sin^{-1}\left(\frac{n_A \times \sin\theta_1}{n_P}\right) \qquad (2)$$

$$\theta_3 = 90^o - \theta_1 + \theta_2 = 90^o - \theta_1 + \sin^{-1}\left(\frac{n_A \times \sin\theta_1}{n_P}\right) \qquad (3)$$

$$\theta_i = \sin^{-1}\left(\frac{n_P}{n_G}\sin\theta_3\right) \qquad (4)$$

$$\theta_i = \sin^{-1}\left[\frac{n_P}{n_G}\sin\left(90^o - \theta_1 + \sin^{-1}\left(\frac{n_A \times \sin\theta_1}{n_P}\right)\right)\right] \qquad (5)$$

where n_A, n_P, n_G are refractive indices of air, PDMS and glass, respectively. Substituting $n_P = 1.46$, $n_G = 1.52$, $n_A = 1$, $\theta_1 = 35.3^0$ into equation (5), we have $\theta_i = 69.5^0$. If the dependence of refractive indices n_P, n_G on wavelength is negligible, then the incident angles θ_i of two excitation laser beams are the same. Clearly, the incident angle $\theta_i = 69.5^0$ is greater than the critical angle $\theta_C = 61^0$ (for glass/water interface), therefore the condition for TIR is satisfied for all excitation lasers.

In multicolor TIRFM systems, it is utmost important to secure the overlap of the evanescent fields or TIR spots generated by two or more excitation wavelengths so that a fluorophore in the field of view can be simultaneously excited and then emitted to the detector. Since the chip itself is symmetric, from Fig. 1(b), it can be observed that the overlap condition will be met if the centers of all evanescent fields fall on the center of the prism projected on the cover

glass. Considering the coaxial overlapping TIR case from Fig. 1(a), the abovementioned condition is geometrically expressed by equation (6). Again due to the symmetry of the chip, this condition also implies the overlap of two, three or four evanescent fields when arbitrarily adding two, three or four excitation optical fibers.

$$l_1 = \frac{l}{2} - \frac{t_3}{\tan 54.7^o} \qquad (6)$$

Again from Fig. 1(a), equations (7) and (8) can be derived:

$$l_2 = t_2 \tan\theta_i \qquad (7)$$

$$t_1 + t_3 = \frac{l_1 - l_2}{\tan\theta_3} \qquad (8)$$

Substituting (7) into (8), then l_1 from (8) into (6) we have:

$$t_1 = \frac{l}{2\tan\theta_3} - \frac{t_2 \tan\theta_i}{\tan\theta_3} - t_3\left(\frac{1}{\tan\theta_3 \tan 54.7^0} + 1\right) \qquad (9)$$

where l is prism size, t_3 is fiber cladding radius, t_1 is chip base thickness and t_2 is cover glass thickness. The cover glass thickness (Matsunami Co., Japan) was fixed at 170 µm, while the fiber cladding radius t_3 was either 62.5 µm or 125 µm. It should be noted that these two values of t_3 are commonly seen in singlemode or multimode optical fibers. Furthermore, t_3 can be easily smaller than prism height t_4, typically in the range of 270-300 µm defined by wet etching time. Finally, l and t_1 were found with the aid of equation (9). Fig. 2 shows

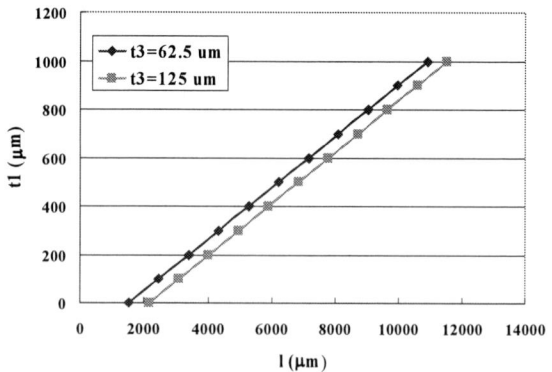

Fig. 2 Linear constraint (9) of chip base thickness t_1 and chip prism size l for two fiber cladding radius t_3, this relationship is to guarantee the center of each evanescent field falls on the center of flow-cell.

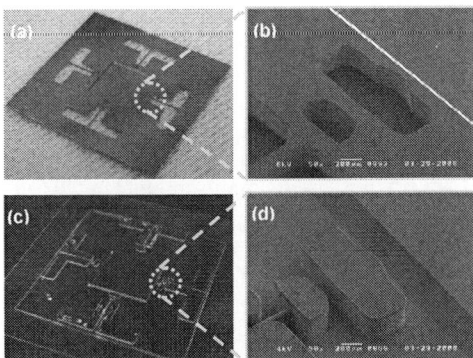

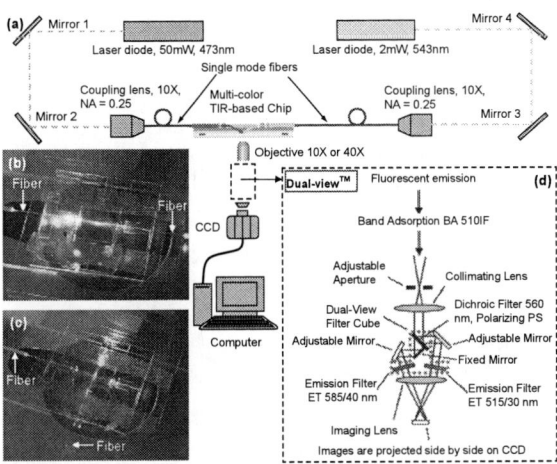

Fig. *3(a) Photo of Si mold of the multicolor TIR chip (b) SEM micrograph of the high-lighted part of the Si mold, (c) Photo of the PDMS multicolor TIR chip (dimensions (L×W×T) of the chip are 20x10x0.9mm³) (d) SEM micrograph of the high-lighted part of the PDMS chip with fiber inserted.*

the linear relationship between chip base thickness and prism size for two fiber cladding radiuses. The smallest prism size required is 2000 μm when the chip base thickness is zero, i.e. when the prism directly attached to the glass slide. However, there should always be a finite value of chip base thickness t_1 to support and fix all components of the chip on the same platform. For our current design, we chose l = 8042 μm, t_3 = 62.5 μm (single mode fiber, Sumitomo, Inc., Japan), t_2 = 170 μm, therefore t_1 = 690 μm which was controlled by the amount of PDMS during casting in the fabrication process.

3. EXPERIMENTAL

Device fabrication

The polymer, multicolor TIR-based chip was fabricated using conventional Si bulk micromachining and PDMS casting technique described in detail elsewhere [14]. As mentioned above, it is particularly important to control the prism length l and the chip base thickness t_1 so as to secure the overlap of the TIR spots. The prism length l was chosen during mask design and the chip base thickness t_1 was controlled by the volume of liquid PDMS mixture during casting. For a Si mold of 27 x 27 mm² in size (Fig. 3 (a)), the volume of liquid PDMS required was 0.4 ml to achieve t_1 approximately 700 μm. Using a single Si mold, the multicolor TIR-based chip can be mass replicated by repeated PDMS casting. The fabrication results are shown in Fig. 3. It is well-known that the DRIE with its cycled etch-passivation nature normally results in the scalloping which leads to severe roughness on the vertical side walls. We are finding ways to reduce this roughness so as to improve the optical smoothness of the microlens.

Experimental setup

Fig. 4(a) shows the experimental setup using the multicolor TIR-based chip which was also adapted from previous work [14]. Two illumination schemes were demonstrated: coaxial (Fig. 4(b)) and perpendicular (Fig. 4(c)) overlapping TIR configurations, respectively. The chip was then placed with the glass flow-cell facing an objective lens (10X or 40X) of an Olympus IX71 microscope equipped

Fig. *4(a) Experimental set-up with Olympus IX71 microscope, Dual-view™, Hamamatsu Orca AG CCD camera and 2 singlemode fibers, fiber coupling optics, Melles Griot (543nm, 2mW) and Photop Suwtech laser diodes (473nm, 50mW); Close-up view of the chip set-up on microscope stage for coaxial (b) and perpendicular (c) TIR configurations; (d) construction of Dual-view™.*

with a deep-cooled Hamamatsu Orca AG CCD camera. A Dual-view™ (Optical Insights Inc., USA) was inserted between the microscope and the CCD camera to split emission larger than 560 nm (mostly from TMR and NR) to ET585/40 while emission smaller than 560 nm (mostly from FI and DG) to the ET515/30 emission filters (Fig. 4(d).

Mixture of TMR (C-1171, Ex/Em = 532 nm/556 nm, Molecular Probes Inc., USA) and FI (CI 45350, Ex/Em = 494 nm/521 nm, Wako, Japan) in DMSO at concentrations of 0.5 μM and 0.5 mM, respectively, were used in observation of TIR spots. Mixture of carboxyl fluorescent NR beads (CFP-0556-2, Ex/Em = 532 nm/556 nm, 400 nm to 600 nm in diameter, Spherotech Inc., USA) and DG beads (CP01F/8206, Ex/Em = 480 nm/520 nm, 510 nm in diameter, Bangs Lab. Inc., USA) dispersed in Tween20®/DI water solution (0.05%), both at 0.05% w/v concentration, were used in observation the Brownian motion of the beads. In each experiment, fluorescent sample with a volume of 10 μL was injected into the flow-cell. Fluorescent images were acquired and analyzed by image processing software (Andor IQ, Andor Technology plc). All experiments were performed in a dark room.

Experimental results

Fig. 5 shows fluorescent images of overlapped evanescent fields on the glass flow-cell deposited with mixture of TMR and FI on two emission channels and overlay for coaxial (a-b) and perpendicular (c-d) overlapping TIR configurations, respectively. Fig. 6 shows fluorescent snapshot of Brownian motion of mixture of NR and DG beads on two emission channels and overlay for coaxial (a-b) and perpendicular (c-d) overlapping TIR configurations, respectively. Due to broad emission spectra, some fluorescent emission from FI and DG beads can go through both ET585/40 and ET515/30 channels as seen in Fig. 5(a, c),

978-1-4244-2977-6/09 $25.00 © 2009 IEEE

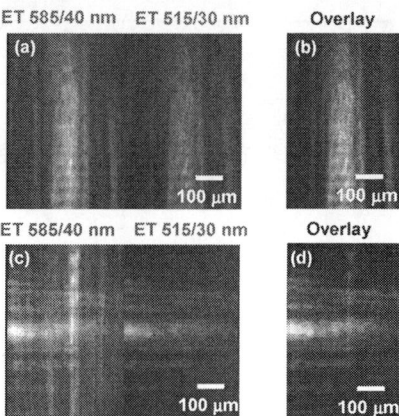

Fig. 5 *Fluorescent images (10X) of overlapped evanescent fields on flow-cell deposited with mixture of TMR dye (0.5 μM) and FI (0.5 mM) on ET585/40, ET515/30 emission channels and overlay (red for TMR and green for FI) for coaxial (a-b, Fig. 4(b)) and perpendicular (c-d, Fig. 4(c)) TIR configurations, respectively.*

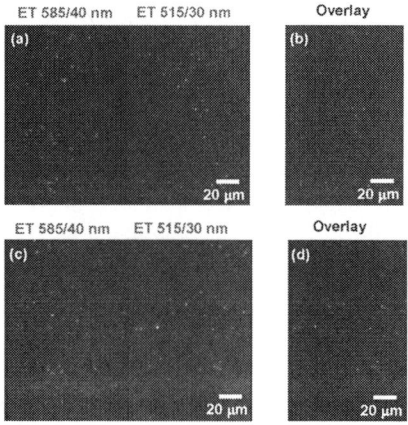

Fig. 6 *Fluorescent images (40X) of mixture of NR 0.5 μm beads (0.0125% w/v) and DG 0.5 μm beads (0.05% w/v) suspended in Tween 20 on ET585/40, ET515/30 emission channels and overlay (red for NR and green for DG) captured with coaxial (a-b, Fig. 4(b)) and perpendicular (c-d, Fig. 4(c)) TIR configuration, respectively.*

Fig. 6(a, c). Nevertheless, in both coaxial and perpendicular overlapping modes, the chip can generate two overlapping evanescent fields in accordance with our design concept. Based on our argument, we can expect the overlapping of even three to four evanescent fields. When more than two excitation fibers are introduced, the filter setup to split the emission wavelengths and the cross talk of emissions from more than two fluorophores will be two significant issues to be addressed.

4. CONCLUSIONS

The design, analysis, fabrication and demonstration of a multicolor TIR-based chip for fluorescent sensing/imaging have been presented. A set of equations governing the conditions for the overlap of up to four evanescent fields has been derived, based on which one could develop different chip configurations that still maintain the functionality. The capability of the chip in highly-sensitive and simultaneous detection and imaging of two fluorophores has been experimentally demonstrated with both coaxial and perpendicular overlapping TIR modes. Our monolithic, miniaturized chip could potentially serve as a multicolor evanescent excitation-based platform integrated into a μ-TAS.

ACKNOWLEDGEMENTS

N. C. H. Le gratefully acknowledges the MEXT of Japan for scholarship and Ritsumeikan University for RA Plus. This study was partially supported by a research grant from CREST, JST, Japan (09-043).

REFERENCES

[1] T. C. Brelje, M. W. Wessendorf and R. L. Sorenson "Multicolor laser scanning confocal immunofluorescence microscopy: practical application and limitations" Methods Cell Biol. Vol. 38, 1993, pp. 97-181.

[2] D. Demandolx and J. Davoust "Multicolor analysis and local image correlation in confocal microscopy" J. Microsc. Vol. 185, 1997, pp. 21-26.

[3] J. G. Dauwerse, J. Wiegant, A. K. Raap, M. H. Breuning and G. J. van Ommen "Multiple colors by fluorescence in situ hybridization using ratio-labelled DNA probes create a molecular karyotype" Hum. Mol. Genet. Vol 1, 1992, pp. 593-598.

[4] T. F. Watson "Fact and artefact in confocal microscopy" Adv. Dent. Res. Vol. 11, 1997, pp. 433-441.

[5] R. Sadr, M. Yoda, Z. Zheng and A. T. Conlisk "An experimental study of electro-osmotic flow in rectangular microchannels" J. Fluid Mech. Vol. 506, 2004, pp. 357–367.

[6] D. Axelrod, T. P. Burghardt and N. L. Thompson "Total internal reflection fluorescence (in biophysics)" Annu. Rev. Biophys. Bioeng. Vol 13, 1984, pp. 247-268.

[7] Y. Sako and T. Yanagida "Single-molecule visualization in cell biology" Nat. Rev. Mol. Cell. Bio. September supplement, 2003, pp. SS1-SS5.

[8] M. Yoda "Nano-Particle Image Velocimetry: A near-wall velocimetry technique with submicron spatial resolution" in BioMEMS and Biomedical Nanotechnology, Springer US, 2007, pp. 331-348.

[9] K-H. Ikuko, R. Ken, F. Takahiro, I. Ryota, M. Hideji, S. K. Rinshi and A. Kusumi "Fluorescence imaging for monitoring the colocalization of two single molecules in living cells" Biophys. J. Vol 88, 2005, pp. 2126–2136.

[10] J. Schmoranzer and S. M. Simon "Role of microtubules in fusion of post-Golgi vesicles to the plasma membrane" Mol. Biol. Cell Vol 14, 2003, pp. 1558–1569.

[11] M. Leutenegger, H. Blom, J. Widengren, C. Eggeling, M. Gosch, R. A. Leitgeb and T. Lasser "Dual-color total internal reflection fluorescence cross-correlation spectroscopy" J. Biomed. Opt.. Vol 11, 2006, pp. 040502 1-3.

[12] S. H. Kang, Y-J. Kim and E. S. Yeung "Detection of single-molecule DNA hybridization by using dual-color total internal reflection fluorescence microscopy" Anal. Bioanal. Chem. Vol. 387, 2007, pp. 2663–2671.

[13] N. C. H. Le, R. Yokokawa, D. V. Dao, T. D. Nguyen, J. Wells and S. Sugiyama "Versatile microfluidic total internal reflection (TIR)-based devices: Application to microbeads velocity measurement and single molecule detection with upright and inverted microscope" Lab Chip DOI. 10.1039/B807408A (to appear).

[14] N. C. H. Le, D. V. Dao, R. Yokokawa, J. Wells and S. Sugiyama "Highly-sensitive dual-fluorescence detection and imaging with integrated dual-color total internal reflection (TIR)-based chip" in Proc. of IEEE MHS 2008 (Micro-Nano Mechatronics and Human Science) (Nagoya, Japan, 6–9 November 2008) (to appear).

CMOS-COMPATIBLE 2-AXIS SELF-ALIGNED VERTICAL COMB-DRIVEN MICROMIRROR FOR LARGE FIELD-OF-VIEW MICROENDOSCOPES

K. Kumar, and X.J. Zhang

Department of Biomedical Engineering, University of Texas at Austin, TX, USA

ABSTRACT

A CMOS-compatible 3-mask process for 2-axis self-aligned vertical comb-driven micromirror fabrication is described. Our 1024μm diameter mirrors exhibit resonance at 2.81kHz, 669Hz, and maximum scan angles of 22°, 12° and 5.0°, 4.5° for resonant and static voltage operation on inner and outer axes. Reflectance confocal images of USAF1951 resolution target and epithelial breast tissue obtained at 3.0fps with 0.49μm, 4.18μm lateral and axial resolution over 200×125μm field of view indicate the potential of these devices for large field-of-view microendoscopes.

INTRODUCTION

Fast-scanning micromirrors are an enabling technology for numerous applications including optical communication switching [1], display [2], and point-by-point image acquisition systems [3-5]. Of particular interest is the ability to integrate such micromirrors into compact laser-scanning micro-endoscopes. *In vivo* microendoscopy is an important tool for biopsy-free disease diagnosis, guided precision surgery, and minimally invasive assessment of the treatment [6]. This miniaturized instrumentation can greatly improve patient prognosis, especially in time-critical applications (such as staging of tumors), while reducing screening costs, treatment delay, and occurrences of unnecessary and potentially harmful treatment.

Current endoscopic approaches to beam deflection across the sample usually involve sideways-imaging proximal scanning techniques, wherein a fiber-fused graded-index lens and micro-prism assembly is rotated and/or translated within a protective sheath to obtain 2D or 3D images [7, 8]. Although such techniques provide reduced form factor, their scans are slow, exhibits poor repeatability, and are useful only in tubular human organs such as the esophagus or gastrointestinal tract. Forward-imaging techniques are better suited to imaging in non-tubular organs, but have been difficult to miniaturize thus far. A fiber-optic bundle of closely-spaced single-mode fibers packaged with an objective lens has been used for *in vivo* imaging by directing illumination through each fiber sequentially [9]. Though compact, this method suffers from pixilation artifacts in imaging and low spatial resolution due to finite fiber core spacing. Microelectromechanical system (MEMS) technologies offer the unique capability to package micro-optical elements with actuators at the distal end for *in vivo* imaging. Distal scanning has been achieved by fiber/objective translation [10, 11], or by micromirror-based angular beam deflection [3-5]. Electrostatic [10] and piezoelectric [11] actuators have been used to translate the fiber or objective lens. These techniques suffer from slow scan rates and the field-of-view is limited either by demagnification of the objective lens or optical aberrations.

Micromirrors actuated in one and two dimensions by parallel-plate and vertical comb drives have been used in optical coherence tomography [5], and confocal [3] and multi-photon microscopy [4]. Both angular [12] and staggered [13] vertical comb-driven micromirrors have been shown to be useful for building forward-imaging probes, providing large actuation torque and scanning angles with smooth optical surfaces and low dynamic mirror deformation. Unfortunately, these micromirrors do not offer perfectly linear transformation between input voltage and mechanical scan angle, and can often experience scanning instabilities due to pull-in phenomena. Complicated comb self-alignment procedures have therefore been adopted in scanner microfabrication to mitigate this problem. The present research attempts to addresses these drawbacks by introducing a simple 3-mask comb self-aligned micromirror fabrication process compatible with traditional CMOS processing in the semiconductor industry. The trend of merging standard COMS process and special post-CMOS processes with MEMS have been summarized [14]. The major advantage is the monolithic integration of the IC and MEMS components toward multifunctional and intelligent microsystems. Our process only utilizes conventional silicon processing tools which operate at temperatures low enough to allow pre-fabrication of CMOS circuitry on the wafer prior to commencing micromirror fabrication. For microendoscopes, this can enable CMOS-MEMS integration [14, 15] of control electronics and sensors to adaptively correct for aberrations in beam scanning along with power amplifier drive electronics, while significantly reducing fabrication costs and lowering the barriers towards clinical applications .

FABRICATION PROCESS

The micromirror fabrication process is shown in Figure 1. Double SOI wafers, having device and buried oxide layers of thickness 30μm and 1μm respectively, serve as the starting material. Silicon dioxide (LTO) is deposited by low-temperature chemical vapor deposition to create a hard etch mask. Coarse features (mask 1) of micromirror stator structures are etched into the LTO by reactive ion etching (RIE). Photolithography of exact micromirror features (mask 2) is performed, aligned to the coarse features in LTO. Self-alignment of the rotor comb structures to the stator occurs during this alignment process. The misalignment tolerance for this lithography step is half the comb gap spacing. Since the alignment procedure is extremely simple, the minimum comb spacing is determined solely by the aspect ratio achievable by the silicon deep reactive ion etching steps following creation of the mask. Oxide RIE to remove exposed silicon dioxide completes hard mask formation. The photoresist remaining after oxide RIE is not removed and, in combination with the underlying silicon dioxide, forms the mask for deep etching. Deep reactive ion etching (DRIE), stopped on the first buried oxide layer, creates features of both stator and rotor combs in the upper device

978-1-4244-2977-6/09 $25.00 © 2009 IEEE

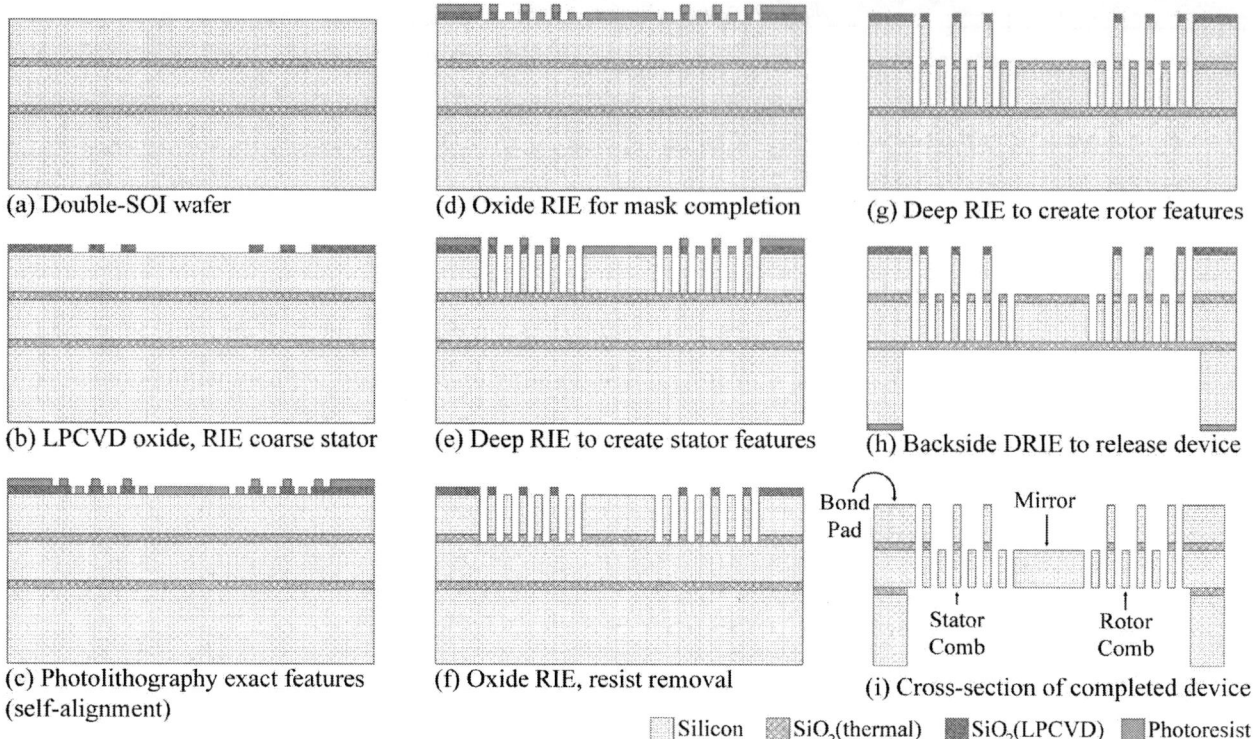

(a) Double-SOI wafer

(b) LPCVD oxide, RIE coarse stator

(c) Photolithography exact features (self-alignment)

(d) Oxide RIE for mask completion

(e) Deep RIE to create stator features

(f) Oxide RIE, resist removal

(g) Deep RIE to create rotor features

(h) Backside DRIE to release device

(i) Cross-section of completed device

Silicon   SiO₂(thermal)   SiO₂(LPCVD)   Photoresist

Figure 1: CMOS-compatible fabrication process for 2-axis self-aligned vertical comb-driven micromirror

layer. After RIE of the first intermediate buried oxide layer the photoresist component of the mask is removed, leaving the rotor features in the upper device layer unprotected by any masking element, while the stator features are still protected by a layer of LTO (Figure 1(f)). A second DRIE etch stopped on the second buried oxide layer removes the upper device layer silicon above the rotor combs while simultaneously defining the rotor features in the lower device layer. After etch completion, rotor layer features reside only in the lower layer of device silicon, while the stator layer extends through both device layers. The lower section of the stator features is redundant from an actuation perspective, but does not affect micromirror operation. Backside substrate DRIE (mask 3) followed by oxide RIE on front and back sides to remove any remaining oxide in the mask and second buried oxide layers releases the device and completes the fabrication process.

DEVICE CHARACTERIZATION

For initial experiments, we fabricated micromirrors

with 1024µm diameter on wafers without pre-fabricated CMOS circuitry. Each micromirror was actuated by two sets of combdrive actuators for each rotation axis (Figure 2). Comb finger thickness and spacing between rotor and stator combs was fixed at 8µm, and the distance between rotor tip and central axis of rotation was maintained at 200µm or 250µm. Chip size is 2.8mm².

A critical requirement of the fabrication process is to ensure optically smooth mirror surface in the completed device. We measured the mirror surface roughness using a Zygo white-light interferometry-based 3D surface profiler

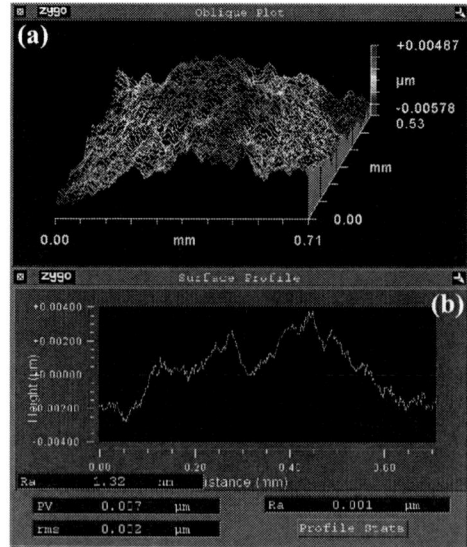

Figure 3: Sample mirror surface roughness measurement by white-light interferometry 3D surface profiler. (a) 2D map of surface roughness over 700×500µm area. (b) 1D slice plot of surface profile across mirror.

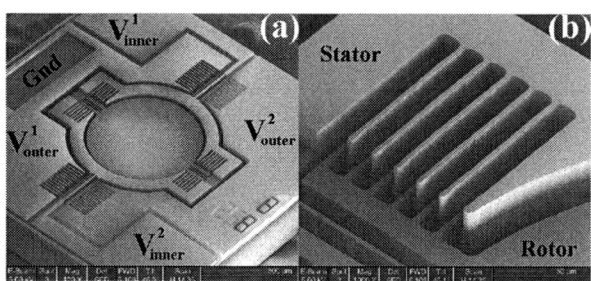

Figure 2: Scanning electron micrographs of fabricated device. (a) Oblique view showing mirror, combdrives, and electrical bond pads. (b) Close-in view of vertical combdrive actuators arrangement.

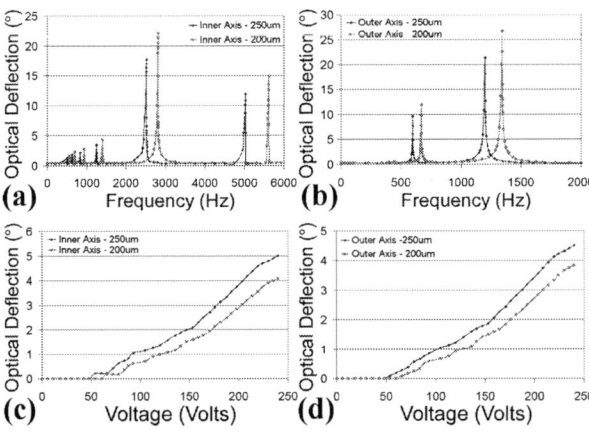

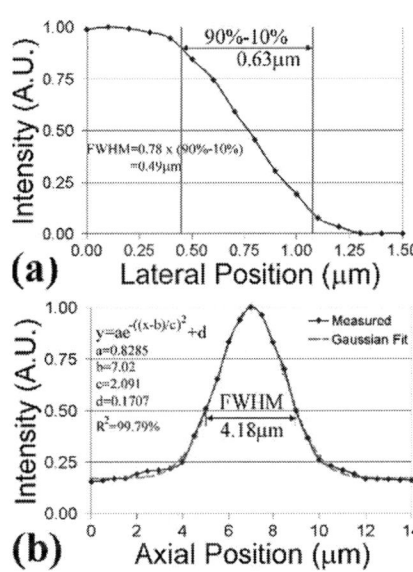

(a)

(b)

Figure 6: Experimental measurements of (a) Lateral and (b) Axial resolution of MEMS-based reflectance confocal microscope instrument.

Figure 4: Operating curves for devices having rotor tip-axis separation of 200/250µm. (a)-(b) Frequency response of inner and outer axes. (c)-(d) Static voltage characteristic (using 1 comb set) of inner and outer axes.

to be 8nm RMS on average (Figure 3).

Micromirror operating characteristics are depicted in Figure 4. The devices exhibit primary resonances (ω_0) around 2.81kHz and 670Hz, with optical scan angles of 22° and 12° on inner and outer axes respectively. Secondary resonances are observed at $2\omega_0$ and $\omega_0/2$, $\omega_0/3$, $\omega_0/4$ etc., as reported previously for micromirrors of similar configuration [16]. Scan angles of 5° and 4.5° for inner and outer axes respectively are observed on applying up to 240V of static voltage to one comb of each rotation axis. Raster scan pattern for imaging is achieved by operating the inner axis on resonance (using one inner comb bank) while actuating the outer axis using both comb banks at low frequency in non-resonant mode. Reducing the comb gap spacing and optimizing comb finger length will reduce operating voltages required to achieve a specified scan angle, while also matching the voltage dynamic range of the high voltage amplifier with the maximum scan angle of the micromirror.

IMAGING EXPERIMENTS

The devices were incorporated into a laser-scanning reflectance confocal microscope instrument, depicted in Figure 5. Linearly polarized light from a 635nm diode laser is coupled into a single-mode polarization maintaining (PM) fiber. Light exiting the fiber is collimated to 1mm beam diameter through a zero-order quarter wave-plate whose axis is oriented at 45° to the incident polarization angle, converting the illumination to circular polarization. After reflection off a stationary mirror, the illumination is incident on the micromirror at

22.5° to the micromirror normal. The micromirror scans the illumination across an objective system consisting of a 3X Keplerian beam expander and a high-NA aspheric objective lens, providing effective numerical aperture of 0.48 at the sample. Reflected light is converted into linear polarization orthogonal to the initial illumination polarization, isolated using a walk-off polarizer and offset mirror, and directed through a spatial filter into an avalanche photodetector. Image acquisition is performed through a National Instruments PCI-6111 card controlled by Matlab® software. Experimental measurements of resolution (Figure 6) by translating a mirror edge laterally across the beam spot in the focal plane, and a mirror

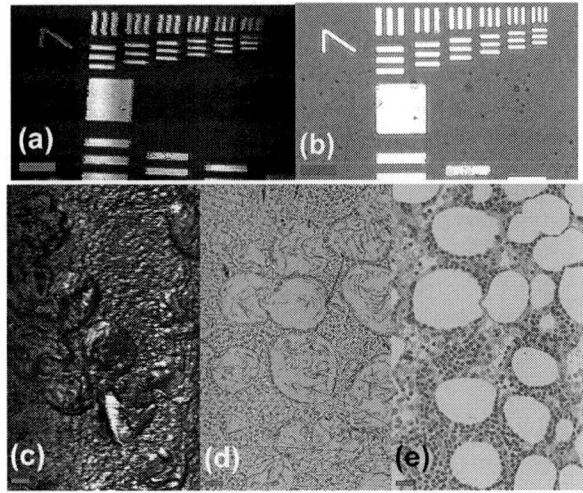

Figure 7: Imaging results. (a) USAF1951 target group 7. (b) Comparison image from Olympus BX51 microscope (20X, reflectance). (c). Human epithelial breast tissue (4 fields stitched). (d) Comparison image from Olympus BX51 microscope (20X, bright-field) (e) Image of H/E stained slice of epithelial breast tissue sectioned from just above slice used in (d) using Olympus BX51 microscope (20X, bright-field). Field of view (unstitched images) is 200×125µm. Scale bar: 25µm.

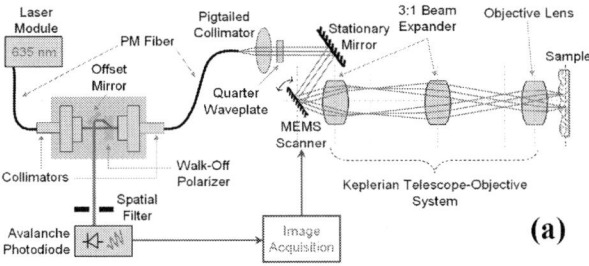

(a)

Figure 5: CMOS-MEMS micromirror-based laser scanning confocal microscope instrumentation set-up.

978-1-4244-2977-6/09 $25.00 © 2009 IEEE

axially through the focal plane showed that lateral and axial resolutions of the system are 0.49μm and 4.18μm respectively, which is comparable with conventional histology.

The instrument provided images of a 200×125μm field of view at 3.0 frames per second. The number of resolvable points in the images (408×255), which is proportional to the product of mirror diameter and optical scan angle, is a 4X improvement over previous results for similar micromirrors. Images of the elements of group 7 of a standard USAF 1951 resolution target are depicted in Figure 7(a), compared against an image acquired by an Olympus BX51 bright-field microscope with a 20X objective. The smallest lines in group 7 are of 2.2μm width. We also obtained images of fixed human epithelial breast tissue without use of contrast agents. The result of stitching four such fields together (Figure 7(c)) is compared against bright-field images using an Olympus BX51microscope with 20X objective of the unstained slice, and the slice stained with hematoxylin/eosin stain used for conventional histology.

CONCLUSIONS

A simple 3-mask CMOS-compatible process for 2-axis self-aligned vertical comb-driven micromirror fabrication is demonstrated. Reflectance confocal imaging using the device is shown with resolution comparable to conventional histology, and 4X improvement in number of resolvable points per image over previous micromirrors of similar configuration. Minimization of comb gap spacing can reduce operating voltages required. The monolithically integrated CMOS-MEMS design further enables pre-fabrication of control electronics and sensors for adaptive beam scan control, and power amplifier drive electronics in compact form suitable for next generation microendoscopes.

ACKNOWLEDGEMENT

The authors gratefully acknowledge financial support for this research from the Wallace H Coulter Foundation Early Career Award 2006–08. The micromirrors were fabricated at Stanford Nanofabrication Facility (supported by National Science Foundation grant no. 9731293) and University of Texas–Austin Microelectronics Research Center (supported by National Science Foundation grant no. 0335765) under the National Nanofabrication Infrastructure Network. Epithelial breast tissue samples were provided by Prof. Tse-Kuan Yu, Department of Radiation Oncology, University of Texas M.D. Anderson Cancer Center, Houston, TX.

REFERENCES

[1] S.-S. Lee, L.-S. Huang, C.-J. Kim, and M. C. Wu, "Free-Space Fiber-Optic Switches Based on MEMS Vertical Torsion Mirrors," *J. Lightwave Technol.*, vol. 17, pp. 7, 1999.

[2] J. Yan, S. T. Kowel, H. J. Cho, and C. H. Ahn, "Real-time full-color three-dimensional display with a micromirror array," *Optics Letters*, vol. 26, pp. 1075-1077, 2001.

[3] K. Kumar, K. Hoshino, and X. Zhang, "Handheld subcellular-resolution single-fiber confocal microscope using high-reflectivity two-axis vertical combdrive silicon microscanner," *Biomedical Microdevices*, vol. 10, pp. 653-660, 2008.

[4] W. Piyawattanametha, R. P. J. Barretto, T. H. Ko, B. A. Flusberg, E. D. Cocker, H. Ra, D. Lee, O. Solgaard, and M. J. Schnitzer, "Fast-scanning two-photon fluorescence imaging based on a microelectromechanical systems two- dimensional scanning mirror," *Opt. Lett.*, vol. 31, pp. 2018-2020, 2006.

[5] T. Xie, H. Xie, G. K. Fedder, and Y. Pan, "Endoscopic optical coherence tomography with a modified microelectromechanical systems mirror for detection of bladder cancers," *Applied Optics*, vol. 42, pp. 6422-6426, 2003.

[6] Z. Yaqoob, J. Wu, E. J. McDowell, X. Heng, and C. Yang, "Methods and application areas of endoscopic optical coherence tomography," *Journal of Biomedical Optics*, vol. 11, pp. 063001, 2006.

[7] J. Su, J. Zhang, L. Yu, and Z. Chen, "In vivo three-dimensional microelectromechanical endoscopic swept source optical coherence tomography," *Opt. Express*, vol. 15, pp. 10390-10396, 2007.

[8] P. H. Tran, D. S. Mukai, M. Brenner, and Z. Chen, "In vivo endoscopic optical coherence tomography by use of a rotational microelectromechanical system probe," *Opt. Lett.*, vol. 29, pp. 1236-1238, 2004.

[9] A. F. Gmitro and D. Aziz, "Confocal microscopy through a fiber-optic imaging bundle," *Opt. Lett.*, vol. 18, pp. 565, 1993.

[10] S. Kwon and L. P. Lee, "Micromachined transmissive scanning confocal microscope," *Optics Letters*, vol. 29, pp. 706-708, 2004.

[11] S. A. Boppart, B. E. Bouma, C. Pitris, G. J. Tearney, and J. G. Fujimoto, "Forward-imaging instruments for optical coherence tomography," *Optics Letters*, vol. 22, pp. 1618-1620, 1997.

[12] D. Hah, P. R. Patterson, H. D. Nguyen, H. Toshiyoshi, and M. C. Wu, "Theory and Experiments of Angular Vertical Comb-Drive Actuators for Scanning Micromirrors," *IEEE Journal of Selected Topics in Quantum Electronics*, vol. 10, pp. 505-513, 2004.

[13] C. Hyuck, D. Garmire, J. Demmel, and R. S. Muller, "Simple Fabrication Process for Self-Aligned, High-Performance Microscanners— Demonstrated Use to Generate a 2-D Ablation Pattern," *Microelectromechanical Systems, Journal of*, vol. 16, pp. 260-268, 2007.

[14] G. K. Fedder, R. T. Howe, L. Tsu-Jae King, and E. P. Quevy, "Technologies for Cofabricating MEMS and Electronics," *Proceedings of the IEEE*, vol. 96, pp. 306-322, 2008.

[15] C.-L. Dai, F.-Y. Xiao, Y.-Z. Juang, and C.-F. Chiu, "An approach to fabricating microstructures that incorporate circuits using a post-CMOS process," *Journal of Micromechanics and Microengineering*, vol. 15, pp. 98-103, 2005.

[16] D. Lee, "Design and fabrication of SOI-based micromirrors for optical applications," *Ph.D. Dissertation, Stanford University*, 2007.

TOPOLOGY OPTIMIZATION FOR MICRO ROTATIONAL MIRROR DESIGN AND SAFE MANUFACTURING

T. Chen[1], Z. Liu[2], J. G. Korvink[2], S. Krausse[1], U. Wallrabe[1]

University of Freiburg – IMTEK, Department of Microsystems Engineering, Freiburg, Germany

[1]Laboratory for Micro-actuators, [2] Laboratory for Simulation

ABSTRACT

We present a new design procedure for a functional MEMS design and, simultaneously, for safe manufacturing. In order to verify our approach we chose a 2.5 D compliant rotational mirror as an example, which is fabricated in single crystal silicon. The design of this compliant mechanism is based on structural topology optimization [1] with subsequent modification by parameter optimization with a pseudo-rigid-body mode analysis [2]. The fabricated compliant mechanism has a linear input at the load point, which is pushed by a piezoelectric actuator, and a rotational output at the mirror section. This single crystal silicon mechanism achieves a rotational angle of 5° with a stationary rotational center at low frequency up to 50 Hz.

INTRODUCTION

The advantages of compliant mechanisms include the simplification of assembly processes and the avoidance of thin hinges required for rotational motion [1], therefore, they are popular in MEMS. A well-designed compliant mechanism uses its flexibility to store strain energy and deliver kinetic energy [2]. Moreover, the mechanism relies on its stiffness to stabilize the kinematic motion. Hence, the main issues in compliant mechanism design are to balance the stiffness and the flexibility, so as to avoid that the maximum stress exceeds the fracture strength, as well as to conserve the maximum energy during motion.

In an initial stage of a traditional design method, finding a suitable mechanism usually consumes most of the time. If one could find a prototype without an initial guess, the design process will be more efficient. For this reason, we use the topology optimization method in the first stage of mechanism design. By this advanced optimization method, we can simultaneously find the topology, the shape and the size of a typically static structure. With the defined design domain, geometry constraints and material parameters the method generates an optimized structure without an initial guess. For dynamic application, through choosing a reasonable in-out stiffness ratio we can also design a compliant mechanism.

The mechanism obtained from topology optimization usually has a complicated topology which in addition is typically hard to fabricate. Moreover, to obtain a smooth structural boundary and to analyze the accurate nonlinear deformation directly from the topology optimization procedure is time consuming. Considering a pseudo-rigid-body model to analyze the kinematic motion can speed up the process. Based on the hypothesis of using common equations for rigid-link mechanisms, the large deflection and nonlinear behavior of a compliant mechanism can be analyzed directly. Furthermore, by using a pseudo-rigid-body model we can also simplify the initially complicated mechanism which was obtained from topology optimization, but maintain the same kinematic motion.

For implementing this design procedure a novel and simple compliant mechanism for a micro-rotational mirror is designed, which is driven by a linear actuator, transferring a horizontal input displacement at the load point to a rotational movement at the mirror section as an output. After that, we verify that the designed device can successfully be fabricated from single crystal silicon and rotate through a specific angle due to the input of a linear piezoelectric actuator. This device has an almost linear relationship between the input force, input displacement and the output rotational angle.

The paper mainly focuses on the design and the modification procedure as well as the problems that are met at the optimization stage. The description of fabrication and measurement in this paper are used to verify the success of our design procedure. We do not focus on the optical properties of this fabricated device.

DESIGN PROCEDURE

First structural layout

We used the topology optimization method to implement the first-step design. The objective of the optimization step is to obtain a reasonably large rotational angle as output with a limited linear horizontal displacement as input. A standard mathematical optimization model needed to design a compliant rotational mirror is:

$$Max: \quad \theta_{out}$$
$$s.t. \quad \nabla \cdot (\rho\sigma) = f, \quad 0 \le \rho \le 1$$
$$\int_{\Omega} \rho \, d\Omega \le Vol^*, \quad 0 \le \rho \le 1 \tag{1}$$

where θ_{out} is the output angle of the mirror, ρ is the density of the material, which is also the structural design variable. The solution of $\nabla \cdot (\rho\sigma) = f$ represents the stress field of the material to be in a static equilibrium. The mechanical equilibrium equation is used to calculate the deformation of the mechanism. Ω is the design domain and Vol^* is the material volume constraint, i.e. the maximum available volume.

For the procedure we have to define the rotational center of the mirror. It is clear that the predictable minimum displacement could be obtained if the rotational center is located at the midpoint of the mirror. By solving the mechanical equilibrium equation, the rotational angle can be expressed directly via the displacement of two points on the mirror. After that, a multi-objective and multi-constraint optimization problem has to be solved, since there are two displacement output points (point 1 and 2 in Fig. 1a). Meanwhile, it is necessary that the compliant mechanism is able to generate a specific rotation angle with a horizontal input load. Because we do not pre-specify any symmetrical constraint for the design domain, the linear input displacement also includes multi-constraints for the structural topology optimization. Point A is defined

978-1-4244-2977-6/09 $25.00 © 2009 IEEE

as the input point where the input force and displacement are along the x direction, then a mathematical expression of the constraint is:

$$\left\{ F_A \le F^*_{in}, \ U_{AX} \le U^*_{in}, \ U_{AY} = 0 \right\} \qquad (2)$$

where F^*_{in} and U^*_{in} are the possible maximum input force and displacement. U_{AX} and U_{AY} are the displacement of the input point along the X and Y axis. Based on the above analysis, the mirror design problem is:

$$
\begin{aligned}
Max: & \quad U_{2X} \\
s.t. & \quad \nabla \cdot (\rho \sigma) = f, \ 0 \le \rho \le 1 \\
& \quad \int_\Omega \rho d\Omega \le Vol^*, \ 0 \le \rho \le 1 \\
& \quad U_{1X} = -U_{2X}, \ U_{1Y} = -U_{2Y} \\
& \quad U_{inX} \le U^*_{in}, f \le F^*_{in}
\end{aligned}
\qquad (3)
$$

To solve this single objective, multi-constraint problem, the "Method of Moving Asymptotes" (MMA) algorithm [3] is used to satisfy the constraints iteratively. The square initial design domain is discretized using 50 × 50 square finite elements. The load point is specified at half the height of the left side, and the position of the mirror is kept fixed at the upper right corner (Fig. 1a). After a multiple-subdivision of the finite elements the first structural layout generated from topology optimization is shown in Fig. 1b. This prototype, generated from a linear model did not have tight stress constraints. Considering a kinematic motion with a large deformation a FEM analysis is necessary subsequently which helps to transfer the structural layout into a real design.

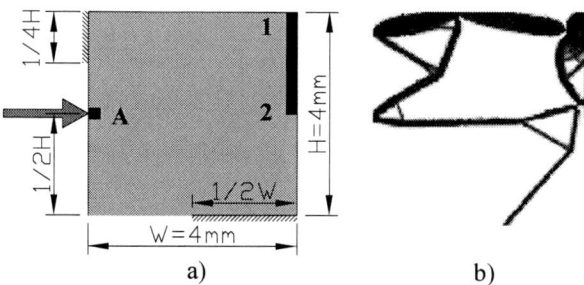

a)   b)

Figure 1: The initial design domain a) and the fist structural layout generated by topology optimization b). The load point is specified at half the height of the left side and the position of the mirror is fixed at the upper right corner.

First Design

In order to smooth the structural layout and to eliminate the stress concentration, a lower ρ with multi-iteration is implemented. Meanwhile, to improve the accuracy of the rotational center of the mirror, an ε-constraint strategy is added to constrain the tolerance in the x and y direction respectively. After an iterative FEM analysis the resulting design is shown in Fig. 2. Compared to the structural layout the load point has moved towards the mirror. The number of beams is reduced, and the beams feature a more homogeneous thickness. The actuating force needed to generate 20° of rotation angle is 0.06 N, with a maximum stress of 1.3 GPa. The dimension of this

compliant mechanism is 4 mm × 4.8 mm × 150 μm, and the minimum beam width is 12 μm.

This design is manufactured from single crystal silicon. However, in the silicon etching process we found that the very thin beams (12-15 μm) are too fragile and that they hardly survive the DRIE and sacrificial layer etching. Moreover, the high flexibility of the mechanism results in an unstable kinematic motion. The rotational center of the mirror can not be kept stationary. Thus we specify a new load point and design domain.

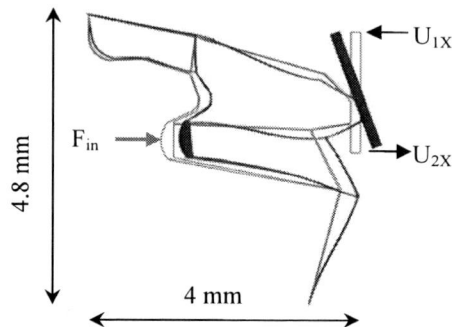

Figure 2: The simulation of kinematics of the initial compliant mechanism design. The actuating force for 20° of rotational angle is 0.06 N and the horizontal displacement needed at the input point is 300 μm. The maximum intrinsic stress is 1.3 GPa.

Second structural layout

The purpose of the second design is to balance the stiffness and the flexibility, as well as to join an amplification mechanism to a fully compliant mechanism to achieve the desired rotational angle. The dimension of this design domain with 50 × 70 square finite elements is 2.2 mm × 3 mm (Fig. 3a).

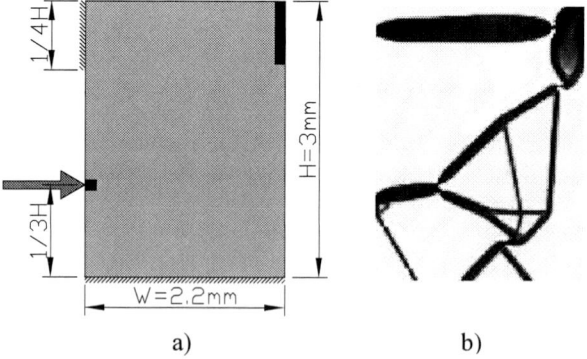

a)   b)

Figure 3: Second design domain a) and the structural layout generated from topology optimization b). The load point is specified at one-third the height of the left side and the mirror is placed at the upper right corner.

In practice, we enlarge the lever effect to decrease the displacement of input, as well as increase the stiffness to stabilize the kinematic motion of our compliant mechanism. The load point is located at one-third the height of the left side. The position of the mirror is fixed at the upper right corner. The structural layout generated by topology optimization is shown in Fig. 3b.

978-1-4244-2977-6/09 $25.00 © 2009 IEEE

Second design

As for the first design, the maximum stress was exceeded. The structural layout is also quite complicated and increases the possibility of failure during fabrication. In order to maintain the same kinematic motion, and to balance the stiffness and flexibility, and in order to avoid a large amount of numerical calculation, we use a pseudo-rigid body model to define and modify the generated mechanism for the first global modification.

Through mechanism analysis we find that the mechanism is composed of three pairs of rotational chains but with a single and accurate rotational output, see Fig. 4a. That means, by relying on the pseudo-rigid-body model we can simplify the structure layout but maintain the same kinematic motion.

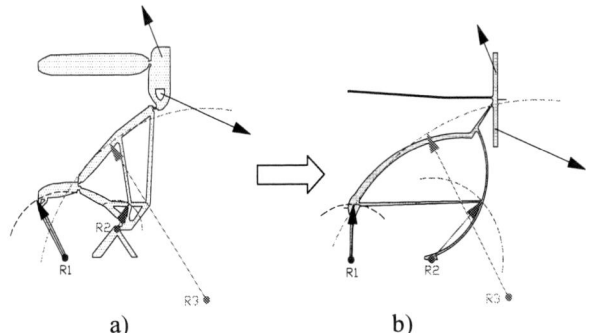

a)   b)

Figure 4: Geometry modification with the analysis of the pseudo-rigid-body mode. The purpose of this modification are decreasing the intrinsic stress and simplifying the structural construction and maintaining the same kinematic mechanism.

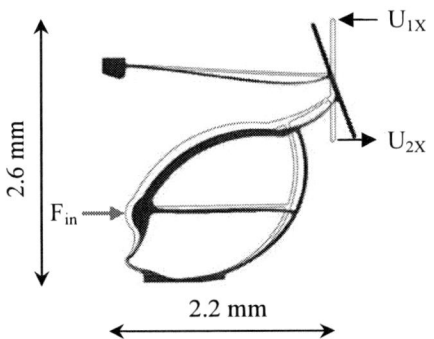

Figure 5: The simulation of kinematic movement of the new compliant mechanism design. The actuating force for 20° of rotational angle is 0.9 N and the horizontal displacement needed at the input point is 65 μm. The maximum intrinsic stress is 2.4 GPa.

To satisfy the stress requirement we use parameter optimization to modify the relative positions between the three chain pairs. The structural modifications are shown in Fig. 4b. The maximum stress is shifted from the joints to the beams and the modified structure becomes more compliant.

After that, two local modifications are implemented: for the beam below the mirror and for the beam below the load point. By FEM analysis we find that the maximum stress occurs in the beam below the mirror, with a value of up to 2.9 GPa when the mirror rotates through 20°. From the references we know that the fracture strength of bulk silicon is about 6.8 GPa [4]. In order to comply with our safety considerations we decide to set the safety factor at 3 resulting in the design of Fig. 5. Therefore, the width and angle of this beam is now modified yet retaining the same rotational angle. After modification the stress of the beam decreased by 27 % down to 2.1 GPa. Meanwhile, a problem with the beam below the load point is that a stress concentration always occurs at its end point, no matter whether it is loaded statically or dynamically. Therefore, a meander structure is added. The biggest advantage of this meander is to avoid fracture after the structure is released by ICP. During the whole modification procedure, the three pairs of rotational kinematic chains are always retained. With a larger stiffness and a minimum beam width (20 μm), this new design is now in a mechanical stability and suitable for fabrication from single crystal silicon.

EXPERIMENTAL VERIFICATION

We use conventional cleanroom processes to fabricate this MEMS structure. First of all, 10 μm of photoresist (AZ9260) is used as etch mask for dry etching but the hard bake process is skipped so as to avoid defects at top edges of sidewalls. PECVD deposited SiO_2 is used as the stop layer in the etch process. The two sides of the wafer are etched by an STS Multiplex Inductively Coupled Plasma (ICP) to shape the structure and cavity sequentially. The etch parameters are modified from the uniform recipe [5]. Most important is the handle wafer which is used to overcome thermal problems and hold the structure during the ICP etching. After the ICP etching, the photoresist and the deposited SiO_2 are removed in acetone and by an Hydrofluoric acid (HF)-dip separately.

The dimension of the fabricated compliant micro mechanism is 2.2 mm × 2.6 mm × 150 μm (Fig. 6a), which is pushed by a piezoelectric actuator whose small displacement is amplified by a precision engineered brass lever amplification mechanism. Because the sidewall roughness of our etched structure is not ideal, small scratches and nanopores are a source of fracture [6]. Hence, in our measurements the maximum input displacement is kept below 30 μm to avoid breakage of the structure. A unipolar 120 V square wave voltage is applied to the piezoelectric actuator. The measured displacement shows a linear dependence on the force and rotational angle of the mirror in one cycle of stroke (Fig. 7).

The sidewall slopes of the etched structures vary from device to device leading to different stiffness. The investigated device has a stiffness of 8.9×10^3 N/m. In a response-speed test we find that the dynamic behavior of the mirror directly follows the piezoelectric actuator at 10 Hz (Fig. 8).

The stiffness of the mechanism suppressed damping and allows for a fast response. The rotational angle of the mirror can thus be controlled very accurately. At low frequency up to 50 Hz the mirror rotated up to 5° of rotational angle with a stationary rotational center (Fig. 6b). Since the power source of the piezoelectric actuator did not supply enough current at high frequency, we observed the rotational angle of the mirror to gradually reduce at frequencies above 50 Hz. Nevertheless, the stable frequency range of the compliant mechanism for precise kinematic motion reaches up to 1 kHz.

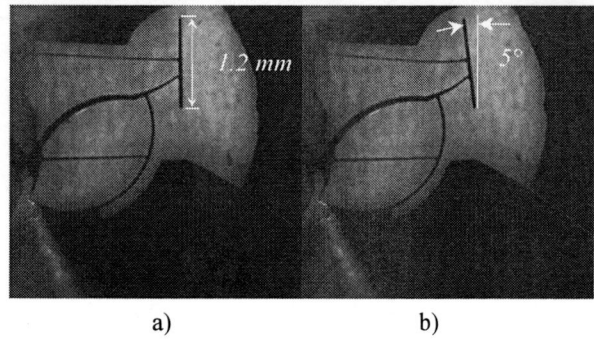

a)   b)

Figure 6: Optical micrograph showing 5° of rotational angle with a fixed rotational center. The compliant micro-rotational mirror is made of single crystal silicon. The dimension is 2.2 mm × 2.6 mm × 150 μm and the minimum beam width is 20 μm.

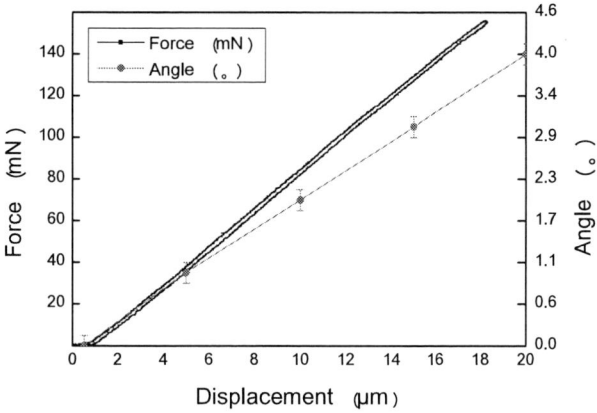

Figure 7: Measured result of the relation between displacement, force and rotational angle of the mirror. The one cycle of forward-and-backward stroke shows a linear behavior and the spring constant of the mechanism is determined as 8.9×10^3 N/m.

CONCLUSION

Topology optimization is a systematic and effective method for the design of hinge-free compliant kinematic mechanisms. To ensure manufacturability, subsequent pseudo-rigid-body mode analysis is recommended, which ensures that the kinematic and the material requirements are satisfied.

We demonstrate the application to a mirror design and show that the kinematic movement of our novel 2.5 D compliant micro rotational mirror is stable. This single crystal silicon mechanism achieves 5° rotational angle with a fixed rotational center at low frequency up to 50 Hz,

moreover, the frequency range for precise kinematic motion is 1 kHz.

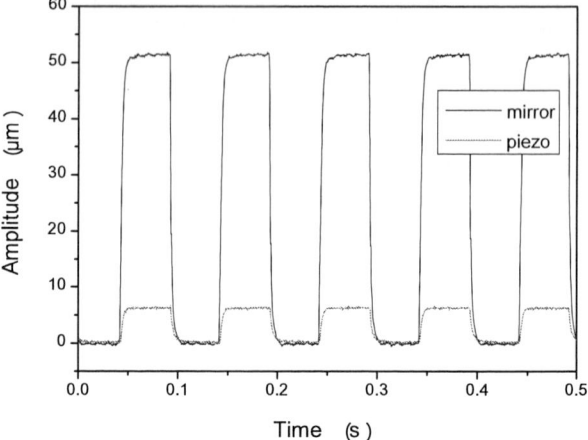

Figure 8: The dynamic behavior comparison between the piezoelectric actuator and the micro-mechanism. The device is driven by 120 V of unipolar electromotive force at 10 Hz of frequency and measured at an arbitrary point on the mirror. The deformation response of the mechanism always follows the piezoelectric actuator. The rotational angle could be controlled exactly from 0° to 5°.

ACKNOWLEDGEMENT

We appreciate technical help from the IMTEK clean room, especially from Mr. Armin Baur, and from other IMTEK scientists, namely Dr. Patrick Ruther, Dr. Joao Gaspar and Mr. Jens Brunne. The work was supported by the Ministry of Education, Republic of China (Taiwan).

REFERENCES

[1] O. Sigmund et al, Topology Optimization 2nd Edition, Springer, Berlin, 2003.

[2] Larry L. Howell, Compliant mechanisms, John Wiley & Sons, Inc., New York, 2001.

[3] K. Svanberg, "The method of moving asymptotes - a new method for structural optimization", Int. J. Num. Meth. Engrg., 24, pp. 359-373, 1987.

[4] K. Petersen, "Silicon as a Mechanical Material", Proceedings of the IEEE, vol. 70, pp. 420, 1982.

[5] H. Ashraf et al, "Advances in deep anisotropic silicon etch processing for MEMS", Sensors and Microsystems: Proceedings of the 5th Conference, Italy, February 12-16, 2000, pp. 322-329.

[6] J. Bagdahn et al, "Fracture Strength of Polysilicon at Stress Concentrations", J. Microelectromech. Syst.,vol 12, pp. 302, 2003.

STRETCHABLE YARN OF DISPLAY ELEMENTS
Seiichi Takamatsu, Kiyoshi Matsumoto, and Isao Shimoyama
The University of Tokyo, Tokyo, Japan
e-mail: (takamatsu, matsu, isao)@leopard.t.u-tokyo.ac.jp

ABSTRACT

We developed the processing technique to form three-dimensionally coiled yarns of display elements from microfabricated organic displays on flat plastic foils in order to realize stretchable electronics. Three dimensional fabrics like sweaters have been much attention as elastic substrates for stretchable electronic devices, but the deposition and patterning of functional materials on the fabric are incompatible to planar process technique of MEMS. To solve this problem, we propose the fabrication technique where firstly display components are constructed on flat plastic foils and the foils are, then, processed to form them into coiled yarns. In detail of the process, organic electrochromic displays are fabricated on the releasable parylene films above the glass. The foils with displays are cut into yarns with dicing cutters and released from the glass. To make yarns stretchable, the resultant foils are spirally rolled around elastic core fibers, forming helical coiled structure. The stretch ratio of the fabricated yarn is tuned with its pitch angle, ranging from 1 to 2. Finally, we demonstrate the projection of the selected one pixel out of 4 pixels both before and after the yarn is stretched at its ratio of 1.3.

INTRODUCTION

Recently, stretchable electronic devices like stretchable transistors[1,2] or image sensors[3] have been of much interest for the applications to wearable computers or sensor suits for robots. Their key technologies are the microstructures such as two dimensional springs or wavy PDMS rubber which enable devices to stretch themselves without destructive strain. As elastic microstructure, very attractive are the fabrics like sweater or socks which are woven with three-dimensionally coiled springs of yarns. Therefore, in MEMS 2008[4], we patterned and dyed the fabric with conductive polymer, but the direct processing of three-
of three dimensional fabric are incompatible to planer MEMS techniques. For example, the deposition and patterning of functional materials are not applicable to 3D

coiled yarn with the display elements is proposed (figure 1). Firstly, the display elements are fabricated on the planar parylene films. The foils are split and rolled around the elastic rubber core, forming them into helical coils. Their mechanical and electrical characteristics are examined.

STRETCH MECHANICS AND DESIGN OF COILED YARN

Most of the materials, including thin metal films, increase in length with regard to elasticity in tension until the yield point is marked. On the other hand, helical coiled springs are capable of sustaining a larger strain because elongation is caused by the untwisting of the coiled fibers. In the proposed yarn, the thin film yarns with electronic components are helically coiled to form helical springs. The length of the fiber is defined by the equation:

$$L = 2\pi rn \tan\theta \qquad \text{------(1),}$$

where L, r, θ, and n are the fiber length, radius, pitch angle, and number of coils. When the initial pitch angle is θ_0, the maximum strain *Stretch Ratio* is defined as:

$$StretchRatio = \frac{2\pi rn/\cos\theta_0}{2\pi rn\tan\theta_0} = \frac{1}{\sin\theta_0} \text{------(2)}$$

because the maximum length is the length of the fiber itself. For example, if θ_0 is 45°, *Stretch Ratio* is 1.41. Namely, by pulling on the structure, the coiled yarn can be stretched without fracture until the ratio of 1.41. Above the stretch ratio, the fiber is plastically deformed. For designing the stretchability of the yarn, the pitch angle is a key parameter. In the experimental section, the angles are changed, ranging from 30 to 90°.

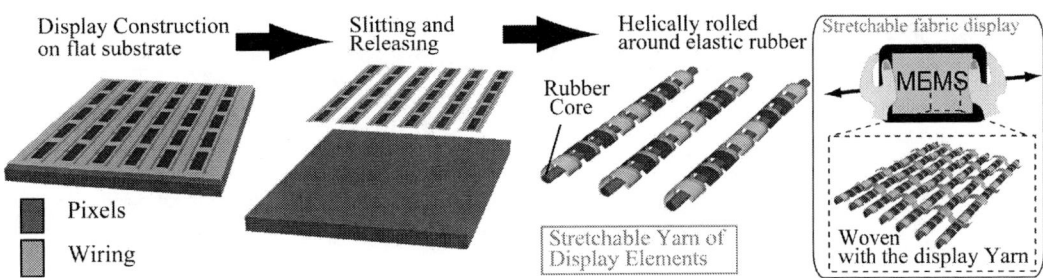

Figure 1 Schematic view of out of plain fabrication process of stretchable yarn of display elements

FABRICATION PROCESS

The construction of the helical coiled yarns with display components demands three fabrication steps which consist of the display formation on the thin plastic membrane, splitting the membrane and releasing the slit yarns from the substrates, and coiling the resultant yarns.

The display elements construction starts with Cytop and parylene coating on the glass substrates. Approximately 1 μm Cytop films are spun on the glass substrate as a dry releasing layer for parylene films. 5 μm parylene films are deposited on the Cytop as a substrate of display elements. Display elements are composed of the wiring and pixel elements with a conductive polymer of PEDOT:PSS, insulation layer and ionic liquid-based gel electrolyte. PEDOT:PSS is patterned on the parylene film with the mechanical peel-off process[4]. Cytop and 1.5 μm sacrificial parylene layers are deposited on the 5 μm parylene film. The top parylene layer is patterned and etched with an AZP4620 photoresist and an oxygen plasma etcher. PEDOT:PSS is coated on the resultant substrate and patterned with the mechanical peel-off of the top parylene layer. For insulating the wiring area of PEDOT:PSS, Cytop film is coated and patterned in the same manner as the described above. As an electrolyte for the electrochromic display, the gel electrolyte consisting of ionic liquid and co-polymer of PVDF-HFP (Kynar 2801) are placed on the substrate. Finally, parylene film is deposited on the gel for sealing. The fabricated organic electrochromic displays on the parylene films are shown in the figure 4 (a).

The thread shape of display elements is formed through the slit and release of fabricated thin film displays. The parylene foils on the glass substrate are cut by a dicing cutter at the intervals of 2.1mm which is the width of the yarn. The slit films are mechanically released from the glass substrates, by using Cytop film as a dry sacrificial layer. This is because a fluorocarbon polymer of Cytop is weakly bonded with parylene films. The figure 4 (b) exhibits the release of the yarns from the glass substrates.

The coiling of the slit films is made by helically rolling them around thermally shrinkable tubes. To define the pitch angle, the threads and the tubes are placed at the certain cross angles and the threads are reeled around the tubes. The yarns are heated at 110 C° for reducing the size of tubes. After the shrunk tubes are pulled out, the core fibers of polyurethane rubber wires are put in.

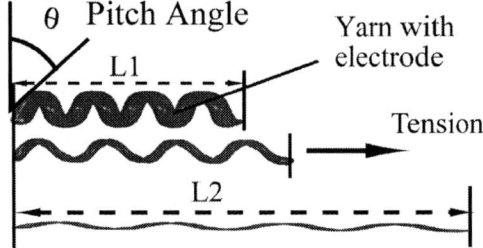

Figure 2 Stretch Ratio of helical coiled structure

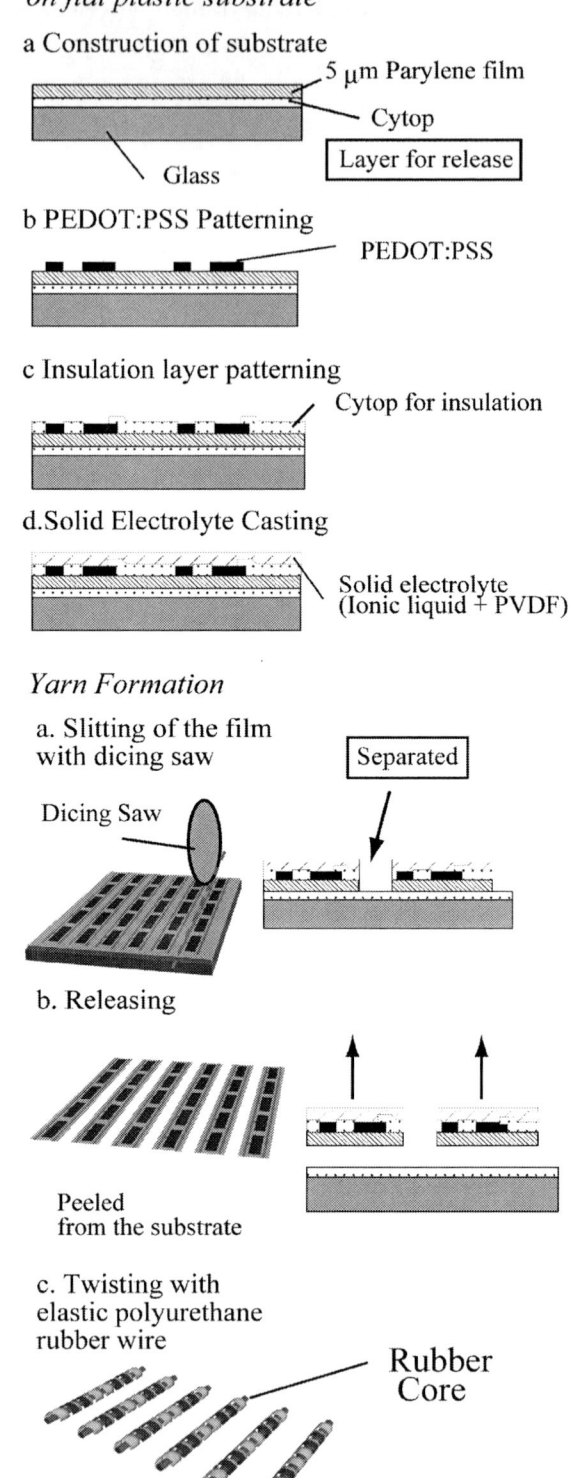

Figure 3 Fabrication process of helical coiled yarn of display elements

(a)

(b)

(c)

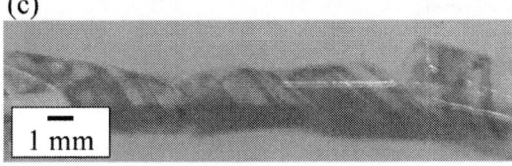

1 mm

Figure 4 Photographs of display elements (a) electrochromic display on flat plastic substrate (b) slit and released display elements (c) coiled yarn of display elements

(a)

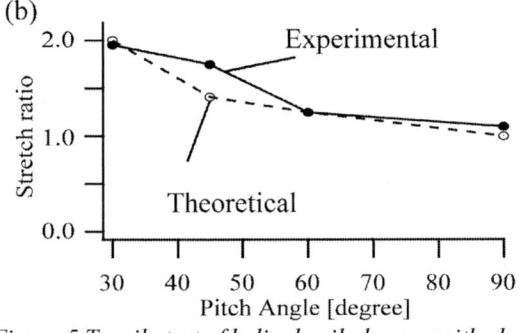

Figure 5 Tensile test of helical coiled yarns with electrodes

The helically coiled yarn with a core fiber is shown in the figure 4 (c).

TENSILE TEST OF COILED YARNS

To verify the stretchability of the fabricated yarn, tensile tests are produced. The stretch ratio of the yarns is defined by the pitch angle as described in the section of stretch mechanics. The PEDOT:PSS-coated coiled yarns with different pitch angles ranging from 30 to 90° are prepared. The both edges of the yarns are fixed on the optical stages and connected to electric resistance meter of Keithley 2400. By moving the stages, applied strains to the yarns are changed and their corresponding electric resistances are measured. Figure 5 (a) shows the diagram of the electric resistances of the yarn at the pitch angle of 45°. In the relation ship with the stretch ratios, the electric resistance does not increase before reaching the ratio of 1.41 which is the calculated limit. This is because the coils are stretched without elongation of the structural materials and finally the yarns are untwisted up. Above the ratio of 1.41, not the coils but the structural materials are stretched and fractured. The limit stretch ratio of the yarn is consistent with the equation (2).

Figure 5 (b) shows the theoretical and experimental tolerant stretch ratio of the coiled yarn with the different pitch angles. The theoretical limits are defined by the equation (2), while the experimental ones are measured in the same manner as the figure 5 (a). The theoretical and experimental limits are well fitted. Therefore, the stretchability of the yarn can be designed according to the formula of (2).

STRETCH OF DISPLAY ELEMTNTS
Electrochromic display

In the yarn with display elements, the pixels are presented by electrochromic effect of PEDOT:PSS. Electrochromism is a phenomenon where the color change is caused electrically[5]. The PEDOT:PSS exhibits the coloration from light blue to dark blue. Its electrochromic effect is induced by the reversible electrochemical reaction consisting of oxidation and reduction.

Figure 6 (a) shows the configuration of the display elements which are of a PEDOT:PSS film for pixels, electrolyte and a PEDOT:PSS film for counter electrodes. The display on plane parylene film is the all solid type of electrochromic display, replacing liquid electrolyte with the ionic liquid solidified by PVDF-HFP polymer. The color is changed if the voltage of -3V applied to the PEDOT:PSS film for pixels. Figure 6 (b) shows the coloration by the electrochromic effect.

Stretch of the yarn

Finally, we demonstrated the stretch of the fabricated yarn. 4 pixels are contained in one yarn and its dimensions are 2.1 mm in width and 20 mm in length while the pixels on the yarn are 1.5 mm in width and 4.1 mm in length. The pitch angle of the yarn is 45 degree. The selected one pixel change its color form pale to dark blue through the voltage application. The one side of it is also pulled at its stretch ratio of 1.3. The yarn is operated even if stretched, as shown in figure 7.

(a)

Parylene (sealing) Solid electrolyte
(Ionic liquid + PVDF)

PEDOT:PSS for pixels

PEDOT:PSS for counter electrode

(b)

Without Voltage biased With Voltage biased

Counter Electrode

Pixel

Wiring

Figure 6 Configuration of Electorchromic Display and its Operation

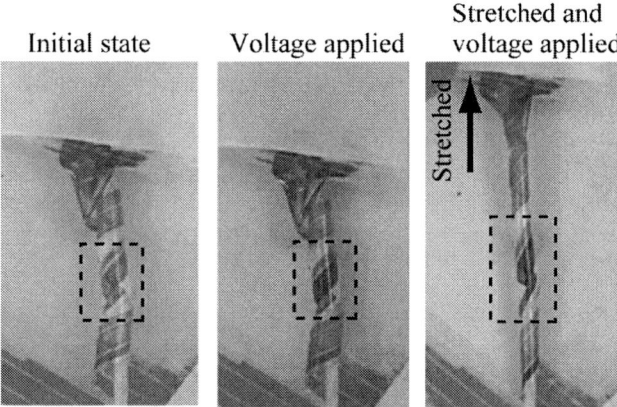

Initial state Voltage applied Stretched and voltage applied

Stretched

Figure 7 Demonstration of mechanical stretch of the yarn with display elements

CONCLUSION

The concept on out of plane fabrication technique for stretchable coiled yarns of display elements is proposed. As a display elements, a PEDOT:PSS-based electrochromic display is fabricated on thin parylene film. With dicing saw, the thin plastic display is slit to form yarns. The yarns are helically coiled by rolling them around the thermally shrinkable tubes and removing them after their shrinkage. The stretchablility of the yarn is defined by the formula of $1/\sin\theta$. The theoretical and experimental limit stretch ratios are fitted. Finally, we demonstrate the stretch of the display yarns. The coloration of selected one-pixel out of 4 pixels is observed before and after being expanded at its ratio of 1.3.

REFERENCES

[1] T. Sekitani, Y. Noguchi, K. Hata, T. Fukushima, T. Aida and T. Someya, "A rubberlike stretchable active matrix using elastic conductors," *Science*, Vol. 12, pp.1468-1472, 2008.

[2] D. H. Kim, J. H. Ahn, W. M. choi, H. S. Kim, T. H . kim, J. Song, Y. Y. Huang, A. Liu, C. Lu and J. A. Rogers, "Stretchable and foldable silicon integrated circuits," *Science*, Vol. 25, pp507-511, 2008.

[3] H. C. Ko, M. P. Stoykovich, J. Song, V. Malyarchuk, W. M. Choi, C. J. Yu, J. B. Geddes, J. Xiao, S. Wang, Y. Huang and J. A. Rogers, "A hemispherical electronic eye camera based on compressible silicon optoelectronics," *Nature*, Vol. 454, pp. 748-754, 2008.

[3] S. Takamatsu, K. Matsumoto and I. Shimoyama, "Mechanically flexible and expandable display with conductive-polymer-coated Nylon fabric," in proceedings of *MEMS2008*.

[4] S. Takamatsu, M. Nikolou, D. A. Bernards, J. DeFranco, G. G. Malliaras, K. Matsumoto and I. Shimoyama, "Flexible, organic light-pen input device with integrated display," *Sensors and Actuators B*, in press.

978-1-4244-2977-6/09 $25.00 © 2009 IEEE

A MICROMACHINED THERMO-OPTIC TUNABLE LASER

H. Cai[1], B. Liu[1], X. M. Zhang[2] W. M. Zhu[1], J. Tamil[1], W. Zhang[1], Q. X. Zhang[3] and A. Q. Liu[1]

[1] School of Electrical & Electronic Engineering,
Nanyang Technological University, Singapore 639798
[2] Department of Mechanical Engineering, University of Maryland, USA 20742
[3] Institute of Microelectronics, Singapore 117685
([†]Corresponding author: A.Q. Liu; Email: eaqliu@ntu.edu.sg; Tel: +65-6790 4336)

ABSTRACT

The paper presents a thermo-optic tunable laser that makes use of a micromachined etalon to form the external cavity. The wavelength tuning is obtained by the thermo-optic effect of the silicon material. In experiment, a wavelength tuning range of 14 nm is demonstrated by applying a heating current of 18.7 mA to a deep-etched silicon etalon of 206 μm wide. In the dynamic test, this laser measures a tuning speed of 3.2 μs, which is much faster than the typical speed of 1 ms as given by the previous MEMS tunable lasers that rely on the motion of mirrors or gratings. Since this laser is based on a different tuning mechanism of thermo-optic effect and requires no mechanical movement, it possesses many advantages such as fast speed, simple configuration and planar structure, and will broaden the applications of MEMS tunable lasers.

1. INTRODUCTION

Miniaturized and reliable tunable lasers have attracted increasing interests of research and development due to their simplicity, compatibility, and capability of spectral control. Meanwhile, tunable lasers have been intriguing because of its wide application range from sources for fiber-optic telecommunication systems to broad band sensors. In this vision, significant efforts have been devoted to the development of tunable lasers, resulting in various types of tunable lasers architectures. One of the prominent progresses is the miniaturized external-cavity tunable lasers that utilize the advanced fabrication and integration technologies of microelectromechanical systems (MEMS) [1-9]. Compared with the conventional mechanical tunable lasers, the MEMS lasers have demonstrated drastic improvements in the laser specifications such as wide tuning range, enhanced tuning speed, small size, batch fabrication, easy integration, low power consumption and so on. However, they are mostly based on mechanical movement of micromirrors and/or gratings, and thus suffer from the mechanics-related problems such as limited tuning speed (typically 1 ms), complicated movement control and unproven mechanical reliability and repeatability. To solve these problems, a fundamental solution is to utilize new mechanisms that require no mechanical movement.

It is well known that the silicon (Si) material has excellent electronic and mechanical properties. It is also a very good optical material and attractive for constructing various optical devices. Especially, it is transparent at $\lambda > 1.2$ μm, making it an exceptional waveguide medium for the 1.3- and 1.55-μm fiber-optic transmission wavelengths. Usually, silicon optoelectronic devices are based on the free carrier dispersion effect[10]. An alternative to this effect is the thermo-optic effect (TOE), namely the variation of the refractive index with the temperature of the material. The TOE of Si has a thermo-optic coefficient $(dn/dT)_{Si}$ around 1.8×10^{-4} K^{-1} at $\lambda=1.5$ μm [11], which is particular strong, about three times larger than those of the other transparent materials such as glasses. Moreover, the TOE requires only the heating (or cooling) and has no mechanical movement. Based on these understandings, this paper proposes to utilize the TOE of Si material for wavelength tuning of the micromachined tuning lasers and presents the experimental demonstration using the MEMS technology.

2. THEORY AND DESIGN

Design of tunable laser

The design of the micromachined thermo-optic tunable laser is illustrated in Fig. 1. The laser cavity consists of a Fabry-Pérot (FP) laser chip, a thermo-optic silicon etalon and a flat mirror. The Si etalon is positioned close to the FP chip as shown in Fig. 1. It has a curved shaped front surface for the collimation and coupling of the laser light. The mirror is intentionally attached on the end surface of the silicon etalon. This avoids the complication of extra cavities if there is a gap between the etalon and the mirror. Narrow serpentine wires on top of the silicon etalon are designed for resistive heating by running the electric current. As a result, a uniform temperature distribution can be obtained over the region of interest in the Si etalon during the heating. When the refractive index n_{Si} is changed, the effective external cavity length (L_{ext}) and the effective reflectance (R_{eff}) are changed accordingly. Consequently, the output wavelength of the FP chip can be tuned.

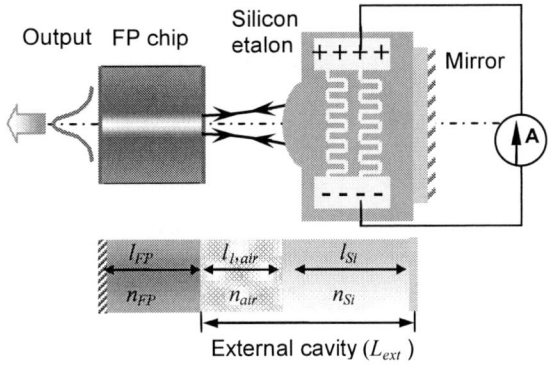

Figure 1: Schematic diagram of the MEMS tunable laser.

Optical phase of silicon etalon

When the Si etalon is heated up, it is subjected to a change of the refractive index. The corresponding optical path (ΔL) is varied as described by

$$\Delta L = \left(\frac{dn}{dT}\right)_{Si} L_h \Delta T \qquad (1)$$

where ΔT is the change in the temperature and L_h is the heater length. As a result, the peak transmission wavelength λ is anticipated to vary as determined by

$$N\lambda = 2n_{eff}L_{eff} = 2(n_{FP}l_{FP} + n_{air}l_{1,air} + n_{Si}l_{Si}) \qquad (2)$$

The light intensity I_t transmitted across the Si etalon can be expressed by

$$I_t = I_i \cdot \frac{(1-R_1)(1-R_2)}{\left(1-\sqrt{R_1 R_2}\right)^2 \cdot \left(1+\frac{4F_R^2}{\pi^2}\sin^2\phi\right)} \qquad (3)$$

where I_i is the incident light intensity; $\phi = 2\pi n_{Si}l_{Si}\cos\theta/\lambda$ is the phase coefficient, whose variation takes into account the tuning of the cavity; l_{Si} is the etalon length; θ is the incidence angle (here $\theta = 0$); and n_{Si} is the refractive index of Si (at room temperature and 1550 nm wavelength, $n_{Si} = 3.42$ [12]). F_R is the reflecting finesse of the etalon as defined by the relationship $F_R = \pi(R_1 R_2)^{1/4}/\left[1-(R_1 R_2)^{1/2}\right]$, here R_1 and R_2 are the reflectance of the two facets of the Si etalon, respectively. In Eq. (3), the absorption losses in silicon and the scattering losses induced by the roughness of the facets are neglected. With this assumption, the maximum of I_t is obtained when the resonance condition $\phi = m\pi$ is satisfied, here m is integer.

On the other hand, the phase coefficient is a function of the temperature T through n_{Si} and l_{Si}, thus its variation is given by

$$\delta\phi = \frac{2\pi l_{Si}\cos\theta}{\lambda}\delta n + \frac{2\pi n_{Si}\cos\theta}{\lambda}\delta l \qquad (4)$$

where δn is the variation of n_{Si} due to TOE, and δl is the thermal expansion of the cavity as given by $\delta l = kl\Delta T$, with k being the thermal expansion coefficient ($k = 2.6\times10^{-6}$ K^{-1} for Si). Therefore, a heating of the etalon structure induces a variation of the phase coefficient, which results in a shift of the transmission comb of the Si etalon.

The TOE of the silicon material is characterized by the empirical expression based on single oscillator model [11], which is expressed as

$$dn/dT = 1.8\times10^{-4} + 3.47\times10^{-7}T - 1.98\times10^{-7}\times T^2 \quad (K^{-1}) \quad (5)$$

Wavelength tuning of FP chip

The concept of wavelength tuning is illustrated in Fig. 2, where the tunable laser system is modeled by an effective cavity with two mirrors (R_{FP} and $R_{eff}(n_{Si})$), which are corresponding to R_1 and R_2 in Eq (1), respectively). The effective reflectance R_{eff} can be expressed as

$$R_{eff} = \frac{R_{FP} - g_{Si}R_{2,Si}}{1-R_{1,ext}R_{2,ext}g_{Si}} \qquad (6)$$

The roundtrip gain within the Si etalon g_{Si} is given by

$$g_{Si}(l_{Si}, n_{Si}, \lambda) = \exp(-2n_{Si}\alpha_{Si}l_{Si} - j4\pi n_{Si}l_{Si}/\lambda) \qquad (7)$$

where α_{Si} denotes the optical loss coefficient. Through the effective reflectance (R_{eff}), the FP chip of the MEMS laser is modulated as a function of refractive index of the Si etalon. Therefore, the lasing mode of the proposed MEMS tunable laser is determined by the coincidence of the resonant mode of the FP chip and the transmission mode of the Si etalon. In this tuning scheme, the effective cavity

could initially have many possible modes for lasing (Fig. 2a). However, at a specific reflectivity R_{eff} (e.g. $R_{2,eff}$), only the mode of λ_2 can be amplified by the cavity for lasing, the rest modes are filtered (Fig. 2b). Once the Si etalon is heated up by the metal heater, R_{eff} is changed (e.g. $R_{3,eff}$). Consequently, the mode of λ_3 (instead of λ_2) is supported for lasing. In the other word, the mode selectivity of the laser cavity is dependent on the refractive index change of the Si etalon, which is in turn determined by the temperature of the etalon.

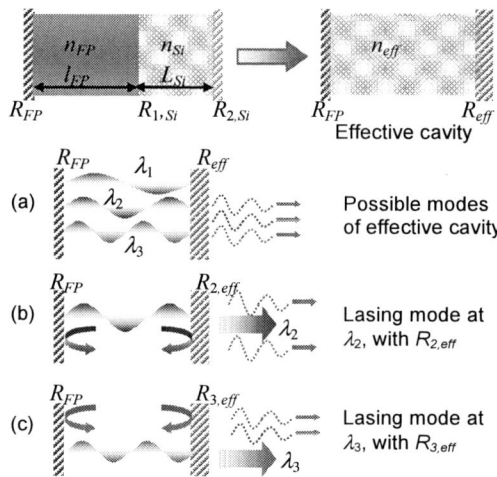

Figure 2. Mechanism of the wavelength tuning based on TOE, with different lasing modes corresponding to the different reflectivity controlled by the Si etalon.

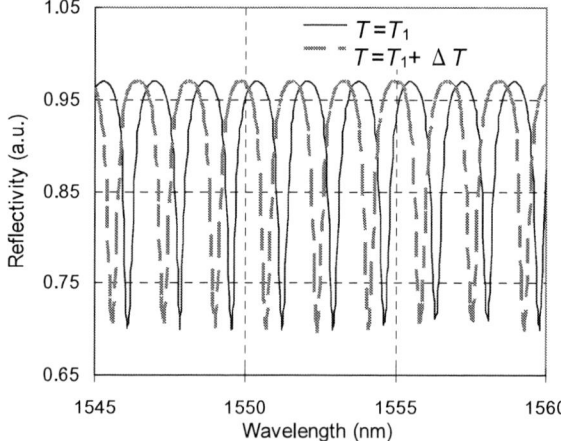

Figure 3. Calculated wavelength shift and reflectivity as a function of the wavelength for a Si etalon at two different temperatures, with the temperature difference $\Delta T = 6.5$ K.

Figure 3 gives the simulation results of the etalon reflectivity and the shift of the transmission peaks in response to the temperature change. When the temperature is increased by 6.5 K, the peaks of the reflectivity are shifted by 0.8 nm. In such case, the thermal expansion effect is neglected since its contribution to the tuning of the cavity is very small.

3. FABRICATION PROCESS

The overview of the fabricated MEMS tunable laser is shown in Fig. 4 by the scanning electron micrograph. All the MEMS structures, including the Si etalon, are fabricated on a silicon-on-insulator (SOI) wafer by deep-reactive-ion-etching (DRIE). The structure layer is 75 μm and the oxide buffer layer is 2 μm thick. Before the etching process, an aluminum layer of 0.5-μm is deposited and patterned for Si etalon heating and electrical connections. After the etching process, the device is undercut with buffered oxide etch (BOE) in order to fully electrically isolate the FP chip bottom contacts and the Si etalon. After that, another layer of 0.2-μm thick aluminum is coated on the end flat facet of the etalon by thermal evaporation. Following that, the FP chip with the size of 250 μm long by 250 μm wide is assembled. Finally, the output fiber is inserted and fixed to the MEMS substrate through the self alignment to the etched tapered fiber trench.

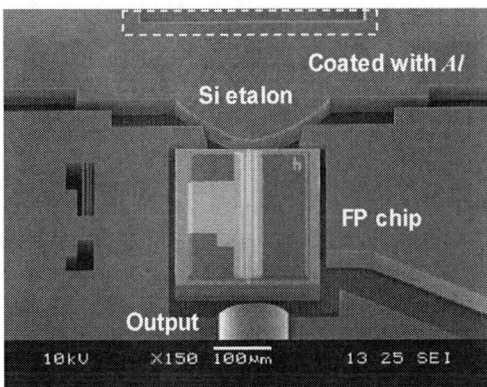

Figure 4. Scanning electron micrograph of the integrated tunable laser

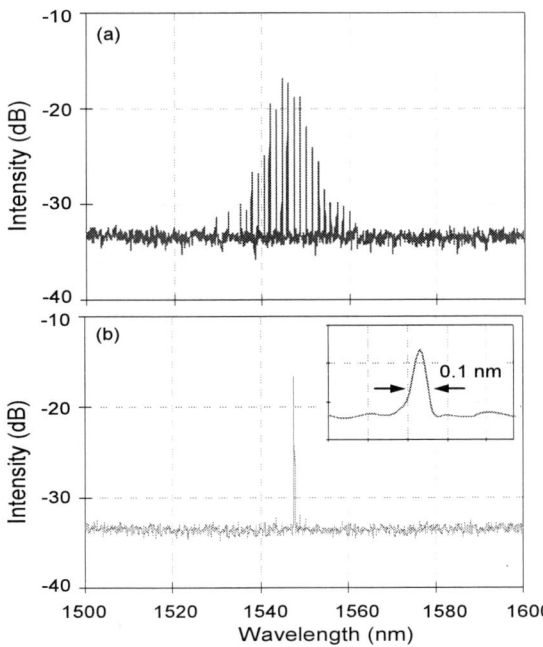

Figure 5. Comparison of the output spectra of the micromachined laser. (a) Original multimode output of the single FP chip; and (b) single mode output spectrum of the micromachined laser.

The Si etalon is design with a curvature profile of front surface, while the rear surface is flat. To further reduce the coupling loss between the FP chip and the Si etalon, the curved front surface of the etalon has been optimized, with a curvature radius of 175.08 μm. On the other hand, to improve the reflectivity of the rear surface,

an aluminum metal coating is applied. As a result, the flat facet acts as a mirror. The long narrow heating elements are formed by serpentine aluminum wires of 2 μm wide on the top of the Si etalon.

4. EXPERIMENTAL RESULTS

Before evaluating the optical performance of the micromachined laser, the FP chip is characterized before assembly. The original spectrum output of the FP chip is shown in Fig. 5(a), with an injection current of 14.1 mA (its threshold is 12 mA). It is observed that the single chip operates in a multimode regime. After the chip is assembled onto the MEMS substrate to build the micromachined laser, the output spectrum is then characterized. During the experiment, the output spectrum is changed to single longitudinal mode as shown in Fig. 5(b) thanks to the wavelength selection property of the Si etalon. The inset exemplifies a close-up of the measured output spectrum of the device, showing a narrow linewidth of ~ 0.1 nm.

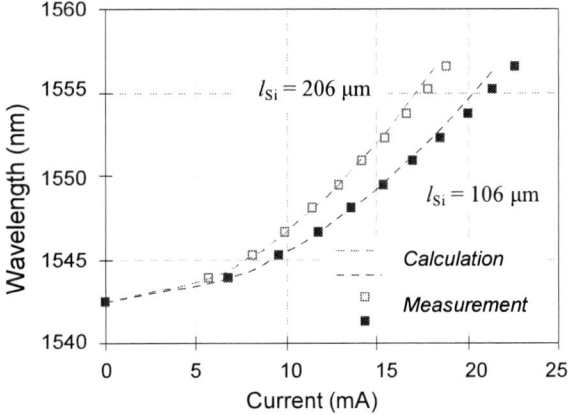

Figure 6. Wavelength tuning as a function of the tuning current for different Si etalons.

The static and dynamic performances of wavelength tuning are both important. In the former case, the Si etalons with different cavity lengths (l_{Si}) are fabricated for different tuning characteristics. Fig. 6 plots the wavelength shift versus the tuning current for two laser devices in which the Si etalons have the lengths of 206 μm and 106 μm, respectively. Experimental results show that both devices can obtain the same wavelength tuning range of 1542.5 to 1556.5 nm. However, the etalon with the length of 206 μm requires 18.7 mA while the other needs 22.7 mA. It shows that the laser with shorter Si etalon requires more current variation to achieve same wavelength tuning. The theoretical results are also plotted in Fig. 6. The discrepancy between theoretical and experimental modulation patterns is mainly due to those effects inherent in the experiments such as the nonideal characteristic of the output (which is not a plane wave) from the FP chip, coupling loss within the external cavity, diffraction effects and the roughness of the semi-reflective Si-air interfaces [13]. Additionally, the output power level observed for cavity length of 106 μm experiences a decrease as compared to that of the length 206 μm. It might be because that higher current induces more temperature increment but it causes a decay of the performance of the FP chip. In the dynamic response experiment, the tuning speed measures approximately 3.2

978-1-4244-2977-6/09 $25.00 © 2009 IEEE   1025

μs, which is comparable with the thermo-optic switching speed demonstrated in our previous work [12]. It is about 1000 times faster than the typical tuning speed of 1 ms in the other micromechanical tunable lasers [2-6].

Figure 7 shows the superimposed output spectra and the wavelength tuning relationship. In Fig. 7 (a), the laser output maintains a single longitudinal mode with the increase of the tuning current, but the wavelength peak moves to longer wavelength. According to Fig. 7 (b), the laser wavelength increases almost linearly when the tuning current applied to the Si etalon is increased from 0 to 18.7 mA. At the same time, the side-mode-suppression-ratio (SMSR) is also improved, with an average value of 22 dB.

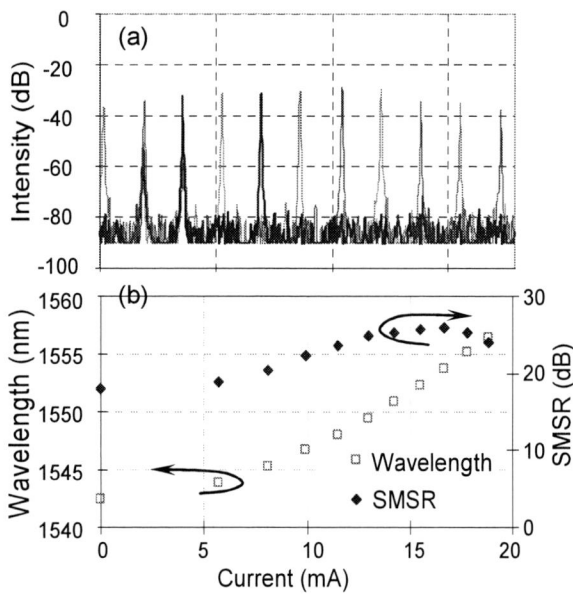

Figure 7. Measured wavelength tuning. (a) Superimposed spectra of the laser output, and (b) wavelength shift as a function of the heating current .

5. CONCLUSIONS

In conclusion, this work demonstrates a MEMS tunable laser that utilizes the thermo-optic effect rather than the mechanical movement for wavelength tuning. The external cavity is constructed by a deep-etched silicon etalon, whose far end is coated with aluminum to serve as a flat mirror, and aluminum wires are patterned on top of the etalon as the microheater. Two devices are fabricated and tested; one has an etalon width of 106 μm while the other is of 206 μm. Both of them demonstrate the wave tuning range of 14 nm, but the longer etalon requires only 18.7 mA as compared to the 22.7 mA of the shorter etalon. The laser devices measure a fast response speed of 3.2 μs. This type of TOE-based MEMS tunable laser requires no mechanical movement for wavelength tuning and thus solves the fundamental problems in the previous demonstrated micromechanical tunable lasers such as limited tuning speed (typically 1 ms), complicated movement control and mechanical stability and reliability issues. In addition, this laser has simple configuration and planar structure, making it suitable for the integration with other semiconductor electronics and high-density photonic integrated systems.

ACKNOWLEDGMENTS

The authors gratefully acknowledge the support by the Agency for Science, Technology and Research (A* STAR) of Singapore under grant no 042 108 0097. Special thanks go to Dr. Yu Aibin and Dr. Selin Teo Hwee Gee for their support to this research work.

REFERENCES

[1] H. Ukita, Y, Uenishi and Y, Katagiri, "Applications of an extremely short strong-feedback configuration of an external-cavity laser diode system fabricated with GaAs-based integration technology," *Appl. Opt.* vol. 33, pp.5557–5563, 1994.

[2] Y, Uenishi, K, Honma and S, Nagaoka, "Tunable laser diode using a nickel micromachined external mirror," *IEE Electron. Lett.* vol. 32, pp.1207–1208, 1996.

[3] X. M. Zhang, A. Q. Liu, D. Y. Tang, and C. Lu, "Discrete wavelength tunable laser using microelectromechanical systems technology," *Appl. Phys. Lett.* vol. 84, pp.329–331, 2004.

[4] A. Q. Liu , X. M. Zhang, D. Y. Tang and C. Lu, "Tunable laser using micromachined grating with continuous wavelength tuning," *Appl. Phys. Lett.* vol. 85, pp.3684–3686, 2004.

[5] X. M. Zhang, A. Q. Liu, C. Lu, and D. Y. Tang, "Continuous wavelength tuning in micromachined Littrow external-cavity lasers," *IEEE J. Quantum Electron.* vol. 41, pp.187–197, 2005.

[6] X. M. Zhang, H. Cai, C. Lu, C. K. Chen and A. Q. Liu, "Design and experiment of 3-dimensional micro-optical system for MEMS tunable lasers," *Proc. MEMS'2006,* Turkey, 22–26, Jan.2006, pp. 830–833.

[7] Heikkinen V, Aikio J K, Alojoki T, Hiltunen J, Mattila A-J, Ollila J and Karioja P, "Single-mode tuning of a 1540-nm diode laser using a Fabry-P´erot interferometer," *IEEE Photon. Technol. Lett.* vol. 16 pp. 1164–1166, 2004.

[8] H. Cai, A. Q. Liu, X. M. Zhang, J. Tamil, Q. X. Zhang, D. Y. Tang, C. Lu, "A miniature tunable coupled-cavity laser constructed by micromachining technology," *Appl. Phys. Lett.* vol. 92, (03), 031105, 2008.

[9] H. Cai, A. Q. Liu, X. M. Zhang, J. Tamil, D. Y. Tang, J. Wu and Q. X. Zhang, "Tunable dual-wavelength laser constructed by silicon micromachining," *Appl. Phys. Lett.* vol. 92, (05), 051113, 2008.

[10] R. A. Soref and J. P. Lorenzo, "All-silicon active and passive guided-wave components for λ = 1.3 and 1.6um," *IEEE. Quantum Electron.,* vol. EQ-22, pp. 873-879, 1986.

[11] F. D. G. Corte, M.E. Montefusco, L. Moretti, I. Rendina and G. Cocorullo, "Temperature dependence analysis of the thermo-optic effect in silicon by single and double oscillator models," *J. Appl. Phys.* vol. 88, pp.7115-7119, 2000.

[12] T. Zhong, X. M. Zhang, A. Q. Liu, J. Li, C. Lu, And D. Y. Tang, "Thermal-optic switch by total-internal reflection of micromachined silicon prism," *IEEE J. Select. Topics Quantum Electron.* vol. 12, no. 2, pp. 348-258, 2007.

[13] S. Solimeno, B. Crosignani, P. D. Porto, *Guiding, diffraction and confinement of the optical radiation*, Academic Press, Orlando, pp. 522-529, 1986.

MASS-ANALYSIS SCANNING FORCE MICROSCOPY WITH ELECTROSTATIC SWITCHING MECHANISM

C.Y. Shao[1], Y. Kawai[1], T. Ono[1], and M. Esashi[2]

1Graduate school of Engineering, Tohoku University, Japan
2The World Premier International Research Center Initiative for Atom Molecule Materials,
Tohoku University, Japan

ABSTRACT

A time-of-flight scanning force microscope (SFM) probe with an ability to switch the measurement mode using the electrostatically switching mechanism is designed, fabricated and demonstrated the performance of it. In order to achieve surface observation and chemical analysis simultaneously for imaging the chemical property of a material surface, this probe can switch the positions of the cantilever between SFM mode and time-of-flight mass analysis (TOF-MA) mode by integrating a couple of electrostatic actuator with curved electrode. This mechanism will be applied to pick up an atom or molecule under SFM mode, then emits them to TOF mass analyzer using field evaporation for analyzing its mass in TOF-MA mode. To switch the cantilever position at TOF-MA mode, the fabricated probe generated the 255μm of maximum displacement at the end of cantilever at an actuation voltage of 180V. The cantilever is attracted with the electrode according to the curved shape using electrostatically pull-in effect. The front edge of the cantilever was aligned in front of integrated extraction electrode for emitting chemical species. In SFM mode, the cantilever was also attracted to another electrode. The fundamental resonant frequency of the cantilever is increased from 1.8kHz to 6.8kHz before and after actuation. A calculated spring constant is changed from 0.05N/m to 0.34N/m.

INTRODUCTION

In nano-science, scanning probe microscopy is playing an important role to investigate of surface properties for a large variety of science field [1]. A lot of powerful analytical techniques are invented using this instrumentation with a high spatial resolution. Scanning Force microscopy (SFM) techniques enable us to obtain various physical and chemical properties on a surface of conductive and insulative material in nanometer scale. The SFM opened up entirely new possibility for studying the structure and dynamics of individual molecules and atoms in surface science. SFM family has been wildly used as an essential tool not only in surface science but also in nano-science, with application in nano-materials, bioscience [2] and nano-engineering. One of the drawbacks of the SFM technique is that it did not have capability for chemical identification of surface. To satisfy this requirement, Time of flight scanning force microscopy (TOF-SFM), which is a scanning force microscopy (SFM) combining with a time-of-flight mass analyzer (TOF-MA), is demonstrated to obtain topographical image of a surface with atomic resolution and identify individual molecule or atom on surface for chemical information simultaneously [3]. In this TOF-SFM, a cantilever probe picks up a single molecule or atom and emits it using field-assisted evaporation toward the TOF mass analyzer. In order to

emit the atom, an extract electrode is placed near the tip of the probe. The cantilever is bended by the actuator for directing its tip position in front of the extract electrode. The integration of the extract electrode into the probe provides low voltage operation in the field-assisted emission. The cantilever should be integrated with an actuator for fast iteration analysis and better position repeatability. This operation requires a large actuation of the probe. However, the power of miniaturized actuator is limited; there for, its stiffness must be reduced to satisfy this requirement. On the other hand, the tip of the probe with a small stiffness often sticks on surface by van der waals force, which causes instability in imaging and gives damage to the tip.

Regarding to the actuator for TOF-SFM, a non-thermal type actuator has benefit for obtaining a high resolution because of less thermal noise. The different types of micro-fabricated probes, which had turning mechanism for stiffening spring constant of the probe using integrated piezoelectric unimorph actuator, has been developed to apply TOF-SFM[4]. However, in TOF mode, the actuator cannot define the position of the probe precisely due to the instability of itself such as hysteresis of piezoelectric material. To fix the cantilever positions of TOF mode and SFM-mode, the position stopper, which is well aligned against extraction electrode and sample surface, is strongly required to switch the probe modes.

In this study, we developed a novel time of flight scanning force microscope probe with an ability to switch the measurement modes between SFM-mode and TOF-mode using an electrostatic switching mechanism with curved electrodes as shown in Fig.1. To change the measurement mode, the integrated electrostatic actuator can bent the cantilever according to the shape of the each curved electrodes. Electrostatic actuator integrated with

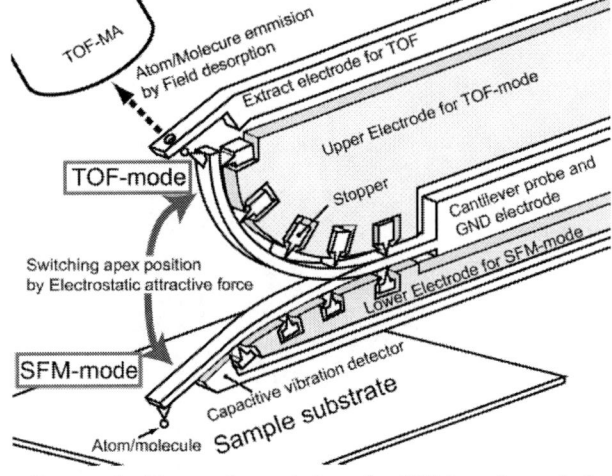

Figure 1: The probe switches the SFM mode and the TOF-MA mode.

978-1-4244-2977-6/09 $25.00 © 2009 IEEE

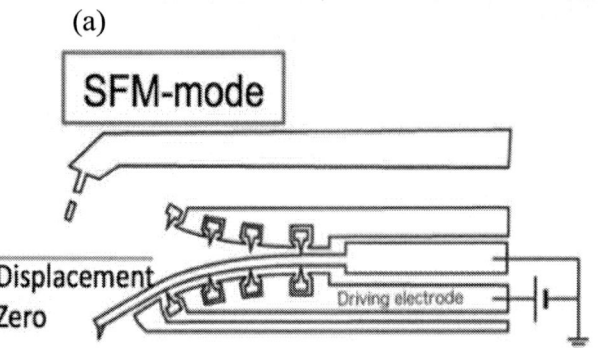

(a)

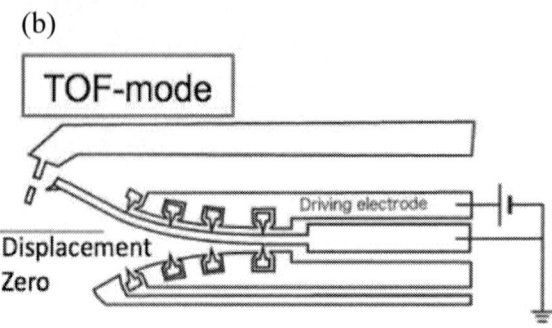

(b)

Figure 2: Schematic figure of the switchable cantilever with curved electrode
(a)In SFM mode, the cantilever is bended by electrostatic force for surface observation and atom manipulation.
(b)In TOF mode, the cantilever bends up and emit the manipulated to TOF mass spectrometer via fixed extract electrode

curved electrodes is simple and easy to achieve large displacement to change position in MEMS application [5-7]. The location of the end of the probe is fixed with lined up stoppers in upper electrode in TOF-mode. In SFM-mode, the actuator also attracts the cantilever on contact with lower stoppers in electrode. The attraction increases the spring constant of the cantilever due to shortening the length of vibration part of cantilever. We present the design, fabrication and characterization of the probe regarding to a static displacement and frequency shift induced by the attraction using integrated curved electrostatic actuator.

DESIGN AND FABRICATION

The TOF-SFM with electrostatic switching mechanism is designed and fabricated. This switching mechanism consist of a cantilever, a couple of curved electrodes for actuating the cantilever and there are many stoppers distributed along the curved electrodes to avoid electrical shortage, capacitive vibration sensor and extraction electrode are place for a force detecting resonate frequency shift in SFM-mode and field evaporation in TOF-mode. The Figure 2 shows the basic working principle of this switching mechanism of the probe. In Figure 2 (a), In SFM mode, the cantilever is attracted onto the lower curved electrode by electrically pull-in phenomenon using electrostatic actuator for measuring the topography and manipulate atom or molecule from the sample surface. Figure 2(b) shows TOF-mode, the cantilever is also attracted onto the upper curved electrode by pull-in effect for field-emission of atoms or molecules toward the

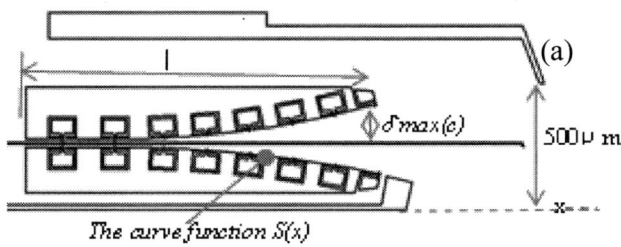

(a)

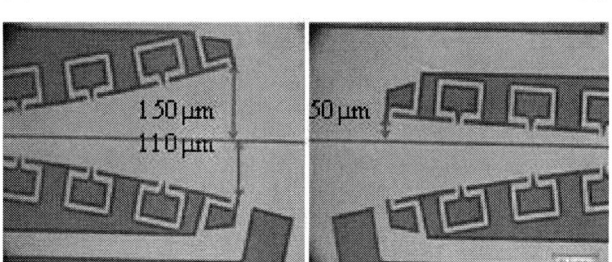

(b)   (c)

Figure 3: (a) Schematic layout of the switching cantilever (b) and (c):Optical photograph of three

TOF-MA. In order to generate large displacement, actuators use curved structures with a specific shape are employing. Figure 3 shows the basic design of the cantilever and curved electrode, for admissible trial function of the deflection profile of the cantilever beam, the following expression has been used for the shape of the curved electrodes.

$$S(x) = \delta_{max}\left[\frac{x}{l}\right]^{n}$$

Where x is the position along the x-axis, and the maximum gap distance of the curved electrode is δmax, l is the length of the beam. n is the polynomial order of the curve ($n \geq 0$).There are three different δmax which correspond to different curve slope two of three is shown in Figure 3 (b) (c). The fabrication process is shown in Figure 4. The staring material is a 20mm x 20mm Pyrex glass with 200 μm thickness as a handling base substrate. To separate each completed device easily, both sides of the Pyrex glass were formed holes and grooves by sand blast (a). A Si substrate with 300μm thickness was bonded with the processed Pyrex glass using anodic bonding (b). the Si

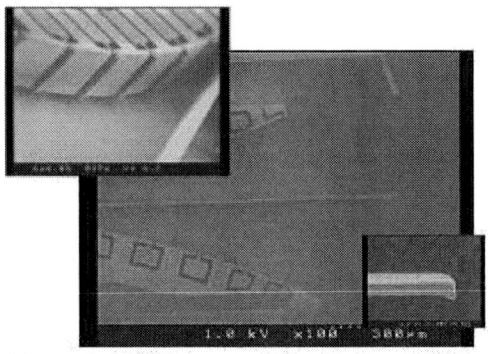

Figure 5 SEM image of the fabricated TOF-SFM probe.

978-1-4244-2977-6/09 $25.00 © 2009 IEEE   1028

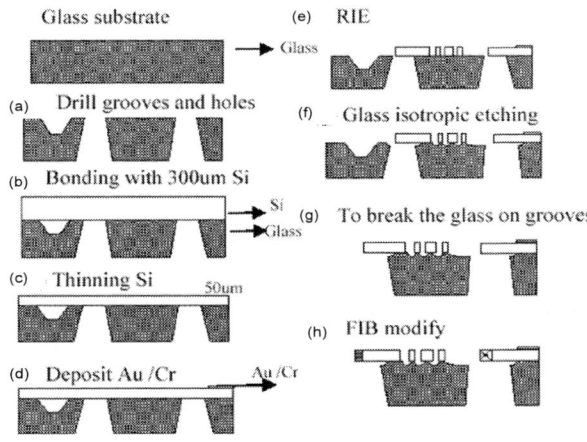

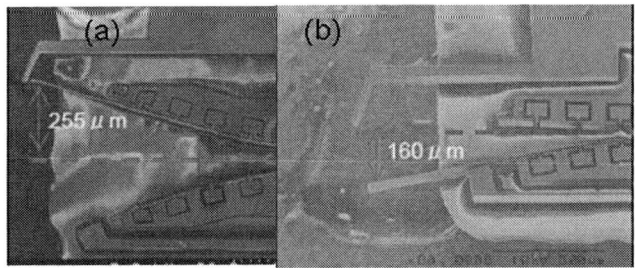

Figure 4 Process sequence of key microfabrication steps

layer is polished from 300μm down to 50 μm in thickness(c). A Cr-Au film with a total thickness of 230 nm as electrical contact pads was formed by lift off technique (d). The actuator structures were defined by reactive ion etching (RIE) dry etching (e). Next, the glass isotropic wet etching using HF is employed to release the cantilever structure from the glass substrate (f). The devices were mechanically diced along the grooves, which are formed by sand blast (g). At final step (h), the sharpened tip at the end of the cantilever and the extraction electrode pattern were formed using focused ion beam (FIB). Figure 5 shows the SEM images of the fabricated probe. There are eight stoppers in the both electrodes. Top of the stopper is protruded with 2μm from the surface of the electrode.

DEVICE EVALUATION

The fabricated actuator can switch the position of the cantilever electrostatically. The three types of curved electrode with different maximum gap have been evaluated. The driving voltage is applied from direct current voltage source using the contact probe with fabricated contact pad of the probe. According to the optical image as shown in Figure 6 (a) (b), the maximum deflection of the different slope curved electrodes are obtain. Figure 6 (a) shows the strongest slope curve of electrode with δmax = 150 μm. The probe can generates largest displacement of 225μm when 180V actuation voltage is applied to upper electrode. In Figure 6 (b), another type of probe with δmax = 110um generates the displacement of 157μm when 135 V of actuation voltage is applied to the lower electrode. To investigate the pull-in voltage, the characteristics of static actuation are measured with the three types of probes. The driving voltage was slowly increased until that all stoppers contact the actuated cantilever. Figure 6 (c) indicates the changes of displacement against applying voltage. The maximum displacement of the cantilevers, which have the δmax = 50μm, 110μm, 150μm, are 55μm, 157μm, 150μm each when the cantilever was attracted with stoppers completely. In three probes, the displacement was increased discrete and cantilever contacts stoppers from support side sequentially as applying voltage increases. The pull-in voltage is increased by stronger slope of electrode curve and by increasing gap distance. The

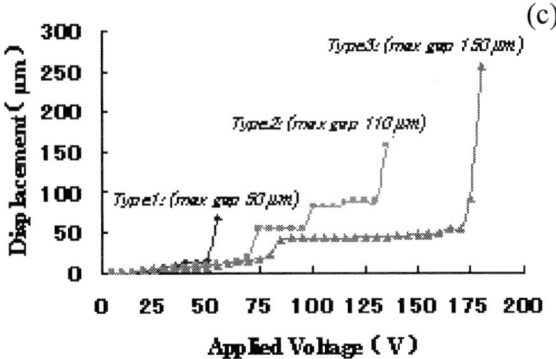

Figure 6: Demonstrated electrostatic actuation. (a) applying upper electrode (b) applying lower electrode (c)actuation

number of pulled-in steps at electrode with the gently slope is smaller than that of steep one. In higher voltage, all probes generate no more displacement.

In order to demonstrate the possibility for TOF- SFM, the fabricated probe is driven with switching mechanism and spring stiffening. Figure 7 shows the characteristics of switching actuation of the probe, which has the gap of δmax 50μm in upper side and 110μm in lower side. Total displacement of the switching mechanism is 230μm whenthe applying voltage is switched from 180V with lower electrode to 180V with upper electrode. In spring stiffening evaluation, the resonant frequency of the cantilever in SFM mode is measured for explaining relevance between pull-in phenomenon and stiffening effect. The electrostatic actuation was supplied by DC voltage using DC source. The vibration spectrum of the cantilever was measured using laser Doppler vibrometer and network analyzer. Figure 8 shows the fundamental resonant frequency of first flexural mode depending on the applying voltage to the lower electrode. The resonant frequency of the cantilever is shifted up according to above pull-in step of displacement. Before each pull-in steps the resonant frequency is little decreased due to as increasing electrostatic attractive force. In SFM-mode, which the cantilever is attracted with the stoppers completely, 1.8 kHz of the initial resonant frequency of the cantilever rerating with the vibration can be changed to 6.8 kHz of that with this switching operation. This change due to switching can increase the spring constant to obtain stable SFM image. The calculated spring constant is changed from 0.05N/m to 0.34N/m. Form the obtained large displacement and hardening effect, it is expected that this probe can be applied to TOF-SFM for combination of image and mass spectroscopy of atoms and molecules.

978-1-4244-2977-6/09 $25.00 © 2009 IEEE

for atomic or molecule desorption.

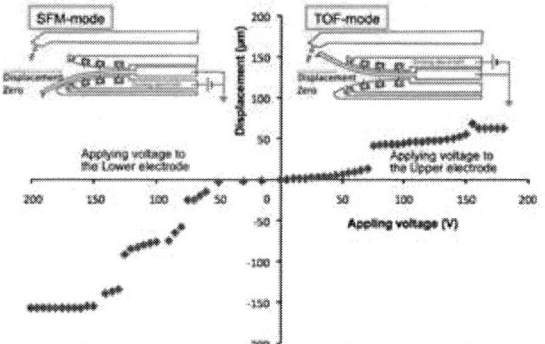

Figure 7: Measured displacement of the electrostatic actuator as a function of the applied voltage.

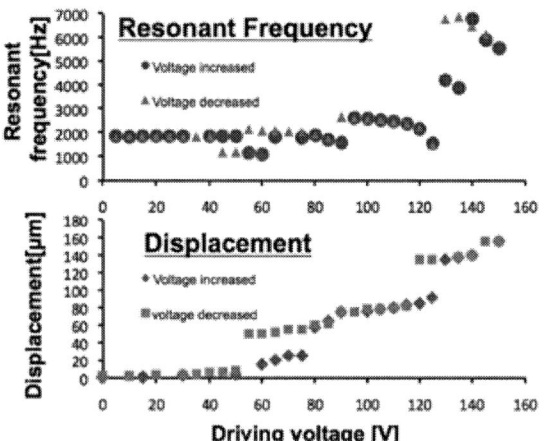

Figure 8: Characteristic of the hardening effect of the cantilever

ACKNOWLEDGEMENT

This study was partially supported by Special Coordination Funds for Promoting Science and Technology, Formation of Innovation Center for Fusion of Advanced Technologys.

CONCLUSION

We have reported an electrostatic switching mechanism, which consists of a switchable micro cantilever and a couple of curved electrode to apply for switching TOF-SFM. This actuator with curved electrode have been characterized and fabricated by micro electro mechanical system (MEMS) technique. According to that actuation evaluation result; The end position of switchable cantilever depends on the slope of curved electrode. The fabricated probes with maximum gap 150µm, 110µm, 50µm, which is from the stopper to cantilever, can generates different maximum static displacement up to 255µm, 157µm, 66µm corresponding with an applied voltage of 180V, 135 V, and 55 V. the resonant frequency of the probe is increased from 1.8 kHz to 6.8 kHz in SFM mode. These evaluations of experiments show the large displacement actuation and required hardening spring effect (0.05N/m to 0.34N/m) are obtained. In order to apply in high resolution of surface observation, optimization of the switching mechanism will improve the performance much more and future work will focus on optimization and sharpening the tip of cantilever

REFERENCE

[1] R. Wiesendanger "Scanning Probe Microscopy Analytical Methods" Springer, 1998

[2] T. Ando, N. Kodera, D. Maruyama, E. Takai, K. Saito, and A. Toda, "A High-Speed Atomic Force Microscope for Studying Biological Macromolecules in Action" Jap. J. Appl. 41, pp. 4851-4856, 2002

[3] D.W. Lee, A. Wetez, R. Bennewitz, E. Meyer, M. Despont, P. Vettiger, C. Gerber, "Switchable cantilever fabrication for a time-of-flight scanning force microscope" J. Applied Physics Letters , vo.l 84, No. 9 pp.1558-1560, 2004

[4] Y. Kawai, T. Ono, M. Esashi, E. Meyer C. Gerber, "Resonator combined with a piezoelectric actuator for chemical analysis by force microscopy" Review Of Scientific Instruments, vol. 78, pp. 063709-1-063709-4, 2007

[5] H. Sasaki, M. Shikida, K. Sato, "A Novel Type Mechanical Power Transmission Arroay For Switching Densely-Arrayed Actuator" in Digest Tech. Papers MEMS 2006 Conference, Istanbul Turkey, January 22-26, pp. 790-793, 2006.

[6] R. Legtenberg, E. Berenschot, M Elwenspoek and J. Fluitman, "Eletrostatic curved electrode actuators " J. Microelectromechanical Syst., vol. 6, NO.3, pp.257-265, 1997

[7] R. McIntosh, P. Mauger, S. Patterson "Capacitive Transducers with Curved Electrodes" Sensors Journal, 6, NO.1, pp. 125-138, 2006

A MULTIPLE DEGREES OF FREEDOM ELECTROTHERMAL ACTUATOR FOR A VERSATILE MEMS GRIPPER

Dian-Sheng Chen, Chun-Yi Yin, Ren-Jie Lai, and Jui-che Tsai

Graduate Institute of Photonics and Optoelectronics and Department of Electrical Engineering
National Taiwan University, Taipei 10617, TAIWAN
E-mail: jctsai@cc.ee.ntu.edu.tw

ABSTRACT

In this paper, we present a versatile MEMS gripper capable of two-dimensional manipulation. This is accomplished by a multiple degrees of freedom electrothermal actuator which can achieve independent in-plane and out-of-plane motions. The device is fabricated with the MetalMUMPs process, which offers silicon nitride, polysilicon, and nickel as the major structural materials. The electrothermal actuator achieves large movement. The in-plane and out-of-plane tip displacements are 83.7 µm at 0.6 V and 98 µm at 87.2 V, respectively.

INTRODUCTION

MEMS (micro-electro-mechanical systems) grippers possess the ability to grasp and manipulate tiny objects such as micro particles or biological cells [1-3]. They can also be used for micro-assembly [4] and micro/nano robotics [5]. With a proper design and fabrication process, a MEMS gripper can further boost up its functionality – for example, operating in solution [3] or being equipped with force sensors [6].

MEMS electrothermal actuators have drawn a great deal of attention due to their ability to achieve large displacements. They are popularly used as the driving elements of micro grippers [1-6]. The electrothermal actuation normally relies on the thermal expansion difference between structural arms or layers, which in general is induced by two major approaches - mismatch between the thermal expansion coefficients of different materials [7]; hot-arm-cold-arm mechanism that employs selective heating [2-3,8]. The former approach generally involves proper selection of materials and determination of layer thicknesses. The latter one induces a temperature difference between structural arms during actuation. The arm with the higher temperature (hot arm) expands more than the lower-temperature one (cold arm), resulting in bending motion. This temperature contrast can be established by the geometric difference between arms [3], which leads to disparities in the resistance and therefore heating, or by selective current feedthrough [2,8].

Until now, most electrothermal actuators only have the capability of one-dimensional (1-D) actuation, either in-plane [2] or out-of-plane [7,8]. For this reason, the majority of MEMS electrothermal grippers are limited to simple grasp-and-release operation. To enhance the gripper versatility, actuators with multiple degrees of freedom are of great interest. An actuator employing both electrostatic and electrothermal mechanisms to achieve out-of-plane and in-plane motions respectively were previously reported [9]. The in-plane horizontal movement was implemented by the hot-arm-cold-arm architecture. However, the vertical displacement was restricted to 3.2 µm (at 100 V) due to the limitation by electrostatic actuation. A purely-electrothermal actuator combining both motions was demonstrated as well [10]. Nevertheless, the in-plane and out-of-plane modes were inherently coupled, making it impossible to accomplish independent control in each direction.

Recently, hybrid polymer-silicon micro structures have been made into in-plane [6,11-12,14] and out-of-plane electrothermal actuators [13,14]. An orthogonal arrangement of the in-plane actuators led to a two-dimensional (2-D) gripper which was not only able to grab and release, but could also move and spin the object in-plane [12]. On the other hand, combining both the in-plane and out-of-plane actuators resulted in a 2-D gripper with clamp and lift motions [14].

In this paper, we present an electrothermal actuator with multiple degrees of freedom. Independent in-plane and out-of-plane motions are achieved. The in-plane tip displacement is 83.7 µm at 0.6 V whereas the out-of-plane movement is 98 µm at 87.2 V. This actuator is used to implement a versatile MEMS gripper with grasp and lift motions.

DESIGN AND OPERATION PRINCIPLE

Figure 1 shows the 3-D schematic of the MEMS gripper implemented with a pair of 2-D electrothermal actuators capable of in-plane and out-of-plane motions. The primary materials used include silicon nitride, polysilicon, and nickel. The use of these multiple materials provides great design flexibility. Silicon nitride also offers insulation, enabling layered electrical routing with polysilicon and nickel. The polysilicon is embedded in silicon nitride but is intentionally revealed in the drawing of Figure 1 for clearer interpretation. The actuator's in-plane motion is achieved by the hot-arm-cold-arm architecture, in which nickel is used as the structural as well as conducting material. The wider cold arm possesses a larger volume and lower electrical resistance, making itself much less heated than the hot arm during electrical current feeding. This results in a temperature contrast and a greater expansion of the hot arm, leading to in-plane bending. For the vertical out-of-plane motion, a structure with separated upper nickel and lower silicon nitride beams are employed. When carrying an electrical current, the embedded serpentine polysilicon line heats up the silicon nitride beam, bending the structure upward. The flexures, essentially silicon nitride-polysilicon stacks, not only mechanically connect the in-plane and out-of-plane moving parts, but also function as soft bending joints for the out-of-plane motion. The white arrowed lines in Figure 1 indicate the independent current flows for controlling the in-plane and out-of-plane modes.

978-1-4244-2977-6/09 $25.00 © 2009 IEEE

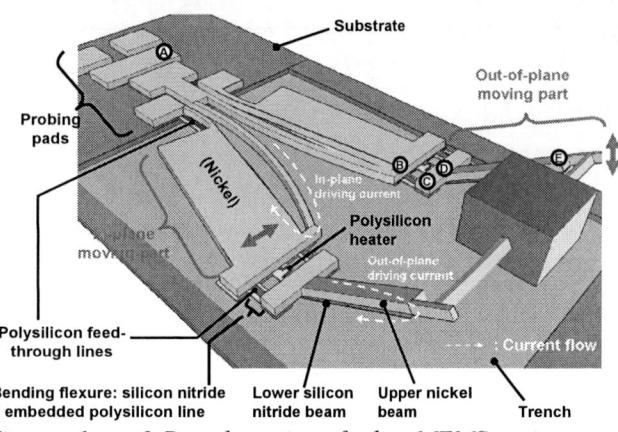

Figure 1: 3-D schematic of the MEMS gripper implemented with a pair of 2-D electrothermal actuators capable of in-plane and out-of-plane motions.

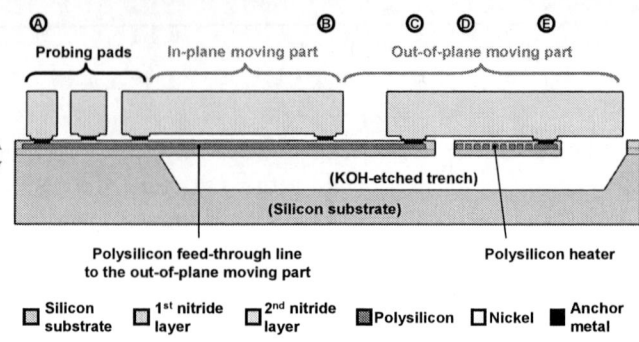

Figure 2: Simplified cross-section view along the A-B-C-D-E direction of Figure 1. The structure details and the corresponding materials/layers are shown.

FABRICATION

Figure 2 is the simplified cross-section view along the A-B-C-D-E direction of Figure 1, depicting the structure details and the corresponding materials/layers. The device is fabricated with the MetalMUMPs process offered by MEMSCAP, Inc [15]. The fabrication flow is summarized as follows. First, a 2-µm thick isolation oxide (not shown in Figure 2) is grown on the n-type (100) silicon substrate, followed by the deposition and patterning of a sacrificial PSG (phosphosilicate glass) layer (0.5 µm thick). This patterned PSG layer defines the trench area. After that, the first silicon nitride and polysilicon layers (0.35 µm and 0.7 µm thick, respectively) are deposited. Reactive ion etching (RIE) is used to pattern the polysilicon, followed by the deposition of the second silicon nitride layer (0.35 µm thick). Then, RIE removes the unwanted nitride. A second sacrificial PSG (1.1 µm thick) is deposited and opening areas are etched for anchoring the metal. Later on, the 20-µm thick nickel is electroplated on the designated areas. On the top of the nickel, a 0.5-µm gold layer is always deposited in the fixed foundry process; however, its contribution is insignificant and it is not shown in the drawing of Figure 2. 49% HF solution is then used to etch away all the sacrificial PSG, leaving a gap between the nickel and nitride beams. Isolation oxide in the trench area is also removed in this step. Finally, a 25 µm deep trench on the silicon substrate is formed using KOH etching. The large clearance provided by this trench prevents the device from sticking to the substrate during the drying or operating stage.

Figure 3 shows the optical microscope images of the device. In the close-up views, the embedded polysilicon is clearly seen. The dashed squares are locations where the suspended nickel and nitride structures are anchored to each other. The total length of the gripper (excluding the probing pads) is 2.15 mm. The initial gap spacing between the gripper tips is 8 µm, which is determined by the MetalMUMPs design rules. When electrical currents are fed into the in-plane moving parts, the gripper opens for slotting an object. The gripper closes and grabs the object after the current is reduced or removed.

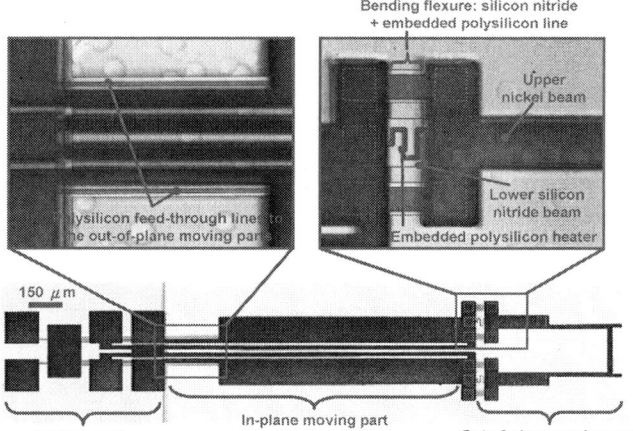

Figure 3: The optical microscope images of the device. The embedded polysilicon is clearly seen in the close-up views.

EXPERIMENTS AND RESULTS

The schematic of the experimental setup is shown in Figure 4. The device is characterized using a high-quality high-magnification optical microscope equipped with an optical scale which has a resolution of 1 µm. The power supply, current meter, and the device are connected in series. For the measurement of in-plane motion, the displacement is observed and measured directly through the microscope. Under out-of-plane actuation, the microscope height is tuned during the measurement so that it remains focused on the gripper's tip. The amount of tuning, which is equivalent to the vertical displacement of the tip, can be read from the optical scale. The experimental dc characteristics of the in-plane and out-of-plane modes are shown in Figures 5(a) and 5(b), respectively. The in-plane tip displacement is 83.7 µm at 0.6 V whereas the out-of-plane movement is 98 µm at 87.2 V.

978-1-4244-2977-6/09 $25.00 © 2009 IEEE   1032

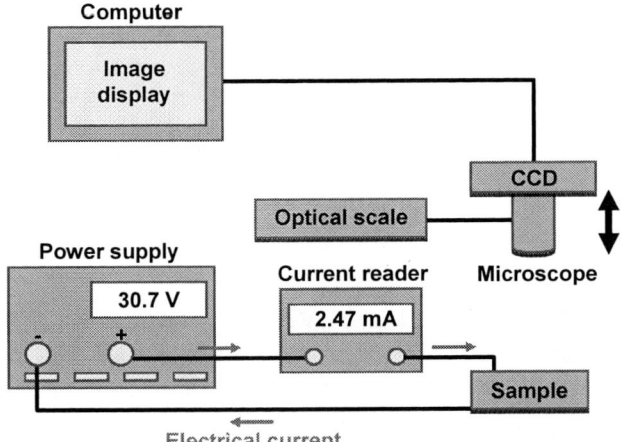

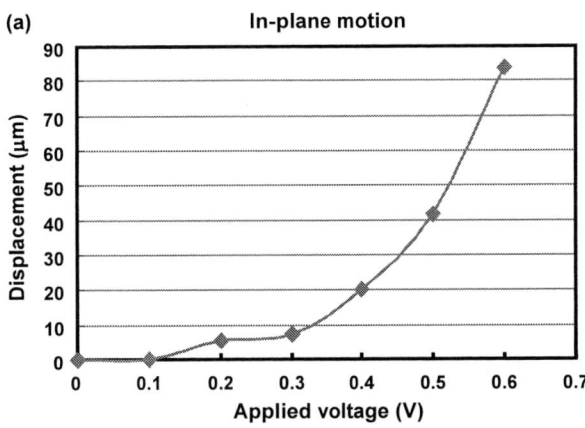

Figure 4: Schematic of the experimental setup. The microscope is used to measure both the in-plane and out-of-plane tip displacements of the electrothermal MEMS gripper.

During the out-of-plane actuation, we observe some undesired induced in-plane motion. For instance, the induced in-plane tip displacement is 35 μm at 43-μm out-of-plane tip movement. This is due to the arrangement of the electrical routing for the out-of-plane moving part and can be explained by Figure 6. The current feed-through strip to the out-of-plane moving part, i.e. polysilicon lines plus the surrounding silicon nitride (also see the upper-left close-up view in Figure 3 for reference), and possibly the metal above are also heated up and expand during the current feeding for the vertical motion. This strip is anchored to the far end of the in-plane moving part. The thermal expansion of the strip plus possibly the metal above could bend the in-plane moving part inward during out-of-plane actuation. This induced horizontal movement can be compensated by the feeding of the in-plane driving current which bends the gripper arm outward. Therefore, independent out-of-plane and in-plane motions are still achievable despite this induced horizontal excursion.

Figure 7 shows the snapshots of one gripper arm during its 2-D movement. The out-of-plane moving part of the other arm is intentionally removed for clearer observation. Figure 7(a) is the photo taken at the initial state with no voltage/current. In Figure 7(b) where a current for the vertical motion is applied, the "fore arm" (out-of-plane moving part) is lifted up while the "upper arm" (in-plane moving part) is induced to bend inward. We later reduce the vertical driving voltage/current so that the "fore arm" is lowered slightly [Figure 7(c)]. The voltage is then increased once more and the "fore arm" tip is up again [Figure 7(d)]. In Figure 7(e), an in-plane driving voltage/current is fed to move the "upper arm" outward toward its original position.

CONCLUSIONS

A novel multiple degrees of freedom electrothermal actuator for a versatile MEMS gripper is presented. Independent in-plane and out-of-plane motions are achieved. The in-plane tip displacement of the gripper is 83.7 μm at 0.6 V whereas the out-of-plane movement is 98 μm at 87.2 V.

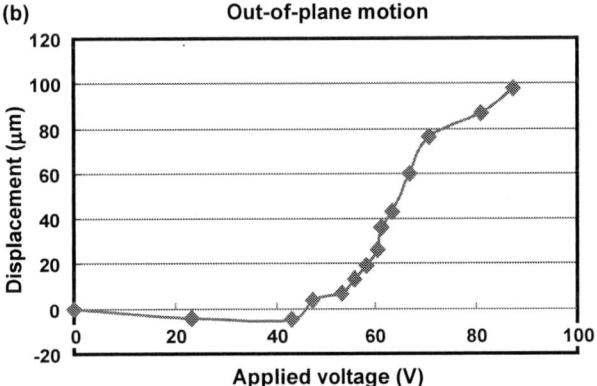

Figure 5: The experimental dc characteristics of the (a) in-plane and (b) out-of-plane modes.

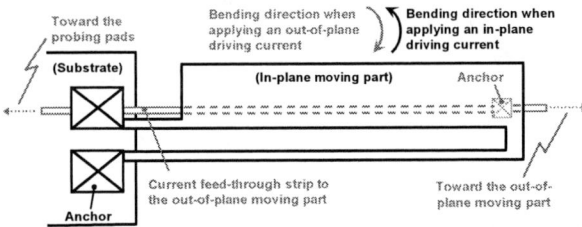

Figure 6: Simplified schematic explaining the induced inward in-plane motion, which is in the opposite direction of the normal outward in-plane bending.

ACKNOWLEDGEMENT

This work was supported by National Science Council of Taiwan under grants NSC 95-2221-E-002-053 and NSC 96-2221-E-002-198-MY2, and Excellent Research Projects of National Taiwan University, 95R0062-AE00-06 and 97R0062-07.

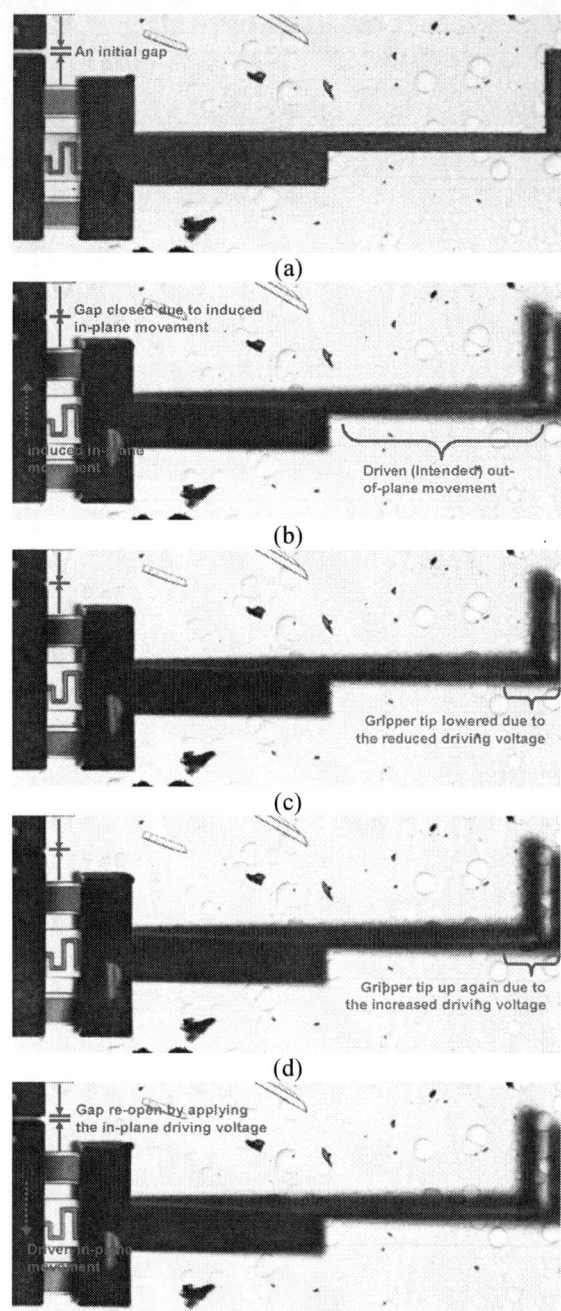

Figure 7: Gripper's snapshots during 2-D movement. (a) initial state with no voltage/current; (b) out-of-plane driving voltage/current applied; (c) out-of-plane voltage/current reduced; (d) out-of-plane voltage/current increased again; (e) in-plane driving voltage/current applied.

REFERENCES

[1] J. Lee, D. S. Park, A. K. Nallani, Y. Cui, A. Skoyles, and J.-B. Lee, "High-aspect ratio metallic nano grippers," in *Proc. of the 1st IEEE International Conference on Nano/Micro Engineered and Molecular Systems*, Jan. 18-21, 2006, Zhuhai, China, pp. 682-686.

[2] B. Solano and D. Wood, "Design and testing of a polymeric microgripper for cell manipulation," *Microelectronic Engineering*, Vol. 84, pp. 1219-1222,

2007.

[3] N. Chronis and L. P. Lee, "Electrothermally activated SU-8 microgripper for single cell manipulation in solution," *Journal of Microelectromechanical Systems*, Vol. 14, No. 4, pp. 857-863, 2005.

[4] K. Ivanova et al., "Thermally driven microgripper as a tool for micro assembly," *Microelectronic Engineering*, Vol. 83, pp. 1393-1395, 2006.

[5] V. Eichhorn, K. Carlson, K. N. Andersen, S. Fatikow, and P. Bøggild, "Nanorobotic manipulation setup for pick-and-place handling and nondestructive characterization of carbon nanotubes," in *Proc. of the 2007 IEEE/RSJ International Conference on Intelligent Robots and Systems*, San Diego, CA, USA, Oct. 29-Nov. 2, 2007, pp 291-296.

[6] T. Chu Duc, G. K. Lau, J. F. Creemer, and P. M. Sarro, "Electrothermal microgripper with large jaw displacement and integrated force sensors," in *Proc. of IEEE MEMS 2008*, Tucson, AZ, USA, Jan. 13-17, 2008, pp. 519-522.

[7] S. T. Todd and H. Xie, "An electrothermomechanical lumped element model of an electrothermal bimorph actuator," *Journal of Microelectromechanical Systems*, Vol. 17, No. 1, pp. 213-225, 2008.

[8] D. Yan, A. Khajepour, and R. Mansour, "Design and modeling of a MEMS bidirectional vertical thermal actuator," *J. Micromech. Microeng.*, Vol. 14, pp. 841-850, 2004.

[9] K. M. Liao, C. C. Chueh, and R. Chen, "A novel electro-thermally driven bi-directional microactuator," in *Proc. of 2002 International Symposium on Micromechatronics and Human Science*, pp. 267-274.

[10] S. C. Chen and M. L. Culpepper, "Design of a six-axis micro-scale nanopositioner - μHexFlex," *Precision Engineering*, Vol. 30, pp. 314-324, 2006.

[11] G. K. Lau, J. F. L. Goosen, F. van Keulen, T. Chu Duc, and P. M. Sarro, "Powerful polymeric thermal microactuator with embedded silicon microstructure," *Applied Physics Letters*, Vol. 90, 214103, 2007.

[12] T. Chu Duc, G. K. Lau, and P. M. Sarro, "Polymeric thermal microactuator with embedded silicon skeleton: part II - fabrication, characterization, and application for 2-DOF microgripper," *Journal of Microelectromechanical Systems*, Vol. 17, No. 4, pp. 823-831, 2008.

[13] J. Wei, T. Chu Duc, G. K. Lau, and P.M. Sarro, "Novel electrothermal bimorph actuator for large out-of-plane displacement and force," in *Proc. of IEEE MEMS 2008*, Tucson, AZ, USA, Jan. 13-17, 2008, pp. 46-49.

[14] J. Wei, T.C. Duc, and P. M. Sarro, "An electro-thermal silicon-polymer micro-gripper for simultaneous in-plane and out-of-plane motions," in *Proc. of Eurosensors 2008*, Dresden, Germany, Sep. 7-10, 2008.

[15] A. Cowen, R. Mahadevan, S. Johnson, and B. Hardy, *MetalMUMPs Design Handbook, Rev. 2.0*, MEMSCAP, Inc.

BIDIRECTIONAL ELECTROTHERMAL ELECTROMAGNETIC TORSIONAL MICROACTUATORS

Youngkee Eun, Hyungjoo Na and Jongbaeg Kim
School of Mechanical Engineering, Yonsei University, Seoul, REPUBLIC OF KOREA

ABSTRACT

This paper presents a novel design of bidirectional torsional micromirror utilizing vertically driven electrothermal electromagnetic silicon beam actuators to generate large static angular motion. The microactuators are fabricated on silicon-on-insulator (SOI) wafer using three photo masks in order to form two different thicknesses of single crystal silicon (SCS) device layer and backside cavities. When the driving bias is applied to the device, four buckle beams placed alongside of the torsion bars are subjected to thermal expansion and buckle in vertical direction generating torsional displacement of the micromirror with respect to two torsion bars placed at the center. The direction of buckle is controlled by Lorentz force caused by the magnetic field applied and the amount of current flowing through the micro beam to be buckled, enabling the bidirectional motion of the torsional micromirror. The maximum static angular displacement of the torsional actuator is up to 14.35° (28.7°, optical angle) under driving DC voltage of 7.5V. For resonance mode operation, the measured angular displacement is 8.46° (16.92°, optical angle) at 10.94 kHz under sinusoidal driving voltage of 0 to 3V.

INTRODUCTION

Torsional microactuators have been applied to various fields such as micromirrors for optical communications [1], scanning devices for projection display devices [2], medical imaging applications [3] and variable capacitors [4]. There are various kinds of torsional actuation methods for microdevices including electrostatic actuation [5], electrothermal actuation [6] and electromagnetic actuation [7]. Even though electrostatic torsional actuators are frequently adopted to drive micromirrors with large angular motion at resonance mode, they are not suitable for static mode operation in most cases due to relatively high operational voltages necessary for large motion. As an alternative, electrothermal actuators have been used for large static angular motion at lower driving voltages, such as bimorph type micromirror actuators [8, 9]. However, these cantilever-like bimorph actuators are very sensitive to shock and vibration due to the excessive compliance for larger motion and asymmetry, deteriorating the quality as an optical component and limiting their usage. More robust design for thermally driven torsional micromirror was introduced previously [10], but it could be operated only at resonance mode.

In this paper, we demonstrate bidirectionally driven electrothermal electromagnetic actuators to achieve large static angular displacement at low operational voltages, maintaining symmetry and high stiffness for mechanical robustness. The fundamental mechanism for bidirectional operation of laterally driven electrothermal electromagnetic actuators was introduced in our previous publication [11], and we adopted and modified this principle for torsional micromirror drive. The bidirectional capability of the presented torsional micromirror enhances the maximum angular displacement range. The novel design combining elastic torsion bar and buckle beam as mechanical flexures results in unique dynamic behavior such as fast recovery from the heated state and resonance at higher than 10 kHz, enabling not only the large angle static mode operation, but also the resonance mode operation at high frequency unlike bimorph thermal actuators.

DESIGN AND PRINCIPLE

The schematic diagram of bidirectional electrothermal electromagnetic torsional actuator is shown in Figure 1. SCS micromirror is supported by four buckle beams and two torsion bars on both sides symmetrically. All the components of the torsional actuator are fabricated as a monolithic structure on a single crystalline silicon-on-insulator layer. When electrical current flows through the buckle beams, the buckle beams thermally expand by Joule heating effect and buckle in an arbitrary direction. In our design, the cross-section of buckle beams has larger width than height in order to induce buckle in vertical direction, towards or away from the substrate, and therefore generate torsional motion of the mirror. The exact direction of buckle is then controlled by the direction of current flow through the buckle beam in the presence of transverse magnetic field.

Figure 2 shows the schematic diagram of bidirectional electrothermal electromagnetic torsional actuator with the arrangement of electrodes, corresponding direction of current, direction of magnetic flux and the direction of Lorentz force generated. When the current flowing through the upper side buckle beams is in opposite direction to that of lower side beams, the Lorentz force (F_1 and F_2) generated decides the buckling direction of the actuation buckle beams to be in an opposite vertical direction leading to a net torsional motion of the mirror with respect to the torsion bar.

When the electrical current flow direction is as in Figure 2(a), the Lorentz force applied to the upper side buckle beam pair guides them to buckle downward,

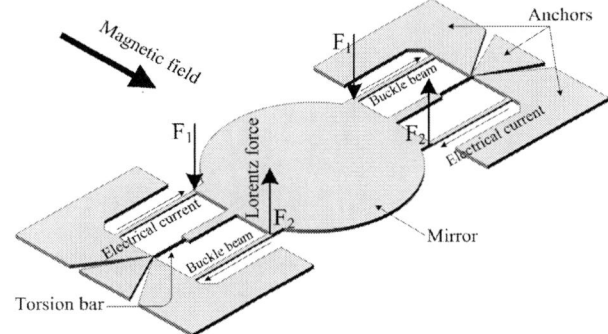

Figure 1: Schematic diagram showing the structures and geometry of the bidirectional electrothermal electromagnetic torsional actuator and the direction of magnetic flux, electrical current and corresponding Lorentz forces.

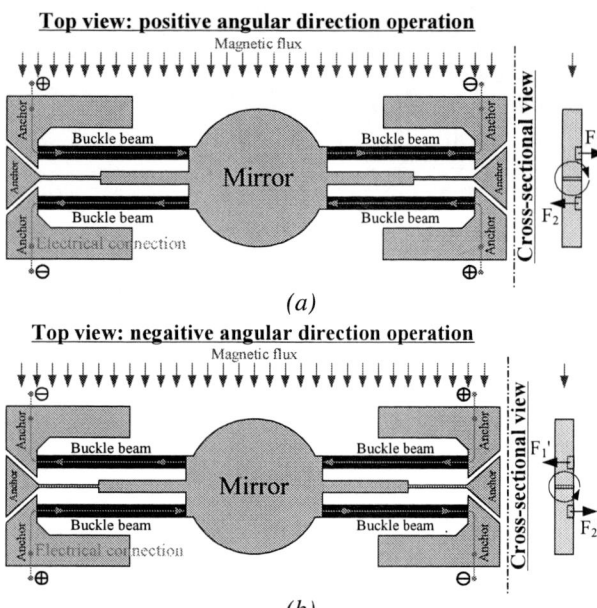

Top view: positive angular direction operation

(a)

Top view: negaitive angular direction operation

(b)

Figure 2: Schematic diagram showing the bidirectional actuation principle. Under the influence of external magnetic field, by controlling the current flow directions, the device can either generate positive (a) or negative (b) angular motion.

pushing down the mirror towards the substrate. The buckle beam pair in the opposite side pushes up the mirror due to the opposite direction of current, inducing positive angular motion of the mirror. Negative angular motion is generated by reversing the current flow direction to result in opposite directional Lorentz forces as shown in Figure 2(b). In order to increase the electrical resistance and reduce the vertical stiffness relative to the lateral stiffness, the buckle beams are designed and fabricated to be thinner than other structures.

A pair of vertically stiff torsion bars are placed on both side of the mirror, and the existence of this torsion bar is very critical for controllable bidirectional tilting motion. When both pairs of buckle beams are operated without the presence of the center torsion bars, they tend to buckle in the same direction resulting only piston motion in our alternative designs tested. This could be due to the fact that two pairs of buckle beams do not initiate buckling at the exactly same instant. Dimensional fabrication errors in width and layer thickness variation of 10μm-thick buckle beams could lead to unequal electrical resistance for different buckle beams and therefore uneven heating, making simultaneous buckling of multiple beams very difficult. Thus the buckle beams heated later tends to follow the buckling direction of the beams buckled ahead even with the Lorentz force generated, resulting in vertical piston motion. On the other hand, with the presence of vertically stiff torsion bar that maintains the vertical position of the mirror center, the buckling direction of the beams heated later is decided not by those buckled ahead, but by the direction of Lorentz force.

FABRICATION PROCESS

SOI wafer with 20μm thickness device layer, 2μm buried oxide layer and 400μm of handle silicon layer is patterned using three photomasks and deep-etched as

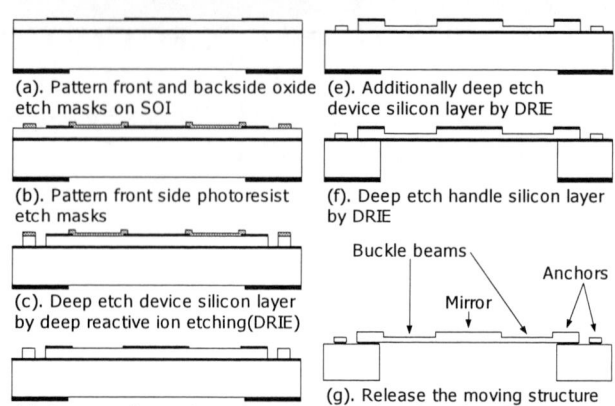

(a). Pattern front and backside oxide etch masks on SOI

(b). Pattern front side photoresist etch masks

(c). Deep etch device silicon layer by deep reactive ion etching(DRIE)

(d). Remove photoresist etch masks

(e). Additionally deep etch device silicon layer by DRIE

(f). Deep etch handle silicon layer by DRIE

(g). Release the moving structure by etching buried oxide layer

Figure 3: Fabrication process of the microactuator. The device is fabricated on SOI wafer with three photomasks to realize two different silicon device layer thickness and backside holes.

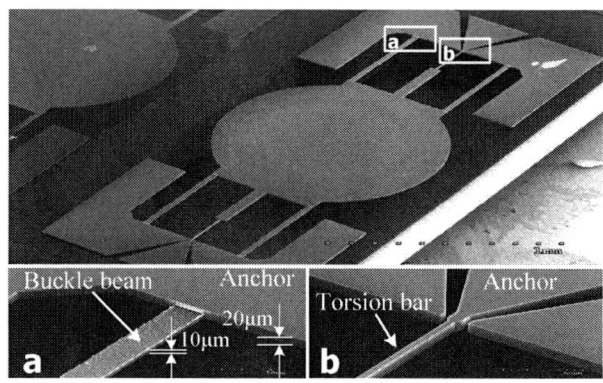

Figure 4: SEM image of bidirectional electrothermal electromagnetic torsional actuator. The diameter of the mirror is 1mm. (a) The buckle beams are thinned down and (b) the torsion bar and anchor are separated from the buckle beams.

shown in Figure 3. First, the silicon oxide masks on both front and backside are patterned (a). Next, front side photoresist is coated and patterned again to give height difference on the buckle beams in device silicon layer (b). After defining all the required etch masks, 20μm depth of silicon device layer is etched by Deep Reactive Ion Etching (DRIE) process (c). Removal of the photoresist etch mask (d) is followed by additional 10μm depth of silicon etch by DRIE (e). After twice of DRIE steps are performed on the device layer to get two different device layer thicknesses, the handle substrate beneath the mirror is fully etched to give sufficient space for the large motion of mirror (f). Finally, oxide etch mask and buried oxide layers are wet-etched to release the moveable structures (g).

Figure 4 shows Scanning Electron Microscope (SEM) images of the fabricated torsional actuator together with the thinned down buckle beams for higher electrical resistance and mechanical compliance. The anchors for buckle beams and torsion bars are separated to induce Joule heating only on the buckle beams.

EXPERIMENTAL RESULTS

For the experimental evaluation, the external magnetic field is applied by two permanent magnets placed on both sides of the device such that the direction of magnetic flux

978-1-4244-2977-6/09 $25.00 © 2009 IEEE

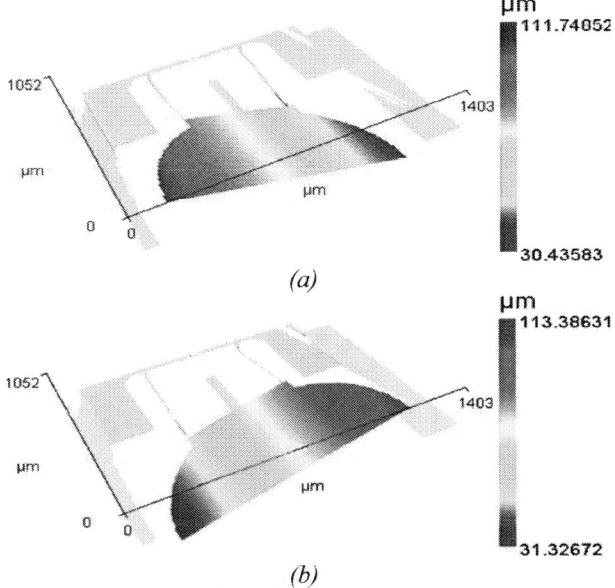

(a)

(b)

Figure 5: White light interferometric measurement of torsional actuators at (a) positive and (b) negative angular direction operations. Both buckle beam pairs are operated by the applied voltage of $5V_{DC}$ resulting in mechanical angular displacement of $4.43°$ and $4.36°$ in positive and negative directions, respectively.

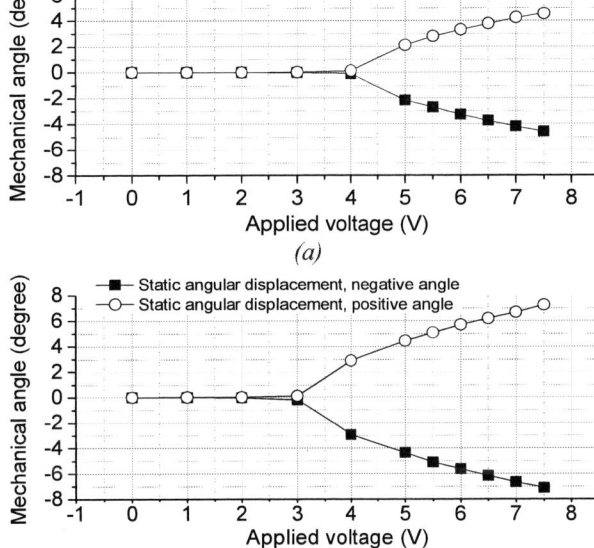

(a)

(b)

Figure 6: Static angular displacement of the device with respect to applied DC voltages. (a) Static angle for one pair of buckle beams actuated is $9.15°$ ($18.3°$, optical scan angle) under $7.5V_{DC}$ and (b) for both pairs of buckle beams actuated is $14.35°$ ($28.7°$, optical scan angle) under $7.5V_{DC}$.

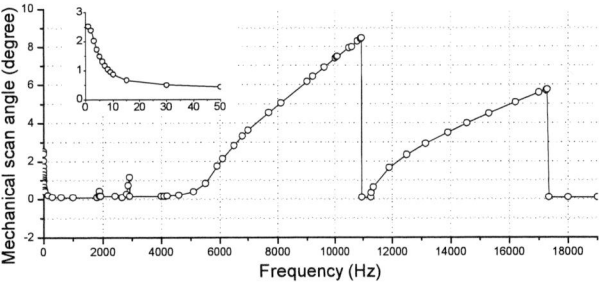

Figure 7: Frequency response of the torsional actuator when 0 to 3V sinusoidal driving voltage is applied to both side buckle beams. The mechanical scan angle was up to $8.46°$ ($16.92°$, optical scan angle) at 10.94 kHz.

is perpendicular to the direction of current flow in buckle beams. White light interferometric measurement images of the actuated device in positive and negative directions under the operation voltage of $5V_{DC}$ are presented in Figure 5. As described earlier, the rotational direction of the torsional actuator is controllable by altering the electrical current flow direction in the buckle beams. Figure 5(a) shows positive angular displacement of $4.43°$ when the actuation voltage of $5V_{DC}$ is applied to both pairs of buckle beams. When the electrical current flow direction is reversed with the same amount of bias difference, negative angular displacement of $4.36°$ is generated as presented in Figure 5(b).

In Figure 6 is shown the mechanical tilt angle with respect to DC voltages applied. The bidirectional torsional actuator can achieve large static angular displacement even with single pair of buckle beam operated, while the other buckle beam pair are inactive. When only two buckle beams on one-side of the torsion bar are actuated, the static tilt angle is $9.15°$ ($18.3°$, optical angle) for $7.5V_{DC}$ applied as shown in Figure 6(a). When all of four buckle beams are actuated, the static tilt angle is increased to $14.35°$ ($28.7°$, optical angle) at the same DC voltage, resulting in 57% increment in angular displacement compared to single side actuation, as shown in Figure 6(b).

Frequency response of the torsional actuator is presented in Figure 7 when sinusoidal alternating voltage of 0 to 3V is applied to all of the four buckle beams. Reduction of angular displacement is observed as the frequency of input voltage increases from the DC mode operation until the resonance is reached. At resonant frequency of 10.94 kHz, the scan angle is $8.46°$ ($16.92°$, optical scan angle), showing jump phenomenon typically observed in nonlinear hardening spring system.

Finite element modal analysis result of the torsional actuator is shown in Figure 8. The first mode shape with calculated modal frequency of 11.07 kHz is the torsional mode which corresponds to the proper intended operational mode for the experimentally tested torsional actuator with the resonant frequency of 10.94 kHz.

The transient responses of the torsional actuator for step input voltages of 0~2V and 0~2.3V are measured by Laser Doppler Vibrometer (LDV) as shown in Figure 9. Unlike conventional thermal actuators, residual vibration is observed whenever the actuation voltage is switched on and off, and the vibration amplitude tends to be increased as the amplitude of the step input voltage increases. Another interesting aspect is the fast restoration to the original position when the actuation voltage is turned off. Both residual vibration and fast restoration are distinct characteristics compared to typical thermally driven microactuators that have relatively long heating and cooling time, and therefore respond to the input signal slowly compared to electrostatic or electromagnetic actuators. Moreover, since the cooling process takes longer than heating, the response of general thermal actuators is

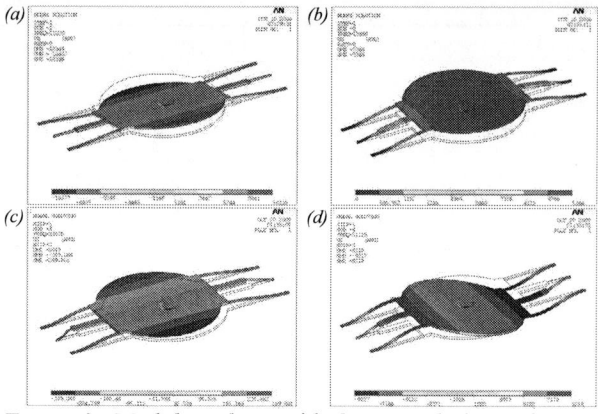

Figure 8: Modal analysis of bidirectional electrothermal electromagnetic torsional actuator. (a) 1st mode: 11.07kHz, torsion. (b) 2nd mode: 15.36kHz, out-of-plane translation. (c) 3rd mode: in-plane translation and torsion combined. (d) 4th mode: 51.23kHz, out-of plane rocking.

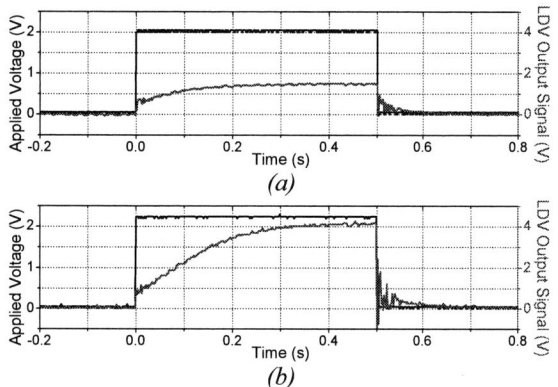

Figure 9: Transient response of the torsional actuator when square wave function of 1Hz with amplitude 2V (a) and 2.3V (b) is applied to both side buckle beams.

even slower for restoration process. The fast restoring characteristic of our torsional actuator is mainly due to the elastic restoring force of the torsion bars placed between the buckle beams.

CONCLUSION

A noble bidirectional torsional actuator based on thermally buckled beams has been successfully designed, fabricated and tested. Two pairs of vertically driven bidirectional electrothermal electromagnetic buckle beams placed alongside of the torsion bars generate torsional motion of the micromirror. It is experimentally assured that the design of unheated elastic torsion bars between buckle beams is essential for bidirectional and rotational motion. The bidirectionality is achieved by switching the direction of electrical current flowing through the buckle beams. Maximum static mode angular displacement of the torsional actuator is 14.35° (28.7°, optical angle) under the operational voltage of $7.5V_{DC}$. At resonance mode,

measured angular displacement is 8.46° (16.92°, optical angle) at 10.94 kHz when the sinusoidal actuation voltage of 0 to 3V is applied.

ACKNOWLEDGMENT

This work was supported by Seoul R&BD Program (11032).

REFERENCES

[1] H. Toshiyoshi and H. Fujita, "Electrostatic micro torsion mirrors for an optical switch matrix," *J. Microelectromech. Syst.*, vol. 5, pp. 231-237, 1996.

[2] K. Ju-Nan, L. Gwo-Bin, and P. Wen-Fung, "Projection display technique utilizing three-color-mixing waveguides and microscanning devices," *Photonics Technol. Lett.*, IEEE, vol. 17, pp. 217-219, 2005.

[3] J. T. W. Yeow, V. X. D. Yang, A. Chahwan, M. L. Gordon, B. Qi, I. A. Vitkin, B. C. Wilson, and A. A. Goldenberg, "Micromachined 2-D scanner for 3-D optical coherence tomography," *Sens. Actuator A-Phys.*, vol. 117, pp. 331-340, 2005.

[4] H. D. Nguyen, D. Hah, P. R. Patterson, R. Chao, W. Piyawattanametha, E. K. Lau, and M. C. Wu, "Angular vertical comb-driven tunable capacitor with high-tuning capabilities," *J. Microelectromech. Syst.*, vol. 13, pp. 406-413, 2004.

[5] J. Kim, H. Choo, L. Lin, and R. S. Muller, "Microfabricated Torsional Actuators Using Self-Aligned Plastic Deformation of Silicon," *J. Microelectromech. Syst.*, vol. 15, pp. 553-562, 2006.

[6] Y. Eun, J. Kim, "1-DOF and 2-DOF Torsional Micromirrors Driven by In-plane Thermal Actuators," in *Asia-Pacific Conference of Transducers and Micro-Nano Technology 2008. APCOT 2008*, Tainan, June 22-25, 2008.

[7] H. Miyajima, N. Asaoka, M. Arima, Y. Minamoto, K. Murakami, K. Tokuda, and K. Matsumoto, "A durable, shock-resistant electromagnetic optical scanner with polyimide-based hinges," *J. Microelectromech. Syst.*, vol. 10, pp. 418-424, 2001.

[8] A. Jain, H. Qu, S. Todd, and H. Xie, "A thermal bimorph micromirror with large bi-directional and vertical actuation," *Sens. Actuator A-Phys.*, vol. 122, pp. 9-15, 2005.

[9] G. Lammel, S. Schweizer, S. Schiesser, and P. Renaud, "Tunable optical filter of porous silicon as key component for a MEMS spectrometer," *J. Microelectromech. Syst.*, vol. 11, pp. 815-828, 2002.

[10] D. Elata and R. Mahameed, "A Temperature-Gradient Driven Micromirror with Large Angles and High Frequencies," in *19th IEEE International Conference on Micro Electro Mechanical Systems, 2006. MEMS 2006*, Istanbul, January 22-26, 2006, pp. 850-853.

[11] A. Cao, J. Kim, and L. Lin, "Bi-directional electrothermal electromagnetic actuators," *J. Micromech. Microeng.*, vol. 17, pp. 975, 2007.

A 2D ELECTRET-BASED RESONANT MICRO ENERGY HARVESTER

U. Bartsch, J. Gaspar, and O. Paul
Department of Microsystems Engineering (IMTEK),
University of Freiburg, GERMANY

ABSTRACT

This paper reports on the integration of several technologies to realize, to our knowledge, the first two-dimensional (2D) electrostatic resonant micro energy generator. In the generator, ambient vibrations lead to the oscillation of the movable parts of two micromachined and assembled substrates. Charged Cytop-based electret structures patterned on these parts and arranged in a variable capacitor configuration translate the oscillations into a displacement current whose energy is harvested. The device results from (i) the design, fabrication and characterization of low-frequency 2D resonators to extract energy from ambient vibrations, (ii) processing, charging and long-term characterization of Cytop as an electret material and (iii) development of a packaging technology based on adhesive bonding using Cytop as well.

INTRODUCTION

Resonators are commonly used in many energy scavengers that exploit electrostatic, piezoelectric or electromagnetic harvesting principles [1]-[3]. The use of resonators is a simple possibility to extract energy from ambient motion and transform it into the kinetic energy of an oscillating seismic mass. Energy is extracted from the moving mass by damping its oscillation using the selected harvesting principle. The generating principle of a one-dimensional (1D) electret-based integrated micro energy generator is schematically shown in Fig. 1. Patterned electret lines and movable counter-electrodes form a variable capacitor ΔC that is charged and discharged as the seismic mass oscillates back and forth. The displacement current through an external load resistor yields the harvested power.

Resonators used in previous work amplify the surrounding vibration around their resonance frequency in a single direction only [2]-[4]. For optimal harvesting, the external acceleration has to be aligned with the direction of the motion of the resonator. This is often delicate since the external vibration needs to be characterized beforehand. Furthermore, a shift of the acceleration direction or an irregular acceleration behavior along several directions prevent an efficient energy conversion and harvesting. To overcome this issue, the seismic mass of the integrated resonators can be allowed to move in two in-plane directions, enabling the extraction of energy from ambient accelerations with arbitrary motion directions in the device plane. This work reports on the fabrication of such resonators and the integration of several technologies

necessary to realize 2D electrostatic resonant micro energy generators.

EXAMPLES OF 2D VIBRATIONS

To illustrate possible applications of the 2D generators presented here, the body vibrations of an autonomous intelligent unmanned quadrocopter and a mechanical vacuum pump (*Trivac D65BCS, Oerlikon Leybold, Germany*) were characterized exemplarily. The setup for the characterization of the quadrocopter acceleration is shown in Fig. 2. It consists of a three-dimensional (3D) accelerometer fixed to the investigated object, a DC sensor supply, and an oscilloscope. The measured data is the average value of 100 measurements of the oscilloscope, to suppress spurious vibration components. The body vibration of the vacuum pump is measured in the same manner. The obtained acceleration spectra are shown in Fig. 3. The inset in each graph shows the time-dependent 2D acceleration for a duration of 1 sec. The measured spectrum of the quadrocopter in Fig. 3 (a) has several peaks around 90 Hz, with accelerations of about $0.08g$, where $g = 9.81$ m/s². The motion shows no preference for either x- or y-direction. In contrast, the spectrum of the vacuum pump shows one dominant peak at 50 Hz resulting from the power grid frequency in use in central Europe. The accelerations are $0.35g$ and $0.14g$ in the x- and y-direction, respectively. The inset for the vacuum pump data shows indeed a direction of maximum acceleration approximately along the x-axis. In case of a suboptimal alignment of a conventional 1D vibration scavenger, a large fraction of the acceleration of the application would be wasted.

FABRICATION

The fabrication process flow of the proposed device is schematically shown in Fig. 4. It combines a Pyrex electret wafer bonded to a micromachined substrate using the flip-chip technique. The Cytop layer on the electret wafer is structured as shown in Figs. 4 (a) and (b) using plasma etching with O₂ as the reactive gas. Photoresist is used as the mask layer. The metallization, a Cr/Au/Cr sandwich, is structured by wet etching. The second photolithography and subsequent plasma etch step shown in Fig. 4 (c) remove the Cytop on top of connection lines

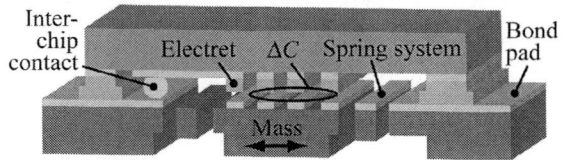

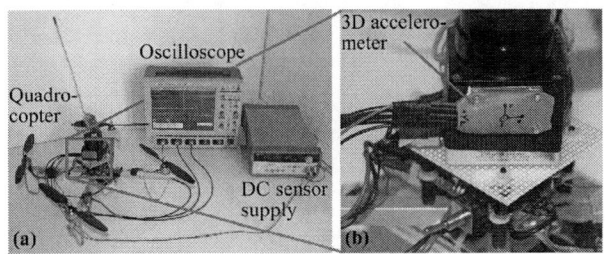

Figure 2: (a) General and (b) detailed views of the measurement setup to characterize the body vibrations of an autonomous intelligent unmanned quadrocopter using a 3D accelerometer.

Figure 1: Schematic view of an integrated vibration micro energy harvester.

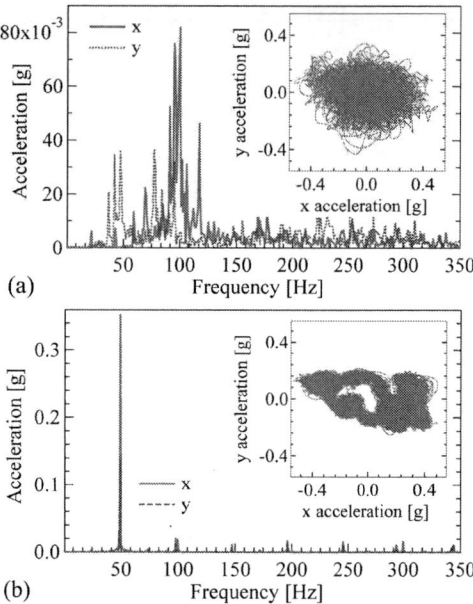

(a)

(b)

Figure 3: 2D acceleration spectrum of (a) an unmanned flying quadrocopter and (b) a vacuum pump.

and bond pads. In Fig. 4 (d), Au stud bumps are placed on the bond pads to establish an electrical inter-chip connection in the subsequent flip-chip bonding process. The Cytop layer is charged using a corona discharge setup.

The back of the oxidized Si resonator wafer is structured as shown in Fig. 4 (e) and (f) and the height of the suspending springs is defined using deep reactive ion etching (DRIE). The metallization is evaporated on the wafer front and a Cytop layer is spun on. This layer is used as an adhesive bond layer in the flip-chip process and therefore no hardbake is performed. Further, the layer thickness defines the gap distance between the metal counter-electrode on the Si side and the moving electret of the Pyrex side. This Cytop layer thickness is defined by the spin-coating process and can range from sub-μm

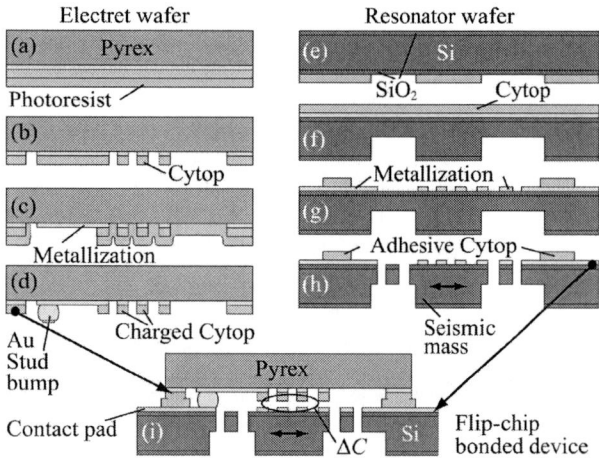

Figure 4: (a)-(c) The Cytop and metallization layers are structured on the electret wafer. (d) Gold stud bumps are placed and electret is corona-charged. The back of (e) an oxidized resonator Si wafer is (f) structured using DRIE. (g) Cytop and metal patterning, (h) subsequent DRIE step to release moving parts before (i) flip-chip bonding.

thickness to several 10 μm. The metallization and the Cytop layers in Fig. 4 (g) are structured in the same way as described above for the electret wafer. The front DRIE step in Fig. 4 (h) releases the moving parts.

After dicing, both chips are flip-chip bonded applying pressure and ultrasonic (US) power at only 106°C to form a variable capacitance ΔC as shown in Fig. 4 (i). Aspects of the different technologies are detailed in the following.

2D RESONATORS

The fabrication of low-frequency resonators using micromachining technology is usually relatively delicate, since both a large mass and a low spring stiffness have to be combined. To address these issues, the full wafer thickness is used for the seismic mass, while the suspension springs have a different height as shown in Fig. 4 (h). 2D resonators with resonance frequencies as low as 100 Hz have thus been realized. Scanning electron micrographs (SEM) of a fabricated structure with the selected coordinate system are shown in Fig. 5. A seismic mass with a diameter of 4 mm is suspended by a circular spring system. The width of each spring element is 10 μm and its height is 235 μm, resulting in an aspect ratio of about 24. The maximum allowed travel range of the seismic mass is 150 μm. The resonators are excited using a piezo-shaker and the resulting in-plane motion is characterized using a stroboscopic video analysis system (MSA 400, Polytec, Germany) as shown in Fig. 6. The observed seismic mass oscillation shows two modes of motion, one along the x- direction at 370.5 Hz and one along the y-direction at 373.7 Hz. The device does not show isotropic behavior because of the anisotropic Young's modulus of Si. The amplitudes of each oscillation mode depend on the excitation angle. The peak-to-peak motion amplitude for the x- and y-mode for different excitation directions is plotted in Fig. 7. The

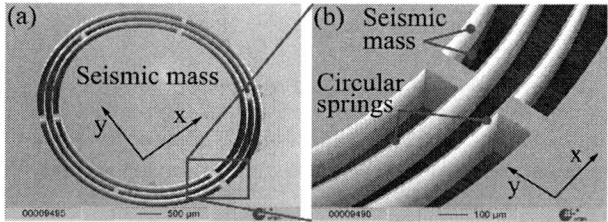

Figure 5: (a), (b) SEM graphs of realized 2D resonator with mass and spring height of 525 μm and 235 μm, respectively.

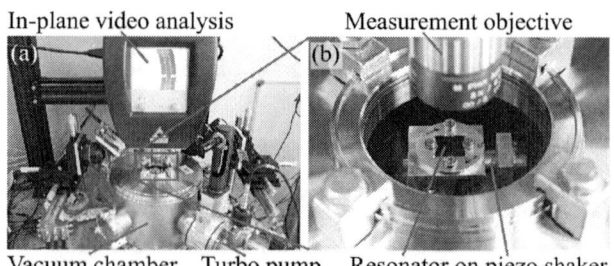

Figure 6: (a) Measurement setup to characterize the 2D motion of the realized resonators at ambient pressure and in vacuum using a stroboscopic video analyzing system. (b) The device is actuated by a piezo-shaker.

978-1-4244-2977-6/09 $25.00 © 2009 IEEE   1040

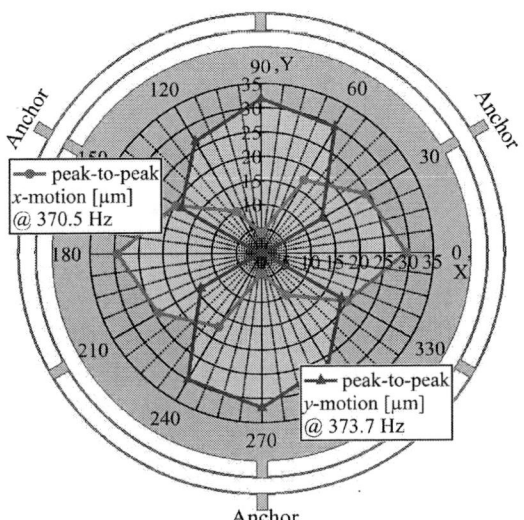

Figure 7: Peak-to-peak motion amplitudes for the x- and y-mode for different excitation directions and a peak-to-peak driving amplitude of 21.4 nm. The device layout is shown in the background.

device layout is shown in the background. The maximum motion is about 30 μm for both the x- and y-directions for an excitation peak-to-peak amplitude of 21.4 nm and occurs at 0° and 90° and equivalently at 180° and 270°. The quality factor Q for both directions is about 1400. Figure 7 also illustrates the disadvantage of a 1D resonator compared to a 2D device when used for the extraction of ambient vibrations. Only the x- or the y-mode would be accessible for the extraction of external vibration. In the worst case, if the direction of motion is orthogonal to the input motion, only a small fraction of the available kinetic energy from the environment would be used. In contrast, for a 2D resonator, both in-plane directions can be accessed for harvesting applications. Therefore, kinetic energy from all directions in the device plane can in principle be harvested.

ELECTRET TECHNOLOGY

The MEMS-compatible polymer electret material Cytop (*CTL-809M, Asahi Glass, Japan*) was charged using the customized corona charging station shown in Fig. 8 (a). The charging station implements a typical corona needle-grid setup. A negative DC voltage of −5 kV at the tip ionizes the surrounding air and lets ions drift through the grid to the electret sample. The resulting surface potential of the charged Cytop layer was measured with the setup in Fig. 8 (b). The setup combines an electrostatic voltmeter (*Model 279, Monroe Electronics, USA*) for a contactless detection of the surface potential of the electret layers and two linear motorized stages for automated surface scans. The effect of the corona grid voltage on the surface potential of a 4×4 mm² electret layer over a period of 143 days is shown in Fig. 8 (c). A maximum surface potential of approximately 80% of the applied corona grid voltage is obtained. Measurements performed 143 days after charging the electret layers show no significant change in the surface potential of the samples, if the layers were stored at 23°C. However, the storage temperature plays an important role for the charge stability and thus for the lifetime of electret-based devices.

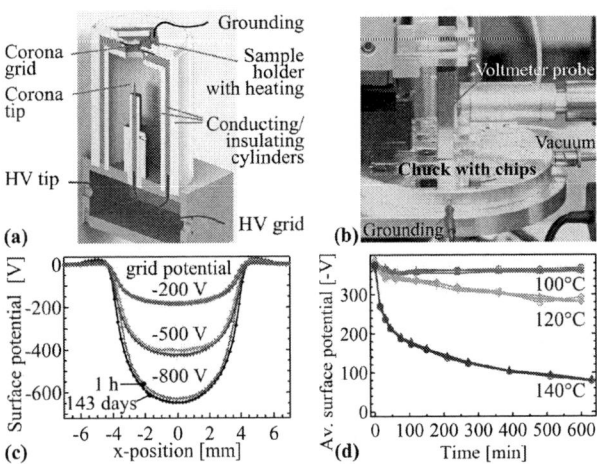

Figure 8: (a) Corona charging station and (b) surface potential measurement setup for electret chips. (c) Long-term measurement of surface potential of 4×4 mm² chips for different grid voltages and (d) temperature-induced charge leakage of Cytop layers.

To investigate the temperature induced discharge of the electret material the samples were heated up. Figure 8 (d) indicates that the stored charge is stable for at least 10 hours at 100°C, which is sufficiently for subsequent processing. By increasing the storage temperature, the surface potential averaged over the electret area, thus the lifetime of the electret decreases rapidly.

ASSEMBLY AND BONDING TECHNIQUE

The assembly of 2D electret-based energy scavengers is a crucial step of the device fabrication. The assembly method combining the resonator chip and the chip with the patterned electret layer needs to be robust and durable. It should offer the possibility of defining a controlled gap between the two chips. Electrical connections between the two chips are also necessary. As the electret layer is corona-charged prior to assembly, the assembly has to be performed at low temperatures to ensure the stability of the electret charges.

In this work, the Cytop polymer is not only used as electret layer, but also as an adhesive bond material for joining chips with excellent mechanical stability and well-controlled gaps. The duration and temperature of the flip-chip bond process are 70 sec and 106°C, respectively, guaranteeing the stability of the electret charges. In order to establish electrical interconnections between the two chips, Au stud bumps are placed at a temperature of 100°C on the Au layer of the Pyrex chips before bonding. A flip-chip bond procedure using heat, pressure and

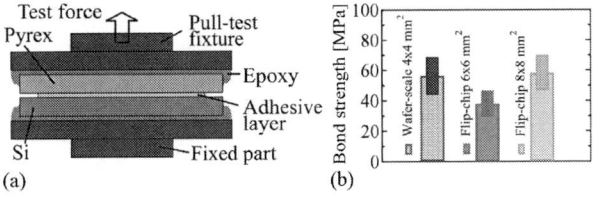

Figure 9: (a) Schematics of the pull-test setup used for evaluating (b) the bond strength of Cytop layers and corresponding standard deviations for wafer-scale and flip-chip bonding processes.

978-1-4244-2977-6/09 $25.00 © 2009 IEEE

ultrasonic power was applied [5]. This process enables one to adhesively bond chips using Cytop and to establish electrical stud bump interconnects at the same time. The adhesive bond strength was characterized using the pull-test schematically illustrated in Fig. 9 (a), showing the strength values presented in Fig. 9 (b) that are higher by a factor of 10 compared to those in Ref. 6. For a bond area of 8×8 mm^2 an excellent bond strength of 58.3 ± 11.8 MPa is achieved. The ohmic resistance of an individual stud bump connection is measured to be 0.5 ± 0.2 Ω.

RESULTING 2D GENERATOR

An optical micrograph taken from the Pyrex side of a 2D harvester fabricated using the technologies described above is shown in Fig. 10. The bonded Si chip with the integrated 2D resonator is larger than the Pyrex electret chip, thus allowing an easy electrical contacting of the device by wire bonding to a chip carrier. Circular electrodes are patterned on the seismic mass as well as on the Pyrex substrate to form a variable capacitor which changes its capacitance in both in-plane directions. Adhesive bond pads and metal pads used for the developed flip-chip bonding technology are also visible in Fig. 10. The Cytop layer spun on the Si resonator substrate defining the gap between the chips has a thickness of 9 µm. For smaller gaps, sticking of the movable parts is more likely to occur. The measured resonance peak of the generator and the electrical power peak of an integrated micro generator, excited along one direction, are shown in Fig. 10. Both peaks have their maximum at the same frequency. The first devices fabricated with the presented technologies show resonance frequencies and extracted harvested powers depending on the design of the circular spring system and capacitor dimensions within the 0.3-2 kHz and 1-100 pW ranges.

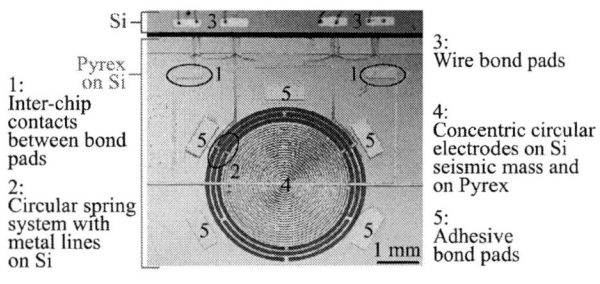

1: Inter-chip contacts between bond pads

2: Circular spring system with metal lines on Si

3: Wire bond pads

4: Concentric circular electrodes on Si seismic mass and on Pyrex

5: Adhesive bond pads

Figure 10: Top view optical micrograph of the fabricated 2D generator (from the Pyrex side) using the developed flip-chip bonding technology along with integrated charged electret layers and 2D mechanical resonator.

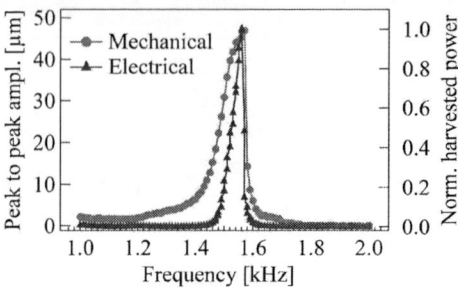

Figure 11: Measured mechanical and harvested power resonance peaks of an integrated micro generator fabricated with the technology presented in this paper.

CONCLUSIONS

The complete integration process needed to fabricate 2D electrostatic harvesters was developed and characterized in this paper. 2D resonators able to extract vibrational energy from all directions of the device plane were integrated together with stable electrets. For the overall integration, a flip-chip process was developed that allows a mechanical robust bond and electrical interconnections from one chip to the other using the same process step. Although 2D harvesting is possible with this structure the level of harvested power leaves room for improvement.

REFERENCES

[1] S. Roundy, "On the Effectiveness of Vibration-based Energy Harvesting", *J. Intell. Mater. Syst. Struct.* **16**, 2005, pp. 809-823.

[2] T. Sterken, P. Fiorini, K. Baert, R. Puers, and G. Borghs, "An Electret-based Electrostatic µ-Generator", *Dig. Tech. Papers Transducers'03*, pp. 1291-1294.

[3] U. Bartsch, A. Trautmann, P. Ruther, J. Gaspar, O. Paul, "Electrostatic Transducers for Micro Energy Harvesting Based on SOI Technology", *Dig. Tech. Papers Transducers'07*, 2007, pp. 141-144.

[4] S. Roundy, P. K. Wright, K. S. J. Pister, "Micro-Electrostatic Vibration-to-Electricity Converters", *Proc. 2002 ASME Int. Mechan. Eng. Congress and Exposition*, Nov. 2002.

[5] U. Bartsch, T. Huesgen, J. Gaspar, P. Woias, and O. Paul, "Low-temperature Adhesive Flip-chip Bonding using Cytop Combined with Electrical Stud Bump Interconnects", *Tech. Digest MicroMechanics Europe Workshop 2008*, pp. 411-414.

[6] K. W. Oh, A. Han, S. Bhansali, and C. H. Ahn, "A Low-temperature Bonding Technique using Spin-on Fluorocarbon Polymers to Assemble Microsystems", *J. Micromech. Microeng.* 12, 2002, pp. 187-191.

PZT MEMS ACTUATED FLAPPING WINGS FOR INSECT-INSPIRED ROBOTICS

J.R. Bronson[1], J.S. Pulskamp[2], R.G. Polcawich[2], C.M. Kroninger[2], and E.D. Wetzel[2]
[1]Oak Ridge Associated Universities, USA
[2]US Army Research Laboratory, USA

ABSTRACT

This paper presents microfabricated flapping-wings for insect-inspired flying robots. To the authors' knowledge, this is the first report of any such MEMS devices for this application. This work is a component of on-going efforts by the US Army Research Laboratory (ARL) in developing millimeter-scale robotic platforms using thin-film lead zirconium titanate (PZT) actuators. Micromachined wing-structures have been successfully fabricated and this report presents experimental results of flapping frequency and stroke-amplitude from a set of wing designs varying in material composition and dimensions. These results establish the feasibility of achieving stroke amplitudes comparable to insects with PiezoMEMS technology.

INTRODUCTION

With recent discoveries regarding the aerodynamics of insect-flight [1], [2], robotics researchers are attempting to create highly-maneuverable, small-scale vehicles with insect-inspired flight capabilities. Insects have the enviable abilities to fly at low speeds and to hover in confined spaces with a skill that has not yet been duplicated in man-made robotic systems. Bio-inspired vehicles with similar capabilities have many potential applications including search and rescue missions and operations in hazardous environments.

Nature has developed multiple strategies that allow insect species with varying wing morphologies and flight mechanics to take flight. The majority of insect species have wingspans ranging from millimeters to centimeters with an average body mass on the order of a mg [3]. The kinematics of insect flight is generally described by three degrees of freedom, the stoke amplitude, the angle of attack (or pitch), and the deviation angle [4]. The stroke amplitude in nature is species dependant, ranging from 66 degrees for hovering syrphid flies to greater than 180 degrees for some species of beetles and moths [3]. The angle of attack also varies in nature, and typically ranges from 20 to 40 degrees. The flapping frequency is generally fixed for a given insect and varies among species from as low as 5 Hz for some butterflies to 1 kHz for the tiny ceratopogonid fly. However, insects generally flap their wings at frequencies on the order of 100 Hz [6].

Previous attempts to develop flapping-wing robots have mostly focused on larger scale flying vehicles. While these vehicles are often referred to as being micro-air vehicles (MAVS,), these devices can have dimensions as large as 15 cm, and lack the hovering, low-speed, high maneuverability performance of insects [5]. The most notable work on small scale insect-inspired devices has been done by Fearing, et al. [6], and Wood [7], who have developed micromechanical flying insects that are several cm in size and operate using bulk-PZT actuators. The bulk-PZT actuators have been successfully demonstrated to provide sufficient lift to achieve tethered flight [7], however they require large voltages (200 V) to operate and produce limited displacements of only a few microns that require a mechanical transmission to achieve the large stroke amplitudes necessary for flight. These approaches utilize a transmission system to amplify the stroke and are analogous to the indirect flight mechanism employed by most insects. With this mechanism, the insect's thorax is driven at resonance and the motion is then coupled to the wings through a hinge. The bulk actuators are also heavy, making up a significant portion of the vehicle weight and limiting payload capabilities. These attributes limit the use of bulk-PZT actuators to these larger cm-scale systems.

This work is the first attempt to construct a flying robotic insect at this scale. The length of the wings investigated in this work range from 1.5 to 2.5 mm, a similar scale to that of a fruit-fly (*Drosophila*). The advantages of using PZT-thin films are numerous at the mm-scale, and include the ability to generate large angular displacements with large actuator compliance and low operating voltages (5-20 V). The ability to fabricate in a MEMS process allows for smaller feature sizes and flexibility in determining actuator size, shape, and placement, and opens possibilities for integrating electronics and sensors. Another key difference to the design of the thin-film actuators as opposed to the bulk-PZT actuators used in [6] and [7] is that the thin-film PZT actuators are able to directly drive the flapping motion of the wing with large displacements, eliminating the need for a larger mechanical transmission. This approach is analogous to the direct flight mechanism employed by dragonflies where the flight muscles attach directly to and drive the wing. The actuator placement and shape can be seen in Figure 1 in which there are three separate actuators at the base of the wing.

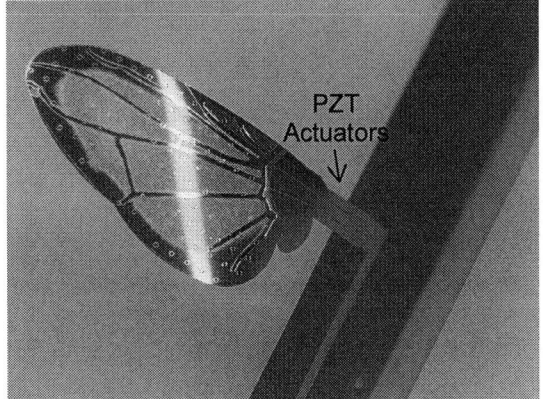

Figure. 1: Optical image of a micro fabricated wing structure (1.5 mm length) with three thin-film PZT actuators at the base of the wing and gold vein structures deposited on the wing surface.

The goal of this work is to establish the feasibility of utilizing a piezoelectric MEMS approach to the creation of mm-scale flapping wings for an insect-inspired flying robot. This paper presents the design, fabrication process, and experimental results of prototype device performance under static and resonant operating conditions.

DESIGN AND FABRICATION

Given the complexity of the aeromechanical design problem, the design of this robot begins with the presumption that the aerodynamic forces required for flight can be produced by mimicking the kinematic patterns of insect lift, using structures of similar mechanical properties to those of insects. This effort attempts to demonstrate stroke amplitude and frequencies that are within the range of actual insect characteristics.

A purely bio-mimetic design approach for these prototype wing structures dictated the choice of the wing shape. Additional design variables include the wing length and wing composition using materials currently available in the microfabrication facilities at ARL, including thin film silicon dioxide, silicon nitride, PZT, and several common evaporated metals (Ti, Au, Pt, Ni, Al). The design features three PZT unimorph mechanically coupled actuators at the base of the wing that are used to directly drive the stroke angle. Altering the phase of the drive signal to the individual actuators permits a limited drive capability of the angle of attack. The actuator performance is a strong function of the underlying structural dielectric layer (see Figure 2). Gold or silicon vein-structures with patterns similar to those found on real insect wings were included in some designs and may be used to decouple the properties of the actuator from the stiffness of the wing. The fabrication process and materials are similar to the process used for PZT RF MEMS switches [9] with an additional backside release etch.

Figure 2 shows a cross-section view of the micro-wing actuator taken longitudinally along the wing-span. Starting with a bare Si wafer (Figure 2b), a thick composite elastic layer comprised of silicon dioxide and silicon nitride is deposited using plasma enhanced chemical vapor deposition (PECVD). The thickness of this layer was varied from one to three microns. The film is then annealed at 700°C in a nitrogen atmosphere. An adhesion/diffusion barrier layer of titanium is sputtered on the PECVD oxide, immediately followed by a sputtered platinum thin film to serve as the bottom electrode layer of the actuator. The sol–gel PZT is deposited in several layers to achieve the desired thickness of 0.5 μm. The top electrode of platinum is sputter deposited and patterned with Argon ion-milling (Figure 2c). The PZT for the actuators and the bottom electrode is defined by ion milling (Figure 2d). With the PZT actuators patterned, the wing itself is lithographically defined and patterned using a reactive ion etch (RIE) of the composite elastic layer. A thin bi-layer of titanium-gold (200/11800 Å) is deposited and patterned using a lift-off process to provide metal to the bond pads and wing surface (Figure 2e). Next, a 2 μm gold layer is deposited via electron beam evaporation and patterned by liftoff (Figure 2f). The wing structure is released using a backside deep reactive ion etch (DRIE) process that creates a window of free space around

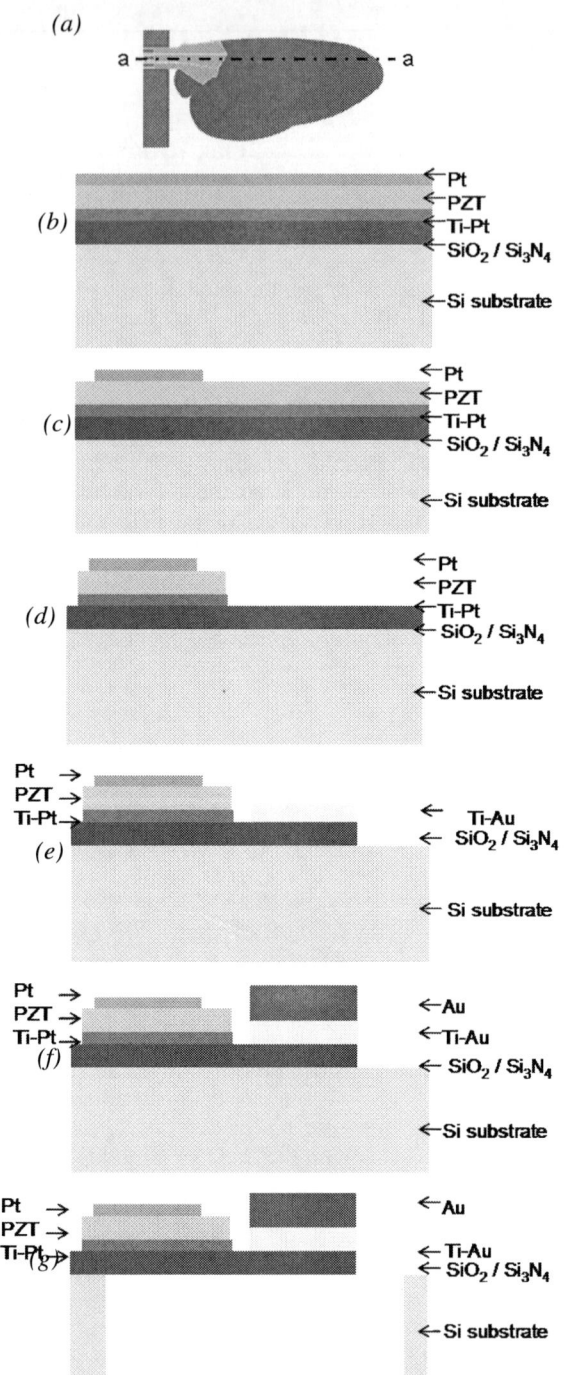

Figure 2: Schematic diagram of fabrication (not to scale). (a) Top view of a wing. (b)-(g)Process flow of wing cross-sectional [a-a].

Table 1: Materials and layer thicknesses of the mm-scale wings.

Device Design #	Elastic layer (μm)	Wing Materials		
		PZT (0.5 μm)	Au full (2 μm)	Au veins
1	1	x		x
2	1	x		
3	1	x	x	
4	3		x	
5	3	x		x
6	3	x		

the wing (Figure 2g). Once the wings are released, the remaining photoresist is ashed in an oxygen plasma.

Table 1 summarizes the different combinations of materials and thicknesses used for the wings including the thickness of the elastic layer, as well as the composition of the wing materials. Variations of this process flow have been explored and include using silicon-on-insulator (SOI) substrates and using XeF_2 etch to ensure that all of the backside silicon is removed from the wing structure.

One concern with the existing devices is stress deformation (see Figure 5) in the released wing structure that varies depending on the wing composition. The stress of the material can be controlled, as demonstrated in [10], and such compensation techniques are being employed to reduce deformation in next generation devices.

EXPERIMENTAL RESULTS

The prototype wings have thus far been tested to determine the static displacement of the actuators, resonant frequency spectrum, and displacement during resonant operation.

Static Displacement Measurement

The static displacement of the PZT-actuators is measured under a DC voltage bias on a Veeco NT1100 optical profilometer. These measurements (Figure 3) reveal that the actuator-tip for a 2.5 mm long wing (*Design 1*) rotates 6.7 degrees at 20 V. This rather modest actuator displacement can be increased significantly by operating the actuators at resonance.

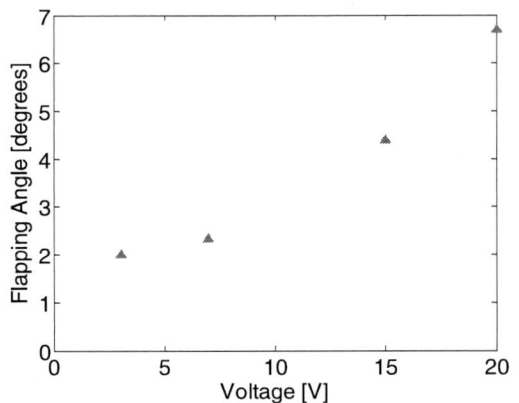

Figure 3: Static displacement measured at the actuator tip as a function of voltage (Design 1, 2.5 mm length).

Frequency Measurement

The wings have been tested to determine the fundamental resonant frequency using a Polytec laser Doppler vibrometer (LDV). These results have been compared to a finite element (FE) modal analysis. The resonant frequencies and mode shapes measured using the LDV show that the fundamental resonant frequency results in the expected longitudinal flexural mode desirable for large flapping strokes, while the second mode, torsional, rotates the wing about a longitudinal axis (see Figure 4a). The fundamental resonant frequencies for a set of devices are listed in Table 2, along with theoretically predicted values from modal analysis (Figure 4b). The FE model was constructed in ANSYS and consists of a structural model of the wing and actuators.

The boundary conditions for the FE model designate that the base of the actuators are fixed in all degrees-of-freedom. The theoretical approximations do not currently account for all of the geometry, such as vein structures and residual backside silicon, which contributes to the error between the theoretical and experimental results.

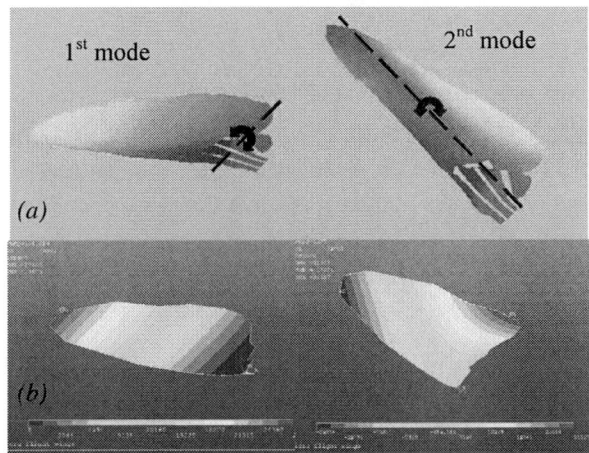

Figure 4: First two resonant mode shapes of the PiezoMEMS micro-wing (a) measured with an LDV and (b) finite element model.

Table 2. Fundamental resonant frequencies for PiezoMEMS micro-wings.

Device Design #	Wing Length (mm)	Resonant Frequency (Hz)	
		experimental	*theoretical*
1	1.5	650	728
2	1.5	681	666
1	2	249	323
3	2.5	173	157
4	1.5	1213	1129
5	2	531	769
6	2.5	656	510

Displacement Measurements at Resonance

Just as many insects flap their wings at a resonant frequency determined by their body size and structure, these wings are intended to operate at resonance to achieve larger stroke amplitudes. Measurement of the wing-tip displacement during flapping at the fundamental resonant frequency is done using optical images and measuring the total flapping angle. The still images captured during resonant operation using a sinusoidal voltage input are shown in Figure 5. Using a 5 V_{p-p} signal, the actuator displacement is 19 degrees, while the displacement of the wing tip is amplified to 44 degrees due to dynamic deformation of the wing. Operating this wing with higher voltages will result in large stroke amplitudes of 70 degrees or greater. Additional testing is currently in progress.

Testing performed under resonant operation a 2.5 mm wing (Design 4) for operating voltages from 5 to 20 V shows that the stroke amplitude in flapping increases as the operating voltage is increased. The results, shown in Figure 6, show that operation at 5 V causes a displacement of 6 degrees. By increasing the operating voltage to 20 V, the displacement increases to 36 degrees.

978-1-4244-2977-6/09 $25.00 © 2009 IEEE

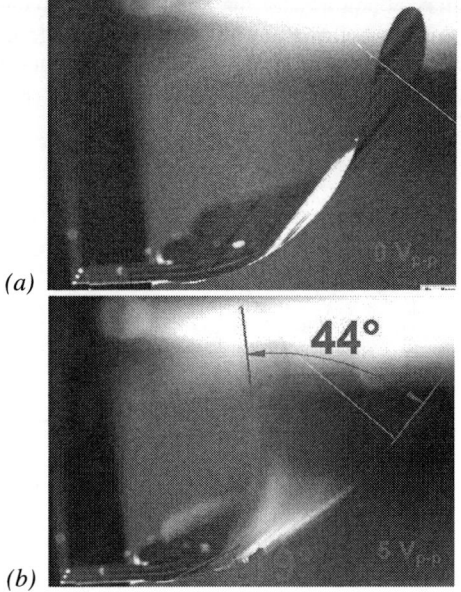

(a)

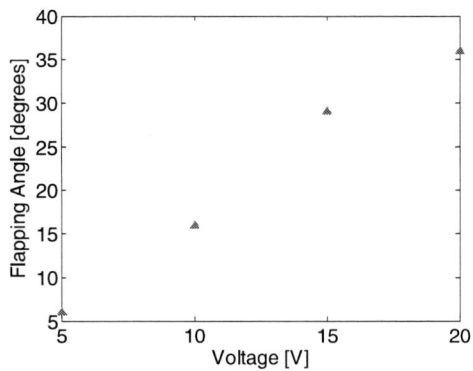

(b)

Figure 5: Images of a wing (Design 3, 2.5 mm length) (a) at 0 V and (b) at resonance (156 Hz, 5 Vp-p), showing that resonant behavior amplifies the stroke.

Figure 6: Dynamic measurement results for stroke amplitude as a function of operating voltage (peak-to-peak) (Design 4, 2.5 mm length).

It is important to take note that the results reported in Figures 3, 5, and 6 show data collected from three different wing designs, each with different material layers (see Table 1). The results in Figure 6 are taken from a wing with a thicker elastic layer of 3 μm, as opposed to the results in Figures 3 and 4 which are from wings with and elastic layer 1 μm thick. Increasing the thickness of the elastic layer and the wing materials causes a significant change in the bending stiffness, which means lower displacements for a given voltage. Taken together, these results illustrate that large flapping angles can be achieved with the appropriate wing design.

CONCLUSIONS AND FUTURE WORK

The investigation reveals that it is possible to fabricate MEMS flapping-wings that can achieve stroke-amplitude and flapping frequencies similar to insects, and that this technology has great potential to produce mm-scale flying robots. Many additional challenges still need to be overcome before a fully operational vehicle can be produced. Solutions will be needed for full system integration of sensors, electronics, and power supply and

pose significant challenges to the long term success of these devices.

In the short term, additional characterization of the current devices is needed including testing the wing material properties and stiffness. Additionally, the wing design should be optimized to have the structural rigidity values that more closely mimic those of insects. Another area of interest is reducing the stiffness of the wings to more closely match properties of insect wings while maintaining structural integrity and robustness. This effort has established the feasibility of achieving insect comparable stroke amplitudes; however, force production in insects is due in large part to the unsteady aerodynamic effects associated with the full 3D wing kinematics that includes the angle of attack. Our next step is to control the angle of attack of the wing during the wingbeat. The aerodynamic characterization is also underway to determine the flight forces generated by the wings.

REFERENCES

[1] C.P. Ellington, C. van der Berg, A.P. Willmott, and A.L.R. Thomas, "Leading-edge vortices in insect flight," *Nature*, vol. 384, pp. 626–630, 1996.

[2] M.H. Dickinson, F.-O. Lehmann, and S.P. Sane, "Wing Rotation and the Aerodynamic Basis of Insect Flight", *Science*, vol. 284, pp. 1954-1960, 1999.

[3] R. Dudley, *"The Biomechanics of Insect Flight"*, Princeton, N.J: Princeton University Press, 2000.

[4] S. P. Sane, M. H. Dickinson, "The Control of Flight Force by a Flapping Wing: Lift and Drag Production", *J. Exp. Biology*, vol. 204, pp. 2607-2626, 2001.

[5] T.J. Mueller, "Fixed and Flapping Wing Aerodynamics for Micro Air Vehicles", *AIAA Progress in Astronautics and Aeronautics*, vol., 195, 2001.

[6] R.S. Fearing, K.H. Chiang, M.H. Dickinson, D.L.M. Sitti, and J. Yan, "Wing Transmission for a Micromechanical Flying Insect", in *IEEE Int. Conf. on Robotics and Automation*, San Fransisco, CA, April 2000, pp. 1509-1516.

[7] R. Wood, "The First Takeoff of a Biologically Inspired At-Scale Robotic Insect", *IEEE Trans. on Robotics*, vol. 24, pp. 341-347, 2008.

[8] D. L. Altshuler, W.B. Dickson, J.T. Vance, S.P. Roberts, and M.H. Dickenson, "Short Amplitude High-Frequency Wing Strokes Determine the Aerodynamics of Honeybee Flight", *Proc. National Academy of Science*, vol. 102, pp. 18213-18318, 2005.

[9] R. Polcawich, J. Pulskamp, D. Judy, P.Ranade, S. Trolier-McKinstry, and M. Dubey, "Surface Micromachined Microelectromechancial Ohmic Series Switch Using Thin-Film Piezoelectric Actuators", *IEEE Trans. Microwave Theory and Tech.*, vol. 55, pp. 2642-2654, 2007.

[10] J. Pulskamp, A. Wickenden, R. Polcawich, B. Peikarski, M. Dubey, and G. Smith, "Mitigation of Residual Film Stress Deformation in Multilayer Microelectromechanical Systems Cantilever Devices", *J. Vac. Sci. Technol. B*, vol. 21, pp. 2482-8486, 2003.

A NOVEL UNDERWATER ACTUATOR DRIVEN BY MAGNETIZATION REPULSION/ATTRACTION

Han-Tang Su[1], Tsung-Lin Tang[2], and Weileun Fang[1,2]
[1]Institute of NEMS, [2]Department of Power Mechanical Engineering,
National Tsing Hua University, HsinChu, TAIWAN

ABSTRACT

This study has successfully designed and implemented a novel magnetic actuator. There are three merits of this magnetic actuator: (1) apply magnetic actuation force by magnetizing soft magnetic material, (2) both attractive and repulsive magnetic forces are available by changing the direction of magnetic field, and (3) the present magnetic actuator can be easily operated underwater. In applications, the test structure successfully demonstrates the repulsive force act as magnetic bearing to prevent contact of micro structures. The magnetization actuation mechanism which is suitable for underwater application is also demonstrated by micro gripper.

INTRODUCTION

Micro actuators are extensively used in micro system. For instance, electrostatic and thermal actuations are very common approaches to drive devices in micro scale. The electrostatic actuators such as the parallel electrode actuator [1], and comb drive actuator [2] are driven by applying different electric potential which causes the attractive electrostatic force. However, the gap spacing issues due to pull-in effect is the main operational limit of electrostatic actuations. The thermal actuators like bimorph thermal actuator [3], and hot-cold arm actuator [4], usually exploit the different thermal expansion of thin film material or micromachined structure to drive the actuator. But the inherent problem of high operating temperature and high power consumption can not be easily prevented. There are also many other micro actuators have been reported [5,6]. Presently, new micro actuators are still under development to meet the requirement of performances. In addition, new actuators are also required to overcome the operational limits of existing devices for different applications.

In the case of micro robotic gripper system, the system may need to mount a micro actuator to operate in biological solution for cell grasping or manipulation. Unfortunately, electrostatic and thermal actuators are designed for operating in dry environment. Both of them will face some problems in underwater application. For instance, the electrolysis problem of electrostatic actuator and the efficiency decay of thermal actuator [7] are important issues for underwater application in Bio-MEMS.

This study employs the magnetization approach to drive micro actuator. The magnetization actuation technique can provide both attractive and repulsive magnetic actuation force, and can be operated in both air and liquid environment. This study will discuss the principle of magnetization actuation technique. In addition, the test structures will also be fabricated and tested to demonstrate the feasibility of the present approach.

DESIGN

The proposed actuator is driven by magnetizing soft magnetic materials. When applying magnetic field on the soft magnetic material, it will be magnetized and have corresponding magnetic pole distribution. The force between two magnetic poles can be expressed as,

$$F_m = \frac{\mu_0 q_1 q_2}{4\pi r^2} \quad (1)$$

where μ_0 is the permeability constant, q_i is the magnetic pole strength, and r is the distance between poles. As shown in Fig.1, the magnetic field will induce corresponding magnetic pole distribution on soft magnetic material. Thus, this study exploits pole distributions to induce different actuation forces. As magnetic field is perpendicular to parallel soft magnetic materials (named magnetization units), the actuation force between two soft magnetic materials is attraction (Fig.1a). On the other hand, as magnetic field is horizontal to magnetization units, the actuation force will become repulsion (Fig.1b).

The actuation force is the summation of force between each couple of magnetic poles. The pole strength q_i can be expressed as,

$$q_i = \frac{B_i A_i}{\mu_0} \quad (2)$$

where B_i is the magnetic flux density and A_i is the cross-section area (along the magnetizing direction) of magnetization units. Finally, for the magnetization units of equal cross-section area, the total actuation force can be expressed as,

$$F_{m\_total} = \frac{B^2 A^2}{4\pi\mu_0} \sum \frac{1}{r^2} \quad (3)$$

Finally, the displacement of actuator is determined by the balance of magnetic force and restoring force of structure.

To verify the actuation concept, two test devices are designed and implemented for following tests. As shown in Fig.2, the magnetization units of test devices are employed to induce magnetic force. As shown in Fig.2a, the micro

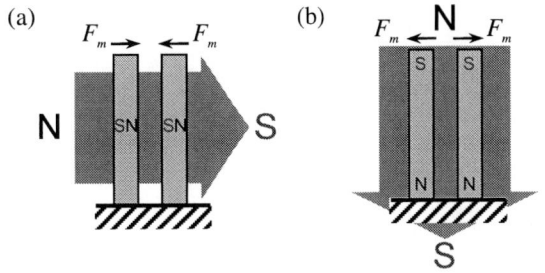

Figure 1: The concept of magnetization actuation: (a) attractive actuation, and (b) repulsive actuation.

978-1-4244-2977-6/09 $25.00 © 2009 IEEE

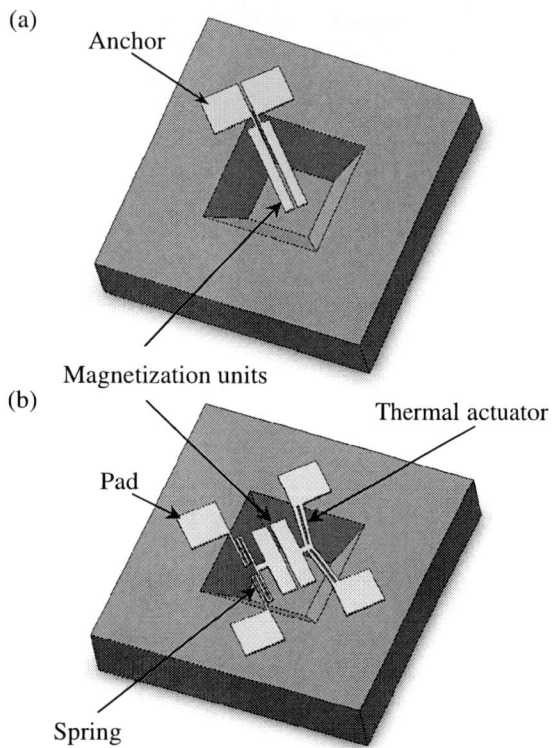

Figure 2: Two designed test devices: (a) micro gripper; (b) repulsive force test device.

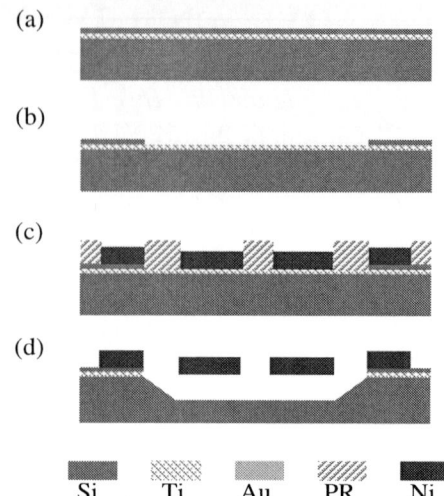

Figure 3: The fabrication steps for test devices.

gripper which contains magnetization units, flexure, and anchors is designed to characterize the actuator displacement vs. magnetic field strength H of both attractive and repulsive actuation. Furthermore, the micro gripper is also used for underwater test of magnetization actuation.

Eq.(3) indicates that the repulsive force becomes extremely large when r is very small. The feature of repulsive force can act as a magnetic bearing to prevent contact of micro structures. The repulsive force test device is shown in Fig.2b which contains magnetization units, V-beam thermal actuator, springs, and pads. The right magnetization unit is driven by the thermal actuator so as to approach the left one supported by springs. The contact of magnetization units can be detected by resistance measurement, and to further demonstrate the existing of magnetic bearing.

FABRICATION

In application, the test devices in Fig.2 were implemented by using the processes shown in Fig.3. In Fig.3a, the adhesion layer *Ti* and seed layer *Au* were deposited on silicon wafer by E-gun evaporator. Following that, the *Au* layer was patterned to define bulk silicon etching window, as shown in Fig.3b. After the *AZ4620* thick photo resist defined by photolithography, a 10μm thick *Ni* was then electroplated and patterned, as illustrated in Fig.3c. After stripping the photoresist, the *Ti* film without protecting by *Au* layer was removed using wet chemical etching. Finally, the *Ni* structure was released from substrate after bulk silicon etching, as shown in Fig.3d.

Fig.4 shows the SEM photos of two fabricated test

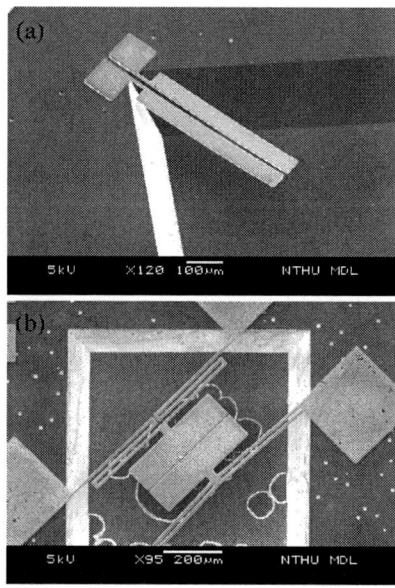

Figure 4: SEM micrographs of two typical test devices, (a) micro-gripper, and (b) repulsive-force test device.

devices. As the above mentioned, both two test devices including micro gripper (Fig.4a) and repulsive force test device (Fig.4b) were designed to verify the characteristic of the presented magnetization actuation. The fabrication processes in Fig.3 have successfully implemented these two test devices for further testing.

TESTING AND RESULTS
B-H Curve

The magnetic property of electroplated *Ni* in fabricated micro gripper is measured by Vibrating Sample Magnetometer (VSM). Fig.5 shows measured B-H curves of the micro gripper in Fig.4a. As discussed in Fig.1, to magnetize the magnetization units along different directions will respectively lead to attractive and repulsive actuations. The B-H curves of these two distinct directions are measured. Since shape anisotropy, different magnetizing directions have distinct B-H characteristics. The measurement results in the zoom-in B-H curve of

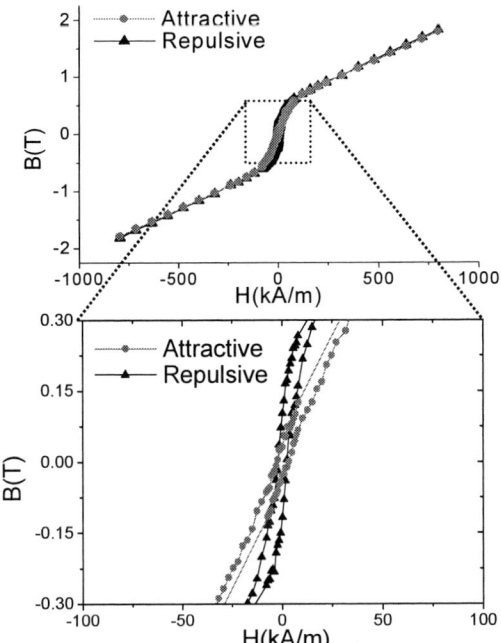

Figure 5: B-H curves of different magnetizing directions.

Figure 6: Experiment setup for actuator driving test.

Fig.5 indicate that magnetizing along long axis (the direction of repulsive actuation) will induce higher magnetic flux density.

Tip Displacement of Gripper Arms vs H

The experiment setup in Fig.6 has been established to provide magnetic field to drive the test devices. The solenoid is used to generate magnetic field. In order to increase the magnetic field strength, the iron cores are placed at the center of the solenoid.

Fig.7 shows the photos of both attractive and repulsive actuation tests of micro gripper. Fig.7a-b shows the attractive actuation test. In Fig.7a, the initial tip gap of micro gripper is G_{tip}=15μm, and the pull-in occurs (G_{tip}=0μm) when the external magnetic field strength H>30.6kA/m, as shown in Fig.7b. Thus, each gripper arm has a tip displacement of D(=ΔG_{tip}/2)=7.5μm when H>30.6kA/m. Fig.7c-d shows the repulsive actuation test. In Fig.7c, the initial tip gap of micro gripper is G_{tip}=3.9μm; and the tip gap shown in Fig.7d becomes G_{tip}=4.82μm when H=50.8kA/m. As a result, each gripper arm has a tip displacement of D=0.46μm when H=50.8kA/m.

Fig.8 summarizes the tip displacement D vs. magnetic field strength H for both attractive and repulsive actuations. This study also established a simplified parallel plate model to predict the tip displacement of gripper, as shown in the inset of Fig.8a. In this simplified model, the parallel plates had a uniform gap of G_{tip}. However, in the real case, the gap between the bent gripper arms is not a constant. As depicted in the inset of Fig.8a, the G_{tip} is the minimum gap between gripper arms for the attraction case. On the other hand, as depicted in the inset of Fig.8b, the G_{tip} is the maximum gap between gripper arms for the repulsion case. Thus, despite the measured and predicted results show the same tendency, the predicted tip displacement by attraction is much higher than the measured one at a given H.

Non-contact by Repulsive Force Test

Fig.9 shows the non-contact characteristic of the repulsive force test device in Fig.4b. The device has two separate magnetization units respectively supported by springs and thermal actuator. The contact of the magnetization units is detected by the on/off of the circuit

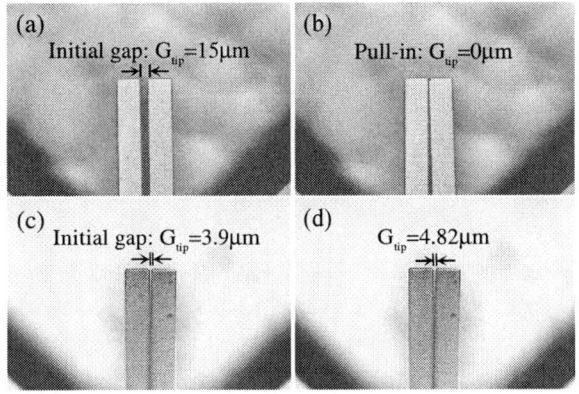

Figure 7: Photos of micro gripper during driving test: (a, c) initial state, (b) pull-in by attraction, and (d) separation by repulsion.

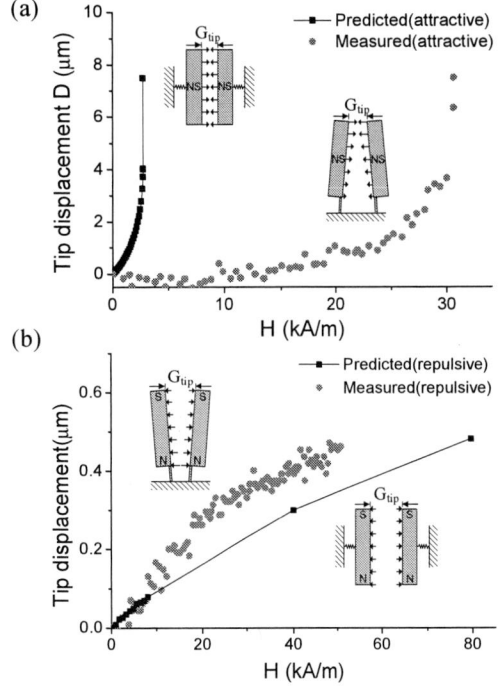

Figure 8: Predicted and measured displacement-H diagram: (a) attractive, and (b) repulsive, actuations.

(a)

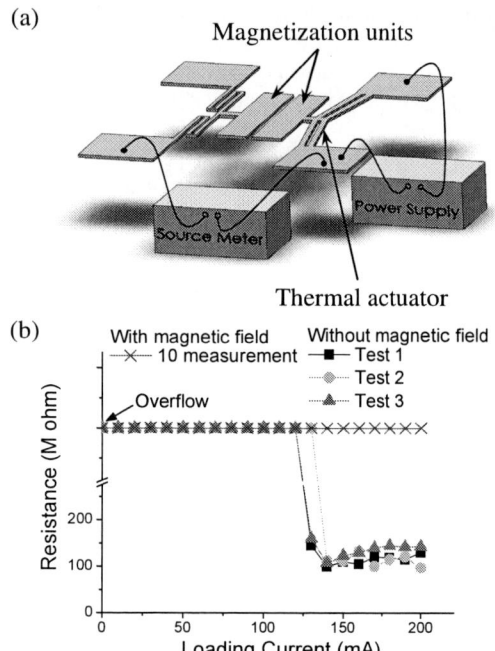

(b)

Figure 9: Non-contact by repulsive force test, (a) the testing setup, and (b) the testing results.

indicated in Fig.9a using source meter. According to the magnetic repulsion, magnetization units will not contact when applying a magnetic field. As demonstrated by cross-dots in Fig. 9b, the circuit remains off when driving current of thermal actuator (horizontal axis) rising from 0 to 200mA. While removing the external magnetic field, the resistance of circuit was measured (solid-dots in Fig. 9b) as loading current I>120mA, which means the magnetization units contact. Thus, the repulsive force can act as a magnetic bearing to prevent the contact of micro components.

Underwater Test

To verify the underwater operation ability of magnetization actuation, the micro gripper was driven in the water, as shown in Fig. 10. Fig. 10a shows the experiment setup for underwater test. The magnetic field source is the same as that in Fig. 6. However, in this test, the micro gripper was placed into a plastic container filled with water. The zoom-in photo in Fig. 10b shows the test chip inside the container with water. Fig. 10c shows the micro gripper in water before actuation. Fig.10d further shows the clipping of micro gripper after attractive actuation using the present approach. The test result successfully demonstrates that the magnetization actuation approach is suitable for underwater application.

CONCLUSIONS

This study presents the concept of driving micro actuator by magnetization. The magnetization actuation approach can provide both attractive and repulsive magnetic force on micro actuators. The test devices are designed and implemented to characterize the actuation characteristic of displacement vs. H. Different from electrostatic actuator, the testing result shows the pull-in

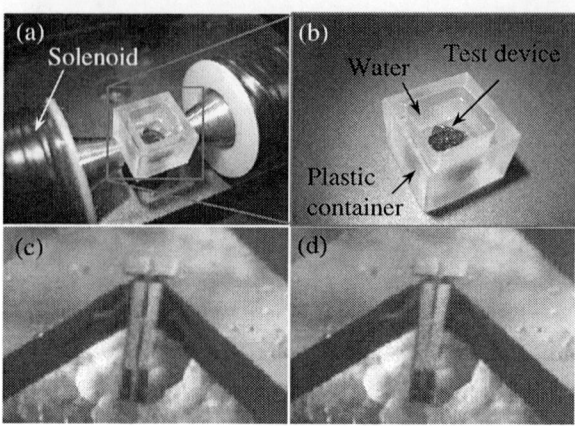

Figure 10: Underwater test: (a) experiment setup, (b) chip inside the water, (c) initial state, and (d) pull-in when magnetic field applied.

effect in magnetization actuation will not damage the micro devices. The test device also successfully demonstrates the magnetic bearing resulted from the repulsive force. Furthermore, the underwater test of micro gripper shows the possibility of operating in liquid by proposed actuation method.

The electroplated *Ni* was employed as the material to realize the test devices as well as magnetization units in this study. For future applications, the actuation efficiency can be improved by using higher susceptibility soft magnetic material like Fe, Ni-Fe, Co, etc. It is also possible to treat the magnetization units during or after the fabrication process for better magnetic anisotropy to improve the actuation efficiency.

ACKNOWLEDGEMENTS

This paper is based (in part) upon work supported by the National Science Council, Taiwan under Grant NSC 95-2221-E-007-068-MY3. The authors would like to express his appreciation to the center of the Nano science and tech. of National Tsing Hua University, and Nano Facility Center of National Chiao Tung University in providing the fabrication facilities.

REFERENCES

[1] K. E. Petersen, *IEEE Transactions on Electron Devices*, vol. 25, pp.1241-1250, 1978
[2] W.C. Tang, T.C.H. Nguyen and R.T. Howe, *Sensors and Actuators*, vol. 20, pp 25-32, 1989
[3] W. Riethmuller, and W. Benecke, *IEEE Transactions on Electron Device*, vol. 35, pp. 758-763, 1988.
[4] J.H. Comtois, V.M. Bright, and M.W. Phipps, *Proc. of SPIE*, vol. 2642, Austin, TX, 1995, pp.10-21.
[5] E. Hashimoto, H. Tanaka, Y. Suzuki, Y. Uenishi, and A. Watabe, *IEEE MEMS'94*, Oiso, Japan, Jan., 1994, pp.108-113.
[6] T. Ueno, T. Higuchi, International Symp. on *Micro-NanoMechatronics and Human Science*, Nagoya, Japan, Nov., 2007, pp.460-465.
[7] D. Sameoto, T. Hubbard, and M. Kujath, *J. of Micromechanics and Microengineering*, vol. 14, pp.1359-1366, 2004.

978-1-4244-2977-6/09 $25.00 © 2009 IEEE

DEVELOPMENT OF THE FORWARD-LOOKING ACTIVE MICRO-CATHETER ACTUATED BY TI-NI SHAPE MEMORY ALLOY SPRINGS

M. Komatsubara[1], T. Namazu[1], H. Nagasawa[2], T. Miki[2], T. Tsurui[2], and S. Inoue[1]

[1]Department of Mechanical and Systems Engineering, University of Hyogo, JAPAN
[2]Kobe Material Testing Laboratory, JAPAN
E-mails: namazu@eng.u-hyogo.ac.jp (T. Namazu); hiroyuki_nagasawa@kmtl.co.jp (H. Nagawasa)

ABSTRACT

This paper describes fabrication, characterization, and demonstration of a novel active micro-catheter with Ti-Ni shape memory alloy (SMA) actuators for wide range of forward-looking area to remedy diseased area inside a blood vessel. The developed active micro-catheter consists of eight Ti-Ni SMA spring actuators for catheter actuation, an ultrasonic piezoelectric transducer for forward-looking, a guide wire, a polyurethane tube for catheter coating, and spiral wirings for various flexure motions of catheter by Ti-Ni SMA actuators. The size of the catheter is 3.5 mm in diameter and 60 mm in length of the sum of transducer and actuator sections. Ti-Ni SMA springs were fabricated from a Ti-50at.%Ni sheet by electrochemical etching with a mixed solution of ethanol and lithium chloride. The micro-catheter was assembled by hand under a stereomicroscope. The tip of the produced micro-catheter was able to bend toward at least eight directions by controlling an applied current to Ti-Ni SMA springs. We have demonstrated the effectiveness of the active micro-catheter by carrying out experiments on blood vessels in a small animal.

1. INTRODUCTION

In the medical field, the micro-catheter is a promising medical tool and can be widely used for minimally invasive treatment, such as intravascular treatment [1]. In order to remedy a diseased area accurately inside a blood vessel, doctors must examine the exact condition of the area. It is, however, difficult to examine the area because doctors must control the catheter outside of the patient so the inside a blood vessel can not be seen directly.

Several techniques have been developed for observing inside blood vessel [1]-[3]. For instance, intravascular ultrasound (IVUS) is used to obtain images inside the blood vessel, but it can only obtain the cross sectional image of the blood vessel. So it cannot be used if the blood vessel is stock with thrombus. In this case, doctors have no choice but to depend on their experience. Thus, a novel technique for examining thrombus in front of the catheter should be developed for a more definitive treatment. Furthermore, even if the thrombus in front of the catheter is observed, obtainable information can be few. Therefore, a technique that can obtain much information from a blood vessel is needed.

The objective of this study is to develop and characterize a forward-looking active micro-catheter for observing diseased area such as thrombus inside a blood vessel. We suggest the catheter with an ultrasonic piezoelectric transducer at the catheter tip and eight Ti-Ni SMA actuators around the catheter tube. Ti-Ni SMA was employed for the actuator as Ti-Ni SMA yields large force output per unit

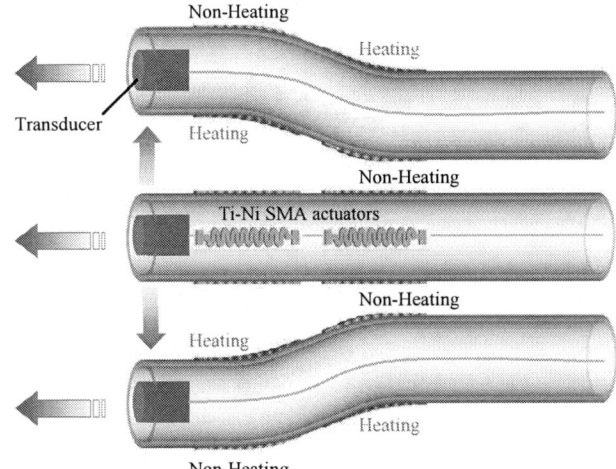

Fig. 1 A concept of the forward-looking active micro-catheter controlled by Ti-Ni SMA spring-actuators.

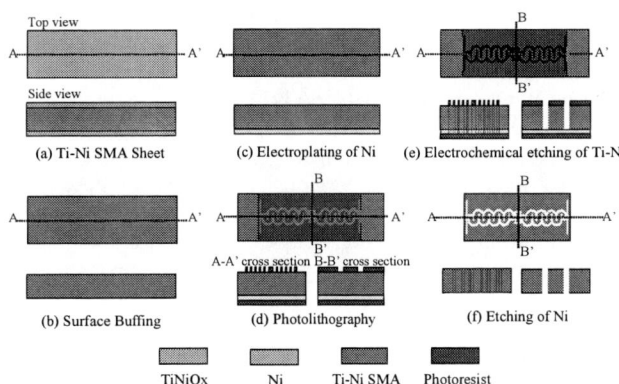

Fig. 2 Process flow for fabricating the Ti-Ni SMA spring actuator.

volume and also is biocompatible materials. With the forward-looking active micro-catheter, the invasive remedy by catheter can be more effective for accurate examination of the diseased area. We have fabricated the catheter and demonstrated the possibility by conducting imaging and actuation tests on blood vessels in a small animal.

2. EXPERIMENTAL PROCEDURE

Forward-looking active micro-catheter

Fig. 1 illustrates the concept of the forward-looking active micro-catheter. The designed micro-catheter has an ultrasonic piezoelectric transducer at the tip of the catheter

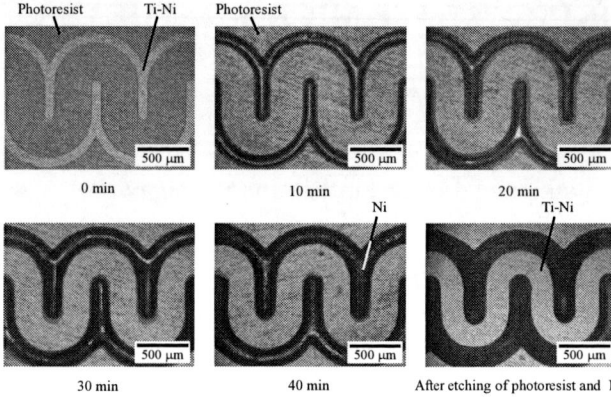

Fig. 3 Snapshot of electrochemical etching of Ti-50at.%Ni sheet.

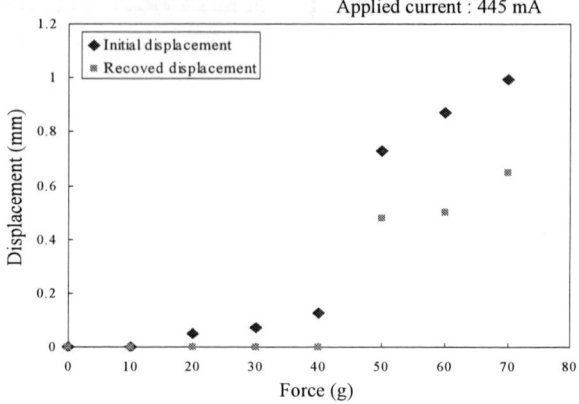

Fig. 4 Relationship between applied force and displacement of the fabricated Ti-Ni SMA springs.

for observing in front of the catheter, and eight Ti-Ni SMA zigzag spring actuators around a catheter tube to bend the tip section of the catheter. The zigzag shape has been adopted for the higher output of recovery force and displacement from the actuator. The zigzag spring SMA actuators, which are in a martensitic phase, are tensioned by external force and fixed both ends onto the catheter tube. Then, by Joule's heating, SMA actuators are transformed from the martensitic into the austenitic phase, so each actuator shrinks to the initial shape. If only one SMA actuator is heated, the catheter tube is bent by transformation because of both ends fixed. Therefore, each application of a current to the Ti-Ni SMA actuator enables the catheter to be actuated towards at least eight directions with an arbitrary angle within 360 degrees. In this case, however, the catheter tip is bent with certain angle. The occurred angle makes the observation by transducer quite difficult. On the other hand, if two SMA actuators which are fixed to diagonally opposite positions are heated, the tip of the catheter is driven without changing the tip angle. Thus, the observation of the diseased area in front of the catheter with wide range is possible using this novel micro-catheter.

Ti-Ni SMA spring actuator

Fig. 2 illustrates the fabrication process for the Ti-Ni SMA actuator. The process is classified broadly into five steps. At first, the surface of the Ti-50at.%Ni sheet, 200 μm in thickness, is buffed well to get rid of an oxidation layer, which disturbs to etch the Ti-Ni sheet, on the surface(Step 1). Then, Ni electroplating is conducted (Step 2). Ni layer of 20 μm in thickness is formed on the back side of the Ti-Ni sheet as a sacrificial layer for electrochemical etching. The sacrificial layer avoids the influence of current concentration during electrochemical etching of Ti-Ni SMA sheet. Photolithography is then carried out to form zigzag-shape spring-like structures made of a photoresist on the surface of the Ti-Ni sheet as an etching mask (Step 3). In this step, the photoresist of a positive type is hard-baked at 473 K for 1.5 hours to get tolerant to chemicals during electrochemical etching. After that, Ti-Ni sheet is selectively etched with a mixed solution of ethanol and lithium chloride to fabricate the Ti-Ni SMA zigzag springs (Step 4). During the electrochemical etching, a voltage of 10 V is applied between electrodes. The photoresist and the Ni layer are removed with a mixed solution of sulfuric acid and hydrogen peroxide solution at 363 K for 10 minutes and Ni etchant at 343 K for 3 hours, respectively. Then, the stand-alone Ti-Ni SMA spring structure is finally formed.

3. RESULTS AND DISCUSSIONS
Characterization of the Ti-Ni SMA spring

Fig. 3 shows the snapshot in the process of the electrochemical etching of the Ti-Ni spring actuator. It appears that the etching is proceeded. After 40 minutes etching, some penetrating spots are observed on the etching surface. Then, photoresist was removed, followed by dipping the sheet in Ni etchant. The Ti-Ni spring was successfully formed without any damages, so the electrochemical etching technique using sacrificial Ni layer on the backside was effective for forming Ti-Ni microstructure. In order to obtain the characteristics of the produced Ti-Ni SMA actuator, we have conducted a uniaxial tensile test using the actuator under various tensile forces. In the experiment, we have determined the relationships between applied force, initial displacement, and recovered displacement when the current of 445 mA is applied to the SMA actuator. Fig. 4 shows the characteristics between applied forces and change on displacements. The initial displacement increased along with the increment of the applied forces to the actuator. On the other hand, the initial displacement was reduced when the SMA spring was actuated by Joule's heating. Although certain residual displacements have been obtained at each force level, the actuator generates recovering force enough to bend the catheter. Note that the significant change between 40 and 50 grams was observed because of static friction of the employed tester.

Fabrication of the active micro-catheter

Fig. 5 shows the fabrication process of the active micro-catheter. We have used a flexible polyethylene double rumen tube. At first, wiring with single-core cable was conducted on the catheter. In order to connect Ti-Ni SMA

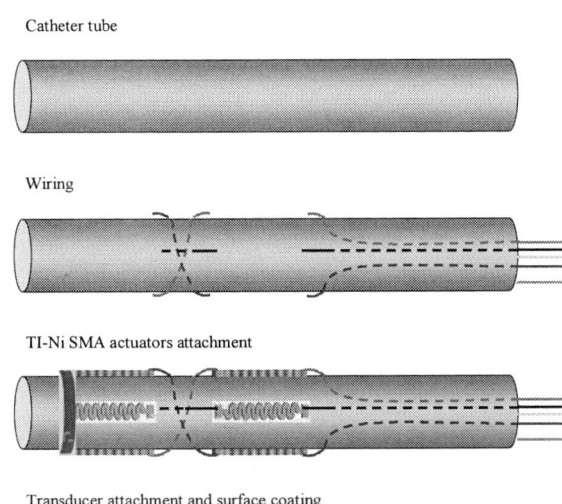

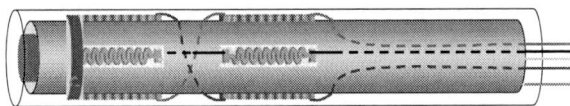

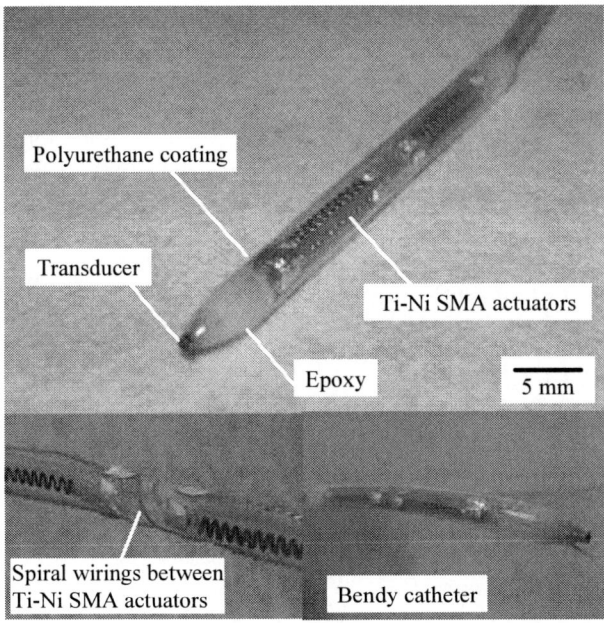

Fig. 5 Fabrication process of the Ti-Ni SMA spring actuated micro-catheter with transducer.

Fig. 6 The produced active micro-catheter with Ti-Ni SMA actuators and transducer for forward-looking.

actuators fixed to diagonally opposite positions, spiral wiring was adopted. After that, eight Ti-Ni SMA actuators are attached onto the catheter tube and connected with the single-core cable using conductive polymer paste. Note that the four actuators at the tip are connected using aluminum film, so the current can be applied to each pair of actuator independently. Then, the transducer was fixed using epoxy and the surface of the catheter was coated with polyurethane

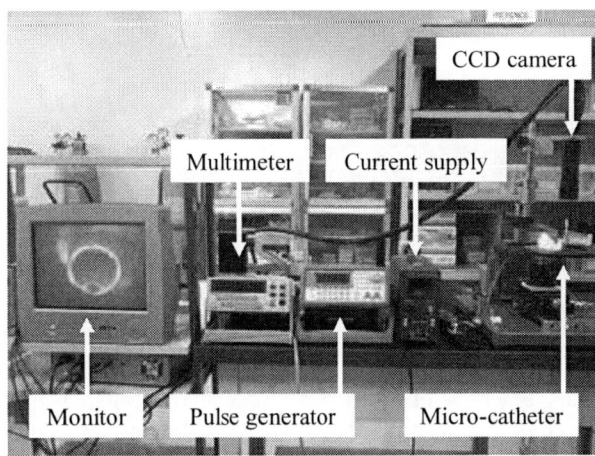

Fig. 7 The catheter actuation testing system.

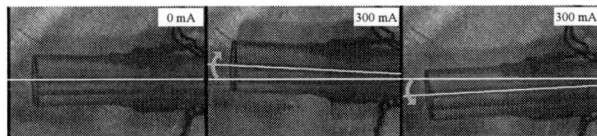

(a) Only one actuator is driven.

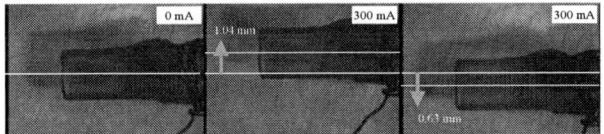

(b) A pair of actuators is driven.

Fig. 8 Novel actuation system using a pair of actuators.

to avoid any contact in a blood vessel. The produced active micro-catheter with Ti-Ni SMA actuators and transducer is shown in Fig. 6. It is 3.5 mm in diameter and 60 mm in length of the sum of transducer and actuator sections. The micro-catheter has good characteristics, such as bendy, biomedical, and can be actuated lithely.

Actuation test

We have conducted the actuation test using the test system, as seen in Fig. 7. The system consists of a current supply for applying constant current to the actuators, multimeter for measuring the current value, pulse generator for converting the current into pulse signal, CCD camera and monitor for observing the catheter motion.

Fig. 8 (a) shows the motion of the tip of the catheter when a current of 300 mA is applied to only one actuator. The catheter tip has moved to indicated direction, but it is clear that the tip of the catheter moves with certain angle toward the moving direction. Observing in front of the catheter is difficult with the angle occurred by the actuation. Moreover, it could skin the inside wall of a blood vessel, so that the motion does not suit on the catheter use. On the other hand, the proposed catheter tip moves without any angle when a pair of actuators is driven at the same time, as seen in

978-1-4244-2977-6/09 $25.00 © 2009 IEEE   1053

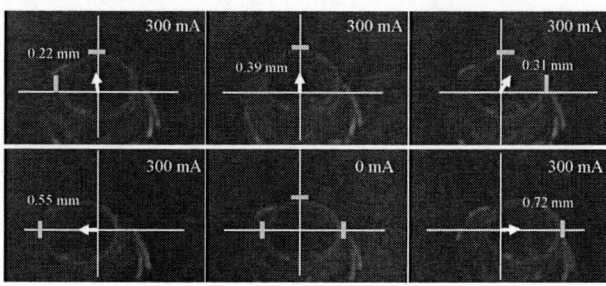

Fig. 9 Actuation test results of active micro-catheter.

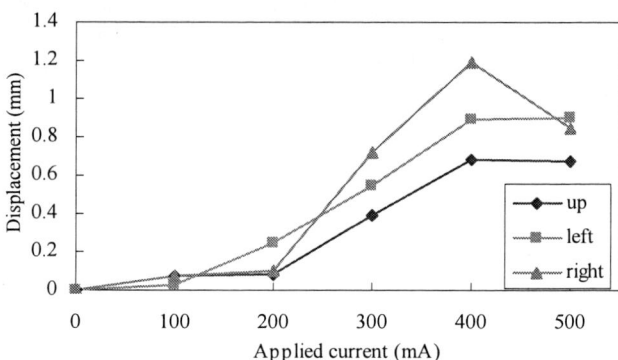

Fig. 10 The relationship between applied current and resultant displacement.

Fig. 8 (b). As the result, it appears that the image in front of the catheter can be obtained in wide range using this novel actuation system. Then, the catheter has been driven in several directions by applying a current to several actuators. Fig. 9 shows the actuation test results of the active micro-catheter. By applying a current to a set of Ti-Ni SMA actuators, the catheter was able to move to the intended directions toward at least eight directions. The displacement of 0.22 - 0.72 mm toward several directions were obtained when the current of 300 mA was applied. Although the amount of displacements differs in each direction, it is due to the individual difference between the actuators and can be controlled by changing the applied current. The relationship between the applied current and resultant displacement is shown in Fig. 10. As applied current increases, the resultant displacement increases. Although the displacement does not change or even reduces under the current of 500 mA, it is because the distance of actuators are too close and the adjacent actuator was also heated. On the other hand, the obtained displacement is enough for operation, so that the tip motion can be fully controlled less than 400 mA.

As the catheter was successfully driven in several directions, the demonstration test was performed in a thick blood vessel of a small animal, though the data, such as photographs in testing, are skipped here. A doctor controlled the catheter, and confirmed that the catheter was bent by Joule's heating. Also, both a bone and silicone thrombus that were ploughed in the blood vessel in the animal were observed using the transducer. The doctor has evaluated that the catheter possesses good operability and potential for the practical use. Thus, the forward-looking catheter with Ti-Ni SMA actuators is effective for wide range examination of the diseased area inside of the blood vessel, and it leads to the accurate and expeditious remedy.

4. CONCLUSIONS

We have designed and developed the forward-looking active micro-catheter for the accurate observation of a diseased area in a blood vessel. We have firstly formed Ti-Ni SMA spring actuators by electrochemical etching. Then, in order to characterize the Ti-Ni SMA spring actuator, the tensile test has been carried out under various forces and confirmed that the spring actuator generates enough force to bend the catheter tube. As forming actuator has been succeeded, the forward-looking active micro-catheter having eight actuators and a transducer has been fabricated. The tip motion without any angle was obtained by applying a certain current to a pair of actuators whereas the clear tilt was observed by applying current to only an actuator. The drive directions and distance was able to be controlled by current value. The tip motion was observed in a thick blood vessel of a small animal. The developed catheter would be useful for the accurate and expeditious remedy in a blood vessel.

ACKNOWLEDGEMENTS

The authors would like to thank members of New Business Department, Kobe Material Testing Laboratory, for fruitful discussion on this study.

REFERENCES

[1] M. Komatsubara, T. Namazu, H. Nagasawa, T. Tsurui, and S. Inoue, "Cylindrical film deposition system for three-dimensional titanium-nickel shape memory alloy microstructure", *Vacuum*, Vol. 83, pp. 703-707, 2009.

[2] K. L. Gentry and S. W. Smith, "Integrated Catheter for 3-D Intracardiac Echocardiography and Ultrasound Ablation", *IEEE Trans Ultrason Ferroelectr Freq Control*, Vol.51, No.7, pp. 800-808, 2004.

[3] Y. Haga, E. Masayoshi, "Biomedical Microsystems for Minimally invasive Diagnosis and Treatment", *Proc. IEEE*, pp. 98-114, 2004.

LOW-RESONANT-FREQUENCY MICRO ELECTRET GENERATOR FOR ENERGY HARVESTING APPLICATION

M. Edamoto[1], Y. Suzuki[1], N. Kasagi[1], K. Kashiwagi[2], Y. Morizawa[2], T. Yokoyama[3], T. Seki[3], and M. Oba[3]
[1]Department of Mechanical Engineering, The University of Tokyo, Tokyo, Japan
[2]Research Center, Asahi Glass Corporation, Kanagawa, Japan
[3]Core Technology Center, OMRON Corporation, Kyoto, Japan

ABSTRACT

A vibration-driven electret generator has been developed for energy harvesting applications. By using parylene as the spring material, a low-resonant-frequency MEMS generator is realized. Electrostatic levitation is adopted for the gap control. Large in-plane amplitude of 0.5 mm at the resonant frequency as low as 21 Hz has been achieved. We also demonstrate electret-powered operation of LED using a low-power-consumption impedance conversion circuit.

INTRODUCTION

Recently, micro power generation attracts significant attention as mobile power source. Energy harvesting from environmental vibration has potential to replace button batteries used for low power applications such as RFIDs and automotive sensors [1-3]. Since the frequency range of vibration existing in the environment is below 100 Hz, electret power generators [4-8] should have higher performance than electromagnetic counterparts.

We recently discover that CYTOP® (Asahi Glass), which is MEMS-friendly amorphous perfluoro-polymer, offers very high surface charge density [6]. Sakene et al. [9] has developed a new high-performance polymer electret material by doping aminosilane into CYTOP, and demonstrated that up to 0.69 mW power can be obtained at an oscillation frequency as low as 20 Hz with the amplitude of 1.2 mm.

In the present study, we microfabricate a prototype electret generator using the CYTOP electret and parylene high-aspect-ratio spring [10], which allows low resonant frequency and large amplitude. Electret-based non-contact bearing [11] is employed as the gap control method between electret and the counter electrode to prevent electrostatic pull-in. In addition, we also develop a novel energy managing circuit for impedance conversion and examine its performance systematically.

ELECTRET POWER GENERATOR

For electret generators, power output is increased with decreasing the gap between the electret and the counter electrode. However, since electrostatic attraction force in the vertical direction is also increased, the gap control is crucial to avoid pull-in. Tsurumi et al. [11] found that electrostatic repulsion force can be obtained between opposed patterned electrets. Thus, in the present design, patterned electrets are formed both on the seismic mass and the bottom substrate to keep the gap constant.

In order to reduce the amplitude of horizontal eletrostatic force, two-phase configuration [12] is adopted, in which two separate generator circuits 180° out-of-phase each other are integrated in a single seismic mass. A ceramic package filled with insulation gas is employed to prevent discharge between electrets and electrodes.

Figure 1 shows a schematic of the micro electret generator designed in the present study. The top substrate consists of a Si proof mass supported with parylene high-aspect-ratio springs [10]. Patterned electrets and electrodes are formed both on the Si mass and the bottom Pyrex substrate. Designed values of the resonant frequency, the amplitude and the power output estimated with the VDRG (velocity-damped resonant generator) model [3] are respectively 74 Hz, 1.0 mm$_{p-p}$ and 2.0 µW. When 1.4 g mass is glued onto the seismic mass, the resonant frequency and the theoretical power output are respectively 20 Hz and 100 µW.

FABRICATION

Fabrication process of the electret generator is shown in Fig. 2. For the top Si substrate with a seismic mass, the process starts with a 400 µm-thick 4" Si wafer with 1.5 µm-thick thermal oxide. The oxide layer on the front side is patterned with BOE for the etch mask of DRIE, and 350 µm-deep trenches are etched into the substrate (Fig. 2a). The trenches are used as the parylene molds. Some of the trenches also define boundaries of Si islands to be left. Then, bottom Cr/Au/Cr electrodes are evaporated on the backside and patterned with standard lithography process, followed by spun-on 15 µm-thick CYTOP (CTL-809M) films and curing at 185 °C for 1.5 hours. A metal mask for parylene etching is evaporated and patterned (Fig. 2b). Next, 15

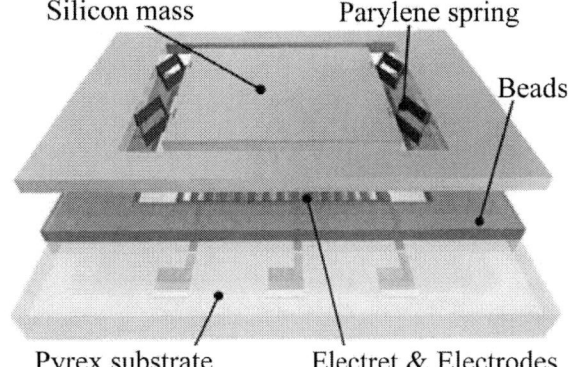

Figure 1. Schematic of the electret generator.

978-1-4244-2977-6/09 $25.00 © 2009 IEEE

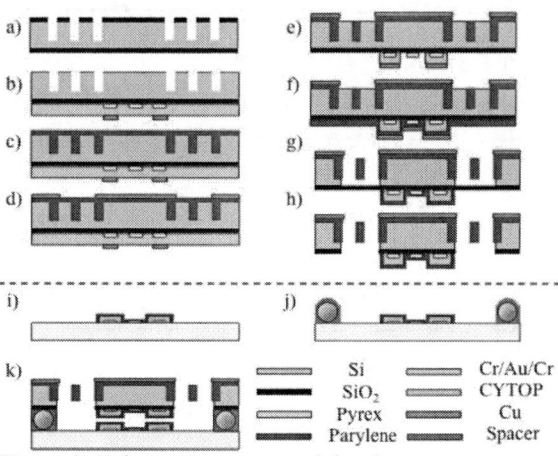

Figure 2. Fabrication process of the electret generator.

Figure 3. Electret generator prototype assembled in ceramic package.

μm-thick parylene-C is deposited on the front side and etched back with O_2 plasma. This is followed by the second parylene-C deposition to fully refill the trenches (Fig. 2c). After the metal mask for the parylene etching is patterned (Fig. 2d), the parylene and CYTOP films are etched with O_2 plasma (Fig. 2e). Another 5 μm-thick parylene layer is then deposited on the top of CYTOP and patterned with O_2 plasma (Fig. 2f). Finally, the Si substrate surrounding the Si mass is etched away with XeF_2 (Fig. 2g), and the structures are released with BOE (Fig. 2h). For the bottom substrate, after patterning Cr/Au/Cr electrodes on with a 700 μm-thick 4" Pyrex wafer, CYTOP and parylene films are deposited and patterned (Fig. 2i).

After these processes, charges are implanted into CYTOP electret using corona charging for 3 minutes at 120 °C, which is slightly higher than the glass transition temperature of CYTOP. The needle and grid voltages are respectively -8 kV and -600 V. Then, micro beads, of which diameter is well defined, are mixed with epoxy adhesive and applied to the Pyrex substrate (Fig. 2j) as the spacer. Finally,

the top Si substrate and the bottom Pyrex substrate are aligned and bonded in SF_6 atmosphere.

Figures 3-5 show photographs of the generator prototype thus fabricated. The dimensions of the device are 18.5 x 16.5 mm², while the size of the mass is 14.6 x 16 mm².

The seismic mass is supported by 25 μm-wide parylene springs (Figs. 4b, d). On the backside of the top substrate, 14 poles of patterned electrets are formed for each phase of the generator (Fig. 4c). The width of the patterned electret and electrode is 150 μm. On the bottom substrate, wire bonding pads for the two phase electrodes and the common ground are located.

MECHANICAL RESPONSE

Firstly, the mechanical response of the spring-mass system is examined. The seismic mass supported with the parylene springs is fixed on an electromagnetic shaker (APS-113, APS Dynamics), and the in-plane amplitude of the mass is measured with a digital microscope (CA-MN80, Keyence). Figure 6 shows the frequency response of the seismic mass. Its resonant frequency is 51 Hz with a quality factor of 7.49. When 1.4 g mass is added, the resonant

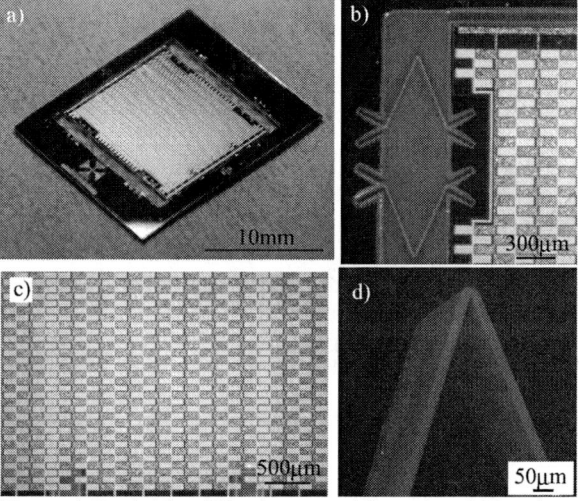

Figure 4. Backside of Si substrate. a)Overview, b)Parylene high-aspect-ratio spring and patterned electret, c)Magnified view of electret and electrodes in the two-phase configuration, d)SEM image of the parylene spring.

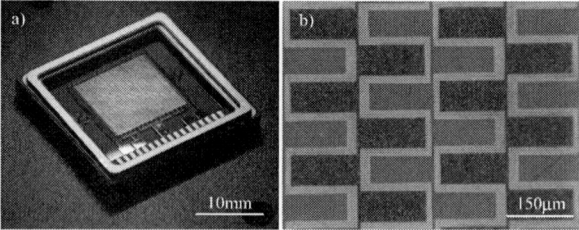

Figure 5. Bottom Pyrex substrate. a)Overview, b)Magnified view of checker-board-patterned electret and electrodes.

frequency becomes 21 Hz with a quality factor of 5.23. The maximum in-plane amplitude at the resonance is as large as 1.0 mm_p-p. Although the quality factor should be improved, the resonant frequency of the present seismic structure is sufficiently low for energy harvesting applications.

Figure 7 shows the experimental setup for the measurement of vertical electrostatic force between the top and bottom substrates. One of the substrates is fixed on an electric balance, and the other is mounted on a 5-axis alignment stage. The gap is measured with a high-precision laser displacement meter (LT-9500, Keyence). Figure 8 shows the electrostatic force. Repulsive force is obtained for the gap smaller than 150 μm when the both electrets are charged, while attractive force is induced when only one electret is charged. Thus, electrostatic levitation can be achieved with the present electret/electrode patterns.

Figure 9 shows the electrostatic force with alignment error of translational and rotational direction. With the present design, repulsive force is obtained even with translational alignment error of 1.2 mm and rotational alignment error of 1.4°. Therefore, the repulsive force in the present design is robust for misalignment, which is inevitable during operation.

IMPEDANCE CONVERSION CIRCUIT

Since the output impedance of electret generators is inherently very high, an impedance conversion circuit is necessary to drive actual circuits. Figure 10 shows the

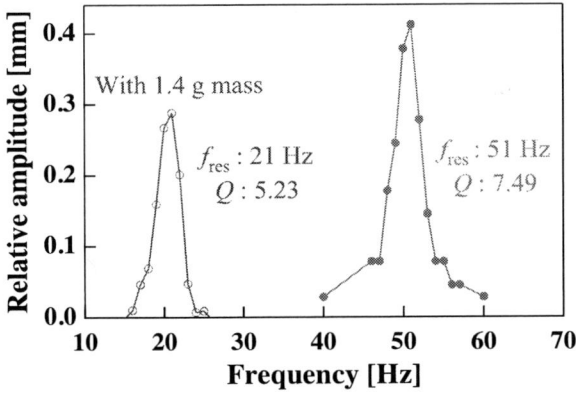

Figure 6. Frequency response of the electret generator.

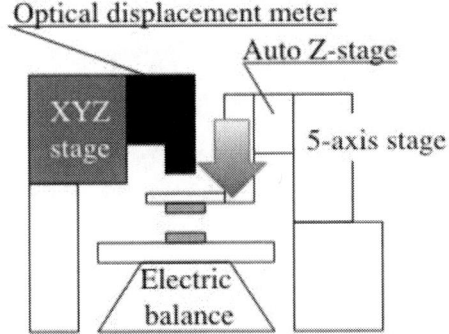

Figure 7. Schematic of the force measurement setup.

circuit diagram of the energy managing circuit including rectifier, smoothing capacitor, and switching circuit. An array of transistors serves as a switch, which drives a LED intermittently. The present circuit requires no external power for wake-up, and consumes only 80 nA.

Figure 11a shows the output voltage of an electret generator with an output power of 12.5 μW and an output impedance of 8.2 MΩ. As shown in Fig. 11b, charges are continuously stored in the capacitor, and intermittently delivered to the LED. In this experiment, peak power

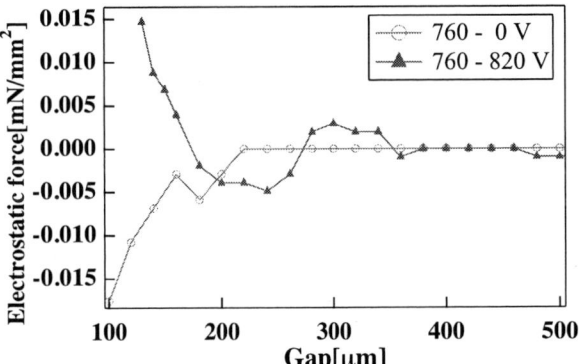

Figure 8. Electrostatic force between electret substrates versus the gap.

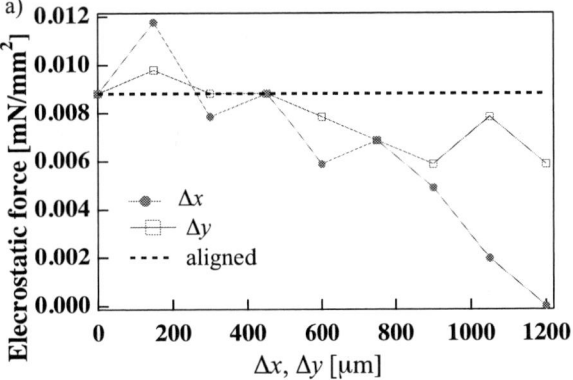

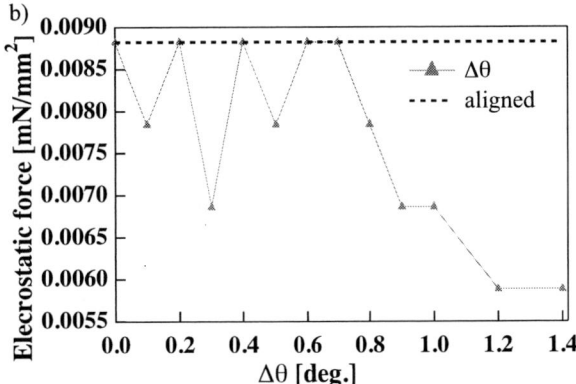

Figure 9. Electrostatic force between electret substrates with finite misalignment. a) Translational displacement, b) Rotational displacement.

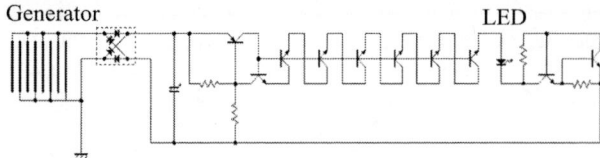

Figure 10. Power managing circuit for rectifying, smoothing and impedance conversion.

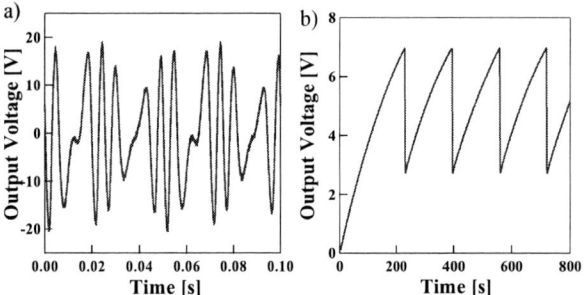

Figure 11. Time trace of the output voltage. a)Electret generator, b)Power managing circuit.

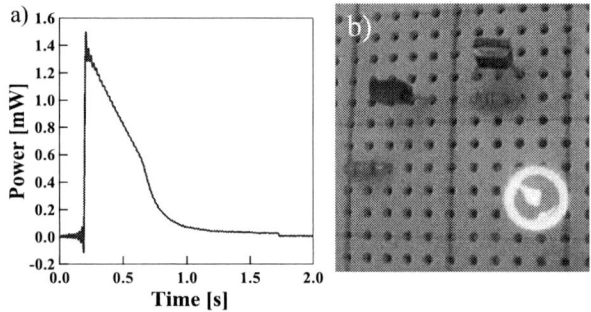

Figure 12. Intermittent operation of a LED. a)Output power, b)Photo. About 0.54 mJ is delivered to the LED for about 1 s with an interval of 165 s.

delivered to the LED is about 1.4 mW, and intermittent operation for about 1 s with an interval of 165 s is accomplished (Fig. 12). Since the output impedance of the present circuit is about 20 Ω, the impedance conversion ratio is as large as 4×10^5.

CONCLCUSIONS

Vibration-driven electret generator for energy harvesting has been developed. Parylene high-aspect-ratio spring is successfully microfabricated to support an in-plane seismic mass. Patterned CYTOP electret is formed on both the top and bottom substrates in order to keep the gap using electrostatic repulsive force. Resonant frequency as low as 21 Hz has been achieved with a large in-plane amplitude of 0.5 mm. With the aid of the patterned electrets on the both side of substrates, robust electrostatic levitation is achieved. We have also demonstrated that, with the present power management circuit, the impedance can be lowered by 400,000 times, and a LED can be intermittently operated

using the power output of an electret generator with 12 µW output.

This work is supported by the New Energy and Industrial Technology Development Organization (NEDO) of Japan. Photomasks are made using the University of Tokyo VLSI Design and Education Center (VDEC)'s 8-inch EB writer F5112+VD01 donated by ADVANTEST Corporation.

REFERENCES

[1] S. P. Beeby, M. J. Tudor, and N. M. White, "Energy harvesting vibration sources for micro systems applications," Meas. Sci. Technol., Vol. 17, pp.175- 195, 2006.

[2] J. A. Paradiso, and T. Starner, "Energy scavenging for mobile and wireless electronics," IEEE Pervasive Comp., Vol. 4, pp. 18-27, 2005.

[3] P. D. Micheson, T. C. Green, E. M. Yeatman, and A. S. Holmes, "Architectures for vibration-driven micropower generators," J. Microelectromech. Syst., Vol. 13, pp. 429-440, 2004.

[4] Y. Tada, "Experimental characteristics of electret generator using polymer film electrets," Jpn. J. Appl. Phys. Vol. 31, pp. 846-851, 1992.

[5] J. Boland, Y.-H. Chao, Y. Suzuki, and Y.-C. Tai, "Micro electret power generator," Proc. 16th IEEE Int. Conf. MEMS, Kyoto, pp. 538-541, 2003.

[6] T. Tsutsumino, Y. Suzuki, N. Kasagi, and Y. Sakane, "Seismic power generator using high-performance polymer electret," Proc. 19th IEEE Int. Conf. MEMS, Istanbul, pp. 98-101, 2006.

[7] T. Sterken, P. Fiorini, G. Altena, C. Van Hoof, and R. Puers, "Harvesting energy from vibrations by a micromachined electret generator," Int. Conf. Solid-State Sensors, Actuators and Microsystems (Transducers'07), pp 129-132, 2007.

[8] H.-W. Lo, and Y.-C. Tai, "Parylene-based electret power generators," J. Micromech. Microeng., Vol. 18, 104011, 2008.

[9] Y. Sakane, Y. Suzuki, and N. Kasagi, "Development of high-performance perfuluoriented polymer electret film and its application to micro power generation," J. Micromech. Microeng., Vol. 18, 104011, 2008.

[10] Y. Suzuki, and Y. -C. Tai, "Micromachined high-aspect- ratio parylene spring and its application to low- frequency accelerometers," J. Microelectromech. Syst., Vol. 15, pp.1364-1370, 2006.

[11] Y. Tsurumi, Y. Suzuki, and N. Kasagi, "Non-contact electrostatic micro-bearing using polymer electret," Proc. 21th IEEE Int. Conf. MEMS, Tucson, pp. 511-514, 2008.

[12] C. Marboutin, Y. Suzuki, and N. Kasagi, "Optimal design of micro electret generator for energy harvesting," 7th Int. Workshop Micro and Nanotechnology for Power Generation and Energy Conversion Applications (PowerMEMS 2007), Freiburg, pp. 141-144, 2007.

DEVELOPMENT OF THE NOVEL ELONGATION-MEASUREMENT DEVICE WITH IN-PLANE BIMORPH ACTUATOR FOR THE TENSILE TEST

Hiroki Fujii, Takahiro Namazu, and Shozo Inoue

Department of Mechanical and Systems Engineering, University of Hyogo, JAPAN

2167 Shosha, Himeji, Hyogo 671-2201, Japan: namazu@eng.u-hyogo.ac.jp

ABSTRACT

This paper describes novel elongation-measurement device utilized for uniaxial tensile test. The device consists of two sets of micro probes with piezoresistive sensor for detection of gauge mark position on specimen, and multiple pairs of bimorph actuators for in-plane motion to scan the gauge mark. The voltage from the sensor is put out by the cantilever deflection when the cantilevers climb the gauge marks. By Joule's heating, movement of the bimorph actuator along one axis was achieved, and detection of gauge mark position was also realized. In the uniaxial tensile test of single-crystal silicon (SCS) film specimen, the mean Young's modulus of 164.0 GPa was able to be measured by using the device. The measured modulus agreed with that derived from CCD-image analysis. The developed device would be useful for measuring elongation of film specimen during the uniaxial tensile test.

1. INTRODUCTION

The uniaxial tensile test is one of the most common methods for investigating mechanical characteristics of a macroscopic specimen. In the case that a target material is a thin film, the tensile test is very popular as with the macro specimen [1]. For such the small specimen, however, technical requirements in tensile testing rise drastically. Especially, since elongation of specimen during the test is sure to be very small, the measurement with high accuracy becomes difficult. To date lots of researchers are developing the elongation measurement technique used for the tensile test of a film specimen. The representative methods for elongation measurement are as follows: CCD-image analysis [2], optical interference [3], and atomic force microscopy [4]. In the former two techniques, elongation of specimen is able to be obtained without contact to specimen. All these techniques that need a gauge mark on specimen to measure the elongation enable us to accurately measure the elongation in the tensile test. However, these experimental setups for measurement are very tricky and costly. Therefore, in order to achieve the ease of the tensile test for a film specimen, a new simple and low cost technique for accurately measuring the elongation is required.

The objective of this study is to propose a new elongation measurement technique for uniaxial tensile test. We have developed the novel MEMS device to achieve the proposed idea. The sensing and actuation tests have been firstly conducted in order to demonstrate the ability of the device for measuring elongation. Then the uniaxial tensile test of single-crystal silicon (SCS) film has been performed to investigate the Young's modulus that is the significant material constant necessary for the structural design of MEMS.

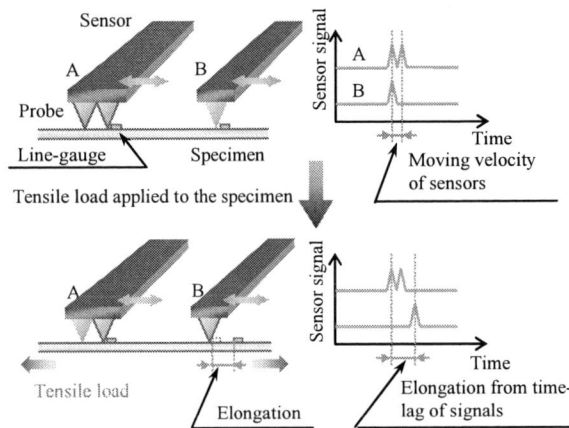

Fig. 1 Elongation-measurement principle.

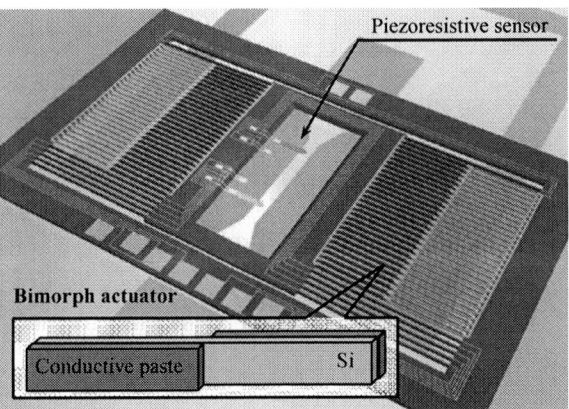

Fig. 2 Schematic of the elongation measurement device.

2. EXPERIMENTAL PROCEDURE

Elongation-measurement principle

Fig. 1 shows the principle of elongation measurement. Two micro-probes (Probe A and B) having piezoresistive sensor scan the two convex line gauges fabricated on specimen. The distance between the probes is the same as that between the gauges. One probe (Probe A) has a two-forked shape to measure moving velocity of sensor head, whereas the other (Probe B) has one sharp probe. While a tensile load is applied to the specimen, the distance between the gauges is elongated. The elongation is able to be calculated from the time-lag of sensor signals arising from the change of the gauge positions.

The designed device

Fig. 2 illustrates the designed elongation-measurement device. The device is composed of piezoresistive sensor mounted two micro-probes, multiple pairs of bimorph

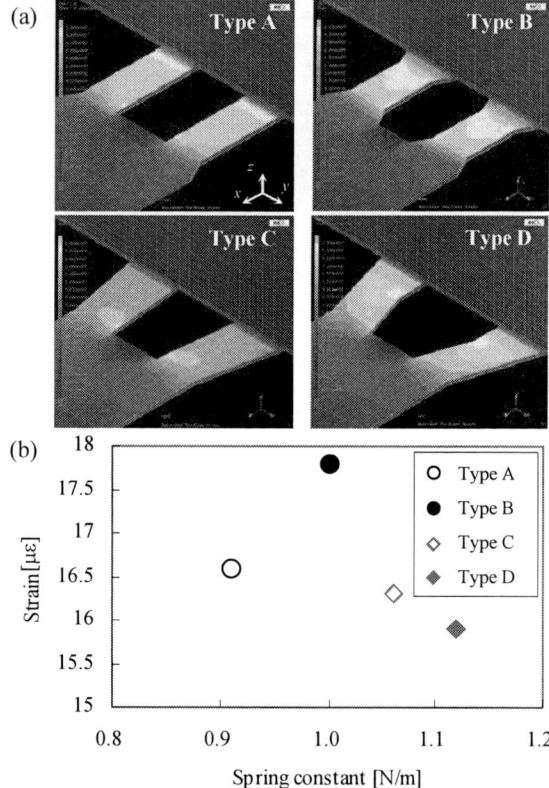

Fig. 3 FEA results of the sensor part: (a) strain distribution, (b) relation between spring constant and strain.

actuator, six pairs of beams for wiring, and frame. The piezoresistive sensor with a temperature compensating circuit is adapted on micro-cantilever probe. The shape was designed using finite element analysis (FEA), as shown in Fig. 3. Several shapes of the cantilever root were designed, and their spring constants were obtained by applying the displacement of 1 μm to their tips. As shown in Fig. 3(b), the strain at the cantilever root for Type B was the largest. This indicates that the spring constant of Type B is comparatively low as compared with other cantilevers. The low spring constant is suitable because damaging the tensile test specimen has to be avoided and a high sensitivity of piezoresistive sensor is better for detection of gauge mark position. On the other hand, the bimorph structure consisting of Si beam and conductive polymer paste is able to actuate by the difference in thermal expansion coefficient between Si and the paste. The actuator section consists of twenty-nine pairs of the bimorph structure. All the actuators have a fixed-fixed beam configuration with rigid ends, so that they are able to cancel out the rotation and out-of-plane deformation of the sensor section. With a small ratio of width to thickness, the actuation beams are able to deform along the scanning axis easier than other axes. In addition, the actuator is able to generate a large force due to the turned two bimorph structures in a beam. The actuator was also designed using FEA, as shown in Fig. 4. The sensor part moved along the scanning axis only by operating the in-plane bimorph

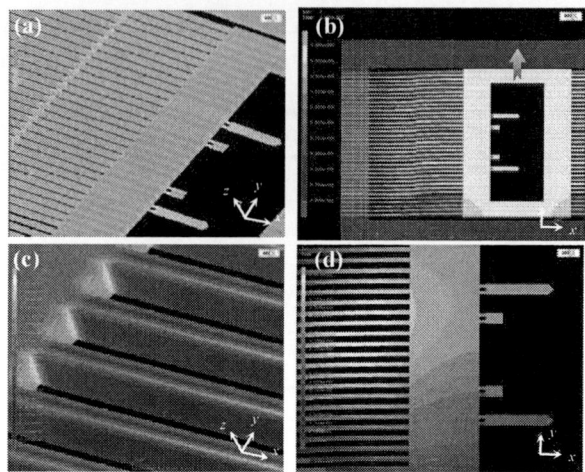

Fig. 4 FEA results of the device: (a) FEA model, (b) displacement distribution, (c) stress distribution, (d) temperature distribution.

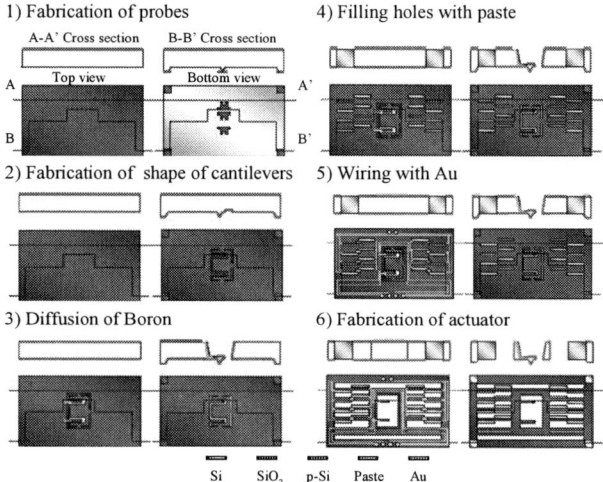

Fig. 5 Fabrication process of the device.

actuator. A little expanse between two cantilever probes was shown because of thermal expansion of Si by Joule's heating.

Fig. 5 depicts the process flow for producing the device. At first, micro-probes were formed by anisotropic wet etching with KOH solution. The piezoresistive sensor was then fabricated by SCS wet etching and thermal diffusion of boron. For the in-plane bimorph actuator, the rectangle holes were etched by deep reactive ion etching (D-RIE), followed by filling a conductive polymer paste. Au wirings were formed through sputtering and wet etching. As the final step, D-RIE was carried out to form the bimorph actuator shape.

Tensile test system

An in-house uniaxial tensile tester was used for evaluating the performance of the produced device. The tester consists of piezoelectric actuator for applying a tensile load to specimen, a load cell for force measurement, and specimen holder. The tester was set under a CCD-image analysis system for specimen elongation measurement with a

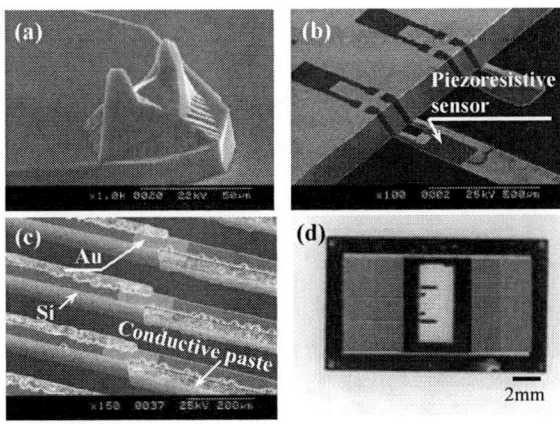

Fig. 6 Photographs of the fabricated device: (a) two-forked probe, (b) sensor section (c) bimorph actuator beams, (d) the entire device.

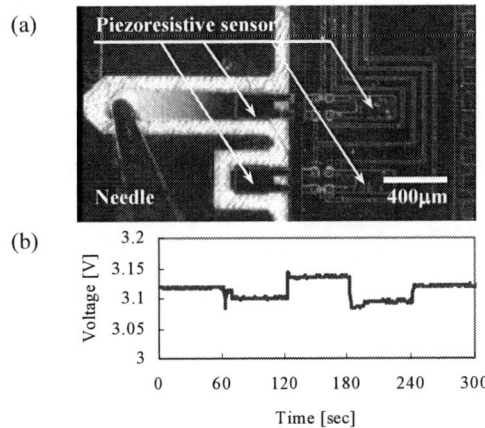

Fig. 7 The sensing test result: (a) snapshot in bending a cantilever, (b) output voltage signal.

resolution of 50 nm in order to check the measurement result derived from the device. The tensile test specimen employed in this study consists of a 15 μm-thick SCS film specimen section with line-shape gauge marks for elongation measurement, chucking holes, SCS bridges, and frame. The specimen has fabricated through conventional micromachining technologies.

The tensile test with elongation measurement using the produced device was conducted through the following steps. At first, the SCS specimen was set onto the tester, and SCS bridges of the specimen were cut with a needle. After that, the device was set onto the gauge section of the specimen. Then, a tensile force was applied by the tensile tester. When a tensile force was kept constant, the device was operated by applying a current for inducing Joule's heating, and scanned the line gauges on the specimen. The specimen elongation was calculated after the test. At the same time, image processing with a CCD camera was conducted to measure the elongation of the specimen.

3. RESULTS AND DISCUSSIONS

The produced device and its performances

Fig. 6 shows the photographs of the fabricated elongation measurement device. The contact probes were formed to a triangular-pyramid-like shape by anisotropic wet etching with KOH (Fig. 6(a)). The piezoresistive sensor was fabricated on the cantilever and base sections (Fig. 6 (b)). The bimorph actuator with SCS and conductive polymer paste was precisely formed by D-RIE (Fig. 6(c)). The fabrication of the entire device was finely completed (Fig. 6(d)).

The performances of cantilever sensor and actuator in the produced device were evaluated. Fig. 7 shows the result of sensor evaluation. The voltage of 5.0 V was supplied to piezoresistive sensors with the Wheatstone bridge circuit. The output voltages from the sensors were measured when the cantilever was deflected with a needle. As a result, the change in voltage was obtained by cantilever's deformation. Therefore, the produced piezoresistive sensors formed on cantilevers were able to be utilized for gauge mark detection

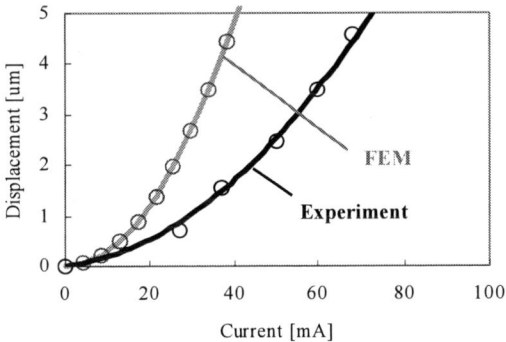

Fig. 8 The relationship between input current and displacement in the actuation test.

in testing.

Fig. 8 shows the result of actuation test. The displacement of the device was measured using a CCD-image analysis system. When a current of 100 mA was applied to actuators, the displacement of 9.0 μm along the scanning axis was observed. The measured value of displacement per unit current was smaller than that from FEA, due to unknown elasticity and resistance of the polymer paste employed. We confirmed that the bimorph structure designed was able to yield the actuation along an axis only.

Uniaxial tensile test for evaluation of the device

The fabricated device was set onto a SCS tensile test specimen, as shown in Fig. 9. The tensile tester was settled onto a stage in the probing tester so as to exchange an electric signal to the device. A tensile force was applied to the specimen by a voltage control to piezoelectric actuator built in the tensile tester. Fig. 10 shows the results of sensor signal measurement when the tensile stresses of 0, 10, and 85 MPa were applied to the specimen. The voltage peaks from two sensor probes (Probe A and B) were observed when respective stresses were applied. However Probe A has two triangle-pyramid shape probes, only one peak from Probe A was observed. This would be caused by poor fabrication precision of the probe tips. The moving velocity of sensor

978-1-4244-2977-6/09 $25.00 © 2009 IEEE   1061

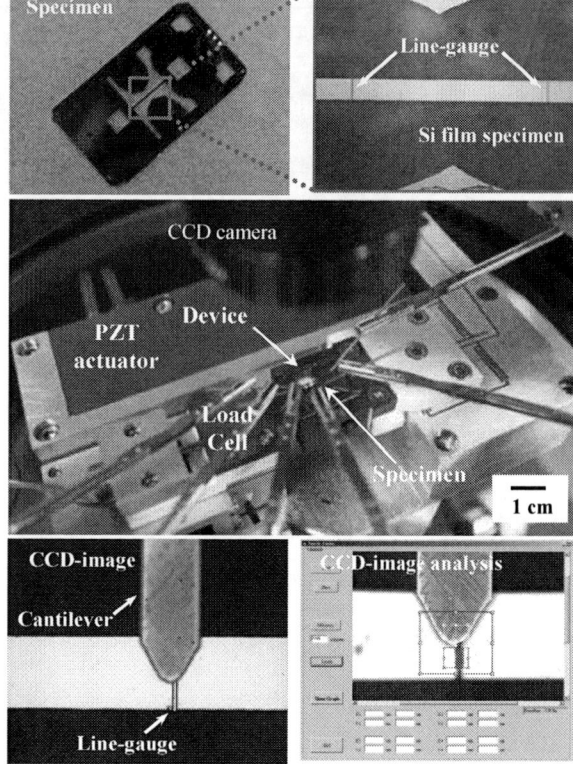

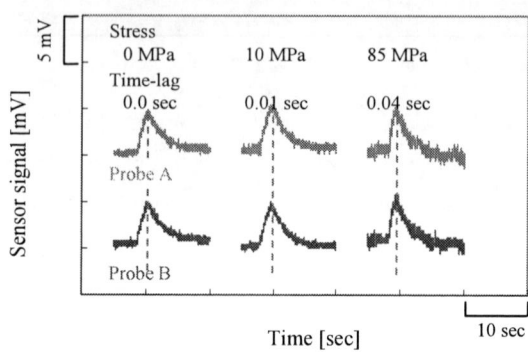

Fig. 10 The obtained sensor signals when the specimen was stretched under various tensile stresses.

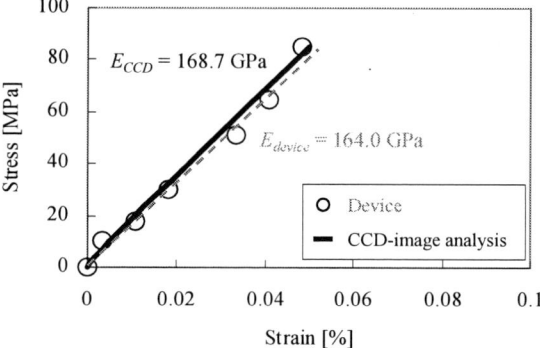

Fig. 11 Typical tensile stress-strain curves of SCS obtained from the device measurement and CCD-image analysis.

Fig. 9 Application of the device toward the compact tensile tester for elongation measurement.

head was measured using a CCD-image analysis system. The calculation of the elongation of specimen was succeeded by the moving velocity and time-lag between voltage peak positions of Probe A and B.

Fig. 11 shows the relationship between tensile stress and strain derived from elongation measurement using the produced device. The open plots represent the measured data using the device, and the solid line is representative of those using a CCD image analysis system. The Young's modulus derived from the device measurement was found to be 164.0 GPa, which was almost the same as that from CCD image analysis. This agreement suggests that the proposed technique using the device is effective for accurately measuring the elongation of a film specimen during the tensile test.

4. CONCLUSIONS

We have designed and produced the novel elongation-measurement device used for uniaxial tensile test. The device includes two micro-probes with piezoresistive sensor, and multiple pairs of in-plane bimorph actuator consisting of SCS and conductive polymer paste. By Joule's heating, movement of the bimorph actuator along a scanning axis was observed. The fabricated device was set onto a SCS specimen, and elongation of the specimen was measured during the tensile test. The mean Young's modulus of 164.0 GPa was obtained from the device measurement.

ACKNOWLEDGEMENTS

The authors would like to express their gratitude to members of SyncMEMS Laboratory in Kagawa University, Japan, for extensive discussion on device fabrication. The authors would like to thank Prof. Tsuchiya in Kyoto University, Japan, for precious ideas regarding the device concept.

REFERENCES

[1] T. Namazu, *et al.*, "Characterization of Single Crystal Silicon and Electroplated Nickel Films by Uniaxial Tensile Test with in-situ X-ray Diffraction Measurement", *Fract. & Fati. Eng. Mater. & Struct.*, Vol. 30, No. 1, pp. 13-20, 2007.

[2] H. Fujii, *et al.*, "Mechanical Characteristics of Al-Si-Cu Structural Films by Uniaxial Tensile Test with Elongation Measurement Image Analysis", *Proc. MRS 2008 Fall Meeting*, 2008.

[3] W. N. Sharpe, Jr., *et al.*, "A New Technique for Measuring The Mechanical Properties of Thin Films", *J. Microelectromech. Syst.*, Vol. 6, No. 3, pp. 193-200, 1997.

[4] Y. Isono, *et al.*, "Development of AFM Tensile Test Technique for Evaluating Mechanical Properties of Sub-Micron Thick DLC Films", *J. Microelectromech. Syst.*, Vol. 15, No. 1, pp. 169-180, 2006.

978-1-4244-2977-6/09 $25.00 © 2009 IEEE

MICROACTUATION UTILIZING WAFER-LEVEL INTEGRATED SMA WIRES

D. Clausi[1], H. Gradin[2], S. Braun[2], J. Peirs[1], G. Stemme[2],
D. Reynaerts[1] and W. van der Wijngaart[2]

[1]Div. PMA, Dep. of Mechanical Engineering, K.U.Leuven, BELGIUM
[2]Microsystem Technology Lab, KTH-Royal Institute of Technology, Stockholm, SWEDEN

ABSTRACT

This paper reports on the wafer-scale integration of pre-strained SMA wires to microstructured silicon devices and the performance of the microactuator prototypes.

The overall goal is to obtain low cost microactuators having high work densities and a mass production compatible manufacturing, without having to deal with the inherently high costs of a pick-and-place approach or with the complex composition control and annealing process of sputtered NiTi films.

Testing above the SMA transformation temperature shows repeatability in actuation of the fabricated structures, with net strokes of 170 µm for the double cantilever actuators.

INTRODUCTION

Microelectromechanical systems (MEMS) technology offers a large range of microactuators for specific applications, based on different actuation principles such as electrostatic, piezoelectric, thermal and magnetic actuation [1], [2]. In a comparison of the different actuation mechanisms, shape memory alloy (SMA) actuators offer the highest work density, which exceeds that of other actuation principles by at least an order of magnitude [3]. Potential issues of SMA actuators, such as low operation frequencies due to limited heat transfer rates, high power consumption and small efficiency, are less pronounced upon miniaturization of the structures. Furthermore, in contrast to other actuation principles, the work density of SMA actuators remains constant upon miniaturization [3]. However, despite its advantages, shape memory alloy is not a standard MEMS material, partially due to the lack of cost efficient and batch-compatible integration methods of such a material into MEMS structures.

So far, there have been two different approaches to integrate SMA material into microsystems. In the hybrid integration method, the SMA components and the MEMS structure are fabricated separately and then assembled on a per-device level. The bias spring is provided by a mechanical obstruction, which deforms the SMA during the assembly of the SMA element and the MEMS structure [4]. This approach features the advantage of using bulk SMA, which is commercially available in a wide thickness range and therefore allows for adjustable mechanical robustness and reduced material cost. However, the required per-device assembly results in high component costs.

The other integration method, monolithic integration, is based on sputter deposition of thin SMA films [5] directly on the MEMS structure, which allows for batch compatible processing. The bias spring is provided by the built-in film stress. However, sputtering of SMA is complicated by the tight requirements on composition control and annealing of the material at high temperatures is necessary, which results in unwanted interdiffusion processes of the SMA with the substrate. A recent report states that in case of TiNi-based films sputtering is mostly feasible for thicknesses less than 10 µm [6], thus setting an upper boundary for the mechanical performance of structures actuated by SMA films.

The present research investigates a novel SMA integration and actuation method for microstructures.

SMA WIRE ACTUATION

We previously demonstrated the per-device integration of thin SMA wires to brass cantilevers to realize miniature actuators using conventional machining techniques [7]. In this configuration the SMA is strained under pure tension, resulting in work efficiencies an order of magnitude larger as compared to torsion or bending loads [8].

The present work extends the previous approach to standard silicon micromachining processes and wafer level integration of the wires on arrays of microstructures.

Pre-strained SMA wires are integrated on top of silicon cantilevers that act as bias springs. In the cold state, the silicon cantilevers stretch the SMA wires and bend slightly upwards as result of the static equilibrium with the detwinning stress in the SMA wires (Figure 1a). In the hot state, the SMA wires contract and bend the cantilevers upwards (Figure 1b).

The availability on the market of SMA wires in a wide range of diameters and material characteristics, together with the possibility of varying geometrical parameters of the silicon microstructures, offers a high flexibility in designing the actuators.

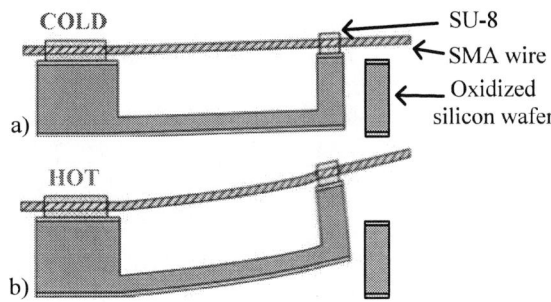

Figure 1: Schematic cross sectional view of the device in the two operational states.

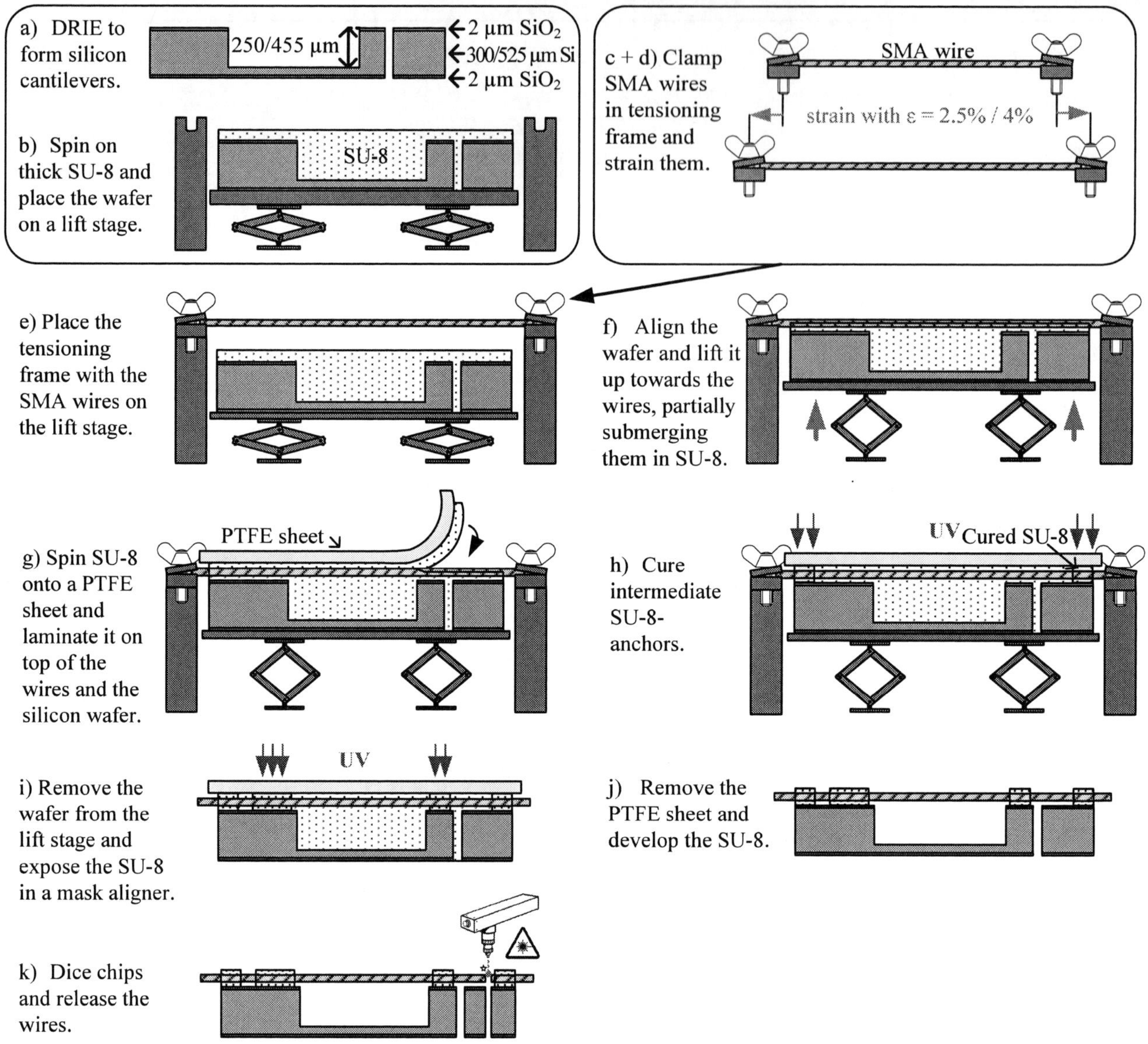
a) DRIE to form silicon cantilevers.
250/455 µm
2 µm SiO2
300/525 µm Si
2 µm SiO2
b) Spin on thick SU-8 and place the wafer on a lift stage.
SU-8
c + d) Clamp SMA wires in tensioning frame and strain them.
SMA wire
strain with ε = 2.5% / 4%
e) Place the tensioning frame with the SMA wires on the lift stage.
f) Align the wafer and lift it up towards the wires, partially submerging them in SU-8.
PTFE sheet
g) Spin SU-8 onto a PTFE sheet and laminate it on top of the wires and the silicon wafer.
UV
Cured SU-8
h) Cure intermediate SU-8-anchors.
UV
i) Remove the wafer from the lift stage and expose the SU-8 in a mask aligner.
j) Remove the PTFE sheet and develop the SU-8.
k) Dice chips and release the wires.

Figure 2: Process flow

FABRICATION

The fabrication of test structures is schematically illustrated in Figure 2.

The actuator is fabricated from two parts. One part consists of the silicon cantilever structures and the other part consists of pre-strained SMA wires. The two parts are finally assembled using an adapted adhesive bonding process with SU-8 (Microchem SU-8 2025) as adhesive layer.

A 7 x 7 array of silicon cantilever structures was fabricated starting with a thermally oxidized 300 µm or 525 µm Silicon wafer. The thermal oxide was patterned to be used as a hard mask. U-shaped cantilevers as shown in Figure 3 were then deep reactive ion etched, first from the back side and then from the front side (Figure 2a). For mechanical support and easier handling, the processed wafer was bonded with photoresist to a 525 µm thick carrier wafer before the front side etch (not shown in the illustration). A thick SU-8 adhesive layer was applied onto the silicon wafer in a standard spin-on step, completely filling the etched structures; subsequently, the wafer was transferred onto a lift stage (Figure 2b).

Si cantilever (50/70 µm thick)
SU-8
SMA wire
300 µm
500 µm
ø 37.5 µm
1 mm
2 mm
SU-8

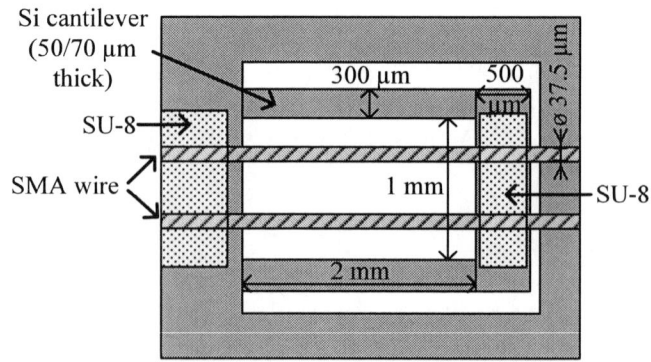

Figure 3: Schematic top view of the device and its dimensions

The SMA wires (Dynalloy Flexinol® HT, nominal diameter φ37.5 μm) were oriented and clamped with the desired pitch on a dedicated tensioning frame (Figure 2c) and strained to the desired value, 2.5% and 4% respectively (Figure 2d). During assembly, the wires remained clamped on the frame to maintain both their orientation and imposed strain. After cleaning the wires with acetone, the tensioning frame was secured to the lift stage on top of the wafer (Figure 2e). Then, the wafer was aligned to the wires and lifted, partially submerging the SMA elements in the adhesive layer (Figure 2f). Finally, the stack was placed for 3 hours in a convection oven at 65°C to soft bake the SU-8.

To fully submerge the SMA wires, a second SU8-layer was applied on top of the stack. However, the tensioning frame and the lift stage inhibit a spin-on process. Therefore the second SU8-layer was spun onto a 125 μm thick PTFE foil, soft-baked on a hotplate (3 hours at 60°C) and laminated onto the stack (Figure 4). Finally, the stack was placed in a convection oven at 65°C for 45 min to reflow both SU-8 layers and achieve proper wetting of the SMA fibers.

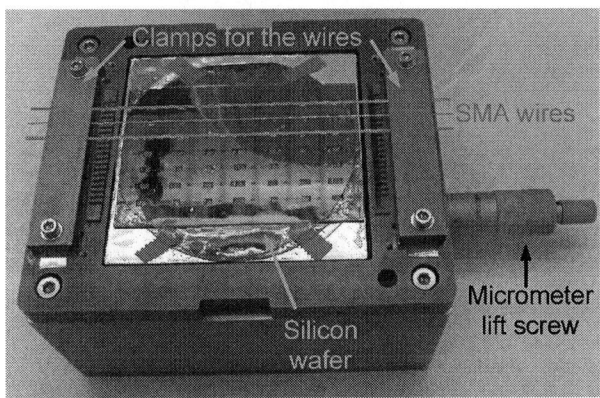

Figure 4: Wafer onto lift stage after lamination of PTFE foil. (The SMA wires are here enhanced.)

Next, the SU-8 was patterned to form the anchors for the SMA wires. First, intermediate SU-8 anchors were defined across the wafer using a course manual curing step to hold the wires in place when removing the wafer from the wire frame and the lift stage. Thereafter a standard mask aligner was used to precisely define the SU-8 anchor pads for each cantilever. The PTFE foil was easily peeled off after the SU-8 was cured on a hotplate (20 min at 65°C, then 40 min at 70°C). Finally, the SU-8 was developed to remove the unexposed material and the wafer was laser-diced into single chips.

The presented integration process allows for batch production at temperatures below the SMA transformation temperature, which ideally eliminates the need for constraining the SMA recovery during processing and thus potentially allows automatic placing of the SMA wires.

Two 7 x 7 arrays of micro devices were successfully fabricated, one on a 300 μm thick silicon wafer with 2.5% prestrained SMA wires and the other on a 525 μm thick wafer with 4% prestrained SMA wires. The prestrain value relative to the cantilever thickness was selected in order to achieve similar deflections between the different wafers and maximum stresses in the of 150-175 MPa in

the SMA wire, as suggested in the data sheet of the SMA wires [9].

Design values obtained with linear beam theory and assuming the prestrain in the SMA completely recovered upon actuation are shown in Table 1.

Table 1: Design values

Wafer thickness	Cantilever thickness	Prestrain of SMA	Stress in SMA	Expected stroke
300 μm	50 μm	2.5%	148 MPa	169 μm
525 μm	70 μm	4%	160 MPa	166 μm

By visual inspection it was concluded that 80% of devices on the wafer were successfully fabricated. A photograph of one of the final devices can be seen in Figure 5.

Figure 5: Photo of one of the final devices

EVALUATION

Two measurement setups based on contactless instruments were used to evaluate the response of the actuators.

The first setup consisted of a hot plate to control the temperature of the chip being tested, and of an optical profilometer to scan the cantilevers at fixed temperatures along the heating-cooling cycles.

Figure 6 shows the color coded surface scans of one of the actuators for both the cold and the hot state.

Figure 6: Color coded optical profilometer images of the device in the cold state (left) and in the hot state (right). Measurements on a device with 50 μm thick etched out cantilevers in a 300 μm thick silicon wafer and 2.5% prestrained SMA wires.

Using these scans from the optical profilometer, both the actuator's initial deflection in the cold state and its maximal vertical displacement upon heating were determined to 28 μm and 188 μm respectively, for a resulting net stroke of 160 μm (Figure 7).

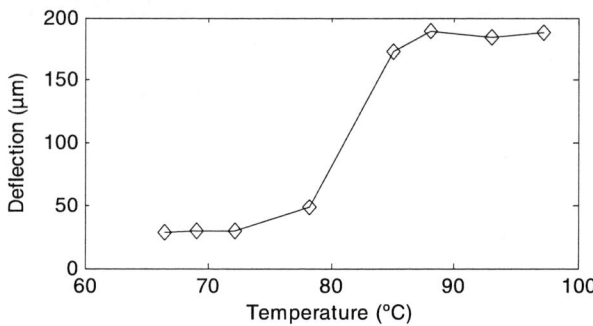

Figure 7: Measurement of deflection vs. temperature on heating a device with 50 μm thick cantilevers in a 300 μm thick silicon wafer and 2.5% prestrained SMA wires.

However, this setup did not allow dynamic measurements: during a time consuming single scan the temperature of the actuator must be kept constant, which proved difficult to achieve and thus hindered the possibility of recording any hysteresis in the actuator's response.

Therefore, a second measurement was performed using a hotplate and a 3 D CNC Vision Measuring Machine, which provides a much faster response and measures the difference in height between one focal point on the moving cantilever tip and another focal point on the surface of the substrate. The temperature was recorded with a thermocouple and slowly cycled between temperatures below and above the transformation temperatures of the SMA without rest steps. In each measurement point, both the height difference between the two focal points and the current temperature of the hotplate were measured.

Figure 8 shows the measured stroke of the actuator in function of temperature, calculated by subtracting the two focal points and thereby eliminating the error introduced by thermal expansion of the hotplate. This measurement method allows for a much higher temperature resolution, when compared with the profilometer measurements, and provides detailed data about the actuator performance between the cold and the hot state.

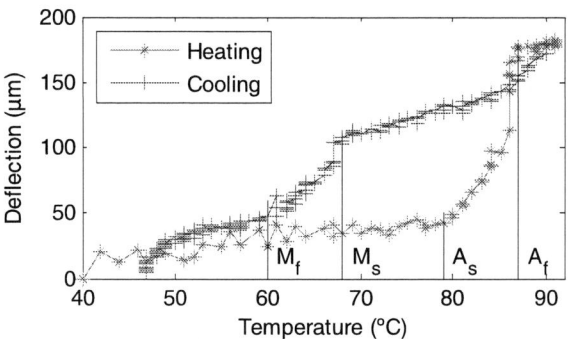

Figure 8: Measurement of deflection vs temperature on a device with 70 μm thick etched out cantilevers in a 525 μm thick silicon wafer and 4% prestrained SMA wires. First measurement point set to zero deflection.

The measurements show transformation temperatures (M_f=60°C, M_s=69°C, A_s=79°C, A_f=88°C) matching the ones provided by the manufacturer of the SMA wires [9]. A hysteresis of about 20°C is visible between the heating and the cooling curves in Figure 8.

The heating curve in figure 8 corresponds well to the optical profilometer measurements in figure 7. Also the measured stroke for this second actuator is in the order of 170 μm, in agreement with the value shown previously. These preliminary results show consistent performances between different batches of actuators and are in good agreement with design values.

CONCLUSION

A novel wafer-scale method for integration of SMA wires to silicon microstructures has been presented, along with evaluation of the actuator performance.

The integration is carried out on a wafer level and potentially suites low cost batch fabrication of the actuators.

Two test Si wafers hosting 7 x 7 arrays of actuators each have been successfully fabricated.

Preliminary measurement results indicate that this 4 mm^2 footprint area actuator features one of the highest displacements reported [2].

The fabrication process hereby presented permits a wide flexibility in designing the actuators, whose performance can therefore be optimized for a broad range of applications.

ACKNOWLEDGMENT

This work is part of the Q2M project and receives research funding from the European Commission through the 6th Framework Programme.

REFERENCES

[1] E. Thielicke and E. Obermeier, "Microactuators and their technologies", *Mechatronics, vol. 10, no. 4/5,* pp. 431–455, Jun. 2000.

[2] D. J. Bell, T J Lu, N. A. Fleck, and S M Spearing, "MEMS actuators and sensors: observations on their performance and selection for purpose", *J. Micromech. Microeng. 15 (2005)* S153–S164

[3] M. Kohl, "Shape Memory Microactuators", *Springer, 2004.*

[4] K. Skrobanek, M. Kohl, and S. Miyazaki, "Stress-optimized shape memory microvalves", *Proc. IEEE. MEMS 1997,* pp. 256–261

[5] P. Krulevitch, A. Lee, P. Ramsey, J. Trevino, J. Hamilton, and M. Northrup, "Thin film shape memory alloy microactuators," *J. Microelectromech. Syst., vol. 5, no. 4,* pp. 270 – 82, Dec. 1996.

[6] S. Miyazaki, M. Tomozawa, and H. Y. Kim, "Development of High-Speed Microactuators utilizing sputter-deposited TiNi-Base Shape Memory Alloy Thin Films", *Proc. Actuator 08, Bremen, Germany,* pp. 372-377

[7] D. Clausi, J. Peirs, and D. Reynaerts, "Towards Batch Integration of SMA into Microsystems: An Actuator Prototype", *Proc.,4M 2008, Cardiff, UK,* pp.50-53.

[8] J. van Humbeeck, D. Reynaerts, and R. Stalmans, "Shape memory alloys: functional and smart", *Proc. Actuator 94, Bremen, Germany,* pp. 312-316.

[9] Dynalloy, Inc., Technical Characteristics of FLEXINOL Actuator Wires. Updated information available at www.dynalloy.com

AN INTEGRATED SOLUTION FOR WAFER-LEVEL PACKAGING AND ELECTROSTATIC ACTUATION OF OUT-OF-PLANE DEVICES.

Kuan-Lin Chen, Renata Melamud, Shasha Wang and Thomas W. Kenny
Stanford University, Stanford, California, USA

ABSTRACT

This paper presents an innovative way to integrate electrodes into our hermetic wafer-level epi-polysilicon encapsulation for out-of-plane actuated devices. The epitaxial-grown silicon encapsulation process was modified to integrate electrodes in the out-of-plane direction and realize ultra-compact packaging. The integration allows reduction of parasitic resistance by more than 180X relative to other conventional interconnects to out-of-plane electrodes. A 200 KHz "see-saw" mode resonator is demonstrated with parasitic resistance of approximately 1 Ω.

INTRODUCTION

Electrostatic MEMS resonators have been a promising technological candidate to replace conventional quartz crystal resonators due to the potential for smaller size, lower power consumption and low-cost silicon manufacturing. However, large motional-impedance (generally > 1 KΩ) limits the range of applications.

MEMS devices operating in the out-of-plane direction have the advantage of large transduction areas on the top and bottom surfaces, resulting smaller motional-impedances (R_x). Therefore, there have been a lot of great inventions based on out-of-plane actuation, such as DMD and IMod, and many of them have also been commercialized [1, 2] In addition to actuator, small R_x (close to 50 Ω) is an important requirement for RF applications and therefore many research projects have been focused on reducing R_x of the resonator.

To explain how to reduce R_x, the equation of R_x is written as:

$$R_x = \frac{c_r}{\eta^2} \quad ; \quad \eta = V \frac{\partial C}{\partial g} = \frac{\varepsilon_0 A V}{g^2} \quad (1)$$

, where c_r, η, g, A, V are the effective damping constant of resonator, transduction efficiency, gap between electrodes, transduction area and bias voltage, respectively. To decrease R_x, η needs to be increased. For in-plane devices, A is $H*L$, where H is the height and L is the length of beam. Due to geometric process constraints, H/g is limited by the etching aspect ratio, normally 20:1, and therefore η is limited to be proportion to $1/g$. However, for out-of-plane devices, A and g are set independently so η is proportion to $1/g^2$. Therefore, out-of-plane devices potentially have better transduction efficiency and excellent R_x of 160 Ω has been reported. [3]

Despite these advantages, the difficulty of implementing electrodes in the out-of-plane direction hinders its development. Packaging is challenging for out-of-plane devices because out-of-plane electrodes can possibly be damaged during packaging processes.

MEMS resonators share some common shortcomings for introducing an out-of-plane electrode. [3-5] First, they require deposition and isolation of another layer of conductive material to create electrodes within the fabrication process. Furthermore, due to the thin and long

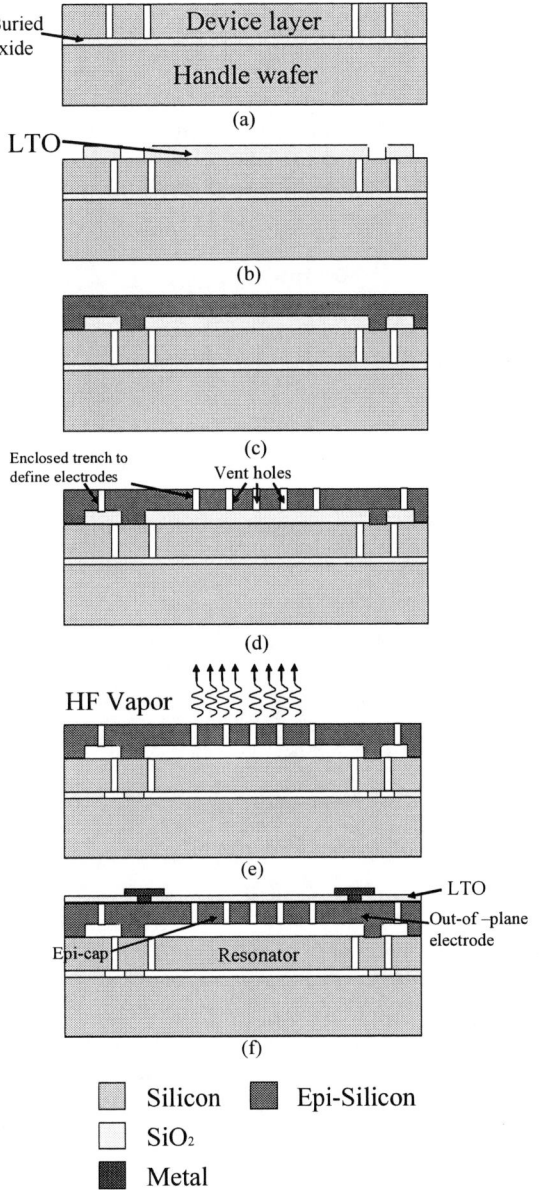

Fig. 1. *Process flow of this work. (a) Process starts with a SOI wafer. Device layer is etched to define the resonator and anchors for top electrodes. (b) Low temperature oxide (LTO) is deposited to seal the trench in device layer. It is also patterned to allow electrical contact. (c) Highly phosphorus doped epi-silicon is deposited and fill in the top surface. (d) Vent holes and enclosed trenches are defined and etched. (e) HF vapor is used for releasing the structure. (f) LTO is used to seal all the trenches on the epi-silicon layer. Aluminum is used as the material for probe pads.*

geometries of electrodes, large electrode resistances, for example 180 Ω in [4], would add up with R_x. Meanwhile, MEMS resonators usually require a vacuum chamber to operate at high Q, so a vacuum encapsulation process is

978-1-4244-2977-6/09 $25.00 © 2009 IEEE

EPITAXIAL SILICON ENCAPSULATION WITH ELECTRODES

This work aims to provide a novel solution for out-of-plane devices with a stable package environment and low impedance electrodes. The packaging is based on the wafer-level epitaxial-grown silicon encapsulation technology developed in collaboration with Bosch at Stanford University.[6] It provides a stable, clean, hermetic vacuum environment for various MEMS devices and it has the merits of superb long-term stability, thermal isolation, high quality factor, and pressure tunability.[7-10]

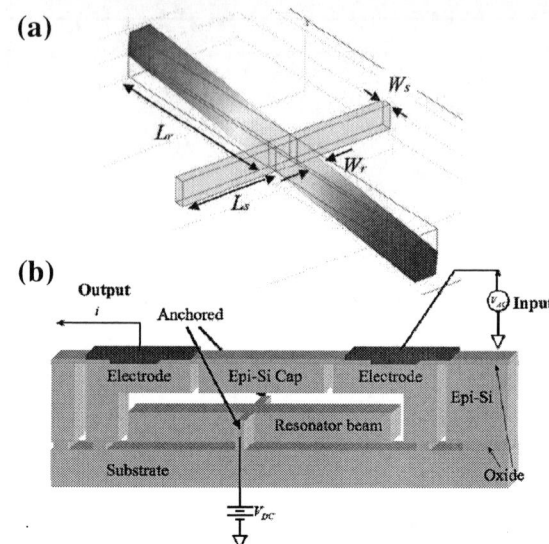

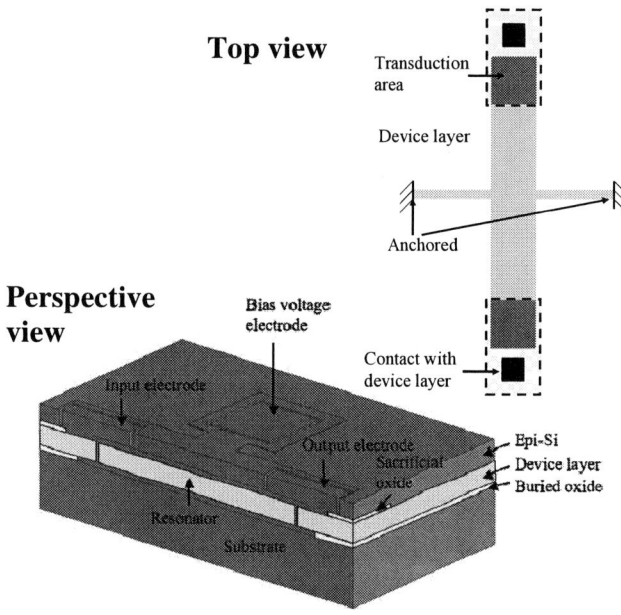

Fig. 2. *Top view of top electrodes on a "see-saw" resonator and the cross-sectional view. The electrode is anchored to device layer as a cantilever. The dash line in top view shows the enclosed trench that is etched in epi-silicon to define the out-of-plane electrode.*

The process flow of this work is shown in Fig. 1. The process starts with a silicon-on-oxide (SOI) wafer with a device layer of 20 μm. The resonator was then patterned and DRIE etched through the device layer. Low temperature oxide (LTO) was deposited as the sacrificial layer to create cavity on top of the resonator. Highly phosphorus doped epi-silicon was then deposited on the top of the wafer to encapsulate the resonators and then the epi-silicon layer was then CMP polished to a flat surface. Top electrodes and vent holes for releasing the device were then defined and etched to the sacrificial LTO. A custom made HF vapor etcher in the Stanford Nanofabrication Facility was used to release the resonator. After releasing, all the trenches were again sealed by LTO in a low pressure CVD vacuum-encapsulating the resonators. Finally, an aluminum layer was used for electrical interconnect and bond pads.

Fig. 3. *(a) COMSOL simulation result of the silicon cross structure shows the displacement of structure at resonance. (b) Schematics shows the "see-saw" resonator encapsulated inside of a epi-silicon packaging and the measurement setup for measuring the resonant frequency.*

The key innovation of the work reported in this paper relative to the work from the original process [4] is that the out-of-plane electrode is defined with other vent holes in step (d). The electrode is defined by the enclosed trench etched in the epi-silicon layer and stopped on the sacrificial oxide. The enclosed trench provides electrical isolation from the ground-shielded epi-silicon packaging. Also, the mechanical anchoring of the electrode is defined in the sacrificial oxide layer to provide stability.

Fig.2 shows the top view of electrode design and its prospective picture, where the over-hanging electrode is incorporated in the epitaxy-silicon layer and can capacitively sense and actuate out-of-plane devices. This compact and highly integrated electrode design solves the common problems that other out-of-plane devices have and further shrink the size of device. Furthermore, the low resistivity (~10 mΩ-cm) of the epi-silicon film gives the resistance of electrodes to be less than 1 Ω.

"SEE-SAW" RESONATOR

A simple torsional-mode resonator is used to demonstrate the out-of-plane actuation. Fig.3 shows the COMSOL simulation result and the schematic diagram of measurement setup. The resonator was anchored at both ends of it tethers and electrically biased with V_{DC} through the tethers.

Fig.4 shows the SEM pictures of fabricated resonators and top electrodes in the package. As shown in the zoom-in picture of Fig. 4 (b), the overhanging electrode does not interfere with the movement of the resonator and it also keeps the integrity of the packaging. The final LTO sealing is also shown to offer a hermitically-sealed environment for the resonator.

978-1-4244-2977-6/09 $25.00 © 2009 IEEE

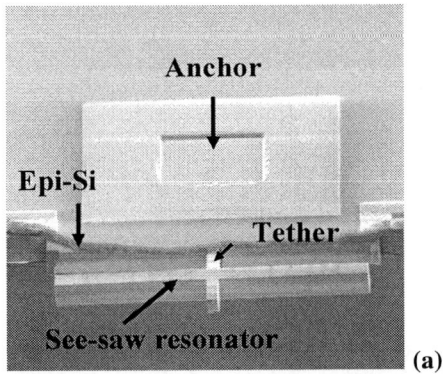

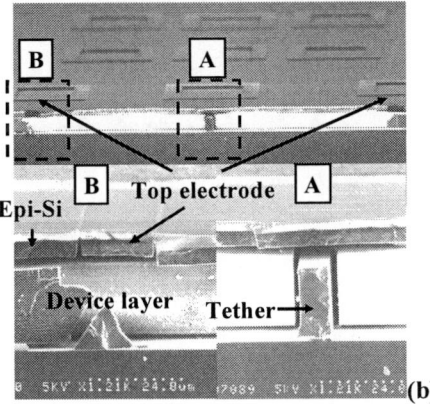

Fig. 4. *SEM pictures of the 200 KHz resonator. (a) An exposed resonator structure packaged inside of the Epi-silicon layer. (b) Section A shows the zoom-in picture of a broken tether. Section B shows the zoom-in picture of a type B electrode suspended above the device layer.*

As with other resonators, the resonance is measured by applying V_{AC} to the input electrode and V_{DC} to the structure and measuring the current from the output electrode. However, as shown in Fig. 7, although the simple two-port measurement shows the resonant peak, the amplitude of the peak is too small to measure its quality factor.

Fig.6(a) shows the electrical equivalent circuit of the resonator. The parasitic capacitance (C_p) is expressed as a lumped component factoring in all of the capacitance components from the packaging. The feedthrough current is based on the value of C_p and the effect can be observed on Fig. 6(b). When the feedthrough current is around the same order of magnitude as the peak amplitude, the parasitic current will mask the motional current coming out from the resonator.

A differential measurement similar to [11] was used to eliminate parasitic feedthrough. As shown in Fig. 5, the RF signal, $V_{AC,}$ from network analyzer is divided into two signals, one with the same phase as the V_{AC} and one with 180 degree phase-shift. The resonator under test is applied with V_{DC} to induce resonance and the dummy resonator is only grounded. Since there is no V_{DC} applied in the second case, the resonator serves only as a parasitic capacitor. By mixing the two signals together, the parasitic current from the parasitic capacitances cancel each other out and reveal the resonant peak hidden by the parasitic current. Fig. 6

shows the measured resonant frequency of 194 KHz with quality factor of 970.

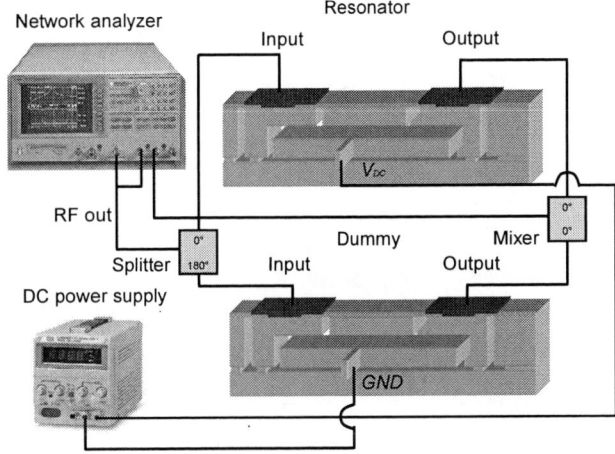

Fig. 5. *The differential measurement setup for the "see-saw" resonator. The upper resonator is applied with V_{DC} to induce resonance. The lower one is grounded so only its parasitic terms will be seen from the circuit point of view.*

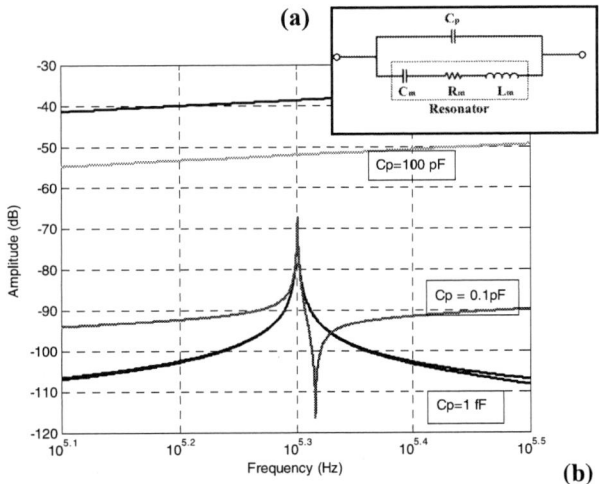

Fig. 6. *(a)The electrical representation of the mechanical resonator. C_m, R_m, L_m, and C_p represent equivalent capacitance, resistance, inductance, and parasitic capacitance, respectively. (b) The plot of resonator responses when different C_p is presented. Blue line shows similar response to differential measurement and red line shows similar response to simple 2-port measurement.*

Based on Equation (2) (3), the motional impedance and equivalent capacitance and inductance can be extracted from the experimental data. The equivalent R_m, L_m, C_m values are 112 KΩ, 86.37 H, and 7.33 fF, respectively. C_p is estimated to be approximately 100 pF using the simple 2-port measurement. The differential setup reduces the parasitic capacitance to less than 0.1 pF, showing more than 1000 X of reduction.

$$R_m = 50 \times 10^{\frac{I.L.}{20}} \qquad (2)$$

$$Q = \frac{\sqrt{L_m/C_m}}{R_m} = \frac{\omega L_m}{R_m} = \frac{1}{\omega R_m C_m} \qquad (3)$$

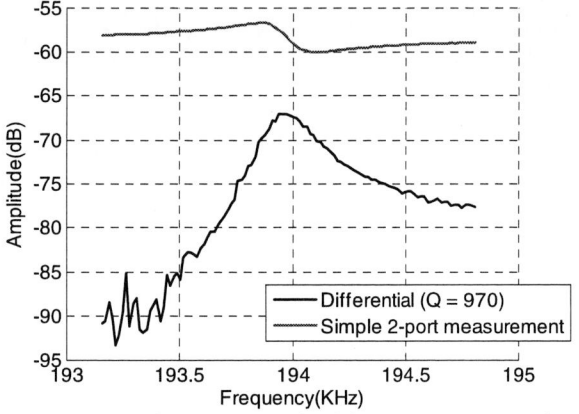

Fig. 7. *Measured resonant peak of the test structure with $L_r = 158\ \mu m$, $W_r = 18\ \mu m$, $L_s = 80\ \mu m$, $W_s = 8\ \mu m$ and $H = 20\ \mu m$. The applied $V_{DC} = 32V$.*

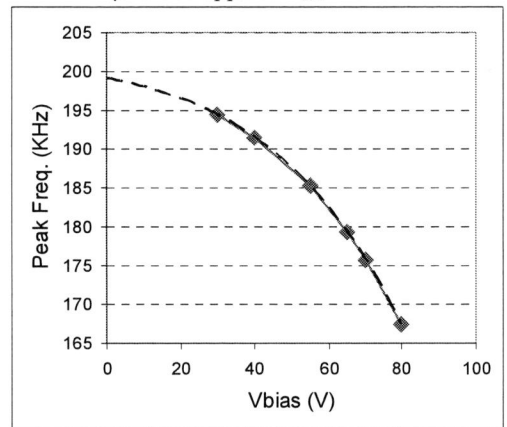

Fig. 8. *Resonant frequencies vs. VDC. The dash line shows the fitting curve, which extrapolates the frequency at zero bias to be 199 KHz.*

Finite-element simulation was used to verify the measurement and showed the resonant frequency of 201.4 KHz. The lower measured frequency is due to the "spring softening" effect from electrostatic actuation. Since larger V_{DC} is required to induce resonance, mostly because of relatively large gap between resonator to electrode, the equivalent spring constant of the resonator is "electrically softened", resulting the lower resonant frequency. The frequency-V_{DC} dependency is plotted in Fig. 7 and the extrapolated frequency at zero voltage is 199 KHz, which confirms the simulation result. Fig. 7 also shows a more than 15% of frequency tunabilty, which is an important feature of out-of-plane resonators. The frequency tunability can be used to compensate for frequency shifts caused by temperature, pressure, and acceleration.[12]

CONCLUSION

This paper presented the concept, design, and demonstration of packaged integrated electrodes for out-of-plane devices. This technology further extends the possibility of reducing the insertion loss, motional impedance and size of the MEMS resonator. It can also be applied to other capacitive MEMS devices that require large transduction area, eg. accelerometer, gyroscope and pressure sensor.

ACKNOLEDGEMENT

The authors would like to thank the staff in Stanford Nanofabrication Facility for their technical support. The authors would also like to thank Maryam Ziaei-Moayyed for helpful discussion.

REFERENCES:

[1] Hornbeck, L.J., "The DMD (TM) projection display chip: A MEMS-based technology." Mrs Bulletin, 2001. 26(4): p. 325-327.

[2] M.W. Miles, "Digital Paper: Reflective Displays Using Interferometric Modulation", SID Digest 5.3, 2000.

[3] John R. Clark, et al., "Parallel-Coupled Square Resonantor Micromachined Filter Arrays". 2006.

[4] Demirci, M.U. and C.T.-C. Nguyen, "Mechanically Corner-Coupled Square Microresonator Array for Reduced Series Motional Resistance", *J. Microelectromech. Syst*, 2006. 15(6): p. 1419-1436.

[5] Wang, K., A.C. Wong, and C.T.C. Nguyen, "VHF free-free beam high-Q micromech. resonators." *J. Microelectromech. Syst*, 2000. 9(3): p. 347-360.

[6] Candler, R.N., et al., "Single wafer encapsulation of MEMS devices". IEEE Transactions on Advanced Packaging, 2003. 26(3): p. 227-232.

[7] Kim, B., et al., "Frequency stability of wafer-scale film encapsulated silicon based MEMS resonators." Sensors and Actuators a-Physical, 2007. 136(1): p. 125-131.

[8] Park, W.T., et al., "Encapsulated submillimeter piezoresistive accelerometers" *J. Microelectromech. Syst*, 2006. 15(3): p. 507-514.

[9] C. M. Jha, M. A. Hopcroft, S. A. Chandorkar, J. Salvia, M. Agarwal, R. N. Candler, R. Melamud, B. Kim, and T. W. Kenny, "Thermal Isolation of Encapsulated MEMS Resonators," *J. Microelectromech. Syst*, In Press, 2007.

[10] B. Kim, R. N. Candler, R. Melamud, S. Yoneoka, H. K. Lee, G. Yama, and T. W. Kenny, "Identification and Management of Diffusion Pathways in Polysilicon Encapsulation for MEMS Devices," accepted for presentation, MEMS 2008.

[11] P. Rantakari et al, "Reducing the effect of parasitic capacitance on MEMS measurements," Transducers'01, pp. 1556-1559

[12] M. Agarwal, K. K. Park, S. A. Chandorkar, R. N. Candler, B. Kim, M. A. Hopcroft, R. Melamud, T.W. Kenny, and B. Murmann, "Acceleration Sensitivity in Beam-type Electrostatic Microresonators," Applied Physics Letters, vol. 90, no. 1, 014103, 2007.

AN ELECTROMAGNETIC MICRO POWER GENERATOR FOR LOW FREQUENCY ENVIRONMENTAL VIBRATIONS BASED ON THE FREQUENCY UP-CONVERSION TECHNIQUE

Ibrahim Sari[1], Tuna Balkan[1], and Haluk Kulah[2,3]

[1]Department of Mechanical Engineering, METU, Ankara, TURKEY
[2]Department of Electrical and Electronics Engineering, METU, Ankara, TURKEY
[3]METU MEMS Center, Ankara, TURKEY

ABSTRACT

This paper presents an electromagnetic (EM) vibration-to-electrical power generator, which can efficiently harvest energy from low-frequency external vibrations by using frequency up-conversion. The generator can effectively scavenge energy from low frequency environmental vibrations of 70-150 Hz and generates 0.57 mV voltage with 0.25 nW power from *a single cantilever* at a vibration frequency of 95 Hz. The fabricated generator size is 8.5 x 7 x 2.5 mm³ and a total number of 20 serially connected cantilevers have been used to multiply the generated voltage and power. The performance of the generator is also compared with a same sized traditional magnet-coil type generator to prove its effectiveness.

INTRODUCTION

MEMS based energy harvesting from environmental vibrations has been an attractive topic and is extensively investigated since 1995 [1]. The maximum generated power for the traditional techniques is proportional to the cube of the vibration frequency [2] and drops dramatically at low frequencies (1-100Hz). However, it is at these low frequencies where most ambient vibration exists. For this reason, vibration based resonant generators are effective at frequencies of several kHz, and at lower frequencies they are ineffective [3]. The proposed electromagnetic generator solves this problem by mechanically up-converting the low frequency vibrations to a higher frequency. This technique has been firstly proposed by Kulah et al. with a milli-scale implementation [4]. In this paper, a micro scale implementation with experimental results is presented for the first time in the literature. The effectiveness of the energy generation of the proposed design has been experimentally verified through comparative tests with a same-sized traditional magnet-coil type generator.

DESIGN

Figure 1 shows the proposed system, which is composed of two mechanical structures: 1) the upper diaphragm and 2) the array of cantilevers located right below the diaphragm. The diaphragm is made of parylene C and holds a NdFeB magnet for both frequency up-conversion and power generation by means of electromagnetic induction. The diaphragm-magnet assembly resonates by vibrations in the range of 1-100 Hz. The cantilevers have a higher resonance frequency of 2-3 kHz and each of them has a coil for induction. Also, at the tip of each cantilever, nickel is electroplated for interaction with the magnet. As the diaphragm resonates in response to external vibrations, it gets closer to the cantilever array. The distance between them is adjusted

such that the magnet *catches* the cantilevers at a certain instance of its movement, pulls them up, and releases at another point. Afterwards, the released cantilevers start resonating at their damped natural frequency with the given initial condition, and hence the frequency up-conversion is realized. The motion of the released cantilevers exponentially decays out and before it completely dies, the cycle starts again.

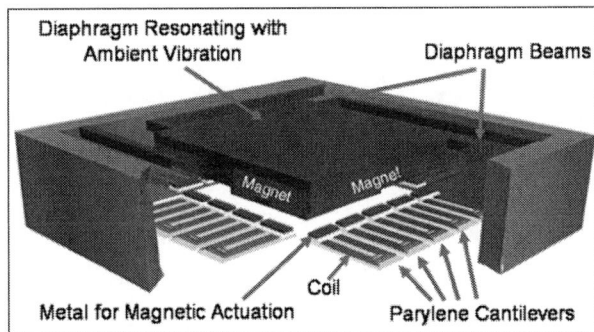

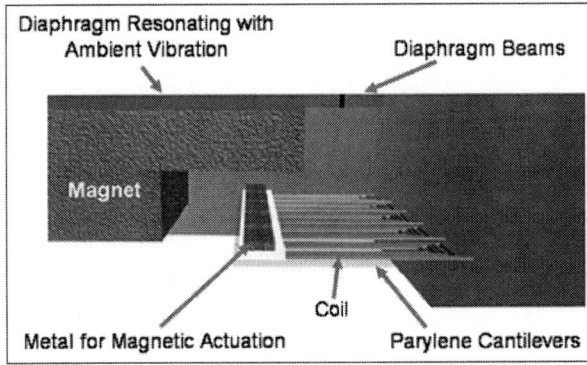

Figure 1: Schematic view of the proposed design: isometric (top) and side (bottom) views.

The equivalent mechanical model of the system is shown in Figure 2, where m, k, and b denote the equivalent mass, stiffness, and damping constants, respectively. The position coordinates are defined by z and δ, with z corresponding to the relative displacements with respect to the base, and δ to the initial offsets with respect to the base. The system is excited by environmental vibrations of frequency ω and displacement amplitude of y. The excitation frequency is close to the natural frequency of the diaphragm-magnet assembly and much smaller than the natural frequency of the cantilevers. This causes only the diaphragm-magnet to be excited by environmental vibrations, and the cantilevers to be remained unaffected. Using this model, the magnet-diaphragm and the cantilever dynamics can be expressed with the following two equations. These equations are the second-order, linear differential equations of motion for the magnet-diaphragm assembly, and the cantilevers and are obtained by following an equivalent modeling approach.

$$m_m \ddot{z}_m + b_m \dot{z}_m + k_m z_m = -m_m \ddot{y} \qquad (1)$$
$$m_c \ddot{z}_c + b_c \dot{z}_c + k_c z_c = 0 \qquad (2)$$

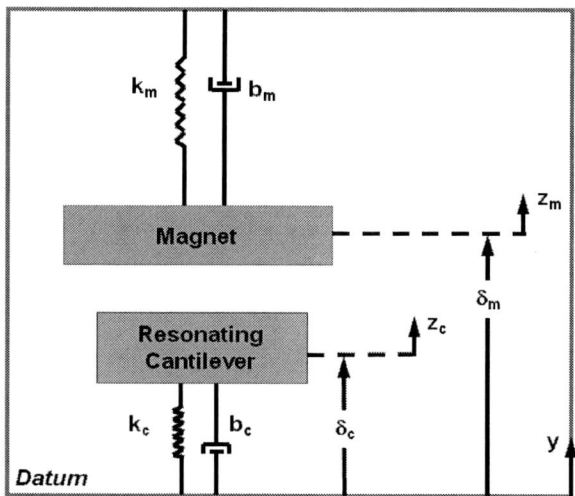

Figure 2. Equivalent mechanical model of the system.

The dynamic behavior of the magnet can be obtained by the steady-state solution of the first equation that is already derived in the literature [2]. The cantilevers are basically excited with their release at a certain instance of their motion. Thus, their dynamics is revealed with the transient solution of Equation (2), by assuming that they are released at a vertical distance of z_o.

$$z_c(t) = \frac{z_o e^{-\zeta_c \omega_n t}}{\sqrt{1-\zeta_c^2}} \left(\zeta_c \sin(\omega_d t) + \sqrt{1-\zeta_c^2} \cos(\omega_d t) \right) \qquad (3)$$

In the last equation, ζ_c and ω_d correspond to the overall damping ratio and the damped natural frequency of the cantilevers. After obtaining the equations representing the dynamic behavior of the system, the generated voltage from a single cantilever can be obtained as:

$$\varepsilon = -BL_p \dot{z}_c \qquad (4)$$

In this equation, B is the magnetic flux density and L_P is the practical coil length. These terms are derived and investigated in detail in the previous work of Sari *et al.* [2]. By using (3) and (4), the output voltage generated by a single cantilever can be derived as:

$$\varepsilon = BL_p z_o \omega_n \frac{e^{-\zeta_c \omega_n t}}{\sqrt{1-\zeta_c^2}} \sin(\omega_d t) \qquad (5)$$

The power output can be obtained using the following equation,

$$P = \frac{1}{2} \frac{R_L}{(R_L + R_c)^2} \varepsilon^2 \qquad (6)$$

where R_L and R_c are the load and coil resistances, respectively. After making necessary substitutions, the power terms can be defined in expanded form as,

$$P = \frac{1}{2} \frac{R_L}{(R_L + R_c)^2} \left(BL_p z_o \omega_n \frac{e^{-\zeta_c \omega_n t}}{\sqrt{1-\zeta_c^2}} \sin(\omega_d t) \right)^2 \qquad (7)$$

The power and voltage terms defined above are optimized using a Pattern Search algorithm in Matlab® to have maximum output from the proposed device. The optimized generator parameters are tabulated in Table 1. According to the optimized results, a maximum voltage of 0.67 mV and a power of 0.33 nW are estimated to be obtained from a single cantilever of the generator. Figure 3 shows the simulated absolute positions of the cantilevers and the magnet obtained using the parameters of Table 1.

Table 1: Optimized parameters of the generator.

Input vibration frequency	70–150 Hz
Diaphragm natural frequency	113 Hz
Input displacement	0.44–2 mm
Device dimensions	8.5 x 7 x 2.5 mm³
Magnetic flux density	0.19 Tesla
Magnet size	3.8x3.8x1.5 mm³
Total number of cantilevers	20
Cantilevers natural frequency	2 kHz
Cantilever size	1000 x 430 x 15 μm³
Magnetic actuation area	430 x 225 x 9 μm³
Nickel thickness	9 μm
Coil width	20 μm
Release height of the cantilevers	200 μm
Practical coil length	1.4 mm
Number of coil turns	6
Collected energy in 1s (20 cantilevers)	610 pJ (30.5 pJ/cantilever)
Peak power output (20 cantilevers)	6.6 nW (0.33 nW/cantilever)
Peak voltage output (20 cantilevers)	13.5 mV (0.67 mV/cantilever)

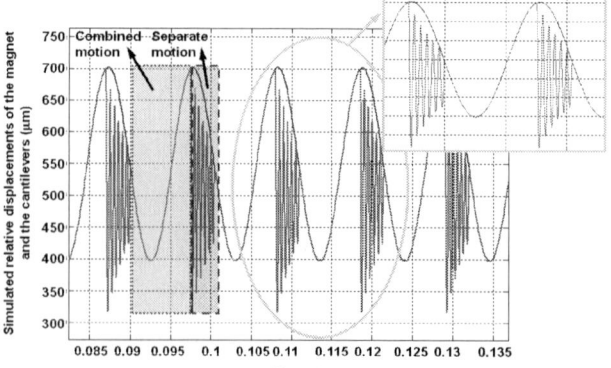

Figure 3: Simulated relative positions of the magnet and the cantilevers.

The system has three different dynamics: (1) cantilever, (2) magnet-diaphragm, and (3) combined motion of the cantilevers and magnet-diaphragm. During the simulation, the model is switched between these three dynamic modes, by continuously checking the effective conditions at each sample time step. The shaded area on the left (bordered by the dotted line) shows the combined low–frequency motion of the cantilevers and the magnet, whereas the shaded area on the right (bordered by the long dashed line) shows the separate motion of the cantilevers (high frequency plot) and the magnet (low frequency plot). The generated voltage from a single cantilever is also plotted in Figure 4, which has an exponentially decaying out form similar to the motion of the cantilevers. It can be seen that voltage generation is realized during the high frequency oscillatory motion of the cantilevers after their release.

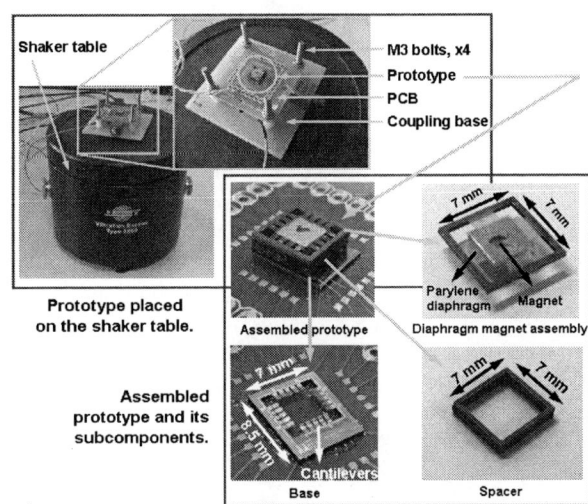

Figure 5: Fabricated prototype prepared for testing.

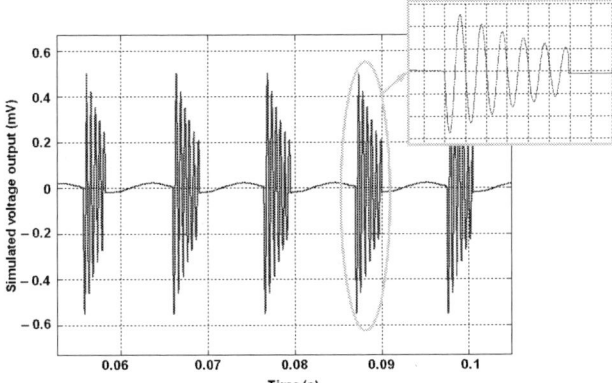

Figure 4: Simulated voltage output from a single cantilever.

IMPLEMENTATION AND TEST RESULTS

Fabrication of the micro generator requires 5 masks and is explained in detail in [4]. Parylene C is used as the structural material for the cantilevers and the diaphragm as it allows much larger deflections before mechanic failure compared to silicon [5].

Figure 5 shows the fabricated prototype placed on the shaker table (upper left) and the assembled prototype with its subcomponents prepared for testing (bottom right). The tests are carried out by sweeping the frequency from 70 to 150 Hz. Table 2 shows the test results indicating that the estimated and measured values for a single cantilever are in good agreement. The main source of minor deviation is due to the error in the estimation of the magnetic flux density, which is a complicated parameter for accurate calculation. Figure 6 shows the simulated and measured output voltages plotted on the same set of axis. The plot on the upper-left hand side is a zoomed out view of the generated voltage and it shows that the cantilevers are continuously caught and released, proving the realization of the up-conversion mechanism. The plots on the right hand side are close-up views of the generated voltage, and they show that the simulation and test results are in close agreement. The bottom plot indicates the catch and release points together with the excitation and up-converted frequencies.

Table 2: Comparison of the simulation and the test results of the frequency up-conversion (FU) design.

	Simulation	Test
Input vibration frequency	70–150 Hz	70–150 Hz
Input displacement	0.44–2 mm	0.44–2 mm
Diaphragm natural frequency	113 Hz	113 Hz
Cantilevers natural frequency	2 kHz	2 kHz
Magnet size (mm^3)	3.8x3.8x1.5	3.8x3.8x1.5
Release height of the cantilevers	200 μm	200 μm
Damping ratio of the cantilevers	0.02	0.02
Coil resistance	160 Ω	150 Ω
Energy in 1 s (single cantilever)	30.5 pJ	20 pJ
Peak power (single cantilever)	0.33 nW	0.25 nW
Peak voltage (single cantilever)	0.67 mV	0.57 mV

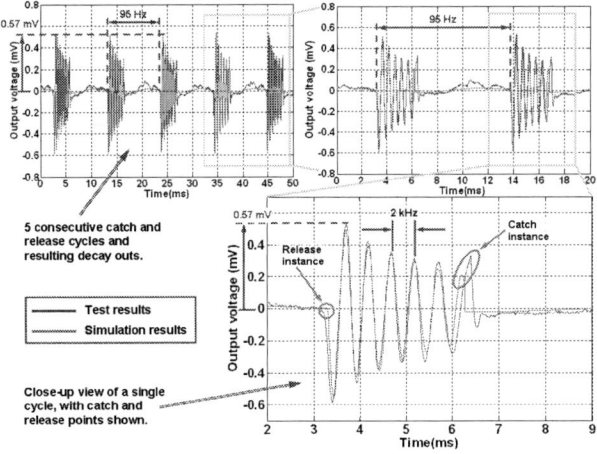

Figure 6: Estimated and measured voltages.

In order to prove the effectiveness of the proposed generator, a same-sized traditional large magnet-coil (MC) type generator has also been fabricated and tested for performance. MC type generator is one of the most popular traditional EM energy scavengers and it has been extensively investigated in the literature [1]. Figure 7 shows the fabricated traditional generator. The base shown on the lower left hand corner is stationary whereas the magnet-diaphragm assembly shown on the upper right hand side is moveable to induce voltage at the terminals of the coil. Table 3 compares the performance of the FU and MC designs.

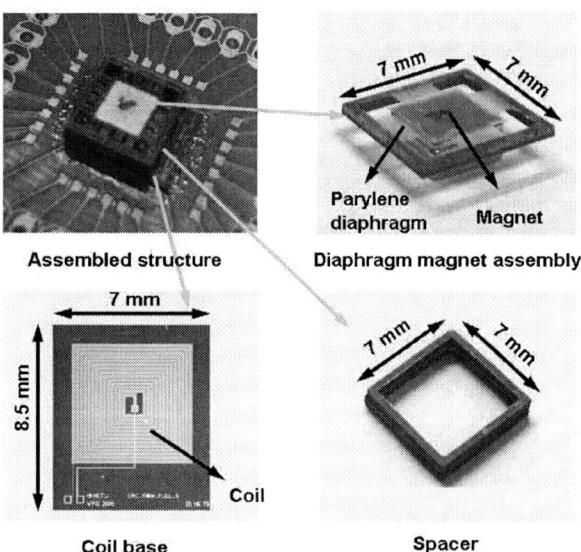

Figure 7: The traditional magnet-coil design fabricated for comparison with the proposed design.

Table 3: Performance comparison of the frequency up-conversion (FU) and the magnet-coil (MC) designs.

	MC design	FU design (single cant.)	FU design (20 cants.)
Size (mm³)	8.5x7x2.5	8.5x7x2.5	8.5x7x2.5
Effective frequency	91-99 Hz	70-150 Hz	70-150 Hz
Excitation frequency	95 Hz	95 Hz	95 Hz
Magnet size (mm³)	3.8x3.8x1.5	3.8x3.8x1.5	3.8x3.8x1.5
Coil resistance	257 Ω	150 Ω	3 kΩ
Energy (in 1s)	30 pJ	20 pJ	400 pJ
Maximum power	0.04 nW	0.25 nW	5 nW
Maximum voltage	0.30 mV	0.57 mV	11.4 mV

As Table 3 shows, a single cantilever of the FU design can generate about 2 times the voltage and 6 times the power that the MC design can provide. These values are multiplied by 20 when the overall performances of the generators are compared. The generated outputs from the generators have different wave forms; sinusoidal for the MC design and decaying out for the FU design. Thus, it is more meaningful to compare the generated energies in a certain amount of time. Measurements show that a single cantilever of the FU design can collect about 0.7 times the energy that the MC design can do. However, when the overall performance of the 20 cantilevers is considered this value goes up to 13.

CONCLUSIONS

In this paper, an EM energy generator that up-converts low frequency environmental vibrations to a higher frequency is presented. The micro scale implementation of this design has been firstly shown here and the concept is proven to work in micro scale with actual test results for the first time. It has been shown that a voltage and power output of 0.57 mV and 0.25 nW can be obtained from a single cantilever of the micro-scale design, respectively. The aim of this study is to show that the proposed generator concept works in micro scale and is more efficient compared to a same sized traditional micro generator operating under the same conditions. For this purpose, the device parameters are optimized in a conservative manner for the verification of the concept. Indeed, it has been shown that the proposed generator performs much better than a traditional MC design. However, it is also possible to further improve the generated voltage and power by decreasing the coil width to increase the coil turns or by increasing the number of cantilevers. For example, in this design, a coil width of 20 µm has been used and it can be decreased to 2 µm to increase the generated outputs. Initial calculations show that this improvement leads to a 6.5 fold increase in the voltage and power output levels. With further improvements in the design parameters, it is possible to improve the performance of the proposed generator.

ACKNOWLEDGMENTS

This work was supported by The Scientific and Technological Research Council of Turkey (TUBITAK) under Grant Number 104E119.

REFERENCES

[1] C.B. Williams and R.B. Yates, "Analysis of a micro-electric generator for microsystems," in *Digest Tech. Papers Transducers'05 Conference*, Sweden, June 25-29, pp. 369-372, 1995.

[2] I. Sari, T. Balkan and H. Kulah, "An Electromagnetic Micro Power Generator for Wideband Environmental Vibrations," *Sensors and Actuators A: Physical*, vol. 145-146, pp. 405-413, 2008.

[3] P.D. Mitcheson, P. Miao, B.H. Stark, E.M. Yeatman, A.S. Holmes, and T.C. Green, "MEMS electrostatic micropower generator for low frequency operation," *Sensors and Actuators A: Physical*, vol. 115, pp. 523-529, 2004.

[4] H. Kulah and K. Najafi, "An electromagnetic micro power generator for low-frequency environmental vibrations," in *Digest Tech. Papers* MEMS 2004, Netherlands, pp. 237-240, 2004.

[5] Product specifications, Parylene knowledge, Specialty Coating Systems, Inc. IN, USA, 1-800-356-8260.

ELECTROSTATIC ROTARY STEPPER MICROMOTOR FOR SKEW ANGLE COMPENSATION IN HARD DISK DRIVE

E. Sarajlic, C. Yamahata, M. Cordero and H. Fujita

CIRMM, Institute of Industrial Science, The University of Tokyo, JAPAN

ABSTRACT

Circular data tracks in present-day hard disk drives (HDD) are accessed by a read/write head mounted on a support arm, which is swung by a voice coil drive. The orientation of the head relative to a data track varies with the radial position of the track, causing an increase in data track misregistration and limiting the performance of HDD. We present a rotary micromotor which can be used as a secondary stage actuator to maintain a constant orientation between the head and the tracks during disk operation. This electrostatic stepper micromotor, bulk micromachined in a standard monocrystalline silicon wafer, uses flexure pivots to avoid any frictional contact of the rotor, providing precise, repeatable and reliable bidirectional stepping motion without feedback control. The experimental characterization of a prototype having a diameter of 1.4 mm has demonstrated a rotational range of 26° (+/– 13°) at 75 V, a resolution of 1/6° in a coarse stepping mode and a maximum speed of 1.67°/ms.

INTRODUCTION

In modern hard disk drives (HDD), read/write heads are swung on an arc by means of a voice coil drive, as illustrated in Figure 1. The skew angle between the head support arm and the track varies with the radial position of the head on the disk. Large variations in the skew angle, which are around 30° for 3.5'' drives, cause an increase in data track misregistration, limiting the performance of HDD. A modified design of the support arm can minimize these variations at the expense of a reduced servo bandwidth [1].

An alternative solution that may not sacrifice the bandwidth is a dual-stage servo system in which a rotary micromotor carrying the read/write head is used to maintain a constant skew angle. Secondary stage rotary micromotors have already been proposed and employed in HDD to provide rapid position correction of the read/write head [2, 3]. However, the secondary micromotors developed to date suffer from a severely limited angular range, which is typically less than 0.1°. This limited range makes them unsuitable for active skew angle compensation, which requires large angular displacements of about 30°. Furthermore, the micromotors aimed to maintain a constant skew angle during disk operation must also be able to provide fast (speeds of about 10°/ms), precise, repeatable and reliable motion without feedback control. These demands are far beyond those reached by existing micromotors. Here, we present a rotary micromotor with exceptionally high performances as a potential candidate to perform this demanding task.

MOTOR DESIGN

Working principle

The working principle of the motor is schematically illustrated in Figure 2. It is a 3-phase electrostatic stepper micromotor, also known as a variable capacitance motor [4, 5]. The motor consists of a rotor with grounded poles and a stator with active poles that are controlled with integrated 3-phase electrical connections. The stator poles are symmetrically located around the rotor (not shown in

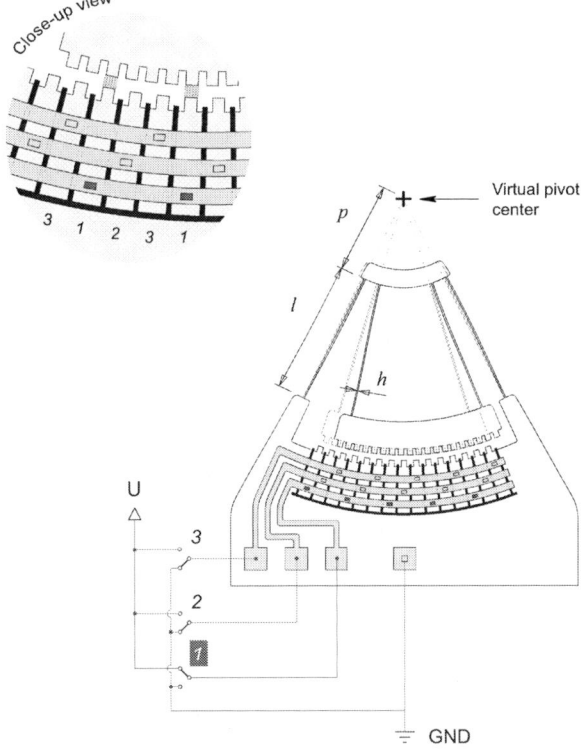

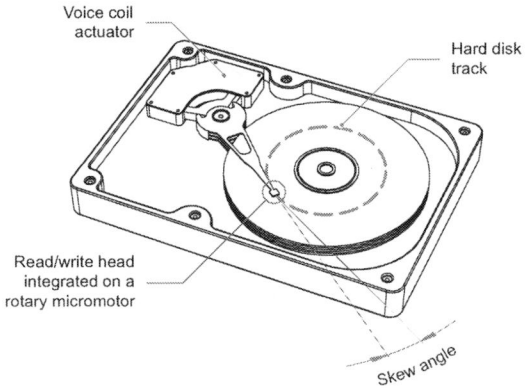

Figure 1: Schematic of a hard disk drive where the read/write head is integrated on a rotary micromotor to maintain a constant skew angle during disk operation. In 3.5'' disks, the skew angle varies typically between +/– 15°.

Figure 2: Working principle of the 3-phase electrostatic stepper motor. The rotor is suspended by a flexure mechanism and the 3-phase electrical connections are integrated on the stator. By applying an appropriate voltage sequence, bidirectional stepping motion can be achieved.

978-1-4244-2977-6/09 $25.00 © 2009 IEEE

Figure 2). The actuation relies on the electrostatic force generated by applying a voltage between the grounded rotor poles and the stator poles of a given phase. While the radial component of the electrostatic force acting on the rotor is canceled due to the symmetry, the tangential component generates a global torque realigning the rotor with the stator poles of the active phase. When the voltage is applied to each phase successively, a stepwise motion of the rotor can be achieved either clockwise or counterclockwise, ensuring a bidirectional rotation of the motor.

Rotor suspension

In our design, the rotor is suspended by flexures instead of the frictional bearings that are commonly employed for this class of micromotors [4, 5]. As a result, any frictional contact between the rotor and the stator is avoided during operation of the motor. Absence of the contact friction – and the resulting wear – ensures precise, repeatable and reliable stepping motion without feedback control. In addition, the flexures serve as an electrical connection through which the rotor is electrically grounded. Furthermore, a read/write head, which should be placed on the rotor, can be wired by using the flexures as a support for the wiring. Naturally, the flexible suspension limits rotational range of the motor. In order to enable large angular displacement required for skew angle compensation we have employed the 'butterfly pivot' suspension proposed by Henein *et al.* [6]. This suspension is composed of four elementary flexure pivots connected in series. A key advantage of the butterfly pivot is a low rotational stiffness and a very low translation of the rotor which accompanies the rotation. Large translation, also called parasitic center shift, can cause pull-in instability of the rotor, resulting in operational failure.

MICROFABRICATION
Vertical Trench Isolation

The motor was fabricated in a standard single-crystal silicon wafer using vertical trench isolation technology [7, 8]. This technology employs trenches refilled with a dielectric material to create electrical insulation between mechanically linked components in a single device layer. It allows fabrication of high-aspect-ratio monocrystalline silicon components with an integrated 3-phase electrical network. High aspect ratio is important for electrostatic micromotors because, for a given gap between the rotor and the stator poles, the output torque increases linearly with the pole height. Furthermore, the ratio between the vertical and the rotational stiffness of the flexible suspension increases quadratically with the aspect ratio, which improves the mechanical stability of the motor. Figure 3 illustrates the layout of the stator. It shows how insulating trenches and interconnects are combined to obtain two-level cross-connections required for a 3-phase electrical network. As depicted, the stator poles, bulk micromachined in the silicon substrate, are enclosed by insulating trenches. Each pole is electrically insulated from its neighbors while remaining mechanically embedded in the bulk silicon. In order to connect the stator poles of a certain phase, a conductive interconnect is employed to access the poles through the vias in a dielectric layer.

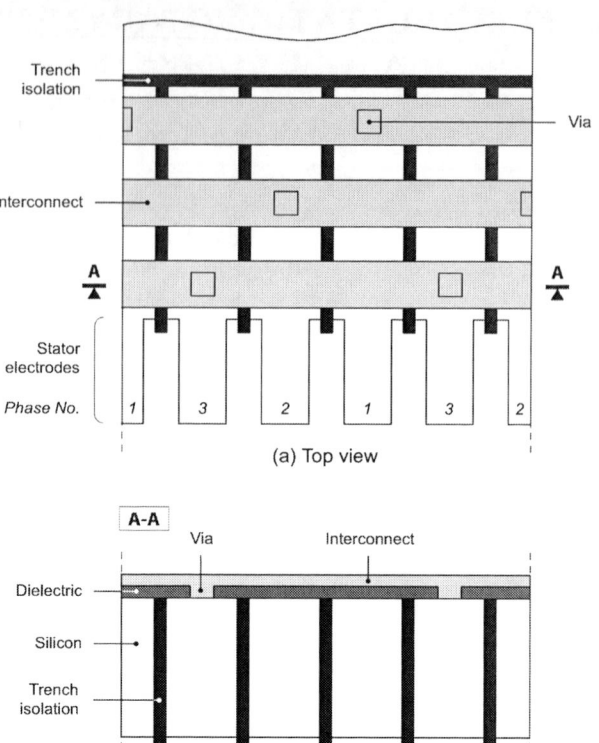

Figure 3: The motor was fabricated in the bulk of a standard silicon wafer using vertical trench isolation technology. Top and cross-sectional views of the stator showing the layout of vertical trench isolation.

Fabrication Process

The 5-mask fabrication process was performed on a standard 3'' silicon wafer having a thickness of 200 µm. To assure electrical conductivity of the substrate, a highly conductive wafer was selected. In this wafer, insulating trenches, 2 µm wide and 40 µm deep, were etched and refilled using a combination of silicon oxide and undoped polysilicon. After the definition of insulting trenches, a 300 nm thick silicon oxide layer was deposited and the vias were etched by BHF etching. In the following fabrication step, interconnects were patterned in a highly doped polysilicon layer deposited by Low Pressure Chemical Vapor Deposition (LPCVD). Next, a deep plasma etch was performed from the backside of the wafer until the bottom of the insulating trenches were visible, leaving a 37 µm thick silicon membrane on the other side. In the last fabrication step, the motor layout was etched in the silicon membrane by deep plasma etching. The motors etched in the silicon membrane were automatically released.

Fabrication Results

Figure 4 shows front and backside SEM micrographs of a completed micromotor. The device has a diameter of 1.4 mm and a height of 37 µm. Poles on the stator and rotor are 10 µm long, 4 µm wide and are separated by a gap of 1.25 µm. The pitch of the poles is 1° and 4/3° on the rotor and stator, respectively. The stator poles are symmetrically located around the rotor in order to cancel

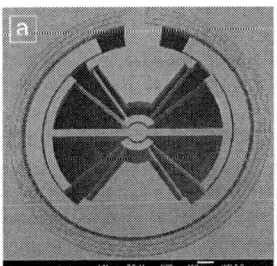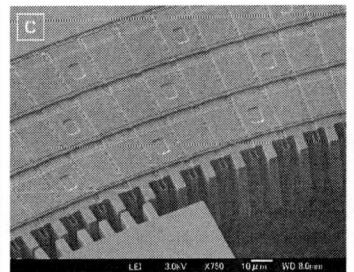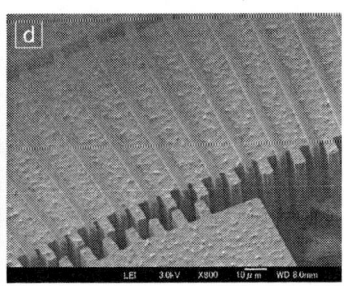

Figure 4: SEM micrographs of the whole motor. (a, b) Overview and (c, d) close-up views from (a, c) the front and (b, d) backside.

Table 1: Design parameters of the 3-phase electrostatic stepper motor.

Parameters	Values
Motor diameter	1400 μm
Pole width	5 μm
Pole height	37 μm
Gap (rotor/stator)	1.25 μm
Rotor pole pitch	1°
Stator pole pitch	4/3°
No. of active poles per phase	64
Flexure length	400 μm
Flexure width	3 μm
Flexure height	37 μm

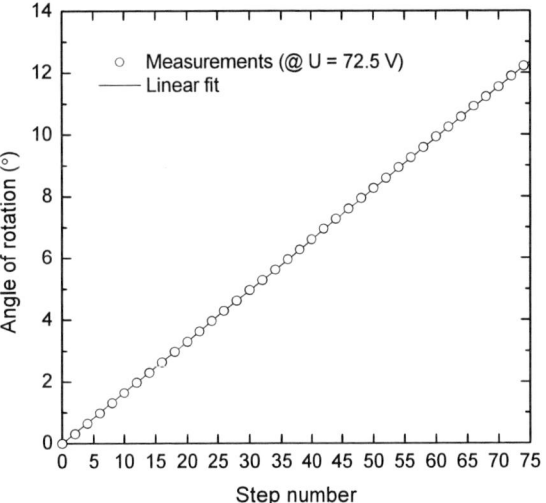

Figure 6: Rotational motion of the 3-phase rotary stepper motor as a function of the number of steps. The measurements were performed with square voltages of 72.5 V using a half stepping sequence

the radial component of the electrostatic force acting on the rotor. There are 85 poles in each phase, giving a total of 255 poles on the stator. During operation of the motor, only 64 stator poles in each phase face the opposite poles on the rotor. Therefore, only 64 pairs of poles are actively involved in torque generation. The motor is suspended with a 'butterfly' flexure pivot, which consists of 8 beam flexures that are 400 μm long, 3 μm wide and 37 μm high. The main dimensions of the fabricated motor are listed in Table 1.

EXPERIMENTAL RESULTS
Motor Operation

We have successfully generated rotational stepping motion of our motor by using an appropriate 3-phase voltage sequence. By reversing the voltage sequence, we could achieve bidirectional motion of the motor, as shown in Figure 5. The 3-phase voltage sequence was generated

using a LabVIEW® control interface, a multichannel analog output card and multichannel high voltage amplifiers. The motor was driven in half-stepping mode, using square wave excitation with overlapping phases. The rotation of the motor was measured by an image processing technique based on the Fourier transform. This technique allows in-plane sub-pixel displacement measurements with a resolution of a few nanometers. Figure 6 shows the measured rotational angle (in a single direction) versus the number of steps at a driving voltage of 72.5 V. One can see the nearly linear increase of the rotational angle with the number of steps. A stepping

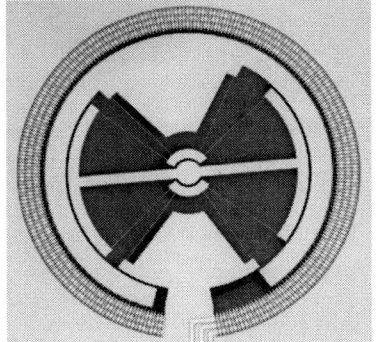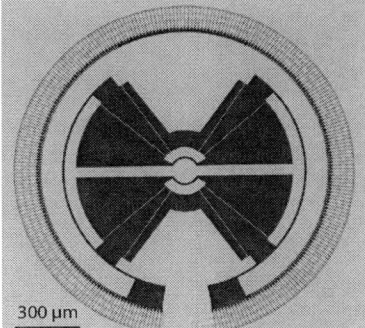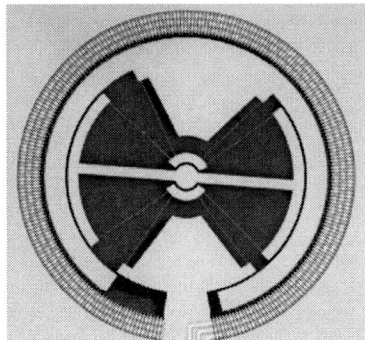

Figure 5: Operation of the 3-phase rotary stepper motor: counterclockwise rotation (left), initial position (middle) and clockwise rotation (right).

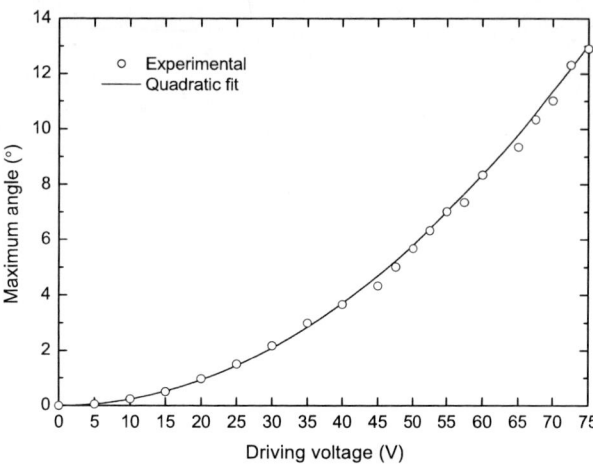

Figure 7: Maximum angular displacement of the 3-phase rotary stepper motor as a function of the driving voltage. The measurements are for one direction only. The angular range is double.

resolution of 1/6°, achieved with the half-stepping sequence, results from the spatial periods of the rotor (1°) and stator poles (4/3°).

Output Range

The rotational range of the motor depends on the stiffness of the mechanical suspension and on the generated torque. The maximum torque in electrostatic motors is a function of the actuation voltage. Figure 7 shows that the angle of rotation is quadratically dependent on the driving voltage, as is expected for an electrostatic motor. We could demonstrate a maximum rotational travel of 26° (+/– 13°) for driving voltages of 75 V. It should be noted that a rotation of 26° corresponds with movement of 317 μm of the rotor edge. The maximum driving voltage of 75 V was limited by electrostatic discharge occurring between the polysilicon interconnects and the grounded substrate. Therefore, higher driving voltages have resulted in operational failure.

Speed

The rotational speed of the stepper motor depends on the driving frequency and on the step size, which is directly related with the driving voltage sequence. We have tested proper operation of the motor at high driving frequencies using two different stepping modes: half-stepping mode with square wave excitation and micro-stepping mode with ramp signals. When operating in the half-stepping mode with the step size of 1.67°, we could observe improper operation (skipping of steps) at high driving frequencies. This is due to abrupt changes of the driving torque, which are inherent with square wave excitation. When driving the motor in a smooth way by using the micro-stepping mode with the step size of 1/48°, we could achieve rotational speeds up to 1.67°/ms.

OUTLOOK

We have devised, fabricated and experimentally tested a monolithic silicon rotary micromotor aimed at skew angle compensation in hard disk drives. With the fabricated motor, we have successfully demonstrated proper operation with a large rotation range of 26° and a high rotational speed of 1.67°/ms. Although the specifications required for skew angle compensation have not yet been fulfilled, the first prototype has demonstrated exceptional performance characteristics, which can be further improved. In our opinion, there is a large yet unexplored room for improvement, such us: further miniaturization of the motor to fit the size of a picoslider (1 mm × 1.25 mm) while preserving or even increasing the number of poles without new fabrication challenges; design optimization of the rotor to reduce its moment of inertia; 'pole shaping' to increase the output torque and development of a model to predict dynamical performance of the motor. We also believe that a motor with such performance characteristics is suitable for a wide range of MEMS applications in which open-loop operation, large angular range, high speed and repeatability are necessary (e.g. optical switches).

ACKNOWLEDGEMENTS

This work is supported by the Storage Research Consortium (SRC) in Japan. The authors gratefully acknowledge the valuable discussions and encouragement from members of the Mechano-Servo Group.

The photolithography mask was made using the 8-inch EB writer F5112+VD01 donated by ADVANTEST Corporation to the VLSI Design and Education Center (VDEC, The University of Tokyo).

REFERENCES

[1] Z. He, E.H. Ong, G. Guo "Optimization of magnetic disk drive actuator with small skew actuation", *J. Appl. Phys.* 91 (1), pp. 8709-8711, 2002.

[2] L.S. Fan, H.H. Otessen, T.C Reiley, R.W. Wood, "Magnetic recording head positioning at very high track densities using a microactuator-based two stage servo system", *IEEE Trans. Indust. Electron.* 42 (3), pp. 222 -233, 1995.

[3] D.A. Horsley, M.B. Cohn, A. Singh, R. Horowitz, A.P. Pisano, "Design and fabrication of an angular microactuator for magnetic disk drives", *J. Microelectromech. Syst.* 7 (2), pp. 141-148, 1998.

[4] L.S. Fan, Y.C. Tai, R.S. Muller, "IC-Processed electrostatic micro-motors", *Proc. IEEE International Electron Devices Meeting*, pp. 666-669, 1988.

[5] M. Mehregany, Y.C. Tai, "Surface micromachined mechanisms and micromotors", *J. Micromech. Microeng.* 1, pp. 73–85, 1991.

[6] S. Henein, P. Spanoudakis, S. Droz, L.I. Myklebust, "Flexure pivot for aerospace mechanisms", *Proc. 10th European Space Mechanisms & Tribology Symposium*, San Sebastian, 2003.

[7] T.J. Brosnihan, J.M. Bustillo, A.P. Pisano, R.T. Howe, "Embedded interconnect and electrical isolation for high-aspect ratio, SOI inertial instruments", *Digest Tech. Papers Transducers'97*, Chicago, 1997, pp. 637–640.

[8] E. Sarajlic, E. Berenschot, G. Krijnen, M. Elwenspoek, "Versatile trench isolation technology for the fabrication of microactuators", *Microelect. Eng.* 67-68 pp. 430-437, 2003.

A SURFACE-TENSION-DRIVEN PROPULSION AND ROTATION PRINCIPLE FOR WATER-FLOATING MINI/MICRO ROBOTS

Sang Kug Chung, Kyungjoo Ryu, and Sung Kwon Cho

Department of Mechanical Engineering and Materials Science, University of Pittsburgh, USA

ABSTRACT

This paper describes development and experimental verifications of a novel propulsion and rotation technique for water-floating objects. As opposed to mechanical paddling, this technique electrically controls surface tension forces acting on water-floating objects without any moving parts (so-called electrowetting-on-dielectric, EWOD). As a proof of concept, a mini water-floating boat of centimeter size is fabricated from a thin plastic foil of which outer surfaces are covered with microfabricated EWOD electrodes. Applying a voltage to the electrodes changes the symmetric configuration of surface tension on the boat, resulting in generation of propulsion and rotation in the boat. By energizing the frontal or rear EWOD electrode, linear propulsion of the boat is achieved. The maximum speed is measured to be 5 mm/s. In addition, energizing two diagonal side EWOD electrodes on the boat surface generates rotational motion at the maximum rotational speed of 20 rpm. Finally, by combining the linear propulsion and rotational actuation, a curvilinear motion with controllability is demonstrated. This novel propulsion and rotation mechanism is simple yet efficient possibly being applied to propel and maneuver water-floating mini/micro robots and boats.

INTRODUCTION

The development of new propulsion mechanisms becomes an important task in applications of micro/nano robots. Recently, various propulsion mechanisms for terrestrial as well as underwater micro/mini robots have been inspired from many small insects. For example, Kim et al. [1] and Guo et al. [2] demonstrated micro underwater robots inspired from fishes that can sense and actuate objects while swimming. These robots implicate the application of microsurgery within blood vessels for minimally invasive medicine. However, the development of propulsion mechanism for water-floating objects has been rarely found. Recently, Hu et al. [3] and Gao & Jiang [4] investigated the propulsion mechanisms of water striders. Song and Sitti [4] developed and demonstrated a mini robot which can walk on the water surface without breaking the water surface. Kim et al. [1] developed a reprogrammable and controllable water strider using ionic polymer-metal composite actuators. In these small insects or robots, the legs are generally hydrophobic such that they do not penetrate into the water surface in many cases but support body weight. The dominant force on the legs is surface tension overwhelming their own weight. As a result, their own body is supported over the water surface without drowning. In addition, paddling the legs generates propulsion, as a result providing mobility to the insects and robots. Compared to underwater swimming, driving on the water surface experiences much less drag force.

However, the above paddling mechanisms in the water-floating insects and robots are so complicated involving many moving parts, joints, and energy (or force) transfer mechanisms. Mimicking and miniaturizing of such complicated mechanisms becomes a challenging issue, even with the state-of-the-art microfabrication technology.

With this regard, we develop a novel propulsion mechanism for water-floating elements that does not require any mechanical moving parts. By simply applying an electric voltage to the electrodes covering the outer surface of water-floating objects, the water-floating objects can be propelled and rotated in a controlled manner. This paper presents its principle, model fabrication of mini boat, and proof of the principle, demonstrating linear, rotational, and curvilinear motions in the model boat.

PRINCIPLE OF PROPULSION AND ROTATION

Figure 1: The larva of Pyrrhalta is able to ascend the inclined meniscus. The larva deforms the water surface by arching its back, thus generate asymmetric forces between on its head and tail that generate the desired capillary thrust to move [6].

The present propulsion and rotation principle for water-floating objects is bio-mimicked from the propelling skill of some water-floating insects (e.g., *Pyrrhalta* larva [6]). The *Pyrrhalta* larva is partially submerged in water due to hydrophilicity of its skin but is a poor swimmer in water; however, it can propel itself by controlling and reorienting the configuration of surface tension without paddling when it needs to ascend inclined menisci. When the larva is stationary, the surface tension forces acting on its head and tail are symmetrically balanced (Fig. 1a). When the larva needs to ascend a water surface, it bends its back (Fig. 1b). In this case, the symmetry of surface tension forces acting on its head and tail breaks. As a result, the asymmetrical reorientation of the surface tension forces produces a net horizontal force. This net force drives the larva through the water, which is called "capillary thrust." This asymmetric force unbalance is also utilized for propulsion by some other insects of which legs are hydrophobic [6]. In this paper, we mimic this thrusting principle.

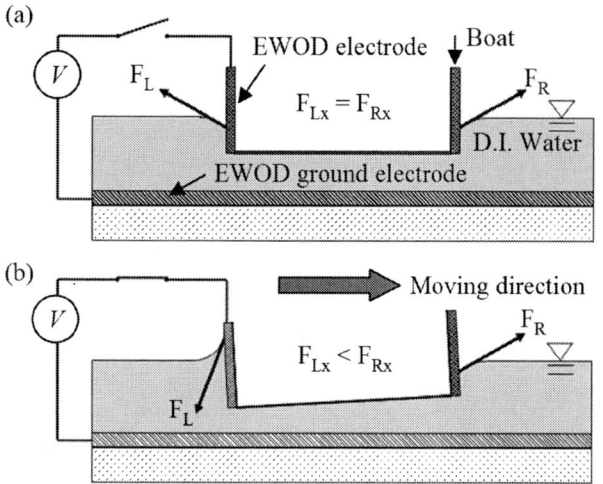

Figure 2: EWOD (electrowetting on dielectrics) propulsion scheme in a mini-boat: (a) Initial state: horizontal force components of surface tension in the left and right are equal to each other ($F_{Lx} = F_{Rx}$). (b) Under voltage application to the left electrode, the contact line rises and the contact angle is decreased (EWOD actuation). That is, $F_{Lx} < F_{Rx}$. The boat would move from left to right.

In the above insect thrust, the asymmetric force unbalance is generated by mechanical displacements in the inset's body or legs, which require mechanical moving parts. To avoid this necessity for mechanical moving parts, we control the configuration of surface tension forces *electrically*, not mechanically, so called electrowetting on dielectrics (EWOD) [7]. EWOD provides the capability of changing the wettability of water on the surface. In other words, the direction of surface tension force can be re-oriented by applying an electric voltage to the electrode buried underneath the surface. Figure 2 illustrates the concept of the present bio-mimicked propulsion in a simplified rectangle mini-boat. The boat has EWOD electrodes on the left and right sides to break force symmetry. Initially, the surface tension forces (F_L and F_R) on the boat are symmetric (Fig. 2a): the boat is stationary. To generate an asymmetric force configuration and thus a horizontal net force, a voltage is applied to the EWOD electrode in the left side of the boat (Fig. 2b). Since the contact angle on the electrode is reduced by the voltage, F_L becomes pointed downward while F_R does not change significantly: the resultant of horizontal forces would direct right, and the boat would move right. Similarly, energizing the EWOD electrode in the right side of the boat would generate a left-directed net horizontal force, thrusting the boat to the left. Furthermore, if a pair of electrodes is diagonally located on the side surfaces (one on each side surface), a net torque would be generated, resulting in

rotational motion in the boat.

The present propulsion mechanism does not have any mechanical moving parts. Thus, the overall system is so simple yet efficient facilitating miniaturization and realization of the propulsion mechanism for such small objects. In addition, it turns out that this mechanism is effective even on horizontal water surfaces, whereas it was reported that the aforementioned water insects use the propulsion skills only when they climb sloped water meniscus [6]. It has not been reported that they can swim on a horizontal water surface using the above skills.

FABRICATION OF TESTING DEVICES

In order to experimentally prove the present propulsion and rotation principle, a centimeter-sized model boat is fabricated with attachment of microfabricated flexible EWOD electrodes to its outer surfaces. A box-type mini-boat with top open ($2.5 \times 1 \times 1$ cm^3) is made of a polymer film. The microfabricated flexible EWOD electrodes are glued on the front and rear faces of the boat for linear propulsion or on the side faces for rotation. The EWOD electrodes are connected to an external, controllable voltage source via a thin copper wire. The wire is chosen as thin (75 μm in diameter) as possible to minimize any effects of wire bending elasticity on propelling and rotating. The EWOD electrodes are fabricated using the standard micro fabrication technology. The main fabrication process consists of (i) patterning a 2 μm thickness Cu layer on the polyimide film (Fig. 3a); (ii) depositing a dielectric layer (Fig. 3b); (iii) depositing a hydrophobic layer (Fig. 3c). For a flexible electrode, Pyralux® flexible circuit material which is commercially available from Dupont is used, and the electrode is patterned and diced. For the EWOD dielectric layer, a 3 μm thick photoresist (AZ4210, Clariant Co.) layer is spin-coated on the Cu electrode layer. Finally, the EWOD electrode is coated with a hydrophobic Teflon layer (2000 Å in thickness).

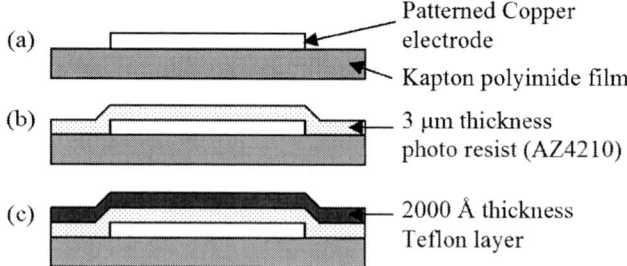

Figure 3: Microfabrication process of flexible EWOD electrodes to be attached onto the boat surfaces: (a) patterning of EWOD electrodes; (b) deposition of dielectric layer (photoresist); (c) deposition of hydrophobic Teflon layer.

EXPERIMENTAL RESULTS

Linear propulsion

The fabricated model boat is tested in a water chamber for linear propulsion. Figure 4 shows sequentially snapshot

978-1-4244-2977-6/09 $25.00 © 2009 IEEE   1080

top-views of mini-boat propelled in the water chamber. For EWOD grounding, a copper foil is placed on the bottom of the chamber. When 160 V (at 1 kHz) is applied to the rear EWOD electrode (red broken circle), the boat is propelled and moved forward at ~ 4 mm/s. In this case, the rear side sinks a little more than the front side while in motion. Similarly, energizing the front electrode generate a reverse motion (not shown). It was rarely observed that the moving direction is opposite to the above ones. The opposite movements may be caused by low contact angle reduction due to contact angle hysteresis on the EWOD electrode surfaces. Almost consistently, however, the boat moves in the direction opposite to where the energized EWOD electrode is placed.

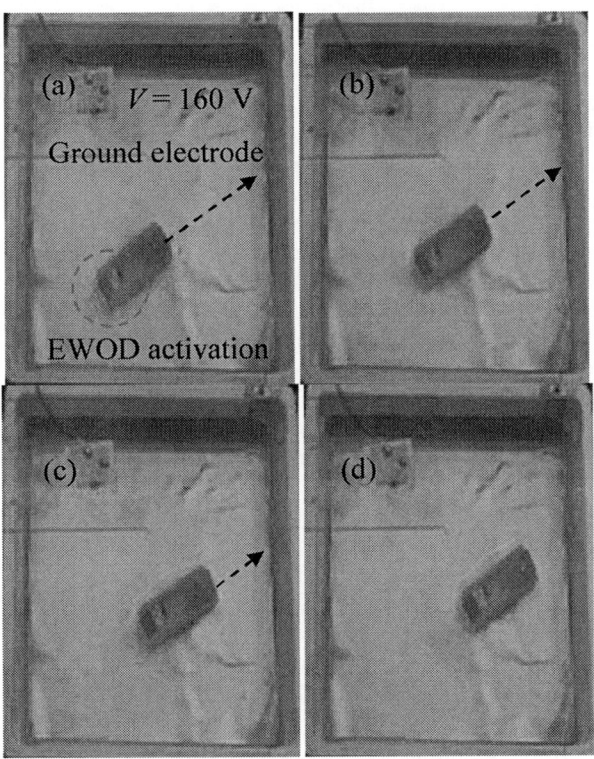

Figure 4: Sequential snapshots of mini-boat propulsion: (a-d) while a voltage of 160 V is applied to the rear electrode (red broken circle), the boat is moving forward as indicated by a dotted arrow. The moving speed is measured at ~4 mm/s.

Contact angle measurement on EWOD electrode

To confirm the working principle described in Fig. 2, the meniscus change on the energized electrode is visualized, as figure 5 shows snapshots of water meniscus (white broken lines) around the EWOD electrode when EWOD is off (Fig. 5a) and on (Fig. 5b). The experiment is conducted in a top opened transparent acryl water chamber ($5 \times 5 \times 2$ cm^3). A ground copper electrode is attached on the bottom of the chamber as shown in Fig. 4 and an EWOD electrode is attached to a glass plate which is placed perpendicularly to the horizontal water level. When the voltage is applied to the electrode, the water meniscus is moved up due to reduced contact angle. The direction of surface tension force is changed from nearly horizontal to downward (yellow arrows), confirming the force

configuration in Fig. 2. However, in this case, the EWOD electrode is fixed, not exactly the same condition as that on the moving boat. Currently, we are visualizing the meniscus motion on the EWOD electrode of moving boats.

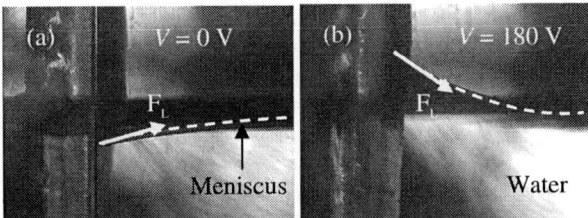

Figure 5: Side view snapshots of the water meniscus on the EWOD electrode before (a) and after (b) applying the voltage. The surface tension force points downward after voltage application, resulting in reduction in the horizontal force component. The broken lines and the arrows denote water menisci near the EWOD electrode of the boat and surface tension force, respectively.

Rotation

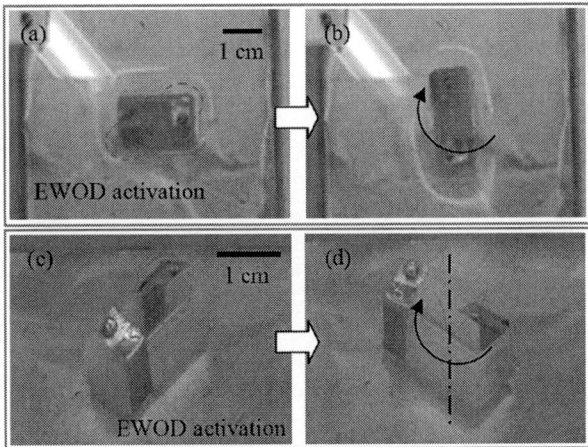

Figure 6: Sequential snapshots of mini-boat rotation: two EWOD electrodes (broken circles) are attached to the side faces of the boat to generate a torque. When a voltage (140 V) is applied to the two EWOD electrodes, the boat is rotated at ~20 rpm: top view (a) and (b); perspective view (c) and (d).

To generate a torque on the boat, one EWOD electrode is attached to one of the side faces while the other to the opposite face. They are diagonally aligned to each other (broken circles). Figure 6 shows sequential snapshots of boat rotation. The experiment is carried out in the same water chamber as shown in Fig. 4. When 140 V is applied to both electrodes, the boat rotates at ~20 rpm in a clockwise direction. The rotation direction is consistent with the result in Fig. 4. The horizontal force component on the energized surface is weaker than that on the non-energized surface. This configuration can be interpreted as the non-energized surface is pulled by resultant surface tension. If this argument is applied to rotation in Fig. 6, the two side areas of non-energized electrode are pulled in opposite directions by two equal but off-centered surface tension forces, generating a clockwise torque. Note that when the applied voltage is off the boat is rotated in a counterclockwise direction due to the wire elasticity.

Dependence of applied voltage on speeds

It is well known that as the applied voltage is increased the contact angle is parabolically decreased until saturated. This fact makes us expect that increasing the voltage leads to faster motion in the boat until the contact angle is saturated. To investigate the effect of the applied voltage on the propulsion and rotation speeds, similar experiments to the ones in Figs. 4 and 6 are carried out with varying the applied voltage. Figure 7 shows the quantitative results of transportation and rotation speeds vs. applied voltage. At lower voltages below the threshold voltages (100 and 80 V for transportation and rotation, respectively), the boat neither moves nor rotates. This may be due to insufficient unbalanced force or contact angle hysteresis. Once the voltage goes over the threshold voltages, the transportation and rotation speeds are increased in proportion to the applied voltage. Finally, the speeds are saturated at ~20 rpm for rotation and 5 mm/s for transportation. This is due to the phenomenon of contact angle saturation which is well known in the EWOD society: above certain voltages, the contact angle is not decreased anymore.

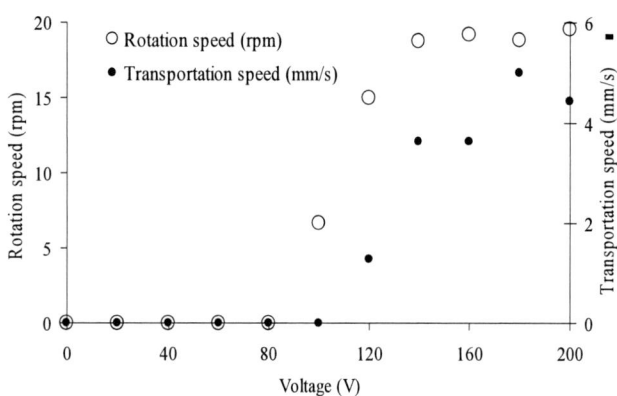

Figure 7: Quantifications of propulsion and rotation speeds vs. applied voltage. Note that there are thresholds to initiate motions. Once passed thresholds, the speeds of rotation and propulsion are increased in proportion to the applied voltage until saturated.

Curved motion

By combining linear actuation and rotational actuation, curvilinear motions can be generated with a steering capability. Figure 8 shows sequential snapshots of a curved motion of the boat. The experiment is carried out in the same water chamber as Fig. 4. To generate curved motions, one electrode covers the front corner of the boat (blue broken-circle) while the other is attached to the rear face (red broken-circle). Upon EWOD voltage application, the forward and side resultant forces (red arrows) are produced, leading to clockwisely curved motion in the boat. Based on the results in Fig. 7, it is expected that changing the voltages to the corner and rear electrodes allows us to control the curvature of the curvilinear motion.

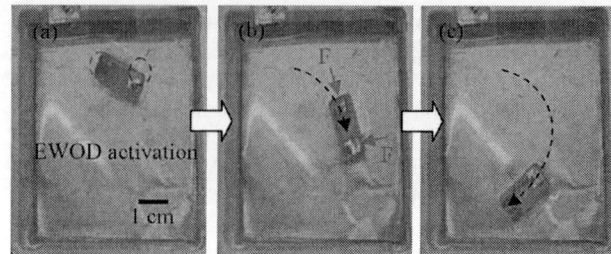

Figure 8: Sequential snapshots of steered propulsion: the boat is driven in a curved path. Note that the front electrode (blue broken circle) covers the front corner of the boat while the rear electrode covers only the rear face (red broken circle).

CONCLUSIONS

In this paper, inspired by the swimming skill of water-floating insects, we present a novel propulsion and rotation mechanism by electrically controlling and reorienting the configuration of surface tension forces acting on water-floating objects (so called electrowetting on dielectrics or EWOD). Using a centimeter-sized model boat with attachment of microfabricated EWOD electrodes on its outer surfaces, linearly-, rotationally-, and curvilinearly driven motions in the boat are experimentally achieved and demonstrated. By observing the dynamic motion of the meniscus on the EWOD electrode, it is confirmed that unbalance in the horizontal force components generates capillary thrust. The present propulsion mechanism is simple yet efficient and does not need any mechanical moving parts, thus being expected to facilitate its application to mini/micro robots.

REFERENCES

[1] B. Kim, D. H. Kim, J. Jung, and J. O. Park, "A biomimetic undulatory tadpole robot using ionic polymer-metal composite actuators," Smart Mater. Struct., vol. 14, pp. 1579–1585, 2005.

[2] S. Guo, Y. Okuda, and K. Asaka, "Hybrid type of underwater micro biped robot with walking and swimming motions," Proc. IEEE Int. Conf. Mech. Autom., pp. 1604–1609, 2005.

[3] D. L. Hu, B. Chan, and J. W. M. Bush, "The hydrodynamics of water strider locomotion," Nature, vol. 424, pp. 663–666, 2003.

[4] X. Gao and L. Jiang, "Water-repellent legs of water striders," Nature, vol. 432, pp. 36, 2004.

[5] Y. S. Song and M. Sitti, "Surface-tension-driven biologically inspired water strider robots: Theory and experiments," IEEE Transactions on robotics, vol. 23, pp. 578-589, 2007.

[6] D. L. Hu and J. W. M. Bush, "Meniscus-climbing insects," Nature, vol. 437, pp. 733–736, 2005.

[7] S. K. Cho and H. Moon, "Electrowetting on dielectric (EWOD): new tool for bio/micro fluid handling," The BioChip Journal, Vol. 2, No. 2, pp. 79-96, 2008.

DEVELOPMENT OF SKEWED DRIE PROCESS AND ITS APPLICATION TO ELECTROSTATIC TILT MIRROR

M.Nakada[1], K. Takahashi[1], T. Takahashi[1], A. Higo[2], H. Fujita[1], and H. Toshiyoshi[1]
[1]Institute of Industrial Science, the University of Tokyo, Tokyo, Japan
[2]Research Center for Advanced Science and Technology, the University of Tokyo, Tokyo, Japan

ABSTRACT

We present a newly developed skewed-DRIE (deep reactive ion etch) process through a silicon wafer. Maximum skew angle of 25 degrees with respect to the wafer surface was obtained with a 360 micron thick silicon substrate. This technique was used to build the slanted counter electrode of 60 degrees for an electrostatic torsion mirror. Pull-in voltage was 100 volts, which was lower than that of a vertically etched counter electrode.

INTRODUCTION

Parallel plate electrostatic actuator

Several different types of electrostatic torsion mirrors have been developed by silicon micromachining techniques mainly for optical MEMS applications. Fig. 1 shows four kinds of such scanners. The torsion mirror in Fig. 1 (a) uses a narrow space of few microns between the movable mirror and the substrate; usually the mirror is electrically biased with respect to the substrate to cause electrostatic torque. SOI (silicon on insulator) micromachining is widely used to develop such structure with a few microns thick of BOX (buried oxide). Deep-RIE (Reactive Ion Etching) and sacrificial release of the SOI microstructure are employed [1]. It is the simplest process among the four types. Fig. 1 (b) used the vertical sidewall of the substrate as a fixed electrode [2]. The scanner is made by etching the both sides of an SOI wafer with DRIE processes. Fig. 1 (c) uses a relatively large space underneath the scanner plate, compared with the scanner shown in Fig. 1 (a). Therefore, it can steer the reflected light with a large optical angle. Wafer bonding process is usually used for fabrication [3]; a recessed wafer as the bottom substrate is bonded with the upper substrate which is later DRIE processed. Apart from these structures, we pursue a new structure shown in Fig. 1 (d), which has a

slanted sidewall as a fixed electrode for large scan angle. The angled etch was produced not by the anisotropic wet etching of silicon but with a newly improved "skewed DRIE process."

The original version of skewed DRIE process was developed by the group at AIST (advanced industrial science and technology)[4]. In their processes, they used a tilted wafer holder in the DRIE chamber and used the directed plasma gas to etch the slanted wall in the silicon substrate. Nevertheless, the angle-etched area was limited near the edge of the wafer. On the other hand, our new version of skewed DRIE exhibited extended process area. Thanks to the skewed DRIE process, we could successfully develop an electrostatic torsion mirror structure that operated at lower actuation voltages.

FABRICATION

Conventional skewed Deep-RIE

As a control of experiment, we first examined the skewed-DRIE method developed by the AIST's group [4]. Normal DRIE processes use a flat orientation of wafer in the substrate holder to make vertically etched holes. On the other hand, the AIST's process used a metal block of slanted wall, onto which a cut wafer was mounted. The metal block height was limited to the maximum slot opening of the DRIE chamber, and the block was also exposed to the plasma during DRIE processes. Thanks to the directed plasma incoming to the silicon surface, tilted trenches and holes were made in the wafer. We first copied the method and home made a metal block tool with a tilted surface of 35.6 degrees with a 16.5 mm long slope length, a 9.6 mm height, and a 20.3 mm in depth in Fig. 2 (b). A sample was put on the tilt surface of the tool, and with the metal block on, the DRIE process was operated (Fig. 2(a)). Cross-sectional image of the sample processed for an hour is shown in Fig. 2 (c). The right-hand side of the SEM (scanning electron microscope) image corresponds to the slope top of the tool, and the left-hand side to the slope bottom. Fig. 2(d) and (e) show the close-up images of the DRIE results measured at the slope top and the bottom, respectively. It was found that the skew angle was marginal and that the etched profile was nearly as vertical as those of usual flat-placed DRIE. It was possibly because of the etching plasma impinging nearly normal to the wafer surface, regardless tilt angle of the wafer. It was also plausible that the process results depended on the tooling factor of DRIE machines used.

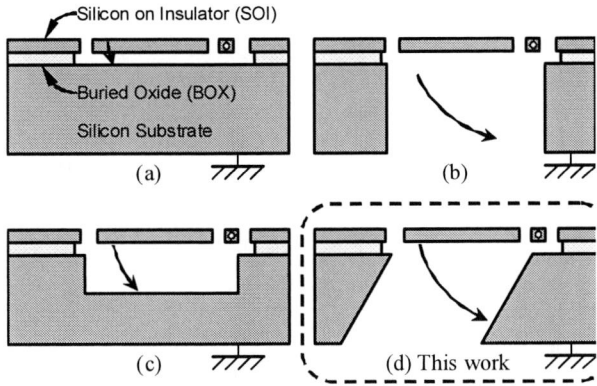

• Fig. 1: Schematic models of electrostatic torsion mirrors (a) parallel plate actuator having the gap of few microns, (b) electrostatic actuator with vertical fixed electrode, (c) parallel plate actuator with the large gap and (d) electrostatic actuator with angled electrode

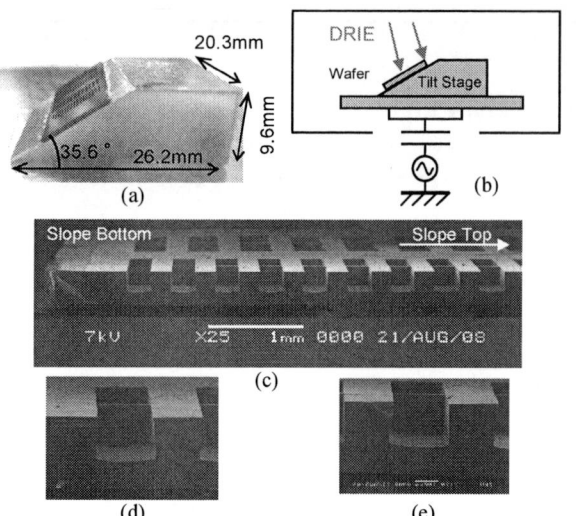

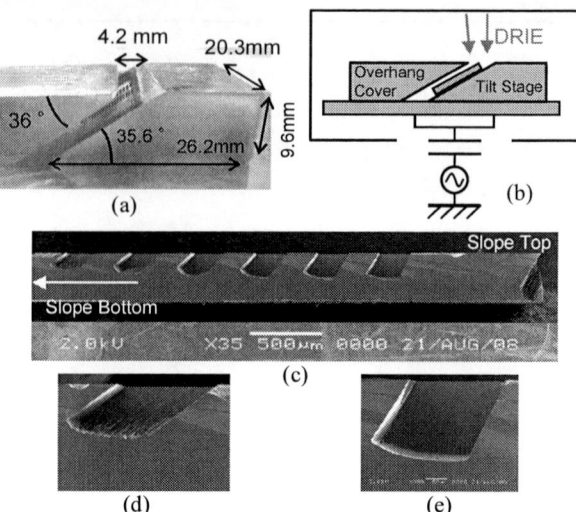

Fig. 2 Conventional angled DRIE process and results. (a) tilt stage block and mounted silicon wafer, (b) DRIE chamber view, (c) cross sectional image of DRIE'd wafer, (d) close up image of slope bottom and (e) close up image of slope top

Fig. 3 Newly developed angled DRIE process and results. (a) tilt stage block and mounted silicon wafer, (b) DRIE chamber view, (c) cross sectional image of DRIE'd wafer, (d) close up image of slope bottom and (e) close up image of slope top

Developed skewed Deep-RIE

In contrast to the previously reported skewed-DRIE tool, we newly developed another type of method by using an overhanging metal block to shield the slanted slope from the directed vertical plasma flux. The overhang cover had a tilt surface of 36 degrees, and it was placed with its face at a 4.2 mm gap, as shown in Fig. 3(a). The DRIE process of the identical recipe was performed with the special tool, as shown in Fig. 3(b). Fig. 3(c) shows the results of cross-sectional image processed for an hour. In the similar manner as before, the right-hand side of the image corresponds to the slope top of the sample, while the left-hand side to the bottom. It was found that the skewed angles were larger than those of the conventional one. Fig. 3(d) and (e) show the close-up images of the slope top and the bottom, respectively. It was found that the slope top had fast etch rate and more skewed angle with respect to the substrate normal, while the slope bottom exhibited slower etch and more verticalness. Etch angle was found to increase with insertion from the slope top aperture, however, the etch depth decreased.

Fig. 4 plots the etch depth D and the etch angle θ as a function of distance from the edge of slope top. The triangles in Fig. 4 indicate the etch depth, while the circles show etch angle. It indicates the etch angle at 1 mm from the edge is 61 degrees and the etch depth is 210 μm after one hour etching. On the other hand, etch angle at 5.25 mm is 25 degrees and the etch depth is 125 μm. It was found that etch angle decreased with increasing distance from edge, while etch depth decreased with distance. From the result of Fig. 4, the overhang cover has been found to direct more angled plasma flux to the substrate surface rather than the conventional method, and such effect was more significant near the bottom of the aperture. Etch rate decreased with insertion, because the reaction was limited by the plasma supply.

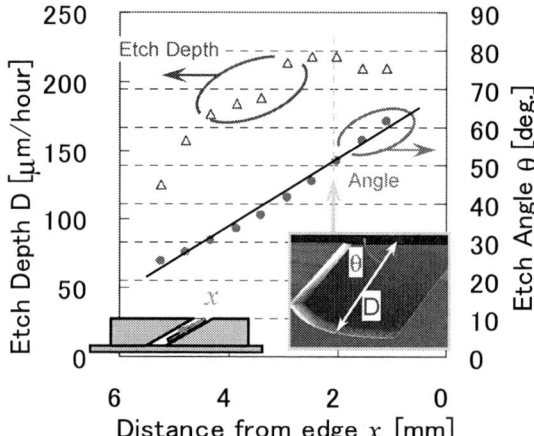

Fig.4 Position dependence of etch depth and tilt angle. Etch rate and tilt angle are maximum near the tilt stage aperture.

Fabrication process for electrostatic actuator with tilted electrode

We used the skewed DRIE process to develop an electrostatic torsion mirror actuator. An SOI wafer of 1.4 μm thick active layer and with a 0.5 micron thick BOX was used. The handle wafer of the SOI was 360 μm thick. First, we patterned the SOI layer by using the usual DRIE recipe and wafer orientation. Next, we patterned the backside of the wafer by using the skewed DRIE process. The aluminum etching mask on the back side was carefully designed and was intentionally shifted by 210 microns with respect to the structures engraved on the SOI front, because the skew angle of 60 degrees was expected (Fig. 5 (a)). The patterned sample was mounted on the slope surface of the tool (Fig. 5 (b)). They were loaded into the process chamber of the DRIE and processed.

978-1-4244-2977-6/09 $25.00 © 2009 IEEE   1084

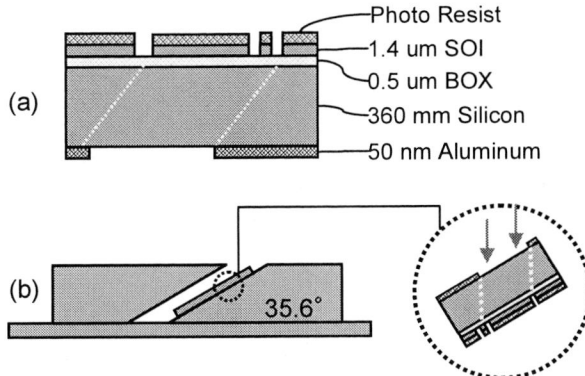

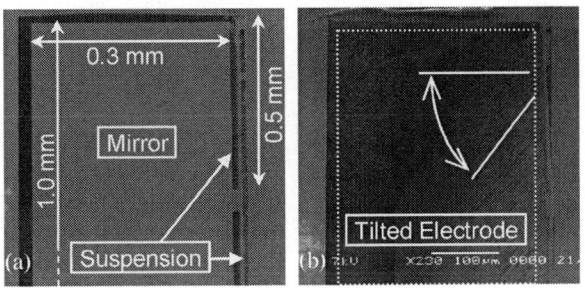

Fig. 5 Process flow of the actuator with tilted counter electrode. (a) Mask patterning for skewed-DRIE and (b) mounted sample on the slope surface of the tool

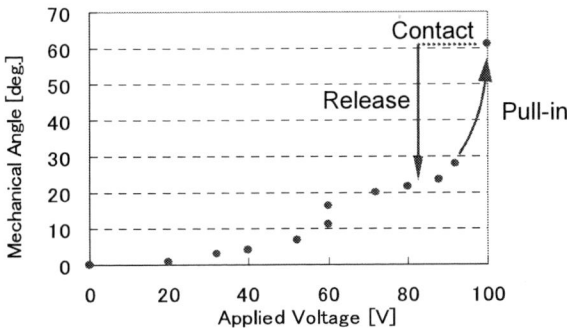

Fig. 7 Electrostatic tilt angle as a function of applied voltage.

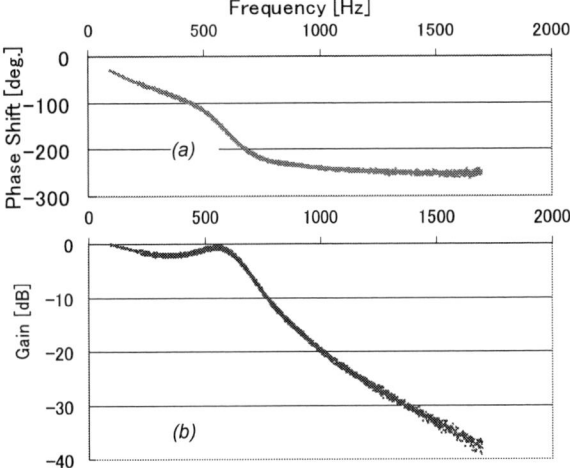

Fig. 6 SEM images of electrostatic torsion mirror. (a) before mirror plate removal and (b) after.

Due to the distribution of the skewed angle on the surface, only a limited area near 1 mm from the wafer edge was found to meet the 60 degree alignment. Later in the process, the BOX was sacrificially removed to complete the torsion mirror release.

The SEM images of the fabricated actuator are shown in Fig. 6. The scanner had a 1.0 mm x 0.3 mm mirror as a movable electrode, suspended with a pair of suspensions of 0.5 mm long and 3.2 μm wide. The slanted electrode made in the substrate is shown in Fig. 6 (b) after manually removing the mirror plate. Slanted wall of 60 degrees was thus visually confirmed by the SEM observation.

RESULTS

The fabricated scanner was operated with a direct current from 0 to 100 volts. The operation voltage was applied to the fixed electrode (tilted electrode), on the other hand, the movable electrode (mirror) was electrically grounded. The rotation angles of the mirror as a function of DC voltage are plotted in Fig. 7. It was found that the pull-in voltage was approximately 100 volts with the pull-in angle of 30 degrees. After the pull-in to full contact of 60 degrees, the torsion mirror was released at 80 volts from 60 to 20 degrees.

The frequency response characteristics of the scanner were measured. The Bode diagram is shown in Fig. 8. Fig. 8(a) indicates a phase shift between the applied voltage and the motion of the mirror. The motion had the lagging phase from 45 degrees (100 Hz) to 260 degrees (1.7 kHz).

Fig. 8 Frequency characteristics of the actuator (a) phase shift (b) gain of rotation angle.

Fig. 8 (b) shows the ratio (gain) of rotation angle to applied voltage from 100 Hz to 1.7 kHz. The figure was normalized with a value at 100 Hz. The gain had a peak at 580 Hz, and the value decreased as the frequency increased. In addition, around 580 Hz the phase delayed 180 degrees because of its resonance.

DISCUSSION

We made an analytical simulation model for the mirror angle as a function of applied voltage. It was assumed that the electrostatic actuator had a shape illustrated in Fig. 9, which is a schematic model of electrostatic scanner with the same shapes of fabricated scanner. Fig. 9 (a) shows the top view and (b) shows the side view. The electrostatic torque between the movable and fixed electrode was described as [5]:

$$T_E = \frac{1}{2}\varepsilon_0 V^2 L_{mirror} \int_{root}^{W_{mirror}+root} x\left[x\left(\frac{\pi}{3}-\theta\right)+g\right]^{-2} dx \quad (1)$$

where ε_0 is the dielectric constant of the air, L_{mirror} and W_{mirror} denote the shapes of mirror (movable electrode), g is the height of the torsion bar axis measured from the counter electrode surface, root is the space between torsion bar and end of the mirror. The torque was a function of rotation angle θ and applied voltage V. The mechanical torque (T_M) is derived from the elastic restoring torque and can be written as [5]:

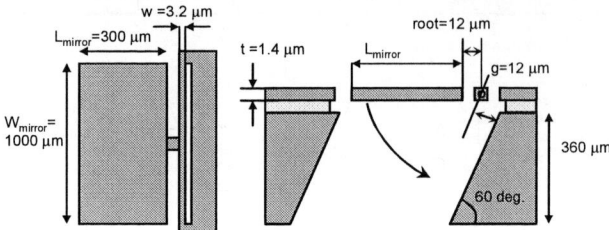

Fig. 9 Simulation model of the actuator (a) top view and (b) side view.

$$T_M = 2 \frac{Gw^3 t}{3l} \left\{ 1 - \frac{192w}{\pi^5 t} \tanh\left(\frac{\pi t}{2w}\right) \right\} \theta \qquad (2)$$

where w, t, l represent the width, thickness, length of torsion bar, respectively, and G is the shear modulus of the torsion bar material (silicon). At the equilibrium condition, electrostatic torque (1) equals to the mechanical torque (2). We solved the equilibrium condition and calculated the angle and voltage. Fig. 10 shows simulation result of rotation angles of the electrostatic actuators with a tilt electrode of 60 degrees and vertical electrode. The solid line shows the angles operated by the actuator with electrode of 60 degrees, while dashed line shows the angles of actuator with vertical electrode.

It was found that the tilt actuator had a pull-in voltage of 112 volts at pull-in angle of 23.4 degrees. On the other hand, actuator with vertical electrode had a pull-in voltage of 190 volts and pull-in angle of 34.4 degrees. In addition, when the torsion mirror was set to 23.4 degrees, the operation voltage of 175 volts was needed. It was higher than the operation voltage of actuator with the electrode of 60 degrees.

CONCLUSION

We have developed a new skewed DRIE process by using a specially tuned etching metal block; the processed wafer was inserted between the slanted gap of the metal block, and directed plasma flux was introduced to the substrate surface to cause slanted DRIE process. By using this process, we created a unique electrostatic actuator which had a slanted-wall electrode of 60 degrees with respect to the substrate surface. The actuator had a pull-in voltage of 100 volts and it operated to a pull-in angle of 60 degrees. According to the simulation developed in this work, the scanner with the electrode of 60 degrees had a lower pull-in voltage compared with a model with vertical electrode. In conclusion, the developed torsion mirror scanner was more beneficial to spatially scan the reflection with larger angle and with lower operation voltages, compared with those made by the conventional vertical DRIE processes.

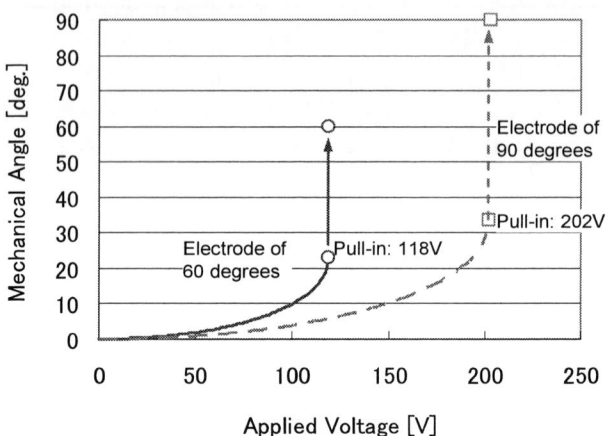

Fig. 10 Simulation result of electrostatic actuators with a electrode of 60 degrees (solid line) and with a vertical electrode (dashed line)

ACKNOWLEDGMENTS

The authors appreciate the VLSI Design and Education Center (VDEC) of the University of Tokyo for making the photo-masks. This research has been supported in part by the G-COE Program of the Electrical Engineering and Electronics, the University of Tokyo.

REFERENCES

[1] K. Isamoto, K. Kato, A. Morosawa, C. H. Chong, H. Fujita, and H. Toshiyoshi, "A 5-V operated MEMS variable optical attenuator by SOI bulk micromachining," IEEE Journal of Selected Topics in Quantum Electronics, vol. 10, pp. 570-578, May-Jun 2004.

[2] H. Toshiyoshi and H. Fujita, "Electrostatic micro torsion mirrors for an optical switch matrix," Journal of Microelectromechanical Systems, vol. 5, pp. 231-237, Dec 1996.

[3] H. Miyajima, K. Murakami, and M. Katashiro, "MEMS optical scanners for microscopes," IEEE Journal of Selected Topics in Quantum Electronics, vol. 10, pp. 514-527, May-Jun 2004.

[4] Y. Ando, T. Ikehara, and S. Matsumoto, "Development of three-dimensional microstages using inclined deep-reactive ion etching," *Journal of Microelectromechanical Systems,* vol. 16, pp. 613-621, Jun 2007.

[5] T. Takahashi, M. Mita, K. Motohara, N. Kobayashi, N. Kashikawa, H. Fujita, and H. Toshiyoshi, "Electrostatically Addressable Gatefold Micro-shutter Arrays for Astronomical Infrared Spectrograph," Technical Digest of Asia Pacific Conference on Transducers and Micro-Nano Technology, Marina-Mandarin Hotel, Singapore, June 25-28 2006.

DESIGN OF A PASSIVE AND PORTABLE DMFC OPERATING IN ALL ORIENTATIONS

N. Paust[1], S. Krumbholz[2], S. Munt[1], C. Müller[1], R. Zengerle[1], C. Ziegler[1], and P. Koltay[1]
[1]University of Freiburg - IMTEK, Georges-Koehler-Allee 106, 79110 Freiburg, Germany
[2]Fraunhofer IZM, Gustav-Meyer-Allee 25, 13355 Berlin, Germany

ABSTRACT

A microfluidic layout concept for passive and portable Direct Methanol fuel Cells (DMFCs) is presented. We proofed this concept by developing a DMFC that continuously runs for 40 hours in all orientations without the need for any active components such as pumps or valves. In contrast to our previous work [1], the system now is truly portable. In any orientation of the DMFC, a bubble driven self regulating supply mechanism safely removes carbon dioxide and transports at least 3.5 times more methanol to the anode than critically needed to sustain DMFC operation. On the cathode, diffusive oxygen supply and the transport of the reaction product water along a capillary gradient out of the DMFC ensures a stable performance. Compared to our previous work [1], the power output was increased by a factor of 2.5 and reached $p = 5.5$ mW cm^{-2}. A stable power output for 40 hours of $p = 4$ mW cm^{-2} was achieved for the preferred vertical position with bubbles moving against buoyancy forces. In the most challenging horizontal position with the anode facing downwards, a power output of at least $p = 3.1$ mW cm^{-2} was reached for the same period of time.

INTRODUCTION

Passive concepts for fuel supply are very attractive for miniaturized Direct Methanol Fuel Cells (DMFCs) [2-4]. Passive in this context means that the DMFC operates fully autonomously without any support of any kind. The abandonment of ancillary pumps and valves makes the system small and compact, increasing the overall energy density. Thus, the passive DMFC is a promising candidate for power supply of small portable appliances. However, a stable long-term performance in all orientations of the fuel cell poses a significant challenge [5]. Methanol supply and the removal of the reaction product carbon dioxide (CO_2) on the anode, as well as oxygen supply and water removal on the cathode must be achieved in any orientation independent of gravity. In our passive DMFC, effects dominant on the micro-scale such as diffusion and capillary force induced transport are applied to account for this challenge.

DESIGN AND OPERATING PRINCIPLE

A schematic of the passive and portable DMFC is depicted in Fig. 1. In the present work, our previously presented passive fuel supply mechanism on the anode [1], propelled by capillary forces of deformed CO_2 bubbles in tapered channel structures, is embedded into a closed design concept. First, the main channel parameters for the anode flow field are determined by analytical calculations ensuring that the capillary pressure gradients are sufficiently large to transport bubbles along the channel even against buoyancy forces. Secondly, the liquid flow rates induced by the bubbly CO_2 gas flow is studied by experiments or Computational Fluid Dynamics (CFD) simulations ensuring that a sufficient amount of methanol is supplied for DMFC operation. In a third step, the channel parameters of the cathode flowfield are determined by CFD-simulations considering diffusive oxygen transport. Subsequently, water formation and transport on the cathode have to be accounted for and the flowfield design is refined to achieve capillary transport of water out of the DMFC by a continuous capillary gradient, hence, the oxygen supply is never blocked.

An exploded view of the current version of our passive DMFC is depicted in Fig. 2. The anode flow field consists of six parallel channels fabricated by micro milling in transparent COC. Each channel has an inlet height of $h_{in} = 0.2$mm, is $w = 1.2$mm wide, has an opening angle of $\alpha = 4°$ and is $l = 14.3$mm long. The channels are coated with a hydrophilic polymer by dip coating with DMAAm-MaBP [6] which results in contact angles at the walls of $\theta = 10°$. At the end of the channels, the CO_2 gas phase is separated from the liquid by a hydrophobic membrane (Fig. 1 (2)) (Whatman PTFE TE36) that holds back the liquid by capillary forces and a bubble fence (Fig. 1 (3)) that allows for the liquid phase to pass but holds back the gas, again, by capillary forces. The bubble fence consists of 20 hydrophilic coated COC poles with a diameter of $d = 300$µm arranged with gaps of $D = 200$µm. The bubbles moving along the tapered channels pump the methanol solution through the bubble fence and the tubing into the reservoir and leave the system through the hydrophobic membrane.

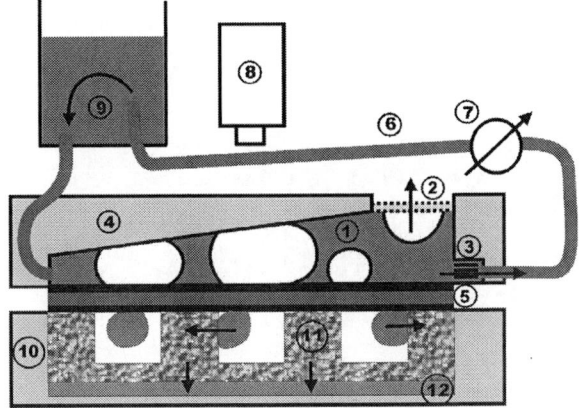

Figure 1: Schematic of the passive DMFC. 1) Tapered channel; 2) Hydrophobic membrane; 3) Bubble fence; 4) Transparent anode flow field; 5) GDL and MEA; 6) Tubing; 7) Flow sensor; 8) Camera; 9) Reservoir; 10) Cathode mount; 11) Cathode flow field; 12) Non woven material for capillary water transport.

Simultaneously, the bubble movement sucks liquid from the reservoir into the narrow channel end. Thus, a net circulation of methanol is induced that supplies the DMFC with fuel as described in detail in [1].

The channels feature an active are of $A = 2.6$ cm^2. Since the flowfield consists of non-conductive COC, a current collector is added (Fig. 2 (4)) that was manufactured by laser cutting of stainless steal and electroplated with a 5μm gold layer. The current collector is separated from the Membrane Electrode Assembley (MEA) (catalyst coated ®Nafion N115; Anode: 3mg cm^{-2} PtRu; Cathode: 1.3 mg cm^{-2} Pt manufactured by balticFuelCells) by a gas diffusion layer GDL (SGL, ®SIGRACET GDL31-BA) that was made hydrophilic by plasma activation in order to increase the bubble mobility in the flow field channels. The anode is sealed by a vacuum casted PDMS layer.

At the cathode side, the GDL (SGL, ®SIGRACET S10CC) is hydrophobic containing 10% weight of PTFE in order to keep the GDL free from liquid water which would hinder oxygen supply. The cathode flow field consists of porous carbon (SGL, PGP) (Fig. 2 (9)) that was also made hydrophilic by plasma activation. Channels in cross direction to the methanol channels on the anode are milled into the carbon. The cross section of these channels is h_1 x w_1 =1mm x 1mm. In order to enhance diffusion of oxygen, on the back side of the porous carbon channels are milled with a cross section of h_2 x w_2 =2mm x 1.5mm in cross direction to the channels on the MEA facing side. A droplet emerging from or forming on the hydrophobic GDL moves into the bulk material of the porous carbon as soon as it touches the channel walls. The water is further transported out of the DMFC by non woven material where it finally evaporates. Thus, the channels are never blocked and diffusive oxygen supply is guaranteed in all orientations of the operating DMFC.

RESULTS

The passive fuel delivery on the anode was measured with a flow sensor and is evaluated by the pump efficiency p_{eff}, defined as the ratio of the measured volumetric liquid flow rate to the bubbly CO_2 gas flow rate. During passive operation, it is assumed that the liquid is saturated with CO_2, thus all CO_2 is generated in the gaseous phase and the volumetric gas flow rate density Φ_{gas} per unit area of the MEA can be calculated by the current density i, the molar weight M_{CO2} and the density ρ_{CO2} as follows:

$$\Phi_{gas} = \frac{M_{CO2}i}{\rho_{CO2}6F} \qquad (1)$$

where F is the Faraday constant.

Fig. 3 shows the liquid flow rate density induced by moving gas bubbles and the electrical current for two load cycles plotted against time. The load cycle was performed in a horizontal position with galvanostatic steps ($\Delta t = 10$s; $i_{step} = 7.7$ mA cm^{-2}). Comparison with pressure signals showed that each peak of the flow rate density is actually caused by single bubbles that move through the channel and pump methanol. The floating average with a time span of 10s shows peaks and minima at the same time as the electrical load. At the first peak, $p_{eff} = 22\%$ was determined and for the second peak $p_{eff} = 20\%$. Considering the average current density and the average flow rate density over the two complete cycles, the pump efficiency yields $p_{eff} = 19\%$. The liquid flow rate density closely follows the electrical load and the methanol supply is regulated by the passive anode flow field with a response time of less than 10 seconds whereas a response time of several minutes would be sufficient due to the storage of methanol in the channel and the GDL.

In Fig. 4, the average pump efficiency of the anode of the DMFC is plotted against the current density for different DMFC orientations. Each data point was acquired by averaging the flow rates over 10 minutes and repeating the measurements 3 times on different days.

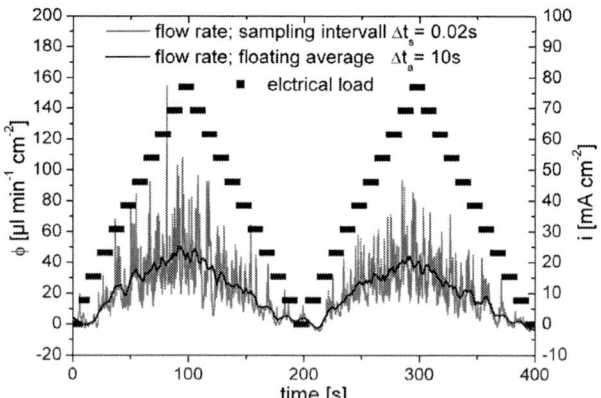

Figure 3: Current density and flow rate density plotted against time for two galvanostatic load cycles; T=23°C.

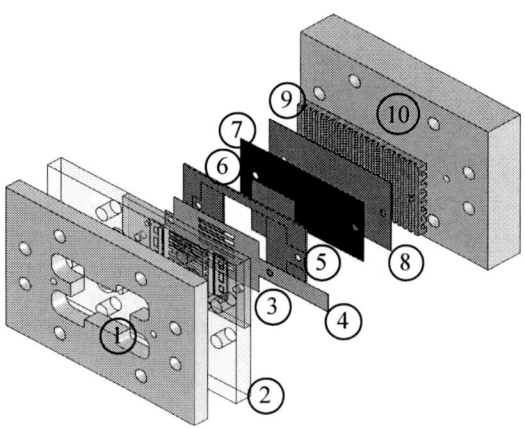

Figure 2: Exploded view on the passive DMFC.
1) Aluminum support plate 2)Transparent end plate;
3) Anode flow field; 4) Current collector; 5) Vacuum casted PDMS sealing; 6) Anode GDL 7) MEA; 8) Cathode GDL; 9) Cathode flow field; 10) Conductive end plate.

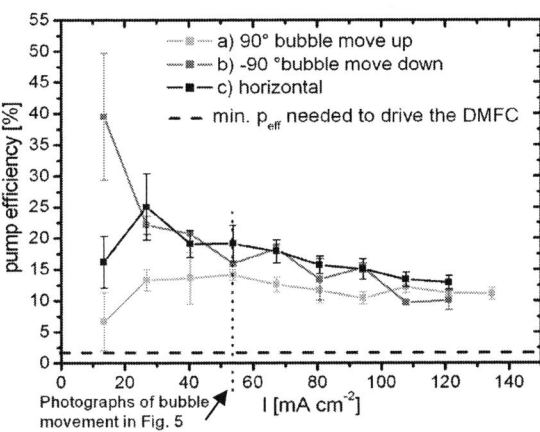

Figure 4: Pump efficiency p_{eff} plotted against the electrical load for different cell orientations; $T = 50°C$.

The smallest p_{eff} was measured at a current density of $i = 13.5$ mA cm^{-2} in a vertical position with the bubbles moving upwards driven by buoyancy forces (Fig. 4 light grey line). The reason for the small p_{eff} is the dominating non-blocking pumping mode in this configuration which can be understood as follows: As depicted in Fig. 5 (a), driven by buoyancy forces, comparably small bubbles that do not fill the cross section of the tapered channel move along the channel from the smaller to the larger cross section. As discussed in our previous work [5], in the non-blocking pumping mode, the major part of the liquid displaced by moving bubbles circulates directly around the bubbles and only a minor part contributes to the pump mechanism. As the bubble generation frequency increases with increasing current densities more liquid is dragged in pump direction by the numerous bubbles moving in the non blocking mode and p_{eff} first increases with an increasing electrical load. As the load increases further p_{eff} remains roughly constant.

The largest p_{eff} occurs at low current densities with the capillary movement of bubbles against buoyancy forces (Fig. 4 dark grey line and Fig. 5 (b)). In this orientation, driven by buoyancy forces, small bubbles move in the opposite direction of the overall pump direction. In the narrower part of the channel these bubbles are caught either by the upper and lower channel wall or by a bubble that has been formed previously. Bubbles grow from the narrow part of the channel towards the increasing cross section. When these bubbles exceed a certain size, the capillary forces of the bubbles that are deformed by the tapered channel walls overcome the buoyancy forces and the bubbles move in the blocking pumping mode towards the larger cross section of the channel. Since these large gas bubbles first form at the inlet and then move through the complete channel, a comparably large amount of liquid is transported by these bubbles. With increasing current densities, blocking bubbles also form in the middle of the channel, thus, less liquid is transported by these bubbles and p_{eff} decreases with an increasing electrical load.

In the horizontal orientation (Fig. 4 black line and Fig. 5 (c)), small bubbles remain fixed due to contact line pinning [1] and large bubbles move in pump direction in the blocking mode. The size of the liquid segments enclosed by moving bubbles in the blocking mode decreases with increasing current densities, thus, a slight decline of p_{eff} can be observed with an increasing electrical load.

In all orientations, the measured pump efficiency was between $p_{eff} = 7\%$ and $p_{eff} = 40\%$. The minimum pump efficiency for a 2 molar methanol solution as considered in this work is $p_{eff,min} = 2\%$, hence, 3.5 up to 20 times more fuel than critically needed to operate the DMFC was supplied to the anode which proves the mechanism to be suitable for fuel supply in portable DMFCs.

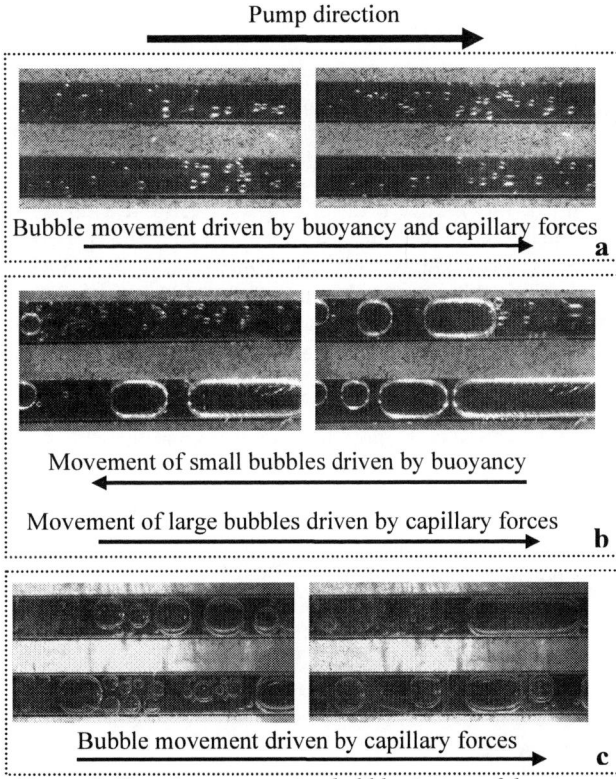

Figure 5: Top views on moving bubbles in two of the six anode channels; Electrical load: $I = 53.8$ mA cm^{-2}. The difference in time between the left and the right pictures is $\Delta t = 5.7s$. a) Vertical orientation with the bubbles moving upwards; b) Vertical orientation with the bubbles moving downwards against buoyancy forces; c) Horizontal orientation.

In Fig. 6 the polarization curve of the passive DMFC operating at ambient temperature and pressure is depicted. For each plotted current density, measurements were performed every 10s for galvanostatic intervals of one hour. The complete measurement lasted ten hours. The maximum power density was reached at $i = 35$ mA cm^{-2} with a power output of $p = 5.5$ mW cm^{-2}. Comparably low fluctuations (less than 12 mV) of the voltage were measured for current densities between $i = 15$ mA cm^{-2} and $i = 35$ mA cm^{-2}. This operation range shows a stable performance where a sufficient amount of fuel is transported to both, the anode and the cathode of the DMFC. The long term performance of the passive DMFC for a current density of $i = 17.3$ mA cm^{-2} is depicted in Fig. 7.

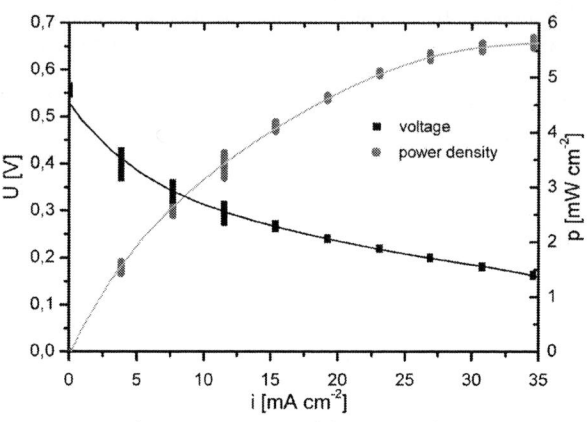

Figure 6: Polarization curve of the passively operating DMFC with galvanostatic intervals of one hour; T=23°C.

First, as a control experiment, measurements were performed for the vertical orientation using a cathode flowfield without capillary water removal. The power density declined rapidly from $p = 4.3$ mW cm^{-2} to less than 2 mW cm^{-2} within the first 5 hours of operation (Fig. 7 light grey line). The reason for the comparably bad performance is that water accumulates in the channels (see Fig. 8 (a)) and blocks the oxygen supply to the active area of the cathode. In contrast, when using the new cathode flowfield water is transported through the porous carbon and the non woven material out of the DMFC where the water finally evaporates. This feature allows for a stable long-term performance, thus, in the same vertical orientation, a continuous operation of more than 40 hours was observed with a power decline of less than 12 %.

Finally, the DMFC was tested in the most challenging position with the anode facing downwards. In this position the buoyancy of the CO_2 gas pushes the bubbles against the active area of the anode and on the cathode gravity does the same to the water. The power output declines from initially $p = 4.5$ mW cm^{-2} to $p = 3.1$ mW cm^{-2}. We assume that this decline is due to the formation of a water film between the catalyst layer and GDL. When taking apart the DMFC after this measurement, the catalyst on the cathode showed comparably large wet spots. Passive management of fluids between the GDL and MEA has not been addressed in this study but might improve the performance of the DMFC.

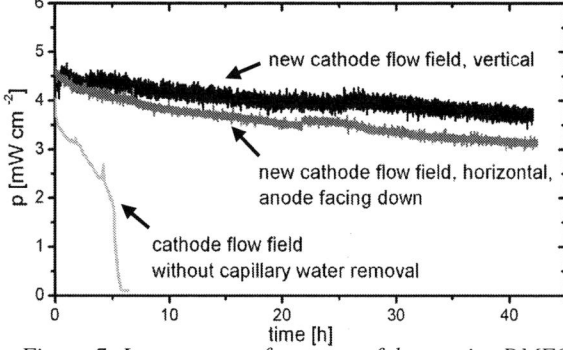

Figure 7: Long term performance of the passive DMFC for different orientations and different cathode flow fields: i =17.3 mA cm^{-2}:T = 23°C.

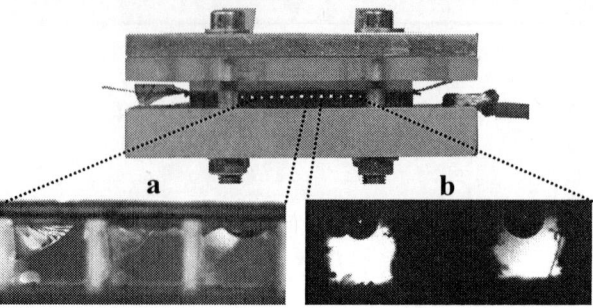

Figure 8: Front view photograph of the passive DMFC with enlarged cathode channels. a) Water accumulates in a conventional cathode flow field; b) The porous carbon flow field removes water as soon as droplets touch the side walls of the channels. Small droplets remain.

CONCLUSIONS

The presented layout concept enables the design of a passive, compact, and portable DMFC that operates in all orientations. On the anode between 3.5 and 20 times more methanol than critically needed to supply the DMFC is transported by the supply mechanism driven by capillary movement of CO_2 bubbles. The flow rates induced by the moving bubbles closely follow the electrical load, thus, the mechanism can be considered as self regulating. On the cathode, oxygen is supplied by diffusion and abundant water is transported out of the DMFC along a continuous capillary gradient. With these features, the DMFC exhibited a stable operation for more than 40 hours in all orientations. This is a major step towards the usage of passive DMFCs in portable systems. To further improve the performance of the DMFC, the fluid management should be extended to the area between the GDL and the MEA.

ACKNOWLEDGEMENTS

This work was supported by the German Ministry of Science and Education within Project 03SF0311B.

REFERENCES

[1]   N. Paust, et. al., "Fully Passive Degassing and Fuel Supply in Direct Methanol Fuel Cells," in Proc. of the 21st IEEE MEMS, Tucson, USA 2008, pp. 34-37

[2]   B. Bae et. al., "Performance evaluation of passive DMFC single cells," in *J. Power Sources*, vol. 158, no. 2, pp. 1256-1261, Aug.2006

[3]   T. Shimizu et. al., "Design and fabrication of pumpless small direct methanol fuel cells for portable applications," in *J. Power Sources*, vol. 137, no. 2, pp. 277-283, Oct.2004

[4]   Y. H. Chan, et. al., "A small mono-polar direct methanol fuel cell stack with passive operation," in *J. Power Sources*, vol. 176, pp. 183-190, Dec.2007

[5]   R. Chen, et. al., "Effect of cell orientation on the performance of a passive direct methanol fuel cell," in *J. Power Sources*, vol. 157, pp. 351-357, Oct.2005

[6]   R. Toomey, et. al., "Swelling Behavior of Thin, Surface-Attached Polymer Networks," in *Macromolecules*, vol. 37, pp. 882-887, 2004

A ROTARY MICROACTUATOR SUPPORTED ON ENCAPSULATED MICROBALL BEARINGS USING AN ELECTRO-PNEUMATIC THRUST BALANCE

M. McCarthy[1], C. M. Waits[2], M. I. Beyaz[1], and R. Ghodssi[1]
[1]University of Maryland, College Park, MD, USA
[2]U.S. Army Research Laboratory, Adelphi, MD, USA

ABSTRACT

The development of a rotary microactuator supported on encapsulated microball bearings and driven by electro-pneumatic actuation is reported. The encapsulated bearing provides full support to an encased rotor, while an electro-pneumatic thrust balance is used to minimize normal load, and therefore bearing friction. Experimental results show excellent agreement with predictions, demonstrating the ability to operate at optimal normal loads up to 2000rpm. This is the first demonstration of a ball bearing supported electrostatic microactuator with a fully encased rotor, capable of being integrated within practical microsystems. The fully-supported rotor allows for direct mechanical attachment or reliable interaction with external media and is a necessity for useful implementation.

INTRODUCTION

Rotary microactuators capable of continuous motion require a robust bearing mechanism able to provide low-friction support over a range of operating speeds and torques. Our group has pioneered the development of microfabricated ball bearings and their use as support mechanisms for rotary MEMS devices [1-7]. Ghalichechian *et al.* demonstrated the first rotary microactuator supported on MEMS-fabricated ball bearings using variable capacitance actuation of a silicon disk supported by microballs housed in a circular trench [3]. While, the microballs allow for continuous rotary motion and stable operation, a key disadvantage of this device was the unrestrained nature of the rotor. The rotating element simply rested on the microballs and is held in place with electrostatic forces. Useful implementation of a microball bearing supported actuator will require direct mechanical attachment and/or interaction with the rotor. Accordingly, encapsulated microball bearings have been developed using a notably more complicated fabrication process and demonstrated in micro-turbomachinery [4-7]. In this design, the rotor is fully supported by microballs around its entire periphery.

Using microturbine actuation, the encapsulated microball bearings have been actuated at speeds of up to 50,000 rpm and tribologically characterized through spin-down testing [7]. Previous publications on the development [4,5] and characterization [6,7] of encapsulated microball bearings have shown the importance of minimizing the thrust load supported by the microballs. Empirical modeling of bearing friction has shown a strong relationship between bearing friction and normal load [7]. In addition, smooth repeatable rotation was observed for small loads while large wear and a deterioration in performance was seen at higher loads.

Accordingly, this work presents the development of a novel bottom-drive electrostatic microactuator supported on microball bearings using a pneumatic thrust balance to reduce normal load and increase speed and performance. The rotor is fully supported using an encapsulated microball bearing design, allowing this device to be implemented in rotary micropumps and directional sensor systems requiring rotational positioning.

MICROACTUATOR

Using the previously reported encapsulated microball bearing design [7] and variable-capacitance electrode design [3], a rotary microactuator has been fabricated and characterized in this work. Figure 1 shows optical photos of the fabricated components and device. An encased rotor (Ø=10mm) is fabricated with approximately ninety stainless steel microballs (Ø=285μm) encapsulated around its periphery. Silicon carbide is conformally sputtered into the bearing housing to act as a friction-reducing hard-coating. Tracking marks are etched into the topside (Fig.1a), while 104 salient poles are etched into the bottomside (Fig.1b). The stator consists of an array of 156 electrodes embedded in benzocyclobutene, a low-k spin-on dielectric (Fig.1c), and fabricated in a six-phase design. The rotor layer sits on top of the stator (Fig.1d) and is aligned using etched ball pits containing a single microball; a central through-hole for thrust balance flow is etched in the stator.

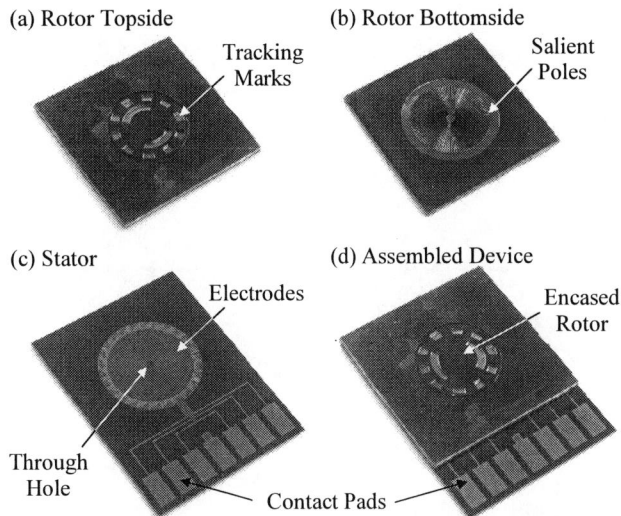

Figure 1: Photographs of the encased rotor (a) topside showing tracking marks and (b) bottomside showing salient poles, (c) the stator with embedded electrodes and through-hole, and (d) the assembled device.

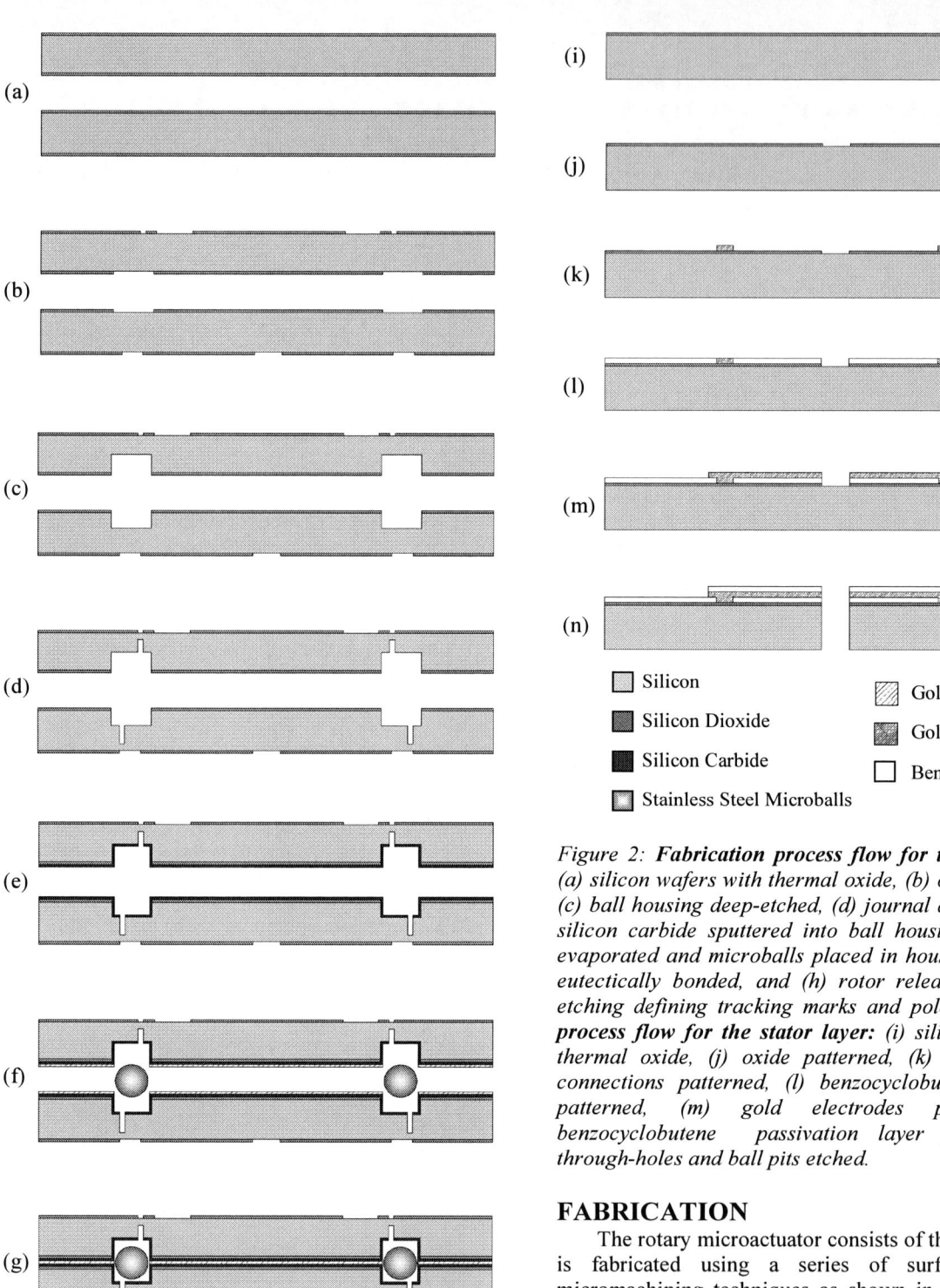

Silicon

Silicon Dioxide

Silicon Carbide

Stainless Steel Microballs

Gold/Tin

Gold

Benzocyclobutene

Figure 2: **Fabrication process flow for the rotor layer:** *(a) silicon wafers with thermal oxide, (b) oxide patterned, (c) ball housing deep-etched, (d) journal deep-etched, (e) silicon carbide sputtered into ball housing, (f) gold/tin evaporated and microballs placed in housing, (g) wafers eutectically bonded, and (h) rotor released with deep-etching defining tracking marks and poles.* **Fabrication process flow for the stator layer:** *(i) silicon wafer with thermal oxide, (j) oxide patterned, (k) gold electrical connections patterned, (l) benzocyclobutene spun and patterned, (m) gold electrodes patterned, (n) benzocyclobutene passivation layer patterned and through-holes and ball pits etched.*

FABRICATION

The rotary microactuator consists of three wafers, and is fabricated using a series of surface and bulk micromachining techniques as shown in Figure 2. The rotor layer fabrication (Fig. 2a-h) begins with two double-side polished four-inch silicon wafers with a total thickness variation of less than 2μm. Low thickness variation wafers are necessary to maintain a uniform gap over the electrostatic drive elements while the rotor spins. Thermal oxide is patterned to define hard masks for the tracking marks, salient poles, and bearing housings on each side of the two wafers (Fig. 2b). Circular trenches are etched in both wafers using a deep reactive ion etching (DRIE) process to define the bearing housing (Fig. 2c). The rotor's journal gaps are etched using spray-coated photoresist (Fig. 2d); these etches are used to release the rotor in the final fabrication step. One micron of sputtered silicon carbide is conformally coated on both wafers filling the bearing housing with the friction-reducing hard film (Fig. 2e). A gold/tin bonding layer is evaporated through a shadow mask on both wafers and approximately ninety stainless steel microballs are placed in each bearing housing (Fig. 2f). The wafers are then eutectically bonded at 315°C in a H_2N_2 environment under an applied load (Fig. 2g). Finally, the rotor is released by etching through the previously patterned hard masks to define the rotor periphery, tracking marks on the topside, and salient poles on the bottom side (Fig. 2h).

The stator layer fabrication (Fig. 2i-n) begins with a four-inch silicon wafer with thermal oxide (Fig. 2i). The oxide is patterned and dry etched to define the central though-hole for thrust balance flow as well as ball pits to be used for alignment (Fig. 2j). Gold lines for electrical connection to the various electrode phases are deposited and patterned using wet etching (Fig. 2k). Benzocyclobutene, a photo-definable low-k dielectric polymer, is spun to a thickness of 3μm and patterned to define electrical vias, isolating the electrical connections from the electrodes (Fig. 2l). Gold is deposited and patterned through a lift-off process to define the stator electrodes (Fig. 2m). Finally, a 3μm passivation layer of benzocyclobutene is patterned over the electrodes and the central through hole is etched along with ball pits for alignment to the rotor layer (Fig. 2n).

The rotor and stator layers are diced into individual dies and assembled using a microball alignment technique, as seen in Figure 3. Four etched ball pits are fabricated on the bottom side of the rotor layer and topside of the stator layer. A single microball is placed in each pit and the two layers are assembled manually. Based on the fabrication tolerances of photolithography, deep etching, and microball manufacturing, the estimated alignment accuracy is ± 2.5μm.

TESTING AND RESULTS

Figure 4 shows a schematic representation of the microactuator cross section and thrust balance operation. The rotor is lifted into contact with the microballs using pressurized nitrogen, and actuated via in-plane electrostatic forces in the form of six-phase sinusoidal signals. The downward electrostatic force is balanced by the upward pneumatic force which is controlled externally with a high-sensitivity flow control valve. A photograph of the packaged device being tested is shown in Figure 5. The device is sandwiched between two plastic manifolds using o-rings for mechanical support and thrust balance flow delivery. The six electrode phases and a ground connection are made using an array of electrical probes. An optical displacement sensor measures rotational speed by monitoring the tracking marks on the rotor topside.

The electrostatic and pneumatic forces counteract each other, while still providing the minimal upward force necessary for reliable operation. By minimizing the net force supported by the microballs, friction and wear can be greatly reduced. Figure 6 shows the maximum stable operating voltage as a function of thrust pressure, compared against theoretical predictions of snap-down voltage.

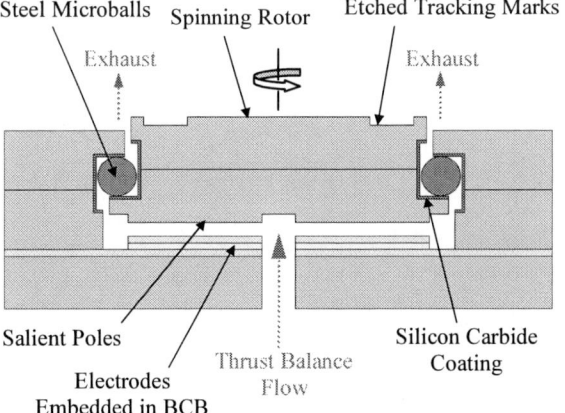

Figure 4: Schematic of the rotary microactuator showing electro-pneumatic thrust balance operation.

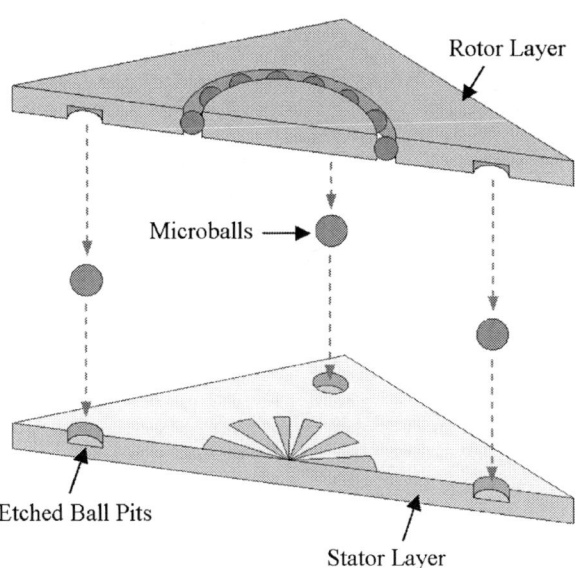

Figure 3: Schematic of the microball alignment technique between the rotor and stator layers.

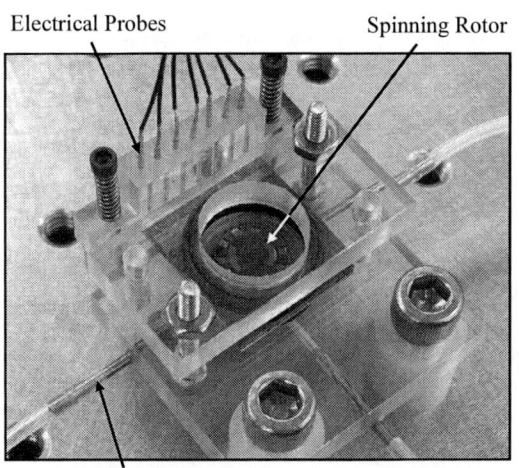

Figure 5: Photograph of a packaged device being tested.

978-1-4244-2977-6/09 $25.00 © 2009 IEEE   1093

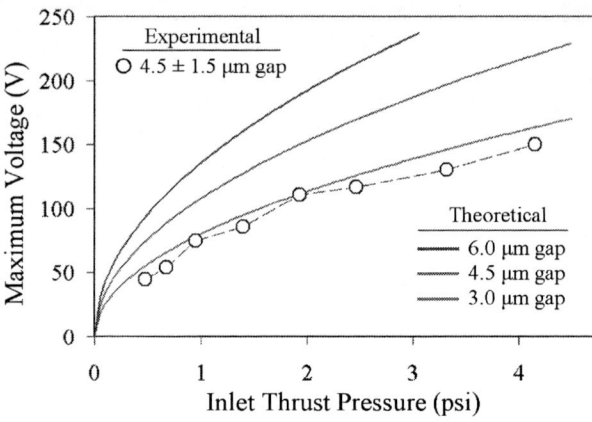

Figure 6: Theoretical model of snap-down voltage compared to experimental results for maximum operating voltage as a function of thrust pressure.

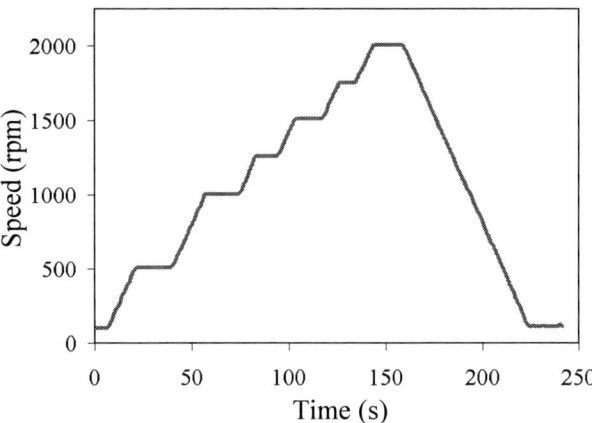

Figure 7: Microactuator performance, showing controlled accelerations and decelerations up to 2000rpm at 150V peak actuation voltage and 4psi of pneumatic thrust pressure.

The tested device has a pole-to-electrode gap of 4.5±1.5μm varying across the rotor and shows excellent agreement with predictions at the minimum gap. The theoretical predictions are based on numerical and analytic models of the electrostatic and pneumatic forces, respectively. Figure 7 shows the controlled acceleration and deceleration of the rotor up to 2000rpm at an optimal loading of 150V peak sinusoidal excitation and 4psi of thrust pressure. This represents a four-fold increase in maximum speed as compared to previous devices [3]. In addition, this device demonstrates a substantial increase in stability and reliability, showing indefinite operation at speeds as low as 5rpm and as high as 2000rpm, with no loss of synchronization. The actuator is capable of abruptly starting and stopping, as well as switching directions, at speeds as high as 50rpm.

The results shown here demonstrate the feasibility of microball bearing support mechanisms for use in next-generation microsystems. This is the first MEMS-fabricated ball bearing supported device capable of useful attachment or interaction with external elements.

This technology is being implemented in the development of directional sensor systems and electrically actuated rotary micropumps.

CONCLUSIONS

This work presents the fabrication and operation of a variable-capacitance rotary microactuator supported on encapsulated microball bearings using an electro-pneumatic thrust balance. The thrust balance, along with a friction-reducing silicon carbide hard-coating, has been used to minimize bearing friction and increase speed and performance as compared to previous microball bearing supported devices. The fully supported rotating element is capable of continuous motion and has demonstrated repeatable operation at speeds from 5-2000rpm.

ACKNOWLEDGEMENTS

This work was supported by the U.S. Army Research Laboratory under Grant No. CA#W911NF-05-2-0026.

REFERENCES

[1] A. Modafe, N. Ghalichechian, A. Frey, J. H. Lang, R. Ghodssi, "Microball-Bearing-Supported Electrostatic Micromachines with Polymer Dielectric Films for Electromechanical Power Conversion," *Journal of Micromechanics and Microengineering (JMM)*, vol. 16, pp. S182-S190, September 2006.

[2] N. Ghalichechian, A. Modafe, J. H. Lang, R. Ghodssi, "Dynamic Characterization of a Linear Electrostatic Micromotor Supported on Microball Bearings," *Sensors and Actuators: A. Physical*, vol. 136 (2), pp. 416-503, May 2007.

[3] N. Ghalichechian, A. Modafe, M. Beyaz, R. Ghodssi, "Design, Fabrication, and Characterization of a Rotary Micromotor Supported on Microball Bearings," *Journal of Microelectromechanical Systems (JMEMS)*, vol. 17, No. 3, pp. 632-642, June 2008.

[4] C. M. Waits, B. Geil, R. Ghodssi, "Encapsulated Ball Bearings for Rotary Micro Machines," *Journal of Micromechanics and Microengineering (JMM)*, vol. 17, pp. S224-S229, August 2007.

[5] C.M. Waits, N. Jankowski, B. Geil, R. Ghodssi, "MEMS Rotary Actuator using an Integrated Ball Bearing and Air Turbine", *The 14th International Conference on Solid-State Sensors, and Microsystems (Transducers '07)*, Lyon, France, June 10-14, 2007.

[6] M. McCarthy, C. M. Waits, R. Ghodssi, "Development of a Hybrid Gas / Ball Bearing Support Mechanism for Microturbomachinery," *7th International Workshop on Micro and Nanotechnology for Power Generation and Energy Conversion Application (Power MEMS '07)*, Freiburg, Germany, November 28-29, 2007.

[7] M. McCarthy, C.M. Waits, R. Ghodssi, "Dynamic Friction and Microturbine Performance Using a Planar-Contact Encapsulated Microball Bearing", *2008 Solid-State Sensor, Actuator, and Microsystems Workshop*, Hilton Head Island, SC, June 1-5, 2008.

PASSIVATED ELECTRODE ACTUATOR WITH STABLE RESONANCE AMPLITUDE

G. Bahl[1,2], R. G. Walmsley[2], B. E. DeMartini[3], K. L. Turner[3], P. G. Hartwell[2]

[1]Stanford University, Stanford, California, USA
[2]Hewlett Packard Laboratories, Palo Alto, California, USA
[3]University of California Santa Barbara, Santa Barbara, California USA

ABSTRACT

In this paper we demonstrate how parametric resonant excitation of a surface drive actuator with dielectric coated electrodes makes the response amplitude independent of the driving force. This behavior can be beneficially utilized to maintain constant excitation amplitude in spite of time-varying charge screening on dielectric passivated electrodes and the associated actuation force decay. Such operation makes the device suitable for applications where constant amplitude operation is critical, such as driving vibratory rate gyroscopes.

INTRODUCTION

Passivated electrodes enhance long-term reliability and can permit less rigorous packaging as the coating prevents electrode corrosion. However, dielectric coatings are typically avoided in electrostatic actuators as dielectric charging can develop shielding potentials affecting device operation. For instance, such shielding behavior is visible in RF capacitive switches where the pull-in voltage changes as the charge builds up [1], and in resonators where the frequency drifts as the potential changes [2].

Although the effects of time-variation of charge are often observable and verifiable, charging is not well modeled. Identical devices from the same wafer may have variations of dielectric defect and trap density, amount of mobile ions, and humidity inside the encapsulation. For commercial devices a better strategy is to design devices that are inherently less sensitive to charging. One could also use actuation schemes where devices are immune or less sensitive to the effects of time-variant charge. This paper discusses one such actuation technique to maintain constant excitation amplitude on a surface drive actuator in the presence of dielectric charging.

DEVICE DESCRIPTION

A surface electrode actuator with individually addressable electrodes has been described previously in [3]. On a similarly structured device, we address stator electrodes (Fig 1) with analog voltages in order to create a periodic sinusoidal potential on the stator surface. This potential profile can be electrically translated across the stator, relative to the rotor, by varying the individual electrode voltages.

The rotor electrodes are driven with an AC actuation voltage at frequency f with 0 and 180 deg phasing indicated by +D and –D signals shown in Fig 1. These voltages interact with the sinusoidal stator potential and produce in-plane force (explained through energy minimization). The direction of this force depends on the relative polarity of the stator and rotor voltages. Thus, as the polarity of this drive signal D alternates the in-plane force also alternates direction. When the driving

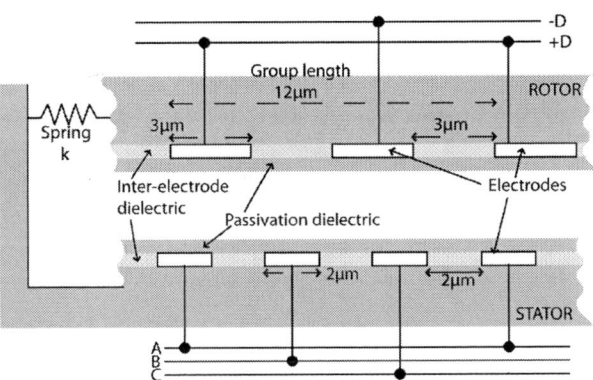

Figure 1: Device cross-section. Electrode groups A, B, C carry the DC bias voltages while groups +D and –D carry the AC actuation signals.

frequency f is near the in-plane resonance frequency (around 1000Hz), the motion of the device due to this actuation signal is large. Displacement can be measured using a physically separate but mechanically connected set of electrodes described previously in [4] that form a capacitive position sensor.

THEORETICAL CONSIDERATIONS
Alignment of potentials

One can analytically obtain the force-displacement curves shown in Fig 2 for any given relative alignment of the stator and rotor potentials using the techniques demonstrated in [5]. At specific alignments the magnitude of the in-plane force experienced by the rotor at the

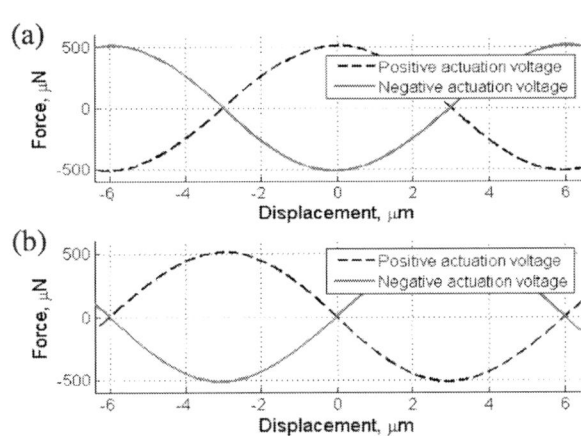

Figure 2: Force-displacement curves for two alignments of the stator-rotor potentials and for D>0 and D<0. (a) High force at zero displacement gives best harmonic resonance excitation (b) Instability at zero displacement for D<0 allows parametric resonance excitation.

978-1-4244-2977-6/09 $25.00 © 2009 IEEE

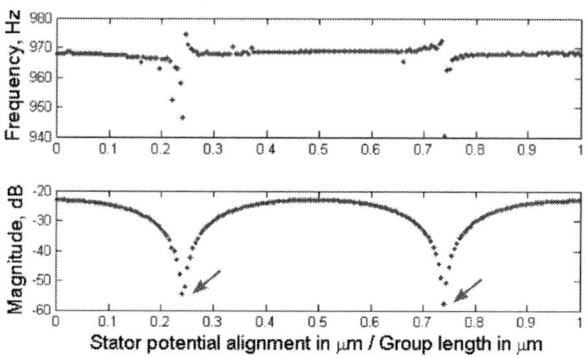

Figure 3: Mapping peak harmonic resonance amplitude as a function of stator-rotor potential alignment helps determine optimal parametric resonance points (the two minima indicated with arrows). This data is taken from an unpassivated device.

harmonic resonance frequency is maximized for zero displacement (Fig 2a). This is a favorable driving condition for harmonic resonant excitation. Varying this alignment of the rotor and stator potentials can produce large harmonic resonance amplitude or in the case shown in Fig 2b reduce the zero displacement force amplitude to zero.

This excitation amplitude can be mapped as a function of electrical alignment (Fig 3). Since physical electrode alignment between the rotor and stator can differ with fabrication tolerances, the identification of these maxima and minima determines the true physical alignment.

Prediction of parametric resonance

Although the force-displacement curves shown in Fig 2b suggest that the force exerted on the device is zero when it is at rest, it should be noted that there is position instability apparent in that graph. A slight displacement on the device when the negative actuation voltage force-displacement curve is valid results in the forces acting to further displace the mass away from the zero position.

The instability can be turned on and off by toggling the polarity of rotor voltage D. It is possible to exploit this instability in order to parametrically excite the device if the rotor voltages are alternated at $2f_o$ where f_o is the mechanical resonance frequency. Thus, electrical alignments where the harmonic resonance amplitude decays to zero are predicted to be points where the system can be operated in a parametric resonance due to this instability.

Model for parametric excitation

A preliminary model has been developed to help understand the oscillator dynamics. The force provided by mechanical springs is modeled as a linear and cubic nonlinear function of displacement, $F_s(x) = k_1 x + k_3 x^3$, where k_1 and k_3 are the mechanical stiffnesses. The net force produced by the surface electrode actuator to facilitate parametric excitation is

$$F_s(x,t) = AV_s \sin(2\pi x / g_L)(V_{DC} + V_r \sin(\omega t))$$

where A is the forcing amplitude, g_L is the group length

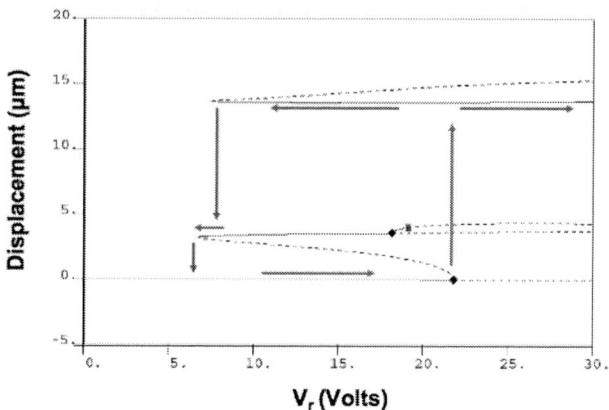

Figure 4: Numerically generated parametric resonance dependence on drive voltage at selected frequencies. (Note, red and blue arrows represent sweeping up and down in rotor voltage respectively, solid lines present the stable oscillatory state, dashed lines the unstable oscillatory state, and diamonds represent bifurcations).

(12 μm), V_s and V_r are the stator and rotor potentials respectively, and V_{DC} is a tuning potential. For this system the scaled equation of motion is,

$$x'' + \frac{1}{Q}x' + x + \frac{k_3}{k_1}x^3 + \frac{AV_s}{k_1}\sin\left(\frac{2\pi x}{g_L}\right)\left(V_{DC} + V_r \sin(\Omega \tau)\right) = 0$$

where Q is the quality factor, time is scaled by the purely elastic natural frequency $\left(\omega_0 = \sqrt{k_1 / m}\right)$ as $\tau = \omega_0 t$, and $\Omega = \omega / \omega_0$.

Using AUTO [8], numerical bifurcation analysis was used to gain a better understanding of the system. Figure 4 shows the parametric resonance amplitude dependence on the driving voltage V_r for A = 0.476 μN/V^2, V_s = 35 V, V_{DC} = 19 V, f = 2020 Hz, f_0 = 1010 Hz, Q = 22, k_1 = 34 μN/μm, and k_3 = 0.0023 μN/μm^3. The important feature of Fig 4 is that the upper stable resonance branch is relatively insensitive to the forcing amplitude as long as there is enough force to maintain parametric excitation. Similar behavior is demonstrated experimentally and discussed in the following section.

EXPERIMENTS AND RESULTS
Drive force effects on harmonic excitation

When driven at the mechanical resonance frequency the amplitude of linear harmonically excited resonance is proportional to the driving force amplitude (verified with an unpassivated device in Fig 5).

For dielectric coated electrodes, the maximum force exerted on the rotor deteriorates over time as charge builds up to screen the electrode. This reduction in force also lowers the excitation amplitude as measured by the position sensor (Fig 6a).

Drive force effects on parametric excitation

For a linear parametric resonance the system is driven at $2f$, mechanically responds at frequency f, and motion amplitude is independent of driving force [6] as long as the force is high enough to create the instability. However, as charge builds up in the dielectric coating, it

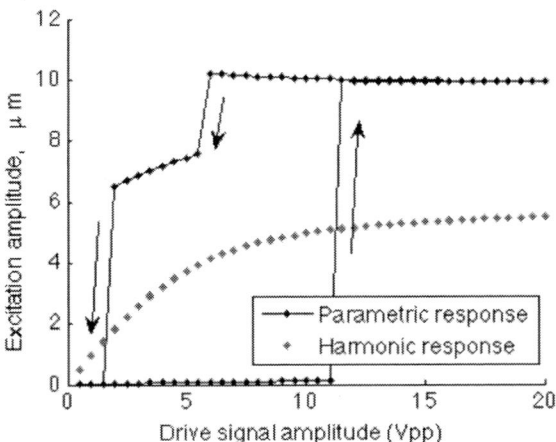

Figure 5: Harmonic and parametric resonance dependence on drive voltage at selected frequencies on an unpassivated device. It can be see that for harmonic resonance the relationship is fairly linear until electromechanical design limits the amplitude. Parametric resonance shows much insensitivity to force amplitude as long as there is enough force to maintain parametric excitation.

acts to screen the applied stator potential and diminishes force magnitude. If the applied force drops below a threshold (Fig 5), the instability will be lost and motion will cease (Fig 6b).

Our passivated devices show non-linear parametric resonance effects as seen in the model and are thus not entirely insensitive to time-varying force changes. They do, however, demonstrate reduced sensitivity to the deterioration of the forcing function over time initially as the driving force magnitude is sufficiently high to sustain the parametric excitation at full amplitude.

Bipolar biasing

To prevent excessive charge screening, often an alternating polarity bias is employed [7]. Such a scheme arrests and reverses the movement or build-up of charge by reversing the electric field applied on the device. However, even under alternating polarity bias, excitation at harmonic resonance shows that time-varying charge screening still occurs in our device. The effects of this can be seen in Fig 7a. This experiment is used to verify that charging is occurring in this device.

This device can also be excited parametrically. Even though oxide and charge behaviors are the same as that of the harmonic case, alternating bias polarity allows us to arrest and reverse the charge before the force decays to a point below the threshold required for parametric excitation. This works successfully as shown in Fig 7b by maintenance of constant response amplitude.

CONCLUSIONS

We have demonstrated surface actuated devices that can be driven harmonically or parametrically using different biasing configurations. Using harmonic resonant excitation it has been shown that these devices are sensitive to charging when a passivation dielectric is coated on the electrodes.

Parametric excitation is relatively insensitive to changes in actuation force and can be used to excite

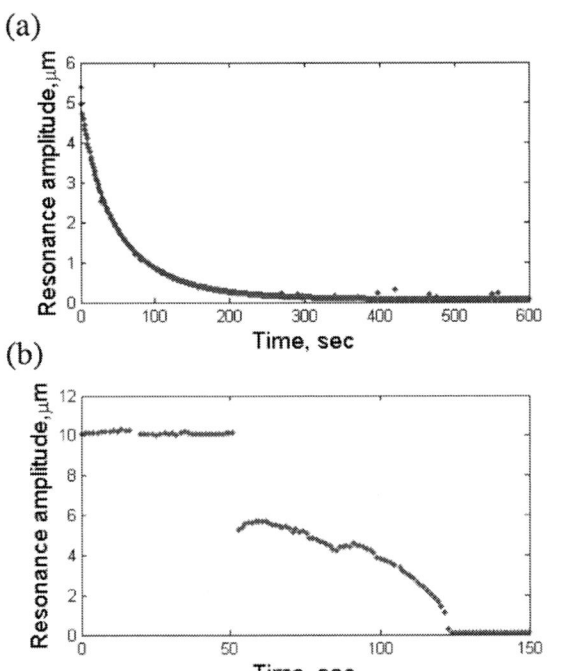

Figure 6: Difference in decay trends on a passivated device while the driving stimulus is kept constant (a) Linear harmonic resonance shows monotonic amplitude decay (b) parametric resonance shows sudden drop after sustained operation exhibiting the force threshold for parametric excitation.

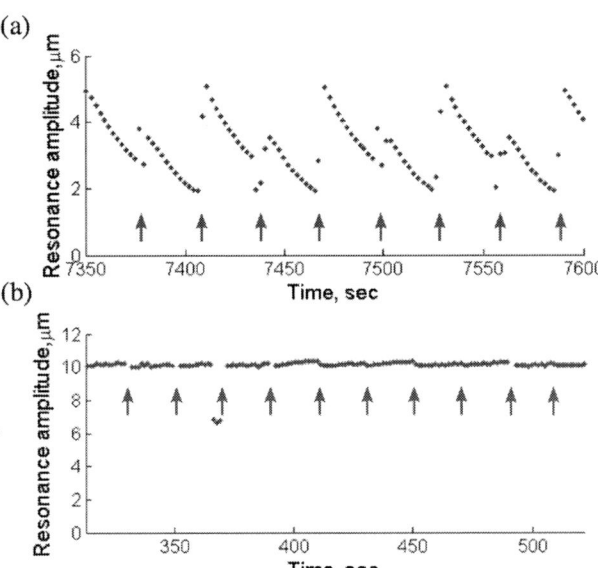

Figure 7: Operation of a passivated device (same device in Fig 6) under bipolar biasing for harmonic and parametric excitation. Arrows indicate instants where bias polarity was inverted (a) Linear harmonic resonance consistently decays in amplitude due to charging. (b) Parametric resonance with bipolar biasing allows for sustained amplitude operation even though the dielectric is charging.

978-1-4244-2977-6/09 $25.00 © 2009 IEEE   1097

dielectric coated devices that are susceptible to charging. The periodic reversal of the bias polarity maintains the force above the threshold required for parametric resonance, thus preserving oscillation amplitude and providing insulation from force amplitude decay resulting from dielectric charging. This parametric driving approach is extensible to systems where constant response amplitude is critical.

ACKNOWLEDGEMENTS

Authors would like to thank Lennie Kiyama and Dr. R. Stanley Williams for their support. Additional thanks go out to the Micro Structures and Sensors Laboratory at Stanford University, and to Dr. Matthew A. Hopcroft for making the Matlab based instrument control software KPIB publicly available.

REFERENCES

[1] W. M. van Spengen, et al, "A comprehensive model to predict the charging and reliability of capacitive RF MEMS switches," J. Micromech. Microeng., 14, pp. 514, 2004

[2] G. Bahl, et al, "Observations of fixed and mobile charge in composite MEMS resonators," Proceedings Hilton Head Workshop 2008

[3] S. Hoen, et al, "A high-performance dipole surface drive for large travel and force," IEEE Transducers 2003, Boston, June 2003

[4] P. G. Hartwell, et al, "Integrated position sensing for control of XY actuator," Proceedings IEEE Sensors 2004, pp. 1407

[5] S. Hoen, et al, "Electrostatic Surface Drives: theoretical considerations and fabrication," Transducers '97, Chicago, June 1997

[6] K. L. Turner, et al, "Five parametric resonances in a microelectromechanical system," Letters to Nature, Vol 396, November 1998

[7] Z. Peng, et al, "To demonstrate the effectiveness of the present scheme, the Dielectric Charging of RF MEMS Capacitive Switches under Bipolar Control-Voltage Waveforms," IEEE Microwave Symp., pp. 1817-1820, 2007

[8] E. Doedel, et al, AUTO 97: Continuation and Bifurcation Software for Ordinary Differential Equations. 1997

FAST POSITIONING AND IMPACT MINIMIZING OF MEMS DEVICES BY SUPPRESSION MOTION-INDUCED VIBRATION BY COMMAND SHAPING METHOD

K.-S. Chen and K.-S. Ou
Department of Mechanical Engineering, National Cheng-Kung University
Tainan, Taiwan, 70101, R.O.C.

ABSTRACT

This work presents a command-shaping based scheme for both fast positioning and reducing contact impact of MEMS devices by suppression motion-induced vibrations. The scheme was developed by applying energy conservation, force equilibrium, and elliptical integrals. Simulink simulations indicated that both of the impact force and settling time can be effectively reduced. Specimen fabricated using Su-8 and a test bed was designed to further demonstrate the performance of the proposed scheme and the test results indicated that the proposed approach can effectively enhance the dynamic performance of MEMS devices such as RF switches.

INTRODUCTION

The dynamic performance and longevity of electrostatically actuated MEMS devices such as RF contact switches have long been an important concern and have been individually performed by many researchers. For example, one can see [1-3] for detail. In general, a higher applied voltage can result in a fast response but the induced higher impact force between the switch and its corresponding substrate could significantly reduce the device reliability [2,3]. As a result, it is important to develop a scheme to enhance the device longevity (by minimizing impact force) without sacrificing or even with improvement on the dynamic performance. Traditionally, this goal could be accomplished by incorporating feedback control. However, due to the inherent nonlinearities and conditionally stability between the controlled electrical potential and the generated electrostatic force, as well as the complexity in sensors/controller/actuator integration, feedback control approach may not be an appropriate choice. Instead, for these devices operated in two destinations (e.g., on-off states), command shaping methods have been proved to be effective for large scale mechatronics systems. However, the traditional linear command-shaping methods may not be proper for MEMS devices due to three important dynamics characteristics of electrostatically actuated MEMS devices, i.e., nonlinear actuating force, non-negligible damping, and the pull-in instability. These characteristics effectively cease the possibility of directly copying those shaping schemes commonly used in mechatronics systems into MEMS. As a result, the major focus and contribution of our work is in developing novel nonlinear command-shaping schemes to overcome the above mentioned difficulties. The developed nonlinear command-shaping schemes would then be useful for enhancing the longevity and bandwidth of various electrostatically actuated systems such as RF MEMS switches and GLVs. In our previous works, a novel nonlinear shaping scheme has been proposed (schematically shown in Figure 1) and verified on a macroscale test rig [4,5]. In this work, we would like to further investigate the method by directly using microfabricated MEMS structures to demonstrate and to evaluate the performance of the proposed shaping scheme for MEMS applications. Due to the restriction of paper length, the discussion would focus on the MEMS switch related applications.

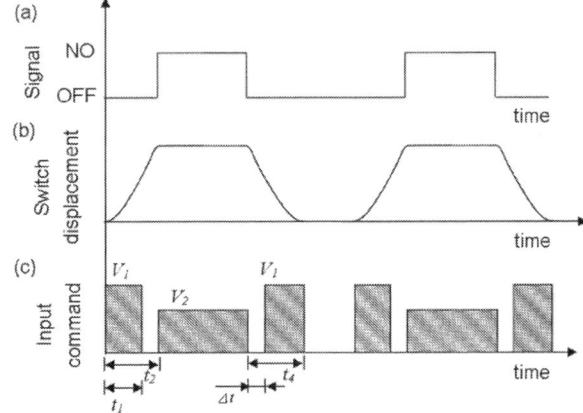

Figure 1: The proposed shaping command for RF MEMS switching operation.

MODELING

A MEMS switch can be modeled as a single degree of freedom vibration system. In most situations, the structure is lightly damped and this implies significant structural vibration would occur with a forcing input. For contact type switches, the contact may also result in bounces and high impact forces.

An energy approach is adapted to develop the proposed nonlinear shaper with an optimal electrical potential waveform to ensure a fast response with a very small impact. By using energy conservation, force equilibrium, and elliptical integral computation, the applied voltage levels and timing are precisely determined. Instead of using a continuous step command or a series of square waves, the proposed command uses a series of pulse trains with precisely calculated applied *voltages* and *time intervals*, schematically shown in Figure 1. The entire operation can be classified into two phases, i.e., the *Pull-In* phase and the *Release* phase, for ease of discussion. The *pull-in* phase mainly concerns the actuation of a contact switch from its original null position to the final contact position and is the most challenging phase. A proper design must

978-1-4244-2977-6/09 $25.00 © 2009 IEEE

achieve a rapid, low impact response. The release phase concerns a fast settling once the switch is released from its contact position back to the null position. Without careful design, the release operation could generate considerable residual vibration at the null position before settling down and this reduces the switching rate during a repeating operation. The developing of the shaping scheme can be found in our previous work [5,6] and the results are briefly addressed below for illustration. The proposed waveform shown in Figure 1 contains a few key parameters, i.e., V_1, V_2, t_1, and t_2. First, at $t=0$, a constant voltage V_1 (user specified, which must exceed the corresponding pull-in voltage) is applied to actuate the switch. The switch then accelerates and moves toward the substrate. This input voltage is then ceased when $t=t_1$ and the switch would then decelerate but continuously move toward the substrate. At the moment of $t=t_2$, the switch would contact with the substrate with a negligible speed. At this moment, an input $V=V_2$ immediately kicks in to counteract the elastic restoring force and to ensure the final contact. As a result, V_2 must be greater than the specified pull-back voltage V_{back}, where

$$V_{back} = \sqrt{\frac{kx_m G_r^2}{\varepsilon_0 A}} \qquad (1)$$

On the other hand, t_1 and t_2 represent two specific times to turn off and turn on the switches. Where

$$t_1 = \int_0^{x_1} \left(\frac{2}{m}\left(\frac{H_1 x}{G_0(G_0 - x)} - \frac{1}{2}kx^2\right)\right)^{-\frac{1}{2}} dx \qquad (2)$$

and

$$t_2 = t_1 + \int_{x_1}^{m} \left(\frac{2}{m}\left(\frac{H_1 x_1}{G_0(G_0 - x_1)} - \frac{1}{2}kx^2\right)\right)^{-\frac{1}{2}} dx \qquad (3)$$

where

$$H_1 = \frac{A\varepsilon_0}{2}V_1^2(t) \qquad (4)$$

For the release phase, by energy and force equilibrium argument, the required input waveform and necessary parameters are identical to these developed in the pull-in phase. The symmetrical nature of commands for pull-in and release phases indicate that the subsequent experimental validation can be performed in releasing phase, which is considered easier in designing experiments for validation.

SIMULATION

A SIMULINK model was constructed to validate the proposed methodology. Note that the impact between the switch and the substrate can be realized in the model by incorporating extra spring and damper elements, in together with proper switching blocks to switch the nominal stiffness/damping to much higher values during contact. Figure 2a shows the simulation results for the

ideal situation with dynamic parameters listed in the figure. The behavior is typical and is similar to the response previously reported [3]. It is obvious that significant bounces are induced for a pure step voltage command and a long time is required before settling down. This implies the switch has significant residual kinetic energy and that a strong impact between the switch and the substrate would be inevitable. On the other hand, the response of the shaped input shows a completely different scenario. That is, although the rising time is slightly longer, its settling time can be significantly reduced by eliminating the possible bouncing. In addition, as shown in Figure 2b, significant impact forces are generated during contact for a step input. These may be eliminated by replacing the step command by a shaped input. This result indicates that for an undamped situation, the proposed shaping scheme works perfectly. However, it is also important to address here that the proposed scheme assumes the damping is negligible. For systems with small damping ratio ς (e.g., ς <1%), the assumption is reasonable and the performance of the scheme is still acceptable (see Figure 2a,b). Many MEMS structures operated in low pressure or vacuum environments belong to this category. On the other hand, for system with higher damping (such as switches operated in ambient air environment), the proposed scheme must be modified. A modified scheme based on assumed mode approach has been developed and validated by the authors. Due to the relatively complicated mathematical development, it would not be presented in this article. Please refer to [6] for the detail.

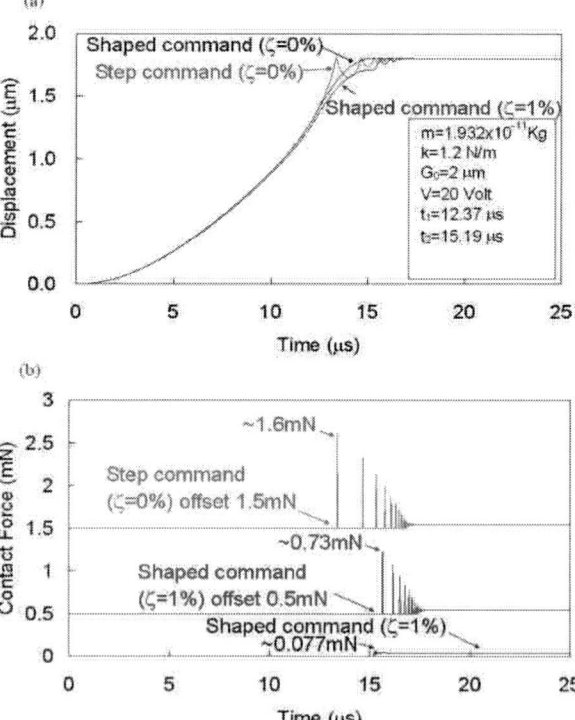

Figure 2: Simulink simulation of time responses (a): Displacement and (b) Impact force.

In addition to the performance demonstration of reduced bouncing and impact force during the operation, the robustness of the shaping scheme should also be investigated. Since the physical system dynamics cannot be completely modeled and the dtnamic parameter uncertainties in measurements always exist, the proposed shaper must be capable to resist a certain level of parameter uncertainties. Otherwise, it would be unrealistic for real applications. The robustness studies were performed systematically. For example, the relationship between the possible variations of t_l with respect to the system stiffness uncertainty is shown in Figure 3. The timing error in t_l would result in excess residual vibrations. Nevertheless, this result still indicates that for a reasonable parameter variation, the performance does not degrade rapidly and this implies that the robustness is acceptable. For the influences of other important system parameters such as air gap uncertainties and damping, please refer to [6] for detail.

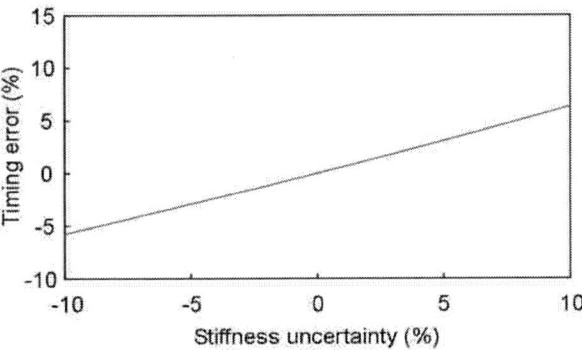

Figure 3: The influence of stiffness uncertainty to the timing error.

FABRICATION AND SYSTEM SETUP

In order to validate the proposed concept, SU-8 based test specimens were design and fabricated. The fabrication process and specimen is outlined in Figure 4. A testing system, schematically shown in Figure 5, consists of a MEMS specimen, a laser displacement sensor (MTI MicotrakTM II, Head model: LTC-025-04, with a linearity of 2μm and resolution of 0.1μm), a self-designed high voltage amplifier with a capability of ±400 V, a precision positioning system, and a data acquisition system, were used to perform the experiment and to analyze the response. It is important to emphasis here that a major design trade off exists between the experimental apparatus and the specimen design. For example, in order to provide a sufficient signal-to-noise ratio (SNR) for the position sensor, the air gap had to be enlarged. However, this action resulted in a requirement of a higher pull-in voltage and therefore a more compliant specimen design would be required. A more compliant design could potentially suffer more fabrication challenges such as warpage, handling damage, and curling during release. As a result, the specimen design was strongly limited by the capability of current experimental apparatus and it took a few trial and error

runs before the final specimen design was determined.

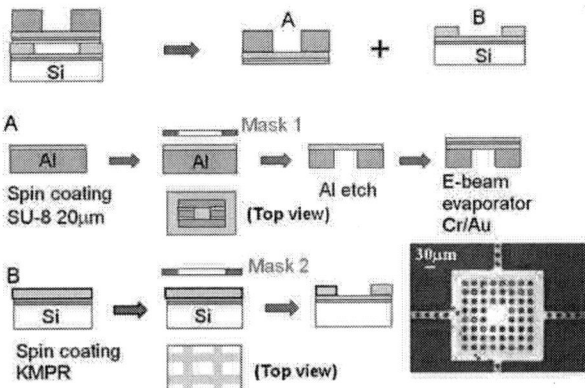

Figure 4: An image of the testing specimen and the corresponding fabrication procedure.

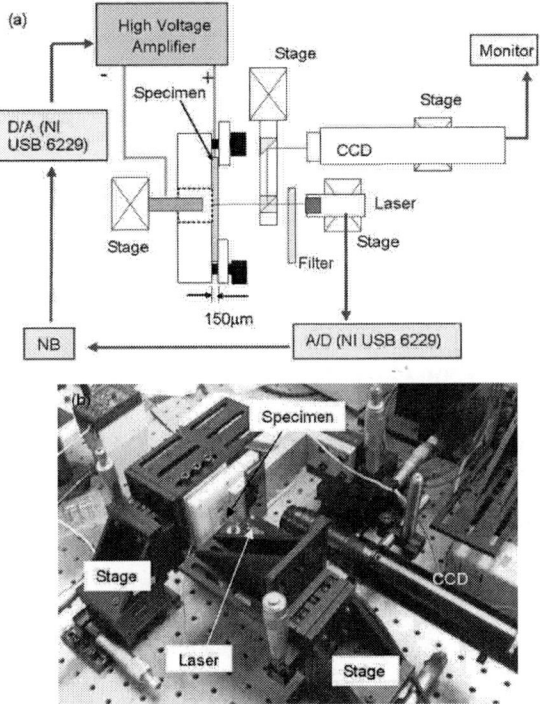

Figure 5: The experimental setup (a) schematic plot and (b) the test rig.

EXPERIMENTAL RESULTS

The test were performed in both the stable regime for testing the positioning performance and the pull in regime for contact performance. As shown in Figure 6, the step command induces significant motion-induced residual vibration and this increases the settling time. The traditional Zero-Vibration (ZV) shaping can only partially reduce the residual vibration. On the other hand, the response subjected to proposed command shaping scheme shows a much superior result. With much less vibration level, the system can potentially be operated in a much faster rate and this certainly improves the performance of the associate applications such as MEMS precision positioning systems.

978-1-4244-2977-6/09 $25.00 © 2009 IEEE   1101

On the other hand, it is more challenging for demonstrating the performance of the proposed shaper in the pull-in regimes. Due to lacking of micro force sensors embedded in the specimen, the impact force cannot be measured currently. However, it is still possible to evaluate the possible improvement by monitoring the contact vibration history. Since the model developed in this work indicates that the modified command is symmetrical and due to the difficulty in performing experiments in pull-in phase, the test could be conducted in release phase for the purpose of feasibility demonstration. The specimen was initially pulled-in with a sufficient applied voltage and was released by reducing the voltage to zero immediately (i.e., step input) or by using an input after shaping. As shown in Figure 7, the step command results in significant residual vibration and it would take approximately 150 *ms* for settling down. On the other hand, the residual vibration of the structure subjected to the shaped input is much less and the settling time is much smaller and this indicates a much fast switching rate could be achieved. Such a performance improvement would be very useful in MEMS devices operated in an on-off manner (such as RF switches) for enhancing their dynamic performance without extra costs or efforts in developing feedback control design.

Finally, both the sensitivity studies for examining the robustness of the proposed scheme and the modified command shaping scheme to incorporate the influence of non-negligible damping commonly existed in many MEMS devices were also performed and developed. Detail results can be found in [6]. We believe that the proposed scheme should be very important for achieving a successful design of electrostatically actuated MEMS actuators with both improved dynamic performance and sufficient longevity.

SUMMARY AND CONCLUSION

Fast positioning and device reliability are always critical concerns of electrosttaically actuated MEMS devices such as RF switches. An optimal command has been proposed in this work to shape the system responses. The scheme was developed by applying energy conservation, force equilibrium, and elliptical integrals. Simulink simulations indicate that the both the impact force and settling time can be effectively reduced. A Su-8-based specimen and a test bed were designed to further demonstrate the performance of the proposed scheme. With this approach, the dynamic performance of many electrostatic-based MEMS actuators could be effectively improved.

ACKNOWLEDGEMENT

This work is supported by National Science Council (NSC94-2212-E006-046) and ITRI (97C073). Many suggestions from Professors T.-S. Yang and S.-Y. Lee of NCKU are also greatly acknowledged.

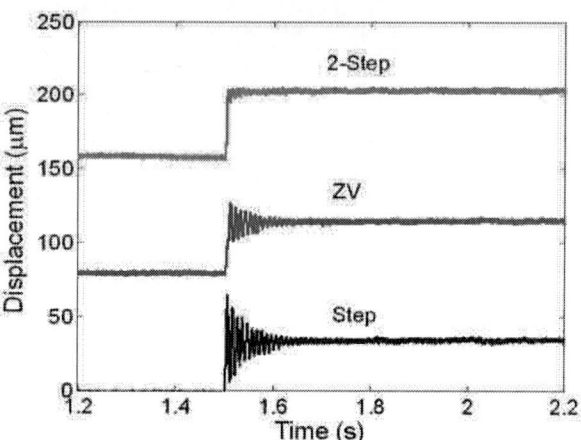

Figure 6: Experiment results for positioning in stable regime.

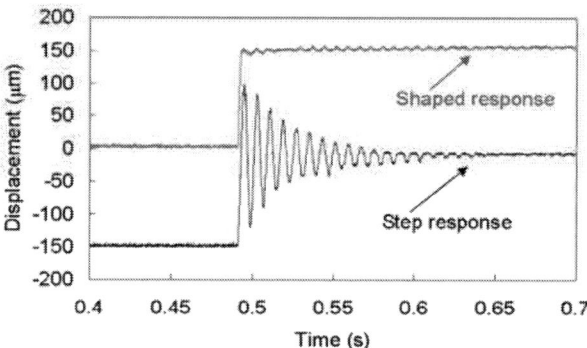

Figure 7: The experimental result in pull-in regime.

REFERENCES

[1] L. M. Castaner and S. D. Senturia, "Speed-energy Optimization of Electrostatic Actuators Based on Pull-in," *J. Microeletromech. Syst.*, vol. 8, pp. 290-298, 1999

[2] G. M. Rebeiz, *RF MEMS, Theory, Design, and Technology*, Wiley, 2003.

[3] B. McCarthy et al, "A Dynamic Model, Including Contact bounce, of an Electrostatically Actuated Microswitch," *J. Microelectromech. Syst.*, vol. 11, pp. 276-283, 2002.

[4] K-S Chen, et al, "Residual Vibration Suppression for Duffing Nonlinear Systems with Electromagnetical Actuation Using Nonlinear Command Shaping techniques," *ASME J. Vibration and Acoustics*, vol.128, pp.778-789, 2006.

[5] K-S Chen and K-S Ou, "Command-shaping Techniques for Electrostatic MEMS Actuation: Analysis and Simulation," *J. Microelectromech. Syst.*, vol. 16, pp.537 -549, 2007.

[6] K-S Ou, *Optimal Motion Planning for Suppressing Vibrations in both Stable and Pull-in Regimes for Electrostatic Actuators Using Nonlinear Command-shaping techniques*, Ph.D. Thesis, National Cheng-Kung University, 2008.

978-1-4244-2977-6/09 $25.00 © 2009 IEEE   1102

A MICRO DIRECT METHANOL FUEL CELL INTEGRATED WITH A TEMPERATURE CONTROL SYSTEM FOR EXTREME ENVIRONMENTS

Qian Zhang[1], Xiaohong Wang[1]*, Yiming Zhu[1], Yan'an Zhou[1], Xinping Qiu[2], and Litian Liu[1]

[1]Institute of Microelectronics, Tsinghua University, Beijing, China

[2]Department of Chemistry, Tsinghua University, Beijing, China

*Phone: +86-10-62789147ext317, Fax: +86-10-62771130, Email: wxh-ime@mail.tsinghua.edu.cn

ABSTRACT

This paper reports a micro direct methanol fuel cell (μDMFC) integrated with a heater and a temperature sensor to realize temperature control. A thermal model for the μDMFC is set up based on heat transfer and emission mechanisms. The μDMFC with a temperature control system is fabricated by using MEMS technologies, assembled with polydimethylsiloxane (PDMS) materials and Polymethylmethacrylate (PMMA) holders, and characterized in two methods, one with different currents applied and another with different methanol velocities. This work would make it possible for a μDMFC to enhance the performance by adjusting to an optimal temperature and employ in extreme environments, such as server winter, polar region, outer space, desert and deep sea area.

INTRODUCTION

With the increasing functionalities of portable devices, the demand for more reliable, longer lasting and clean power sources is rising rapidly. A micro direct methanol fuel cell (μDMFC) is a very promising power source for military equipments and portable devices due to its high energy density, easy recharging, low pollution, etc [1]. Most of the previous researches have focused on structure design and fabrication process of a μDMFC to improve the performance. Great progress has been made in fuel cell stack, reducing methanol crossover, management of heat and water, removal of CO_2, fuel transport, etc. [2].

However, a μDMFC has not optimal performance at lower temperature, even does not work below 0°C, although it has a significant advantage of which it can work at room temperature. Characteristics of polymer electrolyte membrane fuel cell (PEMFC) were examined with thermal cycles during which the temperature of the environment was cycled from 80°C to -10°C by EunAe Cho et al. [3]. The cell performance was degraded by deteriorating the structure of the membrane electrode assemblies (MEAs) due to the phase transformation and volume changes of water. This is particularly problematic in application for portable devices because such devices may be used in cold environments, such as server winter, polar region, outer space, desert and deep sea area. In addition, the proton exchange membrane (PEM) is only stable for a narrow temperature range because its conductivity is affected by humidification. Too high temperature may lead to PEM dry-out or even the sulfonate group decomposition [4]. Therefore it is necessary for a μDMFC to maintain an appropriate operating temperature, especially in extreme environments.

Temperature is an important factor affecting the performance of a μDMFC. A number of papers on thermal management have been published, however, most of them focus on modeling and simulation [5-6]. A temperature control system which consists of a heater and a temperature sensor, is needed for a μDMFC to stably operate, especially when it is placed in extreme climates or unconditioned environments. A thin film temperature sensor was developed for application in an operating PEMFC by Suhao He et al. [7]. They used a structured gold film as a temperature sensor on the Nafion membrane to monitor the operating temperature. However, none of the reported works implemented the temperature control system.

This paper, for the first time, presents a μDMFC integrated with a heater and a temperature sensor to realize temperature control. The μDMFC is fabricated by using micromachining technologies and assembled with polydimethylsiloxane (PDMS) materials and Polymethylmethacrylate (PMMA) holders. The μDMFC with a temperature control system fabricated would show great potential application to enhance the performance by adjusting to an optimal temperature and employ in extreme environments.

DESIGN

Figure 1 shows the schematic of the μDMFC we designed which consists of the MEA sandwiched between two silicon plates, integrated with the temperature control system. During the operation, an aqueous methanol solution is fed to anode and is oxidized to carbon dioxide, protons, and electrons at anodic catalyst layer. The protons migrate directly through the PEM to the cathode. The electrons flow through an external circuit to the cathode where they combine with oxygen and protons to produce water, forming an electrical current as a result.

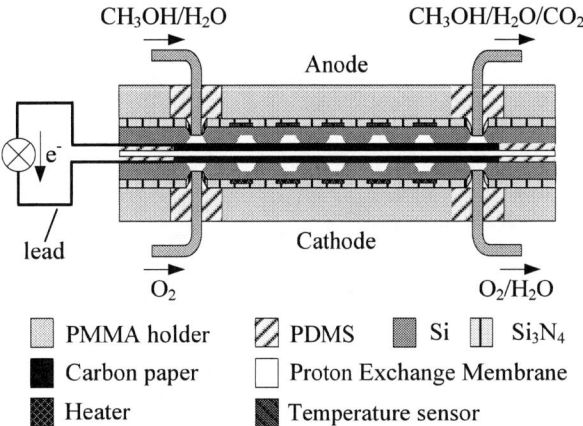

Figure 1: Schematic of the μDMFC integrated with a heater and a temperature sensor.

The operation of a μDMFC is stopped when liquid freezes at temperature below 0°C, although it has a significant advantage of which it can work at room temperature. And the cell performance is improved with an increase of the temperature in a certain range because the conductivity of the PEM and the reaction kinetics at both anode and cathode are enhanced. Therefore, a heater and a temperature sensor, integrated into the µDMFC, are used to control the operating temperature. The flow field structure and the temperature control device are designed on the two sides of the silicon respectively due to the higher thermal conductivity of the silicon and the thinness of the dielectric membrane, as shown in Figure 2. The heat is released from the heater applied by a current and emitted in several ways for a µDMFC, including heat taken away by the methanol solution, thermal convection, radiation, etc. The steady temperature is measured by the temperature sensor when the balance of heat generation and emission is achieved. Platinum (Pt) resistance, with a linear temperature response in a wide temperature range, remains stable and does not undergo any significant physical or chemical changes. Therefore, Pt is a very suitable material as the temperature sensor.

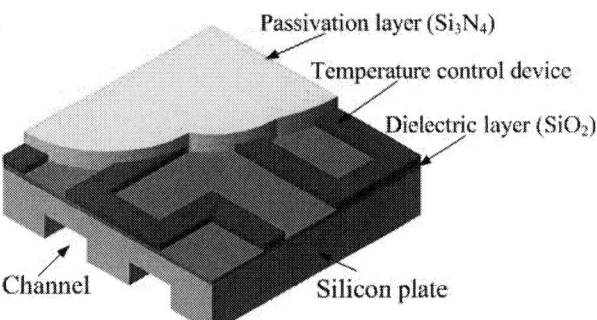

Figure 2: Cross-section schematic of the silicon plate with the temperature control device.

A thermal model of the µDMFC is set up to estimate thermal behavior. The heat generation by electrochemical reactions occurring in the catalyst layers and by the heater is considered. The thermal loss that makes a great negative impact on maintaining an optimal temperature is studied to increase temperature effectively. The heat taken away by the methanol solution, thermal convection, radiation, and heat emission through the lead are considered as main thermal loss ways, as shown in Figure 3.

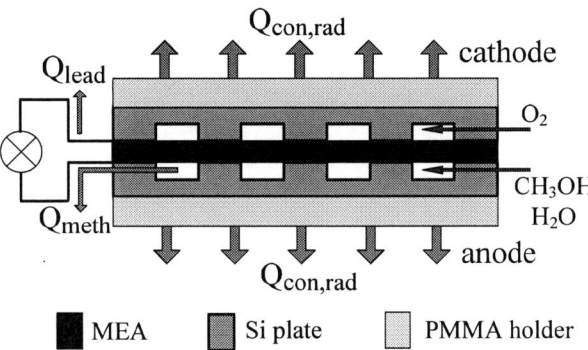

■ MEA **■** Si plate **□** PMMA holder

Figure 3: Schematic of the thermal loss in the µDMFC. The heat is emitted in three main ways: (1) Q_{meth} represents the heat taken away by the methanol solution, (2) $Q_{con,rad}$ represents the heat loss by thermal convection and radiation, and (3) Q_{lead} represents the heat emission through the lead.

The heat loss amount by the methanol solution (Q_{meth}) is calculated by

$$Q_{meth} = \rho v C_P (T - T_0) \qquad (1)$$

where T is the operating temperature, T_0 is the environment temperature, and v, C_P, and ρ are flow velocity, heat capacity, and density of the methanol solution, respectively. The heat loss amount by thermal convection and radiation ($Q_{con,rad}$) is given by

$$Q_{con,rad} = hA(T - T_0) + \varepsilon \sigma A(T^4 - T_0^4) \qquad (2)$$

where h is convective heat transfer coefficient, A is the surface area, ε is emissivity, and δ is Stefan-Boltzmann constant. The heat loss amount through the lead (Q_{lead}) is expressed by

$$Q_{lead} = \sqrt{hPkA_S} (T - T_0) \qquad (3)$$

where P, A_S, and k are cross-section perimeter, cross-section area, and thermal conductivity coefficient of the lead, respectively.

It is necessary to invoke some assumptions to make the complex system simpler. The fuel cell is assumed to operate under steady-state conditions. Temperature gradients inside the cell, especially between the anode and cathode, are negligible. Joule heat caused by the current flow through each component is ignored. Heat exchange with the oxygen in the cathode channel is negligible. The heat capacity of the methanol solution is much larger than that of the oxygen and thus the contribution of gas cooling to the overall heat balance is small.

FABRICATION AND ASSEMBLY

The fabrication process of the anode and cathode plates is presented in Figure 4 and described in detail as follows. (a) Thermal oxide and LPCVD Si₃N₄ are deposited on both sides of a 400µm thick 3 in. double-polished <1 0 0> silicon wafer as the mask layer. (b) Double-side lithography technology is introduced to form complicated flow structure of the anode and cathode plates. (c) KOH-timed etching is employed to anisotropically etch the silicon wafer until the flow patterns and the feeding hole are formed. (d) A thermal oxide of 0.1µm is grown as the dielectric membrane. (e) The patterns of the heater and the temperature sensor are formed by lithography technology. (f) Pt with an adhesive layer of Ti (0.2µm thick) is sputtered to pattern using a lift-off process for making the heater and the temperature sensor on the back. (g) The temperature control devices are covered with a 0.3µm PECVD Si₃N₄ as the passivation layer. (h) Ti/Pt is sputtered to form the current collecting layer on the front.

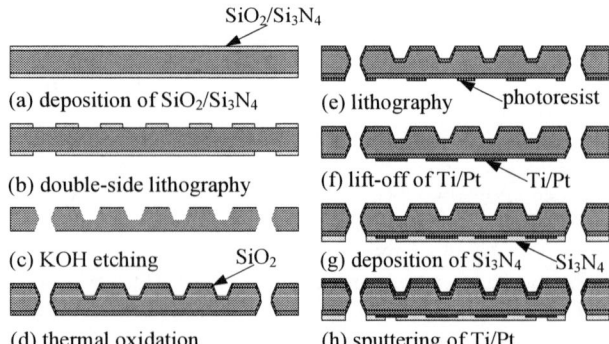

Figure 4: Microfabrication process of the µDMFC anode and cathode plates.

The anode and cathode plates are fabricated simultaneously on the same wafer with identical process. Double-side lithography halves KOH etching time as mentioned in our previous work [8]. Figure 5 shows the two sides of the cathode plate, the fabricated cathode flow structure on the front and the heater on the back. The heater is located on the active area of 6.8mm×6.8mm to enhance the heating efficiency.

The MEA consists of two (anode: 4.0mg/cm^2 Pt-Ru, cathode: 1.5mg/cm^2 Pt) hot pressing wet-proof carbon papers which are used as the diffusion layer, and a layer of Nafion117 in between.

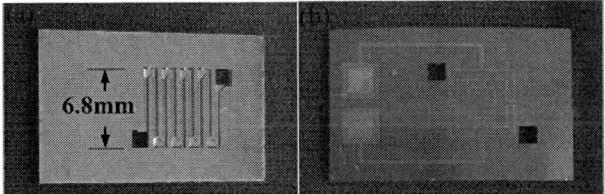

Figure 5: Photos of the cathode plate: (a) the flow pattern on the front of the cathode plate, and (b) the heater on the back of the cathode plate.

PDMS and PMMA holders are used to keep the fuel cell from leakage and protect fragile silicon plates. The flexibility and chemical inertness of PDMS exactly satisfy the assembly requirements of the μDMFC. As previous work has mentioned, the prefabricated PDMS pieces have functions in the packaging: the sealing gaskets around both sides of MEA, the buffering layers between plates and corresponding holders, and the fixture of feeding tubes [8]. The PMMA holders are employed to provide a uniform pressure over the whole plate area and improve the contact between components as the aluminum holders used in our previous work. Moreover, the PMMA, as the material of holders, is advantageous to maintaining the cell operating temperature and reducing the thermal loss, especially the thermal convection and radiation, due to its lower thermal conductivity coefficient than the aluminum. No leakage happens during the daylong testing, indicating that this assembly method is effective and reliable. This assembly method has a special feature of reassembly that brings convenience to replacing components. The μDMFC we developed is shown in Figure 6, and the detailed parameters are listed in Table 1.

RESULTS AND DISCUSSION

Since the temperature sensor is not a standard component, a calibration is determined before the cell performance test. The plate with the temperature sensor is placed in an adjustable oven. The calibration is completed during the temperature changing process.

Calibration results are shown in Figure 7. A plot of the temperature sensor resistance versus temperature during calibration is illustrated by using the method of fitting a straight line to data points. The temperature sensor resistance in Figure 7 is calculated as

$$R = R_0[1 + \alpha(T - T_0)] \tag{4}$$

where R_0 is the sensor resistance at room temperature (T_0=25°C), α is the temperature coefficient of the resistance. The curve shows that the temperature sensor has a linear response of α=0.0025°C^{-1} in the range of 25°C

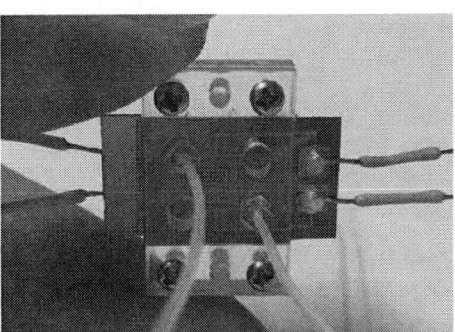

Figure 6: The assembled μDMFC integrated with temperature control devices.

Table 1: Parameters of the μDMFC integrated with a heater and a temperature sensor.

Total size, including holders	23mm×15mm×6.1mm
Plate area	20mm× 12.8mm
Active area	6.8mm×6.8mm
Channel/rib width	0.4mm
Channel depth	0.2mm
Heater resistance(25°C)	239Ω
Sensor resistance(25°C)	169Ω

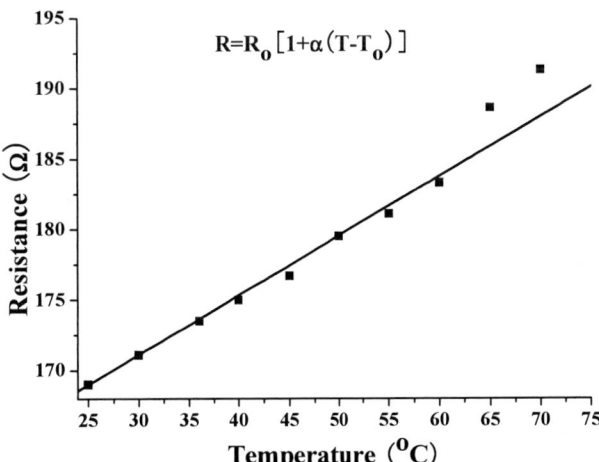

Figure 7: Plot of temperature sensor resistance versus temperature during calibration.

to 70°C temperature. When the μDMFC operates at a certain temperature, the sensor resistance (R) is measured and the temperature is calculated as

$$T = T_0 + \frac{R - R_0}{\alpha R_0} \tag{5}$$

The performance of the μDMFC integrated with the temperature control system is measured on an electrochemical interface, Solartron SI1287. The methanol solution is driven by a horizontal pump (2PB00C Serial, Beijing Satellite Manufacturing Factory). A constant current is applied and the output voltage is monitored for a period (50s in this paper) until the final steady-state value is recorded.

Figure 8 depicts the performance of the μDMFC when the currents applied to the heater are 53.9, 41.5, and 0mA, respectively. The corresponding operating temperatures measured by the sensor are 58.2°C, 41.6°C and room temperature (25°C). Obviously, the cell temperature is raised with the increasing current applied to the heater for more heat released from the heater. The experimental results show that the prototype has the maximum power density of 5.55mW/cm^2 at 58.2°C, which is more twice higher than that of the μDMFC at the value of 2.37mW/cm^2 at room temperature, using 2M methanol solution under 0.1mL/min methanol flow rate.

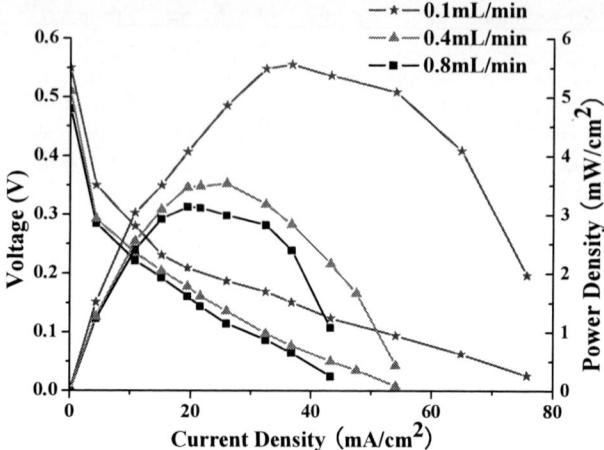

Figure 9: Performance of the μDMFC under different methanol velocities with 2M methanol solution, when a current of 53.9mA is applied to the heater.

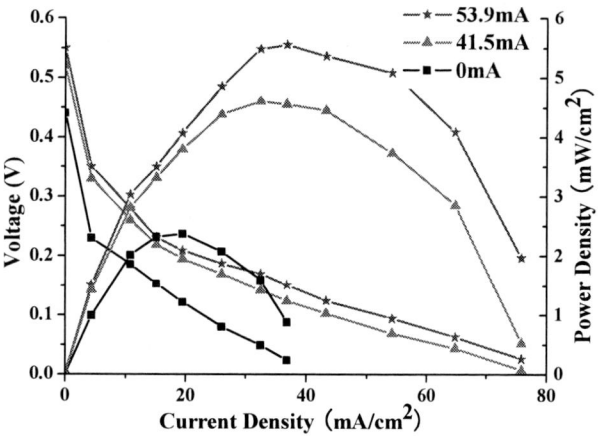

Figure 8: Performance of the μDMFC with 2M methanol solution under 0.1mL/min methanol flow rate, when different currents are applied to the heater.

Moreover, the thermal loss is studied by experiments to verify the great negative impact on maintaining an optimal temperature. As mentioned above, the heat taken away by the methanol solution, accounting for a major proportion in the total thermal loss, is proportional to the flow velocity. Figure 9 illustrates the performance of the μDMFC under the methanol solution flow velocities of 0.1, 0.4, 0.8mL/min with 2M methanol solution, when a current of 53.9mA is applied to the heater. The corresponding operating temperatures measured by the sensor are 58.2°C, 34.5°C and 29.7°C. The maximum power density is 5.55mW/cm^2 under 0.1mL/min, 3.52mW/cm^2 under 0.4mL/min, and 3.13mW/cm^2 under 0.8mL/min. The performance is improved with a decrease of the flow velocity as low velocity contributes to less thermal loss and high operating temperature in the μDMFC.

CONCLUSIONS

A silicon-based μDMFC integrated with a heater and a temperature sensor is presented. A thermal model of the μDMFC is developed based on heat transfer and emission mechanisms in order to estimate the thermal behavior. The μDMFC with a temperature control system is fabricated and assembled by using MEMS technologies and PDMS materials. The PMMA holders are used to not only protect the fragile silicon plates, but also reduce the thermal loss due to its lower thermal conductivity coefficient. The μDMFC is characterized in two methods, one with different currents applied and another with different methanol velocities. This work would make it possible for a μDMFC to enhance the performance by adjusting to an optimal temperature and employ in extreme environments.

ACKNOWLEDGEMENTS

The authors would like to thank the National Natural Science Foundation of China (No. 90607014), 973 program (No. 2009CB320304), and Tsinghua National Laboratory for Information Science and Technology.

REFERENCES

[1] Eiichi Sakaue, "Micromachining/Nanotechnology in direct methanol fuel cell", MEMS'05, pp. 600-605.

[2] S.K. Kamarudin, W.R.W. Daud, S.L. Ho, U.A. Hasran, "Overview on the challenges and developments of micro-direct methanol fuel cells (DMFC)", J. Power Sources 163 (2007) 743-754.

[3] EunAe Cho, Jae-Joon Ko, Heung Yong Ha, Seong-Ahn Hong, Kwan-Young Lee, Tae-Won Lim, In-Hwan Oh, "Characteristics of the PEMFC repetitively brought to temperatures below 0°C", Journal of The Electrochemical Society, 150 (12) A1667-A1670 (2003).

[4] Vladimir Neburchilov, Jonathan Martin, Haijiang Wang, Jiujun Zhang, "A review of polymer electrolyte membranes for direct methanol fuel cells", J. Power Sources 169 (2007) 221-238.

[5] H. Dohle, J. Mergel, D. Stolten, Heat and power management of a direct-methanol-fuel-cell (DMFC) system, J. Power Sources 111 (2002) 268–282.

[6] V.B. Oliveira, D.S. Falcão, C.M. Rangel, A.M.F.R. Pinto, "Heat and mass transfer effects in a direct methanol fuel cell: A 1D model", International Journal of Hydrogen Energy, 33 (2008) 3818-3828.

[7] Suhao He, Matthew M. Mench, Srinivas Tadigadapa, "Thin film temperature sensor for real-time measurement of electrolyte temperature in a polymer electrolyte fuel cell", Sensor and Actuators A 125 (2006) 170-177.

[8] Yingqi Jiang, Xiaohong Wang, Lingyan Zhong, et al., "Design, fabrication and testing of a silicon-based air-breathing micro direct methanol fuel cell", J. Micromech. Microeng. 16, 2006, pp.233-239.

LEVER-BASED CMOS-MEMS PROBES FOR RECONFIGURABLE RF IC'S

J. Liu, M. Noman, J.A. Bain, T.E. Schlesinger and G.K. Fedder

Carnegie Mellon University, Pittsburgh, Pennsylvania, USA

ABSTRACT

We report on new lever-based CMOS-MEMS electrothermal probes for Memory-Intensive Self-Configuring Integrated Circuits (MISCICs). The MISCIC vision is to use MEMS conductive probes to reconfigure ICs, mainly RF ICs such as inductors, by mechanically addressing and passing current through resistance change (RC) vias embedded within the chip circuitry. The lever-based actuation causes the probe tips to move away from the probe substrate and toward the RC via substrate, which is in contrast to the previous design where actuation caused the tips to move away from the RC via substrate. Two independently actuated probes are located 3 μm apart to provide two simultaneous contacts that form a reconfiguration current return path. Thermal management such as dummy heater beams and thermal isolation structures are employed to improve the temperature uniformity and drive efficiency. A statistic semi-sphere model is employed to model the contact.

1. INTRODUCTION

Since the introduction of the STM by Binnig and Rohrer in 1981, probes have been widely used in surface and friction measurements, SPMs [1], AFM nanolithography [2], IC chip testing [3] and high areal density data storage [4-5]. Prior work on MEMS probe-based data storage has explored thermal [4], magnetic [5] and phase change storage modalities [6]. In these applications, probes are either stationary or actuated by piezoelectric, electrostatic, electrothermal, or magnetic mechanisms. Compared with other mechanisms, electrothermal actuation generally achieves larger stroke and higher stiffness but with the possible penalties of slower tuning and greater power dissipation.

In the MISCIC reconfigurable-via concept, MEMS probes are employed to mechanically address 1 μm²-order pads on a CMOS chip, as illustrated in Fig. 1. Current delivered from the probes through the pads heats RC vias to switch their resistance state. This technology is being first applied in reconfigurable RF inductors.

Probe cards in IC chip testing and RF switches explore electrical contact with relatively large electrodes (> 100 μm²) and mN-order forces, with 0.1 Ω-order contact resistance [7]. These mN-order forces are necessary to break through the oxide layer, mainly aluminum oxide or copper oxide in CMOS chips, formed on the chip surface when exposed to air. Contacts with much smaller area are viable, as evidenced by conductive AFMs and RF MEMS switches, which contact on surface asperities and achieve contact resistance typically on the order of 1 Ω. However, we know of no comprehensive studies of MEMS actuated conductive contacts with 1 μm²-order electrodes.

Though the spring constant of the active probes may be as high as 1's of N/m, it is impractical for sub-mm sized active probes to achieve mm-order strokes and mN-order forces to break through the oxide layer. Therefore, non-oxidizing metal such as nickel (Ni) and platinum (Pt) must be deposited on the tips to enable contact forces to decrease to values between 1 to 50 μN. These values of force can be implemented with compact MEMS springs. With contact forces of the order of 1 μN, contact resistance of 10 - 100 Ω was reported with sub-1 μm² contact in our previous Ni-plated push-down probes [8]. However, these probes were soft (0.25 N/m) making comprehensive studies of the contact difficult. Here, we present new probes that are much stiffer (5 N/m) to achieve 1 - 10 contact resistance with 1 μm²-order tips.

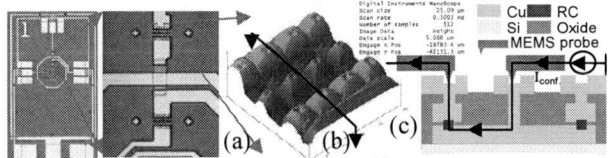

Fig. 1: The MISCIC concept on RF inductors.
(a) RF inductors made with embedded RC vias (a sub-array of vias creates small resistance required for RF).
(b) AFM image showing surface topography of via pads.
(c) Via cross section and schematic of the current flow.

2. PROBE DESIGN

The probe mechanisms are made in a CMOS-MEMS process using the metal-dielectric layers of the CMOS as the composite micromechanical material [9]. The push-up actuation is achieved by the lever structure shown in Fig. 2. The thermally actuated flexures are made from 8 parallel 50 μm-long, 2 μm-wide, 1.5 μm-thick "metal-1" (M1) beams comprising the lowest CMOS metal and oxide layers. The central "metal-3" (M3) flexure folds back toward the M1 anchor and employs 8 parallel 80 μm-long, 2 μm-wide, 5 μm-thick beams made from the metal 1-2-3 stack to help meet the stiff 5 N/m probe spring constant requirement.

Due to the residual stress gradient in the CMOS-MEMS beams, upon release, the M1 flexures self-assemble upward slightly, and the M3 flexure and tip are levered downward. Upon actuation, due to the thermal bimorph effect, the M1 flexures move downward and the M3 flexure and tip move upward. The thinner M1 beams are chosen for the thermal bimorph actuator to reduce the bending moment of inertia, which helps provide high drive efficiency. Polysilicon resistors forming actuation heaters are placed at each end of the M1 flexures. Dummy heater beams are added to eliminate the thermal buckling caused by temperature non-uniformity across the parallel beams [8]. They have the same features as the M1 flexure beams in every respect except that they are mechanically disconnected from the M1 flexure beams, which ensures they do not affect the M1 flexure actuation. Slotted metal thermal isolation structures are placed on both ends of the M1 flexure to increase the temperature uniformity and drive efficiency. This design prevents heat from directly passing through aluminum interconnect to the anchors, instead having to pass through the CMOS tungsten vias. The simulation in Fig. 3 demonstrates the temperature uniformity improvement across the actuation beams. The relative temperature difference at the cantilevered end of the flexure at 165°C was decreased from 4.8% to 0.6% through the addition of the dummy heater beams and thermal isolation.

3. PROBE PROCESS

The particular CMOS MEMS process employed here begins with a custom-designed CMOS chip, followed by a directional oxide etch to define dielectric/metal microstructures [9]. A Si deep-reactive-ion etch (DRIE) is used to set the etch pit depth followed by a timed silicon isotropic etch for structural release.

Probe tips are designed by stacking CMOS vias and metal layers as shown in Fig. 4. The aluminum top layer must be coated with a non-oxidizing metal to be effective as a dc contact. Our initial metallization studies used a derivative of the electroless Ni plating flow previously used for chemiresistors [1] [10]. Only the top M4 layer in the CMOS process is plated with Ni by first performing an oxide etch to expose the M4 layer, then plating, then resuming the MEMS oxide etch.

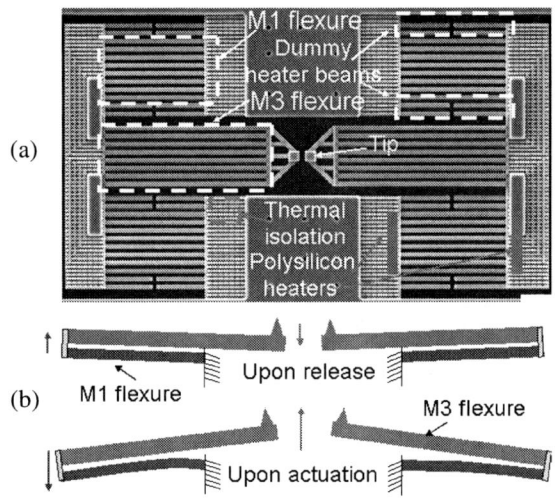

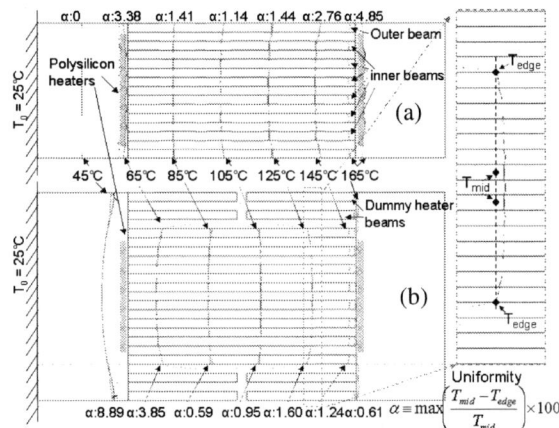

Fig. 2: Lever-based MEMS dual probes.
(a) SEM and (b) Cross-section of lever mechanism.

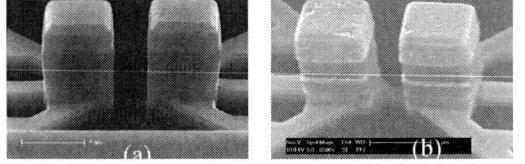

Fig. 3: Simulated uniformity in metal-1 flexure without (a) and with (b) dummy heater beams with 10 mW drive

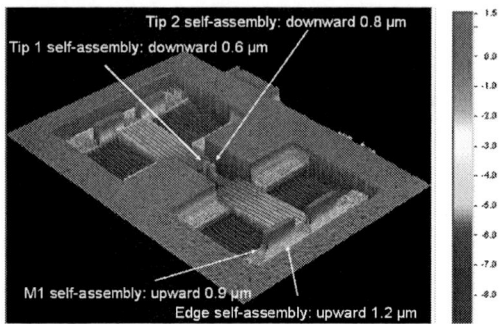

Fig. 4: Probe tips without (a) and with (b) Ni plating

The M1 flexures self-assemble upward by 0.9 μm and the tips are levered downward by 0.6 - 0.8 μm (Fig. 5). The probe achieves an upward stroke of 10.1 μm at a 14 V, 63.7 mW drive with a linear 0.16 μm/mW sensitivity (Fig. 6a). Thermal cross-talk is 6% to the adjacent probe located 3 μm away. The thermal cut-off frequency is 153 Hz, the mechanical resonant frequency is 40920 Hz, and the quality factor is 221 (Fig. 6(b)-(d)). No thermal buckling is observed up to 14 V.

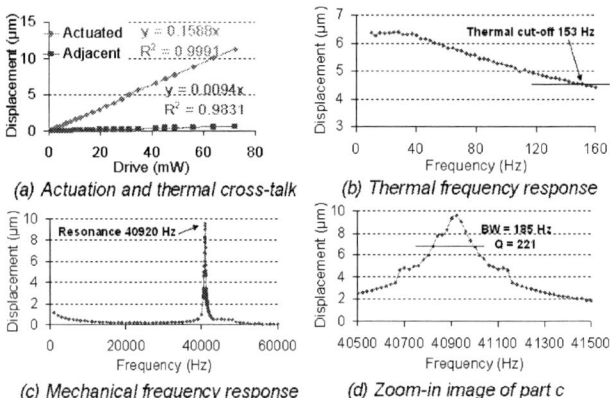

Fig. 5: White-light interference image measuring the self-assembled displacement after release.

(a) Actuation and thermal cross-talk

(b) Thermal frequency response

(c) Mechanical frequency response

(d) Zoom-in image of part c

Fig. 6: Actuation characterization from lever-based probe

4. CONTACT CHARACTERIZATION

A contact test setup was created by carefully positioning a 25 μm-radius tungsten (W) probe over the MEMS probe tip using a custom 6-DOF x-y-z-tilt test-stand. When off, the tips are 0.6 - 0.8 μm lower than the chip surface. A 7 V, 17.9 mW "background" drive is employed to move the tips upward by 2.8 μm to enable contact testing. Electrical contact was not observed immediately upon the physical contact of the W probe and the MEMS probe tip. The physical contact was judged by the reflection change under an optical camera. The W probe was then pushed downward further or the MEMS tip was driven upward to achieve a force strong enough to establish electrical contact.

For a 3×3 μm² tip, a 0.8 μm further displacement was needed to establish electrical contact (Fig. 7a). A 0.7 μm hysteretic displacement is observed, corresponding to a 3.5 μN force. Average contact resistance from 50 up/down cycles was 11 Ω with a standard deviation of 0.2 Ω. For a 4×4 μm² tip, a 0.5 μm further displacement was needed to establish contact, and the hysteresis was 0.3 μm (Fig. 7b).

At a 1 μm further displacement after the electrical contact point, I-V scanning was performed on the

4×4 μm² tip to test the electrical contact linearity. For 10 measurements at each point, the variation was within the measurement error. A match between voltage-increase and voltage-decrease curves was observed as expected (Fig. 8). The contact resistance slightly shifted with time. For 7 loops with 0.5 mA scanning current, the extracted contact resistance, excluding the 8 Ω wiring resistance, shifted from 16 to 20 Ω; for 20 loops with 5 mA current, the resistance shifted from 10 to 14 Ω.

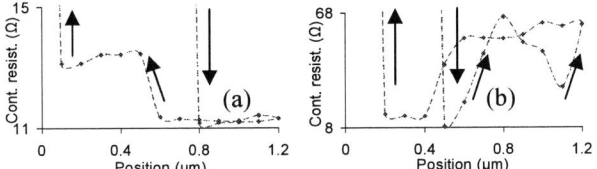

Fig. 7: Hysteresis in 3×3 μm² (a) and 4×4 μm² (b) tips. Physical contact point is at position = 0 μm.

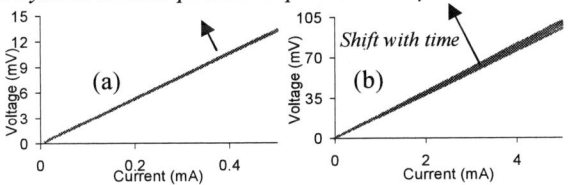

Fig. 8: I-V curves with 0.5 mA (a) and 5 mA (b) current.

At the physical contact point, drive power was applied to the probes. For the 3×3 μm² tip, the relation between the contact resistance and the drive power is shown in Fig. 9. Generally, the contact resistance decreased with the drive power increase as expected due to the normal contact force increase. The "abnormal" resistance increase with the drive power increase may be caused by the overall pressure decrease from the contact sliding caused by the mechanical bending of the flexures. The minimum contact resistance achieved was 10.7 Ω. A relative 4.55 mW hysteretic drive power was observed, which corresponds to a 0.73 μm displacement and 3.65 μN force. Statistical values of 400 measurements yielded a mean of 11.5 Ω and a standard deviation of 0.2 Ω. For the 4×4 μm² tip, the minimum contact resistance achieved was 17.1 Ω. A relative 2.46 mW hysteretic drive power was observed, which corresponds to a 0.39 μm displacement and 1.95 μN force.

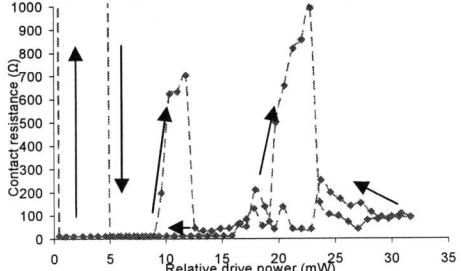

Fig. 9: Contact resistance measurement from a 3×3 μm² tip beginning at physical contact point

At the electrical contact points, drive power was applied to the probes. For the 3×3 μm² tip, the relation between contact resistance and the drive power is shown in Fig. 10. The minimum contact resistance achieved was 9.2 Ω. A relative 9.35 mW hysteretic drive power was observed, which corresponds to a 1.50 μm displacement and 7.50 μN force. For the 4×4 μm² tip, the minimum contact resistance achieved was 14.3 Ω. A relative 4.91 mW hysteretic drive power was observed, corresponding to a 0.79 μm displacement and 3.95 μN force.

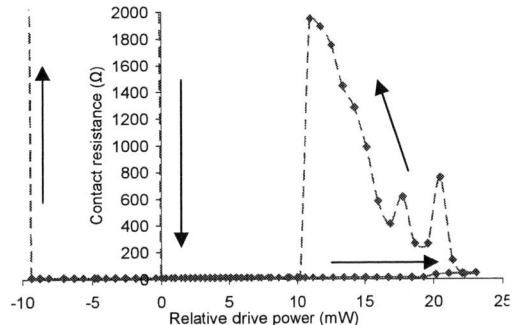

Fig. 10: Contact resistance measurement from a 3×3 μm² tip beginning at electrical contact point

Table 1 shows that the hysteresis displacements and forces at two physical contact conditions match; the larger the drive power, the larger the contact area is, and thus the larger the hysteresis displacement and force is. The hysteresis appears to be caused by an adhesive force between the surfaces of the W probes and the probe tips.

Table 1: Hysteresis displacements and forces

Tip size (μm²)	Contact condition	Hysteresis displacement (μm)	Hysteresis force (μN)
3×3	Physical contact*	0.7	3.5
	Physical contact	0.73	3.65
	1 μm after electrical contact	1.34	6.70
	Electrical contact	1.50	7.50
4×4	Physical contact*	0.3	1.5
	Physical contact	0.39	1.95
	1μm after electrical contact	0.77	3.85
	Electrical contact	0.79	3.95

*Electrical contact achieved by mechanical pushing.

With the 7 V "background" drive being removed, a 10 Hz 10 V sinusoid drive (i.e., 20 Hz contact frequency) was applied to characterize the contact repeatability and reliability. For the 4×4 μm² tip, with a 0.5 mA current passing contact, at the first 3,000 contact cycles, the contact resistance was stable at 8.9 Ω. It jumped in three discrete steps to 13.1 Ω at 6,000 cycles, then it degraded more slowly. Compared to the contact resistance at 6,000 cycles, a 9% degradation was observed after 36,000 cycles (Fig. 11).

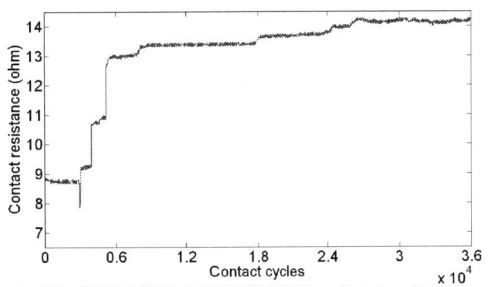

Fig. 11: Repeatability and reliability test on a 4×4 μm² tip

5. CONTACT MODEL

Due to the existence of surface roughness, contact models are built statistically. In view of the grain characteristics of the plated Ni, a semi-sphere model is employed using the 3×3 μm² tip as an example (Fig. 12).

Contact media is assumed to be flat in this model. All the modeled semi-spheres are uniform with a radius, R, of 300 nm, the measured Ni thickness, and a standard deviation of R, σ, of 40 nm, half of the Ni roughness. The tips are tilted in self-assembly and actuation, so only the grains located on one side (6 grains) contribute to contact.

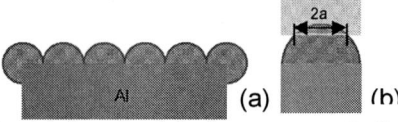

Fig. 12: Semi-sphere model of contact on a $3 \times 3 \ \mu m^2$ tip. (a) Cross-section and (b) Contact on a grain.

With very small force load F, contact resistance on one grain R_{cont} is dominated by elastic deformation [11].

$$R_{cont} = R_{constriction} + R_{ox} = \frac{\rho_{tip}}{4a} + \frac{\rho_{media}}{4a} + \frac{R_{s\_ox}}{A} \quad (1)$$

where ρ_{tip} and ρ_{media} are resistivity of the tip and media, R_{s_ox} is the resistance per m^2 of the possible median film which may be formed by the local oxidization on the tip grain and/or media surface,

$$a = \sqrt{R\sigma} \quad (2)$$

is the contact radius, and

$$A = \frac{F}{E_c} \sqrt{\frac{\pi R}{\sigma}} \quad (3)$$

is area of the grain. E_c is the composite Young's modulus of the two contacting materials. From (1), (2) and (3), $R_{constriction}$ is independent of F, and R_{ox} is inversely proportional to F.

Plastic deformation dominates when the contact load is larger than the nN-scale critical force, which is easily achieved with only nm-scale displacement. The relation between contact area and force in plastic deformation is

$$A = \frac{F}{H} \quad (4)$$

and the relation between contact radius and area is

$$a = \sqrt{\frac{A}{\pi}} \quad (5)$$

where the H is the material hardness. The contact resistance is then

$$R_{cont} = \frac{\rho_{tip}}{4} \sqrt{\frac{\pi H_{tip}}{F_{tip}}} + \frac{\rho_{media}}{4} \sqrt{\frac{\pi H_{media}}{F_{media}}} + \frac{R_{s\_ox} H_{ox}}{F_{ox}} \quad (6)$$

where F_{tip}, F_{media}, F_{ox}, H_{tip}, H_{media} and H_{ox} are the force load and hardness of the tip grain, media and median film, respectively. From (4), (5) and (6), both $R_{constriction}$ and R_{ox} are dependent on F. The total contact resistance R_{total} is

$$R_{total} = \frac{R_{cont}}{(L/2R)}$$
$$= \frac{\rho_{tip}}{4} \sqrt{\frac{2\pi R H_{tip}}{F_{total} L}} + \frac{\rho_{media}}{4} \sqrt{\frac{2\pi R H_{media}}{F_{total} L}} + \frac{2R_{s\_ox} H_{ox}}{F_{total} L} \quad (7)$$

where F_{total} is the total force load applied at the tip and L is the contact length, i.e., 1-D tip size, of the tip.

From (7), different tips can achieve the identical contact resistance with the same product $F_{total}L$. This concept is verified in Table 2: in 0.25 N/m or 5 N/m probes, the products of (Force $\times$ 1-D tip size) are similar in value for different tips. Due to the force load variation, the final contact resistance (under large force load) does not exactly match the quantitative ratios given by (7) though it decreases with the increase of tip size.

All contacts were made on newly Ni- plated probe tips, and it is assumed that there is no median oxide layer on top of the tips. Using typical literature values for ρ and H, this model predicts 0.1 Ω-order contact resistance, which was occasionally observed in our testing with

10 μN forces. A very thin oxide layer may exist on the probe tip and/or W probe. Our experiment showed no electrical contact was observed in a Ni-Au contact after the Ni-plated tip was exposed to air for a couple of weeks, which verified that an oxide layer was formed on Ni surface. As an example, an oxide layer with R_{s_ox} of 10^{-13} Ω-m^2 can contribute 20 Ω to contact resistance at a 10 μN force, which explains the discrepancy between the modeled and experimental values.

Table 2: Force needed to achieve electrical contact

Tip size (μm^2)	Displa-cement (μm)	Force (μN)	Force$\times$1-D tip size (pN-m)	Final cont. resist. (Ω)$^\Theta$	Modeled cont. resist. (Ω)§
$1 \times 1^\Delta$	22	5.50	5.50	22.19	0.35
$2 \times 2^\Delta$	13	3.25	6.50	10.88	0.30
$3 \times 3^\Delta$	10	2.00	6.00	10.17	0.25
$4 \times 4^\Delta$	7	1.75	7.00	9.64	0.22
$5 \times 5^\Delta$	5	1.25	6.25	7.34	0.20
$3 \times 3^*$	0.8	4.0	12.00	9.20	0.25
$4 \times 4^*$	0.5	2.5	12.50	8.45	0.22

$^\Delta$From [8] with spring constant of 0.25 N/m.
*From lever-based probes with spring constant of 5 N/m.
$^\Theta$Final contact resistance under variable large force load.
§Modeled contact resistance at 10 μN force load.

6. CONCLUSIONS AND FUTURE WORK

Lever-based CMOS-MEMS probes are demonstrated in 1 μm^2-order contact to achieve 1 - 10 Ω contact resistance. A semi-sphere contact model is employed. Pt-sputtering is under development to make reliable long-term non-oxidizing contact. The probe application requires sensing mechanism to track the position and force of the tips. Future designs will include capacitive position sensors and piezoresistive force sensors.

ACKNOWLEDGMENTS

We gratefully acknowledge the support of the Center for Circuit and System Solutions (C2S2) and the DARPA N/MEMS MISCIC Center. We thank Dr. Suresh Santhanam and the Carnegie Mellon Nanofabrication Facility for processing support.

REFERENCES

[1] B.P. Jena et al., *Force Microscopy: Applications in Biology and Medicine*, Wiley, NY, 2006.

[2] S.C. Minne et al., *Appl. Phys. Lett.*, vol. 73, no. 12, pp. 1742-1744, 1998.

[3] C. Tsou et al., *J. Micromech. Microeng.*, vol. 16, pp. 2197-2202, 2006.

[4] http://www.zurich.ibm.com/st/index.html.

[5] http://uspam.el.utwente.nl/joomla/.

[6] S. Gidon et al., *Appl. Phys. Lett.*, vol. 85, no. 26, pp. 6392-6394, 2004.

[7] S. Majumder et al., *IEEE MTT-S Int'l Microwave Symposium Digest 2003*, pp. 1935-1938, 2003.

[8] J. Liu et al., *Proc. MEMS 2008*, pp. 515-518.

[9] G.K. Fedder et al., *Sen. Actuators A*, vol. 57, pp. 103-110, 1996.

[10] S. Bedair et al., *Proc. Sensors 2006*, pp. 1074-1077.

[11] R. Holm, *Electrical Contacts*, 4th ed., Springer Verlag, Berlin, 1967.

978-1-4244-2977-6/09 $25.00 © 2009 IEEE

A NEW EFFICIENT METHOD FOR SIMULATING THE DYNAMIC RESPONSE OF ELECTROSTATIC SWITCHES

V. Leus and D. Elata
Technion – Israel Institute of Technology

ABSTRACT

This paper presents a new efficient technique for simulating the dynamic response of electrostatic micro switches. It is demonstrated that the dynamic response of a switch can be well predicted based on only a few *static* states of the system. Using energy methods the deformation of the system is approximated by a single adaptive mode, such that time integration is performed on a scalar equation of motion. Due to the high computational efficiency of the proposed method, it can be used to motivate the functional form of damping forces and extract damping parameters, by fitting experimentally measured dynamic responses.

INTRODUCTION

Electrostatically actuated micro switches for RF applications are becoming potential candidates for replacing traditional solid-state switches (e.g. FETs and pin-diodes), due to their low loss and very low power consumption [1]. In many applications (e.g. phased-array antennas and radars), time-response parameters of RF-MEMS switches, such as switching time and release time, are essential for performance. Better performance of switches requires computationally intensive design optimization. To accurately predict the time-response of such switches, transient analysis of coupled electro-structural system must be performed. This analysis requires full time-integration of nonlinear momentum equations, which is very time-consuming, especially when a 3D model is considered. Consequently, time-response simulation of a specific switch due to a specific voltage, may take from several hours to several days of computation. This makes parametric design of switches very cumbersome. Efficient modeling tools are required to simplify time-response simulation.

The proposed modeling method is capable of simulating the dynamic response of switches with 3-D geometry, including nonlinear effects such as fringing fields and stress stiffening. Using energy methods the distributed deformation of the system is reduced to a single adaptive degree of freedom. Instead of time-integration of a set of momentum equations (e.g. an equation for each node in a finite element mesh), the problem is reduced to a single scalar momentum equation based on the deformation of only four, accurately computed *static* states of the system. High computational efficiency of the new method may be utilized to extract system properties from experimental data (e.g. fitting of functional form of damping forces, or evaluation of structural parameters)

In addition, the proposed method enables effortless extraction of the dynamic pull-in parameters (pull-in voltage and pull-in displacement), which are important performance parameters of electromechanical switches.

THE NEW METHOD

In this section the details of the proposed modeling method are explained using the clamped-clamped beam actuator as a model problem (Fig.1). This actuator consists of a double-clamped beam of length L, width W, and thickness h, which is suspended over a fixed bottom electrode. The electrode is coated by a thin dielectric layer of thickness d, and the nominal gap between the beam and the dielectric layer is g. The system is actuated by application of a step-function voltage to the bottom electrode, while the beam is grounded.

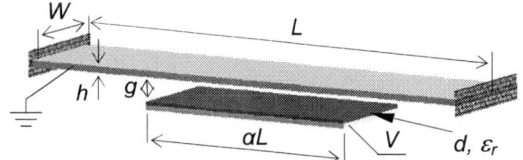

Figure 1: Model of a clamped-clamped (C-C) beam actuator ($L=240\mu m$, $W=40\mu m$, $g=2\mu m$, $h=1\mu m$, $\alpha=0.5$, $d=0.1\mu m$, $\varepsilon_r=4$; $E=70GPa$, $v=0.35$, $\rho=2700 \, kg/m3$).

In the proposed modeling approach, each deformed state during the dynamic response is assumed to be identical to a deformed static state. To uniquely identify any dynamic state with a related static state, we consider a single scalar measure of the deformed state. In the case of the clamped-clamped beam actuator this scalar measure is the deflection of the beam center.

In the following we model the response of the actuator using energy considerations. The total energy of the system at any state is given by its Hamiltonian (1), which is the sum of the kinetic energy T, the elastic strain energy U_S, and the total electrostatic potential energy U_E, which includes the electrostatic potential of the deformable capacitor and of the voltage source.

$$H(t) = T + U_S + U_E \qquad (1)$$

The total electrostatic potential energy is given by $U_E = -\frac{1}{2}CV^2$, where C is the capacitance of the deformable capacitor at any given state, and V is the applied voltage.

For the initial state at $t=0$ the velocity and the deflection of the beam are both zero. Therefore the Hamiltonian is equal to the electrostatic potential energy of the system, $H_0 = -\frac{1}{2}C_0V^2$, where C_0 is the capacitance at the non-deformed state.

The total energy of the system can be dissipated during switch actuation due to viscous damping which results from ambient pressure [2]. The work W_D done by damping force from the initial state up to a specific deformed state 'j' is equal to the difference between the Hamiltonian at the initial state and the Hamiltonian at the deformed state 'j'. This leads to the following energy constraint

978-1-4244-2977-6/09 $25.00 © 2009 IEEE

$$\tfrac{1}{2}\Delta C^j V^2 - T^j - U_S^j = W_D^j \tag{2}$$

where $\Delta C^j = C^j - C_0$.

When the beam is released with zero voltage from the specific static state, the energy constraint is given by

$$\Delta U_S^j - T^j = W_D^j \tag{3}$$

where $\Delta U_S^j = U_{S0}^j - U_S^j$, and U_{S0}^j is the strain energy stored in the beam at the deformed static state just before release.

Kinetic energy

The kinetic energy of the beam at any deformed state 'j' is defined by

$$T^j = \int_0^L \int_0^W \frac{\gamma}{2} \left(\frac{dy^j(x,z)}{dt^j} \right)^2 dxdz \tag{4}$$

Here y is the deflection, x is the coordinate along the beam length, z is the coordinate along the beam width, dy/dt is the local velocity of the beam, $\gamma = \rho h$ where ρ is the density of the material and h is the beam thickness.

The kinetic energy can be rewritten in terms of the velocity of the representative node, which may be extracted from the integral (in the case of the C-C beam actuator this is the velocity of the beam-center)

$$T^j = \frac{\gamma}{2} \int_0^L \int_0^W \left(\frac{dy^j(x,z)}{dy_C^j} \cdot \frac{dy_C^j}{dt^j} \right)^2 dxdz \tag{5}$$

where $dy^j(x,z) = y^{j+1}(x,z) - y^j(x,z)$ is the incremental deflection between two adjacent deflections of the beam during the time interval dt^j, and dy_C^j is the incremental deflection of the beam-center between the two adjacent deflections. In the limit case $dt^j \rightarrow 0$, the velocity of the beam-center is a scalar which we denote by $\dot{y}_C^j = dy_C^j / dt^j$, and extract from the integral. The remaining integral which accounts for the distributed deformation of the beam, is independent of time and denoted by β

$$\beta^j \equiv \frac{\gamma}{2} \int_0^L \int_0^W \left(\frac{dy^j(x,z)}{dy_C^j} \right)^2 dxdz \tag{6}$$

Now, (4) may be rewritten as

$$T^j = \beta^j (\dot{y}_C^j)^2 \tag{7}$$

Dissipated energy

The energy dissipated due to damping from the initial state at $t=0$ up to any specific deformed state at time t is given by

$$W_D = \int_0^t \left[\int_0^L \int_0^W F_D \frac{dy(x,z)}{dt} dxdz \right] dt \tag{8}$$

where F_D is the damping force. The functional form of the distributed damping force can be calibrated using experimental data. In this work we assume that damping is a nonlinear function of displacement, but is a linear function of velocity: $F_D = b(y) \frac{dy}{dt}$, where $b(y)$ is the local gap-dependent damping coefficient.

Now, by extracting the velocity of the beam center using (5), equation (8) takes the form

$$W_D^j = \delta^j \int_0^{t^j} \dot{y}_C^2 dt = \delta^j \int_0^{y_C^j} \dot{y}_C dy_C = \delta^j \sum_{i=1}^j \dot{y}_C^i$$
$$= \delta^j \sum_{i=1}^{j-1} \dot{y}_C^i + \delta^j \dot{y}_C^j \tag{9}$$

where δ is a parameter which accounts for the relative velocity distribution long the beam

$$\delta^j = \int_0^L \int_0^W b^j \left(\frac{dy^j(x,z)}{dy_C^j} \right)^2 dxdz \tag{10}$$

Beam velocity and time-response

The velocity of the beam-center for any applied voltage can be calculated from the quadratic equation obtained by substituting (7) and (9) into (2)

$$\beta^j (\dot{y}_C^j)^2 + \delta^j \dot{y}_C^j = \tfrac{1}{2} \Delta C^j V^2 - U_S^j - \delta^j \sum_{i=1}^{j-1} \dot{y}_C^i \tag{11}$$

In the case of switch release, the beam-center velocity can be found by solving the following quadratic equation

$$\beta^j (\dot{y}_C^j)^2 + \delta^j \dot{y}_C^j = \Delta U_S^j - \delta^j \sum_{i=1}^{j-1} \dot{y}_C^i \tag{12}$$

For the un-damped case where $\delta = 0$, equations (11) and (12) are simplified such that the beam-center velocity for step-function voltage actuation, and for switch release, are respectively

$$\dot{y}_C^j = \sqrt{(\tfrac{1}{2} \Delta C^j V^2 - U_S^j) / \beta^j} \tag{13}$$

$$\dot{y}_C^j = \sqrt{\Delta U_S^j / \beta^j} \tag{14}$$

For a given incremental center-beam deflection dy_C^j, the time required for transition between two adjacent deflection states is given by

$$dt^j = \frac{dy_C^j}{\dot{y}_C^j} \tag{15}$$

Finally, the time-response of the actuator can be reconstructed for any applied step-function voltage V by consistent summation of the time increments dt^j as function of the corresponding center-beam deflection y_C^j. For time-response computation, we have to know the elastic strain energy U_S and the capacitance C at several deflection states from the non-deformed state up to contact. The accuracy of the derived response can be expected to increase if a higher number of deflection states is considered. In the following section we show that the elastic energy, capacitance, and the parameters β and δ may be interpolated and extrapolated such that only four static states must be computed in detail.

SIMULATED RESULTS

The 3-D model of the C-C beam actuator (Fig. 1) was constructed and simulated using static electro-structural analysis in ANSYS to compute four equilibrium deflections of the beam. Nonlinear stress stiffening was included in the analysis. The deflections were computed using displacement iterations (DIPIE) algorithm [3]. Accordingly, the center node of the beam is forced to a specific deflection within the gap. Then the voltage required to nullify the resulting reactive force at the beam center is iteratively computed until the solution converges to an equilibrium static state. Four static deflections of the beam were computed corresponding to four prescribed

978-1-4244-2977-6/09 $25.00 © 2009 IEEE

center-beam deflections: 0.2, 0.4, 0.6 and 0.8 of the nominal gap g. For each distributed deflection, the capacitance and the elastic energy were extracted. The obtained results can be interpolated throughout the whole gap as follows:

By exploiting analogy to parallel-plates actuator, the elastic energy U_S for any deflection throughout the gap can be very well approximated by a third order polynomial, which is optimally fitted by the least square method (Fig. 2a).

The change in capacitance ΔC associated with the beam center is well described by the following formula which is motivated from analogy to the parallel-plates actuator

$$\Delta C = \frac{a\varepsilon_0 A y_c}{(g + d/\varepsilon_r)(g + d/\varepsilon_r - by_c)} \quad (16)$$

where $A = \alpha LW$ is the area of the actuation electrode, and the coefficients a and b are optimally fitted by the least square method (Fig. 2b).

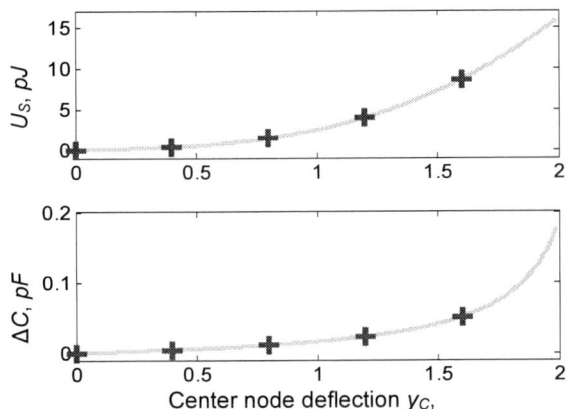

Figure 2: Elastic energy of the beam (top), and change in capacitance (bottom) as function of the center node deflection (the '+' markers represent the extracted values from ANSYS, solid line represents the least square fit).

To compute the parameters β and δ for any deflection throughout the air gap, the deflection of the beam was linearly interpolated and extrapolated, based on the four simulated static deflections.

The incremental deflections between adjacent replicated states were chosen to be sufficiently small to ensure accuracy of the simulated time-response. These results can be then used for reconstructing the dynamic response of the actuator for any applied step-function voltage, and for the release response. For the applied voltage which is higher than the dynamic pull-in voltage V_{Dpi}, a switching response is obtained. In contrast, if the applied voltage is lower than the dynamic pull-in voltage, a periodic response occurs [4].

To verify the accuracy of the method, the same finite element model that is used to simulate the four static states is also used to simulate the dynamic response by performing a regular transient analysis (full time-integration of the momentum equations).

Figure 3 presents the undamped ($\delta = 0$) switching responses of the C-C beam actuator for four different step-function voltages, all larger than the dynamic pull-in

voltage. It can be seen, that new method simulations are in very good agreement with the transient simulations (the relative errors of time-to-contact are less than 1%).

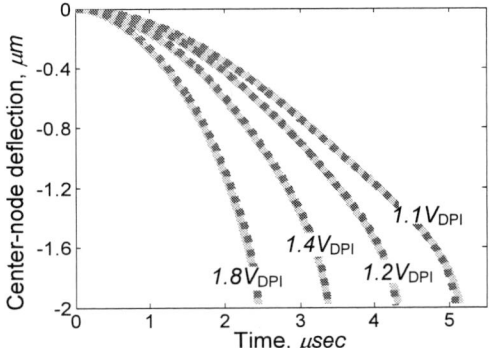

Figure 3: Time-trajectories of the C-C beam actuator for four different applied voltages. For each applied voltage the time response is simulated using transient analysis (dotted lines: computation time is around nine hours) and also using the novel algorithm (solid lines: computation time is ten minutes).

Computation efficiency

The proposed algorithm can be divided into three steps. Step 1 is computation of four static states using ANSYS. This is the main computation effort. The simulation of the considered 3-D model of the C-C beam actuator, including stress-stiffening, took 10 minutes. The next step is fitting and interpolation of relevant parameters which took 15 seconds. And the final step is the time-response reconstruction. The computation time for simulation of each switching response is in the range of 1-2 seconds. Consequently, the overall computation time is around 10 minutes (Table 1).

In the transient analysis the time step should be validated to be sufficiently small to ensure accuracy of the numerical computation. This requires performing time-integration in several iterations, where extraction of each switching response for a specific applied voltage takes at least one and a half hours (depending on the applied voltage).

Table 1: Computation time comparison.

	Transient analysis	New method
Four time-trajectories	≈ 9 hours	10min
Dynamic pull-in voltage extraction	At least 8 hours	
Relative error	-	<1%

Another important parameter of the electromechanical response of the switch is the dynamic pull-in voltage. This parameter can be effortlessly extracted from energy considerations [5], after the second step of the new algorithm is completed. The dynamic pull-in voltage extracted using the new algorithm is found to be 18.83V, whereas the transient analysis result is 18.9V, but at least 8 hours were required for that computation.

978-1-4244-2977-6/09 $25.00 © 2009 IEEE

EXPERIMENTAL VERIFICATION

In the following section we verify the accuracy of the simulation method by fitting experimental measurements, and show that the high time-efficiency of the proposed method can be used to estimate the functional form of damping forces. First, the electrostatic test-switch shown in Fig.4 was experimentally characterized, and was simulated using both the new algorithm and transient analysis.

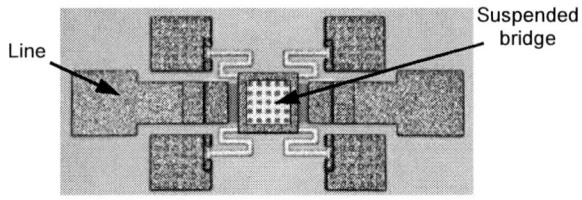

Figure 4: Micro photo of the electrostatic test-switch.

The switch was fabricated using surface micromachining with two structural layers of electroplated Gold with thickness of 1.9μm for the main membrane layer, and 3.2μm for the stiffening frame (Fig. 4). The dielectric is a 100nm thick low temperature oxide with a relative dielectric constant of $\varepsilon_r = 4$.

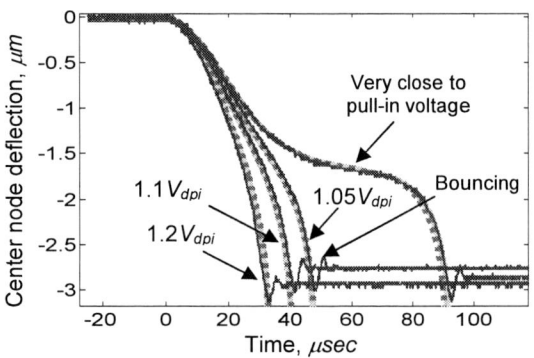

Figure 5: Simulated and measured results for the test-switch (thick sold lines - new method, dotted lines - transient analysis, thin solid lines – measurements).

The nominal gap was measured to be 3μm. The dynamic measurements of the switch were performed at low pressure of $6\cdot10^{-3}$ Torr and at atmospheric conditions. The out-of-plane motion of the switch was measured by a Polytech laser vibrometer.

Figure 5 presents the simulation and measurement results for four undamped time-responses of the switch. Simulations with the new method are presented by thick solid lines (11 minutes of computation), full-time integration simulations are presented by dotted lines (10 hours of computation), and measurements are presented by thin solid lines. The dynamic pull-in voltage of the switch as extracted from simulations is 28.65V by the new method, and 28.75V by transient analysis. The measurement result is around 28.6V. It can be seen that good agreement between the two different simulation methods and measurements is obtained.

The release of the switch from the bottom position at atmospheric conditions is demonstrated in Fig.6. The reconstruction of the release response using the new

method is performed within 2 sec. This enables to efficiently fit a functional form to the damping force (8). The best fitted functional form of the distributed drug force for the test switch is $F_D = b(y)\dot{y}$, where $b(y) = b_0/(g-y)^3$ is the local gap-dependent damping coefficient, and b_0 was fitted to be $0.22\cdot10^{-12} N\cdot\sec$. This fit of damping force is also very good for the release responses from other static states (pre pull-in).

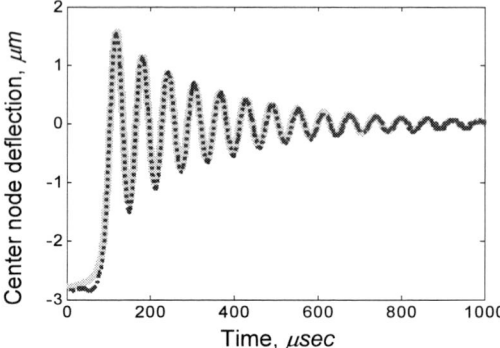

Figure 6: Simulated and measured damped release of the test-switch (thick sold line - new method simulation, dotted line - measurements).

CONCLUSIONS

A new efficient method for simulating the dynamic response of electrostatic micro switches is presented. The method enables to reconstruct the time-response of switches for any applied voltage based on four static states only. It is shown that the proposed technique is very time-efficient relative to common transient analysis, while demonstrating high accuracy. The measurement results are in very good agreement with the simulations. It is also shown that due to high computational efficiency of the new method, it may be used for fitting the functional form of damping forces based on experimental data. Consequently, the proposed method may be an effective tool for parametric design and for characterization and parameter extraction of electrostatic switches.

REFERENCES

[1] G. M. Rebeiz, *RF MEMS: Theory, Design, and Technology* Hoboken, N.J.: Wiley-Interscience, 2003.

[2] T. Veijola, "Compact models for squeezed-film dampers with inertial and rarefied gas effects", *J. Micromech. Microeng.*, Vol. 14, pp. 1109-1118, 2004..

[3] O. Bochobza-Degani, D. Elata, and Y. Nemirovsky, "An efficient DIPIE algorithm for CAD of electrostatically actuated MEMS devices," *J. of Microelectromech. Syst.*, Vol. 11, pp. 612-620, 2002.

[4] V. Leus, D. Elata, "On the dynamic response of electrostatic MEMS switches," *J. of Microelectromech. Syst.*, Vol. 17, pp. 236-243, 2008.

[5] D. Elata and H. Bamberger, "On the dynamic pull-in of electrostatic actuators with multiple degrees of freedom and multiple voltage sources," *J. of Microelectromech. Syst.*, Vol. 15, pp. 131-140, 2006.

978-1-4244-2977-6/09 $25.00 © 2009 IEEE

AUTHOR INDEX

Abbaspour-Tamijani, A..............935

Abe, T..............904

Acquaviva, D...............132

Agah, M...............288

Aïd, M...............619

Aiyar, A. R...............447

Akbarian, M...............160

Akgul, M...............19

Akin, T...............813, 817

Akiyama, S...............567

Aljasem, K...............999

Allen, M. G...............447

Alper, S. E...............813

Amaya, S...............709

Ambacher, O...............923

An, J. Y...............951

Ando, T...............661

Arai, F...............51, 375

Armbruster, S...............693

Arnold, D. P...............665

Autizi, E...............630

Averitt, R. D...............108

Ayazi, F...............749, 888, 931

Baca, A. I...............638

Baek, C.-W...............725

Baghoomian, E...............216

Bahl, G...............801, 1095

Bain, J. A...............1107

Baker, M...............841

Balkan, T...............1071

Barry, T...............356

Bartsch, U...............1039

Baskaran, R...............777

Baumann, W...............220

Baygents, J. C...............431

Bearda, T...............136

Beardslee, L. A...............284

Bedair, S...............268

Bedani, M...............860

Benito-Lopez, F...............439

Benzel, H...............693

Berenschot, E...............152

Berenschot, J. W...............324

Bergaud, C...............567

Berry, C. W...............216

Berry, H...............829

Bever, T...............880

Beyaz, M. I...............1091

Bhagavat, M...............88

Bhave, S. A...............23, 896

Bielen, J. A...............916

Binh-Khiem, N...............168, 963

Birkle, G...............43

Blattmann, R...............669

Boga, B...............817

Bogaerts, L...............136

Böhringer, K. F...............531

Bourouina, T...............975

Brama, R...............860

Brand, O...............284

Braun, S...............1063

Bright, V. M...............120, 638

Bronson, J. R...............1043

Brown, J. J...............638

Brueckner, K...............923

Brugger, J...............144

Bruinink, C. M...............152

Buchaillot, L...............868, 919

Burger, R...............443

Butler, D. P...............833

Byun, D...............523

Cai, H...............967, 975, 1023

AUTHOR INDEX

Candler, R. N. 801

Carlborg, C. F. 39

Carraro, C. 611

Casavant, B. P. 304

Casinovi, G. 931

Castaneda, J. 841

Cellere, G. 630

Ceyssens, F. 451

Chae, J. 96, 935

Chaitanya, C. R. Alla 701

Chandorkar, S. 657

Chandrahalim, H. 23, 896

Chang, C.-T. 296

Chang, C.-W. 200

Chang, J. 128

Chang, P.-Z. 204

Chang, Y. J. 120

Chang, Y.-C. 200

Chao, F.-S. 204

Chen, D.-S. 1031

Chen, H. 264

Chen, J.-S. 627

Chen, J.-W. P. 927

Chen, K. L. 801, 1067

Chen, K.-L. 23

Chen, K.-S. 1099

Chen, M.-H. 296, 455

Chen, P.-J. 196, 244

Chen, R. 200

Chen, T. 1015

Chen, Y. 360

Chen, Z. G. 399

Cheng, M.-Y. 92, 292

Cheng, Y. T. 567

Cheung, L. S. L. 379, 431

Chi, X. Z. 104, 809

Chia, B. T. 292

Chiang, A.-S. 100

Chiao, M. 995

Chicherin, D. 892

Chiravuri, S. 252

Chiu, N.-F. 204

Cho, J. 749

Cho, S. K. 1079

Cho, Y. H. 399

Cho, Y.-H. 64

Choi, C.-A. 793

Choi, D. 360

Choi, H.-C. 939

Choi, W. 160

Choi, Y.-K. 475

Chu, L.-A. 100

Chung, S. K. 1079

Cimalla, V. 923

Cismak, A. 220

Clausi, D. 1063

Collard, D. 607

Collins, S. D. 571

Cooper, J. R. 208

Cordero, M. 1075

Coronel, Ph. 132

Couderc, S. 642

Craighead, H. 646

Creemer, J. F. 76, 844

Cristman, P. 757

Cui, J. 104, 809

Currano, L. 591

Czarnecki, P. 852

Dai, Z. 543

Dainesi, P. 634

Dao, D. V. 709, 1007

Davidson, B. D. 120

AUTHOR INDEX

De Boer, M. J. 152, 463, 479
De Moor, P. 136
De Nooijer, C. 136
De Volder, M. 451
Defay, E. 619
Demartini, B. E. 1095
Demello, A. J. 523
Demirci, K. S. 284
Despont, M. 603
Dhull, R. K. 991
Diamond, D. 439
Dijkstra, M. 479
Dimov, I. K. 356
Ding, H. T. 809
Ding, Z. 407
Draghi, L. 713
Drechsler, U. 603
Drittenbass, S. 575
Ducree, J. 356, 439, 443, 539
Dueck, M. 304
Duncombe, T. A. 531
Durrer, L. 575
Dy, E. 391
Edamoto, M. 1055
Ekeom, D. 919
Ekkels, P. 852
Elata, D. 753, 927, 1111
Elwenspoek, M. C. 324, 479
Esashi, M. 705, 1027
Eskinazi, I. 665
Etzkorn, J. R. 713
Eun, Y. 164, 1035
Evans, A. T. 252
Fan, K. 108
Fan, L.-S. 627
Fang, L. 595

Fang, W. 100, 172, 200, 805, 1047
Fattaccioli, J. 567, 677
Fedder, G. K. 268, 1107
Fernández-Bolaños, M. 634
Fiorini, P. 623
Fischer, M. 693
Flood, A. H. 595
Fonseca, J. G. 443
Foulds, I. G. 583
Frazier, A. B. 399
French, P. J. 140, 844
Fujii, H. 1059
Fujii, M. 519
Fujii, T. 864
Fujita, H. 403, 531, 607, 697, 884, 1075, 1083
Fujita, M. 184
Fujiwara, N. 435
Fujiyoshi, M. 733
Fukuda, M. 260
Fukuoka, M. 124
Fuller, L. 991
Funatsu, N. 507
Fung, A. 360
Ganzerli, M. 860
Gao, X. 931
Garcia-Cordero, J. L. 356, 439
Garipov, G. 979
Garrell, R. L. 280
Gaspar, J. 7, 220, 599, 1039
Gehner, A. 1003
George, S. M. 120
Ghodssi, R. 1091
Gianchandani, Y. B. 212, 252, 837
Gieschke, P. 765
Gilmour, R. 336
Gnudi, A. 860

AUTHOR INDEX

Gojo, R. 256

Goodman, M. B. 188

Goryu, A. 224

Gradin, H. 1063

Graham, A. B. 23, 741, 801

Grasshoff, T. 947

Green, S. R. 212

Gruenberger, R. 880

Gudipaty, T. 379

Guo, T.-F. 415

Guo, Z. Y. 104, 809

Gurau, B. 35

Guzman, R. 431

Hagleitner, C. 603

Hagner, M. 220

Halder, S. 136

Han, A. 399

Han, J. 80

Han, K.-H. 308

Han, S.-I. 308

Haneveld, J. 463

Hao, Y. L. 673

Hartwel, P. 741

Hartwell, P. G. 1095

Hasegawa, Y. 769

Hashimoto, S. 60

Haspeslagh, L. 136

Hassanin, H. S. 717

Hatakeyama, Y. 705

Heilmann, A. 220

Heimark, R. L. 431

Hein, M. A. 923

Helbling, T. 575

Held, J. 7, 220, 599

Helveg, S. 76

Heo, Y. J. 68

Herfst, R. W. 916

Herrmann, M. 765

Herwik, S. 232

Hibert, C. 144

Hida, H. 403, 908

Hierold, C. H. 300, 575, 669

Higgs, G. 407

Higo, A. 697, 1083

Hill, T. A. 841

Hilton, J. P. 344

Hirshberg, A. 753

Ho, C. M. 391

Ho, D. 360

Hoffmann, J. 539

Homma, T. 499

Honda, H. 908

Hong, C.-T. 100

Hong, J. 523

Hoveling, G. H. 76

Howe, R. T. 23, 741, 927

Hsiao, S.-Y. 172

Hsieh, C.-Y. 627

Hsieh, J. 927

Hsu, H. Y. 579

Hsu, H.-Y. 411

Hsu, T.-H. 455

Huang, R. 745

Huang, S.-B. 515

Huang, T. J. 332, 595

Huang, W.-L. 19

Huang, X. 352

Huang, X.-J. 475

Humayun, M. 196

Humayun, M. S. 244

Hwang, G. 793

Hwang, J. 387, 487, 491

AUTHOR INDEX

Hwang, Y.-S.	725
Hyeon, I.-J.	725
Iannacci, J.	860
Ihara, K.	864
Ikeda, A.	677
Ikedo, A.	224, 721
Ikeuchi, M.	124, 348
Ikuta, K.	11, 124, 348
Ilic, B.	646
Im, M.	475
Imai, N.	499
Inoue, N.	507
Inoue, S.	653, 1051, 1059
Ionescu, A. M.	132, 634
Irieda, T.	864
Ishida, M.	192, 224, 503, 721
Ishihara, H.	367
Ito, Y.	419
Itoh, T.	797
Iwai, K.	371
Iwase, E.	68, 176, 551, 943, 963
Izadi, H.	583
Jaganatharaja, R. K.	152
Jakobs, F.	773
Jalabert, L.	607
Jamshidi, A.	411, 579
Jang, W. G.	959
Jansen, H. V.	324
Je, S.-S.	96
Jee, H.	729
Jeon, J. A.	979
Jeong, B.	164
Jeong, D.-H.	793
Jiang, K.	717
Jiang, L.	208, 379
Jiang, Y. Q.	587
Jin, J. Y.	979
Jo, K. W.	959
Johnson, A. T.	320
Jones, D.	841
Jones, T. S.	320
Judy, D. C.	591, 896
Judy, M.	88
Juluri, B. K.	595
Jun, M. H.	951
Jung, H.-K.	725
Jung, K.-D.	156
Junge, S.	773
Kaanta, B. C.	264
Kamezawa, C.	615
Kamiya, S.	599
Kan, T.	551, 963
Kaneko, H.	192
Kang, B.-G.	487
Kanzaki, R.	180
Karim, K. S.	583
Karle, M.	276
Kasagi, N.	60, 615, 1055
Kashiwagi, K.	1055
Kashiwagi, S.	653
Katahira, K.	419
Kato, D.	260
Kato-Negishi, M.	423
Kaul, R.	896
Kawai, K.	519
Kawai, Y.	1027
Kawano, K.	955
Kawano, T.	192, 224, 721
Kawasaki, S.	884
Kawashima, T.	192, 721
Ke, F.	856, 872
Keller, M.	232

AUTHOR INDEX

Kenny, T. W. 23, 657, 741, 801, 1067

Khamis, S. M. 320

Khrenov, B. 979

Khuri-Yakub, B. T. 757

Kim, B. 19, 741

Kim, B. J. 567, 642, 677

Kim, C.-H. 156

Kim, C.-J. 160, 280, 471, 900

Kim, D.-H. 475

Kim, D.-S. 939

Kim, E. S. 316, 761

Kim, H. 483

Kim, H. S. 399

Kim, J. 96, 164, 523, 789, 1035

Kim, J. E. 979

Kim, J. G. 677

Kim, J. K. 729

Kim, J.-K. 84

Kim, J.-M. 27

Kim, K. H. 511

Kim, M. 979

Kim, M. G. 951, 959

Kim, S. 563

Kim, S. G. 563

Kim, S. H. 447

Kim, S.-Y. 487, 491

Kim, W. 156

Kim, Y. K. 979

Kim, Y.-H. 387

Kim, Y.-J. 272, 387, 487, 487, 491, 491

Kim, Y.-K. 725

King, W. P. 543, 848

Kinoshita, Y. 260

Kishi, H. 904

Kitajima, T. 419

Klose, T. 947

Knese, K. 693

Kobayashi, K. 11

Kobayashi, T. 260, 797

Koester, P. J. 220

Koga, H. 260

Kolster, M. L. 152

Koltay, P. 43, 1087

Komatsubara, M. 653, 1051

Kondoh, T. 507

Konishi, S. 260, 435, 499, 737

Kooyman, P. J. 76

Korvink, J. G. 1015

Kotake, N. 256

Kozicki, M. N. 96

Kratt, K. 777

Krausse, S. 1015

Krijnen, G. J. M. 152

Kroninger, C. M. 1043

Krumbholz, S. 1087

Krylov, S. 646

Kuhl, M. 765

Kühne, S. 669

Külah, H. 817, 1071

Kumar, A. S. 595

Kumar, K. 1011

Kumar, S. 833

Kumemura, M. 531, 607

Kuo, P.-L. 340

Kuo, S.-M. 983

Kupnik, M. 757

Kuribayashi, K. 367

Kuypers, F. A. 64

Kwon, S. 387

Kwon, T. H. 785

Lackey, C. 841

Lai, R.-J. 1031

AUTHOR INDEX

Lai, Y.-T. 92
Lal, A. 336, 547
Lambertus, G. 264
Lammerink, T. S. J. 152, 463, 479
Lang, W. 773
Lapisa, M. 1003
Larcher, L. 860
Lauria, J. 136
Lazarus, N. 268
Le Rhun, G. 619
Le, N. C. H. 1007
Lee, B.-K. 148, 785
Lee, C.-C. 172
Lee, C.-I. 272
Lee, D.-W. 84
Lee, G.-B. 47, 340, 415, 515
Lee, H. 308
Lee, H. J. 757
Lee, H. K. 657, 801
Lee, H. S. 785
Lee, H. W. 563
Lee, J. 848, 979
Lee, J. H. 951, 959
Lee, J.-H. 475, 793
Lee, J.-I. 164
Lee, J.-S. 487, 491
Lee, L. P. 304
Lee, Luke P. 55
Lee, M.-L. 793
Lee, S. K. 951
Lee, S.-H. 749
Lee, S.-W. 749
Lee, T.-M. 939
Lee, W. C. 64
Lee, Y. K. 395
Lee, Y.-T. 200

Lei, U. 204
Lemke, B. 777
Leung, M. 360
Leus, V. 1111
Li, C.-M. 312
Li, S. 352
Li, W. 248
Li, W.-C. 19
Li, X. 328, 555
Liao, C. Y. 689
Liao, H.-H. 204
Lien, K.-Y. 340
Lin, A. 316
Lin, C.-H. 983
Lin, C.-L. 292
Lin, C.-W. 200, 204, 292
Lin, J. C.-H. 196
Lin, L. 31, 72, 128, 304, 587, 864
Lin, L. T. 104, 809
Lin, Q. 344, 352
Lin, W.-Y. 47
Lin, Y. 19
Lin, Y.-H. 47, 415
Liu, A. Q. 967, 975, 1023
Liu, B. 967, 975, 1023
Liu, C.-J. 340
Liu, F. 611
Liu, J. 1107
Liu, L. 1103
Liu, L. J. 619
Liu, M. C. 427
Liu, P. 527
Liu, P.-C. 204
Liu, X. D. 527
Liu, X. S. 809
Liu, Y. 595

AUTHOR INDEX

Liu, Z. 1015

Liu, Z. B. 395

Lo, C. 268

Lo, R. 236

Loo, J. A. 280

Lopez, G. 184

Lord, S. F. 657

Löw, P. 567

Lu, S.-S. 204

Lu, Y. W. 459, 991

Luzinova, Y. 284

Maboudian, R. 611

Mabuchi, K. 256

Mader, D. 987

Maeda, M. 184

Mahajerin, A. 304

Maharbiz, M. M. 216

Mao, H. Y. 673

Margesin, B. 860

Mark, D. 539

Maruyama, H. 375

Masel, R. I. 35

Massieu, J. L. 947

Masuda, T. 797

Matsumoto, K. 68, 168, 176, 551, 781, 821, 943, 963, 971, 1019

Matsumoto, Y. 467

McCarthy, M. 1091

McCormick, D. T. 864

Melamud, R. 657, 741, 1067

Meneghesso, G. 630

Meng, E. 236

Messana, M. 741

Metz, M. 693

Metz, T. 43

Miao, J. 312, 856, 872

Miki, N. 467

Miki, T. 1051

Miller, M. 619

Millet, O. 868

Misawa, N. 180

Mita, M. 697

Mitsuno, H. 180

Miwa, J. 60, 276

Mizaikoff, B. 284

Mogulkoc, B. 324

Molenbroek, A. M. 76

Momose, T. 681

Moon, H. 360

Morimoto, K. 260

Morimoto, Y. 56

Morita, N. 507

Morizawa, Y. 1055

Mukerjee, E. V. 848

Müller, C. 1087

Munt, S. 1087

Murase, H. 419

Murata, M. 507

Murmann, B. 657

Na, G. W. 979

Na, H. 1035

Nagamori, E. 403

Nagasawa, H. 1051

Nagaura, Y. 499

Naito, J. 495

Najafi, K. 483, 749

Naka, N. 653

Nakada, M. 697, 1083

Nakai, A. 176

Nakamura, K. 864

Nam, S. 979

Namazu, T. 653, 1051, 1059

Nawaz, M. 880

AUTHOR INDEX

Nayve, R. 507

Neale, S. 411, 579

Nelson, W. 280

Neves, H. 232

Nguyen, C. T.-C. 19

Nguyen, T. 344

Niebelschuetz, F. 923

Niklaus, F. 1003

Ninomiya, T. 467

Nishimura, T. 60

Noda, K. 781

Noge, H. 955

Noh, J.-H. 939

Noman, M. 1107

Nonomura, Y. 733

Norman, J. J. 407

Nurcahyo, Y. 232, 765

Oba, M. 1055

Oberhammer, J. 15, 856, 892

Obinata, K. 403

Ocak, I. E. 817

O'Grady, J. 356

Ogura, D. 769

Ohkubo, T. 681

Ohmori, H. 419

Ohno, K. 507

Ohta, A. T. 411, 579

Ohtsuki, K. 653

Okada, H. 797

Okandan, M. 841

Okayama, Y. 467

Okitsu, T. 423

Okochi, M. 908

Okuda, Y. 348

Okumura, R. 943

Olsson, R. 841

Onda, K. 51

Ono, T. 1027

Onoe, H. 176

Oralkan, Ö. 757

Osawa, K. 467

Osten, W. 7

Ota, K. 864

Ou, K.-S. 1099

Ou, Y.-C. 204

Paccagnella, A. 630

Pacheco, S. 619

Padilla, W. J. 108

Pakula, L. S. 140

Palanivelu, R. 208

Parameswaran, M. 583

Park, D. 387

Park, H. G. 363

Park, I. H. 979

Park, J. 487

Park, J. H. 979

Park, K. 543

Park, K. H. 363

Park, K. K. 757

Park, S. 164

Park, S. E. 491

Park, S.-J. 188

Park, S.-W. 685

Park, Y. S. 959

Park, Y.-S. 979

Pärr, G. 535

Parra, E. 31

Parviz, B. A. 713

Paul, O. 7, 220, 232, 599, 765, 777, 1039

Paust, N. 1087

Pearson, S. 841

Pedrini, G. 7

AUTHOR INDEX

Peeri, Y. 216
Pei, R. 344
Peirs, J. 1063
Pellegrin, S. 829
Peng, I. 280
Peretti, V. 630
Perlin, G. E. 228
Perruchot, F. 619
Petzold, B. 188
Pham, H. T. M. 140, 383
Piazza, G. 320, 912
Pillans, B. 876
Pisano, A. P. 64, 685
Polcawich, R. G. 591, 896, 1043
Pollock, C. R. 547
Presser, N. 360
Provine, J. 741
Pruitt, B. L. 188, 407
Puchades, I. 991
Puers, R. 451, 852
Pulskamp, J. S. 591, 896, 1043
Qin, Y. 208
Qiu, X. 1103
Qu, Y. Q. 657
Quévy, E. P. 927
Raberg, W. 880
Räisenen, A. V. 892
Rajaraman, V. 140
Rebeiz, G. M. 27, 876
Reines, I. 876
Reis, N. 443
Ren, Z. 19
Resnick, P. 841
Rey, P. 619
Reynaerts, D. 451, 1063
Ricco, A. J. 356, 439

Riegger, L. 539
Rinaldi, M. 320, 912
Robin, R. 868
Rodger, D. C. 248
Roman, C. 575
Roper, C. 611
Roth, G. 276
Rothuizen, H. 603
Rottenberg, X. 852
Rubtsov, V. 160
Ruther, P. 220, 232, 765
Ryu, K. 1079
Saati, S. 244
Saeedi, E. 713
Sahin, K. 813
Saif, T. 80
Saito, T. 681
Sakuma, S. 51, 375
Salvia, J. 657, 801
Samarao, A. K. 888
Sammoura, F. 88
Samukawa, Seiji 112
Sandner, T. 947
Santagata, F. 844
Santagata-Iervolino, E. 535
Sarajlic, E. 607, 1075
Sari, I. 1071
Sarro, P. M. 76, 140, 383, 844
Sasaki, H. 403
Sasao, T. 348
Sato, H. 216, 499
Sato, K. 495, 661, 769, 908
Sato, R. 184
Sawada, K. 192, 224, 503, 721
Sawano, S. 435
Sawyer, W. 88

AUTHOR INDEX

Schenk, H. 947

Schlatmann, B. 136

Schlesinger, T. E. 1107

Schmid, S. 300

Schmidt, M. E. 599

Schoen, F. 880

Schroeder, J. 431

Schuck, P. J. 128

Schultz, J. 352

Schüttler, M. 232

Sebastian, A. 603

Sedaghat-Pisheh, H. 27

Seghete, D. 120

Segueni, K. 868

Seidel, H. 693

Seidl, K. 232

Seifert, A. 987, 999

Seita, H. 884

Seki, T. 1055

Sekiguchi, T. 499

Sen, P. 900

Senda, H. 733

Senesky, D. G. 685

Seo, J. H. 284

Seo, Y. H. 511

Seto, S. 507

Shah, G. J. 471

Shannon, M. A. 35

Shao, C. Y. 1027

Shen, C. 383

Shen, C. J. 336

Shetye, S. B. 665

Shi, J. 332

Shih, W.-P. 204

Shikida, M. 403, 495, 769, 908

Shimizu, K. 403, 737

Shimogaki, Y. 681

Shimoyama, I. 68, 168, 176, 551, 781, 821, 943, 963, 971, 1019

Shin, D.-M. 399

Shoji, S. 499, 519

Simard, K. 583

Singh, P. 304

Siu, C. P. B. 995

Skotnicki, T. 132

Smith, R. L. 571

Sochol, R. D. 304

Someya, T. 642

Somjit, N. 15

Song, C. 447

Song, E.-S. 725

Song, Y.-H. 148

Sosnowchik, B. D. 128

Spadaccini, C. M. 848

Sparks, A. 88

Spinney, P. S. 571

Stamm, M. T. 379

Steinecker, W. H. 264

Stemme, G. 15, 39, 892, 1003, 1063

Stephan, R. 923

Sterner, M. 892

Stieglitz, T. 232

Stoddart, J. F. 595

Stojanovic, M. 344

Stopa, A. 431

Strikwerda, A. 108

Stulemeijer, J. 916

Stupian, G. 360

Su, H.-T. 1047

Su, Y. C. 689

Sugano, K. 184

Sugiura, M. 503

Sugiyama, M. 681

AUTHOR INDEX

Sugiyama, S. 709, 1007

Suhm, A. 619

Sun, C.-M. 805

Sun, W. 884

Suzuki, T. 256

Suzuki, Y. 60, 615, 769, 1055

Swaminathan, V. V. 35

Tabata, O. 184

Tachibana, H. 955

Tai, Y. C. 248, 427, 745

Tai, Y.-C. 196, 244

Takahashi, H. 821

Takahashi, K. 697, 1083

Takahashi, T. 1083

Takahashi, Y. 459

Takahata, K. 701

Takahata, T. 971

Takama, N. 567, 677

Takamatsu, S. 943, 1019

Takao, H. 503

Takei, A. 963

Takei, K. 192, 224

Takei, Y. 551

Takeuchi, S. 56, 180, 256, 363, 367, 371, 423

Takumi, T. 661

Tamil, J. 967, 975, 1023

Tan, C. W. 872

Tanaka, K. 507

Tanaka, N. 503

Tanaka, S. 705

Tane, R. 124

Tanemura, T. 184

Tang, T.-L. 1047

Tang, Z. 825

Tao, H. 108

Tautorat, C. 220

Taylor, R. 407

Tazzoli, A. 630

Teh, K. S. 459

Tilmans, H. A. C. 136, 852

Tonisch, K. 923

Tonomura, W. 737

Torfs, T. 232

Toshiyoshi, H. 697, 884, 1083

Trott, W. 841

Truax, S. 284

Tsai, J.-C. 1031

Tsai, M.-H. 805

Tsai, Y.-C. 204

Tsamados, D. 132, 634

Tsang, S.-H. 583

Tsao, C.-M. 92

Tseng, F.-G. 296, 455

Tseng, Y.-T. 296

Tsou, W.-A. 627

Tsuchiya, T. 184

Tsuda, Y. 56, 423

Tsukuni, H. 507

Tsurui, T. 1051

Turner, K. L. 1095

Uchic, M. 80

Ueda, H. 955

Uejima, T. 681

Ullmann, S. 76

Umezu, S. 419

Unnikrishnan, S. 324

Ur, S. C. 761

Usami, H. 507

Valley, J. K. 411, 579

Van Beek, J. T. M. 650

Van Bommel, M. 136

Van Der Avoort, C. 650

AUTHOR INDEX

Van Der Wijngaart, W. 39, 1063

Van Drieenhuizen, B. 136

Van Herwaarden, A. W. 535

Van Hoof, C. 623

Van Wingerden, J. 650

Vanneer, R. 136

Varma, R. 244

Vazquez-Mena, O. 144

Vellekoop, M. J. 535

Vigna, Benedetto 1

Villanueva, G. 144

Villareal, G. 360

Von Stetten, F. 276, 539

Wägli, P. 300

Waguespack, R. P. 829

Waibel, P. 987

Waits, C. M. 1091

Wakui, D. 499

Wallrabe, U. 1015

Walmsley, R. G. 1095

Wang, H. 995

Wang, J. 825

Wang, Q. 352

Wang, S. 1067

Wang, W. 415

Wang, X. 1103

Wang, Z. 312, 623

Watamura, K. 348

Weber, H. 693

Weber, W. 880

Wei, P. 407

Weigel, R. 880

Weiss, P. S. 595

Wells, J. 1007

Wen, K.-A. 627

Wetzel, E. D. 1043

Wiegerink, R. J. 152, 463, 479

Wien, W. 535

Wiesmann, D. 603

Wilson, C. G. 829

Winkler, B. 880

Wise, K. D. 228

Witvrouw, A. 136

Wright, S. A. 837

Wu, D. 673

Wu, M. C. 411, 579

Wu, W. G. 673

Xia, X. 328

Xu, B. J. 395

Xu, J. 673

Xu, T. 312

Xu, W. 935

Yagi, T. 507

Yama, G. 801

Yamahata, C. 607, 1075

Yamane, D. 697, 884

Yamanishi, Y. 51, 375

Yan, G. Z. 104, 809

Yang, C. K. 844

Yang, K. 88

Yang, L.-J. 204

Yang, M. 395

Yang, S. 951

Yang, S.-A. 292

Yang, Y. 328, 555

Yang, Y. W. 595

Yang, Y.-J. 92, 204, 292

Yang, Z. C. 104, 809

Ye, T. 595

Yeh, Y.-T. 627

Yi, S. H. 761

Yin, C.-Y. 1031

AUTHOR INDEX

Yokokawa, R. 1007

Yokota, T. 495

Yokoyama, T. 1055

Yoneoka, S. 741, 801

Yoo, B. W. 979

Yoo, K. 789

Yoon, J.-B. 148, 475

Yoon, S.-I. 272

Yoon, Y. K. 729

Yoshida, H. 260

Yoshihata, Y. 963

Yoshimizu, N. 547

Yu, B. 196

Yu, J. J. 395

Yu, L. 312

Yun, S.-S. 793

Yun, T.-S. 729

Zaman, M. F. 749

Zandbergen, H. W. 76

Zappe, H. 987, 999

Zareian-Jahromi, M. A. 288

Zeng, H. 240, 559, 995

Zengerle, R. 43, 276, 539, 1087

Zhang, F. X. 527

Zhang, H. X. 673

Zhang, Q. 1103

Zhang, Q. X. 967, 1023

Zhang, W. 975, 1023

Zhang, X. 108, 264, 935

Zhang, X. J. 1011

Zhang, X. M. 967, 975, 1023

Zhang, Y. 84

Zhao, Q. C. 104, 809

Zhao, Y. 240, 559

Zhdaneev, O. 264

Zheng, X. J. 431

Zheng, Y. B. 332

Zhou, Q. 72, 587

Zhou, Y. 1103

Zhou, Z. Y. 527

Zhu, L. 35

Zhu, R. 527

Zhu, W. M. 975, 1023

Zhu, Y. 1103

Ziaei-Moayyed, M. 927

Ziaie, B. 407

Ziegler, C. 1087

Zimmer, F. 1003

Zohar, Y. 208, 379, 431

Zuniga, C. 320, 912

CURRAN ASSOCIATES INC.
proceedings
.com

9781424429776

2024 IEEE 17th International Conference on Solid-State & Integrated Circuit Technology (ICSICT 2024)

Zhuhai, China
22-25 October 2024

Pages 1-443

IEEE Catalog Number: CFP24829-POD
ISBN: 979-8-3503-6184-1